Recommended Dietary Allowances (RDA) and Adequate Intakes (AI) for Vitamins

Age (yr)	Thiamin RDA (mg/day)	Riboflavin RDA (mg/day)	Niacin RDA (mg/day)[a]	Biotin AI (μg/day)	Pantothenic acid AI (mg/day)	Vitamin B_6 RDA (mg/day)	Folate RDA (μg/day)[b]	Vitamin B_{12} RDA (μg/day)	Choline AI (mg/day)	Vitamin C RDA (mg/day)	Vitamin A RDA (μg/day)[c]	Vitamin D RDA (IU/day)[d]	Vitamin E RDA (mg/day)[e]	Vitamin K AI (μg/day)
Infants														
0–0.5	0.2	0.3	2	5	1.7	0.1	65	0.4	125	40	400	400 (10 μg)	4	2.0
0.5–1	0.3	0.4	4	6	1.8	0.3	80	0.5	150	50	500	400 (10 μg)	5	2.5
Children														
1–3	0.5	0.5	6	8	2	0.5	150	0.9	200	15	300	600 (15 μg)	6	30
4–8	0.6	0.6	8	12	3	0.6	200	1.2	250	25	400	600 (15 μg)	7	55
Males														
9–13	0.9	0.9	12	20	4	1.0	300	1.8	375	45	600	600 (15 μg)	11	60
14–18	1.2	1.3	16	25	5	1.3	400	2.4	550	75	900	600 (15 μg)	15	75
19–30	1.2	1.3	16	30	5	1.3	400	2.4	550	90	900	600 (15 μg)	15	120
31–50	1.2	1.3	16	30	5	1.3	400	2.4	550	90	900	600 (15 μg)	15	120
51–70	1.2	1.3	16	30	5	1.7	400	2.4	550	90	900	600 (15 μg)	15	120
>70	1.2	1.3	16	30	5	1.7	400	2.4	550	90	900	800 (20 μg)	15	120
Females														
9–13	0.9	0.9	12	20	4	1.0	300	1.8	375	45	600	600 (15 μg)	11	60
14–18	1.0	1.0	14	25	5	1.2	400	2.4	400	65	700	600 (15 μg)	15	75
19–30	1.1	1.1	14	30	5	1.3	400	2.4	425	75	700	600 (15 μg)	15	90
31–50	1.1	1.1	14	30	5	1.3	400	2.4	425	75	700	600 (15 μg)	15	90
51–70	1.1	1.1	14	30	5	1.5	400	2.4	425	75	700	600 (15 μg)	15	90
>70	1.1	1.1	14	30	5	1.5	400	2.4	425	75	700	800 (20 μg)	15	90
Pregnancy														
≤18	1.4	1.4	18	30	6	1.9	600	2.6	450	80	750	600 (15 μg)	15	75
19–30	1.4	1.4	18	30	6	1.9	600	2.6	450	85	770	600 (15 μg)	15	90
31–50	1.4	1.4	18	30	6	1.9	600	2.6	450	85	770	600 (15 μg)	15	90
Lactation														
≤18	1.4	1.6	17	35	7	2.0	500	2.8	550	115	1200	600 (15 μg)	19	75
19–30	1.4	1.6	17	35	7	2.0	500	2.8	550	120	1300	600 (15 μg)	19	90
31–50	1.4	1.6	17	35	7	2.0	500	2.8	550	120	1300	600 (15 μg)	19	90

NOTE: For all nutrients, values for infants are AI. The glossary on the inside back cover defines units of nutrient measure.

[a]Niacin recommendations are expressed as niacin equivalents (NE), except for recommendations for infants younger than 6 months, which are expressed as preformed niacin.

[b]Folate recommendations are expressed as dietary folate equivalents (DFE).

[c]Vitamin A recommendations are expressed as retinol activity equivalents (RAE).

[d]Vitamin D recommendations are expressed as cholecalciferol and assume an absence of adequate exposure to sunlight.

[e]Vitamin E recommendations are expressed as α-tocopherol.

Recommended Dietary Allowances (RDA) and Adequate Intakes (AI) for Minerals

Age (yr)	Sodium AI (mg/day)	Chloride AI (mg/day)	Potassium AI (mg/day)	Calcium RDA (mg/day)	Phosphorus RDA (mg/day)	Magnesium RDA (mg/day)	Iron RDA (mg/day)	Zinc RDA (mg/day)	Iodine RDA (μg/day)	Selenium RDA (μg/day)	Copper RDA (μg/day)	Manganese AI (mg/day)	Fluoride AI (mg/day)	Chromium AI (μg/day)	Molybdenum RDA (μg/day)
Infants															
0–0.5	120	180	400	200	100	30	0.27	2	110	15	200	0.003	0.01	0.2	2
0.5–1	370	570	700	260	275	75	11	3	130	20	220	0.6	0.5	5.5	3
Children															
1–3	1000	1500	3000	700	460	80	7	3	90	20	340	1.2	0.7	11	17
4–8	1200	1900	3800	1000	500	130	10	5	90	30	440	1.5	1.0	15	22
Males															
9–13	1500	2300	4500	1300	1250	240	8	8	120	40	700	1.9	2	25	34
14–18	1500	2300	4700	1300	1250	410	11	11	150	55	890	2.2	3	35	43
19–30	1500	2300	4700	1000	700	400	8	11	150	55	900	2.3	4	35	45
31–50	1500	2300	4700	1000	700	420	8	11	150	55	900	2.3	4	35	45
51–70	1300	2000	4700	1000	700	420	8	11	150	55	900	2.3	4	30	45
>70	1200	1800	4700	1200	700	420	8	11	150	55	900	2.3	4	30	45
Females															
9–13	1500	2300	4500	1300	1250	240	8	8	120	40	700	1.6	2	21	34
14–18	1500	2300	4700	1300	1250	360	15	9	150	55	890	1.6	3	24	43
19–30	1500	2300	4700	1000	700	310	18	8	150	55	900	1.8	3	25	45
31–50	1500	2300	4700	1000	700	320	18	8	150	55	900	1.8	3	25	45
51–70	1300	2000	4700	1200	700	320	8	8	150	55	900	1.8	3	20	45
>70	1200	1800	4700	1200	700	320	8	8	150	55	900	1.8	3	20	45
Pregnancy															
≤18	1500	2300	4700	1300	1250	400	27	12	220	60	1000	2.0	3	29	50
19–30	1500	2300	4700	1000	700	350	27	11	220	60	1000	2.0	3	30	50
31–50	1500	2300	4700	1000	700	360	27	11	220	60	1000	2.0	3	30	50
Lactation															
≤18	1500	2300	5100	1300	1250	360	10	13	290	70	1300	2.6	3	44	50
19–30	1500	2300	5100	1000	700	310	9	12	290	70	1300	2.6	3	45	50
31–50	1500	2300	5100	1000	700	320	9	12	290	70	1300	2.6	3	45	50

NOTE: For all nutrients, values for infants are AI. The glossary on the inside back cover defines units of nutrient measure.

Tolerable Upper Intake Levels (UL) for Vitamins

Age (yr)	Niacin (mg/day)[a]	Vitamin B_6 (mg/day)	Folate (μg/day)[a]	Choline (mg/day)	Vitamin C (mg/day)	Vitamin A (IU/day)[b]	Vitamin D (IU/day)	Vitamin E (mg/day)[c]
Infants								
0–0.5	—	—	—	—	—	600	1000 (25 μg)	—
0.5–1	—	—	—	—	—	600	1500 (38 μg)	—
Children								
1–3	10	30	300	1000	400	600	2500 (63 μg)	200
4–8	15	40	400	1000	650	900	3000 (75 μg)	300
9–13	20	60	600	2000	1200	1700	4000 (100 μg)	600
Adolescents								
14–18	30	80	800	3000	1800	2800	4000 (100 μg)	800
Adults								
19–70	35	100	1000	3500	2000	3000	4000 (100 μg)	1000
>70	35	100	1000	3500	2000	3000	4000 (100 μg)	1000
Pregnancy								
≤18	30	80	800	3000	1800	2800	4000 (100 μg)	800
19–50	35	100	1000	3500	2000	3000	4000 (100 μg)	1000
Lactation								
≤18	30	80	800	3000	1800	2800	4000 (100 μg)	800
19–50	35	100	1000	3500	2000	3000	4000 (100 μg)	1000

[a]The UL for niacin and folate apply to synthetic forms obtained from supplements, fortified foods, or a combination of the two.
[b]The UL for vitamin A applies to the preformed vitamin only.
[c]The UL for vitamin E applies to any form of supplemental α-tocopherol, fortified foods, or a combination of the two.

Tolerable Upper Intake Levels (UL) for Minerals

Age (yr)	Sodium (mg/day)	Chloride (mg/day)	Calcium (mg/day)	Phosphorus (mg/day)	Magnesium (mg/day)[d]	Iron (mg/day)	Zinc (mg/day)	Iodine (μg/day)	Selenium (μg/day)	Copper (μg/day)	Manganese (mg/day)	Fluoride (mg/day)	Molybdenum (μg/day)	Boron (mg/day)	Nickel (mg/day)	Vanadium (mg/day)
Infants																
0–0.5	—	—	1000	—	—	40	4	—	45	—	—	0.7	—	—	—	—
0.5–1	—	—	1500	—	—	40	5	—	60	—	—	0.9	—	—	—	—
Children																
1–3	1500	2300	2500	3000	65	40	7	200	90	1000	2	1.3	300	3	0.2	—
4–8	1900	2900	2500	3000	110	40	12	300	150	3000	3	2.2	600	6	0.3	—
9–13	2200	3400	3000	4000	350	40	23	600	280	5000	6	10	1100	11	0.6	—
Adolescents																
14–18	2300	3600	3000	4000	350	45	34	900	400	8000	9	10	1700	17	1.0	—
Adults																
19–50	2300	3600	2500	4000	350	45	40	1100	400	10,000	11	10	2000	20	1.0	1.8
51–70	2300	3600	2000	4000	350	45	40	1100	400	10,000	11	10	2000	20	1.0	1.8
>70	2300	3600	2000	3000	350	45	40	1100	400	10,000	11	10	2000	20	1.0	1.8
Pregnancy																
≤18	2300	3600	3000	3500	350	45	34	900	400	8000	9	10	1700	17	1.0	—
19–50	2300	3600	2500	3500	350	45	40	1100	400	10,000	11	10	2000	20	1.0	—
Lactation																
≤18	2300	3600	3000	4000	350	45	34	900	400	8000	9	10	1700	17	1.0	—
19–50	2300	3600	2500	4000	350	45	40	1100	400	10,000	11	10	2000	20	1.0	—

[d]The UL for magnesium applies to synthetic forms obtained from supplements or drugs only.
NOTE: An upper Limit was not established for vitamins and minerals not listed and for those age groups listed with a dash (—) because of a lack of data, not because these nutrients are safe to consume at any level of intake. All nutrients can have adverse effects when intakes are excessive.

Understanding Nutrition

for Pima Community College

Ellie Whitney | Sharon Rady Rolfes

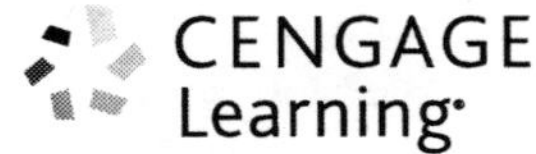

Australia • Brazil • Japan • Korea • Mexico • Singapore • Spain • United Kingdom • United States

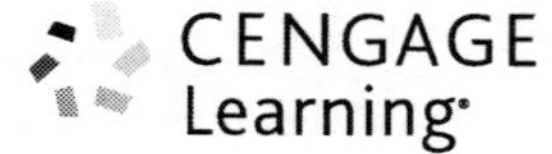

Understanding Nutrition for Pima Community College

Senior Project Development Manager:
Linda deStefano

Market Development Manager:
Heather Kramer

Senior Production/Manufacturing Manager:
Donna M. Brown

Production Editorial Manager:
Kim Fry

Sr. Rights Acquisition Account Manager:
Todd Osborne

Printed in the United States of America

Understanding Nutrition
Ellie Whitney | Sharon Rady Rolfes

ISBN-13: 978-1-285-88770-8

ISBN-10: 1-285-88770-0

Cengage Learning
5191 Natorp Boulevard
Mason, Ohio 45040
USA

Cengage Learning is a leading provider of customized learning solutions with office locations around the globe, including Singapore, the United Kingdom, Australia, Mexico, Brazil, and Japan. Locate your local office at: **international.cengage.com/region.**

Cengage Learning products are represented in Canada by Nelson Education, Ltd.
For your lifelong learning solutions, visit **www.cengage.com/custom.**
Visit our corporate website at **www.cengage.com.**

Brief Contents

Table of Contents

© HSNphotography/Shutterstock.com

© Artem Rudik/Shutterstock.com

© liza1979/Shutterstock.com

© studiogi/Shutterstock.com

Preface

Nutrition is a science. The details of a nutrient's chemistry or a cell's biology can be overwhelming and confusing to some, but it needn't be. When the science is explained step by step and the facts are connected one by one, the details become clear and understandable. By telling stories about fat mice, using analogies of lamps, and applying guidelines to groceries, we make the science of nutrition meaningful and memorable. That has been our mission since the first edition—to reveal the fascination of science and share the excitement of nutrition with readers. We have learned from the hundreds of professors and more than a million students who have used this book through the years that readers want an *understanding* of nutrition so they can make healthy choices in their daily lives. We hope that this book serves you well.

A Book Tour of This Edition

Understanding Nutrition presents the core information of an introductory nutrition course. The early chapters introduce the nutrients and their work in the body, and the later chapters apply that information to people's lives—describing the role of foods and nutrients in energy balance and weight control, in physical activity, in the life cycle, in disease prevention, in food safety, and in hunger.

The Chapters Chapter 1 begins by exploring why we eat the foods we do and continues with a brief overview of the nutrients, the science of nutrition, recommended nutrient intakes, assessment, and important relationships between diet and health. Chapter 2 describes the diet-planning principles and food guides used to create diets that support good health and includes instructions on how to read a food label. In Chapter 3 readers follow the journey of digestion and absorption as the body breaks down foods into nutrients. Chapters 4, 5, and 6 describe carbohydrates, fats, and proteins—their chemistry, roles in the body, and places in the diet. Then Chapter 7 shows how the body derives energy from these three nutrients. Chapters 8 and 9 continue the story with a look at energy balance, the factors associated with overweight and underweight, and the benefits and dangers of weight loss and weight gain. Chapters 10, 11, 12, and 13 complete the introductory lessons by describing the vitamins, the minerals, and water—their roles in the body, deficiency and toxicity symptoms, and sources.

The next seven chapters weave that basic information into practical applications, showing how nutrition influences people's lives. Chapter 14 describes how physical activity and nutrition work together to support fitness. Chapters 15, 16, and 17 present the special nutrient needs of people through the life cycle—pregnancy and lactation; infancy, childhood, and adolescence; and adulthood and the later years. Chapter 18 focuses on the dietary risk factors and recommendations associated with chronic diseases, and Chapter 19 addresses consumer concerns about the safety of the food and water supply. Chapter 20 closes the book by examining hunger and the global environment.

The Highlights Every chapter is followed by a highlight that provides readers with an in-depth look at a current, and often controversial, topic that relates to its companion chapter. For example, Highlight 4 examines the scientific evidence behind some of the current controversies surrounding carbohydrates and their role in weight gain and weight loss.

Special Features The art and layout in this edition have been carefully designed to be inviting while enhancing student learning. For example, numbered steps have been added to several figures to clarify sequences and processes. In addition, special features help readers identify key concepts and apply nutrition knowledge. For example, when a new term is introduced, it is printed in bold type, and a **definition** is provided. These definitions often include pronunciations and derivations to facilitate understanding. The glossary at the end of the book includes all defined terms.

definition (DEF-eh-NISH-en): the meaning of a word.
- **de** = from
- **finis** = boundary

LEARNING GPS

The opening page of each chapter provides a Learning GPS that serves as an outline and directs readers to the main heads (and subheads) within the chapter. Each main head is followed by a Learn It—a learning objective for the content covered in that section. The Learn It also appears within the text at the start of each main section as well as at the start of each Review It. After reading the chapter, students will be able to demonstrate competency in the Learn It objectives.

Below each chapter-opening photo is a suggestion for research using Global Nutrition Watch, the online resource center that provides access to thousands of trusted nutrition sources.

Nutrition in Your Life

The opening paragraph of each chapter—called Nutrition in Your Life—introduces the chapter's content in a friendly and familiar way. This short paragraph closes with a preview of how readers might apply that content to their daily lives by inviting them to use the Nutrition Portfolio section at the end of the chapter.

Nutrition Portfolio

The Nutrition Portfolio section at the end of each chapter prompts readers to consider whether their personal choices are meeting the goals presented in the chapter. Most of these assignments include instructions that use the Diet Analysis Plus program. Such tools help students assess their current choices and make informed decisions about healthy options.

REVIEW IT

Each major section within a chapter concludes with a Review It paragraph that summarizes key concepts. Similarly, Review It tables cue readers to important summaries.

Also featured in this edition are the 2010 *Dietary Guidelines for Americans,* which are introduced in Chapter 2 and presented throughout the text whenever their subjects are discussed. Look for the following design.

> DIETARY GUIDELINES FOR AMERICANS 2010

These guidelines provide science-based advice to promote health and to reduce the risk of chronic disease through diet and physical activity.

>How To

Many of the chapters include "How To" features that guide readers through problem-solving tasks. For example, a "How To" in Chapter 1 presents the steps in calculating energy intake from the grams of carbohydrate, fat, and protein in a food.

TRY IT Each "How To" feature ends with a "Try It" activity that gives readers an opportunity to practice these new lessons.

STUDY CARD > 1

STUDY IT To review the key points of this chapter and take a practice quiz, go to the Study Cards at the end of the book.

Study cards appear at the back of the text—one for each chapter. The Study It side of each card presents a review of the chapter's core concepts, and perhaps a table or figure to remind readers of key points. The Test It side of the study card provides essay and multiple-choice questions to help prepare students for exams.

The Appendixes

The appendixes are valuable references for a number of purposes. Appendix A summarizes background information on the hormonal and nervous systems, complementing Appendixes B and C on basic chemistry, the chemical structures of nutrients, and major metabolic pathways. Appendix D describes measures of protein quality. Appendix E provides detailed coverage of nutrition assessment, and Appendix F presents the estimated energy requirements for men and women at various levels of physical activity. Appendix G presents the 2008 *Choose Your Foods: Exchange List for Diabetes.* Appendix H is a 4000-item food composition table. Appendix I lists nutrition recommendations from the World Health Organization (WHO), and Appendix J presents the 2020 Healthy People nutrition-related objectives. Appendix K features aids to calculations, a short tutorial on converting metric measures and handling basic math problems commonly found in the world of nutrition.

The Inside Covers

The inside covers put commonly used information at your fingertips. The inside front covers (pp. A–C) present the current nutrient recommendations; the inside back covers feature the Daily Values used on food labels and a glossary of nutrient measures (p. Y on the left) as well as suggested weight ranges for various heights (p. Z on the right).

Notable Changes in This Edition

Because nutrition is an active science, staying current is paramount. Just as nutrition research continuously adds to and revises the accepted body of knowledge, this edition builds on the science of previous editions with the latest in nutrition research. Much has changed in the world of nutrition and in our daily lives since the first edition. The number of foods has increased dramatically—even as we spend less time than ever in the kitchen preparing meals. The connections between diet and disease have become more apparent—and consumer interest in making smart health choices has followed. More people are living longer and healthier lives. The science of nutrition has grown rapidly, with new "facts" emerging daily. In this edition, as with all previous editions, every chapter has been revised to enhance learning by presenting current information accurately and attractively. For all chapters and highlights we have:

- Reviewed and updated content
- Created several new figures and tables and revised others to enhance learning
- Updated the *Dietary Guidelines for Americans 2010* feature
- Added learning objectives

Chapter 1

- Enhanced the content of the figure on examples of research designs and deleted the table on strengths and weaknesses of research designs
- Enhanced the content of the figure inaccurate versus accurate view of nutrient intakes and deleted the figure on using the DRI to assess the dietary intake of a healthy individual
- Created a new table for Healthy People 2020 and presented updates on how well Healthy People 2010 goals were met

Chapter 2

- Revised text, tables, and figures to reflect *Dietary Guidelines for Americans 2010*
- Added text and created new figure for MyPlate
- Simplified discussion on health claims on food labels and created new figure comparing health, nutrient, and structure-function claims on food labels
- Revised vegetarian highlight to reflect *Dietary Guidelines for Americans 2010* and include a new table for protein foods subgroups
- Added section on vegetarian eating patterns protecting against diabetes

Chapter 3

- Removed figure on peristalsis and segmentation (available in Instructors' Resources)
- Added a short section on celiac disease to the highlight

Chapter 4

- Consolidated the figures introducing the chemical structure of glucose into one figure
- Consolidated the figures introducing the chemical structures of each of the monosaccharides into one figure
- Created a table summarizing the carbohydrate family
- Clarified the discussions of carbohydrate digestion in the stomach and monosaccharide metabolism in the liver
- Added a new section on obesity and chronic diseases to discussion on health effects of sugars
- Created a new feature on How To Reduce the Intake of Added Sugars
- Tightened the focus in the highlight

Chapter 5

- Removed the figure on acetic acid (available in Instructors' Resources)
- Created a table summarizing the lipid family
- Removed the figure on bile acid (available in Instructors' Resources)
- Reorganized health effects and recommended intake sections into two sections—one for saturated fats, *trans* fats, and cholesterol and another for monounsaturated and polyunsaturated fats
- Removed the figure comparing butter and margarine (available in Instructors' Resources)
- Expanded the discussion on the Mediterranean diet and added a figure of its pyramid
- Removed the figure on relationships among dietary saturated fatty acids, LDL cholesterol, and heart disease risk from the highlight (available in Instructors' Resources)

Chapter 6

- Created new section on protein deficiency and moved full discussion of malnutrition (formerly protein-energy malnutrition) to Chapter 20
- Created new section From Guidelines to Groceries (as in Chapters 4 and 5)

Chapter 7

- Renamed the chapter title and several heads to better reflect the content
- Reworded some sentences and deleted others; rewrote some paragraphs and moved others; created more reviews and simplified figures—all in a never-ending effort to present the science of metabolism clearly and concisely
- Removed the figure on the many carbons of a typical triglyceride providing only three carbons to make glucose (available in Instructors' Resources)

Chapter 8

- Added a figure of BMI silhouettes
- Added a table comparing percent body fat of physically fit people of healthy weight with the national averages; removed the figure on body compositions
- Removed the figure on the declining weight of Miss America (available in Instructors' Resources)

Chapter 9

- Simplified the genetic discussion to focus primarily on leptin and ghrelin
- Simplified the drug discussion and created a table of FDA-approved drugs
- Refocused public health discussion to community programs; created new table on community strategies to prevent obesity
- Removed several tables including the proteins involved in regulation of food intake and energy requirements, the weight-loss consumer bill of rights, recommendations for a weight-loss diet, guidelines for identifying fad diets and other weight loss scams, and the claims and truths of fad diets (available in Instructors' Resources)
- Created new How To on identifying fad diets and weight-loss scams
- Added new diets to table of popular diets compared

Chapter 10

- Revised figure on prevalence of neural tube defects since folate fortification
- Created new figure of red blood cell production (normal compared with folate deficiency)

Chapter 11

- Revised and updated information on vitamin D as per the 2011 DRI
- Added short section on vitamin D supplements
- Revised figure on the synthesis and activation of vitamin D

Chapter 12

- Simplified figure on how the body regulates water volume and blood pressure through the renin-angiotensin-aldosterone system
- Revised and updated information on calcium as per the 2011 DRI

Chapter 13

- Created new figure of red blood cell production (normal compared with iron deficiency)
- Added new foods and examples to the figure of phytochemicals in the highlight
- Created a new table in the highlight featuring the colors of foods rich in phytochemicals

Chapter 14

- Revised section on fat use during physical activity
- Removed How To Carbohydrate Load (available in Instructors' Resources)
- Created new How To on calculating the carbohydrate concentration of sports drinks
- Added a new table of nutrient-dense snacks for athletes
- Created a margin list of when sports drinks may be beneficial
- Reorganized the presentation on ergogenic aids in the highlight

Chapter 15

- Added a little more on the benefits of WIC
- Removed the figure of prenatal supplements (available in Instructors' Resources)

Chapter 16

- Created new table of aerobic and muscle- and bone-strengthening physical activities for children
- Updated and revised figure on trends in childhood obesity
- Updated standards for school lunches

Chapter 17

- Expanded the section on depression in older adults
- Added a section on malnutrition in older adults
- Deleted the pyramid for older adults
- Created a new figure illustrating brain anatomy in Alzheimer's disease
- Created a new table comparing Alzheimer's symptoms with those typical of aging
- Added a section on alcohol abuse in older adults

Chapter 18

- Expanded the discussion on nutrition and immunity
- Deleted the table on the effects of PEM on body defenses
- Revised and updated the table on lifestyle modifications to reduce blood pressure
- Updated the figure on the prevalence of diabetes among adults in the United States
- Revised and updated the table on dietary guidelines and recommendations for chronic diseases
- Deleted the Healthy Eating Pyramid and replaced it with the Healthy Eating Plate
- Deleted the table on advice and precautions on alternative therapies for cancer

Chapter 19

- Removed the figure of increasing number of imported foods (available in Instructors' Resources)

- Limited table on foodborne illnesses to those responsible for 90 percent of illnesses
- Added new paragraph on aflatoxins contaminating grains improperly stored in tropical countries
- Added new paragraph on bisphenol A leaching into water from hard-plastic bottles
- Removed table on preventing H1N1 infection (available in Instructors' Resources)

Chapter 20

- Expanded discussion on obesity paradox
- Added a section on food waste
- Reorganized and revised discussion on malnutrition
- Added new figure on water crisis areas of the world
- Added new How To feature on determining a person's food footprint

Student and Instructor Resources

CengageNOW: An intelligent, Web-based study system and course management tool, CengageNOW provides a completely integrated package of diagnostic quizzes, personalized study, animations, videos, case studies, and more—along with an Instructor Grade Book that automatically captures and tracks student progress. Pretests assess students' understanding of what they have read and direct them to further study of concepts they have not yet mastered using the interactive eBook and other online tools such as Pop-up Tutors, Nutrition Tutorials, and BBC video clips. The self-grading pretests and posttests are also ideal for homework assignments, as results flow automatically into the built-in Instructor Grade Book.

Study Guide: This full-featured guide includes chapter outlines and practice tests, fill-in-the-blank chapter reviews, short-answer questions and calculations, and crossword and matching puzzles for vocabulary review.

Student Course Guide for Nutrition Pathways: Wadsworth, a part of Cengage Learning, is pleased to partner with Dallas TeleLearning and the LeCroy Center for Educational Telecommunication by publishing a text-specific Student Course Guide for the *Nutrition Pathways* Telecourse. The Student Course Guide features chapter and video assignments, lesson overviews, chapter learning objectives, key lesson concepts, and a practice test for each lesson.

Power Lecture DVD-ROM: This one-stop course preparation and presentation resource makes it easy for you to assemble, edit, publish, and present custom lectures for your course, using PowerPoint®. The PowerLecture includes PowerPoint® with stepped art, animations, BBC video clips, the Instructor's Manual, the test bank, "clicker" content, and ExamView computerized testing.

Test Bank: The test bank features a large assortment of multiple-choice questions, essay questions, and matching exercises, now categorized by learning objective as well as difficulty level and text page location.

Instructor's Manual: New to this edition are several food choice–focused case studies, critical-thinking questions, meal comparison activities for in-class discussion, and crossword puzzles, all with keys or grading rubrics. This comprehensive manual also includes expanded chapter objectives, annotated lecture presentation outlines, answer keys for the "Try It" exercises and study card questions from the text, and other assignments, handouts, and classroom activity suggestions.

Transparency Acetates: This set of colorful transparencies includes illustrations from the text. Included is a correlation guide to assist you in utilizing the transparencies from the tenth through thirteenth editions together for the most comprehensive coverage.

CourseMate: This feature brings course concepts to life with interactive learning, study, and exam preparation tools that support the printed textbook, or the included eBook. With CourseMate, professors can use the included Engagement Tracker to assess student preparation and engagement. Use the tracking tools to see progress for the class as a whole or for individual students. Students can access an interactive eBook, chapter-specific interactive learning tools, including flashcards, quizzes, Pop-up Tutors, Nutrition Tutorials, and BBC video clips, and more in their Nutrition CourseMate.

WebTutor™: Provides customizable, text-specific content that allows instructors to edit, reorganize, or delete content to meet their course needs. It offers quizzes, videos, animations, Pop-up Tutors, and test bank materials along with direct access to Diet Analysis Plus, Global Nutrition Watch, and an interactive eBook.

Diet Analysis Plus™: Diet Analysis Plus enables you to track and assess your diet and physical activity online. You can create a personal profile based on height, weight, age, sex, and activity level, and use this tool to easily analyze the nutritional value of the food you eat, adjust your diet to meet your personal health goals, and gain a better understanding of how nutrition relates to your life. Diet Analysis Plus includes a 20,000+ food database, 10 reports for analysis, a food recipe feature, the latest Dietary References, and goals and actual percentages of essential nutrients, vitamins, and minerals. Diet Analysis Plus is a valuable tool that you can use in your nutrition course and then continue to use after the course is over.

Global Nutrition Watch: Bring currency to the classroom with Global Nutrition Watch from Cengage Learning. This user-friendly website provides convenient access to thousands of trusted sources, including academic journals, newspapers, videos, and podcasts, for you to use for research projects or classroom discussion. Global Nutrition Watch is updated daily to offer the most current news about topics related to nutrition.

Closing Comments

We have taken great care to provide accurate information and have included many references at the end of each chapter and highlight. To keep the number of references manageable, however, many statements that appeared in previous editions with references now appear without them. All statements reflect current nutrition knowledge, and the authors will supply references upon request. In addition to supporting text statements, the end-of-chapter references provide readers with resources for finding a good overview or more details on the subject. Nutrition is a fascinating subject, and we hope our enthusiasm for it comes through on every page.

Ellie Whitney
Sharon Rady Rolfes
June 2012

Acknowledgments

To produce a book requires the coordinated effort of a team of people—and, no doubt, each team member has another team of support people as well. We salute, with a big round of applause, everyone who has worked so diligently to ensure the quality of this book.

We thank our partners and friends, Linda DeBruyne and Fran Webb, for their valuable consultations and contributions; working together over the past 30 years has been a most wonderful experience. We especially appreciate Linda's research assistance on several chapters. Special thanks to our colleagues Kathy Pinna for her insightful comments and Sylvia Crews for her careful review of the math explanations in Appendix K. Thank you to Alex Rodriguez for preparing manuscript, Suzie Dorner for clarifying content and providing a student's perspective, Chelsea Mackenzie for assisting in numerous office tasks, and Marni Jay Rolfes for offering behind-the-scenes editorial suggestions.

We also thank the many professors who prepared the ancillaries that accompany this text: Harry Sitren for writing and enhancing the test bank; Carrie King, Melissa Langone, Barbara Quinn, and Daryle Wane for contributing to the Instructor's Manual; and Lori Turner for organizing the Student Study Guide. Thanks also to Miriam Myers, Shelley Ryan, Elesha Feldman, and the folks at Axxya Systems for their assistance in creating the food composition appendix and developing the computerized Diet Analysis Plus program that accompanies this book.

Our heartfelt thanks to our editorial team for their efforts in creating an outstanding nutrition textbook—Peggy Williams for gently pushing through resistance for changes that enhance this edition in amazing ways; Nedah Rose for her calming presence, delightful humor, and thoughtful suggestions; Carol Samet for her management of this project; Janet del Mundo for her energetic efforts in marketing; Miriam Myers for her dedication in developing online animations and study tools; Dean Dauphinais for his assistance in obtaining permissions; and Elesha Feldman for her competent editing of ancillaries.

We also thank Gary Hespenheide for creatively designing these pages; Joan Keyes for her diligent attention to the innumerable details involved in production; Scott Rosen for selecting photographs that deliver nutrition messages attractively; Susan Gall for copyediting more than 2000 manuscript pages; Greg Teague for proofreading close to 1000 final text pages; and Leoni McVey for composing a thorough and useful index. To the hundreds of others involved in production and sales, we tip our hats in appreciation.

We are especially grateful to our friends and families for their continued encouragement and support. We also thank our many reviewers for their comments and contributions.

Reviewers of *Understanding Nutrition*

Becky Alejandre
American River College

Janet B. Anderson
Utah State University

Sandra D. Baker
University of Delaware

Angelina Boyce
Hillsborough Community College

Lynn S. Brann
Syracuse University

Shalon Bull
Palm Beach Community College

Dorothy A. Byrne
University of Texas, San Antonio

Angela Caldwell
Black River Technical College

John R. Capeheart
University of Houston, Downtown

Leah Carter
Bakersfield College

James F. Collins
University of Florida

Diane Curis
Los Rios Community College District

Lisa K. Diewald
Montgomery County Community College

Kelly K. Eichmann
Fresno City College

Mary Flynn
Brown University

Betty J. Forbes
West Virginia University

Sue Fredstrom
Minnesota State University, Mankato

Trish Froehlich
Palm Beach Community College

Stephen P. Gagnon
Hillsborough Community College

Leonard E. Gerber
University of Rhode Island

Jill Golden
Orange Coast College

Barbara J. Goldman
Palm Beach State College

Kathleen Gould
Towson University

Margaret Gunther
Palomar College

Charlene Hamilton
University of Delaware

D. J. Hennager
Kirkwood Community College

Catherine Hagen Howard
Texarkana College

Jasminka Z. Ilich
Florida State University

Ernest B. Izevbigie
Jackson State University

Craig Kasper
Hillsborough Community College

Younghee Kim
Bowling Green State University

Rebecca A. Kleinschmidt
University of Alaska Southeast

Vicki Kloosterhouse
Oakland Community College

Donna M. Kopas
Pennsylvania State University

Susan M. Krueger
University of Wisconsin, Eau Claire

Melissa Langone
Pasco-Hernando Community College

Darlene M. Levinson
Oakland Community College, Orchard Ridge

Kimberly Lower
Collin County Community College

Melissa B. McGuire
Maple Woods Community College

Diane L. McKay
Tufts University

Anne Miller
De Anza College

Anahita M. Mistry
Eastern Michigan University

Mithia Mukutmoni
Sierra College

Steven Nizielski
Grand Valley State University

Yvonne Ortega
Santa Monica College

Jane M. Osowski
University of Southern Mississippi

Sarah Panarello
Yakima Valley Community College

Ryan Paruch
Tulsa Community College

Jill Patterson
Pennsylvania State University

Gina Pazzaglia
West Chester University

Julie Priday
Centralia College

Barbara F. Rabsatt
Brevard Community College

Kathy L. Sedlet
Collin County Community College

Melissa Shock
University of Central Arkansas

Tiffany Shurtz
University of Central Oklahoma

LuAnn Soliah
Baylor University

Bernice Gales Spurlock
Hinds Community College

Kenneth Strothkamp
Lewis & Clark College

Robin Sytsma
Solano Community College

Andrea Villarreal
Phoenix College

Terry Weideman
Oakland Community College, Highland Lake

H. Garrison Wilkes
University of Massachusetts, Boston

Lauri Wright
University of South Florida

Lynne C. Zeman
Kirkwood Community College

Maureen Zimmerman
Mesa Community College

Understanding Nutrition

© Blanche Branton/Shutterstock.com

CENGAGE LEARNING'S GLOBAL NUTRITION WATCH — GLOBAL NUTRITION RESOURCE CENTER

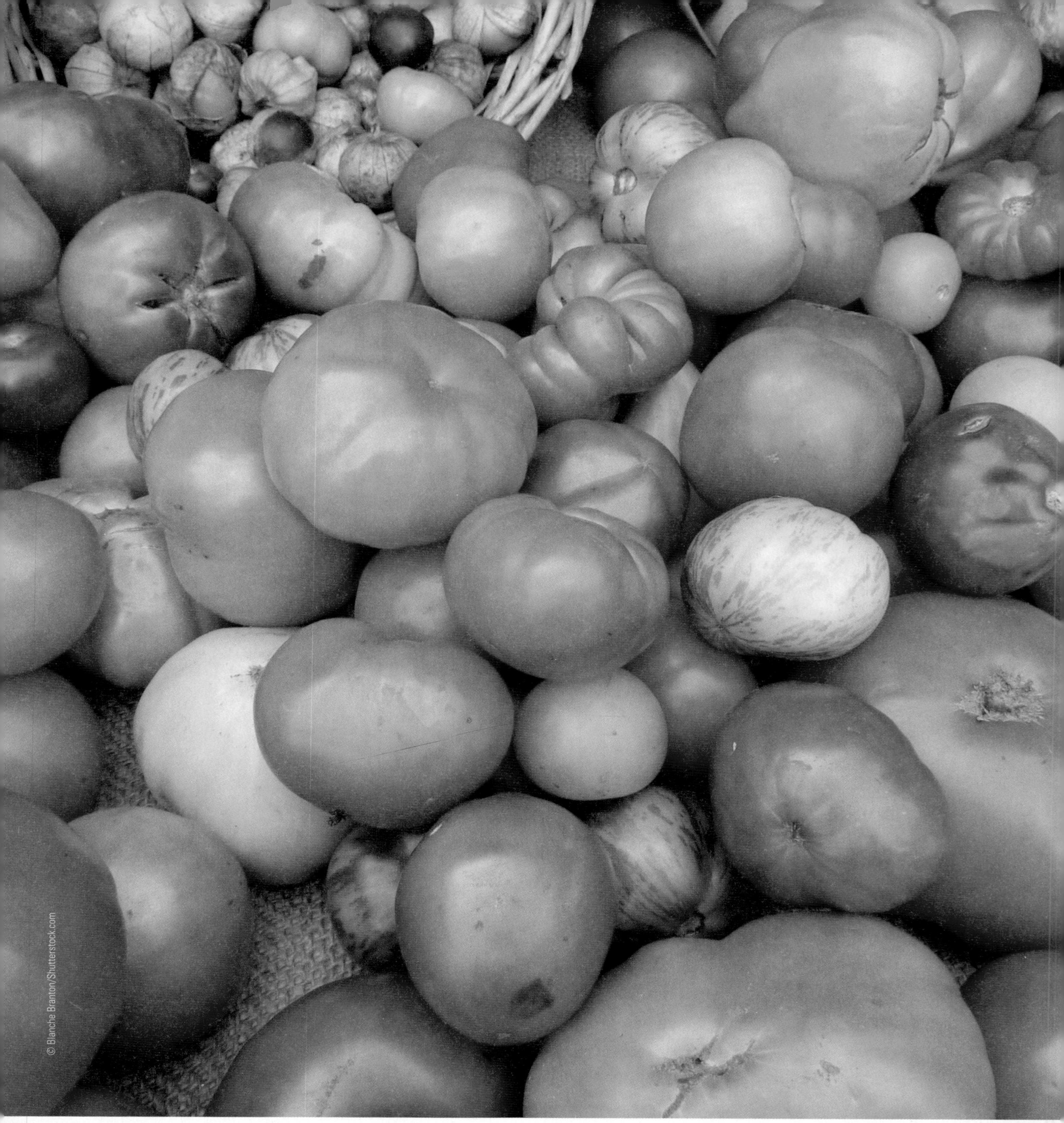

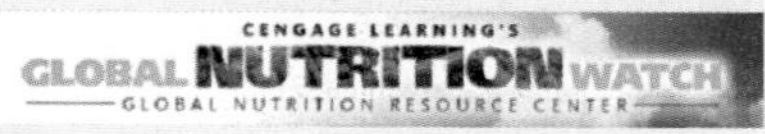

The science of nutrition depends on many other fields of research. Access Global Nutrition Watch and browse some of the recent articles listed on your dashboard. What fields of science are at the foundation of this nutrition research?

1 An Overview of Nutrition

LEARNING GPS

Nutrition in Your Life

Believe it or not, you have probably eaten at least 20,000 meals in your life. Without any conscious effort on your part, your body uses the nutrients from those foods to make all its components, fuel all its activities, and defend itself against diseases. How successfully your body handles these tasks depends, in part, on your food choices. Nutritious food choices support healthy bodies. In the Nutrition Portfolio at the end of this chapter, you can see how your current food choices are influencing your health and risk of chronic diseases.

Although you may not always have been aware of it, **nutrition** has played a significant role in your life. Every day, several times a day, you select **foods** that influence your body's health for better or worse. Each day's food choices may benefit or harm your health only a little, but when these choices are repeated over years and decades, the rewards or consequences become major. That being the case, paying close attention to good eating habits now supports health benefits later. Conversely, carelessness about food choices can contribute to many **chronic diseases,** including heart disease, diabetes, and cancer. Of course, some people will become ill or die young no matter what choices they make, and others will live long lives despite making poor choices. For the majority of us, however, the food choices we make each and every day will benefit or impair our health in proportion to the wisdom of those choices.

Although most people realize that their food habits affect their health, they often choose foods for other reasons. After all, foods bring to the table a variety of pleasures, traditions, and associations as well as nourishment. The challenge, then, is to combine favorite foods and fun times with a nutritionally balanced **diet.**

nutrition: the science of the nutrients in foods and their actions within the body. A broader definition includes the study of human behaviors related to food and eating.

foods: products derived from plants or animals that can be taken into the body to yield energy and nutrients for the maintenance of life and the growth and repair of tissues.

chronic diseases: diseases characterized by slow progression and long duration. Examples include heart disease, diabetes, and some cancers.
- **chronos** = time

diet: the foods and beverages a person eats and drinks.

Before learning more about creating a healthful diet, take a moment to review the definition of *diet*. Note that *diet* does *not* mean a restrictive food plan designed for weight loss. It simply refers to the foods and beverages a person consumes. Whether it's a vegetarian diet, a weight-loss diet, or any other kind of diet depends on the types of foods and beverages a person chooses.

1.1 Food Choices

LEARN IT Describe how various factors influence personal food choices.

People decide what to eat, when to eat, how much to eat, and even whether to eat in highly personal ways based on a complex interaction of genetic, behavioral, or social factors rather than on an awareness of nutrition's importance to health.[1] A variety of food choices can support good health, and an understanding of human nutrition helps you make sensible selections more often.

Preferences As you might expect, the number one reason most people choose certain foods is taste—they like the flavor. Two widely shared preferences are for the sweetness of sugar and the savoriness of salt. Liking high-fat foods also appears to be a universally common preference. Other preferences might be for the hot peppers common in Mexican cooking or the curry spices of Indian cuisine. Research suggests that genetics may influence taste perceptions and therefore food likes and dislikes.[2] Similarly, the hormones of pregnancy seem to influence food cravings and aversions (see Chapter 15).

Habit People sometimes select foods out of habit. They eat cereal every morning, for example, simply because they have always eaten cereal for breakfast. Eating a familiar food and not having to make any decisions can be comforting.

Ethnic Heritage and Regional Cuisines Among the strongest influences on food choices are ethnic heritage and regional cuisines. People tend to prefer the foods they grew up eating. Every country, and in fact every region of a country, has its own typical foods and ways of combining them into meals. These cuisines reflect a unique combination of local ingredients and cooking styles. Chowder in New England is made with clams, but in the Florida Keys conch is the featured ingredient. The Pacific Northwest is as famous for its marionberry pie as Georgia is for its peach cobbler. Philly has its cheesesteaks and New Orleans has its oyster po'boys. The "American diet" includes many ethnic foods and regional styles, all adding variety to the diet.

© Corbis Premium RF/Alamy

An enjoyable way to learn about a culture is to taste the ethnic foods.

Enjoying traditional **ethnic foods** provides an opportunity to celebrate a person's heritage. People offering ethnic foods share a part of their culture with others, and those accepting the foods learn about another's way of life. Developing **cultural competence** honors individual preferences and is particularly important for professionals who help others plan healthy diets.[3]

Social Interactions Most people enjoy companionship while eating. It's fun to go out with friends for a meal or share a snack when watching a movie together. Meals are often social events, and sharing food is part of hospitality. Social customs invite people to accept food or drink offered by a host or shared by a group—regardless of hunger signals. Chapter 9 describes how people tend to eat more food when socializing with others.

Availability, Convenience, and Economy People often eat foods that are accessible, quick and easy to prepare, and within their financial means. Consumers who value convenience frequently eat out, bring home ready-to-eat meals, or have food delivered. Even when they venture into the kitchen, they want to prepare a meal in 15 to 20 minutes, using less than a half dozen ingredients—and those "ingredients" are often semiprepared foods, such as canned soups.

Consumer emphasis on convenience limits food choices to the selections offered on menus and products designed for quick preparation. Whether decisions

ethnic foods: foods associated with particular cultural groups.

cultural competence: having an awareness and acceptance of cultures and the ability to interact effectively with people of diverse cultures.

based on convenience meet a person's nutrition needs depends on the choices made. Eating a banana or a candy bar may be equally convenient, but the fruit provides more vitamins and minerals and less sugar and fat.

Rising food costs have shifted some consumers' priorities and changed their shopping habits. They are less likely to buy higher priced convenience foods and more likely to buy less-expensive store brand items and prepare home-cooked meals.[4] In fact, more than 80 percent of US consumers are eating home-cooked meals at least three times a week.[5] Those who frequently prepare their own meals eat fast food less often and are more likely to meet dietary guidelines for fat, calcium, fruits, vegetables, and whole grains. Not surprisingly, when eating out, consumers choose low-cost fast-food outlets over more expensive fine-dining restaurants.[6]

Positive and Negative Associations People tend to like particular foods associated with happy occasions—such as hot dogs at ball games or cake and ice cream at birthday parties. By the same token, people can develop aversions and dislike foods that they ate when they felt sick or that they were forced to eat in negative situations. Similarly, children learn to like and dislike certain foods when their parents use foods as rewards or punishments. Negative experiences can have long-lasting influences on food preferences. More than 50 years after World War II, veterans who had experienced intense combat in the Pacific dislike Chinese and Japanese food significantly more than their peers who were not engaged in battle or those who fought elsewhere.[7]

Emotions Some people cannot eat when they are emotionally upset. Others may eat in response to a variety of emotional stimuli—for example, to relieve boredom or depression or to calm anxiety. A depressed person may choose to eat rather than to call a friend. A person who has returned home from an exciting evening out may unwind with a late-night snack. These people may find emotional comfort, in part, because foods can influence the brain's chemistry and the mind's response. Carbohydrates and alcohol, for example, tend to calm, whereas proteins and caffeine are more likely to stimulate. Eating in response to emotions and stress can easily lead to overeating and obesity, but it may be helpful at times.[8] For example, sharing food at times of bereavement serves both the giver's need to provide comfort and the receiver's need to be cared for and to interact with others as well as to take nourishment.

Values Food choices may reflect people's religious beliefs, political views, or environmental concerns. For example, some Christians forgo meat on Fridays during Lent (the period prior to Easter), Jewish law includes an extensive set of dietary rules that govern the use of foods derived from animals, and Muslims fast between sunrise and sunset during Ramadan (the ninth month of the Islamic calendar). Some vegetarians select foods based on their concern for animal rights. A concerned consumer may boycott fruit picked by migrant workers who have been exploited. People may buy vegetables from local farmers to save the fuel and environmental costs of foods shipped from far away. They may also select foods packaged in containers that can be reused or recycled. Some consumers accept or reject foods that have been irradiated, grown organically, or genetically modified, depending on their approval of these processes (see Chapter and Highlight 19 for a complete discussion).

To enhance your health, keep nutrition in mind when selecting foods. To protect the environment, shop at local markets and reuse cloth shopping bags.

Body Weight and Image Sometimes people select certain foods and supplements that they believe will improve their physical appearance and avoid those they believe might be detrimental. Such decisions can be beneficial when based on sound nutrition and fitness knowledge, but decisions based on fads or carried to extremes undermine good health, as pointed out in later discussions of eating disorders (Highlight 8) and dietary supplements commonly used by athletes (Highlight 14).

Nutrition and Health Benefits Finally, of course, many consumers make food choices they believe will improve their health.[9] Food manufacturers and restaurant chefs have responded to scientific findings linking health with nutrition by offering an abundant selection of health-promoting foods and beverages. Foods that provide health benefits beyond their nutrient contributions are called **functional foods.**[10] Functional foods may include whole foods, modified foods, or fortified foods. Whole foods—as natural and familiar as oatmeal or tomatoes—are the simplest functional foods. In other cases, foods have been modified to provide health benefits, perhaps by lowering the fat contents. In still other cases, manufacturers have fortified foods by adding nutrients or **phytochemicals** that provide health benefits (see Highlight 13). Examples of these functional foods include orange juice fortified with calcium to help build strong bones and margarine made with a plant sterol that lowers blood cholesterol.

Consumers typically welcome new foods into their diets, provided that these foods are reasonably priced, clearly labeled, easy to find in the grocery store, and convenient to prepare. These foods must also taste good—as good as the traditional choices. Of course, a person need not eat any "special" foods to enjoy a healthy diet; many "regular" foods provide numerous health benefits as well. In fact, "regular" foods such as whole grains; vegetables and legumes; fruits; seafood, meats, poultry, eggs, nuts, and seeds; and milk products are among the healthiest choices a person can make.

REVIEW IT Describe how various factors influence personal food choices.
A person selects foods for a variety of reasons. Whatever those reasons may be, food choices influence health. Individual food selections neither make nor break a diet's healthfulness, but the balance of foods selected over time can make an important difference to health.[11] For this reason, people are wise to think "nutrition" when making their food choices.

Foods bring pleasure—and nutrients.

1.2 The Nutrients

LEARN IT Name the six major classes of nutrients and identify which are organic and which yield energy.

Biologically speaking, people eat to receive nourishment. Do you ever think of yourself as a biological being made of carefully arranged atoms, molecules, cells, tissues, and organs? Are you aware of the activity going on within your body even as you sit still? The atoms, molecules, and cells of your body continuously move and change, even though the structures of your tissues and organs and your external appearance remain relatively constant. The ongoing growth, maintenance, and repair of the body's tissues depend on the **energy** and the **nutrients** received from foods.

Nutrients in Foods and in the Body

Amazingly, our bodies can derive all the energy, structural materials, and regulating agents we need from the foods we eat. This section introduces the nutrients that foods deliver and shows how they participate in the dynamic processes that keep people alive and well.

Nutrient Composition of Foods Chemical analysis of a food such as a tomato shows that it is composed primarily of water (95 percent). Most of the solid materials are carbohydrates, lipids (fats), and proteins. If you could remove these materials, you would find a tiny residue of vitamins, minerals, and other compounds. Water, carbohydrates, lipids, proteins, vitamins, and some of the minerals found in foods represent the six classes ♦ of nutrients—substances the body uses for the growth, maintenance, and repair of its tissues.

This book focuses mostly on the nutrients, but foods contain other compounds as well—fibers, phytochemicals, pigments, additives, alcohols, and others. Some are beneficial, some are neutral, and a few are harmful. Later sections of the book touch on these compounds and their significance.

Nutrient Composition of the Body A chemical analysis of your body would show that it is made of materials similar to those found in foods (see Figure 1-1). A healthy

functional foods: foods that contain bioactive components that provide health benefits beyond their nutrient contributions.

phytochemicals (FIE-toe-KEM-ih-cals): nonnutrient compounds found in plants. Some phytochemicals have biological activity in the body.

- **phyto** = plant

energy: the capacity to do work. The energy in food is chemical energy. The body can convert this chemical energy to mechanical, electrical, or heat energy.

nutrients: chemical substances obtained from food and used in the body to provide energy, structural materials, and regulating agents to support growth, maintenance, and repair of the body's tissues. Nutrients may also reduce the risks of some diseases.

♦ Six classes of nutrients:
- Carbohydrates
- Lipids (fats)
- Proteins
- Vitamins
- Minerals
- Water

> FIGURE 1-1 Body Composition of Healthy-Weight Men and Women

The human body is made of compounds similar to those found in foods—mostly water (60 percent) and some fat (18 to 21 percent for young men, 23 to 26 percent for young women), with carbohydrate, protein, vitamins, minerals, and other minor constituents making up the remainder. (Chapter 8 describes the health hazards of too little or too much body fat.)

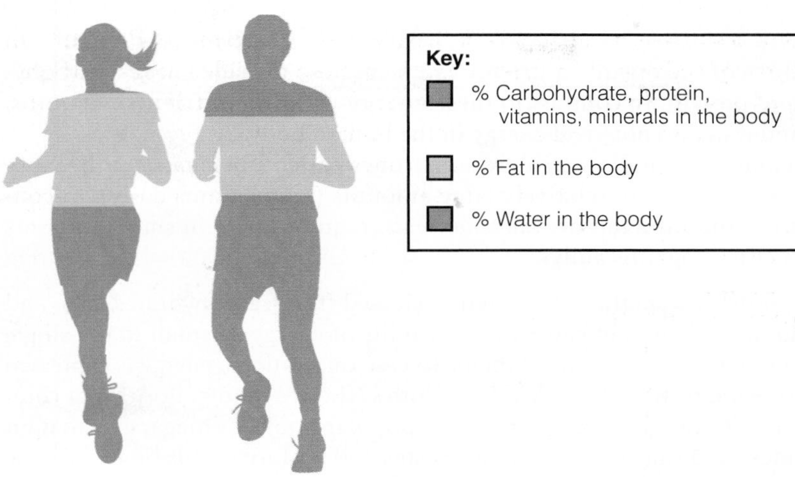

150-pound body contains about 90 pounds of water and about 20 to 45 pounds of fat. The remaining pounds are mostly protein, carbohydrate, and the major minerals of the bones. Vitamins, other minerals, and incidental extras constitute a fraction of a pound.

Chemical Composition of Nutrients The simplest of the nutrients are the minerals. Each mineral is a chemical element; its atoms are all alike. As a result, its identity never changes. For example, iron may have different electrical charges, but the individual iron atoms remain the same when they are in a food, when a person eats the food, when the iron becomes part of a red blood cell, when the cell is broken down, and when the iron is lost from the body by excretion. The next simplest nutrient is water, a compound made of two elements—hydrogen and oxygen. Minerals and water are **inorganic** nutrients—which means they do not contain carbon.

The other four classes of nutrients (carbohydrates, lipids, proteins, and vitamins) are more complex. In addition to hydrogen and oxygen, they all contain carbon, an element found in all living things. They are therefore called **organic** compounds (meaning, literally, "alive"). This chemical definition of *organic* differs from the agricultural definition. As Chapter 19 explains, organic farming refers to growing crops and raising livestock according to standards set by the US Department of Agriculture (USDA). Protein and some vitamins also contain nitrogen and may contain other elements such as sulfur as well (see Table 1-1).

TABLE 1-1 Elements in the Six Classes of Nutrients

Notice that organic nutrients contain carbon.

	Carbon	Hydrogen	Oxygen	Nitrogen	Minerals
Inorganic Nutrients					
Minerals					✓
Water		✓	✓		
Organic Nutrients					
Carbohydrate	✓	✓	✓		
Lipid (fat)	✓	✓	✓		
Protein[a]	✓	✓	✓	✓	
Vitamins[b]	✓	✓	✓		

[a]Some proteins also contain the mineral sulfur.
[b]Some vitamins contain nitrogen; some contain minerals.

inorganic: not containing carbon or pertaining to living things.
- **in** = not

organic: in chemistry, a substance or molecule containing carbon-carbon bonds or carbon-hydrogen bonds. This definition excludes coal, diamonds, and a few carbon-containing compounds that contain only a single carbon and no hydrogen, such as carbon dioxide (CO_2), calcium carbonate ($CaCO_3$), magnesium carbonate ($MgCO_3$), and sodium cyanide (NaCN).

Essential Nutrients The body can make some nutrients, but it cannot make all of them. Also, it makes some in insufficient quantities to meet its needs and, therefore, must obtain these nutrients from foods. The nutrients that foods must supply are **essential nutrients.** When used to refer to nutrients, the word *essential* means more than just "necessary"; it means "needed from outside the body"—normally, from foods.

The Energy-Yielding Nutrients: Carbohydrate, Fat, and Protein In the body, three of the organic nutrients can be used to provide energy: carbohydrate, fat, and protein. In contrast to these **energy-yielding nutrients,** vitamins, minerals, and water do not yield energy in the human body.

Carbohydrate, fat, and protein are sometimes called *macronutrients* because the body requires them in relatively large amounts (many grams daily). In contrast, vitamins and minerals are *micronutrients,* required only in small amounts (milligrams or micrograms daily).

TABLE 1-2 kCalorie Values of Energy Nutrients

Nutrients	Energy (kcal/g)
Carbohydrate	4
Fat	9
Protein	4

NOTE: Alcohol contributes 7 kcalories per gram that can be used for energy, but it is not considered a nutrient because it interferes with the body's growth, maintenance, and repair.

Energy Measured in kCalories The energy released from carbohydrates, fats, and proteins can be measured in **calories**—tiny units of energy so small that a single apple provides tens of thousands of them. To ease calculations, energy is expressed in 1000-calorie metric units known as kilocalories (shortened to kcalories, but commonly called "calories"). When you read in popular books or magazines that an apple provides "100 calories," it actually means 100 kcalories. This book uses the term *kcalorie* and its abbreviation *kcal* throughout, as do other scientific books and journals. The "How To" on p. 9 provides a few tips on "thinking metric."

Energy from Foods The amount of energy a food provides depends on how much carbohydrate, fat, and protein it contains. ♦ When completely broken down in the body, a gram of carbohydrate yields about 4 kcalories of energy; a gram of protein also yields 4 kcalories; and a gram of fat yields 9 kcalories (see Table 1-2).* The "How To" on p. 10 explains how to calculate the energy available from foods.

♦ Energy-yielding nutrients:
- Carbohydrate
- Fat
- Protein

Because fat provides more energy per gram, it has a greater **energy density** than either carbohydrate or protein. Figure 1-2 compares the energy density of two breakfast options, and later chapters describe how foods with a high energy

> **FIGURE 1-2 Energy Density of Two Breakfast Options Compared**

Gram for gram, ounce for ounce, and bite for bite, foods with a high energy density deliver more kcalories than foods with a low energy density. Both of these breakfast options provide 500 kcalories, but the cereal with milk, fruit salad, scrambled egg, turkey sausage, and toast with jam offers three times as much food as the doughnuts (based on weight); it has a lower energy density than the doughnuts. Selecting a variety of foods also helps to ensure nutrient adequacy.

LOWER ENERGY DENSITY
This 450-gram breakfast delivers 500 kcalories, for an energy density of 1.1 (500 kcal ÷ 450 g = 1.1 kcal/g).

HIGHER ENERGY DENSITY
This 144-gram breakfast delivers 500 kcalories, for an energy density of 3.5 (500 kcal ÷ 144 g = 3.5 kcal/g).

*For those using kilojoules: 1 g carbohydrate = 17 kJ; 1 g protein = 17 kJ; 1 g fat = 37 kJ; and 1 g alcohol = 29 kJ.

essential nutrients: nutrients a person must obtain from food because the body cannot make them for itself in sufficient quantity to meet physiological needs; also called *indispensable nutrients.* About 40 nutrients are currently known to be essential for human beings.

energy-yielding nutrients: the nutrients that break down to yield energy the body can use:
- Carbohydrate
- Fat
- Protein

calories: units by which energy is measured. Energy provided by foods and beverages is measured in *kilocalories* (1000 calories equal 1 kilocalorie), abbreviated *kcalories* or *kcal.* One kcalorie is the amount of heat necessary to raise the temperature of 1 kilogram (kg) of water 1°C. The scientific use of the term *kcalorie* is the same as the popular use of the term *calorie.*

energy density: a measure of the energy a food provides relative to the weight of the food (kcalories per gram).

density help with weight *gain*, whereas those with a low energy density help with weight *loss*.

One other substance contributes energy—alcohol. Alcohol, however, is not considered a nutrient. Unlike the essential nutrients, alcohol does not sustain life. In fact, it interferes with the growth, maintenance, and repair of the body. Its only common characteristic with nutrients is that it yields energy (7 kcalories per gram) when metabolized in the body.

>How To

Think Metric

Like other scientists, nutrition scientists use metric units of measure. They measure food energy in kilocalories, people's height in centimeters, people's weight in kilograms, and the weights of foods and nutrients in grams, milligrams, or micrograms. For ease in using these measures, it helps to remember that the prefixes on the grams imply 1000. For example, a *kilo*gram is 1000 grams, a *milli*gram is 1/1000 of a gram, and a *micro*gram is 1/1000 of a milligram.

Most food labels and many recipes provide "dual measures," listing both household measures, such as cups, quarts, and teaspoons, and metric measures, such as milliliters, liters, and grams. This practice gives people an opportunity to gradually learn to "think metric."

A person might begin to "think metric" by simply observing the measure—by noticing the amount of soda in a 2-liter bottle, for example. Through such experiences, a person can become familiar with a measure without having to do any conversions.

The international unit for measuring food energy is the *joule*—the amount of energy expended when 1 kilogram is moved 1 meter by a force of 1 newton. The joule is thus a measure of *work* energy, whereas the kcalorie is a measure of *heat* energy. While many scientists and journals report their findings in kilojoules (kJ), many others, particularly those in the United States, use kcalories (kcal). To convert energy measures from kcalories to kilojoules, multiply by 4.2; to convert kilojoules to kcalories, multiply by 0.24. For example, a 50-kcalorie cookie provides 210 kilojoules:

$$50 \text{ kcal} \times 4.2 = 210 \text{ kJ}$$

Appendix K provides assistance and conversion factors for these and other units of measure.

Volume: Liters (L)

1 L = 1000 milliliters (mL)
0.95 L = 1 quart
1 mL = 0.03 fluid ounces
240 mL = 1 cup

A liter of liquid is approximately one US quart. (Four liters are only about 5 percent more than a gallon.)

One cup is about 240 milliliters; a half-cup of liquid is about 120 milliliters.

Weight: Grams (g)

1 g = 1000 milligrams (mg)
1 g = 0.04 ounce (oz)
1 oz = 28.35 g (or 30 g)
100 g = 3½ oz
1 kilogram (kg) = 1000 g
1 kg = 2.2 pounds (lb)
454 g = 1 lb

A kilogram is slightly more than 2 lb; conversely, a pound is about ½ kg.

A half-cup of vegetables weighs about 100 grams; one pea weighs about ½ gram.

A 5-pound bag of potatoes weighs about 2 kilograms, and a 176-pound person weighs 80 kilograms.

TRY IT Convert your body weight from pounds to kilograms and your height from inches to centimeters.

>How To

Calculate the Energy Available from Foods

To calculate the energy available from a food, multiply the number of grams of carbohydrate, protein, and fat by 4, 4, and 9, respectively. Then add the results together. For example, 1 slice of bread with 1 tablespoon of peanut butter on it contains 16 grams carbohydrate, 7 grams protein, and 9 grams fat:

$$
\begin{aligned}
16 \text{ g carbohydrate} \times 4 \text{ kcal/g} &= 64 \text{ kcal} \\
7 \text{ g protein} \times 4 \text{ kcal/g} &= 28 \text{ kcal} \\
9 \text{ g fat} \times 9 \text{ kcal/g} &= 81 \text{ kcal} \\
\text{Total} &= 173 \text{ kcal}
\end{aligned}
$$

From this information, you can calculate the percentage of kcalories each of the energy nutrients contributes to the total. To determine the percentage of kcalories from fat, for example, divide the 81 fat kcalories by the total 173 kcalories:

$$81 \text{ fat kcal} \div 173 \text{ total kcal} = 0.468$$
(rounded to 0.47)

Then multiply by 100 to get the percentage:

$$0.47 \times 100 = 47\%$$

Dietary recommendations that urge people to limit fat intake to 20 to 35 percent of kcalories refer to the day's total energy intake, not to individual foods. Still, if the proportion of fat in each food choice throughout a day exceeds 35 percent of kcalories, then the day's total surely will, too. Knowing that this snack provides 47 percent of its kcalories from fat alerts a person to the need to make lower-fat selections at other times that day.

TRY IT Calculate the energy available from a bean burrito with cheese (55 grams carbohydrate, 15 grams protein, and 12 grams fat). Determine the percentage of kcalories from each of the energy nutrients.

Most foods contain all three energy-yielding nutrients, as well as vitamins, minerals, water, and other substances. For example, meat contains water, fat, vitamins, and minerals as well as protein. Bread contains water, a trace of fat, a little protein, and some vitamins and minerals in addition to its carbohydrate. Only a few foods are exceptions to this rule, the common ones being sugar (pure carbohydrate) and oil (essentially pure fat).

Energy in the Body When the body uses carbohydrate, fat, or protein to fuel its activities, the bonds between the nutrient's atoms break. As the bonds break, they release energy. Some of this energy is released as heat, but some is used to send electrical impulses through the brain and nerves, to synthesize body compounds, and to move muscles. Thus the energy from food supports every activity from quiet thought to vigorous sports.

If the body does not use these nutrients to fuel its current activities, it converts them into storage compounds (such as body fat), to be used between meals and overnight when fresh energy supplies run low. If more energy is consumed than expended, the result is an increase in energy stores and weight gain. Similarly, if less energy is consumed than expended, the result is a decrease in energy stores and weight loss.

When consumed in excess of energy needs, alcohol, too, can be converted to body fat and stored. When alcohol contributes a substantial portion of the energy in a person's diet, the harm it does far exceeds the problems of excess body fat. (Highlight 7 describes the effects of alcohol on health and nutrition.)

Other Roles of Energy-Yielding Nutrients In addition to providing energy, carbohydrates, fats, and proteins provide the raw materials for building the body's tissues and regulating its many activities. In fact, protein's role as a fuel source is relatively minor compared with both the other two energy-yielding nutrients and its other roles. Proteins are found in structures such as the muscles and skin and help to regulate activities such as digestion and energy metabolism. (Chapter 6 presents a full discussion on proteins.)

The Vitamins The **vitamins** are also organic, but they do not provide energy. Instead, they facilitate the release of energy from carbohydrate, fat, and protein and participate in numerous other activities throughout the body.

Each of the 13 vitamins has its own special roles to play.* One vitamin enables the eyes to see in dim light, another helps protect the lungs from air pollution, and still another helps make the sex hormones—among other things. When you cut yourself, one vitamin helps stop the bleeding and another helps repair the skin. Vitamins busily help replace old red blood cells and the lining of the digestive tract. Almost every action in the body requires the assistance of vitamins.

Vitamins can function only if they are intact, but because they are complex organic molecules, they are vulnerable to destruction by heat, light, and chemical agents. This is why the body handles them carefully, and why nutrition-wise cooks do, too. The strategies of cooking vegetables at moderate temperatures for short times and using small amounts of water help to preserve the vitamins.

The Minerals In the body, some **minerals** are put together in orderly arrays in such structures as bones and teeth. Minerals are also found in the fluids of the body, which influences fluid balance and distribution. Whatever their roles, minerals do not yield energy.

Only 16 minerals are known to be essential in human nutrition.** Others are being studied to determine whether they play significant roles in the human body. Still other minerals, such as lead, are environmental contaminants that displace the nutrient minerals from their workplaces in the body, disrupting body functions. The problems caused by contaminant minerals are described in Chapter 13.

Because minerals are inorganic, they are indestructible and need not be handled with the special care that vitamins require. Minerals can, however, be bound by substances that interfere with the body's ability to absorb them. They can also be lost during food-refining processes or during cooking when they leach into water that is discarded.

Water Water provides the environment in which nearly all the body's activities are conducted. It participates in many metabolic reactions and supplies the medium for transporting vital materials to cells and carrying waste products away from them. Water is discussed fully in Chapter 12, but it is mentioned in every chapter. If you watch for it, you cannot help but be impressed by water's participation in all life processes.

Water is an essential nutrient and naturally carries varying amounts of several minerals.

REVIEW IT Name the six major classes of nutrients and identify which are organic and which yield energy.

Foods provide nutrients—substances that support the growth, maintenance, and repair of the body's tissues. The six classes of nutrients include:

- Carbohydrates
- Lipids (fats)
- Proteins
- Vitamins
- Minerals
- Water

Foods rich in the energy-yielding nutrients (carbohydrate, fat, and protein) provide the major materials for building the body's tissues and yield energy for the body's use or storage. Energy is measured in kcalories—a measure of heat energy. Vitamins, minerals, and water do not yield energy; instead they facilitate a variety of activities in the body.

*The water-soluble vitamins are vitamin C and the eight B vitamins: thiamin, riboflavin, niacin, vitamins B_6 and B_{12}, folate, biotin, and pantothenic acid. The fat-soluble vitamins are vitamins A, D, E, and K. The water-soluble vitamins are the subject of Chapter 10 and the fat-soluble vitamins, of Chapter 11.

**The major minerals are calcium, phosphorus, potassium, sodium, chloride, magnesium, and sulfate. The trace minerals are iron, iodine, zinc, chromium, selenium, fluoride, molybdenum, copper, and manganese. Chapters 12 and 13 are devoted to the major and trace minerals, respectively.

vitamins: organic, essential nutrients required in small amounts by the body for health.

minerals: inorganic elements. Some minerals are essential nutrients required in small amounts by the body for health.

Without exaggeration, nutrients provide the physical and metabolic basis for nearly all that we are and all that we do. The next section introduces the science of nutrition with emphasis on the research methods scientists have used in uncovering the wonders of nutrition.

1.3 The Science of Nutrition

LEARN IT Explain the scientific method and how scientists use various types of research studies and methods to acquire nutrition information.

The science of nutrition is the study of the nutrients and other substances in foods and the body's handling of them. Its foundation depends on several other sciences, including biology, biochemistry, and physiology. As sciences go, nutrition is young, but as you can see from the size of this book, much has happened in nutrition's short life. And it is currently experiencing a tremendous growth spurt as scientists apply knowledge gained from sequencing the human **genome.** The integration of nutrition, genomics, and molecular biology has opened a whole new world of study called **nutritional genomics**—the science of how nutrients affect the activities of genes and how genes affect the interactions between diet and disease. Highlight 6 describes how nutritional genomics is shaping the science of nutrition, and examples of nutrient–gene interactions appear throughout later sections of the book.

Conducting Research Consumers sometimes depend on personal experience or reports from friends to gather information on nutrition. Such a personal account of an experience or event is known as an **anecdote** and is not accepted as reliable scientific information. In contrast, researchers use the scientific method to guide their work (see Figure 1-3). As the figure shows, research always begins with a problem or a question. For example, "What foods or nutrients might protect against the common cold?" In search of an answer, scientists make an educated guess **(hypothesis),** such as "foods rich in vitamin C reduce the number of common colds." Then they systematically conduct research studies to collect data that will test the hypothesis (see the glossary on p. 13 for definitions of research terms). Some examples of various types of research designs are presented in Figure 1-4 on p. 14. Because each type of study has strengths and weaknesses, some provide stronger evidence than others, as Figure 1-4 explains.

In attempting to discover whether a nutrient relieves symptoms or cures a disease, researchers deliberately manipulate one variable (for example, the amount of vitamin C in the diet) and measure any observed changes (perhaps the number of colds). As much as possible, all other conditions are held constant. The following paragraphs illustrate how this is accomplished.

Controls In studies examining the effectiveness of vitamin C, researchers typically divide the **subjects** into two groups. One group (the **experimental group**) receives a vitamin C supplement, and the other (the **control group**) does not. Researchers observe both groups to determine whether one group has fewer, milder, or shorter colds than the other. The following discussion describes some of the pitfalls inherent in an experiment of this kind and ways to avoid them.

In sorting subjects into two groups, researchers must ensure that each person has an equal chance of being assigned to either the experimental group or the control group. This is accomplished by **randomization;** that is, the subjects are chosen randomly from the same population by flipping a coin or some other method involving chance. Randomization helps to ensure that the two groups are "equal" and that observed differences reflect the treatment and not other factors.

Importantly, the two groups of people must be similar and must have the same track record with respect to colds to rule out the possibility that observed differences in the rate, severity, or duration of colds might have occurred anyway. If, for example, the control group would normally catch twice as many colds as the experimental group, then the findings prove nothing.

genome (GEE-nome): the complete set of genetic material (DNA) in an organism or a cell. The study of genomes is called *genomics*.

nutritional genomics: the science of how nutrients affect the activities of genes (*nutrigenomics*) and how genes affect the interactions between diet and disease (*nutrigenetics*).

> FIGURE 1-3 The Scientific Method

Research scientists follow the scientific method. Note that most research generates new questions, not final answers. Thus the sequence begins anew, and research continues in a somewhat cyclical way.

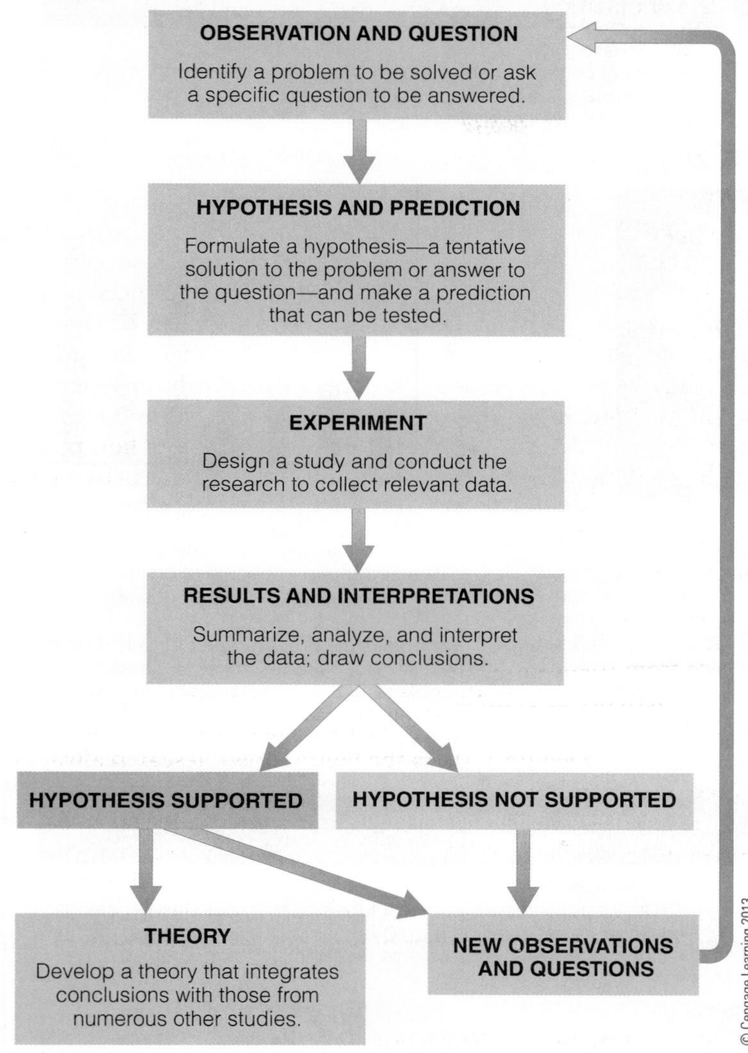

GLOSSARY OF RESEARCH TERMS

anecdote: a personal account of an experience or event; not reliable scientific information.

blind experiment: an experiment in which the subjects do not know whether they are members of the experimental group or the control group.

control group: a group of individuals similar in all possible respects to the experimental group except for the treatment. Ideally, the control group receives a placebo while the experimental group receives a real treatment.

correlation (CORE-ee-LAY-shun): the simultaneous increase, decrease, or change in two variables. If A increases as B increases, or if A decreases as B decreases, the correlation is *positive.* (This does not mean that A causes B or vice versa.) If A increases as B decreases, or if A decreases as B increases, the correlation is *negative.* (This does not mean that A prevents B or vice versa.) Some third factor may account for both A and B.

double-blind experiment: an experiment in which neither the subjects nor the researchers know which subjects are members of the experimental group and which are serving as control subjects, until after the experiment is over.

experimental group: a group of individuals similar in all possible respects to the control group except for the treatment. The experimental group receives the real treatment.

hypothesis (hi-POTH-eh-sis): an unproven statement that tentatively explains the relationships between two or more variables.

peer review: a process in which a panel of scientists rigorously evaluates a research study to assure that the scientific method was followed.

placebo (pla-SEE-bo): an inert, harmless medication given to provide comfort and hope; a sham treatment used in controlled research studies.

placebo effect: a change that occurs in response to expectations about the effectiveness of a treatment that actually has no pharmaceutical effects.

randomization (RAN-dom-ih-ZAY-shun): a process of choosing the members of the experimental and control groups without bias.

replication (REP-lih-KAY-shun): repeating an experiment and getting the same results.

subjects: the people or animals participating in a research project.

theory: a tentative explanation that integrates many and diverse findings to further the understanding of a defined topic.

validity (va-LID-ih-tee): having the quality of being founded on fact or evidence.

variables: factors that change. A variable may depend on another variable (for example, a child's height depends on his age), or it may be independent (for example, a child's height does not depend on the color of her eyes). Sometimes both variables correlate with a third variable (a child's height and eye color both depend on genetics).

> FIGURE 1-4 Examples of Research Designs

EPIDEMIOLOGICAL STUDIES research the incidence, distribution, and control of disease in a population. Epidemiological studies include cross-sectional, case-control, and cohort studies.

Strengths:
- Can narrow down the list of possible causes
- Can raise questions to pursue through other research

Weaknesses:
- Cannot control variables that may influence the development or the prevention of a disease
- Cannot prove cause and effect

CROSS-SECTIONAL STUDIES

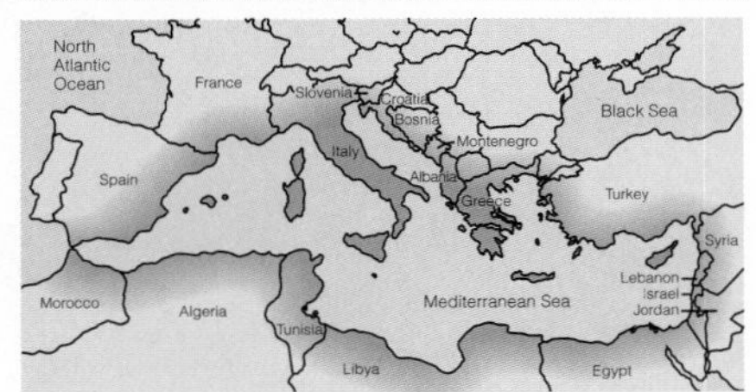

Researchers observe how much and what kinds of foods a group of people eat and how healthy those people are. Their findings identify factors that might influence the incidence of a disease in various populations.

Example. Many people in the Mediterranean region drink more wine, eat more fat from olive oil, and yet have a lower incidence of heart disease than northern Europeans and North Americans.

CASE-CONTROL STUDIES

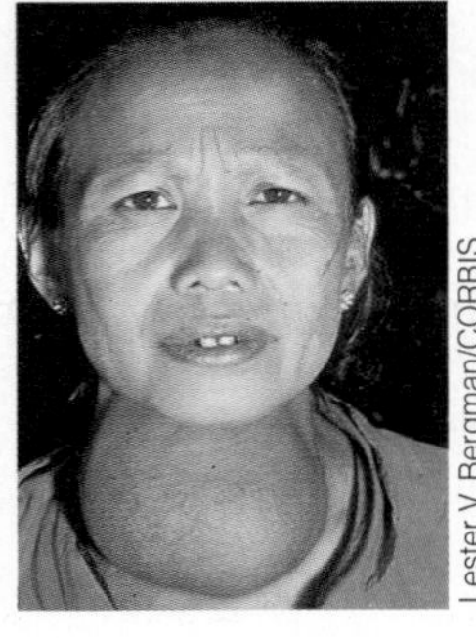
Lester V. Bergman/CORBIS

Researchers compare people who do and do not have a given condition such as a disease, closely matching them in age, gender, and other key variables so that differences in other factors will stand out. These differences may account for the condition in the group that has it.

Example. People with goiter lack iodine in their diets.

COHORT STUDIES

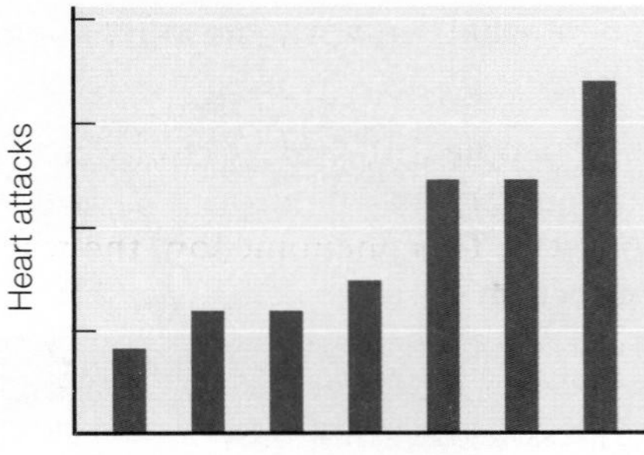

Researchers analyze data collected from a selected group of people (a cohort) at intervals over a certain period of time.

Example. Data collected periodically over the past several decades from more than 5000 people randomly selected from the town of Framingham, Massachusetts, in 1948 have revealed that the risk of heart attack increases as blood cholesterol increases.

EXPERIMENTAL STUDIES test cause-and-effect relationships between variables. Experimental studies include laboratory-based studies—on animals or in test tubes (in vitro)—and human intervention (or clinical) trials.

Strengths:
- Can control conditions (for the most part)
- Can determine effects of a variable
- Can apply some findings on human beings to some groups of human beings

Weaknesses:
- Cannot apply results from test tubes or animals to human beings
- Cannot generalize findings on human beings to all human beings
- Cannot use certain treatments for clinical or ethical reasons

LABORATORY-BASED ANIMAL STUDIES

© R. Benali/Getty Images

Researchers feed animals special diets that provide or omit specific nutrients and then observe any changes in health. Such studies test possible disease causes and treatments in a laboratory where all conditions can be controlled.

Example. Mice fed a high-fat diet eat less food than mice given a lower-fat diet, so they receive the same number of kcalories—but the mice eating the fat-rich diet become severely obese.

LABORATORY-BASED IN VITRO STUDIES

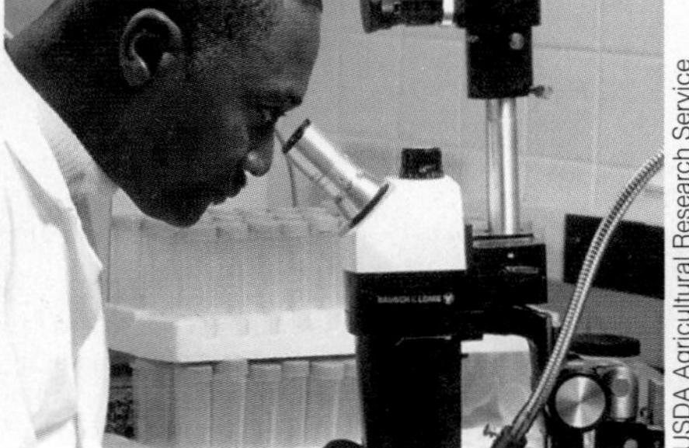
USDA Agricultural Research Service

Researchers examine the effects of a specific variable on a tissue, cell, or molecule isolated from a living organism.

Example. Laboratory studies find that fish oils inhibit the growth and activity of the bacteria implicated in ulcer formation.

HUMAN INTERVENTION (OR CLINICAL) TRIALS

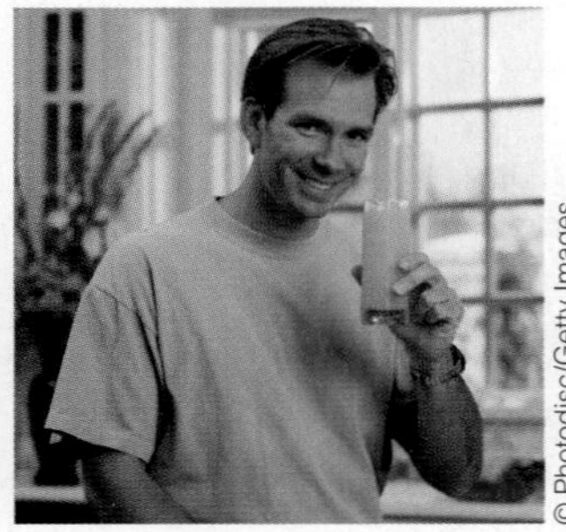
© Photodisc/Getty Images

Researchers ask people to adopt a new behavior (for example, eat a citrus fruit, take a vitamin C supplement, or exercise daily). These trials help determine the effectiveness of such interventions on the development or prevention of disease.

Example. Heart disease risk factors improve when men receive fresh-squeezed orange juice daily for 2 months compared with those on a diet low in vitamin C—even when both groups follow a diet high in saturated fat.

In experiments involving a nutrient, the diets of both groups must also be similar, especially with respect to the nutrient being studied. If those in the experimental group were receiving less vitamin C from their usual diet, then any effects of the supplement may not be apparent.

Sample Size To ensure that chance variation between the two groups does not influence the results, the groups must be large. For example, if one member of a group of five people catches a bad cold by chance, he will pull the whole group's average toward bad colds; but if one member of a group of 500 catches a bad cold, she will not unduly affect the group average. Statistical methods are used to determine whether differences between groups of various sizes support a hypothesis.

Placebos If people who take vitamin C for colds *believe* it will cure them, their chances of recovery may improve. Taking anything believed to be beneficial may hasten recovery. This phenomenon, the result of expectations, is known as the **placebo effect.** In experiments designed to determine vitamin C's effect on colds, this mind-body effect must be rigorously controlled. Severity of symptoms is often a subjective measure, and people who believe they are receiving treatment may report less severe symptoms.

One way experimenters control for the placebo effect is to give pills to all participants. Those in the experimental group, for example, receive pills containing vitamin C, and those in the control group receive a **placebo**—pills of similar appearance and taste containing an inactive ingredient. This way, the expectations of both groups will be equal. It is not necessary to convince all subjects that they are receiving vitamin C, but the extent of belief or unbelief must be the same in both groups. A study conducted under these conditions is called a **blind experiment**—that is, the subjects do not know (are blind to) whether they are members of the experimental group (receiving treatment) or the control group (receiving the placebo).

Double Blind When both the subjects and the researchers do not know which subjects are in which group, the study is called a **double-blind experiment.** Being fallible human beings and having an emotional and sometimes financial investment in a successful outcome, researchers might record and interpret results with a bias in the expected direction. To prevent such bias, the pills are coded by a third party, who does not reveal to the experimenters which subjects are in which group until all results have been recorded.

Analyzing Research Findings

Research findings must be analyzed and interpreted with an awareness of each study's limitations. Scientists must be cautious about drawing any conclusions until they have accumulated a body of evidence from multiple studies that have used various types of research designs. As evidence accumulates, scientists begin to develop a **theory** that integrates the various findings and explains the complex relationships.

Correlations and Causes Researchers often examine the relationships between two or more **variables**—for example, daily vitamin C intake and the number of colds or the duration and severity of cold symptoms. Importantly, researchers must be able to observe, measure, or verify the variables selected. Findings sometimes suggest no **correlation** between variables (regardless of the amount of vitamin C consumed, the number of colds remains the same). Other times, studies find either a **positive correlation** (the more vitamin C, the more colds) or a **negative correlation** (the more vitamin C, the fewer colds). Notice that in a positive correlation, both variables change in the same direction, regardless of whether the direction is "more" or "less"—"the more vitamin C, the more colds" is a positive correlation, just as is "the less vitamin C, the fewer colds." In a negative correlation, the two variables change in opposite directions: "the less vitamin C, the more colds" or "the more vitamin C, the fewer colds." Also notice that a positive correlation does not necessarily reflect a desired outcome, nor does a negative correlation always reflect an unwanted outcome.

Correlational evidence proves only that variables are associated, not that one is the cause of the other. To actually prove that A causes B, scientists have to find evidence of the *mechanism*—that is, an explanation of how A might cause B.

Cautious Conclusions When researchers record and analyze the results of their experiments, they must exercise caution in their interpretation of the findings. For example, in an epidemiological study, scientists may use a specific segment of the population—say, men 18 to 30 years old. When the scientists draw conclusions, they are careful not to generalize the findings to men and women of all ages. Similarly, scientists performing research studies using animals are cautious in applying their findings to human beings. Conclusions from any one research study are always tentative and take into account findings from studies conducted by other scientists as well. As evidence accumulates, scientists gain confidence about making recommendations that affect people's health and lives. Still, their statements are worded cautiously, such as "A diet high in fruits and vegetables *may* protect against *some* cancers."

Quite often, as scientists approach an answer to one research question, they raise several more questions, so future research projects are never lacking. Further scientific investigation then seeks to answer questions, such as "What substance or substances within fruits and vegetables provide protection?" If those substances turn out to be the vitamins found so abundantly in fresh produce, then "How much is needed to offer protection?" "How do these vitamins protect against cancer?" "Is it their action as antioxidant nutrients?" "If not, might it be another action or even another substance that accounts for the protection fruits and vegetables provide against cancer?" (Highlight 11 explores the answers to these questions and reviews recent research on antioxidant nutrients and disease.)

Publishing Research The findings from a research study are submitted to a board of reviewers composed of other scientists who rigorously evaluate the study to assure that the scientific method was followed—a process known as **peer review.** The reviewers critique the study's hypothesis, methodology, statistical significance, and conclusions. They also note the funding source, recognizing that financial support may bias scientific conclusions.[12] If the reviewers consider the conclusions to be well supported by the evidence—that is, if the research has **validity**—they endorse the work for publication in a scientific journal where others can read it. This raises an important point regarding information found on the Internet: much gets published without the rigorous scrutiny of peer review. Consequently, readers must assume greater responsibility for examining the data and conclusions presented—often without the benefit of journal citations. Highlight 1 offers guidance in determining whether website information is reliable. Table 1-3 describes the parts of a typical research article.

Even when a new finding is published or released to the media, it is still only preliminary and not very meaningful by itself. Other scientists will need to confirm or disprove the findings through **replication.** To be accepted into the body of nutrition knowledge, a finding must stand up to rigorous, repeated testing in

TABLE 1-3 Parts of a Research Article

- *Abstract.* The abstract provides a brief overview of the article.
- *Introduction.* The introduction clearly states the purpose of the current study.
- *Review of literature.* A comprehensive review of the literature reveals all that science has uncovered on the subject to date.
- *Methodology.* The methodology section defines key terms and describes the instruments and procedures used in conducting the study.
- *Results.* The results report the findings and may include tables and figures that summarize the information.
- *Conclusions.* The conclusions drawn are those supported by the data and reflect the original purpose as stated in the introduction. Usually, they answer a few questions and raise several more.
- *References.* The references reflect the investigator's knowledge of the subject and should include an extensive list of relevant studies (including key studies several years old as well as current ones).

experiments performed by several different researchers. What we "know" in nutrition results from years of replicating study findings. Communicating the latest finding in its proper context without distorting or oversimplifying the message is a challenge for scientists and journalists alike. For a helpful scientific overview of current topics in nutrition, look for review articles in scholarly journals such as *Nutrition Reviews.* Similar to a review article, a meta-analysis study uses the power of a computer to combine and reanalyze the results of many previously published studies on a single topic.

With each report from scientists, the field of nutrition changes a little—each finding contributes another piece to the whole body of knowledge. People who know how science works understand that single findings, like single frames in a movie, are just small parts of a larger story. Over years, the picture of what is "true" in nutrition gradually changes, and dietary recommendations change to reflect the current understanding of scientific research. Highlight 5 provides a detailed look at how dietary fat recommendations have evolved over the past several decades as researchers have uncovered the relationships between the various kinds of fat and their roles in supporting or harming health.

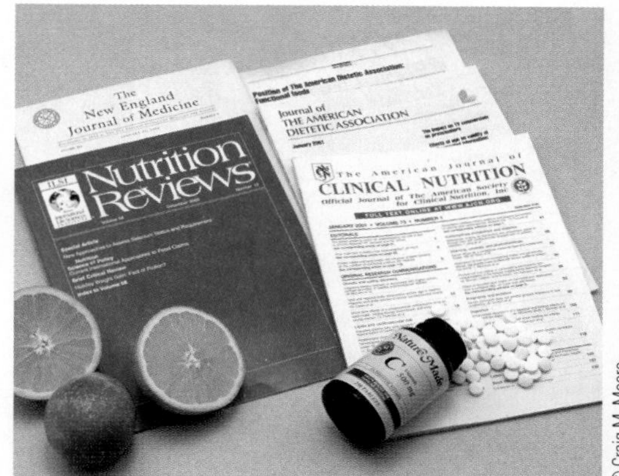

Knowledge about the nutrients and their effects on health comes from scientific studies.

REVIEW IT Explain the scientific method and how scientists use various types of research studies and methods to acquire nutrition information.

Scientists learn about nutrition by conducting experiments that follow the protocol of scientific research. In designing their studies, researchers randomly assign control and experimental groups, seek large sample sizes, provide placebos, and remain blind to treatments. Their findings must be reviewed and replicated by other scientists before being accepted as valid.

The characteristics of well-designed research have enabled scientists to study the actions of nutrients in the body. Such research has laid the foundation for quantifying how much of each nutrient the body needs.

1.4 Dietary Reference Intakes

LEARN IT Define the four categories of the DRI and explain their purposes.

Using the results of thousands of research studies, nutrition experts have produced a set of standards that define the amounts of energy, nutrients, other dietary components, and physical activity that best support health. These recommendations are called **Dietary Reference Intakes (DRI)**, and they reflect the collaborative efforts of researchers in both the United States and Canada.*[13] The inside front cover of this book provides a handy reference for DRI values.

Establishing Nutrient Recommendations The DRI Committee consists of highly qualified scientists who base their estimates of nutrient needs on careful examination and interpretation of scientific evidence. These recommendations apply to healthy people and may not be appropriate for people with diseases that increase or decrease nutrient needs. The next several paragraphs discuss specific aspects of how the committee goes about establishing the values that make up the DRI:

- Estimated Average Requirements (EAR)
- Recommended Dietary Allowances (RDA)
- Adequate Intakes (AI)
- Tolerable Upper Intake Levels (UL)

Estimated Average Requirements (EAR) The committee reviews hundreds of research studies to determine the **requirement** for a nutrient—how much is needed in the diet. The committee selects a different criterion for each nutrient

*The DRI reports are produced by the Food and Nutrition Board, Institute of Medicine of the National Academies, with active involvement of scientists from Canada.

Dietary Reference Intakes (DRI): a set of nutrient intake values for healthy people in the United States and Canada. These values are used for planning and assessing diets and include:
- Estimated Average Requirements (EAR)
- Recommended Dietary Allowances (RDA)
- Adequate Intakes (AI)
- Tolerable Upper Intake Levels (UL)

requirement: the lowest continuing intake of a nutrient that will maintain a specified criterion of adequacy.

based on its roles in supporting various activities in the body and in reducing disease risks.[14]

An examination of all the available data reveals that each person's body is unique and has its own set of requirements. Men differ from women, and needs change as people grow from infancy through old age. For this reason, the committee clusters its recommendations for people into groups based on gender and age. Even so, the exact requirements for people of the same gender and age are likely to be different. Person A might need 40 units of a particular nutrient each day; person B might need 35; and person C might need 57. Looking at enough people might reveal that their individual requirements fall into a symmetrical distribution, with most near the midpoint and only a few at the extremes (see the left side of Figure 1-5). Using this information, the committee determines an **Estimated Average Requirement (EAR)** for each nutrient—the average amount that appears sufficient for half of the population. In Figure 1-5, the EAR is shown as 45 units.

Recommended Dietary Allowances (RDA) Once a nutrient *requirement* is established, the committee must decide what intake to *recommend* for everybody—the **Recommended Dietary Allowance (RDA).** As you can see by the distribution in Figure 1-5, the EAR (shown in the figure as 45 units) is probably closest to everyone's need. If people consumed exactly the average requirement of a given nutrient each day, however, approximately half of the population would develop deficiencies of that nutrient—in Figure 1-5, for example, person C would be among them. Recommendations are therefore set greater than the EAR to meet the needs of most healthy people.

Small amounts greater than the daily requirement do no harm, whereas amounts less than the requirement may lead to health problems. When people's nutrient intakes are consistently **deficient** (less than the requirement), their nutrient stores decline, and over time this decline leads to poor health and deficiency symptoms. Therefore, to ensure that the nutrient RDA meet the needs of as many people as possible, the RDA are set near the top end of the range of the population's estimated requirements.

In this example, a reasonable RDA might be 63 units a day (see the right side of Figure 1-5). Such a point can be calculated mathematically so that the needs of about 98 percent of a population are included. Almost everybody—including person C whose needs were more substantial than the average—would

> FIGURE 1-5 Estimated Average Requirements (EAR) and Recommended Dietary Allowances (RDA) Compared

Each square in the graphs below represents a person with unique nutritional requirements. (The text discusses three of these people—A, B, and C.) Some people require only a small amount of nutrient X and some require a lot. Most people, however, fall somewhere in the middle.

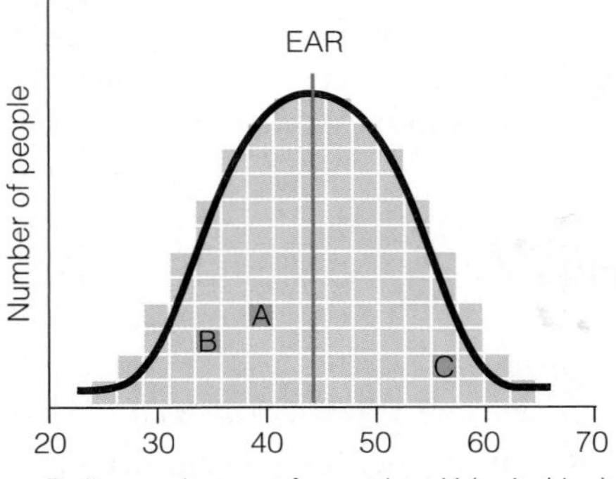

The Estimated Average Requirement (EAR) for a nutrient is the amount that meets the needs of about half of the population (shown here by the red line).

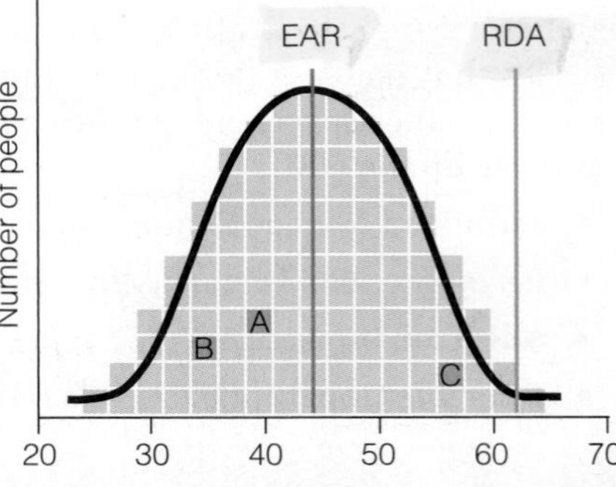

The Recommended Dietary Allowance (RDA) for a nutrient (shown here in green) is set well above the EAR, meeting the needs of about 98% of the population.

Estimated Average Requirement (EAR): the average daily amount of a nutrient that will maintain a specific biochemical or physiological function in half the healthy people of a given age and gender group.

Recommended Dietary Allowance (RDA): the average daily amount of a nutrient considered adequate to meet the known nutrient needs of practically all healthy people; a goal for dietary intake by individuals.

deficient: inadequate; a nutrient amount that fails to meet the body's needs and eventually results in deficiency symptoms.

consume enough of the nutrient if they met this dietary goal. Relatively few people's requirements would exceed this recommendation, and even then, they wouldn't exceed by much.

Adequate Intakes (AI) For some nutrients, such as vitamin K, there is insufficient scientific evidence to determine an EAR (which is needed to set an RDA). In these cases, the committee establishes an **Adequate Intake (AI)** instead of an RDA. An AI reflects the average amount of a nutrient that a group of healthy people consumes. Like the RDA, the AI may be used as nutrient goals for individuals.

Although both the RDA and the AI serve as nutrient intake goals for individuals, their differences are noteworthy. An RDA for a given nutrient is based on enough scientific evidence to expect that the needs of almost all healthy people will be met. An AI, on the other hand, must rely more heavily on scientific judgments because sufficient evidence is lacking. For this reason, AI values are more tentative than RDA values. The table on the inside front cover identifies which nutrients have an RDA and which have an AI. Later chapters present the RDA and AI values for vitamins and minerals.

Tolerable Upper Intake Levels (UL) As mentioned earlier, the recommended intakes for nutrients are generous, yet they may not be sufficient for every individual for every nutrient. Nevertheless, it is probably best not to exceed these recommendations by very much or very often. Individual tolerances for high doses of nutrients vary, and somewhere beyond the recommended intake is a point beyond which a nutrient is likely to become toxic. This point is known as the **Tolerable Upper Intake Level (UL).** It is naïve—and inaccurate—to think of recommendations as minimum amounts. A more accurate view is to see a person's nutrient needs as falling within a range, with marginal and danger zones at each end for intakes that are either inadequate or excessive (see Figure 1-6).

Paying attention to upper levels is particularly useful in guarding against the overconsumption of nutrients, which may occur when people use large-dose dietary supplements and fortified foods regularly. Later chapters discuss the dangers associated with excessively high intakes of vitamins and minerals, and the inside front cover (p. C) presents tables of upper levels for selected nutrients.

Adequate Intake (AI): the average daily amount of a nutrient that appears sufficient to maintain a specified criterion; a value used as a guide for nutrient intake when an RDA cannot be determined.

Tolerable Upper Intake Level (UL): the maximum daily amount of a nutrient that appears safe for most healthy people and beyond which there is an increased risk of adverse health effects.

> **FIGURE 1-6 Inaccurate versus Accurate View of Nutrient Intakes**

The RDA (or AI) for a given nutrient represents a point that lies within a range of appropriate and reasonable intakes between toxicity and deficiency. Both of these recommendations are high enough to provide reserves in times of short-term dietary inadequacies, but not so high as to approach toxicity. Nutrient intakes above or below this range may be equally harmful.

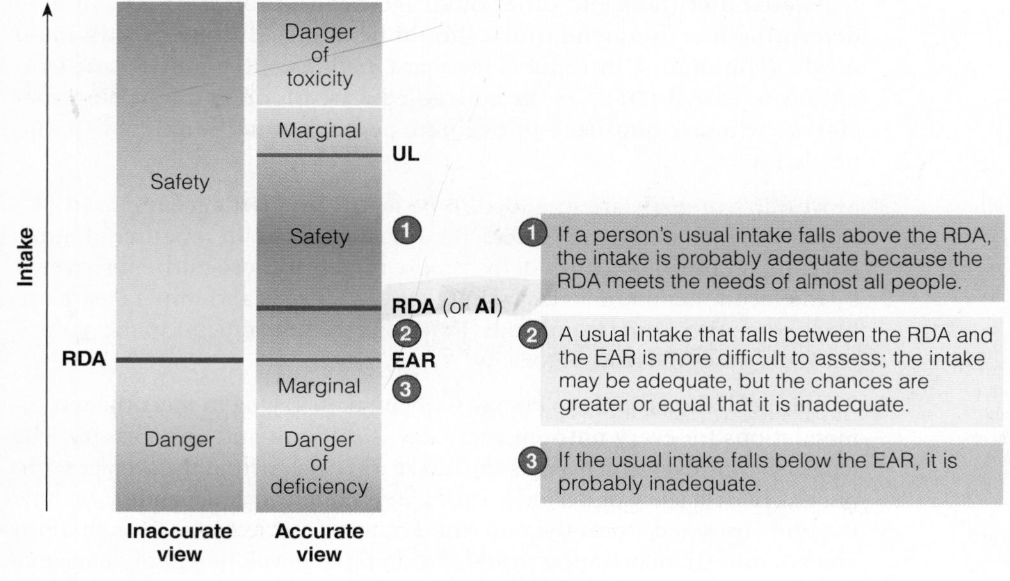
Intake
Safety
RDA
Danger
Inaccurate view
Danger of toxicity
Marginal
UL
Safety
1
RDA (or AI)
2
EAR
Marginal
3
Danger of deficiency
Accurate view
1 If a person's usual intake falls above the RDA, the intake is probably adequate because the RDA meets the needs of almost all people.
2 A usual intake that falls between the RDA and the EAR is more difficult to assess; the intake may be adequate, but the chances are greater or equal that it is inadequate.
3 If the usual intake falls below the EAR, it is probably inadequate.

Establishing Energy Recommendations In contrast to the RDA and AI values for nutrients, the recommendation for energy is not generous. Excess energy cannot be readily excreted and is eventually stored as body fat. These reserves may be beneficial when food is scarce, but they can also lead to obesity and its associated health consequences.

Estimated Energy Requirement (EER) The energy recommendation—called the **Estimated Energy Requirement (EER)**—represents the average dietary energy intake (kcalories per day) that will maintain energy balance in a person who has a healthy body weight and level of physical activity. Balance is key to the energy recommendation. Enough food energy is needed to sustain a healthy and active life, but too much can lead to weight gain and obesity. Because *any* amount in excess of energy needs will result in weight gain, no upper level for energy has been determined.

Acceptable Macronutrient Distribution Ranges (AMDR) People don't eat energy directly; they derive energy from foods containing carbohydrates, fats, and proteins. Each of these three energy-yielding nutrients contributes to the total energy intake, and those contributions vary in relation to one another. The DRI committee has determined that the composition of a diet that provides adequate energy and nutrients and reduces the risk of chronic diseases is:

- 45 to 65 percent kcalories from carbohydrate
- 20 to 35 percent kcalories from fat
- 10 to 35 percent kcalories from protein

These values are known as **Acceptable Macronutrient Distribution Ranges (AMDR).**

Using Nutrient Recommendations Although the intent of nutrient recommendations seems simple, they are the subject of much misunderstanding and controversy. Perhaps the following facts will help put them in perspective:

1. Estimates of adequate energy and nutrient intakes apply to *healthy* people. They need to be adjusted for malnourished people or those with medical problems who may require supplemented or restricted dietary intakes.
2. *Recommendations* are not minimum requirements, nor are they necessarily optimal intakes for all individuals. Recommendations can target only "most" of the people and cannot account for individual variations in nutrient needs—yet. Given the recent explosion of knowledge about genetics, the day may be fast approaching when nutrition scientists will be able to determine an individual's optimal nutrient needs. Until then, registered dietitians and other qualified health professionals can help determine if recommendations should be adjusted to meet individual needs. (Highlight 1 introduces *registered dietitians [RD]* and *dietetic technicians registered [DTR]* as the college-educated food and nutrition specialists who are qualified to evaluate people's nutritional health and needs.)
3. Most nutrient goals are intended to be met through diets composed of a variety of *foods* whenever possible. Because foods contain mixtures of nutrients and nonnutrients, they deliver more than just those nutrients covered by the recommendations. Excess intakes of vitamins and minerals are unlikely when they come from foods. Using dietary supplements to meet nutrient goals raises the risks of toxicity.
4. Recommendations apply to *average* daily intakes. Trying to meet the recommendations for every nutrient every day is difficult and unnecessary. The length of time over which a person's intake can deviate from the average without risk of deficiency or overdose varies for each nutrient, depending on how the body uses and stores the nutrient. For most nutrients (such as thiamin and vitamin C), deprivation would lead to rapid development of deficiency

Estimated Energy Requirement (EER): the average dietary energy intake that maintains energy balance and good health in a person of a given age, gender, weight, height, and level of physical activity.

Acceptable Macronutrient Distribution Ranges (AMDR): ranges of intakes for the energy nutrients that provide adequate energy and nutrients and reduce the risk of chronic diseases.

symptoms (within days or weeks); for others (such as vitamin A and vitamin B_{12}), deficiencies would develop more slowly (over months or years).

5. Each of the DRI categories serves a unique purpose. For example, the EAR are most appropriately used to develop and evaluate nutrition programs for *groups* such as schoolchildren or military personnel. The RDA (or AI if an RDA is not available) can be used to set goals for *individuals*. The UL serve as a reminder to keep nutrient intakes less than amounts that increase the risk of toxicity—not a common problem when nutrients derive from foods, but a real possibility for some nutrients if supplements are used regularly. With these understandings, professionals can use the DRI for a variety of purposes.[15]

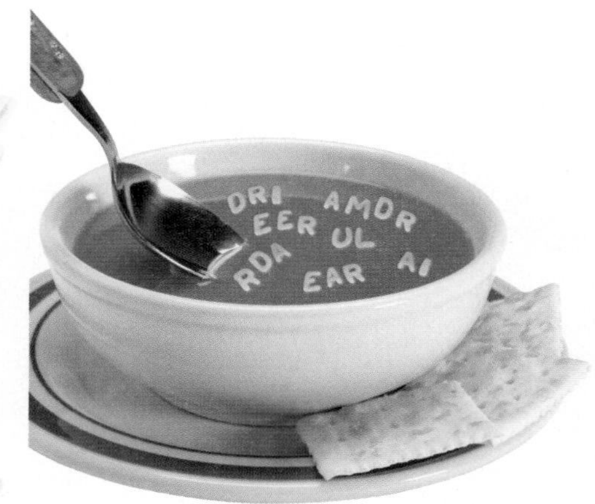

Don't let the DRI "alphabet soup" of nutrient intake standards confuse you. Their names make sense when you learn their purposes.

Comparing Nutrient Recommendations At least 40 different nations and international organizations have published nutrient standards similar to those used in the United States and Canada. Slight differences may be apparent, reflecting differences both in the interpretation of the data from which the standards were derived and in the food habits and physical activities of the populations they serve.

Many countries use the recommendations developed by two international groups: FAO (Food and Agriculture Organization) and WHO (World Health Organization). The FAO/WHO nutrient recommendations are considered sufficient to maintain health in nearly all healthy people worldwide and are provided in Appendix I.

REVIEW IT Define the four categories of the DRI and explain their purposes.
The Dietary Reference Intakes (DRI) are a set of nutrient intake values that can be used to plan and evaluate diets for healthy people. The Estimated Average Requirement (EAR) defines the amount of a nutrient that supports a specific function in the body for half of the population. The Recommended Dietary Allowance (RDA) is based on the Estimated Average Requirement and establishes a goal for dietary intake that will meet the needs of almost all healthy people. An Adequate Intake (AI) serves a similar purpose when an RDA cannot be determined. The Estimated Energy Requirement (EER) defines the average amount of energy intake needed to maintain energy balance, and the Acceptable Macronutrient Distribution Ranges (AMDR) define the proportions contributed by carbohydrate, fat, and protein to a healthy diet. The Tolerable Upper Intake Level (UL) establishes the highest amount that appears safe for regular consumption.

1.5 Nutrition Assessment

LEARN IT Explain how the four assessment methods are used to detect energy and nutrient deficiencies and excesses.

What happens when a person doesn't consume enough or consumes too much of a specific nutrient or energy? If the deficiency or excess is significant over time, the person experiences symptoms of **malnutrition.** With a deficiency of energy, the person may display the symptoms of **undernutrition** by becoming extremely thin, losing muscle tissue, and becoming prone to infection and disease. With a deficiency of a nutrient, the person may experience skin rashes, depression, hair loss, bleeding gums, muscle spasms, night blindness, or other symptoms. Similarly, over time, regular intakes in excess of needs may also have adverse effects. With an excess of energy, the person may become obese and vulnerable to diseases associated with **overnutrition,** such as heart disease and diabetes. With a sudden nutrient overdose, the person may experience hot flashes, yellowing skin, a rapid heart rate, low blood pressure, or other symptoms.

Malnutrition symptoms—such as diarrhea, skin rashes, and fatigue—are easy to miss because they resemble the symptoms of other diseases. But a person who has learned how to use assessment techniques to detect malnutrition can identify when these conditions are caused by poor nutrition and can recommend steps to

malnutrition: any condition caused by excess or deficient food energy or nutrient intake or by an imbalance of nutrients.
- **mal** = bad

undernutrition: deficient energy or nutrients.

overnutrition: excess energy or nutrients.

correct it. This discussion presents the basics of nutrition assessment; many more details are offered in later chapters and in Appendix E.

Nutrition Assessment of Individuals To prepare a **nutrition assessment,** a registered dietitian, dietetic technician registered, or other trained health-care professional uses:

- Historical information
- Anthropometric measurements
- Physical examinations
- Laboratory tests

Each of these methods involves collecting data in various ways and interpreting each finding in relation to the others to create a total picture.

Historical Information One step in evaluating nutrition status is to obtain information about a person's history with respect to health status, socioeconomic status, drug use, and diet. The health history reflects a person's medical record and may reveal a disease that interferes with the person's ability to eat or the body's use of nutrients. The person's family history of major diseases is also noteworthy, especially for conditions such as heart disease that have a genetic tendency to run in families. Economic circumstances may show a financial inability to buy foods or inadequate kitchen facilities in which to prepare them. Social factors such as marital status, ethnic background, and educational level also influence food choices and nutrition status. A drug history, including all prescribed and over-the-counter medications, may highlight possible interactions that lead to nutrient deficiencies (as described in Highlight 17). A diet history that examines a person's intake of foods, beverages, and dietary supplements may reveal either a surplus or inadequacy of nutrients or energy.

To take a diet history, the assessor collects data about the foods a person eats. The data may be collected by recording the foods the person has eaten over a period of 24 hours, 3 days, or a week or more or by asking what foods the person typically eats and how much of each. The days in the record must be fairly typical of the person's diet, and portion sizes must be recorded accurately. To determine the amounts of nutrients consumed, the assessor usually enters the foods and their portion sizes into a computer using a diet analysis program. This step can also be done manually by looking up each food in a table of food composition such as Appendix H in this book. The assessor then compares the calculated nutrient intakes with the DRI to determine the probability of adequacy. Alternatively, the diet history might be compared against standards such as the USDA Food Patterns or *Dietary Guidelines* (described in Chapter 2).

An estimate of energy and nutrient intakes from a diet history, when combined with other sources of information, can help confirm or rule out the *possibility* of suspected nutrition problems. A sufficient intake of a nutrient does not guarantee adequacy, and an insufficient intake does not always indicate a deficiency. Such findings, however, warn of possible problems.

Anthropometric Measurements A second technique that may help to reveal nutrition problems is taking **anthropometric** measures such as height and weight. The assessor compares a person's measurements with standards specific for gender and age or with previous measures on the same individual. (Chapter 8 presents information on body weight and its standards, and Appendix E includes growth charts for children.)

Measurements taken periodically and compared with previous measurements reveal patterns and indicate trends in a person's overall nutrition status, but they provide little information about specific nutrients. Instead, measurements out of line with expectations may reveal such problems as growth failure in children, wasting or swelling of body tissues in adults, and obesity—conditions that may reflect energy or nutrient deficiencies or excesses.

Physical Examinations A third nutrition assessment technique is a physical examination looking for clues to poor nutrition status. Visual inspection of the

nutrition assessment: a comprehensive analysis of a person's nutrition status that uses health, socioeconomic, drug, and diet histories; anthropometric measurements; physical examinations; and laboratory tests.

anthropometric (AN-throw-poe-MET-rick): relating to measurement of the physical characteristics of the body, such as height and weight.

- **anthropos** = human
- **metric** = measuring

hair, eyes, skin, posture, tongue, and fingernails can provide such clues. In addition, information gathered from an interview can help identify symptoms. The examination requires skill because many physical signs and symptoms reflect more than one nutrient deficiency or toxicity—or even nonnutrition conditions. Like the other assessment techniques, a physical examination alone does not yield firm conclusions. Instead, physical examinations reveal possible imbalances that must be confirmed by other assessment techniques, or they confirm results from other assessment measures.

Laboratory Tests A fourth way to detect a developing deficiency, imbalance, or toxicity is to take samples of blood or urine, analyze them in the laboratory, and compare the results with normal values for a similar population. Laboratory tests are most useful in uncovering early signs of malnutrition before symptoms appear. In addition, they can confirm suspicions raised by other assessment methods.

Iron, for Example The mineral iron can be used to illustrate the stages in the development of a nutrient deficiency and the assessment techniques useful in detecting them. The **overt,** or outward, signs of an iron deficiency appear at the end of a long sequence of events. Figure 1-7 describes what happens in the body as a nutrient deficiency progresses and shows which assessment methods can reveal those changes.

> FIGURE 1-7 **Stages in the Development of a Nutrient Deficiency**

Internal changes precede outward signs of deficiencies. Outward signs of sickness, however, need not appear before a person takes corrective measures. Laboratory tests can help determine nutrient status in the early stages.

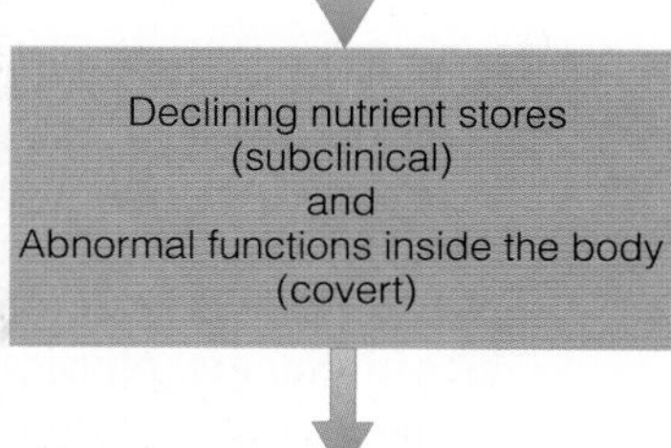

First, the body has too little iron—either because iron is lacking in the person's diet (a **primary deficiency**) or because the person's body doesn't absorb enough, excretes too much, or uses iron inefficiently (a **secondary deficiency**). A diet history provides clues to primary deficiencies; a health history provides clues to secondary deficiencies.

Next, the body begins to use up its stores of iron. At this stage, the deficiency might be described as a **subclinical deficiency.** It exists as a **covert** condition, and although it might be detected by laboratory tests, outward signs are not yet apparent.

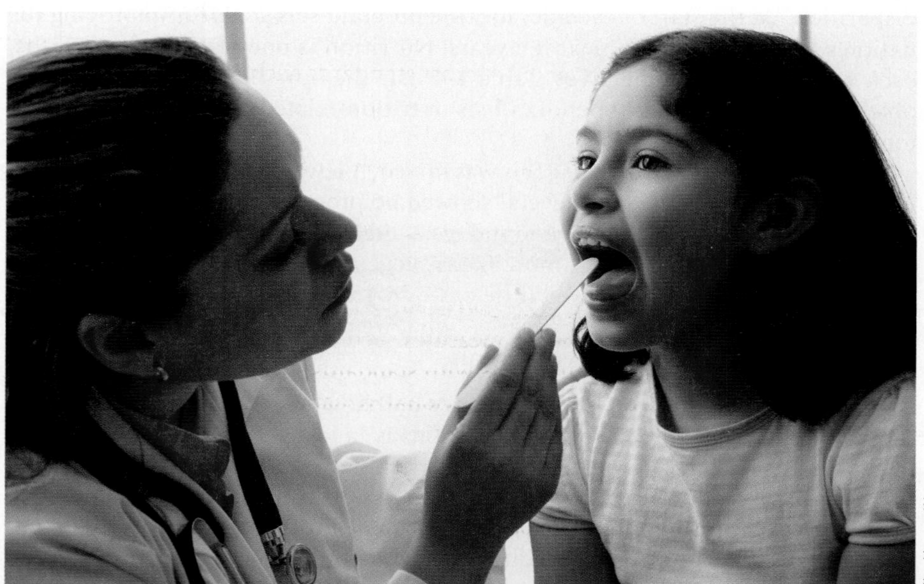

A peek inside the mouth provides clues to a person's nutrition status. An inflamed tongue may indicate a deficiency of one of the B vitamins, and mottled teeth may reveal fluoride toxicity, for example.

overt (oh-VERT): out in the open and easy to observe.

- **ouvrir** = to open

primary deficiency: a nutrient deficiency caused by inadequate dietary intake of a nutrient.

secondary deficiency: a nutrient deficiency caused by something other than an inadequate intake such as a disease condition or drug interaction that reduces absorption, accelerates use, hastens excretion, or destroys the nutrient.

subclinical deficiency: a deficiency in the early stages, before the outward signs have appeared.

covert (KOH-vert): hidden, as if under covers.

- **couvrir** = to cover

Finally, the body's iron stores are exhausted. Now, it cannot make enough iron-containing red blood cells to replace those that are aging and dying. Iron is needed in red blood cells to carry oxygen to all the body's tissues. When iron is lacking, fewer red blood cells are made, the new ones are pale and small, and every part of the body feels the effects of oxygen shortage. At this point in time, the overt symptoms of deficiency appear—weakness, fatigue, pallor, and headaches, reflecting the iron-deficient state of the blood. A physical examination and interview will reveal these symptoms.

Nutrition Assessment of Populations To assess a population's nutrition status, researchers conduct surveys using techniques similar to those used on individuals. The data collected are then used by various agencies for numerous purposes, including the development of national health goals.

National Nutrition Surveys The National Nutrition Monitoring program coordinates the many nutrition-related surveys and research activities of various federal agencies. The integration of two major national surveys provides comprehensive data efficiently. One survey collects data on the kinds and amounts of foods people eat.* The other survey examines the people themselves, using anthropometric measurements, physical examinations, and laboratory tests.** The data provide valuable information on several nutrition-related conditions, such as growth retardation, heart disease, and nutrient deficiencies. National nutrition surveys often oversample high-risk groups (low-income families, pregnant women, adolescents, the elderly, African Americans, and Mexican Americans) to glean an accurate estimate of their health and nutrition status.

The resulting wealth of information from the national nutrition surveys is used for a variety of purposes. For example, Congress uses this information to establish public policy on nutrition education, food assistance programs, and the regulation of the food supply. Scientists use the information to establish research priorities. The food industry uses these data to guide decisions in public relations and product development. The Dietary Reference Intakes and other major reports that examine the relationships between diet and health depend on information collected from these nutrition surveys. These data also provide the basis for developing and monitoring national health goals.

National surveys provide valuable information about the kinds of foods people eat.

National Health Goals The **Healthy People** program sets priorities and guides policies that "increase the quality and years of healthy life" and "eliminate health disparities." At the start of each decade, the program sets goals for improving the nation's health during the next ten years. Nutrition is one of many topic areas, each with numerous objectives. Table 1-4 lists the nutrition and weight status objectives for 2020, and Appendix J lists nutrition-related objectives from other topic areas.

Progress in meeting the 2010 goals was mixed. A few objectives were met, about half made some progress, and several showed no progress—or even moved in the wrong direction.[16] The objective to reduce average blood cholesterol levels was achieved, but objectives to eat more fruits, vegetables, and whole grains and to increase physical activity showed little or no improvement. Trends in overweight and obesity actually worsened. Clearly, "what we eat in America" must change if we hope to meet the Healthy People goals.

National Trends What do we eat in America and how has it changed over the past 40 years? The short answer to both questions is "a lot." We eat more meals away from home, particularly at fast-food restaurants. We eat larger portions. We drink more sweetened beverages and eat more energy-dense, nutrient-poor foods such as candy and chips. We snack frequently. As a result of these dietary habits, our energy intake has risen and, consequently, so has the incidence of overweight and obesity. Overweight and obesity, in turn, profoundly influence our health—as the next section explains.

Healthy People: a national public health initiative under the jurisdiction of the US Department of Health and Human Services (DHHS) that identifies the most significant preventable threats to health and focuses efforts toward eliminating them.

---

*This survey is called *What We Eat in America.*

**This survey is known as the National Health and Nutrition Examination Survey (NHANES).

TABLE 1-4 Healthy People 2020 Nutrition and Weight Status Objectives

- Increase the proportion of adults who are at a healthy weight
- Reduce the proportion of adults who are obese
- Reduce iron deficiency among young children and females of childbearing age
- Reduce iron deficiency among pregnant females
- Reduce the proportion of children and adolescents who are overweight or obese
- Increase the contribution of fruits to the diets of the population aged 2 years and older
- Increase the variety and contribution of vegetables to the diets of the population aged 2 years and older
- Increase the contribution of whole grains to the diets of the population aged 2 years and older
- Reduce consumption of saturated fat in the population aged 2 years and older
- Reduce consumption of sodium in the population aged 2 years and older
- Increase consumption of calcium in the population aged 2 years and older
- Increase the proportion of worksites that offer nutrition or weight management classes or counseling
- Increase the proportion of physician office visits that include counseling or education related to nutrition or weight
- Eliminate very low food security among children in US households
- Prevent inappropriate weight gain in youth and adults
- Increase the proportion of primary care physicians who regularly measure the body mass index of their patients
- Reduce consumption of kcalories from solid fats and added sugars in the population aged 2 years and older
- Increase the number of states that have state-level policies that incentivize food retail outlets to provide foods that are encouraged by the *Dietary Guidelines*
- Increase the number of states with nutrition standards for foods and beverages provided to preschool-aged children in childcare
- Increase the percentage of schools that offer nutritious foods and beverages outside of school meals

NOTE: Nutrition and Weight Status is one of 38 topic areas, each with numerous objectives. Several of the other topic areas have nutrition-related objectives, and these are presented in Appendix J.
SOURCE: www.healthypeople.gov

REVIEW IT Explain how the four assessment methods are used to detect energy and nutrient deficiencies and excesses.

People become malnourished when they get too little or too much energy or nutrients. Deficiencies, excesses, and imbalances of nutrients lead to malnutrition diseases. To detect malnutrition in individuals, health-care professionals use a combination of four nutrition assessment methods. Reviewing historical information on diet and health may suggest a possible nutrition problem. Laboratory tests may detect a possible nutrition problem in its earliest stages, whereas anthropometric measurements and physical examinations pick up on the problem only after it causes symptoms. National surveys use similar assessment methods to measure people's food consumption and to evaluate the nutrition status of populations.

1.6 Diet and Health

LEARN IT Identify several risk factors and explain their relationships to chronic diseases.

Foods play a vital role in supporting health. Early nutrition research focused on identifying the nutrients in foods that would prevent such common diseases as rickets and scurvy, the vitamin D– and vitamin C–deficiency diseases. With this knowledge, developed countries have successfully defended against nutrient deficiency diseases. World hunger and nutrient deficiency diseases still pose a major health threat in developing countries, however, but not because of a lack of nutrition knowledge (as Chapter 20 explains). More recently, nutrition research has focused on chronic diseases associated with energy and nutrient excesses. Chronic diseases are responsible for 7 out of 10 deaths among US adults. Once thought to be "rich countries' problems," chronic diseases have now become epidemic in developing countries as well—contributing to three out of five deaths worldwide.[17]

TABLE 1-5 Leading Causes of Death in the United States

	Percentage of Total Deaths
1. **Heart disease**	24.1
2. **Cancers**	23.3
3. Chronic lung diseases	5.6
4. **Strokes**	5.2
5. Accidents	4.8
6. Alzheimer's disease	3.4
7. **Diabetes mellitus**	2.8
8. Pneumonia and influenza	2.0
9. Kidney disease	2.0
10. Suicide	1.5

NOTE: The diseases highlighted in bold have relationships with diet.
SOURCE: Deaths: Preliminary data for 2010, *National Vital Statistics Reports*, January 11, 2012, Centers for Disease Control and Prevention, www.cdc.gov/nchs.

Chronic Diseases Table 1-5 lists the ten leading causes of death in the United States. These "causes" are stated as if a single condition such as heart disease caused death, but most chronic diseases arise from multiple factors over many years. A person who died of heart disease may have been overweight, had high blood pressure, been a cigarette smoker, and spent years eating a diet high in saturated fat and getting too little exercise.

Of course, not all people who die of heart disease fit this description, nor do all people with these characteristics die of heart disease. People who are overweight might die from the complications of diabetes instead, or those who smoke might die of cancer. They might even die from something totally unrelated to any of these factors, such as an automobile accident. Still, statistical studies have shown that certain conditions and behaviors are linked to certain diseases.

Table 1-5 highlights four of the top seven causes of death as having a link with diet. Notice that these four diseases—heart disease, cancers, strokes, and diabetes—account for more than half of the deaths each year.

Risk Factors for Chronic Diseases Factors that increase or reduce the *risk* of developing chronic diseases can be identified by analyzing statistical data. A strong association between a **risk factor** and a disease means that when the factor is present, the *likelihood* of developing the disease increases. It does not mean that all people with the risk factor will develop the disease. Similarly, a lack of risk factors does not guarantee freedom from a given disease. On the average, though, the more risk factors in a person's life, the greater that person's chances of developing the disease. Conversely, the fewer risk factors in a person's life, the better the chances for good health.

Risk Factors Persist Risk factors tend to persist over time. Without intervention, a young adult with high blood pressure will most likely continue to have high blood pressure as an older adult, for example. Thus, to minimize the damage, early intervention is most effective.

Risk Factors Cluster Risk factors tend to cluster. For example, a person who is obese may be physically inactive, have high blood pressure, and have high blood cholesterol—all risk factors associated with heart disease. Multiple risk factors act synergistically to increase the risk of disease dramatically. Intervention that focuses on one risk factor often benefits the others as well. For example, physical activity can help reduce weight. Physical activity and weight loss will, in turn, help to lower blood pressure and blood cholesterol.

TABLE 1-6 Factors Contributing to Deaths in the United States

Factors	Percentage of Deaths
Tobacco	18
Poor diet/inactivity	15
Alcohol	4
Microbial agents	3
Toxic agents	2
Motor vehicles	2
Firearms	1
Sexual behavior	1
Illicit drugs	1

SOURCE: A. H. Mokdad and coauthors, Actual causes of death in the United States, 2000, *Journal of the American Medical Association* 291 (2004): 1238–1245, with corrections from *Journal of the American Medical Association* 293 (2005): 298.

Risk Factors in Perspective The most prominent factor—contributing to one of every five deaths each year in the United States—is tobacco use, followed closely by diet and activity patterns, and then alcohol use (see Table 1-6).[18] Risk factors such as smoking, poor dietary habits, physical inactivity, and alcohol consumption are personal behaviors that can be changed. Decisions to not smoke, to eat a well-balanced diet, to engage in regular physical activity, and to drink alcohol in moderation (if at all) improve the likelihood that a person will enjoy good health. Other risk factors, such as genetics, gender, and age, also play important roles in the development of chronic diseases, but they cannot be changed. Health recommendations acknowledge the influence of such factors on the development of disease, but they must focus on the factors that are changeable.

Health Behaviors in the United States Despite evidence linking certain behaviors with chronic diseases, many Americans continue to engage in unhealthy behaviors. An estimated 20 percent of US adults consume five or more drinks in a single day at least once a year; 20 percent are cigarette smokers; 40 percent are physically inactive; 60 percent are either overweight or obese; and 30 percent average 6 hours or less of sleep per day.[19] For the two out of three Americans who do not smoke or drink alcohol excessively, the one choice that can influence long-term health prospects more than any other is diet.

REVIEW IT Identify several risk factors and explain their relationships to chronic diseases.

Within the range set by genetics, a person's choice of diet influences long-term health. Diet has no influence on some diseases but is linked closely to others. Personal life choices, such as engaging in physical activity and using tobacco or alcohol, also affect health for the better or worse.

risk factor: a condition or behavior associated with an elevated frequency of a disease but not proved to be causal. Leading risk factors for chronic diseases include obesity, cigarette smoking, high blood pressure, high blood cholesterol, physical inactivity, and a diet high in saturated fats and low in vegetables, fruits, and whole grains.

The next several chapters provide many more details about nutrients and how they support health. Whenever appropriate, the discussion shows how diet influences each of today's major diseases. Dietary recommendations appear again and again, as each nutrient's relationships with health are explored. Most people who follow the recommendations will benefit and can enjoy good health into their later years.

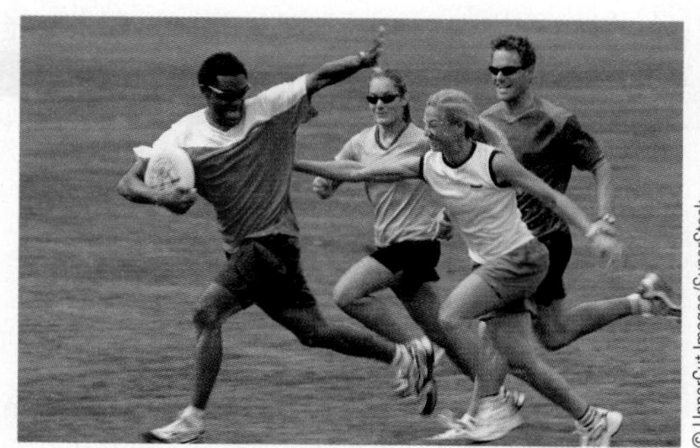

Physical activity can be both fun and beneficial.

Nutrition Portfolio

Each chapter in this book ends with simple Nutrition Portfolio activities that invite you to review key messages and consider whether your personal choices are meeting the dietary goals introduced in the text. By using the information you are recording in Diet Analysis Plus, the dietary tracking software that accompanies this text, and keeping a journal of these Nutrition Portfolio assignments, you can examine how your knowledge and behaviors change as you progress in your study of nutrition.

Your food choices play a key role in keeping you healthy and reducing your risk of chronic diseases. After you have recorded at least one day's foods in Diet Analysis Plus, look at that day's choices and record your answers to the following in your journal:

- Identify the factors that most influence your food choices for meals and snacks.
- List the chronic disease risk factors and conditions (listed in the definition of *risk factor,* p. 26) that you have.
- Describe lifestyle changes you can make to improve your chances of enjoying good health.

To complete this exercise, go to your Diet Analysis Plus at **www.cengagebrain.com.**

STUDY IT To review the key points of this chapter and take a practice quiz, go to the study cards at the end of the book.

REFERENCES

1. E. R. Grimm and N. I. Steinle, Genetics of eating behavior: Established and emerging concepts, *Nutrition Reviews* 69 (2011): 52–60.
2. B. J. Tepper, Nutritional implications of genetic taste variation: The role of PROP sensitivity and other taste phenotypes, *Annual Review of Nutrition* 28 (2008): 367–388.
3. K. Stein, Navigating cultural competency: In preparation for an expected standard in 2010, *Journal of the American Dietetic Association* 109 (2009): 1676–1688.
4. J. Tillotson, Americans' food shopping in today's lousy economy (part 3), *Nutrition Today* 44 (2009): 265–268.
5. Food Marketing Institute, www.fmi.org/research, accessed 19 January 2011.
6. J. Tillotson, Americans' food shopping in today's lousy economy (part 2), *Nutrition Today* 44 (2009): 218–221.
7. B. Wansink, K. van Ittersum, and C. Werle, How negative experiences shape long-term food preferences. Fifty years from the World War II combat front, *Appetite* 52 (2009): 750–752.
8. A. D. Ozier and coauthor, Overweight and obesity are associated with emotion- and stress-related eating as measured by the Eating and Appraisal Due to Emotions and Stress Questionnaire, *Journal of the American Dietetic Association* 108 (2008): 49–56.
9. International Food Information Council Foundation, *2010 Food & Health Survey,* www.foodinsight.org.
10. Position of the American Dietetic Association: Functional foods, *Journal of the American Dietetic Association* 104 (2004): 814–826.
11. Position of the American Dietetic Association: Total diet approach to communicating food and nutrition information, *Journal of the American Dietetic Association* 107 (2007): 1224–1232.
12. S. Rowe and coauthors, Funding food science and nutrition research: Financial conflicts and scientific integrity, *American Journal of Clinical Nutrition* 89 (2009): 1285–1291.
13. Committee on Dietary Reference Intakes, *Dietary Reference Intakes for Calcium and Vitamin D* (Washington, D.C.: National Academies Press, 2011); Committee on Dietary Reference Intakes, *Dietary Reference Intakes for Water, Potassium, Sodium, Chloride, and Sulfate* (Washington, D.C.: National Academies Press, 2005); Committee on Dietary Reference Intakes, *Dietary Reference Intakes for Energy, Carbohydrate, Fiber, Fat, Fatty Acids, Cholesterol, Protein, and Amino Acids* (Washington, D.C.: National Academies Press, 2005); Committee on Dietary Reference Intakes, *Dietary Reference Intakes for Vitamin A, Vitamin K, Arsenic, Boron, Chromium, Copper, Iodine, Iron, Manganese, Molybdenum, Nickel, Silicon, Vanadium, and Zinc* (Washington, D.C.: National Academies Press, 2001); Committee on Dietary Reference Intakes, *Dietary Reference Intakes for Vitamin C, Vitamin E, Selenium, and Carotenoids* (Washington, D.C.: National Academies Press, 2000); Committee on Dietary Reference Intakes, *Dietary Reference Intakes for Thiamin, Riboflavin, Niacin, Vitamin B_6, Folate, Vitamin B_{12}, Pantothenic Acid, Biotin, and Choline* (Washington, D.C.: National Academies Press, 1998); Committee on Dietary Reference Intakes, *Dietary Reference Intakes for Calcium, Phosphorus, Magnesium, Vitamin D, and Fluoride* (Washington, D.C.: National Academies Press, 1997).
14. R. M. Russell, Current framework for DRI development: What are the pros and cons? *Nutrition Reviews* 66 (2008): 455–458.
15. Practice paper of the American Dietetic Association: Using the Dietary Reference Intakes, *Journal of the American Dietetic Association* 111 (2011): 762–770.
16. E. J. Sondik and coauthors, Progress toward the Healthy People 2010 goals and objectives, *Annual Review of Public Health* 31 (2010): 271–281.
17. K. M. V. Narayan, M. K. Ali, and J. P. Koplan, Global noncommunicable diseases: Where worlds meet, *New England Journal of Medicine* 363 (2010): 1196–1198.
18. A. H. Mokdad and coauthors, Actual causes of death in the United States, 2000, *Journal of the American Medical Association* 291 (2004): 1238–1245.
19. C. A. Schoenborn and P. F. Adams, Health behaviors of adults: United States, 2005–2007, National Center for Health Statistics, *Vital and Health Statistics,* 2010, www.cdc.gov/nchs/data/series/sr_10/sr10_245.pdf.

LEARN IT Recognize misinformation and describe how to identify reliable nutrition information.

Nutrition Information and Misinformation

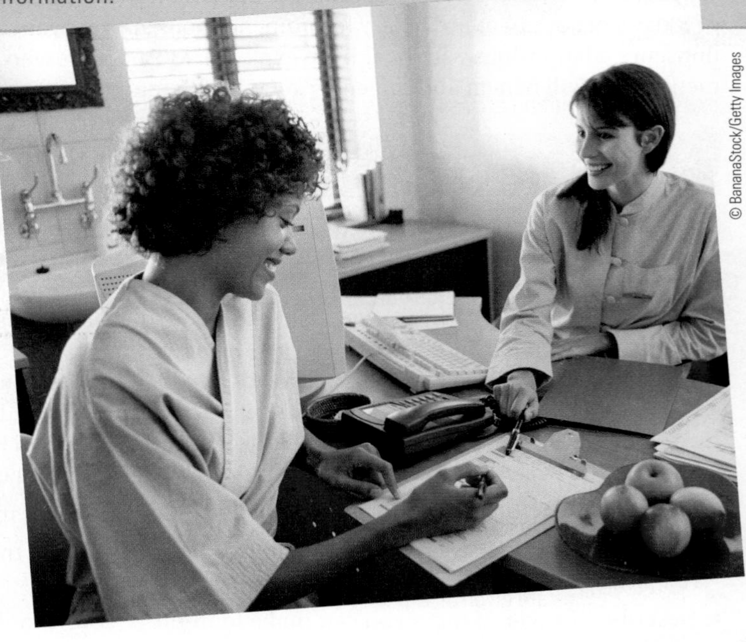

How can people distinguish valid nutrition information from misinformation? One excellent approach is to notice *who* is providing the information. The "who" behind the information is not always evident, though, especially in the world of electronic media. Keep in mind that *people* create websites on the Internet, just as people write books and report the news. In all cases, consumers need to determine whether the person is qualified to provide nutrition information.

This highlight begins by examining the unique potential as well as the problems of relying on the Internet and the media for nutrition information. It continues with a discussion of how to identify reliable nutrition information that applies to all resources, including the Internet and the news. (The accompanying glossary defines related terms.)

Nutrition on the Internet

Got a question? The **Internet** has an answer. An estimated two out of three US adults use the Internet to look up health information or learn about health topics in online chat groups.[1] The Internet offers endless opportunities to obtain high-quality information, but it also delivers an abundance of incomplete, misleading, and inaccurate information.[2] Simply put: anyone can publish anything.

With hundreds of millions of **websites,** searching for nutrition information can be an overwhelming experience, with no guarantees of finding accurate information. When using the Internet, keep in mind that the quality of health-related information available covers a broad range. You must evaluate websites for their accuracy, just as you would any other source. The accompanying "How To" provides tips for determining whether a website is reliable.

One of the most trustworthy sites used by scientists and others is the US National Library of Medicine's PubMed, which provides free access to more than 20 million abstracts (short descriptions) of research papers published in scientific journals around the world. Many abstracts provide links to the full articles. Figure H1-1 (p. 30) introduces this valuable resource.

Did you receive an e-mail warning of the health dangers associated with aspartame? If so, you've been a victim of urban scarelore. When

GLOSSARY

Academy of Nutrition and Dietetics: the professional organization of dietitians in the United States; formerly the American Dietetic Association. The Canadian equivalent is Dietitians of Canada, which operates similarly.

accredited: approved; in the case of medical centers or universities, certified by an agency recognized by the US Department of Education.

certified nutritionist or **certified nutritional consultant** or **certified nutrition therapist:** a person who has been granted a document declaring his or her authority as a nutrition professional.

dietetic technician: a person who has completed a minimum of an associate's degree from an accredited university or college and an approved dietetic technician program that includes a supervised practice experience. See also *dietetic technician, registered.*

dietetic technician, registered (DTR): a dietetic technician who has passed a national examination and maintains registration through continuing professional education.

dietitian: a person trained in nutrition, food science, and diet planning. See also *registered dietitian.*

diploma mills: entities without valid accreditation that provide worthless degrees.

DTR: see *dietetic technician, registered.*

fraudulent: the promotion, for financial gain, of devices, treatments, services, plans, or products (including diets and supplements) that alter or claim to alter a human condition without proof of safety or effectiveness.

Internet (the Net): a worldwide network of millions of computers linked together to share information.

license to practice: permission under state or federal law, granted on meeting specified criteria, to use a certain title (such as dietitian) and offer certain services. *Licensed dietitians* may use the initials *LD* after their names.

misinformation: false or misleading information.

public health dietitians: dietitians who specialize in providing nutrition services through organized community efforts.

RD: see *registered dietitian.*

registered dietitian (RD): a person who has completed a minimum of a bachelor's degree from an accredited university or college, has completed approved course work and a supervised practice program, has passed a national examination, and maintains registration through continuing professional education.

registration: listing; with respect to health professionals, listing with a professional organization that requires specific course work, experience, and passing of an examination.

websites: Internet resources composed of text and graphic files, each with a unique URL (Uniform Resource Locator) that names the site (for example, www.usda.gov).

nutrition information arrives in unsolicited e-mails, be suspicious if:

- The person sending it to you didn't write it and you cannot determine who did or if that person is a nutrition expert
- The phrase "Forward this to everyone you know" appears
- The phrase "This is not a hoax" appears because chances are good that it is
- The news is sensational and you've never heard about it from legitimate sources
- The language is emphatic and the text is sprinkled with capitalized words and exclamation marks
- No references are given or, if present, are of questionable validity when examined
- The message has been debunked on websites such as **www.quackwatch.org, www.snopes.com,** or **www.urbanlegends.com.**

>How To

Determine whether a Website Is Reliable

To determine whether a website offers reliable nutrition information, ask the following questions:

- **Who?** Who is responsible for the site? Is it staffed by qualified professionals? Look for the authors' names and credentials. Have experts reviewed the content for accuracy?
- **When?** When was the site last updated? Because nutrition is an ever-changing science, sites need to be dated and updated frequently.
- **Where?** Where is the information coming from? The three letters following the dot in a Web address identify the site's affiliation. Addresses ending in "gov" (government), "edu" (educational institute), and "org" (organization) generally provide reliable information; "com" (commercial) sites represent businesses and, depending on their qualifications and integrity, may or may not offer dependable information.
- **Why?** Why is the site giving you this information? Is the site providing a public service or selling a product? Many commercial sites provide accurate information, but some do not. When money is the prime motivation, be aware that the information may be biased.

If you are satisfied with the answers to all of the previous questions, then ask this final question:

- **What?** What is the message, and is it in line with other reliable sources? Information that contradicts common knowledge should be questioned. Many reliable sites provide links to other sites to facilitate your quest for knowledge, but this provision alone does not guarantee a reputable intention. Be aware that any site can link to any other site without permission.

TRY IT Visit a nutrition website and answer the five "W" questions to determine whether it is a reliable resource.

Nutrition in the News

Consumers get much of their nutrition information from Internet websites, television news, and magazine articles, which have heightened awareness of how diet influences the development of diseases. Consumers benefit from news coverage of nutrition when they learn to make lifestyle changes that will improve their health. Sometimes, however, popular reports mislead consumers and create confusion. They often tell a lopsided story quickly instead of presenting the integrated results of research studies or a balance of expert opinions.

Tight deadlines and limited understanding sometimes make it difficult to provide a thorough report. Hungry for the latest news, the media often report scientific findings quickly and prematurely—without benefit of careful interpretation, replication, or peer review. Usually, the reports present findings from a single, recently released study, making the news current and controversial. Consequently, the public receives diet and health news fast, but not always in perspective. Reporters may twist inconclusive findings into "meaningful discoveries" when pressured to write catchy headlines and sensational stories.

As a result "surprising new findings" sometimes seem to contradict one another, and consumers may feel frustrated and betrayed. Occasionally, the reports are downright false, but more often the apparent contradictions are simply the normal result of science at work. A single study contributes to the big picture, but when viewed alone, it can easily distort the image. To be meaningful the conclusions of any study must be presented cautiously within the context of other research findings.

Identifying Nutrition Experts

Regardless of whether the medium is electronic, print, or video, consumers need to ask whether the person behind the information is qualified to speak on nutrition. If the creator of an Internet website recommends eating three pineapples a day to lose weight, a trainer at the gym praises a high-protein diet, or a health-food store clerk suggests an herbal supplement, should you believe these people? Can you distinguish between accurate news reports and infomercials on television? Have you noticed that many televised nutrition messages are presented by celebrities, athletes, psychologists, food editors, and chefs—that is, almost anyone except a **dietitian**? When you are confused or need sound dietary advice, whom should you ask?

Physicians and Other Health-Care Professionals

Many people turn to physicians or other health-care professionals for dietary advice, expecting them to know about all health-related matters. But are they the best sources of accurate and current information on nutrition? Only about 30 percent of all medical schools in the United

> **FIGURE H1-1 PubMed (www.pubmed.gov): Internet Resource for Scientific Nutrition References**

The US National Library of Medicine's PubMed website offers tutorials to help teach beginners to use the search system effectively. Often, simply visiting the site, typing a query in the "Search for" box, and clicking "Go" will yield satisfactory results.

For example, to find research concerning calcium and bone health, typing "calcium bone" nets more than 30,000 results. Try setting limits on dates, types of articles, languages, and other criteria to obtain a more manageable number of abstracts to peruse.

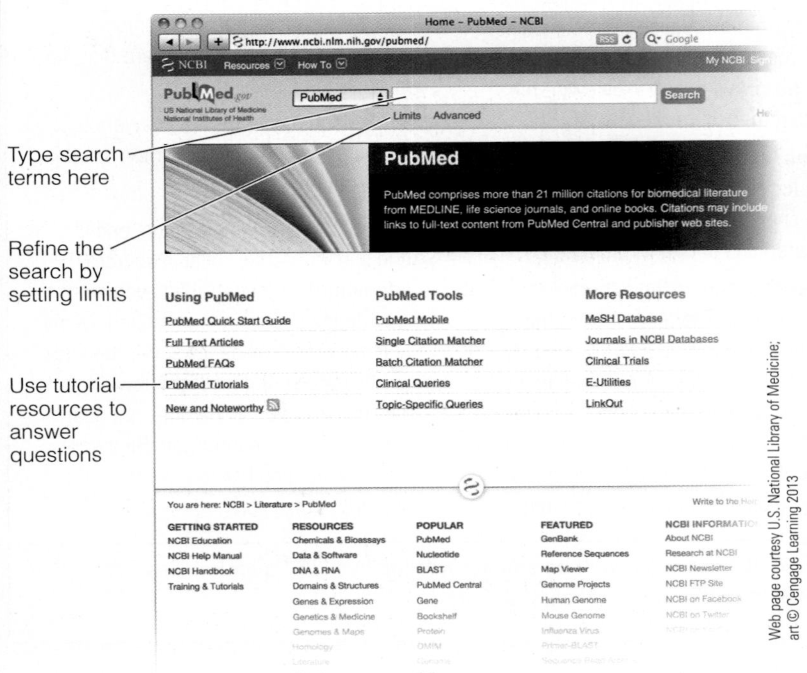

States require students to take a separate nutrition course; less than half require the minimum 25 hours of nutrition instruction recommended by the National Academy of Sciences. By comparison, most students reading this text are taking a nutrition class that provides an average of 45 hours of instruction.

The **Academy of Nutrition and Dietetics** (formerly the American Dietetic Association) asserts that standardized nutrition education should be included in the curricula for all health-care professionals: physicians, nurses, physician's assistants, dental hygienists, physical and occupational therapists, social workers, psychologists, and all others who provide services directly to clients. When these professionals understand the relevance of nutrition in the treatment and prevention of diseases and have command of reliable nutrition information, then all the people they serve will also be better informed.

Most health-care professionals appreciate the connections between health and nutrition. Those who have specialized in clinical nutrition are especially well qualified to speak on the subject. Few, however, have the time or experience to develop diet plans and provide detailed diet instructions for clients. Often they wisely refer clients to a qualified nutrition expert—a **registered dietitian (RD).**

Registered Dietitian (RD)

A registered dietitian (RD) has the educational background necessary to deliver reliable nutrition advice and care.[3] To become an RD, a person must earn an undergraduate degree requiring about 60 credit hours in nutrition, food science, and other related subjects; complete a year's clinical internship or the equivalent; pass a national examination administered by the Academy of Nutrition and Dietetics; and maintain up-to-date knowledge and **registration** by participating in required continuing education activities, such as attending seminars, taking courses, or conducting research.

Some states allow anyone to use the title dietitian or nutritionist, but others allow only an RD or people with specified qualifications to call themselves dietitians. Many states provide a further guarantee: a state registration, certification, or **license to practice.** In this way, states identify people who have met minimal standards of education and experience. Still, these state standards may fall short of those defining an RD. Similarly, some alternative educational programs qualify a graduate as a **certified nutritionist, certified nutritional consultant,** or **certified nutrition therapist**—terms that sound authoritative but lack the credentials of an RD. In fact, even Eddie, an English cocker spaniel, was able to obtain a certificate of membership from the American Association of Nutritional Consultants.[4]

Dietitians perform a multitude of duties in many settings in most communities. They work in the food industry, pharmaceutical companies, home health agencies, long-term care institutions, private practice, public health departments, research centers, education settings, fitness centers, and hospitals. Depending on their work settings, dietitians can assume a number of different job responsibilities and positions. In hospitals, administrative dietitians manage the foodservice system; clinical dietitians provide client care; and nutrition support team dietitians coordinate nutrition care with other health-care professionals. In the food industry, dietitians conduct research, develop products, and market services.

Public health dietitians who work in government-funded agencies such as health departments or clinics play a key role in delivering nutrition services to people in the community. Among their many

Eddie displays his membership certificate from an association of nutritional consultants. His human companion, Connie Diekman, is a registered dietitian and past president of the Academy of Nutrition and Dietetics (formerly the American Dietetic Association).

roles, public health dietitians help plan, coordinate, and evaluate food assistance programs; act as consultants to other agencies; manage finances; and much more.

Dietetic Technician, Registered (DTR)

In some facilities, a **dietetic technician** assists registered dietitians in both administrative and clinical responsibilities. A dietetic technician has been educated and trained to work under the guidance of a registered dietitian; upon passing a national examination, the title changes to **dietetic technician, registered (DTR).**

Other Dietary Employees

In addition to the dietetic technician, other dietary employees may include clerks, aides, cooks, porters, and assistants. These dietary employees do not have extensive formal training in nutrition, and their ability to provide accurate information may be limited.

Identifying Fake Credentials

In contrast to registered dietitians, thousands of people obtain fake nutrition degrees and claim to be nutrition consultants or doctors of "nutrimedicine." These and other such titles may sound meaningful, but most of these people lack the established credentials and training of an RD. If you look closely, you can see signs of their fake expertise.

Consider educational background, for example. The minimum standards of education for a dietitian specify a bachelor of science (BS) degree in food science and human nutrition or related fields from an **accredited** college or university.* Such a degree generally requires 4 to 5 years of study. Similarly, minimum standards of education for a dietetic technician specify an associate degree that typically requires 2 years of study. In contrast, a fake nutritionist may display a degree from a 6-month course. Such a degree simply falls short. In some cases, businesses posing as schools offer even less—they sell certificates to anyone who pays the fees. To obtain these "degrees," a candidate need not attend any classes, read any books, or pass any examinations.

To safeguard educational quality, an accrediting agency recognized by the US Department of Education (DOE) certifies that certain schools meet criteria established to ensure that an institution provides complete and accurate schooling. Unfortunately, fake nutrition degrees are available from schools "accredited" by phony accrediting agencies. Acquiring false degrees and credentials is especially easy today, with **diploma mills** and **fraudulent** businesses operating via the Internet.

Knowing the qualifications of someone who provides nutrition information can help you determine whether that person's advice might be harmful or helpful. Don't be afraid to ask for credentials. Table H1-1 on p. 32 lists credible sources of nutrition information.

Red Flags of Nutrition Quackery

Figure H1-2 on p. 32 features eight red flags consumers can use to identify nutrition **misinformation.** Sales of unproven and dangerous products have always been a concern, but the Internet now provides merchants with an easy and inexpensive way to reach millions of customers around the world. Because of the difficulty in regulating the Internet, fraudulent and illegal sales of medical products have hit a bonanza. As is the case with the air, no one owns the Internet, and similarly, no one has control over the pollution. Countries have different laws regarding sales of drugs, dietary supplements, and other health products, but applying these laws to the Internet marketplace is almost impossible. Even if illegal activities could be defined and identified, finding the person responsible for a particular website is not always possible. Websites can appear and disappear in a blink of a cursor. Now, more than ever, consumers must heed the caution "Buyer beware."

In summary, when you hear nutrition news, consider its source. Ask yourself these two questions: Is the person providing the information qualified to speak on nutrition? Is the information based on valid scientific research? If not, find a better source. After all, your health depends on it.

*To ensure the quality and continued improvement of nutrition and dietetics education programs, an agency known as the Commission on Accreditation for Dietetics Education (CADE) establishes and enforces eligibility requirements and accreditation standards for programs preparing students for careers as registered dietitians or dietetics technicians. Programs meeting those standards are accredited by CADE.

TABLE H1-1 Credible Sources of Nutrition Information

Government agencies, volunteer associations, consumer groups, and professional organizations provide consumers with reliable health and nutrition information. Credible sources of nutrition information include:

- Nutrition and food science departments at a university or community college
- Local agencies such as the health department or County Cooperative Extension Service
- Government resources such as:

Centers for Disease Control and prevention (CDC)	**www.cdc.gov**
Department of Agriculture (USDA)	**www.usda.gov**
Department of Health and Human Services (DHHS)	**www.hhs.gov**
Dietary Guidelines for Americans	**www.cnpp.usda.gov/dietaryguidelines.htm**
Food and Drug Administration (FDA)	**www.fda.gov**
Health Canada	**www.hc-sc.gc.ca/index-eng.php**
Healthy People	**www.healthy people.gov**
Let's Move!	**www.letsmove.gov**
MyPlate	**www.choosemyplate.gov**
National Institutes of Health	**www.nih.gov**
Physical Activity Guidelines for Americans	**www.health.gov/paguidelines**

- Volunteer health agencies such as:

American Cancer Society	**www.cancer.org**
American Diabetes Association	**www.diabetes.org**
American Heart Association	**www.americanheart.org**

- Reputable consumer groups such as:

American Council on Science and Health	**www.acsh.org**
Federal Citizen Information Center	**www.usa.gov**
International Food Information Council	**www.foodinsight.org**

- Professional health organizations such as:

Academy of Nutrition and Dietetics	**www.eatright.org**
American Medical Association	**www.ama-assn.org**
Dietitians of Canada	**www.dietitians.ca**

- Journals such as:

American Journal of Clinical Nutrition	**www.ajcn.org**
Journal of the Academy of Nutrition and Dietetics	**www.adajournal.org**
New England Journal of Medicine	**www.nejm.org**
Nutrition Reviews	**www.ilsi.org**

> FIGURE H1-2 Red Flags of Nutrition Quackery

Satisfaction guaranteed
Marketers may make generous promises, but consumers won't be able to collect on them.
One product does it all
No one product can possibly treat such a diverse array of conditions.
Time tested or newfound treatment
Such findings would be widely publicized and accepted by health professionals.
Paranoid accusations
And this product's company doesn't want money? At least the drug company has scientific research proving the safety and effectiveness of its products.
Quick and easy fixes
Even proven treatments take time to be effective.
Personal testimonials
Hearsay is the weakest form of evidence.
Natural
Natural is not necessarily better or safer; any product that is strong enough to be effective is strong enough to cause side effects.
Meaningless medical jargon
Phony terms hide the lack of scientific proof.
Guaranteed! OR your money back!
"Cures gout, ulcers, diabetes and cancer"
Instant recovery, back to your everyday schedule
Wonder Pills
W
"Best pills around"
The natural way to becoming a better you
Revolutionary product, based on ancient medicine
Super Trim
S
Money grabbing drug companies further corporate means
"My friends feel good as new!"
Beats the hunger stimulation point (HSP)

REFERENCES

1. R. A. Cohen and P. F. Adams, Use of the Internet for health information: United States, 2009, *NCHS Data Brief,* July 2011.
2. Position of the American Dietetic Association: Food and nutrition misinformation, *Journal of the American Dietetic Association* 106 (2006): 601–607.
3. Position of the American Dietetic Association: The roles of registered dietitians and dietetic technicians, registered in health promotion and disease prevention, *Journal of the American Dietetic Association* 106 (2006): 1875–1884.
4. Who's dishing out your nutrition advice? Consumers beware: Make sure your source is a registered dietitian, www.eatright.org, accessed March 3, 2008.

© HSNphotography/Shutterstock.com

CENGAGE LEARNING'S GLOBAL NUTRITION WATCH — GLOBAL NUTRITION RESOURCE CENTER

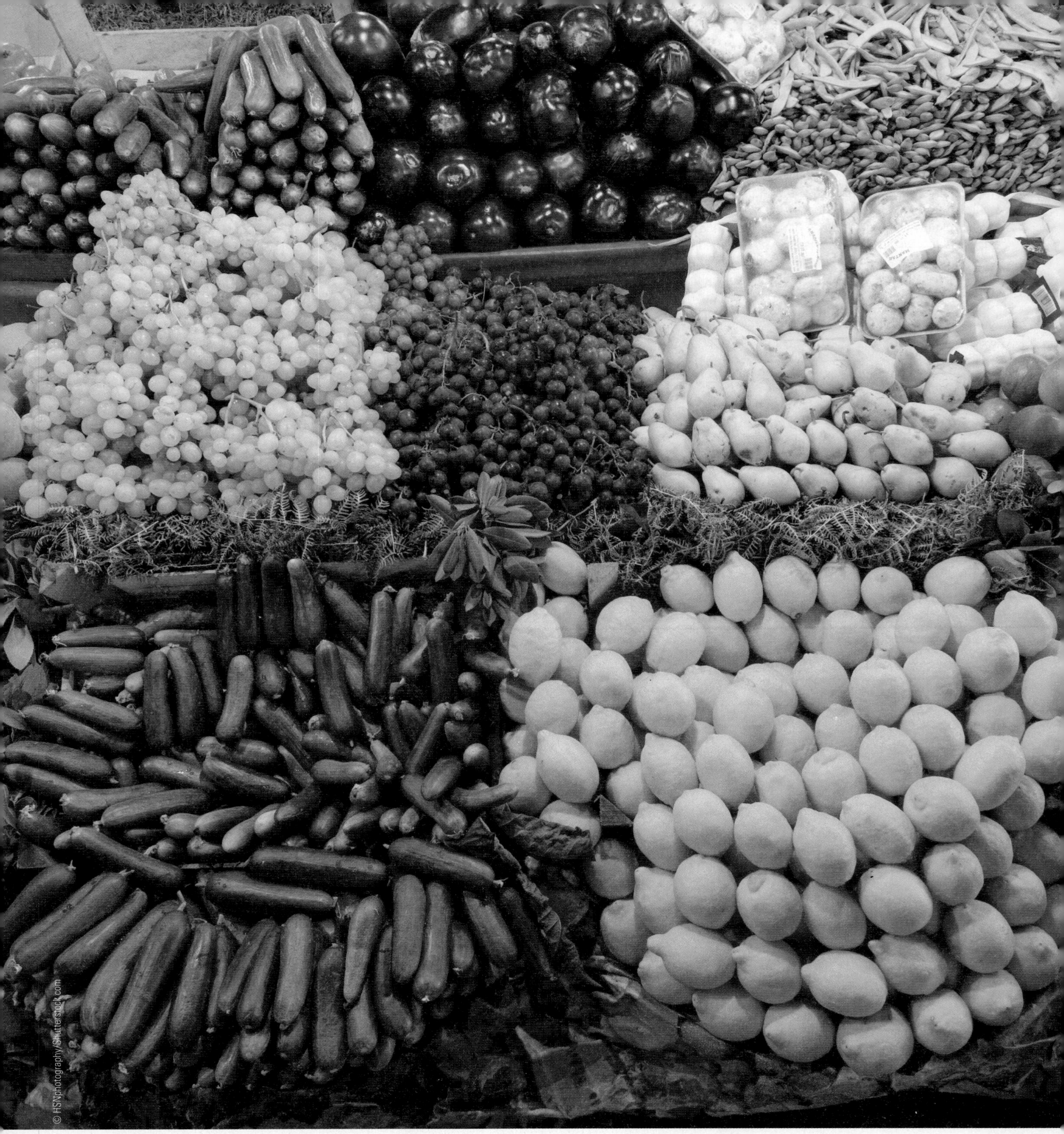

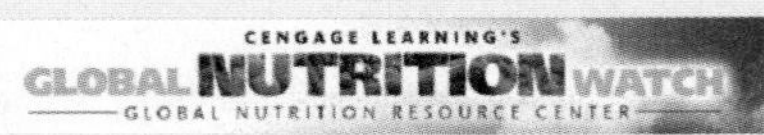

The USDA recently updated many of its guidelines, including the introduction of MyPlate. Go to Nutrition Basics and Tools → Nutrition Standards & Guidelines to learn more about these updates. How do these updates help consumers create a well-balanced diet?

2

Planning a Healthy Diet

LEARNING GPS

Nutrition in Your Life

You make food choices—deciding what to eat and how much to eat—more than 1000 times every year. We eat so frequently that it's easy to choose a meal without giving any thought to its nutrient contributions or health consequences. Even when we want to make healthy choices, we may not know which foods to select or how much to consume. With a few tools and tips, you can learn to plan a healthy diet. In the Nutrition Portfolio exercise at the end of this chapter, you can compare your current diet against a healthy eating plan.

Chapter 1 explained that the body's many activities are supported by the nutrients delivered by the foods people eat. Food choices made over years influence the body's health, and consistently poor choices increase the risks of developing chronic diseases. This chapter shows how a person can select from the tens of thousands of available foods to create a diet that supports good health. Fortunately, most foods provide several nutrients, so one trick for wise diet planning is to select a combination of foods that deliver a full array of nutrients. This chapter begins by introducing the diet-planning principles and dietary guidelines that assist people in selecting foods that will deliver sufficient nutrients without excess energy (kcalories).

2.1 Principles and Guidelines

LEARN IT Explain how each of the diet-planning principles can be used to plan a healthy diet.

How well you nourish yourself does not depend on the selection of any one food. Instead, it depends on the overall **eating pattern**—the combination of many different foods and beverages at numerous meals over days, months, and years.[1] Diet-planning principles and dietary guidelines are key concepts to keep in mind

eating pattern: customary intake of foods and beverages over time.

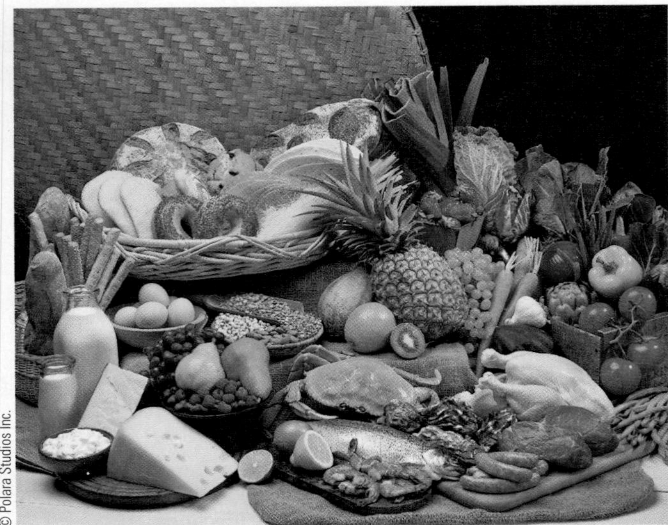

© Polara Studios Inc.

To ensure an adequate and balanced diet, eat a variety of foods daily, choosing different foods from each group.

whenever you are selecting foods—whether shopping at the grocery store, choosing from a restaurant menu, or preparing a home-cooked meal.

Diet-Planning Principles

Diet planners have developed several ways to select foods. Whatever plan or combination of plans they use, though, they keep in mind the six basic diet-planning principles ♦ listed in the margin.

♦ Diet-planning principles:
- Adequacy
- Balance
- kCalorie (energy) control
- Nutrient density
- Moderation
- Variety

Adequacy **Adequacy** reflects a diet that provides sufficient energy and enough of all the nutrients to meet the needs of healthy people. Take the essential nutrient iron, for example. Because the body loses some iron each day, people have to replace it by eating foods that contain iron. A person whose diet fails to provide enough iron-rich foods may develop the symptoms of iron-deficiency anemia: the person may feel weak, tired, and listless; have frequent headaches; and find that even the smallest amount of muscular work brings disabling fatigue. To prevent these deficiency symptoms, a person must include foods that supply adequate iron. The same is true for all the other essential nutrients introduced in Chapter 1.

Balance **Balance** in the diet helps to ensure adequacy. The art of balancing the diet involves consuming enough—but not too much—of different types of foods in proportion to one another. In a balanced diet, foods rich in some nutrients do not crowd out foods that are rich in other nutrients. The essential minerals calcium and iron, taken together, illustrate the importance of dietary balance. Meat is rich in iron but poor in calcium. Conversely, milk is rich in calcium but poor in iron. Use some meat for iron; use some milk for calcium; and save some space for other foods, too, because a diet consisting of milk and meat alone would not be adequate. For the other nutrients, people need to eat other protein foods, whole grains, vegetables, and fruits.

kCalorie (Energy) Control Designing an adequate diet within a reasonable kcalorie allowance requires careful planning. Once again, balance plays a key role. The amount of energy coming into the body from foods should balance with the amount of energy being used by the body to sustain its metabolic and physical activities. Upsetting this balance leads to gains or losses in body weight. The discussion of energy balance and weight control in Chapters 8 and 9 examines this issue in more detail, but one key to **kcalorie control** is to select foods of high **nutrient density.**

Nutrient Density Nutrient density promotes adequacy and kcalorie control. To eat well without overeating, select nutrient-dense foods—that is, foods that deliver the most nutrients for the least food energy.[2] Consider foods containing calcium, for example. You can get about 300 milligrams of calcium from either 1½ ounces of cheddar cheese or 1 cup of fat-free milk, but the cheese delivers about twice as much food energy (kcalories) as the milk. The fat-free milk, then, is twice as calcium dense as the cheddar cheese; it offers the same amount of calcium for half the kcalories. Both foods are excellent choices for adequacy's sake alone, but to achieve adequacy while controlling kcalories, the fat-free milk is the better choice. (Alternatively, a person could select a low-fat cheddar cheese with its kcalories comparable to fat-free milk.) The accompanying "How To" describes how to compare foods based on nutrient density.

Just as a financially responsible person pays for rent, food, clothes, and tuition on a limited budget, healthy people obtain iron, calcium, and all the other essential nutrients on a limited energy (kcalorie) allowance. Success depends on getting many nutrients for each kcalorie "dollar." As Figure 2-1 illustrates, a breakfast of cereal, fruit, egg, and sausage delivers many more nutrients than a couple of doughnuts—even though they both provide about the same number of kcalories. A person who makes nutrient-dense choices can meet daily nutrient needs on a lower energy budget. Such choices support good health.

adequacy (dietary): providing all the essential nutrients, fiber, and energy in amounts sufficient to maintain health.

balance (dietary): providing foods in proportion to one another and in proportion to the body's needs.

kcalorie (energy) control: management of food energy intake.

nutrient density: a measure of the nutrients a food provides relative to the energy it provides. The more nutrients and the fewer kcalories, the higher the nutrient density.

>How To

Compare Foods Based on Nutrient Density

One way to evaluate foods is simply to notice their nutrient contribution *per serving*: 1 cup of milk provides about 300 milligrams of calcium, and ½ cup of fresh, cooked turnip greens provides about 100 milligrams. Thus a serving of milk offers three times as much calcium as a serving of turnip greens. To get 300 milligrams of calcium, a person could choose either 1 cup of milk or 1½ cups of turnip greens.

Another valuable way to evaluate foods is to consider their nutrient density—their nutrient contribution *per kcalorie*. Fat-free milk delivers about 85 kcalories with its 300 milligrams of calcium. To calculate the nutrient density, divide milligrams by kcalories:

$$\frac{300 \text{ mg calcium}}{85 \text{ kcal}} = 3.5 \text{ mg per kcal}$$

Do the same for the fresh turnip greens, which provide 15 kcalories with the 100 milligrams of calcium:

$$\frac{100 \text{ mg calcium}}{15 \text{ kcal}} = 6.7 \text{ mg per kcal}$$

The more milligrams per kcalorie, the greater the nutrient density. Turnip greens are more calcium dense than milk. They provide more calcium *per kcalorie* than milk, but milk offers more calcium *per serving*. Both approaches offer valuable information, especially when combined with a realistic appraisal. What matters most is which are you more likely to consume—1½ cups of turnip greens or 1 cup of milk? You can get 300 milligrams of calcium from either, but the greens will save you about 40 kcalories (the savings would be even greater if you usually use whole milk).

Keep in mind, too, that calcium is only one of the many nutrients that foods provide. Similar calculations for protein, for example, would show that fat-free milk provides more protein both *per kcalorie* and *per serving* than turnip greens—that is, milk is more protein dense. Combining variety with nutrient density helps to ensure the adequacy of all nutrients.

TRY IT Compare the thiamin density of 3 ounces of lean t-bone steak (174 kcalories, 0.09 milligrams thiamin) with ½ cup of fresh cooked broccoli (27 kcalories, 0.05 milligrams thiamin).

> FIGURE 2-1 **Nutrient Density of Two Breakfast Options Compared**

Chapter 1 presented this figure, illustrating energy density—that for the same number of kcalories, the breakfast on the left delivered less energy per gram of food, which benefits weight management. These two breakfasts also illustrate nutrient density—that for the same number of kcalories, the breakfast on the left delivers more nutrients per kcalorie.

Nutrient-dense breakfast

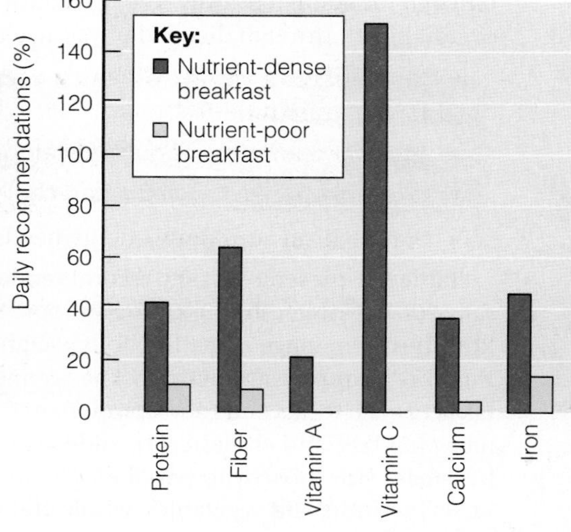

Nutrient-poor breakfast

Foods that are notably low in nutrient density—such as potato chips, candy, and colas—are called **empty-kcalorie foods.** The kcalories these foods provide are called "empty" because they deliver energy (from added sugars, solid fats, or both) with little, or no, protein, vitamins, or minerals.

The concept of nutrient density is relatively simple when examining the contributions of one nutrient to a food or diet. With respect to calcium, milk ranks high and meats rank low. With respect to iron, meats rank high and milk ranks low. But it is a more complex task to answer the question, which food is more nutritious? To answer that question, we need to consider several nutrients—including both nutrients that may harm health as well as those that may be beneficial.[3] Ranking foods based on their overall nutrient composition is known as **nutrient profiling.**[4] Researchers have yet to agree on an ideal way to rate foods based on the nutrient profile, but when they do, nutrient profiling will be quite useful in helping consumers identify nutritious foods and plan healthy diets.[5]

Moderation **Moderation** contributes to adequacy, balance, and kcalorie control. Foods rich in fat and sugar often provide enjoyment and energy but relatively few nutrients; in addition, they promote weight gain when eaten in excess. A person practicing moderation eats such foods only on occasion and regularly selects foods low in **solid fats** and **added sugars**, a practice that automatically improves nutrient density. Returning to the example of cheddar cheese versus fat-free milk, the fat-free milk not only offers the same amount of calcium for less energy, but it also contains far less fat than the cheese.

Variety **Variety** improves nutrient adequacy. A diet may have all of the virtues just described and still lack variety, if a person eats the same foods day after day. People should select foods from each of the food groups daily and vary their choices within each food group from day to day for several reasons. First, different foods within the same group contain different arrays of nutrients. Among the fruits, for example, strawberries are especially rich in vitamin C while apricots are rich in vitamin A. Second, no food is guaranteed entirely free of substances that, in excess, could be harmful. The strawberries might contain trace amounts of one contaminant, the apricots another. By alternating fruit choices, a person will ingest very little of either contaminant. (Contamination of foods is discussed in Chapter 19.) Third, as the adage goes, variety is the spice of life. A person who eats beans frequently can enjoy pinto beans in Mexican burritos today, garbanzo beans in a Greek salad tomorrow, and baked beans with barbecued chicken on the weekend. Eating nutritious meals need never be boring.

Dietary Guidelines for Americans

What should a person eat to stay healthy? The answers can be found in the *Dietary Guidelines for Americans*. These guidelines translate the *nutrient* recommendations of the DRI (presented in Chapter 1) into *food* recommendations.[6] The result is evidence-based advice designed to help people attain and maintain a healthy weight, reduce the risk of chronic diseases, and promote overall health through diet and physical activity. In general, a healthy diet:

- Emphasizes a variety of fruits, vegetables, whole grains, and fat-free and low-fat milk products.
- Includes lean meats, poultry, seafood, legumes, eggs, seeds, and nuts.
- Is low in saturated and *trans* fats, cholesterol, salt (sodium), and added sugars.
- Stays within your daily energy needs for your recommended body weight.

Table 2-1 presents the key recommendations of the *Dietary Guidelines for Americans 2010*, clustered into four major topic areas. The first area focuses on balancing kcalories to manage a healthy body weight by improving eating habits and engaging in regular physical activity. The second area advises people to reduce their intakes of such foods and food components as sodium, solid fats (with their saturated fats, *trans* fats, and cholesterol), added sugars, refined grain products, and alcoholic beverages (for those who partake). The third area encourages people to consume a variety of fruits and vegetables, whole grains, and low-fat milk products and protein

empty-kcalorie foods: a popular term used to denote foods that contribute energy but lack protein, vitamins, and minerals.

nutrient profiling: ranking foods based on their nutrient composition.

moderation (dietary): providing enough but not too much of a substance.

solid fats: fats that are not usually liquid at room temperature; commonly found in most foods derived from animals and vegetable oils that have been hydrogenated. Solid fats typically contain more saturated and *trans* fats than most oils (Chapter 5 provides more details).

added sugars: sugars and other kcaloric sweeteners that are added to foods during processing, preparation, or at the table. Added sugars do not include the naturally occurring sugars found in fruits and milk products.

variety (dietary): eating a wide selection of foods within and among the major food groups.

TABLE 2-1 Key Recommendations of the *Dietary Guidelines for Americans 2010*

Balancing kCalories to Manage Weight

- Prevent and/or reduce overweight and obesity through improved eating and physical activity behaviors (see Chapter 9).
- Control total kcalorie intake to manage body weight. For people who are overweight or obese, this will mean consuming fewer kcalories from foods and beverages (see Chapter 8).
- Increase physical activity and reduce time spent in sedentary behaviors (see Chapter 14).
- Maintain appropriate kcalorie balance during each stage of life—childhood, adolescence, adulthood, pregnancy and breastfeeding, and older age (see Chapters 14–16).

Foods and Food Components to Reduce

- Reduce daily sodium intake to less that 2300 milligrams and further reduce intake to 1500 milligrams among persons who are 51 and older and those of any age who are African American or have hypertension, diabetes, or chronic kidney disease (see Chapter 12).
- Consume less than 10 percent of kcalories from saturated fatty acids by replacing them with monounsaturated and polyunsaturated fatty acids (see Chapter 5).
- Consume less than 300 milligrams per day of dietary cholesterol (see Chapter 5).
- Keep *trans* fatty acid consumption as low as possible by limiting foods that contain synthetic sources of *trans* fats, such as partially hydrogenated oils, and by limiting other solid fats (see Chapter 5).
- Reduce the intake of kcalories from solid fats and added sugars (see Chapters 4–5).
- Limit the consumption of foods that contain refined grains, especially refined grain foods that contain solid fats, added sugars, and sodium (see Chapters 4, 5, and 12).
- If alcohol is consumed it should be consumed in moderation—up to one drink per day for women and two drinks per day for men—and only by adults of legal drinking age (see Highlight 7).

Foods and Nutrients to Increase

- Increase vegetable and fruit intake.
- Eat a variety of vegetables, especially dark-green and red and orange vegetables and beans and peas.
- Consume at least half of all grains as whole grains. Increase whole-grain intake by replacing refined grains with whole grains.
- Increase intake of fat-free or low-fat milk and milk products, such as milk, yogurt, cheese, or fortified soy beverages.
- Choose a variety of protein foods, which include seafood, lean meat and poultry, eggs, beans and peas, soy products, and unsalted nuts and seeds.
- Increase the amount and variety of seafood consumed by choosing seafood in place of some meat and poultry.
- Replace protein foods that are higher in solid fats with choices that are lower in solid fats and kcalories and/or are sources of oils.
- Use oils to replace solid fats where possible (see Highlight 5).
- Choose foods that provide more potassium, dietary fiber, calcium, and vitamin D, which are nutrients of concern in American diets (see Chapters 4, 11, and 12). These foods include vegetables, fruits, whole grains, and milk and milk products.

Building Healthy Eating Patterns

- Select an eating pattern that meets nutrient needs over time at an appropriate kcalorie level.
- Account for all foods and beverages consumed and assess how they fit within a total healthy eating pattern.
- Follow food safety recommendations when preparing and eating foods to reduce the risk of foodborne illnesses (see Chapter 19).

NOTE: These guidelines are intended for adults and healthy children ages 2 and older.
SOURCE: The *Dietary Guidelines for Americans*, available at www.healthierus.gov/dietaryguidelines.

foods (including seafood). The fourth area helps consumers build healthy eating patterns that meet energy and nutrient needs while reducing the risk of foodborne illnesses. Together, the *Dietary Guidelines for Americans 2010* point the way toward longer, healthier, and more active lives. These key recommendations, along with additional recommendations for specific population groups, appear throughout the text as their subjects are discussed.

By law, the *Dietary Guidelines for Americans* are reviewed and revised as needed every five years. Each edition shares some similarities with previous editions but also sets precedent in new ways. Perhaps most noteworthy to the 2010 edition is

© AISPIX/Shutterstock.com

The *Dietary Guidelines* encourage Americans to increase the energy (kcalories) they expend through physical activity.

the overarching focus on curbing the obesity epidemic and improving the health of the American population.[7]

Some people might wonder why *dietary* guidelines include recommendations for physical activity. The simple answer is that most people who maintain a healthy body weight do more than eat right. They also exercise—the equivalent of 30 to 60 minutes or more of moderately intense physical activity on most days.[8] As you will see repeatedly throughout this text, food and physical activity choices are integral partners in supporting good health.

REVIEW IT Explain how each of the diet-planning principles can be used to plan a healthy diet.

A well-planned diet delivers adequate nutrients, a balanced array of nutrients, and an appropriate amount of energy. It is based on nutrient-dense foods, moderate in substances that can be detrimental to health, and varied in its selections. The *Dietary Guidelines* apply these principles, offering practical advice on how to eat for good health.

2.2 Diet-Planning Guides

LEARN IT Use the USDA Food Patterns to develop a meal plan within a specified energy allowance.

To plan a diet that achieves all of the dietary ideals just outlined, a person needs tools as well as knowledge. Among the most widely used tools for diet planning are **food group plans** that build a diet from clusters of foods that are similar in nutrient content. Thus each food group represents a set of nutrients that differs somewhat from the nutrients supplied by the other groups. Selecting foods from each of the groups eases the task of creating an adequate and balanced diet.

USDA Food Patterns The *Dietary Guidelines* encourage consumers to adopt a balanced eating pattern, using the USDA's Food Patterns. The USDA Food Patterns assign foods to five major groups ♦ and recommend daily amounts of foods from each group to meet nutrient needs. Figure 2-2 (pp. 42–43) presents the food groups, the most notable nutrients of each group, the serving equivalents, and the foods within each group. Chapter 16 provides a food guide for young children.

♦ Five food groups:
- Fruits
- Vegetables
- Grains
- Protein foods
- Milk and milk products

Recommended Amounts All food groups offer valuable nutrients, and people should make selections from each group daily. Table 2-2 (p. 44) specifies the amounts of foods from each group needed daily to create a healthful diet for several energy (kcalorie) levels. A person needing 2000 kcalories a day, for example, would select 2 cups of fruit; 2½ cups of vegetables; 6 ounces of grain foods; 5½ ounces of protein foods; and 3 cups of milk or milk products.* Additionally, a small amount of unsaturated oil, such as vegetable oil, or the oils of nuts, olives, or fatty fish, is required to supply needed nutrients. Table 2-3 (p. 44) presents estimated daily kcalorie needs for adults.

> DIETARY GUIDELINES FOR AMERICANS 2010

Select an eating pattern that meets nutrient needs over time at an appropriate kcalorie level.

All vegetables provide an array of nutrients, but some vegetables are especially good sources of certain vitamins, minerals, and beneficial phytochemicals. For this reason, the vegetable group is sorted into five subgroups. The dark-green vegetables deliver the B vitamin folate; the red and orange vegetables provide

food group plans: diet-planning tools that sort foods into groups based on nutrient content and then specify that people should eat certain amounts of foods from each group.

*Milk and milk products also can be referred to as dairy products.

vitamin A; legumes supply iron and protein; the starchy vegetables contribute carbohydrate energy; and the other vegetables fill in the gaps and add more of these same nutrients.

In a 2000-kcalorie diet, then, the recommended 2½ cups of daily vegetables should be varied among the subgroups over a week's time. In other words, consuming 2½ cups of potatoes or even nutrient-rich spinach every day for seven days does *not* meet the recommended amounts for vegetables. Potatoes and spinach make excellent choices when consumed in balance with vegetables from the other subgroups. One way to help ensure selections for all of the subgroups is to eat vegetables of various colors—for example, green broccoli, orange sweet potatoes, black beans, yellow corn, and white cauliflower. Intakes of vegetables are appropriately averaged over a week's time—it is not necessary to include every subgroup every day.

For similar reasons, the protein foods group is sorted into three subgroups. Perhaps most notably, each of these subgroups contributes a different assortment of fats. Table 2-4 (p. 44) presents the recommended *weekly* amounts for each of the subgroups for vegetables and protein foods.

Notable Nutrients As Figure 2-2 notes, each food group contributes key nutrients. This feature provides flexibility in diet planning because a person can select any food from a food group (or its subgroup) and receive similar nutrients. For example, a person can choose milk, cheese, or yogurt and receive the same key nutrients. Importantly, foods provide not only these key nutrients, but small amounts of other nutrients and phytochemicals as well.

Legumes contribute the same key nutrients—notably, protein, iron, and zinc—as meats, poultry, and seafood. They are also excellent sources of fiber, folate, and potassium, which are commonly found in vegetables. To encourage frequent consumption of these nutrient-rich foods, legumes are included as a subgroup of both the vegetable group and the protein foods group. Thus legumes can be counted in either the vegetable group or the protein foods group.[9] In general, people who regularly eat meat, poultry, and seafood count legumes as a vegetable, and vegetarians and others who seldom eat meat, poultry, or seafood count legumes in the protein foods group.

The USDA Food Patterns encourage greater consumption from certain food groups to provide the nutrients most often lacking in the diets of Americans. ♦ In general, most people need to eat:

- *More* vegetables, fruits, whole grains, seafood, and milk and milk products.
- *Less* sodium, saturated fat, *trans* fat, cholesterol, and *fewer* refined grains and foods and beverages with solid fats and added sugars.

♦ Nutrients of concern:
- Dietary fiber
- Vitamin D
- Calcium
- Potassium

Nutrient-Dense Choices A healthy eating pattern emphasizes nutrient-dense options within each food group. By consistently selecting nutrient-dense foods, a person can obtain all the nutrients needed and still keep kcalories under control. In contrast, eating foods that are low in nutrient density makes it difficult to get enough nutrients without exceeding energy needs and gaining weight. For this reason, consumers should select low-fat foods from each group and foods without solid fats or added sugars—for example, fat-free milk instead of whole milk, baked chicken without the skin instead of hot dogs, green beans instead of french fries, orange juice instead of fruit punch, and whole-wheat bread instead of biscuits. Notice that Figure 2-2 indicates which foods *within each group* contain solid fats and/or added sugars and therefore should be limited. Oil is a notable exception: even though oil is pure fat and therefore rich in kcalories, a small amount of oil from sources such as nuts, fish, or vegetable oils is necessary every day to provide nutrients lacking from other foods. Consequently, these high-fat foods are listed among the nutrient-dense foods (see Highlight 5 to learn why).

> DIETARY GUIDELINES FOR AMERICANS 2010

Consume foods from all food groups in nutrient-dense forms and in recommended amounts. Reduce the intake of kcalories from solid fats and added sugars.

legumes (lay-GYOOMS, LEG-yooms): plants of the bean and pea family, with seeds that are rich in protein compared with other plant-derived foods.

> FIGURE 2-2 USDA Food Patterns: Food Groups and Subgroups

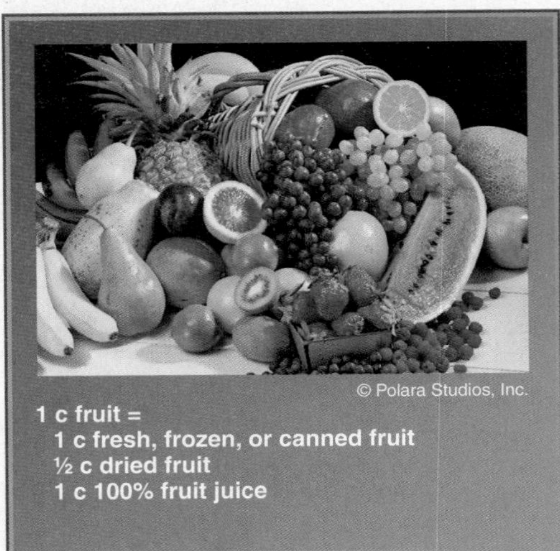

© Polara Studios, Inc.

1 c fruit =
1 c fresh, frozen, or canned fruit
½ c dried fruit
1 c 100% fruit juice

Fruits contribute folate, vitamin A, vitamin C, potassium, and fiber.

Consume a variety of fruits, and choose whole or cut-up fruits more often than fruit juice.

Apples, apricots, avocados, bananas, blueberries, cantaloupe, cherries, grapefruit, grapes, guava, honeydew, kiwi, mango, nectarines, oranges, papaya, peaches, pears, pineapples, plums, raspberries, strawberries, tangerines, watermelon; dried fruit (dates, figs, prunes, raisins); 100% fruit juices

Limit these fruits that contain solid fats and/or added sugars:
Canned or frozen fruit in syrup; juices, punches, ades, and fruit drinks with added sugars; fried plantains

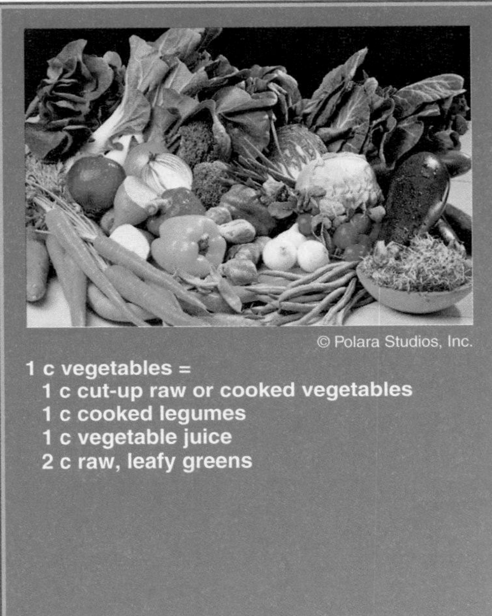

© Polara Studios, Inc.

1 c vegetables =
1 c cut-up raw or cooked vegetables
1 c cooked legumes
1 c vegetable juice
2 c raw, leafy greens

Vegetables contribute folate, vitamin A, vitamin C, vitamin K, vitamin E, magnesium, potassium, and fiber.

Consume a variety of vegetables each day, and choose from all five subgroups several times a week.

Dark-green vegetables: Broccoli and leafy greens such as arugula, beet greens, bok choy, collard greens, kale, mustard greens, romaine lettuce, spinach, turnip greens, watercress

Red and orange vegetables: Carrots, carrot juice, pumpkin, red bell peppers, sweet potatoes, tomatoes, tomato juice, vegetable juice, winter squash (acorn, butternut)

Legumes: Black beans, black-eyed peas, garbanzo beans (chickpeas), kidney beans, lentils, navy beans, pinto beans, soybeans and soy products such as tofu, split peas, white beans

Starchy vegetables: Cassava, corn, green peas, hominy, lima beans, potatoes

Other vegetables: Artichokes, asparagus, bamboo shoots, bean sprouts, beets, brussels sprouts, cabbages, cactus, cauliflower, celery, cucumbers, eggplant, green beans, green bell peppers, iceberg lettuce, mushrooms, okra, onions, seaweed, snow peas, zucchini

Limit these vegetables that contain solid fats and/or added sugars:
Baked beans, candied sweet potatoes, coleslaw, french fries, potato salad, refried beans, scalloped potatoes, tempura vegetables

© Polara Studios, Inc.

1 oz grains =
1 slice bread
½ c cooked rice, pasta, or cereal
1 oz dry pasta or rice
1 c ready-to-eat cereal
3 c popped popcorn

Grains contribute folate, niacin, riboflavin, thiamin, iron, magnesium, selenium, and fiber.

Make most (at least half) of the grain selections whole grains.

Whole grains: amaranth, barley, brown rice, buckwheat, bulgur, cornmeal, millet, oats, quinoa, rye, wheat, wild rice and whole-grain products such as breads, cereals, crackers, and pastas; popcorn

Enriched refined products: bagels, breads, cereals, pastas (couscous, macaroni, spaghetti), pretzels, white rice, rolls, tortillas

Limit these grains that contain solid fats and/or added sugars:
Biscuits, cakes, cookies, cornbread, crackers, croissants, doughnuts, fried rice, granola, muffins, pastries, pies, presweetened cereals, taco shells

Art © Cengage Learning 2013

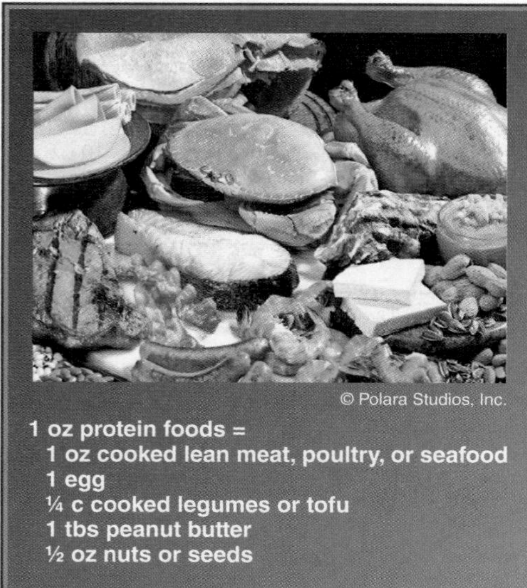

© Polara Studios, Inc.

1 oz protein foods =
1 oz cooked lean meat, poultry, or seafood
1 egg
¼ c cooked legumes or tofu
1 tbs peanut butter
½ oz nuts or seeds

Protein foods contribute protein, essential fatty acids, niacin, thiamin, vitamin B_6, vitamin B_{12}, iron, magnesium, potassium, and zinc.

Choose a variety of protein foods from the three subgroups, including seafood in place of meat or poultry twice a week.

Seafood: Fish (catfish, cod, flounder, haddock, halibut, herring, mackerel, pollock, salmon, sardines, sea bass, snapper, trout, tuna), shellfish (clams, crab, lobster, mussels, oysters, scallops, shrimp)

Meats, poultry, eggs: Lean or low-fat meats (fat-trimmed beef, game, ham, lamb, pork, veal), poultry (no skin), eggs

Nuts, seeds, soy products: Unsalted nuts (almonds, cashews, filberts, pecans, pistachios, walnuts), seeds (flaxseeds, pumpkin seeds, sesame seeds, sunflower seeds), legumes, soy products (textured vegetable protein, tofu, tempeh), peanut butter, peanuts

Limit these protein foods that contain solid fats and/or added sugars:
Bacon; baked beans; fried meat, seafood, poultry, eggs, or tofu; refried beans; ground beef; hot dogs; luncheon meats; marbled steaks; poultry with skin; sausages; spare ribs

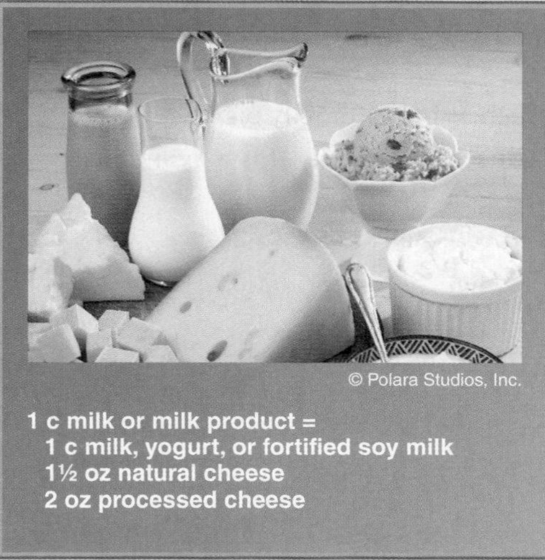

© Polara Studios, Inc.

1 c milk or milk product =
1 c milk, yogurt, or fortified soy milk
1½ oz natural cheese
2 oz processed cheese

Milk and milk products contribute protein, riboflavin, vitamin B_{12}, calcium, potassium, and, when fortified, vitamin A and vitamin D.

Make fat-free or low-fat choices. Choose other calcium-rich foods if you don't consume milk.

Fat-free or 1% low-fat milk and fat-free or 1% low-fat milk products such as buttermilk, cheeses, cottage cheese, yogurt; fat-free fortified soy milk

Limit these milk products that contain solid fats and/or added sugars:
2% reduced-fat milk and whole milk; 2% reduced-fat and whole-milk products such as cheeses, cottage cheese, and yogurt; flavored milk with added sugars such as chocolate milk, custard, frozen yogurt, ice cream, milk shakes, pudding, sherbet; fortified soy milk

© Matthew Farruggio

1 tsp oil =
1 tsp vegetable oil
1 tsp soft margarine
1 tbs low-fat mayonnaise
2 tbs light salad dressing

Oils are not a food group, but are featured here because they contribute vitamin E and essential fatty acids.

Use oils instead of solid fats, when possible.

Liquid vegetable oils such as canola, corn, flaxseed, nut, olive, peanut, safflower, sesame, soybean, sunflower oils; mayonnaise, oil-based salad dressing, soft *trans*-free margarine; unsaturated oils that occur naturally in foods such as avocados, fatty fish, nuts, olives, seeds (flaxseeds, sesame seeds), shellfish

Limit these solid fats:
Butter, animal fats, stick margarine, shortening

TABLE 2-2 USDA Food Patterns: Recommended Daily Amounts from Each Food Group

	1600 kcal	1800 kcal	2000 kcal	2200 kcal	2400 kcal	2600 kcal	2800 kcal	3000 kcal
Fruits	1½ c	1½ c	2 c	2 c	2 c	2 c	2½ c	2½ c
Vegetables	2 c	2½ c	2½ c	3 c	3 c	3½ c	3½ c	4 c
Grains	5 oz	6 oz	6 oz	7 oz	8 oz	9 oz	10 oz	10 oz
Protein foods	5 oz	5 oz	5½ oz	6 oz	6½ oz	6½ oz	7 oz	7 oz
Milk and milk products	3 c	3 c	3 c	3 c	3 c	3 c	3 c	3 c
Oils	5 tsp	5 tsp	6 tsp	6 tsp	7 tsp	8 tsp	8 tsp	10 tsp
Discretionary kcalories	121 kcal	161 kcal	258 kcal	266 kcal	330 kcal	362 kcal	395 kcal	459 kcal

TABLE 2-3 Estimated Daily kCalorie Needs for Adults

	Sedentary[a]	Active[b]
Women		
19–30 yr	1900	2400
31–50 yr	1800	2200
51+ yr	1600	2100
Men		
19–30 yr	2500	3000
31–50 yr	2300	2900
51+ yr	2100	2600

[a]Sedentary describes a lifestyle that includes only the activities typical of day-to-day life.

[b]Active describes a lifestyle that includes physical activity equivalent to walking more than 3 miles per day at a rate of 3 to 4 miles per hour, in addition to the activities typical of day-to-day life.

NOTE: kCalorie values reflect the midpoint of the range appropriate for age and gender, but within each group, older adults may need fewer kcalories and younger adults may need more. In addition to gender, age, and activity level, energy needs vary with height and weight (see Chapter 8 and Appendix F).

Discretionary kCalories At each kcalorie level, people who consistently choose nutrient-dense foods may be able to meet most of their nutrient needs without consuming their full allowance of kcalories. The difference between the kcalories needed to supply nutrients and those needed to maintain weight might be considered **discretionary kcalories** (see Figure 2-3).

Discretionary kcalories allow a person to choose whether to:

- Eat additional nutrient-dense foods, such as an extra serving of skinless chicken or a second ear of corn.
- Select a few foods with fats or added sugars, such as reduced-fat milk or sweetened cereal.
- Add a little fat or sugar to foods, such as butter or jelly on toast.
- Consume some alcohol. (Highlight 7 explains why this may not be a good choice for some individuals.)

Alternatively, a person wanting to lose weight might choose to:

- *Not* use discretionary kcalories.

> DIETARY GUIDELINES FOR AMERICANS 2010

For most people, no more than about 5 to 15 percent of kcalories from solid fats and added sugars (empty kcalories) can be reasonably accommodated in the USDA Food Patterns, which are designed to meet nutrient needs within kcalorie limits.

Serving Equivalents Recommended serving amounts for fruits, vegetables, and milk are measured in cups, and those for grains and protein foods, in ounces. Figure 2-2 provides the **serving sizes** and equivalent measures for foods in each group specifying, for example, that 1 ounce of grains is equivalent to 1 slice of bread or ½ cup of cooked rice.

discretionary kcalories: the kcalories remaining in a person's energy allowance after consuming enough nutrient-dense foods to meet all nutrient needs for a day.

serving sizes: the standardized quantity of a food; such information allows comparisons when reading food labels and consistency when following the *Dietary Guidelines*.

TABLE 2-4 USDA Food Patterns: Recommended Weekly Amounts from the Vegetable and Protein Foods Subgroups

Table 2-2 specifies the recommended amounts of total vegetables and protein foods per *day*. This table shows those amounts dispersed among five vegetable and three protein foods subgroups per *week*.

	1600 kcal	1800 kcal	2000 kcal	2200 kcal	2400 kcal	2600 kcal	2800 kcal	3000 kcal
Vegetable Subgroups								
Dark green	1½ c	1½ c	1½ c	2 c	2 c	2½ c	2½ c	2½ c
Red and orange	4 c	5½ c	5½ c	6 c	6 c	7 c	7 c	7½ c
Legumes	1 c	1½ c	1½ c	2 c	2 c	2½ c	2½ c	3 c
Starchy	4 c	5 c	5 c	6 c	6 c	7 c	7 c	8 c
Other	3½ c	4 c	4 c	5 c	5 c	5½ c	5½ c	7 c
Protein Foods Subgroups								
Seafood	8 oz	8 oz	8 oz	9 oz	10 oz	10 oz	11 oz	11 oz
Meats, poultry, eggs	24 oz	24 oz	26 oz	29 oz	31 oz	31 oz	34 oz	34 oz
Nuts, seeds, soy products	4 oz	4 oz	4 oz	4 oz	5 oz	5 oz	5 oz	5 oz

Consumers using the USDA Food Patterns can learn how standard serving sizes compare with their personal **portion sizes** ♦ by determining the answers to questions such as these: What portion of a cup is a small handful of raisins? Is a "helping" of mashed potatoes more or less than a half cup? How many ounces of cereal do you typically pour into the bowl? How many ounces is the steak at your favorite restaurant? How many cups of milk does your glass hold?

Ethnic Food Choices People can use the USDA Food Patterns and still enjoy a diverse array of culinary styles by sorting ethnic foods into their appropriate food groups. For example, a person eating Mexican foods would find tortillas in the grains group, jicama in the vegetable group, and guava in the fruit group. Table 2-5 (p. 46) features some ethnic food choices.

> FIGURE 2-3 **Discretionary kCalories in a 2000-kCalorie Diet**

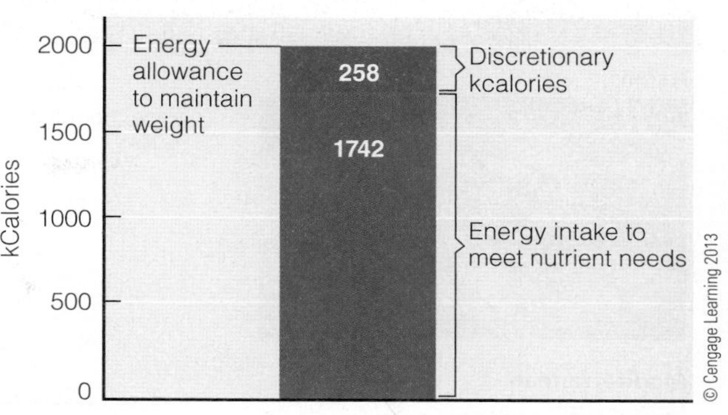

Vegetarian Food Guide Vegetarian diets are plant-based eating patterns that rely mainly on grains, vegetables, legumes, fruits, seeds, and nuts. Some vegetarian diets include eggs, milk products, or both. People who do not eat meats or milk products can still use the USDA Food Patterns to create an adequate diet.[10] The subgroups for protein foods have been reorganized to eliminate meats, poultry, and seafood (see Table H2-1 on p. 64). The other food groups and the recommended daily amounts for each food group remain the same. Highlight 2 defines vegetarian terms and provides details on planning healthy vegetarian diets.

♦ For quick and easy estimates, visualize each portion as being about the size of a common object:

- 1 c fruit or vegetables = a baseball
- ¼ c dried fruit or nuts = a golf ball
- 3 oz meat = a deck of cards
- 2 tbs peanut butter = a ping pong ball
- 1 oz cheese = 4 stacked dice
- ½ c ice cream = a racquetball
- 4 small cookies = 4 poker chips

Mixtures of Foods Some foods—such as casseroles, soups, and sandwiches—fall into two or more food groups. With a little practice, consumers can learn to see these mixtures of foods as items from various food groups. For example, from the USDA Food Patterns point of view, a taco represents four different food groups: the taco shell from the grains group; the onions, lettuce, and tomatoes from the vegetables group; the ground beef from the protein foods group; and the cheese from the milk group.

MyPlate The USDA created an educational tool called MyPlate to illustrate the five food groups. Figure 2-4 (p. 46) shows the MyPlate icon, which was designed to remind consumers to make healthy food choices.

The MyPlate icon divides a plate into four sections, each representing a food group—fruits, vegetables, grains, and protein. The sections vary in size, indicating relative proportion each food group contributes to a healthy diet. A circle next to the plate represents the milk group (dairy).

The MyPlate icon does not stand alone as an educational tool. A wealth of information can be found at the website (**www.choosemyplate.gov**). Consumers can choose the kinds and amounts of foods they need to eat each day based on their height, weight, age, gender, and activity level. Information is also available for children, pregnant and lactating women, and for vegetarians. In addition to creating a personal plan, consumers can find daily tips to help them improve their diet and increase physical activity. A key message of the website is to enjoy food, but eat less by avoiding oversized portions.

Most bagels today weigh in at 4 ounces or more—meaning that a person eating one of these large bagels for breakfast is actually getting four or more grain servings, not one.

Recommendations versus Actual Intakes The USDA Food Patterns and MyPlate were developed to help people choose a balanced and healthful diet. Are consumers actually eating according to these recommendations? The short answer is "not really." In general, consumers are not selecting the most nutrient-dense items from the food groups. Instead, they are consuming too many foods high in solid fats and added sugars—soft drinks, desserts, whole milk products, and fatty meats.[11] They are also not selecting the suggested quantities from each of the food groups, typically eating too few fruits, vegetables, whole grains, and milk products (see Figure 2-5, p. 47).

An assessment tool, called the **Healthy Eating Index,** can be used to measure how well a diet meets the recommendations of the *Dietary Guidelines*.[12] Various components of the diet are given scores that reflect the quantities consumed per

portion sizes: the quantity of a food served or eaten at one meal or snack; *not* a standard amount.

Healthy Eating Index: a measure that assesses how well a diet meets the recommendations of the *Dietary Guidelines for Americans*.

TABLE 2-5 **Ethnic Food Choices**

	Grains	Vegetables	Fruits	Protein Foods	Milk and Milk Products
Asian © Josh Resnick/Shutterstock.com	Rice, noodles, millet	Amaranth, baby corn, bamboo shoots, chayote, bok choy, mung bean sprouts, sugar peas, straw mushrooms, water chestnuts, kelp	Carambola, guava, kumquat, lychee, persimmon, melons, mandarin orange	Soybeans and soy products such as soy milk and tofu, squid, duck eggs, pork, poultry, fish and other seafood, peanuts, cashews	Usually excluded
Mediterranean © PhotoDisc/Getty Images	Pita pocket bread, pastas, rice, couscous, polenta, bulgur, focaccia, Italian bread	Eggplant, tomatoes, peppers, cucumbers, grape leaves	Olives, grapes, figs	Fish and other seafood, gyros, lamb, chicken, beef, pork, sausage, lentils, fava beans	Ricotta, provolone, parmesan, feta, mozzarella, and goat cheeses; yogurt
Mexican © PhotoDisc/Getty Images	Tortillas (corn or flour), taco shells, rice	Chayote, corn, jicama, tomato salsa, cactus, cassava, tomatoes, yams, chilies	Guava, mango, papaya, avocado, plantain, bananas, oranges	Refried beans, fish, chicken, chorizo, beef, eggs	Cheese, custard

1000-kcalorie intake. For most components, higher intakes result in higher scores. For example, selecting at least 3 ounces of grains with at least half of them whole grains gives a score of 10 points, whereas selecting no grains gives a score of 0 points. For a few components, lower intakes provide higher scores. For example, less than 7 percent kcalories from saturated fat receives 10 points, but more than 15 percent gets 0 points. An assessment of recent nutrition surveys using the Healthy Eating Index reports that the American diet scores 58 out of a possible 100 points.[13] To improve this score, the American diet needs to decrease kcalories from solid fats and added sugars by about 60 percent; increase fruits by 100 percent and vegetables and milk products by 70 percent; maintain the quantity of grains but shift the quality to four times as many whole grains; and reduce salt by more than half.[14]

> FIGURE 2-4 **MyPlate**

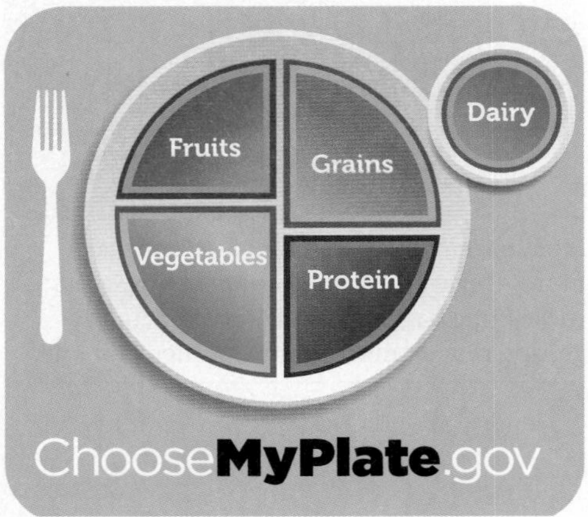

SOURCE: USDA, www.choosemyplate.gov.

MyPlate Shortcomings MyPlate is not perfect and critics are quick to point out its flaws.[15] The first main criticism is that MyPlate fails to convey enough information to help consumers choose a healthy diet. MyPlate contains few words and depends on its website to provide key information—which is helpful for those who have Internet access and are willing to take the time to become familiar with its teachings. The second main criticism is that MyPlate fails to recognize that some foods within a food group are healthier choices than others. For example, MyPlate does not distinguish between fish sticks and salmon or between broccoli and french fries. Many of the upcoming chapters examine the links between diet and health, and Chapter 18 presents a complete summary, including a look at an alternative Healthy Eating Plate created by the faculty members in the Harvard School of Public Health.

exchange lists: diet-planning tools that organize foods by their proportions of carbohydrate, fat, and protein. Foods on any single list can be used interchangeably.

Exchange Lists

Food group plans are particularly well suited to help a person achieve dietary adequacy, balance, and variety. **Exchange lists** provide additional help in achieving kcalorie control and moderation. Originally developed as a

> FIGURE 2-5 **Recommended and Actual Intakes Compared**

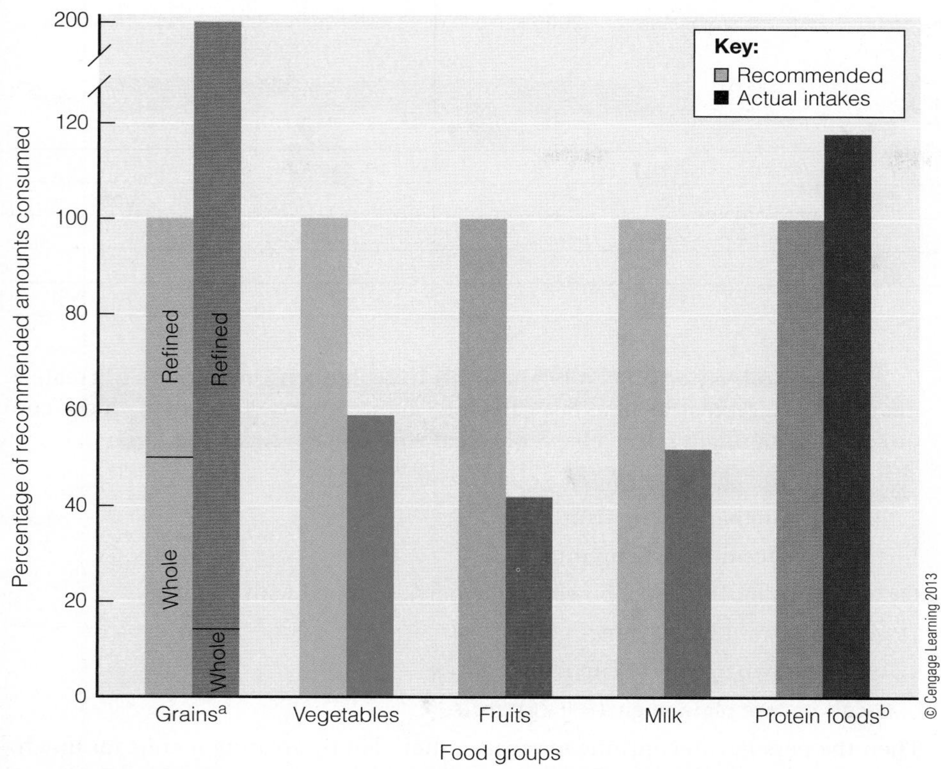

[a]At least half of the grain selections should be whole grains.
[b]On average, actual intakes of all protein foods is close to recommended levels, but actual intakes of the seafood subgroup is only 44 percent of recommended levels.

meal planning guide for people with diabetes, exchange lists have proved useful for general diet planning as well.

Unlike the USDA Food Patterns, which sort foods primarily by their vitamin and mineral contents, the exchange system sorts foods according to their energy-nutrient contents. Consequently, foods do not always appear on the exchange list where you might first expect to find them. For example, cheeses are grouped with meats because, like meats, cheeses contribute energy from protein and fat but provide negligible carbohydrate. (In the USDA Food Patterns presented earlier, cheeses are grouped with milk because they are milk products with similar calcium contents.)

For similar reasons, starchy vegetables such as corn, green peas, and potatoes are listed with grains on the starch list in the exchange system, rather than with the vegetables. Likewise, olives are not classed as a "fruit" as a botanist would claim; they are classified as a "fat" because their fat content makes them more similar to oil than to berries. Cream cheese, bacon, and nuts are also on the fat list to remind users of their high fat content. These groupings highlight the characteristics of foods that are significant to energy intake. To learn more about this useful diet-planning tool, study Appendix G, which gives details of the exchange system used in the United States.

Putting the Plan into Action Familiarizing yourself with each of the food groups is the first step in diet planning. Table 2-6 (p. 48) shows how to use the 2000-kcalorie USDA Food Pattern to plan a diet. The amounts listed from each of the food groups (see the second column of the table) were taken from Table 2-2. The next step is to assign the food groups to meals (and snacks), as shown in the remaining columns of Table 2-6.

TABLE 2-6 Diet Planning Using the 2000-kCalorie USDA Food Pattern

This diet plan is one of many possibilities. It follows the amounts of foods suggested for a 2000-kcalorie diet as shown in Table 2-2 on p. 44 (with a little less oil).

Food Group	Amounts	Breakfast	Lunch	Snack	Dinner	Snack
Fruits	2 c	½ c		½ c	1 c	
Vegetables	2½ c		1 c		1½ c	
Grains	6 oz	1 oz	2 oz	½ oz	2 oz	½ oz
Protein foods	5½ oz		2 oz		3½ oz	
Milk and milk products	3 c	1 c		1 c		1 c
Oils	6 tsp		1½ tsp		4 tsp	

At this point, a person can begin to fill in a plan with real foods to create a menu. For example, the breakfast calls for 1 ounce grain, ½ cup fruit, and 1 cup milk. A person might select a bowl of cereal with banana slices and milk:

1 cup cereal = 1 ounce grain

½ large banana = ½ cup fruit

1 cup fat-free milk = 1 cup milk

Or ½ English muffin and a bowl of strawberries topped with yogurt:

½ English muffin = 1 ounce grain

½ cup strawberries = ½ cup fruit

1 cup fat-free plain yogurt = 1 cup milk

Then the person can continue to create a diet plan by creating menus for lunch, dinner, and snacks. The final menu might look like the one presented in Table 2-7.

TABLE 2-7 A Sample Menu

This sample menu provides about 1850 kcalories and meets the recommendations to provide 45 to 65 percent of kcalories from carbohydrate, 20 to 35 percent from fat, and 10 to 35 percent from protein.

Amounts	Sample Menu	Energy (kcal)
	Breakfast	
1 oz whole grains	1 c whole-grain cereal	108
1 c milk	1 c fat-free milk	100
½ c fruit	1 medium banana (sliced)	105
	Lunch	
2 oz meats, 2 oz whole grains	1 turkey sandwich on whole-wheat roll	272
1½ tsp oils	1½ tbs low-fat mayonnaise	71
1 c vegetables	1 c vegetable juice	50
	Snack	
½ oz whole grains	4 whole-wheat reduced-fat crackers	86
1 c milk	1½ oz low-fat cheddar cheese	74
½ c fruit	1 medium apple	72
	Dinner	
½ c vegetables	1 c raw spinach leaves	8
¼ c vegetables	¼ c shredded carrots	11
1 oz meats	¼ c garbanzo beans	71
2 tsp oils	2 tbs oil-based salad dressing and olives	76
¾ c vegetables, 2½ oz meat, 2 oz enriched grains	Spaghetti with meat and tomato sauce	425
½ c vegetables	½ c green beans	22
2 tsp oils	2 tsp soft margarine	67
1 c fruit	1 c strawberries	49
	Snack	
½ oz whole grains	3 graham crackers	90
1 c milk	1 c fat-free milk	100

As you can see, we all make countless food-related decisions daily—whether we have a plan or not. Following an eating pattern, such as the USDA Food Patterns, that incorporates health recommendations and diet-planning principles helps a person make wise nutrition decisions.

From Guidelines to Groceries Dietary recommendations emphasize nutrient-rich foods such as whole grains, fruits, vegetables, lean meats, poultry, seafood, and low-fat milk products. You can design such a diet for yourself, but how do you begin? Start with the foods you regularly enjoy eating and then try to make a few improvements.[16] For most people that will mean eating less red meat, cheeses, and salted snacks and more fruits, vegetables, whole grains, legumes, nuts, milk products, and seafood. Such small changes can dramatically improve the diet. When shopping, think of the food groups, and choose nutrient-dense foods within each group.

Be aware that many of the tens of thousands of food options available today are **processed foods** that have lost valuable nutrients and gained sugar, fat, and salt as they were transformed from farm-fresh foods to those found in the bags, boxes, and cans that line grocery-store shelves. Their value in the diet depends on the original food and how it was prepared or processed. By eating more fresh foods and fewer processed foods, consumers can reduce their intakes of added sugars, solid fats, and sodium for relatively little effort. Sometimes processed foods have been **fortified** to improve their nutrient contents, which can be helpful in increasing dietary intake of specific vitamins and minerals.

Grains When shopping for grain products, you will find them described as *refined, enriched,* or *whole grain*. These terms refer to the milling process and the making of grain products, and they have different nutrition implications (see Figure 2-6, p. 50). **Refined** grains have lost many nutrients during processing; **enriched** grains have had some nutrients added back; and **whole-grain** products have all the nutrients and fiber found in the original grain. As such, whole-grain products support good health and should account for at least half of the grains daily. Adding more whole grains to the diet can be as easy as eating oatmeal for breakfast and popcorn for a snack or substituting brown rice for white rice and whole-wheat bread for enriched white bread. To find whole-grain products, read food labels and select those that name a whole-grain ♦ first in the ingredient list. Products described as "multi-grain," "stone-ground," or "100% wheat" are usually *not* whole-grain products. Brown color is also not a useful hint, but fiber content often is.

When it became a common practice to refine the wheat flour used for bread by milling it and throwing away the bran and the germ, consumers suffered a tragic loss of many nutrients. As a consequence, in the early 1940s Congress passed legislation requiring that all grain products that cross state lines be enriched with iron, thiamin, riboflavin, and niacin. In 1996 this legislation was amended to include folate, a vitamin considered essential in the prevention of some birth defects. Most grain products that have been refined, such as rice, pastas such as macaroni and spaghetti, and cereals (both cooked and ready-to-eat types), have subsequently been enriched. ♦ Food labels must specify that products have been enriched and include the enrichment nutrients in the ingredients list.

♦ Examples of whole grains:
- Amaranth
- Barley
- Buckwheat
- Bulgur
- Corn (and popcorn)
- Millet
- Oats (and oatmeal)
- Quinoa
- Rice (brown or wild)
- Whole rye
- Whole wheat

♦ Grain enrichment nutrients:
- Iron
- Thiamin
- Riboflavin
- Niacin
- Folate

processed foods: foods that have been treated to change their physical, chemical, microbiological, or sensory properties.

fortified: the addition to a food of nutrients that were either not originally present or present in insignificant amounts. Fortification can be used to correct or prevent a widespread nutrient deficiency or to balance the total nutrient profile of a food.

refined: the process by which the coarse parts of a food are removed. When wheat is refined into flour, the bran, germ, and husk are removed, leaving only the endosperm.

enriched: the addition to a food of specific nutrients to replace losses that occur during processing so that the food will meet a specified standard.

whole grain: a grain that maintains the same relative proportions of starchy endosperm, germ, and bran as the original (all but the husk); not refined.

When shopping for bread, look for the descriptive words *whole grain* or *whole wheat* and check the fiber content on the Nutrition Facts panel of the label—the more fiber, the more likely the bread is a whole-grain product.

> FIGURE 2-6 **A Wheat Plant**

The protective coating of **bran** around the kernel of grain is rich in nutrients and fiber.

The **endosperm** contains starch and proteins.

The **germ** is the seed that grows into a wheat plant, so it is especially rich in vitamins and minerals to support new life.

The outer **husk** (or **chaff**) is the inedible part of a grain.

Whole-grain products contain much of the germ and bran, as well as the endosperm; that is why they are so nutritious. Refined grain products contain only the endosperm. Even with nutrients added back, they are not as nutritious as whole-grain products, as the next figure shows.

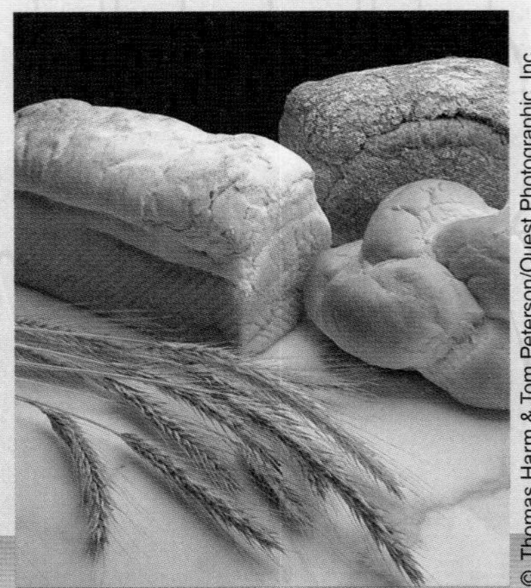

© Thomas Harm & Tom Peterson/Quest Photographic, Inc.

Common types of flour:

- **Refined flour:** finely ground endosperm that is usually enriched with nutrients and bleached for whiteness; sometimes called *white flour.*
- **Wheat flour:** any flour made from the endosperm of the wheat kernel.
- **Whole-wheat flour:** any flour made from the entire wheat kernel.

The difference between *white flour* and *white wheat* is noteworthy. Typically, *white flour* refers to refined flour (as defined above). Most flour—whether refined, white, or whole wheat—is made from red wheat. Whole-grain products made from red wheat are typically brown and full flavored.

To capture the health benefits of whole grains for consumers who prefer white bread, manufacturers use an albino variety of wheat called *white wheat.* Whole-grain products made from white wheat provide the nutrients and fiber of a whole grain with a light color and natural sweetness. Read labels carefully—white bread is a whole-grain product only if it is made from whole white wheat.

Enrichment doesn't make a slice of bread rich in these added nutrients, but people who eat several slices a day obtain significantly more of these nutrients than they would from unenriched bread. Even though the enrichment of flour helps to prevent deficiencies of these nutrients, it fails to compensate for losses of many other nutrients and fiber. As Figure 2-7 (p. 51) shows, whole-grain items deliver many more nutrients than the enriched ones. Only *whole-grain* flour contains all of the nutritive portions of the grain. Whole-grain products, such as brown rice and oatmeal, provide more nutrients and fiber and contain less salt, sugar, and fat than refined grain products. This helps to explain why diet quality tends to be better for consumers who eat more whole grains.[17]

> **DIETARY GUIDELINES FOR AMERICANS 2010**

Increase whole-grain intake. Consume at least half of all grains as whole grains. Whenever possible, replace refined grains with whole grains.

Speaking of processed foods, ready-to-eat breakfast cereals are the most highly fortified foods on the market. Like an enriched food, a *fortified* food has had nutrients added during processing, but in a fortified food, the added nutrients may not have been present in the original product. (The terms *fortified* and *enriched* may be used interchangeably.[18]) Some breakfast cereals made from refined flour and fortified with high doses of vitamins and minerals are actually more like dietary supplements disguised as cereals than they are like whole grains. They may be nutritious—with respect to the nutrients added—but they still may fail to convey the full spectrum of nutrients that a whole-grain food or a mixture of

> FIGURE 2-7 Nutrients in Bread

Whole-grain bread is more nutritious than other breads, even enriched bread. For iron, thiamin, riboflavin, niacin, and folate, enriched bread provides about the same quantities as whole-grain bread and significantly more than unenriched bread. For fiber and the other nutrients (those shown here as well as those not shown), enriched bread provides less than whole-grain bread.

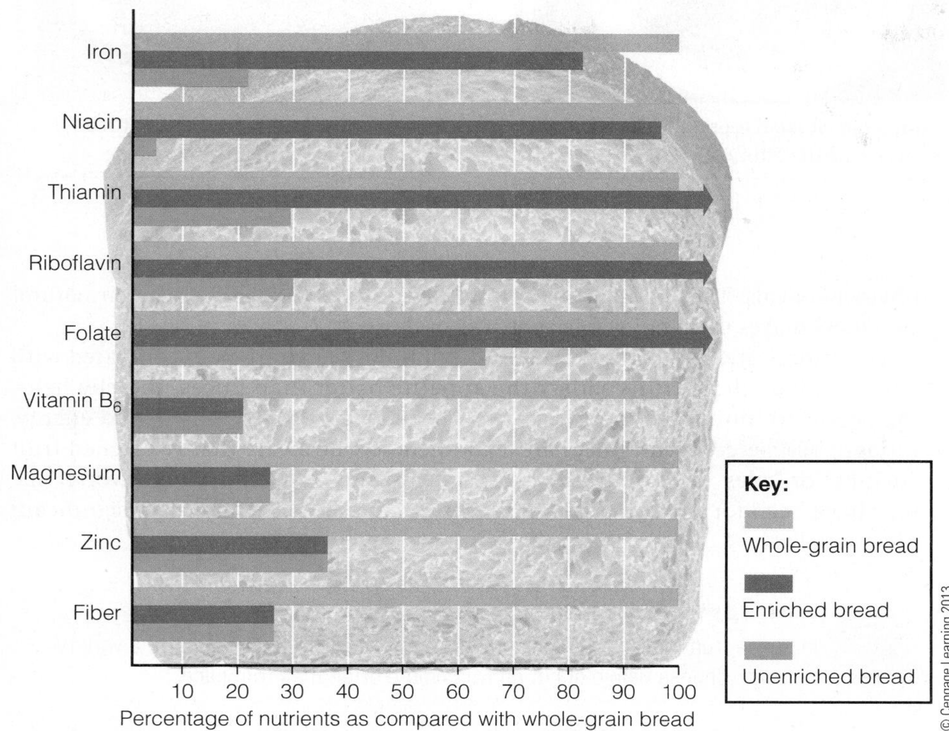

such foods might provide. Still, fortified foods help people meet their vitamin and mineral needs.[19]

Vegetables Posters in the produce section of grocery stores encourage consumers to "think variety, think color." Such efforts are part of a national educational campaign to increase fruit and vegetable consumption.

Choose fresh vegetables often, especially dark-green leafy and red and orange vegetables such as spinach, broccoli, tomatoes, and sweet potatoes. Cooked or raw, vegetables are good sources of vitamins, minerals, and fiber. Frozen and canned vegetables without added salt are acceptable alternatives to fresh. To control fat, energy, and sodium intakes, limit butter and salt on vegetables.

Choose often from the variety of legumes available. ♦ They are an economical, low-fat, nutrient- and fiber-rich food choice. Combining legumes with foods from other food groups creates delicious meals.

♦ Legumes include a variety of beans and peas:

- Adzuki beans
- Black beans
- Black-eyed peas
- Fava beans
- Garbanzo beans
- Great northern beans
- Kidney beans
- Lentils
- Lima beans
- Navy beans
- Peanuts
- Pinto beans
- Soybeans
- Split peas

> DIETARY GUIDELINES FOR AMERICANS 2010

Increase vegetable intake. Eat recommended amounts of vegetables and include a variety of vegetables, especially dark-green vegetables, red and orange vegetables, and legumes.

Fruit Choose fresh fruits often. Frozen, dried, and canned fruits without added sugar are acceptable alternatives to fresh. Fruits supply valuable vitamins, minerals, fibers, and

Consumers can remember to eat a variety of fruits and vegetables every day by selecting from each of five colors.

© PhotoDisc/Getty Images

Add rice to red beans for a hearty meal.

© Felicia Martinez/Photo Edit

Enjoy a Greek salad topped with garbanzo beans for a little ethnic diversity.

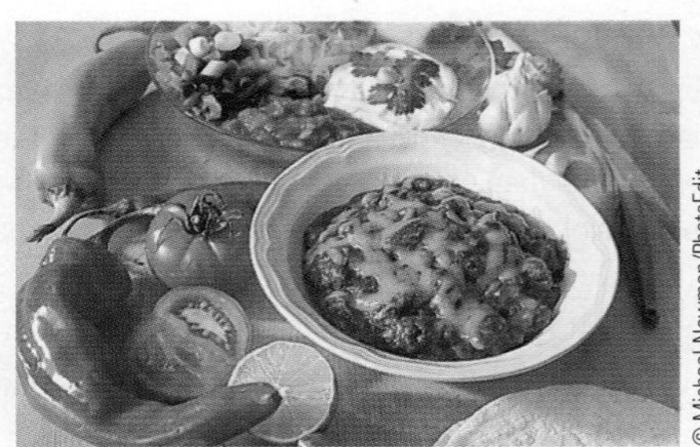
© Michael Newman/PhotoEdit

A bit of meat and lots of spices turn kidney beans into chili con carne.

phytochemicals. They add flavors, colors, and textures to meals, and their natural sweetness makes them enjoyable as snacks or desserts.

Fruit juices are healthy beverages but contain little dietary fiber compared with whole fruits. Whole fruits satisfy the appetite better than juices, thereby helping people to limit food energy intakes. For people who need extra food energy, though, 100 percent fruit juices are a good choice. Be aware that sweetened fruit "drinks" or "ades" contain mostly water, sugar, and a little juice for flavor. Some may have been fortified with vitamin C or calcium but lack any other significant nutritional value.

> DIETARY GUIDELINES FOR AMERICANS 2010

Increase fruit intake. Eat recommended amounts of fruits and choose a variety of fruits. Choose whole or cut-up fruits more often than fruit juice.

Protein Foods Protein foods include seafood, meats, poultry, eggs, legumes, soy products, nuts, and seeds. In addition to protein, these foods provide B vitamins, vitamin E, iron, zinc, and magnesium. To buy and prepare these foods without adding excess energy, fat, and sodium takes a little knowledge and planning.

When shopping in the meat department, choose lean cuts of beef and pork named "round" or "loin" (as in top round or pork tenderloin). As a guide, "prime" and "choice" cuts generally have more fat than "select" cuts. Restaurants usually serve prime cuts. Ground beef, even "lean" ground beef, derives most of its food energy from fat. Have the butcher trim and grind a lean round steak instead. Alternatively, soy products such as **textured vegetable protein** can be used instead of ground beef in a casserole, spaghetti sauce, or chili, saving fat kcalories. Because nuts and seeds are energy dense, they need to be consumed in small quantities and in place of—not in addition to—other protein foods. To lower sodium intake, choose unsalted nuts and seeds.

Serving sizes for meats, poultry, and seafood reflect weight after cooking and without bones. In general, 4 ounces of raw meat is equal to about 3 ounces of cooked meat. Some examples of 3-ounce portions include 1 medium pork chop, ½ chicken breast, or 1 steak or fish filet about the size of a deck of cards. To keep fat intake moderate, bake, roast, broil, grill, or braise meats, poultry, and seafood (but do not fry them in fat); remove the skin from poultry after cooking; trim visible fat before cooking; and drain fat after cooking. Chapter 5 offers many additional strategies for moderating fat intake.

> DIETARY GUIDELINES FOR AMERICANS 2010

Choose a variety of protein foods, which include seafood, lean meats and poultry, eggs, legumes, soy products, and unsalted nuts and seeds. Increase the amount and variety of seafood consumed by choosing seafood in place of some meat and poultry.

textured vegetable protein: processed soybean protein used in vegetarian products such as soy burgers.

Milk and Milk Products Shoppers find a variety of fortified foods in the dairy case. Examples are milk, to which vitamins A and D have been added, and soy milk, to which calcium, vitamin D, and vitamin B_{12} have been added. Be aware that not all soy beverages have been fortified. Read labels carefully.

In addition, shoppers may find **imitation foods** (such as cheese products), **food substitutes** (such as egg substitutes), and functional foods (such as margarine with added plant sterols). As food technology advances, many such foods offer alternatives to traditional choices that may help people reduce their saturated fat and cholesterol intakes. Chapter 5 provides other examples.

When shopping, choose fat-free ♦ or low-fat milk, yogurt, and cheeses. Such selections help consumers meet their vitamin and mineral needs within their energy and fat allowances. Milk products are important sources of calcium but can provide too much sodium and fat if not selected with care.

♦ Milk descriptions:
- Fat-free milk = nonfat, skim, zero-fat, or no-fat
- Low-fat milk = 1% milk
- Reduced-fat milk = 2% milk or less-fat

> DIETARY GUIDELINES FOR AMERICANS 2010
Increase intake of fat-free or low-fat milk and milk products—such as milk, yogurt, cheese, or fortified soy milk—and replace whole milk products with fat-free or low-fat options.

REVIEW IT Use the USDA Food Patterns to develop a meal plan within a specified energy allowance.

Food group plans such as the USDA Food Patterns help consumers select the types and amounts of foods to provide adequacy, balance, and variety in the diet. They make it easier to plan a diet that includes a balance of grains, vegetables, fruits, protein foods, and milk and milk products. In making any food choice, remember to view the food in the context of the total diet. The combination of many different foods provides the array of nutrients that is so essential to a healthy diet.

2.3 Food Labels

LEARN IT Compare and contrast the information on food labels to make selections that meet specific dietary and health goals.

Many consumers, especially those interested in preventing chronic diseases, read food labels to help them make healthy choices.[20] Food labels appear on virtually all packaged foods, and posters or brochures provide similar nutrition information for fresh meats, fruits, and vegetables (see Figure 2-8, p. 54). A few foods need not carry nutrition labels: those contributing few nutrients, such as plain coffee, tea, and spices; those produced by small businesses; and those prepared and sold in the same establishment. Producers of some of these items, however, voluntarily use food labels. Even markets selling nonpackaged items voluntarily present nutrient information, either in brochures or on signs posted at the point of purchase.

Restaurants with 20 or more locations must provide menu listings of an item's kcalories, grams of saturated fat, and milligrams of sodium.[21] Other restaurants need not supply nutrition information for menu items unless claims such as "low-fat" or "heart healthy" have been made. When ordering such items, keep in mind that restaurants tend to serve extra-large portions—two to three times standard serving sizes. A "low-fat" ice cream, for example, may have only 3 grams of fat per ½ cup, but you may be served 2 cups for a total of 12 grams of fat and all their accompanying kcalories.

The Ingredient List *All* packaged foods must list *all* ingredients—including additives used to preserve or enhance foods, such as vitamins and minerals added to enrich or fortify products. The ingredients are listed on the label in descending order of predominance by weight. Knowing that the first ingredient predominates

imitation foods: foods that substitute for and resemble another food, but are nutritionally inferior to it with respect to vitamin, mineral, or protein content. If the substitute is not inferior to the food it resembles and if its name provides an accurate description of the product, it need not be labeled "imitation."

food substitutes: foods that are designed to replace other foods.

> **FIGURE 2-8 Example of a Food Label**

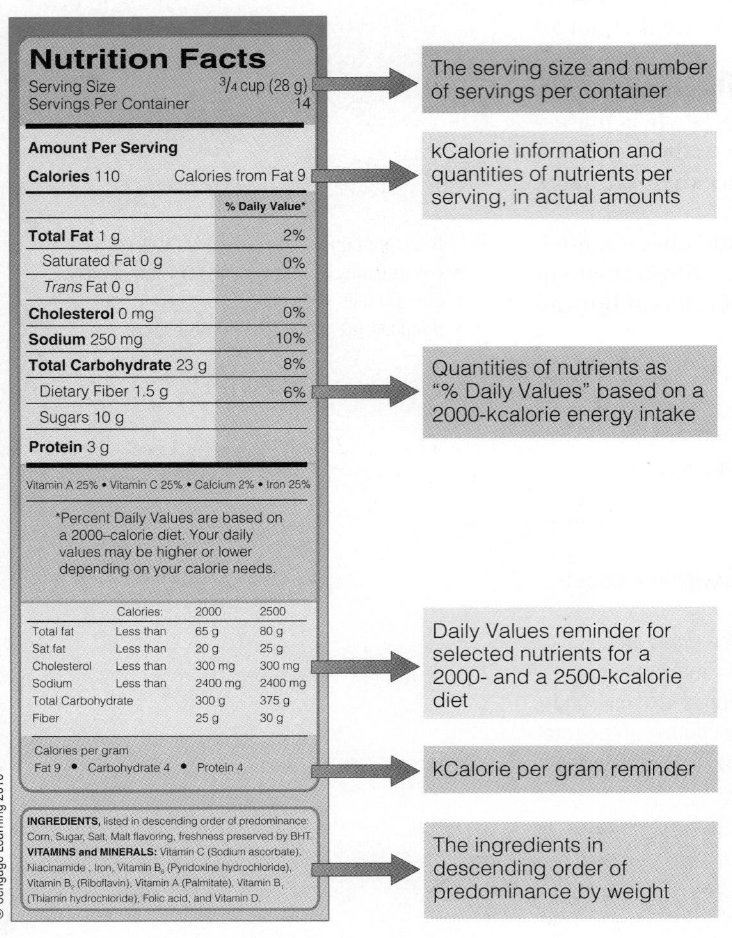

Nutrition Facts
Serving Size 3/4 cup (28 g)
Servings Per Container 14
Amount Per Serving
Calories 110 Calories from Fat 9
% Daily Value*
Total Fat 1 g 2%
Saturated Fat 0 g 0%
Trans Fat 0 g
Cholesterol 0 mg 0%
Sodium 250 mg 10%
Total Carbohydrate 23 g 8%
Dietary Fiber 1.5 g 6%
Sugars 10 g
Protein 3 g
Vitamin A 25% • Vitamin C 25% • Calcium 2% • Iron 25%
*Percent Daily Values are based on a 2000-calorie diet. Your daily values may be higher or lower depending on your calorie needs.
Calories: 2000 2500
Total fat Less than 65 g 80 g
Sat fat Less than 20 g 25 g
Cholesterol Less than 300 mg 300 mg
Sodium Less than 2400 mg 2400 mg
Total Carbohydrate 300 g 375 g
Fiber 25 g 30 g
Calories per gram
Fat 9 • Carbohydrate 4 • Protein 4
INGREDIENTS, listed in descending order of predominance: Corn, Sugar, Salt, Malt flavoring, freshness preserved by BHT. VITAMINS and MINERALS: Vitamin C (Sodium ascorbate), Niacinamide, Iron, Vitamin B6 (Pyridoxine hydrochloride), Vitamin B2 (Riboflavin), Vitamin A (Palmitate), Vitamin B1 (Thiamin hydrochloride), Folic acid, and Vitamin D.
The serving size and number of servings per container
kCalorie information and quantities of nutrients per serving, in actual amounts
Quantities of nutrients as "% Daily Values" based on a 2000-kcalorie energy intake
Daily Values reminder for selected nutrients for a 2000- and a 2500-kcalorie diet
kCalorie per gram reminder
The ingredients in descending order of predominance by weight

♦ Household and metric measures:

- 1 teaspoon (tsp) = 5 milliliters (mL)
- 1 tablespoon (tbs) = 15 mL
- 1 cup (c) = 240 mL
- 1 fluid ounce (fl oz) = 30 mL
- 1 ounce (oz) = 28 grams (g)

by weight, consumers can glean much information. Compare these products, for example:

- A beverage powder that contains "sugar, citric acid, natural flavors . . ." versus a juice that contains "water, tomato concentrate, concentrated juices of carrots, celery . . ."
- A cereal that contains "puffed milled corn, sugar, corn syrup, molasses, salt . . ." versus one that contains "100 percent rolled oats"
- A canned fruit that contains "sugar, apples, water" versus one that contains simply "apples, water"

In each of these comparisons, consumers can see that the second product is more nutrient dense.

Nutrition Facts Panel The FDA requires labels to include key nutrition facts. The Nutrition Facts panel provides such information as serving sizes, nutrient quantities, and Daily Values.

Serving Sizes Because labels present nutrient information based on one serving, they must identify the size of the serving. The Food and Drug Administration (FDA) has established specific serving sizes for various foods and requires that all labels for a given product use the same serving size. For example, the standard serving size for all ice creams is ½ cup and for all beverages it is 8 fluid ounces. This facilitates comparison shopping. Consumers can see at a glance which brand has more or fewer kcalories or grams of fat, for example. Standard serving sizes are expressed in both common household measures, ♦ such as cups, and metric measures, such as milliliters, to accommodate users of both types of measures.

When examining the nutrition facts on a food label, consumers need to compare the serving size on the label with how much they actually eat and adjust their calculations accordingly. For example, if the serving size is four cookies and you eat only two, then you need to cut the nutrient and kcalorie values in half; similarly, if you eat eight cookies, then you need to double the values. Notice, too, that small bags or individually wrapped items, such as chips or candy bars, may contain more than a single serving. The total number of servings per container is listed just below the serving size.

Be aware that serving sizes on food labels are not always the same as those of the USDA Food Patterns. For example, a serving of rice on a food label is 1 cup, whereas in the USDA Food Patterns it is ½ cup. Unfortunately, this discrepancy, coupled with each person's own perception (oftentimes misperception) of standard serving sizes, sometimes creates confusion for consumers trying to follow recommendations.

Nutrient Quantities In addition to the serving size and the servings per container, the FDA requires that the Nutrition Facts panel on food labels present nutrient information in two ways—in quantities (such as grams) and as percentages of standards called the **Daily Values.** The Nutrition Facts panel must provide the nutrient amount, **percent Daily Value,** or both for the following:

- Total food energy (kcalories)
- Food energy from fat (kcalories)
- Total fat (grams and percent Daily Value)
- Saturated fat (grams and percent Daily Value)

Daily Values (DV): reference values developed by the FDA specifically for use on food labels.

percent Daily Value (%DV): the percentage of a Daily Value recommendation found in a specified serving of food for key nutrients based on a 2000-kcalorie diet.

- *Trans* fat (grams)
- Cholesterol (milligrams and percent Daily Value)
- Sodium (milligrams and percent Daily Value)
- Total carbohydrate, which includes starch, sugar, and fiber (grams and percent Daily Value)
- Dietary fiber (grams and percent Daily Value)
- Sugars, which includes both those naturally present in and those added to the food (grams)
- Protein (grams)

The labels must also present nutrient content information as a percent Daily Value for the following vitamins and minerals:

- Vitamin A
- Vitamin C
- Iron
- Calcium

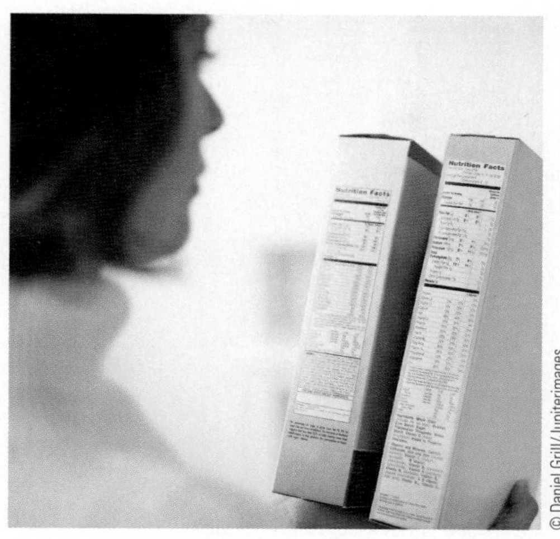

Consumers read food labels to learn about the nutrient contents of a food or to compare similar foods.

The Daily Values Table 2-8 presents the Daily Value standards for nutrients that are required to provide this information. Food labels list the amount of some nutrients in a product as a percentage of its Daily Value, which makes the numbers more meaningful to consumers. A person reading a food label might wonder, for example, whether 1 milligram of iron or calcium is a little or a lot. As Table 2-8 shows, the Daily Value for iron is 18 milligrams, so 1 milligram of iron is enough to notice—it is more than 5 percent, and that is what the food label will say. But because the Daily Value for calcium on food labels is 1000 milligrams, 1 milligram of calcium is insignificant, and the food label will read "0%."

The Daily Values reflect dietary recommendations for nutrients and dietary components that have important relationships with health. For example, for heart health, consumers are advised to limit saturated fat to 10 percent of energy intake. For a 2000-kcalorie diet, 10 percent is 200 kcalories, or 22 grams of fat. (Remember that fats deliver 9 kcalories per gram.) As Table 2-8 shows, the Daily Value for saturated fat has been rounded down to 20 grams.

The "% Daily Value" column on a label provides a ballpark estimate of how individual foods contribute to the total diet. It compares key nutrients in a serving of food with the goals of a person consuming 2000 kcalories per day. A 2000-kcalorie diet is considered about right for sedentary younger women, active older women, and sedentary older men. Young children and sedentary older women may need fewer kcalories. Most labels list, at the bottom, Daily Values for both a 2000-kcalorie and a 2500-kcalorie diet, but the "% Daily Value" column on all labels applies only to a 2000-kcalorie diet. A 2500-kcalorie diet is considered about right for many men, teenage boys, and active younger women. People who are exceptionally active may have still higher energy needs. Labels may also provide a reminder of the kcalories in a gram of carbohydrate, fat, and protein just below the Daily Value information (review Figure 2-8).

People who consume 2000 kcalories a day can simply add up all of the "% Daily Values" for a particular nutrient to see if their diet for the day fits recommendations. People who require more or less than 2000 kcalories daily must do some calculations to see how foods compare with their personal nutrition goals. Those interested can use the Calculation Factors column in Table 2-8 or the suggestions presented in the "How To" feature on p. 56.

TABLE 2-8 Daily Values for Food Labels

Food labels must present the "% Daily Value" for these nutrients.

Food Component	Daily Value	Calculation Factors
Fat	65 g	30% of kcalories
Saturated fat	20 g	10% of kcalories
Cholesterol	300 mg	—
Carbohydrate (total)	300 g	60% of kcalories
Fiber	25 g	11.5 g per 1000 kcalories
Protein	50 g	10% of kcalories
Sodium	2400 mg	—
Potassium	3500 mg	—
Vitamin C	60 mg	—
Vitamin A	1500 μg	—
Calcium	1000 mg	—
Iron	18 mg	—

NOTE: Daily Values were established for adults and children more than 4 years old. The values for energy-yielding nutrients are based on 2000 kcalories a day. For fiber, the Daily Value was rounded up from 23.

>How To

Calculate Personal Daily Values

The Daily Values on food labels are designed for a 2000-kcalorie intake, but you can calculate a personal set of Daily Values based on your energy allowance. Consider a 1500-kcalorie intake, for example. To calculate a daily goal for fat, multiply energy intake by 30 percent:

$$1500 \text{ kcal} \times 0.30 \text{ kcal from fat} = 450 \text{ kcal from fat}$$

The "kcalories from fat" are listed on food labels, so you can add all the "kcalories from fat" values for a day, using 450 as an upper limit. A person who prefers to count grams of fat can divide this 450 kcalories from fat by 9 kcalories per gram to determine the goal in grams:

$$450 \text{ kcal from fat} \div 9 \text{ kcal/g} = 50 \text{ g fat}$$

Alternatively, a person can calculate that 1500 kcalories is 75 percent of the 2000-kcalorie intake used for Daily Values:

$$1500 \text{ kcal} \div 2000 \text{ kcal} = 0.75$$
$$0.75 \times 100 = 75\%$$

Then, instead of trying to achieve 100 percent of the Daily Value, a person consuming 1500 kcalories will aim for 75 percent. Similarly, a person consuming 2800 kcalories would aim for 140 percent:

$$2800 \text{ kcal} \div 2000 \text{ kcal} = 1.40 \text{ or } 140\%$$

Table 2-8 includes a calculation column that can help you estimate your personal daily value for several nutrients.

TRY IT Calculate the Daily Values for a 1800-kcalorie diet and revise the Daily Value percentages on the cereal label found on p. 54.

♦ % Daily Values:

- ≥20% = high or excellent source
- 10–19% = good source
- < 5% = low source

Daily Values help consumers readily see whether a food contributes "a little" or "a lot" of a nutrient. ♦ For example, the "% Daily Value" column on a package of frozen macaroni and cheese may say 20 percent for fat. This tells the consumer that each serving of this food contains about 20 percent of the day's allotted 65 grams of fat. Be aware that for some nutrients (such as fat and sodium) you will want to select foods with a low "% Daily Value" and for others (such as calcium and fiber) you will want a high "% Daily Value." To determine whether a particular food is a wise choice, a consumer needs to consider its place in the diet among all the other foods eaten during the day.

Daily Values also make it easy to compare foods. For example, a consumer might discover that frozen macaroni and cheese has a Daily Value for fat of 20 percent, whereas macaroni and cheese prepared from a boxed mix has a Daily Value of 15 percent. By comparing labels, consumers who are concerned about their fat intakes can make informed decisions.

Claims on Labels

In addition to the Nutrition Facts panel, consumers may find various claims on labels. These claims include nutrient claims, health claims, and structure-function claims.

Nutrient Claims Have you noticed phrases such as "good source of fiber" on a box of cereal or "rich in calcium" on a package of cheese? These and other **nutrient claims** may be used on labels so long as they meet FDA definitions, which include the conditions under which each term can be used. For example, in addition to having less than 2 milligrams of cholesterol, a "cholesterol-free" product may not contain more than 2 grams of saturated fat and *trans* fat combined per serving. The accompanying glossary defines nutrient terms on food labels, including criteria for foods described as "low," "reduced," and "free." When nutrients have been added to enriched or fortified products, they must appear in the ingredients list.

nutrient claims: statements that characterize the quantity of a nutrient in a food.

Some descriptions *imply* that a food contains, or does not contain, a nutrient. Implied claims are prohibited unless they meet specified criteria. For example, a claim that a product "contains no oil" *implies* that the food contains no fat. If the product is truly fat-free, then it may make the no-oil claim, but if it contains another source of fat, such as butter, it may not.

Health Claims **Health claims** describe a relationship between a food (or food component) and a disease or health-related condition. In some cases, the FDA authorizes health claims based on an extensive review of the scientific literature.[22] For example, the health claim that "Diets low in sodium may reduce the risk of high blood pressure" is based on enough scientific evidence to establish a clear link between diet and health. In cases where there is emerging—but not established—evidence for a relationship between a food or food component and disease, the FDA allows the use of *qualified* health claims that must use specific language indicating that the evidence supporting the claim is limited. A qualified health claim might claim that "Very limited and preliminary research suggests that eating one-half to one cup of tomatoes and/or tomato sauce a week may reduce the risk of prostate cancer. The FDA concludes that there is little scientific evidence supporting the claim."

Structure-Function Claims Unlike health claims, which require food manufacturers to collect scientific evidence and petition the FDA, **structure-function claims** can be made without any FDA approval. Product labels can claim to "slow aging," "improve memory," and "build strong bones" without any proof. The only

health claims: statements that characterize the relationship between a nutrient or other substance in a food and a disease or health-related condition.

structure-function claims: statements that characterize the relationship between a nutrient or other substance in a food and its role in the body.

GLOSSARY OF TERMS ON FOOD LABELS

GENERAL TERMS

free: "nutritionally trivial" and unlikely to have a physiological consequence; synonyms include *without, no,* and *zero.* A food that does not contain a nutrient naturally may make such a claim, but only as it applies to all similar foods (for example, "applesauce, a fat-free food").

good source of: the product provides between 10 and 19 percent of the Daily Value for a given nutrient per serving.

healthy: a food that is low in fat, saturated fat, cholesterol, and sodium and that contains at least 10 percent of the Daily Values for vitamin A, vitamin C, iron, calcium, protein, or fiber.

high: 20 percent or more of the Daily Value for a given nutrient per serving; synonyms include *rich in* or *excellent source.*

less: at least 25 percent less of a given nutrient or kcalories than the comparison food (see individual nutrients); synonyms include *fewer* and *reduced.*

light or **lite:** one-third fewer kcalories than the comparison food; 50 percent or less of the fat or sodium than the comparison food; any use of the term other than as defined must specify what it is referring to (for example, "light in color" or "light in texture").

low: an amount that would allow frequent consumption of a food without exceeding the Daily Value for the nutrient. A food that is naturally low in a nutrient may make such a claim, but only as it applies to all similar foods (for example, "fresh cauliflower, a low-sodium food"); synonyms include *little, few,* and *low source of.*

more: at least 10 percent more of the Daily Value for a given nutrient than the comparison food; synonyms include *added* and *extra.*

organic: on food labels, that at least 95 percent of the product's ingredients have been grown and processed according to USDA regulations defining the use of fertilizers, herbicides, insecticides, fungicides, preservatives, and other chemical ingredients (see Chapter 19).

ENERGY

kcalorie-free: fewer than 5 kcalories per serving.

low kcalorie: 40 kcalories or less per serving.

reduced kcalorie: at least 25 percent fewer kcalories per serving than the comparison food.

FAT AND CHOLESTEROL[a]

percent fat-free: may be used only if the product meets the definition of *low fat* or *fat-free* and must reflect the amount of fat in 100 grams (for example, a food that contains 2.5 grams of fat per 50 grams can claim to be "95 percent fat-free").

fat-free: less than 0.5 gram of fat per serving (and no added fat or oil); synonyms include *zero-fat, no-fat,* and *nonfat.*

low fat: 3 grams or less fat per serving.

less fat: 25 percent or less fat than the comparison food.

saturated fat-free: less than 0.5 gram of saturated fat and 0.5 gram of *trans* fat per serving.

low saturated fat: 1 gram or less saturated fat and less than 0.5 gram of *trans* fat per serving.

less saturated fat: 25 percent or less saturated fat and *trans* fat combined than the comparison food.

***trans* fat-free:** less than 0.5 gram of *trans* fat and less than 0.5 gram of saturated fat per serving.

cholesterol-free: less than 2 milligrams cholesterol per serving and 2 grams or less saturated fat and *trans* fat combined per serving.

low cholesterol: 20 milligrams or less cholesterol per serving and 2 grams or less saturated fat and *trans* fat combined per serving.

less cholesterol: 25 percent or less cholesterol than the comparison food (reflecting a reduction of at least 20 milligrams per serving), and 2 grams or less saturated fat and *trans* fat combined per serving.

extra lean: less than 5 grams of fat, 2 grams of saturated fat and *trans* fat combined, and 95 milligrams of cholesterol per serving and per 100 grams of meat, poultry, and seafood.

lean: less than 10 grams of fat, 4.5 grams of saturated fat and *trans* fat combined, and 95 milligrams of cholesterol per serving and per 100 grams of meat, poultry, and seafood. For mixed dishes such as burritos and sandwiches, less than 8 grams of fat, 3.5 grams of saturated fat, and 80 milligrams of cholesterol per reference amount customarily consumed.

CARBOHYDRATES: FIBER AND SUGAR

high fiber: 5 grams or more of fiber per serving. A high-fiber claim made on a food that contains more than 3 grams of fat per serving and per 100 grams of food must also declare total fat.

sugar-free: less than 0.5 gram of sugar per serving.

SODIUM

sodium-free and salt-free: less than 5 milligrams of sodium per serving.

low sodium: 140 milligrams or less per serving.

very low sodium: 35 milligrams or less per serving.

[a]Foods containing more than 13 grams total fat per serving or per 50 grams of food must indicate those contents immediately after a cholesterol claim. As you can see, all cholesterol claims are prohibited when the food contains more than 2 grams saturated fat and *trans* fat combined per serving.

> FIGURE 2-9 **Label Claims**

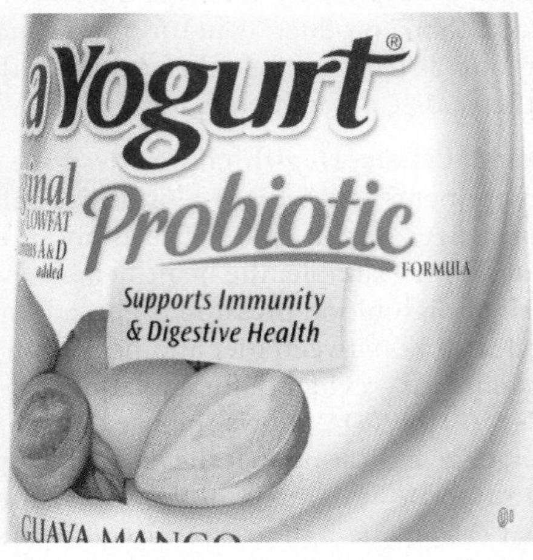

© Sam Kolich/Bill Smith Group/Cengage Learning

Nutrient claims characterize the level of a nutrient in the food—for example, "fat free" or "less sodium."

Health claims characterize the relationship of a food or food component to a disease or health-related condition—for example, "soluble fiber from oatmeal daily in a diet low in saturated fat and cholesterol may reduce the risk of heart disease" or "a diet low in total fat may reduce the risk of some cancers."

Structure/function claims describe the effect that a substance has on the structure or function of the body and do not make reference to a disease—for example, "supports immunity and digestive health" or "calcium builds strong bones."

criterion for a structure-function claim is that it must not mention a disease or symptom. Unfortunately, structure-function claims can be deceptively similar to health claims, and most consumers do not distinguish between these two types of claims. Consider these statements:

- "May reduce the risk of heart disease"
- "Promotes a healthy heart"

The first is a health claim that requires FDA approval and the second is an unproven, but legal, structure-function claim. Figure 2-9 compares label claims.

Consumer Education Food labels are a primary source of information for consumers trying to make healthy diet choices, but label messages are often confusing.[23] Because labels are valuable only if people know how to use them, the FDA has designed several programs to educate consumers. Consumers who understand how to read labels are best able to apply the information to achieve and maintain healthful dietary practices. Table 2-9 shows how the messages from the *Dietary Guidelines*, the USDA Food Patterns, and food labels coordinate with one another.

REVIEW IT Compare and contrast the information on food labels to make selections that meet specific dietary and health goals.
Food labels provide consumers with information they need to select foods that will help them meet their nutrition and health goals. When labels contain relevant information presented in a standardized, easy-to-read format, consumers are well prepared to plan and create healthful diets.

This chapter provides the links to go from dietary guidelines to buying groceries and offers helpful tips for selecting nutritious foods. For additional information on foods, including organic foods, irradiated foods, genetically modified foods, and more, turn to Chapter 19.

TABLE 2-9 From Guidelines to Groceries

Dietary Guidelines	USDA Food Patterns/MyPlate	Food Labels
Balancing kcalories to manage weight	Enjoy your food, but eat less. Select the recommended amounts from each food group at the energy level appropriate for your energy needs; meet, but do not exceed, energy needs. Limit foods and beverages with solid fats and added sugars. Use appropriate portion sizes; avoid oversized portions. Increase physical activity and reduce time spent in sedentary behaviors.	Read the Nutrition Facts to see how many kcalories are in a serving and the number of servings that are in a package. Look for foods that describe their kcalorie contents as *free, low, reduced, light*, or *less*.
Foods and food components to reduce	Choose foods within each group that are low in salt or sodium. Choose foods within each group that are lean, low fat, or fat free and have little solid fat (sources of saturated and *trans* fats); use unsaturated oils instead of solid fats whenever possible. Choose foods and beverages within each group that have little added sugars; drink water instead of sugary beverages. If alcohol is consumed by adults, use in moderation (no more than one drink a day for women and two drinks a day for men).	Read the Nutrition Facts to see how much sodium, saturated fat, *trans* fat, and cholesterol is in a serving of food. Look for foods that describe their salt and sodium contents as *free, low*, or *reduced*; foods that describe their fat, saturated fat, *trans* fat, and cholesterol contents as *free, less, low, light, reduced, lean*, or *extra lean*; foods that describe their sugar contents as *free* or *reduced*. Look for foods that provide no more than 5 percent of the Daily Value for sodium, fat, saturated fat, and cholesterol. A food may be high in solid fats if its ingredients list begins with or contains several of the following: *beef fat (tallow, suet), butter, chicken fat, coconut oil, cream, hydrogenated oils, palm kernel oil, palm oil, partially hydrogenated oils, pork fat (lard), shortening*, or *stick margarine*. A food most likely contains *trans* fats if its ingredients list includes: *partially hydrogenated oils*. A food may be high in added sugars if its ingredients list begins with or contains several of the following: *brown sugar, confectioner's powdered sugar, corn syrup, dextrin, fructose, high-fructose corn syrup, honey, invert sugar, lactose, malt syrup, maltose, molasses, nectars, sucrose, sugar*, or *syrup*. Light beverages contain fewer kcalories and less alcohol than regular versions.
Foods and nutrients to increase	Make half your plate fruits and vegetables. Choose a variety of vegetables from all five subgroups (dark green, red and orange, legumes, starchy vegetables, and other vegetables) several times a week. Choose a variety of fruits; consume whole or cut-up fruits more often than fruit juice. Choose potassium-rich foods such as fruits and vegetables often. Choose fiber-rich fruits, vegetables, and whole grains often. Choose whole grains; make at least half of the grain selections whole grains by replacing refined grains with whole grains whenever possible. Choose fat-free or low-fat milk and milk products. Choose a variety of protein foods; increase the amount and variety of seafood by choosing seafood in place of some meat and poultry.	Look for foods that describe their fiber, calcium, potassium, and vitamin D contents as *good, high*, or *excellent*. Look for foods that provide at least 10 percent of the Daily Value for fiber, calcium, potassium, and vitamin D from a variety of sources. A food may be a good source of whole grains if its ingredients list begins with or contains several of the following: *barley, brown rice, buckwheat, bulgur, corn, millet, oatmeal, popcorn, quinoa, rolled oats, rye, sorghum, triticale, whole wheat*, or *wild rice*.
Building healthy eating patterns	Select nutrient-dense foods and beverages within and among the food groups. Keep food safe.	Look for foods that describe their vitamin, mineral, or fiber contents as a *good source* or *high*. Follow the *safe handling instructions* on packages of meat and other safety instructions, such as *keep refrigerated*, on packages of perishable foods.

Nutrition Portfolio

The secret to making healthy food choices is learning to incorporate the *Dietary Guidelines for Americans* and the USDA Food Patterns into your decision-making process.

Go to Diet Analysis Plus and choose one of the days on which you have tracked your diet for the entire day. Choose the MyPlate Report and, looking at it, record in your journal the answers to the following:

- How do the foods you consumed on the day you have chosen stack up with the daily goals (the percentages) in the MyPlate breakdown? Which food groups are over- or under-represented?
- Think about your choices within each food group for the day you recorded. Are they typical of the foods you choose from day to day? Are there simple and realistic ways to enhance the variety in your diet?
- Write yourself a letter describing the dietary changes you can make to improve your chances of enjoying good health.

Diet Analysis PLUS To complete this exercise, go to your Diet Analysis Plus at **www.cengagebrain.com.**

STUDY IT To review the key points of this chapter and take a practice quiz, go to the study cards at the end of the book.

REFERENCES

1. Position of the American Dietetic Association: Total diet approach to communicating food and nutrition information, *Journal of the American Dietetic Association* 107 (2007): 1224–1232.
2. Practice paper of the American Dietetic Association: Nutrient density: Meeting nutrient goals within calorie needs, *Journal of the American Dietetic Association* 107 (2007): 860–869.
3. G. D. Miller and coauthors, It is time for a positive approach to dietary guidance using nutrient density as a basic principle, *Journal of Nutrition* 139 (2009): 1198–1202.
4. A. Drewnowski and V. Fulgoni III, Nutrient profiling of foods: Creating a nutrient-rich food index, *Nutrition Reviews* 66 (2008): 23–39.
5. The science behind current nutrition profiling systems to promote consumer intake of nutrient-dense foods, *American Journal of Clinical Nutrition* 91 (2010): entire supplement; N. Darmon and coauthors, Nutrient profiles discriminate between foods according to their contribution to nutritionally adequate diets: A validation study using linear programming and the SAIN, LIM system, *American Journal of Clinical Nutrition* 89 (2009): 1227–1236; E. Kennedy, Food rating systems, diet quality, and health, *Nutrition Reviews* 66 (2008): 21–22.
6. D. Mozaffarian and D. S. Ludwig, Dietary Guidelines in the 21st century: A time for food, *Journal of the American Medical Association* 304 (2010): 681–682.
7. ˙L. V. Horn, Development of the 2010 US Dietary Guidelines Advisory Committee Report: Perspectives from a registered dietitian, *Journal of the American Dietetic Association* 110 (2010): 1638–1645.
8. US Department of Health and Human Services, *2008 Physical Activity Guidelines for Americans*, www.health.gov/paguidelines; US Department of Agriculture and US Department of Health and Human Services, *Dietary Guidelines for Americans, 2010*, www.dietaryguidelines.gov.
9. www.choosemyplate.gov/food-groups/vegetables-beans-peas.html, accessed February 15, 2012.
10. Position of the American Dietetic Association: Vegetarian diets, *Journal of the American Dietetic Association* 109 (2009): 1266–1282.
11. J. L. Bachman and coauthors, Sources of food group intakes among the US population, 2001–2002, *Journal of the American Dietetic Association* 108 (2008): 804–814.
12. P. M. Guenther, J. Reedy, and S. M. Krebs-Smith, Development of the Healthy Eating Index—2005, *Journal of the American Dietetic Association* 108 (2008): 1896–1901; Center for Nutrition Policy and Promotion, Healthy Eating Index—2005, fact sheet revised June 2008, www.cnpp.usda.gov.
13. Center for Nutrition Policy and Promotion, Diet quality of Americans in 1994–1996 and 2001–02 as measured by the Healthy Eating Index—2005, nutrition insight 37 revised August 2008, www.cnpp.usda.gov.
14. S. M. Krebs-Smith, J. Reedy, and C. Bosire, Healthfulness of the US food supply: Little improvement despite decades of dietary guidance, *American Journal of Preventive Medicine* 38 (2010): 472–477.
15. Harvard School of Public Health, *The Nutrition Source: Healthy Eating Plate vs. USDA's MyPlate*, www.hsph.harvard.edu, accessed October 4, 2011.
16. M. Maillot and coauthors, Individual diet modeling translates nutrient recommendations into realistic and individual-specific food choices, *American Journal of Clinical Nutrition* 91 (2010): 421–430.
17. C. E. O'Neil and coauthors, Whole-grain consumption is associated with diet quality and nutrient intake in adults: The National Health and Nutrition Examination Survey, 1999–2004, *Journal of the American Dietetic Association* 110 (2010): 1461–1468.
18. As cited in 21 Code of Federal Regulations—Food and Drugs, Section 104.20, 45 *Federal Register* 6323, January 25, 1980, as amended in 58 *Federal Register* 2228, January 6, 1993.
19. Position of the American Dietetic Association: Fortification and nutritional supplements, *Journal of the American Dietetic Association* 105 (2005): 1300–1311.
20. J. E. Lewis and coauthors, Food label use and awareness of nutritional information and recommendations among persons with chronic disease, *American Journal of Clinical Nutrition* 90 (2009):1351–1357; C. L. Taylor and

V. L. Wilkening, How the nutrition food label was developed, part 1: The nutrition facts panel, *Journal of the American Dietetic Association* 108 (2008): 437–442.

21. L. Marr, National restaurant menu labeling legislation: Public nutrition education and professional opportunities, *Journal of the American Dietetic Association* 111 (2011): S7; K. Stein, A national approach to restaurant menu labeling: The Patient Protection and Affordable Health Care Act, section 4205, *Journal of the American Dietetic Association* 110 (2010): 1280–1286.

22. C. L. Taylor and V. L. Wilkening, How the nutrition food label was developed, part 2: The purpose and promise of nutrition claims, *Journal of the American Dietetic Association* 108 (2008): 618–623.

23. N. J. Ollberding, R. L. Wolf, and I. Contento, Food label use and its relation to dietary intake among US adults, *Journal of the American Dietetic Association* 111 (2011): S47–S51; J. M. Wills and coauthors, Exploring global consumer attitudes toward nutrition information on food labels, *Nutrition Reviews* 67 (2009): S102–S106.

 LEARN IT Develop a well-balanced vegetarian meal plan.

Vegetarian Diets

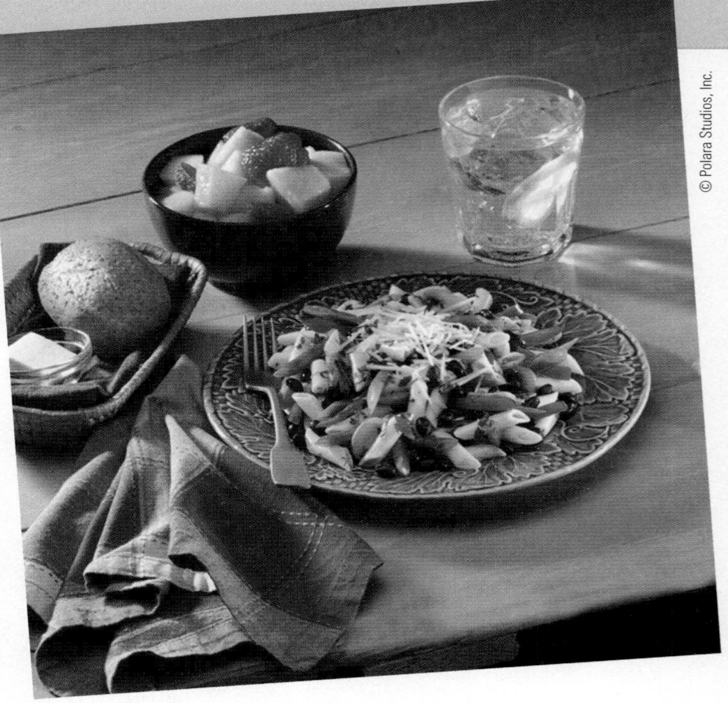
© Polara Studios, Inc.

The waiter presents this evening's specials: a fresh spinach salad topped with mandarin oranges, raisins, and sunflower seeds, served with a bowl of pasta smothered in a mushroom and tomato sauce and topped with grated parmesan cheese. Then this one: a salad made of chopped parsley, scallions, celery, and tomatoes mixed with bulgur wheat and dressed with olive oil and lemon juice, served with a spinach and feta cheese pie. Do these meals sound good to you? Or is something missing . . . a pork chop or chicken breast, perhaps?

Would vegetarian fare be acceptable to you some of the time? Most of the time? Ever? The health benefits of a primarily vegetarian diet seem to have encouraged many people to eat more plant-based meals. The popular press sometimes refers to individuals who eat small amounts of meat, seafood, or poultry from time to time as "flexitarians."

People who choose to exclude meat and other animal-derived foods from their diets today do so for many of the same reasons the Greek philosopher Pythagoras cited in the sixth century B.C.: physical health, ecological responsibility, and philosophical concerns. They might also cite world hunger issues, economic reasons, ethical concerns, or religious beliefs as motivating factors. Whatever their reasons—and even if they don't have a particular reason—people who exclude meat will be better prepared to plan well-balanced meals if they understand the nutrition and health implications of their choices.

Vegetarians generally are categorized, not by their motivations but by the foods they choose to exclude (see the accompanying glossary). Some people exclude red meat only; some also exclude poultry or seafood; others also exclude eggs; and still others exclude milk and milk products as well. In fact, finding agreement on the definition of the term *vegetarian* is a challenge.

As you will see, though, the foods a person *excludes* are not nearly as important as the foods a person *includes* in the diet. Vegetarian diets that include a variety of whole grains, vegetables, legumes, fruits, and nuts and seeds offer abundant complex carbohydrates and fibers, an assortment of vitamins and minerals, a mixture of phytochemicals, and little fat—characteristics that reflect current dietary recommendations aimed at maintaining good health and an appropriate body weight. Each of these foods—whole grains, vegetables, legumes, fruits, and nuts and seeds—independently reduces the risk for several chronic diseases. This highlight examines the health benefits and potential problems of vegetarian diets and shows how to plan a well-balanced vegetarian diet. Highlight 20 includes a discussion of the environmental benefits of a plant-based diet.[1]

Health Benefits of Vegetarian Diets

Research findings suggest that well-planned vegetarian diets offer sound nutrition and health benefits to adults.[2] Eating patterns that include very little, if any, meat may even increase life expectancy.

Obesity

Vegetarians tend to maintain a lower and healthier body weight than nonvegetarians. In general, those who eat meat have higher energy intakes and body weights.[3] Vegetarians' lower body weights correlate with their high intakes of fiber and low intakes of fat. Because obesity impairs health in a number of ways, this gives vegetarians a health advantage.

GLOSSARY

lactovegetarians: people who include milk and milk products, but exclude meat, poultry, seafood, and eggs from their diets.
- **lacto** = milk

lacto-ovo-vegetarians: people who include milk, milk products, and eggs, but exclude meat, poultry, and seafood from their diets.
- **ovo** = egg

macrobiotic diet: a philosophical approach of eating mostly plant-based foods such as whole grains, legumes, and vegetables, with small amounts of fish, fruits, nuts, and seeds.
- **macro** = large, great
- **biotic** = life

meat replacements: products formulated to look and taste like meat, seafood, or poultry; usually made of textured vegetable protein.

omnivorous: an eating pattern that includes foods derived from both animals and plants.
- **omni** = all
- **vores** = to eat

tempeh (TEM-pay): a fermented-soybean food, rich in protein and fiber.

plant-based diets: an eating pattern that derives most of its protein from plant products (although some animal products may be included).

tofu (TOE-foo): a curd made from soybeans, rich in protein and often fortified with calcium; used in many Asian and vegetarian dishes in place of meat.

vegans (VEE-gans): people who exclude all animal-derived foods (including meat, poultry, fish, eggs, and dairy products); also called *pure vegetarians, strict vegetarians,* or *total vegetarians.*

vegetarians: a general term used to describe people who exclude meat, poultry, fish, or other animal-derived foods from their diets.

Diabetes

Obesity and weight gains are strong risk factors for diabetes, which partially explains why **omnivorous** diets are more often associated with diabetes than vegetarian diets.[4] Even when body weight and lifestyle factors are taken into account, vegetarian eating patterns seem to protect against diabetes.[5]

Hypertension

Vegetarians tend to have lower blood pressure and lower rates of hypertension than nonvegetarians. Appropriate body weight helps to maintain a healthy blood pressure, as does a diet low in total fat and saturated fat and high in fiber, fruits, vegetables, and soy protein. Lifestyle factors also influence blood pressure: smoking and alcohol intake raise blood pressure, and physical activity lowers it.

Heart Disease

The incidence of heart disease and related deaths and the concentrations of blood cholesterol are lower for vegetarians than for nonvegetarians, which can partly be explained by their avoidance of meat.[6] The dietary factor most directly related to heart disease is saturated animal fat, and in general, vegetarian diets are lower in total fat, saturated fat, and cholesterol than typical meat-based diets. The fats common in **plant-based diets**—the monounsaturated fats of olives, seeds, and nuts and the polyunsaturated fats of vegetable oils—are associated with a decreased risk of heart disease. Furthermore, vegetarian diets are generally higher in dietary fiber, antioxidant vitamins, and phytochemicals—all factors that help control blood lipids and protect against heart disease.

Many vegetarians include soy products such as **tofu** in their diets. Soy products—with their polyunsaturated fats, fibers, vitamins, and minerals, and little saturated fat—may help to protect against heart disease.[7]

Cancer

Vegetarians have a lower rate of some cancers than the general population. Their low cancer rates may be due to their high intakes of fruits and vegetables (as Highlight 11 explains). In fact, the ratio of vegetables to meat may be the most relevant dietary factor responsible for cancer prevention.

Some scientific findings indicate that vegetarian diets are associated not only with lower cancer mortality in general, but also with lower incidence of cancer at specific sites as well, most notably, colon cancer. People with colon cancer seem to eat more meat. Some cancer experts recommend limiting "consumption of red meat to no more than 11 ounces a week, with very little (if any) processed meat."[8]

Other Diseases

In addition to obesity, diabetes, hypertension, heart disease, and some cancers, vegetarian diets may help prevent osteoporosis, diverticular disease, gallstones, cataracts, and rheumatoid arthritis. Health benefits of a vegetarian diet depend on wise diet planning.

Vegetarian Diet Planning

The vegetarian has the same meal-planning task as everyone else—using a variety of foods to deliver all the needed nutrients within an energy allowance that maintains a healthy body weight (as discussed in Chapter 2). Vegetarians who include milk, milk products, and eggs can meet recommendations for most nutrients about as easily as nonvegetarians. Such eating patterns may rely on some fortified foods, but generally provide enough energy, protein, and other nutrients to support the health of adults and the growth of children and adolescents. The USDA Food Patterns for vegetarians are flexible enough that a variety of people can use it: people who have adopted various vegetarian diets, those who want to make the transition to a vegetarian diet, and those who simply want to include more plant-based meals in their diets.

Vegans who exclude milk, milk products, and eggs can select protein foods such as legumes, nuts, and seeds as well as foods made from them, such as peanut butter, **tempeh**, and tofu. Those who do not use milk can use soy "milk"—a product made from soybeans that provides similar nutrients if fortified with calcium, vitamin D, and vitamin B_{12} (see Figure H2-1). Similarly, "milks" made from rice, almonds, and oats are reasonable alternatives, if adequately fortified. Vegan eating patterns must include fortified foods or supplements to provide adequate intakes of all essential nutrients.

MyPlate includes tips for planning vegetarian diets using an adaptation of the USDA Food Patterns. The recommended daily amounts

> **FIGURE H2-1 Low-Fat Milk and Soy Milk Compared**

A comparison of low-fat milk and enriched soy milk shows that they provide similar amounts of key nutrients.

Low-Fat Milk

Nutrition Facts
Serving Size 1 cup (240mL)
Servings Per Container About 8

Amount Per Serving	
Calories 110 Calories from Fat 25	
	% Daily Value*
Total Fat 2.5g	4%
Saturated Fat 1.5g	8%
Trans Fat 0g	
Polyunsaturated Fat 0.5g	
Monounsaturated Fat 0.5g	
Cholesterol 15mg	4%
Sodium 130mg	5%
Potassium 380mg	11%
Total Carbohydrate 13g	4%
Dietary Fiber 0g	0%
Sugars 12g	
Protein 8g	
Vitamin A 10% • Vitamin C 0%	
Calcium 30% • Iron 0%	
Vitamin D 25%	

Soy Milk

Nutrition Facts
Serving Size 1 cup (240mL)
Servings Per Container About 8

Amount Per Serving	
Calories 100 Calories from Fat 35	
	% Daily Value*
Total Fat 4g	6%
Saturated Fat 0.5g	3%
Trans Fat 0g	
Polyunsaturated Fat 2.5g	
Monounsaturated Fat 1g	
Cholesterol 0mg	0%
Sodium 120mg	5%
Potassium 300mg	8%
Total Carbohydrate 8g	3%
Dietary Fiber 1g	4%
Sugars 6g	
Protein 7g	
Vitamin A 10% • Vitamin C 0%	
Calcium 30% • Iron 6%	
Vitamin D 30% • Riboflavin 30%	
Folate 6% • Vitamin B12 50%	

TABLE H2-1 USDA Food Patterns: Recommended Weekly Amounts of Protein Foods for Vegetarians and Vegans

The daily amounts for protein foods is the same for both vegetarians and nonvegetarians, but the subgroups and weekly amounts for protein foods differ. The recommended daily amounts from each of the other food groups—fruits, vegetables, grains, and milk products—is the same (see Table 2-2).

Protein Foods	1600 kcal	1800 kcal	2000 kcal	2200 kcal	2400 kcal	2600 kcal	2800 kcal	3000 kcal
Daily Amounts	5 oz	5 oz	5½ oz	6 oz	6½ oz	6½ oz	7 oz	7 oz
Vegetarian Subgroups								
Eggs	4 oz	4 oz	4 oz	4 oz	5 oz	5 oz	5 oz	5 oz
Legumes	9 oz	9 oz	10 oz	10 oz	11 oz	11 oz	12 oz	12 oz
Soy products	11 oz	11 oz	12 oz	13 oz	14 oz	14 oz	15 oz	15 oz
Nuts and seeds	12 oz	12 oz	13 oz	15 oz	16 oz	16 oz	17 oz	17 oz
Vegan Subgroups								
Legumes	12 oz	12 oz	13 oz	15 oz	16 oz	16 oz	17 oz	17 oz
Soy products	9 oz	9 oz	10 oz	11 oz	11 oz	11 oz	12 oz	12 oz
Nuts and seeds	14 oz	14 oz	15 oz	17 oz	18 oz	18 oz	20 oz	20 oz

NOTE: Total recommended amounts for legumes include the sum of both the vegetables and protein foods. An ounce-equivalent of legumes in the protein foods group is ¼ cup. For a 2000-kcal vegan diet, that's 3¼ cups legumes for protein foods plus 1½ cups legumes for vegetables (see Table 2-4), or about almost 5 cups legumes weekly.

from the food groups is the same for both vegetarians and nonvegetarians (see Table 2-2 on p. 44). Selections from within the food groups may differ, of course. For example, the milk group features fortified soy milks for those who do not use milk, cheese, or yogurt. When selecting from the vegetable and fruit groups, vegetarians may want to emphasize particularly good sources of calcium and iron. Green leafy vegetables provide almost five times as much calcium per serving as other vegetables. Similarly, dried fruits deserve special notice in the fruit group because they deliver six times as much iron as other fruits. The protein foods group includes eggs (for those who use them), legumes, soy products, and nuts and seeds. Table H2-1 provides recommended *weekly* amounts of protein food subgroups for both vegetarians and vegans. The use of vegetable oils rich in unsaturated fats provides essential omega-3 fatty acids.

Most vegetarians easily obtain large quantities of the nutrients that are abundant in plant foods, including carbohydrate, fiber, thiamin, folate, vitamin B_6, vitamin C, vitamin A, and vitamin E. Well-planned vegetarian eating patterns help to ensure adequate intakes of the nutrients vegetarian diets might otherwise lack, including protein, iron, zinc, calcium, vitamin B_{12}, vitamin D, and omega-3 fatty acids. Table H2-2 presents good vegetarian sources of these key nutrients.

Protein

The protein RDA for vegetarians is the same as for others, although some have suggested that it should be higher because plant proteins are not digested as completely. **Lacto-ovo-vegetarians**, who use animal-derived foods such as milk and eggs, receive high-quality proteins and are likely to meet their protein needs. Even those who adopt only plant-based diets are likely to meet protein needs provided that their energy intakes are adequate and the protein sources varied.[9] The proteins of whole grains, vegetables, legumes, and nuts and seeds can provide adequate amounts of all the amino acids. An advantage of many vegetarian sources of protein is that they are generally lower in saturated fat than meats and are often higher in fiber and richer in some vitamins and minerals.

Vegetarians sometimes use **meat replacements** made of textured vegetable protein (soy protein). These foods are formulated to look and taste like meat, seafood, or poultry. Many of these products are fortified to provide the vitamins and minerals found in animal sources of protein. Some may be high in salt, sugars, and saturated fats. A wise vegetarian learns to read labels and use a variety of whole, unrefined foods often and commercially prepared foods less frequently. Vegetarians may also use soy products such as tofu to bolster protein intake.

Iron

Getting enough iron can be a problem even for meat eaters, and those who eat no meat must pay special attention to their iron intake. The iron in plant foods such as legumes, dark-green leafy vegetables, iron-fortified cereals, and whole-grain breads and cereals is poorly absorbed. Because iron absorption from a vegetarian diet is low, the iron RDA for vegetarians is higher than for others (see Chapter 13 for more details).

Fortunately, the body seems to adapt to a low-iron vegetarian diet by increasing iron absorption and decreasing iron losses. Furthermore, iron absorption is enhanced by vitamin C, and vegetarians typically eat many vitamin C–rich fruits and vegetables. Consequently, vegetarians are no more iron deficient than other people.

Zinc

Zinc is similar to iron in that meat is its richest food source, and zinc from plant sources is not well absorbed. In addition, phytates, fiber, and calcium, which are common in vegetarian diets, interfere with zinc absorption. Nevertheless, most vegetarian adults are not zinc

TABLE H2-2 **Good Vegetarian Sources of Key Nutrients**

Nutrients	Grains	Vegetables	Fruits	Protein Foods	Milk	Oils
Protein	Whole grains[a]			Legumes, seeds, nuts, soy products (tempeh, tofu, veggie burgers)[a] Eggs (for ovo-vegetarians)	Milk, cheese, yogurt (for lactovegetarians)	
Iron	Fortified cereals, enriched and whole grains	Dark green leafy vegetables (spinach, turnip greens)	Dried fruits (apricots, prunes, raisins)	Legumes (black-eyed peas, kidney beans, lentils)		
Zinc	Fortified cereals, whole grains			Legumes (garbanzo beans, kidney beans, navy beans), nuts, seeds (pumpkin seeds)	Milk, cheese, yogurt (for lactovegetarians)	
Calcium	Fortified cereals	Dark green leafy vegetables (bok choy, broccoli, collard greens, kale, mustard greens, turnip greens, watercress)	Fortified juices, figs	Fortified soy products, nuts (almonds), seeds (sesame seeds)	Milk, cheese, yogurt (for lactovegetarians) Fortified soy milk	
Vitamin B_{12}	Fortified cereals			Eggs (for ovo-vegetarians) Fortified soy products	Milk, cheese, yogurt (for lactovegetarians) Fortified soy milk	
Vitamin D					Milk, cheese, yogurt (for lactovegetarians) Fortified soy milk	
Omega-3 fatty acids				Flaxseed, walnuts, soybeans		Flaxseed oil, walnut oil, soybean oil

[a]As Chapter 6 explains, many plant proteins do not contain all the essential amino acids in the amounts and proportions needed by human beings. To improve protein quality, vegetarians can eat grains and legumes together, for example, although it is not necessary if protein intake is varied and energy intake is sufficient.

deficient. Perhaps the best advice to vegetarians regarding zinc is to eat a variety of nutrient-dense foods; include whole grains, nuts, and legumes such as black-eyed peas, pinto beans, and kidney beans; and maintain an adequate energy intake. For those who include seafood in their diets, oysters, crabmeat, and shrimp are rich in zinc.

Calcium

The calcium intakes of **lactovegetarians** are similar to those of the general population, but vegans who use no milk or milk products may risk inadequate intakes. To ensure adequate intakes, vegans can select calcium-rich foods, such as calcium-fortified juices, soy milk, and breakfast cereals, in ample quantities regularly. This advice is especially important for children and adolescents. Soy formulas for infants are fortified with calcium and can be used in cooking, even for adults. Other good calcium sources include figs, some legumes, some green vegetables such as broccoli and turnip greens, some nuts such as almonds, certain seeds such as sesame seeds, and calcium-set tofu.* The choices should be varied because calcium absorption from some plant foods may be limited (as Chapter 12 explains).

*Calcium salts are often added during processing to coagulate the tofu.

Vitamin B_{12}

The requirement for vitamin B_{12} is small, but this vitamin is found only in animal-derived foods. Consequently, vegetarians, in general, and vegans who eat no foods of animal original, in particular, may not get enough vitamin B_{12} in their diets.[10] Fermented soy products such as tempeh may contain some vitamin B_{12} from the bacteria, but unfortunately, much of the vitamin B_{12} found in these products may be an inactive form. Seaweeds such as nori and chlorella supply some vitamin B_{12}, but not much, and excessive intakes of these foods can lead to iodine toxicity. To defend against vitamin B_{12} deficiency, vegans must rely on vitamin B_{12}-fortified sources (such as soy milk or breakfast cereals) or supplements. Without vitamin B_{12}, the nerves suffer damage, leading to such health consequences as loss of vision.

Vitamin D

The vitamin D status of vegetarians is similar to that of nonvegetarians.[11] People who do not use vitamin D–fortified foods and do not receive enough exposure to sunlight to synthesize adequate vitamin D may need supplements to defend against bone loss. This is particularly important for infants, children, and older adults. In northern climates during winter months, young children on vegan diets can readily develop rickets, the vitamin D–deficiency disease.

Omega-3 Fatty Acids

Both Chapter 5 and Highlight 5 describe the health benefits of unsaturated fats, most notably the omega-3 fatty acids commonly found in fatty fish. A diet that includes some meat, fish, and eggs provides much more omega-3 fatty acids than a vegetarian diet, but the *blood* differences between those eating fish and others is relatively small.[12] Researchers speculate that the smaller-than-expected differences may reflect a more efficient conversion of plant-derived fats to omega-3 fats in non–fish eaters. Vegetarians can obtain sufficient amounts of omega-3 fatty acids from plant sources such as flaxseed, walnuts, soy, and canola oil.[13]

Healthy Food Choices

Later chapters provide details on how vegetarian diets can meet nutrient needs for various stages of the life cycle, including pregnancy, lactation, infancy, childhood, and adolescence. In general, well-planned vegetarian eating patterns may lower the risk of mortality and several chronic diseases, including obesity, diabetes, high blood pressure, heart disease, and some cancers.[14] But there is nothing mysterious or magical about a vegetarian eating pattern. The quality of the diet depends not on whether it includes meat but on whether the other food choices are nutritionally sound. A plant-based eating pattern that includes ample fruits, vegetables, whole grains, legumes, nuts, and seeds is higher in fiber, antioxidant vitamins, and phytochemicals and lower in saturated fats and cholesterol than meat-based diets. Variety is key to nutritional adequacy in a vegetarian diet. Restrictive plans that limit selections to a few grains and vegetables cannot possibly deliver a full array of nutrients.

Vegetarianism is not a religion like Buddhism or Hinduism, but merely an eating pattern that selects plant foods to deliver needed nutrients. That said, some vegetarians choose to follow a **macrobiotic diet.** Those following a macrobiotic diet select natural, organic foods and embrace a Zen-like spirituality. In other words, a macrobiotic diet represents a way of life, not just a meal plan. A macrobiotic diet emphasizes whole grains, legumes, and vegetables, with small amounts of fish, fruits, nuts, and seeds. Practices include selecting locally grown foods, eating foods in their most natural state, and balancing cold, sweet, and passive foods with hot, salty, and aggressive ones. Some items, such as processed foods, alcohol, hot spices, and potatoes are excluded from the diet. Early versions of the macrobiotic diet followed a progression that ended with the "ultimate" diet of brown rice and water—a less-than nutritiously balanced diet. Today's version reflects a modified vegetarian approach with an appreciation of how foods can enhance health. With careful planning, a macrobiotic diet can provide an array of nutrients that support good health.

If not properly balanced, any diet—vegetarian, macrobiotic, or otherwise—can lack nutrients. Poorly planned vegetarian diets typically lack iron, zinc, calcium, vitamin B_{12}, and vitamin D; without planning, meat-based diets may lack vitamin A, vitamin C, folate, and fiber, among others. Quite simply, the negative health aspects of any diet, including vegetarian diets, reflect poor diet planning. Careful attention to energy intake and specific nutrients of concern can ensure adequacy.

Keep in mind, too, that diet is only one factor influencing health. Whatever a diet consists of, its context is also important: no smoking, alcohol consumption in moderation (if at all), regular physical activity, adequate rest, and medical attention when needed all contribute to good health. Establishing these healthy habits early in life seems to be the most important step one can take to reduce the risks of chronic diseases later in life (as Highlight 16 explains).

REFERENCES

1. B. M. Popkin, Reducing meat consumption has multiple benefits for the world's health, *Archives of Internal Medicine* 169 (2009): 543–545.
2. D. R. Jacobs and coauthors, Food, plant food, and vegetarian diets in the US dietary guidelines: Conclusions of an expert panel, *American Journal of Clinical Nutrition* 89 (2009): 1549S–1552S; S. E. Berkow and N. Barnard, Vegetarian diets and weight status, *Nutrition Review* 64 (2006): 175–188; T. J. Key, P. N. Appleby, and M. S. Rosell, Health effects of vegetarian and vegan diets, *Proceedings of the Nutrition Society* 65 (2006): 35–41; Position of the American Dietetic Association: Vegetarian diets, *Journal of the American Dietetic Association* 109 (2009): 1266–1282.
3. Y. Wang and M. A. Beydoun, Meat consumption is associated with obesity and central obesity among US adults, *International Journal of Obesity* 33 (2009): 621-628; S. Tonstad and coauthors, Type of vegetarian diet, body weight, and prevalence of type 2 diabetes, *Diabetes Care* 32 (2009): 791–796.
4. A. Vang and coauthors, Meats, processed meats, obesity, weight gain and occurrence of diabetes among adults: Findings from Adventist Health Studies, *Annals of Nutrition and Metabolism* 52 (2008): 96–104.
5. S. Tonstad and coauthors, Type of vegetarian diet, body weight, and prevalence of type 2 diabetes, *Diabetes Care* 32 (2009): 791–796.
6. H. R. Ferdowsian and N. D. Barnard, Effects of plant-based diets on plasma lipids, *American Journal of Cardiology* 104 (2009): 947–956.
7. M. Messina, Insights gained from 20 years of soy research, *Journal of Nutrition* 140 (2010): 2289S–2295S.
8. World Cancer Research Fund/American Institute for Cancer Research, *Policy and Action for Cancer Prevention. Food, Nutrition, and Physical Activity: A Global Perspective* (Washington, D.C.: AICR, 2009), p. 26.
9. Position of the American Dietetic Association, 2009.
10. I. Elmadfa and I. Singer, Vitamin B-12 and homocysteine status among vegetarians: A global perspective, *American Journal of Clinical Nutrition* 89 (2009): 1693S–1698S.
11. J. Chan, K. Jaceldo-Siegl, and Gary E. Fraser, Serum 25-hydroxyvitamin D status of vegetarians, partial vegetarians, and nonvegetarians: The Adventist Health Study, *American Journal of Clinical Nutrition* 89 (2009): 1686S–1692S.
12. A. A. Welch and coauthors, Dietary intake and status of n-3 polyunsaturated fatty acids in a population of fish-eating and

non-fish-eating meat-eaters, vegetarians, and vegans and the precursor-product ratio of α-linolenic acid to long-chain n-3 polyunsaturated fatty acids: Results from the EPIC-Norfolk cohort, *American Journal of Clinical Nutrition* 92 (2010): 1040–1051.
13. I. Mangat, Do vegetarians have to eat fish for optimal cardiovascular protection? *American Journal of Clinical Nutrition* 89 (2009): 1597S–1601S.
14. W. J. Craig, Nutrition concerns and health effects of vegetarian diets, *Nutrition in Clinical Practice* 25 (2010): 613–620; G. E. Fraser, Vegetarian diets: What do we know of their effects on common chronic diseases? *American Journal of Clinical Nutrition* 89 (2009): 1607S–1612S.

© Istockphoto.com/Svetlana Prikhodko

CENGAGE LEARNING'S GLOBAL NUTRITION WATCH — GLOBAL NUTRITION RESOURCE CENTER

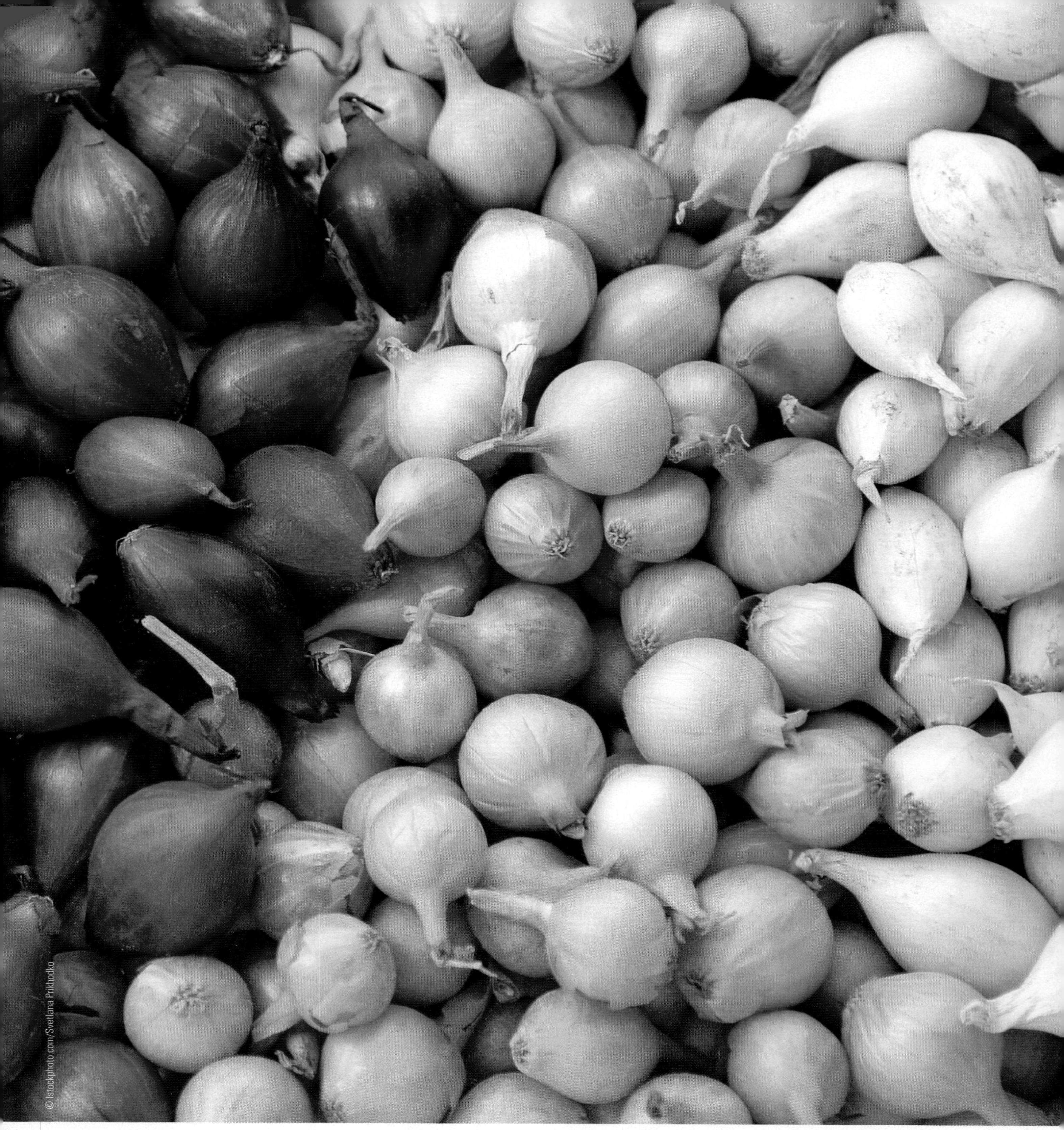

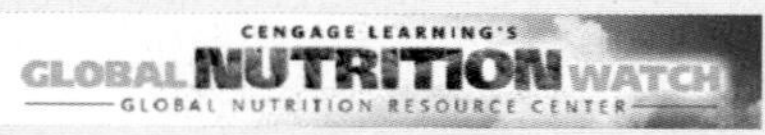

Probiotics offer a wide array of digestive benefits, but some studies suggest the benefits may go beyond the GI tract. Go to Nutrition Basics and Tools → Probiotics to learn more about the ways probiotics support good health.

3 Digestion, Absorption, and Transport

LEARNING GPS

Nutrition in Your Life

Have you ever wondered what happens to the food you eat after you swallow it? Or how your body extracts nutrients from food? Have you ever marveled at how it all just seems to happen? Follow foods as they travel through the digestive system. Learn how a healthy digestive system takes whatever food you give it—whether sirloin steak and potatoes or tofu and brussels sprouts—and extracts the nutrients that will nourish the cells of your body. In the Nutrition Portfolio at the end of the chapter, you can determine whether your current eating habits are supporting a healthy digestive system.

This chapter follows the journey that breaks down foods into the nutrients featured in the later chapters. Then it follows the nutrients as they travel through the intestinal cells and into the body to do their work. This introduction presents a general overview of the processes common to all nutrients; later chapters discuss the specifics of digesting and absorbing individual nutrients.

3.1 Digestion

LEARN IT Explain how foods move through the digestive system, describing the actions of the organs, muscles, and digestive secretions along the way.

Digestion is the body's ingenious way of breaking down foods into nutrients in preparation for **absorption.** In the process, the body overcomes many challenges without any conscious effort. Consider these challenges:

1. Human beings breathe, eat, and drink through their mouths. Air taken in through the mouth must go to the lungs; food and liquid must go to the stomach. The throat must be arranged so that swallowing and breathing don't interfere with each other.
2. Below the lungs lies the diaphragm, a dome of muscle that separates the upper half of the major body cavity from the lower half. The body needs a

digestion: the process by which food is broken down into absorbable units.

- **digest** = take apart

absorption: the uptake of nutrients by the cells of the small intestine for transport into either the blood or the lymph.

- **absorb** = suck in

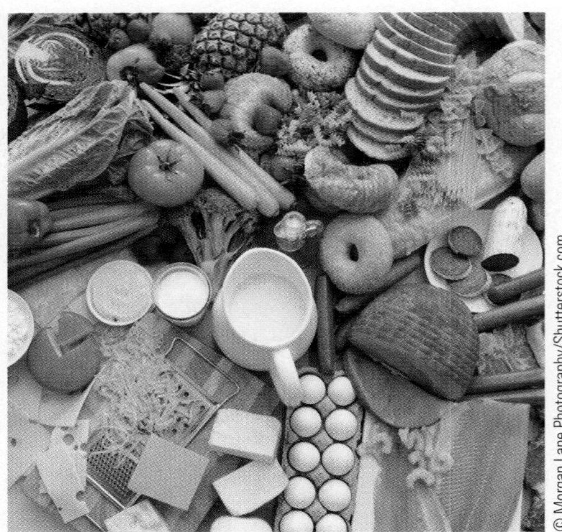

The process of digestion breaks down all kinds of *foods* into *nutrients.*

passageway that will allow food from the mouth to pass through the diaphragm to reach the stomach below.

3. The contents of the digestive tract should be kept moving forward, slowly but steadily, at a pace that permits all reactions to reach completion.
4. To move through the system, food must be lubricated with fluids. Too much would form a liquid that would flow too rapidly; too little would form a paste too dry and compact to move at all. The amount of fluids must be regulated to keep the intestinal contents at the right consistency to move along smoothly.
5. For digestive enzymes to work, foods must be broken down into small particles and suspended in enough liquid so that every particle is accessible. Once digestion is complete and nutrients have been absorbed from the GI tract into the body, the remaining waste must be excreted. Excreting all the water along with the solid residue, however, would be both wasteful and messy. Some water must be withdrawn, leaving a solid waste product that is easy to pass.
6. The digestive enzymes are designed to digest carbohydrate, fat, and protein. The cells of the GI tract are also made of carbohydrate, fat, and protein. These cells must be protected against the powerful digestive juices that they secrete.
7. Once waste matter has reached the end of the GI tract, it must be excreted, but it would be inconvenient and embarrassing if this function occurred continuously. Evacuation needs to occur periodically.

The following sections show how the body elegantly and efficiently handles these challenges. Each section follows the GI tract from one end to the other—first describing its anatomy, then its muscular actions, and finally its secretions.

Anatomy of the Digestive Tract The **gastrointestinal (GI) tract** is a flexible muscular tube that extends from the mouth, through the esophagus, stomach, small intestine, large intestine, and rectum to the anus. Figure 3-1 traces the path followed by food from one end to the other. In a sense, the human body surrounds the GI tract. The inner space within the GI tract, called the **lumen,** is continuous from one end to the other. (GI anatomy terms appear in boldface type and are defined in the accompanying glossary.) Only when a nutrient or other substance finally penetrates the GI tract's wall does it enter the body proper; many materials pass through the GI tract without being digested or absorbed.

gastrointestinal (GI) tract: the digestive tract. The principal organs are the stomach and intestines.
- **gastro** = stomach
- **intestinalis** = intestine

GLOSSARY OF GI ANATOMY TERMS

anus (AY-nus): the terminal outlet of the GI tract.

appendix: a narrow blind sac extending from the beginning of the colon that contains bacteria and lymph cells.

duodenum (doo-oh-DEEN-um, doo-ODD-num): the top portion of the small intestine (about "12 fingers' breadth" long in ancient terminology).
- **duodecim** = twelve

epiglottis (epp-ih-GLOTT-iss): cartilage in the throat that guards the entrance to the trachea and prevents fluid or food from entering it when a person swallows.
- **epi** = upon (over)
- **glottis** = back of tongue

esophageal (ee -SOF-ah-GEE-al) **sphincter:** a sphincter muscle at the upper or lower end of the esophagus. The *lower esophageal sphincter* is also called the *cardiac sphincter* because of its proximity to the heart.

esophagus (ee-SOFF-ah-gus): the food pipe; the conduit from the mouth to the stomach.

gallbladder: the organ that stores and concentrates bile. When it receives the signal that fat is present in the duodenum, the gallbladder contracts and squirts bile through the bile duct into the duodenum.

ileocecal (ill-ee-oh-SEEK-ul) **valve:** the sphincter separating the small and large intestines.

ileum (ILL-ee-um): the last segment of the small intestine.

jejunum (je-JOON-um): the first two-fifths of the small intestine beyond the duodenum.

large intestine or colon (COAL-un): the lower portion of intestine that completes the digestive process. Its segments are the *ascending colon*, the *transverse colon*, the *descending colon*, and the *sigmoid colon*.
- **sigmoid** = shaped like the letter S (sigma in Greek)

lumen (LOO-men): the space within a vessel, such as the intestine.

mouth: the oral cavity containing the tongue and teeth.

pancreas: a gland that secretes digestive enzymes and juices into the duodenum. (The pancreas also secretes hormones into the blood that help to maintain glucose homeostasis.)

pharynx (FAIR-inks): the passageway leading from the nose and mouth to the larynx and esophagus, respectively.

pyloric (pie-LORE-ic) **sphincter:** the circular muscle that separates the stomach from the small intestine and regulates the flow of partially digested food into the small intestine; also called *pylorus* or *pyloric valve.*
- **pylorus** = gatekeeper

rectum: the muscular terminal part of the intestine, extending from the sigmoid colon to the anus.

small intestine: a 10-foot length of small-diameter intestine that is the major site of digestion of food and absorption of nutrients. Its segments are the *duodenum, jejunum,* and *ileum.*

sphincter (SFINK-ter): a circular muscle surrounding, and able to close, a body opening. Sphincters are found at specific points along the GI tract and regulate the flow of food particles.
- **sphincter** = band (binder)

stomach: a muscular, elastic, saclike portion of the digestive tract that grinds and churns swallowed food, mixing it with acid and enzymes to form chyme.

> FIGURE 3-1 **The Gastrointestinal Tract**

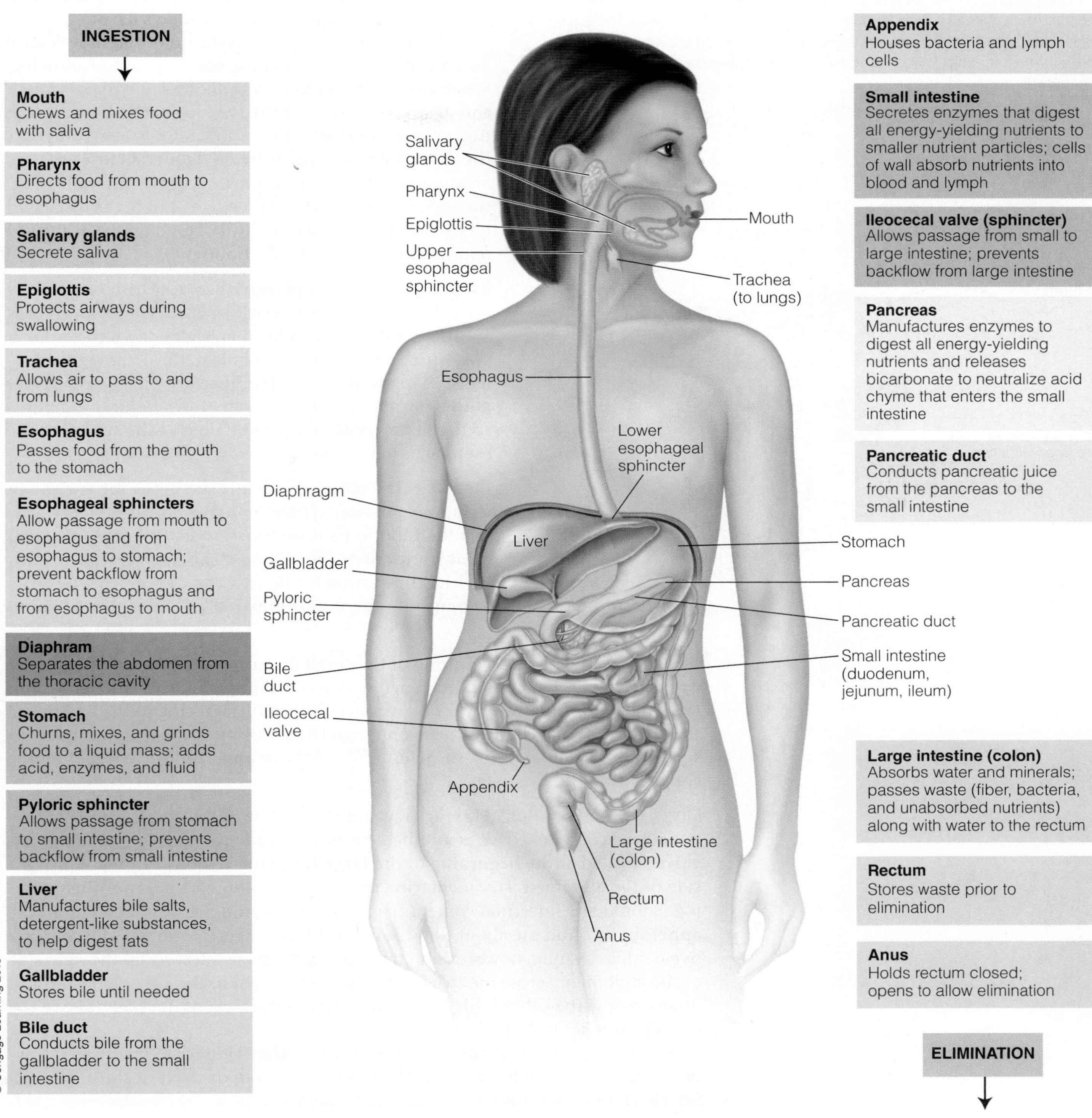

INGESTION
Mouth
Chews and mixes food with saliva
Pharynx
Directs food from mouth to esophagus
Salivary glands
Secrete saliva
Epiglottis
Protects airways during swallowing
Trachea
Allows air to pass to and from lungs
Esophagus
Passes food from the mouth to the stomach
Esophageal sphincters
Allow passage from mouth to esophagus and from esophagus to stomach; prevent backflow from stomach to esophagus and from esophagus to mouth
Diaphram
Separates the abdomen from the thoracic cavity
Stomach
Churns, mixes, and grinds food to a liquid mass; adds acid, enzymes, and fluid
Pyloric sphincter
Allows passage from stomach to small intestine; prevents backflow from small intestine
Liver
Manufactures bile salts, detergent-like substances, to help digest fats
Gallbladder
Stores bile until needed
Bile duct
Conducts bile from the gallbladder to the small intestine
Salivary glands
Pharynx
Epiglottis
Upper esophageal sphincter
Mouth
Trachea (to lungs)
Esophagus
Lower esophageal sphincter
Diaphragm
Liver
Gallbladder
Pyloric sphincter
Bile duct
Ileocecal valve
Appendix
Stomach
Pancreas
Pancreatic duct
Small intestine (duodenum, jejunum, ileum)
Large intestine (colon)
Rectum
Anus
Appendix
Houses bacteria and lymph cells
Small intestine
Secretes enzymes that digest all energy-yielding nutrients to smaller nutrient particles; cells of wall absorb nutrients into blood and lymph
Ileocecal valve (sphincter)
Allows passage from small to large intestine; prevents backflow from large intestine
Pancreas
Manufactures enzymes to digest all energy-yielding nutrients and releases bicarbonate to neutralize acid chyme that enters the small intestine
Pancreatic duct
Conducts pancreatic juice from the pancreas to the small intestine
Large intestine (colon)
Absorbs water and minerals; passes waste (fiber, bacteria, and unabsorbed nutrients) along with water to the rectum
Rectum
Stores waste prior to elimination
Anus
Holds rectum closed; opens to allow elimination
ELIMINATION

Mouth The process of digestion begins in the **mouth.** During chewing, teeth crush large pieces of food into smaller ones, and fluids from foods, beverages, and salivary glands blend with these pieces to ease swallowing.* Fluids also help dissolve the food so that the tongue can taste it; only particles in solution can react with taste buds. When stimulated, the taste buds detect one, or a combination, of the four basic taste sensations: sweet, sour, bitter, and salty. Some scientists also include the flavor associated with monosodium glutamate, sometimes called

*The process of chewing is called *mastication* (mass-tih-KAY-shun).

> FIGURE 3-2 The Colon

The colon begins with the ascending colon rising upward toward the liver. It becomes the transverse colon as it turns and crosses the body toward the spleen. The descending colon turns downward and becomes the sigmoid colon, which extends to the rectum. Along the way, the colon mixes the intestinal contents, absorbs water and salts, and forms stools.

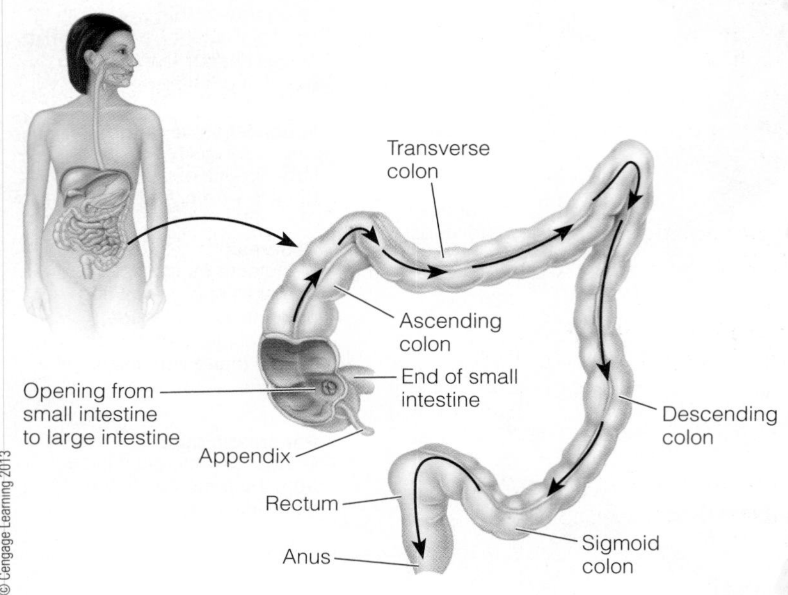

savory or its Asian name, *umami* (oo-MOM-ee).[1] In addition to these chemical triggers, aroma, appearance, texture, and temperature also affect a food's flavor.

The tongue provides taste sensations and moves food around the mouth, facilitating chewing and swallowing. When a mouthful of food is swallowed, it passes through the **pharynx,** a short tube that is shared by both the **digestive system** and the respiratory system. To bypass the entrance to the lungs, the **epiglottis** closes off the airway so that choking doesn't occur when swallowing, thus resolving the first challenge. (Choking is discussed on pp. 88–89.) After a mouthful of food has been chewed and swallowed, it is called a **bolus.**

Esophagus The **esophagus** has a **sphincter** muscle at each end. During a swallow, the upper **esophageal sphincter** opens. The bolus then slides down the esophagus, which passes through a hole in the diaphragm (challenge 2) to the stomach. The lower esophageal sphincter at the entrance to the stomach closes behind the bolus so that it proceeds forward and doesn't slip back into the esophagus (challenge 3).

Stomach The **stomach** retains the bolus for a while in its upper portion. Little by little, the stomach transfers the food to its lower portion, adds juices to it, and grinds it to a semiliquid mass called **chyme.** Then, bit by bit, the stomach releases the chyme through the **pyloric sphincter,** which opens into the **small intestine** and then closes behind the chyme.

Small Intestine At the beginning of the small intestine, the chyme bypasses the opening from the common bile duct, which is dripping fluids (challenge 4) into the small intestine from two organs outside the GI tract—the **gallbladder** and the **pancreas.** The chyme travels on down the small intestine through its three segments—the **duodenum,** the **jejunum,** and the **ileum**—almost 10 feet of tubing coiled within the abdomen.*

Large Intestine (Colon) Having traveled the length of the small intestine, the remaining contents arrive at another sphincter (challenge 3 again): the **ileocecal valve,** located at the beginning of the **large intestine (colon)** in the lower right side of the abdomen. Upon entering the colon, the contents pass another opening. Should any intestinal contents slip into this opening, it would end up in the **appendix,** a blind sac about the size of your little finger. Normally, the contents bypass this opening, however, and travel along the large intestine up the right side of the abdomen, across the front to the left side, down to the lower left side, and finally below the other folds of the intestines to the back of the body, above the **rectum** (see Figure 3-2).

As the intestinal contents pass to the rectum, the colon withdraws water, leaving semisolid waste (challenge 5). The strong muscles of the rectum and anal canal hold back this waste until it is time to defecate. Then the rectal muscles relax (challenge 7), and the two sphincters of the **anus** open to allow passage of the waste.

The Muscular Action of Digestion

In the mouth, chewing, the addition of saliva, and the action of the tongue transform food into a coarse mash that can be swallowed. After swallowing, all the activity that follows occurs without much conscience thought. As is the case with so much else that happens in the body, the muscles of the digestive tract meet internal needs without any concerted effort on

*The small intestine is almost two and a half times shorter in living adults than it is at death, when muscles are relaxed and elongated.

digestive system: all the organs and glands associated with the ingestion and digestion of food.

bolus (BOH-lus): a portion; with respect to food, the amount swallowed at one time.

- **bolos** = lump

chyme (KIME): the semiliquid mass of partly digested food expelled by the stomach into the duodenum.

- **chymos** = juice

your part. They keep things moving at just the right pace, slow enough to get the job done and fast enough to make progress.*

Peristalsis The entire GI tract is ringed with circular muscles. Surrounding these rings of muscle are longitudinal muscles. When the rings tighten and the long muscles relax, the tube is constricted. When the rings relax and the long muscles tighten, the tube bulges. This action—called **peristalsis**—occurs continuously and pushes the intestinal contents along (challenge 3 again). (If you have ever watched a lump of food pass along the body of a snake, you have a good picture of how these muscles work.)

The waves of contraction normally ripple along the GI tract at varying rates and intensities depending on the part of the GI tract and on whether food is present. Factors such as stress, medicines, and medical conditions may interfere with normal GI tract contractions.[2]

> FIGURE 3-3 **Stomach Muscles**

The stomach has three layers of muscles.

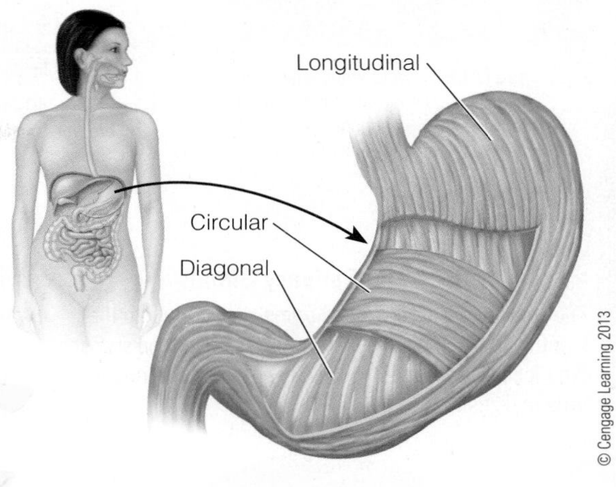

Stomach Action The stomach has the thickest walls and strongest muscles of all the GI tract organs. In addition to the circular and longitudinal muscles, it has a third layer of diagonal muscles that also alternately contract and relax (see Figure 3-3). These three sets of muscles work to force the chyme downward, but the pyloric sphincter usually remains tightly closed, preventing the chyme from passing into the duodenum of the small intestine. As a result, the chyme is churned and forced down, hits the pyloric sphincter, and remains in the stomach. Meanwhile, the stomach wall releases gastric juices. When the chyme is completely liquefied with gastric juices, the pyloric sphincter opens briefly, about three times a minute, to allow small portions of chyme to pass through. At this point, the chyme no longer resembles food in the least.

Segmentation The circular muscles of the intestines rhythmically contract and squeeze their contents. These contractions, called **segmentation,** mix the chyme and promote close contact with the digestive juices and the absorbing cells of the intestinal walls before letting the contents move slowly along.

Sphincter Contractions Sphincter muscles periodically open and close, allowing the contents of the GI tract to move along at a controlled pace (challenge 3 again). At the top of the esophagus, the upper esophageal sphincter opens in response to swallowing. At the bottom of the esophagus, the lower esophageal sphincter (sometimes called the cardiac sphincter because of its proximity to the heart) prevents **reflux** of the stomach contents. At the bottom of the stomach, the pyloric sphincter, which stays closed most of the time, holds the chyme in the stomach long enough for it to be thoroughly mixed with gastric juice and liquefied. The pyloric sphincter also prevents the intestinal contents from backing up into the stomach. At the end of the small intestine, the ileocecal valve performs a similar function, allowing the contents of the small intestine to empty into the large intestine. Finally, the tightness of the rectal muscle acts as a kind of safety device; together with the two sphincters of the anus, it prevents continuous elimination (challenge 7). Figure 3-4 illustrates how sphincter muscles contract and relax to close and open passageways.

peristalsis (per-ih-STALL-sis): wavelike muscular contractions of the GI tract that push its contents along.

- **peri** = around
- **stellein** = wrap

segmentation (SEG-men-TAY-shun): a periodic squeezing or partitioning of the intestine at intervals along its length by its circular muscles.

reflux: a backward flow.

- **re** = back
- **flux** = flow

> FIGURE 3-4 **An Example of a Sphincter Muscle**

When the circular muscles of a sphincter contract, the passage closes; when they relax, the passage opens.

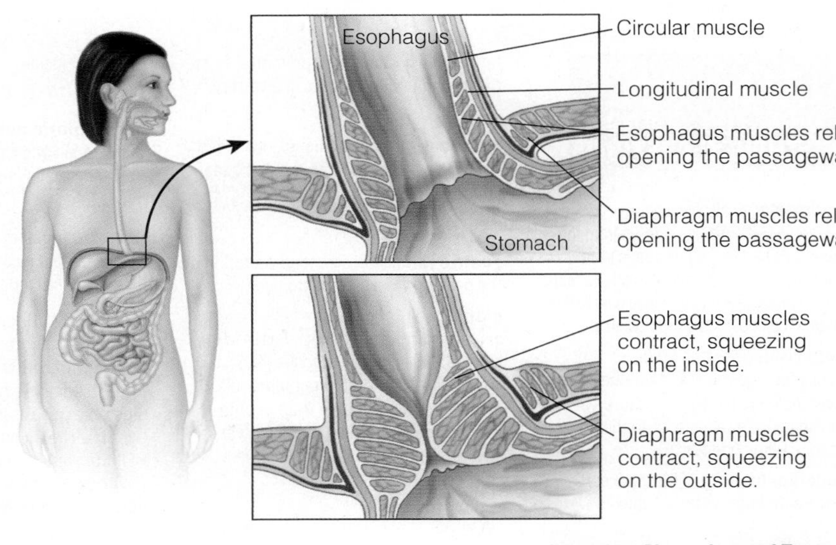

Esophagus
Stomach
Circular muscle
Longitudinal muscle
Esophagus muscles relax, opening the passageway.
Diaphragm muscles relax, opening the passageway.
Esophagus muscles contract, squeezing on the inside.
Diaphragm muscles contract, squeezing on the outside.
© Cengage Learning 2013

The Secretions of Digestion

The breakdown of food into nutrients requires secretions from five different organs: the salivary glands, the stomach, the pancreas, the liver (via the gallbladder), and the small

*The ability of the GI tract muscles to move is called *motility* (moh-TIL-ih-tee).

GLOSSARY OF DIGESTIVE ENZYMES

-ase (ACE): a word ending denoting an enzyme. The word beginning often identifies the compounds the enzyme works on. Examples include:

- *carbohydrase* (KAR-boe-HIGH-drase), an enzyme that hydrolyzes carbohydrates.
- *lipase* (LYE-pase), an enzyme that hydrolyzes lipids (fats).
- *protease* (PRO-tee-ase), an enzyme that hydrolyzes proteins.

digestive enzymes: proteins found in digestive juices that act on food substances, causing them to break down into simpler compounds.

hydrolysis (high-DROL-ih-sis): a chemical reaction in which one molecule is split into two molecules, with hydrogen (H) added to one and a hydroxyl group (OH) to the other (from water, H_2O). (The noun is *hydrolysis*; the verb is *hydrolyze*.)

- **hydro** = water
- **lysis** = breaking

> FIGURE 3-5 **The Salivary Glands**

The salivary glands secrete enzyme-rich saliva into the mouth and begin the digestive process. Given the short time food is in the mouth, salivary enzymes contribute little to digestion.

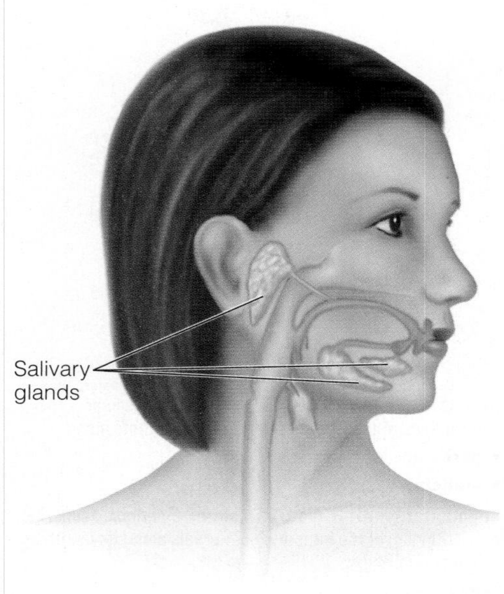

catalyst (CAT-uh-list): a compound that facilitates chemical reactions without itself being changed in the process.

intestine. These secretions enter the GI tract at various points along the way, bringing an abundance of water (challenge 4) and a variety of enzymes.

Enzymes are formally introduced in Chapter 6, but for now a simple definition will suffice. An enzyme is a protein that facilitates a chemical reaction—making a molecule, breaking a molecule apart, changing the arrangement of a molecule, or exchanging parts of molecules. As a **catalyst**, the enzyme itself remains unchanged. The enzymes involved in digestion facilitate a chemical reaction known as **hydrolysis**—the addition of water *(hydro)* to break *(lysis)* a molecule into smaller pieces. The glossary above describes how to identify some of the common **digestive enzymes** and related terms; later chapters introduce specific enzymes. When learning about enzymes, it helps to know that the word ending *-ase* denotes an enzyme. Enzymes are often identified by the organ they come from and the compounds they work on. *Gastric lipase,* for example, is a stomach enzyme that acts on lipids, whereas *pancreatic lipase* comes from the pancreas (and also works on lipids).

Saliva The **salivary glands,** shown in Figure 3-5, squirt just enough **saliva** to moisten each mouthful of food so that it can pass easily down the esophagus (challenge 4). (Digestive **glands** and their secretions are defined in the glossary below.) The saliva contains water, salts, mucus, and enzymes that initiate the digestion of carbohydrates. Saliva also protects the teeth and the linings of the mouth, esophagus, and stomach from substances that might cause damage.

Gastric Juice In the stomach, **gastric glands** secrete **gastric juice,** a mixture of water, enzymes, and **hydrochloric acid,** which acts primarily in protein digestion. The acid is so strong that it causes the sensation of heartburn if it happens to reflux into the esophagus. Highlight 3, following this chapter, discusses heartburn, ulcers, and other common digestive problems.

The strong acidity of the stomach prevents bacterial growth and kills most bacteria that enter the body with food. It would destroy the cells of the stomach as well, but for their natural defenses. To protect themselves from gastric juice, the cells of

GLOSSARY OF DIGESTIVE GLANDS AND THEIR SECRETIONS

bicarbonate: an alkaline compound with the formula HCO_3 that is secreted from the pancreas as part of the pancreatic juice. (Bicarbonate is also produced in all cell fluids from the dissociation of carbonic acid to help maintain the body's acid-base balance.)

bile: an emulsifier that prepares fats and oils for digestion; an exocrine secretion made by the liver, stored in the gallbladder, and released into the small intestine when needed.

emulsifier (ee-MUL-sih-fire): a substance with both water-soluble and fat-soluble portions that promotes the mixing of oils and fats in a watery solution.

gastric glands: exocrine glands in the stomach wall that secrete gastric juice into the stomach.

- **gastro** = stomach

gastric juice: the digestive secretion of the gastric glands of the stomach.

glands: cells or groups of cells that secrete materials for special uses in the body. Glands may be *exocrine* (EKS-oh-crin) *glands*, secreting their materials "out" (into the digestive tract or onto the surface of the skin), or *endocrine* (EN-doe-crin) *glands*, secreting their materials "in" (into the blood).

- **exo** = outside
- **endo** = inside
- **krine** = to separate

hydrochloric acid: an acid composed of hydrogen and chloride atoms (HCl) that is normally produced by the gastric glands.

liver: the organ that manufactures bile. (The liver's many other functions are described in Chapter 7.)

mucus (MYOO-kus): a slippery substance secreted by cells of the GI lining (and other body linings) that protects the cells from exposure to digestive juices (and other destructive agents). The lining of the GI tract with its coat of mucus is a *mucous membrane*. (The noun is *mucus*; the adjective is *mucous*.)

pancreatic (pank-ree-AT-ic) **juice:** the exocrine secretion of the pancreas that contains both enzymes for the digestion of carbohydrate, fat, and protein as well as bicarbonate, a neutralizing agent. The juice flows from the pancreas into the small intestine through the pancreatic duct. (The pancreas also has an endocrine function, the secretion of insulin and other hormones.)

saliva: the secretion of the salivary glands. Its principal enzyme begins carbohydrate digestion.

salivary glands: exocrine glands that secrete saliva into the mouth.

the stomach wall (in fact, of the entire gastrointestinal lining) secrete **mucus,** a thick, slippery, white substance that coats the cells, protecting them from the acid, enzymes, and disease-causing bacteria that might otherwise cause harm (challenge 6).

Figure 3-6 shows how the strength of acids is measured—in **pH** units. Note that the acidity of gastric juice registers below 2 on the pH scale—stronger than vinegar. The stomach enzymes work most efficiently in the stomach's strong acid, but the salivary enzymes, which are swallowed with food, do not work in acid this strong. Consequently, the salivary digestion of carbohydrates gradually ceases when the stomach acid penetrates each newly swallowed bolus of food. Once in the stomach, salivary enzymes simply become other proteins to be digested.

Pancreatic Juice and Intestinal Enzymes By the time food leaves the stomach, digestion of all three energy nutrients (carbohydrates, fats, and proteins) has begun, and the action gains momentum in the small intestine. There the pancreas contributes digestive juices by way of ducts leading into the duodenum. The **pancreatic juice** contains enzymes that act on all three energy nutrients, and the cells of the intestinal wall also possess digestive enzymes on their surfaces.

In addition to enzymes, the pancreatic juice contains sodium **bicarbonate,** which is basic or alkaline—the opposite of the stomach's acid (review Figure 3-6). The pancreatic juice thus neutralizes the acidic chyme arriving in the small intestine from the stomach. From this point on, the chyme remains at a neutral or slightly alkaline pH. The enzymes of both the intestine and the pancreas work best in this environment.

Bile **Bile** also flows into the duodenum. The **liver** continuously produces bile, which is then concentrated and stored in the gallbladder. The gallbladder squirts bile into the duodenum of the small intestine when fat arrives there. Bile is not an enzyme; it is an **emulsifier** that brings fats into suspension in water so that enzymes can break them down into their component parts. A summary of digestive secretions and their actions is presented in Table 3-1.

The Final Stage

At this point, the three energy-yielding nutrients—carbohydrate, fat, and protein—have been digested and are ready to be absorbed. Some vitamins and minerals are altered slightly during digestion, but most are absorbed as they are. Undigested residues, such as some fibers, are not absorbed. Instead, they continue through the digestive tract, carrying some minerals, bile acids, additives, and contaminants out of the body. This semisolid mass helps exercise the GI muscles and keep them strong enough to perform peristalsis efficiently. Fiber also retains water, accounting for the consistency of **stools.**

> FIGURE 3-6 The pH Scale

A substance's acidity or alkalinity is measured in pH units. The pH is the negative logarithm of the hydrogen ion concentration. Each increment represents a tenfold increase in concentration of hydrogen ions, meaning, for example, that a pH of 2 is 1000 times stronger than a pH of 5.

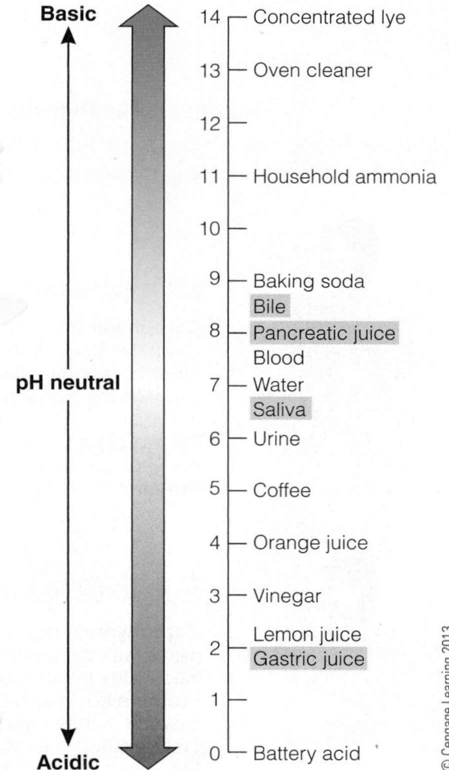

TABLE 3-1 Summary of Digestive Secretions and Their Major Actions

Organ or Gland	Target Organ	Secretion	Action
Salivary glands	Mouth	Saliva	Fluid eases swallowing; salivary enzyme breaks down some **carbohydrate.***
Gastric glands	Stomach	Gastric juice	Fluid mixes with bolus; hydrochloric acid uncoils **proteins;** enzymes break down proteins; mucus protects stomach cells.*
Pancreas	Small intestine	Pancreatic juice	Bicarbonate neutralizes acidic gastric juices; pancreatic enzymes break down **carbohydrates, fats,** and **proteins.**
Liver	Gallbladder	Bile	Bile is stored until needed.
Gallbladder	Small intestine	Bile	Bile emulsifies **fat** so that enzymes can have access to break it down.
Intestinal glands	Small intestine	Intestinal juice	Intestinal enzymes break down **carbohydrate, fat,** and **protein** fragments; mucus protects the intestinal wall.

*Saliva and gastric juice also contain lipases, but most fat breakdown occurs in the small intestine.

pH: the unit of measure expressing a substance's acidity or alkalinity. The lower the pH, the higher the H^+ ion concentration and the stronger the acid. A pH above 7 is alkaline, or base (a solution in which OH^- ions predominate).

stools: waste matter discharged from the colon; also called *feces* (FEE-seez).

By the time the contents of the GI tract reach the end of the small intestine, little remains but water, a few dissolved salts and body secretions, and undigested materials such as fiber (with some fat, cholesterol, and a few minerals bound to it). All of this remaining matter enters the large intestine (colon).

In the colon, intestinal bacteria ferment some fibers, producing water, gas, and small fragments of fat that provide energy for the cells of the colon. The colon itself retrieves all materials that the body can recycle—water and dissolved salts. The waste that is finally excreted has little or nothing of value left in it. The body has extracted all that it can use from the food. Figure 3-7 summarizes digestion by following a sandwich through the GI tract and into the body.

> FIGURE 3-7 ***Animated*** **The Digestive Fate of a Sandwich**

To review the digestive processes, follow a peanut butter and banana sandwich on whole-wheat, sesame seed bread through the GI tract. As the graph on the right illustrates, digestion of the energy nutrients begins in different parts of the GI tract, but all are ready for absorption by the time they reach the end of the small intestine.

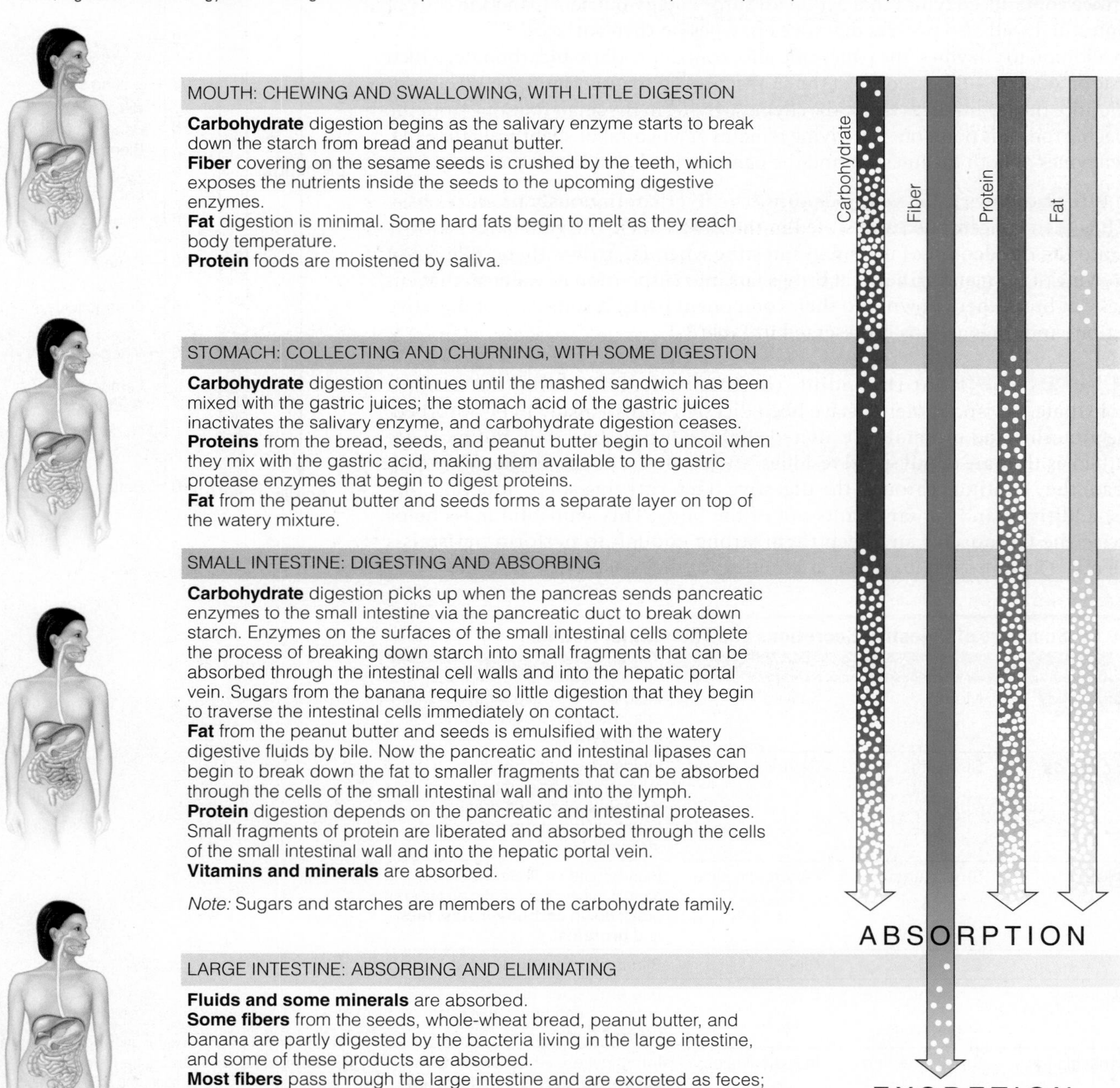

MOUTH: CHEWING AND SWALLOWING, WITH LITTLE DIGESTION
Carbohydrate digestion begins as the salivary enzyme starts to break down the starch from bread and peanut butter.
Fiber covering on the sesame seeds is crushed by the teeth, which exposes the nutrients inside the seeds to the upcoming digestive enzymes.
Fat digestion is minimal. Some hard fats begin to melt as they reach body temperature.
Protein foods are moistened by saliva.
STOMACH: COLLECTING AND CHURNING, WITH SOME DIGESTION
Carbohydrate digestion continues until the mashed sandwich has been mixed with the gastric juices; the stomach acid of the gastric juices inactivates the salivary enzyme, and carbohydrate digestion ceases.
Proteins from the bread, seeds, and peanut butter begin to uncoil when they mix with the gastric acid, making them available to the gastric protease enzymes that begin to digest proteins.
Fat from the peanut butter and seeds forms a separate layer on top of the watery mixture.
SMALL INTESTINE: DIGESTING AND ABSORBING
Carbohydrate digestion picks up when the pancreas sends pancreatic enzymes to the small intestine via the pancreatic duct to break down starch. Enzymes on the surfaces of the small intestinal cells complete the process of breaking down starch into small fragments that can be absorbed through the intestinal cell walls and into the hepatic portal vein. Sugars from the banana require so little digestion that they begin to traverse the intestinal cells immediately on contact.
Fat from the peanut butter and seeds is emulsified with the watery digestive fluids by bile. Now the pancreatic and intestinal lipases can begin to break down the fat to smaller fragments that can be absorbed through the cells of the small intestinal wall and into the lymph.
Protein digestion depends on the pancreatic and intestinal proteases. Small fragments of protein are liberated and absorbed through the cells of the small intestinal wall and into the hepatic portal vein.
Vitamins and minerals are absorbed.
Note: Sugars and starches are members of the carbohydrate family.
LARGE INTESTINE: ABSORBING AND ELIMINATING
Fluids and some minerals are absorbed.
Some fibers from the seeds, whole-wheat bread, peanut butter, and banana are partly digested by the bacteria living in the large intestine, and some of these products are absorbed.
Most fibers pass through the large intestine and are excreted as feces; some fat, cholesterol, and minerals bind to fiber and are also excreted.
Carbohydrate
Fiber
Protein
Fat
ABSORPTION
EXCRETION

REVIEW IT Explain how foods move through the digestive system, describing the actions of the organs, muscles, and digestive secretions along the way.

As Figure 3-1 shows, food enters the mouth and travels down the esophagus and through the upper and lower esophageal sphincters to the stomach, then through the pyloric sphincter to the small intestine, on through the ileocecal valve to the large intestine, past the appendix to the rectum, ending at the anus. The wavelike contractions of peristalsis and the periodic squeezing of segmentation keep things moving at a reasonable pace. Along the way, secretions from the salivary glands, stomach, pancreas, liver (via the gallbladder), and small intestine deliver fluids and digestive enzymes.

3.2 Absorption

LEARN IT Describe the anatomical details of the intestinal cells that facilitate nutrient absorption.

Within three or four hours after a person has eaten a dinner of beans and rice (or spinach lasagna, or steak and potatoes) with vegetable, salad, beverage, and dessert, the body must find a way to absorb the molecules derived from carbohydrate, protein, and fat digestion—and the vitamin and mineral molecules as well. Most absorption takes place in the small intestine, one of the most elegantly designed organ systems in the body. Within its 10-foot length, which provides a surface area equivalent to a tennis court, the small intestine traps and absorbs the nutrient molecules. To remove the absorbed molecules rapidly and provide room for more to be absorbed, a rush of circulating blood continuously washes the underside of this surface, carrying the absorbed nutrients away to the liver and other parts of the body. Figure 3-8 describes how most nutrients are absorbed by simple diffusion, facilitated diffusion, or active transport. Later chapters provide details on specific nutrients. Before following nutrients through the body, we must look more closely at the anatomy of the absorptive system.

Foods must first be digested and nutrients must be absorbed before the body can use them.

Anatomy of the Absorptive System

The inner surface of the small intestine looks smooth and slippery, but when viewed through a microscope, it turns out to be wrinkled into hundreds of folds. Each fold is contoured into thousands

> **FIGURE 3-8 Absorption of Nutrients**

Absorption of nutrients into intestinal cells typically occurs by simple diffusion, facilitated diffusion, or active transport. Occasionally, a large molecule is absorbed by *endocytosis*—a process in which the cell membrane engulfs the molecule, forming a sac that separates from the membrane and moves into the cell.

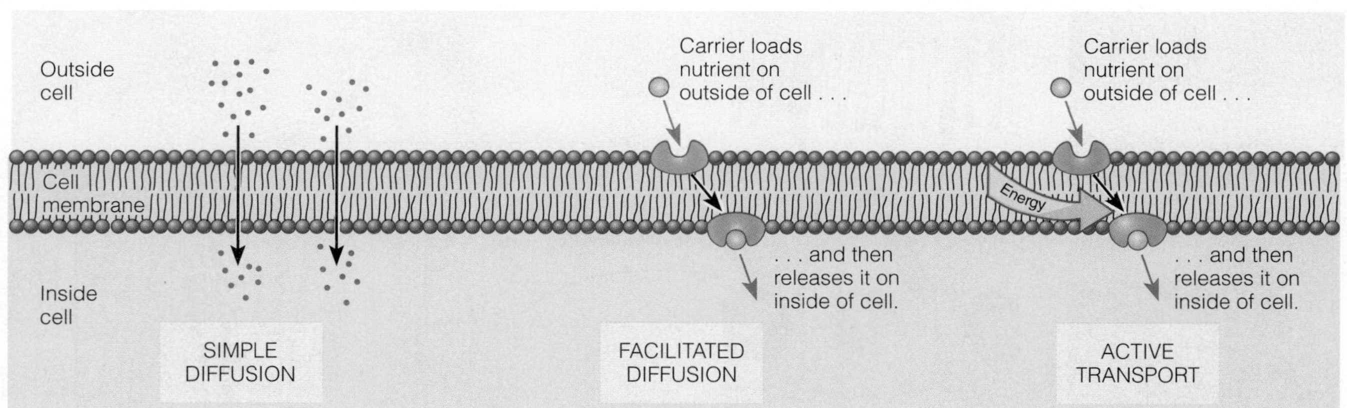

Some nutrients (such as water and small lipids) are absorbed by simple diffusion. They cross into intestinal cells freely.

Some nutrients (such as the water-soluble vitamins) are absorbed by facilitated diffusion. They need a specific carrier to transport them from one side of the cell membrane to the other. (Alternatively, facilitated diffusion may occur when the carrier changes the cell membrane in such a way that the nutrients can pass through.)

Some nutrients (such as glucose and amino acids) must be absorbed actively. These nutrients move against a concentration gradient, which requires energy.

villi (VILL-ee, VILL-eye): fingerlike projections from the folds of the small intestine; singular *villus.*

microvilli (MY-cro-VILL-ee, MY-cro-VILL-eye): tiny, hairlike projections on each cell of every villus that can trap nutrient particles and transport them into the cells; singular *microvillus.*

crypts (KRIPTS): tubular glands that lie between the intestinal villi and secrete intestinal juices into the small intestine.

goblet cells: cells of the GI tract (and lungs) that secrete mucus.

of fingerlike projections, as numerous as the hairs on velvet fabric. These small intestinal projections are called **villi.** A single villus, magnified still more, turns out to be composed of hundreds of cells, each covered with its own microscopic hairs, called **microvilli** (see Figure 3-9). In the crevices between the villi lie the **crypts**—tubular glands that secrete the intestinal juices into the small intestine. Nearby **goblet cells** secrete mucus.

The villi are in constant motion. Each villus is lined by a thin sheet of muscle, so it can wave, squirm, and wriggle like the tentacles of a sea anemone. Any nutrient molecule small enough to be absorbed is trapped among the microvilli and then drawn into the cells. Some partially digested nutrients are caught in the microvilli, digested further by enzymes there, and then absorbed into the cells.

> **FIGURE 3-9 The Small Intestinal Villi**

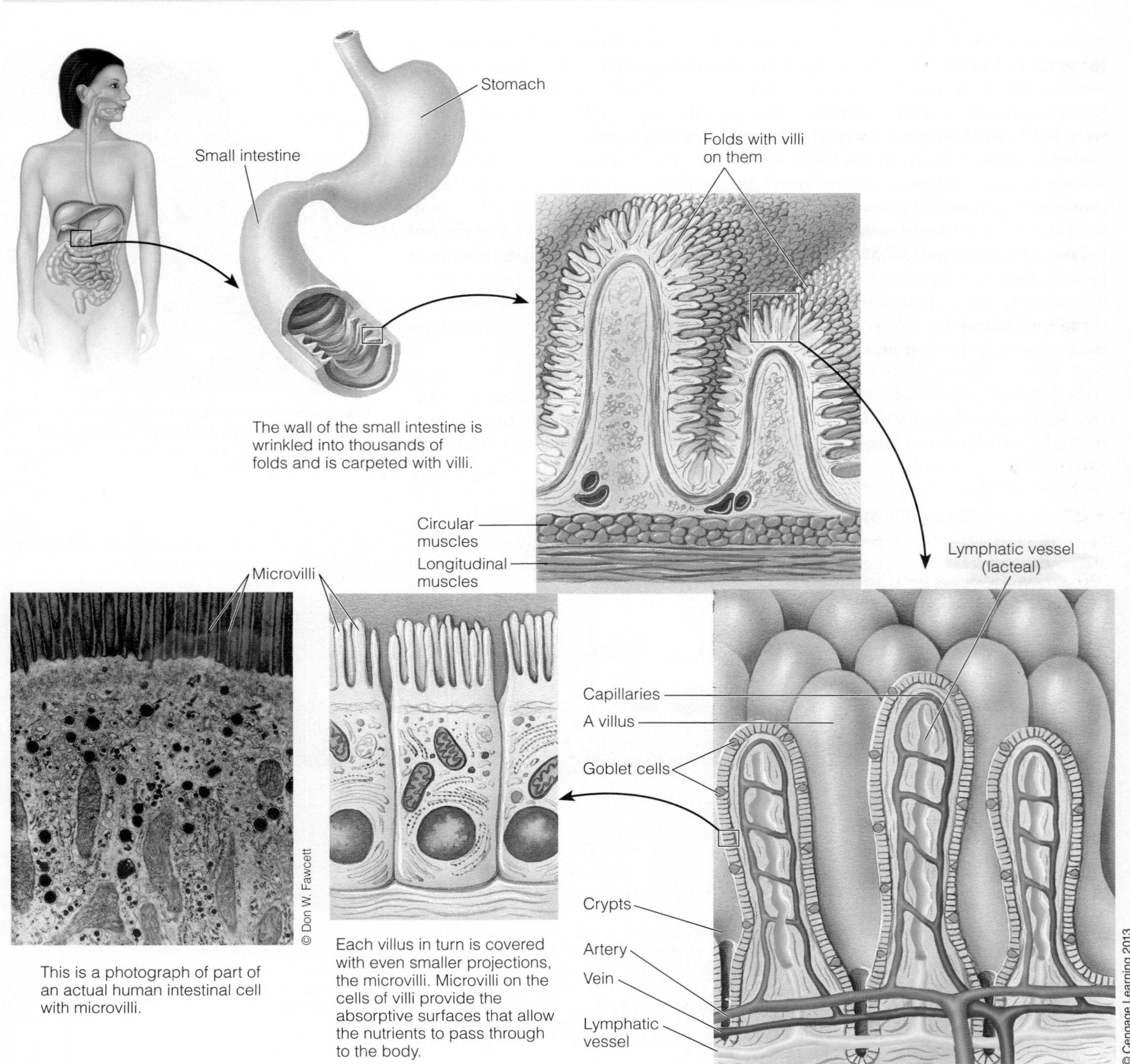

Stomach
Small intestine
Folds with villi on them
The wall of the small intestine is wrinkled into thousands of folds and is carpeted with villi.
Circular muscles
Longitudinal muscles
Microvilli
Lymphatic vessel (lacteal)
Capillaries
A villus
Goblet cells
Crypts
Artery
Vein
Lymphatic vessel
This is a photograph of part of an actual human intestinal cell with microvilli.
Each villus in turn is covered with even smaller projections, the microvilli. Microvilli on the cells of villi provide the absorptive surfaces that allow the nutrients to pass through to the body.

A Closer Look at the Intestinal Cells

The cells of the villi are among the most amazing in the body, for they recognize and select the nutrients the body needs and regulate their absorption. As already described, each cell of a villus is coated with thousands of microvilli, which project from the cell's membrane (review Figure 3-9). In these microvilli, and in the membrane, lie hundreds of different kinds of enzymes and "pumps," which recognize and act on different nutrients. Descriptions of specific enzymes and "pumps" for each nutrient are presented in later chapters where appropriate; the point here is that the cells are equipped to handle all kinds and combinations of foods and their nutrients.

If you have ever watched a sea anemone with its fingerlike projections in constant motion, you have a good picture of how the intestinal villi move.

Specialized Cells A further refinement of the system is that the cells of successive portions of the intestinal tract are specialized to absorb different nutrients. The nutrients that are ready for absorption early are absorbed near the top of the GI tract; those that take longer to be digested are absorbed farther down. Registered dietitians and medical professionals who treat digestive disorders learn the specialized absorptive functions of different parts of the GI tract so that if one part becomes dysfunctional, the diet can be adjusted accordingly.

Food Combining The idea that people should not eat certain food combinations (for example, fruit and meat) at the same meal, because the digestive system cannot handle more than one task at a time, is a myth. The art of "food combining"—which actually emphasizes "food separating"—is based on this myth, and it represents faulty logic and a gross underestimation of the body's capabilities. In fact, the contrary is often true; foods eaten together can enhance each other's use by the body. For example, vitamin C in a pineapple or other citrus fruit can enhance the absorption of iron from a meal of chicken and rice or other iron-containing foods. Many other instances of mutually beneficial interactions are presented in later chapters.

Preparing Nutrients for Transport When a nutrient molecule has crossed the cell of a villus, it enters either the bloodstream or the lymphatic system. Both transport systems supply vessels to each villus, as shown in Figure 3-9. The water-soluble nutrients and the smaller products of fat digestion are released directly into the bloodstream and guided directly to the liver where their fate and destination will be determined.

The larger fats and the fat-soluble vitamins are insoluble in water, however, and blood is mostly water. The intestinal cells assemble many of the products of fat digestion into larger molecules. These larger molecules cluster together with special proteins, forming chylomicrons. Chylomicrons (kye-lo-MY-cronz) are defined and described in more detail in Chapter 5. For now, keep in mind that because chylomicrons carry fats, they are released into the lymphatic system. They move through the lymph until they can enter the bloodstream at a point near the heart. Consequently, chylomicrons bypass the liver at first. Details follow.

REVIEW IT Describe the anatomical details of the intestinal cells that facilitate nutrient absorption.

The many folds and villi of the small intestine dramatically increase its surface area, facilitating nutrient absorption. Nutrients pass through the cells of the villi and enter either the blood (if they are water soluble or small fat fragments) or the lymph (if they are fat soluble).

3.3 The Circulatory Systems

LEARN IT Explain how nutrients are routed in the circulatory systems from the GI tract into the body and identify which nutrients enter the blood directly and which must first enter the lymph.

Once a nutrient has entered the bloodstream, it may be transported to any of the cells in the body, from the tips of the toes to the roots of the hair. The circulatory systems deliver nutrients wherever they are needed.

The Vascular System The vascular, or blood circulatory, system is a closed system of vessels through which blood flows continuously, with the heart serving as the pump (see Figure 3-10). As the blood circulates through this system, it picks up and delivers materials as needed.

All the body tissues derive nutrients and oxygen from the blood and deposit carbon dioxide and other wastes back into the blood. The digestive system supplies the nutrients. The lungs exchange oxygen (which enters the blood to be delivered to all cells) and carbon dioxide (which leaves the blood to be exhaled). The kidneys filter wastes other than carbon dioxide out of the blood to be excreted in the urine.

> FIGURE 3-10 *Animated* **The Vascular System**

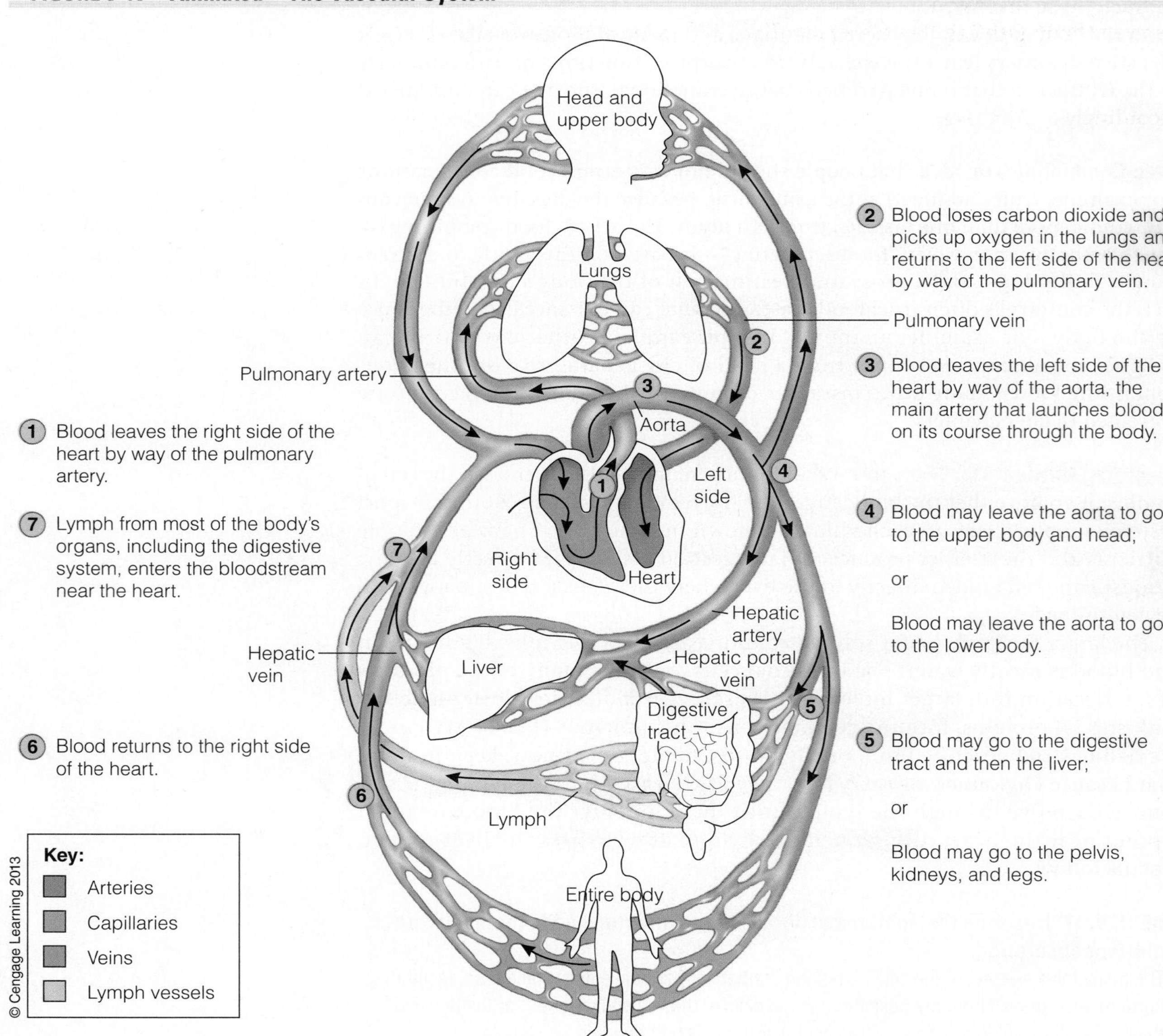
Head and upper body
Lungs
Pulmonary vein
Pulmonary artery
Aorta
Left side
Right side
Heart
Hepatic artery
Hepatic portal vein
Hepatic vein
Liver
Digestive tract
Lymph
Entire body
1 Blood leaves the right side of the heart by way of the pulmonary artery.
2 Blood loses carbon dioxide and picks up oxygen in the lungs and returns to the left side of the heart by way of the pulmonary vein.
3 Blood leaves the left side of the heart by way of the aorta, the main artery that launches blood on its course through the body.
4 Blood may leave the aorta to go to the upper body and head;
or
Blood may leave the aorta to go to the lower body.
5 Blood may go to the digestive tract and then the liver;
or
Blood may go to the pelvis, kidneys, and legs.
6 Blood returns to the right side of the heart.
7 Lymph from most of the body's organs, including the digestive system, enters the bloodstream near the heart.
Key:
Arteries
Capillaries
Veins
Lymph vessels
© Cengage Learning 2013

Blood leaving the right side of the heart circulates through the lungs and then back to the left side of the heart. The left side of the heart then pumps the blood out of the **aorta** through **arteries** to all systems of the body. The blood circulates in the **capillaries,** where it exchanges material with the cells and then collects into **veins,** which return it again to the right side of the heart. In short, blood travels this simple route:

Heart to arteries to capillaries to veins to heart

The routing of the blood leaving the digestive system has a special feature. The blood is carried to the digestive system (as to all organs) by way of an artery, which (as in all organs) branches into capillaries to reach every cell. Blood leaving the digestive system, however, goes by way of a vein. The **hepatic portal vein** directs blood not back to the heart but to another organ, the liver. This vein branches into a network of large capillaries so that every cell of the liver has access to the blood. Blood leaving the liver then collects into the **hepatic vein,** which returns blood to the heart. The route is:

Heart to arteries to capillaries (in intestines) to hepatic portal vein to capillaries (in liver) to hepatic vein to heart

Figure 3-11 shows the liver's key position in nutrient transport. An anatomist studying this system knows there must be a reason for this special arrangement. The liver's placement ensures that it will be first to receive the nutrients absorbed from the GI tract. In fact, the liver has many jobs to do in preparing the absorbed nutrients for use by the body. Of all the body's organs, the liver is the most metabolically active.

aorta (ay-OR-tuh): the large, primary artery that conducts blood from the heart to the body's smaller arteries.

arteries: vessels that carry blood from the heart to the tissues.

capillaries (CAP-ill-aries): small vessels that branch from an artery. Capillaries connect arteries to veins. Exchange of oxygen, nutrients, and waste materials takes place across capillary walls.

veins (VANES): vessels that carry blood to the heart.

hepatic portal vein: the vein that collects blood from the GI tract and conducts it to the liver.

- **portal** = gateway

hepatic vein: the vein that collects blood from the liver and returns it to the heart.

- **hepatic** = liver

> **FIGURE 3-11 The Liver**

1. Vessels gather up nutrients from the digestive tract.

Not shown here:
Parallel to these vessels (veins) are other vessels (arteries) that carry oxygen-rich blood from the heart to the intestines.

2. The vessels merge into the hepatic portal vein, which conducts all absorbed materials to the liver.

3. The hepatic artery brings a supply of freshly oxygenated blood (not loaded with nutrients) from the lungs to supply oxygen to the liver's own cells.

4. A network of large capillaries branch all over the liver, making nutrients and oxygen available to all its cells and giving the cells access to blood from the digestive system.

5. The hepatic vein gathers up blood in the liver and returns it to the heart.

In contrast, nutrients absorbed into lymph do not go to the liver first. They go to the heart, which pumps them to all the body's cells. The cells remove the nutrients they need, and the liver then has to deal only with the remnants.

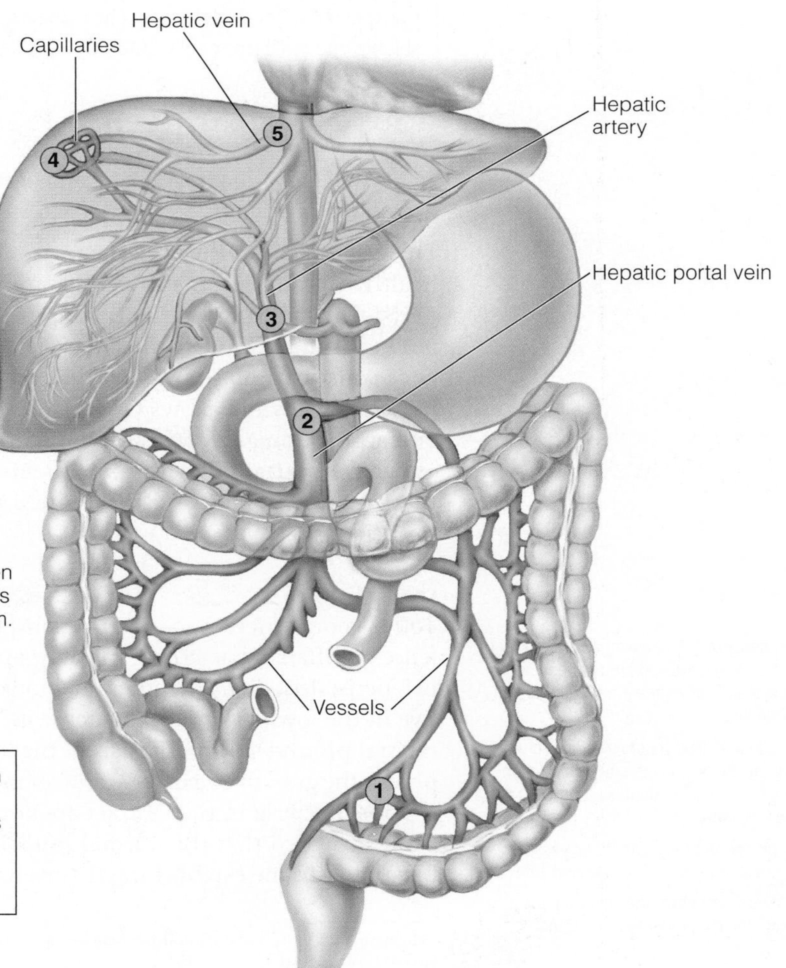

In addition, the liver defends the body by detoxifying substances that might cause harm and preparing waste products for excretion. This is why, when people ingest poisons that succeed in passing the first barrier (the intestinal cells), the liver quite often suffers the damage—from viruses such as hepatitis, from drugs such as barbiturates or alcohol, from toxins such as pesticide residues, and from contaminants such as mercury. Perhaps, in fact, you have been undervaluing your liver, not knowing what heroic tasks it quietly performs for you.

The Lymphatic System The **lymphatic system** provides a one-way route for fluid from the tissue spaces to enter the blood. Unlike the vascular system, the lymphatic system has no pump; instead, **lymph** circulates *between* the cells of the body and collects into tiny vessels. The fluid moves from one portion of the body to another as muscles contract and create pressure here and there. Ultimately, much of the lymph collects in the **thoracic duct** behind the heart. The thoracic duct opens into the **subclavian vein,** where the lymph enters the bloodstream. Thus nutrients from the GI tract that enter lymphatic vessels (large fats and fat-soluble vitamins) ultimately enter the bloodstream, circulating through arteries, capillaries, and veins like the other nutrients, with a notable exception—they bypass the liver at first.*

Once inside the vascular system, the nutrients can travel all over the body where they can be taken into cells and used as needed. What becomes of them is described in later chapters.

REVIEW IT Explain how nutrients are routed in the circulatory systems from the GI tract into the body and identify which nutrients enter the blood directly and which must first enter the lymph.

Nutrients leaving the digestive system via the blood are routed directly to the liver before being transported to the body's cells. Those leaving via the lymphatic system (large fats and fat-soluble vitamins) eventually enter the vascular system but bypass the liver at first.

3.4 The Health and Regulation of the GI Tract

LEARN IT Describe how bacteria, hormones, and nerves influence the health and activities of the GI tract.

This section describes the bacterial conditions and hormonal regulation of a healthy GI tract, but many factors can influence normal GI function. For example, peristalsis and sphincter action are poorly coordinated in newborns, so infants tend to "spit up" during the first several months of life. Older adults often experience constipation, in part because the intestinal wall loses strength and elasticity with age, which slows GI motility. Diseases can also interfere with digestion and absorption and often lead to malnutrition. Lack of nourishment, in general, and lack of certain dietary constituents such as fiber, in particular, alter the structure and function of GI cells. Quite simply, GI tract health depends on adequate nutrition.

Gastrointestinal Bacteria An estimated 10 trillion bacteria representing some 400 or more different species and subspecies live in a healthy GI tract.** The prevalence of different bacteria in various parts of the GI tract depends on such factors as pH, peristalsis, diet, and other microorganisms.[3] Relatively few microorganisms can live in the low pH of the stomach with its relatively rapid peristalsis, whereas the neutral pH and slow peristalsis of the lower small intestine and the large intestine permit the growth of a diverse and abundant bacterial population.

Most bacteria in the GI tract are not harmful; in fact, they are actually beneficial.[4] Provided that the normal intestinal **flora** are thriving, infectious bacteria have a hard time establishing themselves to launch an attack on the system.

*The lymphatic vessels of the intestine that take up nutrients and pass them to the lymph circulation are called *lacteals* (LACK-tee-als).

**Bacteria in the intestines are sometimes referred to as *flora*.

lymphatic (lim-FAT-ic) **system:** a loosely organized system of vessels and ducts that convey fluids toward the heart. The GI part of the lymphatic system carries the products of fat digestion into the bloodstream.

lymph (LIMF): a clear yellowish fluid that is similar to blood except that it contains no red blood cells or platelets. Lymph from the GI tract transports fat and fat-soluble vitamins to the bloodstream via lymphatic vessels.

thoracic (thor-ASS-ic) **duct:** the main lymphatic vessel that collects lymph and drains into the left subclavian vein.

subclavian (sub-KLAY-vee-an) **vein:** the vein that provides passageway from the lymphatic system to the vascular system.

flora: bacteria in the intestines.

Diet is one of several factors that influence the body's bacterial population and environment. Consider **yogurt**, for example. Yogurt contains *Lactobacillus* and other living bacteria. These microorganisms are considered **probiotics** because they change the conditions and native bacterial colonies in the GI tract in ways that seem to benefit health.[5] The potential GI health benefits of probiotics include helping to alleviate diarrhea, constipation, inflammatory bowel disease, ulcers, allergies, lactose intolerance, and infant colic; enhance immune function; and protect against colon cancer. Some probiotics may have adverse effects under certain circumstances.[6] Research studies continue to explore how diet influences GI bacteria and which foods—with their probiotics—affect GI health. In addition, research studies are beginning to reveal several health benefits beyond the GI tract—such as improving blood pressure and immune responses.[7]

GI bacteria also digest fibers and complex proteins. These food components are called **prebiotics** because they encourage the growth and activity of bacteria. In the process, the bacteria produce nutrients such as short fragments of fat that the cells of the colon use for energy. Bacteria in the GI tract also produce several vitamins, ♦ although the amount is insufficient to meet the body's total need for these vitamins.

Eaten regularly, yogurt can alleviate common digestive problems.

♦ Vitamins produced by bacteria include:

- Biotin
- Folate
- Pantothenic acid
- Riboflavin
- Thiamin
- Vitamin B_6
- Vitamin B_{12}
- Vitamin K

Gastrointestinal Hormones and Nerve Pathways The ability of the digestive tract to handle its ever-changing contents illustrates an important physiological principle that governs the way all living things function—the principle of **homeostasis.** Simply stated, survival depends on body conditions staying about the same; if they deviate too far from the norm, the body must "do something" to bring them back to normal. The body's regulation of digestion is one example of homeostatic regulation. The body also regulates its temperature, its blood pressure, and all other aspects of its blood chemistry in similar ways.

Two intricate and sensitive systems coordinate all the digestive and absorptive processes: the hormonal (or endocrine) system and the nervous system. Even before the first bite of food is taken, the mere thought, sight, or smell of food can trigger a response from these systems. Then, as food travels through the GI tract, it either stimulates or inhibits digestive secretions by way of messages that are carried from one section of the GI tract to another by both **hormones** and nerve pathways. (Appendix A presents a brief summary of the body's hormonal system and nervous system.)

Notice that the kinds of regulation described next are all examples of *feedback* mechanisms. A certain condition demands a response. The response changes that condition, and the change then cuts off the response. Thus the system is self-correcting. Examples follow:

- *The stomach normally maintains a pH between 1.5 and 1.7. How does it stay that way?* Food entering the stomach stimulates cells in the stomach wall to release the hormone **gastrin.** Gastrin, in turn, stimulates the stomach glands to secrete the components of hydrochloric acid. When pH 1.5 is reached, the acid itself turns off the gastrin-producing cells; they stop releasing gastrin, and the glands stop producing hydrochloric acid. Thus the system adjusts itself, as Figure 3-12 (p. 84) shows.

Nerve receptors in the stomach wall also respond to the presence of food and stimulate the gastric glands to secrete juices and the muscles to contract. As the stomach empties, the receptors are no longer stimulated, the flow of juices slows, and the stomach quiets down.

- *The pyloric sphincter opens to let out a little chyme, then closes again. How does it know when to open and close?* When the pyloric sphincter relaxes, acidic chyme slips through. The cells of the pyloric muscle on the intestinal side sense the acid, causing the pyloric sphincter to close tightly. Only after the chyme has been neutralized by pancreatic bicarbonate and the juices surrounding the pyloric sphincter have become alkaline can the muscle relax again. This process ensures that the chyme will be released slowly enough to be neutralized as it flows through the

yogurt: milk product that results from the fermentation of lactic acid in milk by *Lactobacillus bulgaricus* and *Streptococcus thermophilus.*

probiotics: living microorganisms found in foods and dietary supplements that, when consumed in sufficient quantities, are beneficial to health.

- **pro** = for
- **bios** = life

prebiotics: food components (such as fibers) that are not digested by the human body but are used as food by the GI bacteria to promote their growth and activity.

homeostasis (HOME-ee-oh-STAY-sis): the maintenance of constant internal conditions (such as blood chemistry, temperature, and blood pressure) by the body's control systems. A homeostatic system is constantly reacting to external forces to maintain limits set by the body's needs.

- **homeo** = the same
- **stasis** = staying

hormones: chemical messengers. Hormones are secreted by a variety of glands in response to altered conditions in the body. Each hormone travels to one or more specific target tissues or organs, where it elicits a specific response to maintain homeostasis.

gastrin: a hormone secreted by cells in the stomach wall. Target organ: the glands of the stomach. Response: secretion of gastric acid.

> FIGURE 3-12 **An Example of a Negative Feedback Loop**

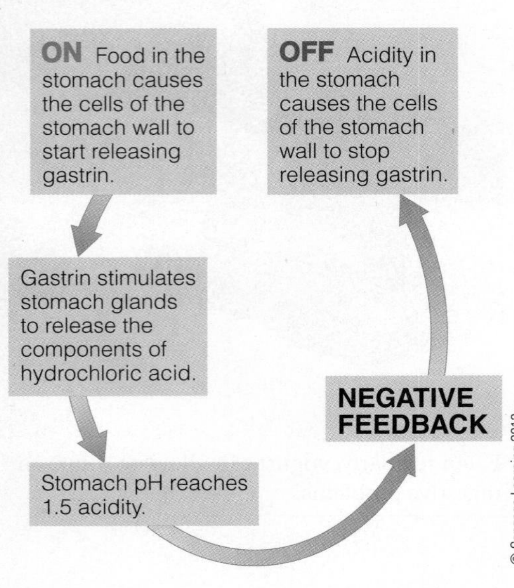

small intestine. This is important because the small intestine has less of a mucous coating than the stomach does and so is not as well protected from acid.

ON Food in the stomach causes the cells of the stomach wall to start releasing gastrin.
OFF Acidity in the stomach causes the cells of the stomach wall to stop releasing gastrin.
Gastrin stimulates stomach glands to release the components of hydrochloric acid.
NEGATIVE FEEDBACK
Stomach pH reaches 1.5 acidity.
© Cengage Learning 2013

- *As the chyme enters the small intestine, the pancreas adds bicarbonate to it so that the intestinal contents always remain at a slightly alkaline pH. How does the pancreas know how much to add?* The presence of chyme stimulates the cells of the duodenum wall to release the hormone **secretin** into the blood. When secretin reaches the pancreas, it stimulates the pancreas to release its bicarbonate-rich juices. Thus, whenever the duodenum signals that acidic chyme is present, the pancreas responds by sending bicarbonate to neutralize it. When the need has been met, the cells of the duodenum wall are no longer stimulated to release secretin, the hormone no longer flows through the blood, and the pancreas no longer receives the message and stops sending pancreatic juice. Nerves also regulate pancreatic secretions.

- *Pancreatic secretions contain a mixture of enzymes to digest carbohydrate, fat, and protein. How does the pancreas know how much of each type of enzyme to provide?* This is one of the most interesting questions physiologists have asked. Clearly, the pancreas does know what its owner has been eating, and it secretes enzyme mixtures tailored to handle the food mixtures that have been arriving recently (over the last several days). Enzyme activity changes proportionately in response to the amounts of carbohydrate, fat, and protein in the diet. If a person has been eating mostly carbohydrates, the pancreas makes and secretes mostly carbohydrases; if the person's diet has been high in fat, the pancreas produces more lipases; and so forth. Hormones from the GI tract, secreted in response to meals, keep the pancreas informed as to its digestive tasks. The day or two lag between the time a person's diet changes dramatically and the time digestion of the new diet becomes efficient explains why dietary changes can "upset digestion" and should be made gradually.

- *Why don't the digestive enzymes damage the pancreas?* The pancreas protects itself from harm by producing an inactive form of the enzymes.* It releases these proteins into the small intestine, where they are activated to become enzymes. In pancreatitis, the digestive enzymes become active within the infected pancreas, causing inflammation and damaging the delicate pancreatic tissues.

- *When fat is present in the intestine, the gallbladder contracts to squirt bile into the intestine to emulsify the fat. How does the gallbladder get the message that fat is present?* Fat in the intestine stimulates cells of the intestinal wall to release the hormone **cholecystokinin (CCK).** This hormone travels by way of the blood to the gallbladder and stimulates it to contract, which releases bile into the small intestine. Cholecystokinin also travels to the pancreas and stimulates it to secrete its juices, which releases bicarbonate and enzymes into the small intestine. Once the fat in the intestine is emulsified and enzymes have begun to work on it, the fat no longer provokes release of the hormone, and the message to contract is canceled. (By the way, fat emulsification can continue even after a diseased gallbladder has been surgically removed because the liver can deliver bile directly to the small intestine.)

- *Fat and protein take longer to digest than carbohydrate does. When fat or protein is present, intestinal motility slows to allow time for its digestion. How does the intestine know when to slow down?* Cholecystokinin is released in response to fat or protein in the small intestine. In addition to its role in fat emulsification and digestion, cholecystokinin slows GI tract motility. Slowing the digestive process helps to maintain a pace that allows all reactions to reach completion. Hormonal and nervous mechanisms like these account for much of the body's ability to adapt to changing conditions.

Table 3-2 summarizes the actions of these three GI hormones. Gastrin, secretin, and cholecystokinin are among the most studied GI hormones, but the GI tract releases more than 20 hormones. In addition to assisting with

secretin (see-CREET-in): a hormone produced by cells in the duodenum wall. Target organ: the pancreas. Response: secretion of bicarbonate-rich pancreatic juice.

cholecystokinin (COAL-ee-SIS-toe-KINE-in), or **CCK**: a hormone produced by cells of the intestinal wall. Target organ: the gallbladder. Response: release of bile and slowing of GI motility.

*The inactive precursor of an enzyme is called a *zymogen* (ZYE-mo-jen).

TABLE 3-2 The Primary Actions of Selected GI Hormones

Hormone	Responds to	Secreted from	Stimulates	Response
Gastrin	Food in the stomach	Stomach wall	Stomach glands	Hydrochloric acid secreted into the stomach
Secretin	Acidic chyme in the small intestine	Duodenal wall	Pancreas	Bicarbonate-rich juices secreted into the small intestine
Cholecystokinin	Fat or protein in the small intestine	Intestinal wall	Gallbladder	Bile secreted into the duodenum
			Pancreas	Bicarbonate- and enzyme-rich juices secreted into the small intestine

digestion and absorption, many of these hormones regulate food intake and influence satiation—the feeling of satisfaction and fullness that occurs during a meal and halts eating. Current research is focusing on the roles these hormones may play in the development of obesity and its treatments (more details provided in Chapter 9).

Discovering the answers to questions like these has led some people to devote their whole lives to the study of physiology. For now, however, these few examples illustrate how all the processes throughout the digestive system are precisely and automatically regulated without any conscious effort.

The System at Its Best This chapter describes the anatomy of the digestive tract on several levels: the sequence of digestive organs, the cells and structures of the villi, and the selective machinery of the cell membranes. The intricate architecture of the digestive system makes it sensitive and responsive to conditions in its environment. Several different kinds of GI tract cells confer specific immunity against intestinal diseases such as inflammatory bowel disease. In addition, secretions from the GI tract—saliva, mucus, gastric acid, and digestive enzymes—not only help with digestion, but also defend against foreign invaders. Together the GI's team of bacteria, cells, and secretions defend the body against numerous challenges.

One indispensable condition is good health of the digestive system itself. Like all the other organs of the body, the GI tract depends on a healthy supply of blood. The cells of the GI tract become weak and inflamed when blood flow is diminished, as may occur in heart disease when arteries become clogged or blood clots form. Just as a diminished blood flow to the heart or brain can cause a heart attack or stroke, respectively, too little blood to the intestines can also be

Nourishing foods and pleasant conversations support a healthy digestive system.

damaging—or even fatal. A diminished blood flow to the intestines—called **intestinal ischemia**—is characterized by abdominal pain, forceful bowel movements, and blood in the stool.

The health of the digestive system is also affected by such lifestyle factors as sleep, physical activity, and state of mind. Adequate sleep allows for repair and maintenance of tissue and removal of wastes that might impair efficient functioning. Activity promotes healthy muscle tone. Stress influences the activity of regulatory nerves and hormones.[8] For healthy digestion, mealtimes should be relaxed and tranquil. Pleasant conversations and peaceful environments during meals ease the digestive process.

Another factor in GI health is the kind of foods eaten. Among the characteristics of meals that promote optimal absorption of nutrients are those mentioned in Chapter 2: balance, moderation, variety, and adequacy. Balance and moderation require having neither too much nor too little of anything. For example, too much fat can be harmful, but some fat is beneficial in slowing down intestinal motility and providing time for absorption of some of the nutrients that are slow to be absorbed.

Variety is important for many reasons, but one is that some food constituents interfere with nutrient absorption. For example, some compounds common in high-fiber foods such as whole-grain cereals, certain leafy green vegetables, and legumes bind with minerals. To some extent, then, the minerals in those foods may become unavailable for absorption. These high-fiber foods are still valuable, but they need to be balanced with a variety of other foods that can provide the minerals.

As for adequacy—in a sense, this entire book is about dietary adequacy. A diet must provide all the essential nutrients, fiber, and energy in amounts sufficient to maintain health. But here, at the end of this chapter, is a good place to emphasize the interdependence of the nutrients. It could almost be said that every nutrient depends on every other. All the nutrients work together, and all are present in the cells of a healthy digestive tract. To maintain health and promote the functions of the GI tract, make balance, moderation, variety, and adequacy features of every day's meals.

REVIEW IT Describe how bacteria, hormones, and nerves influence the health and activities of the GI tract.

A diverse and abundant bacteria population supports GI health. The regulation of GI processes depends on the coordinated efforts of the hormonal system and the nervous system. Together, digestion and absorption break down foods into nutrients for the body's use. To function optimally, a healthy GI tract needs a balanced diet, adequate rest, and regular physical activity.

Nutrition Portfolio

A digestive system that is well cared for most of the time can adjust to handle almost any diet or combination of foods with ease on occasion. Go to Diet Analysis Plus and choose one of the days on which you have tracked your diet for the entire day. Choose the day you thought you ate most poorly, and looking at it, record in your journal answers to the following:

- Describe the physical and emotional environment that typically surrounds your meals, including how it affects you and how it might be improved.
- Did you experience any GI discomfort on that day? Do you experience any GI discomfort regularly? If so, which of the foods that you ate might have contributed to your discomfort? What can you do to prevent or alleviate GI problems in the future? Use Table H3-1 (p. 93) as a guide.
- List any changes you can make in your eating habits to promote overall GI health.

To complete this exercise, go to your Diet Analysis Plus at **www.cengagebrain.com**.

STUDY IT To review the key points of this chapter and take a practice quiz, go to the study cards at the end of the book.

intestinal ischemia (is-KEY-me-ah): a diminished blood flow to the intestines that is characterized by abdominal pain, forceful bowel movements, and blood in the stool.

REFERENCES

1. 100th Anniversary Symposium of Umami Discovery, *American Journal of Clinical Nutrition* 90 (2009): entire supplement.
2. A. Stengel and Y. Taché, Neuroendocrine control of the gut during stress: Corticotropin-releasing factor signaling pathways in the spotlight, *Annual Review of Physiology* 71 (2009): 219–239; P. Holzer, Opioid receptors in the gastrointestinal tract, *Regulatory Peptides* 155 (2009): 11–17; J. López-Herce, Gastrointestinal complications in critically ill patients: What differs between adults and children? *Current Opinion in Clinical Nutrition and Metabolic Care* 12 (2009): 180–185.
3. E. M. Bik, Composition and function of the human-associated microbiota, *Nutrition Reviews* 67 (2009): S164–S171.
4. P. Hunter, The secret garden. The human digestive system is teeming with microbiotic life, but just how important are these interlopers for health and disease? *European Molecular Biology Organization Reports* 10 (2009): 1082–1086.
5. S. C. Bischoff and M. Zeitz, Scientific evidence for the medical use of probiotics, *Annals of Nutrition and Metabolism* 57 (2010): S1–S5; N. T. Williams, Probiotics, *American Journal of Health System Pharmacy* 15 (2010): 449–458; M. E. Sanders, How do we know when something called "probiotic" is really a probiotic? A guideline for consumers and health care professionals, *Functional Food Reviews* 1 (2009): 3–12.
6. K. Whelan and C. E. Myers, Safety of probiotics in patients receiving nutritional support: A systematic review of case reports, randomized controlled trials, and nonrandomized trials, *American Journal of Clinical Nutrition* 91 (2010): 687–703.
7. I. Trebichavsky and coauthors, Cross-talk of human gut with bifidobacteria, *Nutrition Reviews* 67 (2009): 77–82; N. G. Hord, Eukaryotic-microbiota crosstalk: Potential mechanisms for health benefits of prebiotics and probiotics, *Annual Review of Nutrition* 28 (2008): 215–231.
8. A. Stengel and Y. Taché, Neuroendocrine control of the gut during stress: Corticotropin-releasing factor signaling pathways in the spotlight, *Annual Review of Physiology* 71 (2009): 219–239.

HIGHLIGHT > 3

LEARN IT Outline strategies to prevent or alleviate common GI problems.

© Corbis Super RF/Alamy

Common Digestive Problems

The facts of anatomy and physiology presented in Chapter 3 permit easy understanding of some common problems that occasionally arise in the digestive tract. Food may slip into the airways instead of the esophagus, causing choking. Bowel movements may be loose and watery, as in diarrhea, or painful and hard, as in constipation. Some people complain about belching, while others are bothered by intestinal gas. Sometimes people develop medical problems such as ulcers. This highlight describes some of the symptoms of these common digestive problems and suggests strategies for preventing them (the accompanying glossary defines related terms).

Choking

Sometimes a sip of a beverage or a tiny bit of food "slips down the wrong pipe." The body's first response is to cough, and quite often coughing clears the passage. When someone is truly choking, however, food has slipped into the **trachea** and completely blocked the air passageways (see Figure H3-1). Thus the person cannot cough—or even breathe. Without oxygen, the person may suffer permanent brain damage within 5 minutes—or die. For this reason, it is imperative that everyone learn to recognize the universal distress signal for choking (shown in Figure H3-2) and act promptly.

Because the **larynx** is in the trachea and makes sounds only when air is pushed across it, a person choking will be unable to speak. For this reason, to help a person who is choking, first ask "Can you speak?" If the person is coughing, breathing adequately, or able to speak, do not interfere. Whatever you do, do not hit him on the back as the particle may become lodged more firmly in his air passageway. If the person cannot speak or cough, shout for help and perform the **Heimlich maneuver** (described in Figure H3-2). Almost any food can cause choking, although some are cited more often than others: chunks of meat, hot dogs, nuts, whole grapes, raw carrots, marshmallows, hard or sticky candies, gum, popcorn, and peanut butter. These foods are particularly difficult for young children (especially those 4 years of

GLOSSARY

acid controllers: medications used to prevent or relieve indigestion by suppressing production of acid in the stomach; also called *H2 blockers*. Common brands include Pepcid AC, Tagamet HB, Zantac 75, and Axid AR.

antacids: medications used to relieve indigestion by neutralizing acid in the stomach. Common brands include Alka-Seltzer, Maalox, Rolaids, and Tums.

belching: the release of air or gas from the stomach through the mouth.

bloating: uncomfortable abdominal fullness or distention.

celiac disease: an intestinal disorder in which the inability to absorb the protein portion of gluten results in an immune response that damages intestinal cells, also called *celiac sprue* or *gluten-sensitive enteropathy*.

colitis (ko-LYE-tis): inflammation of the colon.

colonic irrigation: the popular, but potentially harmful practice of "washing" the large intestine with a powerful enema machine.

constipation: the condition of having infrequent or difficult bowel movements.

defecate (DEF-uh-cate): to move the bowels and eliminate waste.

- **defaecare** = to remove dregs

diarrhea: the frequent passage of watery bowel movements.

diverticula (dye-ver-TIC-you-la): sacs or pouches that develop in the weakened areas of the intestinal wall (like bulges in an inner tube where the tire wall is weak).

- **divertir** = to turn aside

diverticulitis (DYE-ver-tic-you-LYE-tis): infected or inflamed diverticula.

- **itis** = infection or inflammation

diverticulosis (DYE-ver-tic-you-LOH-sis): the condition of having diverticula. Diverticulosis affects more than 50 percent of adults in later life.

- **osis** = condition

enema: solution inserted into the rectum and colon to stimulate a bowel movement and empty the lower large intestine.

flatulence: passage of excessive amounts of intestinal gas.

gastroesophageal reflux: the backflow of stomach acid into the esophagus, causing damage to the cells of the esophagus and the sensation of heartburn; commonly known as *heartburn* or *acid indigestion. Gastroesophageal reflux disease (GERD)* is characterized by symptoms of reflux occurring two or more times a week.

Heimlich (HIME-lick) **maneuver (abdominal thrusts):** a technique for dislodging an object from the trachea of a choking person (see Figure H3-2); named for the physician who developed it.

hemorrhoids (HEM-oh-royds): painful swelling of the veins surrounding the rectum.

indigestion: incomplete or uncomfortable digestion, usually accompanied by pain, nausea, vomiting, heartburn, intestinal gas, or belching.

- **in** = not

irritable bowel syndrome: an intestinal disorder of unknown cause. Symptoms include abdominal discomfort and cramping, diarrhea, constipation, or alternating diarrhea and constipation.

larynx (LAIR-inks): the entryway to the trachea that contains the vocal cords; also called the *voice box* (see Figure H3-1).

laxatives: substances that loosen the bowels and thereby prevent or treat constipation.

mineral oil: a purified liquid derived from petroleum and used to treat constipation.

peptic ulcer: a lesion in the mucous membrane of either the stomach (a *gastric ulcer*) or the duodenum (a *duodenal ulcer*).

- **peptic** = concerning digestion

trachea (TRAKE-ee-uh): the air passageway from the larynx to the lungs; also called the *windpipe*.

ulcer: a lesion of the skin or mucous membranes characterized by inflammation and damaged tissues. See also *peptic ulcer*.

vomiting: expulsion of the contents of the stomach up through the esophagus to the mouth.

> **FIGURE H3-1 Normal Swallowing and Choking**

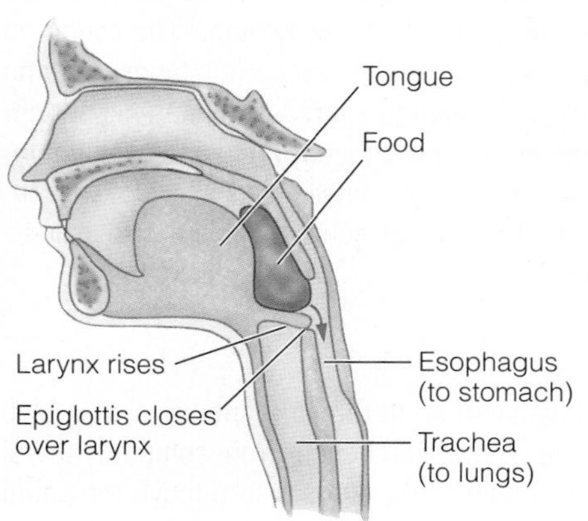

Swallowing. The epiglottis closes over the larynx, blocking entrance to the lungs via the trachea. The red arrow shows that food is heading down the esophagus normally.

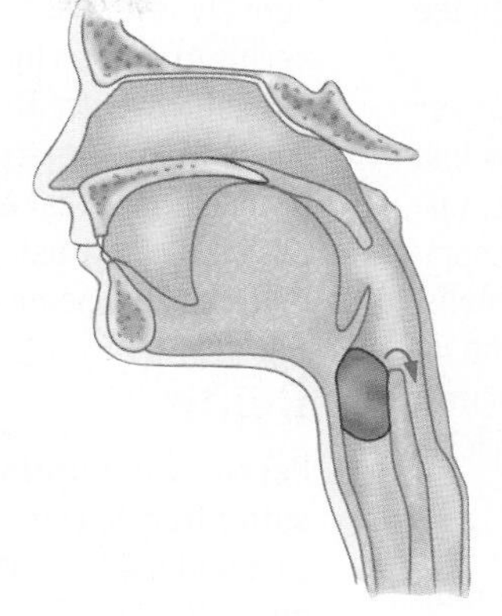

Choking. A choking person cannot speak or gasp because food lodged in the trachea blocks the passage of air. The red arrow points to where the food should have gone to prevent choking.

age and younger) to safely chew and swallow. Each year more than 10,000 children (14 years old or younger) in the United States choke; more than half choke on food. Every 5 days, a child in the United States chokes to death on food.[1] An adult should be present and alert to the dangers of choking whenever young children are eating. To prevent choking, cut food into small pieces, chew thoroughly before swallowing, don't talk or laugh with food in your mouth, and don't eat when breathing hard.

Vomiting

Vomiting can be a symptom of many different diseases or may arise in situations that upset the body's equilibrium, such as air or sea travel. For whatever reason, the contents of the stomach are propelled up through the esophagus to the mouth and expelled. Sometimes the muscular contractions will extend beyond the stomach and carry the contents of the duodenum, with its green bile, into the stomach and then up the esophagus. Although certainly unpleasant and wearying for the nauseated person,

> **FIGURE H3-2 First Aid for Choking**

First aid for choking relies on abdominal thrusts, sometimes called the Heimlich maneuver. If abdominal thrusts are not successful and the person loses consciousness, lower him to the floor, call 911, remove the object blocking the airway if possible, and begin CPR. Because there is no time for hesitation when called upon to perform this death-defying act, you would do well to take a life-saving course to learn these techniques.

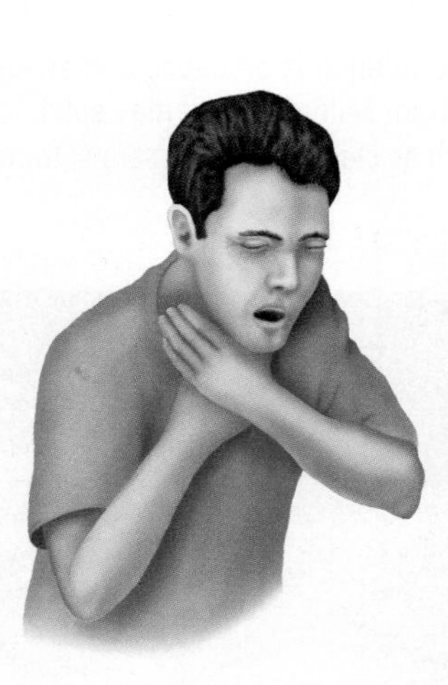

The universal signal for choking alerts others to the need for assistance.

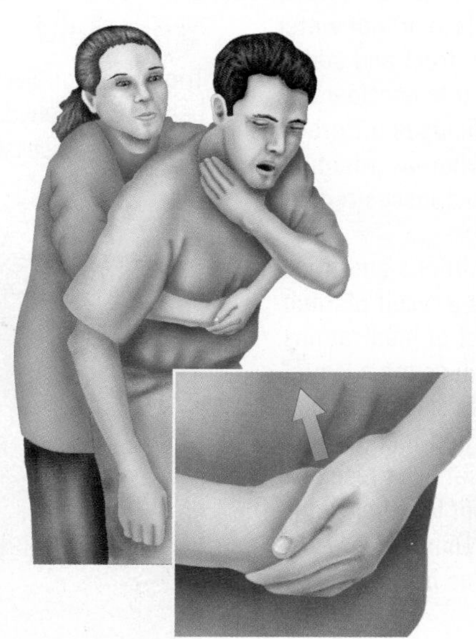

Stand behind the person with your arms wrapped around him. Make a fist with one hand and place the thumb side snugly against the body, slightly above the navel and below the breastbone.

Grasp the fist with your other hand and make a quick upward and inward thrust. Repeat thrusts until the object is dislodged.

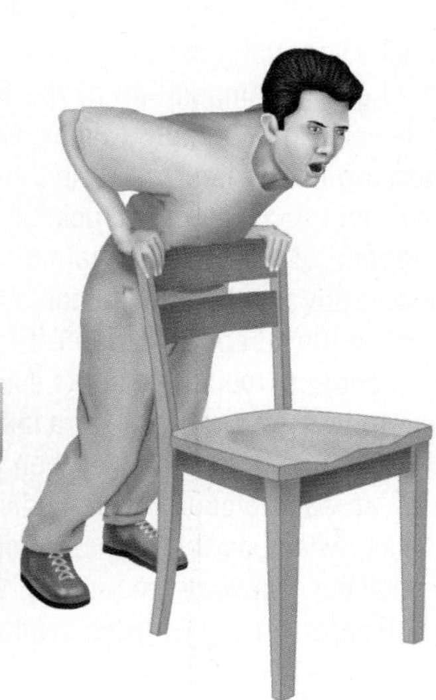

To perform abdominal thrusts on yourself, make a fist and place the thumb below your breastbone and above your navel. Grasp your fist with your other hand and press inward with a quick upward thrust. Alternatively, quickly thrust your upper body against a table edge, chair, or railing.

vomiting is often not a cause for alarm. Vomiting is one of the body's adaptive mechanisms to rid itself of something irritating. The best advice is to rest and drink small amounts of liquids as tolerated until the nausea subsides.

A physician's care may be needed, however, if vomiting causes such large losses of fluid as to threaten dehydration. As fluid is lost from the GI tract, the body's other fluids redistribute themselves, taking fluid from every cell of the body. Fluid leaving the cells is accompanied by salts that are absolutely essential to the life of the cells. Replacing salts and fluid is difficult if the vomiting continues, and intravenous feedings of saline and glucose may be necessary. Vomiting and dehydration are especially serious in an infant, and a physician should be contacted without delay.

Self-induced vomiting, such as occurs in bulimia nervosa, also has serious consequences. In addition to fluid and salt imbalances, repeated vomiting can cause irritation and infection of the pharynx, esophagus, and salivary glands; erosion of the teeth and gums; and dental caries. The esophagus may rupture or tear, as may the stomach. Sometimes the eyes become red from pressure during vomiting. Bulimic behavior reflects underlying psychological problems that require intervention. (Bulimia nervosa is discussed fully in Highlight 8.)

Diarrhea

Diarrhea is characterized by frequent, loose, watery stools. Such stools indicate that the intestinal contents have moved too quickly through the intestines for fluid absorption to take place or that water has been drawn from the cells lining the intestinal tract and added to the food residue. Like vomiting, diarrhea can lead to considerable fluid and salt losses, but the composition of the fluids is different. Stomach fluids lost in vomiting are highly acidic, whereas intestinal fluids lost in diarrhea are nearly neutral. When fluid losses require medical attention, correct replacement is crucial.

Diarrhea is a symptom of various medical conditions and treatments. It may occur abruptly in a healthy person as a result of infections (such as foodborne illness) or as a side effect of medications. When used in large quantities, food ingredients such as the sugar alternative sorbitol and the fat alternative olestra may also cause diarrhea in some people. If a food is responsible, then that food must be omitted from the diet, at least temporarily. If medication is responsible, a different medicine, when possible, or a different form (injectable versus oral, for example) may alleviate the problem. Diarrhea may also occur as a result of disorders of the GI tract, such as irritable bowel syndrome or colitis.

Irritable Bowel Syndrome

Irritable bowel syndrome is one of the most common GI disorders and is characterized by frequent or severe abdominal discomfort and a disturbance in the motility of the GI tract.[2] In most cases, GI contractions are stronger and last longer than normal, forcing intestinal contents through quickly and causing gas, **bloating**, and diarrhea. In some cases, however, GI contractions are weaker than normal, slowing the passage of intestinal contents and causing constipation. The exact cause of irritable bowel syndrome is not known, but researchers are actively investigating the role of the nervous system.[3] The condition seems to worsen for some people when they eat certain foods or during stressful events. These triggers seem to aggravate symptoms but not cause them. Dietary treatment hinges on identifying and avoiding individual foods that aggravate symptoms; small meals may also be beneficial.[4] Other treatments that may be effective include antispasmodic drugs and peppermint oil.[5]

Colitis

People with **colitis**, an inflammation of the large intestine, may also suffer from severe diarrhea. They often benefit from complete bowel rest and medication. If treatment fails, surgery to remove the colon and rectum may be necessary.

Celiac Disease

Celiac disease is an autoimmune disease characterized by inflammation of the small intestine that occurs in response to foods that contain gluten, a protein commonly found in wheat, barley, rye, and possibly oats. In people with celiac disease, gluten triggers an immune system reaction in the small intestine that causes inflammation, which damages the villi and decreases nutrient absorption. Common symptoms include abdominal pains, bloating and gas, and diarrhea—making it commonly misdiagnosed as irritable bowel syndrome. Treatment focuses on a gluten-free diet.[6]

Treatment

Treatment for diarrhea depends on cause and severity, but it always begins with rehydration. Mild diarrhea may subside with simple rest and extra liquids (such as clear juices and soups) to replace fluid losses.

Personal hygiene (such as regular hand washing with soap and water) and safe food preparation (as described in Chapter 19) are easy and effective steps to take in preventing diarrheal diseases.

If diarrhea is bloody or if it worsens or persists—especially in an infant, young child, elderly person, or person with a compromised immune system—call a physician. Severe diarrhea can be life threatening.

Constipation

Like diarrhea, **constipation** describes a symptom, not a disease. Each person's GI tract has its own cycle of waste elimination, which depends on its owner's health, the type of food eaten, when it was eaten, and when the person takes time to **defecate.** What's normal for some people may not be normal for others. Some people have bowel movements three times a day; others may have them three times a week. The symptoms of constipation include straining during bowel movements, hard stools, and infrequent bowel movements (fewer than three per week). Abdominal discomfort, headaches, backaches, and the passing of gas sometimes accompany constipation.

Often a person's lifestyle may cause constipation. Being too busy to respond to the defecation signal is a common complaint. If a person receives the signal to defecate and ignores it, the signal may not return for several hours. In the meantime, fluids continue to be withdrawn from the fecal matter, so when the person does defecate, the stools are dry and hard. In such a case, a person's daily regimen may need to be revised to allow time to have a bowel movement when the body sends its signal.

Although constipation usually reflects lifestyle habits, in some cases it may be a side effect of medication or a medical problem such as bowel obstruction. If discomfort is associated with passing fecal matter, seek medical advice to rule out disease. Once this has been done, simple treatments, such as increased fiber, fluids, and exercise, are recommended before the use of medications.

One dietary measure that may be appropriate is to increase dietary fiber to 20 to 25 grams per day gradually over the course of a week or two. Fibers found in fruits, vegetables, and whole grains help to prevent constipation by increasing fecal mass. In the GI tract, fiber attracts water, creating soft, bulky stools that stimulate bowel contractions to push the contents along. These contractions strengthen the intestinal muscles. The improved muscle tone, together with the water content of the stools, eases elimination, reducing the pressure in the rectal veins and helping to prevent **hemorrhoids**. Chapter 4 provides more information on fiber's role in maintaining a healthy colon and reducing the risks of colon cancer and diverticulosis. **Diverticulosis** is a condition in which the intestinal walls develop bulges in weakened areas, most commonly in the colon (see Figure H3-3). These bulging pockets, known as **diverticula**, can worsen constipation, entrap feces, and become painfully infected and inflamed **(diverticulitis)**. Treatment may require hospitalization, antibiotics, or surgery.

Drinking plenty of water in conjunction with eating high-fiber foods also helps to prevent constipation. The increased bulk physically stimulates the upper GI tract, promoting peristalsis throughout. Similarly, physical activity improves the muscle tone and motility of the digestive tract. As little as 30 minutes of physical activity a day can help prevent or alleviate constipation.

Eating prunes—or "dried plums" as some have renamed them—can also be helpful. Prunes are high in fiber and also contain a laxative substance.* If a morning defecation is desired, a person can drink prune juice at bedtime; if the evening is preferred, the person can drink prune juice with breakfast.

> FIGURE H3-3 **Diverticula in the Colon**

Diverticula may develop anywhere along the GI tract, but they are most common in the colon.

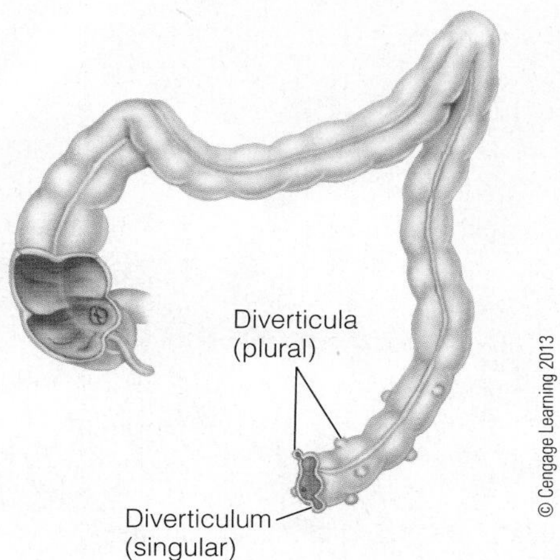

If these suggested changes in lifestyle or diet do not correct constipation, then a physician might recommend the use of stool softeners, **laxatives,** or **mineral oil**. These products are best used for brief periods. If needed for extended times, they should be used under physician supervision. Frequent use of laxatives can lead to dependency and upset the body's fluid, salt, and mineral balances. Mineral oil interferes with the absorption of fat-soluble vitamins.

One potentially harmful but currently popular practice is **colonic irrigation**—the internal washing of the large intestine with a powerful **enema** machine. Such an extreme cleansing is not only unnecessary, but it can be hazardous, causing illness and death from equipment contamination, electrolyte depletion, and intestinal perforation. Less extreme practices can cause problems, too.

Belching and Gas

Many people complain of problems that they attribute to excessive gas. For some, belching is the complaint. Others blame intestinal gas for abdominal discomforts and embarrassment.

Belching

Belching results from swallowing air. Everyone swallows a little bit of air with each mouthful of food, but people who eat too fast may swallow too much air. Ill-fitting dentures, carbonated beverages, and chewing gum can also contribute to the swallowing of air with resultant belching. The best advice for belching seems to be to eat slowly, chew thoroughly, and relax while eating.

Intestinal Gas

Although **flatulence** can be an embarrassing experience, it is quite normal. (People who experience painful bloating from malabsorption diseases,

*This laxative substance is dihydroxyphenyl isatin.

People troubled by intestinal gas need to determine which foods bother them and then eat those foods in moderation.

however, require medical treatment.) Healthy people expel several hundred milliliters of intestinal gas several times a day. Almost all (99 percent) of the gases expelled—nitrogen, oxygen, hydrogen, methane, and carbon dioxide—are odorless. The remaining "volatile" gases are the infamous ones.

Foods that produce gas usually must be determined individually. The most common offenders are foods rich in the carbohydrates—sugars, starches, and fibers. When partially digested carbohydrates reach the large intestine, bacteria digest them, giving off gas as a by-product. People can test foods suspected of forming gas by omitting them individually for a trial period to see if there is any improvement.

Gastroesophageal Reflux

Almost everyone has experienced heartburn at one time or another, usually soon after eating a meal. Medically known as **gastroesophageal reflux**, heartburn is the painful sensation a person feels behind the breastbone when the lower esophageal sphincter allows the stomach contents to reflux into the esophagus (see Figure H3-4).[7] This may happen if a person eats or drinks too much (or both). Tight clothing and even changes of position (lying down, bending over) can cause it, too, as can some medications and smoking. Weight gain and overweight increase the frequency, severity, and duration of heartburn symptoms. A defect of the sphincter muscle itself is a possible, but less common, cause.

If heartburn is not caused by an anatomical defect, treatment is fairly simple. To avoid such misery in the future, the person needs to learn to eat less at a sitting, chew food more thoroughly, and eat more slowly. Additional strategies are presented in Table H3-1 at the end of this highlight.

People who overeat or eat too quickly are likely to suffer from **indigestion.** The muscular reaction of the stomach to unchewed lumps or to being overfilled may be so intense that it upsets normal peristalsis. When this happens, overeaters may taste the stomach acid and feel pain. Over-the-counter **antacids** and **acid controllers** may provide relief but should be used only infrequently for occasional heartburn; they may mask or cause problems if used regularly. If problems continue, people who suffer from frequent and regular bouts of heartburn and indigestion may need to see a physician, who can prescribe specific medication to control gastroesophageal reflux. Without treatment, the repeated splashes of acid can severely damage the cells of the esophagus, creating a condition known as Barrett's esophagus. At that stage, the risk of cancer in the throat or esophagus increases dramatically.[8] To repeat, if symptoms persist, see a doctor—don't self-medicate.

Ulcers

Ulcers are another common digestive problem, affecting an estimated 1 out of every 12 adults in the United States. An **ulcer** is a lesion (a sore), and a **peptic ulcer** is a lesion in the lining of the stomach (gastric ulcers) or the duodenum of the small intestine (duodenal ulcers). The compromised lining is left unprotected and exposed to gastric juices, which can be painful. In some cases, ulcers can cause internal bleeding. If GI bleeding is excessive, iron deficiency may develop. Ulcers that perforate the GI lining can pose life-threatening complications.

> FIGURE H3-4 **Gastroesophageal Reflux**

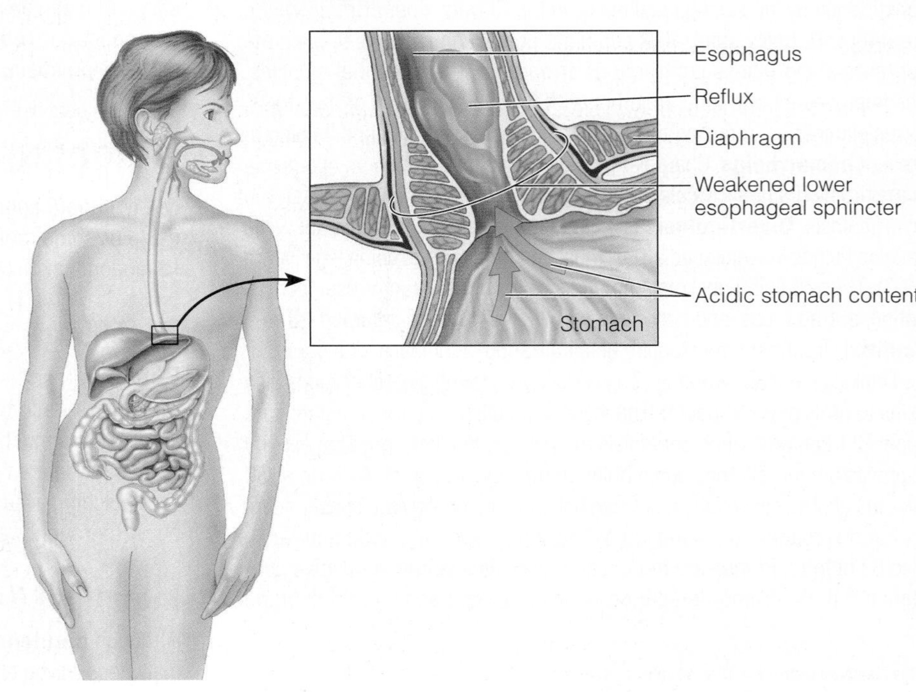

TABLE H3-1 Strategies to Prevent or Alleviate Common GI Problems

GI Problem	Strategies
Choking	• Take small bites of food. • Chew thoroughly before swallowing. • Don't talk or laugh with food in your mouth. • Don't eat when breathing hard.
Diarrhea	• Avoid strenuous activity. • Rest. • Drink fluids to replace losses. • Call for medical help if diarrhea persists.
Constipation	• Eat a high-fiber diet. • Drink plenty of fluids. • Exercise regularly. • Respond promptly to the urge to defecate.
Belching	• Eat slowly. • Chew thoroughly. • Relax while eating.
Intestinal gas	• Eat bothersome foods in moderation.
Heartburn	• Eat small meals. • Drink liquids between meals. • Sit up while eating; elevate your head when lying down. • Wait 3 hours after eating before lying down. • Wait 2 hours after eating before exercising. • Refrain from wearing tight-fitting clothing. • Avoid foods, beverages, and medications that aggravate your heartburn. • Refrain from smoking cigarettes or using tobacco products. • Lose weight if overweight.
Ulcer	• Take medicine as prescribed by your physician. • Avoid coffee and caffeine- and alcohol-containing beverages. • Avoid foods that aggravate your ulcer. • Minimize aspirin, ibuprofen, and naproxen use. • Refrain from smoking cigarettes.

Many people naively believe that an ulcer is caused by stress or spicy foods, but this is not the case. The stomach lining in a healthy person is well protected by its mucous coat. What, then, causes ulcers to form?

Three major causes of ulcers have been identified: bacterial infection with *Helicobacter pylori* (commonly abbreviated *H. pylori*); the use of certain anti-inflammatory drugs such as aspirin, ibuprofen, and naproxen; and disorders that cause excessive gastric acid secretion. Most commonly, ulcers develop in response to *H. pylori* infection. The cause of the ulcer dictates the type of medication used in treatment. For example, people with ulcers caused by infection receive antibiotics, whereas those with ulcers caused by medicines discontinue their use.[9] In addition, all treatment plans aim to relieve pain, heal the ulcer, and prevent recurrence.

The regimen for ulcer treatment is to treat for infection, eliminate any food that routinely causes indigestion or pain, and avoid coffee and caffeine- and alcohol-containing beverages. Both regular and decaffeinated coffee stimulate acid secretion and so aggravate *existing* ulcers.

Ulcers and their treatments highlight the importance of not self-medicating when symptoms persist. People with *H. pylori* infection often take over-the-counter acid controllers to relieve the pain of their ulcers when, instead, they need physician-prescribed antibiotics. Suppressing gastric acidity not only fails to heal the ulcer, but it also actually worsens inflammation during an *H. pylori* infection. Furthermore, *H. pylori* infection has been linked with stomach cancer, making prompt diagnosis and appropriate treatment essential.

Table H3-1 summarizes strategies to prevent or alleviate common GI problems. Many of these problems reflect hurried lifestyles. For this reason, many of their remedies require that people slow down and take the time to eat leisurely; chew food thoroughly to prevent choking, heartburn, and acid indigestion; rest until vomiting and diarrhea subside; and heed the urge to defecate. In addition, people must learn how to handle life's day-to-day problems and challenges without overreacting and becoming upset; learn how to relax, get enough sleep, and enjoy life. Remember, "what's eating you" may cause more GI distress than what you eat.

REFERENCES

1. American Academy of Pediatrics, Policy statement: Prevention of choking among children, *Pediatrics* 125 (2010): 601–607.
2. E. A. Mayer, Irritable bowel syndrome, *New England Journal of Medicine* 358 (2008): 1692–1699.
3. C. M. Surawicz, Mechanisms of diarrhea, *Current Gastroenterology Reports* 12 (2010): 236–241.
4. W. D. Heizer, S. Southern, and S. McGovern, The role of diet in symptoms of irritable bowel syndrome in adults: A narrative review, *Journal of the American Dietetic Association* 109 (2009): 1204–1214.
5. American College of Gastroenterology IBS Task Force, An evidence-based position statement on the management of irritable bowel syndrome, *American Journal of Gastroenterology* 104 (2009): S1–S35; A. C. Ford and coauthors, Effect of fibre, antispasmodics, and peppermint oil in the treatment of irritable bowel syndrome: Systematic review and meta-analysis, *British Journal of Medicine* 337 (2008): a2313.
6. P. Fric, D. Gabrovska, and J. Nevoral, Celiac disease, gluten-free diet, and oats, *Nutrition Reviews* 69 (2011): 107–115.
7. P. J. Kahrilas, Gastroesophageal reflux disease, *New England Journal of Medicine* 359 (2008): 1700–1707.
8. P. Sharma, Barrett's esophagus, *New England Journal of Medicine* 361 (2009): 2548–2556.
9. N. Vakil and D. Vaira, Sequential therapy for *Helicobacter pylori:* Time to consider making the switch? *Journal of the American Medical Association* 300 (2008): 1346–1347.

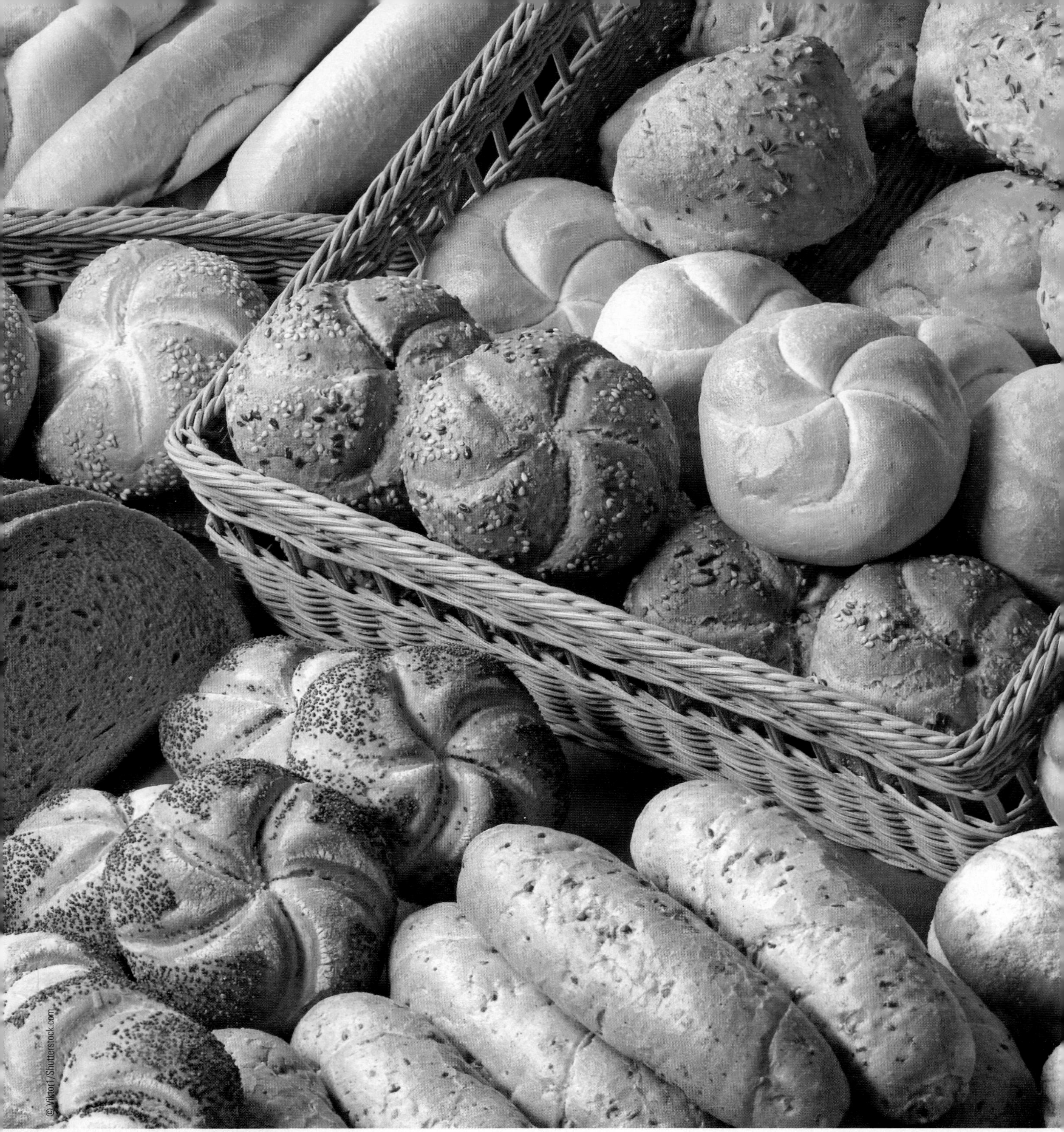

© Viktor1/Shutterstock.com

CENGAGE LEARNING'S GLOBAL NUTRITION WATCH — GLOBAL NUTRITION RESOURCE CENTER

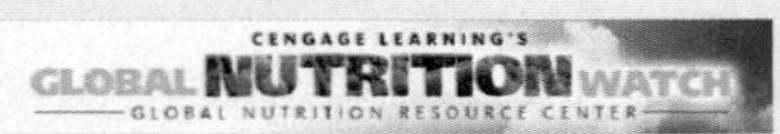

Carbohydrates support the quiet work of the brain and the active movement of the muscles. Go to Nutrition Basics and Tools → Carbohydrates to learn how they provide energy to the body. What are the risks and benefits of a low-carbohydrate diet? What other diet choices are needed to support a balanced diet?

4

The Carbohydrates: Sugars, Starches, and Fibers

LEARNING GPS

Nutrition in Your Life

Whether you are studying for an exam or daydreaming about your next vacation, your brain needs carbohydrate to power its activities. Your muscles need carbohydrate to fuel their work, too, whether you are racing up the stairs to class or moving on the dance floor to your favorite music. Where can you get carbohydrate? Are some foods healthier choices than others? As you will learn from this chapter, whole grains, vegetables, legumes, and fruits naturally deliver ample carbohydrate and fiber with valuable vitamins and minerals and little or no fat. Milk products typically lack fiber, but they also provide carbohydrate along with an assortment of vitamins and minerals. In the Nutrition Portfolio at the end of this chapter, you can examine whether your current carbohydrate choices are meeting dietary goals.

A student, quietly studying a textbook, is seldom aware that within his brain cells, billions of glucose molecules are splitting to provide the energy that permits him to learn. Yet glucose provides nearly all of the energy the human brain uses daily. Similarly, a marathon runner, bursting across the finish line in an explosion of sweat and triumph, seldom gives credit to the glycogen fuel her muscles have devoured to help her finish the race. Yet, together, these two **carbohydrates**—glucose and its storage form glycogen—provide about half of all the energy muscles and other body tissues use. The other half of the body's energy comes mostly from fat.

People don't eat glucose and glycogen directly. When they eat foods rich in carbohydrates, their bodies receive glucose for immediate energy and convert some glucose into glycogen for reserve energy. All plant foods—whole grains, vegetables, legumes, and fruits—provide ample carbohydrate. Milk also contains carbohydrate.

Many people mistakenly think of carbohydrates as "fattening" and avoid them when trying to lose weight. Such a strategy may be helpful if the

carbohydrates: compounds composed of carbon, oxygen, and hydrogen arranged as monosaccharides or multiples of monosaccharides. Most, but not all, carbohydrates have a ratio of one carbon molecule to one water molecule: $(CH_2O)_n$.

- **carbo** = carbon (C)
- **hydrate** = with water (H_2O)

carbohydrates are the concentrated sugars of soft drinks, candies, and cookies, but it is counterproductive if the carbohydrates are from whole grains, vegetables, and legumes. As the next section explains, not all carbohydrates are created equal.

4.1 The Chemist's View of Carbohydrates

LEARN IT Identify the monosaccharides, disaccharides, and polysaccharides common in nutrition by their chemical structures and major food sources.

The dietary carbohydrate family includes:

- Monosaccharides: single sugars
- Disaccharides: sugars composed of pairs of monosaccharides
- Polysaccharides: large molecules composed of chains of monosaccharides

Monosaccharides and disaccharides (the sugars) are sometimes called *simple carbohydrates,* and polysaccharides (starches and fibers) are sometimes called *complex carbohydrates.*

To understand the structure of carbohydrates, look at the atoms within them. Each atom can form a certain number of chemical bonds with other atoms:

- Carbon atoms, four
- Nitrogen atoms, three
- Oxygen atoms, two
- Hydrogen atoms, only one

Chemists represent the bonds as lines between the chemical symbols (such as C, N, O, and H) that stand for the atoms (see Figure 4-1).

Atoms form molecules in ways that satisfy the bonding requirements of each atom. Figure 4-1 includes the structure of ethyl alcohol, the active ingredient of alcoholic beverages, as an example. The two carbons each have four bonds represented by lines; the oxygen has two; and each hydrogen has one bond connecting it to other atoms. Chemical structures always bond according to these rules.

The following list of the most important **sugars** in nutrition symbolizes them as hexagons and pentagons of different colors.* Three are monosaccharides:

- Glucose
- Fructose
- Galactose

Three are disaccharides:

- Maltose (glucose + glucose)
- Sucrose (glucose + fructose)
- Lactose (glucose + galactose)

> FIGURE 4-1 Atoms and Their Bonds

The four main types of atoms found in nutrients are hydrogen (H), oxygen (O), nitrogen (N), and carbon (C).

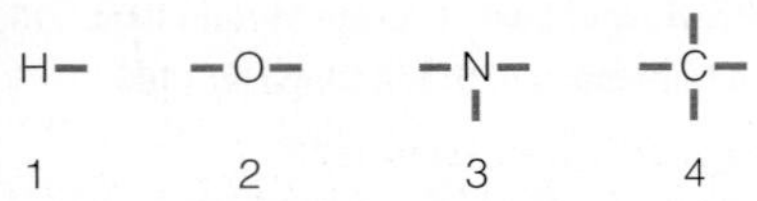

H– (1) –O– (2) –N– (3) –C– (4)

Each atom has a characteristic number of bonds it can form with other atoms.

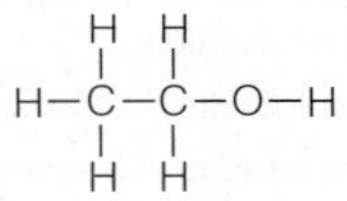

H–C–C–O–H (with H above and below each C)

Notice that in this simple molecule of ethyl alcohol, each H has one bond, O has two, and each C has four.

Monosaccharides

The three **monosaccharides** most important in nutrition all have the same numbers and kinds of atoms—each contains 6 carbon atoms, 12 hydrogens, and 6 oxygens (written in shorthand as $C_6H_{12}O_6$). The monosaccharides differ in their arrangements of the atoms. These chemical differences account for the differing sweetness of the monosaccharides. A pinch of purified glucose on the tongue gives only a mild sweet flavor, and galactose hardly tastes sweet at all. Fructose, however, is as intensely sweet as honey and, in fact, is the sugar primarily responsible for honey's sweetness.

Glucose Chemically, **glucose** is a larger and more complicated molecule than the ethyl alcohol shown in Figure 4-1, but it obeys the same rules of chemistry: each

sugars: simple carbohydrates composed of monosaccharides or disaccharides.

monosaccharides (mon-oh-SACK-uh-rides): carbohydrates of the general formula $C_nH_{2n}O_n$ that typically form a single ring. The monosaccharides important in nutrition are *hexoses,* sugars with six atoms of carbon and the formula $C_6H_{12}O_6$. See Appendix C for the chemical structures of the monosaccharides.

- **mono** = one
- **saccharide** = sugar
- **hex** = six

glucose (GLOO-kose): a monosaccharide; sometimes known as *blood sugar* in the body or *dextrose* in foods.

- **ose** = carbohydrate
- ⬢ = glucose

*Fructose is shown as a pentagon, but like the other monosaccharides, it has six carbons (as you will see in Figure 4-3). The disaccharides are illustrated with a simple bond, but actual linkages differ (as shown in Appendix C).

> FIGURE 4-2 Chemical Structure of Glucose

The diagram of a glucose molecule on the left shows all the bonds between the 6 carbon (C), 12 hydrogen (H), and 6 oxygen (O) atoms. It proves simple on examination, but chemists have adopted shortcuts to depict chemical structures. The middle and right diagrams also present the chemical structure of glucose, but as simplified versions with fewer symbols and bonds showing.

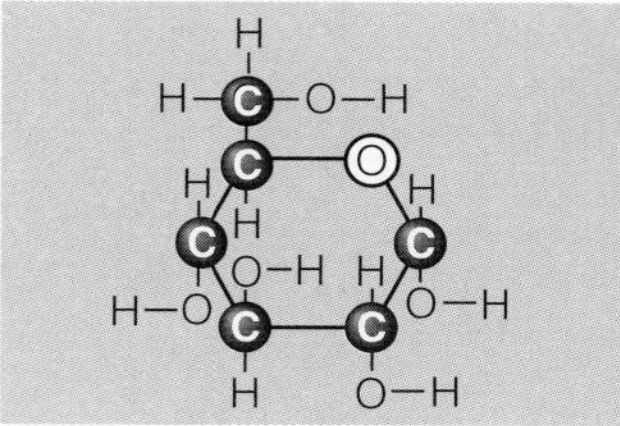

On paper, the structure of glucose has to be drawn flat, but in nature the five carbons and oxygen are roughly in a plane. The atoms attached to the ring carbons extend above and below the plane.

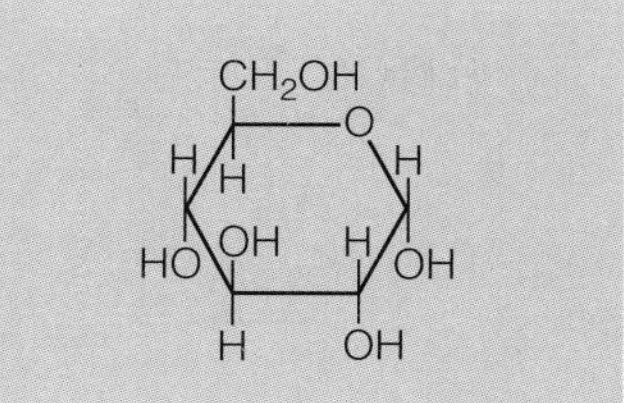

The lines representing some of the bonds and the carbons at the corners are not shown.

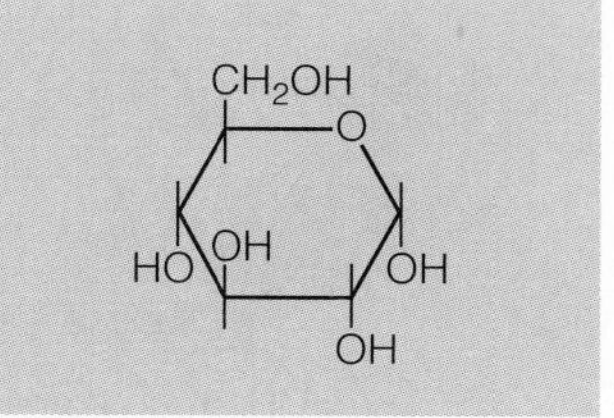

Now the single hydrogens are not shown, but lines still extend upward or downward from the ring to show where they belong.

carbon atom has four bonds; each oxygen, two bonds; and each hydrogen, one bond. Figure 4-2 illustrates the chemical structure of a glucose molecule.

Commonly known as blood sugar, glucose serves as an essential energy source for all the body's activities. Its significance to nutrition is tremendous. Later sections explain that glucose is one of the two sugars in every disaccharide and the unit from which the polysaccharides are made almost exclusively. One of these polysaccharides, starch, is the chief food source of energy for all the world's people; another, glycogen, is an important storage form of energy in the body. Glucose reappears frequently throughout this chapter and all those that follow.

Fructose **Fructose** is the sweetest of the sugars. Curiously, fructose has exactly the same chemical *formula* as glucose—$C_6H_{12}O_6$—but its *structure* differs (see Figure 4-3). The arrangement of the atoms in fructose stimulates the taste buds on the tongue to produce the sweet sensation. Fructose occurs naturally in fruits and honey; other sources include products such as soft drinks, ready-to-eat cereals, and desserts that have been sweetened with high-fructose corn syrup (defined on p. 109).

Galactose The monosaccharide **galactose** occurs naturally in foods as a single sugar only in very small amounts. Galactose has the same numbers and kinds of atoms as glucose and fructose in yet another arrangement. Figure 4-3 shows galactose beside a molecule of glucose for comparison.

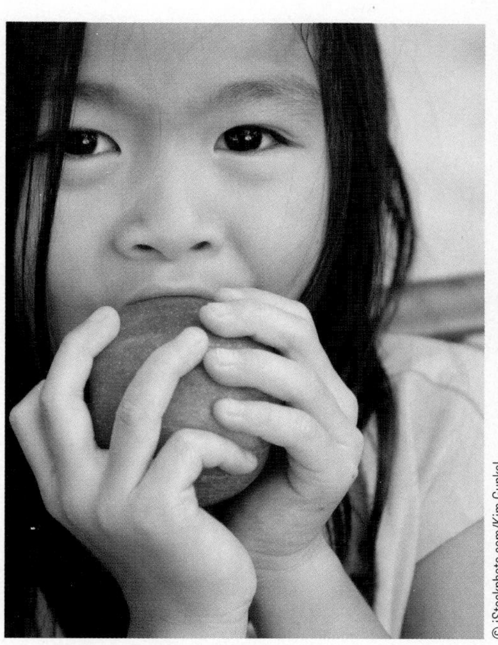

Fruits package their sugars with fibers, vitamins, and minerals, making them a sweet and healthy snack.

> FIGURE 4-3 The Monosaccharides

Notice the similarities—all three monosaccharides have 6 carbons (those shown plus one in each corner), 12 hydrogens (those shown plus one at the end of each single line), and 6 oxygens (all shown). Also notice the differences compared with glucose—in fructose, the ring is five-sided and in galactose, the position of one OH group differs slightly.

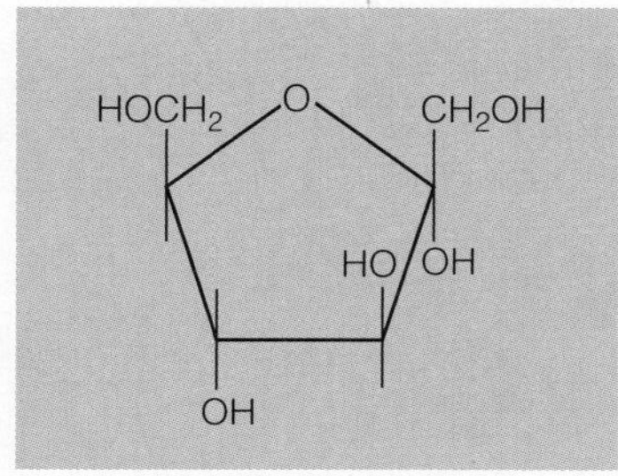

Fructose

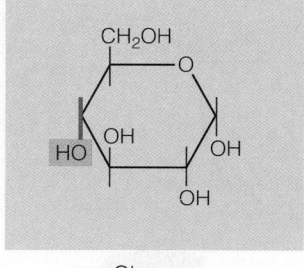

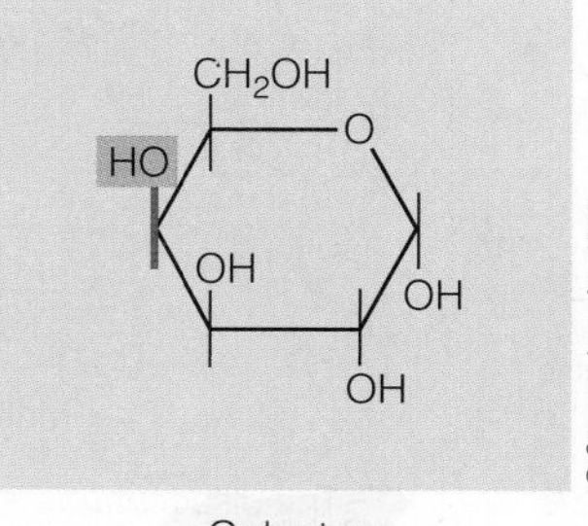

Glucose

Galactose

fructose (FRUK-tose or FROOK-tose): a monosaccharide; sometimes known as *fruit sugar* or *levulose:* Fructose is found abundantly in fruits, honey, and saps.

- **fruct** = fruit
- ⬟ = fructose

galactose (ga-LAK-tose): a monosaccharide; part of the disaccharide lactose.

- ⬢ = galactose

> FIGURE 4-4 Condensation of Two Monosaccharides to Form a Disaccharide

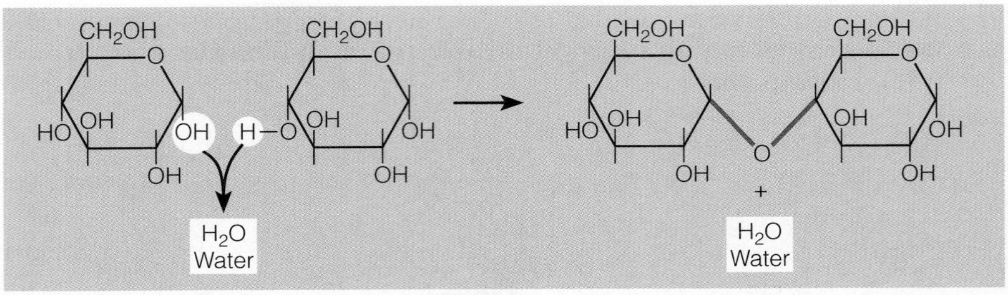

Glucose + glucose → Maltose

An OH group from one glucose and an H atom from another glucose combine to create a molecule of H_2O.

The two glucose molecules bond together with a single O atom to form the disaccharide maltose.

© Cengage Learning 2013

Disaccharides The **disaccharides** are pairs of the three monosaccharides just described. Glucose occurs in all three; the second member of the pair is fructose, galactose, or another glucose. These carbohydrates—and all the other energy nutrients—are put together and taken apart by similar chemical reactions: condensation and hydrolysis.

Condensation To make a disaccharide, a chemical reaction known as **condensation** links two monosaccharides together (see Figure 4-4). A hydroxyl (OH) group from one monosaccharide and a hydrogen atom (H) from the other combine to create a molecule of water (H_2O). The two originally separate monosaccharides link together with a single oxygen (O).

Hydrolysis To break a disaccharide in two, a chemical reaction known as **hydrolysis** occurs (see Figure 4-5). A molecule of water (H_2O) splits to provide the H and OH needed to complete the resulting monosaccharides. Hydrolysis reactions commonly occur during digestion.

Maltose The disaccharide **maltose** consists of two glucose units. Maltose is produced whenever starch breaks down—as happens in human beings during carbohydrate digestion. It also occurs during the fermentation process that yields alcohol. Maltose is only a minor constituent of a few foods, most notably barley.

Sucrose Fructose and glucose together form the disaccharide **sucrose**. Sucrose is the sweetest of the disaccharides because it contains fructose, the sweetest of the monosaccharides. These sugars account for the natural sweetness of fruits, vegetables, and grains. To make table sugar, sucrose is refined from the juices of sugarcane and sugar beets, then granulated. Depending on the extent to which it

> FIGURE 4-5 Hydrolysis of a Disaccharide

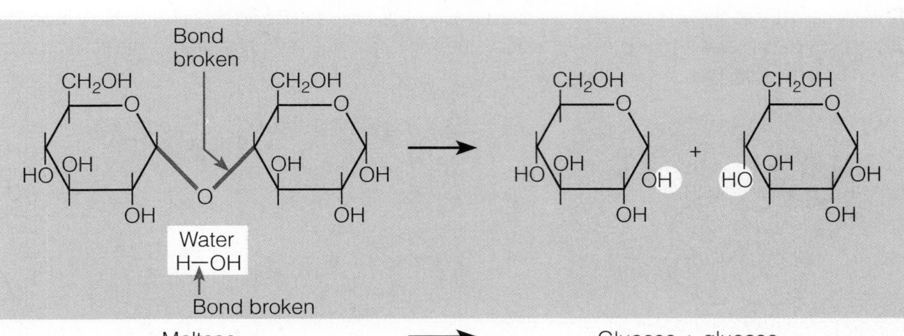

The disaccharide maltose splits into two glucose molecules with H added to one and OH to the other (from the water molecule).

disaccharides (dye-SACK-uh-rides): pairs of monosaccharides linked together. See Appendix C for the chemical structures of the disaccharides.

- **di** = two

condensation: a chemical reaction in which water is released as two molecules combine to form one larger product.

hydrolysis (high-DROL-ih-sis): a chemical reaction in which one molecule is split into two molecules, with hydrogen (H) added to one and a hydroxyl group (OH) to the other (from water, H_2O). (The noun is *hydrolysis;* the verb is *hydrolyze.*)

- **hydro** = water
- **lysis** = breaking

maltose (MAWL-tose): a disaccharide composed of two glucose units; sometimes known as malt sugar.

- 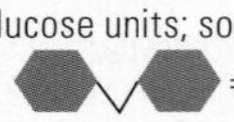 = maltose

sucrose (SUE-krose): a disaccharide composed of glucose and fructose; commonly known as *table sugar, beet sugar,* or *cane sugar.* Sucrose also occurs in many fruits and some vegetables and grains.

- **sucro** = sugar
- 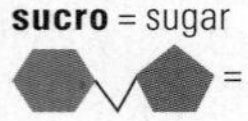 = sucrose

is refined, the product becomes the familiar brown, white, and powdered sugars available at grocery stores.

Lactose The combination of galactose and glucose makes the disaccharide **lactose**, the principal carbohydrate of milk. Known as milk sugar, lactose contributes half of the energy (kcalories) provided by fat-free milk.

Polysaccharides

In contrast to the simple carbohydrates just mentioned—the monosaccharides glucose, fructose, and galactose and the disaccharides maltose, sucrose, and lactose—the **polysaccharides** are slightly more complex, containing many glucose units and, in some cases, a few other monosaccharides strung together. Three types of polysaccharides are important in nutrition: glycogen, starches, and fibers.

Glycogen is a storage form of energy in the body; starch is the storage form of energy in plants; and fibers provide structure in stems, trunks, roots, leaves, and skins of plants. Both glycogen and starch are built of glucose units; fibers are composed of a variety of monosaccharides and other carbohydrate derivatives.

Glycogen **Glycogen** is found to only a limited extent in meats and not at all in plants.* For this reason, food is not a significant source of glycogen. Glycogen performs an important role in the body, however: it stores glucose for future use. Glycogen is made of many glucose molecules linked together in highly branched chains (see the left side of Figure 4-6). When the hormonal message "release energy" arrives at a liver or muscle cell, enzymes respond by attacking the many branches of glycogen simultaneously, making a surge of glucose available.**

Starches The human body stores glucose as glycogen, but plant cells store glucose as **starches**—long, branched or unbranched chains of hundreds or thousands of glucose molecules linked together (see the middle and right side of Figure 4-6). These giant starch molecules are packed side by side in grains such as wheat or rice, in root crops and tubers such as yams and potatoes, and in legumes such as peas and beans. When you eat the plant, your body hydrolyzes the starch to glucose and uses the glucose for its own energy purposes.

All starchy foods come from plants. Grains are the richest food source of starch, providing much of the food energy for people all over the world—rice in Asia; wheat in Canada, the United States, and Europe; corn in much of Central and

*Glycogen in animal muscles rapidly hydrolyzes after slaughter.

**Normally, liver cells produce glucose from glycogen to be sent directly to the blood; muscle cells can also produce glucose from glycogen, but must use it themselves. Muscle cells can restore the blood glucose level indirectly, however, as Chapter 7 explains.

Major sources of starch include grains (such as rice, wheat, millet, rye, barley, and oats), legumes (such as kidney beans, black-eyed peas, pinto beans, navy beans, and garbanzo beans), tubers (such as potatoes), and root crops (such as yams and cassava).

lactose (LAK-tose): a disaccharide composed of glucose and galactose; commonly known as *milk sugar*.

- **lact** = milk
- [two hexagons] = lactose

polysaccharides: compounds composed of many monosaccharides linked together. An intermediate string of 3 to 10 monosaccharides is an *oligosaccharide*.

- **poly** = many
- **oligo** = few

glycogen (GLY-ko-jen): an animal polysaccharide composed of glucose; a storage form of glucose manufactured and stored in the liver and muscles. Glycogen is not a significant food source of carbohydrate and is not counted as a dietary carbohydrate in foods.

- **glyco** = glucose
- **gen** = gives rise to

starches: plant polysaccharides composed of many glucose molecules.

> **FIGURE 4-6 Glycogen and Starch Compared**

For details of the chemical structures, see Appendix C.

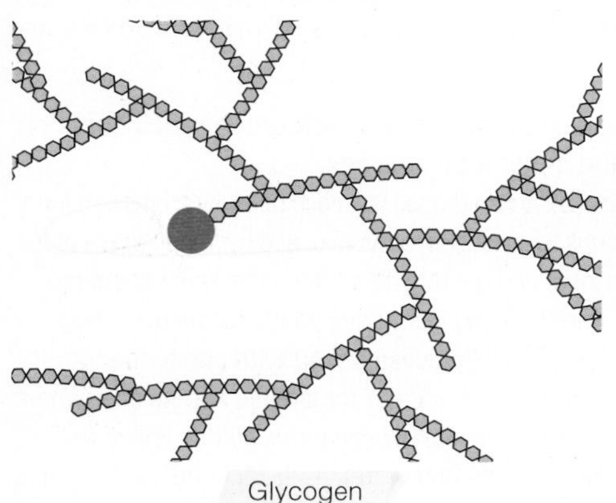

Glycogen

A glycogen molecule contains hundreds of glucose units in highly branched chains. Each new glycogen molecule needs a special protein (shown here in red) for the attachment of the first glucose.

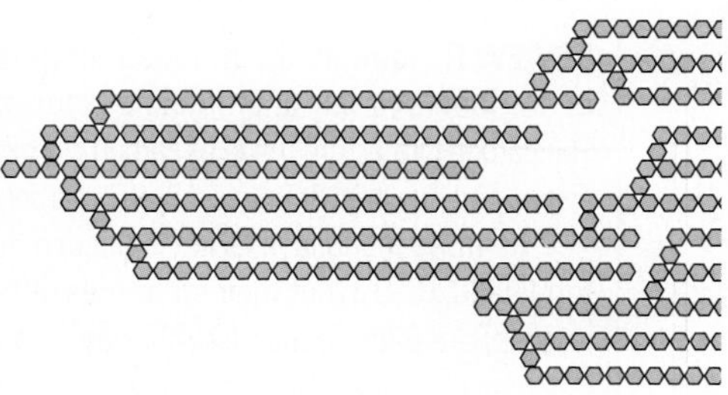

Starch (amylopectin)

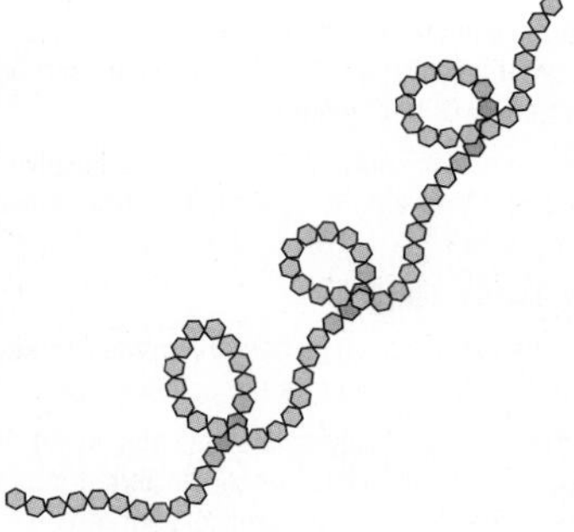

Starch (amylose)

A starch molecule contains hundreds of glucose molecules in either occasionally branched chains (amylopectin) or unbranched chains (amylose).

> FIGURE 4-7 The Bonds of Starch and Cellulose Compared

Human enzymes can digest starch but they cannot digest cellulose because the bonds that link the glucose molecules together are different. See Appendix C for chemical structures and descriptions of linkages.

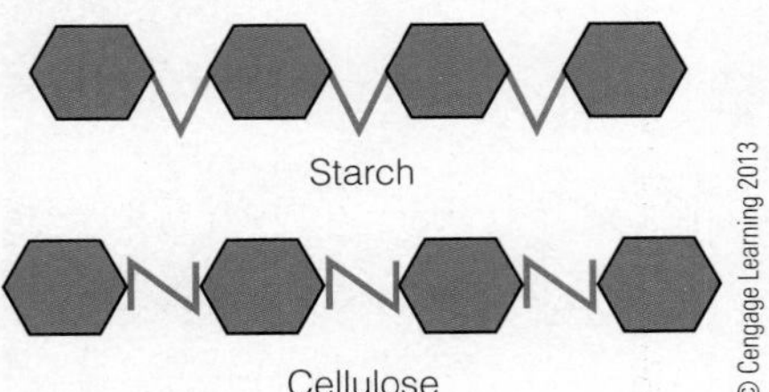

South America, and millet, rye, barley, and oats elsewhere. Legumes and tubers are also important sources of starch.

Fibers **Dietary fibers** are the structural parts of plants and thus are found in all plant-derived foods—vegetables, fruits, whole grains, and legumes. Most dietary fibers are polysaccharides. As mentioned earlier, starches are also polysaccharides, but dietary fibers differ from starches in that the bonds between their monosaccharides cannot be broken down by digestive enzymes in the body. For this reason, dietary fibers are often described as *nonstarch polysaccharides*.* Figure 4-7 illustrates the difference in the bonds that link glucose molecules together in starch with those found in the fiber cellulose. Because dietary fibers pass through the body undigested, they contribute no monosaccharides, and therefore little or no energy.

Even though most foods contain a variety of fibers, researchers often sort dietary fibers into two groups according to their solubility. Such distinctions help to explain their actions in the body.

Some dietary fibers dissolve in water **(soluble fibers)**, form gels **(viscous)**, and are easily digested by bacteria in the colon **(fermentable)**.** Commonly found in oats, barley, legumes, and citrus fruits, soluble fibers are most often associated with protecting against heart disease and diabetes by lowering blood cholesterol and glucose levels, respectively.[1]

Other fibers do not dissolve in water **(insoluble fibers)**, do not form gels (nonviscous), and are less readily fermented. Found mostly in whole grains (bran) and vegetables, insoluble fibers promote bowel movements, alleviate constipation, and prevent diverticular disease.[2]

As mentioned, *dietary fibers* occur naturally in plants. When these fibers have been extracted from plants or are manufactured and then added to foods or used in supplements, they are called *functional fibers*—if they have beneficial health effects. Cellulose in cereals, for example, is a dietary fiber, but when consumed as a supplement to alleviate constipation, cellulose is considered a functional fiber. *Total fiber* refers to the sum of dietary fibers and functional fibers.

A few starches are classified as dietary fibers. Known as **resistant starches**, these starches escape digestion and absorption in the small intestine. Starch may resist digestion for several reasons, including the body's efficiency in digesting starches and the food's physical properties. Resistant starch is common in whole or partially milled grains, legumes, and just-ripened bananas. Cooked potatoes, pasta, and rice that have been chilled also contain resistant starch. Similar to insoluble fibers, resistant starch may support a healthy colon.[3]

Phytic acid is not a dietary fiber, but it is often found in fiber-rich foods. Because of this close association, researchers have been unable to determine whether it is the dietary fiber, the phytic acid, or both, that binds with minerals, preventing their absorption. This binding presents a risk of mineral deficiencies, but the risk is minimal when total fiber intake is reasonable (less than 40 grams a day) and mineral intake is adequate. The nutrition consequences of mineral losses are described further in Chapters 12 and 13.

REVIEW IT Identify the monosaccharides, disaccharides, and polysaccharides common in nutrition by their chemical structures and major food sources.

The carbohydrates are made of carbon (C), oxygen (O), and hydrogen (H). Each of these atoms can form a specified number of chemical bonds: carbon forms four, oxygen forms two, and hydrogen forms one.

The three monosaccharides (glucose, fructose, and galactose) all have the same chemical formula ($C_6H_{12}O_6$), but their structures differ. The three disaccharides (maltose, sucrose, and lactose) are pairs of monosaccharides, each containing a glucose paired with one of the three monosaccharides. The sugars derive primarily from plants, except for lactose and its component galactose, which come from milk and milk products. Two monosaccharides can be linked together by a condensation reaction to form a disaccharide and water. A disaccharide, in turn, can be broken into its two monosaccharides by a hydrolysis reaction using water.

dietary fibers: in plant foods, the *nonstarch polysaccharides* that are not digested by human digestive enzymes, although some are digested by GI tract bacteria.

soluble fibers: nonstarch polysaccharides that dissolve in water to form a gel. An example is pectin from fruit, which is used to thicken jellies.

viscous: a gel-like consistency.

fermentable: the extent to which bacteria in the GI tract can break down fibers to fragments that the body can use.

insoluble fibers: nonstarch polysaccharides that do not dissolve in water. Examples include the tough, fibrous structures found in the strings of celery and the skins of corn kernels.

resistant starches: starches that escape digestion and absorption in the small intestine of healthy people.

phytic (FYE-tick) **acid:** a nonnutrient component of plant seeds; also called *phytate* (FYE-tate). Phytic acid occurs in the husks of grains, legumes, and seeds and is capable of binding minerals such as zinc, iron, calcium, magnesium, and copper in insoluble complexes in the intestine, which the body excretes unused.

*The nonstarch polysaccharide fibers include cellulose, hemicelluloses, pectins, gums, and mucilages. Fibers also include some *nonpolysaccharides* such as lignins, cutins, and tannins.

**Dietary fibers are fermented by bacteria in the colon to short-chain fatty acids, which are absorbed and metabolized by cells in the GI tract and liver (Chapter 5 describes fatty acids).

The polysaccharides are chains of monosaccharides and include glycogen, starches, and dietary fibers. Both glycogen and starch are storage forms of glucose—glycogen in the body, and starch in plants—and both yield energy for human use. The dietary fibers also contain glucose (and other monosaccharides), but their bonds cannot be broken by human digestive enzymes, so they yield little, if any, energy. Table 4-1 summarizes the carbohydrate family of compounds.

TABLE 4-1 The Carbohydrate Family

Monosaccharides
Glucose
Fructose
Galactose
Disaccharides
Maltose (glucose + glucose)
Sucrose (glucose + fructose)
Lactose (glucose + galactose)
Polysaccharides
Glycogen[a]
Starches (amylose and amylopectin)
Fibers (soluble and insoluble)

[a]Glycogen is a polysaccharide, but not a common dietary source of carbohydrate.

4.2 Digestion and Absorption of Carbohydrates

LEARN IT Summarize carbohydrate digestion and absorption.

The ultimate goal of digestion and absorption of sugars and starches is to break them into small molecules—chiefly glucose—that the body can absorb and use. The large starch molecules require extensive breakdown; the disaccharides need only be broken once and the monosaccharides not at all. The details follow.

Carbohydrate Digestion Figure 4-8 (p. 102) traces the digestion of carbohydrates through the GI tract. When a person eats foods containing starch, enzymes hydrolyze the long chains to shorter chains, the short chains to disaccharides, and, finally, the disaccharides to monosaccharides.* This process begins in the mouth.

In the Mouth In the mouth, thoroughly chewing high-fiber foods slows eating and stimulates the flow of saliva. The salivary enzyme **amylase** starts to work, hydrolyzing starch to shorter polysaccharides and to the disaccharide maltose. In fact, you can taste the change if you chew a piece of starchy food like a cracker and hold it in your mouth for a few minutes without swallowing it—the cracker begins tasting sweeter as the enzyme acts on it. Because food is in the mouth for a relatively short time, very little carbohydrate digestion takes place there; it begins again in the small intestine.

In the Stomach Carbohydrate digestion ceases in the stomach. The activity of salivary amylase diminishes as the stomach's acid and protein-digesting enzymes inactivate the enzyme. The stomach's digestive juices contain no enzymes to break down carbohydrates. Fibers are not digested, but because they linger in the stomach, they delay gastric emptying, thereby providing a feeling of fullness and **satiety**.

In the Small Intestine The small intestine performs most of the work of carbohydrate digestion. A major carbohydrate-digesting enzyme, pancreatic amylase, enters the intestine via the pancreatic duct and continues breaking down the polysaccharides to shorter glucose chains and maltose. The final step takes place on the outer membranes of the intestinal cells. There specific enzymes break down specific disaccharides:

- **Maltase** breaks maltose into two glucose molecules.
- **Sucrase** breaks sucrose into one glucose and one fructose molecule.
- **Lactase** breaks lactose into one glucose and one galactose molecule.

At this point, all polysaccharides and disaccharides have been broken down to monosaccharides—mostly glucose molecules, with some fructose and galactose molecules as well.

In the Large Intestine Within 1 to 4 hours after a meal, all the sugars and most of the starches have been digested. Only the fibers remain in the digestive tract. Fibers in the large intestine attract water, which softens the stools for passage

When a person eats carbohydrate-rich foods, the body receives a valuable commodity—glucose.

amylase (AM-ih-lace): an enzyme that hydrolyzes amylose (a form of starch). Amylase is a *carbohydrase,* an enzyme that breaks down carbohydrates.

satiety (sah-TIE-eh-tee): the feeling of fullness and satisfaction that occurs after a meal and inhibits eating until the next meal. Satiety determines how much time passes between meals.
- **sate** = to fill

maltase: an enzyme that hydrolyzes maltose.

sucrase: an enzyme that hydrolyzes sucrose.

lactase: an enzyme that hydrolyzes lactose.

*The short chains of glucose units that result from the breakdown of starch are known as *dextrins.* The word sometimes appears on food labels because dextrins can be used as thickening agents in processed foods.

> FIGURE 4-8 Carbohydrate Digestion in the GI Tract

STARCH

Mouth and salivary glands
The salivary glands secrete saliva into the mouth to moisten the food. The salivary enzyme amylase begins digestion:

Starch —Amylase→ Small polysaccharides, maltose

Stomach
Stomach acid inactivates salivary enzymes, halting starch digestion.

Small intestine and pancreas
The pancreas produces an amylase that is released through the pancreatic duct into the small intestine:

Starch —Pancreatic amylase→ Small polysaccharides, maltose

Then disaccharidase enzymes on the surface of the small intestinal cells hydrolyze the disaccharides into monosaccharides:

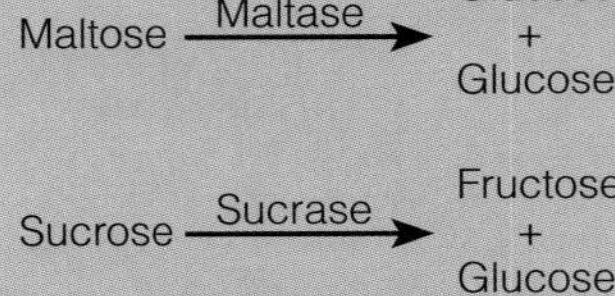

Maltose —Maltase→ Glucose + Glucose

Sucrose —Sucrase→ Fructose + Glucose

Lactose —Lactase→ Galactose + Glucose

Intestinal cells absorb these monosaccharides.

FIBER

Mouth
The mechanical action of the mouth crushes and tears fiber in food and mixes it with saliva to moisten it for swallowing.

Stomach
Fiber is not digested, and it delays gastric emptying.

Small intestine
Fiber is not digested, and it delays absorption of other nutrients.

Large intestine
Most fiber passes intact through the digestive tract to the large intestine. Here, bacterial enzymes digest fiber:

Some fiber —Bacterial enzymes→ Short-chain fatty acids, gas

Fiber holds water; regulates bowel activity; and binds substances such as bile, cholesterol, and some minerals, carrying them out of the body.

without straining. Also, bacteria in the GI tract ferment some fibers. This process generates water, gas, and short-chain fatty acids (described in Chapter 5).* The cells of the colon use these small fat molecules for energy. Metabolism of short-chain fatty acids also occurs in the cells of the liver. Fibers, therefore, can contribute some energy (1.5 to 2.5 kcalories per gram), depending on the extent to which they are broken down by bacteria and the fatty acids are absorbed. How much energy fiber contributes to a person's daily intake remains unclear.[4]

*The short-chain fatty acids produced by GI bacteria are primarily acetic acid, propionic acid, and butyric acid.

Carbohydrate Absorption Glucose is unique in that it can be absorbed to some extent through the lining of the mouth, but for the most part, nutrient absorption takes place in the small intestine. Glucose and galactose enter the cells lining the small intestine by active transport; fructose is absorbed by facilitated diffusion.

As the blood from the small intestine circulates through the liver, cells there take up fructose and galactose and most often convert them to compounds within the same metabolic pathways as glucose. Figure 4-9 shows that fructose and galactose are mostly metabolized in the liver, whereas glucose is sent out to the body's cells for energy. In the end, all disaccharides provide at least one glucose molecule directly, and they can provide the equivalent of another one indirectly—through the metabolism of fructose and galactose in the liver.

Lactose Intolerance Normally, the intestinal cells produce enough of the enzyme lactase to ensure that the disaccharide lactose found in milk is both digested and absorbed efficiently. Lactase activity is highest immediately after birth, as befits an infant whose first and only food for a while will be breast milk or infant formula. In the great majority of the world's populations, lactase activity declines dramatically during childhood and adolescence to about 5 to 10 percent of the activity at birth. Only a relatively small percentage (about 30 percent) of the people in the world retain enough lactase to digest and absorb lactose efficiently throughout adult life.

Symptoms When more lactose is consumed than the available lactase can handle, lactose molecules remain in the intestine undigested, attracting water and causing bloating, abdominal discomfort, and diarrhea—the symptoms of **lactose intolerance**. The undigested lactose becomes food for intestinal bacteria, which multiply and produce irritating acid and gas, further contributing to the discomfort and diarrhea.

Causes As mentioned, lactase activity commonly declines with age. **Lactase deficiency** may also develop when the intestinal villi are damaged by disease, certain medicines, prolonged diarrhea, or malnutrition. Depending on the extent of the intestinal damage, lactose malabsorption may be temporary or permanent. In extremely rare cases, an infant is born with a lactase deficiency, making feeding a challenge.

Prevalence The prevalence of lactose intolerance varies widely among ethnic groups, indicating that the trait has a genetic component. The prevalence of

lactose intolerance: a condition that results from the inability to digest the milk sugar lactose; characterized by bloating, gas, abdominal discomfort, and diarrhea. Lactose intolerance differs from milk allergy, which is caused by an immune reaction to the protein in milk.

lactase deficiency: a lack of the enzyme required to digest the disaccharide lactose into its component monosaccharides (glucose and galactose).

> FIGURE 4-9 **Absorption of Monosaccharides**

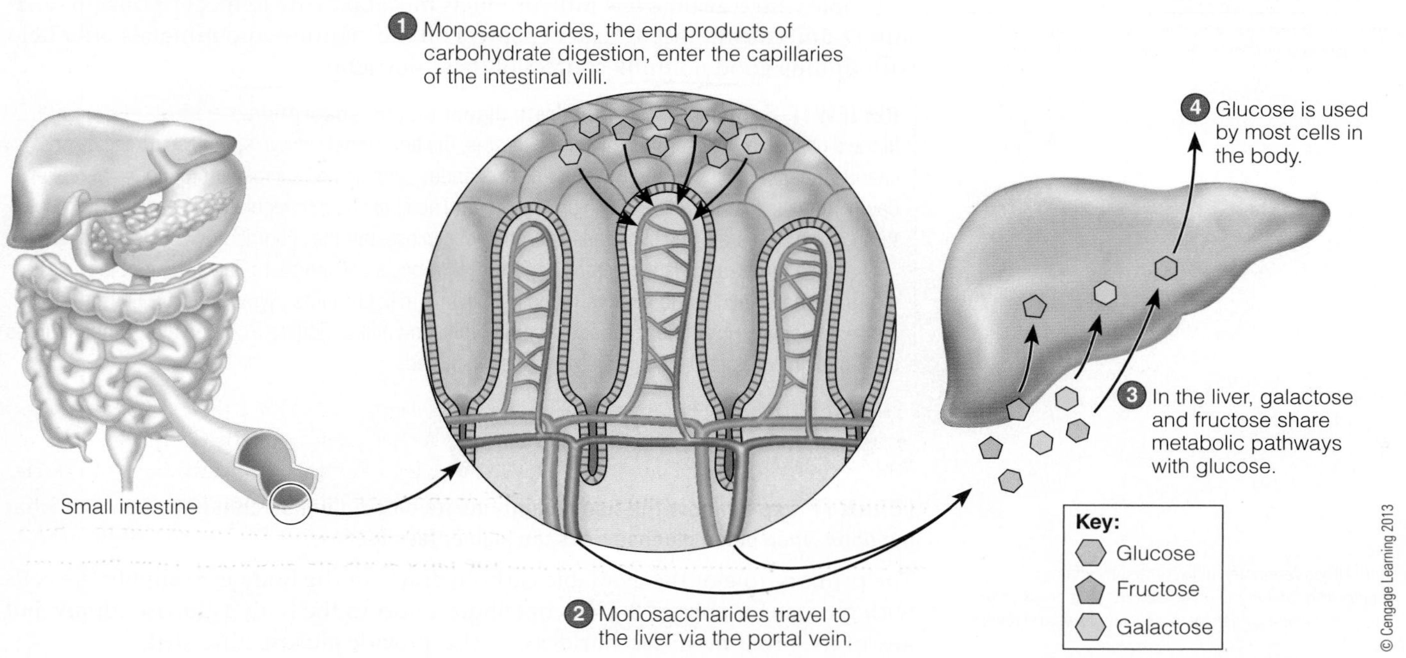
1 Monosaccharides, the end products of carbohydrate digestion, enter the capillaries of the intestinal villi.
Small intestine
2 Monosaccharides travel to the liver via the portal vein.
3 In the liver, galactose and fructose share metabolic pathways with glucose.
4 Glucose is used by most cells in the body.
Key:
Glucose
Fructose
Galactose
© Cengage Learning 2013

lactose intolerance is lowest among Scandinavians and other northern Europeans and highest among native North Americans and Southeast Asians. An estimated 30 to 50 million people in the United States are lactose intolerant.

Dietary Changes Managing lactose intolerance requires some dietary changes, although total elimination of milk products usually is not necessary.[5] Excluding all milk products from the diet can lead to nutrient deficiencies because these foods are a major source of several nutrients, notably the mineral calcium, vitamin D, and the B vitamin riboflavin. Fortunately, many people with lactose intolerance can consume foods containing up to 6 grams of lactose (½ cup milk) without symptoms. The most successful strategies are to increase intake of milk products gradually, consume them with other foods in meals, and spread their intake throughout the day. In addition, yogurt containing live bacteria seems to improve lactose intolerance. A change in the type, number, and activity of GI bacteria—not the reappearance of the missing enzyme—accounts for the ability to adapt to milk products.[6] Importantly, most lactose-intolerant individuals need to *manage* their dairy consumption rather than *restrict* it.

In many cases, lactose-intolerant people can tolerate fermented milk products such as yogurt and **kefir.** The bacteria in these products digest lactose for their own use, thus reducing the lactose content. Even when the lactose content is equivalent to milk's, yogurt produces fewer symptoms. Hard cheeses, such as cheddar, and cottage cheese are often well tolerated because most of the lactose is removed with the whey during manufacturing. Lactose continues to diminish as cheese ages.

Many lactose-intolerant people use commercially prepared milk products (such as Lactaid) that have been treated with an enzyme that breaks down the lactose. Alternatively, they take enzyme tablets with meals or add enzyme drops to their milk. The enzyme hydrolyzes much of the lactose in milk to glucose and galactose, which lactose-intolerant people can absorb without ill effects.

Because people's tolerance to lactose varies widely, lactose-restricted diets must be highly individualized. A completely lactose-free diet can be difficult because lactose appears not only in milk and milk products but also as an ingredient in many nondairy foods such as breads, cereals, breakfast drinks, salad dressings, and cake mixes (see Table 4-2). People on strict lactose-free diets need to read labels and avoid foods that include milk, milk solids, whey (milk liquid), and casein (milk protein, which may contain traces of lactose). They also need to check all medications with the pharmacist because 20 percent of prescription drugs and 5 percent of over-the-counter drugs contain lactose as a filler.

People who consume few milk products must take care to meet riboflavin, vitamin D, and calcium needs. Later chapters on the vitamins and minerals offer help with finding good nonmilk sources of these nutrients.

TABLE 4-2 Lactose in Selected Foods

Foods	Lactose (g)
Whole-wheat bread, 1 slice	0.5
Dinner roll, 1	0.5
Cheese, 1 oz	
Cheddar or American	0.5
Parmesan or cream	0.8
Doughnut (cake type), 1	1.2
Chocolate candy, 1 oz	2.3
Sherbet, 1 c	4.0
Cottage cheese (low-fat), 1 c	7.5
Ice cream, 1 c	9.0
Milk, 1 c	12.0
Yogurt (low-fat), 1 c	15.0

NOTE: Yogurt is often enriched with nonfat milk solids, which increase its lactose content to a level higher than milk's.

REVIEW IT Summarize carbohydrate digestion and absorption.

In the digestion and absorption of carbohydrates, the body breaks down starches into the disaccharide maltose. Maltose and the other disaccharides (lactose and sucrose) from foods are broken down into monosaccharides, which are absorbed. The fibers help to regulate the passage of food through the GI system and slow the absorption of glucose, but they contribute little, if any, energy.

Lactose intolerance is a common condition that occurs when there is insufficient lactase to digest the disaccharide lactose found in milk and milk products. Symptoms are limited to GI distress. Because treatment requires limiting milk and milk products from the diet, other sources of riboflavin, vitamin D, and calcium must be included.

4.3 Glucose in the Body

LEARN IT Explain how the body maintains its blood glucose concentration and what happens when blood glucose rises too high or falls too low.

The primary role of the available carbohydrates in the body is to supply the cells with glucose for energy. Starch contributes most to the body's glucose supply, but any of the other monosaccharides can also provide glucose if needed.

kefir (keh-FUR): a fermented milk created by adding *Lactobacillus acidophilus* and other bacteria that break down lactose to glucose and galactose, producing a sweet, lactose-free product.

Scientists have long known that providing energy is glucose's primary role in the body, but they have recently uncovered additional roles that glucose and other sugars perform in the body.* When sugar molecules adhere to the body's protein and fat molecules, the consequences can be dramatic. Sugars attached to a protein change the protein's shape and function; when they bind to lipids in a cell's membranes, sugars alter the way cells recognize one another.**

A Preview of Carbohydrate Metabolism Glucose plays the central role in carbohydrate metabolism. This brief discussion provides just enough information about carbohydrate metabolism to illustrate that the body needs and uses glucose as a chief energy nutrient. Chapter 7 provides a full description of energy metabolism, and Chapter 10 shows how the B vitamins participate.

Storing Glucose as Glycogen After a meal, blood glucose rises, and liver cells link excess glucose molecules by condensation reactions into long, branching chains of glycogen (review Figure 4-6, p. 99). When blood glucose falls, the liver cells break down glycogen by hydrolysis reactions into single molecules of glucose and release them into the bloodstream. Thus glucose becomes available to supply energy to the brain and other tissues regardless of whether the person has eaten recently.

The liver stores about one-third of the body's total glycogen and releases glucose into the bloodstream as needed. Muscle cells can also store glucose as glycogen (the other two-thirds), but muscles hoard most of their supply, using it just for themselves during exercise. The brain maintains a small amount of glycogen, which is thought to provide an emergency energy reserve during times of severe glucose deprivation.

Glycogen holds water and, therefore, is rather bulky. The body can store only enough glycogen to provide energy for relatively short periods of time—less than a day during rest and a few hours at most during exercise. For its long-term energy reserves, for use over days or weeks of food deprivation, the body uses its abundant, water-free fuel, fat, as Chapter 5 describes.

Using Glucose for Energy Glucose fuels the work of most of the body's cells. Inside a cell, a series of reactions can break glucose into smaller compounds that yield energy when broken down completely to carbon dioxide and water (see Chapter 7).

Making Glucose from Protein As mentioned, the liver's glycogen stores last only for hours, not for days. Glucose is the preferred energy source for brain cells, other nerve cells, and developing red blood cells. To keep providing glucose to meet the body's energy needs, a person has to eat carbohydrate-rich foods frequently. Yet people who do not always attend faithfully to their bodies' carbohydrate needs still survive. How do they manage without glucose from dietary carbohydrate? Do they simply draw energy from the other two energy-yielding nutrients, fat and protein? They do draw energy from them, but not simply.

Fat cannot make glucose to any significant extent. The amino acids of protein can be used to make glucose to some extent, but amino acids and proteins have jobs of their own that no other nutrient can perform. Still, when a person does not replenish glucose by eating carbohydrate, body proteins are broken down to make glucose to fuel the brain and other special cells. These body proteins derive primarily from the liver and skeletal muscles.

The conversion of protein to glucose is called **gluconeogenesis**—literally, the making of new glucose. Only adequate dietary carbohydrate can prevent this use of protein for energy, and this role of carbohydrate is known as its **protein-sparing action**.

Making Ketone Bodies from Fat Fragments An inadequate supply of carbohydrate can shift the body's energy metabolism in a precarious direction. With less carbohydrate providing glucose to meet the brain's energy needs, fat takes an alternative

*The study of sugars and their derivatives is known as *glycobiology*.

**These combination molecules are known as *glycoproteins* and *glycolipids*, respectively.

© Brian Leatart/Getty Images

The carbohydrates of grains, vegetables, fruits, and legumes supply most of the energy in a healthful diet.

gluconeogenesis (gloo-ko-nee-oh-JEN-ih-sis): the making of glucose from a noncarbohydrate source (described in more detail in Chapter 7).

- **gluco** = glucose
- **neo** = new
- **genesis** = making

protein-sparing action: the action of carbohydrate (and fat) in providing energy that allows protein to be used for other purposes.

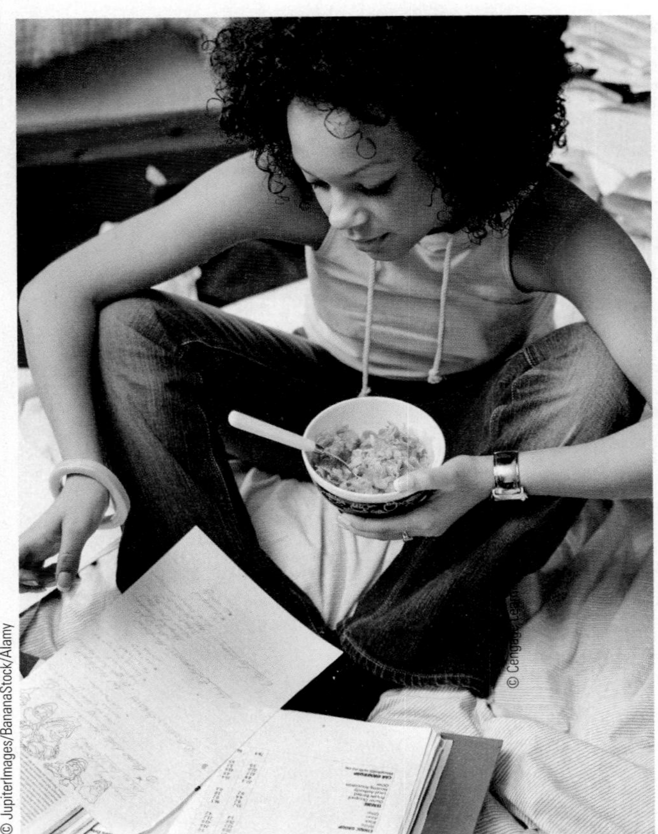

The brain uses glucose as its primary fuel for energy.

metabolic pathway; instead of entering the main energy pathway, fat fragments combine with one another, forming **ketone bodies**. Ketone bodies provide an alternate fuel source during starvation, but when their production exceeds their use, they accumulate in the blood, causing **ketosis**. Because most ketone bodies are acidic, ketosis disturbs the body's normal **acid-base balance**. (Chapter 7 explores ketosis and the metabolic consequences of low-carbohydrate diets further.)

To spare body protein and prevent ketosis, the body needs at least 50 to 100 grams of carbohydrate a day. Dietary recommendations urge people to select abundantly from carbohydrate-rich foods to provide for considerably more.

Using Glucose to Make Fat After meeting its immediate energy needs and filling its glycogen stores to capacity, the body must find a way to handle any extra glucose. When glucose is abundant, energy metabolism shifts to use more glucose instead of fat. If that isn't enough to restore glucose balance, the liver breaks glucose into smaller molecules and puts them together into the more permanent energy-storage compound—fat. Thus when carbohydrate is abundant, fat is either conserved (by using more carbohydrate in the fuel mix) or created (by using excess carbohydrate to make body fat). The fat then travels to the fatty tissues of the body for storage. Unlike the liver cells, which can store only enough glycogen to meet less than a day's energy needs, fat cells can store seemingly unlimited quantities of fat.

The Constancy of Blood Glucose

Every body cell depends on glucose for its fuel to some extent, and the cells of the brain and the rest of the nervous system depend almost exclusively on glucose for their energy. The activities of these cells never cease, and they have limited ability to store glucose. Day and night, they continually draw on the supply of glucose in the fluid surrounding them. To maintain the supply, a steady stream of blood moves past these cells bringing more glucose from either the small intestine (food) or the liver (via glycogen breakdown or gluconeogenesis).

Maintaining Glucose Homeostasis To function optimally, the body must maintain blood glucose within limits that permit the cells to nourish themselves. If blood glucose falls below normal, ♦ a person may become dizzy and weak; if it rises above normal, a person may become fatigued. Left untreated, fluctuations to the extremes—either high or low—can be fatal.

♦ Normal blood glucose (fasting): 70 to 100 mg/dL (published values vary slightly)

The Regulating Hormones Blood glucose homeostasis is regulated primarily by two hormones: *insulin,* which moves glucose from the blood into the cells, and *glucagon,* which brings glucose out of storage when necessary. Figure 4-10 depicts these hormonal regulators at work.

After a meal, as blood glucose rises, special cells of the pancreas respond by secreting **insulin** into the blood.* In general, the amount of insulin secreted corresponds with the rise in glucose. As the circulating insulin contacts the body's cells, receptors respond by ushering glucose from the blood into the cells. Most of the cells take only the glucose they can use for energy right away, but the liver and muscle cells can assemble the small glucose units into long, branching chains of glycogen for storage. The liver cells can also convert extra glucose to fat for export to other cells. Thus elevated blood glucose returns to normal levels as excess glucose is stored as glycogen and fat.

When blood glucose falls (as occurs between meals), other special cells of the pancreas respond by secreting **glucagon** into the blood.** Glucagon raises blood glucose by signaling the liver to break down its glycogen stores and release glucose into the blood for use by all the other body cells.

*The *beta* (BAY-tuh) *cells,* one of several types of cells in the pancreas, secrete insulin in response to elevated blood glucose concentration.

**The *alpha cells* of the pancreas secrete glucagon in response to low blood glucose concentration.

ketone (KEE-tone) **bodies:** compounds produced during the incomplete breakdown of fat when glucose is not available in the cells.

ketosis (kee-TOE-sis): an undesirably high concentration of ketone bodies in the blood and urine.

acid-base balance: the equilibrium in the body between acid and base concentrations (see Chapter 12).

insulin (IN-suh-lin): a hormone secreted by special cells in the pancreas in response to (among other things) elevated blood glucose concentration. Insulin controls the transport of glucose from the bloodstream into the muscle and fat cells.

glucagon (GLOO-ka-gon): a hormone secreted by special cells in the pancreas in response to low blood glucose concentration. Glucagon elicits release of glucose from liver glycogen stores.

> FIGURE 4-10 Maintaining Blood Glucose Homeostasis

1 When a person eats, blood glucose rises.

2 High blood glucose stimulates the pancreas to release insulin into the bloodstream.

3 Insulin stimulates the uptake of glucose into cells and storage as glycogen in the liver and muscles. Insulin also stimulates the conversion of excess glucose into fat for storage.

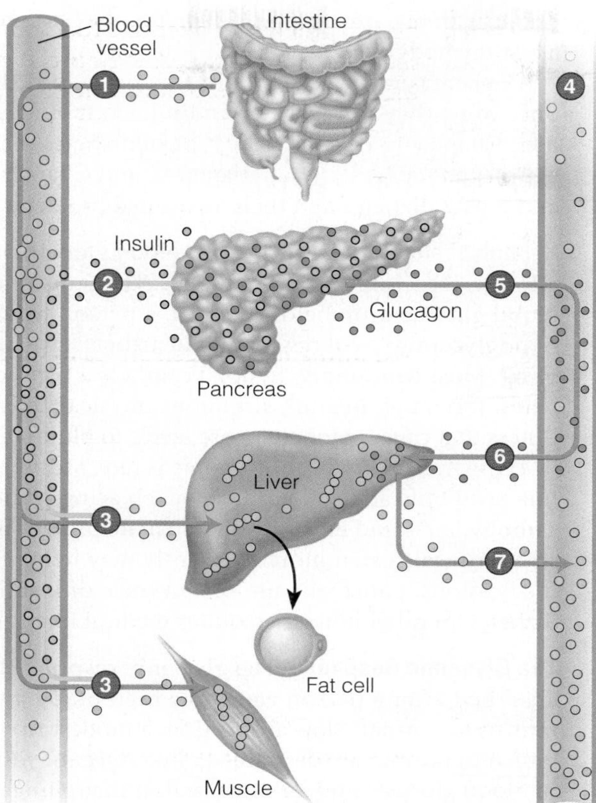

4 As the body's cells use glucose, blood levels decline.

5 Low blood glucose stimulates the pancreas to release glucagon into the bloodstream.

6 Glucagon stimulates liver cells to break down glycogen and release glucose into the blood.[a]

7 Blood glucose begins to rise.

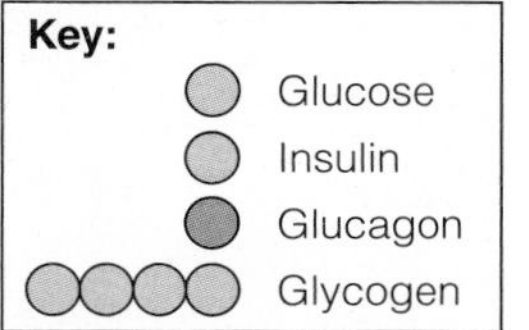

[a]The stress hormone epinephrine and other hormones also bring glucose out of storage.

Another hormone that signals the liver cells to release glucose is the "fight-or-flight" hormone, **epinephrine**. When a person experiences stress, epinephrine acts quickly to ensure that all the body cells have energy fuel in emergencies. Among its many roles in the body, epinephrine works to release glucose from liver glycogen to the blood.

Balancing within the Normal Range The maintenance of normal blood glucose depends on foods and hormones. When blood glucose falls below normal, food can replenish it, or in the absence of food, glucagon can signal the liver to break down glycogen stores. When blood glucose rises above normal, insulin can signal the cells to take in glucose for energy. Eating balanced meals that provide abundant carbohydrates, including fibers, and a little fat help to slow down the digestion and absorption of carbohydrate so that glucose enters the blood gradually. Eating at regular intervals also helps the body maintain a balance between the extremes.

Falling outside the Normal Range In some people, blood glucose regulation fails. When this happens, either of two conditions can result: diabetes or hypoglycemia. People with these conditions need to plan their diets and physical activities to help maintain their blood glucose within a normal range.

Diabetes In **diabetes**, blood glucose rises after a meal and remains above normal levels ♦ because insulin is either inadequate or ineffective. Notice that *blood* glucose is central to diabetes, but *dietary* carbohydrate does not cause diabetes.

There are two main types of diabetes. In **type 1 diabetes**, the less common type, the pancreas fails to produce insulin. Although the exact cause is unclear, some research suggests that in genetically susceptible people, certain viruses activate the immune system to attack and destroy cells in the pancreas as if they were foreign cells. In **type 2 diabetes,** the more common type of diabetes, the cells fail to respond to insulin. This condition tends to occur as a consequence of obesity. As the incidence of obesity in the United States has risen in recent decades, the incidence

♦ **Blood glucose (fasting):**
- Prediabetes: 100 to 125 mg/dL
- Diabetes: ≥126 mg/dL

epinephrine (EP-ih-NEFF-rin): a hormone of the adrenal gland that modulates the stress response; formerly called *adrenaline.* When administered by injection, epinephrine counteracts anaphylactic shock by opening the airways and maintaining heartbeat and blood pressure.

diabetes (DYE-uh-BEET-eez): a chronic disorder of carbohydrate metabolism, usually resulting from insufficient or ineffective insulin. When blood glucose levels are higher than normal, but below the diagnosis of diabetes, the condition is called *prediabetes.*

type 1 diabetes: the less common type of diabetes in which the pancreas produces little or no insulin.

type 2 diabetes: the more common type of diabetes in which the cells fail to respond to insulin.

of diabetes has followed. This trend is most notable among children and adolescents as obesity among the nation's youth reaches epidemic proportions. Because obesity can precipitate type 2 diabetes, the best preventive measure is to maintain a healthy body weight.

Concentrated sweets are not strictly excluded from the diabetic diet as they once were; they can be eaten in limited amounts with meals as part of a healthy diet. Chapter 15 describes the type of diabetes that develops in some women during pregnancy (gestational diabetes), and Chapter 18 gives full coverage to type 1 and type 2 diabetes and their associated problems.

Hypoglycemia In healthy people, blood glucose rises after eating and then gradually falls back into the normal range. The transition occurs without notice. Should blood glucose drop below normal, a person would experience the symptoms of **hypoglycemia:** weakness, rapid heartbeat, sweating, anxiety, hunger, and trembling. Most commonly, hypoglycemia is a consequence of poorly managed diabetes: too much insulin, strenuous physical activity, inadequate food intake, or illness that causes blood glucose levels to plummet.

Hypoglycemia in healthy people is rare. Most people who experience hypoglycemia need only adjust their diets by replacing refined carbohydrates with fiber-rich carbohydrates and ensuring an adequate protein intake at each meal. In addition, smaller meals eaten more frequently may help. Hypoglycemia caused by certain medications, pancreatic tumors, overuse of insulin, alcohol abuse, uncontrolled diabetes, or other illnesses requires medical intervention.

The Glycemic Response The **glycemic response** refers to how quickly glucose is absorbed after a person eats, how high blood glucose rises, and how quickly it returns to normal. Slow absorption, a modest rise in blood glucose, and a smooth return to normal are desirable (a low glycemic response). Fast absorption, a surge in blood glucose, and an overreaction that plunges glucose below normal are less desirable (a high glycemic response). The glycemic response may be particularly important to people with diabetes, who may benefit from limiting foods that produce too great a rise, or too sudden a fall, in blood glucose.[7]

Different foods elicit different glycemic responses; the **glycemic index** classifies foods accordingly. ♦ Some studies have shown that selecting foods with a low glycemic index is a practical way to improve glucose control.[8] Lowering the glycemic index of the diet may improve blood lipids and reduce the risk of heart disease as well.[9] A low glycemic diet may also help with weight management, although research findings are mixed.

♦ Glycemic index generalizations:
- Low: Legumes, milk and milk products
- Moderate: Whole grains, fruits
- High: Processed foods made from refined flour such as snack foods, breads, ready-to-eat cereals

Researchers debate whether selecting foods based on the glycemic index is practical or offers any real health benefits.[10] Those opposing the use of the glycemic index argue that it is not sufficiently supported by scientific research. The glycemic index has been determined for relatively few foods, and when the glycemic index has been established, it is based on an average of multiple tests with wide variations in their results. Values vary because of differences in the physical and chemical characteristics of foods, testing methods of laboratories, and digestive processes of individuals.[11]

Furthermore, the practical utility of the glycemic index is limited because this information is neither provided on food labels nor intuitively apparent. Indeed, a food's glycemic index is not always what one might expect. Ice cream, for example, is a high-sugar food but produces less of a glycemic response than baked potatoes, a high-starch food. Perhaps most relevant to real life, a food's glycemic effect differs depending on plant variety, food processing, cooking method, and whether it is eaten alone or with other foods. Most people eat a variety of foods, cooked and raw, that provide different amounts of carbohydrate, fat, and protein—all of which influence the glycemic index of a meal.

Paying attention to the glycemic index may be unnecessary because current guidelines already suggest many low and moderate glycemic index choices: whole grains, legumes, vegetables, fruits, and milk and milk products. In addition, eating frequent, small meals spreads glucose absorption across the day and thus

hypoglycemia (HIGH-po-gly-SEE-me-ah): an abnormally low blood glucose concentration.

glycemic (gly-SEEM-ic) **response:** the extent to which a food raises the blood glucose concentration and elicits an insulin response.

glycemic index: a method of classifying foods according to their potential for raising blood glucose.

offers similar metabolic advantages to eating foods with a low glycemic response. People wanting to follow a low glycemic diet should be careful not to adopt a low carbohydrate diet as well. Highlight 4 explores the controversies surrounding low-carbohydrate and high-glycemic diets.

REVIEW IT Explain how the body maintains its blood glucose concentration and what happens when blood glucose rises too high or falls too low.

Dietary carbohydrates provide glucose that can be used by the cells for energy, stored by the liver and muscles as glycogen, or converted into fat if intakes exceed needs. All of the body's cells depend on glucose; those of the central nervous system are especially dependent on it. Without glucose, the body is forced to break down its protein tissues to make glucose and to alter energy metabolism to make ketone bodies from fats. Blood glucose regulation depends primarily on two pancreatic hormones: insulin to move glucose from the blood into the cells when levels are high and glucagon to free glucose from glycogen stores and release it into the blood when levels are low.

4.4 Health Effects and Recommended Intakes of Sugars

LEARN IT Describe how added sugars can contribute to health problems.

Almost everyone finds pleasure in sweet foods—after all, the taste preference for sweets is inborn. To a child, the sweeter the food, the better. In adults, this preference is somewhat diminished, but most adults still enjoy an occasional sweet food or beverage.

In the United States, the natural sugars of milk, fruits, vegetables, and grains account for about half of the sugar intake; the other half consists of concentrated sugars that have been refined and added to foods for a variety of purposes. ♦ The use of added sugars has risen steadily over the past several decades, both in the United States and around the world, with soft drinks and sugared fruit drinks accounting for most of the increase. These added sugars assume various names on food labels: sucrose, invert sugar, corn sugar, corn syrups and solids, high-fructose corn syrup, and honey. A food is likely to be high in added sugars if its ingredient list starts with any of the sugars named in the accompanying glossary or if it includes several of them.

♦ As an additive, sugar:
- Enhances flavor
- Supplies texture and color to baked goods
- Provides fuel for fermentation, causing bread to rise or producing alcohol
- Acts as a bulking agent in ice cream and baked goods
- Acts as a preservative in jams
- Balances the acidity of tomato- and vinegar-based products

Health Effects of Sugars In moderate amounts, sugars add pleasure to meals without harming health.[12] In excess, however, sugars can be detrimental, and the average American diet currently delivers excessive amounts. The *Dietary Guidelines* caution that added sugars may increase the risk of certain chronic diseases—even in the absence of overweight or obesity.[13]

GLOSSARY OF ADDED SUGARS

brown sugar: refined white sugar crystals to which manufacturers have added molasses syrup with natural flavor and color; 91 to 96 percent pure sucrose.

confectioners' sugar: finely powdered sucrose, 99.9 percent pure.

corn sweeteners: corn syrup and sugars derived from corn.

corn syrup: a syrup made from cornstarch that has been treated with acid, high temperatures, and enzymes to produce glucose, maltose, and dextrins. It may be dried and used as *corn syrup solids*. See also *high-fructose corn syrup (HFCS)*.

dextrose: the name food manufacturers use for the sugar that is chemically the same as glucose.

high-fructose corn syrup (HFCS): a syrup made from cornstarch that has been treated with an enzyme that converts some of the glucose to the sweeter fructose; made especially for use in processed foods and beverages, where it is the predominant sweetener. With a chemical structure similar to sucrose, HFCS has a fructose content of 42, 55, or 90 percent, with glucose making up the remainder.

honey: sugar (mostly sucrose) formed from nectar gathered by bees. Composition and flavor vary, but honey always contains a mixture of sucrose, fructose, and glucose.

invert sugar: a mixture of glucose and fructose formed by the hydrolysis of sucrose in a chemical process; sold only in liquid form and sweeter than sucrose. Invert sugar is used as a food additive to help preserve freshness and prevent shrinkage.

levulose: an older name for fructose.

maple sugar: a sugar (mostly sucrose) purified from the concentrated sap of the sugar maple tree.

molasses: the thick brown syrup produced during sugar refining. Molasses retains residual sugar and other by-products and a few minerals; blackstrap molasses contains significant amounts of calcium and iron.

raw sugar: the first crop of crystals harvested during sugar processing. Raw sugar cannot be sold in the United States because it contains too much filth (dirt, insect fragments, and the like). Sugar sold as "raw sugar" domestically has actually gone through more than half of the refining steps.

turbinado (ter-bih-NOD-oh) **sugar:** sugar produced using the same refining process as white sugar, but without the bleaching and anticaking treatment. Traces of molasses give turbinado its sandy color.

white sugar: granulated sucrose or "table sugar," produced by dissolving, concentrating, and recrystallizing raw sugar.

© Polara Studios, Inc.

Almost half of the added sugars in our diet come from sugar-sweetened beverages, but baked goods, ice cream, candy, and breakfast cereals also make substantial contributions.

Obesity and Chronic Diseases Over the past several decades, as obesity rates increased sharply, consumption of added sugars reached an all-time high—much of it because high-fructose corn syrup use, especially in beverages, surged.[14] High-fructose corn syrup is composed of fructose and glucose in a ratio of roughly 50:50. Compared with sucrose, high-fructose corn syrup is less expensive, easier to use, and more soluble. Manufacturers prefer high-fructose corn syrup because it retains moisture, resists drying out, controls crystallization, prevents microbial growth and blends easily with other sweeteners, acids, and flavorings. In addition to being used in beverages, high-fructose corn syrup sweetens candies, baked goods, and hundreds of other foods.

In general, the energy intake of people who drink soft drinks, fruit punches, and other sugary beverages is greater than those who choose differently.[15] Adolescents, for example, who drink as much as 26 ounces or more (about two cans) of sugar-sweetened soft drinks daily, consume 400 more kcalories a day than teens who don't. Not too surprisingly, they also tend to weigh more.[16] Overweight children and adolescents consume more sweet desserts and soft drinks than their normal-weight peers. Research confirms that consuming sugary beverages correlates with increases in energy intake, body weight, and associated diseases.[17]

Some research suggests that added sugars in general, and fructose in particular, favor the fat-making pathways and impair the fat-clearing pathways in the body.[18] The resulting blood lipid profile increases the risk of heart disease.[19] As the liver busily makes lipids, its handling of glucose becomes unbalanced and insulin resistance develops—an indicator of prediabetes.[20] All in all, research is finding links between added sugars and the risk of diabetes, inflammation, hypertension, and heart disease.[21]

Nutrient Deficiencies Foods such as whole grains, vegetables, legumes, and fruits that contain some natural sugars and lots of starches and fibers provide protein, vitamins, and minerals. By comparison, foods and beverages that contain lots of added sugars such as cakes, candies, and sodas provide the body with glucose and energy, but few, if any, other essential nutrients or fiber. The more added sugars (and solid fats) in the diet, the more difficult it is to meet recommendations for dietary fiber, vitamins, and minerals and still stay within kcalorie limits.

A person spending 200 kcalories of a day's energy allowance on a 16-ounce soda gets little of value for those kcalories. In contrast, a person using 200 kcalories on three slices of whole-wheat bread gets 9 grams of protein, 6 grams of fiber, plus several of the B vitamins with those kcalories. For the person who wants something sweet, a reasonable compromise might be two slices of bread with a teaspoon of jam on each. The amount of sugar a person can afford to eat depends on how many discretionary kcalories are available beyond those needed to deliver indispensable vitamins and minerals.

By following the USDA Food Pattern and making careful food selections, a typical adult can obtain all the needed nutrients within an allowance of about 1500 kcalories. An inactive older woman who is limited to fewer than 1500 kcalories a day can afford to eat only the most nutrient-dense foods—with few, or no, discretionary kcalories available. In contrast, an active teenage boy may need as many as 3000 kcalories a day. If he chooses wisely, then he may use discretionary kcalories for nutrient-dense foods that contain added sugars—or even indulge in empty kcalorie choices such as cola beverages. Examples of nutrient-dense foods containing some added sugars include whole-grain breakfast cereals and vanilla yogurt.

© Matthew Farruggio

You receive about the same amount and kinds of sugars from an orange as from a tablespoon of honey, but the packaging makes a big nutrition difference.

Some people believe that because honey is a natural food, it is nutritious—or, at least, more nutritious than sugar.* A look at their chemical structures reveals the truth. Honey, like table sugar, contains glucose and fructose. The primary difference is that in table sugar the two monosaccharides are bonded together as the disaccharide sucrose, whereas in honey some of the monosaccharides are free. Whether a person eats monosaccharides individually, as in honey, or linked together, as in table sugar, they end up the same way in the body: as glucose and fructose.

*Honey should never be fed to infants because of the risk of botulism. Chapters 16 and 19 provide more details.

TABLE 4-3 **Sample Nutrients in Sugar and Other Foods**

The indicated portion of any of these foods provides approximately 100 kcalories. Notice that for a similar number of kcalories and grams of carbohydrate, foods such as milk, legumes, fruits, grains, and vegetables offer more of the other nutrients than do the sugars.

	Size of 100 kcal Portion	Carbohydrate (g)	Protein (g)	Calcium (mg)	Iron (mg)	Vitamin A (µg)	Vitamin C (mg)
Foods							
Milk, 1% low-fat	1 c	12	8	300	0.1	144	2
Kidney beans	½ c	20	7	30	1.6	0	2
Apricots	6	24	2	30	1.1	554	22
Bread, whole-wheat	1½ slices	20	4	30	1.9	0	0
Broccoli, cooked	2 c	20	12	188	2.2	696	148
Sugars							
Sugar, white	2 tbs	24	0	trace	trace	0	0
Molasses, blackstrap	2½ tbs	28	0	343	12.6	0	0.1
Cola beverage	1 c	26	0	6	trace	0	0
Honey	1½ tbs	26	trace	2	0.2	0	trace

Honey does contain a few vitamins and minerals, but not many. Honey is denser than crystalline sugar, too, so it provides more energy per spoonful. Table 4-3 shows that honey and white sugar are similar nutritionally—and both fall short of milk, legumes, fruits, grains, and vegetables.

Although the body cannot distinguish whether fructose and glucose derive from honey or table sugar, this is not to say that all sugar sources are alike. Some sugar sources are more nutritious than others. Consider a fruit, say, an orange. The fruit may give you the same amounts of fructose and glucose and the same number of kcalories as a spoonful of sugar or honey, but the packaging is more valuable nutritionally. The fruit's sugars arrive in the body diluted in a large volume of water, packaged in fiber, and mixed with essential vitamins, minerals, and phytochemicals.

As these comparisons illustrate, the significant difference between sugar sources is not between "natural" honey and "purified" sugar but between concentrated sugars and the dilute, naturally occurring sugars that sweeten foods. You can suspect an exaggerated nutrition claim when someone asserts that one product is more nutritious than another because it contains honey.

Added sugars contribute to nutrient deficiencies by displacing nutrients. For nutrition's sake, the appropriate attitude to take is not that sugar is "bad" and must be avoided, but that nutritious foods must come first. If nutritious foods crowd sugar out of the diet, that is fine—but not the other way around. As always, balance, variety, and moderation guide healthy food choices.

Dental Caries Both naturally occurring and added sugars from foods and from the breakdown of starches in the mouth can contribute to tooth decay. Bacteria in the mouth ferment the sugars and, in the process, produce an acid that erodes tooth enamel (see Figure 4-11), causing **dental caries,** or tooth decay. People can eat sugar without this happening, though. Much depends on how long foods stay in the mouth. Sticky foods stay on the teeth longer and continue to yield acid longer than foods that are readily cleared from the mouth. For that reason, sugar in a juice consumed quickly, for example, is less likely to cause dental caries than sugar in a pastry. By the same token, the sugar in sticky foods such as raisins can be more detrimental than the quantity alone would suggest.

Another concern is how often people eat sugar. Bacteria produce acid for 20 to 30 minutes after each exposure. If a person eats three pieces of candy at one time, the teeth will be exposed to approximately 30 minutes of acid destruction. But, if the person eats three pieces at half-hour intervals, the time of exposure increases to 90 minutes. Likewise, slowly sipping a sugary sports beverage may be more harmful

> FIGURE 4-11 **Dental Caries**

Dental caries begins when acid dissolves the enamel that covers the tooth. If not repaired, the decay may penetrate the dentin and spread into the pulp of the tooth, causing inflammation, abscess, and possible loss of the tooth.

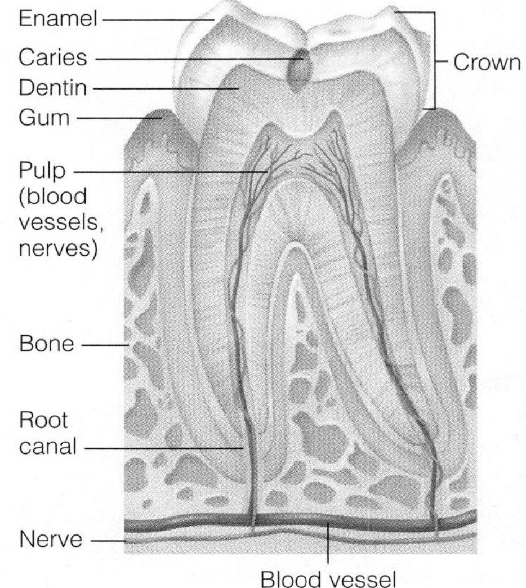

dental caries: decay of teeth.

- **caries** = rottenness

than drinking quickly and clearing the mouth of sugar. Nonsugary foods can help remove sugar from tooth surfaces; hence, it is better to eat sugar with meals than between meals. Foods such as milk and cheese may be particularly helpful in protecting against dental caries by neutralizing acids, stimulating salivary flow, inhibiting bacterial activity, and promoting remineralization of damaged enamel.

Beverages such as soft drinks, orange juice, and sports drinks not only contain sugar but also have a low pH. These acidic drinks can erode tooth enamel and may explain why the prevalence of dental erosion is growing steadily.

The development of caries depends on several factors: the bacteria that reside in **dental plaque**, the saliva that cleanses the mouth, the minerals that form the teeth, and the foods that remain after swallowing. For most people, good oral hygiene will prevent ♦ dental caries. In fact, regular brushing (twice a day, with a fluoride toothpaste) and flossing may be more effective in preventing dental caries than restricting sugary foods. Still nutrition is a key component of dental health.[22] *The Dietary Guidelines for Americans* recommend a combined approach to prevent dental caries—practicing good oral hygiene, drinking fluoridated water, and consuming sugar- and starch-containing foods and beverages less frequently.

♦ To prevent dental caries:
- Limit between-meal juices and snacks containing sugars and starches.
- Brush with a fluoride toothpaste and floss teeth regularly.
- If brushing and flossing are not possible, at least rinse with water.
- Get a dental checkup regularly.
- Drink fluoridated water.

Recommended Intakes of Sugars Estimates indicate that, on average, each person in the United States consumes about 30 teaspoons (about 120 grams) of sugars a day.[23] Most of the sugars in the average American diet are added to foods and beverages by manufacturers during processing; major sources of added sugars include sugar-sweetened beverages (sodas, energy drinks, sports drinks, fruit drinks), desserts, and candy. Some sugars are also added by consumers during food preparation and at the table. Because added sugars deliver kcalories, but few or no nutrients or fiber, the *Dietary Guidelines for Americans* urge consumers to "reduce the intake of kcalories from added sugars." These added sugar kcalories (and those from solid fats and alcohol) are considered discretionary kcalories—and most people need to limit their intake. By reducing the intake of foods and beverages with added sugars, consumers can lower the kcalorie content of the diet without compromising the nutrient content. The accompanying "How To" provides strategies for reducing the intake of added sugars.

> **DIETARY GUIDELINES FOR AMERICANS 2010**
Reduce the intake of kcalories from added sugars.

Estimating the *added* sugars in a diet is not always easy for consumers. Food labels list the *total* grams of sugar a food provides, but this total reflects both added sugars and those occurring naturally in foods. To help estimate sugar and energy intakes accurately, the list in the margin ♦ shows the amounts of concentrated sweets that are equivalent to 1 teaspoon of white sugar. These sugars all provide *about* 5 grams of carbohydrate and *about* 20 kcalories per teaspoon. Some are lower (16 kcalories for table sugar) and others are higher (22 kcalories for honey), but a 20-kcalorie average is an acceptable approximation. For a person who uses ketchup liberally, it may help to remember that 1 tablespoon of ketchup supplies about 1 teaspoon of sugar.

♦ 1 tsp white sugar =
- 1 tsp brown sugar
- 1 tsp candy
- 1 tsp corn syrup
- 1 tsp honey or agave nectar
- 1 tsp jam or jelly
- 1 tsp maple sugar or maple syrup
- 1 tsp molasses
- 1½ oz carbonated soda
- 1 tbs ketchup

The DRI committee did not publish a Tolerable Upper Intake Level (UL) for sugar, but as mentioned, excessive intakes can interfere with sound nutrition and good health. Few people can eat lots of sugary treats and still meet all of their nutrient needs without exceeding their kcalorie allowance. Specifically, the DRI suggests that added sugars should account for no more than 25 percent of the day's total energy intake.[24] One out of eight in the US population exceeds this maximum intake.[25] When added sugars occupy this much of a diet, intakes from the five food groups usually fall below recommendations. For a person consuming 2000 kcalories a day, 25 percent represents 500 kcalories ♦ (that is, 125 grams, or 31 teaspoons) from concentrated sugars—and that's a lot of sugar. Perhaps an athlete in training whose energy needs are high can afford the added

♦ For perspective, each of these concentrated sugars provides about 500 kcal:
- 40 oz cola
- ½ c honey
- 125 jelly beans
- 23 marshmallows
- 30 tsp sugar

How many kcalories from sugar does your favorite beverage or snack provide?

dental plaque: a gummy mass of bacteria that grows on teeth and can lead to dental caries and gum disease.

>How To

Reduce the Intake of Added Sugars

- Use less table sugar when preparing meals and at the table.
- Use your sugar kcalories to sweeten nutrient-dense foods (such as oatmeal) instead of consuming empty kcalorie foods and beverages (such as candy and soda).
- Drink fewer regular sodas, sports drinks, energy drinks, and fruit drinks; choose water, fat-free milk, 100 percent fruit juice, or unsweetened tea or coffee instead. If you do drink sugar-sweetened beverages, have a small portion.
- Select fruit for dessert. Eat less cake, cookies, ice cream, other desserts, and candy. If you do eat these foods, have a small portion.
- Read the Nutrition Facts on a label to choose foods with less sugar. Compare the unsweetened version of a food (such as corn flakes) with the sweetened version (such as frosted corn flakes) to estimate the quantity of added sugars. The quantity of sugars listed in the Nutrition Facts for foods containing little or no milk or fruit are a good estimate of added sugars per serving.
- Read the ingredients list to identify foods with little or no added sugars. A food is likely to be high in added sugars if its ingredient list starts with any of the sugars named in the glossary on p. 109, or if it includes several of them.

TRY IT Compare the energy contents and ingredients lists of 1 cup of the following foods: fruit-flavored yogurt and plain yogurt, sugar-frosted corn flakes and plain corn flakes, orange soda and orange juice.

sugars from sports drinks without compromising nutrient intake, but most people do better by limiting their use of added sugars. Added sugars contribute an average of 16 percent of the total energy in the typical American diet.[26] The World Health Organization (WHO) and the Food and Agriculture Organization (FAO) suggest restricting consumption of added sugars to less than 10 percent of total energy.

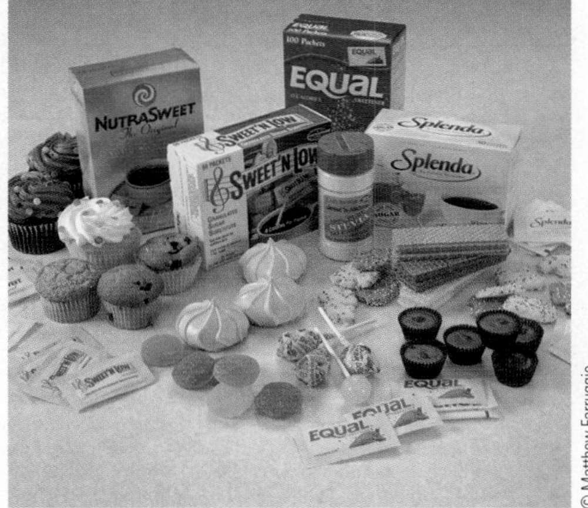

© Matthew Farruggio

Consumers use artificial sweeteners to help them limit kcalories and minimize sugar intake.

Alternative Sweeteners To control weight gain, blood glucose, and dental caries, many consumers turn to alternative sweeteners to help them limit kcalories and minimize sugar intake. In doing so, they encounter three sets of alternative sweeteners: artificial sweeteners, an herbal sweetener, and sugar alcohols.

Artificial Sweeteners **Artificial sweeteners** are sometimes called **nonnutritive sweeteners** because they provide virtually no energy. Table 4-4 (p. 114) provides general details about each of the sweeteners, including their **Acceptable Daily Intakes (ADI).** Chapter 9 includes a discussion of their use in weight control and Chapter 19 focuses on some of the safety issues surrounding their use. Considering that all substances are toxic at some dose, it is little surprise that large doses of artificial sweeteners (or their components or metabolic by-products) may have adverse effects. The question to ask is whether their ingestion is safe for human beings in quantities people normally use (and potentially abuse).

Stevia—An Herbal Sweetener The herb stevia derives from a plant whose leaves have long been used by the people of South America to sweeten their beverages. The FDA has granted stevia the status of "generally recognized as safe," and it can be used as an additive in a variety of foods and beverages.

Sugar Alcohols Some "sugar-free" or reduced-kcalorie products contain **sugar alcohols.** ♦ The sugar alcohols (or polyols) occur naturally in fruits and vegetables; manufacturers also use sugar alcohols in many processed foods to add bulk and

♦ Examples of sugar alcohols:
- Erythritol
- Isomalt
- Lactitol
- Maltitol
- Mannitol
- Sorbitol
- Xylitol

artificial sweeteners: sugar substitutes that provide negligible, if any, energy; sometimes called *nonnutritive sweeteners.*

nonnutritive sweeteners: sweeteners that yield no energy (or insignificant energy in the case of aspartame).

Acceptable Daily Intake (ADI): the estimated amount of a sweetener that individuals can safely consume each day over the course of a lifetime without adverse effect.

sugar alcohols: sugarlike compounds that can be derived from fruits or commercially produced from dextrose; also called *polyols.* Sugar alcohols are absorbed more slowly than other sugars and metabolized differently in the human body; they are not readily utilized by ordinary mouth bacteria.

TABLE 4-4 **Alternative Sweeteners**

Sweetener	Chemical Composition	Body's Response	Relative Sweetness[a]	Energy (kcal/g)	Acceptable Daily Intake (ADI) and (Estimated Equivalent)[b]	Approval Status
			Artificial Sweeteners			
Acesulfame potassium or Acesulfame K[c] (AY-sul-fame)	Potassium salt	Not digested or absorbed	200	0	15 mg/kg body weight[d] (30 cans diet soda)	Approved for use in the United States and Canada
Aspartame[e] (ah-SPAR-tame or ASS-par-tame)	Amino acids (phenylalanine and aspartic acid) and a methyl group	Digested and absorbed	200	4[f]	50 mg/kg body weight[g] (18 cans diet soda)	Approved for use in the United States and Canada; warning for PKU
Cyclamate (SIGH-kla-mate)	Sodium or calcium salt of cyclamic acid	Incompletely absorbed; absorbed cyclamate is excreted unchanged; unabsorbed cyclamate may be metabolized by bacteria in the GI tract	30	0	11 mg/kg body weight (8 cans of diet soda)	Approval pending in the United States; approved for use in Canada
Neotame (NEE-oh-tame)	Aspartame with an additional side group attached	Not digested or absorbed	8000	0	18 mg/day	Approved for use in the United States; no warning for PKU
Saccharin[h] (SAK-ah-ren)	Benzoic sulfimide	Rapidly absorbed and excreted	450	0	5 mg/kg body weight (10 packets of sweetener)	Approved for use in the United States; restricted use as a tabletop sweetener in Canada
Sucralose[i] (SUE-kra-lose)	Sucrose with Cl atoms instead of OH groups	Not digested or absorbed	600	0	5 mg/kg body weight (6 cans diet soda)	Approved for use in the United States and Canada
Tagatose[j] (TAG-ah-tose)	Monosaccharide similar in structure to fructose; naturally occurring or derived from lactose	Mostly not absorbed; some short-chain fatty acids absorbed	0.8	1.5	7.5 g/day	GRAS[k] approved; does not promote dental caries and may carry a health claim
			Herbal Sweetener			
Stevia[l] (STEE-vee-ah)	Glycosides found in the leaves of the *Stevia rebaudiana* herb	Digested and absorbed	300	0	4 mg/kg body weight	GRAS[k] approved

[a]Relative sweetness is determined by comparing the approximate sweetness of a sugar substitute with the sweetness of pure sucrose, which has been defined as 1.0. Chemical structure, temperature, acidity, and other flavors of the foods in which the substance occurs all influence relative sweetness.
[b]The Acceptable Daily Intake (ADI) is the estimated amount of a sweetener that individuals can safely consume each day over the course of a lifetime without adverse effects. The Estimated Equivalent is based on a person weighing 70 kg (154 lb).
[c]Marketed under the trade names Sunett, Sweet One.
[d]Recommendations from the World Health Organization limit acesulfame-K intake to 9 mg per kilogram of body weight per day.
[e]Marketed under the trade names NutraSweet, Equal, NatraTaste, Canderel.
[f]Aspartame provides 4 kcal per gram, as does protein, but because so little is used, its energy contribution is negligible. In powdered form, it is sometimes mixed with lactose, however, so a 1 g packet may provide 4 kcal.
[g]Recommendations from the World Health Organization and in Europe and Canada limit aspartame intake to 40 mg per kilogram of body weight per day.
[h]Marketed under the trade names Sweet'N Low, Necta Sweet.
[i]Marketed under the trade names Splenda, SucraPlus.
[j]Marketed under the trade names Nutralose, Nutrilatose, Tagatesse.
[k]GRAS = food additives that are generally recognized as safe. The GRAS list is subject to revision as new facts become known.
[l]Marketed under the trade names Sweetleaf, Purevia, Truvia, Honey Leaf.

texture, to provide a cooling effect or sweet taste, to inhibit browning from heat, and to retain moisture. These products may claim to be "sugar-free" on their labels, but in this case, "sugar-free" does not mean free of kcalories. Sugar alcohols do provide kcalories (0.2 to 2.6 kcalories per gram), but fewer than the sugars. Because sugar alcohols yield energy, they are sometimes referred to as **nutritive sweeteners.**

Sugar alcohols evoke a low glycemic response. The body partially absorbs some sugar alcohols and absorbs others slowly; consequently, they are slower to enter the bloodstream than other sugars. Unabsorbed sugar alcohols may be metabolized by bacteria in the GI tract, producing side effects such as intestinal gas, abdominal discomfort, and diarrhea. For this reason, regulations require food labels to state "Excess consumption may have a laxative effect" if reasonable consumption of that food could result in the daily ingestion of 50 grams of a sugar alcohol. ♦

The real benefit of using sugar alcohols is that they do not contribute to dental caries. Bacteria in the mouth cannot metabolize sugar alcohols as rapidly as sugar. Sugar alcohols are therefore valuable in chewing gums, breath mints, and other

♦ For perspective, a low-carbohydrate energy bar or shake may contain 10 to 15 grams of a sugar alcohol.

nutritive sweeteners: sweeteners that yield energy, including both sugars and sugar alcohols.

> FIGURE 4-12 **Sugar Alternatives on Food Labels**

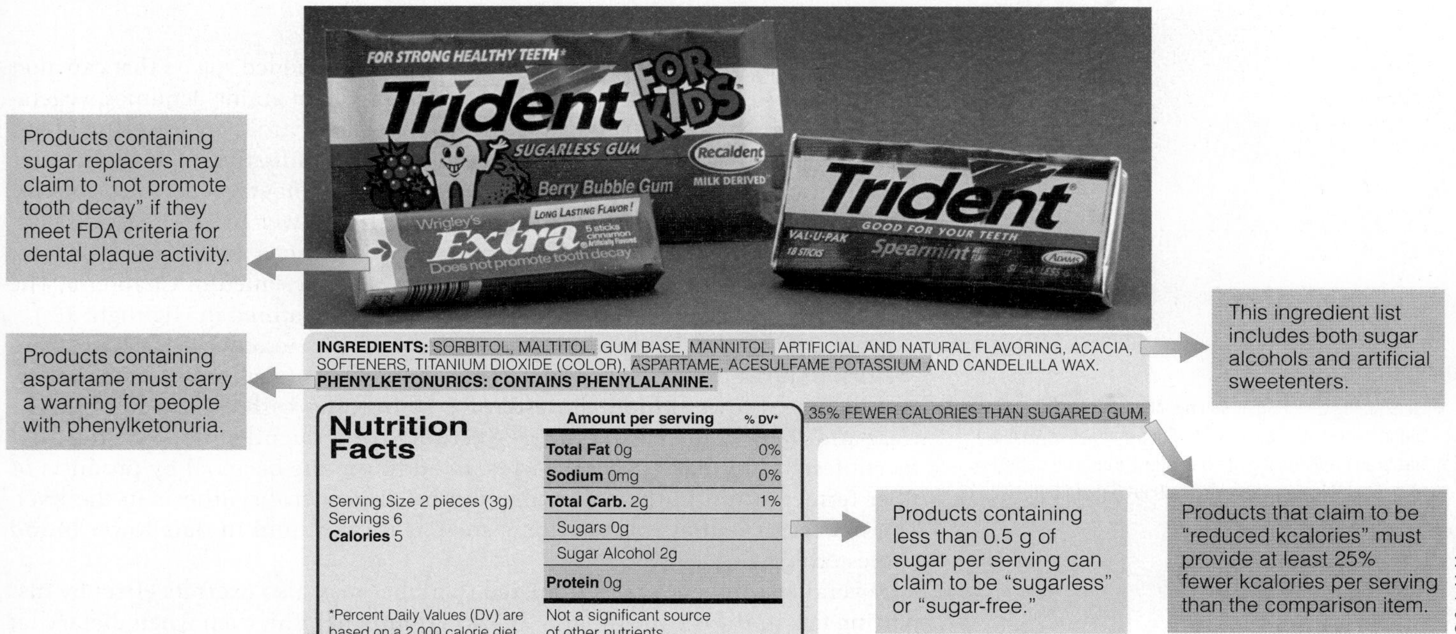

products that people keep in their mouths for a while. Figure 4-12 presents labeling information for products using sugar alternatives.

FOR STRONG HEALTHY TEETH*
Trident FOR KIDS
SUGARLESS GUM
Recaldent MILK DERIVED
Berry Bubble Gum
Wrigley's Extra
LONG LASTING FLAVOR!
Does not promote tooth decay
Trident
GOOD FOR YOUR TEETH
VAL-U-PAK
18 STICKS
Spearmint
Products containing sugar replacers may claim to "not promote tooth decay" if they meet FDA criteria for dental plaque activity.
Products containing aspartame must carry a warning for people with phenylketonuria.
INGREDIENTS: SORBITOL, MALTITOL, GUM BASE, MANNITOL, ARTIFICIAL AND NATURAL FLAVORING, ACACIA, SOFTENERS, TITANIUM DIOXIDE (COLOR), ASPARTAME, ACESULFAME POTASSIUM AND CANDELILLA WAX.
PHENYLKETONURICS: CONTAINS PHENYLALANINE.
This ingredient list includes both sugar alcohols and artificial sweetenters.
Nutrition Facts
Serving Size 2 pieces (3g)
Servings 6
Calories 5
Amount per serving % DV*
Total Fat 0g 0%
Sodium 0mg 0%
Total Carb. 2g 1%
Sugars 0g
Sugar Alcohol 2g
Protein 0g
*Percent Daily Values (DV) are based on a 2,000 calorie diet.
Not a significant source of other nutrients.
35% FEWER CALORIES THAN SUGARED GUM.
Products containing less than 0.5 g of sugar per serving can claim to be "sugarless" or "sugar-free."
Products that claim to be "reduced kcalories" must provide at least 25% fewer kcalories per serving than the comparison item.

For consumers choosing to use alternative sweeteners, the Academy of Nutrition and Dietetics wisely advises that they be used in moderation and only as part of a well-balanced nutritious diet.[27] When used in moderation, these sweeteners will do no harm. In fact, they may even help, by providing an alternative to sugar for people with diabetes, by inhibiting caries-causing bacteria, and by limiting energy intake. People may find it appropriate to choose from among any of the sweeteners at times: artificial sweeteners, an herbal sweetener, sugar alcohols, and sugar itself.

REVIEW IT Describe how added sugars can contribute to health problems.
Sugars increase the risk of dental caries; excessive intakes displace needed nutrients and fiber and contribute to obesity when energy intake exceeds needs. A person deciding to limit daily sugar intake should recognize that not all sugars need to be restricted, just concentrated sweets, which are relatively empty of other nutrients and high in kcalories. Sugars that occur naturally in fruits, vegetables, and milk are acceptable. Alternative sweeteners may help to limit kcalories and sugar intake.

4.5 Health Effects and Recommended Intakes of Starch and Fibers

LEARN IT Indentify the health benefits of, and recommendations for, starches and fibers.

Carbohydrates and fats are the two major sources of energy in the diet. When one is high, the other is usually low—and vice versa. A diet that provides abundant carbohydrate (45 to 65 percent of energy intake) and some fat (20 to 35 percent of energy intake) within a reasonable energy allowance best supports good health. To increase carbohydrates in the diet, focus on whole grains, vegetables, legumes, and fruits—foods noted for their starch, fibers, and naturally occurring sugars.

Health Effects of Starch and Fibers In addition to starch, fibers, and natural sugars, remember that whole grains, vegetables, legumes, and fruits supply valuable vitamins and minerals and little or no fat. The following paragraphs describe some of the health benefits of diets that include a variety of these foods daily.

Heart Disease Unlike high-carbohydrate diets rich in added sugars that can alter blood lipids to favor heart disease, those rich in whole grains, legumes, vegetables, and fruits may protect against heart attack and stroke by lowering blood pressure, improving blood lipids, and reducing inflammation.[28] Such diets are low in animal fat and cholesterol and high in dietary fibers, vegetable proteins, and phytochemicals—all factors associated with a lower risk of heart disease. (The role of animal fat and cholesterol in heart disease is discussed in Chapter 5. The role of vegetable proteins in heart disease is presented in Chapter 6. The benefits of phytochemicals in disease prevention are featured in Highlight 13.)

Oatmeal was the first food recognized for its ability to reduce cholesterol and the risk of heart disease.[29] Foods rich in soluble fibers (such as oat bran, barley, and legumes) lower blood cholesterol ♦ by binding with bile acids in the GI tract and thereby increasing their excretion. Consequently, the liver must use its cholesterol to make new bile acids. In addition, the bacterial by-products of fiber fermentation in the colon also inhibit cholesterol synthesis in the liver. The net result is that soluble fibers such as those found in oats lower blood cholesterol.[30]

♦ Consuming 5 to 10 g of soluble fiber daily reduces blood cholesterol by 3 to 5%. For perspective, ½ c dry oat bran provides 8 g of fiber, and 1 c cooked barley or ½ c cooked legumes provides about 6 g of fiber.

Several researchers have speculated that fiber may also exert its effect by displacing fats in the diet. Although this is certainly helpful, even when dietary fat is low, fibers exert a separate and significant cholesterol-lowering effect. In other words, a high-fiber diet helps to decrease the risk of heart disease independent of fat intake.

Diabetes High-fiber foods—especially whole grains—play a key role managing and preventing type 2 diabetes.[31] When soluble fibers trap nutrients and delay their transit through the GI tract, glucose absorption is slowed, which helps to prevent glucose surge and rebound.

GI Health Dietary fibers also enhance the health of the large intestine. The healthier the intestinal walls, the better they can block absorption of unwanted constituents. Taken with ample fluids, insoluble fibers such as cellulose (as in cereal brans, fruits, and vegetables) increase stool weight, ease passage, and reduce transit time.

Large, soft stools ease elimination for the rectal muscles and reduce pressure in the lower bowel, preventing constipation and making it less likely that rectal veins will swell (hemorrhoids). Fiber prevents compaction of the intestinal contents, which could obstruct the appendix and permit bacteria to invade and infect it (appendicitis). In addition, fiber stimulates the GI tract muscles so that they retain their strength and resist bulging out into pouches known as diverticula (illustrated in Figure H3-3 on p. 91).[32]

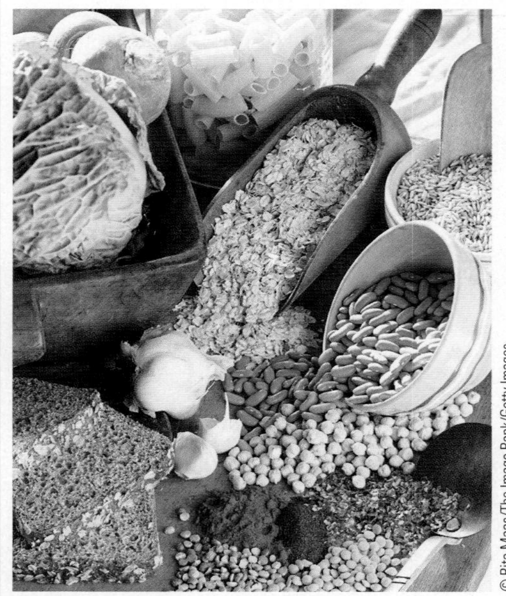

Foods rich in starch and fiber offer many health benefits.

Cancer Research studies suggest that a high-fiber diet protects against colon cancer.[33] When a large study of diet and cancer examined the diets of more than a half million people in ten countries for several years, the researchers found an inverse association between dietary fiber and colon cancer.[34] People who ate the most dietary fiber (35 grams per day) reduced their risk of colon cancer by 40 percent compared with those who ate the least fiber (15 grams per day). Importantly, the study focused on dietary fiber, not fiber supplements or additives, which lack valuable nutrients and phytochemicals that also help protect against cancer. Plant foods—vegetables, fruits, and whole-grain products—reduce the risks of colon and rectal cancers.

Fibers may help prevent colon cancer by diluting, binding, and rapidly removing potential cancer-causing agents from the colon. In addition, soluble fibers stimulate bacterial fermentation of resistant starch and fiber in the colon, a process that produces short-chain fatty acids that lower the pH. These small fat molecules activate cancer-killing enzymes and inhibit inflammation in the colon.[35]

Weight Management High-fiber and whole-grain foods may help a person to maintain a healthy body weight.[36] Foods rich in fiber tend to be low in fat and added sugars and can therefore prevent weight gain and promote weight loss by delivering less energy per bite.[37] In addition, as fibers absorb water from the digestive juices, they swell, creating feelings of fullness, lowering food intake, and delaying hunger.[38]

Many weight-loss products on the market today contain bulk-inducing fibers such as methylcellulose, but buying pure fiber compounds like this is neither necessary nor advisable. Instead of fiber supplements, consumers should select whole grains, legumes, fruits, and vegetables. High-fiber foods not only add bulk to the diet but are economical and nutritious as well.

Dietary fiber provides numerous health benefits.[39] Table 4-5 summarizes fiber characteristics, food sources, actions in the body, and their health benefits.

Harmful Effects of Excessive Fiber Intake Despite fibers' benefits to health, a diet excessively high in fiber also has a few drawbacks. A person who has a small capacity and eats mostly high-fiber foods may not be able to eat enough food to meet energy or nutrient needs. The malnourished, the elderly, and young children adhering to all-plant (vegan) diets are especially vulnerable to this problem.

Switching from a low-fiber diet to a high-fiber diet suddenly can cause temporary bouts of abdominal discomfort, gas, and diarrhea and, more seriously, can obstruct the GI tract. To prevent such complications, a person adopting a high-fiber diet can take the following precautions:

- Increase fiber intake gradually over several weeks to give the GI tract time to adapt.
- Drink plenty of liquids to soften the fiber as it moves through the GI tract.
- Select fiber-rich foods from a variety of sources—fruits, vegetables, legumes, and whole-grain breads and cereals.

TABLE 4-5 Characteristics, Sources, and Health Effects of Fibers

	Major Food Sources	Types of Fibers	Actions in the Body	Probable Health Benefits
	Viscous, Soluble, More Fermentable			
© Brian Leatart/Getty Images	• Barley, oats, oat bran, rye, fruits (apples, citrus), legumes (especially young green peas and black-eyed peas), seaweeds, seeds and husks, many vegetables, fibers used as food additives	• Gums • Pectins • Psyllium[a] • Some hemicellulose	• Lower blood cholesterol by binding bile • Slow glucose absorption • Slow transit of food through upper GI tract • Hold moisture in stools, softening them • Yield small fat molecules after fermentation that the colon can use for energy • Increase satiety	• Lower risk of heart disease • Lower risk of diabetes • Lower risk of colon and rectal cancer • Increased satiety, and may help with weight management
	Nonviscous, Insoluble, Less Fermentable			
© Stockbyte/Getty Images	• Brown rice, fruits, legumes, seeds, vegetables (cabbage, carrots, brussels sprouts), wheat bran, whole grains, extracted fibers used as food additives	• Cellulose • Lignins • Resistant starch • Hemicellulose	• Increase fecal weight and speed fecal passage through colon • Provide bulk and feelings of fullness	• Alleviate constipation • Lower risk of diverticulosis, hemorrhoids, and appendicitis • Lower risk of colon and rectal cancer

[a]Psyllium, a soluble fiber derived from seeds, is used as a laxative and food additive.

Some fibers can limit the absorption of nutrients by speeding the transit of foods through the GI tract and by binding to minerals. When mineral intake is adequate, however, a *reasonable* intake of high-fiber foods (less than 40 grams a day) does not compromise mineral balance.

Clearly, fiber is like all nutrients in that "more" is "better" only up to a point. Again, the key dietary goals are balance, moderation, and variety.

Recommended Intakes of Starch and Fibers The DRI suggest that carbohydrates provide about half (45 to 65 percent) of the energy requirement. ♦ A person consuming 2000 kcalories a day should therefore have 900 to 1300 kcalories of carbohydrate, or about 225 to 325 grams. (Appendix K explains how to solve such problems.) This amount is more than adequate to meet the RDA for carbohydrate, which is set at 130 grams per day, based on the average minimum amount of glucose used by the brain.[40]

♦ Acceptable Macronutrient Distribution Ranges (AMDR):
- Carbohydrate: 45 to 65%
- Fat: 20 to 35%
- Protein: 10 to 35%

On food labels, the Food and Drug Administration (FDA) uses a 60 percent of kcalories guideline in setting the Daily Value for carbohydrate at 300 grams per day. To meet this goal, the *Dietary Guidelines* encourage people to choose a variety of whole grains, vegetables, fruits, and legumes daily.

> DIETARY GUIDELINES FOR AMERICANS 2010

Choose foods that provide more dietary fiber, a nutrient of concern in American diets. Dietary fiber is found in plants, notably legumes, vegetables, fruits, whole grains, and nuts.

Recommendations for fiber ♦ suggest the same foods just mentioned: whole grains, vegetables, fruits, and legumes, which also provide minerals and vitamins. The FDA sets the Daily Value for fiber at 25 grams, rounding up from the recommended 11.5 grams per 1000 kcalories for a 2000-kcalorie intake. The DRI recommendation is slightly higher, at 14 grams per 1000-kcalorie intake—roughly 25 to 35 grams of dietary fiber daily. These recommendations are about two times higher than the usual intake in the United States.[41] An effective way to add fiber while lowering fat is to substitute plant sources of proteins (legumes) for animal sources (meats). Table 4-6 presents a list of fiber sources.

♦ To increase your fiber intake:
- Eat raw vegetables.
- Eat fresh and dried fruit for snacks.
- Add legumes to soups, salads, and casseroles.
- Eat whole-grain breads that contain ≥3 g fiber per serving.
- Eat whole-grain cereals that contain ≥5 g fiber per serving.
- Eat fruits (such as pears) and vegetables (such as potatoes) with their skins.

Because high-fiber foods are so filling, they are not likely to be eaten in excess. Too much fiber can cause GI problems for some people, but it generally does not have adverse effects in most healthy people. For these reasons, the DRI committee did not set an Upper Level for fiber.

WHOLE GRAIN 8g or more per serving WholeGrainsCouncil.org EAT 48g OR MORE OF WHOLE GRAINS DAILY

100% WHOLE GRAIN 16g or more per serving WholeGrainsCouncil.org EAT 48g OR MORE OF WHOLE GRAINS DAILY

Some food labels use a "whole-grain stamp" to help consumers identify whole-grain foods.

From Guidelines to Groceries A diet following the USDA Food Patterns, which include several servings of fruits, vegetables, and whole grains daily, can easily supply the recommended amount of carbohydrates and fiber. In selecting high-fiber foods, keep in mind the principle of variety. The fibers in oats lower cholesterol, whereas those in bran help promote GI tract health. (Review Table 4-5, p. 117, to see the diverse health effects of various fibers.)

Grains An ounce-equivalent of most foods in the grain group (for example, one slice of bread) provides about 15 grams of carbohydrate, mostly as starch. Be aware that some foods in this group, especially snack crackers and baked goods such as biscuits, croissants, and muffins, contain added sugars, solid fats, and sodium. When selecting from the grain group, limit refined grains and be sure to include at least half as whole-grain products (see Figure 4-13, p. 120). The "three are key" message may help consumers to remember to

TABLE 4-6 Fiber in Selected Foods

Grains

Whole-grain products provide about 1 to 2 g (or more) of fiber per serving:

- 1 slice whole-wheat, pumpernickel, rye bread
- 1 oz ready-to-eat cereal (100% bran cereals contain 10 g or more)
- ½ c cooked barley, bulgur, grits, oatmeal

© Polara Studios, Inc.

Vegetables

Most vegetables contain about 2 to 3 g of fiber per serving:

- 1 c raw bean sprouts
- ½ c cooked broccoli, brussels sprouts, cabbage, carrots, cauliflower, collards, corn, eggplant, green beans, green peas, kale, mushrooms, okra, parsnips, potatoes, pumpkin, spinach, sweet potatoes, swiss chard, winter squash
- ½ c chopped raw carrots, peppers

© Polara Studios, Inc.

Fruit

Fresh, frozen, and dried fruits have about 2 g of fiber per serving:

- 1 medium apple, banana, kiwi, nectarine, orange, pear
- ½ c applesauce, blackberries, blueberries, raspberries, strawberries
- Fruit juices contain very little fiber

© Polara Studios, Inc.

Legumes

Many legumes provide about 6 to 8 g of fiber per serving:

- ½ c cooked baked beans, black beans, black-eyed peas, kidney beans, navy beans, pinto beans

Some legumes provide about 5 g of fiber per serving:

- ½ c cooked garbanzo beans, great northern beans, lentils, lima beans, split peas

© Polara Studios, Inc.

NOTE: Appendix H provides fiber grams for more than 2000 foods.

choose a whole-grain cereal for breakfast, a whole-grain bread for lunch, and a whole-grain pasta or rice for dinner. Consumers who eat more whole grains tend to have healthier diets.[42]

> DIETARY GUIDELINES FOR AMERICANS 2010

Limit the consumption of foods that contain refined grains, especially refined-grain foods that contain solid fats, added sugars, and sodium. Replace refined grains with whole grains.

Vegetables The amount of carbohydrate a serving of vegetables provides depends primarily on its starch content. Starchy vegetables—corn, peas, or potatoes—provide about 15 grams of carbohydrate per half-cup serving. A serving of most other *nonstarchy* vegetables—such as a half-cup of broccoli, green beans, or tomatoes—provides about 5 grams.

Fruits A typical fruit serving—a small banana, apple, or orange or a half-cup of most canned or fresh fruit—contains an average of about 15 grams of carbohydrate, mostly as sugars, including the fruit sugar fructose. Fruits vary greatly in their water and fiber contents and, therefore, in their sugar concentrations.

Milks and Milk Products A serving (a cup) of milk or yogurt provides about 12 grams of carbohydrate. Cottage cheese provides about 6 grams of carbohydrate per cup, but most other cheeses contain little, if any, carbohydrate.

> **FIGURE 4-13 Bread Labels Compared**

Although breads may appear similar, their ingredients vary widely. Breads made mostly from whole-grain flours provide more benefits to the body than breads made of enriched, refined, wheat flours.

Some "high-fiber" breads may contain purified cellulose or more nutritious whole grains. "Low-carbohydrate" breads may be regular white bread, thinly sliced to reduce carbohydrates per serving, or may contain soy flour, barley flour, or flaxseed to reduce starch content.

A trick for estimating a bread's content of a nutritious ingredient, such as whole-grain flour, is to read the ingredients list (ingredients are listed in order of predominance). Bread recipes generally include one teaspoon of salt per loaf. Therefore, when a bulky nutritious ingredient, such as whole grain, is listed after the salt, you'll know that less than a teaspoonful of the nutritious ingredient was added to the loaf—not enough to significantly improve the nutrient value of one slice of bread.

Natural Wheat Bread

Multi-fiber Low carb

	Whole Grain Whole Wheat	Natural Wheat Bread	Multi-fiber Low carb
Nutrition Facts			
Serving size	1 slice (30g)	1 slice (30g)	1 slice (30g)
Servings Per Container	18	15	21
Amount per serving			
Calories	90	90	60
Calories from Fat	14	14	15
	% Daily Value*	**% Daily Value***	**% Daily Value***
Total Fat	1.5g 2%	1.5g 2%	1.5g 2%
***Trans* Fat**	0g	0g	0g
Sodium	135mg 6%	220mg 9%	135mg 6%
Total Carbohydrate	15g 5%	15g 5%	9g 3%
Dietary fiber	2g 8%	less than 1g 2%	3g 12%
Sugars	2g	2g	0g
Protein	4g	4g	5g

Whole Grain Whole Wheat — MADE FROM: UNBROMATED STONE GROUND 100% WHOLE WHEAT FLOUR, WATER, CRUSHED WHEAT, HIGH FRUCTOSE CORN SYRUP, PARTIALLY HYDROGENATED VEGETABLE SHORTENING (SOYBEAN AND COTTONSEED OILS), RAISIN JUICE CONCENTRATE, WHEAT GLUTEN, YEAST, WHOLE WHEAT FLAKES, UNSULPHURED MOLASSES, SALT, HONEY, VINEGAR, ENZYME MODIFIED SOY LECITHIN, CULTURED WHEY, UNBLEACHED WHEAT FLOUR AND SOY LECITHIN.

Natural Wheat Bread — INGREDIENTS: UNBLEACHED ENRICHED WHEAT FLOUR [MALTED BARLEY FLOUR, NIACIN, REDUCED IRON, THIAMIN MONONITRATE (VITAMIN B1), RIBOFLAVIN (VITAMIN B2), FOLIC ACID], WATER, HIGH FRUCTOSE CORN SYRUP, MOLASSES, PARTIALLY HYDROGENATED SOYBEAN OIL, YEAST, CORN FLOUR, SALT, GROUND CARAWAY, WHEAT GLUTEN, CALCIUM PROPIONATE (PRESERVATIVE), MONOGLYCERIDES, SOY LECITHIN.

Multi-fiber Low carb — INGREDIENTS: UNBLEACHED ENRICHED WHEAT FLOUR, WATER, WHEAT GLUTEN, CELLULOSE, YEAST, SOYBEAN OIL, CRACKED WHEAT, SALT, BARLEY, NATURAL FLAVOR PRESERVATIVES, MONOCALCIUM PHOSPHATE, MILLET, CORN, OATS, SOYBEANS, BROWN RICE, FLAXSEED, SUCRALOSE.

Protein Foods With two exceptions, protein foods deliver almost no carbohydrate to the diet. The exceptions are nuts, which provide a little starch and fiber along with their abundant fat, and legumes, which provide an abundance of both starch and fiber. Just a half-cup serving of legumes provides about 20 grams of carbohydrate, a third from fiber.

Read Food Labels Food labels list the amount, in grams, of *total* carbohydrate—including starch, fibers, and sugars—per serving (review Figure 4-13). Fiber grams are also listed separately, as are the grams of sugars. (With this information, you can calculate starch grams ♦ by subtracting the grams of fibers and sugars from the total carbohydrate.) Sugars reflect both added sugars and those that occur naturally in foods. Total carbohydrate and dietary fiber are also expressed as "% Daily Values" for a person consuming 2000 kcalories; there is no Daily Value for sugars.

♦ To calculate starch grams using the first label in Figure 4-13:
- 15 g total – 4 g (dietary fiber + sugars) = 11 g starch

REVIEW IT Identify the health benefits of, and recommendations for, starches and fibers.

Clearly, a diet rich in starches and fibers supports efforts to control body weight and prevent heart disease, some cancers, diabetes, and GI disorders. For these reasons, recommendations urge people to eat plenty of whole grains, vegetables, legumes, and fruits—enough to provide 45 to 65 percent of the daily energy intake from carbohydrate.

In today's world, there is one other reason why plant foods rich in complex carbohydrates and natural sugars are a better choice than animal foods or foods high in concentrated sugars: in general, less energy and fewer resources are required to grow and process plant foods than to produce sugar or foods derived from animals. Chapter 20 takes a closer look at the environmental impacts of food production and use.

Nutrition Portfolio

Foods that derive from plants—whole grains, vegetables, legumes, and fruits—naturally provide ample carbohydrates and fiber with little or no fat. Refined foods often contain added sugars and fat.

Go to Diet Analysis Plus and choose one of the days on which you have tracked your diet for the entire day. Go to the Intake Spreadsheet report. Scroll down until you see: carb (g).

- Which of your foods for this day were highest in carbohydrate? Which of these foods also contain added sugars and fats? List better alternatives.
- List the types and amounts of grain products you ate on that day, making note of which are whole-grain or refined foods and how your choices could include more whole-grain options.
- List the types and amounts of fruits and vegetables you ate on that day, making note of how many are dark green, red and orange, or deep yellow, how many are starchy or legumes, and how your choices could include more of these options.
- Describe choices you can make in selecting and preparing foods and beverages to lower your intake of added sugars.

Diet Analysis PLUS

To complete this exercise, go to your Diet Analysis Plus at **www.cengagebrain.com.**

STUDY IT To review the key points of this chapter and take a practice quiz, go to the study cards at the end of the book.

REFERENCES

1. K. C. Maki and coauthors, Whole-grain ready-to-eat oat cereal, as part of a dietary program for weight loss, reduces low-density lipoprotein cholesterol in adults with overweight and obesity more than a dietary program including low-fiber control foods, *Journal of the American Dietetic Association* 110 (2010): 205–214; Z. Wei and coauthors, Time- and dose-dependent effect of psyllium on serum lipids in mild-to-moderate hypercholesterolemia: A meta-analysis of controlled clinical trials, *European Journal of Clinical Nutrition* 63 (2009): 821–827; L. Van Horn and coauthors, The evidence for dietary prevention and treatment of cardiovascular disease, *Journal of the American Dietetic Association* 108 (2008): 287–331; M. O. Weickert and A. F. Pfeiffer, Metabolic effects of dietary fiber consumption and prevention of diabetes, *Journal of Nutrition* 138 (2008): 439–442.
2. V. Vuksan and coauthors, Using cereal to increase dietary fiber intake to the recommended level and the effect of fiber on bowel function in healthy persons consuming North American diets, *American Journal of Clinical Nutrition* 88 (2008): 1256–1262.
3. S. S. Dronamraju and coauthors, Cell kinetics and gene expression in colorectal cancer patients given resistant starch: A randomized controlled trial, *Gut* 58 (2009): 413–420; K. C. Maki and coauthors, Beneficial effects of resistant starch on laxation in healthy adults, *International Journal of Food Sciences & Nutrition* 60 (2009): 296–305; S. J. D. O'Keefe and coauthors, Products of the colonic microbiota mediate the effects of diet on colon cancer risk, *Journal of Nutrition* 139 (2009): 2044–2048.
4. Committee on Dietary Reference Intakes, *Dietary Reference Intakes: Energy, Carbohydrate, Fiber, Fat, Fatty Acids, Cholesterol, Protein, and Amino Acids* (Washington, D.C.: National Academies Press, 2005).
5. NIH Consensus Development Conference: Lactose intolerance and health, http://consensus.nih.gov/2010/lactosestatement.htm.
6. T. He and coauthors, Effects of yogurt and bifidobacteria supplementation on the colonic microbiotia in lactose-intolerant subjects, *Journal of Applied Microbiology* 104 (2008): 595–604.
7. G. Riccardi, A. A. Rivellese, and R. Giacco, Role of glycemic index and glycemic load in the healthy state, in prediabetes, and in diabetes, *American Journal of Clinical Nutrition* 87 (2008): 269S–274S.
8. G. Livesey and coauthors, Glycemic response and health—A systematic review and meta-analysis: Relations between dietary glycemic properties and health outcomes, *American Journal of Clinical Nutrition* 87 (2008): 258S–268S.
9. A. W. Barclay and coauthors, Glycemic index, glycemic load, and chronic disease risk: A meta-analysis of observational studies, *American Journal of Clinical Nutrition* 87 (2008): 627–637; J. Howlett and M. Ashwell, Glycemic response and health: Summary of a workshop, *American Journal of Clinical Nutrition* 87 (2008): 212S–216S.
10. H. Hare-Bruun and coauthors, Should glycemic index and glycemic load be considered in dietary recommendations? *Nutrition Reviews* 66 (2008): 569–590.

11. T. M. S. Wolever and coauthors, Measuring the glycemic index of foods: Interlaboratory study, *American Journal of Clinical Nutrition* 87 (2008): 247S–257S.
12. S. W. Rizkalla, Health implications of fructose consumption: A review of recent data, *Nutrition and Metabolism* 4 (2010): 82–98.
13. US Department of Agriculture and US Department of Health and Human Services, *Dietary Guidelines for Americans, 2010*, www.dietaryguidelines.gov; R. E. Kavey, How sweet it is: Sugar-sweetened beverage consumption, obesity, and cardiovascular risk in childhood, *Journal of the American Dietetic Association* 110 (2010): 1456–1460.
14. O. I. Bermudez and X. Gao, Greater consumption of sweetened beverages and added sugars is associated with obesity among US young adults, *Annals of Nutrition and Metabolism* 57 (2010): 211–218; S. N. Bleich and coauthors, Increasing consumption of sugar-sweetened beverages among US adults: 1988–1994 to 1999–2004, *American Journal of Clinical Nutrition* 89 (2009): 372–381.
15. Bermudez and Gao, 2010.
16. S. Harrington, The role of sugar-sweetened beverage consumption in adolescent obesity: A review of the literature, *Journal of School Nursing* 24 (2008): 3–12.
17. D. I. Jalal and coauthors, Increased fructose associates with elevated blood pressure, *Journal of the American Society of Nephrology* 21 (2010): 1543–1549; R. K. Johnson and B. A. Yon, Weighing in on added sugars and health, *Journal of the American Dietetic Association* 110 (2010): 1296–1299; V. S. Malik and coauthors, Sugar-sweetened beverages and risk of metabolic syndrome and type 2 diabetes: A meta-analysis, *Diabetes Care* 33 (2010): 2477–2483; L. Tappy and coauthors, Fructose and metabolic diseases: New findings, new questions, *Nutrition* 26 (2010): 1044–1049; T. T. Fung and coauthors, Sweetened beverage consumption and risk of coronary heart disease in women, *American Journal of Clinical Nutrition* 89 (2009): 1037–1042.
18. M. J. Dekker and coauthors, Fructose: A highly lipogenic nutrient implicated in insulin resistance, hepatic steatosis, and the metabolic syndrome, *American Journal of Physiology, Endocrinology and Metabolism* 299 (2010): E685–E694; M. E. Bocarsly and coauthors, High-fructose corn syrup causes characteristics of obesity in rats: Increased body weight, body fat and triglyceride levels, *Pharmacology Biochemistry and Behavior* 97 (2010): 101–106; K. L. Stanhope and coauthors, Consuming fructose-sweetened, not glucose-sweetened, beverages increases visceral adiposity and lipids and decreases insulin sensitivity in overweight/obese humans, *Journal of Clinical Investigation* 119 (2009): 1322–1334; K. A. Lê and coauthors, Fructose overconsumption causes dyslipidemia and ectopic lipid deposition in healthy subjects with and without a family history of type 2 diabetes, *American Journal of Clinical Nutrition* 89 (2009): 1760–1765; M. M. Swarbrick and coauthors, Consumption of fructose sweetened beverages for 10 weeks increases postprandial triacylglycerol and apolipoprotein-B concentrations in overweight and obese women, *British Journal of Nutrition* 100 (2008): 947–952; E. J. Parks and coauthors, Dietary sugars stimulate fatty acid synthesis in adults, *Journal of Nutrition* 138 (2008): 1039–1046.
19. J. A. Welsh and coauthors, Consumption of added sugars and indicators of cardiovascular disease risk among US adolescents, *Circulation* 123 (2011): 249–257; J. A. Welsh and coauthors, Caloric sweetener consumption and dyslipidemia among US adults, *Journal of the American Medical Association* 303 (2010): 1490–1497.
20. Lê and coauthors, 2009.
21. L. de Koning and coauthors, Sugar-sweetened and artificially sweetened beverage consumption and risk of type 2 diabetes in men, *American Journal of Clinical Nutrition* 93 (2011): 1321–1327; K. J. Duffey and coauthors, Drinking caloric beverages increases the risk of adverse cardiometabolic outcomes in the Coronary Artery Risk Development in Young Adults (CARDIA) Study, *American Journal of Clinical Nutrition* 92 (2010): 954–959; F. B. Hu and V. S. Malik, Sugar-sweetened beverages and risk of obesity and type 2 diabetes: Epidemiologic evidence, *Physiology and Behavior* 100 (2010): 47–54; R. K. Johnson and coauthors, Dietary sugars intake and cardiovascular health. A scientific statement from the American Heart Association, *Circulation* 120 (2009): 1011–1020; V. S. Malik and coauthors, 2010; A. Miller and K. Adeli, Dietary fructose and the metabolic syndrome, *Current Opinion in Gastroenterology* 24 (2008): 204–209.
22. Position of the American Dietetic Association: Oral health and nutrition, *Journal of the American Dietetic Association* 107 (2007): 1418–1428.
23. US Department of Agriculture, Agricultural Research Service, Beltsville Human Nutrition Research Center, Food Surveys Research Group (Beltsville, MD) and US Department of Health and Human Services, Centers for Disease Control and Prevention, National Center for Health Statistics (Hyattsville, MD), *What We Eat in America*, NHANES 2007–2008, www.ars.usda.gov/ ba/bhnrc/fsrg, published 2010.
24. Committee on Dietary Reference Intakes, 2005.
25. B. P. Marriott and coauthors, Intake of added sugars and selected nutrients in the United States, National Health and Nutrition Examination Survey (NHANES) 2003–2006, *Critical Reviews in Food Science and Nutrition* 50 (2010): 228–258.
26. *Dietary Guidelines for Americans, 2010.*
27. Position of the American Dietetic Association: Use of nutritive and nonnutritive sweeteners, *Journal of the American Dietetic Association* 104 (2004): 255–275.
28. M. U. Jakobsen and coauthors, Intake of carbohydrates compared with intake of saturated fatty acids and risk of myocardial infarction: Importance of the glycemic index, *American Journal of Clinical Nutrition* 91 (2010): 1764–1768; M. T. Streppel and coauthors, Dietary fiber intake in relation to coronary heart disease and all-cause mortality over 40 y: The Zutphen Study, *American Journal of Clinical Nutrition* 88 (2008): 1119–1125.
29. M. B. Andon and J. W. Anderson, State of the art reviews: The oatmeal-cholesterol connection: 10 Years later, *American Journal of Lifestyle Medicine* 2 (2008): 51–57.
30. R. A. Othman, M. H. Moghadasian, and P. J. H. Jones, Cholesterol-lowering effects of oat β-glucan, *Nutrition Reviews* 69 (2011): 299–309.
31. H. Kim and coauthors, Glucose and insulin responses to whole grain breakfasts varying in soluble fiber, beta-glucan: A dose response study in obese women with increased risk for insulin resistance, *European Journal of Nutrition* 48 (2009): 170–175; V. Vuksan and coauthors, Fiber facts: Benefits and recommendations for individuals with type 2 diabetes, *Current Diabetes Reports* 9 (2009): 405–411.
32. S. Tarleton and J. K. Dibaise, Low-residue diet in diverticular disease: Putting an end to a myth, *Nutrition in Clinical Practice* 26 (2011): 137–142; A. Rocco and coauthors, Treatment options for uncomplicated diverticular disease of the colon, *Journal of Clinical Gastroenterology* 43 (2009): 803–808.
33. C. C. Dahm and coauthors, Dietary fiber and colorectal cancer risk: A nested case-control study using food diaries, *Journal of the National Cancer Institute* 102 (2010): 614–626; L. B. Sansbury and coauthors, The effect of strict adherence to a high-fiber, high-fruit and -vegetable, and low-fat eating pattern on adenoma recurrence, *American Journal of Epidemiology* 170 (2009): 576–584.
34. N. Slimani and B. Margetts, Nutrient intakes and patterns in the EPIC cohorts from ten European countries, *European J ournal of Clinical Nutrition* 63 (2009): S1–S274.
35. M. H. Pan and coauthors, Molecular mechanisms for chemoprevention of colorectal cancer by natural dietary compounds, *Molecular Nutrition and Food Research* 55 (2011): 32–45.
36. H. Du and coauthors, Dietary fiber and subsequent changes in body weight and waist circumference in European men and women, *American Journal of Clinical Nutrition* 91 (2010): 329–336; J. N. Davis and coauthors, Inverse relation between dietary fiber intake and visceral adiposity in overweight Latino youth, *American Journal of Clinical Nutrition* 90 (2009): 1160–1166.
37. L. A. Tucker and K. S. Thomas, Increasing total fiber intake reduces risk of weight and fat gains in women, *Journal of Nutrition* 139 (2009): 567–581.
38. P. Vitaglione and coauthors, β-Glucan-enriched bread reduces energy intake and modifies plasma ghrelin and peptide YY concentrations in the short term, *Appetite* 53 (2009): 338–344; A. Hamedani and coauthors, Reduced energy intake at breakfast is not compensated for at lunch if a high-insoluble-fiber cereal replaces a low-fiber cereal, *American Journal of Clinical Nutrition* 89 (2009): 1343–1349; N. Schroeder and coauthors, Influence of whole grain barely, whole grain wheat, and refined

rice-based foods on short-term satiety and energy intake, *Appetite* 53 (2009): 363–369.
39. J. W. Anderson and coauthors, Health benefits of dietary fiber, *Nutrition Reviews* 67 (2009): 188–205.
40. Committee on Dietary Reference Intakes, 2005.
41. *What We Eat in America*, 2010; Position of the American Dietetic Association: Health implications of dietary fiber, *Journal of the American Dietetic Association* 108 (2008): 1716–1731.
42. C. E. O'Neil and coauthors, Consumption of whole grains is associated with improved diet quality and nutrient intake in children and adolescents: The National Health and Nutrition Examination Survey 1999–2004, *Public Health Nutrition* 14 (2011): 347–355; C. E. O'Neil and coauthors, Whole-grain consumption is associated with diet quality and nutrient intake in adults: The National Health and Nutrition Examination Survey, 1999–2004, *Journal of the American Dietetic Association* 110 (2010): 1461–1468.

LEARN IT Summarize the key scientific evidence behind some of the current controversies surrounding carbohydrates and their kcalories.

Carbs, kCalories, and Controversies

Carbohydrate-rich foods are easy to like. Mashed potatoes, warm muffins, blueberry pancakes, freshly baked bread, and tasty rice and pasta dishes tempt most people's palates. In recent years, such homey foods have been blamed for causing weight gain and harming health. Popular writers have persuaded consumers that carbohydrates are "bad."[1] In contrast, the *Dietary Guidelines* urge people to consume plenty of fruits, vegetables, legumes, and whole grains—all carbohydrate-rich foods.

Do carbohydrate-rich foods cause obesity and related health problems?[2] Should people "cut carbs" to lose weight and protect their health? Many popular diet books espouse a carbohydrate-restricted or carbohydrate-modified diet. Some claim that all or some types of carbohydrates are bad. Some go so far as to equate carbohydrates with toxic poisons or addictive drugs. "Bad" carbohydrates—such as sugar, white flour, and potatoes—are considered evil because they are absorbed easily and raise blood glucose. The pancreas then responds by secreting insulin—and insulin is touted as the real villain responsible for our nation's obesity epidemic. Whether restricting overall carbohydrate intake or replacing certain "bad" carbohydrates with "good" carbohydrates, many of these popular diets tend to distort the facts. This highlight examines the scientific evidence behind some of the current controversies surrounding carbohydrates and their kcalories.

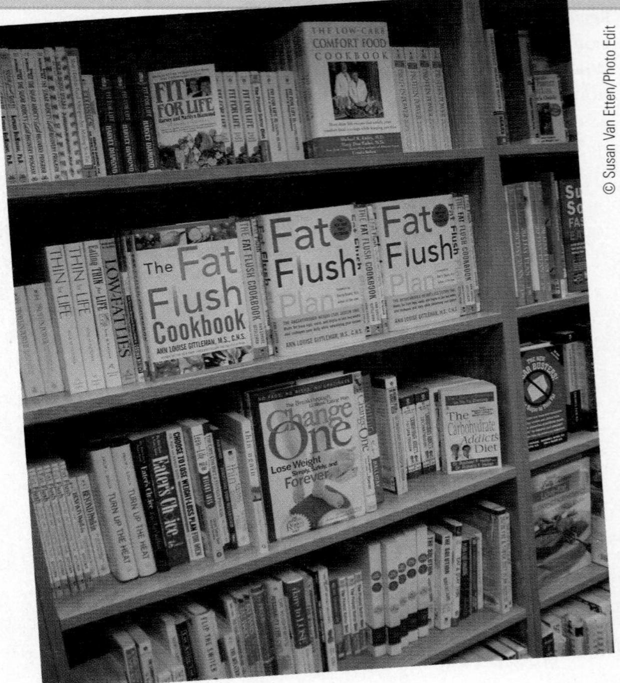

© Susan Van Etten/Photo Edit

Carbohydrates' kCalorie Contributions

The incidence of obesity in the United States has risen dramatically over the past several decades.[3] Popular diet books often blame carbohydrates for this increase in obesity. One way researchers can explore whether the amount of carbohydrate in the diet contributes to increases in body weight over time is by reviewing national food intake survey records, such as NHANES (introduced in Chapter 1). Figure H4-1 presents a summary of energy nutrient data over the past three decades. Since the 1970s, kcalories from carbohydrates increased from 42 percent to 49 percent today.[4] At the same time, kcalories from fat dropped from 41 percent to 34 percent. The percentage of protein intake stayed about the same.

A closer look at the data reveals that, as the percentage of kcalories from the three energy nutrients shifted slightly, total daily energy intake increased significantly. In general, as food became more readily available in this nation, consumers began to eat more than they had in the past.[5] Since the 1970s, total energy intakes have increased by about 200 to 300 kcalories a day (see Figure H4-2).[6] All of the increase in kcalories came from an increase in carbohydrate kcalories. At the same time, most people were not active enough to use up those extra kcalories; in fact, activity levels declined.[7] Consequently, the average body weight for adults increased over these decades by about 20 pounds (see Figure H4-3).

> FIGURE H4-1 **Energy Nutrients over Time**

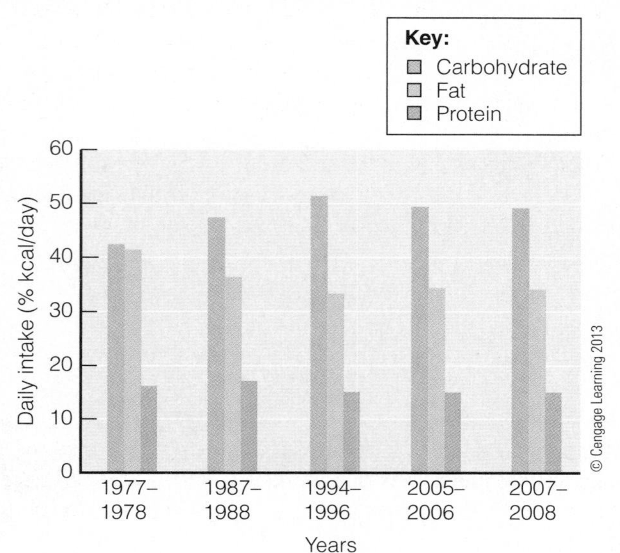

> FIGURE H4-2 **Daily Energy Intake over Time**

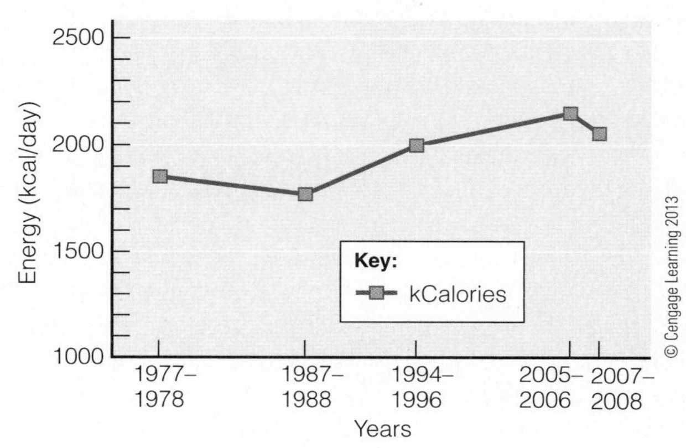

> FIGURE H4-3 **Increases in Adult Body Weight over Time**

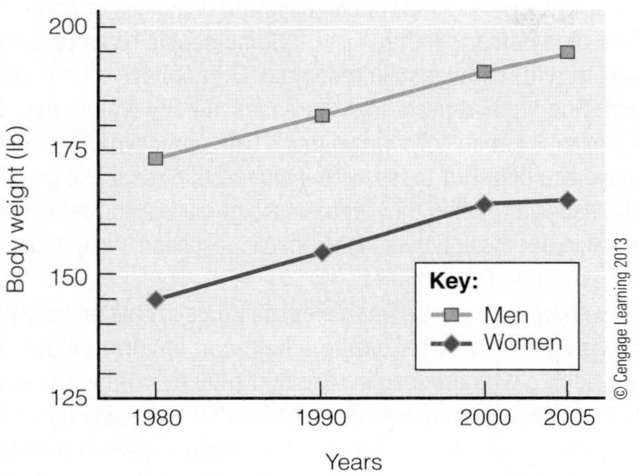

Might too many carbohydrates in the diet be to blame for weight gains? Interestingly, epidemiological studies find an *inverse* relationship between carbohydrate intake and body weight.[8] Those with the highest carbohydrate intake have the lowest body weight and vice versa. Dietary fiber, which favors a healthy body weight, explains some but not all of this relationship.

Might a low-carbohydrate diet support weight losses? For the most part, weight loss is similar for people following either a low-carbohydrate diet or a low-fat diet.[9] This is an important point. Weight losses reflect restricted kcalories—not the proportion of energy nutrients in the diet.[10] Any diet can produce weight loss, at least temporarily, if energy intake is restricted.

Sugars' Share in the Problem

As the chapter mentioned, the use of high-fructose corn syrup sweetener parallels unprecedented increases in the incidence of obesity, but does it mean that the increasing sugar intakes are responsible for the increase in body fat and its associated health problems?[11] Excess sugar in the diet is associated with more fat on the body. When eaten in excess of need, energy from added sugars contributes to body fat stores, just as excess energy from other sources does. Added sugars provide excess energy, raising the risk of weight gain. When total energy intake is controlled, however, *moderate* amounts of sugar do not *cause* obesity. In other words, foods containing added sugars are no more likely to contribute to weight gain than any other foods. Yet moderating sugar intake can be a challenge.

The liquid form of sugar in soft drinks makes it especially easy to overconsume kcalories. Swallowing liquid kcalories requires little effort. The sugar kcalories of sweet beverages also cost less than many other energy sources, and they are widely available. Also, beverages are energy-dense, providing more than 150 kcalories per 12-ounce can, and many people drink several cans a day. The convenience, economy, availability, and flavors of sugary foods and beverages make overconsumption especially likely.

Limiting selections of foods and beverages high in added sugars can be an effective weight-loss strategy, especially for people whose excess kcalories come primarily from added sugars. Replacing a can of cola with a glass of water every day, for example, can help a person lose a pound (or at least not gain a pound) in 1 month.[12] That may not sound like much, but it adds up to more than 10 pounds a year, for very little effort.

Cravings and Addictions

Some people blame their excessive sugar intakes on cravings and addictions. Foods in general, and carbohydrates and sugars more specifically, are not physically addictive in the ways that drugs are. Yet some people describe themselves as having "carbohydrate cravings" or being "sugar addicts." One frequently noted theory is that people seek carbohydrates as a way to increase their levels of the brain neurotransmitter serotonin, which elevates mood. Interestingly, when those with self-described carbohydrate cravings indulge, they tend to eat more of everything; the percentage of energy from carbohydrates remains unchanged.

One reasonable explanation for the carbohydrate cravings that some people experience involves the self-imposed labeling of a food as both "good" and "bad"—that is, one that is desirable but should be eaten with restraint. Restricting intake heightens the desire further (a "craving"). Then "addiction" is used to explain why resisting the food is so difficult and, sometimes, even impossible. But the "addiction" is not physiological or pharmacological.

Appetite Control

Recall from Chapter 4 that glucose stimulates the release of insulin from the pancreas. Insulin, in turn, sets off a sequence of hormonal actions that suppress the appetite.[13] (Appetite regulation is discussed fully in Chapter 8.) Fructose, in contrast, does not stimulate the release of insulin, and therefore does not suppress appetite. Theoretically, then, eating lots of fructose would never satisfy a person's appetite. In fact, fructose metabolism tends to promote hunger and thereby increase food intake.[14] Although, this idea sounds plausible, a major flaw exists: people don't typically eat pure fructose. They eat sucrose or high-fructose corn syrup, and both of these sugars contain sufficient glucose to stimulate the release of insulin and suppress appetite accordingly.

Whether the meal or snack is liquid or solid may also affect appetite.[15] Even when kcaloric intake is the same, a fresh apple suppresses appetite more than apple juice.[16] Consequently, beverages can influence weight gains both by providing energy and by not satisfying hunger.

Insulin's Response

Several popular diet books hold insulin responsible for the obesity problem and a low-glycemic diet as the weight-loss solution. Yet, among nutrition researchers, controversy continues to surround the questions of whether insulin promotes weight gain or a low-glycemic diet fosters weight loss.[17]

Recall that just after a meal, blood glucose rises and insulin responds. How high insulin levels surge may influence whether the body stores or uses its glucose and fat supplies. What does insulin do? Among its roles, insulin facilitates the transport of glucose into the cells, the storage of fatty acids as fat, and the synthesis of cholesterol. It is an anabolic hormone that builds and stores. True—but there's more to the story. Insulin is only one of many factors involved in the body's metabolism of nutrients and regulation of body weight.

Furthermore, as Chapter 4's discussion of the glycemic index pointed out, the glycemic effect of a particular food varies; diet books often mislead people by claiming that each food has a set glycemic effect. The glycemic effect of a food depends on how the food is ripened, processed, and cooked; the time of day the food is eaten; the other foods eaten with it; and the presence or absence of certain diseases such as type 2 diabetes in the person eating the food.[18]

Most importantly, insulin is critical to maintaining health, as any person with type 1 diabetes can attest. Insulin causes problems only when a person develops insulin resistance—that is, when the body's cells do not respond to the large quantities of insulin that the pancreas continues to pump out in an effort to get a response. Insulin resistance is a major health problem—but it is not caused by carbohydrate, or by protein, or by fat. It most often results from being obese. Importantly, when a person loses weight, insulin response usually improves.

The Glycemic Index and Body Weight

As Chapter 4 mentions, the glycemic index identifies foods that raise blood glucose and stimulate insulin secretion. What is the relationship between a diet's glycemic index and body weight? In general, studies find that diets with a high glycemic index are positively associated with body weight.[19]

Might a low-glycemic diet foster weight loss?[20] In general, research examining the use of low glycemic diets for weight loss finds inconsistent results.[21] Still, a low-glycemic diet may offer other advantages. A low-glycemic meal seems to curb appetite and limit energy intake of the next meal. Low-glycemic diets are also more likely to be rich in nutrients and fiber than high-glycemic diets.

Clearly, if kcalories are low, obese people on either a low-glycemic diet or a traditional low-fat diet can lose weight. Overweight people can lose as much or more weight by emphasizing low-glycemic foods as they can by following a typical low-fat, portion-controlled weight-loss diet.

The Individual's Response to Foods

The body's insulin response to carbohydrate depends not only on a food, but also on a person's metabolism.[22] Some people react to dietary carbohydrate with a low insulin response. Others have a high insulin response. One study reports that increases in body weight over 6 years were similar in people following either a high-carbohydrate diet or a low-carbohydrate diet. But those with a high insulin response gained more weight, especially when they were on a high-carbohydrate diet.[23] By the same token, for those with a higher insulin response, weight loss may be greater on a low-glycemic diet.

How energy is stored after a meal depends in part on how the body responds to insulin. After eating a high-carbohydrate meal, normal-weight people who are insulin resistant tend to synthesize about half as much glycogen in muscles and make about twice as much fat in the liver as people who are insulin sensitive. Some research suggests that restricting carbohydrate intake may improve glucose control, insulin response, and blood lipids.[24]

In Summary

As might be expected given the similarity in their chemical composition, high-fructose corn syrup and sucrose produce similar effects in appetite control and energy metabolism.[25] In fact, high-fructose corn syrup is more like sucrose than it is like pure fructose. As mentioned, people don't eat pure fructose; they eat foods and drink beverages that contain added sugars—either high-fructose corn syrup or sucrose. Limiting these sugars is a helpful strategy when trying to control body weight, but restricting all carbohydrates would be unwise.

The quality of the diet suffers when carbohydrates are restricted. Without fruits, vegetables, and whole grains, low-carbohydrate diets lack not only carbohydrate, but fiber, vitamins, minerals, and phytochemicals as well—all dietary factors protective against disease. The DRI recommends that carbohydrates contribute between 45 and 65 percent of daily energy intake. Intakes within this range can support healthy body weight and do not contribute to obesity—when added sugar intake is moderate and total energy intake is appropriate. Similarly, added sugars increase energy intake, but need not contribute to obesity—when added sugar intake is moderate and total energy intake is appropriate. When choosing carbohydrates, emphasize a variety of naturally occurring carbohydrates—such as whole grains, legumes, vegetables, and fruits—and limit foods and beverages with added sugars.

REFERENCES

1. G. A. Bray, *Viewpoint: Good kCalories, Bad kCalories* by Gary Taubes, *Obesity Reviews* 9 (2008): 251–263.
2. D. B. Allison and R. D. Mattes, Nutritively sweetened beverage consumption and obesity: The need for solid evidence on a fluid issue, *Journal of the American Medical Association* 301 (2009): 318–320.
3. K. M. Flegal and coauthors, Prevalence and trends in obesity among US adults, 1999–2008, *Journal of the American Medical Association* 303 (2010): 235–241.
4. US Department of Agriculture, Agricultural Research Service, Beltsville Human Nutrition Research Center, Food Surveys Research Group (Beltsville, MD) and US Department of Health and Human Services, Centers for Disease Control and Prevention, National Center for Health Statistics (Hyattsville, MD), *What We Eat in America,* NHANES 2007–2008, www.ars.usda.gov/ba/bhnrc/fsrg, published 2010.
5. B. A. Swinburn and coauthors, Estimating the changes in energy flux that characterize the rise in obesity prevalence, *American Journal of Clinical Nutrition* 89 (2009): 1723–1728.
6. *What We Eat in America*, 2010.
7. US Department of Health and Human Services, *2008 Physical Activity Guidelines for Americans* Summary, www.health.gov/PAGuidelines/guidelines/summary.aspx.

8. A. T. Merchant and coauthors, Carbohydrate intake and overweight and obesity among healthy adults, *Journal of the American Dietetic Association* 109 (2009): 1165–1172.
9. G. D. Foster and coauthors, Weight and metabolic outcomes after 2 years on a low-carbohydrate versus low-fat diet: A randomized trial, *Annals of Internal Medicine* 153 (2010): 147–157; I. Shai and coauthors, Weight loss with a low-carbohydrate, Mediterranean, or low-fat diet, *New England Journal of Medicine* 359 (2008): 229–241; R. F. Kushner and B. Doerfier, Low-carbohydrate, high-protein diets revisited, *Current Opinion in Gastroenterology* 24 (2008): 198–203.
10. F. M. Sacks and coauthors, Comparison of weight-loss diets with different compositions of fat, protein, and carbohydrates, *New England Journal of Medicine* 360 (2009): 859–873; R. M. van Dam and J. C. Seidell, Carbohydrate intake and obesity, *European Journal of Clinical Nutrition* 61 (2007): S75–S99.
11. R. D. Mattes and coauthors, Nutritively sweetened beverage consumption and body weight: A systematic review and meta-analysis of randomized experiments, *Obesity Reviews* 12 (2011): 346–365; S. M. Moeller and coauthors, The effects of high fructose syrup, *Journal of the American College of Nutrition* 28 (2009): 619–626.
12. L. Chen and coauthors, Reduction in consumption of sugar-sweetened beverages is associated with weight loss: The PREMIER trial, *American Journal of Clinical Nutrition* 89 (2009): 1299–1306.
13. K. J. Melanson and coauthors, High-fructose corn syrup, energy intake, and appetite regulation, *American Journal of Clinical Nutrition* 88 (2008): 1738S–1744S.
14. M. D. Lane and S. H. Cha, Effect of glucose and fructose on food intake via malonyl-CoA signaling in the brain, *Biochemical and Biophysical Research Communications* 382 (2009): 1–5.
15. R. D. Mattes and W. W. Campbell, Effects of food form and timing of ingestion on appetite and energy intake in lean young adults and in young adults with obesity, *Journal of the American Dietetic Association* 109 (2009): 430–437.
16. Mattes and Campbell, 2009.
17. G. Radulian and coauthors, Metabolic effects of low glycaemic index diets, *Nutrition Journal* 8 (2009): 5; J. Brand-Miller and coauthors, Carbohydrates: The good, the bad and the whole grain, *Asia Pacific Journal of Clinical Nutrition* 17 (2008): 16–19; A. W. Barclay and coauthors, Glycemic index, glycemic load, and chronic disease risk: A meta-analysis of observational studies, *American Journal of Clinical Nutrition* 87 (2008): 627–637; D. J. A. Jenkins and coauthors, Effect of a low-glycemic index or a high-cereal fiber diet on type 2 diabetes, *Journal of the American Medical Association* 300 (2008): 2742–2753.
18. P. Small and J. Brand-Miller, From complex carbohydrate to glycemic index, *Nutrition Today* 44 (2009) 236–243; G. Riccardi, A. A. Rivellese, and R. Giacco, Role of glycemic index and glycemic load in the healthy state, in prediabetes, and in diabetes, *American Journal of Clinical Nutrition* 87 (2008): 269S–274S.
19. H. Hare-Bruun, A. Flint, and B. L. Heitmann, Glycemic index and glycemic load in relation to changes in body weight, body fat distribution, and body composition in adult Danes, *American Journal of Clinical Nutrition* 84 (2006): 871–879.
20. Brand-Miller and coauthors, 2008.
21. A. Esfahani and coauthors, The application of the glycemic index and glycemic load in weight loss: A review of the clinical evidence, *International Union of Biochemistry and Molecular Biology* 63 (2011): 7–13; S. Vega-López and S. N. Mayol-Kreiser, Use of the glycemic index for weight loss and glycemic control: A review of recent evidence, *Current Diabetes Reports* 9 (2009): 379–388.
22. W. J. Whelan and coauthors, The glycemic response is a personal attribute, *International Union of Biochemistry and Molecular Biology* 62 (2010): 637–641.
23. J. P. Chaput and coauthors, A novel interaction between dietary composition and insulin secretion: Effects on weight gain in the Quebec Family Study, *American Journal of Clinical Nutrition* 87 (2008): 303–309.
24. R. J. Wood and M. L. Fernandez, Carbohydrate-restricted versus low-glycemic-index diets for the treatment of insulin resistance and metabolic syndrome, *Nutrition Reviews* 67 (2009): 179–183.
25. K. L. Stanhope and P. J. Havel, Fructose consumption: Recent results and their potential implications, *Annals of the New York Academy of Sciences* 1190 (2010): 15–24; K. J. Melanson and coauthors, 2008.

CENGAGE LEARNING'S GLOBAL NUTRITION WATCH — GLOBAL NUTRITION RESOURCE CENTER

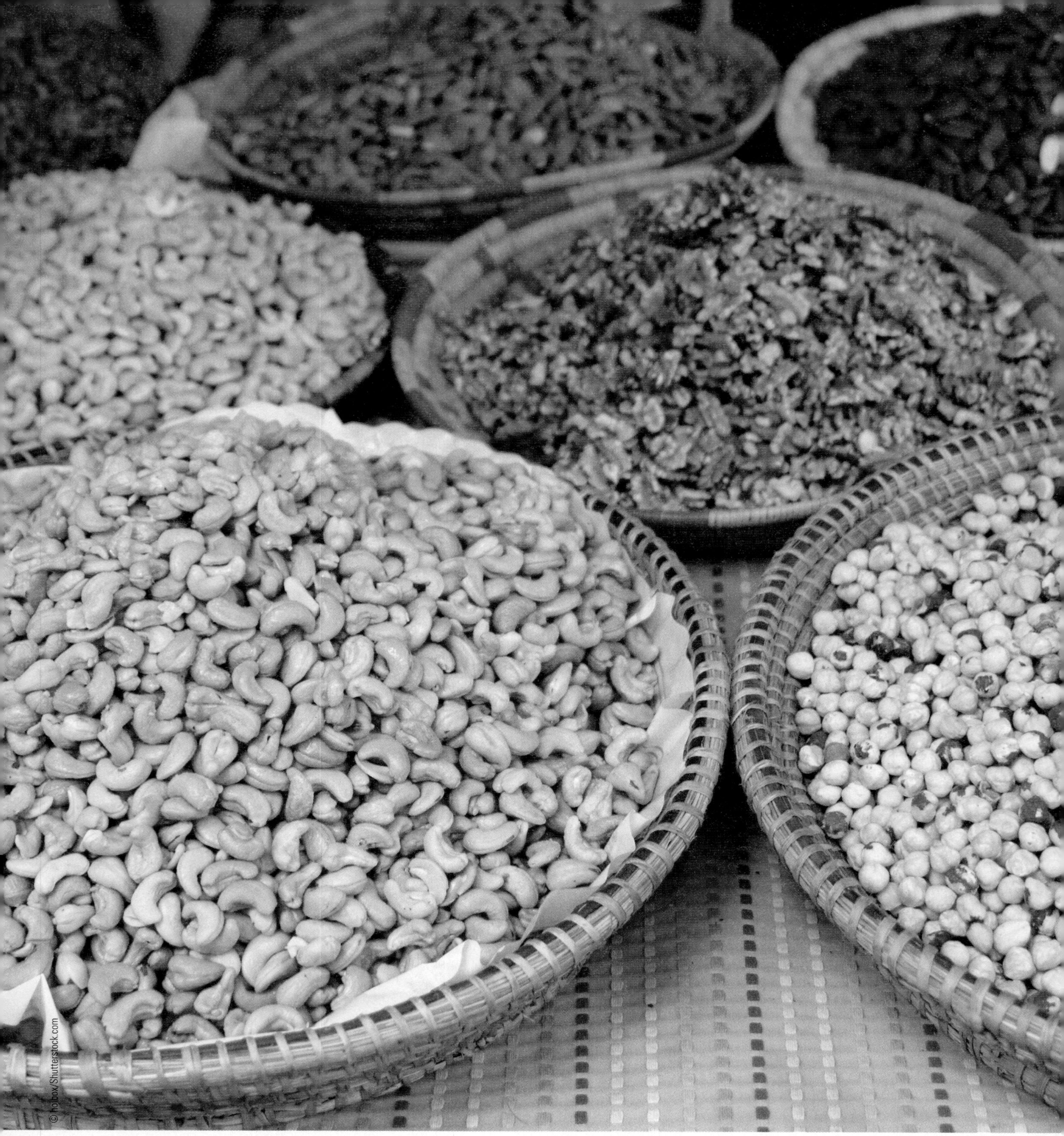

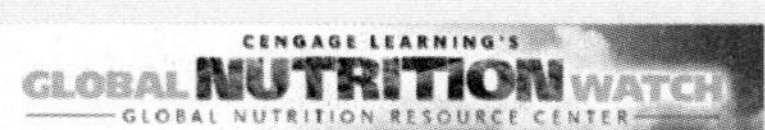

A recent study found a strong correlation between *trans* fatty acids and an increase in allergies in adolescents. Go to Nutrition Basics and Tools → Fats to learn about other adverse effects *trans* fatty acids may have on health. What food choices might be healthier alternatives to foods high in *trans* fats?

5

The Lipids: Triglycerides, Phospholipids, and Sterols

LEARNING GPS

Nutrition in Your Life

Most likely, you know what you don't like about body fat, but do you appreciate how it insulates you against the cold or powers your hike around a lake? And what about food fat? You're right to credit fat for providing the delicious flavors and aromas of buttered popcorn and fried chicken—and to criticize it for contributing to the weight gain and heart disease so common today. The challenge is to strike a healthy balance of enjoying some fat, but not too much. Learning which kinds of fats are beneficial and which are most harmful will help you make wise decisions. In the Nutrition Portfolio at the end of this chapter, you can examine whether your current fat choices are meeting dietary goals.

Most people are surprised to learn that fat has some virtues and that a well-balanced diet needs at least a little fat. Getting enough fat is rarely a problem. In our society of abundance, people are more likely to consume too much fat, or too much of some kinds of fat—with consequent health problems.

Fat refers to the class of nutrients known as lipids. The lipid family includes triglycerides (fats and oils), phospholipids, and sterols. Triglycerides are most abundant, both in foods and in the body.

5.1 The Chemist's View of Fatty Acids and Triglycerides

LEARN IT Recognize the chemistry of fatty acids and triglycerides and differences between saturated and unsaturated fats.

Like carbohydrates, lipids are composed of carbon (C), hydrogen (H), and oxygen (O). Because lipids have many more carbons and hydrogens in proportion to their oxygens, they can supply more energy per gram than carbohydrates can (Chapter 7 provides details).

The many names and relationships in the lipid family can seem overwhelming—like meeting a friend's extended family for the first time. To ease the introductions, this chapter first presents each of the lipids from a chemist's point of view using both words and diagrams. Then the chapter follows the **lipids** through digestion and absorption and into the body to examine their roles in health and disease. For people who think more easily in words than in chemical symbols, this *preview* of the upcoming chemistry may be helpful:

1. Every triglyceride contains one molecule of glycerol and three fatty acids (basically, chains of carbon atoms).
2. Fatty acids may be 4 to 24 (even numbers of) carbons long, the 18-carbon ones being the most common in foods and especially noteworthy in nutrition.
3. Fatty acids may be saturated or unsaturated. Unsaturated fatty acids may have one or more points of unsaturation—that is, they may be *mono*unsaturated or *poly*unsaturated.
4. Of special importance in nutrition are the polyunsaturated fatty acids known as omega-3 fatty acids and omega-6 fatty acids.
5. The 18-carbon polyunsaturated fatty acids are linolenic acid (omega-3) and linoleic acid (omega-6). Both are essential fatty acids that the body cannot make. Each is the primary member of a family of longer-chain fatty acids that help to regulate blood pressure, blood clotting, and other body functions important to health.

The paragraphs, definitions, and diagrams that follow present this information again in much more detail.

lipids: a family of compounds that includes triglycerides, phospholipids, and sterols. Lipids are characterized by their insolubility in water. (Lipids also include the fat-soluble vitamins, described in Chapter 11.)

Fatty Acids

All **fatty acids** have the same basic structure—a chain of carbon and hydrogen atoms with an acid group (COOH) at one end and a methyl group (CH_3) at the other end. Fatty acids may differ from one another, however, in the length of their carbon chains and in the number and location of their double bonds, as the following paragraphs describe. (Fatty acids and related terms are defined in the accompanying glossary.)

The Length of the Carbon Chain

Most naturally occurring fatty acids contain even numbers of carbons in their chains—up to 24 carbons in length. This discussion begins with the 18-carbon fatty acids, which are abundant in our food supply. Stearic acid is the simplest of the 18-carbon fatty acids; the bonds between its carbons are all alike:

Stearic acid, an 18-carbon saturated fatty acid

```
  H H H H H H H H H H H H H H H H H O
  | | | | | | | | | | | | | | | | | ‖
H-C-C-C-C-C-C-C-C-C-C-C-C-C-C-C-C-C-C-O-H
  | | | | | | | | | | | | | | | | |
  H H H H H H H H H H H H H H H H H
```

GLOSSARY OF FATTY ACID TERMS

fatty acids: organic compounds composed of a carbon chain with hydrogens attached and an acid group (COOH) at one end and a methyl group (CH_3) at the other end.

monounsaturated fatty acid (MUFA): a fatty acid that lacks two hydrogen atoms and has one double bond between carbons—for example, oleic acid. A *monounsaturated fat* is composed of triglycerides in which most of the fatty acids are monounsaturated.

- **mono** = one

point of unsaturation: the double bond of a fatty acid, where hydrogen atoms can easily be added to the structure.

polyunsaturated fatty acid (PUFA): a fatty acid that lacks four or more hydrogen atoms and has two or more double bonds between carbons—for example, linoleic acid (two double bonds) and linolenic acid (three double bonds). A *polyunsaturated fat* is composed of triglycerides in which most of the fatty acids are polyunsaturated.

- **poly** = many

saturated fatty acid: a fatty acid carrying the maximum possible number of hydrogen atoms—for example, stearic acid. A *saturated fat* is composed of triglycerides in which most of the fatty acids are saturated.

unsaturated fatty acid: a fatty acid that lacks hydrogen atoms and has at least one double bond between carbons (includes monounsaturated and polyunsaturated fatty acids). An *unsaturated fat* is composed of triglycerides in which most of the fatty acids are unsaturated.

As you can see, stearic acid is 18 carbons long, and each atom meets the rules of chemical bonding described in Figure 4-1 on p. 96. The following structure also depicts stearic acid, but in a simpler way, with each "corner" on the zigzag line representing a carbon atom with two attached hydrogens:

Stearic acid (simplified structure)

As mentioned, the carbon chains of fatty acids vary in length. The long-chain (12 to 24 carbons) fatty acids of meats, seafood, and vegetable oils are most common in the diet. Smaller amounts of medium-chain (6 to 10 carbons) and short-chain (fewer than 6 carbons) fatty acids also occur, primarily in dairy products. (Tables C-1 and C-2 in Appendix C provide the names, chain lengths, and sources of fatty acids commonly found in foods.)

The Number of Double Bonds Stearic acid (described and shown previously) is a **saturated fatty acid.** A saturated fatty acid is fully loaded with all its hydrogen atoms and contains only single bonds between its carbon atoms. If two hydrogens were missing from the middle of the carbon chain, the remaining structure might be:

An impossible chemical structure

Notice that in this impossible chemical structure, two of the carbons have only three bonds each. Such a compound cannot exist because every carbon must have four bonds. To satisfy this rule, the two carbons form a double bond:

Oleic acid, an 18-carbon monounsaturated fatty acid

The same structure drawn more simply looks like this:*

Oleic acid (simplified structure)

Although drawn straight here, the actual shape bends at the double bond. The double bond is a **point of unsaturation.** A fatty acid like this—with two hydrogens missing and a double bond—is an **unsaturated fatty acid.** This one is the 18-carbon **monounsaturated fatty acid** oleic acid, which is abundant in olive oil and canola oil.

A **polyunsaturated fatty acid** has two or more carbon-to-carbon double bonds. **Linoleic acid**, the 18-carbon fatty acid common in vegetable oils, lacks four hydrogens and has two double bonds:

Linoleic acid, an 18-carbon polyunsaturated fatty acid

*Remember that each "corner" on the zigzag line represents a carbon atom with two attached hydrogens.

linoleic (lin-oh-LAY-ick) **acid:** an essential fatty acid with 18 carbons and two double bonds.

TABLE 5-1 18-Carbon Fatty Acids

Name	Number of Carbon Atoms	Number of Double Bonds	Saturation	Common Food Sources
Stearic acid	18	0	Saturated	Most animal fats
Oleic acid	18	1	Monounsaturated	Olive and canola oils
Linoleic acid	18	2	Polyunsaturated	Sunflower, safflower, corn, and soybean oils
Linolenic acid	18	3	Polyunsaturated	Soybean and canola oils, flaxseed, walnuts

NOTE: Chemists use a shorthand notation to describe fatty acids. The first number indicates the number of carbon atoms; the second, the number of the double bonds. For example, the notation for stearic acid is 18:0.

Drawn more simply, linoleic acid looks like this (though the actual shape would bend at the double bonds):

Linoleic acid (simplified structure)

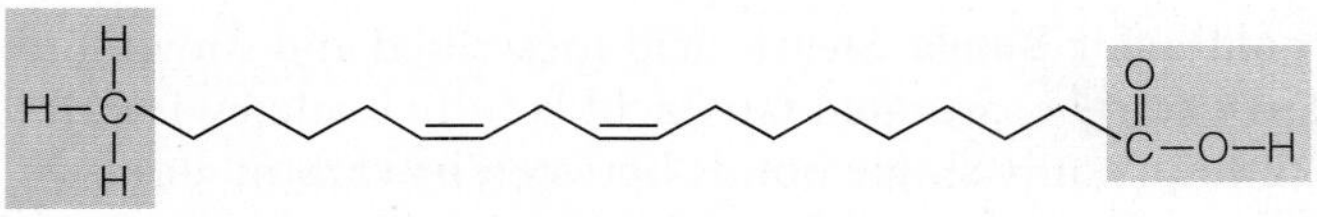

A fourth 18-carbon fatty acid is **linolenic acid,** which has three double bonds. Table 5-1 presents the 18-carbon fatty acids.

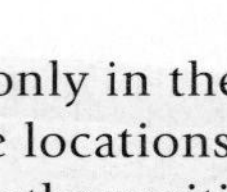

The Location of Double Bonds Fatty acids differ not only in the length of their chains and their degree of saturation, but also in the locations of their double bonds. Chemists identify polyunsaturated fatty acids by the position of the double bond closest to the methyl (CH_3) end of the carbon chain, which is described by an **omega** number. A polyunsaturated fatty acid with its closest double bond three carbons away from the methyl end is an **omega-3 fatty acid.** Similarly, an **omega-6 fatty acid** is a polyunsaturated fatty acid with its closest double bond six carbons away from the methyl end. Figure 5-1 compares two 18-carbon fatty acids—linolenic acid (an omega-3 fatty acid) and linoleic acid (an omega-6 fatty acid).

Monounsaturated fatty acids tend to belong to the omega-9 group, with their closest (and only) double bond nine carbons away from the methyl end. Oleic acid—the 18 carbon monounsaturated fatty acid common in olive oil mentioned earlier—is an omega-9 fatty acid. It is also the most predominant monounsaturated fatty acid in the diet.

> FIGURE 5-1 Omega-3 and Omega-6 Fatty Acids Compared

The omega number indicates the position of the double bond closest to the methyl (CH_3) end. The fatty acids of an omega family may have different lengths and different numbers of double bonds, but the location of the double bond closest to the methyl end is the same in all of them. These structures are drawn linearly here to ease counting carbons and locating double bonds, but their shapes actually bend at the double bonds.

Linolenic acid, an 18-carbon, omega-3 fatty acid

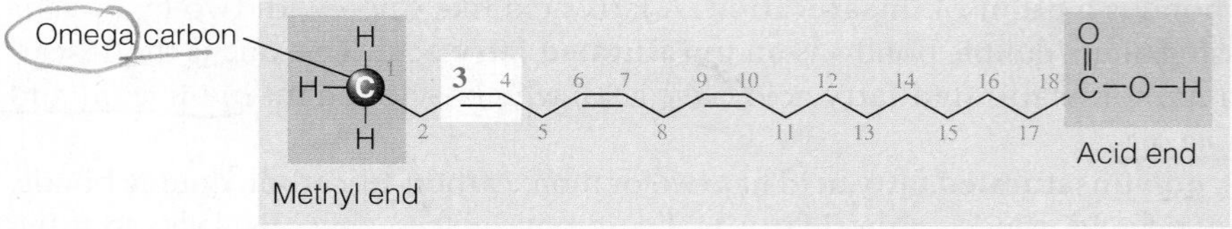

Linoleic acid, an 18-carbon, omega-6 fatty acid

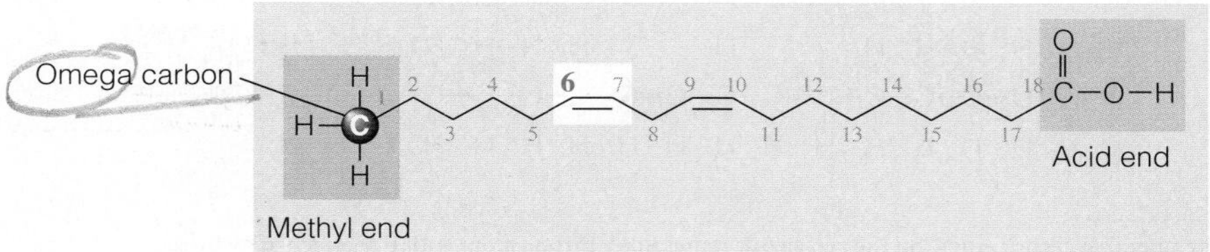

linolenic (lin-oh-LEN-ick) **acid:** an essential fatty acid with 18 carbons and three double bonds.

omega: the last letter of the Greek alphabet (ω), used by chemists to refer to the position of the closest double bond to the methyl (CH_3) end of a fatty acid.

omega-3 fatty acid: a polyunsaturated fatty acid in which the closest double bond to the methyl (CH_3) end of the carbon chain is three carbons away.

omega-6 fatty acid: a polyunsaturated fatty acid in which the closest double bond to the methyl (CH_3) end of the carbon chain is six carbons away.

Triglycerides Few fatty acids occur free in foods or in the body. Most often, they are incorporated into **triglycerides**—lipids composed of three fatty acids attached to a **glycerol**.* Figure 5-2 presents a glycerol molecule.

To make a triglyceride, a series of **condensation** reactions combine a hydrogen atom (H) from the glycerol and a hydroxyl (OH) group from a fatty acid, forming a molecule of water (H_2O) and leaving a bond between the two molecules (see Figure 5-3). Most triglycerides contain a mixture of more than one type of fatty acid (as shown on the right side of Figure 5-3).

Characteristics of Solid Fats and Oils

The chemistry of a fatty acid—whether it is short or long, saturated or unsaturated, with its closest double bond at carbon 3 or carbon 6—influences the characteristics of foods and the health of the body. A section later in this chapter explains how these features affect health; this section describes how the chemistry influences the fats and oils in foods.

Firmness The degree of unsaturation influences the firmness of fats at room temperature (see Figure 5-4). Generally speaking, most polyunsaturated vegetable oils are liquid at room temperature, and the more saturated animal fats are solid. Some oils—notably, cocoa butter, palm oil, palm kernel oil, and coconut oil—are

*Research scientists commonly use the term *triacylglycerols;* this book continues to use the more familiar term *triglycerides,* as do many other health and nutrition books and journals.

> FIGURE 5-2 **Glycerol**

When glycerol is free, an OH group is attached to each carbon. When glycerol is part of a triglyceride, each carbon is attached to a fatty acid (as shown in the next figure).

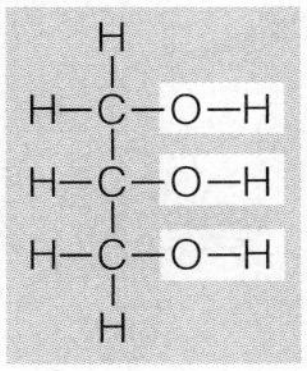

H
H–C–O–H
H–C–O–H
H–C–O–H
H

triglycerides (try-GLISS-er-rides): the chief form of fat in the diet and the major storage form of fat in the body; composed of a molecule of glycerol with three fatty acids attached; also called *triacylglycerols* (try-ay-seel-GLISS-er-ols).

- **tri** = three
- **glyceride** = of glycerol

glycerol (GLISS-er-ol): an alcohol composed of a three-carbon chain, which can serve as the backbone for a triglyceride.

condensation: a chemical reaction in which water is released as two molecules combine to form one larger product.

> FIGURE 5-3 **Condensation of Glycerol and Fatty Acids to Form a Triglyceride**

To make a triglyceride, three fatty acids attach to glycerol in condensation reactions.

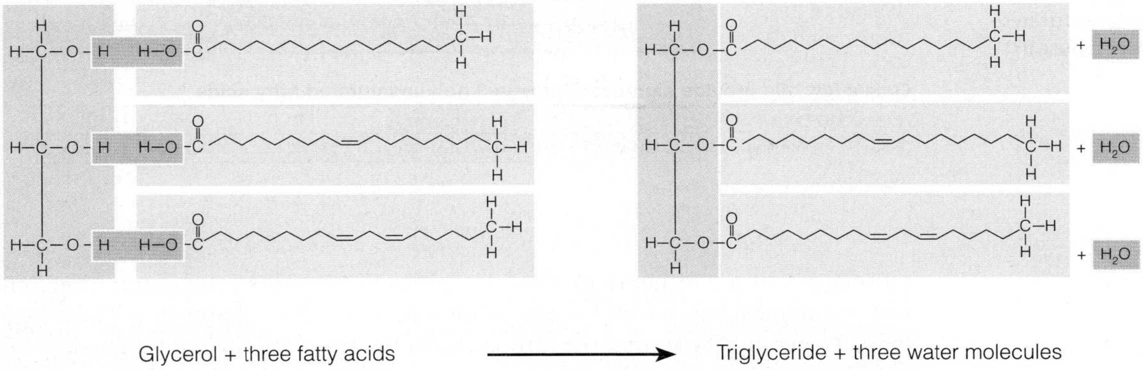

Glycerol + three fatty acids → Triglyceride + three water molecules

An H atom from glycerol and an OH group from a fatty acid combine to create water, leaving the O on the glycerol and the C at the acid end of each fatty acid to form a bond.

Three fatty acids attached to a glycerol form a triglyceride and yield water. In this example, the triglyceride includes a saturated fatty acid, a monounsaturated fatty acid, and a polyunsaturated fatty acid, respectively.

> FIGURE 5-4 **Diagram of Saturated and Unsaturated Fatty Acids Compared**

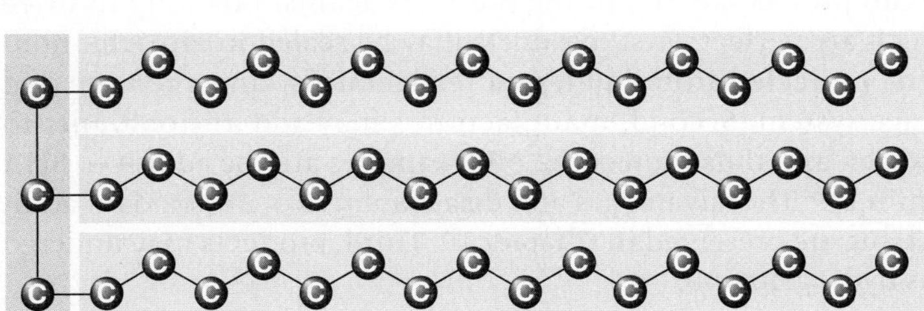

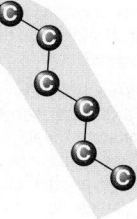

Saturated fatty acids tend to stack together. Consequently, saturated fats tend to be solid (or more firm) at room temperature.

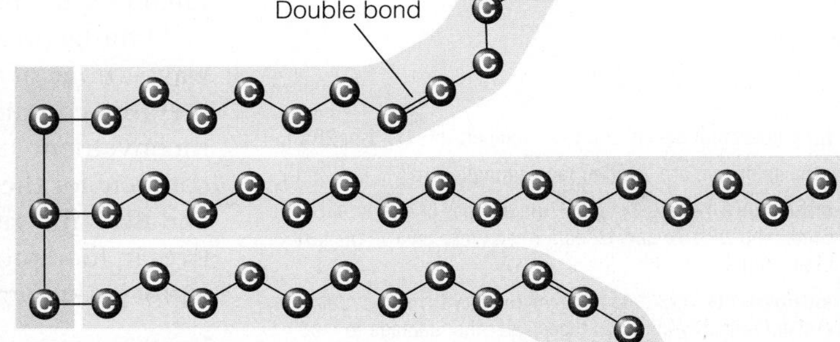

This mixture of saturated and unsaturated fatty acids does not stack neatly because unsaturated fatty acids bend at the double bond(s). Consequently, unsaturated fats tend to be liquid (or less firm) at room temperature.

At room temperature, saturated fats (such as those commonly found in butter and other animal fats) are solid, whereas unsaturated fats (such as those found in vegetable oils) are usually liquid.

> FIGURE 5-5 Fatty Acid Composition of Common Food Fats

Most fats are a mixture of saturated, monounsaturated, and polyunsaturated fatty acids.

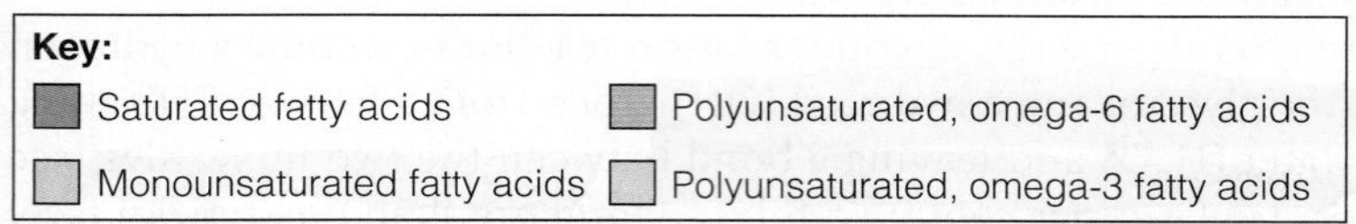

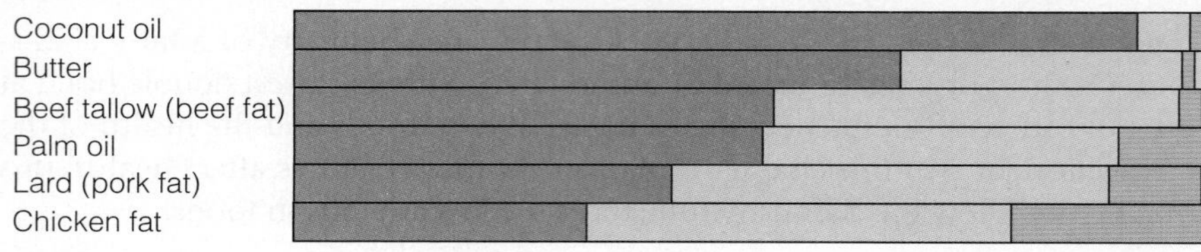

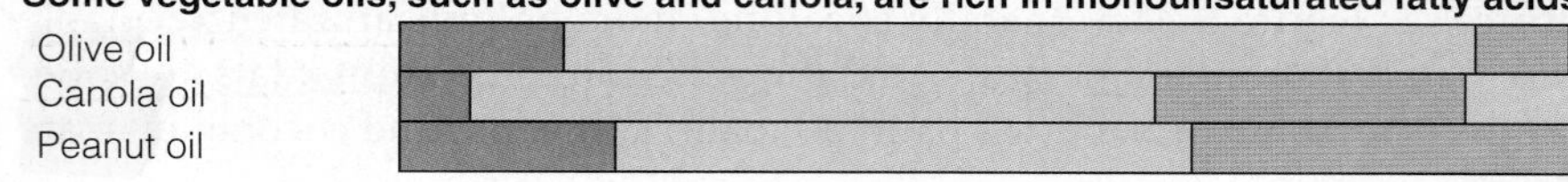

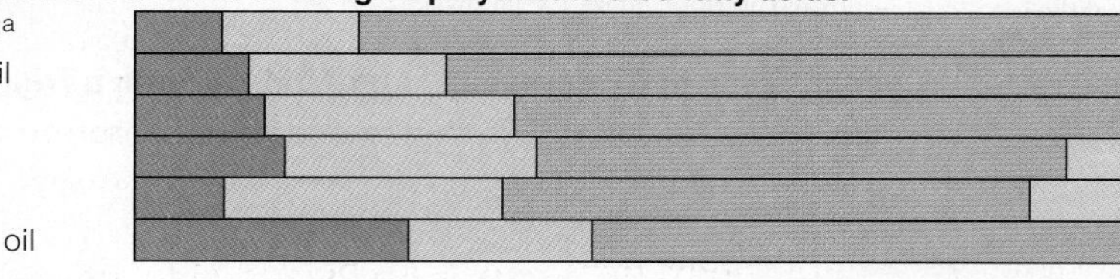

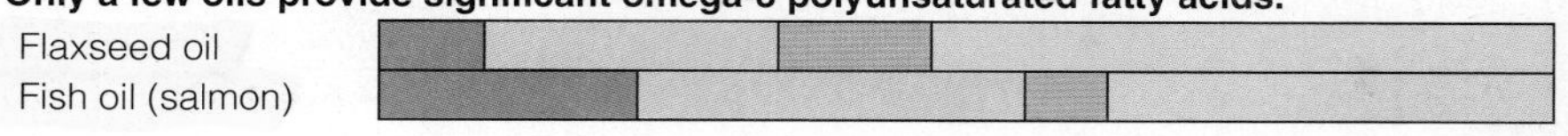

[a]*Salad or cooking type over 70% linoleic acid.*

saturated; they are firmer than most vegetable oils because of their saturation, but softer than most animal fats because of their shorter carbon chains (8 to 14 carbons long). Generally, the shorter the carbon chain, the softer the fat is at room temperature. Fatty acid compositions of selected **fats** and **oils** are shown in Figure 5-5, and Appendix H provides the fat and fatty acid contents of many other foods.

Stability The degree of unsaturation also influences stability. All fats become spoiled when exposed to oxygen. The **oxidation** of fats produces a variety of compounds that smell and taste rancid. (Other types of spoilage can occur due to microbial growth.) Polyunsaturated fats spoil most readily because their double bonds are unstable; monounsaturated fats are slightly less susceptible. Saturated fats are most resistant to oxidation and thus least likely to become rancid.

Manufacturers can protect fat-containing products against rancidity in three ways—none of which are perfect. First, products may be sealed in air-tight, nonmetallic containers, protected from light, and refrigerated—an expensive and inconvenient storage system. Second, manufacturers may add **antioxidants** to compete for the oxygen and thus protect the oil (examples are the additives BHA and BHT and vitamin E).* The advantages and disadvantages of antioxidant additives in food processing are presented in Chapter 19. Third, products may undergo a process known as hydrogenation.

Hydrogenation During **hydrogenation**, some or all of the points of unsaturation are saturated by adding hydrogen molecules. Hydrogenation offers two advantages. First, it protects against oxidation (thereby prolonging shelf life) by making polyunsaturated fats more saturated. Second, it alters the texture of foods

*BHA is butylated hydroxyanisole; BHT is butylated hydroxytoluene.

fats: lipids that are solid at room temperature (77°F or 25°C).

oils: lipids that are liquid at room temperature (77°F or 25°C).

oxidation (OKS-ee-day-shun): the process of a substance combining with oxygen; oxidation reactions involve the loss of electrons.

antioxidants: as a food additive, preservatives that delay or prevent rancidity of fats in foods and other damage to food caused by oxygen.

hydrogenation (HIGH-dro-jen-AY-shun or high-DROJ-eh-NAY-shun): a chemical process by which hydrogens are added to monounsaturated or polyunsaturated fatty acids to reduce the number of double bonds, making the fats more saturated (solid) and more resistant to oxidation (protecting against rancidity). Hydrogenation produces *trans*-fatty acids.

> **FIGURE 5-6 Hydrogenation**

Double bonds carry a slightly negative charge and readily accept positively charged hydrogen atoms, creating a saturated fatty acid. Most often, fat is *partially* hydrogenated, creating a *trans*-fatty acid (shown in Figure 5-7).

Polyunsaturated fatty acid → Hydrogenated (saturated) fatty acid

by making liquid vegetable oils more solid (as in margarine and shortening). Hydrogenated fats improve the texture of foods, making margarines spreadable, pie crusts flaky, and puddings creamy.

Figure 5-6 illustrates the *total* hydrogenation of a polyunsaturated fatty acid to a saturated fatty acid. Total hydrogenation rarely occurs during food processing. Most often, a fat is *partially* hydrogenated, and some of the double bonds that remain after processing change their configuration from *cis* to *trans*.

***Trans*-Fatty Acids** In nature, most double bonds are *cis*—meaning that the hydrogens next to the double bonds are on the same side of the carbon chain. Only a few fatty acids (notably a small percentage of those found in milk and meat products) naturally occur as ***trans*-fatty acids**—meaning that the hydrogens next to the double bonds are on opposite sides of the carbon chain (see Figure 5-7). In the body, *trans*-fatty acids behave more like saturated fats, increasing blood cholesterol and the risk of heart disease (as a later section describes).[1]

Some research suggests that both naturally occurring and commercially created *trans* fats change blood lipids similarly; other research suggests that the negative effects are specific to only the commercial *trans* fats.[2] In any case, the important distinction is that a relatively small amount of *trans* fat in the diet comes from natural sources.* At current levels of consumption, natural *trans* fats have little, if any, effect on blood lipids. Some naturally occurring *trans* fatty acids, known as **conjugated linoleic acids,** may even have health benefits.[3] Conjugated linoleic acids are not counted as *trans* fats on food labels.

***cis*:** on the near side of; refers to a chemical configuration in which the hydrogen atoms are located on the same side of a double bond.

***trans*:** on the other side of; refers to a chemical configuration in which the hydrogen atoms are located on opposite sides of a double bond.

***trans*-fatty acids:** fatty acids with hydrogens on opposite sides of the double bond.

conjugated linoleic acids: several fatty acids that have the same chemical formula as linoleic acid (18 carbons, two double bonds) but with different configurations (the double bonds occur on adjacent carbons).

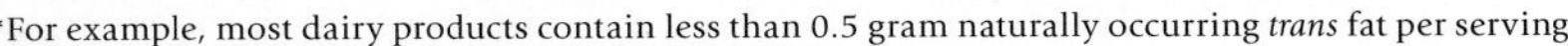

*For example, most dairy products contain less than 0.5 gram naturally occurring *trans* fat per serving.

> **FIGURE 5-7 *Cis*- and *Trans*-Fatty Acids Compared**

This example compares the *cis* configuration for an 18-carbon monounsaturated fatty acid (oleic acid) with its corresponding *trans* configuration (elaidic acid).

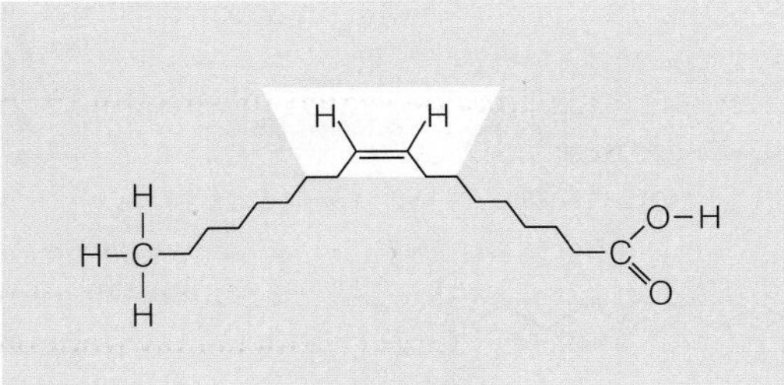

cis-fatty acid

A *cis*-fatty acid has its hydrogens on the same side of the double bond; *cis* molecules bend into a U-like formation. Most naturally occuring unsaturated fatty acids in foods are *cis*.

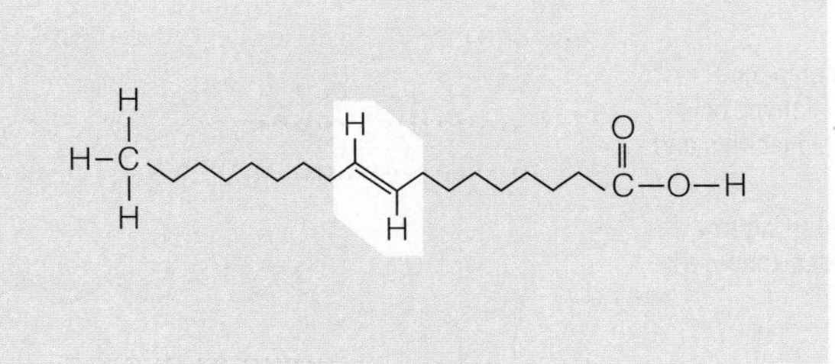

trans-fatty acid

A *trans*-fatty acid has its hydrogens on the opposite sides of the double bond; *trans* molecules are more linear. The *trans* form typically occurs in partially hydrogenated foods when hydrogen atoms shift around some double bonds and change the configuration from *cis* to *trans*.

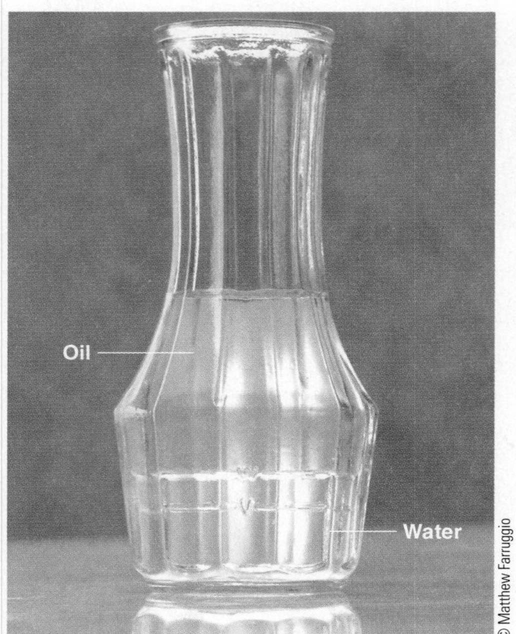

© Matthew Farruggio

Without help from emulsifiers, fats and water don't mix.

REVIEW IT Recognize the chemistry of fatty acids and triglycerides and differences between saturated and unsaturated fats.

The predominant lipids both in foods and in the body are triglycerides: a molecule of glycerol with three fatty acids attached. Fatty acids vary in the length of their carbon chains, their degrees of unsaturation (number of double bonds), and the location of their double bond(s). Those that are fully loaded with hydrogens are saturated; those that are missing hydrogens and therefore have double bonds are unsaturated (monounsaturated or polyunsaturated). The vast majority of triglycerides contain more than one type of fatty acid. Fatty acid saturation affects fats' physical characteristics and storage properties. Hydrogenation, which converts polyunsaturated fats to saturated fats, protects fats from oxidation and alters the texture by making liquid vegetable oils more solid. In the process, hydrogenation creates *trans*-fatty acids that damage health in ways similar to those of saturated fatty acids.

5.2 The Chemist's View of Phospholipids and Sterols

LEARN IT Describe the chemistry, food sources, and roles of phospholipids and sterols.

The preceding pages have been devoted to one of the classes of lipids, the triglycerides, and their component parts, glycerol and the fatty acids. The other lipids, the phospholipids and sterols, make up only 5 percent of the lipids in the diet.

Phospholipids

The best-known **phospholipid** is **lecithin** (see Figure 5-8). Notice that lecithin has one glycerol with two of its three attachment sites occupied by fatty acids like those in triglycerides. The third site is occupied by a phosphate group and a molecule of **choline.** The **hydrophobic** fatty acids make phospholipids soluble in fat; the **hydrophilic** phosphate group allows them to dissolve in water. Such versatility enables the food industry to use phospholipids as **emulsifiers** to mix fats with water in such products as mayonnaise, salad dressings, and candy bars.

Phospholipids in Foods In addition to the phospholipids used by the food industry as emulsifiers, phospholipids are also found naturally in foods. The richest food sources of lecithin are eggs, liver, soybeans, wheat germ, and peanuts.

Roles of Phospholipids Lecithin and other phospholipids are constituents of cell membranes (see Figure 5-9). Because phospholipids are soluble in both water and fat, they can help fat-soluble substances, including vitamins and hormones, to pass

> FIGURE 5-9 Phospholipids of a Cell Membrane

A cell membrane is made of phospholipids assembled into an orderly formation called a bilayer. The fatty acid "tails" orient themselves away from the watery fluid inside and outside of the cell. The glycerol and phosphate "heads" are attracted to the watery fluid.

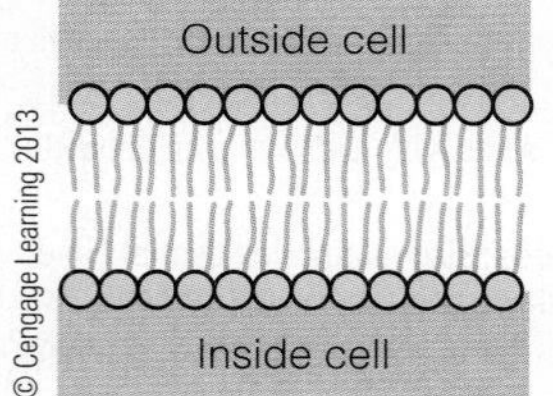

Watery fluid
Glycerol heads
Fatty acid tails
Watery fluid

© Cengage Learning 2013

phospholipid (FOS-foe-LIP-id): a compound similar to a triglyceride but having a phosphate group (a phosphorus-containing salt) and choline (or another nitrogen-containing compound) in place of one of the fatty acids.

lecithin (LESS-uh-thin): one of the phospholipids. Both nature and the food industry use lecithin as an emulsifier to combine water-soluble and fat-soluble ingredients that do not ordinarily mix, such as water and oil.

choline (KOH-leen): a nitrogen-containing compound found in foods and made in the body from the amino acid methionine. Choline is part of the phospholipid lecithin and the neurotransmitter acetylcholine.

hydrophobic (high-dro-FOE-bick): a term referring to water-fearing, or non-water-soluble, substances; also known as *lipophilic* (fat loving).

- **hydro** = water
- **phobia** = fear
- **lipo** = lipid
- **phile** = love

hydrophilic (high-dro-FIL-ick): a term referring to water-loving, or water-soluble, substances.

emulsifiers: substances with both water-soluble and fat-soluble portions that promote the mixing of oils and fats in watery solutions.

> FIGURE 5-8 Lecithin

Lecithin is one of the phospholipids. Notice that a molecule of lecithin is similar to a triglyceride but contains only two fatty acids. The third position is occupied by a phosphate group and a molecule of choline. Other phospholipids have different fatty acids at the upper two positions and different groups attached to phosphate.

From 2 fatty acids

The plus charge on the N is balanced by a negative ion—usually chloride.

From choline

From glycerol

From phosphate

© Cengage Learning 2013

easily in and out of cells. Phospholipids also act as emulsifiers in the body, helping to keep fats suspended in the blood and body fluids.

Sterols In addition to triglycerides and phospholipids, the lipids include the **sterols,** compounds with a multiple-ring structure.* The most famous sterol is **cholesterol;** Figure 5-10 shows its chemical structure.

Sterols in Foods Foods derived from both plants and animals contain sterols, but only those from animals contain significant amounts of cholesterol—meats, eggs, seafood, poultry, and dairy products. Some people, confused about the distinction between dietary cholesterol and blood cholesterol, have asked which foods contain the "good" cholesterol. "Good" cholesterol is not a type of cholesterol found in foods, but it refers to the way the body transports cholesterol in the blood, as explained in a later section of this chapter.

Sterols other than cholesterol are naturally found in plants. Being structurally similar to cholesterol, plant sterols interfere with cholesterol absorption. By inhibiting cholesterol absorption, a diet rich in plant sterols lowers blood cholesterol levels.[4] Food manufacturers have fortified foods such as margarine with plant sterols, creating a functional food that helps to reduce blood cholesterol.

Roles of Sterols Many vitally important body compounds are sterols. Among them are bile acids, the sex hormones (such as testosterone), the adrenal hormones (such as cortisol), and vitamin D, as well as cholesterol itself. Cholesterol in the body can serve as the starting material for the synthesis of these compounds ♦ or as a structural component of cell membranes; more than 90 percent of all the body's cholesterol is found in the cells. Despite popular impressions, cholesterol is not a villain lurking in some evil foods—it is a compound the body makes and uses. The chemical structure is the same, but cholesterol that is made in the body is called **endogenous,** whereas cholesterol from outside the body (from foods) is called **exogenous.** Right now, as you read, your liver is manufacturing cholesterol from fragments of carbohydrate, protein, and fat. In fact, the liver makes about 800 to 1500 milligrams of cholesterol per day, thus contributing much more to the body's total than does the diet. For perspective, the Daily Value on food labels for cholesterol is 300 milligrams per day.

Cholesterol's harmful effects in the body occur when it accumulates in the artery walls and contributes to the formation of **plaque.** These plaque deposits lead to **atherosclerosis,** a disease that causes heart attacks and strokes. Chapter 18 provides many more details.

REVIEW IT Describe the chemistry, food sources, and roles of phospholipids and sterols.

Phospholipids, including lecithin, have a unique chemical structure that allows them to be soluble in both water and fat. The food industry uses phospholipids as emulsifiers, and in the body, phospholipids are part of cell membranes. Sterols have a multiple-ring structure that differs from the structure of other lipids. In the body, sterols include cholesterol, bile, vitamin D, and some hormones. Animal-derived foods are rich sources of cholesterol. Table 5-2 (p. 138) summarizes the lipid family of compounds.

5.3 Digestion, Absorption, and Transport of Lipids

LEARN IT Summarize fat digestion, absorption, and transport.

Each day, the GI tract receives, on average from the food we eat, 50 to 100 grams of triglycerides, 4 to 8 grams of phospholipids, and 200 to 350 milligrams of cholesterol. These lipids are hydrophobic, whereas the digestive enzymes are hydrophilic. As you read, notice how the body elegantly meets the challenges of keeping

*The four-ring core structure identifies a steroid; sterols are alcohol derivatives with a steroid ring structure.

> **FIGURE 5-10 Cholesterol**

Notice how different cholesterol is from the triglycerides and phospholipids. The fat-soluble vitamin D is synthesized from cholesterol; notice the many structural similarities. The only difference is that cholesterol has a closed ring (highlighted in red), whereas vitamin D's is open, accounting for its vitamin activity.

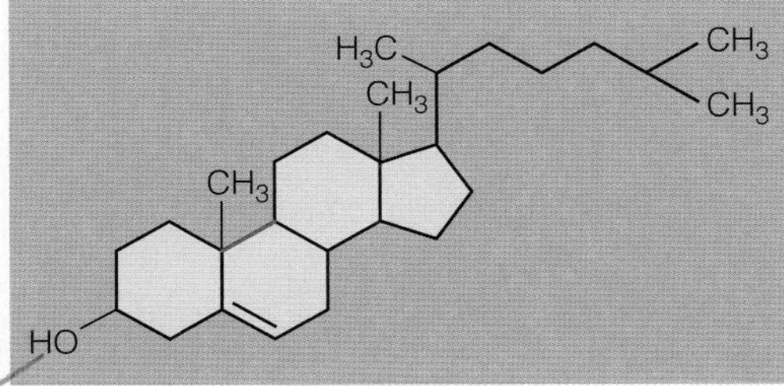

Cholesterol

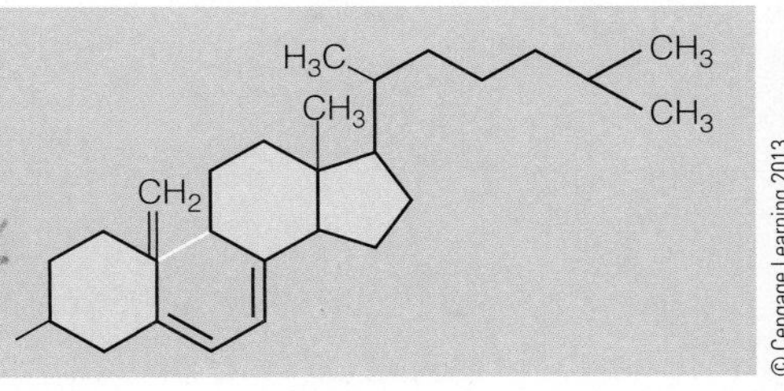

Vitamin D_3

♦ Compounds made from cholesterol:

- Bile acids
- Steroid hormones (testosterone, androgens, estrogens, progesterones, cortisol, cortisone, and aldosterone)
- Vitamin D

sterols (STARE-ols or STEER-ols): compounds containing a four-ring carbon structure with any of a variety of side chains attached.

cholesterol (koh-LESS-ter-ol): one of the sterols containing a four-ring carbon structure with a carbon side chain.

endogenous (en-DODGE-eh-nus): from within the body.

- **endo** = within
- **gen** = arising

exogenous (eks-ODGE-eh-nus): from outside the body.

- **exo** = outside

plaque (PLACK): an accumulation of fatty deposits, smooth muscle cells, and fibrous connective tissue that develops in the artery walls in atherosclerosis. Plaque associated with atherosclerosis is known as *atheromatous* (ATH-er-OH-ma-tus) *plaque.*

atherosclerosis (ATH-er-oh-scler-OH-sis): a type of artery disease characterized by plaques (accumulations of lipid-containing material) on the inner walls of the arteries (see Chapter 18).

TABLE 5-2 The Lipid Family

Triglycerides ■ 1 Glycerol (per triglyceride) and ■ 3 Fatty acids (per triglyceride); depending on the number of double bonds, fatty acids may be: • *Saturated* (no double bonds) • *Monounsaturated* (one double bond) • *Polyunsaturated* (more than one double bond); depending on the location of the double bonds, polyunsaturated fatty acids may be: ◆ *Omega-3* (double bond closest to methyl end is 3 carbons away) ◆ *Omega-6* (double bond closest to methyl end is 6 carbons away)
Phospholipids (such as lecithin)
Sterols (such as cholesterol)

the lipids mixed in the watery fluids of the GI tract and facilitating the work of the **lipases.**

Lipid Digestion

Figure 5-11 traces the digestion of fat through the GI tract. The goal of fat digestion is to dismantle triglycerides into small molecules that the body can absorb and use—namely, **monoglycerides,** fatty acids, and glycerol. The following paragraphs provide the details.

In the Mouth Fat digestion starts off slowly in the mouth, with some hard fats beginning to melt when they reach body temperature. A salivary gland at the base of the tongue releases an enzyme (lingual lipase) that plays an active role in fat digestion in infants, but a relatively minor role in adults. In infants, this enzyme efficiently digests the short- and medium-chain fatty acids found in milk.

In the Stomach In a quiet stomach, fat would float as a layer above the watery components of swallowed food. But whenever food is present, the stomach becomes active. The strong muscle contractions of the stomach propel its contents toward the pyloric sphincter. Some chyme passes through the pyloric sphincter periodically, but the remaining partially digested food is propelled back into the body of the stomach. This churning grinds the solid pieces to finer particles, mixes the chyme, and disperses the fat into small droplets. These actions help to expose the fat for attack by the gastric lipase enzyme—an enzyme that performs best in the acidic environment of the stomach. Still, little fat digestion takes place in the stomach; most of the action occurs in the small intestine.

In the Small Intestine When fat enters the small intestine, it triggers the release of the hormone cholecystokinin (CCK), which signals the gallbladder to release its stores of bile. (Remember that the liver makes bile, and the gallbladder stores it until it is needed.) Among bile's many ingredients are bile acids, which are made in the liver from cholesterol and have a similar structure. In addition, bile acids often pair up with an amino acid (a building block of protein). The amino acid end is hydrophilic, and the sterol end is hydrophobic. This structure enables bile to act as an emulsifier, drawing fat molecules into the surrounding watery fluids. There, the fats are fully digested as they encounter lipase enzymes from the pancreas and small intestine. The process of emulsification is diagrammed in Figure 5-12 (p. 140).

Most of the hydrolysis of triglycerides occurs in the small intestine. The major fat-digesting enzymes are pancreatic lipases; some intestinal lipases are also active. These enzymes remove each of a triglyceride's outer fatty acids one at a

lipases: enzymes that hydrolyze lipids. *Lingual lipase* refers to the fat-digesting enzyme secreted from the salivary gland at the base of the tongue.

monoglycerides: molecules of glycerol with one fatty acid attached. A molecule of glycerol with two fatty acids attached is a *diglyceride.*
- **mono** = one
- **di** = two

> **FIGURE 5-11 Fat Digestion in the GI Tract**

FAT

Mouth and salivary glands
Some hard fats begin to melt as they reach body temperature. The sublingual salivary gland in the base of the tongue secretes lingual lipase. The degree of hydrolysis by lingual lipase is slight for most fats but may be appreciable for milk fats.

Stomach
The stomach's churning action mixes fat with water and acid. A gastric lipase accesses and hydrolyzes (only a very small amount of) fat.

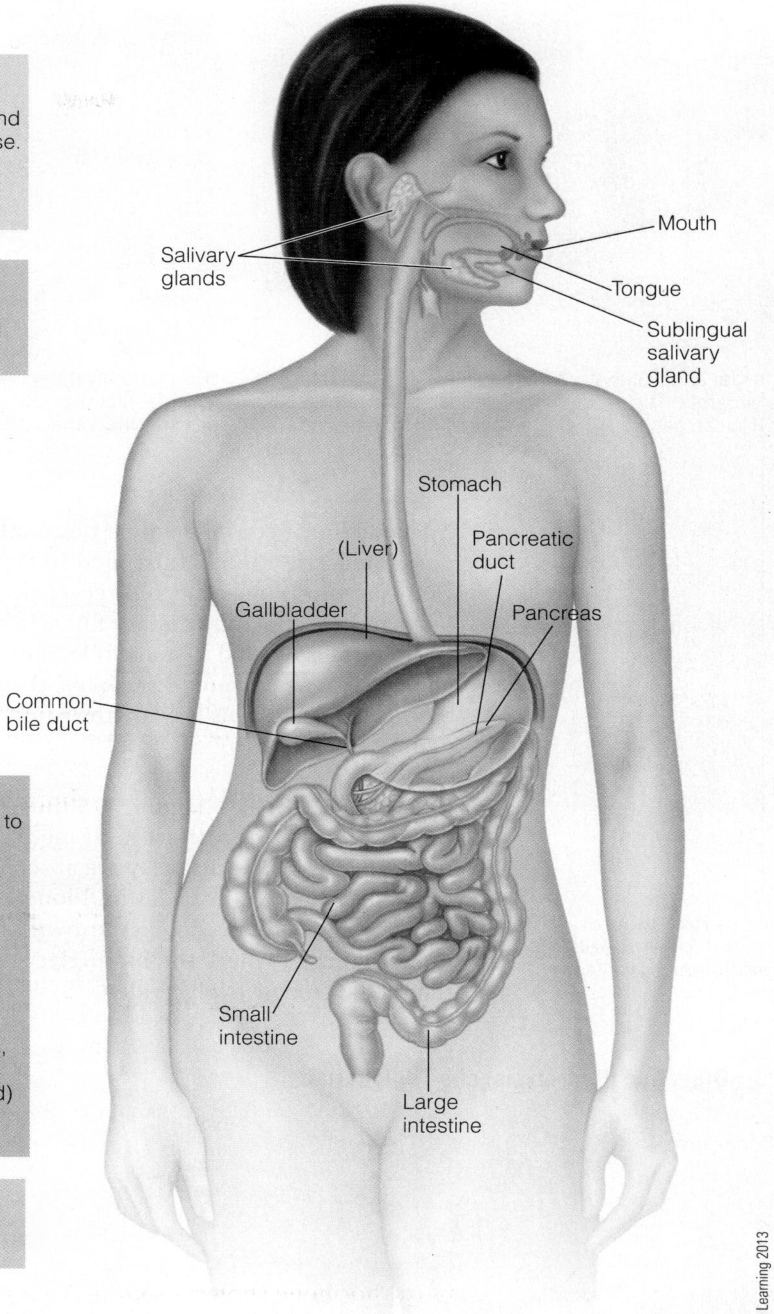

Small intestine
Cholecystokinin (CCK) signals the gallbladder to release bile (via the common bile duct):

Fat —Bile→ Emulsified fat

Pancreatic lipase flows in from the pancreas (via the pancreatic duct):

Emulsified fat (triglycerides) —Pancreatic (and intestinal) lipase→ Monoglycerides, glycerol, fatty acids (absorbed)

Large intestine
Some fat and cholesterol, trapped in fiber, exit in feces.

time, leaving a monoglyceride. Occasionally, enzymes remove all three fatty acids, leaving a free molecule of glycerol. Hydrolysis of a triglyceride is shown in Figure 5-13 (p. 140).

Phospholipids are digested similarly—that is, their fatty acids are removed by hydrolysis. The two fatty acids and the remaining glycerol and phosphate fragments are then absorbed. Most sterols can be absorbed as is; if any fatty acids are attached, they are first hydrolyzed off.

> **FIGURE 5-12 Emulsification of Fat by Bile**

Like bile, detergents are emulsifiers and work the same way, which is why they are effective in removing grease spots from clothes. Molecule by molecule, the grease is dissolved out of the spot and suspended in the water, where it can be rinsed away.

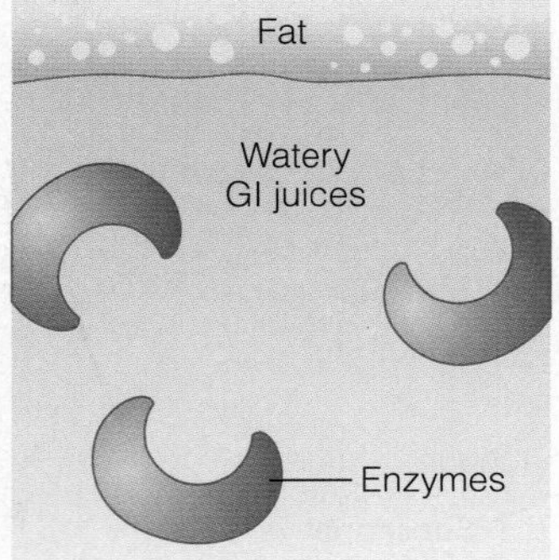

In the stomach, the fat and watery GI juices tend to separate. The enzymes in the GI juices can't get at the fat.

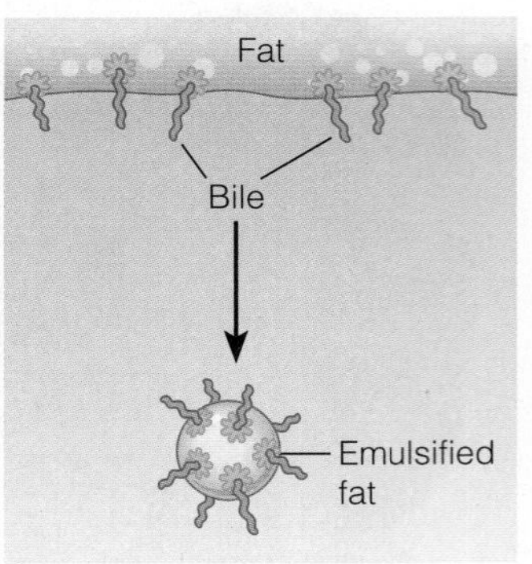

When fat enters the small intestine, the gallbladder secretes bile. Bile has an affinity for both fat and water, so it can bring the fat into the water.

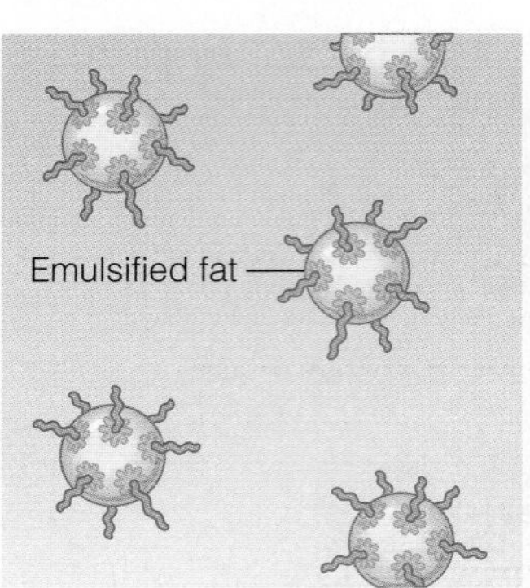

Bile's emulsifying action converts large fat globules into small droplets that repel one another.

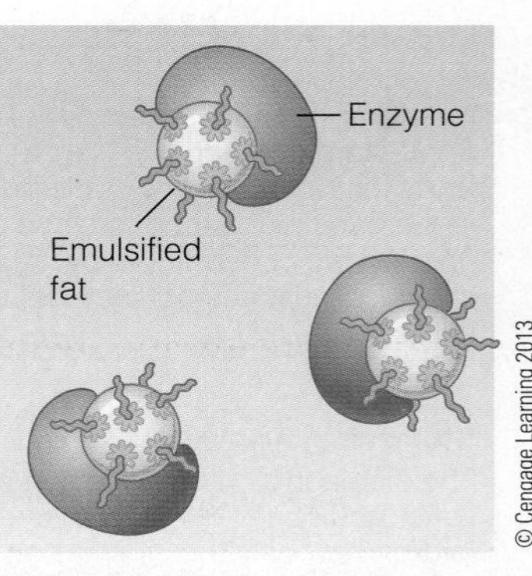

After emulsification, more fat is exposed to the enzymes, making fat digestion more efficient.

Bile's Routes After bile enters the small intestine and emulsifies fat, it has two possible destinations, illustrated in Figure 5-14. Most of the bile is reabsorbed from the small intestine and recycled. The other possibility is that some of the bile can be trapped by dietary fibers in the large intestine and excreted. Because cholesterol is needed to make bile, the excretion of bile effectively reduces blood cholesterol. As Chapter 4 explains, the dietary fibers most effective at lowering blood cholesterol this way are the soluble fibers commonly found in fruits, whole grains, and legumes.

Lipid Absorption

Figure 5-15 illustrates the absorption of lipids. Small molecules (glycerol and short- and medium-chain fatty acids) can diffuse easily into the intestinal cells; they are absorbed directly into the bloodstream. Larger molecules (monoglycerides and long-chain fatty acids) are emulsified by bile, forming spherical complexes known as **micelles**. The micelles diffuse into the intestinal cells where the monoglycerides and long-chain fatty acids are reassembled into new triglycerides.

micelles (MY-cells): tiny spherical complexes of emulsified fat that arise during digestion; most contain bile salts and the products of lipid digestion, including fatty acids, monoglycerides, and cholesterol.

> **FIGURE 5-13 Digestion (Hydrolysis) of a Triglyceride**

Bonds break

Bonds break

Triglyceride

The triglyceride and two molecules of water are split. The H and OH from water complete the structures of two fatty acids and leave a monoglyceride.

Monoglyceride + two fatty acids

These products may pass into the intestinal cells, but sometimes the monoglyceride is split with another molecule of water to give a third fatty acid and glycerol. Fatty acids, monoglycerides, and glycerol are absorbed into intestinal cells.

Within the intestinal cells, the newly made triglycerides and other lipids (cholesterol and phospholipids) are packed with protein into transport vehicles known as chylomicrons. The intestinal cells then release the chylomicrons into the lymphatic system. The chylomicrons glide through the lymph until they reach a point of entry into the bloodstream at the thoracic duct near the heart. (Recall from Chapter 3 that nutrients from the GI tract that enter the lymph system initially bypass the liver.) The blood carries these lipids to the rest of the body for immediate use or storage. A look at these lipids in the body reveals the kinds of fat the diet has been delivering. The blood, fat stores, and muscle cells of people who eat a diet rich in unsaturated fats, for example, contain more unsaturated fats than those of people who select a diet high in saturated fats.

Lipid Transport The chylomicrons are only one of several clusters of lipids and proteins that are used as transport vehicles for fats. As a group, these vehicles are known as **lipoproteins**, and they solve the body's challenge of transporting fat through the watery bloodstream. The body makes four main types of lipoproteins, distinguished by their size and density.* Each type contains different kinds and amounts of lipids and proteins. The more lipids, the less dense; the more proteins, the more dense. Figure 5-16 (p. 142) shows the relative compositions and sizes of the lipoproteins.

Chylomicrons The **chylomicrons** are the largest and least dense of the lipoproteins. They transport *diet*-derived lipids (mostly triglycerides) from the

*Chemists can identify the various lipoproteins by their density. They place a blood sample below a thick fluid in a test tube and spin the tube in a centrifuge. The most buoyant particles (highest in lipids) rise to the top and have the lowest density; the densest particles (highest in proteins) remain at the bottom and have the highest density. Others distribute themselves in between.

> **FIGURE 5-14 Enterohepatic Circulation**

Most of the bile released into the small intestine is reabsorbed and sent back to the liver to be reused. This cycle is called the *enterohepatic circulation* of bile. Some bile is excreted.

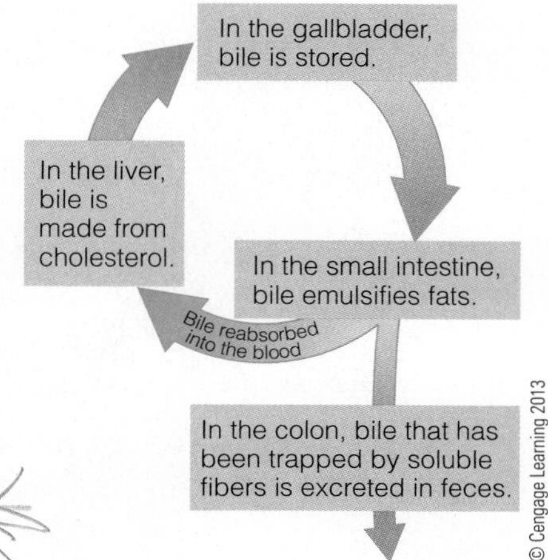

In the gallbladder, bile is stored.
In the liver, bile is made from cholesterol.
In the small intestine, bile emulsifies fats.
Bile reabsorbed into the blood
In the colon, bile that has been trapped by soluble fibers is excreted in feces.

lipoproteins (LIP-oh-PRO-teenz): clusters of lipids associated with proteins that serve as transport vehicles for lipids in the lymph and blood.

chylomicrons (kye-lo-MY-cronz): the class of lipoproteins that transport lipids from the intestinal cells to the rest of the body.

> **FIGURE 5-15** ***Animated*** **Absorption of Fat**

The end products of fat digestion are mostly monoglycerides, some fatty acids, and very little glycerol. Their absorption differs depending on their size. (In reality, molecules of fatty acid are too small to see without a powerful microscope, whereas villi are visible to the naked eye.)

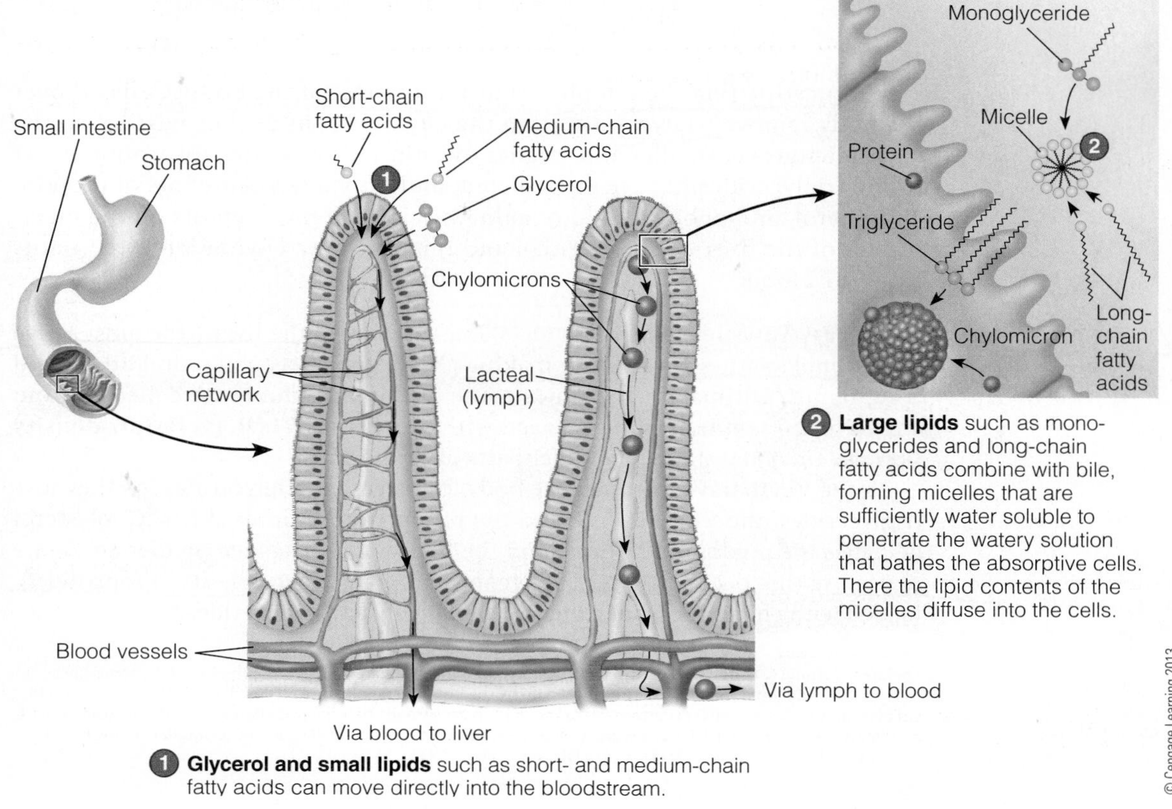

② **Large lipids** such as monoglycerides and long-chain fatty acids combine with bile, forming micelles that are sufficiently water soluble to penetrate the watery solution that bathes the absorptive cells. There the lipid contents of the micelles diffuse into the cells.

① **Glycerol and small lipids** such as short- and medium-chain fatty acids can move directly into the bloodstream.

> FIGURE 5-16 **Sizes and Compositions of the Lipoproteins**

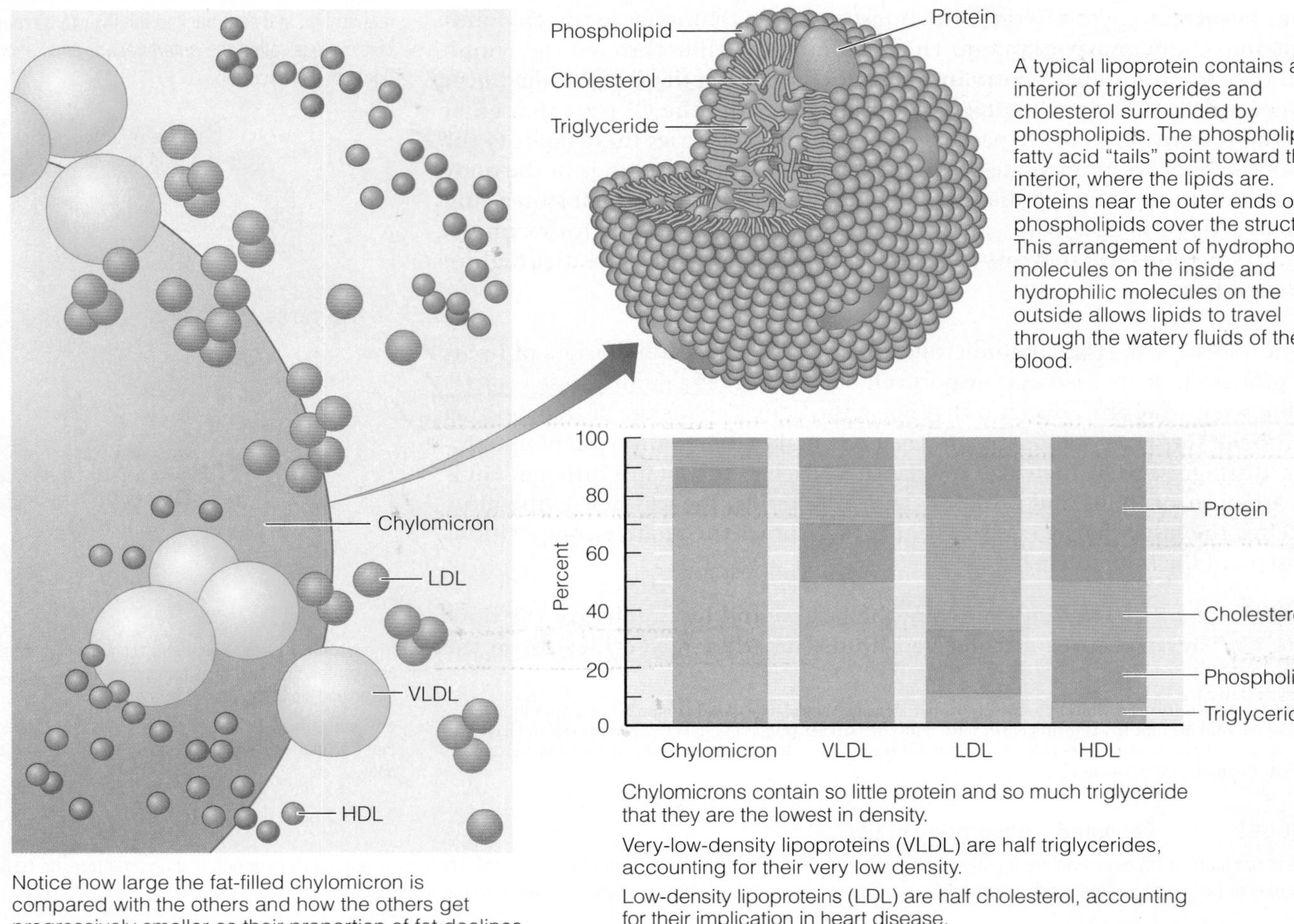

Phospholipid
Protein
Cholesterol
Triglyceride
A typical lipoprotein contains an interior of triglycerides and cholesterol surrounded by phospholipids. The phospholipids' fatty acid "tails" point toward the interior, where the lipids are. Proteins near the outer ends of the phospholipids cover the structure. This arrangement of hydrophobic molecules on the inside and hydrophilic molecules on the outside allows lipids to travel through the watery fluids of the blood.
Chylomicron
LDL
VLDL
HDL
Percent
100
80
60
40
20
0
Chylomicron
VLDL
LDL
HDL
Protein
Cholesterol
Phospholipid
Triglyceride

Notice how large the fat-filled chylomicron is compared with the others and how the others get progressively smaller as their proportion of fat declines and protein increases.

Chylomicrons contain so little protein and so much triglyceride that they are the lowest in density.

Very-low-density lipoproteins (VLDL) are half triglycerides, accounting for their very low density.

Low-density lipoproteins (LDL) are half cholesterol, accounting for their implication in heart disease.

High-density lipoproteins (HDL) are half protein, accounting for their high density.

small intestine (via the lymph system) to the rest of the body. Cells all over the body remove triglycerides from the chylomicrons as they pass by, so the chylomicrons get smaller and smaller. Within 14 hours after absorption, most of the triglycerides have been depleted, and only a few remnants of protein, cholesterol, and phospholipid remain. Special protein receptors on the membranes of the liver cells recognize and remove these chylomicron remnants from the blood.

VLDL (Very-Low-Density Lipoproteins) Meanwhile, in the liver—the most active site of lipid synthesis—cells are making cholesterol, fatty acids, and other lipid compounds. Ultimately, the lipids made in the liver and those collected from chylomicron remnants are packaged with proteins as **VLDL (very-low-density lipoproteins)** and shipped to other parts of the body.

As the VLDL travel through the body, cells remove triglycerides. As they lose triglycerides, the VLDL shrink and the proportion of lipids shifts. Cholesterol becomes the predominant lipid, and the lipoprotein becomes smaller and more dense. As this occurs, the VLDL becomes an **LDL (low-density lipoprotein)**, loaded with cholesterol, but containing relatively few triglycerides.*

VLDL (very-low-density lipoprotein): the type of lipoprotein made primarily by liver cells to transport lipids to various tissues in the body; composed primarily of triglycerides.

LDL (low-density lipoprotein): the type of lipoprotein derived from very-low-density lipoproteins (VLDL) as triglycerides are removed and broken down; composed primarily of cholesterol.

*Before becoming LDL, the VLDL are first transformed into intermediate-density lipoproteins (IDL), sometimes called VLDL remnants. Some IDL may be picked up by the liver and rapidly broken down; those IDL that remain in circulation continue to deliver triglycerides to the cells and eventually become LDL. Researchers debate whether IDL are simply transitional particles or a separate class of lipoproteins; normally, IDL do not accumulate in the blood. Measures of blood lipids include IDL with LDL.

LDL (Low-Density Lipoproteins) The LDL circulate throughout the body, making their contents available to the cells of all tissues—muscles (including the heart muscle), fat stores, the mammary glands, and others. The cells take triglycerides, cholesterol, and phospholipids to use for energy, make hormones or other compounds, or build new membranes. Special LDL receptors on the liver cells play a crucial role in the control of blood cholesterol concentrations by removing LDL from circulation.

HDL (High-Density Lipoproteins) The liver makes **HDL (high-density lipoprotein)** to remove cholesterol from the cells and carry it back to the liver for recycling or disposal. By efficiently clearing cholesterol, HDL lowers the risk of heart disease.[5] In addition, HDL have anti-inflammatory properties that seem to keep artery-clogging plaque from breaking apart and causing heart attacks. Figure 5-17 summarizes lipid transport via the lipoproteins.

Health Implications The distinction between LDL and HDL has implications for the health of the heart and blood vessels. The blood lipid linked most directly to heart disease is LDL cholesterol. As mentioned, HDL also carry cholesterol, but elevated HDL represent cholesterol returning from the rest of the body to the liver for breakdown and excretion. The transport of cholesterol from the tissues back to the liver is sometimes called *reverse cholesterol transport* or the *scavenger pathway*.

High LDL cholesterol is associated with a high risk of heart attack, whereas high HDL cholesterol seems to have a protective effect. ♦ This explains why some people refer to LDL as "bad," and HDL as "good," cholesterol. Keep in mind that the cholesterol itself is the same and that the differences between LDL and HDL

♦ Think of **H**DL as **H**ealthy and **L**DL as **L**ess healthy.

HDL (high-density lipoprotein): the type of lipoprotein that transports cholesterol back to the liver from the cells; composed primarily of protein.

> **FIGURE 5-17 Lipid Transport via Lipoproteins**

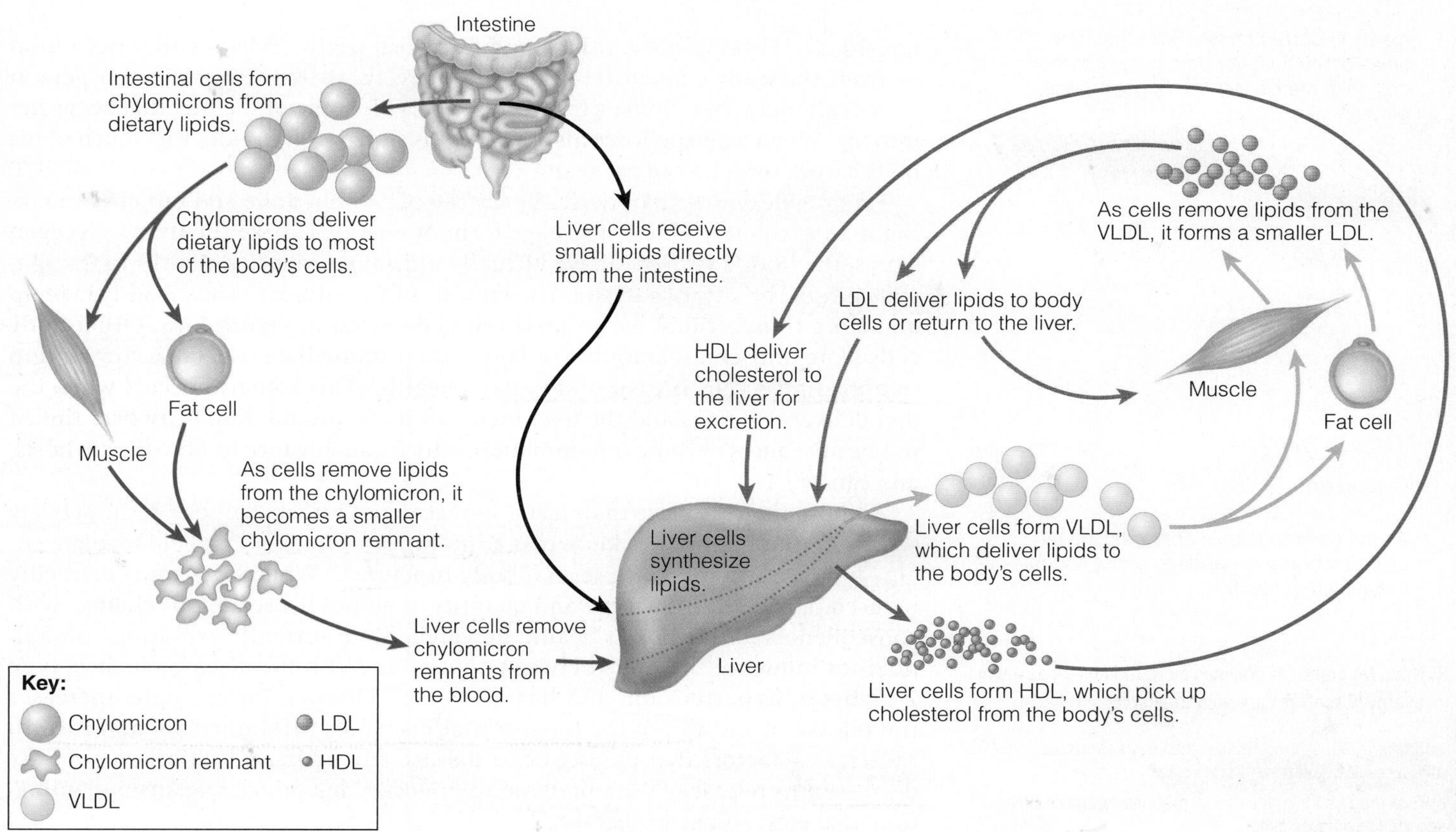
Intestine
Intestinal cells form chylomicrons from dietary lipids.
Chylomicrons deliver dietary lipids to most of the body's cells.
Liver cells receive small lipids directly from the intestine.
Muscle
Fat cell
As cells remove lipids from the chylomicron, it becomes a smaller chylomicron remnant.
Liver cells remove chylomicron remnants from the blood.
Liver cells synthesize lipids.
Liver
HDL deliver cholesterol to the liver for excretion.
LDL deliver lipids to body cells or return to the liver.
As cells remove lipids from the VLDL, it forms a smaller LDL.
Muscle
Fat cell
Liver cells form VLDL, which deliver lipids to the body's cells.
Liver cells form HDL, which pick up cholesterol from the body's cells.
Key:
Chylomicron
Chylomicron remnant
VLDL
LDL
HDL

◆ Factors that lower LDL and/or raise HDL:

- Weight control
- Monounsaturated or polyunsaturated, instead of saturated, fat in the diet
- Soluble dietary fibers (see Chapter 4)
- Phytochemicals (see Highlight 13)
- *Moderate* alcohol consumption
- Physical activity

reflect the *proportions* and *types* of lipids and proteins within them—not the type of cholesterol. The margin ◆ lists factors that influence LDL and HDL, and Chapter 18 provides many more details.

Not too surprisingly, numerous genes influence how the body handles the synthesis, transport, and degradation of lipids and lipoproteins. Much current research is focused on how nutrient-gene interactions may direct the progression of heart disease.

REVIEW IT Summarize fat digestion, absorption, and transport.
The body makes special arrangements to digest and absorb lipids. It provides the emulsifier bile to make them accessible to the fat-digesting lipases that dismantle triglycerides, mostly to monoglycerides and fatty acids, for absorption by the intestinal cells. The intestinal cells assemble absorbed lipids into chylomicrons, lipid packages with protein escorts, for transport so that cells all over the body may select needed lipids from them. Lipoproteins transport lipids around the body. All four types of lipoproteins carry all classes of lipids (triglycerides, phospholipids, and cholesterol), but the chylomicrons are the largest and contain mostly triglycerides from the diet; VLDL are smaller and are about half triglycerides; LDL are smaller still and contain mostly cholesterol; and HDL are the densest and are rich in protein. High LDL cholesterol indicates increased risk of heart disease, whereas high HDL cholesterol has a protective effect.

5.4 Lipids in the Body

LEARN IT Outline the major roles of fats in the body, including a discussion of essential fatty acids and the omega fatty acids.

In the body, lipids provide energy, insulate against temperature extremes, protect against shock, and maintain cell membranes. This section provides an overview of the roles of triglycerides and fatty acids and then of the metabolic pathways they can follow within the body's cells.

Roles of Triglycerides First and foremost, triglycerides—either from food or from the body's fat stores—provide the cells with energy. When a person dances all night, her dinner's triglycerides provide some of the fuel that keeps her moving. When a person loses his appetite, his stored triglycerides fuel much of his body's work until he can eat again.

Fat provides more than twice the energy of carbohydrate and protein,◆ making it an extremely efficient storage form of energy. Unlike the liver's glycogen stores, the body's fat stores have virtually unlimited capacity, thanks to the special cells of the **adipose tissue.** The fat cells of the adipose tissue readily take up and store triglycerides. An adipose cell is depicted in Figure 5-18. Other body cells store only small amounts of fat for their immediate use; fat accumulation in nonadipose cells is toxic and impairs health.[6] This scenario occurs when the diet delivers excesses and the liver increases its fat production. Fatty liver linked to obesity causes chronic inflammation, which can advance to fibrosis, cirrhosis, and cancer.[7]

> **FIGURE 5-18 An Adipose Cell**
Newly imported triglycerides first form small droplets at the periphery of the cell, then merge with the large, central globule.

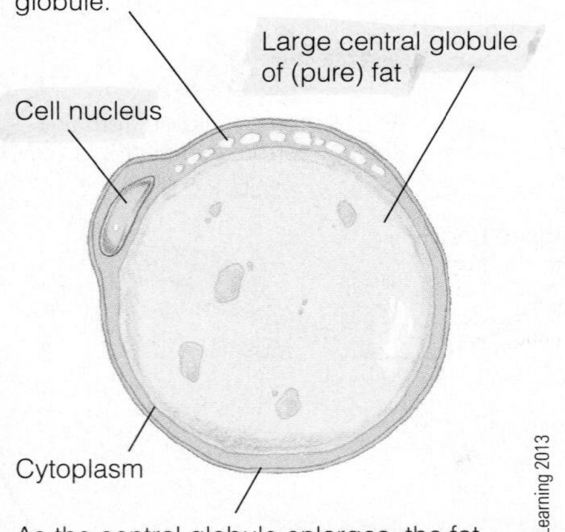
Newly imported triglycerides first form small droplets at the periphery of the cell, then merge with the large, central globule.
Large central globule of (pure) fat
Cell nucleus
Cytoplasm
As the central globule enlarges, the fat cell membrane expands to accommodate its swollen contents.
© Cengage Learning 2013

Adipose tissue is more than just a storage depot for fat. Adipose tissue actively secretes several hormones known as **adipokines**—proteins that help regulate energy balance and influence several body functions.[8] When body fat is markedly reduced or excessive, the type and quantity of adipokine secretions change, with consequences for the body's health. Researchers are currently exploring how adipokines influence the links between obesity and chronic diseases such as type 2 diabetes, hypertension, and heart disease.[9] Obesity, for example, increases the release of the adipokine resistin that promotes inflammation and insulin resistance—factors that predict heart disease and diabetes.[10] Similarly, obesity decreases the release of the adipokine adiponectin that protects against inflammation, diabetes, and heart disease.

◆ Gram for gram, fat provides more than twice as much energy (9 kcal) as carbohydrate or protein (4 kcal).

adipose (ADD-ih-poce) **tissue:** the body's fat tissue; consists of masses of triglyceride-storing cells.

adipokines (ADD-ih-poe-kines): proteins synthesized and secreted by adipose cells.

Fat serves other roles in the body as well. Because fat is a poor conductor of heat, the layer of fat beneath the skin insulates the body from temperature extremes. Fat pads also serve as natural shock absorbers, providing a cushion for the bones and vital organs. Fat provides the structural material for cell membranes and participates in cell signaling pathways.

> **FIGURE 5-19 The Pathway from One Omega-6 Fatty Acid to Another**

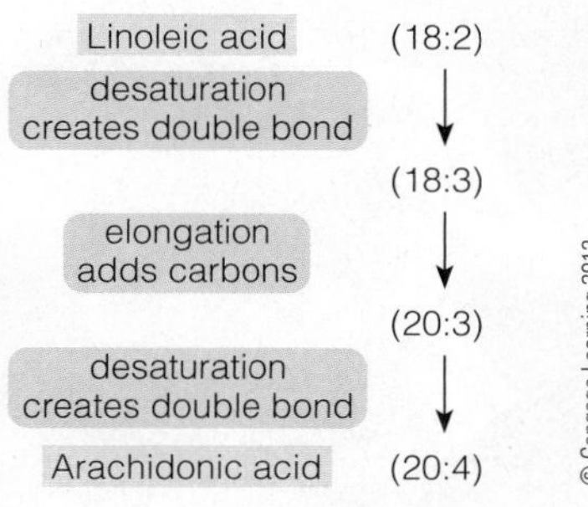

The first number indicates the number of carbons and the second, the number of double bonds. Similar reactions occur when the body makes the omega-3 fatty acids EPA and DHA from linolenic acid.

Essential Fatty Acids The human body needs fatty acids, and it can make all but two of them—linoleic acid (the 18-carbon omega-6 fatty acid) and linolenic acid (the 18-carbon omega-3 fatty acid). These two fatty acids must be supplied by the diet and are therefore **essential fatty acids.** The cells do not possess the enzymes to make any of the omega-6 or omega-3 fatty acids from scratch, nor can they convert an omega-6 fatty acid to an omega-3 fatty acid or vice versa. Cells *can,* however, use the 18-carbon member of an omega family from the diet to make the longer fatty acids of that family by forming double bonds (desaturation) and lengthening the chain two carbons at a time (elongation), as shown in Figure 5-19. This is a slow process because the omega-3 and omega-6 families compete for the same enzymes. Too much of a fatty acid from one family can create a deficiency of the other family's longer fatty acids, which becomes critical only when the diet fails to deliver adequate supplies. Therefore, the most effective way to maintain body supplies of all the omega-6 and omega-3 fatty acids is to obtain them directly from foods—most notably, from vegetable oils, seeds, nuts, fish, and other seafoods.

Linoleic Acid and the Omega-6 Family Linoleic acid is the primary member of the omega-6 fatty acid family. When the body receives linoleic acid from the diet, it can make other members of the omega-6 family—such as the 20-carbon polyunsaturated fatty acid, **arachidonic acid** (as shown in Figure 5-19). Should a linoleic acid deficiency develop, arachidonic acid, and all other omega-6 fatty acids that derive from linoleic acid, would also become essential and have to be obtained from the diet. A nonessential nutrient (such as arachidonic acid) that must be supplied by the diet in special circumstances (as in a linoleic acid deficiency) is considered a **conditionally essential nutrient.** Normally, vegetable oils and meats supply enough omega-6 fatty acids to meet the body's needs.

Linolenic Acid and the Omega-3 Family Linolenic acid is the primary member of the omega-3 fatty acid family.* Like linoleic acid, linolenic acid cannot be made in the body and must be supplied by foods. Given the 18-carbon linolenic acid, the body can make small amounts of the 20- and 22-carbon members of the omega-3 family, **eicosapentaenoic acid (EPA)** and **docosahexaenoic acid (DHA),** respectively. These omega-3 fatty acids are found in the eyes and brain and are essential for normal growth, visual acuity, and cognitive development.[11] They may also play an important role in the prevention and treatment of heart disease, as later sections explain.[12]

Eicosanoids The body uses the longer omega-3 and omega-6 fatty acids to make substances known as **eicosanoids.** Eicosanoids are a diverse group of compounds that are sometimes described as "hormonelike," but they differ from hormones in important ways. For one, hormones are secreted in one location and travel to affect cells all over the body, whereas eicosanoids appear to affect only the cells in which they are made or nearby cells in the same localized environment. For another, hormones elicit the same response from all their target cells, whereas eicosanoids often have different effects on different cells.

The actions of various eicosanoids sometimes oppose one another. For example, one causes muscles to relax and blood vessels to dilate, whereas another causes muscles to contract and blood vessels to constrict. Certain eicosanoids

---

*This omega-3 linolenic acid is known as alpha-linolenic acid and is the fatty acid referred to in this chapter. Another fatty acid, also with 18 carbons and three double bonds, belongs to the omega-6 family and is known as gamma-linolenic acid.

essential fatty acids: fatty acids needed by the body but not made by it in amounts sufficient to meet physiological needs.

arachidonic (a-RACK-ih-DON-ic) **acid:** an omega-6 polyunsaturated fatty acid with 20 carbons and four double bonds; present in small amounts in meat and other animal products and synthesized in the body from linoleic acid.

conditionally essential nutrient: a nutrient that is normally nonessential, but must be supplied by the diet in special circumstances when the need for it exceeds the body's ability to produce it.

eicosapentaenoic (EYE-cossa-PENTA-ee-NO-ick) **acid (EPA):** an omega-3 polyunsaturated fatty acid with 20 carbons and five double bonds; present in fatty fish and synthesized in limited amounts in the body from linolenic acid.

docosahexaenoic (DOE-cossa-HEXA-ee-NO-ick) **acid (DHA):** an omega-3 polyunsaturated fatty acid with 22 carbons and six double bonds; present in fatty fish and synthesized in limited amounts in the body from linolenic acid.

eicosanoids (eye-COSS-uh-noyds): derivatives of 20-carbon fatty acids; biologically active compounds that help to regulate blood pressure, blood clotting, and other body functions. They include *prostaglandins* (PROS-tah-GLAND-ins), *thromboxanes* (throm-BOX-ains), and *leukotrienes* (LOO-ko-TRY-eens).

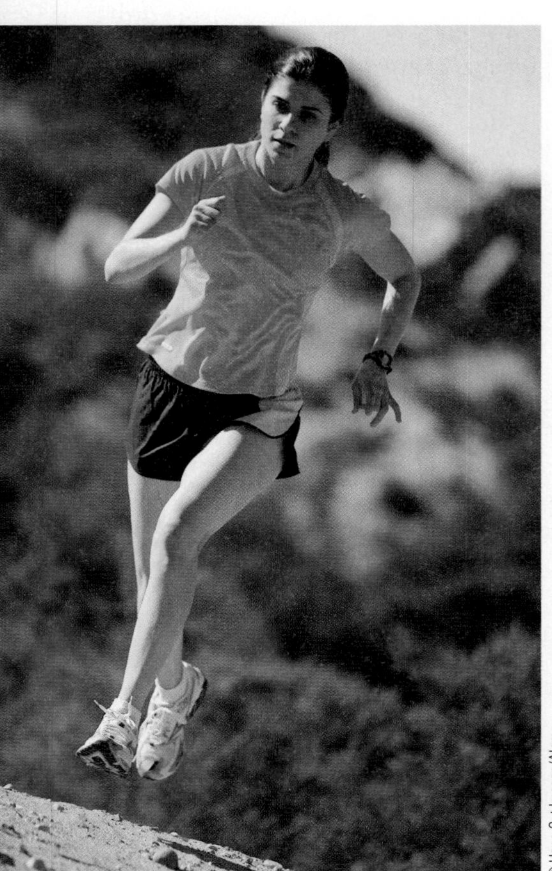

© UpperCut Images/Alamy

Fat supplies most of the energy during a long-distance run.

participate in the immune response to injury and infection, producing fever, inflammation, and pain. One of the ways aspirin relieves these symptoms is by slowing the synthesis of these eicosanoids.

Eicosanoids that derive from omega-3 fatty acids differ from those that derive from omega-6 fatty acids, with the omega-3 family providing greater health benefits. The omega-3 eicosanoids help lower blood pressure, prevent blood clot formation, protect against irregular heartbeats, and reduce inflammation, whereas the omega-6 eicosanoids tend to promote clot formation, inflammation, and blood vessel constriction.[13]

Omega-6 to Omega-3 Ratio Because omega-6 and omega-3 fatty acids compete for the same enzymes and their actions often oppose each other, researchers have studied whether there is an ideal ratio that best supports cardiovascular health. Suggested ratios range from 4:1 to 10:1; while some researchers support such recommendations, others find the ratio of little value in improving health or predicting risk.[14] Increasing the amount of omega-3 fatty acids in the diet is clearly beneficial, but reducing the amount of omega-6 fatty acids in the diet to improve the ratio may not be helpful. Omega-6 fatty acids protect heart health by lowering LDL cholesterol and improving insulin resistance.[15]

Fatty Acid Deficiencies Most diets in the United States and Canada meet the minimum essential fatty acid requirement adequately. Historically, deficiencies have developed only in infants and young children who have been fed fat-free milk and low-fat diets or in hospital clients who have been mistakenly fed formulas that provided no polyunsaturated fatty acids for long periods of time. Classic deficiency symptoms include growth retardation, reproductive failure, skin lesions, kidney and liver disorders, and subtle neurological and visual problems.

A Preview of Lipid Metabolism

This preview of fat metabolism describes how the cells store and release energy from fat. Chapter 7 provides details.

Storing Fat as Fat After meals, the blood delivers chylomicrons and VLDL loaded with triglycerides to the adipose cells for storage. An enzyme—**lipoprotein lipase (LPL)**—hydrolyzes triglycerides from these lipoproteins, releasing fatty acids, diglycerides, and monoglycerides into the adipose cells. Enzymes inside the adipose cells reassemble these fatty acids, diglycerides, and monoglycerides into triglycerides again for storage. As Figure 5-18 (p. 144) shows, triglycerides fill the adipose cells, storing a lot of energy in a relatively small space.

Using Fat for Energy Efficient energy metabolism depends on the energy nutrients—carbohydrate, fat, and protein—supporting one another. Glucose fragments combine with fat fragments during energy metabolism, and fat and carbohydrate help spare protein, providing energy so that protein can be used for other important tasks.

Fat supplies about 60 percent of the body's ongoing energy needs during rest. During prolonged light to moderately intense exercise or extended periods of food deprivation, fat stores may make a slightly greater contribution to energy needs.

During energy deprivation, several lipase enzymes (most notably **hormone-sensitive lipase**) inside the adipose cells respond by dismantling stored triglycerides and releasing the glycerol and fatty acids directly into the blood. Energy-hungry cells anywhere in the body can then capture these compounds and take them through a series of chemical reactions to yield energy, carbon dioxide, and water.

A person who fasts (drinking only water) will rapidly metabolize body fat. A pound of body fat provides 3500 kcalories, ♦ so you might think a fasting person who expends 2000 kcalories a day could lose more than half a pound of body fat each day.* Actually, the person has to obtain some energy from lean tissue because the brain, nerves, and red blood cells need glucose—and only the small glycerol molecule can be converted to glucose; fatty acids cannot be. (As Chapter 7 explains, only 3 of the 50 or

*The reader who knows that 1 pound = 454 grams and that 1 gram of fat = 9 kcalories may wonder why a pound of body fat does not equal 4086 (9 × 454) kcalories. The reason is that body fat contains some cell water and other materials; it is not quite pure fat.

♦ 1 lb body fat = 3500 kcal

lipoprotein lipase (LPL): an enzyme that hydrolyzes triglycerides passing by in the bloodstream and directs their parts into the cells, where they can be metabolized for energy or reassembled for storage.

hormone-sensitive lipase: an enzyme inside adipose cells that responds to the body's need for fuel by hydrolyzing triglycerides so that their parts (glycerol and fatty acids) escape into the general circulation and thus become available to other cells for fuel. The signals to which this enzyme responds include epinephrine and glucagon, which oppose insulin (see Chapter 4).

so carbon atoms in a molecule of fat can yield glucose.) Also, the complete breakdown of fat requires carbohydrate or protein. So, even on a total fast, a person cannot lose more than half a pound of pure fat per day. Still, in times of severe hunger and starvation, a fatter person can survive longer than a thinner person thanks to this energy reserve. But as Chapter 7 explains, fasting for too long will eventually cause death, even if the person still has ample body fat.

REVIEW IT Outline the major roles of fats in the body, including a discussion of essential fatty acids and the omega fatty acids.

In the body, triglycerides provide energy, insulate against temperature extremes, protect against shock, provide structural material for cell membranes, and participate in cell signaling pathways. Linoleic acid (18 carbons, omega-6) and linolenic acid (18 carbons, omega-3) are essential nutrients. They serve as structural parts of cell membranes and as precursors to the longer fatty acids that can make eicosanoids—powerful compounds that participate in blood pressure regulation, blood clot formation, and the immune response to injury and infection, among other functions. Because essential fatty acids are common in the diet and stored in the body, deficiencies are unlikely. The body can easily store unlimited amounts of fat if given excesses, and this body fat is used for energy when needed.

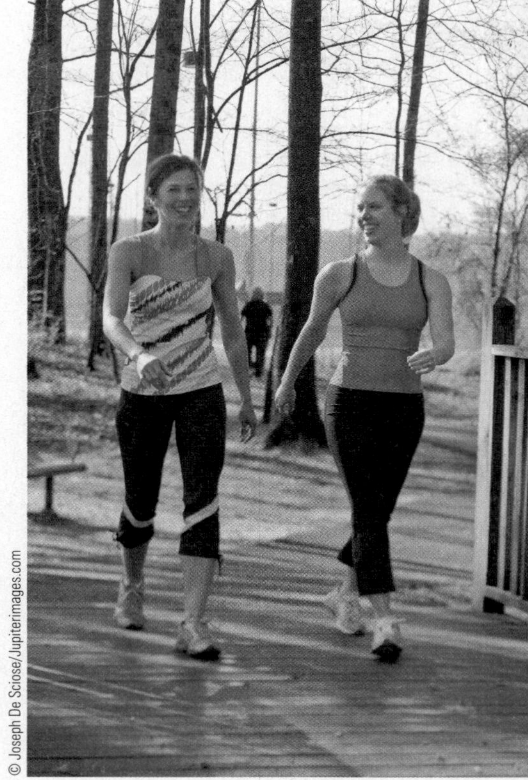
© Joseph De Sciose/Jupiterimages.com

Double thanks: The body's fat stores provide energy for a walk, and fat pads on the heels provide cushion against the hard pavement.

5.5 Health Effects and Recommended Intakes of Saturated Fats, *Trans* Fats, and Cholesterol

LEARN IT Explain the relationships among saturated fats, *trans* fat, and cholesterol and chronic diseases, noting recommendations.

Some fat in the diet is essential for good health. The current American diet, however, delivers excessive amounts of solid fats, representing an average of almost one-fifth of the day's total kcalories. Major sources of solid fats in the American diet include desserts, pizza, cheese, and processed and fatty meats (sausages, hot dogs, bacon, ribs). Because foods made with solid fats provide abundant energy, but few if any essential nutrients, they contribute to weight gain and make it difficult to meet nutrient needs. Solid fats also provide abundant saturated fat, *trans* fat, and cholesterol. Even without overweight or obesity, high intakes of solid fats increase the risk of some chronic diseases. The easiest way to control saturated fat, *trans* fat, cholesterol, and kcalories is to limit solid fats in the diet.

Health Effects of Saturated Fats, *Trans* Fats, and Cholesterol

Hearing a physician say, "Your blood lipid profile looks fine," is reassuring. The **blood lipid profile** ♦ reveals the concentrations of various lipids in the blood, notably triglycerides and cholesterol, and their lipoprotein carriers (VLDL, LDL, and HDL). This information alerts people to possible disease risks and perhaps to a need for changing their physical activity and eating habits. Both the amounts and types of fat in the diet influence the risk for disease.

♦ Desirable blood lipid profile:
- Total cholesterol: <200 mg/dL
- LDL cholesterol: <100 mg/dL
- HDL cholesterol: ≥60 mg/dL
- Triglycerides: <150 mg/dL

Heart Disease As mentioned earlier, elevated LDL cholesterol is a major risk factor for **cardiovascular disease (CVD)**.[16] As LDL cholesterol accumulates in the arteries, blood flow becomes restricted and blood pressure rises. The consequences are deadly; in fact, heart disease is the nation's number-one killer of adults. LDL cholesterol is often used to predict the likelihood of a person's suffering a heart attack or stroke; the higher the LDL, the earlier and more likely the tragedy. Much of the effort to prevent heart disease focuses on lowering LDL cholesterol.

Saturated fats are most often implicated in raising LDL cholesterol. In general, the more saturated fat in the diet, the more LDL cholesterol in the blood. Not all saturated fats have the same cholesterol-raising effect, however. Most notable among the saturated fatty acids that raise blood cholesterol are lauric, myristic, and palmitic acids (12, 14, and 16 carbons, respectively). In contrast, stearic acid (18 carbons) seems to have little or no effect on blood cholesterol.[17] Making such

blood lipid profile: results of blood tests that reveal a person's total cholesterol, triglycerides, and various lipoproteins.

cardiovascular disease (CVD): diseases of the heart and blood vessels throughout the body. Atherosclerosis is the main cause of CVD. When the arteries that carry blood to the heart muscle become blocked, the heart suffers damage known as *coronary heart disease (CHD)*.
- **cardio** = heart
- **vascular** = blood vessels

distinctions may be impractical in diet planning, however, because these saturated fatty acids typically appear together in the same foods. In addition to raising blood cholesterol, saturated fatty acids contribute to heart disease by promoting blood clotting.[18] Fats from animal sources (meats, milk, and milk products) are the main sources of saturated fats ♦ in most people's diets.[19] Selecting lean cuts of meat, skinless poultry, and fat-free milk products helps to lower saturated fat intake and the risk of heart disease.

♦ Major sources of saturated fats:
- Whole milk, cream, butter, cheese, ice cream
- Fatty cuts of beef and pork
- Coconut, palm, and palm kernel oils (the tropical oils and products containing them such as candies, pastries, pies, doughnuts, and cookies)

Research also suggests an association between dietary *trans* fats and heart disease.[20] In the body, *trans* fats alter blood cholesterol the same way some saturated fats do: they raise LDL cholesterol and, at high intakes, lower HDL cholesterol. Limiting the intake of *trans* fats ♦ can improve blood cholesterol and lower the risk of heart disease. To that end, many restaurants and manufacturers have taken steps to eliminate or greatly reduce *trans* fats in foods.[21]

♦ Major sources of *trans* fats:
- Cakes, cookies, doughnuts, pastry, crackers
- Margarine
- Deep-fried foods (vegetable shortening)
- Snack chips

Although its effect is not as strong as that of saturated fat or *trans* fat, dietary cholesterol ♦ may also raise blood cholesterol. Less clear is its role in heart disease.[22]

♦ Major sources of cholesterol:
- Eggs
- Milk and milk products
- Meat, poultry, shellfish

Cancer The links between dietary fats and cancer are not as evident as they are for heart disease. Dietary fat does not seem to *initiate* cancer development but, instead, may *promote* cancer once it has arisen. Stronger risk factors for cancer include smoking, alcohol, and environmental contaminants. (Chapter 18 provides many more details about these risk factors and the development of cancer.)

The relationship between dietary fat and the risk of cancer differs for various types of cancers. In the case of breast cancer, evidence has been weak and inconclusive. Some studies indicate an association between dietary fat and breast cancer; more convincing evidence indicates that body fatness contributes to the risk. In the case of colon cancer, limited evidence suggests a harmful association with foods containing animal fats.

The relationship between dietary fat and the risk of cancer differs for various types and combinations of fats as well. The increased risk in cancer from fat appears to be due primarily to saturated fats or dietary fat from meats (which is mostly saturated). Fat from milk or fish has not been implicated in cancer risk.

Obesity Fat contributes more than twice as many kcalories ♦ per gram as either carbohydrate or protein. Consequently, people who eat high-fat diets regularly may exceed their energy needs and gain weight, especially if they are inactive.[23] Because fat boosts energy intake, cutting fat from the diet can be an effective strategy in cutting kcalories. In some cases, though, choosing a fat-free food offers no kcalorie savings. Fat-free frozen desserts, for example, often have so much sugar added that the kcalorie count can be as high as in the regular-fat product. In this case, cutting fat and adding carbohydrate offers no kcalorie savings or weight-loss advantage. In fact, it may even raise energy intake and exacerbate weight problems. Later chapters revisit the role of dietary fat in the development of obesity.

♦ 1 g fat = 9 kcal
1 g carbohydrate or protein = 4 kcal

Recommended Intakes of Saturated Fat, *Trans* Fat, and Cholesterol

Defining the exact amount of saturated fat, *trans* fat, or cholesterol that begins to harm health is difficult.[24] For this reason, no RDA or Upper Level has been set. Instead, the DRI and *Dietary Guidelines* suggest a diet that provides 20 to 35 percent of the daily energy intake from fat, ♦ less than 10 percent of daily energy intake from saturated fat, as little *trans* fat as possible, and less than 300 milligrams cholesterol. These recommendations recognize that diets with up to 35 percent of kcalories from fat can be compatible with good health if energy intake is reasonable and saturated fat, *trans* fat, and cholesterol intakes are low. When total fat exceeds 35 percent, however, saturated fat usually rises to unhealthy levels.[25] For a 2000-kcalorie diet, 20 to 35 percent represents 400 to 700 kcalories from fat (roughly 45 to 75 grams).

♦ DRI and *Dietary Guidelines* for fat:
- 20 to 35% of energy intake

> **DIETARY GUIDELINES FOR AMERICANS 2010**
Consume less than 10 percent of kcalories from saturated fat. Consume less than 300 milligrams per day of dietary cholesterol. Keep *trans*-fat consumption as low as possible by limiting foods that contain synthetic sources of *trans* fats, such as partially hydrogenated oils, and by limiting other solid fats.

According to surveys, diets in the United States provide about 34 percent of their total energy from fat, with saturated fat contributing about 11 percent of the total.[26] The average daily intake of *trans*-fatty acids in the United States is about 6 grams per day—mostly from products that have been hydrogenated. Cholesterol intakes in the United States average 230 milligrams a day for women and 362 for men.

Although it is very difficult to do, some people actually manage to eat too little fat—to their detriment. Among them are people with eating disorders, described in Highlight 8, and athletes. Athletes following a diet too low in fat (less than 20 percent of total kcalories) fall short on energy, vitamins, minerals, and essential fatty acids as well as on performance.[27] As a practical guideline, it is wise to include the equivalent of at least a teaspoon of fat in every meal—a little peanut butter on toast or mayonnaise in tuna salad, for example. Dietary recommendations that limit fat are designed for healthy people older than age 2; Chapter 16 discusses the fat needs of infants and young children.

REVIEW IT Explain the relationships among saturated fats, *trans* fat, and cholesterol and chronic diseases, noting recommendations.
Although some fat in the diet is necessary, too much fat adds kcalories without nutrients, which leads to obesity and nutrient inadequacies. Too much saturated fat, *trans* fat, and cholesterol increases the risk of heart disease and possibly cancer. For these reasons, health authorities recommend a diet moderate in total fat and low in saturated fat, *trans* fat, and cholesterol.

5.6 Health Effects and Recommended Intakes of Monounsaturated and Polyunsaturated Fats

LEARN IT Explain the relationships between monounsaturated and polyunsaturated fats and health, noting recommendations.

Whereas saturated fats, *trans* fats, and cholesterol are implicated in chronic diseases, monounsaturated and polyunsaturated fats seem to offer health benefits. For this reason, dietary recommendations suggest replacing sources of saturated fats, *trans* fats, and cholesterol with foods rich in monounsaturated and polyunsaturated fats—foods such as seafood, nuts, seeds, and vegetable oils.

Well-balanced, healthy meals provide some fat with an emphasis on monounsaturated and polyunsaturated fats.

Health Effects of Monounsaturated and Polyunsaturated Fats Researchers examining eating patterns from around the world have noted that some diets support good health despite being high in fat. As Highlight 5 explains, the *type* of fat may be more important than the *amount* of fat.[28]

Heart Disease Replacing saturated fats with unsaturated fats ♦ reduces LDL cholesterol and lowers the risk of heart disease.[29] To replace saturated fats with unsaturated fats, sauté foods in olive oil instead of butter, garnish salads with sunflower seeds instead of bacon, snack on mixed nuts instead of potato chips, use avocado instead of cheese on a sandwich, and eat salmon instead of steak. Table 5-3 (p. 150) shows how these simple substitutions can lower the saturated fat and raise the unsaturated fat in a meal. Highlight 5 provides more details about the benefits of healthy fats in the diet.

Research on the different types of fats has spotlighted the many beneficial effects of the omega-3 ♦ polyunsaturated fatty acids.[30] Regular consumption of omega-3 fatty acids helps to prevent blood clots, protect against irregular heartbeats, improve blood lipids, and lower blood pressure, especially in people with hypertension or atherosclerosis.[31] In addition, omega-3 fatty acids support a healthy immune system and suppress inflammation.[32]

♦ Major sources of unsaturated fats:

Monounsaturated fats
- Olive oil, canola oil, peanut oil, safflower oil
- Avocados

Polyunsaturated fats
- Vegetable oils (sesame, soy, corn, sunflower)
- Nuts and seeds

♦ Major sources of omega-3 fats:
- Vegetable oils (canola, soybean, flaxseed)
- Walnuts, flaxseeds
- Fatty fish (mackerel, salmon, sardines)

TABLE 5-3 Replacing Saturated Fat with Unsaturated Fat

Portion sizes have been adjusted so that each of these foods provides approximately 100 kcalories. Notice that for a similar number of kcalories and grams of fat, the second choices offer less saturated fat and more unsaturated fat.

Replace these foods . . .

	Saturated Fat (g)	Unsaturated Fat (g)	Total Fat (g)
Butter (1 tbs)	7	4	11
Bacon (2 slices)	3	6	9
Potato chips (10 chips)	2	5	7
Cheese (1 slice)	4	4	8
Steak (1½ oz)	2	3	5
Totals	**18**	**22**	**40**

. . . with these foods.

	Saturated Fat (g)	Unsaturated Fat (g)	Total Fat (g)
Olive oil (1 tbs)	2	9	11
Sunflower seeds (2 tbs)	1	7	8
Mixed nuts (2 tbs)	1	8	9
Avocado (6 slices)	2	8	10
Salmon (2 oz)	1	3	4
Totals	**7**	**35**	**42**

Cancer The omega-3 fatty acids of fatty fish may protect against some cancers as well, perhaps by suppressing inflammation.[33] Even when omega-3 fats do not protect against cancer development, there seems to be a significant reduction in cancer-related deaths.[34] Thus dietary advice to reduce cancer risks parallels that given to reduce heart disease risks: reduce saturated fats and increase omega-3 fatty acids. Evidence does not support omega-3 supplementation.

Omega-3 Supplements Omega-3 fatty acids are available in capsules of fish oil supplements, although routine supplementation is not recommended.[35] High intakes of omega-3 polyunsaturated fatty acids may increase bleeding time, interfere with wound healing, raise LDL cholesterol, and suppress immune function.* Such findings reinforce the concept that too much of a good thing can sometimes be harmful. People with heart disease, however, may benefit from doses greater than can be achieved through diet alone. Because high intakes of omega-3 fatty acids can cause excessive bleeding, supplements should be used only under close medical supervision.[36]

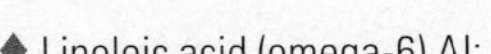

♦ Linoleic acid (omega-6) AI:

Men:
- 19–50 yr: 17 g/day
- 51+ yr: 14 g/day

Women:
- 19–50 yr: 12 g/day
- 51+ yr: 11 g/day

♦ Linolenic acid (omega-3) AI:
- Men: 1.6 g/day
- Women: 1.1 g/day

Recommended Intakes of Monounsaturated and Polyunsaturated Fats

The 20 to 35 percent of kcalories from fat recommendation provides for the essential fatty acids—linoleic acid and linolenic acid—and Adequate Intakes (AI) have been established for these two fatty acids. The DRI suggest that linoleic acid ♦ provide 5 to 10 percent of the daily energy intake and linolenic acid ♦ 0.6 to 1.2 percent.[37]

Beware of fast-food meals delivering too much fat, especially saturated fat. This double bacon cheeseburger, fries, and milkshake provide more than 1600 kcalories, with almost 90 grams of fat and more than 30 grams of saturated fat—far exceeding dietary fat guidelines for the entire day.

From Guidelines to Groceries

Fats accompany protein in foods derived from animals such as meat, seafood, poultry, and eggs, and fats accompany carbohydrate in foods derived from plants such as avocados and coconuts. Fats carry with them the four fat-soluble vitamins—A, D, E, and K—together with many of the compounds that give foods their flavor, texture, and palatability. Fat is responsible for the delicious aromas associated with sizzling bacon, hamburgers on the grill, onions being sautéed, and vegetables in a stir-fry. The essential oils of many spices are fat-soluble. Of course, these wonderful characteristics lure people into eating too much from time to time. With careful selections, a diet can support good health and still meet fat recommendations.

As the photos in Figure 5-20 show, fat accounts for much of the energy in foods, and removing the fat from foods cuts energy and saturated fat intakes dramatically. To reduce dietary fat, eliminate fat as a seasoning and in cooking; remove the fat from high-fat foods; replace high-fat foods with low-fat alternatives; and emphasize whole grains, fruits, and vegetables. The "How To" feature on p. 152 suggests additional heart-healthy choices by food group.

In general, except for seafood, animal fats tend to have a higher proportion of saturated fatty acids. Except for the tropical oils, plant foods tend to have a higher proportion of monounsaturated and polyunsaturated fatty acids. Consumers can find

*Suppressed immune function is seen with daily intake of 0.9 to 9.4 gram EPA and 0.6 to 6.0 gram DHA for 3 to 24 weeks.

an abundant array of fresh, unprocessed foods that are naturally low in saturated fat, *trans* fat, cholesterol, and total fat. In addition, many familiar foods have been processed to provide less fat. For example, fat can be removed by skimming milk or trimming meats. Manufacturers can dilute fat by adding water or whipping in air. They can use fat-free milk in creamy desserts and lean meats in frozen entrées. Sometimes manufacturers simply prepare the products differently. For example, fat-free potato chips may be baked instead of fried. Such choices make healthy eating easy.

Protein Foods The fats in seafood, nuts, and seeds are considered oils, whereas the fats in meat, poultry, and eggs are considered solid fats because of their high fat, saturated fat, and cholesterol content. Because these meats provide high-quality protein and valuable vitamins and minerals, however, they can be included in a healthy diet if a person makes lean choices, ♦ prepares them using the suggestions outlined in the "How To" feature, and eats small portions. Selecting wild game or grass-fed cattle or bison instead of grain-fed livestock offers the nutrient advantages of being lower in fat and higher in omega-3 polyunsaturated fatty acids.[38] Another strategy to lower blood cholesterol is to prepare meals using soy protein instead of animal protein.[39] When preparing meat, fish, or poultry, consider grilling, baking, or broiling, but not frying. Fried fish does not benefit heart disease.[40] Fried fish from fast-food restaurants and frozen fried fish products are often low in omega-3 fatty acids and high in *trans*- and saturated fatty acids.

Table 5-4 (p. 154) provides sources of omega-3 and omega-6 fatty acids. Fatty fish are among the best sources of omega-3 fatty acids, and Highlight 5 features their role in supporting heart health.[41] The American Heart Association recommends two servings of fish a week, with an emphasis on fatty fish (salmon, herring, and mackerel, for example).[42] Two servings of fatty fish provide about 500 milligrams of omega-3 fatty acids. Eating fish supports heart health, especially when combined with physical activity. Fish provides many minerals (except iron) and vitamins.

♦ Protein foods
- Very lean options:
  Chicken (white meat, no skin); cod, flounder, trout; tuna (canned in water); legumes
- Lean options:
  Beef or pork "round" or "loin" cuts; chicken (dark meat, no skin); herring or salmon; tuna (canned in oil)
- Medium-fat options:
  Ground beef, eggs, tofu
- High-fat options:
  Sausage, bacon, luncheon meats, hot dogs, peanut butter, nuts

> **FIGURE 5-20 Cutting Fat Cuts kCalories—and Saturated Fat**

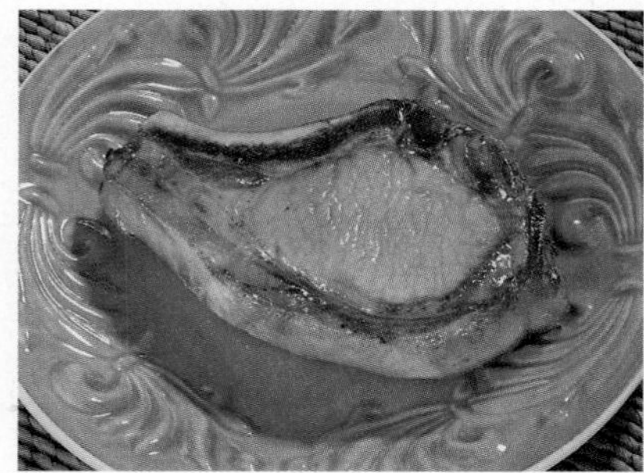

Pork chop with fat (340 kcal, 19 g fat, 7 g saturated fat)

Potato with 1 tbs butter and 1 tbs sour cream (350 kcal, 14 g fat, 10 g saturated fat)

Whole milk, 1 c (150 kcal, 8 g fat, 5 g saturated fat)

Pork chop with fat trimmed off (230 kcal, 9 g fat, 3 g saturated fat)

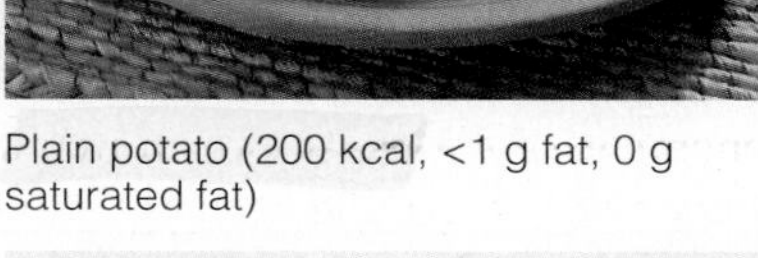

Plain potato (200 kcal, <1 g fat, 0 g saturated fat)

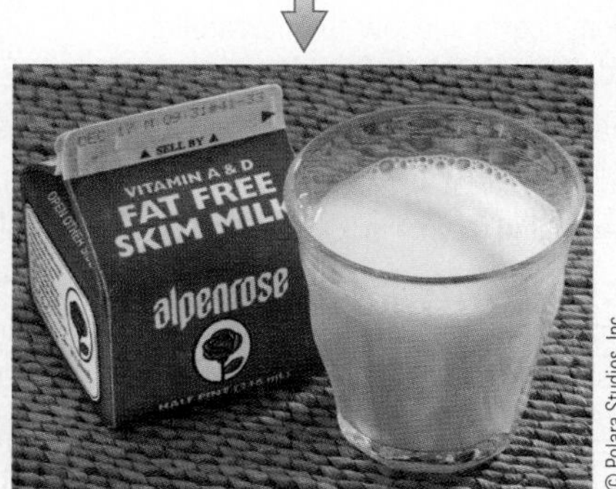

Fat-free milk, 1 c (90 kcal, <1 g fat, <1 g saturated fat)

Savings:
110 kcal, 10 g fat, 4 g saturated fat

Savings:
150 kcal, 13 g fat, 10 g saturated fat

Savings:
60 kcal, 7 g fat, 4 g saturated fat

>How To

Make Heart-Healthy Choices—by Food Group

In General

- Select the most nutrient-dense foods from all food groups.
- Consume fewer and smaller portions of foods and beverages that contain solid fats.
- Check the Nutrition Facts label to choose foods with little or no saturated fat and no *trans* fat.

Grains

- Select breads, cereals, and crackers that are low in saturated and *trans* fat (for example, bagels instead of croissants).
- Prepare pasta with a tomato sauce instead of a cheese or cream sauce.
- Limit intake of cookies, doughnuts, pastries, and croissants.

Vegetables and Fruits

- Enjoy the natural flavor of steamed or roasted vegetables (without butter) for dinner and fruits for dessert.
- Eat at least two vegetables (in addition to a salad) with dinner.
- Snack on raw vegetables or fruits instead of high-fat items like potato chips.
- Buy frozen vegetables without sauce.

Milk and Milk Products

- Switch from whole milk to reduced-fat, from reduced-fat to low-fat, and from low-fat to fat-free (nonfat).
- Use fat-free and low-fat cheeses (such as part-skim ricotta and low-fat mozzarella) instead of regular cheeses.
- Use fat-free or low-fat yogurt or sour cream instead of regular sour cream.
- Use evaporated fat-free milk instead of cream.
- Enjoy fat-free frozen yogurt, sherbet, or ice milk instead of ice cream.

Protein Foods

- Fat adds up quickly, even with lean meat; limit intake to about 6 ounces (cooked weight) daily.
- Eat at least two servings of fish per week (particularly fish such as mackerel, lake trout, herring, sardines, and salmon).
- Choose fish, poultry, or lean cuts of pork or beef; look for unmarbled cuts named *round* or *loin* (eye of round, top round, bottom round, round tip, tenderloin, sirloin, center loin, and top loin).
- Choose processed meats such as lunch meats and hot dogs that are low in saturated fat and cholesterol.
- Trim the fat from pork and beef; remove the skin from poultry.
- Grill, roast, broil, bake, stir-fry, stew, or braise meats; don't fry. When possible, place food on a rack so that fat can drain.
- Use lean ground turkey or lean ground beef in recipes; brown ground meats without added fat, then drain off fat.
- Select tuna, sardines, and other canned meats packed in water; rinse oil-packed items with hot water to remove much of the fat.
- Fill kabob skewers with lots of vegetables and slivers of meat; create main dishes and casseroles by combining a little meat, fish, or poultry with a lot of pasta, rice, or vegetables.
- Use legumes often.
- Eat a meatless meal or two daily.
- Use egg substitutes in recipes instead of whole eggs or use two egg whites in place of each whole egg.

Fats and Oils

- Use small amounts of vegetable oils in place of solid fats.
- Use butter or stick margarine sparingly; select soft margarines instead of hard margarines.
- Use fruit butters, reduced-kcalorie margarines, or butter replacers instead of butter.
- Use low-fat or fat-free mayonnaise and salad dressing instead of regular.
- Limit use of lard and meat fat.
- Limit use of products made with coconut oil, palm kernel oil, and palm oil (read labels on bakery goods, processed foods, popcorn oils, and nondairy creamers).
- Reduce use of hydrogenated shortenings and stick margarines and products that contain them (read labels on crackers, cookies, and other commercially prepared baked goods); use vegetable oils instead.

Miscellaneous

- Use a nonstick pan or coat the pan lightly with vegetable oil.
- Refrigerate soups and stews; when the fat solidifies, remove it before reheating.
- Use wine; lemon, orange, or tomato juice; herbs; spices; fruits; or broth instead of butter or margarine when cooking.
- Stir-fry in a small amount of oil; add moisture and flavor with broth, tomato juice, or wine.
- Use variety to enhance enjoyment of the meal: vary colors, textures, and temperatures—hot cooked versus cool raw foods—and use garnishes to complement food.
- Omit high-fat meat gravies and cheese sauces.
- Order pizzas with lots of vegetables, a little lean meat, and half the cheese.

SOURCE: Adapted from *Third Report of the National Cholesterol Education Program (NCEP) Expert Panel on Detection, Evaluation, and Treatment of High Blood Cholesterol in Adults (Adult Treatment Panel III)*, NIH publication no. 02-5215 (Bethesda, MD.: National heart, Lung, and Blood Institute, 2002), pp. V-25–V-27.

TRY IT Compare the total kcalories, grams of fat, and percent kcalories from fat for 1 cup of whole milk, reduced-fat milk, low-fat milk, and nonfat milk.

TABLE 5-4 Sources of Omega-3 and Omega-6 Fatty Acids

Omega-3	
Linolenic acid	Oils (canola, flaxseed, soybean, walnut, wheat germ, liquid or soft margarine made from canola or soybean oil) Nuts and seeds (flaxseeds, walnuts, soybeans) Vegetables (soybeans)
EPA and DHA	Human milk Fish and seafood: *>500 mg per 3.5 oz serving:* European seabass (bronzini), herring (Atlantic and Pacific), mackerel, oyster (Pacific wild), salmon (wild and farmed), sardines, toothfish (includes Chilian seabass), trout (wild and farmed) *150–500 mg per 3.5 oz serving:* black bass, catfish (wild and farmed), clam, cod (Atlantic), crab (Alaskan king), croakers, flounder, haddock, hake, halibut, oyster (eastern and farmed), perch, scallop, shrimp (mixed varieties), sole, swordfish, tilapia (farmed) *<150 mg per 3.5 oz serving:* cod (Pacific), grouper, lobster, mahimahi, monkfish, red snapper, skate, triggerfish, tuna, wahoo
Omega-6	
Linoleic acid	Seeds, nuts, vegetable oils (corn, cottonseed, safflower, sesame, soybean, sunflower), poultry fat

Source for fish data: K. L. Weaver and coauthors. The content of favorable and unfavorable polyunsaturated fatty acids found in commonly eaten fish, *Journal of the American Dietetic Association* 108 (2008): 1178–1185; P. M. Kris-Etherton, W. S. Harris, and L. J, Appel , Fish consumption, fish oil, omega-3 fatty acids, and cardiovascular disease, *Circulation* 106 (2002): 2747–2757.

Because fish is leaner than most other animal-protein sources it can help with weight-loss efforts. The combination of losing weight and eating fish improves blood lipids even more effectively than can be explained by either the weight loss or the omega-3 fats of the fish. Chapter 19 discusses the adverse consequences of mercury, ♦ an environmental contaminant common in some fish, and suggests that most healthy people who eat two servings of fish a week can maximize the health benefits while incurring minimal risks.

♦ Mercury in fish
- Relatively high:
  Tilefish (also called golden snapper or golden bass), swordfish, king mackerel, shark
- Relatively low:
  Cod, haddock, pollock, salmon, sole, tilapia
  Most shellfish

Recall that cholesterol is found in all foods derived from animals. Consequently, eating less fat from meats, eggs, milk, and milk products helps lower dietary cholesterol intake (as well as total and saturated fat intakes). Most foods that are high in cholesterol are also high in saturated fat, but eggs are an exception. An egg contains only 1 gram of saturated fat but has a little more than 200 milligrams of cholesterol—roughly two-thirds of the recommended daily limit. For people with a healthy lipid profile, eating one egg a day is not detrimental. People with high blood cholesterol, however, may benefit from limiting daily cholesterol intake to less than 200 milligrams.[43] When eggs are included in the diet, other sources of cholesterol may need to be limited on that day. Eggs are a valuable part of the diet because they are inexpensive, useful in cooking, and a source of high-quality protein, other nutrients, and phytochemicals.[44] To help consumers improve their omega-3 fatty acid intake, hens fed flaxseed produce eggs rich in omega-3 fatty acids (about 150 milligrams per egg). Including even one enriched egg in the diet daily can significantly increase a person's intake of omega-3 fatty acids. Food manufacturers have produced several fat-free, cholesterol-free egg substitutes.

> DIETARY GUIDELINES FOR AMERICANS 2010

Replace protein foods that are higher in solid fats (meat, poultry, and eggs) with choices that are lower in solid fats and higher in oils (seafood, nuts, and seeds).

Milk and Milk Products Like meats, milk and milk products ♦ should also be selected with an awareness of their fat, saturated fat, and cholesterol contents. Keep in mind that the fat in milk is a solid fat; it is apparent as butter, but less so when suspended in homogenized milk. Fat-free and low-fat milk products provide as much or more protein, calcium, and other nutrients as their whole-milk versions—but with little or no saturated fat. Selecting fermented milk products, such as yogurt, may also help to lower blood cholesterol. These foods increase the population and activity of bacteria in the colon that use cholesterol.

♦ Milk and milk products
- Fat-free and low-fat options:
  Fat-free or 1% milk or yogurt (plain); fat-free and low-fat cheeses
- Reduced-fat options:
  2% milk, low-fat yogurt (plain)
- High-fat options:
  Whole milk, regular cheeses

Vegetables, Fruits, and Grains Most vegetables and fruits naturally contain little or no fat. Although avocados and olives are exceptions, most of their fat is unsaturated, which is not harmful to heart health. Most grains contain only small amounts of fat. Consumers need to read food labels, though, because many refined grain products such as fried taco shells, croissants, and biscuits are high in saturated fat, and pastries, crackers, and cookies may be high in *trans* fats. Similarly, many people add butter, margarine, or cheese sauce to grains and vegetables, which raises the saturated- and *trans*-fat contents. Because fruits are often eaten without added fat, a diet that includes several servings of fruit daily can help a person meet the dietary recommendations for fat.

A diet rich in vegetables, fruits, whole grains, and legumes also offers abundant vitamin C, folate, vitamin A, vitamin E, and dietary fiber—all important in supporting health. Consequently, such a diet protects against disease by reducing saturated fat, cholesterol, and total fat as well as by increasing nutrients. It also provides valuable phytochemicals, which help defend against heart disease.

Solid Fats and Oils Because solid fats ♦ deliver an abundance of saturated fatty acids, they are considered discretionary kcalories. The fats of fish, nuts, and vegetable oils are *not* counted as discretionary kcalories because they provide valuable omega-3 fatty acids, essential fatty acids, and vitamin E. When discretionary kcalories are available, they may be used to add fats in cooking or at the table or to select higher fat items from the food groups.

♦ Solid fats include:
- Meat and poultry fats (as in poultry skin, luncheon meats, sausage)
- Milk fat (as in whole milk, cheese, butter)
- Shortening (as in fried foods and baked goods)
- Hard margarines

> DIETARY GUIDELINES FOR AMERICANS 2010

Reduce intake of solid fats (major sources of saturated and *trans* fats). Replace solid fats with oils (major sources of polyunsaturated and monounsaturated fats) when possible.

Some solid fats, such as butter and the fat trimmed from meat, are easy to see. Others—such as the fat that "marbles" a steak or is hidden in foods such as cheese—are less apparent and can be present in foods in surprisingly high amounts. Any *fried* food contains abundant solid fats—potato chips, french fries, fried wontons, and fried fish. Many *baked* goods, too, are high in solid fats—pie crusts, pastries, crackers, biscuits, cornbread, doughnuts, sweet rolls, cookies, and cakes.

Reports on *trans*-fatty acids raise the question whether margarine or butter is a better choice for heart health. The American Heart Association has stated that because butter is rich in both saturated fat and cholesterol whereas margarine is made from vegetable fat with no dietary cholesterol, margarine is still preferable to butter. Be aware that soft margarines (liquid or tub) ♦ are less hydrogenated and relatively lower in *trans*-fatty acids; consequently, they do not raise blood cholesterol as much as the saturated fats of butter or the *trans* fats of hard (stick) margarines do. Many manufacturers are now offering nonhydrogenated margarines that are "*trans*-fat free." In addition, manufacturers have developed margarines fortified with plant sterols that lower blood cholesterol. (Highlight 13 explores these and other functional foods designed to support health.) Whichever you decide to use, remember to use them sparingly.

♦ When selecting margarine, look for:
- Soft (liquid or tub) instead of hard (stick)
- ≤2 g saturated fat
- Liquid vegetable oil (not hydrogenated or partially hydrogenated) as first ingredient
- "*Trans*-fat free"

Read Food Labels Labels list total fat, saturated fat, *trans* fat, and cholesterol contents of foods in addition to fat kcalories per serving. Because each package provides information for a single serving and because serving sizes are standardized, consumers can easily compare similar products.

Total fat, saturated fat, and cholesterol are also expressed as "% Daily Values" for a person consuming 2000 kcalories, ♦ using 30 percent of energy intake as the guideline for fat and 10 percent for saturated fat. The Daily Value for cholesterol is 300 milligrams regardless of energy intake. There is no Daily Value for *trans* fat, but consumers should try to keep intakes as low as possible and within the 10 percent allotted for saturated fat. People who consume more or less than 2000 kcalories daily can calculate their personal Daily Value for fat as described in the "How To" feature.

♦ Daily Values:
- 65 g fat (based on 30% of 2000 kcal diet)
- 20 g saturated fat (based on 10% of 2000 kcal diet)
- 300 mg cholesterol

>How To

Calculate a Personal Daily Value for Fat

The % Daily Value for fat on food labels is based on 65 grams. To know how your intake compares with this recommendation, you can either count grams until you reach 65 or add the "% Daily Values" until you reach 100 percent—if your energy intake is 2000 kcalories a day. If your energy intake is more or less, you can calculate your personal daily fat allowance in grams. Suppose your energy intake is 1800 kcalories per day and your goal is 30 percent kcalories from fat. Multiply your total energy intake by 30 percent, then divide by 9:

$$1800 \text{ total kcal} \times 0.30 \text{ from fat} = 540 \text{ fat kcal}$$
$$540 \text{ fat kcal} \div 9 \text{ kcal/g} = 60 \text{ g fat}$$

(In familiar measures, 60 grams of fat is about the same as ⅔ stick of butter or ¼ cup of oil.)

The accompanying table shows the numbers of grams of fat allowed per day for various energy intakes. With one of these numbers in mind, you can quickly evaluate the number of fat grams in foods you are considering eating.

Energy (kcal/day)	20% kCal from Fat	35% kCal from Fat	Fat (g/day)
1200	240	420	27–47
1400	280	490	31–54
1600	320	560	36–62
1800	360	630	40–70
2000	400	700	44–78
2200	440	770	49–86
2400	480	840	53–93
2600	520	910	58–101
2800	560	980	62–109
3000	600	1050	67–117

TRY IT Calculate a personal daily fat allowance for a person with an energy intake of 2100 kcalories and a goal of 25 percent kcalories from fat.

Be aware that the "% Daily Value" for fat is not the same as "% kcalories from fat." This important distinction is explained in the "How To" feature on p. 156. Because recommendations apply to average daily intakes rather than individual food items, food labels do not provide "% kcalories from fat." Still, you can get an idea of whether a particular food is high or low in fat.

Fat Replacers Some foods are made with **fat replacers**—ingredients that provide some of the taste and texture of fats, but with fewer kcalories. Because the body may digest and absorb some of these fat replacers, they may contribute energy, although significantly less energy than fat's 9 kcalories per gram.

Some fat replacers are derived from carbohydrate, protein, or fat. Carbohydrate-based fat replacers are used primarily as thickeners or stabilizers in foods such as soups and salad dressings. Protein-based fat replacers provide a creamy feeling in the mouth and are often used in foods such as ice creams and yogurts. Fat-based replacers act as emulsifiers and are heat stable, making them most versatile in shortenings used in cake mixes and cookies.

Fat replacers offering the sensory and cooking qualities of fats but none of the kcalories are called **artificial fats.** A familiar example of an artificial fat that has been approved for use in snack foods such as potato chips, crackers, and tortilla chips is **olestra.** Olestra's chemical structure is similar to that of a regular fat (a triglyceride) but with important differences. A triglyceride is composed of a glycerol molecule with three fatty acids attached, whereas olestra is made of a sucrose molecule with six to eight fatty acids attached. Enzymes in the digestive tract cannot break the bonds of olestra, so unlike sucrose or fatty acids, olestra passes through the digestive system unabsorbed.

The FDA's evaluation of olestra's safety addressed two questions. First, is olestra toxic? Research on both animals and human beings supports the safety of olestra as a partial replacement for dietary fats and oils, with no reports of cancer or birth defects. Second, does olestra affect either nutrient absorption or the health

fat replacers: ingredients that replace some or all of the functions of fat and may or may not provide energy.

artificial fats: zero-energy fat replacers that are chemically synthesized to mimic the sensory and cooking qualities of naturally occurring fats but are totally or partially resistant to digestion.

olestra: a synthetic fat made from sucrose and fatty acids that provides 0 kcalories per gram; also known as *sucrose polyester.*

>How To

Understand "% Daily Value" and "% kCalories from Fat"

The "% Daily Value" that is used on food labels to describe the amount of fat in a food is not the same as the "% kcalories from fat" that is used in dietary recommendations to describe the amount of fat in the diet. They may appear similar, but their difference is worth understanding. Consider, for example, a piece of lemon meringue pie that provides 140 kcalories and 12 grams of fat. Because the Daily Value for fat is 65 grams for a 2000-kcalorie intake, 12 grams represent about 18 percent:

$$12 \text{ g} \div 65 \text{ g} = 0.18$$
$$0.18 \times 100 = 18\%$$

The pie's "% Daily Value" is 18 percent, or almost one-fifth, of the day's fat allowance.

Uninformed consumers may mistakenly believe that this food meets recommendations to limit fat to "20 to 35 percent kcalories," but it doesn't—for two reasons. First, the pie's 12 grams of fat contribute 108 of the 140 kcalories, for a total of 77 percent kcalories from fat:

$$12 \text{ g fat} \times 9 \text{ kcal/g} = 108 \text{ kcal}$$
$$108 \text{ kcal} \div 140 \text{ kcal} = 77\%$$

Second, the "percent kcalories from fat" guideline applies to a day's total intake, not to an individual food. Of course, if every selection throughout the day exceeds 35 percent kcalories from fat, you can be certain that the day's total intake will, too.

© PhotoDisc/Getty Images

Whether a person's energy and fat allowance can afford a piece of a lemon meringue pie depends on the other food and activity choices made that day.

TRY IT Calculate the percent Daily Value and the percent kcalories from fat for ½ cup frozen yogurt that provides 115 kcalories and 4 grams of fat.

of the digestive tract? When olestra passes through the digestive tract unabsorbed, it binds with some of the fat-soluble vitamins, A, D, E, and K, and carries them out of the body, robbing the person of these valuable nutrients. To compensate for these losses, the FDA requires the manufacturer to fortify olestra with vitamins A, D, E, and K. Saturating olestra with these vitamins does not make the product a good source of vitamins, but it does block olestra's ability to bind with the vitamins from other foods. An asterisk in the ingredients list informs consumers that these added vitamins are "dietarily insignificant."

Consumers need to keep in mind that low-fat and fat-free foods still deliver kcalories. Alternatives to fat can help to lower energy intake and support weight loss only when they actually *replace* fat and energy in the diet.[45]

REVIEW IT Explain the relationships between monounsaturated and polyunsaturated fats and health, noting recommendations.

Some fat in the diet has health benefits, especially the monounsaturated and polyunsaturated fats that protect against heart disease and possibly cancer. For this reason, *Dietary Guidelines* recommend replacing saturated fats with monounsaturated and polyunsaturated fats, particularly omega-3 fatty acids from foods such as fatty fish, not from supplements. Many selection and preparation strategies can help bring these goals within reach, and food labels help to identify foods consistent with these guidelines.

If people were to make only one change in their diets, they would be wise to replace saturated fat with unsaturated fat. Sometimes these choices can be difficult, though, because fats make foods taste delicious. To maintain good health, must a person give up all high-fat foods forever—never again to eat marbled steak, hollandaise sauce, or go oey chocolate cake? Not at all. These foods bring pleasure to a meal and can be enjoyed as part of a healthy diet when eaten occasionally in small quantities; but they should not be everyday foods. The key dietary principle for fat is *moderation,* not *deprivation.* Appreciate the energy and enjoyment that fat provides, but take care not to exceed your needs.

Nutrition Portfolio

To maintain good health, eat enough, but not too much, fat and select the right kinds. Go to Diet Analysis Plus and choose one of the days on which you have tracked your diet for the entire day. Go to the Intake Spreadsheet report. Scroll down until you see: fat (g), sat fat (g), mono fat (g), poly fat (g), and chol (g), which stand for grams of total fat, saturated fat, monounsaturated fat, polyunsaturated fat, and cholesterol, respectively. Use these columns to answer the following questions:

- List the types and amounts of fats and oils you ate on that day, making note of which are saturated, monounsaturated, or polyunsaturated and how your choices could include fewer saturated options.
- List the types and amounts of milk and milk products, meats, fish, and poultry you eat daily, noting how your choices could include more low-fat options.
- Describe choices you can make in selecting and preparing foods to lower your intake of solid fats.

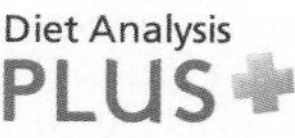

To complete this exercise, go to your Diet Analysis Plus at **www.cengagebrain.com.**

STUDY IT To review the key points of this chapter and take a practice quiz, go to the study cards at the end of the book.

REFERENCES

1. V. Remig and coauthors, *Trans* fats in America: A review of their use, consumption, health implications, and regulation, *Journal of the American Dietetic Association* 110 (2010): 585–592.
2. I. A. Brouwer, A. J. Wanders, M. B. Katan, Effect of animal and industrial *trans* fatty acids on HDL and LDL cholesterol levels in humans—A quantitative review, *PLos One* 5 (2010): e9434; A. Motard-Bélanger and coauthors, Study of the effect of *trans* fatty acids from ruminants on blood lipids and other risk factors for cardiovascular disease, *American Journal of Clinical Nutrition* 87 (2008): 593–599; J. M. Chardigny and coauthors, Do *trans* fatty acids from industrially produced sources and from natural sources have the same effect on cardiovascular disease risk factors in healthy subjects? Results of the *trans* Fatty Acids Collaboration (TRABSFACT) Study, *American Journal of Clinical Nutrition* 87 (2008): 558–566.
3. S. W. Ing and M. A. Belury, Impact of conjugated linoleic acid on bone physiology: Proposed mechanism involving inhibition of adipogenesis, *Nutrition Reviews* 69 (2011): 123–131.
4. S. Klingberg and coauthors, Inverse relation between dietary intake of naturally occurring plant sterols and serum cholesterol in northern Sweden, *American Journal of Clinical Nutrition* 87 (2008): 993–1001.
5. A. V. Khera and coauthors, Cholesterol efflux capacity, high-density lipoprotein function, and atherosclerosis, *New England Journal of Medicine* 364 (2011): 127–135.
6. D. M. Muoio, Metabolism and vascular fatty acid transport, *New England Journal of Medicine* 363 (2010): 291–293.
7. M. Krawczyk, L. Bonfrate, and P. Portincasa, Nonalcoholic fatty liver disease, *Best Practice and Research, Clinical Gastroenterology* 24 (2010): 695–708; G. Tarantino, S. Savastano, and A. Colao, Hepatic steatosis, low-grade chronic inflammation and hormone/growth factor/adipokine imbalance, *World Journal of Gastroenterology* 16 (2010): 4773–4783.
8. Y. Deng and P. E. Scherer, Adipokines as novel biomarkers and regulators of the metabolic syndrome, *Annals of the New York Academy of Sciences* 1212 (2010): E1–E19.
9. N. Ouchi and coauthors, Adipokines in inflammation and metabolic disease, *Nature Reviews, Immunology* 11 (2011): 85–97; C. Stryjecki and D. M. Mutch, Fatty acid-gene interactions, adipokines and obesity, *European Journal of Clinical Nutrition* 65 (2011): 285–297; G. Govindarajan, M. A. Alpert, and L. Tejwani, Endocrine and metabolic effects of fat: Cardiovascular implications, *American Journal of Medicine* 121 (2008): 366–370.
10. R. N. Redinger, The physiology of adiposity, *Journal of the Kentucky Medical Association* 106 (2008): 53–62.
11. N. G. Bazan, M. F. Molina, and W. C. Gordon, Docosahexaenoic acid signalolipidomics in nutrition: Significance in aging, neuroinflammation, macular degeneration, Alzheimer's and other neurodegenerative diseases, *Annual Review of Nutrition* 31 (2011): 321–351; E. E. Birch and coauthors, The DIAMOND (DHA Intake and Measurement of Neural Development) Study: A double-masked, randomized controlled clinical trial of the maturation of infant visual acuity as a function of the dietary level of docosahexaenoic acid, *American Journal of Clinical Nutrition* 91 (2010): 848–859; R. K. McNamara and coauthors, Docosahexaenoic acid supplementation increases prefrontal cortex activation during sustained attention in healthy boys: A placebo-controlled, dose-ranging, functional magnetic resonance imaging study, *American Journal of Clinical Nutrition* 91 (2010): 1060–1067; S. E. Carlson, Early determinants of development: A lipid perspective, *American Journal of Clinical Nutrition* 89 (2009): 1523S–1529S.
12. A. H. Stark, M. A. Crawford, and R. Reifen, Update on alpha-linolenic acid, *Nutrition Reviews* 66 (2008): 326–332.
13. R. Wall and coauthors, Fatty acids from fish: The anti-inflammatory potential of long-chain omega-3 fatty acids, *Nutrition Reviews* 68 (2010): 280–289.
14. K. Zelman, The great fat debate: A closer look at the controversy—Questioning the validity of age-old dietary guidance, *Journal of the American Dietetic Association* 111 (2011): 655–658
15. P. Kris-Etherton, J. Fleming, W. S. Harris, The debate about n-6 polyunsaturated fatty acid recommendations for cardiovascular health, *Journal of the American Dietetic Association* 110 (2010): 201–204; W.S. Harris and coauthors, Omega-6 fatty acids and risk for cardiovascular disease: A Science Advisory from the American Heart Association Nutrition Subcommittee of the Council of Nutrition, Physical Activity, and Metabolism; Council on Cardiovascular Nursing, and Council on Epidemiology and Prevention, *Circulation* 108 (2009): 902–907; B. A. Griffin, How relevant is the ratio of dietary n-6 to n-3 polyunsaturated fatty acids to cardiovascular disease risk? Evidence from the OPTILIP study, *Current Opinion in Lipidology* 19 (2008): 57–62.

16. L. H. Kuller, The great fat debate: Reducing cholesterol, *Journal of the American Dietetic Association* 111 (2011): 663–664.
17. J. E. Hunter, J. Zhang, and P. M. Kris-Etherton, Cardiovascular disease risk of dietary stearic acid compared with *trans*, other saturated, and unsaturated fatty acids: A systematic review, *American Journal of Clinical Nutrition* 91 (2010): 46–63.
18. J. Delgado-Lista and coauthors, Chronic dietary fat intake modifies the postprandial response of hemostatic markers to a single fatty test meal, *American Journal of Clinical Nutrition* 87 (2008): 317–322.
19. Position of the American Dietetic Association and Dietitians of Canada: Dietary fatty acids, *Journal of the American Dietetic Association* 107 (2007): 1599–1611.
20. S. K. Wallace and D. Mozaffarian, *Trans*-fatty acids and nonlipid risk factors, *Current Atherosclerosis Reports* 11 (2009): 423–433.
21. M. J. Albers and coauthors, 2006 Marketplace survey of *trans*–fatty acid content of margarines and butters, cookies and snack cakes, and savory snacks, *Journal of the American Dietetic Association* 108 (2008): 367–370.
22. A. M. Brownawell and M. C. Falk, Cholesterol: Where science and public health policy intersect, *Nutrition Reviews* 68 (2010): 355–364; M. L. Fernandez and M. Calle, Revisiting dietary cholesterol recommendations: Does the evidence support a limit of 300 mg/d? *Current Atherosclerosis Reports* 12 (2010): 377–383.
23. Committee on Dietary Reference Intakes, *Dietary Reference Intakes for Energy, Carbohydrate, Fiber, Fat, Fatty Acids, Cholesterol, Protein, and Amino Acids* (Washington, D.C.: National Academies Press, 2005).
24. P. R. Trumbo and T. Shimakawa, Tolerable upper intake levels for *trans* fat, saturated fat, and cholesterol, *Nutrition Reviews* 69 (2011): 270–278.
25. Committee on Dietary Reference Intakes, 2005.
26. US Department of Agricultural Research Service, Nutrient intakes from food: Mean amounts consumed per individual, 2007–2008, www.ars.usda.gov/ba/bhnrc/fsrg, updated August 2010.
27. Position of the American Dietetic Association, Dietitians of Canada, and the American College of Sports Medicine: Nutrition and athletic performance, *Journal of the American Dietetic Association* 100 (2000): 1543–1556.
28. J. M. Lecerf, Fatty acids and cardiovascular disease, *Nutrition Reviews* 67 (2009): 273–283.
29. A. Astrup, The role of reducing intakes of saturated fats in the prevention of cardiovascular disease: Where does the evidence stand in 2010? *American Journal of Clinical Nutrition* 93 (2011): 684–688.
30. Z. Makhoul and coauthors, Associations of very high intakes of eicosapentaenoic and docosahexaenoic acids with biomarkers of chronic disease risk among Yup'ik Eskimos, *American Journal of Clinical Nutrition* 91 (2010): 777–785; P. P. Dimitrow and M. Jawien, Pleiotropic, cardioprotective effects of omega-3 polyunsaturated fatty acids, *Mini Reviews in Medicinal Chemistry* 9 (2009): 1030–1039; N. D. Riediger and coauthors, A systemic review of the roles of n-3 fatty acids in health and disease, *Journal of the American Dietetic Association* 109 (2009): 668–679.
31. A. C. Skulas-Ray and coauthors, Dose-response effects of omega-3 fatty acids on triglycerides, inflammation, and endothelial function in healthy persons with moderate hypertriglyceridemia, *American Journal of Clinical Nutrition* 93 (2011): 243–252; K. Musa-Velosa and coauthors, Long-chain omega-3 fatty acids eicosapentaenoic acid and docosahexaenoic acid dose-dependently reduce fasting serum triglycerides, *Nutrition Reviews* 68 (2010): 155–167; K. R. Motoyama and coauthors, Association of serum n-6 and n-3 polyunsaturated fatty acids with lipids in 3 populations of middle-aged men, *American Journal of Clinical Nutrition* 90 (2009): 49–55; S. Rajaram and coauthors, Walnuts and fatty fish influence different serum lipid fractions in normal to mildly hyperlipidemic individuals: A randomized controlled study, *American Journal of Clinical Nutrition* 89 (2009): 1657S–1663S; P. J. Smith and coauthors, Association between n-3 fatty acid consumption and ventricular ectopy after myocardial infarction, *American Journal of Clinical Nutrition* 89 (2009): 1315–1320.
32. M. Bouwens and coauthors, Fish-oil supplementation induces antiinflammatory gene expression profiles in human blood mononuclear cells, *American Journal of Clinical Nutrition* 90 (2009): 415–424; M. K. Duda and coauthors, Fish oil, but not flaxseed oil, decreases inflammation and prevents pressure overload-induced cardiac dysfunction, *Cardiovascular Research* 81 (2009): 319–327.
33. T. M. Brasky and coauthors, Specialty supplements and breast cancer risk in the VITamins And Lifestyle (VITAL) Cohort, *Cancer Epidemiology, Biomarkers and Prevention* 19 (2010): 1696–1708.
34. K. M. Szymanski, D. C. Wheeler, and L. A. Mucci, Fish consumption and prostate cancer risk: A review and meta-analysis, *American Journal of Clinical Nutrition* 92 (2010): 1223–1233; J. E. Chavarro and coauthors, A 22-y prospective study of fish intake in relation to prostate cancer incidence and mortality, *American Journal of Clinical Nutrition* 88 (2008): 1297–1303.
35. P. M. Kris-Etherton and A. M. Hill, n-3 Fatty acids: Foods or supplements? *Journal of the American Dietetic Association* 108 (2008): 1125–1130.
36. Fish and omega-3 fatty acids, www.heart.org, updated September 2010.
37. Committee on Dietary Reference Intakes, 2005.
38. A. J. McAfee and coauthors, Red meat from animals offered a grass diet increases plasma and platelet *n*-3 PUFA in healthy consumers, *British Journal of Nutrition* 105 (2011): 80–89.
39. C. W. Xiao, J. Mei, and C. M. Wood, Effect of soy proteins and isoflavones on lipid metabolism and involved gene expression, *Frontiers in Bioscience* 13 (2008): 2660–2673.
40. K. He and coauthors, Intakes of long-chain n-3 polyunsaturated fatty acids and fish in relation to measurement of subclinical atherosclerosis, *American Journal of Clinical Nutrition* 88 (2008): 1111–1118.
41. E. B. Levitan, A. Wolk, and M. A. Mittleman, Fish consumption, marine omega-3 fatty acids, and incidence of heart failure: A population-based prospective study of middle-aged and elderly men, *European Heart Journal* 30 (2009): 1495–1500; A. Philibert and coauthors, Fish intake and serum fatty acid profiles from freshwater fish, *American Journal of Clinical Nutrition* 84 (2006): 1299–1307.
42. AHA Scientific statement: Diet and lifestyle recommendations revision 2006, *Circulation* 114 (2006): 82–96.
43. Expert Panel on Detection, Evaluation, and Treatment of High Blood Cholesterol in Adults (Adult Treatment Panel III), *Third Report of the National Cholesterol Education Program* (NCEP), NIH publication no. 02-5215 (Bethesda, Md.: National Heart, Lung, and Blood Institute, 2002), pp. v–10.
44. Fernandez and Calle, 2010.
45. Position of the American Dietetic Association: Fat replacers, *Journal of the American Dietetic Association* 105 (2005): 266–275.

 LEARN IT Identify which fats support health and which impair it.

© John A. Rizzo/Jupiterimages

High-Fat Foods—Friend or Foe?

Eat less fat. Eat more fatty fish. Give up butter. Use margarine. Give up margarine. Use olive oil. Steer clear of saturated. Seek out omega-3. Stay away from *trans*. Stick with monounsaturated and polyunsaturated. Keep fat intake moderate. Today's fat messages seem to be forever multiplying and changing. No wonder some people feel confused about dietary fat. The confusion stems in part from the complexities of fat and in part from the nature of recommendations. As Chapter 5 explained, "dietary fat" refers to several kinds of fats. Some fats support health whereas others impair it, and foods typically provide a mixture of fats in varying proportions. Researchers have spent decades sorting through the relationships among the various kinds of fat and their roles in supporting or harming health. Translating these research findings into dietary recommendations is challenging. Too little information can mislead consumers, but too much detail can overwhelm them. As research findings accumulate, recommendations slowly evolve and become more refined. Fortunately, that's where we are with fat recommendations today—refining them from the general to the specific. Though they may seem to be "forever multiplying and changing," in fact, they are becoming more meaningful.

This highlight begins with a look at the dietary guidelines for fat intake. It continues by identifying which foods provide which fats and presenting the Mediterranean diet, an example of an eating pattern that embraces the heart-healthy fats. It closes with strategies to help consumers choose the right amounts of the right kinds of fats for a healthy diet.

Guidelines for Fat Intake

Dietary recommendations for fat have shifted emphasis from lowering total fat, in general, to limiting saturated and *trans* fat, specifically. Instead of urging people to cut back on all fats, recommendations suggest carefully replacing the "bad" saturated fats with the "good" unsaturated fats and enjoying them in moderation. The goal is to create a diet moderate in kcalories that provides enough of the fats that support good health, but not too much of those that harm health. (Turn to pp. 147–150 for a review of the health consequences of each type of fat.)

With these findings and goals in mind, the Dietary Reference Intakes (DRI) committee suggests a healthy range of 20 to 35 percent of energy intake from fat. This range appears to be compatible with low rates of heart disease, diabetes, obesity, and cancer. Heart-healthy recommendations suggest that within this range, consumers should try to minimize their intakes of saturated fat, *trans* fat, and cholesterol and use monounsaturated and polyunsaturated fats instead.

Asking consumers to limit their total fat intake is less than perfect advice, but it is straightforward—find the fat and cut back. Asking consumers to keep their intakes of saturated fats, *trans* fats, and cholesterol low and to use monounsaturated and polyunsaturated fats instead is more on target with heart health, but it also makes diet planning a bit more challenging. To make appropriate selections, consumers must first learn which foods contain which fats.

High-Fat Foods and Heart Health

Avocados, bacon, walnuts, potato chips, and mackerel are all high-fat foods, yet some of these foods have detrimental effects on heart health when consumed in excess, whereas others seem neutral or even beneficial. This section presents some of the accumulating evidence that help distinguish which high-fat foods belong in a healthy diet and which ones need to be kept to a minimum. As you will see, fat in the diet can be compatible with heart health, but only if most of it is unsaturated.

Cook with Olive Oil

As it turns out, the traditional diets of Greece and other countries in the Mediterranean region offer an excellent example of eating patterns that use "good" fats liberally. The primary fat in these diets is olive oil, which seems to play a key role in providing health benefits.[1] A classic study of the world's people, the Seven Countries Study, found that death rates from heart disease were strongly associated with diets high in saturated fats but only weakly linked with total fat.[2] In fact, the two countries with the highest fat intakes, Finland and the Greek island of Crete, had the highest (Finland) and lowest (Crete) rates of heart disease deaths. In both countries, the people consumed 40 percent or more of their kcalories from fat. Clearly, a high-fat diet is not the primary problem.[3] When researchers refocused their attention on the *type* of fat, they began to notice the benefits of olive oil.

A diet that uses olive oil instead of other cooking fats, especially butter, stick margarine, and meat fats, may offer numerous health benefits. Olive oil and other oils rich in monounsaturated fatty acids help to protect against heart disease by:[4]

- Lowering total and LDL cholesterol and not lowering HDL cholesterol or raising triglycerides
- Lowering LDL cholesterol susceptibility to oxidation

- Lowering blood-clotting factors
- Providing phytochemicals that act as antioxidants (see Highlight 11)
- Lowering blood pressure
- Interfering with the inflammatory response

When compared with other fats, olive oil seems to be a wise choice, but controlled clinical trials are too scarce to support population-wide recommendations to switch to a high-fat diet rich in olive oil. Importantly, olive oil is not a magic potion; drizzling it on foods does not make them healthier. Like other fats, olive oil delivers 9 kcalories per gram, which can contribute to weight gain in people who fail to balance their energy intake with physical activity. Its role in a healthy diet is to *replace* the saturated fats. Other vegetable oils, such as canola or safflower oil, are also generally low in saturated fats and high in unsaturated fats. For this reason, heart-healthy diets use these unsaturated vegetable oils to replace the more saturated fats of butter, hydrogenated stick margarine, lard, or shortening. (Remember that the tropical oils—coconut, palm, and palm kernel—are too saturated to be included with the heart-healthy vegetable oils.)

Nibble on Nuts

Tree nuts and peanuts are traditionally excluded from low-fat diets, and for good reasons. Nuts provide up to 80 percent of their kcalories from fat, and a quarter cup (about an ounce) of mixed nuts provides more than 200 kcalories. Frequent nut consumption (1-ounce serving of nuts on five or more days a week), however, protects against heart disease.[5] Benefits are seen for a variety of nuts commonly eaten in the United States: almonds, Brazil nuts, cashews, hazelnuts, macadamia nuts, pecans, pistachios, walnuts, and even peanuts. On average, these nuts contain mostly monounsaturated fat (59 percent), some polyunsaturated fat (27 percent), and little saturated fat (14 percent). Nuts also provide valuable fiber, vegetable protein, vitamin E, minerals, and phytochemicals.

Olives and their oil may benefit heart health.

© Matthew Farruggio

For heart health, snack on a few nuts instead of potato chips. Because nuts are energy dense (high in kcalories per ounce), it is especially important to keep portion size in mind when eating them.

© Matthew Farruggio

Including nuts may be a wise diet strategy against heart disease. Nuts may protect against heart disease by:[6]

- Lowering blood cholesterol
- Lowering blood pressure
- Limiting oxidative stress and inflammation

Some research suggests that a diet that includes nuts may benefit other diseases as well.[7]

Before advising consumers to include nuts in their diets, however, a caution is in order. As mentioned, most of the energy nuts provide comes from fats. Consequently, they deliver many kcalories per bite. Incorporating nuts in the diet, however, does not necessarily lead to weight gains and may even help with weight control.[8] Consumers can enjoy nuts without increasing total kcalories by using nuts *instead of, not in addition to,* other foods (such as meats, potato chips, oils, margarine, and butter).

Feast on Fish

Research into the health benefits of the long-chain omega-3 polyunsaturated fatty acids began with a simple observation: the native peoples of Alaska, northern Canada, and Greenland, who eat a traditional diet rich in omega-3 fatty acids, notably EPA (eicosapentaenoic acid) and DHA (docosaheaenoic acid), have a remarkably low rate of heart disease even though their diets are relatively high in fat.[9] These omega-3 fatty acids help to protect against heart disease by:[10]

- Reducing blood triglycerides
- Stabilizing plaque
- Lowering blood pressure
- Reducing inflammation
- Serving as precursors to eicosanoids

For people with hypertension or atherosclerosis, these actions can be life saving.

Research studies have provided strong evidence that increasing omega-3 fatty acids in the diet supports heart health and lowers the rate of deaths from heart disease. For this reason, the American Heart Association recommends including fish in a heart-healthy diet. People who eat some fish each week can lower their risks of heart attack and stroke.

Fish is the best source of EPA and DHA in the diet, but it is also a source of mercury, an environmental contaminant. Most fish contain at least trace amounts of mercury, but some have especially high levels. For this reason, the FDA advises pregnant and lactating women, women of childbearing age who may become pregnant, and young children to include fish in their diets, but to avoid:

- Tilefish (also called golden snapper or golden bass), swordfish, king mackeral, marlin, and shark

And to limit average weekly consumption of:

- A variety of ocean fish and shellfish to 12 ounces (cooked or canned)
- White (albacore) tuna to 6 ounces (cooked or canned)

Commonly eaten seafood relatively low in mercury include shrimp, catfish, pollock, salmon, and canned light tuna.

In addition to the direct toxic effects of mercury, some (but not all) research suggests that mercury may diminish the health benefits of omega-3 fatty acids. Such findings serve as a reminder that our health depends on the health of our planet. The protective effect of fish in the diet is available, provided that the fish and their surrounding waters are not heavily contaminated. (Chapter 19 discusses the adverse consequences of mercury, and Chapter 20 presents the relationships between diet and the environment in more detail.)

In an effort to limit exposure to pollutants, some consumers choose farm-raised fish. Compared with fish caught in the wild, farm-raised fish tend to be lower in mercury, but they are also lower in omega-3 fatty acids. When selecting fish, keep the diet strategies of variety and moderation in mind. Varying choices and eating moderate amounts helps to limit the intake of contaminants such as mercury.

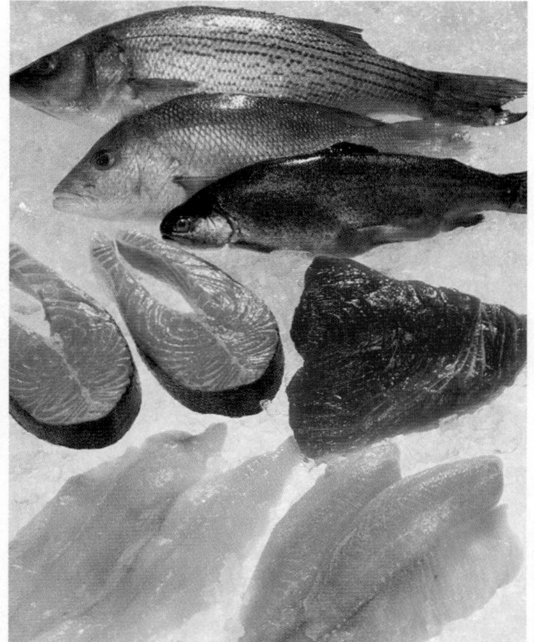
© www.comstock.com

Fish is a good source of the omega-3 fatty acids.

High-Fat Foods and Heart Disease

The number-one dietary determinant of LDL cholesterol is saturated fat. Each 1 percent increase in energy from saturated fatty acids in the diet produces a 2 percent jump in heart disease risk by elevating LDL cholesterol. Conversely, reducing saturated fat intake by 1 percent can be expected to produce a 2 percent drop in heart disease risk by the same mechanism. Even a 2 percent drop in LDL represents a significant improvement for heart health.[11] Like saturated fats, *trans* fats also raise heart disease risk by elevating LDL cholesterol. A heart-healthy diet limits foods rich in these two types of fat.

Limit Fatty Meats, Whole-Milk Products, and Tropical Oils

The major sources of saturated fats in the US diet are fatty meats, whole milk, tropical oils, and products made from any of these foods. To limit saturated fat intake, consumers must choose carefully among these high-fat foods. More than a third of the fat in most meats is saturated. Similarly, more than half of the fat is saturated in whole milk and other high-fat milk products, such as cheese, butter, cream, half-and-half, cream cheese, sour cream, and ice cream. The tropical oils of palm, palm kernel, and coconut, which are rarely used by consumers in the kitchen, are used heavily by food manufacturers, and are commonly found in many commercially prepared foods.

When choosing meats, milk products, and commercially prepared foods, look for those lowest in saturated fat. Labels provide a useful guide for comparing products in this regard, and Appendix H lists the saturated fat in several thousand foods.

Even with careful selections, a nutritionally adequate diet will provide some saturated fat. Zero saturated fat is not possible even when experts design menus with the mission to keep saturated fat as low as possible.[12] Because most saturated fats come from animal foods, vegetarian diets can, and usually do, deliver fewer saturated fats than mixed diets.

Limit Hydrogenated Foods

Chapter 5 explained that solid shortening and margarine are made from vegetable oil that has been hardened through hydrogenation. This process both saturates some of the unsaturated fatty acids and introduces *trans*-fatty acids. Many convenience foods contain *trans* fats, including:

- Fried foods such as french fries, chicken, and other commercially fried foods
- Commercial baked goods such as cookies, doughnuts, pastries, breads, and crackers
- Snack foods such as chips
- Imitation cheeses

To keep *trans*-fat intake low, use these foods sparingly.

Table H5-1 summarizes which foods provide which fats. Substituting unsaturated fats for saturated fats at each meal and snack can help protect against heart disease. Figure H5-1 compares two meals and shows how such substitutions can lower saturated fat and raise unsaturated fat—even when total fat and kcalories remain unchanged.

The Mediterranean Diet

The links between good health and traditional Mediterranean eating patterns of the mid-1900s were introduced earlier with regard to olive oil. For people who eat these diets, the incidence of heart disease, some cancers, diabetes, and other chronic inflammatory diseases is low, and life expectancy is high.[13] Some research suggests that the health benefits of the Mediterranean eating pattern are partially due to its favorable effects on body weight.[14]

Although each of the many countries that border the Mediterranean Sea has its own culture, traditions, and dietary habits, their similarities are much greater than the use of olive oil alone. In fact, no one factor alone can be credited with reducing disease risks—the association holds true only when the overall eating pattern is present. Apparently, each of the foods contributes small benefits that harmonize to produce either a substantial cumulative or synergistic effect.

The Mediterranean eating pattern features fresh, whole foods. The people select crusty breads, whole grains, potatoes, and pastas; a variety of vegetables (including wild greens) and legumes; feta and mozzarella cheeses and yogurt; nuts; and fruits (especially grapes and figs). They eat some fish, other seafood, poultry, a few eggs, and little meat. Along with olives and olive oil, their principal sources of fat are nuts and fish; they rarely use butter or encounter hydrogenated fats. They commonly use herbs and spices instead of salt. Consequently, traditional Mediterranean diets are:

- Low in saturated fat
- Very low in *trans* fat
- Rich in monounsaturated and polyunsaturated fat
- Rich in complex carbohydrate and fiber
- Rich in nutrients and phytochemicals that support good health

As a result, lipid profiles improve, inflammation diminishes, and the risk of heart disease declines.

People following the traditional Mediterranean diet can receive as much as 40 percent of a day's kcalories from fat, but their limited consumption of milk and milk products and meats provides less than 10 percent from saturated fats. In addition, because the animals in the Mediterranean region pasture-graze, the meat, milk and milk products, and eggs are richer in omega-3 fatty acids than those from animals fed grain.

TABLE H5-1 Major Sources of Various Fatty Acids

Healthful Fatty Acids Monounsaturated	Omega-6 Polyunsaturated	Omega-3 Polyunsaturated
Avocado	Margarine (nonhydrogenated)	Fatty fish (herring, mackerel, salmon, tuna)
Oils (canola, olive, peanut, sesame)	Oils (corn, cottonseed, safflower, soybean)	Flaxseed
Nuts (almonds, cashews, filberts, hazelnuts, macadamia nuts, peanuts, pecans, pistachios)	Nuts (pine nuts, walnuts)	Nuts (walnuts)
Olives	Mayonnaise	
Peanut butter	Salad dressing	
Seeds (sesame)	Seeds (pumpkin, sunflower)	
Harmful Fatty Acids Saturated	***Trans***	
Bacon	Fried foods (hydrogenated shortening)	
Butter	Margarine (hydrogenated or partially hydrogenated)	
Chocolate	Nondairy creamers	
Coconut	Many fast foods	
Cream cheese	Shortening	
Cream, half-and-half	Commercial baked goods (including doughnuts, cakes, cookies)	
Lard	Many snack foods (including microwave popcorn, chips, crackers)	
Meat		
Milk and milk products (whole)		
Oils (coconut, palm, palm kernel)		
Shortening		
Sour cream		

NOTE: Keep in mind that foods contain a mixture of fatty acids.

> FIGURE H5-1 **Two Meals Compared: Replacing Saturated Fat with Unsaturated Fat**

Examples of ways to replace saturated fats with unsaturated fats include sautéing vegetables in olive oil instead of butter, garnishing salads with avocado and sunflower seeds instead of bacon and blue cheese, and eating salmon instead of steak. Each of these meals provides roughly the same number of kcalories and grams of fat, but the one on the left has almost four times as much saturated fat and only half as many omega-3 fatty acids.

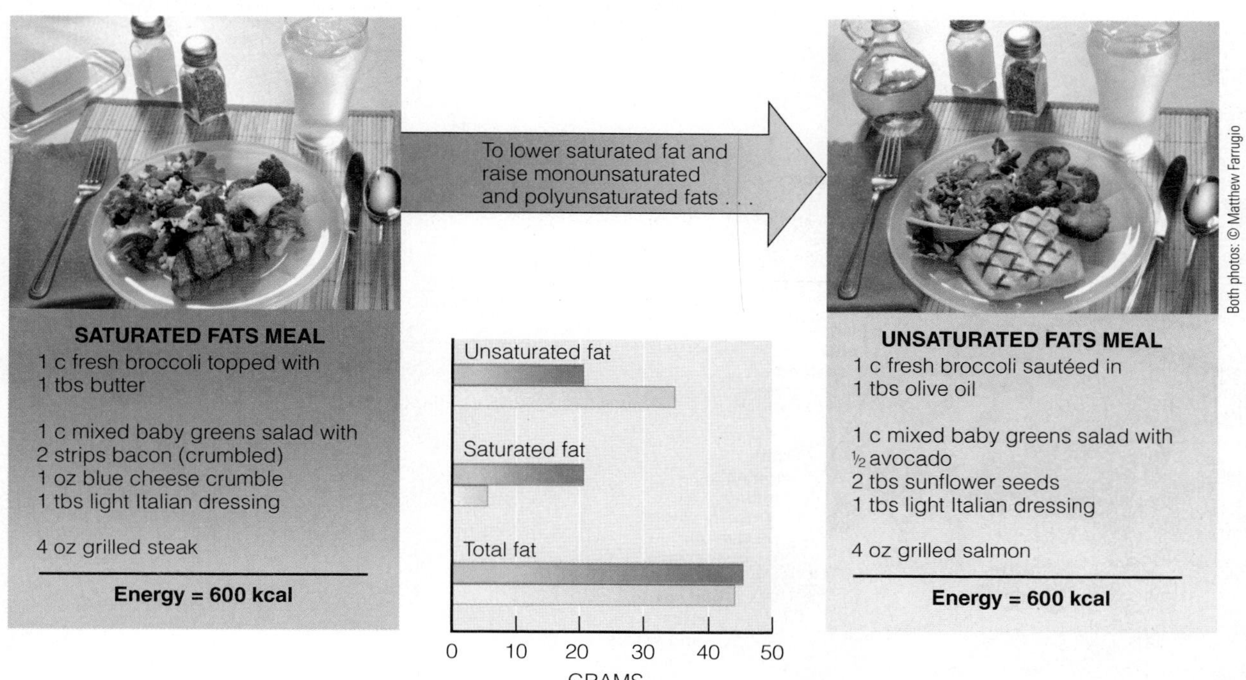

Other foods typical of the Mediterranean, such as wild plants and snails, provide omega-3 fatty acids as well. All in all, the traditional Mediterranean diet has gained a reputation for its health benefits as well as its delicious flavors. By following a Mediterranean eating pattern, consumers improve their blood lipid profile, insulin resistance, blood pressure, and body weight.[15] Consumers need to beware that the typical Mediterranean-style cuisine available in US restaurants, however, has been adjusted to popular tastes. Quite often, these meals are much higher in saturated fats and meats—and much lower in the potentially beneficial constituents—than the traditional fare. Unfortunately, it appears that people in the Mediterranean region who are replacing some of their traditional dietary habits with those of the United States are losing the health benefits previously enjoyed.[16] Figure H5-2 (p. 164) presents a Mediterranean Diet Pyramid. Notice the emphasis on abundant plant foods.

Conclusion

Are some fats "good," and others "bad" from the body's point of view? The saturated and *trans* fats indeed seem mostly bad for the health of the heart. Aside from providing energy, which unsaturated fats can do equally well, *saturated* and *trans* fats bring no indispensable benefits to the body. Furthermore, no harm can come from consuming diets low in them. Still, foods rich in these fats are often delicious, giving them an occasional place in the diet.

In contrast, the unsaturated fats are mostly good for the heart health when consumed in moderation. To date, their one proven fault seems to be that they, like all fats, provide abundant energy to the body and so may promote obesity if they drive kcalorie intakes higher than energy needs.[17] Obesity, in turn, often begets many body ills, as Chapter 8 describes.

Clearly, different fatty acids have different actions in the body and risks of chronic diseases.[18] When judging foods by their fatty acids, keep in mind that the fat in foods is a mixture of "good" and "bad," providing both unsaturated and saturated fatty acids. Even predominantly monounsaturated olive oil delivers some saturated fat. Consequently, even when a person chooses foods with mostly unsaturated fats, saturated fat can still add up if total fat is high.

Adopting the Mediterranean eating pattern may serve those who enjoy a little more fat in the diet. Including vegetables, fruits, whole grains, and legumes as part of a balanced daily diet is a good idea, as is *replacing* saturated fats such as butter, shortening, and meat fat with unsaturated fats such as olive oil and the oils from nuts and fish.[19] These foods provide beneficial fatty acids, fiber, vitamins, minerals, and phytochemicals as well as little salt, saturated fat, and *trans* fat—all valuable in protecting the body's health. In addition, take care to select portion sizes that will best meet energy needs. And enjoy some physical activity daily. Remember that even a healthy eating pattern can be detrimental if eaten in excess.[20]

> FIGURE H5-2 **Mediterranean Diet Pyramid**

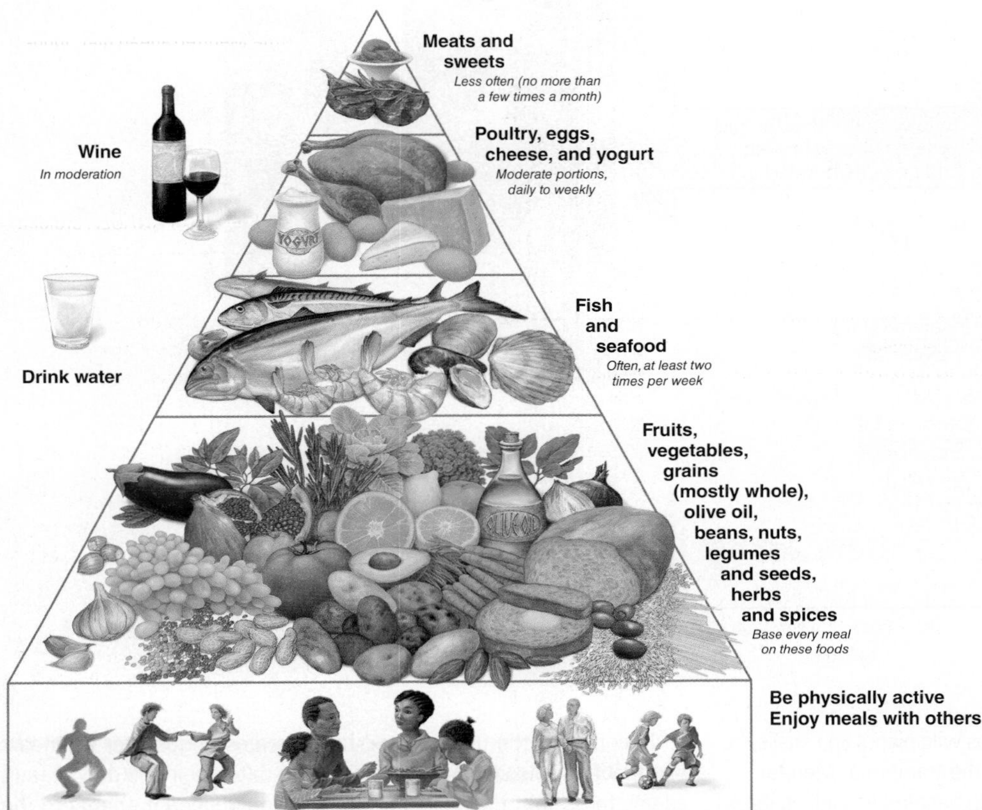

REFERENCES

1. B. Bendinelli and coauthors, Fruit, vegetables, and olive oil and risk of coronary heart disease in Italian women: The EPICOR Study, *American Journal of Clinical Nutrition* 93 (2011): 275–283; M. I. Covas, V. Konstantinidou, and M. Fitó, Olive oil and cardiovascular health, *Journal of Cardiovascular Pharmacology* 54 (2009): 477–482.
2. A. Keys, *Seven Countries: A Multivariate Analysis of Death and Coronary Heart Disease* (Cambridge: Harvard University Press, 1980).
3. W. C. Willet, The great fat debate: Total fat and health, *Journal of the American Dietetic Association* 111 (2011): 660–662.
4. D. Bester and coauthors, Cardiovascular effects of edible oils: A comparison between four popular edible oils, *Nutrition Research Reviews* 23 (2010): 334–348.
5. J. Sabaté and M. Wien, Nuts, blood lipids and cardiovascular disease, *Asia Pacific Journal of Clinical Nutrition* 19 (2010): 131–136.
6. C. E. Berryman and coauthors, Effects of almond consumption on the reduction of LDL-cholesterol: A discussion of potential mechanisms and future research directions, *Nutrition Reviews* 69 (2011): 171–185; E. Ros, L. C. Tapsell, and J. Sabaté, Nuts and berries for heart health, *Current Atherosclerosis Reports* 12 (2010): 397–406; J. Sabaté, K. Oda, and E. Ros, Nut consumption and blood lipid levels: A pooled analysis of 25 intervention trials, *Archives of Internal Medicine* 170 (2010): 821–827; P. López-Uriarte and coauthors, Nuts and oxidation: A systematic review, Nutrition Reviews 67 (2009): 497–508; D. K. Banel and F. B. Hu, Effects of walnut consumption on blood lipids and other cardiovascular risk factors: A meta-analysis and systematic review, *American Journal of Clinical Nutrition* 90 (2009): 56–63.
7. J. Sabaté and Y. Ang, Nuts and health outcomes: New epidemiologic evidence, *American Journal of Clinical Nutrition* 89 (2009): 1643S–1648S.
8. M. Bes-Rastrollow and coauthors, Prospective study of nut consumption, long-term weight change, and obesity risk in women, *American Journal of Clinical Nutrition* 89 (2009): 1913–1919.
9. A. Bersamin and coauthors, Westernizing diets influence fat intake, red blood cell fatty acid composition, and health in remote Alaskan native communities in the Center for Alaska Native Health Study, *Journal of the American Dietetic Association* 108 (2008): 266–273.
10. A. C. Skulas-Ray and coauthors, Dose-response effects of omega-3 fatty acids on triglycerides, inflammation, and endothelial function in healthy persons with moderate hypertriglyceridemia, *American Journal of Clinical Nutrition* 93 (2011): 243–252; P. C. Calder and P. Yaqoob, Omega-3 (n-3) fatty

acids, cardiovascular disease and stability of atherosclerotic plaques, *Journal of Molecular Cell Biology* 56 (2010): 28–37; F. Dangardt and coauthors, Omega-3 fatty acid supplementation improves vascular function and reduces inflammation in obese adolescents, *Atherosclerosis* 212 (2010): 580–585; M. N. DiMinno and coauthors, Exploring newer cardioprotective strategies: ω-3 Fatty acids in perspective, *Journal of Thrombosis and Haemostasis* 104 (2010): 664–680.

11. *Third Report of the National Cholesterol Education Program (NCEP) Expert Panel on Detection, Evaluation, and Treatment of High Blood Cholesterol in Adults (Adult Treatment Panel III)*, publication NIH no. 02-5215 (Bethesda, Md.: National Heart, Lung, and Blood Institute, 2002), pp. v–8.
12. *Third Report of the National Cholesterol Education Program (NCEP) Expert Panel on Detection, Evaluation, and Treatment of High Blood Cholesterol in Adults (Adult Treatment Panel III)*, 2002; Committee on Dietary Reference Intakes, *Dietary Reference Intakes for Energy, Carbohydrate, Fiber, Fat, Fatty Acids, Cholesterol, Protein, and Amino Acids* (Washington, D.C.: National Academies Press, 2005).
13. K. Esposito and coauthors, Prevention and control of type 2 diabetes by Mediterranean diet: A systematic review, *Diabetes Research and Clinical Practice* 89 (2010): 97–102; P. P. McKeown and coauthors, Session 4: CVD, diabetes and cancer: Evidence for the use of the Mediterranean diet in patients with CHD, *Proceedings of the Nutrition Society* 69 (2010): 45–60; F. Sofi and coauthors, Accruing evidence on benefits of adherence to the Mediterranean diet on health: An updated systematic review and meta-analysis, *American Journal of Clinical Nutrition* 92 (2010): 1189–1196; L. Verberne and coauthors, Association between the Mediterranean diet and cancer risk: A review of observational studies, *Nutrition and Cancer* 62 (2010): 860–870; G. Buckland and coauthors, Adherence to the Mediterranean diet and risk of coronary heart disease in the Spanish EPIC Cohort Study, *American Journal of Epidemiology* 170 (2009): 1518–1529; N. Babio, M. Bulló, and J. Salas-Salvadó, Mediterranean diet and metabolic syndrome: The evidence, *Public Health Nutrition* 12 (2009): 1607–1617; M. A. Martinez-Gonzalez and coauthors, Mediterranean food pattern and the primary prevention of chronic disease: Recent developments, *Nutrition Reviews* 67 (2009): S111–S116; M. E. Rumawas and coauthors, Mediterranean-style dietary pattern, reduced risk of metabolic syndrome traits, and incidence in the Framingham Offspring Cohort, *American Journal of Clinical Nutrition* 90 (2009): 1608–1614; P. Sjögren and coauthors, Mediterranean and carbohydrate-restricted diets and mortality among elderly men: A cohort study in Sweden, *American Journal of Clinical Nutrition* 92 (2010): 967–974; C. LaVecchia, Association between Mediterranean dietary patterns and cancer risk, *Nutrition Reviews* 67 (2009): S126–S129.
14. J. J. Beunza and coauthors, Adherence to the Mediterranean diet, long-term weight change, and incident overweight or obesity: The Seguimiento Universidad de Navarra (SUN) cohort, *American Journal of Clinical Nutrition* 92 (2010): 1484–1493; C. M. Kastorini and coauthors, Mediterranean diet and coronary heart disease: Is obesity a link?: A systematic review, *Nutrition, Metabolism & Cardiovascular Diseases* 20 (2010): 536–551; D. Romaguera and coauthors, Mediterranean dietary patterns and prospective weight change in participants of the EPIC-PANACEA project, *American Journal of Clinical Nutrition* 92 (2010): 912–921; Babio, Bulló, and Salas-Salvadó, 2009.
15. R. Estruch, Anti-inflammatory effects of the Mediterranean diet: The experience of the PREDIMED study, *Proceedings of the Nutrition Society* 69 (2010): 333–340; Babio, Bulló, and Salas-Salvadó, 2009; S. Piscopo, The Mediterranean diet as a nutrition education, health promotion and disease prevention tool, *Public Health Nutrition* 12 (2009): 1648–1655.
16. P. A. Gilbert and S. Khokhar, Changing dietary habits of ethnic groups in Europe and implications for health, *Nutrition Reviews* 66 (2008): 203–215.
17. Committee on Dietary Reference Intakes, 2005, pp. 796–797.
18. J. M. Lecerf, Fatty acids and cardiovascular disease, *Nutrition Reviews* 67 (2009): 273–283; J. Y. Lee, L. Zhao, and D. H. Hwang, Modulation of pattern recognition receptor-mediated inflammation and risk of chronic diseases by dietary fatty acids, *Nutrition Reviews* 68 (2009): 38–61.
19. D. Mozaffarian, The great fat debate: Taking the focus off of saturated fat, *Journal of the American Dietetic Association* 111 (2011): 665–666.
20. A. H. Lichtenstein, The great fat debate: The importance of message translation, *Journal of the American Dietetic Association* 111 (2011): 667–670.

© nito/Shutterstock.com

CENGAGE LEARNING'S GLOBAL NUTRITION WATCH — GLOBAL NUTRITION RESOURCE CENTER

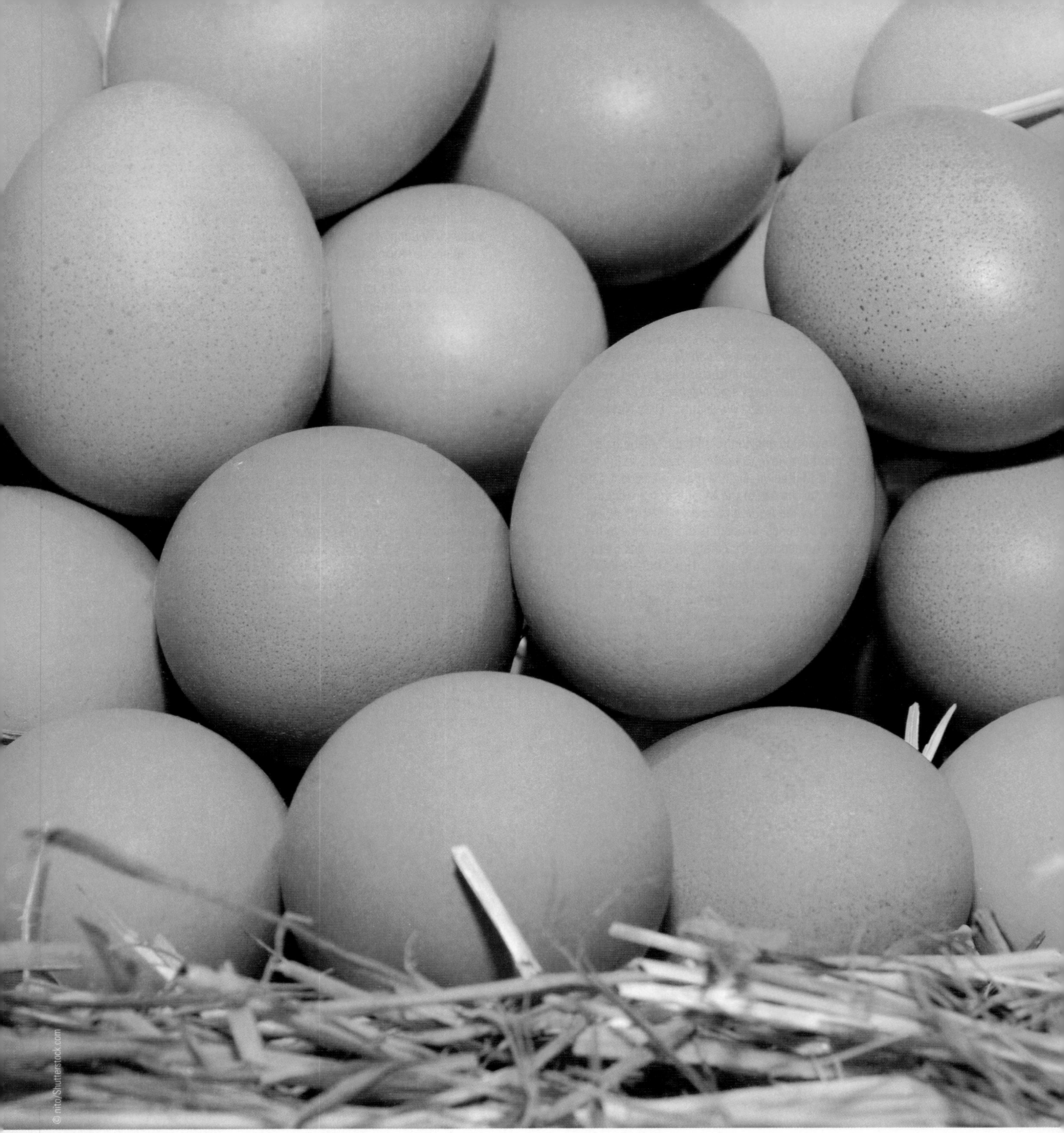

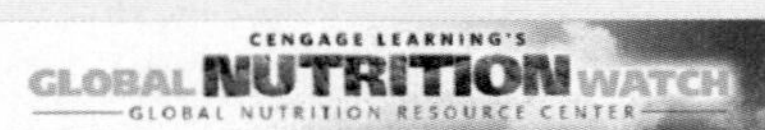

The quality of proteins derived from animal sources versus plant sources differs greatly. Go to Nutrition Basics and Tools → Proteins to learn about protein quality. How can the protein quality of vegetarian diets be improved?

6

Protein: Amino Acids

LEARNING GPS

Nutrition in Your Life

The versatility of proteins in the body is impressive. They help your muscles to contract, your blood to clot, and your eyes to see. They keep you alive and well by facilitating chemical reactions and defending against infections. Without them, your bones, skin, and hair would have no structure. No wonder they were named *proteins,* meaning "of prime importance." Does that mean proteins deserve top billing in your diet as well? Are the best sources of protein beef, beans, or broccoli? Learn which foods will supply you with enough, but not too much, high-quality protein. In the Nutrition Portfolio at the end of this chapter, you can determine whether your diet is meeting your protein needs.

A few misconceptions surround the roles of protein in the body and the importance of protein in the diet. For example, people who associate meat with protein and protein with strength may eat steak to build muscles. Their thinking is only partly correct, however. Protein is a vital structural and working substance in all cells—not just muscle cells. To build strength, muscles cells need physical activity and all the nutrients—not just protein. Furthermore, protein is found in milk, eggs, legumes, and many grains and vegetables—not just meat. By overvaluing protein and overemphasizing meat in the diet, a person may mistakenly crowd out other, equally important nutrients and foods. As this chapter describes the various roles of protein in the body and food sources in the diet, keep in mind that protein is one of many nutrients needed to maintain good health.

6.1 The Chemist's View of Proteins

LEARN IT Recognize the chemical structures of amino acids and proteins.

Chemically, **proteins** contain nitrogen (N) atoms in addition to the same atoms as carbohydrates and lipids—carbon (C), hydrogen (H), and oxygen (O). These nitrogen atoms give the name *amino* (nitrogen containing) to the amino acids that make the links in the chains of proteins.

Amino Acids

All **amino acids** have the same basic structure—a central carbon (C) atom with a hydrogen atom (H), an amino group (NH_2), and an acid group (COOH) attached to it. Remember, however, that carbon atoms must have four bonds, ♦ so a fourth attachment is necessary. This fourth site distinguishes each amino acid from the others. Attached to the central carbon at the fourth bond is a distinct atom, or group of atoms, known as the *side group* or *side chain* (see Figure 6-1).

♦ Chemical bonds:
- H forms one bond
- O forms two bonds
- N forms three bonds
- C forms four bonds

Unique Side Groups The side groups on the central carbon vary from one amino acid to the next, making proteins more complex than either carbohydrates or lipids. A polysaccharide (starch, for example) may be several thousand units long, but each unit is a glucose molecule just like all the others. A protein, on the other hand, is made up of about 20 different amino acids, each with a different side group. Table 6-1 lists the amino acids most common in proteins.*

The simplest amino acid, glycine, has a hydrogen atom as its side group. A slightly more complex amino acid, alanine, has an extra carbon with three hydrogen atoms. Other amino acids have more complex side groups (see Figure 6-2 for examples). Thus, although all amino acids share a common structure, they differ in size, shape, electrical charge, and other characteristics because of differences in these side groups.

> FIGURE 6-1 Amino Acid Structure

All amino acids have a central carbon with an amino group (NH_2), an acid group (COOH), a hydrogen (H), and a side group attached. The side group is a unique chemical structure that differentiates one amino acid from another.

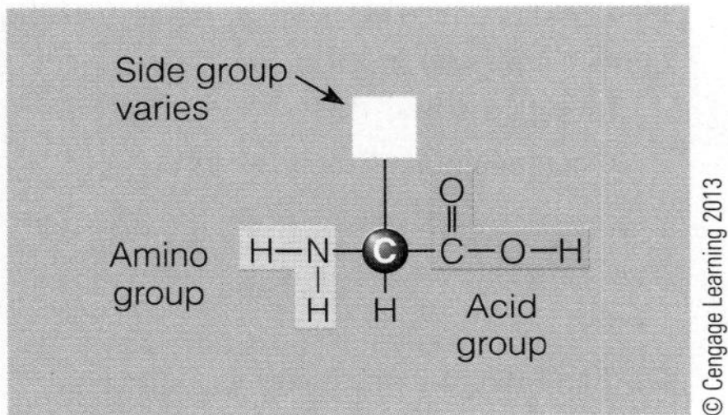

Nonessential Amino Acids More than half of the amino acids are *nonessential,* meaning that the body can synthesize them for itself. Proteins in foods usually deliver these amino acids, but it is not essential that they do so. The body can make all **nonessential amino acids,** given nitrogen to form the amino group and fragments from carbohydrate or fat to form the rest of the structure.

TABLE 6-1 Amino Acids

Proteins are made up of about 20 common amino acids. The first column lists the *essential amino acids* for human beings (those the body cannot make—that must be provided in the diet). The second column lists the *nonessential amino acids.* In special cases, some nonessential amino acids may become *conditionally essential.* In a newborn, for example, only five amino acids are truly nonessential; the other nonessential amino acids are conditionally essential until the metabolic pathways are developed enough to make those amino acids in adequate amounts.

Essential Amino Acids		Nonessential Amino Acids	
Histidine	(HISS-tuh-deen)	Alanine	(AL-ah-neen)
Isoleucine	(eye-so-LOO-seen)	Arginine	(ARJ-ih-neen)
Leucine	(LOO-seen)	Asparagine	(ah-SPAR-ah-geen)
Lysine	(LYE-seen)	Aspartic acid	(ah-SPAR-tic acid)
Methionine	(meh-THIGH-oh-neen)	Cysteine	(SIS-teh-een)
Phenylalanine	(fen-il-AL-ah-neen)	Glutamic acid	(GLU-tam-ic acid)
Threonine	(THREE-oh-neen)	Glutamine	(GLU-tah-meen)
Tryptophan	(TRIP-toe-fan, TRIP-toe-fane)	Glycine	(GLY-seen)
Valine	(VAY-leen)	Proline	(PRO-leen)
		Serine	(SEER-een)
		Tyrosine	(TIE-roe-seen)

*These 20 amino acids can all be commonly found in proteins. In addition, other amino acids do not occur in proteins but can be found individually (for example, taurine and ornithine). Some amino acids occur in related forms (for example, proline can acquire an OH group to become hydroxyproline).

proteins: compounds composed of carbon, hydrogen, oxygen, and nitrogen atoms, arranged into amino acids linked in a chain. Some amino acids also contain sulfur atoms.

amino (a-MEEN-oh) **acids:** building blocks of proteins. Each contains an amino group, an acid group, a hydrogen atom, and a distinctive side group, all attached to a central carbon atom.
- **amino** = containing nitrogen

nonessential amino acids: amino acids that the body can synthesize (see Table 6-1).

> **FIGURE 6-2 Examples of Amino Acids**

Note that all amino acids have a common chemical structure but that each has a different side group. Appendix C presents the chemical structures of the 20 amino acids most common in proteins.

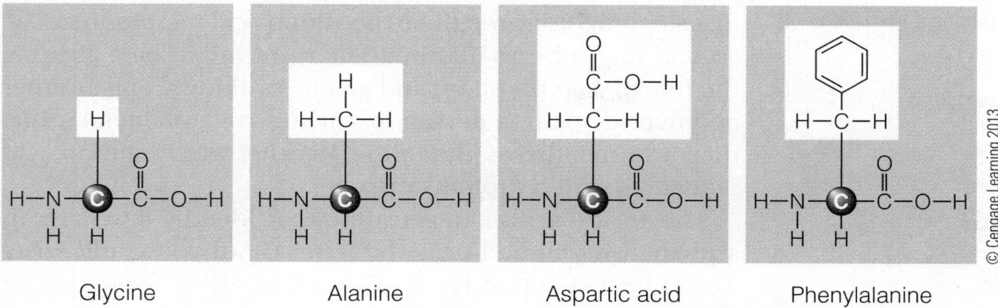

Essential Amino Acids There are nine amino acids that the human body either cannot make at all or cannot make in sufficient quantity to meet its needs. These nine amino acids must be supplied by the diet; they are *essential.* The first column in Table 6-1 presents the **essential amino acids.** Some researchers refer to essential amino acids as *indispensable* and to nonessential amino acids as *dispensable.*

Conditionally Essential Amino Acids Sometimes a nonessential amino acid becomes essential under special circumstances. For example, the body normally uses the essential amino acid phenylalanine to make tyrosine (a nonessential amino acid). But if the diet fails to supply enough phenylalanine, or if the body cannot make the conversion for some reason (as happens in the inherited disease phenylketonuria, described in Highlight 6), then tyrosine becomes a **conditionally essential amino acid.**

Proteins

Cells link amino acids end-to-end in a variety of sequences to form thousands of different proteins. A **peptide bond** unites each amino acid to the next.

Amino Acid Chains Condensation reactions connect amino acids, just as they combine two monosaccharides to form a disaccharide and three fatty acids with a glycerol to form a triglyceride. Two amino acids bonded together form a **dipeptide** (see Figure 6-3). By another such reaction, a third amino acid can be added to the chain to form a **tripeptide.** As additional amino acids join the chain, a **polypeptide** is formed. Most proteins are a few dozen to several hundred amino acids long. Figure 6-4 (p. 170) illustrates the protein insulin.

Primary Structure—Amino Acid Sequence The primary structure of a protein is determined by the sequence of amino acids. If a person could walk along a carbohydrate molecule like starch, the first stepping stone would be a glucose. The

> **FIGURE 6-3 Condensation of Two Amino Acids to Form a Dipeptide**

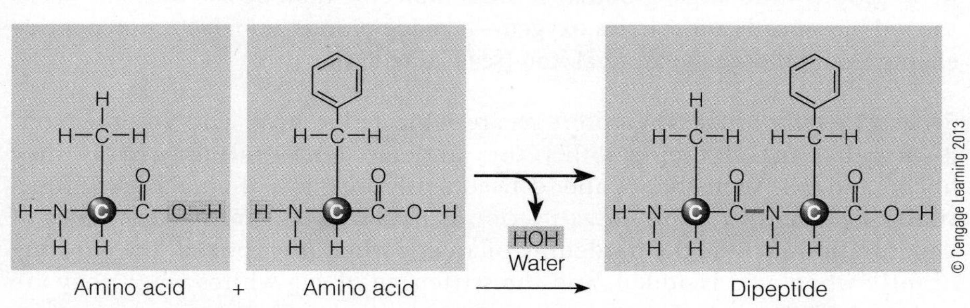

An OH group from the acid end of one amino acid and an H atom from the amino group of another join to form a molecule of water.

A peptide bond (highlighted in red) forms between the two amino acids, creating a dipeptide.

essential amino acids: amino acids that the body cannot synthesize in amounts sufficient to meet physiological needs (see Table 6-1).

conditionally essential amino acid: an amino acid that is normally nonessential, but must be supplied by the diet in special circumstances when the need for it exceeds the body's ability to produce it.

peptide bond: a bond that connects the acid end of one amino acid with the amino end of another, forming a link in a protein chain.

dipeptide (dye-PEP-tide): two amino acids bonded together.

- **di** = two
- **peptide** = amino acid

tripeptide: three amino acids bonded together.

- **tri** = three

polypeptide: many (10 or more) amino acids bonded together.

- **poly** = many

> FIGURE 6-4 Amino Acid Sequence of Human Insulin

Human insulin is a relatively small protein that consists of 51 amino acids in two short polypeptide chains. (For amino acid abbreviations, see Appendix C.) Two bridges link the two chains. A third bridge spans a section within the short chain. Known as *disulfide bridges*, these links form between the cysteine (Cys) amino acids, whose side group contains sulfur (S).

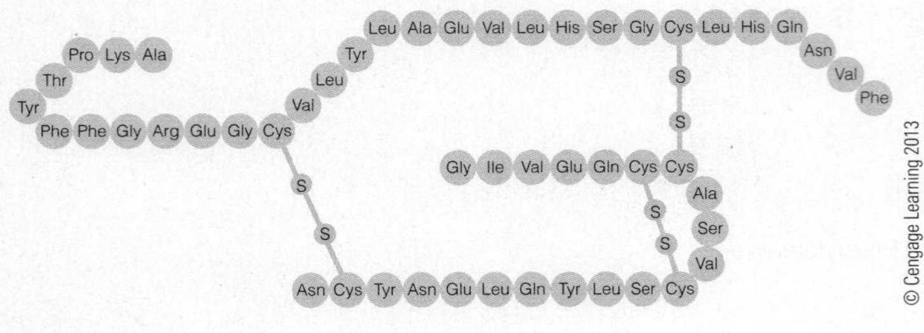

next stepping stone would also be a glucose, and it would be followed by a glucose, and yet another glucose. But if a person were to walk along a polypeptide chain, each stepping stone would be one of 20 different amino acids. The first stepping stone might be the amino acid methionine. The second might be an alanine. The third might be a glycine, the fourth a tryptophan, and so on. Walking along another polypeptide path, a person might step on a phenylalanine, then a valine, then a glutamine. In other words, amino acid sequences within proteins vary.

The amino acids can act somewhat like the letters in an alphabet. If you had only the letter G, all you could write would be a string of Gs: G–G–G–G–G–G–G. But with 20 different letters available, you can create poems, songs, and novels. Similarly, the 20 amino acids can be linked together in a variety of sequences—even more than are possible for letters in a word or words in a sentence. Thus the variety of possible sequences for polypeptide chains is tremendous.

Secondary Structure—Polypeptide Shapes The secondary structure of proteins is determined not by chemical bonds as between the amino acids but by weak electrical attractions within the polypeptide chain. As positively charged hydrogens attract nearby negatively charged oxygens, sections of the polypeptide chain twist into a helix or fold into a pleated sheet, for example. These shapes give proteins strength and rigidity.

Tertiary Structure—Polypeptide Tangles The tertiary structure of proteins occurs as long polypeptide chains twist and fold into a variety of complex, tangled shapes. The unique side group of each amino acid gives it characteristics that attract it to, or repel it from, the surrounding fluids and other amino acids. Some amino acid side groups are attracted to water molecules; they are *hydrophilic*. Other side groups are repelled by water; they are *hydrophobic*. As amino acids are strung together to make a polypeptide, the chain folds so that its hydrophilic side groups are on the outer surface near water; the hydrophobic groups tuck themselves inside, away from water. Similarly, the disulfide bridges in insulin (see Figure 6-4) determine its tertiary structure. The extraordinary and unique shapes of proteins enable them to perform their various tasks in the body. Some form globular or spherical structures that can carry and store materials within them, and some, such as those of tendons, form linear structures that are more than 10 times as long as they are wide. The intricate shape a protein finally assumes gives it maximum stability.

Cooking an egg denatures its proteins.

Quaternary Structure—Multiple Polypeptide Interactions Some polypeptides are functioning proteins just as they are; others need to associate with other polypeptides to form larger working complexes. The quaternary structure of proteins involves the interactions between two or more polypeptides. One molecule of **hemoglobin**—the large, globular protein molecule that, by the billions, packs the red blood cells and carries oxygen—is made of four associated polypeptide chains, each holding the mineral iron (see Figure 6-5).

Protein Denaturation When proteins are subjected to heat, acid, or other conditions that disturb their stability, they undergo **denaturation**—that is, they uncoil and lose their shapes and, consequently, also lose their ability to function. Past a certain point, denaturation is irreversible. Familiar examples of denaturation include the hardening of an egg when it is cooked, the curdling of milk when acid is added, and the stiffening of egg whites when they are whipped. In the body, proteins are denatured when they are exposed to stomach acid.

hemoglobin (HE-moh-GLO-bin): the globular protein of the red blood cells that carries oxygen from the lungs to the cells throughout the body.

- **hemo** = blood
- **globin** = globular protein

denaturation (dee-NAY-chur-AY-shun): the change in a protein's shape and consequent loss of its function brought about by heat, agitation, acid, base, alcohol, heavy metals, or other agents.

REVIEW IT Recognize the chemical structures of amino acids and proteins.
Chemically speaking, proteins are more complex than carbohydrates or lipids; they are made of some 20 different amino acids, 9 of which the body cannot make (the essential amino acids). Each amino acid contains an amino group, an acid group, a hydrogen atom, and a distinctive side group, all attached to a central carbon atom. Peptide bonds link amino acids together in a series of condensation reactions to create proteins. The distinctive sequence of amino acids in each protein determines its unique shape and function.

> **FIGURE 6-5 The Structure of Hemoglobin**

The globular hemoglobin protein is made of four polypeptide chains (quaternary structure).

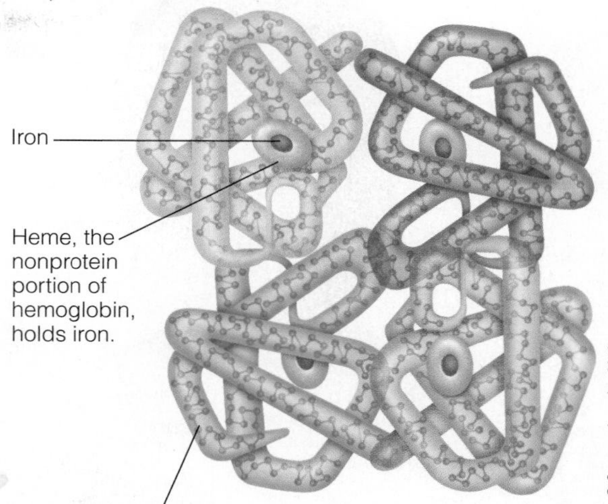

Iron
Heme, the nonprotein portion of hemoglobin, holds iron.

The shape of each polypeptide chain is determined by an amino acid sequence (primary structure) that twists into a helix (secondary structure) and bends itself into a ball shape (tertiary structure).

6.2 Digestion and Absorption of Proteins

LEARN IT Summarize protein digestion and absorption.

Proteins in foods do not become body proteins directly. Instead, foods supply the amino acids from which the body makes its own proteins. When a person eats foods containing protein, enzymes break the long polypeptides into short polypeptides, the short polypeptides into tripeptides and dipeptides, and, finally, the tripeptides and dipeptides into amino acids.

Protein Digestion

Figure 6-6 (p. 172) illustrates the digestion of protein through the GI tract. Proteins are crushed and moistened in the mouth, but the real action begins in the stomach.

In the Stomach The major event in the stomach is the partial breakdown (hydrolysis) of proteins. Hydrochloric acid uncoils (denatures) each protein's tangled strands so that digestive enzymes can attack the peptide bonds. The hydrochloric acid also converts the inactive form of the enzyme pepsinogen to its active form, **pepsin.*** Pepsin cleaves proteins—large polypeptides—into smaller polypeptides and some amino acids.

In the Small Intestine When polypeptides enter the small intestine, several pancreatic and intestinal **proteases** hydrolyze them further into short peptide chains, tripeptides, dipeptides, and amino acids.** Then **peptidase** enzymes on the membrane surfaces of the intestinal cells split most of the dipeptides and tripeptides into single amino acids. Only a few peptides escape digestion and enter the blood intact. Figure 6-6 (p. 172) includes names of the digestive enzymes for protein and describes their actions.

Protein Absorption

A number of specific carriers transport amino acids (and some dipeptides and tripeptides) into the intestinal cells. Once inside the intestinal cells, amino acids may be used for energy or to synthesize needed compounds. Amino acids that are not used by the intestinal cells are transported across the cell membrane into the surrounding fluid where they enter the capillaries on their way to the liver.

Consumers lacking nutrition knowledge may fail to realize that most proteins are broken down to amino acids before absorption. They may be misled by advertisements urging them to "Take this enzyme supplement to help you digest your food." Or "Don't eat this food that contains these enzymes that will digest cells in your body." In reality, though, enzymes in supplements and foods are proteins that are digested to amino acids, just as all proteins are. Even the digestive enzymes—which function optimally at their specific pH—are denatured and digested when the pH of their environment changes. The enzyme pepsin, for example, which works best in the low pH of the stomach becomes inactive and digested when it enters the higher pH of the small intestine.

Another misconception is that eating predigested proteins (amino acid supplements) saves the body from having to digest proteins and keeps the digestive system from "overworking." Such a belief grossly underestimates the body's abilities. As a matter of fact, the digestive system handles whole proteins *better* than predigested ones because it dismantles and absorbs the amino acids at rates that are optimal for the body's use. (The last section of this chapter discusses protein and amino acid supplements further.)

*The inactive form of an enzyme is called a *proenzyme* or a *zymogen* (ZYE-moh-jen).
**A string of four to nine amino acids is an *oligopeptide* (OL-ee-go-PEP-tide); *oligo* means few.

pepsin: a gastric enzyme that hydrolyzes protein. Pepsin is secreted in an inactive form, *pepsinogen*, which is activated by hydrochloric acid in the stomach.

proteases (PRO-tee-aces): enzymes that hydrolyze protein.

peptidase: a digestive enzyme that hydrolyzes peptide bonds. *Tripeptidases* cleave tripeptides; *dipeptidases* cleave dipeptides. *Endopeptidases* cleave peptide bonds within the chain to create smaller fragments, whereas *exopeptidases* cleave bonds at the ends to release free amino acids.

- **tri** = three
- **di** = two
- **endo** = within
- **exo** = outside

> FIGURE 6-6 *Animated* Protein Digestion in the GI Tract

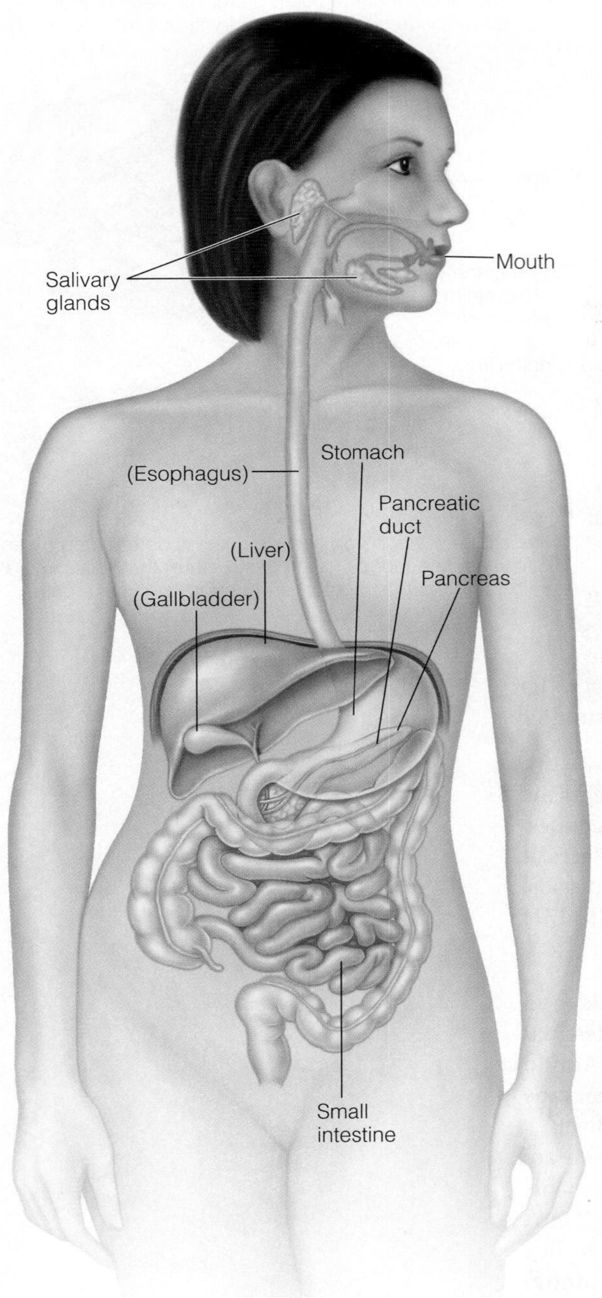

PROTEIN

Mouth and salivary glands
Chewing and crushing moisten protein-rich foods and mix them with saliva to be swallowed

HYDROCHLORIC ACID AND THE DIGESTIVE ENZYMES

PROTEIN	HYDROCHLORIC ACID AND THE DIGESTIVE ENZYMES
Stomach Hydrochloric acid (HCl) uncoils protein strands and activates stomach enzymes: Protein —(Pepsin, HCl)→ Smaller polypeptides	**In the stomach:** **Hydrochloric acid (HCl)** • Denatures protein structure • Activates pepsinogen to pepsin **Pepsin** • Cleaves proteins to smaller polypeptides and some free amino acids • Inhibits pepsinogen synthesis
Small intestine and pancreas Pancreatic and small intestinal enzymes split polypeptides further: Poly-peptides —(Pancreatic and intestinal proteases)→ Tripeptides, dipeptides, amino acids Then enzymes on the surface of the small intestinal cells hydrolyze these peptides and the cells absorb them: Peptides —(Intestinal tripeptidases and dipeptidases)→ Amino acids (absorbed)	**In the small intestine:** **Enteropeptidase** • Converts pancreatic trypsinogen to trypsin **Trypsin** • Inhibits trypsinogen synthesis • Cleaves peptide bonds next to the amino acids lysine and arginine • Converts pancreatic procarboxypeptidases to carboxypeptidases • Converts pancreatic chymotrypsinogen to chymotrypsin **Chymotrypsin** • Cleaves peptide bonds next to the amino acids phenylalanine, tyrosine, tryptophan, methionine, asparagine, and histidine **Carboxypeptidases** • Cleave amino acids from the acid (carboxyl) ends of polypeptides **Elastase and collagenase** • Cleave polypeptides into smaller polypeptides and tripeptides **Intestinal tripeptidases** • Cleave tripeptides to dipeptides and amino acids **Intestinal dipeptidases** • Cleave dipeptides to amino acids **Intestinal aminopeptidases** • Cleave amino acids from the amino ends of small polypeptides (oligopeptides)

REVIEW IT *Summarize protein digestion and absorption.*
Digestion is facilitated mostly by the stomach's acid and enzymes, which first denature dietary proteins, then cleave them into smaller polypeptides and some amino acids. Pancreatic and intestinal enzymes split these polypeptides further, to oligopeptides, tripeptides, and dipeptides, and then split most of these to single amino acids. Then carriers in the membranes of intestinal cells transport the amino acids into the cells, where they are released into the bloodstream.

6.3 Proteins in the Body

LEARN IT Describe how the body makes proteins and uses them to perform various roles.

The human body has an estimated 20,000 to 25,000 genes that code for hundreds of thousands of proteins. Relatively few proteins have been studied in detail, although this number is growing rapidly with the surge in knowledge gained from sequencing the human genome. The relatively few proteins described in this chapter illustrate the versatility, uniqueness, and importance of proteins. As you will see, each protein has a specific function, and that function is determined during protein synthesis.

Protein Synthesis Each human being is unique because of small differences in the body's proteins. These differences are determined by the amino acid sequences of proteins, which, in turn, are determined by genes. The following paragraphs describe in words the ways cells synthesize proteins; Figure 6-7 (p. 174) provides a pictorial description. Protein synthesis depends on a diet that provides adequate protein and essential amino acids.

The instructions for making every protein in a person's body are transmitted by way of the genetic information received at conception. This body of knowledge, which is filed in the DNA (deoxyribonucleic acid) within the nucleus of every cell, never leaves the nucleus.

Delivering the Instructions Transforming the information in DNA into the appropriate sequence of amino acids needed to make a specific protein requires two major steps. ♦ In the first step, known as **transcription,** a stretch of DNA is used as a template to make messenger RNA. Messenger RNA then carries the code across the nuclear membrane into the body of the cell, where it seeks out and attaches itself to one of the ribosomes (a protein-making machine, which is itself composed of RNA and protein). There the second step, known as **translation,** takes place. Situated on a ribosome, messenger RNA specifies the sequence in which the amino acids line up for the synthesis of a protein.

♦ DNA $\xrightarrow{\text{transcription}}$ RNA $\xrightarrow{\text{translation}}$ protein

Lining Up the Amino Acids Other forms of RNA, called transfer RNA, collect amino acids from the cell fluid and take them to messenger RNA. Each of the 20 amino acids has a specific transfer RNA. Thousands of transfer RNA, each carrying its amino acid, cluster around the ribosomes, awaiting their turn to unload. When the messenger RNA calls for a specific amino acid, the transfer RNA carrying that amino acid moves into position. Then the next loaded transfer RNA moves into place and then the next and the next. In this way, the amino acids line up in the sequence that is genetically determined, and enzymes bind them together. Finally, the completed protein strand is released, and the transfer RNA are freed to return for another load of amino acids.

Sequencing Errors The sequence of amino acids in each protein determines its shape, which supports a specific function. An error in the amino acid sequence results in an altered protein—sometimes with dramatic consequences. The protein hemoglobin offers one example of such a genetic variation. In a person with **sickle-cell anemia,** two of hemoglobin's four polypeptide chains (described earlier on pp. 170–171) have the normal sequence of amino acids, but the other two chains do not—they have the amino acid valine in a position that is normally occupied by glutamic acid (see Figure 6-8, p. 175). This single alteration in the amino acid sequence changes the characteristics and shape of hemoglobin so much that it loses its ability to carry oxygen effectively. The red blood cells filled with this abnormal hemoglobin stiffen into elongated sickle, or crescent, shapes instead of maintaining their normal pliable disc shape—hence the name, sickle-cell anemia. Sickle-cell anemia raises energy needs, causes many medical problems, and can be fatal.[1] Caring for children with sickle-cell anemia includes diligent attention to their water needs because dehydration can trigger a crisis.

transcription: the process of messenger RNA being made from a template of DNA.

translation: the process of messenger RNA directing the sequence of amino acids and synthesis of proteins.

sickle-cell anemia: a hereditary form of anemia characterized by abnormal sickle- or crescent-shaped red blood cells. Sickled cells interfere with oxygen transport and blood flow. Symptoms are precipitated by dehydration and insufficient oxygen (as may occur at high altitudes) and include hemolytic anemia (red blood cells burst), fever, and severe pain in the joints and abdomen.

> FIGURE 6-7 *Animated* **Protein Synthesis**

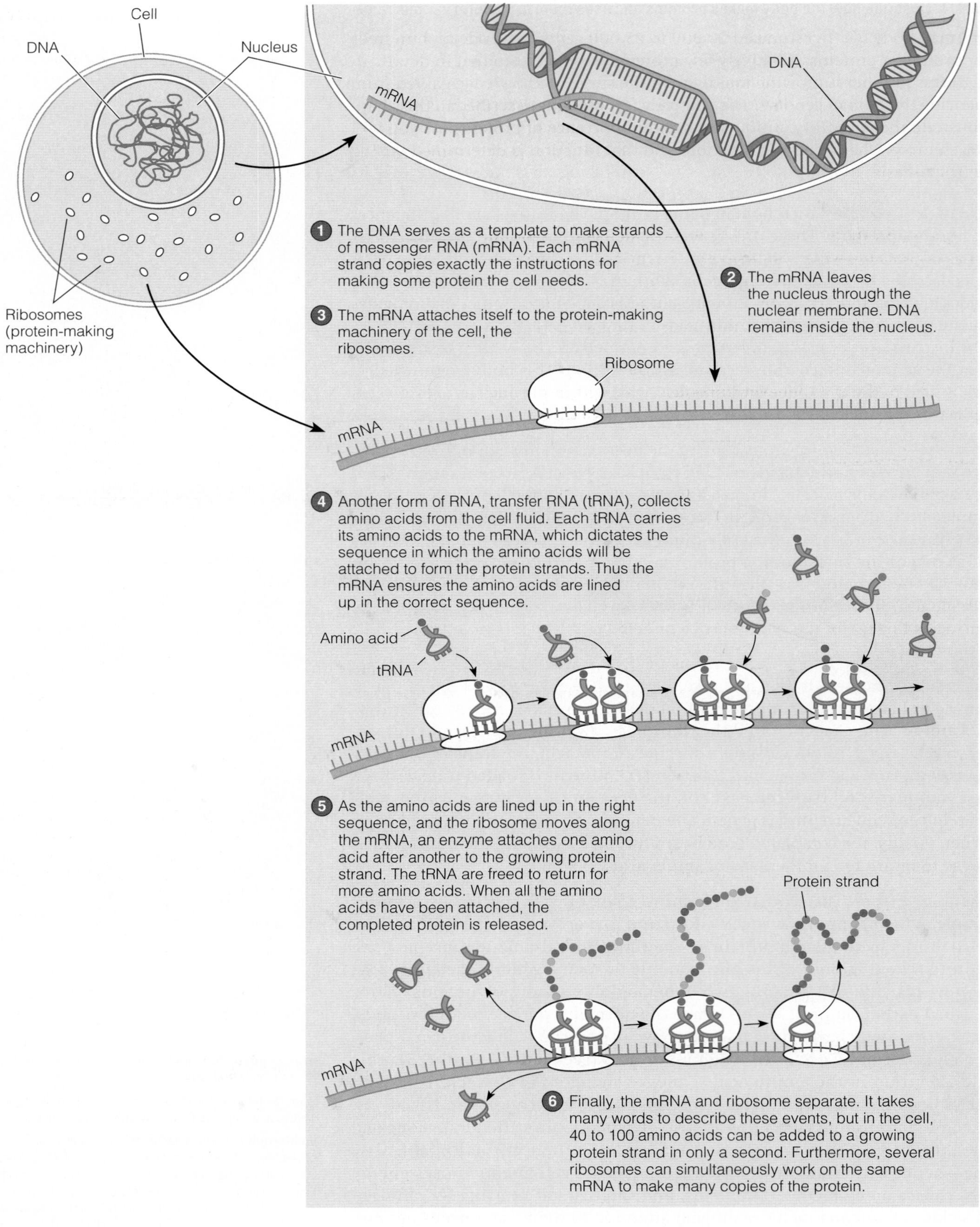
Cell
DNA
Nucleus
Ribosomes (protein-making machinery)
mRNA
DNA
1 The DNA serves as a template to make strands of messenger RNA (mRNA). Each mRNA strand copies exactly the instructions for making some protein the cell needs.
2 The mRNA leaves the nucleus through the nuclear membrane. DNA remains inside the nucleus.
3 The mRNA attaches itself to the protein-making machinery of the cell, the ribosomes.
Ribosome
mRNA
4 Another form of RNA, transfer RNA (tRNA), collects amino acids from the cell fluid. Each tRNA carries its amino acids to the mRNA, which dictates the sequence in which the amino acids will be attached to form the protein strands. Thus the mRNA ensures the amino acids are lined up in the correct sequence.
Amino acid
tRNA
mRNA
5 As the amino acids are lined up in the right sequence, and the ribosome moves along the mRNA, an enzyme attaches one amino acid after another to the growing protein strand. The tRNA are freed to return for more amino acids. When all the amino acids have been attached, the completed protein is released.
Protein strand
mRNA
6 Finally, the mRNA and ribosome separate. It takes many words to describe these events, but in the cell, 40 to 100 amino acids can be added to a growing protein strand in only a second. Furthermore, several ribosomes can simultaneously work on the same mRNA to make many copies of the protein.

Gene Expression When a cell makes a protein as described earlier, scientists say that the gene for that protein has been "expressed." Cells can regulate **gene expression** to make the type of protein, in the amounts and at the rate, they need. Nearly all of the body's cells possess the genes for making all human proteins, but each type of cell makes only the proteins it needs. For example, cells of the pancreas express the gene for insulin; in other cells, that gene is idle. Similarly, the cells of the pancreas do not make the protein hemoglobin, which is needed only by the red blood cells.

Recent research has unveiled some of the fascinating ways nutrients regulate gene expression and protein synthesis (see Highlight 6). Because diet plays an ongoing role in our lives from conception to death, it has a major influence on gene expression and disease development. The benefits of polyunsaturated fatty acids in defending against heart disease, for example, are partially explained by their role in influencing gene expression for lipid enzymes. Later chapters provide additional examples of relationships among nutrients, genes, and disease development.

Roles of Proteins

Whenever the body is growing, repairing, or replacing tissue, proteins are involved. Sometimes their role is to facilitate or to regulate; other times it is to become part of a structure. Versatility is a key feature of proteins.

As Structural Materials From the moment of conception, proteins form the building blocks of muscles, blood, and skin—in fact, protein is the major structural component of all the body's cells. To build a bone or a tooth, for example, cells first lay down a **matrix** of the protein **collagen** and then fill it with crystals of calcium, phosphorus, magnesium, fluoride, and other minerals.

Collagen also provides the material of ligaments and tendons and the strengthening "glue" between the cells of the artery walls that enables the arteries to withstand the pressure of the blood surging through them with each heartbeat. Also made of collagen are scars that knit the separated parts of torn tissues together.

Proteins are also needed for replacing dead or damaged cells. The average life span of a skin cell is only about 30 days. As old skin cells are shed, new cells made largely of protein grow from underneath to replace them. Cells in the deeper skin layers synthesize new proteins to form hair and fingernails. Muscle cells make new proteins to grow larger and stronger in response to exercise. Cells of the GI tract are replaced every few days. Both inside and outside, the body continuously uses protein to create new cells that replace those that have been lost.

As Enzymes Some proteins act as **enzymes.** Digestive enzymes have appeared in every chapter since Chapter 3, but digestion is only one of the many processes facilitated by enzymes. Enzymes not only break down substances, but they also build substances (such as bone) and transform one substance into another (amino acids into glucose, for example). Breaking down reactions are *catabolic,* whereas building up reactions are *anabolic.* (Chapter 7 provides more details.) Figure 6-9 diagrams a synthesis reaction.

An analogy may help to clarify the role of enzymes. Enzymes are comparable to the clergy and judges who make and dissolve marriages. When a minister marries two people, they become a couple, with a new bond between them. They are joined together—but the minister remains unchanged. The minister represents enzymes that synthesize large compounds from smaller ones. One minister can

> FIGURE 6-8 ***Animated*** **Sickle Cell Compared with Normal Red Blood Cell**

Normally, red blood cells are disc-shaped, but in the inherited disorder sickle-cell anemia, red blood cells are sickle- or crescent-shaped. This alteration in shape occurs because valine replaces glutamic acid in the amino acid sequence of two of hemoglobin's polypeptide chains. As a result of this one alteration, the hemoglobin has a diminished capacity to carry oxygen.

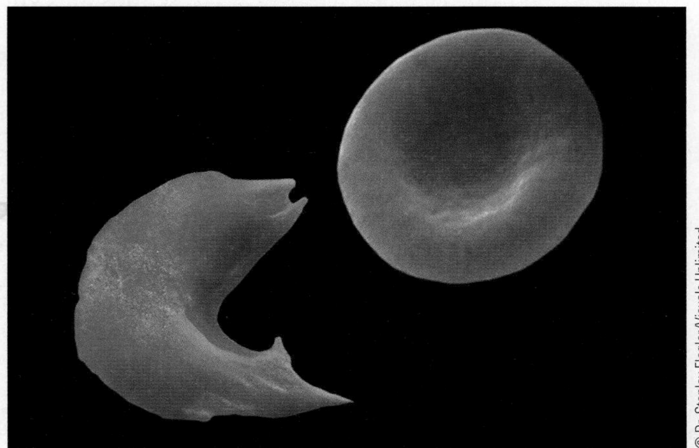

© Dr. Stanley Flegler/Visuals Unlimited

Sickle-shaped blood cell — Normal red blood cell

Amino acid sequence of normal hemoglobin:
Val—His—Leu—Thr—Pro—Glu—Glu

Amino acid sequence of sickle-cell hemoglobin:
Val—His—Leu—Thr—Pro—Val—Glu

gene expression: the process by which a cell converts the genetic code into RNA and protein.

matrix (MAY-tricks): the basic substance that gives form to a developing structure; in the body, the formative cells from which teeth and bones grow.

collagen (KOL-ah-jen): the structural protein from which connective tissues such as scars, tendons, ligaments, and the foundations of bones and teeth are made.

enzymes: proteins that facilitate chemical reactions without being changed in the process; protein catalysts.

> FIGURE 6-9 **Enzyme Action**

Each enzyme facilitates a specific chemical reaction. In this diagram, an enzyme enables two compounds to make a more complex structure, but the enzyme itself remains unchanged.

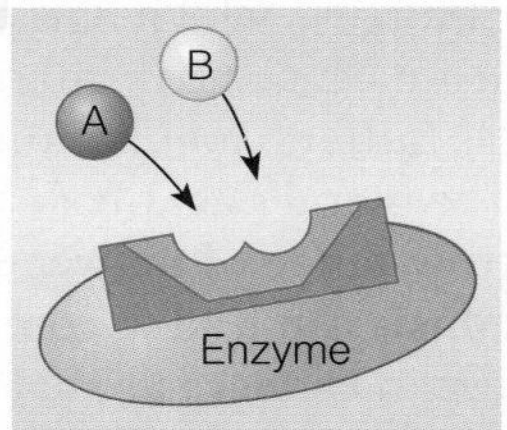

The separate compounds, A and B, are attracted to the enzyme's active site, making a reaction likely.

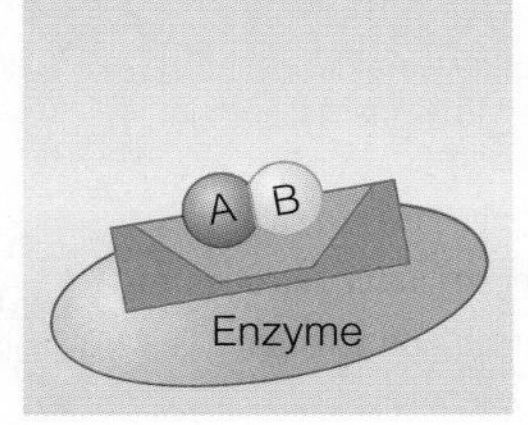

The enzyme forms a complex with A and B.

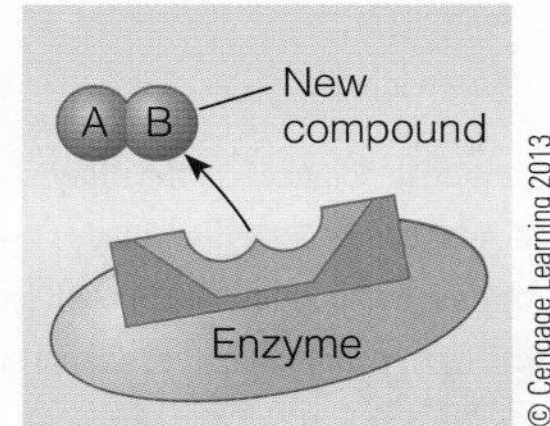

The enzyme is unchanged, but A and B have formed a new compound, AB.

© Cengage Learning 2013

TABLE 6-2 Examples of Hormones and Their Actions

Hormones	Actions
Growth hormone	Promotes growth
Insulin and glucagon	Regulate blood glucose (see Chapter 4)
Thyroxin	Regulates the body's metabolic rate (see Chapter 8)
Calcitonin and parathyroid hormone	Regulate blood calcium (see Chapter 12)
Antidiuretic hormone	Regulates fluid and electrolyte balance (see Chapter 12)

NOTE: Hormones are chemical messengers that are secreted by endocrine glands in response to altered conditions in the body. Each travels to one or more specific target tissues or organs, where it elicits a specific response. For descriptions of many hormones important in nutrition, see Appendix A.

perform thousands of marriage ceremonies, just as one enzyme can expedite billions of reactions.

Similarly, a judge who lets married couples separate may decree many divorces before retiring. The judge represents enzymes that hydrolyze larger compounds to smaller ones; for example, the digestive enzymes. The point is that, like the minister and the judge, enzymes themselves are not altered by the reactions they facilitate. They are catalysts, permitting reactions to occur more quickly and efficiently than if substances depended on chance encounters alone.

As Hormones The body's many hormones are messenger molecules, and *some* hormones are proteins. (Recall from Chapter 5 that some hormones, such as estrogen and testosterone, are made from the lipid cholesterol.) Various endocrine glands in the body release hormones in response to changes that challenge the body. The blood carries the hormones from these glands to their target tissues, where they elicit the appropriate responses to restore and maintain normal conditions.

The hormone insulin provides a familiar example. After a meal, when blood glucose rises, the pancreas releases insulin. Insulin stimulates the transport proteins of the muscles and adipose tissue to pump glucose into the cells faster than it can leak out. After acting on the message, the cells destroy the insulin. As blood glucose falls, the pancreas slows its release of insulin. Many other proteins act as hormones, regulating a variety of actions in the body (see Table 6-2 for examples).

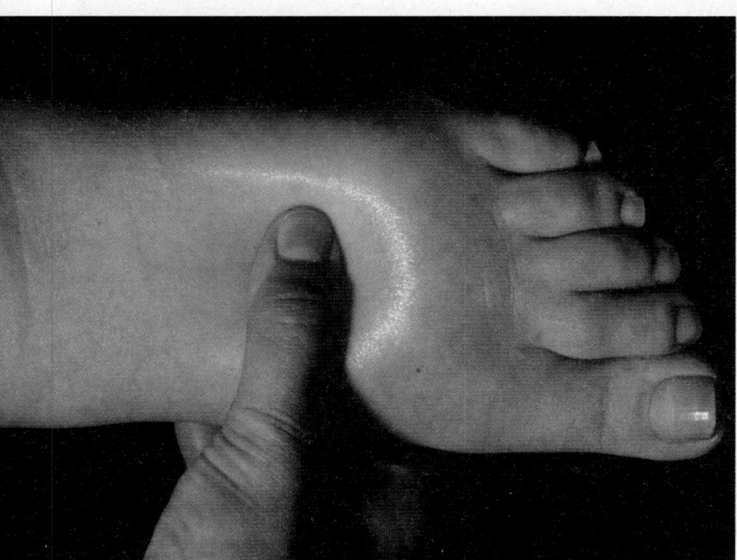

In critical illness and protein malnutrition, blood vessels become "leaky" and allow plasma proteins to move into the tissues. Because proteins attract water, the tissues swell, causing edema.

As Regulators of Fluid Balance Proteins help to maintain the body's **fluid balance.** Normally, proteins are found primarily within the cells and in the plasma (essentially blood without its red blood cells). Being large, proteins do not normally cross the walls of the blood vessels. During times of critical illness or protein malnutrition, however, plasma proteins leak out of the blood vessels into the spaces between the cells. Because proteins attract water, fluid accumulates and causes swelling. Swelling due to an excess of fluid in the tissues is known as **edema.** The protein-related causes of edema include:

- Excessive protein losses caused by inflammation and critical illnesses
- Inadequate protein synthesis caused by liver disease
- Inadequate dietary intake of protein

Whatever the cause of edema, the result is the same: a diminished capacity to deliver nutrients and oxygen to the cells and to remove wastes from them. As a consequence, cells fail to function adequately.

As Acid-Base Regulators Proteins also help to maintain the balance between **acids** and **bases** within the body fluids. Normal body processes continuously produce acids and bases, which the blood carries to the kidneys and lungs for excretion. The challenge is to maintain acid-base balance as conditions continually change.

An acid solution contains an abundance of hydrogen ions (H^+); the greater the concentration of hydrogen ions, the more acidic the solution and the lower the pH. Proteins, which have negative charges on their surfaces, attract hydrogen ions, which have positive charges. By accepting and releasing hydrogen ions, proteins act as **buffers,** maintaining the acid-base balance of the blood and body fluids.

The blood's acid-base balance is tightly controlled to maintain pH within the narrow range of between 7.35 and 7.45. Outside this range, either **acidosis** or **alkalosis** can lead to coma and death, largely by denaturing proteins. Denaturing a protein changes its shape and renders it useless. To give just one example, denatured hemoglobin loses its capacity to carry oxygen.

As Transporters Some proteins move about in the body fluids, carrying nutrients and other molecules. The protein hemoglobin carries oxygen from the lungs to

fluid balance: maintenance of the proper types and amounts of fluid in each compartment of the body fluids (see also Chapter 12).

edema (eh-DEEM-uh): the swelling of body tissue caused by excessive amounts of fluid in the interstitial spaces; seen in protein deficiency (among other conditions).

acids: compounds that release hydrogen ions in a solution.

bases: compounds that accept hydrogen ions in a solution.

buffers: compounds that keep a solution's pH constant when acids or bases are added.

acidosis (assi-DOE-sis): higher-than-normal acidity in the blood and body fluids.

alkalosis (alka-LOE-sis): higher-than-normal alkalinity (base) in the blood and body fluids.

the cells. The lipoproteins transport lipids around the body. Special transport proteins carry vitamins and minerals.

The transport of the mineral iron provides an especially good illustration of these proteins' specificity and precision. When iron is absorbed, it is captured in an intestinal cell by a protein. Before leaving the intestinal cell, iron is attached to another protein that carries it through the bloodstream to the cells. Once iron enters a cell, it is attached to a storage protein that will hold the iron until it is needed. When it is needed, iron is incorporated into proteins in the red blood cells and muscles that assist in oxygen transport and use. (Chapter 13 provides more details on how these protein carriers transport and store iron.)

Some transport proteins reside in cell membranes and act as "pumps," picking up compounds on one side of the membrane and releasing them on the other as needed. Each transport protein is specific for a certain compound or group of related compounds. Figure 6-10 illustrates how a membrane-bound transport protein helps to maintain the sodium and potassium concentrations in the fluids inside and outside cells. The balance of these two minerals is critical to nerve transmissions and muscle contractions; imbalances can cause irregular heartbeats, muscular weakness, kidney failure, and even death.

As Antibodies Proteins also defend the body against disease. A virus—whether it is one that causes flu, smallpox, measles, or the common cold—enters the cells and multiplies there. One virus may produce 100 replicas of itself within an hour or so. Each replica can then burst out and invade 100 different cells, soon yielding 10,000 viruses, which invade 10,000 cells. Left free to do their worst, they will soon overwhelm the body with disease.

Fortunately, when the body detects these invading **antigens,** it manufactures **antibodies,** giant protein molecules designed specifically to combat them. The antibodies work so swiftly and efficiently that in a healthy individual, most diseases never get started. Without sufficient protein, though, the body cannot maintain its army of antibodies to resist infectious diseases.

Each antibody is designed to destroy a specific antigen. Once the body has manufactured antibodies against a particular antigen (such as the measles virus), it "remembers" how to make them. Consequently, the next time the body encounters that same antigen, it produces antibodies even more quickly. In other words, the body develops a molecular memory, known as **immunity.** (Chapter 16 describes food allergies—the immune system's response to food antigens.)

As a Source of Energy and Glucose Without energy, cells die; without glucose, the brain and nervous system falter. Even though proteins are needed to do the

antigens: substances that elicit the formation of antibodies or an inflammation reaction from the immune system. A bacterium, a virus, a toxin, and a protein in food that causes allergy are all examples of antigens.

antibodies: large proteins of the blood and body fluids, produced by the immune system in response to the invasion of the body by foreign molecules (usually proteins called *antigens*). Antibodies combine with and inactivate the foreign invaders, thus protecting the body.

immunity: the body's ability to defend itself against diseases (see also Chapter 18).

> **FIGURE 6-10** ***Animated*** **An Example of a Transport Protein**

This transport protein resides within a cell membrane and acts as a two-door passageway. Molecules enter on one side of the membrane and exit on the other, but the protein doesn't leave the membrane. This example shows how the transport protein moves sodium and potassium in opposite directions across the membrane to maintain a high concentration of potassium and a low concentration of sodium within the cell. This active transport system requires energy.

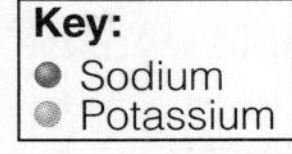

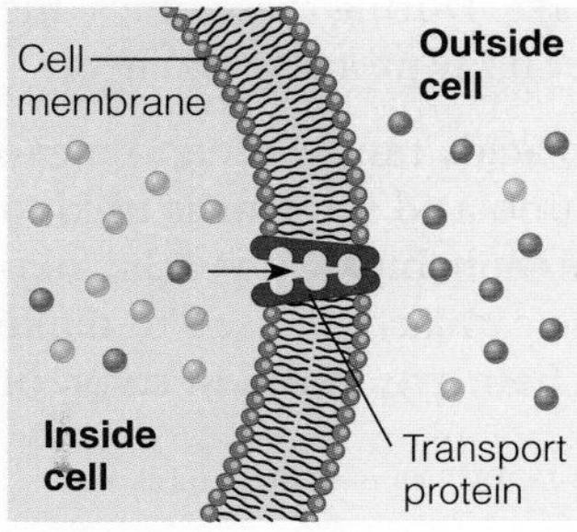

The transport protein picks up sodium from inside the cell.

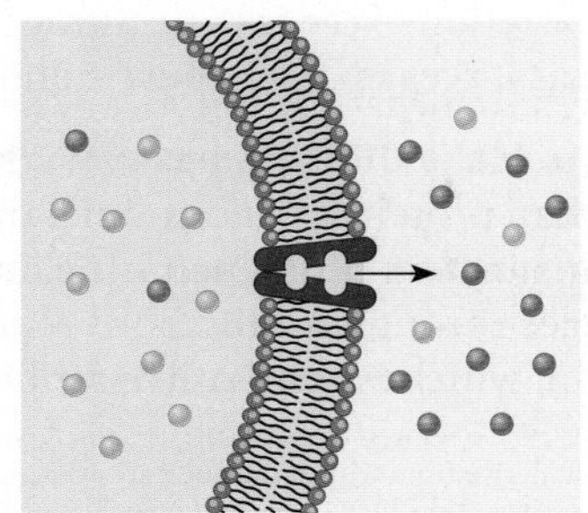

The protein changes shape and releases sodium outside the cell.

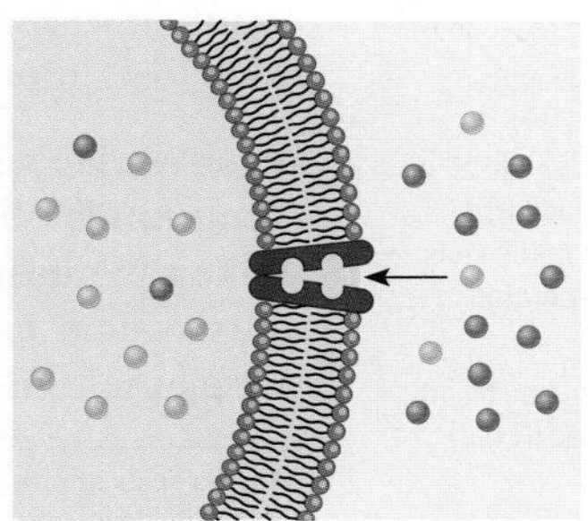

The transport protein picks up potassium from outside the cell.

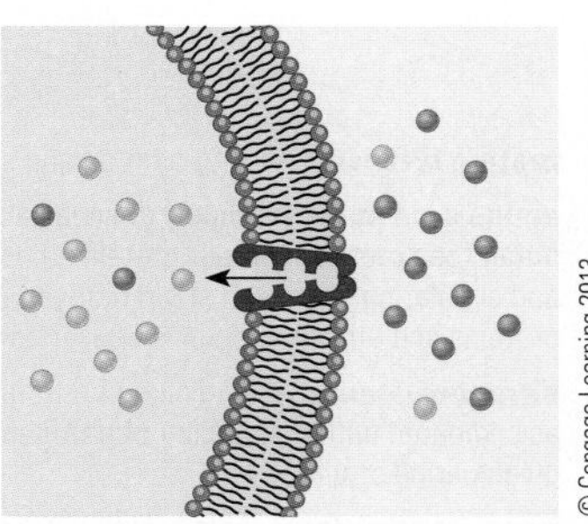

The protein changes shape and releases potassium inside the cell.

work that only they can perform, they will be sacrificed to provide energy and glucose during times of starvation or insufficient carbohydrate intake. The body will break down its tissue proteins to make amino acids available for energy or glucose production (a process known as *gluconeogenesis*). In this way, protein can maintain blood glucose levels, but at the expense of losing lean body tissue. Chapter 7 provides many more details on energy metabolism.

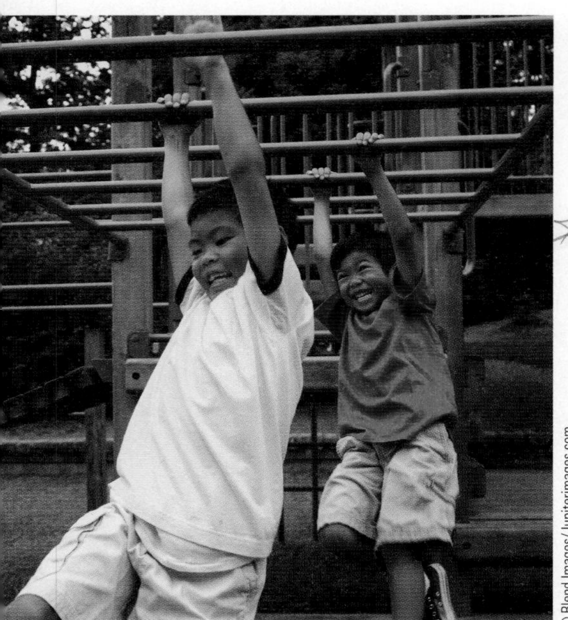

© Blend Images/Jupiterimages.com

Growing children end each day with more bone, blood, muscle, and skin cells than they had at the beginning of the day.

Other Roles As mentioned earlier, proteins form integral parts of most body structures such as skin, muscles, and bones. They also participate in some of the body's most amazing activities such as blood clotting and vision. When a tissue is injured, a rapid chain of events leads to the production of fibrin, a stringy, insoluble mass of protein fibers that forms a solid clot from liquid blood. Later, more slowly, the protein collagen forms a scar to replace the clot and permanently heal the wound. The light-sensitive pigments in the cells of the eye's retina are molecules of the protein opsin. Opsin responds to light by changing its shape, thus initiating the nerve impulses that convey the sense of sight to the brain.

The amino acids are as versatile as the proteins. In addition to serving as building blocks for proteins in the body, amino acids have multiple roles in regulating pathways that support growth, reproduction, metabolism, and immunity.[2]

A Preview of Protein Metabolism

This section previews protein metabolism; Chapter 7 provides a full description. Cells have several metabolic options, depending on their protein and energy needs.

Protein Turnover and the Amino Acid Pool Within each cell, proteins are continually being made and broken down, a process known as **protein turnover.** Protein breakdown releases amino acids.* These amino acids mix with amino acids from dietary protein to form an **"amino acid pool"** within the cells and circulating blood. The rate of protein degradation and the amount of protein intake may vary, but the pattern of amino acids within the pool remains fairly constant. Regardless of their source, any of these amino acids can be used to make body proteins or other nitrogen-containing compounds, or they can be stripped of their nitrogen and used for energy (either immediately or stored as fat for later use).

◆ Nitrogen balance:

- Nitrogen equilibrium (zero nitrogen balance): N in = N out
- Positive nitrogen: N in > N out
- Negative nitrogen: N in < N out

Nitrogen Balance Protein turnover and **nitrogen balance** go hand in hand. ◆ In healthy adults, protein synthesis balances with degradation, and protein intake from food balances with nitrogen excretion in the urine, feces, and sweat. When nitrogen intake equals nitrogen output, the person is in nitrogen equilibrium, or zero nitrogen balance. Researchers use nitrogen balance studies to estimate protein requirements.**

If the body synthesizes more than it degrades, then protein is added and nitrogen status becomes positive. Nitrogen status is positive in growing infants, children, adolescents, pregnant women, and people recovering from protein deficiency or illness; their nitrogen intake exceeds their nitrogen excretion. They are retaining protein in new tissues as they add blood, bone, skin, and muscle cells to their bodies.

If the body degrades more than it synthesizes, then protein is being lost and nitrogen status becomes negative. Nitrogen status is negative in people who are starving or suffering other severe stresses such as burns, injuries, infections, and fever; their nitrogen excretion exceeds their nitrogen intake. During these times, the body loses nitrogen as it breaks down muscle and other body proteins for energy.

Using Amino Acids to Make Other Compounds Amino acids can be used to make compounds other than proteins. For example, the amino acid tyrosine is used to make the **neurotransmitters** norepinephrine and epinephrine, which relay nervous system messages throughout the body. Tyrosine can also be used to make the pigment melanin, which is responsible for brown hair, eye, and skin color, or

protein turnover: the degradation and synthesis of protein.

amino acid pool: the supply of amino acids derived from either food proteins or body proteins that collect in the cells and circulating blood and stand ready to be incorporated in proteins and other compounds or used for energy.

nitrogen balance: the amount of nitrogen consumed (N in) as compared with the amount of nitrogen excreted (N out) in a given period of time.

neurotransmitters: chemicals that are released at the end of a nerve cell when a nerve impulse arrives there. They diffuse across the gap to the next cell and alter the membrane of that second cell to either inhibit or excite it.

*Amino acids or proteins that derive from within the body are *endogenous* (en-DODGE-eh-nus). In contrast, those that derive from foods are *exogenous* (eks-ODGE-eh-nus).

**The genetic materials DNA and RNA contain nitrogen, but the quantity is insignificant compared with the amount in protein. Protein is 16 percent nitrogen. Said another way, the average protein weighs about 6.25 times as much as the nitrogen it contains, so scientists can estimate the amount of protein in a sample of food, body tissue, or other material by multiplying the weight of the nitrogen in it by 6.25.

the hormone thyroxin, which helps to regulate the metabolic rate. For another example, the amino acid tryptophan serves as a precursor for the vitamin niacin and for **serotonin,** a neurotransmitter important in sleep regulation, appetite control, and sensory perception.

Using Amino Acids for Energy and Glucose As mentioned earlier, when glucose or fatty acids are limited, cells are forced to use amino acids for energy and glucose. The body does not have a specialized storage site for protein as it does for carbohydrate and fat. Glucose is stored as glycogen in the liver and fat as triglycerides in adipose tissue, but protein is not stored as such. When the need arises, the body breaks down its working and structural proteins and uses the amino acids for energy or glucose. Thus, over time, energy deprivation (fasting or starvation) always causes wasting of lean body tissue as well as fat loss. An adequate supply of carbohydrates and fats spares amino acids from being used for energy and allows them to perform their unique roles.

Using Amino Acids to Make Fat Amino acids may be converted to fat when energy and protein intakes exceed needs and carbohydrate intake is adequate. In this way, protein-rich foods can contribute to weight gain.

Deaminating Amino Acids When amino acids are broken down (as occurs when they are used for energy or to make glucose or fat), they are first deaminated—stripped of their nitrogen-containing amino groups (see Figure 6-11). Two products result from **deamination:** one is **ammonia** (NH_3); the other product is the carbon structure without its amino group—often a **keto acid.** Keto acids may enter a number of metabolic pathways—for example, they may be used for energy or for the production of glucose, ketones, cholesterol, or fat.* They may also be used to make nonessential amino acids.

Using Amino Acids to Make Proteins and Nonessential Amino Acids As mentioned, cells can assemble amino acids into the proteins they need to do their work. If an essential amino acid is missing, the body may break down some of its own proteins to obtain it. If a particular nonessential amino acid is not readily available, cells can make it from a keto acid—if a nitrogen source is available. Cells can also make a nonessential amino acid by transferring an amino group from one amino acid to its corresponding keto acid, as shown in Figure 6-12. Through many such **transamination** reactions, involving many different keto acids, the liver cells can synthesize the nonessential amino acids.

Converting Ammonia to Urea As mentioned earlier, deamination produces ammonia. Ammonia is a toxic compound chemically identical to the strong-smelling ammonia in bottled cleaning solutions. Because ammonia is a base, excessive

> **FIGURE 6-11 Deamination and Synthesis of a Nonessential Amino Acid**

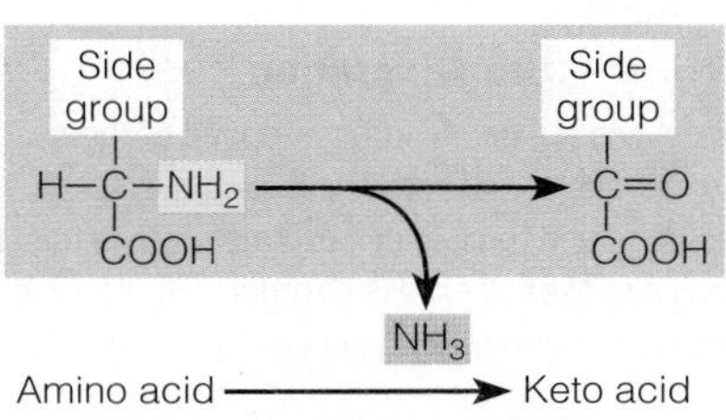

The deamination of an amino acid produces ammonia (NH_3) and a keto acid.

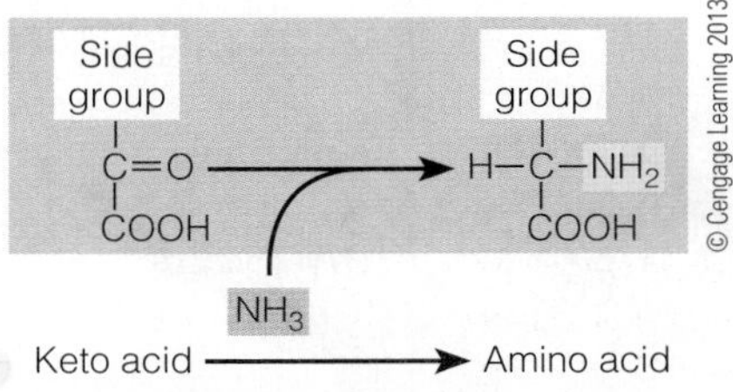

Given a source of NH_3, the body can make nonessential amino acids from keto acids.

> **FIGURE 6-12 Transamination and Synthesis of a Nonessential Amino Acid**

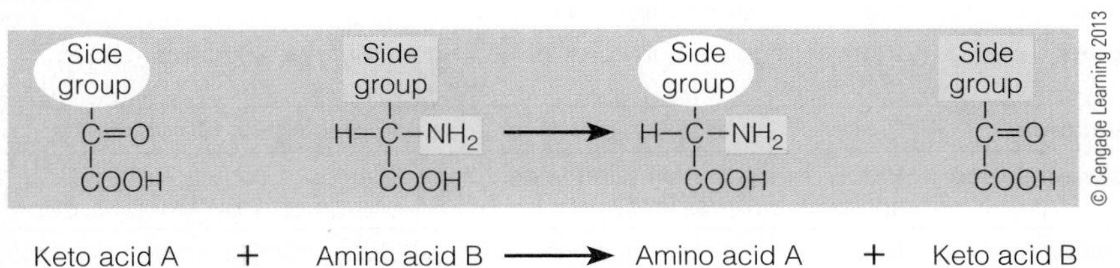

Keto acid A + Amino acid B → Amino acid A + Keto acid B

The body can transfer amino groups (NH_2) from an amino acid to a keto acid, forming a new *nonessential* amino acid and a new keto acid. Transamination reactions require the vitamin B_6 coenzyme.

*Chemists sometimes classify amino acids according to the destinations of their carbon fragments after deamination. If the fragment leads to the production of glucose, the amino acid is called *glucogenic;* if it leads to the formation of ketone bodies, fats, and sterols, the amino acid is called *ketogenic.* There is no sharp distinction between glucogenic and ketogenic amino acids, however. A few are both, most are considered glucogenic, only one (leucine) is clearly ketogenic.

serotonin: a neurotransmitter important in sleep regulation, appetite control, and sensory perception, among other roles. Serotonin is synthesized in the body from the amino acid tryptophan with the help of vitamin B_6.

deamination (dee-AM-ih-NAY-shun): removal of the amino (NH_2) group from a compound such as an amino acid.

ammonia: a compound with the chemical formula NH_3; produced during the deamination of amino acids.

keto (KEY-toe) **acid:** an organic acid that contains a carbonyl group (C=O).

transamination (TRANS-am-ih-NAY-shun): the transfer of an amino group from one amino acid to a keto acid, producing a new nonessential amino acid and a new keto acid.

> **FIGURE 6-13 Urea Synthesis**

Ammonia is produced when amino acids are deaminated. The liver detoxifies ammonia by combining it with another waste product, carbon dioxide, to produce urea. See Appendix C for details.

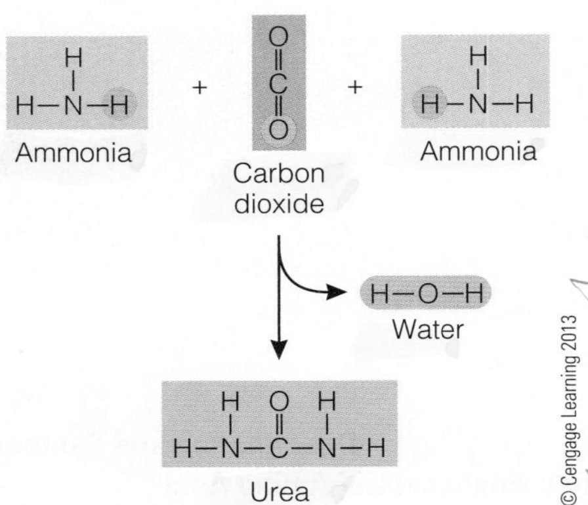

> **FIGURE 6-14 Urea Excretion**

When amino acids are deaminated (stripped of their nitrogen), ammonia is released. The liver converts ammonia to urea, and the kidneys excrete urea. In this way the body disposes of excess nitrogen. (Figure 12-2 provides details of how the kidneys work.)

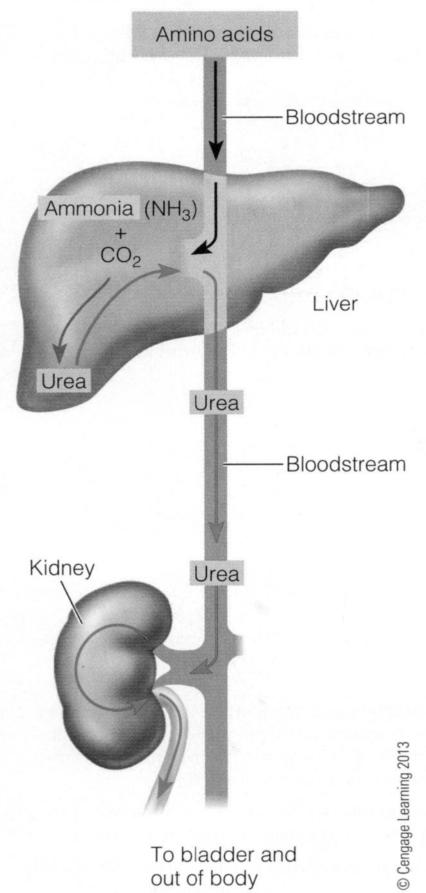

urea (you-REE-uh): the principal nitrogen-excretion product of protein metabolism. Two ammonia fragments are combined with carbon dioxide to form urea.

quantities upset the blood's critical acid-base balance. To prevent such a crisis, the liver combines ammonia with carbon dioxide to make **urea,** a much less toxic compound. Figure 6-13 provides a greatly oversimplified diagram of urea synthesis; details are shown in Appendix C. The production of urea increases as dietary protein increases, until production hits its maximum rate at intakes approaching 250 grams per day. (For perspective, the average daily intake of protein in the United States is 78 grams.[3])

Excreting Urea Liver cells release urea into the blood, where it circulates until it passes through the kidneys (see Figure 6-14). The kidneys then filter urea out of the blood for excretion in the urine. Normally, the liver efficiently captures all the ammonia, makes urea from it, and releases the urea into the blood; then the kidneys clear all the urea from the blood. This division of labor allows easy diagnosis of diseases of both organs. In liver disease, blood ammonia is high; in kidney disease, blood urea is high.

Urea is the body's principal vehicle for excreting unused nitrogen, and the amount of urea produced increases with protein intake. To keep urea in solution, the body needs water. For this reason, a person who regularly consumes a high-protein diet (say, 100 grams a day or more) must drink plenty of water to dilute and excrete urea from the body. Without extra water, a person on a high-protein diet risks dehydration because the body uses its water to rid itself of urea. This explains some of the water loss that accompanies high-protein diets. Such losses may make high-protein diets *appear* to be effective, but water loss, of course, is of no value to the person who wants to lose body fat (as Highlight 8 explains).

REVIEW IT Describe how the body makes proteins and uses them to perform various roles.

Cells synthesize proteins according to genetic information that dictates the sequence in which amino acids are linked together. Each protein plays a specific role. Table 6-3 summarizes some of the many roles proteins play and conveys a sense of the immense variety and importance of proteins in the body.

Proteins are constantly being synthesized and broken down as needed. The body's assimilation of amino acids into proteins and its release of amino acids via protein breakdown and excretion can be tracked by measuring nitrogen balance, which should be positive during growth and steady in adulthood. An energy deficit or an inadequate protein intake may force the body to use amino acids as fuel, creating a negative nitrogen balance. Protein eaten in excess of need is broken down and stored as body fat.

TABLE 6-3 Protein Functions in the Body

Structural materials	Proteins form integral parts of most body tissues and provide strength and shape to skin, tendons, membranes, muscles, organs, and bones.
Enzymes	Proteins facilitate chemical reactions.
Hormones	Proteins regulate body processes. (Some, but not all, hormones are proteins.)
Fluid balance	Proteins help to maintain the volume and composition of body fluids.
Acid-base balance	Proteins help to maintain the acid-base balance of body fluids by acting as buffers.
Transportation	Proteins transport substances, such as lipids, vitamins, minerals, and oxygen, around the body.
Antibodies	Proteins inactivate foreign invaders, thus protecting the body against diseases.
Energy and glucose	Proteins provide some fuel, and glucose if needed, for the body's energy needs.
Other	The protein fibrin creates blood clots; the protein collagen forms scars; the protein opsin participates in vision.

6.4 Protein in Foods

LEARN IT Explain the differences between high-quality and low-quality proteins, including notable food sources of each.

In the United States and other countries where nutritious foods are abundant, most people eat protein in such large quantities that they receive all the amino acids they need. In countries where food is scarce and the people eat only marginal amounts of protein-rich foods, however, the *quality* of the protein becomes crucial.

Protein Quality The protein quality of the diet determines, in large part, how well children grow and how well adults maintain their health. Put simply, **high-quality proteins** provide enough of all the essential amino acids needed to support the body's work, and low-quality proteins don't. Two factors influence protein quality—the protein's digestibility and its amino acid composition.

Digestibility As explained earlier, proteins must be digested before they can provide amino acids. **Protein digestibility** depends on such factors as the protein's source and the other foods eaten with it. The digestibility of most animal proteins is high (90 to 99 percent); plant proteins are less digestible (70 to 90 percent for most, but more than 90 percent for soy and legumes).

Amino Acid Composition To make proteins, a cell must have all the needed amino acids available simultaneously. The liver can make any nonessential amino acid that may be in short supply so that the cells can continue linking amino acids into protein strands. If an essential amino acid is missing, though, a cell must dismantle its own proteins to obtain it. Therefore, to prevent protein breakdown in the body, dietary protein must supply at least the nine essential amino acids plus enough nitrogen-containing amino groups and energy for the synthesis of the nonessential ones. If the diet supplies too little of any essential amino acid, protein synthesis will be limited. The body makes whole proteins only; if one amino acid is missing, the others cannot form a "partial" protein. An essential amino acid supplied in less than the amount needed to support protein synthesis is called a **limiting amino acid.**

Reference Protein The quality of a food protein is determined by comparing its amino acid composition with the essential amino acid requirements of preschool-age children. Such a standard is called a **reference protein.** The rationale behind using the requirements of this age group is that if a protein will effectively support a young child's growth and development, then it will meet or exceed the requirements of older children and adults.

High-Quality Proteins As mentioned earlier, a high-quality protein contains all the essential amino acids in relatively the same amounts and proportions that human beings require; it may or may not contain all the nonessential amino acids. Proteins that are low in an essential amino acid cannot, by themselves, support protein synthesis. Generally, foods derived from animals (meat, seafood, poultry, eggs, and milk and milk products) provide high-quality proteins, although gelatin is an exception. Gelatin lacks tryptophan and cannot support growth and health as a diet's sole protein. Proteins from plants (vegetables, nuts, seeds, grains, and legumes) have more diverse amino acid patterns and tend to be limiting in one or more essential amino acids. Some plant proteins are notoriously low quality (for example, corn protein). A few others are high quality (for example, soy protein).

Researchers have developed several methods for evaluating the quality of food proteins and identifying high-quality proteins. Appendix D provides details.

Complementary Proteins In general, plant proteins are lower quality than animal proteins, and plants also offer less protein (per weight or measure of food). For this reason, many vegetarians improve the quality of proteins in their diets by combining plant-protein foods that have different but complementary amino acid

high-quality proteins: dietary proteins containing all the essential amino acids in relatively the same amounts that human beings require. They may also contain nonessential amino acids.

protein digestibility: a measure of the amount of amino acids absorbed from a given protein intake.

limiting amino acid: the essential amino acid found in the shortest supply relative to the amounts needed for protein synthesis in the body. Four amino acids are most likely to be limiting:

- Lysine
- Methionine
- Threonine
- Tryptophan

reference protein: a standard against which to measure the quality of other proteins.

© Polara Studios, Inc.

Black beans and rice, a favorite Hispanic combination, together provide a balanced array of amino acids.

patterns. This strategy yields **complementary proteins** that together contain all the essential amino acids in quantities sufficient to support health. The protein quality of the combination is greater than either food alone (see Figure 6-15).

Many people have long believed that combining plant proteins at every meal is critical to protein nutrition. For most healthy vegetarians, though, it is *not* necessary to balance amino acids at each meal if protein intake is varied and energy intake is sufficient.[4] Vegetarians can receive all the amino acids they need over the course of a day by eating a variety of whole grains, legumes, seeds, nuts, and vegetables. Protein deficiency will develop, however, when fruits and certain vegetables make up the core of the diet, severely limiting both the *quantity* and *quality* of protein. Highlight 2 describes how to plan a nutritious vegetarian diet.

REVIEW IT Explain the differences between high-quality and low-quality proteins, including notable food sources of each.

A diet that supplies all of the essential amino acids in adequate amounts ensures protein synthesis. The best guarantee of amino acid adequacy is to eat foods containing high-quality proteins or mixtures of foods containing complementary proteins that can each supply the amino acids missing in the other. In addition to its amino acid content, the quality of protein is measured by its digestibility and its ability to support growth. Such measures are of great importance in dealing with malnutrition worldwide, but in countries where protein deficiency is not common, the protein quality of individual foods deserves little emphasis.

> FIGURE 6-15 **Complementary Proteins**

In general, legumes provide plenty of isoleucine (Ile) and lysine (Lys) but fall short in methionine (Met) and tryptophan (Trp). Grains have the opposite strengths and weaknesses, making them a perfect match for legumes.

	Ile	Lys	Met	Trp
Legumes	✓	✓		
Grains			✓	✓
Together	✓	✓	✓	✓

© Cengage Learning 2013

6.5 Health Effects and Recommended Intakes of Protein

LEARN IT Identify the health benefits of, and recommendations for, protein.

As you know by now, protein is indispensable to life. This section examines the health effects and recommended intakes of protein.

Health Effects of Protein

It should come as no surprise that protein deficiency can have devastating effects on people's health. But like the other nutrients, protein in excess can also be harmful. High-protein diets have been implicated in several chronic diseases, including heart disease, cancer, osteoporosis, obesity, and kidney stones, but evidence is insufficient to establish an Upper Level (UL).[5]

Protein Deficiency In protein deficiency, as occurs when the diet supplies too little protein or lacks a specific essential amino acid relative to the others (a limiting amino acid), the synthesis of body proteins decreases and degradation increases to provide the needed amino acids. Without proteins to perform their critical roles, many of the body's activities come to a halt. The consequences of protein deficiency include slowed growth, impaired brain and kidney functions, poor immunity, and inadequate nutrient absorption.

The term *protein-energy malnutrition* has traditionally been used to describe the condition that develops when the diet delivers too little protein, too little energy, or both. The causes and consequences are complex, but clearly, such malnutrition reflects insufficient food intake. Importantly, not only are protein and energy inadequate, but so are many, if not all, of the vitamins and minerals. For this reason, severe malnutrition and its clinical forms—marasmus and kwashiorkor—are included in Chapter 20's discussion of world hunger.

Heart Disease In the United States and other developed countries, protein is so abundant that problems of excess are more common than deficiency. A high-protein diet may contribute to the progression of heart disease. As Chapter 5 mentions, foods rich in animal protein also tend to be rich in saturated fats. Consequently, it is not surprising to find a correlation between animal-protein intake (red meats and milk products) and heart disease.[6] On the other hand, substituting vegetable protein for animal protein may improve blood pressure and blood lipids and decrease heart disease mortality.[7]

complementary proteins: two or more dietary proteins whose amino acid assortments complement each other in such a way that the essential amino acids missing from one are supplied by the other.

Research suggests that elevated levels of the amino acid homocysteine may be an independent risk factor for heart disease, heart attacks, and sudden death in patients with heart disease.[8] Researchers do not yet fully understand the many factors—including a diet high in saturated fatty acids—that can raise homocysteine in the blood or whether elevated levels are a cause or an effect of heart disease.[9] Elevated homocysteine is associated with increased oxidative stress and inflammation. Until researchers can determine the exact role homocysteine plays in heart disease, they are following several leads in pursuit of the answers. Coffee's role in heart disease has been controversial, but research suggests it is among the most influential factors in raising homocysteine, which may explain some of the adverse health effects of heavy consumption. Elevated homocysteine levels are among the many adverse health consequences of smoking cigarettes and drinking alcohol as well. Homocysteine is also elevated with inadequate intakes of B vitamins and can usually be lowered with fortified foods or supplements of vitamin B_{12}, vitamin B_6, and folate.[10] Lowering homocysteine, however, may not help in lowering risks or preventing heart attacks.[11] Supplements of the B vitamins do not always benefit those with heart disease and, in fact, may actually increase risks.[12]

In contrast to homocysteine, the amino acid arginine may help protect against heart disease by lowering blood pressure and homocysteine levels.[13] Additional research is needed to confirm the benefits of arginine. In the meantime, it is unwise for consumers to use supplements of arginine, or any other amino acid for that matter (as pp. 186–187 explain). Physicians, however, may consider the benefits of adding arginine supplements to their heart patients' treatment plan.

Cancer Protein does not seem to increase the risk of cancer, but some protein-rich foods do. For example, evidence suggests a strong correlation between high intakes of red meat and processed meats with cancer of the colon.[14]* Chapter 18 discusses dietary links with cancer, and Chapter 19 presents food safety issues of processed meats.

Adult Bone Loss (Osteoporosis) Chapter 12 presents calcium metabolism, and Highlight 12 elaborates on the main factors that influence osteoporosis. This section briefly describes the relationships between protein intake and bone loss. When protein intake is high, calcium excretion increases. Whether excess protein depletes the bones of their chief mineral may depend upon the ratio of calcium intake to protein intake. After all, bones need both protein and calcium. An ideal ratio has not been determined, but a young woman whose intake meets recommendations for both nutrients has a calcium-to-protein ratio of more than 20 to 1 (milligrams to grams), which probably provides adequate protection for the bones. For most women in the United States, however, average calcium intakes are lower and protein intakes are higher, yielding a 9-to-1 ratio, which may produce calcium losses significant enough to compromise bone health. In other words, the problem may reflect too little calcium, not too much protein. In establishing recommendations, the DRI Committee considered protein's effect on calcium metabolism and bone health, but it did not find sufficient evidence to warrant an adjustment for calcium or a UL for protein.[15]

Some (but not all) research suggests that animal protein may be more detrimental to calcium metabolism and bone health than vegetable protein. Importantly, adequate protein does not harm bones and may even improve bone mineral density, whereas *inadequate* intakes of protein may compromise bone health.[16] Osteoporosis is particularly common in elderly women and in adolescents with anorexia nervosa—groups who typically receive less protein than they need. For these people, increasing protein intake may be just what they need to protect their bones.[17]

Weight Control Fad weight-loss diets that encourage a high-protein, low-carbohydrate diet may be effective, but only because they are low-kcalorie diets. Diets that provide adequate protein (at least 65 to 70 grams a day), moderate fat, and sufficient energy

*Processed meats include ham, bacon, pastrami, salami, sausage, bratwurst, and hot dogs that have been preserved by smoking, curing, salting, or adding preservatives.

from carbohydrates can better support weight loss and good health. Including protein at each meal may help with weight loss by providing satiety.[18] Selecting too many protein-rich foods may crowd out fruits, vegetables, and whole grains, making the diet inadequate in other nutrients.

Kidney Disease Excretion of the end products of protein metabolism depends, in part, on an adequate fluid intake and healthy kidneys. A high protein intake does not cause kidney disease, but it does increase the work of the kidneys.[19] It may also accelerate kidney deterioration in people with chronic kidney disease. Restricting dietary protein may help to slow the progression of kidney disease in people who have this condition.

Recommended Intakes of Protein

As mentioned earlier, the body continuously breaks down and loses some protein and it cannot store proteins or amino acids. To replace protein, the body needs dietary protein for two reasons. First, dietary protein is the only source of the *essential* amino acids, and second, it is the only practical source of *nitrogen* with which to build the nonessential amino acids and other nitrogen-containing compounds the body needs.

Given recommendations that fat should contribute 20 to 35 percent of total food energy and carbohydrate should contribute 45 to 65 percent, that leaves 10 to 35 percent for protein. In a 2000-kcalorie diet, that represents 200 to 700 kcalories from protein, or 50 to 175 grams. Average intakes in the United States and Canada fall within this range.

♦ RDA for protein:
- 0.8 g/kg/day
- 10 to 35% of energy intake

Protein RDA The protein RDA ♦ for adults is 0.8 grams per kilogram of healthy body weight per day. For infants and children, the RDA is slightly higher. The table on the inside front cover lists the RDA for males and females at various ages in two ways—grams per day based on reference body weights and grams per kilogram of body weight per day. Some evidence suggests that intakes greater than the protein RDA may be beneficial.[20]

The RDA covers the needs for replacing worn-out tissue, so it increases for larger people; it also covers the needs for building new tissue during growth, so it increases for infants, children, adolescents, and pregnant and lactating women. The accompanying "How To" feature explains how to calculate your RDA for protein.

♦ Protein recommendations for athletes: 1.2–1.7 g/kg/day

The protein RDA is the same for athletes as for others, even though athletes may need more protein and many fitness authorities recommend a higher range of protein intakes for athletes pursuing different activities ♦ (see Table 14-5 in Chapter 14 for details).[21] Most athletes in training typically don't need to actually increase their protein intakes, however, because the additional foods they eat to meet their high energy needs deliver protein as well. Importantly, these higher recommendations still fall within the 10 to 35 percent Acceptable Macronutrient Distribution Range (AMDR).

In setting the RDA, the DRI Committee assumes that people are healthy and do not have unusual metabolic needs for protein, that the protein eaten will be of mixed quality (from both high- and low-quality sources), and that the body will use the protein efficiently. In addition, the committee assumes that the protein is consumed along with sufficient carbohydrate and fat to provide adequate energy and that other nutrients in the diet are also adequate.

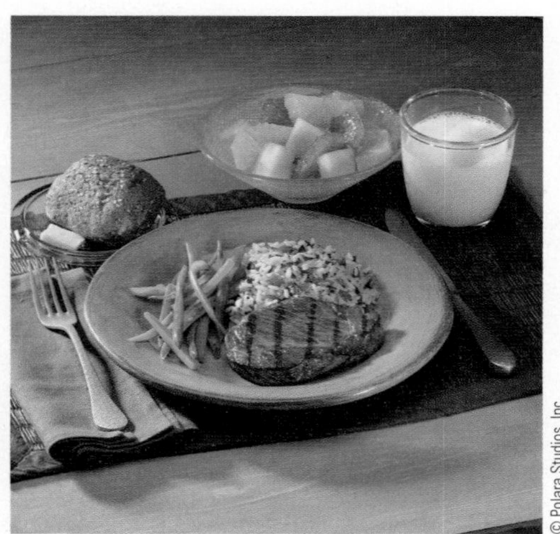

For many people, this 5-ounce steak provides almost all of the meat and much of the protein recommended for a day's intake.

Adequate Energy Note the qualification "adequate energy" in the preceding statement, and consider what happens if energy intake falls short of needs. An intake of 50 grams of protein provides 200 kcalories, which represents 10 percent of the total energy from protein, if the person receives 2000 kcalories a day. But if the person cuts energy intake drastically—to, say, 800 kcalories a day—then an intake of 200 kcalories from protein is suddenly 25 percent of the total; yet it's still the same amount of protein (number of grams). The protein intake is reasonable, but the energy intake is not. The low energy intake forces the body to use the protein to meet energy needs rather than to replace lost body protein. Similarly, if the person's energy intake is high—say, 4000 kcalories—the 50 gram

>How To

Calculate Recommended Protein Intakes

To figure your protein RDA:

- Look up the healthy weight for a person of your height (inside back cover). If your present weight falls within that range, use it for the following calculations. If your present weight falls outside the range, use the midpoint of the healthy weight range as your reference weight.
- Convert pounds to kilograms, if necessary (pounds divided by 2.2 equals kilograms).
- Multiply kilograms by 0.8 to get your RDA in grams per day. (Teens 14 to 18 years old, multiply by 0.85.) Example:

Weight = 150 lb

150 lb ÷ 2.2 lb/kg = 68 kg (rounded off)

68 kg × 0.8 g/kg = 54 g protein (rounded off)

TRY IT Calculate your protein RDA.

protein intake represents only 5 percent of the total; yet it *still* is a reasonable protein intake. Again, the energy intake is unreasonable for most people, but in this case, it permits the protein to be used to meet the body's needs.

Be careful when judging protein (or carbohydrate or fat) intake as a percentage of energy. Always consider the number of grams as well, and compare it with the RDA or another standard stated in grams. A recommendation stated as a percentage of energy intake is useful only if the energy intake is within reason.

From Guidelines to Groceries A diet following the USDA Food Patterns can easily supply the recommended amount of protein. In selecting foods for protein, keep in mind the principles of variety and moderation.

Protein Foods An ounce of most protein foods delivers about 7 grams of protein. The USDA Food Patterns encourage a variety by sorting protein foods into three subgroups (review Figure 2-2, p. 43 and Table 2-4, p. 44). Over a week's time, the total recommended intake of protein foods should be about 20 percent from seafood; almost 70 percent from meat, poultry, and eggs; and 10 percent from nuts, seeds, and legumes.

Either plant or animal sources of protein can support a healthy eating pattern, but some protein foods—notably those derived from animals—may be high in saturated fat. To minimize saturated fat intake, select lean meats and poultry. Trim fat from meats before cooking and drain fat from meat after cooking. Remove skin from poultry before eating. Include plant sources of protein as well. By select a variety of protein foods, consumers can improve their nutrient intake and incur health benefits. For example, Highlight 5 describes how nuts and fish can reduce the risks of heart disease when consumed in place of other protein foods.

> DIETARY GUIDELINES FOR AMERICANS 2010

Choose a variety of foods from the protein foods group, which includes seafood, lean meat and poultry, eggs, beans and peas, soy products, and unsalted nuts and seeds. Increase the amount and variety of seafood consumed by choosing seafood in place of some meat and poultry. Replace protein foods that are higher in solid fats with choices that are lower in solid fats and kcalories.

© Polara Studios, Inc.

Vegetarians obtain their protein from whole grains, legumes, nuts, vegetables, and, in some cases, eggs and milk products.

Milk and Milk Products The only other food group to provide significant amounts of protein per serving is the milk and milk products group. A serving (a cup) of milk or yogurt provides about 8 grams of protein.

Fruits, Vegetables, and Grains Fruits do not contain protein. A serving of vegetables or grains provides 2 to 3 grams of protein, respectively.

◆ Daily Value:
- 50 g protein (based on 10% of 2000 kcal diet)

Read Food Labels

Food labels state the quantity of protein in grams. The "% Daily Value" ◆ for protein is not mandatory on all labels but is required whenever a food makes a protein claim or is intended for consumption by children younger than 4 years old.* Whenever the Daily Value percentage is declared, researchers must determine the *quality* of the protein. Thus, when a % Daily Value is stated for protein, it reflects both quantity and quality.

To illustrate how easy it is to get enough protein, consider the amounts recommended by the USDA Food Pattern for a 2000-kcalorie diet. Six ounces of grains provide about 18 grams of protein; 2½ cups of vegetables deliver about 10 grams; 3 cups of milk offer 24 grams; and 5½ ounces of protein foods supply 38 grams. This totals 90 grams of protein—higher than the protein RDA for most people and yet still lower than the average intake of people in the United States.

People in the United States and Canada typically get more protein than they need. If they have an adequate *food* intake, they have a more-than-adequate protein intake. The key diet-planning principle to emphasize for protein is moderation. Even though most people receive plenty of protein, some feel compelled to take supplements as well, as the next section describes.

Protein and Amino Acid Supplements

Websites, health-food stores, and popular magazine articles advertise a wide variety of protein supplements, and people take these supplements for many different reasons. Athletes take protein powders to build muscle. Dieters take them to spare their bodies' protein while losing weight. Women take them to strengthen their fingernails. People take individual amino acids, too—to cure herpes, to make themselves sleep better, to lose weight, and to relieve pain and depression.** Like many other magic solutions to health problems, protein and amino acid supplements don't work these miracles.

Protein Powders Because the body builds muscle protein from amino acids, many athletes take protein powders with the false hope of stimulating muscle growth. Muscle work builds muscle; protein supplements do not, and athletes do not need them. Getting enough protein to support protein synthesis in the muscles is certainly important, but ingesting "more than enough" protein does not further enhance muscle growth or function.[22] (Highlight 14 presents more information on other supplements athletes commonly use.) Protein powders can supply amino acids to the body, but nature's protein sources—lean meat, milk, eggs, and legumes—supply all these amino acids and more.

Whey protein appears to be particularly popular among athletes hoping to achieve greater muscle gains. A waste product of cheese manufacturing, whey protein is a common ingredient in many low-cost protein powders. When combined with strength training, whey supplements may increase protein synthesis slightly, but they do not seem to enhance athletic performance. To build stronger muscles, athletes need to eat food with adequate energy and protein to support the weight-training work that does increase muscle mass. Those who still think they need more whey can drink a glass of milk; one cup provides 1.5 grams of whey.

Amino Acid Supplements Single amino acids do not occur naturally in foods and offer no benefit to the body; in fact, they may be harmful. The body was not designed to handle the high concentrations and unusual combinations of amino acids found in supplements. Large doses of amino acids cause diarrhea. An excess of

whey protein: a by-product of cheese production; falsely promoted as increasing muscle mass. Whey is the watery part of milk that separates from the curds.

*For labeling purposes, the Daily Values for protein are as follows: for infants, 14 grams; for children younger than age 4, 16 grams; for older children and adults, 50 grams; for pregnant women, 60 grams; and for lactating women, 65 grams.

**Canada requires single amino acid supplements to be sold as drugs; they are, however, allowed to be used as food additives.

one amino acid can create such a demand for a carrier that it limits the absorption of another amino acid, presenting the possibility of a deficiency. Those amino acids winning the competition enter in excess, creating the possibility of toxicity.[23] Toxicity of single amino acids in animal studies raises concerns about their use in human beings. Anyone considering taking amino acid supplements should be cautious not to exceed levels normally found in foods.[24]

Most healthy athletes eating well-balanced diets do not need amino acid supplements. Advertisers point to research that identifies the **branched-chain amino acids** as the main ones used as fuel by exercising muscles. What the ads leave out is that compared to glucose and fatty acids, branched-chain amino acids provide very little fuel and that ordinary foods provide them in abundance anyway. Large doses of branched-chain amino acids can raise plasma ammonia concentrations, which can be toxic to the brain. Branched-chain amino acid supplements may be beneficial in conditions such as liver disease, but otherwise, they are not routinely recommended.[25]

In two cases, recommendations for single amino acid supplements have led to widespread public use—lysine to prevent or relieve the infections that cause herpes cold sores on the mouth or genital organs, and tryptophan to relieve depression and insomnia. In both cases, enthusiastic popular reports preceded careful scientific experiments and health recommendations. Research is insufficient to determine whether lysine suppresses herpes infections, but it appears safe (up to 3 grams per day) when taken in divided doses with meals.

Tryptophan may be effective with respect to inducing drowsiness, but caution is still advised. About 25 years ago, more than 1500 people who elected to take tryptophan supplements developed a rare blood disorder known as eosinophilia-myalgia syndrome (EMS). EMS is characterized by severe muscle and joint pain, extremely high fever, and, in more than three dozen cases, death. Treatment for EMS usually involves physical therapy and low doses of corticosteroids to relieve symptoms temporarily. The Food and Drug Administration implicated impurities in the supplements, issued a recall of all products containing manufactured tryptophan, and warned that high-dose supplements of tryptophan might provoke symptoms of EMS even in the absence of impurities.

REVIEW IT Identify the health benefits of, and recommendations for, protein.
Protein deficiency impairs the body's ability to grow and function optimally. Excesses of protein offer no advantage; in fact, overconsumption of protein-rich foods may incur health problems as well. The optimal diet is adequate in energy from carbohydrate and fat and delivers 0.8 grams of protein per kilogram of healthy body weight each day. US and Canadian diets are typically more than adequate in this respect. Normal, healthy people do not need protein or amino acid supplements.

As is true for the other nutrients as well, it is safest to obtain amino acids and protein from foods, eaten with abundant carbohydrate and some fat to facilitate their use in the body. With all that we know about science, it is hard to improve on nature.

Nutrition Portfolio

Foods that derive from animals—meats, fish, poultry, eggs, and milk products—provide plenty of protein but are often accompanied by fat. Those that derive from plants—whole grains, vegetables, and legumes—may provide less protein but also less fat.

Go to Diet Analysis Plus and choose one of the days on which you have tracked your diet for the entire day. Go to the Intake Spreadsheet report. Scroll down until you see: protein (g).

- Which of your food choices provided you with the most protein on that day? Does that food also have a lot of fat? Refer to the fat (g) column for this information.

branched-chain amino acids: the essential amino acids leucine, isoleucine, and valine, which are present in large amounts in skeletal muscle tissue; falsely promoted as fuel for exercising muscles.

- Describe your dietary sources of proteins and whether you use mostly plant-based or animal-based protein foods in your diet.

Now take a look at the Intake vs. Goals report.

- How do your protein needs compare with your protein intake? Consider whether you receive enough, but not too much, protein daily. Remember, 100 percent means your intake is meeting your needs based on your intake and profile information.
- If your protein intake exceeds 100 percent, consider the possible negative consequences of a high protein intake over many years.
- Debate the risks and benefits of taking protein or amino acid supplements.

Diet Analysis PLUS To complete this exercise, go to your Diet Analysis Plus at **www.cengagebrain.com.**

STUDY IT To review the key points of this chapter and take a practice quiz, go to the study cards at the end of the book.

REFERENCES

1. M. T. Gladwin and E. Vichinsky, Pulmonary complications of sickle cell disease, *New England Journal of Medicine* 359 (2008): 2254–2265.
2. S. Tesseraud and coauthors, Role of sulfur amino acids in controlling nutrient metabolism and cell functions: Implications for nutrition, *British Journal of Nutrition* 101 (2009): 1132–1139; G. Wu, Amino acids: Metabolism, functions, and nutrition, *Amino Acids* 37 (2009): 1–17.
3. US Department of Agriculture, Agricultural Research Service, Beltsville Human Nutrition Research Center, Food Surveys Research Group (Beltsville, MD) and US Department of Health and Human Services, Centers for Disease Control and Prevention, National Center for Health Statistics (Hyattsville, MD), *What We Eat in America*, NHANES 2007–2008, accessed at www.ars.usda.gov/ba/bhnrc/fsrg, published 2010.
4. Position of the American Dietetic Association and Dietitians of Canada: Vegetarian diets, *Journal of the American Dietetic Association* 109 (2009): 1266–1282.
5. Committee on Dietary Reference Intakes, *Dietary Reference Intakes: Energy, Carbohydrate, Fiber, Fat, Fatty Acids, Cholesterol, Protein, and Amino Acids* (Washington, D.C.: National Academies Press, 2005), p. 694.
6. S. R. Preis and coauthors, Dietary protein and risk of ischemic heart disease in middle-aged men, *American Journal of Clinical Nutrition* 92 (2010): 1265–1272.
7. D. G. Hackam and coauthors, The 2010 Canadian Hypertension Education Program recommendations for the management of hypertension: Part 2—Therapy, *Canadian Journal of Cardiology* 26 (2010): 249–258.
8. L. L. Humphrey and coauthors, Homocysteine level and coronary heart disease incidence: A systematic review and meta-analysis, *Mayo Clinic Proceedings* 83 (2008): 1203–1212.
9. R. Clarke and coauthors, Homocysteine and vascular disease: Review of published results of the homocysteine-lowering trials, *Journal of Inherited Metabolic Disease* 34 (2011): 83–91.
10. P. Tighe and coauthors, A dose-finding trial of the effect of long-term folic acid intervention: Implications for food fortification policy, *American Journal of Clinical Nutrition* 93 (2011): 11–18.
11. Clarke and coauthors, 2011; S. Eilat-Adar and U. Goldbourt, Nutritional recommendations for preventing coronary heart disease in women: Evidence concerning whole foods and supplements, *Nutrition, Metabolism, and Cardiovascular Disease* 20 (2010): 459–466; A. J. Martí-Carvajal and coauthors, Homocysteine lowering interventions for preventing cardiovascular events, *Cochrane Database of Systematic Reviews* 4 (2009): CD006612.
12. L. Chao-Qiang, *MAT1A* variants are associated with hypertension, stroke, and markers of DNA damage and are modulated by plasma vitamin B-6 and folate, *American Journal of Clinical Nutrition* 91 (2010): 1377–1386; J. M. Armitage and coauthors, Effects of homocysteine-lowering with folic acid plus vitamin B_{12} vs placebo on mortality and major morbidity in myocardial infarction survivors: A randomized trial, *Journal of the American Medical Association* 303 (2010): 2486–2494; M. Ebbing and coauthors, Cancer incidence and mortality after treatment with folic acid and vitamin B_{12}, *Journal of the American Medical Association* 302 (2009): 2119–2126.
13. U. N. Das and coauthors, L-arginine, NO and asymmetrical dimethylarginine in hypertension and type 2 diabetes, *Frontiers in Bioscience* 16 (2011): 13–20; D. Tousoulis and coauthors, Novel therapeutic strategies targeting vascular endothelium in essential hypertension, *Expert Opinion on Investigational Drugs* 19 (2010): 1395–1412.
14. A. T. Chan and E. L. Giovannucci, Primary prevention of colorectal cancer, *Gastroenterology* 138 (2010): 2029–2043.
15. Committee on Dietary Reference Intakes, 2005, p. 841; Committee on Dietary Reference Intakes, *Dietary Reference Intakes for Calcium and Vitamin D* (Washington, D.C.: National Academies Press, 2011).
16. M. P. Thorpe and E. M. Evans, Dietry protein and bone health: Harmonizing conflicting theories, *Nutrition Reviews* 9 (2011): 215–230; J. J. Cao, L. K. Johnson, and J. R. Hunt, A diet high in meat protein and potential renal acid load increases fractional calcium absorption and urinary calcium excretion without affecting markers of bone resorption or formation in postmenopausal women, *Journal of Nutrition* 141 (2011): 391–397; J. M. Beasley and coauthors, Is protein intake associated with bone mineral density in young women? *American Journal of Clinical Nutrition* 91 (2010): 1311–1316; A. L. Darling and coauthors, Dietary protein and bone health: A systematic review and meta-analysis, *American Journal of Clinical Nutrition* 90 (2009): 1674–1692; M. S. Westerterp-Plantenga and coauthors, Dietary protein, weight loss, and weight maintenance, *Annual Review of Nutrition* 29 (2009): 21–41.
17. J. J. Cao and F. H. Nielsen, Acid diet (high-meat protein) effects on calcium metabolism and bone health, *Current Opinion in Clinical Nutrition and Metabolic Care* 13 (2010): 698–702; A. D. Conigrave, E. M. Brown, and R. Rizzoli, Dietary protein and bone health: Roles of amino acid-sensing receptors in the control of calcium metabolism and bone homeostasis, *Annual Review of Nutrition* 28 (2008): 131–155.
18. Westerterp-Plantenga and coauthors, 2009; R. F. Kushner and B. Doerfler, Low-carbohydrate, high-protein diets revisited, *Current Opinion in Gastroenterology* 24 (2008): 198–203.
19. H. Frank and coauthors, Effect of short-term high-protein compared with normal-protein diets on renal hemodynamics and associated variables in healthy young men, *American Journal of Clinical Nutrition* 90 (2009): 1509–1516.
20. R. R. Wolfe and S. L. Miller, The recommended dietary allowance of protein: A misunderstood concept, *Journal of the American Medical Association* 299 (2008): 2891–2893.

21. Position of the American Dietetic Association, Dietitians of Canada, and the American College of Sports Medicine, Nutrition and athletic performance, *Journal of the American Dietetic Association* 109 (2009): 509–527.
22. T. B. Symons and coauthors, A moderate serving of high-quality protein maximally stimulates skeletal muscle protein synthesis in young and elderly subjects, *Journal of the American Dietetic Association* 109 (2009): 1582–1586.
23. Wu, 2009.
24. Committee on Dietary Reference Intakes, *Dietary Reference Intakes: The Essential Guide to Nutrient Requirements* (Washington, D.C.: National Academies Press, 2006), p. 152.
25. S. Takeshita and coauthors, A snack enriched with oral branched-chain amino acids prevents a fall in albumin in patients with liver cirrhosis undergoing chemoembolization for hepatocellular carcinoma, *Nutrition Research* 29 (2009): 89–93; T. Kawaguchi and coauthors, Branched-chain amino acid–enriched supplementation improves insulin resistance in patients with chronic liver disease, *International Journal of Molecular Medicine* (2008): 105–112.

 LEARN IT Explain how nutrients influence gene activity (nutrigenomics) and how genes influence the activities of nutrients (nutrigenetics).

Nutritional Genomics

Imagine this scenario: A physician scrapes a sample of cells from inside your cheek and submits it to a **genomics** lab. The lab returns a report based on your genetic profile that reveals which diseases you are most likely to develop, and your physician recommends specific lifestyle changes and medical treatments that can help you maintain good health. You may also be given a prescription for an individualized diet and dietary supplements that will best meet your personal nutrient requirements. This scenario may one day become a common reality as scientists uncover the relationships among **genetics,** diet, and disease.[1] Such genetic testing holds great promise, but consumers need to know that current genetic test kits commonly available to the public are unproven and may create more problems than they resolve.[2]

Figure H6-1 introduces **nutritional genomics,** a new field of study that examines how nutrients influence gene activity *(nutrigenomics)* and how **genes** influence the activities of nutrients *(nutrigenetics)*.[3] The accompanying glossary defines related terms.

The recent surge in genomics research grew from the Human Genome Project, an international effort by industry and government scientists to identify and describe all of the genes in the **human genome**—that is, all the genetic information contained within a person's cells. Completed in 2003, this project developed many of the research technologies needed to study genes and genetic variation. Scientists are now working on the human **proteome** and hope to identify each of the proteins made by the genes, the genes associated with aging and diseases, and the dietary and lifestyle choices that most influence the expression of those genes. Such information will have major implications for society in general, and for health care in particular.[4]

A Genomics Primer

Figure H6-2 (p. 192) shows the relationships among the materials that comprise the genome. As Chapter 6's discussion of protein synthesis points out, genetic information is encoded in DNA molecules within the nucleus of cells. The **DNA (deoxyribonucleic acid)** molecules and associated proteins are packed within 46 **chromosomes.** The genes are segments of a DNA strand that can eventually be translated into one or more proteins. The sequence of **nucleotide bases** within each gene determines the amino acid sequence of a particular protein. Scientists currently estimate that there are between 20,000 and 25,000 protein-coding genes in the human genome.

As Figure 6-7 (p. 174) explains, when cells make proteins, a DNA sequence is used to make messenger **RNA (ribonucleic acid).** The **nucleotide** sequence in messenger RNA then determines the amino acid sequence to make a protein. This process—from genetic information to protein synthesis—is known as **gene expression.** Gene expression can be determined by measuring the amounts of messenger RNA in a tissue sample. **Microarray technology** (see photo above) allows researchers to detect messenger RNA and analyze the

GLOSSARY

chromosomes: structures within the nucleus of a cell made of DNA and associated proteins. Human beings have 46 chromosomes in 23 pairs. Each chromosome has many genes.

DNA (deoxyribonucleic acid): the double helix molecules of which genes are made.

epigenetics: the study of heritable changes in gene function that occur without a change in the DNA sequence.

gene expression: the process by which a cell converts the genetic code into RNA and protein.

genes: sections of chromosomes that contain the instructions needed to make one or more proteins.

genetics: the study of genes and inheritance.

genomics: the study of all the genes in an organism and their interactions with environmental factors.

human genome (GEE-nome): the complete set of genetic material (DNA) in a human being.

methylation: the addition of a methyl group (CH_3).

microarray technology: research tools that analyze the expression of thousands of genes simultaneously and search for particular gene changes associated with a disease. DNA microarrays are also called *DNA chips.*

mutations: permanent changes in the DNA that can be inherited.

nucleotide bases: the nitrogen-containing building blocks of DNA and RNA—cytosine (C), thymine (T), uracil (U), guanine (G), and adenine (A). In DNA, the base pairs are A–T and C–G and in RNA, the base pairs are A–U and C–G.

nucleotides: the subunits of DNA and RNA molecules, composed of a phosphate group, a 5-carbon sugar (deoxyribose for DNA and ribose for RNA), and a nitrogen-containing base.

nutritional genomics: the science of how nutrients affect the activities of genes (*nutrigenomics*) and how genes affect the activities of nutrients (*nutrigenetics*).

phenylketonuria (FEN-il-KEY-toe-NEW-ree-ah) or **PKU:** an inherited disorder characterized by failure to metabolize the amino acid phenylalanine to tyrosine.

proteome: all proteins in a cell. The study of all proteins produced by a species is called *proteomics.*

RNA (ribonucleic acid): a compound similar to DNA, but RNA is a single strand with a ribose sugar instead of a deoxyribose sugar and uracil instead of thymine as one of its bases.

> **FIGURE H6-1 Nutritional Genomics**

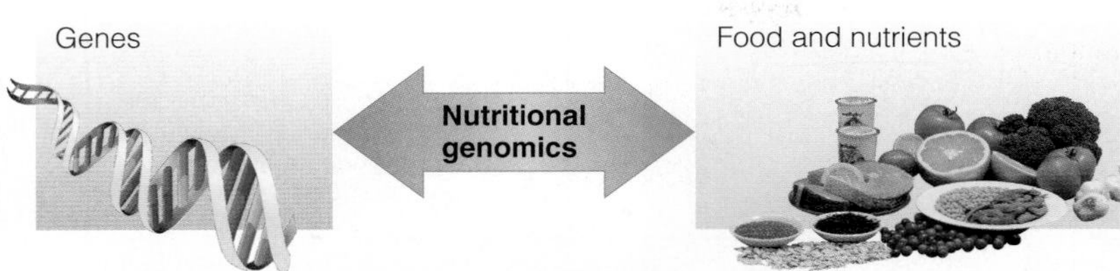

Nutritional genomics examines the interactions of genes and nutrients. These interactions include both nutrigenetics and nutrigenomics.

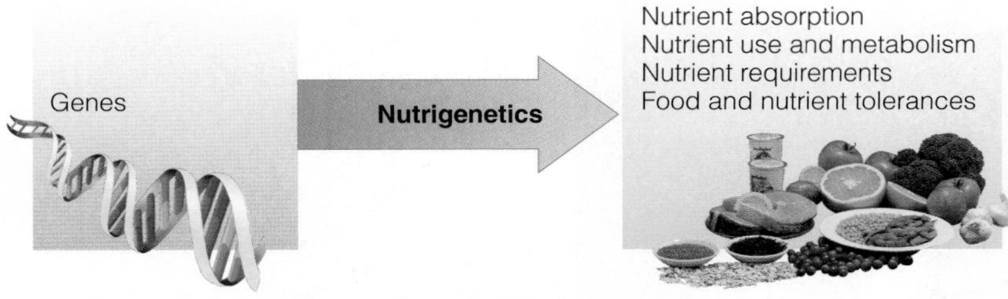

Nutrigenetics (or nutritional genetics) examines how genes influence the activities of nutrients.

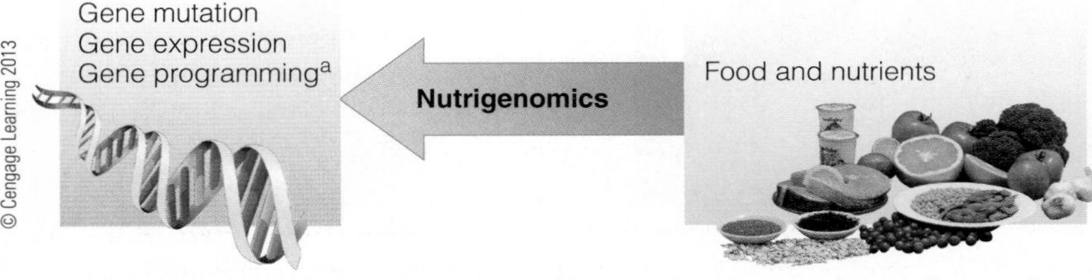

Nutrigenomics, which includes epigenetics, examines how nutrients influence the activities of genes.

[a]Chapter 15 introduces programming and describes how a mother's nutrition can permanently change gene expression in the fetus with consequences for future generations.

expression of thousands of genes simultaneously. These patterns of gene expression help to explain the development of diseases and relationships between diseases.[5]

Simply having a certain gene does not determine that its associated trait will be expressed; the gene has to be activated. (Similarly, owning lamps does not ensure you will have light in your home unless you turn them on.) Nutrients are among many environmental factors that play key roles in either activating or silencing genes. Switching genes on and off does not change the DNA itself, but it can have dramatic consequences for a person's health.

The area of study that examines how environmental factors influence gene expression without changing the DNA is known as **epigenetics.**[6] To turn genes on, enzymes attach proteins near the beginning of a gene. If enzymes attach a methyl group (CH_3) instead, the protein is blocked from binding to the gene and the gene remains switched off. Other factors influence gene expression as well, but methyl groups are currently the most well understood.[7] They also are known to have dietary connections.

The accompanying photo of two mice illustrates epigenetics and how diet can influence genetic traits such as hair color and body weight. Both mice have a gene that tends to produce fat, yellow pups, but their mothers were given different diets during pregnancy. The mother of the mouse on the right was given a dietary supplement containing the B vitamins folate and vitamin B_{12}. These nutrients silenced the gene for "yellow and fat," resulting in brown pups with normal appetites. As Chapter 10 explains, one of the main roles of these B vitamins is to transfer methyl groups. In the case of the supplemented mice, methyl groups migrated onto DNA and silenced several genes, thus producing brown coats and protecting against the development of obesity (and consequently, some related diseases). Keep in mind that these changes occurred epigenetically. In other words, the DNA sequence within the genes of the mice remained the same. Nutrition and other environmental factors can influence genes in a way that creates inheritable changes in the body's metabolism and susceptibility to disease.[8] In this way, the dietary habits of parents, and even grandparents, can influence future generations.

Many nutrients and phytochemicals regulate gene expression and influence health through their involvement in DNA **methylation.**[9] Some, such as folate, silence genes and protect against some cancers by providing methylation.[10] Others, such as a phytochemical found in green tea, activate genes and protect against some cancers by inhibiting methylation activity.[11] Whether silencing or activating a gene is beneficial or harmful depends on what the gene does. Silencing a gene that stimulates cancer growth, for example, would be beneficial, but silencing a gene that suppresses cancer growth would be harmful. Similarly, activating a gene that defends against obesity would be beneficial, but activating a gene that promotes obesity would

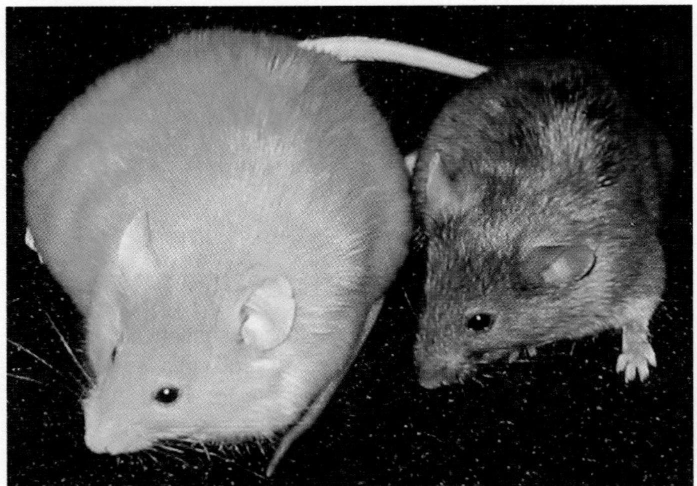

Both of these mice have the gene that tends to produce fat, yellow pups, but their mothers had different diets. The mother of the mouse on the right received a dietary supplement, which silenced the gene, resulting in brown pups with normal appetites.

> FIGURE H6-2 **The Human Genome**

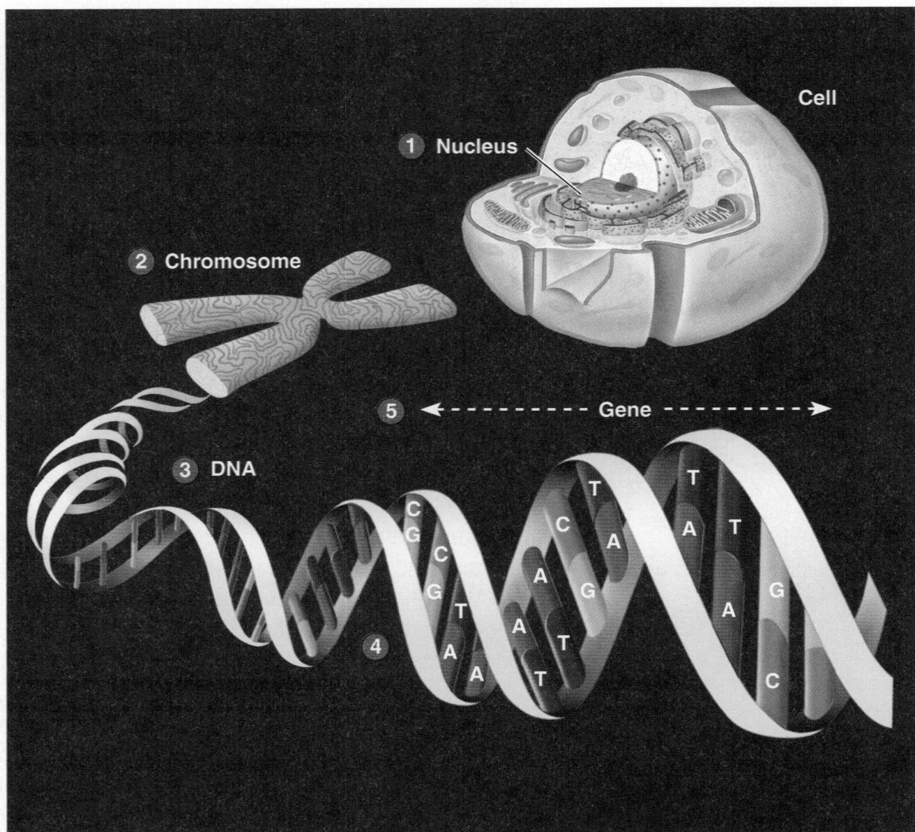

Cell
1 Nucleus
2 Chromosome
5 Gene
3 DNA
4
C G C G T A A A C A T A G T T A A T G A C
1 The human genome is a complete set of genetic material organized into 46 chromosomes, located within the nucleus of a cell.
2 A chromosome is made of DNA and associated proteins.
3 The double helical structure of a DNA molecule is made up of two long chains of nucleotides. Each nucleotide is composed of a phosphate group, a 5-carbon sugar, and a base.
4 The sequence of nucleotide bases (C, G, A, T) determines the amino acid sequence of proteins. These bases are connected by hydrogen bonding to form base pairs—adenine (A) with thymine (T) and guanine (G) with cytosine (C).
5 A gene is a segment of DNA that includes the information needed to synthesize one or more proteins.

SOURCE: Adapted from "A Primer: From DNA to Life," Human Genome Project, US Department of Energy Office of Science, accessed at www.orn.gov/sci/techresources/human_genome/primer-pic.shtml.

be harmful. Figure H6-3 illustrates how nutrient regulation of gene expression can influence a person's health. Much research is under way to determine which nutrients activate or silence which genes. Such knowledge is expected to help researchers reverse the epigenetic changes that lead to cancer.[12] Similarly, researchers exploring how kcalorie-restricted diets influence DNA methylation are gaining new insights on the regulation of appetite and the metabolism of weight loss.[13]

Genetic Variation and Disease

Except for identical twins, no two persons are genetically identical. Even then, a particular gene may become active in one twin and silenced in the other because of epigenetic changes.[14]

The variation in the genomes of any two persons is only about 0.1 percent, a difference of only one nucleotide base in every 1000. Yet it is this incredibly small difference that makes each of us unique and explains why, given the same environmental influences, some of us develop certain diseases and others do not. Similarly, genetic variation explains why some of us respond to interventions such as diet and others do not. For example, following a diet low in saturated fats will significantly lower LDL cholesterol for most people, but the degree of change varies dramatically among individuals, with some people having only a small decrease or even a slight increase. In other words, dietary factors may be more helpful or more harmful depending on a person's particular genetic variations. Such findings help to explain some of the conflicting results from research studies. One of the goals of nutritional genomics is to custom design *specific* recommendations that fit the needs of *each* individual. Such personalized recommendations are expected to provide more effective disease prevention and treatment solutions.

Diseases characterized by a single-gene disorder are genetically predetermined, usually exert their effects early in life, and greatly affect those touched by them; such diseases are relatively rare. The cause and effect of single-gene disorders is clear—those with the genetic defect get the disease and those without it don't. In contrast, the more common diseases, such as heart disease and cancer, are influenced by many genes and typically develop over several decades. These chronic diseases have multiple genetic components that *predispose* the prevention or development of a disease, depending on a variety of environmental factors (such as smoking, diet, and physical activity).[15] Both types of diseases are of interest to researchers studying nutritional genomics.

> **FIGURE H6-3 Nutrient Regulation of Gene Expression**

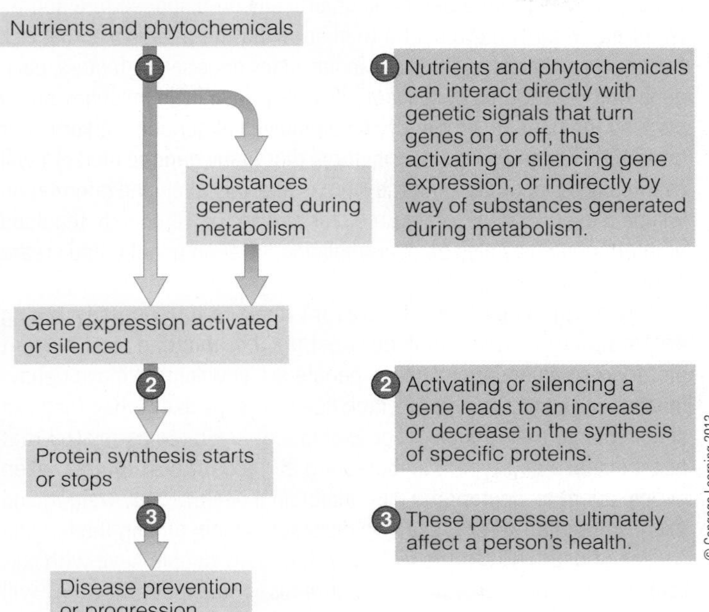
Nutrients and phytochemicals
1
Substances generated during metabolism
Gene expression activated or silenced
2
Protein synthesis starts or stops
3
Disease prevention or progression
1 Nutrients and phytochemicals can interact directly with genetic signals that turn genes on or off, thus activating or silencing gene expression, or indirectly by way of substances generated during metabolism.
2 Activating or silencing a gene leads to an increase or decrease in the synthesis of specific proteins.
3 These processes ultimately affect a person's health.
© Cengage Learning 2013

Single-Gene Disorders

Some disorders are caused by **mutations** in single genes that are inherited at birth. The consequences of a missing or malfunctioning protein can seriously disrupt metabolism and may require significant dietary or medical intervention. A classic example of a diet-related, single-gene disorder is **phenylketonuria,** or **PKU.**

Approximately one in every 15,000 infants in the United States is born with PKU. PKU arises from mutations in the gene that codes for the enzyme that converts the essential amino acid phenylalanine to the amino acid tyrosine. Without this enzyme, phenylalanine and its metabolites accumulate and damage the nervous system, resulting in mental retardation, seizures, and behavior abnormalities. At the same time, the body cannot make tyrosine or compounds made from it (such as the neurotransmitter epinephrine). Consequently, tyrosine becomes an essential amino acid: because the body cannot make it, the diet must supply it.

Although the most debilitating effect is on brain development, other symptoms of PKU become evident if the condition is left untreated. Infants with PKU may have poor appetites and grow slowly. They may be irritable or have tremors or seizures. Their bodies and urine may have a musty odor. Their skin coloring may be unusually pale, and they may develop skin rashes.

The effect of nutrition intervention in PKU is remarkable. In fact, the only current treatment for PKU is a diet that restricts phenylalanine and supplies tyrosine to maintain blood levels of these amino acids within safe ranges. Because all foods containing protein provide phenylalanine, the diet must depend on a special formula to supply a phenylalanine-free source of energy, amino acids, vitamins, and minerals. If the restricted diet is conscientiously followed, the symptoms can be prevented. Because phenylalanine is an essential amino acid, the diet cannot exclude it completely. Children with PKU need phenylalanine to grow, but they cannot handle excesses without detrimental effects. Therefore, their diets must provide enough phenylalanine to support normal growth and health but not enough to cause harm. The diet must also provide tyrosine. To ensure that blood concentrations of phenylalanine and tyrosine are close to normal, children and adults who have PKU must have blood tests periodically and adjust their diets as necessary.

Multigene Disorders

In multigene disorders, several genes can influence the progression of a disease, but no single gene causes the disease on its own.[16] For this reason, genomics researchers must study the expression and interactions of *multiple* genes. Because multigene disorders are often sensitive to interactions with environmental influences, they are not as straightforward as single-gene disorders.

Heart disease provides an example of a chronic disease with multiple gene and environmental influences. Consider that major risk factors for heart disease include elevated blood cholesterol levels, obesity, diabetes, and hypertension. Each of these risk factors has multiple underlying genetic and environmental causes, many of which are not completely understood. Research in nutritional genomics involves coordinating multiple findings on each of these risk factors and explaining the interactions among several genes, biological pathways, and nutrients in relatively little time.[17] Studies have been quite successful in examining the genome and identifying multiple pathways in the development of complex diseases.[18] This information could then guide physicians and dietitians to prescribe the most appropriate medical and dietary interventions from among many possible solutions.[19] Finding the best option for each person is a challenge given the many possible interactions between genes and environmental factors and the millions of possible gene variations in the human genome that make each individual unique.

The results of genomic research are helping to explain findings from previous nutrition research. Consider dietary fat and heart disease, for example. As Highlight 5 explains, epidemiological and clinical studies have found that a diet high in omega-3 polyunsaturated fatty acids benefits heart health. Now genetic studies offer an underlying explanation of this relationship: diets rich in omega-3 polyunsaturated fatty acids alter gene expression of immune cells to suppress inflammation and inhibit plaque build-up.[20] Both actions support a healthy heart.

To learn more about how individuals respond to diet, researchers examine the genetic differences among people. The most common genetic differences involve a change in a single nucleotide base located in a particular region of a DNA strand—thymine replacing cytosine, for example. Such variations are called single nucleotide polymorphisms (SNPs), and they commonly occur throughout the genome. Many SNPs (commonly pronounced "snips") have no effect on cell activity. In fact, SNPs are significant only if they affect the amino acid sequence of a protein in a way that alters its function *and* if that function is critical to the body's well-being. In these cases, SNPs may reveal fascinating answers to previously unexplained findings. Consider that research on a gene that plays a key role in lipid metabolism reveals differences in a person's response to diet depending on whether the gene has a common SNP. People with the SNP have lower LDL when eating a diet rich in polyunsaturated fatty acids—and higher LDL with a low intake—than those without the SNP.[21] These findings clearly show how diet (in this case, polyunsaturated fat) interacts with a gene (in this case, a fat metabolism gene with a SNP) to influence the development of a disease (changing blood lipids implicated in heart disease).

Clinical Concerns

Because multigene, chronic diseases are common, an understanding of the human genome will have widespread ramifications for health care.[22] This new understanding of the human genome is expected to change health care by:

- Providing knowledge of an individual's genetic predisposition to specific diseases[23]
- Allowing physicians to develop "designer" therapies—prescribing the most effective schedule of screening, behavior changes (including diet), and medical interventions based on each individual's genetic profile
- Enabling manufacturers to create new medications for each genetic variation so that physicians can prescribe the best medicine in the exact dose and frequency to enhance effectiveness and minimize the risks of side effects[24]
- Providing a better understanding of how nutrition influences the biological pathways of diseases[25]

Enthusiasm surrounding genomic research needs to be put into perspective, however, in terms of the present status of clinical medicine as well as people's willingness to make difficult lifestyle choices. Critics have questioned whether genetic markers for disease would be more useful than simple family history and clinical measurements, which reflect both genetic *and* environmental influences. In other words, knowing that a person is genetically predisposed to diabetes is not necessarily more useful than knowing the person's actual risk factors.[26] Furthermore, if a disease has many genetic risk factors, each gene that contributes to susceptibility may have little influence on its own, so the benefits of identifying an individual genetic marker might be small. The long-range possibility is that many genetic markers will eventually be identified, and the hope is that the combined information will be a useful and accurate predictor of disease. Of course, the flood of information may also be overwhelming, offer no benefit, and create anxiety.[27]

Having the knowledge to prevent disease and actually taking action do not always coincide. Despite the abundance of current dietary recommendations, many people are unwilling to make behavior changes known to improve their health—especially when they can simply blame their genes.[28] For example, it has been estimated that heart disease and type 2 diabetes are 90 percent preventable when people adopt an appropriate diet, maintain a healthy body weight, and exercise regularly. Yet these two diseases remain among the leading causes of death. Given the difficulty that many people have with current recommendations, it may be unrealistic to expect that they will enthusiastically adopt an even more detailed list of lifestyle modifications. Then again, compliance may be better when it is supported by information based on a person's own genetic profile and the knowledge that the epigenetic profile can be changed.

The debate over nature versus nurture—whether genes or the environment are more influential—has quieted. The focus has shifted. Scientists acknowledge the important roles of each and understand the real answers lie within the myriad interactions. Current research is sorting through how nutrients and other dietary factors interact with genes to confer health benefits or risks. Answers from genomic research may not become apparent for years to come, but the opportunities and rewards may prove well worth the efforts.

REFERENCES

1. W. G. Feero, A. E. Guttmacher, and F. S. Collins, Genomic medicine: An updated primer, *New England Journal of Medicine* 362 (2010): 2001–2011.
2. Government Accountability Office, *Direct-to-consumer genetic tests: Misleading test results are further complicated by deceptive marketing and other questionable practices*, GAO-10-847T (Washington D.C.: July 22, 2010); J. P. Annes, M. A. Giovanni, and M. F. Murray, Risks of presymptomatic direct-to-consumer genetic testing, *New England Journal of Medicine* 363 (2010): 1100–1101; L. Esserman and V. Kaklamani, Lessons learned from genetic testing, *Journal of the American Medical Association* 304 (2010): 1011–1012; J. P. Evans, D. C. Dale, and C. Fomous, Preparing for a consumer-driven genomic age, *New England Journal of Medicine* 363 (2010): 1099–1103; A. L. McGuire and W. Burke, An unwelcome side effect of direct-to-customer personal genome testing: Raiding the medical commons, *Journal of the American Medical Association* 300 (2008): 2669–2671.
3. P. J. Stover and M. A. Caudill, Genetic and epigenetic contributions to human nutrition and health: Managing genome-diet interactions, *Journal of the American Dietetic Association* 108 (2008): 1480–1487.
4. H. Varmus, Ten years on: The human genome and medicine, *New England Journal of Medicine* 362 (2010): 2028–2029; R. Raqib and A. Cravioto, Nutrition, immunology, and genetics: Future perspectives, *Nutrition Reviews* 67 (2009): S227–S236; A. P. Feinberg, Epigenetics at the epicenter of modern medicine, *Journal of the American Medical Association* 299 (2008): 1345–1350.
5. M. P. Keller and A. D. Attie, Physiological insights gained from gene expression analysis in obesity and diabetes, *Annual Review of Nutrition* 30 (2010): 341–364; M. I. McCarthy, Genomics, type 2 diabetes, and obesity, *New England Journal of Medicine* 363 (2010): 2339–2350.
6. S. W. Choi and S. Friso, Epigenetics: A new bridge between nutrition and health, *Advances in Nutrition* 1 (2010): 8–16; G. P. Kauwell, Epigenetics: What it is and how it can affect dietetics practice, *Journal of the American Dietetic Association* 108 (2008): 1056–1059.
7. M. Esteller, Epigenetics in cancer, *New England Journal of Medicine* 358 (2008): 1148–1159.
8. S. A. Ross and coauthors, Introduction: Diet, epigenetic events and cancer prevention, *Nutrition Reviews* 66 (2008): S1–S6.
9. M. P. Lee and B. K. Dunn, Influence of genetic inheritance on global epigenetic states and cancer risk prediction with DNA methylation signature: Challenges in technology and data analysis, *Nutrition Reviews* 66 (2008): S69–S72.

10. C. M. Ulrich, M. C. Reed, and H. F. Nijhout, Modeling folate, one-carbon metabolism, and DNA methylation, *Nutrition Reviews* 66 (2008): S27–S30.
11. C. S. Yang and coauthors, Reverse of hypermethylation and reactivation of genes by dietary polyphenolic compounds, *Nutrition Reviews* 66 (2008): S18–S20.
12. S. Sharma, T. K. Kelly, and P. A. Jones, Epigenetics in cancer, *Carcinogenesis* 31 (2010): 27–36.
13. L. Bouchard and coauthors, Differential epigenomic and transcriptomic responses in subcutaneous adipose tissue between low and high responders to caloric restriction, *American Journal of Clinical Nutrition* 91 (2010): 309–320; M. den Hoed and coauthors, Postprandial responses in hunger and satiety are associated with the rs9939609 single nucleotide polymorphism in *FTO*, *American Journal of Clinical Nutrition* 90 (2009): 1426–1432.
14. Z. A. Kaminsky and coauthors, DNA methylation profiles in monozygotic and dizygotic twins, *Nature Genetics* 41 (2009): 240–245.
15. L. R. Ferguson and M. Philpott, Nutrition and mutagenesis, *Annual Review of Nutrition* 28 (2008): 313–329.
16. J. Hardy and A. Singleton, Genomewide association studies and human disease, *New England Journal of Medicine* 360 (2009): 1759–1768.
17. Hardy and Singleton, 2009.
18. T. A. Manolio, Genomewide association studies and assessment of the risk of disease, *New England Journal of Medicine* 363 (2010): 166–176.
19. R. DeBusk, Diet-related disease, nutritional genomics, and food and nutrition professionals, *Journal of the American Dietetic Association* 109 (2009): 410–413.
20. M. Bouwens and coauthors, Fish-oil supplementation induces anti-inflammatory gene expression profiles in human blood mononuclear cells, *American Journal of Clinical Nutrition* 90 (2009): 415–424.
21. E. S. Tai and coauthors, Polyunsaturated fatty acids interact with PPARA–L162V polymorphism to affect plasma triglyceride apolipoprotein C-III concentrations in the Framingham Heart Study, *Journal of Nutrition* 135 (2005): 397–403.
22. M. M. Bergmann, U. Görman, and J. C. Mathers, Bioethical considerations for human nutrigenomics, *Annual Review of Nutrition* 28 (2008): 447–467.
23. R. P. Lifton, Individual genomes on the horizon, *New England Journal of Medicine* 362 (2010): 1235–1236.
24. S. B. Shurin and E. G. Nabel, Pharmacogenomics: Ready for prime time? *New England Journal of Medicine* 358 (2008): 1061–1063.
25. J. N. Hirschhorn, Genomewide association studies: Illuminating biologic pathways, *New England Journal of Medicine* 360 (2009): 1699–1701; G. Panagiotou and J. Nielsen, Nutritional systems biology: Definitions and approaches, *Annual Review of Nutrition* 29 (2009): 329–339.
26. N. P. Paynter and coauthors, Association between a literature-based genetic risk score and cardiovascular events in women, *Journal of the American Medical Association* 303 (2010): 631–637; P. Kraft and D. J. Hunter, Genetic risk prediction: Are we there yet? *New England Journal of Medicine* 360 (2009): 1701–1703; J. B. Meigs and coauthors, Genotype score in addition to common risk factors for prediction of type 2 diabetes, *New England Journal of Medicine* 359 (2008): 2208–2219.
27. R. A. Kane and R. L. Kane, Effect of genetic testing for risk of Alzheimer's disease, *New England Journal of Medicine* 361 (2009): 298–299; P. M. Visscher and G. W. Montgomery, Genome-wide association studies and human disease: From trickle to flood, *Journal of the American Medical Association* 302 (2009): 2028–2029.
28. S. C. O'Neill and coauthors, Preferences for genetic and behavioral health information: The impact of risk factors and disease attributions, *Annals of Behavioral Medicine* 40 (2010): 127–173.

CENGAGE LEARNING'S GLOBAL NUTRITION WATCH — GLOBAL NUTRITION RESOURCE CENTER

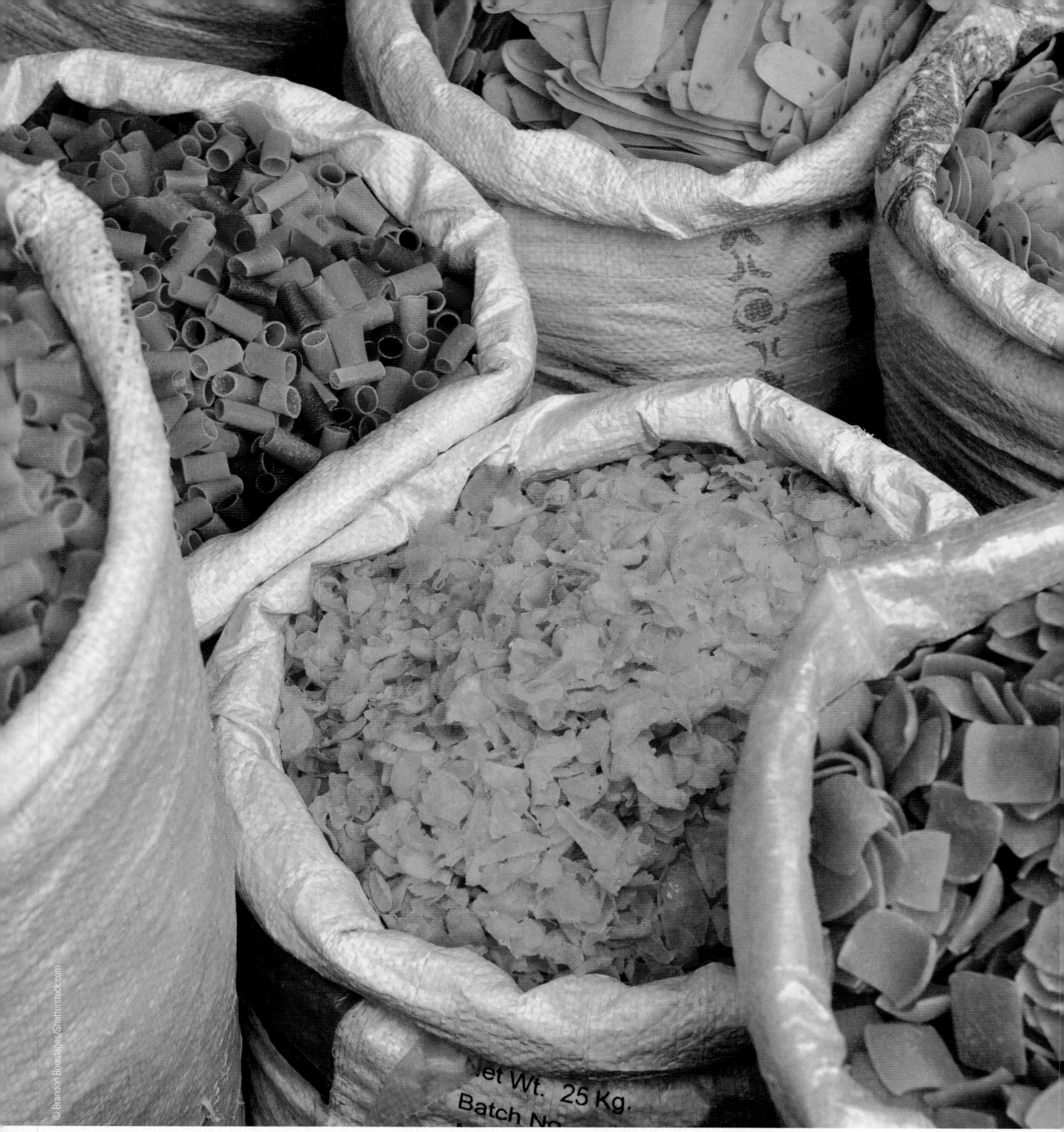

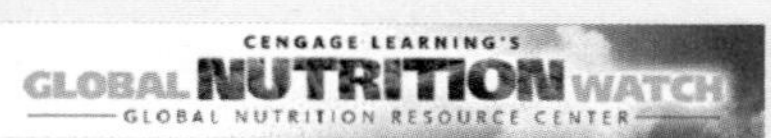

Most of the body's metabolic work involves either anabolic or catabolic reactions. Go to Nutrition and Health/Disease → Metabolism to review anabolic and catabolic reactions. In what ways are anabolic and catabolic reactions similar and how do they differ?

7 Energy Metabolism

LEARNING GPS

Nutrition in Your Life

You eat breakfast and hustle off to class. After lunch, you study for tomorrow's exam. Dinner is followed by an evening of dancing. Do you ever think about how the food you eat powers the activities of your life? What happens when you don't eat—or when you eat too much? Learn how the cells of your body transform carbohydrates, fats, and proteins into energy—and what happens when you give your cells too much or too little of any of these nutrients. Discover the metabolic pathways that lead to body fat and those that support physical activity. It's really quite fascinating. In the Nutrition Portfolio at the end of this chapter, you can determine whether your diet provides a healthy balance of the energy nutrients.

Energy makes it possible for people to breathe, ride bicycles, compose music, and do everything else they do. As Chapter 1 explains, *energy* is the capacity to do work. Although every aspect of our lives depends on energy, the concept of energy can be difficult to grasp because it cannot be seen or touched, and it manifests in various forms, including heat, mechanical, electrical, and chemical energy. In the body, heat energy maintains a constant body temperature, mechanical energy moves muscles, and electrical energy sends nerve impulses. Energy is stored in foods and in the body as chemical energy. This chemical energy powers the myriad activities of all cells.

All the energy that sustains human life initially comes from the sun—the ultimate source of energy. During **photosynthesis,** plants make simple sugars from carbon dioxide and capture the sun's light energy in the chemical bonds of those sugars. Then human beings eat either the plants or animals that have eaten the plants. These foods provide energy, but how does the body obtain that energy from foods? This chapter answers that question by following the nutrients that provide the body with **fuel** through a series of reactions that

photosynthesis: the process by which green plants use the sun's energy to make carbohydrates from carbon dioxide and water.

- **photo** = light
- **synthesis** = put together (making)

fuel: compounds that cells can use for energy. The major fuels include glucose, fatty acids, and amino acids; other fuels include ketone bodies, lactate, glycerol, and alcohol.

release energy from their chemical bonds. As the bonds break, they release energy in a controlled version of the same process by which wood burns in a fire. Both wood and food have the potential to provide energy. When wood burns in the presence of oxygen, it generates heat and light (energy), steam (water), and some carbon dioxide and ash (waste). Similarly, during **metabolism,** the body releases energy, water, and carbon dioxide (and other waste products).

By studying metabolism, you will understand how the body uses foods to meet its needs and why some foods meet those needs better than others. Readers who are interested in weight control will discover which foods contribute most to body fat and which to select when trying to gain or lose weight safely. Readers who are physically active will discover which foods best support endurance activities and which to select when trying to build lean body mass.

7.1 Chemical Reactions in the Body

LEARN IT Identify the nutrients involved in energy metabolism and the high-energy compound that captures the energy released during their breakdown.

Earlier chapters introduced some of the body's chemical reactions: the making and breaking of the bonds in carbohydrates, fats, and proteins. Metabolism is the sum of these and all the other chemical reactions that go on in living cells; *energy metabolism* includes all the ways the body obtains and uses energy from food.

The Site of Metabolic Reactions—Cells The human body is made up of trillions of cells, and each cell busily conducts its metabolic work all the time. (Appendix A presents a brief summary of the structure and function of the cell.) Figure 7-1 depicts a typical cell and shows where the major reactions of energy metabolism take place. The type and extent of metabolic activities vary depending on the type of cell, but of all the body's cells, the liver cells are the most versatile and metabolically active. Table 7-1 offers insights into the liver's work.

metabolism: the sum total of all the chemical reactions that go on in living cells. *Energy metabolism* includes all the reactions by which the body obtains and expends the energy from food.

- **metaballein** = change

> FIGURE 7-1 **A Typical Cell (Simplified Diagram)**

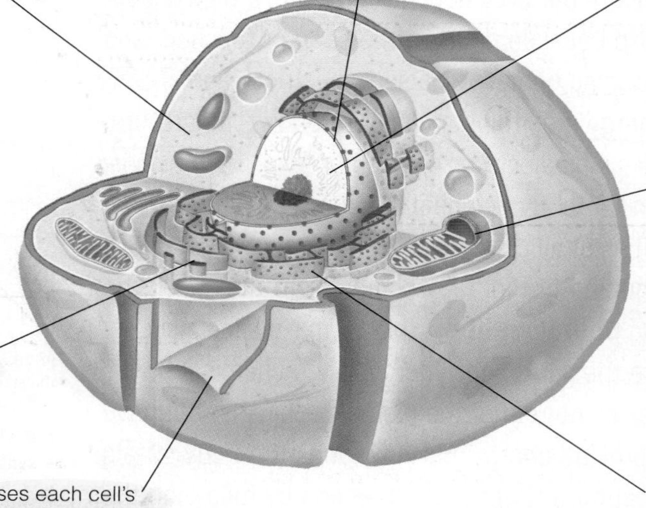

Inside the cell membrane lies the cytoplasm, a lattice-type structure that supports and controls the movement of the cell's structures. A protein-rich jelly-like fluid called cytosol fills the spaces within the lattice. The cytosol contains the enzymes involved in glucose breakdown (glycolysis).
A separate inner membrane encloses the cell's nucleus.
Inside the nucleus are the chromosomes, which contain the genetic material DNA.
Known as the "powerhouses" of the cells, the mitochondria are intricately folded membranes that house all the enzymes involved in the conversion of pyruvate to acetyl CoA, fatty acid oxidation, the TCA cycle, and the electron transport chain (described later in the chapter).
This network of membranes is known as smooth endoplasmic reticulum—the site of lipid synthesis.
A membrane encloses each cell's contents and regulates the passage of molecules in and out of the cell.
Rough endoplasmic reticulum is dotted with ribosomes—the site of protein synthesis (described in Chapter 6).

TABLE 7-1 Metabolic Work of the Liver

The liver is the most active processing center in the body. When nutrients enter the body from the digestive tract, the liver receives them first; then it metabolizes, packages, stores, or ships them out for use by other tissues. When alcohol, drugs, or poisons enter the body, they are also sent directly to the liver; here they are detoxified and their by-products shipped out for excretion. An enthusiastic anatomy and physiology professor once remarked that given the many vital activities of the liver, we should express our feelings for others by saying, "I love you with all my liver" instead of "with all my heart." Granted, this declaration lacks romance, but it makes a valid point. Here are just some of the many jobs performed by the liver. To renew your appreciation for this remarkable organ, review Figure 3-11 (p. 81).

Carbohydrates

- Metabolizes fructose, galactose, and glucose
- Makes and stores glycogen
- Breaks down glycogen and releases glucose
- Breaks down glucose for energy when needed
- Makes glucose from some amino acids and glycerol when needed
- Converts excess glucose and fructose to fatty acids

Lipids

- Builds and breaks down triglycerides, phospholipids, and cholesterol as needed
- Breaks down fatty acids for energy when needed
- Packages lipids in lipoproteins for transport to other body tissues
- Manufactures bile to send to the gallbladder for use in fat digestion
- Makes ketone bodies when necessary

Proteins

- Manufactures nonessential amino acids that are in short supply
- Removes from circulation amino acids that are present in excess of need and converts them to other amino acids or deaminates them and converts them to glucose or fatty acids
- Removes ammonia from the blood and converts it to urea to be sent to the kidneys for excretion
- Makes other nitrogen-containing compounds the body needs (such as bases used in DNA and RNA)
- Makes many proteins

Other

- Detoxifies alcohol, other drugs, and poisons; prepares waste products for excretion
- Helps dismantle old red blood cells and captures the iron for recycling
- Stores most vitamins and many minerals
- Activates vitamin D

The Building Reactions—Anabolism Earlier chapters describe how condensation reactions combine molecules to build body compounds. Glucose molecules may be joined together to make glycogen chains. Glycerol and fatty acids may be assembled into triglycerides. Amino acids may be linked together to make proteins. Each of these reactions starts with small, simple compounds and uses them as building blocks to form larger, more complex structures. Because such reactions involve doing work, they require energy. The building up of body compounds is known as **anabolism**. Anabolic reactions are represented in this book, wherever possible, with "up" arrows in chemical diagrams (such as those shown at the top of Figure 7-2, p. 200).

The Breakdown Reactions—Catabolism The breaking down of body compounds is known as **catabolism**; catabolic reactions release energy and are represented, wherever possible, by "down" arrows in chemical diagrams (as in the bottom of Figure 7-2, p. 200). Earlier chapters describe how hydrolysis reactions break down glycogen to glucose, triglycerides to fatty acids and glycerol, and proteins to amino acids. When the body needs energy, it breaks down any or all of these molecules further.

The Transfer of Energy in Reactions—ATP Some of the energy released during the breakdown of glucose, glycerol, fatty acids, and amino acids is captured in the high-energy compound **ATP (adenosine triphosphate).** ATP, as its name indicates, contains three phosphate groups (see Figure 7-3, p. 200). ♦ The negative charges on the phosphate groups make ATP vulnerable to hydrolysis. When the bonds between the phosphate groups are hydrolyzed, they readily break, splitting off one or two phosphate groups and releasing energy. In this way, ATP provides the energy that powers all the activities of living cells. Figure 7-4 (p. 201) describes how the body captures and releases energy in the bonds of ATP.

Quite often, the hydrolysis of ATP occurs simultaneously with reactions that will use that energy—a metabolic duet known as **coupled reactions.** In essence, the body uses ATP to transfer the energy released during catabolic reactions to power anabolic reactions. The body converts the chemical energy of food to the chemical energy of ATP with about 50 percent efficiency, radiating the rest as

♦ ATP = A–P~P~P.
Each ~ denotes a "high-energy" bond.

anabolism (an-AB-o-lism): reactions in which small molecules are put together to build larger ones. Anabolic reactions require energy.
- **ana** = (build) up

catabolism (ca-TAB-o-lism): reactions in which large molecules are broken down to smaller ones. Catabolic reactions release energy.
- **kata** = (break) down

ATP or **adenosine** (ah-DEN-oh-seen) **triphosphate** (try-FOS-fate): a common high-energy compound composed of a purine (adenine), a sugar (ribose), and three phosphate groups.

coupled reactions: pairs of chemical reactions in which some of the energy released from the breakdown of one compound is used to create a bond in the formation of another compound.

> FIGURE 7-2 Anabolic and Catabolic Reactions Compared

ANABOLIC REACTIONS

Anabolic reactions include the making of glycogen, triglycerides, and protein; these reactions require differing amounts of energy.

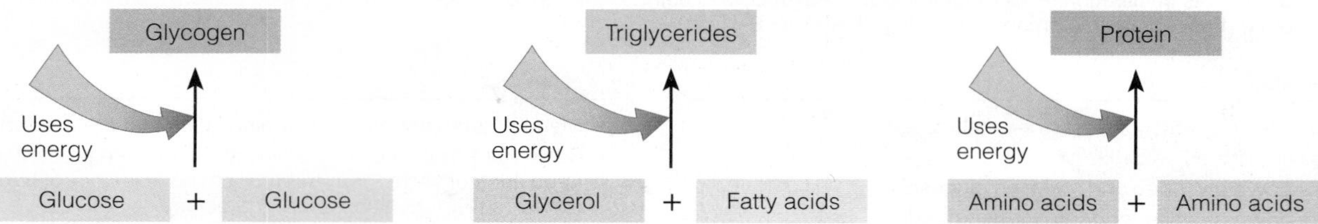

CATABOLIC REACTIONS

Catabolic reactions include the breakdown of glycogen, triglycerides, and protein; the further catabolism of glucose, glycerol, fatty acids, and amino acids releases differing amounts of energy. Much of the energy released is captured in the bonds of adenosine triphosphate (ATP).

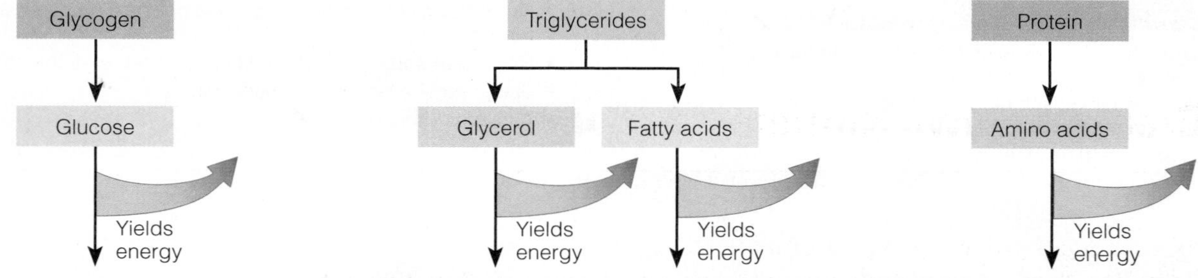

NOTE: You need not memorize a color code to understand the figures in this chapter, but you may find it helpful to know that blue is used for carbohydrates, yellow for fats, and red for proteins.

> FIGURE 7-3 ATP (Adenosine Triphosphate)

Notice that the bonds connecting the three phosphate groups have been drawn as wavy lines, indicating a high-energy bond. When these bonds are broken, energy is released.

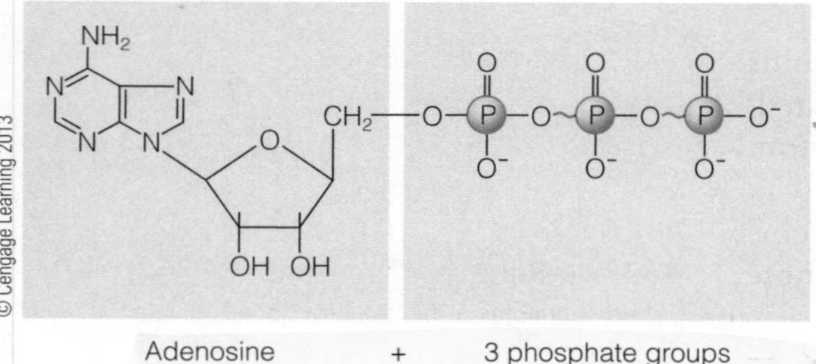

heat. Some energy is lost as heat again when the body uses the chemical energy of ATP to do its work—moving muscles, synthesizing compounds, or transporting nutrients, for example.

The Helpers in Metabolic Reactions—Enzymes and Coenzymes Metabolic reactions almost always require **enzymes** to facilitate their action. In many cases, the enzymes need assistants to help them. Enzyme helpers are called **coenzymes.***

Coenzymes are complex organic molecules that associate closely with enzymes but are not proteins themselves. The relationships between various coenzymes and their respective enzymes may differ in detail, but one thing is true of all: without its coenzyme, an enzyme cannot function. Some of the B vitamins serve as coenzymes that participate in the energy metabolism of glucose, glycerol, fatty acids, and amino acids. (Chapter 10 provides more details.)

REVIEW IT Identify the nutrients involved in energy metabolism and the high-energy compound that captures the energy released during their breakdown.

During digestion the energy-yielding nutrients—carbohydrates, fats, and proteins—are broken down to glucose (and other monosaccharides), glycerol, fatty acids, and amino acids. With the help of enzymes and coenzymes, the cells use these molecules to build more complex compounds (anabolism) or break them down further to release energy (catabolism). High-energy compounds such as ATP may capture the energy released during catabolism and provide the energy needed for anabolism.

enzymes: proteins that facilitate chemical reactions without being changed in the process; protein catalysts.

coenzymes: complex organic molecules that work with enzymes to facilitate the enzymes' activity. Many coenzymes have B vitamins as part of their structures. (Figure 10-2 on p. 301 illustrates coenzyme action.)

- **co** = with

*The general term for substances that facilitate enzyme action is *cofactors;* they include both organic coenzymes made from vitamins and inorganic substances such as minerals.

> FIGURE 7-4 The Capture and Release of Energy by ATP

It may help to think of ATP as a rechargeable battery—capturing and releasing energy as it does the body's work.

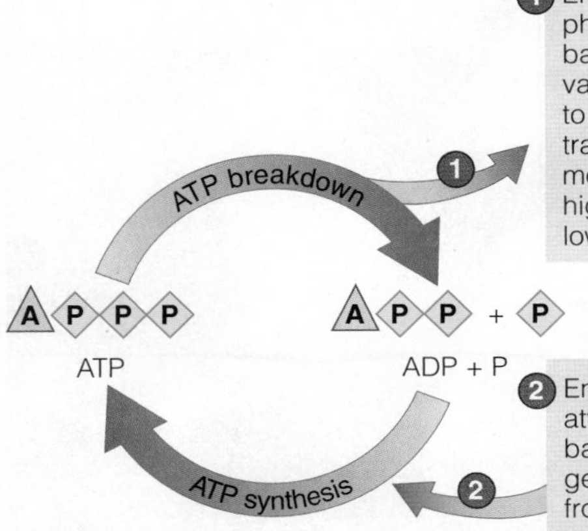

1 Energy is released when a high-energy phosphate bond in ATP is broken. Just as a battery can be used to provide energy for a variety of uses, the energy from ATP can be used to do most of the body's work—contract muscles, transport compounds, make new molecules, and more. With the loss of a phosphate group, high-energy ATP (charged battery) becomes low-energy ADP (used battery).

2 Energy is required when a phosphate group is attached to ADP, making ATP. Just as a used battery needs energy from an electrical outlet to get recharged, ADP (used battery) needs energy from the breakdown of carbohydrate, fat, and protein to make ATP (recharged battery).

7.2 Breaking Down Nutrients for Energy

LEARN IT Summarize the main steps in the energy metabolism of glucose, glycerol, fatty acids, and amino acids.

Chapters 4, 5, and 6 provide previews of metabolism; a brief review may be helpful. During digestion, the body breaks down the three energy-yielding nutrients—carbohydrates, fats, and proteins—into smaller molecules that can be absorbed:

- From carbohydrates—glucose (and other monosaccharides)
- From fats (triglycerides)—glycerol and fatty acids
- From proteins—amino acids

Each molecule of glucose, glycerol, fatty acids, and amino acids is composed of atoms—carbons, nitrogens, oxygens, and hydrogens. During catabolism, the bonds between these atoms break, releasing energy. To follow this action, recall how many carbons are in each of these molecules:

- Glucose has 6 carbons:

- Glycerol has 3 carbons:

- A fatty acid usually has an even number of carbons, commonly 16 or 18 carbons:*

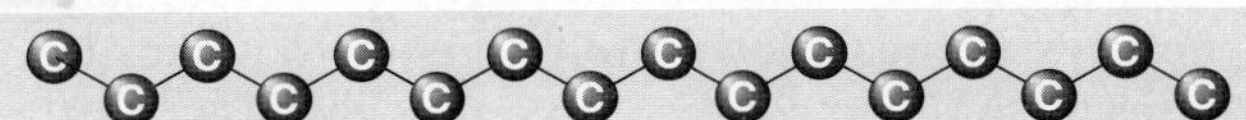

- An amino acid has 2, 3, or more carbons with a nitrogen attached:**

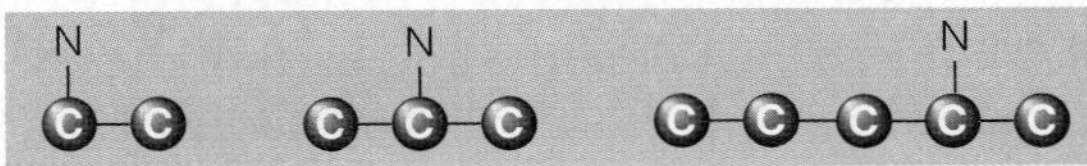

*The figures in this chapter show 16- or 18-carbon fatty acids. Fatty acids may have 4 to 20 or more carbons, with chain lengths of 16 and 18 carbons most prevalent.

**The figures in this chapter usually show amino acids as compounds of 2, 3, or 5 carbons arranged in a straight line, but in reality amino acids may contain other numbers of carbons and assume other structural shapes (see Appendix C).

All the energy used to keep the heart beating, the brain thinking, and the body moving comes from the carbohydrates, fats, and proteins in foods.

Full chemical structures and reactions appear both in the earlier chapters and in Appendix C; this chapter diagrams the reactions using just the compounds' carbon and nitrogen atoms.

As you will see, each of these molecules—glucose, glycerol, fatty acids, and amino acids—starts down a different path, but they all can end up in the same place. (Similarly, three people entering an interstate highway at three different locations can all travel to the same destination.) Along the way, two new names appear—**pyruvate** (a 3-carbon structure) and **acetyl CoA** (a 2-carbon structure with a coenzyme, **CoA**, attached)—and the rest of the story falls into place around them.* Two major points to notice in the following discussion:

- Pyruvate can be used to make glucose.
- Acetyl CoA cannot be used to make glucose.

Learning which fuels can be converted to glucose and which cannot is a major key to understanding energy metabolism. Amino acids and glycerol can be converted to pyruvate and therefore *can* provide glucose for the body. Fatty acids are converted to acetyl CoA and therefore *cannot* make glucose. Acetyl CoA can readily make fat. Whereas most of the body's cells can use glucose, fat, or both for energy, the body *must* have glucose to fuel the activities of the central nervous system and red blood cells. Without glucose from food, the body will break down its own lean (protein-containing) tissue to get the amino acids needed to make glucose. To protect this protein tissue, the body needs foods that provide glucose—primarily carbohydrate. Eating only fat provides abundant acetyl CoA, but forces the body to break down protein tissue to make glucose. Eating only protein requires the body to convert protein to glucose. Clearly, the best diet ♦ provides ample carbohydrate, adequate protein, and some fat.

Figure 7-5 provides a simplified overview of the energy-yielding pathways. Upcoming sections of the chapter describe how each of the energy-yielding nutrients follows its pathway as it is broken down to acetyl CoA. Their paths merge at acetyl CoA, where the real action begins. Acetyl CoA enters the **TCA cycle,** and energy is harnessed through the **electron transport chain.** The TCA cycle and electron transport chain have central roles in energy metabolism and receive

*The term *pyruvate* means a salt of *pyruvic acid*. (Throughout this book, the ending *-ate* is used interchangeably with *-ic acid;* for our purposes they mean the same thing.)

♦ A healthy diet provides:
- 45–65% kcalories from carbohydrate
- 10–35% kcalories from protein
- 20–35% kcalories from fat

pyruvate (PIE-roo-vate): a 3-carbon compound that plays a key role in energy metabolism.

$$\begin{array}{c} CH_3 \\ | \\ C{=}O \\ | \\ COOH \end{array}$$

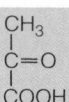

acetyl CoA (ASS-eh-teel or ah-SEET-il, coh-AY): a 2-carbon compound (acetate or acetic acid) to which a molecule of CoA is attached.

CoA (coh-AY): coenzyme A; the coenzyme derived from the B vitamin pantothenic acid and central to energy metabolism.

TCA cycle or **tricarboxylic** (try-car-box-ILL-ick) **acid cycle:** a series of metabolic reactions that break down molecules of acetyl CoA to carbon dioxide and hydrogen atoms; also called the *citric acid cycle* or the *Kreb's cycle* after the biochemist who elucidated its reactions.

electron transport chain: the final pathway in energy metabolism that transports electrons from hydrogen to oxygen and captures the energy released in the bonds of ATP; also called the *respiratory chain*.

> FIGURE 7-5 **Simplified Overview of Energy-Yielding Pathways**

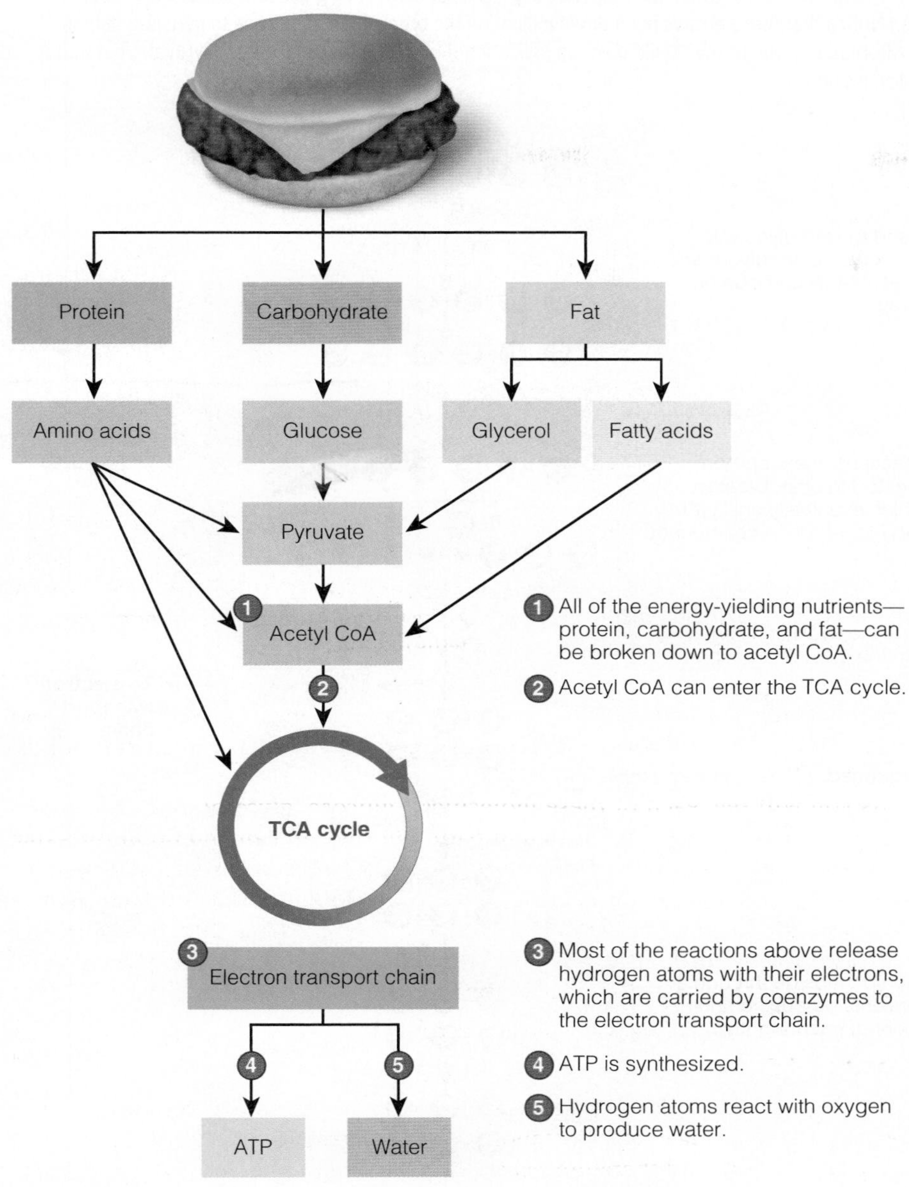
Protein
Carbohydrate
Fat
Amino acids
Glucose
Glycerol
Fatty acids
Pyruvate
1 Acetyl CoA
2
TCA cycle
3 Electron transport chain
4
5
ATP
Water
1 All of the energy-yielding nutrients—protein, carbohydrate, and fat—can be broken down to acetyl CoA.
2 Acetyl CoA can enter the TCA cycle.
3 Most of the reactions above release hydrogen atoms with their electrons, which are carried by coenzymes to the electron transport chain.
4 ATP is synthesized.
5 Hydrogen atoms react with oxygen to produce water.

full attention later in the chapter—after following each of the energy nutrient pathways to acetyl CoA.

Glucose What happens to glucose, glycerol, fatty acids, and amino acids during energy metabolism can best be understood by starting with glucose. This discussion features glucose because of its central role in carbohydrate metabolism and because liver cells can convert the monosaccharides fructose and galactose to compounds that enter the same energy pathways.

Glucose-to-Pyruvate The first pathway glucose takes on its way to yield energy is called **glycolysis** (glucose splitting).* Figure 7-6 (p. 204) shows a simplified drawing of glycolysis. (This pathway actually involves several steps and several enzymes, which are detailed in Appendix C.) In a series of reactions, the 6-carbon glucose is converted to similar 6-carbon compounds before being split in half, forming two 3-carbon compounds. These 3-carbon compounds continue along the pathway until they are converted to pyruvate. Thus the net yield of one glucose molecule is two pyruvate molecules.

*Glycolysis takes place in the cytosol of the cell (see Figure 7-1, p. 198).

glycolysis (gly-COLL-ih-sis): the metabolic breakdown of glucose to pyruvate. Glycolysis does not require oxygen (anaerobic).

- **glyco** = glucose
- **lysis** = breakdown

> **FIGURE 7-6 Glycolysis: Glucose-to-Pyruvate**

This simplified overview of glycolysis illustrates the steps in the process of converting glucose to pyruvate. (Appendix C provides more details.) Notice that these arrows point down indicating the breakdown of glucose to pyruvate during energy metabolism. (Alternatively, the arrows could point up indicating the making of glucose from pyruvate, but that is not the focus of this discussion.)

1. A little ATP is used to start glycolysis. Galactose and fructose enter glycolysis at different places, but all continue on the same pathway.
2. In a series of reactions, the 6-carbon glucose is converted to other 6-carbon compounds, which eventually split into two interchangeable 3-carbon compounds.
3. Coenzymes carry the hydrogens and their electrons to the electron transport chain.
4. A little ATP is produced.
5. The 3-carbon compounds go through a series of conversions, producing another 3-carbon compound, each slightly different.
6. Eventually, the 3-carbon compounds are converted to pyruvate. Glycolysis of one molecule of glucose produces two molecules of pyruvate.

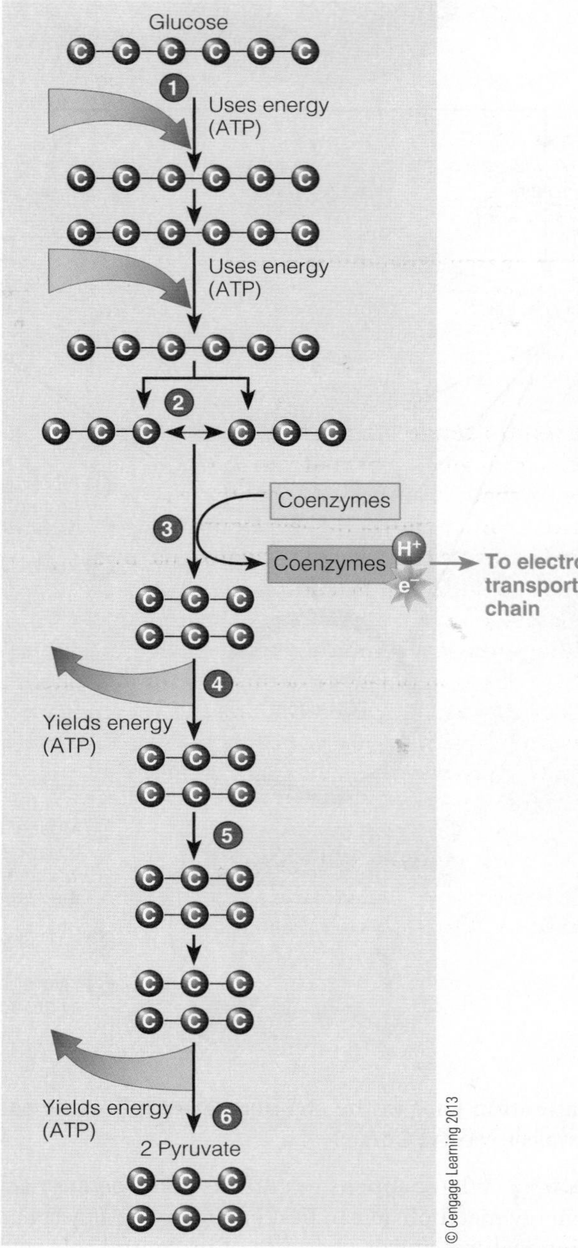

NOTE: The cell uses a little energy (–2 ATP) to begin the breakdown of glucose to pyruvate, but then it gains a little more energy (4 ATP) for a small net gain (of 2 ATP).

The net yield of energy at this point is small; to start glycolysis, the cell uses a little energy and then produces only a little more than it invested initially.* In addition, as glucose breaks down to pyruvate, hydrogen atoms with their electrons are released and carried to the electron transport chain by coenzymes made from the B vitamin niacin. A later section of the chapter explains how oxygen accepts the electrons and combines with the hydrogens to form water and how the process captures energy in the bonds of ATP.

This discussion focuses primarily on the breakdown of glucose for energy, but if needed, cells in the liver (and to some extent, the kidneys) can make glucose again

*The cell uses two ATP to begin the breakdown of glucose to pyruvate, but it then gains four ATP for a net gain of two ATP.

from pyruvate in a process similar to the reversal of glycolysis. Making glucose requires energy, however, and a few different enzymes. Still, glucose can be made from pyruvate, so the arrows between glucose and pyruvate could point up as well as down. ♦

♦ Glucose may go "down" to make pyruvate, or pyruvate may go "up" to make glucose, depending on the cell's needs.

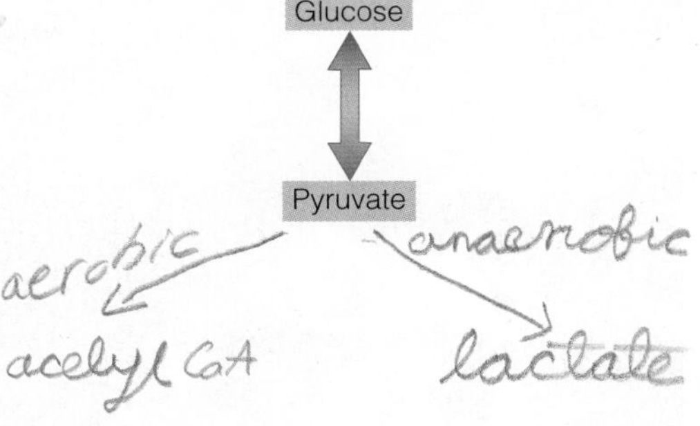

Pyruvate's Options—Anaerobic or Aerobic Whenever carbohydrates, fats, or proteins are broken down to provide energy, oxygen is always ultimately involved in the process. The role of oxygen in metabolism is worth noticing, for it helps our understanding of physiology and metabolic reactions. Chapter 14 describes the body's use of the energy nutrients to fuel physical activity, but the facts presented here offer a sneak preview.

When the body needs energy quickly—as occurs when you run a quarter mile as fast as you can—pyruvate is converted to lactate. The breakdown of glucose-to-pyruvate-to-lactate proceeds without oxygen—it is **anaerobic.** This anaerobic pathway yields energy quickly, but it cannot be sustained for long—a couple of minutes at most.

When energy expenditure proceeds at a slower pace—as occurs when you jog around the track for an hour—pyruvate breaks down to acetyl CoA in an **aerobic** pathway. Aerobic pathways produce energy more slowly, but because they can be sustained for a long time, their total energy yield is greater. The following paragraphs provide more details.

Pyruvate-to-Lactate (Anaerobic) As mentioned earlier, coenzymes carry the hydrogens from glucose breakdown to the electron transport chain. If the electron transport chain is unable to accept these hydrogens, as may occur when cells lack sufficient **mitochondria** (review Figure 7-1, p. 198) or in the absence of sufficient oxygen, pyruvate can accept the hydrogens. By accepting the hydrogens, pyruvate becomes **lactate,** and the coenzymes are freed to return to glycolysis to pick up more hydrogens (see the left side of Figure 7-7). In this way, glucose can continue providing energy anaerobically for a while.

The production of lactate occurs to a limited extent even at rest. During high-intensity exercise, however, the muscles rely heavily on anaerobic glycolysis to produce ATP quickly, and the concentration of lactate increases dramatically. The rapid rate of glycolysis produces abundant pyruvate and releases hydrogen-carrying coenzymes more rapidly than the mitochondria can handle. To enable exercise to

anaerobic (AN-air-ROE-bic): not requiring oxygen.

- **an** = not

aerobic (air-ROE-bic): requiring oxygen.

mitochondria (my-toh-KON-dree-uh): the cellular organelles responsible for producing ATP aerobically; made of membranes (lipid and protein) with enzymes mounted on them. (The singular is *mitochondrion.*)

- **mitos** = thread (referring to their slender shape)
- **chondros** = cartilage (referring to their external appearance)

lactate: a 3-carbon compound produced from pyruvate during anaerobic metabolism.

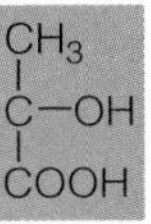

> **FIGURE 7-7 Pyruvate-to-Lactate**

Because muscle cells lack the enzyme to convert lactate to glucose, lactate must first travel to the liver. The process of converting lactate from the muscles to glucose in the liver that can be returned to the muscles is known as the Cori cycle.

1 Working muscles break down most of their glucose molecules to pyruvate.

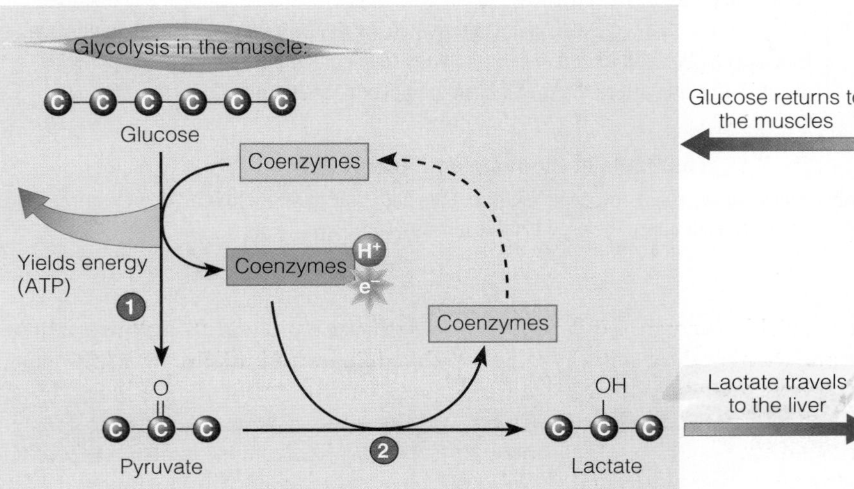

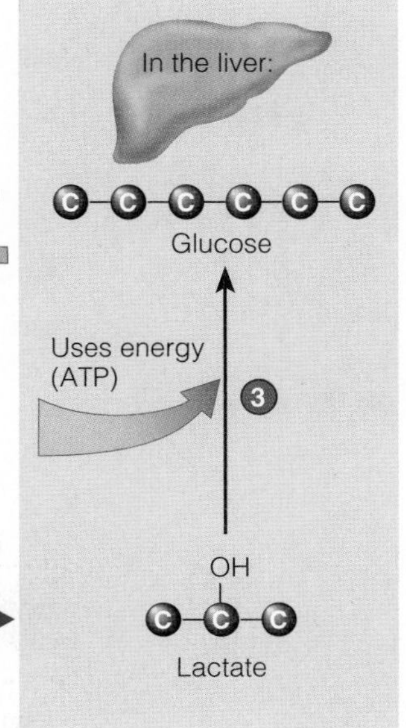

2 If the cells lack sufficient mitochondria or in the absence of sufficient oxygen, pyruvate can accept the hydrogens from glucose breakdown and become lactate. This conversion frees the coenzymes so that glycolysis can continue.

3 Liver enzymes can convert lactate to glucose, but this reaction requires energy.

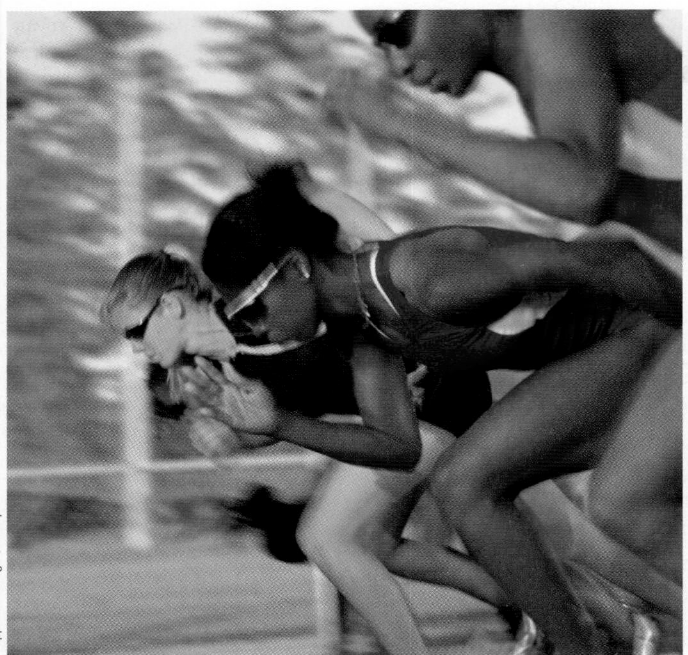

The anaerobic breakdown of glucose-to-pyruvate-to-lactate is the major source of energy for short, intense exercise.

continue at this intensity, pyruvate is converted to lactate and coenzymes are released, which allows glycolysis to continue. The accumulation of lactate in the muscles coincides with—but does not seem to be the cause of—the subsequent drop in blood pH, burning pain, and fatigue that are commonly associated with intense exercise. In fact, making lactate from pyruvate consumes two hydrogen ions, which actually diminishes acidity and improves the performance of tired muscles. A person performing the same exercise following endurance training actually experiences less discomfort—in part because the number of mitochondria in the muscle cells has increased. This adaptation improves the mitochondria's ability to keep pace with the muscles' demand for energy.

One possible fate of lactate is to be transported from the muscles to the liver. There the liver can convert the lactate produced in muscles to glucose, which can then be returned to the muscles. (Muscle cells cannot convert lactate to glucose because they lack the necessary enzyme.) This recycling process is called the **Cori cycle** (see the right side of Figure 7-7, p. 205).

Pyruvate-to-Acetyl CoA (Aerobic) If a cell needs energy and oxygen is available, pyruvate molecules enter the mitochondria of the cell. There a carbon group (COOH) from the 3-carbon pyruvate is removed to produce a 2-carbon compound that bonds with a molecule of CoA, becoming acetyl CoA. The carbon group from pyruvate becomes carbon dioxide (CO_2), which is released into the blood, circulated to the lungs, and breathed out. Figure 7-8 diagrams the pyruvate-to-acetyl CoA reaction.

Figure 7-9 shows that many of the body's metabolic pathways are reversible, but the step from pyruvate to acetyl CoA is not one of them. A cell cannot retrieve the carbons from carbon dioxide to remake pyruvate and then glucose. It is one way only.

The story of acetyl CoA continues on p. 208 after a discussion of how fat and protein arrive at the same crossroads. For now, know that when acetyl CoA continues on its energy-yielding pathway, much more ATP is produced than during glycolysis.

REVIEW IT

The glucose-to-energy pathway begins with glycolysis—the breakdown of glucose to pyruvate. Pyruvate may be converted to lactate anaerobically or to acetyl CoA aerobically. The pathway from pyruvate to acetyl CoA is irreversible. Once the commitment to acetyl CoA is made, glucose is not retrievable; acetyl CoA cannot go back to glucose. Glucose can be synthesized only from pyruvate or compounds earlier in the pathway. Figure 7-10 summarizes the metabolism of glucose for energy.

> FIGURE 7-8 Pyruvate-to-Acetyl CoA

The pyruvate-to-acetyl CoA reaction is not reversible.

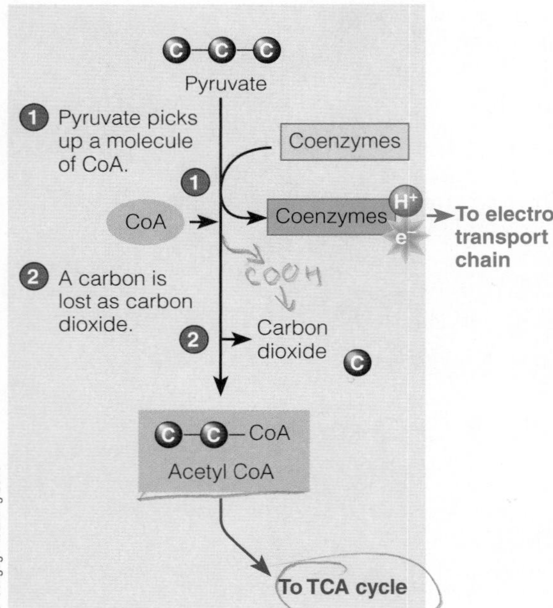

> FIGURE 7-9 The Paths of Pyruvate and Acetyl CoA

Pyruvate may follow several reversible paths, but the path from pyruvate to acetyl CoA is irreversible. Notice that fatty acids cannot be used to make glucose.

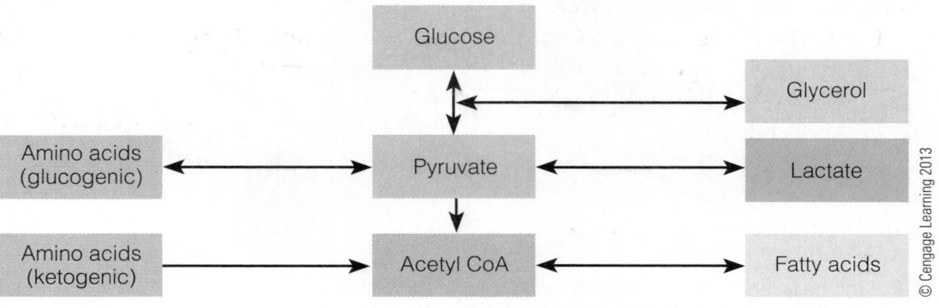

NOTE: Amino acids that can be used to make glucose are called *glucogenic;* amino acids that are converted to acetyl CoA are called *ketogenic*.

Cori cycle: the pathway in which glucose is metabolized to lactate (by anaerobic glycolysis) in the muscle, lactate is converted back to glucose in the liver, and then glucose is returned to the muscle; named after the scientist who elucidated this pathway.

Glycerol and Fatty Acids Recall that triglycerides can break down to glycerol and fatty acids. They enter energy metabolism via different pathways.

Glycerol-to-Pyruvate Glycerol is a 3-carbon compound like pyruvate but with a different arrangement of H and OH on the C. As such, glycerol can easily be converted to another 3-carbon compound that can go either "up" to glucose or "down" to pyruvate and then to acetyl CoA (review Figure 7-9, p. 206).

Fatty Acids-to-Acetyl CoA Fatty acids are taken apart two carbons at a time in a series of reactions known as **fatty acid oxidation.*** Figure 7-11 (p. 208) illustrates fatty acid oxidation and shows that in the process, each 2-carbon fragment splits off and combines with a molecule of CoA to make acetyl CoA. As each 2-carbon fragment breaks off, hydrogens and their electrons are released and carried to the electron transport chain by coenzymes made from the B vitamins riboflavin and niacin.

Fatty Acids Cannot Make Glucose As mentioned earlier, red blood cells and the brain and nervous system depend primarily on glucose as fuel. When carbohydrate is unavailable, liver cells can make glucose from pyruvate and other 3-carbon compounds, such as glycerol. Importantly, cells cannot make glucose from the 2-carbon fragments of fatty acids.

Remember that almost all dietary fats are triglycerides and that triglycerides contain only one small molecule of glycerol with three fatty acids. The glycerol can yield glucose, but that represents only 3 of the 50 or so carbons in a triglyceride—about 5 percent of its weight. The other 95 percent cannot be used to make glucose.

REVIEW IT

The body can convert the small glycerol portion of a triglyceride to either pyruvate (and then glucose) or acetyl CoA. The fatty acids of a triglyceride, on the other hand, cannot make glucose, but they can provide abundant acetyl CoA. Acetyl CoA may then enter the TCA cycle to release energy or combine with other molecules of acetyl CoA to make body fat. Figure 7-12 (p. 209) summarizes the metabolism of fats for energy.

> FIGURE 7-10 **Glucose Enters the Energy Pathway**

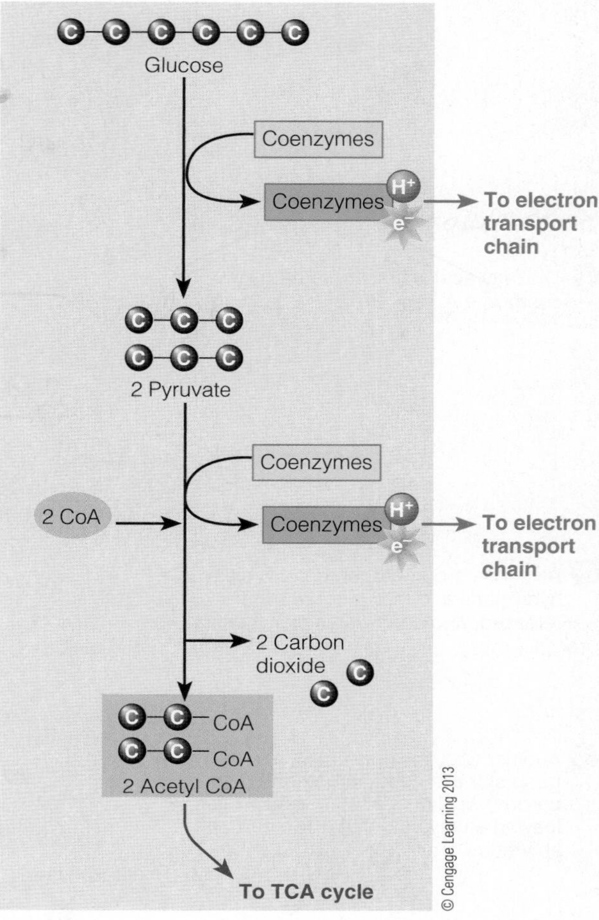

REVIEW IT

1 glucose yields 2 pyruvate, which yield 2 acetyl CoA.

Amino Acids The preceding two sections have described how the breakdown of carbohydrate and fat produces acetyl CoA. One energy-yielding nutrient remains: protein or, rather, the amino acids of protein.

Amino Acid Deamination Before entering the metabolic pathways, amino acids are deaminated (that is, they lose their nitrogen-containing amino group). Chapter 6 describes how deamination produces ammonia (NH_3), which provides the nitrogen needed to make nonessential amino acids and other nitrogen-containing compounds. Any remaining ammonia is cleared from the body via urea synthesis in the liver and excretion in the kidneys.

Amino Acid Pathways Amino acids can enter the energy pathways in several ways. Some amino acids can be converted to pyruvate, others are converted to acetyl CoA, and still others enter the TCA cycle directly as compounds other than acetyl CoA.

As you might expect, amino acids that are used to make pyruvate can provide glucose, whereas those used to make acetyl CoA can provide additional energy or make body fat but cannot make glucose. Amino acids entering the TCA cycle directly can continue in the cycle and generate energy; alternatively, they can generate glucose.** Thus protein, unlike fat, is a fairly good source of glucose when carbohydrate is not available.

*Oxidation of fatty acids occurs in the mitochondria of the cells (review Figure 7-1, p. 198).

**Amino acids that can make glucose via either pyruvate or TCA cycle intermediates are *glucogenic;* amino acids that are degraded to acetyl CoA are *ketogenic.*

fatty acid oxidation: the metabolic breakdown of fatty acids to acetyl CoA; also called *beta oxidation.*

> **FIGURE 7-11 Fatty Acid-to-Acetyl CoA**

Fatty acids are broken apart into 2-carbon fragments that combine with CoA to make acetyl CoA.

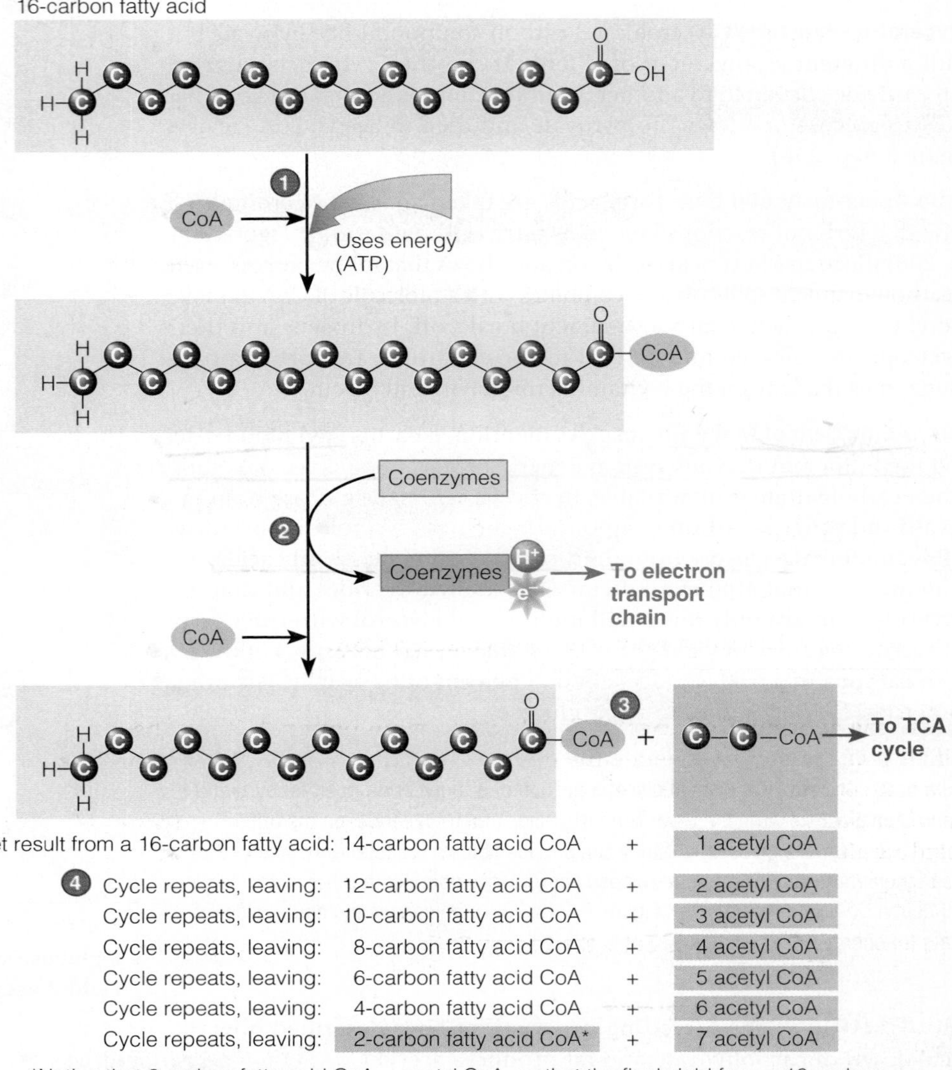

1. The fatty acid is first activated by coenzyme A.
2. As each carbon-carbon bond is cleaved, hydrogens and their electrons are released, and coenzymes pick them up.
3. Another CoA joins the chain, and the bond at the second carbon (the beta-carbon) weakens. Acetyl CoA splits off, leaving a fatty acid that is two carbons shorter.
4. The shorter fatty acid enters the pathway and the cycle repeats, releasing more hydrogens with their electrons to coenzymes and producing more acetyl CoA. The molecules of acetyl CoA enter the TCA cycle, and the coenzymes carry the hydrogens and their electrons to the electron transport chain.

REVIEW IT

The body can use some amino acids to make glucose, whereas others can be used either to provide energy or to make fat. Before an amino acid enters any of these metabolic pathways, its nitrogen-containing amino group must be removed through deamination. Figure 7-13 summarizes the metabolism of amino acids for energy.

Table 7-2 (p. 210) reviews the ways the body can use the energy-yielding nutrients. To obtain energy, the body uses glucose and fatty acids as its primary fuels and amino acids to a lesser extent. To make glucose, the body can use all carbohydrates and most amino acids, but it can convert only 5 percent of fat (the glycerol portion) to glucose. Fatty acids cannot make glucose. To make proteins, the body needs amino acids. It can use glucose and glycerol to make some nonessential amino acids when nitrogen is available; it cannot use fatty acids to make body proteins. Finally, when energy intake exceeds the body's needs, all three energy-yielding nutrients can contribute to body fat stores.

The Final Steps of Energy Metabolism Thus far the discussion has followed each of the energy-yielding nutrients down three different pathways,

> **FIGURE 7-12 Fats Enter the Energy Pathway**

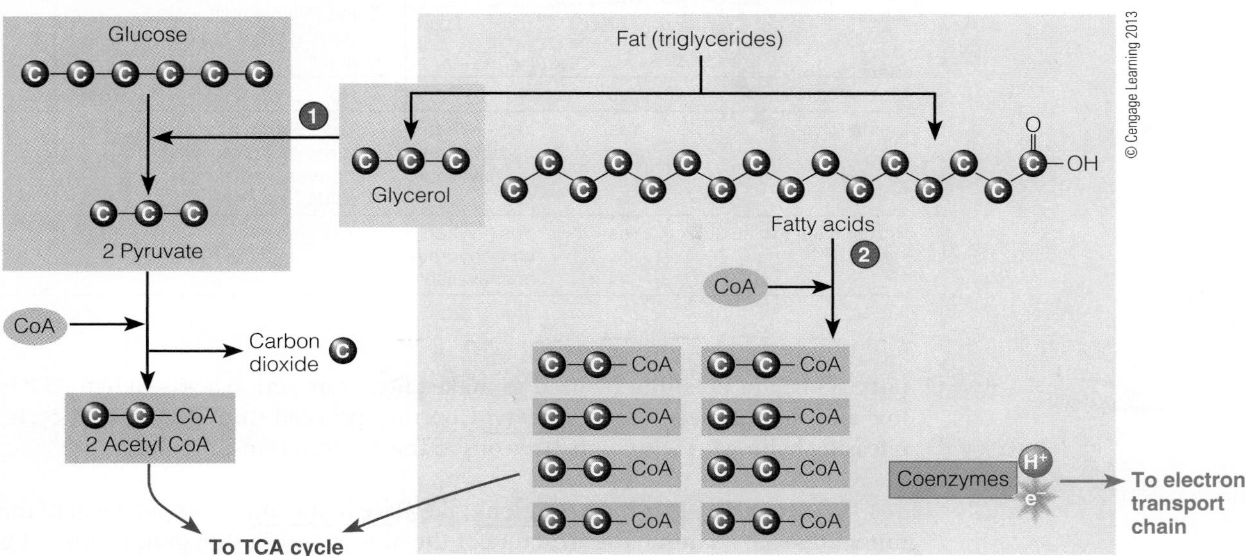

REVIEW IT 16-carbon fatty acid yields 8 acetyl CoA.

all arriving at acetyl CoA. Acetyl CoA has two main options—it may be used to synthesize fats or to generate the high-energy compound ATP. When ATP is abundant, acetyl CoA makes fat, the most efficient way to store energy for later use when energy may be needed. Thus any molecule that can make acetyl CoA—including glucose, glycerol, fatty acids, and amino acids—can make fat. In reviewing Figure 7-9 (p. 206) notice that acetyl CoA can be used as a building block for

> **FIGURE 7-13 Amino Acids Enter the Energy Pathway**

Amino acids

1 Most amino acids can be converted to pyruvate, which can be used to make glucose; they are glucogenic.

2 Some amino acids are converted directly to acetyl CoA; they are ketogenic.

3 Some amino acids can enter the TCA cycle directly; they are glucogenic.

NH_2 · NH_3 · NH_3 · 1 · C-C-C Pyruvate · CoA · Coenzymes · Coenzymes · H^+ · e^- · To electron transport chain · Carbon dioxide · NH_2 · NH_3 · 2 · C-C-CoA Acetyl CoA · NH_2 · NH_3 · NH_3 · 3 · To TCA cycle

© Cengage Learning 2013

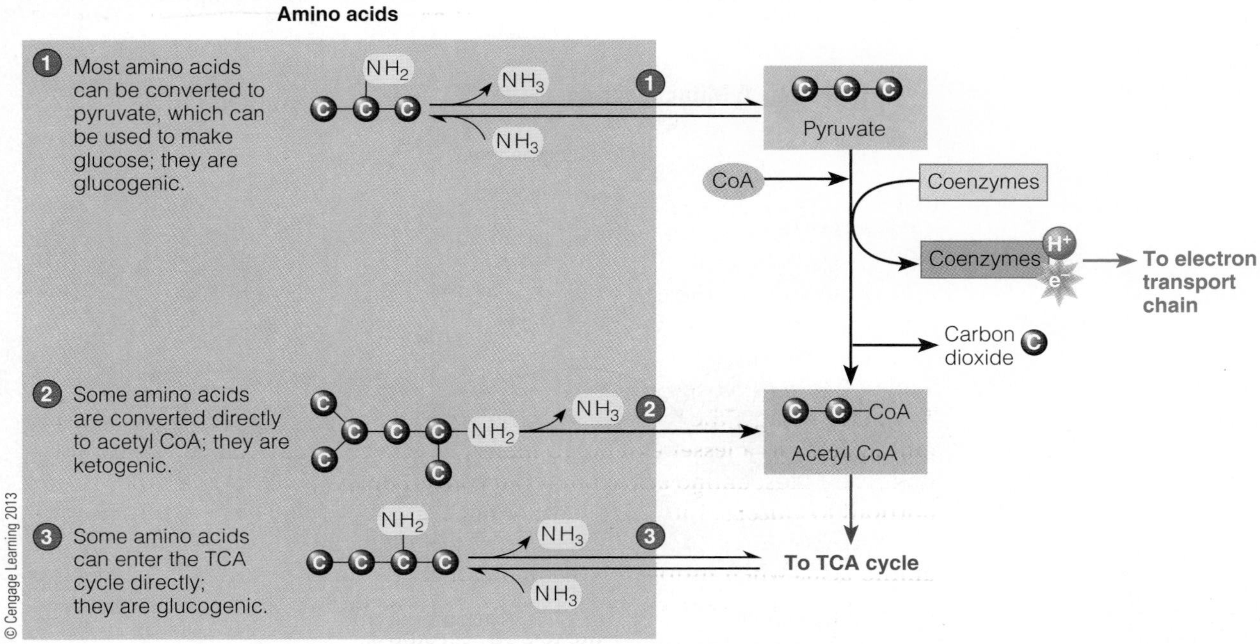

NOTE: Deamination and the synthesis of urea are discussed and illustrated in Chapter 6, Figure 6-13 (p. 180). The arrows from pyruvate and the TCA cycle to amino acids are possible only for *nonessential* amino acids; remember, the body cannot make essential amino acids.

TABLE 7-2 Review of Energy-Yielding Nutrient Endpoints

Nutrient	Yields energy?	Yields glucose?	Yields amino acids and body proteins?	Yields fat stores?
Carbohydrates (glucose)	Yes	Yes	Yes—when nitrogen is available, can yield *nonessential* amino acids	Yes
Lipids (fatty acids)	Yes	No	No	Yes
Lipids (glycerol)	Yes	Yes—when carbohydrate is unavailable	Yes—when nitrogen is available, can yield *nonessential* amino acids	Yes
Proteins (amino acids)	Yes	Yes—when carbohydrate is unavailable	Yes	Yes

acetil CoA →

fatty acids, but it cannot be used to make glucose or amino acids. When ATP is low and the cell needs energy, acetyl CoA may proceed through the TCA cycle, releasing hydrogens with their electrons to the electron transport chain.

The TCA Cycle The TCA cycle reactions take place in the inner compartment of the mitochondria. Examine the structure of the mitochondria shown in Figure 7-14. The significance of its structure will become evident as details unfold.

When cells need energy, acetyl CoA enters the TCA cycle, a busy metabolic traffic center. The TCA cycle is a circular path, but that doesn't mean it regenerates acetyl CoA. Acetyl CoA goes one way only—down to two carbon dioxide molecules and a coenzyme (CoA). The TCA cycle is a circular path because a 4-carbon compound known as **oxaloacetate** is needed in the first step and it is synthesized in the last step.

Oxaloacetate's role in replenishing the TCA cycle is critical. When oxaloacetate is insufficient, the TCA cycle slows down, and the cells face an energy crisis. Oxaloacetate is made primarily from pyruvate, although it can also be made from certain amino acids. Importantly, oxaloacetate cannot be made from fat. That oxaloacetate must be available for acetyl CoA to enter the TCA cycle underscores the importance of carbohydrates in the diet. A diet that provides ample carbohydrate

> **FIGURE 7-14 A Mitochondrion**

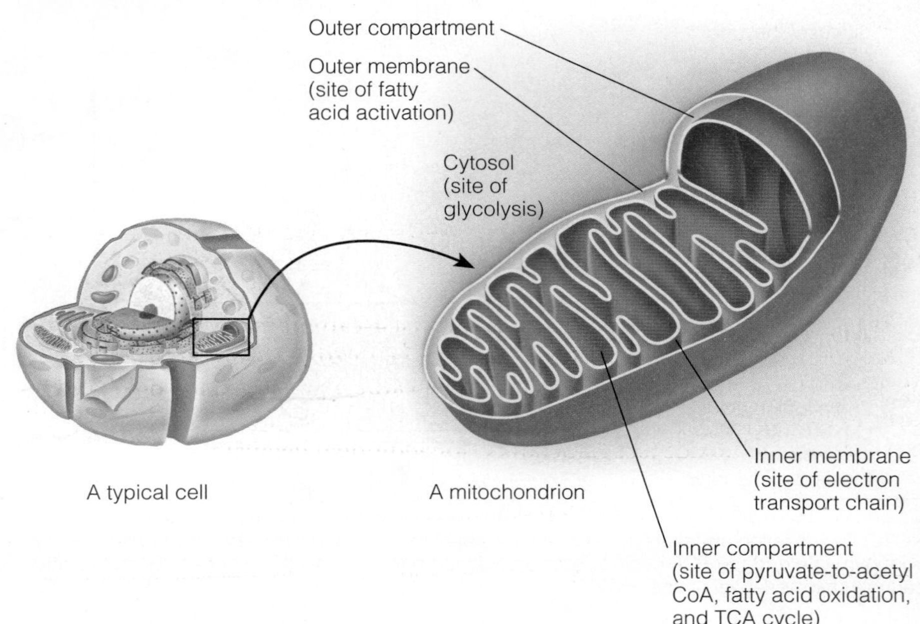

oxaloacetate (OKS-ah-low-AS-eh-tate): a carbohydrate intermediate of the TCA cycle.

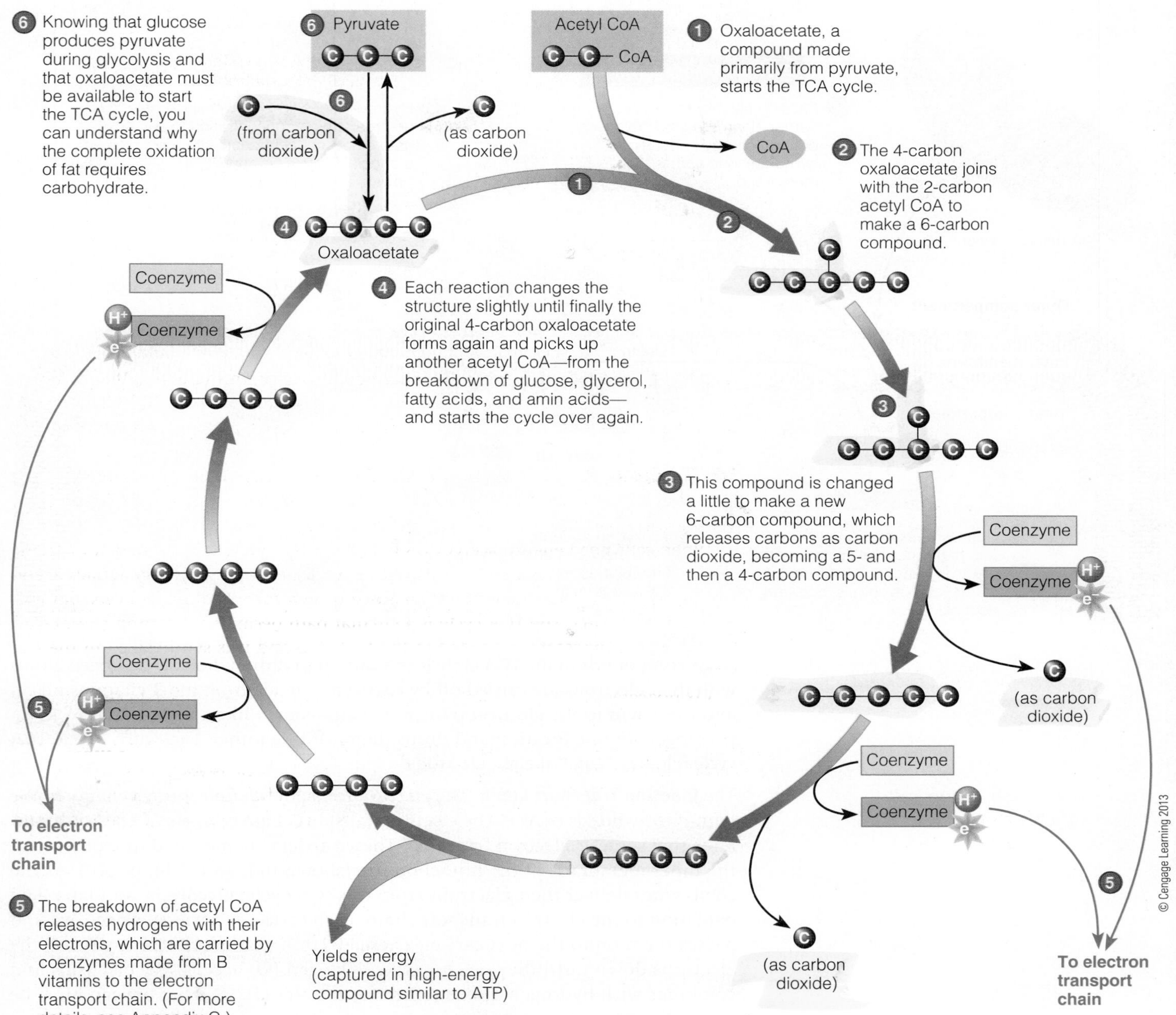

Knowing that glucose produces pyruvate during glycolysis and that oxaloacetate must be available to start the TCA cycle, you can understand why the complete oxidation of fat requires carbohydrate.
Pyruvate
Acetyl CoA
CoA
Oxaloacetate, a compound made primarily from pyruvate, starts the TCA cycle.
(from carbon dioxide)
(as carbon dioxide)
CoA
The 4-carbon oxaloacetate joins with the 2-carbon acetyl CoA to make a 6-carbon compound.
Oxaloacetate
Each reaction changes the structure slightly until finally the original 4-carbon oxaloacetate forms again and picks up another acetyl CoA—from the breakdown of glucose, glycerol, fatty acids, and amin acids—and starts the cycle over again.
Coenzyme
Coenzyme
This compound is changed a little to make a new 6-carbon compound, which releases carbons as carbon dioxide, becoming a 5- and then a 4-carbon compound.
(as carbon dioxide)
To electron transport chain
The breakdown of acetyl CoA releases hydrogens with their electrons, which are carried by coenzymes made from B vitamins to the electron transport chain. (For more details, see Appendix C.)
Yields energy (captured in high-energy compound similar to ATP)
(as carbon dioxide)
To electron transport chain
© Cengage Learning 2013

FIGURE 7-15 The TCA Cycle

ensures an adequate supply of oxaloacetate—because glucose produces pyruvate during glycolysis. (This chapter closes with a discussion of the consequences of low-carbohydrate diets.)

As Figure 7-15 shows, oxaloacetate is the first 4-carbon compound to enter the TCA cycle. Oxaloacetate picks up acetyl CoA (a 2-carbon compound), drops off one carbon (as carbon dioxide), then another carbon (as carbon dioxide), and returns to pick up another acetyl CoA. As for the acetyl CoA, its carbons go only one way—to carbon dioxide (see Appendix C for additional details).*

*Actually, the carbons that enter the cycle in acetyl CoA may not be the exact ones that are given off as carbon dioxide. In one of the steps of the cycle, a 6-carbon compound of the cycle becomes symmetrical, both ends being identical. Thereafter it loses carbons to carbon dioxide at one end or the other. Thus only half of the carbons from acetyl CoA are given off as carbon dioxide in any one turn of the cycle; the other half become part of the compound that returns to pick up another acetyl CoA. It is true to say, though, that for each acetyl CoA that enters the TCA cycle, two carbons are given off as carbon dioxide. It is also true that with each turn of the cycle, the energy equivalent of one acetyl CoA is released.

> FIGURE 7-16 **Electron Transport Chain and ATP Synthesis**

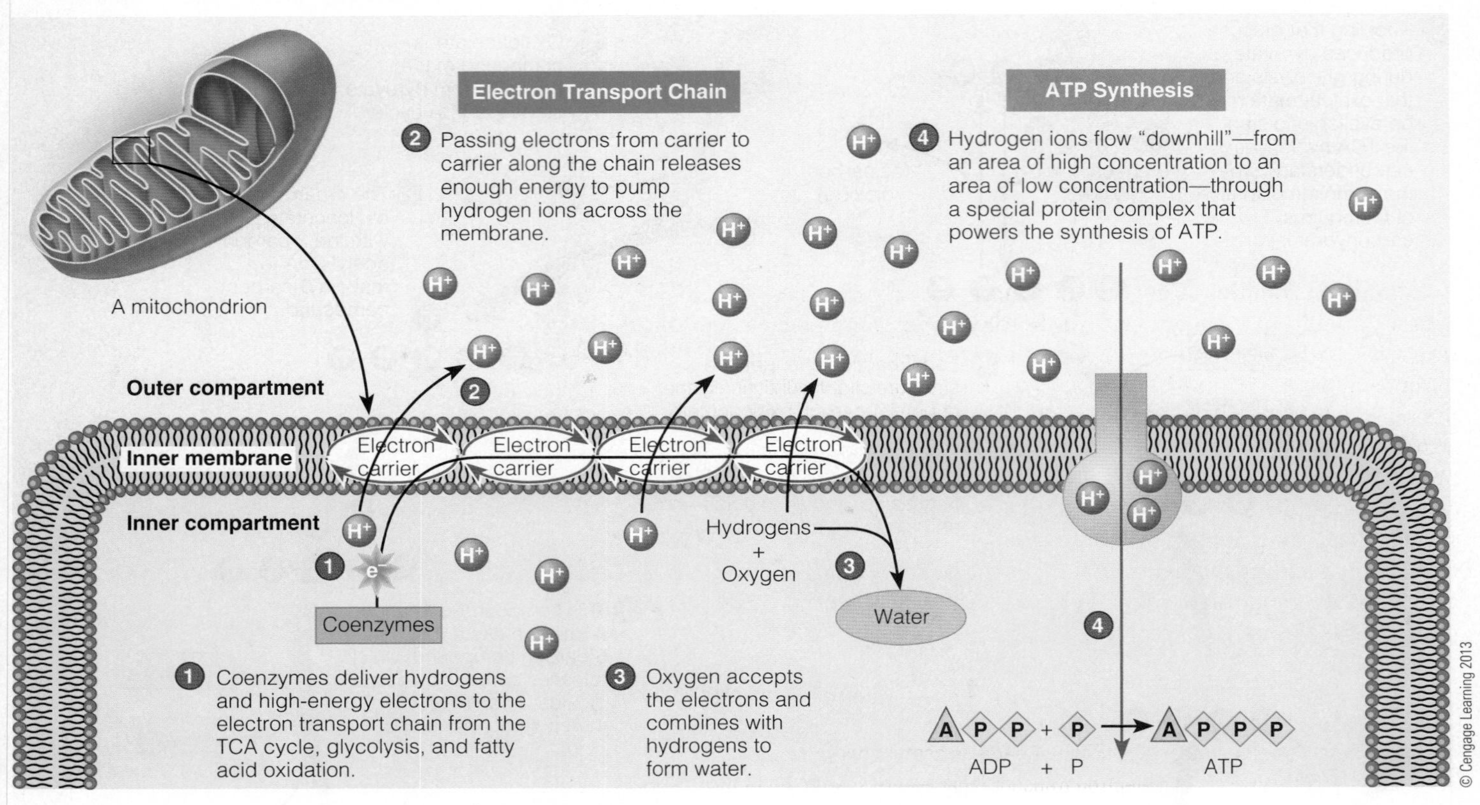
Electron Transport Chain
ATP Synthesis
2 Passing electrons from carrier to carrier along the chain releases enough energy to pump hydrogen ions across the membrane.
4 Hydrogen ions flow "downhill"—from an area of high concentration to an area of low concentration—through a special protein complex that powers the synthesis of ATP.
A mitochondrion
Outer compartment
Inner membrane
Inner compartment
Electron carrier
Electron carrier
Electron carrier
Electron carrier
Hydrogens + Oxygen
Water
Coenzymes
1 Coenzymes deliver hydrogens and high-energy electrons to the electron transport chain from the TCA cycle, glycolysis, and fatty acid oxidation.
3 Oxygen accepts the electrons and combines with hydrogens to form water.
ADP + P
ATP
© Cengage Learning 2013

As compounds in the TCA cycle lose a carbon to carbon dioxide, hydrogen atoms with their electrons are carried off by coenzymes made from the B vitamins niacin and riboflavin to the electron transport chain—much like a taxicab that picks up passengers in one location and drops them off in another. Each turn of the TCA cycle releases a total of eight electrons.

The Electron Transport Chain The electron transport chain captures energy in the high-energy bonds of ATP. The electron transport chain consists of a series of proteins that serve as electron "carriers." These carriers are mounted in sequence on the inner membrane of the mitochondria (shown in Figure 7-14, p. 210). As the coenzymes deliver their electrons from the TCA cycle, glycolysis, and fatty acid oxidation to the electron transport chain, each carrier receives the electrons and passes them on to the next carrier. These electron carriers continue passing the electrons down until they reach oxygen. Oxygen (O) accepts the electrons and combines with hydrogen atoms (H) to form water (H_2O). That oxygen must be available for energy metabolism explains why it is essential to life.

As electrons are passed from carrier to carrier, hydrogen ions are pumped across the membrane to the outer compartment of the mitochondria. The rush of hydrogen ions back into the inner compartment powers the synthesis of ATP. In this way, energy is captured in the bonds of ATP. The ATP leaves the mitochondria and enters the cytoplasm, where it can be used for energy. Figure 7-16 provides a simple diagram of the electron transport chain (see Appendix C for details).

The kCalories-per-Gram Secret Revealed Of the three energy-yielding nutrients, fat provides the most energy per gram. ♦ The reason may be apparent in Figure 7-17, which compares a fatty acid with a glucose molecule. Notice that nearly all the bonds in the fatty acid are between carbons and hydrogens. Oxygen can be added to all of them—forming carbon dioxide (CO_2) with the carbons and water (H_2O) with the hydrogens. As this happens, hydrogens are released to coenzymes heading for the electron transport chain. In glucose, on the other hand, an oxygen is already bonded to each carbon. Thus there is less potential for oxidation, and fewer hydrogens are released when the remaining bonds are broken.

♦ Fat = 9 kcal/g
Carbohydrate = 4 kcal/g
Protein = 4 kcal/g

> **FIGURE 7-17 Chemical Structures of a Fatty Acid and Glucose Compared**

To ease comparison, the structure shown here for glucose is not the ring structure shown in Chapter 4, but an alternative way of drawing its chemical structure.

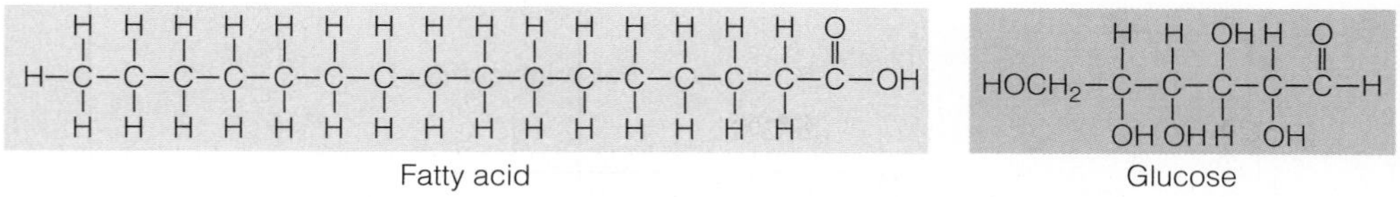

Fatty acid

Glucose

Because fat contains many carbon-hydrogen bonds that can be readily oxidized, it sends numerous coenzymes with their hydrogens and electrons to the electron transport chain where that energy can be captured in the bonds of ATP. This explains why fat yields more kcalories per gram than carbohydrate or protein. (Remember that each ATP holds energy and that kcalories measure energy; thus the more ATP, the more kcalories.) For example, one glucose molecule will yield 30 to 32 ATP when completely oxidized. In comparison, one 16-carbon fatty acid molecule will yield 129 ATP when completely oxidized. Fat is a more efficient fuel source. Gram for gram, fat can provide much more energy than either of the other two energy-yielding nutrients, making it the body's preferred form of energy storage. (Similarly, you might prefer to fill your car with a fuel that provides 130 miles per gallon versus one that provides 30 miles per gallon.)

REVIEW IT Summarize the main steps in the energy metabolism of glucose, glycerol, fatty acids, and amino acids.

Carbohydrate, fat, and protein take different paths to acetyl CoA, but once there, the final pathways—the TCA cycle and electron transport chain—are shared. All of the pathways, which are shown as a simplified overview in Figure 7-5 (p. 203), are shown again in more detail in Figure 7-18 (p. 214). Instead of dismissing this figure as "too busy," take a few moments to appreciate the busyness of it all. Consider that this figure is merely an overview of energy metabolism, and then imagine how busy a living cell really is during the metabolism of hundreds of compounds, each of which may be involved in several reactions, each requiring an enzyme.

7.3 Feasting and Fasting

LEARN IT Explain how an excess of any of the three energy-yielding nutrients contributes to body fat and how an inadequate intake of any of them shifts metabolism.

Every day, a healthy diet delivers more than a thousand kcalories of energy—from carbohydrate, fat, and protein—to fuel the physical activity and metabolic work of the body. The details of energy metabolism have already been described; this discussion examines what happens when energy intake is excessive or inadequate and how metabolism shifts when the three energy-yielding nutrients are out of balance.

Feasting—Excess Energy When a person eats too much, metabolism favors fat formation. Fat cells enlarge regardless of whether the excess in kcalories derives from protein, carbohydrate, or fat. The pathway from dietary fat to body fat, however, is the most direct (requiring only a few metabolic steps) and the most efficient (costing only a few kcalories). To convert a dietary triglyceride to a triglyceride in adipose tissue, the body removes two of the fatty acids from the glycerol, absorbs the parts, and puts them (and others) together again. By comparison, to convert a molecule of sucrose, the body has to split glucose from fructose, absorb them, dismantle them to pyruvate and acetyl CoA, assemble many acetyl CoA molecules into fatty acid chains, and finally attach fatty acids to a glycerol molecule to make a triglyceride for storage in adipose tissue. Quite simply, the body uses much less energy to convert dietary fat to body fat than it does to convert dietary carbohydrate to body fat. On average, storing excess energy from dietary fat as body fat uses only 5 percent of the ingested energy intake, but storing excess energy from dietary carbohydrate as body fat requires 25 percent of the ingested energy intake.

> **FIGURE 7-18 The Central Pathways of Energy Metabolism**

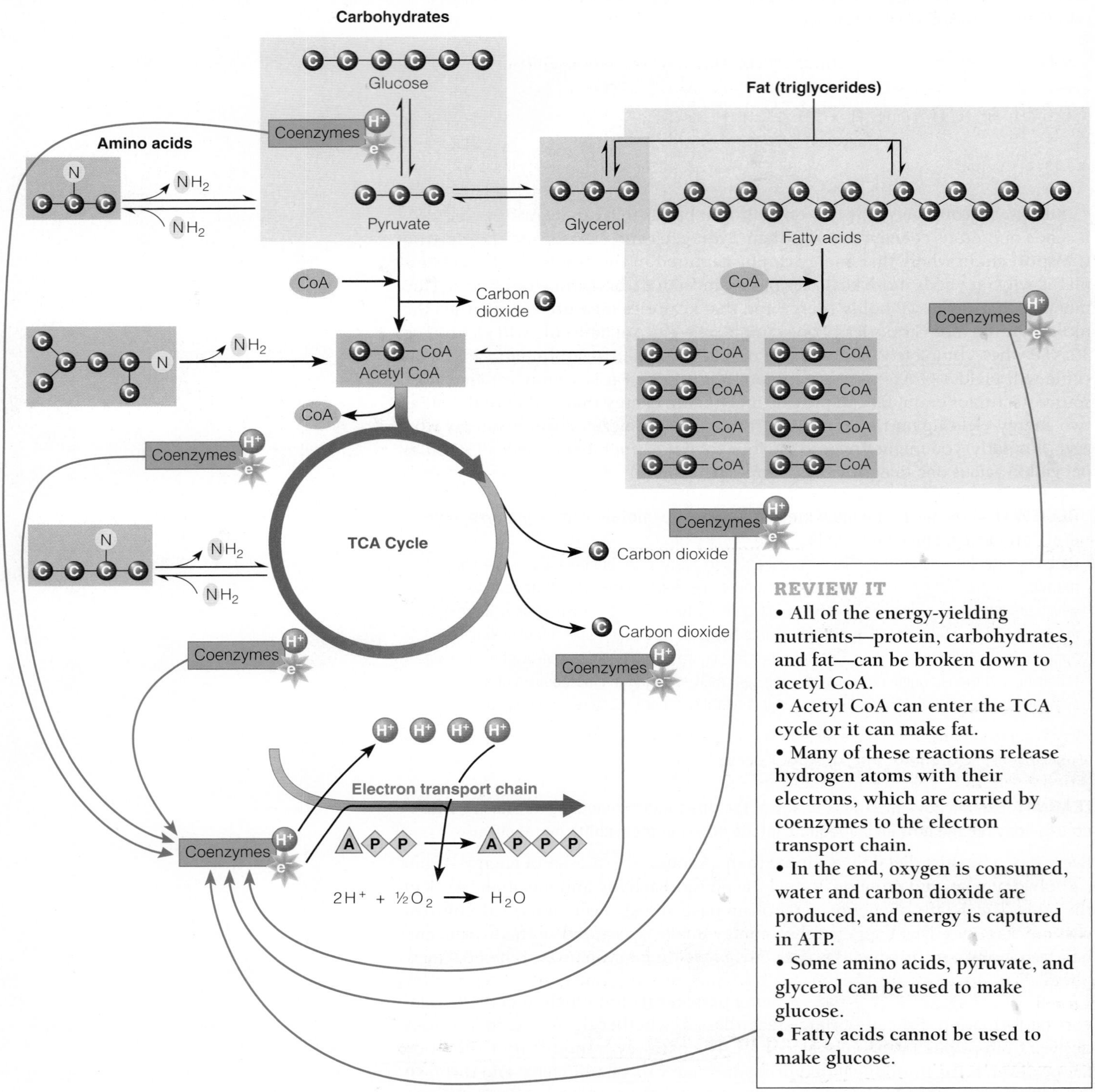

REVIEW IT

- All of the energy-yielding nutrients—protein, carbohydrates, and fat—can be broken down to acetyl CoA.
- Acetyl CoA can enter the TCA cycle or it can make fat.
- Many of these reactions release hydrogen atoms with their electrons, which are carried by coenzymes to the electron transport chain.
- In the end, oxygen is consumed, water and carbon dioxide are produced, and energy is captured in ATP.
- Some amino acids, pyruvate, and glycerol can be used to make glucose.
- Fatty acids cannot be used to make glucose.

The pathways from excess protein and excess carbohydrate to body fat are not only indirect and inefficient, but they are also less preferred by the body (having other priorities for using these nutrients). Before entering fat storage, protein must first tend to its many roles in the body's lean tissues, and carbohydrate must fill the glycogen stores. Simply put, using these two nutrients to make fat is a low priority for the body. Still, if eaten in abundance, any of the energy-yielding nutrients will be converted to fat for storage.

This chapter has described each of the energy-yielding nutrients individually, but cells use a mixture of these fuels. How much of which nutrient is in the fuel mix depends, in part, on its availability from the diet.[1] (The proportion of each fuel also depends on physical activity, as Chapter 14 explains.) Usually, protein's contribution to the fuel mix is relatively minor and fairly constant, but protein oxidation does increase when protein is eaten in excess. Similarly, carbohydrate eaten in excess significantly enhances carbohydrate oxidation. In contrast, fat oxidation does *not* respond to dietary fat intake. The more protein or carbohydrate in the fuel mix, the less fat contributes to the fuel mix. Instead of being oxidized, fat accumulates in storage. Details follow.

Excess Protein Recall from Chapter 6 that the body cannot store excess amino acids as such; it has to convert them to other compounds. Contrary to popular opinion, a person cannot grow muscle simply by overeating protein. Lean tissue such as muscle develops in response to a stimulus such as hormones or physical activity. When a person overeats protein, the body uses the surplus first by replacing normal daily losses and then by increasing protein oxidation. An increase in protein oxidation uses some protein excess, but it displaces fat in the fuel mix. If excess protein is still available, the amino acids are deaminated and the remaining carbons are used to make fatty acids, which are stored as triglycerides in adipose tissue. Thus a person can grow fat by eating too much protein.

People who eat huge portions of meat and other protein-rich foods may wonder why they have weight problems. Not only does the fat in those foods lead to body fat, but the protein can, too, when energy intake exceeds energy needs. Many fad weight-loss diets encourage high protein intakes based on the false assumption that protein builds only muscle, not fat.

Excess Carbohydrate Compared with protein, the proportion of carbohydrate in the fuel mix changes more dramatically when a person overeats. The body handles abundant carbohydrate by first storing it as glycogen, but glycogen storage areas are limited and fill quickly. Because maintaining glucose balance is critical, the body uses glucose frugally when the diet provides only small amounts and freely when supplies are abundant. In other words, glucose oxidation rapidly adjusts to the dietary intake of carbohydrate.

Like protein, excess glucose can also be converted to fat directly.[2] This pathway is relatively minor, however. As mentioned earlier, converting glucose to fat is energetically expensive and does not occur until after glycogen stores have been filled. Even then, only a little, if any, new fat is made from carbohydrate.

Nevertheless, excess dietary carbohydrate can displace fat in the fuel mix.[3] When this occurs, carbohydrate spares both dietary fat and body fat from oxidation—an effect that may be more pronounced in overweight people than in lean people. The net result: excess carbohydrate contributes to obesity or at least to the maintenance of an overweight body.

Excess Fat Unlike excess protein and carbohydrate, which both increase oxidation, eating too much fat does not promote fat oxidation. Instead, excess dietary fat moves efficiently into the body's fat stores; almost all of the excess is stored.

The Transition from Feasting to Fasting

Figure 7-19 (p. 216) shows the metabolic pathways operating in the body as it shifts from feasting (part A) to fasting (parts B and C). After a meal, glucose, glycerol, and fatty acids from foods are used as needed and then stored. Later, as the body shifts from a fed state to a fasting one, it begins drawing on these stores. Glycogen and fat are released from storage to provide more glucose, glycerol, and fatty acids for energy.

Energy is needed all the time. Even when a person is asleep and totally relaxed, the cells are hard at work. In fact, this work—the cells' work that maintains all life processes without any conscious effort—represents about two-thirds of the total energy a person expends in a day.* The small remainder is the work that a person's muscles perform voluntarily during waking hours.

© Pixland/Jupiterimages.com

People can enjoy bountiful meals such as this without storing body fat, provided they expend as much energy as they take in.

*The cells' work that maintains all life processes refers to the body's *basal metabolism,* which is described in Chapter 8.

> **FIGURE 7-19 Feasting and Fasting**

A Feasting: When a person eats in excess of energy needs, the body stores a small amount of glycogen and much larger quantities of fat.

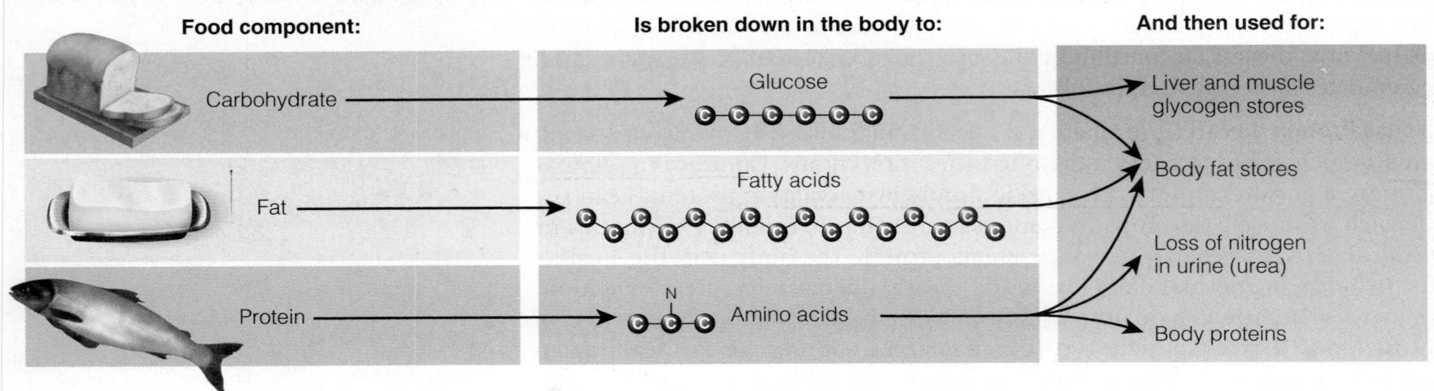

B Fasting: When nutrients from a meal are no longer available to provide energy (about 2 to 3 hours after a meal), the body draws on its glycogen and fat stores for energy.

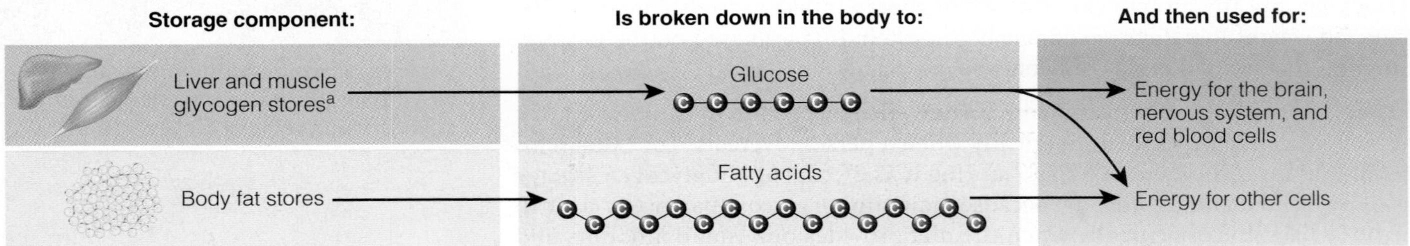

C Fasting beyond glycogen depletion: As glycogen stores dwindle (after about 24 hours of starvation), the body begins to break down its protein (muscle and lean tissue) to amino acids to synthesize glucose needed for brain and nervous system energy. In addition, the liver converts fats to ketone bodies, which serve as an alternative energy source for the brain, thus slowing the breakdown of body protein.

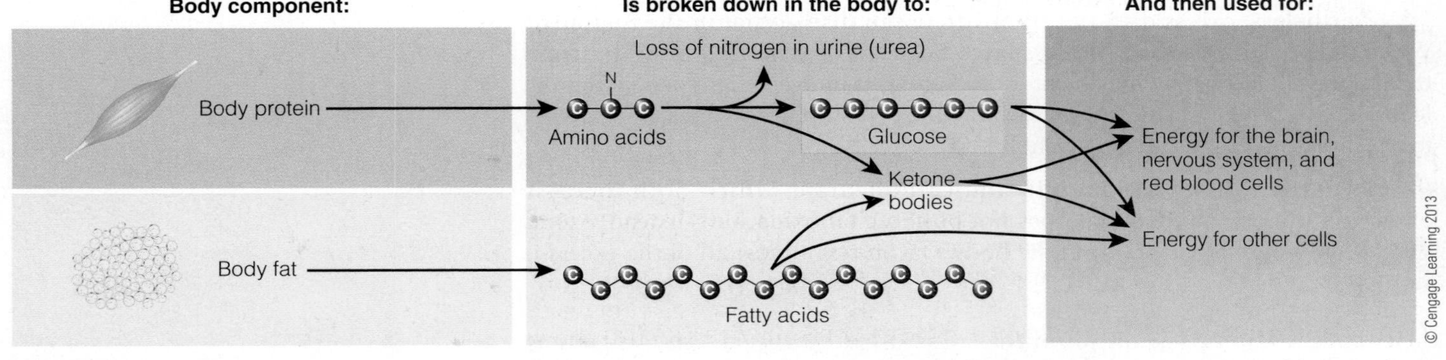

NOTE: Alcohol is not included because it is a toxin and not a nutrient, but it does contribute energy to the body. After detoxifying the alcohol, the body uses the remaining two carbon fragments to build fatty acids and stores them as fat.

[a]The muscles' stored glycogen provides glucose only for the muscle in which the glycogen is stored.

The body's top priority is to meet the cells' needs for energy, and it normally does this by periodic refueling—that is, by eating several times a day. When food is not available, the body turns to its own tissues for fuel. If people choose not to eat, we say they are fasting; if they have no choice, we say they are starving. The body makes no such distinction. In either case, the body must draw on its reserves of carbohydrate and fat and, within a day or so, on its vital protein tissues as well.

Fasting—Inadequate Energy During fasting, carbohydrate, fat, and protein are all eventually used for energy—fuel must be delivered to every cell. As the fast begins, glucose from the liver's stored glycogen and fatty acids from the adipose tissue's stored fat travel to the cells. As described earlier, these molecules are broken down to acetyl CoA, which enters the energy pathways that power the cells' work. Several hours later, however, liver glycogen is depleted and blood glucose begins to fall. The body must adjust its normal metabolism to survive without food. Starvation demands cells to degrade their components for fuel.[4]

Adaptation: Making Glucose At this point, most cells are using fatty acids for their fuel. But, as mentioned earlier, red blood cells and the cells of the nervous system need glucose. Glucose is their primary energy fuel. Normally, the brain and nerve cells—which weigh only about three pounds—consume about half of the total *glucose* used each day (about 500 kcalories' worth). About one-fourth of the *energy* the adult body uses when it is at rest is spent by the brain.

During a fast, the need for glucose poses a major problem. The body can use its stores of fat, which may be quite generous, to furnish most of its cells with energy, but the red blood cells are completely dependent on glucose, and the brain and nerves prefer energy in the form of glucose.* Amino acids that yield pyruvate can be used for **gluconeogenesis**—the making of glucose from noncarbohydrate sources. The liver is the major site of gluconeogenesis, but the kidneys become increasingly involved under certain circumstances, such as starvation.

The glycerol portion of a triglyceride and most amino acids can be used to make glucose (review Figure 7-9, p. 206). To obtain the amino acids, body proteins must be broken down. ♦ For this reason, protein tissues such as muscle and liver always break down to some extent during fasting. The amino acids that cannot be used to make glucose are used as an energy source for other body cells.

The breakdown of body protein is an expensive way to obtain glucose. In the first few days of a fast, body protein provides about 90 percent of the needed glucose; glycerol, about 10 percent. If body protein losses were to continue at this rate, death would follow within three weeks, regardless of the quantity of fat a person had stored. Fortunately, fat breakdown also increases with fasting—in fact, fat breakdown almost doubles, providing energy for other body cells and glycerol for glucose production.

Adaptation: Creating an Alternate Fuel As the fast continues, the body finds a way to use its fat to fuel the brain. It adapts by combining acetyl CoA fragments derived from fatty acids to produce an alternate energy source, **ketone bodies** (see Figure 7-20, p. 218). Normally produced and used only in small quantities, ketone bodies can efficiently provide fuel for brain cells. Ketone body production rises until, after about 10 days of fasting, it is meeting much of the nervous system's energy needs. Still, many areas of the brain rely exclusively on glucose, and to produce it, the body continues to sacrifice protein—albeit at a slower rate than in the early days of fasting.

A ketone body that contains an acid group (COOH) is called a **keto acid.** Small amounts of keto acids are a normal part of the blood chemistry, but when their concentration rises, the pH of the blood drops. This is ketosis, a sign that the body's chemistry is going awry. Acidic blood denatures proteins, leaving them unable to function. Elevated blood ketones (ketonemia) are excreted in the urine (ketonuria). A fruity odor on the breath (known as acetone breath) develops, reflecting the presence of the ketone acetone.

Ketosis induces a loss of appetite. As starvation continues, this loss of appetite becomes an advantage to a person without access to food. When food becomes available again and the person eats, the body shifts out of ketosis and appetite returns.

*Red blood cells contain no mitochondria. Review Figure 7-1 (p. 198) to fully appreciate why red blood cells must depend on glucose for energy.

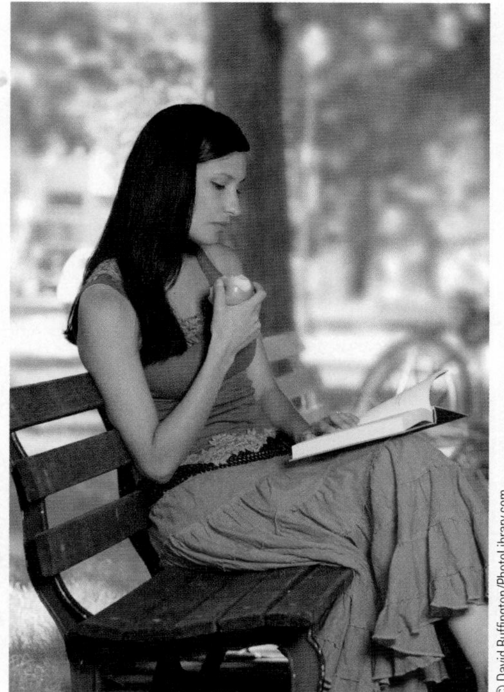

The brain and nerve cells depend on glucose—either directly from carbohydrates or indirectly from proteins (through gluconeogenesis). Importantly, fatty acids cannot provide glucose.

♦ 1 g protein can make ½ g glucose

gluconeogenesis (gloo-ko-nee-oh-JEN-ih-sis): the making of glucose from a noncarbohydrate source such as amino acids or glycerol.

- **gluco** = glucose
- **neo** = new
- **genesis** = making

ketone (KEE-tone) **bodies:** compounds produced during the incomplete breakdown of fat when glucose is not available in the cells.

keto (KEY-toe) **acid:** an organic acid that contains a carbonyl group (C=O).

FIGURE 7-20 Ketone Body Formation

1 The first step in the formation of ketone bodies is the condensation of two molecules of acetyl CoA and the removal of the CoA to form a compound that is converted to the first ketone body, acetoacetate.

Acetyl CoA + Acetyl CoA + H_2O

→ 2 CoA

A ketone body, acetoacetate

2 Acetoacetate may lose a molecule of carbon dioxide to become another ketone body, acetone.

→ CO_2

A ketone body, acetone

3 Or, acetoacetate may add two hydrogens, becoming another ketone body (beta-hydroxybutyrate). See Appendix C for more details.

Adaptation: Conserving Energy In an effort to conserve body tissues for as long as possible, the hormones of fasting slow metabolism. As the body shifts to the use of ketone bodies, it simultaneously reduces its energy output and conserves both its fat and its lean tissue. Still the lean protein tissues shrink and perform less metabolic work, reducing energy expenditures. As the muscles waste, they can do less work and so demand less energy, reducing expenditures further. Although fasting may promote dramatic *weight* loss, a low-kcalorie diet and physical activity better support *fat* loss while retaining lean tissue.

These adaptations of fasting—slowing of energy output and reduction in fat loss—occur in the starving child, the hungry homeless adult, the fasting religious person, the adolescent with anorexia nervosa, and the malnourished hospital patient. Such adaptations help to prolong their lives and explain the physical symptoms of starvation: wasting; slowed heart rate, respiration, and metabolism; lowered body temperature; impaired vision; organ failure; and reduced resistance to disease. Psychological effects of food deprivation include depression, anxiety, and food-related dreams.

The body's adaptations to fasting are sufficient to maintain life for a long time—up to 2 months. Mental alertness need not be diminished, and even some physical energy may remain unimpaired for a surprisingly long time. These remarkable adaptations, however, should not prevent anyone from recognizing the very real hazards that fasting presents.

Low-Carbohydrate Diets When a person consumes a low-carbohydrate diet, a metabolism similar to that of fasting prevails. With little dietary carbohydrate coming in, the body uses its glycogen stores to provide glucose for the cells of the brain, nerves, and blood. Once the body depletes its glycogen reserves, it begins making glucose from the amino acids of protein (gluconeogenesis). A low-carbohydrate diet may provide abundant protein from food, but the body still uses some protein from body tissues.

Dieters can know glycogen depletion has occurred and gluconeogenesis has begun by monitoring their urine. Whenever glycogen or protein is broken down, water is released and urine production increases. Low-carbohydrate diets also induce ketosis, and ketones can be detected in the urine. Ketones form whenever glucose is lacking and fat breakdown is incomplete.

Many fad diets regard ketosis as the key to losing weight, but studies comparing weight-loss diets find no relation between ketosis and weight loss. People in ketosis may experience a loss of appetite and a dramatic weight loss within the first few days.[5] They should know that much of this weight loss reflects the loss of glycogen and protein together with large quantities of body fluids and important minerals. They need to appreciate the difference between loss of *fat* and loss of *weight*. Fat losses on ketogenic diets are no greater than on other diets providing the same number of kcalories. Once the dieter returns to well-balanced meals that provide adequate energy, carbohydrate, fat, protein, vitamins, and minerals, the body avidly retains these needed nutrients.

Low-carbohydrate meals overemphasize meat, fish, poultry, eggs, and cheeses, and shun breads, pastas, fruits, and starchy vegetables.

The weight will return, quite often to a level higher than the starting point. Table 7-3 lists other consequences of a ketogenic diet.

TABLE 7-3 Adverse Side Effects of Low-Carbohydrate, Ketogenic Diets

- Nausea
- Fatigue (especially if physically active)
- Constipation
- Low blood pressure
- Elevated uric acid (which may exacerbate kidney disease and cause inflammation of the joints in those predisposed to gout)
- Stale, foul taste in the mouth (bad breath)
- In pregnant women, fetal harm and stillbirth

REVIEW IT Explain how an excess of any of the three energy-yielding nutrients contributes to body fat and how an inadequate intake of any of them shifts metabolism.
When energy intake exceeds energy needs, the body makes fat—regardless of whether the excess intake is from protein, carbohydrate, or fat. The only difference is that the body is much more efficient at storing energy when the excess derives from dietary fat.

When fasting, the body makes a number of adaptations: increasing the breakdown of fat to provide energy for most of the cells, using glycerol and amino acids to make glucose for the red blood cells and central nervous system, producing ketones to fuel the brain, suppressing the appetite, and slowing metabolism. All of these measures conserve energy and minimize losses. Low-carbohydrate diets incur similar changes in metabolism.

This chapter has probed the intricate details of metabolism at the level of the cells. Several upcoming chapters and highlights build on this information. The highlight that follows this chapter focuses on how alcohol disrupts metabolism. Chapter 8 describes how a person's intake and expenditure of energy are reflected in body weight and body composition. Chapter 9 examines the consequences of unbalanced energy budgets—overweight and underweight. Chapter 10 shows the vital roles the B vitamins play as coenzymes assisting in all the metabolic pathways described here. And Chapter 14 revisits metabolism to show how it supports the work of physically active people and how athletes can best apply that information in their choices of foods to eat.

Nutrition Portfolio

All day, every day, your cells dismantle carbohydrates, fats, and proteins, with the help of vitamins, minerals, and water, releasing energy to meet your body's immediate needs or storing it as fat for later use. Go to Diet Analysis Plus and choose one of the days on which you have tracked your diet for the entire day. Go to the Intake vs. Goals report and answer the following questions. Keep in mind that in this report 100 percent means you are meeting your needs perfectly.

- How close were you to 100 percent for: carbohydrates, fats, proteins, vitamins, minerals, and water? In general, which category was lowest? Which category was highest?
- Describe what types of foods best support aerobic and anaerobic activities.
- Consider whether you eat more protein, carbohydrate, or fat than your body needs.
- Explain how a low-carbohydrate diet forces your body into ketosis.

Diet Analysis PLUS

To complete this exercise, go to your Diet Analysis Plus at **www.cengagebrain.com.**

STUDY IT To review the key points of this chapter and take a practice quiz, go to the study cards at the end of the book.

REFERENCES

1. A. Wise, Transcriptional switches in the control of macronutrient metabolism, *Nutrition Reviews* 66 (2008): 321–325.
2. M. F. Chong and coauthors, Parallel activation of de novo lipogenesis and stearoyl-CoA desaturase activity after 3 d of high-carbohydrate feeding, *American Journal of Clinical Nutrition* 87 (2008): 817–823.
3. R. Roberts and coauthors, Reduced oxidation of dietary fat after a short-term high-carbohydrate diet, *American Journal of Clinical Nutrition* 87 (2008): 824–831.
4. M. H. Stipanuk, Macroautophaghy and its role in nutrient homeostasis, *Nutrition Reviews* 67 (2009): 677–689.
5. A. M. Johnstone and coauthors, Effects of a high-protein ketogenic diet on hunger, appetite, and weight loss in obese men feeding ad libitum, *American Journal of Clinical Nutrition* 87 (2008): 44–55.

LEARN IT Describe how alcohol disrupts metabolism and impairs health.

Alcohol in the Body

© iStockphoto.com/Mark Bowden

With the understanding of metabolism gained from Chapter 7, you are in a position to understand how the body handles alcohol, how alcohol interferes with metabolism, and how alcohol impairs health and nutrition. Before examining alcohol's damaging effects, it may be appropriate to mention that drinking alcohol in *moderation* may have some health benefits, including reduced risks of heart disease, diabetes, and osteoporosis.[1] Moderate alcohol consumption may lower mortality from all causes, but only in adults aged 35 and older.[2] No health benefits are evident before middle age. Similarly, health benefits begin to disappear in older age, as metabolism changes and organs become more sensitive to toxic substances.[3] Importantly, any benefits of moderate alcohol use must be weighed against the many harmful effects of excessive alcohol use described in this highlight, as well as the possibility of alcohol abuse.[4]

Alcohol in Beverages

To the chemist, **alcohol** refers to a class of organic compounds containing hydroxyl (OH) groups (the accompanying glossary defines *alcohol* and related terms). The glycerol to which fatty acids are attached in triglycerides is an example of an alcohol to a chemist. To most people, though, *alcohol* refers to the intoxicating ingredient in **beer, wine,** and **liquor (distilled spirits).** The chemist's name for this particular alcohol is *ethyl alcohol,* or **ethanol.** Glycerol has three carbons with three hydroxyl groups attached; ethanol has only two carbons and one hydroxyl group (see Figure H7-1). The remainder of this highlight talks about the particular alcohol ethanol but refers to it simply as *alcohol.*

Alcohols affect living things profoundly, partly because they act as lipid solvents. Their ability to dissolve lipids out of cell membranes allows alcohols to penetrate rapidly into cells,

GLOSSARY

acetaldehyde (ass-et-AL-duh-hide): an intermediate in alcohol metabolism.

alcohol: a class of organic compounds containing hydroxyl (OH) groups.

- **ol** = alcohol

alcohol abuse: a pattern of drinking that includes failure to fulfill work, school, or home responsibilities; drinking in situations that are physically dangerous (as in driving while intoxicated); recurring alcohol-related legal problems (as in aggravated assault charges); or continued drinking despite ongoing social problems that are caused by or worsened by alcohol.

alcohol dehydrogenase (dee-high-DROJ-eh-nayz): an enzyme active in the stomach and the liver that converts ethanol to acetaldehyde.

alcoholism: a pattern of drinking that includes a strong craving for alcohol, a loss of control and an inability to stop drinking once begun, withdrawal symptoms (nausea, sweating, shakiness, and anxiety) after heavy drinking, and the need for increasing amounts of alcohol to feel "high."

antidiuretic hormone (ADH): a hormone produced by the pituitary gland in response to dehydration (or a high sodium concentration in the blood) that stimulates the kidneys to reabsorb more water and therefore to excrete less. In addition to its antidiuretic effect, ADH elevates blood pressure and so is also called *vasopressin* (VAS-oh-PRES-in).

beer: an alcoholic beverage traditionally brewed by fermenting malted barley and adding hops for flavor.

binge drinking: pattern of drinking that raises blood alcohol concentration to 0.08 percent or higher, usually corresponds to four or more drinks for women and five or more drinks for men on a single occasion, generally within a couple of hours.

cirrhosis (seer-OH-sis): advanced liver disease in which liver cells turn orange, die, and harden, permanently losing their function; often associated with alcoholism.

- **cirrhos** = an orange

drink: a dose of any alcoholic beverage that delivers ½ ounce of pure ethanol:

- 5 ounces of wine
- 10 ounces of wine cooler
- 12 ounces of beer
- 1½ ounces of liquor (80 proof whiskey, scotch, rum, or vodka)

drug: a substance that can modify one or more of the body's functions.

ethanol: a particular type of alcohol found in beer, wine, and liquor; also called *ethyl alcohol* (see Figure H7-1). Ethanol is the most widely used—and abused—drug in our society. It is also the only legal, nonprescription drug that produces euphoria.

excessive drinking: heavy drinking, binge drinking, or both.

fatty liver: an early stage of liver deterioration seen in several diseases, including kwashiorkor and alcoholic liver disease. Fatty liver is characterized by an accumulation of fat in the liver cells.

fibrosis (fye-BROH-sis): an intermediate stage of liver deterioration seen in several diseases, including viral hepatitis and alcoholic liver disease. In fibrosis, the liver cells lose their function and assume the characteristics of connective tissue cells (fibers).

heavy drinking: consuming an average of more than one drink per day for women and more than two drinks per day for men.

liquor or **distilled spirits:** an alcoholic beverage traditionally made by fermenting and distilling a carbohydrate source such as molasses, potatoes, rye, beets, barley, or corn.

MEOS or **microsomal** (my-krow-SO-mal) **ethanol-oxidizing system:** a system of enzymes in the liver that oxidize not only alcohol but also several classes of drugs.

moderation: up to one drink per day for women with no more than three drinks on any single day and up to two drinks per day for men with no more than four drinks on any single day.

narcotic (nar-KOT-ic): a drug that dulls the senses, induces sleep, and becomes addictive with prolonged use.

proof: a way of stating the percentage of alcohol in distilled liquor. Liquor that is 100 proof is 50 percent alcohol; 90 proof is 45 percent, and so forth.

Wernicke-Korsakoff (VER-nee-key KORE-sah-kof) **syndrome:** a neurological disorder typically associated with chronic alcoholism and caused by a deficiency of the B vitamin thiamin; also called *alcohol-related dementia.*

wine: an alcoholic beverage traditionally made by fermenting a sugar source such as grape juice.

> **FIGURE H7-1 Two Alcohols: Glycerol and Ethanol**

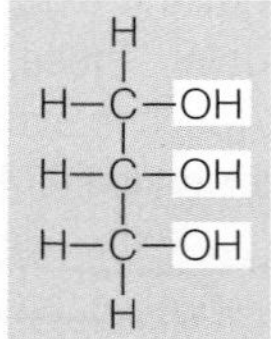

Glycerol is the alcohol used to make triglycerides.

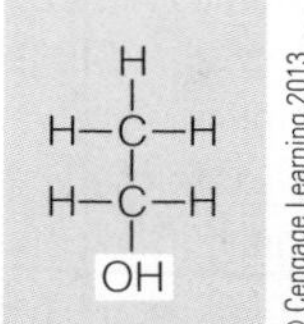

Ethanol is the alcohol in beer, wine, and liquor.

destroying cell structures and thereby killing the cells. For this reason, most alcohols are toxic in relatively small amounts; by the same token, because they kill microbial cells, they are useful as skin disinfectants.

Ethanol is less toxic than the other alcohols. Sufficiently diluted and taken in small enough doses, its action in the brain produces an effect that people seek—not with zero risk, but with a low enough risk (if the doses are low enough) to be tolerable. Used in this way, alcohol is a **drug**—that is, a substance that modifies body functions. Like all drugs, alcohol both offers benefits and poses hazards. The *Dietary Guidelines for Americans* advise "if alcohol is consumed, it should be consumed in moderation."

> **DIETARY GUIDELINES FOR AMERICANS 2010**

- If alcohol is consumed, it should be consumed in moderation—up to one drink per day for women and two drinks per day for men—and only by adults of legal drinking age.
- Alcoholic beverages should not be consumed by some individuals, including those who cannot restrict their alcohol intake to moderate levels, women of childbearing age who may become pregnant, pregnant and lactating women, children and adolescents, individuals taking medications that can interact with alcohol, and those with specific medical conditions.
- Alcoholic beverages should not be consumed by individuals engaging in activities that require attention, skill, or coordination, such as driving or swimming.

The term **moderation** is important when describing alcohol use. How many drinks constitute moderate use, and how much is "a drink"? First, a **drink** is any alcoholic beverage that delivers ½ ounce of *pure ethanol:*

- 5 ounces of wine
- 10 ounces of wine cooler
- 12 ounces of beer
- 1½ ounces of liquor (80 proof whiskey, scotch, rum, or vodka)

As a practical tip, prevent overpouring by measuring liquids and using tall, narrow glasses.

Beer, wine, and liquor deliver different amounts of alcohol. The amount of alcohol in liquor is stated as **proof:** 100 proof liquor is 50 percent alcohol, 80 proof is 40 percent alcohol, and so forth. Wine and beer have less alcohol than liquor, although some fortified wines and beers have more alcohol than the regular varieties (see photo caption below).

Second, because people have different tolerances for alcohol, it is impossible to name an exact daily amount of alcohol that is appropriate for everyone. Authorities have attempted to identify amounts that are acceptable for most healthy people. An accepted definition of *moderation* is up to two drinks per day for men and up to one drink per day for women. (Pregnant women are advised to abstain from alcohol, as Highlight 15 explains.) Notice that this advice is stated as a maximum, not as an average; seven drinks one night a week would not be considered moderate, even though one a day would be. Doubtless,

Each of these servings equals one drink. Moderation is up to one drink per day for women and two drinks per day for men.

Wines contain 7 to 24 percent alcohol by volume; those containing 14 percent or more must state their alcohol content on the label, whereas those with less than 14 percent may simply state "table wine" or "light wine." Beers typically contain less than 5 percent alcohol by volume and malt liquors, 5 to 8 percent; regulations vary, with some states requiring beer labels to show the alcohol content and others prohibiting such statements.

some people could consume slightly more; others could not handle nearly so much without risk. The amount a person can drink safely is highly individual, depending on genetics, health, gender, body composition, age, and family history.

Alcohol's Influence

From the moment an alcoholic beverage enters the body, alcohol is treated as if it has special privileges. Its influence is most apparent in the GI tract, the liver, and the brain.

In the GI Tract

Unlike foods, which require time for digestion, alcohol needs no digestion and is quickly absorbed across the walls of an empty stomach, reaching the brain within a few minutes. Consequently, a person can immediately feel euphoric when drinking, especially on an empty stomach.

When the stomach is full of food, alcohol has less chance of touching the walls and diffusing through, so its influence on the brain is slightly delayed. This information leads to another practical tip: eat snacks when drinking alcoholic beverages. Carbohydrate snacks slow alcohol absorption and high-fat snacks slow peristalsis, keeping the alcohol in the stomach longer. Salty snacks make a person thirsty; to quench thirst, drink water instead of more alcohol.

The stomach begins to break down alcohol with its **alcohol dehydrogenase** enzyme. Women produce less of this stomach enzyme than men; consequently, more alcohol reaches the intestine for absorption into the bloodstream. As a result, women absorb more alcohol than men of the same size who drink the same amount of alcohol. Consequently, they are more likely to become more intoxicated on less alcohol than men. Such differences between men and women help explain why women have a lower alcohol tolerance and a lower guideline for moderate intake.

In the small intestine, alcohol is rapidly absorbed. From this point on, alcohol receives priority treatment: it gets absorbed and metabolized before most nutrients. Alcohol's priority status helps to ensure a speedy disposal and reflects two facts: alcohol cannot be stored in the body, and it is potentially toxic.

In the Liver

As Chapter 3 explains, the capillaries of the digestive tract merge into veins that carry blood first to the liver. These veins branch and rebranch into a capillary network that touches every liver cell. Consequently, liver cells are the first to receive alcohol-laden blood. Liver cells are also the only other cells in the body that can make enough of the alcohol dehydrogenase enzyme to oxidize alcohol at an appreciable rate. The routing of blood through the liver cells gives them the chance to dispose of some alcohol before it moves on.

Alcohol affects every organ of the body, but the most dramatic evidence of its disruptive behavior appears in the liver. If liver cells could talk, they would describe alcohol as demanding, egocentric, and disruptive of the liver's efficient way of running its business. For example, liver cells normally prefer fatty acids as their fuel, and they like to package excess fatty acids into triglycerides and ship them out to other tissues. When alcohol is present, however, the liver cells metabolize alcohol first and let the fatty acids accumulate, sometimes in huge stockpiles. Alcohol metabolism can also permanently change liver cell structure, impairing the liver's ability to metabolize fats. As a result, heavy drinkers develop fatty livers.

The liver is the primary site of alcohol metabolism. It can process about ½ ounce of *ethanol* per hour (the amount defined as a drink), depending on the person's body size, previous drinking experience, food intake, and general health. This maximum rate of alcohol breakdown is determined by the amount of alcohol dehydrogenase available. If more alcohol arrives at the liver than the enzymes can handle, the extra alcohol travels around the body, circulating again and again until liver enzymes are finally available to process it. Another practical tip derives from this information: drink slowly enough to allow the liver to keep up—no more than one drink per hour.

The amount of alcohol dehydrogenase enzyme present in the liver varies with individuals, depending on the genes they have inherited and on how recently they have eaten. Fasting for as little as a day prompts the body to degrade its proteins, including the alcohol-processing enzymes, and this can slow the rate of alcohol metabolism by half. Drinking after not eating all day thus causes the drinker to feel the effects more promptly for two reasons: rapid absorption and slowed breakdown.

Figure H7-2 provides a simplified diagram of alcohol metabolism; Appendix C provides the chemical details. The alcohol dehydrogenase enzyme breaks down alcohol by removing hydrogens in two steps. In the first step, alcohol dehydrogenase oxidizes alcohol to

> FIGURE H7-2 **Alcohol Metabolism**

The conversion of alcohol to acetyl CoA requires the B vitamin niacin in its role as a coenzyme. When the enzymes oxidize alcohol, they remove H atoms and attach them to the niacin coenzyme.

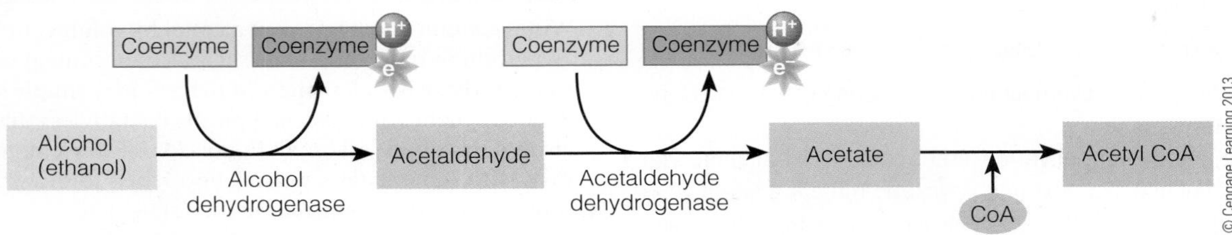

acetaldehyde—a highly reactive and toxic compound. High concentrations of acetaldehyde in the brain and other tissues are responsible for many of the damaging effects of **alcohol abuse.**

In the second step, a related enzyme, acetaldehyde dehydrogenase, converts acetaldehyde to acetate, which is then converted to either carbon dioxide (CO_2) or acetyl CoA—the compound that plays such a central role in energy metabolism, as described in Chapter 7. The reactions from alcohol to acetaldehyde to acetate produce hydrogens (H^+) and electrons. The B vitamin niacin, in its role as a coenzyme, helpfully picks up these hydrogens and electrons and escorts them to the electron transport chain. (Chapter 10 presents information on the coenzyme roles of the B vitamins.)

During alcohol metabolism, the multitude of other metabolic processes for which the niacin coenzyme is required, including glycolysis, the TCA cycle, and the electron transport chain, falter. Its presence is sorely missed in these energy pathways because it is the chief carrier of the hydrogens that travel with their electrons along the electron transport chain. Without adequate coenzymes, these energy pathways cannot function. Traffic either backs up or an alternate route is taken.

Such changes in the normal flow of energy pathways have striking metabolic consequences. For one, the accumulation of hydrogen ions during alcohol metabolism shifts the body's acid-base balance toward acid. For another, alcohol's interference with energy metabolism promotes the making of lactate from pyruvate. The conversion of pyruvate to lactate uses some of the excess hydrogens, but a lactate build-up has serious consequences of its own—it adds still further to the body's acid burden and interferes with the excretion of another acid, uric acid, causing inflammation of the joints.

Alcohol alters both amino acid and protein metabolism. Synthesis of proteins important in the immune system slows down, weakening the body's defenses against infections. Evidence of protein deficiency becomes apparent, both from a diminished synthesis of proteins and from a poor diet. Normally, the cells would at least use the amino acids from the protein foods a person eats, but the drinker's liver deaminates the amino acids and uses the carbon fragments primarily to make fat or ketone bodies. Eating well does not protect the drinker from protein depletion; a person has to stop drinking alcohol.

The accumulation of coenzymes with their hydrogens and electrons slows the TCA cycle, so pyruvate and acetyl CoA build up. Excess acetyl CoA then takes the pathway to fatty acid synthesis (as Figure H7-3 illustrates), and fat clogs the liver. As you might expect, a liver overburdened with fat cannot function properly. Liver cells become less efficient at performing a number of tasks. Much of this inefficiency impairs a person's nutritional health in ways that cannot be corrected by diet alone. For example, the liver has difficulty activating vitamin D, as well as producing and releasing bile. The fatty liver has difficulty making glucose from protein. Without gluconeogenesis, blood glucose can plummet, leading to irreversible damage to the central nervous system. The lack of glucose together with the overabundance of acetyl CoA sets the stage for ketosis. The body uses excess acetyl CoA to make ketone bodies; their acidity pushes the acid-base balance further toward acid and suppresses nervous system activity. To overcome such problems, a person needs to stop drinking alcohol.

The synthesis of fatty acids accelerates with exposure to alcohol. Fat accumulation can be seen in the liver after a single night of heavy drinking. **Fatty liver,** the first stage of liver deterioration seen in heavy drinkers, interferes with the distribution of nutrients and oxygen to the liver cells. Fatty liver is reversible with abstinence from alcohol. If fatty liver lasts long enough, however, the liver cells will die and form fibrous scar tissue. This second stage of liver deterioration is called **fibrosis.** Some liver cells can regenerate with good nutrition and abstinence from alcohol, but in the most advanced stage, **cirrhosis,** damage is the least reversible.

The liver's priority treatment of alcohol affects its handling of drugs as well as nutrients. In addition to the dehydrogenase enzyme already described, the liver possesses an enzyme system that metabolizes *both* alcohol and several other types of drugs. Called the **MEOS (microsomal ethanol-oxidizing system),** this system handles

> **FIGURE H7-3 Alternate Route for Acetyl CoA: To Fat**

Acetyl CoA molecules are blocked from getting into the TCA cycle by the low level of coenzymes. Instead of being used for energy, the acetyl CoA molecules become building blocks for fatty acids.

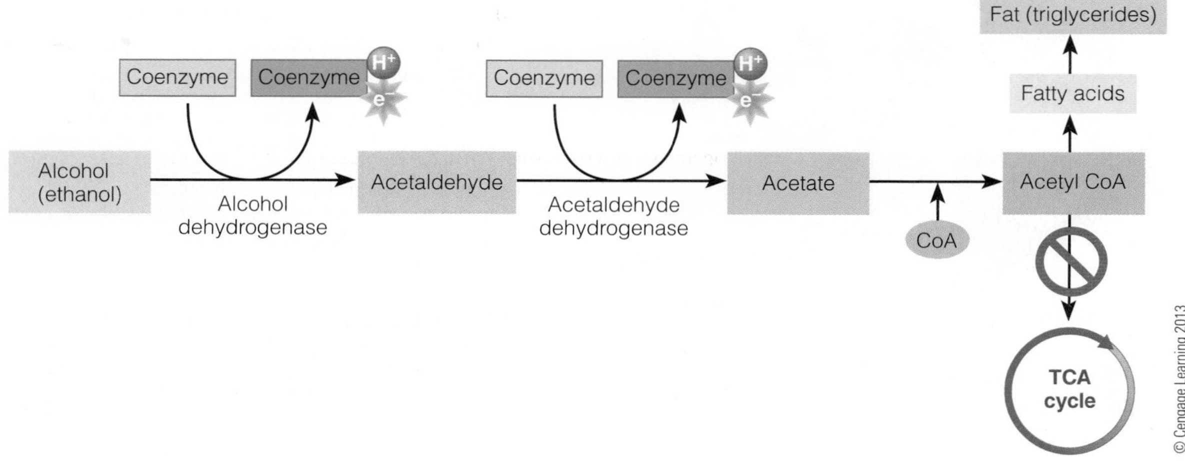

about one-fifth of the total alcohol a person consumes. At high blood concentrations or with repeated exposures, alcohol stimulates the synthesis of enzymes in the MEOS. The result is a more efficient metabolism of alcohol and tolerance to its effects.

As a person's blood alcohol rises, alcohol competes with—and wins out over—other drugs whose metabolism also relies on the MEOS. If a person drinks and uses another drug at the same time, the MEOS will dispose of alcohol first and metabolize the drug more slowly. While the drug waits to be handled later, the dose may build up so that its effects are greatly amplified—sometimes to the point of being fatal. Many drug labels provide warnings to avoid alcohol while taking the drug.

In contrast, once a heavy drinker stops drinking and alcohol is no longer competing with other drugs, the enhanced MEOS metabolizes drugs much faster than before. As a result, determining the correct dosages of medications can be challenging.

This discussion has emphasized the major way that the blood is cleared of alcohol—metabolism by the liver—but there is another way. About 10 percent of the alcohol leaves the body through the breath and in the urine. This is the basis for the breath and urine tests for drunkenness. The amounts of alcohol in the breath and in the urine are in proportion to the amount still in the bloodstream and brain. In nearly all states, legal drunkenness is set at 0.10 percent or less, reflecting the relationship between alcohol use and traffic and other accidents.

In the Brain

Figure H7-4 describes alcohol's effects on the brain. Alcohol is a **narcotic.** People used it for centuries as an anesthetic because it can deaden pain. But alcohol was a poor anesthetic because one could never be sure how much a person would need and how much would be a fatal dose. Today's anesthetics provide a more predictable response. Alcohol continues to be used socially to help people relax or to relieve anxiety. People think that alcohol is a stimulant because it seems to relieve inhibitions. Actually, though, it accomplishes this by sedating *inhibitory* nerves, which are more numerous than excitatory nerves. Ultimately, alcohol acts as a depressant and affects all the nerve cells.

It is lucky that the brain centers respond to a rising blood alcohol concentration in the order described in Figure H7-4 because a person usually passes out before managing to drink a lethal dose. It is possible, though, to drink so fast that the effects of alcohol continue to accelerate after the person has passed out. Occasionally, a person drinks so much as to stop breathing and die. Table H7-1 shows the blood alcohol levels that correspond to progressively greater intoxication, and Table H7-2 shows the brain responses that occur at these blood levels.

Like liver cells, brain cells die with excessive exposure to alcohol. Liver cells may be replaced, but not all brain cells can regenerate.

> FIGURE H7-4 **Alcohol's Effects on the Brain**

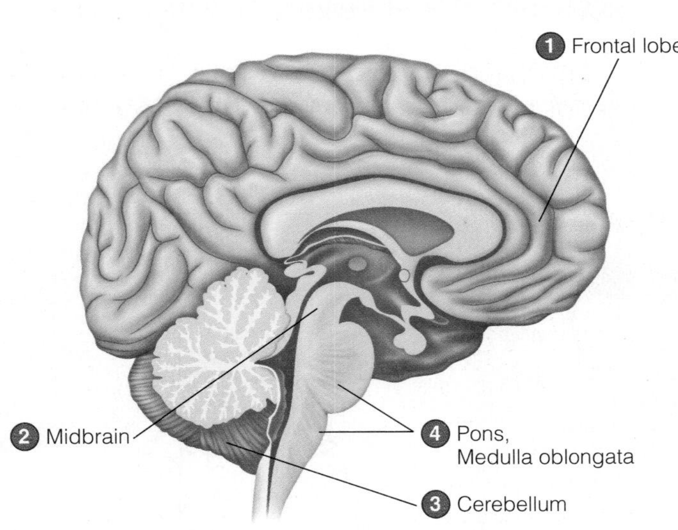

1. Judgment and reasoning centers are most sensitive to alcohol. When alcohol flows to the brain, it first sedates the frontal lobe, the center of all conscious activity. As alcohol diffuses into the cells of these lobes, it interferes with reasoning and judgment.

2. Speech and vision centers in the midbrain are affected next. If the drinker drinks faster than the rate at which the liver can oxidize the alcohol, blood alcohol concentrations rise: the speech becomes challenging and vision becomes blurry.

3. Voluntary muscular control is then affected. At still higher concentrations, the cells in the cerebellum responsible for coordination of voluntary muscles are affected, including those used in speech, eye-hand coordination, and limb movements. At this point people under the influence stagger or weave when they try to walk, or they may slur their speech.

4. Respiration and heart action are the last to be affected. Finally, the conscious brain is completely subdued, and the person passes out. Now the person can drink no more; this is fortunate because higher doses would anesthetize the deepest brain centers that control breathing and heartbeat, causing death.

TABLE H7-1 Alcohol Doses and Approximate Blood Level Percentages for Men and Women

Drinks[a]	Body Weight in Pounds—Men								
	100	120	140	160	180	200	220	240	
	00	00	00	00	00	00	00	00	ONLY SAFE DRIVING LIMIT
1	.04	.03	.03	.02	.02	.02	.02	.02	IMPAIRMENT BEGINS
2	.08	.06	.05	.05	.04	.04	.03	.03	
3	.11	.09	.08	.07	.06	.06	.05	.05	DRIVING SKILLS SIGNIFICANTLY AFFECTED
4	.15	.12	.11	.09	.08	.08	.07	.06	
5	.19	.16	.13	.12	.11	.09	.09	.08	
6	.23	.19	.16	.14	.13	.11	.10	.09	
7	.26	.22	.19	.16	.15	.13	.12	.11	LEGALLY INTOXICATED
8	.30	.25	.21	.19	.17	.15	.14	.13	
9	.34	.28	.24	.21	.19	.17	.15	.14	
10	.38	.31	.27	.23	.21	.19	.17	.16	

Drinks[a]	Body Weight in Pounds—Women									
	90	100	120	140	160	180	200	220	240	
	00	00	00	00	00	00	00	00	00	ONLY SAFE DRIVING LIMIT
1	.05	.05	.04	.03	.03	.03	.02	.02	.02	IMPAIRMENT BEGINS
2	.10	.09	.08	.07	.06	.05	.05	.04	.04	DRIVING SKILLS SIGNIFICANTLY AFFECTED
3	.15	.14	.11	.10	.09	.08	.07	.06	.06	
4	.20	.18	.15	.13	.11	.10	.09	.08	.08	
5	.25	.23	.19	.16	.14	.13	.11	.10	.09	
6	.30	.27	.23	.19	.17	.15	.14	.12	.11	LEGALLY INTOXICATED
7	.35	.32	.27	.23	.20	.18	.16	.14	.13	
8	.40	.36	.30	.26	.23	.20	.18	.17	.15	
9	.45	.41	.34	.29	.26	.23	.20	.19	.17	
10	.51	.45	.38	.32	.28	.25	.23	.21	.19	

NOTE: Driving under the influence is proved when an adult's blood contains 0.08 percent alcohol. Many states have adopted a "zero-tolerance" policy for drivers under age 21, using 0.00 to 0.02 percent as the limit.

[a]Taken within an hour or so; each drink equivalent to ½ ounce pure ethanol.

SOURCE: National Clearinghouse for Alcohol and Drug Information.

Thus some heavy drinkers suffer permanent brain damage. Whether alcohol impairs cognition in moderate drinkers is unclear.

Alcohol's Damage

As alcohol busily disrupts cellular activities, the physical consequences become apparent. People become dehydrated and malnourished; their alcohol use brings both short- and long-term effects.

Dehydration

People who drink alcoholic beverages may notice that they urinate more, but they may be unaware of the vicious cycle that results. Alcohol depresses production of **antidiuretic hormone (ADH),** a hormone produced by the pituitary gland that retains water—consequently, with less ADH, more water is lost. Loss of body water leads to thirst, and thirst leads to more drinking. Water will relieve dehydration, but the thirsty drinker may drink alcohol instead, which only worsens the problem. Such information provides another practical tip: drink water when thirsty and before each alcoholic drink. Drink an extra glass or two before going to bed. This strategy will help lessen the effects of a hangover.

Water loss is accompanied by the loss of important minerals. As Chapters 12 and 13 explain, these minerals are vital to the body's fluid balance and to many chemical reactions in the cells, including muscle action. Detoxification treatment includes restoration of mineral balance as quickly as possible.

Malnutrition

For some light-to-moderate drinkers, alcohol may suppress food intake and prevent weight gains.[5] For others, however, alcohol may actually stimulate appetite. Moderate drinkers usually consume alcohol as *added* energy—on top of their normal food intake. In addition, alcohol in moderate doses is efficiently metabolized. Consequently, alcohol can contribute to body fat and weight gain—either by inhibiting oxidation or by being converted to fat. Metabolically, alcohol is almost as efficient as fat in promoting obesity; each ounce of alcohol represents about a half-ounce of fat. Alcohol's contribution to body fat is most evident in the central obesity that commonly accompanies alcohol consumption, popularly known as the "beer belly."[6]

Alcohol in heavy doses, though, is not efficiently metabolized, generating more heat than fat. Heavy drinkers usually consume alcohol

TABLE H7-2 Alcohol Blood Levels and Brain Responses

Blood Alcohol Concentration	Effect on Brain
0.05	Impaired judgment, relaxed inhibitions, altered mood, increased heart rate
0.10	Impaired coordination, delayed reaction time, exaggerated emotions, impaired peripheral vision, impaired ability to operate a vehicle
0.15	Slurred speech, blurred vision, staggered walk, seriously impaired coordination and judgment
0.20	Double vision, inability to walk
0.30	Uninhibited behavior, stupor, confusion, inability to comprehend
0.40 to 0.60	Unconsciousness, shock, coma, death (cardiac or respiratory failure)

NOTE: Blood alcohol concentration depends on a number of factors, including alcohol in the beverage, the rate of consumption, the person's gender, and body weight. For example, a 100-pound female can become legally drunk (≥0.10 concentration) by drinking three beers in an hour, whereas a 220-pound male consuming that amount at the same rate would have a 0.05 blood alcohol concentration.

as *substituted* energy—instead of their normal food intake. Diet quality declines as alcohol consumption increases.[7] Consequently, many heavy drinkers suffer malnutrition.

Alcohol is rich in energy (7 kcalories per gram), but as with pure sugar or fat, the kcalories are empty of nutrients. The more alcohol people drink, the less likely that they will eat enough food to obtain adequate nutrients. The more kcalories used for alcohol, the fewer kcalories available to use from nutritious foods. Table H7-3 shows the kcalorie amounts of typical alcoholic beverages.

Chronic alcohol abuse not only displaces nutrients from the diet, but it also interferes with the body's metabolism of nutrients. Most dramatic is alcohol's effect on the B vitamin folate. The liver loses its ability to retain folate, and the kidneys increase their excretion of it. Alcohol abuse creates a folate deficiency that devastates digestive system function. The small intestine normally releases and retrieves folate continuously, but it becomes damaged by folate deficiency and alcohol toxicity, so it fails to retrieve its own folate and misses any that may trickle in from food as well. Alcohol also interferes with the action of folate in converting the amino acid homocysteine to methionine. The result is an excess of homocysteine, which has been linked to heart disease, and an inadequate supply of methionine, which slows the production of new cells, especially the rapidly dividing cells of the intestine and the blood. The combination of poor folate status and alcohol consumption has also been implicated in promoting colorectal cancer.

The inadequate food intake and impaired nutrient absorption that accompany chronic alcohol abuse frequently lead to a deficiency of another B vitamin—thiamin. In fact, the cluster of thiamin-deficiency symptoms commonly seen in chronic **alcoholism** has its own name—**Wernicke-Korsakoff syndrome.**[8] This syndrome is characterized by paralysis of the eye muscles, poor muscle coordination, impaired memory, and damaged nerves; it and other alcohol-related memory problems may respond to thiamin supplements.

Acetaldehyde, an intermediate in alcohol metabolism (review Figure H7-2), interferes with nutrient use, too. For example, acetaldehyde dislodges vitamin B_6 from its protective binding protein so that it is destroyed, causing a vitamin B_6 deficiency and, thereby, lowered production of red blood cells.

Malnutrition occurs not only because of lack of intake and altered metabolism but because of direct toxic effects as well. Alcohol causes stomach cells to oversecrete both gastric acid and histamine, an immune system agent that produces inflammation. Beer in particular stimulates gastric acid secretion, irritating the linings of the stomach and esophagus and making them vulnerable to ulcer formation.

Overall, nutrient deficiencies are virtually inevitable in alcohol abuse, not only because alcohol displaces food but also because alcohol directly interferes with the body's use of nutrients, making them ineffective even if they are present. Intestinal cells fail to absorb B vitamins, notably, thiamin, folate, and vitamin B_{12}. Liver cells lose efficiency in activating vitamin D. Cells in the retina of the eye, which normally process the alcohol form of vitamin A (retinol) to the aldehyde form needed in vision (retinal), find themselves processing ethanol to acetaldehyde instead. Likewise, the liver cannot convert the aldehyde form of vitamin A to its acid form (retinoic acid), which is needed to support the growth of its (and all) cells. Regardless of dietary intake,

TABLE H7-3 kCalories in Alcoholic Beverages and Mixers

Beverage	Amount (oz)	Energy (kcal)	Alcohol (g)
Beer			
Regular	12	153	14
Light	12	103	11
Nonalcoholic	12	32	0
Cocktails			
Daiquiri, canned	6.8	259	20
Daiquiri, from recipe	4.5	223	28
Piña colada, canned	6.8	526	20
Piña colada, from recipe	4.5	245	14
Tequila sunrise, canned	6.8	232	20
Whiskey sour, canned	6.8	249	20
Liquor (gin, rum, vodka, whiskey)			
80 proof	1.5	97	14
86 proof	1.5	105	15
90 proof	1.5	110	16
94 proof	1.5	116	17
100 proof	1.5	124	18
Sake	1.5	58	7
Liqueurs			
Coffee and cream liqueur, 34 proof	1.5	154	7
Coffee liqueur, 53 proof	1.5	170	11
Coffee liqueur, 63 proof	1.5	160	14
Crème de menthe, 72 proof	1.5	186	15
Mixers			
Club soda	12	0	0
Cola	12	136	0
Cranberry juice cocktail	4	72	0
Ginger ale or tonic water	12	124	0
Grapefruit juice	4	48	0
Orange juice	4	56	0
Tomato or vegetable juice	4	21	0
Wine			
Champagne	5	105	13
Cooking	5	72	5
Dessert, dry	5	224	23
Dessert, sweet	5	236	23
Red or rosé	5	125	16
White	5	121	15
Wine cooler	10	150	11

excessive drinking over a lifetime creates deficits of all the nutrients mentioned in this discussion and more. No diet can compensate for the damage caused by heavy alcohol consumption.

Short-Term Effects

The effects of abusing alcohol may be apparent immediately, or they may not become evident for years to come. Among the immediate consequences, all of the following involve alcohol use:[9]

- 20 percent of all boating fatalities
- 23 percent of all suicides
- 39 percent of all traffic fatalities
- 40 percent of all residential fire fatalities
- 47 percent of all homicides
- 65 percent of all domestic violence incidents

These statistics are sobering. The consequences of heavy drinking touch all races and all segments of society—men and women, young and old, rich and poor. One group particularly hard hit by heavy drinking is college students—not because they are prone to alcoholism, but because they live in an environment and are in a developmental stage of life in which risk-taking behaviors are common and heavy drinking is acceptable.

Excessive drinking—including both **heavy drinking** and **binge drinking**—is widespread on college campuses and poses serious health and social consequences to drinkers and nondrinkers alike. In fact, binge drinking can kill: the respiratory center of the brain becomes anesthetized, and breathing stops. It can also cause coronary artery spasms, leading to a heart attack and death.[10]

Binge drinking is especially common among college students, especially males.[11] Compared with nondrinkers or moderate drinkers, people who frequently binge drink (at least three times within two weeks) are more likely to engage in unprotected sex, have multiple sex partners, damage property, and assault others. On average, *every day* alcohol is involved in the:

- Death of 5 college students
- Sexual assault of 266 college students
- Injury of 1641 college students
- Assault of 1907 college students

Binge drinkers skew the statistics on college students' alcohol use. The median number of drinks consumed by college students is 1.5 per week, but for binge drinkers, it is 14.5. Nationally, only 20 percent of all students are frequent binge drinkers; yet they account for two-thirds of all the alcohol students report consuming and most of the alcohol-related problems.

The dangers of binge drinking have been amplified by the use of beverages that contain caffeine as an additive. The caffeine seems to mask the sensory cues that an individual normally relies on to determine intoxication. Consequently, individuals drinking these beverages typically consume more alcohol and become more intoxicated than they realize. The Food and Drug Administration (FDA) has warned manufacturers of packaged caffeinated alcoholic beverages to stop sales. The combination of alcohol and added caffeine has not been approved because these products are associated with risky behaviors that may lead to hazardous and life-threatening situations.[12] For the same reasons, individuals should not mix alcohol with high-energy drinks.

Binge drinking is not limited to college campuses, of course, but it is most common among 18 to 24 year olds.[13] That age group and campus environment seem most accepting of such behavior despite its problems. Social acceptance may make it difficult for binge drinkers to recognize themselves as problem drinkers. For this reason, interventions must focus both on educating individuals and on changing the campus social environment. The damage alcohol causes becomes worse if the pattern is not broken. Alcohol abuse sets in much more quickly in young people than in adults. Those who start drinking at an early age more often suffer from alcoholism than others. Table H7-4 lists the key signs of alcoholism.

Long-Term Effects

The most devastating long-term effect of alcohol is the damage done to a child whose mother abused alcohol during pregnancy. The effects of alcohol on the unborn and the message that pregnant women should not drink alcohol are presented in Highlight 15.

For nonpregnant adults, a drink or two sets in motion many destructive processes in the body, but the next day's abstinence reverses them. As long as the doses are moderate, the time between them is ample, and nutrition is adequate, recovery is probably complete.

If the doses of alcohol are heavy and the time between them short, complete recovery cannot take place. Repeated onslaughts of alcohol gradually take a toll on all parts of the body and increase the risks of several chronic diseases (see Table H7-5, p. 228).[14] Compared with nondrinkers and moderate drinkers, heavy drinkers have significantly greater risks of dying from all causes. Excessive alcohol consumption is the third leading preventable cause of death in the United States.

TABLE H7-4 Signs of Alcoholism

- Tolerance: the person needs higher and higher intakes of alcohol to achieve intoxication.
- Withdrawal: the person who stops drinking experiences anxiety, agitation, increased blood pressure, or seizures, or seeks alcohol to relieve these symptoms.
- Impaired control: the person intends to have 1 or 2 drinks, but has 9 or 10 instead, or the person tries to control or quit drinking, but fails.
- Disinterest: the person neglects important social, family, job, or school activities because of drinking.
- Time: the person spends a great deal of time obtaining and drinking alcohol or recovering from excessive drinking.
- Impaired ability: the person's intoxication or withdrawal symptoms interfere with work, school, or home.
- Problems: the person continues drinking despite physical hazards or medical, legal, psychological, family, employment, or school problems.

The presence of three or more of these conditions is required to make a diagnosis.

SOURCE: Adapted from *Diagnostic and Statistical Manual of Mental Disorders,* 4th ed. (Washington, D.C.: American Psychiatric Association, 1994).

TABLE H7-5 Health Effects of Heavy Alcohol Consumption

Health Problem	Effects of Alcohol
Arthritis	Increases the risk of inflamed joints.
Cancer	Increases the risk of cancer of the liver, breast, mouth, pharynx, larynx, esophagus, colon, and rectum.
Fetal alcohol syndrome	Causes physical and behavioral abnormalities in the fetus (see Highlight 15).
Heart disease	In heavy drinkers, raises blood pressure, blood lipids, and the risk of stroke and heart disease; when compared with those who abstain, heart disease risk is generally lower in light-to-moderate drinkers.
Hyperglycemia	Raises blood glucose.
Hypoglycemia	Lowers blood glucose, especially in people with diabetes.
Infertility	Increases the risks of menstrual disorders and spontaneous abortions (in women); suppresses luteinizing hormone (in women) and testosterone (in men).
Kidney disease	Enlarges the kidneys, alters hormone functions, and increases the risk of kidney failure.
Liver disease	Causes fatty liver, alcoholic hepatitis, and cirrhosis.
Malnutrition	Increases the risk of malnutrition; low intakes of protein, calcium, iron, vitamin A, vitamin C, thiamin, vitamin B_6, and riboflavin; and impaired absorption of calcium, phosphorus, vitamin D, and zinc.
Nerve disorders	Causes neuropathy and dementia; impairs balance and memory.
Obesity	Increases energy intake, but is not a primary cause of obesity.
Psychological disturbances	Causes depression, anxiety, and insomnia.

NOTE: This list is by no means all-inclusive. Alcohol has direct toxic effects on all body systems.

Personal Strategies

One obvious option available to people attending social gatherings is to enjoy the conversation, eat the food, and drink nonalcoholic beverages. Several nonalcoholic beverages are available that mimic the look and taste of their alcoholic counterparts. For those who enjoy champagne or beer, sparkling ciders and beers without alcohol are available. Instead of drinking a cocktail, a person can sip tomato juice with a slice of lime and a stalk of celery or just a plain cola beverage. The person who chooses to drink alcohol should sip each drink slowly accompanied by food and water.

If you want to help sober up a friend who has had too much to drink, don't bother walking arm in arm around the block. Walking muscles have to work harder, but muscle cells don't have the enzymes to metabolize alcohol; only liver cells can clear alcohol from the blood. Remember that each person has a limited amount of the alcohol dehydrogenase enzyme, which clears the blood at a steady rate. Time alone will do the job. Nor will it help to give your friend a cup of coffee. Caffeine is a stimulant, but it won't speed up alcohol metabolism. Table H7-6 presents other alcohol myths.

People who have passed out from drinking need 24 hours to sober up completely. Let them sleep, but watch over them. Encourage them to lie on their sides, instead of their backs. That way, if they vomit, they won't choke.

TABLE H7-6 Myths and Truths Concerning Alcohol

Myth:	Liquors such as rum, vodka, and tequila are more harmful than wine and beer.
Truth:	The damage caused by alcohol depends largely on the *amount* consumed. Compared with liquor, beer and wine have relatively low percentages of alcohol, but they are often consumed in larger quantities.
Myth:	Consuming alcohol with raw seafood diminishes the likelihood of getting hepatitis.
Truth:	People have eaten contaminated oysters while drinking alcoholic beverages and not gotten as sick as those who were not drinking. But do not be misled: hepatitis is too serious an illness for anyone to depend on alcohol for protection.
Myth:	Alcohol stimulates the appetite.
Truth:	For some people, alcohol may stimulate appetite, but it seems to have the opposite effect in heavy drinkers. Heavy drinkers tend to eat poorly and suffer malnutrition.
Myth:	Drinking alcohol is healthy.
Truth:	Moderate alcohol consumption is associated with a lower risk for heart disease. Higher intakes, however, raise the risks for high blood pressure, stroke, heart disease, some cancers, accidents, violence, suicide, birth defects, and deaths in general. Furthermore, excessive alcohol consumption damages the liver, pancreas, brain, and heart. No authority recommends that nondrinkers begin drinking alcoholic beverages to obtain health benefits.
Myth:	Wine increases the body's absorption of minerals.
Truth:	Wine may increase the body's absorption of potassium, calcium, phosphorus, magnesium, and zinc, but the alcohol in wine also promotes the body's excretion of these minerals, so no benefit is gained.
Myth:	Alcohol is legal and, therefore, not a drug.
Truth:	Alcohol is legal for adults 21 years old and older, but it is also a drug—a substance that alters one or more of the body's functions.
Myth:	A shot of alcohol warms you up.
Truth:	Alcohol diverts blood flow to the skin making you *feel* warmer, but it actually cools the body.
Myth:	Wine and beer are mild; they do not lead to alcoholism.
Truth:	Alcoholism is not related to the kind of beverage, but rather to the quantity and frequency of consumption.
Myth:	Mixing different types of drinks gives you a hangover.
Truth:	Too much alcohol in any form produces a hangover.
Myth:	Alcohol is a stimulant.
Truth:	People think alcohol is a stimulant because it seems to relieve inhibitions, but it does so by depressing the activity of the brain. Alcohol is medically defined as a depressant drug.
Myth:	Beer is a great source of carbohydrate, vitamins, minerals, and fluids.
Truth:	Beer does provide some carbohydrate, but most of its kcalories come from alcohol. The few vitamins and minerals in beer cannot compete with rich food sources. And the diuretic effect of alcohol causes the body to lose more fluid in urine than is provided by the beer.

Don't drive too soon after drinking. The lack of glucose for the brain to function and the length of time to clear the blood of alcohol make alcohol's adverse effects linger long after its blood concentration has fallen. Driving coordination is still impaired the morning *after* a night of drinking, even if the drinking was moderate. Responsible aircraft pilots know that they must allow 24 hours for their bodies to clear alcohol completely, and they do not fly any sooner. The Federal Aviation Administration and major airlines enforce this rule.

Look again at the drawing of the brain in Figure H7-4 (p. 224), and note that when someone drinks, judgment fails first. Judgment might tell a person to limit alcohol consumption to two drinks at a party, but if the first drink takes judgment away, many more drinks may follow. The failure to stop drinking as planned, on repeated occasions, is a warning sign that the person should not drink at all.

Ethanol interferes with a multitude of metabolic reactions in the body—many more than have been enumerated here. With heavy alcohol consumption, the potential for harm is great. If you drink alcoholic beverages, do so with care, and in moderation.

REFERENCES

1. U. Benedetto and coauthors, Alcohol intake and outcomes following coronary artery bypass grafting, *Circulation* 122 (2010): A14440; D. A. Boggs and coauthors, Coffee, tea, and alcohol intake in relation to risk of type 2 diabetes in African American women, *American Journal of Clinical Nutrition* 92 (2010): 960–966; M. M. Joosten and coauthors, Combined effect of alcohol consumption and lifestyle behaviors on risk of type 2 diabetes, *American Journal of Clinical Nutrition* 91 (2010): 1777–1783; D. O. Baliunas and coauthors, Alcohol as a risk factor for type 2 diabetes: A systematic review and meta-analysis, *Diabetes Care* 32 (2009): 2123–2132; K. L. Tucker and coauthors, Effects of beer, wine, and liquor intakes on bone mineral density in older men and women, *American Journal of Clinical Nutrition* 89 (2009): 1188–1196.
2. M. P. Ferreira and D. Willoughby, Alcohol consumption: The good, the bad, and the indifferent, *Applied Physiology, Nutrition, and Metabolism* 33 (2008): 12–20.
3. A. J. Barnes and coauthors, Prevalence and correlates of at-risk drinking among older adults: The project SHARE study, *Journal of General Internal Medicine* 25 (2010): 840–846; P. Meier and H. K. Seitz, Age, alcohol metabolism and liver disease, *Current Opinion in Clinical Nutrition and Metabolic Care* 11 (2008): 21–26.
4. K. J. Mukamal, A 42-year-old man considering whether to drink alcohol for his health, *Journal of the American Medical Association* 303 (2010): 2065–2073.
5. C. Sayon-Orea, M. A. Martinez-Gonzalez, and M. Bes-Rastrollo, Alcohol consumption and body weight: A systematic review, *Nutrition Reviews* 69 (2011): 419–431; L. Wang and coauthors, Alcohol consumption, weight gain, and risk of becoming overweight in middle-aged and older women, *Archives of Internal Medicine* 170 (2010): 453–461.
6. M. Schütze and coauthors, Beer consumption and the "beer belly": Scientific basis or common belief? *European Journal of Clinical Nutrition* 63 (2009): 1143–1149.
7. R. A. Breslow and coauthors, Alcoholic beverage consumption, nutrient intakes, and diet quality in the US adult population, 1999-2006, *Journal of the American Dietetic Association* 110 (2010): 551–562.
8. A. D. Thomson and coauthors, Wernicke's encephalopathy: "Plus ça change, plus c'est la mème chose," *Alcohol and Alcoholism* 43 (2008): 180–186.
9. Centers for Disease Control and Prevention, National Center for Injury Prevention and Control (NCIPC), www.cdc.gov, accessed June 23, 2008.
10. M. Roerecke and J. Rehm, Irregular heavy drinking occasions and risk of ischemic heart disease: A systematic review and meta-analysis, *American Journal of Epidemiology* 171 (2010): 633–644.
11. R. W. Hingson, Z. Wenxing, and E. R. Weitzman, Magnitude of and trends in alcohol-related mortality and morbidity among US college students ages 18–24, 1998-2005, *Journal of Studies on Alcohol and Drugs* Supplement No. 16 (2009): 12–20.
12. *Questions and Answers: Caffeinated Alcoholic Beverages*, www.fda.gov/Food/FoodIngredientsPackaging/ucm233726.htm, accessed November 17, 2010.
13. Centers for Disease Control and Prevention, Vital signs: Binge drinking among high school students and adults—United States, 2009, *Morbidity and Mortality Weekly Report* 59 (2010): 1274–1279.
14. A. Z. Fan and coauthors, Patterns of alcohol consumption and the metabolic syndrome, *Journal of Clinical Endocrinology and Metabolism* 93 (2008): 3833–3838.

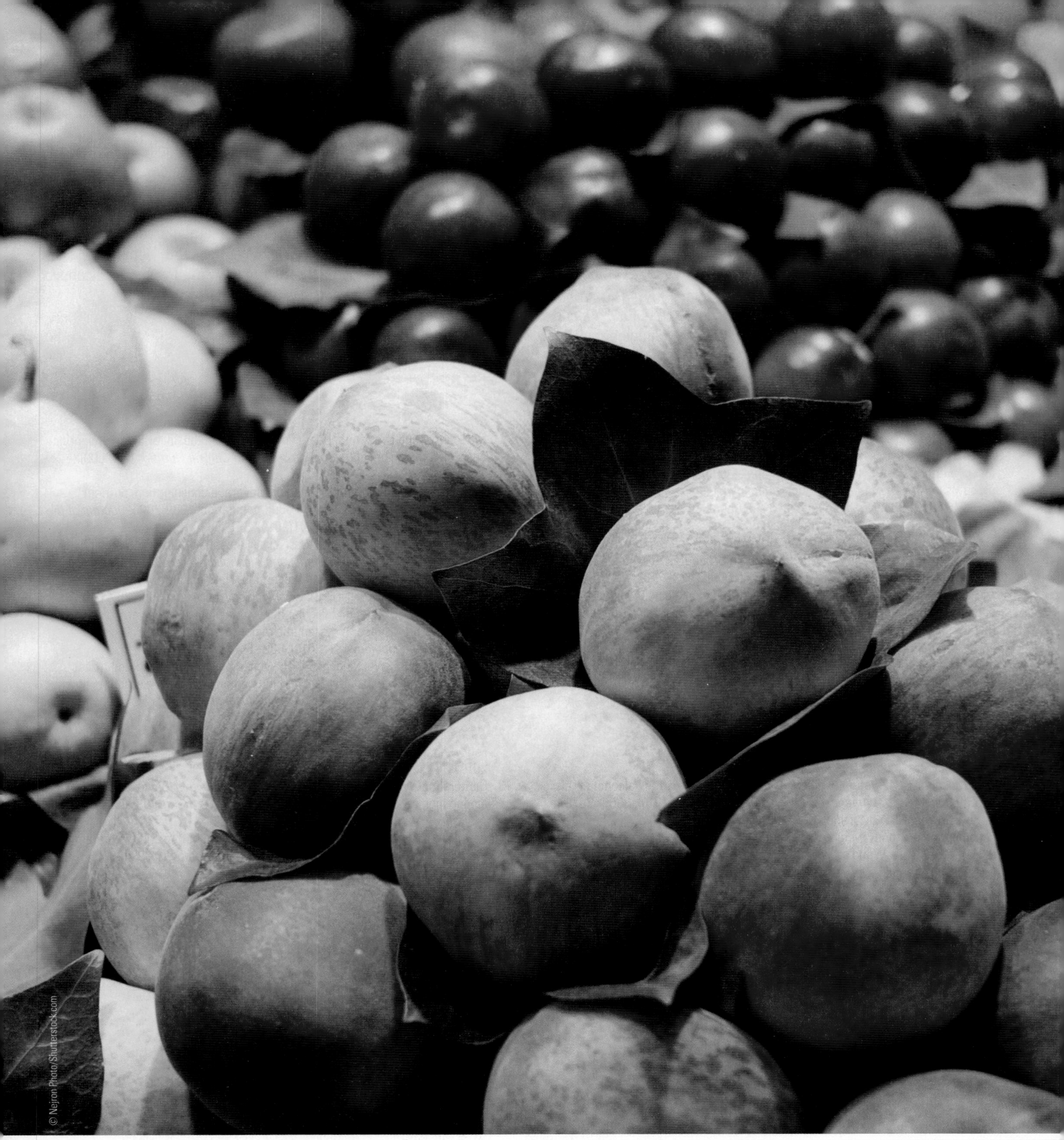

© Nejron Photo/Shutterstock.com

Body mass index is a measure of a person's weight relative to height. Go to Weight Management/Diets → BMI to review the calculation and how to interpret the data. When would it be appropriate to use BMI in research and when would it be best to use another measure of body composition?

8 Energy Balance and Body Composition

LEARNING GPS

Nutrition in Your Life

It's a simple equation: energy in + energy out = energy balance. The reality, of course, is much more complex. One day you may devour a dozen doughnuts at midnight and sleep through your morning workout—tipping the scales toward weight gain. Another day you may snack on veggies and train for this weekend's 10k race—shifting the balance toward weight loss. Your body weight—especially as it relates to your body fat—and your level of fitness have consequences for your health. So, how are you doing? In the Nutrition Portfolio at the end of this chapter, you can see how your "energy in" and "energy out" balance and whether your body weight and fat measures are consistent with good health.

As Chapter 7 explains, the body's remarkable metabolism can cope with variations in the diet. When the diet delivers too little energy, carbohydrate, or protein, the body uses its fat to meet energy needs and degrades its lean tissue to meet glucose and protein needs. When the diet delivers too much energy—whether from excess carbohydrate, excess protein, or excess fat—the body stores fat.

Both excessive and deficient body fat result from an energy imbalance. The simple picture is as follows. People who consume more food energy than they expend store the surplus as body fat. To reduce body fat, they need to expend more energy than they take in from food. In contrast, people who consume too little food energy to support their bodies' activities must rely on their bodies' fat stores and possibly some of their lean tissues as well. To gain weight, these people need to take in more food energy than they expend. As you will see, though, the details of energy balance and weight regulation are quite complex.[1] This chapter describes energy balance and body composition and examines the health problems associated with having too much or too little body fat. The next chapter presents strategies toward resolving these problems.

When energy in balances with energy out, a person's body weight is stable.

8.1 Energy Balance

LEARN IT Describe energy balance and the consequences of not being in balance.

People expend energy continuously and eat periodically to refuel. Ideally, their energy intakes cover their energy expenditures with little, or no, excess. Excess energy is stored as fat, and stored fat is used for energy between meals. The fat stores of even a healthy-weight adult represent an ample reserve of energy—50,000 to 200,000 kcalories.

The amount of body fat a person deposits in, or withdraws from, storage on any given day depends on the **energy balance** for that day—the amount consumed (energy in) versus the amount expended (energy out). When a person is maintaining weight, energy in equals energy out. When the balance shifts, weight changes. For each 3500 kcalories eaten in excess, a pound of body fat is stored; similarly, a pound of fat is lost for each 3500 kcalories expended beyond those consumed.*

Quick changes in body weight are not simple changes in fat stores. Weight gained or lost rapidly includes some fat, large amounts of fluid, and some lean tissues such as muscle proteins and bone minerals. Because water constitutes about 60 percent of an adult's body weight, retention or loss of water can greatly influence body weight. Even over the long term, the composition of weight gained or lost is normally about 75 percent fat and 25 percent lean. During starvation, losses of fat and lean are about equal. (Recall from Chapter 7 that without adequate carbohydrate, protein-rich lean tissues break down to provide glucose.) Invariably, though, *fat* gains and losses are gradual. The next two sections examine the two sides of the energy-balance equation—energy in and energy out. As you read, keep in mind that changes on one side usually affect the other.[2]

REVIEW IT Describe energy balance and the consequences of not being in balance.
When energy consumed equals energy expended, a person is in energy balance and body weight is stable. If more energy is taken in than is expended, a person gains weight. If more energy is expended than is taken in, a person loses weight.

> **FIGURE 8-1 Bomb Calorimeter**
When food is burned, energy is released in the form of heat. Heat energy is measured in kcalories.

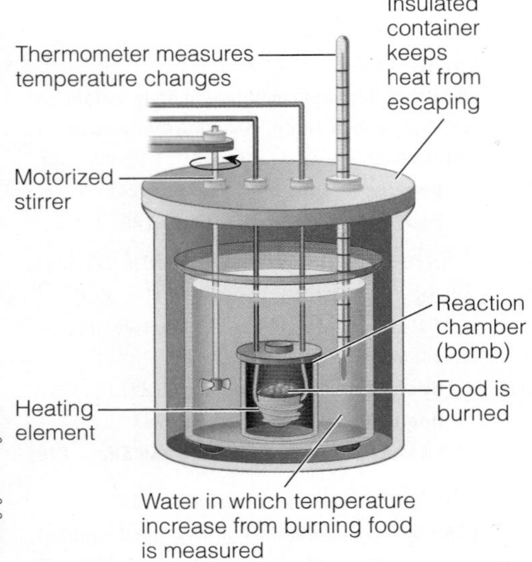

8.2 Energy In: The kCalories Foods Provide

LEARN IT Discuss some of the physical, emotional, and environmental influences on food intake.

Foods and beverages provide the "energy in" part of the energy-balance equation. How much energy a person receives depends on the composition of the foods and beverages and on the amount the person eats and drinks.

Food Composition To find out how many kcalories a food provides, a scientist can burn the food in a **bomb calorimeter** (see Figure 8-1). When the food burns, energy is released in the form of heat. The amount of heat given off provides a *direct* measure of the food's energy value (remember that kcalories are units of heat energy).** In addition to releasing heat, these reactions generate carbon dioxide and water—just as the body's cells do when they metabolize the energy-yielding nutrients from foods. Details of the chemical reactions in a calorimeter and in the body differ, but the overall process is similar: when the food burns and the chemical bonds break, the carbons (C) and hydrogens (H) combine with oxygens (O) to form carbon dioxide (CO_2) and water (H_2O). The amount of oxygen consumed gives an *indirect* measure ♦ of the amount of energy released.

A bomb calorimeter measures the available energy in foods but overstates the **physiological fuel value**—the amount of energy that the human body derives from foods. The body is less efficient than a calorimeter and cannot metabolize all of the energy-yielding nutrients in a food completely. Researchers can correct

♦ Food energy values can be determined by:
- *Direct calorimetry,* which measures the amount of heat released.
- *Indirect calorimetry,* which measures the amount of oxygen consumed.

energy balance: the energy (kcalories) consumed from foods and beverages compared with the energy expended through metabolic processes and physical activities.

bomb calorimeter (KAL-oh-RIM-eh-ter): an instrument that measures the heat energy released when foods are burned, thus providing an estimate of the potential energy of the foods.
- **calor** = heat
- **metron** = measure

physiological fuel value: the number of kcalories that the body derives from a food, in contrast to the number of kcalories determined by calorimetry.

*Body fat, or adipose tissue, is composed of a mixture of mostly fat, some protein, and water. A pound of body fat (454 g) is approximately 87 percent fat, or (454 x 0.87) 395 g, and 395 g x 9 kcal/g = 3555 kcal.
**As Chapter 1 mentions, many scientists measure food energy in *kilojoules* (a measure of work energy). Conversion factors for these and other measures can be found in Appendix K.

for this discrepancy mathematically to create useful tables of the energy values of foods (such as Appendix H). These values provide reasonable estimates, but they do not reflect the *precise* amount of energy a person will derive from the foods consumed.

The energy values of foods can also be computed from the amounts of carbohydrate, fat, and protein (and alcohol, if present) in the foods.* For example, a food ♦ containing 12 grams of carbohydrate, 5 grams of fat, and 8 grams of protein will provide 48 carbohydrate kcalories, 45 fat kcalories, and 32 protein kcalories, for a total of 125 kcalories. (To review how to calculate the energy foods provide, turn to p. 9.)

♦ Energy values:

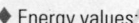

- 1 g carbohydrate = 4 kcal
- 1 g fat = 9 kcal
- 1 g protein = 4 kcal
- 1 g alcohol = 7 kcal

Food Intake To achieve energy balance, the body must meet its needs without taking in too much or too little energy. **Appetite** prompts a person to eat—or not to eat. Somehow the body decides how much and how often to eat—when to start eating and when to stop. As you will see, many signals—from both the environment and genetics—initiate or delay eating.[3]

Hunger People eat for a variety of reasons, most obviously (although not necessarily most commonly) because they are hungry. Most people recognize **hunger** as an irritating feeling that prompts thoughts of food and motivates them to start eating. In the body, hunger is the physiological response to a need for food triggered by nerve signals and chemical messengers originating and acting in the brain, primarily in the **hypothalamus.** Hunger can be influenced by the presence or absence of nutrients in the bloodstream, the size and composition of the preceding meal, customary eating patterns, climate (heat reduces food intake; cold increases it), physical activity, hormones, and illnesses. Hunger determines what to eat, when to eat, and how much to eat.

The stomach is ideally designed to handle periodic batches of food, and people typically eat meals at roughly 4-hour intervals. Four hours after a meal, most, if not all, of the food has left the stomach. Most people do not feel like eating again until the stomach is either empty or almost so. Even then, a person may not feel hungry for quite a while.

Satiation During the course of a meal, as food enters the GI tract and hunger diminishes, **satiation** occurs. As receptors in the stomach stretch and hormones such as cholecystokinin become active, the person begins to feel full. The response: satiation, which prompts the person to stop eating.

Satiety After a meal, the feeling of **satiety** continues to suppress hunger and allows a person to not eat again for a while. Whereas *satiation* tells us to "stop eating," *satiety* reminds us to "not start eating again." Figure 8-2 (p. 234) summarizes the relationships among hunger, satiation, and satiety. Of course, people can override these signals, especially when presented with stressful situations or favorite foods.

Overriding Hunger and Satiety Not surprisingly, eating can be triggered by signals other than hunger, even when the body does not need food. Some people experience food cravings when they are bored or anxious. In fact, they may eat in response to any kind of stress, negative or positive. ("What do I do when I'm grieving? Eat. What do I do when I'm celebrating? Eat!") Not too surprisingly, repeatedly eating to relieve chronic stress can lead to overeating and weight gain.

Many people respond to external cues such as the time of day ("It's time to eat") or the availability, sight, and taste of food ("I'd love a piece of chocolate even though I'm full"). Environmental influences such as large portion sizes, favorite foods, or an abundance or variety of foods stimulate eating and increase energy intake. Cognitive influences—such as perceptions, memories, intellect, and social interactions—can easily lead to weight gain. Those who are overweight or obese may be especially susceptible to external cues that trigger hunger and the desire to eat.[4]

appetite: the integrated response to the sight, smell, thought, or taste of food that initiates or delays eating.

hunger: the painful sensation caused by a lack of food that initiates food-seeking behavior.

hypothalamus (high-po-THAL-ah-mus): a brain center that controls activities such as maintenance of water balance, regulation of body temperature, and control of appetite.

satiation (say-she-AY-shun): the feeling of satisfaction and fullness that occurs during a meal and halts eating. Satiation determines how much food is consumed during a meal.

satiety: the feeling of fullness and satisfaction that occurs after a meal and inhibits eating until the next meal. Satiety determines how much time passes between meals.

*Some of the food energy values in the table of food composition in Appendix H were derived by bomb calorimetry, and many were calculated from their energy-yielding nutrient contents.

> **FIGURE 8-2 Hunger, Satiation, and Satiety**

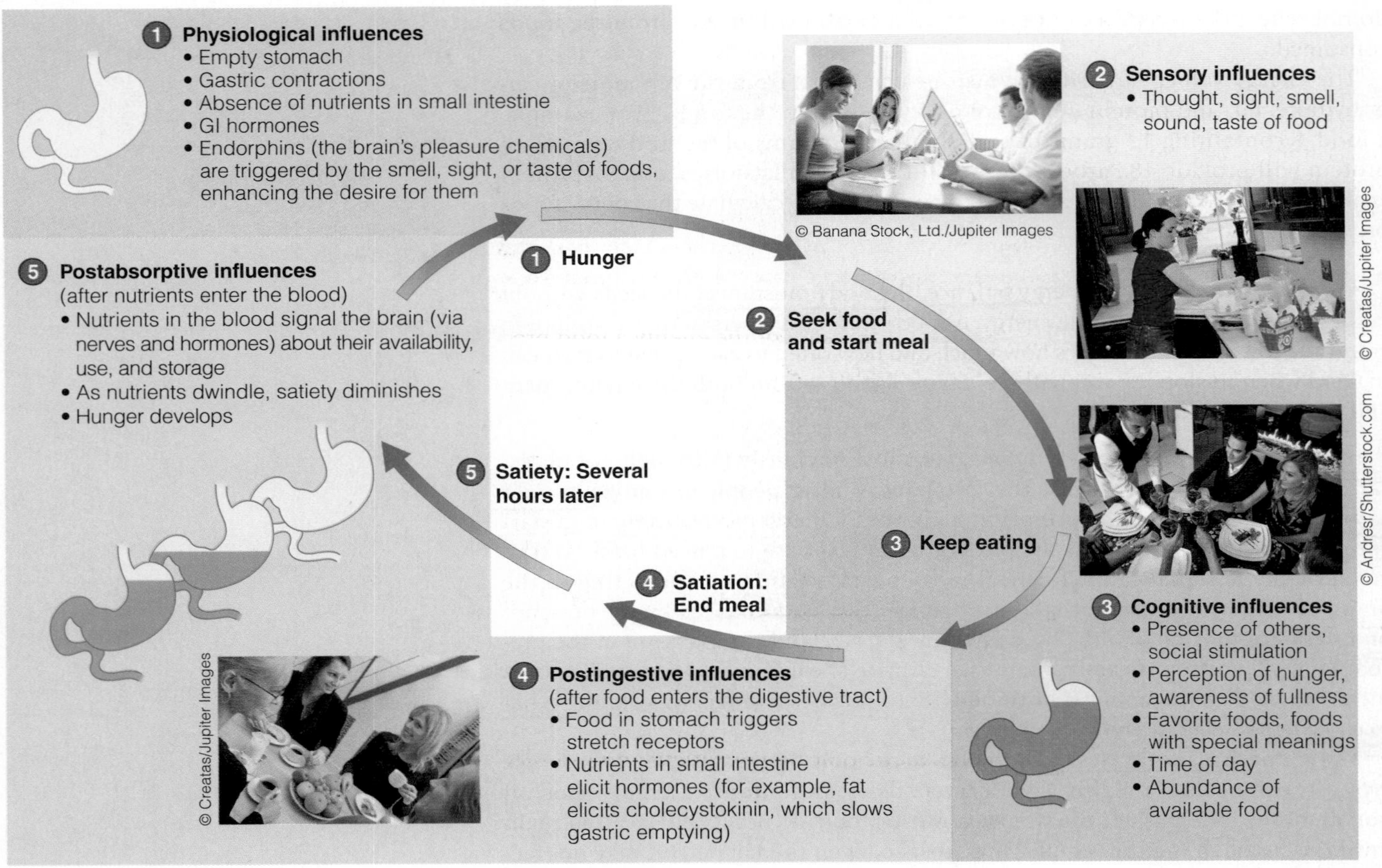

1 Physiological influences
- Empty stomach
- Gastric contractions
- Absence of nutrients in small intestine
- GI hormones
- Endorphins (the brain's pleasure chemicals) are triggered by the smell, sight, or taste of foods, enhancing the desire for them

1 Hunger

2 Sensory influences
- Thought, sight, smell, sound, taste of food

© Banana Stock, Ltd./Jupiter Images

© Creatas/Jupiter Images

2 Seek food and start meal

3 Keep eating

© Andresr/Shutterstock.com

3 Cognitive influences
- Presence of others, social stimulation
- Perception of hunger, awareness of fullness
- Favorite foods, foods with special meanings
- Time of day
- Abundance of available food

4 Satiation: End meal

4 Postingestive influences
(after food enters the digestive tract)
- Food in stomach triggers stretch receptors
- Nutrients in small intestine elicit hormones (for example, fat elicits cholecystokinin, which slows gastric emptying)

© Creatas/Jupiter Images

5 Satiety: Several hours later

5 Postabsorptive influences
(after nutrients enter the blood)
- Nutrients in the blood signal the brain (via nerves and hormones) about their availability, use, and storage
- As nutrients dwindle, satiety diminishes
- Hunger develops

Eating can also be suppressed by signals other than satiety, even when a person is hungry. People with the eating disorder anorexia nervosa, for example, use tremendous discipline to ignore the pangs of hunger. Some people simply cannot eat during times of stress, negative or positive. ("I'm too sad to eat." "I'm too excited to eat!") Why some people overeat in response to stress and others cannot eat at

© Jupiterimages/Getty Images

Regardless of hunger, people typically overeat when offered the abundance and variety of a buffet. To limit unhealthy weight gains, listen to hunger and satiety signals.

all remains a bit of a mystery, although researchers are beginning to understand the connections between stress hormones, brain activity, and "comfort foods." Factors that appear to be involved include how the person perceives the stress and whether usual eating behaviors are restrained. (Highlight 8 features anorexia nervosa and other eating disorders.)

Sustaining Satiation and Satiety The extent to which foods produce satiation and sustain satiety depends in part on the nutrient composition of a meal. Of the three energy-yielding nutrients, protein is considered the most **satiating.**[5] In fact, too little protein in the diet can leave a person feeling hungry. Including some protein—such as drinking milk—provides satiety and decreases energy intake at the next meal.[6] In contrast, fructose in a sugary fruit drink seems to stimulate appetite and increase food intake.[7]

Chapter 1 explains that energy density is a measure of the energy a food provides relative to the amount of food (kcalories per gram). Foods with a high energy density provide more kcalories, and those with low energy density provide fewer kcalories, for the same amount of food. Foods low in energy density are also more satiating. High-fiber foods effectively provide satiation by filling the stomach and delaying the absorption of nutrients. For this reason, eating a large salad as a first course helps a person eat less during the meal. In contrast, fat has a weak effect on satiation; consequently, eating high-fat foods may lead to passive overconsumption. High-fat foods are flavorful, which stimulates the appetite and entices people to eat more. High-fat foods are also energy dense; consequently, they deliver more kcalories per bite. (Chapter 1 introduces the concept of energy density, and Chapter 9 describes how considering a food's energy density can help with weight management.) Although fat provides little satiation during a meal, it produces strong satiety signals once it enters the intestine. Fat in the intestine triggers the release of cholecystokinin—a hormone that signals satiety and inhibits food intake.

Eating high-fat foods while trying to limit energy intake requires small portion sizes, which can leave a person feeling unsatisfied. Portion size correlates directly with a food's satiety. Instead of eating small portions of high-fat foods and feeling deprived, a person can feel satisfied by eating large portions of low-fat, high-fiber, and low energy density foods. Figure 8-3 illustrates how fat influences portion size.

Message Central—The Hypothalamus As you can see, eating is a complex behavior controlled by a variety of psychological, social, metabolic, and physiological factors. The hypothalamus appears to be the control center, integrating messages

satiating: having the power to suppress hunger and inhibit eating.

> FIGURE 8-3 **How Fat Influences Portion Sizes**

For the same size portion, peanuts deliver more than 15 times the kcalories and 20 times the fat of popcorn.

For the same number of kcalories, a person can have a few high-fat peanuts or almost 2 cups of high-fiber popcorn. (This comparison used oil-based popcorn; using air-popped popcorn would double the amount of popcorn in this example.)

about energy intake, expenditure, and storage from other parts of the brain and from the mouth, GI tract, and liver. Some of these messages influence satiation, which helps control the size of a meal; others influence satiety, which helps determine the frequency of meals.

♦ Gastrointestinal hormones that regulate food intake:
- Amylin
- Cholecystokinin (CCK)
- Enterostatin
- Ghrelin
- Glucagon-like peptide-1 (GLP-1)
- Oxyntomodulin
- Pancreatic polypeptide (PP)
- Peptide YY (PYY)

Dozens of gastrointestinal hormones ♦ influence appetite control and energy balance.[8] By understanding the action of these hormones, researchers may one day be able to develop anti-obesity treatments. The greatest challenge now is to sort out the many actions of these brain chemicals. For example, one of these chemicals, **neuropeptide Y,** causes carbohydrate cravings, initiates eating, decreases energy expenditure, and increases fat storage—all factors favoring a positive energy balance and weight gain.

REVIEW IT Discuss some of the physical, emotional, and environmental influences on food intake.

A mixture of signals governs a person's eating behaviors. Hunger and appetite initiate eating, whereas satiation and satiety stop and delay eating, respectively. Each responds to messages from the nervous and hormonal systems. Superimposed on these signals are complex factors involving emotions, habits, and other aspects of human behavior.

8.3 Energy Out: The kCalories the Body Expends

LEARN IT List the components of energy expenditure and factors that might influence each.

Chapter 7 explains that heat is released whenever the body breaks down carbohydrate, fat, or protein for energy and again when that energy is used to do work. The generation of heat, known as **thermogenesis,** can be measured to determine the amount of energy expended. ♦ The total energy a body expends reflects three main categories of thermogenesis:

♦ Energy expenditure, like food energy, can be determined by:
- *Direct calorimetry,* which measures the amount of heat released
- *Indirect calorimetry,* which measures the amount of oxygen consumed and carbon dioxide expelled

- Energy expended for basal metabolism
- Energy expended for physical activity
- Energy expended for food consumption

A fourth category is sometimes involved:

- Energy expended for adaptation

Components of Energy Expenditure People expend energy when they are physically active, of course, but they also expend energy when they are resting quietly. In fact, quiet metabolic activities account for the largest share of most people's energy expenditures, as Figure 8-4 shows.

♦ Quick and easy estimates for basal energy needs:
- Men: Slightly >1 kcal/min (1.1 to 1.3 kcal/min) or 24 kcal/kg/day
- Women: Slightly <1 kcal/min (0.8 to 1.0 kcal/min) or 23 kcal/kg/day

For perspective, a burning candle or a 75-watt light bulb releases about 1 kcal/min.

Basal Metabolism About two-thirds of the energy the average person expends in a day supports the body's **basal metabolism.** Metabolic activities include the lungs inhaling and exhaling air, the bone marrow making new red blood cells, the heart beating 100,000 times a day, and the kidneys filtering wastes—in short, they support all the basic processes of life.

♦ BMR equations use actual weight in kilograms, height in centimeters, and age in years:
- Men: $(10 \times \text{wt}) + (6.25 \times \text{ht}) - (5 \times \text{age}) + 5$
- Women: $(10 \times \text{wt}) + (6.25 \times \text{ht}) - (5 \times \text{age}) - 161$

The **basal metabolic rate (BMR)** is the rate at which the body expends energy for these life-sustaining activities. ♦ The rate may vary from person to person and may vary for the same individual with a change in circumstance or physical condition. The rate is slowest when a person is sleeping undisturbed, but it is usually measured in a room with a comfortable temperature when the person is awake, but lying still, after a restful sleep and an overnight (12 to 14 hour) fast. A similar measure of energy output—called the **resting metabolic rate (RMR)**—is slightly higher than the BMR because its criteria for recent food intake and physical activity are not as strict. When energy needs cannot be measured, equations ♦ can provide reasonably accurate estimates.

In general, the more a person weighs, the more *total* energy is expended on basal metabolism, but the amount of energy *per pound* of body weight may be lower. For example, an adult's BMR might be 1500 kcalories per day and an

neuropeptide Y: a chemical produced in the brain that stimulates appetite, diminishes energy expenditure, and increases fat storage.

thermogenesis: the generation of heat; used in physiology and nutrition studies as an index of how much energy the body is expending.

basal metabolism: the energy needed to maintain life when a body is at complete digestive, physical, and emotional rest.

basal metabolic rate (BMR): the rate of energy use for metabolism under specified conditions: after a 12-hour fast and restful sleep, without any physical activity or emotional excitement, and in a comfortable setting. It is usually expressed as kcalories per kilogram body weight per hour.

resting metabolic rate (RMR): similar to the basal metabolic rate (BMR), a measure of energy use for a person at rest in a comfortable setting, but with less stringent criteria for recent food intake and physical activity. Consequently, the RMR is slightly higher than the BMR.

infant's only 500, but compared to body weight, the infant's BMR is more than twice as fast. Similarly, a normal-weight adult may have a metabolic rate one and a half times that of an obese adult when compared to body weight because lean tissue is metabolically more active than body fat.

Table 8-1 summarizes the factors that raise and lower the BMR. For the most part, the BMR is highest in people who are growing (children, adolescents, and pregnant women) and in those with considerable **lean body mass** (physically fit people and males). One way to increase the BMR, then, is to participate in endurance and strength-training activities regularly to maximize lean body mass. The BMR is also high in people with fever or under stress and in people with highly active thyroid glands. The BMR slows down with a loss of lean body mass and during fasting and malnutrition.

Physical Activity The second component of a person's energy output is physical activity: voluntary movement of the skeletal muscles and support systems. Physical activity is the most variable—and the most changeable—component of energy expenditure. Consequently, its influence on both weight gain and weight loss can be significant.

During physical activity, the muscles need extra energy to move, and the heart and lungs need extra energy to deliver nutrients and oxygen and dispose of wastes. The amount of energy needed for any activity, whether playing tennis or studying for an exam, depends on three factors: muscle mass, body weight, and activity. The larger the muscle mass and the heavier the weight of the body part being moved, the more energy is expended. Table 8-2 (p. 238) gives average energy expenditures for various activities. The activity's duration, frequency, and intensity also influence energy expenditure: the longer, the more frequent, and the more intense the activity, the more kcalories expended. (Chapter 14 describes how an activity's duration, frequency, and intensity also influence the body's use of the energy-yielding nutrients.)

> FIGURE 8-4 **Components of Energy Expenditure**

The amount of energy expended in voluntary physical activities has the greatest variability, depending on a person's activity patterns. For a sedentary person, physical activities may account for less than half as much energy as basal metabolism, whereas an extremely active person may expend as much on activity as for basal metabolism.

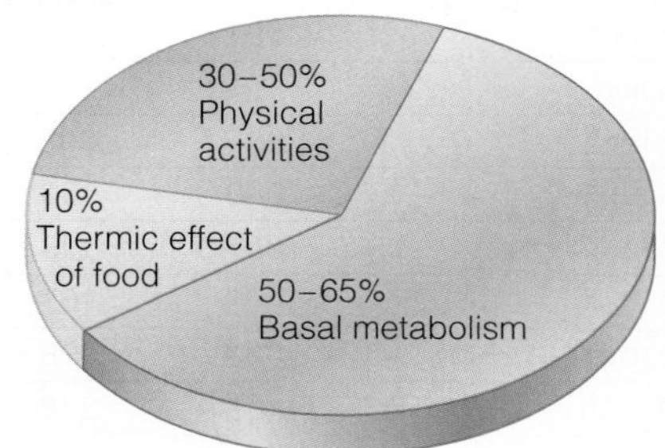

The amount of energy expended in a day differs for each individual, but in general, basal metabolism is the largest component of energy expenditure and thermic effect of food is the smallest.

TABLE 8-1 Factors that Affect the BMR

Factor	Effect on BMR
Age	Lean body mass diminishes with age, slowing the BMR.[a]
Height	In tall, thin people, the BMR is higher.[b]
Growth	In children, adolescents, and pregnant women, the BMR is higher.
Body composition (gender)	The more lean tissue, the higher the BMR (which is why males usually have a higher BMR than females). The more fat tissue, the lower the BMR.
Fever	Fever raises the BMR.[c]
Stresses	Stresses (including many diseases and certain drugs) raise the BMR.
Environmental temperature	Both heat and cold raise the BMR.
Fasting/starvation	Fasting/starvation lowers the BMR.[d]
Malnutrition	Malnutrition lowers the BMR.
Hormones (gender)	The thyroid hormone thyroxin, for example, can speed up or slow down the BMR.[e] Premenstrual hormones slightly raise the BMR.
Smoking	Nicotine increases energy expenditure.
Caffeine	Caffeine increases energy expenditure.
Sleep	BMR is lowest when sleeping.

[a]The BMR begins to decrease in early adulthood (after growth and development cease) at a rate of about 2 percent/decade. A reduction in voluntary activity as well brings the total decline in energy expenditure to about 5 percent/decade.
[b]If two people weigh the same, the taller, thinner person will have the faster metabolic rate, reflecting the greater skin surface, through which heat is lost by radiation, in proportion to the body's volume (see Figure 8-5, p. 239).
[c]Fever raises the BMR by 7 percent for each degree Fahrenheit.
[d]Prolonged starvation reduces the total amount of metabolically active lean tissue in the body, although the decline occurs sooner and to a greater extent than body losses alone can explain. More likely, the neural and hormonal changes that accompany fasting are responsible for changes in the BMR.
[e]The thyroid gland releases hormones that travel to the cells and influence cellular metabolism. Thyroid hormone activity can speed up or slow down the rate of metabolism by as much as 50 percent.

lean body mass: the body minus its fat.

TABLE 8-2 Energy Expended on Various Activities

The values listed in this table reflect both the energy expended in physical activity *and* the amount used for BMR. To calculate kcalories spent per minute of activity for your own body weight, multiply kcal/lb/min (or kcal/kg/min) by your exact weight and then multiply that number by the number of minutes spent in the activity. For example, if you weigh 142 pounds, and you want to know how many kcalories you spent doing 30 minutes of vigorous aerobic dance: 0.062 × 142 = 8.8 kcalories per minute; 8.8 × 30 minutes = 264 total kcalories spent.

Activity	kCal/lb min	kCal/kg min
Aerobic dance (vigorous)	.062	.136
Basketball (vigorous, full court)	.097	.213
Bicycling		
13 mph	.045	.099
15 mph	.049	.108
17 mph	.057	.125
19 mph	.076	.167
21 mph	.090	.198
23 mph	.109	.240
25 mph	.139	.306
Canoeing, flat water, moderate pace	.045	.099
Cross-country skiing 8 mph	.104	.229
Gardening	.045	.099
Golf (carrying clubs)	.045	.099
Handball	.078	.172
Horseback riding (trot)	.052	.114
Rowing (vigorous)	.097	.213
Running		
5 mph	.061	.134
6 mph	.074	.163
7.5 mph	.094	.207
9 mph	.103	.227
10 mph	.114	.251
11 mph	.131	.288
Soccer (vigorous)	.097	.213
Studying	.011	.024
Swimming		
20 yd/min	.032	.070
45 yd/min	.058	.128
50 yd/min	.070	.154
Table tennis (skilled)	.045	.099
Tennis (beginner)	.032	.070
Vacuuming and other household tasks	.030	.066
Walking (brisk pace)		
3.5 mph	.035	.077
4.5 mph	.048	.106
Weight lifting		
light-to-moderate	.024	.053
vigorous	.048	.106
Wheelchair basketball	.084	.185
Wheeling self in wheelchair	.030	.066
Wii games		
bowling	.021	.046
boxing	.021	.047
tennis	.022	.048

Thermic Effect of Food When a person eats, the GI tract muscles speed up their rhythmic contractions, the cells that manufacture and secrete digestive juices become active, and some nutrients require energy to be absorbed. This acceleration of activity requires energy and produces heat; it is known as the **thermic effect of food (TEF)**.

It feels like work and it may make you tired, but studying requires only one or two kcalories per minute.

thermic effect of food (TEF): an estimation of the energy required to process food (digest, absorb, transport, metabolize, and store ingested nutrients); also called the *specific dynamic effect (SDE)* of food or the *specific dynamic activity (SDA)* of food. The sum of the TEF and any increase in the metabolic rate due to overeating is known as *diet-induced thermogenesis (DIT).*

The thermic effect of food is proportional to the food energy taken in and is usually estimated at 10 percent of energy intake. Thus a person who ingests 2000 kcalories probably expends about 200 kcalories on the thermic effect of food. The proportions vary for different foods, however, and are also influenced by factors such as meal size and frequency. In general, the thermic effect of food is greater for high-protein foods than for high-fat foods ♦ and for a meal eaten all at once rather than spread out over a couple of hours. For most purposes, however, the thermic effect of food can be ignored when estimating energy expenditure because its contribution to total energy output is smaller than the probable errors involved in estimating overall energy intake and output.

♦ Thermic effect of foods:
- Carbohydrate: 5–10%
- Fat: 0–5%
- Protein: 20–30%
- Alcohol: 15–20%

Percentages are calculated by dividing the energy expended during digestion and absorption (above basal) by the energy content of the food.

Adaptive Thermogenesis Additional energy is expended when circumstances in the body are dramatically changed. A body challenged to physical conditioning, extreme cold, overfeeding, starvation, trauma, or other types of stress must adapt; it has extra work to do and uses extra energy to build the tissues and produce the enzymes and hormones necessary to cope with the demand. This energy is known as **adaptive thermogenesis**, and in some circumstances, it makes a considerable difference in the total energy expended. Because this component of energy expenditure is so variable and specific to individuals, it is not included when calculating energy requirements.

Estimating Energy Requirements

In estimating energy requirements, the DRI Committee developed equations based on research measuring total daily energy expenditure. These equations consider how the following factors influence BMR and consequently energy expenditure:

- *Gender.* In general, women have a lower BMR than men, in large part because men typically have more lean body mass. Two sets of energy equations—one for men and one for women—were developed to accommodate the influence of gender on energy expenditure (provided on the next page).
- *Growth.* The BMR is high in people who are growing. For this reason, pregnant and lactating women, infants, children, and adolescents have their own sets of energy equations (provided in Appendix F).
- *Age.* The BMR declines during adulthood as lean body mass diminishes. This change in body composition occurs, in part, because some hormones that influence appetite, body weight, and metabolism become more, or less, active with age. Physical activities tend to decline as well, bringing the average reduction in energy expenditure to about 5 percent per decade. The decline in BMR that occurs when a person becomes less active reflects the loss of lean body mass and may be minimized with ongoing physical activity. Because age influences energy expenditure, it is also factored into the energy equations.
- *Physical activity.* Using individual values for various physical activities (as in Table 8-2) is time-consuming and impractical for estimating the energy needs of a population. Instead, various activities are clustered according to the typical intensity of a day's efforts. Energy equations include a physical activity factor for various levels of intensity for each gender.
- *Body composition and body size.* The BMR is high in people who are tall and so have a large surface area, as illustrated in Figure 8-5. Similarly, the more a person weighs, the more energy is expended on basal metabolism. For these reasons, energy equations include a factor for both height and weight.

> FIGURE 8-5 **How Body Size Influences BMR**

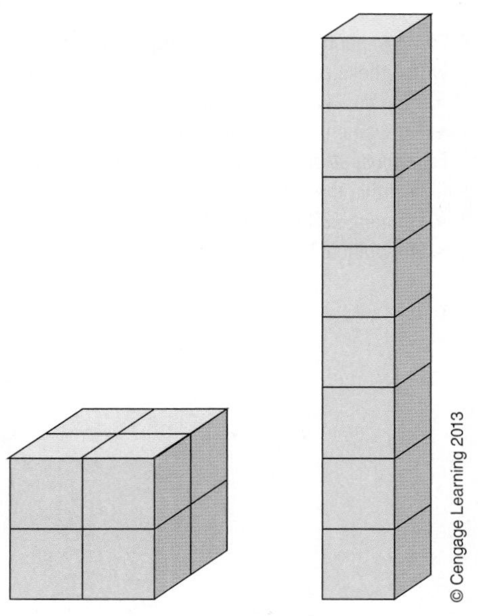

Each of these structures is made of eight blocks. They weigh the same, but they are arranged differently. The short, wide structure has 24 sides exposed and the tall, thin one has 34. Because the tall, thin structure has a greater surface area, it will lose more heat (expend more energy) than the short, wide one. Similarly, two people of different heights might weigh the same, but the taller, thin one will have a higher BMR (expending more energy) because of the greater skin surface.

As just explained, energy needs vary between individuals depending on such factors as gender, growth, age, physical activity, and body size and composition. Even when two people are similarly matched, however, their energy needs still differ because of genetic differences. Perhaps one day genetic research will reveal how to estimate requirements for each individual. For now, the "How To" feature on p. 240 provides instructions on calculating your estimated energy requirements using the DRI equations and physical activity factors. Appendix F presents DRI tables

adaptive thermogenesis: adjustments in energy expenditure related to changes in environment such as extreme cold and to physiological events such as overfeeding, trauma, and changes in hormone status.

>How To

Estimate Energy Requirements

To determine your estimated energy requirement (EER), use the appropriate equation, inserting your age in years, weight (wt) in kilograms, height (ht) in meters, and physical activity (PA) factor from the accompanying table. (To convert pounds to kilograms, divide by 2.2; to convert inches to meters, divide by 39.37.)

- For men 19 years and older:
  $EER = [662 - (9.53 \times age)] + PA \times [(15.91 \times wt) + (539.6 \times ht)]$
- For women 19 years and older:
  $EER = [354 - (6.91 \times age)] + PA \times [(9.36 \times wt) + (726 \times ht)]$

For example, consider an active 30-year-old male who is 5 feet 11 inches tall and weighs 178 pounds. First, he converts his weight from pounds to kilograms and his height from inches to meters, if necessary:

$$178 \text{ lb} \div 2.2 = 80.9 \text{ kg}$$
$$71 \text{ in} \div 39.37 = 1.8 \text{ m}$$

Next, he considers his level of daily physical activity and selects the appropriate PA factor from the accompanying table. (In this example, 1.25 for an active male.) Then, he inserts his age, PA factor, weight, and height into the appropriate equation:

$$EER = [662 - (9.53 \times 30)] + 1.25 \times [(15.91 \times 80.9) + (539.6 \times 1.8)]$$

(A reminder: Do calculations within the parentheses first.) He calculates:

$$EER = [662 - 286] + 1.25 \times [1287 + 971]$$

(Another reminder: Do calculations within the brackets next.)

$$EER = 376 + 1.25 \times 2258$$

(One more reminder: Do multiplication before addition.)

$$EER = 376 + 2823$$
$$EER = 3199$$

The estimated energy requirement for an active 30-year-old male who is 5 feet 11 inches tall and weighs 178 pounds is about 3200 kcalories/day. His actual requirement probably falls within a range ♦ of 200 kcalories above and below this estimate.

NOTE: Appendix F provides tables of energy expenditure for adults at various levels of activity and various heights and weights. It also includes ERR equations for infants, children, adolescents, and pregnant women.

♦ For *most* people, the actual energy requirement falls within these ranges:
- For men, EER ± 200 kcal
- For women, EER ± 160 kcal

For *almost all* people, the actual energy requirement falls within these ranges:
- For men, EER ± 400 kcal
- For women, EER ± 320 kcal

Physical Activity (PA) Factors for EER Equations

	Men	Women	Physical Activity
Sedentary	1.0	1.0	Typical daily living activities
Low active	1.11	1.12	plus 30–60 min moderate activity
Active	1.25	1.27	plus ≥ 60 min moderate activity
Very active	1.48	1.45	plus ≥ 60 min moderate activity and 60 min vigorous or 120 min moderate activity

NOTE: Moderate activity is equivalent to walking at 3 to 4½ mph.

TRY IT Estimate your energy requirement based on your current age, weight, height, and activity level.

that provide a shortcut to estimating total energy expenditure and instructions to help you determine the appropriate physical activity factor to use in the equation.

REVIEW IT List the components of energy expenditure and factors that might influence each.

A person in energy balance takes in energy from food and expends much of it on basal metabolic activities, some of it on physical activities, and a little on the thermic effect of food. Energy requirements vary from person to person, depending on such factors as gender, age, weight, and height as well as the intensity and duration of physical activity. All of these factors must be considered when estimating energy requirements.

8.4 Body Weight and Body Composition

LEARN IT Distinguish between body weight and body composition, including methods to assess each.

A person 5 feet 10 inches tall who weighs 150 pounds may carry only about 30 of those pounds as fat.* The rest is mostly water and lean tissues—muscles, organs such as the heart and liver, and the bones of the skeleton. Direct measures of **body composition** are impossible in living human beings; instead, researchers assess body composition indirectly based on the following assumption:

$$\text{Body weight} = \text{fat} + \text{lean tissue (including water)}$$

Weight gains and losses tell us nothing about how the body's composition may have changed, yet weight is the measure most people use to judge their "fatness." For many people, overweight is overfat, but this is not always the case. Athletes with dense bones and well-developed muscles may be overweight by some standards but have little body fat. Conversely, inactive people may seem to have acceptable weights, when, in fact, they may have too much body fat.

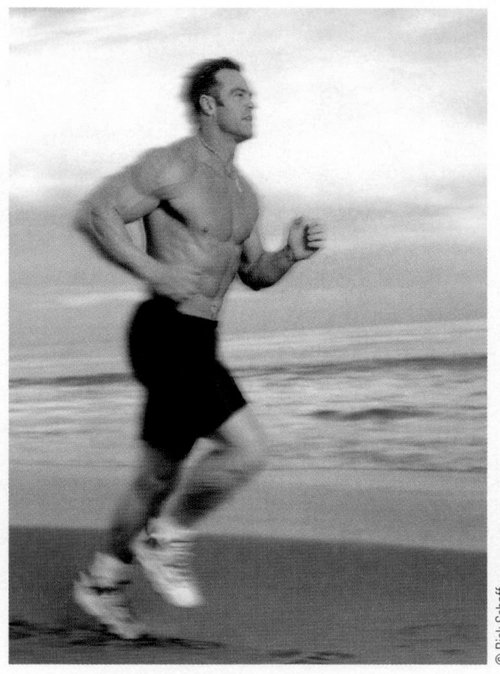

© Rick Schaff

At 6 feet 4 inches tall and 250 pounds (1.93 meters and 113 kilograms), this runner would be considered over*weight* by most standards. Yet he is clearly not over*fat*.

Defining Healthy Body Weight

How much should a person weigh? How can a person know if her weight is appropriate for her height? How can a person know if his weight is jeopardizing his health? Such questions seem so simple, yet the answers can be complex—and quite different depending on whom you ask.

The Criterion of Fashion In asking what is ideal, people often mistakenly turn to friends and fashion for the answer and judge body weight by appearances. No doubt our society sets unrealistic ideals for body weight, especially for women. Magazines, movies, and television all convey the message that to be thin is to be beautiful and happy. As a result, the media have a great influence on the weight concerns and dieting patterns of people of all ages, but most tragically on young, impressionable children and adolescents.[9]

Importantly, perceived body image may have little to do with actual body weight or size. People of all shapes, sizes, and ages—including extremely thin fashion models with anorexia nervosa and fitness instructors with ideal body composition—have learned to be unhappy with their "overweight" bodies. Such dissatisfaction can lead to damaging behaviors, such as starvation diets, diet pill abuse, and health-care avoidance. The first step toward making healthy changes may be self-acceptance. Keep in mind that fashion is fickle; the body shapes valued by our society change with time. Furthermore, body shapes valued by one society differ from those of other societies. The standards defining "ideal" are subjective and may have little in common with health. Table 8-3 offers some tips for adopting health as an ideal.

body composition: the proportions of muscle, bone, fat, and other tissue that make up a person's total body weight.

TABLE 8-3 Tips for Accepting a Healthy Body Weight

- Value yourself and others for human attributes other than body weight. Realize that prejudging people by weight is as harmful as prejudging them by race, religion, or gender.
- Use positive, nonjudgmental descriptions of your body.
- Accept positive comments from others.
- Focus on your whole self including your intelligence, social grace, and professional and scholastic achievements.
- Accept that no magic diet exists.
- Stop dieting to lose weight. Adopt a lifestyle of healthy eating and physical activity permanently.
- Follow the USDA Food Patterns. Never restrict food intake below the minimum levels that meet nutrient needs.
- Become physically active, not because it will help you get thin but because it will make you feel good and improve your health.
- Seek support from loved ones. Tell them of your plan for a healthy life in the body you have been given.
- Seek professional counseling, *not* from a weight-loss counselor, but from someone who can help you make gains in self-esteem without weight as the primary focus.
- Appreciate body weight for its influence on health, not appearance.

*In metric terms, a person 1.78 meters tall who weights 68 kilograms may carry only about 14 of those kilograms as fat.

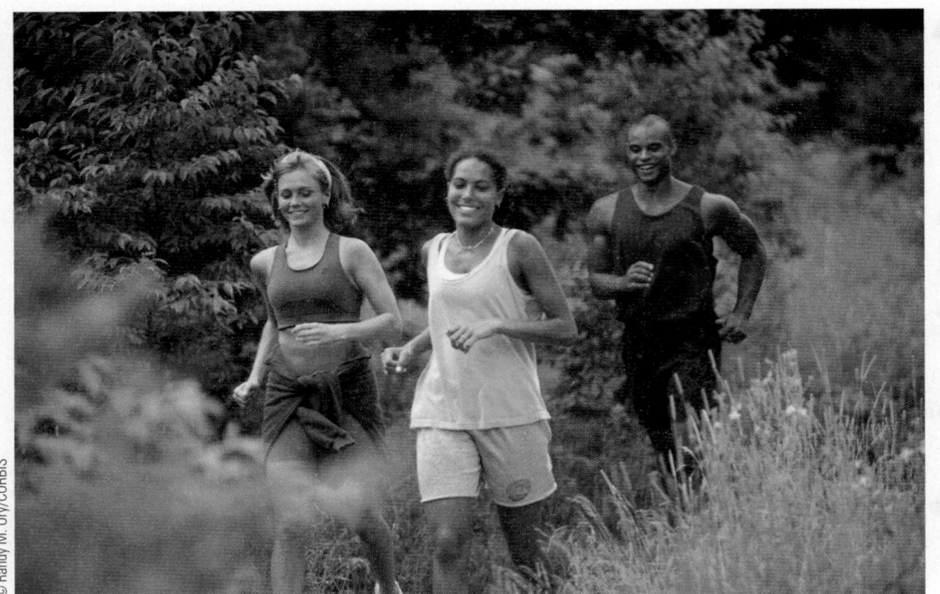

© Randy M. Ury/CORBIS

A healthy body contains enough lean tissue to support health and the right amount of fat to meet body needs.

The Criterion of Health Even if our society were to accept fat as beautiful, obesity is still a major risk factor for several life-threatening diseases, including heart disease, type 2 diabetes, and some cancers.[10] For this reason, the most important criterion for determining how much a person should weigh and how much body fat a person needs is not appearance but good health and longevity. Ideally, a person has enough fat to meet basic needs ♦ but not so much as to incur health risks. This range of healthy body weights has been identified using a common measure of weight and height—the body mass index.

Body Mass Index The **body mass index (BMI)** describes relative weight for height: ♦

$$\text{BMI} = \frac{\text{weight (kg)}}{\text{height (m)}^2} \text{ or } \frac{\text{weight (lb)}}{\text{height (in)}^2} \times 703$$

♦ Fat in the body:

- Provides energy
- Insulates against temperature extremes
- Protects against physical shock
- Forms cell membranes
- Makes compounds such as hormones, vitamin D, and bile

♦ To convert pounds to kilograms:
lb ÷ 2.2 lb/kg = kg
To convert inches to meters:
in ÷ 39.37 in/m = m

Weight classifications based on BMI are presented in Table 8-4. Notice that healthy weight falls between a BMI of 18.5 and 24.9, with **underweight** below 18.5, **overweight** above 25, and **obese** above 30. Figure 8-6 shows examples of body shapes with different BMI. More than two-thirds of adults in the United States have a BMI greater than 25, as Figure 8-7 shows.[11]

Obesity-related diseases become evident beyond a BMI of 25. For this reason, a BMI of 25 for adults represents a healthy goal for overweight people and an upper limit for others. The lower end of the healthy range may be a reasonable target for severely underweight people. BMI values slightly below the healthy range may be compatible with good health if food intake is adequate, but signs of illness, reduced work capacity, and poor reproductive function

> FIGURE 8-6 BMI and Body Shapes

Standard silhouette figures such as those shown below are commonly used in research studies (without the BMI numbers) to determine how accurately people perceive their body size.

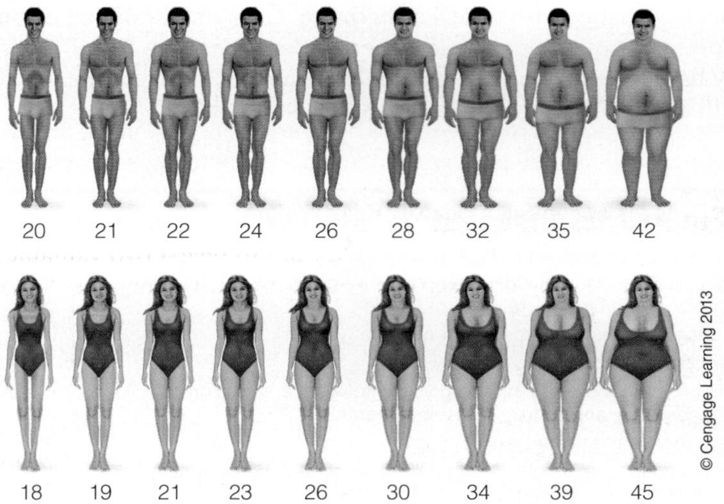

Source: A. J. Stunkard, T. Sorensen, and F. Schulsinger, Use of the Danish Adoption Register for the study of obesity and thinness, *Research Publications: Association for Research in Nervous and Mental Disorders* 60 (1983): 115–120.

body mass index (BMI): a measure of a person's weight relative to height; determined by dividing the weight (in kilograms) by the square of the height (in meters).

underweight: body weight lower than the weight range that is considered healthy; BMI less than 18.5.

overweight: body weight greater than the weight range that is considered healthy; BMI 25 to 29.9.

obese: too much body fat with adverse health effects; BMI 30 or more.

TABLE 8-4 Body Mass Index (BMI)

	Underweight (<18.5)	Healthy Weight (18.5–24.9)						Overweight (25–29.9)					Obese (≥30)										
	18	19	20	21	22	23	24	25	26	27	28	29	30	31	32	33	34	35	36	37	38	39	40
Height	***Body weight (pounds)***																						
4′10″	86	91	96	100	105	110	115	119	124	129	134	138	143	148	153	158	162	167	172	177	181	186	191
4′11″	89	94	99	104	109	114	119	124	128	133	138	143	148	153	158	163	168	173	178	183	188	193	198
5′0″	92	97	102	107	112	118	123	128	133	138	143	148	153	158	163	168	174	179	184	189	194	199	204
5′1″	95	100	106	111	116	122	127	132	137	143	148	153	158	164	169	174	180	185	190	195	201	206	211
5′2″	98	104	109	115	120	126	131	136	142	147	153	158	164	169	175	180	186	191	196	202	207	213	218
5′3″	102	107	113	118	124	130	135	141	146	152	158	163	169	175	180	186	191	197	203	208	214	220	225
5′4″	105	110	116	122	128	134	140	145	151	157	163	169	174	180	186	192	197	204	209	215	221	227	232
5′5″	108	114	120	126	132	138	144	150	156	162	168	174	180	186	192	198	204	210	216	222	228	234	240
5′6″	112	118	124	130	136	142	148	155	161	167	173	179	186	192	198	204	210	216	223	229	235	241	247
5′7″	115	121	127	134	140	146	153	159	166	172	178	185	191	198	204	211	217	223	230	236	242	249	255
5′8″	118	125	131	138	144	151	158	164	171	177	184	190	197	203	210	216	223	230	236	243	249	256	262
5′9″	122	128	135	142	149	155	162	169	176	182	189	196	203	209	216	223	230	236	243	250	257	263	270
5′10″	126	132	139	146	153	160	167	174	181	188	195	202	209	216	222	229	236	243	250	257	264	271	278
5′11″	129	136	143	150	157	165	172	179	186	193	200	208	215	222	229	236	243	250	257	265	272	279	286
6′0″	132	140	147	154	162	169	177	184	191	199	206	213	221	228	235	242	250	258	265	272	279	287	294
6′1″	136	144	151	159	166	174	182	189	197	204	212	219	227	235	242	250	257	265	272	280	288	295	302
6′2″	141	148	155	163	171	179	186	194	202	210	218	225	233	241	249	256	264	272	280	287	295	303	311
6′3″	144	152	160	168	176	184	192	200	208	216	224	232	240	248	256	264	272	279	287	295	303	311	319
6′4″	148	156	164	172	180	189	197	205	213	221	230	238	246	254	263	271	279	287	295	304	312	320	328
6′5″	151	160	168	176	185	193	202	210	218	227	235	244	252	261	269	277	286	294	303	311	319	328	336
6′6″	155	164	172	181	190	198	207	216	224	233	241	250	259	267	276	284	293	302	310	319	328	336	345

become apparent when BMI is below 17. The "How To" feature on p. 244 describes how to determine your BMI and how to find a goal weight based on a desired BMI.

Keep in mind that BMI reflects height and weight measures and not body composition. Consequently, muscular athletes may be classified as over*weight* by BMI standards and not be over*fat*. At the peak of his bodybuilding career, Arnold Schwarzenegger won the Mr. Olympia competition with a BMI of 31; the runner on p. 241 also has a BMI greater than 30. Yet neither would be considered obese. Striking differences in body composition are also apparent among people of different ages and various ethnic and racial groups, making standard BMI guidelines inappropriate for some populations.[12] For example, blacks tend to have a greater bone density and protein content than whites; consequently, using BMI as the standard may overestimate the prevalence of overweight and obesity among blacks.

> FIGURE 8-7 Distribution of Body Weights in US Adults

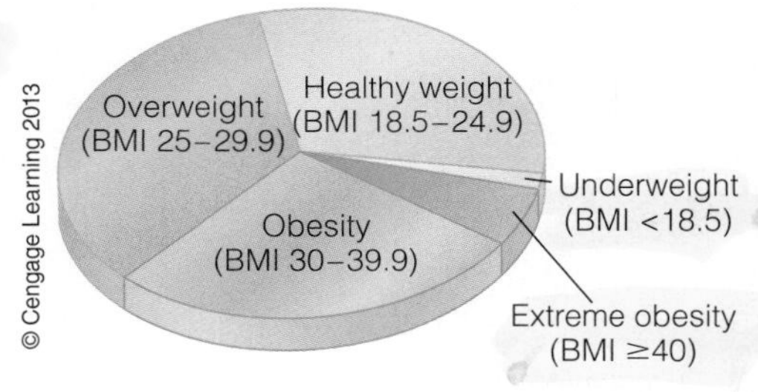

Body Fat and Its Distribution

Although weight measures are inexpensive, easy to take, and highly accurate, they fail to reveal two valuable pieces of information in assessing disease risk: how much of the weight is fat and where the fat is located. The ideal amount of body fat depends partly on the person. Table 8-5 (p. 244) compares percent body fat values of healthy weight, average fitness individuals with averages from national surveys.

Some People Need Less Body Fat

For many athletes, a lower percentage of body fat may be ideal—just enough fat to provide fuel, insulate and protect the body, assist in nerve impulse transmissions, and support normal hormone activity, but not so much as to burden the body with excess bulk. Percent body fat for athletes, then, might be 7 to 16 percent for young men and 15 to 22 percent for young

TABLE 8-5 Percent Body Fat

Age (yr)	Ideal (Healthy weight, average fitness)	Actual (US average)
Male		
20–39	18–21	26
40–59	22–25	29
60+	24–27	31
Female		
20–39	23–26	38
40–59	28–32	41
60+	31–34	42

SOURCE: L. G. Borrud and coauthors, Body composition data for individuals 8 years of age and older: US population, 1999–2004, *Vital and Health Statistics* 11 (2010): 1–87; *ACSM's Health-Related Physical Fitness Assessment Manual*, 2nd ed. (Baltimore, M.D.: Lippincott Williams & Wilkins, 2008), p. 59.

>How To

Determine BMI

To calculate your body mass index (BMI), use one of the following equations:

$$BMI = \frac{\text{weight (lb)}}{\text{height (in)}^2} \times 703$$

or

$$BMI = \frac{\text{weight (kg)}}{\text{height (m)}^2}$$

Consider, for example, a person who is 5′5″ (1.65 m) tall and weighs 174 lb (79 kg):

$$BMI = \frac{174 \text{ lb}}{65 \text{ in}^2} \times 703 = 29$$

or

$$BMI = \frac{79 \text{ kg}}{1.65 \text{ m}^2} = 29$$

This person has a BMI of 29 and is considered overweight.

You could also use Table 8-4 to determine your BMI. Locate your height in the first column (in this example, 5′5″). Then look across the row until you find the number that is closest to your weight (in this example, 174). The number at the top of that column identifies your BMI (in this example, 29).

A reasonable initial target for most overweight people is a BMI 2 units below their current one. To determine a goal weight based on a desired BMI, locate your height in the first column and then look across the row until you reach the column with the desired BMI at the top. In this example, to reach a BMI of 27, this person's goal weight is 162 pounds, which represents a 12-pound weight loss. Such a determination can help a person set realistic weight goals using health risk as a guide.

TRY IT Calculate your BMI and determine whether you are underweight, healthy weight, overweight, or obese. If your BMI is less than 18.5 or greater than 25, identify a weight that takes your BMI 2 units closer to the healthy weight range.

women.[13] (Review the photo on p. 241 to appreciate what 8 percent body fat looks like—even with a BMI greater than 30.)

Some People Need More Body Fat For an Alaska fisherman, a higher percentage of body fat is probably beneficial because fat provides an insulating blanket to prevent excessive loss of body heat in cold climates. A woman starting a pregnancy needs sufficient body fat to support conception and fetal growth. Below a certain threshold for body fat, hormone synthesis falters, and individuals may become infertile, develop depression, experience abnormal hunger regulation, or become unable to keep warm. These thresholds differ for each function and for each individual; much remains to be learned about them.

Fat Distribution The location of fat on the body may influence health as much, or more than, total fat alone. **Visceral fat** that is stored around the organs of the abdomen is referred to as **central obesity** or upper-body fat (see Figure 8-8). Much research supports the widely held belief that central obesity—significantly and independently of BMI—contributes to heart disease and related deaths.[14] Some research, however, casts doubt and suggests that *all* types of obesity are linked to heart disease; central obesity is no riskier than other shapes.[15]

Visceral fat is most common in men and to a lesser extent in women past menopause. Even when total body fat is similar, men have more visceral fat than women. **Subcutaneous fat** around the hips and thighs, sometimes referred to as lower-body fat, is most common in women during their reproductive years. Figure 8-9 compares the body shapes of people with upper-body fat and lower-body fat.

Waist Circumference A person's **waist circumference** is a good indicator of central obesity. In general, women with a waist circumference of greater than 35 inches (88 centimeters) and men with a waist circumference of greater than 40 inches (102 centimeters) have a high risk of central obesity–related

visceral fat: fat stored within the abdominal cavity in association with the internal abdominal organs; also called *intra-abdominal fat.*

central obesity: excess fat around the trunk of the body; also called *abdominal fat* or *upper-body fat.*

subcutaneous fat: fat stored directly under the skin.
- **sub** = beneath
- **cutaneous** = skin

waist circumference: an anthropometric measurement used to assess a person's abdominal fat.

> FIGURE 8-8 **Abdominal Fat**

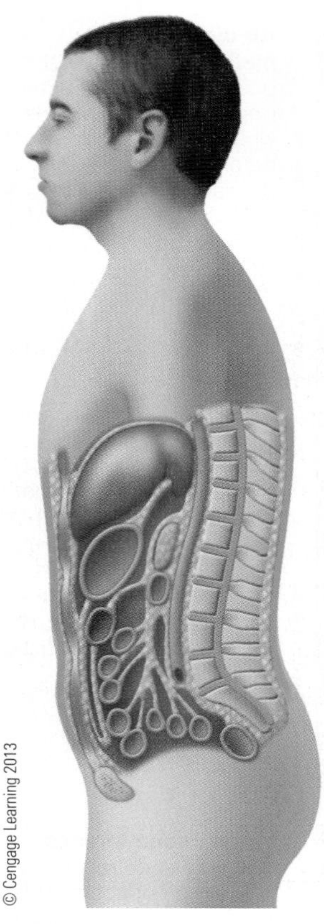

In healthy-weight people, some fat is stored around the organs of the abdomen.

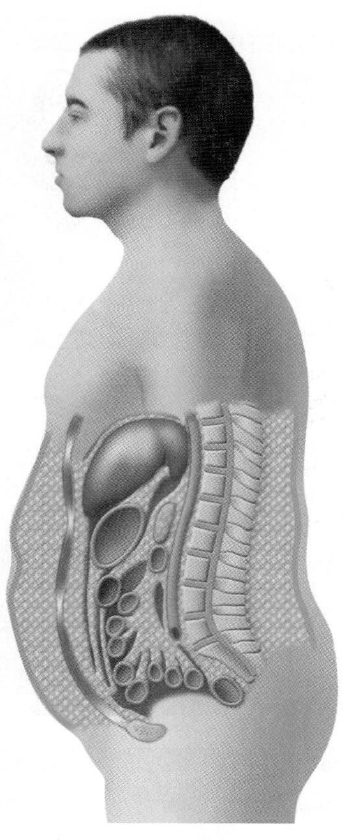

In overweight people, excess abdominal fat increases the risks of diseases.

health problems. As waist circumference increases, disease risks increase. Appendix E includes instructions for measuring waist circumference and assessing abdominal fat.

Some researchers use the waist-to-hip ratio as an indicator of disease risks. The ratio requires another step or two (measuring the hips and comparing that measurement to the waist measurement), but it does not provide any additional information. Therefore, waist circumference alone is the preferred method for assessing abdominal fat in a clinical setting.*

Other Measures of Body Composition Health-care professionals commonly use BMI and waist circumference measurements because they are relatively easy and inexpensive. Together, these two measurements prove most valuable in assessing a person's health risks and monitoring changes over time.[16] Researchers needing more precise measures of body composition may choose any of several other techniques to estimate body fat and its distribution (see Figure 8-10). Mastering these techniques requires proper instruction and practice to ensure reliability. In addition to the methods shown in Figure 8-10, researchers sometimes estimate body composition using these methods: total body water, radioactive potassium count, near-infrared spectrophotometry, ultrasound, computed tomography, and magnetic resonance imaging. Each method has advantages and disadvantages with respect to cost, technical difficulty, and precision of estimating body fat. Appendix E provides additional details and includes many of the tables and charts routinely used in assessment procedures.

*The National Heart, Lung, and Blood Institute recommends using the waist circumference instead of the waist-to-hip ratio to assess obesity health risks.

> FIGURE 8-9 **"Apple" and "Pear" Body Shapes Compared**

Popular articles sometimes call bodies with upper-body fat "apples" and those with lower-body fat, "pears." Researchers sometimes refer to upper-body fat as "android" (manlike) obesity and to lower-body fat as "gynoid" (womanlike) obesity.

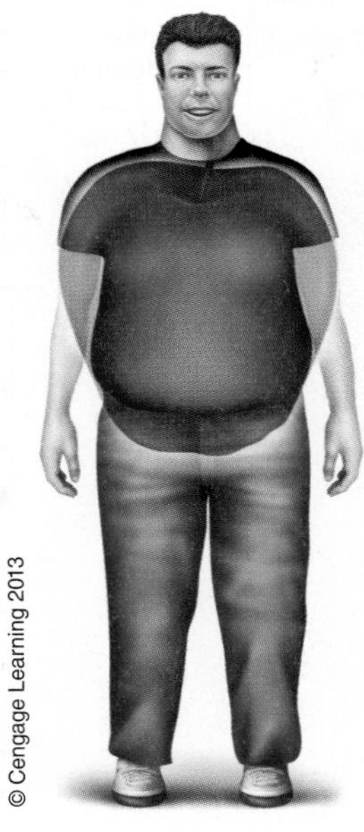

Upper-body fat is more common in men than in women and may be more closely associated with chronic diseases.

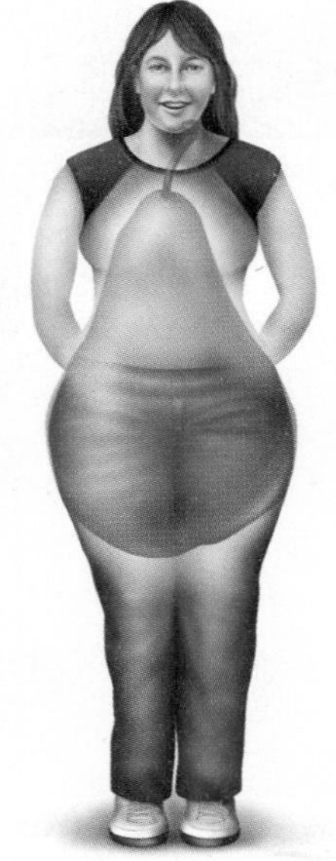

Lower-body fat is more common in women than in men and is not usually associated with chronic diseases.

SOURCE: R.E.C. Wildman and D. M. Medeiros, *Advanced Human Nutrition* (Boca Raton, FL: CRC Press, 2000), pp. 321–323. Copyright © 2000 Taylor and Francis Books LLC. Reprinted with permission.

> FIGURE 8-10 **Common Methods Used to Assess Body Fat**

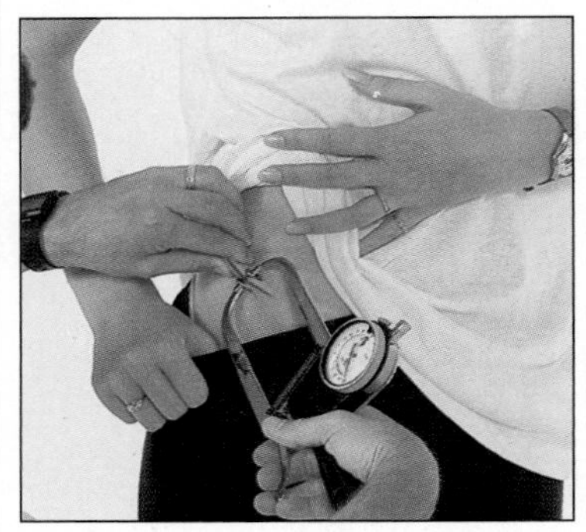

© Fitness & Wellness, Boise, Idaho

Skinfold measures estimate body fat by using a caliper to gauge the thickness of a fold of skin on the back of the arm (over the triceps), below the shoulder blade (subscapular), and in other places (including lower-body sites), and then comparing these measurements with standards.

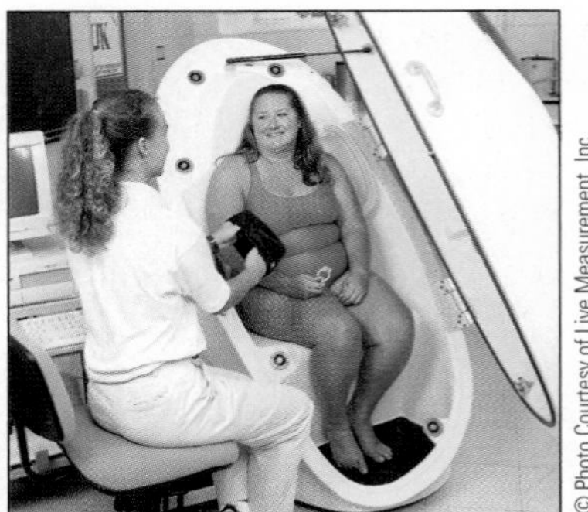

© Photo Courtesy of Live Measurement, Inc.

Air displacement plethysmography estimates body composition by having a person sit inside a chamber while computerized sensors determine the amount of air displaced by the person's body.

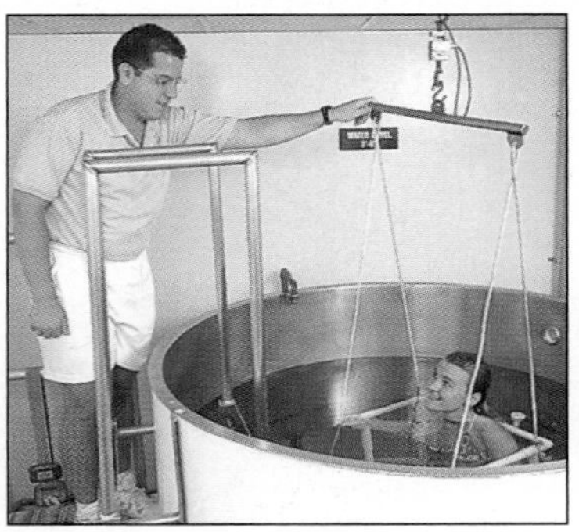

© David Young-Wolff/PhotoEdit

Hydrodensitometry measures body density by weighing the person first on land and then again while submerged in water. The difference between the person's actual weight and underwater weight provides a measure of the body's volume. A mathematical equation using the two measurements (volume and actual weight) determines body density, from which the percentage of body fat can be estimated.

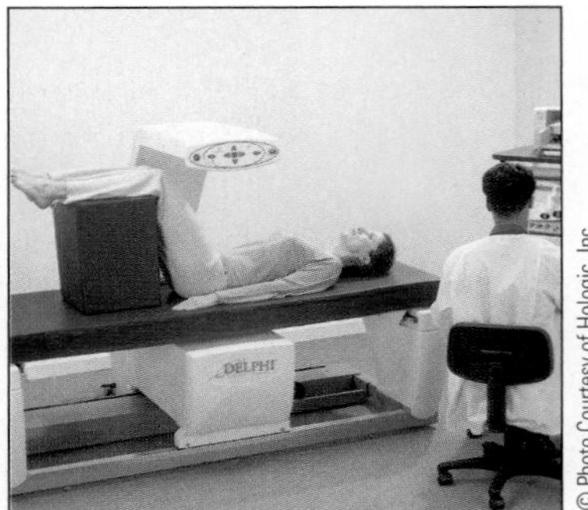

© Photo Courtesy of Hologic, Inc.

Dual energy X-ray absorptiometry (DEXA) uses two low-dose X-rays that differentiate among fat-free soft tissue (lean body mass), fat tissue, and bone tissue, providing a precise measurement of total fat and its distribution in all but extremely obese subjects.

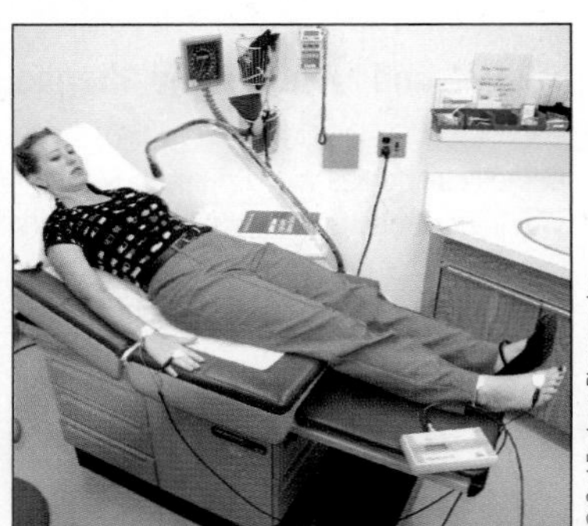

© Geri Engberg Photography

Bioelectrical impedance measures body fat by using a low-intensity electrical current. Because electrolyte-containing fluids, which readily conduct an electrical current, are found primarily in lean body tissues, the leaner the person, the less resistance to the current. The measurement of electrical resistance is then used in a mathematical equation to estimate the percentage of body fat.

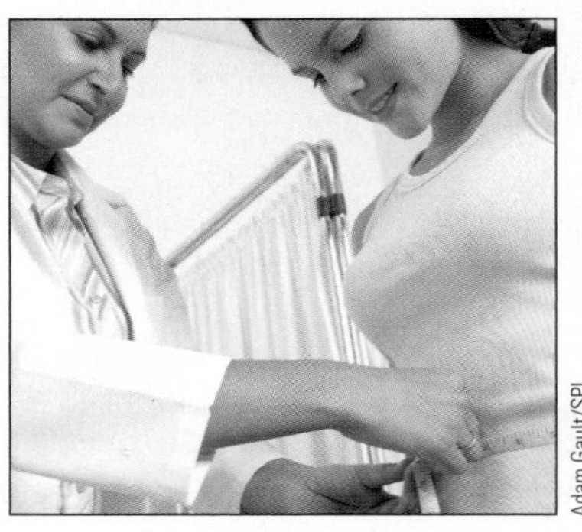

Adam Gault/SPL

Waist circumference measures central obesity by placing a nonstretchable measuring tape around the waist just above the bony crest of the hip. The tape runs parallel to the floor and is snug, but does not compress the skin.

REVIEW IT Distinguish between body weight and body composition, including methods to assess each.

The body mass index (BMI) is based on weight in kilograms (kg) relative to height in meters (m) squared. As a useful measure of a person's weight status, the BMI serves as a reliable indicator of chronic disease risks, but it says little about body composition. The ideal amount of body fat varies from person to person, but researchers have found that body fat in excess of 22 percent for young men and 27 percent for young women (the levels rise slightly with age) poses health risks. Central obesity is measured by waist circumference and indicates excess abdominal fat distributed around the trunk of the body. Central obesity contributes to chronic diseases, but whether those risks are greater than fat elsewhere in the body is uncertain.

8.5 Health Risks Associated with Body Weight and Body Fat

LEARN IT Identify relationships between body weight and chronic diseases.

Body weight and body fat correlate with disease risks and life expectancy. The correlation suggests a greater *likelihood* of developing a chronic disease and shortening life expectancy for those with a higher BMI.[17] Not all overweight and underweight people will get sick and die before their time nor will all normal-weight people live long healthy lives. *Correlations* are not *causes*. For the most part though, people with a BMI between 18.5 and 24.9 have relatively few weight-related health risks; risks increase as BMI falls below or rises above this range, indicating that both too little and too much body fat impair health.[18] Epidemiological data show a J- or U-shaped relationship between body weights and mortality (see Figure 8-11).[19] People who are extremely underweight or extremely obese carry higher risks of early deaths ♦ than those whose weights fall within the healthy, or even the slightly overweight, range.[20] These mortality risks decline with age.

Independently of BMI, factors such as smoking habits raise health risks, and physical fitness lowers them. A man with a BMI of 22 who smokes two packs of cigarettes a day is jeopardizing his health, whereas a woman with a BMI of 32 who walks briskly for an hour a day is improving her health.

> FIGURE 8-11 **BMI and Mortality**

This J-shaped curve describes the relationship between body mass index (BMI) and mortality and shows that both underweight and overweight present risks of a premature death.

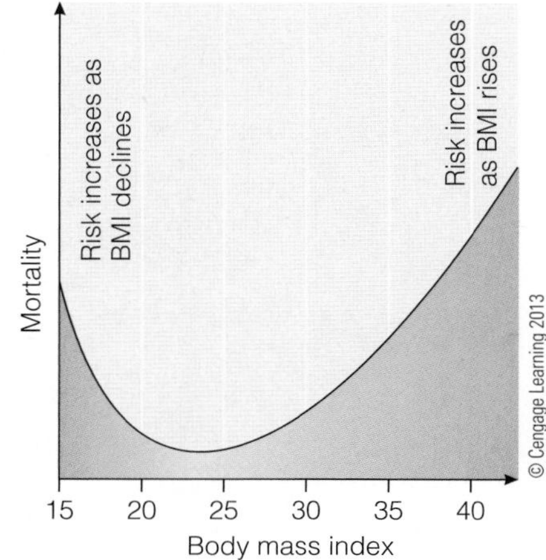

♦ BMI and mortality:
- BMI 22.5–25 = optimal survival
- BMI 30–35 = 3 years' loss of life
- BMI >40 = 10 years' loss of life (equivalent to lifetime of smoking)

Health Risks of Underweight Some underweight people enjoy an active, healthy life, but others are underweight because of malnutrition, smoking habits, substance abuse, or illnesses. Weight and fat measures alone would not reveal these underlying causes, but a complete assessment that includes a diet and medical history, physical examination, and biochemical analysis would.

An underweight person, especially an older adult, may be unable to preserve lean tissue during the fight against a wasting disease such as cancer or a digestive disorder, especially when the disease is accompanied by malnutrition. Without adequate nutrient and energy reserves, an underweight person will have a particularly tough battle against such medical stresses. Underweight women develop menstrual irregularities and become infertile. Those who do conceive may give birth to unhealthy infants. An underweight woman can improve her chances of having a healthy infant by gaining weight prior to conception, during pregnancy, or both. Underweight and significant weight loss are also associated with osteoporosis and bone fractures. For all these reasons, underweight people may benefit from enough of a weight gain to provide an energy reserve and protective amounts of all the nutrients.

Health Risks of Overweight As for excessive body fat, the health risks are so many that it has been designated a disease—obesity. Among the health risks associated with obesity are diabetes, hypertension, cardiovascular disease, sleep apnea (abnormal ceasing of breathing during sleep), osteoarthritis, some cancers, gallbladder disease, kidney stones, respiratory problems (including Pickwickian syndrome, a breathing blockage linked with sudden death), infertility, and complications in pregnancy and surgery. Obese people are more likely to be disabled in their later years. Each year, these obesity-related illnesses cost our nation $147 billion—in fact, as much as, or more than, the medical costs of smoking.[21] An additional $72 billion is estimated in a loss of productivity at work due to mortality and disability.[22]

The cost in terms of lives is also great. In fact, obesity is second only to tobacco in causing premature deaths.[23] Mortality is lowest when BMI is in the optimal range of 20.0 to 24.9; it increases as excess weight increases.[24]

Equally important, both central obesity and weight gains of more than 20 pounds (9 kilograms) between early and middle adulthood correlate with increased disease risks. Fluctuations in body weight, as typically occur with "yo-yo" dieting, may also increase the risks of chronic diseases and premature death. In contrast, sustained weight loss improves physical well-being, reduces disease risks, and increases life expectancy.

Cardiovascular Disease The relationship between obesity and cardiovascular disease risk is strong, with links to both elevated blood cholesterol and hypertension.[25] Central obesity may raise the risk of heart attack and stroke as much as the three leading risk factors (high LDL cholesterol, hypertension, and smoking) do. ♦ In addition to body fat, weight gain also increases the risk of cardiovascular

♦ Cardiovascular disease risk factors associated with obesity:
- High LDL cholesterol
- Low HDL cholesterol
- High blood pressure (hypertension)
- Diabetes

© Craig Stephen / Alamy

Smoking is the leading cause of preventable illnesses and early deaths. Obesity is a close second.

disease. Weight loss, on the other hand, can effectively reverse atherosclerosis and lower both blood cholesterol and blood pressure in overweight and obese people.[26] Of course, lean and normal-weight people may also have high blood cholesterol and blood pressure, and these factors are just as dangerous in lean people as in obese people.

Type 2 Diabetes Most adults with type 2 diabetes are overweight or obese.[27] Type 2 diabetes is three times more likely to develop in an obese person than in a nonobese person. Furthermore, the person with type 2 diabetes often has central obesity. Central-body fat cells appear to be larger and more insulin-resistant than lower-body fat cells. The association between **insulin resistance** and obesity is strong, and both are major risk factors for the development of type 2 diabetes.

Diabetes appears to be influenced by weight gains as well as by body weight. A weight gain of more than 10 pounds (4.5 kilograms) after the age of 18 doubles the risk of developing diabetes, even in adults of average weight. In contrast, weight loss is effective in improving glucose tolerance and insulin resistance.

Inflammation and the Metabolic Syndrome Chronic **inflammation** accompanies obesity, and inflammation contributes to chronic diseases.[28] As a person grows fatter, lipids first fill the adipose tissue and then migrate into other tissues such as the muscles and liver. This accumulation of fat, especially in the abdominal region, changes the body's metabolism, resulting in insulin resistance, low HDL, high triglycerides, and high blood pressure.[29] This cluster of symptoms—collectively known as the metabolic syndrome—increases the risks for diabetes, hypertension, and atherosclerosis. ♦ Fat accumulation, especially in the abdominal region, activates genes that code for proteins (adipokines) involved in inflammation.[30] Furthermore, although relatively few immune cells are commonly found in adipose tissue, weight gain significantly increases their number and their role in inflammation. Elevated blood lipids—whether due to obesity or to a high-fat diet—also promote inflammation. Together, these factors help to explain why chronic inflammation accompanies obesity and how obesity contributes to the metabolic syndrome and the progression of chronic diseases.[31] Even in healthy youngsters, body fat correlates positively with chronic inflammation. As might be expected, weight loss reduces

♦ Metabolic syndrome is a cluster of at least three of the following risk factors:
- High blood pressure
- High blood glucose
- High blood triglycerides
- Low HDL cholesterol
- High waist circumference

Chapter 18 provides many more details.

insulin resistance: the condition in which a normal amount of insulin produces a subnormal effect in muscle, adipose, and liver cells, resulting in an elevated fasting glucose; a metabolic consequence of obesity that precedes type 2 diabetes.

inflammation: an immunological response to cellular injury characterized by an increase in white blood cells.

the number of immune cells in adipose tissue and changes gene expression to reduce inflammation.

Cancer The risk of some cancers increases with both body weight and weight gain, but researchers do not fully understand the relationships. One possible explanation may be that obese people have elevated levels of hormones that could influence cancer development. For example, adipose tissue is the major site of estrogen synthesis in women, obese women have elevated levels of estrogen, and estrogen has been implicated in the development of cancers of the female reproductive system—cancers that account for half of all cancers in women.

© Steven Frame/Alamy

Being active—even if overweight—is healthier than being sedentary.

Fit and Fat versus Sedentary and Slim Importantly, BMI and weight gains and losses do not tell the whole story. Cardiorespiratory and muscular fitness play major roles in health and longevity, independently of body weight.[32] Normal-weight people who are fit have a lower risk of mortality than normal-weight people who are unfit. Furthermore, overweight but fit people have lower risks than normal-weight, unfit ones.[33] Fit people are also likely to gain less weight over the years. Clearly, a healthy body weight is good, but it may not be good enough. Fitness, in and of itself, offers many health benefits, as Chapter 14 confirms.

REVIEW IT Identify relationships between body weight and chronic diseases.

The weight appropriate for an individual depends largely on factors specific to that individual, including body fat distribution, family health history, and current health status. At the extremes, both overweight and underweight carry clear risks to health.

This chapter has described energy balance and body composition with a focus on the health problems associated with too much or too little body weight and body fat. Highlight 8 examines the health problems that arise when efforts to control body weight become eating disorders. The next chapter continues the discussion with a look at weight management and the benefits of choosing nutritious foods and exercising regularly.

Nutrition Portfolio

When combined with fitness, a healthy body weight will help you to defend against chronic diseases. Go to Diet Analysis Plus and choose one of the days on which you have tracked your diet for the entire day. Go to the Energy Balance report; use this report to help you answer the following questions:

- Describe how your daily food intake and physical activity balance with each other.
- What did the diet analysis program estimate as your daily energy requirement? What information was this based on?
- Describe any health risks that may be of concern for a person who continuously has very high "net kcalories" or very low "net kcalories" for many years?

Diet Analysis PLUS — To complete this exercise, go to your Diet Analysis Plus at **www.cengagebrain.com**.

STUDY IT To review the key points of this chapter and take a practice quiz, go to the study cards at the end of the book.

REFERENCES

1. M. B. Katan and D. S. Ludwig, Extra calories cause weight gain: But how much? *Journal of the American Medical Association* 303 (2010): 65–66.
2. D. A. Schoeller, The energy balance equation: Looking back and looking forward are two very different views, *Nutrition Reviews* 67 (2009): 249–254.
3. E. R. Grimm and N. I. Steinle, Genetics of eating behavior: Established and emerging concepts, *Nutrition Reviews* 69 (2011): 52–60; M. de Krom and coauthors, Genetic variation and effects on human eating behavior, *Annual Review of Nutrition* 29 (2009): 283–304.
4. D. Ferriday and J. M. Brunstrom, "I just can't help myself": Effects of food-cue exposure in overweight and lean individuals, *International Journal of Obesity* 35 (2011): 142–149; L. B. Shomaker and coauthors, Eating in the absence of hunger in adolescents: Intake after a large-array meal compared with that after a standardized meal, *American Journal of Clinical Nutrition* 92 (2010): 697–703.
5. M. S. Westerterp-Plantenga and coauthors, Dietary protein, weight loss, and weight maintenance, *Annual Review of Nutrition* 29 (2009): 21–41.
6. J. A. Gilbert and coauthors, Milk supplementation facilitates appetite control in obese women during weight loss: A randomized, single-blind, placebo-controlled trial, *British Journal of Nutrition* 105 (2011): 133–143; E. R. Dove and coauthors, Skim milk compared with a fruit drink acutely reduces appetite and intake in overweight men and women, *American Journal of Clinical Nutrition* 90 (2009): 70–75.
7. M. D. Lane and S. H. Cha, Effect of glucose and fructose on food intake via malonyl-CoA signaling in the brain, *Biochemical and Biophysical Research Communications* 382 (2009): 1–5.
8. H. Schloegl and coauthors, Peptide hormones regulating appetite: Focus on neuroimaging studies in humans, *Diabetes/Metabolism Research and Reviews* 27 (2011): 104–112; K. A. Simpson and S. R. Bloom, Appetite and hedonism: Gut hormones and the brain, *Endocrinology and Metabolism Clinics of North America* 39 (2010): 729–743; S. Zac-Varghese, T. Tan, and S. R. Bloom, Hormonal interactions between gut and brain, *Discovery Medicine* 10 (2010): 543–552; E. Valassi, M. Scacchi, and F. Cavagnini, Neuroendocrine control of food intake, *Nutrition, Metabolism, and Cardiovascular Disease* 18 (2008): 158–168.
9. D. Anschutz and coauthors, Watching your weight? The relations between watching soaps and music television and body dissatisfaction and restrained eating in young girls, *Psychology and Health* 24 (2009): 1035–1050; M. J. Hogan and V. C. Strasburger, Body image, eating disorders, and the media, *Adolescent Medicine: State of the Art Reviews* 19 (2008): 521–546.
10. W. V. Brown and coauthors, Obesity: Why be concerned? *The American Journal of Medicine* 122 (2009): S4–S11; D. P. Guh and coauthors, The incidence of co-morbidities related to obesity and overweight: A systematic review and meta-analysis, *BioMed Central Public Health* 25 (2009): 88.
11. K. M. Flegal and coauthors, Prevalence and trends in obesity among US adults, 1999–2008, *Journal of the American Medical Association* 303 (2010): 235–241.
12. R. Huxley and coauthors, Ethnic comparisons of the cross-sectional relationships between measures of body size with diabetes and hypertension, *Obesity Reviews* 9 (2008): 53–61.
13. ACSM's *Health-Related Physical Fitness Assessment Manual,* 2nd ed., (Baltimore, M.D.: Lippincott Williams & Wilkins, 2008), p. 59.
14. B. J. Arsenault and coauthors, Physical inactivity, abdominal obesity and risk of coronary heart disease in apparently healthy men and women, *International Journal of Obesity* 34 (2010): 340–347; E. J. Jacobs and coauthors, Waist circumference and all-cause mortality in a large US cohort, *Archives of Internal Medicine* 170 (2010): 1293–1301; S. S. Dhaliwal and T. A. Welborn, Central obesity and multivariable cardiovascular risk as assessed by the Framingham prediction scores, *American Journal of Cardiology* 103 (2009): 1403–1407; E. B. Levitan and coauthors, Adiposity and incidence of heart failure hospitalization and mortality: A population-based prospective study, *Circulation Heart Failure* 2 (2009): 202–208; J. P. Després and coauthors, Abdominal obesity and the metabolic syndrome: Contribution to global cardiometabolic risk, *Arteriosclerosis, Thrombosis, and Vascular Biology* 28 (2008): 1039–1049.
15. D. Wormser and coauthors, Separate and combined associations of body-mass index and abdominal adiposity with cardiovascular disease: Collaborative analysis of 58 prospective studies, *The Lancet* 377 (2011): 1085–1095; K. M. Flegal and B. I. Graubard, Estimates of excess deaths associated with body mass index and other anthropometric variables, *American Journal of Clinical Nutrition* 89 (2009): 1213–1219.
16. Position of the American Dietetic Association: Weight management, *Journal of the American Dietetic Association* 109 (2009): 330–346.
17. S. T. Stewart, D. M. Cutler, and A. B. Rosen, Forecasting the effects of obesity and smoking on US life expectancy, *New England Journal of Medicine* 361 (2009): 2252–2260.
18. Flegal and Graubard, 2009.
19. A. Berrington de Gonzalez and coauthors, Body-mass index and mortality among 1.46 million white adults, *New England Journal of Medicine* 363 (2010): 2211–2219; G. Whitlock and coauthors, Body-mass index and cause-specific mortality in 900,000 adults: Collaborative analyses of 57 prospective studies, *Lancet* 373 (2009): 1083–1096; T. Pischon and coauthors, General and abdominal adiposity and risk of death in Europe, *New England Journal of Medicine* 359 (2008): 2105–2120.
20. W. Zheng and coauthors, Association between body-mass index and risk of death in more than 1 million Asians, *New England Journal of Medicine* 364 (2011): 719–729.
21. E. A. Finkelstein and coauthors, Annual medical spending attributable to obesity: Payer- and service-specific estimates, *Health Affairs* 28 (2009): W822–W831.
22. Society of Actuaries, Obesity and its relation to mortality and morbidity costs, www.soa.org/files/pdf/research-2011-obesity-relation-mortality.pdf.
23. H. Jia and E. I. Lubetkin, Trends in quality-adjusted life-years lost contributed by smoking and obesity, *American Journal of Preventive Medicine* 38 (2010): 138–144.
24. Berrington de Gonzalez and coauthors, 2010; Whitlock and coauthors, 2009.
25. C. J. Lavie , R. V. Milani, and H. O. Ventura, Obesity and cardiovascular disease: Risk factor, paradox, and impact of weight loss, *Journal of the American College of Cardiology* 53 (2009): 1925–1932; R. P. Bogers and coauthors, Association of overweight with increased risk of coronary heart disease partly independent of blood pressure and cholesterol levels: A meta-analysis of 21 cohort studies including more than 300,000 persons, *Archives of Internal Medicine* 167 (2007): 1720–1728.
26. I. Shai and coauthors, Dietary intervention to reverse carotid atherosclerosis, *Circulation* 121 (2010): 1200–1208.
27. M. L. Biggs and coauthors, Association between adiposity in midlife and older age and risk of diabetes in older adults, *Journal of the American Medical Association* 303 (2010): 2504–2512.
28. B. B. Aggarwal, Targeting inflammation-induced obesity and metabolic diseases by curcumin and other nutraceuticals, *Annual Review of Nutrition* 30 (2010): 173–199; J. A. Alvarez and coauthors, Fasting and postprandial markers of inflammation in lean and overweight children, *American Journal of Clinical Nutrition* 89 (2009): 1138–1144; A. Cartier and coauthors, Sex differences in inflammatory markers: What is the contribution of visceral adiposity? *American Journal of Clinical Nutrition* 89 (2009): 1307–1314; M. Hamer and A. Steptoe, Prospective study of physical fitness, adiposity, and inflammatory markers in healthy middle-aged men and women, *American Journal of Clinical Nutrition* 89 (2009): 85–89; A. W. Fogarty and coauthors, A prospective study of weight change and systemic inflammation over 9 y, *American Journal of Clinical Nutrition* 87 (2008): 30–35.
29. E. W. Demerath, Causes and consequences of human variation in visceral adiposity, *American Journal of Clinical Nutrition* 91 (2010): 1–2.

30. E. Dalmas and coauthors, Variations in circulating inflammatory factors are related to changes in calorie and carbohydrate intakes early in the course of surgery-induced weight reduction, *American Journal of Clinical Nutrition* 94 (2011): 450–458; J. Korner, S. C. Woods, and K. A. Woodworth, Regulation of energy homeostasis and health consequences in obesity, *The American Journal of Medicine* 122 (2009): S12–S18.
31. P. Calabro and E. T. Yeh, Intra-abdominal adiposity, inflammation, and cardiovascular risk: New insight into global cardiometabolic risk, *Current Hypertension Reports* 10 (2008): 32–38; V. Z. Rocha and P. Libby, The multiple facets of the fat tissue, *Thyroid* 18 (2008): 175–183.
32. V. Hainer, H. Toplak, and V. Stich, Fat or fit: What is more important? *Diabetes Care* 32 (2009): S392–S397.
33. D. E. Larson-Meyer and coauthors, Caloric restriction with or without exercise: The fitness versus fatness debate, *Medicine and Science in Sports and Exercise* 42 (2010): 152–159; R. P. Wildman and coauthors, The obese without cardiometabolic risk factor clustering and the normal weight with cardiometabolic risk factor clustering, *Archives of Internal Medicine* 168 (2008): 1617–1624.

HIGHLIGHT > 8

LEARN IT Compare and contrast the diagnoses, characteristics, and treatments of the different eating disorders.

© BananaStock/SuperStock

Eating Disorders

For some people, the struggle with body weight manifests itself as an **eating disorder.** (The accompanying glossary defines this and related terms.) Three eating disorders—anorexia nervosa, bulimia nervosa, and binge eating disorder—are relatively uncommon, but present real concerns because of their health consequences. Findings from a large national survey suggest that 0.9 percent of women and 0.3 percent of men suffer from anorexia nervosa at some time in their lives.[1] Prevalence of bulimia nervosa is slightly higher, with 1.5 percent of women and 0.5 percent of men. Binge eating disorder is higher still, with 3.5 percent of women and 2 percent of men. Many more suffer from other unspecified conditions that, even though they do not meet the strict criteria for an eating disorder, imperil a person's well-being.

Why do so many people in our society suffer from eating disorders? Most experts agree that the causes include multiple factors: sociocultural, psychological, and perhaps neurochemical. Excessive pressure to be thin is at least partly to blame. Family attitudes concerning body shape and eating habits can have profound effects.[2] Young people may have learned to identify discomforts such as anger, jealousy, or disappointment with "feeling fat." They often have other psychological issues such as depression, anxiety, or substance abuse. As weight issues become more of a focus, psychological problems worsen, and the likelihood of developing eating disorders intensifies. Unfortunately, few seek health care for eating disorders. Athletes are among those most likely to develop eating disorders.

The Female Athlete Triad

At age 14, Suzanne was a top contender for a spot on the state gymnastics team. Each day her coach reminded team members that they must weigh no more than their assigned weights to qualify for competition. The coach chastised gymnasts who gained weight, and Suzanne was terrified of being singled out. Convinced that the less she weighed the better she would perform, Suzanne weighed herself several times a day to confirm that she had not exceeded her 80-pound limit. Driven to excel in her sport, Suzanne kept her weight down by eating very little and training very hard. Unlike many of her friends, Suzanne never began to menstruate. A few months before her fifteenth birthday, Suzanne's coach dropped her back to the second-level team. Suzanne blamed her poor performance on a slow-healing stress fracture. Mentally stressed and physically exhausted, she quit gymnastics and began overeating between periods of self-starvation. Suzanne had

GLOSSARY

amenorrhea (ay-MEN-oh-REE-ah): the absence of or cessation of menstruation. *Primary amenorrhea* is menarche delayed beyond 16 years of age. *Secondary amenorrhea* is the absence of three to six consecutive menstrual cycles.

anorexia (an-oh-RECK-see-ah) **nervosa:** an eating disorder characterized by a refusal to maintain a minimally normal body weight and a distortion in perception of body shape and weight.

- **an** = without
- **orex** = mouth
- **nervos** = of nervous origin

binge-eating disorder: an eating disorder with criteria similar to those of bulimia nervosa, excluding purging or other compensatory behaviors.

bulimia (byoo-LEEM-ee-ah) **nervosa:** an eating disorder characterized by repeated episodes of binge eating usually followed by self-induced vomiting, misuse of laxatives or diuretics, fasting, or excessive exercise.

- **buli** = ox

cathartic (ka-THAR-tik): a strong laxative.

disordered eating: eating behaviors that are neither normal nor healthy, including restrained eating, fasting, binge eating, and purging.

eating disorders: disturbances in eating behavior that jeopardize a person's physical or psychological health.

emetic (em-ETT-ic): an agent that causes vomiting.

female athlete triad: a potentially fatal combination of three medical problems—disordered eating, amenorrhea, and osteoporosis.

muscle dysmorphia (dis-MORE-fee-ah): a psychiatric disorder characterized by a preoccupation with building body mass.

stress fractures: bone damage or breaks caused by stress on bone surfaces during exercise.

unspecified eating disorders: eating disorders that do not meet the defined criteria for specific eating disorders.

> **FIGURE H8-1 The Female Athlete Triad**

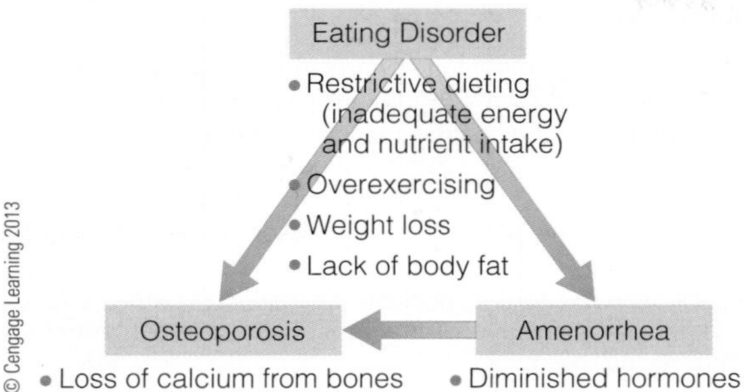

developed the dangerous combination of problems that characterize the **female athlete triad**—disordered eating, amenorrhea, and osteoporosis (see Figure H8-1).[3]

Disordered Eating

Part of the reason many athletes engage in **disordered eating** behaviors may be that they and their coaches have embraced unsuitable weight standards. An athlete's body must be heavier for a given height than a nonathlete's body because the athlete's body is dense, containing more healthy bone and muscle and less fat. When athletes rely only on the scales, they may mistakenly believe they are too fat because weight standards, such as the BMI, do not provide adequate information about body composition.

Many young athletes severely restrict energy intakes to improve performance, enhance the aesthetic appeal of their performance, or meet the weight guidelines of their specific sports. They fail to realize that the loss of lean tissue that accompanies energy restriction actually impairs their physical performance. The increasing incidence of abnormal eating habits among athletes is cause for concern. Male athletes, especially wrestlers and gymnasts, are affected by these disorders as well, but females are most vulnerable. Risk factors for eating disorders among athletes include:

- Young age (adolescence)
- Pressure to excel at a chosen sport
- Focus on achieving or maintaining an "ideal" body weight or body fat percentage
- Participation in sports or competitions that emphasize a lean appearance or judge performance on aesthetic appeal such as gymnastics, wrestling, figure skating, or dance
- Weight-loss dieting at an early age
- Unsupervised dieting

Amenorrhea

The prevalence of **amenorrhea** among premenopausal women in the United States is about 2 to 5 percent overall, but among female athletes, it may be as high as 66 percent. Contrary to previous notions, amenorrhea is *not* a normal adaptation to strenuous physical training: it is a symptom of something going wrong. Amenorrhea is characterized by low blood estrogen, infertility, and often bone mineral losses. Excessive training, depleted body fat, low body weight, and inadequate nutrition all contribute to amenorrhea. However amenorrhea develops, it threatens the integrity of the bones. Bone losses remain significant even after recovery. (Women with bulimia frequently have menstrual irregularities, but because they rarely cease menstruating, they may be spared this loss of bone integrity.)

Osteoporosis

For most people, weight-bearing physical activity, dietary calcium, and (for women) the hormone estrogen protect against the bone loss of osteoporosis. For young women with disordered eating and amenorrhea, strenuous activity can increase bone turnover and impair bone health.[4] Vigorous training combined with inadequate food intake disrupt metabolic and hormonal balances. These disturbances compromise bone health, greatly increasing the risks of **stress fractures.** Stress fractures commonly occur among dancers and other competitive athletes with amenorrhea, low calcium intakes, and disordered eating. Many underweight young athletes have bones like those of postmenopausal women, and they may never recover their lost bone even after diagnosis and treatment—which makes prevention critical. Young athletes should be encouraged to consume 1300 milligrams of calcium each day, to eat nutrient-dense foods, and to obtain enough energy to support both a healthy body weight and the energy expended in physical activity. Nutrition is critical to bone recovery.

Other Dangerous Practices of Athletes

Only females face the threats of the female athlete triad, of course, but many male athletes face pressure to achieve an unrealistic body weight and may develop eating disorders.[5] Each week throughout the season, David drastically restricts his food and fluid intake before a wrestling match in an effort to "make weight." Competitors in judo and wrestling and their coaches believe that competing in a lower weight class will give them a competitive advantage over smaller opponents. To that end, David intensifies his exercise, skips meals, restricts fluids, practices in plastic suits, and trains in heated rooms to lose 4 to 7 pounds rapidly.[6] He does this several times during the competitive season. These athletes hope to replenish the lost fluids, glycogen, and lean tissue during the hours between weigh-in and competition, but the body needs days to correct this metabolic mayhem. Reestablishing fluid and electrolyte balances may take 1 to 2 days, replenishing glycogen stores may take 2 to 3 days, and replacing lean tissue may take even longer.

Ironically, the combination of food deprivation and dehydration impairs physical performance by reducing muscle strength, decreasing anaerobic power, and reducing endurance capacity. For optimal performance, athletes need to first achieve their competitive weight during the off-season and then eat well-balanced meals and drink plenty of fluids during the competitive season.

Some athletes, usually males, go to extreme measures to bulk up and *gain* weight. People afflicted with **muscle dysmorphia** eat high-protein diets, take dietary supplements, weight train for hours at a time, and often abuse steroids in an attempt to increase muscle mass. Their bodies are large and muscular, yet they see themselves as puny 90-pound weaklings. They are preoccupied with the idea that their bodies are too small or inadequately muscular. Like others with distorted body images, people with muscle dysmorphia weigh themselves frequently and center their lives on diet and exercise. Paying attention to diet and pumping iron for fitness is admirable, but obsessing over it can cause serious social, occupational, and physical problems.

Preventing Eating Disorders in Athletes

To prevent eating disorders in athletes and dancers, the performers, their coaches, and their parents must learn about inappropriate body weight ideals, improper weight-loss techniques, eating disorder development, proper nutrition, and safe weight-control methods. Young people naturally search for identity and will often follow the advice of a person in authority without question. Therefore, coaches and dance instructors should never encourage unhealthy weight loss to qualify for competition or to conform to distorted artistic ideals. Athletes who truly need to lose weight should try to do so during the off-season and under the supervision of a health-care professional. Frequent weigh-ins can push young people who are striving to lose weight into a cycle of starving to confront the scale, then bingeing uncontrollably afterward. The erosion of self-esteem that accompanies this cycle can interfere with normal psychological development and set the stage for serious problems later on.

Table H8-1 includes suggestions to help athletes and dancers protect themselves against developing eating disorders. The remaining sections describe eating disorders that anyone, athlete or nonathlete, may experience.

TABLE H8-1 Tips for Combating Eating Disorders

General Guidelines

- Never restrict food amounts to below those suggested for adequacy by the USDA Food Patterns (see Table 2-2 on p. 44).
- Eat frequently. Include healthy snacks between meals. The person who eats frequently never gets so hungry as to allow hunger to dictate food choices.
- If not at a healthy weight, establish a reasonable weight goal based on a healthy body composition.
- Allow a reasonable time to achieve the goal. A reasonable loss of excess fat can be achieved at the rate of about 10 percent of body weight in 6 months.
- Establish a weight-maintenance support group with people who share interests.

Specific Guidelines for Athletes and Dancers

- Replace weight-based goals with performance-based goals.
- Restrict weight-loss activities to the off-season.
- Remember that eating disorders impair physical performance. Seek professional help in obtaining treatment if needed.
- Focus on proper nutrition as an important facet of your training, as important as proper technique.

Anorexia Nervosa

Julie, 18 years old, is a superachiever in school. She watches her diet with great care, and she exercises daily, maintaining a rigorous schedule of self-discipline. She is thin, but she is determined to lose more weight. She is 5 feet 6 inches tall and weighs 104 pounds (roughly 1.68 meters and 47 kilograms). Her BMI is less than 17. She has **anorexia nervosa.**

Characteristics of Anorexia Nervosa

Julie is unaware that she is undernourished, and she sees no need to obtain treatment. She developed amenorrhea several months ago and has become moody and chronically depressed. She views normal healthy body weight as too fat and insists that she needs to lose weight, although her eyes are sunk in deep hollows in her face. Julie denies that she is ever tired, although she is close to physical exhaustion and no longer sleeps easily. Her family is concerned, and though reluctant to push her, they have finally insisted that she see a psychiatrist. Julie's psychiatrist has diagnosed anorexia nervosa using specific criteria that describe such characteristics as minimal body weight, distorted body image, and the absence of menstrual cycles.[7] She is prescribed group therapy as a start and if she does not begin to gain weight soon, she may need to enter a residential program or be hospitalized.

Central to the diagnosis of anorexia nervosa is a distorted body image that overestimates personal body fatness. When Julie looks at herself in the mirror, she sees a "fat" 104-pound body. The more Julie overestimates her body size, the more resistant she is to treatment, and the more unwilling she is to examine her faulty values and misconceptions. In fact, she finds value in her condition. Malnutrition and weight loss affect brain functioning and judgment in this way, causing lethargy, confusion, and delirium and influencing mood, anxiety, and emotions.[8]

Anorexia nervosa cannot be self-diagnosed. Many people in our society are engaged in the pursuit of thinness, and denial runs high among people with anorexia nervosa. Some women have all the attitudes and behaviors associated with the condition, but without the dramatic weight loss.

How can a person as thin as Julie continue to starve herself? Julie uses tremendous discipline against her hunger to strictly limit her portions of low-fat, high-fiber, low-kcalorie foods.[9] She will deny her hunger, and having adapted to eating so little food, she feels full after nibbling on a few carrot sticks. She knows the kcalorie intake of various foods and the kcalorie expenditure of different physical activities. If she feels that she has gained an ounce of weight, she

People with anorexia nervosa see themselves as fat, even when they are dangerously underweight.

runs or jumps rope until she is sure she has exercised it off. If she fears that the food she has eaten outweighs her physical activity, she may take laxatives to hasten the passage of food from her system. She drinks water incessantly to fill her stomach, risking dangerous mineral imbalances. She is desperately hungry. In fact, she is starving, but she doesn't eat because her need for self-control dominates.

Many people, on learning of this disorder, say they wish they had "a touch" of it to get thin. They mistakenly think that people with anorexia nervosa feel no hunger. They also fail to recognize the pain of the associated psychological and physical trauma.

The starvation of anorexia nervosa damages the body just as the starvation of war and poverty does. In fact, most people with anorexia nervosa are malnourished. Their bodies have been depleted of both body fat and protein. Victims are dying to be thin—quite literally. In young people, growth ceases and normal development falters. They lose so much lean tissue that their basal metabolic rate slows. In addition, the heart pumps inefficiently and irregularly, the heart muscle becomes weak and thin, the chambers diminish in size, and the blood pressure falls. Minerals that help to regulate heartbeat become unbalanced. Many deaths occur due to multiple organ system failure when the heart, kidneys, and liver cease to function.

Starvation brings other physical consequences as well, such as loss of brain tissue, impaired immune response, anemia, and a loss of digestive functions that worsen malnutrition. Peristalsis becomes sluggish, the stomach empties slowly, and the lining of the intestinal tract atrophies. The pancreas slows its production of digestive enzymes. The deteriorated GI tract fails to provide sufficient digestive enzymes and absorptive surfaces for handling any food that is eaten. The person may suffer from diarrhea, further worsening malnutrition.

Other effects of starvation include altered blood lipids, high blood vitamin A and vitamin E, low blood proteins, dry thin skin, abnormal nerve functioning, reduced bone density, low body temperature, low blood pressure, and the development of fine body hair (the body's attempt to keep warm). The electrical activity of the brain becomes abnormal, and insomnia is common. Both women and men lose their sex drives.

Women with anorexia nervosa develop amenorrhea. (It is one of the diagnostic criteria.) In young girls, the onset of menstruation is delayed. Menstrual periods typically resume with recovery, although some women never restart even after they have gained weight. Should an underweight woman with anorexia nervosa become pregnant, she is likely to give birth to an underweight baby—and low-birthweight babies face many health problems (as Chapter 15 explains). Mothers with anorexia nervosa may underfeed their children, who then fail to grow and may also suffer the other consequences of starvation.

Treatment of Anorexia Nervosa

Treatment of anorexia nervosa requires a multidisciplinary approach.[10] Teams of physicians, nurses, psychiatrists, family therapists, and dietitians work together to resolve two sets of issues and behaviors: those relating to food and weight and those involving relationships with oneself and others.

The first dietary objective is to stop weight loss while establishing regular eating patterns. Appropriate diet is crucial to recovery and must be tailored to each individual's needs. Because body weight is low and fear of weight gain is high, initial food intake may be small—perhaps only 1200 kcalories per day. A variety of foods and foods with a higher energy density help to ensure greater success.[11] As eating becomes more comfortable, clients should gradually increase energy intake. Initially, clients may be unwilling to eat for themselves. Those who do eat will have a good chance of recovering without additional interventions. Even after recovery, however, energy intakes and eating behaviors may not fully return to normal. Furthermore, weight gains may be slow because energy needs may be slightly elevated due to anxiety, abdominal pain, and cigarette smoking.

Because anorexia nervosa is like starvation physically, health-care professionals classify clients based on indicators of malnutrition. Low-risk clients need nutrition counseling. Intermediate-risk clients may need supplements such as high-kcalorie, high-protein formulas in addition to regular meals. High-risk clients may require hospitalization and may need to be fed by tube at first to prevent death.[12] Residential programs that provide intensive behavioral treatment may be most appropriate for those who do not respond to less intensive approaches.[13]

Denial runs high among those with anorexia nervosa. Few seek treatment on their own. About half of the women who are treated can maintain their body weight at 85 percent or more of a healthy weight, and at that weight, many of them may begin menstruating again. The other half have poor to fair treatment outcomes, relapse into abnormal eating behaviors, or die. Anorexia nervosa has one of the highest mortality rates among psychiatric disorders—most commonly from cardiac complications or by suicide.[14] Much like treatment for drug addictions, treatment for eating disorders engages family members. Therapists help family members to understand how their past interactions have enabled the client to continue destructive behaviors and how new ways of interacting can support change.[15]

Before drawing conclusions about someone who is extremely thin or who eats very little, remember that diagnosis requires professional assessment. Several national organizations offer information for people who are seeking help with anorexia nervosa, either for themselves or for others.

Bulimia Nervosa

Kelly is a charming, intelligent, 30-year-old flight attendant of normal weight who thinks constantly about food. She alternates between starving herself and secretly bingeing, and when she has eaten too much, she makes herself vomit. Most readers recognize these symptoms as those of **bulimia nervosa.**

Characteristics of Bulimia Nervosa

Bulimia nervosa is distinct from anorexia nervosa and is more prevalent, although the true incidence is difficult to establish because bulimia nervosa is not as physically apparent. More men suffer from bulimia nervosa than from anorexia nervosa, but bulimia nervosa is still more common in women than in men. The secretive nature of bulimic behaviors makes recognition of the problem difficult, but once it is recognized, diagnosis is based on such criteria as number and frequency of binge eating episodes, inappropriate compensatory behaviors, and distorted body image.[16]

Like the typical person with bulimia nervosa, Kelly is single, female, and white. She is well educated and close to her ideal body weight, although her weight fluctuates over a range of 10 pounds or so every few weeks. She prefers to weigh less than the weight that her body maintains naturally.

Kelly seldom lets her eating disorder interfere with work or other activities, although a third of all bulimics do. From early childhood, she has been a high achiever and emotionally dependent on her parents. As a young teen, Kelly frequently followed severely restricted diets but could never maintain the weight loss. Kelly feels anxious at social events and cannot easily establish close personal relationships. She is usually depressed, is often impulsive, and has low self-esteem. When crisis hits, Kelly responds by replaying events, worrying excessively, and blaming herself but never asking for help—behaviors that interfere with effective coping.

Like the person with anorexia nervosa, the person with bulimia nervosa spends much time thinking about body image and food. The preoccupation with food manifests itself in secret binge-eating episodes, which usually progress through several emotional stages: anticipation and planning, anxiety, urgency to begin, rapid and uncontrollable consumption of food, relief and relaxation, disappointment, and finally shame or disgust.

A bulimic binge is characterized by a sense of no control over eating. During a binge, the person consumes food for its emotional comfort and cannot stop eating or control what or how much is eaten. A typical binge occurs periodically, in secret, usually at night, and lasts an hour or more. Because a binge frequently follows a period of restrictive dieting, eating is accelerated by intense hunger. Energy restriction followed by bingeing can set in motion a pattern of weight cycling, which may make weight loss and maintenance more difficult over time.

During a binge, Kelly consumes thousands of kcalories of easy-to-eat, low-fiber, high-fat, and, especially, high-carbohydrate foods. Typically, she chooses cookies, cakes, and ice cream—and she eats the entire bag of cookies, the whole cake, and every last spoonful in a carton of ice cream. After the binge, Kelly pays the price with swollen hands and feet, bloating, fatigue, headache, nausea, and pain.

To purge the food from her body, Kelly may use a **cathartic**—a strong laxative that can injure the lower intestinal tract. Or she may induce vomiting, with or without the use of an **emetic**—a drug intended as first aid for poisoning. These purging behaviors are often accompanied by feelings of shame or guilt. Hence a vicious cycle develops: negative self-perceptions followed by dieting, bingeing, and purging, which in turn lead to negative self-perceptions (see Figure H8-2).

On first glance, purging seems to offer a quick and easy solution to the problems of unwanted kcalories and body weight. Many people perceive such behavior as neutral or even positive, when, in fact, binge eating and purging have serious physical consequences. Signs of subclinical malnutrition are evident in a compromised immune system. Fluid and mineral imbalances caused by vomiting or diarrhea can lead to abnormal heart rhythms and injury to the kidneys. Urinary tract infections can lead to kidney failure. Vomiting causes irritation and infection of the pharynx, esophagus, and salivary glands; erosion of the teeth; and dental caries. The esophagus may rupture or tear, as may the stomach. Sometimes the eyes

> **FIGURE H8-2 The Vicious Cycle of Restrictive Dieting and Binge Eating**

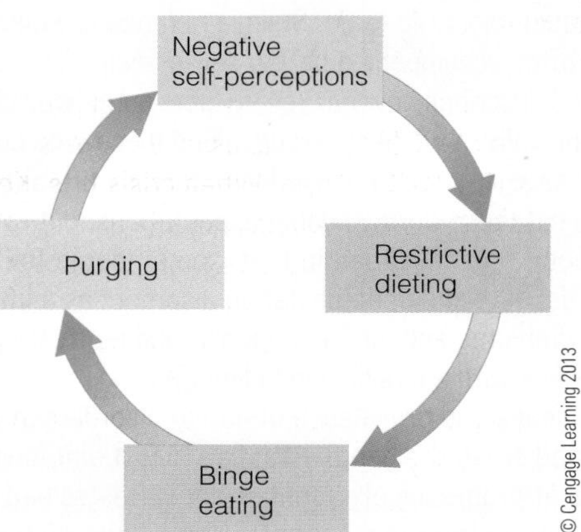

become red from pressure during vomiting. The hands may be calloused or cut by the teeth while inducing vomiting. Overuse of emetics depletes potassium concentrations and can lead to death by heart failure.

Unlike Julie, Kelly is aware that her behavior is abnormal, and she is deeply ashamed of it. She wants to recover, and this makes recovery more likely for her than for Julie, who clings to denial. Feeling inadequate ("I can't even control my eating"), Kelly tends to be passive and to look to others for confirmation of her sense of worth. When she experiences rejection, either in reality or in her imagination, her bulimia nervosa becomes worse. If Kelly's depression deepens, she may seek solace in drug or alcohol abuse or in other addictive behaviors. Clinical depression is common in people with bulimia nervosa, and the rates of substance abuse are high.

Treatment of Bulimia Nervosa

Kelly needs to establish regular eating patterns. She may also benefit from a regular exercise program. Weight maintenance, rather than cyclic weight gains and losses, is the treatment goal. Major steps toward recovery include discontinuing purging and restrictive dieting habits and learning to eat three meals a day plus snacks. Initially, energy intake should provide enough food to satisfy hunger and maintain body weight. Table H8-2 offers diet strategies to correct the eating problems of bulimia nervosa. Most women diagnosed with bulimia nervosa recover within 5 to 10 years, with or without treatment, but treatment probably speeds the recovery process. Cognitive behavioral therapy may be more effective than other types of treatment.[17] A mental health professional should be on the treatment team to help clients with their depression and addictive behaviors.

Anorexia nervosa and bulimia nervosa are distinct eating disorders, yet they sometimes overlap in important ways. Anorexia victims may purge, and victims of both disorders are overly concerned with body image and have a tendency to drastically undereat. Many perceive foods as "forbidden" and "give in" to an eating binge. The two disorders can also appear in the same person, or one can lead to the other. Treatment is challenging and relapses are not unusual. Other people have **unspecified eating disorders** that fall short of the criteria for anorexia nervosa or bulimia nervosa but share several of the same features. One such condition is **binge-eating disorder**.

TABLE H8-2 Diet Strategies for Combating Bulimia Nervosa

Planning Principles

- Plan meals and snacks; record plans in a food diary prior to eating.
- Plan meals and snacks that require eating at the table and using utensils.
- Refrain from finger foods.
- Refrain from "dieting" or skipping meals.

Nutrition Principles

- Eat a well-balanced diet and regularly timed meals consisting of a variety of foods.
- Include raw vegetables, salad, or raw fruit at meals to prolong eating times.
- Choose whole-grain, high-fiber breads, pasta, rice, and cereals to increase bulk.
- Consume adequate fluid, particularly water.

Other Tips

- Choose foods that provide protein and fat for satiety and bulky, fiber-rich carbohydrates for immediate feelings of fullness.
- Try including soups and other water-rich foods for satiety.
- Choose portions that meet the definition of "a serving" according to the USDA Food Patterns (pp. 42–43).
- For convenience (and to reduce temptation) select foods that naturally divide into portions. Select one potato, rather than rice or pasta that can be overloaded onto the plate; purchase yogurt and cottage cheese in individual containers; look for small packages of precut steak or chicken; choose frozen dinners with measured portions.
- Include 30 minutes of physical activity every day—exercise may be an important tool in defeating bulimia.

Binge-Eating Disorder

Charlie is a 40-year-old schoolteacher who has been overweight all his life. His friends and family are forever encouraging him to lose weight, and he has come to believe that if he only had more willpower, dieting would work. He periodically gives dieting his best shot—restricting energy intake for a day or two only to succumb to uncontrollable cravings, especially for high-fat foods. Like Charlie, up to half of the obese people who try to lose weight periodically binge; unlike people with bulimia nervosa, however, they typically do not purge. Such an eating disorder does not meet the criteria for either anorexia nervosa or bulimia nervosa—yet such compulsive overeating is a problem and occurs in people of normal weight as well as those who are severely overweight. Obesity alone is not an eating disorder.

Clinicians note differences between people with bulimia nervosa and those with binge-eating disorder. People with binge-eating disorder consume less during a binge, rarely purge, and exert less restraint during times of dieting. Similarities also exist, including feeling out of control, disgusted, depressed, embarrassed, guilty, or distressed because of their self-perceived gluttony.

There are also differences between obese binge eaters and obese people who do not binge. Those with binge-eating disorder report higher rates of self-loathing, disgust about body size, depression, and anxiety. Their eating habits differ as well. Obese binge eaters tend to consume more kcalories and more dessert and snack-type foods during regular meals and binges than obese people who do not binge. Binge eating may incur health risks greater than those of obesity alone.[18]

Binge eating is a behavioral disorder that can be resolved with treatment. Even a simple Internet-based treatment program can help.[19] Reducing binge eating makes participation in weight-control programs easier. It also improves physical health, mental health, and the chances of success in breaking the cycle of rapid weight losses and gains.

Eating Disorders in Society

Society plays a central role in eating disorders. Adolescent girls who read magazine articles on dieting and weight loss are likely to engage in unhealthy eating habits. Further proof of society's influence is found in the demographic distribution of eating disorders—they are known only in developed nations, and they become more prevalent as wealth increases and food becomes plentiful. Some people point to the vomitoriums of ancient times and claim that bulimia nervosa is not new, but the two are actually distinct. Ancient people were eating for pleasure, without guilt, and in the company of others; they vomited so that they could rejoin the feast. Bulimia nervosa is a disorder of isolation and is often accompanied by low self-esteem.

Chapter 8 describes how our society sets unrealistic ideals for body weight, especially in women, and devalues those who do not conform to them. Anorexia nervosa and bulimia nervosa are not a form of rebellion against these unreasonable expectations, but rather an exaggerated acceptance of them. In fact, some people fail to recognize the health dangers and endorse eating disorders as a lifestyle choice. Some 200 websites encourage, support, and motivate users to continue their lives with anorexia and bulimia.[20]

The incidence and prevalence of eating disorders in young people has increased steadily since the 1950s.[21] Most alarming is the rising prevalence at progressively younger ages. Restrained eating, fasting, binge eating, purging, fear of fatness, and distortion of body image are extraordinarily common among children and adolescents. Most are "on diets," and many are poorly nourished. Some eat too little food to support normal growth, thus they miss out on their adolescent growth spurts and may never catch up. Many eat so little that hunger propels them into binge-purge cycles. Disordered eating behaviors set a pattern that likely continues into young adulthood.[22]

Perhaps a person's best defense against these disorders is to learn to appreciate his or her own uniqueness. When people discover and honor their body's real physical needs, they become unwilling to sacrifice health for conformity. To respect and value oneself may be lifesaving.

REFERENCES

1. J. I. Hudson and coauthors, The prevalence and correlates of eating disorders in the National Comorbidity Survey Replication, *Biological Psychiatry* 61 (2007): 348–358.
2. R. Rodgers and H. Chabrol, Parental attitudes, body image disturbance and disordered eating amongst adolescents and young adults: A review, *European Eating Disorders Review* 17 (2009): 137–151.
3. D. L. Nichols, C. F. Sanborn, and E. V. Essery, Bone density and young athletic women: An update, *Sports Medicine* 37 (2007): 1001–1014; American College of Sports Medicine, Position stand: The female athlete triad, *Medicine and Science in Sports and Exercise* 39 (2007): 1867–1882.
4. M. T. Barrack and coauthors, Physiologic and behavioral indicators of energy deficiency in female adolescent runners with elevated bone turnover, *American Journal of Clinical Nutrition* 92 (2010): 652–659; M. T. Barrack and coauthors, Dietary restraint and low bone mass in female adolescent endurance runners, *American Journal of Clinical Nutrition* 87 (2008): 36–43.
5. J. L. Glazer, Eating disorders among male athletes, *Current Sports Medicine Reports* 7 (2008): 332–337.
6. G. G. Artioli and coauthors, Prevalence, magnitude, and methods of rapid weight loss among judo competitors, *Medicine and Science in Sports and Exercise* 42 (2010): 436–442.
7. American Psychiatric Association, *Diagnostic and Statistical Manual of Mental Disorders,* (Washington, D.C.: American Psychiatric Association, 2000).
8. W. Kaye, Neurobiology of anorexia and bulimia nervosa, *Physiology and Behavior* 94 (2008): 121–135.
9. M. Misra and coauthors, Nutrient intake in community-dwelling adolescent girls with anorexia nervosa and in healthy adolescents, *American Journal of Clinical Nutrition* 84 (2006): 698–706.
10. Position of the American Dietetic Association: Nutrition intervention in the treatment of anorexia nervosa, bulimia nervosa, and other eating disorders, *Journal of the American Dietetic Association* 106 (2006): 2073–2082.
11. J. E. Schebendach and coauthors, Dietary energy density and diet variety as predictors of outcome in anorexia nervosa, *American Journal of Clinical Nutrition* 87 (2008): 810–816.
12. E. Attia and B. T. Walsh, Behavioral management for anorexia nervosa, *New England Journal of Medicine* 360 (2009): 500–506.
13. Attia and Walsh, 2009.
14. J. M. Holm-Denoma and coauthors, Deaths by suicide among individuals with anorexia as arbiters between competing explanations of the anorexia-suicide link, *Journal of Affective Disorders* 107 (2008): 231–236.
15. J. Treasure and coauthors, The assessment of the family of people with eating disorders, *European Eating Disorders Review* 16 (2008): 247–255.
16. American Psychiatric Association, *Diagnostic and Statistical Manual of Mental Disorders* (Washington, D.C.: American Psychiatric Association, 2000).
17. ECRI Institute, *Bulimia Nervosa: Comparative Efficacy of Available Psychological and Pharmacological Treatments*, as cited in M. Mitka, Reports weighs options for bulimia nervosa treatment, *Journal of the American Medical Association* 305 (2011): 875.
18. J. I. Hudson and coauthors, Longitudinal study of the diagnosis of components of the metabolic syndrome in individuals with binge-eating disorder, *American Journal of Clinical Nutrition* 91 (2010): 1568–1573.

19. M. Jones and coauthors, Randomized, controlled trial of an Internet-facilitated intervention for reducing binge eating and overweight in adolescents, *Pediatrics* 121 (2008): 453–462.
20. D. L. G. Borzekowski and coauthors, e-Ana and e-Mia: A content analysis of pro-eating disorder web sites, *American Journal of Public Health* 100 (2010): 1526–1534.
21. D. S. Rosen and the Committee on Adolescence, Clinical report: Identification and management of eating disorders in children and adolescents, *Pediatrics* 126 (2010): 1240–1253.
22. D. Neumark-Sztainer and coauthors, Dieting and disordered eating behaviors from adolescence to young adulthood: Findings from a 10-year longitudinal study, *Journal of the American Dietetic Association* 111 (2011): 1004–1011.

© Artem Rudik/Shutterstock.com

CENGAGE LEARNING'S GLOBAL NUTRITION WATCH
GLOBAL NUTRITION RESOURCE CENTER

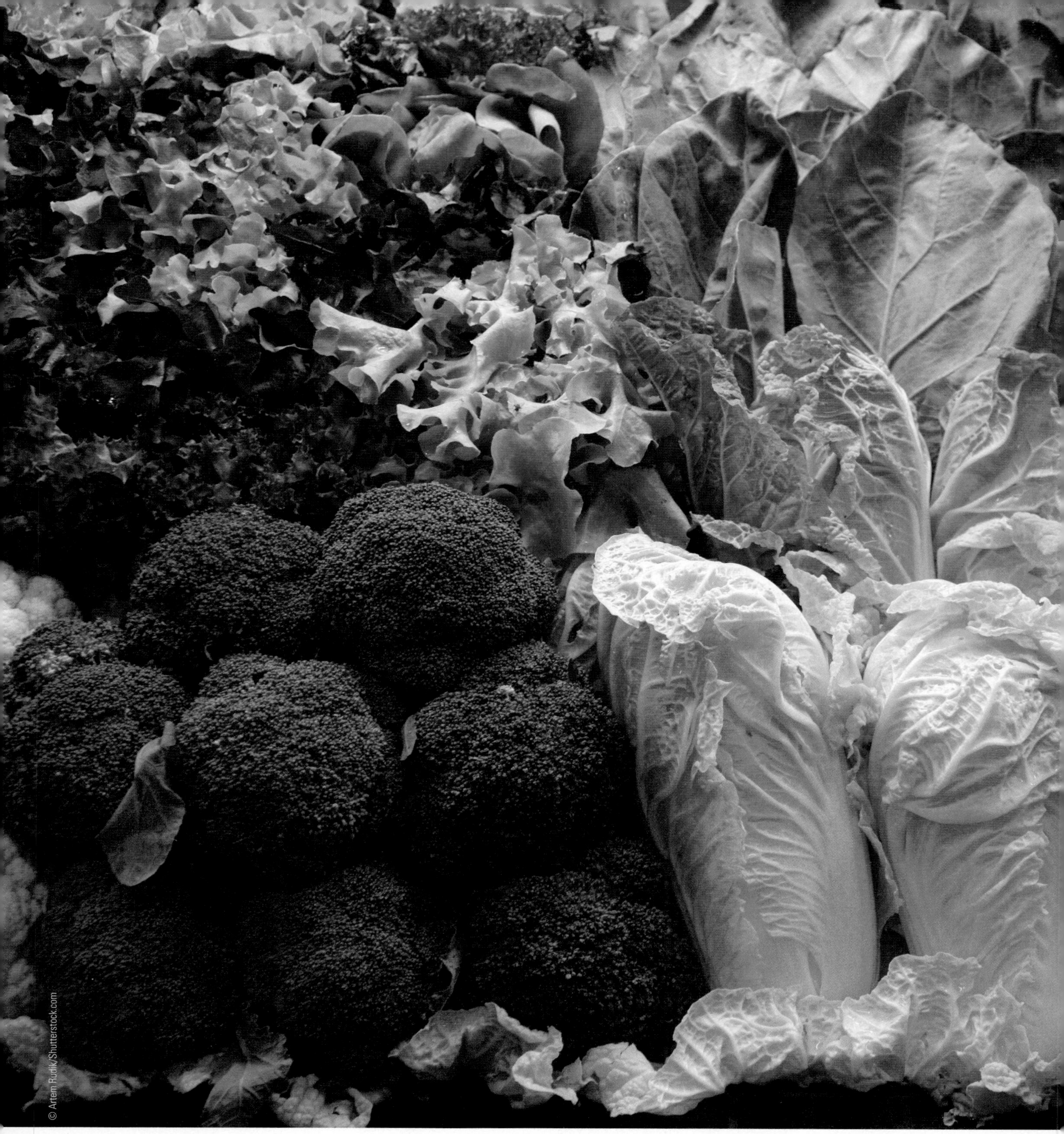

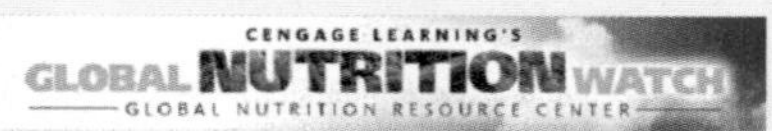

Gastric surgery is an option for some cases of clinically severe obesity. There are several surgical options available, each with benefits and risks. Go to Weight Management/Diets → Obesity to compare the risks and benefits of surgery with those of other weight-loss strategies. Under what circumstances would surgery be the best decision?

9 Weight Management: Overweight, Obesity, and Underweight

LEARNING GPS

Nutrition in Your Life

Are you pleased with your body weight? If so, you are a rare individual. Most people in our society think they should weigh more or less (mostly less) than they do. Usually, their primary concern is appearance, but they often understand that physical health is also somehow related to body weight. One does not necessarily cause the other—that is, an ideal body weight does not ensure good health. Instead, both depend on diet and physical activity. A well-balanced diet and active lifestyle support good health—and help maintain body weight within a reasonable range. In the Nutrition Portfolio at the end of this chapter, you can consider whether your eating habits and physical activities are supporting good health and a reasonable body weight.

The previous chapter described how body weight is stable when energy in equals energy out. Weight gains occur when energy intake exceeds energy expended, and conversely, weight losses occur when energy expended exceeds energy intake. At the extremes, both overweight and underweight present health risks. **Weight management** is a key component of good health.

This chapter emphasizes overweight and obesity, partly because they have been more intensively studied and partly because they represent a major health problem in the United States and a growing concern worldwide. Information on underweight is presented at the end of the chapter. The highlight that follows this chapter examines fad diets.

9.1 Overweight and Obesity

LEARN IT Describe how body fat develops and why it can be difficult to maintain weight gains and losses.

Despite our preoccupation with body image and weight loss, the prevalence of overweight and obesity in the United States continues to be high.[1] In the past

weight management: maintaining body weight in a healthy range by preventing gradual weight gains over time and losing weight if overweight, and by preventing weight losses and gaining weight if underweight.

> FIGURE 9-1 **Increasing Prevalence of Obesity (BMI ≥ 30) among US Adults**

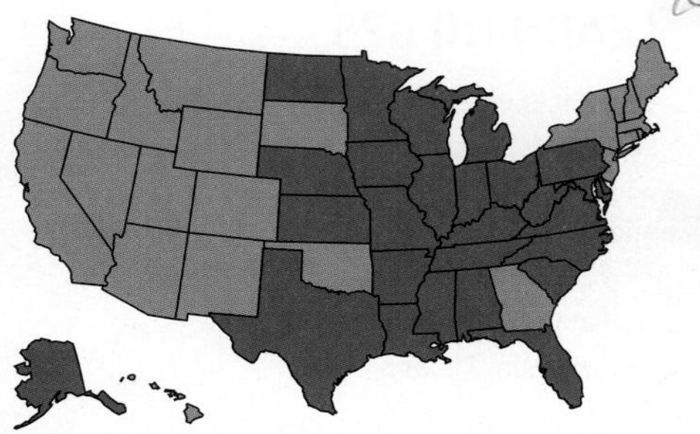

1995: No state had a prevalence rate less than 10%, all states had prevalence rates between 10–19%, and no state had a prevalence rate greater than or equal to 20%.

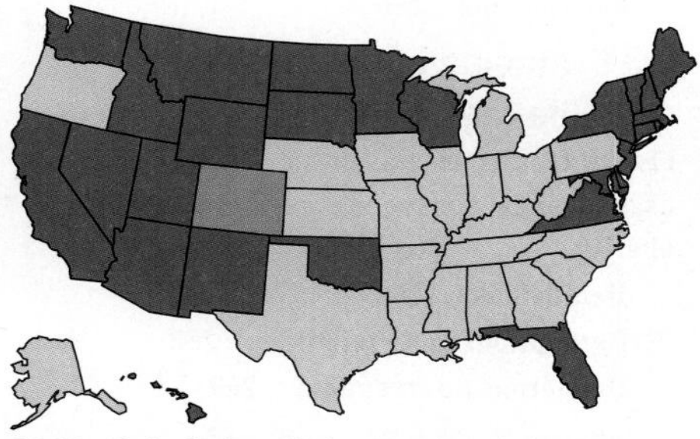

2000: Only Colorado had a prevalence rate less than 15%, the other states had prevalence rates between 15–24%, and no state had a prevalence rate greater than or equal to 25%.

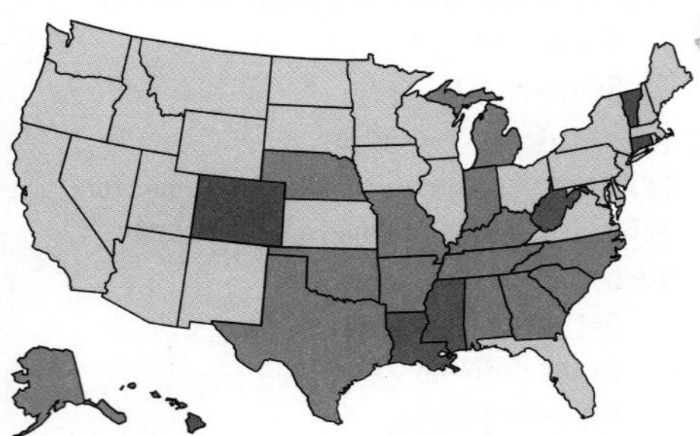

2005: No state had a prevalence rate less than 15%, most states had prevalence rates between 20–29%, and 3 states had a prevalence rate greater than or equal to 30%.

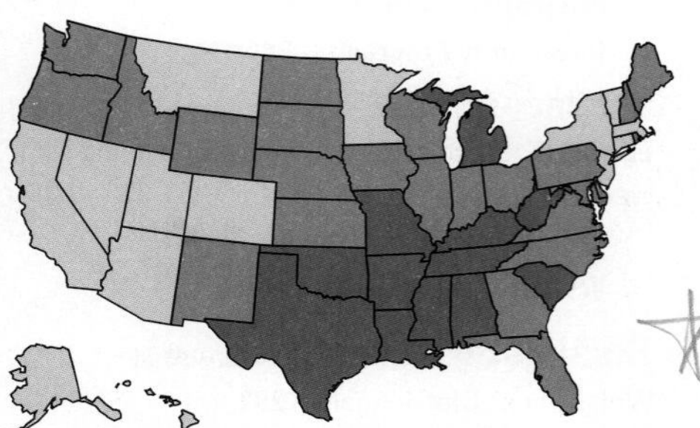

2010: No state had a prevalence rate less than 20%, most states had prevalence rates between 20–29%, and 12 states had a prevalence rate greater than or equal to 30%.

Key:
10%–14%
15%–19%
20%–24%
25%–29%
≥ 30%

SOURCE: www.cdc.gov/obesity/data/trends.html

three decades, obesity increased in every state, in both genders, and across all ages, races, and educational levels (see Figure 9-1).[2] An estimated 68 percent of the adults in the United States are now considered overweight or obese, as defined by a BMI of 25 or greater.[3] ♦ The prevalence of overweight is especially high among women, the poor, African Americans, and Mexican Americans.

The prevalence of overweight among children in the United States has also risen at an alarming rate. An estimated 34 percent of children and adolescents aged 2 to 19 years are either overweight or obese.[4] Chapter and Highlight 16 present information on overweight during childhood and adolescence.

Obesity in the United States is widespread. Prevalence increased rapidly over the past four decades, but seems to have leveled out in recent years. This **epidemic** of obesity has spread worldwide, affecting 1.5 billion adults and 43 million children younger than age 5.[5] Contrary to popular opinion, obesity is not limited to industrialized nations; people in developing countries also suffer from obesity-related problems. Before examining the suspected causes of obesity and the various strategies used to treat it, it is helpful to understand the development and metabolism of body fat.

Fat Cell Development When "energy in" exceeds "energy out," much of the excess energy is stored in the fat cells of adipose tissue. The amount of fat in adipose tissue reflects both the *number* and the *size* of the fat cells.* The number of fat cells increases most rapidly during the growing years of late childhood and early puberty. After growth ceases, fat cell number may continue to increase whenever energy balance is positive.[6] Obese people have more fat cells than healthy-weight people; their fat cells are also larger.

As fat cells accumulate triglycerides, they expand in size (review Figure 5-18 on p. 144). When the cells enlarge, they stimulate cell proliferation so that their numbers increase again. Thus obesity develops when a person's fat cells increase in number, in size, or quite often both. Figure 9-2 illustrates fat cell development.

When "energy out" exceeds "energy in," the size of fat cells dwindles, but not their number. People with extra fat cells tend to regain lost weight rapidly; with weight gain, their many fat cells readily fill. In contrast, people with an average number of enlarged fat cells may be more successful in maintaining weight losses; when their cells shrink, both cell size and number are normal. Prevention of obesity is most critical, then, during the growing years of childhood and adolescence when fat cells increase in number.[7] Researchers are exploring ways to induce fat cell death—which would decrease the number.[8]**

As mentioned, excess fat first fills the body's natural storage site—adipose tissue. If fat is still abundant, the excess is deposited in organs such as the heart and liver and plays a key role in the development of diseases such as heart failure and fatty liver, respectively.[9]*** As adipose tissue produces adipokines, metabolic changes that indicate disease risk—such as insulin resistance—become apparent and chronic inflammation develops.[10] The adipokine profile begins to improve with as little as a 5 percent weight loss and a decrease in fat cell size, suggesting that other metabolic changes might also occur at that time to improve disease risks.[11]

Fat Cell Metabolism The enzyme **lipoprotein lipase (LPL)** is a major determinant in the development of obesity.[12] Its role is to remove triglycerides from the blood for storage in both adipose tissue and muscle cells. Obese people generally have much more LPL activity in their adipose cells than lean people do (their muscle cell LPL activity is similar, though). This high LPL activity makes fat storage especially efficient. Consequently, even modest excesses in energy intake have a more dramatic impact on obese people than on lean people. When obese people lose weight, their LPL activity diminishes.[13]

The activity of LPL in different regions of the body is partially influenced by gender. In women, fat cells in the breasts, hips, and thighs produce abundant LPL,

---

*Obesity due to an increase in the *number* of fat cells is *hyperplastic obesity.* Obesity due to an increase in the *size* of fat cells is *hypertrophic obesity.*

**Cell death is known as as *apoptosis.*

***The adverse effect of fat in nonadipose tissue is known as *lipotoxicity.*

> FIGURE 9-2 **Fat Cell Development**

Fat cells are capable of increasing their size by 20-fold and their number by several thousandfold.

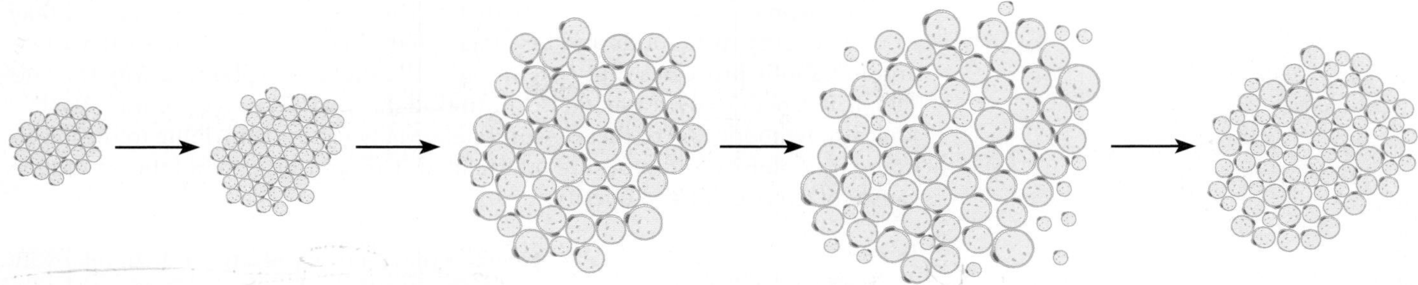

storing fat in those body sites; in men, fat cells in the abdomen produce abundant LPL. This enzyme activity explains why men tend to develop central obesity around the abdomen (apple-shaped) whereas women more readily develop lower-body fat around the hips and thighs (pear-shaped).

During growth, fat cells increase in number.
When energy intake exceeds expenditure, fat cells increase in size.
When fat cells have enlarged and energy intake continues to exceed energy expenditure, fat cells may increase in number again.
With fat loss, the size of the fat cells shrinks but not the number.

Gender differences are also apparent in the activity of the lipase enzymes controlling the release and breakdown of fat in various parts of the body. The release of lower-body fat is less active in women than in men, whereas the release of upper-body fat is similar. Furthermore, the rate of fat breakdown is lower in women than in men. Consequently, women may have a more difficult time losing fat in general, and from the hips and thighs in particular.

Enzyme activity may also explain why some people who lose weight regain it so easily. After weight loss, adipose LPL activity increases. Apparently, weight loss serves as a signal to the gene that produces the LPL enzyme, saying "Make more of the enzyme that stores fat." People easily regain weight after having lost it because they are battling against enzymes that want to store fat. Not only is fat storage efficient, but fat oxidation is not. Dietary fat oxidation correlates negatively with body fatness: obese people have the least activity.[14] The activities of these and other proteins provide an explanation for the observation that some inner mechanism seems to set a person's body weight or composition at a fixed point; the body will make adjustments to restore that **set point** if the person tries to change it.

♦ BMI:

- Underweight: <18.5
- Healthy weight: 18.5–24.9
- Overweight: 25.0–29.9
- Obese: ≥30

Set-Point Theory

Many physiological variables, such as blood glucose, blood pH, and body temperature, remain fairly stable under a variety of conditions. The hypothalamus and other regulatory centers constantly monitor and delicately adjust conditions to maintain homeostasis. The stability of such complex systems may depend on set-point regulators that maintain variables within specified limits.

Researchers have confirmed that after weight gains or losses, the body adjusts its metabolism to restore the original weight. Energy expenditure increases after weight gain and decreases after weight loss. These changes in energy expenditure differ from those that would be expected based on body composition alone, and they help to explain why it can be difficult for an underweight person to maintain weight gains and an overweight person to maintain weight losses.

REVIEW IT Describe how body fat develops and why it can be difficult to maintain weight gains and losses.

Fat cells develop by increasing in number and size. Obesity prevention depends on maintaining a reasonable number of fat cells. With weight gains or losses, the body adjusts in an attempt to return to its set-point weight.

epidemic (ep-ih-DEM-ick): the appearance of a disease (usually infectious) or condition that attacks many people at the same time in the same region.

- **epi** = upon
- **demos** = people

lipoprotein lipase (LPL): an enzyme that hydrolyzes triglycerides passing by in the bloodstream and directs their parts into the cells, where they can be metabolized or reassembled for storage.

set point: the point at which controls are set (for example, on a thermostat). The set-point theory that relates to body weight proposes that the body tends to maintain a certain weight by means of its own internal controls.

9.2 Causes of Overweight and Obesity

LEARN IT Review some of the causes of obesity.

Why do people accumulate excess body fat? The obvious answer is that they take in more energy from foods and beverages than they expend in physical activity and metabolic processes. But that answer falls short of explaining why they do this. Is it genetic? Environmental? Cultural? Behavioral? Socioeconomic? Psychological? Metabolic? All of these? Most likely. Many factors contribute to the development of obesity and most are interrelated. This section reviews these possible causes of overweight and obesity.

Genetics and Epigenetics Genetics plays a true causative role in relatively few cases of obesity, for example, in Prader-Willi syndrome—a genetic disorder characterized by excessive appetite, massive obesity, short stature, and often mental retardation. Most cases of obesity, however, do not stem from a single gene, yet genetic influences do seem to be involved.[15] Highlight 6 describes epigenetics—the influence of environmental factors, such as diet and physical activity, on gene expression. Obesity provides a classic example.[16]

Researchers have found that adopted children tend to be more similar in weight to their biological parents than to their adoptive parents.[17] Studies of twins yield similar findings: compared with fraternal twins, identical twins are twice as likely to weigh the same.[18] These findings suggest an important role for genetics in determining a person's *predisposition* to obesity.[19] In other words, genes interact with the diet and activity patterns that lead to obesity and the metabolic pathways that influence satiety and energy balance. Even identical twins with identical genes become different over the years as epigenetic changes accumulate. This raises an important point: you cannot change the genome you inherit, but you can influence the epigenome. Physical activity, for example, can minimize the genetic influences on BMI.[20] Likewise, high-fat diets and low physical activity can accentuate the genetic influences on obesity.[21]

Clearly, something genetic makes a person more or less likely to gain or lose weight when overeating or undereating. Some people gain more weight than others on comparable energy intakes. Given an extra 1000 kcalories a day for 100 days, some pairs of identical twins gain less than 10 pounds while others gain up to 30 pounds. Within each pair, the amounts of weight gained, percentages of body fat, and locations of fat deposits are similar. Similarly, some people lose more weight than others following comparable exercise routines.

Researchers have been examining the human genome in search of genetic and epigenetic answers to obesity questions.[22] As the section on protein synthesis in Chapter 6 describes, each cell expresses only the genes for the proteins it needs, and each protein performs a unique function. The following paragraphs describe only a couple of the proteins that help explain appetite control, energy regulation, and obesity development.[23]

Leptin Researchers have identified an obesity gene, called *ob*, that is expressed primarily in the adipose tissue and codes for the protein **leptin.** Leptin acts as a hormone, primarily in the hypothalamus. Leptin maintains homeostasis by regulating food intake and energy expenditure in response to adipose tissue.[24] When body fat increases, leptin increases—which suppresses appetite. When body fat decreases, leptin decreases—which stimulates appetite and suppresses energy expenditure.[25]

Mice with a defective *ob* gene do not produce leptin and can weigh up to three times as much as normal mice and have five times as much body fat (see Figure 9-3). When injected with a synthetic form of leptin, the mice rapidly lose body fat. (Because leptin is a protein, it would be destroyed during digestion if given orally; consequently, it must be given by injection.) The fat cells not only lose fat, but they self-destruct (reducing cell number), which may explain why weight gains are delayed when the mice are fed again.

leptin: a protein produced by fat cells under direction of the *ob* gene that decreases appetite and increases energy expenditure.

- **leptos** = thin

> FIGURE 9-3 Mice with and without Leptin Compared

Both of these mice have a defective *ob* gene. Consequently, they do not produce leptin. They both became obese, but the one on the right received daily injections of leptin, which suppressed food intake and increased energy expenditure, resulting in weight loss.

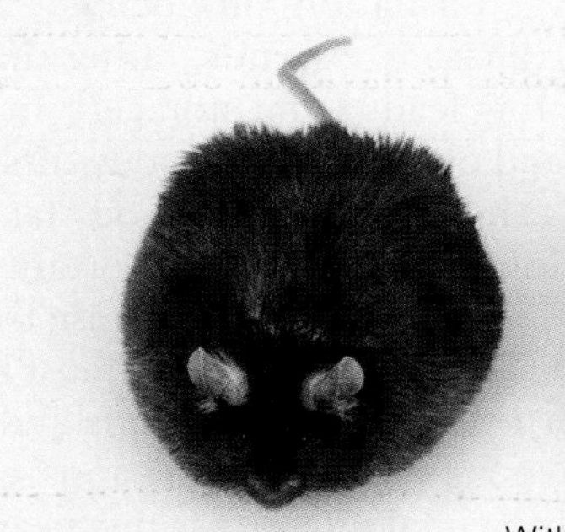
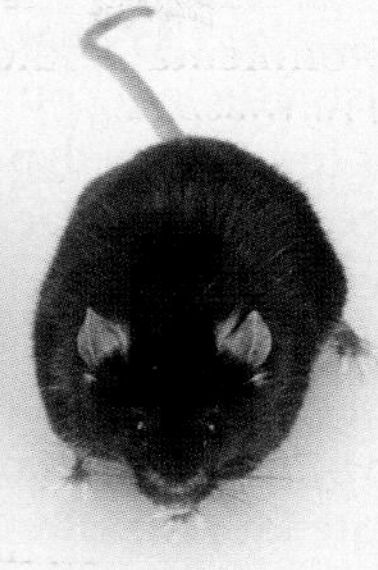

Without leptin, this mouse weighs almost three times as much as a normal mouse.
With leptin treatment, this mouse lost a significant amount of weight but still weighs almost one and a half times as much as a normal mouse.

Although extremely rare, a genetic deficiency of leptin or genetic mutation of its receptor has been identified in human beings as well.[26] Extremely obese children with barely detectable blood levels of leptin have little appetite control; they are constantly hungry and eat considerably more than their siblings or peers. Given daily injections of leptin, these children lose a substantial amount of weight, confirming leptin's role in regulating appetite and body weight.

Not too surprisingly, leptin injections are effective in suppressing appetite and supporting weight loss only when overeating and obesity are the result of a leptin deficiency. Very few obese people have a leptin deficiency, however. In fact, leptin levels increase as BMI increases. Leptin rises but fails to suppress appetite or enhance energy expenditure—a condition researchers describe as leptin resistance.[27] Interestingly, excessive fructose consumption seems to induce leptin resistance and accelerate fat storage.[28]

Ghrelin Another protein, known as **ghrelin,** also acts as a hormone primarily in the hypothalamus. In contrast to leptin, ghrelin is secreted primarily by the stomach cells and promotes weight gain by stimulating appetite and promoting efficient energy storage.[29]

Ghrelin triggers the desire to eat. Blood levels of ghrelin typically rise before and fall after a meal in proportion to the kcalories ingested—reflecting the hunger and satiety that precede and follow eating. On average, ghrelin levels are high whenever the body is in negative energy balance, as occurs during low-kcalorie diets, for example. This response may help explain why weight loss is so difficult to maintain. Weight loss is more successful with exercise and after gastric bypass surgery, in part because ghrelin levels are relatively low. Ghrelin levels decline again whenever the body is in positive energy balance, as occurs with weight gains.

Some research indicates that ghrelin also promotes sleep. Interestingly, a lack of sleep increases the hunger hormone ghrelin and decreases the satiety hormone leptin—which may help to explain the association between inadequate sleep and overweight.[30]

These two proteins—leptin and ghrelin—illustrate some of the complex factors involved in the regulation of food intake and energy homeostasis. Scientists have identified numerous proteins expressed by dozens of genes linked to obesity and several others associated with fat distribution in the body.[31] Each of these genes have slight variations that differ among individuals.[32] Furthermore, these genes interact with one another and with the environment. The complexity of it all creates a multitude of possible genetic explanations.[33]

Uncoupling Proteins Genes also code for proteins involved in energy metabolism. These proteins may influence the storing or expending of energy with different

ghrelin (GRELL-in): a protein produced by the stomach cells that enhances appetite and decreases energy expenditure.
- **ghre** = growth

The food industry spends billions of dollars a year on advertising. The message? "Eat more."

efficiencies or in different types of fat. The body has two types of fat: white and **brown adipose tissue.**[34] White adipose tissue stores fat for other cells to use for energy; brown adipose tissue releases stored energy as heat. Recall from Chapter 7 that when fat is oxidized, some of the energy is released in heat and some is captured in ATP. In brown adipose tissue, oxidation is uncoupled from ATP formation, producing heat only.[35]* By radiating energy away as heat, the body expends, rather than stores, energy. In contrast, efficient coupling leads to fat storage.[36] In other words, weight gains or losses may depend on whether the body dissipates the energy from an ice cream sundae as heat or stores it in body fat.

Brown fat and heat production is particularly important in newborns and in animals exposed to cold weather, especially those that hibernate. They have plenty of brown adipose tissue. In contrast, most human adults have little brown fat—less than 1 percent of all fat cells and interspersed among the white fat cells. Brown fat activity is most apparent during exposure to cold.[37] Importantly, brown fat quantity is inversely related with BMI; overweight and obese individuals have less brown fat activity than others.[38] The role of brown fat in body weight regulation is not yet understood, but such an understanding may prove most useful in developing obesity treatments.[39]

Uncoupling proteins are active not only in brown fat, but also in white fat and many other tissues. Their actions seem to influence the basal metabolic rate (BMR) and oppose the development of obesity. Animals with abundant amounts of these uncoupling proteins resist weight gain, whereas those with minimal amounts gain weight easily. Similarly, people with a genetic variant of an uncoupling protein have lower metabolic rates and are more overweight than others.

Environment With obesity rates rising and the **gene pool** remaining relatively unchanged, environment must also play a role in obesity. Obesity reflects the interactions between genes and the environment.[40] An **obesogenic environment** includes all of the circumstances that we encounter daily that push us toward fatness. Over the past 4 decades, the demand for physical activity has decreased as the abundance of food has increased.[41]

Keep in mind that genetic and environmental factors are not mutually exclusive; in fact, their *interactions* create the epigenetics that provide a greater understanding of obesity and related diseases.[42] Genes can influence eating behaviors, for example, and food and activity behaviors influence the genes that regulate body weight. Interestingly, even social relationships can influence the development of obesity.[43] The likelihood that a person will become obese increases when a friend, sibling, or spouse becomes obese.

Overeating One explanation for obesity is that overweight people overeat, although diet histories may not always reflect high intakes. Diet histories are not always accurate records of actual intakes; both normal-weight and obese people commonly misreport their dietary intakes. Most importantly, current dietary intakes may not reflect the eating habits that led to obesity. Obese people who had a positive energy balance for years and accumulated excess body fat may not currently have a positive energy balance. This reality highlights an important point: the energy-balance equation must consider time. Both present *and* past eating and activity patterns influence current body weight.

We live in an environment that exposes us to an abundance of high-kcalorie, high-fat foods that are readily available, relatively inexpensive, heavily advertised, and reasonably tasty. Food is available everywhere, all the time—thanks largely to fast food. Our highways are lined with fast-food restaurants. Convenience stores and service stations offer fast food and snacks as well. Fast food is available in our schools, malls, and airports. The mere proximity of fast food increases the risk of obesity.[44] It's convenient and it's available morning, noon, and night—and all

*In *coupled reactions,* the energy released from the breakdown of one compound is used to create a bond in the formation of another compound. In *uncoupled reactions,* the energy is released as heat.

brown adipose tissue: masses of specialized fat cells packed with pigmented mitochondria that produce heat instead of ATP.

gene pool: all the genetic information of a population at a given time.

obesogenic (oh-BES-oh-JEN-ick) **environment:** all the factors surrounding a person that promote weight gain, such as increased food intake, especially of unhealthy choices, and decreased physical activity.

times in between. Consequently, we are eating more meals more frequently than in decades past—and energy intake has risen accordingly.[45]

"Want fries with that?" A supersize portion delivers more than 600 kcalories.

Most alarming are the extraordinarily large portions and ready-to-go combo-meals. Eating large portion sizes multiple times a day accounts for much of the weight increase seen over the decades.[46] People buy the large portions and combinations, perceiving them to be a good value, but then they eat more than they need—a bad deal. In fact, one research study calculated that for the 67 cents extra to upsize a meal, consumers receive an extra 400 kcalories, an extra 36 grams of body fat, and an extra \$1 to \$7 in health-care costs.[47]

Simply put, large portion sizes deliver more kcalories.[48] And portion sizes of virtually all foods and beverages have increased markedly in the past several decades, most notably at fast-food restaurants. Not only have portion sizes increased over time, but they are now two to eight times larger than standard serving sizes. The trend toward large portion sizes parallels the increasing prevalence of overweight and obesity in the United States, beginning in the 1970s, increasing sharply in the 1980s, and continuing today.

Restaurant food, especially fast food, contributes significantly to the development of obesity. Fast food is often energy-dense food, which increases energy intake, BMI, and body fatness. The combination of large portions and energy-dense foods is a double whammy. Reducing portion sizes is somewhat helpful, but the real kcalorie savings come from lowering the energy density. Low-energy density foods such as fruits and vegetables can help with weight loss.

Physical Inactivity Our environment fosters physical inactivity as well. Life requires little exertion—escalators carry us up stairs, automobiles take us across town, and remote controls change television channels from a distance. Modern technology has replaced physical activity at home, at work, and in transportation.[49] Inactivity contributes to weight gain and poor health. Most physical inactivity occurs when watching television, playing video games, and using the computer. The more time people spend in these sedentary activities, the more likely they are to be overweight—and to incur the metabolic risk factors of heart disease (high blood lipids, high blood pressure, and high blood glucose).[50]

Sedentary activities contribute to weight gain in several ways. First, they require little energy beyond the resting metabolic rate. Second, they replace time spent in more vigorous activities. Third, watching television influences food purchases and correlates with between-meal snacking on the high-kcalorie, solid fat and added sugars foods and beverages most heavily advertised.

Some obese people are so extraordinarily inactive that even when they eat less than lean people, they still have an energy surplus. Reducing their food intake further would incur nutrient deficiencies and jeopardize health. Physical activity is a necessary component of nutritional health. People must be physically active if they are to eat enough food to deliver all the nutrients they need without unhealthy weight gain. In fact, *to prevent weight gain*, the DRI ♦ suggests an accumulation of 60 minutes of moderately intense physical activities every day in addition to the less intense activities of daily living. Recommendations *to lose weight* encourage even greater duration, intensity, or frequency of physical activity (as a later section of the chapter discusses).

♦ DRI for physical activity: 60 min/day (moderate intensity)

People may be obese, therefore, not because they eat too much, but because they move too little—both in purposeful exercise and in the activities of daily life. Studies report that the differences in the time obese and lean people spend lying, sitting, standing, and moving accounts for about 350 kcalories a day. In general, lean people tend to be more spontaneously active in their occupations and their leisure time.[51] The energy expended in these everyday spontaneous activities—called *nonexercise activity thermogenesis (NEAT)*—plays a pivotal role in energy balance and weight management.

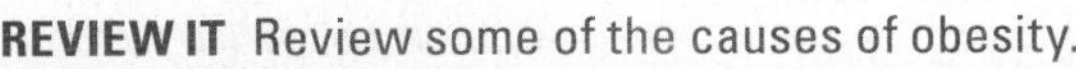

REVIEW IT Review some of the causes of obesity.

Obesity has many causes and most interact, creating a complex scenario. Environmental factors, such as overeating and physical inactivity, may influence a person's genetic predisposition to obesity.

Lack of physical activity fosters obesity.

9.3 Problems of Overweight and Obesity

LEARN IT Discuss the physical, social, and psychological consequences of overweight and obesity.

An estimated 59 percent of US adults are trying to lose weight at any given time.[52] Some of these people may not even need to lose weight. Others may benefit from weight loss, but they will not be successful. Relatively few people succeed in losing weight, and even fewer succeed permanently. For many, improving diet and activity habits to simply prevent further weight gains may be sufficient.[53] Whether a person will benefit from weight loss is a question of health.

Health Risks

Chapter 8 describes some of the health problems that commonly accompany obesity. In evaluating the risks to health from obesity, health-care professionals use three indicators:

- Body mass index ♦
- Waist circumference ♦
- Disease risk profile

♦ BMI:
- 25.0–29.9 = overweight
- ≥30 = obese

♦ Waist circumference:
- Men: >40 in (>102 cm)
- Women: >35 in (>88 cm)

Importantly, the disease risk profile takes into account family history, life-threatening diseases, and common risk factors for chronic diseases (such as blood lipid profile). The higher the BMI, the greater the waist circumference, and the more risk factors—the greater the urgency to treat obesity.

People can best decide whether weight loss might be beneficial by considering their health status and motivation. People who are overweight by BMI standards, but otherwise in good health, might not benefit from losing weight; they might focus on preventing further weight gains instead. In contrast, those who are obese and suffering from a life-threatening disease such as diabetes might improve their health substantially by adopting a diet and activity plan that supports weight loss. Motivation is a key component; to lose weight, a person needs to be ready and willing to make lifestyle changes for a lifetime.

Overweight in Good Health

Often a person's motivations for weight loss have nothing to do with health. A healthy young woman with a BMI of 26 might want to lose a few pounds for spring break, but doing so might not improve her health. In fact, if she opts for a starvation diet or diet pills, she would be healthier *not* trying to lose weight.

Obese or Overweight with Risk Factors

Weight loss is recommended for people who are obese and those who are overweight (or who have a high waist circumference) with two or more risk factors for chronic diseases. ♦ A 50-year-old man with a BMI of 28 who has high blood pressure and a family history of heart disease can improve his health by adopting a diet low in saturated fat and a regular exercise plan.

♦ Obese people and overweight people with two or more of these risk factors require aggressive treatment:
- Hypertension
- Cigarette smoking
- High LDL
- Low HDL
- Impaired glucose tolerance
- Family history of heart disease
- Men ≥45 yr; women ≥55 yr

Obese or Overweight with Life-Threatening Condition

Weight loss is also recommended for a person who is either obese or overweight and suffering from a life-threatening condition such as heart disease, diabetes, or sleep apnea. ♦ The health benefits of weight loss are clear. For example, a 30-year-old man with a BMI of 40 might be able to prevent or control diabetes by losing 75 pounds. Although the effort required to do so may be great, it may be no greater than the effort and consequences of living with diabetes.

♦ Obese people and overweight people with any of these diseases require aggressive treatment:
- Heart disease
- Diabetes (type 2)
- Sleep apnea (interruption of breathing during sleep)

Perceptions and Prejudices

Many people assume that every obese person can achieve slenderness and should pursue that goal. First consider that most obese people do not—for whatever reason—successfully lose weight and maintain their losses. Then consider the prejudice involved in that assumption. People come with varying weight tendencies, just as they come with varying potentials for height and physical talents, yet we do not expect tall people to shrink or fast runners to slow down in an effort to become "normal."

Social Consequences Large segments of our society place such enormous value on thinness that obese people face prejudice and discrimination on the job, at school, and in social situations: they are judged on their appearance more than on their character.[54] Socially, obese people are negatively stereotyped as lazy and lacking in self-control.[55] Such a critical view of overweight is not prevalent in many other cultures, including segments of our own society. Instead, overweight is simply accepted or even embraced as a sign of robust health and beauty. To free society of its obsession with body weight and prejudice against obesity, people must first learn to judge others—and themselves—for who they are and not for what they weigh.

Psychological Problems Psychologically, obese people may suffer embarrassment when others treat them with hostility and contempt, and many have come to view their own bodies as flawed. Feelings of rejection, shame, and depression are common among obese people. Anxiety and depression, in turn, may contribute to the development of obesity, which perpetuates the problem.[56]

Most weight-loss programs assume that the problem can be solved simply by applying willpower and hard work. If determination were the only factor involved, though, the success rate would be far greater than it is. Overweight people may readily assume blame for failure to lose weight and maintain the losses when, in fact, it is the programs that have failed. Ineffective treatment and its associated sense of failure add to a person's psychological burden. Figure 9-4 illustrates how the devastating psychological effects of obesity and dieting perpetuate themselves.

> **FIGURE 9-4 The Psychology of Weight Cycling**

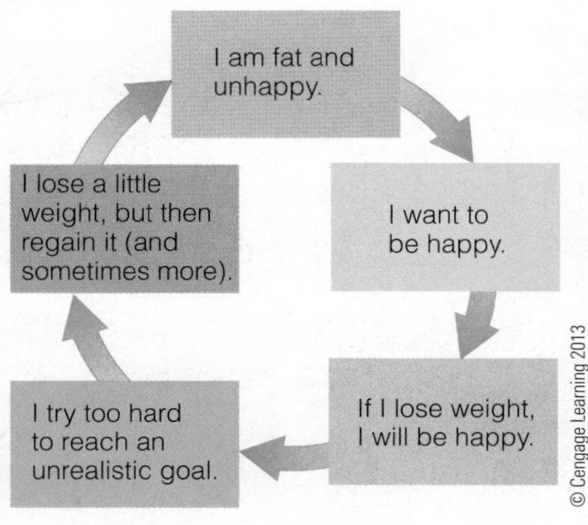

Dangerous Interventions

Some people attach so many dreams of happiness to weight loss that they willingly risk huge sums of money for the slightest chance of success. As a result, weight-loss schemes flourish. Of the tens of thousands of claims, treatments, and theories for losing weight, few are effective—and many are downright dangerous. The negative effects must be carefully considered before embarking on any weight-loss program. Some interventions entail greater dangers than the risk of being overweight. Physical, metabolic, and psychosocial problems may arise from fad diets and "yo-yo" dieting.[57] Wise consumers scrutinize fad diets, magic potions, and wonder gizmos with a healthy dose of skepticism.

Some of the nation's most popular diet books and weight-loss programs have misled consumers with unsubstantiated claims and deceptive testimonials. Furthermore, they fail to provide an assessment of the short- and long-term results of their treatment plans, even though such evaluations are possible and would permit consumers to make informed decisions. Of course, some weight-loss programs are better than others in terms of cost, approach, and customer satisfaction. Reputable weight-loss programs will explain the risks associated with their plans and provide honest predictions of success.

Fad Diets **Fad diets** often sound good, but they typically fall short of delivering on their promises. They espouse exaggerated or false theories of weight loss and advise consumers to follow inadequate diets. Some fad diets are hazardous to health as Highlight 9 explains. Adverse reactions can be as minor as headaches, nausea, and dizziness or as serious as death. The "How To" on p. 295 offers guidelines for identifying unsound weight-loss schemes and fad diets.

Weight-Loss Products Millions of people in the United States use nonprescription weight-loss products. Most of them are women, especially young overweight women, but almost 10 percent are of normal weight.

In their search for weight-loss magic, some consumers turn to "natural" herbal products and dietary supplements, even though few have proved to be effective and many have proved to be harmful.[58] For example, in addition to the many cautions that accompany the use of all herbal remedies, consumers should be aware that St. John's wort is often prepared in combination with the herbal stimulant ephedrine. Ephedrine-containing supplements promote modest short-term weight loss (about 2 pounds a month), but with great risks. These supplements

fad diets: popular eating plans that promise quick weight loss. Most fad diets severely limit certain foods or overemphasize others (for example, never eat potatoes or pasta or eat cabbage soup daily).

So many promises, so little success.

© Richard Levine / Alamy

have been implicated in numerous heart attacks and seizures, resulting in about 100 deaths. For this reason, the FDA has banned the sale of ephedrine-containing supplements, but they are still readily available on the Internet.* Similarly, the FDA has issued warnings for another herbal weight loss supplement called Que She, which contains not only ephedrine but two weight-loss drugs that have been withdrawn from the market and another drug used to treat heart conditions.

Highlight 18 explores the possible benefits and potential dangers of herbal products and other alternative therapies. As it explains, dietary supplements do not need to be approved by the FDA, and manufacturers do not need to test the safety or effectiveness of any product. In other words, consumers cannot assume that an herbal product or dietary supplement is safe or effective just because it is available on the market. In fact, the FDA has identified more than 70 products that contain undeclared, active pharmaceutical ingredients that can have serious consequences such as seizures and heart attacks.[59] These ingredients are not listed on the labels, and consumers have no way of knowing what the products actually contain. Anyone considering whether to use dietary supplements for weight loss should consult with a physician and research the product with the FDA (**www.fda.gov**).

Other Gimmicks Other gimmicks don't help with weight loss either. Hot baths do not speed up metabolism so that pounds can be lost in hours. Steam and sauna baths do not melt the fat off the body, although they may dehydrate people so that they lose water weight. Brushes, sponges, wraps, creams, and massages intended to move, burn, or break up fat do nothing of the kind.

REVIEW IT Discuss the physical, social, and psychological consequences of overweight and obesity.

The question of whether a person should lose weight depends on many factors: among them are the extent of overweight, age, health, and genetic makeup. Not all obesity will cause disease or shorten life expectancy. Just as there are unhealthy, normal-weight people, there are healthy, overweight people. Some people may risk more in the process of losing weight than in remaining overweight. Fad diets and weight-loss supplements can be as physically and psychologically damaging as excess body weight.

9.4 Aggressive Treatments for Obesity

LEARN IT Explain the risks and benefits, if any, of several aggressive ways to treat obesity.

The appropriate strategies for weight reduction depend on the degree of obesity and the risk of disease. An overweight person in good health may need only to improve eating habits and increase physical activity, but someone with **clinically severe obesity** may need more aggressive treatment options—drugs or surgery. Drugs appear to be modestly effective and safe, at least in the short term; surgery appears to be dramatically effective but can have severe complications, at least for some people.

Drugs

Based on new understandings of obesity's genetic basis and its classification as a chronic disease, much research effort has focused on drug treatments for obesity. Experts reason that if obesity is a chronic disease, it should be treated as such—and the treatment of most chronic diseases includes drugs. The challenge, then, is to develop an effective drug—or more likely, a combination of drugs—that

clinically severe obesity: a BMI of 40 or greater or a BMI of 35 or greater with additional medical problems. A less preferred term used to describe the same condition is *morbid obesity*.

*Ephedrine is an amphetamine-like substance extracted from the Chinese ephedra herb *ma huang*. The FDA has banned the sale of *ma huang* in the United States, and it is illegal in Canada.

TABLE 9-1 FDA Approved Drugs for Weight Loss

Product	Action	Side Effects
Orlistat (OR-leh-stat)	Inhibits pancreatic lipase activity in the GI tract, thus blocking digestion and absorption of dietary fat and limiting energy intake	Cramping, diarrhea, gas, frequent bowel movements, reduced absorption of fat-soluble vitamins; rare cases of liver injury
Phentermine (FEN-ter-mean), diethylpropion (DYE-eth-ill-PRO-pee-on), phendimetrazine (FEN-dye-MEH-tra-zeen)	Enhances the release of the neurotransmitter norepinephrine, which suppresses appetite	Increased blood pressure and heart rate, insomnia, nervousness, dizziness, headache

NOTE: Weight-loss drugs are most effective when taken as directed and used in combination with a reduced-kcalorie diet and increased physical activity.

can be used over time without adverse side effects or the potential for abuse.[60] Weight-loss drugs should be prescribed only to those with medical risks—not for cosmetic reasons—and in tandem with a healthy diet and activity program.

Several drugs for weight loss have been tried over the years, with varying degrees of effectiveness and safety.[61] When used as part of a long-term, comprehensive weight-loss program, drugs can help with modest weight loss. Because weight regain commonly occurs with the discontinuation of drug therapy, treatment must be long term. Yet the long-term use of drugs poses risks. We don't yet know whether a person would be harmed more from maintaining a 100-pound excess or from taking a drug for a decade to keep the 100 pounds off. Physicians must prescribe drugs appropriately, inform consumers of the potential risks, and monitor side effects carefully. Table 9-1 presents the FDA-approved drugs to treat obesity.

Some physicians prescribe drugs that have not been approved for weight loss, a practice known as "off-label" use. These drugs have been approved for other conditions (such as seizures) and incidentally cause modest weight loss. Physicians using off-label drugs must be well-informed of the drugs' use and effects and monitor their patients' responses closely.

Surgery The prevalence of clinically severe obesity (BMI >40) is increasing at an incredibly rapid rate. At this level of obesity, lifestyle changes and modest weight losses can improve disease risks a little, but the most effective treatment is surgery.[62] ♦ More than 200,000 such surgeries are performed in the United States annually. As Figure 9-5 (p. 272) shows, the two most common surgical procedures effectively limit food intake by reducing the capacity of the stomach. In addition, gastric bypass suppresses hunger by reducing production of gastrointestinal hormones.[63] The results are significant: depending on the type of surgery, initial weight loss is 20 to 32 percent of body weight and 14 to 25 percent after 10 years. Importantly, most people experience dramatic improvements in their diabetes, blood lipids, and blood pressure.[64] Improvements in depression and anxiety are not as likely.[65] Whether surgery is a reasonable option for obese teens is the subject of much debate among pediatricians and bariatric surgeons (see Chapter 16).[66]

♦ Surgery may be an option for people with all of the following conditions:
- Unable to achieve adequate weight loss with diet and exercise
- BMI ≥40 or BMI ≥35 with weight-related health problems (such as diabetes or hypertension)
- No medical or psychological contraindications
- Understanding of risks and strong motivation to comply with post-surgery treatment plan

Because the long-term safety and effectiveness of surgery depend, in large part, on compliance with dietary instructions, nutrition care plays an important role in follow-up treatment.[67] Vitamin and mineral deficiencies are common, and dietary supplements are routinely prescribed.[68] Weight regain may occur and psychological problems—such as disordered eating behaviors—may also develop.[69] Lifelong medical supervision is necessary, but the possible health benefits of weight loss—improved blood lipid profile, blood pressure, and insulin sensitivity—may outweigh the risks. Overall risk of death and health problems is lower for obese people after successful surgery than for obese people who do not pursue surgery.[70]

Another surgical procedure removes some fat deposits by liposuction. This cosmetic procedure has little effect on body weight (less than 10 pounds), but can alter body shape slightly in specific areas. Liposuction is a popular procedure in part because of its perceived safety, but immediate and delayed complications can arise.[71] Furthermore, removing adipose tissue by way of liposuction does not provide the health benefits that typically accompany weight loss. In other words, liposuction does not improve blood pressure, inflammation, blood lipid profile, or insulin sensitivity. Perhaps most surprisingly, a year after liposuction, body fat returns and redistributes itself from the thighs to the abdomen.[72]

> **FIGURE 9-5 Gastric Surgery Used in the Treatment of Clinically Severe Obesity**

Both of these surgical procedures limit the amount of food that can be comfortably eaten.

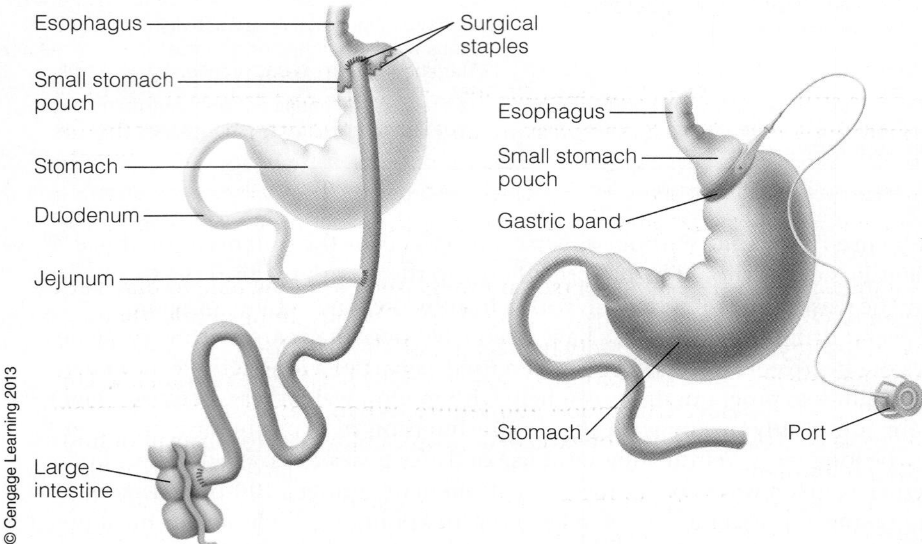

In gastric bypass, the surgeon constructs a small stomach pouch and creates an outlet directly to the small intestine, bypassing most of the stomach, the entire duodenum, and some of the jejunum. (Dark areas highlight the flow of food through the GI tract; pale areas indicate bypassed sections.)

In gastric banding, the surgeon uses a gastric band to reduce the opening from the esophagus to the stomach. The size of the opening can be adjusted by inflating or deflating the band by way of a port placed in the abdomen just beneath the skin.

Gastric bypass	Gastric banding
Advantages: • No foreign object in abdomen or need for adjustments • More durable, reliable, and effective	Advantages: • No malabsorption • More flexible, less invasive, safer

REVIEW IT Explain the risks and benefits, if any, of several aggressive ways to treat obesity.

Overweight and obese people may benefit most from improving eating and activity habits. Those with high risks of medical problems may need more aggressive treatment, including drugs or surgery. Such treatments may offer benefits, but also incur some risks.

9.5 Weight-Loss Strategies

LEARN IT Outline reasonable strategies for achieving and maintaining a healthy body weight.

From the bustling activity of a cell making fat to the inactivity of a person watching television, the factors contributing to obesity are numerous and complex. Each interacts with many others. Efforts to combat obesity must integrate healthy eating patterns, physical activities, supportive environments, and psychosocial support.[73]

Set Reasonable Goals Successful weight-loss strategies embrace small changes, moderate losses, and reasonable goals. People who lose 10 to 20 pounds in a year by consistently choosing nutrient-dense foods and engaging in regular physical activity are much more likely to maintain the loss and reap health benefits than if they were to lose more weight in less time by adopting a radical fad diet. In keeping with this philosophy, the *Dietary Guidelines for Americans* advise those who need to lose weight to "consume fewer kcalories from foods and beverages, increase physical activity, and reduce time in sedentary behaviors."

> **DIETARY GUIDELINES FOR AMERICANS 2010**
Strive to achieve and maintain a healthy body weight through improved eating and physical activity behaviors.

Even modest weight loss brings health benefits. Modest weight loss, even when a person is still overweight, can improve blood glucose and reduce the risks of heart disease by lowering blood pressure and blood cholesterol, especially for those with central obesity.[74] Improvements in physical capabilities and quality of life become evident with even a 5- to 10-pound weight loss.[75] For these reasons, parameters such as blood pressure, blood cholesterol, or even vitality are more useful than body weight in marking success. People less concerned with disease risks may prefer to set goals for personal fitness, such as being able to play with children or climb stairs without becoming short of breath. Importantly, they can focus on healthy eating and activity habits instead of weight loss.

Whether the goal is health or fitness, expectations need to be reasonable. Unreachable targets ensure frustration and failure. When realistic yet moderately challenging goals are achieved or exceeded, people enjoy rewards instead of finding disappointment.

Research findings highlight the great disparity between lofty expectations and reasonable success.[76] Before beginning a weight-loss program, obese women identified the weights they would describe as "dream," "happy," "acceptable," and "disappointing" (see Figure 9-6). All of these weights were below their starting weight. Their goals far exceeded the 5 to 10 percent weight loss recommended by experts, or even the 15 percent reported by the most successful weight-loss studies. Even their "disappointing" weights exceeded recommended goals. Close to a year later, and after an average loss of 35 pounds, almost half of the women did not achieve even their "disappointing" weights. They did, however, experience more physical, social, and psychological benefits than they had predicted for that weight. Still, in a culture that overvalues thinness, these women were not satisfied with a 16 percent reduction in weight—not because their efforts were unsuccessful, but because their goals and expectations were unrealistic.

Depending on initial body weight, a reasonable rate of weight loss for overweight adults is ½ to 2 pounds a week, ♦ or 10 percent of body weight over 6 months. For a person weighing 250 pounds, a 10 percent loss is 25 pounds, or about 1 pound a week for 6 months. Such gradual weight losses are more likely to be maintained than rapid losses. Keep in mind that pursuing good health is a lifelong journey.

♦ Safe rate for weight loss:
- ½ to 2 lbs/week (0.2 to 0.9 kg)
- 10% body weight/6 mo

For a person weighing 110 kg, a 10% loss is 11 kg, or about 0.5 kg a week for 6 months.

> FIGURE 9-6 **Reasonable Weight Goals vs Unrealistic Expectations Compared**

Obese women achieved remarkable success during a year's weight-loss program, but they were disappointed because they had set unrealistic expectations at the beginning.

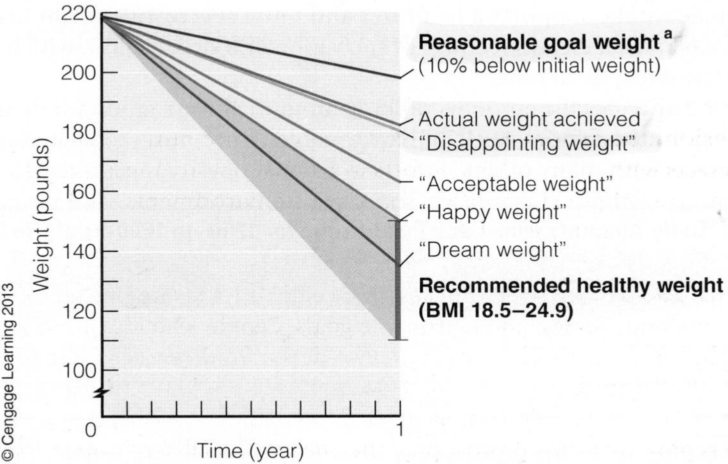

[a]Reasonable goal weights reflect pounds lost over time. Given more time, reasonable goals may eventually fall within the recommended weight range.

Most adults are keenly aware of their body weights and shapes and realize that what they eat and what they do can make a difference to some extent. Those who are most successful at weight management seem to have fully incorporated healthful eating and physical activity into their daily lives.

Eating Patterns Contrary to the claims of fad diets, no single food plan is magical, and no specific food must be included or avoided in a weight-management program. In designing an eating pattern, people need only consider foods that they like or can learn to like, that are available, and that are within their means. Creating a healthful eating pattern is the first step. The important next step is following it for the rest of one's life. Achieving and maintaining a healthy weight requires permanent lifestyle changes.

Be Realistic about Energy Intake The main characteristic of a weight-loss diet is that it provides less energy than the person needs to maintain present body weight. If food energy is restricted too severely, dieters may not receive sufficient nutrients and may lose lean tissue. Rapid weight loss usually means excessive loss of lean tissue, a lower BMR, and a rapid weight gain to follow. In addition, restrictive eating may create stress or foster unhealthy behaviors of eating disorders as described in Highlight 8.[77]

Energy intake should provide nutritional adequacy without excess—that is, somewhere between deprivation and complete freedom to eat whatever, whenever. A reasonable suggestion is to increase activity and reduce food intake enough to create a deficit of 500 to 1000 kcalories per day for adults with a BMI of 35 or greater and 300 to 500 kcalories per day for adults with a BMI of 27 to 35. Such a deficit produces a weight loss of 1 to 2 pounds per week—a rate that supports the loss of fat efficiently while retaining lean tissue.[78] In general, weight-loss diets need to provide about 1200 kcalories per day for women and 1600 kcalories a day for men.

Some people skip meals, typically breakfast, in an effort to reduce energy intake, but research suggests such a strategy may be counterproductive. Breakfast frequency is inversely associated with obesity—that is, people who frequently eat breakfast have a lower BMI than those who tend to skip breakfast.[79] Furthermore, when people eat breakfast, their overall diet quality is better and daily energy density is lower—two factors that support healthy body weight.[80]

> DIETARY GUIDELINES FOR AMERICANS 2010

Control total kcalorie intake to manage body weight. For people who are overweight or obese, this will mean consuming fewer kcalories from foods and beverages.

Emphasize Nutritional Adequacy Healthy diet plans make nutritional adequacy a priority. Nutritional adequacy is difficult to achieve on fewer than 1200 kcalories a day, and most healthy adults need never consume any less. A plan that provides an adequate intake supports a healthier and more successful weight loss than a restrictive plan that creates feelings of starvation and deprivation, which can lead to an irresistible urge to binge.

Table 9-2 specifies the amounts of foods from each food group for diets providing 1200 to 1600 kcalories. Such an intake would allow most people to lose weight

TABLE 9-2 Daily Amounts from Each Food Group for 1200- to 1600-kCalorie Diets

Food Group	1200 kCalories	1400 kCalories	1600 kCalories
Fruit	1 c	1½ c	1½ c
Vegetables	1½ c	1½ c	2 c
Grains	4 oz	5 oz	5 oz
Protein foods	3 oz	4 oz	5 oz
Milk and milk products	2½ c	2½ c	3 c
Oils	4 tsp	4 tsp	5 tsp

and still meet their nutrient needs with careful, nutrient-dense food selections. Keep in mind, too, that well-balanced diets that emphasize fruits, vegetables, whole grains, lean protein foods, and low-fat milk products offer many health rewards even when they don't result in weight loss. A dietary supplement providing vitamins and minerals—especially iron and calcium for women—at or below 100 percent of the Daily Values can help people following low-kcalorie diets to achieve nutrient adequacy.

Eat Small Portions As mentioned earlier, portion sizes at markets, at restaurants, and even at home have increased dramatically over the years. We have come to expect large portions, and we have learned to clean our plates. Many of us pay more attention to these external cues defining how much to eat than to our internal cues of hunger and satiety. For health's sake, we may need to learn to eat less food at each meal—one piece of chicken for dinner instead of two, a teaspoon of butter on vegetables instead of a tablespoon, and one cookie for dessert instead of six. The goal is to eat enough food for adequate energy, abundant vitamins and minerals, and some pleasure, but not more. This amount should leave a person feeling satisfied—not stuffed.

Keep in mind that even fat-free and low-fat foods can deliver a lot of kcalories when a person eats large quantities. A low-fat cookie or two can be a sweet treat even on a weight-loss diet, but larger portions defeat the savings.

People who have difficulty making low-kcalorie selections or controlling portion sizes may find it easier to use prepared meal plans. Prepared meals that provide low-kcalorie, nutritious meals or snacks can support weight loss while easing the task of diet planning.[81] Ideally, those using a prepared meal plan will also receive counsel from a registered dietitian to learn how to select appropriately from conventional food choices as well.

Slow Down Eating can be a pleasurable experience, and taking the time to savor the flavors can help with weight management. Eating slowly, taking small bites, and chewing thoroughly all help to decrease food intake.[82] A person who slows down and savors each bite eats less before hormones signal satiety and the end of a meal.[83] Consequently, energy intake is lower when meals are eaten slowly.[84] Savoring each bite also activates the pleasure centers of the brain. Some research suggests that people may overeat when the brain doesn't sense enough gratification from food.[85] Faster eating correlates with higher weights.[86]

Lower Energy Density Most people take their cues about how much to eat based on portion sizes, and the larger the portion size, the more they eat. To lower energy intake, a person can either reduce the portion size or reduce the energy density. Reducing energy density while maintaining food quantity, especially by including fruits and vegetables, seems to be a successful strategy to control hunger and lose weight.[87] Figure 9-7 illustrates how water, fiber, and fat influence energy density, and the accompanying "How To" feature (p. 276) compares foods based

> FIGURE 9-7 Energy Density

Decreasing the energy density (kcal/g) of foods allows a person to eat satisfying portions while still reducing energy intake. To lower energy density, select foods high in water or fiber and low in fat.

Selecting grapes with their high water content instead of raisins increases the volume and cuts the energy intake.

Even at the same weight and similar serving sizes, the fiber-rich broccoli delivers twice the fiber for about one-third the energy.

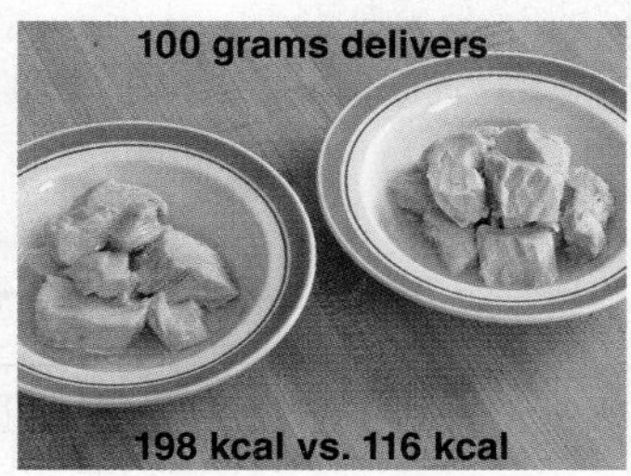

By selecting the water-packed tuna (on the right) instead of the oil-packed tuna (on the left), a person can enjoy the same amount for fewer kcalories.

>How To

Compare Foods Based on Energy Density

Chapter 2 describes how to evaluate foods based on their nutrient density—their nutrient contribution per kcalorie. Another way to evaluate foods is to consider their energy density—their energy contribution per gram. This example compares carrot sticks with french fries. The conclusion is no surprise, but understanding the mathematics may offer valuable insight into the concept of energy density. A carrot weighing 72 grams delivers 31 kcalories. To calculate the energy density, divide kcalories by grams:

$$\frac{31 \text{ kcal}}{72 \text{ g}} = 0.43 \text{ kcal/g}$$

Do the same for french fries weighing 50 grams and contributing 167 kcalories:

$$\frac{167 \text{ kcal}}{50 \text{ g}} = 3.34 \text{ kcal/g}$$

The more kcalories per gram, the greater the energy density. French fries are more energy dense than carrots. They provide more energy per gram—and per bite. Considering a food's energy density is especially useful in planning diets for weight management. Foods with a high energy density help with weight gain, whereas foods with a low energy density help with weight loss.

© Matthew Farruggio

TRY IT Compare the energy density of a hard-boiled egg (50 grams and 78 kcalories) with light tuna canned in water (57 grams and 66 kcalories).

on their energy density. Foods containing water, those rich in fiber, and those low in fat help to lower energy density, providing more satiety for fewer kcalories. Because a low-energy-density diet is a low-fat, high-fiber diet rich in many vitamins and minerals, it supports good health in addition to weight loss.

Remember Water In addition to lowering the energy density of foods, water helps with weight management in other ways. For one, foods with high water content (such as broth-based soups) increase fullness, reduce hunger, and consequently reduce energy intake. For another, drinking a large glass of water before a meal eases hunger, fills the stomach, and consequently reduces energy intake.[88] Importantly, water adds no kcalories. The average US diet delivers an estimated 75 to 150 kcalories a day from sweetened beverages. Simply replacing nutrient-poor, energy-dense beverages with water could save a person up to 15 pounds a year. Water also helps the GI tract adapt to a high-fiber diet.

Focus on Fiber High-fiber foods such as fresh fruits, vegetables, legumes, and whole grains may help with weight management.[89] By offering abundant vitamins, minerals, and fiber but little fat, these foods tend to be relatively low in energy and high in nutrients. Eating high-fiber foods also takes time, which eases hunger and promotes satiety.[90]

Choose Fats Sensibly One way to lower energy intake is to lower fat intake. Lowering the fat content of a food lowers its energy density—for example, selecting fat-free milk instead of whole milk. That way, a person can consume the usual amount (say, a cup of milk) at a lower energy intake (85 instead of 150 kcalories).

Fat has a weak satiating effect, and satiation plays a key role in determining food intake during a meal. Consequently, a person eating a high-fat meal increases energy intake in two ways—more food and more fat kcalories. For these reasons, measure fat with extra caution. (Review p. 152 for strategies to lower fat in the diet.) Be careful not to take this advice to extremes, however; too little fat incurs health risks as well, as Chapter 5 explains.

Whether a low-fat diet is the best option for weight loss is the subject of some controversy and much debate. An important point to notice in any discussion on weight-loss diets is total energy intake. A low-fat diet supports weight loss only when energy intake is less than energy expenditure.

Select Carbohydrates Carefully Another popular way to lower energy intake is to lower carbohydrate intake. Highlight 4's discussion of carbohydrate-restricted and carbohydrate-modified diets reaches the same conclusion as the previous paragraph on low-fat diets: they work only when energy intake is less than energy expenditure.

Chapter 4 describes how foods with added sugars increase energy intake and contribute to weight gain. Limiting consumption of foods with added sugars can help with weight management. One way people try to control weight is to use foods and beverages sweetened with artificial sweeteners. Using artificial sweeteners instead of sugars can lower energy intake and may support modest weight loss, although evidence is lacking.[91] In fact, some research indicates a relationship between artificial sweetener use and weight gain.[92]

To what extent artificial sweeteners can help someone lose weight depends in part on the person's motivations and actions. For example, one person might drink an artificially sweetened beverage now so as to be able to eat a high-kcalorie food later. This person's energy intake might stay the same or increase. A person trying to control energy intake might drink an artificially sweetened beverage now and choose a low-kcalorie food later. This plan would help reduce the person's total energy intake. Using artificial sweeteners will not automatically lower energy intake. To control energy intake successfully, a person needs to make informed diet and activity decisions throughout the day.

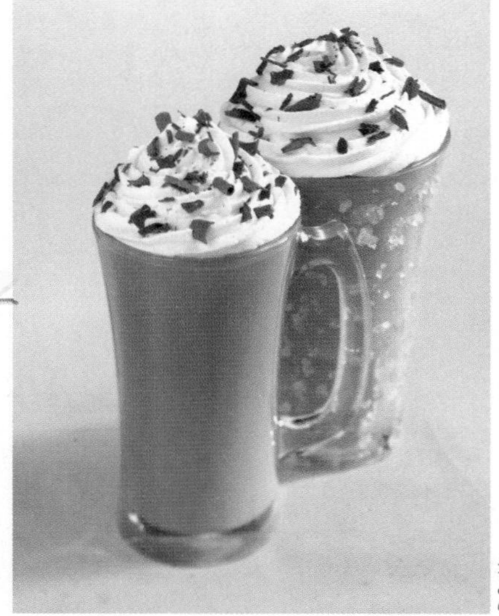

If you want to lose weight, steer clear of the empty kcalories in fancy coffee drinks. A 16-ounce café mocha delivers 400 kcalories—half of them from fat.

Watch for Other Empty kCalories A person trying to achieve or maintain a healthy weight needs to pay attention not only to fat and sugar, but to alcohol too. Not only does alcohol add kcalories, but accompanying mixers can also add both kcalories and fat, especially in creamy drinks such as piña coladas (review Table H7-3 on p. 226). Furthermore, drinking alcohol reduces a person's inhibitions, which can sabotage weight-control efforts—at least temporarily.

A person who adopts a lifelong "eating plan for good health" rather than a "diet for weight loss" will be more likely to keep the lost weight off. Table 9-3 provides several tips for successful weight management.

Physical Activity

Whether trying to minimize weight gains or support weight losses, the best approach includes physical activity.[93] To prevent weight gains and

TABLE 9-3 Weight-Loss Strategies

Food	Activities
• To maintain weight, consume foods and drinks to meet, not exceed, kcalorie needs. To lose weight, energy out should exceed energy in by about 500 kcalories/day. • Emphasize foods with a low energy density and a high nutrient density; make legumes, whole grains, vegetables, and fruits central to your diet plan. • Eat slowly. • Drink water before you eat and while you eat; drink plenty of water throughout the day. • Track food and kcalorie intake. • Plan ahead to make better food choices. • Limit kcalorie intake from solid fats and added sugars. • Reduce portions, especially of high-kcalorie foods. • Cook and eat more meals at home, instead of eating out. When eating out, think about choosing healthy options.	• Limit screen time. • Increase physical activity. • Choose moderate- or vigorous-intensity physical activities. • Avoid inactivity. Some physical activity is better than none. • Slowly build up the amount of physical activity you choose.

support weight losses, current recommendations advise 200 to 300 minutes of moderately intense physical activity a week in addition to activities of daily life.[94] People who combine diet and exercise typically lose more fat, retain more muscle, and regain less weight than those who only follow a weight-loss diet. Even when they do not lose more weight, they seem to follow their diet plans more closely and maintain their losses better than those who do not exercise. Consequently, they benefit from taking in a little less energy from the diet as well as from expending a little more energy in physical activity. Importantly, those who exercise reap important health benefits—reduced abdominal obesity and improved blood pressure, insulin resistance, and cardiorespiratory fitness—regardless of weight loss.[95] Fitness benefits—such as strength and balance—also improve when exercise is part of a weight-loss program.[96] Chapter 14 presents the many benefits of physical activity; the focus here is on its role in weight management.

> DIETARY GUIDELINES FOR AMERICANS 2010
Increase physical activity and reduce time spent in sedentary behaviors.

Activity and Energy Expenditure Table 8-2 (p. 238) shows how much energy each of several activities uses. The number of kcalories spent in an activity depends on body weight, intensity, and duration. For example, a person who weighs 150 pounds and walks 3½ miles in 60 minutes expends about 315 kcalories. That same person running 3 miles in 30 minutes uses a similar amount. By comparison, a 200-pound person running 3 miles in 30 minutes expends an additional 100 kcalories or so. The goal is to expend as much energy as your time allows. The greater the energy deficit created by exercise, the greater the fat loss. And be careful not to compensate for the energy expended in exercise by eating more food.[97] Otherwise, energy balance won't shift, and fat loss will be less significant.

Activity and Discretionary kCalories Chapter 2 introduced the concept of discretionary kcalories as the difference between the kcalories needed to supply nutrients and those needed to maintain energy balance. Because exercise expends energy, the energy allowance to maintain weight increases with increased physical activity—yet the energy needed to deliver needed nutrients remains about the same. In this way, physical activity increases discretionary kcalories (see Figure 9-8). Having more discretionary kcalories puts a little wiggle room in a weight-loss diet for such options as second helpings, sweet treats, or alcoholic beverages on occasion. Of course, selecting nutrient-dense foods and *not* using discretionary kcalories will maximize weight loss.

Activity and Metabolism Activity also contributes to energy expenditure in an indirect way—by speeding up metabolism. It does this both immediately and over

> FIGURE 9-8 ***Animated*** **Influence of Physical Activity on Discretionary kCalories**

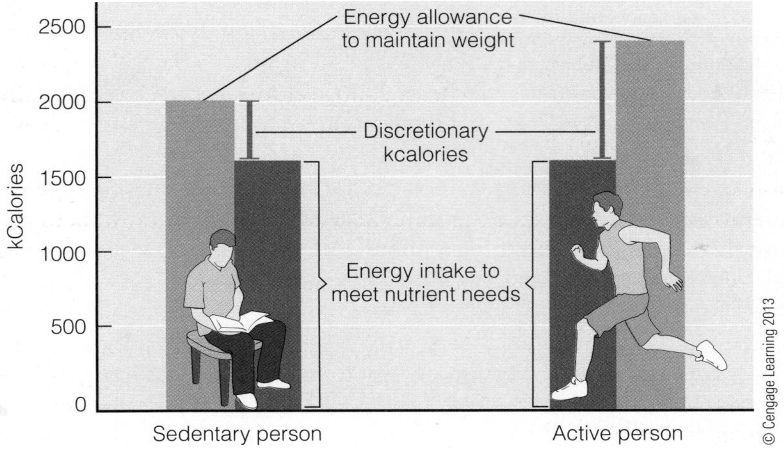

the long term. On any given day, metabolism remains elevated for several hours after vigorous and prolonged exercise.[98] This postexercise effect may raise the energy expenditure of exercise up to 15 percent. Over the long term, a person who engages in daily vigorous activity gradually develops more lean tissue. Metabolic rate rises accordingly, and this supports continued weight loss or maintenance.[99]

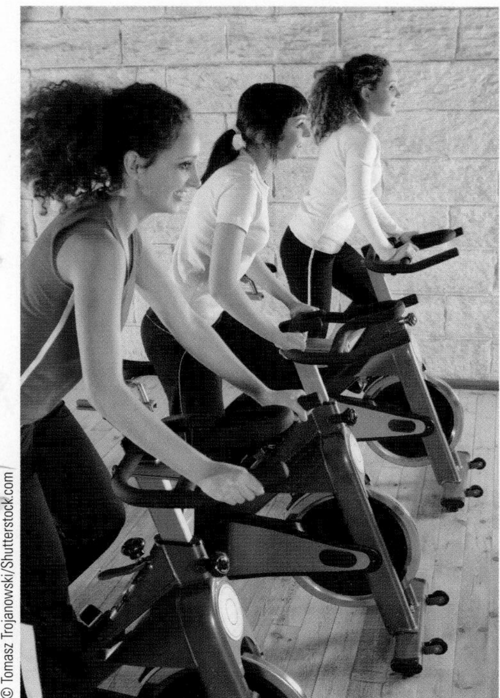

© Tomasz Trojanowski/Shutterstock.com

The key to good health is to combine sensible eating with regular exercise.

Activity and Body Composition Physically active people have less body fat than sedentary people do—even if they have the same BMI. Physical activity, even without weight loss, changes body composition: body fat decreases and lean body mass increases. Furthermore, strength training exercises specifically prevent increases in body fat and abdominal fat.[100]

Activity and Appetite Control Many people think that exercising will increase hunger, but not necessarily.[101] Active people do have healthy appetites, but appetite is suppressed after a workout and satiation during a meal and satiety between meals is enhanced.[102] The body has released fuels from storage to support the exercise, so glucose and fatty acids are abundant in the blood. At the same time, the body has suppressed its digestive functions. Hard physical work and eating are not compatible. A person must calm down, put energy fuels back in storage, and relax before eating. At that time, a physically active person may eat more than a sedentary person, but not so much as to fully compensate for the energy expended in exercise.

Exercise may also help curb the inappropriate appetite that accompanies boredom, anxiety, or depression. Weight-management programs encourage people who feel the urge to eat when not particularly hungry to exercise instead. The activity passes time, relieves anxiety, and prevents inappropriate eating.

Activity and Psychological Benefits Activity also helps reduce stress, which is especially helpful for people who respond to stress with inappropriate eating. In addition, a fit person looks and feels healthy and, as a result, gains self-esteem. High self-esteem motivates a person to continue seeking good health and fitness, which keeps the beneficial ♦ cycle going. Chapter 14 presents additional benefits of physical activity.

♦ Benefits of physical activity in a weight-management program:

- Short-term increase in energy expenditure (from exercise and from a slight rise in metabolism)
- Long-term increase in BMR (from an increase in lean tissue)
- Improved body composition
- Appetite control
- Stress reduction and control of stress eating
- Physical, and therefore psychological, well-being
- Improved self-esteem

Choosing Activities Clearly, physical activity is a plus in a weight-management program. What kind of physical activity is best? People should choose activities that they enjoy and are willing to do regularly. What schedule of physical activity is best? It doesn't matter; a person can benefit from either several short bouts of exercise or one continuous workout. Any activity is better than being sedentary. For an active life, limit sedentary activities, engage in strength and flexibility activities, enjoy leisure activities often, engage in vigorous activities regularly, and be as active as possible every day.

Health-care professionals frequently advise people to engage in activities of low-to-moderate intensity for a long duration, such as an hour-long, fast-paced walk. The reasoning behind such advice is that walking offers the health benefits of aerobic physical activity with low risk of injury. It can be done almost anywhere at any time. A person who stays with an activity routine long enough to enjoy the rewards will be less inclined to give it up and will, over the long term, reap many health benefits. A regular walking program can prevent or slow the weight gain that commonly occurs in most adults.[103] An average of 60 minutes a day of moderate-intensity activity ♦ or an expenditure of at least 2000 kcalories per week is especially helpful for weight management.[104] Higher levels of duration, frequency, or intensity produce greater losses.

♦ Estimated energy expended when walking at a moderate pace = 1 kcal/mi/kg body weight

In addition to exercise, a person can incorporate hundreds of energy-expending activities into daily routines: take the stairs instead of the elevator, walk to the neighbor's apartment instead of making a phone call, and rake the leaves instead of using a blower. Remember that sitting uses more kcalories than lying down, standing uses more kcalories than sitting, and moving uses more kcalories than standing. A 175-pound person who replaces a 30-minute television program with a 2-mile walk a day can expend enough energy to lose (or at least not gain) 18 pounds in a year. Meeting an activity goal of 10,000 steps a day is another way to support a healthy BMI. By wearing a pedometer, a person can easily increase physical activity, lose weight, and lower blood pressure without measuring miles or watching the clock. The point is to be active. Walk. Run. Swim. Dance. Cycle. Climb. Skip. Do whatever you enjoy doing—and do it often.

Spot Reducing People sometimes ask about "spot reducing." Unfortunately, muscles do not "own" the fat that surrounds them. Fat cells all over the body release fat in response to the demand of physical activity for use by whatever muscles are active. Specific exercises—whether moderate or intense—do not influence the site of adipose tissue loss.[105]

Exercise can help with trouble spots in another way, though. The "trouble spot" for most men is the abdomen, their primary site of fat storage. During aerobic exercise, abdominal fat readily releases its stores, providing fuel to the physically active body. With regular exercise and weight loss, men will deplete these abdominal fat stores before those in the lower body. Women may also deplete abdominal fat with exercise, but their "trouble spots" are more likely to be their hips and thighs.

In addition to aerobic activity, strength training can help to improve the tone of muscles in a trouble area, and stretching to gain flexibility can help with associated posture problems. A combination of aerobic, strength, and flexibility workouts best improves fitness and physical appearance.

Environmental Influences Chapter 8 describes how hormones regulate hunger, satiety, and satiation, but people don't always pay close attention to such internal signals. Instead, their eating behaviors are often dictated by environmental factors—those surrounding the eating experience as well as those pertaining to the food itself. Changing any of these factors can influence how much a person eats.

Atmosphere The environment surrounding a meal or snack influences its duration. When the lighting, décor, aromas, and sounds of an environment are pleasant and comfortable, people tend to spend more time eating and thus eat more. A person needn't eat under neon lights with offensive music to eat less, of course. Instead, after completing a meal, remove food from the table and enjoy the ambience—without the presence of visual cues to stimulate additional eating.

Accessibility Among the strongest influences on how much we eat are the accessibility, ease, and convenience of obtaining food. In general, the less effort needed to obtain food, the more likely food will be eaten. Think about it. Are you more likely to eat if half a leftover pizza is in your refrigerator or if you have to drive to the grocery store, buy a frozen pizza, and bake it for 45 minutes? Having food nearby and visible encourages eating—regardless of hunger. The message is clear. For people wanting to eat fewer empty-kcalorie foods, keep them out of sight in an inconvenient place, or better yet, don't even bring them home. In contrast, a bowl of fruit on the counter and vegetables in the refrigerator promote healthy eating options.

Socializing People tend to eat more when socializing with others. Pleasant conversations extend the duration of a meal, allowing a person more time to eat more, and the longer the meal, the greater the consumption. In addition, by taking a visual cue from companions, a person might eat more when others at the table eat large portions or go to the buffet line for seconds.[106] One way to eat less is to pace yourself with the person who seems to be eating the least and slowest.

Social interactions also distract a person from paying attention to how much has been eaten. In some cases, socializing with friends during a meal may provide comfort and lower a person's motivation to limit consumption. In other cases, socializing with unfamiliar people during a meal—during a job interview or blind date, for example—may create stress and reduce food consumption. To eat less while socializing, pay attention to portion size.

Distractions Distractions influence food intake by initiating eating, interfering with internal controls to stop eating, and extending the duration of eating. Some people start eating dinner when a favorite television program comes on, regardless of hunger. Other people continue eating breakfast until they finish reading the newspaper. Such mindless eating can easily become overeating. Distractions interfere with a person's ability to monitor and regulate how much is consumed. If distractions are a part of the eating experience, extra care is needed to control portion sizes.

Multiple Choices When offered a large assortment of foods, or several flavors of the same food, people tend to eat more. To limit intake, then, focus on a limited number of foods per meal. Be careful not to misunderstand and abandon variety in diet planning. Eating a variety of nutrient-dense foods from each of the food groups is still a healthy plan.

© moodboard / SuperStock

Eating from the package while distracted by television is a weight-gaining combination.

Package and Portion Sizes As noted earlier, the sizes of packages in grocery stores as well as portion sizes at restaurants and at home have increased dramatically in recent decades, contributing to the increase in obesity in the United States. Put simply, we tend to clean our plates and finish the package. The larger the bag of potato chips, the greater the intake. To keep from overeating, repackage snacks into smaller containers or eat a measured portion from a plate, not directly from the package.

Serving Containers We often use plates, utensils, and glasses as visual cues to guide our decisions on how much to eat and drink. If you plan to eat a bowl of ice cream, it matters whether the bowl you select holds 8 ounces or 24 ounces. Large dinner plates and wide glasses create illusions and misperceptions about quantities consumed. A scoop of mashed potatoes on a small plate looks larger than the same-size scoop on a large plate, leading a person to underestimate the amount of food eaten. To control portion sizes, use small bowls and plates, small serving spoons, and tall, narrow glasses. Of course, using a small plate will not result in less food eaten if multiple servings are taken.

Behavior and Attitude

Changes in behavior and attitude can be very effective in supporting efforts to achieve and maintain appropriate body weight and composition. **Behavior modification** focuses on how to change behaviors to increase energy expenditure and decrease energy intake. A person must commit to taking action. Adopting a positive, matter-of-fact attitude helps to ensure success. Healthy eating and activity choices are an essential part of healthy living and should simply be incorporated into the day—much like brushing one's teeth or wearing a safety belt.

Become Aware of Behaviors To solve a problem, a person must first identify all the behaviors that created the problem. Keeping a record will help to identify eating and exercise behaviors that may need changing (see Figure 9-9, p. 282). Such self-monitoring raises awareness and establishes a baseline against which to measure future progress.[107]

In this era of technology, many companies have developed weight-loss applications for smartphones to help users manage their daily food and physical activity behaviors. Applications include diet analysis tools that can track eating habits, scanning devices that can quickly enter food data, customized activity and meal plans that can be sent to users, and support programs that deliver encouraging messages and helpful tips. Social media sites allow users to upload progress reports and receive texts. Using these applications can help a person become more aware of behaviors that lead to weight gains and losses.

Change Behaviors Behavior modification strategies ♦ focus on learning desired eating and activity behaviors and eliminating unwanted behaviors. With so many possible behavior changes, a person can feel overwhelmed. Start with small time-specific goals for each behavior—for example, "I'm going to take a 30-minute walk after dinner every evening" instead of "I'm going to run in a marathon someday." Practice desired behaviors until they become routine. Addressing multiple behaviors that focus on a common goal simultaneously may better support changes

♦ Examples of behavioral strategies to support weight change:

- Do not grocery shop when hungry.
- Eat slowly (pause during meals, chew thoroughly, put down utensils between bites).
- Exercise when watching television.

behavior modification: the changing of behavior by the manipulation of antecedents (cues or environmental factors that trigger behavior), the behavior itself, and consequences (the penalties or rewards attached to behavior).

> FIGURE 9-9 Food Record

The entries in a food record should include the times and places of meals and snacks, the types and amounts of foods eaten, and a description of the individual's feelings when eating. The diary should also record physical activities: the kind, the intensity level, the duration, and the person's feelings about them.

Time	Place	Activity or food eaten	People present	Mood
10:30–10:40	School vending machine	6 peanut butter crackers and 12 oz. cola	by myself	Starved
12:15–12:30	Restaurant	Sub sandwich and 12 oz. cola	friends	relaxed & friendly
3:00–3:45	Gym	Weight training	work out partner	tired
4:00–4:10	Snack bar	Small frozen yogurt	by myself	OK

than taking on one at a time. Using a reward system also seems to effectively support weight-loss efforts.[108]

Cognitive Skills Successful behavior changes depend in part on two cognitive skills—problem solving and cognitive restructuring. Problem solving skills enable a person to identify the problem, generate potential solutions, list the pros and cons of each, implement the most feasible solution, and evaluate whether behaviors should be continued or abandoned.[109] Cognitive restructuring requires a person to replace negative thoughts that derail success with positive thoughts that support behavior change.

The effectiveness of cognitive behavioral therapy in weight-loss extends to other health behaviors as well. Overweight smokers who participated in a cognitive program for weight management lost weight, made healthy food choices, increased their confidence to manage their eating and smoking habits, decreased the number of cigarettes smoked, and increased their readiness to quit smoking.[110] Such research highlights the need to include dietary strategies in smoking cessation programs. Smoking a cigarette overrides feelings of hunger. When smokers receive a hunger signal, they can quiet it with cigarettes instead of food. Such behavior ignores body signals and postpones energy and nutrient intake. Indeed, smokers tend to weigh less than nonsmokers and to gain weight when they stop smoking. People contemplating giving up cigarettes should know that the average weight gain is about 10 pounds in the first year. Smokers wanting to quit should prepare for the possibility of weight gain and adjust their diet and activity habits so as to maintain weight during and after quitting.

Personal Attitude For many people, overeating and being overweight have become an integral part of their identity. Those who fully understand their personal relationships with food are best prepared to make healthful changes in eating and activity behaviors.

Sometimes habitual behaviors that are hazardous to health, such as smoking or drinking alcohol, contribute positively by helping people adapt to stressful situations. Similarly, many people overeat to cope with the stresses of life. Weight gains, in turn, contribute to psychosocial stress, thus creating an unhealthy cycle.[111] To break out of that pattern, they must first identify the particular stressors that trigger the urge to overeat. Then, when faced with these situations, they must learn and practice problem-solving skills that will help them to respond appropriately. Learning to reduce episodes of emotional eating can lead to weight loss.[112]

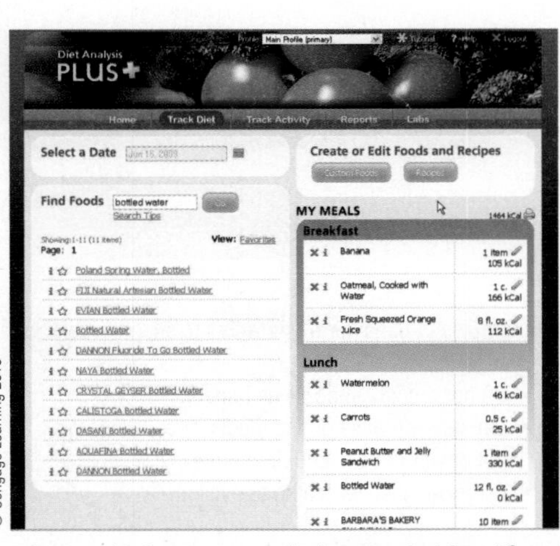

Diet analysis programs help people identify high-kcalorie foods and monitor their eating habits.

All this is not to imply that psychotherapy holds the magic answer to a weight problem. Still, efforts to improve one's general well-being may result in healthy eating and activity habits even when weight loss is not the primary goal. When the problems that trigger the urge to overeat are resolved in alternative ways, people may find they eat less. They may begin to respond appropriately to internal cues of hunger rather than inappropriately to external cues of stress. Sound emotional health supports a person's ability to take care of physical health in all ways—including nutrition, weight management, and fitness.

Support Groups Group support can prove helpful when making life changes. Some people find it useful to join a group such as Take Off Pounds Sensibly (TOPS), Weight Watchers (WW), Overeaters Anonymous (OA), or others. Some dieters prefer to form their own self-help groups or find support online. The Internet offers numerous opportunities for weight-loss education and counseling that may be effective alternatives to face-to-face or telephone counseling programs.[113] As always, consumers need to choose wisely and avoid rip-offs.

Weight Maintenance People who are successful often experience much of their weight loss within half a year and then reach a plateau. This slowdown can be disappointing, but it should be recognized as an opportunity for the body to adjust to its new weight. Reaching a plateau provides a little relief from the distraction of weight-loss dieting. An appropriate goal at this point is to continue the eating and activity behaviors that will maintain weight. Attempting to lose additional weight at this point would require major effort and would almost certainly meet with failure.

The prevalence of **successful weight-loss maintenance** is difficult to determine, in part because researchers have used different criteria. Some look at success after 1 year and others after 5 years; some quantify success as 10 or more pounds lost and others as 5 or 10 percent of initial body weight lost. Furthermore, most research studies examine the success of one episode of weight loss in a structured program, but this scenario does not necessarily reflect the experiences of the general population. In reality, most people have lost weight several times in their lifetimes and did so on their own, not in a formal program. An estimated one out of every six overweight adults in the United States has successfully maintained at least a 10 percent loss for at least a year.[114]

Those who are successful in maintaining their weight loss have established regular exercise regimens and careful eating patterns, taking in less energy than the

© Blend Images/Dan Bannister/Jupiterimages

Maintaining a healthy body weight requires maintaining the vigorous physical activities and careful eating habits that supported weight loss.

successful weight-loss maintenance: achieving a weight loss of at least 10 percent of initial body weight and maintaining the loss for at least 1 year.

national average. Because these people are more efficient at storing fat, they do not have the same flexibility in their food and activity habits as their friends who have never been overweight.[115] With weight loss, metabolism shifts downward so that formerly overweight people require less energy than might be expected given their current body weight and body composition. This decrease in energy expenditure persists over time.[116] Consequently, to keep weight off, they must either eat less or exercise more than people the same size who have never been obese. Put simply, it takes more to prevent weight *regain* than to prevent weight gain.

Physical activity plays a key role in preventing weight gains and maintaining weight losses.[117] Those who exercise vigorously are far more successful than those who are inactive. Weight maintenance may require a person to expend at least 2500 kcalories in physical activity per week.[118] To accomplish this, a person might exercise either moderately (such as brisk walking at 4 miles per hour) for 60 minutes a day or vigorously (such as fast bicycling at 18 miles per hour) for 30 minutes a day, for example. Being active during both work hours and leisure time also helps a person expend more energy and maintain weight loss.[119]

In addition to limiting energy intake and exercising regularly, one other strategy helps with weight maintenance: frequent self-monitoring. People who weigh themselves periodically and monitor their eating and exercise habits regularly can detect weight gains in the early stages and promptly initiate changes to prevent relapse.

Losing weight and maintaining the loss may not be easy, but it is possible. Strategies of those who have been successful may differ in the details, but in general, most do the following:[120]

- Eat a low-kcalorie diet (usually small portions four to five times a day).
- Follow a diet that is high in nutrient density and low in energy density.
- Eat breakfast (curbs hunger).
- Be physically active regularly (at least 60 minutes of moderate activity daily).
- Monitor weight frequently (take prompt action with small gains).
- Limit television time (less than 10 hours a week).
- Consult a registered dietitian, physician, or other support person (or group).

Importantly, people who are successful losing weight find that it gets easier with time—the changes in diet and activity patterns become permanent.

Prevention Given the information presented up to this point in the chapter, the adage "An ounce of prevention is worth a pound of cure" seems particularly apropos. Obesity is a major risk factor for numerous diseases, and losing weight is challenging and often temporary. Strategies for preventing weight gain ♦ are very similar to those for losing weight, with one exception: they begin early. Over the years, they become an integral part of a person's life. It is much easier for a person to resist doughnuts for breakfast if he rarely eats them. Similarly, a person will have little trouble walking each morning if she has always been active.

♦ To prevent weight gain:

- Eat regular meals and limit snacking
- Drink water instead of high-kcalorie beverages
- Select sensible portion sizes and limit daily energy intake to no more than energy expended
- Become physically active and limit sedentary activities

> DIETARY GUIDELINES FOR AMERICANS 2010

Maintain appropriate kcalorie balance during each stage of life—childhood, adolescence, adulthood, pregnancy and lactation, and older age.

Community Programs Reversing the US obesity epidemic is a challenge in an environment of abundant food and physical inactivity. Success may depend on community actions to promote healthy lifestyle choices.[121] Table 9-4 lists health strategies to prevent obesity in the United States.[122] Whether changes in public policy—such as a tax on sugared beverages and snack foods—will influence diet habits or simply generate revenues remains to be seen.[123] Clearly, effective strategies will need to reach beyond individuals to address social networks, community institutions, and government policies.[124]

TABLE 9-4 Community Strategies to Prevent Obesity

- Promote the availability of affordable healthy food and beverages.
- Support healthy food and beverage choices.
- Encourage breastfeeding.
- Encourage physical activity and limit sedentary activity, especially among children and youth.
- Create safe communities that support physical activity.
- Encourage communities to organize for change.

REVIEW IT Outline reasonable strategies for achieving and maintaining a healthy body weight.

A surefire remedy for obesity has yet to be found, although many people find a combination of approaches to be most effective. Diet and exercise shift energy balance so that more energy is expended than is taken in. Behavior modification and cognitive restructuring retrain habits to support a healthy eating and activity plan. Such a plan requires time, individualization, and sometimes the assistance of a registered dietitian or support group.

9.6 Underweight

LEARN IT Summarize strategies for gaining weight.

Underweight is a far less prevalent problem than overweight, affecting no more than 2 percent of US adults (review Figure 8-7 on p. 243).[125] Whether an underweight person needs to gain weight is a question of health and, like weight loss, a highly individual matter. There are no compelling reasons for people who are healthy at their present weight to try to gain weight. Those who are thin because of malnourishment or illness, however, might benefit from a diet that supports weight gain. Medical advice can help make the distinction.

Thin people may find gaining weight difficult. Unlike the genes expressed in obesity, the genes in lean people protect against energy excesses.[126] Those who wish to gain weight for appearance's sake or to improve their athletic performance need to be aware that healthful weight gains can be achieved only by physical conditioning combined with high energy intakes. On a high-kcalorie diet alone, a person may gain weight, but it will be mostly fat. Even if the gain improves appearance, it can be detrimental to health and might impair athletic performance. Therefore, in weight gain, as in weight loss, physical activity and energy intake are essential components of a sound plan.

Problems of Underweight

The causes of underweight may be as diverse as those of overweight—genetic tendencies; hunger, appetite, and satiety irregularities; psychological traits; and metabolic factors. Habits learned early in childhood, especially food aversions, may perpetuate themselves.

The high demand for energy to support physical activity and growth may contribute to underweight. An active, growing boy may need more than 4000 kcalories a day to maintain his weight and may be too busy to take time to eat adequately. In addition, underweight people may find it hard to gain weight because they are expending energy in adaptive thermogenesis. So much energy may be expended adapting to a higher food intake that at first as many as 750 to 800 extra kcalories a day may be needed to gain a pound a week. Like those who want to lose weight, people who want to gain must learn new habits and learn to like new foods. They are also similarly vulnerable to potentially harmful schemes.

As described in Highlight 8, the underweight condition anorexia nervosa sometimes develops in people who employ self-denial to control their weight. They go to such extremes that they become severely undernourished, achieving final body weights of 70 pounds or even less. One difference between a person with anorexia nervosa and other underweight people is that starvation is intentional. (See Highlight 8 for a review of anorexia nervosa and other eating disorders.)

underweight: a body weight so low as to have adverse health effects; generally defined as BMI <18.5.

Weight-Gain Strategies Adequacy and balance are the key diet planning strategies for weight gain. Meals focus on energy-dense foods to provide many kcalories in a small volume and exercise to build muscle. By using the USDA Food Pattern recommendations for the higher kcalorie levels (see Table 2-2 on p. 44), a person can gain weight while meeting nutrient needs.

Energy-Dense Foods Energy-dense foods (the very ones eliminated from a successful weight-loss diet) hold the key to weight gain. Pick the highest-kcalorie items from each food group—that is, milk shakes instead of fat-free milk, salmon instead of snapper, avocados instead of cucumbers, a cup of grape juice instead of a small apple, and whole-wheat muffins instead of whole-wheat bread. Because fat provides more than twice as many kcalories per teaspoon as sugar does, fat adds kcalories without adding much bulk.

Although eating high-kcalorie, high-fat foods is not healthy for most people, it may be essential for an underweight individual who needs to gain weight. An underweight person who is physically active and eating a nutritionally adequate diet can afford a few extra kcalories from fat. For health's sake, it is wise to select foods with monounsaturated and polyunsaturated fats instead of those with saturated or *trans* fats: for example, sautéing vegetables in olive oil instead of butter or hydrogenated margarine.

Regular Meals Daily People who are underweight need to make meals a priority and take the time to plan, prepare, and eat each meal. They should eat at least three healthy meals every day. Another suggestion is to eat meaty appetizers or the main course first and leave the soup or salad until later.

Large Portions Underweight people need to learn to eat more food at each meal. For example, they can add extra slices of ham and cheese on a sandwich for lunch, drink milk from a larger glass, and eat cereal from a larger bowl.

The person should expect to feel full. Most underweight individuals are accustomed to small quantities of food. When they begin eating significantly more, they feel uncomfortable. This is normal and passes over time.

Extra Snacks Because a substantially higher energy intake is needed each day, in addition to eating more food at each meal, it is necessary to eat more frequently. Between-meal snacks can readily lead to weight gains. For example, a student might make three sandwiches in the morning and eat them between classes in addition to the day's three regular meals. Snacking on dried fruit, nuts, and seeds is also an easy way to add kcalories.

Juice and Milk Beverages provide an easy way to increase energy intake. Consider that 6 cups of cranberry juice add almost 1000 kcalories to the day's intake. kCalories can be added to milk by mixing in powdered milk or packets of instant breakfast.

For people who are underweight due to illness, liquid dietary supplements are often recommended because a weak person can swallow them easily. Used in addition to regular meals, these high-protein, high-kcalorie formulas can help an underweight person maintain or gain weight easily.

Exercising to Build Muscles To gain weight, use strength training primarily, and increase energy intake to support that exercise. Eating extra food to provide an additional 500 to 1000 kcalories a day above normal energy needs is enough to support the exercise as well as to build muscle.[127]

> **REVIEW IT** Summarize strategies for gaining weight.
> Both the incidence of underweight and the health problems associated with it are less prevalent than overweight and its associated problems. To gain weight, a person must train physically and increase energy intake by selecting energy-dense foods, eating regular meals, taking larger portions, and consuming extra snacks and beverages. Table 9-5 includes a summary of weight-gain strategies.

Achieving and maintaining a healthy weight requires vigilant attention to diet and physical activity. Taking care of oneself is a lifelong responsibility.

TABLE 9-5 Weight-Gain Strategies

- Energy in should exceed energy out by at least 500 kcalories/day. Eat enough to store more energy than you expend in exercise. Exercise and eat to build muscles.
- Expect weight gain to take time (1 pound per month would be reasonable).
- Emphasize energy-dense foods.
- Eat at least three meals a day.
- Eat large portions of foods and expect to feel full.
- Eat snacks between meals.
- Drink plenty of juice and milk.

Nutrition Portfolio

To enjoy good health and maintain a reasonable body weight, combine sensible eating habits and regular physical activity. Go to Diet Analysis Plus and choose one of the days on which you have tracked your diet for the entire day. Go to the Energy Balance and Intake vs. Goals reports.

- Calculate your BMI and consider whether you need to lose or gain weight for the sake of good health. If you do need to gain or lose weight, do the Diet Analysis reports give you insight into why you may be overweight or underweight?
- Reflect on your weight over the past year or so and explain any weight gains or losses. Using the Intake vs. Goals report, can you identify areas in which you need to adjust your food intake, perhaps eating more or less?
- Describe the potential risks and possible benefits of fad diets and over-the-counter weight-loss drugs or herbal supplements.

Diet Analysis PLUS To complete this exercise, go to your Diet Analysis Plus at **www.cengagebrain.com.**

STUDY IT To review the key points of this chapter and take a practice quiz, go to the study cards at the end of the book.

REFERENCES

1. K. M. Flegal and coauthors, Prevalence and trends in obesity among US adults, 1999–2008, *Journal of the American Medical Association* 303 (2010): 235–241.
2. Y. Wang and M. A. Beydoun, The obesity epidemic in the United States—Gender, age, socioeconomic, racial/ethnic, and geographic characteristics: A systematic review and meta-regression analysis, *Epidemiological Reviews* 29 (2007): 6–28.
3. C. L. Ogden and M. D. Carroll, Prevalence of overweight, obesity, and extreme obesity among adults: United States, trends 1960–1962 through 2007–2008, *NCHS Health E-Stats*, www.cdc.gov/nchs/data/hestat/obesity_adult_07_08/obesity_adult_07_08.htm, accessed June 2010.
4. Wang and Beydoun, 2007.
5. World Health Organization, www.who.int/features/factfiles/obesity, updated February 2010.
6. Y. D. Tchoukalova and coauthors, Regional differences in cellular mechanisms of adipose tissue gain with overfeeding, *Proceedings of the National Academic of Sciences of the United States of America* 107 (2010): 18226–18231; R. Drolet and coauthors, Hypertrophy and hyperplasia of abdominal adipose tissues in women, *International Journal of Obesity* 32 (2008): 283–291.
7. K. L. Spalding and coauthors, Dynamics of fat cell turnover in humans, *Nature* 453 (2008): 783–787.
8. C. A. Baile and coauthors, Effect of resveratrol on fat mobilization, *Annals of the New York Academy of Sciences* 1215 (2011): 40–47.
9. M. Krawczyk, L. Bonfrate, and P. Portincasa, Nonalcoholic fatty liver disease, *Best Practice and Research, Clinical Gastroenterology* 24 (2010): 695–708; D. M. Muoio, Metabolism and vascular fatty acid transport, *New England Journal of Medicine* 363 (2010): 291–293; G. Tarantino, S. Savastano, and A. Colao, Hepatic steatosis, low-grade chronic inflammation and hormone/growth factor/adipokine imbalance, *World Journal of Gastroenterology* 16 (2010): 4773–4783.
10. N. Ouchi and coauthors, Adipokines in inflammation and metabolic disease, *Nature Reviews Immunology* 11 (2011): 85–97; C. Stryjecki and D. M. Mutch, Fatty acid-gene interactions, adipokines and obesity, *European Journal of Clinical Nutrition* 65 (2011): 285–297; G. Govindarajan, M. A. Alpert, and L. Tejwani, Endocrine and metabolic effects of fat: Cardiovascular implications, *American Journal of Medicine* 121 (2008): 366–370.
11. K. A. Varady and coauthors, Degree of weight loss required to improve adipokine concentrations and decrease fat cell size in severely obese women, *Metabolism Clinical and Experimental* 58 (2009): 1096–1101.
12. H. Wang and R. H. Eckel, Lipoprotein lipase: From gene to obesity, *American Journal of Physiology: Endocrinology and Metabolism* 297 (2009): E271–E288; P. J. Voshol and coauthors, Effect of plasma triglyceride metabolism on lipid storage in adipose tissue: Studies using genetically engineered mouse models, *Biochimica et Biophysica Acta* 1791 (2009): 479–485.
13. E. Pardina and coauthors, Lipoprotein lipase but not hormone-sensitive lipase activities achieve normality after surgically induced weight loss in morbidly obese patients, *Obesity Surgery* 19 (2009): 1150–1158.
14. K. R. Westerterp and coauthors, Dietary fat oxidation as a function of body fat, *American Journal of Clinical Nutrition* 87 (2008): 132–135.

15. E. Lebenthal, Leptin: Weight management and beyond. Introduction to the symposium, *American Journal of Clinical Nutrition* 89 (2009): 971S–972S.
16. R. A. Waterland, Epigenetic epidemiology of obesity: Application of epigenomic technology, *Nutrition Reviews* 66 (2008): S21–S23.
17. K. Silventoinen and coauthors, The genetic and environmental influences on childhood obesity: A systematic review of twin and adoption studies, *International Journal of Obesity* 34 (2010): 29–40.
18. J. Wardle and coauthors, Evidence for a strong genetic influence on childhood adiposity despite the force of the obesogenic environment, *American Journal of Clinical Nutrition* 87 (2008): 398–404.
19. C. M. Lindgren and M. I. McCarthy, Mechanisms of disease: Genetic insights into the etiology of type 2 diabetes and obesity, *Nature Clinical Practice. Endocrinology & Metabolism* 4 (2008): 156–163.
20. S. Li and coauthors, Physical activity attenuates the genetic predisposition to obesity in 20,000 men and women from EPIC-Norfolk prospective population study, *PLoS Medicine* 7 (2010): e1000331; J. M. McCaffery and coauthors, Gene X environment interaction of vigorous exercise and body mass index among male Vietnam-era twins, *American Journal of Clinical Nutrition* 89 (2009): 1011–1018; K. S. Vimaleswaran and coauthors, Physical activity attenuates the body mass index-increasing influence of genetic variation in the *FTO* gene, *American Journal of Clinical Nutrition* 90 (2009): 425–428.
21. E. Sonestedt and coauthors, Fat and carbohydrate intake modify the association between genetic variation in the *FTO* genotype and obesity, *American Journal of Clinical Nutrition* 90 (2009): 1418–1425.
22. R. J. Loos and C. Bouchard, FTO: The first gene contributing to common forms of human obesity, *Obesity Reviews* 9 (2008): 246–250; N. J. Timpson and coauthors, The fat mass- and obesity-associated locus and dietary intake in children, *American Journal of Clinical Nutrition* 88 (2008): 971–978; A. Körner and coauthors, Polygenic contribution to obesity: Genome-wide strategies reveal new targets, *Frontiers of Hormone Research* 36 (2008): 12–36; R. L. Leibel, Energy in, energy out, and the effects of obesity-related genes, *New England Journal of Medicine* 359 (2008): 2603–2604.
23. J. Hall, R. Roberts, and N. Vora, Energy homeostasis: The roles of adipose tissue-derived hormones, peptide YY and ghrelin, *Obesity Facts* 2 (2009): 117–125.
24. J. M. Friedman, Leptin at 14 y of age: An ongoing story, *American Journal of Clinical Nutrition* 89 (2009): 973S–979S; M. Rosenbaum and coauthors, Leptin reverses weight loss–induced changes in regional neural activity responses to visual food stimuli, *Journal of Clinical Investigation* 118 (2008): 2583–2591.
25. K. W. Williams, M. M. Scott, and J. K. Elmquist, From observation to experimentation: Leptin action in the mediobasal hypothalamus, *American Journal of Clinical Nutrition* 89 (2009): 985S–990S; Rosenbaum and coauthors, 2008.
26. I. S. Farooqi and S. O'Rahilly, Leptin: A pivotal regulator of human energy homeostasis, *American Journal of Clinical Nutrition* 89 (2009): 980S–984S.
27. M. G. Myers, M. A. Cowley, and H. Münzberg, Mechanisms of leptin action and leptin resistance, *Annual Review of Physiology* 70 (2008): 537–556.
28. A. Shapiro and coauthors, Fructose-induced leptin resistance exacerbates weight gain in response to subsequent high fat feeding, *American Journal of Physiology. Regulatory, Integrative and Comparative Physiology* 295 (2008): R1370–R1375.
29. T. R. Castañeda and coauthors, Ghrelin in the regulation of body weight and metabolism, *Frontiers in Neuroendocrinology* 31 (2010): 44–60.
30. P. Lyytikäinen and coauthors, Sleep problems and major weight gain: A follow-up study, *International Journal of Obesity* 35 (2011): 109–114; L. Brondel and coauthors, Acute partial sleep deprivation increases food intake in healthy men, *American Journal of Clinical Nutrition* 91 (2010): 1550–1559; A. V. Nedeltcheva and coauthors, Sleep curtailment is accompanied by increased intake of calories from snacks, *American Journal of Clinical Nutrition* 89 (2009): 126–133.
31. H. H. Chen and coauthors, Severe obesity is associated with novel single nucleotide polymorphisms of the *ESR1* and *PPARγ* locus in Han Chinese, *American Journal of Clinical Nutrition* 90 (2009): 255–262; M. Hofker and C. Wijmenga, A supersized list of obesity genes, *Nature Genetics* 41 (2009): 139–140.
32. S. Li and coauthors, Cumulative effects and predictive value of common obesity-susceptibility variants identified by genome-wide association studies, *American Journal of Clinical Nutrition* 91 (2010): 184–190.
33. M. M. Hetherington and J. E. Cecil, Gene-environment interactions in obesity, *Forum of Nutrition* 63 (2010): 195–203; C. Bouchard, Defining the genetic architecture of the predisposition to obesity: A challenging but not insurmountable task, *American Journal of Clinical Nutrition* 91 (2010): 5–6; J. Hebebrand and A. Hinney, Environmental and genetic risk factors in obesity, *Child and Adolescent Psychiatric Clinics of North America* 18 (2009): 83–94.
34. S. Enerbäck, The origins of brown adipose tissue, *New England Journal of Medicine* 360 (2009): 2021–2023.
35. F. S. Celi, Brown adipose tissue: When it pays to be inefficient, *New England Journal of Medicine* 360 (2009): 1553–1556; G. Wolf, Brown adipose tissue: The molecular mechanism of its formation, *Nutrition Reviews* 67 (2009): 167–171.
36. M. Harper, K. Green, and M. D. Brand, The efficiency of cellular energy transduction and its implications for obesity, *Annual Review of Nutrition* 28 (2008): 13–33.
37. M. Saito, High incidence of metabolically active brown adipose tissue in healthy adult humans, *Diabetes* 58 (2009): 1526–1531; W. D. van Marken Lichtenbelt and coauthors, Cold-activated brown adipose tissue in healthy men, *New England Journal of Medicine* 360 (2009): 1500–1508.
38. A. M. Cypess and coauthors, Identification and importance of brown adipose tissue in adult humans, *New England Journal of Medicine* 360 (2009): 1509–1517.
39. E. Ravussin and J. E. Galgani, The implication of brown adipose tissue for humans, *Annual Review of Nutrition* 31 (2011): 33–47; P. Seale, S. Kajimura, and B. M. Spiegelman, Transcriptional control of brown adipocyte development and physiological function: Of mice and men, *Genes and Development* 23 (2009): 788–797; K. A. Virtanen and coauthors, Functional brown adipose tissue in healthy adults, *New England Journal of Medicine* 360 (2009): 1518–1525.
40. L. Qi and Y. A. Cho, Gene-environment interaction and obesity, *Nutrition Reviews* 66 (2008): 684–694.
41. B. A. Swinburn, G. Sacks, and E. Ravussin, Increased food energy supply is more than sufficient to explain the US epidemic of obesity, *American Journal of Clinical Nutrition* 90 (2009): 1453–1456; B. A. Swinburn and coauthors, Estimating the changes in energy flux that characterizes the rise in obesity prevalence, *American Journal of Clinical Nutrition* 89 (2009): 1723–1728; W. P. James, The fundamental drivers of the obesity epidemic, *Obesity Reviews* 9 (2008): S6–S13.
42. I. Romao and J. Roth, Genetic and environmental interactions in obesity and type 2 diabetes, *Journal of the American Dietetic Association* 108 (2008): S24–S28.
43. J. M. McCaffery and coauthors, Effects of social contact and zygosity on 21-y weight change in male twins, *American Journal of Clinical Nutrition* 94 (2011): 404–409.
44. J. Currie and coauthors, The effect of fast food restaurants on obesity and weight gain, *American Economic Journal* 2 (2009): 32–63.
45. B. M. Popkin and K. J. Duffey, Does hunger and satiety drive eating anymore? Increasing eating occasions and decreasing time between eating occasions in the United States, *American Journal of Clinical Nutrition* 91 (2010): 1342–1347; K. J. Duffey and coauthors, Regular consumption from fast food establishments relative to other restaurants is differentially associated with metabolic outcomes in young adults, *Journal of Nutrition* 139 (2009): 2113–2118.
46. K. J. Duffey and B. M. Popkin, Energy density, portion size, and eating occasions: Contributions to increased energy intake in the United States, 1977–2006, *PLoS Medicine* 8 (2011): e1001050.
47. R. N. Close and D. A. Schoeller, The financial reality of overeating, *Journal of the American College of Nutrition* 25 (2006): 203–209.
48. I. H. Steenhuis and W. M. Vermeer, Portion size: Review and framework for interventions, *International Journal of Behavioral Nutrition and Physical Activity* 6 (2009): 58–68.
49. T. S. Church and coauthors, Trends over 5 decades in US occupation-related physical activity and their association with obesity, *PLoS ONE* 6 (2011): e19657.

50. K. Wijndaele and coauthors, Increased cardiometabolic risk is associated with increased TV viewing time, *Medicine and Science in Sports and Exercise* 42 (2010): 1511–1518.
51. J. A. Teske, C. J. Billington, and C. M. Kotz, Neuropeptidergic mediators of spontaneous physical activity and non-exercise activity thermogenesis, *Neuroendocrinology* 87 (2008): 71–90.
52. American on the move: Steps to a healthier way of life, press release, September 10, 2007.
53. J. O. Hill, Can a small-changes approach help address the obesity epidemic? A report of the Joint Task Force of the American Society for Nutrition, Institute of Food Technologists, and International Food Information Council, *American Journal of Clinical Nutrition* 89 (2009): 477–484.
54. K. E. Giel and coauthors, Weight bias in work settings: A qualitative review, *Obesity Facts* 3 (2010): 33–40; C. Greenleaf, S. B. Martin, and D. Rhea, Fighting fat: How do fat stereotypes influence beliefs about physical education? *Obesity* 16 (2008): S53–S59.
55. G. Horsburgh-McLeod, J. D. Latner, and K. S. O'Brien, Unprompted generation of obesity stereotypes, *Eating and Weight Disorders* 14 (2009): e153–e157.
56. H. Konttinen and coauthors, Emotional eating and physical activity self-efficacy as pathways in the association between depressive symptoms and adiposity indicators, *American Journal of Clinical Nutrition* 92 (2010): 1031–1039; B. Blaine, Does depression cause obesity?: A meta-analysis of longitudinal studies of depression and weight control, *Journal of Health Psychology* 13 (2008): 1190–1197.
57. K. Stohacker and B. K. McFarlin, Influence of obesity, physical inactivity, and weight cycling on chronic inflammation, *Frontiers in Bioscience (Elite Edition)* 2 (2010): 98–104; I. Strychar and coauthors, Anthropometric, metabolic, psychosocial, and dietary characteristics of overweight/obese postmenopausal women with a history of weight cycling: A MONET (Montreal Ottawa New Emerging Team) Study, *Journal of the American Dietetic Association* 109 (2009): 718–724.
58. S. Hasani-Ranjbar and coauthors, A systematic review of the efficacy and safety of herbal medicines used in the treatment of obesity, *World Journal of Gastroenterology* 15 (2009): 3073–3085; C. K. Biesemeier and coauthors, Ethics opinion: Weight loss products and medications, *Journal of the American Dietetic Association* 108 (2008): 2109–2113.
59. P. A. Cohen, American Roulette: Contaminated dietary supplements, *New England Journal of Medicine* 361 (2009): 1523–1525.
60. R. F. Kushner, Anti-obesity drugs, *Expert Opinion on Pharmacotherapy* 9 (2008): 1339–1350.
61. A. Astrup, Drug management of obesity: Efficacy versus safety, *New England Journal of Medicine* 363 (2010): 288–290.
62. A. Nagle, Bariatric surgery: A surgeon's perspective, *Journal of the American Dietetic Association* 110 (2010): 520–523; G. L. Blackburn, S. Wollner, and S. B. Heymsfield, Lifestyle interventions for the treatment of class III obesity: A primary target for nutrition medicine in the obesity epidemic, *American Journal of Clinical Nutrition* 91 (2010): 289S–292S.
63. L. M. Beckman, T. R. Beckman, and C. P. Earthman, Changes in gastrointestinal hormones and leptin after Roux-en-Y gastric bypass procedure: A review, *Journal of the American Dietetic Association* 110 (2010): 571–584.
64. T. H. Inge and coauthors, Reversal of type 2 diabetes mellitus and improvements in cardiovascular risk factors after surgical weight loss in adolescents, *Pediatrics* 123 (2009): 214–222; J. B. Dixon and coauthors, Adjustable gastric banding and conventional therapy for type 2 diabetes: A randomized controlled trial, *Journal of the American Medical Association* 299 (2008): 316–323.
65. C. C. Wee, A 52-year-old woman with obesity, *Journal of the American Medical Association* 302 (2009): 1097–1104.
66. E. H. Livingston, Surgical treatment of obesity in adolescence, *Journal of the American Medical Association* 303 (2010): 559–560.
67. Y. Chen, Acute bariatric surgery complications: Managing parenteral nutrition in the morbidly obese, *Journal of the American Dietetic Association* 110 (2010): 1734–1737; D. Kulick, L. Hark, and D. Deen, The bariatric surgery patient: A growing role for registered dietitians, *Journal of the American Dietetic Association* 110 (2010): 593–599; G. Snyder-Marlow, D. Taylor, and J. Lenhard, Nutrition care for patients undergoing laparoscopic sleeve gastrectomy for weight loss, *Journal of the American Dietetic Association* 110 (2010): 600–607.
68. Snyder-Marlow, Taylor, and Lenhard, 2010; E. T. Aasheim and coauthors, Vitamin status after bariatric surgery: A randomized study of gastric bypass and duodenal switch, *American Journal of Clinical Nutrition* 90 (2009): 15–22; M. Ruz and coauthors, Iron absorption and iron status are reduced after Roux-en-Y gastric bypass, *American Journal of Clinical Nutrition* 90 (2009): 527–532.
69. M. Kruseman and coauthors, Dietary, weight, and psychological changes among patients with obesity, 8 years after gastric bypass, *Journal of the American Dietetic Association* 110 (2010): 527–534.
70. D. R. Flum and the Longitudinal Assessment of Bariatric Surgery (LABS) Consortium, Perioperative safety in the longitudinal assessment of bariatric surgery, *New England Journal of Medicine* 361 (2009): 445–454.
71. P. J. Stephan and J. M. Kenkel, Updates and advances in liposuction, *Aesthetic Surgery Journal* 30 (2010) 83–97.
72. T. L. Hernandez and coauthors, Fat redistribution following suction lipectomy: Defense of body fat and patterns of restoration, *Obesity* 19 (2011): 1388–1395.
73. D. Heber, An integrative view of obesity, *American Journal of Clinical Nutrition* 91 (2010): 280S–283S.
74. D. R. Jacobs and coauthors, Association of 1-y changes in diet pattern with cardiovascular disease risk factors and adipokines: Results from the 1-y randomized Oslo Diet and Exercise Study, *American Journal of Clinical Nutrition* 89 (2009): 509–517.
75. D. R. Young and coauthors, Effects of the PREMIER interventions on health-related quality of life, *Annals of Behavioral Medicine* 40 (2010): 302–312.
76. G. D. Foster and coauthors, Obese patients' perceptions of treatment outcomes and the factors that influence them, *Archives of Internal Medicine* 161 (2001): 2133–2139; G. D. Foster and coauthors, What is a reasonable weight loss? Patients' expectations and evaluations of obesity treatment outcomes, *Journal of Consulting and Clinical Psychology* 65 (1997): 79–85.
77. A. J. Tomiyama and coauthors, Low calorie dieting increases cortisol, *Psychosomatic Medicine* 72 (2010): 357–364.
78. H. M. Seagle and coauthors, Position of the American Dietetic Association: Weight management, *Journal of the American Dietetic Association* 109 (2009): 330–346.
79. M. T. Timlin and coauthors, Breakfast eating and weight change in a 5-year prospective analysis of adolescents: Project EAT (Eating Among Teens), *Pediatrics* 121 (2008): e638.
80. L. Dubois and coauthors, Breakfast skipping is associated with differences in meal patterns, macronutrient intakes and overweight among pre-school children, *Public Health Nutrition* 12 (2009): 19–28; A. K. Kant and coauthors, Association of breakfast energy density with diet quality and body mass index in American adults: National Health and Nutrition Examination Surveys, 1999–2004, *American Journal of Clinical Nutrition* 88 (2008): 1396–1404.
81. C. L. Rock and coauthors, Effect of a free prepared meal and incentivized weight loss program on weight loss and weight loss maintenance in obese and overweight women: A randomized controlled trial, *Journal of the American Medical Association* 304 (2010): 1803–1810; Position of the American Dietetic Association: Weight management, *Journal of the American Dietetic Association* 109 (2009): 330–346.
82. N. Zijlstra and coauthors, Effect of bite size and oral processing time of a semisolid food on satiation, *American Journal of Clinical Nutrition* 90 (2009): 269–275.
83. A. Kokkinos and coauthors, Eating slowly increases the postprandial response of the anorexigenic gut hormones, peptide YY and glucagon-like peptide-1, *Journal of Clinical Endocrinology and Metabolism* 95 (2010): 333–337.
84. A. M. Andrade and coauthors, Eating slowly led to decreases in energy intake within meals in healthy women, *Journal of the American Dietetic Association* 108 (2008): 1186–1191.
85. E. Stice and coauthors, Relation between obesity and blunted striatal response to food is moderated by TaqIA A1 allele, *Science* 322 (2008): 449–452.
86. C. H. Llewellyn and coauthors, Eating rate is a heritable phenotype related to weight in children, *American Journal of Clinical Nutrition* 88 (2008): 1560–1566.

87. B. J. Rolls, Plenary lecture 1: Dietary strategies for the prevention and treatment of obesity, *Proceedings of the Nutrition Society* 3 (2009): 1–10; K. E. Leahy, L. L. Birch, and B. J. Rolls, Reducing the energy density of multiple meals decreases the energy intake of preschool-age children, *American Journal of Clinical Nutrition* 88 (2008): 1459–1468.
88. M. C. Daniels and B. M. Popkin, Impact of water intake on energy intake and weight status: A systematic review, *Nutrition Reviews* 68 (2010): 505–521; B. M. Davy and coauthors, Water consumption reduces energy intake at a breakfast meal in obese older adults, *Journal of the American Dietetic Association* 108 (2008): 1236–1239.
89. B. Buijsse and coauthors, Fruit and vegetable intakes and subsequent changes in body weight in European populations: Results from the project on Diet, Obesity, and Genes (DiOGenes), *American Journal of Clinical Nutrition* 90 (2009): 202–209.
90. N. Schroeder and coauthors, Influence of whole grain barley, whole grain wheat, and refined rice-based foods on short-term satiety and energy intake, *Appetite* 53 (2009): 363–369; P. Vitaglione and coauthors, β-Glucan-enriched bread reduces energy intake and modifies plasma ghrelin and peptide YY concentrations in the short term, *Appetite* 53 (2009): 338–344.
91. US Department of Agriculture and US Department of Health and Human Services, *Dietary Guidelines for Americans, 2010,* available at www.dietaryguidelines.gov; R. D. Mattes and B. M. Popkin, Nonnutritive sweetener consumption in humans: Effects on appetite and food intake and their putative mechanisms, *American Journal of Clinical Nutrition* 89 (2009): 1–14.
92. S. P. Fowler and coauthors, Fueling the obesity epidemic? Artificially sweetened beverage use and long-term weight gain, *Obesity* 16 (2008): 1894–1900.
93. B. H. Goodpaster and coauthors, Effects of diet and physical activity interventions on weight loss and cardiometabolic risk factors in severely obese adults: A randomized study, *Journal of the American Medical Association* 304 (2010): 1795–1802; A. L. Hankinson and coauthors, Maintaining a high physical activity level over 20 years and weight gain, *Journal of the American Medical Association* 304 (2010): 2603–2610.
94. J. E. Donnelly and coauthors, American College of Sports Medicine Position Stand: Appropriate physical activity intervention strategies for weight loss and prevention of weight regain for adults, *Medicine and Science in Sports and Exercise* 41 (2009): 459–471; Committee on Dietary Reference Intakes, *Dietary Reference Intakes for Energy, Carbohydrate, Fiber, Fat, Fatty Acids, Cholesterol, Protein, and Amino Acids* (Washington, D.C.: National Academies Press, 2005).
95. Goodpaster and coauthors, 2010; M. Hamer and G. O'Donovan, Cardiorespiratory fitness and metabolic risk factors in obesity, *Current Opinion in Lipidology* 21 (2010): 1–7; D. E. Larson-Meyer and coauthors, Caloric restriction with or without exercise: The fitness versus fatness debate, *Medicine and Science in Sports and Exercise* 42 (2010): 152–159.
96. D. T. Villareal and coauthors, Weight loss, exercise, or both and physical function in obese older adults, *New England Journal of Medicine* 364 (2011): 1218–1229.
97. J. E. Turner and coauthors, Nonprescribed physical activity energy expenditure is maintained with structured exercise and implicates a compensatory increase in energy intake, *American Journal of Clinical Nutrition* 92 (2010): 1009–1016.
98. A. M. Knab and coauthors, A 45-minute vigorous exercise bout increases metabolic rate for 14 hours, *Medicine and Science in Sports and Exercise* 43 (2011): 1643–1648; K. Ohkawara and coauthors, Twenty-four-hour analysis of elevated energy expenditure after physical activity in a metabolic chamber: Models of daily total energy expenditure, *American Journal of Clinical Nutrition* 87 (2008): 1268–1276.
99. E. P. Kirk and coauthors, Minimal resistance training improves daily energy expenditure and fat oxidation, *Medicine and Science in Sports and Exercise* 41 (2009): 1122–1129.
100. J. W. Bea and coauthors, Resistance training predicts 6-yr body composition change in postmenopausal women, *Medicine and Science in Sports and Exercise* 42 (2010): 1286–1295.
101. J. A. King and coauthors, Influence of brisk walking on appetite, energy intake, and plasma acylated ghrelin, *Medicine and Science in Sports and Exercise* 42 (2010): 485–492.
102. N. A. King and coauthors, Dual process action of exercise on appetite control: Increase in orexigenic drive but improvement in meal-induced satiety, *American Journal of Clinical Nutrition* 90 (2009): 921–927.
103. P. Gordon-Larsen and coauthors, Fifteen-year longitudinal trends in walking patterns and their impact on weight change, *American Journal of Clinical Nutrition* 89 (2009): 19–26.
104. I. M. Lee and coauthors, Physical activity and weight gain prevention, *Journal of the American Medical Association* 303 (2010): 1173–1179.
105. B. J. Nicklas and coauthors, Effect of exercise intensity on abdominal fat loss during calorie restriction in overweight and obese postmenopausal women: A randomized, controlled trial, *American Journal of Clinical Nutrition* 89 (2009): 1043–1052.
106. B. McFerran and coauthors, I'll have what she's having: Effects of social influence and body type on the food choices of others, *Journal of Consumer Research* 36 (2010): 915–929.
107. L. E. Burke, J. Wang, and M. A. Sevick, Self-monitoring in weight loss: A systematic review of the literature, *Journal of the American Dietetic Association* 111 (2011): 92–102.
108. K. G. Volpp and coauthors, Financial incentive–based approaches for weight loss: A randomized trial, *Journal of the American Medical Association* 300 (2008): 2631–2637.
109. M. E. Murawski and coauthors, Problem solving, treatment adherence, and weight-loss outcome among women participating in lifestyle treatment for obesity, *Eating Behaviors* 10 (2009): 146–151.
110. J. Sallit, M. Cuiccazzo, and Z. Dixon, A cognitive-behavioral weight control program improves eating and smoking behaviors in weight-concerned female smokers, *Journal of the American Dietetic Association* 109 (2009): 1398–1405.
111. J. P. Block and coauthors, Psychosocial stress and change in weight among US adults, *American Journal of Epidemiology* 170 (2009): 181–192.
112. G. M. Manzoni and coauthors, Can relaxation training reduce emotional eating in women with obesity? An exploratory study with 3 months of follow-up, *Journal of the American Dietetic Association* 109 (2009): 1427–1432.
113. S. Kodama and coauthors, Effect of web-based lifestyle modification on weight control: A meta-analysis, *International Journal of Obesity* (2011): doi:10.1038/ijo.2011.121; A. G. Digenio and coauthors, Comparison of methods for delivering a lifestyle modification program for obese patients: A randomized trial, *Annals of Internal Medicine* 150 (2009): 255–262; L. P. Svetkey and coauthors, Comparison of strategies for sustaining weight loss: The weight loss maintenance randomized controlled trial, *Journal of the American Medical Association* 299 (2008): 1139–1148.
114. J. L. Kraschnewski and coauthors, Long-term weight loss maintenance in the United States, *International Journal of Obesity* 34 (2010): 1644–1654.
115. S. Phelan and coauthors, Use of artificial sweeteners and fat-modified foods in weight loss maintainers and always-normal weight individuals, *International Journal of Obesity* 33 (2009): 1183–1190.
116. M. Rosenbaum and coauthors, Long-term persistence of adaptive thermogenesis in subjects who have maintained a reduced body weight, *American Journal of Clinical Nutrition* 88 (2008): 906–912.
117. Donnelly and coauthors, 2009; Hankinson and coauthors, 2010; J. L. Unick, J. M. Jakicic, and B. H. Marcus, Contribution of behavior intervention components to 24-month weight loss, *Medicine and Science in Sports and Exercise* 42 (2010): 745–753.
118. V. A. Catenacci and coauthors, Physical activity patterns in the National Weight Control Registry, *Obesity* 16 (2008): 153–161.
119. E. Manthou and coauthors, Behavioral compensatory adjustments to exercise training in overweight women, *Medicine and Science in Sports and Medicine* 42 (2010): 1221–1228.
120. S. N. Grief and R. L. Miranda, Weight loss maintenance, *American Family Physician* 82 (2010): 630–634.
121. L. K. Khan and coauthors, Recommended community strategies and measurements to prevent obesity in the United States, *MMWR Recommendations and Reports* 58 (2009): 1–26; A. M. Wolf and

K. A. Woodworth, Obesity prevention: Recommended strategies and challenges, *The American Journal of Medicine* 122 (2009): S19–S23.

122. Khan and coauthors, 2009.
123. E. A. Finkelstein and coauthors, Impact of targeted beverage taxes on higher- and lower-income households, *Archives of Internal Medicine* 170 (2010): 2028–2034; K. D. Brownell and T. R. Frieden, Ounces of prevention: The public policy case for taxes on sugared beverages, *New England Journal of Medicine* 360 (2009): 1805–1808.
124. T. T. Huang and T. A. Glass, Transforming research strategies for understanding and preventing obesity, *Journal of the American Medical Association* 300 (2008): 1811–1813.
125. C. D. Fryar and C. L. Ogden, Prevalence of underweight among adults aged 20 years and over: United States, 2007–2008, *NCHS Health E-Stats,* www.cdc.gov/nchs/data/hestat/underweight/underweight_adults.htm.
126. J. Shea and coauthors, Changes in the transcriptome of abdominal subcutaneous adipose tissue in response to short-term overfeeding in lean and obese men, *American Journal of Clinical Nutrition* 89 (2009): 407–415.
127. Position Paper: Nutrition and athletic performance: Position of the American Dietetic Association, Dietitians of Canada, and the American College of Sports Medicine, *Journal of the American Dietetic Association* 100 (2000): 1543–1556.

HIGHLIGHT > 9

LEARN IT Contrast the differences between popular fad diets and weight-loss diets based on sound nutrition.

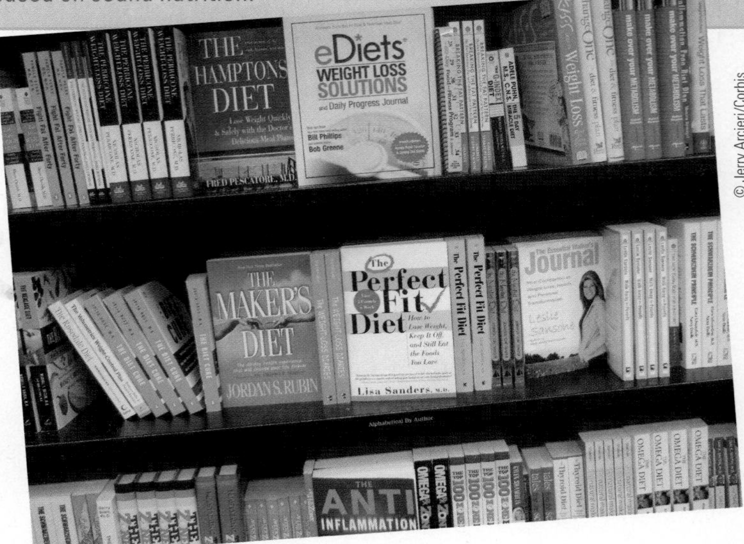

The Latest and Greatest Weight-Loss Diet—Again

To paraphrase William Shakespeare, "A fad diet by any other name would still be a fad diet." Year after year, "new and improved" diets appear on bookstore shelves and circulate among friends.* People of all sizes eagerly try the best diet ever on the market, hoping that this one will really work. Sometimes these diets seem to work for a while, but more often than not, their success is short-lived. Then another diet takes the spotlight. Here's how Dr. K. Brownell, an obesity researcher at Yale University, describes this phenomenon: "When I get calls about the latest diet fad, I imagine a trick birthday cake candle that keeps lighting up and we have to keep blowing it out."

Realizing that fad diets do not offer a safe and effective long-term plan for weight loss, health professionals speak out, but they never get the candle blown out permanently. New fad diets can keep making outrageous claims because no one requires their advocates to prove what they say. Fad diet gurus do not have to conduct credible research on the benefits or dangers of their diets. They can simply make recommendations and then later, if questioned, search for bits and pieces of research that support the conclusions they have already reached. That's backward. Diet and health recommendations should *follow* years of sound scientific research *before* being offered to the public.

Because anyone can publish anything—in books or on the Internet—peddlers of fad diets can make unsubstantiated statements that fall far short of the truth but sound impressive to the uninformed. They often offer distorted bits of legitimate research. They may start with one or more actual facts but then leap from one erroneous conclusion to the next. Anyone who wants to believe these claims has to wonder how the thousands of scientists working on obesity research over the past century could possibly have missed such obvious connections.

Fad diets come in almost as many shapes and sizes as the people who search them out. Some restrict fats or carbohydrates, some limit portion sizes, some focus on food combinations, and some claim that a person's genetic type or blood type determines the foods best suited to manage weight and prevent disease. Table H9-1 compares some of today's popular diets.

*The Academy of Nutrition and American Dietetics offers evaluations of popular diets for your review. Look for Popular Diet Reviews at their website, **www.eatright.org/media.**

TABLE H9-1 Popular Diets Compared

Diet	Major Premise Promoted	Strong Point(s)	Weak Point(s)
The 4-Hour Body	• Less is more, and small, simple changes produce long-lasting effects.	• Quick results.	• Restricts carbohydrates. • All fruit and milk (except cottage cheese) are excluded and vegetables are limited. • Encourages eating the same small meals repeatedly. • Recommends a weekly binge.
The 17 Day Diet	• Changing the way you eat every few days creates "body confusion," which prevents metabolism from settling into homeostasis. • You can boost metabolism by "eating clean," which means no sugar, no processed food, and no fried foods.	• Quick results. • Prevents boredom by alternating between cycles. • Fairly well-balanced diet that promotes healthy eating.	• No scientific evidence that changing the diet creates "body confusion." • No individualized kcalorie goals. • Promotes its own processed foods.
Atkins Diet	• People are overweight or obese because they have metabolic imbalances caused by eating too many carbohydrates; by restricting carbohydrates, these imbalances can be corrected. • You can lose weight without lowering kcalorie intake.	• Quick, short-term weight loss is achieved.	• Restricts carbohydrates to a level that induces ketosis which can cause nausea, light-headedness, and fatigue and can worsen existing medical problems such as kidney disease. • A diet high in fat such as Atkins can increase the risk of heart disease and some cancers.

TABLE H9-1 Popular Diets Compared (*continued*)

Diet	Major Premise Promoted	Strong Point(s)	Weak Point(s)
Cheater's Diet	• Successful weight loss depends on eliminating boredom and allowing indulgences. • Cheating on weekends "stokes your metabolism."	• Meals are proportioned one-half fruit or vegetables, one-fourth lean protein, and one-fourth whole grains. • Encourages as much exercise as possible.	• No scientific data on cheating boosting metabolism or supporting weight loss.
Cinch!	• A nutrient-dense diet composed mainly of plant-based foods will help you lose weight and lower the risk of disease.	• Plant-based, nutrient-dense diet. • Stresses the importance of exercise.	• A little confusing and dense with facts.
The Dukan Diet	• A high-protein, low-kcalorie diet promotes rapid weight loss and will keep it off for good.	• Encourages daily exercise, moderate salt intake, and lifelong weight management. • Provides a highly structured plan.	• Restricts carbohydrates to a level that induces ketosis which can cause nausea, light-headedness, and fatigue and can worsen medical problems such as kidney disease. • Not suited for vegetarians and others who prefer not to emphasize animal proteins.
Glucose Revolution	• Low glycemic index foods satisfy hunger, control blood glucose, and promote weight loss.	• Emphasizes fiber-rich vegetables, legumes, fruits, and whole grains. • Minimizes saturated fat intake.	• Difficult to know the glycemic index of some foods.
New Sonoma Diet	• Enjoying portion-controlled Coastal California style foods supports weight loss and promotes good health.	• Emphasizes nutrient-dense foods. • Limits processed foods.	• No individualized kcalorie plans.
Ornish Diet	• By strictly limiting fat (both animal and vegetable), you eat fewer kcalories without eating less food.	• High-fiber, low-fat foods in this plan can lower blood cholesterol and blood pressure.	• So little fat that essential fatty acids may be lacking. • Limits fish, nuts, and olive oil, which may protect against heart disease.
South Beach Diet	• Eating "good carbohydrates" such as vegetables, whole-wheat pastas, and brown rice will maintain satiety and resist cravings for "bad carbohydrates" such as white rice and potatoes.	• Encourages consumption of vegetables, lean meats, and fish, and the use of unsaturated oils when cooking. • Restricts fatty meats and cheeses as well as sweets.	• Starchy carbohydrates and all fruits are completely excluded during the first two weeks.
Ultimate Weight Solution Diet	• Foods that require great effort to prepare and eat are nutrient-dense; eating these kinds of foods (raw vegetables, vegetable soups, whole grains, beans, meats, poultry, and fish) will lead to weight loss. • Foods that take little effort to prepare and eat provide excess kcalories relative to nutrients; eating these kinds of foods (fast foods, puddings, high-kcalorie convenience foods, processed foods) leads to uncontrolled eating and weight gain.	• Encourages consumption of lean meats and fish; whole grains; vegetables; fruit; and low-fat milk, yogurt, and cheese. • Restricts fatty meats and cheeses as well as sweets. • Encourages exercise.	• Confusing as to exactly what to eat or how much.
Zone Diet	• Eating the correct proportions of carbohydrates, fat, and protein leads to hormonal balance, weight loss, disease prevention, and increased vitality.	• Promotes weight loss because it is a low-kcalorie diet.	• The diet is rigid, restrictive, and complicated, making it difficult for most people to follow accurately. • The overblown health claims of the diet's proponents are based on misinterpreted science and remain unsubstantiated.

Fad Diets' Appeal

With more than half of our nation's adults overweight and many more concerned about their weight, the market for a weight-loss book, product, or program is huge (no pun intended). Americans spend an estimated $33 billion a year on weight-loss books and products. Even a plan that offers only minimal weight-loss success easily attracts a following.

Perhaps the greatest appeal of fad diets is that they tend to ignore dietary recommendations. Foods such as meats and milk products that need to be selected carefully to limit saturated fat can be eaten with abandon. Whole grains, legumes, vegetables, and fruits that should be eaten in abundance can now be bypassed. For some people, this is a dream come true: steaks without the potatoes, ribs without the coleslaw, and meatballs without the pasta. Who can resist the promise of weight loss while eating freely from a list of favorite foods?

Dieters are also lured into fad diets by sophisticated—yet often erroneous—explanations of the metabolic consequences of eating certain foods. Terms such as *eicosanoids* and *de novo lipogenesis* are scattered about, often intimidating readers into believing that the authors must be right given their brilliance in understanding the body.

If fad diets were as successful as some people claim, then consumers who tried them would lose weight, and their obesity problems would be solved. But this is not the case. Similarly, if fad diets were as worthless as others claim, then consumers would eventually stop pursuing them. Clearly, this is not happening either. Most fad diets have enough going for them that they work for some people at least for a short time, but they fail to produce long-lasting results for most people.

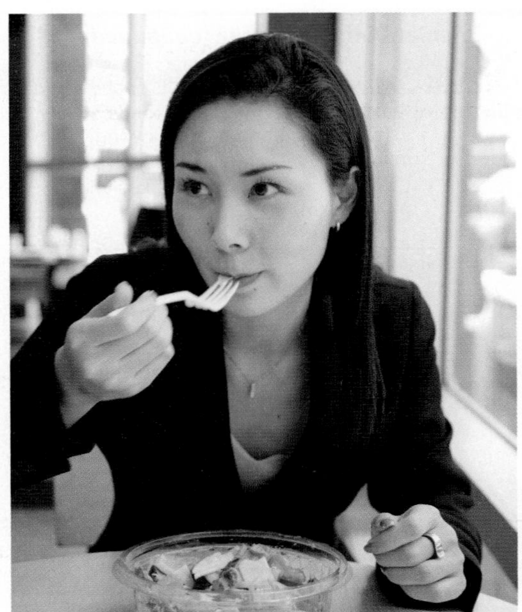

The wise consumer seeks a diet that supports not only weight loss, but health gains.

Don't Count kCalories

Who wants to count kcalories? Even experienced dieters find counting kcalories burdensome, not to mention timeworn. They want a new, easy way to lose weight, and fad diet plans seem to offer this boon. But, though fad diets often claim to disregard kcalories, their design typically ensures a low energy intake. Most of the sample menu plans, especially in the early stages, are designed to deliver an average of 1200 kcalories a day.

Even when counting kcalories is truly not necessary, total kcalories tend to be low simply because food intake is so limited. Diets that omit hundreds of foods and several food groups limit a person's options and lack variety. Chapter 2 praises variety as a valuable way to ensure an adequate intake of nutrients, but variety also entices people to eat more food and gain more weight. Without variety, some people lose interest in eating, which further reduces energy intake. Even if the allowed foods are favorites, eating the same foods week after week can become monotonous.

Without its refried beans, tortilla wrapping, and chopped vegetables, a burrito is reduced to a pile of ground beef. Without the baked potato, there's no need for butter and sour cream. Weight loss occurs because of the low energy intake. This is an important point. Any diet can produce weight loss, at least temporarily, if intake is restricted. The real value of a diet is determined by its ability to maintain weight loss and support good health over the long term. The goal is not simply weight loss, but health gains—and most fad diets cannot support optimal health over time. In fact, some weight-loss diets can create or exacerbate health problems.[1]

When food choices are limited, nutrient intakes may be inadequate. To help shore up some of these inadequacies, fad diets often recommend a dietary supplement. Conveniently, many of the companies selling fad diets also peddle these supplements. But as Highlights 10 and 11 explain, foods offer many more health benefits than any supplement can provide. Quite simply, if the diet is inadequate, it needs to be improved.

Follow a Plan

Most people need specific instructions and examples to make dietary changes. Popular diets offer dieters a plan. The user doesn't have to decide what foods to eat, how to prepare them, or how much to eat. Unfortunately, these instructions serve only short-term weight-loss needs. They do not provide for long-term changes in lifestyle that will support weight maintenance or health goals.

The success of any weight-loss diet depends on the person adopting the plan and sticking with it. People who prefer a high-protein, low-carbohydrate diet over a high-carbohydrate, low-fat diet, for example, may have more success at sticking with it.[2] Keep in mind, though, that weight loss occurs because of the duration of a low-kcalorie plan—not the proportion of energy nutrients.[3]

The Real Deal

Fad diets attribute magical powers to their weight-loss plans, but in reality, the magic is in tipping the energy balance so that metabolic and physical activities expend more kcalories than foods bring in. Because new diets emerge in the market regularly, it can be challenging to sort the fad diets from the healthy options. Furthermore, it can be difficult determining how a diet's overall quality rates and how it compares with others.

Keep in mind that healthy weight loss requires long-term lifestyle changes in eating and activity habits—not quick, short-term fixes. A healthy plan may not be quick, but it allows for flexibility and a variety of foods, including some favorite treats on occasion.

Some currently popular diet plans offer a sensible approach to weight loss and healthy eating. The challenge is sorting through "the good, the bad, and the ugly." The accompanying "How To" feature offers tips for identifying fad diets and other weight-loss scams. Fad diets may not harm healthy people if used for only a little while, but they cannot support optimal health for long. Chapter 9 includes reasonable approaches to weight management and concludes that the ideal diet is one you can live with for the rest of your life. Keep that criterion in mind when you evaluate the next "latest and greatest weight-loss diet" that comes along.

>How To

Identify a Fad Diet or Weight-Loss Scam

It may be a fad diet or weight-loss scam if it:

- Sounds too good to be true.
- Recommends using a single food consistently as the key to the program's success.
- Promises quick and easy weight loss with no effort. "Lose weight while you sleep!"
- Eliminates an entire food group such as grains or milk and milk products.
- Guarantees an unrealistic outcome in an unreasonable time period. "Lose 10 pounds in 2 days!"
- Bases evidence for its effectiveness solely on anecdotal stories.
- Requires you to buy special products that are not readily available in the marketplace at affordable prices.
- Specifies a proportion for the energy nutrients that falls outside the recommended ranges—carbohydrate (45 to 65 percent), fat (20 to 35 percent), and protein (10 to 35 percent).
- Claims to alter your genetic code or reset your metabolism.
- Fails to mention potential risks or additional costs.
- Promotes products or procedures that have not been proven safe and effective.
- Neglects plans for weight maintenance following weight loss.

TRY IT Review an advertisement for a popular weight-loss plan and explain why you think it might—or might not—be a fad diet.

REFERENCES

1. P. Sjögren and coauthors, Mediterranean and carbohydrate-restricted diets and mortality among elderly men: A cohort study in Sweden, *American Journal of Clinical Nutrition* 92 (2010): 967–974; T. D. Barnett, N. D. Barnard, and T. L. Radak, Development of symptomatic cardiovascular disease after self-reported adherence to the Atkins diet, *Journal of the American Dietetic Association* 109 (2009): 1263–1265; M. Miller and coauthors, Comparative effects of three popular diets on lipids, endothelial function, and c-reactive protein during weight maintenance, *Journal of the American Dietetic Association* 109 (2009): 713–717.
2. T. M. Larsen and coauthors, Diets with high or low protein content and glycemic index for weight-loss maintenance, *New England Journal of Medicine* 363 (2010): 2102–2113.
3. E. A. Delbridge and coauthors, One-year weight maintenance after significant weight loss in healthy overweight and obese subjects: Does diet composition matter? *American Journal of Clinical Nutrition* 90 (2009): 1203–1214; G. D. Foster and coauthors, Weight and metabolic outcomes after 2 years on a low-carbohydrate versus low-fat diet: A randomized trial, *Annals of Internal Medicine* 153 (2010): 147–157; W. S. Yancy and coauthors, A randomized trial of a low-carbohydrate diet vs orlistat plus a low-fat diet for weight loss, *Archives of Internal Medicine* 170 (2010): 136–145; F. M. Sacks and coauthors, Comparison of weight-loss diets with different compositions of fat, protein, and carbohydrates, *New England Journal of Medicine* 360 (2009): 859–873.

© Rocky33/Shutterstock.com

CENGAGE LEARNING'S GLOBAL NUTRITION WATCH — GLOBAL NUTRITION RESOURCE CENTER

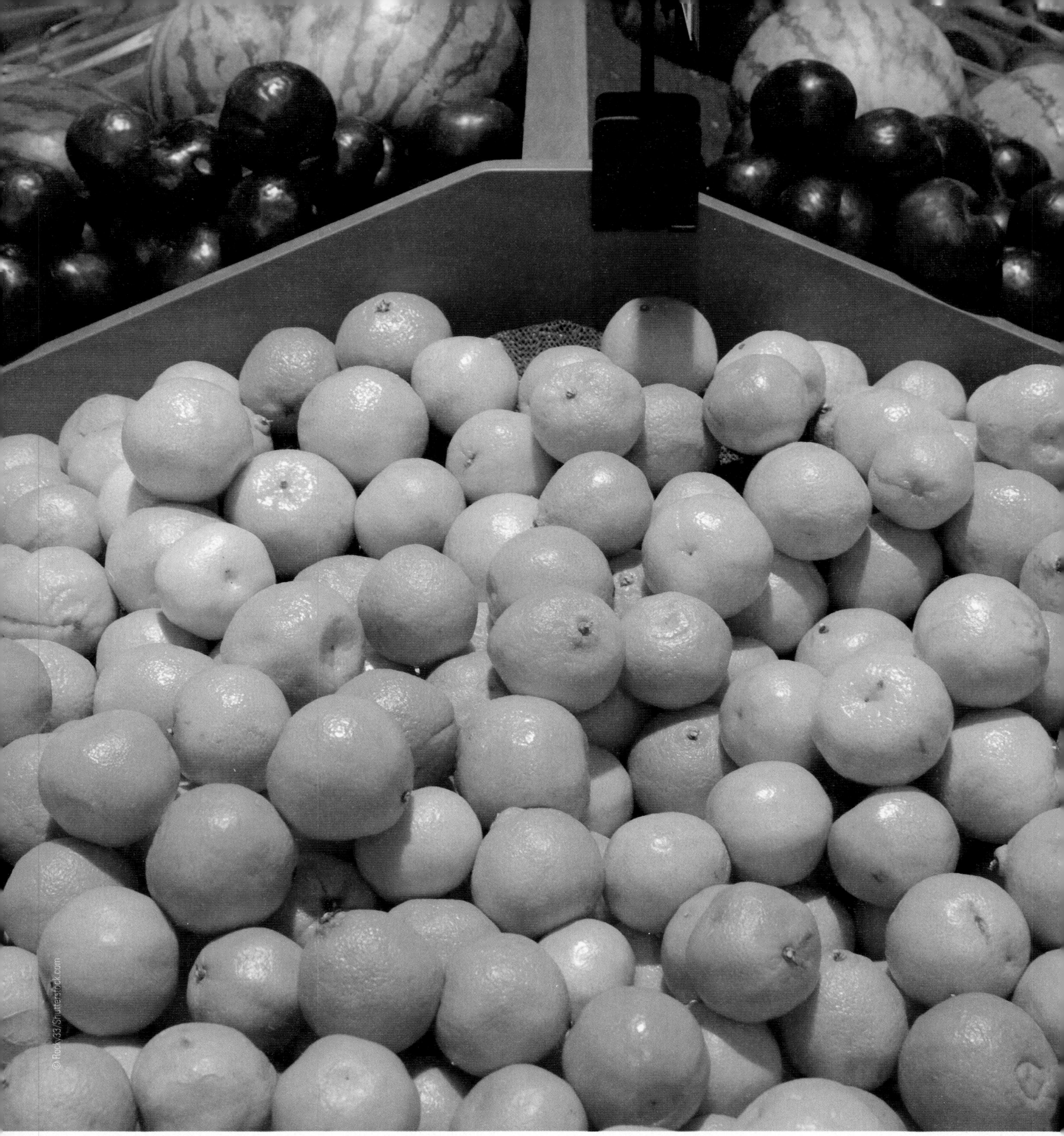

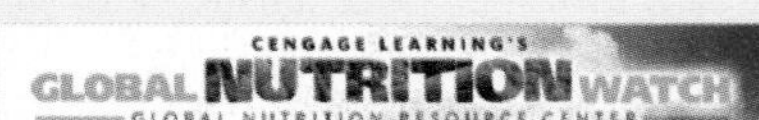

Vitamin C's role in strengthening the immune system has led it to be touted as a remedy for many ailments, including the common cold. Go to Nutrition Basics and Tools → Vitamins, Minerals, and Water to review its other benefits. What benefit or harm could come from increasing intake of water-soluble vitamins beyond DRI guidelines?

10

The Water-Soluble Vitamins: B Vitamins and Vitamin C

LEARNING GPS

Nutrition in Your Life

If you were playing a word game and your partner said "vitamins," how would you respond? If "pills" and "supplements" immediately come to mind, you may be missing the main message of the vitamin story—that hundreds of foods deliver more than a dozen vitamins that participate in thousands of activities throughout your body. Quite simply, foods supply vitamins to support all that you are and all that you do—and supplements of any one of them, or even a combination of them, can't compete with foods in keeping you healthy. In the Nutrition Portfolio at the end of this chapter, you can determine whether the foods you are eating are meeting your water-soluble vitamin needs.

Earlier chapters focused on the energy-yielding nutrients—carbohydrates, fats, and proteins. This chapter begins with an overview of the vitamins and then examines each of the water-soluble vitamins; the next chapter features the fat-soluble vitamins.

10.1 The Vitamins—An Overview

LEARN IT Describe how vitamins differ from the energy nutrients and how fat-soluble vitamins differ from water-soluble vitamins.

Researchers first recognized that foods contain substances that are "vital to life" in the early 1900s.[1] Since then, the world of **vitamins** has opened up dramatically. The vitamins are powerful substances, as their *absence* attests. Vitamin A deficiency can cause blindness; a lack of the B vitamin niacin can cause dementia; and a lack of vitamin D can retard bone growth. The consequences of deficiencies are so dire, and the effects of restoring the needed vitamins so dramatic, that people spend billions of dollars every year in the belief that vitamin supplements will cure a host of ailments (see Highlight 10). Vitamins certainly support sound nutritional health, but they do not cure all ills. Furthermore, vitamin supplements do not offer the many benefits that come from vitamin-rich foods.

vitamins: organic, essential nutrients required in small amounts by the body for health. Vitamins regulate body processes that support growth and maintain life.

- **vita** = life
- **amine** = containing nitrogen (the first vitamins discovered contained nitrogen)

The *presence* of the vitamins also attests to their power. The B vitamin folate helps to prevent birth defects and vitamin K allows blood to clot. As you will see, the vitamins' roles in supporting optimal health extend far beyond preventing deficiency diseases. In fact, some of the credit given to low-fat diets in preventing disease actually belongs to the vitamins found in vegetables, fruits, and whole grains (see Highlight 11 for more on vitamins in disease prevention).

The vitamins differ from carbohydrates, fats, and proteins in the following ways:

- *Structure.* Vitamins are individual units; they are not linked together (as are molecules of glucose or amino acids). Appendix C presents the chemical structure for each of the vitamins.
- *Function.* Vitamins do not yield energy when metabolized; many of them do, however, assist the enzymes that participate in the release of energy from carbohydrates, fats, and proteins.
- *Food contents.* The amounts of vitamins people ingest from foods and the amounts they require daily are measured in *micrograms* (μg) or *milligrams* (mg), rather than grams (g). ◆

The vitamins are similar to the energy-yielding nutrients, though, in that they are vital to life, organic, and available from foods.

◆ For perspective, a dollar bill weighs about 1 g.
1 g = 1000 mg
1 mg = 1000 μg
Appendix K explains how to convert a measurement from one unit of measure to another.

Bioavailability The amount of vitamins available from foods depends not only on the quantity provided by a food but also on the amount absorbed and used by the body—referred to as the vitamins' **bioavailability.** The quantity of vitamins in a food can be determined relatively easily. Researchers analyze foods to determine the vitamin contents and publish the results in tables of food composition such as Appendix H. Determining the bioavailability of a vitamin is a more complex task because it depends on many factors, including:

- Efficiency of digestion and time of transit through the GI tract
- Previous nutrient intake and nutrition status
- Method of food preparation (raw, cooked, or processed)
- Source of the nutrient (synthetic, fortified, or naturally occurring)
- Other foods consumed at the same time

Chapters 10 through 13 describe factors that inhibit or enhance the absorption of individual vitamins and minerals. Experts consider these factors when estimating recommended intakes.

Precursors Some of the vitamins are available from foods in inactive forms known as **precursors,** or provitamins. Once inside the body, the precursor is converted to an active form of the vitamin. For example, beta-carotene, a red-orange pigment found in fruits and vegetables, is a precursor to vitamin A. Thus, in measuring a person's vitamin intake, it is important to count both the amount of the active vitamin and the potential amount available from its precursors. The discussions and summary tables throughout this chapter and the next indicate which vitamins have precursors.

© Polara Studios, Inc.

To minimize vitamin losses, wrap cut fruits and vegetables or store them in airtight containers.

Organic Nature Fresh foods naturally contain vitamins, but because they are organic, vitamins can be readily destroyed during processing.[2] Therefore, processed foods should be used sparingly, and fresh foods should be handled with care during storage and in cooking. Prolonged heating may destroy much of the thiamin in food. Because riboflavin can be destroyed by the ultraviolet rays of the sun or by fluorescent light, foods stored in transparent glass containers are most likely to lose riboflavin. Oxygen destroys vitamin C, so losses occur when foods are cut, processed, and stored; these losses may be enough to reduce its action in the body. Table 10-1 summarizes ways to minimize nutrient losses in the kitchen, and Chapter 19 provides more details.

Solubility As you may recall, carbohydrates and proteins are hydrophilic and lipids are hydrophobic. The vitamins divide along the same lines—the hydrophilic, water-soluble ones are the B vitamins and vitamin C; the hydrophobic, fat-soluble ones are vitamins A, D, E, and K. As each vitamin was discovered, it was given a name and sometimes a letter and number as well. Many of the vitamins have multiple names, which has led to some confusion. The margin lists the standard

bioavailability: the rate at and the extent to which a nutrient is absorbed and used.

precursors: substances that precede others; with regard to vitamins, compounds that can be converted into active vitamins; also known as *provitamins.*

TABLE 10-1 Minimizing Nutrient Losses

- To slow the degradation of vitamins, refrigerate (most) fruits and vegetables.
- To minimize the oxidation of vitamins, store fruits and vegetables that have been cut in airtight wrappers, and store juices that have been opened in closed containers (and refrigerate them).
- To prevent vitamin losses during washing, rinse fruits and vegetables before cutting (not after).
- To minimize vitamin losses during cooking, use a microwave oven or steam vegetables in a small amount of water. Add vegetables after water has come to a boil. Use the cooking water in mixed dishes such as casseroles and soups. Avoid high temperatures and long cooking times.

names, ♦ and summary tables throughout this chapter and the next provide the common alternative names.

Solubility is apparent in the food sources of the different vitamins, and it affects their absorption, transport, storage, and excretion by the body. The water-soluble vitamins are found in the watery compartments of foods; the fat-soluble vitamins usually occur together in the fats and oils of foods. On being absorbed, the water-soluble vitamins move directly into the blood. Like fats, the fat-soluble vitamins must first enter the lymph, then the blood. Once in the blood, many of the water-soluble vitamins travel freely, whereas many of the fat-soluble vitamins require transport proteins. Upon reaching the cells, water-soluble vitamins freely circulate in the water-filled compartments whereas fat-soluble vitamins are held in fatty tissues and the liver until needed. The kidneys, monitoring the blood that flows through them, detect and remove small excesses of water-soluble vitamins; large excesses, however, may overwhelm the system, creating adverse effects. Fat-soluble vitamins tend to remain in fat-storage sites in the body rather than being excreted, and so are more likely to reach toxic levels when consumed in excess.

Because the body stores fat-soluble vitamins, they can be eaten in large amounts once in a while and still meet the body's needs over time. Water-soluble vitamins are retained for varying lengths of time in the body. The water-soluble vitamins must be eaten more regularly than the fat-soluble vitamins, although a single day's omission from the diet does not create a deficiency.

♦ **Water-soluble vitamins:**
- B vitamins:
  Thiamin
  Riboflavin
  Niacin
  Biotin
  Pantothenic acid
  Vitamin B_6
  Folate
  Vitamin B_{12}
- Vitamin C

Fat-soluble vitamins:
- Vitamin A
- Vitamin D
- Vitamin E
- Vitamin K

Toxicity Knowledge about some of the amazing roles of vitamins has prompted many people to take vitamin supplements, assuming that "more is better." Just as an inadequate intake can cause harm, so can an excessive intake. Even some of the water-soluble vitamins have adverse effects when taken in large doses.

That a vitamin can be both essential and harmful may seem surprising, but the same is true of most nutrients. The effects of every substance depend on its dose, and this is one reason consumers should not self-prescribe supplements. Figure 10-1 shows three possible relationships between dose levels and effects.

> FIGURE 10-1 Dose Levels and Effects

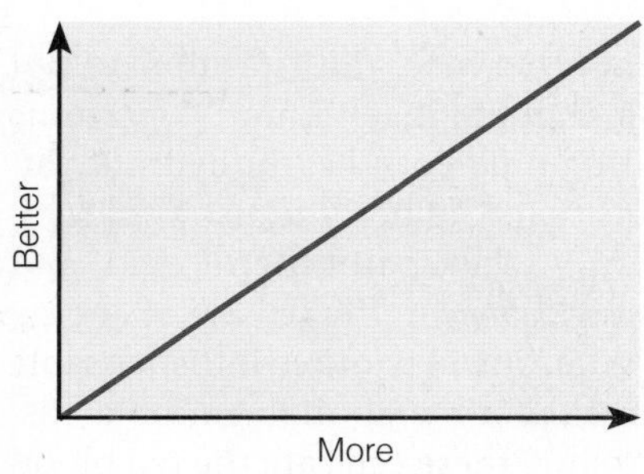

As you progress in the direction of more, the effect gets better and better, with no end in sight (real life is seldom, if ever, like this).

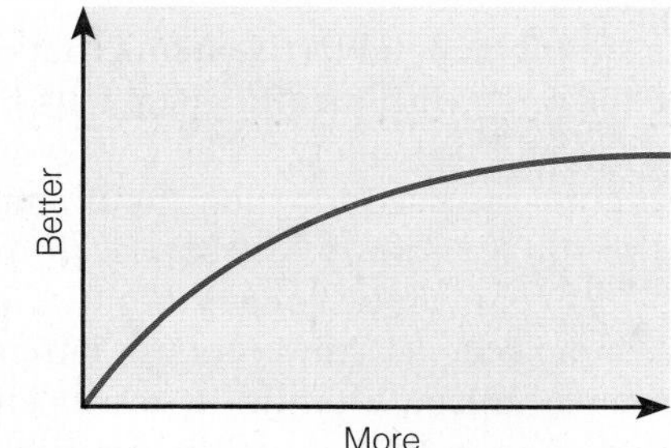

As you progress in the direction of more, the effect reaches a maximum and then a plateau, becoming no better with higher doses.

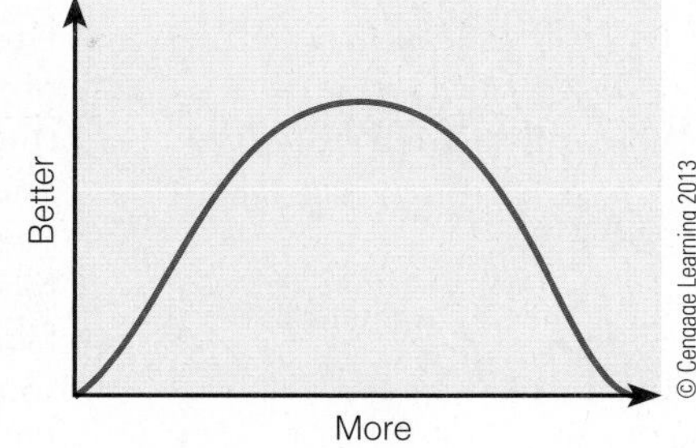

As you progress in the direction of more, the effect reaches an optimum at some intermediate dose and then declines, showing that more is better up to a point and then harmful. That too much can be as harmful as too little represents the situation with most nutrients.

TABLE 10-2 Water-Soluble and Fat-Soluble Vitamins Compared

	Water-Soluble Vitamins: B Vitamins and Vitamin C	Fat-Soluble Vitamins: Vitamins A, D, E, and K
Absorption	Directly into the blood	First into the lymph, then the blood
Transport	Travel freely	Many require transport proteins
Storage	Circulate freely in water-filled parts of the body	Stored in the cells associated with fat
Excretion	Kidneys detect and remove excess in urine	Less readily excreted; tend to remain in fat-storage sites
Toxicity	Possible to reach toxic levels when consumed from supplements	Likely to reach toxic levels when consumed from supplements
Requirements	Needed in frequent doses (perhaps 1 to 3 days)	Needed in periodic doses (perhaps weeks or even months)

NOTE: Exceptions occur, but these differences between the water-soluble and fat-solule vitamins are valid generalizations.

The third diagram in Figure 10-1 represents the situation with nutrients—more is better up to a point, but beyond that point, still more can be harmful.

The Committee on Dietary Reference Intakes (DRI) addresses the possibility of adverse effects from high doses of nutrients by establishing Tolerable Upper Intake Levels (UL). The UL defines the highest amount of a nutrient that is likely not to cause harm for most healthy people when consumed daily. The risk of harm increases as intakes rise above the UL. Of the nutrients discussed in this chapter, niacin, vitamin B_6, folate, choline, and vitamin C have UL, and these values are presented in their respective summary tables. Data are lacking to establish UL for the remaining B vitamins, but this does not mean that excessively high intakes would be without risk. (The inside front cover pages present UL for the vitamins and minerals.)

REVIEW IT Describe how vitamins differ from the energy nutrients and how fat-soluble vitamins differ from water-soluble vitamins.

The vitamins are essential nutrients needed in tiny amounts in the diet both to prevent deficiency diseases and to support optimal health. The water-soluble vitamins are the B vitamins and vitamin C; the fat-soluble vitamins are vitamins A, D, E, and K. Table 10-2 summarizes the differences between the water-soluble and fat-soluble vitamins.

The discussion of B vitamins that follows begins with a brief description of each of them, then offers a look at the ways they work together. Thus, a preview of the individual vitamins is followed by a discussion of their interactions.

10.2 The B Vitamins

LEARN IT Identify the main roles, deficiency symptoms, and food sources for each of the B vitamins.

Despite supplement advertisements that claim otherwise, the vitamins do not provide the body with fuel for energy. It is true, though, that without B vitamins the body would lack energy. The energy-yielding nutrients—carbohydrate, fat, and protein—are used for fuel; the B vitamins help the body to use that fuel. Several of the B vitamins—thiamin, riboflavin, niacin, pantothenic acid, and biotin—form part of the **coenzymes** that assist enzymes in the release of energy from carbohydrate, fat, and protein. Other B vitamins play other indispensable roles in metabolism. Vitamin B_6 assists enzymes that metabolize amino acids. Folate and vitamin B_{12} help cells to multiply. Among these cells are the red blood cells and the cells lining the GI tract—cells that deliver energy to all the others.

The vitamin portion of a coenzyme allows a chemical reaction to occur; the remaining portion of the coenzyme binds to the enzyme. Without its coenzyme, an enzyme cannot function. Thus symptoms of B vitamin deficiencies directly reflect the disturbances of metabolism caused by a lack of coenzymes. Figure 10-2 illustrates coenzyme action.

coenzymes: complex organic molecules that work with enzymes to facilitate the enzymes' activity. Many coenzymes have B vitamins as part of their structures.

> **FIGURE 10-2 Coenzyme Action**

Some vitamins form part of the coenzymes that enable enzymes either to synthesize compounds (as illustrated by the lower enzymes in this figure) or to dismantle compounds (as illustrated by the upper enzymes).

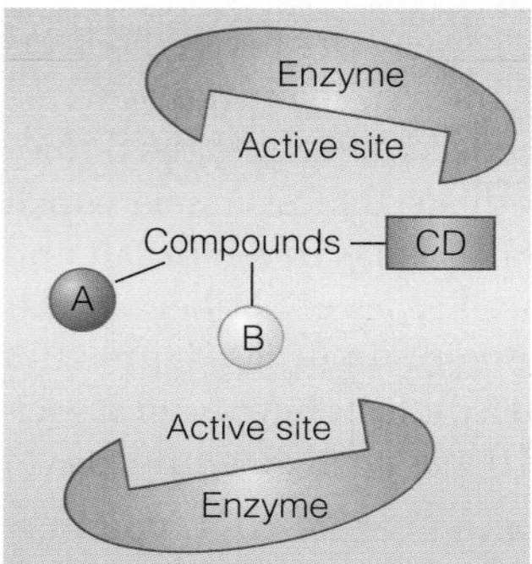

Without coenzymes, compounds A, B, and CD don't respond to their enzymes.

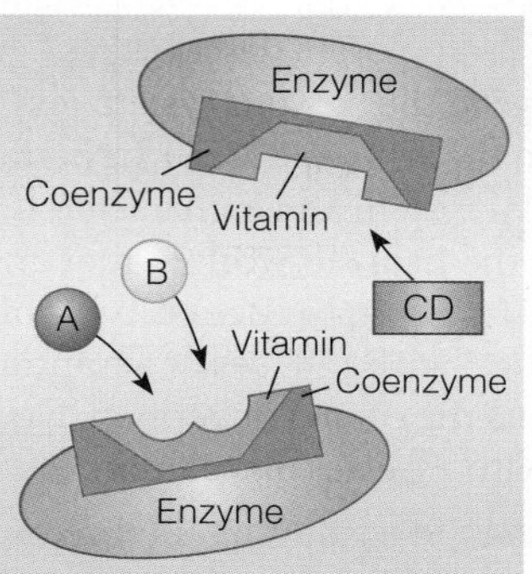

With the coenzymes in place, compounds are attracted to their sites on the enzymes . . .

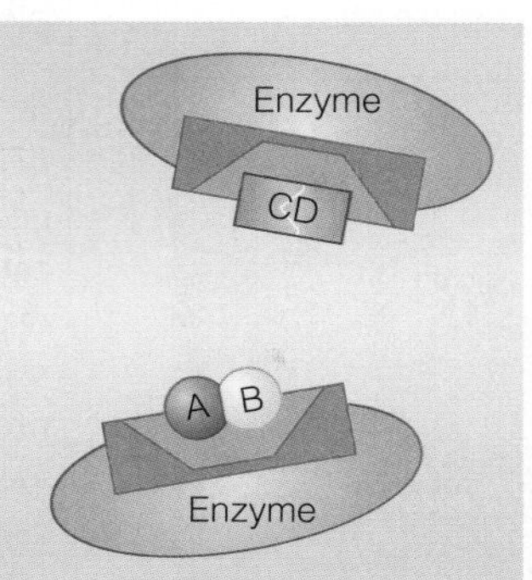

. . . and the reactions proceed instantaneously. The coenzymes often donate or accept electrons, atoms, or groups of atoms.

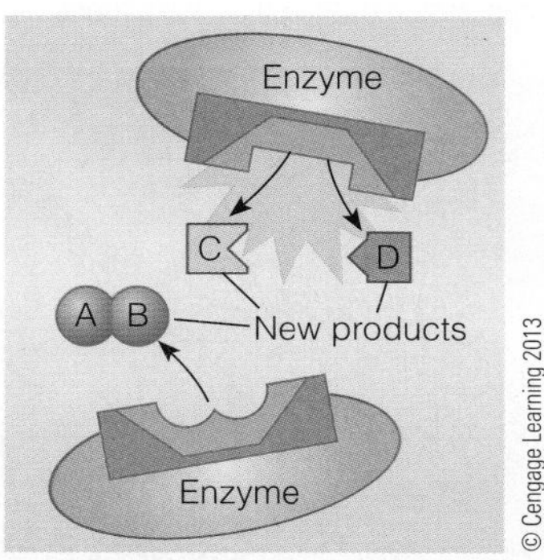

The reactions are completed with either the formation of a new product, AB, or the breaking apart of a compound into two new products, C and D, and the release of energy.

The following sections describe the roles of individual B vitamins and note many coenzymes and metabolic pathways. Keep in mind that a later discussion assembles these pieces of information into a whole picture. The following sections also present the recommendations, deficiency and toxicity symptoms, and food sources for each vitamin. For thiamin, riboflavin, niacin, vitamin B_6, folate, vitamin B_{12}, and vitamin C, sufficient data were available to establish an RDA; for biotin, pantothenic acid, and choline, an Adequate Intake (AI) was set; only niacin, vitamin B_6, folate, choline, and vitamin C have Tolerable Upper Intake Levels (UL).[3] These values appear in the summary tables and figures that follow and on the pages of the inside front cover.

Thiamin **Thiamin** is the vitamin part of the coenzyme TPP (thiamin pyrophosphate) that assists in energy metabolism. The TPP coenzyme participates in the conversion of pyruvate to acetyl CoA (described in Chapter 7). The reaction removes 1 carbon from the 3-carbon pyruvate to make the 2-carbon acetyl CoA and carbon dioxide (CO_2). In a similar step in the TCA cycle, TPP helps convert a 5-carbon compound to a 4-carbon compound. Besides playing these pivotal roles in energy metabolism, thiamin occupies a special site on the membranes of nerve cells. Consequently, nerve activity and muscle activity in response to nerves depend heavily on thiamin.

Thiamin Recommendations Dietary recommendations are based primarily on thiamin's role in enzyme activity. Generally, thiamin needs will be met if a person eats enough food to meet energy needs—if that energy comes from nutritious foods. The average thiamin intake in the United States and Canada meets or exceeds recommendations.

Thiamin Deficiency and Toxicity People who fail to eat enough food to meet energy needs risk nutrient deficiencies, including thiamin deficiency. Inadequate thiamin intakes have been reported among the nation's malnourished and homeless people. Similarly, people who derive most of their energy from empty-kcalorie foods and beverages risk thiamin deficiency. Alcohol provides a good example of how empty kcalories can lead to thiamin deficiency. Alcohol contributes energy but provides few, if any, nutrients and often displaces food. In addition, alcohol impairs thiamin absorption and enhances thiamin excretion in the urine, doubling the risk of deficiency. An estimated four out of five alcoholics are thiamin deficient.*

Prolonged thiamin deficiency can result in the disease **beriberi,** which was first observed in Indonesia when the custom of polishing rice became widespread.[4] Rice

thiamin (THIGH-ah-min): a B vitamin. The coenzyme form is *TPP* (*thiamin pyrophosphate*).

beriberi: the thiamin-deficiency disease.

- **beri** = weakness
- **beriberi** = "I can't, I can't"

*Severe thiamin deficiency in alcohol abusers is called the *Wernicke-Korsakoff* (VER-nee-key KORE-sah-kof) *syndrome.* Symptoms include disorientation, loss of short-term memory, jerky eye movements, and staggering gait.

> FIGURE 10-3 **Thiamin-Deficiency Symptom—The Edema of Beriberi**

Physical examination confirms that this person has wet beriberi. Notice how the impression of the physician's thumb remains on the leg.

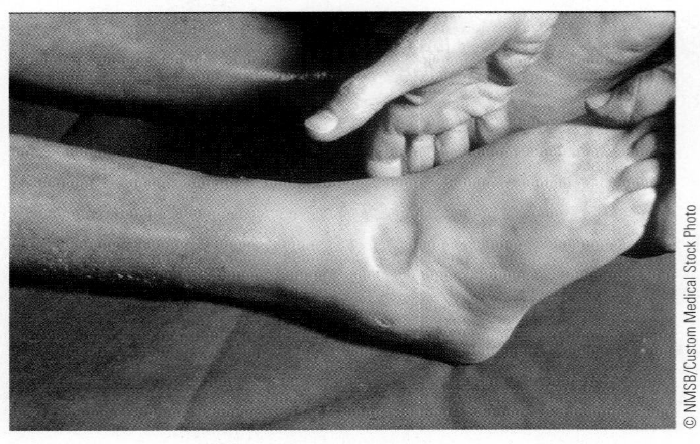

© NMSB/Custom Medical Stock Photo

provided 80 percent of the energy intake of the people of that area, and the germ and bran of the rice grain was their principal source of thiamin. When the germ and bran were removed in the preparation of white rice, beriberi became rampant.

Beriberi is often described as "dry" or "wet." Dry beriberi reflects damage to the nervous system and is characterized by muscle weakness in the arms and legs. Wet beriberi reflects damage to the cardiovascular system and is characterized by dilated blood vessels, which cause the heart to work harder and the kidneys to retain salt and water, resulting in edema. Typically, both types of beriberi appear together, with one set of symptoms predominating. Figure 10-3 presents the edema of beriberi. No adverse effects have been associated with excesses of thiamin, and no UL has been determined.

Thiamin Food Sources Before examining Figure 10-4, you may want to read the accompanying "How To" feature, which describes the content in this and similar figures found in this chapter and the next three chapters. When you look at Figure 10-4, notice that thiamin occurs in small quantities in many nutritious foods. The long red bar near the bottom of the graph shows that meats in the pork family are exceptionally rich in thiamin. Yellow bars confirm that grains—whole grains or enriched—are a reliable source of thiamin.

As mentioned earlier, prolonged cooking can destroy thiamin. Also, like other water-soluble vitamins, thiamin leaches into water when foods are boiled or blanched. Cooking methods that require little or no water such as steaming and microwave heating conserve thiamin and other water-soluble vitamins. The accompanying table provides a summary of thiamin.

riboflavin (RYE-boh-flay-vin): a B vitamin. The coenzyme forms are *FMN (flavin mononucleotide)* and *FAD (flavin adenine dinucleotide)*.

Riboflavin

Like thiamin, **riboflavin** serves as a coenzyme in many reactions, most notably in energy metabolism. The coenzyme forms of riboflavin are FMN (flavin mononucleotide) and FAD (flavin adenine dinucleotide); both can accept and then donate two hydrogens (see Figure 10-5, p. 304). During energy metabolism,

>How To

Evaluate Foods for Their Nutrient Contributions

Figure 10-4 is the first of a series of figures in this and the next three chapters that present the vitamins and minerals in foods. Each figure presents the same 24 foods, which were selected to ensure a variety of choices representative of each of the food groups as suggested by the USDA Food Patterns. For example, a bread, a cereal, and a pasta were chosen from the grain group. The suggestion to include a variety of vegetables was also considered: dark green vegetables (broccoli); orange and red vegetables (carrots); starchy vegetables (potatoes); legumes (pinto beans) and other vegetables (tomato juice). The selection of fruits followed suggestions to use whole fruits (bananas); citrus fruits (oranges); melons (watermelon); and berries (strawberries). Items were selected from the milk group and protein foods in a similar way. In addition to the 24 foods that appear in all of the figures, three different foods were selected for each of the nutrients to add variety and often reflect excellent, and sometimes unusual, sources.

Notice that the figures list the food, the serving size, and the food energy (kcalories) on the left. The amount of the nutrient per serving is presented in the graph on the right along with the RDA (or AI) for adults, so you can see how many servings would be needed to meet recommendations.

The colored bars show at a glance which food groups best provide a nutrient: yellow for grains; green for vegetables; purple for fruits; white for milk and milk products; brown for legumes; and red for protein foods. Because the USDA Food Patterns include legumes with both the protein foods group and the vegetable group and because legumes are especially rich in many vitamins and minerals, they have been given their own color to highlight their nutrient contributions.

Notice how the bar graphs shift in the various figures. Careful study of all of the figures taken together will confirm that variety is the key to nutrient adequacy.

Another way to evaluate foods for their nutrient contributions is to consider their nutrient density (their thiamin *per 100 kcalories*, for example). Quite often, vegetables rank higher on a nutrient-per-kcalorie list than they do on a nutrient-per-serving list (see p. 37 to review how to evaluate foods based on nutrient density). The left column in the figure highlights about five foods that offer the best nutrient density. Notice how many of them are vegetables.

Realistically, people cannot eat for single nutrients. Fortunately, most foods deliver more than one nutrient, allowing people to combine foods into nourishing meals.

TRY IT Calculate which food provides more riboflavin per 1-ounce serving—a pork chop (3 oz, 291 kcal, 0.25 mg riboflavin) or cheddar cheese (1½ oz, 165 kcal, 0.11 mg riboflavin). Which food is more nutrient dense with respect to riboflavin?

> FIGURE 10-4 Thiamin in Selected Foods

See the "How To" feature on p. 302 for more information on using this figure.

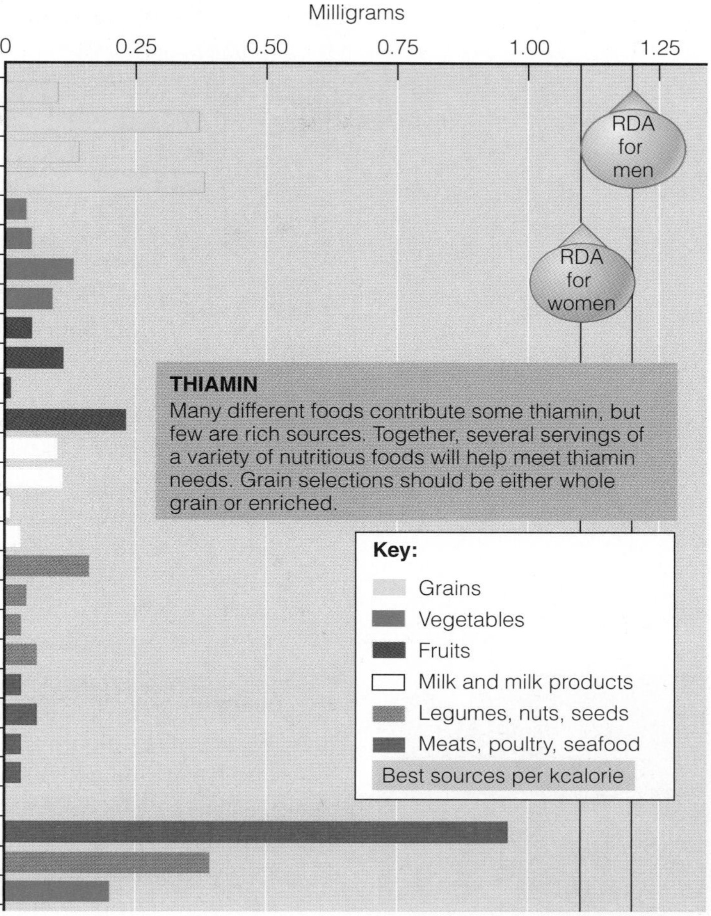

Milligrams: 0, 0.25, 0.50, 0.75, 1.00, 1.25

Food	Serving size (kcalories)
Bread, whole wheat	1 oz slice (70 kcal)
Cornflakes, fortified	1 oz (110 kcal)
Spaghetti pasta	½ c cooked (99 kcal)
Tortilla, flour	1 10"-round (234 kcal)
Broccoli	½ c cooked (22 kcal)
Carrots	½ c shredded raw (24 kcal)
Potato	1 medium baked w/skin (133 kcal)
Tomato juice	¾ c (31 kcal)
Banana	1 medium raw (109 kcal)
Orange	1 medium raw (62 kcal)
Strawberries	½ c fresh (22 kcal)
Watermelon	1 slice (92 kcal)
Milk	1 c reduced-fat 2% (121 kcal)
Yogurt, plain	1 c low-fat (155 kcal)
Cheddar cheese	1½ oz (171 kcal)
Cottage cheese	½ c low-fat 2% (101 kcal)
Pinto beans	½ c cooked (117 kcal)
Peanut butter	2 tbs (188 kcal)
Sunflower seeds	1 oz dry (165 kcal)
Tofu (soybean curd)	½ c (76 kcal)
Ground beef, lean	3 oz broiled (244 kcal)
Chicken breast	3 oz roasted (140 kcal)
Tuna, canned in water	3 oz (99 kcal)
Egg	1 hard cooked (78 kcal)
Excellent, and sometimes unusual, sources:	
Pork chop, lean	3 oz broiled (169 kcal)
Soy milk	1 c (81 kcal)
Squash, acorn	½ c baked (69 kcal)

RDA for men

RDA for women

THIAMIN
Many different foods contribute some thiamin, but few are rich sources. Together, several servings of a variety of nutritious foods will help meet thiamin needs. Grain selections should be either whole grain or enriched.

Key:
- Grains
- Vegetables
- Fruits
- Milk and milk products
- Legumes, nuts, seeds
- Meats, poultry, seafood
- Best sources per kcalorie

FAD picks up two hydrogens (with their electrons) from the TCA cycle and delivers them to the electron transport chain (described in Chapter 7).

Riboflavin Recommendations Like thiamin's RDA, riboflavin's RDA is based primarily on its role in enzyme activity. Most people in the United States and Canada meet or exceed riboflavin recommendations.

Pork is the richest source of thiamin, but enriched or whole-grain products typically make the greatest contribution to a day's intake because of the quantities eaten. Legumes such as split peas are also valuable sources of thiamin.

REVIEW IT Thiamin

Other Names

Vitamin B_1

RDA

Men: 1.2 mg/day

Women: 1.1 mg/day

Chief Functions in the Body

Part of coenzyme TPP (thiamin pyrophosphate) used in energy metabolism

Significant Sources

Whole-grain, fortified, or enriched grain products; moderate amounts in all nutritious food; pork

Easily destroyed by heat

Deficiency Disease

Beriberi (wet, with edema; dry, with muscle wasting)

Deficiency Symptoms[a]

Enlarged heart, cardiac failure; muscular weakness; apathy, poor short-term memory, confusion, irritability; anorexia, weight loss

Toxicity Symptoms

None reported

[a]Severe thiamin deficiency is often related to heavy alcohol consumption with limited food consumption (Wernicke-Korsakoff syndrome).

> FIGURE 10-5 **Riboflavin Coenzyme, Accepting and Donating Hydrogens**

This figure shows the chemical structure of the riboflavin portion of the coenzyme only; the remainder of the coenzyme structure is represented by dotted lines (see Appendix C for the complete chemical structures of FAD and FMN). The reactive sites that accept and donate hydrogens are highlighted in white.

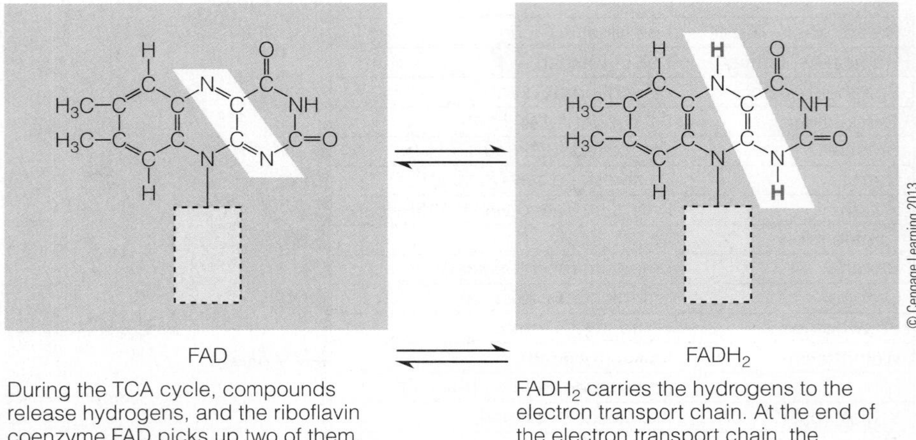

During the TCA cycle, compounds release hydrogens, and the riboflavin coenzyme FAD picks up two of them. As it accepts two hydrogens, FAD becomes $FADH_2$.

$FADH_2$ carries the hydrogens to the electron transport chain. At the end of the electron transport chain, the hydrogens are accepted by oxygen, creating water, and $FADH_2$ becomes FAD again. For every $FADH_2$ that passes through the electron transport chain, two ATP are generated.

© Polara Studios, Inc.

All of these foods are rich in riboflavin, but milk and milk products provide much of the riboflavin in the diets of most people.

Riboflavin Deficiency and Toxicity Riboflavin deficiency most often accompanies other nutrient deficiencies.* Lack of the vitamin causes inflammation of the membranes of the mouth, skin, eyes, and GI tract. Excesses of riboflavin appear to cause no harm, and no UL has been established.

Riboflavin Food Sources The greatest contributions of riboflavin come from milk and milk products (see Figure 10-6). Whole-grain or enriched grains are also valuable sources because of the quantities typically consumed. When riboflavin sources are ranked by nutrient density (per kcalorie), many dark green, leafy vegetables (such as broccoli, turnip greens, asparagus, and spinach) appear high on the list. Vegans and others who don't use milk must rely on ample servings of dark greens and enriched grains for riboflavin. Nutritional yeast is another good source.

Ultraviolet light and irradiation destroy riboflavin. For these reasons, milk is sold in cardboard or opaque plastic containers, instead of clear glass bottles. In contrast, riboflavin is stable to heat, so cooking does not destroy it. The accompanying table provides a summary of riboflavin.

REVIEW IT Riboflavin

Other Names

Vitamin B_2

RDA

Men: 1.3 mg/day

Women: 1.1 mg/day

Chief Functions in the Body

Part of coenzymes FMN (flavin mononucleotide) and FAD (flavin adenine dinucleotide) used in energy metabolism

Significant Sources

Milk products (yogurt, cheese); whole-grain, fortified, or enriched grain products; liver

Easily destroyed by ultraviolet light and irradiation

Deficiency Disease

Ariboflavinosis (ay-RYE-boh-FLAY-vin-oh-sis)

Deficiency Symptoms

Sore throat; cracks and redness at corners of mouth;[a] painful, smooth, purplish red tongue;[b] inflammation characterized by skin lesions covered with greasy scales

Toxicity Symptoms

None reported

[a]Cracks at the corners of the mouth are called *angular stomatitis* or *cheilosis* (kye-LOH-sis or kee-LOH-sis).
[b]Smoothness of the tongue is caused by loss of its surface structures and is termed *glossitis* (gloss-EYE-tis).

*Riboflavin deficiency is called *ariboflavinosis* (ay-RYE-boh-FLAY-vin-oh-sis).

> FIGURE 10-6 Riboflavin in Selected Foods

See the "How To" feature on p. 302 for more information on using this figure.

Food	Serving size (kcalories)
Bread, whole wheat	1 oz slice (70 kcal)
Cornflakes, fortified	1 oz (110 kcal)
Spaghetti pasta	½ c cooked (99 kcal)
Tortilla, flour	1 10"-round (234 kcal)
Broccoli	½ c cooked (22 kcal)
Carrots	½ c shredded raw (24 kcal)
Potato	1 medium baked w/skin (133 kcal)
Tomato juice	¾ c (31 kcal)
Banana	1 medium raw (109 kcal)
Orange	1 medium raw (62 kcal)
Strawberries	½ c fresh (22 kcal)
Watermelon	1 slice (92 kcal)
Milk	1 c reduced-fat 2% (121 kcal)
Yogurt, plain	1 c low-fat (155 kcal)
Cheddar cheese	1½ oz (171 kcal)
Cottage cheese	½ c low-fat 2% (101 kcal)
Pinto beans	½ c cooked (117 kcal)
Peanut butter	2 tbs (188 kcal)
Sunflower seeds	1 oz dry (165 kcal)
Tofu (soybean curd)	½ c (76 kcal)
Ground beef, lean	3 oz broiled (244 kcal)
Chicken breast	3 oz roasted (140 kcal)
Tuna, canned in water	3 oz (99 kcal)
Egg	1 hard cooked (78 kcal)
Excellent, and sometimes unusual, sources:	
Liver	3 oz fried (184 kcal)
Clams, canned	3 oz (126 kcal)
Mushrooms	½ c cooked (21 kcal)

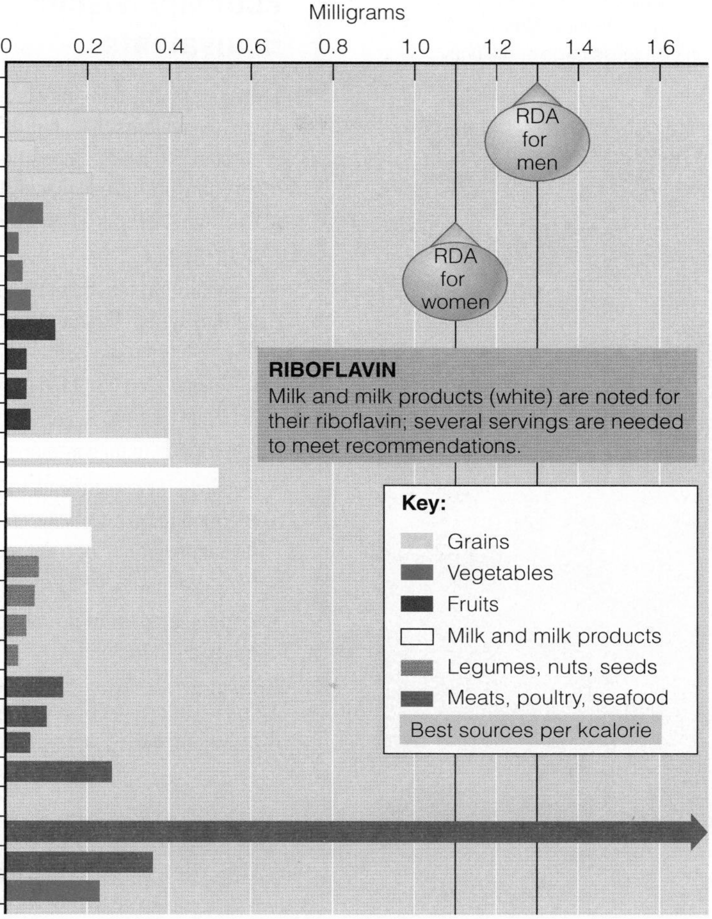

Niacin The name **niacin** describes two chemical structures: nicotinic acid and nicotinamide (also known as niacinamide). The body can easily convert nicotinic acid to nicotinamide, which is the major form of niacin in the blood.

The two coenzyme forms of niacin, NAD (nicotinamide adenine dinucleotide) and NADP (the phosphate form), participate in numerous metabolic reactions. They are central in energy-transfer reactions, especially the metabolism of glucose, fat, and alcohol. NAD is similar to the riboflavin coenzymes in that it carries hydrogens (and their electrons) during metabolic reactions, including the pathway from the TCA cycle to the electron transport chain. NAD also protects against neurological degeneration.[5]

Niacin Recommendations Niacin is unique among the B vitamins in that the body can make it from the amino acid tryptophan. This use of tryptophan occurs only after protein synthesis needs have been met.[6] Approximately 60 milligrams of dietary tryptophan is needed to make 1 milligram of niacin. For this reason, recommended intakes are stated in **niacin equivalents (NE).** ♦ A food containing 1 milligram of niacin and 60 milligrams of tryptophan provides the equivalent of 2 milligrams of niacin, or 2 niacin equivalents. The RDA for niacin allows for this conversion and is stated in niacin equivalents; average niacin intakes in the United States and Canada exceed recommendations. The "How To" feature on p. 306 shows how to estimate niacin equivalents from both tryptophan and preformed niacin in the diet.

♦ 1 NE = 1 mg niacin or 60 mg tryptophan

niacin (NIGH-a-sin): a B vitamin. The coenzyme forms are NAD (*nicotinamide adenine dinucleotide*) and NADP (*the phosphate form of NAD*). Niacin can be eaten preformed or made in the body from its precursor, tryptophan, an essential amino acid.

niacin equivalents (NE): the amount of niacin present in food, including the niacin that can theoretically be made from its precursor, tryptophan, present in the food.

pellagra (pell-AY-gra): the niacin-deficiency disease.
- **pellis** = skin
- **agra** = rough

Niacin Deficiency The niacin-deficiency disease, **pellagra,** produces the symptoms of diarrhea, dermatitis, dementia, and eventually death (often called "the four Ds"). Figure 10-7 (p. 306) illustrates the dermatitis of pellagra.

In the early 1900s, pellagra caused widespread misery and some 87,000 deaths in the US South, where many people subsisted on a low-protein diet centered on

> FIGURE 10-7 **Niacin-Deficiency Symptom—The Dermatitis of Pellagra**

In the dermatitis of pellagra, the skin darkens and flakes away as if it were sunburned. Skin lesions typically develop only on those parts of the body exposed to the sun.

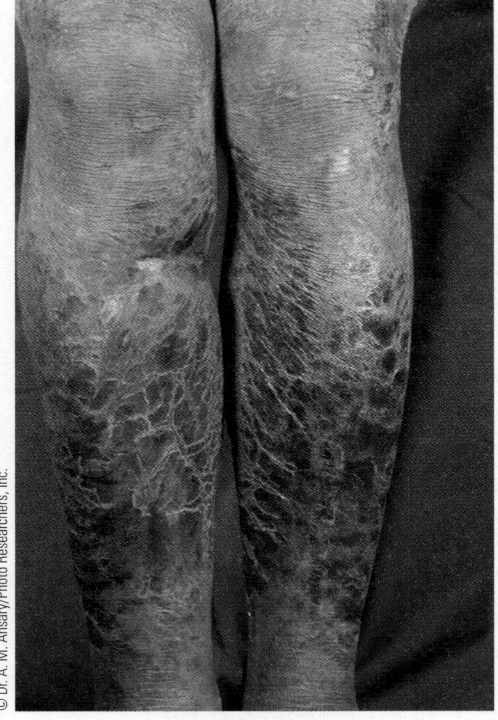

>How To

Estimate Niacin Equivalents

Niacin recommendations are expressed as niacin equivalents (NE), but diet analysis programs and food composition tables report only preformed niacin. To estimate niacin equivalents from the tryptophan in dietary protein:

- Assume that most dietary proteins contain about 1 percent tryptophan. To determine the amount of tryptophan in protein, divide grams of protein by 100.
- Multiply by 1000 to convert grams of tryptophan to milligrams.
- Because it takes 60 milligrams of tryptophan to make 1 milligram of niacin, divide milligrams of tryptophan by 60 to get niacin equivalents.
- Add the amount of preformed niacin obtained in the diet.

Consider, for example, a person who consumes 80 grams of protein and 5 milligrams of preformed niacin.

- Estimate the amount of tryptophan in 80 grams of protein and convert to milligrams:

  80 g protein ÷ 100 = 0.8 g tryptophan

  0.8 g tryptophan × 1000 = 800 mg tryptophan

- Convert milligrams of tryptophan to niacin equivalents:

  800 mg tryptophan ÷ 60 = 13 mg NE

To determine the total amount of niacin available from the diet, add the amount available from tryptophan to the amount preformed in the diet.

13 mg NE + 5 mg preformed niacin = 18 mg NE

TRY IT Calculate how many niacin equivalents a person receives from a diet that delivers 60 grams protein and 6 milligrams niacin.

corn. This diet supplied neither enough niacin nor enough tryptophan. At least 70 percent of the niacin in corn is bound to complex carbohydrates and small peptides, making it unavailable for absorption. Furthermore, corn is high in the amino acid leucine, which interferes with the tryptophan-to-niacin conversion, thus further contributing to the development of pellagra.

Pellagra was originally believed to be caused by an infection. Medical researchers spent many years and much effort searching for infectious microbes until they realized that the problem was not what was *present* in the food but what was *absent* from it. That a disease such as pellagra could be caused by diet—and not by pathogens—was a groundbreaking discovery. It contradicted commonly held medical opinions that diseases were caused only by infectious agents. By carefully following the scientific method (as described in Chapter 1), researchers advanced the science of nutrition dramatically.

Niacin Toxicity When a normal dose of a nutrient (levels commonly found in foods) provides a normal blood concentration, the nutrient is having a *physiological* effect. When a large dose (levels commonly available only from supplements) overwhelms the body and raises blood concentrations to abnormally high levels, the nutrient is acting like a drug and having a *pharmacological* effect. Naturally occurring niacin from foods has a physiological effect that causes no harm. Large doses of nicotinic acid from supplements or drugs, however, produce a variety of pharmacological effects, most notably **"niacin flush."** Niacin flush occurs when nicotinic acid is taken in doses only three to four times the RDA. It dilates the capillaries and causes a tingling sensation that can be painful. The nicotinamide form does not produce this effect.

Large doses of nicotinic acid have been used to lower LDL cholesterol, raise HDL cholesterol, and increase adiponectin levels—all factors that help to protect against heart disease.[7] Such therapy must be closely monitored. People with the following conditions may be particularly susceptible to the toxic effects of niacin: liver disease, diabetes, peptic ulcers, gout, irregular heartbeats, inflammatory

niacin flush: a temporary burning, tingling, and itching sensation that occurs when a person takes a large dose of nicotinic acid; often accompanied by a headache and reddened face, arms, and chest.

> **FIGURE 10-8 Niacin in Selected Foods**

See the "How To" feature on p. 302 for more information on using this figure.

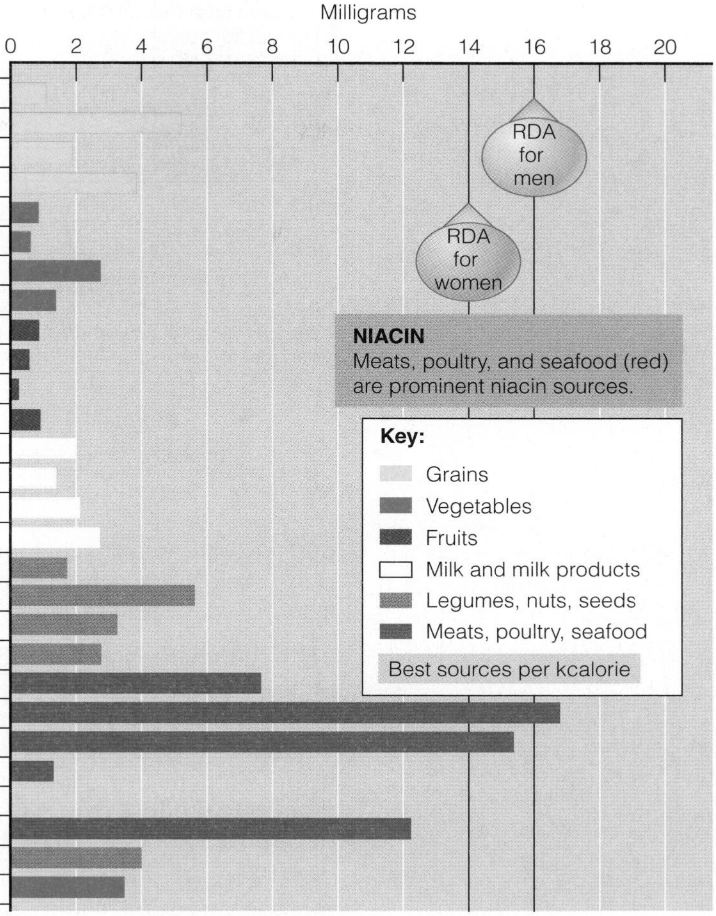

Milligrams: 0, 2, 4, 6, 8, 10, 12, 14, 16, 18, 20

Food	Serving size (kcalories)
Bread, whole wheat	1 oz slice (70 kcal)
Cornflakes, fortified	1 oz (110 kcal)
Spaghetti pasta	½ c cooked (99 kcal)
Tortilla, flour	1 10"-round (234 kcal)
Broccoli	½ c cooked (22 kcal)
Carrots	½ c shredded raw (24 kcal)
Potato	1 medium baked w/skin (133 kcal)
Tomato juice	¾ c (31 kcal)
Banana	1 medium raw (109 kcal)
Orange	1 medium raw (62 kcal)
Strawberries	½ c fresh (22 kcal)
Watermelon	1 slice (92 kcal)
Milk	1 c reduced-fat 2% (121 kcal)
Yogurt, plain	1 c low-fat (155 kcal)
Cheddar cheese	1½ oz (171 kcal)
Cottage cheese	½ c low-fat 2% (101 kcal)
Pinto beans	½ c cooked (117 kcal)
Peanut butter	2 tbs (188 kcal)
Sunflower seeds	1 oz dry (165 kcal)
Tofu (soybean curd)	½ c (76 kcal)
Ground beef, lean	3 oz broiled (244 kcal)
Chicken breast	3 oz roasted (140 kcal)
Tuna, canned in water	3 oz (99 kcal)
Egg	1 hard cooked (78 kcal)
Excellent, and sometimes unusual, sources:	
Liver	3 oz fried (184 kcal)
Peanuts	1 oz roasted (165 kcal)
Mushrooms	½ c cooked (21 kcal)

RDA for men

RDA for women

NIACIN
Meats, poultry, and seafood (red) are prominent niacin sources.

Key:
- Grains
- Vegetables
- Fruits
- Milk and milk products
- Legumes, nuts, seeds
- Meats, poultry, seafood
- Best sources per kcalorie

bowel disease, migraine headaches, and alcoholism. The nicotinamide form does not improve blood cholesterol levels.

Niacin Food Sources Tables of food composition typically list preformed niacin only, but as mentioned, niacin can also be made in the body from the amino acid tryptophan. Dietary tryptophan could meet about half the daily niacin need for most people, but the average diet easily supplies enough preformed niacin.

Figure 10-8 presents niacin in selected foods. Meat, poultry, fish, legumes, and enriched and whole grains contribute about half the niacin people consume. Mushrooms, potatoes, and tomatoes are among the richest vegetable sources, and they can provide abundant niacin when eaten in generous amounts.

Niacin is less vulnerable to losses during food preparation and storage than other water-soluble vitamins. Being fairly heat resistant, niacin can withstand reasonable cooking times, but like other water-soluble vitamins, it will leach into cooking water. The table on p. 308 provides a summary of niacin.

Biotin

Biotin plays an important role in metabolism as a coenzyme that carries activated carbon dioxide. This role is critical in the TCA cycle: biotin delivers a carbon to 3-carbon pyruvate, thus replenishing oxaloacetate, the 4-carbon compound needed to combine with acetyl CoA to keep the TCA cycle turning (review Figure 7-15 on p. 211). The biotin coenzyme also participates in gluconeogenesis, fatty acid synthesis, and the breakdown of certain fatty acids and amino acids.

Protein-rich foods such as meat, fish, poultry, and peanut butter contribute much of the niacin in people's diets. Enriched breads and cereals and a few vegetables are also rich in niacin.

biotin (BY-oh-tin): a B vitamin that functions as a coenzyme in metabolism.

REVIEW IT Niacin

Other Names

Nicotinic acid, nicotinamide, niacinamide, vitamin B_3; precursor is dietary tryptophan (an amino acid)

RDA

Men: 16 mg NE/day

Women: 14 mg NE/day

UL[a]

Adults: 35 mg/day

Chief Functions in the Body

Part of coenzymes NAD (nicotinamide adenine dinucleotide) and NADP (its phosphate form) used in energy metabolism

Significant Sources

Milk, eggs, meat, poultry, fish; whole-grain, fortified, and enriched grain products; nuts and all protein-containing foods

Deficiency Disease

Pellagra

Deficiency Symptoms

Diarrhea, abdominal pain, vomiting; inflamed, swollen, smooth, bright red tongue;[b] depression, apathy, fatigue, loss of memory, headache; bilateral symmetrical rash on areas exposed to sunlight

Toxicity Symptoms

Painful flush, hives, and rash ("niacin flush"); nausea and vomiting; liver damage, impaired glucose tolerance

[a]The UL applies to synthetic forms obtained from supplements, fortified foods, or a combination.
[b]Smoothness of the tongue is caused by loss of its surface structures and is termed *glossitis* (gloss-EYE-tis).

Biotin Recommendations Biotin is needed in very small amounts. Because there is insufficient research on biotin requirements, an Adequate Intake (AI) has been determined, instead of an RDA.

Biotin Deficiency and Toxicity Biotin deficiencies rarely occur. Researchers can induce a biotin deficiency in animals or human beings by feeding them raw egg whites, which contain a protein that binds biotin and thus prevents its absorption.* Biotin-deficiency symptoms include skin rash, hair loss, and neurological impairment. More than two dozen raw egg whites must be consumed daily for several months to produce these effects; cooking eggs denatures the binding protein. Because no adverse effects have been reported from high biotin intakes, a UL has not been set.

Biotin Food Sources Biotin is widespread in foods (including egg yolks), so eating a variety of foods protects against deficiencies. Some biotin is also synthesized by GI tract bacteria, but this amount does not contribute much to the biotin absorbed. The accompanying table provides a summary of biotin.

REVIEW IT Biotin

Adequate Intake (AI)

Adults: 30 µg/day

Chief Functions in the Body

Part of a coenzyme used in energy metabolism, fat synthesis, amino acid metabolism, and glycogen synthesis

Significant Sources

Widespread in foods; liver, egg yolks, soybeans, fish, whole grains; also produced by GI bacteria

Deficiency Symptoms

Depression, lethargy, hallucinations, numb or tingling sensation in the arms and legs; red, scaly rash around the eyes, nose, and mouth; hair loss

Toxicity Symptoms

None reported

Pantothenic Acid

Pantothenic acid is part of the chemical structure of coenzyme A—the same CoA that forms acetyl CoA, the "crossroads" compound in several metabolic pathways, including the TCA cycle. (Appendix C presents the chemical structures of these two molecules and shows that coenzyme A is made up in part of pantothenic acid.) As such, it is involved in more than 100 different steps in the synthesis of lipids, neurotransmitters, steroid hormones, and hemoglobin.

pantothenic (PAN-toe-THEN-ick) **acid:** a B vitamin. The principal active form is part of coenzyme A, called "CoA" throughout Chapter 7.
- **pantos** = everywhere

*The protein *avidin* (AV-eh-din) in egg whites binds biotin.

Pantothenic Acid Recommendations An Adequate Intake (AI) for pantothenic acid has been set. It reflects the amount needed to replace daily losses.

Pantothenic Acid Deficiency and Toxicity Pantothenic acid deficiency is rare. Its symptoms involve a general failure of all the body's systems and include fatigue, GI distress, and neurological disturbances. The "burning feet" syndrome that affected prisoners of war in Asia during World War II is thought to have been caused by pantothenic acid deficiency. No toxic effects have been reported, and no UL has been established.

Pantothenic Acid Food Sources Pantothenic acid is widespread in foods, and typical diets seem to provide adequate intakes. Beef, poultry, whole grains, potatoes, tomatoes, and broccoli are particularly good sources. Losses of pantothenic acid during food production can be substantial because it is readily destroyed by the freezing, canning, and refining processes. The accompanying table provides a summary of pantothenic acid.

REVIEW IT Pantothenic Acid

Adequate Intake (AI)	Deficiency Symptoms
Adults: 5 mg/day	Vomiting, nausea, stomach cramps; insomnia, fatigue, depression, irritability, restlessness, apathy; hypoglycemia, increased sensitivity to insulin; numbness, muscle cramps, inability to walk
Chief Functions in the Body	
Part of coenzyme A, used in energy metabolism	
Significant Sources	**Toxicity Symptoms**
Widespread in foods; chicken, beef, potatoes, oats, tomatoes, liver, egg yolk, broccoli, whole grains	None reported
Easily destroyed by food processing	

Vitamin B_6

Vitamin B_6 occurs in three forms—pyridoxal, pyridoxine, and pyridoxamine. All three can be converted to the coenzyme PLP (pyridoxal phosphate), which is active in amino acid metabolism. Because PLP can transfer amino groups (NH_2) from an amino acid to a keto acid, the body can make nonessential amino acids (review Figure 6-12 on p. 179). The ability to add and remove amino groups makes PLP valuable in protein and urea metabolism as well. The conversions of the amino acid tryptophan to niacin or to the neurotransmitter serotonin also depend on PLP. In addition, PLP participates in the synthesis of heme (the nonprotein portion of hemoglobin), nucleic acids (such as DNA and RNA), and lecithin (a phospholipid).

Vitamin B_6 Recommendations The RDA for vitamin B_6 is based on the amounts needed to maintain adequate levels of its coenzymes. Unlike other water-soluble vitamins, vitamin B_6 is stored extensively in muscle tissue. Research does not support claims, however, that large doses of vitamin B_6 enhance muscle strength or physical endurance.

Vitamin B_6 Deficiency Without adequate vitamin B_6, synthesis of key neurotransmitters diminishes, and abnormal compounds produced during tryptophan metabolism accumulate in the brain. Early symptoms of vitamin B_6 deficiency include depression and confusion; advanced symptoms include abnormal brain wave patterns and convulsions. Low levels of vitamin B_6 are associated with increased risks of some cancers and cardiovascular disease.[8]

Alcohol contributes to the destruction and loss of vitamin B_6 from the body. As Highlight 7 describes, when the body breaks down alcohol, it produces acetaldehyde. If allowed to accumulate, acetaldehyde dislodges the PLP coenzyme from its enzymes; once loose, PLP breaks down and is excreted.

Another drug that acts as a vitamin B_6 **antagonist** is isoniazid, a medication that inhibits the growth of the tuberculosis bacterium.* This drug has saved countless lives, but because isoniazid binds and inactivates vitamin B_6, it can induce a deficiency. Whenever isoniazid is used to treat tuberculosis, vitamin B_6 supplements must be given to protect against deficiency.

*Isoniazid (eye-so-NYE-uh-zid) is also known as INH (isonicotinic acid hydrazide).

vitamin B_6: a family of compounds—pyridoxal, pyridoxine, and pyridoxamine. The primary active coenzyme form is *PLP* (*pyridoxal phosphate*).

antagonist: a competing factor that counteracts the action of another factor. When a drug displaces a vitamin from its site of action, the drug renders the vitamin ineffective and thus acts as a vitamin antagonist.

> FIGURE 10-9 Vitamin B_6 in Selected Foods

See the "How To" feature on p. 302 for more information on using this figure.

Food	Serving size (kcalories)
Bread, whole wheat	1 oz slice (70 kcal)
Cornflakes, fortified	1 oz (110 kcal)
Spaghetti pasta	½ c cooked (99 kcal)
Tortilla, flour	1 10"-round (234 kcal)
Broccoli	½ c cooked (22 kcal)
Carrots	½ c shredded raw (24 kcal)
Potato	1 medium baked w/skin (133 kcal)
Tomato juice	¾ c (31 kcal)
Banana	1 medium raw (109 kcal)
Orange	1 medium raw (62 kcal)
Strawberries	½ c fresh (22 kcal)
Watermelon	1 slice (92 kcal)
Milk	1 c reduced-fat 2% (121 kcal)
Yogurt, plain	1 c low-fat (155 kcal)
Cheddar cheese	1½ oz (171 kcal)
Cottage cheese	½ c low-fat 2% (101 kcal)
Pinto beans	½ c cooked (117 kcal)
Peanut butter	2 tbs (188 kcal)
Sunflower seeds	1 oz dry (165 kcal)
Tofu (soybean curd)	½ c (76 kcal)
Ground beef, lean	3 oz broiled (244 kcal)
Chicken breast	3 oz roasted (140 kcal)
Tuna, canned in water	3 oz (99 kcal)
Egg	1 hard cooked (78 kcal)
Excellent, and sometimes unusual, sources:	
Prune juice	¾ c (137 kcal)
Bluefish	3 oz baked (135 kcal)
Squash, acorn	½ c baked (69 kcal)

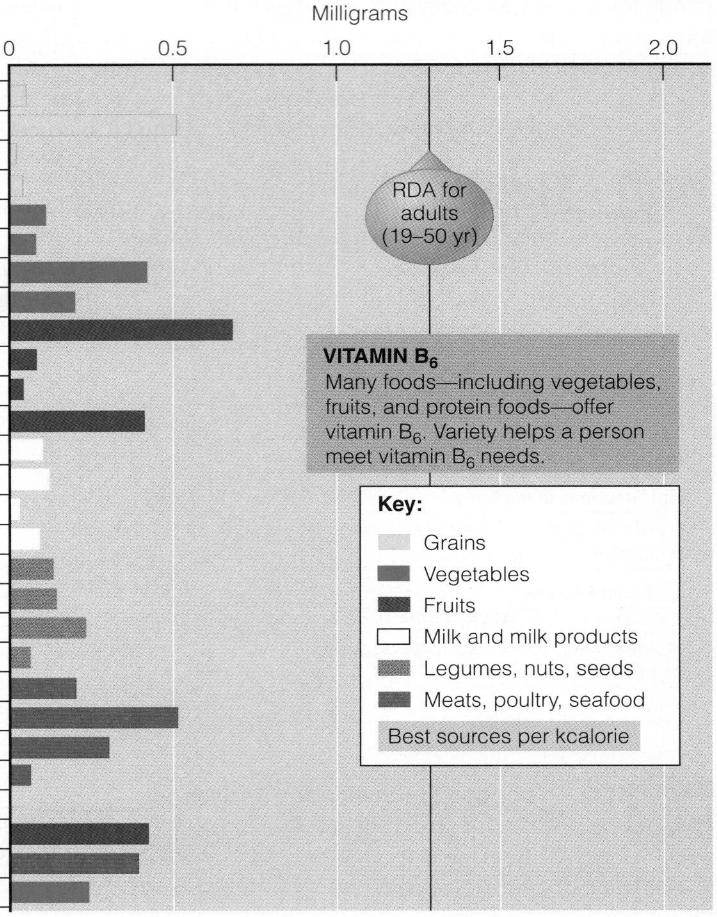

Milligrams: 0, 0.5, 1.0, 1.5, 2.0

RDA for adults (19–50 yr)

VITAMIN B_6
Many foods—including vegetables, fruits, and protein foods—offer vitamin B_6. Variety helps a person meet vitamin B_6 needs.

Key:
- Grains
- Vegetables
- Fruits
- Milk and milk products
- Legumes, nuts, seeds
- Meats, poultry, seafood
- Best sources per kcalorie

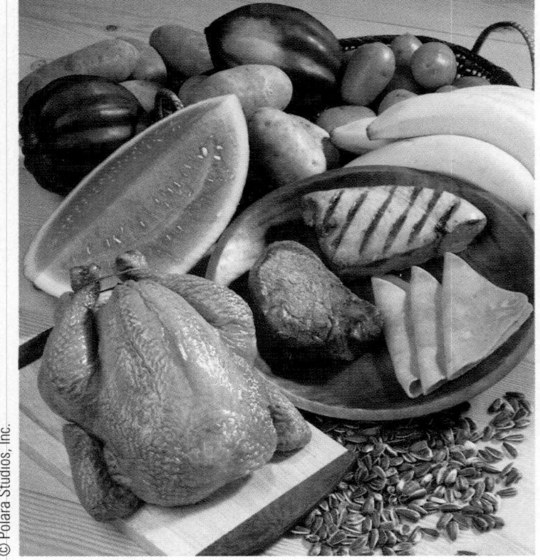

Most protein-rich foods such as meat, fish, and poultry provide ample vitamin B_6; some vegetables and fruits are good sources too.

folate (FOLE-ate): a B vitamin; also known as folic acid, folacin, or pteroylglutamic (tare-o-EEL-glue-TAM-ick) acid (PGA). The coenzyme forms are *DHF (dihydrofolate)* and *THF (tetrahydrofolate)*.

Vitamin B_6 Toxicity The first major report of vitamin B_6 toxicity appeared in the early 1980s. Until that time, most researchers and dietitians believed that, like the other water-soluble vitamins, vitamin B_6 could not reach toxic concentrations in the body. The report described neurological damage in people who had been taking more than 2 *grams* of vitamin B_6 daily (20 times the current UL of 100 *milligrams* per day) for 2 months or more.

Vitamin B_6 Food Sources As you can see from the colors in Figure 10-9, meats, fish, and poultry (red bars), potatoes and a few other vegetables (green bars), and fruits (purple bars) offer vitamin B_6. As is true of most of the other vitamins, fruits and vegetables rank considerably higher when foods are judged by nutrient density (vitamin B_6 per kcalorie). Several servings of vitamin B_6–rich foods are needed to meet recommended intakes.

Foods lose vitamin B_6 when heated. Information is limited, but vitamin B_6 bioavailability from plant-derived foods seems to be lower than from animal-derived foods. Fiber does not appear to interfere with vitamin B_6 absorption. The accompanying table provides a summary of vitamin B_6.

Folate

Folate, also known as folacin or folic acid, has a chemical name that would fit a flying dinosaur: pteroylglutamic acid (PGA for short). Its primary coenzyme form, THF (tetrahydrofolate), serves as part of an enzyme complex that transfers 1-carbon compounds that arise during metabolism.[9] This action converts vitamin B_{12} to one of its coenzyme forms, synthesizes the DNA required for all rapidly growing cells, and regenerates the amino acid methionine from homocysteine.

REVIEW IT Vitamin B_6

Other Names	Significant Sources
Pyridoxine, pyridoxal, pyridoxamine	Meats, fish, poultry, potatoes and other starchy vegetables, legumes, noncitrus fruits, fortified cereals, liver, soy products
	Easily destroyed by heat

RDA	
Adults (19–50 yr): 1.3 mg/day	

UL	Deficiency Symptoms
Adults: 100 mg/day	Scaly dermatitis; anemia (small-cell type);[a] depression, confusion, convulsions

Chief Functions in the Body	Toxicity Symptoms
Part of coenzymes PLP (pyridoxal phosphate) and PMP (pyridoxamine phosphate) used in amino acid and fatty acid metabolism; helps to convert tryptophan to niacin and to serotonin; helps to make red blood cells	Depression, fatigue, irritability, headaches, nerve damage causing numbness and muscle weakness leading to an inability to walk and convulsions; skin lesions

[a]Small-cell-type anemia is called *microcytic anemia.*

Figure 10-10 summarizes folate's absorption, activation, and relationship with vitamin B_{12}. It explains that foods deliver folate mostly in the "bound" form—that is, combined with a string of amino acids (all glutamate), known as polyglutamate. (See Appendix C for the chemical structure.) Enzymes on the intestinal cell surfaces hydrolyze the polyglutamate to monoglutamate—folate with only one glutamate attached—and several single glutamates. The monoglutamate is then attached to a methyl group (CH_3) and delivered to the liver and

> **FIGURE 10-10 Folate's Absorption and Activation**

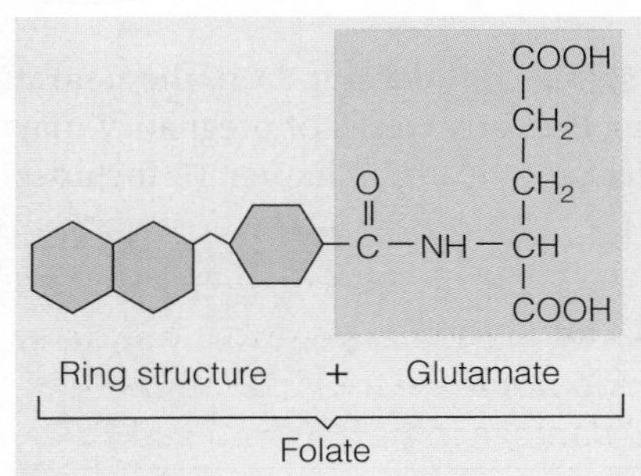

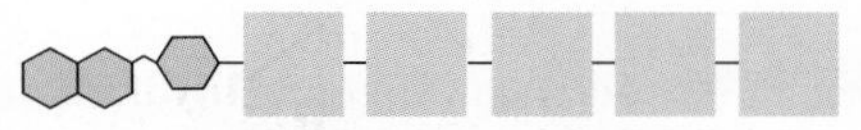

In foods, folate naturally occurs as polyglutamate. (Folate occurs as monoglutamate in fortified foods and supplements.)

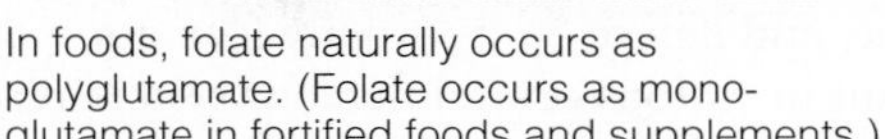

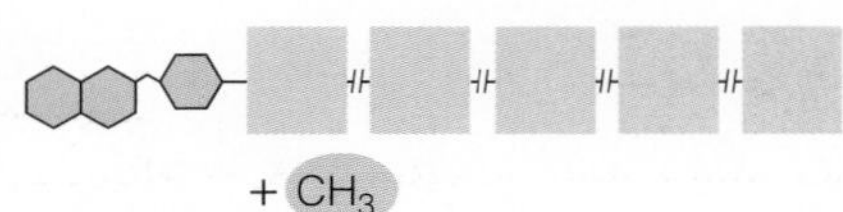

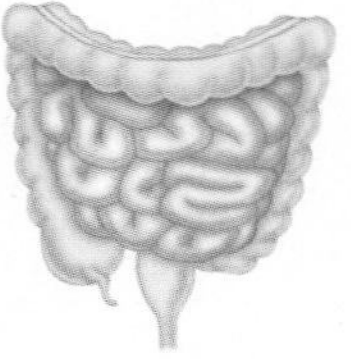

In the intestine, digestion breaks glutamates off . . . and adds a methyl group. Folate is absorbed and delivered to cells.

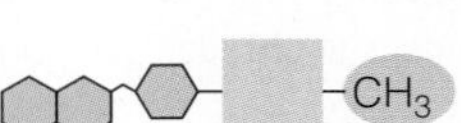

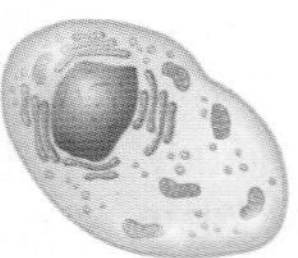

In the cells, folate is trapped in its inactive form.

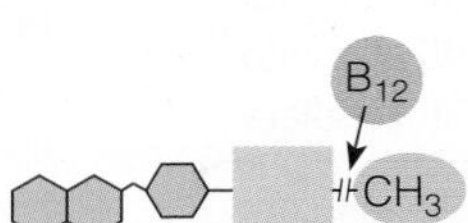

To activate folate, vitamin B_{12} removes and keeps the methyl group, which activates vitamin B_{12}.

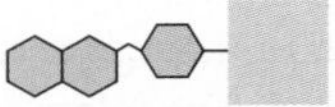

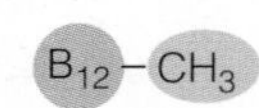

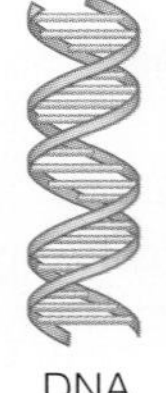

Both the folate coenzyme and the vitamin B_{12} coenzyme are now active and available for DNA synthesis.

other body cells. To activate folate, the methyl group must be removed by an enzyme that requires the help of vitamin B_{12}. Without that help, folate becomes trapped inside cells in its methyl form, unavailable to support DNA synthesis and cell growth.

The liver incorporates excess folate into bile that is then sent to the gallbladder and GI tract. Thus folate travels in the same enterohepatic circulation as bile (review Figure 5-14 on p. 141).

This complicated system for handling folate is vulnerable to GI tract injuries. Because folate is actively secreted back into the GI tract with bile, it can be reabsorbed repeatedly. If the GI tract cells are damaged, then folate is lost. Such is the case in alcohol abuse; folate deficiency rapidly develops and, ironically, further damages the GI tract. Remember, folate is active in cell multiplication—and the cells lining the GI tract are among the most rapidly replaced cells in the body. When unable to make new cells, the GI tract deteriorates and not only loses folate, but fails to absorb other nutrients as well.

Folate Recommendations The bioavailability of folate ranges from 50 percent for foods to 100 percent for supplements taken on an empty stomach. These differences in bioavailability must be considered when establishing folate recommendations.[10] The DRI committee gives naturally occurring folate from foods full credit. Synthetic folate from fortified foods and supplements is given extra credit because, on average, it is 1.7 times more available than naturally occurring food folate. Thus a person consuming 100 micrograms of folate from foods and 100 micrograms from a supplement receives 270 **dietary folate equivalents (DFE).** ♦ (The "How To" feature below describes how to estimate dietary folate equivalents.) The need for folate rises considerably during pregnancy and whenever cells are multiplying, so the recommendations for pregnant women are considerably higher than for other adults.

♦ To calculate DFE:
DFE = μg food folate + (1.7 × μg synthetic folate)
Using the example in the text:
100 μg food
+ 170 μg supplement (1.7 × 100 μg)
270 μg DFE

dietary folate equivalents (DFE): the amount of folate available to the body from naturally occurring sources, fortified foods, and supplements, accounting for differences in the bioavailability from each source.

neural tube: the embryonic tissue that forms the brain and spinal cord. The two main types of neural tube defects are *spina bifida* (literally "split spine") and *anencephaly* ("no brain").

Folate and Neural Tube Defects The brain and spinal cord develop from the **neural tube,** and defects in its orderly formation during the early weeks of pregnancy may result in various central nervous system disorders and death. (Chapter 15 includes an illustration of a neural tube defect.)

>How To

Estimate Dietary Folate Equivalents

Folate is expressed in terms of DFE (dietary folate equivalents) because synthetic folate from supplements and fortified foods is absorbed at almost twice (1.7 times) the rate of naturally occurring folate from other foods. Use the following equation to calculate:

DFE = μg food folate + (1.7 × μg synthetic folate)

Consider, for example, a pregnant woman who takes a supplement and eats a bowl of fortified cornflakes, 2 slices of fortified bread, and a cup of fortified pasta. From the supplement and fortified foods, she obtains synthetic folate:

Supplement	100 μg folate
Fortified cornflakes	100 μg folate
Fortified bread	40 μg folate
Fortified pasta	60 μg folate
	300 μg folate

To calculate the DFE, multiply the amount of synthetic folate by 1.7:

300 μg × 1.7 = 510 μg DFE

Now add the naturally occurring folate from the other foods in her diet—in this example, another 90 μg of folate.

510 μg DFE + 90 μg = 600 μg DFE

Notice that if we had not converted synthetic folate from supplements and fortified foods to DFE, then this woman's intake would appear to fall short of the 600 μg recommendation for pregnancy (300 μg + 90 μg = 390 μg). But as our example shows, her intake does meet the recommendation. At this time, supplement and fortified food labels list folate in μg only, not μg DFE, making such calculations necessary.

TRY IT Calculate how many dietary folate equivalents a person receives from 200 μg folate from a supplement, 75 μg folate from fortified cereal, and 120 μg folate from other foods.

Folate supplements taken 1 month before conception and continued throughout the first trimester of pregnancy can help prevent **neural tube defects.** For this reason, all women of childbearing age ♦ who are capable of becoming pregnant should consume 0.4 milligram (400 micrograms) of folate daily. Because half of the pregnancies each year are unplanned and because neural tube defects occur early in development before most women realize they are pregnant, the Food and Drug Administration (FDA) has mandated that grain products be fortified to deliver folate to the US population.* Labels on fortified products may claim that "adequate intake of folate has been shown to reduce the risk of neural tube defects." Fortification has improved folate status in women of childbearing age and lowered the prevalence rate of neural tube defects, as Figure 10-11 shows.

Some research suggests that folate may also prevent other congenital birth defects, such as cleft lip and cleft palate.[11] Such findings strengthen recommendations for women to pay attention to their folate needs.

Folate fortification raises safety concerns as well. Because high intakes of folate can mask a vitamin B_{12} deficiency, folate consumption should not exceed 1 milligram daily without close medical supervision.[12] The risks and benefits of folate fortification continue to be a topic of current debate.[13]

Folate and Heart Disease The FDA's decision to fortify grain products with folate was strengthened by research suggesting a role for folate in protecting against heart disease.[14] One of folate's key roles in the body is to break down the amino acid homocysteine. Without folate, homocysteine accumulates, which seems to enhance formation of blood clots and atherosclerotic lesions. Fortified foods and folate supplements raise blood folate and reduce blood homocysteine, but do not seem to reduce the risk of heart attacks, strokes, or death from cardiovascular causes.[15]

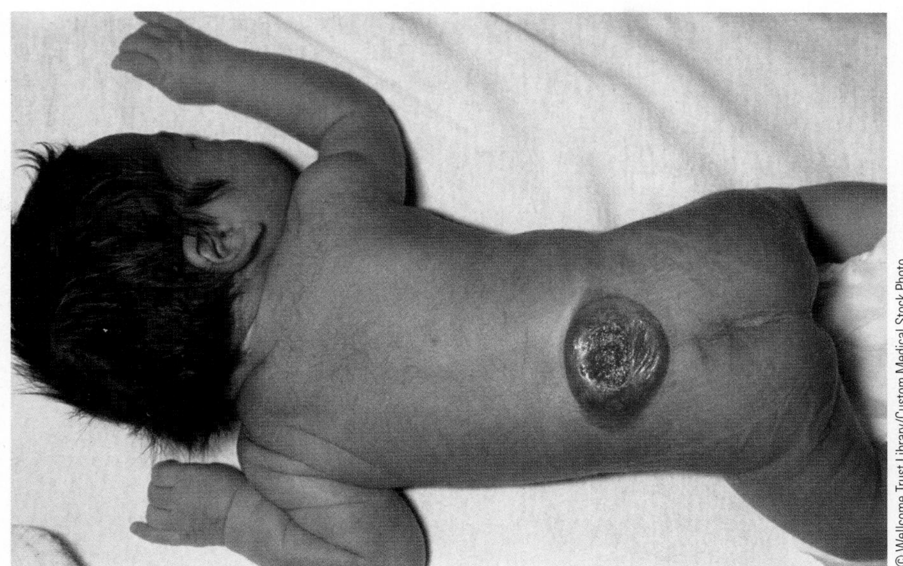

Folate helps to protect against spina bifida, a neural tube defect characterized by the incomplete closure of the spinal cord and its bony encasement.

> **FIGURE 10-11 Decreasing Prevalence of Neural Tube Defects since Folate Fortification**

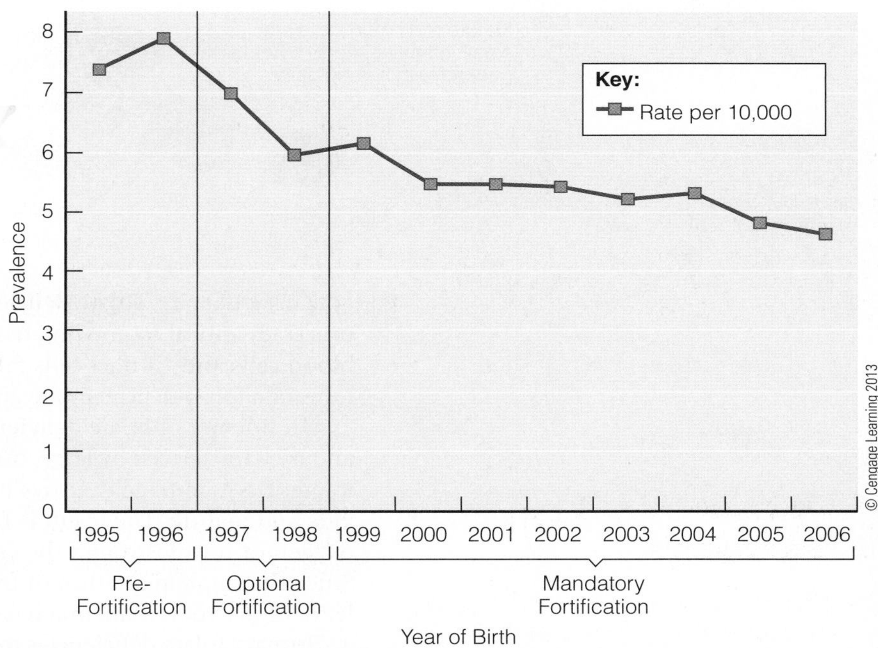

SOURCE: National Center for Health Statistics, Centers for Disease Control and Prevention, www.cdc.gov, updated January 2010.

Folate and Cancer Because the synthesis of DNA and the transfer of methyl groups depend on folate, its relationships with cancer are complex, depending on the type of cancer and the timing of folate supplementation.[16] Some research suggests that sufficient folate may protect against the initiation of cancer, whereas other studies report that high intakes may enhance progression once cancer has begun.[17] In general, foods containing folate probably reduce the risk of pancreatic cancer.[18] Limited evidence suggests folate may also reduce the risk of esophageal and colorectal cancer.[19]

♦ Women of childbearing age (15 to 45 yr) should:
- Eat folate-rich foods
- Eat folate-fortified foods
- Take a multivitamin daily (most provide 400 µg folate)

*Bread products, flour, corn grits, cornmeal, farina, rice, macaroni, and noodles must be fortified with 140 micrograms of folate per 100 grams of grain. For perspective, 100 grams is roughly 3 slices of bread; 1 cup of flour; ½ cup of corn grits, cornmeal, farina, or rice; or ¾ cup of macaroni or noodles.

neural tube defects: malformations of the brain, spinal cord, or both during embryonic development that often result in lifelong disability or death.

> FIGURE 10-12 **Normal Blood Cells and Blood Cells in Macrocytic Anemia Compared**

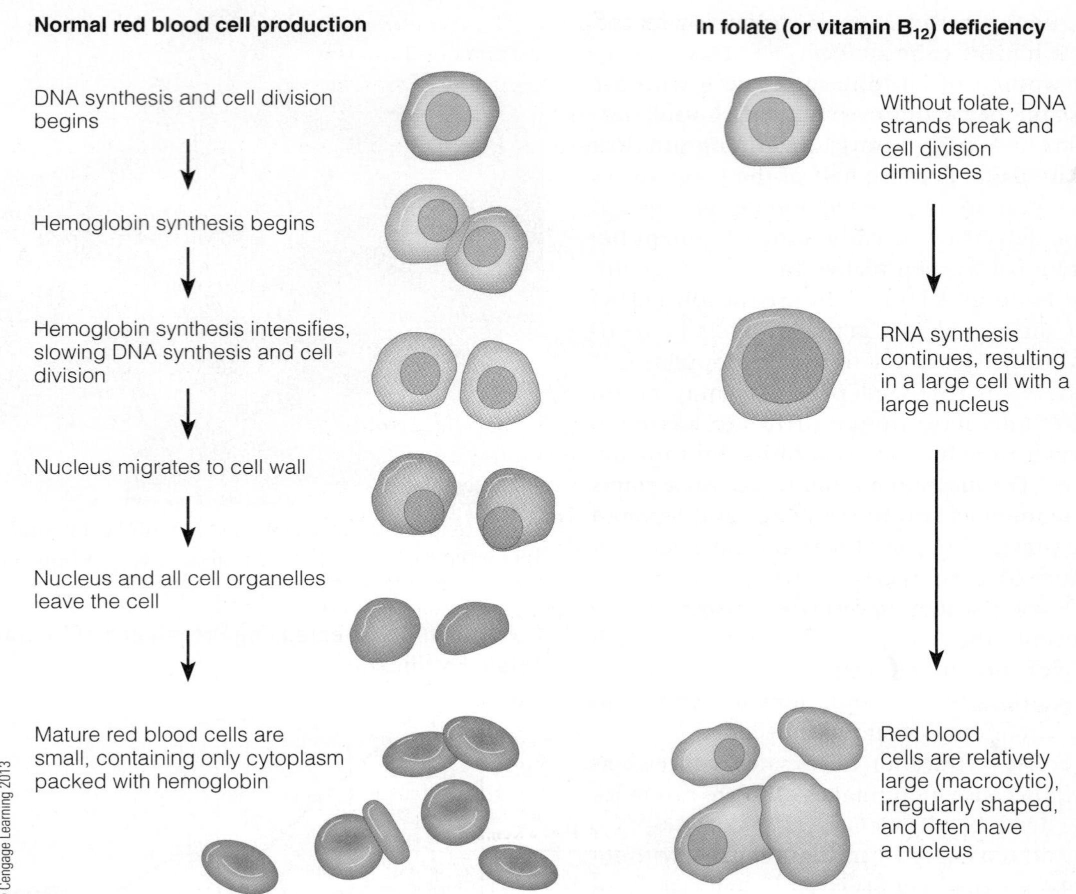
Normal red blood cell production
DNA synthesis and cell division begins
Hemoglobin synthesis begins
Hemoglobin synthesis intensifies, slowing DNA synthesis and cell division
Nucleus migrates to cell wall
Nucleus and all cell organelles leave the cell
Mature red blood cells are small, containing only cytoplasm packed with hemoglobin
In folate (or vitamin B12) deficiency
Without folate, DNA strands break and cell division diminishes
RNA synthesis continues, resulting in a large cell with a large nucleus
Red blood cells are relatively large (macrocytic), irregularly shaped, and often have a nucleus

Folate Deficiency Folate deficiency impairs cell division and protein synthesis—processes critical to growing tissues. In a folate deficiency, the replacement of red blood cells and GI tract cells falters. Not surprisingly, then, two of the first symptoms of a folate deficiency are **anemia** and GI tract deterioration.

The anemia of folate deficiency is known as *macrocytic* or *megaloblastic anemia* and is characterized by large, immature red blood cells (see Figure 10-12). Without folate, DNA damage destroys many of the red blood cells as they attempt to divide and mature. The result is fewer, but larger, red blood cells that cannot carry oxygen or travel through the capillaries as efficiently as normal red blood cells. Since the implementation of folate fortification in the United States, the prevalence of macrocytic anemia has decreased.[20]

Primary folate deficiencies may develop from inadequate intake and have been reported in infants who were fed goat's milk, which is notoriously low in folate. Secondary folate deficiencies may result from impaired absorption or an unusual metabolic need for the vitamin. Metabolic needs increase in situations where cell multiplication must speed up, such as pregnancies involving twins and triplets; cancer; skin-destroying diseases such as chicken pox and measles; and burns, blood loss, GI tract damage, and the like.

Of all the vitamins, folate appears to be most vulnerable to interactions with drugs, which can also lead to a secondary deficiency. Some medications, notably anticancer drugs, have a chemical structure similar to folate's structure and can displace the vitamin from enzymes and interfere with normal metabolism. Like all cells, cancer cells need the real vitamin to multiply—without it, they die. Unfortunately, anticancer drugs affect both cancerous cells and healthy cells, creating a folate deficiency for all cells. (Highlight 17 discusses nutrient-drug interactions,

anemia (ah-NEE-me-ah): literally, "too little blood." Anemia is any condition in which too few red blood cells are present, or the red blood cells are immature (and therefore large) or too small or contain too little hemoglobin to carry the normal amount of oxygen to the tissues. Anemia is not a disease itself but can be a consequence of many different disease conditions, including many nutrient deficiencies, bleeding, excessive red blood cell destruction, and defective red blood cell formation.

- **an** = without
- **emia** = blood

and Figure H17-1 illustrates the similarities between the vitamin folate and the anticancer drug methotrexate.)

Aspirin and antacids also interfere with the body's folate status: aspirin inhibits the action of folate-requiring enzymes, and antacids limit the absorption of folate. Healthy adults who use these drugs to relieve an occasional headache or upset stomach need not be concerned, but people who rely heavily on aspirin or antacids should be aware of the nutrition consequences.

© Polara Studios, Inc.

Dark green and leafy vegetables (such as spinach and broccoli), legumes (such as black beans, kidney beans, and black-eyed peas), liver, and some fruits (notably citrus fruits and juices) are naturally rich in folate.

Folate Toxicity A UL has been established for folate from fortified foods or supplements (see the inside front cover). Commonly consumed amounts of folate from both natural sources and fortified foods appear to cause no harm. The small percentage of adults who also take high-dose folate supplements, however, can reach levels that are high enough to obscure a vitamin B_{12} deficiency and delay diagnosis of neurological damage.[21]

Folate Food Sources Figure 10-13 shows that folate is especially abundant in legumes, fruits, and vegetables. The vitamin's name suggests the word *foliage,* and indeed, dark green, leafy vegetables are outstanding sources. With fortification, grain products also contribute folate. The small red and white bars in Figure 10-13 indicate that meats and milk products are poor folate sources. Heat and oxidation during cooking and storage can destroy as much as half of the folate in foods. The accompanying table provides a summary of folate.

REVIEW IT Folate

Other Names

Folic acid, folacin, pteroylglutamic acid (PGA)

RDA

Adults: 400 μg/day

UL[a]

Adults: 1000 μg/day

Chief Functions in the Body

Part of coenzymes THF (tetrahydrofolate) and DHF (dihydrofolate) used in DNA synthesis and therefore important in new cell formation

Significant Sources

Fortified grains, leafy green vegetables, legumes, seeds, liver

Easily destroyed by heat and oxygen

Deficiency Symptoms

Anemia (large-cell type);[b] smooth, red tongue;[c] mental confusion, weakness, fatigue, irritability, headache; shortness of breath; elevated homocysteine

Toxicity Symptoms

Masks vitamin B_{12}–deficiency symptoms

[a]The UL applies to synthetic forms obtained from supplements, fortified foods, or a combination.
[b]Large-cell-type anemia is known as either *macrocytic* or *megaloblastic anemia.*
[c]Smoothness of the tongue is caused by loss of its surface structures and is termed *glossitis* (gloss-EYE-tis).

Vitamin B_{12}

Vitamin B_{12} and folate are closely related: each depends on the other for activation. Recall that vitamin B_{12} removes a methyl group to activate the folate coenzyme. When folate gives up its methyl group, the vitamin B_{12} coenzyme becomes activated (review Figure 10-10 on p. 311).

The regeneration of the amino acid methionine and the synthesis of DNA and RNA depend on both folate and vitamin B_{12}.* In addition, without any help from folate, vitamin B_{12} maintains the sheath that surrounds and protects nerve fibers and promotes their normal growth. Bone cell activity and metabolism also depend on vitamin B_{12}.

The digestion and absorption of vitamin B_{12} depends on several steps. In the stomach, hydrochloric acid and the digestive enzyme pepsin release vitamin B_{12} from the proteins to which it is attached in foods. Then as vitamin B_{12} passes from the stomach to the small intestine, it binds with a stomach secretion called **intrinsic factor.** Bound together, intrinsic factor and vitamin B_{12} travel to the end of the small intestine, where receptors recognize the complex. Importantly, the receptors do not recognize vitamin B_{12} without intrinsic

vitamin B_{12}: a B vitamin characterized by the presence of cobalt (see Figure 13-2). The active forms of coenzyme B_{12} are methylcobalamin and deoxyadenosylcobalamin.

intrinsic factor: a glycoprotein (a protein with short polysaccharide chains attached) secreted by the stomach cells that binds with vitamin B_{12} in the small intestine to aid in the absorption of vitamin B_{12}.

- **intrinsic** = on the inside

*In the body, methionine serves as a methyl (CH_3) donor. In doing so, methionine can be converted to other amino acids. Some of these amino acids can regenerate methionine, but methionine is still considered an essential amino acid that is needed in the diet.

> **FIGURE 10-13 Folate in Selected Foods**

See the "How To" feature on p. 302 for more information on using this figure.

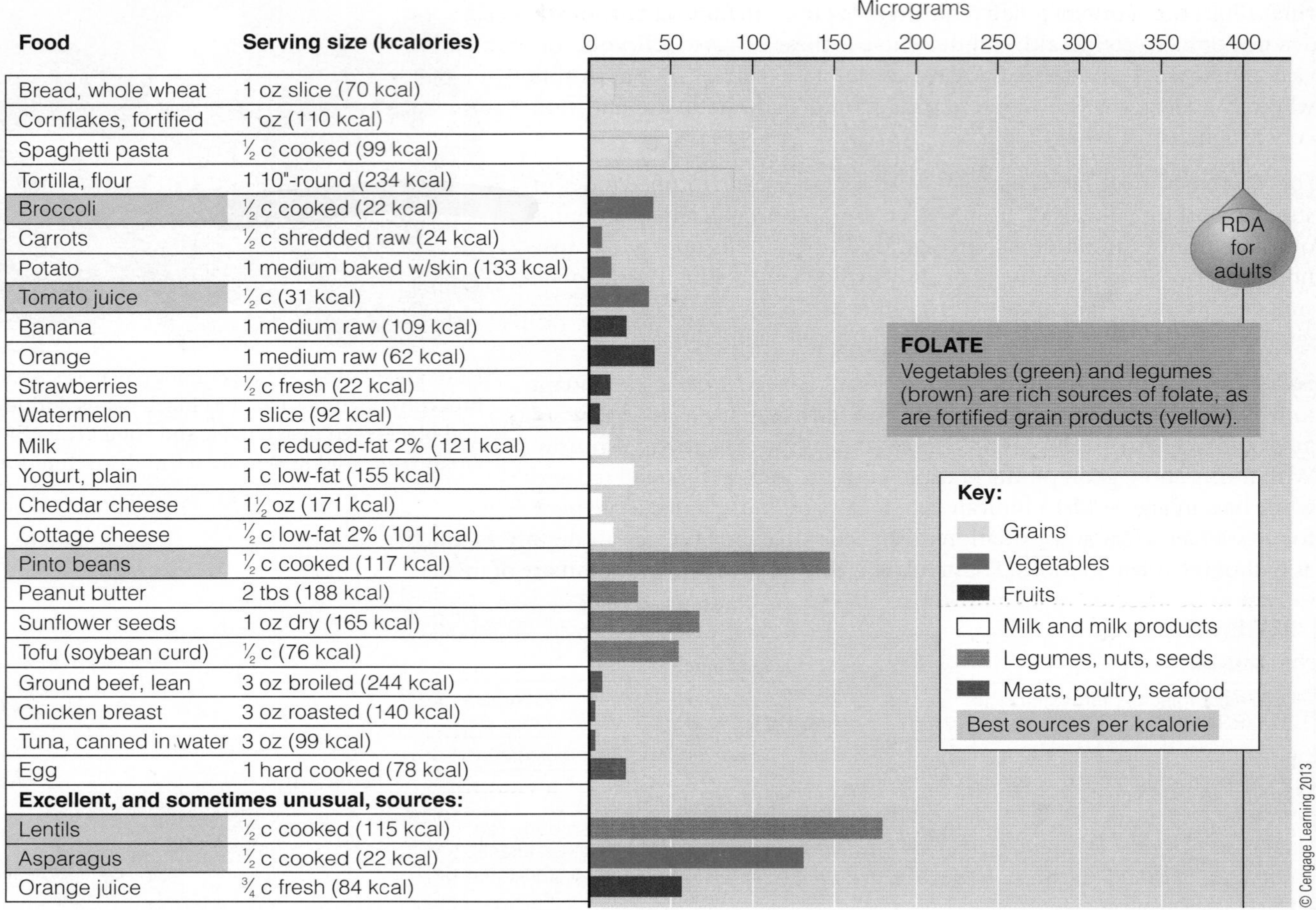

factor. The vitamin is gradually absorbed into the bloodstream as the intrinsic factor is degraded. Transport of vitamin B_{12} in the blood depends on specific binding proteins.

Like folate, vitamin B_{12} enters the enterohepatic circulation—continuously being secreted into bile and delivered to the intestine where it is reabsorbed. Because most vitamin B_{12} is reabsorbed, healthy people rarely develop a deficiency even when their intake is minimal.

Vitamin B_{12} Recommendations The RDA for adults is only 2.4 micrograms of vitamin B_{12} a day—just over two-millionths of a gram. The ink in the period at the end of this sentence may weigh about 2.4 micrograms. As tiny as this amount appears to the human eye, it contains billions of molecules of vitamin B_{12}, enough to provide coenzymes for all the enzymes that need its help.

Vitamin B_{12} Deficiency and Toxicity Most vitamin B_{12} deficiencies reflect inadequate absorption, not poor intake. Inadequate absorption typically occurs for one of two reasons: a lack of hydrochloric acid or a lack of intrinsic factor. Without hydrochloric acid, the vitamin is not released from the dietary proteins and so is not available for binding with the intrinsic factor. Without the intrinsic factor, the vitamin cannot be absorbed.

Vitamin B_{12} deficiency is common among the elderly.[22] Many older adults develop **atrophic gastritis,** a condition that damages the cells of the stomach. Atrophic gastritis may also develop in response to iron deficiency or infection with *Helicobacter pylori,* the bacterium implicated in ulcer formation. Without healthy stomach cells, production of hydrochloric acid and intrinsic factor diminishes. Even

atrophic (a-TRO-fik) **gastritis** (gas-TRY-tis): chronic inflammation of the stomach accompanied by a diminished size and functioning of the mucous membrane and glands. This condition is also characterized by inadequate hydrochloric acid and intrinsic factor—two substances needed for vitamin B_{12} absorption.

- **atrophy** = wasting
- **gastro** = stomach
- **itis** = inflammation

with an adequate intake from foods, vitamin B_{12} status suffers. The vitamin B_{12} deficiency caused by atrophic gastritis and a lack of intrinsic factor is known as **pernicious anemia.**

Some people inherit a defective gene for the intrinsic factor. In such cases, or when the stomach has been injured and cannot produce enough of the intrinsic factor, vitamin B_{12} must be given by injection to bypass the need for intestinal absorption. Alternatively, the vitamin may be delivered by nasal spray; absorption is rapid, high, and well tolerated.

Because vitamin B_{12} is found primarily in foods derived from animals, people who follow a vegan diet may develop a vitamin B_{12} deficiency. It may take several years for people who stop eating animal-derived foods to develop deficiency symptoms because the body recycles much of its vitamin B_{12}, reabsorbing it over and over again. Even when the body fails to absorb vitamin B_{12}, deficiency may take up to 3 years to develop because the body conserves its supply. Neurological degeneration, a sign of vitamin B_{12} deficiency, appears more rapidly in infants born to mothers with unsupplemented vegan diets or untreated pernicious anemia.[23]

Because vitamin B_{12} is required to convert folate to its active form, one of the most obvious vitamin B_{12}–deficiency symptoms is the anemia commonly seen in folate deficiency. This anemia is characterized by large, immature red blood cells, which indicate slow DNA synthesis and an inability to divide (see Figure 10-12, p. 314). When folate is trapped in its inactive (methyl folate) form due to vitamin B_{12} deficiency or is unavailable due to folate deficiency itself, DNA synthesis slows.

First to be affected in a vitamin B_{12} or folate deficiency are the rapidly growing blood cells. Either vitamin B_{12} or folate will clear up the anemia, but if folate is given when vitamin B_{12} is needed, the result is disastrous: devastating neurological symptoms. Remember that vitamin B_{12}, but not folate, maintains the sheath that surrounds and protects nerve fibers and promotes their normal growth. Folate "cures" the *blood* symptoms of a vitamin B_{12} deficiency, but cannot stop the *nerve* symptoms from progressing. By doing so, folate "masks" a vitamin B_{12} deficiency.[24]

Marginal vitamin B_{12} deficiency impairs cognition.[25] Advanced neurological symptoms include a creeping paralysis that begins at the extremities and works inward and up the spine. Early detection and correction are necessary to prevent permanent nerve damage and paralysis. With sufficient folate in the diet, the neurological symptoms of vitamin B_{12} deficiency can develop without evidence of anemia.[26] Such interactions between folate and vitamin B_{12} highlight some of the safety issues surrounding the use of supplements and the fortification of foods.[27] No adverse effects have been reported for excess vitamin B_{12}, and no UL has been set.

Vitamin B_{12} Food Sources Vitamin B_{12} is unique among the vitamins in being found almost exclusively in foods derived from animals. Its bioavailability is greatest from milk and fish.[28] Anyone who eats reasonable amounts of animal-derived foods is most likely to have an adequate intake, including vegetarians who use milk products or eggs. Vegans, who restrict all foods derived from animals, need a reliable source, such as vitamin B_{12}–fortified soy milk or vitamin B_{12} supplements. Yeast grown on a vitamin B_{12}–enriched medium and mixed with that medium provides some vitamin B_{12}, but yeast itself does not contain active vitamin B_{12}. Similarly, neither fermented soy products such as miso (a soybean paste) nor sea algae such as spirulina provide active vitamin B_{12}. Extensive research shows that the amounts listed on the labels of these plant products are inaccurate and misleading because the vitamin B_{12} is in an inactive, unavailable form.

As mentioned earlier, the water-soluble vitamins are particularly vulnerable to losses in cooking. For most of these nutrients, microwave heating minimizes losses as well as, or better than, traditional cooking methods. Such is not the case for vitamin B_{12}, however. Microwave heating inactivates vitamin B_{12}. To preserve this vitamin, use the oven or stovetop instead of a microwave to cook meats and milk products (major sources of vitamin B_{12}). The table on p. 318 provides a summary of vitamin B_{12}.

pernicious (per-NISH-us) **anemia:** a blood disorder that reflects a vitamin B_{12} deficiency caused by lack of intrinsic factor and characterized by abnormally large and immature red blood cells. Other symptoms include muscle weakness and irreversible neurological damage.

- **pernicious** = destructive

REVIEW IT Vitamin B_{12}

Other Names

Cobalamin (and related forms)

RDA

Adults: 2.4 μg/day

Chief Functions in the Body

Part of coenzymes methylcobalamin and deoxyadenosylcobalamin used in new cell synthesis; helps to maintain nerve cells; reforms folate coenzyme; helps to break down some fatty acids and amino acids

Significant Sources

Foods of animal origin (meat, fish, poultry, shellfish, milk, cheese, eggs), fortified cereals

Easily destroyed by microwave cooking

Deficiency Disease

Pernicious anemia[a]

Deficiency Symptoms

Anemia (large-cell type);[b] fatigue, degeneration of peripheral nerves progressing to paralysis; sore tongue, loss of appetite, constipation

Toxicity Symptoms

None reported

[a]The name *pernicious anemia* refers to the vitamin B_{12} deficiency caused by atrophic gastritis and a lack of intrinsic factor, but not to that caused by inadequate dietary intake.
[b]Large-cell-type anemia is known as either *macrocytic* or *megaloblastic anemia*.

Choline Although not defined as a vitamin, choline is an essential nutrient that is commonly grouped with the B vitamins.[29] The body uses choline to make the neurotransmitter acetylcholine and the phospholipid lecithin. During fetal development, choline supports the structure and function of the brain and spinal cord, by supporting neural tube closure and enhancing learning performance.

Choline Recommendations The body can make choline from the amino acid methionine, but without dietary choline, synthesis alone appears to be insufficient to meet the body's needs. For this reason, the DRI Committee established an Adequate Intake (AI) for choline.

Choline Deficiency and Toxicity Average choline intakes fall below the AI, but the impact of deficiencies are not fully understood.[30] The UL for choline is based on its critical effect in lowering blood pressure.

Choline Food Sources Choline is found in a variety of common foods such as milk, eggs, and peanuts and as part of lecithin, a food additive commonly used as an emulsifying agent (review Figure 5-8 on p. 136). The accompanying table provides a summary of choline.

Nonvitamins Some substances have been mistaken for vitamins, but they are not essential nutrients. Among them are the compounds **inositol** and **carnitine,** which can be made by the body. Inositol is a part of cell membrane structures,

REVIEW IT Choline

Adequate Intake (AI)

Men: 550 mg/day

Women: 425 mg/day

UL

Adults: 3500 mg/day

Chief Functions in the Body

Needed for the synthesis of the neurotransmitter acetylcholine and the phospholipid lecithin

Deficiency Symptoms

Liver damage

Toxicity Symptoms

Body odor, sweating, salivation, reduced growth rate, low blood pressure, liver damage

Significant Sources

Milk, liver, eggs, peanuts

inositol (in-OSS-ih-tall): a nonessential nutrient that can be made in the body from glucose. Inositol is a part of cell membrane structures.

carnitine (CAR-neh-teen): a nonessential, nonprotein amino acid made in the body from lysine that helps transport fatty acids across the mitochondrial membrane.

and carnitine transports long-chain fatty acids from the cytosol to the mitochondria for oxidation. Other nonvitamins include PABA (para-aminobenzoic acid, a component of folate's chemical structure), the bioflavonoids (vitamin P or hesperidin), pyrroloquinoline quinone (methoxatin), orotic acid, lipoic acid, and ubiquinone (coenzyme Q_{10}). Other names erroneously associated with vitamins are "vitamin O" (oxygenated saltwater), "vitamin B_5" (another name for pantothenic acid), "vitamin B_{15}" (also called "pangamic acid," a hoax), and "vitamin B_{17}" (laetrile, an alleged "cancer cure" and not a vitamin or a cure by any stretch of the imagination—in fact, laetrile is a potentially dangerous substance).

Interactions among the B Vitamins This chapter has described some of the impressive ways that vitamins work individually, as if their many actions in the body could easily be disentangled. In fact, it is often difficult to tell which vitamin is truly responsible for a given effect because the nutrients are interdependent; the presence or absence of one affects another's absorption, metabolism, and excretion. You have already seen this interdependence with folate and vitamin B_{12}.

Riboflavin and vitamin B_6 provide another example. One of the riboflavin coenzymes, FMN, assists the enzyme that converts vitamin B_6 to its coenzyme form PLP. Consequently, a severe riboflavin deficiency can impair vitamin B_6 activity. Thus a deficiency of one nutrient may alter the action of another. Furthermore, a deficiency of one nutrient may create a deficiency of another. For example, both riboflavin and vitamin B_6 (as well as iron) are required for the conversion of tryptophan to niacin. Consequently, an inadequate intake of either riboflavin or vitamin B_6 can diminish the body's niacin supply. These interdependent relationships are evident in many of the roles B vitamins play in the body.

B Vitamin Roles Figure 10-14 (p. 320) summarizes the metabolic pathways introduced in Chapter 7 and conveys an *impression* of the many ways B vitamins assist in metabolic pathways. Metabolism is the body's work, and the B vitamin coenzymes are indispensable to every step. In scanning the pathways of metabolism depicted in the figure, note the many abbreviations for the coenzymes that keep the processes going.

Look at the now-familiar pathway of glucose breakdown. To break down glucose to pyruvate, the cells must have certain enzymes. For the enzymes to work, they must have the niacin coenzyme NAD. Cells can make NAD, but only if they have enough niacin (or enough of the amino acid tryptophan to make niacin).

The next step is the breakdown of pyruvate to acetyl CoA. The enzymes involved in this step require both NAD and the thiamin and riboflavin coenzymes TPP and FAD, respectively. The cells can manufacture the enzymes they need from the vitamins, if the vitamins are in the diet.

Another coenzyme needed for this step is CoA. Predictably, the cells can make CoA except for an essential part that must be obtained in the diet—pantothenic acid. Another coenzyme requiring biotin serves the enzyme complex involved in converting pyruvate to oxaloacetate, the compound that combines with acetyl CoA to start the TCA cycle.

These and other coenzymes participate throughout all the metabolic pathways. Vitamin B_6 is an indispensable part of PLP—a coenzyme required for many amino acid conversions, for a crucial step in the making of the iron-containing portion of hemoglobin for red blood cells, and for many other reactions. Folate becomes THF—the coenzyme required for the synthesis of new genetic material and therefore new cells. The vitamin B_{12} coenzyme, in turn, regenerates THF to its active form; thus vitamin B_{12} is also necessary for the formation of new cells.

Thus each of the B vitamin coenzymes is involved, directly or indirectly, in energy metabolism. Some facilitate the energy-releasing reactions themselves; others help build new cells to deliver the oxygen and nutrients that allow the energy reactions to occur.

B Vitamin Deficiencies Now suppose the body's cells lack one of these B vitamins—niacin, for example. Without niacin, the cells cannot make NAD. Without NAD, the enzymes involved in every step of the glucose-to-energy pathway cannot function.

> **FIGURE 10-14 Metabolic Pathways Involving B Vitamins**

These metabolic pathways are introduced in Chapter 7 and are presented here to highlight the many coenzymes that facilitate the reactions. These coenzymes depend on the following vitamins:

- NAD and NADP: niacin
- TPP: thiamin
- CoA: pantothenic acid
- B_{12}: vitamin B_{12}
- FMN and FAD: riboflavin
- THF: folate
- PLP: vitamin B_6
- Biotin

Pathways leading toward acetyl CoA and the TCA cycle are catabolic, and those leading toward amino acids, glycogen, and fat are anabolic. For further details, see Appendix C.

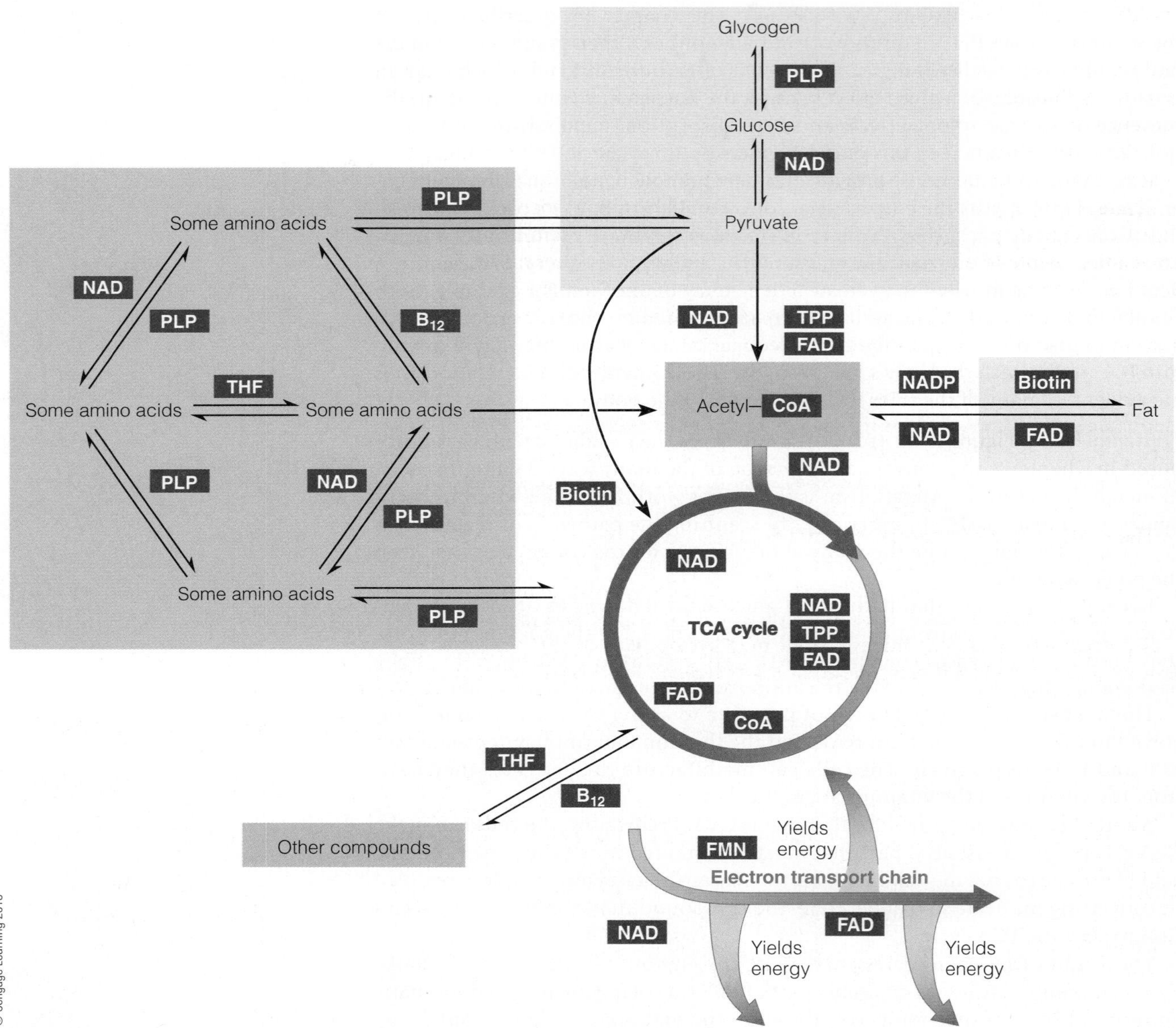

Then, because all the body's activities require energy, literally everything begins to grind to a halt. This is no exaggeration. The deadly disease pellagra, caused by niacin deficiency, produces the "devastating four *D*s": dermatitis, which reflects a failure of the skin; dementia, a failure of the nervous system; diarrhea, a failure of digestion and absorption; and eventually, as would be the case for any severe nutrient deficiency, death. These symptoms are the obvious ones, but a niacin deficiency affects all other organs, too, because all are dependent on the energy pathways.

All the vitamins are as essential as niacin. With any B vitamin deficiency, many body systems become deranged, and similar symptoms may appear. A lack of any of them can have disastrous and far-reaching effects.

Deficiencies of single B vitamins seldom show up in isolation, however. After all, people do not eat nutrients singly; they eat foods, which contain mixtures of nutrients. Only in two cases described earlier—beriberi and pellagra—have dietary deficiencies associated with single B vitamins been observed on a large scale in human populations. Even in these cases, several vitamins were lacking even though one vitamin stood out above the rest. When foods containing the vitamin known to be needed were provided, the other vitamins that were in short supply came as part of the package.

Major deficiency diseases of epidemic proportions such as pellagra and beriberi are no longer seen in the United States and Canada, but lesser deficiencies of nutrients, including the B vitamins, sometimes occur in people whose food choices are poor because of poverty, ignorance, illness, or poor health habits like alcohol abuse. (Review Highlight 7 to fully appreciate how alcohol induces vitamin deficiencies and interferes with energy metabolism.) Remember from Chapter 1 that deficiencies can arise not only from deficient intakes (primary causes), but also for other (secondary) reasons.

In identifying nutrient deficiencies, it is important to realize that a particular sign or symptom may not always have the same cause. The skin and the tongue (shown in Figure 10-15) appear to be especially sensitive to B vitamin deficiencies, but focusing on these body parts gives them undue emphasis. Both the skin and the tongue are readily visible in a physical examination.* The physician sees and reports the deficiency's outward signs, but the full impact of a vitamin deficiency occurs inside the cells of the body. If the skin develops a rash or lesions, other tissues beneath it may be degenerating too. Similarly, the mouth and tongue are the visible part of the digestive system; if they are abnormal, most likely the rest of the GI tract is as well.

Keep in mind that the cause of a sign or symptom is not always apparent. The summary tables in this chapter show that deficiencies of riboflavin, niacin, biotin, and vitamin B_6 can all cause skin rashes. So can a deficiency of protein, linoleic acid, or vitamin A. Because skin is on the outside and easy to see, it is a useful indicator of "things going wrong inside cells." By itself, a skin condition says nothing about its possible cause.

The same is true of anemia. Anemia is often caused by iron deficiency, but it can also be caused by a folate or vitamin B_{12} deficiency; by digestive tract failure to absorb any of these nutrients; or by such nonnutritional causes as infections, parasites, cancer, or loss of blood. No single nutrient will always cure a given symptom.

A person who feels chronically tired may be tempted to self-diagnose iron-deficiency anemia and self-prescribe an iron supplement. But this will relieve tiredness

> **FIGURE 10-15 B Vitamin–Deficiency Symptoms—The Smooth Tongue of Glossitis and the Skin Lesions of Cheilosis**

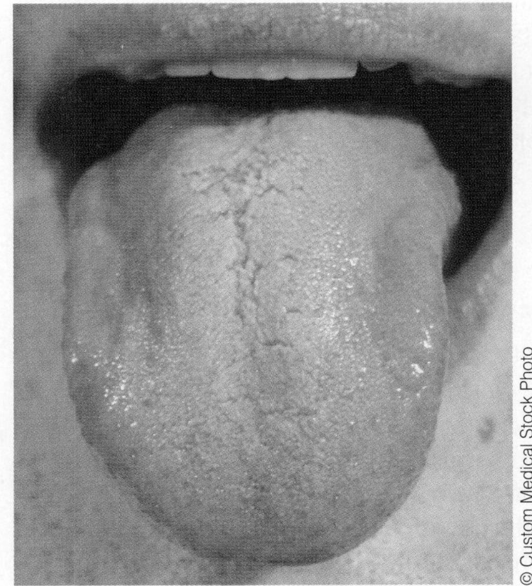

A healthy tongue has a rough and somewhat bumpy surface.

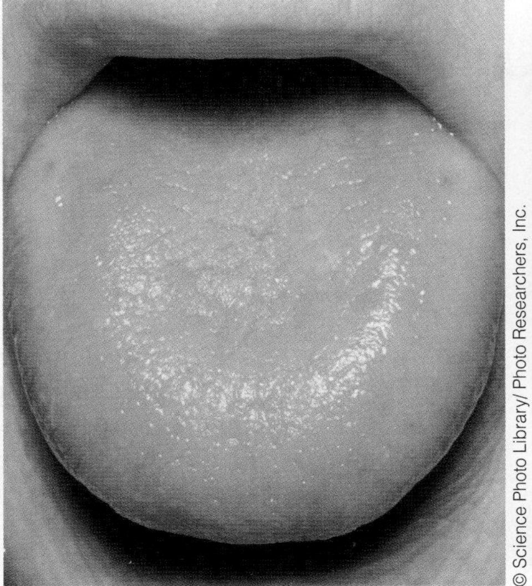

In a B vitamin deficiency, the tongue becomes smooth and swollen due to atrophy of the tissue (glossitis).

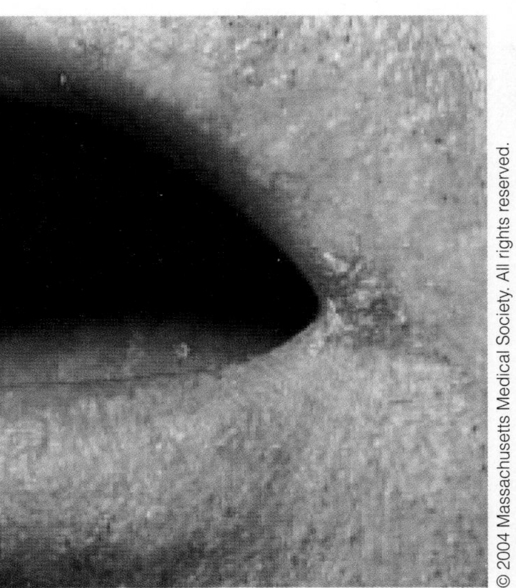

In a B vitamin deficiency, the corners of the mouth become irritated and inflamed (cheilosis).

*The two common signs of B vitamin deficiencies are *glossitis* (gloss-EYE-tis), an inflammation of the tongue, and *cheilosis* (kye-LOH-sis or kee-LOH-sis), a condition of reddened lips with cracks at the corners of the mouth.

only if the cause is indeed iron-deficiency anemia. If the cause is a folate deficiency, taking iron will only prolong the fatigue. A person who is better informed may decide to take a vitamin supplement with iron, covering the possibility of a vitamin deficiency. But the symptom may have a nonnutritional cause. If the cause of the tiredness is actually hidden blood loss due to cancer, the postponement of a diagnosis may be fatal. When fatigue is caused by a lack of sleep, of course, no nutrient or combination of nutrients can replace a good night's rest. A person who is chronically tired should see a physician rather than self-prescribe. If the condition is nutrition related, a registered dietitian should be consulted as well.

B Vitamin Toxicities Toxicities of the B vitamins from foods alone are unknown, but they can occur when people overuse dietary supplements. With supplements, the quantities can quickly overwhelm the cells. Consider that one small capsule can easily deliver 2 milligrams of vitamin B_6, but it would take more than 3000 bananas, 6600 cups of rice, or 3600 chicken breasts to supply an equivalent amount. When the cells become oversaturated with a vitamin, they must work to eliminate the excess. The cells dispatch water-soluble vitamins to the urine for excretion, but sometimes they cannot keep pace with the onslaught. Homeostasis becomes disturbed and symptoms of toxicity develop.

B Vitamin Food Sources Significantly, deficiency diseases, such as beriberi and pellagra, were eliminated by providing foods. Dietary supplements advertise that vitamins are indispensable to life, but human beings obtained their nourishment from foods for centuries before supplements existed. If the diet lacks a vitamin, the first solution is to adjust food intake to obtain that vitamin.

The bar graphs of selected foods in this chapter, taken together, sing the praises of a balanced diet. The grains deliver thiamin, riboflavin, niacin, and folate. The fruit and vegetable groups excel in folate. Protein foods serve thiamin, niacin, vitamin B_6, and vitamin B_{12} well. The milk group stands out for riboflavin and vitamin B_{12}. A diet that offers a variety of foods from each group, prepared with reasonable care, serves up ample B vitamins.

REVIEW IT Identify the main roles, deficiency symptoms, and food sources for each of the B vitamins.

The B vitamins serve as coenzymes that facilitate the work of every cell. They are active in carbohydrate, fat, and protein metabolism and in the making of DNA and thus new cells. Historically famous B vitamin–deficiency diseases are beriberi (thiamin), pellagra (niacin), and pernicious anemia (vitamin B_{12}). Pellagra can be prevented by adequate protein because the amino acid tryptophan can be converted to niacin in the body. A high intake of folate can mask the blood symptoms of a vitamin B_{12} deficiency, but it will not prevent the associated nerve damage. Vitamin B_6 participates in amino acid metabolism and can be harmful in excess. Biotin and pantothenic acid serve important roles in energy metabolism and are common in a variety of foods. Many substances that people claim as B vitamins are not. Fortunately, a variety of foods from each of the food groups provides an adequate supply of all of the B vitamins.

10.3 Vitamin C

LEARN IT Identify the main roles, deficiency symptoms, and food sources for vitamin C.

For many centuries, any man who joined the crew of a seagoing ship knew he had at best a 50–50 chance of returning alive—not because he might be slain by pirates or die in a storm, but because he might contract **scurvy.**[31] As many as two-thirds of a ship's crew could die of scurvy during a long voyage. Only men on short voyages, especially around the Mediterranean Sea, were free of scurvy. No one knew the reason: that on long ocean voyages, the ship's cook used up the fresh fruits and vegetables early and then served only cereals and meats until the return to port.

In the mid-1700s, James Lind, a British physician serving in the navy, devised an experiment to find a cure for scurvy. He divided 12 sailors with scurvy into 6 pairs. Each pair received a different supplemental ration: cider, vinegar, sulfuric acid, seawater, oranges and lemons, or a strong laxative. Those receiving the citrus fruits quickly

scurvy: the vitamin C–deficiency disease.

recovered, but sadly, it was almost 50 years before the British navy required all vessels to provide every sailor with lemon or lime juice daily. The tradition of providing British sailors with citrus juice daily to prevent scurvy gave them the nickname "limeys."

The antiscurvy "something" in citrus and other foods was dubbed the **antiscorbutic factor.** Nearly 200 years later, the factor was isolated and found to be a 6-carbon compound similar to glucose; it was named **ascorbic acid.**

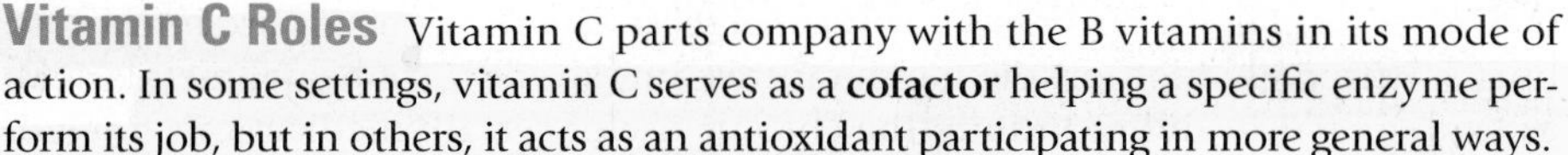

Vitamin C Roles

Vitamin C parts company with the B vitamins in its mode of action. In some settings, vitamin C serves as a **cofactor** helping a specific enzyme perform its job, but in others, it acts as an antioxidant participating in more general ways.

As an Antioxidant Vitamin C loses electrons easily, a characteristic that allows it to perform as an antioxidant. ♦ In the body, **antioxidants** defend against free radicals. Free radicals are discussed fully in Highlight 11, but for now, a simple definition will suffice. A **free radical** is a molecule with one or more unpaired electrons, which makes it unstable and highly reactive. Antioxidants can neutralize free radicals by donating an electron or two. In doing so, antioxidants protect other substances from free radical damage. Figure 10-16 illustrates how vitamin C can give up electrons and then accept them again to become reactivated. This recycling of vitamin C is key to limiting losses and maintaining a reserve of antioxidants in the body.

♦ Key antioxidant nutrients:
- Vitamin C, vitamin E, beta-carotene
- Selenium

Vitamin C is like a bodyguard for water-soluble substances; it stands ready to sacrifice its own life to save theirs. In the cells and body fluids, vitamin C protects tissues from the **oxidative stress** of free radicals and thus may play an important role in preventing diseases. In the intestines, vitamin C enhances iron absorption by protecting iron from oxidation. (Chapter 13 provides more details about the relationship between vitamin C and iron.)

As a Cofactor in Collagen Formation Vitamin C helps to form the fibrous structural protein of connective tissues known as **collagen.** Collagen serves as the matrix on which bones and teeth are formed. When a person is wounded, collagen glues the separated tissues together, forming scars. Cells are held together largely by collagen; this is especially important in the walls of the blood vessels, which must withstand the pressure of blood surging with each beat of the heart.

Chapter 6 describes how the body makes proteins by stringing together chains of amino acids. During the synthesis of collagen, each time a proline or lysine is added to the growing protein chain, an enzyme hydroxylates it (adds an OH group), making the amino acid hydroxyproline or hydroxylysine, respectively. These two special amino acids facilitate the binding together of collagen fibers to make strong, rope-like structures. The conversion of proline to hydroxyproline requires both vitamin C and iron. Iron works as a cofactor in the reaction, and vitamin C protects iron from oxidation, thereby allowing iron to perform its duty. Without vitamin C and iron, the hydroxylation step does not occur.

As a Cofactor in Other Reactions Vitamin C also serves as a cofactor in the synthesis of several other compounds. As in collagen formation, vitamin C helps in the

> **FIGURE 10-16 Active Forms of Vitamin C**

The two hydrogens highlighted in yellow give vitamin C its acidity and its ability to act as an antioxidant.

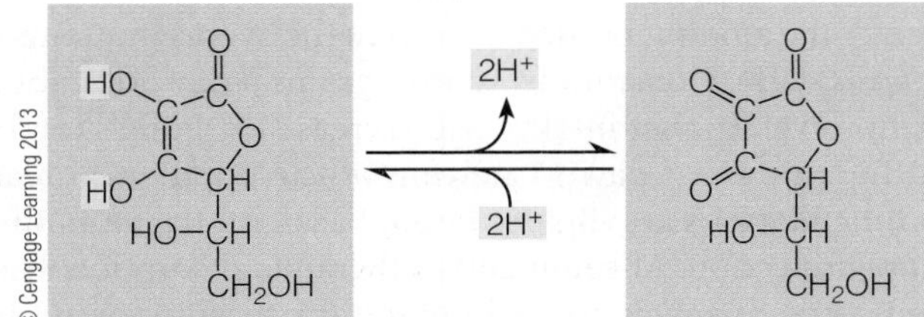

Ascorbic acid protects against oxidative damage by donating its two hydrogens with their electrons to free radicals (molecules with unpaired electrons). In doing so, ascorbic acid becomes dehydroascorbic acid.

Dehydroascorbic acid can readily accept hydrogens to become ascorbic acid. The reversibility of this reaction is key to vitamin C's role as an antioxidant.

antiscorbutic (AN-tee-skor-BUE-tik) **factor:** the original name for vitamin C.
- **anti** = against
- **scorbutic** = causing scurvy

ascorbic acid: one of the two active forms of vitamin C (see Figure 10-16). Many people refer to vitamin C by this name.
- **a** = without
- **scorbic** = having scurvy

cofactor: a small, inorganic or organic substance that facilitates the action of an enzyme.

antioxidants: in the body, substances that significantly decrease the adverse effects of free radicals on normal physiological functions.

free radical: an unstable molecule with one or more unpaired electrons.

oxidative stress: a condition in which the production of oxidants and free radicals exceeds the body's ability to handle them and prevent damage.

collagen: the structural protein from which connective tissues such as scars, tendons, ligaments, and the foundations of bones and teeth are made.

> FIGURE 10-17 Vitamin C Intake (mg/day)

Recommendations for vitamin C are set generously above the minimum requirement and well below the toxicity level.

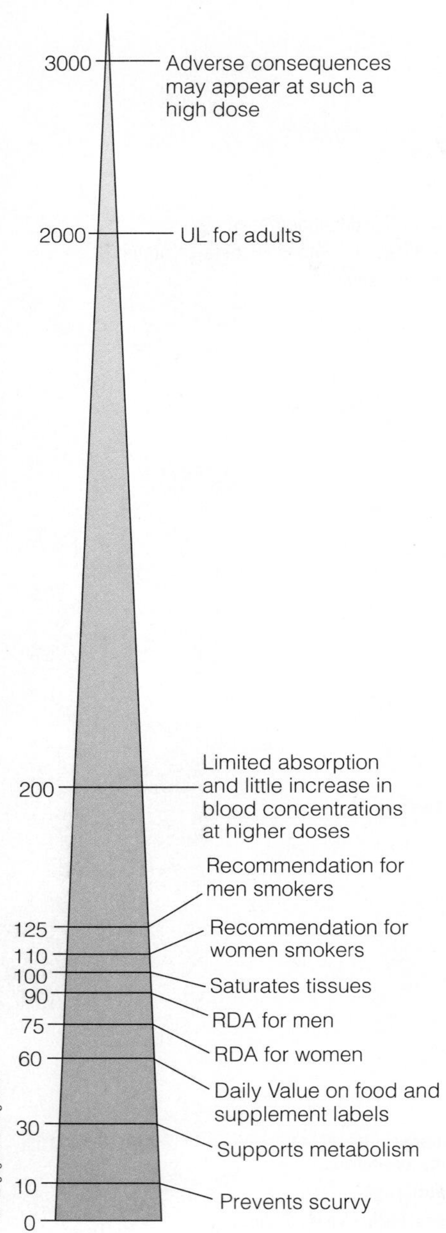

♦ Vitamin C is found in:

- High amounts: Adrenal glands, pituitary glands
- Medium amounts: Liver, spleen, heart, kidneys, lungs, pancreas, white blood cells
- Small amounts: Muscles, red blood cells

♦ For perspective, 1 c orange juice provides >100 mg vitamin C.

histamine (HISS-tah-mean or HISS-tah-men): a substance produced by cells of the immune system as part of a local immune reaction to an antigen.

hydroxylation of carnitine, a compound that transports fatty acids, especially long-chain fatty acids, across the inner membrane of mitochondria in cells. It also participates in the conversions of the amino acids tryptophan and tyrosine to the neurotransmitters serotonin and norepinephrine, respectively. Vitamin C also assists in the making of hormones, including thyroxin, which regulates the metabolic rate; when metabolism speeds up in times of extreme physical stress, the body's use of vitamin C increases.

In Stress Among the stresses known to increase vitamin C needs are infections; burns; extremely high or low temperatures; intakes of toxic heavy metals such as lead, mercury, and cadmium; the chronic use of certain medications, including aspirin, barbiturates, and oral contraceptives; and cigarette smoking. During stress, the adrenal glands—which contain more vitamin C than any other organ in the body—release vitamin C ♦ and hormones into the blood.

When immune system cells are called into action, they use a great deal of oxygen and produce free radicals. In this case, free radicals are helpful. They act as ammunition in an "oxidative burst" that demolishes the offending viruses and bacteria and destroys the damaged cells. Vitamin C steps in as an antioxidant to control this oxidative activity.

In the Prevention and Treatment of the Common Cold Vitamin C has been a popular option for the prevention and treatment of the common cold for decades, but research supporting such claims has been conflicting and controversial. Some studies find no relationship between vitamin C and the occurrence of the common cold, whereas others report modest benefits—fewer colds, fewer days, and shorter duration of severe symptoms, especially for those exposed to physical and environmental stresses.[32] A review of the research on vitamin C in the treatment and prevention of the common cold reveals a slight, but consistent reduction (of 8 percent) in the duration of the common cold in favor of those taking a daily dose of at least 200 milligrams of vitamin C.[33] The question for consumers to consider is, "Is this enough to warrant routine daily supplementation?" Findings from one study show that consumers want their colds to be at least 25 percent less severe to justify the costs of taking vitamin C supplements regularly.[34]

Discoveries about how vitamin C works in the body provide possible links between the vitamin and the common cold. Anyone who has ever had a cold knows the discomfort of a runny or stuffed-up nose. Nasal congestion develops in response to elevated blood **histamine,** and people commonly take antihistamines for relief. Like an antihistamine, vitamin C comes to the rescue and deactivates histamine.

In Disease Prevention Whether vitamin C may help in preventing or treating cancer, heart disease, cataract, and other diseases is still being studied, and findings are presented in Highlight 11's discussion on antioxidants. Conducting research in the United States and Canada can be difficult, however, because diets typically contribute enough vitamin C to provide optimal health benefits.

Vitamin C Recommendations

For decades, vitamin C ranked at the top of dietary supplement sales. How much vitamin C does a person need? As is true of all the vitamins, recommendations are set generously above the minimum requirement to prevent deficiency disease and well below the toxicity level (see Figure 10-17).[35]

The requirement—the amount needed to prevent the overt symptoms of scurvy—is only 10 milligrams daily. Consuming 10 milligrams a day does not saturate all the body tissues, however; higher intakes will increase the body's total vitamin C. At about 100 milligrams ♦ per day, 95 percent of the population reaches tissue saturation. Recommendations are slightly lower, based on the amounts needed to provide antioxidant protection. At about 200 milligrams, absorption reaches a maximum, and there is little, if any, increase in blood concentrations at higher doses. Excess vitamin C is readily excreted.

As mentioned earlier, cigarette smoking increases the need for vitamin C. Cigarette smoke contains oxidants, which greedily deplete this potent antioxidant. Exposure to cigarette smoke, especially when accompanied by low dietary intakes of vitamin C, depletes the body's vitamin C in both active and passive smokers. People who chew tobacco also have low levels of vitamin C. Because people who smoke cigarettes regularly suffer significant oxidative stress, their requirement for vitamin C is increased an additional 35 milligrams; nonsmokers regularly exposed to cigarette smoke should also be sure to meet their RDA for vitamin C. Smokers are among those most likely to suffer vitamin C deficiency.[36]

Vitamin C Deficiency Early signs of nutrient deficiencies can be difficult to recognize.[37] Two of the most notable signs of a vitamin C deficiency reflect its role in maintaining the integrity of blood vessels. The gums bleed easily around the teeth, and capillaries under the skin break spontaneously, producing pinpoint hemorrhages (see Figure 10-18).

When vitamin C concentrations fall to about a fifth of optimal levels (this may take more than a month on a diet lacking vitamin C), scurvy symptoms begin to appear. Inadequate collagen synthesis causes further hemorrhaging. Muscles, including the heart muscle, degenerate. The skin becomes rough, brown, scaly, and dry. Wounds fail to heal because scar tissue will not form. Bone rebuilding falters; the ends of the long bones become softened, malformed, and painful, and fractures develop. The teeth become loose as the cartilage around them weakens. Anemia and infections are common. There are also characteristic psychological signs, including hysteria and depression. Sudden death is likely, caused by massive internal bleeding.

Once diagnosed, scurvy is readily resolved by vitamin C. Moderate doses in the neighborhood of 100 milligrams per day are sufficient, curing the scurvy within about 5 days. Such an intake is easily achieved by including vitamin C–rich foods in the diet.

Vitamin C Toxicity The availability of vitamin C supplements and the publication of books recommending vitamin C to prevent colds and cancer have led many people to take large doses of vitamin C. Not surprisingly, side effects of vitamin C supplementation such as gastrointestinal distress and diarrhea have been reported. The UL for vitamin C was established based on these symptoms.

Several instances of interference with medical regimens are also known. Large amounts of vitamin C excreted in the urine obscure the results of tests used to detect glucose or ketones in the diagnosis of diabetes. In some instances, excess vitamin C gives a **false positive** result; in others, a **false negative.** People taking anticlotting medications may unwittingly counteract the effect if they also take massive doses of vitamin C. Those with kidney disease, a tendency toward gout, or a genetic abnormality that alters vitamin C's breakdown to its excretion products are prone to forming kidney stones if they take large doses of vitamin C.* Vitamin C supplements may adversely affect people with iron overload. As Chapter 13 explains, vitamin C enhances iron absorption and releases iron from body stores; too much free iron causes the kind of cellular damage typical of free

> FIGURE 10-18 Vitamin C–Deficiency Symptoms—Scorbutic Gums and Pinpoint Hemorrhages

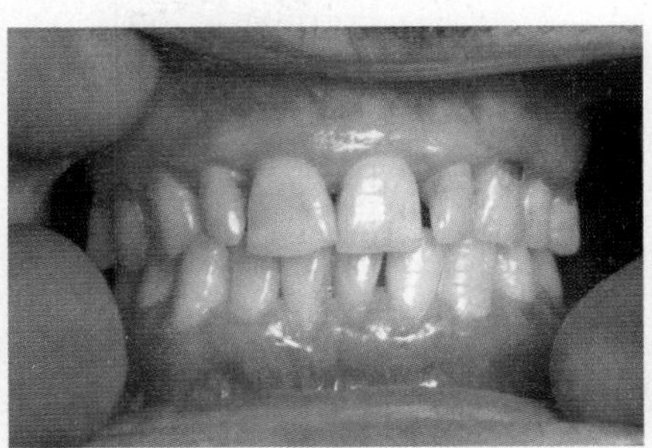

Biophoto Associates/Photo Researchers, Inc.

Scorbutic gums. Unlike other lesions of the mouth, scurvy presents a symmetrical appearance without infection.

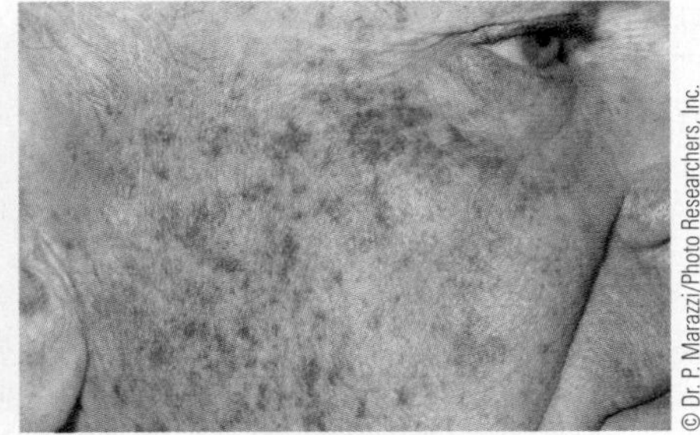

© Dr. P. Marazzi/Photo Researchers, Inc.

Pinpoint hemorrhages. Small red spots appear in the skin, indicating spontaneous bleeding internally.

false positive: a test result indicating that a condition is present (positive) when in fact it is not present (therefore false).

false negative: a test result indicating that a condition is not present (negative) when in fact it is present (therefore false).

*Vitamin C is inactivated and degraded by several routes, and sometimes oxalate, which can form kidney stones, is produced along the way. People may also develop oxalate crystals in their kidneys regardless of vitamin C status.

radicals. These adverse consequences illustrate how vitamin C can act as a *pro*oxidant when quantities exceed the body's needs.

Vitamin C Food Sources Fruits and vegetables can easily provide a generous amount of vitamin C. A cup of orange juice at breakfast, a salad for lunch, and a stalk of broccoli and a potato for dinner alone provide more than 300 milligrams. (For perspective, review Figure 10-17, p. 324) Clearly, a person making such food choices does not need vitamin C supplements.

Figure 10-19 shows the amounts of vitamin C in various common foods. The overwhelming abundance of purple and green bars reveals not only that the citrus fruits are justly famous for being rich in vitamin C, but that other fruits and vegetables are in the same league. A half cup of broccoli, bell pepper, or strawberries provides more than 50 milligrams of the vitamin (and an array of other nutrients). Because vitamin C is vulnerable to heat, raw fruits and vegetables usually have a higher nutrient density than their cooked counterparts. Similarly, because vitamin C is readily destroyed by oxygen, foods and juices should be stored properly and consumed within a week of opening.

The potato is an important source of vitamin C, not because one potato by itself meets the daily need, but because potatoes are such a common staple that they make significant contributions. In fact, scurvy was unknown in Ireland until the potato blight of the mid-1840s when some 2 million people died of malnutrition and infection.

The lack of yellow, white, brown, and red bars in Figure 10-19 confirms that grains, milk and milk products (except breast milk), and most protein foods are notoriously poor sources of vitamin C. Organ meats (liver, kidneys, and others)

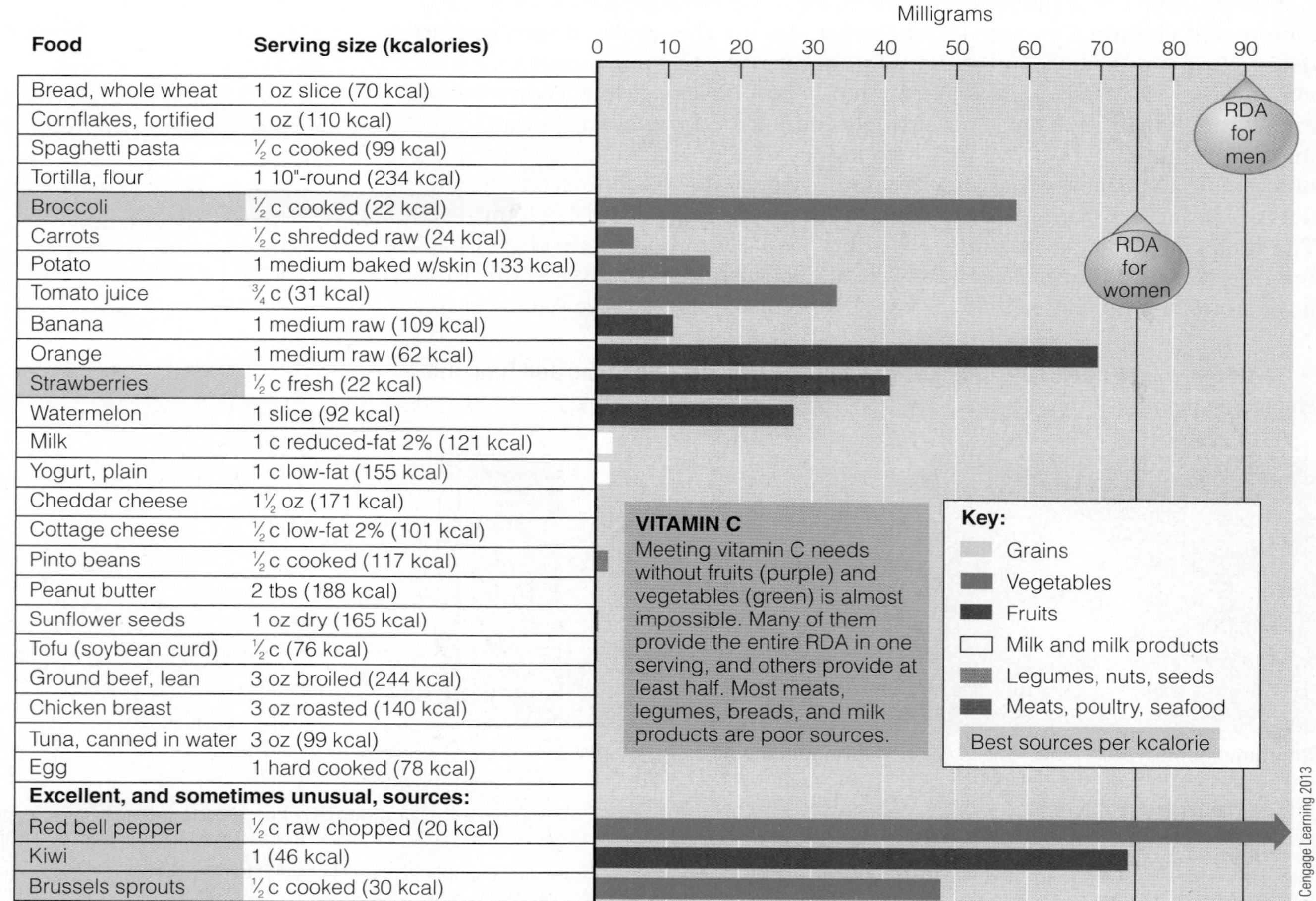

> FIGURE 10-19 **Vitamin C in Selected Foods**

See the "How To" feature on p. 302 for more information on using this figure.

© Photodisc/ Getty Images

When dietitians say "vitamin C," people think "citrus fruits" . . .

© Polara Studios, Inc.

. . . but these foods are also rich in vitamin C.

and raw meats contain some vitamin C, but most people don't eat large quantities of these foods. Raw meats and fish contribute enough vitamin C to be significant sources in parts of Alaska, Canada, and Japan, but elsewhere fruits and vegetables are necessary to supply sufficient vitamin C.

Because of vitamin C's antioxidant property, food manufacturers sometimes add a variation of vitamin C to some beverages and most cured meats, such as luncheon meats, to prevent oxidation and spoilage. This compound safely preserves these foods, but it does not have vitamin C activity in the body. Simply put, "Ham and bacon cannot replace fruits and vegetables."

REVIEW IT Identify the main roles, deficiency symptoms, and food sources for vitamin C.

Vitamin C acts primarily as an antioxidant and a cofactor. Recommendations are set well above the amount needed to prevent the deficiency disease scurvy. A variety of fruits and vegetables—most notably citrus fruits—provide generous amounts of vitamin C. The accompanying table provides a summary of vitamin C.

Vitamin C

Other Names
Ascorbic acid

RDA
Men: 90 mg/day
Women: 75 mg/day
Smokers: +35 mg/day

UL
Adults: 2000 mg/day

Chief Functions in the Body
Collagen synthesis (strengthens blood vessel walls, forms scar tissue, provides matrix for bone growth), antioxidant, thyroxin synthesis, amino acid metabolism, strengthens resistance to infection, helps in absorption of iron

Significant Sources
Citrus fruits, cabbage-type vegetables (such as brussels sprouts and cauliflower), dark green vegetables (such as bell peppers and broccoli), cantaloupe, strawberries, lettuce, tomatoes, potatoes, papayas, mangoes
Easily destroyed by heat and oxygen

Deficiency Disease
Scurvy

Deficiency Symptoms
Anemia (small-cell type),[a] atherosclerotic plaques, pinpoint hemorrhages; bone fragility, joint pain; poor wound healing, frequent infections; bleeding gums, loosened teeth; muscle degeneration, pain, hysteria, depression; rough skin, blotchy bruises

Toxicity Symptoms
Nausea, abdominal cramps, diarrhea; headache, fatigue, insomnia; hot flashes; rashes; interference with medical tests, aggravation of gout symptoms, urinary tract problems, kidney stones[b]

[a]Small-cell-type anemia is *microcytic anemia*.

[b]People with kidney disease, a tendency toward gout, or a genetic abnormality that alters the breakdown of vitamin C are prone to forming kidney stones. Vitamin C is inactivated and degraded by several routes, sometimes producing oxalate, which can form stones in the kidneys.

REVIEW IT The Water-Soluble Vitamins

Vitamin and Chief Functions	Deficiency Symptoms	Toxicity Symptoms	Food Sources
Thiamin Part of coenzyme TPP in energy metabolism	Beriberi (edema or muscle wasting), anorexia and weight loss, neurological disturbances, muscular weakness, heart enlargement and failure	None reported	Enriched, fortified, or whole-grain products; pork
Riboflavin Part of coenzymes FAD and FMN in energy metabolism	Inflammation of the mouth, skin, and eyelids	None reported	Milk products; enriched, fortified, or whole-grain products; liver
Niacin Part of coenzymes NAD and NADP in energy metabolism	Pellagra (diarrhea, dermatitis, and dementia)	Niacin flush, liver damage, impaired glucose tolerance	Protein-rich foods
Biotin Part of coenzyme in energy metabolism	Skin rash, hair loss, neurological disturbances	None reported	Widespread in foods; GI bacteria synthesis
Pantothenic acid Part of coenzyme A in energy metabolism	Digestive and neurological disturbances	None reported	Widespread in foods
Vitamin B_6 Part of coenzymes used in amino acid and fatty acid metabolism	Scaly dermatitis, depression, confusion, convulsions, anemia	Nerve degeneration, skin lesions	Protein-rich foods
Folate Activates vitamin B_{12}; helps synthesize DNA for new cell growth	Anemia, glossitis, neurological disturbances, elevated homocysteine	Masks vitamin B_{12} deficiency	Legumes, vegetables, fortified grain products
Vitamin B_{12} Activates folate; helps synthesize DNA for new cell growth; protects nerve cells	Anemia; nerve damage and paralysis	None reported	Foods derived from animals
Vitamin C Synthesis of collagen, carnitine, hormones, neurotransmitters; antioxidant	Scurvy (bleeding gums, pinpoint hemorrhages, abnormal bone growth, and joint pain)	Diarrhea, GI distress	Fruits and vegetables

Vita means life. After this discourse on the vitamins, who could dispute that they deserve their name? Their regulation of metabolic processes makes them vital to the normal growth, development, and maintenance of the body. The accompanying table condenses the information provided in this chapter for a quick review. The remarkable roles of the vitamins continue in the next chapter.

Nutrition Portfolio

To obtain all the vitamins you need each day, be sure to select from a variety of foods from all the food groups. Go to Diet Analysis Plus and choose one of the days on which you have tracked your diet for the entire day. Go to the Intake vs. Goals report. Near the bottom of this report, you will see all of the vitamins grouped together; using this section of the report for reference, answer the following questions:

- How was your vitamin intake overall? Did you consume too much or too little of any vitamin? Which vitamins concerned you most?

Next go to the Intake Spreadsheet report, and looking at each of the vitamins, answer the following questions:

- Which of your foods provided high intakes of vitamins?
- Which of your foods provided few or no vitamins?
- Examine your daily choices of whole or enriched grains, dark green vegetables, citrus fruits, and legumes, then evaluate their contributions to your vitamin intakes.

- If you are a woman of childbearing age, calculate the dietary folate equivalents you receive from folate-rich foods, fortified foods, and supplements, then compare that to your RDA.
- Compare your vitamin intakes from supplements with their UL.

To complete this exercise, go to your Diet Analysis Plus at **www.cengagebrain.com.**

STUDY IT To review the key points of this chapter and take a practice quiz, go to the study cards at the end of the book.

REFERENCES

1. A. Piro and coauthors, Casimir Funk: His discovery of the vitamins and their deficiency disorders, *Annals of Nutrition and Metabolism* 57 (2010): 85–88.
2. M. N. Riaz, M. Asif, and R. Ali, Stability of vitamins during extrusion, *Critical Reviews in Food Science and Nutrition* 49 (2009): 361–368.
3. Committee on Dietary Reference Intakes, *Dietary Reference Intakes for Vitamin C, Vitamin E, Selenium, and Carotenoids* (Washington, D.C.: National Academies Press, 2000); Committee on Dietary Reference Intakes, *Dietary Reference Intakes for Thiamin, Riboflavin, Niacin, Vitamin B_6, Folate, Vitamin B_{12}, Pantothenic Acid, Biotin, and Choline* (Washington, D.C.: National Academies Press, 1998).
4. K. J. Carpenter, *Beriberi, White Rice, and Vitamin B: A Disease, a Cause, and a Cure* (Berkeley: University of California Press, 2000).
5. K. L. Bogan and C. Brenner, Nicotinic acid, nicotinamide, and nicotinamide riboside: A molecular evaluation of NAD+ precursor vitamins in human nutrition, *Annual Review of Nutrition* 28 (2008): 115–130.
6. Committee on Dietary Reference Intakes, 1998, pp. 128–129.
7. J. M. Backes, R. J. Padley, and P. M. Moriarty, Important considerations for treatment with dietary supplement versus prescription niacin products, *Postgraduate Medicine* 123 (2011): 70–83.
8. S. C. Larsson, N. Orsini, and A. Wolk, Vitamin B_6 and risk of colorectal cancer: A meta-analysis of prospective studies, *Journal of the American Medical Association* 303 (2010): 1077–1083; J. Shen and coauthors, Association of vitamin B-6 status with inflammation, oxidative stress, and chronic inflammatory conditions: The Boston Puerto Rican Health Study, *American Journal of Clinical Nutrition* 91 (2010): 337–342; J. H. Page and coauthors, Plasma vitamin B(6) and risk of myocardial infarction in women, *Circulation* 120 (2009): 649–655.
9. A. S. Tibbetts and D. R. Appling, Compartmentalization of mammalian folate-mediated one-carbon metabolism, *Annual Review of Nutrition* 30 (2010): 57–81.
10. M. A. Caudill, Folate bioavailability: Implications for establishing dietary recommendations and optimizing status, *American Journal of Clinical Nutrition* 91 (2010): 1455S–1460S.
11. S. H. Blanton and coauthors, Folate pathway and nonsyndromic cleft lip and palate, *Birth Defects Research, Part A, Clinical and Molecular Teratology* 91 (2011): 50–60.
12. Committee on Dietary Reference Intakes, 1998.
13. O. Dary, Nutritional interpretation of folic acid interventions, *Nutrition Reviews* 67 (2009): 235–244; J. Selhub and coauthors, Folate-vitamin B-12 interaction in relation to cognitive impairment, anemia, and biochemical indicators of vitamin B-12 deficiency, *American Journal of Clinical Nutrition* 89 (2009): 702S–706S; M. D. Thompson, D. E. C. Cole, and J. G. Ray, Vitamin B-12 and neural tube defects: The Canadian experience, *American Journal of Clinical Nutrition* 89 (2009): 697S–701S; G. Varela-Moreiras, M. M. Murphy, and J. M. Scott, Cobalamin, folic acid, and homocysteine, *Nutrition Reviews* 67 (2009): S69–S72; A. D. Smith, Y. I. Kim, and H. Refsum, Is folic acid good for everyone? *American Journal of Clinical Nutrition* 87 (2008): 517–533.
14. R. Cui and coauthors, Dietary folate and vitamin B6 and B12 intake in relation to mortality from cardiovascular diseases: Japan collaborative cohort study, *Stroke* 41 (2010): 1285–1289; A. Imamura and coauthors, Low folate levels may be an atherogenic factor regardless of homocysteine levels in young healthy nonsmokers, *Metabolism* 59 (2010): 728–733; A. Mente and coauthors, A systematic review of the evidence supporting a causal link between dietary factors and coronary heart disease, *Archives of Internal Medicine* 169 (2009): 659–669.
15. J. M. Artmitage and the Study of Effectiveness of Additional Reductions in Cholesterol and Homocysteine (SEARCH) Collaborative Group, Effects of homocysteine-lowering with folic acid plus vitamin B_{12} vs placebo on mortality and major morbidity in myocardial infarction survivors: A randomized trial, *Journal of the American Medical Association* 303 (2010): 2486–2494; P. Tighe and coauthors, A dose-finding trial of the effect of the long-term folic acid intervention: Implications for food fortification policy, *American Journal of Clinical Nutrition* 93 (2011): 11–18; C. M. Albert and coauthors, Effect of folic acid and B vitamins on risk of cardiovascular events and total mortality among women at high risk for cardiovascular disease: A randomized trial, *Journal of the American Medical Association* 299 (2008): 2027–2036; M. Ebbing and coauthors, Mortality and cardiovascular events in patients treated with homocysteine-lowering B vitamins after coronary angiography, *Journal of the American Medical Association* 300 (2008): 795–801; E. Lonn, Homocysteine-lowering B vitamin therapy in cardiovascular prevention: Wrong again? *Journal of the American Medical Association* 299 (2008): 2086–2087.
16. J. W. Crott and coauthors, Moderate folate depletion modulates the expression of selected genes involved in cell cycle, intracellular signaling and folate uptake in human colonic epithelial cell lines, *Journal of Nutritional Biochemistry* 19 (2008): 328–335.
17. V. L. Stevens and coauthors, Folate and other one-carbon metabolism-related nutrients and risk of postmenopausal breast cancer in the Cancer Prevention Study II Nutrition Cohort, *American Journal of Clinical Nutrition* 91 (2010): 1708–1715; M. Ebbing and coauthors, Cancer incidence and mortality after treatment with folic acid and vitamin B_{12}, *Journal of the American Medical Association* 302 (2009): 2119–2126; U. C. Ericson and coauthors, Increased breast cancer risk at high plasma folate concentrations among women with the *MTHFR 677T* allele, *American Journal of Clinical Nutrition* 90 (2009): 1380–1389; S. S. Maruti, C. M. Ulrich, and E. White, Folate and one-carbon metabolism nutrients from supplements and diet in relation to breast cancer risk, *American Journal of Clinical Nutrition* 89 (2009): 624–633; J. B. Mason, Folate, cancer risk, and the Greek god, Proteus: A tale of two chameleons, *Nutrition Reviews* 67 (2009): 206–212; J. Sauer, J. B. Mason, and S. W. Choi, Too much folate: A risk factor for cancer and cardiovascular disease? *Current Opinion in Clinical Nutrition and Metabolic Care* 12 (2009): 30–36.
18. B. M. Oaks and coauthors, Folate intake, post-folic acid grain fortification, and pancreatic cancer risk in the Prostate, Lung, Colorectal, and Ovarian Cancer Screening Trial, *American Journal of Clinical Nutrition* 91 (2010): 449–455; A. R. Hart, H. Kennedy, and I. Harvey, Pancreatic

cancer: A review of the evidence on causation, *Clinical Gastroenterology Hepatology* 6 (2008): 275–282; World Cancer Research Fund and American Institute for Cancer Research, *Policy and Action for Cancer Prevention—Food, Nutrition, Physical Activity: A Global Perspective* (Washington, D.C.: AICR, 2009), p. 23.

19. Z. Liu and coauthors, Mild depletion of dietary folate combined with other B vitamins alters multiple components of the Wnt pathway in mouse colon, *Journal of Nutrition* 137 (2007): 2701–2708.
20. V. Ganji and M. R. Kafai, Hemoglobin and hematocrit values are higher and prevalence of anemia is lower in the post-folic acid fortification period than in the pre-folic acid fortification period in US adults, *American Journal of Clinical Nutrition* 89 (2009): 363–371.
21. R. L. Bailey and coauthors, Total folate and folic acid intake from foods and dietary supplements in the United States: 2003–2006, *American Journal of Clinical Nutrition* 91 (2010): 231–237; Q. Yang and coauthors, Folic acid source, usual intake, and folate and vitamin B-12 status in US adults: National Health and Nutrition Examination Survey (NHANES) 2003–2006, *American Journal of Clinical Nutrition* 91 (2010): 64–72.
22. L. H. Allen, How common is vitamin B-12 deficiency? *American Journal of Clinical Nutrition* 89 (2009): 693S–696S.
23. D. K. Dror and L. H. Allen, Effect of vitamin B_{12} deficiency on neurodevelopment in infants: Current knowledge and possible mechanisms, *Nutrition Reviews* 66 (2008): 250–255.
24. Varela-Moreiras, Murphy, and Scott, 2009.
25. L. Feng and coauthors, Vitamin B-12, apolipoprotein E genotype, and cognitive performance in community-living older adults: Evidence of a gene-micronutrient interaction, *American Journal of Clinical Nutrition* 89 (2009): 1263–1268.
26. Varela-Moreiras, Murphy, and Scott, 2009.
27. Allen, 2009; Dary, 2009; R. Green, Is it time for vitamin B-12 fortification? What are the questions? *American Journal of Clinical Nutrition* 89 (2009): 712S–716S; J. W. Miller and coauthors, Metabolic evidence of vitamin B-12 deficiency, including high homocysteine and methylmalonic acid and low holotranscobalamin, is more pronounced in older adults with elevated plasma folate, *American Journal of Clinical Nutrition* 90 (2009): 1586–1592; J. Selhub and coauthors, Folate-vitamin B-12 interaction in relation to cognitive impairment, anemia, and biochemical indicators of vitamin B-12 deficiency, *American Journal of Clinical Nutrition* 89 (2009): 702S–706S; Thompson, Cole, and Ray, 2009.
28. A. Vogiatzoglou and coauthors, Dietary sources of vitamin B-12 and their association with plasma vitamin B-12 concentrations in the general population: The Hordaland Homocysteine Study, *American Journal of Clinical Nutrition* 89 (2009): 1078–1087.
29. S. H. Zeisel and K. A. da Costa, Choline: An essential nutrient for public health, *Nutrition Reviews* 67 (2009): 615–623.
30. Zeisel and da Costa, 2009.
31. J. H. Baron, Sailors' scurvy before and after James Lind—A reassessment, *Nutrition Reviews* 67 (2009): 315–332.
32. M. Simasek and D. A. Blandino, Treatment of the common cold, *American Family Physician* 75 (2007): 515–520.
33. R. M. Douglas and coauthors, Vitamin C for preventing and treating the common cold, *Cochrane Database of Systematic Reviews* 3 (2007): CD000980.
34. B. Barrett and coauthors, Sufficiently important difference for common cold: Severity reduction, *Annals of Family Medicine* 5 (2007): 216–223.
35. Committee on Dietary Reference Intakes, 2000.
36. R. L. Schleicher and coauthors, Serum vitamin C and the prevalence of vitamin C deficiency in the United States: 2003–2004 National Health and Nutrition Examination Survey (NHANES), *American Journal of Clinical Nutrition* 90 (2009): 1252–1263.
37. D. Léger, Scurvy: Reemergence of nutritional deficiencies, *Canadian Family Physician* 54 (2008): 1403–1406.

LEARN IT Present arguments for and against the use of dietary supplements.

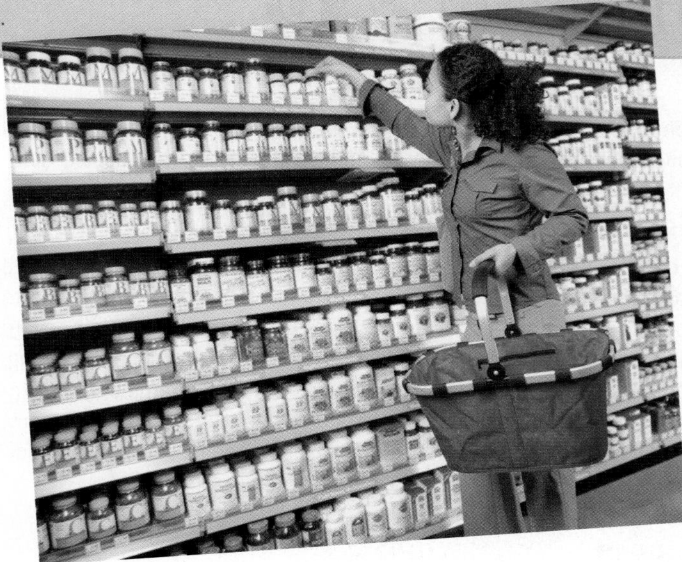
© Tanya Constantine/Jupiterimages

Vitamin and Mineral Supplements

An estimated 75,000 supplements are currently on the market. More than half of the adults in the United States take a **dietary supplement** regularly, spending almost $24 billion each year.[1] Many people take supplements as dietary insurance—in case they are not meeting their nutrient needs from foods alone. Others take supplements as health insurance—to protect against certain diseases.

One out of every three people takes multivitamin-mineral supplements daily. Others take large doses of single nutrients, most commonly, vitamin C, B vitamins, vitamin D, and calcium. In many cases, taking supplements is a costly but harmless practice; sometimes, it is both costly and harmful to health.[2]

For the most part, people self-prescribe supplements, taking them on the advice of friends, advertisements, websites, or books that may or may not be reliable. Sometimes, they take supplements on the recommendation of a physician. When such advice follows a valid nutrition assessment, supplementation may be warranted, but even then the preferred course of action is to improve food choices and eating habits.[3] Without an assessment, the advice to take supplements may be inappropriate. A registered dietitian can help with the decision.

When people think of dietary supplements, they often think of vitamins, but a diet that lacks vitamins probably lacks several minerals as well. This highlight asks several questions related to vitamin-mineral supplements. (The accompanying glossary defines dietary supplements and related terms.) What are the arguments *for* taking supplements? What are the arguments *against* taking them? Finally, if people do take supplements, how can they choose the appropriate ones? (Amino acid supplements and herbal supplements are discussed in Chapter 6 and Highlight 18, respectively.)

Arguments for Supplements

Vitamin-mineral supplements may be appropriate in some circumstances. In some cases, they can prevent or correct deficiencies; in others, they can reduce the risk of diseases. Consumers should discuss supplement use with their health-care professionals, who can help monitor for adverse effects or nutrient-drug interactions.

Correct Overt Deficiencies

In the United States and Canada, adults rarely suffer nutrient deficiency diseases such as scurvy, pellagra, and beriberi, but nutrient deficiencies do still occur. To correct an overt deficiency disease, a physician may prescribe therapeutic doses two to ten times the RDA (or AI) of a nutrient. At such high doses, the supplement is acting as a drug.

Support Increased Nutrient Needs

As Chapters 15 through 17 explain, nutrient needs increase during certain stages of life, making it difficult to meet some of those needs without supplementation. For example, women who lose a lot of blood and therefore a lot of iron during menstruation each month may need an iron supplement. Women of childbearing age need folate supplements to reduce the risks of neural tube defects. Similarly, pregnant women and women who are breastfeeding their infants have exceptionally high nutrient needs and so usually need special supplements. Newborns routinely receive a single dose of vitamin K at birth to prevent abnormal bleeding. Infants may need other supplements as well, depending on whether they are breastfed or receiving formula, and on whether their water contains fluoride.

Improve Nutrition Status

In contrast to the classical deficiencies, which present a multitude of symptoms and are relatively easy to recognize, subclinical deficiencies are subtle and easy to overlook—and they are also more likely to occur. People who do not eat enough food to deliver the needed amounts of nutrients, such as habitual dieters and the elderly, risk developing subclinical deficiencies. Similarly, vegetarians who restrict their use of entire food groups without appropriate substitutions may fail to fully meet their nutrient needs. If there is no way for these people to eat

GLOSSARY

dietary supplement: any pill, capsule, tablet, liquid, or powder that contains vitamins, minerals, herbs, or amino acids; intended to increase dietary intake of these substances.

FDA (Food and Drug Administration): a part of the Department of Health and Human Services' Public Health Service that is responsible for ensuring the safety and wholesomeness of all dietary supplements and food processed and sold in interstate commerce except meat, poultry, and eggs (which are under the jurisdiction of the USDA); inspecting food plants and imported foods; and setting standards for food composition and product labeling.

high potency: 100% or more of the Daily Value for the nutrient in a single supplement and for at least two-thirds of the nutrients in a multinutrient supplement.

nanotechnology: a manufacturing technology that manipulates atoms to change the structure of matter.

nanoceuticals: substances with extremely small particles that have been manufactured by nanotechnology.

enough nutritious foods to meet their needs, then vitamin-mineral supplements may be appropriate to help prevent nutrient deficiencies.

Improve the Body's Defenses

Health-care professionals may provide special supplementation to people being treated for addictions to alcohol or other drugs and to people with prolonged illnesses, extensive injuries, or other severe stresses such as surgery. Illnesses that interfere with appetite, eating, or nutrient absorption impair nutrition status. For example, the stomach condition atrophic gastritis often creates a vitamin B_{12} deficiency. In addition, nutrient needs are often heightened by diseases or medications. In all these cases, supplements are appropriate.

Reduce Disease Risks

Few people consume the optimal amounts of all the vitamins and minerals by diet alone. Inadequate intakes have been linked to chronic diseases such as heart disease, some cancers, and osteoporosis. For this reason, some physicians recommend that all adults take vitamin-mineral supplements. Such regular supplementation would provide an optimum intake to enhance metabolic harmony and prevent disease at relatively little cost. Others recognize the lack of conclusive evidence and the potential harm of supplementation and advise against such a recommendation. The most recent statement from the National Institutes of Health acknowledges that evidence is insufficient to recommend either for or against the use of supplements to prevent chronic diseases.

Highlight 11 reviews the relationships between supplement use and disease prevention. It describes some of the accumulating evidence suggesting that intakes of certain nutrients at levels much higher than can be attained from foods alone may be beneficial in reducing some disease risks. It also presents research confirming the associated risks. Clearly, consumers must be cautious in taking supplements to prevent disease.

Who Needs Supplements?

In summary, the following list acknowledges that in these specific conditions, these people may need to take supplements:

- People with specific nutrient deficiencies may need specific nutrient supplements.
- People whose energy intakes are particularly low (fewer than 1600 kcalories per day) may need multivitamin-mineral supplements.
- Vegetarians who eat all-plant diets (vegans) and older adults with atrophic gastritis may need vitamin B_{12}.
- People who have lactose intolerance or milk allergies or who otherwise do not consume enough milk products to forestall extensive bone loss may need calcium.
- People in certain stages of the life cycle who have increased nutrient requirements may need specific nutrient supplements. For example, infants may need vitamin D, iron, and fluoride; women of childbearing age and pregnant women may need folate and iron; and the elderly may need vitamin B_{12} and vitamin D.
- People who have inadequate intakes of milk or milk products, limited sun exposure, or heavily pigmented skin may need vitamin D.
- People who have diseases, infections, or injuries or who have undergone surgery that interferes with the intake, absorption, metabolism, or excretion of nutrients may need specific nutrient supplements.
- People taking medications that interfere with the body's use of specific nutrients may need specific nutrient supplements.

Except for people in these circumstances, most adults can get all the nutrients they need by eating a variety of nutrient-dense foods. Even athletes can meet their nutrient needs without the help of supplements, as Chapter 14 explains.

Arguments against Supplements

Foods rarely cause nutrient imbalances or toxicities, but supplements can. The higher the dose, the greater the risk of harm. People's tolerances for high doses of nutrients vary, just as their risks of deficiencies do. Amounts that some can tolerate may be harmful for others, and no one knows who falls where along the spectrum. It is difficult to determine just how much of a nutrient is enough—or too much. The Tolerable Upper Intake Levels of the DRI answer the question "How much is too much?" by defining the highest amount that appears safe for most healthy people. Table H10-1 presents Upper Levels and Daily Values for selected vitamins and minerals and the quantities typically found in supplements.

Who Should Not Take Supplements?

The following list recognizes that in certain circumstances, these people may need to avoid specific supplements:[4]

- Men and postmenopausal women should not take iron supplements given that excess iron is harmful and generally more likely than inadequacies.
- Smokers should not take beta-carotene supplements given that high doses have been associated with increased lung cancer and mortality.
- Postmenopausal women should not take vitamin A supplements given that excess retinol has been associated with increased risk of hip fractures and reduced bone density.
- Surgery patients should not take vitamin E supplements during the week before surgery because vitamin E acts as a blood thinner.

Toxicity

Supplement users are more likely to have excessive intakes of certain nutrients—notably iron, zinc, vitamin A, and niacin. The extent and severity of supplement toxicity remain unclear. Only a few alert health-care professionals can recognize toxicity, even when it is acute. When it is chronic, with the effects developing subtly and progressing slowly, it often goes unrecognized. In view of the potential hazards, some authorities believe supplements should bear warning labels, advising consumers that large doses may be toxic.

TABLE H10-1 **Vitamin and Mineral Intakes for Adults**

Nutrient	Tolerable Upper Intake Levels[a]	Daily Values	Typical Multivitamin-Mineral Supplement	Average Single-Nutrient Supplement
Vitamins				
Vitamin A	3000 µg (10,000 IU)	5000 IU	5000 IU	8000 to 10,000 IU
Vitamin D	100 µg (4000 IU)	400 IU	400 IU	1000 to 5,000 IU[b]
Vitamin E	1000 mg (1500 to 2200 IU)[c]	30 IU	30 IU	100 to 1000 IU
Vitamin K	—[d]	80 µg	40 µg	—[f]
Thiamin	—[d]	1.5 mg	1.5 mg	50 mg
Riboflavin	—[d]	1.7 mg	1.7 mg	25 mg
Niacin (as niacinamide)	35 mg[c]	20 mg	20 mg	100 to 500 mg
Vitamin B_6	100 mg	2 mg	2 mg	100 to 200 mg
Folate	1000 µg[c]	400 µg	400 µg	400 µg
Vitamin B_{12}	—[d]	6 µg	6 µg	100 to 1000 µg
Pantothenic acid	—[d]	10 mg	10 mg	100 to 500 mg
Biotin	—[d]	300 µg	30 µg	300 to 600 µg
Vitamin C	2000 mg	60 mg	10 mg	500 to 2000 mg
Choline	3500 mg	—	10 mg	250 mg
Minerals				
Calcium	2500 mg	1000 mg	160 mg	250 to 600 mg
Phosphorus	4000 mg	1000 mg	110 mg	—[f]
Magnesium	350 mg[e]	400 mg	100 mg	250 mg
Iron	45 mg	18 mg	18 mg	18 to 30 mg
Zinc	40 mg	15 mg	15 mg	10 to 100 mg
Iodine	1100 µg	150 µg	150 µg	—[f]
Selenium	400 µg	70 µg	10 µg	50 to 200 µg
Fluoride	10 mg	—	—	—[f]
Copper	10 mg	2 mg	0.5 mg	—[f]
Manganese	11 mg	2 mg	5 mg	—[f]
Chromium	—[d]	120 µg	25 µg	200 to 400 µg
Molybdenum	2000 µg	75 µg	25 µg	—[f]

[a]Unless otherwise noted, Upper Levels represent total intakes from food, water, and supplements.
[b]Most commonly, single supplements are of vitamin D_3 (cholecalciferol); single supplements of vitamin D_2 (ergocalciferol) may provide up to 50,000 IU per dose.
[c]Upper Levels represent intakes from supplements, fortified foods, or both.
[d]These nutrients have been evaluated by the DRI Committee for Tolerable Upper Intake Levels, but none were established because of insufficient data. No adverse effects have been reported with intakes of these nutrients at levels typical of supplements, but caution is still advised, given the potential for harm that accompanies excessive intakes.
[e]Upper Levels represent intakes from supplements only.
[f]Available as a single supplement by prescription.

Toxic overdoses of vitamins and minerals in children are more readily recognized and, unfortunately, fairly common. Fruit-flavored, chewable vitamins shaped like cartoon characters entice young children to eat them like candy in amounts that can cause poisoning. Iron supplements (30 milligrams of iron or more per tablet) are especially toxic and are the leading cause of accidental ingestion fatalities among children. Even mild overdoses cause GI distress, nausea, and black diarrhea, which reflects gastric bleeding. Severe overdoses result in bloody diarrhea, shock, liver damage, coma, and death.

Life-Threatening Misinformation

Another problem arises when people who are ill come to believe that high doses of vitamins or minerals can be therapeutic. Not only can high doses be toxic, but the person may take them instead of seeking medical help. Furthermore, there are no guarantees that the supplements will be effective. Taking vitamin supplements instead of medication may sound appealing, but they do not protect against the progression of heart disease or cancers.[5] In some cases, supplements may even be harmful.[6] Supplements of beta-carotene and vitamin A increased the risk of lung cancer and mortality, especially among smokers.[7] Similarly, supplements of vitamin E increased the risk of prostate cancer among healthy men.[8]

Marketing materials for supplements often make health statements that are required to be "truthful and not misleading," but they often fall far short of both. Highlight 18 revisits this topic and includes a discussion of herbal preparations and other alternative therapies.

Unknown Needs

Another argument against the use of supplements is that no one knows exactly how to formulate the "ideal" supplement. What nutrients should be included? Which, if any, of the phytochemicals should be included? How much of each? On whose needs should the choices be based? Surveys have repeatedly shown little relationship between the supplements people take and the nutrients they actually need.

False Sense of Security

Another argument against supplement use is that it may lull people into a false sense of security. A person might eat irresponsibly, thinking, "My supplement will ensure my needs are met." Or, experiencing a warning symptom of a disease, a person might postpone seeking a diagnosis, thinking, "I probably just need a supplement to make this go away." Such self-diagnosis is potentially dangerous.

Other Invalid Reasons

Other invalid reasons people might use for taking supplements include:

- The belief that the food supply or soil contains inadequate nutrients
- The belief that supplements can provide energy
- The belief that supplements can enhance athletic performance or build lean body tissues without physical work or faster than work alone (see Highlight 14)
- The belief that supplements will help a person cope with stress
- The belief that supplements can prevent, treat, or cure conditions ranging from the common cold to cancer

Ironically, people with health problems are more likely to take supplements than other people, yet today's health problems are more likely to be due to overnutrition and poor lifestyle choices than to nutrient deficiencies. The truth—that most people would benefit from improving their eating and activity patterns—is harder to swallow than a supplement pill.

Bioavailability and Antagonistic Actions

In general, the body absorbs nutrients best from foods in which the nutrients are diluted and dispersed among other substances that may facilitate their absorption. Taken in pure, concentrated form, nutrients are likely to interfere with one another's absorption or with the absorption of nutrients in foods eaten at the same time. Documentation of these effects is particularly extensive for minerals: zinc hinders copper and calcium absorption, iron hinders zinc absorption, calcium hinders magnesium and iron absorption, and magnesium hinders the absorption of calcium and iron. Similarly, binding agents in supplements limit mineral absorption.

Although minerals provide the most-familiar and best-documented examples, interference among vitamins is now being seen as supplement use increases. The vitamin A precursor beta-carotene, long thought to be nontoxic, interferes with vitamin E metabolism when taken over the long term as a dietary supplement. Vitamin E, on the other hand, antagonizes vitamin K activity and so should not be used by people being treated for blood-clotting disorders. Consumers who want the benefits of optimal absorption of nutrients should eat foods selected for nutrient density and variety.

Whenever the diet is inadequate, the person should first attempt to improve it so as to obtain the needed nutrients from foods. If that is truly impossible, then the person needs a multivitamin-mineral supplement that supplies between 50 and 150 percent of the Daily Value for each of the nutrients. These amounts reflect the ranges commonly found in foods and therefore are compatible with the body's normal handling of nutrients (its physiologic tolerance). The next section provides some pointers to assist in the selection of an appropriate supplement.

Selection of Supplements

Whenever a physician or registered dietitian recommends a supplement, follow the directions carefully. When selecting a supplement yourself, look for a single, balanced vitamin-mineral supplement. Supplements with a USP verification logo have been tested by the US Pharmacopeia (USP) to assure that the supplement:

- Contains the declared ingredients and amounts listed on the label
- Does not contain harmful levels of contaminants
- Will disintegrate and release ingredients in the body
- Was made under safe and sanitary conditions

If you decide to take a vitamin-mineral supplement, ignore the eye-catching art and meaningless claims. Pay attention to the form the supplements are in, the list of ingredients, and the price. Here's where the truth lies, and from it you can make a rational decision based on facts. You have two basic questions to answer.

Form

The first question: What form do you want—chewable, liquid, or pills? If you'd rather drink your supplements than chew them, fine. If you choose a chewable form, though, be aware that chewable vitamin C can dissolve tooth enamel. If you choose pills, look for statements about the disintegration time. The USP suggests that supplements should completely disintegrate within 30 to 45 minutes. Obviously, supplements that don't dissolve have little chance of entering the bloodstream, so look for a brand that claims to meet USP disintegration standards.

Contents

The second question: What vitamins and minerals do you need? Generally, an appropriate supplement provides vitamins and minerals in amounts that do not exceed recommended intakes. Avoid supplements that, in a daily dose, provide more than the Upper Level for *any* nutrient. Avoid preparations with more than 10 milligrams of iron per dose, except as prescribed by a physician. Iron is hard to get rid of once it's in the body, and an excess of iron can cause problems, just as a deficiency can (see Chapter 13).

Misleading Claims

Manufacturers of *organic* or natural vitamins boast that their pills are purified from real foods rather than synthesized in a laboratory. These supplements are no more effective than others and often cost more. The word *synthetic* may sound like "fake," but to synthesize just means to put together. Think back on the course of human evolution; it is not natural to take any kind of pill. In reality, the finest, most natural vitamin "supplements" available are whole grains, vegetables, fruits, meat, fish, poultry, eggs, legumes, nuts, and milk and milk products.

Avoid products that make **"high potency"** claims. More is not better (review Figure 10-1 on p. 299). Remember that foods are also providing these nutrients. Nutrients can build up and cause unexpected problems. For example, a man who takes vitamins and begins to lose his hair may think his hair loss means he needs *more* vitamins, when in fact it may be the early sign of a vitamin A overdose. (Of course, it may be completely unrelated to nutrition as well.)

Be aware that fake vitamins and preparations that contain items not needed in human nutrition, such as carnitine and inositol, reflect a marketing strategy aimed at your pocket, not at your health. The manufacturer wants you to believe that its pills contain the latest "new" nutrient that other brands omit, but in reality, these substances are not known to be needed by human beings.

Realize that the claim that supplements "relieve stress" is another marketing ploy. If you give even passing thought to what people mean by "stress," you'll realize manufacturers could never design a supplement to meet everyone's needs. Is it stressful to take an exam? Well, yes. Is it stressful to survive a major car wreck with third-degree burns and multiple bone fractures? Definitely, yes. The body's responses to these stresses are different. The body does use vitamins and minerals in mounting a stress response, but a body fed a well-balanced diet can meet the needs of most minor stresses. For the major ones, medical intervention is needed. In any case, taking a dietary supplement won't make life any less stressful.

Other marketing tricks to sidestep are "green" pills that contain dehydrated, crushed parsley, alfalfa, and other fruit and vegetable extracts. The nutrients and phytochemicals advertised can be obtained from a serving of vegetables more easily and for less money. Such pills may also provide enzymes, but enzymes are inactivated in the stomach during protein digestion.

Recognize the latest nutrition buzzwords. Manufacturers were marketing "antioxidant" supplements before the print had time to dry on the first scientific reports of antioxidant vitamins' action in the body. Remember, too, that high doses can alter a nutrient's action in the body. An antioxidant in physiological quantities may be beneficial, but in pharmacological quantities, it may act as a prooxidant and cause harm. Highlight 11 explores antioxidants and supplement use in more detail.

Similarly, manufacturers began making dietary supplements using **nanotechnology** before the FDA had created guidelines defining their use in consumer products.[9] These **nanoceuticals** promise enhanced nutrient absorption and activity. Such claims may sound good, but again, more does not always mean better.

Finally, be aware that advertising on the Internet is cheap and not closely regulated. Promotional e-mails can be sent to millions of people in an instant. Internet messages can easily cite references and provide links to other sites, implying an endorsement when in fact none has been given. Be cautious when examining unsolicited information and search for a balanced perspective.

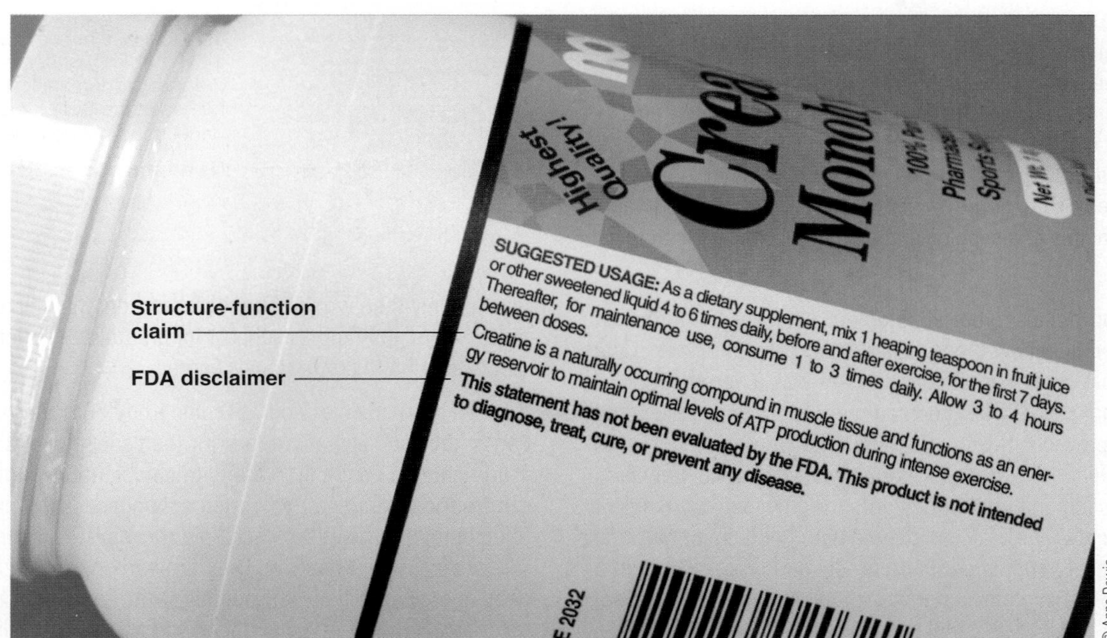

Structure-function claims do not need FDA authorization, but they must be accompanied by a disclaimer.

Cost

When shopping for supplements, remember that local or store brands may be just as good as nationally advertised brands. If they are less expensive, it may be because the price does not have to cover the cost of national advertising.

Regulation of Supplements

Dietary supplements are regulated by the **FDA (Food and Drug Administration)** as foods. Details of supplement regulation are defined in the Dietary Supplement Health and Education Act of 1994, which was intended to enable consumers to make informed choices about nutrient supplements. The act subjects supplements to the same general labeling requirements that apply to foods. Specifically:

- Nutrition labeling for dietary supplements is required.
- Labels may make nutrient claims (as "high" or "low") according to specific criteria (for example, "an excellent source of vitamin C").
- Labels may claim that the lack of a nutrient can cause a deficiency disease, but if they do, they must also include the prevalence of that deficiency disease in the United States.
- Labels may make health claims that are supported by significant scientific agreement and are not brand specific (for example, "folate protects against neural tube defects").
- Labels may claim to diagnose, treat, cure, or relieve common complaints such as menstrual cramps or memory loss, but may *not* make claims about specific diseases (except as noted previously).
- Labels may make structure-function claims about the role a nutrient plays in the body, how the nutrient performs its function, and how consuming the nutrient is associated with general well-being. The manufacturer is responsible for ensuring that the claims are truthful and not misleading. Claims must be accompanied by an FDA disclaimer statement: "This statement has not been evaluated by the Food and Drug Administration. This product is not intended to diagnose, treat, cure, or prevent any disease." Figure H10-1 provides an example of a supplement label that complies with the requirements.

The multibillion-dollar-a-year supplement industry spends much money and effort influencing these regulations. The net effect of the Dietary Supplement Health and Education Act was a deregulation of the supplement industry. Unlike food additives or drugs, supplements do not need to be proved safe and effective, nor do they need the FDA's approval before being marketed. Furthermore, there are no standards for potency or dosage and no requirements for providing warnings of potential side effects. The FDA can only require good manufacturing practices: that dietary supplements be produced and packaged in a quality manner, do not contain contaminants or impurities, and are accurately labeled to reflect the actual contents.

Should a problem arise, the burden falls to the FDA to prove that the supplement poses a "significant or unreasonable risk of illness or injury." Only then would it be removed from the market. When asked, most Americans express support for greater regulation of dietary supplements. Health professionals agree.

> FIGURE H10-1 **An Example of a Supplement Label**

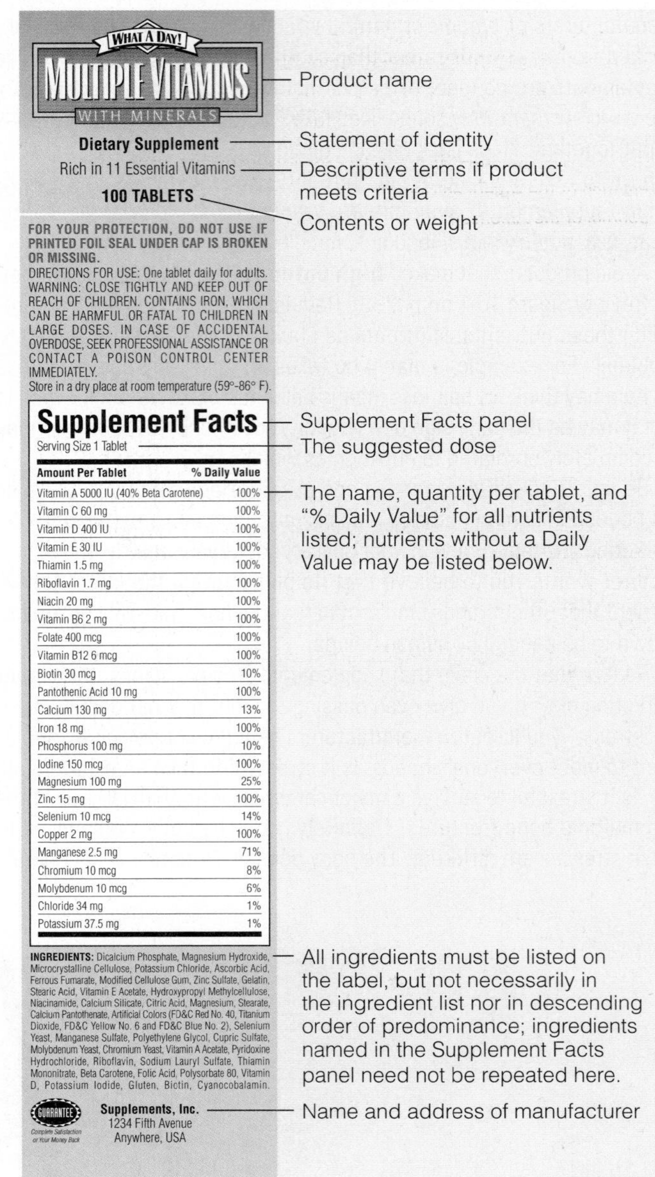

If all the nutrients we need can come from food, why not just eat food? Foods have so much more to offer than supplements do. Nutrients in foods come in an infinite variety of combinations with a multitude of different carriers and absorption enhancers. They come with water, fiber, and an array of beneficial phytochemicals. Foods stimulate the GI tract to keep it healthy. They provide energy, and as long as you need energy each day, why not have nutritious foods deliver it? Foods offer pleasure, satiety, and opportunities for socializing while eating. Quite simply, foods meet human health needs far better than dietary supplements.[10] For further proof, read Highlight 11.

REFERENCES

1. J. Gahche and coauthors, Dietary supplement use among US adults has increased since NHANES III (1988–1994), *National Center for Health Statistics: Data Brief* 61 (2011): 1–8; Capitol health call: Dietary supplements, *Journal of the American Medical Association* 301 (2009): 1427.
2. D. B. McCormick, Vitamin/mineral supplements: Of questionable benefit for the general population, *Nutrition Reviews* 68 (2010): 207–213.
3. Position of the American Dietetic Association: Nutrient supplementation, *Journal of the American Dietetic Association* 109 (2009): 2073–2085.
4. Position of the American Dietetic Association, 2009.
5. M. G. O'Doherty and coauthors, Effect of supplementation with B vitamins and antioxidants on levels of asymmetric dimethylarginine (ADMA) and C-reactive protein (CRP): A double-blind, randomized, factorial design, placebo-controlled trial, *European Journal of Nutrition* 49 (2010): 483–492; G. J. Hankey and VITATOPS Trial Study Group, B vitamins in patients with recent transient ischaemic attack or stroke in the VITAmins TO Prevent Stroke (VITATOPS) trial: A randomized, double-blind, parallel, placebo-controlled trial, *The Lancet Neurology* 9 (2010): 855–865; S. Czernichow and coauthors, Effects of long-term antioxidant supplementation and association of serum antioxidant concentrations with risk of metabolic syndrome in adults, *American Journal of Clinical Nutrition* 90 (2009): 329–335; A. M. Hill, J. A. Fleming, and P. M. Kris-Etherton, The role of diet and nutritional supplements in preventing and treating cardiovascular disease, *Current Opinion in Cardiology* 24 (2009): 433–441; M. L. Neuhouser and coauthors, Multivitamin use and risk of cancer and cardiovascular disease in the Women's Health Initiative Cohorts, *Archives of Internal Medicine* 169 (2009): 294–304; G. Pocobelli and coauthors, Use of supplements of multivitamins, vitamin C, and vitamin E in relation to mortality, *American Journal of Epidemiology* 170 (2009): 472–483; H. D. Sesso and coauthors, Vitamins E and C in the prevention of cardiovascular disease in men, *Journal of the American Medical Association* 300 (2008): 2123–2133.
6. G. Bjelakovic and C. Gluud, Vitamin and mineral supplement use in relation to all-cause mortality in the Iowa Women's Health Study, *Archives of Internal Medicine* 171 (2011): 1633–1634.
7. G. Bjelakovic and coauthors, Antioxidant supplements for prevention of mortality in healthy participants and patients with various diseases, *Cochrane Database of Systematic Reviews* 16 (2008): CD007176; J. A. Satia and coauthors, Long-term use of beta-carotene, retinol, lycopene, and lutein supplements and lung cancer risk: Results from the VITamins And Lifestyle (VITAL) Study, *American Journal of Epidemiology* 169 (2009): 1409.
8. E. A. Klein and coauthors, Vitamin E and the risk of prostate cancer: The Selenium and Vitamin E Cancer Prevention Trial (SELECT), *Journal of the American Medical Association* 306 (2011): 1549–1556.
9. FDA opens dialogue on "nano" regulation, www.fda.gov/forconsumers/consumerupdates/ucm258462.html; B. E. Erickson, Dietary supplements made with little government oversight, accessed February 2009, http://pubs.acs.org/cen/government/87/8706gov3.html.
10. D. R. Jacobs Jr, M. D. Gross, and L. C. Tapsell, Food synergy: An operational concept for understanding nutrition, *American Journal of Clinical Nutrition* 89 (2009): 1543S–1548S.

CENGAGE LEARNING'S GLOBAL NUTRITION WATCH — GLOBAL NUTRITION RESOURCE CENTER

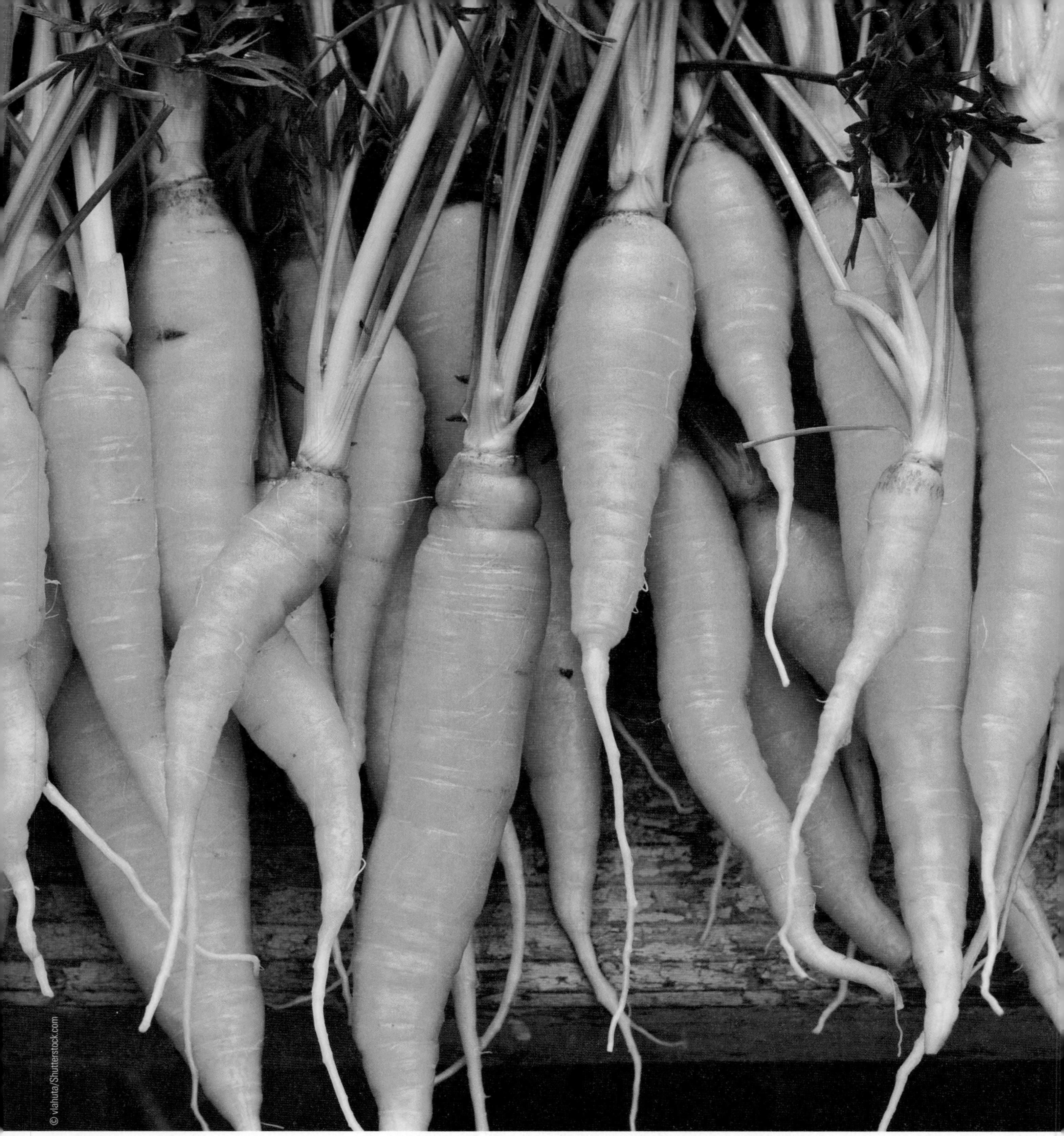

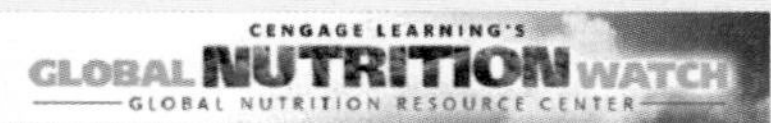

The fat-soluble vitamins are intricately bound through shared receptors. Learn more about their relationships and how scientists continue to debate appropriate recommendations for each by going to Nutrition Basics and Tools → Vitamins, Minerals, and Water.

11

The Fat-Soluble Vitamins: A, D, E, and K

LEARNING GPS

Nutrition in Your Life

Realizing that vitamin A from vegetables participates in vision, a mom encourages her children to "eat your carrots" because "they're good for your eyes." A dad takes his children outside to "enjoy the fresh air and sunshine" because they need the vitamin D that is made with the help of the sun. A physician recommends that a patient use vitamin E to slow the progression of heart disease. Another physician gives a newborn a dose of vitamin K to protect against life-threatening blood loss. These common daily occurrences highlight some of the heroic work of the fat-soluble vitamins. In the Nutrition Portfolio at the end of this chapter, you can determine whether the foods you are eating are meeting your fat-soluble vitamin needs.

The fat-soluble vitamins A, D, E, and K differ from the water-soluble vitamins in several significant ways (review Table 10-2 on p. 300). Being insoluble in the watery juices of the GI tract, the fat-soluble vitamins require bile for their digestion and absorption. Upon absorption, fat-soluble vitamins travel through the lymphatic system within chylomicrons before entering the bloodstream, where many of them require protein carriers for transport. The fat-soluble vitamins participate in numerous activities throughout the body, but excesses are stored primarily in the liver and adipose tissue. The body maintains blood concentrations by retrieving these vitamins from storage as needed; thus people can eat less than their daily need for days, weeks, or even months or years without ill effects. They need only ensure that, over time, *average* daily intakes approximate recommendations. By the same token, because fat-soluble vitamins are not readily excreted, the risk of toxicity is greater than it is for the water-soluble vitamins.

11.1 Vitamin A and Beta-Carotene

LEARN IT Identify the main roles, deficiency symptoms, and food sources for vitamin A.

Vitamin A was the first fat-soluble vitamin to be recognized. More than a century later, vitamin A and its precursor, **beta-carotene,** continue to intrigue researchers with their diverse roles and profound effects on health.

Three different forms of vitamin A are active in the body: **retinol, retinal,** and **retinoic acid.** Collectively, these compounds are known as **retinoids.** Foods derived from animals provide compounds (retinyl esters) that are readily digested and absorbed as retinol in the intestine. Foods derived from plants provide **carotenoids,** some of which can be converted to vitamin A.* The most studied of the carotenoids with **vitamin A activity** is beta-carotene, which can be split to form retinol in the intestine and liver.[2] Beta-carotene's absorption and conversion are significantly less efficient than those of the retinoids. Figure 11-1 illustrates the structural similarities and differences of these vitamin A compounds and the cleavage of beta-carotene.

The cells can convert retinol and retinal to the other active forms of vitamin A as needed. The conversion of retinol to retinal is reversible, but the further conversion of retinal to retinoic acid is irreversible (see Figure 11-2). This irreversibility is significant because each form of vitamin A performs a function that the others cannot.

Several proteins participate in the digestion and absorption of vitamin A. After absorption via the lymph system, vitamin A eventually arrives at the liver, where it is stored. There, a special transport protein, **retinol-binding protein (RBP),** picks up vitamin A from the liver and carries it in the blood. Cells that use vitamin A have special protein receptors for it, and its action within each cell may differ depending on the receptor. For example, retinoic acid can stimulate cell growth in the skin and inhibit cell growth in tumors.[3]

Roles in the Body

Vitamin A is a versatile vitamin, known to regulate the expression of several hundred genes. Its major roles include:

- Promoting vision
- Participating in protein synthesis and cell differentiation, thereby maintaining the health of epithelial tissues and skin
- Supporting reproduction and regulating growth

As mentioned, each form of vitamin A performs specific tasks. Retinol supports reproduction and is the major transport and storage form of the vitamin.

vitamin A: all naturally occurring compounds with the biological activity of *retinol* (RET-ih-nol), the alcohol form of vitamin A.

beta-carotene (BAY-tah KARE-oh-teen): one of the carotenoids; an orange pigment and vitamin A precursor found in plants.

retinol (RET-ih-nol): the alcohol form of vitamin A.

retinal (RET-ih-nal): the aldehyde form of vitamin A.

retinoic (RET-ih-NO-ick) **acid:** the acid form of vitamin A.

retinoids (RET-ih-noyds): chemically related compounds with biological activity similar to that of retinol; metabolites of retinol.

carotenoids (kah-ROT-eh-noyds): pigments commonly found in plants and animals, some of which have vitamin A activity. The carotenoid with the greatest vitamin A activity is beta-carotene.

vitamin A activity: a term referring to both the active forms of vitamin A and the precursor forms in foods without distinguishing between them.

retinol-binding protein (RBP): the specific protein responsible for transporting retinol.

*Carotenoids with vitamin A activity include alpha-carotene, beta-carotene, and beta-cryptoxanthin; carotenoids with no vitamin A activity include lycopene, lutein, and zeaxanthin.

> **FIGURE 11-1 Forms of Vitamin A**

In this diagram, corners represent carbon atoms, as in all previous diagrams in this book. A further simplification here is that methyl groups (CH_3) are understood to be at the ends of the lines extending from corners. (See Appendix C for complete structures.)

Retinol, the alcohol form

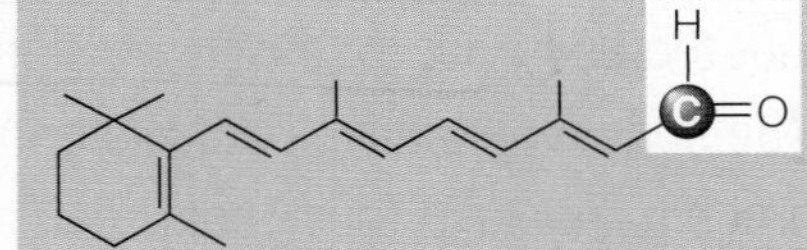

Retinal, the aldehyde form

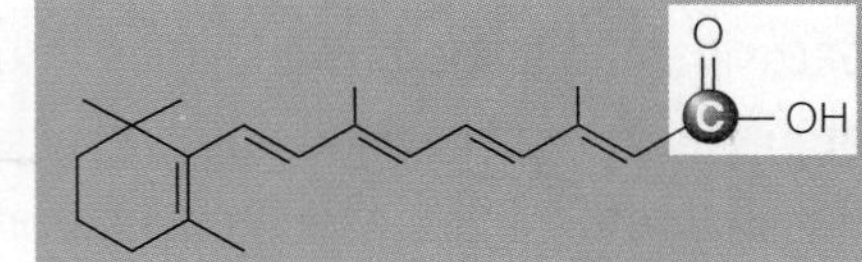

Retinoic acid, the acid form

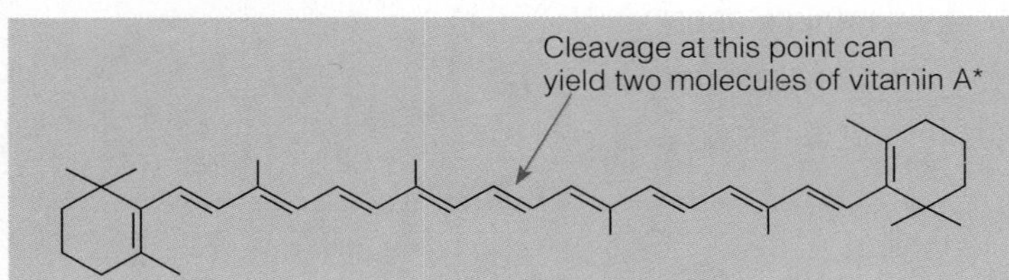
Cleavage at this point can yield two molecules of vitamin A*

Beta-carotene, a precursor

*Sometimes cleavage occurs at other points as well, so that one molecule of beta-carotene may yield only one molecule of vitamin A. Furthermore, not all beta-carotene is converted to vitamin A, and absorption of beta-carotene is not as efficient as that of vitamin A. For these reasons, 12 μg of beta-carotene are equivalent to 1 μg of vitamin A. Conversion of other carotenoids to vitamin A is even less efficient.

> **FIGURE 11-2 Conversion of Vitamin A Compounds**

Notice that the conversion from retinol to retinal is reversible, whereas the pathway from retinal to retinoic acid is not.

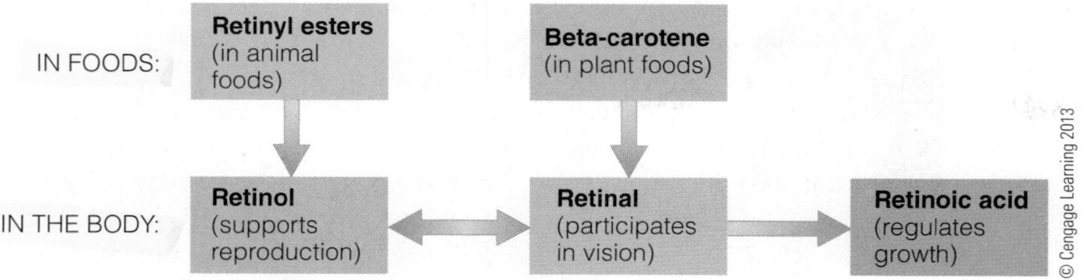

Retinal is active in vision and is also an intermediate in the conversion of retinol to retinoic acid (review Figure 11-2). Retinoic acid acts like a hormone, regulating cell differentiation, growth, and embryonic development. Animals raised on retinoic acid as their sole source of vitamin A can grow normally, but they become blind because retinoic acid cannot be converted to retinal (review Figure 11-2).

Vitamin A in Vision Vitamin A plays two indispensable roles in the eye: it helps maintain a crystal-clear outer window, the **cornea,** and it participates in the conversion of light energy into nerve impulses at the **retina** (see Figure 11-3 for details). Some of the photosensitive cells of the retina contain **pigment** molecules called **rhodopsin;** each rhodopsin molecule is composed of a protein called **opsin** bonded to a molecule of retinal. When light passes through the cornea of the eye and strikes the retina, rhodopsin responds by changing shape and becoming bleached. As it does, the retinal shifts from a *cis* to a *trans* configuration, just as fatty acids do during hydrogenation (see pp. 134–135). The bleached *trans*-retinal cannot remain bonded to opsin. When retinal is released, opsin changes shape, thereby disturbing the membrane of the cell and generating an electrical impulse that travels along the cell's length. At the other end of the cell, the impulse is transmitted to a nerve cell, which conveys the message to the brain. Much of the retinal is then converted back to its active *cis* form and combined with the opsin protein to regenerate the pigment rhodopsin. Some retinal, however, may be oxidized to retinoic acid, a biochemical dead end for the visual process. Visual activity leads to repeated small losses of retinal, necessitating its constant replenishment either directly from foods or indirectly from retinol stores.

Vitamin A in Protein Synthesis and Cell Differentiation Despite its important role in vision, only one-thousandth of the body's vitamin A is in the retina. Much more is in the cells lining the body's surfaces. There, the vitamin participates in

cornea (KOR-nee-uh): the transparent membrane covering the outside of the eye.

retina (RET-in-uh): the innermost membrane of the eye, composed of several layers including one that contains the rods and cones.

pigment: a molecule capable of absorbing certain wavelengths of light so that it reflects only those that we perceive as a certain color.

rhodopsin (ro-DOP-sin): a light-sensitive pigment of the retina that contains the retinal form of vitamin A and the protein opsin.

- **rhod** = red (pigment)
- **opsin** = visual protein

opsin (OP-sin): the protein portion of visual pigment molecules.

> **FIGURE 11-3** ***Animated*** **Vitamin A's Role in Vision**

More than 100 million photosensitive cells reside in the retina, and each contains about 30 million molecules of vitamin A-containing visual pigments. The rods contain the rhodopsin pigment and respond to faint light; the cones contain the iodopsin pigment and function in color vision.

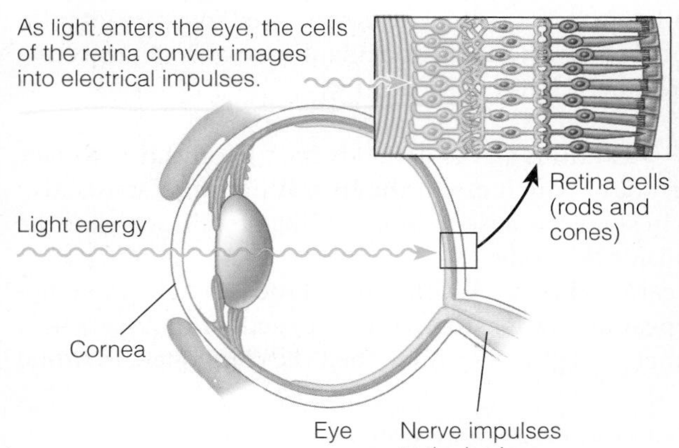

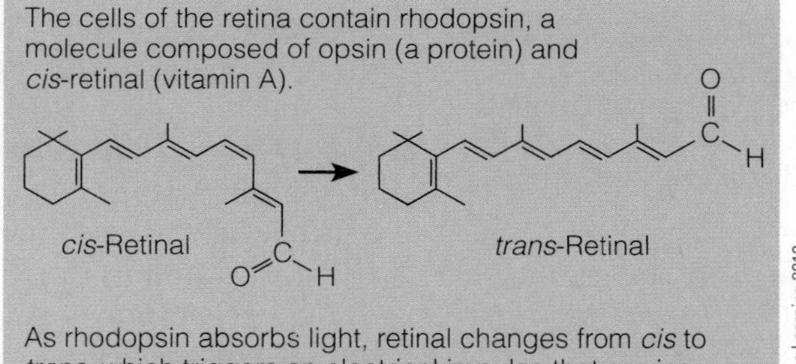

The cells of the retina contain rhodopsin, a molecule composed of opsin (a protein) and cis-retinal (vitamin A).
cis-Retinal
trans-Retinal
As rhodopsin absorbs light, retinal changes from cis to trans, which triggers an electrical impulse that carries visual information to the brain through the optic nerve.
© Cengage Learning 2013

> FIGURE 11-4 **Mucous Membrane Integrity**

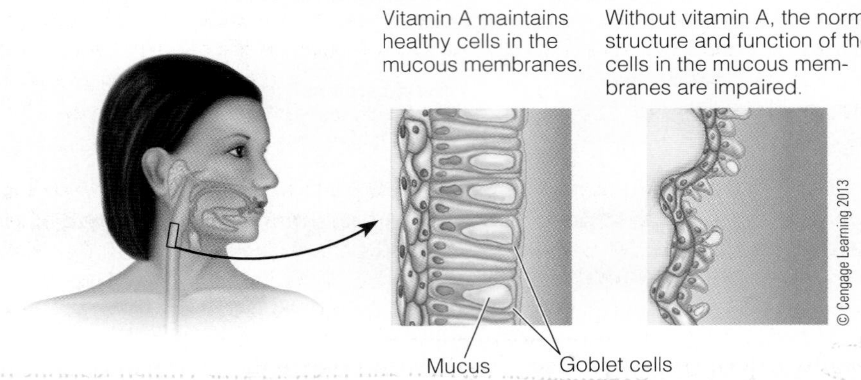

Vitamin A maintains healthy cells in the mucous membranes.
Without vitamin A, the normal structure and function of the cells in the mucous membranes are impaired.
Mucus
Goblet cells
© Cengage Learning 2013

protein synthesis and **cell differentiation,** a process by which each type of cell develops to perform a specific function.

All body surfaces, both inside and out, are covered by layers of cells known as **epithelial cells.** The **epithelial tissue** on the outside of the body is, of course, the skin—and vitamin A helps to protect against skin damage from sunlight. The epithelial tissues that line the inside of the body are the **mucous membranes:** the linings of the mouth, stomach, and intestines; the linings of the lungs and the passages leading to them; the linings of the urinary bladder and urethra; the linings of the uterus and vagina; and the linings of the eyelids and sinus passageways. Within the body, the mucous membranes of the GI tract alone line an area larger than a quarter of a football field, and vitamin A helps to maintain their integrity (see Figure 11-4).

Vitamin A promotes differentiation of epithelial cells and goblet cells, one-celled glands that synthesize and secrete mucus. Mucus coats and protects the epithelial cells from invasive microorganisms and other potentially damaging substances, such as gastric juices.

Vitamin A in Reproduction and Growth As mentioned, vitamin A also supports reproduction and regulates growth.[4] In men, retinol participates in sperm development, and in women, vitamin A supports normal fetal development during pregnancy. Children lacking vitamin A fail to grow; given vitamin A supplements, these children gain weight and grow taller.

The growth of bones illustrates that growth is a complex phenomenon of **remodeling.** To convert a small bone into a large bone, the bone-remodeling cells must "undo" some parts of the bone as they go, and vitamin A participates in the dismantling.* The cells that break down bone contain sacs of degradative enzymes.** With the help of vitamin A, these enzymes destroy selected sites in the bone, removing the parts that are not needed.

Beta-Carotene as an Antioxidant In the body, beta-carotene serves primarily as a vitamin A precursor.[5] Not all dietary beta-carotene is converted to active vitamin A, however. Some beta-carotene may act as an antioxidant ♦ capable of protecting the body against disease. (Highlight 11 provides details.)

Vitamin A Deficiency

Vitamin A status depends mostly on the adequacy of vitamin A stores, 90 percent of which are in the liver. Vitamin A status also depends on a person's protein status because retinol-binding protein serves as the vitamin's transport carrier inside the body.

If a person were to stop eating vitamin A–containing foods, deficiency symptoms would not begin to appear until after stores were depleted—1 to 2 years for a healthy adult but much sooner for a growing child. Then the consequences would

♦ Key antioxidant nutrients:
- Vitamin C, vitamin E, beta-carotene
- Selenium

cell differentiation (DIF-er-EN-she-AY-shun): the process by which immature cells develop specific functions different from those of the original that are characteristic of their mature cell type.

epithelial (ep-i-THEE-lee-ul) **cells:** cells on the surface of the skin and mucous membranes.

epithelial tissue: the layer of the body that serves as a selective barrier between the body's interior and the environment. Examples are the cornea of the eyes, the skin, the respiratory lining of the lungs, and the lining of the digestive tract.

mucous (MYOO-kus) **membranes:** the membranes, composed of mucus-secreting cells, that line the surfaces of body tissues.

remodeling: the dismantling and re-formation of a structure.

*The cells that destroy bone during growth are *osteoclasts;* those that build bone are *osteoblasts.*

**The sacs of degradative enzymes are *lysosomes* (LYE-so-zomes).

be profound and severe.[6] Vitamin A deficiency is uncommon in the United States, but it is a major nutrition problem in many developing countries. An estimated 250 million children worldwide have some degree of vitamin A deficiency and thus are vulnerable to infectious diseases and blindness. About 1 to 2 percent of them become blind every year, half of them dying within a year of losing their sight. Routine vitamin A supplementation and food fortification can be a life-saving intervention.[7]

Infectious Diseases Vitamin A supports immune function and inhibits replication of the measles virus.[8] In developing countries around the world, measles is a devastating infectious disease, killing 450 children each day.[9] The severity of the illness often correlates with the degree of vitamin A deficiency; deaths are usually due to related infections such as pneumonia and severe diarrhea. Providing large doses of vitamin A reduces the risk of dying from these infections by half.

The World Health Organization (WHO) and UNICEF (the United Nations International Children's Emergency Fund) have made the control of vitamin A deficiency a major goal in their quest to improve child health and survival throughout the developing world. They recommend two doses of vitamin A supplements, given 24 hours apart, for all children with measles. In the United States, the American Academy of Pediatrics recommends vitamin A supplements for certain groups of measles-infected infants and children. Vitamin A supplements also protect against blindness and the complications of other life-threatening infections, including malaria, lung diseases, and HIV (human immunodeficiency virus, the virus that causes AIDS).

Night Blindness **Night blindness** is one of the first detectable signs of vitamin A deficiency and permits early diagnosis. In night blindness, the retina does not receive enough retinal to regenerate the visual pigments bleached by light. The person loses the ability to recover promptly from the temporary blinding that follows a flash of bright light at night or to see after dark. In many parts of the world, after the sun goes down, vitamin A–deficient people become night-blind. They often cling to others or sit still, afraid that they may trip and fall or lose their way if they try to walk alone. Figure 11-5 shows the eyes' slow recovery in response to a flash of bright light in night blindness.

Blindness (Xerophthalmia) Beyond night blindness is total blindness—failure to see at all. Night blindness is caused by a lack of vitamin A at the back of the eye, the retina; total blindness is caused by a lack of vitamin A at the front of the eye, the cornea. Severe vitamin A deficiency is the leading cause of preventable blindness in the world, causing half a million preschool children to lose their sight each year. Blindness due to vitamin A deficiency, known as **xerophthalmia,** develops in stages. At first, the cornea becomes dry and hard because of inadequate mucous

night blindness: slow recovery of vision after flashes of bright light at night or an inability to see in dim light; an early symptom of vitamin A deficiency.

xerophthalmia (zer-off-THAL-mee-uh): progressive blindness caused by inadequate mucus production due to severe vitamin A deficiency.

- **xero** = dry
- **ophthalm** = eye

> FIGURE 11-5 Vitamin A–Deficiency Symptom—Night Blindness

These photographs illustrate the eyes' slow recovery in response to a flash of bright light at night. In animal research studies, the response rate is measured with electrodes.

In dim light, you can make out the details in this room. You are using your rods for vision.

A flash of bright light momentarily blinds you as the pigment in the rods is bleached.

You quickly recover and can see the details again in a few seconds.

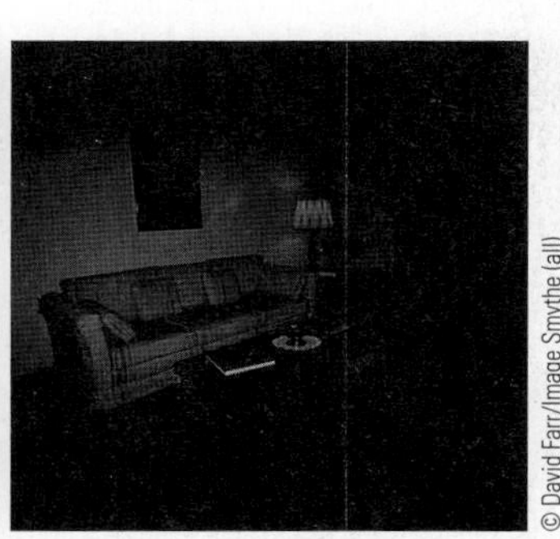

With inadequate vitamin A, you do not recover but remain blinded for many seconds.

production—a condition known as **xerosis.** Then xerosis quickly progresses to **keratomalacia,** the softening of the cornea that leads to irreversible blindness.

Keratinization Elsewhere in the body, vitamin A deficiency affects other surfaces. On the body's outer surface, the epithelial cells change shape and begin to secrete the protein **keratin**—the hard, inflexible protein of hair and nails. As Figure 11-6 shows, the skin becomes dry, rough, and scaly as lumps of keratin accumulate **(keratinization).** Without vitamin A, the goblet cells in the GI tract diminish in number and activity, limiting the secretion of mucus. With less mucus, normal digestion and absorption of nutrients falter, and this, in turn, worsens malnutrition by limiting the absorption of whatever nutrients the diet may deliver. Similar changes in the cells of other epithelial tissues weaken defenses, making infections of the respiratory tract, the GI tract, the urinary tract, the vagina, and inner ear likely.

> FIGURE 11-6 Vitamin A–Deficiency Symptom—The Rough Skin of Keratinization

© H. Sanstead/ U. of Texas/Galveston

In vitamin A deficiency, the epithelial cells secrete the protein keratin in a process known as *keratinization.* (Keratinization doesn't occur in the GI tract, but mucus-producing cells dwindle and mucus production declines.) The extreme of this condition is *hyperkeratinization* or *hyperkeratosis.* When keratin accumulates around hair follicles, the condition is known as *follicular hyperkeratosis.*

Vitamin A Toxicity

Just as a deficiency of vitamin A affects all body systems, so does a toxicity. Symptoms of toxicity begin to develop when all the binding proteins are loaded, and vitamin A is free to damage cells. Such effects are unlikely when a person depends on a balanced diet for nutrients, but toxicity is a real possibility when concentrated amounts of **preformed vitamin A** in foods derived from animals, fortified foods, or supplements is consumed.[10] Children are most vulnerable to toxicity because they need less vitamin A and are more sensitive to overdoses. An Upper Level (UL) has been set for preformed vitamin A (see inside front cover). Even multivitamin supplements ♦ provide more vitamin A than most people need.

♦ Multivitamin supplements typically provide:
- 750 µg (2500 IU)
- 1500 µg (5000 IU)

For perspective, the RDA for vitamin A is 700 mg for women and 900 µg for men.

Beta-carotene, which is found in a wide variety of fruits and vegetables, is not converted efficiently enough in the body to cause vitamin A toxicity; instead, it is stored in the fat just under the skin. Although overconsumption of beta-carotene from foods may turn the skin yellow, this is not harmful (see Figure 11-7). In contrast, overconsumption of beta-carotene from supplements may be quite harmful. In excess, this antioxidant may act as a prooxidant (as Highlight 11 explains). Adverse effects of beta-carotene supplements are most evident in people who drink alcohol and smoke cigarettes.

Bone Defects Excessive intake of vitamin A over the years may weaken the bones and contribute to fractures and osteoporosis.[11] Vitamin A suppresses bone-building activity, stimulates bone-dismantling activity, and interferes with vitamin D's ability to maintain normal blood calcium.

Birth Defects Excessive vitamin A during pregnancy leads to abnormal cell death in the spinal cord, which increases the risk of birth defects.[12] In such cases, vitamin A is considered a **teratogen.** High intakes (10,000 IU ♦ of supplemental vitamin A daily) before the seventh week of pregnancy appear to be the most damaging. For this reason, vitamin A is not given as a supplement in the first trimester of pregnancy without specific evidence of deficiency, which is rare.

♦ For perspective, 10,000 IU ≈ 3000 µg vitamin A, roughly four times the RDA for women.

Not for Acne Adolescents need to know that massive doses of vitamin A have no beneficial effect on **acne.** The prescription medicine Accutane is made from vitamin A but is chemically different.* Taken orally, Accutane is effective against the deep lesions of cystic acne. It is highly toxic, however, especially during growth, and has caused birth defects in infants when women have taken it during their pregnancies. For this reason, women taking Accutane must agree to pregnancy testing and to using contraception from at least 1 month before taking the drug through at least 1 month after discontinuing its use.[13] Should they become pregnant, they need to stop taking Accutane immediately and notify their physician.

Another vitamin A relative, Retin-A, fights acne, the wrinkles of aging, and other skin disorders.** Applied topically, this ointment smooths and softens skin; it also lightens skin that has become darkly pigmented after inflammation. During treatment, the skin becomes red and tender and peels.

xerosis (zee-ROW-sis): abnormal drying of the skin and mucous membranes; a sign of vitamin A deficiency.

keratomalacia (KARE-ah-toe-ma-LAY-shuh): softening of the cornea that leads to irreversible blindness; a sign of severe vitamin A deficiency.

keratin (KARE-uh-tin): a water-insoluble protein; the normal protein of hair and nails.

keratinization: accumulation of keratin in a tissue; a sign of vitamin A deficiency.

preformed vitamin A: dietary vitamin A in its active form.

teratogen (ter-AT-oh-jen): a substance that causes abnormal fetal development and birth defects.

acne: a chronic inflammation of the skin's follicles and oil-producing glands, which leads to an accumulation of oils inside the ducts that surround hairs; usually associated with the maturation of young adults.

*The generic name for Accutane is *isotretinoin.*

**The generic name for Retin-A is *tretinoin topical.*

>How To

Convert International Units (IU) to Weight Measurements

Supplement labels often list the amount of fat-soluble vitamins in International Units (IU), a universally accepted measure of a vitamin's biological effect. Such a measure allows scientists to compare the potency of substances and was most useful decades ago before chemicals could be purified and weighed accurately.

Because IU measures biological activity and not weight, the conversion factors differ for each fat-soluble vitamin:

For vitamin A (retinol):

- 1 IU = 0.3 µg
- 1 µg = 3.33 IU

For vitamin D (cholecalciferol):

- 1 IU = 0.025 µg
- 1 µg = 40 IU

For vitamin E (natural α-tocopherol):

- 1 IU = 0.67 mg
- 1 mg = 1.49 IU

To convert from IU to a weight measurement, multiply IU by the appropriate equivalent. For example, for a supplement listing 5000 IU vitamin A, 400 IU vitamin D, and 30 IU vitamin E:

5000 IU X 0.3 µg/IU = 1500 µg retinol

400 IU X 0.025 µg/IU = 10 µg cholecalciferol

30 IU X 0.67 mg/IU = 20 mg α-tocopherol

TRY IT Convert these values on a supplement label from IU to weight measurements: 4000 IU vitamin A, 600 IU vitamin D, and 12 IU vitamin E.

> FIGURE 11-7 **Symptom of Beta-Carotene Excess—Discoloration of the Skin**

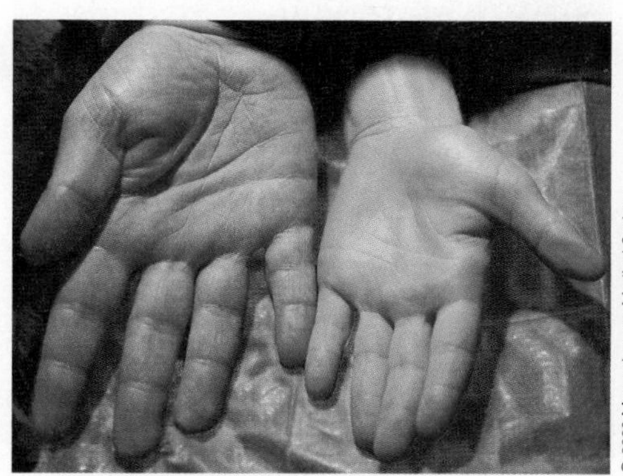

The hand on the right shows the skin discoloration that occurs when blood levels of beta-carotene rise in response to a low-kcalorie diet that features carrots, pumpkins, and orange juice. (The hand on the left belongs to someone else and is shown here for comparison.)

Vitamin A Recommendations Because the body can derive vitamin A from various retinoids and carotenoids, its content in foods and its recommendations are expressed as **retinol activity equivalents (RAE).** One microgram of retinol counts as 1 RAE, ♦ as does 12 micrograms of dietary beta-carotene. Most food and supplement labels report their vitamin A contents using International Units (IU), ♦ a measure of vitamin activity used before direct chemical analysis was possible. The accompanying "How To" feature explains how to convert IU to a weight measurement.

♦ 1 mg RAE
= 1 µg retinol
= 2 µg beta-carotene (supplement)
= 12 µg beta-carotene (dietary)
= 24 µg other vitamin A precursor carotenoids

♦ 1 IU retinol = 0.3 µg retinol or 0.3 µg RAE
1 IU beta-carotene (supplement) = 0.5 IU retinol or 0.15 µg RAE
1 IU beta-carotene (dietary) = 0.165 IU retinol or 0.05 µg RAE
1 IU other vitamin A precursor carotenoids = 0.025 µg RAE

Vitamin A in Foods The richest sources of the retinoids are foods derived from animals—liver, fish liver oils, milk and milk products, butter, and eggs. Because vitamin A is fat soluble, it is lost when milk is skimmed. To compensate, reduced-fat, low-fat, and fat-free milks are often fortified so as to supply 6 to 10 percent of the Daily Value per cup.* Margarine is usually fortified to provide the same amount of vitamin A as butter.

Plants contain no retinoids, but many vegetables and some fruits contain vitamin A precursors—the carotenoids. Only a few carotenoids have vitamin A activity; the carotenoid with the greatest vitamin A activity is beta-carotene. Beta-carotene is a rich, deep yellow, almost orange compound. The beta-carotene in dark green, leafy vegetables is abundant, but masked by large amounts of the green pigment **chlorophyll.** Attractive meals that include colorful fruits and vegetables are likely to provide vitamin A.

The Colors of Vitamin A Foods Dark leafy greens (like spinach—not celery or cabbage) and rich yellow or deep orange vegetables and fruits (such as cantaloupe, carrots, and sweet potatoes—not corn or bananas) help people meet their vitamin A needs (see Figure 11-8 on p. 346). A diet including several servings of such carotene-rich sources helps to ensure a sufficient intake.

Bright color is not always a sign of vitamin A activity, however. Beets and corn, for example, derive their colors from the red and yellow **xanthophylls,** which have no vitamin A activity. As for white plant foods such as potatoes, cauliflower,

*Vitamin A fortification of milk in the United States is required to a level found in whole milk (1200 IU per quart), but many manufacturers commonly fortify to a higher level (2000 IU per quart). Similarly, in Canada all milk that has had fat removed must be fortified with vitamin A.

retinol activity equivalents (RAE): a measure of vitamin A activity; the amount of retinol that the body will derive from a food containing preformed retinol or its precursor, beta-carotene.

chlorophyll (KLO-row-fil): the green pigment of plants, which absorbs light and transfers the energy to other molecules, thereby initiating photosynthesis.

xanthophylls (ZAN-tho-fills): pigments found in plants responsible for the color changes seen in autumn leaves.

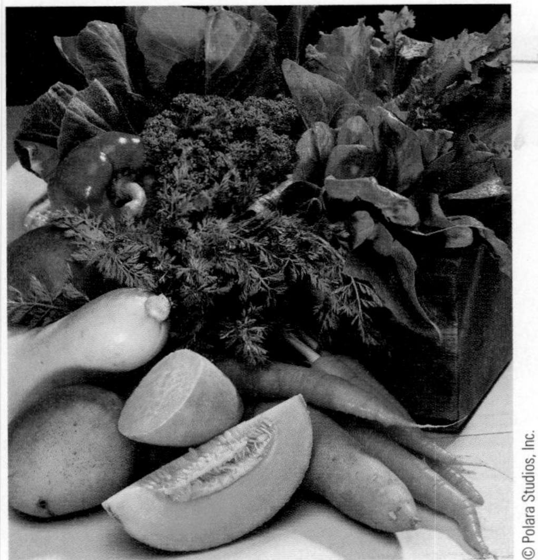

The carotenoids in foods bring colors to meals; the retinoids in our eyes allow us to see them.

pasta, and rice, they also offer little or no vitamin A. Similarly, fast foods often lack vitamin A. Anyone who dines frequently on hamburgers, french fries, and colas is wise to emphasize colorful vegetables and fruits at other meals.

Vitamin A–Rich Liver People sometimes wonder if eating liver too frequently can cause vitamin A toxicity. Liver is a rich source because vitamin A is stored in the livers of animals, just as in humans.* Arctic explorers who have eaten large quantities of polar bear liver have become ill with symptoms suggesting vitamin A toxicity. Liver offers many nutrients, and eating it periodically may improve a person's nutrition status, but caution is warranted not to eat too much too often, especially for pregnant women. With 1 ounce of beef liver providing more than three times the RDA for vitamin A, intakes can rise quickly.

REVIEW IT Identify the main roles, deficiency symptoms, and food sources for vitamin A.
Vitamin A is found in the body in three forms: retinol, retinal, and retinoic acid. Together, they are essential to vision, healthy epithelial tissues, and growth. Vitamin A deficiency is a major health problem worldwide, leading to infections, blindness, and keratinization. Toxicity can also cause problems and is most often associated with supplement abuse. Animal-derived foods such as liver and whole or fortified milk provide retinoids, whereas brightly colored plant-derived foods such as spinach, carrots, and pumpkins provide beta-carotene and other carotenoids. In addition to serving as a precursor for vitamin A, beta-carotene acts as an antioxidant in the body. The accompanying table provides a summary of vitamin A.

> FIGURE 11-8 Vitamin A in Selected Foods

See the "How To" feature on p. 302 for more information on using this figure.

Food	Serving size (kcalories)
Bread, whole wheat	1 oz slice (70 kcal)
Cornflakes, fortified	1 oz (110 kcal)
Spaghetti pasta	½ c cooked (99 kcal)
Tortilla, flour	1 10"-round (234 kcal)
Broccoli	½ c cooked (22 kcal)
Carrots	½ c shredded raw (24 kcal)
Potato	1 medium baked w/skin (133 kcal)
Tomato juice	¾ c (31 kcal)
Banana	1 medium raw (109 kcal)
Orange	1 medium raw (62 kcal)
Strawberries	½ c fresh (22 kcal)
Watermelon	1 slice (92 kcal)
Milk, fortified	1 c reduced-fat 2% (121 kcal)
Yogurt, plain	1 c low-fat (155 kcal)
Cheddar cheese	1½ oz (171 kcal)
Cottage cheese	½ c low-fat 2% (101 kcal)
Pinto beans	½ c cooked (117 kcal)
Peanut butter	2 tbs (188 kcal)
Sunflower seeds	1 oz dry (165 kcal)
Tofu (soybean curd)	½ c (76 kcal)
Ground beef, lean	3 oz broiled (244 kcal)
Chicken breast	3 oz roasted (140 kcal)
Tuna, canned in water	3 oz (99 kcal)
Egg	1 hard cooked (78 kcal)
Excellent, and sometimes unusual, sources:	
Beef liver	3 oz fried (184 kcal)
Sweet potatoes	½ c cooked (116 kcal)
Mango	1 (135 kcal)

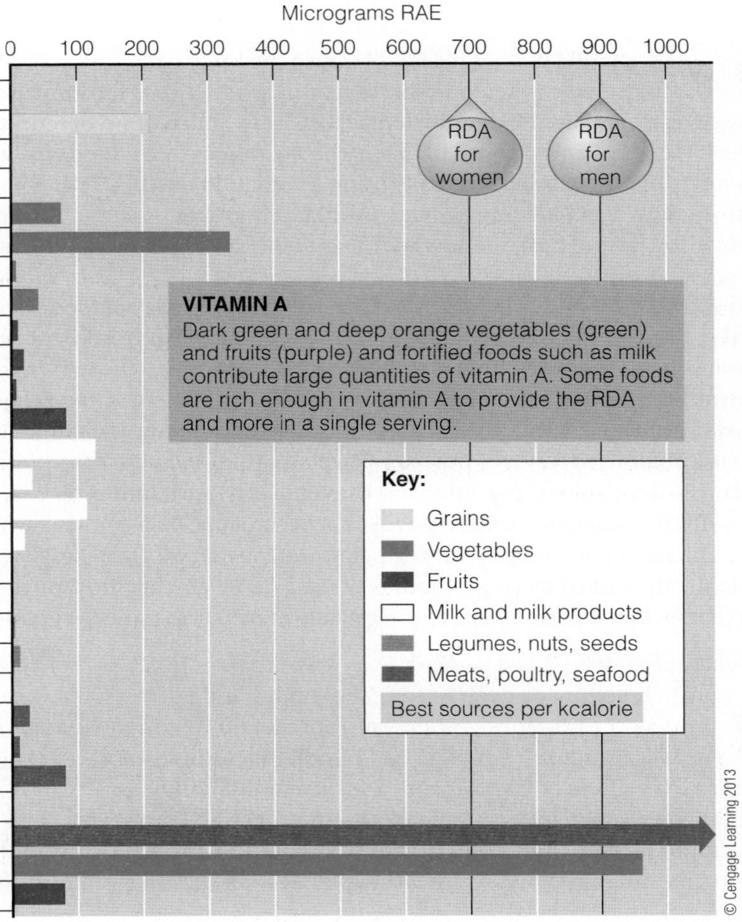

Micrograms RAE
0 100 200 300 400 500 600 700 800 900 1000

RDA for women
RDA for men

VITAMIN A
Dark green and deep orange vegetables (green) and fruits (purple) and fortified foods such as milk contribute large quantities of vitamin A. Some foods are rich enough in vitamin A to provide the RDA and more in a single serving.

Key:
Grains
Vegetables
Fruits
Milk and milk products
Legumes, nuts, seeds
Meats, poultry, seafood
Best sources per kcalorie

*The liver is not the only organ that stores vitamin A. The kidneys, adrenal glands, and other organs do, too, but the liver stores the most and is the most commonly eaten organ meat.

REVIEW IT Vitamin A (*continued*)

Other Names	Deficiency Disease
Retinol, retinal, retinoic acid; precursors are carotenoids such as beta-carotene	Hypovitaminosis A
2001 RDA	**Deficiency Symptoms**
Men: 900 μg RAE/day Women: 700 μg RAE/day	Night blindness, corneal drying (xerosis), triangular gray spots on eye (Bitot's spots), softening of the cornea (keratomalacia), and corneal degeneration and blindness (xerophthalmia); impaired immunity (infectious diseases); plugging of hair follicles with keratin, forming white lumps (hyperkeratosis)
Upper Level	**Toxicity Disease**
Adults: 3000 μg/day	Hypervitaminosis A[a]
Chief Functions in the Body	**Chronic Toxicity Symptoms**
Vision; maintenance of cornea, epithelial cells, mucous membranes, skin; bone and tooth growth; reproduction; immunity	Increased activity of osteoclasts[b] causing reduced bone density; liver abnormalities; birth defects
Significant Sources	**Acute Toxicity Symptoms**
Retinol: fortified milk, cheese, cream, butter, fortified margarine, eggs, liver Beta-carotene: spinach and other dark green, leafy vegetables, broccoli, deep orange fruits (apricots, cantaloupe) and vegetables (squash, carrots, sweet potatoes, pumpkin)	Blurred vision, nausea, vomiting, vertigo; increase of pressure inside skull, mimicking brain tumor; headaches; muscle incoordination

[a]A related condition, *hypercarotenemia,* is caused by the accumulation of too much of the vitamin A precursor beta-carotene in the blood, which turns the skin noticeably yellow. Hypercarotenemia is not, strictly speaking, a toxicity symptom.
[b]*Osteoclasts* are the cells that destroy bone during its growth. Those that build bone are *osteoblasts.*

11.2 Vitamin D

LEARN IT Identify the main roles, deficiency symptoms, and sources for vitamin D.

Vitamin D differs from the other nutrients in that the body can synthesize it, with the help of sunlight, from a precursor that the body makes from cholesterol. Therefore, vitamin D is not an essential nutrient; given enough time in the sun, people need no vitamin D from foods.

Also known as **calciferol,** vitamin D comes in two major forms. **Vitamin D_2** (or **ergocalciferol**) derives primarily from plant foods in the diet. **Vitamin D_3** (or **cholecalciferol**) derives from animal foods in the diet and from synthesis in the skin. These two forms of vitamin D are similar and both must be activated before they can fully function.

Figure 11-9 (p. 348) diagrams the pathway for making and activating vitamin D in the body. To make vitamin D, ultraviolet rays from the sun hit a precursor in the skin and convert it to previtamin D_3, which is converted to vitamin D_3 with the help of the body's heat. To activate vitamin D—whether made in the body or consumed from the diet—two hydroxylation reactions must occur. First, the liver adds an OH group, and then the kidneys add another OH group to produce the active vitamin. As you might expect, diseases affecting either the liver or the kidneys can interfere with the activation of vitamin D and produce symptoms of deficiency.

Roles in the Body Though called a vitamin, the active form of vitamin D is actually a hormone—a compound manufactured by one part of the body that travels through the blood and causes another body part to respond.[14] Like vitamin A, vitamin D has a binding protein that carries it to the target organs—most notably, the intestines, the kidneys, and the bones. All respond to vitamin D by making the minerals needed for bone growth and maintenance available.

Vitamin D in Bone Growth Vitamin D is a member of a large and cooperative bone-making and maintenance team ♦ composed of nutrients and other compounds, including vitamins A and K; the hormones parathyroid hormone and

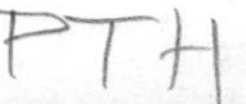

♦ Key bone nutrients:
- Vitamin D, vitamin K, vitamin A
- Calcium, phosphorus, magnesium, fluoride

calciferol (kal-SIF-er-ol): vitamin D.

vitamin D_2 or **ergocalciferol** (ER-go-kal-SIF-er-ol): vitamin D derived from plants in the diet.

vitamin D_3 or **cholecalciferol** (KO-lee-kal-SIF-er-ol): vitamin D derived from animals in the diet or made in the skin from 7-dehydrocholesterol, a precursor of cholesterol, with the help of sunlight.

> **FIGURE 11-9 Vitamin D Synthesis and Activation**

The final activation step in the kidneys is tightly regulated by hormones.

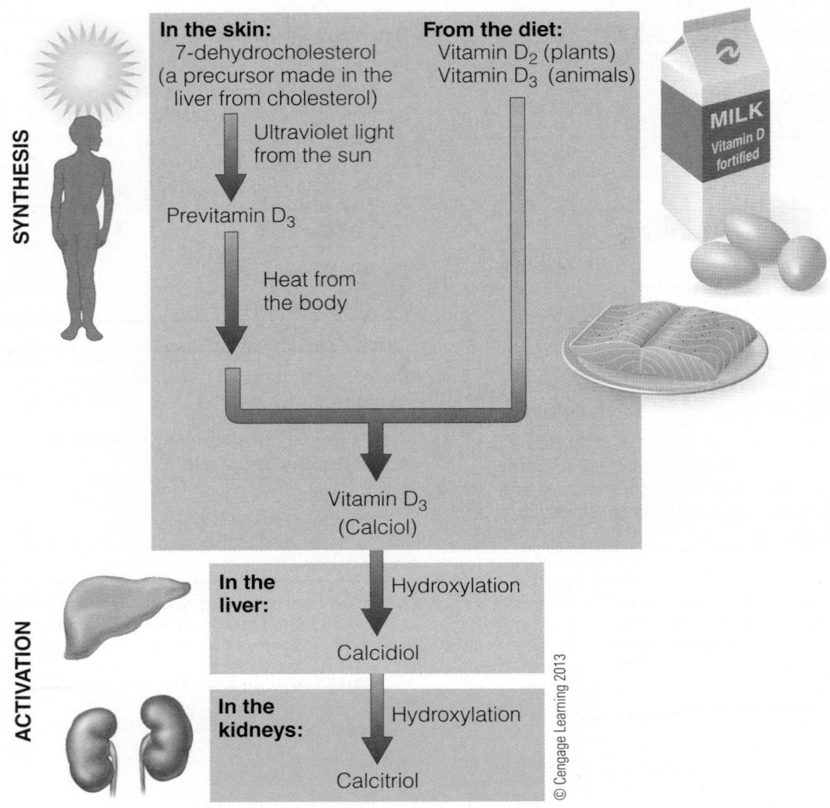

calcitonin; the protein collagen; and the minerals calcium, phosphorus, magnesium, and fluoride. Vitamin D's special role in bone health is to assist in the absorption of calcium and phosphorus, thus helping to maintain blood concentrations of these minerals.[15] The bones grow denser and stronger as they absorb and deposit these minerals. Details of calcium balance and mineral deposition appear in Chapter 12.

Vitamin D raises blood concentrations of bone minerals in three ways. When the diet is sufficient, vitamin D enhances their absorption from the GI tract. When the diet is insufficient, vitamin D provides the needed minerals from other sources: reabsorption by the kidneys and mobilization from the bones into the blood.[16] The vitamin may work alone, as it does in the GI tract, or in combination with parathyroid hormone, as it does in the bones and kidneys.[17]

Vitamin D in Other Roles Scientists have discovered many other tissues that respond to vitamin D, including cells of the immune system, brain and nervous system, pancreas, skin, muscles and cartilage, and reproductive organs. In many cases, vitamin D enhances or suppresses the activity of genes that regulate cell growth.[18] As such, it may be valuable in treating a number of diseases.[19] Recent research suggests that vitamin D may protect against tuberculosis, inflammation, multiple sclerosis, macular degeneration, hypertension, and some cancers.[20] Even so, evidence does not support vitamin D supplementation to improve health beyond correcting deficiencies.[21]

Vitamin D Deficiency

Overt signs of vitamin D deficiency are relatively rare, but vitamin D insufficiency is remarkably common.[22] Almost 10 percent of the US population is deficient and another 25 percent are marginal.[23] Factors that contribute to vitamin D deficiency include dark skin, breastfeeding without supplementation, lack of sunlight, and not using fortified milk. In vitamin D deficiency, production of **calbindin,** a protein that binds calcium in the intestinal

calbindin: a calcium-binding transport protein that requires vitamin D for its synthesis.

> **FIGURE 11-10 Vitamin D–Deficiency Symptoms—Bowed Legs and Beaded Ribs of Rickets**

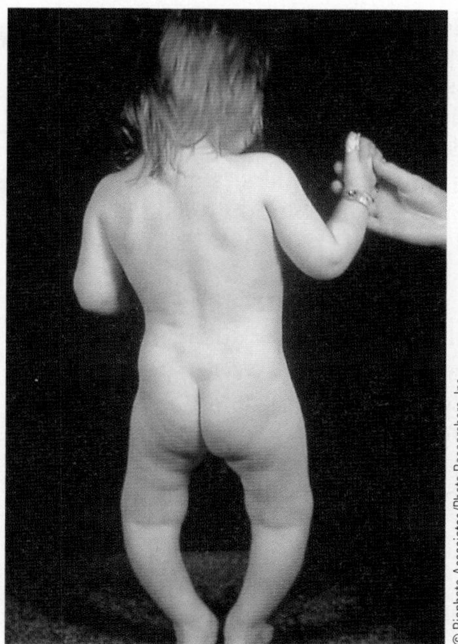

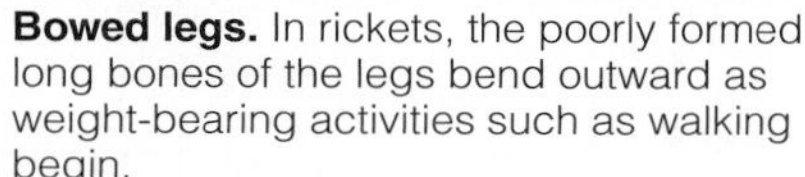

Bowed legs. In rickets, the poorly formed long bones of the legs bend outward as weight-bearing activities such as walking begin.

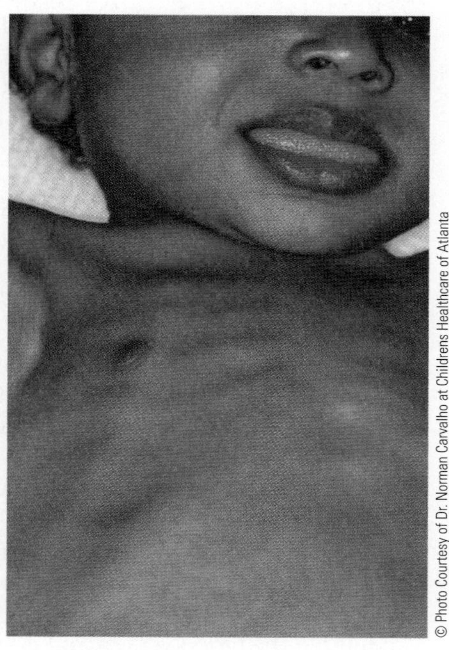

Beaded ribs. In rickets, a series of "beads" develop where the cartilages and bones attach.

cells, slows. Thus, even when calcium in the diet is adequate, it passes through the GI tract unabsorbed, leaving the bones undersupplied. Consequently, a vitamin D deficiency creates a calcium deficiency and increases the risks of several chronic diseases and osteoporosis.[24] Vitamin D–deficient adolescents do not reach their peak bone mass.[25]

Rickets Worldwide, the prevalence of the vitamin D–deficiency disease **rickets** is extremely high, affecting more than half of the children in countries such as Mongolia, Tibet, and the Netherlands.[26] In the United States, rickets is not common, but when it occurs, black children and adolescents—especially females and overweight teens—are the ones most likely to be affected.[27] To prevent rickets, the American Academy of Pediatrics recommends a supplement for all infants, children, and adolescents who do not receive enough vitamin D.[28]

In rickets, the bones fail to calcify normally, causing growth retardation and skeletal abnormalities. The bones become so weak that they bend when they have to support the body's weight (see Figure 11-10). A child with rickets who is old enough to walk characteristically develops bowed legs, often the most obvious sign of the disease. Another sign is the beaded ribs that result from the poorly formed attachments of the bones to the cartilage.*

Osteomalacia In adults, the poor mineralization of bone results in the painful bone disease **osteomalacia.** The bones become increasingly soft, flexible, brittle, and deformed.

Osteoporosis Any failure to synthesize adequate vitamin D or obtain enough from foods sets the stage for a loss of calcium from the bones, which can result in fractures.[29] Highlight 12 describes the many factors that lead to osteoporosis, a condition of reduced bone density.

---

*Because the poorly formed rib attachments resemble rosary beads, this symptom is commonly known as *rachitic* (ra-KIT-ik) *rosary* ("the rosary of rickets").

rickets: the vitamin D–deficiency disease in children characterized by inadequate mineralization of bone (manifested in bowed legs or knock-knees, outward-bowed chest, and "beads" on ribs). A rare type of rickets, not caused by vitamin D deficiency, is known as *vitamin D–refractory rickets.*

osteomalacia (OS-tee-oh-ma-LAY-shuh): a bone disease characterized by softening of the bones. Symptoms include bending of the spine and bowing of the legs. The disease occurs most often in adult women.

- **osteo** = bone
- **malacia** = softening

© Westend61 GmbH/Alamy

A cold glass of milk refreshes as it replenishes vitamin D and other bone-building nutrients.

The Elderly Vitamin D deficiency is especially likely in older adults for several reasons. For one, the skin, liver, and kidneys lose their capacity to make and activate vitamin D with advancing age. For another, older adults typically drink little or no milk—the main dietary source of vitamin D. And finally, older adults typically spend much of the day indoors, and when they do venture outside, many of them cautiously wear protective clothing or apply sunscreen to all sun-exposed areas of their skin. Dark-skinned adults living in northern regions are particularly vulnerable. All of these factors increase the likelihood of vitamin D deficiency and its consequences: bone losses and fractures. Vitamin D supplementation helps to raise blood levels, reduce bone loss, improve muscle performance, and lower the risks of falls and fractures in elderly persons.[30]

Vitamin D Toxicity

Vitamin D clearly illustrates how nutrients in optimal amounts support health, but both inadequacies and excesses create harm. Vitamin D is among the most likely of the vitamins to have toxic effects when consumed in excessive amounts. The amounts of vitamin D made by the skin and found in foods are well within the safe limits set by the UL, but supplements containing the vitamin in concentrated form should be kept out of the reach of children and used cautiously, if at all, by adults.

Excess vitamin D raises the concentration of blood calcium.[31]* Excess blood calcium tends to precipitate in the soft tissue, forming stones, especially in the kidneys where calcium is concentrated in an effort to excrete it. Calcification may also harden the blood vessels and is especially dangerous in the major arteries of the brain, heart, and lungs, where it can cause death.

Vitamin D Recommendations and Sources

◆ Vitamin D RDA for adults <71: 15 μg/day (600 IU/day)

Only a few foods contain vitamin D naturally. Fortunately, the body can make vitamin D with the help of a little sunshine. In setting dietary recommendations, however, the DRI Committee assumed that no vitamin D was available from skin synthesis. In order to reach sufficient levels of vitamin D in the blood without contributions from the sun, dietary recommendations were recently increased.[32] ◆ Some research suggests that vitamin D recommendations should be higher still.[33]

Vitamin D in Foods Most adults, especially in sunny regions, need not make special efforts to obtain vitamin D from food. People who are not outdoors much or who live in northern or predominantly cloudy or smoggy areas are advised to drink at least 2 cups of vitamin D–fortified milk a day. The fortification of milk and other foods with vitamin D is the best guarantee that people will meet their needs.[34]** Despite vitamin D fortification, the average intake in the United States falls short of recommendations. Egg yolks and oily fish such as salmon, mackerel, and sardines are the best natural sources of vitamin D.

Meeting vitamin D needs is difficult without adequate sunshine, fortification, or supplementation. Vegetarians who do not include milk in their diets may use vitamin D–fortified soy milk and cereals. Importantly, feeding infants and young children nonfortified "health beverages" instead of milk or infant formula can create severe nutrient deficiencies, including rickets.

Vitamin D from the Sun Most of the world's population relies on natural exposure to sunlight to maintain adequate vitamin D nutrition. The sun imposes no risk of vitamin D toxicity; prolonged exposure to sunlight degrades the vitamin D precursor in the skin, preventing its conversion to the active vitamin.

Prolonged exposure to sunlight can, however, prematurely wrinkle the skin and cause skin cancer. Sunscreens help reduce these risks, but sunscreens with a sun protection factor (SPF) of 8 and higher can also reduce vitamin D synthesis.

*High blood calcium is known as *hypercalcemia* and may develop from a variety of disorders, including vitamin D toxicity. It does *not* develop from a high calcium intake.

**Vitamin D fortification of milk in the United States is 400 IU per quart; in Canada, 350 to 450 IU per liter.

> **FIGURE 11-11 Vitamin D Synthesis and Latitude**

Above 40° north latitude (and below 40° south latitude in the southern hemisphere), vitamin D synthesis essentially ceases for the 4 months of winter. Synthesis increases as spring approaches, peaks in summer, and declines again in the fall. People living in regions of extreme northern (or extreme southern) latitudes may miss as much as 6 months of vitamin D production.

The sunshine vitamin—vitamin D.

Still, even with an SPF 15 to 30 sunscreen, sufficient vitamin D synthesis can be obtained in 10 to 20 minutes of sun exposure.[35] Alternatively, a person could apply sunscreen after enough time has elapsed to provide sufficient vitamin D synthesis. For most people, exposing hands, face, and arms on a clear summer day for 5 to 10 minutes two or three times a week should be sufficient to maintain vitamin D nutrition.

The pigments of dark skin provide some protection from the sun's damage, but they also reduce vitamin D synthesis. Dark-skinned people require longer sunlight exposure than light-skinned people: heavily pigmented skin achieves the same amount of vitamin D synthesis in 3 hours as fair skin in a half hour. Latitude, season, and time of day ♦ also have dramatic effects on vitamin D synthesis and status (see Figure 11-11). Heavy clouds, smoke, or smog block the ultraviolet (UV) rays of the sun that promote vitamin D synthesis. Differences in skin pigmentation, latitude, and smog may account for the finding that African American people, especially those in northern, smoggy cities, are most likely to be vitamin D deficient. Vitamin D deficiency is especially prevalent in the winter and in the Arctic and Antarctic regions of the world.[36] To ensure an adequate vitamin D status, supplements may be needed. The body's vitamin D supplies from summer synthesis alone are insufficient to meet winter needs.[37]

♦ Factors that may limit sun exposure and, therefore, vitamin D synthesis:

- Geographic location (latitude)
- Season of the year
- Time of day
- Air pollution and cloud cover
- Clothing
- Tall buildings
- Indoor living
- Sunscreens

> **DIETARY GUIDELINES FOR AMERICANS 2010**

Choose foods that provide more vitamin D, a nutrient of concern in American diets. Most dietary vitamin D derives from fortified foods such as milk, yogurt, and breakfast cereals; natural dietary sources of vitamin D include oily fish and egg yolks.

Depending on the radiation used, the UV rays from tanning lamps and tanning beds may also stimulate vitamin D synthesis and increase bone density. The potential hazards of skin damage, however, may outweigh any possible benefits.* The Food and Drug Administration (FDA) warns that if the lamps are not properly filtered, people using tanning booths risk burns, damage to the eyes and blood vessels, and skin cancer.

*The best wavelengths for vitamin D synthesis are UV-B rays between 290 and 310 nanometers. Some tanning parlors advertise "UV-A rays only, for a tan without the burn," but UV-A rays can damage the skin.

Vitamin D from Supplements As mentioned, some people may benefit from taking vitamin D supplements. Vitamin D can be found in multivitamin-mineral supplements as well as a high-dose single supplement. As a single supplement, vitamin D_3 is less expensive, more commonly available, and more potent than vitamin D_2.[38] Taking vitamin D supplements with the largest meal of the day improves absorption, resulting in a 50 percent increase in blood levels.[39]

REVIEW IT Identify the main roles, deficiency symptoms, and sources for vitamin D.
Vitamin D can be synthesized in the body with the help of sunlight or obtained from some foods, most notably fortified milk. Vitamin D sends signals to three primary target sites: the GI tract to absorb more calcium and phosphorus, the bones to release more, and the kidneys to retain more. These actions maintain blood calcium concentrations and support bone formation. A deficiency causes rickets in childhood and osteomalacia in later life. The accompanying table provides a summary of vitamin D.

Vitamin D

Other Names

calciferol (vitamin D)

ergocalciferol (vitamin D_2): vitamin D derived from plants in the diet and made from the yeast and plant sterol ergosterol.

cholecalciferol (vitamin D_3 or **calciol):** vitamin D derived from animals in the diet or made in the skin from 7-dehydrocholesterol, a precursor of cholesterol, with the help of sunlight.

calcidiol (25-hydroxyvitamin D): vitamin D found in the blood that is made from the hydroxylation of calciol in the liver.

calcitriol (1,25-dihydroxyvitamin D): vitamin D that is made from the hydroxylation of calcidiol in the kidneys; the biologically active hormone, sometimes called *active vitamin D.*

2011 RDA

Adults: 15 µg/day or 600 IU/day (19–70 yr)
20 µg/day or 800 IU/day (>70 yr)

Upper Level

Adults: 100 µg/day or 4000 IU/day

Chief Functions in the Body

Mineralization of bones (raises blood calcium and phosphorus by increasing absorption from digestive tract, withdrawing calcium from bones, stimulating retention by kidneys)

Significant Sources

Synthesized in the body with the help of sunlight; fortified milk, margarine, butter, juices, cereals, and chocolate mixes; veal, beef, egg yolks, liver, fatty fish (herring, salmon, sardines) and their oils

Deficiency Diseases

Rickets, osteomalacia

Deficiency Symptoms

Rickets in children:

Inadequate calcification, resulting in misshapen bones (bowing of legs); enlargement of ends of long bones (knees, wrists); deformities of ribs (bowed, with beads or knobs);[a] delayed closing of fontanel, resulting in rapid enlargement of head (see figure below); lax muscles resulting in protrusion of abdomen; muscle spasms

Osteomalacia or osteoporosis in adults:

Loss of calcium, resulting in soft, flexible, brittle, and deformed bones; progressive weakness; pain in pelvis, lower back, and legs

Toxicity Disease

Hypervitaminosis D

Toxicity Symptoms

Elevated blood calcium; calcification of soft tissues (blood vessels, kidneys, heart, lungs, tissues around joints)

Fontanel
A fontanel is an open space in the top of a baby's skull before the bones have grown together. In rickets, closing of the fontanel is delayed.

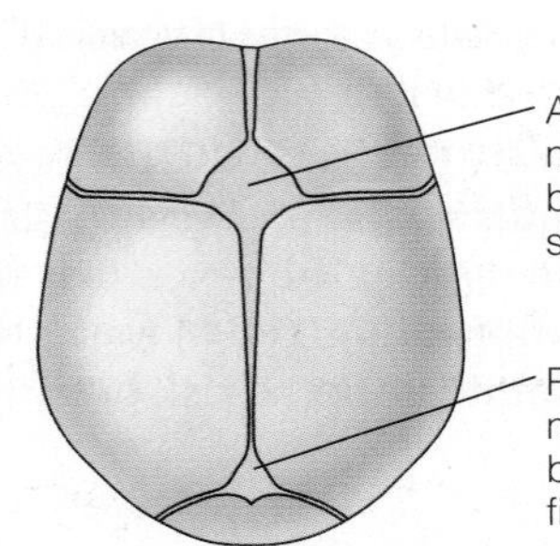

Anterior fontanel normally closes by the end of the second year.
Posterior fontanel normally closes by the end of the first year.

[a]Bowing of the ribs causes the symptoms known as *pigeon breast.* The beads that form on the ribs resemble rosary beads; thus this symptom is known as *rachitic* (ra-KIT-ik) *rosary* ("the rosary of rickets").

11.3 Vitamin E

LEARN IT Identify the main roles, deficiency symptoms, and food sources for vitamin E.

Researchers discovered a component of vegetable oils necessary for reproduction in rats and named this antisterility factor **tocopherol,** which means "to bring forth offspring." When chemists first isolated tocopherol compounds, they designated them by letters of the Greek alphabet: alpha, beta, gamma, and delta. All tocopherols consist of a complex ring structure and a long saturated side chain. (Appendix C provides the chemical structures.) The positions of methyl groups (CH_3) on the side chain and their chemical rotations distinguish one tocopherol from another. **Alpha-tocopherol** is currently the only one recognized as having vitamin E activity in the human body.[40] The other tocopherols are not readily converted to alpha-tocopherol in the body, nor do they perform the same roles. Whether these other tocopherols might be beneficial in other ways is the subject of current research. Gamma-tocopherol, for example, appears to be most effective in inhibiting inflammation and cancer growth.[41]

Vitamin E as an Antioxidant Vitamin E is a fat-soluble antioxidant ♦ and one of the body's primary defenders against the adverse effects of free radicals. Its main action is to stop the chain reaction of free radicals from producing more free radicals (see Highlight 11). In doing so, vitamin E protects the vulnerable components of the cells and their membranes from destruction. Most notably, vitamin E prevents the oxidation of the polyunsaturated fatty acids, but it protects other lipids and related compounds (for example, vitamin A) as well.

♦ Key antioxidant nutrients:
- Vitamin C, vitamin E, beta-carotene
- Selenium

Accumulating evidence suggests that vitamin E may reduce the risk of heart disease by protecting low-density lipoproteins (LDL) against oxidation and reducing inflammation. The oxidation of LDL and inflammation have been implicated as key factors in the development of heart disease. Highlight 11 explains how vitamin E and other antioxidants might protect against chronic diseases, such as heart disease and cancer, and explores whether foods or supplements might be most helpful—or harmful.

Vitamin E Deficiency A primary deficiency of vitamin E (from poor dietary intake) is rare; deficiency is usually associated with diseases of fat malabsorption such as cystic fibrosis. Without vitamin E, the red blood cells break and spill their contents, probably due to oxidation of the polyunsaturated fatty acids in their membranes. This classic sign of vitamin E deficiency, known as **erythrocyte hemolysis,** is seen in premature infants born before the transfer of vitamin E from the mother to the infant that takes place in the last weeks of pregnancy. Vitamin E treatment corrects **hemolytic anemia.**

Prolonged vitamin E deficiency also causes neuromuscular dysfunction involving the spinal cord and retina of the eye. Common symptoms include loss of muscle coordination and reflexes and impaired vision and speech. Vitamin E treatment corrects these neurological symptoms of vitamin E deficiency.

Two other conditions seem to respond to vitamin E treatment, although results are inconsistent. One is **fibrocystic breast disease,** a nonmalignant breast disease. The other is **intermittent claudication,** an abnormality of blood flow that causes cramping in the legs.

Vitamin E Toxicity Vitamin E supplement use has risen in recent years as its protective actions against chronic diseases have been recognized. Fortunately, the liver carefully regulates vitamin E concentrations. Toxicity is rare, and vitamin E appears safe across a broad range of intakes. The UL for vitamin E (1000 milligrams) is more than 65 times greater than the recommended intake for adults (15 milligrams). Extremely high doses of vitamin E may interfere with the blood-clotting action of vitamin K and enhance the effects of drugs used to oppose blood clotting, causing hemorrhage. Additional research is needed to determine whether vitamin E supplements increase the risk of hemorrhagic stroke.

Vitamin E Recommendations The RDA for vitamin E is based on the alpha-tocopherol form only. As mentioned earlier, the other tocopherols cannot be converted to alpha-tocopherol, nor can they perform the same metabolic roles

tocopherol (tuh-KOFF-er-ol): a general term for several chemically related compounds, one of which has vitamin E activity. (See Appendix C for chemical structures.)

alpha-tocopherol: the active vitamin E compound.

erythrocyte (eh-RITH-ro-cite) **hemolysis** (he-MOLL-uh-sis): the breaking open of red blood cells (erythrocytes); a symptom of vitamin E–deficiency disease in human beings.
- **erythro** = red
- **cyte** = cell
- **hemo** = blood
- **lysis** = breaking

hemolytic (HE-moh-LIT-ick) **anemia:** the condition of having too few red blood cells as a result of erythrocyte hemolysis.

fibrocystic (FYE-bro-SIS-tik) **breast disease:** a harmless condition in which the breasts develop lumps, sometimes associated with caffeine consumption. In some, it responds to abstinence from caffeine; in others, it can be treated with vitamin E.
- **fibro** = fibrous tissue
- **cyst** = closed sac

intermittent claudication (klaw-dih-KAY-shun): severe calf pain caused by inadequate blood supply. It occurs when walking and subsides during rest.
- **intermittent** = at intervals
- **claudicare** = to limp

© Craig, M. Moore

Fat-soluble vitamin E is found predominantly in vegetable oils, seeds, and nuts.

in the body. A person who consumes large quantities of polyunsaturated fatty acids needs more vitamin E. Fortunately, vitamin E and polyunsaturated fatty acids tend to occur together in the same foods.

Vitamin E in Foods Vitamin E is widespread in foods. Much of the vitamin E in the diet comes from vegetable oils and products made from them, such as margarine and salad dressings. Wheat germ oil is especially rich in vitamin E.

Because vitamin E is readily destroyed by heat processing (such as deep-fat frying) and oxidation, fresh or lightly processed foods are preferable sources. Most processed and convenience foods do not contribute enough vitamin E to ensure an adequate intake.

Prior to 2000, values of the vitamin E in foods reflected all of the tocopherols and were expressed in "milligrams of tocopherol equivalents."* These measures overestimated the amount of alpha-tocopherol. To estimate the alpha-tocopherol content of foods stated in tocopherol equivalents, multiply by 0.8.[42]

REVIEW IT Identify the main roles, deficiency symptoms, and foods sources for vitamin E.

Vitamin E acts as an antioxidant, defending lipids and other components of the cells against oxidative damage. Deficiencies are rare, but they do occur in premature infants, the primary symptom being erythrocyte hemolysis. Vitamin E is found predominantly in vegetable oils and appears to be one of the least toxic of the fat-soluble vitamins. The accompanying table provides a summary of vitamin E.

Vitamin E

Other Names	**Significant Sources**
Alpha-tocopherol	Polyunsaturated plant oils (margarine, salad dressings), dark green, leafy vegetables (spinach, turnip greens, collard greens, broccoli), wheat germ, whole grains, liver, egg yolks, nuts, seeds, fatty meats
2000 RDA	Easily destroyed by heat and oxygen
Adults: 15 mg/day	**Deficiency Symptoms**
Upper Level	Red blood cell breakage,[a] nerve damage
Adults: 1000 mg/day	**Toxicity Symptoms**
Chief Functions in the Body	Augments the effects of anticlotting medication
Antioxidant (stabilization of cell membranes, regulation of oxidation reactions, protection of polyunsaturated fatty acids [PUFA] and vitamin A)	

[a]The breaking of red blood cells is called *erythrocyte hemolysis*.

11.4 Vitamin K

LEARN IT Identify the main roles, deficiency symptoms, and sources for vitamin K.

Like vitamin D, vitamin K can be obtained both from foods and from a nonfood source. Bacteria in the GI tract synthesize vitamin K that the body can absorb. Vitamin K appropriately gets its name from the Danish word *koagulation* ("coagulation" or "clotting"). Its primary action is blood clotting, where its presence can make the difference between life and death.[43] Blood has a remarkable ability to remain liquid, but it can clot within seconds when the integrity of that system is disturbed.

Roles in the Body More than a dozen different proteins and the mineral calcium are involved in making a blood clot. Vitamin K is essential for the activation of several of these proteins, among them prothrombin, made by the

*Appendix H accurately presents vitamin E data in milligrams of alpha-tocopherol.

> FIGURE 11-12 Blood-Clotting Process

Vitamin K is essential for the synthesis of prothrombin and several other clotting factors. Blood clots are formed by a cascade of reactions, with each step creating a compound that activates the next step.

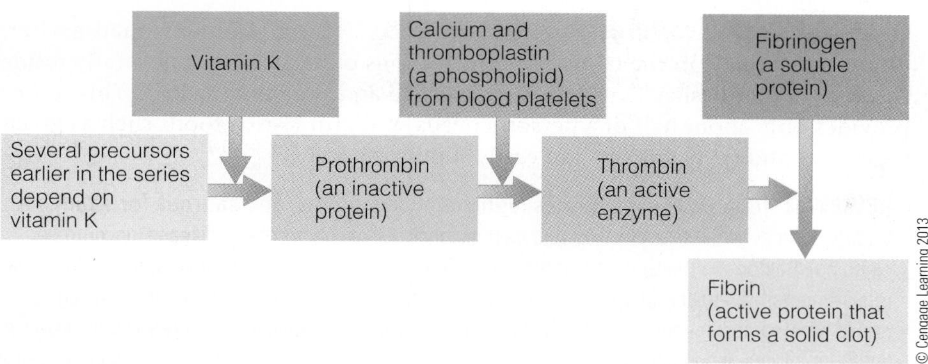

liver as a precursor of the protein thrombin (see Figure 11-12). When any of the blood-clotting factors is lacking, **hemorrhagic disease** results. If an artery or vein is cut or broken, bleeding goes unchecked. Of course, this is not to say that hemorrhaging is always caused by vitamin K deficiency. Another cause is the genetic disorder **hemophilia,** which is neither caused nor cured by vitamin K.

Vitamin K also participates in the metabolism of bone proteins, most notably **osteocalcin.** Without vitamin K, osteocalcin cannot bind to the minerals that normally form bones, resulting in low bone density.[44]* An adequate intake of vitamin K helps to decrease bone turnover and protect against fractures.[45] The effectiveness of vitamin K supplements on bone health is inconclusive.

Vitamin K is historically known for its role in blood clotting, and more recently for its participation in bone building, but researchers continue to discover proteins needing vitamin K's assistance.[46] These proteins have been identified in the plaques of atherosclerosis, the kidneys, and the nervous system.

Vitamin K Deficiency Chapter 1 explains that a *primary deficiency* develops in response to an inadequate dietary intake whereas a *secondary deficiency* occurs for other reasons. A primary deficiency of vitamin K is rare, but a secondary deficiency may occur in two circumstances. First, whenever fat absorption falters, as occurs when bile production fails, vitamin K absorption diminishes. Second, some drugs disrupt vitamin K's synthesis and action in the body: antibiotics kill the vitamin K–producing bacteria in the intestine, and anticoagulant drugs interfere with vitamin K metabolism and activity. Excessive bleeding due to a vitamin K deficiency can be fatal.

Newborn infants present a unique case of vitamin K nutrition because they are born with a **sterile** intestinal tract, and the vitamin K–producing bacteria take weeks to establish themselves. At the same time, plasma prothrombin concentrations are low. This reduces the likelihood of fatal blood clotting during the stress of birth. To prevent hemorrhagic disease in the newborn, a single dose of vitamin K is given at birth by intramuscular injection.** Concerns that vitamin K given at birth raises the risks of childhood cancer are unproved and unlikely.

Vitamin K Toxicity Toxicity is not common, and no adverse effects have been reported with high intakes of vitamin K. Therefore, a UL has not been established. High doses of vitamin K can, however,

hemorrhagic (hem-oh-RAJ-ik) **disease:** a disease characterized by excessive bleeding.

hemophilia (HE-moh-FEEL-ee-ah): a hereditary disease in which the blood is unable to clot because it lacks the ability to synthesize certain clotting factors.

osteocalcin (os-teo-KAL-sen): a calcium-binding protein in bones, essential for normal mineralization.

sterile: free of microorganisms, such as bacteria.

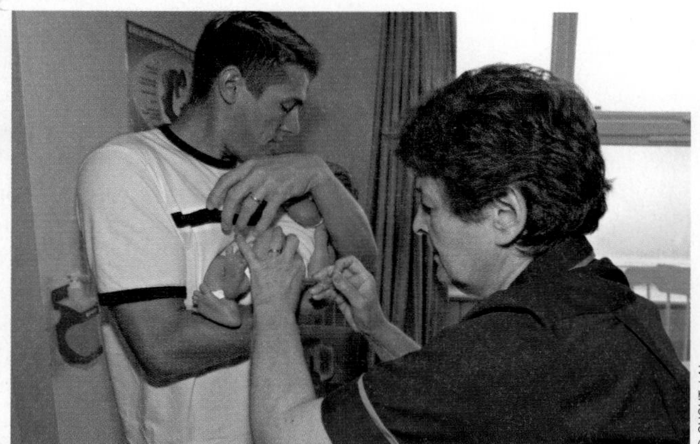

Soon after birth, newborn infants receive a dose of vitamin K to prevent hemorrhagic disease.

*Vitamin K is a cofactor for a carboxylase enzyme. When vitamin K is inadequate, osteocalcin is undercarboxylated and therefore less effective in binding calcium.

**Vitamin K is usually given in the naturally occurring form known as *phylloquinone* (FILL-oh-KWIN-own); the synthetic form of vitamin K is *menadione* (men-uh-DYE-own). See Appendix C for the chemistry of these structures.

reduce the effectiveness of anticoagulant drugs used to prevent blood clotting. People taking these drugs can continue eating their usual diets. Their blood clotting times should be monitored closely and drug dosages adjusted accordingly.[47]

© Matthew Farruggio

Notable food sources of vitamin K include green vegetables such as collards, spinach, bib lettuce, brussels sprouts, and cabbage and vegetable oils such as soybean oil and canola oil.

Vitamin K Recommendations and Sources As mentioned earlier, vitamin K is made in the GI tract by the billions of bacteria that normally reside there. Once synthesized, vitamin K is absorbed and stored in the liver. This source provides only about half of a person's needs. Vitamin K–rich foods such as green vegetables and vegetable oils can easily supply the rest.

REVIEW IT Identify the main roles, deficiency symptoms, and sources for vitamin K.
Vitamin K helps with blood clotting, and its deficiency causes hemorrhagic disease (uncontrolled bleeding). Bacteria in the GI tract can make the vitamin; people typically receive about half of their requirements from bacterial synthesis and half from foods such as green vegetables and vegetable oils. Because people depend on bacterial synthesis for vitamin K, deficiency is most likely in newborn infants and in people taking antibiotics. The accompanying table provides a summary of vitamin K.

Vitamin K

Other Names	**Significant Sources**
Phylloquinone, menaquinone, menadione, naphthoquinone	Bacterial synthesis in the digestive tract;[a] liver; dark green, leafy vegetables, cabbage-type vegetables; milk
2001 Adequate Intake (AI)	**Deficiency Symptoms**
Men: 120 μg/day Women: 90 μg/day	Hemorrhaging
Chief Functions in the Body	**Toxicity Symptoms**
Synthesis of blood-clotting proteins and bone proteins	None known

[a]Vitamin K needs cannot be met from bacterial synthesis alone; however, it is a potentially important source in the small intestine, where absorption efficiency ranges from 40 to 70 percent.

The four fat-soluble vitamins play many specific roles in the growth and maintenance of the body. Their presence affects the health and function of the eyes, skin, GI tract, lungs, bones, teeth, nervous system, and blood; their deficiencies become apparent in these same areas. Toxicities of the fat-soluble vitamins are possible, especially when people use supplements, because the body stores excesses.

As with the water-soluble vitamins, the function of one fat-soluble vitamin often depends on the presence of another. Recall that vitamin E protects vitamin A from oxidation. In vitamin E deficiency, vitamin A absorption and storage are impaired. Three of the four fat-soluble vitamins—A, D, and K—play important roles in bone growth and remodeling. As mentioned, vitamin K helps synthesize a specific bone protein, and vitamin D regulates that synthesis. Vitamin A, in turn, may control which bone-building genes respond to vitamin D. Vitamin E and vitamin K share some metabolic pathways, which can create problems, especially in blood clotting.[48]

Fat-soluble vitamins also interact with minerals. Vitamin D and calcium cooperate in bone formation, and zinc is required for the synthesis of vitamin A's transport protein, retinol-binding protein. Zinc also assists the enzyme that regenerates retinal from retinol in the eye. Vitamin A deficiency and iron deficiency often occur together and each seems to worsen the other's metabolism.[49]

The roles of the fat-soluble vitamins differ from those of the water-soluble vitamins, and they appear in different foods—yet they are just as essential to life. The need for them underlines the importance of eating a wide variety of nourishing foods daily. The accompanying table provides a summary of the fat-soluble vitamins.

REVIEW IT The Fat-Soluble Vitamins

Vitamin and Chief Functions	Deficiency Symptoms	Toxicity Symptoms	Significant Sources
Vitamin A			
Vision; maintenance of cornea, epithelial cells, mucous membranes, skin; bone and tooth growth; reproduction; immunity	Infectious diseases, night blindness, blindness (xerophthalmia), keratinization	Reduced bone mineral density, liver abnormalities, birth defects	Retinol: milk and milk products Beta-carotene: dark green, leafy and deep yellow/orange vegetables
Vitamin D			
Mineralization of bones (raises blood calcium and phosphorus by increasing absorption from digestive tract, withdrawing calcium from bones, stimulating retention by kidneys)	Rickets, osteomalacia	Calcium imbalance (calcification of soft tissues and formation of stones)	Synthesized in the body with the help of sunshine; fortified milk
Vitamin E			
Antioxidant (stabilization of cell membranes, regulation of oxidation reactions, protection of polyunsaturated fatty acids [PUFA] and vitamin A)	Erythrocyte hemolysis, nerve damage	Hemorrhagic effects	Vegetable oils
Vitamin K			
Synthesis of blood-clotting proteins and bone proteins	Hemorrhage	None known	Synthesized in the body by GI bacteria; dark green, leafy vegetables

Nutrition Portfolio

For the fat-soluble vitamins, select colorful fruits and vegetables, fortified milk or soy products, and vegetable oils; use supplements with caution, if at all. Go to Diet Analysis Plus and choose one of the days on which you tracked your diet for an entire day. Select the MyPlate report and then consider the following questions:

- How was your overall intake in the vegetable group? Do you need improvement in this area? If so, what are some changes you could make?

Now look at the report titled Intake Spreadsheet to answer the following questions:

- Examine your weekly choices of vegetables and evaluate whether you meet the recommendations for dark green or orange and deep yellow vegetables.
- Consider whether you drink enough vitamin D–fortified milk or go outside in the sunshine regularly.
- Describe the vegetable oils you use when you cook and their vitamin contributions.

Diet Analysis PLUS To complete this exercise, go to your Diet Analysis Plus at **www.cengagebrain.com.**

STUDY IT To review the key points of this chapter and take a practice quiz, go to the study cards at the end of the book.

REFERENCES

1. G. Tang, Bioconversion of dietary provitamin A carotenoids to vitamin A in humans, *American Journal of Clinical Nutrition* 91 (2010): 1468S–1473S.
2. J. von Lintig, Colors with functions: Elucidating the biochemical and molecular basis of carotenoid metabolism, *Annual Review of Nutrition* 30 (2010): 35–56.
3. G. Wolf, Retinoic acid as cause of cell proliferation or cell growth inhibition depending on activation of one of two different nuclear receptors, *Nutrition Reviews* 66 (2008): 55–59.
4. N. Noy, Between death and survival: Retinoic acid in regulation of apoptosis, *Annual Review of Nutrition* 30 (2010): 201–217.
5. Committee on Dietary Reference Intakes, *Dietary Reference Intakes for Vitamin C, Vitamin E, Selenium, and Carotenoids* (Washington, D.C.: National Academies Press, 2000).
6. A. Sommer, Vitamin A deficiency and clinical disease: An historical overview, *Journal of Nutrition* 138 (2008): 1835–1839.
7. E. Mayo-Wilson and coauthors, Vitamin A supplements for preventing mortality, illness, and blindness in children aged under 5: Systematic review and meta-analysis, *British Medical Journal* 343 (2011): d5094; S. A. Abrams and D. C. Hilmers, Postnatal vitamin A supplementation in developing countries: An intervention whose time has come? *Pediatrics*

122 (2008): 180–181; K. Kraemer and coauthors, Are low tolerable upper intake levels for vitamin A undermining effective food fortification efforts? *Nutrition Reviews* 66 (2008): 517–525.

8. S. M. Ahmad and coauthors, Vitamin A status is associated with T-cell responses in Bangladeshi men, *British Journal of Nutrition* 102 (2009): 797–802; C. Trottier and coauthors, Retinoids inhibit measles virus through a type I IFN-dependent bystander effect, *The FASEB Journal* 23 (2009): 3203–3212.
9. www.who.int/mediacentre/factsheets, f286, updated December 2009.
10. A. Sheth, R. Khurana, and V. Khurana, Potential liver damage associated with over-the-counter vitamin supplements, *Journal of the American Dietetic Association* 108 (2008): 1536–1537.
11. S. L. Morgan, Nutrition and bone: It is more than calcium and vitamin D, *Women's Health* 5 (2009): 727–737.
12. J. Zhao and coauthors, Retinoic acid downregulates microRNAs to induce abnormal development of spinal cord in spina bifida rat model, *Child's Nervous System* 24 (2008): 485–492.
13. T. L. Schonfeld, N. J. Amoura, and C. J. Kratochvil, iPLEDGE allegiance to the pill: Evaluation of year 1 of a birth defect prevention and monitoring system, *Journal of Law, Medicine, and Ethics* 37 (2009): 104–117.
14. H. F. Deluca, Evolution of our understanding of vitamin D, *Nutrition Reviews* 66 (2008): S73–S87; A. W. Norman, From vitamin D to hormone D: Fundamentals of the vitamin D endocrine system essential for good health, *American Journal of Clinical Nutrition* 88 (2008): 491S–499S.
15. R. P. Heaney, Vitamin D and calcium interactions: Functional outcomes, *American Journal of Clinical Nutrition* 88 (2008): 541S–544S.
16. Deluca, 2008; M. R. Haussler and coauthors, Vitamin D receptor: Molecular signaling and actions of nutritional ligands in disease prevention, *Nutrition Reviews* 66 (2008): S98–S112.
17. R. C. Khanal and I. Nemere, Regulation of intestinal calcium transport, *Annual Review of Nutrition* 28 (2008): 179–196.
18. S. Samual and M. D. Sitrin, Vitamin D's role in cell proliferation and differentiation, *Nutrition Reviews* 66 (2008): S116–S124.
19. E. van Etten and coauthors, Regulation of vitamin D homeostasis: Implications for the immune system, *Nutrition Reviews* 66 (2008): S125–S134.
20. E. M. Mowry, Vitamin D: Evidence for its role as a prognostic factor in multiple sclerosis, *Journal of Neurological Science* 311 (2011): 19–22; M. H. Hopkins and coauthors, Effects of supplemental vitamin D and calcium on biomarkers of inflammation in colorectal adenoma patients: A randomized, controlled clinical trial, *Cancer Prevention Research* 4 (2011): 1645–1654; K. Luong and L. T. Nguyen, Impact of vitamin D in the treatment of tuberculosis, *American Journal of Medical Sciences* 341 (2011): 493–498; A. E. Millen and coauthors, Vitamin D status and early age-related macular degeneration in postmenopausal women, *Archives of Ophthalmology* 129 (2011): 481–489; N. Parekh, Protective role of vitamin D against age-related macular degeneration: A hypothesis, *Topics in Clinical Nutrition* 25 (2010): 290–301; A. G. Pittas and coauthors, Systematic review: Vitamin D and cardiometabolic outcomes, *Annals of Internal Medicine* 152 (2010): 307–314; C. D. Toner, C. D. Davis, and J. A. Milner, The vitamin D and cancer conundrum: Aiming at a moving target, *Journal of the American Dietetic Association* 110 (2010): 1492–1500; A. Zitterman and coauthors, Vitamin D supplementation enhances the beneficial effects of weight loss on cardiovascular disease risk markers, *American Journal of Clinical Nutrition* 89 (2009): 1321–1327; M. T. Cantorna, Vitamin D and multiple sclerosis: An update, *Nutrition Reviews* 66 (2008): S135–S138; M. F. Holick, Vitamin D: A d-lightful health perspective, *Nutrition Reviews* 66 (2008): S182–S194; S. E. Judd and coauthors, Optimal vitamin D status attenuates the age-associated increase in systolic blood pressure in white Americans: Results from the third National Health and Nutrition Examination Survey, *American Journal of Clinical Nutrition* 87 (2008): 136–141.
21. A. Grey and M. Bolland, Vitamin D: A place in the sun? *Archives of Internal Medicine* 170 (2010): 1099–1100.
22. C. J. Rosen, Vitamin D insufficiency, *New England Journal of Medicine* 364 (2011): 248–254; A. A. Ginde , M. C. Liu, and C. A. Camargo, Demographic differences and trends of vitamin D insufficiency in the US population, 1988–2004, *Archives of Internal Medicine* 169 (2009): 626–632; S. A. Bowden and coauthors, Prevalence of vitamin D deficiency and insufficiency in children with osteopenia or osteoporosis referred to a pediatric metabolic bone clinic, *Pediatrics* 121 (2008): e1585–e1590; M. L. Neuhouser and coauthors, Vitamin D insufficiency in a multiethnic cohort of breast cancer survivors, *American Journal of Clinical Nutrition* 88 (2008): 133–139.
23. A. C. Looker and coauthors, Vitamin D status: United States, 2001–2006, *NCHS Data Brief* 59 (2011): 1–8.
24. L. A. Martini and R. J. Wood, Vitamin D and blood pressure connection: Update on epidemiologic, clinical, and mechanistic evidence, *Nutrition Reviews* 66 (2008): 291–297.
25. K. D. Cashman and coauthors, Low vitamin D status adversely affects bone health parameters in adolescents, *American Journal of Clinical Nutrition* 87 (2008): 1039–1044.
26. A. Prentice, Vitamin D deficiency: A global perspective, *Nutrition Reviews* 66 (2008): S153–S164.
27. C. M. Lenders and coauthors, Relation of body fat indexes to vitamin D status and deficiency among obese adolescents, *American Journal of Clinical Nutrition* 90 (2009): 459–467; S. Saintonge, H. Bang, and L. M. Gerber, Implications of a new definition of vitamin D deficiency in a multiracial US adolescent population: The National Health and Nutrition Examination Survey III, *Pediatrics* 123 (2009): 797–803; F. R. Greer, 25-Hydroxyvitamin D: Functional outcome in infants and young children, *American Journal of Clinical Nutrition* 88 (2008): 529S–533S; S. Y. Huh and C. M. Gordon, Vitamin D deficiency in children and adolescents: Epidemiology, impact and treatment, *Reviews in Endocrine and Metabolic Disorders* 9 (2008): 161–170; M. Misra and coauthors, Vitamin D deficiency in children and its management: Review of current knowledge and recommendations, *Pediatrics* 122 (2008): 398–417.
28. C. L. Wagner, F. R. Greer, and the Section on Breastfeeding and Committee on Nutrition, Prevention of rickets and vitamin D deficiency in infants, children, and adolescents, *Pediatrics* 122 (2008): 1142–1152.
29. K. Ukine, Severe osteomalacia presenting with multiple vertebral fractures: A case report and review of the literature, *Endocrine* 36 (2009): 30–36.
30. P. Lips and coauthors, Once-weekly dose of 8400 IU vitamin D_3 compared with placebo: Effects on neuromuscular function and tolerability in older adults with vitamin D insufficiency, *American Journal of Clinical Nutrition* 91 (2010): 985–991; B. Dawson-Hughes, Serum 25-hydroxyvitamin D and functional outcomes in the elderly, *American Journal of Clinical Nutrition* 88 (2008): 537S–540S.
31. G. Jones, Pharmacokinetics of vitamin D toxicity, *American Journal of Clinical Nutrition* 88 (2008): 582S–586S.
32. Committee on Dietary Reference Intakes, *Dietary Reference Intakes for Calcium and Vitamin D*, (Washington, D.C.: National Academies Press, 2011), pp. 75–124.
33. L. M. Hall and coauthors, Vitamin D intake needed to maintain target serum 25-hydroxyvitamin D concentrations in participants with low sun exposure and dark skin pigmentation is substantially higher than current recommendations, *Journal of Nutrition* 140 (2010): 542–550; K. D. Cashman and coauthors, Estimation of the dietary requirement for vitamin D in healthy adults, *American Journal of Clinical Nutrition* 88 (2008): 1535–1542; J. F. Aloia and coauthors, Vitamin D intake to attain a desired serum 25-hydroxyvitamin D concentration, *American Journal of Clinical Nutrition* 87 (2008): 1952–1958.
34. R. M. Biancuzzo and coauthors, Fortification of orange juice with vitamin D_2 or vitamin D_3 is as effective as an oral supplement in maintaining vitamin D status in adults, *American Journal of Clinical Nutrition* 91 (2010): 1621–1626; V. Mocanu and coauthors, Long-term effects of giving nursing home residents bread fortified with 125 μg (5000 IU) vitamin D_3 per daily serving, *American Journal of Clinical Nutrition* 89 (2009): 1132–1137; S. O'Donnell and coauthors, Efficacy of food fortification on serum 25-hydroxyvitamin D concentrations: Systematic review, *American Journal of Clinical Nutrition* 88 (2008): 1528–1534.
35. B. A. Gilchrest, Sun exposure and vitamin D sufficiency, *American Journal of Clinical Nutrition* 88 (2008): 570S–577S.
36. S. Sharma and coauthors, Vitamin D deficiency and disease risk among aboriginal Arctic populations, *Nutrition Reviews* 69 (2011): 468–478; Hall and coauthors, 2010; S. M. Smith and coauthors, Vitamin D supplementation during Antarctic winter, *American Journal of Clinical Nutrition* 89 (2009): 1092–1098.

37. L. A. Houghton and coauthors, Predictors of vitamin D status and its association with parathyroid hormone in young New Zealand children, *American Journal of Clinical Nutrition* 92 (2010): 69–76.
38. R. P. Heaney and coauthors, Vitamin D(3) is more potent than vitamin D(2) in humans, *Journal of Clinical Endocrinology and Metabolism* 96 (2011): E447–E452.
39. G. B. Mulligan and A. Licata, Taking vitamin D with the largest meal improves absorption and results in higher serum levels of 25-hydroxyvitamin D, *Journal of Bone and Mineral Research* 25 (2010): 928–930.
40. Committee on Dietary Reference Intakes, 2000.
41. C. S. Yang and coauthors, Inhibition of inflammation and carcinogenesis in the lung and colon by tocopherols, *Annals of the New York Academy of Sciences* 1203 (2010): 29–34; J. Ju and coauthors, Cancer-preventive activities of tocopherols and tocotrienols, *Carcinogenesis* 31 (2010): 533–542; S. R. Wells and coauthors, α-, γ-, and δ-tocopherols reduce inflammatory angiogenesis in human microvascular endothelial cells, *Journal of Nutritional Biochemistry* 21 (2009): 589–597.
42. Committee on Dietary Reference Intakes, 2000.
43. G. J. Merli and J. Fink, Vitamin K and thrombosis, *Vitamins and Hormones* 78 (2008): 265–279.
44. H. M. Macdonald and coauthors, Vitamin K_1 intake is associated with higher bone mineral density and reduced bone resorption in early postmenopausal Scottish women: No evidence of gene-nutrient interaction with apolipoprotein E polymorphisms, *American Journal of Clinical Nutrition* 87 (2008): 1513–1520.
45. S. Cockayne and coauthors, Vitamin K and the prevention of fractures: Systematic review and meta-analysis of randomized controlled trials, *Archives of Internal Medicine* 166 (2006): 1256–1261; J. Iwamoto, T. Takeda, and Y. Sato, Menatetrenone (vitamin K_2) and bone quality in the treatment of postmenopausal osteoporosis, *Nutrition Reviews* 64 (2006): 509–517; K. D. Cashman, Vitamin K status may be an important determinant of childhood bone health, *Nutrition Reviews* 63 (2005): 284–293.
46. S. L. Booth, Roles for vitamin K beyond coagulation, *Annual Review of Nutrition* 29 (2009): 89–110.
47. Committee on Dietary Reference Intakes, *Dietary Reference Intakes for Vitamin A, Vitamin K, Arsenic, Boron, Chromium, Copper, Iodine, Iron, Manganese, Molybdenum, Nickel, Silicon, Vanadium, and Zinc,* (Washington, D.C.: National Academies Press, 2001), pp. 174–175.
48. M. G. Traber, B. Frei, and J. S. Beckman, Vitamin E revisited: Do new data validate benefits for chronic disease prevention? *Current Opinion in Lipidology* 19 (2008): 30–38; M. G. Traber, Vitamin E and K interactions—a 50-year-old problem, *Nutrition Reviews* 66 (2008): 624–629.
49. J. M. Oliveira and coauthors, Influence of iron on vitamin A nutritional status, *Nutrition Reviews* 66 (2008): 141–147.

HIGHLIGHT > 11

LEARN IT Describe how antioxidants defend against free radicals that contribute to diseases.

Antioxidant Nutrients in Disease Prevention

Count on supplement manufacturers to exploit the day's hot topics in nutrition. The moment bits of research news surface, new supplements appear—and terms such as *antioxidants* and *lycopene* become household words. Friendly faces in TV commercials try to persuade us that these supplements hold magic in the fight against aging and disease. New supplements hit the market and cash registers ring. Vitamin C, for years the leading single nutrient supplement, gains new popularity, and sales of lutein, beta-carotene, and vitamin E supplements soar as well.

In the meantime, scientists and medical experts around the world continue their work to clarify and confirm the roles of antioxidants in preventing chronic diseases. This highlight summarizes some of the accumulating evidence. It also revisits the advantages of foods over supplements. But first it is important to introduce the troublemaker—an unstable molecule known as a **free radical.** (The accompanying glossary defines free radicals and related terms.)

Free Radicals and Disease

Chapter 7 describes how the body's cells use oxygen in metabolic reactions. In the process, oxygen reacts with body compounds and produces highly unstable molecules known as free radicals.[1] In addition to normal body processes, environmental factors such as ultraviolet radiation, air pollution, and tobacco smoke generate free radicals.

A free radical is a molecule with one or more unpaired electrons.* An electron without a partner is unstable and highly reactive. To regain its stability, the free radical quickly finds a stable but vulnerable compound from which to steal an electron.

With the loss of an electron, the formerly stable molecule becomes a free radical itself and steals an electron from another nearby molecule. Thus an electron-snatching chain reaction is under way with free radicals producing more free radicals. **Antioxidants** neutralize free radicals by donating one of their own electrons, thus ending the chain reaction. When they lose electrons, antioxidants do not become free radicals because they are stable in either form. (Review Figure 10-16 on p. 323 to see how ascorbic acid can give up two hydrogens with their electrons and become dehydroascorbic acid.)

Free radicals attack. Occasionally, these free-radical attacks are helpful. For example, cells of the immune system use free radicals as ammunition in an "oxidative burst" that demolishes disease-causing viruses and bacteria. Most often, however, free-radical attacks cause widespread damage. They commonly damage the polyunsaturated fatty acids in lipoproteins and in cell membranes, disrupting the transport of substances into and out of cells. Free radicals also alter DNA, RNA, and proteins, creating excesses and deficiencies of specific proteins, impairing cell functions, and eliciting an inflammatory response. All of these actions contribute to cell damage, disease progression, and aging (see Figure H11-1).

The body's natural defenses and repair systems try to control the destruction caused by free radicals, but these systems are not 100 percent effective. In fact, they become less effective with age, and the unrepaired damage accumulates. To some extent, dietary antioxidants defend the body against **oxidative stress**, but if antioxidants are unavailable or if free-radical production becomes excessive, health problems may develop.[2] Oxygen-derived free radicals may cause diseases, not only by indiscriminately destroying the valuable components of cells, but also by serving as signals for specific activities within the cells. Scientists have identified oxidative stress as a causative factor and antioxidants as a protective factor in cognitive performance and the aging process as well as in the development of diseases such as cancer, arthritis, cataracts, diabetes, hypertension, and heart disease.[3]

*Many free radicals exist, but oxygen-derived free radicals are most common in the human body. Examples of oxygen-derived free radicals include superoxide radical ($O_2^{\cdot -}$), hydroxyl radical ($OH\cdot$), and nitric oxide ($NO\cdot$). (The dots in the symbols represent the unpaired electrons.) Technically, hydrogen peroxide (H_2O_2) and singlet oxygen are not free radicals because they contain paired electrons, but the unstable conformation of their electrons makes radical-producing reactions likely. Scientists sometimes use the term *reactive oxygen species (ROS)* to describe all of these compounds.

GLOSSARY

antioxidants: in the body, substances that significantly decrease the adverse effects of free radicals on normal physiological functions.

free radical: an unstable molecule with one or more unpaired electrons. (See Appendix B for a review of basic chemistry concepts.)

oxidants (OKS-ih-dants): compounds (such as oxygen itself) that oxidize other compounds. Compounds that prevent oxidation are called *antioxidants,* whereas those that promote it are called *prooxidants.*

- **anti** = against
- **pro** = for

oxidative stress: a condition in which the production of oxidants and free radicals exceeds the body's ability to handle them and prevent damage.

phytochemicals: nonnutrient compounds found in plant-derived foods that have biological activity in the body.

prooxidants: substances that significantly induce oxidative stress.

> FIGURE H11-1 **Free-Radical Damage**

Free radicals are highly reactive. They might attack the polyunsaturated fatty acids in a cell membrane, which generates lipid radicals that damage cells and accelerate disease progression. Free radicals might also attack and damage DNA, RNA, and proteins, which interferes with the body's ability to maintain normal cell function, causing disease and premature aging.

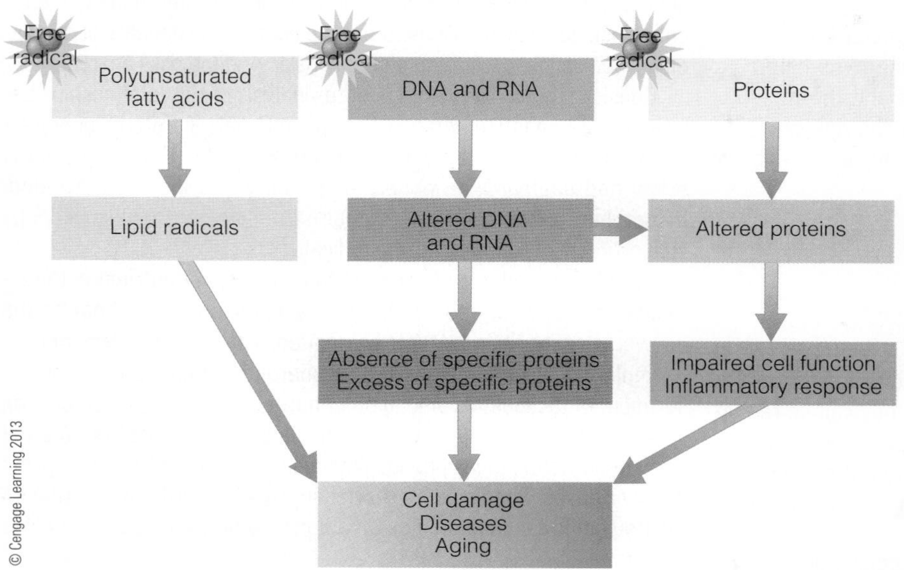

Defending against Free Radicals

The body maintains a couple lines of defense against free-radical damage. A system of enzymes disarms the most harmful **oxidants.*** The action of these enzymes depends on the minerals selenium, copper, manganese, and zinc. If the diet fails to provide adequate supplies of these minerals, this line of defense weakens. The body also uses the antioxidant vitamins—vitamin E, beta-carotene, and vitamin C. Vitamin E defends the body's lipids (cell membranes and lipoproteins, for example) by efficiently stopping the free-radical chain reaction. Beta-carotene also acts as an antioxidant in lipid membranes. Vitamin C protects other tissues, such as the skin and fluid of the blood, against free-radical attacks. Vitamin C seems especially adept at neutralizing free radicals from polluted air and cigarette smoke; it also restores oxidized vitamin E to its active state.[4]

Dietary antioxidants also include some of the **phytochemicals** (featured in Highlight 13). Together, nutrients and phytochemicals with antioxidant activity minimize damage and prevent disease in the following ways:[5]

- Limiting free-radical formation
- Destroying free radicals or their precursors
- Stimulating antioxidant enzyme activity
- Repairing oxidative damage
- Stimulating repair enzyme activity
- Supporting a healthy immune system

These actions play key roles in defending the body against chronic diseases such as cancer and heart disease.

*These enzymes include *glutathione peroxidase, thioredoxin reductase, superoxide dismutase,* and *catalase.*

Defending against Cancer

Cancers arise when cellular DNA is damaged—sometimes by free-radical attacks. Antioxidants may reduce cancer risks by protecting DNA from this damage. Many researchers have reported low rates of cancer in people whose diets include abundant vegetables and fruits, rich in antioxidants. Preliminary reports suggest an inverse relationship between DNA damage and vegetable intake and a positive relationship with beef and pork intake.[6]

Foods rich in vitamin C seem to protect against certain types of cancers, especially those of the esophagus. Such a correlation may reflect the benefits of a diet rich in fruits and vegetables and low in fat; it does not necessarily support taking vitamin C supplements to treat or prevent cancer. At high doses vitamin C acts as a **prooxidant,** generating free radicals. Limited research suggests this action may be useful in destroying cancer cells.[7] Evidence that vitamin C supplements reduce the risk of cancer is lacking.[8]

Researchers hypothesize that vitamin E might inhibit cancer formation by attacking free radicals that damage DNA. Evidence that vitamin E supplements help guard against cancer, however, is lacking.[9]

Several studies report a cancer-preventing benefit of vegetables and fruits rich in beta-carotene and the other carotenoids as well. Carotenoids may protect against oxidative damage to DNA. Some research suggests that high concentrations of beta-carotene are associated with lower rates of some cancers, but intervention studies do not find a reduction in cancer risk with beta-carotene supplementation.[10] Benefits most likely reflect a healthy diet abundant in fruits and vegetables.[11]

Defending against Heart Disease

Decades of research have contributed to our understanding of how oxidative stress contributes to atherosclerosis and how antioxidants might protect against heart disease, yet questions remain.[12] High blood cholesterol carried in LDL (low-density lipoproteins) is a major risk factor for cardiovascular disease, but how do LDL exert their damage? One scenario is that free radicals within the arterial walls oxidize LDL, changing their structure and function. The oxidized LDL then accelerate the formation of artery-clogging plaques. These free radicals also oxidize the polyunsaturated fatty acids of the cell membranes, sparking additional changes in the arterial walls, which impede the flow of blood. Susceptibility to such oxidative damage within the arterial walls is heightened by a diet high in saturated fat and by cigarette smoke. In contrast, diets that include plenty of fruits and vegetables, especially when combined with little saturated fat, strengthen antioxidant defenses against LDL oxidation.

Antioxidants, especially vitamin E, may protect against hypertension and cardiovascular disease.[13] Epidemiological studies suggest that people who eat foods rich in vitamin E have relatively few atherosclerotic plaques and low rates of death from heart disease. Among its many protective roles, vitamin E defends against LDL oxidation, inflammation, arterial injuries, and blood clotting. Whether vitamin E supplements may slow the progression of heart disease is less clear.[14]

Some studies suggest that vitamin C protects against LDL oxidation, raises HDL, lowers total cholesterol, and improves blood pressure. Vitamin C may also minimize inflammation and the free-radical action within the arterial wall. Like vitamin E, the role of vitamin C supplements in reducing the risk of heart disease remains uncertain.[15]

Foods, Supplements, or Both?

In the process of scavenging and quenching free radicals, antioxidants themselves become oxidized. To some extent, they can be regenerated, but losses still occur and free radicals attack continuously. To maintain defenses, a person must replenish dietary antioxidants regularly. But should antioxidants be replenished from foods or from supplements?[16]

Foods—especially fruits and vegetables—offer not only antioxidants, but an array of other valuable vitamins and minerals as well. Importantly, deficiencies of these nutrients can damage DNA as readily as free radicals can. Eating fruits and vegetables in abundance protects against both deficiencies and diseases—and may protect against DNA damage.[17] A major review of the evidence gathered from metabolic studies, epidemiologic studies, and dietary intervention trials identified three dietary strategies most effective in preventing heart disease:

- Use unsaturated fats (that have not been hydrogenated) instead of saturated or *trans* fats (see Highlight 5).
- Select foods rich in omega-3 fatty acids (see Chapter 5).
- Consume a diet high in fruits, vegetables, nuts, and whole grains and low in refined grain products.

Such a diet combined with exercise, weight control, and not smoking serves as the best prescription for health. Notably, taking supplements is not among these disease-prevention recommendations.

Diets that deliver sufficient quantities of antioxidant vitamins may protect against cancer and heart disease—but only a small fraction of the US population consumes recommended amounts.[18] Some research suggests a protective effect from as little as a daily glass of orange juice or carrot juice (rich sources of vitamin C and beta-carotene, respectively). Other intervention studies, however, have used levels of nutrients that far exceed current recommendations and can be achieved only by taking supplements. In making their recommendations for the antioxidant nutrients, members of the DRI Committee considered whether these studies support substantially higher intakes to help protect against chronic diseases. They did raise the recommendations for vitamins C and E, but they do not support taking supplements over eating a healthy diet.

Though fruits and vegetables containing many antioxidant nutrients and phytochemicals have been associated with a diminished risk of many chronic diseases, supplements have not always proved beneficial.[19] In fact, sometimes the benefits are more apparent when the vitamins come from foods rather than from supplements. In other words, the antioxidant actions of fruits and vegetables are greater than their nutrients alone can explain. Without data to confirm the benefits of supplements, we cannot accept the potential risks. And the risks are real.

Consider the findings from a meta-analysis of the relationships between supplements of vitamin A, vitamin E, beta-carotene, or combinations and total mortality. Researchers concluded that supplements either had *no benefit* or *increased* mortality and should be avoided. In fact, beta-carotene *increases* the risk of lung cancer in smokers by supporting the formation of free radicals.[20]

Even if research clearly proves that a particular nutrient is the ultimate protective ingredient in foods, supplements would not be the answer because their contents are limited. Vitamin E supplements, for example, usually contain alpha-tocopherol, but foods provide an assortment of tocopherols among other nutrients, many of which provide valuable protection against free-radical damage. In addition to a full array of nutrients, foods provide phytochemicals that also fight against many diseases. Supplements shortchange users. Furthermore, supplements should be used only as an adjunct to other measures such as

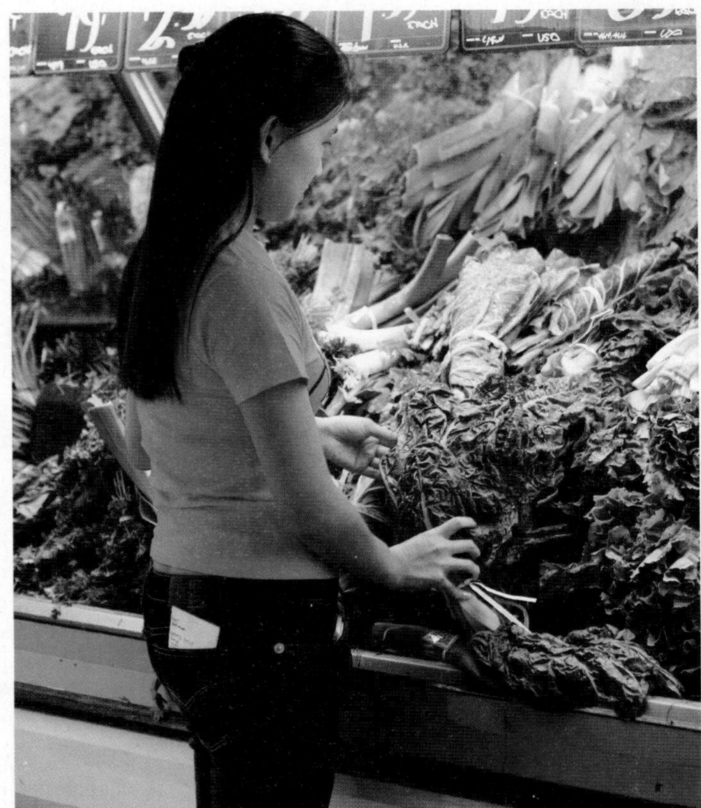

Many cancer-fighting products are available now at your local produce counter.

TABLE H11-1 Antioxidants and Chronic Disease Risk

Antioxidant	Disease	Risk from Foods	Risk from Supplements
Vitamin C	Coronary heart disease	Inconsistent results	Inconsistent results
	Breast cancer	Inconsistent results	—
	Colorectal cancer	Inconsistent results	—
	Gastrointestinal cancer	—	Not known
	Lung cancer	No effect	Not known
Vitamin E	Coronary heart disease	Inconsistent results	No effect or possible increased risk
	Breast cancer	—	No effect
	Colorectal cancer	Inconsistent results	—
	Gastrointestinal cancer	—	No effect
	Lung cancer	No effect	No effect
	Prostate cancer	Decreased risk	Decreased risk in smokers
Beta-carotene	Coronary heart disease	Decreased risk	No effect in nonsmokers, increased risk in smokers
	Lung cancer	Inconsistent results	No effect in nonsmokers, increased risk in smokers
	Colorectal cancer	Decreased risk	—
	Gastrointestinal cancer	—	No effect
	Prostate cancer	No effect	—
Other carotenoids	Lung cancer	Decreased risk for beta-cryptoxanthin	—
	Colorectal cancer	Decreased risk	—
	Prostate cancer	Decreased risk for lycopene	—
Fruits and vegetables	Coronary heart disease	Decreased risk	
	Breast cancer	No effect	
	Colorectal cancer	Inconsistent results	
	Gastric and esophageal cancer	Decreased risk	
	Lung cancer	Decreased risk for fruits, no effect for vegetables	
	Prostate cancer	No effect	
Supplement containing a combination of antioxidants	Coronary heart disease		Possibly increased risk
	Gastrointestinal cancer		Possibly increased risk
	Lung cancer		No effect in nonsmokers, increased risk in smokers

SOURCE: Adapted from H. Verhagen and coauthors, The state of antioxidant affairs, *Nutrition Today* 41 (2006): 244–249.

smoking cessation, weight control, physical activity, and medication as needed.

Clearly, much more research is needed to define optimal and harmful levels of intake. This much we know: antioxidants behave differently under various conditions. At physiological levels typical of a healthy diet, they act as antioxidants, but at pharmacological doses typical of supplements, they may act as prooxidants, stimulating the production of free radicals and altering metabolism in a way that may promote disease. A high intake of vitamin C from supplements, for example, may *increase* the risk of heart disease in women with diabetes. Until the optimum intake of antioxidant nutrients can be determined, the risks of supplement use remain unclear. Table H11-1 presents a summary of the relationships between antioxidants and chronic diseases—sorted by foods or supplements. As you can see, many studies report either no effect or inconsistent results. Any decrease in risk is attributed to foods 9 out of 10 times. Any increase in risk is always from supplements, and often in smokers. Clearly, the best way to add antioxidants to the diet is to eat generous servings of fruits and vegetables daily.

It should be clear by now that we cannot know the identity and action of every chemical in every food. Even if we did, why create a supplement to replicate a food? Why not eat foods and enjoy the pleasure, nourishment, and health benefits they provide? The beneficial constituents in foods are widespread among plants. Among the fruits, pomegranates, berries, and citrus rank high in antioxidants; top antioxidant vegetables include kale, spinach, and brussels sprouts; millet and oats contain the most antioxidants among the grains; pinto beans and soybeans are outstanding legumes; and walnuts outshine the other nuts. But don't try to single out one particular food for its "magical" nutrient, antioxidant, or phytochemical. Instead, eat a wide variety of fruits, vegetables, grains, legumes, and nuts every day—and get *all* the benefits these foods have to offer.

REFERENCES

1. D. Orsucci and coauthors, Electron transfer mediators and other metabolites and cofactors in the treatment of mitochondrial dysfunction, *Nutrition Reviews* 67 (2009): 427–438.
2. Z. Zadák and coauthors, Antioxidants and vitamins in clinical conditions, *Physiological Research* 58 (2009): S13–S17.
3. A. Whaley-Connell, P. A. McCullough, and J. R. Sowers, The role of oxidative stress in the metabolic syndrome, *Reviews in Cardiovascular Medicine* 12 (2011): 21–29.
4. Committee on Dietary Reference Intakes, *Dietary Reference Intakes for Vitamin C, Vitamin E, Selenium, and Carotenoids* (Washington, D.C.: National Academies Press, 2000), p. 225.
5. H. Yao and coauthors, Dietary flavonoids as cancer prevention agents, *Environmental Carcinogenesis and Ecotoxicology Reviews* 29 (2011): 1–31.
6. P. Riso and coauthors, DNA damage and repair activity after broccoli intake in young healthy smokers, *Mutagenesis* 25 (2010): 595–602; A. Brevik and coauthors, Polymorphisms in base excision repair genes as colorectal cancer risk factors and modifiers of the effects of diets high in red meat, *Cancer Epidemiology, Biomarkers and Prevention* 19 (2010): 3167–3173; L. C. Yong and coauthors, High dietary antioxidant intakes are associated with decreased chromosome translocation frequency in airline pilots, *American Journal of Clinical Nutrition* 90 (2009): 1402–1410.
7. Q. Chen and coauthors, Pharmacologic doses of ascorbate act as a prooxidant and decrease growth of aggressive tumor xenografts in mice, *Proceedings of the National Academy of Sciences* 105 (2008): 11105–11109.
8. J. M. Gaziano and coauthors, Vitamins E and C in the prevention of prostate and total cancer in men: The Physicians' Health Study II Randomized Controlled Trial, *Journal of the American Medical Association* 301 (2009): 52–62.
9. Gaziano and coauthors, 2009; S. M. Lippman and coauthors, Effect of selenium and vitamin E on risk of prostate cancer and other cancers: The Selenium and Vitamin E Cancer Prevention Trial (SELECT), *Journal of the American Medical Association* 301 (2009): 39–51.
10. K. Musa-Veloso and coauthors, Influence of observational study design on the interpretation of cancer risk reduction by carotenoids, *Nutrition Reviews* 67 (2009): 527–545.
11. L. Gallicchio and coauthors, Carotenoids and the risk of developing lung cancer: A systematic review, *American Journal of Clinical Nutrition* 88 (2008): 372–383.
12. B. J. Willcox, J. D. Curb, and B. L. Rodriguez, Antioxidants in cardiovascular health and disease: Key lessons from epidemiologic studies, *American Journal of Cardiology* 101 (2008): 75D-86D.
13. E. L. Schiffrin, Antioxidants in hypertension and cardiovascular disease, *Molecular Interventions* 10 (2010): 354–362.
14. S. Czernichow and coauthors, Effects of long-term antioxidant supplementation and association of serum antioxidant concentrations with risk of metabolic syndrome in adults, *American Journal of Clinical Nutrition* 90 (2009): 329–335; J. Zingg, A. Azzi, and M. Meydani, Genetic polymorphisms as determinants for disease-preventive effects of vitamin E, *Nutrition Reviews* 66 (2008): 406–414.
15. Czernichow and coauthors, 2009.
16. E. Herrera and coauthors, Aspects of antioxidant foods and supplements in health and disease, *Nutrition Reviews* 67 (2009): S140–S144.
17. M. K. Shanmugam, R. Kannaiyan, and G. Sethi, Targeting cell signaling and apoptotic pathways by dietary agents: Role in the prevention and treatment of cancer, *Nutrition and Cancer* 63 (2011): 161–173; Yong and coauthors, 2009.
18. M. G. Traber, B. Frei, and J. S. Beckman, Vitamin E revisited: Do new data validate benefits for chronic disease prevention? *Current Opinion in Lipidology* 19 (2008): 30–38.
19. B. Halliwell, Free radicals and antioxidants: Quo vadis? *Trends in Pharmacological Sciences* 32 (2011): 125–130.
20. T. Tanvetyanon and G. Bepler, Beta-carotene in multivitamins and the possible risk of lung cancer among smokers versus former smokers: A meta-analysis and evaluation of national brands, *Cancer* 113 (2008): 150–157; Y. G. J. van Helden and coauthors, β-Carotene metabolites enhance inflammation-induced oxidative DNA damage in lung epithelial cells, *Free Radical Biology and Medicine* 46 (2009): 299–304.

CENGAGE LEARNING'S GLOBAL NUTRITION WATCH
GLOBAL NUTRITION RESOURCE CENTER

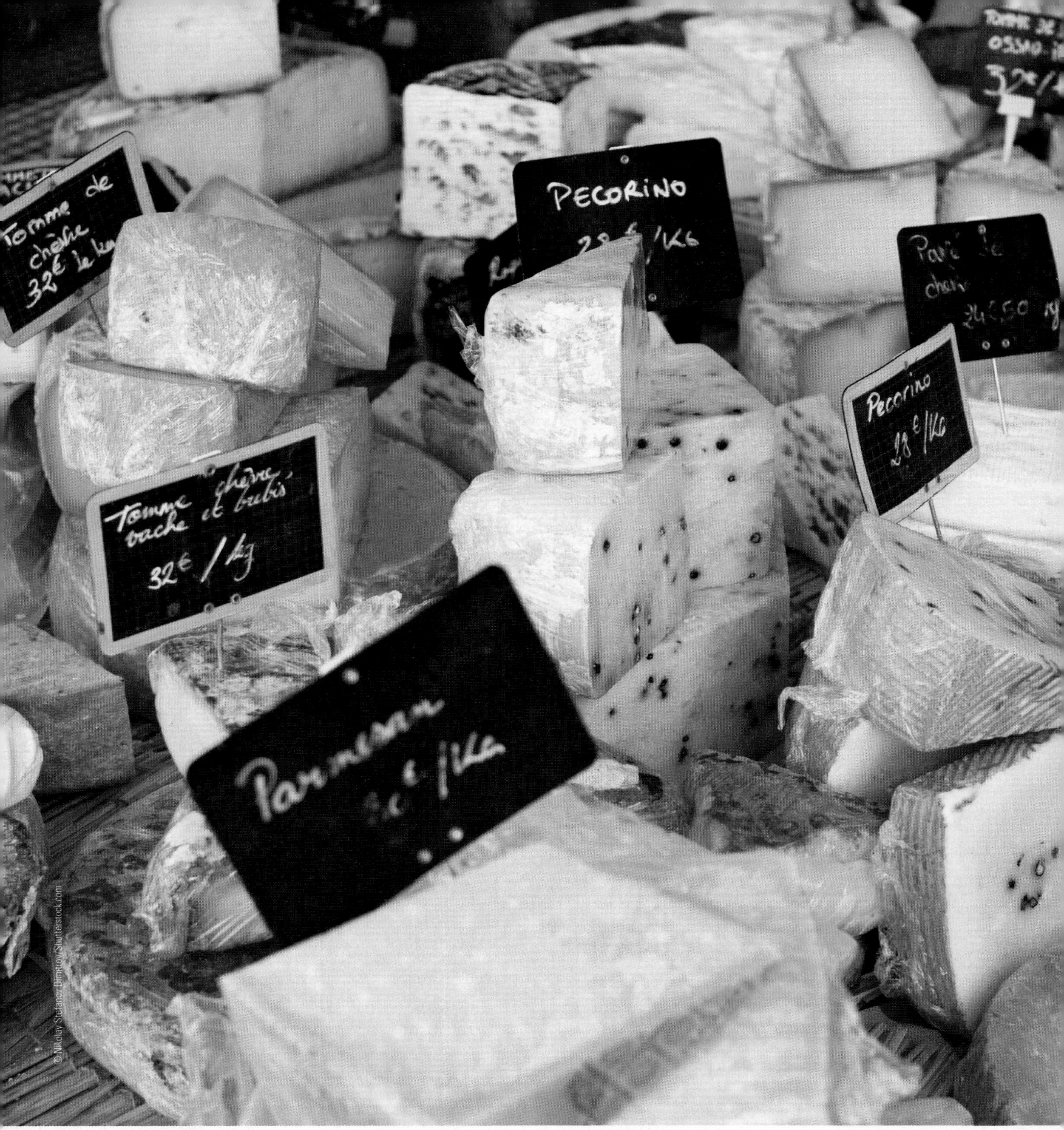

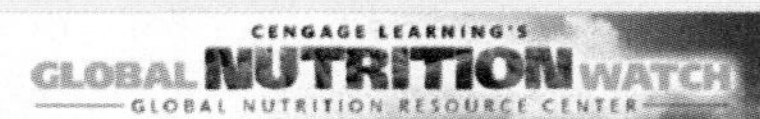

Water is essential for many body functions. Go to Nutrition Basics and Tools → Hydration to review recommendations on proper hydration. In what ways would inadequate hydration affect your daily activities?

12

Water and the Major Minerals

LEARNING GPS

Nutrition in Your Life

What's your beverage of choice? If you said water, then congratulate yourself for recognizing its importance in maintaining your body's fluid balance. If you answered milk, then pat yourself on the back for taking good care of your bones. Without water, you would realize within days how vital it is to your survival. The consequences of a lack of milk (or other calcium-rich foods) are also dramatic, but may not become apparent for decades. Water, calcium, and all the other major minerals support fluid balance and bone health. Before getting too comfortable reading this chapter, pour yourself a glass of water or milk. Your body will thank you. In the Nutrition Portfolio at the end of this chapter, you can determine whether the foods you are eating are meeting your water and major mineral needs.

Water is an essential nutrient, more important to life than any of the others. The body needs more water each day than any other nutrient. Furthermore, you can survive only a few days without water, whereas a deficiency of the other nutrients may take weeks, months, or even years to develop.

This chapter begins with a look at water and the body's fluids. The body maintains an appropriate balance and distribution of fluids with the help of another class of nutrients—the minerals. In addition to introducing the minerals that help regulate body fluids, this chapter describes many of the other important functions minerals perform in the body. Chapter 19 revisits water as a beverage and addresses consumer concerns about its safety.

Water is the most indispensable nutrient.

> FIGURE 12-1 **One Cell and Its Associated Fluids**

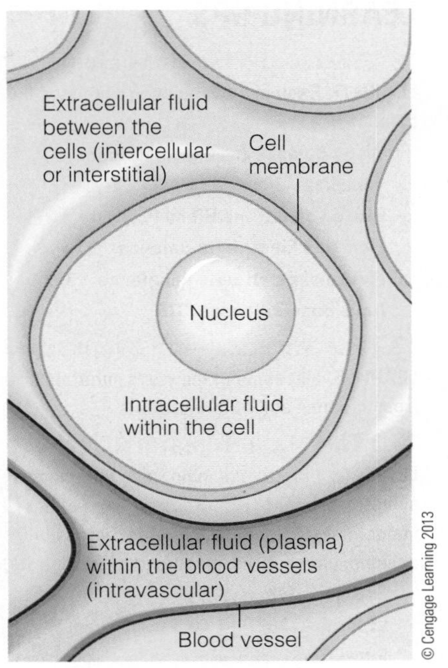

12.1 Water and the Body Fluids

LEARN IT Explain how the body regulates fluid balance.

Water constitutes about 60 percent of an adult's body weight and a higher percentage of a child's (see Figure 1-1, p. 7). Because water makes up about 75 percent of the weight of lean tissue and less than 25 percent of the weight of fat, a person's body composition influences how much of the body's weight is water. The proportion of water is generally smaller in females, obese people, and the elderly because of their smaller proportion of lean tissue.

In the body, water is the fluid in which all life processes occur. The water in the body fluids:

- Carries nutrients and waste products throughout the body
- Maintains the structure of large molecules such as proteins and glycogen
- Participates in metabolic reactions
- Serves as the solvent for minerals, vitamins, amino acids, glucose, and many other small molecules so that they can participate in metabolic activities
- Acts as a lubricant and cushion around joints and inside the eyes, the spinal cord, and, in pregnancy, the amniotic sac surrounding the fetus in the womb
- Aids in the regulation of normal body temperature, as the evaporation of sweat from the skin removes excess heat from the body
- Maintains blood volume

To support these and other vital functions, the body actively maintains an appropriate **water balance.** ♦

♦ Water balance: intake = output

Water Balance and Recommended Intakes

Every cell contains fluid of the exact composition that is best for that cell. Fluid inside cells is called **intracellular fluid,** whereas fluid outside cells is called **extracellular fluid.** ♦ The extracellular fluid that surrounds each cell is called **interstitial fluid.** Figure 12-1 illustrates a cell and its associated fluids. The compositions of intercellular and extracellular fluids differ from each other. They continuously lose and replace their components, yet the composition in each compartment remains remarkably constant under normal conditions. Because imbalances can be devastating, the body quickly responds by adjusting both water intake and excretion as needed. Consequently, the entire system of cells and fluids remains in a delicate, but controlled, state of homeostasis.

♦ Fluids in the body:
- Intracellular (inside cells)
- Extracellular (outside cells)
- Interstitial (between cells)
- Intravascular (inside blood vessels)

Water Intake

Thirst and satiety influence water intake, apparently in response to changes sensed by the mouth, **hypothalamus,** and nerves.[1] When water intake is inadequate, the blood becomes concentrated (having lost water but not the dissolved substances within it), the mouth becomes dry, and the hypothalamus initiates drinking behavior. When water intake is excessive, the stomach expands and stretch receptors send signals to stop drinking. Similar signals are sent from receptors in the heart as blood volume increases.

When too much water is lost from the body and not replaced, **dehydration** develops. A first sign of dehydration is thirst, the signal that the body has lost some fluid. If a person is unable to obtain water or, as in many elderly people, fails to perceive the thirst message, the symptoms of dehydration may progress rapidly from thirst to weakness, exhaustion, and delirium—and end in death if not corrected (see Table 12-1). Dehydration develops with either water deprivation or excessive water losses. (Chapter 14 revisits dehydration and the fluid needs of athletes.)

Water intoxication, on the other hand, is rare but can occur with excessive water intake and kidney disorders that reduce urine production. The symptoms may include confusion, convulsions, and even death in extreme cases. Excessive water ingestion (10 to 20 liters) within a few hours dilutes the sodium concentration of the blood and contributes to a dangerous condition known as **hyponatremia.** For this reason, guidelines suggest limiting fluid intake during times of heavy sweating to between 1 and 1.5 liters per hour. (Chapter 14 revisits hyponatremia as sometimes seen in endurance athletes.)

water balance: the balance between water intake and output (losses).

intracellular fluid: fluid within the cells, usually high in potassium and phosphate. Intracellular fluid accounts for approximately two-thirds of the body's water.
- **intra** = within

extracellular fluid: fluid outside the cells. Extracellular fluid includes two main components—the interstitial fluid between cells and the intravascular fluid of plasma. Extracellular fluid accounts for approximately one-third of the body's water.
- **extra** = outside

interstitial (IN-ter-STISH-al) **fluid:** fluid between the cells (intercellular), usually high in sodium and chloride. Interstitial fluid is a large component of extracellular fluid.
- **inter** = in the midst, between

thirst: a conscious desire to drink.

hypothalamus: a brain center that controls activities such as maintenance of water balance, regulation of body temperature, and control of appetite.

dehydration: the condition in which body water output exceeds water input. Symptoms include thirst, dry skin and mucous membranes, rapid heartbeat, low blood pressure, and weakness.

water intoxication: the rare condition in which body water contents are too high in all body fluid compartments.

hyponatremia (HIGH-po-na-TREE-me-ah): a decreased concentration of sodium in the blood.

TABLE 12-1 Signs of Dehydration

Body Weight Lost (%)	Symptoms
1–2	Thirst, fatigue, weakness, vague discomfort, loss of appetite
3–4	Impaired physical performance, dry mouth, reduction in urine, flushed skin, impatience, apathy
5–6	Difficulty concentrating, headache, irritability, sleepiness, impaired temperature regulation, increased respiratory rate
7–10	Dizziness, spastic muscles, loss of balance, delirium, exhaustion, collapse

NOTE: The onset and severity of symptoms at various percentages of body weight lost depend on the activity, fitness level, degree of acclimation, temperature, and humidity. If not corrected, dehydration can lead to death.

Water Sources The obvious dietary sources of water are water itself and other beverages, but nearly all foods also contain water. Most fruits and vegetables contain up to 90 percent water, and many meats and cheeses contain at least 50 percent. See Table 12-2 for selected foods and Appendix H for many more. Also, **metabolic water** is generated as an end product during condensation reactions and the oxidation of energy-yielding nutrients. Recall from Chapter 7 that when the energy-yielding nutrients break down, their carbons and hydrogens combine with oxygen to yield carbon dioxide (CO_2) and water (H_2O). As Table 12-3 (p. 370) shows, the water derived daily from these three sources—beverages, foods, and metabolism—averages about 2500 milliliters (roughly 2.5 quarts or 10.5 cups).

Water Losses At the very least, the body must excrete enough water to carry away the waste products generated by a day's metabolic activities. This **obligatory water excretion** is a minimum of about 500 milliliters (about 2 cups) of water each day. Above this amount, excretion adjusts to balance intake. If a person drinks more water, the kidneys excrete more urine, and the urine becomes more dilute. In addition to urine, water is lost from the lungs as vapor and from the skin as sweat; some is also lost in feces.* The amount of fluid lost from each source varies, depending on the environment (such as heat or humidity) and physical conditions (such as exercise or fever). On average, daily losses total about 2500 milliliters. Table 12-3 (p. 370) shows how daily water losses and intakes balance; maintaining this balance requires healthy kidneys and an adequate intake of fluids.

Water Recommendations Because water needs vary depending on diet, activity, environmental temperature, and humidity, a general water requirement is difficult to establish. Recommendations ♦ are sometimes expressed in proportion to the amount of energy expended under average environmental conditions. The recommended

♦ Water recommendation:
- 1.0 to 1.5 mL/kcal expended (adults)*
- 1.5 mL/kcal expended (infants and athletes)

Conversion factors:
- 1 mL = 0.03 fluid oz
- 125 mL ≈ ½ c

Easy estimation: ½ c per 100 kcal expended

*For those using kilojoules: 4.2 to 6.3 mL/kJ expended.

TABLE 12-2 Percentage of Water in Selected Foods

100%	Water
90–99%	Fat-free milk, strawberries, watermelon, lettuce, cabbage, celery, spinach, broccoli
80–89%	Fruit juice, yogurt, apples, grapes, oranges, carrots
70–79%	Shrimp, bananas, corn, potatoes, avocados, cottage cheese, ricotta cheese
60–69%	Pasta, legumes, salmon, ice cream, chicken breast
50–59%	Ground beef, hot dogs, feta cheese
40–49%	Pizza
30–39%	Cheddar cheese, bagels, bread
20–29%	Pepperoni sausage, cake, biscuits
10–19%	Butter, margarine, raisins
1–9%	Crackers, cereals, pretzels, taco shells, peanut butter, nuts
0%	Oils, sugars

*Water lost from the lungs and skin accounts for almost one-half of the daily losses even when a person is not visibly perspiring; these losses are commonly referred to as *insensible water losses*.

metabolic water: water generated during metabolism.

obligatory (ah-BLIG-ah-TORE-ee) **water excretion:** the minimum amount of water the body has to excrete each day to dispose of its wastes—about 500 mL (about 2 cups, or 1 pint).

TABLE 12-3 Water Balance

Water Sources	Amount (mL)	Water Losses	Amount (mL)
Beverages	550 to 1500	Kidneys (urine)	500 to 1400
Foods	700 to 1000	Skin (sweat)	450 to 900
Metabolism	200 to 300	Lungs (breath)	350
		GI tract (feces)	150
Total	1450 to 2800	Total	1450 to 2800

NOTE: For perspective, 100 milliliters is a little less than ½ cup and 1000 milliliters is a little more than 1 quart (1 mL = 0.03 oz).

water intake for a person who expends 2000 kcalories a day, for example, is 2 to 3 liters of water (about 8 to 12 cups). This recommendation is in line with the Adequate Intake (AI) for *total* water set by the DRI Committee. ♦ Total water includes not only drinking water, but water in other beverages and in foods as well. Adults in the United States report drinking about 3 liters of water and other beverages a day.[2]

♦ AI for *total* water:
- Men: 3.7 L/day
- Women: 2.7 L/day

Conversion factors:
- 1 L ≈ 1 qt ≈ 32 oz ≈ 4 c

Because a wide range of water intakes will prevent dehydration and its harmful consequences, the AI is based on average intakes. People who are physically active or who live in hot environments may need more.

Which beverages are best? Any beverage can readily meet the body's fluid needs, but those with few or no kcalories do so without contributing to weight gain. Given that obesity is a major health problem and that beverages currently represent more than 20 percent of the total energy intake in the United States, water is the best choice for most people. Other choices include tea, coffee, nonfat and low-fat milk and soymilk, artificially sweetened beverages, fruit and vegetable juices, sports drinks, and lastly, sweetened nutrient-poor beverages.

Some research indicates that people who drink caffeinated beverages lose a little more fluid than when drinking water because caffeine acts as a diuretic. The DRI Committee considered such findings in their recommendations for water intake and concluded: "Caffeinated beverages contribute to the daily total water intake similar to that contributed by non-caffeinated beverages."[3] In other words, it doesn't seem to matter whether people rely on caffeine-containing beverages or other beverages to meet their fluid needs.

As Highlight 7 explains, alcohol acts as a diuretic and can impair a person's health. Alcohol should not be used to meet fluid needs.

Health Effects of Water Water supports good health.[4] Physical and mental performances depend on it, as does the optimal functioning of the GI tract, kidneys, heart, and other body systems.

The kind of water a person drinks may also make a difference to health. Water is usually either hard or soft. **Hard water** has high concentrations of calcium and magnesium; the principal mineral of **soft water** is sodium or potassium. (See the accompanying glossary for other common terms used to describe water.) In practical terms, soft water makes more bubbles with less soap; hard water leaves a ring on the tub, a crust of rocklike crystals in the teakettle, and a gray residue in the laundry.

Soft water may seem more desirable around the house, and some homeowners purchase water softeners that replace magnesium and calcium with sodium. In the body, however, soft water with sodium may aggravate hypertension and heart disease. In contrast, the minerals in hard water may benefit these conditions.

Soft water also more easily dissolves certain contaminant minerals, such as cadmium and lead, from old plumbing pipes. As Chapter 13 explains, these contaminant minerals harm the body by displacing the nutrient minerals from their normal sites of action. People who live in buildings with old plumbing should run the cold water tap a minute or two to flush out harmful minerals whenever the water faucet has been off for more than 6 hours.[5]

Many people select **bottled water**, believing it to be safer than tap water and therefore worth its substantial cost. Chapter 19 offers a discussion of bottled water safety and regulations.

hard water: water with a high calcium and magnesium content.

soft water: water with a high sodium or potassium content.

bottled water: drinking water sold in bottles.

> FIGURE 12-2 ***Animated*** **A Nephron, One of the Kidney's Many Functioning Units**

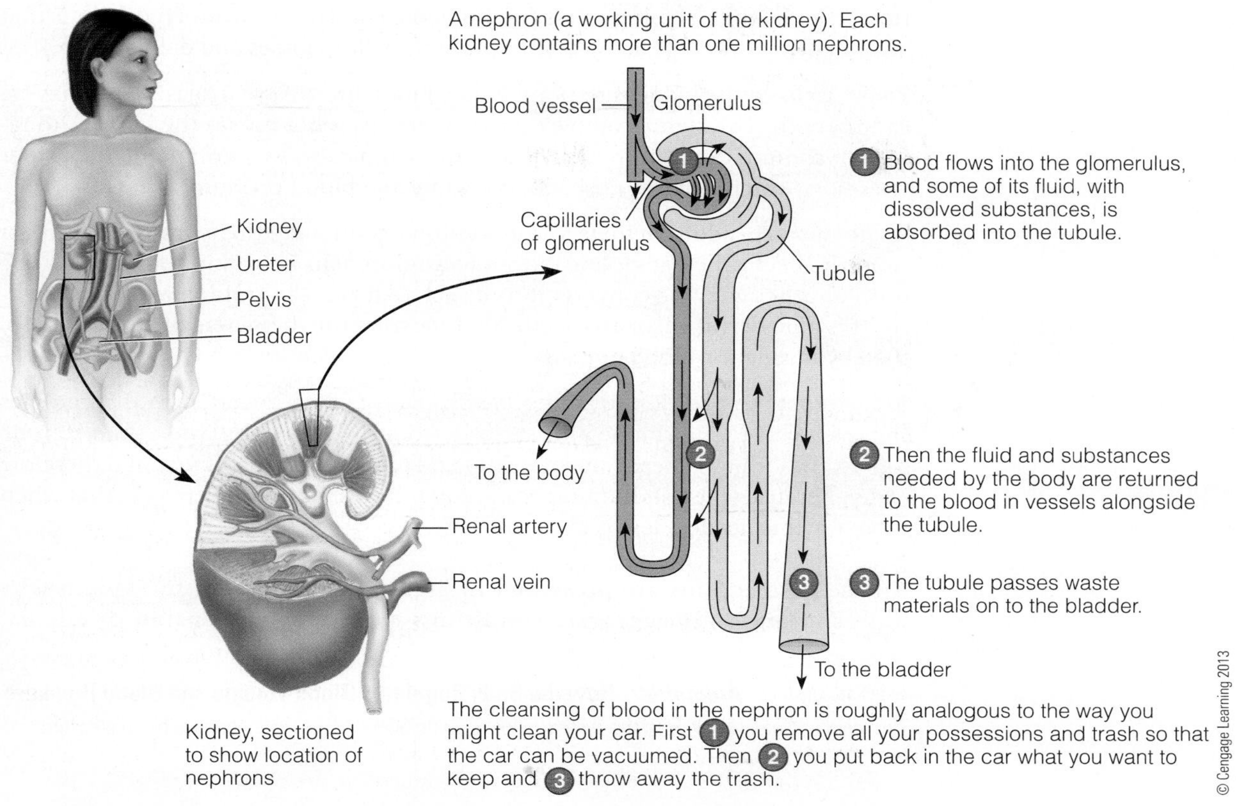
A nephron (a working unit of the kidney). Each kidney contains more than one million nephrons.
Blood vessel
Glomerulus
Capillaries of glomerulus
Tubule
Kidney
Ureter
Pelvis
Bladder
To the body
Renal artery
Renal vein
To the bladder
1 Blood flows into the glomerulus, and some of its fluid, with dissolved substances, is absorbed into the tubule.
2 Then the fluid and substances needed by the body are returned to the blood in vessels alongside the tubule.
3 The tubule passes waste materials on to the bladder.
Kidney, sectioned to show location of nephrons

The cleansing of blood in the nephron is roughly analogous to the way you might clean your car. First (1) you remove all your possessions and trash so that the car can be vacuumed. Then (2) you put back in the car what you want to keep and (3) throw away the trash.

Blood Volume and Blood Pressure Fluids maintain the blood volume, which in turn influences blood pressure. The kidneys are central to the regulation of blood volume and blood pressure. All day, every day, the kidneys reabsorb needed substances and water and excrete wastes with some water in the urine (see Figure 12-2). The kidneys meticulously adjust the volume and the concentration of the urine to accommodate changes in the body, including variations in the day's food and beverage intakes. Instructions on whether to retain or release substances or water come from ADH, renin, angiotensin, and aldosterone.

ADH Whenever blood volume or blood pressure falls too low, or whenever the extracellular fluid becomes too concentrated, the hypothalamus signals the pituitary gland to release **antidiuretic hormone (ADH).** ADH is a water-conserving hormone that stimulates the kidneys to reabsorb water. Consequently, the more

antidiuretic hormone (ADH): a hormone produced by the pituitary gland in response to dehydration (or a high sodium concentration in the blood) that stimulates the kidneys to reabsorb more water and therefore to excrete less. In addition to its antidiuretic effect, ADH elevates blood pressure and so is also called *vasopressin* (VAS-oh-PRES-in).
- **vaso** = vessel
- **press** = pressure

GLOSSARY OF TYPES OF WATER

artesian water: water drawn from a well that taps a confined aquifer in which the water is under pressure.

carbonated water: water that contains carbon dioxide gas, either naturally occurring or added, that causes bubbles to form in it; also called *bubbling* or *sparkling water.* The FDA defines seltzer, soda, and tonic waters as soft drinks; they are not regulated as water.

distilled water: water that has been vaporized and recondensed, leaving it free of dissolved minerals.

filtered water: water treated by filtration, usually through *activated carbon filters* that reduce the lead in tap water, or by *reverse osmosis* units that force pressurized water across a membrane removing lead, arsenic, and some microorganisms from tap water.

mineral water: water from a spring or well that naturally contains at least 250 parts per million (ppm) of minerals. Minerals give water a distinctive flavor. Many mineral waters are high in sodium.

natural water: water obtained from a spring or well that is certified to be safe and sanitary. The mineral content may not be changed, but the water may be treated in other ways such as with ozone or by filtration.

public water: water from a municipal or county water system that has been treated and disinfected.

purified water: water that has been treated by distillation or other physical or chemical processes that remove dissolved solids. Because purified water contains no minerals or contaminants, it is useful for medical and research purposes.

spring water: water originating from an underground spring or well. It may be bubbly (carbonated), or "flat" or "still," meaning not carbonated. Brand names such as "Spring Pure" do not necessarily mean that the water comes from a spring.

well water: water drawn from groundwater by tapping into an aquifer.

water you need, the less your kidneys excrete. These events also trigger thirst. Drinking water and retaining fluids raise the blood volume and dilute the concentrated fluids, thus helping to restore homeostasis. (Recall from Highlight 7 that alcohol depresses ADH activity, thus promoting fluid losses and dehydration.)

Renin Cells in the kidneys respond to low blood pressure by releasing an enzyme called **renin.** Through a complex series of events, renin causes the kidneys to reabsorb sodium. Sodium reabsorption, in turn, is always accompanied by water retention, which helps to raise blood volume and blood pressure.

Angiotensin In addition to its role in sodium retention, renin hydrolyzes a protein from the liver called angiotensinogen to **angiotensin I.** Angiotensin I is inactive until another enzyme converts it to its active form—**angiotensin II.** Angiotensin II is a powerful **vasoconstrictor** that narrows the diameters of blood vessels, thereby raising the blood pressure.

Aldosterone In addition to acting as a vasoconstrictor, angiotensin II stimulates the release of the hormone **aldosterone** from the **adrenal glands.** Aldosterone signals the kidneys to excrete potassium and to retain more sodium, and therefore water, because when sodium moves, water follows. Again, the effect is that when more water is needed, less is excreted.

All of these actions are presented in Figure 12-3 and help to explain why high-sodium diets aggravate conditions such as hypertension or edema.

> FIGURE 12-3 ***Animated*** **How the Body Regulates Blood Volume and Blood Pressure**

The renin-angiotensin-aldosterone system helps regulate blood volume and therefore blood pressure.

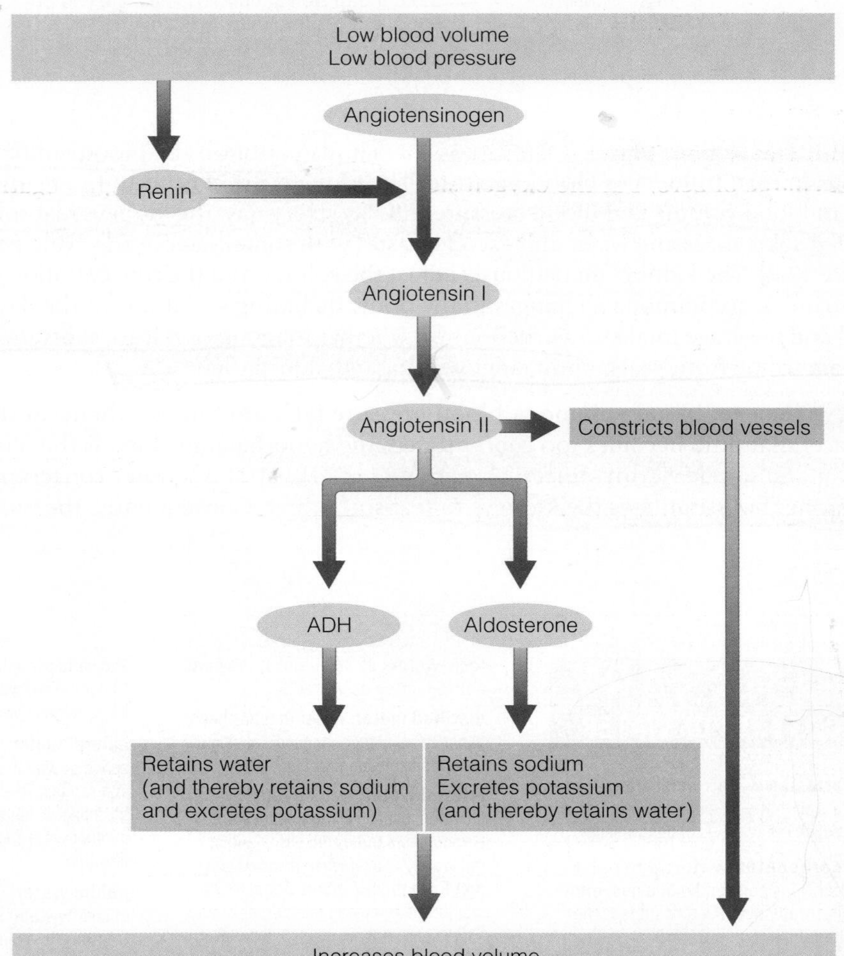

renin (REN-in): an enzyme from the kidneys that hydrolyzes the protein angiotensinogen to angiotensin I, which results in the kidneys reabsorbing sodium.

angiotensin I (AN-gee-oh-TEN-sin): an inactive precursor that is converted by an enzyme to yield active angiotensin II.

angiotensin II: a hormone involved in blood pressure regulation.

vasoconstrictor (VAS-oh-kon-STRIK-tor): a substance that constricts or narrows the blood vessels.

aldosterone (al-DOS-ter-own): a hormone secreted by the adrenal glands that regulates blood pressure by increasing the reabsorption of sodium by the kidneys. Aldosterone also regulates chloride and potassium concentrations.

adrenal glands: glands adjacent to, and just above, each kidney.

Too much sodium causes water retention and an accompanying rise in blood pressure or swelling in the interstitial spaces. Chapter 18 discusses hypertension in detail.

Fluid and Electrolyte Balance Maintaining a balance of about two-thirds of the body fluids inside the cells and one-third outside is vital to the life of the cells. If too much water were to enter the cells, they might rupture; if too much water were to leave, they would collapse. To control the movement of water, the cells direct the movement of the major minerals. ♦

Dissociation of Salt in Water When a mineral **salt** such as sodium chloride (NaCl) dissolves in water, it separates **(dissociates)** into **ions**—positively and negatively charged particles (Na^+ and Cl^-). The positive ions are **cations;** the negative ones are **anions.** ♦ Unlike pure water, which conducts electricity poorly, ions dissolved in water carry electrical current. For this reason, salts that dissociate into ions are called **electrolytes,** and fluids that contain them are **electrolyte solutions.**

In all electrolyte solutions, anion and cation concentrations are balanced (the number of negative and positive charges are equal). If a fluid contains 1000 negative charges, it must contain 1000 positive charges too. If an anion enters the fluid, a cation must accompany it or another anion must leave so that electrical neutrality will be maintained. Thus, whenever sodium (Na^+) ions leave a cell, potassium (K^+) ions enter, for example. In fact, it's a good bet that whenever Na^+ and K^+ ions are moving, they are going in opposite directions.

Table 12-4 shows that, indeed, the positive and negative charges inside and outside cells are perfectly balanced even though the numbers of each kind of ion differ over a wide range. Inside the cells, the positive charges total 202 and the negative charges balance these perfectly. Outside the cells, the amounts and proportions of the ions differ from those inside, but again the positive and negative charges balance. Scientists count these charges in **milliequivalents per liter (mEq/L)**.

Electrolytes Attract Water Electrolytes attract water. Each water molecule has a net charge of zero, but the oxygen side of the molecule has a slight negative

♦ The major minerals:
- Sodium
- Chloride
- Potassium
- Calcium
- Phosphorus
- Magnesium
- Sulfur

♦ To remember the difference between cations and anions, think of the "t" in cations as a "plus" (+) sign and the "n" in anions as a "negative."

TABLE 12-4 Important Body Electrolytes

Electrolytes	Intracellular (inside cells) Concentration (mEq/L)	Extracellular (outside cells) Concentration (mEq/L)
Cations (positively charged ions)		
Sodium (Na^+)	10	142
Potassium (K^+)	150	5
Calcium (Ca^{++})	2	5
Magnesium (Mg^{++})	40	3
	202	155
Anions (negatively charged ions)		
Chloride (Cl^-)	2	103
Bicarbonate (hCO_3^-)	10	27
Phosphate ($hpO_4^=$)	103	2
Sulfate ($SO_4^=$)	20	1
Organic acids (lactate, pyruvate)	10	6
Proteins	57	16
	202	155

NOTE: The numbers of positive and negative charges in a given fluid are the same. For example, in extracellular fluid, the cations and anions both equal 155 milliequivalents per liter (mEq/L). Of the cations, sodium ions make up 142 mEq/L; and potassium, calcium, and magnesium ions make up the remainder. Of the anions, chloride ions number 103 mEq/L; bicarbonate ions number 27; and the rest are provided by phosphate ions, sulfate ions, organic acids, and protein.

salt: a compound composed of a positive ion other than H^+ and a negative ion other than OH^-. An example is sodium chloride ($Na^+ Cl^-$).
- **Na** = sodium
- **Cl** = chloride

dissociates (dis-SO-see-aites): physically separates.

ions (EYE-uns): atoms or molecules that have gained or lost electrons and therefore have electrical charges. Examples include the positively charged sodium ion (Na^+) and the negatively charged chloride ion (Cl^-). For a closer look at ions, see Appendix B.

cations (CAT-eye-uns): positively charged ions.

anions (AN-eye-uns): negatively charged ions.

electrolytes: salts that dissolve in water and dissociate into charged particles called ions.

electrolyte solutions: solutions that can conduct electricity.

milliequivalents per liter (mEq/L): the concentration of electrolytes in a volume of solution. Milliequivalents reveal characteristics about the solution that are not evident when the concentration is expressed in terms of weight.

> **FIGURE 12-4 Water Dissolves Salts and Follows Electrolytes**

The structural arrangement of the two hydrogen atoms and one oxygen atom enables water to dissolve salts. Water's role as a solvent is one of its most valuable characteristics.

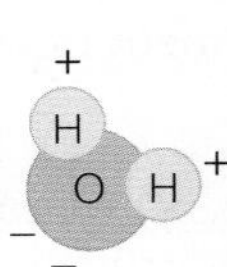

The negatively charged electrons that bond the hydrogens to the oxygen spend most of their time near the oxygen atom. As a result, the oxygen is slightly negative, and the hydrogens are slightly positive (see Appendix B).

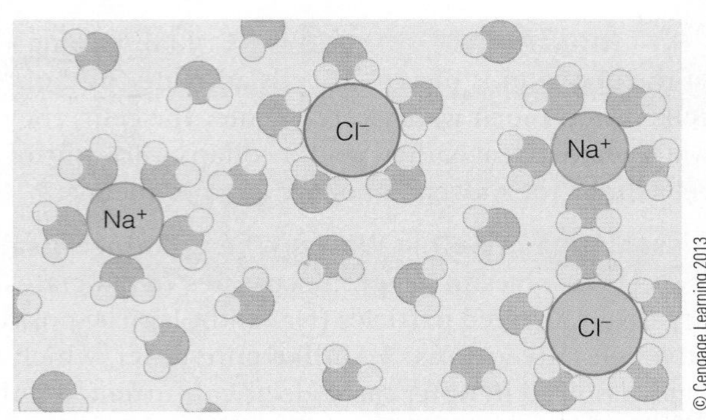

In an electrolyte solution, water molecules are attracted to both anions and cations. Notice that the negative oxygen atoms of the water molecules are drawn to the sodium cation (Na^+), whereas the positive hydrogen atoms of the water molecules are drawn to the chloride ions (Cl^-).

> **FIGURE 12-5 A Cell and Its Electrolytes**

All of these electrolytes are found both inside and outside the cells, but each can be found mostly on one side or the other of the cell membrane.

Outside the cells

Cell membrane

K

Mg

P

S

Na

Cl

Within the cell

Blood vessel

Chemical symbols:
K = potassium
P = phosphorus
Mg = magnesium
S = sulfate
Na = sodium
Cl = chloride

Key:
Cations
Anions

charge, and the hydrogens have a slight positive charge. Figure 12-4 shows the result in an electrolyte solution: both positive and negative ions attract clusters of water molecules around them. This attraction dissolves salts in water and enables the body to move fluids into appropriate compartments.

Water Follows Electrolytes As Figure 12-5 shows, some electrolytes reside primarily outside the cells (notably, sodium and chloride), whereas others reside predominantly inside the cells (notably, potassium, magnesium, phosphate, and sulfate). Cell membranes are *selectively permeable,* meaning that they allow the passage of some molecules, but not others. Whenever electrolytes move across the membrane, water follows.

The movement of water across a membrane toward the more concentrated **solutes** is called **osmosis.** The amount of pressure needed to prevent the movement of water across a membrane is called the **osmotic pressure.** Figure 12-6 presents osmosis, and the photos of salted eggplant and rehydrated raisins provide familiar examples.

Proteins Regulate Flow of Fluids and Ions Chapter 6 describes how proteins attract water and help to regulate fluid movement. In addition, transport proteins in the cell membranes regulate the passage of positive ions and other substances from one side of the membrane to the other. Negative ions follow positive ions, and water flows toward the more concentrated solution.

An example of a protein that regulates the flow of fluids and ions in and out of cells is the sodium-potassium pump. The pump actively exchanges sodium for potassium across the cell membrane, using ATP as an energy source. Figure 6-10 on p. 177 illustrates this action.

Regulation of Fluid and Electrolyte Balance The amounts of various minerals in the body must remain nearly constant. Regulation occurs chiefly at two sites: the GI tract and the kidneys.

Minerals in foods enter the body by way of the GI tract. In addition, the digestive juices of the GI tract contain minerals. These minerals and those from foods are absorbed in the large intestine or excreted as needed. Each day, 8 liters of fluids and associated minerals are recycled this way, providing ample opportunity for the regulation of electrolyte balance.

solutes (SOLL-yutes): the substances that are dissolved in a solution. The number of molecules in a given volume of fluid is the *solute concentration.*

osmosis: the movement of water across a membrane *toward* the side where the solutes are more concentrated.

osmotic pressure: the amount of pressure needed to prevent the movement of water across a membrane.

> **FIGURE 12-6 Osmosis**

Water flows in the direction of the more highly concentrated solution.

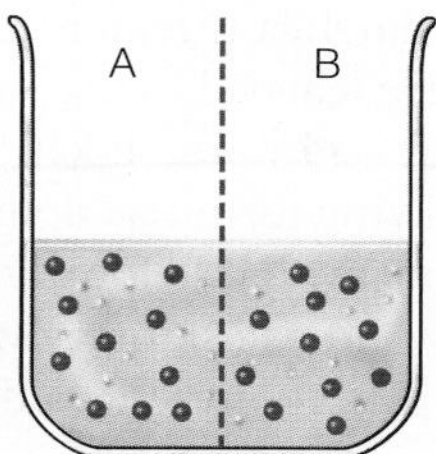

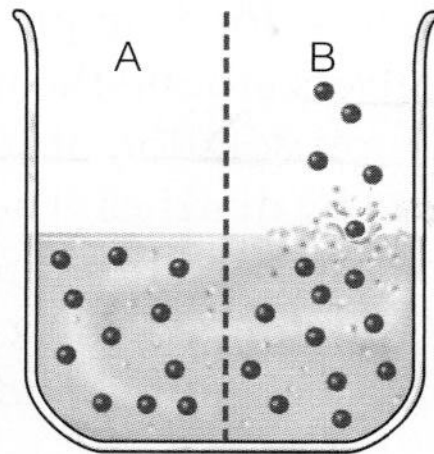

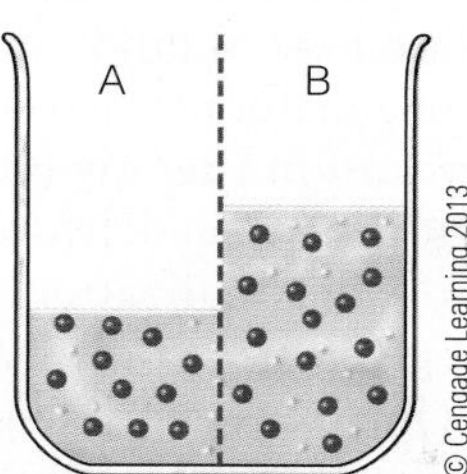

1. With equal numbers of solute particles on both sides of the semi-permeable membrane, the concentrations are equal, and the tendency of water to move in either direction is about the same.
2. Now additional solute is added to side B. Solute cannot flow across the divider (in the case of a cell, its membrane).
3. Water can flow both ways across the divider, but has a greater tendency to move from side A to side B, where there is a greater concentration of solute. The volume of water becomes greater on side B, and the concentrations on side A and B become equal.

The kidneys' control of the body's *water* content by way of the hormone ADH has already been described (see pp. 371–372). The kidneys regulate the *electrolyte* contents by responding to the hormone aldosterone (also explained on p. 372). If the body's sodium is low, aldosterone stimulates sodium reabsorption from the kidneys. As sodium is reabsorbed, potassium (another positive ion) is excreted in accordance with the rule that total positive charges must remain in balance with total negative charges.

Fluid and Electrolyte Imbalance Normally, the body defends itself successfully against fluid and electrolyte imbalances. Certain situations and some medications, however, may overwhelm the body's ability to compensate. Severe, prolonged vomiting and diarrhea as well as heavy sweating, burns, and traumatic wounds may incur such great fluid and electrolyte losses as to precipitate a medical emergency.

Different Solutes Lost by Different Routes Different solutes are lost depending on why fluid is lost. If fluid is lost by vomiting or diarrhea, sodium is lost indiscriminately. If the adrenal glands oversecrete aldosterone, as may occur when they develop a tumor, the kidneys may excrete too much potassium. A person with uncontrolled diabetes may lose glucose, a solute not normally excreted, and large amounts of fluid with it. Each situation results in dehydration, but drinking water alone will not restore electrolyte balance. Medical intervention is required.

When immersed in water, raisins become plump because water moves toward the higher concentration of sugar inside the raisins.

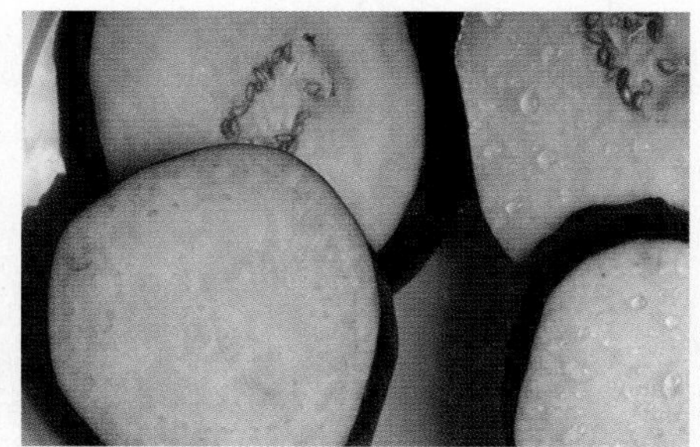

When sprinkled with salt, vegetables "sweat" because water moves toward the higher concentration of salt outside the eggplant.

© Helder Almeida/Shutterstock.com

Physically active people must remember to replace their body fluids.

♦ A simple ORT recipe (cool before giving):

- ½ L boiling water
- A small handful of sugar (4 tsp)
- 3 pinches of salt (½ tsp)

Replacing Lost Fluids and Electrolytes In many cases, people can replace the fluids and minerals lost in sweat or in a temporary bout of diarrhea by drinking plain cool water and eating regular foods. Some cases, however, demand rapid replacement of fluids and electrolytes—for example, when diarrhea threatens the life of a malnourished child. Caregivers around the world have learned to use **oral rehydration therapy (ORT)**—a simple solution of sugar, salt, and water, taken by mouth—to treat dehydration caused by diarrhea. These lifesaving formulas do not require hospitalization and can be prepared from ingredients available locally. ♦ Caregivers need only learn to measure ingredients carefully and use sanitary water. Once rehydrated, a person can begin eating foods. (Chapter 14 presents a discussion of sport drinks.)

Acid-Base Balance

The body uses its ions not only to help maintain fluid and electrolyte balance, but also to regulate the acidity (pH) of its fluids. The pH scale introduced in Chapter 3 is repeated here, in Figure 12-7, with the normal and abnormal pH ranges of the blood added. As you can see, the body must maintain the pH within a narrow range to avoid life-threatening consequences. Slight deviations in either direction can denature proteins, rendering them useless. Enzymes couldn't catalyze reactions and hemoglobin couldn't carry oxygen—to name just two examples.

The acidity of the body's fluids is determined by the concentration of hydrogen ions (H^+).* A high concentration of hydrogen ions is very acidic. Normal energy metabolism generates hydrogen ions, as well as many other acids, that must be

> FIGURE 12-7 The pH Scale

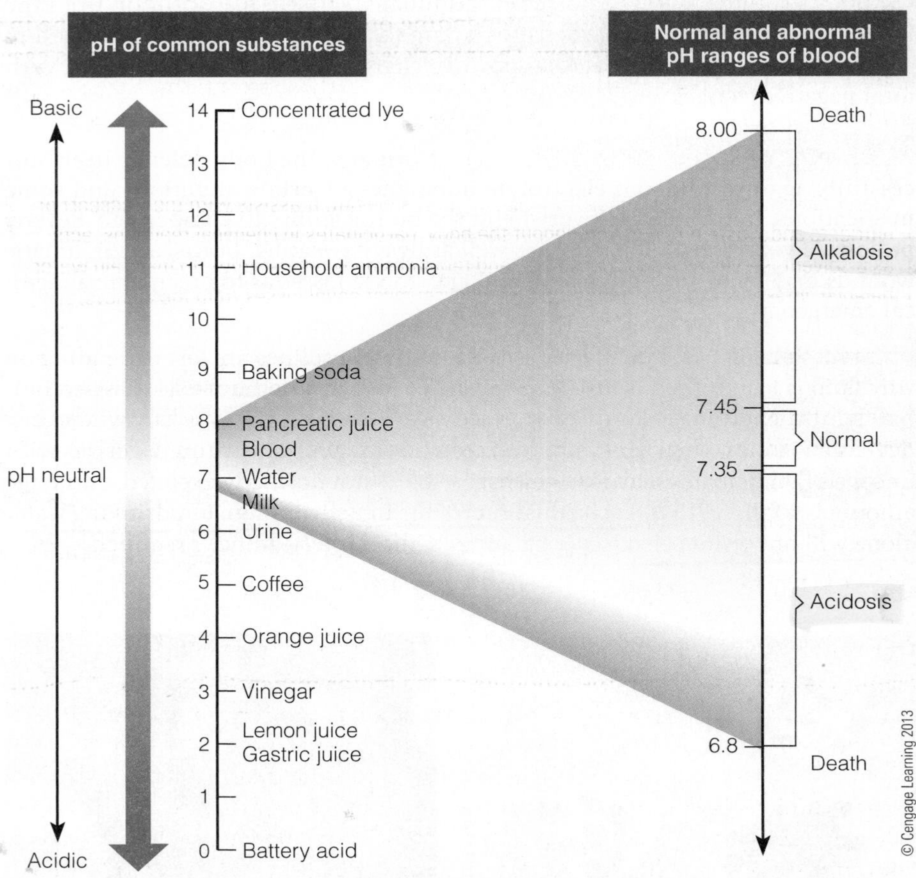

NOTE: Each step is 10 times as concentrated in base (1/10 as much as acid, or H^+) as the one below it.

oral rehydration therapy (ORT): the administration of a simple solution of sugar, salt, and water, taken by mouth, to treat dehydration caused by diarrhea.

*The lower the pH, the higher the H^+ ion concentration and the stronger the acid. A pH above 7 is alkaline, or base—a solution in which OH^- ions predominate.

neutralized. Three systems defend the body against fluctuations in pH—buffers in the blood, respiration in the lungs, and excretion in the kidneys.

Regulation by the Buffers **Bicarbonate** (a base) and **carbonic acid** (an acid) in the body fluids, as well as some proteins, protect the body against changes in acidity by acting as **buffers**—substances that can neutralize acids or bases. Carbon dioxide, which is formed all the time during energy metabolism, dissolves in water to form carbonic acid in the blood. Carbonic acid, in turn, dissociates to form hydrogen ions and bicarbonate ions. The appropriate balance between carbonic acid and bicarbonate is essential to maintaining optimal blood pH. Figure 12-8 presents the chemical reactions of this buffer system, which is primarily under the control of the lungs and kidneys.

> **FIGURE 12-8 Bicarbonate–Carbonic Acid Buffer System**

The reversible reactions of the bicarbonate–carbonic acid buffer system help to regulate the body's pH. Recall from Chapter 7 that carbon dioxide and water are formed during energy metabolism.

Carbon dioxide (CO_2) is a volatile gas that quickly dissolves in water (H_2O), forming carbonic acid (H_2CO_3):

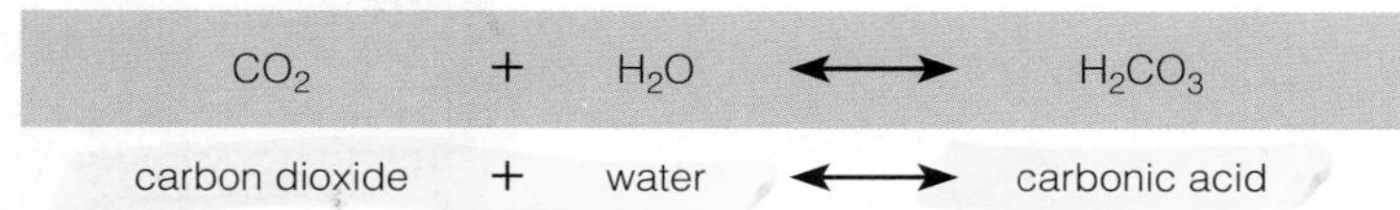

Carbonic acid readily dissociates to a hydrogen ion (H^+) and a bicarbonate ion (HCO_3^-):

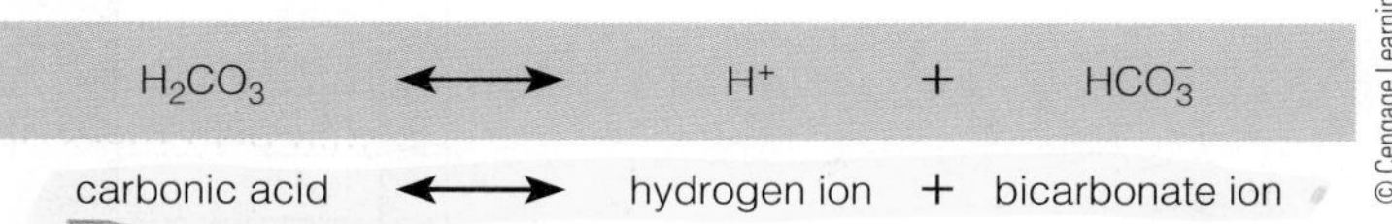

Respiration in the Lungs The lungs control the concentration of carbonic acid by raising or slowing the respiration rate, depending on whether the pH needs to be increased or decreased. If too much carbonic acid builds up, the respiration rate speeds up; this hyperventilation increases the amount of carbon dioxide exhaled, thereby lowering the carbonic acid concentration and restoring homeostasis. Conversely, if bicarbonate builds up, the respiration rate slows; carbon dioxide is retained and forms more carbonic acid. Again, homeostasis is restored.

Excretion in the Kidneys The kidneys control the concentration of bicarbonate by either reabsorbing or excreting it, depending on whether the pH needs to be increased or decreased, respectively. Their work is complex, but the net effect is easy to sum up. The *body's* total acid burden remains nearly constant; the acidity of the *urine* fluctuates to accommodate that balance.

REVIEW IT Explain how the body regulates fluid balance.
Water makes up about 60 percent of the adult body's weight. It assists with the transport of nutrients and waste products throughout the body, participates in chemical reactions, acts as a solvent, serves as a shock absorber, and regulates body temperature. To maintain water balance, intake from liquids, foods, and metabolism must equal losses from the kidneys, skin, lungs, and GI tract. Whenever the body experiences low blood volume, low blood pressure, or highly concentrated body fluids, the actions of ADH, renin, angiotensin, and aldosterone restore homeostasis. Electrolytes (charged minerals) in the fluids help distribute the fluids inside and outside the cells, thus ensuring the appropriate water balance and acid-base balance to support all life processes. Excessive losses of fluids and electrolytes upset these balances, and the kidneys play a key role in restoring homeostasis.

12.2 The Minerals—An Overview

LEARN IT List some of the ways minerals differ from vitamins and other nutrients.

Figure 12-9 (p. 378) shows the amounts of the **major minerals** found in the body and, for comparison, some of the **trace minerals.** The distinction between the major and trace minerals does not mean that one group is more important than the other—all minerals are vital. The major minerals are so named because they are present, and needed, in larger amounts in the body. They are shown at the top of the figure and are discussed in this chapter. The trace minerals, shown at the bottom, are discussed in Chapter 13. A few generalizations pertain to all of the minerals and distinguish them from the vitamins. Especially notable is their chemical nature.

Inorganic Elements Unlike the organic vitamins, which are easily destroyed, minerals are inorganic elements that always retain their chemical identity. Once minerals enter the body proper, they remain there until excreted; they cannot be changed into anything else. Iron, for example, may temporarily combine with other charged

bicarbonate: an alkaline compound with the formula HCO_3 that is produced in all cell fluids from the dissociation of carbonic acid to help maintain the body's acid-base balance. Bicarbonate is also secreted from the pancreas as part of the pancreatic juice.

carbonic acid: a compound with the formula H_2CO_3 that results from the combination of carbon dioxide (CO_2) and water (H_2O); of particular importance in maintaining the body's acid-base balance.

buffers: compounds that keep a solution's pH constant when acids or bases are added.

major minerals: essential mineral nutrients the human body requires in relatively large amounts (greater than 100 milligrams per day); sometimes called *macrominerals.*

trace minerals: essential mineral nutrients the human body requires in relatively small amounts (less than 100 milligrams per day); sometimes called *microminerals.*

> FIGURE 12-9 Minerals in a 60-kilogram (132-pound) Human Body

Not only are the major minerals needed by the body in larger amounts, but they are also present in the body in larger amounts than the trace minerals.

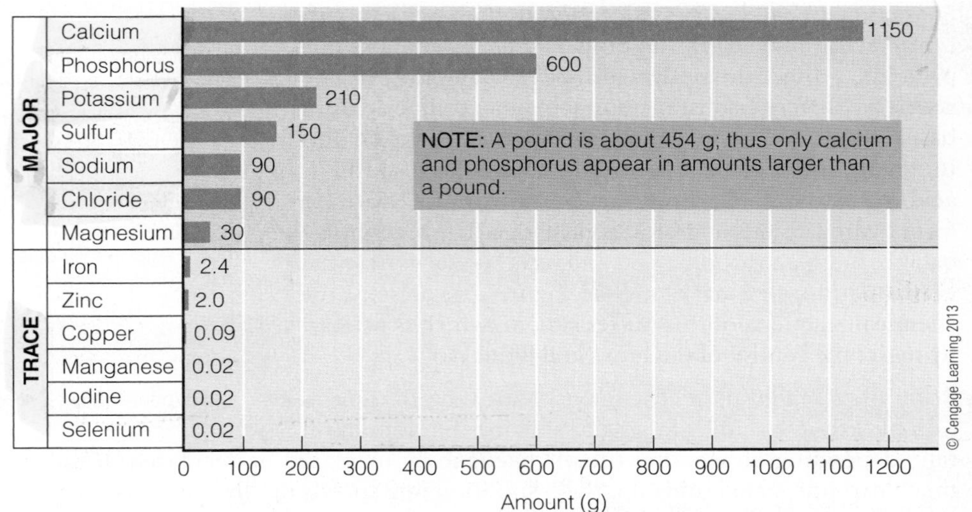

elements in salts, but it is always iron. Neither can minerals be destroyed by heat, air, acid, or mixing. Consequently, little care is needed to preserve minerals during food preparation. In fact, the ash that remains when a food is burned contains all the minerals that were in the food originally. Minerals can be lost from food only when they leach into cooking water that is then poured down the drain.

The Body's Handling of Minerals The minerals also differ from the vitamins in the amounts the body can absorb and in the extent to which they must be specially handled. Some minerals, such as potassium, are easily absorbed into the blood, transported freely, and readily excreted by the kidneys, much like the water-soluble vitamins. Other minerals, such as calcium, are more like fat-soluble vitamins in that they must have carriers to be absorbed and transported. And, like some of the fat-soluble vitamins, minerals taken in excess can be toxic.

Variable Bioavailability The **bioavailability** of minerals varies. Some foods contain **binders** that combine chemically with minerals, preventing their absorption and carrying them out of the body with other wastes. Examples of binders include phytates, which are found primarily in legumes and grains, and oxalates, which are present in rhubarb and spinach, among other foods. These foods contain more minerals than the body actually receives for use.

Nutrient Interactions Chapter 10 describes how the presence or absence of one vitamin can affect another's absorption, metabolism, and excretion. The same is true of the minerals. The interactions between sodium and calcium, for example, cause both to be excreted when sodium intakes are high. Phosphorus binds with magnesium in the GI tract, so magnesium absorption is limited when phosphorus intakes are high. These are just two examples of the interactions involving minerals featured in this chapter. Discussions in both this chapter and the next point out additional problems that arise from such interactions. Notice how often they reflect an excess of one mineral creating an inadequacy of another and how supplements—not foods—are most often to blame.

REVIEW IT List some of the ways minerals differ from vitamins and other nutrients.
Compared with the trace minerals, major minerals are found, and needed, in larger quantities in the body. Unlike vitamins and the energy-yielding nutrients, minerals are inorganic elements that retain their chemical identities. Minerals usually receive special handling and regulation in the body, and they may bind with other substances or interact with other minerals, thus limiting their absorption.

bioavailability: the rate at and the extent to which a nutrient is absorbed and used.

binders: chemical compounds in foods that combine with nutrients (especially minerals) to form complexes the body cannot absorb. Examples include *phytates* (FYE-tates) and *oxalates* (OCK-sa-lates).

12.3 The Major Minerals

LEARN IT Identify the main roles, deficiency symptoms, and food sources for each of the major minerals (sodium, chloride, potassium, calcium, phosphorus, magnesium, and sulfate).

Although all the major minerals help to maintain the body's fluid balance as described earlier, sodium, chloride, and potassium are most noted for that role. ♦ For this reason, these three minerals are discussed first here. Later sections describe the minerals most noted for their roles in bone growth and health—calcium, phosphorus, and magnesium. The chapter closes with a brief discussion on sulfate, a mineral required for the synthesis of several sulfur-containing compounds.

♦ Key fluid balance nutrients:
- Sodium, potassium, chloride

Sodium

People have held salt (sodium chloride) in high regard throughout recorded history. We describe someone we admire as "the salt of the earth" and people we consider worthless as "not worth their salt." Even the word *salary* comes from the Latin word for salt.

Cultures vary in their use of salt, but most people find its taste innately appealing. Salt brings its own tangy taste and enhances other flavors, most likely by suppressing the bitter flavors. You can taste this effect for yourself: tonic water with its bitter quinine tastes sweeter with a little salt added.

Sodium Roles in the Body **Sodium** is the principal cation of the extracellular fluid and the primary regulator of its volume. Sodium also helps maintain acid-base balance and is essential to nerve impulse transmission and muscle contraction.*

Sodium is readily absorbed by the intestinal tract and travels freely in the blood until it reaches the kidneys, which filter all the sodium out of the blood. Then, with great precision, the kidneys return to the blood the exact amount of sodium the body needs. Normally, the amount excreted is approximately equal to the amount ingested on a given day. When blood sodium rises, as when a person eats salted foods, thirst signals the person to drink until the appropriate sodium-to-water concentration is restored. Then the kidneys excrete both the excess water and the excess sodium together.

Sodium Recommendations Diets rarely lack sodium, and even when intakes are low, the body adapts by reducing sodium losses in urine and sweat, thus making deficiencies unlikely. Sodium recommendations ♦ are set low enough to protect against high blood pressure, but high enough to allow an adequate intake of other nutrients with a typical diet. Because high sodium intakes correlate with high blood pressure, the Upper Level (UL) for adults is set at 2300 milligrams per day, slightly lower than the Daily Value used on food labels (2400 milligrams). The average sodium intake for adults in the United States is 3400 milligrams, which exceeds the UL—and most adults will develop hypertension at some point in their lives.[6]

♦ AI for sodium:
- 1500 mg/day (19–50 yr)
- 1300 mg/day (51–70 yr)
- 1200 mg/day (>70 yr)

Sodium and Hypertension For years, a high *sodium* intake was considered the primary factor responsible for high blood pressure. Then research pointed to *salt* (sodium chloride) ♦ as the dietary culprit. Salt has a greater effect on blood pressure than either sodium or chloride alone or in combination with other ions. The increase in blood pressure in response to a high salt intake may be immediate and is reversible with salt restriction. The elevation of blood pressure in response to a high-salt diet over years is progressive, and the damage caused to blood vessels is irreversible.[7]

♦ Salt (sodium chloride) is about 40% sodium.
- 1 g salt contributes about 400 mg sodium
- 6 g salt = 1 tsp
- 1 tsp salt contributes about 2300 mg sodium

Blood pressure increases in response to excesses in salt intake—most notably for those with hypertension, African Americans, and people older than 40 years of age.[8] **Salt sensitivity** is apparent in about 25 percent of those with normal blood pressure and in about 50 percent of those with high blood pressure.** For them, a high salt intake correlates strongly with heart disease, and salt restriction (to no more than 1500 milligrams of sodium per day) helps to lower blood pressure.

*One of the ways the kidneys regulate acid-base balance is by excreting hydrogen ions (H^+) in exchange for sodium ions (Na^+).

**Compared with others, salt-sensitive individuals have elevated concentrations of renin in their blood.

sodium: the principal cation in the extracellular fluids of the body; critical to the maintenance of fluid balance, nerve impulse transmissions, and muscle contractions.

salt sensitivity: a characteristic of individuals who respond to a high salt intake with an increase in blood pressure or to a low salt intake with a decrease in blood pressure.

>How To

Cut Salt (and Sodium) Intake

Most people eat more salt (and therefore sodium) than they need. Some people can lower their blood pressure by avoiding highly salted foods and removing the salt shaker from the table. Foods eaten without salt may seem less tasty at first, but with repetition, people can learn to enjoy the natural flavors of many unsalted foods. Strategies to cut salt intake include:

- Select fresh, unprocessed foods.
- Cook with little or no added salt.
- Prepare foods with sodium-free spices such as basil, bay leaves, curry, garlic, ginger, mint, oregano, pepper, rosemary, and thyme; lemon juice; vinegar; or wine.
- Add little or no salt at the table; taste foods before adding salt.
- Read labels with an eye open for sodium. (See the glossary on p. 57 for terms used to describe the sodium contents of foods on labels.)
- Select low-salt or salt-free products when available.

Use these foods sparingly:

- Foods prepared in brine, such as pickles, olives, and sauerkraut
- Salty or smoked meats, such as bologna, corned or chipped beef, bacon, frankfurters, ham, lunchmeats, salt pork, sausage, and smoked tongue
- Salty or smoked fish, such as anchovies, caviar, salted and dried cod, herring, sardines, and smoked salmon
- Snack items such as potato chips, pretzels, salted popcorn, salted nuts, and crackers
- Condiments such as bouillon cubes; seasoned salts; MSG; soy, teriyaki, Worcestershire, and barbeque sauces; prepared horseradish, ketchup, and mustard
- Cheeses, especially processed types
- Canned and instant soups

TRY IT Compare the sodium contents of 1 ounce of the following foods: a plain bagel, potato chips, and animal crackers.

A salt-restricted diet lowers blood pressure and improves blood vessel dilation in people without hypertension as well.[9] Because reducing salt intake causes no harm and diminishes the risk of hypertension and heart disease, the *Dietary Guidelines for Americans* advise limiting daily *salt* intake to about 1 teaspoon (the equivalent of about 2.3 grams or 2300 milligrams of *sodium*).[10] The American Heart Association goal ♦ is to lower blood pressure by reducing sodium intake to less than 1500 milligrams a day.[11] The accompanying "How To" feature offers strategies for cutting salt (and therefore sodium) intake.

♦ AHA goals:
- Blood pressure: <120/80 mmHg
- Sodium intake: <1500 mg/day

> DIETARY GUIDELINES FOR AMERICANS 2010

Choose foods low in sodium and prepare foods with little salt. Reduce daily sodium intake to less than 2300 milligrams and further reduce intake to 1500 milligrams among persons who are 51 and older and those of any age who are African American or have hypertension, diabetes, or chronic kidney disease.

One eating pattern, known as the DASH (Dietary Approaches to Stop Hypertension) Eating Plan, is effective in lowering blood pressure. Like other USDA Food Patterns, the DASH Eating Plan reflects the *Dietary Guidelines* and allows people to stay within their energy allowance, meet nutrient needs, and reduce chronic disease risk. The DASH approach emphasizes potassium-rich fruits, vegetables, and low-fat milk products; includes whole grains, nuts, poultry, and fish; and calls for reduced intakes of sodium, red and processed meats, sweets, and sugar-containing beverages. In combination with a reduced sodium intake, DASH is even more effective at lowering blood pressure than either strategy alone. Chapter 18 offers a complete discussion of hypertension and the dietary recommendations ♦ for its prevention and treatment.

♦ Hypertension

Beneficial factors:
- Dietary fiber
- Physical activity
- Potassium

Harmful factors:
- Overweight/obesity
- Sodium in excess
- Alcohol in excess

Sodium and Bone Loss (Osteoporosis) A high salt intake is also associated with increased calcium excretion, but its influence on bone loss is less clear. In addition, potassium may prevent the calcium excretion caused by a high-salt diet. For these reasons, dietary advice to prevent bone loss parallels that suggested for hypertension—a DASH eating pattern that is low in sodium and abundant in potassium-rich fruits and vegetables and calcium-rich low-fat milk.

Fresh herbs add flavor to a recipe without adding salt.

Sodium in Foods In general, processed foods have the most sodium, whereas unprocessed foods such as fresh fruits and vegetables have the least. In fact, as much as 75 percent of the sodium in people's diets comes from salt added to foods by manufacturers; about 15 percent comes from salt added during cooking and at the table; and only 10 percent comes from the natural content in foods. To help consumers limit their intake, public health organizations and policymakers worldwide are calling for manufacturers to reduce sodium in the food supply.[12] Reducing the sodium content in processed foods could prevent an estimated 100,000 deaths and save up to $24 billion in health care costs in the United States annually.[13]

Because processed foods may contain sodium without chloride, as in additives such as sodium bicarbonate or sodium saccharin, they do not always taste salty. Most people are surprised to learn that 1 ounce of some cereals contains more sodium than 1 ounce of salted peanuts—and that ½ cup of instant chocolate pudding contains still more. The peanuts taste saltier because the salt is all on the surface, where the tongue's taste receptors immediately pick it up.

Figure 12-10 shows that processed foods not only contain more sodium than their less-processed counterparts but also have less potassium. Low potassium may be as significant as high sodium when it comes to blood pressure regulation, so processed foods have two strikes against them.

Sodium Deficiency Sodium deficiency does not develop from an inadequate diet. The body needs so little and diets provide enough. Blood sodium may drop with vomiting, diarrhea, or heavy sweating, and in these cases, both sodium and water must be replenished. Under normal conditions of sweating due to physical activity, salt losses can easily be replaced later in the day with ordinary foods. Salt tablets are not recommended because too much salt, especially if taken with too little water, can induce dehydration. During intense activities, such as ultra-endurance events, athletes can lose so much sodium and drink so much water that they develop hyponatremia—the dangerous condition ♦ of having too little sodium in the blood. Hyponatremia is caused by excessive sodium losses, not from inadequate sodium intake.[14] (Chapter 14 offers details about hyponatremia and guidelines for ultra-endurance athletes.)

♦ Symptoms of hyponatremia:
- Headache, confusion, stupor
- Seizures, coma

> **FIGURE 12-10 What Processing Does to the Sodium and Potassium Contents of Foods**

People who eat foods high in salt often happen to be eating fewer potassium-containing foods at the same time. Notice how potassium is lost and sodium is gained as foods become more processed, causing the potassium-to-sodium ratio to fall dramatically. Even when potassium isn't lost, the addition of sodium still lowers the potassium-to-sodium ratio. Selecting fresh, unprocessed foods lowers blood pressure in two ways, then—by lowering sodium intakes and by raising potassium intakes.

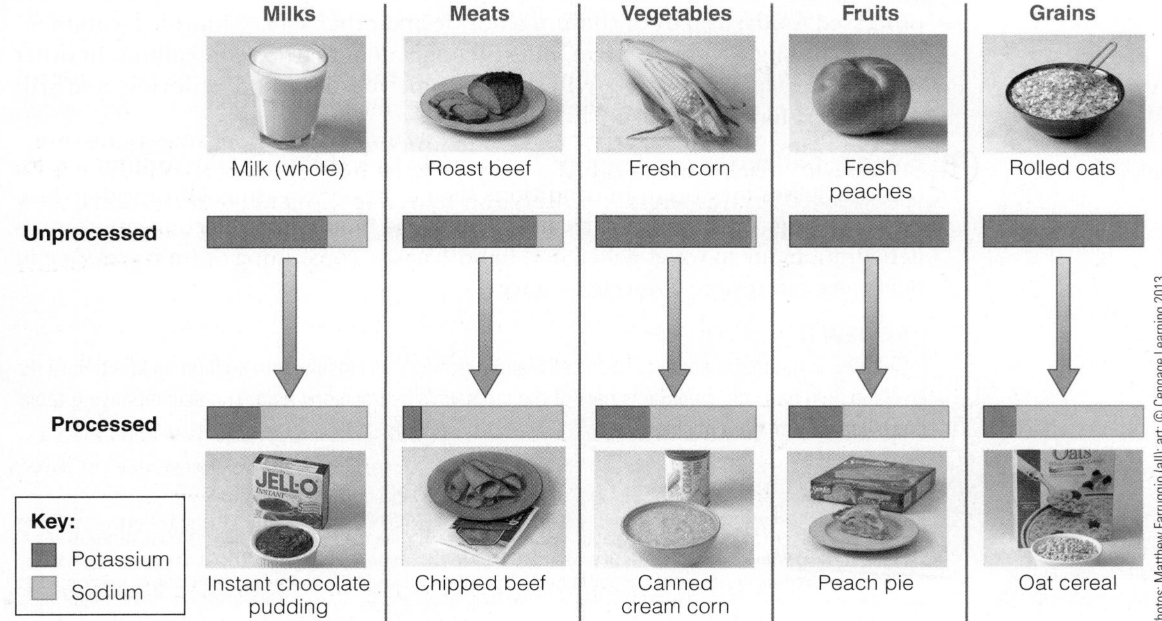

Sodium Toxicity and Excessive Intakes The immediate symptoms of acute sodium toxicity are edema and high blood pressure. Prolonged excessive sodium intake may also contribute to hypertension in some people, as explained earlier.

REVIEW IT

Sodium is the main cation outside cells and one of the primary electrolytes responsible for maintaining fluid balance. Dietary deficiency is unlikely, and excesses raise blood pressure in many people. For this reason, health professionals advise a diet moderate in salt and sodium. The accompanying table provides a summary of sodium.

Sodium

Adequate Intake (AI)

Adults: 1500 mg/day (19–50 yr)
1300 mg/day (51–70 yr)
1200 mg/day (>70 yr)

Upper Level

Adults: 2300 mg/day

Chief Functions in the Body

Maintains normal fluid and electrolyte balance; assists in nerve impulse transmission and muscle contraction

Deficiency Symptoms

Not from inadequate intakes
Hyponatremia from excessive losses

Toxicity Symptoms

Edema, acute hypertension

Significant Sources

Table salt, soy sauce; moderate amounts in meats, milks, breads, and vegetables; large amounts in processed foods

Chloride

The element *chlorine* (Cl_2) is a poisonous gas. When chlorine reacts with sodium or hydrogen, however, it forms the negative chloride ion (Cl^-). *Chloride,* an essential nutrient, is required in the diet.

Chloride Roles in the Body **Chloride** is the major anion of the extracellular fluids (outside the cells), where it occurs mostly in association with sodium. Chloride moves passively across membranes through channels and so also associates with potassium inside cells. Like sodium and potassium, chloride maintains fluid and electrolyte balance.

In the stomach, the chloride ion is part of hydrochloric acid, which maintains the strong acidity of gastric juice. One of the most serious consequences of vomiting is the loss of this acid from the stomach, which upsets the acid-base balance.* Such imbalances are commonly seen in bulimia nervosa, as described in Highlight 8.

Chloride Recommendations and Intakes Chloride is abundant in foods (especially processed foods) as part of sodium chloride and other salts. Chloride recommendations are slightly higher than, but still equivalent to, those of sodium. In other words, ¾ teaspoon of salt ♦ will deliver some sodium, more chloride, and still meet the AI for both.

♦ Salt (sodium chloride) is about 60% chloride.
- 1 g salt contributes about 600 mg chloride
- 6 g salt = 1 tsp
- 1 tsp salt contributes about 3700 mg chloride

Chloride Deficiency and Toxicity Diets rarely lack chloride. Like sodium losses, chloride losses may occur in conditions such as heavy sweating, chronic diarrhea, and vomiting. The only known cause of elevated blood chloride concentrations is dehydration due to water deficiency. In both cases, consuming ordinary foods and beverages can restore chloride balance.

REVIEW IT

Chloride is the major anion outside cells, and it associates closely with sodium. In addition to its role in fluid balance, chloride is part of the stomach's hydrochloric acid. The accompanying table provides a summary of chloride.

chloride (KLO-ride): the major anion in the extracellular fluids of the body. Chloride is the ionic form of chlorine, Cl^-. See Appendix B for a description of the chlorine-to-chloride conversion.

*Hydrochloric acid secretion into the stomach involves the addition of bicarbonate ions (base) to the plasma. These bicarbonate ions (HCO_3^-) are neutralized by hydrogen ions (H^+) from the gastric secretions that are reabsorbed into the plasma. When hydrochloric acid is lost during vomiting, these hydrogen ions are no longer available for reabsorption, and so, in effect, the concentrations of bicarbonate ions in the plasma are increased. In this way, excessive vomiting of acidic gastric juices leads to *metabolic alkalosis*—an above-normal alkalinity in the blood and body fluids.

REVIEW IT Chloride (*continued*)

Adequate Intake (AI)

Adults: 2300 mg/day (19–50 yr)
2000 mg/day (51–70 yr)
1800 mg/day (>70 yr)

Upper Level

Adults: 3600 mg/day

Chief Functions in the Body

Maintains normal fluid and electrolyte balance; part of hydrochloric acid found in the stomach, necessary for proper digestion

Deficiency Symptoms

Do not occur under normal circumstances

Toxicity Symptoms

Vomiting

Significant Sources

Table salt, soy sauce; moderate amounts in meats, milks, eggs; large amounts in processed foods

Potassium

Like sodium, **potassium** is a positively charged ion. In contrast to sodium, potassium is the body's principal intracellular cation, *inside* the body cells.

Potassium Roles in the Body Potassium plays a major role in maintaining fluid and electrolyte balance and cell integrity. During nerve impulse transmissions and muscle contractions, potassium and sodium briefly trade places across the cell membrane. The cell then quickly pumps them back into place. Controlling potassium distribution is a high priority for the body because it affects many aspects of homeostasis, including a steady heartbeat.

Potassium Recommendations and Intakes Potassium is abundant in all living cells. Because cells remain intact unless foods are processed, the richest sources of potassium are *fresh* foods—as Figure 12-11 (p. 384) shows. In contrast, most processed foods such as canned vegetables, ready-to-eat cereals, and luncheon meats contain less potassium—and more sodium (recall Figure 12-10, p. 381). To meet the AI for potassium, most people need to increase their intake of fruits and vegetables to five to nine servings daily.

> DIETARY GUIDELINES FOR AMERICANS 2010

Choose foods that provide more potassium, a nutrient of concern in American diets. Potassium is found in all food groups, notably vegetables, fruits, and milk and milk products.

Potassium and Hypertension Diets low in potassium, especially when combined with high sodium intakes, raise blood pressure and increase the risk of death from heart disease.[15] In contrast, high potassium intakes, especially when combined with low sodium intakes, appear to both prevent and correct hypertension. Recall that the DASH eating pattern described earlier is used to lower blood pressure and emphasizes potassium-rich foods such as fruits and vegetables. Potassium-rich fruits and vegetables also appear to reduce the risk of stroke—more so than can be explained by the reduction in blood pressure alone.

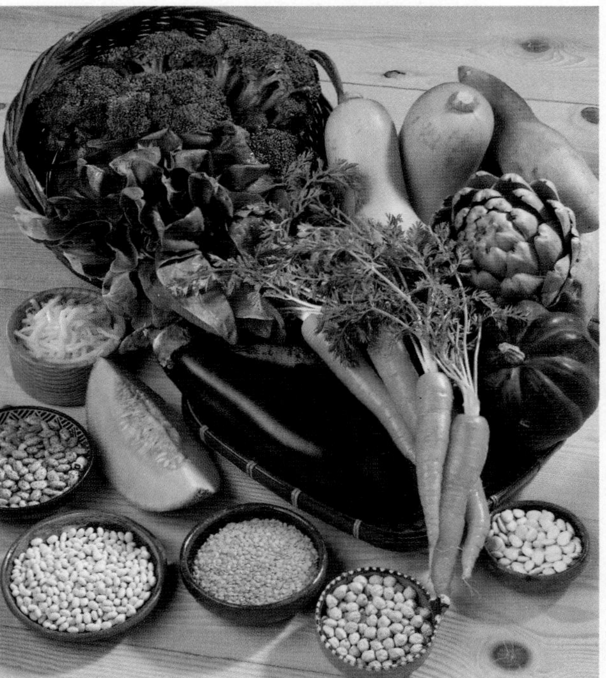

Fresh foods, especially fruits and vegetables, provide potassium in abundance.

Potassium Deficiency Potassium deficiency is characterized by an increase in blood pressure, salt sensitivity, kidney stones, and bone turnover. As deficiency progresses, symptoms include irregular heartbeats, muscle weakness, and glucose intolerance.

Potassium Toxicity Potassium toxicity does not result from overeating foods high in potassium; therefore a UL has not been set. It can result from overconsumption of potassium salts or supplements (including some "energy fitness shakes") and from certain diseases or treatments. Given more potassium than the body needs, the kidneys accelerate excretion. If potassium is injected directly into a vein, however, it can stop the heart.

potassium: the principal cation within the body's cells; critical to the maintenance of fluid balance, nerve impulse transmissions, and muscle contractions.

> FIGURE 12-11 Potassium in Selected Foods

See the "How To" feature on p. 302 for more information on using this figure.

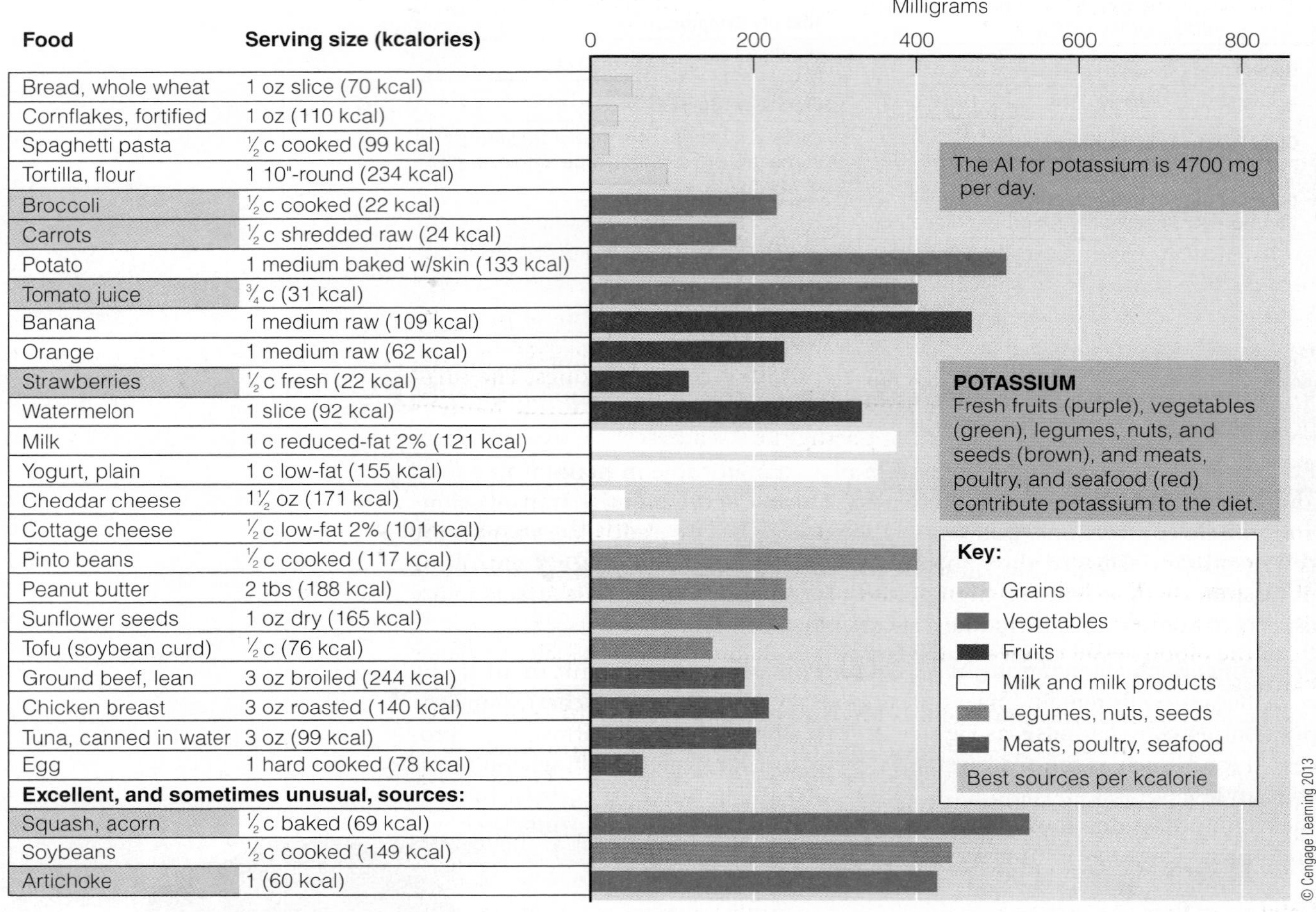

Food	Serving size (kcalories)
Bread, whole wheat	1 oz slice (70 kcal)
Cornflakes, fortified	1 oz (110 kcal)
Spaghetti pasta	½ c cooked (99 kcal)
Tortilla, flour	1 10"-round (234 kcal)
Broccoli	½ c cooked (22 kcal)
Carrots	½ c shredded raw (24 kcal)
Potato	1 medium baked w/skin (133 kcal)
Tomato juice	¾ c (31 kcal)
Banana	1 medium raw (109 kcal)
Orange	1 medium raw (62 kcal)
Strawberries	½ c fresh (22 kcal)
Watermelon	1 slice (92 kcal)
Milk	1 c reduced-fat 2% (121 kcal)
Yogurt, plain	1 c low-fat (155 kcal)
Cheddar cheese	1½ oz (171 kcal)
Cottage cheese	½ c low-fat 2% (101 kcal)
Pinto beans	½ c cooked (117 kcal)
Peanut butter	2 tbs (188 kcal)
Sunflower seeds	1 oz dry (165 kcal)
Tofu (soybean curd)	½ c (76 kcal)
Ground beef, lean	3 oz broiled (244 kcal)
Chicken breast	3 oz roasted (140 kcal)
Tuna, canned in water	3 oz (99 kcal)
Egg	1 hard cooked (78 kcal)
Excellent, and sometimes unusual, sources:	
Squash, acorn	½ c baked (69 kcal)
Soybeans	½ c cooked (149 kcal)
Artichoke	1 (60 kcal)

REVIEW IT

Potassium, like sodium and chloride, is an electrolyte that plays an important role in maintaining fluid balance. Potassium is the primary cation inside cells; fresh foods, notably fruits and vegetables, are its best sources. The accompanying table provides a summary of potassium.

Potassium

Adequate Intake (AI)

Adults: 4700 mg/day

Chief Functions in the Body

Maintains normal fluid and electrolyte balance; facilitates many reactions; supports cell integrity; assists in nerve impulse transmission and muscle contractions

Deficiency Symptoms[a]

Irregular heatbeat, muscular weakness, glucose intolerance

Toxicity Symptoms

Muscular weakness; vomiting; if given into a vein, can stop the heart

Significant Sources

All whole foods: meats, milks, fruits, vegetables, grains, legumes

[a]Deficiency accompanies dehydration.

Calcium Calcium is the most abundant mineral in the body. It receives much emphasis in this chapter and in the highlight that follows because an adequate intake helps grow a healthy skeleton in early life and minimize bone loss in later life.

calcium: the most abundant mineral in the body; found primarily in the body's bones and teeth.

Calcium Roles in the Body Ninety-nine percent of the body's calcium is in the bones (and teeth), where it plays two roles. First, it is an integral part of bone structure, providing a rigid frame that holds the body upright and serves as attachment points for muscles, making motion possible. Second, it serves as a calcium bank, offering a readily available source of calcium to the body fluids should a drop in blood calcium occur. The remaining 1 percent of the body's calcium is in the body fluids.

As bones begin to form, calcium salts form crystals, called **hydroxyapatite,** on a matrix of the protein collagen. During **mineralization,** as the crystals become denser, they give strength and rigidity to the maturing bones. As a result, the long leg bones of children can support their weight by the time they have learned to walk.

Many people have the idea that once a bone is built, it is inert like a rock. Actually, the bones are gaining and losing minerals continuously in an ongoing process of remodeling. Growing children gain more bone than they lose, and healthy adults maintain a reasonable balance. When withdrawals substantially exceed deposits, problems such as osteoporosis develop (as described in Highlight 12).

The formation of teeth follows a pattern similar to that of bones. The turnover of minerals in teeth is not as rapid as in bone, however; fluoride hardens and stabilizes the crystals of teeth, opposing the withdrawal of minerals from them.

Although only 1 percent of the body's calcium circulates in the extracellular and intracellular fluids, its presence there is vital to life. Cells throughout the body can detect calcium in the extracellular fluids and respond accordingly. Many of calcium's actions help to maintain normal blood pressure, perhaps by stabilizing the smooth muscle cells of the blood vessels or by releasing relaxing factors from the blood vessel cell walls. Extracellular calcium also participates in blood clotting.

The calcium in intracellular fluids binds to proteins within the cells and activates them. For example, when the protein **calmodulin** binds with calcium, it activates the enzymes involved in breaking down glycogen, which releases energy for muscle contractions. Many such proteins participate in the regulation of muscle contractions, the transmission of nerve impulses, the secretion of hormones, and the activation of some enzyme reactions.

Calcium in Disease Prevention Calcium may protect against some chronic diseases, including hypertension.[16] Considering the success of DASH in lowering blood pressure, restricting sodium to treat hypertension may be narrow advice. The DASH eating pattern is rich in calcium, as well as in magnesium and potassium. As mentioned earlier, the combination of DASH with a reduced sodium intake is more effective in lowering blood pressure than either strategy alone.

Some research also suggests protective relationships between dietary calcium and blood cholesterol, diabetes, and colon cancer.[17] Highlight 12 explores calcium's role in preventing osteoporosis.

Calcium from dairy foods, but *not* from supplements, may play a role in maintaining a healthy body weight.[18] Some epidemiological studies suggest an inverse relationship between calcium intake and body weight: the higher the calcium intake, the lower the prevalence of overweight. Teenagers who drink little or no milk gain more weight than their peers who do drink milk.[19]

An adequate dietary calcium intake may help prevent excessive fat accumulation by stimulating hormonal action that targets the breakdown of stored fat. Alternatively, milk may weaken the motivation to eat that typically accompanies weight loss.[20] Not all research suggests that consumption of calcium from dairy foods alters fat metabolism, energy expenditure, or body weight.[21] But even when body weight does not improve, dairy foods seem to help in another way—by suppressing the inflammation commonly associated with overweight.[22]

Calcium Balance Calcium homeostasis involves a system of hormones and vitamin D.[23] Whenever blood calcium falls too low or rises too high, three organ systems respond: the intestines, bones, and kidneys. Figure 12-12 (p. 386) illustrates how vitamin D and two hormones—**parathyroid hormone** and **calcitonin**—return blood calcium to normal.

hydroxyapatite (high-drox-ee-APP-ah-tite): crystals made of calcium and phosphorus.

mineralization: the process in which calcium, phosphorus, and other minerals crystallize on the collagen matrix of a growing bone, hardening the bone.

calmodulin (cal-MOD-you-lin): a calcium-binding protein that regulates such cell activities as muscle contractions.

parathyroid hormone: a hormone from the parathyroid glands that regulates blood calcium by raising it when levels fall too low; also known as *parathormone* (PAIR-ah-THOR-moan).

calcitonin (KAL-seh-TOE-nin): a hormone secreted by the thyroid gland that regulates blood calcium by lowering it when levels rise too high.

> **FIGURE 12-12** ***Animated*** **Calcium Balance**

Blood calcium is regulated by vitamin D and two hormones—calcitonin and parathyroid hormone. Bones serve as a reservoir when blood calcium is high and as a source of calcium when blood calcium is low. Osteoclasts break down bone and release calcium into the blood; osteoblasts build new bone using calcium from the blood.

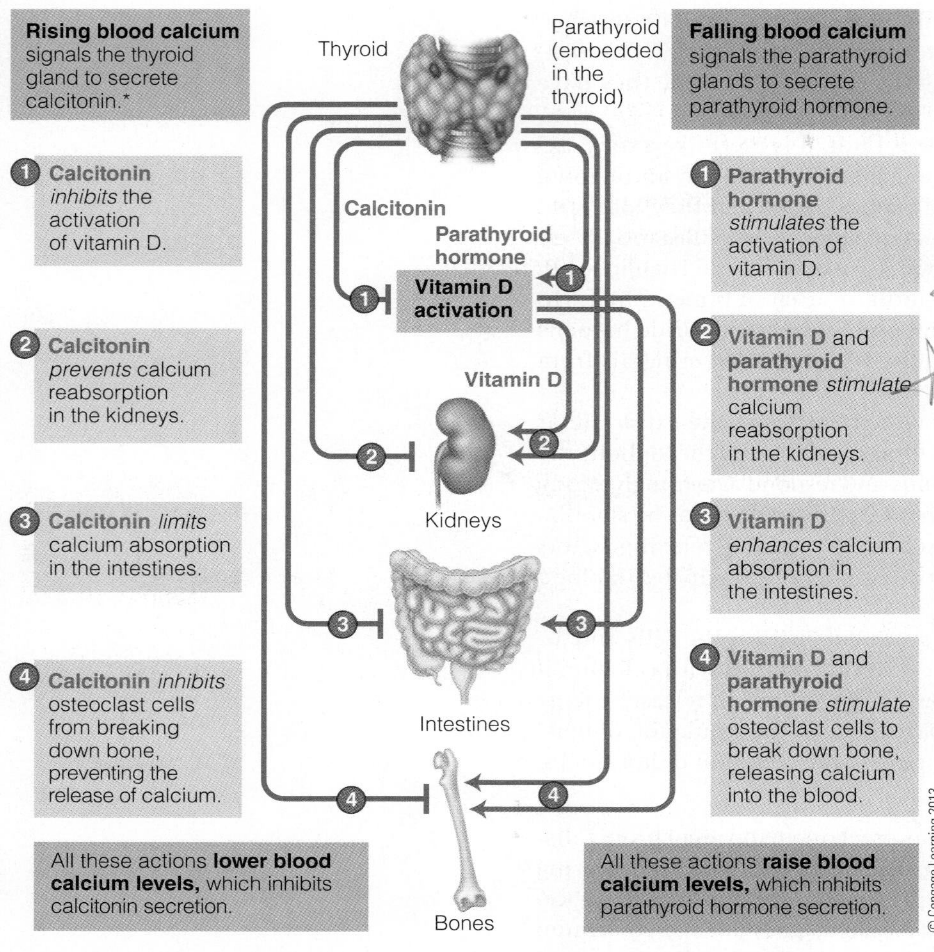

Rising blood calcium signals the thyroid gland to secrete calcitonin.*
Thyroid
Parathyroid (embedded in the thyroid)
Falling blood calcium signals the parathyroid glands to secrete parathyroid hormone.
1 Calcitonin inhibits the activation of vitamin D.
Calcitonin
Parathyroid hormone
Vitamin D activation
1 Parathyroid hormone stimulates the activation of vitamin D.
2 Calcitonin prevents calcium reabsorption in the kidneys.
Vitamin D
2 Vitamin D and parathyroid hormone stimulate calcium reabsorption in the kidneys.
Kidneys
3 Calcitonin limits calcium absorption in the intestines.
3 Vitamin D enhances calcium absorption in the intestines.
4 Calcitonin inhibits osteoclast cells from breaking down bone, preventing the release of calcium.
Intestines
4 Vitamin D and parathyroid hormone stimulate osteoclast cells to break down bone, releasing calcium into the blood.
All these actions lower blood calcium levels, which inhibits calcitonin secretion.
Bones
All these actions raise blood calcium levels, which inhibits parathyroid hormone secretion.

*Calcitonin plays a major role in defending infants and young children against the dangers of rising blood calcium that can occur when regular feedings of milk deliver large quantities of calcium to a small body. In contrast, calcitonin plays a relatively minor role in adults because their absorption of calcium is less efficient and their bodies are larger, making elevated blood calcium unlikely.

The calcium in bones provides a nearly inexhaustible bank of calcium for the blood. The blood borrows and returns calcium as needed so that even with an inadequate diet, *blood* calcium remains normal—even as *bone* calcium diminishes (see Figure 12-13). Blood calcium changes only in response to abnormal regulatory control, not to diet. A person can have an inadequate calcium intake for years and have no noticeable symptoms. Only later in life does it become apparent that bone integrity has been compromised.

Blood calcium above normal results in **calcium rigor:** the muscles contract and cannot relax. Similarly, blood calcium below normal causes **calcium tetany**—also characterized by uncontrolled muscle contraction. These conditions do *not* reflect a *dietary* excess or lack of calcium; they are caused by a lack of vitamin D or by abnormal secretion of the regulatory hormones. A chronic *dietary* deficiency of calcium, or a chronic deficiency due to poor absorption over the years, depletes the bones. Again: the *bones,* not the blood, are robbed by a calcium deficiency.

Calcium Absorption Because many factors affect calcium absorption, the most effective way to ensure adequacy is to increase calcium intake.[24] On average, adults absorb about 30 percent of the calcium they ingest. The stomach's acidity helps to keep calcium soluble, and vitamin D helps to make the **calcium-binding protein** needed for absorption. This relationship explains why calcium-rich milk is a good choice for vitamin D fortification.

Whenever calcium is needed, the body increases its calcium absorption. The result is obvious in the case of a newborn infant, whose calcium absorption is 55 to 60 percent. Similarly, a pregnant woman doubles her absorption of calcium. Growing children and teens absorb up to 50 percent of the calcium they consume. Then, when bone growth slows or stops, absorption falls to the adult level of about 30 percent. In addition, absorption becomes more efficient during times of inadequate intakes.

Many of the conditions that enhance calcium absorption limit its absorption when they are absent. For example, sufficient vitamin D supports absorption, and a deficiency impairs it. In addition, fiber in general, and the binders phytate and oxalate in particular, interfere with calcium absorption, but their effects are relatively minor in typical US diets. Vegetables with oxalates and whole grains with phytates are nutritious foods, of course, but they are not useful calcium sources. The margin note ♦ presents factors that influence calcium balance.

♦ *Enhance* calcium absorption:
- Stomach acid
- Vitamin D
- Lactose (in infants only)

Limit calcium absorption:
- Lack of stomach acid
- Vitamin D deficiency
- High phosphorus intake
- Phytates (in seeds, nuts, grains)
- Oxalates (in beet greens, rhubarb, spinach, sweet potatoes)

calcium rigor: hardness or stiffness of the muscles caused by high blood calcium concentrations.

calcium tetany (TET-ah-nee): intermittent spasm of the extremities due to nervous and muscular excitability caused by low blood calcium concentrations.

calcium-binding protein: a protein in the intestinal cells, made with the help of vitamin D, that facilitates calcium absorption.

Calcium Recommendations Calcium is unlike most other nutrients in that hormones maintain its *blood* concentration regardless of dietary intake. As Figure 12-13 shows, when calcium intake is high, the *bones* benefit; when intake is low, the *bones* suffer. Calcium recommendations are therefore based on the amount needed to retain the most calcium in bones. By retaining the most calcium possible, the bones can develop to their fullest potential in size and density—their **peak bone mass**—within genetic limits.

Calcium recommendations have been set high enough to accommodate a 30 percent absorption rate. Because obtaining enough calcium during growth helps to ensure that the skeleton will be strong and dense, the recommendation for adolescents to the age of 18 years is 1300 milligrams daily. Between the ages of 19 and 50, recommendations are lowered to 1000 milligrams a day; for women older than 50 and all adults older than 70, recommendations are raised again to 1200 milligrams a day to minimize the bone loss that tends to occur later in life. Some authorities advocate as much as 1500 milligrams a day for women older than 50. Most people in the United States have calcium intakes below current recommendations.[25] Those meeting recommendations for calcium are likely to be using calcium supplements.[26] High intakes of calcium from supplements may have adverse effects such as kidney stone formation. For this reason, a UL has been established. ♦

High intakes of both dietary protein and sodium increase calcium losses, but whether these losses impair bone development remains unclear. In the case of protein, any effects of high intakes are offset by the beneficial effects of accompanying nutrients—for example, by potassium in legumes and calcium in milk.[27] In fact, protein may even improve calcium absorption and bone strength.[28] The DRI Committee considered these nutrient interactions in establishing the RDA for calcium and did not adjust dietary recommendations based on this information.[29]

Calcium Food Sources Figure 12-14 (p. 388) shows that calcium is found most abundantly in a single class of foods—milk and milk products. ♦ The person who doesn't like to drink milk may prefer to eat cheese or yogurt. Alternatively, milk and milk products can be concealed in foods. Powdered fat-free milk can be added to casseroles, soups, and other mixed dishes during preparation; 5 heaping tablespoons offer the equivalent of 1 cup of milk. This simple step is an excellent way for older women to obtain not only extra calcium, but more protein, vitamins, and minerals as well.

It is especially difficult for children who don't drink milk to meet their calcium needs. The consequences of drinking too little milk during childhood and adolescence persist into adulthood. Women who seldom drank milk as children or teenagers have lower bone density and greater risk of fractures than those who drank milk regularly. It is possible for people who do not drink milk to obtain adequate calcium, but only if they carefully select other calcium-rich foods.

> **DIETARY GUIDELINES FOR AMERICANS 2010**
> Choose foods that provide more calcium, a nutrient of concern in American diets. The best sources of calcium are milk and milk products.

Many people, for a variety of reasons, cannot or do not drink milk. Some cultures do not use milk in their cuisines; some vegetarians exclude milk as well as meat; and some people are allergic to milk protein or are lactose intolerant. Others simply do not enjoy the taste of milk. These people need to find other foods to help meet their calcium needs. Some brands of tofu, corn tortillas, some nuts (such as almonds), and some seeds (such as sesame seeds) can supply calcium for the person who doesn't use milk products. A slice of most breads contains only about 5 to 10 percent of the calcium found in milk, but it can be a major source for people who eat many slices because the calcium is well absorbed. Oysters are also a rich source of calcium, as are small fish eaten with their bones, such as canned sardines.

Among the vegetables, mustard and turnip greens, bok choy, kale, parsley, watercress, and broccoli are good sources of available calcium. So are some seaweeds such as the nori popular in Japanese cooking. Some dark green, leafy vegetables—notably spinach and Swiss chard—appear to be calcium-rich but actually provide little, if any, calcium because they contain binders that limit absorption. It would

> **FIGURE 12-13 Maintaining Blood Calcium from the Diet and from the Bones**

With an adequate intake of calcium-rich food, blood calcium remains normal . . .

With a dietary deficiency, blood calcium still remains normal . . .

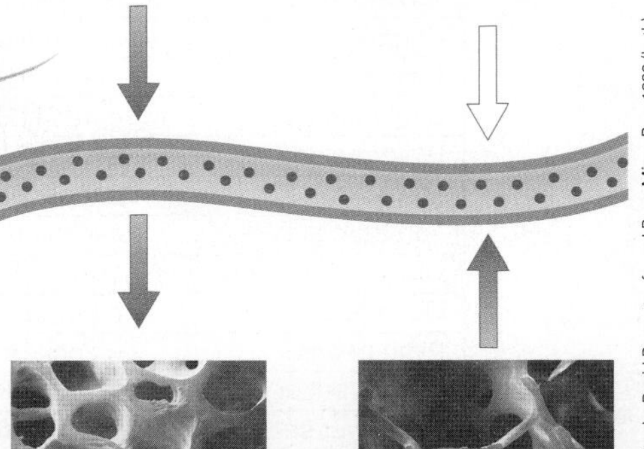

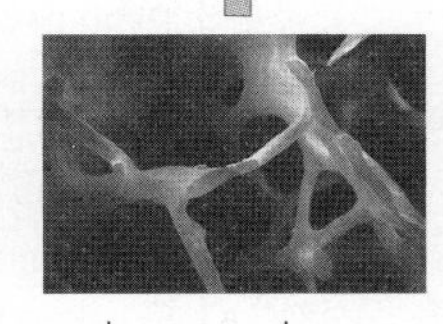

. . . and bones deposit calcium. The result is strong, dense bones.

. . . because bones give up calcium to the blood. The result is weak, osteoporotic bones.

♦ UL for calcium:
- 2500 mg/day (adults 19–50 yr)
- 2000 mg/day (adults >50 yr)

♦ Suggested daily milk amounts:
- Young children (2 to 8 yr): 2 c
- Older children, teenagers, and all adults: 3 c

peak bone mass: the highest attainable bone density for an individual, developed during the first three decades of life.

> FIGURE 12-14 Calcium in Selected Foods

See the "How To" feature on p. 302 for more information on using this figure.

Food	Serving size (kcalories)
Bread, whole wheat	1 oz slice (70 kcal)
Cornflakes, fortified	1 oz (110 kcal)
Spaghetti pasta	½c cooked (99 kcal)
Tortilla, flour	1 10"-round (234 kcal)
Broccoli	½c cooked (22 kcal)
Carrots	½c shredded raw (24 kcal)
Potato	1 medium baked w/skin (133 kcal)
Tomato juice	¾c (31 kcal)
Banana	1 medium raw (109 kcal)
Orange	1 medium raw (62 kcal)
Strawberries	½c fresh (22 kcal)
Watermelon	1 slice (92 kcal)
Milk	1 c reduced-fat 2% (121 kcal)
Yogurt, plain	1 c low-fat (155 kcal)
Cheddar cheese	1½ oz (171 kcal)
Cottage cheese	½c low-fat 2% (101 kcal)
Pinto beans	½c cooked (117 kcal)
Peanut butter	2 tbs (188 kcal)
Sunflower seeds	1 oz dry (165 kcal)
Tofu (soybean curd)[a]	½c (76 kcal)
Ground beef, lean	3 oz broiled (244 kcal)
Chicken breast	3 oz roasted (140 kcal)
Tuna, canned in water	3 oz (99 kcal)
Egg	1 hard cooked (78 kcal)
Excellent, and sometimes unusual, sources:	
Sardines, with bones[b]	3 oz canned (176 kcal)
Bok choy (Chinese cabbage)	½ c cooked (10 kcal)
Almonds	1 oz (167 kcal)

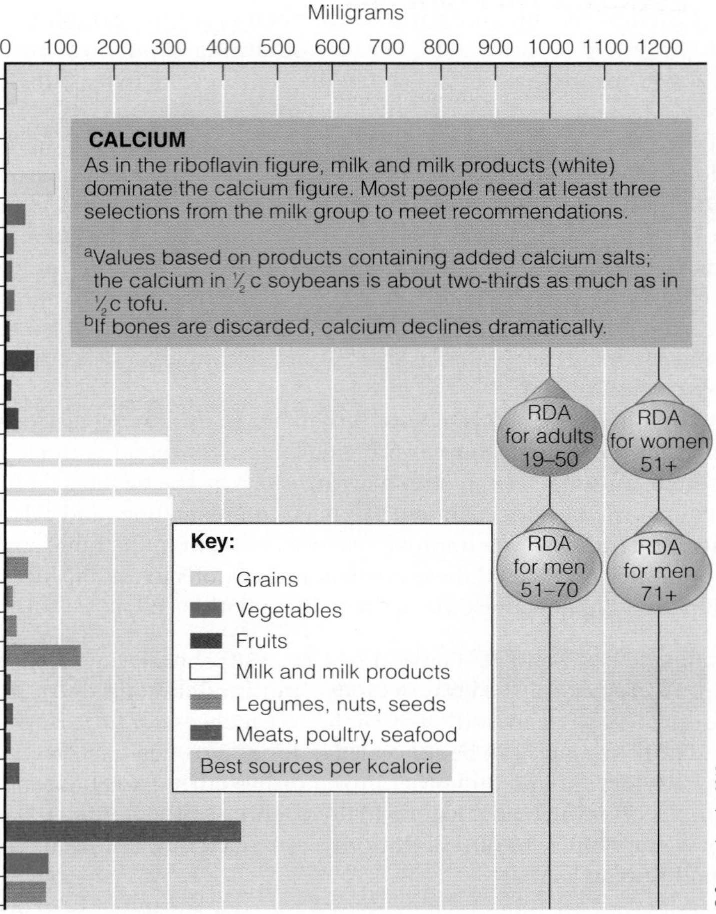

Milligrams
0 100 200 300 400 500 600 700 800 900 1000 1100 1200
CALCIUM
As in the riboflavin figure, milk and milk products (white) dominate the calcium figure. Most people need at least three selections from the milk group to meet recommendations.
[a]Values based on products containing added calcium salts; the calcium in ½ c soybeans is about two-thirds as much as in ½c tofu.
[b]If bones are discarded, calcium declines dramatically.
RDA for adults 19–50
RDA for women 51+
RDA for men 51–70
RDA for men 71+
Key:
Grains
Vegetables
Fruits
Milk and milk products
Legumes, nuts, seeds
Meats, poultry, seafood
Best sources per kcalorie

take 8 cups of spinach—containing six times as much calcium as 1 cup of milk—to deliver the equivalent in *absorbable* calcium.

With the exception of foods such as spinach that contain calcium binders, the calcium content of foods is usually more important than bioavailability. Consequently, recognizing that people eat a variety of foods containing calcium, the DRI Committee did not adjust for calcium bioavailability when setting recommendations. Figure 12-15 ranks selected foods according to their calcium bioavailability.

Some mineral waters provide as much as 500 milligrams of calcium per liter, offering a convenient way to meet both calcium and water needs. Similarly, calcium-fortified orange juice and other fruit and vegetable juices allow a person to obtain both calcium and vitamins easily. Other examples of calcium-fortified foods include high-calcium milk (milk with extra calcium added) and calcium-fortified cereals. Fortified juices and foods help consumers increase calcium intakes, but depending on the calcium sources, the bioavailability may be significantly less than quantities listed on food labels. The accompanying "How To" feature describes a quick way to estimate calcium intake. Highlight 12 discusses calcium supplements.

A generalization that has been gaining strength throughout this book is supported by the information given here about calcium. A balanced diet that supplies a variety of foods is the best plan to ensure adequacy for all essential nutrients. All food groups should be included, and none should be overemphasized. In our culture, calcium intake is usually inadequate wherever milk is lacking in the diet. By contrast, iron is usually lacking whenever milk is overemphasized, as Chapter 13 explains.

Calcium Deficiency A low calcium intake during the growing years limits the bones' ability to reach their optimal mass and density. Most people achieve a

CALCIUM ADDED
ORGANIC VALLEY
ORGANIC ORANGE JUICE

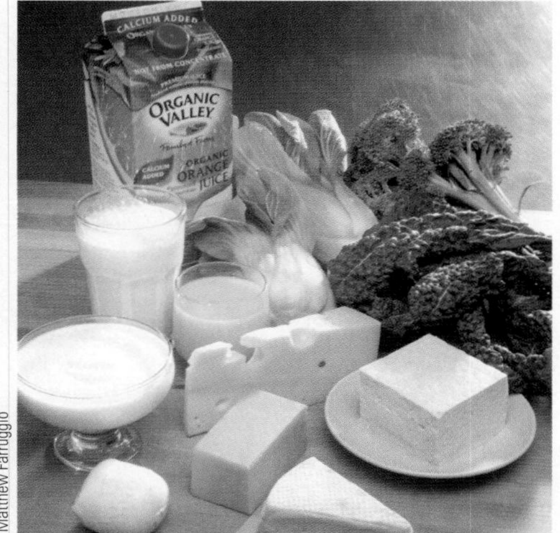

Milk and milk products are well known for their calcium, but calcium-set tofu, bok choy, kale, calcium-fortified orange juice, and broccoli are also rich in calcium.

>How To

Estimate Your Calcium Intake

Most dietitians have developed useful shortcuts to help them estimate nutrient intakes and "see" inadequacies in the diet. They can tell at a glance whether a day's meals fall short of calcium recommendations, for example.

To estimate calcium intakes, keep two bits of information in mind:

- A cup of milk provides about 300 milligrams of calcium.
- Adults need between 1000 and 1200 milligrams of calcium per day, which represents 3 to 4 cups of milk—or the equivalent:

$$1000 \text{ mg} \div 300 \text{ mg/c} = 3\tfrac{1}{3} \text{ c}$$
$$1200 \text{ mg} \div 300 \text{ mg/c} = 4 \text{ c}$$

If a person drinks 3 to 4 cups of milk a day, it's easy to see that calcium needs are being met. If not, it takes some detective work to identify the other sources and estimate total calcium intake.

To estimate a person's daily calcium intake, use this shortcut, which compares the calcium in calcium-rich foods to the calcium content of milk. The calcium in a cup of milk is assigned 1 point, and the goal is to attain 3 to 4 points per day. Foods are given points as follows:

- 1 c milk, yogurt, or fortified soy milk or 1½ oz cheese = 1 point
- 4 oz canned fish with bones (sardines) = 1 point
- 1 c ice cream, cottage cheese, or calcium-rich vegetable (see the text) = ½ point

Then, because other foods also contribute small amounts of calcium, together they are given a point.

- Well-balanced diet containing a variety of foods = 1 point

Now consider a day's meals with calcium in mind. Cereal with 1 cup of milk for breakfast (1 point for milk), a ham and cheese sub sandwich for lunch (1 point for cheese), and a cup of broccoli and lasagna for dinner (½ point for calcium-rich vegetable and 1 point for cheese in lasagna)—plus 1 point for all other foods eaten that day—adds up to 4½ points. This shortcut estimate indicates that calcium recommendations have been met, and a diet analysis of these few foods reveals a calcium intake of more than 1000 milligrams. By knowing the best sources of each nutrient, you can learn to scan the day's meals and quickly see if you are meeting your daily goals.

TRY IT Compare the calcium contents of ½ cup of the following foods: almonds, broccoli, and yogurt.

> **FIGURE 12-15 Bioavailability of Calcium from Selected Foods**

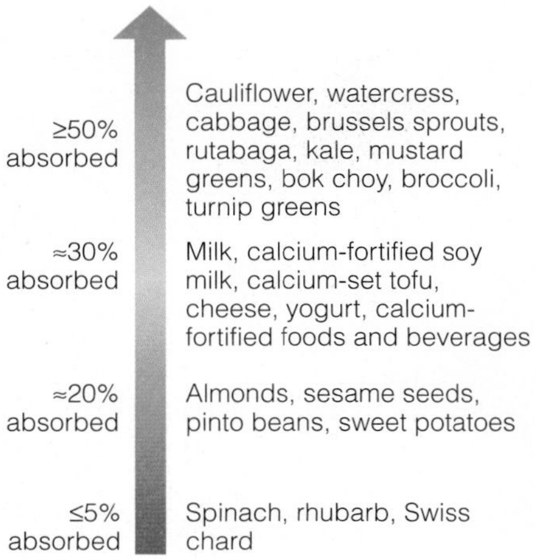

≥50% absorbed	Cauliflower, watercress, cabbage, brussels sprouts, rutabaga, kale, mustard greens, bok choy, broccoli, turnip greens
≈30% absorbed	Milk, calcium-fortified soy milk, calcium-set tofu, cheese, yogurt, calcium-fortified foods and beverages
≈20% absorbed	Almonds, sesame seeds, pinto beans, sweet potatoes
≤5% absorbed	Spinach, rhubarb, Swiss chard

peak bone mass by their late 20s, and dense bones best protect against age-related bone loss and fractures (see Figure 12-16). All adults lose bone as they grow older, beginning between the ages of 30 and 40. When bone losses reach the point of causing fractures under common, everyday stresses, the condition is known as **osteoporosis.** Osteoporosis and low bone mass (osteopenia) affect an estimated 44 million people in the United States, mostly older women.[30]

Unlike many diseases that make themselves known through symptoms such as pain, shortness of breath, skin lesions, tiredness, and the like, osteoporosis is

> **FIGURE 12-16 Phases of Bone Development throughout Life**

The active growth phase occurs from birth to approximately age 20. The phase of peak bone mass development occurs between the ages of 12 and 30. The final phase, when bone resorption exceeds formation, begins between the ages of 30 and 40 and continues through the remainder of life.

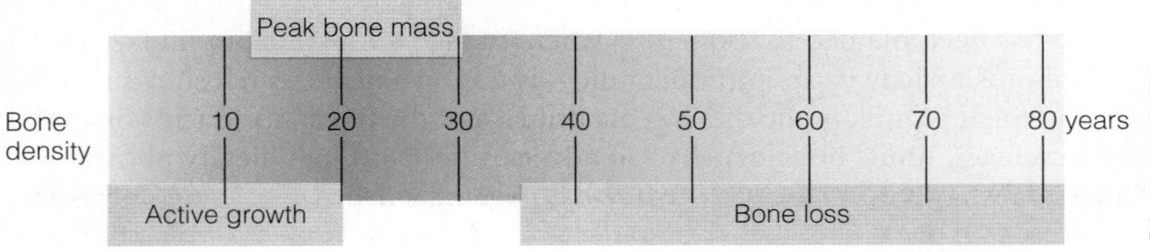

osteoporosis (OS-tee-oh-pore-OH-sis): a disease in which the bones become porous and fragile due to a loss of minerals; also called *adult bone loss.*

- **osteo** = bone
- **porosis** = porous

silent. The body sends no signals saying bones are losing their calcium and, as a result, their integrity. Blood samples offer no clues because blood calcium remains normal regardless of bone content, and measures of bone density may not be routinely taken until later in life. Highlight 12 suggests strategies to protect against bone loss, of which eating calcium-rich foods is only one.

REVIEW IT

Most of the body's calcium is in the bones, where it provides a rigid structure and a reservoir of calcium for the blood. Blood calcium participates in muscle contraction, blood clotting, and nerve impulses, and it is closely regulated by a system of hormones and vitamin D. Calcium is found predominantly in milk and milk products. Even when calcium intake is inadequate, blood calcium remains normal, but at the expense of bone loss, which can lead to osteoporosis. The accompanying table provides a summary of calcium.

Calcium

2011 RDA	**Deficiency Symptoms**
Adults: 1000 mg/day (adults, 19–50 yr) 1000 mg/day (men, 51–70 yr) 1200 mg/day (men, ≥71 yr) 1200 mg/day (women, ≥51 yr)	Stunted growth in children; bone loss (osteoporosis) in adults
Upper Level	**Toxicity Symptoms**
Adults: 2500 mg/day (adults, 19–50 yr) 2000 mg/day (adults, ≥51 yr)	Constipation; increased risk of urinary stone formation and kidney dysfunction; interference with absorption of other minerals
Chief Functions in the Body	**Significant Sources**
Mineralization of bones and teeth; also involved in muscle contraction and relaxation, nerve functioning, blood clotting, blood pressure	Milk and milk products, small fish (with bones), calcium-set tofu (bean curd), greens (bok choy, broccoli, chard, kale), legumes

Phosphorus

Phosphorus is the second most abundant mineral in the body. About 85 percent of it is found combined with calcium in the hydroxyapatite crystals of bones and teeth.

Phosphorus Roles in the Body Phosphorus is found not only in bones and teeth, but in all body cells as part of a major buffer system. Phosphorus is also part of DNA and RNA and is therefore necessary for all growth.

Phosphorus assists in energy metabolism. The high-energy compound ATP uses three phosphate groups to do its work. Many enzymes and the B vitamins become active only when a phosphate group is attached.

Phosphorus provides structure to the phospholipid vehicles that help to transport lipids in the blood. Phospholipids are also the major structural components of cell membranes, where they control the transport of nutrients into and out of the cells. Some proteins, such as the casein in milk, contain phosphorus as part of their structures (phosphoproteins).

Phosphorus Recommendations and Intakes Because phosphorus is commonly found in almost all foods, dietary deficiencies are unlikely. As Figure 12-17 shows, foods rich in proteins are the best sources of phosphorus. Milk and cheese contribute about 25 percent of the phosphorus in the US diet.

Over the years, researchers have emphasized the importance of an ideal calcium-to-phosphorus ratio to support calcium metabolism, but there is little or no evidence to support this concept. The quantities of calcium and phosphorus in the diet are far more important than their ratio to each other. A high phosphorus intake has been blamed for bone loss when, in fact, a low calcium intake—not a phosphorus toxicity or an improper ratio—is responsible. Research shows that the displacement of milk in the diet by cola drinks, not the phosphoric acid content of the beverages, limits bone density. No adverse effects of high dietary phosphorus intakes have been reported; still, a UL ♦ has been established.

♦ Phosphorus UL:
- 4000 mg/day (adults 19–70 yr)

phosphorus: a major mineral found mostly in the body's bones and teeth.

> FIGURE 12-17 Phosphorus in Selected Foods

See the "How To" feature on p. 302 for more information on using this figure.

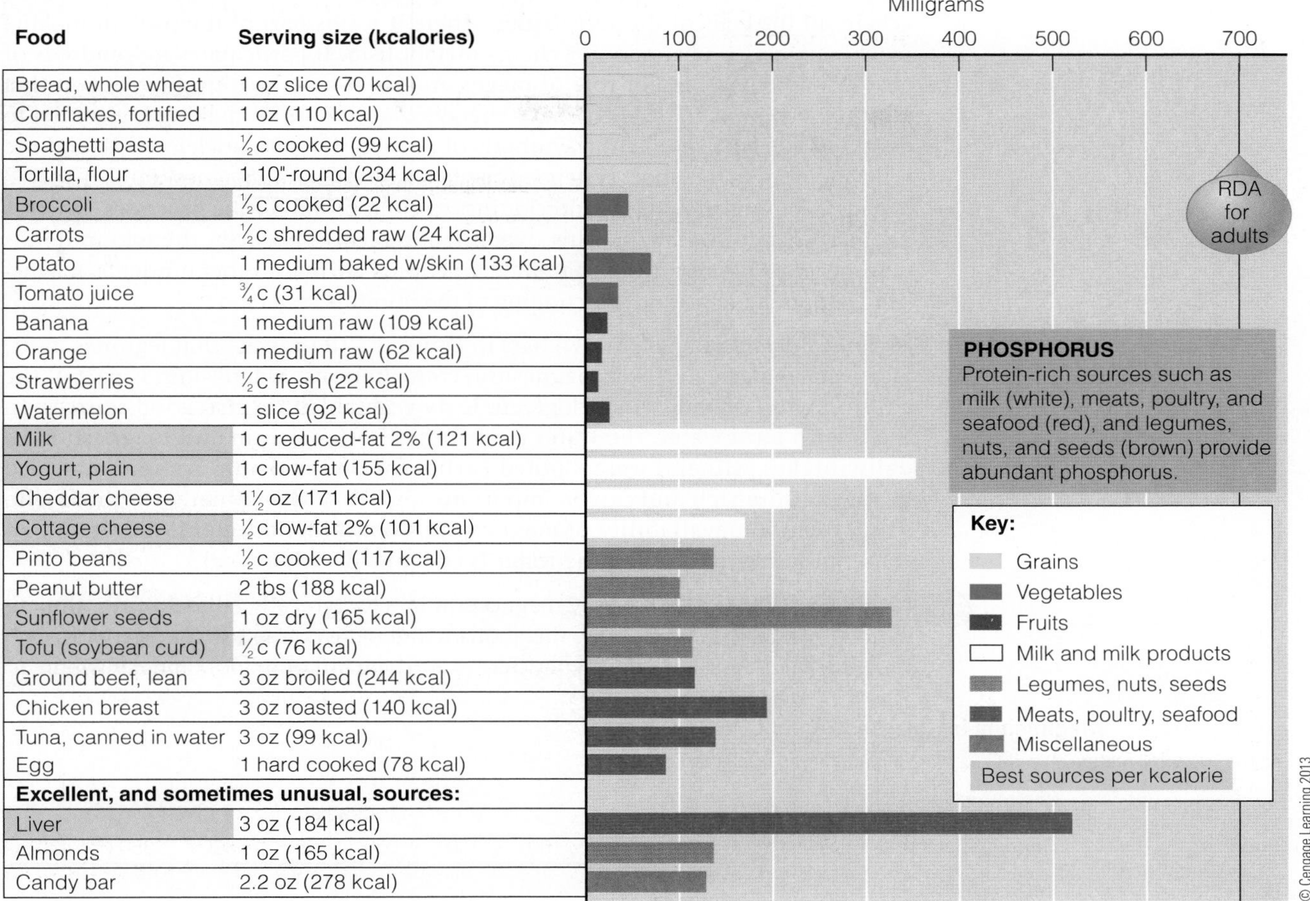

REVIEW IT

Phosphorus accompanies calcium both in the crystals of bone and in many foods such as milk. Phosphorus is also important in energy metabolism as part of ATP, in lipid transport as part of phospholipids, and in genetic materials as part of DNA and RNA. The accompanying table provides a summary of phosphorus.

Phosphorus

RDA

Adults: 700 mg/day

Upper Level

Adults (19–70 yr): 4000 mg/day

Chief Functions in the Body

Mineralization of bones and teeth; part of every cell; important in genetic material, part of phospholipids, used in energy transfer and in buffer systems that maintain acid-base balance

Deficiency Symptoms

Muscular weakness, bone pain[a]

Toxicity Symptoms

Calcification of nonskeletal tissues, particularly the kidneys

Significant Sources

Foods derived from animals (meat, fish, poultry, eggs, milk)

[a]Dietary deficiency rarely occurs, but some drugs can bind with phosphorus making it unavailable and resulting in bone loss that is characterized by weakness and pain.

Magnesium Only about 1 ounce of **magnesium** is present in the body of a 132-pound person (review Figure 12-9, p. 378). More than half of the body's magnesium is in the bones. Much of the rest is in the muscles and soft tissues, with

magnesium: a cation within the body's cells, active in many enzyme systems.

only 1 percent in the extracellular fluid. As with calcium, bone magnesium may serve as a reservoir to ensure normal blood concentrations.

Magnesium Roles in the Body In addition to maintaining bone health, magnesium acts in all the cells of the soft tissues, where it forms part of the protein-making machinery and is necessary for energy metabolism. It participates in hundreds of enzyme systems. A major role of magnesium is as a catalyst in the reaction that adds the last phosphate to the high-energy compound ATP, making it essential to the body's use of glucose; the synthesis of protein, fat, and nucleic acids; and the cells' membrane transport systems. Together with calcium, magnesium is involved in muscle contraction and blood clotting: calcium promotes the processes, whereas magnesium inhibits them. This dynamic interaction between the two minerals helps regulate blood pressure and lung function. Like many other nutrients, magnesium supports the normal functioning of the immune system.

Magnesium Intakes The brown bars in Figure 12-18 indicate that legumes, nuts, and seeds make significant magnesium contributions. Magnesium is part of the chlorophyll molecule, so dark green, leafy vegetables are also good sources. In areas with hard water, the water contributes both calcium and magnesium to daily intakes. Mineral waters noted earlier for their calcium content may also be magnesium-rich and can be important sources of this mineral for those who drink them. Bioavailability of magnesium from mineral water is about 50 percent, but it improves when the water is consumed with a meal.

Magnesium Deficiency Average magnesium intakes typically fall below recommendations, which may upset bone metabolism and increase the risk for osteoporosis.[31] In addition, magnesium deficiency may exacerbate inflammation and contribute to

> FIGURE 12-18 **Magnesium in Selected Foods**

See the "How To" feature on p. 302 for more information on using this figure.

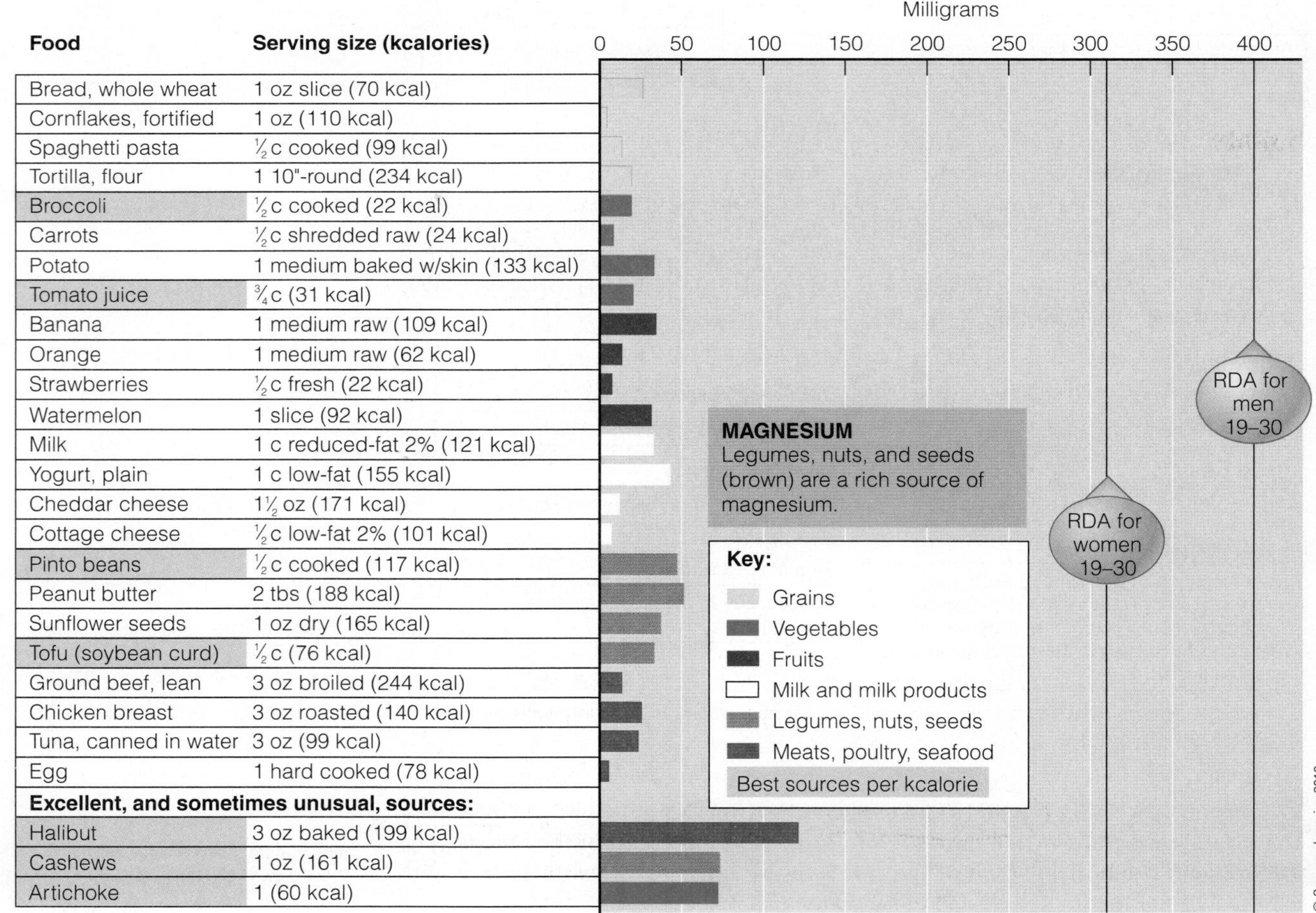

Food	Serving size (kcalories)
Bread, whole wheat	1 oz slice (70 kcal)
Cornflakes, fortified	1 oz (110 kcal)
Spaghetti pasta	½ c cooked (99 kcal)
Tortilla, flour	1 10"-round (234 kcal)
Broccoli	½ c cooked (22 kcal)
Carrots	½ c shredded raw (24 kcal)
Potato	1 medium baked w/skin (133 kcal)
Tomato juice	¾ c (31 kcal)
Banana	1 medium raw (109 kcal)
Orange	1 medium raw (62 kcal)
Strawberries	½ c fresh (22 kcal)
Watermelon	1 slice (92 kcal)
Milk	1 c reduced-fat 2% (121 kcal)
Yogurt, plain	1 c low-fat (155 kcal)
Cheddar cheese	1½ oz (171 kcal)
Cottage cheese	½ c low-fat 2% (101 kcal)
Pinto beans	½ c cooked (117 kcal)
Peanut butter	2 tbs (188 kcal)
Sunflower seeds	1 oz dry (165 kcal)
Tofu (soybean curd)	½ c (76 kcal)
Ground beef, lean	3 oz broiled (244 kcal)
Chicken breast	3 oz roasted (140 kcal)
Tuna, canned in water	3 oz (99 kcal)
Egg	1 hard cooked (78 kcal)
Excellent, and sometimes unusual, sources:	
Halibut	3 oz baked (199 kcal)
Cashews	1 oz (161 kcal)
Artichoke	1 (60 kcal)

chronic diseases such as heart disease, hypertension, diabetes, and cancer.[32] A severe magnesium deficiency causes a tetany similar to the calcium tetany described earlier. Magnesium deficiencies also impair central nervous system activity and may be responsible for the hallucinations experienced during alcohol withdrawal.

Magnesium and Hypertension Magnesium is critical to heart function and seems to protect against hypertension and heart disease. Interestingly, people living in areas of the country with hard water, which contains high concentrations of calcium and magnesium, tend to have low rates of heart disease. With magnesium deficiency, the walls of the arteries and capillaries tend to constrict—a possible explanation for the hypertensive effect.

Magnesium Toxicity Magnesium toxicity is rare, but it can be fatal. The UL for magnesium applies only to nonfood sources such as supplements or magnesium salts.

REVIEW IT

Like calcium and phosphorus, magnesium supports bone mineralization. Magnesium is also involved in numerous enzyme systems and in heart function. It is found abundantly in legumes and dark green, leafy vegetables and, in some areas, in water. The accompanying table offers a summary of magnesium.

Magnesium

RDA	**Deficiency Symptoms**
Men (19–30 yr): 400 mg/day Women (19–30 yr): 310 mg/day	Weakness; confusion; if extreme, convulsions, bizarre muscle movements (especially of eye and face muscles), hallucinations, and difficulty in swallowing; in children, growth failure[a]
Upper Level	**Toxicity Symptoms**
Adults: 350 mg nonfood magnesium/day	From nonfood sources only; diarrhea, alkalosis, dehydration
Chief Functions in the Body	**Significant Sources**
Bone mineralization, building of protein, enzyme action, normal muscle contraction, nerve impulse transmission, maintenance of teeth, and functioning of immune system	Nuts, legumes, whole grains, dark green vegetables, seafood, chocolate, cocoa

[a]A still more severe deficiency causes tetany, an extreme, prolonged contraction of the muscles similar to that caused by low blood calcium.

Sulfate **Sulfate** is the oxidized form of the mineral **sulfur**, as it exists in food and water. The body's need for sulfate is easily met by a variety of foods and beverages. In addition, the body receives sulfate from the amino acids methionine and cysteine, which are found in dietary proteins. These sulfur-containing amino acids help determine the contour of protein molecules. The sulfur-containing side chains in cysteine molecules can link to each other via disulfide bridges, which stabilize the protein structure. (See the drawing of insulin with its disulfide bridges on p. 170.) Skin, hair, and nails contain some of the body's more rigid proteins, which have a high sulfur content.

Because the body's sulfate needs are easily met with normal protein intakes, there is no recommended intake for sulfate. Deficiencies do not occur when diets contain protein. Only when people lack protein to the point of severe deficiency will they lack the sulfur-containing amino acids.

REVIEW IT Identify the main roles, deficiency symptoms, and food sources for each of the major minerals (sodium, chloride, potassium, calcium, phosphorus, magnesium, and sulfate).

Like the other nutrients, minerals' actions are coordinated to get the body's work done. The major minerals, especially sodium, chloride, and potassium, influence the body's fluid balance; whenever an anion moves, a cation moves—always maintaining homeostasis. Sodium, chloride, potassium, calcium, and magnesium are key members of the team of nutrients that direct nerve impulse transmission and muscle contraction. They are also the primary nutrients involved in regulating blood pressure. Phosphorus and magnesium participate in many reactions involving glucose, fatty acids, amino acids, and the vitamins. Calcium, phosphorus, and magnesium combine to form the structure of the bones and teeth. Each major mineral also plays other specific roles in the body. The table on p. 394 provides a summary of the major minerals.

sulfate: a salt produced from the oxidation of sulfur.

sulfur: a mineral present in the body as part of some proteins.

REVIEW IT The Major Minerals

Chief Functions	Deficiency Symptoms	Toxicity Symptoms	Significant Sources
Sodium Maintains normal fluid and electrolyte balance; assists in nerve impulse transmission and muscle contraction	Muscle cramps, mental apathy, loss of appetite	Edema, acute hypertension	Table salt, soy sauce; moderate amounts in meats, milks, breads, and vegetables; large amounts in processed foods
Chloride Maintains normal fluid and electrolyte balance; part of hydrochloric acid found in the stomach, necessary for proper digestion	Do not occur under normal circumstances	Vomiting	Table salt, soy sauce; moderate amounts in meats, milks, eggs; large amounts in processed foods
Potassium Maintains normal fluid and electrolyte balance; facilitates many reactions; supports cell integrity; assists in nerve impulse transmission and muscle contractions	Irregular heartbeat, muscular weakness, glucose intolerance	Muscular weakness; vomiting; if injected into a vein, can stop the heart	All whole foods; meats, milks, fruits, vegetables, grains, legumes
Calcium Mineralization of bones and teeth; also involved in muscle contraction and relaxation, nerve functioning, blood clotting, and blood pressure	Stunted growth in children; bone loss (osteoporosis) in adults	Constipation; increased risk of urinary stone formation and kidney dysfunction; interference with absorption of other minerals	Milk and milk products, small fish (with bones), tofu, greens (bok choy, broccoli, chard), legumes
Phosphorus Mineralization of bones and teeth; part of every cell; important in genetic material, part of phospholipids, used in energy transfer and in buffer systems that maintain acid-base balance	Muscular weakness, bone pain[a]	Calcification of nonskeletal tissues, particularly the kidneys	All animal tissues (meat, fish, poultry, eggs, milk)
Magnesium Bone mineralization, building of protein, enzyme action, normal muscle contraction, nerve impulse transmission, maintenance of teeth, and functioning of immune system	Weakness; confusion; if extreme, convulsions, bizarre muscle movements (especially of eye and face muscles), hallucinations, and difficulty in swallowing; in children, growth failure[b]	From nonfood sources only; diarrhea, alkalosis, dehydration	Nuts, legumes, whole grains, dark green vegetables, seafood, chocolate, cocoa
Sulfate As part of proteins, stabilizes their shape by forming disulfide bridges; part of the vitamins biotin and thiamin and the hormone insulin	None known; protein deficiency would occur first	Toxicity would occur only if sulfur-containing amino acids were eaten in excess; this (in animals) suppresses growth	All protein-containing foods (meats, fish, poultry, eggs, milk, legumes, nuts)

[a]Dietary deficiency rarely occurs, but some drugs can bind with phosphorus making it unavailable and resulting in bone loss that is characterized by weakness and pain.
[b]A still more severe deficiency causes tetany, an extreme, prolonged contraction of the muscles similar to that caused by low blood calcium.

With all of the tasks these minerals perform, they are of great importance to life. Consuming enough of each of them every day is easy, given a variety of foods from each of the food groups. Whole-grain breads supply magnesium; fruits, vegetables, and legumes provide magnesium and potassium too; milk products offer calcium and phosphorus; meats, poultry, and seafood offer phosphorus and sulfate as well; all foods provide sodium and chloride, with excesses being more problematic than inadequacies. The message is quite simple and has been repeated throughout this text: for an adequate intake of all the nutrients, including the major minerals, choose a variety of foods from each of the five food groups. And drink plenty of water.

Nutrition Portfolio

Many people may miss the mark when it comes to drinking enough water to keep their bodies well hydrated or obtaining enough calcium to promote strong bones; in contrast, sodium intakes often exceed those recommended for health. Go to Diet Analysis Plus and choose one of the days on which you tracked your diet for an entire day. Select the Intake vs. Goals report and then consider the following questions. Remember that scoring 100 percent on this report means you met your goal.

- Did you exceed, fail to meet, or meet your goal for water intake? Was that a typical day for you? Describe your strategy for ensuring that you drink plenty of water—about eight glasses—every day.
- Take a look at your sodium intake in this report. Most people in the United States exceed the UL. Did you? Explain the importance of selecting and preparing foods with less salt.
- How was your intake of calcium for that day? If you are not getting enough calcium, consult Chapter 12 for ideas to help you get more, then list at least three foods or beverages you would be willing to eat or drink that would improve your intake.

Diet Analysis PLUS

To complete this exercise, go to your Diet Analysis Plus at **www.cengagebrain.com.**

STUDY IT To review the key points of this chapter and take a practice quiz, go to the study cards at the end of the book.

REFERENCES

1. A. K. Johnson, The sensory psychobiology of thirst and salt appetite, *Medicine and Science in Sports and Exercise* 39 (2007): 1388–1400.
2. A. K. Kant, B. I. Graubard, and E. A. Atchison, Intakes of plain water, moisture in foods and beverages, and total water in the adult US population—Nutritional, meal pattern, and body weight correlates: National Health and Nutrition Examination Surveys 1999–2006, *American Journal of Clinical Nutrition* 90 (2009): 655–663.
3. Committee on Dietary Reference Intakes, *Dietary Reference Intakes for Water, Potassium, Sodium, Chloride, and Sulfate* (Washington, D.C.: National Academies Press, 2004), p. 67.
4. B. M. Popkin, K. E. D'Anci, and I. H. Rosenberg, Water, hydration, and health, *Nutrition Reviews* 68 (2010): 439–458.
5. Actions You Can Take to Reduce Lead in Drinking Water, www.epa.gov/ogwdw/lead/lead1.html, updated April 2008.
6. Centers for Disease Control and Prevention, Application of lower sodium intake recommendations to adults: United States, 1999–2006, *Morbidity and Mortality Weekly Report* 58 (2009): 281–283.
7. B. N. Van Vliet and J. P. Montani, The time course of salt-induced hypertension, and why it matters, *International Journal of Obesity* 32 (2008): S35–S47.
8. Centers for Disease Control and Prevention, 2009.
9. K. M. Dickinson, J. B. Keogh, and P. M. Clifton, Effects of a low-salt diet on flow-mediated dilation in humans, *American Journal of Clinical Nutrition* 89 (2009): 485–490.
10. US Department of Agriculture and US Department of Health and Human Services, *Dietary Guidelines for Americans, 2010*, www.dietaryguidelines.gov.
11. L. J. Appel and coauthors, The importance of population-wide sodium reduction as a means to prevent cardiovascular disease and stroke: A call to action from the American Heart Association, *Circulation* 123 (2011): 1138–1143.
12. C. N. Mhurchu and coauthors, Sodium content of processed foods in the United Kingdom: Analysis of 44,000 foods purchased by 21,000 households, *American Journal of Clinical Nutrition* 93 (2011): 594–600; C.A.M. Anderson and coauthors, Dietary sources of sodium in China, Japan, the United Kingdom, and the United States, women and men aged 40 to 59 years: The INTERMAP Study, *Journal of the American Dietetic Association* 110 (2010): 736–745; J. L. Webster, E. K. Dunford, and B. C. Neal, *American Journal of Clinical Nutrition* 91 (2010): 413–420; F. J. He and G. A. MacGregor, A comprehensive review on salt and health and current experience of worldwide salt reduction programmes, *Journal of Human Hypertension* 23 (2009): 363–384; R. A. Forshee, Innovative regulatory approaches to reduce sodium consumption: Could a cap-and-trade system work? *Nutrition Reviews* 66 (2008): 280–285.
13. K. Bibbins-Domingo and coauthors, Projected effect of dietary salt reductions on future cardiovascular disease, *New England Journal of Medicine* 362 (2010): 590–599; Institute of Medicine (US) Committee on Strategies to Reduce Sodium Intake, *Strategies to Reduce Sodium Intake in the United States* (Washington, D.C.: National Academies Press, 2010).
14. Committee on Dietary Reference Intakes, *Dietary Reference Intakes for Water, Potassium, Sodium, Chloride, and Sulfate* (Washington, D.C.: National Academies Press, 2005), p. 281.
15. M. Umesawa and coauthors, Relations between dietary sodium and potassium intakes and mortality from cardiovascular disease: The Japan Collaborative Cohort Study for Evaluation of Cancer Risks, *American Journal of Clinical Nutrition* 88 (2008): 195–202.
16. J. Kaluza and coauthors, Dietary calcium and magnesium intake and mortality: A prospective study of men, *American Journal of Epidemiology* 171 (2010): 801–807; I. R. Reid and coauthors, Effects of calcium supplementation on lipids, blood pressure, and body composition in healthy older men: A randomized controlled trial, *American Journal of Clinical Nutrition* 91 (2010): 131–139; V. Centeno and coauthors, Molecular mechanisms triggered by low-calcium diets, *Nutrition Research Reviews* 22 (2009): 163–174; L. Wang and coauthors, Dietary intake of dairy products, calcium, and vitamin D and the risk of hypertension in middle-aged and older women, *Hypertension* 51 (2008): 1073–1079.
17. M. Huncharek, J. Muscat, and B. Kupelnick, Colorectal cancer risk and dietary intake of calcium, vitamin D, and dairy products: A meta-analysis of 26,335 cases from 60 observational studies, *Nutrition and Cancer* 61 (2009): 47–69; Y. Park and coauthors, Dairy food, calcium, and risk of cancer in the NIH-AARP Diet and Health Study, *Archives of Internal Medicine* 169 (2009): 391–401; J. Ishihara and coauthors, Dietary calcium, vitamin D, and the risk of colorectal cancer, *American Journal of Clinical Nutrition* 88 (2008): 1576 –1583.
18. D. R. Shahar and coauthors, Dairy calcium intake, serum vitamin D, and successful weight loss, *American Journal of Clinical Nutrition* 92 (2010): 1017–1022; R. P. Heaney and K. Rafferty, Preponderance of the evidence: An example from the issue of calcium intake and body composition, *Nutrition Reviews* 67 (2009): 32–39; G. C. Major and coauthors, Recent developments in calcium-related obesity research, *Obesity Reviews* 9 (2008): 428–445.
19. M. S. Vanselow and coauthors, Adolescent beverage habits and changes in weight over time: Findings from Project EAT, *American Journal of Clinical Nutrition* 90 (2009): 1489–1495.
20. J. A. Gilbert and coauthors, Milk supplementation facilitates appetite control in obese women during weight loss: A randomized, single-blind, placebo-controlled trial, *British Journal of Nutrition* 105 (2011): 133–143.
21. M. Bortolotti and coauthors, Dairy calcium supplementation in overweight or obese persons: Its effect on markers of fat metabolism, *American Journal of Clinical Nutrition* 88 (2008): 877–885; A. J. Lanou and N. D. Barnard, Dairy and weight loss hypothesis: An evaluation of the clinical trials, *Nutrition Reviews* 66 (2008): 272–279; D. Teegarden and C. W. Gunther, Can the controversial relationship between dietary calcium and body weight be mechanistically explained by alterations in appetite and food intake? *Nutrition Reviews* 66 (2008): 601–605.
22. M. B. Zemel and coauthors, Effects of dairy compared with soy on oxidative and inflammatory stress in overweight and obese subjects, *American Journal of Clinical Nutrition* 91 (2010): 16–22.
23. R. C. Khanal and I. Nemere, Regulation of intestinal calcium transport, *Annual Review of Nutrition* 28 (2008): 179–196.
24. F. Bronner, Recent developments in intestinal calcium absorption, *Nutrition Reviews* 67 (2009): 109–113.
25. C. M. Weaver, Closing the gap between calcium intake and requirements, *Journal of the American Dietetic Association* 109 (2009): 812–813.
26. R. L. Bailey and coauthors, Estimation of total usual calcium and vitamin D intakes in the United States, *Journal of Nutrition* 140 (2010): 817–822.
27. J. R Hunt, L. K. Johnson, and Z.K.F. Roughead, Dietary protein and calcium interact to influence calcium retention: A controlled feeding study, *American Journal of Clinical Nutrition* 89 (2009): 1357–1365; K. Rafferty and R. P. Heaney, Nutrient effects on the calcium economy: Emphasizing the potassium controversy, *Journal of Nutrition* 138 (2008): 166S–171S.
28. M. P. Thorpe and E. M. Evans, Dietary protein and bone health: Harmonizing conflicting theories, *Nutrition Reviews* 69 (2011): 215–230; Hunt, Johnson, and Roughead, 2009.
29. S. A. Abrams, Setting Dietary Reference Intakes with the use of bioavailability data: Calcium, *American Journal of Clinical Nutrition* 91 (2010): 1474S–1477S.
30. National Osteoporosis Foundation, **www.nof.org**, accessed July 2011.
31. R. K. Rude, F. R. Singer, and H. E. Gruber, Skeletal and hormonal effects of magnesium deficiency, *Journal of the American College of Nutrition* 28 (2009): 131–141.
32. F. H. Nielsen, Magnesium, inflammation, and obesity in chronic disease, *Nutrition Reviews* 68 (2010): 333–340.

HIGHLIGHT > 12

LEARN IT Describe factors that contribute to the development of osteoporosis and strategies to prevent it.

Osteoporosis and Calcium

© ONOKY-Photononstop/Alamy

Osteoporosis becomes apparent during the later years, but it develops much earlier—and without warning. Few people are aware that their bones are being robbed of their strength. The problem often first becomes evident when someone's hip suddenly gives way. People say, "She fell and broke her hip," but in fact the hip may have been so fragile that it broke *before* she fell. Even bumping into a table may be enough to shatter a porous bone into fragments so numerous and scattered that they cannot be reassembled. Removing them and replacing them with an artificial joint requires major surgery. More than 400,000 people in the United States are hospitalized each year because of fractures related to osteoporosis. About a third die of complications within a year; many more will never walk or live independently again.[1] Their quality of life slips downward.

This highlight examines low bone density and osteoporosis, one of the most prevalent diseases of aging, affecting an estimated 44 million people in the United States—most of them women older than 50.[2] It reviews the many factors that contribute to the 2 million fractures in the bones of the hips, vertebrae, wrists, arms, and ankles each year. And it presents strategies to reduce the risks, paying special attention to the role of dietary calcium.

Bone Development and Disintegration

Bone has two compartments: the outer, hard shell of **cortical bone** and the inner, lacy matrix of **trabecular bone.** (The glossary defines these and other bone-related terms.) Both can lose minerals, but in different ways and at different rates. The first photograph in Figure H12-1 shows a human leg bone sliced lengthwise, exposing the lacy, calcium-containing crystals of trabecular bone. These crystals give up calcium to the blood when the diet runs short, and they take up calcium again when the supply is plentiful (review Figure 12-13 on p. 387). For people who have eaten calcium-rich foods throughout the bone-forming years of their youth, these deposits make bones dense and provide a rich reservoir of calcium.

Surrounding and protecting the trabecular bone is a dense, ivorylike exterior shell—the cortical bone. Cortical bone composes the shafts of the long bones, and a thin cortical shell caps the ends of the bones too. Both compartments confer strength on bone: cortical bone provides the sturdy outer wall, and trabecular bone provides support along the lines of stress.

The two types of bone handle calcium in different ways. Supplied with blood vessels and metabolically active, trabecular bone is sensitive to hormones that govern day-to-day deposits and withdrawals of calcium. It readily gives up minerals whenever blood calcium needs replenishing. Losses of trabecular bone start becoming significant for men and women in their 30s, although losses can occur whenever calcium withdrawals exceed deposits. Cortical bone also gives up calcium, but slowly and at a steady pace. Cortical bone losses typically begin at about age 40 and continue slowly but surely thereafter.

As bone loss continues, bone density declines, and osteoporosis becomes apparent (see Figure H12-1). Bones become so fragile that even the body's own weight can overburden the spine—vertebrae may suddenly disintegrate and crush down, painfully pinching major nerves.[3] Or the vertebrae may compress into wedge shapes, forming what is often called a "dowager's hump," the posture many older people assume as they "grow shorter." Figure H12-2 (p. 398) shows the effect of compressed spinal bone on a woman's height and posture. Because both the cortical shell and the trabecular interior weaken, breaks most often occur in the hip, as mentioned in the introductory paragraph.

Physicians can determine bone loss and diagnose osteoporosis by measuring **bone density** using dual-energy X-ray absorptiometry (DEXA scan).[4] They also consider risk factors for osteoporosis, including age, personal and family history of fractures, and physical inactivity.

GLOSSARY

antacids: medications used to relieve indigestion by neutralizing acid in the stomach. Calcium-containing preparations (such as Tums) contain available calcium. Antacids with aluminum or magnesium hydroxides (such as Rolaids) can accelerate calcium losses.

bone meal or **powdered bone:** crushed or ground bone preparations intended to supply calcium to the diet. Calcium from bone is not well absorbed and is often contaminated with toxic minerals such as arsenic, mercury, lead, and cadmium.

bone density: a measure of bone strength. When minerals fill the bone matrix (making it dense), they give it strength.

cortical bone: the very dense bone tissue that forms the outer shell surrounding trabecular bone and comprises the shaft of a long bone.

dolomite: a compound of minerals (calcium magnesium carbonate) found in limestone and marble. Dolomite is powdered and is sold as a calcium-magnesium supplement. However, it may be contaminated with toxic minerals, is not well absorbed, and interacts adversely with absorption of other essential minerals.

osteoporosis: a disease characterized by porous and fragile bones.

oyster shell: a product made from the powdered shells of oysters that is sold as a calcium supplement, but it is not well absorbed by the digestive system.

trabecular (tra-BECK-you-lar) **bone:** the lacy inner structure of calcium crystals that supports the bone's structure and provides a calcium storage bank.

> FIGURE H12-1 Healthy and Osteoporotic Trabecular Bones

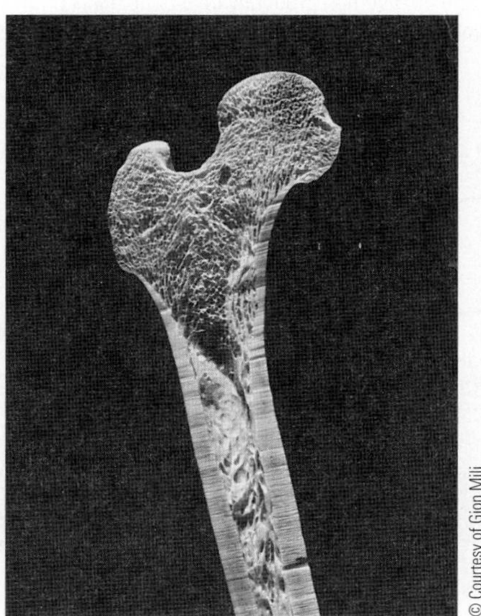
© Courtesy of Gjon Mili

Trabecular bone is the lacy network of calcium-containing crystals that fills the interior. Cortical bone is the dense, ivorylike bone that forms the exterior shell.

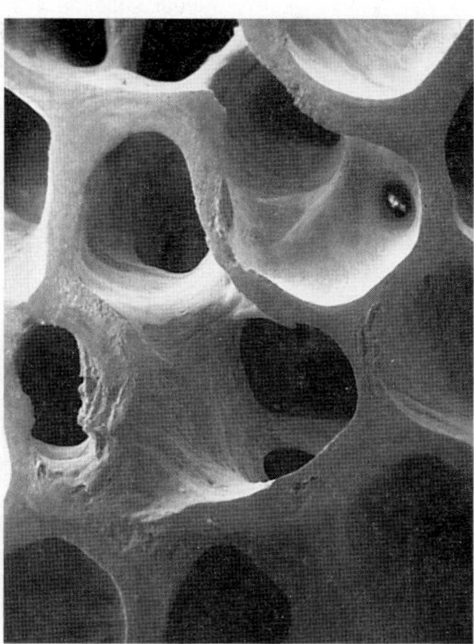

Electron micrograph of healthy trabecular bone.

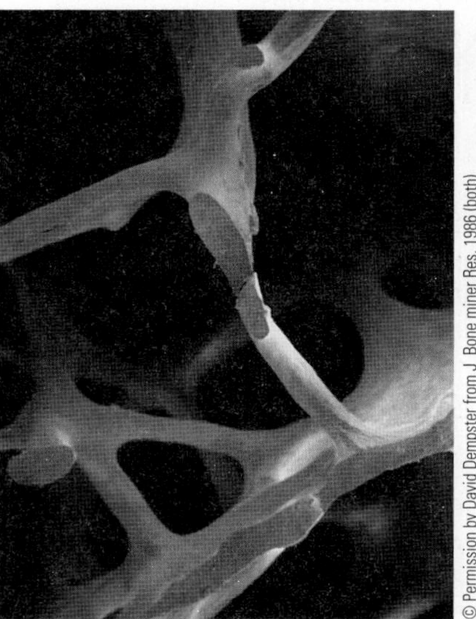
© Permission by David Dempster from J Bone miner Res, 1986 (both)

Electron micrograph of trabecular bone affected by osteoporosis.

Table H12-1 summarizes the major risk factors for osteoporosis. The more risk factors that apply to a person, the greater the chances of bone loss. Notice that several risk factors that are influential in the development of osteoporosis—such as age, gender, and genetics—cannot be changed. Other risk factors—such as diet, physical activity, body weight, smoking, and alcohol use—are personal behaviors that can be changed. By eating a calcium-rich, well-balanced diet, being physically active, abstaining from smoking, and drinking alcohol in moderation (if at all), people can defend themselves against osteoporosis. These decisions are particularly important for those with other risk factors that cannot be changed.

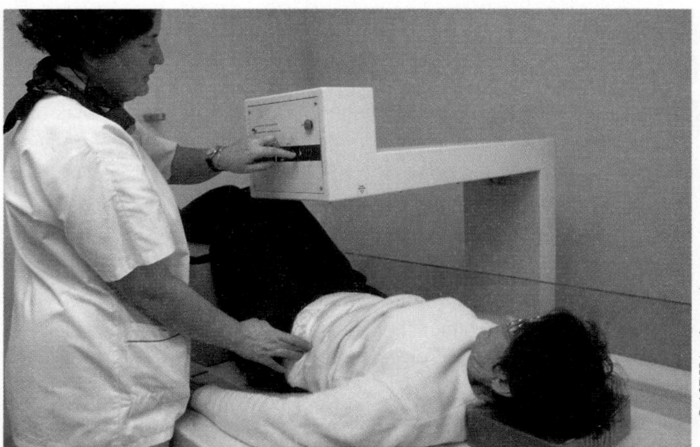
© ASTIER/agefotostock

Using a DEXA (dual-energy X-ray absorpiometry) test to measure bone mineral density identifies osteoporosis, determines risks for fractures, and tracks responses to treatment.

Whether a person develops osteoporosis seems to depend on the interactions of several factors, including nutrition. The strongest predictor of bone density is age.

Age and Bone Calcium

Two major stages of life are critical in the development of osteoporosis. The first is the bone-acquiring stage of childhood and adolescence. The second is the bone-losing decades of late adulthood, especially in women after menopause. The bones gain strength and density all through the growing years and into young adulthood. As people age, the cells that build bone gradually become less active, but those that dismantle bone continue working. The result is that bone loss exceeds bone formation. Some bone loss is inevitable, but losses can be curtailed by maximizing bone mass.

TABLE H12-1 Risk Factors for Osteoporosis

Nonmodifiable	Modifiable
• Female gender • Older age • Small frame • Caucasian, Asian, or Hispanic/Latino • Family history of osteoporosis or fractures • Personal history of fractures • Estrogen deficiency in women (amenorrhea or menopause, especially early or surgically induced); testosterone deficiency in men	• Sedentary lifestyle • Diet inadequate in calcium and vitamin D • Diet excessive in protein, sodium, caffeine • Cigarette smoking • Alcohol abuse • Low body weight • Certain medications, such as glucocorticoids and anticonvulsants

> **FIGURE H12-2 Loss of Height in a Woman Caused by Osteoporosis**

The woman on the left is about 50 years old. On the right, she is 80 years old. Her legs have not grown shorter. Instead, her back has lost length due to collapse of her spinal bones (vertebrae). Collapsed vertebrae cannot protect the spinal nerves from pressure that causes excruciating pain.

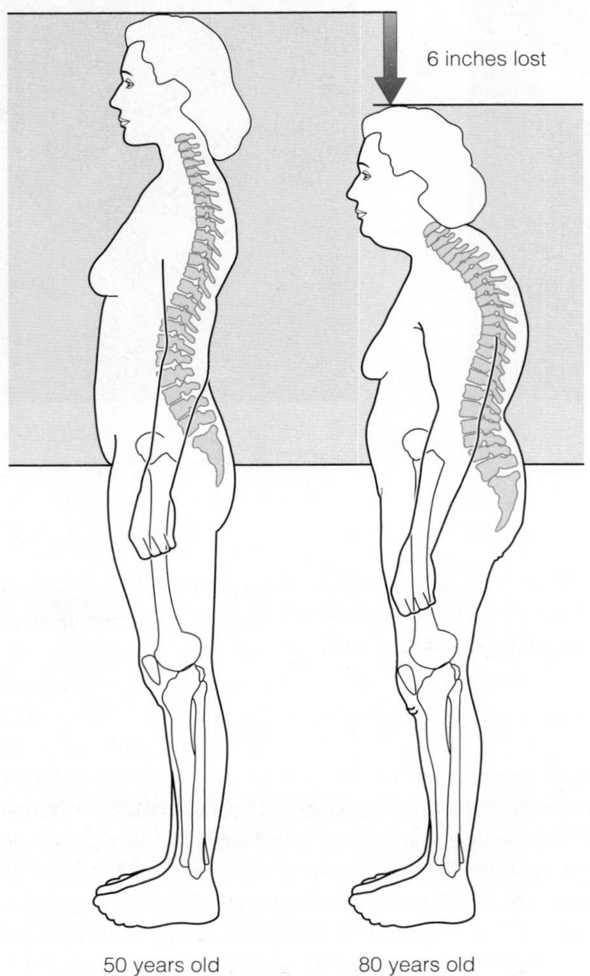
6 inches lost
50 years old
80 years old

> **FIGURE H12-3 Bone Losses over Time Compared**

Peak bone mass is achieved by age 30. Women gradually lose bone mass until menopause, when losses accelerate dramatically and then gradually taper off.

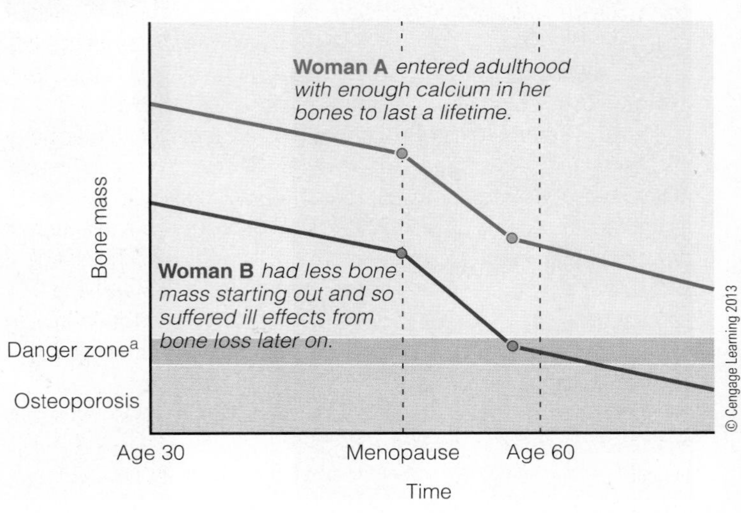
Woman A entered adulthood with enough calcium in her bones to last a lifetime.
Bone mass
Woman B had less bone mass starting out and so suffered ill effects from bone loss later on.
Danger zone[a]
Osteoporosis
Age 30
Menopause
Age 60
Time
© Cengage Learning 2013

[a]People with a moderate degree of bone mass loss are said to have *osteopenia* and are at increased risk of fractures.

Maximizing Bone Mass

To maximize bone mass, the diet must deliver an adequate supply of calcium during the first three decades of life. Children and teens who consume milk products and get enough calcium have denser bones than those with inadequate intakes.[5] With little or no calcium from the diet, the body must depend on bone to supply calcium to the blood—bone mass diminishes, and bones lose their density and strength. When people reach the bone-losing years of middle age, those who formed dense bones during their youth have the advantage. They simply have more bone starting out and can lose more before suffering ill effects. Figure H12-3 demonstrates this effect.

Minimizing Bone Loss

Not only does dietary calcium build strong bones in youth, but it remains important in protecting against losses in the later years. Unfortunately, calcium intakes of older adults are typically low, and calcium absorption declines after menopause. The kidneys do not activate vitamin D as well as they did earlier (recall that vitamin D enhances calcium absorption). Also, sunlight is needed to form vitamin D, and many older people spend little or no time outdoors in the sunshine. For these reasons, and because intakes of vitamin D are typically low anyway, blood vitamin D declines.

Some of the hormones that regulate bone and calcium metabolism—parathyroid hormone, calcitonin, and estrogen—also change with age and accelerate bone loss. Together, these age-related factors contribute to bone loss: inefficient bone remodeling, reduced calcium intakes, impaired calcium absorption, poor vitamin D status, and hormonal changes that favor bone mineral withdrawal.

Gender and Hormones

After age, gender is the next strongest predictor of osteoporosis. The sex hormones play a major role in regulating the rate of bone turnover.[6] Men have greater bone density than women at maturity, and women have greater losses than men in later life. Consequently, men develop bone problems about 10 years later than women, and women account for two out of three cases of osteoporosis.

Menopause imperils women's bones. Bone dwindles rapidly when the hormone estrogen diminishes and menstruation ceases. The lack of estrogen contributes to the release of cytokines that produce inflammation and accelerate bone loss.[7] Women may lose up to 20 percent of their bone mass during the 6 to 8 years following menopause. Eventually, losses taper off so that women again lose bone at the same rate as men their age. Losses of bone minerals continue throughout the remainder of a woman's lifetime, but not at the free-fall pace of the menopause years (review Figure H12-3).

Rapid bone losses also occur when *young* women's ovaries fail to produce enough estrogen, causing menstruation to cease. In some cases, diseased ovaries are to blame and must be removed; in others, the ovaries fail to produce sufficient estrogen because the women suffer from anorexia nervosa and have unreasonably restricted their body weight (see Highlight 8). The amenorrhea and low body weights explain much of the bone loss seen in these young women, even years after diagnosis and treatment.

Estrogen therapy may help some women prevent further bone loss and reduce the incidence of fractures. Because estrogen therapy may increase the risks for breast cancer, women must carefully weigh any potential benefits against the possible dangers. A combination of drugs or of hormone replacement and a drug may be most beneficial.

Several drug therapies have been developed to inhibit bone loss and enhance bone formation.[8] The FDA has approved the following drugs to prevent or treat osteoporosis: biophosphates, calcitonin, estrogens, estrogen antagonists, and parathyroid hormone.[9]

Some women who choose not to use estrogen therapy turn to soy as an alternative treatment. Interestingly, the phytochemicals commonly found in soybeans mimic the actions of estrogen in the body. When natural estrogen is lacking, as it is after menopause, these phytochemicals may step in to stimulate estrogen-sensitive tissues. By way of this action, soy and its phytochemicals may help to prevent the rapid bone losses of the menopause years.[10] Unfortunately, soy phytochemicals may also stimulate breast cancer cell growth in some women.[11] Because the risks and benefits vary depending on life stage and prior history of breast cancer, women should discuss soy options with their physicians.

As in women, sex hormones—including estrogen—appear to play a key role in men's bone loss as well.[12] Other common causes of osteoporosis in men include corticosteroid use and alcohol abuse.[13]

Genetics

Risks of osteoporosis appear to run along racial lines and reflect genetic differences in bone development. African Americans, for example, seem to use and retain calcium more efficiently than Caucasians. Consequently, even though their calcium intakes are typically lower, black people have denser bones than white people do. Greater bone density expresses itself in less bone loss, fewer fractures, and a lower rate of osteoporosis among blacks.

The exact role of genetics is unclear, and specific genes have yet to be identified.[14] Most likely, genes influence both the peak bone mass achieved during growth and the bone loss incurred during the later years. The extent to which a given genetic potential is realized, however, depends on many outside factors. Diet and physical activity, for example, can maximize peak bone density during growth, whereas alcohol and tobacco abuse can accelerate bone losses later in life. Importantly, these factors are within a person's control.

Physical Activity and Body Weight

Physical activity may be the single most important factor supporting bone growth during adolescence.[15] Muscle strength and bone strength go together.[16] When muscles work, they pull on the bones, stimulating them to grow denser.[17] The hormones that promote new muscle growth also favor the building of bone. As a result, active bones are denser and stronger than sedentary bones.

© Tetra Images/Alamy

Strength training helps to build strong bones.

Both the muscle contraction and the gravitational pull of the body's weight create a load that benefits bone metabolism.[18] To keep bones healthy, a person should engage in weight training or weight-bearing endurance activities (such as tennis and jogging or sprint cycling) regularly.[19] Regular physical activity combined with an adequate calcium intake helps to maximize bone density in adolescence. Adults can also maximize and maintain bone density with a regular program of weight training. Even past menopause, when most women are losing bone, weight training improves bone density.

Heavier body weights and weight gains place a similar stress on the bones and promote their density. The many relationships between bone and fat are beginning to explain how overweight may protect bones.[20] For example, the adipokine leptin may play a key role in the relationship between body weight and bone mass.[21] Obese mice that are deficient in leptin show increases in bone formation. In contrast, weight losses reduce bone density and increase the risk of fractures—in part because energy restriction diminishes calcium absorption and compromises calcium balance. As mentioned in Highlight 8, the combination of underweight, severely restricted energy intake, extreme daily exercise, and amenorrhea reliably predicts bone loss.

Smoking and Alcohol

Add bone damage to the list of ill consequences associated with smoking. The bones of smokers are less dense than those of nonsmokers—even after controlling for differences in age, body weight, and physical activity habits. Fortunately, the damaging effects can be reversed with smoking cessation. Blood indicators of beneficial bone activity are apparent 6 weeks after a person stops smoking. In time, bone density is similar for former smokers and nonsmokers.

People who abuse alcohol often suffer from osteoporosis and experience more bone breaks than others. Several factors appear to be involved. Alcohol enhances fluid excretion, leading to excessive calcium losses in the urine; upsets the hormonal balance required for healthy bones; slows bone formation, leading to lower bone density; stimulates bone breakdown; and increases the risk of falling. Whereas high alcohol intakes are associated with lower bone density, limited research suggests that *moderate* alcohol consumption may benefit bone mineral density.[22]

Dietary Calcium

For older adults, an adequate calcium intake alone cannot protect against bone fractures. Bone strength later in life depends primarily on how well the bones were built during childhood and adolescence. Adequate calcium nutrition during the growing years is essential to achieving optimal peak bone mass. Simply put, growing children who do not get enough calcium do not develop strong bones. For this reason, the DRI Committee recommends 1300 milligrams of calcium per day for everyone 9 through 18 years of age. Unfortunately, few girls meet the recommendations for calcium during these bone-forming years. (Boys generally obtain intakes close to those recommended because they eat more food.) Consequently, most girls start their adult years with less-than-optimal bone density. As adults, women rarely meet their recommended intakes of 1000 to 1200 milligrams from food. Some authorities suggest 1500 milligrams of calcium for postmenopausal women who are not receiving estrogen.

Other Nutrients

Much research has focused on calcium, but other nutrients support bone health too.[23] Adequate protein protects bones and reduces the likelihood of hip fractures.[24] As mentioned earlier, vitamin D is needed to maintain calcium metabolism and optimal bone health.[25] Research suggests that a combination of calcium and vitamin D supplements is the best option for the prevention or treatment of osteoporosis.[26] Daily supplementation with a low dose of vitamin D may reduce bone loss and the risk of fractures, but a single high dose seems to increase the risk of falls and fractures.[27] Vitamin K decreases bone turnover and protects against hip fractures. Vitamin C may slow bone losses.[28] The minerals magnesium and potassium also help to maintain bone mineral density. Vitamin A is needed in the bone-remodeling process, but too much vitamin A may be associated with osteoporosis. Carotenoids may inhibit bone loss.[29] Omega-3 fatty acids may help preserve bone integrity.[30] Additional research points to the bone benefits not of a specific nutrient, but of a diet rich in fruits, vegetables, and whole grains.[31] In contrast, diets containing too much salt are associated with bone losses. Similarly, diets containing too many colas or commercially baked snack and fried foods are associated with low bone mineral density. Clearly, a well-balanced diet that depends on all the food groups to supply a full array of nutrients is central to bone health.

A Perspective on Calcium Supplements

Bone health depends, in part, on calcium. People who do not consume milk products or other calcium-rich foods in amounts that provide even half the recommendation should consider consulting a registered dietitian who can assess the diet and suggest food choices to correct any inadequacies. Calcium from foods may support bone health better than calcium from supplements. For those who are unable to consume enough calcium-rich foods, however, taking calcium supplements—especially in combination with vitamin D—may help to enhance bone density and protect against bone loss and fractures.[32] Because calcium supplements may increase the risk of heart attacks and strokes, women should consult their physicians when making this decision.[33]

An estimated 60 percent of women aged 60 and over take calcium supplements.[34] Selecting a calcium supplement requires a little investigative work to sort through the many options. Before examining calcium supplements, recognize that multivitamin-mineral pills contain little or no calcium. The label may list a few milligrams of calcium, but remember that the recommended intake is a gram or more for adults.

Calcium supplements are typically sold as compounds of calcium carbonate (common in **antacids** and fortified chocolate candies), citrate, gluconate, lactate, malate, or phosphate. These supplements often include magnesium, vitamin D, or both. In addition, some calcium supplements are made from **bone meal, oyster shell,** or **dolomite** (limestone). Many calcium supplements, especially those derived from these natural products, contain lead—which impairs health in numerous ways, as Chapter 13 points out. Fortunately, calcium interferes with the absorption and action of lead in the body.

The first question to ask is how much calcium the supplement provides. Most calcium supplements provide between 250 and 1000 milligrams of calcium. To be safe, total calcium intake from both foods and supplements should not exceed the UL. Read the label to find out how much a dose supplies. Unless the label states otherwise, supplements of calcium carbonate are 40 percent calcium; those of calcium citrate are 21 percent; lactate, 13 percent; and gluconate, 9 percent. Select a low-dose supplement and take it several times a day rather than taking a large-dose supplement all at once. Taking supplements in doses of 500 milligrams or less improves absorption. Small doses also help ease the GI distress (constipation, intestinal bloating, and excessive gas) that sometimes accompanies calcium supplement use.

The next question to ask is how well the body absorbs and uses the calcium from various supplements. Most healthy people absorb calcium equally well from milk and any of these supplements: calcium carbonate, citrate, or phosphate. More important than supplement solubility is tablet disintegration. When manufacturers compress large quantities of calcium into small pills, the stomach acid has difficulty penetrating the pill. To test a supplement's ability to dissolve, drop it into a 6-ounce cup of vinegar, and stir occasionally. A high-quality formulation will dissolve within a half-hour.

Finally, people who choose supplements must take them regularly. Furthermore, consideration should be given to the best time to take the supplements. To circumvent adverse nutrient interactions, take calcium supplements between, not with, meals. (Importantly, do not take calcium supplements with iron supplements or iron-rich meals; calcium inhibits iron absorption.) To enhance calcium absorption, take supplements with meals. If such contradictory advice drives you crazy, reconsider the benefits of food sources of calcium. Most experts agree that foods are the best source of most nutrients.

Some Closing Thoughts

Unfortunately, many of the strongest risk factors for osteoporosis are beyond people's control: age, gender, and genetics. But several strategies are effective for prevention. First, ensure an optimal peak bone mass during childhood and adolescence by eating a balanced diet rich in calcium and vitamin D and by engaging in regular physical activity. Then, maintain that bone mass in early adulthood by continuing those healthy diet and activity habits, abstaining from cigarette smoking and using alcohol moderately, if at all. Finally, minimize bone loss in later life by maintaining an adequate nutrition and exercise regimen, and, especially for older women, consult a physician about bone density tests, calcium supplements, or other drug therapies that may be effective both in preventing bone loss and in restoring lost bone.[35] The reward is the best possible chance of preserving bone health throughout life.

REFERENCES

1. A. Leboime and coauthors, Osteoporosis and mortality, *Joint Bone Spine* 77 (2010): S107–S112; C. A. Brauer and coauthors, Incidence and mortality of hip fractures in the United States, *Journal of the American Medical Association* 302 (2009): 1573–1579.
2. National Osteoporosis Foundation, www.nof.org, accessed July 2011; R. Nuti and coauthors, Bone fragility in men: Where are we? *Journal of Endocrinological Investigation* 33 (2010): 33–38.
3. A. M. Cheung and A. S. Detsky, Osteoporosis and fractures: Missing the bridge? *Journal of the American Medical Association* 299 (2008): 1468–1470.
4. R. Lorente-Ramos and coauthors, Dual-energy x-ray absorptiometry in the diagnosis of osteoporosis: A practical guide, *American Journal of Roentgenology* 196 (2011): 897–904.
5. L. L. Moore and coauthors, Effects of average childhood dairy intake on adolescent bone health, *Journal of Pediatrics* 153 (2008): 667–673.
6. B. Frenkel and coauthors, Regulation of adult bone turnover by sex steroids, *Journal of Cellular Physiology* 224 (2010): 305–310.
7. Y. Imai and coauthors, Minireview: Osteoprotective action of estrogens is mediated by osteoclastic estrogen receptor-alpha, *Molecular Endocrinology* 24 (2010): 877–885.
8. T. D. Rachner, S. Khosla, and L. C. Hofbauer, Osteoporosis: Now and the future, *Lancet* 377 (2011): 1276–1287.
9. National Osteoporosis Foundation, *Clinician's Guide to Prevention and Treatment of Osteoporosis* (Washington, D.C.: National Osteoporosis Foundation, 2010), pp. 21–24.
10. A. Atmaca and coauthors, Soy isoflavones in the management of postmenopausal osteoporosis, *Menopause* 15 (2008): 748–757; R. C. Poulsen and M. C. Kruger, Soy phytoestrogens: Impact on postmenopausal bone loss and mechanisms of action, *Nutrition Reviews* 66 (2008): 359–374.
11. W. G. Helferich, J. E. Andrade, and M. S. Hoagland, Phytoestrogens and breast cancer: A complex story, *Inflammopharmacology* 16 (2008): 219–226; Y. Zhang and coauthors, Soy isoflavones and their bone protective effect, *Inflammopharmacology* 16 (2008): 213–215.
12. E. Gielen and coauthors, Osteoporosis in men, *Best Practice and Research: Clinical Endocrinology and Metabolism* 25 (2011): 321–335; N. Ducharme, Male osteoporosis, *Clinics in Geriatric Medicine* 26 (2010): 301-309; S. Khosla, Update in male osteoporosis, *Journal of Clinical Endocrinology and Metabolism* 95 (2010): 3–10.
13. P. R. Ebeling, Osteoporosis in men, *New England Journal of Medicine* 358 (2008): 1474–1482.
14. B. D. Mitchell and L. M. Yerges-Armstrong, The genetics of bone loss: Challenges and prospects, *Journal of Clinical Endocrinology and Metabolism* 96 (2011): 1258–1268.
15. K. F. Janz and coauthors, Early physical activity provides sustained bone health benefits later in childhood, *Medicine and Science in Sports and Exercise* 42 (2010): 1072–1078; A. Guadalupe-Grau and coauthors, Exercise and bone mass in adults, *Sports Medicine* 39 (2009): 439–468.
16. B. R. Beck, Muscle forces or gravity—What predominates mechanical loading on bone?: Introduction, *Medicine and Science in Sports and Exercise* 41 (2009): 2033–2036.
17. S. Judex and K. J. Carlson, Is bone's response to mechanical signals dominated by gravitational loading? *Medicine and Science in Sports and Exercise* 41 (2009): 2037–2043.
18. W. M. Kohrt, D. W. Barry, and R. S. Schwartz, Muscle forces or gravity: What predominates mechanical loading on bone? *Medicine and Science in Sports and Exercise* 41 (2009): 2050–2055.
19. A. J. Smock and coauthors, Bone volumetric density, geometry, and strength in female and male collegiate runners, *Medicine and Science in Sports and Exercise* 41 (2009): 2026–2032; D. C. Wilks, S. F. Gilliver, and J. Rittweger, Forearm and tibial bone measures of distance- and sprint-trained master cyclists, *Medicine and Science in Sports and Exercise* 41 (2009): 566–573; American College of Sports Medicine Position Stand, Physical activity and bone health, *Medicine and Science in Sports and Exercise* 36 (2004): 1985–1996.
20. C. J. Rosen and A. Klibanski, Bone, fat, and body composition: Evolving concepts in the pathogenesis of osteoporosis, *The American Journal of Medicine* 122 (2009): 409–414.
21. G. Wolf, Energy regulation by the skeleton, *Nutrition Reviews* 66 (2008): 229–233.
22. K. L. Tucker and coauthors, Effects of beer, wine, and liquor intakes on bone mineral density in older men and women, *American Journal of Clinical Nutrition* 89 (2009): 1188–1196.
23. J. P. Bonjour and coauthors, Minerals and vitamins in bone health: The potential value of dietary enhancement, *British Journal of Nutrition* 101 (2009): 1581–1596.
24. A. L. Darling and coauthors, Dietary protein and bone health: A systematic review and meta-analysis, *American Journal of Clinical Nutrition* 90 (2009): 1674–1692; A. D. Conigrave, E. M. Brown, and R. Rizzoli, Dietary protein and bone health: Roles of amino acid–sensing receptors in the control of calcium metabolism and bone homeostasis, *Annual Review of Nutrition* 28 (2008): 131–155.
25. K. D. Cashman and coauthors, Low vitamin D status adversely affects bone health parameters in adolescents, *American Journal of Clinical Nutrition* 87 (2008): 1039–1044.
26. J. M. Quesada Gómez and coauthors, Calcium citrate and vitamin D in the treatment of osteoporosis, *Clinical Drug Investigation* 31 (2011): 285–298.
27. K. M. Sanders and coauthors, Annual high-dose oral vitamin D and falls and fractures in older women, *Journal of the American Medical Association* 303 (2010): 1815–1822.
28. S. Sahni and coauthors, High vitamin C intake is associated with lower 4-year bone loss in elderly men, *Journal of Nutrition* 138 (2008): 1931–1938.
29. S. Sahni and coauthors, Inverse association of carotenoid intakes with 4-y change in bone mineral density in elderly men and women: The Framingham Osteoporosis Study, *American Journal of Clinical Nutrition* 89 (2009): 416–424.
30. E. P. Weiss and coauthors, Dehydroepiandrosterone replacement therapy in older adults: 1- and 2-y effects on bone, *American Journal of Clinical Nutrition* 89 (2009): 1459–1467.
31. L. Langsetmo and coauthors, Dietary patterns and incident low-trauma fractures in postmenopausal women and men aged ≥50 y: A population-based cohort study, *American Journal of Clinical Nutrition* 93 (2011): 192–199.
32. Quesada Gómez and coauthors, 2011; H. A. Bischoff-Ferrari and coauthors, Effect of calcium supplementation on fracture risk: A double-blind randomized controlled trial, *American Journal of Clinical Nutrition* 87 (2008): 1945–1951; R. M. Daly and coauthors, The skeletal benefits of calcium- and vitamin D_3-fortified milk are sustained in older men after withdrawal of supplementation: An 18-mo follow-up study, *American Journal of Clinical Nutrition* 87 (2008): 771–777.
33. M. J. Bolland and coauthors, Calcium supplements with or without vitamin D and risk of cardiovascular events: Reanalysis of the Women's Health Initiative limited access dataset and meta-analysis, *British Medical Journal* 342 (2011): d2040; Z. Sabbagh and H. Vatanparast, Is calcium supplementation a risk factor for cardiovascular disease in older women? *Nutrition Reviews* 67 (2009): 105–108.
34. J. Gahche and coauthors, Dietary supplement use among US adults has increased since NHANES III (1998-1994), *NCHS Data Brief* 61 (2011): 1–8.
35. R. C. Hamdy and coauthors, Algorithm for the management of osteoporosis, *Southern Medical Journal* 103 (2010): 1009–1015.

© anyamay/Shutterstock.com

CENGAGE LEARNING'S GLOBAL NUTRITION WATCH — GLOBAL NUTRITION RESOURCE CENTER

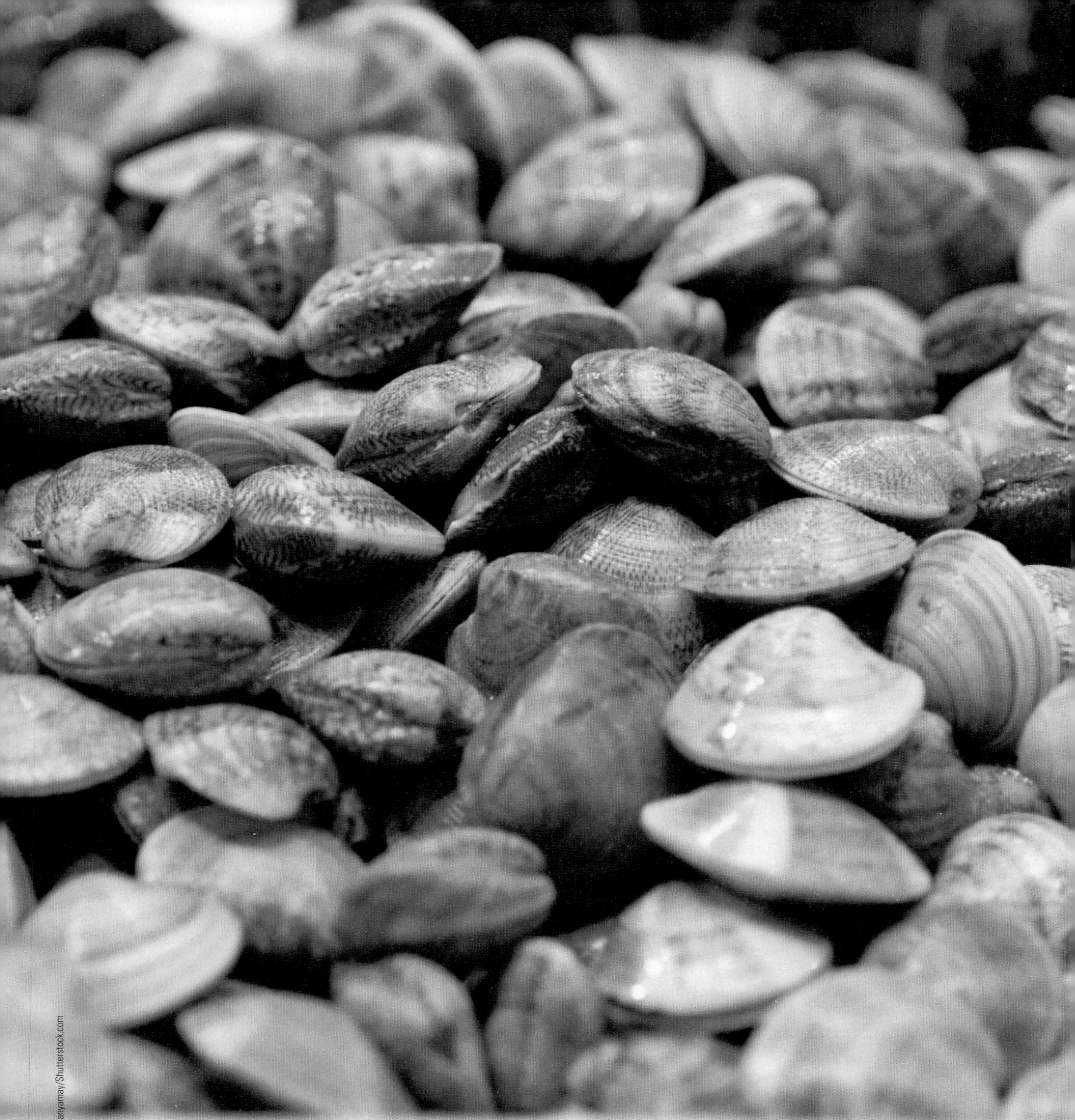

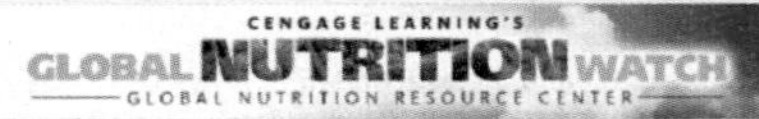

The body requires small amounts of the trace minerals, but that little bit is essential. Go to Nutrition Basics and Tools → Vitamins, Minerals, and Water to review why these minerals are vital in maintaining good health. What individuals are most at risk for deficiency?

13

The Trace Minerals

Nutrition in Your Life

Trace—barely a perceptible amount. But the trace minerals tackle big jobs. Your blood can't carry oxygen without iron, and insulin can't deliver glucose without chromium. Teeth become decayed without fluoride, and thyroid glands develop goiter without iodine. Together, the trace minerals keep you healthy and strong. Where can you get these amazing minerals? A variety of foods, especially those from the protein foods group, sprinkled with a little iodized salt and complemented by a glass of fluoridated water will do the trick. It's remarkable what your body can do with only a few milligrams—or even micrograms—of the trace minerals. In the Nutrition Portfolio at the end of this chapter, you can determine whether the foods you are eating are meeting your trace mineral needs.

Figure 12-9 in Chapter 12 (p. 378) showed the tiny quantities of **trace minerals** in the human body. The trace minerals are so named because they are present, and needed, in relatively small amounts in the body. All together, they would hardly fill a teaspoon. Yet they are no less important than the major minerals or any of the other nutrients. Each of the trace minerals performs a vital role. A deficiency of any of them may be fatal, and excesses are equally deadly. Remarkably, a well-balanced diet supplies enough of these minerals to maintain health.

13.1 The Trace Minerals—An Overview

LEARN IT Summarize key factors unique to the trace minerals.

The body requires the trace minerals in minuscule quantities. They participate in diverse tasks all over the body, each having special duties that only it can perform.

Food Sources The trace mineral contents of foods depend on soil and water composition and on how foods are processed. Furthermore, many factors in the diet and within the body affect the minerals' **bioavailability.** Still, outstanding food

trace minerals: essential mineral nutrients the human body requires in relatively small amounts (less than 100 milligrams per day); sometimes called *microminerals.*

bioavailability: the rate at and the extent to which a nutrient is absorbed and used.

sources for each of the trace minerals, just like those for the other nutrients, include a wide variety of foods.

Deficiencies Assessing trace mineral status is challenging.[1] Severe deficiencies of the better-known minerals are relatively easy to recognize. Deficiencies of the others may be harder to diagnose, and for all minerals, mild deficiencies are easy to overlook. Because the minerals are active in many body systems—digestive, cardiovascular, circulatory, muscular, skeletal, and nervous—deficiencies can have wide-reaching effects and can affect people of all ages. The most common result of a deficiency in children is failure to grow and thrive.

Toxicities Most of the trace minerals are toxic at intakes only two and a half to seven times above the estimated requirements (see Figure 13-1). Thus it is important not to habitually exceed the Upper Level (UL) of recommended intakes (see inside front cover). Many dietary supplements contain trace minerals, making it easy for users to exceed their needs. Highlight 10 discusses supplement use and some of the regulations included in the Dietary Supplement Health and Education Act. As that discussion notes, consumers have demanded the freedom to choose their own doses of nutrients. By law, the Food and Drug Administration (FDA) has no authority to limit the amounts of trace minerals in supplements.* Individuals who take supplements must therefore be aware of the possible dangers and select supplements that contain no more than 100 percent of the Daily Value. It is easier and safer to meet nutrient needs by selecting a variety of foods than by combining an assortment of supplements.

*Canada regulates the amounts of trace minerals in supplements.

> FIGURE 13-1 **RDA (or AI) and UL Compared for Selected Trace Minerals**

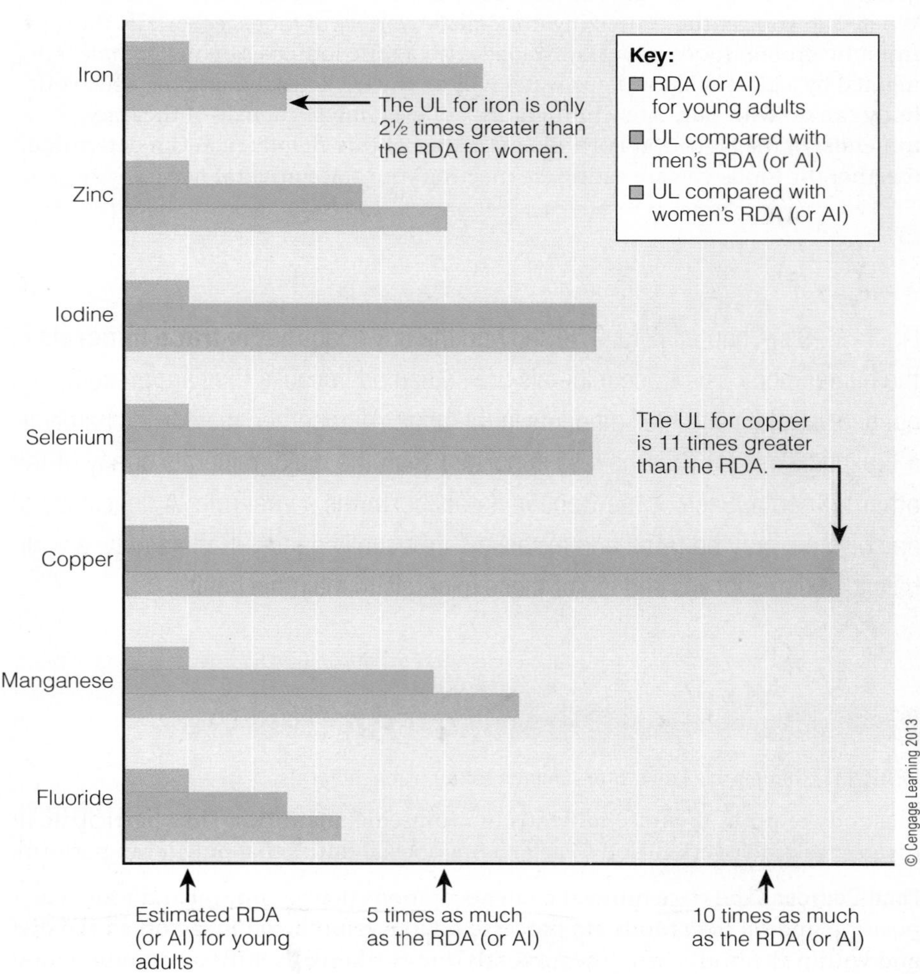

Interactions Interactions among the trace minerals are common and often well coordinated to meet the body's needs. For example, several of the trace minerals support insulin's work, influencing its synthesis, storage, release, and action.

At other times, interactions lead to nutrient imbalances. An excess of one may cause a deficiency of another. (A slight manganese overload, for example, may aggravate an iron deficiency.) A deficiency of one may interfere with the work of another. (A selenium deficiency halts the activation of the iodine-containing thyroid hormones.) A deficiency of a trace mineral may even open the way for a contaminant mineral to cause a toxic reaction. (Iron deficiency, for example, makes the body more vulnerable to lead poisoning.) These examples of nutrient interactions highlight one of the many reasons why people should use supplements conservatively, if at all: supplementation can easily create imbalances.

A good food source of one nutrient may be a poor food source of another, and factors that enhance the action of some trace minerals may interfere with others. Meats, for example, are a good source of iron but a poor source of calcium; vitamin C enhances the absorption of iron but hinders that of copper.

Nonessential Trace Minerals This chapter features the essential trace minerals—iron, zinc, iodine, selenium, copper, manganese, fluoride, chromium, and molybdenum. Research to determine whether other trace minerals are essential is difficult because their quantities in the body are so small and also because human deficiencies are unknown. Guessing their functions in the body can be particularly problematic. Much of the available knowledge comes from research using animals.

Nickel may serve as a cofactor for certain enzymes. Silicon is involved in the formation of bones and collagen. Vanadium, too, is necessary for growth and bone development and for normal reproduction. Cobalt is a key mineral in the large vitamin B_{12} molecule (see Figure 13-2), but it is not an essential nutrient and no recommendation has been established. Boron may play a key role in bone health, brain activities, and immune response.[2]

In the future, we may discover that these and other trace minerals play key nutritional roles. Even arsenic—famous as a poison used by murderers and known to be a carcinogen—may turn out to be essential for human beings in tiny quantities. It has already proved useful in the treatment of some types of leukemia.[3] Research on all the trace minerals is active, suggesting that we have much more to learn about them.

> **FIGURE 13-2 Cobalt in Vitamin B_{12}**

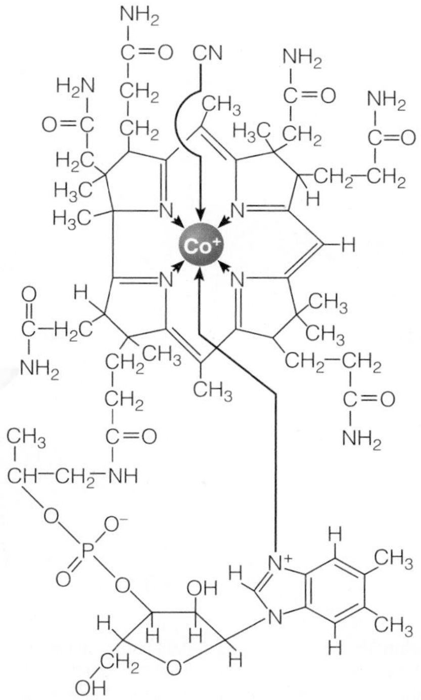

The intricate vitamin B_{12} molecule contains one atom of the mineral cobalt. The alternative name for vitamin B_{12}, cobalamin, reflects the presence of cobalt in its structure.

REVIEW IT Summarize key factors unique to the trace minerals.
Although the body uses only tiny amounts of the trace minerals, they are vital to health. Because so little is required, the trace minerals can be toxic at levels not far above estimated requirements—a consideration for supplement users. Like the other nutrients, the trace minerals are best obtained by eating a variety of foods.

13.2 The Trace Minerals

LEARN IT Identify the main roles, deficiency symptoms, and food sources for each of the trace minerals (iron, zinc, iodine, selenium, copper, manganese, fluoride, chromium, and molybdenum).

Iron **Iron** is an essential nutrient, vital to many of the cells' activities, but it poses a problem for millions of people. Some people simply don't eat enough iron-containing foods to support their health optimally, whereas others absorb so much iron that it threatens their health. Iron exemplifies the principle that both too little and too much of a nutrient in the body can be harmful. In its wisdom, the body has several ways to maintain iron balance, protecting against both deficiency and toxicity.

Iron Roles in the Body Iron has the knack of switching back and forth between two ionic states. ♦ In the reduced state, iron has lost two electrons and therefore has a net positive charge of two; it is known as *ferrous iron*. In the oxidized state, iron has lost a third electron, has a net positive charge of three, and is known as *ferric iron*. Ferrous iron can be oxidized to ferric iron, and ferric iron can be reduced to ferrous iron. By doing so, iron can serve as a **cofactor** to enzymes

♦ Iron's two ionic states:
- Ferrous iron (reduced): Fe^{++}
- Ferric iron (oxidized): Fe^{+++}

iron: an essential trace mineral that is needed for the transport of oxygen and the metabolism of energy nutrients.

cofactor: a small, inorganic or organic substance that facilitates the action of an enzyme.

involved in the numerous oxidation-reduction reactions that commonly occur in all cells. Enzymes involved in making amino acids, collagen, hormones, and neurotransmitters all require iron. (For details about ions, oxidation, and reduction, see Appendix B.)

Iron forms a part of the electron carriers that participate in the electron transport chain (discussed in Chapter 7).* These carriers transfer hydrogens and electrons to oxygen, forming water, and in the process, make ATP for the cells' energy use.

Most of the body's iron is found in two proteins: **hemoglobin** in the red blood cells and **myoglobin** in the muscle cells. In both, iron helps accept, carry, and then release oxygen.

Iron Absorption The body conserves iron. Because it is difficult to excrete iron once it is in the body, balance is maintained primarily through absorption. More iron is absorbed when stores are empty and less is absorbed when stores are full. Special proteins help the body absorb iron from food (see Figure 13-3).[4] The iron-storage protein **ferritin** captures iron from food and stores it in the cells of the small intestine. When the body needs iron, ferritin releases some iron to an iron transport protein called **transferrin.** If the body does not need iron, it is carried out when the intestinal cells are shed and excreted in the feces; intestinal cells are replaced about every 3 to 5 days. By holding iron temporarily, these cells control iron absorption by either delivering iron when the day's intake falls short or disposing of it when intakes exceed needs.

Iron absorption depends in part on its dietary source. Iron occurs in two forms in foods: as **heme iron,** which is found only in foods derived from the flesh of animals, such as meats, poultry, and fish and as **nonheme iron,** which is found in both plant-derived and animal-derived foods (see Figure 13-4). On average, heme iron represents about 10 percent of the iron a person consumes in a day. Even though heme iron accounts for only a small proportion of the intake, it is so well absorbed that it contributes significant iron. About 25 percent of heme iron and 17 percent of nonheme iron is absorbed, depending on dietary factors and the body's iron stores.[5] In iron deficiency, absorption increases.[6] In iron overload, absorption declines.

*The iron-containing electron carriers of the electron transport chain are known as *cytochromes*. See Appendix C for details on the electron transport chain.

hemoglobin: the oxygen-carrying protein of the red blood cells that transports oxygen from the lungs to tissues throughout the body; hemoglobin accounts for 80 percent of the body's iron.

myoglobin: the oxygen-holding protein of the muscle cells.
- **myo** = muscle

ferritin (FAIR-ih-tin): the iron storage protein.

transferrin (trans-FAIR-in): the iron transport protein.

heme (HEEM) **iron:** the iron in foods that is bound to the hemoglobin and myoglobin proteins; found only in meat, fish, and poultry.

nonheme iron: the iron in foods that is not bound to proteins; found in both plant-derived and animal-derived foods.

> **FIGURE 13-3 Iron Absorption**

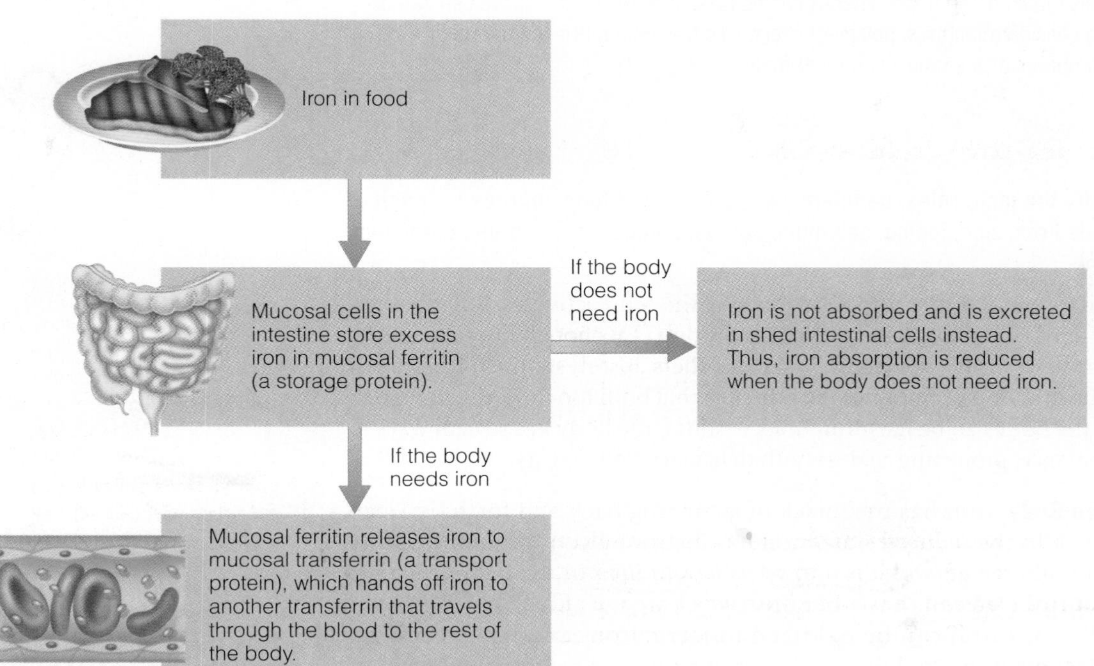

Iron in food
Mucosal cells in the intestine store excess iron in mucosal ferritin (a storage protein).
If the body does not need iron
Iron is not absorbed and is excreted in shed intestinal cells instead. Thus, iron absorption is reduced when the body does not need iron.
If the body needs iron
Mucosal ferritin releases iron to mucosal transferrin (a transport protein), which hands off iron to another transferrin that travels through the blood to the rest of the body.

> FIGURE 13-4 **Heme and Nonheme Iron in Foods**

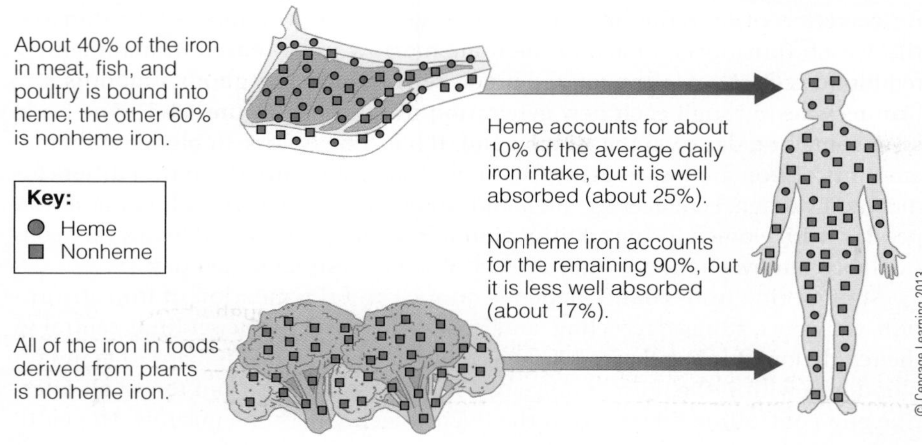

About 40% of the iron in meat, fish, and poultry is bound into heme; the other 60% is nonheme iron.
Key:
Heme
Nonheme
Heme accounts for about 10% of the average daily iron intake, but it is well absorbed (about 25%).
Nonheme iron accounts for the remaining 90%, but it is less well absorbed (about 17%).
All of the iron in foods derived from plants is nonheme iron.
© Cengage Learning 2013

Heme iron has a high bioavailability and is not influenced by dietary factors. In contrast, several dietary factors enhance nonheme iron absorption. ♦ Meat, fish, and poultry contain not only the well-absorbed heme iron, but also a peptide (sometimes called the **MFP factor**) that promotes the absorption of nonheme iron from other foods eaten at the same meal. Vitamin C also enhances nonheme iron absorption from foods eaten at the same meal by capturing the iron and keeping it in the reduced ferrous form, ready for absorption. Some acids and sugars also enhance nonheme iron absorption.

Some dietary factors bind with nonheme iron, inhibiting absorption. ♦ These factors include the phytates in legumes, whole grains, and rice; the vegetable proteins in soybeans, other legumes, and nuts; the calcium in milk; and the polyphenols (such as tannic acid) in tea, coffee, grain products, oregano, and red wine.

The many dietary enhancers, inhibitors, and their combined effects make it difficult to estimate iron absorption. Most of these factors exert a strong influence individually, but not when combined with the others in a meal. Furthermore, the impact of the combined effects diminishes when a diet is evaluated over several days. When multiple meals are analyzed together, three factors appear to be most relevant: MFP factor and vitamin C as enhancers and phytates as inhibitors.

Overall, about 18 percent of dietary iron is absorbed from mixed diets and only about 10 percent from vegetarian diets.[7] As you might expect, vegetarian diets do not have the benefit of easy-to-absorb heme iron or the help of the MFP factor in enhancing absorption. In addition to dietary influences, iron absorption also depends on an individual's health, stage in the life cycle, and iron status. Absorption can be as low as 2 percent in a person with GI disease or as high as 35 percent in a rapidly growing, healthy child. The body adapts to absorb more iron when a person's iron stores fall short or when the need increases for any reason (such as pregnancy). The body makes more ferritin to absorb more iron from the small intestine and more transferrin to carry more iron around the body. Similarly, when iron stores are sufficient, the body adapts to absorb less iron.

Iron Transport and Storage The blood transport protein transferrin delivers iron to the bone marrow and other tissues. The bone marrow uses large quantities of iron to make new red blood cells, whereas other tissues use less. Surplus iron is stored in the protein ferritin, primarily in the liver, but also in the bone marrow and spleen. When dietary iron has been plentiful, ferritin is constantly and rapidly made and broken down, providing an ever-ready supply of iron. When iron concentrations become abnormally high, the liver converts some ferritin into another storage protein called **hemosiderin**. Hemosiderin releases iron more slowly than ferritin does. Storing excess iron in hemosiderin protects the body against the damage that free iron can cause. Free iron acts as a free radical, attacking cell lipids, DNA, and protein. (See Highlight 11 for more information on free radicals and the damage they can cause.)

♦ Factors that *enhance* nonheme iron absorption:
- MFP factor
- Vitamin C (ascorbic acid)
- Acids (citric and lactic)
- Sugars (fructose)

♦ Factors that *inhibit* nonheme iron absorption:
- Phytates (legumes, grains, and rice)
- Vegetable proteins (soybeans, legumes, nuts)
- Calcium (milk)
- Tannic acid (and other polyphenols in tea and coffee)

This chili dinner provides several factors that may enhance iron absorption: heme and nonheme iron and the MFP factor from meat, nonheme iron from legumes, and vitamin C from tomatoes.

MFP factor: a peptide released during the digestion of meat, fish, and poultry that enhances nonheme iron absorption.

hemosiderin (heem-oh-SID-er-in): an iron-storage protein primarily made in times of iron overload.

The average red blood cell lives about 4 months; then the spleen and liver cells remove it from the blood, take it apart, and prepare the degradation products for excretion or recycling. The iron is salvaged: the liver attaches it to transferrin, which transports it back to the bone marrow to be reused in making new red blood cells. Thus, although red blood cells live for only about 4 months, the iron recycles through each new generation of cells (see Figure 13-5). The body loses some iron daily via the GI tract and, if bleeding occurs, in blood. Only tiny amounts of iron are lost in urine, sweat, and shed skin. Iron excretion differs for men and women.[8] On average, men and women lose about 1.0 milligram of iron per day, with women losing additional iron in menses; menstrual losses vary considerably, but over a month, they average about 0.5 milligram per day.

Maintaining iron balance depends on the careful regulation of iron absorption, transport, storage, recycling, and losses. The hormone **hepcidin** is central to the regulation of iron balance.[9] Produced by the liver, hepcidin helps to maintain blood iron within the normal range by limiting absorption from the small intestine and controlling release from the liver, spleen, and bone marrow. Hepcidin production increases in iron overload and decreases in iron deficiency.[10]

♦ High risk for iron deficiency:
- Women in their reproductive years
- Pregnant women
- Infants and young children
- Adolescents

hepcidin: a hormone produced by the liver that regulates iron balance.

iron deficiency: the state of having depleted iron stores.

iron-deficiency anemia: severe depletion of iron stores that results in low hemoglobin and small, pale red blood cells. Anemias that impair hemoglobin synthesis are *microcytic* (small cell).
- **micro** = small
- **cytic** = cell

Iron Deficiency Worldwide, **iron deficiency** is the most common nutrient deficiency, with **iron-deficiency anemia** affecting more than 1.6 billion people—mostly preschool children and pregnant women.[11] In the United States, iron deficiency is less prevalent, but it still affects about 10 percent of toddlers, adolescent girls, and women of childbearing age. Iron deficiency is also relatively common among overweight children and adolescents compared with those who are normal weight.[12] The association between iron deficiency and obesity has yet to be explained, but researchers are currently examining the relationships between the inflammation that develops with excess body fat and reduced iron absorption.[13] The increased production of hepcidin in obesity may also help to explain the relationship between obesity and iron deficiency.[14] Preventing and correcting iron deficiency are high priorities.

Some stages of life ♦ demand more iron but provide less, making deficiency likely.[15] Women in their reproductive years are especially prone to iron deficiency because of repeated blood losses during menstruation. Pregnancy demands additional

> **FIGURE 13-5** ***Animated*** **Iron Recycled in the Body**

Once iron enters the body, most of it is recycled. Some is lost with body tissues and must be replaced by eating iron-containing foods.

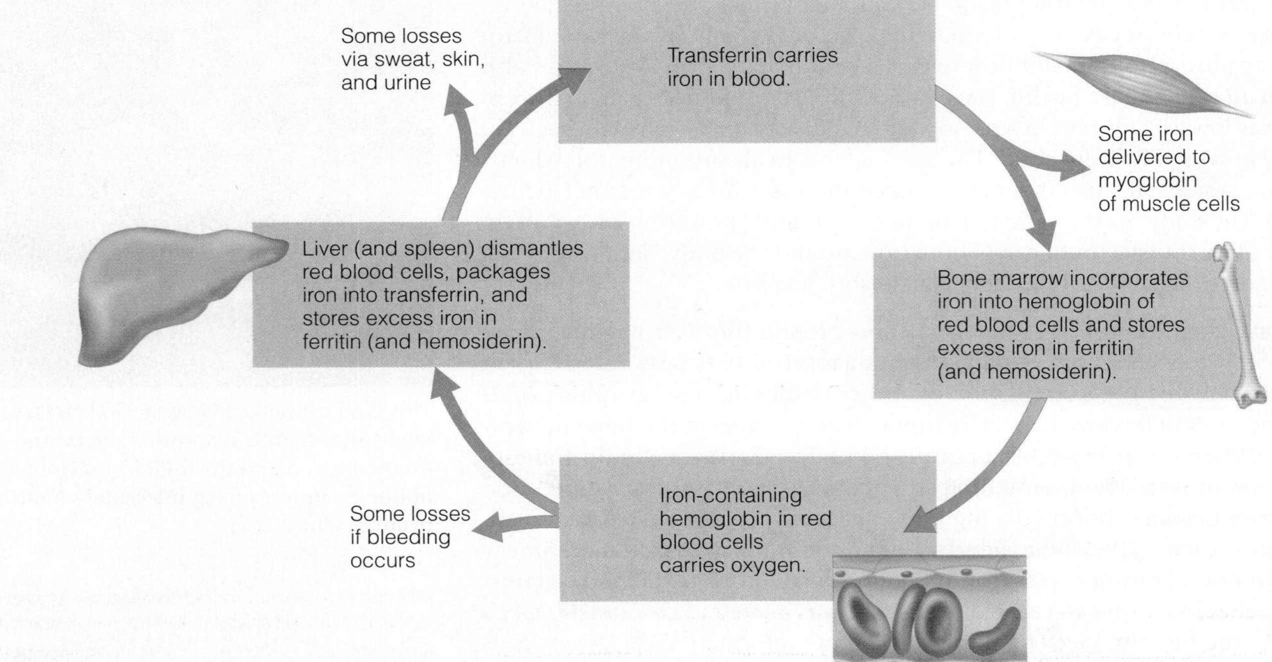
Some losses via sweat, skin, and urine
Transferrin carries iron in blood.
Some iron delivered to myoglobin of muscle cells
Liver (and spleen) dismantles red blood cells, packages iron into transferrin, and stores excess iron in ferritin (and hemosiderin).
Bone marrow incorporates iron into hemoglobin of red blood cells and stores excess iron in ferritin (and hemosiderin).
Some losses if bleeding occurs
Iron-containing hemoglobin in red blood cells carries oxygen.

iron to support the added blood volume, growth of the fetus, and blood loss during childbirth. Infants and young children receive little iron from their high-milk diets, yet need extra iron to support their rapid growth and brain development.[16]* Iron deficiency among toddlers in the United States is common. The rapid growth of adolescence, especially for males, and the menstrual losses of females also demand extra iron that a typical teen diet may not provide. An adequate iron intake is especially important during these stages of life.

Bleeding from any site incurs iron losses.** In some cases, such as an active ulcer, the bleeding may not be obvious, but even small chronic blood losses significantly deplete iron reserves. In developing countries, blood loss is often brought on by malaria and parasitic infections of the GI tract. People who donate blood regularly also incur losses and may benefit from iron supplements. As mentioned, menstrual losses can be considerable as they tap women's iron stores regularly.

Assessment of Iron Deficiency Iron deficiency develops in stages. ♦ This section provides a brief overview of how to detect these stages, and Appendix E provides more details. In the first stage of iron deficiency, iron stores diminish. Measures of serum ferritin (in the blood) reflect iron stores and are most valuable in assessing iron status at this earliest stage.[17] Unfortunately, serum ferritin increases with infections, which interferes with an accurate diagnosis and estimate of prevalence.[18]

♦ Stages of iron deficiency:
- Iron stores diminish
- Transport iron decreases
- Hemoglobin production declines

The second stage of iron deficiency is characterized by a decrease in transport iron: serum iron falls, and the iron-carrying protein transferrin *increases* (an adaptation that enhances iron absorption). Together, measurements of serum iron and transferrin can determine the severity of the deficiency—the more transferrin and the less iron in the blood, the more advanced the deficiency is. Transferrin saturation—the percentage of transferrin that is saturated with iron—decreases as iron stores decline.

The third stage of iron deficiency occurs when the lack of iron limits hemoglobin production. Now the hemoglobin precursor, **erythrocyte protoporphyrin,** begins to accumulate as hemoglobin and **hematocrit** values decline. Hemoglobin and hematocrit tests are easy, quick, and inexpensive, so they are the tests most commonly used in evaluating iron status. Their usefulness in detecting iron deficiency is limited, however, because they are late indicators. Furthermore, other nutrient deficiencies and medical conditions can influence their values.

Iron Deficiency and Anemia Notice that iron deficiency and iron-deficiency anemia are not the same: people may be iron deficient without being anemic. The term *iron deficiency* refers to depleted body iron stores without regard to the degree of depletion or to the presence of anemia. The term *iron-deficiency anemia* refers to the severe depletion of iron stores that results in a low hemoglobin concentration. In iron-deficiency anemia, hemoglobin synthesis decreases, resulting in red blood cells that are pale (hypochromic) and small (microcytic), ♦ as shown in Figure 13-6, p. 410. Without adequate iron, these cells can't carry enough oxygen from the lungs to the tissues. Energy metabolism in the cells falters. The result is fatigue, weakness, headaches, apathy, pallor, and poor resistance to cold temperatures. Because hemoglobin is the bright red pigment of the blood, the skin of a fair person who is anemic may become noticeably pale. In a dark-skinned person, the tongue and eye lining, normally pink, is very pale.

♦ Iron-deficiency anemia is a *microcytic* (my-cro-SIT-ic) *hypochromic* (high-po-KROME-ic) *anemia.*
- **micro** = small
- **cytic** = cell
- **hypo** = too little
- **chrom** = color

The fatigue that accompanies iron-deficiency anemia differs from the tiredness a person experiences from a simple lack of sleep. People with anemia feel fatigue only when they exert themselves. Iron supplementation can relieve the fatigue and improve the body's response to physical activity. (The iron needs of physically active people and the special iron deficiency known as sports anemia are discussed in Chapter 14.)

*The condition of developing iron-deficiency anemia because iron-poor milk displaces iron-rich foods in the diet is called *milk anemia*.

**The iron content of blood is about 0.5 miligram/100 milliliters blood. A person donating a pint of blood (approximately 500 milliliters) loses about 2.5 milligrams of iron.

erythrocyte protoporphyrin (PRO-toe-PORE-fe-rin): a precursor to hemoglobin.

hematocrit (hee-MAT-oh-krit): the percentage of total blood volume that consists of red blood cells.

> FIGURE 13-6 **Normal Blood Cells and Blood Cells in Iron Deficiency Anemia Compared**

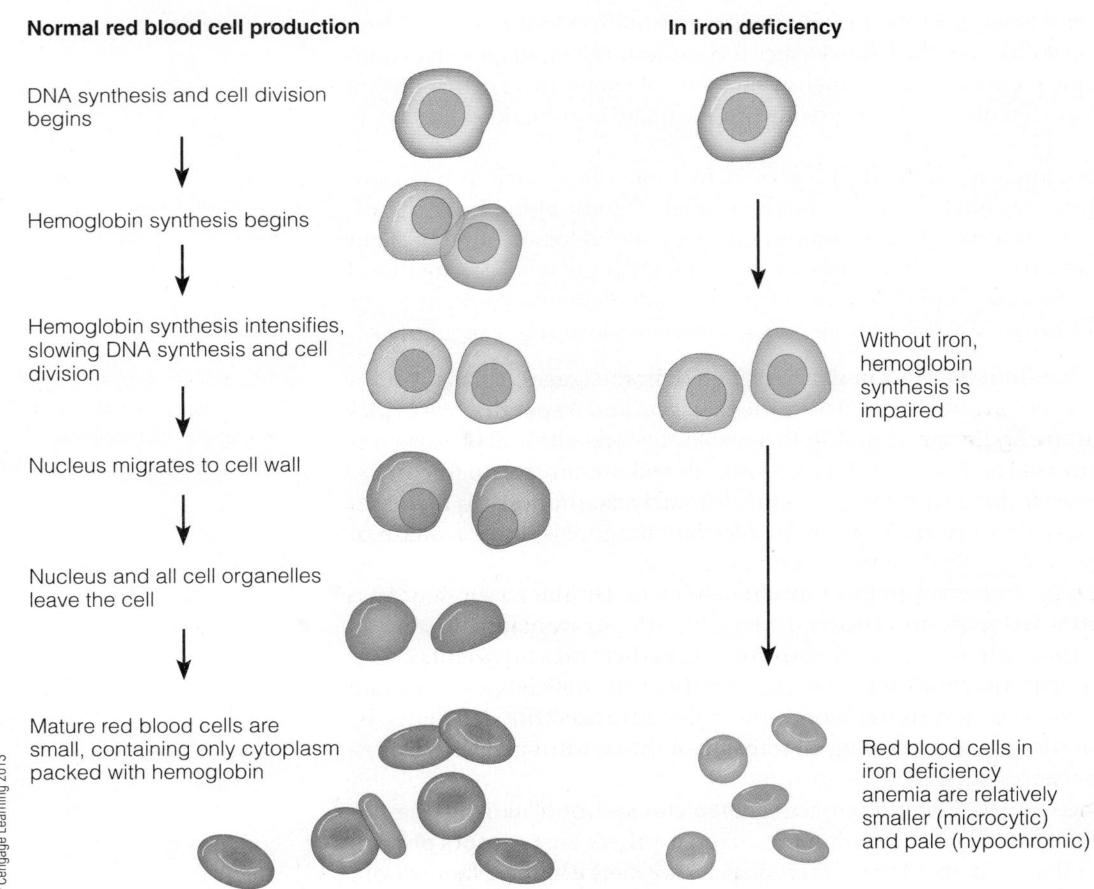
Normal red blood cell production
In iron deficiency
DNA synthesis and cell division begins
Hemoglobin synthesis begins
Hemoglobin synthesis intensifies, slowing DNA synthesis and cell division
Without iron, hemoglobin synthesis is impaired
Nucleus migrates to cell wall
Nucleus and all cell organelles leave the cell
Mature red blood cells are small, containing only cytoplasm packed with hemoglobin
Red blood cells in iron deficiency anemia are relatively smaller (microcytic) and pale (hypochromic)
© Cengage Learning 2013

Iron Deficiency and Behavior Long before the red blood cells are affected and anemia is diagnosed, a developing iron deficiency affects behavior.[19] Even at slightly lowered iron levels, energy metabolism is impaired and neurotransmitter synthesis is altered, reducing physical work capacity and mental productivity.[20] Without the physical energy and mental alertness to work, plan, think, play, sing, or learn, people simply do less. They have no obvious deficiency symptoms; they just appear unmotivated, apathetic, and less physically fit. Work productivity and voluntary activities decline.

Many of the symptoms associated with iron deficiency are easily mistaken for behavioral or motivational problems. A restless child who fails to pay attention in class might be thought contrary. An apathetic homemaker who has let housework pile up might be thought lazy. No responsible dietitian would ever claim that all behavioral problems are caused by nutrient deficiencies, but poor nutrition is always a possible contributor to problems like these. When investigating a behavioral problem, check the adequacy of the diet and seek a routine physical examination before undertaking more expensive, and possibly more harmful, treatment options. Treatment with iron supplements may improve mood, cognitive skills, and physical performance.[21] The effects of iron deficiency on children's behavior are discussed further in Chapter 16.

Iron Deficiency and Pica A curious behavior seen in some iron-deficient people, especially in women and children of low-income groups, is **pica**—the craving and consumption of ice, chalk, starch, and other nonfood substances. These substances contain no iron and cannot remedy a deficiency; in fact, clay actually

pica (PIE-ka): a craving for and consumption of nonfood substances. Pica is known as *geophagia* (gee-oh-FAY-gee-uh) when referring to eating clay, baby powder, chalk, ash, ceramics, paper, paint chips, or charcoal; *pagophagia* (pag-oh-FAY-gee-uh) when referring to eating large quantities of ice; and *amylophagia* (AM-ee-low-FAY-gee-ah) when referring to eating uncooked starch (flour, laundry starch, or raw rice).

inhibits iron absorption, which may explain the iron deficiency that accompanies such behavior. Pica is poorly understood. Its cause is unknown, but researchers hypothesize that it may be motivated by hunger, nutrient deficiencies, or an attempt to protect against toxins or microbes.[22] The consequence of pica is anemia.

Iron Overload As mentioned earlier, iron is closely regulated and absorption normally decreases when iron stores are full. Even a diet that includes fortified foods usually poses no risk for most people, but some individuals are vulnerable to excess iron. Once considered rare, **iron overload** has emerged as an important disorder of iron metabolism and regulation.

The iron overload disorder known as **hemochromatosis** is caused by a genetic failure to prevent unneeded iron in the diet from being absorbed.[23] Research suggests that just as insulin supports normal glucose homeostasis and its absence or ineffectiveness causes diabetes, the hormone hepcidin supports iron homeostasis and its deficiency or (rarely) resistance causes hemochromatosis.[24]

Hereditary hemochromatosis is the most common genetic disorder in the United States, affecting some 1.5 million people. Other causes of iron overload include repeated blood transfusions (which bypass the intestinal defense), massive doses of supplementary iron (which overwhelm the intestinal defense), and other rare metabolic disorders. Excess iron overload is characterized by a toxic accumulation of iron in the liver, heart, joints, and other tissues.[25]

Some of the signs and symptoms of iron overload are similar to those of iron deficiency: apathy, lethargy, and fatigue. Therefore, taking iron supplements before assessing iron status is clearly unwise; hemoglobin tests alone would fail to make the distinction because excess iron accumulates in storage. Iron overload assessment tests measure transferrin saturation and serum ferritin.

Iron overload is characterized by free-radical tissue damage, especially in iron-storing organs such as the liver.[26] Infections are likely because viruses and bacteria thrive on iron-rich blood.[27] Symptoms are most severe in alcohol abusers because alcohol damages the small intestine, further impairing its defenses against absorbing excess iron. Untreated iron overload increases the risks of diabetes, liver cancer, heart disease, and arthritis.[28] Treatment involves chelation therapy, which uses a **chelate** to form a complex with iron and promote its excretion.[29]

Iron overload is much more common in men than in women and is twice as prevalent among men as iron deficiency.[30] The widespread fortification of foods with iron makes it difficult for people with hemochromatosis to follow a low-iron diet, and greater dangers lie in the indiscriminate use of iron and vitamin C supplements. Vitamin C not only enhances iron absorption, but also releases iron from ferritin, allowing free iron to wreak the damage typical of free radicals. Thus vitamin C acts as a *pro*oxidant when taken in high doses. (See Highlight 11 for a discussion of free radicals and their effects on disease development.)

Iron and Chronic Diseases Some research suggests a link between heart disease and excess iron.[31] Limited evidence suggests an association between iron and some cancers. Explanations for how iron might be involved in contributing to these chronic diseases focus on its free-radical activity. One of the benefits of a high-fiber diet may be that the accompanying phytates bind iron, making it less available for such reactions.

Iron Poisoning Large doses of iron supplements cause GI distress, including constipation, nausea, vomiting, and diarrhea. These effects may not be as serious as other consequences of iron toxicity, but they are consistent enough to establish a UL of 45 milligrams per day for adults.

Ingestion of iron-containing supplements is a common cause of accidental poisoning in young children.[32] Symptoms of toxicity include nausea, vomiting, diarrhea, a rapid heartbeat, a weak pulse, dizziness, shock, and confusion. As few as five iron tablets containing as little as 200 milligrams of iron have caused death in young children. The exact cause of death is uncertain, but excessive free-radical damage is thought to play a role in heart failure and respiratory distress. Autopsy reports reveal iron deposits and cell death in the stomach, small intestine, liver,

iron overload: toxicity from excess iron.

hemochromatosis (HE-moh-KRO-ma-toe-sis): a genetically determined failure to prevent absorption of unneeded dietary iron that is characterized by iron overload and tissue damage.

chelate (KEY-late): a substance that can grasp the positive ions of a mineral.

- **chele** = claw

>How To

Estimate the Recommended Daily Intake for Iron

To calculate the recommended daily iron intake, the DRI Committee considers a number of factors. For example, for a woman of childbearing age (19 to 50):

- Losses from feces, urine, sweat, and shed skin: 1.0 milligram
- Losses through menstruation: 0.5 milligram (about 14 milligrams total averaged over 28 days)

These losses reflect an average daily need (total) of 1.5 milligrams of *absorbed iron.*

An estimated average requirement is determined based on the daily need and the assumption that an average of 18 percent of ingested iron is absorbed:

1.5 mg iron (needed)
÷ 0.18 (percent iron absorbed)
= 8 mg iron (estimated average requirement)

Then, a margin of safety is added to cover the needs of essentially all women of childbearing age, and the RDA is set at 18 milligrams.

TRY IT Calculate how many slices of whole-wheat bread, cups of broccoli, ounces of hamburger meat, and cups of milk it takes to provide 18 milligrams of iron.

and blood vessels (which can cause internal bleeding). As with medicines and other potentially toxic substances, keep iron-containing tablets out of the reach of children. If you suspect iron poisoning, call the nearest poison control center or a physician immediately.

Iron Recommendations The usual diet in the United States provides about 6 to 7 milligrams of iron for every 1000 kcalories. The recommended daily intake for men is 8 milligrams, and because most men eat more than 2000 kcalories a day, they can meet their iron needs with little effort. Women in their reproductive years, however, need 18 milligrams a day. The accompanying "How To" feature explains how to calculate the recommended intake.

Because women have higher iron needs and lower energy needs, they sometimes have trouble obtaining enough iron. On average, women receive only 12 to 13 milligrams of iron per day, which is not enough iron for women until after menopause. To meet their iron needs from foods, premenopausal women need to select iron-rich foods at every meal.

> DIETARY GUIDELINES FOR AMERICANS 2010

Women capable of becoming pregnant should choose foods that supply heme iron, which is more readily absorbed by the body, additional iron sources, and enhancers of iron absorption such as vitamin C. If pregnant, women should take an iron supplement, as recommended by a health-care provider.

◆ To calculate the RDA for vegetarians, multiply by 1.8:
- 8 mg × 1.8 = 14 mg/day (vegetarian men)
- 18 mg × 1.8 = 32 mg/day (vegetarian women, 19 to 50 yr)

Vegetarians need 1.8 times as much iron ◆ to make up for the low bioavailability typical of their diets.[33] To maximize iron absorption, vegetarians should incorporate iron-rich foods into a diet that is low in inhibitors (foods such as leavened breads and fermented soy products such as miso and tempeh) and high in enhancers (foods rich in vitamin C and the organic acids found in fruits and vegetables). Good vegetarian sources of iron include soy foods (such as soybeans and tofu), legumes (such as lentils and kidney beans), nuts (such as cashews and almonds), seeds (such as pumpkin seeds and sunflower seeds), cereals (such as cream of wheat and oatmeal), dried fruit (such as apricots and raisins), vegetables (such as mushrooms and potatoes), and blackstrap molasses.

Iron Food Sources To obtain enough iron, people must first select iron-rich foods—both naturally occurring and enriched or fortified—and then take advantage of factors that maximize iron absorption. This discussion begins by identifying iron-rich foods and then reviews the factors affecting absorption. Figure 13-7 shows

> **FIGURE 13-7 Iron in Selected Foods**

See the "How To" feature on p. 302 for more information on using this figure.

Food	Serving size (kcalories)
Bread, whole wheat	1 oz slice (70 kcal)
Cornflakes, fortified	1 oz (110 kcal)
Spaghetti pasta	½ c cooked (99 kcal)
Tortilla, flour	1 10"-round (234 kcal)
Broccoli	½ c cooked (22 kcal)
Carrots	½ c shredded raw (24 kcal)
Potato	1 medium baked w/skin (133 kcal)
Tomato juice	½ c (31 kcal)
Banana	1 medium raw (109 kcal)
Orange	1 medium raw (62 kcal)
Strawberries	½ c fresh (22 kcal)
Watermelon	1 slice (92 kcal)
Milk	1 c reduced-fat 2% (121 kcal)
Yogurt, plain	1 c low-fat (155 kcal)
Cheddar cheese	1½ oz (171 kcal)
Cottage cheese	½ c low-fat 2% (101 kcal)
Pinto beans	½ c cooked (117 kcal)
Peanut butter	2 tbs (188 kcal)
Sunflower seeds	1 oz dry (165 kcal)
Tofu (soybean curd)	½ c (76 kcal)
Ground beef, lean	3 oz broiled (244 kcal)
Chicken breast	3 oz roasted (140 kcal)
Tuna, canned in water	3 oz (99 kcal)
Egg	1 hard cooked (78 kcal)
Excellent, and sometimes unusual, sources:	
Clams, canned	3 oz (126 kcal)
Beef liver	3 oz fried (184 kcal)
Parsley	1 c raw (22 kcal)

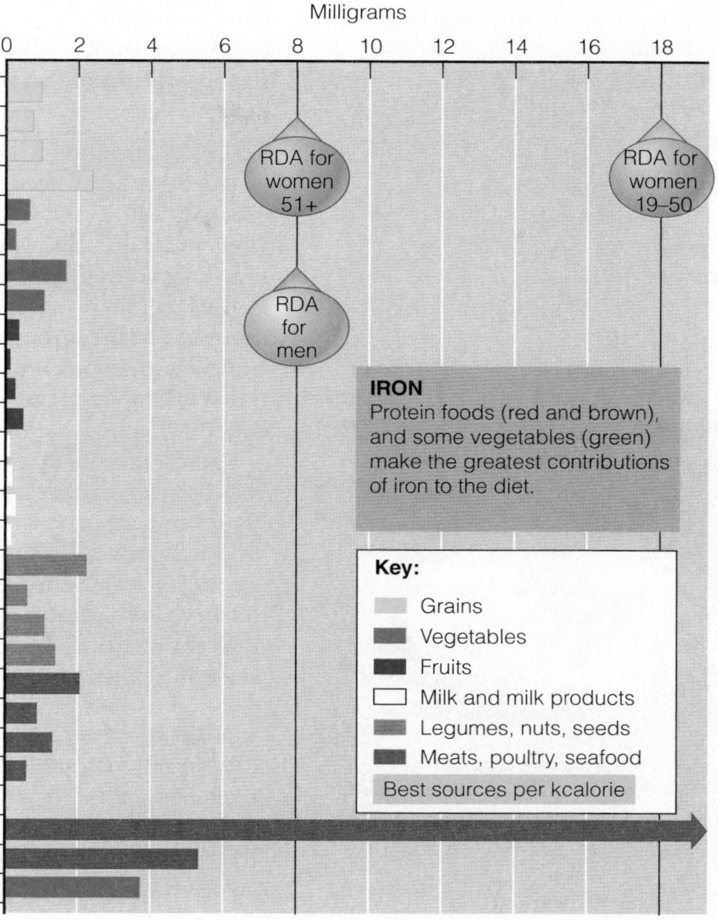

the amounts of iron in selected foods. Meats, fish, and poultry contribute the most iron per serving; other protein-rich foods such as legumes and eggs are also good sources. Although an indispensable part of the diet, foods in the milk group are notoriously poor in iron. Grain products vary, with whole-grain, enriched, and fortified breads and cereals contributing significantly to iron intakes. Finally, dark greens (such as broccoli) and dried fruits (such as raisins) contribute some iron.

The FDA does not mandate iron enrichment, but most states require manufacturers to enrich flour and grain products with iron.* One serving of enriched bread or cereal provides only a little iron, but because people eat many servings of these foods, the contribution can be significant. Iron added to foods is nonheme iron, which is not absorbed as well as heme iron, but when eaten with absorption-enhancing foods, enrichment iron can increase iron stores and reduce iron deficiency. In cases of iron overload, enrichment may exacerbate the problem.

In general, the bioavailability of iron is high in meats, fish, and poultry, intermediate in grains and legumes, and low in most vegetables, especially those containing oxalates such as spinach. As mentioned earlier, the amount of iron ultimately absorbed from a meal depends on the combined effects of several enhancing and inhibiting factors. For maximum absorption of nonheme iron, eat meat for the MFP factor and fruits or vegetables for vitamin C. The iron of baked beans, for example, will be enhanced by the MFP factor in a piece of pork served with them. The iron of bread will be enhanced by the vitamin C in a slice of tomato on a sandwich.

When the label on a grain product says "enriched," it means iron and several B vitamins have been added to meet FDA standards.

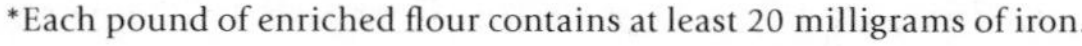

*Each pound of enriched flour contains at least 20 milligrams of iron.

Iron Contamination In addition to the iron from foods, **contamination iron** from nonfood sources of inorganic iron salts can contribute to the day's intakes. Foods cooked in iron cookware take up iron salts. The more acidic the food and the longer it is cooked in iron cookware, the higher the iron content. ♦ The iron content of eggs can triple in the time it takes to scramble them in an iron pan. Admittedly, the absorption of this iron may be poor (perhaps only 1 to 2 percent), but every little bit helps a person who is trying to increase iron intake.

♦ Increase in iron content (mg) for selected foods (3 oz) after cooking in iron skillet:

Beef stew	0.66→3.40
Chili	0.96→6.27
Cornbread	0.67→0.86
Hamburger	1.49→2.29
Pancake	0.63→1.31
Rice	0.67→1.97
Scrambled egg	1.49→4.76
Spaghetti sauce	0.61→5.77

Iron Supplementation People can also get iron from supplements. People who are iron deficient may need supplements as well as an iron-rich, absorption-enhancing diet. Many physicians routinely recommend iron supplements to pregnant women, infants, and young children. Iron from supplements is less well absorbed than that from food, so the doses must be high. The absorption of iron taken as ferrous sulfate is better than that from other iron supplements. Absorption also improves when supplements are taken between meals, at bedtime on an empty stomach, and with liquids (other than milk, tea, or coffee, which inhibit absorption). Taking iron supplements in a single dose instead of several doses per day is equally effective and may improve a person's willingness to take it regularly.

There is no benefit to taking iron supplements with orange juice because vitamin C does not enhance absorption from supplements as it does from foods. Vitamin C enhances iron absorption by converting insoluble ferric iron in foods to the more soluble ferrous iron, and supplemental iron is already in the ferrous form. Constipation is a common side effect of iron supplementation; drinking plenty of water may help to relieve this problem. The best strategy to ensure compliance is to individualize the dose, formulation, and schedule.[34] Most importantly, iron supplements should be taken only when prescribed by a physician who has assessed an iron deficiency.

An old-fashioned iron skillet adds iron to foods.

REVIEW IT

Most of the body's iron is in hemoglobin and myoglobin, where it carries oxygen for use in energy metabolism; some iron is also required for enzymes involved in a variety of reactions. Special proteins assist with iron absorption, transport, and storage—all helping to maintain an appropriate balance—because both too little and too much iron can be damaging. Iron deficiency is most common among infants and young children, teenagers, women of childbearing age, and pregnant women. Symptoms include fatigue and anemia. Iron overload is most common in men. Heme iron, which is found only in meat, fish, and poultry, is better absorbed than nonheme iron, which occurs in most foods. Nonheme iron absorption is improved by eating iron-containing foods with foods containing the MFP factor and vitamin C; absorption is limited by phytates and oxalates. The accompanying table provides a summary of iron.

Iron

RDA

Men: 8 mg/day

Women: 18 mg/day (19–50 yr)
8 mg/day (51+)

Upper Level

Adults: 45 mg/day

Chief Functions in the Body

Part of the protein hemoglobin, which carries oxygen in the blood; part of the protein myoglobin in muscles, which makes oxygen available for muscle contraction; necessary for the utilization of energy as part of the cells' metabolic machinery

Significant Sources

Red meats, fish, poultry, shellfish, eggs, legumes, dried fruits

Deficiency Symptoms

Anemia: weakness, fatigue, headaches; impaired work performance and cognitive function; impaired immunity; pale skin, nail beds, mucous membranes, and palm creases; concave nails; inability to regulate body temperature; pica

Toxicity Symptoms

GI distress

Iron overload: infections, fatigue, joint pain, skin pigmentation, organ damage

Zinc

Zinc Zinc is a versatile trace element required as a cofactor by more than 100 enzymes. Virtually all cells contain zinc, but the highest concentrations are found in muscle and bone.

Zinc Roles in the Body Zinc supports the work of numerous proteins in the body, such as the **metalloenzymes**, ♦ which are involved in a variety of metabolic processes,

♦ Metalloenzymes that require zinc:

- Synthesize parts of the genetic materials DNA and RNA
- Manufacture heme for hemoglobin
- Convert retinol to retinal and alcohol to acetaldehyde
- Influence carbohydrate metabolism
- Regulate protein synthesis
- Digest proteins and folate
- Dispose of damaging free radicals
- Dispose of carbon dioxide

contamination iron: iron found in foods as the result of contamination by inorganic iron salts from iron cookware, iron-containing soils, and the like.

zinc: an essential trace mineral that is part of many enzymes and a constituent of insulin.

metalloenzymes (meh-TAL-oh-EN-zimes): enzymes that contain one or more minerals as part of their structures.

including the regulation of gene expression.* In addition, zinc stabilizes cell membranes and DNA, helping to strengthen antioxidant defenses against free-radical attacks.[35] Zinc also assists in immune function and in growth and development.[36] Zinc participates in the synthesis, storage, and release of the hormone insulin in the pancreas, although it does not appear to play a direct role in insulin's action. Zinc interacts with platelets in blood clotting, affects thyroid hormone function, and influences behavior and learning performance. It is needed to produce the active form of vitamin A (retinal) in visual pigments and the retinol-binding protein that transports vitamin A. It is essential to normal taste perception, wound healing, sperm production, and fetal development. A zinc deficiency impairs all these and other functions, underlining the vast importance of zinc in supporting the body's proteins.

Zinc Absorption The body's handling of zinc resembles that of iron in some ways and differs in others. A key difference is the circular passage of zinc from the small intestine to the body and back again.

The rate of zinc absorption varies from about 15 to 40 percent, depending on the amount of zinc consumed—as zinc intake increases, the rate of absorption decreases, and as zinc intake decreases, the rate of absorption increases.[37] Like iron, dietary factors such as phytates influence absorption, limiting its bioavailability.[38]

Upon absorption into an intestinal cell, zinc has two options. Zinc may participate in the metabolic functions of the intestinal cell itself, or it may be retained within the intestinal cells by **metallothionein** until the body needs zinc. Metallothionein plays a key role in storing and distributing zinc throughout the body.[39]

Zinc Transport Some zinc eventually reaches the pancreas, where it is incorporated into many of the digestive enzymes that the pancreas releases into the small intestine at mealtimes. The small intestine thus receives two doses of zinc with each meal—one from foods and the other from the zinc-rich pancreatic juices. The recycling of zinc in the body from the pancreas to the small intestine and back to the pancreas is referred to as the **enteropancreatic circulation** of zinc. Each time zinc circulates through the small intestine, it may be excreted in shed intestinal cells or reabsorbed into the body (see Figure 13-8). The body loses zinc

*Among the metalloenzymes requiring zinc are carbonic anhydrase, deoxythymidine kinase, DNA and RNA polymerase, and alkaline phosphatase.

metallothionein (meh-TAL-oh-THIGH-oh-neen): a sulfur-rich protein that avidly binds with and transports metals such as zinc.

- **metallo** = containing a metal
- **thio** = containing sulfur
- **ein** = a protein

enteropancreatic (EN-ter-oh-PAN-kree-AT-ik) **circulation:** the circulatory route from the pancreas to the small intestine and back to the pancreas.

> FIGURE 13-8 ***Animated*** **Enteropancreatic Circulation of Zinc**

Some zinc from food is absorbed by the small intestine and sent to the pancreas to be incorporated into digestive enzymes that return to the small intestine. This cycle is called the *enteropancreatic circulation* of zinc.

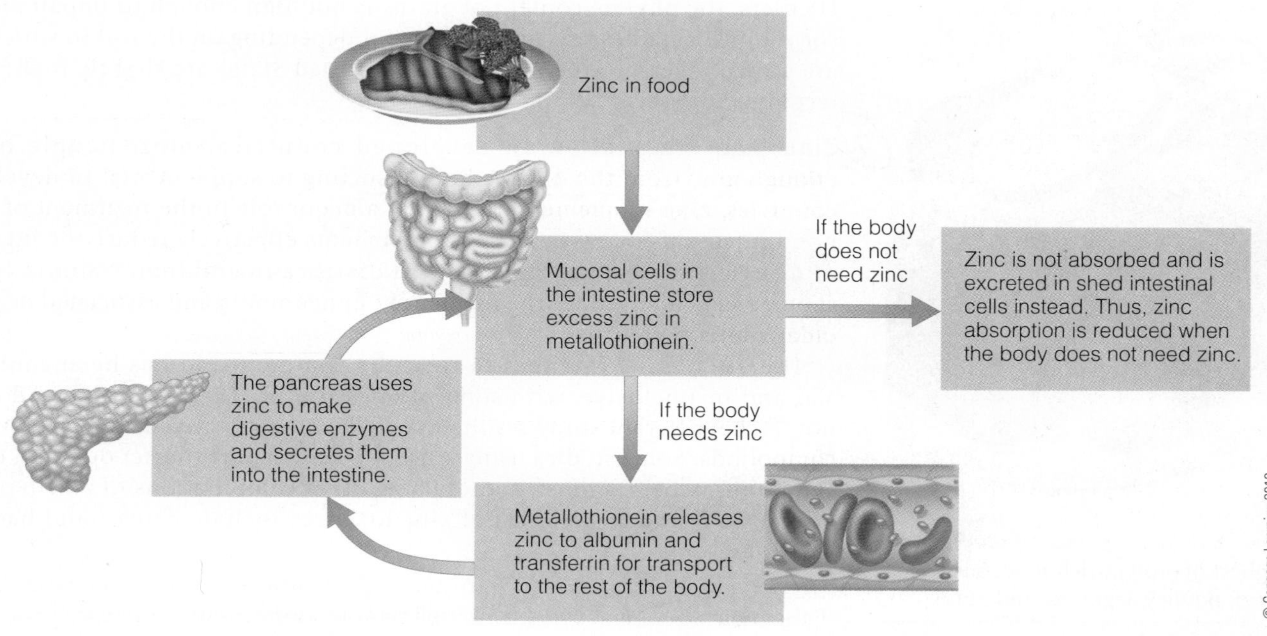

Zinc in food
Mucosal cells in the intestine store excess zinc in metallothionein.
If the body does not need zinc
Zinc is not absorbed and is excreted in shed intestinal cells instead. Thus, zinc absorption is reduced when the body does not need zinc.
If the body needs zinc
Metallothionein releases zinc to albumin and transferrin for transport to the rest of the body.
The pancreas uses zinc to make digestive enzymes and secretes them into the intestine.

> FIGURE 13-9 Zinc-Deficiency Symptom—The Stunted Growth of Dwarfism

The growth retardation, known as dwarfism, is rightly ascribed to zinc deficiency because it is partially reversible when zinc is restored to the diet.

The Egyptian man on the right is an adult of average height. The Egyptian boy on the left is 17 years old but is only 4 feet tall, like a 7-year-old in the United States. His genitalia are like those of a 6-year-old.

primarily in feces. Smaller losses occur in urine, shed skin, hair, sweat, menstrual fluids, and semen.

Numerous proteins participate in zinc transport.[40] Zinc's main transport vehicle in the blood is the protein albumin. Some zinc also binds to transferrin—the same transferrin that carries iron in the blood.

Zinc Deficiency Severe zinc deficiency is not widespread in developed countries, but it is responsible for an estimated 1 out of 20 childhood deaths in Africa, Asia, and Latin America.[41] Human zinc deficiency was first reported in the 1960s in children and adolescent boys in Egypt, Iran, and Turkey. Children have especially high zinc needs because they are growing rapidly and synthesizing many zinc-containing proteins, and the native diets among those populations were not meeting these needs. Middle Eastern diets are traditionally low in the richest zinc source, meats. Furthermore, the staple foods in these diets are legumes, unleavened breads, and other whole-grain foods—all high in fiber and phytates, which inhibit zinc absorption.*

Figure 13-9 shows the severe growth retardation and mentions the immature sexual development characteristic of zinc deficiency.[42] In addition, zinc deficiency hinders digestion and absorption, causing diarrhea, which worsens malnutrition not only for zinc, but for other nutrients as well. It also impairs the immune response, making infections likely—among them, pneumonia and GI tract infections, which worsen malnutrition, including zinc malnutrition (a classic downward spiral of events).[43] Chronic zinc deficiency damages the central nervous system and brain and may lead to poor motor development and cognitive performance. Because zinc deficiency directly impairs vitamin A metabolism, vitamin A–deficiency symptoms often appear. Zinc deficiency also disturbs thyroid function and the metabolic rate. It alters taste, causes loss of appetite, and slows wound healing—in fact, its symptoms are so pervasive that generalized malnutrition and sickness are more likely to be the diagnosis than simple zinc deficiency.

Zinc Toxicity High doses (more than 50 milligrams) of zinc may cause vomiting, diarrhea, headaches, exhaustion, and other symptoms. The UL for adults was set at 40 milligrams based on zinc's interference in copper metabolism—an effect that, in animals, leads to degeneration of the heart muscle.

Zinc Recommendations and Sources Figure 13-10 shows zinc amounts in selected foods per serving. Zinc is highest in protein-rich foods such as shellfish (especially oysters), meats, poultry, milk, and cheese. Legumes and whole-grain products are good sources of zinc if eaten in large quantities; in typical US diets, the phytate content of grains is not high enough to impair zinc absorption. Vegetables vary in zinc content depending on the soil in which they are grown. Average zinc intakes in the United States are slightly higher than recommendations.

Zinc is highest in protein-rich foods such as oysters, beef, poultry, legumes, and nuts.

Zinc Supplementation In developed countries, most people obtain enough zinc from the diet without resorting to supplements. In developing countries, zinc supplementation plays a major role in the treatment of childhood infectious diseases.[44] Zinc supplements effectively reduce the incidence of disease and death associated with diarrhea in children.[45] Similarly, zinc supplements may reduce the incidence of pneumonia and associated deaths in older adults.[46]

The use of zinc lozenges to treat the common cold has been controversial and inconclusive, with some studies finding them effective and others not.[47] The different study results may reflect the effectiveness of various zinc compounds. Some studies using zinc gluconate report shorter duration of cold symptoms, whereas most studies using other combinations of zinc report no effect. Common side effects of zinc lozenges include nausea and bad taste reactions.

*Unleavened bread contains no yeast, which normally breaks down phytates during fermentation.

> **FIGURE 13-10 Zinc in Selected Foods**

See the "How To" feature on p. 302 for more information on using this figure.

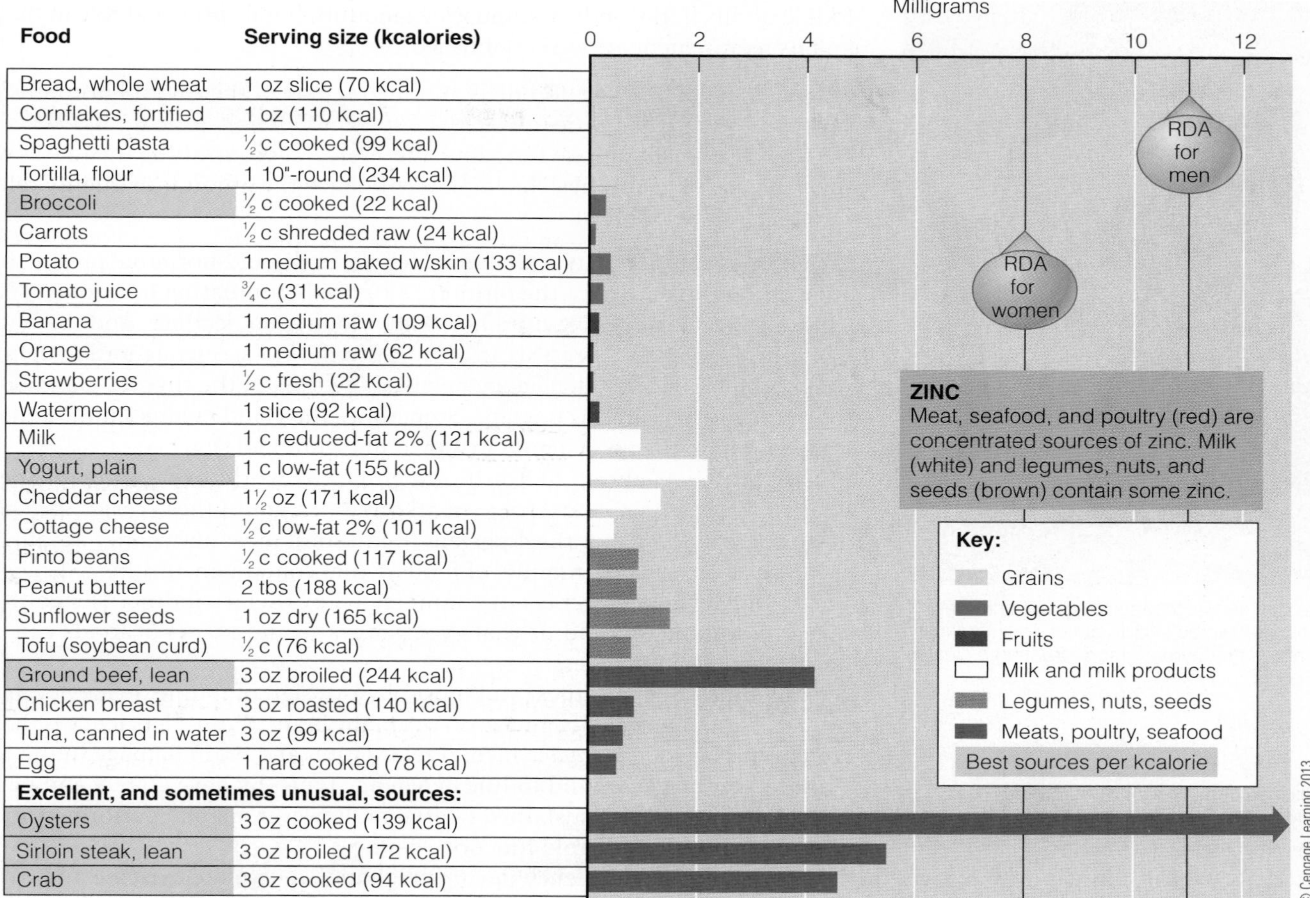

Milligrams
Food
Serving size (kcalories)
0
2
4
6
8
10
12
Bread, whole wheat | 1 oz slice (70 kcal)
Cornflakes, fortified | 1 oz (110 kcal)
Spaghetti pasta | ½ c cooked (99 kcal)
Tortilla, flour | 1 10"-round (234 kcal)
Broccoli | ½ c cooked (22 kcal)
Carrots | ½ c shredded raw (24 kcal)
Potato | 1 medium baked w/skin (133 kcal)
Tomato juice | ¾ c (31 kcal)
Banana | 1 medium raw (109 kcal)
Orange | 1 medium raw (62 kcal)
Strawberries | ½ c fresh (22 kcal)
Watermelon | 1 slice (92 kcal)
Milk | 1 c reduced-fat 2% (121 kcal)
Yogurt, plain | 1 c low-fat (155 kcal)
Cheddar cheese | 1½ oz (171 kcal)
Cottage cheese | ½ c low-fat 2% (101 kcal)
Pinto beans | ½ c cooked (117 kcal)
Peanut butter | 2 tbs (188 kcal)
Sunflower seeds | 1 oz dry (165 kcal)
Tofu (soybean curd) | ½ c (76 kcal)
Ground beef, lean | 3 oz broiled (244 kcal)
Chicken breast | 3 oz roasted (140 kcal)
Tuna, canned in water | 3 oz (99 kcal)
Egg | 1 hard cooked (78 kcal)
Excellent, and sometimes unusual, sources:
Oysters | 3 oz cooked (139 kcal)
Sirloin steak, lean | 3 oz broiled (172 kcal)
Crab | 3 oz cooked (94 kcal)
RDA for men
RDA for women
ZINC
Meat, seafood, and poultry (red) are concentrated sources of zinc. Milk (white) and legumes, nuts, and seeds (brown) contain some zinc.
Key:
Grains
Vegetables
Fruits
Milk and milk products
Legumes, nuts, seeds
Meats, poultry, seafood
Best sources per kcalorie
© Cengage Learning 2013

REVIEW IT

Zinc-requiring enzymes participate in a multitude of reactions affecting growth, vitamin A activity, and pancreatic digestive enzyme synthesis, among others. After a meal, both dietary zinc and zinc-rich pancreatic secretions (via enteropancreatic circulation) are absorbed. Absorption is regulated by a special binding protein (metallothionein) in the small intestine. Protein-rich foods derived from animals are the best sources of bioavailable zinc. Fiber and phytates in cereals bind zinc, limiting absorption. Growth retardation and sexual immaturity are hallmark symptoms of zinc deficiency. The accompanying table provides a summary of zinc.

Zinc

RDA

Men: 11 mg/day

Women: 8 mg/day

Upper Level

Adults: 40 mg/day

Chief Functions in the Body

Part of many enzymes; associated with the hormone insulin; involved in making genetic material and proteins, immune reactions, transport of vitamin A, taste perception, wound healing, the making of sperm, and the normal development of the fetus

Significant Sources

Protein-containing foods: red meats, shellfish, whole grains; some fortified cereals

Deficiency Symptoms[a]

Growth retardation, delayed sexual maturation, impaired immune function, hair loss, eye and skin lesions, loss of appetite

Toxicity Symptoms

Loss of appetite, impaired immunity, low HDL, copper and iron deficiencies

[a]A rare inherited disease of zinc malabsorption, *acrodermatitis* (AK-roh-der-ma-TIE-tis) *enteropathica* (EN-teroh- PATH-ick-ah), causes additional and more severe symptoms.

> FIGURE 13-11 **Iodine-Deficiency Symptom—The Enlarged Thyroid of Goiter**

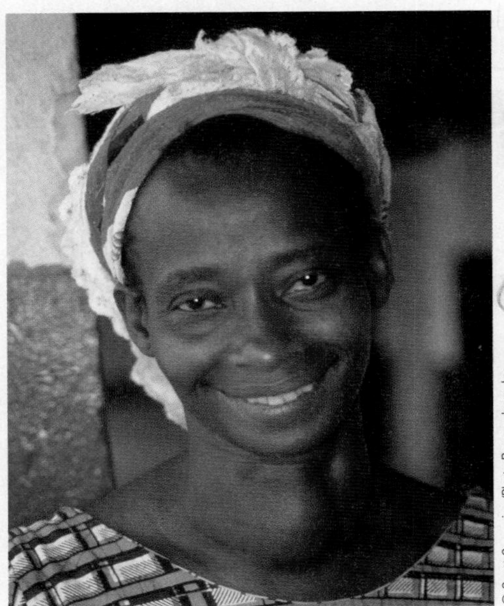

© Scott Camazine/Photo Researchers, Inc.

In iodine deficiency, the thyroid gland enlarges—a condition known as simple goiter. Iodine toxicity also enlarges the thyroid gland, creating a similar-looking goiter.

Iodine Traces of the **iodine** ion (called iodide) are indispensable to life. In the GI tract, iodine from foods becomes iodide. This chapter uses the term *iodine* when referring to the nutrient in foods and *iodide* when referring to it in the body. Iodide occurs in the body in minuscule amounts, but its principal role in the body and its requirement are well established.

Iodide Roles in the Body Iodide is an integral part of the thyroid hormones that regulate body temperature, metabolic rate, reproduction, growth, blood cell production, nerve and muscle function, and more.* By controlling the rate at which the cells use oxygen, these hormones influence the amount of energy released during basal metabolism.

Iodine Deficiency The hypothalamus regulates thyroid hormone production by controlling the release of the pituitary's thyroid-stimulating hormone (TSH).** With iodine deficiency, thyroid hormone production declines, and the body responds by secreting more TSH in a futile attempt to accelerate iodide uptake by the thyroid gland. If a deficiency persists, the cells of the thyroid gland enlarge to trap as much iodide as possible. Sometimes the gland enlarges until it makes a visible lump in the neck, a **goiter** (shown in Figure 13-11).

Goiter afflicts about 200 million people the world over, many of them in South America, Asia, and Africa. In all but 4 percent of these cases, the cause is iodine deficiency. As for the 4 percent (8 million), most have goiter because they regularly eat excessive amounts of foods ♦ that contain an antithyroid substance **(goitrogen)** whose effect is not counteracted by dietary iodine. The goitrogens present in plants remind us that even natural components of foods can cause harm when eaten in excess.

♦ Examples of goitrogen-containing foods:

- Cabbage, spinach, radishes, rutabagas
- Soybeans, peanuts
- Peaches, strawberries

Goiter may be the earliest and most obvious sign of iodine deficiency, but the most tragic and prevalent damage occurs in the brain. Iodine deficiency is the most common cause of *preventable* mental retardation and brain damage in the world. Children with even a mild iodine deficiency typically have goiters and perform poorly in school. With sustained treatment, however, mental performance in the classroom as well as thyroid function improves.[48]

A severe iodine deficiency during pregnancy causes the extreme and irreversible mental and physical retardation known as **cretinism**.*** Cretinism affects approximately six million people worldwide and can be averted by the early diagnosis and treatment of maternal iodine deficiency.[49] A worldwide effort to provide iodized salt to people living in iodine-deficient areas has been dramatically successful. An estimated 70 percent of all households in developing countries have access to iodized salt.[50] Because iron deficiency is common among people with iodine deficiency and because iron deficiency reduces the effectiveness of iodized salt, dual fortification with both iron and iodine may be most beneficial.

Iodine Toxicity Excessive intakes of iodine can interfere with thyroid function and enlarge the gland, just as deficiency can. During pregnancy, exposure to excessive iodine from foods, prenatal supplements, or medications is especially damaging to the developing infant. An infant exposed to toxic amounts of iodine during gestation may develop a goiter so severe as to block the airways and cause suffocation. The UL is 1100 micrograms ♦ per day for an adult—several times higher than average or recommended intakes (review Figure 13-1 on p. 404).

♦ For perspective, most foods provide 3 to 75 μg iodine per serving.

Iodine Recommendations and Sources The ocean is the world's major source of iodine. In coastal areas, kelp, seafood, water, and even iodine-containing sea mist are dependable iodine sources. Further inland, the amount of iodine in foods is variable and generally reflects the amount present in the soil in which plants are grown or on which animals graze. Landmasses that were once under the ocean

iodine: an essential trace mineral that is needed for the synthesis of thyroid hormones.

goiter (GOY-ter): an enlargement of the thyroid gland due to an iodine deficiency, malfunction of the gland, or overconsumption of a goitrogen. Goiter caused by iodine deficiency is sometimes called *simple goiter.*

goitrogen (GOY-troh-jen): a substance that enlarges the thyroid gland and causes *toxic goiter.* Goitrogens occur naturally in such foods as cabbage, kale, brussels sprouts, cauliflower, broccoli, and kohlrabi.

cretinism (CREE-tin-ism): a congenital disease characterized by mental and physical retardation and commonly caused by maternal iodine deficiency during pregnancy.

*The thyroid gland releases tetraiodothyronine (T_4), commonly known as *thyroxine* (thigh-ROCKS-in), to its target tissues. Upon reaching the cells, T_4 loses one iodine, becoming triiodothyronine (T_3), which is the active form of the hormone.

**Thyroid-stimulating hormone is also called *thyrotropin.*

***The underactivity of the thyroid gland is known as *hypothyroidism* and may be caused by iodine deficiency or any number of other causes. Without treatment, an infant with *congenital hypothyroidism* will develop the physical and mental retardation of *cretinism.*

have soils rich in iodine; those in flood-prone areas where water leaches iodine from the soil are poor in iodine. In the United States and Canada, the iodization of salt ♦ has eliminated the widespread misery caused by iodine deficiency during the 1930s, but iodized salt is not available in many parts of the world. Some countries add iodine to bread, fish paste, or drinking water instead. Families who do not use iodized foods have a higher prevalence of child malnutrition and mortality.[51]

♦ Iodized salt contains about 60 μg iodine per gram salt.

Although average consumption of iodine in the United States exceeds recommendations, it falls below toxic levels. Some of the excess iodine in the US diet stems from fast foods, which use iodized salt liberally. Some iodine comes from bakery products and from milk. The baking industry uses iodates (iodine salts) as dough conditioners, and most dairies feed cows iodine-containing medications and use iodine to disinfect milking equipment. Now that these sources have been identified, food industries have reduced their use of these compounds, but the sudden emergence of this problem points to a need for continued surveillance of the food supply. Processed foods in the United States use regular salt, not iodized salt.

The recommended intake of iodine for adults is a minuscule amount. The need for iodine is easily met by consuming seafood, vegetables grown in iodine-rich soil, and iodized salt. ♦ In the United States, labels indicate whether salt is iodized; in Canada, all table salt is iodized.

♦ On average, ½ tsp iodized salt provides the RDA for iodine.

REVIEW IT

Iodide, the ion of the mineral iodine, is an essential component of the thyroid hormones. An iodine deficiency can lead to simple goiter (enlargement of the thyroid gland) and can impair fetal development, causing cretinism. Iodization of salt has largely eliminated iodine deficiency in the United States and Canada. The accompanying table provides a summary of iodine.

Iodine

RDA	**Deficiency Disease**
Adults: 150 μg/day	Simple goiter, cretinism
Upper Level	**Deficiency Symptoms**
1100 μg/day	Underactive thyroid gland, goiter, mental and physical retardation in infants (cretinism)
Chief Functions in the Body	**Toxicity Symptoms**
A component of two thyroid hormones that help to regulate growth, development, and metabolic rate	Underactive thyroid gland, elevated TSH, goiter
Significant Sources	
Iodized salt, seafood, bread, dairy products, plants grown in iodine-rich soil and animals fed those plants	

Only "iodized salt" has had iodine added.

Selenium

Selenium The essential mineral **selenium** shares some of the chemical characteristics of the mineral sulfur. This similarity allows selenium to substitute for sulfur in the amino acids methionine, cysteine, and cystine.

Selenium Roles in the Body Selenium is one of the body's antioxidant nutrients, ♦ working primarily as a part of proteins—most notably, the enzyme glutathione peroxidase.[52] Glutathione peroxidase and vitamin E work in tandem. Glutathione peroxidase prevents free-radical formation, thus blocking the chain reaction before it begins; if free radicals do form and a chain reaction starts, vitamin E stops it. (Highlight 11 describes free-radical formation, chain reactions, and antioxidant action in detail.) Other selenium-containing enzymes selectively activate or inactivate the thyroid hormones.[53]

♦ Key antioxidant nutrients:
- Vitamin C, vitamin E, beta-carotene
- Selenium

Selenium Deficiency Selenium deficiency is associated with **Keshan disease**—a heart disease that is prevalent in regions of China where the soil and foods lack selenium.[54] Although the primary cause of this heart disease is probably a virus or toxin, selenium deficiency appears to predispose people to it, and adequate selenium seems to prevent it.[55]

selenium (se-LEEN-ee-um): an essential trace mineral that is part of an antioxidant enzyme.

Keshan (KESH-an or ka-SHAWN) **disease:** the heart disease associated with selenium deficiency; named for one of the provinces of China where it was first studied. Keshan disease is characterized by heart enlargement and insufficiency; fibrous tissue replaces the muscle tissue that normally composes the middle layer of the walls of the heart.

Selenium and Cancer Limited research suggests that the antioxidant action of selenium may protect against some types of cancers.[56] Selenium supplements, however, have not proved effective in preventing cancer and may in fact damage DNA and cause harm.[57]

Selenium Recommendations and Sources Selenium is found in the soil, and therefore in the crops grown for consumption. People living in regions with selenium-poor soil may still get enough selenium, partly because they eat vegetables and grains transported from other regions and partly because they eat meats, milk, and eggs, which are reliable sources of selenium. Eating as few as two Brazil nuts a day effectively improves selenium status.[58] Average intakes in the United States and Canada exceed the RDA, which is based on the amount needed to maximize glutathione peroxidase activity.

Selenium Toxicity Because high doses of selenium are toxic, a UL has been set. Selenium toxicity causes loss and brittleness of hair and nails, garlic breath odor, and nervous system abnormalities.

REVIEW IT

Selenium is an antioxidant nutrient that works closely with the glutathione peroxidase enzyme and vitamin E. Selenium is found in association with protein in foods. Deficiencies are associated with a predisposition to a type of heart abnormality known as Keshan disease. The accompanying table provides a summary of selenium.

Selenium

RDA

Adults: 55 µg/day

Upper Level

Adults: 400 µg/day

Chief Functions in the Body

Defends against oxidation; regulates thyroid hormone

Significant Sources

Seafood, meat, whole grains, fruits, and vegetables (depending on soil content)

Deficiency Symptoms

Predisposition to heart disease characterized by cardiac tissue becoming fibrous (Keshan disease)

Toxicity Symptoms

Loss and brittleness of hair and nails; skin rash, fatigue, irritability, and nervous system disorders; garlic breath odor

Copper

The body contains about 100 milligrams of **copper** in a variety of cells and tissues. Copper balance and transport depend on a system of proteins.[59]

Copper Roles in the Body Copper serves as a constituent of several enzymes. The copper-containing enzymes have diverse metabolic roles with one common characteristic: all involve reactions that consume oxygen or oxygen radicals. For example, copper-containing enzymes catalyze the oxidation of ferrous iron to ferric iron, which allows iron to bind to transferrin. Copper's role in iron metabolism makes it a key factor in hemoglobin synthesis. Copper- and zinc-containing enzymes participate in the body's natural defenses against the oxidative damage of free radicals. Still other copper enzymes help to manufacture collagen, inactivate histamine, and degrade serotonin. Copper, like iron, is needed in many of the metabolic reactions involved in the release of energy.

Copper Deficiency and Toxicity Typical US diets provide adequate amounts of copper and deficiency is rare. In animals, copper deficiency raises blood cholesterol and damages blood vessels, raising questions about whether low dietary copper might contribute to cardiovascular disease in humans.

Some genetic disorders create a copper toxicity, but excessive intakes from foods are unlikely. Excessive intakes from supplements may cause liver damage, and therefore a UL has been set.

Two rare genetic disorders affect copper status in opposite directions.[60] In **Menkes disease**, the intestinal cells absorb copper, but cannot release it into

copper: an essential trace mineral that is part of many enzymes.

Menkes disease: a genetic disorder of copper transport that creates a copper deficiency and results in mental retardation, poor muscle tone, seizures, brittle kinky hair, and failure to thrive.

circulation, causing a life-threatening deficiency. Treatment involves giving copper intravenously. In **Wilson's disease,** copper accumulates in the liver and brain, creating a life-threatening toxicity. Wilson's disease can be controlled by reducing copper intake, using chelating agents such as penicillamine, and taking zinc supplements, which interfere with copper absorption.

Copper Recommendations and Sources The richest food sources of copper are legumes, whole grains, nuts, shellfish, and seeds. More than half of the copper from foods is absorbed, and the major route of elimination appears to be bile. Water may also provide copper, depending on the type of plumbing pipe and the hardness of the water.

REVIEW IT

Copper is a component of several enzymes, all of which are involved in some way with oxygen or oxidation. Some act as antioxidants; others are essential to iron metabolism. Legumes, whole grains, and shellfish are good sources of copper. The accompanying table provides a summary of copper.

Copper

RDA	Significant Sources
Adults: 900 µg/day	Seafood, nuts, whole grains, seeds, legumes
Upper Level	**Deficiency Symptoms**
Adults: 10,000 µg/day (10 mg/day)	Anemia, bone abnormalities
Chief Functions in the Body	**Toxicity Symptoms**
Necessary for the absorption and use of iron in the formation of hemoglobin; part of several enzymes	Liver damage

Manganese

The human body contains a mere 20 milligrams of **manganese.** Most of it can be found in the bones and metabolically active organs such as the liver, kidneys, and pancreas.

Manganese Roles in the Body Manganese acts as a cofactor for many enzymes that facilitate the metabolism of carbohydrate, lipids, and amino acids. In addition, manganese-containing metalloenzymes assist in bone formation and the conversion of pyruvate to a TCA cycle compound.

Manganese Deficiency and Toxicity Manganese requirements are low, and many plant foods contain significant amounts of this trace mineral, so deficiencies are rare. As is true of other trace minerals, however, dietary factors such as phytates inhibit its absorption. In addition, high intakes of iron and calcium limit manganese absorption, so people who use supplements of those minerals regularly may impair their manganese status.

Toxicity is more likely to occur from an environment contaminated with manganese than from dietary intake. Miners who inhale large quantities of manganese dust on the job over prolonged periods show symptoms of a brain disease, along with abnormalities in appearance and behavior. Still, a UL has been established based on intakes from food, water, and supplements.

Manganese Recommendations and Sources Grain products make the greatest contribution of manganese to the diet. With insufficient information to establish an RDA, an AI was set based on average intakes.

REVIEW IT

Manganese-dependent enzymes are involved in bone formation and various metabolic processes. Because manganese is widespread in plant foods, deficiencies are rare, although regular use of calcium and iron supplements may limit manganese absorption. The table on p. 422 provides a summary of manganese.

Wilson's disease: a genetic disorder of copper metabolism that creates a copper toxicity and results in neurologic symptoms such as tremors, impaired speech, inappropriate behaviors, and personality changes.

manganese: an essential trace mineral that acts as a cofactor for many enzymes.

REVIEW IT *(continued)*

Manganese

AI	Significant Sources
Men: 2.3 mg/day	Nuts, whole grains, leafy vegetables, tea
Women: 1.8 mg/day	**Deficiency Symptoms**
Upper Level	Rare
Adults: 11 mg/day	**Toxicity Symptoms**
Chief Functions in the Body	Nervous system disorders
Cofactor for several enzymes; bone formation	

Fluoride

Fluoride is present in virtually all soils, water supplies, plants, and animals. The body contains only a trace of fluoride, but with this amount, the crystalline deposits in teeth are larger and more perfectly formed.

Fluoride Roles in the Body As Chapter 12 explains, during the mineralization of bones ♦ and teeth, calcium and phosphorus form crystals called hydroxyapatite. Then fluoride replaces the hydroxyl (OH) portions of the hydroxyapatite crystal, forming **fluorapatite,** which makes the teeth stronger and more resistant to decay.

Dental caries ranks as the nation's most widespread health problem: an estimated 95 percent of the population have decayed, missing, or filled teeth. These dental problems can quickly lead to a multitude of nutrition problems by interfering with a person's ability to chew and eat a wide variety of foods. Where fluoride is lacking, dental decay is common.

Drinking water is usually the best source of fluoride, and most of the US population served by public water systems receives optimal levels of fluoride (see Figure 13-12).[61] Fluoridation of drinking water (to raise the concentration to one part fluoride per one million parts water) offers the greatest protection against dental caries at virtually no risk of toxicity.[62] By fluoridating the drinking water, a community offers its residents, particularly the children, a safe, economical, practical, and effective way to defend against dental caries. Most bottled waters lack fluoride.

♦ Key bone nutrients:
- Vitamin D, vitamin K, vitamin A
- Calcium, phosphorus, magnesium, fluoride

♦ To prevent fluorosis:
- Monitor the fluoride content of the local water supply.
- Supervise toddlers when they brush their teeth—using only a little toothpaste (pea-size amount).
- Use fluoride supplements only as prescribed by a physician.

Fluoride Toxicity Too much fluoride can damage the teeth, causing **fluorosis.**[63] For this reason, a UL has been established. In mild cases, the teeth develop small white flecks; in severe cases, the enamel becomes pitted and permanently stained (as shown in Figure 13-13). Fluorosis occurs only during tooth development and cannot be reversed, making its prevention ♦ a high priority. To limit fluoride ingestion, take care not to swallow fluoride-containing dental products such as toothpaste and mouthwash.

> FIGURE 13-12 US Population with Access to Fluoridated Water through Public Water Systems

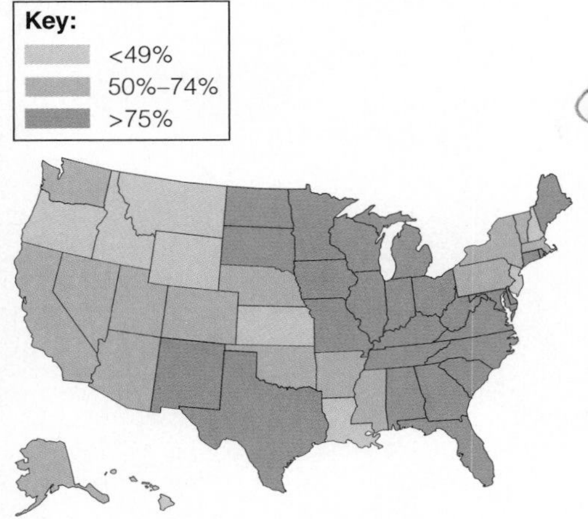

Fluoride Recommendations and Sources As mentioned earlier, much of the US population has access to water with an optimal fluoride concentration, which typically delivers about 1 milligram per person per day. Fish and most teas contain appreciable amounts of natural fluoride.

REVIEW IT

Fluoride makes teeth stronger and more resistant to decay. Fluoridation of public water supplies can significantly reduce the incidence of dental caries, but excess fluoride during tooth development can cause fluorosis—discolored and pitted tooth enamel. The accompanying table provides a summary of fluoride.

Fluoride

AI	Significant Sources
Men: 4 mg/day	Drinking water (if fluoride containing or fluoridated), tea, seafood
Women: 3 mg/day	**Deficiency Symptoms**
Upper Level	Susceptibility to tooth decay
Adults: 10 mg/day	**Toxicity Symptoms**
Chief Functions in the Body	Fluorosis (pitting and discoloration of teeth)
Strengthens teeth; helps to make teeth resistant to decay	

fluoride: an essential trace mineral that makes teeth stronger and more resistant to decay.

fluorapatite (floor-APP-uh-tite): the stabilized form of bone and tooth crystal, in which fluoride has replaced the hydroxyl groups of hydroxyapatite.

fluorosis (floor-OH-sis): discoloration and pitting of tooth enamel caused by excess fluoride during tooth development.

Chromium

Chromium is an essential mineral that participates in carbohydrate and lipid metabolism. Like iron, chromium assumes different charges. In chromium, the Cr^{+++} ion is the most stable and most commonly found in foods.

Chromium Roles in the Body Chromium helps maintain glucose homeostasis by enhancing the activity of the hormone insulin.* When chromium is lacking, a diabetes-like condition may develop with elevated blood glucose and impaired glucose tolerance, insulin response, and glucagon response. Some research suggests that chromium supplements lower blood glucose or improve insulin responses in type 2 diabetes, but findings have not been consistent.[64]

Chromium Recommendations and Sources Chromium is present in a variety of foods. The best sources are unrefined foods, particularly liver, brewer's yeast, and whole grains. The more refined foods people eat, the less chromium they ingest.

Chromium Supplements Supplement advertisements have succeeded in convincing consumers that they can lose fat and build muscle by taking chromium picolinate. Whether chromium supplements (either picolinate or plain) reduce body fat or improve muscle strength remains controversial. (Highlight 14 discusses chromium picolinate and other supplements athletes use in the hopes of improving their performance.)

> FIGURE 13-13 Fluoride-Toxicity Symptom—The Mottled Teeth of Fluorosis

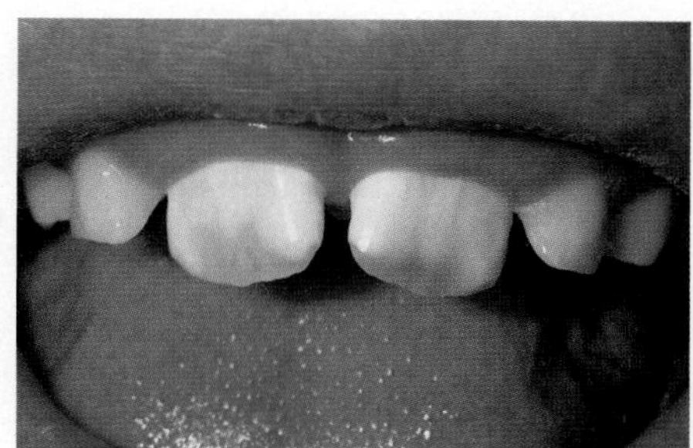

REVIEW IT

Chromium enhances insulin's action. A deficiency can impair glucose homeostasis. Chromium is widely available in unrefined foods including brewer's yeast, whole grains, and liver. The accompanying table provides a summary of chromium.

Chromium

AI	Significant Sources
Men: 35 μg/day Women: 25 μg/day	Meats (especially liver), whole grains, brewer's yeast
Chief Functions in the Body	**Deficiency Symptoms**
Enhances insulin action and may improve glucose tolerance	Diabetes-like condition
	Toxicity Symptoms
	None reported

Molybdenum

Molybdenum acts as a working part of several metalloenzymes. Dietary deficiencies of molybdenum are unknown because the amounts needed are minuscule—as little as 0.1 part per million parts of body tissue. Legumes, breads and other grain products, leafy green vegetables, milk, and liver are molybdenum-rich foods. Average daily intakes fall within the suggested range of intakes.

Molybdenum toxicity in people is rare. It has been reported in animal studies, and a UL has been established. Characteristics of molybdenum toxicity include kidney damage and reproductive abnormalities.

REVIEW IT

Molybdenum is found in a variety of foods and participates in several metabolic reactions. The table on p. 424 provides a summary of molybdenum.

*Small organic compounds that enhance insulin's actions are called *glucose tolerance factors (GTF)*. Some glucose tolerance factors contain chromium.v

chromium (KRO-mee-um): an essential trace mineral that enhances the activity of insulin.

molybdenum (mo-LIB-duh-num): an essential trace mineral that acts as a cofactor for many enzymes.

Molybdenum

RDA	Significant Sources
Adult: 45 μg/day	Legumes, cereals, nuts
Upper Level	**Deficiency Symptoms**
Adults: 2 mg/day	Unknown
Chief Functions in the Body	**Toxicity Symptoms**
Cofactor for several enzymes	None reported; reproductive effects in animals

13.3 Contaminant Minerals

LEARN IT Describe how contaminant minerals disrupt body processes and impair nutrition status.

Chapter 12 and this chapter explain the many ways minerals serve the body—maintaining fluid and electrolyte balance, providing structural support to the bones, transporting oxygen, and assisting enzymes. In contrast to the essential minerals that the body requires, contaminant minerals impair the body's growth, work capacity, and general health. Contaminant minerals include the **heavy metals** lead, mercury, and cadmium, which enter the food supply by way of soil, water, and air pollution. This section focuses on lead poisoning because it is a serious environmental threat to young children and because reducing blood lead levels in children is a goal of the Healthy People initiative. Much of the information on lead applies to the other contaminant minerals as well—they all disrupt body processes and impair nutrition status similarly.

Like other minerals, lead is indestructible; the body cannot change its chemistry. Chemically similar to nutrient minerals such as iron, calcium, and zinc (cations with two positive charges), lead displaces them from some of the metabolic sites they normally occupy so they are then unable to perform their roles. For example, lead competes with iron in heme, but it cannot carry oxygen. Similarly, lead competes with calcium in the brain, but it cannot signal messages from nerve cells. Excess lead in the blood also deranges the structure of red blood cell membranes, making them leaky and fragile. Lead interacts with white blood cells, too, impairing their ability to fight infection, and it binds to antibodies, thwarting their effort to resist disease.

Children with iron deficiency are particularly vulnerable to lead toxicity.[65] Chapter 16 examines the damaging effects of iron deficiency and lead toxicity on a child's growth and development.[66]

REVIEW IT Describe how contaminant minerals disrupt body processes and impair nutrition status.

Lead typifies the ways all heavy metals behave in the body: they interfere with nutrients that are trying to do their jobs. The "good guy" nutrients are shoved aside by the "bad guy" contaminants. Then, when the contaminants cannot perform the roles of the nutrients, health diminishes. To safeguard our health, we must defend ourselves against contamination by eating nutrient-rich foods and preserving a clean environment.

This chapter completes the introductory lessons on the nutrients. Each nutrient from the amino acids to zinc has been described rather thoroughly—its chemistry, roles in the body, sources in the diet, symptoms of deficiency and toxicity, and influences on health and disease. Such a detailed examination is informative, but it can also be misleading. It is important to step back from the detailed study of the individual nutrients to look at them as a whole. After all, people eat foods, not nutrients, and most foods deliver dozens of nutrients. Furthermore, nutrients work cooperatively with one another in the body; their actions are most often *inter*actions. This chapter alone mentioned how iron depends on vitamin C to keep it in its active form and copper to incorporate it into hemoglobin, how zinc is needed to activate and transport vitamin A, and how both iodine and selenium are needed for the synthesis of thyroid hormone. The accompanying table provides a summary of the trace minerals for your review. Highlight 13 explores the benefits of phytochemicals.

heavy metals: mineral ions such as mercury and lead, so called because they are of relatively high atomic weight. Many heavy metals are poisonous.

REVIEW IT The Trace Minerals

Mineral and Chief Functions	Deficiency Symptoms	Toxicity Symptoms[a]	Significant Sources
Iron Part of the protein hemoglobin, which carries oxygen in the blood; part of the protein myoglobin in muscles, which makes oxygen available for muscle contraction; necessary for energy metabolism	Anemia: weakness, fatigue, headaches; impaired work performance; impaired immunity; pale skin, nail beds, mucous membranes, and palm creases; concave nails; inability to regulate body temperature; pica	GI distress; iron overload: infections, fatigue, joint pain, skin pigmentation, organ damage	Red meats, fish, poultry, shellfish, eggs, legumes, dried fruits
Zinc Part of insulin and many enzymes; involved in making genetic material and proteins, immune reactions, transport of vitamin A, taste perception, wound healing, the making of sperm, and normal fetal development	Growth retardation, delayed sexual maturation, impaired immune function, hair loss, eye and skin lesions, loss of appetite	Loss of appetite, impaired immunity, low HDL, copper and iron deficiencies	Protein-containing foods: red meats, fish, shellfish, poultry, whole grains; fortified cereals
Iodine A component of the thyroid hormones that help to regulate growth, development, and metabolic rate	Underactive thyroid gland, goiter, mental and physical retardation (cretinism)	Underactive thyroid gland, elevated TSH, goiter	Iodized salt; seafood; plants grown in iodine-rich soil and animals fed those plants
Selenium Part of an enzyme that defends against oxidation; regulates thyroid hormone	Associated with Keshan disease	Nail and hair brittleness and loss; fatigue, irritability, and nervous system disorders, skin rash, garlic breath odor	Seafoods, organ meats; other meats, whole grains, fruits, and vegetables (depending on soil content)
Copper Helps form hemoglobin; part of several enzymes	Anemia, bone abnormalities	Liver damage	Seafood, nuts, legumes, whole grains, seeds
Manganese Cofactor for several enzymes; bone formation	Rare	Nervous symptom disorders	Nuts, whole grains, leafy vegetables, tea
Fluoride Maintains health of bones and teeth; confers decay resistance on teeth	Susceptibility to tooth decay	Fluorosis (pitting and discoloration) of teeth	Drinking water (if fluoridated), tea, seafood
Chromium Enhances insulin action, may improve glucose intolerance	Diabetes-like condition	None reported	Meats (liver), whole grains, brewer's yeast
Molybdenum Cofactor for several enzymes	Unknown	None reported	Legumes, cereals, nuts

[a]Acute toxicities of many minerals cause abdominal pain, nausea, vomiting, and diarrhea.

Nutrition Portfolio

Trace minerals from a variety of foods, especially those in the protein foods group, support many of your body's activities. Go to Diet Analysis Plus and choose one of the days on which you tracked your diet for an entire day. Select the Intake vs. Goals report and then consider the following questions. Remember that scoring 100 percent on this report means you met your goal.

- Your Intake vs. Goals report may only display your intake for two of the trace minerals: iron and zinc. How was your intake for these two trace minerals?

Now look at the Intake Spreadsheet report and consider the following questions:

- Examine the variety in your food intake, taking particular notice of how often you include meats, seafood, poultry, legumes, and enriched or fortified grain products weekly. These foods often contain trace minerals.

- Describe the advantages of using iodized salt.
- Determine whether your community provides fluoridated water.

Diet Analysis PLUS To complete this exercise, go to your Diet Analysis Plus at **www.cengagebrain.com.**

STUDY IT To review the key points of this chapter and take a practice quiz, go to the study cards at the end of the book.

REFERENCES

1. L. H. Allen, Limitations of current indicators of micronutrient status, *Nutrition Reviews* 67 (2009): S21–S23.
2. F. H. Nielsen, Is boron nutritionally relevant? *Nutrition Reviews* 66 (2008): 183–191.
3. M. S. Tallman, What is the role of arsenic in newly diagnosed APL? *Best Practice and Research Clinical Haematology* 21 (2008): 659–666.
4. M. Muñoz, J. A. García-Erce, and A. F. Remacha, Disorders of iron metabolism. Part 1: Molecular basis of iron homeostasis, *Journal of Clinical Pathology* 64 (2011): 281–286; M. D. Knutson, Iron-sensing proteins that regulate hepcidin and enteric iron absorption, *Annual Review of Nutrition* 30 (2010): 149–171.
5. P. A. Sharp, Intestinal iron absorption: Regulation by dietary & systematic factors, *International Journal for Vitamin and Nutrition Research* 80 (2010): 231–242; Committee on Dietary Reference Intakes, *Dietary Reference Intakes for Vitamin A, Vitamin K, Arsenic, Boron, Chromium, Copper, Iodine, Iron, Manganese, Molybdenum, Nickel, Silicon, Vanadium, and Zinc* (Washington, D.C.: National Academies Press, 2001), p. 315.
6. P. Thankachan and coauthors, Iron absorption in young Indian women: The interaction of iron status with the influence of tea and ascorbic acid, *American Journal of Clinical Nutrition* 87 (2008): 881–886.
7. Committee on Dietary Reference Intakes, 2001, p. 351.
8. J. R. Hunt, C. A. Zito, and L. K. Johnson, Body iron excretion by healthy men and women, *American Journal of Clinical Nutrition* 89 (2009): 1792–1798.
9. T. Ganz and E. Nemeth, Hepcidin and disorders of iron metabolism, *Annual Review of Medicine* 62 (2011): 347–360; M. Wessling-Resnick, Iron homeostasis and the inflammatory response, *Annual Review of Nutrition* 30 (2010): 105–122; D. M. Frazer and G. J. Anderson, Hepcidin compared with prohepcidin: An absorbing story, *American Journal of Clinical Nutrition* 89 (2009): 475–476; M. D. Knutson, Into the matrix: Regulation of the iron regulatory hormone hepcidin by matriptase-2, *Nutrition Reviews* 67 (2009): 284–288; M. A. Roe and coauthors, Plasma hepcidin concentrations significantly predict interindividual variation in iron absorption in healthy men, *American Journal of Clinical Nutrition* 89 (2009): 1088–1091; M. F. Young and coauthors, Serum hepcidin is significantly associated with iron absorption from food and supplemental sources in healthy young women, *American Journal of Clinical Nutrition* 89 (2009): 533–538; M. U. Muckenthaler, B. Galy, and M. W. Hentze, Systemic iron homeostasis and the iron-responsive element/iron-regulatory protein (IRE/IRP) regulatory network, *Annual Review of Nutrition* 28 (2008): 197–213.
10. T. Ganz, Hepcidin and iron regulation: 10 years later, *Blood* 117 (2011): 4425–4433; J. Kaplan, D. M. Ward, and I. De Domenico, The molecular basis of iron overload disorders and iron-linked anemias, *International Journal of Hematology* 93 (2011): 14–20; A. Pietrangelo, Hereditary hemochromatosis: Pathogenesis, diagnosis, and treatment, *Gastroenterology* 139 (2010): 393–408; B. Borch-Iohnsen and coauthors, Regulation of the iron metabolism, *Journal of the Norwegian Medical Association* 129 (2009): 858–862.
11. S. R. Lynch, Why nutritional iron deficiency persists as a worldwide problem, *Journal of Nutrition* 141 (2011): 763S–768S; Worldwide prevalence of anaemia 1993–2005: WHO Global Database on Anaemia, 2008, **www.who.org.**
12. L. M. Tussing-Humphreys and coauthors, Excess adiposity, inflammation, and iron-deficiency in female adolescents, *Journal of the American Dietetic Association* 109 (2009): 297–302.
13. A. C. Cepeda-Lopez and coauthors, Sharply higher rates of iron deficiency in obese Mexican women and children are predicted by obesity-related inflammation rather than by differences in dietary iron intake, *American Journal of Clinical Nutrition* 93 (2011): 975–983; J. P. McClung and J. P. Karl, Iron deficiency and obesity: The contribution of inflammation and diminished iron absorption, *Nutrition Reviews* 67 (2008): 100–104.
14. E. M. del Giudice and coauthors, Hepcidin in obese children as a potential mediator of the association between obesity and iron deficiency, *Journal of Endocrinology and Metabolism* 94 (2009): 5102–5107.
15. N. Milman, Anemia: Still a major health problem in many parts of the world, *Annals of Hematology* 90 (2011): 369–377.
16. J. L. Beard, Why iron deficiency is important in infant development, *Journal of Nutrition* 138 (2008): 2534–2536.
17. Z. Yang and coauthors, Comparison of plasma ferritin concentration with the ratio of plasma transferrin receptor to ferritin in estimating body iron stores: Results of 4 intervention trials, *American Journal of Clinical Nutrition* 87 (2008): 1892–1898.
18. D. I. Thurnham and coauthors, Adjusting plasma ferritin concentrations to remove the effects of subclinical inflammation in the assessment of iron deficiency: A meta-analysis, *American Journal of Clinical Nutrition* 92 (2010): 546–555; M. A. Ayoya and coauthors, α_1-Acid glycoprotein, hepcidin, C-reactive protein, and serum ferritin are correlated in anemic schoolchildren with *Schistosoma haematobium*, *American Journal of Clinical Nutrition* 91 (2010): 1784–1790.
19. L. E. Murray-Kolb, Iron status and neuropsychological consequences in women of reproductive age: What do we know and where are we headed? *Journal of Nutrition* 141 (2011): 747S–755S; K. Kordas, Iron, lead, and children's behavior and cognition, *Annual Review of Nutrition* 30 (2010): 123–148.
20. B. Lozoff, Early iron deficiency has brain and behavior effects consistent with dopaminergic dysfunction, *Journal of Nutrition* 141 (2011): 740S–746S.
21. J. P. McClung and coauthors, Randomized, double-blind, placebo-controlled trial of iron supplementation in female soldiers during military training: Effects on iron status, physical performance, and mood, *American Journal of Clinical Nutrition* 90 (2009): 124–131; L. E. Murray-Kolb and J. L. Beard, Iron treatment normalizes cognitive functioning in young women, *American Journal of Clinical Nutrition* 85 (2007): 778–787.
22. S. L. Young, Pica in pregnancy: New ideas about an old condition, *Annual Review of Nutrition* 30 (2010): 403–422.
23. P. Brissot and coauthors, Molecular diagnosis of genetic iron-overload disorders, *Expert Review of Molecular Diagnostics* 10 (2010): 755–763.
24. C. Camaschella and E. Poggiali, Inherited disorders of iron metabolism, *Current Opinion in Pediatrics* 23 (2011): 14–20.
25. O. K. Fix and K. V. Kowdley, Hereditary hemochromatosis, *Minerva Medica* 99 (2008): 605–617.
26. G. A. Ramm and R. G. Ruddell, Iron homeostasis, hepatocellular injury, and fibrogenesis in hemochromatosis: The role of inflammation in a non-inflammatory liver disease, *Seminars in Liver Disease* 30 (2010): 271–287; S. Lekawanvijit and N. Chattipakorn, Iron overload thalassemic cardiomyopathy: Iron status assessment and mechanisms of mechanical and electrical disturbance due to iron toxicity, *Canadian Journal of Cardiology* 25 (2009): 213–218.
27. H. Drakesmith and A. Prentice, Viral infection and iron metabolism, *Nature Reviews. Microbiology* 6 (2008): 541–552; A. M. Prentice, Iron

metabolism, malaria, and other infections: What is all the fuss about? *Journal of Nutrition* 138 (2008): 2537–2541.

28. Q. Liu and coauthors, Role of iron deficiency and overload in the pathogenesis of diabetes and diabetic complications, *Current Medicinal Chemistry* 16 (2009): 113–129.
29. G. M. Brittenham, Iron-chelating therapy for transfusional iron overload, *New England Journal of Medicine* 364 (2011): 146–156; F. Dreyfus, The deleterious effects of iron overload in patients with myelodysplastic syndromes, *Blood Reviews* 22 (2008): S29–S34.
30. K. J. Allen and coauthors, Iron-overload: Related disease in *HFE* hereditary hemochromatosis, *New England Journal of Medicine* 358 (2008): 221–230.
31. N. Ahluwalia and coauthors, Iron status is associated with carotid atherosclerotic plaques in middle-aged adults, *Journal of Nutrition* 140 (2010): 812–816.
32. A. C. Bronstein and coauthors, 2009 Annual Report of the American Association of Poison Control Centers' National Poison Data System (NPDS): 27th Annual Report, *Clinical Toxicology* 28 (2010): 979–1178.
33. Committee on Dietary Reference Intakes, 2001, p. 351.
34. M. Alleyne, M. K. Horne, and J. L. Miller, Individualized treatment for iron-deficiency anemia in adults, *American Journal of Medicine* 121 (2008): 943–948.
35. Y. Song and coauthors, Dietary zinc restriction and repletion affects DNA integrity in healthy men, *American Journal of Clinical Nutrition* 90 (2009): 321–328.
36. H. Haase and L. Rink, Functional significance of zinc-related signaling pathways in immune cells, *Annual Review of Nutrition* 29 (2009): 133–152.
37. J. R. Hunt, Algorithms for iron and zinc bioavailability: Are they accurate? *International Journal of Vitamin and Nutrition Research* 80 (2010): 257–262; J. C. King, Does zinc absorption reflect zinc status? *International Journal of Vitamin and Nutrition Research* 80 (2010): 300–306.
38. K. M. Hambidge and coauthors, Zinc bioavailability and homeostasis, *American Journal of Clinical Nutrition* 91 (2010): 1478S–1483S; J. R. Hunt, J. M. Beiseigel, and L. K. Johnson, Adaptation in human zinc absorption as influenced by dietary zinc and bioavailability, *American Journal of Clinical Nutrition* 87 (2008): 1336–1345.
39. S. G. Bell and B. L. Vallee, The metallothionein/thionein system: An oxidoreductive metabolic zinc link, *European Journal of Chemical Biology* 10 (2009): 55–62.
40. L. A. Lichten and R. J. Cousins, Mammalian zinc transporters: Nutritional and physiologic regulation, *Annual Review of Nutrition* 29 (2009): 153–176.
41. C.L.F. Walker, M. Ezzati, and R. E. Black, Global and regional child mortality and burden of disease attributable to zinc deficiency, *European Journal of Clinical Nutrition* 63 (2009): 591–597.
42. C. R. Cole and F. Lifshitz, Zinc nutrition and growth retardation, *Pediatric Endocrinology Reviews* 5 (2008): 889–896.
43. J. B. Barnett, D. H. Hamer, and S. N. Meydani, Low zinc status: A new risk factor for pneumonia in the elderly? *Nutrition Reviews* 68 (2010): 30–37.
44. D. E. Roth and coauthors, Acute lower respiratory infections in childhood: Opportunities for reducing the global burden through nutritional interventions, *Bulletin of the World Health Organization* 86 (2008): 321–416.
45. Walker, Ezzati, and Black, 2009; M. Lukacik, R. L. Thomas, and J. V. Aranda, A meta-analysis of the effects of oral zinc in the treatment of acute and persistent diarrhea, *Pediatrics* 121 (2008): 326–336; S. E. Wuehler, F. Sempértegui, and K. H. Brown, Dose-response trial of prophylactic zinc supplements, with or without copper, in young Ecuadorian children at risk of zinc deficiency, *American Journal of Clinical Nutrition* 87 (2008): 723–733.
46. Barnett, Hamer, and Meydani, 2010.
47. T. J. Caruso, C. G. Prober, and J. M. Gwaltney, Jr., Treatment of naturally acquired common colds with zinc: A structured review, *Clinical Infectious Disease* 45 (2007): 569–574.
48. R. C. Gordon and coauthors, Iodine supplementation improves cognition in mildly iodine-deficient children, *American Journal of Clinical Nutrition* 90 (2009): 1264–1271.
49. M. B. Zimmerman, Iodine deficiency in pregnancy and the effects of maternal iodine supplementation on the offspring: A review, *American Journal of Clinical Nutrition* 89 (2009): 668S–672S.
50. GAIN-UNICEF Universal Salt Iodization Partnership Program, 2011, **www.gainhealth.gov/programs/usi.**
51. R. D. Semba and coauthors, Child malnutrition and mortality among families not utilizing adequately iodized salt in Indonesia, *American Journal of Clinical Nutrition* 87 (2008): 438–444.
52. F. P. Bellinger and coauthors, Regulation and function of selenoproteins in human disease, *Biochemical Journal* 422 (2009): 11–22.
53. D. L. St. Germain, V. A. Galton, and A. Hernandez, Minireview: Defining the roles of the iodothyronine deiodinases: Current concepts and challenges, *Endocrinology* 150 (2009): 1097–1107.
54. C. Lei and coauthors, Is selenium deficiency really the cause of Keshan disease? *Environmental Geochemistry and Health* 33 (2011): 183–188; J. Yang and coauthors, Selenium level surveillance for the year 2007 of Keshan disease in endemic areas and analysis on surveillance results between 2003 and 2007, *Biological Trace Element Research* 138 (2010): 53–59.
55. S. Sun, Chronic exposure to cereal mycotoxin likely citreoviridin may be a trigger for Keshan disease mainly through oxidative stress mechanism, *Medical Hypotheses* 74 (2010): 841–842.
56. G. Dennert and coauthors, Selenium for preventing cancer, *Cochrane Database of Systematic Reviews* 5 (2011): CD005195; S. J. Fairweather-Tait and coauthors, Selenium in human health and disease, *Antioxidants and Redox Signaling* 14 (2011): 1337–1383.
57. B. K. Dunn and coauthors, A nutrient approach to prostate cancer prevention: The Selenium and Vitamin E Cancer Prevention Trial (SELECT), *Nutrition and Cancer* 62 (2010): 896–918; J. Brozmanová and coauthors, Selenium: A double-edged sword for defense and offence in cancer, *Archives of Toxicology* 84 (2010): 919–938.
58. C. D. Thomson and coauthors, Brazil nuts: An effective way to improve selenium status, *American Journal of Clinical Nutrition* 87 (2008): 379–384.
59. J. R. Prohaska, Role of copper transporters in copper homeostasis, *American Journal of Clinical Nutrition* 88 (2008): 826S–829S.
60. D. L. de Romaña and coauthors, Risks and benefits of copper in light of new insights of copper homeostasis, *Journal of Trace Elements in Medicine and Biology* 25 (2011): 3–13.
61. 2008 Water Fluoridation Statistics, **www.cdc.gov/fluoridation/statistics/2008stats.htm**, updated October 22, 2010.
62. Position of the American Dietetic Association: The impact of fluoride on health, *Journal of the American Dietetic Association* 105 (2005): 1620–1628.
63. E. D. Beltrán-Aguilar, L. Barker, and B. A. Dye, Prevalence and severity of dental fluorosis in the United States, *NCHS Data Brief* 53 (2010): 1–8.
64. F. C. Lau and coauthors, Nutrigenomic basis of beneficial effects of chromium (III) on obesity and diabetes, *Molecular and Cellular Biochemistry* 317 (2008): 1–10; Z. Q. Wang and W. T. Cefalu, Current concepts about chromium supplementation in type 2 diabetes and insulin resistance, *Current Diabetes Reports* 10 (2010): 145–151; H. E. Bartlett and F. Eperjesi, Nutritional supplements for type 2 diabetes: A systematic review, *Ophthalmic & Physiological Optics* 28 (2008): 503–523.
65. Kordas, 2010.
66. C. Warniment, K. Tsang, and S. S. Galazka, Lead poisoning in children, *American Family Physician* 81 (2010): 751–757.

LEARN IT Define *phytochemicals* and explain how they might defend against chronic diseases.

Phytochemicals and Functional Foods

Chapter 13 completes the introductory lessons on the six classes of nutrients—carbohydrates, lipids, proteins, vitamins, minerals, and water. In addition to these nutrients, foods contain thousands of other compounds, including the **phytochemicals.** Chapter 1 introduces the phytochemicals as compounds found in plant-derived foods (*phyto* means plant) that have biological activity in the body. Research on phytochemicals is unfolding daily, adding to our knowledge of their roles in human health, but there are still many questions and only tentative answers. Just a few of the tens of thousands of phytochemicals have been researched at all, and only a sampling are mentioned in this highlight—enough to illustrate their wide variety of food sources and roles in supporting health.

The concept that foods provide health benefits beyond those of the nutrients emerged from numerous epidemiological studies showing the protective effects of plant-based diets on cancer and heart disease. People have been using foods to maintain health and prevent disease for years, but now these foods have been given a name—they are called **functional foods.**[1] (The accompanying glossary defines this and other terms.) As Chapter 1 explains, functional foods include all foods (whole, fortified, or modified foods) that have a potentially beneficial effect on health.[2] Much of this text touts the benefits of nature's functional foods—whole grains rich in dietary fibers, oily fish rich in omega-3 fatty acids, and fresh fruits rich in phytochemicals, for example. This highlight begins with a look at some of these familiar functional foods, the phytochemicals they contain, and their roles in disease prevention. Then the discussion turns to examine the most controversial of functional foods—novel foods to which phytochemicals have been added to promote health. How these foods fit into a healthy diet is still unclear.

The Phytochemicals

In foods, phytochemicals impart tastes, aromas, colors, and other characteristics. They give hot peppers their burning sensation, garlic its pungent flavor, and tomatoes their red color. In the body, phytochemicals can have profound physiological effects—acting as antioxidants, mimicking hormones, stimulating enzymes, interfering with DNA replication, suppressing inflammation, destroying bacteria, and binding to cell walls. Any of these actions may prevent the development of chronic diseases, depending in part on how genetic factors interact with the phytochemicals.[3] Phytochemicals might also have adverse effects when consumed in excess. Table H13-1 presents the names, possible effects, and food sources of some of the better-known phytochemicals.

Defending against Cancer

A variety of phytochemicals from a variety of foods appear to protect against DNA damage and defend the body against cancer. A few examples follow.

Soy may protect against breast and prostate cancers and reduce the risk of death and recurrence.[4] Soybeans—as well as other legumes, **flaxseeds,** whole grains, fruits, and vegetables—are a rich source of an array of phytochemicals, among them the **phytoestrogens.** Because the chemical structure of phytoestrogens is similar to the steroid hormone estrogen, they can weakly mimic or modulate the effects of estrogen in the body. They also have antioxidant activity that appears to slow the growth of some cancers. Soy foods appear to be most

GLOSSARY

carotenoids (kah-ROT-eh-noyds): pigments commonly found in plants and animals, some of which have vitamin A activity. The carotenoid with the greatest vitamin A activity is beta-carotene.

flavonoids (FLAY-von-oyds): yellow pigments in foods; phytochemicals that may exert physiological effects on the body.

flaxseeds: the small brown seeds of the flax plant; valued in nutrition as a source of fiber, lignans, and omega-3 fatty acids.

functional foods: whole, fortified, or modified foods that contain bioactive compounds that provide health benefits beyond their nutrient contributions; sometimes called *designer foods* and *nutraceuticals*.

lignans: phytochemicals present in flaxseed, that are converted to phytosterols by intestinal bacteria and are under study as possible anticancer agents.

lutein (LOO-teen): a plant pigment of yellow hue; a phytochemical believed to play roles in eye functioning and health.

lycopene (LYE-koh-peen): a pigment responsible for the red color of tomatoes and other red-hued vegetables; a phytochemical that may act as an antioxidant in the body.

phytochemicals: nonnutrient compounds found in plants that confer taste, color, and other characteristics. Some phytochemicals have biological activity in the body.

phytoestrogens: phytochemicals structurally similar to human estrogen that weakly mimic or modulate estrogen's action in the body. Phytoestrogens include the isoflavones *genistein, daidzein,* and *glycitein.*

plant sterols: phytochemicals that have structural similarities to cholesterol and lower blood cholesterol by interfering with cholesterol absorption. Plant sterols include *sterol esters* and *stanol esters.*

TABLE H13-1 Phytochemicals—Their Food Sources and Actions

Name	Possible Effects	Food Sources
Alkylresorcinols (phenolic lipids)	May contribute to the protective effect of grains in reducing the risks of diabetes, heart disease, and some cancers.	Whole-grain wheat and rye
Allicin (organosulfur compound)	Antimicrobial that may reduce ulcers; may lower blood cholesterol.	Chives, garlic, leeks, onions, scallions
Capsaicin	Modulates blood clotting, possibly reducing the risk of fatal clots in heart and artery disease.	Hot peppers
Carotenoids (include beta-carotene, lycopene, lutein, zeaxanthin, and hundreds of related compounds)	Act as antioxidants, possibly reducing risks of cancer and other diseases.	Deeply pigmented fruits and vegetables (apricots, broccoli, cantaloupe, carrots, pink grapefruit, pumpkin, spinach, sweet potatoes, tomatoes, red peppers, watermelon)
Curcumin (polyphenol)	Acts as an antioxidant and anti-inflammatory agent; may reduce blood clot formation; may inhibit enzymes that activate carcinogens.	Turmeric, a yellow-colored spice common in curry powder
Flavonoids (include anthocyanins, flavones, flavonols, isoflavones, catechins, and others)	Act as antioxidants; scavenge carcinogens; bind to nitrates in the stomach, preventing conversion to nitrosamines; inhibit cell proliferation.	Berries, black tea, celery, citrus fruits, green tea, olives, onions, oregano, purple grapes, purple grape juice, soybeans and soy products, vegetables, whole wheat, wine
Genistein and daidzein (isoflavonoids)	Phytoestrogens that inhibit cell replication in GI tract; may reduce risk of breast, colon, ovarian, prostate, and other estrogen-sensitive cancers; may reduce cancer cell survival; may reduce risk of osteoporosis.	Soybeans, soy flour, soy milk, tofu, textured vegetable protein, other legume products
Indoles (organosulfur compound)	May trigger production of enzymes that block DNA damage from carcinogens; may inhibit estrogen action.	Cruciferous vegetables such as bok choy, broccoli, brussels sprouts, cabbage, cauliflower, collard greens, mustard greens, kale, swiss chard, watercress
Isothiocyanates (organosulfur compounds, including sulforaphane)	Act as antioxidants; inhibit enzymes that activate carcinogens; activate enzymes that detoxify carcinogens; may reduce risk of breast cancer, prostate cancer.	Cruciferous vegetables such as bok choy, broccoli, brussels sprouts, cabbage, cauliflower, collard greens, mustard greens, kale, swiss chard, watercress
Lignans (polyphenol)	Phytoestrogens that block estrogen activity in cells possibly reducing the risk of cancer of the breast, colon, ovaries, and prostate.	Flaxseed, whole grains
Monoterpenes (including limonene)	May trigger enzyme production to detoxify carcinogens; inhibit cancer promotion and cell proliferation.	Citrus fruits, cherries
Phenolic acids (including ellagic acid)	May trigger enzyme production to make carcinogens water soluble, facilitating excretion.	Coffee beans, fruits (apples, blueberries, cherries, grapes, oranges, pears, prunes), oats, potatoes, soybeans
Phytic acid (phenolic acid)	Binds to minerals, preventing free-radical formation, possibly reducing cancer risk.	Whole grains
Resveratrol (flavonoid)	Acts as an antioxidant; may inhibit cancer growth; reduce inflammation, LDL oxidation, and blood clot formation.	Red wine, peanuts, grapes, raspberries
Saponins (glucosides)	May interfere with DNA replication, preventing cancer cells from multiplying; stimulate immune response.	Alfalfa sprouts, other sprouts, green vegetables, potatoes, tomatoes
Tannins (flavonoid)	Act as antioxidants; may inhibit carcinogen activation and cancer promotion.	Black-eyed peas, grapes, lentils, red and white wine, tea

effective when consumed in moderation early and throughout life.[5] Importantly, soy concentrates or the use of phytoestrogen supplements are ill-advised—especially for women with breast cancer and those with high risk factors—as phytoestrogens may stimulate the growth of estrogen-dependent cancers (such as breast cancer).[6] The American Cancer Society recommends that women with breast cancer should consume only *moderate* amounts of soy as part of a healthy plant-based diet and should not intentionally ingest high levels of soy or supplements of phytoestrogens.

Limited evidence suggests that tomatoes may offer protection against some cancers.[7] Among the phytochemicals thought to be responsible for this effect is **lycopene,** one of the many **carotenoids**.[8] Lycopene is the pigment that gives apricots, guava, papaya, pink grapefruits, and watermelon their red color—and it is especially abundant in tomatoes and cooked tomato products. Lycopene is a powerful antioxidant that seems to inhibit the growth of cancer cells.[9] Importantly, the benefits of lycopene have been seen when people have eaten *foods* containing lycopene; lycopene supplements may interfere with cancer treatments.[10]

Soybeans and tomatoes are only two of the many fruits and vegetables credited with providing anticancer activity. Strong and convincing evidence shows that the risk of many cancers, and perhaps of cancer in general, decreases when diets include an abundance of fruits and vegetables.[11] To that end, current recommendations urge consumers to eat five to nine servings of fruits and vegetables a day.

Defending against Heart Disease

Diets based primarily on unprocessed foods appear to support heart health better than those founded on highly refined foods—perhaps because of the abundance of nutrients, fiber, or phytochemicals such as the **flavonoids.** Flavonoids, a large group of phytochemicals known for their health-promoting qualities, are found in whole grains, legumes, soy, vegetables, fruits, herbs, spices, teas, chocolate, nuts, olive oil, and red wines.[12] Flavonoids are powerful antioxidants that may help to protect LDL cholesterol against oxidation, minimize inflammation, and reduce blood platelet stickiness, thereby slowing the progression of atherosclerosis and making blood clots less likely.[13] Whereas an abundance of flavonoid-containing *foods* in the diet may lower the risks of chronic diseases, no claims can be made for flavonoids themselves as the protective factor, particularly when they are extracted from foods and sold as supplements. In fact, purified flavonoids may even be harmful.[14]

In addition to flavonoids, fruits and vegetables are rich in carotenoids such as beta-carotene and **lutein.** Studies suggest that a diet rich in carotenoids is associated with a lower risk of hypertension and heart disease.[15]

The **plant sterols** of soybeans and the **lignans** of flaxseed may also protect against heart disease.[16] These cholesterol-like molecules are naturally found in all plants and inhibit cholesterol absorption in the body. As a result, blood cholesterol levels decline.[17] These phytochemicals also seem to protect against heart disease by reducing inflammation and lowering blood pressure.[18]

Defending against Other Diseases

Most research on phytochemicals has focused on cancer and heart disease, but phytochemicals defend against other diseases as well. The orange-yellow pigment curcumin, commonly found in curry powder, may help reverse insulin resistance, inflammation, and other symptoms associated with obesity.[19] The carotenoids lutein and zeaxanthin may protect the eyes and skin from ultraviolet light damage and the bones from mineral loss.[20]

The Phytochemicals in Perspective

Because foods deliver thousands of phytochemicals in addition to dozens of nutrients, researchers must be careful in giving credit for particular health benefits to any one compound. Diets rich in whole grains, legumes, vegetables, fruits, and nuts seem to protect against heart disease and cancer, but identifying *the* specific foods or components of foods that are responsible is difficult. Each food possesses a unique array of phytochemicals—citrus fruits provide monoterpenes; grapes, resveratrol; and flaxseed, lignans. (Review Table H13-1, p. 429, for the possible effects and other food sources of these phytochemicals.) Broccoli may contain as many as 10,000 different phytochemicals—each with the potential to influence some action in the body. Beverages such as wine, spices such as oregano, and oils such as olive oil (especially virgin olive oil) contain many phytochemicals that may explain, in part, why people who eat a traditional Mediterranean diet have reduced risks of heart disease and cancer.[21] Phytochemicals might also explain why the DASH diet is so effective in lowering blood pressure and blood lipids. Even identifying all of the phytochemicals and their effects doesn't answer all the questions because the actions of phytochemicals may be complementary or overlapping—which reinforces the principle of variety in diet planning. For an appreciation of the array of phytochemicals offered by a variety of foods, see Figure H13-1.

Functional Foods

Because foods naturally contain thousands of phytochemicals that are biologically active in the body, virtually all of them have some value in supporting health.[22] In other words, even simple, whole foods, in reality, are functional foods.[23] Cranberries may help prevent urinary tract infections; garlic may lower blood cholesterol; grapes may reduce inflammation; and green tea may inhibit ulcer infections, just to name a few examples.[24] Functional foods rich in phytochemicals are easy to find in the produce section of grocery stores. Just look for the colorful fruits and vegetables (see Table H13-2, p. 432). But food manufacturers continue to create new functional foods as well. The creation of more functional foods has become the fastest-growing trend and the greatest influence transforming the global food supply.[25]

Many processed foods become functional foods when they are fortified with nutrients or enhanced with phytochemicals or herbs (calcium-fortified orange juice, for example). Less frequently, an entirely new food is created, as in the case of a meat substitute made of mycoprotein—a protein derived from a fungus.* This functional food not only provides dietary fiber, polyunsaturated fats, and high-quality protein, but it lowers LDL cholesterol, raises HDL cholesterol, improves glucose response, and prolongs satiety after a meal. Such a novel functional food raises the question—is it a food or a drug?

Foods as Pharmacy

Not too long ago, most of us could agree on what was a food and what was a drug. Today, functional foods blur the distinctions.[26] They have characteristics similar to both foods and drugs, but do not fit neatly into either category. Consider margarine, for example.

Eating nonhydrogenated margarine sparingly instead of butter generously may lower blood cholesterol slightly over several months and clearly falls into the food category. Taking a statin drug, on the other hand, lowers blood cholesterol significantly within weeks and clearly falls into the drug category. But margarine enhanced with a plant sterol that lowers blood cholesterol is in a gray area between the two. The margarine looks and tastes like a food, but it acts like a drug.

The use of functional foods as drugs creates a whole new set of diet-planning challenges. Not only must foods provide an adequate intake of all the nutrients to support good health, but they must also

*This mycoprotein product is marketed under the trade name Quorn (pronounced KWORN).

> FIGURE H13-1 **An Array of Phytochemicals in a Variety of Fruits and Vegetables**

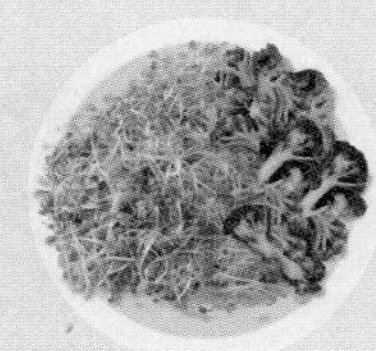

Broccoli and broccoli sprouts (and brussels sprouts, bok choy, cabbage, cauliflower, kale, collard greens, swiss chard, turnips, and watercress) contain an abundance of the cancer-fighting phytochemicals sulforaphane and indoles.

The phytochemical resveratrol found in grapes (and nuts) protects against cancer by inhibiting cell growth and against heart disease by limiting clot formation and inflammation.

The flavonoids in cocoa and chocolate defend against oxidation and reduce the tendency of blood to clot.

Spinach (and collard greens, corn, swiss chard, and winter squash) contains the carotenoids lutein and zeaxanthin, which help protect the eyes against macular degeneration.

An apple a day—rich in phenolic acids—may protect against heart disease.

The ellagic acid of strawberries (and blackberries, blueberries, raspberries, and grapes) may inhibit certain types of cancer and decrease cholesterol levels.

Tomatoes (and pink grapefruit, red peppers, and watermelons), with their abundant lycopene, may defend against cancer and heart disease by protecting DNA from oxidative damage.

Quercetins—commonly found in kale (and onions, pears, and grapes)—reduce inflammation from allergies, inhibit tumor growth, and protect the lungs.

The phytoestrogens of soybeans seem to starve cancer cells and inhibit tumor growth; the plant sterols may lower blood cholesterol and protect cardiac arteries.

The monoterpenes of citrus fruits (and cherries) may protect the lungs.

Colorful foods such as apricots (and cantaloupes, carrots, kale, kiwifruit, mangoes, papaya, pumpkins, spinach, sweet potatoes, and winter squash) contain beta-carotene, which may help slow aging, protect against some cancers, improve lung function, and reduce complications of diabetes.

Garlic (and chives, leeks, onions, and scallions), with its abundant organosulfur compounds, may lower blood cholesterol and blood pressure and protect against stomach cancer.

The flavonoids in black tea may protect against heart disease, whereas those in green tea may defend against cancer.

Blueberries (and cherries, plums, and strawberries), a rich source of anthocyanins, may protect against the effects of aging.

Flaxseed, the richest source of lignans, may prevent the spread of cancer.

TABLE H13-2 **The Colors of Foods Rich in Phytochemicals**

Red	White-Brown	Orange-Yellow	Blue-Purple	Green
Anthocyanins Lycopene	Allicin Allyl sulfides	Beta-carotene Limonene	Anthocyanins Ellagic acid Phenolics	Beta-carotene Lutein Indoles
Beets Cherries Cranberries Pink grapefruit Pomegranates Radicchio Radishes Raspberries Red apples Red peppers Red potatoes Rhubarb Strawberries Tomatoes Watermelon	Bananas Brown pears Cauliflower Chives Dates Garlic Ginger Leeks Mushrooms Onions Parsnips Shallots Turnips	Apricots Cantaloupe Carrots Lemons Mangoes Nectarines Oranges Papayas Peaches Persimmons Pineapple Pumpkin Rutabagas Squash Sweet potatoes Tangerines Yellow peppers	Black currants Blackberries Blueberries Dried plums Eggplant Elderberries Plums Purple figs Purple peppers Raisins Purple cabbage Purple grapes	Artichokes Arugula Asparagus Avocados Broccoli Brussels sprouts Cabbage Celery Cucumbers Endive Green apples Green beans Green grapes Green onions Green pears Green peppers Honeydew melon Kiwifruit Leafy greens Limes Okra Peas Snow peas Spinach Sugar snap peas Watercress Zucchini

Nature offers a variety of functional foods that provide us with many health benefits.

deliver druglike ingredients to protect against disease. Like drugs used to treat chronic diseases, functional foods may need to be eaten several times a day for several months or even years to have a beneficial effect. Sporadic users may be disappointed in the results. Margarine enriched with 2 to 3 grams of plant sterols may reduce cholesterol by up to 15 percent, much more than regular margarine does, but not nearly as much as the more than 30 percent reduction seen with cholesterol-lowering drugs. For this reason, functional foods may be more useful for prevention and mild cases of disease than for intervention and more severe cases.

Foods and drugs differ dramatically in cost as well. Functional foods such as fruits and vegetables incur no added costs, but foods that have been manufactured with added phytochemicals can be expensive, costing up to six times as much as their conventional counterparts.

The price of functional foods typically falls between that of traditional foods and medicines.

Unanswered Questions

To achieve a desired health effect, which is the better choice: to eat a novel functional food created to affect a specific body function or to adjust the diet? Does it make more sense to use a margarine enhanced with a plant sterol that lowers blood cholesterol or simply to limit the amount of butter eaten?* Is it smarter to eat eggs enriched with omega-3 fatty acids or to restrict egg consumption? Might functional foods offer a sensible solution for improving our nation's health—if done correctly? Perhaps so, but the problem is that the food industry is moving too fast for either scientists or the Food and Drug Administration to keep up. Consumers were able to buy soup with St. John's wort that claimed to enhance mood and fruit juice with echinacea that was supposed to fight colds while scientists were still conducting their studies on these ingredients. Research to determine the safety and effectiveness of these substances is still in progress. Until this work is complete, consumers are on their own in finding answers to the following questions:

- *Does it work?* Research is generally lacking and findings are often inconclusive.
- *How much does it contain?* Food labels are not required to list the quantities of added phytochemicals. Even if they were, consumers have no standard for comparison and cannot deduce whether the amounts listed are a little or a lot. Most importantly, until research is complete, food manufacturers do not know what amounts (if any) are most effective—or most toxic.
- *Is it safe?* Functional foods can act like drugs. They contain ingredients that can alter body functions and cause allergies, drug interactions, drowsiness, and other side effects. Yet, unlike drug labels, food labels do not provide instructions for the dosage, frequency, or duration of treatment.
- *Is it healthy?* Adding phytochemicals to a food does not magically make it a healthy choice. A candy bar fortified with phytochemicals is still made mostly of sugar and fat.

Critics suggest that the designation "functional foods" may be nothing more than a marketing tool. After all, even the most experienced researchers cannot yet identify the perfect combination of nutrients and phytochemicals to support optimal health. Yet manufacturers are freely experimenting with various concoctions as if they possessed that knowledge. Is it okay for them to sprinkle phytochemicals on fried snack foods or caramel candies and label them "functional," thus implying health benefits?

*Margarine products that lower blood cholesterol contain either sterol esters from vegetable oils, soybeans, and corn or stanol esters from wood pulp.

© Craig M. Moore

Functional foods currently on the market promise to "enhance mood," "promote relaxation and good karma," "increase alertness," and "improve memory," among other claims.

Future Foods

Nature has elegantly designed foods to provide us with a complex array of dozens of nutrients and thousands of additional compounds that may benefit health—most of which we have yet to identify or understand. Over the years, we have taken those foods, deconstructed them, and then reconstructed them in an effort to "improve" them. With new scientific understandings of how nutrients—and the myriad other compounds in foods—interact with genes, we may someday be able to design *specific* eating patterns to meet the *exact* health needs of *each* individual.[27] Indeed, our knowledge of the human genome and of human nutrition may well merge to allow specific recommendations for individuals based on their predisposition to diet-related diseases.

If the present trend continues, someday physicians may be able to prescribe the perfect foods to enhance your health, and farmers will be able to grow them. As Highlight 19 explains, scientists have already developed gene technology to alter the composition of food crops. They can grow rice enriched with vitamin A and tomatoes containing a hepatitis vaccine, for example. It seems quite likely that foods can be created to meet every possible human need. But then, in a sense, that was largely true 100 years ago when we relied on the bounty of nature.

REFERENCES

1. W. R. Kapsak and coauthors, Functional foods: Consumer attitudes, perceptions, and behaviors in a growing market, *Journal of the American Dietetic Association* 111 (2011): 804–810.
2. Position of the American Dietetic Association: Functional foods, *Journal of the American Dietetic Association* 109 (2009): 735–746.
3. J. W. Lampe, Interindividual differences in response to plant-based diets: Implications for cancer risk, *American Journal of Clinical Nutrition* 89 (2009): 1553S–1557S.
4. P. L. de Souza and coauthors, Clinical pharmacology of isoflavones and its relevance for potential prevention of prostate cancer, *Nutrition Reviews* 68 (2010): 542–555; S. A. Lee and coauthors, Adolescent and adult soy food intake and breast cancer risk: Results from the Shanghai Women's Health Study, *American Journal of Clinical Nutrition* 89 (2009): 1920–1926; X. O. Shu and coauthors, Soy food intake and breast cancer survival, *Journal of the American Medical Association* 302 (2009): 2437–2443; E. Cheung and coauthors, Diet

and prostate cancer risk reduction, *Expert Review of Anticancer Therapy* 8 (2008): 43–50.
5. L. Hilakivi-Clarke, J. E. Andrade, and W. Helferich, Is soy consumption good or bad for the breast? *Journal of Nutrition* 140 (2010): 2326S–2334S.
6. M. Messina and A. H. Wu, Perspectives on the soy-breast cancer relation, *American Journal of Clinical Nutrition* 89 (2009): 1673S–1679S.
7. N. P. Gullet and coauthors, Cancer prevention with natural compounds, *Seminars in Oncology* 37 (2010): 258–281.
8. J. Talvas and coauthors, Differential effects of lycopene consumed in tomato paste and lycopene in the form of a purified extract on target genes of cancer prostatic cells, *American Journal of Clinical Nutrition* 91 (2010): 1716–1724; J. R. Mein, F. Lian, and X. Wang, Biological activity of lycopene metabolites: Implications for cancer prevention, *Nutrition Reviews* 66 (2008): 667–683.
9. N. Khan, F. Afaq, and H. Mukhtar, Cancer chemoprevention through dietary antioxidants: Progress and promise, *Antioxidants and Redox Signaling* 10 (2008): 475–510.
10. B. Cassileth, Lycopene, *Oncology* 24 (2010): 296.
11. T. J. Key, Fruit and vegetables and cancer risk, *British Journal of Cancer* 104 (2011): 6–11; J. M. Matés and coauthors, Anticancer antioxidant regulatory functions of phytochemicals, *Current Medicinal Chemistry* 18 (2011): 2315–2338.
12. O. K. Chun and coauthors, Estimation of antioxidant intakes from diet and supplements in US adults, *Journal of Nutrition* 140 (2010): 317–324.
13. G. Williamson and coauthors, Functional foods for health promotion: State-of-the-science on dietary flavonoids. Extended abstracts from the 12th Annual Conference on Functional Foods for Health Promotion, April 2009, *Nutrition Reviews* 67 (2009): 736–743; R. di Giuseppe and coauthors, Regular consumption of dark chocolate is associated with low serum concentrations of C-reactive protein in a healthy Italian population, *Journal of Nutrition* 138 (2008): 1939–1945; I. Erlund and coauthors, Favorable effects of berry consumption on platelet function, blood pressure, and HDL cholesterol, *American Journal of Clinical Nutrition* 87 (2008): 323–331; L. Hooper and coauthors, Flavonoids, flavonoid-rich foods, and cardiovascular risk: A meta-analysis of randomized controlled trials, *American Journal of Clinical Nutrition* 88 (2008): 38–50; W. M. Loke and coauthors, Pure dietary flavonoids quercetin and (–)-epicatechin augment nitric oxide products and reduce endothelin-1 acutely in healthy men, *American Journal of Clinical Nutrition* 88 (2008): 1018–1025.
14. S. Egert and G. Rimbach, Which sources of flavonoids: Complex diets or dietary supplements? *Advances in Nutrition* 2 (2011): 8–14.
15. A. Hozawa and coauthors, Circulating carotenoid concentrations and incident hypertension: The Coronary Artery Risk Development in Young Adults (CARDIA) study, *Journal of Hypertension* 27 (2009): 237–242; G. Riccioni, Carotenoids and cardiovascular disease, *Current Atherosclerosis Reports* 11 (2009): 434–439.
16. A. Pan and coauthors, Meta-analysis of the effects of flaxseed interventions on blood lipids, *American Journal of Clinical Nutrition* 90 (2009): 288–297.
17. S. B. Racette and coauthors, Dose effects of dietary phytosterols on cholesterol metabolism: A controlled feeding study, *American Journal of Clinical Nutrition* 91 (2010): 32–38; J. H. van Ee, Soy constituents: Modes of action in low-density lipoprotein management, *Nutrition Reviews* 67 (2009): 222–234.
18. R. A. Othman and M. H. Moghadasian, Beyond cholesterol-lowering effects of plant sterols: Clinical and experimental evidence of anti-inflammatory properties, *Nutrition Reviews* 69 (2011): 371–382.
19. B. B. Aggarwal, Targeting inflammation-induced obesity and metabolic diseases by curcumin and other nutraceuticals, *Annual Review of Nutrition* 30 (2010): 173–199; L. Alappat and A. B. Awad, Curcumin and obesity: Evidence and mechanisms, *Nutrition Reviews* 68 (2010): 729–738.
20. R. L. Roberts, J. Green, and B. Lewis, Lutein and zeaxanthin in eye and skin health, *Clinics in Dermatology* 27 (2009): 195–201; S. Shivani and coauthors, Inverse association of carotenoid intakes with 4-y change in bone mineral density in elderly men and women: The Framingham Osteoporosis Study, *American Journal of Clinical Nutrition* 89 (2009): 416–424.
21. S. Granados-Principal and coauthors, Hydroxytyrosol: From laboratory investigations to future clinical trials, *Nutrition Reviews* 68 (2010): 191–206; D. M. Minich and J. S. Bland, Dietary management of the metabolic syndrome beyond macronutrients, *Nutrition Reviews* 66 (2008): 429–444.
22. A. S. Chang, B. Y. Yeong, and W. P. Koh, Symposium on plant polyphenols: Nutrition, health and innovations, June 2009, *Nutrition Reviews* 68 (2010): 246–252.
23. Position of the American Dietetic Association, 2009.
24. C. Chuang and M. K. McIntosh, Potential mechanisms by which polyphenol-rich grapes prevent obesity-mediated inflammation and metabolic diseases, *Annual Review of Nutrition* 31 (2011): 155–176; K. M. Reinhart and coauthors, The impact of garlic on lipid parameters: A systematic review and meta-analysis, *Nutrition Research Reviews* 22 (2009): 39–48; S. Y. Lee, Y. W. Shin, and K. B. Hahm, Phytoceuticals: Mighty but ignored weapons against *Helicobacter pylori* infection, *Journal of Digestive Diseases* 9 (2008): 129–139; Y. Liu and coauthors, Cranberry changes the physicochemical surface properties of *E. coli* and adhesion with uroepithelial cells, *Colloids and Surfaces. B: Biointerfaces* 65 (2008): 35–42.
25. A. E. Sloan, The top functional food trends, *Food Technology* 62 (2008): 24–44; I. Siró and coauthors, Functional food. Product development, marketing and consumer acceptance: A review, *Appetite* 51 (2008): 456–467.
26. P. J. Jones and K. A. Varady, Are functional foods redefining nutritional requirements? *Applied Physiology, Nutrition, and Metabolism* 33 (2008): 118–123.
27. L. R. Ferguson, Nutrigenomics approaches to functional foods, *Journal of the American Dietetic Association* 109 (2009): 452–458.

© Ammit/Shutterstock.com

Good health depends on a well-balanced diet and regular exercise. Go to Fitness → Exercise to learn how a physically active body improves nutrient absorption.

14

Fitness: Physical Activity, Nutrients, and Body Adaptations

LEARNING GPS

Nutrition in Your Life

Every day, you choose whether to be physically active or not, and your choices over time can influence how well you feel and how long you live. Today's world makes it easy to be inactive—too easy in fact—but the many health rewards of being physically active make it well worth the effort. You may even discover how much fun it is to be active, and with a little perseverance, you may become physically fit as well. The choice is yours. In the Nutrition Portfolio at the end of this chapter, you can determine whether your physical activities meet current recommendations and whether your daily food, fluid, and carbohydrate intakes are appropriate to support those activities.

Are you physically fit? If so, the following description applies to you. Your joints are flexible, your muscles are strong, and your body is lean with enough, but not too much, fat. You have the endurance to engage in daily physical activities with enough reserve energy to handle added challenges. Carrying heavy suitcases, opening a stuck window, or climbing four flights of stairs, which might strain an unfit person, is easy for you. What's more, you are prepared to meet mental and emotional challenges too. All these characteristics of **fitness** describe the same wonderful condition of a healthy body.

Or perhaps you are leading a **sedentary** life. Today's world encourages inactivity, and people who go through life exerting minimal physical effort become weak and unfit and may begin to feel unwell. In fact, a sedentary lifestyle fosters the development of several chronic diseases.[1]

Regardless of your level of fitness, this chapter is written for "you," whoever you are and whatever your goals—whether you want to improve your health, lose weight, hone your athletic skills, ensure your position on a sports team, or simply adopt an active lifestyle. This chapter begins by discussing fitness and

fitness: the characteristics that enable the body to perform physical activity; more broadly, the ability to meet routine physical demands with enough reserve energy to rise to a physical challenge; or the body's ability to withstand stress of all kinds.

sedentary: physically inactive (literally, "sitting down a lot").

its benefits, and then goes on to explain how the body uses energy nutrients to fuel physical activity. Finally, it describes diets to support fitness.

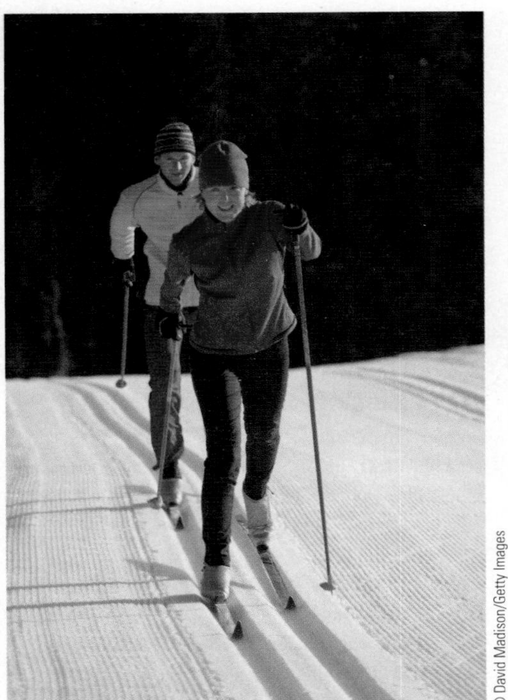

© David Madison/Getty Images

Physical activity, or lack of it, exerts a significant and pervasive influence on everyone's nutrition and overall health.

♦ Each comparison influences the risks associated with chronic disease and death similarly:

- Vigorous exercise vs. minimal exercise
- Healthy weight vs. 20% overweight
- Nonsmoking vs. smoking (one pack a day)

14.1 Fitness

LEARN IT Describe the health benefits of being physically fit and explain how to develop the components of fitness.

Fitness depends on a certain minimum amount of **physical activity** or **exercise.** Both physical activity and exercise involve body movement, muscle contraction, and enhanced energy expenditure, but "exercise" is often used to describe structured, planned physical activity. This chapter focuses on how the active body uses energy nutrients—whether that body is pedaling a bike across campus or pedaling a stationary bike in a gym. Thus, for our purposes, the terms *physical activity* and *exercise* are used interchangeably.

Benefits of Fitness The *Dietary Guidelines for Americans 2010* emphasize the benefits of increasing physical activities and reducing sedentary activities.[2] Extensive evidence confirms that regular physical activity promotes health ♦ and reduces the risk of developing a number of diseases.[3] Yet, despite an increasing awareness of the health benefits that physical activity confers, about 40 percent of adults in the United States are not regularly active, and about 13 percent are completely inactive.[4] Physical inactivity is linked to the major degenerative diseases—heart disease, cancer, stroke, diabetes, and hypertension—the primary killers of adults in developed countries.[5]

> DIETARY GUIDELINES FOR AMERICANS 2010

Increase physical activity and reduce time spent in sedentary behaviors.

As a person becomes physically fit, the health of the entire body improves. In general, physically fit people enjoy:[6]

- *Restful sleep.* Rest and sleep occur naturally after periods of physical activity.
- *Nutritional health.* Physical activity expends energy and thus allows people to eat more food. If they choose wisely, active people will consume more nutrients and be less likely to develop nutrient deficiencies.
- *Improved body composition.* A balanced program of physical activity limits body fat and increases or maintains lean tissue. Thus physically active people have relatively less body fat than sedentary people at the same body weight.[7]
- *Improved bone density.* Weight-bearing physical activity builds bone strength and protects against osteoporosis.[8]
- *Resistance to colds and other infectious diseases.* Fitness enhances immunity.[9]*
- *Low risks of some cancers.* Lifelong physical activity may help to protect against colon cancer, breast cancer, and some other cancers.[10]
- *Strong circulation and lung function.* Physical activity that challenges the heart and lungs strengthens the circulatory system.
- *Low risk of cardiovascular disease.* Physical activity lowers blood pressure, slows resting pulse rate, and lowers blood cholesterol, thus reducing the risks of heart attacks and strokes.[11] Some research suggests that physical

physical activity: bodily movement produced by muscle contractions that substantially increase energy expenditure.

exercise: planned, structured, and repetitive body movements that promote or maintain physical fitness.

*Moderate physical activity can stimulate immune function. Intense, vigorous, prolonged activity such as marathon running, however, may compromise immune function.

activity may reduce the risk of cardiovascular disease in another way as well—by reducing visceral fat stores.[12]

- *Low risk of type 2 diabetes.* Physical activity normalizes glucose tolerance.[13] Regular physical activity reduces the risk of developing type 2 diabetes and improves glucose control.
- *Reduced risk of gallbladder disease.* Regular physical activity reduces the risk of gallbladder disease—perhaps by facilitating weight control and lowering blood lipid levels.[14]
- *Low incidence and severity of anxiety and depression.* Physical activity may improve mood and enhance the quality of life by reducing depression and anxiety.[15]
- *Strong self-image.* The sense of achievement that comes from meeting physical challenges promotes self-confidence.
- *Long life and high quality of life in the later years.* Active people live longer, healthier lives than sedentary people do.[16] Even as little as 15 minutes a day of moderate-intensity activity can add years to a person's life.[17] In addition to extending longevity, physical activity supports independence and mobility in later life by reducing the risk of falls and minimizing the risk of injury should a fall occur.[18]

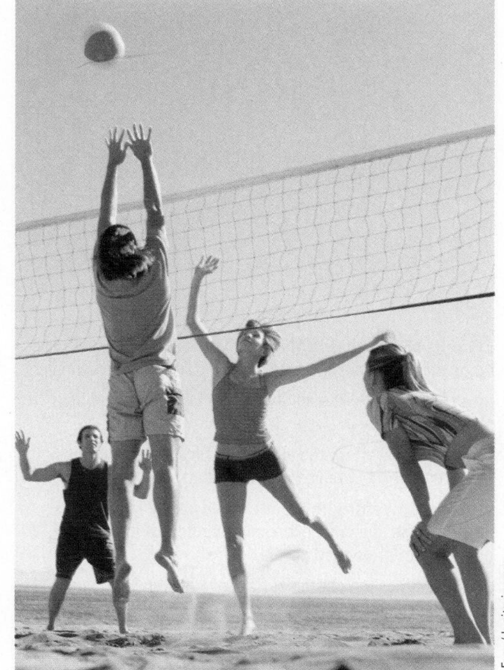

Physical activity helps you look good, feel good, and have fun, and it brings many long-term health benefits as well.

What does a person have to do to reap the health rewards of physical activity? To gain substantial *health* benefits, most guidelines recommend a minimum amount of time performing **aerobic physical activity.**[19] The minimum amount of time depends on whether the activity is **moderate-intensity physical activity** or **vigorous-intensity physical activity.** Table 14-1 compares intensity levels. For clarity and effectiveness, a minimum length of 10 minutes for short bouts of aerobic physical activity is recommended.[20] Of course, more time and greater intensity bring even greater health benefits—such as maintaining a healthy body weight (BMI of 18.5 to 24.9) and further reducing the risk of chronic diseases.

> DIETARY GUIDELINES FOR AMERICANS 2010

Choose moderate- or vigorous-intensity physical activities. Slowly build up the amount of physical activity you choose.

In addition to providing health benefits, physical activity helps to develop and maintain *fitness*. Table 14-2 (p. 440) presents the American College of Sports Medicine (ACSM) guidelines for physical activity.[21] Following these guidelines will help adults improve their cardiorespiratory endurance, body composition, strength, and flexibility. At this level of fitness, a person can reap still greater health benefits (substantially lower risk of premature death compared with those who are inactive, improved cardiorespiratory fitness, and improved body composition, for example).[22] Fitness and health depend on maintaining an active lifestyle every day.

aerobic physical activity: activity in which the body's large muscles move in a rhythmic manner for a sustained period of time. Aerobic activity, also called *endurance activity,* improves cardiorespiratory fitness. Brisk walking, running, swimming, and bicycling are examples.

moderate-intensity physical activity: physical activity that requires some increase in breathing and/or heart rate and expends 3.5 to 7 kcalories per minute. Walking at a speed of 3 to 4.5 miles per hour (about 15 to 20 minutes to walk 1 mile) is an example.

vigorous-intensity physical activity: physical activity that requires a large increase in breathing and/or heart rate and expends more than 7 kcalories per minute. Walking at a very brisk pace (>4.5 miles per hour) or running at a pace of at least 5 miles per hour are examples.

TABLE 14-1 Levels of Physical Activity Intensity Compared

Level of Intensity	Breathing and/or Heart Rate	Perceived Exertion (on a Scale of 0 to 10)	Talk Test	Energy Expenditure	Walking Pace
Light	Little to no increase	<5	Able to sing	<3.5 kcal/min	<3 mph
Moderate	Some increase	5 or 6	Able to have a conversation	3.5 to 7 kcal/min	3 to 4.5 mph
Vigorous	Large increase	7 or 8	Conversation is difficult or "broken"	>7 kcal/min	>4.5 mph

SOURCE: Centers for Disease Control and Prevention, www.cdc.gov/physicalactivity/everyone; updated March 30, 2011, accessed September 12, 2011.

The bottom line is that any physical activity, even moderate activity, *beyond* activities of daily living, provides some health benefits, and these benefits follow a dose-response relationship. In other words, some activity is better than none, and more activity is better still—up to a point. Pursued in excess, intense physical activity, especially when combined with poor eating habits, can undermine health, as Highlight 8, Eating Disorders, explains (pp. 252–259).

Developing Fitness To be physically fit, a person must develop enough flexibility, muscle strength and endurance, and cardiorespiratory endurance to meet the everyday demands of life with some to spare and to achieve a reasonable body weight and body composition. **Flexibility** allows the joints to move freely, reducing the risk of injury. **Muscle strength** and **muscle endurance** enable muscles to work harder and longer without fatigue. **Cardiorespiratory endurance** supports the ongoing activity of the heart and lungs. Physical activity supports lean body tissues and reduces excess body fat. A person who practices a physical activity *adapts* by becoming better able to perform that activity after each session—with more flexibility, more strength, and more endurance.

flexibility: the capacity of the joints to move through a full range of motion; the ability to bend and recover without injury.

muscle strength: the ability of muscles to work against resistance.

muscle endurance: the ability of a muscle to contract repeatedly without becoming exhausted.

cardiorespiratory endurance: the ability to perform large-muscle, dynamic exercise of moderate to high intensity for prolonged periods.

conditioning: the physical effect of training; improved flexibility, strength, and endurance.

training: practicing an activity regularly, which leads to conditioning. (Training is what you do; conditioning is what you get.)

progressive overload principle: the training principle that a body system, in order to improve, must be worked at frequencies, durations, or intensities that gradually increase physical demands.

frequency: the number of occurrences per unit of time (for example, the number of activity sessions per week).

The principles of **conditioning** apply to each component of fitness—flexibility, strength, and endurance. During conditioning, the body adapts microscopically to perform the work it is asked to do. The way to achieve conditioning is by **training,** primarily by applying the **progressive overload principle**—that is, by asking a little more of the body in each training session.

The Overload Principle You can apply the progressive overload principle in several different ways. You can perform the activity more often—that is, increase its **frequency.** You can perform it more strenuously—that is, increase

TABLE 14-2 ACSM Guidelines for Physical Fitness

	Cardiorespiratory	Strength	Flexibility
	© Photodisc/Getty Images	 © David Hanover Photography	 © David Hanover Photography
Type of Activity	Aerobic activity that uses large-muscle groups and can be maintained continuously	Resistance activity that is performed at a controlled speed and through a full range of motion	Stretching activity that uses the major muscle groups
Frequency	5 to 7 days per week	2 to 3 nonconsecutive days per week	2 to 7 days per week
Intensity	Moderate (equivalent to walking at a pace of 3 to 4 miles per hour)[a]	Enough to enhance muscle strength and improve body composition	Enough to feel tightness or slight discomfort
Duration	At least 30 minutes per day	2 to 4 sets of 8 to 12 repetitions involving each major muscle group	2 to 4 repetitions of 15 to 30 seconds per muscle group
Examples	Running, cycling, dancing, swimming, inline skating, rowing, power walking, cross-country skiing, kickboxing, water aerobics, jumping rope; sports activities such as basketball, soccer, racquetball, tennis, volleyball	Pull-ups, push-ups, sit-ups, weightlifting, pilates	Yoga

[a]For those who prefer vigorous-intensity aerobic activity such as walking at a very brisk pace (>4.5 mph) or running (≥5 mph), a minimum of 20 minutes per day, 3 days per week is recommended.

SOURCE: American College of Sports Medicine position stand, Quantity and quality of exercise for developing and maintaining cardiorespiratory, musculoskeletal, and neuromotor fitness in apparently healthy adults: Guidance for prescribing exercise, *Medicine and Science in Sports and Exercise* 43 (2011): 1334–1359; W. L. Haskell and coauthors, Physical activity and public health: Updated recommendation for adults from the American College of Sports Medicine and the American Heart Association, *Medicine & Science in Sports & Exercise* 39 (2007): 1423–1434.

its **intensity.** Or you can do it for longer time periods—that is, increase its **duration.** All three strategies, individually or in combination, work well. The rate of progression depends on individual characteristics such as fitness level, health status, age, and preference. If you want continuous improvements, remember to overload progressively as you reach higher levels of fitness.

When increasing the frequency, intensity, or duration of a workout, however, exercise to a point that only *slightly* exceeds the comfortable capacity to work. It is better to progress slowly than to risk injury by overexertion.

© Mike Powell/Getty Images

Choose an active lifestyle. Use the stairs; walk or bike to work, class, or shops; wash and wax the car; mow the grass; rake the leaves; shovel snow; walk the dog; play with children. Be active and have fun.

The Body's Response to Physical Activity

Fitness develops in response to demand and wanes when demand ceases. Muscles gain size and strength after being made to work repeatedly, a response called **hypertrophy.** Conversely, without activity, muscles diminish in size and lose strength, a response called **atrophy.**

Hypertrophy and atrophy are adaptive responses to the muscles' greater and lesser work demands, respectively. Thus cyclists often have strong, well-developed legs but less arm and chest strength; a tennis player may have one superbly strong arm, while the other is just average. A variety of physical activities produces the best overall fitness, and to this end, people need to work different muscle groups from day to day. This strategy provides a day or two of rest for different muscle groups, allowing time to replenish nutrients and to repair any minor damage incurred by the activity.

Other tips for building fitness and minimizing the risk of overuse injuries are:

- Be active all week, not just on the weekends.
- Use proper equipment and wear proper attire.
- Perform exercises using proper form.
- Include **warm-up** and **cool-down** activities in each session. Warming up helps to prepare muscles, ligaments, and tendons for the upcoming activity and mobilizes fuels to support strength and endurance activities. Cooling down reduces muscle cramping and allows the heart rate to slow gradually.
- Train hard enough to challenge your strength or endurance a few times each week rather than every time you work out. Between challenges, do moderate workouts and include at least one day of rest each week.
- Pay attention to body signals. Symptoms such as abnormal heartbeat, dizziness, lightheadedness, cold sweat, confusion, or pain or pressure in the middle of the chest, teeth, jaw, neck, or arm demand immediate medical attention.
- Work out wisely. Do not start with activities so demanding that pain stops you within a day or two. Learn to enjoy small steps toward improvement. Fitness builds slowly.

Cautions on Starting a Fitness Program Before beginning a fitness program, make sure it is safe for you to do so. Most apparently healthy people can begin a moderate exercise program such as walking or increasing daily activities without a

intensity: the degree of exertion while exercising (for example, the amount of weight lifted or the speed of running).

duration: length of time (for example, the time spent in each activity session).

hypertrophy (high-PER-tro-fee): growing larger; with regard to muscles, an increase in size (and strength) in response to use.

atrophy (AT-ro-fee): becoming smaller; with regard to muscles, a decrease in size (and strength) because of disuse, undernutrition, or wasting diseases.

warm-up: 5 to 10 minutes of light activity, such as easy jogging or cycling, prior to a workout to prepare the body for more vigorous activity.

cool-down: 5 to 10 minutes of light activity, such as walking or stretching, following a vigorous workout to gradually return the body's core to near-normal temperature.

© Kzenon/Shutterstock.com

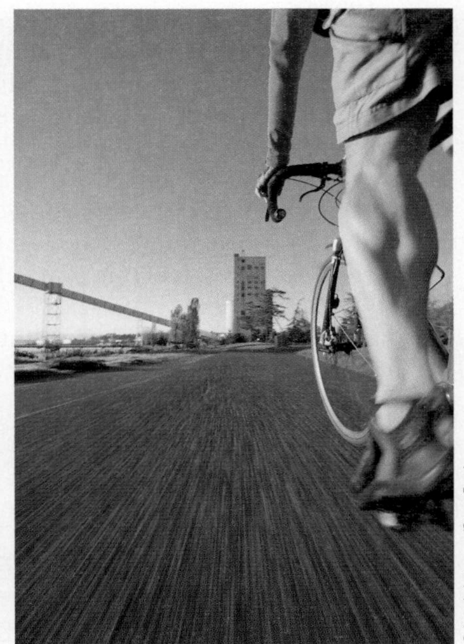

© Istockphoto.com/Steve Froebe

People's bodies are shaped by the activities they perform.

medical examination, but people with any of the risk factors listed in the margin ♦ may need medical advice.[23]

♦ Major coronary risk factors:
- Age (men ≥ 45 yr or women ≥ 55 yr)
- Family history of heart disease
- Cigarette smoking
- Hypertension
- Serum cholesterol ≥200 mg/dL, LDL ≥ 130 mg/dL or HDL <40 mg/dL, or taking lipid-lowering medication
- Prediabetes (fasting plasma glucose ≥ 100 mg/dL)
- Sedentary lifestyle
- Obesity (BMI ≥30)

Cardiorespiratory Endurance The length of time a person can remain active with an elevated heart rate—that is, the ability of the heart, lungs, and blood to sustain a given demand—defines a person's cardiorespiratory endurance. Cardiorespiratory endurance training improves a person's ability to sustain vigorous activities such as running, brisk walking, or swimming. Such training enhances the capacity of the heart, lungs, and blood to deliver oxygen to, and remove waste from, the body's cells.[24] Cardiorespiratory endurance training, therefore, is *aerobic,* meaning oxygen requiring. As the cardiorespiratory system gradually adapts to the demands of aerobic activity, the body delivers oxygen more efficiently. In fact, the accepted measure of a person's cardiorespiratory fitness is maximal oxygen uptake **(VO_2max).** The benefits of cardiorespiratory training are not just physical, though, because all of the body's cells, including the brain cells, require oxygen to function. When the cells receive more oxygen more readily, both the body and the mind benefit.

♦ Cardiorespiratory conditioning:
- Increases cardiac output and oxygen delivery
- Increases blood volume per heart beat (stroke volume)
- Slows resting pulse
- Increases breathing efficiency
- Improves circulation
- Reduces blood pressure

VO_2max: the maximum rate of oxygen consumption by an individual at sea level.

cardiorespiratory conditioning: improvements in heart and lung function and increased blood volume, brought about by aerobic training.

cardiac output: the volume of blood discharged by the heart each minute; determined by multiplying the stroke volume by the heart rate. The stroke volume is the amount of oxygenated blood the heart ejects toward the tissues at each beat. Cardiac output (volume/minute) = stroke volume (volume/beat) × heart rate (beats/minute)

Cardiorespiratory Conditioning **Cardiorespiratory conditioning** ♦ occurs as aerobic workouts improve heart and lung function. **Cardiac output** increases, thus enhancing oxygen delivery. The heart becomes stronger, and each beat pumps more blood. Because the heart pumps more blood with each beat, fewer beats are necessary, and the resting heart rate slows down. The average resting pulse rate for adults is around 70 beats per minute, but people who achieve cardiorespiratory conditioning may have resting pulse rates of 50 or even lower. The muscles that work the lungs become stronger, too, so breathing becomes more efficient. Circulation through the arteries and veins improves. Blood moves easily, and blood pressure falls.[25]

Cardiorespiratory endurance reflects the health of the heart and circulatory system, on which all other body systems depend. To improve your cardiorespiratory endurance, activities must be sustained for 20 minutes or longer and use most of the large-muscle groups of the body (legs, buttocks, and abdomen). The level of training must be intense enough to elevate your heart rate.

A person's own perceived effort is usually a reliable indicator of the intensity of an activity. In general, workouts should be at an intensity that raises your heart

© Jason Maehl/Shutterstock.com

The key to regular physical activity is finding an activity you enjoy.

rate but still leaves you able to talk comfortably. For those who are more competitive and want to work to their limits on some days, a treadmill test can reveal the maximum heart rate. Workouts are safe at up to 85 percent of that rate. Table 14-2 (p. 440) includes the ACSM guidelines for developing and maintaining cardiorespiratory fitness.

Muscle Conditioning One of the benefits of cardiorespiratory training is that fit muscles use oxygen efficiently, reducing the heart's workload. An added bonus is that muscles that use oxygen efficiently can burn fat longer—a plus for body composition and weight control.

A Balanced Fitness Program The intensity and type of physical activities that are best for one person may not be good for another. A person who has been sedentary will initially perform at a dramatically different level of intensity than a fit person.

The type of physical activity that is best for you depends, too, on what you want to achieve and what you enjoy doing. Some people love walking, whereas others prefer to dance or ride a bike. Those who want to be stronger and firmer, lift weights. Keep in mind that muscle is more metabolically active than body fat, so the more muscle you have, the more energy you'll burn.

In a balanced fitness program, aerobic activity improves cardiorespiratory fitness, stretching enhances flexibility, and resistance training develops muscle strength, **muscle power**, and muscle endurance. Table 14-3 provides an example of a balanced fitness program.

TABLE 14-3 A Sample Balanced Fitness Program

Monday, Tuesday, Wednesday, Thursday, Friday:

- 5 minutes of warm-up activity
- 30–60 minutes of aerobic activity
- 10 minutes of cool-down activity and stretching

Tuesday, Thursday, Saturday:

- 5 minutes of warm-up activity
- 30 minutes of resistance training
- 10 minutes of cool-down activity and stretching

Saturday and/or Sunday:

- Sports, walking, hiking, biking, or swimming

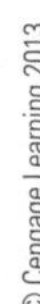

Resistance Training

Resistance training has long been recognized as a means to build muscle mass and develop and maintain muscle strength, muscle power, and muscle endurance. Additional benefits of resistance training, however, have also emerged. Progressive resistance training helps prevent and manage several chronic diseases, including cardiovascular disease, and enhances psychological well-being.[26] Resistance training can also help to maximize and maintain bone mass.[27] Even in women past menopause (when most women are losing bone), resistance training can improve bone density, especially in combination with adequate dietary calcium and vitamin D intake.[28]

By promoting strong muscles in the back and abdomen, resistance training can improve posture and reduce the risk of back injury. Resistance training can also help prevent the decline in physical mobility that often accompanies aging.[29]

muscle power: the product of force generation (strength) and movement velocity (speed); the speed at which a given amount of exertion is completed.

resistance training: the use of free weights or weight machines to provide resistance for developing muscle strength, power, and endurance; also called *weight training*. A person's own body weight may also be used to provide resistance such as when a person does push-ups, pull-ups, or abdominal crunches.

Older adults, even those in their 80s, who participate in resistance training programs not only gain muscle strength but also improve their muscle endurance, which enables them to walk longer before exhaustion. Leg strength and walking endurance are powerful indicators of an older adult's physical abilities.

Resistance training builds muscle strength, muscle power, and muscle endurance. To emphasize muscle strength, combine high resistance (heavy weight) with a low number of repetitions (8 to 12).[30] To emphasize muscle power, combine moderate resistance (light to medium weight) with high velocity (as fast as safely possible). To emphasize muscle endurance, combine less resistance (lighter weight) with more repetitions (15 to 20). Resistance training enhances performance in other sports too. Swimmers can develop a more efficient stroke and tennis players a more powerful serve when they train with weights, for example.

REVIEW IT Describe the health benefits of being physically fit and explain how to develop the components of fitness.
Physical activity brings good health and long life. To develop fitness—whose components are flexibility, muscle strength and endurance, and cardiorespiratory endurance—a person must condition the body, through training, to adapt to the activity performed.

14.2 Energy Systems and Fuels to Support Activity

LEARN IT Identify the factors that influence fuel use during physical activity and the types of activities that depend more on glucose or fat, respectively.

Nutrition and physical activity go hand in hand. Activity demands carbohydrate and fat as fuel, protein to build and maintain lean tissues, vitamins and minerals to support both energy metabolism and tissue building, and water to help distribute the fuels and to dissipate the resulting heat and wastes. This section describes how nutrition supports a person who decides to get up and go.

The Energy Systems of Physical Activity—ATP and CP Muscles contract fast. When called upon, they respond quickly without taking time to metabolize fat or carbohydrate for energy. In the first fractions of a second, muscles starting to move depend on their supplies of quick-energy compounds to power their movements. Exercise physiologists know these compounds by their abbreviations, ATP and CP.

ATP As Chapter 7 describes, all of the energy-yielding nutrients—carbohydrate, fat, and protein—can enter metabolic pathways that make the high-energy compound ATP (adenosine triphosphate). ATP is present in small amounts in all body tissues all the time, and it can deliver energy instantly. In the muscles, ATP provides the chemical energy for contraction. When an ATP molecule is split, its energy is released, and the muscle cells channel some of that energy into mechanical movement and most of it into heat.

CP Immediately after the onset of a demand, before muscle ATP pools dwindle, a muscle enzyme begins to break down another high-energy compound that is stored in the muscle, **CP,** or **creatine phosphate.*** CP is made from creatine, a compound commonly found in muscles, with a phosphate group attached. CP can split anaerobically (not requiring oxygen) to release phosphate, which can be used to replenish ATP. Supplies of CP in a muscle last for only about 10 seconds, producing enough quick energy, without oxygen, for a 100-meter dash.

When activity ceases and the muscles are resting, ATP gives up one of its phosphate groups to creatine. Thus CP is produced during rest by reversing the process that occurs during muscular activity. ♦ (Highlight 14 includes creatine supplements in its discussion of substances commonly used in the pursuit of fitness.)

♦ During rest: ATP + creatine → CP
During activity: CP → ATP + creatine

CP, or **creatine phosphate** (also called **phosphocreatine**): a high-energy compound in muscle cells that acts as a reservoir of energy that can maintain a steady supply of ATP. CP provides the energy for short bursts of activity.

*Creatine phosphate is also called *phosophocreatine (PC).*

© UpperCut Images/Alamy

Sustained muscular efforts as in a long-distance rowing event or a cross-country run involve *aerobic* work.

© Comstock/JupiterImages

Split-second surges of power as in the heave of a barbell or jump of a basketball player involve *anaerobic* work.

The Energy-Yielding Nutrients To meet the more prolonged demands of sustained activity, the muscles generate ATP from the more abundant fuels—carbohydrate, fat, and protein—as described in Chapter 7. The breakdown of these nutrients generates ATP all day every day.

During rest, the body derives more than half of its ATP from fatty acids and most of the rest from glucose, along with a small percentage from amino acids. During physical activity, the body adjusts its mixture of fuels. Muscles always use a mixture of fuels—never just one. How much of which fuel ♦ the muscles use during physical activity depends on an interplay among the fuels available from the diet, the intensity and duration of the activity, and the degree to which the body is conditioned to perform that activity. The next sections explain these relationships by examining each of the energy-yielding nutrients individually, but keep in mind that although one fuel may predominate at a given time, the other two will still be involved. Table 14-4 shows how fuel use changes according to the intensity and duration of the activity.

♦ Fuel mixture during activity depends on:

- Diet
- Intensity and duration of activity
- Training

As you read about each of the energy-yielding nutrients, notice how its contribution to the fuel mixture shifts depending on whether the activity is anaerobic or aerobic. Anaerobic activities are associated with strength, agility, and split-second surges of power. The jump of a slam dunk, the power of a tennis serve, and the heave of a bench press all involve anaerobic work. Such high-intensity, short-duration activities depend mostly on glucose as the chief energy fuel.

Endurance activities of low to moderate intensity and long duration depend more on fat to provide energy aerobically. The ability to continue swimming to the shore, to keep on hiking to the top of the mountain, or to continue pedaling all the way home reflects aerobic capacity. As mentioned earlier, aerobic capacity

TABLE 14-4 Primary Fuels Used for Activities of Different Intensities and Durations

Activity Intensity	Activity Duration	Preferred Fuel Source	Oxygen Needed?	Activity Example
Extreme[a]	8 to 10 sec	ATP-CP (immediate availability)	No (anaerobic)	100-yard dash, shot put
Very high	20 sec to 3 min	ATP from carbohydrate (lactate)	No (anaerobic)	¼-mile run at maximal speed
High	3 min to 20 min	ATP from carbohydrate	Yes (aerobic)	Cycling, swimming, or running
Moderate	More than 20 min	ATP from fat	Yes (aerobic)	Hiking

[a]All levels of activity intensity use the ATP-CP system initially; extremely intense short-term activities rely solely on the ATP-CP system.

© Cengage Learning 2013

is also crucial to maintaining a healthy heart and circulatory system. The relationships among fuels and physical activity bear heavily on what foods best support your chosen activities.

Glucose Use during Physical Activity

Glucose, stored in the liver and muscles as glycogen, is vital to physical activity. During exertion, the liver breaks down its glycogen and releases the glucose into the bloodstream. The muscles use this glucose as well as their own private glycogen stores to fuel their work. Glycogen supplies can easily support everyday activities but are limited to less than 2000 kcalories of energy, enough for about 20 miles of running.[31] The more glycogen the muscles store, the longer the glycogen will last during physical activity, which in turn influences performance. When glycogen is depleted, the muscles become fatigued.

Diet Affects Glycogen Storage and Use How much carbohydrate a person eats influences how much glycogen is stored. A classic study compared fuel use during activity among three groups of runners on different diets.[32] For several days before testing, one group consumed a normal mixed diet, a second group consumed a high-carbohydrate diet, and the third group consumed a no-carbohydrate diet (fat and protein diet). As Figure 14-1 shows, the high-carbohydrate diet allowed the runners to keep going longer before exhaustion. This study and many others that followed have confirmed that high-carbohydrate diets enhance endurance by ensuring ample glycogen stores. Thus, to fill glycogen stores, eat plenty of carbohydrate-rich foods.

Intensity of Activity Affects Glycogen Use How long an exercising person's glycogen will last depends not only on diet, but also on the intensity of the activity. Moderate activities such as jogging, during which breathing is steady and easy, use glycogen slowly. The lungs and circulatory system have no trouble keeping up with the muscles' need for oxygen. The individual breathes easily, and the heart beats steadily—the activity is aerobic. The muscles derive their energy from both glucose and fatty acids. By depending partly on fatty acids, moderate aerobic activity conserves glycogen.

Intense activities—the kind that make it difficult "to catch your breath," such as a quarter-mile race—use glycogen quickly. In such activities, the muscles break down glucose to pyruvate anaerobically, producing ATP quickly.

> **FIGURE 14-1 The Effect of Diet on Physical Endurance**

A high-carbohydrate diet can increase an athlete's endurance. In this study, the fat and protein diet provided 94 percent of kcalories from fat and 6 percent from protein; the normal mixed diet provided 55 percent of kcalories from carbohydrate; and the high-carbohydrate diet provided 83 percent of kcalories from carbohydrate.

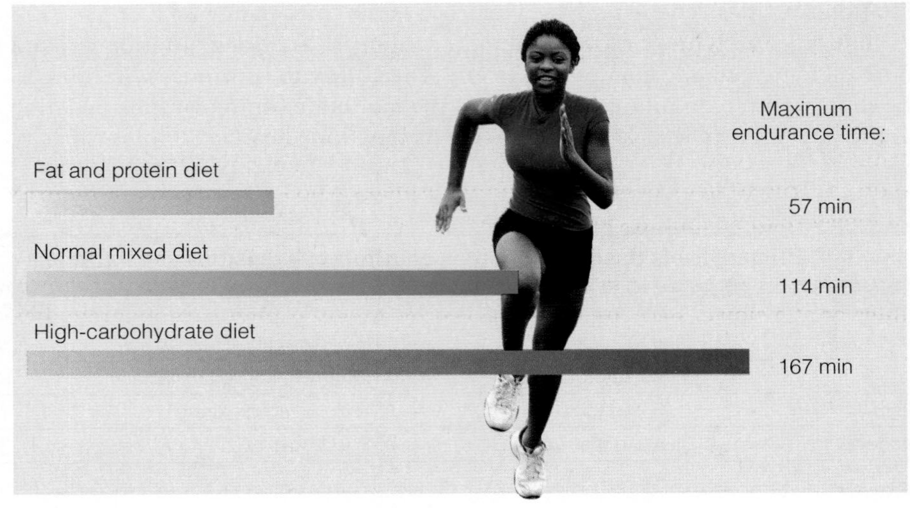

Lactate When the rate of glycolysis exceeds the capacity of the mitochondria to accept hydrogens with their electrons for the electron transport chain, the accumulating pyruvate molecules are converted to lactate. Lactate is the product of anaerobic glycolysis. At low intensities, lactate is readily cleared from the blood, but at higher intensities, lactate accumulates. When the rate of lactate production exceeds the rate of clearance, intense activity can be maintained for only 1 to 3 minutes (as in a 400- to 800-meter race or a boxing match). Lactate has long been blamed for muscle fatigue, but recent research disputes this idea. Working muscles may produce lactate and experience fatigue, but the lactate does not cause the fatigue.[33]

Lactate quickly leaves the muscles and travels in the blood to the liver. There, liver enzymes convert the lactate back into glucose. Glucose can then return to the muscles to fuel additional activity. (The recycling process that regenerates glucose from lactate is known as the *Cori cycle,* as shown in Figure 7-7 on p. 205.)

Duration of Activity Affects Glycogen Use Glycogen use depends not only on the intensity of an activity, but also on its duration. Within the first 20 minutes or so of moderate activity, a person uses mostly glycogen for fuel—about one-fifth of the available glycogen. As the muscles devour their own glycogen, they become ravenous for more glucose, and the liver responds by emptying out its glycogen stores.

After 20 minutes, a person who continues exercising moderately (mostly aerobically) begins to use less and less glycogen and more and more fat for fuel (review Table 14-4 on p. 445). Still, glycogen use continues, and if the activity lasts long enough and is intense enough, blood glucose declines and muscle and liver glycogen stores become depleted. Physical activity can continue for a short time thereafter only because the liver scrambles to produce, from lactate and certain amino acids, the minimum amount of glucose needed to briefly forestall total depletion.

© David Leahy/Taxi/Getty Images

Moderate- to high-intensity aerobic exercises that can be sustained for only a short time (less than 20 minutes) use some fat, but more glucose for fuel.

Glucose Depletion After a couple of hours of strenuous activity, glucose stores are depleted. When depletion occurs, it brings nervous system function to a near halt, making continued exertion at the same intensity almost impossible. Marathon runners refer to this point of glucose exhaustion as "hitting the wall."

To avoid such debilitation, endurance athletes try to maintain their blood glucose for as long as they can. The following guidelines will help endurance athletes maximize glucose supply:

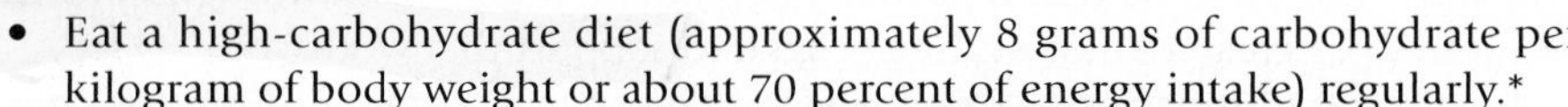

- Eat a high-carbohydrate diet (approximately 8 grams of carbohydrate per kilogram of body weight or about 70 percent of energy intake) regularly.*
- Consume glucose (usually in sports drinks) periodically during activities that last for 1 hour or more.
- Eat carbohydrate-rich foods (approximately 60 grams of carbohydrate) ♦ immediately following activity.

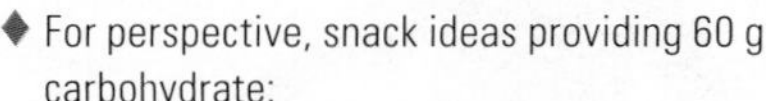

♦ For perspective, snack ideas providing 60 g carbohydrate:
- 6 oz sports drink and 1 small bagel
- 16 oz milk and 4 oatmeal cookies
- 8 oz pineapple juice and 1 granola bar

Another strategy, **carbohydrate loading,** involves training the muscles to store as much glycogen as they can, while supplying the dietary glucose to enable them to do so. This strategy benefits endurance athletes who exercise at high intensity for longer than 90 minutes and who cannot meet their carbohydrate needs during competition. Although carbohydrate-loading can increase an individual's glycogen stores, performance may or may not improve.[34] Those who exercise for shorter times or at a slower pace are better served by a regular high-carbohydrate diet.

*Percentage of energy intake is meaningful only when total energy intake is known. Consider that at high energy intakes (say, 5000 kcalories/day), even a moderate-carbohydrate diet (50 percent of energy intake) supplies 625 grams of carbohydrate—enough for a 165-pound (75 kilogram) athlete in heavy training. By comparison, at a moderate energy intake (2000 kcalories/day), a high carbohydrate intake (70 percent of energy intake) supplies 350 grams—plenty of carbohydrate for most people, but not enough for athletes in heavy training.

carbohydrate loading: a regimen of moderate exercise followed by the consumption of a high-carbohydrate diet that enables muscles to store glycogen beyond their normal capacities; also called *glycogen loading* or *glycogen super compensation.*

The last section of this chapter, "Diets for Physically Active People," discusses how to design a high-carbohydrate diet for performance.

Glucose during Activity Muscles can obtain the glucose they need not only from glycogen stores, but also from foods and beverages consumed during activity. Consuming carbohydrate is especially useful during exhausting endurance activities (lasting more than 1 hour) and during games such as soccer or hockey, which last for hours and demand repeated bursts of intense activity.[35]

Endurance athletes often run short of glucose by the end of competitive events. To ensure optimal carbohydrate intake, sports nutrition experts recommend 30 to 60 grams of carbohydrate per hour during prolonged events.[36] Carbohydrate-based sports drinks offer a convenient way to meet this recommendation and also help replace water and electrolyte losses as a later section explains. Thus, to ensure optimal hydration and carbohydrate intake, endurance athletes are advised to drink one-half to one liter of a 4 to 8 percent carbohydrate-based sports drink per hour, in small, frequent doses during activity.[37] During the last stages of an endurance competition, when glycogen is running low, glucose consumed during the event can slowly make its way from the digestive tract to the muscles and augment the body's supply of glucose enough to forestall exhaustion.

Some researchers have questioned whether adding protein to carbohydrate-containing sports beverages would offer a performance advantage to endurance athletes.[38] Evidence so far suggests that when carbohydrate intake is optimal, protein provides no additional performance benefit.[39]

Glucose after Activity Eating high-carbohydrate foods *after* physical activity also enlarges glycogen stores. Train normally; then, within 2 hours after physical activity, consume a high-carbohydrate meal, such as a glass of orange juice and some graham crackers, toast, or cereal. This method accelerates the rate of glycogen storage. After 2 hours, the rate of glycogen storage declines by almost half. Despite this slower rate of glycogen restoration, muscles continue to accumulate glycogen as long as athletes eat carbohydrate-rich foods within 2 hours following activity.[40] This is particularly important to athletes who train hard more than once a day.

Chapter 4 introduces the glycemic response and discusses the possible health benefits of eating a *low*-glycemic diet. Such a diet may also benefit endurance performance. Some research indicates that foods with a low glycemic response enhance fatty acid availability and use during subsequent activity, thereby reducing reliance on the muscles' own lipid and glycogen stores.[41] More research is needed to confirm these findings.

Training Affects Glycogen Use Training, too, affects how much glycogen muscles will store. Muscle cells that repeatedly deplete their glycogen through hard work adapt to store greater amounts of glycogen to support that work.

Conditioned muscles also rely less on glycogen and more on fat for energy, so glycogen breakdown and glucose use occur more slowly in trained than in untrained individuals at a given work intensity. A person attempting an activity for the first time uses much more glucose than an athlete who is trained to perform the same activity. Oxygen delivery to the muscles by the heart and lungs plays a role, but equally importantly, trained muscles are better equipped to use the oxygen because their cells contain more mitochondria, the structures within a cell responsible for producing ATP (see Figure 7-14 on p. 210). Untrained muscles depend more heavily on anaerobic glucose breakdown, even when physical activity is just moderate.

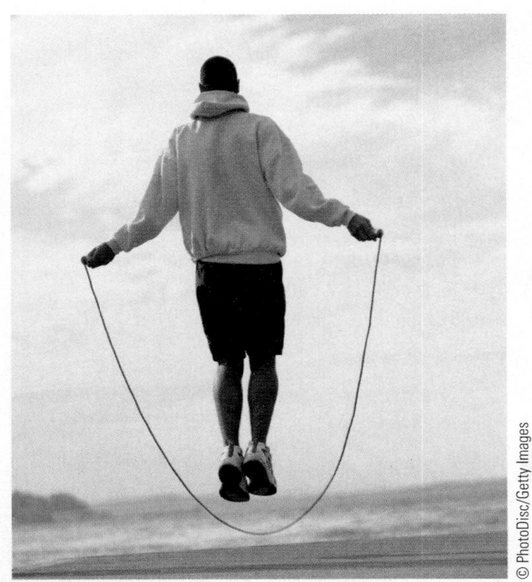

Abundant energy from the breakdown of fat comes from aerobic metabolism.

Fat Use during Physical Activity

Sports nutrition experts recommend that athletes consume 20 to 35 percent of their energy from fat to meet nutrient and energy needs—the same recommendation as for others.[42] Athletes who restrict fat below 20 percent of total energy intake may fail to consume adequate energy and nutrients. Recommendations to include vegetable oils, nuts, olives, fatty fish, and other sources of health-promoting fats in the diet apply to athletes as well as to everyone else.

Diets high in saturated fat carry risks of heart disease. Physical activity offers some protection against cardiovascular disease, but athletes may still suffer heart attacks and strokes. Controlling saturated fat intake is another way athletes can protect themselves from heart disease.

In contrast to *dietary* fat, *body* fat stores are extremely important during physical activity, so long as the activity is not too intense. Unlike glycogen stores, the body's fat stores can usually provide more than 70,000 kcalories and fuel hours of activity without running out.

The fat used in physical activity is liberated as fatty acids from the internal fat stores and from the fat under the skin. Areas that have the most fat to spare donate the greatest amounts to the blood (although they may not be the areas that appear fattiest). Thus "spot reducing" doesn't work because muscles do not "own" the fat that surrounds them. Fat cells release fatty acids into the blood, not into the underlying muscles. Then the blood gives to each muscle the amount of fat that it needs. Proof of this is found in a tennis player's arms—the skinfold measures of fat are the same in both arms, even though the muscles of one arm work much harder and may be larger than those of the other. A balanced fitness program that includes strength training will tighten muscles underneath the fat, improving the overall appearance. Keep in mind that some body fat is essential to good health. (Chapter 8 discusses the health risks of too little body fat.)

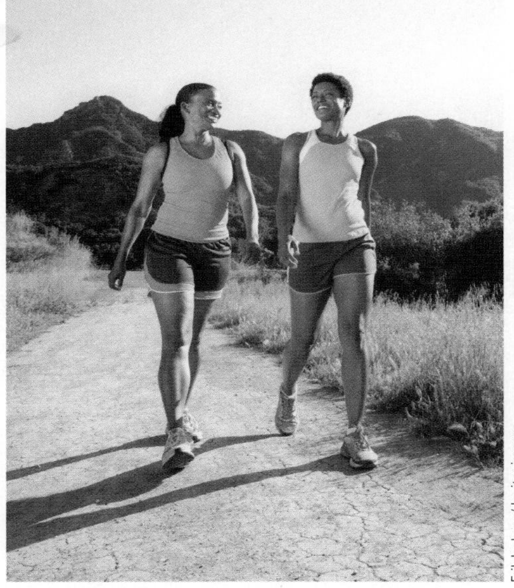

Erik Isakson/Jupiter images

Low- to moderate-intensity aerobic exercises that can be sustained for a long time (more than 20 minutes) use some glucose, but more fat for fuel.

Duration of Activity Affects Fat Use Early in an activity, as the muscles draw on fatty acids, blood levels fall. If the activity continues for more than a few minutes, the hormone epinephrine signals the fat cells to begin breaking down their stored triglycerides and liberating fatty acids into the blood. After about 20 minutes of physical activity, the blood fatty acid concentration surpasses the normal resting concentration. Thereafter, sustained, moderate activity uses body fat stores as its major fuel.

Intensity of Activity Affects Fat Use The intensity of physical activity also affects fat use. As the intensity of activity increases, fat makes less and less of a contribution to the fuel mixture. Remember that fat can be broken down for energy only by aerobic metabolism. For fat to fuel activity, then, oxygen must be abundantly available. If a person is breathing easily during activity, the muscles are getting all the oxygen they need and are able to use more fat in the fuel mixture.

Training Affects Fat Use Training—repeated aerobic activity—produces the adaptations that permit the body to draw more heavily on fat for fuel. Training stimulates the muscle cells to manufacture more and larger mitochondria, the "powerhouse" structures of the cells that produce ATP for energy. Another adaptation: the heart and lungs become stronger and better able to deliver oxygen to muscles at high activity intensities. Still another: hormones in the body of a trained person slow glucose release from the liver and speed up the use of fat instead. These adaptations reward not only trained athletes but all active people; a person who trains in aerobic activities such as distance running or cycling becomes well suited to the activity.

Protein Use during Physical Activity—and between Times

Table 14-4 on p. 445 summarizes the fuel uses discussed so far, but does not include the third energy-yielding nutrient, protein, because protein is not a major fuel for physical activity. Nevertheless, physically active people use protein just as other people do—to build muscle and other lean tissues and, to some extent, to fuel activity. The body does, however, handle protein differently during activity than during rest.

Protein Used in Muscle Building Synthesis of body proteins is suppressed during activity. In the hours of recovery following activity, though, protein synthesis accelerates beyond normal resting levels. As noted earlier, eating high-carbohydrate foods immediately after exercise accelerates muscle glycogen storage. Similarly, research shows that eating high-quality protein, either by itself or together with carbohydrate, enhances muscle protein synthesis.[43] Remember that the body adapts and builds the molecules, cells, and tissues it needs for the next period of activity. Whenever the body remodels a part of itself, it tears down old structures to make way for new ones. Repeated activity, with just a slight overload, triggers the protein-dismantling and protein-synthesizing equipment of each muscle cell to make needed changes—that is, to adapt.

The physical work of each muscle cell acts as a signal to its DNA and RNA to begin producing the kinds of proteins that will best support that work. Take running, for example. In the first difficult sessions, the body is not yet equipped to perform aerobic work easily, but with each session, the cells' genetic material gets the message that an overhaul is needed. In the hours that follow the session, the genes send molecular messages to the protein-building equipment that tell it what old structures to break down and what new structures to build. Within the limits of its genetic potential, the body responds. Running (or any aerobic activity) stimulates synthesis of mitochondria to facilitate efficient aerobic metabolism. Over time, the body adapts and running becomes easier.

The body of a weightlifter responds to training as well, but the response differs from that of aerobic training. Weightlifting stimulates synthesis of muscle fiber protein to enhance muscle mass and strength—with little change in mitochondria. An athlete may add between ¼ ounce and 1 ounce (between 7 and 28 grams) of protein to muscle mass each day during active muscle-building phases of training.

Protein Used as Fuel Not only do athletes retain more protein in their muscles, but they also use more protein as fuel. Muscles speed up their use of amino acids for energy during physical activity, just as they speed up their use of fat and carbohydrate. Still, protein contributes only about 10 percent of the total fuel used, both during activity and during rest. The most active people of all—endurance athletes—use large amounts of all energy fuels, including protein, during performance, but they also eat more food and therefore usually consume enough protein.

Diet Affects Protein Use during Activity The factors that affect how much protein is used during activity seem to be the same three that influence the use of fat and carbohydrate. The first factor is diet. People who consume diets adequate in energy and rich in *carbohydrate* use less protein than those who eat protein- and fat-rich diets. Recall that carbohydrates spare proteins from being broken down to make glucose when needed. Because physical activity requires glucose, a diet lacking in carbohydrate necessitates the conversion of amino acids to glucose. The same is true for a diet high in fat because fatty acids can never provide glucose. In short, to conserve protein, eat a diet adequate in energy and rich in carbohydrate.

Intensity and Duration of Activity Affect Protein Use during Activity A second factor, the intensity and duration of activity, also modifies protein use. Endurance athletes who train for more than an hour a day, engaging in aerobic activity of moderate intensity and long duration, may deplete their glycogen stores by the end of their workouts and become somewhat more dependent on body protein for energy.

In contrast, anaerobic strength training does not use more protein for energy, but it does demand more protein to build muscle. Thus the protein needs of both endurance and strength athletes are slightly higher than those of sedentary people, but certainly not as high as the protein intakes many athletes consume.

Training Affects Protein Use A third factor that influences a person's use of protein during physical activity is the extent of training. Particularly in strength athletes such as weightlifters, the higher the degree of training, the less protein a person uses during an activity.

Protein Recommendations for Active People As mentioned, all active people, and especially athletes in training, probably need more protein than sedentary people do. Endurance athletes, such as long-distance runners and cyclists, use more protein for fuel than strength or power athletes do, and they retain some, especially in the muscles used for their sport. Strength athletes, such as weightlifters, and power athletes, such as football players, use less protein for fuel, but they still use some and retain much more. Therefore, *all* athletes in training should attend to

TABLE 14-5 Recommended Protein Intakes for Athletes

	Recommendations (g/kg/day)	Protein Intakes (g/day)	
		Males	*Females*
RDA for adults	0.8	56	44
Recommended intake for power (strength or speed) athletes	1.2–1.7	84–119	66–94
Recommended intake for endurance athletes	1.2–1.4	84–98	66–77
US average intake		102	70

NOTE: Daily protein intakes are based on a 70-kilogram (154-pound) man and 55-kilogram (121-pound) woman.
SOURCES: Position of the American Dietetic Association, Dietitians of Canada, and the American College of Sports Medicine: Nutrition and athletic performance, *Journal of the American Dietetic Association* 109 (2009): 509–527; US Department of Agriculture, Agricultural Research Service, 2008, Nutrient Intakes from Food: Mean Amounts Consumed per Individual, One Day, 2005–2006, www.ars.usda.gov/ba/bhnrc/fsrg; Committee on Dietary Reference Intakes, *Dietary Reference Intakes for Energy, Carbohydrate, Fiber, Fat, Fatty Acids, Cholesterol, Protein and Amino Acids* (Washington, D.C.: National Academies Press, 2005), pp. 660–661.

protein needs, but they should first meet their energy needs with adequate carbohydrate intakes. Without adequate carbohydrate intake, athletes will burn off as fuel the very protein they wish to retain in muscle.

How much protein, then, should an active person consume? The DRI Committee does not recommend greater than normal protein intakes for athletes, but other authorities do.[44] Recommendations specify different protein intakes for athletes pursuing different activities (see Table 14-5).[45] Even the highest protein recommendations can be met without protein supplements, or even excessive servings of meat. Chapter 6 reviews the potential dangers of using protein and amino acid supplements (see pp. 186–187).

REVIEW IT Identify the factors that influence fuel use during physical activity and the types of activities that depend more on glucose or fat, respectively.

The mixture of fuels the muscles use during physical activity depends on diet, the intensity and duration of the activity, and training. During intense activity, the fuel mix contains mostly glucose, whereas during less intense, moderate activity, fat makes a greater contribution. With endurance training, muscle cells adapt to store more glycogen and to rely less on glucose and more on fat for energy. Athletes in training may need more protein than sedentary people do, but they typically eat more food as well and therefore obtain enough protein without supplements.

14.3 Vitamins and Minerals to Support Activity

LEARN IT List which vitamin and mineral supplements, if any, athletes may need and why.

Many of the vitamins and minerals assist in releasing energy from fuels and in transporting oxygen. This knowledge has led many people to believe, mistakenly, that vitamin and mineral *supplements* offer physically active people both health benefits and athletic advantages. (Review Highlight 10 for a discussion of vitamin and mineral supplements, and see Highlight 14, which explores supplements and other products people use in the hope of enhancing athletic performance.)

Dietary Supplements Nutrient supplements do not enhance the performance of well-nourished people. Deficiencies of vitamins and minerals, however, do impede performance. Regular, strenuous, physical activity increases the demand for energy, and athletes and active people tend to eat more food. If they select nutrient-dense foods to meet those increased energy needs, they will also meet their vitamin and mineral needs.

For perfect functioning, every nutrient is needed.

As Highlight 8 mentions, some athletes who struggle to meet low body-weight requirements may eat so little food that they fail to obtain all the nutrients they need.[46] The practice of "making weight" is opposed by many health and fitness organizations, but for athletes who choose this course of action, a single daily multivitamin-mineral supplement that provides no more than the DRI recommendations for nutrients may be beneficial. In addition, some athletes simply do not eat enough food to maintain a healthy body weight during times of intense training or competition. For these athletes as well, a daily multivitamin-mineral supplement can be helpful.

Some athletes believe that taking vitamin or mineral supplements directly before competition will enhance performance. These beliefs are contrary to scientific reality. Most vitamins and minerals function as small parts of larger working units. After absorption into the blood, these nutrients travel to the cells for assembly with their appropriate other parts before they can do their work. This takes time—hours or days. Vitamins or minerals taken right before an event are useless for improving performance, even if the person is actually suffering deficiencies of them.

Nutrients of Concern In general, then, active people who eat well-balanced meals do not need vitamin or mineral supplements. Two nutrients, vitamin E and iron, do merit special mention here, however, each for a different reason. Vitamin E is discussed because so many athletes take vitamin E supplements. Iron is discussed because some athletes may be unaware that they need iron supplements.

Vitamin E During prolonged, high-intensity physical activity, the muscles' consumption of oxygen increases tenfold or more, which increases the production of free radicals in the body.[47] As Highlight 11 explains, vitamin E is a potent antioxidant that vigorously defends cell membranes against the oxidative damage of free radicals.

Does vitamin E supplementation protect against exercise-induced oxidative stress? Some studies find that it does; others show no effect, and still others report enhanced oxidative stress.[48] Recent research may offer some insight into these inconsistencies. Although free radicals are usually damaging, during repeated episodes of endurance activities, they may actually be beneficial. Free radicals activate powerful antioxidant enzymes, which may enhance the athlete's tolerance to oxidative stresses.[49] Researchers speculate that antioxidant supplements such as vitamin E interfere with this protective adaptation. This may explain why, in some studies, athletes taking vitamin E show signs of increased oxidative stress. Clearly, more research is needed on supplements, ♦ but in the meantime, active people can benefit by using vitamin E–rich vegetable oils and eating generous servings of antioxidant-rich fruits and vegetables regularly.

♦ UL for vitamin E: 1000 mg/day

Iron Deficiency Physically active young women, especially those who engage in endurance activities such as distance running, are prone to iron deficiency.[50] Habitually low intakes of iron-rich foods, high iron losses through menstruation, and the high demands of muscles for the iron-containing electron carriers of the mitochondria and the muscle protein myoglobin can contribute to iron deficiency in physically active young women.

Adolescent female athletes who eat vegetarian diets may be particularly vulnerable to iron deficiency. As Chapter 13 explains, the bioavailability of iron is often poor in vegetarian diets.[51] To protect against iron deficiency, vegetarian athletes need to select good dietary sources of iron (fortified cereals, legumes, nuts, and seeds) and include vitamin C–rich foods with each meal. So long as vegetarian athletes, like all athletes, consume enough nutrient-dense foods, they can perform as well as anyone.

Iron-Deficiency Anemia Iron-deficiency anemia impairs physical performance because iron is an essential component of hemoglobin. Without adequate iron, hemoglobin in red blood cells cannot deliver oxygen to the cells for energy metabolism. Without adequate oxygen, an active person cannot perform aerobic activities and tires easily. Whether iron deficiency without clinical signs of anemia impairs physical performance is less clear.[52]

Sports Anemia Early in training, athletes may develop low blood hemoglobin for a while. This condition, sometimes called **sports anemia,** is not a true iron-deficiency condition. Strenuous aerobic activity promotes destruction of the more fragile, older red blood cells, and the resulting cleanup work reduces the blood's iron content temporarily. Strenuous activity also expands the blood's plasma volume, thereby reducing the red blood cell count per unit of blood. This low hematocrit looks like iron deficiency anemia, but it is not the same. In sports anemia, the red blood cells do not diminish in size or number as in anemia, so the oxygen-carrying capacity is not hindered. Most researchers view sports anemia as an *adaptive,* temporary response to endurance training. Iron-deficiency anemia requires iron supplementation, but sports anemia does not.

Iron Recommendations for Athletes The best strategy for maintaining adequate iron nutrition depends on the individual. Menstruating women may border on iron deficiency even without the iron losses incurred by physical activity. Active teens of both genders have high iron needs because they are growing. Especially for women and teens, then, prescribed supplements may be needed to correct iron deficiencies. Physicians use the results of blood tests to determine whether such supplementation is needed. (Review Chapter 13 for many more details about iron, and see Appendix E for a description of the tests used in assessing its status.)

REVIEW IT List which vitamin and mineral supplements, if any, athletes may need and why.

With the possible exception of iron, well-nourished active people and athletes do not need nutrient supplements. Women and teens may need to pay special attention to their iron needs.

14.4 Fluids and Electrolytes to Support Activity

LEARN IT Identify the factors that influence an athlete's fluid needs and describe the differences between water and sports drinks.

The need for water far surpasses the need for any other nutrient. The body relies on watery fluids as the medium for all of its life-supporting activities, and if it loses too much water, its well-being will be compromised.

Obviously, the body loses water via sweat. Breathing uses water, too, exhaled as vapor. During physical activity, water losses from both routes are significant, and dehydration becomes a threat. Dehydration's first symptom is fatigue: a water loss of greater than 2 percent of body weight can reduce a person's capacity to do muscular work.[53] ♦ With a water loss of about 7 percent, a person is likely to collapse.

© David Vintiner/Corbis

To prevent dehydration and the fatigue that accompanies it, drink liquids before, during, and after physical activity.

Temperature Regulation As Chapter 7 discusses, working muscles produce heat as a by-product of energy metabolism. During intense activity, muscle heat production can be 15 to 20 times greater than at rest. The body cools itself by sweating. Each liter of sweat dissipates almost 600 kcalories of heat, preventing a rise in body temperature of almost 10 degrees on the Celsius scale.* The body routes its blood supply through the capillaries just under the skin, and the skin secretes sweat to evaporate and cool the skin and the underlying blood. The blood then flows back to cool the deeper body chambers.

Hyperthermia In hot, humid weather, sweat doesn't evaporate well because the surrounding air is already laden with water. In **hyperthermia,** body heat builds up and triggers maximum sweating, but without sweat evaporation, little cooling takes place. In such conditions, active people must take precautions to prevent **heat stroke.** To reduce the risk of heat stroke, drink enough fluid before and during the activity, rest in the shade when tired, and wear lightweight clothing

♦ For perspective:
- 2% of 150 lb = 3 lb
- 7% of 150 lb = 10½ lb

sports anemia: a transient condition of low hemoglobin in the blood, associated with the early stages of sports training or other strenuous activity.

hyperthermia: an above-normal body temperature.

heat stroke: a dangerous accumulation of body heat with accompanying loss of body fluid.

*10 degrees on the Celcius scale is about 18 degrees on the Fahrenheit scale.

that allows sweat to evaporate. (Hence the danger of rubber or heavy suits that supposedly promote weight loss during physical activity—they promote profuse sweating, prevent sweat evaporation, and invite heat stroke.) If you ever experience any of the symptoms of heat stroke listed in the margin, ♦ stop your activity, sip fluids, seek shade, and ask for help. Heat stroke can be fatal, young people often die of it, and these symptoms demand attention.

♦ Symptoms of dehydration and heat stroke:
- Headache
- Nausea
- Dizziness
- Clumsiness
- Stumbling
- Sudden cessation of sweating (hot, dry skin)
- Confusion or other mental changes

Hypothermia In cold weather, **hypothermia,** or low body temperature, can be as serious as heat stroke is in hot weather. Inexperienced, slow runners participating in long races on cold or wet, chilly days are especially vulnerable to hypothermia. Slow runners who produce little heat can become too cold if clothing is inadequate. Early symptoms of hypothermia include feeling cold, shivering, apathy, and social withdrawal.[54] As body temperature continues to fall, shivering may stop, and disorientation, slurred speech, and change in behavior or appearance set in. People with these symptoms ♦ soon become helpless to protect themselves from further body heat losses. Even in cold weather, however, the active body still sweats and still needs fluids. The fluids should be warm or at room temperature to help protect against hypothermia.

♦ Symptoms of hypothermia:
- Shivering
- Slurred speech
- Clumsiness
- Confusion
- Slow, shallow breathing
- Poor decision making such as trying to remove warm clothes
- Drowsiness
- Lack of concern about one's condition
- Progressive loss of consciousness

Fluid Replacement via Hydration Endurance athletes can easily lose 1.5 liters or more of fluid during *each hour* of activity. Table 14-6 presents a schedule of hydration for physical activity. To prepare for fluid losses, a person must hydrate before activity; to replace fluid losses, the person must rehydrate during and after activity. Even then, in hot weather, the GI tract may not be able to absorb enough water fast enough to keep up with sweat losses, and some degree of dehydration may be inevitable. Athletes who know their body's **hourly sweat rate** can strive to replace the total amount of fluid lost during activity to prevent dehydration.

Athletes who are preparing for competition are often advised to drink extra fluids in the *days* immediately before the event, especially if they are still training. The extra water is not stored in the body, but drinking extra water ensures maximum hydration at the start of the event. Full hydration is imperative for every athlete both in training and in competition. The athlete who arrives at an event even slightly dehydrated begins with a disadvantage.

What is the best fluid for an exercising body? For noncompetitive, everyday active people, plain, cool water is recommended, especially in warm weather, for two reasons: (1) water rapidly leaves the digestive tract to enter the tissues where it is needed, and (2) it cools the body from the inside out. For endurance athletes, carbohydrate-containing beverages may be appropriate. Fluid ingestion during the event has the dual purposes of replenishing water lost through sweating and providing a source of carbohydrate to supplement the body's limited glycogen stores. Many carbohydrate-containing sports drinks are marketed for active people; a later section compares them with water.

TABLE 14-6 Hydration Schedule for Physical Activity

When to Drink	Amount of Fluid
2 to 3 hr before activity	2 to 3 c
15 min before activity	1 to 2 c
Every 15 min during activity	½ to 1 c (Drink enough to minimize loss of body weight, but don't overdrink.)
After activity	2 c for each pound of body weight lost[a]

[a]Drinking 2 cups of fluid every 20 to 30 minutes after exercise until the total amount required is consumed is more effective for rehydration than drinking the needed amount all at once. Rapid fluid replacement after exercise stimulates urine production and results in less body water retention.

SOURCES: Adapted from American College of Sports Medicine, Position stand, Exercise and fluid replacement, *Medicine and Science in Sports and Exercise* 39 (2007): 377–390; C. K. Seto, D. Way, and N. O'Connor, Environmental illness in athletes, *Clinics in Sports Medicine* 24 (2005): 695–718; R. Murray, Fluid, electrolytes, and exercise in *Sports Nutrition: A Practice Manual for Professionals*. 4th ed., ed. M. Dunford (Chicago: American Dietetic Association, 2006), pp. 94–115; D. J. Casa, P. M. Clarkson, and W. O. Roberts, American College of Sports Medicine Roundtable on Hydration and Physical Activity: Consensus statements, *Current Sports Medicine Reports* 4 (2005): 115–127.

hypothermia: a below-normal body temperature.

hourly sweat rate: the amount of weight lost plus fluid consumed during exercise per hour. One pound equals roughly 2 cups (500 milliliters) of fluid.

Electrolyte Losses and Replacement When a person sweats, small amounts of electrolytes—the electrically charged minerals sodium, potassium, chloride, and magnesium—are lost from the body along with water. Losses are greatest in beginners; training improves electrolyte retention.

To replenish lost electrolytes, a person ordinarily needs only to eat a regular diet that meets energy and nutrient needs. In events lasting more than 1 hour, sports drinks may be needed to replace fluids and electrolytes. Salt tablets can worsen dehydration and impair performance; they increase potassium losses, irritate the stomach, and cause vomiting.

Water is the best fluid for most physically active people, but some consumers prefer the flavors of sports drinks.

Hyponatremia When athletes compete in endurance sports lasting longer than 3 hours, replenishing electrolytes is crucial. If athletes sweat profusely over a long period of time and do not replace lost sodium, a dangerous condition known as **hyponatremia** may result. Research shows that some athletes who sweat profusely may also lose more sodium in their sweat than others—and are prone to debilitating heat cramps.[55] These athletes lose twice as much sodium in sweat as athletes who don't cramp. Depending on individual variation, exercise intensity, and changes in ambient temperature and humidity, sweat rates for these athletes can exceed 2 liters per hour.[56]

Hyponatremia may also occur when endurance athletes drink such large amounts of water over the course of a long event that they overhydrate, diluting the body's fluids to such an extent that the sodium concentration becomes extremely low. During long competitions, when athletes lose sodium through heavy sweating *and* consume excessive amounts of liquids, especially water, hyponatremia becomes likely.

Some athletes may still be vulnerable to hyponatremia even when they consume sports drinks during an event. Sports drinks do contain sodium, but as a later section points out, the sodium content of sports drinks is low and, in some cases, too low to replace sweat losses. Still, sports drinks do offer more sodium than plain water.

To prevent hyponatremia, athletes need to replace sodium during prolonged events. Sports drinks, salty pretzels, and other sodium sources can provide sodium during long competitions. Some athletes may need beverages with higher sodium concentrations than commercial sports drinks. In the days before the event, especially an event in the heat, athletes should not restrict salt in their diets. The symptoms of hyponatremia ♦ are similar to, but not the same as, those of dehydration.

♦ Symptoms of hyponatremia:

- Severe headache
- Vomiting
- Bloating, puffiness from water retention (shoes tight, rings tight)
- Confusion
- Seizure

Sports Drinks

Hydration is critical to optimal performance. As stated earlier, water best meets the fluid needs of most people, yet manufacturers market many good-tasting sports drinks that deliver both fluid and carbohydrate for active people. The term *sports drink* generally refers to beverages that contain carbohydrates and electrolytes in specific concentrations, and they are the focus of this discussion.

Many sports drinks compete for their share of the multi-billion-dollar market. What do sports drinks have to offer?

Fluid Sports drinks offer fluids to help offset the loss of fluids during physical activity, but plain water can do this too. Alternatively, diluted fruit juices or flavored water can be used if preferred to plain water.

Glucose Sports drinks offer simple sugars or **glucose polymers** that help maintain hydration and blood glucose and enhance performance as effectively as, or in some circumstances, even better than, water. Such measures are especially beneficial for strenuous endurance activities lasting longer than 1 hour, during

hyponatremia (HIGH-poe-na-TREE-mee-ah): a decreased concentration of sodium in the blood.

- **hypo** = below
- **natrium** = sodium (Na)
- **emia** = blood

glucose polymers: compounds that supply glucose, not as single molecules, but linked in chains somewhat like starch. The objective is to attract less water from the body into the digestive tract (osmotic attraction depends on the number, not the size, of particles).

>How To

Calculate the Carbohydrate Concentration of Sports Drinks

Sports drinks with a carbohydrate concentration of 6 to 8 percent best support athletes participating in events lasting 1 hour or longer. To calculate the carbohydrate concentration of a sports drink, divide the grams of carbohydrates per serving by the serving size (in milliliters) and multiply by 100.

Consider a sports drink that contains 22 grams of carbohydrates per serving and lists the serving size as 360 milliliters (12 ounces):

$$\frac{\text{22 grams carbohydrate}}{\text{360 milliliters}} = 0.06$$

$$0.06 \times 100 = 6\% \text{ carbohydrates}$$

TRY IT Calculate the carbohydrate concentration of a sports drink that contains 14 grams of carbohydrate per serving (240 milliliters).

intense activities, or during prolonged competitive games that demand repeated intermittent activity.[57] Sports drinks are also suitable for events lasting less than 1 hour, although plain water is appropriate as well.

Fluid transport to the tissues from beverages containing up to 8 percent glucose is rapid. Most sports drinks contain about 7 percent carbohydrate (about half the sugar of ordinary soft drinks, or about 5 teaspoons in each 12 ounces). Less than 6 percent carbohydrate may not enhance performance, and more than 8 percent may cause abdominal cramps, nausea, and diarrhea. The accompanying "How To" feature describes how to calculate the carbohydrate concentration of sports drinks.

Although glucose can enhance endurance and performance in strenuous competitive events, for the moderate exerciser, it can be counterproductive if weight loss is the goal. Glucose is sugar, and like candy, it provides only empty kcalories—no vitamins or minerals. Most sports drinks provide between 50 and 100 kcalories per 8-ounce cup.

Sodium and Other Electrolytes Sports drinks offer sodium and other electrolytes to help replace those lost during physical activity. Sodium in sports drinks also helps to increase the rate of fluid absorption from the GI tract and maintain plasma volume during activity and recovery. Most physically active people do not need to replace the minerals lost in sweat immediately; a meal eaten within hours of competition replaces these minerals soon enough. Most sports drinks are relatively low in sodium, however, so those who choose to use these beverages run little risk of excessive intake.

Good Taste Manufacturers reason that if a drink tastes good, people will drink more, thereby ensuring adequate hydration. For athletes who prefer the flavors of sports drinks over water, it may be worth paying for good taste to replace lost fluids.

For athletes who exercise for 1 hour or more, sports drinks provide an advantage over water, and as the margin shows, there may be other circumstances in which sports drinks may be helpful. ♦ For most physically active people, though, water is the best fluid to replenish the body's lost fluids. The most important thing to do is drink—even if you don't feel thirsty.

♦ Sports drinks may be beneficial for athletes who:

- Exercise on an empty stomach
- Don't eat enough carbohydrate
- Want to load carbohydrates
- Want to gain weight
- Train at altitude or in extreme weather
- Had diarrhea (or vomiting) recently
- Refuse to drink adequate water

Enhanced Water Another beverage often marketed to athletes and active people is **enhanced water.** Enhanced waters are lightly flavored waters with lower carbohydrate and electrolyte contents than traditional sports drinks. Marketers promote these beverages for the added vitamins, minerals, and in some cases, protein, they contain. In fact, most enhanced waters contain small amounts of only a few minerals, some of the B vitamins, and sometimes vitamin C or vitamin E. In the context of daily needs, the vitamins and minerals in these drinks do not add up to much. For example, it takes a quart of most of these beverages to provide only 10 percent of the RDA for iron or calcium. Quite simply, enhanced waters are not a

enhanced water: water that is fortified with ingredients such as vitamins, minerals, protein, oxygen, or herbs. Enhanced water is marketed as *vitamin water, sports water, oxygenated water,* and *protein water.*

substitute for eating nutrient-rich fruits and vegetables. Enhanced waters may not be harmful (except maybe to the wallet), but most people do not need them. Plain water can meet fluid needs. If the flavor of enhanced waters encourages greater fluid intake, then they may offer some advantage. Serious endurance athletes need the carbohydrate-electrolyte sports drinks discussed earlier.

Poor Beverage Choices: Caffeine and Alcohol Athletes, like others, sometimes drink beverages that contain caffeine or alcohol. Each of these substances can influence physical performance.

Caffeine Caffeine is a stimulant, and athletes sometimes use it to enhance performance as Highlight 14 explains. Carbonated soft drinks, with or without caffeine, may not be a wise choice for athletes: bubbles make a person feel full quickly and so limit fluid intake. Some of the increasingly popular beverages, called energy drinks, contain amounts of caffeine equivalent to a cup or two of coffee. When used in excess or in combination with stimulants or other unregulated substances, energy drinks can hinder performance and are potentially dangerous.[58] Another reason energy drinks should not be used for fluid replacement during athletic events is that the carbohydrate concentrations are too high for optimal fluid absorption. The caffeine contents of selected energy drinks are listed in the table at the beginning of Appendix H.

Alcohol Some athletes mistakenly believe that they can replace fluids and load up on carbohydrates by drinking beer. ♦ A 12-ounce beer provides 13 grams of carbohydrate—one-third the amount of carbohydrate in a glass of orange juice the same size. In addition to carbohydrate, beer also contains alcohol, of course. Energy from alcohol breakdown generates heat, but it does not fuel muscle work because alcohol is metabolized in the liver.

Alcohol's diuretic effect impairs the body's fluid balance, making dehydration likely. After physical activity, a person needs to replace fluids, not lose them by drinking beer. Alcohol also impairs the body's ability to regulate its temperature, increasing the likelihood of hypothermia or heat stroke.

It is difficult to overstate alcohol's detrimental effects on physical activity. Alcohol alters perceptions; slows reaction time; reduces strength, power, and endurance; and hinders accuracy, balance, eye-hand coordination, and coordination in general—all opposing optimal athletic performance. In addition, it deprives people of their judgment, thereby compromising safety in sports. Many sports-related fatalities and injuries involve alcohol or other drugs.

Clearly, alcohol impairs performance. For those who do drink, do not drink alcohol before exercising and drink plenty of water after exercising before drinking alcohol.

♦ Beer facts:

- *Beer is not carbohydrate-rich.* Beer is *kcalorie-rich*, but only one-third of its kcalories are from carbohydrates. The other two-thirds are from alcohol.
- *Beer is mineral-poor.* Beer contains a few minerals, but to replace the minerals lost in sweat, athletes need good sources such as fruit juices or sports drinks.
- *Beer is vitamin-poor.* Beer contains traces of some B vitamins, but it cannot compete with food sources.
- *Beer causes fluid losses.* Beer is a fluid, but alcohol is a diuretic and causes the body to lose valuable fluid.

REVIEW IT Identify the factors that influence an athlete's fluid needs and describe the differences between water and sports drinks.

Active people need to drink plenty of water; endurance athletes need to drink both water and carbohydrate-containing beverages, especially during training and competition. During events lasting longer than 3 hours, athletes need to pay special attention to replace sodium losses to prevent hyponatremia.

14.5 Diets for Physically Active People

LEARN IT Discuss an appropriate daily eating pattern for athletes and list one example of a recommended pre- or post-game meal.

No one diet best supports physical performance. Active people who choose foods within the framework of the diet-planning principles presented in Chapter 2 can design many excellent diets.

Choosing a Diet to Support Fitness Above all, keep in mind that water is depleted more rapidly than any other nutrient. A diet to support fitness must provide water, energy, and all the other nutrients.

Water Even casual exercisers must attend conscientiously to their fluid needs. Physical activity blunts the thirst mechanism, especially in cold weather. During activity, thirst signals come too late, so don't wait to feel thirsty before drinking. To find out how much water is needed to replenish activity losses, weigh yourself before and after the activity—the difference is almost all water. One pound equals roughly 2 cups (500 milliliters) of fluid.

Nutrient Density A healthful diet is based on nutrient-dense foods—foods that supply adequate vitamins and minerals for the energy they provide. Active people need to eat both for nutrient adequacy and for energy—and energy needs can be extremely high. For example, during training, meals for some Olympian athletes provide as much as 12,000 kcalories a day. Still, a nutrient-rich diet remains central for adequacy's sake. Though vital, energy alone is not enough to support performance. Table 14-7 offers convenient, nutrient-dense snack ideas for athletes and active people.[59]

♦ Carbohydrate recommendation for athletes in heavy training: 6 to 10 g/kg body weight

Carbohydrate A diet that is high in carbohydrate (60 to 70 percent of total kcalories), moderate in fat (20 to 35 percent), and adequate in protein (10 to 20 percent) ensures full glycogen and other nutrient stores. On two occasions, however, the active person's regular high-carbohydrate, ♦ fiber-rich diet may require temporary adjustment. Both of these exceptions involve training for competition rather than for fitness in general. One special occasion is the pregame meal, when fiber-rich, bulky foods are best avoided. The pregame meal is discussed in a later section.

The other occasion is during intensive training, when energy needs may be so high as to outstrip the person's capacity to eat enough food to meet them. In this case, the athlete can add concentrated carbohydrate foods, such as dried fruits, sweet potatoes, and nectars, and even high-fat foods, such as avocados and nuts. Some athletes use commercial high-carbohydrate liquid supplements to obtain the carbohydrate and energy needed for intense training and top performance. These supplements do not *replace* regular food; they are meant to be used in *addition* to it. Unlike the sports drinks discussed earlier, these high-carbohydrate supplement beverages are too concentrated in carbohydrate to be used for fluid replacement.

Protein In addition to carbohydrate and some fat (and the energy they provide), physically active people need protein. Meats and milk products are rich protein sources, but recommending that active people emphasize these foods is narrow advice. As mentioned repeatedly, active people need diets rich in carbohydrate, and of course, meats have none to offer. Legumes, whole grains, and vegetables provide some protein with abundant carbohydrate.

Meals before and after Competition

No single food improves speed, strength, or skill in competitive events, although some *kinds* of foods do support performance better than others, as already explained. Still, a competitor

TABLE 14-7 Nutrient-Dense Snacks for Athletes and Active People

© silver-john/Shutterstock.com

One ounce of almonds provides protein, fiber, calcium, vitamin E, and healthy unsaturated fats. Similar choices include other nuts or trail mix consisting of dried fruit, nuts, and seeds.

© Robert Gubbins /Shutterstock.com

Low-fat Greek yogurt contains more protein per serving than regular yogurt, but a little less calcium. A small amount of fresh fruit adds fiber and vitamins. A similar choice is low-fat cheese paired with fresh fruit.

© Cheryl E. Davis/Shutterstock.com

Low-fat milk or chocolate milk together with fig bars or oatmeal-raisin cookies offer protein and fiber. A similar choice is whole-grain cereal with low-fat milk.

© Polara Studios, Inc.

Popcorn offers fiber and a fruit smoothie quenches thirst and provides valuable vitamins. A similar choice is pretzels and fruit juice.

may eat a particular food before or after an event for psychological reasons. One eats a steak the night before wrestling. Another eats a spoonful of honey within 5 minutes of diving. So long as these practices remain harmless, they should be respected.

A variety of foods is the best source of nutrients for athletes.

Pregame Meals Science indicates that the pregame meal or snack should include plenty of fluids and be light and easy to digest. It should provide between 300 and 800 kcalories, primarily from carbohydrate-rich foods that are familiar and well tolerated by the athlete. The meal should end 3 to 4 hours before competition to allow time for the stomach to empty before exertion.

Breads, potatoes, pasta, and fruit juices—that is, carbohydrate-rich foods low in fat and fiber—form the basis of the best pregame meal (see Figure 14-2 for some examples). Bulky, fiber-rich foods such as raw vegetables or high-bran cereals, although usually desirable, are best avoided just before competition. Fiber in the digestive tract attracts water and can cause stomach discomfort during performance. Liquid meals ♦ are easy to digest, and many such meals are commercially available. Alternatively, athletes can mix fat-free milk or juice, frozen fruits, and flavorings in a blender.

♦ High-carbohydrate, liquid pregame meal ideas:
- Apple juice, frozen banana, and cinnamon
- Papaya juice, frozen strawberries, and mint
- Fat-free milk, frozen banana, and vanilla

Postgame Meals As mentioned earlier, eating high-carbohydrate foods *after* physical activity enhances glycogen storage. Because people are usually not hungry immediately following physical activity, carbohydrate-containing beverages such as sports drinks or fruit juices may be preferred. If an active person does feel hungry after an event, then foods high in carbohydrate, moderate in protein, and low in fat and fiber are the ones to choose—similar to those recommended prior to competition.

REVIEW IT Discuss an appropriate daily eating pattern for athletes and list one example of a recommended pre- or post-game meal.

The person who wants to excel physically will apply accurate nutrition knowledge along with dedication to rigorous training. A diet that provides ample fluid and includes a variety of nutrient-dense foods in quantities to meet energy needs will enhance not only athletic performance, but overall health as well. Carbohydrate-rich foods that are light and easy to digest are recommended for both the pregame and the postgame meal.

> FIGURE 14-2 Examples of High-Carbohydrate Pregame Meals

Pregame meals should be eaten 3 to 4 hours before the event and provide 300 to 800 kcalories, primarily from carbohydrate-rich foods. Each of these sample meals provides at least 65 percent of total kcalories from carbohydrate.

300-kcalorie meal
1 large apple
4 saltine crackers
1½ tbs reduced-fat peanut butter

500-kcalorie meal
1 large whole-wheat bagel
2 tbs jelly
1½ c low-fat milk

750-kcalorie meal
1 large baked potato
2 tsp margarine
1 c steamed broccoli
1 c mixed carrots and green peas
5 vanilla wafers
1½ c apple or pineapple juice

Training and genetics being equal, who will win a competition. The athlete who habitually consumes inadequate amounts of needed nutrients will lag behind the competitor who arrives at the event with a long history of full nutrient stores and well-met metabolic needs. Some athletes learn that nutrition can support physical performance and turn to pills and powders instead of foods. In case you need further convincing that a healthful diet surpasses such potions, the following highlight addresses this issue.

Nutrition Portfolio

The foods and beverages you eat and drink provide fuel and other nutrients to support your physical activity. Go to Diet Analysis Plus and choose one of the days on which you tracked your diet and activity for an entire day. Select the Activities Spreadsheet report to help you answer the following.

- Describe your daily physical activities and how they compare with recommendations to be physically active for at least 30 minutes, and preferably 60 minutes, a day on most or all days of the week.

Now click on the Intake vs. Goals report and consider the following:

- Estimate your daily fluid intake, making note of whether you drink fluids, especially water, before, during, and after physical activity.
- Evaluate the carbohydrate contents of your diet and consider whether it would meet the needs of a physically active person.

Diet Analysis PLUS To complete this exercise, go to your Diet Analysis Plus at **www.cengagebrain.com.**

STUDY IT To review the key points of this chapter and take a practice quiz, go to the study cards at the end of the book.

REFERENCES

1. A. Grøntved and F. B. Hu, Television viewing and risk of type 2 diabetes, cardiovascular disease, and all-cause mortality, *Journal of the American Medical Association* 305 (2011) 2448–2455; T. Y. Warren and coauthors, Sedentary behaviors increase risk of cardiovascular disease mortality in men, *Medicine and Science in Sports and Exercise* 42 (2010): 879–885; A. V. Patel and coauthors, Leisure time spent sitting in relation to total mortality in a prospective cohort of adults, *American Journal of Epidemiology* 172 (2010): 419–429; P. T. Katzmarzyk and coauthors, Sitting time and mortality from all causes, cardiovascular disease, and cancer, *Medicine and Science in Sports and Exercise* 41 (2009): 998–1005.
2. Centers for Disease Control and Prevention, www.cdc.gov/physicalactivity/everyone; updated February 16, 2011, accessed June 30, 2011.
3. Centers for Disease Control and Prevention, 2011; M. Hamer and coauthors, Physical activity and cardiovascular mortality risk: Possible protective mechanisms, *Medicine and Science in Sports and Exercise* 44 (2012): 84–88; X. Yang and coauthors, The longitudinal effects of physical activity history on metabolic syndrome, *Medicine and Science in Sports and Exercise* 40 (2008): 1424–1431; L. B. Yates and coauthors, Exceptional longevity in men, *Archives of Internal Medicine* 168 (2008): 284–290.
4. Centers for Disease Control and Prevention, US physical activity statistics, National average: Summary of physical activity, 2007, www.cdc.gov/physicalactivity.
5. Warren and coauthors, 2010; Patel and coauthors, 2010; Katzmarzyk and coauthors, 2009; W. Kemmler and coauthors, Exercise decreases the risk of metabolic syndrome in elderly females, *Medicine and Science in Sports and Exercise* 41 (2009): 297–305; K. Y Wolin and coauthors, Physical activity and colon cancer prevention: A meta-analysis, *British Journal of Cancer* 100 (2009): 611–616; P. Anand, and coauthors, Cancer is a preventable disease that requires major lifestyle changes, *Pharmaceutical Research* 25 (2008): 2097–2116; N. Orsini and coauthors, Combined effects of obesity and physical activity in predicting mortality among men, *Journal of Internal Medicine* 264 (2008): 442–451; A. R. Weinstein and coauthors, The joint effects of physical activity and body mass index on coronary heart disease risk in women, *Archives of Internal Medicine* 168 (2008): 884–890; Yang and coauthors, 2008.
6. Centers for Disease Control and Prevention, 2011.
7. American College of Sports Medicine, Position paper: Appropriate physical activity intervention strategies for weight loss and prevention of weight regain for adults, *Medicine and Science in Sports and Exercise* 41 (2009): 459–471; K. S. Vimaleswaran and coauthors, Physical activity attenuates the body mass index-increasing influence of genetic variation in the *FTO* gene, *American Journal of Clinical Nutrition* 90 (2009): 425–428.
8. R. S. Rector and coauthors, Lean body mass and weight-bearing activity in the prediction of bone mineral density in physically active men, *Journal of Strength and Conditioning Research* 23 (2009): 427–435; A. Guadalupe-Grau and coauthors, Exercise and bone mass in adults, *Sports Medicine* 39 (2009): 439–468.
9. J. Romeo and coauthors, Physical activity, immunity and infection, *Proceedings of the Nutrition Society* 69 (2010): 390–399.
10. X. Sui and coauthors, Influence of cardiorespiratory fitness on lung cancer mortality, *Medicine and Science in Sports and Exercise* 42 (2010): 872–878; J. B. Peel and coauthors, A prospective study of cardiorespiratory fitness and breast cancer mortality, *Medicine and Science in Sports and Exercise* 41 (2009): 742–748; S. Y. Pan and M. DesMeules, Energy intake, physical activity energy balance, and cancer: Epidemiologic evidence,

Methods in Molecular Biology 472 (2009): 191–215; World Cancer Research Fund/American Institute for Cancer Research, *Food, Nutrition, Physical Activity, and the Prevention of Cancer: A Global Perspective* (Washington, D.C.: AICR, 2007), pp. 244–321.

11. N. T. Artinian and coauthors, Interventions to promote physical activity and dietary lifestyle changes for cardiovascular risk factor reduction in adults: A scientific statement from the American Heart Association, *Circulation* 122 (2010): 406–441; Hamer and coauthors, 2012; N. L. Chase and coauthors, The association of cardiorespiratory fitness and physical activity with incidence of hypertension in men, *American Journal of Hypertension* 22 (2009): 417–424; P. T. Williams, Reduced diabetic, hypertensive, and cholesterol medication use with walking, *Medicine and Science in Sports and Exercise* 40 (2008): 433–443.
12. T. S. Church and coauthors, Changes in weight, waist circumference and compensatory responses with different doses of exercise among sedentary, overweight postmenopausal women, *PLoS ONE* 4 (2009): e4515; M. Fogelhom, How physical activity can work? *International Journal of Pediatric Obesity* 3 (2008): 10–14; B. A. Irving and coauthors, Effect of exercise training intensity on abdominal visceral fat and body composition, *Medicine and Science in Sports and Exercise* 40 (2008): 1863–1872.
13. J. Ralph and coauthors, Low-intensity exercise reduces the prevalence of hyperglycemia in type 2 diabetes, *Medicine and Science in Sports and Exercise* 42 (2010): 219–225.
14. P. J. Banim and coauthors, Physical activity reduces the risk of symptomatic gallstones: A prospective cohort study, *European Journal of Gastroenterology and Hepatology* 22 (2010): 983–988; P. T. Williams, Independent effects of cardiorespiratory fitness, vigorous physical activity, and body mass index on clinical gallbladder disease risk, *American Journal of Gastroenterology* 103 (2008): 2239–2247.
15. J. C. Sieverdes and coauthors, Association between leisure-time physical activity and depressive symptoms in men, *Medicine and Science in Sports and Exercise* 44 (2012), 260–265; D. B. Nelson and coauthors, Effect of physical activity on menopausal symptoms among urban women, *Medicine and Science in Sports and Exercise* 40 (2008): 50–58.
16. P. Kokkinos and coauthors, Exercise capacity and mortality in black and white men, *Circulation* 117 (2008): 614–622; Yates and coauthors, 2008.
17. C. P. Wen and coauthors, Minimum amount of physical activity for reduced mortality and extended life expectancy: A prospective cohort study, *The Lancet* 378 (2011): 1244–1253.
18. M. E. Nelson and coauthors, Physical activity and public health in older adults: Recommendation from the American College of Sports Medicine and the American Heart Association, *Medicine and Science in Sports and Exercise* 39 (2007): 1435–1445.
19. Centers for Disease Control and Prevention, 2011.
20. American College of Sports Medicine position stand, Quantity and quality of exercise for developing and maintaining cardiorespiratory, musculoskeletal, and neuromotor fitness in apparently healthy adults: Guidance for prescribing exercise, *Medicine and Science in Sports and Exercise* 43 (2011): 1334–1359; W. L. Haskell and coauthors, Physical activity and public health: Updated recommendation for adults from the American College of Sports Medicine and the American Heart Association, *Medicine and Science in Sports and Exercise* 39 (2007): 1423–1434.
21. American College of Sports Medicine position stand, 2011; Haskell and coauthors, 2007.
22. Centers for Disease Control and Prevention, 2011; Haskell and coauthors, 2007.
23. American College of Sports Medicine, *ACSM's Guidelines for Exercise Testing and Prescription*, 8th ed. (Philadelphia, Pa.: Lippincott, Williams, & Wilkins, 2010), pp. 18–39.
24. S. Marwood and coauthors, Faster pulmonary oxygen uptake kinetics in trained versus untrained male adolescents, *Medicine and Science in Sports and Exercise* 42 (2010): 127–134.
25. E. G. Ciolac and coauthors, Effects of high-intensity aerobic interval training vs. moderate exercise on hemodynamic, metabolic and neurohumoral abnormalities of young normotensive women at high familial risk for hypertension, *Hypertension Research* 33 (2010): 836–843; American College of Sports Medicine position stand: Exercise and hypertension, *Medicine and Science in Sports and Exercise* 36 (2004): 533–553.
26. American College of Sports Medicine position stand, 2011; American College of Sports Medicine position stand: Progression models in resistance training for healthy adults, *Medicine and Science in Sports and Exercise* 41 (2009): 687–708; Haskell and coauthors, 2007; Nelson and coauthors, 2007.
27. American College of Sports Medicine position stand, 2011.
28. Guadalupe-Grau and coauthors, 2009; R. Rizzoli and coauthors, Management of osteoporosis in the elderly, *Current Medical Research and Opinion* 25 (2009): 2373–2387.
29. H. Littbrand and coauthors, The effect of a high-intensity functional exercise program on activities of daily living: A randomized controlled trial in residential care facilities, *Journal of the American Geriatric Society* 57 (2009): 1741–1749; M. J. Chin A Paw and coauthors, The functional effects of physical exercise training in frail older people: A systematic review, *Sports Medicine* 38 (2008): 781–793; S. Karikanta and coauthors, A multi-component exercise regimen to prevent functional decline and bone fragility in home-dwelling elderly women: Randomized, controlled trial, *Osteoporosis International* 18 (2007): 453–462; Nelson and coauthors, 2007.
30. American College of Sports Medicine position stand, 2009.
31. R. Beneke and D. Böning, The limits of human performance, *Essays in Biochemistry* 44 (2008): 11–25.
32. J. Bergstrom and coauthors, Diet, muscle glycogen and physical performance, *Acta Physiologica Scandanavica* 71 (1967): 140–150.
33. A. M. Bellinger and coauthors, Remodeling of ryanodine receptor complex causes "leaky" channels: a molecular mechanism for decreased exercise capacity, *Proceedings of the National Academy of Sciences* 105 (2008): 2198–2202.
34. D. A. Sedlock, The latest on carbohydrate loading: A practical approach, *Current Sports Medicine Reports* 7 (2008): 209–213.
35. J. Temesi and coauthors, Carbohydrate ingestion during endurance exercise improves performance in adults, *Journal of Nutrition* 141 (2011): 890–897; L. M. Burke, Fueling strategies to optimize performance: Training high or training low? *Scandinavian Journal of Medicine and Science in Sports* 20 (2010): 48–58; A. Ali and C. Williams, Carbohydrate ingestion and soccer skill performance during prolonged intermittent exercise, *Journal of Sport Science* 27 (2009): 1499–1508; A. Foskett and coauthors, Carbohydrate availability and muscle energy metabolism during intermittent running, *Medicine and Science in Sports and Exercise* 40 (2008): 96–103.
36. American College of Sports Medicine position stand: Exercise and fluid replacement, *Medicine and Science in Sports and Exercise* 39 (2007): 377–390.
37. American College of Sports Medicine position stand: Exercise and fluid replacement, 2007.
38. L. Breen, K. D. Tipton, and A. E. Jeukendrup, No effect of carbohydrate-protein on cycling performance and indices of recovery, *Medicine and Science in Sports and Exercise* 42 (2010): 1140–1148; N. M. Cermak and coauthors, Muscle metabolism during exercise with carbohydrate or protein-carbohydrate ingestion, *Medicine and Science in Sports and Exercise* 41 (2009): 2158–2164.
39. Breen, Tipton, and Jeukendrup, 2010; Cermak and coauthors, 2009; M. Millard-Stafford and coauthors, Recovery nutrition: Timing and composition after endurance exercise, *Current Sports Medicine Reports* 7 (2008): 193–201.
40. Position of the American Dietetic Association, Dietitians of Canada, and the American College of Sports Medicine: Nutrition and athletic performance, *Journal of the American Dietetic Association* 109 (2009): 509–527; G. A. Wallis and coauthors, Postexercise muscle glycogen synthesis with combined glucose and fructose ingestion, *Medicine and Science in Sports and Exercise* 40 (2008): 1789–1794.
41. C. M. Perry, T. L. Perry, and M.C. Rose, Glycemic index and endurance performance, *International Journal of Sport Nutrition and Exercise Metabolism* 20 (2010): 154–165; E. J. Stevenson and coauthors, Dietary glycemic index influences lipid oxidation but not muscles or liver glycogen oxidation during exercise, *American Journal of Physiology, Endocrinology and Metabolism* 296 (2009): E1140–E1147; M. I. Trenell and coauthors, Effect of high and low glycaemic index recovery diets on intramuscular lipid oxidation during aerobic exercise, *British Journal of Nutrition* 99 (2008): 326–332.
42. Position of the American Dietetic Association, Dietitians of Canada, and the American College of Sports Medicine, 2009.
43. A. R. Josse and coauthors, Body composition and strength changes in women with milk and resistance exercise, *Medicine and Science in*

Sports and Exercise 42 (2010): 1122–1130; T. B. Symons and coauthors, a moderate serving of high-quality protein maximally stimulates skeletal muscle protein synthesis in young and elderly subjects, *Journal of the American Dietetic Association* 109 (2009): 1582–1586; K. D. Tipton and A. A. Ferrando, Improving muscle mass: response of muscle metabolism to exercise, nutrition, and anabolic agents, *Essays in Biochemistry* 44 (2008): 85–98.

44. Position of the American Dietetic Association, Dietitians of Canada, and the American College of Sports Medicine, 2009; Committee on Dietary Reference Intakes, *Dietary Reference Intakes for Energy, Carbohydrate, Fiber, Fat, Fatty Acids, Cholesterol, Protein, and Amino Acids* (Washington, D.C.: National Academies Press, 2002), pp. 660–661.
45. Position of the American Dietetic Association, Dietitians of Canada, and the American College of Sports Medicine, 2009.
46. Position of the American Dietetic Association, Dietitians of Canada, and the American College of Sports Medicine, 2009; American College of Sports Medicine position stand: The female athlete triad, *Medicine and Science in Sports and Exercise* 39 (2007): 1867–1882.
47. S. K. Powers, E. E. Talbert, and P. J. Adhihetty, Reactive oxygen and nitrogen species as intracellular signals in skeletal muscle, *Journal of Physiology* 589 (2011): 2129–2138; M. C. Gomez-Cabrera, J. Viña, and L. L. Ji, Interplay of oxidants and antioxidants during exercise: Implications for muscle health, *Physician and Sportsmedicine* 37 (2009): 116–123; M. J. Jackson, Free radicals generated by contracting muscle; By-products of metabolism or key regulators of muscle function? *Free Radical Biology & Medicine* 44 (2008): 132–141; S. Sachdev and K. J. Davies, Production, detection, and adaptive responses to free radicals in exercise, *Free Radical Biology & Medicine* 44 (2008): 215–223.
48. V. H. Teixeira and coauthors, Antioxidants do not prevent postexercise peroxidation and may delay muscle recovery, *Medicine and Science in Sports and Exercise* 41 (2009): 1752–1760.
49. Teixeira and coauthors, 2009; Gomez-Cabrera, Viña, and Ji, 2009; M. C. Gomez-Cabrera, E. Domenech, and J. Viña, Moderate exercise is an antioxidant: Upregulation of antioxidant genes by training, *Free Radical Biology and Medicine* 44 (2008): 126–131; Jackson 2008; Sachdev and Davies, 2008; J. Padilla and T. D. Mickleborough, Does antioxidant supplementation prevent favorable adaptations to exercise training? *Medicine and Science in Sports and Exercise* 39 (2007): 1887.
50. K. Woolf and coauthors, Iron status in highly active and sedentary young women, *International Journal of Sport Nutrition and Exercise Metabolism* 19 (2009): 519–535; J. P. McClung and coauthors, Randomized, double-blind, placebo-controlled trial of iron supplementation in female soldiers during military training: Effects on iron status, physical performace, and mood, *American Journal of Clinical Nutrition* 90 (2009): 124–131.
51. Position of the American Dietetic Association, Dietitians of Canada, and the American College of Sports Medicine, 2009.
52. McClung and coauthors, 2009.
53. American College of Sports Medicine position stand: Exercise and fluid replacement, 2007; Committee on Dietary Reference Intakes, *Dietary Reference Intakes for Water, Potassium, Sodium, Chloride, and Sulfate* (Washington, D.C.: National Academies Press, 2005), pp. 108–110.
54. American College of Sports Medicine position stand: Prevention of cold injuries during exercise, *Medicine and Science in Sports and Exercise* 38 (2006): 2012–2029.
55. American College of Sports Medicine position stand: Exercise and fluid replacement, 2007.
56. Committee on Dietary Reference Intakes, 2005, pp. 108–110.
57. American College of Sports Medicine position stand: Exercise and fluid replacement, 2007.
58. Position of the American Dietetic Association, Dietitians of Canada, and the American College of Sports Medicine, 2009.
59. C. Rosenbloom, Snack attack: Evaluating and rating snacks for athletes, *Nutrition Today* 46 (2011): 106–115.

 LEARN IT Present arguments for and against the use of ergogenic aids.

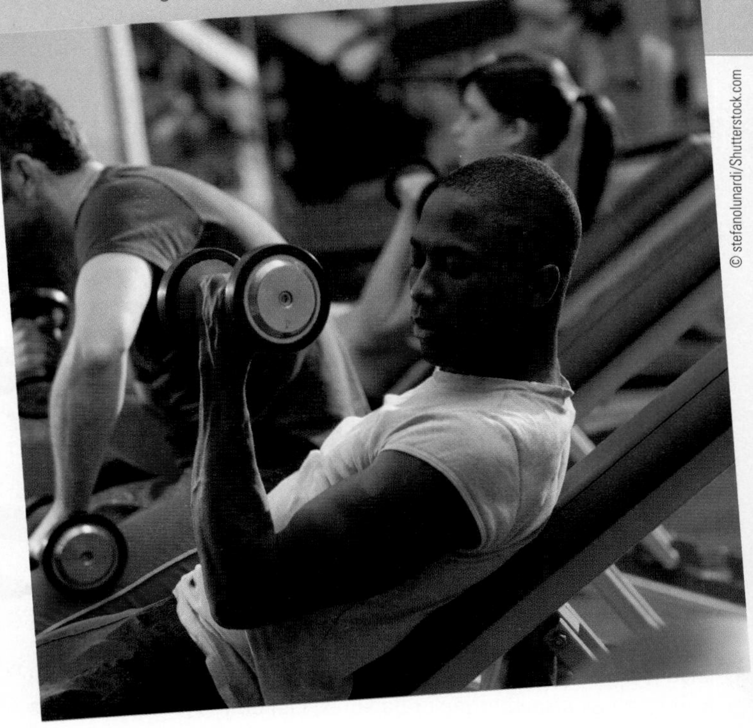
© stefanolunardi/Shutterstock.com

Supplements as Ergogenic Aids

Athletes gravitate to promises that they can enhance their performance by taking pills, powders, or potions. Unfortunately, they often hear such promises from their coaches and peers, who advise them to use dietary supplements, take drugs, or follow procedures that claim to deliver results with little effort.[1] When such performance-enhancing aids are harmless, they are only a waste of money; when they impair performance or harm health, they waste athletic potential and cost lives. This highlight looks at some promises of supplements to enhance physical performance.

Many substances or treatments claim to be *ergogenic,* meaning work enhancing. In connection with athletic performance, **ergogenic aids** are substances or treatments that purportedly improve athletic performance above and beyond what is possible through training. For practical purposes, most ergogenic aids can be categorized as follows:[2]

- Those that perform as claimed
- Those that may perform as claimed but for which there is insufficient evidence at this time
- Those that do not perform as claimed
- Those that are dangerous, banned, or illegal, and therefore should not be used

The accompanying glossary defines several of the commonly used ergogenic aids discussed in this highlight.

For the vast majority of ergogenic aids, research findings do not support the claims made.[3] Athletes who hear that a product is ergogenic should ask who is making the claim and who will profit from the sale. Chapter 6 includes a discussion on protein powders and amino acid supplements (pp. 186–187).

Athletes who supplement their diets with products promoted to improve athletic performance should be aware that some supplements are contaminated with illegal substances such as steroids or stimulants that are not listed on the label.[4] Supplements contaminated with illegal substances pose health risks to those who use them as well as the risk of positive drug testing for athletes subject to such tests.

Dietary Supplements that Perform as Claimed

Among the extensive array of dietary supplements and other ergogenic aids that athletes use, a few seem to live up to the claims made for them, based on research so far. Convenient dietary

GLOSSARY

anabolic steroids: drugs related to the male sex hormone, testosterone, that stimulate the development of lean body mass.

- **anabolic** = promoting growth
- **sterols** = compounds chemically related to cholesterol

androstenedione: *See DHEA.*

beta-hydroxymethylbutyrate (HMB): a metabolite of the amino acid leucine promoted to increase muscle mass and strength.

caffeine: a natural stimulant found in many common foods and beverages, including coffee, tea, and chocolate; may enhance endurance by stimulating fatty acid release. High doses cause headaches, trembling, rapid heart rate, and other undesirable side effects.

carnitine: a nonessential nonprotein amino acid made in the body from lysine that helps transport fatty acids across the mitochondrial membrane. As a supplement, carnitine supposedly "burns" fat and spares glycogen during endurance events, but in reality it does neither.

chromium picolinate (CROW-mee-um pick-oh-LYN-ate): a trace mineral supplement; falsely promoted as building muscle, enhancing energy, and burning fat. *Picolinate* is a derivative of the amino acid tryptophan that seems to enhance chromium absorption.

convenient dietary supplements: liquid meal replacers, energy drinks, energy bars, and energy gels that athletes and active people use to replenish energy and nutrients when time is limited.

creatine (KREE-ah-tin): a nitrogen-containing compound that combines with phosphate to form the high-energy compound creatine phosphate (or phosphocreatine) in muscles.

DHEA (dehydroepiandrosterone) and **androstenedione:** hormones made in the adrenal glands that serve as precursors to the male hormone, testosterone; falsely promoted as burning fat, building muscle, and slowing aging.

ergogenic (ER-go-JEN-ick) **aids:** substances or techniques used in an attempt to enhance physical performance.

- **ergo** = work
- **genic** = gives rise to

hGH (human growth hormone): a hormone produced by the brain's pituitary gland that regulates normal growth and development; also called *somatotropin.*

ribose: a naturally occurring 5-carbon sugar needed for the synthesis of ATP.

sodium bicarbonate (baking soda): a white crystalline powder that is used to buffer acid that accumulates in the muscles and blood during high-intensity exercise.

supplements, caffeine, creatine, and sodium bicarbonate are the examples discussed here.

Convenient Dietary Supplements

Ready to drink supplements such as liquid meal replacers and energy drinks, energy bars, and energy gels are **convenient dietary supplements** for athletes and active people, especially when time is limited. Many such products appeal to athletes by claiming to provide "complete" nutrition. These supplements usually taste good and provide extra carbohydrate and food energy, but they fall short of providing "complete" nutrition. They can be useful as a pregame meal or a between-meal snack, but they should not replace regular meals.

Liquid meal replacers may help a nervous athlete who cannot tolerate solid food on the day of an event. A liquid meal 2 to 3 hours before competition can supply some of the fluid and carbohydrate needed in a pregame meal, but a shake of fat-free milk or juice (such as apple or papaya) and frozen fat-free yogurt or frozen fruit (such as strawberries or bananas) can do the same thing less expensively.

Caffeine

Some research supports the use of **caffeine** to enhance endurance and, to some extent, to enhance short-term, high-intensity exercise performance.[5] Caffeine may stimulate fatty acid release during endurance activity, but in contrast to what was previously thought, caffeine does not slow muscle glycogen use.[6] Light activity before a workout also stimulates fat release, and unlike caffeine, also warms the muscles and connective tissues, making them flexible and resistant to injury.

Caffeine is a stimulant that elicits a number of physiological and psychological effects in the body. Caffeine enhances alertness and reduces fatigue.[7] The possible benefits of caffeine use must be weighed against its adverse effects—stomach upset, nervousness, irritability, headaches, and diarrhea. Caffeine-containing beverages should be used in moderation, if at all, and *in addition* to other fluids, not as a substitute for them.

Caffeine is a *restricted* substance by the National Collegiate Athletic Association, which allows urine concentrations of 15 milligrams per liter or less (equivalent to about 5 cups of coffee consumed within a few hours before testing). Urine tests that detect more caffeine than this disqualify athletes from competition. (The table at the start of Appendix H provides a list of common caffeine-containing items and the doses they deliver.)

Creatine

Interest in—and use of—**creatine** supplements to enhance performance during intense activity has grown dramatically in the last few years. Power athletes such as weightlifters use creatine supplements to enhance stores of the high-energy compound creatine phosphate (CP) in muscles. Theoretically, the more creatine phosphate in muscles, the higher the intensity at which an athlete can train. High-intensity training stimulates the muscles to adapt, which in turn, improves performance.

Research suggests that creatine supplementation does enhance performance of short-term, repetitive, high-intensity activity such as weightlifting or sprinting.[8] Creatine may improve performance by increasing muscle strength and size, cell hydration, or glycogen loading capacity. In contrast, creatine supplementation has not been shown to benefit endurance activity.

The question of whether short-term use (up to a year) of creatine supplements (up to 5 grams per day) is safe continues to be studied, but so far, the supplements are considered safe for healthy adults.[9] More research is needed, however, to confirm the safety of larger doses and long-term use. Creatine supplementation may pose risks to athletes with kidney disease or other conditions.[10] One side effect of creatine supplementation that no one disputes is weight gain. For some athletes, weight gain, especially muscle gain, is beneficial, but for others, it is not.

Some medical and fitness experts voice concern that, like many performance enhancement supplements before it, creatine is being taken in huge doses (up to 30 *grams* per day) before evidence of its value has been ascertained. Even people who eat red meat, which is a creatine-rich food, do not consume nearly the amount supplements provide. (Creatine content varies, but on average, pork, chicken, and beef provide 65 to 180 *milligrams* per ounce.) Despite the uncertainties, creatine supplements are not illegal in international competition. The American Academy of Pediatrics strongly discourages the use of creatine supplements, as well as the use of any performance-enhancing substance in adolescents younger than 18 years old.[11]

Sodium Bicarbonate

During short-term, high-intensity activity, acid and carbon dioxide (CO_2) accumulate in the blood and muscles. **Sodium bicarbonate** (0.3 grams per kilogram of body weight) ingested prior to high-intensity sports performance buffers the acid and neutralizes the carbon dioxide, thereby maintaining muscle pH levels closer to normal and enhancing exercise capacity.[12] Sodium bicarbonate supplementation may cause unpleasant side effects such as diarrhea in some athletes.

Dietary Supplements that May Perform as Claimed

As noted earlier, dozens of supplements are promoted to enhance performance or to improve training adaptations of athletes and active people. For some of these supplements, it is just too early to tell whether they deliver on the promises made for them because research thus far is inconclusive. Examples of supplements that may perform as claimed, but for which there is insufficient evidence of efficacy, include beta-hydroxymethylbutyrate and ribose.

Beta-hydroxymethylbutyrate

Beta-hydroxymethylbutyrate (HMB) is a metabolite of the amino acid leucine. Supplementing the diet with HMB during training has been shown to increase muscle mass and strength, especially among untrained individuals.[13] Additional research is needed, however, to determine whether HMB supplementation in trained athletes enhances training adaptations.

Ribose

Manufacturers of ribose supplements claim that **ribose** increases ATP and improves performance. Ribose, a naturally occurring sugar, does help resynthesize muscle ATP. Based on research thus far, however, ribose supplements do not improve athletic performance.[14]

Dietary Supplements that Do Not Perform as Claimed

Most of the dietary supplements promoted as ergogenic aids fall into the category of "those that do not perform as claimed." Carnitine and chromium picolinate are two examples of ineffective supplements discussed here, but others include bee pollen, boron, coenzyme Q, ginseng, pyruvate, and vanadium.[15]

Carnitine

In the body, **carnitine** facilitates the transfer of fatty acids across the mitochondrial membrane. Supplement manufacturers suggest that with more carnitine available, fat oxidation will be enhanced, but this does not seem to be the case. Carnitine supplements neither raise muscle carnitine concentrations nor enhance exercise performance.[16] Milk and meat products are good sources of carnitine, and supplements are not needed.

Chromium Picolinate

Chapter 13 introduces chromium as an essential trace mineral involved in carbohydrate and lipid metabolism. Claims that **chromium picolinate,** which is more easily absorbed than chromium alone, builds muscle, enhances energy, and burns fat derive from one or two early studies reporting that men who strength-trained while taking chromium picolinate supplements increased lean body mass and reduced body fat. Most subsequent studies, however, show no effects of chromium picolinate on strength, lean body mass, or body fat.[17] Chromium-rich foods include whole grains, liver, and nuts.

Dangerous, Banned, or Illegal Supplements

The dietary supplements discussed thus far may or may not help athletic performance, but in the doses commonly taken, they seem to cause little harm. The remaining discussion features hormonal supplements that are clearly damaging: anabolic steroids, DHEA (dehydroepiandrosterone), androstenedione, and hGH (human growth hormone). All of these ergogenic aids are dangerous to use and are banned by most professional sports leagues and the World Anti-Doping Agency (WADA) established by the International Olympic Committee. The American Academy of Pediatrics and the American College of Sports Medicine also condemn athletes' use of these substances.

Anabolic Steroids

Among the most dangerous and illegal ergogenic practices is the taking of **anabolic steroids.** These drugs are derived from the male sex hormone testosterone, which promotes the development of male characteristics and lean body mass. Athletes who take steroids do so to stimulate muscle bulking.

The known toxic side effects of steroids include, but are not limited to, extreme aggression and hostility, heart disease, and liver damage. Taking these drugs is a form of cheating. Other athletes are put in the difficult position of either conceding an unfair advantage to competitors who use steroids or taking steroids and accepting the risk of harmful side effects. Athletes, especially young athletes, should not be forced to make such a choice.

The price for the potential competitive edge that steroids confer is high—sometimes it is life itself. Steroids are not simple pills that build bigger muscles. They are complex chemicals to which the body reacts in many ways, particularly when bodybuilders and other athletes take large amounts.[18] The safest, most effective way to build muscle has always been through hard training and a sound diet, and—despite popular misconceptions—it still is.

Some manufacturers peddle specific herbs as legal substitutes for steroid drugs. They falsely claim that these herbs contain hormones, enhance the body's hormonal activity, or both. In some cases, an herb may contain plant sterols, such as gamma-oryzanol, but these compounds are poorly absorbed. Even if absorption occurs, the body cannot convert herbal compounds to anabolic steroids. None of these products has any proven anabolic steroid activity, none enhances muscle strength, and some contain natural toxins. In short, "natural" does not mean "harmless."

DHEA and Androstenedione

Some athletes use **DHEA** and **androstenedione** as alternatives to anabolic steroids. DHEA (dehydroepiandrosterone) and androstenedione are hormones made in the adrenal glands that serve as precursors to the male hormone testosterone. Advertisements claim the hormones "burn fat," "build muscle," and "slow aging," but evidence to support such claims is lacking.

Short-term side effects of DHEA and androstenedione may include oily skin, acne, body hair growth, liver enlargement, testicular shrinkage, and aggressive behavior. Long-term effects such as serious liver damage may take years to become evident. The potential for harm from DHEA and androstenedione supplements is great, and athletes, as well as others, should avoid them.

Human Growth Hormone

A wide range of athletes, including weightlifters, baseball players, cyclists, and track and field participants use **hGH** (**human growth hormone**) to build lean tissue and improve athletic performance. The athletes use hGH, believing the injectable hormone will provide the benefits of anabolic steroids without the dangerous side effects.

Taken in large quantities, hGH causes the disease acromegaly, in which the body becomes huge and the organs and bones overenlarge. Other effects include diabetes, thyroid disorder, heart disease, menstrual irregularities, diminished sexual desire, and shortened life span.

Sometimes it is difficult to distinguish valid claims from bogus ones. Fitness magazines and Internet websites are particularly troublesome because many of them present both valid and invalid nutrition information along with slick advertisements for nutrition products. Advertisements often feature colorful anatomical figures, graphs, and tables that appear scientific. Some ads even include references, citing or linking to such credible sources as the *American Journal of Clinical Nutrition* and the *Journal of the American Medical Association.* These ads create the illusion of endorsement and credibility to gain readers' trust. Keep in mind, however, that the ads are created not to teach, but to sell. A careful reading of the cited research might reveal that the ads have presented the research findings out of context. For example, an ad might use a research article to conclude that its human growth hormone supplement "increases lean body mass and bone mineral," when in fact, the researchers would conclude that "its general use now or in the immediate future is not justified." Scientific facts are often exaggerated and twisted to promote sales. Highlight 1 describes ways to recognize misinformation and quackery.

The search for a single food, nutrient, drug, or technique that will safely and effectively enhance athletic performance will no doubt continue as long as people strive to achieve excellence in sports. When athletic performance does improve after use of an ergogenic aid, the improvement can often be attributed to the placebo effect, which is strongly at work in athletes. Even if a reliable source reports a performance boost from a newly tried product, give the effect time to fade away. Chances are excellent that it simply reflects the power of the mind over the body.

The overwhelming majority of performance-enhancing aids sold for athletes are frauds. Wishful thinking will not substitute for talent, hard training, adequate diet, and mental preparedness in competition. But don't discount the power of mind over body for a minute—it is formidable, and sports psychologists dedicate their work to harnessing it. You can use it by imagining yourself a winner and visualizing yourself excelling in your sport. You don't have to buy magic to obtain a winning edge; you already possess it—your physically fit mind and body.

REFERENCES

1. J. R. Hoffman and coauthors, Position stand on androgen and human growth hormone use, *Journal of Strength and Conditioning Research* 23 (2009): S1–S59; A. Petróczi and coauthors, Nutritional supplement use by elite young UK athletes: Fallacies of advice regarding efficacy, *Journal of the International Society of Sports Nutrition* 5 (2008): www.jissn.com/content/5/1/22.
2. Position of the American Dietetic Association, Dietetians of Canada, and the American College of Sports Medicine: Nutrition and athletic performance, *Journal of the American Dietetic Association* 109 (2009): 509–527.
3. R. B. Kreider and coauthors, ISSN exercise and sport nutrition review: Research and recommendations, *Journal of the International Society of Sports Nutrition* 7 (2010): www.jissn.com/content/7/1/7 ; L. Di Luigi, Supplements and the endocrine system in athletes, *Clinics in Sports Medicine* 27 (2008): 131–151.
4. Position of the American Dietetic Association, Dietitians of Canada, and the American College of Sports Medicine, 2009; L. M. Burke, L. M. Castell, and S. J. Stear, BJSM reviews: A–Z of supplements: Dietary supplements, sports nutrition foods and ergogenic aids for health and performance Part 1, *British Journal of Sports Medicine* 43 (2009): 728–729; Di Luigi, 2008.
5. Kreider and coauthors, 2010; E. R. Goldstein and coauthors, International Society of Sports Nutrition position stand: Caffeine and performance, *Journal of the International Society of Sports Nutrition* 7 (2010): www.jissn.com/content/7/1/5; G. L. Warren and coauthors, Effect of caffeine ingestion on muscular strength and endurance: A meta-analysis, *Medicine and Science in Sports and Exercise* 42 (2010): 1375–1387; J. K. Davis and J. M. Green, Caffeine and anaerobic performance: Ergogenic value and mechanisms of action, *Sports Medicine* 39 (2009): 813–832; E. Hogervorst and coauthors, Caffeine improves physical and cognitive performance during exhaustive exercise, *Medicine and Science in Sports and Exercise* 40 (2008): 1841–1851; M. Glaister and coauthors, Caffeine supplementation and multiple sprint running performance, *Medicine and Science in Sports and Exercise* 40 (2008):1835–1840; G. Jones, Caffeine and other sympathomimetic stimulants: Modes of action and effects on sports performance, *Essays in Biochemistry* 44 (2008): 109–123.
6. T. E. Graham and coauthors, Does caffeine alter muscle carbohydrate and fat metabolism during exercise? *Applied Physiology, Nutrition, and Metabolism* 33 (2008): 1311–1318.
7. Goldstein and coauthors, 2010; Jones, 2008.
8. M. A. Tarnopolsky, Caffeine and creatine use in sport, *Annals of Nutrition and Metabolism* 57 (2010): 1–8 (Supplement 2); D. H. Fukuda and coauthors, The effects of creatine loading and gender on anaerobic running capacity, *Journal of Strength and Conditioning Research* 24 (2010): 1826–1833; L. L. Spriet, C. G. Perry, and J. L. Talanian, Legal pre-event nutritional supplements to assist energy metabolism, *Essays in Biochemistry* 44 (2008): 27–43; K. D. Tipton and A. A. Ferrando, Improving muscle mass: response of muscle metabolism to exercise, nutrition, and anabolic agents, *Essays in Biochemistry* 44 (2008): 85–98.
9. Kreider and coauthors, 2010; Position of the American Dietetic Association, Dietetians of Canada, and the American College of Sports Medicine, 2009.
10. Position of the American Dietetic Association, Dietitians of Canada, and the American College of Sports Medicine, 2009.
11. American Academy of Pediatrics, Policy Statement, Committee on Sports Medicine and Fitness, Use of performance-enhancing substances, *Pediatrics* 115 (2005): 1103–1106.
12. J. C. Siegler and D. O. Gleadall-Siddall, Sodium bicarbonate ingestion and repeated swim sprint performance, *Journal of Strength and Conditioning*

Research 24 (2010): 3105–3111; S. L. Cameron and coauthors, Increased blood pH but not performance in elite rugby union players, *International Journal of Sport Nutrition and Exercise Metabolism* 20 (2010): 307–321; J. C. Siegler and K. Hirscher, Sodium bicarbonate ingestion and boxing performance, *Journal of Strength and Conditioning Research* 24 (2010): 103–108; A. M. Lindh and coauthors, Sodium bicarbonate improves swimming performance, *International Journal of Sports Medicine* 29 (2008): 519–523.

13. Kreider and coauthors, 2010; D. S. Rowlands and J. S. Thomson, Effects of beta-hydroxy-beta-methylbutyrate supplementation during resistance training on strength, body composition, and muscle damage in trained and untrained young men: a meta-analysis, *Journal of Strength and Conditioning Research* 23 (2009): 836–846.
14. Kreider and coauthors, 2010; D. Bishop, Dietary supplements and team-sport performance, *Sports Medicine* 40 (2010): 995–1017.
15. Kreider and coauthors, 2010; Position of the American Dietetic Association, Dietetians of Canada, and the American College of Sports Medicine, 2009.
16. Spriet, Perry, and Talanian, 2008.
17. Di Luigi, 2008; M. L. Diaz and coauthors, Chromium picolinate and conjugated linoleic acid do not synergistically influence diet-and exercise-induced changes in body composition and health indexes in overweight women, *Journal of Nutritional Biochemistry* 19 (2008): 61–68.
18. Di Luigi, 2008.

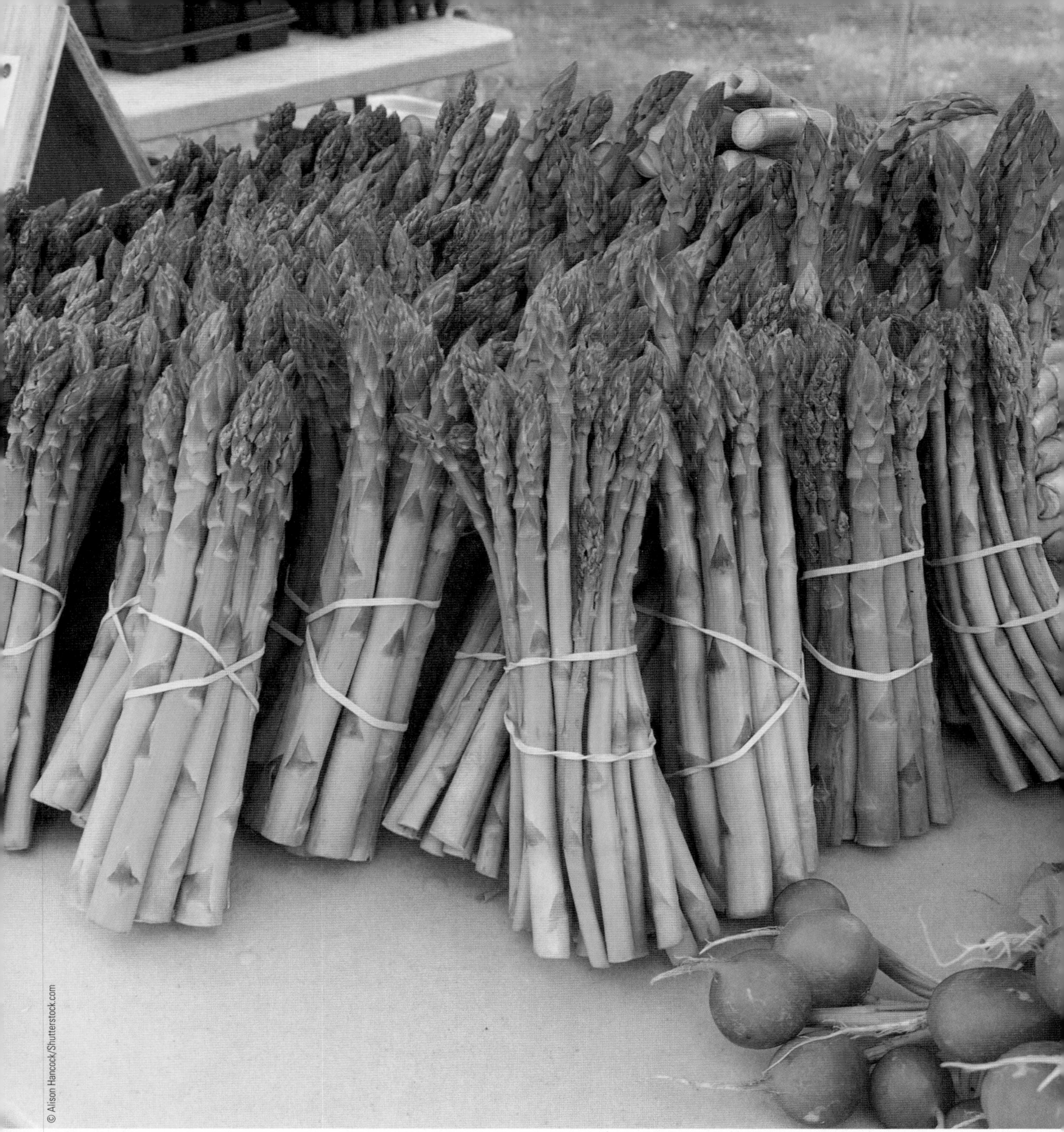

Some women develop gestational diabetes during their pregnancies. Access research on gestational diabetes by going to Weight Management/Diets → Diabetes. What possible effects could gestational diabetes have on both mother and child?

15

Life Cycle Nutrition: Pregnancy and Lactation

LEARNING GPS

Nutrition in Your Life

Food choices have consequences. Sometimes they are immediate, such as when you get heartburn after eating a pepperoni pizza. Other times they sneak up on you, such as when you gain weight after repeatedly overindulging in hot fudge sundaes. Quite often, they are temporary and easily resolved, such as when hunger pangs strike after you skip lunch. During pregnancy, however, the consequences of a woman's food choices are dramatic. They affect not only her health, but also the growth and development of another human being—and not just for today, but for years to come. Making smart food choices is a huge responsibility, but fortunately, it's fairly simple. In the Nutrition Portfolio at the end of this chapter, you can determine how well your current diet might support the needs of a pregnant woman.

All people—pregnant and lactating women, infants, children, adolescents, and adults—need the same nutrients, but the amounts they need vary depending on their stage of life. This chapter focuses on nutrition in preparation for, and support of, pregnancy and lactation. The next two chapters address the needs of infants, children, adolescents, and older adults.

15.1 Nutrition prior to Pregnancy

LEARN IT List the ways men and women can prepare for a healthy pregnancy.

A section on nutrition prior to pregnancy must, by its nature, focus mainly on women. Both a man's and a woman's nutrition may affect **fertility** and possibly the genetic contributions they make to their children, but it is the woman's nutrition that has the most direct influence on the developing fetus. Her body provides the environment for the growth and development of a new human being. Prior to pregnancy, a woman has a unique opportunity to prepare herself physically, mentally,

fertility: the capacity of a woman to produce a normal ovum periodically and of a man to produce normal sperm; the ability to reproduce.

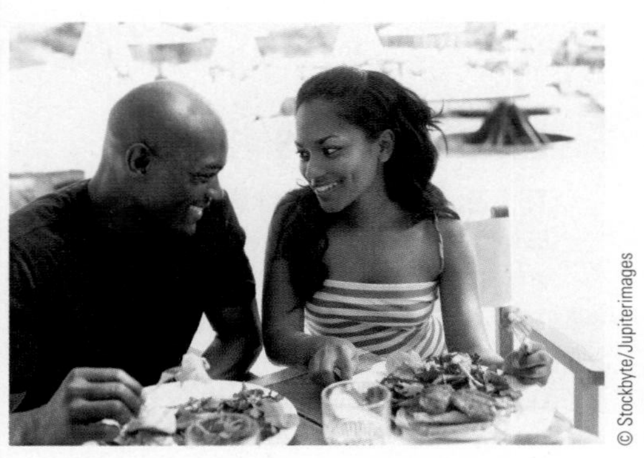

© Stockbyte/Jupiterimages

Young adults can prepare for a healthy pregnancy by taking care of themselves today.

and emotionally for the many changes to come.[1] In preparation for a healthy pregnancy, a woman can establish the following habits:

- *Achieve and maintain a healthy body weight.* Both underweight and overweight are associated with infertility.[2] Overweight and obese men have low sperm counts and hormonal changes that reduce fertility.[3] Excess body fat in women disrupts menstrual regularity and ovarian hormone production.[4] Should a pregnancy occur, mothers, both underweight and overweight, and their newborns, face increased risks of complications.
- *Choose an adequate and balanced diet.* Malnutrition reduces fertility and impairs the early development of an infant should a woman become pregnant. In contrast, a healthy diet that includes a full array of vitamins and minerals can favorably influence fertility.[5] Men with diets rich in antioxidant nutrients have higher sperm numbers and motility.[6]
- *Be physically active.* A woman who wants to be physically active *when* she is pregnant needs to become physically active *beforehand.*
- *Receive regular medical care.* Regular health-care visits help ensure a healthy start to pregnancy.
- *Manage chronic conditions.* Conditions such as diabetes, hypertension, HIV/AIDS, phenylketonuria (PKU), and sexually transmitted diseases can adversely affect a pregnancy and need close medical attention to help ensure a healthy outcome.
- *Avoid harmful influences.* Both maternal and paternal ingestion of, or exposure to, harmful substances (such as cigarettes, alcohol, drugs, or environmental contaminants) can cause miscarriage or abnormalities, alter genes or their expression, and may interfere with fertility.[7]

Young adults who nourish and protect their bodies do so not only for their own sakes, but also for future generations.

> DIETARY GUIDELINES FOR AMERICANS 2010

Women who are capable of becoming pregnant should:

- Achieve and maintain a healthy weight before becoming pregnant.
- Choose foods that supply heme iron, which is more readily absorbed by the body, additional iron sources, and enhancers of iron absorption such as vitamin C–rich foods.
- Consume 400 micrograms per day of synthetic folate from fortified foods and/or supplements in addition to folate from a varied diet.

REVIEW IT List the ways men and women can prepare for a healthy pregnancy.
Prior to pregnancy, the health and behaviors of both men and women can influence fertility and fetal development. In preparation, they can achieve and maintain a healthy body weight, choose an adequate and balanced diet, be physically active, receive regular medical care, manage chronic conditions, and avoid harmful influences.

15.2 Growth and Development during Pregnancy

LEARN IT Describe fetal development from conception to birth and explain how maternal malnutrition can affect critical periods.

A whole new life begins at **conception.** Organ systems develop rapidly, and nutrition plays many supportive roles. This section describes placental development and fetal growth, paying close attention to times of intense developmental activity.

Placental Development In the early days of pregnancy, a spongy structure known as the **placenta** develops in the **uterus.** Two associated structures also form (see Figure 15-1). One is the **amniotic sac,** a fluid-filled balloonlike structure

conception: the union of the male sperm and the female ovum; fertilization.

placenta (plah-SEN-tuh): the organ that develops inside the uterus early in pregnancy, through which the fetus receives nutrients and oxygen and returns carbon dioxide and other waste products to be excreted.

uterus (YOU-ter-us): the muscular organ within which the infant develops before birth.

amniotic (am-nee-OTT-ic) **sac:** the "bag of waters" in the uterus, in which the fetus floats.

> FIGURE 15-1 **The Placenta and Associated Structures**

To understand how placental villi absorb nutrients without maternal and fetal blood interacting directly, think of how the intestinal villi work. The GI side of the intestinal villi is bathed in a nutrient-rich fluid (chyme). The intestinal villi absorb the nutrient molecules and release them into the body via capillaries. Similarly, the maternal side of the placental villi is bathed in nutrient-rich maternal blood. The placental villi absorb the nutrient molecules and release them to the fetus via fetal capillaries.

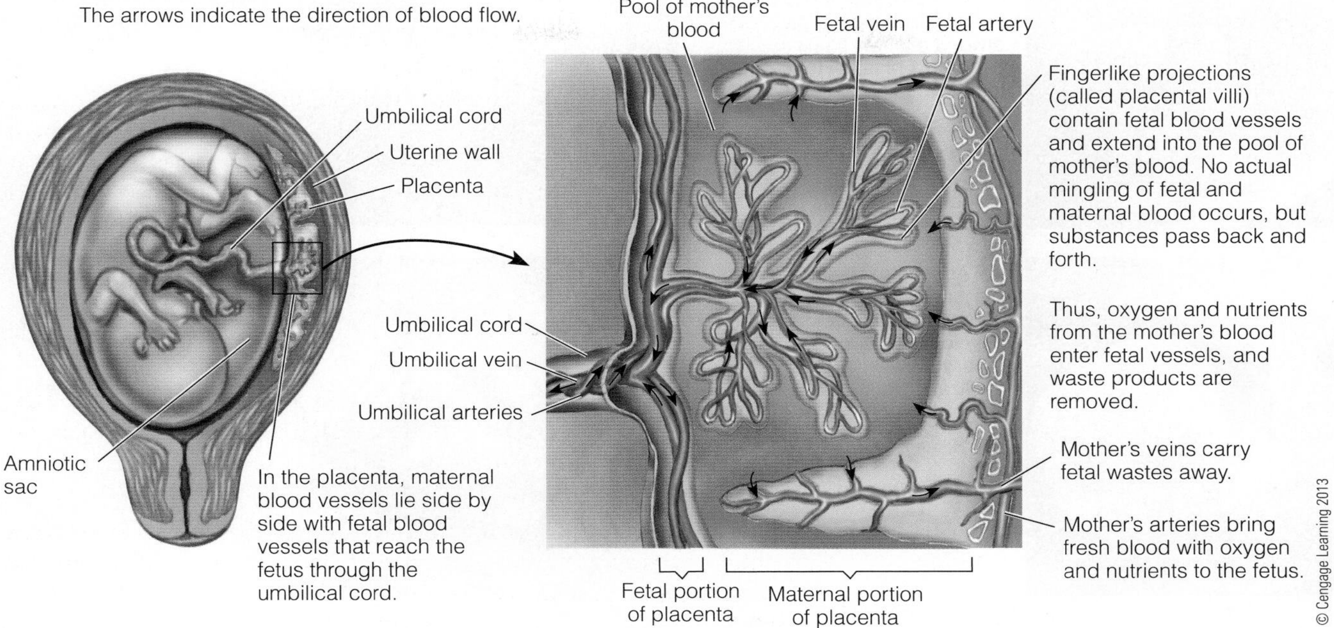

The arrows indicate the direction of blood flow.
Umbilical cord
Uterine wall
Placenta
Amniotic sac
In the placenta, maternal blood vessels lie side by side with fetal blood vessels that reach the fetus through the umbilical cord.
Umbilical cord
Umbilical vein
Umbilical arteries
Pool of mother's blood
Fetal vein
Fetal artery
Fingerlike projections (called placental villi) contain fetal blood vessels and extend into the pool of mother's blood. No actual mingling of fetal and maternal blood occurs, but substances pass back and forth.
Thus, oxygen and nutrients from the mother's blood enter fetal vessels, and waste products are removed.
Mother's veins carry fetal wastes away.
Mother's arteries bring fresh blood with oxygen and nutrients to the fetus.
Fetal portion of placenta
Maternal portion of placenta

that houses the developing fetus. The other is the **umbilical cord,** a ropelike structure containing fetal blood vessels that extends through the fetus's "belly button" (the umbilicus) to the placenta. These three structures play crucial roles during pregnancy, and then are expelled from the uterus during childbirth.

The placenta develops as an interweaving of fetal and maternal blood vessels embedded in the uterine wall. The maternal blood transfers oxygen and nutrients to the fetus's blood and picks up fetal waste products. By exchanging oxygen, nutrients, and waste products, the placenta performs the respiratory, absorptive, and excretory functions that the fetus's lungs, digestive system, and kidneys will provide after birth.

The placenta is a versatile, metabolically active organ. Like all body tissues, the placenta uses energy and nutrients to support its work. It produces an array of hormones that maintain pregnancy and prepare the mother's breasts for lactation (making milk). A healthy placenta is essential for the developing fetus to attain its full potential.[8]

Fetal Growth and Development

Fetal development begins with the fertilization of an **ovum** by a **sperm.** Three stages follow: the zygote, the embryo, and the fetus (see Figure 15-2, p. 472).

The Zygote

The newly fertilized ovum is called a **zygote.** It begins as a single cell and rapidly divides to become a **blastocyst.** During that first week, the blastocyst floats down into the uterus where it will embed itself in the inner uterine wall—a process known as **implantation.** Cell division continues at an amazing rate as each set of cells divides into many other cells.

The Embryo

At first, the number of cells in the **embryo** doubles approximately every 24 hours; later the rate slows, and only one doubling occurs during the final 10 weeks of pregnancy. At 8 weeks, the 1¼-inch embryo has a complete central nervous system, a beating heart, a digestive system, well-defined fingers and toes, and the beginnings of facial features.

umbilical (um-BILL-ih-cul) **cord:** the ropelike structure through which the fetus's veins and arteries reach the placenta; the route of nourishment and oxygen to the fetus and the route of waste disposal from the fetus. The scar in the middle of the abdomen that marks the former attachment of the umbilical cord is the *umbilicus* (um-BILL-ih-cus), commonly known as the "belly button."

ovum (OH-vum): the female reproductive cell, capable of developing into a new organism upon fertilization; commonly referred to as an egg.

sperm: the male reproductive cell, capable of fertilizing an ovum.

zygote (ZY-goat): the initial product of the union of ovum and sperm; a fertilized ovum.

blastocyst (BLASS-toe-sist): the developmental stage of the zygote when it is about 5 days old and ready for implantation.

implantation (IM-plan-TAY-shun): the embedding of the blastocyst in the inner lining of the uterus.

embryo (EM-bree-oh): the developing infant from 2 to 8 weeks after conception.

> **FIGURE 15-2 Stages of Embryonic and Fetal Development**

1 A newly fertilized ovum is called a **zygote** and is about the size of the period at the end of this sentence. Less than 1 week after fertilization, these cells have rapidly divided multiple times to become a blastocyte ready for implantation.

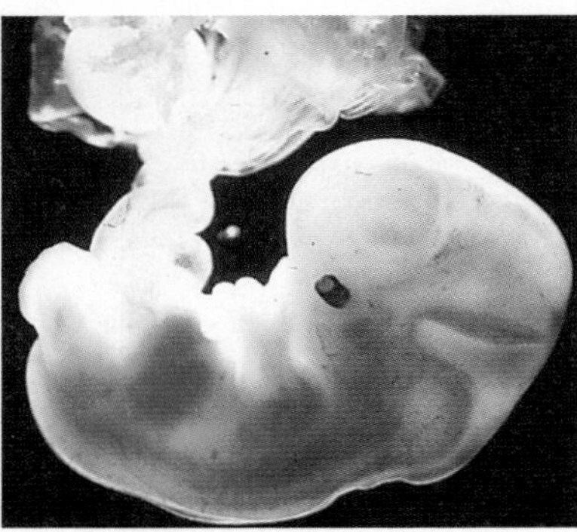

2 After implantation, the placenta develops and begins to provide nourishment to the developing embryo. An **embryo** 5 weeks after fertilization is about ½ inch long.

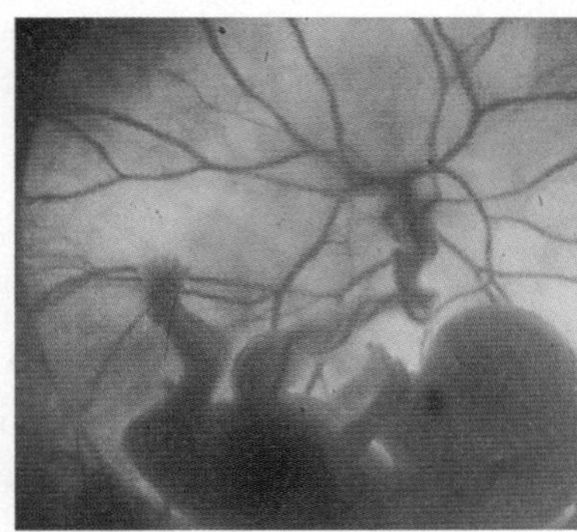

3 A **fetus** after 11 weeks of development is just over an inch long. Notice the umbilical cord and blood vessels connecting the fetus with the placenta.

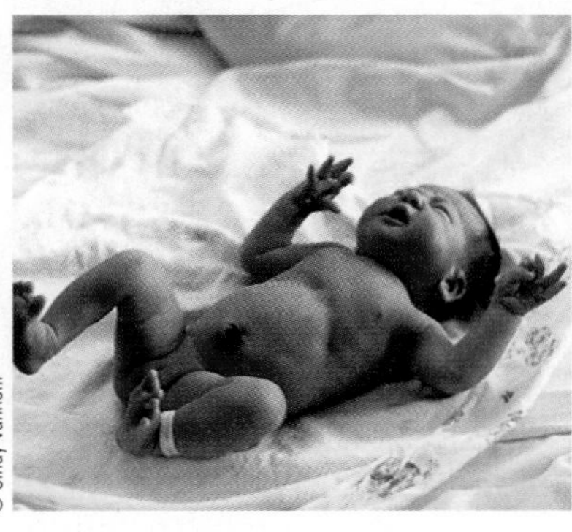

4 A **newborn infant** after 9 months of development measures close to 20 inches in length. From 8 weeks to term, this infant grew 20 times longer and 50 times heavier.

The Fetus The **fetus** continues to grow during the next 7 months. Each organ grows to maturity according to its own schedule, with greater intensity at some times than at others. As Figure 15-2 shows, fetal growth is phenomenal: weight increases from less than an ounce to about 7½ pounds (3500 grams). Most successful pregnancies are **full term**—lasting 38 to 42 weeks—and produce a healthy infant weighing 6½ to 8 pounds.

Critical Periods

Times of intense development and rapid cell division are called **critical periods**—critical in the sense that those cellular activities can occur only at those times. If cell division and number are limited during a critical period, full recovery is not possible (see Figure 15-3). Damage during these critical times of pregnancy has permanent consequences for the life and health of the fetus.[9]

The development of each organ and tissue is most vulnerable to adverse influences (such as nutrient deficiencies or toxins) during its own critical period (see Figure 15-4). The neural tube, for example, is the structure that eventually becomes the brain and the spinal cord, and its critical period of development is from 17 to 30 days **gestation**. Consequently, neural tube development is most vulnerable to nutrient deficiencies, nutrient excesses, or toxins during this critical time—when most women do not yet even realize they are pregnant. Any abnormal development of the neural tube or its failure to close completely can cause a major defect in the central nervous system.

Neural Tube Defects Each year in the United States, approximately 3000 pregnancies are affected by a **neural tube defect**—a malformation of the brain, spinal cord, or both during embryonic development.* Many of these pregnancies end in abortions or stillbirths; an estimated 5 of every 10,000 births results in an infant with a neural tube defect.

The two most common types of neural tube defects are anencephaly (no brain) and spina bifida (split brain). In **anencephaly,** the upper end of the neural

fetus (FEET-us): the developing infant from 8 weeks after conception until term.

full term: between 38 and 42 weeks of pregnancy.

critical periods: finite periods during development in which certain events occur that will have irreversible effects on later developmental stages; usually a period of rapid cell division.

gestation (jes-TAY-shun): the period from conception to birth. For human beings, the average length of a healthy gestation is 40 weeks. Pregnancy is often divided into 3-month periods, called *trimesters.*

neural tube defect: malformations of the brain, spinal cord, or both during embryonic development that often results in lifelong disability or death.

anencephaly (AN-en-SEF-a-lee): an uncommon and always fatal type of neural tube defect, characterized by the absence of a brain.

- **an** = not (without)
- **encephalus** = brain

*Worldwide, some 300,000 pregnancies are affected by neural tube defects each year.

tube fails to close. Consequently, the brain is either missing or fails to develop. Pregnancies affected by anencephaly often end in miscarriage; infants born with anencephaly die shortly after birth.

Spina bifida is characterized by incomplete closure of the spinal cord and its bony encasement (see Figure 15-5, p. 474). The meninges membranes covering the spinal cord often protrude as a sac, which may rupture and lead to meningitis, a life-threatening infection. Spina bifida is accompanied by varying degrees of paralysis, depending on the extent of the spinal cord damage. Mild cases may not even be noticed, but severe cases lead to death. Common problems include clubfoot, dislocated hip, kidney disorders, curvature of the spine, muscle weakness, mental impairments, and motor and sensory losses.

The cause of neural tube defects is unknown, but researchers are examining several gene-gene, gene-nutrient, and gene-environment interactions. A pregnancy affected by a neural tube defect can occur in any woman, but these factors make it more likely:[10]

- A personal or family history of a pregnancy affected by a neural tube defect
- Maternal diabetes
- Maternal use of certain antiseizure medications
- Mutations in folate-related enzymes
- Maternal obesity

Folate supplementation reduces the risk.

Folate Supplementation Chapter 10 describes how folate supplements taken 1 month before conception and continued throughout the first trimester can help support a healthy pregnancy, prevent neural tube defects, and reduce the severity of defects that do occur. For this reason, all women of childbearing age who are

> FIGURE 15-3 The Concept of Critical Periods in Fetal Development

Critical periods occur early in fetal development. An adverse influence felt early in pregnancy can have a much more severe and prolonged impact than one felt later on.

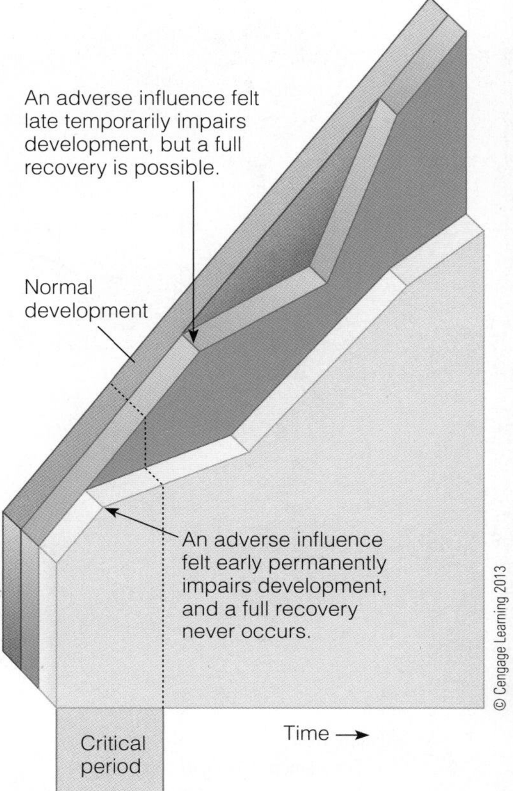

> FIGURE 15-4 Critical Periods of Development

During embryonic development (from 2 to 8 weeks), many of the tissues are in their critical periods (purple area of the bars); events occur that will have irreversible effects on the development of those tissues. In the later stages of development (green area of the bars), the tissues continue to grow and change, but the events are less critical in that they are relatively minor or reversible.

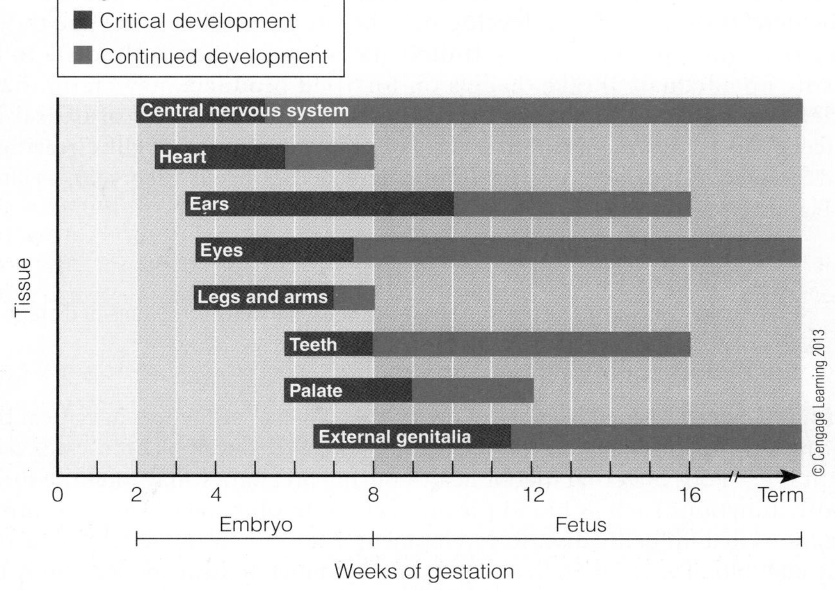

SOURCE: Adapted from *Before We Are Born: Essentials of Embryology and Birth Defects* by K. L. Moore and T.V.N. Persaud (W. B. Saunders, 2003).

spina (SPY-nah) **bifida** (BIFF-ih-dah): one of the most common types of neural tube defects, characterized by the incomplete closure of the spinal cord and its bony encasement.

- **spina** = spine
- **bifida** = split

> FIGURE 15-5 **Spina Bifida**

Spina bifida, a common neural tube defect, occurs when the vertebrae of the spine fail to close around the spinal cord, leaving it unprotected. The B vitamin folate—consumed prior to and during pregnancy—helps prevent spina bifida and other neural tube defects.

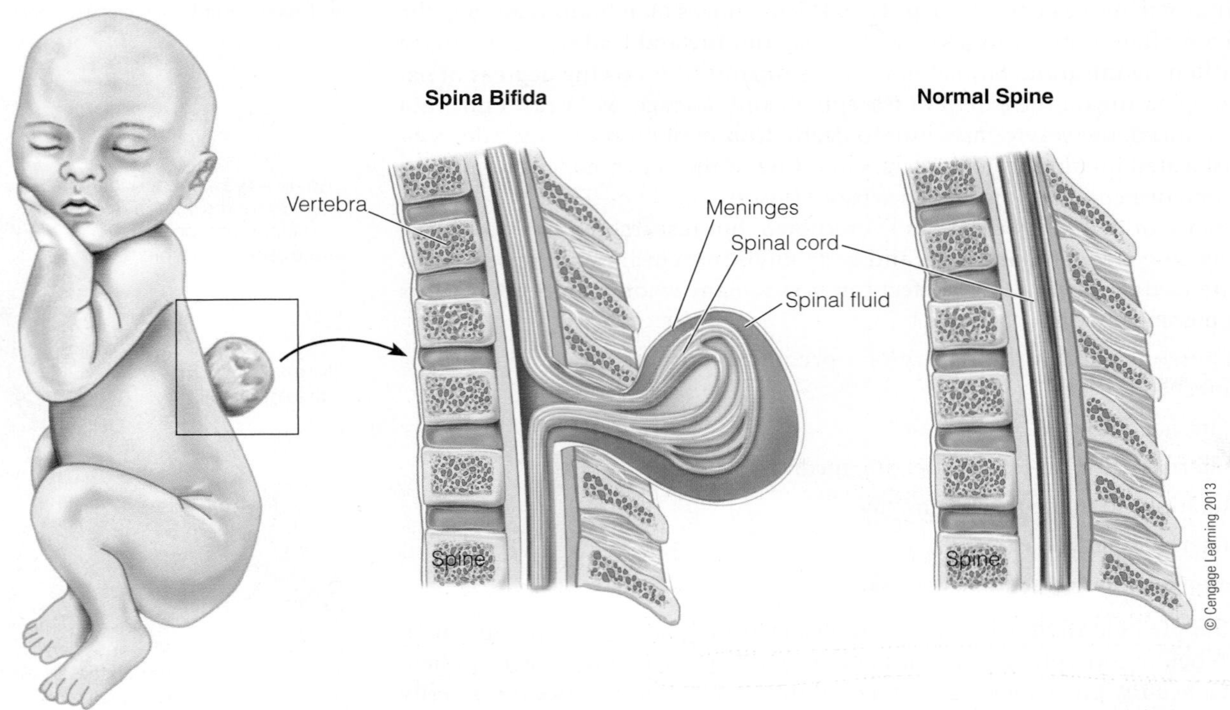

♦ Folate RDA:

- For women: 400 μg (0.4 mg)/day
- During pregnancy: 600 μg (0.6 mg)/day

capable of becoming pregnant should consume 400 micrograms (0.4 milligrams) of folate daily. ♦ A woman who has previously had an infant with a neural tube defect may be advised by her physician to take folate supplements in doses ten times larger—4 milligrams daily. Because high doses of folate can mask the symptoms of pernicious anemia associated with a vitamin B_{12} deficiency, quantities of 1 milligram or more require a prescription. Most over-the-counter multivitamin-mineral supplements contain 400 micrograms of folate; prenatal supplements usually contain 800 micrograms.

Because half of the pregnancies each year are unplanned and because neural tube defects occur early in development before most women realize they are pregnant, grain products in the United States are fortified with folate to help ensure an adequate intake. Labels on fortified products may claim that an "adequate intake of folate has been shown to reduce the risk of neural tube defects." Fortification has improved folate status in women of childbearing age and lowered the number of neural tube defects that occur each year, as shown in Figure 10-11 on p. 313.

> **DIETARY GUIDELINES FOR AMERICANS 2010**

Women who are pregnant are advised to consume 600 micrograms of dietary folate equivalents from all sources.

Chronic Diseases Much research suggests that adverse influences at critical times during fetal development set the stage for the infant to develop chronic diseases in adult life.[11] Poor maternal diet or health during pregnancy may alter the infant's bodily functions such as blood pressure, cholesterol metabolism, and immune functions that influence disease development. For example, prenatal malnutrition may alter blood vessel growth and program lipid metabolism and lean body mass development in such a way that the infant will develop risk factors for cardiovascular disease as an adult.[12]

Malnutrition during the critical period of pancreatic cell growth provides an example of how type 2 diabetes may develop in adulthood. The pancreatic cells responsible for producing insulin (the beta cells) normally increase more than 130-fold between 12 weeks gestation and 5 months after birth. Nutrition is a primary determinant of beta cell growth, and infants who have suffered prenatal malnutrition have significantly fewer beta cells than well-nourished infants. They are also more likely to be low-birthweight infants—and low birthweight and premature birth correlate with insulin resistance later in life.[13] One hypothesis suggests that diabetes may develop from the interaction of inadequate nutrition early in life with abundant nutrition later in life: the small mass of beta cells developed in times of undernutrition during fetal development may be insufficient in times of overnutrition during adulthood when the body needs more insulin.

Hypertension may develop from a similar scenario of inadequate growth during placental and gestational development followed by accelerated growth during early childhood: the small mass of kidney cells developed during malnutrition may be insufficient to handle the excessive demands of later life. As adults, low-birthweight infants may be particularly sensitive to the blood-pressure raising effects of salt.[14]

Fetal Programming Recent genetic research may help to explain how substances such as nutrients during gestation influence the development of obesity and diseases in adulthood. This process is commonly known as **fetal programming,** although "developmental origins of disease" may more appropriately describe the ever-changing interactions involved in disease development.[15] In the case of pregnancy, the mother's nutrition can change gene expression in the fetus.[16] Such epigenetic changes during pregnancy can affect the infant's development of obesity and related adult diseases.[17] Some research suggests that these epigenetic changes during pregnancy may even influence succeeding generations.[18] (See Highlight 6 for further discussion of epigenetics.)

REVIEW IT Describe fetal development from conception to birth and explain how maternal malnutrition can affect critical periods.

Maternal nutrition before and during pregnancy affects both the mother's health and the infant's growth. As the infant develops through its three stages—the zygote, embryo, and fetus—its organs and tissues grow, each on its own schedule. Times of intense development are critical periods that depend on nutrients to proceed smoothly. Without folate, for example, the neural tube fails to develop completely during the first month of pregnancy, prompting recommendations that all women of childbearing age take folate daily.

Because critical periods occur throughout pregnancy, a woman should continuously take good care of her health. That care should include achieving and maintaining a healthy body weight prior to pregnancy and gaining sufficient weight during pregnancy to support a healthy infant.

15.3 Maternal Weight

LEARN IT Explain how both underweight and overweight can interfere with a healthy pregnancy and how weight gain and physical activity can support maternal health and infant growth.

Birthweight is the most reliable indicator of an infant's health. As a later section of this chapter explains, compared with a normal-weight infant, an underweight infant is more likely to have physical and mental abnormalities, suffer illnesses, and die. In general, higher birthweights present fewer risks for infants. Two characteristics of the mother's weight influence an infant's birthweight: her weight *prior* to conception and her weight gain *during* pregnancy.

Weight prior to Conception A woman's weight prior to conception influences fetal growth. ♦ Even with the same weight gain during pregnancy, underweight women tend to have smaller babies than heavier women. Ideally, before a woman

♦ BMI is introduced in Chapter 8.

- Underweight = BMI <18.5
- Normal weight = BMI 18.5 to 24.9
- Overweight = BMI 25 to 29.9
- Obesity = BMI ≥30

fetal programming: the influence of substances during fetal growth on the development of diseases in later life.

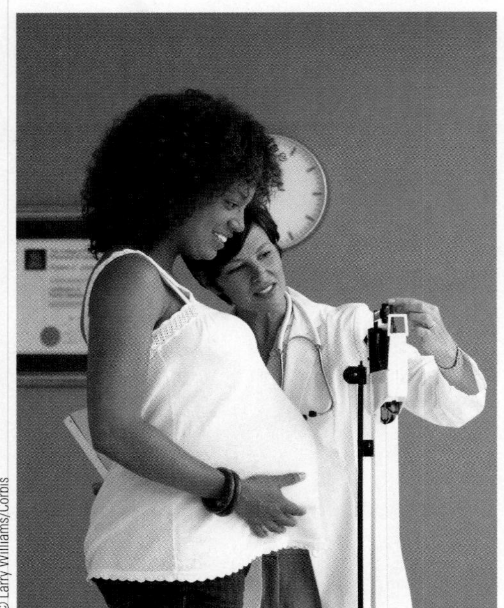

© Larry Williams/Corbis

Fetal growth and maternal health depend on a sufficient weight gain during pregnancy.

becomes pregnant, she will have established diet and activity habits to support an adequate, and not excessive, weight gain during pregnancy.[19]

Underweight An underweight woman has a high risk of having a low-birthweight infant, especially if she is malnourished or unable to gain sufficient weight during pregnancy. In addition, the rates of **preterm** births and infant deaths are higher for underweight women. An underweight woman improves her chances of having a healthy infant by gaining sufficient weight prior to conception or by gaining extra pounds during pregnancy. To gain weight and ensure nutrient adequacy, an underweight woman can follow the dietary recommendations for pregnant women (described on pp. 479–483).

Overweight and Obesity An estimated one-third of all pregnant women in the United States are obese, which can create problems related to pregnancy and childbirth.[20] Obese women have an especially high risk of medical complications such as hypertension, gestational diabetes, and postpartum infections. Compared with other women, obese women are also more likely to have other complications of labor and delivery.[21] Complications in women after gastric bypass surgery and weight loss are lower than in obese women; careful monitoring during pregnancy is advised.[22]

Overweight women have the lowest rate of low-birthweight infants. In fact, infants of overweight women are more likely to be born **post term** and to be large for gestational age, weighing more than 9 pounds. Problems associated with **macrosomia** include increases in the likelihood of a difficult labor and delivery, birth trauma, and **cesarean section.**[23] Consequently, these infants have a greater risk of poor health and death than infants of normal weight.

Of greater concern than infant birthweight is the poor development of infants born to obese mothers.[24] Obesity may double the risk for neural tube defects. Folate's role has been examined, but a more likely explanation seems to be poor glycemic control. Undiagnosed diabetes might also explain why obese women have a greater risk of giving birth to infants with heart defects and other abnormalities.

Health-care providers have traditionally advised against weight-loss dieting during pregnancy. Limited research, however, suggests that obese women who follow a well-balanced, kcalorie-restricted diet and regular exercise program can gain little or no weight without adverse consequences.[25] Ideally, overweight women will achieve a healthy body weight before becoming pregnant and avoid excessive weight gain during pregnancy.[26]

Weight Gain during Pregnancy

Fetal growth and maternal health depend on a sufficient weight gain during pregnancy. Maternal weight gain during pregnancy correlates closely with infant birthweight, which is a strong predictor of the health and subsequent development of the infant.

> DIETARY GUIDELINES FOR AMERICANS 2010

Maintain appropriate kcalorie balance during pregnancy. Pregnant women are encouraged to gain weight within the gestational weight gain guidelines (see Table 15-1).

Recommended Weight Gains Table 15-1 presents recommended weight gains for various prepregnancy weights. The recommended gain for a woman who begins pregnancy at a healthy weight and is carrying a single fetus is 25 to 35 pounds.[27] An underweight woman needs to gain between 28 and 40 pounds; and an overweight woman, between 15 and 25 pounds. About one-third of US women gain weight within these recommended ranges; most gain more than recommended.[28] Appropriate weight gains reduce complications, help women limit weight retention and gains after pregnancy, and help their infants prevent obesity during childhood.[29] To limit excessive weight gains, pregnant women can select foods with a high nutrient density but a low energy density.♦[30] Physical activity also plays a key role in preventing excessive weight gains during pregnancy.[31]

♦ Nutrient density = nutrient/kcal
Energy density = kcal/g

preterm (premature): prior to the thirty-eighth week of pregnancy.

post term: after the forty-second week of pregnancy.

macrosomia (mak-roh-SO-me-ah): abnormally large body size. In the case of infants, a birthweight at the 90th percentile or higher for gestational age (roughly 9 lb—or 4000 g—or more); macrosomia results from prepregnancy obesity, excessive weight gain during pregnancy, or uncontrolled gestational diabetes.

- **macro** = large
- **soma** = body

cesarean (si-ZAIR-ee-un) **section:** a surgically assisted birth involving removal of the fetus by an incision into the uterus, usually by way of the abdominal wall.

TABLE 15-1 Recommended Weight Gains Based on Prepregnancy Weight

Prepregnancy Weight	Recommended Weight Gain: For single birth	For twin birth
Underweight (BMI <18.5)	28 to 40 lb (12.5 to 18.0 kg)	Insufficient data to make recommendation
Healthy weight (BMI 18.5 to 24.9)	25 to 35 lb (11.5 to 16.0 kg)	37 to 54 lb (17.0 to 25.0 kg)
Overweight (BMI 25.0 to 29.9)	15 to 25 lb (7.0 to 11.5 kg)	31 to 50 lb (14.0 to 23.0 kg)
Obese (BMI ≥30)	11 to 20 lb (5.0 to 9.0 kg)	25 to 42 lb (11.0 to 19.0 kg)

SOURCE: Institute of Medicine, *Weight Gain during Pregnancy: Reexamining the Guidelines* (Washington, D.C.: National Academies Press, 2009).

Weight-Gain Patterns For the normal-weight woman, weight gain ideally follows a pattern of 3½ pounds during the first trimester and 1 pound per week thereafter. Health-care professionals monitor weight gain using a prenatal weight-gain grid (see Figure 15-6).

If a woman gains more than is recommended early in pregnancy, she should not restrict her energy intake later in order to lose weight. A large weight gain over a short time, however, indicates excessive fluid retention and may be the first sign of the serious medical complication preeclampsia, which is discussed later.

Components of Weight Gain Women often express concern about the weight gain that accompanies a healthy pregnancy. They may find comfort by remembering that most of the gain supports the growth and development of the placenta, uterus, blood, and breasts, the increase in blood supply and fluid volume, as well as a healthy 7½-pound infant. A small amount goes into maternal fat stores, and even that fat has a special purpose—to provide energy for labor and lactation. Figure 15-7 (p. 478) shows the components of a healthy 30-pound weight gain.

Weight Loss after Pregnancy The pregnant woman loses some weight at delivery. In the following weeks, she loses more as her blood volume returns to normal and she sheds accumulated fluids. The typical woman does not, however, return

FIGURE 15-6 Recommended Prenatal Weight Gain Based on Prepregnancy Weight

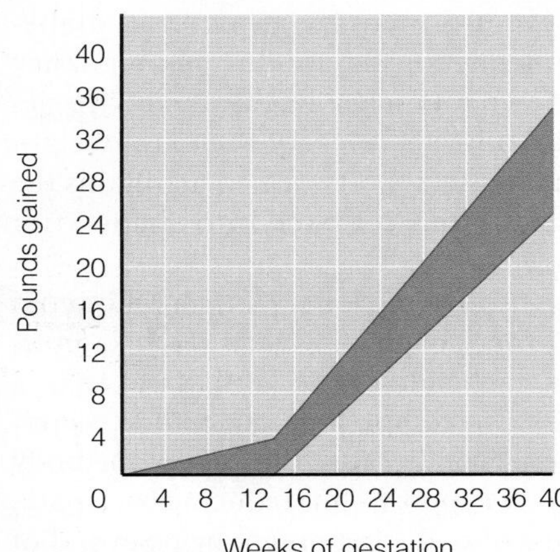

Normal-weight women should gain about 3½ pounds in the first trimester and just under 1 pound/week thereafter, achieving a total gain of 25 to 35 pounds by term.

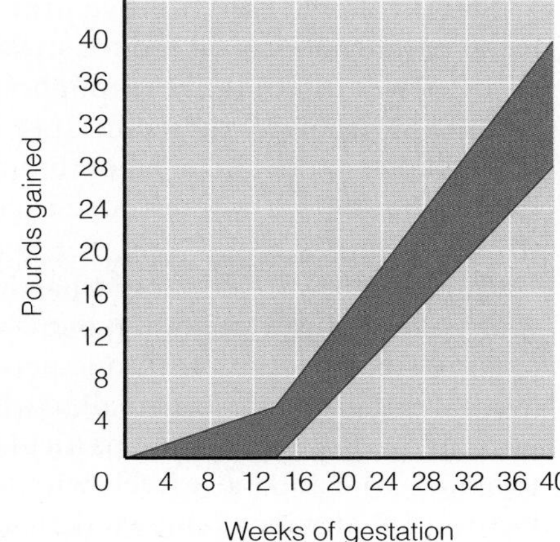

Underweight women should gain about 5 pounds in the first trimester and just over 1 pound/week thereafter, achieving a total gain of 28 to 40 pounds by term.

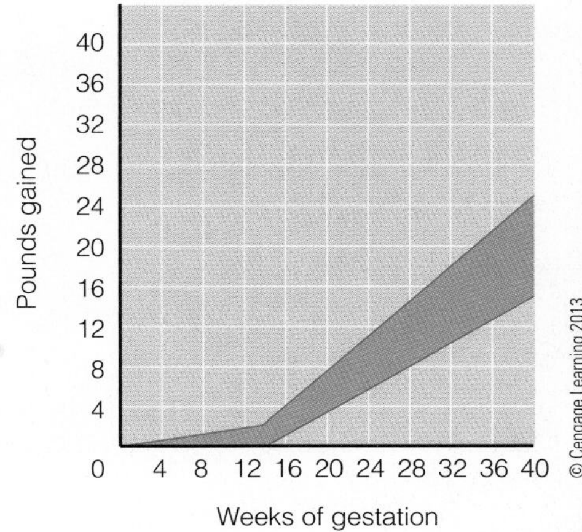

Overweight women should gain about 2 pounds in the first trimester and ⅔ pound/week thereafter, achieving a total gain of 15 to 25 pounds.

> FIGURE 15-7 **Components of Weight Gain during Pregnancy**

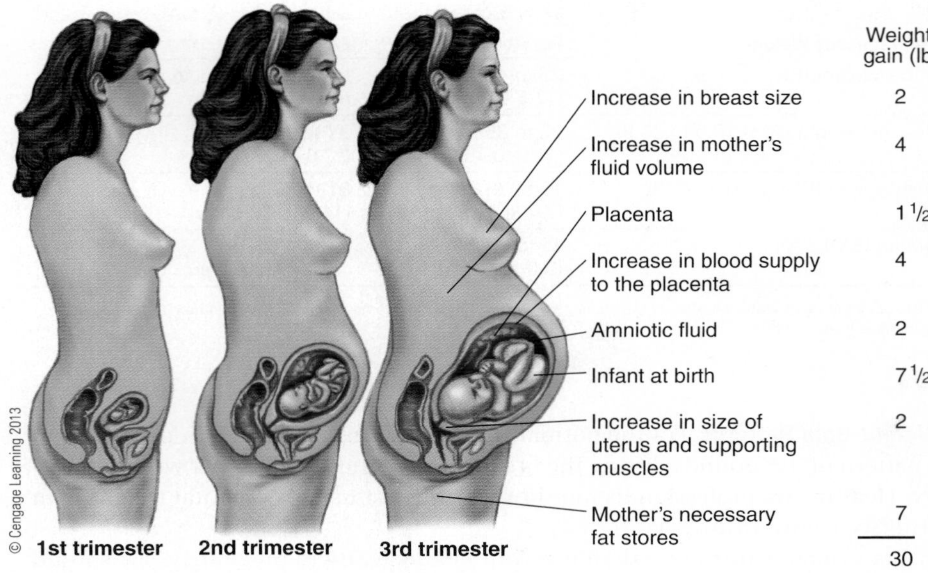

to her prepregnancy weight. In general, the more weight a woman gains beyond the needs of pregnancy, the more she retains and the more likely she will continue to gain.[32] Even with an average weight gain during pregnancy, most women tend to retain a couple of pounds with each pregnancy. When those couple of pounds become 7 or more and BMI increases by a unit or more, complications such as diabetes and hypertension in future pregnancies as well as chronic diseases in later life become more likely—even for women who are not overweight. Those who are successful in losing their pregnancy weight are more likely to limit weight gains through middle adulthood. Eating breakfast regularly supports postpartum weight loss.[33]

Exercise during Pregnancy An active, physically fit woman experiencing a normal pregnancy can continue to exercise throughout pregnancy, adjusting the duration and intensity of activity as the pregnancy progresses.[34] Inactive women and those experiencing pregnancy complications should discuss physical activity options with their health-care provider.

Staying active can improve fitness, prevent or manage gestational diabetes, facilitate labor, and reduce stress. Women who exercise during pregnancy report fewer discomforts throughout their pregnancies. Regular exercise develops the strength and endurance a woman needs to carry the extra weight through pregnancy and to labor through an intense delivery. It also maintains the habits that help a woman lose excess weight and get back into shape after the birth.

A pregnant woman should participate in low-impact activities and avoid sports in which she might fall or be hit by other people or objects. For example, playing singles tennis with one person on each side of the net is safer than a fast-moving game of racquetball in which the two competitors can collide. Swimming and water aerobics are particularly beneficial because they allow the body to remain cool and move freely with the water's support, thus reducing back pain. Figure 15-8 provides some guidelines for exercise during pregnancy. Several of the guidelines are aimed at preventing excessively high internal body temperature and dehydration, both of which can harm fetal development. To this end, pregnant women should also stay out of saunas, steam rooms, and hot tubs or hot whirlpool baths.

> FIGURE 15-8 **Exercise Guidelines during Pregnancy**

DO

Do begin to exercise gradually.

Do exercise regularly (most, if not all, days of the week).

Do warm up with 5 to 10 minutes of light activity.

Do 30 minutes or more of moderate physical activity.

Do cool down with 5 to 10 minutes of slow activity and gentle stretching.

Do drink water before, after, and during exercise.

Do eat enough to support the needs of pregnancy plus exercise.

Do rest adequately.

Pregnant women can enjoy the benefits of exercise.

DON'T

Don't exercise vigorously after long periods of inactivity.

Don't exercise in hot, humid weather.

Don't exercise when sick with fever.

Don't exercise while lying on your back after the 1st trimester of pregnancy or stand motionless for prolonged periods.

Don't exercise if you experience any pain, discomfort, or fatigue.

Don't participate in activities that may harm the abdomen or involve jerky, bouncy movements.

Don't scuba dive.

REVIEW IT Explain how both underweight and overweight can interfere with a healthy pregnancy and how weight gain and physical activity can support maternal health and infant growth.

A healthy pregnancy depends on a sufficient weight gain. Women who begin their pregnancies at a healthy weight need to gain about 30 pounds, which covers the growth and development of the placenta, uterus, blood, breasts, and infant. By remaining active throughout pregnancy, a woman can develop the strength she needs to carry the extra weight and maintain habits that will help her lose weight after the birth.

15.4 Nutrition during Pregnancy

LEARN IT Summarize the nutrient needs of women during pregnancy.

A woman's body changes dramatically during pregnancy. Her uterus and its supporting muscles increase in size and strength; her blood volume increases by half to carry the additional nutrients and other materials; her joints become more flexible in preparation for childbirth; her feet swell in response to high concentrations of the hormone estrogen, which promotes water retention and helps to ready the uterus for delivery; and her breasts enlarge in preparation for lactation. The hormones that mediate all these changes may influence her mood. She can best prepare to handle these changes given a nutritious diet, regular physical activity, plenty of rest, and caring companions. This section highlights the role of nutrition.

In general, the following guidelines will allow most women to enjoy a healthy pregnancy:[35]

- Strive for good nutrition and health prior to pregnancy and get prenatal care during pregnancy.
- Gain a healthy amount of weight.
- Eat a balanced diet, safely prepared, and engage in physical activity regularly.
- Take prenatal vitamin and mineral supplements as prescribed.
- Refrain from cigarettes, alcohol, and drugs (including herbal remedies, unless prescribed by a physician).

An adequate diet may also help a woman manage the challenges and possible depression that can arise *after* the infant arrives.[36] Details follow.

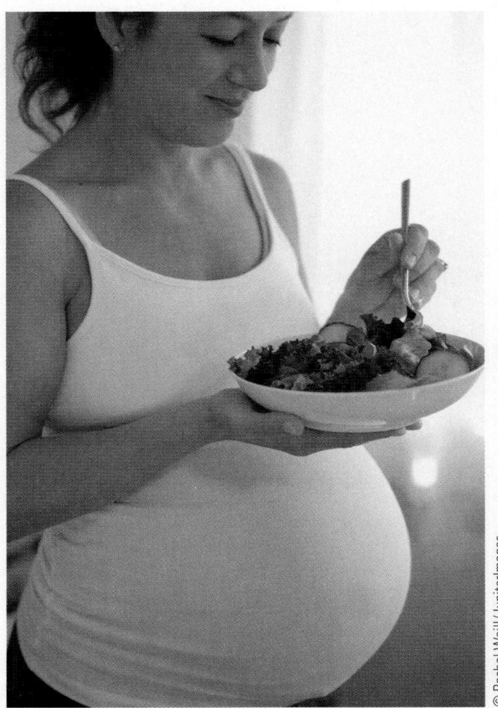

A pregnant woman's food choices support both her health and her infant's growth and development.

Energy and Nutrient Needs during Pregnancy

From conception to birth, all parts of the infant—bones, muscles, blood cells, skin, and all other tissues—are made from nutrients in the foods the mother eats. For most women,

> **FIGURE 15-9 Comparison of Nutrient Recommendations for Nonpregnant, Pregnant, and Lactating Women**

For actual values, turn to the table on the inside front cover. For vitamins and minerals not shown here, the values do not change for pregnant and lactating women.

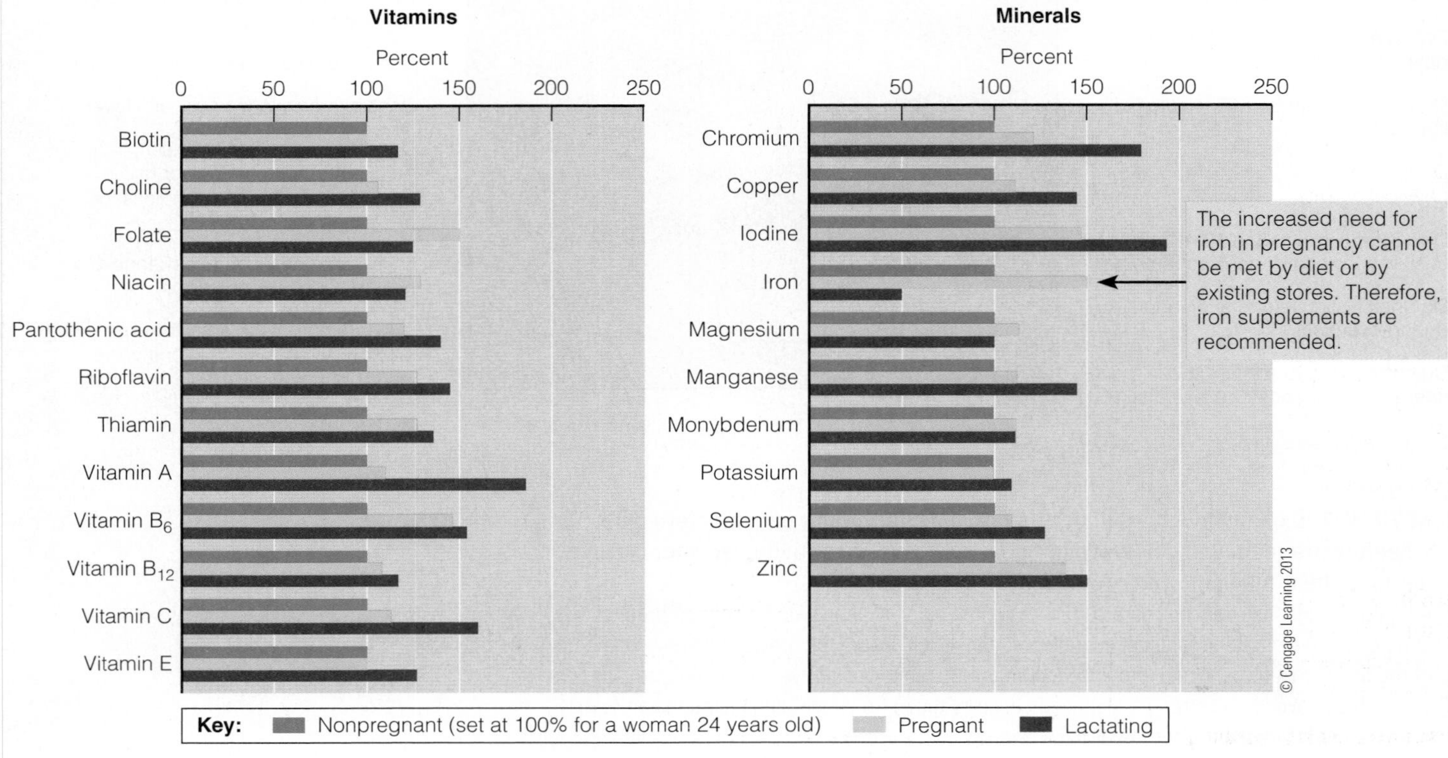

nutrient needs during pregnancy and lactation are higher than at any other time (see Figure 15-9). To meet the high nutrient demands of pregnancy, a woman will need to make careful food choices, but her body will also help by maximizing absorption and minimizing losses. The Dietary Reference Intakes (DRI) table on the inside front cover provides separate listings for women during pregnancy and lactation, reflecting their heightened nutrient needs.

Energy The enhanced work of pregnancy raises the woman's basal metabolic rate dramatically and demands extra energy. Energy needs of pregnant women are greater than those of nonpregnant women—an additional 340 kcalories per day during the second trimester and an extra 450 kcalories per day during the third trimester. ♦ A woman can easily get these added kcalories with nutrient-dense selections from the five food groups. See Table 2-2 (p. 44) for suggested dietary patterns for several kcalorie levels. A sample menu for pregnant and lactating women is presented in Figure 15-10.

♦ Energy requirement during pregnancy:
- 1st trimester: + 0 kcal/day
- 2nd trimester: + 340 kcal/day
- 3rd trimester: + 450 kcal/day

For a 2000-kcalorie daily intake, these added kcalories represent about 15 to 20 percent more food energy than before pregnancy. The increase in nutrient needs is often greater than this, so nutrient-dense foods should be chosen to supply the extra kcalories: foods such as whole-grain breads and cereals, legumes, dark green vegetables, citrus fruits, low-fat milk and milk products, and lean meats, fish, poultry, and eggs.

Carbohydrate Ample carbohydrate (ideally, 175 grams or more per day and certainly no less than 135 grams) is necessary to fuel the fetal brain. Sufficient carbohydrate also ensures that the protein needed for growth will not be broken down and used to make glucose.

Protein The protein RDA for pregnancy is an additional 25 grams per day higher than for nonpregnant women. ♦ Pregnant women can easily meet their protein needs by selecting meats, milk products, and protein-containing plant foods

♦ Protein RDA during pregnancy:
- + 25 g/day

> FIGURE 15-10 **Daily Food Choices for Pregnancy and Lactation**

Food Group	Amount	SAMPLE MENU	
Fruits	2 c	***Breakfast***	***Dinner***
		1 whole-wheat English muffin	Chicken cacciatore
		2 tbs peanut butter	3 oz chicken
Vegetables	2½–3 c	1 c low-fat vanilla yogurt	½ c stewed tomatoes
		½ c fresh strawberries	1 c rice
		1 c orange juice	½ c summer squash
Grains	6–8 oz	***Midmorning snack***	1½ c salad (spinach, mushrooms, carrots)
		½ c cranberry juice	1 tbs salad dressing
		1 oz pretzels	1 slice Italian bread
Protein foods	5½–6½ oz	***Lunch***	2 tsp soft margarine
		Sandwich (tuna salad on whole-wheat bread)	1 c low-fat milk
		½ carrot (sticks)	
Milk	3 c	1 c low-fat milk	

NOTE: The range of recommended amounts reflects the differences of the first trimester versus the second and third trimesters. This sample meal plan provides about 2500 kcalories (55% from carbohydrate, 20% from protein, and 25% from fat) and meets most of the vitamin and mineral needs of pregnant and lactating women.

such as legumes, whole grains, nuts, and seeds. Because use of high-protein supplements during pregnancy may be harmful to the infant's development, it is discouraged unless medically prescribed and carefully monitored to treat fetal growth problems.[37]

Essential Fatty Acids The high nutrient requirements of pregnancy leave little room in the diet for excess fat, but the essential long-chain polyunsaturated fatty acids are particularly important to the growth and development of the fetus.[38] The brain is largely made of lipid material, and it depends heavily on the long-chain omega-3 and omega-6 fatty acids for its growth, function, and structure.[39] (See Table 5-4 on p. 153 for a list of good food sources of the omega fatty acids.)

Nutrients for Blood Production and Cell Growth New cells are laid down at a tremendous pace as the fetus grows and develops. At the same time, the mother's red blood cell mass expands. All nutrients are important in these processes, but for folate, vitamin B_{12}, iron, and zinc, the needs are especially great due to their key roles in the synthesis of DNA and new cells.

The requirement for folate increases dramatically during pregnancy. ♦ It is best to obtain sufficient folate from a combination of supplements, fortified foods, and a diet that includes fruits, juices, green vegetables, and whole grains.[40] The "How To" feature in Chapter 10 on p. 312 describes how folate from each of these sources contributes to a day's intake.

♦ Folate RDA during pregnancy:
- 600 µg/day

The pregnant woman also has a slightly greater need for the B vitamin that activates the folate enzyme—vitamin B_{12}. ♦ Generally, even modest amounts of meat, fish, eggs, or milk products together with body stores easily meet the need for vitamin B_{12}. Vegans who exclude all foods of animal origin, however, need daily supplements of vitamin B_{12} or vitamin B_{12}–fortified foods to prevent the neurological complications of a deficiency.

♦ Vitamin B_{12} RDA during pregnancy:
- 2.6 µg/day

Pregnant women need iron to support their increased blood volume and to provide for placental and fetal needs. ♦ The developing fetus draws on maternal iron stores to create sufficient stores of its own to last through the first 4 to 6 months after birth. Ideally, a woman enters pregnancy with adequate iron stores and maintains sufficient iron nutrition throughout the pregnancy.[41] The transfer of significant amounts of iron to the fetus is regulated by the placenta, which gives the iron needs of the fetus priority over those of the mother.[42] Women with inadequate iron stores are left with too little iron to meet their own health needs.

♦ Iron RDA during pregnancy:
- 27 mg/day

In addition, blood losses are inevitable at birth, especially during a cesarean section, and can further drain the mother's supply.*

During pregnancy, the body makes several adaptations to help meet the exceptionally high need for iron. Menstruation, the major route of iron loss in women, ceases, and iron absorption improves thanks to an increase in transferrin, the body's iron-absorbing and iron-carrying protein. Without sufficient intake, though, iron stores quickly dwindle. Women with iron-deficiency anemia are likely to give birth to low birthweight infants.[43]

Few women enter pregnancy with adequate iron stores, so a daily iron supplement is recommended early in pregnancy, if not before.[44] To enhance iron absorption, the supplement should be taken between meals or at bedtime and with liquids other than milk, coffee, or tea, which inhibit iron absorption. Drinking orange juice does not enhance iron absorption from supplements as it does from foods; vitamin C enhances iron absorption by converting iron from ferric to ferrous, but supplemental iron is already in the ferrous form. Vitamin C is helpful, however, in preventing the premature rupture of amniotic membranes.

> **DIETARY GUIDELINES FOR AMERICANS 2010**
> Women who are pregnant should take an iron supplement as recommended by an obstetrician or other health-care provider.

♦ Zinc RDA during pregnancy:
- 12 mg/day (≤18 yr)
- 11 mg/day (19–50 yr)

Zinc is required for DNA and RNA synthesis and thus for protein synthesis and cell development. ♦ Typical zinc intakes for pregnant women are lower than recommendations, but fortunately, zinc absorption increases when intakes are low.

Nutrients for Bone Development Vitamin D and the bone-building minerals calcium, phosphorus, magnesium, and fluoride are in great demand during pregnancy. Insufficient intakes may produce abnormal fetal bones and teeth.

Vitamin D plays a central role in calcium absorption and utilization. Consequently, severe maternal vitamin D deficiency interferes with normal calcium metabolism, resulting in rickets in the infant and osteomalacia in the mother.[45] Regular exposure to sunlight and consumption of vitamin D–fortified milk are usually sufficient to provide the recommended amount of vitamin D during pregnancy, which is the same as for nonpregnant women.[46] ♦ Pregnant women who do not receive sufficient dietary vitamin D or enough exposure to sunlight may need a supplement.

♦ The RDA for vitamin D does not increase during pregnancy.

Calcium absorption and retention increases dramatically in pregnancy, helping the mother to meet the calcium needs of pregnancy. ♦ During the last trimester, as the fetal bones begin to calcify, more than 300 milligrams a day are transferred to the fetus. This mobilization of calcium from the mother's bones is apparent in a decrease in her bone density.[47] Recommendations to ensure an adequate calcium intake during pregnancy help to conserve maternal bones while meeting fetal needs.

♦ The RDA for calcium does not increase during pregnancy.

Calcium intakes for pregnant women typically fall below recommendations. Because bones are still actively depositing minerals until about age 30, adequate calcium is especially important for young women. Pregnant women younger than age 25 who receive less than 600 milligrams of dietary calcium daily need to increase their intake of milk, cheese, yogurt, and other calcium-rich foods. The USDA Food Patterns suggest consuming 3 cups per day of fat-free or low-fat milk or the equivalent in milk products. Alternatively, and less preferably, they may need a daily supplement of 600 milligrams of calcium, taken with meals.

Other Nutrients The nutrients mentioned here are those most intensely involved in blood production, cell growth, and bone development. Of course, other vitamins and minerals are also needed during pregnancy to support the growth and health of both fetus and mother.[48] Even with adequate nutrition, repeated pregnancies within a short time span can deplete nutrient reserves. When this

*On average, almost twice as much blood is lost during a cesarean delivery as during the average vaginal delivery of a single fetus.

happens, fetal growth may be compromised and maternal health may decline. The optimal interval between pregnancies is 18 to 23 months.

Nutrient Supplements A healthy pregnancy and optimal infant development depend on the mother's diet.[49] Pregnant women who make wise food choices can meet most of their nutrient needs, with the possible exception of iron. Even so, physicians routinely recommend daily multivitamin-mineral supplements for pregnant women.[50] Prenatal supplements typically contain greater amounts of folate, iron, and calcium than regular multivitamin-mineral supplements. These supplements are particularly beneficial for women who do not eat adequately and for those in high-risk groups: women carrying multiple fetuses, cigarette smokers, and alcohol and drug abusers. The use of prenatal supplements may help reduce the risks of preterm delivery, low infant birthweights, and birth defects.[51] Supplement use *prior* to conception also seems to reduce these risks.

Vegetarian Diets during Pregnancy and Lactation In general, a well-planned vegetarian diet can support a healthy pregnancy and successful lactation if it provides adequate energy and contains a wide variety of legumes, whole grains, nuts, seeds, fruits, and vegetables.[52] Many vegetarian women are well nourished, with nutrient intakes from diet alone exceeding the RDA for most vitamins and minerals except iron, which is low for most women. In contrast, vegan women who restrict themselves to an exclusively plant-based diet generally have low energy intakes and are underweight. For pregnant women, this can be a problem. Women with low prepregnancy weights and insufficient weight gains during pregnancy jeopardize a healthy pregnancy.

Vegan diets may require supplementation with vitamin B_{12}, calcium, and vitamin D, or the addition of foods fortified with these nutrients. Infants may suffer spinal cord damage and develop severe psychomotor retardation due to a lack of vitamin B_{12} in the mother's diet during pregnancy. Breastfed infants of vegan mothers have been reported to develop vitamin B_{12} deficiency and severe movement disorders. Giving infants vitamin B_{12} supplements corrects the blood and neurological symptoms of the deficiency, as well as the structural abnormalities, but cognitive and language development delays may persist. A pregnant woman needs a regular source of vitamin B_{12}–fortified foods or a supplement that provides 2.6 micrograms daily.

Common Nutrition-Related Concerns of Pregnancy Nausea, constipation, heartburn, and food sensitivities are common nutrition-related concerns during pregnancy. A few simple strategies can help alleviate maternal discomforts (see Table 15-2).

Nausea and Vomiting Not all women have queasy stomachs in the early months of pregnancy, but about half do. The nausea of "morning sickness" may actually occur anytime and ranges from mild queasiness to debilitating nausea and vomiting. Severe and continued vomiting may require hospitalization if it results in acidosis, dehydration, or excessive weight loss. The hormonal changes of early pregnancy seem to be responsible for a woman's sensitivities to the appearance, texture, or smell of foods.

TABLE 15-2 **Strategies to Alleviate Maternal Discomforts**

To Alleviate the Nausea of Pregnancy	To Prevent or Alleviate Constipation	To Prevent or Relieve Heartburn
• On waking, arise slowly. • Eat dry toast or crackers. • Chew gum or suck hard candies. • Eat small, frequent meals. • Avoid foods with offensive odors. • When nauseated, drink carbonated beverages instead of citrus juice, water, milk, coffee, or tea.	• Eat foods high in fiber (fruits, vegetables, and whole grains). • Exercise regularly. • Drink at least eight glasses of liquids a day. • Respond promptly to the urge to defecate. • Use laxatives only as prescribed by a physician; do not use mineral oil, because it interferes with absorption of fat-soluble vitamins.	• Relax and eat slowly. • Chew food thoroughly. • Eat small, frequent meals. • Drink liquids between meals. • Avoid spicy or greasy foods. • Sit up while eating; elevate the head while sleeping. • Wait 3 hours after eating before lying down. • Wait 2 hours after eating before exercising.

The problem typically peaks at 9 weeks gestation and resolves within a month or two.[53] Traditional strategies for quelling nausea are listed in Table 15-2 (p. 483), but there is little evidence to support such advice.[54] Many women benefit most from simply resting when nauseous and eating the foods they want when they feel like eating. They may also find comfort in a clean, quiet, and temperate environment.

Constipation and Hemorrhoids As the hormones of pregnancy alter muscle tone and the growing fetus crowds intestinal organs, an expectant mother may experience constipation. She may also develop hemorrhoids (swollen veins of the rectum). Hemorrhoids can be painful, and straining during bowel movements may cause bleeding. She can gain relief by following the strategies listed in Table 15-2.

Heartburn Heartburn is another common complaint during pregnancy. The hormones of pregnancy relax the digestive muscles, and the growing fetus puts increasing pressure on the mother's stomach. This combination causes gastroesophageal reflux, the painful sensation a person feels behind the breastbone when stomach acid splashes back up into the lower esophagus (see Highlight 3). Tips to help relieve heartburn are included in Table 15-2.

Food Cravings and Aversions Some women develop cravings for, or aversions to, particular foods and beverages during pregnancy. **Food cravings** and **food aversions** are fairly common, but they do not seem to reflect real physiological needs. In other words, a woman who craves pickles does not necessarily need salt. Similarly, cravings for ice cream are common in pregnancy but do not signify a calcium deficiency. Cravings and aversions that arise during pregnancy are most likely due to hormone-induced changes in sensitivity to taste and smell.

Nonfood Cravings Some pregnant women develop cravings for nonfood items such as freezer frost, laundry starch, clay, soil, or ice—a practice known as **pica.**[55] Pica is a cultural phenomenon that reflects a society's folklore; it is especially common among African American women. Pica is often associated with iron-deficiency anemia, but whether iron deficiency leads to pica or pica leads to iron deficiency is unclear. Eating clay or soil may interfere with iron absorption and displace iron-rich foods from the diet.

REVIEW IT Summarize the nutrient needs of women during pregnancy.
Energy and nutrient needs are high during pregnancy. A balanced diet that includes an extra serving from each of the five food groups can usually meet these needs, with the possible exception of iron and folate (supplements are recommended). The nausea, constipation, and heartburn that sometimes accompany pregnancy can usually be alleviated with a few simple strategies. Food cravings do not typically reflect physiological needs.

15.5 High-Risk Pregnancies

LEARN IT Identify factors predicting low-risk and high-risk pregnancies and describe ways to manage them.

Some pregnancies jeopardize the life and health of the mother and infant. Table 15-3 identifies several characteristics of a **high-risk pregnancy.** A woman with none of these risk factors is said to have a **low-risk pregnancy.** The more factors that apply, the higher the risk. All pregnant women, especially those in high-risk categories, need prenatal medical care, including nutrition advice. ♦

The Infant's Birthweight A high-risk pregnancy is likely to produce an infant with **low birthweight (LBW).** Low-birthweight infants, defined as infants who weigh 5½ pounds or less, are classified according to their gestational age. Preterm infants are born before they are fully developed; they are often underweight and have trouble breathing because their lungs are immature. Preterm infants may be small, but if their size and weight are appropriate for their gestational age, they can catch up in growth given adequate nutrition support. In contrast, small-for-gestational-age infants have suffered growth failure in the uterus

♦ Nutrition advice in prenatal care:
- Eat well-balanced meals.
- Gain enough weight to support fetal growth.
- Take prenatal supplements as prescribed.
- Stop drinking alcohol.

food cravings: strong desires to eat particular foods.

food aversions: strong desires to avoid particular foods.

pica: the general term for eating nonfood items. The specific craving for nonfood items that come from the earth, such as clay or dirt, is known as *geophagia*.

high-risk pregnancy: a pregnancy characterized by risk factors that make it likely the birth will be surrounded by problems such as premature delivery, difficult birth, restricted growth, birth defects, and early infant death.

low-risk pregnancy: a pregnancy characterized by factors that make it likely the birth will be normal and the infant healthy.

low birthweight (LBW): a birthweight of 5½ pounds (2500 grams) or less; indicates probable poor health in the newborn and poor nutrition status in the mother during pregnancy, before pregnancy, or both. Optimal birthweight for a full-term baby is about 6¾ to 8 pounds (about 3100 to 3600 grams).

TABLE 15-3 High-Risk Pregnancy Factors

Factor	Condition that Raises Risk
Maternal weight	
• Prior to pregnancy	Prepregnancy BMI either <18.5 or ≥25
• During pregnancy	Insufficient or excessive pregnancy weight gain (see Table 15-1)
Maternal nutrition	Nutrient deficiencies or toxicities; eating disorders
Socioeconomic status	Poverty, lack of family support, low level of education, limited food available
Lifestyle habits	Smoking, alcohol or other drug use
Age	Teens, especially 15 years or younger; women 35 years or older
Previous pregnancies	
• Number	Many previous pregnancies (3 or more to mothers younger than age 20; 4 or more to mothers age 20 or older)
• Interval	Short or long intervals between pregnancies (<18 months or >59 months)
• Outcomes	Previous history of problems
• Multiple births	Twins or triplets
• Birthweight	Low- or high-birthweight infants
Maternal health	
• High blood pressure	Development of gestational hypertension
• Diabetes	Development of gestational diabetes
• Chronic diseases	Diabetes; heart, respiratory, and kidney disease; certain genetic disorders; special diets and medications

and do not catch up as well. For the most part, survival improves with increased gestational age and birthweight.

Low-birthweight infants are more likely to experience complications during delivery than normal-weight babies. They also have a statistically greater chance of having physical and mental birth defects, becoming ill, and dying early in life. Of infants who die before their first birthdays, about two-thirds were low-birthweight newborns. Very-low-birthweight infants (3½ pounds or less) struggle not only for their immediate physical health and survival, but for their future cognitive development and abilities as well.

A strong association is seen between socioeconomic disadvantage and low birthweight. Low socioeconomic status impairs fetal development by causing stress and by limiting access to medical care and nutritious foods. Low socioeconomic status often accompanies teen pregnancies, smoking, and alcohol and drug abuse—all predictors of low birthweight.

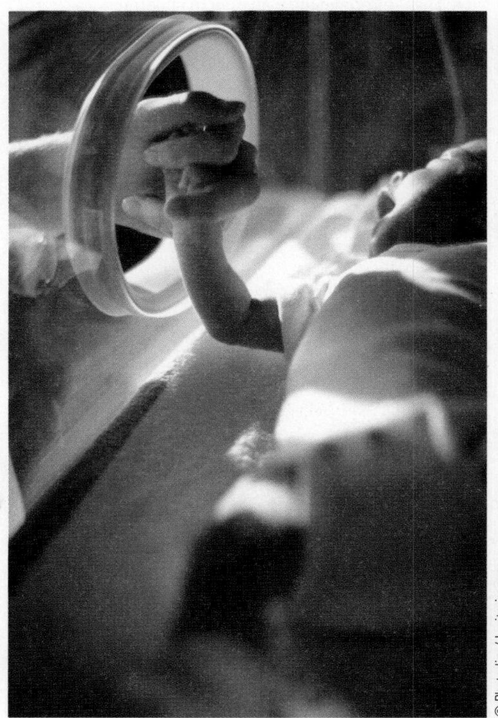

Low-birthweight babies need special care and nourishment.

Malnutrition and Pregnancy

Good nutrition clearly supports a healthy pregnancy. In contrast, malnutrition interferes with the ability to conceive, the likelihood of implantation, and the subsequent development of a fetus should conception and implantation occur.

Malnutrition and Fertility The nutrition habits and lifestyle choices people make can influence the course of a pregnancy they are not even planning at the time. Severe malnutrition and food deprivation can reduce fertility because women may develop amenorrhea—the temporary or permanent absence of menstrual periods.* Men who are malnourished may be unable to produce viable sperm. Furthermore, both men and women lose sexual interest during times of starvation. Starvation arises predictably during famines, wars, and droughts, but it can also occur amid peace and plenty. Many young women who diet excessively are starving and suffering from malnutrition (see Highlight 8).

*Amenorrhea is normal before puberty, after menopause, during pregnancy, and during lactation; otherwise it is abnormal.

Malnutrition and Early Pregnancy If a malnourished woman does become pregnant, she faces the challenge of supporting both the growth of a baby and her own health with inadequate nutrient stores. Malnutrition prior to and around conception prevents the placenta from developing fully. A poorly developed placenta cannot deliver optimum nourishment to the fetus, and the infant will be born small and possibly with physical and cognitive abnormalities. If this small infant is a female, she may develop poorly and have an elevated risk of developing a chronic condition that could impair her ability to give birth to a healthy infant. Thus a woman's malnutrition can adversely affect not only her children but her *grandchildren*.

Malnutrition and Fetal Development Without adequate nutrition during pregnancy, fetal growth and infant health are compromised.[56] In general, consequences of malnutrition during pregnancy include fetal growth restriction, congenital malformations (birth defects), spontaneous abortion and stillbirth, preterm birth, and low infant birthweight. Malnutrition, coupled with low birthweight, is a factor in more than half of all deaths of children younger than 4 years of age worldwide.

Food Assistance Programs

Women in high-risk pregnancies can find assistance from the WIC program—a high-quality, cost-effective health-care and nutrition services program for women, infants, and children in the United States. Formally known as the Special Supplemental Nutrition Program for Women, Infants, and Children, WIC provides nutrition education and nutritious foods to infants, children to age 5, and pregnant and breastfeeding women who qualify financially and have a high risk of medical or nutritional problems. ♦ The program is both remedial and preventive: services include health-care referrals, nutrition education, and food packages or vouchers for specific foods. These foods supply nutrients known to be lacking in the diets of the target population—most notably, protein, calcium, iron, vitamin A, and vitamin C. WIC-sponsored foods include tuna, tofu, fruits, vegetables, eggs, milk, iron-fortified cereal, whole-grain breads, vitamin C–rich juices, cheeses, legumes, peanut butter, and iron-fortified infant formula and cereal.

♦ How WIC helps:
- Earlier prenatal care
- Better diet during pregnancy
- Better weight gain during pregnancy
- Longer duration of pregnancy
- Fewer fetal and infant deaths
- Fewer low birthweight infants
- Better growth in infants and children
- Less iron-deficiency anemia in children
- Better diet for children
- Better medical care for children
- Better preparation for school
- Improved intellectual development

More than 9 million people—most of them young children—receive WIC benefits each month. Prenatal WIC participation can effectively reduce iron deficiency, infant mortality, low birthweight, and maternal and newborn medical costs.[57] In 2011, Congress appropriated more than $6.7 billion for WIC. For every dollar spent on WIC, an estimated $3 in medical costs are saved in the first 2 months after birth.

Maternal Health

Medical disorders can threaten the life and health of both mother and fetus. If diagnosed and treated early, many diseases can be managed to ensure a healthy outcome—another strong argument for early prenatal care. Furthermore, the changes in pregnancy can reveal disease risks, making screening important and early intervention possible.

Preexisting Diabetes The risks of diabetes depend on how well it is managed before and during pregnancy. Without proper management of maternal diabetes, women face high infertility rates, and those who do conceive may experience episodes of severe hypoglycemia or hyperglycemia, preterm labor, and pregnancy-related hypertension. Infants may be large, suffer physical and mental abnormalities, and experience other complications such as severe hypoglycemia or respiratory distress, both of which can be fatal. Signs of fetal health problems are apparent even when maternal glucose is above normal but still below the diagnosis of diabetes.[58] To minimize complications, a woman needs to achieve glucose control before conception and continued glucose control throughout pregnancy.[59]

Gestational Diabetes An estimated 2 to 10 percent of pregnancies in the United States are complicated by a condition known as **gestational diabetes.**[60] Gestational diabetes usually develops during the second half of pregnancy, with subsequent return to normal after childbirth. Some women with gestational diabetes, however, develop diabetes (usually type 2) after pregnancy, especially if they are overweight.[61] For this reason, health-care professionals strongly advise against excessive weight gain during—and after—pregnancy. Weight gains after pregnancy increase the risk of gestational diabetes in the next pregnancy.[62]

The most common consequences of gestational diabetes are complications during labor and delivery and a high infant birthweight.[63] Birth defects associated with gestational diabetes include heart damage, limb deformities, and neural tube defects. To ensure that the problems of gestational diabetes are dealt with promptly, physicians screen for the risk factors listed in the margin and test high-risk women for glucose intolerance immediately and average-risk women between 24 and 28 weeks gestation.[64] ♦

♦ Risk factors for gestational diabetes:
- Age 25 or older
- BMI ≥25 or excessive weight gain
- Complications in previous pregnancies, including gestational diabetes or high-birthweight infant
- Prediabetes or symptoms of diabetes
- Family history of diabetes
- Hispanic American, African American, Native American, Asian American, Pacific Islander

Dietary recommendations should meet the needs of pregnancy and maternal blood glucose goals.[65] Diet and moderate exercise may control gestational diabetes, but if blood glucose fails to normalize, insulin or other drugs may be required. Importantly, treatment reduces birth complications and infant deaths.[66]

Chronic Hypertension Hypertension complicates pregnancy and affects its outcome in different ways, depending on when the hypertension first develops and on how severe it becomes.[67] In addition to the threats hypertension always carries (such as heart attack and stroke), high blood pressure increases the risks of fetal growth restriction, preterm birth, and separation of the placenta from the wall of the uterus before the birth, resulting in stillbirth.[68] To minimize complications, blood pressure needs to be under control before a woman with hypertension becomes pregnant.

Gestational Hypertension Women with chronic hypertension have a greater likelihood of developing **gestational hypertension**—high blood pressure during the second half of pregnancy.* For some women with gestational hypertension, the rise in blood pressure is mild and does not affect the pregnancy adversely. Blood pressure usually returns to normal during the first few weeks after childbirth. For others, gestational hypertension increases the risks of subsequent hypertension and type 2 diabetes.[69] Gestational hypertension is also an early sign of the most serious maternal complication of pregnancy—preeclampsia.

Preeclampsia **Preeclampsia** is a condition characterized not only by gestational hypertension but also by protein in the urine. The cause of preeclampsia remains unclear, but it usually occurs with first pregnancies and most often after 20 weeks gestation. ♦ Symptoms typically regress within 2 days of delivery. Both men and women who were born of pregnancies complicated by preeclampsia are more likely to have a child born of a pregnancy complicated by preeclampsia, suggesting a genetic predisposition. Black women have a much greater risk of preeclampsia than white women.

♦ Signs and symptoms of preeclampsia:
- Hypertension
- Protein in the urine
- Upper abdominal pain
- Severe headaches
- Swelling of hands, feet, and face
- Vomiting
- Blurred vision
- Sudden weight gain (1 lb/day)
- Fetal growth restriction

Preeclampsia affects almost all of the mother's organs—the circulatory system, liver, kidneys, and brain.[70] Blood flow through the vessels that supply oxygen and nutrients to the placenta diminishes. For this reason, preeclampsia often restricts fetal growth. It also seems to increase the risk of epilepsy for the infant.[71] In some cases, the placenta separates from the uterus, resulting in preterm birth or stillbirth.

Preeclampsia can progress rapidly to **eclampsia**—a condition characterized by seizures and coma. Maternal death during pregnancy and childbirth is rare in developed countries, but when it does occur, eclampsia is a common cause.[72] The rate of death for black women with eclampsia is more than four times the rate for white women.

gestational diabetes: glucose intolerance with onset or first recognition during pregnancy.

gestational hypertension: high blood pressure that develops in the second half of pregnancy and resolves after childbirth, usually without affecting the outcome of the pregnancy.

preeclampsia (PRE-ee-KLAMP-see-ah): a condition characterized by high blood pressure and some protein in the urine.

eclampsia (eh-KLAMP-see-ah): a condition characterized by extremely high blood pressure, elevated protein in the urine, seizures, and possibly coma.

*Blood pressure of 140/90 millimeters mercury or greater during the second half of pregnancy in a woman who has not previously exhibited hypertension indicates high blood pressure.

Preeclampsia demands prompt medical attention. Treatment focuses on controlling blood pressure and preventing seizures. If preeclampsia develops early and is severe, induced labor or cesarean section may be necessary, regardless of gestational age. The infant will be preterm, with all of the associated problems, including poor lung development and special care needs. Several dietary factors have been studied, but none have proved beneficial in preventing preeclampsia.[73] Limited research suggests that exercise may protect against preeclampsia by stimulating placenta growth and vascularity and reducing oxidative stress.[74]

The Mother's Age Maternal age also influences the course of a pregnancy. Compared with women of the physically ideal childbearing age of 20 to 25, both younger and older women face more complications of pregnancy.

Pregnancy in Adolescents Many adolescents become sexually active before age 19; more than 400,000 adolescent girls give birth each year in the United States.[75] Nourishing a growing fetus adds to a teenage girl's nutrition burden, especially if her growth is still incomplete. Simply being young and physically immature increases the risks of pregnancy complications. Pregnant teens are less likely to receive early prenatal care and are more likely to smoke during pregnancy—two factors that predict low birthweight and infant death.[76]

The typical energy-dense, but nutrient-poor diet of pregnant adolescents increases the risk of low birthweight infants.[77] Common complications among adolescent mothers include iron-deficiency anemia (which may reflect poor diet and inadequate prenatal care) and prolonged labor (which reflects the mother's physical immaturity). On a positive note, maternal death is lowest for mothers younger than age 20.

The rates of stillbirths, preterm births, and low-birthweight infants is high for teenagers—both for teen moms and for teen dads.[78] Many of these infants suffer physical problems, require intensive care, and die within the first year. The care of teen mothers and their infants costs our society billions of dollars annually.[79] Because teenagers have few financial resources, they cannot pay these costs. Furthermore, their low economic status contributes significantly to the complications surrounding their pregnancies. At a time when prenatal care is most important, it is less accessible. And the pattern of teenage pregnancies continues from generation to generation, with daughters of teenage mothers 66 percent more likely to become teenage mothers themselves.[80] Clearly, teenage pregnancy is a major public health problem.

To support the needs of both mother and fetus, young teenagers (13 to 16 years old) are encouraged to strive for the highest weight gains recommended for pregnancy. For a teen who enters pregnancy at a healthy body weight, a weight gain of approximately 35 pounds is recommended; this amount minimizes the risk of delivering a low-birthweight infant. Pregnant and lactating teenagers can use the food patterns presented in Table 2-2 (p. 44), making sure to select a high enough kcalorie level to support adequate weight gain.

Without the appropriate economic, social, and physical support, a young mother will not be able to care for herself during her pregnancy and for her child after the birth. To improve her chances for a successful pregnancy and a healthy infant, she must seek prenatal care. WIC provides health-care referrals and helps pregnant teenagers obtain adequate food for themselves and their infants. (WIC is introduced on p. 486.)

Pregnancy in Older Women In the last several decades, many women have delayed childbearing while they pursue education and careers. As a result, the number of first births to women 35 and older has increased dramatically. Most of these women, even those older than age 50, have healthy pregnancies.

The few complications associated with later childbearing often reflect chronic conditions such as hypertension and diabetes, which can complicate an otherwise healthy pregnancy. These complications may result in a cesarean section, which is twice as common in women older than 35 as among younger women. For all these reasons, maternal death rates are higher in women older than 35 than in younger women.

The babies of older mothers face problems of their own including higher rates of preterm births and low birthweight. Their rates of birth defects are also high. Because 1 out of 50 pregnancies in older women produces an infant with genetic abnormalities, obstetricians routinely screen women older than 35. For a 40-year-old mother, the risk of having a child with **Down syndrome,** for example, is about 1 in 100 compared with 1 in 300 for a 35 year old and 1 in 10,000 for a 20 year old. In addition, fetal death is twice as high for women 35 years and older than for younger women. Why this is so remains unclear. One possibility is that the uterine blood vessels of older women may not fully adapt to the increased demands of pregnancy.

Practices Incompatible with Pregnancy Besides malnutrition, a variety of lifestyle factors can have adverse effects on pregnancy, and some may be teratogenic, causing abnormal fetal development and birth defects. By practicing healthy behaviors, people who are planning to have children can reduce the risks.

Alcohol One out of 8 pregnant women drinks alcohol at some time during her pregnancy; 1 out of 50 reports binge drinking.[81] Alcohol consumption during pregnancy can cause the irreversible mental and physical retardation of the fetus known as fetal alcohol syndrome (FAS). Of the leading causes of mental retardation, FAS is the only one that is totally *preventable.* To that end, the surgeon general urges all pregnant women to refrain from drinking alcohol. Fetal alcohol syndrome is the topic of Highlight 15, which includes mention of how alcohol consumption by men may also affect fertility and fetal development.

> DIETARY GUIDELINES FOR AMERICANS 2010

Women who are pregnant should not drink alcohol.

Medicinal Drugs Drugs other than alcohol can also cause complications during pregnancy, problems in labor, and serious birth defects. For these reasons, pregnant women should not take any medicines without consulting their physicians, who must weigh the benefits against the risks.

Herbal Supplements Similarly, pregnant women should seek a physician's advice before using herbal supplements. Women sometimes seek herbal preparations during their pregnancies to quell nausea, induce labor, aid digestion, promote water loss, support restful sleep, and fight depression. As Highlight 18 explains, some herbs may be safe, but many others are definitely harmful.

Illicit Drugs The recommendation to avoid drugs during pregnancy also includes illicit drugs, of course. Unfortunately, use of illicit drugs, such as cocaine and marijuana, is common among some pregnant women.

Drugs of abuse, such as cocaine, easily cross the placenta and impair fetal growth and development. Furthermore, they are responsible for preterm births, low-birthweight infants, **perinatal** deaths, and sudden infant deaths. If these newborns survive, central nervous system damage is evident: their cries, sleep, and behaviors early in life are abnormal, and their cognitive development later in life is impaired.[82] They may be hypersensitive or underaroused; those who test positive for drugs suffer the greatest effects of toxicity and withdrawal.[83] Their growth throughout childhood continues at a slow rate.

Smoking and Chewing Tobacco Unfortunately, an estimated 10 to 14 percent of pregnant women in the United States smoke.[84] Smoking cigarettes and chewing tobacco at any time exert harmful effects, and pregnancy dramatically magnifies the hazards of these practices. Smoking restricts the blood supply to the growing fetus and thus limits oxygen and nutrient delivery and waste removal. A mother who smokes is more likely to have a complicated birth and a low-birthweight infant. Indeed, of all preventable causes of low birthweight in the United States, smoking is at the top of the list. Although most infants born to cigarette smokers are low birthweight, some are not, suggesting that the effect of

Down syndrome: a genetic abnormality that causes mental retardation, short stature, and flattened facial features.

perinatal: referring to the time between the twenty-eighth week of gestation and 1 month after birth.

smoking on birthweight also depends, in part, on genes involved in the metabolism of smoking toxins.

In addition to contributing to low birthweight, smoking interferes with heart and lung growth and increases the risks of heart defects, poor lung function, respiratory infections, and childhood asthma.[85] It can also cause death in an otherwise healthy fetus or newborn. A positive relationship exists between **sudden infant death syndrome (SIDS)** and both cigarette smoking during pregnancy and postnatal exposure to passive smoke. Smoking during pregnancy may reduce brain size and impair the intellectual and behavioral development of the child later in life. The margin lists complications of smoking during pregnancy.[86] ♦

♦ Complications associated with smoking during pregnancy:

- Fetal growth restriction
- Preterm birth
- Low birthweight
- Premature separation of the placenta
- Miscarriage
- Stillbirth
- Sudden infant death syndrome (SIDS)
- Congenital malformations

Alternatives to smoking—such as using snuff, chewing tobacco, or nicotine-replacement therapy—are not safe during pregnancy.[87] Any woman who uses nicotine in any form and is considering pregnancy or who is already pregnant needs to quit. Avoiding secondhand smoke is also advised. Pregnant women exposed to secondhand smoke during pregnancy are more likely to experience complications such as stillbirth and the birth of an infant with congenital malformations.[88]

Environmental Contaminants Proving that environmental contaminants cause reproductive damage is difficult, but evidence in wildlife is established and seems likely for human beings. Infants and young children of pregnant women exposed to environmental contaminants such as lead show signs of delayed mental and psychomotor development. During pregnancy, lead readily crosses into the placenta, inflicting severe damage on the developing fetal nervous system. In addition, infants exposed to even low levels of lead during gestation weigh less at birth and consequently struggle to survive. For these reasons, it is particularly important that pregnant women receive foods and beverages grown and prepared in environments free of contamination.

Mercury is another contaminant of concern. As Chapter 5 mentions, fatty fish are a good source of omega-3 fatty acids, but some fish contain large amounts of the pollutant mercury, which can impair fetal growth and harm the developing brain and nervous system. Because the benefits of seafood consumption seem to outweigh the risks, pregnant (and lactating) women need reliable information on which fish are safe to eat.[89] In general, they need to do the following:[90]

- Avoid shark, swordfish, king mackerel, and tilefish (also called golden snapper or golden bass).
- Limit average weekly consumption to 12 ounces (cooked or canned) of seafood *or* to 6 ounces (cooked or canned) of white (albacore) tuna.

Ideally, pregnant (and lactating) women will select fish that are both high in omega-3 fatty acids and low in mercury.[91] ♦ Supplements of fish oil are not recommended because they may contain concentrated toxins and because their effects on pregnancy remain unknown.

♦ Fish relatively high in omega-3 fatty acids and low in mercury:
Anchovies
Bonito
Eel
Herring
Mackerel
Pollock
Salmon
Sardines
Smelt
Tilapia
Trout

> **DIETARY GUIDELINES FOR AMERICANS 2010**
Women who are pregnant or breastfeeding should consume 8 to 12 ounces of seafood per week from a variety of seafood types. Because of the high methyl mercury content of some types of fish, pregnant or breastfeeding women should avoid tilefish, shark, swordfish, and king mackerel and should limit white (albacore) tuna to 6 ounces per week.

Foodborne Illness As Chapter 19 explains, foodborne illnesses arise when people eat foods that contain infectious microbes or microbes that produce toxins. At best, the vomiting and diarrhea associated with these illnesses can leave a pregnant woman exhausted and dehydrated; at worst, foodborne illnesses can cause meningitis, pneumonia, or even fetal death. Pregnant women are about 20 times more likely than other healthy adults to get the foodborne illness

sudden infant death syndrome (SIDS): the unexpected and unexplained death of an apparently well infant; the most common cause of death of infants between the second week and the end of the first year of life; also called *crib death.*

listeriosis. The margin presents tips to prevent listeriosis; ♦ Chapter 19 includes precautions to minimize the risks of other common foodborne illnesses.

> **DIETARY GUIDELINES FOR AMERICANS 2010**
> Women who are pregnant should:
> - Eat foods containing seafood, meat, poultry, or eggs only if cooked to recommended safe minimum internal temperatures.
> - Take special precautions not to consume unpasteurized juice or milk products.
> - Reheat deli and luncheon meats and hot dogs to steaming hot and not eat raw sprouts.

Vitamin-Mineral Megadoses Pregnant women who are trying to eat well may mistakenly assume that more is better when it comes to multivitamin-mineral supplements. This is simply not true; many vitamins and minerals are toxic when taken in excess. Excessive vitamin A is particularly infamous for its role in fetal malformations of the cranial nervous system. Intakes before the seventh week appear to be the most damaging. (Review Figure 15-4 on p. 473 to see how many tissues are in their critical periods prior to the seventh week.) For this reason, vitamin A supplements are not given during pregnancy unless there is specific evidence of deficiency, which is rare. A pregnant woman can obtain all the vitamin A and most of the other vitamins and minerals she needs by making wise food choices. She should take supplements only on the advice of a registered dietitian or physician.

Caffeine Caffeine crosses the placenta, and the developing fetus has a limited ability to metabolize it. Research studies have not proved that caffeine (even in high doses) causes birth defects or preterm births in human infants (as it does in animals), but limited evidence suggests that heavy use increases the risk of hypertension, miscarriage, and stillbirth.[92] (In these studies, heavy caffeine use is defined as the equivalent of 3 or more cups of coffee a day.) Depending on the quantities consumed and the mother's metabolism, caffeine may also interfere with fetal growth.[93] All things considered, it is most sensible to limit caffeine consumption to the equivalent of a cup of coffee or two 12-ounce cola beverages a day. (The caffeine contents of selected beverages, foods, and drugs are listed at the beginning of Appendix H.)

Restrictive Dieting Restrictive dieting, even for short periods, can be hazardous during pregnancy. Low-carbohydrate diets or fasts that cause ketosis deprive the fetal brain of needed glucose and may impair cognitive development. Such diets are also likely to lack other nutrients vital to fetal growth. Regardless of prepregnancy weight, pregnant women need an adequate diet to support healthy fetal development.

Sugar Substitutes Artificial sweeteners have been extensively investigated and found to be acceptable during pregnancy if used within the FDA's guidelines.[94] Still, it is prudent for pregnant women to use sweeteners in moderation and within an otherwise nutritious and well-balanced diet. Women with the inherited disease phenylketonuria (PKU) should not use the artificial sweetener aspartame. Aspartame contains the amino acid phenylalanine, and people with PKU are unable to dispose of any excess phenylalanine. The accumulation of phenylalanine and its by-products is toxic to the developing nervous system, causing irreversible brain damage.

REVIEW IT Identify factors predicting low-risk and high-risk pregnancies and describe ways to manage them.

High-risk pregnancies, especially for teenagers, threaten the life and health of both mother and infant. Proper nutrition and abstinence from smoking, alcohol, and other drugs improve the outcome. In addition, prenatal care includes monitoring pregnant women for gestational diabetes, gestational hypertension, and preeclampsia.

♦ To prevent listeriosis:

- Use only pasteurized juices and dairy products; do not eat soft cheeses such as feta, brie, Camembert, Panela, "queso blanco," "queso fresco," and blue-veined cheeses such as Roquefort; do not drink raw (unpasteurized) milk or eat foods that contain it.
- Thoroughly cook meat, poultry, eggs, and seafood.
- Do not eat hot dogs or luncheon meats unless heated until steaming hot.
- Wash all fruits and vegetables.
- Do not eat refrigerated pâtés or meat spreads.
- Do not eat refrigerated smoked seafood such as salmon or trout, or any fish labeled "nova," "lox," or "kippered," unless prepared in a cooked dish.

listeriosis (lis-TEAR-ee-OH-sis): an infection caused by eating food contaminated with the bacterium *Listeria monocytogenes*, which can be killed by pasteurization and cooking but can survive at refrigerated temperatures; certain ready-to-eat foods, such as hot dogs and deli meats, may become contaminated after cooking or processing, but before packaging.

15.6 Nutrition during Lactation

LEARN IT Summarize the nutrient needs of women during lactation.

Childbirth marks the end of pregnancy and the beginning of a new set of parental responsibilities—including feeding the newborn. Before the end of her pregnancy, a woman needs to consider whether to feed her infant breast milk, infant formula, or both. These options are the only recommended foods for an infant during the first 4 to 6 months of life. The current rate of breastfeeding met the Healthy People goal of 75 percent at birth, but it falls far short of goals at 3 months, 6 months, and 1 year.[95] This section focuses on how the mother's nutrition supports the making of breast milk, and the next chapter describes how the infant benefits from drinking breast milk.

In many countries around the world, a woman breastfeeds her newborn without considering the alternatives or making a conscious decision. In other parts of the world, a woman feeds her newborn formula simply because she knows so little about breastfeeding. She may have misconceptions or feel uncomfortable about a process she has never seen or experienced. Breastfeeding offers many health benefits to both mother and infant, and every pregnant woman should seriously consider it (see Table 15-4).[96] Even so, women's choices are often influenced by factors other than health and science—factors such as culture, politics, religion, and marketing.[97] In any case, keep in mind that mothers may have valid reasons for not breastfeeding and that formula-fed infants grow and develop into healthy children.

Lactation: A Physiological Process **Lactation** naturally follows pregnancy, as the mother's body continues to nourish the infant. The **mammary glands** secrete milk for this purpose. The mammary glands develop during puberty but remain fairly inactive until pregnancy. During pregnancy, hormones promote the growth and branching of a duct system in the breasts and the development of the milk-producing cells.

TABLE 15-4 Benefits of Breastfeeding

For Infants

- Provides the appropriate composition and balance of nutrients with high bioavailability
- Provides hormones that promote physiological development
- Improves cognitive development
- Protects against a variety of infections and illnesses, including diarrhea, ear infections, and pneumonia
- May protect against some chronic diseases—such as diabetes (both types), obesity, atherosclerosis, asthma, and hypertension—later in life
- Protects against food allergies
- Reduces the risk of SIDS
- Supports healthy weight

For Mothers

- Contracts the uterus
- Delays the return of regular ovulation, thus lengthening birth intervals (this is not, however, a dependable method of contraception)
- Conserves iron stores (by prolonging amenorrhea)
- May protect against breast and ovarian cancer and reduce the risk of diabetes (type 2)
- Increases energy expenditure, which may contribute to weight loss

Other

- Cost and time savings from not needing medical treatment for childhood illnesses or leaving work to care for sick infants
- Cost and time savings from not needing to purchase and prepare formula (even after adjusting for added foods in the diet of a lactating mother)[a]
- Environmental savings to society from not needing to manufacture, package, and ship formula and dispose of the packaging
- Convenience of not having to shop for and prepare formula

[a]Estimated savings of $1200–$1500 in the first year.

lactation: production and secretion of breast milk for the purpose of nourishing an infant.

mammary glands: glands of the female breast that secrete milk.

The hormones **prolactin** and **oxytocin** finely coordinate lactation. The infant's demand for milk stimulates the release of these hormones, which signal the mammary glands to supply milk. Prolactin is responsible for milk production. As long as the infant is nursing, prolactin concentrations remain high, and milk production continues.

The hormone oxytocin causes the mammary glands to eject milk into the ducts, a response known as the **let-down reflex.** The mother feels this reflex as a contraction of the breast, followed by the flow of milk and the release of pressure. By relaxing and eating well, the nursing mother promotes easy let-down of milk and greatly enhances her chances of successful lactation.

© LWA/Getty Images

A woman who decides to breastfeed provides her infant with a full array of nutrients and protective factors to support optimal health and development.

Breastfeeding: A Learned Behavior Lactation is an automatic physiological process that virtually all mothers are capable of doing. Breastfeeding, on the other hand, is a learned behavior that not all mothers decide to do. Of women who do breastfeed, those who receive early and repeated information and support breastfeed their infants longer than others. Mothers who are confident and committed are most successful in breastfeeding, especially when challenged by obstacles such as a lack of support from friends and family.[98] Health-care professionals play an important role in providing encouragement and accurate information on breastfeeding. Especially helpful are **certified lactation consultants** who specialize in helping new mothers establish a healthy breastfeeding relationship with their newborn. These consultants are often registered nurses with specialized training in breast and infant anatomy and physiology. Women who have been successful breastfeeding can offer advice and dispel misperceptions about lifestyle issues. Table 15-5 lists tips to promote successful breastfeeding among new mothers.

The mother's partner also plays an important role in encouraging breastfeeding. When partners support the decision, mothers are more likely to start and continue breastfeeding. Clearly, educating those closest to the mother could change attitudes and promote breastfeeding.

Most healthy women who want to breastfeed can do so with a little preparation. Physical obstacles to breastfeeding are rare, although most nursing mothers quit within a few months because of perceived difficulties. Obese mothers seem to have a particularly difficult time because of both biological and sociocultural factors.[99] Successful breastfeeding requires adequate nutrition and rest. This, plus the support of all who care, will help to enhance the well-being of mother and infant.

Maternal Energy and Nutrient Needs during Lactation Ideally, the mother who chooses to breastfeed her infant will continue to eat nutrient-dense foods throughout lactation. An adequate diet is needed to support the stamina, patience, and self-confidence that nursing an infant demands.

Energy Intake and Exercise A nursing mother produces about 25 ounces of milk per day, with considerable variation from woman to woman and in the same woman from time to time, depending primarily on the infant's demand for milk. To produce an adequate supply of milk, a woman needs extra energy—almost 500 kcalories a day above her regular need during the first 6 months of lactation. To meet this energy need, she can eat an extra 330 kcalories of food each day and let the fat reserves she accumulated during pregnancy provide the rest. ♦ Most

♦ Energy requirement during lactation:
- 1st 6 mo: +330 kcal/day
- 2nd 6 mo: +400 kcal/day

TABLE 15-5 Tips for Successful Breastfeeding

- Learn about the benefits of breastfeeding
- Initiate breastfeeding within 1 hour of birth
- Ask a health-care professional to explain how to breastfeed and how to maintain lactation
- Give newborn infants no food or drink other than breast milk, unless medically indicated
- Breastfeed on demand
- Give no artificial nipples or pacifiers to breastfeeding infants[a]
- Find breastfeeding support groups, books, or websites to help troubleshoot breastfeeding problems

[a]Compared with nonusers, infants who use pacifiers breastfeed less frequently and stop breastfeeding at a younger age.

prolactin (pro-LAK-tin): a hormone secreted from the anterior pituitary gland that acts on the mammary glands to promote the production of milk. The release of prolactin is mediated by *prolactin-inhibiting hormone (PIH).*
- **pro** = promote
- **lacto** = milk

oxytocin (OCK-see-TOH-sin): a hormone that stimulates the mammary glands to eject milk during lactation and the uterus to contract during childbirth.

let-down reflex: the reflex that forces milk to the front of the breast when the infant begins to nurse.

certified lactation consultants: health-care providers who specialize in helping new mothers establish a healthy breastfeeding relationship with their newborn. These consultants are often registered nurses with specialized training in breast and infant anatomy and physiology.

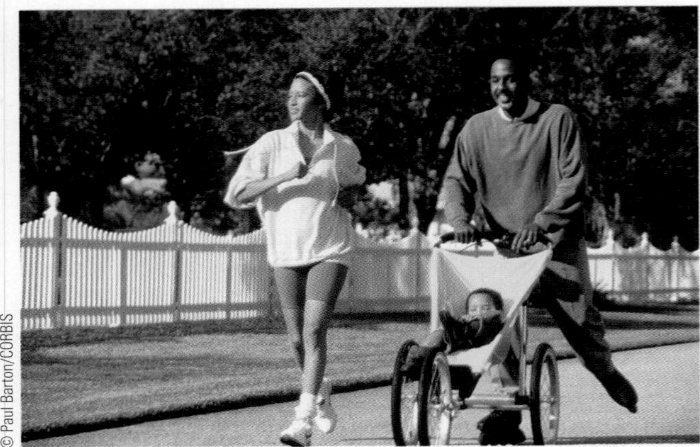
© Paul Barton/CORBIS

A jog through the park provides an opportunity for physical activity and fresh air.

women need at least 1800 kcalories a day to receive all the nutrients required for successful lactation. Severe energy restriction may hinder milk production.

After the birth of the infant, many women actively try to lose the extra weight and body fat they accumulated during pregnancy. How much weight a woman retains after pregnancy depends on her gestational weight gain and the duration and intensity of breastfeeding. Many women who follow recommendations for gestational weight gain and breastfeeding can readily return to prepregnancy weight by 6 months. Neither the quality nor the quantity of breast milk is adversely affected by moderate weight loss, and infants grow normally.

Women often exercise to lose weight and improve fitness, and this is compatible with breastfeeding and infant growth. Because intense physical activity can raise the lactate concentration of breast milk and influence the milk's taste, some infants may prefer milk produced prior to exercise. In these cases, mothers can either breastfeed before exercise or express their milk before exercise for use afterward.

> **DIETARY GUIDELINES FOR AMERICANS 2010**
> Maintain appropriate kcalorie balance during breastfeeding.

Energy Nutrients Recommendations for protein and fatty acids remain about the same during lactation as during pregnancy, but they increase for carbohydrates. Nursing mothers need additional carbohydrate to replace the glucose used to make the lactose in breast milk. The fiber recommendation is 1 gram higher simply because it is based on kcalorie intake, which increases during lactation.

Vitamins and Minerals A question often raised is whether a mother's milk may lack a nutrient if she fails to get enough in her diet. The answer differs from one nutrient to the next, but in general, nutrient inadequacies reduce the *quantity,* not the *quality,* of breast milk. Women can produce milk with adequate protein, carbohydrate, fat, and most minerals, even when their own supplies are limited. For these nutrients and for the vitamin folate as well, milk quality is maintained at the expense of maternal stores. This is most evident in the case of calcium: dietary calcium has no effect on the calcium concentration of breast milk, but maternal bones lose some density during lactation if calcium intakes are inadequate. Exercise may help protect against bone loss during lactation.[100] The nutrients in breast milk that are most likely to decline in response to prolonged inadequate intakes are the vitamins—especially vitamins B_6, B_{12}, A, and D. Review Figure 15-9 (p. 480) to compare a lactating woman's nutrient needs with those of pregnant and nonpregnant women.

Water Despite misconceptions, a mother who drinks more fluid does not produce more breast milk. To protect herself from dehydration, however, a lactating woman needs to drink plenty of fluids. ♦ A sensible guideline is to drink a glass of milk, juice, or water at each meal and each time the infant nurses.

♦ AI for *total* water (including drinking water, other beverages, and foods) during lactation: 3.8 L/day. Because foods provide about 20% of total water intake, beverages—including drinking water—should provide 3.1 L/ day (≈13 c).

Nutrient Supplements Most lactating women can obtain all the nutrients they need from a well-balanced diet without taking multivitamin-mineral supplements. Nevertheless, some may need iron supplements, not to enhance the iron in breast milk, but to refill depleted iron stores. The mother's iron stores dwindle during pregnancy as she supplies the developing fetus with enough iron to last through the first 4 to 6 months of the infant's life. In addition, childbirth may have incurred blood losses. Thus a woman may need iron supplements during lactation even though, until menstruation resumes, her iron requirement is about half that of other nonpregnant women her age.

Food Assistance Programs In general, women most likely to participate in the food assistance program WIC—those who are poor and have little education—are less likely to breastfeed.[101] Furthermore, WIC provides infant formula at no cost.

Because WIC recognizes the many benefits of breastfeeding, efforts are made to overcome this dilemma.[102] In addition to nutrition education and encouragement, breastfeeding mothers receive the WIC incentives listed in the margin. ♦ Together, these efforts help to provide nutrition support and encourage WIC mothers to breastfeed.

© Istockphoto.com/Morgan Lane Studios

Nutritious foods support successful lactation.

Particular Foods Foods with strong or spicy flavors (such as garlic) may alter the flavor of breast milk. A sudden change in the taste of the milk may annoy some infants. Familiar flavors may enhance enjoyment. Flavors in breast milk from the mother's diet can influence the infant's later food preferences.[103]

Current evidence does not support a major role for maternal dietary restrictions during lactation to prevent or delay the onset of food allergy in infants.[104] Infants who develop symptoms of food allergy, however, may be more comfortable if the mother's diet excludes the most common offenders—cow's milk, eggs, fish, peanuts, and tree nuts. Generally, infants with a strong family history of food allergies benefit from breastfeeding.[105]

A nursing mother can usually eat whatever nutritious foods she chooses. If she suspects a particular food is causing the infant discomfort, her physician may recommend a dietary challenge: eliminate the food from the diet to see if the infant's reactions subside, then return the food to the diet and again monitor the infant's reactions. If a food must be eliminated for an extended time, appropriate substitutions must be made to ensure nutrient adequacy.

♦ WIC incentives to breastfeed:

- Higher priority in certification into WIC
- Longer eligibility to participate in WIC
- More foods and larger quantities
- Breast pumps and other support materials

Maternal Health

If a woman has an ordinary cold, she can continue nursing without worry. If susceptible, the infant will catch it from her anyway. Thanks to the immunological protection of breast milk, the baby may be less susceptible than a formula-fed baby would be. With appropriate treatment, a woman who has an infectious disease such as tuberculosis or hepatitis can breastfeed; transmission is rare. Women with HIV (human immunodeficiency virus) infections, however, should consider other options.

HIV Infection and AIDS Mothers with HIV infections can transmit the virus (which causes AIDS) to their infants through breast milk, especially during the early months of breastfeeding. In developed countries such as the United States, where safe alternatives are available, HIV-positive women should *not* breastfeed their infants.[106] In developing countries, where the feeding of inappropriate or contaminated formulas causes more than 1 million infant deaths each year, breastfeeding can be critical to infant survival.[107] Thus, in making the decision of whether to breastfeed, HIV-infected women in developing countries must weigh the potential risks and benefits. The World Health Organization (WHO) recommends exclusive breastfeeding for infants of HIV-infected women for the first six months of life unless formula feeding is acceptable, feasible, affordable, sustainable, and safe before that time.[108] Alternatively, HIV-exposed infants may be protected by receiving antiretroviral treatment while being breastfed.[109]

Diabetes Women with diabetes (type 1) may need careful monitoring and counseling to ensure successful lactation. These women need to adjust their energy intakes and insulin doses to meet the heightened needs of lactation. Maintaining good glucose control helps to initiate lactation and support milk production.

Postpartum Amenorrhea Women who breastfeed experience prolonged **postpartum amenorrhea.** Absent menstrual periods, however, do not protect a woman from pregnancy. To prevent pregnancy, a couple must use some form of contraception. Breastfeeding women who use oral contraceptives should use progestin-only agents for at least the first 6 months. Estrogen-containing oral contraceptives reduce the volume and the protein content of breast milk.

Breast Health Some women fear that breastfeeding will cause their breasts to sag. The breasts do swell and become heavy and large immediately after the birth,

postpartum amenorrhea (ay-MEN-oh-REE-ah): the normal temporary absence of menstrual periods immediately following childbirth.

but even when they produce enough milk to nourish a thriving infant, they eventually shrink back to their prepregnant size. Given proper support, diet, and exercise, breasts often return to their former shape and size when lactation ends. Breasts change their shape as the body ages, but breastfeeding does not accelerate this process.

Whether the physical and hormonal events of pregnancy and lactation protect women from later breast cancer is an area of active research.[110] Some research suggests no association between breastfeeding and breast cancer, whereas other research suggests a protective effect.

Practices Incompatible with Lactation Some substances impair milk production or enter breast milk and interfere with infant development. This section discusses practices that a breastfeeding mother should avoid.

Alcohol Alcohol easily enters breast milk, and its concentration peaks within an hour of ingestion. Infants drink less breast milk when their mothers have consumed even small amounts of alcohol (equivalent to a can of beer). Three possible reasons, acting separately or together, may explain why. For one, the alcohol may have altered the flavor of the breast milk and thereby the infants' acceptance of it. For another, because infants metabolize alcohol inefficiently, even low doses may be potent enough to suppress their feeding and cause sleepiness. Third, the alcohol may have interfered with lactation by inhibiting the hormone oxytocin.

In the past, alcohol has been recommended to mothers to facilitate lactation despite a lack of scientific evidence that it does so. The research summarized here suggests that alcohol actually hinders breastfeeding. An occasional alcoholic beverage may be within safe limits, but breastfeeding should be delayed for at least 2 hours afterward.

Medicinal Drugs Most medicines are compatible with breastfeeding, but some are contraindicated, either because they suppress lactation or because they are secreted into breast milk and can harm the infant. As a precaution, a nursing mother should consult with her physician prior to taking any drug, including herbal supplements.

Illicit Drugs Illicit drugs, of course, are harmful to the physical and emotional health of both the mother and the nursing infant. Breast milk can deliver such high doses of illicit drugs as to cause irritability, tremors, hallucinations, and even death in infants. Women whose infants have overdosed on illicit drugs contained in breast milk have been convicted of murder. Women who use methadone to control withdrawal symptoms for opiate addiction can safely breastfeed their infants.[111]

Smoking About half the women who quit smoking during pregnancy relapse after delivery.[112] Mothers who smoke are likely to not breastfeed and to wean earlier than nonsmoking mothers.[113] Because cigarette smoking reduces milk volume, smokers may produce too little milk to meet their infants' energy needs. The milk they do produce contains nicotine, which alters its smell and flavor. Furthermore, infants of breastfeeding mothers who smoke have an increased risk for SIDS.[114] Breastfeeding helps to protect against SIDS, but infant exposure to passive smoke negates this protective effect.[115]

Environmental Contaminants Chapter 19 discusses environmental contaminants in the food supply. Some of these environmental contaminants, such as DDT, PCBs, and dioxin, can find their way into breast milk. Inuit mothers living in Arctic Québec who eat seal and beluga whale blubber have high concentrations of DDT and PCBs in their breast milk, but the impact on infant development is unclear. Preliminary studies indicate that the children of these Inuit mothers are developing normally. Researchers speculate that the abundant omega-3 fatty acids of the Inuit diet may protect against damage to the central nervous system. Breast milk tainted with dioxin interferes with tooth development during early infancy, producing soft, mottled teeth that are vulnerable to dental caries. To limit mercury intake, lactating women should heed the fish restrictions mentioned previously for pregnant women (see p. 490).

Caffeine Caffeine enters breast milk and may make an infant irritable and wakeful. As during pregnancy, caffeine consumption should be moderate—the equivalent of 1 to 2 cups of coffee a day. Larger doses of caffeine may interfere with the bioavailability of iron from breast milk and impair the infant's iron status.

REVIEW IT Summarize the nutrient needs of women during lactation.
The lactating woman needs extra fluid and enough energy and nutrients to produce about 25 ounces of milk a day. Breastfeeding is contraindicated for those with HIV/AIDS. Alcohol, other drugs, smoking, and contaminants may reduce milk production or enter breast milk and impair infant development.

This chapter has focused on the nutrition needs of the mother during pregnancy and lactation. The next chapter explores the dietary needs of infants, children, and adolescents.

Nutrition Portfolio

The choices a woman makes in preparation for, and in support of, pregnancy and lactation can influence both her health and her infant's development—today and for decades to come. Go to Diet Analysis Plus and choose one of the days on which you tracked your diet and activity for an entire day. Select the Intake vs. Goals report to help you answer the following questions:

- For women of childbearing age, determine whether you consume at least 400 micrograms of dietary folate equivalents daily.
- For women who are pregnant, evaluate whether you are meeting your nutrition needs and gaining the amount of weight recommended.
- For women who are about to give birth, carefully consider all the advantages of breastfeeding your infant and obtain the needed advice to support you.

Diet Analysis PLUS

To complete this exercise, go to your Diet Analysis Plus at **www.cengagebrain.com.**

STUDY IT To review the key points of this chapter and take a practice quiz, go to the study cards at the end of the book.

REFERENCES

1. V. Berghella and coauthors, Preconception care, *Obstetric and Gynecological Survey* 65 (2010): 119–131; P. M. Gardiner and coauthors, The clinical content of preconception care: Nutrition and dietary supplements, *American Journal of Obstetrics and Gynecology* 199 (2008): S345–S356.
2. M. Jokela, M. Elovainio, and M. Kivimäki, Lower fertility associated with obesity and underweight: The US National Longitudinal Survey of Youth, *American Journal of Clinical Nutrition* 88 (2008): 886–893.
3. U. Paasch and coauthors, Obesity and age affect male fertility potential, *Fertility and Sterility* 94 (2010): 2898–2901.
4. C. J. Brewer and A. H. Balen, The adverse effects of obesity on conception and implantation, *Reproduction* 140 (2010): 347–364; S. Wilkes and A. Murdoch, Obesity and female fertility: A primary care perspective, *The Journal of Family Planning and Reproductive Health Care* 35 (2009): 181–185.
5. I. Cetin, C. Berti, and S. Calabrese, Role of micronutrients in the periconceptional period, *Human Reproduction Update* 16 (2010): 80–95.
6. J. Mendiola and coauthors, A low intake of antioxidant nutrients is associated with poor semen quality in patients attending fertility clinics, *Fertility and Sterility* 93 (2010): 1128–1133.
7. J. C. Sadeu and coauthors, Alcohol, drugs, caffeine, tobacco, and environmental contaminant exposure: Reproductive health consequences and clinical implications, *Critical Reviews in Toxicology* 40 (2010): 633–652; S. Cordier, Evidence for a role of paternal exposures in developmental toxicity, *Basic and Clinical Pharmacology and Toxicology* 102 (2008): 176–181.
8. M. Desforges and C. P. Sibley, Placental nutrient supply and fetal growth, *The International Journal of Developmental Biology* 54 (2010): 377–390.
9. D. O. Mook-Kanamori and coauthors, Risk factors and outcomes associated with first-trimester fetal growth restriction, *Journal of the American Medical Association* 303 (2010): 527–534.
10. US Preventive Services Task Force, Folic acid for the prevention of neural tube defects: US Preventive Services Task Force recommendation statement, *Annals of Internal Medicine* 150 (2009): 626–631.
11. P. D. Gluckman and coauthors, Effect of in utero and early-life conditions on adult health and disease, *New England Journal of Medicine* 359 (2008): 61–73.
12. L. H. Lumey and coauthors, Lipid profiles in middle-aged men and women after famine exposure during gestation: The Dutch Hunger Winter

Families Study, *American Journal of Clinical Nutrition* 89 (2009): 1737–1743; F. Lussana and coauthors, Prenatal exposure to the Dutch famine is associated with a preference for fatty foods and a more atherogenic lipid profile, *American Journal of Clinical Nutrition* 88 (2008): 1648–1652.

13. J. Rotteveel and coauthors, Infant and childhood growth patterns, insulin sensitivity, and blood pressure in prematurely born young adults, *Pediatrics* 122 (2008): 313–321.
14. M. M. Perälä and coauthors, The association between salt intake and adult systolic blood pressure is modified by birth weight, *American Journal of Clinical Nutrition* 93 (2011): 422–426.
15. M. L. de Gusmão Correia and coauthors, Developmental origins of health and disease: Experimental and human evidence of fetal programming for metabolic syndrome, *Journal of Human Hypertension* doi: 10.1038/jhh.2011.61; N. W. Solomons, Developmental origins of health and disease: Concepts, caveats, and consequences for public health nutrition, *Nutrition Reviews* 67 (2009): S12–S16.
16. G. C. Burdge and K. A. Lillycrop, Nutrition, epigenetics, and developmental plasticity: Implications for understanding human disease, *Annual Review of Nutrition* 30 (2010): 315–339.
17. J. E. Wiedmeier and coauthors, Early postnatal nutrition and programming of the preterm neonate, *Nutrition Reviews* 69 (2011): 76–82; S. A. Atkinson, Introduction to the workshop, *American Journal of Clinical Nutrition* 89 (2009): 1485S–1487S; M. E. Symonds, T. Stephenson, and H. Budge, Early determinants of cardiovascular disease: The role of early diet in later blood pressure control, *American Journal of Clinical Nutrition* 89 (2009): 1518S–1522S; B. Delage and R. H. Dashwood, Dietary manipulation of histone structure and function, *Annual Review of Nutrition* 28 (2008): 347–366; W. Kiess and coauthors, Adipocytes and adipose tissues, *Best Practice and Research. Clinical Endocrinology and Metabolism* 22 (2008): 135–153.
18. G. K. Swamy, T. Østbye, and R. Skjærven, Association of preterm birth with long-term survival, reproduction, and next-generation preterm birth, *Journal of the American Medical Association* 299 (2008): 1429–1436.
19. Position of the American Dietetic Association and American Society for Nutrition: Obesity, reproduction, and pregnancy outcomes, *Journal of the American Dietetic Association* 109 (2009): 918–927.
20. Position of the American Dietetic Association and American Society for Nutrition, 2009.
21. S. Y. Chu and coauthors, Association between obesity during pregnancy and increased use of health care, *New England Journal of Medicine* 358 (2008): 1444–1453.
22. M. A. Kominiarek, Pregnancy after bariatric surgery, *Obstetrics & Gynecology Clinics of North America* 37 (2010): 305–320; I. Guelinckx, R. Devlieger, and G. Vansant, Reproductive outcome after bariatric surgery: A critical review, *Human Reproduction Update* 15 (2009): 189–201; N. Sapre and coauthors, Pregnancy following gastric bypass surgery: What is the expected course and outcome? *The New Zealand Medical Journal* 122 (2009): 33–42; J. R. Wax and coauthors, Pregnancy following gastric bypass for morbid obesity: Effect of surgery-to-conception interval on maternal and neonatal outcomes, *Obesity Surgery* 18 (2008): 1517–1521; M. A. Maggard and coauthors, Pregnancy and fertility following bariatric surgery: A systematic review, *Journal of the American Medical Association* 300 (2008): 2286–2296.
23. E. A. Nohr and coauthors, Combined associations of prepregnancy body mass index and gestational weight gain with the outcome of pregnancy, *American Journal of Clinical Nutrition* 87 (2008): 1750–1759.
24. J. L. Mills and coauthors, Maternal obesity and congenital heart defects: A population-based study, *American Journal of Clinical Nutrition* 91 (2010): 1543–1549; K. J. Stothard and coauthors, Maternal overweight and obesity and the risk of congenital anomalies: A systematic review and meta-analysis, *Journal of the American Medical Association* 301 (2009): 636–650.
25. Y. S. Thornton and coauthors, Perinatal outcomes in nutritionally monitored obese pregnant women: A randomized clinical trial, *Journal of the National Medical Association* 101 (2009): 569–577.
26. G. A. L. Davies and coauthors, SOGC Clinical Practice Guideline: Obesity in pregnancy, *Journal of Obstetrics and Gynaecology Canada* 110 (2010): 165–173; J. H. Cohen and H. Kim, Sociodemographic and health characteristics associated with attempting weight loss during pregnancy, *Preventing Chronic Disease* 6 (2009): A07.
27. Institute of Medicine, *Weight Gain During Pregnancy: Reexamining the Guidelines* (Washington, D.C.: The National Academies Press, 2009).
28. S. Y. Chu and coauthors, Gestational weight gain by body mass index among US women delivering live births, 2004–2005: Fueling future obesity, *American Journal of Obstetrics and Gynecology* 200 (2009): 271.e1–271.e7; C. M. Olson, Achieving a healthy weight gain during pregnancy, *Annual Review of Nutrition* 28 (2008): 411–423.
29. J. Josefson, The impact of pregnancy nutrition on offspring obesity, *Journal of the American Dietetic Association* 111 (2011): 50–52; S. R. Crozier and coauthors, Weight gain in pregnancy and childhood body composition: Findings from the Southhampton Women's Survey, *American Journal of Clinical Nutrition* 91 (2010): 1745–1751; M. F. Mottola and coauthors, Nutrition and exercise prevent excess weight gain in overweight pregnant women, *Medicine and Science in Sports and Exercise* 42 (2010): 265–272; J. M. Crane and coauthors, The effect of gestational weight gain by body mass index on maternal and neonatal outcomes, *Journal of Obstetrics and Gynaecology Canada* 31 (2009): 28–35; B. H. Wrotniak and coauthors, Gestational weight gain and risk of overweight in the offspring at age 7 y in a multicenter, multiethnic cohort study, *American Journal of Clinical Nutrition* 87 (2008): 1818–1824.
30. A. L. Deierlein, A. M. Siega-Riz, and A. Herring, Dietary energy density but not glycemic load is associated with gestational weight gain, *American Journal of Clinical Nutrition* 88 (2008): 693–699.
31. I. Streuling, A. Beyerlein, and R. von Kries, Can gestational weight gain be modified by increasing physical activity and diet counseling? A meta-analysis of interventional trials, *American Journal of Clinical Nutrition* 92 (2010): 678–687.
32. A. A. Mamun and coauthors, Associations of excess weight gain during pregnancy with long-term maternal overweight and obesity: Evidence from 21 y postpartum follow-up, *American Journal of Clinical Nutrition* 91 (2010): 1336–1341; J. L. Baker and coauthors, Breastfeeding reduces postpartum weight retention, *American Journal of Clinical Nutrition* 88 (2008): 1543–1551.
33. D. Haire-Joshu and coauthors, Postpartum teens' breakfast consumption is associated with snack and beverage intake and body mass index, *Journal of the American Dietetic Association* 111 (2011): 124–130.
34. Position of the American Dietetic Association: Nutrition and lifestyle for a healthy pregnancy outcome, *Journal of the American Dietetic Association* 108 (2008): 553–561; US Department of Health and Human Services, *2008 Physical Activity Guidelines for Americans*, www.health.gov/paguidelines/guidelines/chapter7.aspx accessed 6/26/2009.
35. Position of the American Dietetic Association, 2008.
36. B.M.Y. Leung and B. J. Kaplan, Perinatal depression: Prevalence, risks, and the nutrition link—A review of the literature, *Journal of the American Dietetic Association* 109 (2009): 1566–1575.
37. L. D. Brown and coauthors, Maternal amino acid supplementation for intrauterine growth restriction, *Frontiers in Bioscience* 3 (2011): 428–444.
38. P. Haggarty, Fatty acid supply to the human fetus, *Annual Review of Nutrition* 30 (2010): 237–255; S. M. Innis and R. W. Freisen, Essential n-3 fatty acids in pregnant women and early visual acuity maturation in term infants, *American Journal of Clinical Nutrition* 87 (2008): 548–557; M. von Eijsden and coauthors, Maternal n-3, n-6, and *trans* fatty acid profile early in pregnancy and term birth weight: A prospective cohort study, *American Journal of Clinical Nutrition* 87 (2008): 887–895.
39. A. S. de Souza, F. S. Fernandes, and M. das Graças Tavares do Carmo, Effects of maternal malnutrition and postnatal nutritional rehabilitation on brain fatty acids, learning, and memory, *Nutrition Reviews* 69 (2011): 132–144.
40. Committee on Dietary Reference Intakes, *Dietary Reference Intakes for Thiamin, Riboflavin, Niacin, Vitamin B_6, Folate, Vitamin B_{12}, Pantothenic Acid, Biotin, and Choline* (Washington, D.C.: National Academies Press, 1998), pp. 196–305.
41. F. E. Viteri, Iron endowment at birth: Maternal iron status and other influences, *Nutrition Reviews* 69 (2011): S3–S16.
42. H. J. McArdle and coauthors, Role of the placenta in regulation of fetal iron status, *Nutrition Reviews* 69 (2011): S17–S22.
43. S. Mahajan and coauthors, Nutritional anaemia dysregulates endocrine control of fetal growth, *British Journal of Nutrition* 100 (2008): 408–417.
44. T. O. Scholl, Maternal iron status: Relation to fetal growth, length of gestation, and iron endowment of the neonate, *Nutrition Reviews* 69 (2011): S23–S29; J. Berger and coauthors, Strategies to prevent iron

deficiency and improve reproductive health, *Nutrition Reviews* 69 (2011): S78–S86.
45. P. M. Brannon and M. F. Picciano, Vitamin D in pregnancy and lactation in humans, *Annual Review of Nutrition* 31 (2011): 89–115; D. K. Dror and L. H. Allen, Vitamin D inadequacy in pregnancy: Biology, outcomes, and interventions, *Nutrition Reviews* 68 (2010): 465–477.
46. Committee on Dietary Reference Intakes, *Dietary Reference Intakes for Calcium and Vitamin D* (Washington, D.C.: National Academies Press, 2011).
47. H. Olausson and coauthors, Changes in bone mineral status and bone size during pregnancy and the influences of body weight and calcium intake, *American Journal of Clinical Nutrition* 88 (2008): 1032–1039.
48. P. Christian, Micronutrients, birth weight, and survival, *Annual Review of Nutrition* 30 (2010): 83–104.
49. S. H. Zeisel, Is maternal diet supplementation beneficial? Optimal development of infant depends on mother's diet, *American Journal of Clinical Nutrition* 89 (2009): 685S–687S.
50. M. F. Picciano and M. K. McGuire, Use of dietary supplements by pregnant and lactating women in North America, *American Journal of Clinical Nutrition* 89 (2009): 663S–667S.
51. J. M. Catov and coauthors, Periconceptional multivitamin use and risk of preterm or small-for-gestational-age births in the Danich National Birth Cohort, *American Journal of Clinical Nutrition* 94 (2011): 906–912; T. O. Scholl, Maternal nutrition before and during pregnancy, *Nestlé Nutrition Workshop Series: Pediatric Program* 61 (2008): 79–89.
52. Position of the American Dietetic Association: Vegetarian diets, *Journal of the American Dietetic Association* 109 (2009): 1266–1282.
53. J. R. Niebyl, Nausea and vomiting in pregnancy, *New England Journal of Medicine* 363 (2010): 1544–1550.
54. A. Matthews and coauthors, Interventions for nausea and vomiting in early pregnancy, *Cochrane Database of Systemic Reviews* 8 (2010): CD007575.
55. S. L. Young, Pica in pregnancy: New ideas about an old condition, *Annual Review of Nutrition* 30 (2010): 403–422.
56. F. H. Bloomfield, How is maternal nutrition related to preterm birth? *Annual Review of Nutrition* 31 (2011): 235–261.
57. J. M. Schneider and coauthors, The use of multiple logistic regression to identify risk factors associated with anemia and iron deficiency in a convenience sample of 12–36-mo-old children from low-income families, *American Journal of Clinical Nutrition* 87 (2008): 614–620.
58. The HAPO Study Cooperative Research Group, Hyperglycemia and adverse pregnancy outcomes, *New England Journal of Medicine* 358 (2008): 1991–2002.
59. J. L. Kitzmiller and coauthors, Preconception care for women with diabetes and prevention of major congenital malformations, *Birth Defects Research Part A: Clinical and Molecular Teratology* 88 (2010): 791–803; J. L. Kitzmiller and coauthors, Managing preexisting diabetes for pregnancy: Summary of evidence and consensus recommendations for care, *Diabetes Care* 31 (2008): 1060-1079.
60. Centers for Disease Control and Prevention, National Diabetes Fact Sheet 2011: National estimates and general information on diabetes and prediabetes in the United States, 2011, www.cdc.gov/diabetes/pubs/pdf/ndfs_2011.pdf; National Institute of Child Health and Human Development, www.nichd.nih.gov/health/topics/gestational_diabetes.cfm, updated April 2008.
61. Y. Yogev and G. H. Visser, Obesity, gestational diabetes and pregnancy outcome, *Seminars in Fetal and Neonatal Medicine* 14 (2009): 77–84.
62. S. F. Ehrlich and coauthors, Change in body mass index between pregnancies and the risk of gestational diabetes in a second pregnancy, *Obstetrics and Gynecology* 117 (2011): 1323–1330.
63. V. Seshiah and coauthors, "Abnormal" fasting plasma glucose during pregnancy, *Diabetes Care* 31 (2008): e92.
64. Position of the American Diabetes Association, Diagnosis and classification of diabetes mellitus, *Diabetes Care* 31 (2008): S55–S60.
65. Position of the American Diabetes Association: Nutrition recommendations and interventions for diabetes, *Diabetes Care* 31 (2008): S61–S78.
66. T. Sathyapalan, D. Mellor, and S. L. Atkin, Obesity and gestational diabetes, *Seminars in Fetal and Neonatal Medicine* 15 (2010): 89–93.
67. P. E. Marik, Hypertensive disorders of pregnancy, *Postgraduate Medicine* 121 (2009): 69–76.
68. E. W. Seely and J. Ecker, Chronic hypertension in pregnancy, *New England Journal of Medicine* 365 (2011): 439–446.
69. J. A. Lykke and coauthors, Hypertensive pregnancy disorders and subsequent cardiovascular morbidity and type 2 diabetes mellitus in the mother, *Hypertension* 53 (2009): 944–951.
70. B. E. Vikse and coauthors, Preeclampsia and the risk of end-stage renal disease, *New England Journal of Medicine* 359 (2008): 800–809.
71. C. S. Wu and coauthors, Preeclampsia and risk for epilepsy in offspring, *Pediatrics* 122 (2008): 1072–1078.
72. A. Paxton and T. Wardlaw, Are we making progress in maternal mortality? *New England Journal of Medicine* 364 (2011): 1990–1993.
73. H. Xu and coauthors, Role of nutrition in the risk of preeclampsia, *Nutrition Reviews* 67 (2009): 639–657.
74. C. B. Rudra and coauthors, A prospective analysis of recreational physical activity and preeclampsia risk, *Medicine and Science in Sports and Exercise* 40 (2008): 1581–1588.
75. Vital signs: Teen pregnancy—United States, 1991–2009, *Morbidity and Mortality Weekly Report* 60 (2011): 414–420; J. A. Martin and coauthors, Births: Final data for 2006, *National Vital Statistics Reports* 57 (2009): 1–102.
76. Trends in smoking before, during, and after pregnancy: Pregnancy Risk Assessment Monitoring System (PRAMS), United States, 31 sites, 2000–2005, *Morbidity and Mortality Weekly Report* 58 (2009): 1–29.
77. P. N. Baker and coauthors, A prospective study of micronutrient status in adolescent pregnancy, *American Journal of Clinical Nutrition* 89 (2009): 1114–1124.
78. X. Chen and coauthors, Paternal age and adverse birth outcomes: Teenager or 40+, who is at risk? *Human Reproduction* 23 (2008): 1290–1296.
79. Centers for Disease Control and Prevention, US Teen Birth Rate Fell to Record Low in 2009, www.cdc.gov/media/releases/2011/p0405_vitalsigns.html, updated April 5, 2011; The National Campaign to Prevent Teen and Unplanned Pregnancy, Counting it up: The public costs of teen childbearing, www.thenationalcampaign.org/costs/, updated June 9, 2011
80. C. S. Meade, T. S. Kershaw, and J. R. Ickovics, The intergenerational cycle of teenage motherhood: An ecological approach, *Health Psychology* 27 (2008): 419–429.
81. Alcohol use among pregnant and nonpregnant women of childbearing age: United States, 1991–2005, *Mortality and Morbidity Weekly Report* 58 (2009): 529–532.
82. J. P. Ackerman, T. Riggins, and M. M. Black, A review of the effects of prenatal cocaine exposure among school-aged children, *Pediatrics* 125 (2010): 554–565; M. J. Rivkin and coauthors, Volumetric MRI study of brain in children with intrauterine exposure to cocaine, alcohol, tobacco, and marijuana, *Pediatrics* 121 (2008): 741–750.
83. M. O'Donnell and coauthors, Increasing prevalence of neonatal withdrawal syndrome: Population study of maternal factors and child protection involvement, *Pediatrics* 123 (2009): e614–e621.
84. Trends in smoking before, during, and after pregnancy, 2009; J. A. Martin and coauthors, Annual summary of vital statistics: 2008, *Pediatrics* 121 (2008): 788–801.
85. A. Bjerg and coauthors, A strong synergism of low birth weight and prenatal smoking on asthma in schoolchildren, *Pediatrics* 127 (2011): e905–e912; C. J. Alverson and coauthors, Maternal smoking and congenital heart defects in the Baltimore-Washington infant study, *Pediatrics* 127 (2011): e647–e653.
86. J. M. Rogers, Tobacco and pregnancy: Overview of exposures and effects, *Birth Defects Research (Part C)* 84 (2008): 1–15; E. Jauniaux and G. J. Burton, Morphological and biological effects of maternal exposure to tobacco smoke on the feto-placental unit, *Early Human Development* 83 (2007): 699–706.
87. A. Gunnerbeck and coauthors, Relationship of maternal snuff use and cigarette smoking with neonatal apnea, *Pediatrics* 128 (2011): 503–509.
88. J. Leonardi-Bee, J. Britton, and A. Venn, Secondhand smoke and adverse fetal outcomes in nonsmoking pregnant women: A meta-analysis, *Pediatrics* 127 (2011): 734–741.
89. A. Bloomingdale and coauthors, A qualitative study of fish consumption during pregnancy, *American Journal of Clinical Nutrition* 92 (2010): 1234–1240.
90. E. Oken and coauthors, Maternal fish intake during pregnancy, blood mercury levels, and child cognition at age 3 years in a US cohort, *American Journal of Epidemiology* 167 (2008): 1171–1181.

91. K. R. Mahaffey and coauthors, Balancing the benefits of n-3 polyunsaturated fatty acids and the risks of methylmercury exposure from fish consumption, *Nutrition Reviews* 69 (2011): 493–508.
92. R. Bakker and coauthors, Maternal caffeine intake, blood pressure, and the risk of hypertensive complications during pregnancy: The Generation R Study, *American Journal of Hypertension* 24 (2011): 421–428; E. Maslova and coauthors, Caffeine consumption during pregnancy and risk of preterm birth: A meta-analysis, *American Journal of Clinical Nutrition* 92 (2010): 1120–1132; D. C. Greenwood and coauthors, Caffeine intake during pregnancy, late miscarriage and stillbirth, *European Journal of Epidemiology* 25 (2010): 275–280; X. Weng, R. Odouli, and D. Li, Maternal caffeine consumption during pregnancy and the risk of miscarriage: A prospective cohort study, *American Journal of Obstetrics and Gynecology* 198 (2008): 279.e1–279.e8.
93. R. Bakker and coauthors, Maternal caffeine intake from coffee and tea, fetal growth, and the risks of adverse birth outcomes: The Generation R Study, *American Journal of Clinical Nutrition* 91 (2010): 1691–1698; CARE Study Group, Maternal caffeine intake during pregnancy and risk of fetal growth restriction: A large prospective observational study, *British Medical Journal* 337 (2008): 1334–1338.
94. Position of the American Dietetic Association: Use of nutritive and nonnutritive sweeteners, *Journal of the American Dietetic Association* 104 (2004): 255–275.
95. Centers for Disease Control and Prevention, Breastfeeding report card: United States, 2010, www.cdc.gov/breastfeeding/data/reportcard.htm, updated September 13, 2010.
96. US Department of Health and Human Services, *The Surgeon General's Call to Action to Support Breastfeeding*, (Washington, D.C.: U.S. Department of Health and Human Services, Office of the Surgeon General, 2011); Position of the American Dietetic Association: Promoting and supporting breastfeeding, *Journal of the American Dietetic Association* 109 (2009): 1926–1942; Breastfeeding, *Pediatric Nutrition Handbook*, 6th ed., ed. R. E. Kleinman (Elk Grove Village, IL: American Academy of Pediatrics, 2009), pp. 29–59.
97. D. Thulier, Breastfeeding in America: A history of influencing factors, *Journal of Human Lactation* 25 (2009): 85–94.
98. A. Avery and coauthors, Confident commitment is a key factor for sustained breastfeeding, *Birth* 36 (2009): 141–148.
99. L. A. Nommsen-Rivers and coauthors, Delayed onset of lactogenesis among first-time mothers is related to maternal obesity and factors associated with ineffective breastfeeding, *American Journal of Clinical Nutrition* 92 (2010): 574–584; K. M. Rasmussen, Association of maternal obesity before conception with poor lactation performance, *Annual Review of Nutrition* 27 (2007): 103–121; C. A. Lovelady, Is maternal obesity a cause of poor lactation performance? *Nutrition Reviews* 63 (2005): 352–355.
100. C. A. Lovelady and coauthors, Effect of exercise training on loss of bone mineral density during lactation, *Medicine and Science in Sports and Exercise* 41 (2009): 1902–1907.
101. K. M. Ziol-Guest and D. C. Hernandez, First- and second-trimester WIC participation is associated with lower rates of breastfeeding and early introduction of cow's milk during infancy, *Journal of the American Dietetic Association* 110 (2010): 702–709.
102. M. Murimi and coauthors, Factors that influence breastfeeding decisions among Special Supplemental Nutrition Program for Women, Infants, and Children participants from Central Louisiana, *Journal of the American Dietetic Association* 110 (2010): 624–627.
103. G. K. Beauchamp and J. A. Mennella, Flavor perception in human infants: Development and functional significance, *Digestion* 83 (2011): 1–6.
104. F. R. Greer, S. H. Sicherer, A. Wesley Burks, and the Committee on Nutrition and Section on Allergy and Immunology, Effects of early nutritional interventions on the development of atopic disease in infants and children: The role of maternal dietary restriction, breastfeeding, timing of introduction of complementary foods, and hydrolyzed formulas, *Pediatrics* 121 (2008): 183–191.
105. Greer, Sicherer, Burks, and the Committee on Nutrition and Section on Allergy and Immunology, 2008.
106. P. L. Havens, L. M. Mofenson, and the Committee on Pediatric AIDS, Evaluation and management of the infant exposed to HIV-1 in the United States, *Pediatrics* 123 (2009): 175–187.
107. A. Koyanagi and coauthors, Effect of early exclusive breastfeeding on morbidity among infants born to HIV-negative mothers in Zimbabwe, *American Journal of Clinical Nutrition* 89 (2009): 1375–1382.
108. World Health Organization, *HIV and Infant Feeding*, www.who.int/child_adolescent_health/topics/prevention_care/child/nutrition/hivif/en/, accessed June 1, 2009; M. W. Kline, Early exclusive breastfeeding: Still the cornerstone of child survival, *American Journal of Clinical Nutrition* 89 (2009): 1281–1282.
109. G. E. Gray and H. Saloojee, Breast-feeding, antiretroviral prophylaxis, and HIV, *New England Journal of Medicine* 359 (2008): 189–191; N. I. Kumwenda and coauthors, Extended antiretroviral prophylaxis to reduce breast-milk HIV-1 transmission, *New England Journal of Medicine* 359 (2008): 119–129; W. T. Shearer, Breastfeeding and HIV infection, *Pediatrics* 121 (2008): 1046–1047.
110. L. Yang and K. H. Jacobsen, A systematic review of the association between breastfeeding and breast cancer, *Journal of Women's Health* 17 (2008): 1635–1645.
111. L. M. Jansson and coauthors, Methadone maintenance and breastfeeding in the neonatal period, *Pediatrics* 121 (2008): 106–114.
112. Trends in smoking before, during, and after pregnancy, 2009.
113. T. M. Weiser and coauthors, Association of maternal smoking status with breastfeeding practices: Missouri, 2005, *Pediatrics* 124 (2009): 1603–1610.
114. B. M. Ostfeld and coauthors, Concurrent risks in sudden infant death syndrome, *Pediatrics* 125 (2010): 447–453.
115. F. R. Hauck and coauthors, Breastfeeding and reduced risk of sudden infant death syndrome: A meta-analysis, *Pediatrics* 128 (2011): 103–110; M. M. Vennemann and coauthors, Does breastfeeding reduce the risk of sudden infant death syndrome? *Pediatrics* 123 (2009): e406–e410.

LEARN IT Explain how drinking alcohol endangers the fetus and how women can prevent fetal alcohol syndrome.

Fetal Alcohol Syndrome

As Chapter 15 mentions, drinking alcohol during pregnancy endangers the fetus. Alcohol crosses the placenta freely and deprives the developing fetus of both nutrients and oxygen. The damaging effects of alcohol on the developing fetus cover a range of abnormalities referred to as **fetal alcohol spectrum disorder** (see the accompanying glossary).[1] Those at the most severe end of the spectrum are described as having **fetal alcohol syndrome (FAS),** a cluster of physical, mental, and neurobehavioral symptoms that includes:[2]

- Prenatal and postnatal growth restriction
- Abnormalities of the brain and central nervous system, with consequent impairment in cognition and behavior[3]
- Physical abnormalities of the face and skull that alter normal patterns of symmetry (see Figure H15-1, p. 502)[4]
- Increased frequency of major birth defects such as cleft palate, heart defects, and defects in ears, eyes, genitals, and urinary system

Those with more severe physical abnormalities have more cognitive limitations. Tragically, the damage evident at birth persists: children with FAS never fully recover.

Each year, an estimated 6000 infants are born in the United States with FAS because their mothers drank too much alcohol during pregnancy.[5] In addition, at least three times as many infants have had enough **prenatal alcohol exposure** to result in some symptoms of fetal alcohol spectrum disorder. The cluster of *mental* problems associated with prenatal alcohol exposure is known as **alcohol-related neurodevelopmental disorder (ARND),** and the *physical* malformations are referred to as **alcohol-related birth defects (ARBD).** Some children with ARBD and ARND have no outward signs; others may be short or have only minor facial abnormalities. Diagnosis is often overlooked until after 1 year of age.[6] Children commonly go undiagnosed even when they develop learning difficulties in the early school years. Mood disorders and problem behaviors, such as aggression, are common.[7] These children typically need support and guidance to function and participate in daily activities.[8]

The surgeon general states that pregnant women should abstain from alcohol. Abstinence from alcohol is the best policy for pregnant women both because alcohol consumption during pregnancy has such severe consequences and because FAS can only be prevented—it cannot be treated. Further, because the most severe damage occurs around the time of conception—*before a woman may even realize that she is pregnant*—the warning to abstain includes women who may become pregnant.

> **DIETARY GUIDELINES FOR AMERICANS 2010**
Women who are pregnant should not drink alcohol.

Drinking during Pregnancy

As mentioned in Chapter 15, 1 out of 8 pregnant women drinks alcohol at some time during her pregnancy; 1 out of 50 admits to binge drinking.[9] When a woman drinks during pregnancy, she causes damage in two ways: directly, by intoxication, and indirectly, by malnutrition. Prior to the complete formation of the placenta (approximately 12 weeks), alcohol diffuses directly into the tissues of the developing embryo, causing incredible damage. (Review Figure 15-4 on p. 473 and note that the critical periods for most tissues occur during this time of embryonic development.) Alcohol interferes with the orderly development of tissues during their critical periods, reducing the number of cells and damaging those that are produced. The damage of alcohol toxicity during brain development is apparent in its reduced size and impaired function.

GLOSSARY

alcohol-related birth defects (ARBD): malformations in the skeletal and organ systems (heart, kidneys, eyes, ears) associated with prenatal alcohol exposure.

alcohol-related neurodevelopmental disorder (ARND): abnormalities in the central nervous system and cognitive development associated with prenatal alcohol exposure.

fetal alcohol spectrum disorder: a range of physical, behavioral, and cognitive abnormalities caused by prenatal alcohol exposure.

fetal alcohol syndrome (FAS): a cluster of physical, behavioral, and cognitive abnormalities associated with prenatal alcohol exposure, including facial malformations, growth retardation, and central nervous disorders.

prenatal alcohol exposure: subjecting a fetus to a pattern of excessive alcohol intake characterized by substantial regular use or heavy episodic drinking.

Note: See Highlight 7 for other alcohol-related terms and information.

> FIGURE H15-1 **Typical Facial Characteristics of FAS**

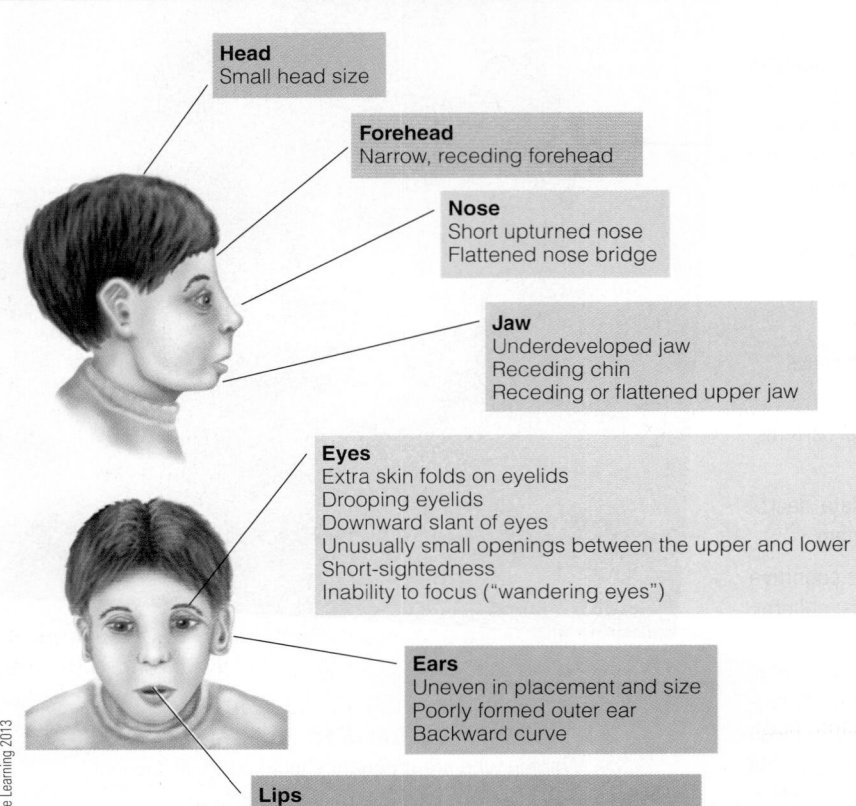

Head
Small head size

Forehead
Narrow, receding forehead

Nose
Short upturned nose
Flattened nose bridge

Jaw
Underdeveloped jaw
Receding chin
Receding or flattened upper jaw

Eyes
Extra skin folds on eyelids
Drooping eyelids
Downward slant of eyes
Unusually small openings between the upper and lower lids
Short-sightedness
Inability to focus ("wandering eyes")

Ears
Uneven in placement and size
Poorly formed outer ear
Backward curve

Lips
Absence of groove in upper lip; flat upper lip
Thin upper lip

When alcohol crosses the placenta, fetal blood alcohol rises until it reaches equilibrium with maternal blood alcohol. The mother may not even appear drunk, but the fetus may be poisoned. The fetus's body is small, its detoxification system is immature, and alcohol remains in fetal blood long after it has been cleared from maternal blood.

A pregnant woman harms her unborn child not only by consuming alcohol but also by not consuming food. This combination enhances the likelihood of malnutrition and a poorly developed infant. It is important to realize, however, that malnutrition is not the cause of FAS. It is true that mothers of FAS children often have unbalanced diets and nutrient deficiencies. It is also true that nutrient deficiencies may exacerbate the clinical signs seen in these children, but it is the *alcohol* that causes the damage. An adequate diet alone will not prevent FAS if alcohol use continues.

How Much Is Too Much?

A pregnant woman need not have an alcohol-abuse problem to give birth to a baby with FAS. She need only drink in excess of her liver's capacity to detoxify alcohol. The damaging effects on the developing fetus are dose dependent, becoming greater as the dose increases.[10] Even one drink a day threatens neurological development and behaviors. Four drinks a day dramatically increase the risk of having an infant with physical malformations.

In addition to total alcohol intake, drinking patterns play an important role. Most FAS studies report their findings in terms of average intake per day, but people often drink more heavily on some days than on others. For example, a woman who drinks an *average* of 1 ounce of alcohol (2 drinks) a day may not drink at all during the week, but then have 10 drinks on Saturday night, exposing the fetus to extremely toxic quantities of alcohol. Whether various drinking patterns incur damage depends on the frequency of consumption, the quantity consumed, and the stage of fetal development at the time of each drinking episode.

An occasional drink may be innocuous, but researchers are unable to say how much alcohol is safe to consume during pregnancy. For this reason, health-care professionals urge women to stop drinking alcohol as soon as they realize they are pregnant, or better, as soon as they *plan* to become pregnant. Why take any risk? The only sure way to protect an infant from alcohol damage is for the mother to abstain.

Characteristic facial features may diminish with time, but children with FAS typically continue to be short and underweight for their age.

When Is the Damage Done?

The type of abnormality observed in an FAS infant depends on the developmental events occurring at the times of alcohol exposure. During the first trimester, developing organs such as the brain, heart,

© 1995 George Steinmetz

Children born with FAS must live with the long-term consequences of prenatal brain damage.

© Matt Faruggio Photography

All containers of beer, wine, and liquor warn women not to drink alcoholic beverages during pregnancy because of the risk of birth defects.

and kidneys may be malformed. During the second trimester, the risk of spontaneous abortion increases. During the third trimester, body and brain growth may be retarded.

The father's alcohol ingestion may also affect fertility and fetal development. Animal studies have found smaller litter sizes, lower birthweights, reduced survival rates, and impaired learning ability in the offspring of males consuming alcohol prior to conception. An association between paternal alcohol intake 1 month prior to conception and low infant birthweight has been reported in human beings. Alcohol use creates epigenetic changes in sperm DNA that may alter gene expression and result in features of fetal alcohol spectrum disorders.[11]

In view of the damage caused by FAS, prevention efforts focus on educating women not to drink during pregnancy. Public service announcements and alcohol beverage warning labels help to raise awareness. Everyone should hear the message loud and clear: don't drink alcohol during pregnancy.

REFERENCES

1. S. N. Mattson, N. Crocker, and T. T. Nguyen, Fetal alcohol spectrum disorders: Neuropsychological and behavioral features, *Neuropsychology Review* 21 (2011): 81–101.
2. E. P. Riley, M. A. Infante, and K. R. Warren, Fetal alcohol spectrum disorders: An overview, *Neuropsychological Review* 21 (2011): 73–80; K. L. Jones and coauthors, Fetal alcohol spectrum disorders: Extending the range of structural defects, *American Journal of Medical Genetics* 152A (2010): 2731–2735.
3. C. Guerri, A. Bazinet, and E. P. Riley, Foetal alcohol spectrum disorders and alterations in brain and behavior, *Alcohol and Alcoholism* 44 (2009): 108–114.
4. C. P. Klingenberg and coauthors, Prenatal alcohol exposure alters the patterns of facial asymmetry, *Alcohol* 44 (2010): 649–657; P. A. May and coauthors, Population differences in dysmorphic features among children with fetal alcohol spectrum disorders, *Journal of Developmental and Behavioral Pediatrics* 31 (2010): 304–316.
5. Centers for Disease Control and Prevention, *Fetal Alcohol Spectrum Disorders,* www.cdc.gov/ncbddd/fasd/data.html, updated October 6, 2010.
6. C. Bower and coauthors, Age at diagnosis of birth defects, *Birth Defects Research Part A: Clinical and Molecular Teratology* 88 (2010): 251–255.
7. E. R. Disney and coauthors, Strengthening the case: Prenatal alcohol exposure is associated with increased risk for conduct disorder, *Pediatrics* 122 (2008): e1225–e1230.
8. T. Jirikowic, D. Kartin, and H. C. Olsen, Children with fetal alcohol spectrum disorders: A descriptive profile of adaptive function, *Canadian Journal of Occupational Therapy* 75 (2008): 238–248.
9. Alcohol use among pregnant and nonpregnant women of childbearing age: United States, 1991–2005, *Mortality and Morbidity Weekly Report* 58 (2009): 529–532.
10. A. Ornoy and Z. Ergaz, Alcohol abuse in pregnant women: Effects on the fetus and newborn, mode of action and maternal treatment, *International Journal of Environmental Research and Public Health* 7 (2010): 364–379.
11. L. A. Ouko and coauthors, Effect of alcohol consumption on CpG methylation in the differentially methylated regions of H19 and IG-DMR in male gametes: Implications for fetal alcohol spectrum disorders, *Alcoholism: Clinical and Experimental Research* 33 (2009): 1615–1627.

CENGAGE LEARNING'S GLOBAL NUTRITION WATCH — GLOBAL NUTRITION RESOURCE CENTER

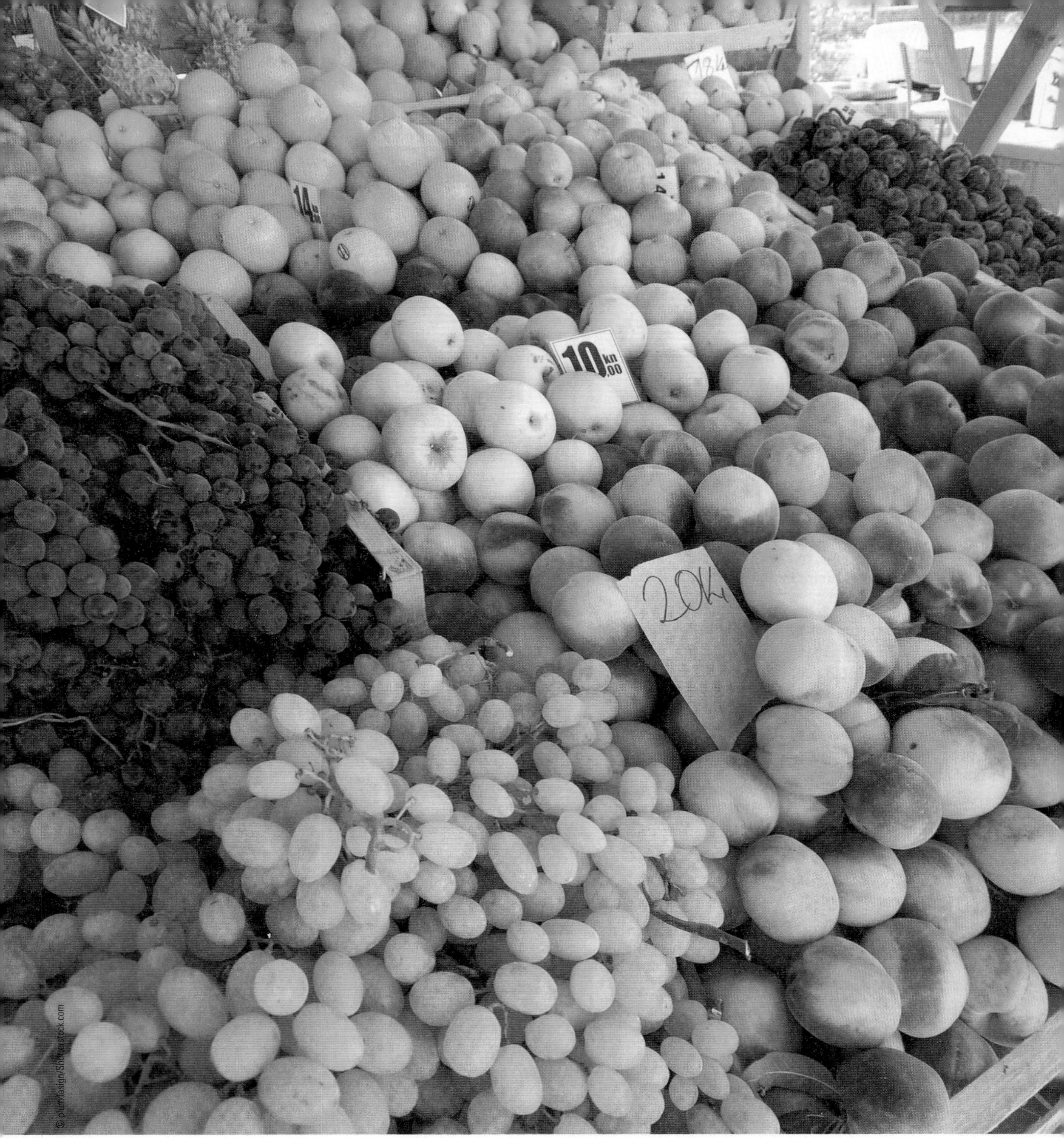

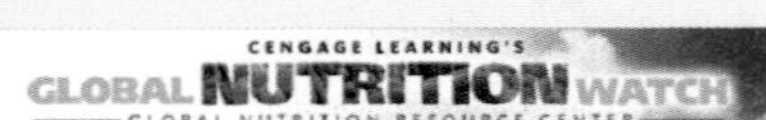

An alarming number of children under 18 years old in the United States are overweight. Visit Weight Management/Diets → Obesity to review how various groups are looking at early prevention methods. What global lifestyle changes can be made to prevent obesity in the early stages of life?

16

Life Cycle Nutrition: Infancy, Childhood, and Adolescence

LEARNING GPS

Nutrition in Your Life

Much of this book has focused on you—your food choices and how they might affect your health. This chapter shifts the focus from you the recipient to you the caregiver. One day (if not already), children may depend on you to feed them well and teach them wisely. The responsibility of nourishing children can seem overwhelming at times, but the job is fairly simple. Offer children a variety of nutritious foods to support their growth and teach them how to make healthy food and activity choices. Presenting foods in a relaxed and supportive environment nourishes both physical and emotional well-being. In the Nutrition Portfolio at the end of this chapter, you can plan a day's menu for a child 4 to 8 years of age and determine whether it meets nutrient requirements to support healthy growth but not so much as to promote obesity.

The first year of life (infancy) is a time of phenomenal growth and development. After the first year, a child continues to grow and change, but more slowly. Still, the cumulative effects over the next decade are remarkable. Then, as the child enters the teen years, the pace toward adulthood accelerates dramatically. This chapter examines the special nutrient needs of infants, children, and adolescents.

16.1 Nutrition during Infancy

LEARN IT List some of the immune factors in breast milk and describe the appropriate foods for infants during the first year of life.

Initially, the infant drinks only breast milk or formula but later begins to eat some foods, as appropriate. Common sense in the selection of infant foods along with a nurturing, relaxed environment support an infant's health and well-being.

> FIGURE 16-1 **Weight Gain of Infants in Their First Five Years of Life**

In the first year, an infant's birthweight may triple, but over the following several years, the rate of weight gain gradually diminishes.

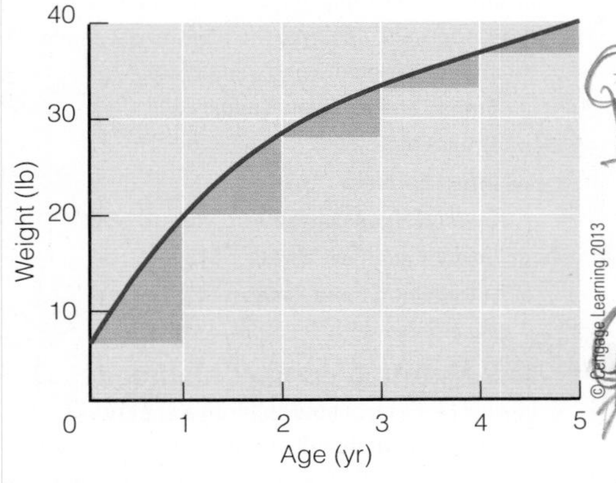

TABLE 16-1 **Infant and Adult Heart Rate, Respiration Rate, and Energy Needs Compared**

	Infants	Adults
Heart rate (beats/minute)	120 to 140	70 to 80
Respiration rate (breaths/minute)	20 to 40	15 to 20
Energy needs (kcal/body weight)	45/lb (100/kg)	<18/lb (<40/kg)

After 6 months, energy saved by slower growth is spent in increased activity.

Energy and Nutrient Needs An infant grows fast during the first year, as Figure 16-1 shows. Growth directly reflects nutrient intake and is an important factor in assessing the nutrition status of infants and children. Health-care professionals measure the height and weight of an infant or child at intervals and compare the measurements with standard growth charts for gender and age and with previous measures of that infant or child (see the accompanying "How To" feature).

Energy Intake and Activity A healthy infant's birthweight doubles by about 5 months of age and triples by 1 year, typically reaching 20 to 25 pounds. The infant's length changes more slowly than weight, increasing about 10 inches from birth to 1 year. By the end of the first year, infant growth slows considerably; during the second year, an infant typically gains less than 10 pounds and grows about 5 inches in length.

Not only do infants grow rapidly, but their energy requirement is remarkably high—about twice that of an adult, based on body weight. A newborn baby requires about 450 kcalories per day, whereas most adults require about 2000 kcalories per day. In terms of body weight, the difference is remarkable. Infants require about 100 kcalories per kilogram of body weight per day, whereas most adults need fewer than 40 (see Table 16-1). If an infant's energy needs were applied to an adult, a 170-pound adult would require more than 7000 kcalories a day. After 6 months, the infant's energy needs decline as the growth rate slows, but some of the energy saved by slower growth is spent in increased activity.

Energy Nutrients Recommendations for the energy nutrients—carbohydrate, fat, and protein—during the first 6 months of life are based on the average intakes of healthy, full-term infants fed breast milk.[1] During the second 6 months of life, recommendations reflect typical intakes from solid foods as well as breast milk.

As Chapter 4 discusses, carbohydrates provide energy to all the cells of the body, but those in the brain depend primarily on glucose to fuel activities. Relative to the size of the body, the size of an infant's brain is greater than that of an adult's. An infant's brain weight is about 12 percent of its body weight, whereas an adult's brain weight is about 2 percent. Thus, an infant's brain uses *relatively* more glucose—about 60 percent of the day's total energy intake.[2]

Fat provides most of the energy in breast milk and standard infant formula. Its high energy density supports the rapid growth of early infancy.

No single nutrient is more essential to growth than protein; it is the basic building material of the body's tissues. All of the body's cells and most of its fluids contain protein. Consequently, inadequate protein has widespread effects, limiting brain function, weakening immune defenses, and disrupting digestion and absorption. The term *failure to thrive* is used to describe the many problems associated with infants and children suffering from protein deficiency. Excess dietary protein can cause problems, too, especially in a small infant. Too much protein stresses the liver and kidneys, which have to metabolize and excrete the excess nitrogen. Signs of protein overload include acidosis, dehydration, diarrhea, elevated blood ammonia, elevated blood urea, and fever. Such problems are not common, but they have been observed in infants fed inappropriate foods, such as fat-free milk or concentrated formula.

Vitamins and Minerals An infant's needs for most nutrients, in proportion to body weight, are more than double those of an adult. Figure 16-2 (p. 508) illustrates this by comparing a 5-month-old infant's needs per unit of body weight with those of an adult man. Some of the differences are extraordinary. Infant recommendations are based on the average amount of nutrients consumed by thriving infants breastfed by well-nourished mothers.

Water One of the most essential nutrients for infants, as for everyone, is water. The younger the infant, the greater the percentage of body weight is water. During early infancy, breast milk or infant formula normally provides enough water to replace fluid losses in a healthy infant. If the environmental temperature is extremely high, however, infants need supplemental water.[3] Because much of

>How To

Plot Measures on a Growth Chart

You can assess the growth of infants and children by plotting their measurements on a percentile graph. Percentile graphs divide the measures of a population into 100 equal divisions so that half of the population falls at or above the 50th percentile and half falls below. Using percentiles allows for comparisons among people of the same age and gender.

To plot measures on a growth chart, follow these steps:

- Select the appropriate chart based on age and gender. For this example, use the accompanying chart, which gives percentiles for weight for girls from birth to 36 months. (Appendix E provides other growth charts for both boys and girls of various ages.)
- Locate the infant's age along the horizontal axis at the bottom of the chart (in this example, 6 months).
- Locate the infant's weight in pounds or kilograms along the vertical axis of the chart (in this example, 17 pounds or 7.7 kilograms).
- Mark the chart where the age and weight lines intersect (shown here with a red dot), and follow the curved line to find the percentile.

This 6-month-old infant is at the 75th percentile. Her pediatrician will weigh her again over the next few months and expect the growth curve to follow the same percentile throughout the first year. In general, dramatic changes or measures much above the 80th percentile or much below the 10th percentile may be cause for concern.

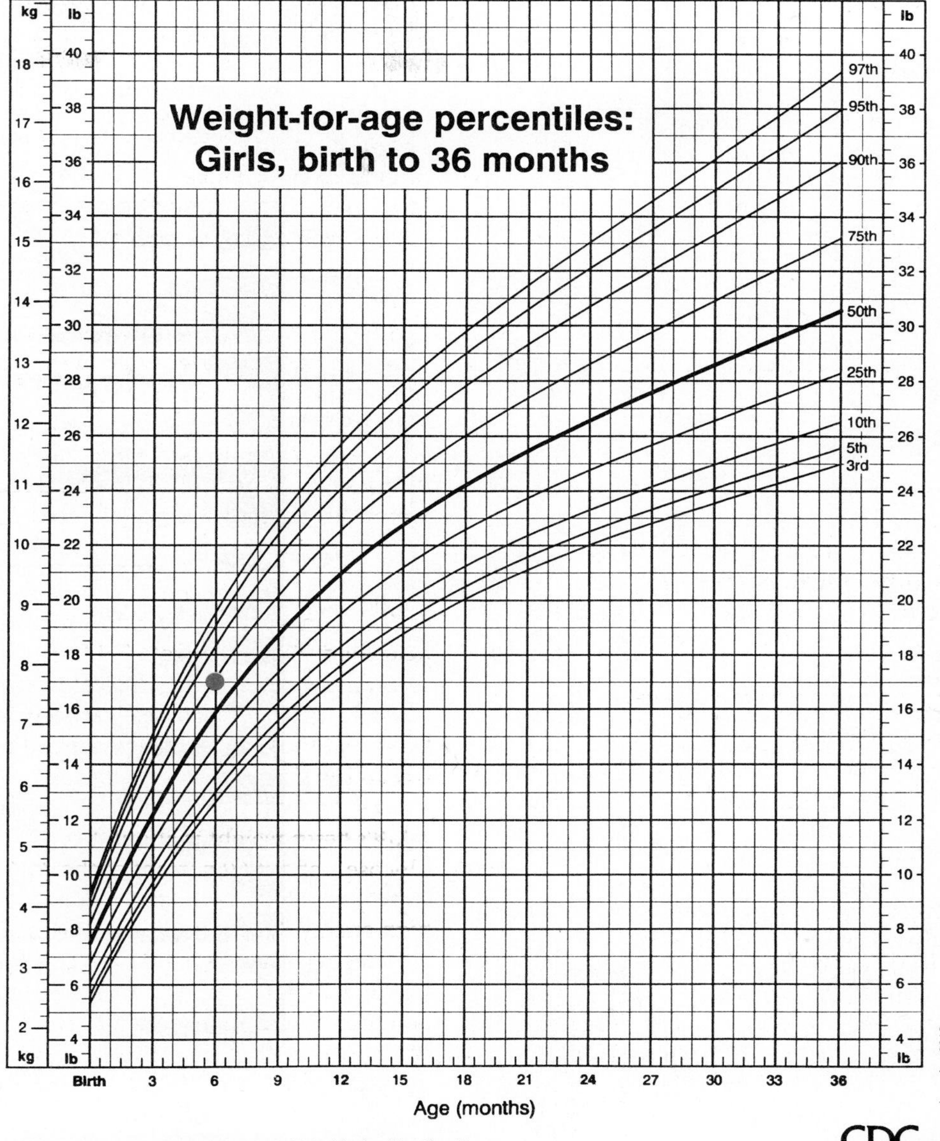

TRY IT Determine the percentile for a 12-month-old girl who weighs 21 pounds.

the fluid in an infant's body is located *outside* the cells—between the cells and in the blood vessels—rapid fluid losses and the resulting dehydration can be life-threatening. Conditions that cause rapid fluid loss, such as diarrhea or vomiting, require prompt treatment with an electrolyte solution designed for infants.

Breast Milk In the United States and Canada, the two dietary practices that have the most significant effect on an infant's nutrition are the milk the infant receives and the age at which solid foods are introduced. A later section discusses the introduction of solid foods, but as to the milk, both the American Academy of Pediatrics and the Canadian Paediatric Society strongly recommend breastfeeding for healthy full-term infants, except where specific contraindications exist.

> FIGURE 16-2 **Recommended Intakes of an Infant and an Adult Compared on the Basis of Body Weight**

Because infants are small, they need smaller total amounts of the nutrients than adults do, but when comparisons are based on body weight, infants need more than twice as much of many nutrients. Infants use large amounts of energy and nutrients, in proportion to their body size, to keep all their metabolic processes going.

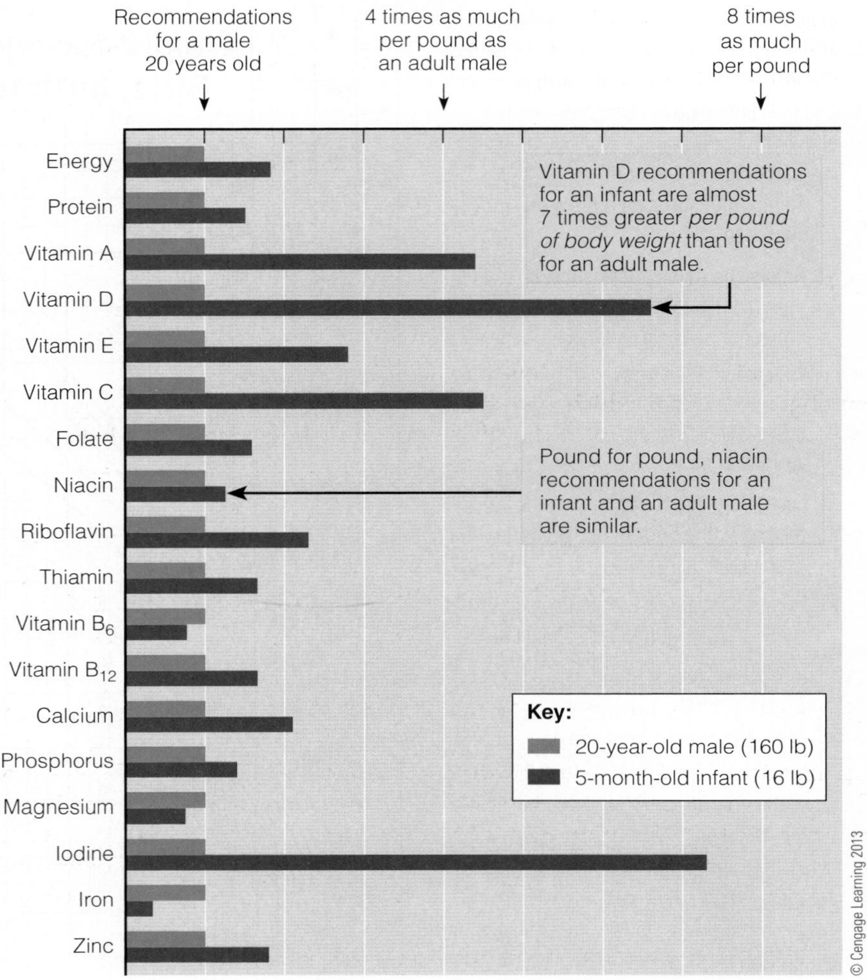

The Academy of Nutrition and Dietetics also advocates breastfeeding for the nutritional health of the infant as well as for the many other benefits it provides both infant and mother (review Table 15-4 on p. 492).[4]

Breast milk excels as a source of nutrients for infants. Its unique nutrient composition and protective factors promote optimal infant health and development throughout the first year of life. Ideally, infants will receive exclusive breastfeeding for 6 months, and breastfeeding with complementary foods for at least 12 months.[5] Experts add, though, that iron-fortified formula, which imitates the nutrient composition of breast milk, is an acceptable alternative. After all, the primary goal is to provide the infant nourishment in a relaxed and loving environment. Chapter 15 discusses breastfeeding, breastfeeding support, reasons why some women choose not to breastfeed, and contraindications to breastfeeding.

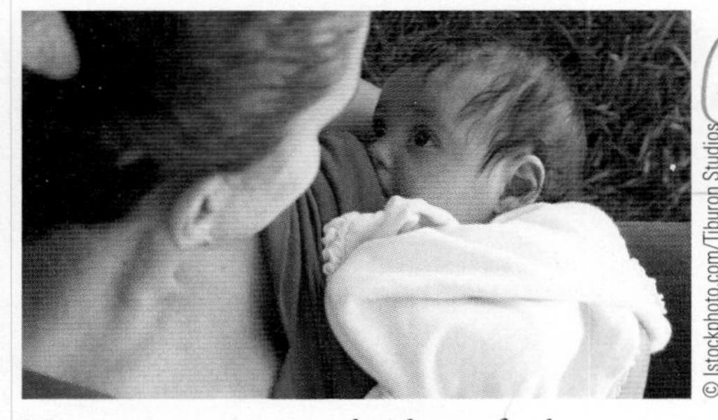

Women are encouraged to breastfeed whenever possible because breast milk provides infants with many nutrient and health advantages.

Frequency and Duration of Breastfeeding Breast milk is more easily and completely digested than formula, so breastfed infants usually need to eat more frequently than formula-fed infants do. During the first few weeks, approximately 8 to 12 feedings a day, on demand, as soon as the infant shows early signs of hunger such as increased alertness, activity, or suckling motions. Such a schedule promotes optimal milk production and infant growth. Crying is a late indicator of hunger. An infant who nurses every 2 to 3 hours and sleeps contentedly between feedings is adequately nourished. As the infant gets older, stomach capacity enlarges and the mother's milk production increases, allowing for longer intervals between feedings.

Even though the infant obtains about half the milk from the breast during the first 2 to 3 minutes of suckling, the infant should be encouraged to breastfeed on the first breast for as long as he or she is actively suckling, before being offered the second breast. Begin each feeding on the breast offered last. The infant's suckling, as well as the complete removal of milk from the breast, stimulates milk production.

Energy Nutrients The energy-nutrient composition of breast milk differs dramatically from that recommended for adult diets (see Figure 16-3). Yet for infants, breast milk is nature's most nearly perfect food, providing the clear lesson that people at different stages of life have different nutrient needs.

The main carbohydrate in breast milk (and standard infant formula) is the disaccharide lactose. In addition to being easily digested, lactose enhances calcium absorption. The carbohydrate component of breast milk also contains abundant oligosaccharides, which are present only in trace amounts in cow's milk and infant formula made from cow's milk.[6] Breast milk oligosaccharides help protect the infant from infection by preventing the binding of pathogens to the infant's intestinal cells.[7]

The amount of protein in breast milk is less than in cow's milk, but this quantity is actually beneficial because it places less stress on the infant's immature kidneys to excrete urea, the major end product of protein metabolism. Much of the protein in breast milk is **alpha-lactalbumin,** which is efficiently digested and absorbed.

As for the lipids, breast milk contains a generous proportion of the essential fatty acids linoleic acid and linolenic acid, as well as their longer-chain derivatives arachidonic acid and DHA (docosahexaenoic acid). As Chapter 5 mentions, DHA is the most abundant fatty acid in the brain and is also present in the retina of the eye, contributing to neural and visual development. DHA accumulation in the brain is greatest during the first year of life and is higher in breastfed infants than formula-fed infants.[8] Research has focused on the mental and visual development of breastfed infants and infants fed standard formula with and without DHA added.[9] Most studies show no beneficial effect of DHA supplementation of formula for term infants.[10] Adding DHA to standard infant formulas has no adverse effects, however, and most standard formulas are currently fortified with both DHA and arachidonic acid.

> FIGURE 16-3 Percentages of Energy-Yielding Nutrients in Breast Milk and in Recommended Adult Diets

The proportions of energy-yielding nutrients in human breast milk differ from those recommended for adults.

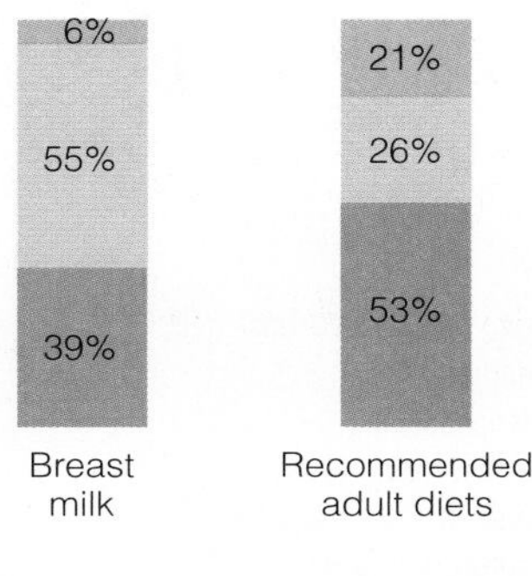

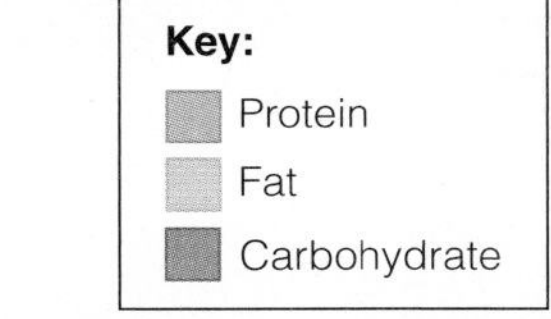

NOTE: The values listed for adults represent approximate midpoints of the acceptable ranges for protein (10 to 35 percent), fat (20 to 35 percent), and carbohydrate (45 to 65 percent).

Vitamins With the exception of vitamin D, the vitamins in breast milk are ample to support infant growth. The vitamin D in breast milk is low, however, and vitamin D deficiency impairs bone mineralization.[11] Vitamin D deficiency is most likely in infants who are not exposed to sunlight daily, have darkly pigmented skin, and receive breast milk without vitamin D supplementation.[12] Reports of infants in the United States developing the vitamin D–deficiency disease rickets and recommendations to keep infants younger than 6 months of age out of direct sunlight prompted revisions in vitamin D guidelines. The American Academy of Pediatrics currently recommends a vitamin D supplement for all infants who are breastfed exclusively and for all infants who do not receive at least 1 liter (1000 milliliters, roughly 1 quart or 32 ounces) of vitamin D–fortified formula daily.[13]

Minerals The calcium content of breast milk is ideal for infant bone growth, and the calcium is well absorbed. Breast milk contains relatively small amounts of iron, but the iron has a high bioavailability. Zinc also has a high bioavailability, thanks to the presence of a zinc-binding protein. Breast milk is low in sodium, another benefit for the infant's immature kidneys. Fluoride promotes the development of strong teeth, but breast milk is not a good source.

alpha-lactalbumin (lact-AL-byoo-min): a major protein in human breast milk, as opposed to *casein* (CAY-seen), a major protein in cow's milk.

TABLE 16-2 Supplements for Full-Term Infants

	Vitamin D[a]	Iron[b]	Fluoride[c]
Breastfed infants			
Birth to 6 months of age	✓		
6 months to 1 year	✓	✓	✓
Formula-fed infants			
Birth to 6 months of age			
6 months to 1 year		✓	✓

[a]Vitamin D supplements are recommended for all infants who are exclusively breastfed and for any infants who do not receive at least 1 liter (1000 milliliters) or 1 quart (32 ounces) of vitamin D–fortified formula per day.
[b]All infants 6 months of age need additional iron, preferably in the form of iron-fortified infant cereal and/or infant meats. Formula-fed infants need iron-fortified infant formula.
[c]At 6 months of age, breastfed infants and formula-fed infants who receive ready-to-use formulas (these are prepared with water low in fluoride) or formula mixed with water that contains little or no fluoride (less than 0.3 ppm) need supplements.
SOURCE: Adapted from Committee on Nutrition, American Academy of Pediatrics, *Pediatric Nutrition Handbook*, 6th ed., ed. R. E. Kleinman (Elk Grove Village, Ill.: American Academy of Pediatrics, 2009).

Supplements Many pediatricians routinely prescribe liquid supplements containing vitamin D, iron, and fluoride as outlined in Table 16-2. In addition, infants receive a single dose of vitamin K at birth to protect them from bleeding to death. (See Chapter 11 for a description of vitamin K's role in blood clotting.)

Immunological Protection In addition to its nutritional benefits, breast milk offers immunological protection. Not only is breast milk sterile, but it actively fights disease and protects infants from illnesses.[14] Such protection is most valuable during the first year, when the infant's immune system is not fully prepared to mount a response against infections.

During the first 2 to 3 days after delivery, the breasts produce **colostrum**, a premilk substance containing mostly serum with antibodies and white blood cells. Colostrum (like breast milk) helps protect the newborn from infections against which the mother has developed immunity. The maternal antibodies in the breast milk inactivate disease-causing bacteria within the infant's digestive tract before they can start infections.[15] This explains, in part, why breastfed infants have fewer intestinal infections than formula-fed infants.

In addition to antibodies, colostrum and breast milk provide other powerful agents that help to fight against bacterial infection. ♦ Among them are the oligosaccharides, described earlier, that prevent pathogens from binding to intestinal cells. Also present are **bifidus factors**, which favor the growth of the "friendly" bacterium *Lactobacillus bifidus* in the infant's digestive tract, so that other, harmful bacteria cannot become established. An iron-binding protein in breast milk, **lactoferrin**, keeps bacteria from getting the iron they need to grow, helps absorb iron into the infant's intestinal cells, and kills some bacteria directly. The protein **lactadherin** in breast milk binds to, and inhibits replication of, the virus that causes most infant diarrhea. Breast milk also protects against other common illnesses of infancy such as ear infections and respiratory illness.[16] In addition, a growth factor present in breast milk stimulates the development and maintenance of the infant's digestive tract and its protective factors. Several breast milk enzymes such as lipase also help protect the infant against infection. Clearly, breast milk is a very special substance.

♦ Protective factors in breast milk:

- Antibodies
- Oligosaccharides
- Bifidus factors
- Lactoferrin
- Lactadherin
- Growth factor
- Lipase enzyme

Allergy and Disease Protection In addition to protection against infections, breast milk may offer protection against the development of allergies.[17] Compared with formula-fed infants, breastfed infants have a lower incidence of allergic reactions, such as recurrent wheezing and skin rashes.[18] This protection is especially noticeable among infants with a family history of allergies.[19] Breastfeeding may also reduce the risk of sudden infant death syndrome (SIDS). In one study, exclusively breastfeeding an infant for the first month reduced the risk of SIDS by half compared to never breastfeeding.[20] Similarly, breast milk may offer protection against the development of cardiovascular disease in adulthood. Compared with formula-fed infants, breastfed infants have lower blood pressure and lower blood cholesterol as adults.[21]

Other Potential Benefits Breastfeeding may offer some protection against excessive weight gain later, although findings are inconsistent.[22] One review reports that various studies have shown a protective effect, a protective effect only in certain groups, or no effect. Some research suggests that the longer the duration of breastfeeding, the lower the risk of overweight in childhood, while other research conflicts with such findings.[23] Researchers note that many other factors—socioeconomic status, other infant and child feeding practices, and especially the mother's weight—strongly predict a child's body weight.[24]

Many studies suggest a beneficial effect of breastfeeding on intelligence, but when subjected to strict standards of methodology (for example, large sample size

colostrum (ko-LAHS-trum): a milklike secretion from the breast, present during the first few days after delivery before milk appears; rich in protective factors.

bifidus (BIFF-id-us, by-FEED-us) **factors:** factors in colostrum and breast milk that favor the growth of the "friendly" bacterium *Lactobacillus* (lack-toh-ba-SILL-us) *bifidus* in the infant's intestinal tract, so that other, less desirable intestinal bacteria will not flourish.

lactoferrin (lack-toh-FERR-in): a protein in breast milk that binds iron and keeps it from supporting the growth of the infant's intestinal bacteria.

lactadherin (lack-tad-HAIR-in): a protein in breast milk that attacks diarrhea-causing viruses.

and appropriate intelligence testing), the evidence is less convincing.[25] Nevertheless, the possibility that breastfeeding may positively affect later intelligence is intriguing. It may be that some specific component of breast milk, such as DHA, stimulates brain development or that certain factors associated with the feeding process itself promote intellect.[26] Most likely, a combination of factors is involved. More large, well-controlled studies are needed to confirm the effects, if any, of breastfeeding on intelligence.

Breast Milk Banks Similar to blood banks that collect blood from individuals to give to others in need, **breast milk banks** receive milk from lactating women who have an abundant supply to give to infants whose own mothers' milk is unavailable or insufficient.[27] The women who donate breast milk are carefully screened to exclude those who smoke cigarettes, use illicit drugs, take medications (including high doses of dietary supplements), drink more than two alcoholic beverages a day, or have communicable diseases. The breast milk from several donors is pooled to ensure an even distribution of all components, pasteurized to destroy bacteria, checked for contamination, and frozen before being shipped overnight to hospitals, where it is dispensed by physician prescription. In the absence of a mother's own breast milk, donor milk may be the life-saving solution for fragile infants, most notably those with very low birthweight or unusual medical conditions.

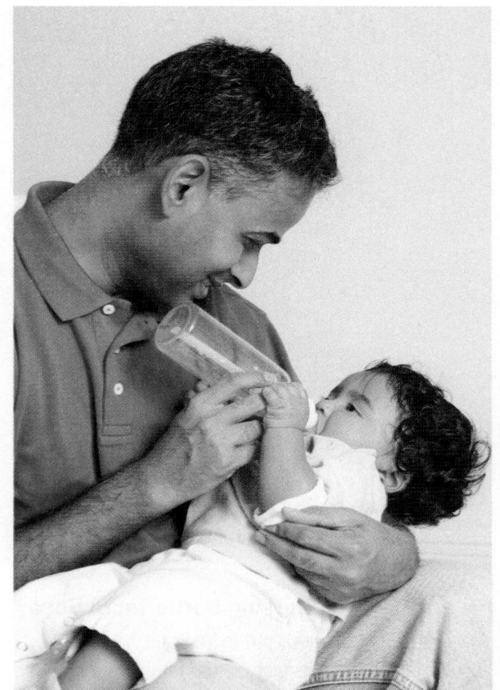

The infant thrives on infant formula offered with affection.

Infant Formula

A woman who breastfeeds for a year can **wean** her infant to cow's milk, bypassing the need for infant formula. A woman who decides to feed her infant formula from birth, to wean to formula after less than a year of breastfeeding, or to substitute formula for breastfeeding on occasion must select an appropriate infant formula and learn to prepare it. Cow's milk is inappropriate during the first year of life.

Infant Formula Composition Formula manufacturers attempt to copy the nutrient composition of breast milk as closely as possible. Figure 16-4 illustrates the energy-nutrient balance of breast milk, infant formula, and cow's milk. All formula-fed infants should be given iron-fortified infant formulas.[28] The increasing use of iron-fortified formulas over the past few decades is responsible for the decline in iron-deficiency anemia among infants in the United States.

> FIGURE 16-4 Percentages of Energy-Yielding Nutrients in Breast Milk, Infant Formula, and Cow's Milk

The average proportions of energy-yielding nutrients in human breast milk and formula differ slightly. In contrast, cow's milk provides too much protein and too little carbohydrate.

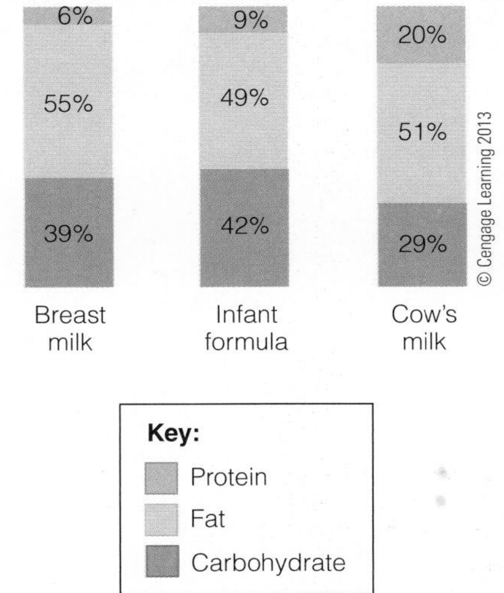

Risks of Formula Feeding Infant formulas contain no protective antibodies for infants, but in general, vaccinations, purified water, and clean environments in developed countries help protect infants from infections. Formulas can be prepared safely by following the rules of proper food handling and by using water that is free of contamination. Of particular concern is lead-contaminated water, a major source of lead poisoning in infants. Because the first water drawn from the tap each day is highest in lead, a person living in a house with old, lead-soldered plumbing should let the water run a few minutes before drinking or using it to prepare formula or food.

In developing countries and in poor areas of the United States, formula may be unavailable, prepared with contaminated water, or overdiluted in an attempt to save money. Contaminated formulas may cause infections, leading to diarrhea, dehydration, and malabsorption. Without sterilization and refrigeration, formula is an ideal breeding ground for bacteria. Whenever such risks are present, breastfeeding can be a life-saving option: breast milk is sterile, and its antibodies enhance an infant's resistance to infections.

Infant Formula Standards National and international standards have been set for the nutrient contents of infant formulas. In the United States, the standard developed by the American Academy of Pediatrics reflects "human milk taken from well-nourished mothers during the first or second month of lactation, when the infant's growth rate is high." The Food and Drug Administration (FDA) mandates the safety and nutritional quality of infant formulas. Formulas meeting these standards have similar nutrient compositions. Small differences among formulas are sometimes confusing, but they are usually unimportant.

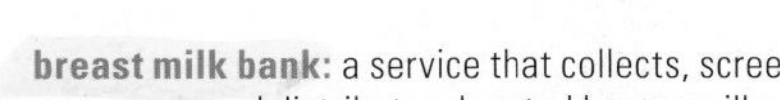

breast milk bank: a service that collects, screens, processes, and distributes donated human milk.

wean: to gradually replace breast milk with infant formula or other foods appropriate to an infant's diet.

Special Formulas Standard formulas are inappropriate for some infants. Special formulas have been designed to meet the dietary needs of infants with specific conditions such as prematurity or inherited diseases. Most infants allergic to milk protein can drink formulas based on soy protein.[29] Soy formulas also use cornstarch and sucrose instead of lactose and so are recommended for infants with lactose intolerance as well. They are also useful as an alternative to milk-based formulas for vegan families. Despite these limited uses, soy formulas account for about 25 percent of the infant formulas sold today. Although soy formulas support the normal growth and development of infants, for infants who don't need them, they offer no advantage over milk formulas.

Some infants who are allergic to cow's milk protein may also be allergic to soy protein.[30] For these infants, special formulas based on hydrolyzed protein are available.

Inappropriate Formulas Caregivers must use only products designed for infants; soy *beverages*, for example, are nutritionally incomplete and inappropriate for infants. Goat's milk is also inappropriate for infants in part because of its low folate content.[31] An infant receiving goat's milk is likely to develop "goat's milk anemia," an anemia characteristic of folate deficiency.

Nursing Bottle Tooth Decay An infant cannot be allowed to sleep with a bottle because of the potential damage to developing teeth. Salivary flow, which normally cleanses the mouth, diminishes as the infant falls asleep. Prolonged sucking on a bottle of formula, milk, or juice bathes the upper teeth in a carbohydrate-rich fluid that nourishes decay-producing bacteria. (The tongue covers and protects most of the lower teeth, but they, too, may be affected.) The result is extensive and rapid tooth decay (see Figure 16-5). To prevent **nursing bottle tooth decay,** no infant should be put to bed with a bottle of nourishing fluid.

> FIGURE 16-5 **Nursing Bottle Tooth Decay**
This child was frequently put to bed sucking on a bottle filled with apple juice, so the teeth were bathed in carbohydrate for long periods of time—a perfect medium for bacterial growth. The upper teeth show signs of decay.

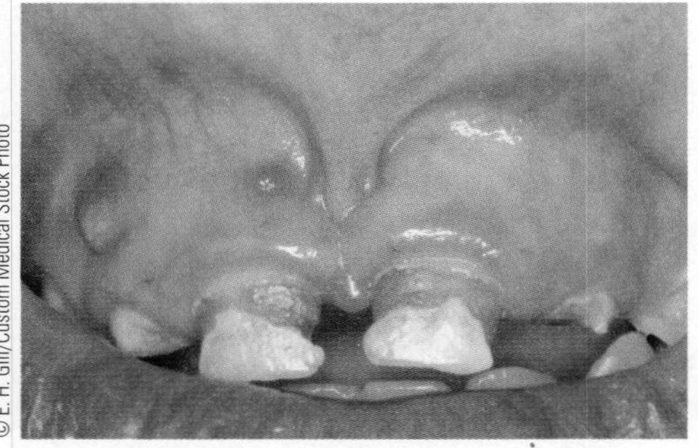

Special Needs of Preterm Infants

An estimated one out of eight pregnancies in the United States results in a preterm birth.[32] The terms *preterm* and *premature* imply incomplete fetal development, or immaturity, of many body systems. As might be expected, preterm birth is a leading cause of infant deaths. Preterm infants face physical independence from their mothers before some of their organs and body tissues are ready. The rate of weight gain in the fetus is greater during the last trimester of gestation than at any other time. Therefore, a preterm infant is most often a low-birthweight infant as well. A premature birth deprives the infant of the nutritional support of the placenta during a time of maximal growth.

The last trimester of gestation is also a time of building nutrient stores. Being born with limited nutrient stores intensifies the already precarious situation for the infant. The physical and metabolic immaturity of preterm infants further compromises their nutrition status. Nutrient absorption, especially of fat and calcium, from an immature GI tract is limited. Consequently, preterm, low-birthweight infants are candidates for nutrient imbalances. Deficiencies of the fat-soluble vitamins, calcium, iron, and zinc are common.

Preterm breast milk is well suited to meet a preterm infant's needs. During early lactation, preterm breast milk contains higher concentrations of protein and is lower in volume than term breast milk. The low milk volume is advantageous because preterm infants consume small quantities of milk per feeding, and the higher protein concentration allows for better growth. In many instances, supplements of nutrients specifically designed for preterm infants are added to the mother's expressed breast milk and fed to the infant from a bottle. When fortified with a preterm supplement, preterm breast milk supports growth at a rate that approximates the growth rate that would have occurred within the uterus.[33]

Introducing Cow's Milk

The age at which cow's milk should be introduced to the infant's diet has long been a source of controversy. The American Academy of Pediatrics advises that cow's milk is not appropriate during the first year.[34] Between the ages of 1 and 2 years, a transition from breast milk

nursing bottle tooth decay: extensive tooth decay due to prolonged tooth contact with formula, milk, fruit juice, or other carbohydrate-rich liquid offered to an infant in a bottle.

or formula to reduced-fat cow's milk can take place, but care should be taken to avoid excessive restriction of dietary fat.[35]

> DIETARY GUIDELINES FOR AMERICANS 2010

Children 2 to 3 years of age should consume 2 cups of milk or milk products per day, and children 4 to 8 years of age should consume 2½ cups per day.

Cow's milk is a poor choice during the first year of life.[36] For some infants, particularly those younger than 6 months of age, cow's milk may cause intestinal bleeding, which can lead to iron deficiency. Cow's milk is also a poor source of iron. Consequently, it both causes iron loss and fails to replace iron. Furthermore, the bioavailability of iron from infant cereal and other foods is reduced when cow's milk replaces breast milk or iron-fortified formula during the first year. Compared with breast milk or iron-fortified formula, cow's milk is higher in calcium and lower in vitamin C, characteristics that further reduce iron absorption. In addition, the higher protein concentration of cow's milk can stress the infant's kidneys. In short, infants need breast milk or iron-fortified infant formula, *not* cow's milk.

Introducing Solid Foods The high nutrient needs of infancy are met first by breast milk or formula only and then by the limited addition of selected foods over time. Infants gradually develop the ability to chew, swallow, and digest the wide variety of foods available to adults. The caregiver's selection of appropriate foods at the appropriate stages of development is prerequisite to the infant's optimal growth and health.

When to Begin In addition to breast milk or formula, an infant can begin eating solid foods between 4 and 6 months.[37] The American Academy of Pediatrics supports exclusive breastfeeding for 6 months but recognizes that infants are often developmentally ready to accept complementary foods between 4 and 6 months of age. The main purpose of introducing solid foods is to provide needed nutrients that are no longer supplied adequately by breast milk or formula alone. The foods chosen must be those that the infant is developmentally capable of handling both physically and metabolically. As digestive secretions gradually increase throughout the first year of life, the digestion of solid foods becomes more efficient. The exact timing depends on the individual infant's needs and developmental readiness (see Table 16-3, p. 514), which vary from infant to infant because of differences in growth rates, activities, and environmental conditions. In addition to the infant's nutrient needs and physical readiness to handle different forms of foods, the need to detect and control allergic reactions should also be considered when introducing solid foods.

Food Allergies To prevent allergy and to facilitate its prompt identification should it occur, experts recommend introducing single-ingredient foods, one at a time, in small portions, and waiting 3 to 5 days before introducing the next new food.[38] For example, rice cereal is usually the first cereal introduced because it is the least allergenic. When it is clear that rice cereal is not causing an allergy, another grain, perhaps barley or oat is introduced. Wheat cereal is offered last because it is the most common offender. If a cereal causes an allergic reaction such as a skin rash, digestive upset, or respiratory discomfort, it should be discontinued before introducing the next food. A later section in this chapter offers more information about food allergies.

Choice of Infant Foods Infant foods should be selected to provide variety, balance, and moderation. Commercial baby foods offer a wide variety of palatable, nutritious foods in a safe and convenient form. Homemade infant foods can be as nutritious as commercially prepared ones, so long as the cook minimizes nutrient losses during preparation. Ingredients for homemade foods should be fresh, whole foods without added salt, sugar, or seasonings. Pureed food can be frozen in ice cube trays, providing convenient-size blocks of food that can be thawed,

TABLE 16-3 Infant Development and Recommended Foods

Because each stage of development builds on the previous stage, the foods from an earlier stage continue to be included in all later stages.

Age (mo)	Feeding Skill	Appropriate Foods Added to the Diet
0–4	Turns head toward any object that brushes cheek. Initially swallows using back of tongue; gradually begins to swallow using front of tongue as well. Strong reflex (extrusion) to push food out during first 2 to 3 months.	Feed breast milk or infant formula.
4–6	Extrusion reflex diminishes, and the ability to swallow nonliquid foods develops. Indicates desire for food by opening mouth and leaning forward. Indicates satiety or disinterest by turning away and leaning back. Sits erect with support at 6 months. Begins chewing action. Brings hand to mouth. Grasps objects with palm of hand.	Begin iron-fortified cereal mixed with breast milk, formula, or water. Begin pureed meats, legumes, vegetables, and fruits.
6–8	Able to self-feed finger foods. Develops pincer (finger to thumb) grasp. Begins to drink from cup.	Begin textured vegetables and fruits. Begin unsweetened, diluted fruit juices from cup.
8–10	Begins to hold own bottle. Reaches for and grabs food and spoon. Sits unsupported.	Begin breads and cereals from table. Begin yogurt. Begin pieces of soft, cooked vegetables and fruit from table. Gradually begin finely cut meats, fish, casseroles, cheese, eggs, and mashed legumes.
10–12	Begins to master spoon, but still spills some.	Add variety. Gradually increase portion sizes.[a]

[a]Portion sizes for infants and young children are smaller than those for an adult. For example, a grain serving might be ½ slice of bread instead of 1 slice, or ¼ cup rice instead of ½ cup.

SOURCE: Adapted in part from Committee on Nutrition, American Academy of Pediatrics, *Pediatric Nutrition Handbook*, 6th ed., ed. R. E. Kleinman (Elk Grove Village, Ill.: American Academy of Pediatrics, 2009), pp. 113–142.

warmed, and fed to the infant. To guard against foodborne illnesses, hands and equipment must be kept clean.

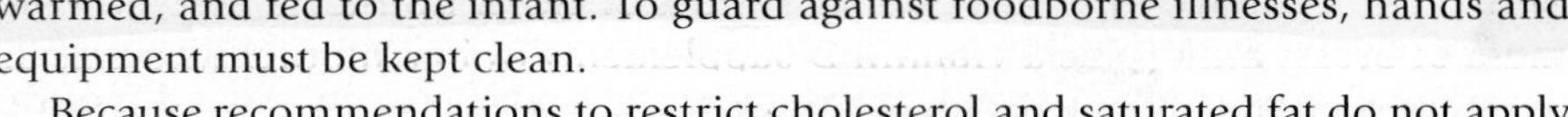

Because recommendations to restrict cholesterol and saturated fat do not apply to children younger than age 2, labels on foods for children younger than 2 (such as infant meats and cereals) only list amounts for total fat. Fat information is limited on infant food labels to prevent parents from overly restricting fat in infants' diets. Fearing that their infant will become overweight, parents may unintentionally malnourish the infant by limiting fat. In fact, infants and young children, because of their rapid growth, need more fat than older children and adults.

Foods to Provide Iron Rapid growth demands iron. At about 4 to 6 months of age, the infant begins to need more iron than body stores plus breast milk or iron-fortified formula can provide. In addition to breast milk or iron-fortified formula, infants can receive iron from iron-fortified cereals and, once they readily accept solid foods, from meats or legumes. Iron-fortified cereals contribute a significant amount of iron to an infant's diet, but the iron's bioavailability is poor.[39] Caregivers can enhance iron absorption from iron-fortified cereals by serving vitamin C–rich foods with meals.

Foods such as iron-fortified cereals and formulas, mashed legumes, and strained meats provide iron.

Foods to Provide Vitamin C The best sources of vitamin C are fruits and vegetables (see pp. 326–327 in Chapter 10). It has been suggested that infants who are introduced to fruits before vegetables may develop a preference for sweets and find the vegetables less palatable, but there is no evidence to support offering these foods in a particular order.[40]

Fruit juice is a good source of vitamin C, but excessive juice intake can lead to diarrhea in infants and young children.[41] Furthermore, too much fruit juice contributes excessive kcalories and displaces other nutrient-rich foods. The American Academy of Pediatrics recommends limiting juice consumption for infants and young children (1 to 6 years of age) to between 4 and 6 ounces per day.[42] Fruit juices should be diluted and served in a cup, not a bottle, once the infant is 6 months of age or older.

> DIETARY GUIDELINES FOR AMERICANS 2010

Monitor intake of 100 percent fruit juice for children, especially those who are overweight or obese.

Foods to Omit Concentrated sweets, including baby food "desserts," have no place in an infant's diet. They convey no nutrients to support growth, and the extra food energy can promote obesity. Products containing sugar alcohols such as sorbitol should also be limited because they may cause diarrhea. Canned vegetables are also inappropriate for infants because they often contain too much sodium. Honey and corn syrup should never be fed to infants because of the risk of **botulism**.* Infants and young children are vulnerable to foodborne illnesses, and the *Dietary Guidelines for Americans* address this risk.

> DIETARY GUIDELINES FOR AMERICANS 2010

Infants and young children should not eat or drink unpasteurized milk, milk products, or juices; raw or undercooked eggs, meat, poultry, fish, or shellfish; or raw sprouts.

Infants and even young children cannot safely chew and swallow any of the foods listed in the margin; ♦ they can easily choke on these foods, a risk not worth taking. Nonfood items may present even greater choking hazards to infants and young children. Parents and caregivers must pay careful attention to eliminate choking hazards in children's environments.

♦ To prevent choking, do not give infants or young children:
- Cherries
- Gum
- Hard or gel-type candies
- Hot dog slices
- Marshmallows
- Nuts
- Peanut butter
- Popcorn
- Raw carrots
- Raw celery
- Whole beans
- Whole grapes

Keep these nonfood items out of their reach:
- Balloons
- Coins
- Pen tops
- Small balls and marbles

Vegetarian Diets during Infancy The newborn infant is a lacto-vegetarian. As long as the infant has access to sufficient quantities of either iron-fortified infant formula or breast milk (plus a vitamin D supplement) from a mother who eats an adequate diet, the infant will thrive during the early months. "Health-food beverages," such as rice milk, are inappropriate choices because they lack the protein, vitamins, and minerals infants and toddlers need; in fact, their use can lead to nutrient deficiencies.

Infants older than about 6 months of age present a greater challenge in terms of meeting nutrient needs by way of vegetarian and, especially, vegan diets. Continued breastfeeding or formula feeding is recommended, but supplementary feedings are necessary to ensure adequate energy and iron intakes. Infants and young children in vegetarian families should be given iron-fortified infant cereals well into the second year. Mashed or pureed legumes, tofu, and cooked eggs can be added to their diets in place of meats.

Infants who receive a well-balanced vegetarian diet that includes milk products and a variety of other foods can easily meet their nutritional requirements for growth. This is not always true for vegan infants; the growth of vegan infants slows significantly when weaning from breast milk to solid foods. Deficiencies of protein, vitamin D, vitamin B_{12}, iron, and calcium have been reported in infants fed vegan diets. Vegan diets that are high in fiber, other complex carbohydrates,

Ideally, a 1-year-old eats many of the same foods as the rest of the family.

*In infants, but not in older individuals, ingestion of *Clostridium botulinum* spores can cause illness when the spores germinate in the intestine and produce a toxin, which is absorbed. Symptoms include poor feeding, constipation, loss of tension in the arteries and muscles, weakness, and respiratory compromise. Infant botulism has been implicated in 5 percent of cases of sudden infant death syndrome (SIDS).

botulism (BOT-chew-lism): an often fatal foodborne illness caused by the ingestion of foods containing a toxin produced by bacteria that grow without oxygen. (See Chapter 19 for details.)

> **FIGURE 16-6 Sample Meal Plan for a 1 Year Old**

SAMPLE MENU	
Breakfast	1 scrambled egg 1 slice whole-wheat toast ½ c reduced-fat milk
Morning snack	½ c yogurt ¼ c fruit[a]
Lunch	½ grilled cheese sandwich: 1 slice whole-wheat bread with ½ slice cheese ½ c vegetables[b] (steamed carrots) ¼ c 100% fruit juice
Afternoon snack	½ c fruit[a] ½ c toasted oat cereal
Dinner	1 oz chopped meat or ¼ c well-cooked mashed legumes ½ c rice or pasta ½ c vegetables[b] (chopped broccoli) ½ c reduced-fat milk

NOTE: This sample menu provides about 1000 kcalories.
[a]Include citrus fruits, melons, and berries.
[b]Include dark-green, leafy, and deep-yellow vegetables.

and water will fill infants' stomachs before meeting energy needs. This problem can be partially alleviated by providing more energy-dense foods, such as mashed legumes, tofu, and avocado. Using soy formulas (or milk) fortified with calcium, vitamin B_{12}, and vitamin D and including vitamin C–containing foods at meals to enhance iron absorption will help prevent some nutrient deficiencies in vegan diets. Parents or caregivers who choose to feed their infants vegan diets should consult with their pediatrician and a registered dietitian frequently to ensure a nutritionally adequate diet that will support growth.

Foods at 1 Year At 1 year of age, reduced-fat or low-fat cow's milk can become a primary source of most of the nutrients an infant needs; 2 to 3 cups a day meets those needs sufficiently. Ingesting more milk than this can displace iron-rich foods, which can lead to **milk anemia.** If powdered milk is used, it should contain some fat.

Other foods—meats, iron-fortified cereals, enriched or whole-grain breads, fruits, and vegetables—should be supplied in variety and in amounts sufficient to round out total energy needs. Ideally, a 1 year old will sit at the table, eat many of the same foods everyone else eats, and drink liquids from a cup, not a bottle. Figure 16-6 shows a meal plan that meets a 1 year old's requirements.

Mealtimes with Toddlers

The nurturing of a young child involves more than nutrition. Those who care for young children are responsible not only for providing nutritious foods, milk, and water, but also a safe, loving, secure environment in which the children may grow and develop. In light of toddlers' developmental and nutrient needs and their often contrary and willful behavior, a few feeding guidelines may be helpful:

- *Discourage unacceptable behavior, such as standing at the table or throwing food.* Be consistent and firm, not punitive. For example, instead of saying "You make me mad when you don't sit down," say "The fruit salad tastes good; please sit down and eat some with me." The child will soon learn to sit and eat.
- *Let toddlers explore and enjoy food, even if this means eating with fingers for a while.* Learning to use a spoon will come in time. Children who are allowed to touch, mash, and smell their food while exploring it are more likely to accept it.
- *Don't force food on children.* Rejecting new foods is normal, and acceptance is more likely as children become familiar with new foods through repeated opportunities to taste them. Instead of saying "You cannot go outside to play until you taste your carrots," say "You can try the carrots again another time."
- *Provide nutritious foods and let children choose which ones, and how much, they will eat.* Gradually, they will acquire a taste for different foods.
- *Limit sweets.* Infants and young children have little room for empty-kcalorie foods in their daily energy allowance. Do not use sweets as a reward for eating meals.
- *Don't turn the dining table into a battleground.* Make mealtimes enjoyable. Teach healthy food choices and eating habits in a pleasant environment. Mealtimes are not the time to fight, argue, or scold.

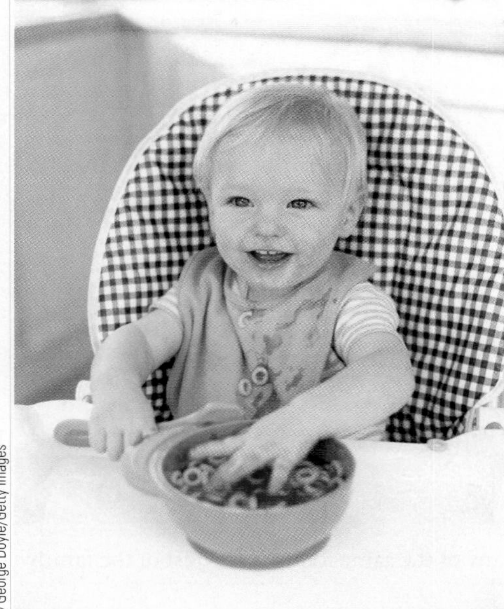

Let toddlers explore and enjoy their food.

REVIEW IT List some of the immune factors in breast milk and describe the appropriate foods for infants during the first year of life.
The primary food for infants during the first 12 months is either breast milk or iron-fortified formula. In addition to nutrients, breast milk also offers immunological protection. At about 4 to 6 months of age, infants should gradually begin eating solid foods. By 1 year, they are drinking from a cup and eating many of the same foods as the rest of the family.

milk anemia: iron-deficiency anemia that develops when an excessive milk intake displaces iron-rich foods from the diet.

16.2 Nutrition during Childhood

LEARN IT Explain how children's appetites and nutrient needs reflect their stage of growth and why iron deficiency and obesity are often concerns during childhood.

Each year from age 1 to adolescence, a child typically grows taller by 2 to 3 inches and heavier by 5 to 6 pounds. Growth charts provide valuable clues to a child's health. Weight gains out of proportion to height gains may reflect overeating and inactivity, whereas measures significantly less than the standard suggest malnutrition.

Increases in height and weight are only two of the many changes growing children experience (see Figure 16-7). At age 1, children can stand alone and are beginning to toddle; by 2, they can walk and are learning to run; and by 3, they can jump and climb with confidence. Bones and muscles increase in mass and density to make these accomplishments possible. Thereafter, lengthening of the long bones and increases in musculature proceed unevenly and more slowly until adolescence.

Energy and Nutrient Needs Children's appetites begin to diminish around 1 year, consistent with the slowing growth. Thereafter, children spontaneously vary their food intakes to coincide with their growth patterns; they demand more food during periods of rapid growth than during slow growth. Sometimes they seem insatiable, and other times they seem to live on air and water.

Children's energy intakes also vary widely from meal to meal. Even so, their total daily intakes remain remarkably constant. If children eat less at one meal, they typically eat more at the next, and vice versa. Overweight children do not always adjust their energy intakes appropriately, however, and may eat in response to external cues, disregarding hunger and satiety signals.[43]

Energy Intake and Activity As mentioned, children's energy needs vary widely, depending on their growth and physical activity. A 1-year-old child needs about 800 kcalories a day; an active 6-year-old child needs twice as many kcalories a day. By age 10, an active child needs about 2000 kcalories a day. Total energy needs increase slightly with age, but energy needs per kilogram of body weight actually decline gradually.

Physically active children of any age need more energy because they expend more, and inactive children can become obese even when they eat less food than the average. Unfortunately, our nation's children are becoming less and

> FIGURE 16-7 **Body Shape of 1 Year Old and 2 Year Old Compared**

The body shape of a 1 year old (left) changes dramatically by age 2 (right). The 2 year old has lost much of the baby fat; the muscles (especially in the back, buttocks, and legs) have firmed and strengthened; and the leg bones have lengthened.

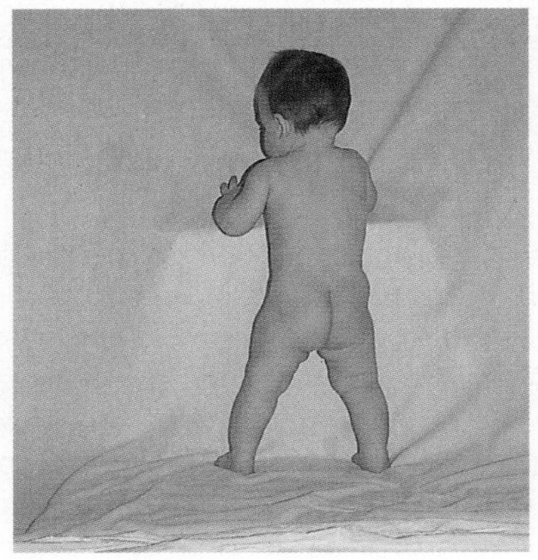

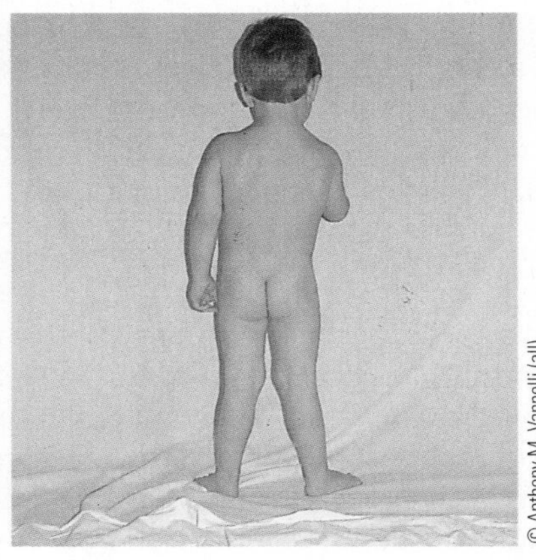

less active; schools would serve our children well by offering more activities to promote physical fitness.[44] Children who learn to enjoy physical play and exercise, both at home and at school, are best prepared to maintain active lifestyles as adults.

> **DIETARY GUIDELINES FOR AMERICANS 2010**
> Children should do 60 minutes or more of physical activity daily. Children are encouraged to spend no more than 1 to 2 hours each day watching television, playing electronic games, or using the computer (other than for homework).

Some children, notably those adhering to a vegan diet, may have difficulty meeting their energy needs. Grains, vegetables, and fruits provide plenty of fiber, adding bulk, but may provide too little energy to support growth. Soy products, other legumes, and nut or seed butters offer more concentrated sources of energy to support optimal growth and development.[45]

Carbohydrate and Fiber Carbohydrate recommendations are based on the brain's glucose needs. After 1 year of age, the brain's use of glucose remains fairly constant and is within the adult range. Carbohydrate recommendations for children older than 1 year are therefore the same as for adults (see inside front cover).[46]

Fiber recommendations derive from adult intakes shown to reduce the risk of heart disease and are based on energy intakes. ♦ Consequently, fiber recommendations for younger children with low energy intakes are less than those for older ones with high energy intakes.[47]

♦ Fiber recommendations for children:

Age (yr)	AI (g)
1–3	19
4–8	25
9–13	31 (boys)
	26 (girls)
14–18	38 (boys)
	26 (girls)

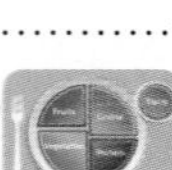

> **DIETARY GUIDELINES FOR AMERICANS 2010**
> Children and adolescents should consume whole-grain products often, and at least half of the grains should be whole grains.

Fat and Fatty Acids No RDA for total fat has been established, but the DRI Committee recommends a fat intake of 30 to 40 percent of energy for children 1 to 3 years of age and 25 to 35 percent for children 4 to 18 years of age.[48] As long as children's energy intakes are adequate, however, fat intakes less than 30 percent of total energy do not impair growth.[49] Children who eat low-fat diets, however, tend to have low intakes of some vitamins and minerals. Recommended intakes of the essential fatty acids are based on average intakes (see inside front cover).

> **DIETARY GUIDELINES FOR AMERICANS 2010**
> Keep total fat intake between 30 to 40 percent of kcalories for children 1 to 3 years of age and between 25 and 35 percent of kcalories for children and adolescents 4 to 18 years of age, with most fats coming from sources of polyunsaturated and monounsaturated fatty acids, such as fish, nuts, and vegetable oils.

Protein Like energy needs, total protein needs increase slightly with age, but when the child's body weight is considered, the protein requirement actually declines slightly (see inside front cover). Protein recommendations must consider the requirements for maintaining nitrogen balance, the quality of protein consumed, and the added needs of growth.

Vitamins and Minerals The vitamin and mineral needs of children increase with age (see inside front cover). A balanced diet of nutritious foods can meet children's needs for these nutrients, with the notable exception of iron, and possibly vitamin D. Iron-deficiency anemia is a major problem worldwide and is prevalent among both US and Canadian children, especially toddlers 1 to 3 years of age.[50] During the second year of life, toddlers progress from a diet of iron-rich infant foods such as breast milk, iron-fortified formula, and iron-fortified infant cereal to a diet of adult foods and iron-poor cow's milk. In addition, their appetites often

TABLE 16-4 Recommended Daily Amounts from Each Food Group (1000 to 1800 kCalories)

Food Group	1000 kcal	1200 kcal	1400 kcal	1600 kcal	1800 kcal
Fruits	1 c	1 c	1½ c	1½ c	1½ c
Vegetables	1 c	1½ c	1½ c	2 c	2½ c
Grains	3 oz	4 oz	5 oz	5 oz	6 oz
Protein foods	2 oz	3 oz	4 oz	5 oz	5 oz
Milk	2 c	2½ c	2½ c	3 c	3 c

fluctuate—some become finicky about the foods they eat, and others prefer milk and juice to solid foods. These situations can interfere with children eating iron-rich foods at a critical time for brain growth and development.

To prevent iron deficiency, children's foods must deliver 7 to 10 milligrams of iron per day. To achieve this goal, snacks and meals should include iron-rich foods, and milk intake should be reasonable so that it will not displace lean meats, fish, poultry, eggs, legumes, and whole-grain or enriched products. (Chapter 13 describes iron-rich foods and ways to maximize iron absorption.)

The DRI committee recently revised their recommendations for vitamin D intakes for healthy Americans.[51] Children typically obtain most of their vitamin D from fortified milk (2.5 micrograms per 1 cup serving) and dry cereals (1 microgram per ½ cup serving). Children who do not meet their RDA from these sources should receive a vitamin D supplement.[52] Remember that sunlight is also a source of vitamin D, especially in warm climates and warm seasons.

Supplements With the exception of specific recommendations for fluoride, iron, and vitamin D during infancy and childhood, the American Academy of Pediatrics and other professional groups agree that well-nourished children do not need vitamin and mineral supplements. Despite this, many children and adolescents take supplements.[53] Ironically, children with poor nutrient intakes typically do not receive supplements, and those who do take supplements typically receive extra nutrients they do not need.[54] Furthermore, researchers are still studying the safety of supplement use by children. The Federal Trade Commission has warned parents about giving supplements advertised to prevent or cure childhood illnesses such as colds, ear infections, or asthma. Dietary supplements on the market today include many herbal products that have not been tested for safety and effectiveness in children.

Planning Children's Meals To provide all the needed nutrients, children's meals should include a variety of foods from each food group—in amounts suited to their appetites and needs. Table 16-4 lists recommended amounts from each food group for several kcalorie levels. Estimated daily kcalorie needs for children of various ages are shown in Table 16-5. Figure 16-8 (p. 520) presents food patterns for children. The figure includes the recommended amounts of food for a 1200-kcalorie intake (appropriate for many younger children) and for an 1800-kcalorie intake (appropriate for many older children).

Children whose diets follow the patterns presented in Figure 16-8 can meet their nutrient needs fully, but few children eat according to these recommendations. One analysis of the quality of children's diets found that most (up to 88 percent) children between 2 and 9 years of age have diets that need substantial improvement.[55] A comprehensive survey, called the Feeding Infants and Toddlers Study (FITS), assessed the food and nutrient intakes of more than 3000 infants and toddlers.[56] The survey found that fruit and vegetable intakes of infants and toddlers are limited, and in fact, about 25 percent of infants and toddlers older than 9 months did not eat a single serving of fruits or vegetables in a day.[57] Not only are nutritious fruits and vegetables lacking, but more than 80 percent of young preschoolers (2 to 3 years of age) consumed nutrient-poor, energy-dense beverages, desserts, and snack foods each day. The most popular vegetable among this age

TABLE 16-5 Estimated Daily Energy Needs for Children

Age	Energy (kcal)
2 to 3 yr	1000
Females	
4 to 8 yr	1200–1400
9 to 13 yr	1600
Males	
4 to 8 yr	1200–1400
9 to 13 yr	1800

NOTE: Active children may need more kcalories.

> FIGURE 16-8 **Food Patterns for Children**

ChooseMyPlate.gov

Fruits, Grains, Dairy, Vegetables, Protein

© U.S. Department of Agriculture

Grains Make half your grains whole	Vegetables Vary your veggies	Fruits Focus on fruits	Milk Get your calcium-rich foods	Meat & Beans Go lean with protein
Start smart with breakfast. Look for whole-grain cereals. Just because bread is brown doesn't mean it's whole grain. Search the ingredients list to make sure the first word is "whole" (like "whole wheat").	Color your plate with all kinds of great-tasting veggies. What's green and orange and tastes good? Veggies! Go dark green with broccoli and spinach, or try orange ones like carrots and sweet potatoes.	Fruits are nature's treats—sweet and delicious. Go easy on juice and make sure it's 100%.	Move to the milk group to get your calcium. Calcium builds strong bones. Look at the carton or container to make sure your milk, yogurt, or cheese is low fat or fat-free. 	Eat lean or low-fat meat, chicken, turkey, and fish. Ask for it baked, broiled, or grilled—not fried. It's nutty, but true. Nuts, seeds, peas, and beans are all great sources of protein, too.
For a 1400-kcalorie diet (suitable for many children ages 4 to 8), include the amounts below from each food group.				
Eat 5 oz. every day; at least half should be whole	Eat 1½ cups every day	Eat 1½ cups every day	Get 2½ cups every day	Eat 4 oz. every day
For a 1800-kcalorie diet (suitable for many children ages 9 to 13), include the amounts below from each food group.				
Eat 6 oz. every day; at least half should be whole	Eat 2½ cups every day	Eat 1½ cups every day	Get 3 cups every day	Eat 5 oz. every day

Oils Oils are not a food group, but you need some for good health. Get your oils from fish, nuts, and liquid oils such as corn oil, soybean oil, and canola oil.

Find your balance between food and fun

- Move more. Aim for at least 60 minutes every day, or most days.
- Walk, dance, bike, rollerblade—it all counts. How great is that!

Fats and sugars—know your limits

- Get your fat facts and sugar smarts from the Nutrition Facts label.
- Limit solid fats as well as foods that contain them.
- Choose food and beverages low in added sugars and other kcaloric sweeteners.

group was french fries. Parents and caregivers of infants and toddlers thus need to offer a much greater variety of nutrient-dense vegetables and fruits at meals and snacks to help ensure adequate nutrition. Among other nutrition concerns for US children are inadequate intakes of vitamin E, potassium, and fiber, and excessive intakes of sodium.[58]

Hunger and Malnutrition in Children Most children in the United States and Canada have access to regular meals, but hunger and malnutrition are not uncommon, especially among children in very low-income families. More than

16 million US children are hungry at least some of the time and are living in poverty.[59] Chapter 20 examines the causes and consequences of hunger in the United States and around the world.

Hunger and Behavior Both short-term and long-term hunger exert negative effects on behavior and health. Short-term hunger, such as when a child misses a meal, impairs the child's ability to pay attention and be productive. Hungry children are irritable, apathetic, and uninterested in their environment. Long-term hunger impairs growth and immune defenses. Food assistance programs such as the WIC program (discussed in Chapter 15) and the School Breakfast and National School Lunch Programs (discussed later in this chapter) are designed to protect against hunger and improve children's health.[60]

Children who eat nutritious breakfasts improve their school performance and are tardy or absent significantly less often than their peers who do not.[61] A nutritious breakfast is a central feature of a diet that meets the needs of children and supports their healthy growth and development.[62] Children who skip breakfast typically do not make up the deficits at later meals—they simply have lower intakes of energy, vitamins, and minerals than those who eat breakfast. Without breakfast, children perform poorly in tasks requiring concentration, their attention spans are shorter, and they even score lower on intelligence tests than their well-fed peers. Malnourished children are particularly vulnerable. Common sense dictates that it is unreasonable to expect anyone to learn and perform without fuel. For the child who hasn't had breakfast, the morning's lessons may be lost altogether. Even if a child has eaten breakfast, discomfort from hunger may become distracting by late morning. Teachers aware of the late-morning slump in their classrooms wisely request that midmorning snacks be provided; snacks improve classroom performance all the way to lunchtime.

Healthy, well-nourished children are alert in the classroom and energetic at play.

> DIETARY GUIDELINES FOR AMERICANS 2010

Eat a nutrient-dense breakfast. Not eating breakfast has been associated with excess body weight, especially among children and adolescents.

Iron Deficiency and Behavior Iron deficiency has well-known and widespread effects on children's behavior and intellectual performance.[63] In addition to carrying oxygen in the blood, iron transports oxygen within cells, which use it during energy metabolism. Iron is also used to make neurotransmitters—most notably, those that regulate the ability to pay attention, which is crucial to learning. Consequently, iron deficiency not only causes an energy crisis, but also directly impairs attention span and learning ability.

Iron deficiency is often diagnosed by a quick, easy, inexpensive hemoglobin or hematocrit test that detects a deficit of iron in the *blood*. A child's *brain*, however, is sensitive to low iron concentrations long before the blood effects appear. Iron deficiency lowers the "motivation to persist in intellectually challenging tasks" and impairs overall intellectual performance. Anemic children perform poorly on tests and are disruptive in the classroom; iron supplementation improves learning and memory. When combined with other nutrient deficiencies, iron-deficiency anemia has synergistic effects that are especially detrimental to learning. Furthermore, children who had iron-deficiency anemia *as infants* continue to perform poorly as they grow older, even if their iron status improves.[64] The long-term damaging effects on mental development make prevention and treatment of iron deficiency during infancy and early childhood a high priority.

Other Nutrient Deficiencies and Behavior A child with any of several nutrient deficiencies may be irritable, aggressive, and disagreeable, or sad and withdrawn. Such a child may be labeled "hyperactive," "depressed," or "unlikable," when in fact these traits may be due to simple, even marginal, malnutrition. Parents and medical practitioners often overlook the possibility that malnutrition may account for abnormalities of appearance and

TABLE 16-6 Physical Signs of Malnutrition in Children

	Well-Nourished	Malnourished	Possible Nutrient Deficiencies
Hair	Shiny, firm in the scalp	Dull, brittle, dry, loose; falls out	Protein
Eyes	Bright, clear pink membranes; adjust easily to light	Pale membranes; spots; redness; adjust slowly to darkness	Vitamin A, the B vitamins, zinc, and iron
Teeth and gums	No pain or caries, gums firm, teeth bright	Missing, discolored, decayed teeth; gums bleed easily and are swollen and spongy	Minerals and vitamin C
Face	Clear complexion without dryness or scaliness	Off-color, scaly, flaky, cracked skin	Protein, vitamin A, and iron
Glands	No lumps	Swollen at front of neck, cheeks	Protein and iodine
Tongue	Red, bumpy, rough	Sore, smooth, purplish, swollen	B vitamins
Skin	Smooth, firm, good color	Dry, rough, spotty; "sandpaper" feel or sores; lack of fat under skin	Protein, essential fatty acids, vitamin A, B vitamins, and vitamin C
Nails	Firm, pink	Spoon-shaped, brittle, ridged	Iron
Internal systems	Regular heart rhythm, heart rate, and blood pressure; no impairment of digestive function, reflexes, or mental status	Abnormal heart rate, heart rhythm, or blood pressure; enlarged liver, spleen; abnormal digestion; burning, tingling of hands, feet; loss of balance, coordination; mental confusion, irritability, fatigue	Protein and minerals
Muscles and bones	Muscle tone; posture, long bone development appropriate for age	"Wasted" appearance of muscles; swollen bumps on skull or ends of bones; small bumps on ribs; bowed legs or knock-knees	Protein, minerals, and vitamin D

behavior. Any departure from normal healthy appearance and behavior is a sign of possible poor nutrition (see Table 16-6). In any such case, inspection of the child's diet by a registered dietitian or other qualified health-care professional is in order. Any suspicion of dietary inadequacies, no matter what other causes may be implicated, should prompt steps to correct those inadequacies immediately.

The Malnutrition-Lead Connection Children who are malnourished are vulnerable to lead poisoning. They absorb more lead if their stomachs are empty; if they have low intakes of calcium, zinc, vitamin C, or vitamin D; and, of greatest concern because it is so common, if they have an iron deficiency. Iron deficiency weakens the body's defenses against lead absorption, and lead poisoning can cause iron deficiency.[65] Common to both iron deficiency and lead poisoning are a low socioeconomic background and a lack of immunizations against infectious diseases. Another common factor is pica—a craving for nonfood items. Many children with lead poisoning eat dirt or chips of old paint, two common sources of lead.

The anemia brought on by lead poisoning may be mistaken for a simple iron deficiency and therefore may be incorrectly treated. Like iron deficiency, mild lead toxicity has nonspecific symptoms, including diarrhea, irritability, and fatigue. Adding iron to the diet does not reverse the symptoms; exposure to lead must stop and treatment for lead poisoning must begin. With further exposure, the symptoms become more pronounced, and children develop learning disabilities and behavioral problems. Still more severe lead toxicity can cause irreversible nerve damage, paralysis, mental retardation, and death.

Childhood lead exposure disrupts normal brain development—a finding that may partially explain the impaired cognitive and behavioral abilities of lead-exposed children. For 6 years, researchers measured blood lead levels at intervals in young children who lived in lead-contaminated houses.[66] Years later, brain images revealed that the higher the blood lead concentrations during childhood, the smaller the brain size as a young adult. The brains of boys were more affected than the brains of girls.

Approximately 250,000 children in the United States—most of them younger than age 6—have blood lead concentrations high enough to cause mental, behavioral, and other health problems.[67] Lead toxicity in young children comes from their own behaviors and activities—putting their hands in their mouths, playing in dirt and dust, and chewing on nonfood items.[68] Unfortunately, the body readily absorbs lead during times of rapid growth and hoards it possessively thereafter. Lead is not easily excreted and accumulates mainly in the bones, but also in the brain, teeth, and kidneys. Tragically, a child's neuromuscular system is also maturing during these first few years of life. No wonder children with elevated lead levels experience impairment of balance, motor development, and the relaying of nerve messages to and from the brain. Deficits in intellectual development are only partially reversed when lead levels decline.

Federal laws mandating reductions in leaded gasoline, lead-based solder, and other products over the past four decades have helped to reduce the amounts of lead in food and in the environment in the United States. As a consequence, the prevalence of lead toxicity in children has declined dramatically for most of the United States, but lead exposure is still a threat in certain communities. The accompanying "How To" feature presents strategies for defending children against lead toxicity.

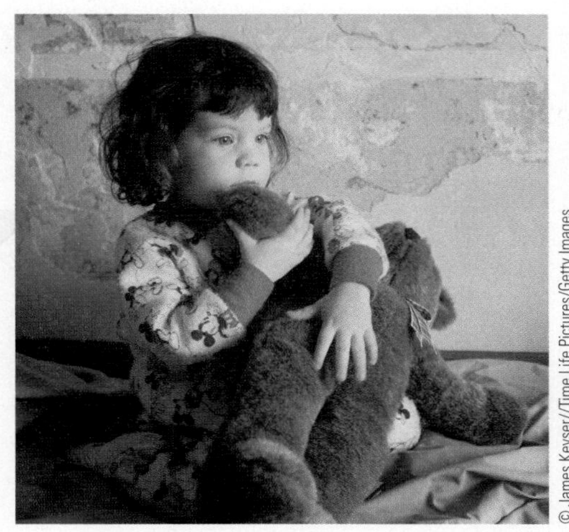

Old, lead-based paint threatens the health of an exploring child.

>How To

Protect against Lead Toxicity

Researchers simultaneously made three major discoveries about lead toxicity: lead poisoning has *subtle* effects, the effects are *permanent*, and they occur at *low levels of exposure*. The amount of lead recognized to cause harm is only 10 micrograms per 100 milliliters of blood. Some research shows that blood lead concentrations *below* this amount may adversely affect children's physical and mental development. Consequently, consumers should take ultraconservative measures to protect themselves, and especially their infants and young children, from lead poisoning. The American Academy of Pediatrics and the Centers for Disease Control recommend screening in communities with a substantial number of houses built before 1950 and in those with a substantial number of children with elevated lead levels. In addition to screening children most likely to be exposed, pediatricians should alert all parents to the possible dangers of lead exposure and explain prevention strategies.

Preventive strategies include:

- In contaminated environments, keep small children from putting dirty or old painted objects in their mouths, and make sure children wash their hands before eating. Similarly, keep small children from eating any nonfood items. Lead poisoning has been reported in young children who have eaten crayons or pool cue chalk.
- Wet-mop floors and damp-sponge walls regularly. Children's blood lead levels decline when the homes they live in are cleaned regularly.
- Be aware that other countries do not have the same regulations protecting consumers against lead. Children have been poisoned by eating crayons made in China and drinking fruit juice canned in Mexico.
- Do not use lead-contaminated water to make infant formula.
- Once you have opened canned food, store it in a lead-free container to prevent lead migration into the food.
- Do not store acidic foods or beverages (such as vinegar or orange juice) in ceramic dishware or alcoholic beverages in pewter or crystal decanters.
- Many manufacturers are now making lead-safe products. Old, handmade, or imported ceramic cups and bowls may contain lead and should not be used to heat coffee or tea or acidic foods such as tomato soup.
- Feed children nutritious meals regularly.
- Before using your newspaper to wrap food, mulch garden plants, or add to your compost, confirm with the publisher that the paper uses no lead in its ink.

The Environmental Protection Agency (EPA) also publishes a booklet, *Lead and Your Drinking Water*, in which the following cautions appear:

- Have the water in your home tested by a competent laboratory.
- Use only cold water for drinking, cooking, and making formula (cold water absorbs less lead).
- When water has been standing in pipes for more than two hours, flush the coldwater pipes by running water through them for 30 seconds before using it for drinking, cooking, or mixing formulas.
- If lead contamination of your water supply seems probable, obtain additional information and advice from the EPA and your local public health agency.

By taking these steps, parents can protect themselves and their children from this preventable danger.

TRY IT Visit the lead section of the website for the Environmental Protection Agency (www.epa.gov/lead) and identify the most common sources of lead poisoning.

Hyperactivity and "Hyper" Behavior All children are naturally active, and many of them become overly active on occasion—for example, in anticipation of a birthday party. Such behavior is markedly different from true **hyperactivity.**

Hyperactivity Hyperactive children have trouble sleeping, cannot sit still for more than a few minutes at a time, act impulsively, and have difficulty paying attention. These behaviors interfere with social development and academic progress. The cause of hyperactivity remains unknown, but it affects about 9 percent of young school-age children.[69] To resolve the problems surrounding hyperactivity, physicians often recommend specific behavioral strategies, special educational programs, and psychological counseling. If these interventions are ineffective, they prescribe medication.[70]

Research on hyperactivity has focused on several nutritional factors as possible causes or treatments.[71] Parents often blame sugar. They mistakenly believe that simply eliminating candy and other sweet treats will solve the problem. This dietary change will not solve the problem, however, and studies have consistently found no convincing evidence that sugar causes hyperactivity or worsens behavior. Such speculation has been based on personal stories. No scientific evidence supports a relationship between sugar and hyperactivity or other misbehaviors.

Food additives have also been blamed for hyperactivity and other behavior problems in children, but scientific evidence to substantiate the connection has been elusive.[72] Limited research suggests that food additives such as artificial colors or sodium benzoate preservative (or both) may exacerbate hyperactive symptoms such as inattention and impulsivity. Additional studies are needed to confirm the findings and to determine which additives might be responsible for specific negative behaviors. A Food and Drug Administration (FDA) review determined that evidence linking color additives to hyperactivity is lacking.[73] The FDA did not rule out the possibility that some food additives, including food colorings, may aggravate hyperactivity and other behavioral problems in some susceptible children.

Misbehaving Even a child who is not truly hyperactive can be difficult to manage at times. Michael may act unruly out of a desire for attention, Jessica may be cranky because of a lack of sleep, Christopher may react violently after watching too much television, and Ashley may be unable to sit still in class due to a lack of exercise. All of these children may benefit from more consistent care—regular hours of sleep, regular mealtimes, and regular outdoor activity.

Food Allergy and Intolerance **Food allergy** is frequently blamed for physical and behavioral abnormalities in children, but only 4 to 8 percent of children younger than 4 years of age are diagnosed with true food allergies.[74] Food allergies diminish with age, until in adulthood they affect less than 4 percent of the population. The prevalence of food allergy, especially peanut allergy, is on the rise, however.[75] Reasons for an increase in peanut allergy are not yet clear, but possible contributing factors include genetics, food preparation methods (roasting peanuts at very high temperatures makes them more allergenic), and exposure to medicinal skin creams containing peanut oil.[76]

A true food allergy occurs when fractions of a food protein or other large molecule are absorbed into the blood and elicit an immunologic response. (Recall that proteins are normally dismantled in the digestive tract to amino acids that are absorbed without such a reaction.) The body's immune system reacts to these large food molecules as it does to other antigens—by producing antibodies, histamines, or other defensive agents.

Detecting Food Allergy Allergies may have one or two components. They always involve antibodies, but they may or may not involve symptoms.* Therefore, allergies cannot be diagnosed from symptoms alone. The National Institute of Allergy and Infectious Diseases (NIAID) has developed clinical guidelines for the

*A person who produces antibodies *without* having any symptoms has an *asymptomatic allergy;* a person who produces antibodies *and* has symptoms has a *symptomatic allergy.*

hyperactivity: inattentive and impulsive behavior that is more frequent and severe than is typical of others a similar age; professionally called *attention-deficit/hyperactivity disorder (ADHD).*

food allergy: an adverse reaction to food that involves an immune response; also called *food-hypersensitivity reaction.*

diagnosis and management of food allergy.[77] Even symptoms exactly like those of an allergy may not be caused by an allergy. The NIAID recommends that food allergy should be considered when an individual experiences symptoms such as skin rash, respiratory difficulties, vomiting, diarrhea, or anaphylactic shock (described later) within minutes to hours of eating food, especially in young children.

Diagnosis of food allergy requires medical testing and food challenges. Once a food allergy has been diagnosed, the required treatment is strict elimination of the offending food. Children with allergies, like all children, need all their nutrients, so it is important to include other foods that offer the same nutrients as the omitted foods.[78]

Allergic reactions to food may be immediate or delayed. In either case, the antigen interacts immediately with the immune system, but the timing of symptoms varies from minutes to 24 hours after consumption of the antigen. Identifying the food that causes an immediate allergic reaction is fairly easy because the symptoms appear shortly after the food is eaten. Identifying the food that causes a delayed reaction is more difficult because the symptoms may not appear until much later. By this time, many other foods may have been eaten, complicating the picture.

These normally wholesome foods may cause life-threatening symptoms in people with allergies.

© Polara Studios, Inc.

Anaphylactic Shock The life-threatening food allergy reaction of **anaphylactic shock** is most often caused by peanuts, tree nuts, milk, eggs, wheat, soybeans, fish, or shellfish. Among these foods, eggs, milk, soy, and peanuts most often cause problems in children.[79] Children are more likely to outgrow allergies to eggs, milk, and soy than allergies to peanuts. Peanuts cause more life-threatening reactions than do all other food allergies combined. Research is currently under way to help people with peanut allergies tolerate small doses, thus saving lives and minimizing reactions.[80] Families of children with a life-threatening food allergy and the school personnel who supervise those children must guard them against any exposure to the allergen. The child must learn to identify which foods pose a problem and then learn and use refusal skills for all foods that may contain the allergen.

Parents of children with allergies can pack safe foods for lunches and snacks and ask school officials to strictly enforce a "no swapping" policy in the lunchroom. The child must be able to recognize the symptoms of impending anaphylactic shock, such as a tingling of the tongue, throat, or skin, or difficulty breathing. ♦ Any person with food allergies severe enough to cause anaphylactic shock should wear a medical alert bracelet or necklace. Finally, the responsible child and the school staff should be prepared with injections of epinephrine, which prevents anaphylactic shock after exposure to the allergen.* Many preventable deaths occur each year when people with food allergies accidentally ingest the allergen but have no epinephrine available.

♦ Symptoms of impending anaphylactic shock:
- Tingling sensation in mouth
- Swelling of the tongue and throat
- Irritated, reddened eyes
- Difficulty breathing, asthma
- Hives, swelling, rashes
- Vomiting, abdominal cramps, diarrhea
- Drop in blood pressure
- Loss of consciousness

Food Labeling Food labels must list the presence of common allergens in plain language, using the names of the eight most common allergy-causing foods. For example, a food containing "textured vegetable protein" must say "soy" on its label. Similarly, "casein" must be identified as "milk," and so forth. Food producers must also prevent cross-contamination during production and clearly label foods in which it is likely to occur. For example, equipment used for making peanut butter cookies must be scrupulously clean before being used to make oatmeal cookies; even then, the oatmeal cookie label warns consumers that this product "may contain peanuts" or "was made in a facility that uses peanuts."

Technology may soon offer new solutions. New drugs are being developed that may interfere with the immune response that causes allergic reactions.

*Epinephrine is a hormone of the adrenal gland that modulates the stress response; formerly called *adrenaline*. When administered by injection, epinephrine counteracts anaphylactic shock by opening the airways and maintaining heartbeat and blood pressure.

anaphylactic (ana-fill-LAC-tic) **shock:** a life-threatening, whole-body allergic reaction to an offending substance.

Also, through genetic engineering, scientists may one day create allergen-free peanuts, soybeans, and other foods to make them safer.

Food Intolerances Not all **adverse reactions** to foods are food allergies, although even physicians may describe them as such. Signs of adverse reactions to foods include stomachaches, headaches, rapid pulse rate, nausea, wheezing, hives, bronchial irritation, coughs, and other such discomforts. Among the causes may be reactions to chemicals in foods, such as the flavor enhancer monosodium glutamate (MSG), the natural laxative in prunes, or the mineral sulfur; digestive diseases, obstructions, or injuries; enzyme deficiencies, such as lactose intolerance; and even psychological aversions. These reactions involve symptoms but no antibody production. Therefore, they are **food intolerances,** not allergies.

Pesticides on produce may also cause adverse reactions. Pesticides that were applied in the fields may linger on foods. Health risks from pesticide exposure may be low for healthy adults, but children are vulnerable. Therefore, government agencies have set a **tolerance level** for each pesticide by first identifying foods that children commonly eat in large amounts and then considering the effects of pesticide exposure during each stage of development. Chapter 19 revisits the issues surrounding the use of pesticides on food crops.

Hunger, lead poisoning, hyperactivity, and allergic reactions can all adversely affect a child's nutrition status and health. Fortunately, each of these problems has solutions. They may not be easy solutions, but at least we have a reasonably good understanding of the problems and ways to correct them. Such is not the case with the most pervasive health problem for children in the United States—obesity.

Childhood Obesity The number of overweight children has increased dramatically over the past four decades (see Figure 16-9).[81] Like their parents, children in the United States are becoming fatter. An estimated 32 percent of US children and adolescents 2 to 19 years of age are overweight and 17 percent are obese.[82] Based on data from the BMI-for-age growth charts, children and adolescents are categorized as *overweight* above the 85th percentile and as *obese* at the 95th percentile and above.[83] There are exceptions to the use of the 85th and 95th percentile cutoff points. For older adolescents, a BMI at the 95th percentile is higher than a BMI of 30, the adult obesity cutoff point. Therefore, obesity is defined as a BMI at the 95th percentile or a BMI of 30 or greater, whichever is lower. For children younger than 2 years of age, BMI values are not available. For this age group, weight-for-height values above the 95th percentile are classified as overweight. Figure 16-10 presents the BMI for children and adolescents, indicating cutoff points for obesity and overweight.

The Expert Committee of the American Medical Association recommends a third cutoff point (99th percentile) to define severe obesity in childhood.[84] Unfortunately, severe obesity in children is becoming more prevalent. Many of these children have multiple risk factors for cardiovascular disease and a high risk of severe obesity in adulthood.[85] The special risks and treatment needs of severely obese children need to be recognized.

The problem of obesity in children is especially troubling because overweight children have the potential of becoming obese adults with all the social, economic, and medical ramifications that often accompany obesity. They have additional problems, too, arising from differences in their growth, physical health, and psychological development. In trying to explain the rise in childhood obesity, researchers point to both genetic and environmental factors.

adverse reactions: unusual responses to food (including intolerances and allergies).

food intolerances: adverse reactions to foods that do not involve the immune system.

tolerance level: the maximum amount of residue permitted in a food when a pesticide is used according to the label directions.

> FIGURE 16-9 **Trends in Childhood Obesity**

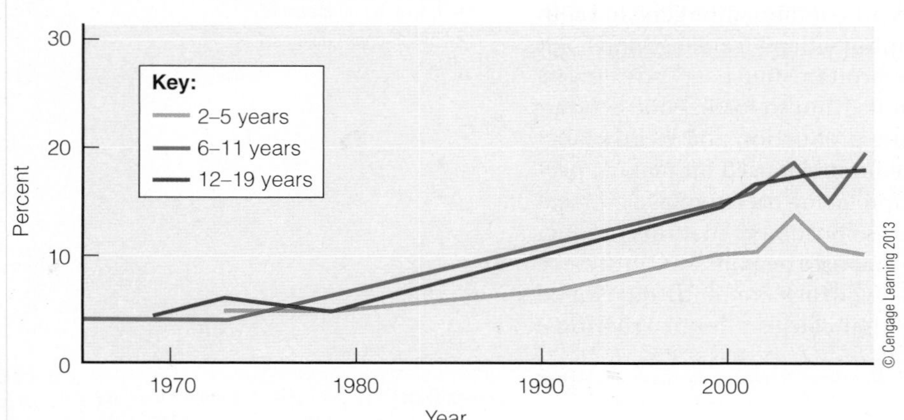

> **FIGURE 16-10 Body Mass Index-for-Age Percentiles: Boys and Girls, Age 2 to 20**

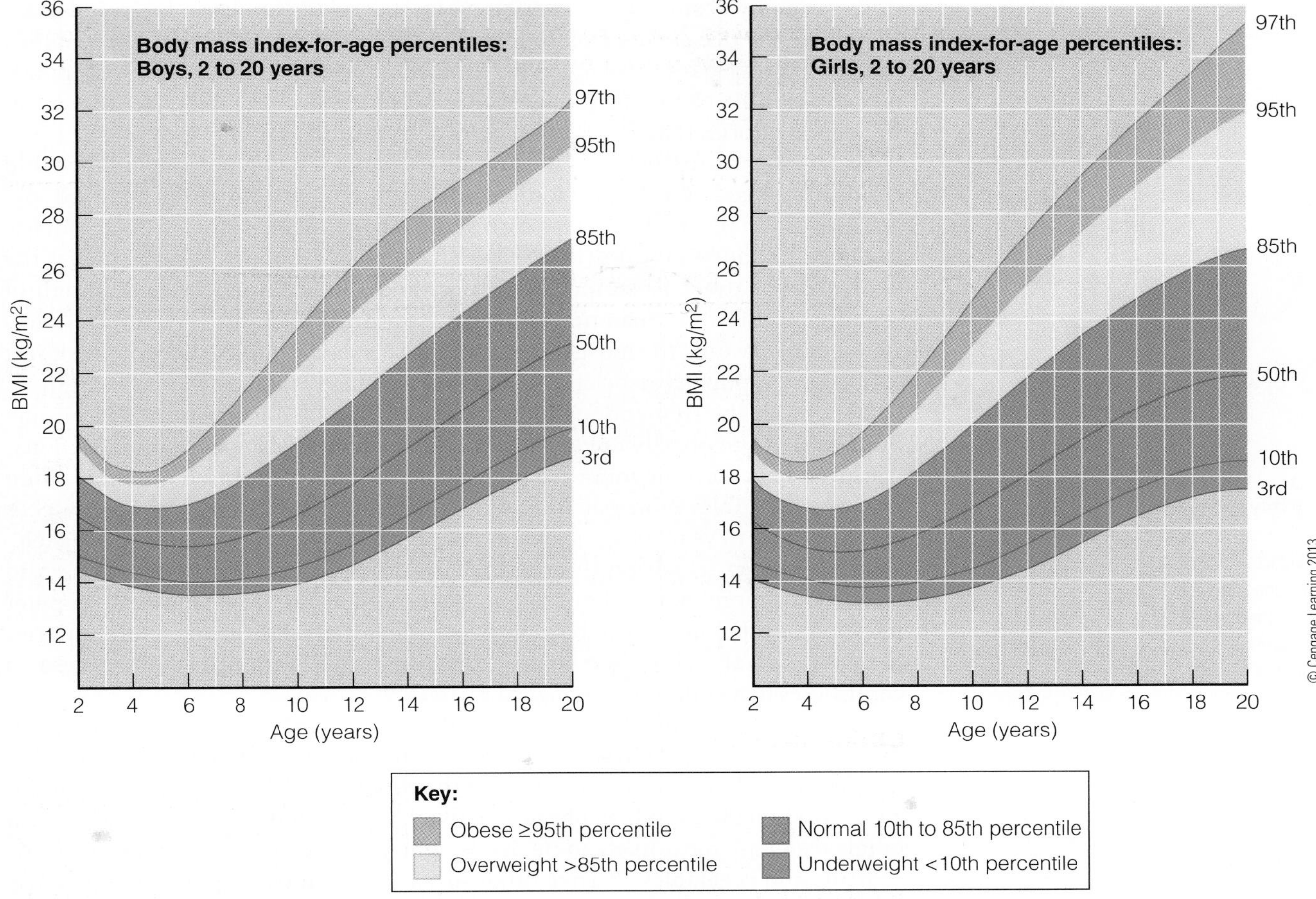

Genetic and Environmental Factors Parental obesity predicts an early increase in a young child's BMI, and it more than doubles the chances that a young child will become an obese adult.[86] Children with neither parent obese have a less than 10 percent chance of becoming obese in adulthood, whereas overweight teens with at least one obese parent have a greater than 80 percent chance of being obese adults. The chances of an obese child becoming an obese adult grow greater as the child grows older.[87] The link between parental and child obesity reflects both genetic and environmental factors (as described in Chapter 9).

Diet and physical inactivity are the two strongest environmental factors explaining why children are heavier today than they were 40 or so years ago. In that time, the prevalence of childhood obesity throughout the United States more than doubled for young children and more than tripled for children 6 to 11 years of age and adolescents. As a society, our eating and activity patterns changed considerably. In many families, both parents work outside the home and work longer hours; more emphasis is placed on convenience foods and foods eaten away from home; meal choices at school are more diverse and often less nutritious; sedentary activities such as watching television and playing video or computer games occupy much of children's free time; and opportunities for physical activity and outdoor play both during and after school have declined.[88] All of these factors—and many others—influence children's eating and activity patterns.

Children learn food behaviors from their families, and research confirms the significant roles parents play in teaching their children about healthy food choices, providing nutrient-dense foods, and serving as role models.[89] When parents eat fruits and vegetables frequently, their children do too.[90] The more fruits and vegetables children eat, the more vitamins, minerals, and fiber, and the less saturated fat, in their diets.[91]

In children 2 to 18 years of age, about 40 percent of total energy intake comes from solid fats and added sugars—in other words—empty kcalories.[92] About half of these empty kcalories are contributed by six specific foods: soda, fruit drinks, dairy desserts (ice cream, frozen yogurt, sorbet, sherbet, pudding, and custard), grain desserts (cakes, cookies, pies, cobblers, donuts, and granola bars), pizza, and whole milk. Not surprisingly, when researchers ask "Are today's children eating more kcalories than those of 40 years ago?" the answer is, "Yes."

As Highlight 4 discusses, as the prevalence of obesity among both children and adults has surged over the past four decades, so has the consumption of added sugars and, especially, high-fructose corn syrup—the easily consumed, energy-dense liquid sugar added to soft drinks.[93] Each 12-ounce can of soft drink provides the equivalent of about 10 teaspoons of sugar and 150 kcalories. More than half of school-age children consume at least one soft drink each day at school; adolescent males consume the most—four or more cans daily.[94] Research shows that soft drink consumption is associated with increased energy intake and body weight.[95]

No doubt, the tremendous increase in soft drink consumption plays a role, but much of the obesity epidemic can be explained by lack of physical activity. Children have become more sedentary, and sedentary children are more often overweight.[96] Television watching may contribute most to physical inactivity. ♦ Children 8 to 18 years of age spend an average of 4.5 hours per day watching television.[97] Longer television time is linked with overweight in children.[98] A child who spends more than an hour or two each day in front of a television, computer monitor, or other media can become overweight even while eating fewer kcalories than a more active child. Too much screen time and not enough activity time also contributes to a child's psychological distress.[99]

♦ TV fosters obesity because it:
- Requires no energy beyond basal metabolism
- Replaces vigorous activities
- Encourages snacking
- Promotes a sedentary lifestyle

Playing video games influences children's activity patterns similarly.

Children who have television sets in their bedrooms spend more time watching TV, less time being physically active, and are more likely to be overweight than children who do not have televisions in their rooms.[100] Watching television influences food intake as well as physical activity.[101] Children who watch a great deal of television are most likely to be overweight and least likely to eat family meals or fruits and vegetables.[102] They often snack on the nutrient-poor, energy-dense foods that are advertised.[103] The average child sees an estimated 40,000 TV commercials a year—many peddling foods high in sugar, saturated fat, and salt such as sugar-coated breakfast cereals, candy bars, chips, fast foods, and carbonated beverages.[104] More than half of all food advertisements are aimed specifically at children and market their products as fun and exciting. Not surprisingly, the more time children spend watching television, the more they request these advertised foods and beverages—and they get their requests about half of the time. The most popular foods and beverages are marketed to children and adolescents on the Internet as well, using "advergaming" (advertised product as part of a game), cartoon characters or "spokes-characters," and designated children's areas.[105]

The physically inactive time spent watching television is second only to time spent sleeping. Children also spend more time playing computer and video games. In most cases, these activities use no more energy than resting, displace participation in more vigorous activities, and foster snacking on high-fat foods.[106] Compared to sedentary screen-time activities, playing active video games does expend a little more energy, but not enough to count toward the 60 minutes of moderate-to-vigorous physical activity recommended for children.[107] Simply reducing the amount of time spent watching television (and playing sedentary video games) can improve a child's BMI. The American Academy of Pediatrics now recommends no television viewing before 2 years of age and thereafter limiting television and video time to 2 hours per day as a strategy to help prevent childhood obesity.[108]

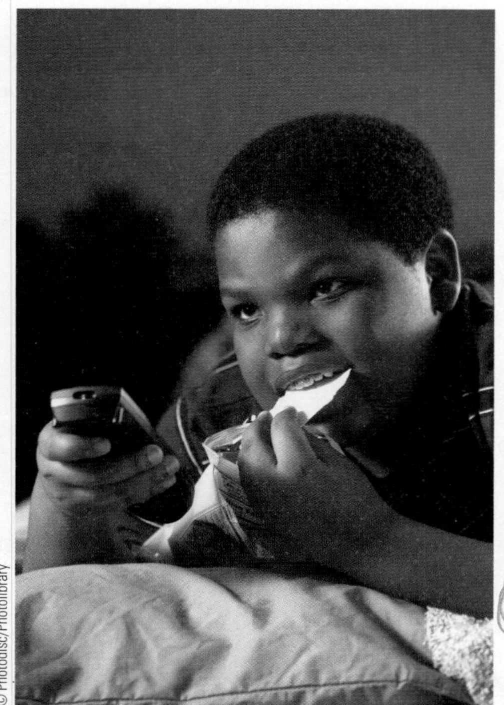

Excessive television watching promotes physical inactivity and poor snacking habits.

Growth Overweight children develop a characteristic set of physical traits. They typically begin puberty earlier and so grow taller than their peers at first, but then they stop growing at a shorter height. They develop greater bone and muscle mass in response to the demand of having to carry more weight—both fat and lean weight. Consequently, they appear "stocky" even when they lose their excess fat.

Physical Health Like overweight adults, overweight children display a blood lipid profile indicating that atherosclerosis is beginning to develop—high levels of total cholesterol, triglycerides, and LDL cholesterol. Overweight children also tend to have high blood pressure; in fact, obesity is a leading cause of pediatric hypertension.[109] Their risks for developing type 2 diabetes and respiratory diseases (such as asthma) are also exceptionally high.[110] These relationships between childhood obesity and chronic diseases are discussed fully in Highlight 16.

Psychological Development In addition to the physical consequences, childhood obesity brings a host of emotional and social problems. Because people frequently judge others on appearance more than on character, overweight children are often victims of prejudice and bullying. Many suffer discrimination by adults and rejection by their peers. They may have poor self-images, a sense of failure, and a passive approach to life. Television shows, which are a major influence in children's lives, often portray the fat person as the bumbling misfit. Overweight children may come to accept this negative stereotype in themselves and in others, which can lead to additional emotional and social problems. Researchers investigating children's reactions to various body types find that both normal-weight and underweight children respond unfavorably to overweight bodies.

Prevention and Treatment of Obesity Medical science has worked wonders in preventing or curing many of even the most serious childhood diseases, but obesity remains a challenge. Once excess fat has been stored, it is difficult to lose. In light of all this, parents are encouraged to make major efforts to prevent childhood obesity, starting at birth, or to begin treatment early—before adolescence. The Expert Committee of the American Medical Association recommends specific eating and physical activity behaviors to prevent obesity, for all children (see Table 16-7).

The main goal of obesity treatment is to improve long-term physical health through permanent changes in lifestyle habits. The most successful approach integrates diet, physical activity, psychological support, and behavioral changes. As a first step, the Expert Committee recommends that overweight and obese children and their families adopt the same healthy eating and activity behaviors presented in Table 16-7 for obesity prevention. The goal for overweight and obese children is to improve BMI. If the child's BMI does not improve after several months, the Expert Committee recommends increasing the intensity of the treatment. The level of intensity depends on treatment response, age, degree of obesity, health risks, and the family's readiness to change. Advanced treatment involves close follow-up monitoring by a health-care provider and greater support and structure for the child.

TABLE 16-7 Recommended Eating and Physical Activity Behaviors to Prevent Obesity

The Expert Committee of the American Medical Association recommends the following healthy habits for children 2 to 18 years of age to help prevent childhood obesity:

- Limit consumption of sugar-sweetened beverages, such as soft drinks and fruit-flavored punches.
- Eat the recommended amounts of fruits and vegetables every day (2 to 4.5 cups per day based on age).
- Learn to eat age-appropriate portions of foods.
- Eat foods low in energy density such as those high in fiber and/or water and modest in fat.
- Eat a nutritious breakfast every day.
- Eat a diet rich in calcium.
- Eat a diet balanced in recommended proportions for carbohydrate, fat, and protein.
- Eat a diet high in fiber.
- Eat together as a family as often as possible.
- Limit the frequency of restaurant meals.
- Limit television watching or other screen time to no more than 2 hours per day and do not have televisions or computers in bedrooms.
- Engage in at least 60 minutes of moderate to vigorous physical activity every day.

SOURCE: S. E. Barlow, Expert Committee recommendations regarding the prevention, assessment, and treatment of child and adolescent overweight and obesity: Summary report, *Pediatrics* 120 (2007): S164–S192.

Diet The initial goal for overweight children is to reduce the rate of weight gain; that is, to maintain weight as the child grows taller. Continued growth will then accomplish the desired change in BMI. Weight loss is usually not recommended because diet restriction can interfere with growth and development. Intervention for some overweight children with accompanying medical conditions may warrant weight loss, but this treatment requires an individualized approach based on the degree of overweight and severity of the medical conditions.[111] Dietary strategies begin with those listed in Table 16-7 and progress to more structured family meal plans when necessary. For example, the child or the parent may be instructed to keep detailed records of dietary intake and physical activity.

> DIETARY GUIDELINES FOR AMERICANS 2010

Children are encouraged to maintain kcalorie balance to support normal growth and development without promoting excess weight gain. Children who are overweight or obese should change their eating and physical activity behaviors so that their BMI-for-age percentile does not increase over time.

Physical Activity The many benefits of physical activity are well known but often are not enough to motivate overweight people, especially children. Yet regular vigorous activity can improve a child's weight, body composition, and physical fitness. Ideally, parents will limit sedentary activities and encourage at least 1 hour of daily physical activity to promote strong skeletal, muscular, and cardiovascular development and instill in their children the desire to be physically active throughout life. Opportunities to be physically active can include team, individual, and recreational activities (see Table 16-8). Most importantly, parents need to set a good example. Physical activity is a natural and lifelong behavior of healthy living. It can be as simple as riding a bike, playing tag, jumping rope, or doing chores. The American Academy of Pediatrics supports the efforts of schools to include more physical activity in the curriculum and encourages parents to support their children's participation.[112]

TABLE 16-8 Examples of Aerobic, Muscle-Strengthening, and Bone-Strengthening Physical Activities for Children and Adolescents

Moderate-to-Vigorous Aerobic Activities	Muscle-Strengthening Activities	Bone-Strengthening Activities
 © Fotokostic/Shutterstock.com	 © Istockphoto.com/ Christopher Futcher	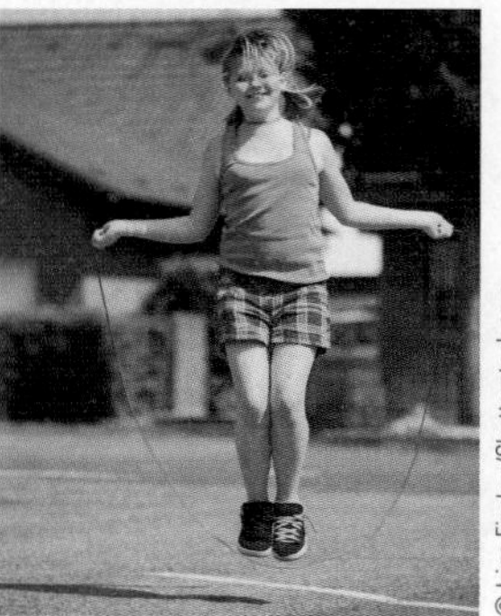 © Irina Fischer/Shutterstock.com
Moderate Active recreation such as hiking, skateboarding, rollerblading Bicycle riding[a] Brisk walking **Vigorous** Active games involving running and chasing, such as tag Bicycle riding[a] Cross-country skiing Jumping rope Martial arts Running Sports such as soccer, ice or field hockey, basketball, swimming, tennis	Games such as tug-of-war Modified push-ups (with knees on the floor) Resistance exercises using body weight, free weights, or resistance bands Rope or tree climbing Sit-ups (curl-ups or crunches) Swinging on playground equipment/bars	Games such as hopscotch Hopping, skipping, jumping Jumping rope Running Sports such as gymnastics, basketball, volleyball, tennis

[a]Some activities, such as bicycling, can be moderate or vigorous, depending on level of effort.

> **PHYSICAL ACTIVITY GUIDELINES FOR AMERICANS 2008**

- Children and adolescents should do 60 minutes (1 hour) or more of physical activity daily.
- *Aerobic.* Most of the 60 or more minutes a day should be either moderate- or vigorous-intensity aerobic physical activity and should include vigorous-intensity physical activity at least 3 days a week.
- *Muscle-strengthening.* As part of their 60 or more minutes of daily physical activity, children and adolescents should include muscle-strengthening physical activity on at least 3 days of the week.
- *Bone-strengthening.* As part of their 60 or more minutes of daily physical activity, children and adolescents should include bone-strengthening physical activity on at least 3 days of the week.

Physical activity is fun—play games in the park, build a sandcastle or a snowman, row a boat, toss a Frisbee, run with the dog, or plant a garden.

Psychological Support Weight-loss programs that involve parents and other caregivers in treatment report greater success than those without parental involvement. Because obesity in parents and their children tends to be positively correlated, both benefit when parents participate in a weight-loss program. Parental attitudes about food greatly influence children's eating behavior, so it is important that the influence be positive. Otherwise, eating problems may become exacerbated.

Behavioral Changes In contrast to traditional weight-loss programs that focus on *what* to eat, behavioral programs focus on *how* to eat. These techniques involve learning new habits that lead a child to make healthy choices.

Drugs The use of weight-loss drugs to treat obesity in children merits special concern because the long-term effects of these drugs on growth and development have not been studied. The drugs may be used in addition to structured lifestyle changes for carefully selected children or adolescents who are at high risk for severe obesity in adulthood. Only orlistat (see Chapter 9) has been approved for limited use in children and adolescents. Alli, the over-the-counter version of orlistat, should not be given to anyone younger than age 18.

Surgery The use of surgery to treat severe obesity in adults (see Chapter 9) has created interest in its use for adolescents. Limited research shows that after surgery extremely obese adolescents lose significant weight and experience improvements in type 2 diabetes and cardiovascular risk factors.[113] The selection criteria ♦ for surgery to treat obesity in adolescents are based on recommendations of a panel of pediatricians and surgeons.[114]

♦ Surgery criteria for adolescents:
- Physically mature
- BMI ≥50 or BMI ≥40 with significant weight-related health problems
- Failure in a formal, 6-month weight-loss program
- Capable of adhering to the long-term lifestyle changes required after surgery

Obesity is prevalent in our society. Because treatment of obesity is frequently unsuccessful, it is most important to prevent its onset. Above all, be sensible in teaching children how to maintain appropriate body weight. Children can easily get the impression that their worth is tied to their body weight. Parents and the media are most influential in shaping self-concept, weight concerns, and dieting practices.[115] Some parents fail to realize that society's ideal of slimness can be perilously close to starvation and that a child encouraged to "diet" cannot obtain the energy and nutrients required for normal growth and development. Weight loss in truly overweight children can be managed without compromising growth, but it should be overseen by a health-care professional.

Mealtimes at Home

Traditionally, parents served as **gatekeepers**, determining what foods and activities were available in their children's lives. Then the children made their own selections. Gatekeepers who wanted to promote nutritious choices and healthful habits provided access to nutrient-dense foods and opportunities for active play at home.

In today's consumer-oriented society, children have greater influence over family decisions concerning food—the fast-food restaurant the family chooses when eating out, the snacks the family eats at home, and the specific brands the family

gatekeepers: with respect to nutrition, key people who control other people's access to foods and thereby exert profound impacts on their nutrition. Examples are the spouse who buys and cooks the food, the parent who feeds the children, and the caregiver in a day-care center.

purchases at the grocery store. Parental guidance in food choices is still necessary, but teaching children consumer skills to help them make informed choices is equally important.

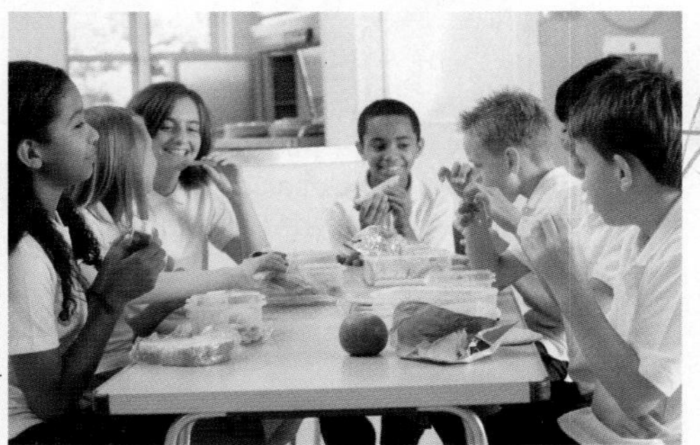

© Istockphoto.com/Catherine Yeulet

Eating is more fun for children when friends are there.

Honoring Children's Preferences Researchers attempting to explain children's food preferences encounter contradictions. Children say they like colorful foods, yet they most often reject green and yellow vegetables in favor of brown peanut butter and white potatoes, apple wedges, and bread. They seem to like raw vegetables better than cooked ones, so it is wise to offer vegetables that are raw or slightly undercooked, served separately, and easy to eat. Foods should be warm, not hot, because a child's mouth is much more sensitive than an adult's. The flavor should be mild because a child has more taste buds, and smooth foods such as mashed potatoes or split-pea soup should contain no lumps (a child wonders, with some disgust, what the lumps might be).

Make mealtimes fun for children. Young children like to eat at little tables and to be served small portions of food. They like sandwiches cut in different geometric shapes and common foods called silly names. They also like to eat with other children, and they tend to eat more when in the company of their friends. Children are also more likely to give up their prejudices against foods when they see their peers eating them.

Learning through Participation Allowing children to help plan and prepare the family's meals provides enjoyable learning experiences and encourages children to eat the foods they have prepared. Vegetables are attractive, especially when fresh, and provide opportunities for children to learn about color, seeds, growing vegetables, and shapes and textures—all of which are fascinating to young children. Measuring, stirring, washing, and arranging foods are skills that even a young child can practice with enjoyment and pride (see Table 16-9).

TABLE 16-9 Food Skills of Preschool Children

Age 2 years, when large muscles develop:

- Uses a spoon
- Helps feed self
- Lifts and drinks from a cup
- Helps scrub fruits and vegetables, tear lettuce or greens, snap green beans, or dip foods
- Wipes table
- Places items in recycle bin or trash

Age 3 years, when medium hand muscles develop:

- Spears food with a fork
- Feeds self independently
- Adds ingredients to pancake batters, cookie recipes, salads or other mixed dishes
- Helps wrap, pour, mix, shake, stir, or spread foods
- Helps crack nuts with supervision

Age 4 years, when small finger muscles develop:

- Uses all utensils and napkin
- Helps roll, juice, or mash foods
- Helps measure dry ingredients
- Cracks egg shells
- Helps make sandwiches and toss salads
- Peels foods such as hard-boiled eggs and bananas

Age 5 years, when fine coordination of fingers and hands develops:

- Measures liquids
- Helps grind, grate, and cut (soft foods with dull knife)
- Uses hand mixer with supervision

NOTE: These ages are approximate. Healthy children develop at their own pace.

Avoiding Power Struggles Problems over food often arise during the second or third year, when children begin asserting their independence. Many of these problems stem from the conflict between children's developmental stages and capabilities and parents who, in attempting to do what they think is best for their children, try to control every aspect of eating. Such conflicts can disrupt children's abilities to regulate their own food intakes and to determine their own likes and dislikes. For example, many people share the misconception that children must be persuaded or coerced to try new foods. In fact, the opposite is true. When children are forced to try new foods, even by way of rewards, they are less likely to try those foods again than are children who are left to decide for themselves. Similarly, when children are restricted from eating their favorite foods, they are more likely to want those foods.[116] Wise parents provide healthful foods and allow their child to determine *how much* and even *whether* to eat.

When introducing new foods, offer them one at a time and only in small amounts such as a small bite at first. The more often a food is presented to a young child, the more likely the child will accept that food.[117] Offer the new food at the beginning of the meal, when the child is hungry, and allow the child to make the decision to accept or reject it. Never make an issue of food acceptance.

Choking Prevention Parents must always be alert to the dangers of choking. A choking child is silent, so an adult should be present whenever a child is eating. Make sure the child sits when eating; choking is more likely when a child is running or falling. (See the margin list on p. 515 for foods and nonfood items most likely to cause choking.)

Playing First Children may be more relaxed and attentive during meals if outdoor play or other fun activities are scheduled before, rather than immediately after, mealtimes. Otherwise children "hurry up and eat" so that they can go play.

Snacking Parents may find that when their children snack, they aren't hungry at mealtimes. Instead of teaching children *not* to snack, parents are wise to teach them *how* to snack. Provide snacks that are as nutritious as the foods served at mealtime. Snacks can even be mealtime foods served individually over time, instead of all at once on one plate. When providing snacks to children, think of the five food groups and offer such snacks as pieces of cheese, tangerine slices, and egg salad on whole-wheat crackers (see Table 16-10). Snacks that are easy to prepare should be readily available to children, especially if they arrive home from school before their parents.

To ensure that children have healthy appetites and plenty of room for nutritious foods when they are hungry, parents and teachers must limit access to candy, soft drinks, and other concentrated sweets. Limiting access includes limiting the amount of pocket money children have to buy such foods themselves. If these foods are permitted in large quantities, the only possible outcomes are nutrient deficiencies, obesity, or both. The preference for sweets is innate; most children do not naturally select nutritious foods on the basis of taste. When children are allowed to create meals freely from a variety of foods, they typically select foods that provide a lot of sugar. When their parents are watching, or even when they only think their parents are watching, children improve their selections.

Sweets need not be banned altogether. Children who are exceptionally active can enjoy high-kcalorie foods such as ice cream or pudding from the milk group or pancakes from the bread group. Sedentary children need to become more active so they can also enjoy some of these foods without unhealthy weight gain.

Preventing Dental Caries Children frequently snack on sticky, sugary foods that stay on the teeth and provide an ideal environment for the growth of bacteria that cause dental caries. Teach children to brush and floss after meals, to brush or

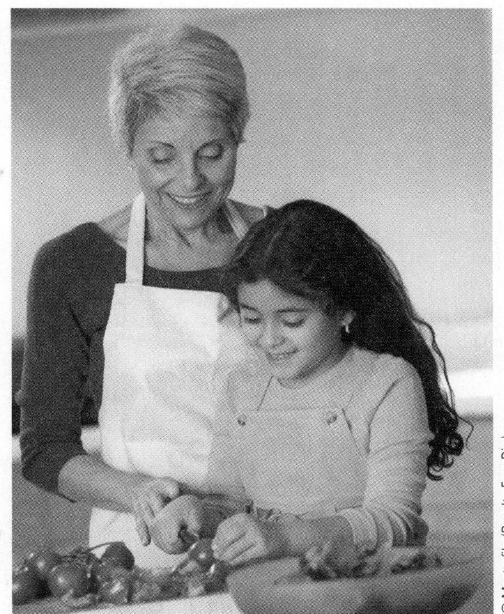

Children enjoy eating the foods they help to prepare.

TABLE 16-10 Healthful Snack Ideas—Think Food Groups, Alone and in Combination

Selecting two or more foods from different food groups adds variety and nutrient balance to snacks. The combinations are endless, so be creative. Whenever possible, choose whole grains, low-fat or reduced-fat milk products, and lean meats.

Grains

Grain products are filling snacks, especially when combined with other foods:

- Cereal with fruit and milk
- Crackers and cheese
- Whole-grain toast with peanut butter
- Popcorn with grated cheese
- Oatmeal raisin cookies with milk

Vegetables

Cut-up, fresh, raw vegetables make great snacks alone or in combination with foods from other food groups:

- Celery with peanut butter
- Broccoli, cauliflower, and carrot sticks with a flavored cottage cheese dip

Fruits

Fruits are delicious snacks and can be eaten alone—fresh, dried, or juiced—or combined with other foods:

- Apples and cheese
- Bananas and peanut butter
- Peaches with yogurt
- Raisins mixed with sunflower seeds or nuts

Protein Foods

Seafood, meat, poultry, eggs, legumes, nuts, seeds, and soy products add protein to snacks:

- Refried beans with nachos and cheese
- Tuna on crackers
- Luncheon meat on whole-grain bread

Milk and Milk Products

Milk can be used as a beverage with any snack, and many other milk products, such as yogurt and cheese, can be eaten alone or with other foods as listed above.

rinse after eating snacks, to avoid sticky foods, and to select crisp or fibrous foods frequently.

Serving as Role Models In an effort to practice these many tips, parents may overlook perhaps the single most important influence on their children's food habits—themselves.[118] Parents who don't eat carrots shouldn't be surprised when their children refuse to eat carrots. Likewise, parents who comment negatively on the smell of brussels sprouts may not be able to persuade children to try them. Children learn much through imitation. It is not surprising that children prefer the foods other family members enjoy and dislike foods that are never offered to them. Parents, older siblings, and other caregivers set an irresistible example by sitting with younger children, eating the same foods, and having pleasant conversations during mealtimes.

While serving and enjoying food, caregivers can promote both physical and emotional growth at every stage of a child's life. They can help their children develop both a positive self-concept and a positive attitude toward food. With good beginnings, children will grow without the conflicts and confusions about food that can lead to nutrition and health problems.

Nutrition at School

While parents are doing what they can to establish good eating habits in their children at home, others are preparing and serving foods to their children at day-care centers and schools. In addition, children begin to learn about food and nutrition in the classroom. Meeting the nutrition and education needs of children is critical to supporting their healthy growth and development.[119]

The Academy of Nutrition and Dietetics has set nutrition standards for child-care programs. Among them, meal plans should:

- Be nutritionally adequate and consistent with the *Dietary Guidelines for Americans.*
- Emphasize fresh fruit, fresh and frozen vegetables, whole grains, and low-fat milk and milk products.
- Limit foods and beverages high in energy, added sugars, solid fats, and sodium, and low in vitamins and minerals.
- Provide foods and beverages in quantities and meal patterns appropriate to ensure optimal growth and development.
- Involve parents in planning.
- Provide furniture and eating utensils that are age appropriate and developmentally suitable to encourage children to accept and enjoy mealtime.

In addition, child-care providers can encourage active play for children by creating opportunities for children to engage in both structured and unstructured activity throughout the day.

Meals at School The US government assists schools financially so that every student can receive nutritious meals at school. Both the School Breakfast Program and the National School Lunch Program provide meals free or at reduced cost to children from low-income families. In addition, schools can obtain food commodities. Nationally, the US Department of Agriculture (USDA) administers the programs; on the state level, state departments of education operate them.* The educational rewards of school meal programs are great. Several studies have reported that children who participate in school food programs perform better in the classroom.[120] Unfortunately, many school districts are finding it more and more difficult to operate the programs because the cost to produce lunches and breakfasts exceeds the reimbursement rate. Furthermore, the number of children qualifying for free and reduced-price school meals continues to rise.[121]

More than 31 million children receive lunches through the National School Lunch Program—more than half of them free or at a reduced price.[122] School lunches

*School lunches in Canada are administered locally and therefore vary from area to area.

TABLE 16-11 School Lunch Patterns

	Grades		
	K–5	6–8	9–12
Food Group	Amount per week (minimum per day)		
Fruits[a] (cups)	2½ (½)	2½ (½)	5 (1)
Vegetables[a] (cups)	3¾ (¾)	3¾ (¾)	5 (1)
Dark green	≥½	≥½	≥½
Red and orange	≥¾	≥¾	≥1¼
Legumes	≥½	≥½	≥½
Starchy	≥½	≥½	≥½
Other	≥½	≥½	≥¾
Any additional vegetables to meet total requirement	1	1	1½
Grains (oz equivalents)	8–9 (1)	8–10 (1)	10–12 (2)
Protein foods (oz equivalents)	8–10 (1)	9–10 (1)	10–12 (2)
Fluid milk[b] (cups)	5 (1)	5 (1)	5 (1)
Other			
kCalories	550–650	600–700	750–850
Saturated fat (% of total kcalories)	<10	<10	<10
Sodium (mg)	≤640	≤710	≤740
Trans fat	0 g per serving		

[a]No more than half of the fruit or vegetable servings may be in the form of juice. All juice must be 100% full strength.
[b]Fluid milk must be low-fat (unflavored) or fat-free (flavored or unflavored).
SOURCE: US Department of Agriculture, *Nutrition Standards in the National School Lunch and School Breakfast Programs*, January 25, 2012.

offer a variety of food choices and help children meet at least one-third of their recommended intakes for energy, protein, vitamin A, vitamin C, iron, and calcium. Table 16-11 shows school lunch patterns for children of different ages and specifies the numbers of servings of milk, protein-rich foods (meat, poultry, fish, cheese, eggs, legumes, or peanut butter), vegetables, fruits, and whole grains. In an effort to help reduce disease risk, all government-funded meals served at schools must follow the *Dietary Guidelines for Americans*.[123] Findings from a comprehensive study of school meal programs indicate that many schools have improved the nutritional quality of the school breakfast and lunch programs, but room for improvement remains.[124] Reimbursable school meals are still too high in saturated fat and sodium, and low in fiber.

◆ The school breakfast must contain at a minimum:
- One serving of fluid milk (either unflavored low-fat or flavored or unflavored fat-free)
- One serving of fruit or vegetable (no more than half of the servings may be 100% full-strength juice)
- One to two servings of whole grains; or one serving of whole grains and one serving of meat or meat alternates

Parents often rely on school lunches to meet a significant part of their children's nutrient needs on school days. Indeed, students who regularly eat school lunches have higher intakes of many nutrients and fiber than students who do not.[125]

The School Breakfast Program ◆ is available in more than 80 percent of the nation's schools that offer school lunch, and close to 12 million children participate in it.[126] Nevertheless, for many children who need it, the School Breakfast Program is either unavailable or the children do not participate in it.[127] The majority of children who eat school breakfasts are from low-income families. As research results continue to emphasize the positive impact breakfast has on school performance and health, vigorous campaigns to expand and improve school breakfast programs are under way.[128]

Another federal program, the Child and Adult Care Food Program (CACFP), operates similarly and provides funds to organized child-care programs. All eligible children, centers, and

School lunches provide children with nourishment at little or no charge.

family day-care homes may participate. Sponsors are reimbursed for most meal costs and may also receive USDA commodity foods.

Competing Influences at School Serving healthful lunches is only half the battle; students need to eat them too. Short lunch periods and long waiting lines prevent some students from eating a school lunch and leave others with too little time to finish their meals.[129] Nutrition efforts at schools are also undermined when students can buy what the USDA labels "competitive" or "nonreimbursable" foods—meals from fast-food restaurants or a la carte foods such as pizza or snack foods and carbonated beverages from snack bars, school stores, and vending machines.[130] When students have access to competitive foods, participation in the school lunch program decreases, nutrient intake from lunch declines, and more food is discarded.[131]

Increasingly, school-based nutrition issues are being addressed by legislation. Some states restrict the sale of competitive foods and have higher rates of participation in school meal programs than the national average. Federal legislation mandates that all school districts that participate in the USDA's National School Lunch Program develop and implement a local wellness policy.[132] By law, wellness policies must:[133]

- Set goals for nutrition education, physical activity, and other school-based activities.
- Establish nutrition guidelines for all foods available on school campuses during the school day.
- Develop a plan to measure policy implementation.

School districts across the nation have made progress toward meeting these goals, but implementation is inconsistent, and because wellness policies are established locally, a great deal of variety exists among them. Some are well defined and detailed, while others are vague and less detailed.[134] To enhance local wellness policies, standards for competitive foods and beverages served in schools define limits for fat, saturated fat, kcalories, sugar, and sodium.[135] Establishing and implementing these nutrition standards for competitive foods and beverages helps to ensure that all foods served in schools are consistent and comply with the *Dietary Guidelines for Americans.*

REVIEW IT Explain how children's appetites and nutrient needs reflect their stage of growth and why iron deficiency and obesity are often concerns during childhood.
Children's appetites and nutrient needs reflect their stage of growth. Those who are chronically hungry and malnourished suffer growth retardation; when hunger is temporary and nutrient deficiencies are mild, the problems are usually more subtle—such as poor academic performance. Iron deficiency is widespread and has many physical and behavioral consequences. "Hyper" behavior is not caused by poor nutrition; misbehavior may be due to lack of sleep, too little physical activity, or too much television, among other factors. Childhood obesity has become a major health problem. Adults at home and at school need to provide children with nutrient-dense foods and teach them how to make healthful diet and activity choices.

16.3 Nutrition during Adolescence

LEARN IT Describe some of the challenges in meeting the nutrient needs of adolescents.

Teenagers make many more choices for themselves than they did as children. They are not fed, they eat; they are not sent out to play, they choose to go. At the same time, social pressures thrust choices at them, such as whether to drink alcoholic beverages and whether to develop their bodies to meet extreme ideals of slimness or athletic prowess. Their interest in nutrition—both valid information and misinformation—derives from personal, immediate experiences. They are concerned with how diet can improve their lives now—they engage in fad dieting in order to fit into a new bathing suit, avoid greasy foods in an effort to clear acne, or eat a plate of spaghetti to prepare for a big sporting event. In presenting information on the nutrition and health of adolescents, this section includes many topics of interest to teens.

Growth and Development With the onset of **adolescence,** the steady growth of childhood speeds up abruptly and dramatically, and the growth patterns of females and males become distinct. Hormones direct the intensity of the adolescent growth spurt, profoundly affecting every organ of the body, including the brain. After 2 to 3 years of intense growth and a few more at a slower pace, physically mature adults emerge.

In general, the adolescent growth spurt begins at age 10 to 11 for females and at 12 to 13 for males. It lasts about 2½ years. Before **puberty,** male and female body compositions differ only slightly, but during the adolescent spurt, differences between the genders become apparent in the skeletal system, lean body mass, and fat stores. In females, fat assumes a larger percentage of total body weight, and in males, the lean body mass—principally muscle and bone—increases much more than in females (review Figure 1-1 on p. 7). During adolescence, males grow an average 8 inches taller, and females, 6 inches taller. Males gain approximately 45 pounds, and females, about 35 pounds.

© Istockphoto.com/Steve Debenport

Nutritious snacks contribute valuable nutrients and energy to an active teen's diet.

Energy and Nutrient Needs Energy and nutrient needs are greater during adolescence than at any other time of life, except pregnancy and lactation. In general, nutrient needs rise throughout childhood, peak in adolescence, and then level off or even diminish as the teen becomes an adult.

Energy Intake and Activity The energy needs of adolescents vary greatly, depending on the current rate of growth, gender, body composition, and physical activity.[136] Boys' energy needs may be especially high; they typically grow faster than girls and, as mentioned, develop a greater proportion of lean body mass. An exceptionally active boy of 15 may need 3500 kcalories or more a day just to maintain his weight. Girls start growing earlier than boys and attain shorter heights and lower weights, so their energy needs peak sooner and decline earlier than those of their male peers. A sedentary girl of 15 whose growth is nearly at a standstill may need fewer than 1800 kcalories a day if she is to avoid excessive weight gain. Thus adolescent girls need to pay special attention to being physically active and selecting foods of high nutrient density so as to meet their nutrient needs without exceeding their energy needs.

The problems of obesity become ever more apparent in adolescence, especially for females of African American descent and Hispanic children of both genders. Without intervention, overweight adolescents face numerous physical and emotional consequences. The consequences of obesity are so dramatic and our society's attitude toward thin people is so positive that even teens of normal or below-normal weight may perceive a need to lose weight. When taken to extremes, restrictive diets bring dramatic physical consequences of their own, as Highlight 8 explains.

> **DIETARY GUIDELINES FOR AMERICANS 2010**

- Adolescents are encouraged to maintain kcalorie balance to support normal growth and development without promoting excess weight gain. Adolescents who are overweight or obese should change their eating and physical activity behaviors so that their BMI-for-age percentile does not increase over time.
- Adolescents should do 60 minutes or more of physical activity daily. Adolescents are encouraged to spend no more than 1 to 2 hours each day watching television, playing electronic games, or using the computer (other than for homework).

Vitamins The RDA (or AI) for most vitamins increases during the adolescent years (see the table on the inside front cover). Several of the vitamin recommendations for adolescents are similar to those for adults, including the recently revised recommendations for vitamin D.[137] Vitamin D is essential for bone growth and development. Recent studies of vitamin D status in adolescents show that many adolescents are vitamin D deficient; blacks, females, and overweight adolescents are most at risk.[138] Adolescents who do not receive enough vitamin D from

adolescence: the period from the beginning of puberty until maturity.

puberty: the period in life in which a person becomes physically capable of reproduction.

fortified foods such as milk and cereals, or from sun exposure each day, may need a vitamin D supplement.[139]

Iron The need for iron increases during adolescence for both females and males, but for different reasons. Iron needs increase for females as they start to lose blood through menstruation and for males as their lean body mass develops. Hence the RDA increases at age 14 for both males and females. For females, the RDA remains high into late adulthood. For males, the RDA returns to preadolescent values in early adulthood.

In addition, iron needs increase when the adolescent growth spurt begins, whether that occurs before or after age 14. Therefore, boys in a growth spurt need an additional 2.9 milligrams of iron per day above the RDA for their age; girls need an additional 1.1 milligrams per day.[140]

Furthermore, iron recommendations for girls before age 14 do not reflect the iron losses of menstruation. The average age of menarche (first menstruation) in the United States is 12.5 years. Therefore, for girls younger than the age of 14 who have started to menstruate, an additional 2.5 milligrams of iron per day is recommended.[141] Thus the RDA for iron depends not only on age and gender but also on whether the individual is in a growth spurt or has begun to menstruate, as listed in the margin. ♦

♦ Iron RDA for males:
- 9–13 yr: 8 mg/day
- 9–13 yr in growth spurt: 10.9 mg/day
- 14–18 yr: 11 mg/day
- 14–18 yr in growth spurt: 13.9 mg/day

Iron RDA for females:
- 9–13 yr: 8 mg/day
- 9–13 yr in menarche: 10.5 mg/day
- 9–13 yr in menarche and growth spurt: 11.6 mg/day
- 14–18 yr: 15 mg/day
- 14–18 yr in growth spurt: 16.1 mg/day

Iron intakes often fail to keep pace with increasing needs, especially for females, who typically consume fewer iron-rich foods such as meat and fewer total kcalories than males. Not surprisingly, iron deficiency is most prevalent among adolescent girls. Iron-deficient children and teens score lower on standardized tests than those who are not iron deficient.

> **DIETARY GUIDELINES FOR AMERICANS 2010**

Adolescent girls should choose foods that supply heme iron, which is more readily absorbed by the body, additional iron sources, and enhancers of iron absorption such as vitamin C–rich foods.

Calcium Adolescence is a crucial time for bone development, and the requirement for calcium reaches its peak during these years.[142] Unfortunately, low calcium intakes among adolescents have reached crisis proportions: 90 percent of females and 70 percent of males aged 12 to 19 years have calcium intakes below recommendations. ♦ Low calcium intakes during times of active growth, especially if paired with physical inactivity, can compromise the development of peak bone mass, which is considered the best protection against adolescent fractures and adult osteoporosis. Increasing milk and milk products in the diet to meet calcium recommendations greatly increases bone density.[143] Once again, however, teenage girls are most vulnerable because their milk—and therefore their calcium—intakes begin to decline at the time when their calcium needs are greatest.[144] Furthermore, women have much greater bone losses than men in later life. In addition to dietary calcium, bones grow stronger with physical activity. Because most high schools do not require students to participate in physical education classes, however, many adolescents are not as physically active as healthy bones demand.

♦ Calcium RDA for males and females:
- 9–18 yr: 1300 mg/day

Bones grow stronger with physical activity.

> **DIETARY GUIDELINES FOR AMERICANS 2010**

Children 9 years of age and older should consume 3 cups per day of fat-free or low-fat milk or equivalent milk products.

Food Choices and Health Habits Teenagers like the freedom to come and go as they choose. They eat what they want if it is convenient and if they have the time. With a multitude of after school, social, and job activities, they almost inevitably fall into irregular eating habits. At any given time on any given day, a teenager may be skipping a meal, eating a snack, preparing a meal, or consuming food prepared by a parent or restaurant. Adolescents who frequently eat meals with their families, however, eat more fruits, vegetables, grains, and calcium-rich foods, and drink fewer soft drinks, than those who seldom eat with their families.[145] Some research shows that the more often teenagers eat dinner with their families, the less likely they are to smoke, drink, or use drugs; other research supports these findings only in teenage girls.[146] Many adolescents also begin to skip breakfast on a regular basis, missing out on important nutrients that are not made up at later meals during the day. Compared with those who skip breakfast, teenagers who do eat breakfast have higher intakes of vitamin A, vitamin C, and riboflavin, as well as calcium, iron, and zinc.[147] Teenagers who eat breakfast are therefore more likely to meet their nutrient recommendations.

Breakfast skipping may also lead to weight gain in adolescents. Research shows a dose-response, inverse relationship between breakfast eating and BMI.[148] As adolescents make the transition to adulthood, not only do they skip breakfast more often, they also eat fast food more often. Both skipping breakfast and eating fast foods lead to weight gain.[149]

Ideally, parents continue to play the role of gatekeepers, providing nutritious, easy-to-grab foods in the refrigerator (meats for sandwiches; low-fat cheeses; fresh, raw vegetables and fruits; fruit juices; and milk) and more in the cabinets (whole-grain breads and crackers, peanut butter, nuts, popcorn, and cereal). In many households today, adults work outside the home and teenagers help with some of the gatekeepers' tasks, such as shopping for groceries or choosing fast or prepared foods.

Snacks Snacks typically provide at least 25 percent of the average teenager's daily food energy intake. Often, favorite snacks are too high in added sugars, saturated fat, and sodium and too low in fiber. A survey of more than 4000 adolescents, however, found that those who ate snacks more often had higher intakes of fruit compared with those who ate snacks less often.[150] Table 16-10 on p. 533 shows how to combine foods from different food groups to create healthy snacks.

Beverages Most frequently, adolescents drink soft drinks instead of fruit juice or milk with lunch, supper, and snacks. About the only time they select fruit juices is at breakfast. When teens drink milk, they are more likely to consume it with a meal (especially breakfast) than as a snack. Because of their greater food intakes, boys are more likely than girls to drink enough milk to meet their calcium needs.[151] Adolescents who drink soft drinks regularly have a higher energy intake and a lower calcium intake than those who do not.

Over the past three decades, teens (especially girls) have been drinking more soft drinks and less milk. When soft drinks displace milk as the primary beverage, bone density is limited.[152]

Soft drinks containing caffeine present another problem if caffeine intake becomes excessive. Caffeine seems to be relatively harmless when used in moderate doses (the equivalent of fewer than three 12-ounce cola beverages a day).* In greater amounts, however, it can cause symptoms associated with anxiety, such as sweating, tenseness, and inability to concentrate.

> DIETARY GUIDELINES FOR AMERICANS 2010

Reduce intake of sugar-sweetened beverages. Children and adolescents who consume more sugar-sweetened beverages have higher body weight compared to those who drink less.

Eating Away from Home Adolescents eat about one-third of their meals away from home, and their nutritional welfare is enhanced or hindered by the choices they make.[153] A lunch consisting of a hamburger, a chocolate shake, and french fries supplies substantial quantities of many nutrients at a kcalorie cost of about 800, an energy intake some adolescents can afford. When they eat this sort of lunch,

*Caffeine-containing soft drinks typically deliver between 30 and 55 mg of caffeine per 12-ounce can. A pharmacologically active dose of caffeine is defined as 200 mg. Appendix H starts with a table listing the caffeine contents of selected foods, beverages, and drugs.

© Monkeybusiness Images/ Dreamstime.com

Because their lunches rarely include fruits, vegetables, or milk, many teens fail to get all the vitamins and minerals they need each day.

teens can adjust their breakfast and dinner choices to include fruits and vegetables for vitamin A, vitamin C, folate, and fiber and lean meats and legumes for iron and zinc. (See Appendix H for the nutrient contents of fast foods.) Fortunately, many fast-food restaurants are offering more nutritious choices than the standard hamburger meal.

Peer Influence Physical maturity and growing independence present adolescents with new choices. The consequences of those choices will influence their health and nutrition status both today and throughout life. Many of the food and health choices adolescents make reflect the opinions and actions of their peers. When others perceive milk as "babyish," a teen may choose soft drinks instead; when others skip lunch and hang out in the parking lot, a teen may join in for the camaraderie, regardless of hunger. Some teenagers begin using drugs, alcohol, and tobacco; others wisely refrain. Adults can set up the environment so that nutritious foods are available and can stand by with reliable information and advice about health and nutrition, but the rest is up to the adolescents. Ultimately, they make the choices. (Highlight 8 examines the influence of social pressures on the development of eating disorders.)

REVIEW IT Describe some of the challenges in meeting the nutrient needs of adolescents.

Nutrient needs rise dramatically as children enter the rapid growth phase of adolescence. Teenagers' busy lifestyles add to the challenge of meeting their nutrient needs, especially for iron and calcium.

The nutrition and lifestyle choices people make as children and adolescents have long-term, as well as immediate, effects on their health. Highlight 16 describes how sound choices and good habits during childhood and adolescence can help prevent chronic diseases later in life.

Nutrition Portfolio

Encouraging children to eat nutritious foods today helps them learn how to make healthy food choices tomorrow.

- If there are children in your life, think about the food they eat and consider whether they receive enough food for healthy growth, but not so much as to lead to obesity.
- Describe the advantages of physical activity to children's health and well-being.
- Plan a day's menu for a child 4 to 8 years of age, making sure to include foods that provide enough calcium and iron.
- Now, go to Diet Analysis Plus and create a profile for a child 4 to 8 years of age. Enter the day's menu you suggested in the previous exercise and see if you met the basic requirements for that child.

To complete this exercise, go to your Diet Analysis Plus at **www.cengagebrain.com**.

STUDY IT To review the key points of this chapter and take a practice quiz, go to the study cards at the end of the book.

REFERENCES

1. Committee on Dietary Reference Intakes, *Dietary Reference Intakes for Energy, Carbohydrate, Fiber, Fat, Fatty Acids, Cholesterol, Protein, and Amino Acids* (Washington, D.C.: National Academies Press, 2005).
2. Committee on Dietary Reference Intakes, 2005, pp. 280–281.
3. Formula feeding of term infants, in *Pediatric Nutrition Handbook*, 6th ed., R. E. Kleinman (Elk Grove Village, Ill.: American Academy of Pediatrics, 2009), pp. 61–78.
4. Position of the American Dietetic Association: Promoting and supporting breastfeeding, *Journal of the American Dietetic Association* 109 (2009): 1926–1942.
5. Breastfeeding, in *Pediatric Nutrition Handbook*, 6th ed., ed. R. E. Kleinman (Elk Grove Village, Ill.: American Academy of Pediatrics, 2009), pp. 29–59; Position of the American Dietetic Association, 2009.

6. L. Bode, Human milk oligosaccharides: Prebiotics and beyond, *Nutrition Reviews* 67 (2009): S183–S191.
7. S. M. Donovan, Human milk oligosaccharides: The plot thickens, *British Journal of Nutrition* 101 (2009): 1267–1269.
8. S. E. Carlson, Docosahexaenoic acid supplementation in pregnancy and lactation, *American Journal of Clinical Nutrition* 89 (2009): 678S–684S; S. M. Innis, Dietary omega 3 fatty acids and the developing brain, *Brain Research* 1237 (2008): 35–43.
9. S. J. Meldrum and coauthors, Achieving definitive results in long-chain polyunsaturated fatty acid supplementation trials of term infants: Factors for consideration, *Nutrition Reviews* 69 (2011): 205– 214; L. G. Smithers, R. A. Gibson, and M. Makrides, Maternal supplementation with docosahexaenoic acid during pregnancy does not affect early visual development in the infant: A randomized controlled trial, *American Journal of Clinical Nutrition* 93 (2011): 1293–1299; E. E. Birch and coauthors, The DIAMOND (DHA Intake and Measurement of Neural Development) Study: A double-masked, randomized controlled clinical trial of the maturation of infant visual acuity as a function the dietary level of docosahexaenoic acid, *American Journal of Clinical Nutrition* 91 (2010): 848–859; K. Simmer, S. K. Patole, and S. C. Rao, Longchain polyunsaturated fatty acid supplementation in infants born at term, *Cochrane Database of Systematic Reviews*, January 23, 2008, CD000376; Innis, 2008.
10. Simmer, Patole, and Rao, 2008.
11. S. A. Abrams, What are the risks and benefits to increasing dietary bone minerals and vitamin D intake in infants and small children? *Annual Review of Nutrition* 31 (2011): 285–297.
12. Fat-soluble vitamins, in *Pediatric Nutrition Handbook*, 6th ed., ed. R. E. Kleinman (Elk Grove Village, Ill.: American Academy of Pediatrics, 2009), pp. 461–474.
13. C. L. Wagner, F. R. Greer, and the Section on Breastfeeding and Committee on Nutrition, Prevention of rickets and vitamin D deficiency in infants, children, and adolescents, *Pediatrics* 122 (2008): 1142–1152.
14. A. Walker, Breast milk as the gold standard for protective nutrients, *Journal of Pediatrics* 156 (2010): S3–S7; L. Duijts and coauthors, Prolonged and exclusive breastfeeding reduces the risk of infectious diseases in infancy, *Pediatrics* 126 (2010): e18–e25; Breastfeeding, 2009; Position of the American Dietetic Association, 2009.
15. Breastfeeding, 2009.
16. Breastfeeding, 2009; Position of the American Dietetic Association, 2009.
17. F. R. Greer, S. H. Sicherer, A. W. Burks, and the Committee on Nutrition and Section on Allergy and Immunology, Effects of early nutritional interventions on the development of atopic disease in infants and children: The role of maternal dietary restriction, breastfeeding, timing of introduction of complementary foods, and hydrolyzed formulas, *Pediatrics* 121 (2008): 183–191.
18. Greer, Sicherer, Burks, and the Committee on Nutrition and Section on Allergy and Immunology, 2008.
19. Greer, Sicherer, Burks, and the Committee on Nutrition and Section on Allergy and Immunology, 2008; A. C. Krakowski and coauthors, Management of atopic dermatitis in the pediatric population, *Pediatrics* 122 (2008): 812–824.
20. M. M. Vennemann and coauthors, Does breastfeeding reduce the risk of sudden infant death syndrome? *Pediatrics* 123 (2009): e406–e410.
21. D. A. Leon and G. Ronalds, Breast-feeding influences on later life: Cardiovascular disease, *Advances in Experimental Medicine and* Biology 639 (2009): 153–166; C. G. Owen and coauthors, Does initial breastfeeding lead to lower blood cholesterol in adult life? A quantitative review of the evidence, *American Journal of Clinical Nutrition* 88 (2008): 305–314.
22. L. Shields and coauthors, Breastfeeding and obesity at 21 years: A cohort study, *Journal of Clinical Nursing* 19 (2010): 1612–1617; L. Schack-Nielsen and coauthors, Late introduction of complementary feeding, rather than duration breastfeeding, may protect against adult overweight, *American Journal of Clinical Nutrition* 91 (2010): 619–627; L. Twells and L. A. Newhook, Can exclusive breastfeeding reduce the likelihood of childhood obesity in some regions of Canada? *Canadian Journal of Public Health* 101 (2010): 36–39; P. Chivers and coauthors, Body mass index, adiposity rebound and early feeding in a longitudinal cohort (Raine Study), *International Journal of Obesity* 34 (2010): 1169–1176.
23. Chivers and coauthors, 2010; Shields and coauthors, 2010.
24. K. L. Whitaker and coauthors, Comparing maternal and paternal intergenerational transmission of obesity risk in a large population-based sample, *American Journal of Clinical Nutrition* 91 (2010): 1560–1567; R. Li, S. B. Fein, and L. M. Grummer-Strawn, Do infants fed from bottles lack self-regulation of milk intake compared with directly breastfed infants? *Pediatrics* 125 (2010): e1386–e1393.
25. W. Jedrychowski and coauthors, Effect of exclusive breastfeeding on the development of children's cognitive function in the Krakow prospective birth cohort study, *European Journal of Pediatrics* 171 (2012): 151–158.
26. M. Guxens and coauthors, Breastfeeding, long-chain polyunsaturated fatty acids in colostrum, and infant mental development, *Pediatrics* 128 (2011): e880–e889.,
27. K. Y. Wojcik and coauthors, Macronutrient analysis of a nationwide sample of donor breast milk, *Journal of the American Dietetic Association* 109 (2009): 137–140.
28. Formula feeding of term infants, in *Pediatric Nutrition Handbook*, 6th ed., ed. R. E. Kleinman (Elk Grove Village, Ill.: American Academy of Pediatrics, 2009), pp. 61–78.
29. Formula feeding of term infants, 2009.
30. Formula feeding of term infants, 2009.
31. S. Basnet and coauthors, Fresh goat's milk for infants: Myths and realities: A review, *Pediatrics* 125 (2010): e973–977.
32. T. J. Mathews and coauthors, Annual summary of vital statistics: 2008, *Pediatrics* 127 (2011): 146–157.
33. D. L. O'Connor and coauthors, Growth and nutrient intakes of human milk-fed preterm infants provided with extra energy and nutrients after hospital discharge, *Pediatrics* 121 (2008): 766–776.
34. Formula feeding of term infants, 2009.
35. Expert panel on integrated guidelines for cardiovascular health and risk reduction in children and adolescents: Summary report, *Pediatrics* 128 (2011): S213–S256.
36. E. E. Ziegler, Consumption of cow's milk as a cause of iron deficiency in infants and toddlers, *Nutrition Reviews* 69 (2011): S37–S42.
37. Complementary feeding, in *Pediatric Nutrition Handbook*, 6th ed., ed. R. E. Kleinman (Elk Grove Village, Ill.: American Academy of Pediatrics, 2009), pp. 113–142.
38. Complementary feeding, in *Pediatric Nutrition Handbook*, 6th ed., ed. R. E. Kleinman (Elk Grove Village, Ill.: American Academy of Pediatrics, 2009), pp. 113–142.
39. E. E. Ziegler, S. E. Nelson, and J. M. Jeter, Iron supplementation of breastfed infants, *Nutrition Reviews* 69 (2011): S71–S77; Iron, in *Pediatric Nutrition Handbook*, 6th ed., ed. R. E. Kleinman (Elk Grove Village, Ill.: American Academy of Pediatrics, 2009), pp. 403–422.
40. Complementary feeding, 2009.
41. Feeding the child, in *Pediatric Nutrition Handbook*, 6th ed., ed. R. E. Kleinman (Elk Grove Village, Ill.: American Academy of Pediatrics, 2009), pp. 145–174.
42. Feeding the child, in *Pediatric Nutrition Handbook*, 6th ed., ed. R. E. Kleinman (Elk Grove Village, Ill.: American Academy of Pediatrics, 2009), pp. 145–174.
43. Position of the American Dietetic Association: Nutrition guidance for healthy children ages 2 to 11 years, *Journal of the American Dietetic Association* 108 (2008): 1038–1047; Position of the American Dietetic Association: Individual-, family-, school-, and community-based interventions for pediatric overweight, *Journal of the American Dietetic Association* 106 (2006): 925–945.
44. J. E. Fulton and coauthors, Physical activity levels of high school students: United States, 2010, *Morbidity and Mortality Weekly Report* 60 (2011): 773–777; D. A. McCarron and coauthors, Community-based priorities for improving nutrition and physical activity in childhood, *Pediatrics* 126 (2010): S73–S89; G. A. Nyberg and coauthors, Physical activity patterns measured by accelerometry in 6- to 10-yr-old children, *Medicine and Science in Sports and Exercise* 41 (2009): 1842–1848; M. Dowda and coauthors, Policies and characteristics of the preschool environment and physical activity of young children, *Pediatrics* 123 (2009): e261–e266; American Academy of Pediatrics, Council on Sports Medicine and Fitness and Council on School Health, Active healthy living: Prevention of childhood obesity through increased physical activity, *Pediatrics* 117 (2006): 1834–1842.
45. Nutritional aspects of vegetarian diets, in *Pediatric Nutrition Handbook*, 6th ed., ed. R. E. Kleinman (Elk Grove Village, Ill.: American Academy of Pediatrics, 2009), pp. 201–224.

46. Committee on Dietary Reference Intakes, 2005, Chapter 6.
47. Committee on Dietary Reference Intakes, 2005, Chapter 7.
48. Committee on Dietary Reference Intakes, 2005, Chapter 11.
49. Committee on Dietary Reference Intakes, 2005, Chapter 8.
50. R. D. Baker, F. R. Greer, and the Committee on Nutrition, Clinical Report: Diagnosis and prevention of iron deficiency and iron-deficiency anemia in infants and young children (0–3 years of age), *Pediatrics* 126 (2010): 1040–1050.
51. Committee on Dietary Reference Intakes, *Dietary Reference Intakes for Calcium and Vitamin D* (Washington, D.C.: National Academies Press, 2011), pp. 345–402.
52. Wagner, Greer, and the Section on Breastfeeding and Committee on Nutrition, 2008.
53. J. T. Dwyer and coauthors, Feeding infants and toddlers study 2008: Progress, continuing concerns, and implications, *Journal of the American Dietetic Association* 110 (2010): S60–S67; Feeding the child, 2009.
54. U. Shaikh, R. S. Byrd, and P. Auinger, Vitamin and mineral supplement use by children and adolescents in the 1999–2004 National Health and Nutrition Examination survey: Relationship with nutrition, food security, physical activity, and health care access, *Archives of Pediatrics and Adolescent Medicine* 163 (2009): 150–157.
55. Position of the American Dietetic Association, 2008.
56. R. R. Briefel and coauthors, The Feeding Infants and Toddlers Study 2008: Study design and methods, *Journal of the American Dietetic Association* 110 (2010): S16–S26.
57. M. K. Fox and coauthors, Food consumption patterns of young preschoolers: Are they starting off on the right path? *Journal of the American Dietetic Association* 110 (2010): S52–S59.
58. Dwyer and coauthors, 2010.
59. Bread for the World, Hunger facts: Domestic, www.bread.org/learn/hunger-basics/hunger-facts-domestic.html, updated February 2008; accessed January 23, 2009.
60. Position of the American Dietetic Association: Child and adolescent nutrition assistance programs, *Journal of the American Dietetic Association* 110 (2010): 791–799.
61. K. Widenhorn-Muller and coauthors, Influence of having breakfast on cognitive performance and mood in 13- to 20-year-old high school students: Results of a crossover trial, *Pediatrics* 122 (2008): 279–284.
62. T.V.E. Kral and coauthors, Effects of eating breakfast compared with skipping breakfast on ratings of appetite and intake at subsequent meals in 8- to 10-y-old children, *American Journal of Clinical Nutrition* 93 (2011): 284–291; K. J. Smith and coauthors, Skipping breakfast: Longitudinal associations with cardiometabolic risk factors in the Childhood Determinants of Adult Health Study, *American Journal of Clinical Nutrition* 92 (2010): 1316–1325; L. Dubois and coauthors, Breakfast skipping is associated with differences in meal patterns, macronutrient intakes and overweight among pre-school children, *Public Health Nutrition* 12 (2009): 19–28.
63. K. Kordas, Iron, lead, and children's behavior and cognition, *Annual Review of Nutrition* 30 (2010): 123–148.
64. M. M. Black and coauthors, Iron deficiency and iron-deficiency anemia in the first two years of life: Strategies to prevent loss of developmental potential, *Nutrition Reviews* 69 (2011): S64–S70; M. K. Georgieff, Long-term brain and behavioral consequences of early iron deficiency, *Nutrition Reviews* 69 (2011): S43–S48; C. S. Wang and coauthors, Iron-deficiency anemia in infancy and social emotional development in preschool-aged Chinese children, *Pediatrics* 127 (2011): e927–e933; F. Corapci and coauthors, Longitudinal evaluation of externalizing and internalizing behavior problems following iron deficiency in infancy, *Journal of Pediatric Psychology* 35 (2010): 296–305.
65. Kordas, 2010.
66. K. M. Cecil and coauthors, Decreased brain volume in adults with childhood lead exposure, *PLoS Medicine* 27 (2008): e112.
67. Centers for Disease Control and Prevention, Lead, www.cdc.gov/nceh/lead/.htm, updated on August 26, 2011 and accessed on September 29, 2011.
68. Centers for Disease control and Prevention, Interpreting and managing blood lead levels <10μg/dL in children and reducing childhood exposures to lead: Recommendations of CDC's Advisory Committee on Childhood Lead Poisoning Prevention, *Morbidity and Mortality Weekly Report* 56/RR-8 (2007): 1–16; Position of the Committee on Environmental Health, American Academy of Pediatrics: Lead exposure in children: Prevention, detection, and management, *Pediatrics* 116 (2005): 1036–1046.
69. L. J. Akinbami and coauthors, Attention Deficit Hyperactivity Disorder among children aged 5–17 years in the United States, 1998–2009, www.cdc.gov/nchs/data/databriefs/db70.htm.
70. Subcommittee on Attention-Deficit/Hyperactivity Disorder, Steering Committee on Quality Improvement and Management, ADHD: Clinical practice guideline for the diagnosis, evaluation, and treatment of attention-deficit/hyperactivity disorder in children and adolescents, *Pediatrics* 128 (2011): 1007–1022; W. B. Brinkman and coauthors, Parental angst making and revisiting decisions about treatment of Attention-Deficit/Hyperactivity Disorder, *Pediatrics* 124 (2009): 580–589.
71. L. F. Martí, Effectiveness of nutritional interventions on the functioning of children with ADHD and/or ASD. An updated review of research evidence, *Bulletin of the Porto Rico Medical Association* 102 (2010): 31–42.
72. R. E. Kleinman and coauthors, A research model for investigating the effects of artificial food colorings on children with ADHD, *Pediatrics* 127 (2011): e1575–1584; A. Connolly and coauthors, Pattern of intake of food additives associated with hyperactivity in Irish children and teenagers, *Food Additives and Contaminants: Part A, Chemistry, Analysis, Control, Exposure and Risk Assessment* 27 (2010): 447–456.
73. Food and Drug Administration, Food ingredients and colors, www.fda.gov. revised April 2010.
74. R. S. Gupta and coauthors, The prevalence, severity, and distribution of childhood food allergy in the United States, *Pediatrics* 128 (2011): e9–e17; Food and Drug Administration, Food allergies: Reducing the risks, January 23, 2009, www.fda.gov/consumer/updates/foodallergies012209.html.
75. Food and Drug Administration, Food allergies; Reducing the risks, January 23, 2009, www.fda.gov/consumer/updates/foodallergies012209.html.
76. A. Boulay and coauthors, A EuroPrevall review of factors affecting incidence of peanut allergy: Priorities for research and policy, *Allergy* 63 (2008): 797–809.
77. J. A. Boyce and coauthors, Guidelines for the diagnosis and management of food allergy in the United States: Summary of the NIAID-Sponsored Expert Panel Report, *Journal of the American Dietetic Association* 111 (2011): 17–27.
78. M. Boguniewicz, N. Moore, and K. Paranto, Allergic diseases, quality of life, and the role of the dietitian, *Nutrition Today* 43 (2008): 6–10.
79. S. Ramesh, Food allergy overview in children, *Clinical Reviews in Allergy & Immunology* 34 (2008): 217–230.
80. M. Pansare and D. Kamat, Peanut allergy, *Current Opinions in Pediatrics* 22 (2010): 642–646; A. T. Clark and coauthors, Successful oral tolerance induction in severe peanut allergy, *Allergy* 64 (2009): 1218–1220.
81. C. L. Ogden and M. D. Carroll, Prevalence of obesity among children and adolescents: United States, trends 1963–1965 through 2007–2008, *NCHS Health E-Stat*, www.cdc.gov/nchs/data/hestat/obesity_child_07_08/obesity_child_07_08.pdf.
82. C. L. Ogden and coauthors, Prevalence of high body mass index in US children and adolescents, 2007–2008, *Journal of the American Medical Association* 303 (2010): 242–249.
83. S. E. Barlow and the Expert Committee, Expert Committee recommendations regarding the prevention, assessment, and treatment of child and adolescent overweight and obesity: Summary report, *Pediatrics* 120 (2007): S164–S192.
84. Barlow and the Expert Committee, 2007.
85. N. S. The and coauthors, Association of adolescent obesity with risk of severe obesity in adulthood, *Journal of the American Medical Association* 304 (2010): 2042–2057; F. M. Biro and M. Wien, Childhood obesity and adult morbidities, *American Journal of Clinical Nutrition* 91 (2010): 1499S–1505S.
86. R. Cooper and coauthors, Associations between parental and offspring adiposity up to midlife: The contribution of adult lifestyle factors in the 1958 British Birth Cohort Study, *American Journal of Clinical Nutrition* 92 (2010): 946–953; Whitaker and coauthors, 2010; L. Li and coauthors, Intergenerational influences on childhood body mass index: The effect of parental body mass index trajectories, *American Journal of Clinical Nutrition* 89 (2009): 551–557.

87. Biro and Wien, 2010.
88. Centers for Disease Control and Prevention, Overweight and obesity: A growing problem, www.cdc.gov/obesity/childhood/problem.html, accessed on September 30, 2011.
89. S. J. Salvy and coauthors, Influence of parents and friends on children's and adolescents' food intake and food selection, *American Journal of Clinical Nutrition* 93 (2011): 87–92; L. Hall and coauthors, Children's intake of fruit and selected energy-dense nutrient-poor foods is associated with fathers' intake, *Journal of the American Dietetic Association* 111 (2011): 1039–1044; H. A. Raynor and coauthors, The relationship between child and parent food hedonics and parent and child food group intake in children with overweight/obesity, *Journal of the American Dietetic Association* 111 (2011): 425–430; C. Sweetman and coauthors, Characteristics of family mealtimes affecting children's vegetable consumption and liking, *Journal of the American Dietetic Association* 111 (2011): 269–273.
90. K. S. Geller and D. A. Dzewaltowski, Longitudinal and cross-sectional influences on youth fruit and vegetable consumption, *Nutrition Reviews* 67 (2009): 65–76; J. Brug and coauthors, Taste preferences, liking and other factors related to fruit and vegetable intakes among schoolchildren: Results from observational studies, *British Journal of Nutrition* 99 (2008): S7–S14.
91. K. E. Leahy, L. L. Birch, and B. Rolls, Reducing the energy density of multiple meals decreases the energy intake of preschool-age children, *American Journal of Clinical Nutrition* 88 (2008): 1459–1468.
92. J. Reedy and S. M. Krebs-Smith, Dietary sources of energy, solid fats, and added sugars among children and adolescents in the United States, *Journal of the American Dietetic Association* 110 (2010): 1477–1484.
93. S. N. Bleich and coauthors, Increasing consumption of sugar-sweetened beverages among US adults: 1988–1994 to 1999–2004, *American Journal of Clinical Nutrition* 89 (2009): 372–381.
94. S. Harrington, The role of sugar-sweetened beverage consumption in adolescent obesity: A review of the literature, *Journal of School Nursing* 24 (2008): 3–12.
95. L. Chen and coauthors, Reduction in consumption of sugar-sweetened beverages is associated with weight loss: The PREMIER Trial, *American Journal of Clinical Nutrition* 89 (2009): 1299–1306.
96. American Academy of Pediatrics, Council on Sports Medicine and Fitness and Council on School Health, Active healthy living: Prevention of childhood obesity through increased physical activity, *Pediatrics* 117 (2006): 1834–1842.
97. Centers for Disease Control and Prevention, 2011.
98. D. M. Jackson and coauthors, Increased television viewing is associated with elevated body fatness but not with lower total energy expenditure in children, *American Journal of Clinical Nutrition* 89 (2009): 1031–1036.
99. M. Hamer, E. Stamsatakis, and G. Mishra, Psychological distress, television viewing, and physical activity in children aged 4 to 12 years, *Pediatrics* 123 (2009): 1263–1268.
100. D. J. Barr-Anderson and coauthors, Characteristics associated with older adolescents who have a television in their bedrooms, *Pediatrics* 121 (2008): 718–724.
101. Jackson and coauthors, 2009.
102. L. Dubois and coauthors, Social factors and television use during meals and snacks is associated with higher BMI among pre-school children, *Public Health Nutrition* 11 (2008): 1267–1279.
103. D. J. Anschutz, R. C.M.E. Engels, and T. Van Strien, Side effects of television food commercials on concurrent nonadvertised sweet snack food intakes in young children, *American Journal of Clinical Nutrition* 89 (2009): 1328–1333.
104. A. Batada and coauthors, Nine out of 10 food advertisements shown during Saturday morning children's television programming are for foods high in fat, sodium, or added sugars, or low in nutrients, *Journal of the American Dietetic Association* 108 (2008): 673–678.
105. A. E. Henry and M. Story, Food and beverage brands that market to children and adolescents on the internet: A content analysis of branded web sites, *Journal of Nutrition Education and Behavior* 41 (2009): 353–359.
106. J. P. Chaput and coauthors, Video game playing increases food intake in adolescents: A randomized crossover study, *American Journal of Clinical Nutrition* 93 (2011): 1196–1203.
107. K. White, G. Schofield, and A. E. Kilding, Energy expended by boys playing active video games, *Journal of Science and Medicine in Sport* 14 (2011): 130–134.
108. Policy statement: Media use by children younger than 2 years, *Pediatrics* 128 (2011): 1040–1045; Barlow and the Expert Committee, 2007.
109. W. Tu and coauthors, Intensified effect of adiposity on blood pressure in overweight and obese children, *Hypertension* 58 (2011): 818–824; M. Salvadori and coauthors, Elevated blood pressure in relation to overweight and obesity among children in a rural Canadian community, *Pediatrics* 122 (2008): e821–e827.
110. S. Cook and coauthors, Metabolic Syndrome rates in United States adolescents, from the National Health and Nutrition Examination Survey, 1999–2002, *Journal of Pediatrics* 152 (2008): 165–170; K. L. Jones, Role of obesity in complicating and confusing the diagnosis and treatment of diabetes in children, *Pediatrics* 121 (2008): 361–368.
111. Barlow and the Expert Committee, 2007.
112. American Academy of Pediatrics, 2006.
113. P. E. O'Brien and coauthors, Laparoscopic adjustable gastric banding in severely obese adolescents, *Journal of the American Medical Association* 303 (2010): 519–526; T. H. Inge and coauthors, Reversal of type 2 diabetes mellitus and improvements in cardiovascular risk factors after surgical weight loss in adolescents, *Pediatrics* 123 (2009): 214–222.
114. B. A. Spear and coauthors, Recommendations for treatment of child and adolescent overweight and obesity, *Pediatrics* 120 (2007): S254–S288.
115. Position of the American Dietetic Association, 2008; Position of the American Dietetic Association, 2006.
116. E. Jansen and coauthors, From the Garden of Eden to the land of plenty: Restriction of fruit and sweets intake leads to increased fruit and sweets consumption in children, *Appetite* 51 (2008): 570–575.
117. Position of the American Dietetic Association, 2008.
118. Position of the American Dietetic Association, 2008.
119. Position of the American Dietetic Association: Benchmarks for nutrition in child care, *Journal of the American Dietetic Association* 111 (2011): 607–615; Position of the American Dietetic Association, School Nutrition Association, and Society of Nutrition Education: Comprehensive school nutrition services, *Journal of the American Dietetic Association* 110 (2010): 1738–1749.
120. Position of the American Dietetic Association, School Nutrition Association, and Society of Nutrition Education, 2010.
121. Position of the American Dietetic Association, School Nutrition Association, and Society of Nutrition Education, 2010.
122. School Lunch Program participation and meals served 2010, www.fns.usda.gov/pd/slsummar.htm; updated September 1, 2011, accessed September 26, 2011; Position of the American Dietetic Association, School Nutrition Association, and Society for Nutrition Education, 2010; Position of the American Dietetic Association: Local support for nutrition integrity in schools, *Journal of the American Dietetic Association* 110 (2010): 1244–1254.
123. US Department of Agriculture, *Nutrition Standards in the National School Lunch and School Breakfast Programs*, January 25, 2012.
124. M. Story, The Third School Nutrition Dietary Assessment Study: Findings and policy implications for improving the health of US children, *Journal of the American Dietetic Association* 109 (2009): S7–S13.
125. Position of the American Dietetic Association, 2008.
126. School Breakfast Program participation and meals served 2010, www.fns.usda.gov/pd/sbsummar.htm; updated September 1, 2011, accessed September 26, 2011; Position of the American Dietetic Association, 2010.
127. Position of the American Dietetic Association, 2010.
128. M. D. Crepinsek and coauthors, Meals offered and served in US public schools: do they meet nutrient standards? *Journal of the American Dietetic Association* 109 (2009): S31–S43; Widenhorn-Muller and coauthors, 2008.
129. Position of the American Dietetic Association, 2010.
130. Position of the American Dietetic Association, 2010; Centers for Disease Control and Prevention, Competitive foods and beverages available for purchase in secondary schools: Selected sites, United States, 2006, *Morbidity and Mortality Weekly Report* 57 (2008): 935–938.

131. M. Kakarala, D. R. Keast, and S. Hoerr, Schoolchildren's consumption of competitive foods and beverages, excluding á la carté, *Journal of School Health* 80 (2010): 429–435; Position of the American Dietetic Association: Local support for nutrition integrity in schools, *Journal of the American Dietetic Association* 110 (2010): 1244–1254.
132. Position of the American Dietetic Association, 2010.
133. Institute of Medicine, Food and Nutrition Board, Committee on Nutrition Standards for Foods in Schools, eds. V. A. Stallings and A. L. Yaktine, *Nutrition Standards for Foods in Schools: Leading the Way toward Healthier Youth* (Washington, D.C.: National Academies Press, 2007).
134. Position of the American Dietetic Association, School Nutrition Association, and Society for Nutrition Education, 2010; Institute of Medicine, Food and Nutrition Board, Committee on Nutrition Standards for Foods in Schools, 2007.
135. Institute of Medicine, Food and Nutrition Board, Committee on Nutrition Standards for Foods in Schools, 2007).
136. L. B. Shomaker and coauthors, Puberty and observed energy intake: Boy, can they eat! *American Journal of Clinical Nutrition* 92 (2010): 123–129; Committee on Dietary Reference Intakes, 2005, Chapter 5.
137. Committee on Dietary Reference Intakes, 2011.
138. S. Saintonge, H. Bang, and L. M. Gerber, Implications of a new definition of vitamin D deficiency in a multiracial US adolescent population: The National Health and Nutrition Examination Survey III, *Pediatrics* 123 (2009): 797–803; Wagner, Greer, and the Section on Breastfeeding and Committee on Nutrition, 2008.
139. Wagner, Greer, and the Section on Breastfeeding and Committee on Nutrition, 2008.
140. Committee on Dietary Reference Intakes, *Dietary Reference Intakes for Vitamin A, Vitamin K, Arsenic, Boron, Chromium, Copper, Iodine, Iron, Manganese, Molybdenum, Nickel, Silicon, Vanadium, and Zinc* (Washington, D.C.: National Academies Press, 2001), pp. 290–393.
141. Committee on Dietary Reference Intakes, 2001.
142. Committee on Dietary Reference Intakes, 2011, pp. 35–74.
143. M. Mesias, I. Seiquer, and M. P. Navarro, Calcium nutrition in adolescence, *Critical Reviews in Food Science and Nutrition* 51 (2011): 195–209; L. Esterie and coauthors, Milk, rather than other foods, is associated with vertebral bone mass and circulating IGF-1 in female adolescents, *Osteoporosis International* 20 (2009): 567–575; M. M. Murphy and coauthors, Drinking flavored or plain milk is positively associated with nutrient intake and is not associated with adverse effects on weight status in US children and adolescents, *Journal of the American Dietetic Association* 108 (2008): 631–639.
144. L. M. Fiorito and coauthors, Girls' early sweetened carbonated beverage intake predicts different patterns of beverage and nutrient intake across childhood and adolescence, *Journal of the American Dietetic Association* 110 (2010): 543–550.
145. A. J. Hammons and B. H. Fiese, Is frequency of shared family meals related to the nutritional health of children and adolescents? *Pediatrics* 127 (2011): e1565–e1574; S. J. Woodruff and R. M. Hanning, Associations between family dinner frequency and specific food behaviors among grade six, seven, and eight students from Ontario and Nova Scotia, *Journal of Adolescent Health* 44 (2009): 431–436.
146. J. White and E. Halliwell, Alcohol and tobacco use during adolescence: the importance of the family mealtime environment, *Journal of Health Psychology* 15 (2010): 526–532; B. Sen, The relationship between frequency of family dinner and adolescent problem behaviors after adjusting for other family characteristics, *Journal of Adolescence* 33 (2010): 187–196; M. E. Eisenberg and coauthors, Family meals and substance use: Is there a long-term association? *Journal of Adolescent Health* 43 (2008): 151–156.
147. R. Priya and coauthors, The relationship of breakfast skipping and type of breakfast consumption with nutrient intake and weight status in children and adolescents: The National Health and Nutrition examination Survey 1999–2006.
148. M. T. Timlin and coauthors, Breakfast eating and weight change in a 5-year prospective analysis of adolescents: Project EAT (Eating Among Teens), *Pediatrics* 121 (2008): e638–e645.
149. Smith and coauthors, 2010; C. M. McDonald and coauthors, Overweight is more prevalent than stunting and is associated with socioeconomic status, maternal obesity, and a snacking dietary pattern in school children from Bogota, Columbia, *Journal of Nutrition* 139 (2009): 370–376.
150. R. S. Sebastian, L. E. Cleveland, and J. D. Goldman, Effect of snacking frequency on adolescents' dietary intakes and meeting national recommendations, *Journal of Adolescent Health* 42 (2008): 503–511.
151. N. D. Brener and coauthors, Beverage consumption among high school students: United States, 2010, *Morbidity and Mortality Weekly Report* 60 (2011): 778–780.
152. L. Libuda and coauthors, Association between long-term consumption of soft drinks and variables of bone modeling and remodeling in a sample of healthy German children and adolescents, *American Journal of Clinical Nutrition* 88 (2008): 1670–1677.
153. L. Mancino and coauthors, How food away from home affects children's diet quality, *Economic Research Report No. (ERR-104)*, October 2010, www.ers.usda.gov/publications/err104.

HIGHLIGHT > 16

LEARN IT Describe the lifestyle factors that can help prevent childhood obesity and the development of type 2 diabetes and heart disease.

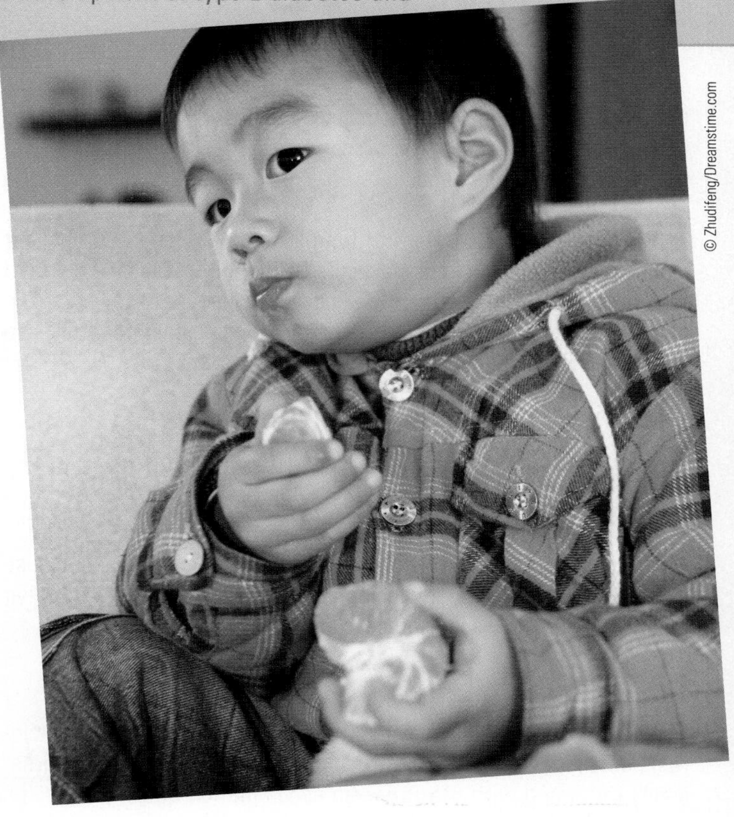

© Zhudifeng/Dreamstime.com

Childhood Obesity and the Early Development of Chronic Diseases

When people think about the health problems of children and adolescents, they typically think of ear infections, colds, and acne—not heart disease, diabetes, or hypertension. Today, however, unprecedented numbers of US children are being diagnosed with obesity and serious "adult diseases," such as **type 2 diabetes,** that accompany overweight.[1] When type 2 diabetes develops before the age of 20, the incidence of diabetic kidney disease and death in middle age increases dramatically, largely because of the long duration of the disease. For children born in the United States in the year 2000, the risk of developing type 2 diabetes sometime in their lives is estimated to be 30 percent for boys and 40 percent for girls. US children are not alone—rapidly rising rates of obesity threaten the health of an alarming number of children around the globe.[2] Without immediate intervention, millions of children are destined to develop type 2 diabetes and hypertension in childhood followed by **cardiovascular disease (CVD)** in early adulthood.[3] (See the accompanying glossary for this and related terms.)

This highlight focuses on efforts to prevent childhood obesity and the development of heart disease and type 2 diabetes, but the benefits extend to other obesity-related diseases as well. The years of childhood (ages 2 to 18) are emphasized here because the earlier in life health-promoting habits become established, the better they will stick. Chapter 18 fills in the rest of the story of nutrition's role in reducing chronic disease risk.

Invariably, questions arise as to what extent genetics is involved in disease development. For heart disease and type 2 diabetes, genetics does not appear to play a *determining* role; that is, a person is not simply destined at birth to develop these diseases. Instead, genetics appears to play a *permissive* role—the potential is inherited and will develop if given a push by poor health choices such as excessive weight gain, poor diet, sedentary lifestyle, and cigarette smoking.[4]

Many experts agree that preventing or treating obesity in childhood will reduce the rate of chronic diseases in adulthood. Without intervention, most overweight children become overweight adolescents who become overweight adults, and being overweight exacerbates every chronic disease that adults face.[5] Fatty liver, a condition that correlates directly with BMI, was not even recognized in pediatric research until recently. Today, fatty liver disease has a high prevalence in obese children.[6]

Early Development of Type 2 Diabetes

In recent years, type 2 diabetes, a chronic disease closely linked with obesity, has been on the rise among children and adolescents as the prevalence of obesity in US youth has increased.[7] Obesity is the most important risk factor for type 2 diabetes; it is most likely to occur in those who are obese and sedentary and have a family history of diabetes. Diagnoses typically are made during puberty, but as younger children become more obese and less active, the trend is shifting to younger ages.

In type 2 diabetes, the cells become insulin-resistant—that is, the cells become less sensitive to insulin, reducing the amount of glucose entering the cells from the blood. The combination of obesity and insulin resistance produces a cluster of symptoms, including high blood cholesterol and high blood pressure, which, in turn, promotes the development of atherosclerosis and the early development of heart

GLOSSARY

atherosclerosis (ATH-er-OH-scler-OH-sis): a type of artery disease characterized by plaques (accumulations of lipid-containing material) on the inner walls of the arteries (see Chapter 18).

- **athero** = porridge or soft
- **scleros** = hard
- **osis** = condition

cardiovascular disease (CVD): a general term for all diseases of the heart and blood vessels. Atherosclerosis is the main cause of CVD. When the arteries that carry blood to the heart muscle become blocked, the heart suffers damage known as *coronary heart disease (CHD).*

- **cardio** = heart
- **vascular** = blood vessels

fatty streaks: accumulations of cholesterol and other lipids along the walls of the arteries.

plaque (PLACK): an accumulation of fatty deposits, smooth muscle cells, and fibrous connective tissue that develops in the artery walls in atherosclerosis. Plaque associated with atherosclerosis is known as *atheromatous* (ATH-er-OH-ma-tus) *plaque.*

type 2 diabetes: a chronic disorder of carbohydrate metabolism, commonly associated with obesity and characterized by insulin resistance.

disease.[8] Other common problems evident by early adulthood include kidney disease, blindness, and miscarriages. The complications of diabetes, especially when encountered at a young age, can shorten life expectancy.

Prevention and treatment of type 2 diabetes depend on weight management, which can be particularly difficult in a child's world of food advertising, video games, and pocket money for candy bars. The activity and dietary suggestions to help defend against heart disease later in this highlight apply to type 2 diabetes as well.

Early Development of Heart Disease

Most people consider heart disease to be an adult disease because its incidence rises with advancing age, and symptoms rarely appear before age 30. The disease process actually begins much earlier.

Atherosclerosis

Most cardiovascular disease involves **atherosclerosis**. Atherosclerosis develops when regions of an artery's walls become progressively thickened with **plaque**—an accumulation of fatty deposits, smooth muscle cells, and fibrous connective tissue. If it progresses, atherosclerosis may eventually block the flow of blood to the heart and cause a heart attack or cut off blood flow to the brain and cause a stroke. Infants are born with healthy, smooth, clear arteries, but within the first decade of life, **fatty streaks** may begin to appear (see Figure H16-1). During adolescence, these fatty streaks may begin to accumulate fibrous connective tissue. By early adulthood, the fibrous plaques may begin to calcify and become raised lesions, especially in boys and young men. As the lesions grow more numerous and enlarge, the heart disease rate begins to rise, most dramatically at about age 45 in men and 55 in women. From this point on, arterial damage and blockage progress rapidly, and heart attacks and strokes threaten life. In short, the consequences of atherosclerosis, which become apparent only in adulthood, have their beginnings in the first decades of life.[9]

Atherosclerosis is not inevitable; people can grow old with relatively clear arteries. Early lesions may either progress or regress, depending on several factors, many of which reflect lifestyle behaviors. Smoking, for example, is strongly associated with the prevalence of fatty streaks and raised lesions, even in young adults.

Blood Cholesterol

As blood cholesterol rises, atherosclerosis worsens. Cholesterol values at birth are similar in all populations; differences emerge in early childhood. Standard values for cholesterol in children and adolescents (ages 2 to 18 years) are listed in Table H16-1. Cholesterol concentrations change with age in children and adolescents, however, and are especially variable during puberty.[10] Thus, use of a single cut-off point for all pediatric age groups has limitations.

In general, blood cholesterol tends to rise as dietary saturated fat intakes increase. Blood cholesterol also correlates with childhood obesity, especially abdominal obesity; LDL cholesterol rises and HDL declines. These relationships are apparent throughout childhood, and their magnitude increases with age.

Children who are both overweight and have high blood cholesterol are likely to have parents who develop heart disease early. For this reason, selective screening is recommended for children and adolescents of any age who are overweight or obese; those whose parents (or grandparents) have premature heart disease (≤55 years of age for men and ≤65 years of age for women); those whose parents have elevated blood cholesterol; those who have other risk factors for heart disease such as hypertension, cigarette smoking, or diabetes; and those whose family history is unavailable. Because blood cholesterol in children is a good predictor of adult values, some experts recommend universal screening for all children aged 9 to 11.[11]

Early—but not advanced—atherosclerotic lesions are reversible, making screening and education a high priority. Both those with family histories of heart disease and those with multiple risk factors need

> FIGURE H16-1 **The Formation of Plaques in Atherosclerosis**

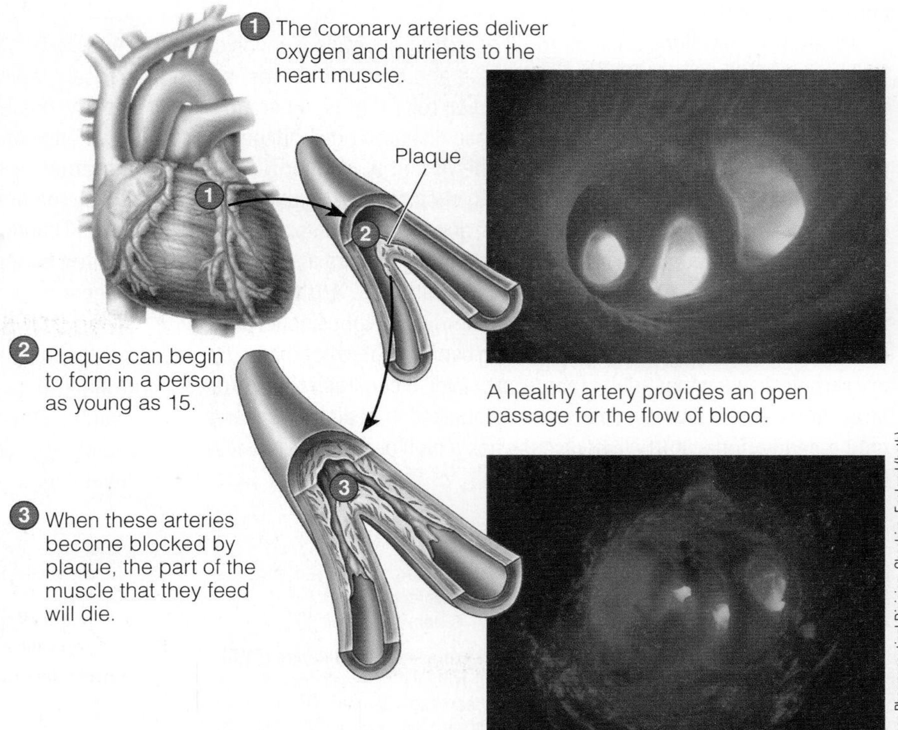

1 The coronary arteries deliver oxygen and nutrients to the heart muscle.
Plaque
2 Plaques can begin to form in a person as young as 15.
3 When these arteries become blocked by plaque, the part of the muscle that they feed will die.

A healthy artery provides an open passage for the flow of blood.

Plaques form along the artery's inner wall, reducing blood flow. Clots can form, aggravating the problem.

TABLE H16-1 Cholesterol Values for Children and Adolescents

Disease Risk	Total Cholesterol (mg/dL)	LDL Cholesterol (mg/dL)
Acceptable	<170	<110
Borderline	170–199	110–129
High	≥200	≥130

NOTE: Adult values appear in Chapter 18.

intervention. Children with the highest risks of developing heart disease are sedentary and obese, with high blood pressure and high blood cholesterol.[12] In contrast, children with the lowest risks of heart disease are physically active and of normal weight, with low blood pressure and favorable lipid profiles. Routine pediatric care should identify these known risk factors and provide intervention when needed.

Blood Pressure

Pediatricians routinely monitor blood pressure in children and adolescents. High blood pressure may signal an underlying disease or the early onset of hypertension. Hypertension accelerates the development of atheroscerlosis.

Like atherosclerosis and high blood cholesterol, hypertension may develop in the first decades of life, especially among obese children, and worsen with time. Children can control their hypertension by participating in regular aerobic activity and by losing weight or maintaining their weight as they grow taller. Restricting dietary sodium also causes an immediate drop in most children's and adolescents' blood pressure.[13]

Physical Activity

Research has also confirmed an association between blood lipids and physical activity in children, similar to that seen in adults. Physically active children have a better lipid profile and lower blood pressure than physically inactive children, and these positive findings often persist into adulthood.[14] The *Physical Activity Guidelines for Americans 2008* recommendations for children and adolescents are listed in Chapter 16 on p. 531.

Just as blood cholesterol and obesity track over the years, so does a child's level of physical activity. Those who are inactive now are likely to still be inactive years later. Similarly, those who are physically active now tend to remain so. Compared with inactive teens, those who are physically active weigh less, smoke less, eat a diet lower in saturated fats, and have better blood lipid profiles. Both obesity and blood cholesterol correlate with the inactive pastime of watching television. The message is clear: physical activity offers numerous health benefits, and children who are active today are most likely to be active for years to come.

Dietary Recommendations for Children

Regardless of family history, experts agree that all children older than age 2 should eat a variety of foods and maintain desirable weight (see Table H16-2). Children (4 to 18 years of age) should receive at least 25 percent and no more than 35 percent of total energy from fat, less than 10 percent from saturated fat, and less than 300 milligrams of cholesterol per day.[15]

TABLE H16-2 American Heart Association Dietary Guidelines and Strategies for Children

- Balance dietary kcalories with physical activity to maintain normal growth.
- Every day, engage in 60 minutes of moderate to vigorous play or physical activity.
- Eat vegetables and fruits daily. Serve fresh, frozen, or canned vegetables and fruits at every meal; limit those with added fats, salt, and sugar.
- Limit juice intake (4 to 6 ounces per day for children 1 to 6 years of age, 8 to 12 ounces for children 7 to 18 years of age).
- Use vegetable oils (canola, soybean, olive, safflower, or other unsaturated oils) and soft margarines low in saturated fat and *trans*-fatty acids instead of butter or most other animal fats in the diet.
- Choose whole-grain breads and cereals rather than refined products; read labels and make sure that "whole grain" is the first ingredient.
- Reduce the intake of sugar-sweetened beverages and foods.
- Consume low-fat and non-fat milk and milk products daily.
- Include two servings of fish per week, especially fatty fish such as broiled or baked salmon.
- Choose legumes and tofu in place of meat for some meals.
- Choose only lean cuts of meat and reduced-fat meat products; remove the skin from poultry.
- Use less salt, including salt from processed foods. Because breads, breakfast cereals, and soups may be high in salt and/or sugar, read food labels and choose high-fiber, low-salt, low-sugar alternatives.
- Limit the intake of high-kcalorie add-ons such as gravy, Alfredo sauce, cream sauce, cheese sauce, and hollandaise sauce.
- Serve age-appropriate portion sizes on appropriately sized plates and bowls.

NOTE: These guidelines are for children 2 years of age and older.
SOURCE: Adapted from American Heart Association, Samuel S. Gidding, and coauthors, Dietary recommendations for children and adolescents: A guide for practitioners, *Pediatrics* 117 (2006): 544–559.

Moderation, Not Deprivation

Healthy children older than age 2 can begin the transition to eating according to recommendations by selecting more fruits and vegetables and fewer foods high in saturated fat. Healthy meals can occasionally include moderate amounts of a child's favorite food, such as ice cream, even if it is high in saturated fat. A steady diet from the children's menus in some restaurants—which feature chicken nuggets, hot dogs, and french fries—easily exceeds a prudent intake of saturated fat, *trans* fat, and kcalories, however, and invites both nutrient shortages and weight gains.[16] Fortunately, most restaurants chains are changing children's menus to include steamed vegetables, fruit cups, and broiled or grilled chicken—additions welcomed by busy parents who often dine out or purchase take-out foods.

Other fatty foods, such as nuts, vegetable oils, and some varieties of fish such as tuna or salmon, contribute essential fatty acids. Low-fat milk and milk products also deserve special attention in a child's diet for the needed calcium and other nutrients they supply.

Parents and caregivers play a key role in helping children establish healthy eating habits. Balanced meals need to provide lean meat, poultry, fish, and legumes; fruits and vegetables; whole grains; and low-fat milk products. Such meals can provide enough energy and nutrients to support growth and maintain blood cholesterol within a healthy range.

Pediatricians warn parents to avoid extremes. Although intentions may be good, excessive food restriction may create nutrient deficiencies and impair growth. Furthermore, parental control over eating may instigate battles and foster attitudes about foods that can lead to inappropriate eating behaviors.

Diet First, Drugs Later

Experts agree that children with high blood cholesterol should first be treated with diet. If high blood cholesterol persists despite dietary intervention in children 10 years of age and older, then drugs may be necessary to lower blood cholesterol. Drugs can effectively lower blood cholesterol without interfering with adolescent growth or development.

Smoking

Even though the focus of this text is nutrition, another risk factor for heart disease that starts in childhood and carries over into adulthood must also be addressed—cigarette smoking. Each day more than 3000 young people between the ages of 12 and 17 light up for the first time, and an estimated 850 become daily cigarette smokers. Among high school students, one in six smokes regularly.[17] Approximately 80 percent of all adult smokers began smoking before the age of 18.

Of those teenagers who continue smoking, half will eventually die of smoking-related causes. Efforts to teach children about the dangers of smoking need to be aggressive. Children are not likely to consider the long-term health consequences of tobacco use. They are more likely to be struck by the immediate health consequences, such as shortness of breath when playing sports, or social consequences, such as having bad breath. Whatever the context, the message to all children and teens should be clear: don't start smoking. If you've already started, quit.

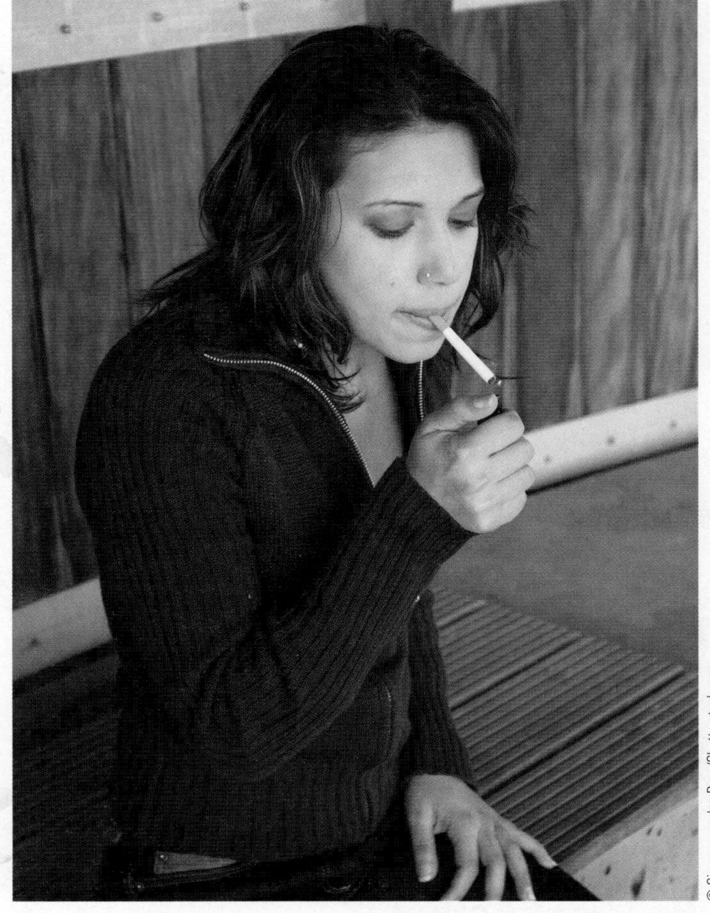

Cigarette smoking is the number one preventable cause of deaths.

In conclusion, *adult* heart disease is a major *pediatric* problem. Without intervention, some 60 million children are destined to suffer its consequences within the next 30 years. Optimal prevention efforts focus on children, especially on those who are overweight. Just as young children receive vaccinations against infectious diseases, they need screening for, and education about, chronic diseases. Many health education programs have been implemented in schools around the country. These programs are most effective when they include education in the classroom, heart-healthy meals in the lunch room, fitness activities on the playground, and parental involvement at home.

REFERENCES

1. A. Tirosh and coauthors, Adolescent BMI trajectory and risk of diabetes versus coronary disease, *New England Journal of Medicine* 364 (2011): 1315–1325; J. C. Han, D. A. Lawlor, and S.Y.S. Kimm, Childhood obesity, *Lancet* 375 (2010): 1737–1748; F. M. Biro and M. Wien, Childhood obesity and adult morbidities, *American Journal of Clinical Nutrition* 91 (2010): 1499S–1505S; C. L. Ogden and coauthors, Prevalences of high body mass index in US children and adolescents, 2007–2008, *Journal of the American Medical Association* 303 (2010): 242–249; D. S. Freedman and coauthors, Risk factors and adult body mass index among overweight children: The Bogalusa Heart Study, *Pediatrics* 123 (2009): 750–757; S. Cook and coauthors, Metabolic syndrome rates in United States adolescents, from the National Health and Nutrition Examination survey, 1999–2002, *Journal of Pediatrics* 152 (2008): 165–170; M. Gardner, D. W. Gardner, and J. R. Sowers, The cardiometabolic syndrome in the adolescent, *Pediatric Endocrinology Reviews* Suppl. 4 (2008): 964–968; K. L. Jones, Role of obesity in complicating and confusing the diagnosis and treatment of diabetes in children, *Pediatrics* 121 (2008): 361–368.
2. C. Bouchard, Childhood obesity: Are genetic differences involved? *American Journal of Clinical Nutrition* 89 (2009): 1494S–1501S; W. Maziak, K. D. Ward,

and M. B. Stockton, Childhood obesity: Are we missing the big picture? *Obesity Reviews* 9 (2008): 35–42.

3. Expert Panel on Integrated Guidelines for Cardiovascular Health and Risk Reduction in Children and Adolescents, Summary report, *Pediatrics* 128 (2011): S213–S256; Tirosh and coauthors, 2011; P. W. Franks and coauthors, Childhood obesity, other cardiovascular risk factors, and premature death, *New England Journal of Medicine* 362 (2010): 485–493; Freedman and coauthors, 2009; H. Zhu and coauthors, Relationships of cardiovascular phenotypes with healthy weight, at risk of overweight, and overweight in US youths, *Pediatrics* 121 (2008): 115–122.
4. Bouchard, 2009; J. M. Ordovas, Genetic influences on blood lipids and cardiovascular disease risk: Tools for primary prevention, *American Journal of Clinical Nutrition* 89 (2009): 1509S–1517S.
5. M. Juonaloa and coauthors, Childhood adiposity, adult adiposity, and cardiovascular risk factors, *New England Journal of Medicine* 365 (2011): 1876–1885; M. Neovius, J. Sundström, and F. Rasmussen, Combined effects of overweight and smoking in late adolescence on subsequent mortality: Nationwide cohort study, *British Medical Journal* 338 (2009): b496.
6. B. G. Koot and coauthors, Lifestyle intervention for non-alcoholic fatty liver disease: Prospective cohort study of its efficacy and factors related to improvement, *Archives of Disease in Childhood* 96 (2011): 669–674.
7. Biro and Wien, 2010.
8. Biro and Wien, 2010.
9. Expert Panel on Integrated Guidelines for Cardiovascular Health and Risk Reduction in Children and Adolescents, 2011; G. Raghuveer, Lifetime cardiovascular risk of childhood obesity, *American Journal of Clinical Nutrition* 91 (2010): 1514S–1519S; A. C. Skinner and coauthors, Multiple markers of inflammation and weight status: Cross-sectional analyses throughout childhood, *Pediatrics* 125 (2010): e801–809; Zhu and coauthors, 2008.
10. S. R. Daniels, F. R. Greer, and the Committee on Nutrition, Lipid screening and cardiovascular health in childhood, *Pediatrics* 122 (2008): 198–208.
11. Expert Panel on Integrated Guidelines for Cardiovascular Health and Risk Reduction in Children and Adolescents, 2011.
12. Centers for Disease Control and Prevention, Prevalence of abnormal lipid levels among youths: United States, 1999–2006, *Morbidity and Mortality Weekly Report* 59 (2010): 29–33.
13. J. Feber and M. Ahmed, Hypertension in children: New trends and challenges, *Clinical Science* 119 (2010): 151–161.
14. N. J. Farpour-Lambert and coauthors, Physical activity reduces systemic blood pressure and improves early markers of atherosclerosis in pre-pubertal obese children, *Journal of the American College of Cardiology* 54 (2009): 2396–2406.
15. Committee on Dietary Reference Intakes, *Dietary Reference Intakes for Energy, Carbohydrate, Fiber, Fat, Fatty Acids, Cholesterol, Protein, and Amino Acids* (Washington, D.C.: National Academies Press, 2005), pp. 769–879.
16. K. N. Boutelle and coauthors, Nutritional quality of lunch meal purchased for children at a fast-food restaurant, *Childhood Obesity* 7 (2011): 316–322; L. Johnson and coauthors, Energy-dense, low-fiber, high-fat dietary pattern is associated with increased fatness in childhood, *American Journal of Clinical Nutrition* 87 (2008): 846–854.
17. Centers for Disease Control and Prevention, Youth and tobacco use, 2009, www.cdc.gov/tobacco/data_statistics/fact_sheets/youth_data/tobacco_use/index.htm.

CENGAGE LEARNING'S GLOBAL NUTRITION WATCH — GLOBAL NUTRITION RESOURCE CENTER

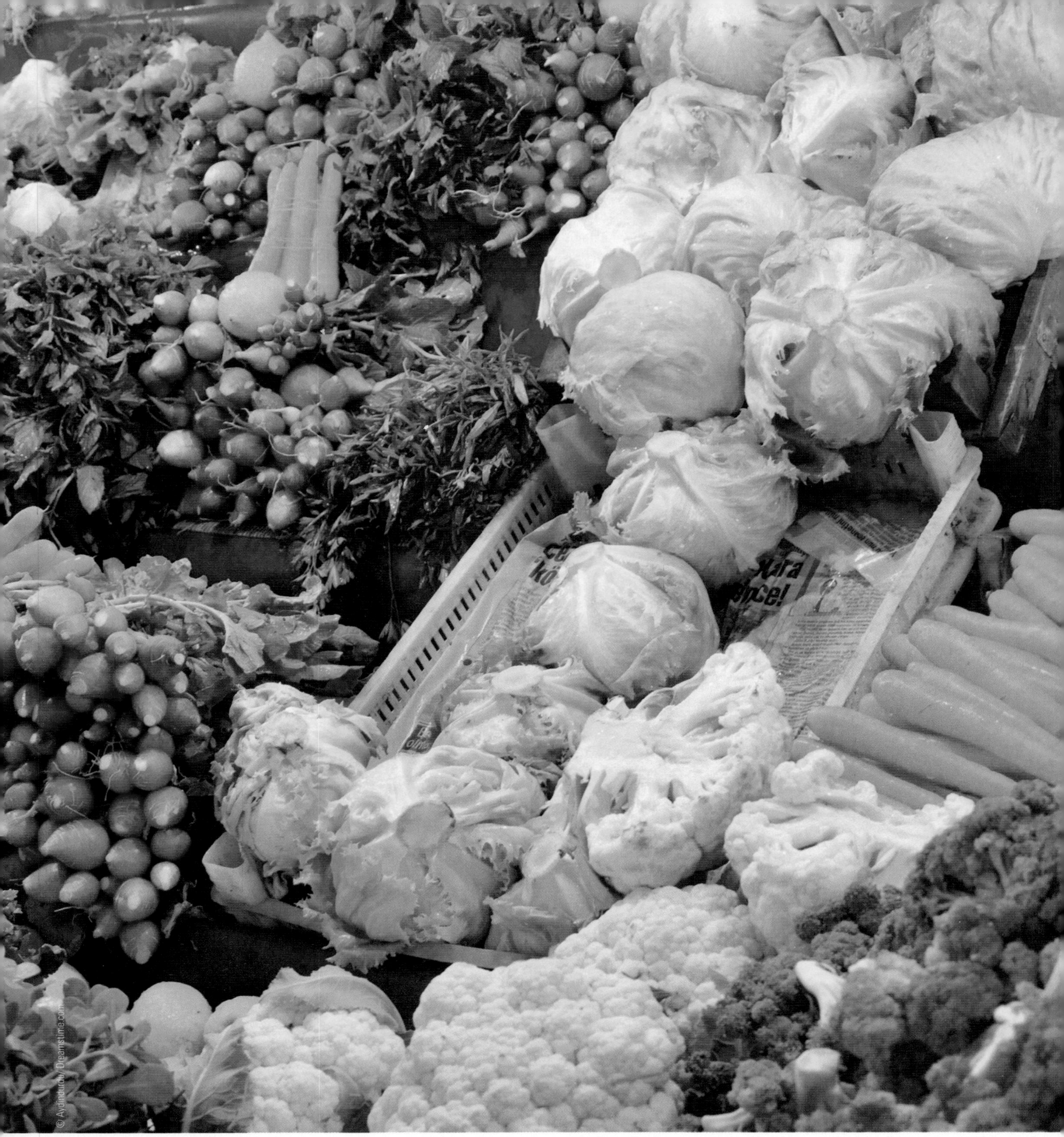

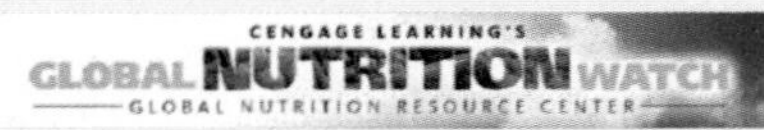

Many adults in the United States take at least one dietary supplement. Go to Nutrition Basics and Tools → Dietary Supplements to learn why older adults believe supplements are beneficial. Why might supplements benefit adults as they age?

17

Life Cycle Nutrition: Adulthood and the Later Years

LEARNING GPS

Nutrition in Your Life

Take a moment to envision yourself at age 60, 75, or even 90. Are you physically fit and healthy? Can you see yourself walking on the beach with friends or tossing a ball with children? Are you able to climb stairs and carry your own groceries? Importantly, are you enjoying life? If you're lucky, you will enjoy old age in good health. Making nutritious foods and physical activities a priority in your life can help bring rewards of continued health and enjoyment throughout life. In the Nutrition Portfolio at the end of this chapter, you can examine the nutritional health and concerns of an older adult.

Wise food choices, made throughout adulthood, can support a person's ability to meet physical, emotional, and mental challenges and to enjoy freedom from disease. Two goals motivate adults to pay attention to their diets: promoting health and slowing aging. Much of this text has focused on nutrition to support health, and Chapter 18 features prevention of chronic diseases such as cancer and heart disease. This chapter focuses on aging and the nutrition needs of older adults. As you will see, the same diet and behaviors that reduce disease risks also slow aging.

The US population is growing older. The majority is now middle-aged, and the ratio of old people to young is increasing, as Figure 17-1 (p. 552) shows. In 1900, only 1 out of 25 people was 65 or older. In 2000, 1 out of 8 had reached age 65. Projections for 2030 are 1 out of 5.

Our society uses the arbitrary age of 65 years ♦ to define the transition point between middle age and old age, but growing "old" happens day by day, with changes occurring gradually over time. Since 1950 the population of those older than 65 has almost tripled. Remarkably, the fastest-growing age group has been

♦ Commonly used age groups:
- Young old (65–74 years)
- Old old (75–84 years)
- Oldest old (≥85 years)

> FIGURE 17-1 The Aging of the US Population

In general, the percentage of older people in the population has increased over the decades whereas the percentage of younger people has decreased.

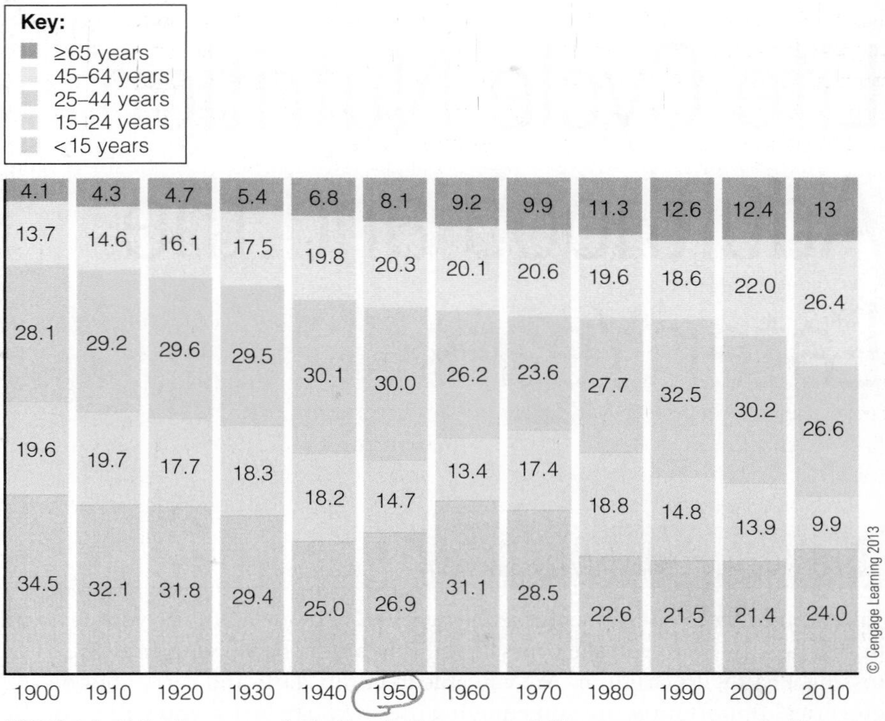

Key	1900	1910	1920	1930	1940	1950	1960	1970	1980	1990	2000	2010
≥65 years	4.1	4.3	4.7	5.4	6.8	8.1	9.2	9.9	11.3	12.6	12.4	13
45–64 years	13.7	14.6	16.1	17.5	19.8	20.3	20.1	20.6	19.6	18.6	22.0	26.4
25–44 years	28.1	29.2	29.6	29.5	30.1	30.0	26.2	23.6	27.7	32.5	30.2	26.6
15–24 years	19.6	19.7	17.7	18.3	18.2	14.7	13.4	17.4	18.8	14.8	13.9	9.9
<15 years	34.5	32.1	31.8	29.4	25.0	26.9	31.1	28.5	22.6	21.5	21.4	24.0

NOTE: Data for 2010 split age groups slightly differently. Blue represents 18–24 years and purple represents <18 years.
SOURCE: US Census Bureau.

people older than 85 years; since 1950 their numbers have increased sevenfold. The number of people in the United States age 100 or older doubled in the last decade. Similar trends are occurring in populations worldwide.

Life expectancy in the United States is 78 years: 81 years for white women and 77 years for black women, 76 years for white men and 71 years for black men.[1] All of these record highs are much higher than the average life expectancy of 47 years in 1900.[2] Women who live to 70 can expect to survive an additional 16 years, on average; men, an additional 14 years. Advances in medical science—antibiotics and other treatments—are largely responsible for almost doubling the life expectancy in the 20th century. Improved nutrition and an abundant food supply have also contributed to lengthening life expectancy. Ironically, an abundant food supply has also jeopardized the chances of lengthening life expectancy as obesity rates increase.

The **life span** has not lengthened as dramatically; human **longevity** appears to have an upper limit. The maximum potential human life span is currently about 130 years. The verifiably oldest person died in 1997 at age 122. With recent advances in medical technology and genetic knowledge, researchers may one day be able to extend the life span even further by slowing, or perhaps preventing, aging and its accompanying diseases.

life expectancy: the average number of years lived by people in a given society.

life span: the maximum number of years of life attainable by a member of a species.

longevity: long duration of life.

17.1 Nutrition and Longevity

LEARN IT Describe the role nutrition plays in longevity.

Research in the field of aging is active—and difficult. Researchers are challenged by the diversity of older adults. When older adults experience health problems, it is hard to know whether to attribute these problems to genetics, aging, or environmental factors such as nutrition. The idea that nutrition can influence the aging process is particularly appealing because people can control and change their eating habits. The questions being asked include:

- To what extent is aging inevitable, and can it be slowed through changes in lifestyle and environment?
- What role does nutrition play in the aging process, and what role can it play in slowing aging?

With respect to the first question, it seems that aging is an inevitable, natural process, programmed into the genes at conception. People can, however, slow the process within genetic limits by adopting healthy lifestyle habits such as eating nutritious foods and engaging in physical activities. In fact, an estimated 70 to 80 percent of the average person's life expectancy may depend on individual health-related behaviors; genes determine the remaining 20 to 30 percent.

With respect to the second question, good nutrition helps to maintain a healthy body and can therefore ease the aging process in many significant ways. Clearly, nutrition can improve the **quality of life** in the later years.

Observation of Older Adults

The strategies adults use to meet the two goals mentioned at the start of this chapter—promoting health and slowing aging—are actually very much the same. What to eat, how physically active to be, and other lifestyle choices greatly influence both physical health and the aging process.

Healthy Habits A person's **physiological age** reflects his or her health status and may or may not reflect the person's **chronological age.** Quite simply, some people seem younger, and others older, than their years. Six lifestyle behaviors seem to have the greatest influence on people's health and therefore on their physiological age:[3]

- Eating well-balanced meals (rich in fruits, vegetables, whole grains, poultry, fish, and low fat milk products)[4]
- Engaging in physical activity regularly
- Not smoking
- Not using alcohol, or using it in moderation
- Maintaining a healthy body weight
- Sleeping regularly and adequately

Over the years, the effects of these lifestyle choices accumulate—that is, people who follow most of these practices live longer and have fewer disabilities as they age. They are in better health, even when older in chronological age, than people who do not adopt these behaviors.[5] Even though people cannot change their birth dates, they may be able to add years to, and enhance the quality of, their lives.[6] Physical activity seems to be most influential in preventing or slowing the many changes that define a stereotypical "old" person. After all, many of the physical limitations that accompany aging occur because people become inactive, not because they become older.

Physical Activity The many remarkable benefits of regular physical activity outlined in Chapter 14 are not limited to the young. Compared with those who are inactive, older adults who are active weigh less; have greater flexibility, more endurance, better balance, and better health; and they live longer.[7] They reap additional benefits from various activities as well: aerobic activities improve cardiorespiratory endurance, blood pressure, and blood lipid concentrations; moderate-endurance activities improve the quality of sleep; and strength training improves posture

quality of life: a person's perceived physical and mental well-being.

physiological age: a person's age as estimated from her or his body's health and probable life expectancy.

chronological age: a person's age in years from his or her date of birth.

and mobility. In fact, regular physical activity is the most powerful predictor of a person's mobility in the later years. Mobility, in turn, is closely associated with longevity.[8] Physical activity also increases blood flow to the brain, thereby preserving mental ability, alleviating depression, supporting independence, and improving quality of life.[9]

Muscle mass and muscle strength tend to decline with aging, making older people vulnerable to falls and immobility. Falls are a major cause of fear, injury, disability, and even death among older adults. Many lose their independence as a result of falls. Regular physical activity tones, firms, and strengthens muscles, helping to improve balance, restore confidence, reduce the risk of falling, and lessen the risk of injury should a fall occur.

Even without a fall, older adults may become so weak that they can no longer perform life's daily tasks, such as climbing stairs, carrying packages, and opening jars. Resistance training helps older adults to maintain independence by improving muscle strength to perform these tasks. Even in frail, elderly people older than 85 years of age, strength training not only improves balance, muscle strength, and mobility, but it also increases energy expenditure and energy intake, thereby enhancing nutrient intakes. This finding highlights another reason to be physically active: a person expending energy can afford to eat more food and thus receives more nutrients. People who are committed to an ongoing fitness program can benefit from higher energy and nutrient intakes and still maintain healthy body weights.

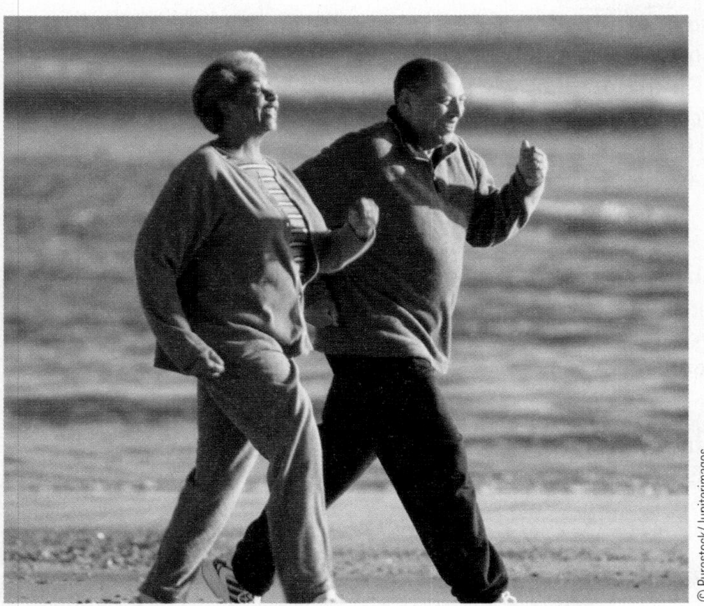
© Purestock/Jupiterimages

Regular physical activity promotes a healthy, independent lifestyle.

Ideally, physical activity should be part of each day's schedule and should be intense enough to prevent muscle atrophy and to speed the heartbeat and respiration rate. Although aging reduces both speed and endurance to some degree, older adults can still train and achieve exceptional performances. Healthy older adults who have not been active can ease into a suitable routine, becoming as physically active as their abilities allow. They can start by walking short distances until they are walking at least 10 minutes continuously, and then gradually increase their distance to a 30- to 60-minute workout at least 5 days a week. Table 17-1 provides exercise goals and guidelines for older adults.[10] Relatively few older adults meet these goals. People with medical conditions should check with a physician before beginning an exercise routine, as should sedentary men older than 40 and sedentary women older than 50 who want to participate in a vigorous program.

> **DIETARY GUIDELINES FOR AMERICANS 2010**
> Older adults should be as physically active as their abilities and conditions will allow.

Manipulation of Diet In their efforts to understand longevity, researchers have not only observed people, but they have also manipulated influencing factors, such as diet, in animals. This research has given rise to some interesting and suggestive findings.

Energy Restriction in Animals Decades of research has revealed that animals live longer and have fewer age-related diseases when their energy intakes are restricted.[11] These life-prolonging benefits become evident when the diet provides enough food to prevent malnutrition and an energy intake of about 70 percent of normal; benefits decline as the age of starting the energy restriction is delayed. Exactly how energy restriction prolongs life remains unexplained, although gene activity appears to play a key role. The genetic activity of old mice differs from that of young mice, with some genes becoming more active with age and others less active. With an energy-restricted diet, many of the genetic activities of older mice revert to those of younger mice. These "slow-aging" genetic changes are apparent in as little as 1 month on an energy-restricted, but still nutritionally adequate, diet.

TABLE 17-1 **Exercise Guidelines for Older Adults**

	Aerobic	Strength	Balance	Flexibility
Examples	© Geoff Manasse/PhotoDisc/Jupiter Images	© Polka Dot Images/Jupiterimages	© IT Stock Free/PictureQuest	© Ron Chapple/Thinkstock/Jupiter Images
Start easy and progress gradually	Be active 5 minutes on most or all days	Using 0- to 2-pound weights, do 1 set of 8–12 repetitions twice a week	Hold onto table or chair with one hand, then with one finger	Hold stretch for 10 seconds; do each stretch 3 times
Frequency	At least 5 days per week of moderate activity or at least 4 days per week of vigorous activity	At least 2 (nonconsecutive) days per week	2 to 3 days each week	At least 2 days per week; preferably on all days that aerobic or strength activities are performed
Intensity[a]	Moderate, vigorous, or combination	Moderate to high; 10 to 15 repetitions per exercise; gradually increase weights		Moderate
Duration	At least 30 minutes of moderate activity in bouts of at least 10 minutes each or at least 20 minutes of continuous vigorous activity	8 to 10 exercises involving the major muscle groups	At least 20 to 30 minutes	Stretch major muscle groups for 10–30 seconds, repeating each stretch 3 to 4 times
Cautions and comments	Stop if you are breathing so hard you can't talk or if you feel dizziness or chest pain	Breathe out as you contract and in as you relax (do not hold breath); use smooth, steady movements	Incorporate balance techniques with strength exercises as you progress	Stretch after strength and endurance exercises for 20 minutes, 3 times a week; use slow, steady movements; bend joints slightly

[a]On a 10-point scale, where sitting = 0 and maximum effort = 10, moderate intensity = 5 to 6 and vigorous intensity = 7 to 8.

NOTE: Activity recommendations are in addition to routine activities of daily living (such as getting dressed, cooking, grocery shopping) and moderate activities lasting less than 10 minutes.

SOURCE: W. J. Chodzko–Zajko and coauthors, Position stand of the American College of Sports Medicine: Exercise and physical activity for older adults, *Medicine & Science in Sports & Exercise* 41 (2009): 1510–1530; C. E. Garber and coauthors, Quantity and quality of exercise for developing and maintaining cardiorespiratory, musculoskeletal, and neuromotor fitness in apparently healthy adults: Guidance for prescribing exercise, *Medicine & Science in Sports & Exercise* 43 (2011): 1334–1359.

The consequences of energy restriction in animals include a delay in the onset, or prevention, of chronic diseases such as cancer and atherosclerosis and age-related conditions such as neuron degeneration; prolonged growth and development; and improved blood glucose, insulin sensitivity, and blood lipids. In addition, energy metabolism slows and body temperature drops—indications of a reduced rate of oxygen consumption. Oxygen consumption is lower in mice prone to obesity, and they benefit more from energy-restricted diets than other mice who tend to maintain normal weight.[12] As Highlight 11 explains, the use of oxygen during energy metabolism produces free radicals, which have been implicated in the aging process. Restricting energy intake in animals not only produces fewer free radicals, but also increases antioxidant activity and enhances DNA repair. Reducing oxidative stress may at least partially explain how restricting energy intake lengthens life expectancy.

Interestingly, longevity appears to depend on restricting energy intake and not on energy balance or body composition.[13] Genetically obese rats live longer when given a restricted diet even though their body fat is similar to that of other rats allowed to eat freely.

Energy Restriction in Human Beings Research on a variety of species—including mice, rats, monkeys, spiders, and fish—confirms the relationship between energy restriction and longevity.[14] Applying the results of animal studies to human beings is problematic, however, and conducting studies on human beings raises numerous questions—beginning with how to define *energy restriction*.[15] Does it mean eating less or just weighing less? Is it less than you want or less than the average? Does eating less have to result in weight loss? Does it matter whether weight loss results from more exercise or from less food? Or whether weight loss is intentional or unintentional? Answers await research.

Extreme starvation to extend life, like any extreme, is rarely, if ever, worth the price. Hunger is persistent when energy is restricted by 30 percent. Furthermore, using animal data to extrapolate to humans, researchers estimate that it would take 30 years of such energy-restricted dieting to increase life expectancy by less than 3 years.

Moderation, on the other hand, may be valuable. Many of the physiological responses to energy restriction seen in animals also occur in people whose intakes are *moderately* restricted. When people cut back on their usual energy intake by 10 to 20 percent, ♦ body weight, body fat, and blood pressure drop, and blood lipids and insulin response improve—favorable changes for preventing chronic diseases such as type 2 diabetes, hypertension, and heart disease.[16] Some research suggests that fasting on alternative days may provide similar benefits.[17]

♦ For perspective, a person with a usual energy intake of 2000 kcalories might cut back to 1600 to 1800 kcalories.

The reduction in oxidative damage that occurs with energy restriction in animals also occurs in people whose diets include antioxidant nutrients and phytochemicals. Diets, such as the Mediterranean diet, which include an abundance of fruits, vegetables, olive oil, and moderate amounts of red wine—with their array of phytochemicals that have antioxidant activity—support good health and long life.[18] Clearly, nutritional adequacy is essential to living a long and healthy life.

REVIEW IT Describe the role nutrition plays in longevity.
Life expectancy in the United States increased dramatically in the 20th century. Factors that enhance longevity include well-balanced meals, regular physical activity, abstinence from smoking, limited or no alcohol use, healthy body weight, and adequate sleep. Energy restriction in animals seems to lengthen their lives. Whether such dietary intervention in human beings is beneficial remains unknown. At the very least, nutrition—especially when combined with regular physical activity—can influence aging and longevity in human beings by supporting good health and preventing disease.

17.2 The Aging Process

LEARN IT Summarize how nutrition interacts with the physical, psychological, economic, and social changes involved in aging.

As people get older, each person becomes less and less like anyone else. The older people are, the more time has elapsed for such factors as nutrition, genetics, physical activity, and everyday **stress** to influence physical and psychological aging.

Stress contributes to a variety of age-related diseases. Both physical **stressors** (such as alcohol abuse, other drug abuse, smoking, pain, and illness) and psychological stressors (such as exams, divorce, moving, and the death of a loved one) elicit the body's **stress response.** The body responds to such stressors with an elaborate series of physiological steps, as the nervous and hormonal systems bring about defensive readiness in every body part. These effects favor physical action—the classic fight-or-flight response. Prolonged or severe stress can drain the body of its reserves and leave it weakened, aged, and vulnerable to illness, especially if physical action is not taken. As people age, they lose their ability to adapt to both external and internal disturbances. When disease strikes, the reduced ability to adapt makes the aging individual more vulnerable to death than a younger person. Measures to preserve health forestall disease, disability, and death.

Because the stress response is mediated by hormones, it differs between men and women. The fight-or-flight response may be more typical of men than of women. Women's reactions to stress more typically follow a pattern of "tend-and-befriend." Women *tend* by nurturing and protecting themselves and their children. These actions promote safety and reduce stress. Women *befriend* by creating and maintaining a social group that can help in the process.

Highlight 11 describes the oxidative stresses and cellular damage that occur when free radicals exceed the body's ability to defend itself. Increased free-radical activity and decreased antioxidant protection are common features of aging—and foods rich in antioxidants may help slow the aging process and improve cognition.[19] Such findings seem to suggest that the fountain of youth may actually be a

stress: any threat to a person's well-being; a demand placed on the body to adapt.

stressors: environmental elements, physical or psychological, that cause stress.

stress response: the body's response to stress, mediated by both nerves and hormones.

cornucopia of fruits and vegetables rich in antioxidants. (Return to Highlight 11 for more details on the antioxidant action of fruits and vegetables in defending against oxidative stress.)

Physiological Changes As aging progresses, inevitable changes in each of the body's organs contribute to the body's declining function. These physiological changes influence nutrition status, just as growth and development do in the earlier stages of the life cycle.

Body Weight Two-thirds of older adults in the United States are now considered overweight or obese. Chapter 8 presents the many health problems that accompany obesity and the BMI guidelines for a healthy body weight (18.5 to 24.9). These guidelines apply to all adults, regardless of age, but they may be too restrictive for older adults. The importance of body weight in defending against chronic diseases differs for older adults.[20] Being *moderately overweight* may not be harmful. For adults older than 65, health risks do not become apparent until BMI reaches at least 27—and the relationship tends to diminish with age until it disappears by age 75. Older adults who are *obese,* however, face serious medical complications and can significantly improve their quality of life with weight loss.

For some older adults, a low body weight may be more detrimental than a high one. Low body weight often reflects malnutrition and the trauma associated with a fall. Many older adults experience unintentional weight loss, in large part because of inadequate food intake. Without adequate body fat and nutrient reserves, an underweight person may be unprepared to fight against diseases.[21] For underweight people, even a slight weight loss (5 percent) increases the likelihood of disease and premature death, making every meal a life-saving event. Snacking between meals can help older adults obtain needed nutrients and energy.

> DIETARY GUIDELINES FOR AMERICANS 2010

Maintain appropriate kcalorie balance during older age. Adults aged 65 and older who are overweight are encouraged to not gain additional weight.

Body Composition In general, older people tend to lose bone and muscle and gain body fat. Many of these changes occur because some hormones that regulate appetite and metabolism become less active with age, whereas others become more active.*

Loss of muscle, known as **sarcopenia,** can be significant in the later years, and its consequences can be quite dramatic (see Figure 17-2).[22] As muscles diminish and weaken, people lose the ability to move and maintain balance—making falls likely. The limitations that accompany the loss of muscle mass and strength play a key role in the diminishing health that often accompanies aging. Optimal nutrition with sufficient protein at each meal along with regular physical activity can help maintain muscle mass and strength and minimize the changes in body composition associated with aging.[23]

> FIGURE 17-2 Sarcopenia

These cross sections of two women's thighs may appear to be about the same size from the outside, but the 20-year-old woman's thigh (left) is dense with muscle tissue. The 64-year-old woman's thigh (right) has lost muscle and gained fat, changes that may be largely preventable with good nutrition and strength-building physical activities.

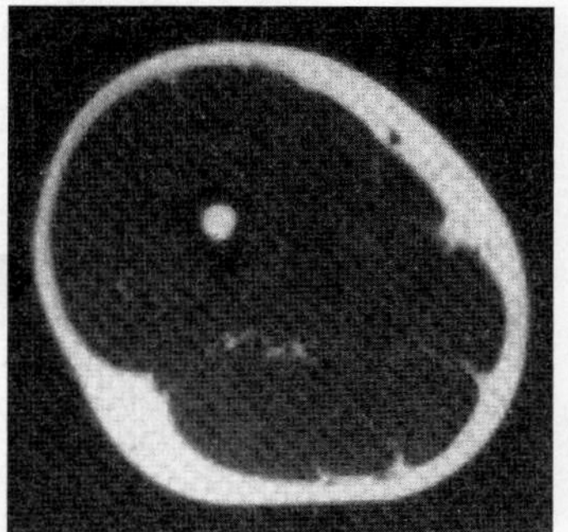

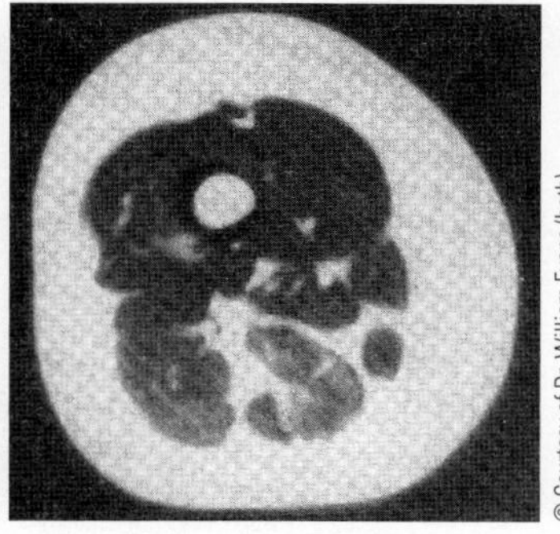

Immunity and Inflammation As people age, the immune system loses function. As they become ill, the immune system becomes overstimulated. The combination of an inefficient and overactive response in aging—known as "inflammaging"—results in a chronic inflammation that accompanies frailty, illness, and death.

*Causes of diminished appetite in older adults include increased cholecystokinin, leptin, and cytokines and decreased ghrelin and testosterone. Additional examples of hormones that change with age include growth hormone and androgens, which decline with advancing age, thus contributing to the decrease in lean body mass, and prolactin, which increases with age, helping to maintain body fat. Insulin sensitivity also diminishes as people grow older, most likely because of increases in body fat and decreases in physical activity.

sarcopenia (SAR-koh-PEE-nee-ah): loss of skeletal muscle mass, strength, and quality.
- **sarco** = flesh
- **penia** = loss or lack

Most diseases common in older adults—such as atherosclerosis, Alzheimer's disease, obesity, and rheumatoid arthritis—are different in obvious ways, but they all reflect an underlying inflammatory process. Because of this association with diseases, inflammation is often perceived as a harmful process, yet it is critical in supporting health as the immune system destroys invading organisms and repairs damaged tissues. Thus inflammation presents a challenge to identify factors that will both protect the beneficial effects and limit the harmful consequences.

In addition to aging and diseases, the immune system is compromised by nutrient deficiencies. Thus the combination of age, illness, and poor nutrition makes older people particularly vulnerable to infectious diseases. Adding insult to injury, antibiotics often are not effective against infections in people with compromised immune systems. Consequently, infectious diseases are a major cause of death in older adults. Older adults may improve their immune system responses with regular physical activity.

GI Tract In the GI tract, the intestinal wall loses strength and elasticity with age, and GI hormone secretions change. All of these actions slow motility. Constipation is much more common in the elderly than in the young. Changes in GI hormone secretions also diminish appetite, leading to decreased energy intake and unintentional weight loss.

Atrophic gastritis, a condition that affects almost one-third of those older than 60, is characterized ♦ by an inflamed stomach, bacterial overgrowth, and a lack of hydrochloric acid and intrinsic factor. All of these factors can impair the digestion and absorption of nutrients, most notably, vitamin B_{12}, but also biotin, folate, calcium, iron, and zinc.

♦ Consequences of atrophic gastritis:
- Inflamed stomach
- Increased bacterial growth
- Reduced hydrochloric acid
- Reduced intrinsic factor
- Increased risk of nutrient deficiencies, notably of vitamin B_{12}

Difficulty swallowing, medically known as **dysphagia**, occurs in all age groups, but especially in the elderly. Being unable to swallow a mouthful of food can be scary, painful, and dangerous. Even swallowing liquids can be a problem for some people. Consequently, the person may eat less food and drink fewer beverages, resulting in weight loss, malnutrition, and dehydration. Dietary intervention for dysphagia is highly individualized based on the person's abilities and tolerances. The diet typically provides moist, soft-textured, tender-cooked, or pureed foods and thickened liquids.

Tooth Loss Regular dental care over a lifetime protects against tooth loss and gum disease, which are common in old age. These conditions make chewing difficult or painful. Dentures, even when they fit properly, are less effective than natural teeth, and inefficient chewing can cause choking. Chewing crushes foods into smaller pieces in preparation for digestion. Inefficient chewing leaves larger pieces of food moving from the stomach into the small intestine, thus limiting enzyme accessibility.

People with tooth loss, gum disease, and ill-fitting dentures tend to limit their food selections to soft foods.* If foods such as corn on the cob and apples are replaced by creamed corn and applesauce, then nutrition status may not be greatly affected. However, when food groups are eliminated and variety is limited, poor nutrition follows. People without teeth typically eat fewer fruits and vegetables and have less variety in their diets.[24] Consequently, they have low intakes of fiber and vitamins, which exacerbates their dental and overall health problems. To determine whether a visit to the dentist is needed, an older adult can check the conditions listed in the margin. ♦

♦ Conditions requiring dental care:
Dry mouth
Eating difficulty
No dental care within 2 years
Tooth or mouth pain
Altered food selections
Lesions, sores, or lumps in mouth

Sensory Losses and Other Physical Problems Sensory losses and other physical problems can also interfere with an older person's ability to obtain adequate nourishment. Failing eyesight, for example, can make driving to the grocery store impossible and shopping for food a frustrating experience. It may become so difficult to read food labels and count money that the person doesn't buy needed foods. Carrying bags of groceries may be an unmanageable task. Similarly, a person with limited mobility may find cooking and cleaning up too hard to do. Not

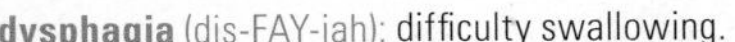
dysphagia (dis-FAY-jah): difficulty swallowing.

*The medical term for lack of teeth is *edentulous* (ee-DENT-you-lus).

too surprisingly, the prevalence of undernutrition is high among those who are home-bound.

Sensory losses can also interfere with a person's ability or willingness to eat. Taste and smell sensitivities tend to diminish with age and may make eating less enjoyable. If a person eats less, then weight loss and nutrient deficiencies may follow. Loss of vision and hearing may contribute to social isolation, and eating alone may lead to poor intake.

Other Changes In addition to the physiological changes that accompany aging, adults change in many other ways that influence their nutrition status.[25] Psychological, economic, and social factors play major roles in a person's ability and willingness to eat.

Psychological Changes Late-life depression is associated with an increased risk of mortality.[26] Although not an inevitable component of aging, depression is common among older adults, especially among those in poor health and those living in long-term nursing homes.[27] Relatively few receive adequate treatment from either antidepressant medication or mental health counseling.[28]

Depressed people, even those without physical disabilities, lose their ability to perform simple physical tasks. They frequently lose their appetite and the motivation to cook or even to eat. An overwhelming sense of grief and sadness at the death of a spouse, friend, or family member may leave a person, especially an elderly person, feeling powerless to overcome depression. When a person is suffering the heartache and loneliness of bereavement, cooking meals may not seem worthwhile. The support and companionship of family and friends, especially at mealtimes, can help overcome depression and enhance appetite.

Several nutrient interventions to relieve depression have been studied, but evidence of effectiveness is inconclusive.[29] A balanced, healthy diet may be the best nutritional approach to reducing symptoms of depression and improving quality of life.[30]

Economic Changes Overall, older adults today have higher incomes than their cohorts of previous generations. Still, 9 percent of the people older than age 65 live in poverty.[31] Factors such as living arrangements and income make significant differences in the food choices, eating habits, and nutrition status of older adults, especially those older than age 80. People of low socioeconomic means are likely to have inadequate food and nutrient intakes. Only about one-third of eligible seniors participate in the Supplemental Nutrition Assistance Program (SNAP).

© MBI / Alamy

Shared meals can brighten the day and enhance the appetite.

Social Changes Malnutrition is most likely to occur among those living alone, especially men; those with the least education; those living in federally funded housing (an indicator of low income); and those who have recently experienced a change in lifestyle. Adults who live alone do not necessarily make poor food choices, but they often consume too little food. Loneliness is directly related to nutritional inadequacies, especially of energy intake.

> **REVIEW IT** **Summarize how nutrition interacts with the physical, psychological, economic, and social changes involved in aging.**
> Many changes that accompany aging can impair nutrition status. Among physiological changes, hormone activity alters body composition, immune system changes raise the risk of infections, atrophic gastritis interferes with digestion and absorption, and tooth loss limits food choices. Psychological changes such as depression, economic changes such as loss of income, and social changes such as loneliness contribute to poor food intake.

17.3 Energy and Nutrient Needs of Older Adults

LEARN IT Explain why the needs for some nutrients increase or decrease during aging.

Knowledge about the nutrient needs and nutrition status of older adults has grown considerably in recent years. The Dietary Reference Intakes (DRI) cluster people older than 50 into two age categories—one group of 51 to 70 years and one of 71 and older.

Setting standards for older people is difficult because individual differences become more pronounced as people grow older. People start out with different genetic predispositions and ways of handling nutrients, and the effects of these differences become magnified with years of unique dietary habits. For example, one person may tend to omit fruits and vegetables from his diet, and by the time he is old, he may have a set of nutrition problems associated with a lack of fiber and antioxidants. Another person may have omitted milk and milk products all her life—her nutrition problems may be related to a lack of calcium. Also, as people age, they suffer different chronic diseases and take various medicines—both of which will affect nutrient needs. For all of these reasons, researchers have difficulty even defining *healthy aging,* a prerequisite to developing recommendations to meet the "needs of practically all healthy persons." The following discussion gives special attention to the nutrients of greatest concern.

© Fancy / Alamy

To ensure adequate hydration, keep a glass of water next to you at home, drink from water fountains whenever you walk by, and put a bottle of water in your car.

Water Despite real fluid needs, many older people do not seem to feel thirsty or notice mouth dryness. Many nursing home employees say it is hard to persuade their elderly clients to drink enough water and fruit juices. Older adults may find it difficult and bothersome to get a drink or to get to a bathroom. Those who have lost bladder control may be afraid to drink too much water.

Dehydration is a risk for older adults. Total body water decreases as people age, so even mild stresses such as fever or hot weather can precipitate rapid dehydration in older adults. Dehydrated older adults seem to be more susceptible to urinary tract infections, pneumonia, **pressure ulcers,** and confusion and disorientation. To prevent dehydration, older adults need to drink *at least* six glasses of water or other beverages every day. ♦ Emphasizing foods with high-water content, such as melons and soups, can also be helpful.

♦ Beverage recommendation for adults 51+ yr:
- Men: 13 c/day
- Women: 9 c/day

Energy and Energy Nutrients On average, energy needs decline an estimated 5 percent per decade. One reason is that people usually reduce their physical activity as they age, although they need not do so. Another reason is that basal metabolic rate declines 1 to 2 percent per decade in part because lean body mass and thyroid hormones diminish.

The lower energy expenditure of older adults means that they need to eat less food to maintain their weights. Accordingly, the estimated energy requirements

pressure ulcers: damage to the skin and underlying tissues as a result of compression and poor circulation; commonly seen in people who are bedridden or chair-bound.

for adults decrease steadily after age 19. The accompanying "How To" feature explains how to estimate energy requirements for older adults.

Older adults need fewer kcalories as they age, but their nutrient needs remain high. For this reason, it is most important that they select mostly nutrient-dense foods. There is little leeway for added sugars, solid fats, or alcohol; such nutrient-poor selections can easily lead to weight gain and malnutrition. The USDA Food Patterns (p. 44) offer a dietary framework for adults of all ages.

© Corbis Super RF / Alamy

Growing old can be enjoyable for people who take care of their health and live each day fully.

Protein Because energy needs decrease, protein must be obtained from low-kcalorie sources of high-quality protein, such as lean meats, poultry, fish, and eggs; fat-free and low-fat milk products; and legumes. Protein is especially important for the elderly to support a healthy immune system, prevent muscle wasting, and optimize bone mass.[32] Maintaining muscles helps to support protein metabolism and immune function.

Underweight or malnourished older adults need protein- and energy-dense snacks such as hard-boiled eggs, tuna salad, peanut butter on wheat toast, and hearty soups. Drinking liquid nutritional formulas between meals can also boost energy and nutrient intakes. Importantly, the diet should provide enjoyment as well as nutrients.[33]

Carbohydrate and Fiber As always, abundant carbohydrate is needed to protect protein from being used as an energy source. Carbohydrate-rich foods such as legumes, vegetables, whole grains, and fruits are also rich in fiber and essential vitamins and minerals. Average fiber intakes among older adults are lower than current recommendations (14 grams per 1000 kcalories). Eating high-fiber foods and drinking water can alleviate constipation—a condition common among older adults, especially nursing home residents. (Physical inactivity and medications also contribute to the high incidence of constipation.)

Fat As is true for people of all ages, fat intake needs to be moderate in the diets of most older adults—enough to enhance flavors and provide valuable nutrients, but not so much as to raise the risks of atherosclerosis and other degenerative diseases. This recommendation should not be taken too far; limiting fat too severely may lead to nutrient deficiencies and weight loss—two problems that carry greater health risks in the elderly than being overweight.

>How To

Estimate Energy Requirements for Older Adults

The "How To" feature on p. 240 describes how to estimate the energy requirements for adults using an equation that accounts for age, physical activity, weight, and height. Alternatively, energy requirements for older adults can be "guesstimated" by using the values listed in the tables in Appendix F for adults 30 years of age and subtracting 7 kcalories for women and 10 kcalories for men per day for each year older than 30.

For example, Table F-4 lists 2556 kcalories per day for a 30-year-old woman who is 5 feet 5 inches tall, weighs 150 pounds, and has a low activity level. To estimate the energy requirements of a similar 50-year-old woman, subtract 7 kcalories per day for each year older than 30:

$$50 - 30 = 20 \text{ yr}$$
$$20 \text{ yr} \times 7 \text{ kcal/day} = 140 \text{ kcal/day}$$
$$2556 \text{ kcal/day (at age 30)} - 140 \text{ kcal/day} = 2416 \text{ kcal/day (at age 50)}$$

Similarly, using Table F-5 to estimate the energy requirements of a sedentary 65-year-old man who is 5 feet 11 inches tall and weighs 250 pounds, subtract 10 kcalories per day for each year older than 30:

$$65 - 30 = 35 \text{ yr}$$
$$35 \text{ yr} \times 10 \text{ kcal/day} = 350 \text{ kcal/day}$$
$$3088 \text{ kcal/day (at age 30)} - 350 \text{ kcal/day} = 2738 \text{ kcal/day (at age 65)}$$

Adults between the ages of 19 and 30 can also use the values listed in the tables in Appendix F by adding 7 kcalories for women and 10 kcalories for men per day for each year younger than 30.

TRY IT Use Appendix F to calculate your energy requirements at your current age and at your age 20 years from now.

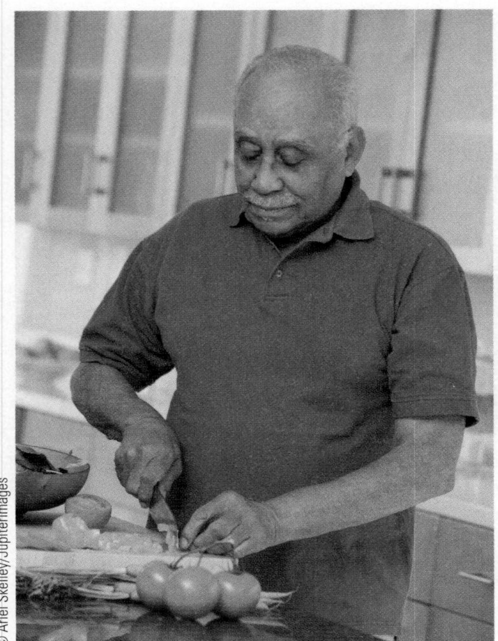

© Ariel Skelley/Jupiterimages

Taking time to nourish your body well is a gift you give yourself.

Vitamins and Minerals Most people can achieve adequate vitamin and mineral intakes simply by including foods from all food groups in their diets, but older adults often omit fruits and vegetables. Similarly, few older adults consume the recommended amounts of milk or milk products.

Vitamin B_{12} An estimated 10 to 30 percent of adults older than 50 have **atrophic gastritis;** as Chapter 10 explains, people with atrophic gastritis are particularly vulnerable to vitamin B_{12} deficiency. The bacterial overgrowth that accompanies this condition uses up the vitamin, and without hydrochloric acid and intrinsic factor, digestion and absorption of vitamin B_{12} are inefficient. Given the poor cognition, anemia, and devastating neurological effects associated with a vitamin B_{12} deficiency, an adequate intake is imperative.[34] The RDA for older adults is the same as for younger adults, but with the added suggestion to obtain most of a day's intake from vitamin B_{12}–fortified foods and supplements.[35] The bioavailability of vitamin B_{12} from these sources is better than from foods.

> DIETARY GUIDELINES FOR AMERICANS 2010
Individuals aged 50 years and older should consume foods fortified with vitamin B_{12}, such as fortified cereals, or take dietary supplements.

Vitamin D Vitamin D deficiency is a problem for many older adults. Vitamin D–fortified milk is the most reliable source of vitamin D, but many older adults drink little or no milk. Further compromising the vitamin D status of many older people, especially those in nursing homes, is their limited exposure to sunlight. Finally, aging reduces the skin's capacity to make vitamin D and the kidneys' ability to convert it to its active form. Not only are older adults not getting enough vitamin D, but they may actually need more to improve both muscle and bone strength. To prevent bone loss and to maintain vitamin D status, especially in those who engage in minimal outdoor activity, adults 51 to 70 years old need 15 micrograms daily, and those 71 and older need 20 micrograms.[36]

> DIETARY GUIDELINES FOR AMERICANS 2010
Choose fortified foods and dietary supplements to provide more vitamin D, a nutrient of concern.

Folate As is true of vitamin B_{12}, folate intakes of older adults typically fall short of recommendations. The elderly are also more likely to have medical conditions or to take medications that can compromise folate status (see Highlight 17).

Calcium Both Chapter 12 and Highlight 12 emphasize the importance of abundant dietary calcium throughout life, especially for women after menopause, to protect against osteoporosis. The DRI Committee recommends 1200 milligrams of calcium daily for women older than 50 and men older than 70, but the calcium intakes of older people in the United States are well below recommendations. Some older adults avoid milk and milk products because they dislike these foods or associate them with stomach discomfort. Simple solutions include using calcium-fortified juices, adding powdered milk to recipes, and taking supplements. Chapter 12 offers many other strategies for including nonmilk sources of calcium for those who do not drink milk.

Iron The iron needs of men remain unchanged throughout adulthood. For women, iron needs decrease substantially at menopause when blood loss through menstruation ceases. Consequently, iron-deficiency anemia is less common in older adults than in younger people. In fact, elevated iron stores are more likely than deficiency in older people, especially those who take iron supplements, eat red meat regularly, and include vitamin C–rich fruits in their daily diet.

atrophic (a-TRO-fik) **gastritis** (gas-TRY-tis): chronic inflammation of the stomach accompanied by a diminished size and functioning of the mucous membrane and glands. This condition is also characterized by inadequate hydrochloric acid and intrinsic factor—two substances needed for vitamin B_{12} absorption.

Nevertheless, iron deficiency may develop in older adults, especially when their food energy intakes are low. Aside from diet, two other factors may lead to iron deficiency in older people: chronic blood loss from diseases and medicines and poor iron absorption due to reduced stomach acid secretion and antacid use. Iron deficiency impairs immunity and leaves older adults vulnerable to infectious diseases. Anyone concerned with older people's nutrition should keep these possibilities in mind.

Zinc Zinc intake is commonly low in older people. Zinc deficiency can depress the appetite and blunt the sense of taste, thereby reducing food intake and worsening zinc status. Many medications that older adults commonly use can impair zinc absorption or enhance its excretion and thus lead to deficiency.

Nutrient Supplements People judge for themselves how to manage their nutrition, and more than half of older adults turn to dietary supplements.[37] When recommended by a physician or registered dietitian, vitamin D and calcium supplements for osteoporosis or vitamin B_{12} for pernicious anemia may be beneficial. Many health-care professionals recommend a daily multivitamin-mineral supplement that provides 100 percent or less of the Daily Value for the listed nutrients. They reason that such a supplement is more likely to be beneficial than to cause harm. Supplement use may help older adults obtain enough of some nutrients, but it may also lead to excessive intakes of others.[38]

People with small energy allowances would do well to become more active so they can afford to eat more food. Food is the best source of nutrients for everybody. Supplements are just that—supplements to foods, not substitutes for them. For anyone who is motivated to obtain the best possible health, it is never too late to learn to eat well, drink water, exercise regularly, and adopt other lifestyle habits such as quitting smoking and moderating alcohol use.

REVIEW IT Explain why the needs for some nutrients increase or decrease during aging. Table 17-2 provides a summary of the nutrient concerns of aging. Although some nutrients need special attention in the diet, supplements are not routinely recommended. The ever-growing number of older people creates an urgent need to learn more about how their nutrient requirements differ from those of others and how such knowledge can enhance their health.

TABLE 17-2 Nutrient Concerns of Aging

Nutrient	Effect of Aging	Comments
Water	Lack of thirst and decreased total body water make dehydration likely.	Mild dehydration is a common cause of confusion. Difficulty obtaining water or getting to the bathroom may compound the problem.
Energy	Need decreases as muscle mass decreases (sarcopenia).	Physical activity moderates the decline.
Fiber	Likelihood of constipation increases with low intakes and changes in the GI tract.	Inadequate water intakes and lack of physical activity, along with some medications, compound the problem.
Protein	Needs may stay the same or increase slightly.	Low-fat, high-fiber legumes and grains meet both protein and other nutrient needs.
Vitamin B_{12}	Atrophic gastritis is common.	Deficiency causes neurological damages; supplements may be needed.
Vitamin D	Increased likelihood of inadequate intake; skin synthesis declines.	Daily sunlight exposure in moderation or supplements may be beneficial.
Calcium	Intakes may be low; osteoporosis is common.	Stomach discomfort commonly limits milk intake; calcium substitutes or supplements may be needed.
Iron	In women, status improves after menopause; deficiencies are linked to chronic blood losses and low stomach acid output.	Adequate stomach acid is required for absorption; antacid or other medicine use may aggravate iron deficiency; vitamin C and meat increase absorption.
Zinc	Intakes are often inadequate and absorption may be poor, but needs may also increase.	Medications interfere with absorption; deficiency may depress appetite and sense of taste.

17.4 Nutrition-Related Concerns of Older Adults

LEARN IT Identify how nutrition might contribute to, or prevent, the development of age-related problems associated with vision, arthritis, the brain, and alcohol use.

Nutrition may play a greater role than has been realized in preventing many changes once thought to be inevitable consequences of growing older. The following discussions of vision, arthritis, the aging brain, and alcohol use show how nutrition interacts with these conditions.

Vision

One key aspect of healthy aging is maintaining good vision. Age-related eye diseases that impair vision, such as cataract and macular degeneration, correlate with poor survival that cannot be explained by other risk factors. Following a healthy diet as described by the *Dietary Guidelines for Americans* is one way to protect against these age-related vision problems. Foods containing phytochemicals that act as antioxidants or anti-inflammatory agents may be especially beneficial.[39]

Cataracts **Cataracts** are age-related clouding of the lenses of the eyes that impairs vision. If not surgically removed, they ultimately lead to blindness. Cataracts may develop as a result of ultraviolet light exposure, oxidative stress, injury, viral infections, toxic substances, and genetic disorders. Most cataracts, however, are vaguely called senile cataracts—meaning "caused by aging." In the United States, more than half of all adults 65 and older have a cataract.

Oxidative stress appears to play a significant role in the development of cataracts, and the antioxidant nutrients may help minimize the damage. Studies have reported an inverse relationship between cataracts and dietary intakes of vitamin C, vitamin E, and carotenoids; taking supplements or eating fruits and vegetables rich in these antioxidant nutrients seems to slow the progression or reduce the risk of developing cataracts.[40] A word of caution, however: vitamin C supplements in high doses (1000 milligrams) and long duration (several years) may *increase* the risk of cataracts.[41]

One other diet-related factor may play a role in the development of cataracts—obesity. Obesity appears to be associated with cataracts, but its role has not been identified. Risk factors that typically accompany overweight, such as inactivity, diabetes, or hypertension, do not explain the association.[42]

Macular Degeneration The leading cause of visual loss among older people is age-related **macular degeneration,** a deterioration of the macular region of the retina.[43] As with cataracts, risk factors for age-related macular degeneration include oxidative stress from sunlight.[44] Preventive factors may include supplements of the omega-3 fatty acid DHA, some B vitamins (folate, vitamin B_6, and vitamin B_{12}), and the carotenoids lutein and zeaxanthin.[45]

Arthritis

Almost 50 million people in the United States have some form of **arthritis.**[46] As the population ages, it is expected that the prevalence will increase to 70 million by 2030.[47] Arthritis pain and fear of further damage limit physical activity.[48]

Osteoarthritis The most common type of arthritis that disables older people is **osteoarthritis,** a painful deterioration of the cartilage in the joints. During movement, the ends of bones are normally protected from wear by cartilage and by small sacs of fluid that act as a lubricant. With age, the cartilage sometimes disintegrates, and the joints become malformed and painful to move.

One known connection between osteoarthritis ♦ and nutrition is overweight. Weight loss may relieve some of the pain for overweight persons with osteoarthritis, partly because the joints affected are often weight-bearing joints that are stressed and irritated by having to carry excess pounds. Interestingly, though, weight loss often relieves much of the pain of arthritis in the hands as well, even

♦ Risk factors for osteoarthritis:
- Age
- Smoking
- High BMI at age 40
- Lack of hormone therapy (in women)

cataracts (KAT-ah-rakts): clouding of the eye lenses that impairs vision and can lead to blindness.

macular (MACK-you-lar) **degeneration:** deterioration of the macular area of the eye that can lead to loss of central vision and eventual blindness. The *macula* is a small, oval, yellowish region in the center of the retina that provides the sharp, straight-ahead vision so critical to reading and driving.

arthritis: inflammation of a joint, usually accompanied by pain, swelling, and structural changes.

osteoarthritis: a painful, degenerative disease of the joints that occurs when the cartilage in a joint deteriorates; joint structure is damaged, with loss of function; also called *degenerative arthritis.*

though they are not weight-bearing joints. Importantly, walking and other weight-bearing exercises do not worsen arthritis. In fact, low-impact aerobic activity and strength training offer improvements in physical performance and pain relief, especially when accompanied by even modest weight loss.

Rheumatoid Arthritis Another type of arthritis known as **rheumatoid arthritis** has possible links to diet through the immune system. In rheumatoid arthritis, the immune system mistakenly attacks the bone coverings as if they were made of foreign tissue. In some individuals, certain foods, notably a Mediterranean-type diet of fish, vegetables, and olive oil, may moderate the inflammatory response and provide some relief.

The omega-3 fatty acids commonly found in fatty fish reduce joint tenderness and improve mobility in some people with rheumatoid arthritis. The same diet recommended for heart health—one low in saturated fat from meats and milk products and high in omega-3 fats from fish—helps prevent or reduce the inflammation in the joints that makes arthritis so painful.

Another possible link between nutrition and rheumatoid arthritis involves the oxidative damage to the membranes within joints that causes inflammation and swelling. The antioxidant vitamins C and E and the carotenoids defend against oxidation, and increased intakes of these nutrients may help prevent or relieve the pain of rheumatoid arthritis.

Gout Another form of arthritis, which most commonly affects men, is **gout**, a condition characterized by deposits of uric acid crystals in the joints. Uric acid derives from the breakdown of **purines**, primarily from those made by the body but also from those found in foods. Recommendations to lower uric acid levels and the risk of gout include limiting alcohol and excessive amounts of meat, seafood, and sugar-sweetened beverages.[49]

Treatment Treatment for arthritis—dietary or otherwise—may help relieve discomfort and improve mobility, but it does not cure the condition. Traditional medical intervention for arthritis includes medication and surgery. Alternative therapies to treat arthritis abound, but none have proved safe and effective in scientific studies. Popular supplements—glucosamine, chondroitin, or a combination—may relieve pain and improve mobility as well as over-the-counter pain relievers, but mixed reports from studies emphasize the need for additional research. Drugs and supplements used to relieve arthritis can impose nutrition risks; many affect appetite and alter the body's use of nutrients, as Highlight 17 explains.

The Aging Brain

The brain, like all of the body's organs, responds to both genetic and environmental factors ♦ that can enhance or diminish its amazing capacities. One of the challenges researchers face when studying the human brain is to distinguish among normal age-related physiological changes, changes caused by diseases, and changes that result from cumulative, environmental factors such as diet.

♦ Factors that protect brain function:
- Physical activities
- Intellectual challenges
- Social interactions
- Balanced diet rich in antioxidants

The brain normally changes in some characteristic ways as it ages. For one thing, its blood supply decreases. For another, the number of **neurons**, the brain cells that specialize in transmitting information, diminishes as people age. When the number of nerve cells in one part of the cerebral cortex diminishes, hearing and speech are affected. Losses of neurons in other parts of the cortex can impair memory and cognitive function. When the number of neurons in the cerebellum diminishes, balance and posture are affected. Losses of neurons in other parts of the brain affect still other functions. Some of the cognitive loss and forgetfulness generally attributed to aging may be due in part to environmental, and therefore controllable, factors—including nutrient deficiencies.

Nutrient Deficiencies and Brain Function Nutrients influence the development and activities of the brain. The ability of neurons to synthesize specific neurotransmitters depends in part on the availability of precursor nutrients that are obtained from the diet. The neurotransmitter serotonin, for example, derives

rheumatoid (ROO-ma-toyd) **arthritis:** a disease of the immune system involving painful inflammation of the joints and related structures.

gout (GOWT): a common form of arthritis characterized by deposits of uric acid crystals in the joints.

purines: compounds of nitrogen-containing bases such as adenine, guanine, and caffeine. Purines that originate from the body are *endogenous* and those that derive from foods are *exogenous*.

neurons: nerve cells; the structural and functional units of the nervous system. Neurons initiate and conduct nerve impulse transmissions.

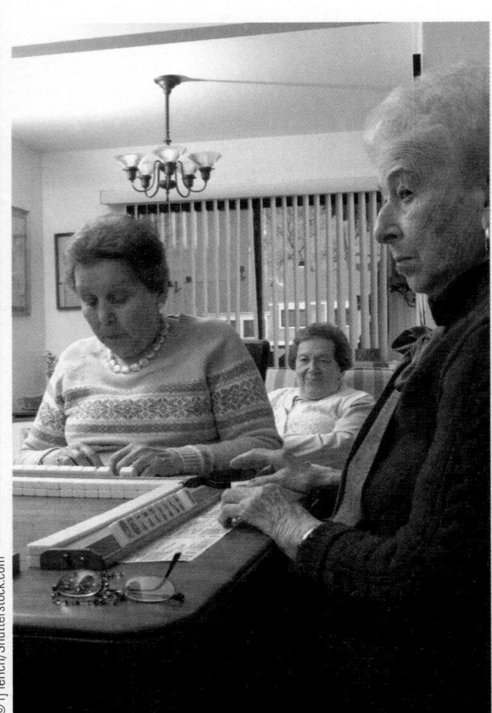

Both foods and mental challenges nourish the brain.

from the amino acid tryptophan. To function properly, the enzymes involved in neurotransmitter synthesis require vitamins and minerals. The essential fatty acid DHA counteracts the cognitive decline commonly seen in elderly adults.[50] Thus nutrient deficiencies may contribute to the loss of memory and cognition that some older adults experience. Such losses may be preventable or at least diminished or delayed through diet and exercise. In some instances, the degree of cognitive loss is extensive. Some **senile dementia** may be attributable to a specific disorder such as a brain tumor or Alzheimer's disease.

Alzheimer's Disease Much attention has focused on the *abnormal* deterioration of the brain called **Alzheimer's disease,** which affects one out of eight US adults older than age 65 and almost half of adults aged 85 and older.[51] Nerve cells in the brain die, and communication between the cells breaks down. Diagnosis of Alzheimer's disease depends on its characteristic symptoms: the victim gradually loses memory and reasoning, the ability to communicate, physical capabilities, and eventually life itself.[52] Table 17-3 compares the signs of Alzheimer's disease with typical age-related changes.

The primary risk factor for Alzheimer's disease is age, but the exact cause remains uknown.[53] Clearly, genetic factors are involved.[54] Free radicals and oxidative stress also seem to be involved. Nerve cells in the brains of people with Alzheimer's disease show evidence of free-radical attack—damage to DNA, cell membranes, and proteins.[55] They also show evidence of the minerals that trigger free-radical attacks—iron, copper, zinc, and aluminum. Increasing evidence also suggests that overweight and obesity in middle age are associated with dementia in general, and with Alzheimer's disease in particular.[56]

In Alzheimer's disease, the brain develops **senile plaques** and **neurofibrillary tangles** (see Figure 17-3). Senile plaques are clumps of a protein fragment called beta-amyloid, whereas neurofibrillary tangles are snarls of the fibers that extend from the nerve cells. Both seem to occur in response to oxidative stress.[57]

TABLE 17-3 Signs of Alzheimer's and Typical Age-Related Changes Compared

Signs of Alzheimer's	Typical Age-Related Changes
Memory loss that disrupts daily life such as asking for the same information repeatedly or asking others to handle tasks of daily living.	Forgetting a name or missing an appointment
Challenges in planning or solving problems such as following a recipe or paying monthly bills	Missing a monthly payment or making an error when balancing the checkbook
Difficulty completing familiar tasks at home such as using the microwave, at work such as preparing a report, or at leisure such as playing a game	Needing help recording a television program
Confusion with time or place including current season and location	Not knowing today's date
Trouble understanding visual images and spatial relationships such as judging distances and recognizing self in a mirror	Experiencing visual changes due to cataracts
New problems with words in speaking or writing such as knowing the name of a common object	Being unable to find the right word to use
Misplacing things and losing the ability to retrace steps such as putting the milk in the closet and having no idea when or where the milk was last seen	Misplacing a pair of glasses or the car keys
Decreased or poor judgment such as giving large sums of money to strangers	Making a bad decision on occasion
Withdrawal from work projects or social activities	Feeling too tired to participate in work, family, or social activities
Changes in mood and personality such as confusion, suspicion, depression, and anxiety especially when in unfamiliar places or with unfamiliar people	Becoming irritable when routines are disrupted

SOURCE: Adapted from Alzheimer's Association, www.alz.org/alzheimers_disease_10_signs_of_alzheimers.asp.

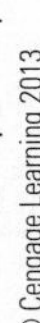

senile dementia: the loss of brain function beyond the normal loss of physical adeptness and memory that occurs with aging.

Alzheimer's (AHLZ-high-merz) **disease:** a degenerative disease of the brain involving memory loss and major structural changes in neuron networks; also known as *senile dementia of the Alzheimer's type (SDAT), primary degenerative dementia of senile onset,* or *chronic brain syndrome.*

senile plaques: clumps of the protein fragment beta-amyloid on the nerve cells, commonly found in the brains of people with Alzheimer's dementia.

neurofibrillary tangles: snarls of the threadlike strands that extend from the nerve cells, commonly found in the brains of people with Alzheimer's dementia.

> **FIGURE 17.3 Alzheimer's and Healthy Brains Compared**

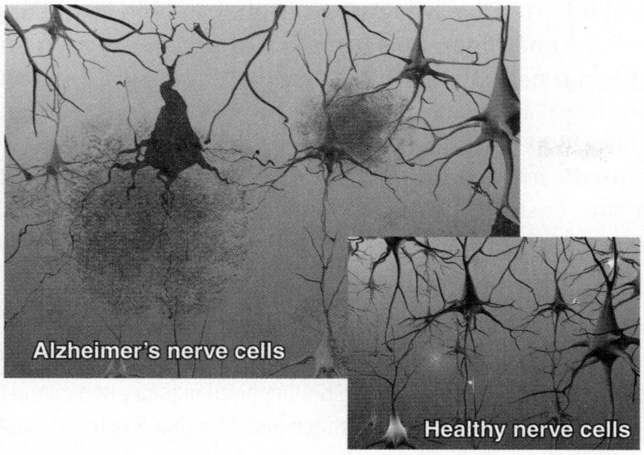

Alzheimer's nerve cells

Healthy nerve cells

© 2012 Alzheimer's Association; illustration by Stacy Janis

Plaques—clumps of beta-amyloid protein pieces—block cell-to-cell synapse signals. Tangles—twisted strands of protein—destroy the cell transport system. As plaques and tangles block essential nutrients from reaching the nerve cells, they eventually die.

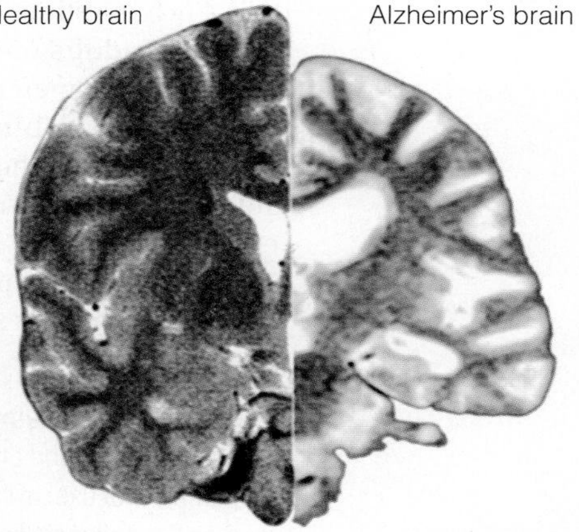

Healthy brain

Alzheimer's brain

Medical Body Scans/Jessica Wilson/Photo Researchers, Inc.

As nerve cells die, the brain shrinks and loses its ability to think, plan, remember, and form new memories. The fluid-filled spaces within the brain grow larger.

© Cengage Learning 2013

SOURCE: Alzheimer's Association, www.alz.org/research/science/alzheimers_brain_tour.asp.

Researchers question whether these characteristics are the cause or the result of Alzheimer's disease.[58] In fact, scientists are unsure whether these plaques and tangles are causing the damage, serving as markers, or even protecting by sequestering the proteins that begin the dementia process. In any case, treatment research focuses on lowering beta-amyloid levels.

Late in the course of the disease there is a decline in the activity of the enzyme that assists in the production of the neurotransmitter acetylcholine from choline and acetyl CoA. Acetylcholine is essential to memory, but supplements of choline (or of lecithin, which contains choline) have no effect on memory or on the progression of the disease. Drugs that inhibit the breakdown of acetylcholine, on the other hand, have proved beneficial.

Research suggests that cardiovascular disease risk factors such as high blood pressure, diabetes, and elevated levels of homocysteine may be related to the development of Alzheimer's disease. Heart healthy diets that include the omega-3 fatty acid DHA may benefit brain health as well.[59] Similarly, physical activity supports heart health and slows the cognitive decline of Alzheimer's disease.[60]

Treatment for Alzheimer's disease involves providing care to clients and support to their families. Drugs may be used to improve or at least to slow the loss of short-term memory and cognition, but they do not cure the disease. Other drugs may be used to control depression, anxiety, and behavior problems.

Maintaining appropriate body weight may be the most important nutrition concern for the person with Alzheimer's disease. Depression and forgetfulness can lead to changes in eating behaviors and poor food intake. Furthermore, changes in the body's weight-regulation system may contribute to weight loss. Perhaps the best that a caregiver can do nutritionally for a person with Alzheimer's disease is to supervise food planning and mealtimes. Providing well-liked and well-balanced meals and snacks in a cheerful atmosphere encourages food consumption. To minimize confusion, offer a few ready-to-eat foods, in bite-size pieces, with seasonings and sauces. To avoid mealtime disruptions, control distractions such as music, television, children, and the telephone.

Alcohol Highlight 7 presented information on alcohol metabolism and some of the health consequences of excessive use. Among the consequences of chronic alcohol use are impaired memory and cognition, which can complicate the diagnosis and treatment of age-related dementia.[61]

A variety of tools can be used to diagnose alcohol abuse, but simply asking a question or two can identify hazardous drinking behaviors and potential problems in the elderly.[62] "In the past year, how often did you drink four (for women, and five for men) or more drinks? What is the maximum number of drinks you consumed on any given day?" Such questions help to identify regular heavy use of alcohol and binge drinking.

Among adults 65 years of age and older, an estimated 13 percent of men and 8 percent of women report regular heavy use, and 14 percent of men and 3 percent of women report binge drinking.[63] Excessive alcohol use among elderly adults is associated with other risk factors as well, including illicit drug use, tobacco use, and misuse of prescription medications—all factors exacerbating overall health, independence, and health-care costs.[64] Fortunately, alcohol-dependent elderly adults seeking treatment seem to experience less intense alcohol cravings during withdrawal therapy than younger adults.[65]

REVIEW IT Identify how nutrition might contribute to, or prevent, the development of age-related problems associated with vision, arthritis, the brain, and alcohol use.
Senile dementia and other losses of brain function, including the impaired memory and cognition of alcohol use, afflict millions of older adults, and others face loss of vision due to cataracts or macular degeneration or cope with the pain of arthritis. As the number of people older than age 65 continues to grow, the need for solutions to these problems becomes urgent. Some problems may be inevitable, but others are preventable and good nutrition may play a key role.

17.5 Food Choices and Eating Habits of Older Adults

LEARN IT Instruct an adult on how to shop for groceries and prepare healthy meals for one person on a tight budget.

Older people are an incredibly diverse group, and for the most part, they are independent, socially sophisticated, mentally lucid, fully participating members of society who report themselves to be happy and healthy. In fact, the quality of life among the elderly has improved, and their chronic disabilities have declined dramatically in recent years. By practicing stress-management skills, maintaining physical fitness, participating in activities of interest, and cultivating spiritual health, as well as obtaining adequate nourishment, people can support a high quality of life into old age (see Table 17-4 for some strategies).

Compared with other age groups, older people spend more money per person on foods to eat at home and less money on foods away from home. Manufacturers would be wise to cater to the preferences of older adults by providing good-tasting, nutritious foods in easy-to-open, single-serving packages with labels that are easy to read.

TABLE 17-4 Strategies for Growing Old Healthfully

- Choose nutrient-dense foods.
- Be physically active. Walk, run, dance, swim, bike, or row for aerobic activity. Lift weights, do calisthenics, or pursue some other activity to tone, firm, and strengthen muscles. Practice balancing on one foot or doing simple movements with your eyes closed. Modify activities to suit changing abilities and preferences.
- Maintain appropriate body weight.
- Reduce stress—cultivate self-esteem, maintain a positive attitude, manage time wisely, know your limits, practice assertiveness, release tension, and take action.
- For women, discuss with a physician the risks and benefits of estrogen replacement therapy.
- For people who smoke, discuss with a physician strategies and programs to help you quit.
- Expect to enjoy sex, and learn new ways of enhancing it.
- Use alcohol only moderately, if at all; use drugs only as prescribed.
- Take care to prevent accidents.
- Expect good vision and hearing throughout life; obtain glasses and hearing aids if necessary.
- Take care of your teeth; obtain dentures if necessary.
- Be alert to confusion as a disease symptom, and seek diagnosis.
- Take medications as prescribed; see a physician before self-prescribing medicines or herbal remedies and a registered dietitian before self-prescribing supplements.
- Control depression through activities and friendships; seek professional help if necessary.
- Drink six to eight glasses of water every day.
- Practice mental skills. Keep on solving math problems and crossword puzzles, playing cards or other games, reading, writing, imagining, and creating.
- Make financial plans early to ensure security.
- Accept change. Work at recovering from losses; make new friends.
- Cultivate spiritual health. Cherish personal values. Make life meaningful.
- Go outside for sunshine and fresh air as often as possible.
- Be socially active—play bridge, join an exercise or dance group, take a class, teach a class, eat with friends, volunteer time to help others.
- Stay interested in life—pursue a hobby, spend time with grandchildren, take a trip, read, grow a garden, or go to the movies.
- Enjoy life.

Such services enable older adults to maintain their independence and to feel a sense of control and involvement in their own lives. Another way older adults can take care of themselves is by remaining or becoming physically active. As mentioned earlier, physical activity helps preserve one's ability to perform daily tasks and so promotes independence.

Familiarity, taste, and health beliefs are most influential on older people's food choices. Eating foods that are familiar, especially ethnic foods that recall family meals and pleasant times, can be comforting. People 65 and older are less likely to diet to lose weight than younger people are, but they are more likely to diet in pursuit of medical goals such as controlling blood glucose and cholesterol.

Malnutrition As mentioned, most older adults are adequately nourished, but an estimated one out of six is malnourished. Chronic illnesses, medications, depression, and social isolation can all contribute to malnutrition. Malnutrition limits a person's ability to function and diminishes quality of life by:[66]

- Impairing muscle function
- Decreasing bone mass
- Limiting immune defenses
- Reducing cognitive abilities[67]
- Delaying wound healing
- Slowing recovery from surgery
- Increasing hospitalizations

Healthy snacks or liquid nutrition supplements between meals enhance energy and nutrient intakes, which improves body weight and body composition as well as physical and cognitive functioning.[68]

The Nutrition Screening Initiative is part of a national effort to identify and treat nutrition problems in older adults; it uses a screening checklist. To *determine* the risk of malnutrition, health-care professionals can keep in mind the characteristics and questions listed in Table 17-5. Providing access to safe, adequate food and nutrition programs and services can help ensure healthful aging.[69]

Food Assistance Programs An integral component of the Older Americans Act (OAA) is the OAA Nutrition Program. Its services are designed to improve older people's nutrition status and enable them to avoid medical problems, continue living in communities of their own choice, and stay out of institutions. Its specific goals are to provide low-cost, nutritious meals; opportunities for social interaction; homemaker education and shopping assistance; counseling and referral to social

TABLE 17-5 Risk Factors for Malnutrition in Older Adults

	These questions help *determine* the risk of malnutrition in older adults:
Disease	• Do you have an illness or condition that changes the types or amounts of foods you eat?
Eating poorly	• Do you eat fewer than two meals a day? Do you eat fruits, vegetables, and milk products daily?
Tooth loss or mouth pain	• Is it difficult or painful to eat?
Economic hardship	• Do you have enough money to buy the food you need?
Reduced social contact	• Do you eat alone most of the time?
Multiple medications	• Do you take three or more different prescribed or over-the-counter medications daily?
Involuntary weight loss or gain	• Have you lost or gained 10 pounds or more in the last 6 months?
Needs assistance	• Are you physically able to shop, cook, and feed yourself?
Elderly person	• Are you older than 80?

NOTE: A complete description of DETERMINE and its scoring system are available online from the American Academy of Family Physicians: www.aafp.org/afp/980301ap/edits.html.

© Thinkstock/Jupiterimages

Social interactions at a congregate meal site can be as nourishing as the foods served.

services; and transportation. The program's mission has always been to provide "more than a meal."

The OAA Nutrition Program provides for **congregate meals** at group settings such as community centers. Administrators try to select sites for congregate meals where as many eligible people as possible can participate. Volunteers may also deliver meals to those who are home-bound either permanently or temporarily; these home-delivered meals are known as **Meals on Wheels.** Although the home-delivery program ensures nutrition, its recipients miss out on the social benefits of the congregate meals. Therefore, every effort is made to persuade older people to come to the shared meals, if they can. All persons aged 60 years and older and their spouses are eligible to receive meals from these programs, regardless of their income. Priority is given to those who are economically and socially needy. An estimated 3 million of our nation's older adults benefit from these meals.

These programs provide at least one meal a day that meets one-third of the RDA for this age group, and they operate five or more days a week. Many programs voluntarily offer additional services designed to appeal to older adults: provisions for special diets (to meet medical needs or religious preferences), food pantries, ethnic meals, and delivery of meals to the homeless. Adding breakfast to the service increases energy and nutrient intakes, which helps to relieve hunger and depression.

Older adults can also take advantage of the Senior Farmers Market Nutrition Program, which provides low-income older adults with coupons that can be exchanged for fresh fruits, vegetables, and herbs at community-supported farmers' markets and roadside stands. This program increases fresh fruit and vegetable consumption, provides nutrition information, and even reaches the home-bound elderly, a group of people who normally do not have access to farmers' markets.

Older adults can learn about the available programs in their communities by contacting the Eldercare section of the Department of Health and Human Services.* In addition, the local senior center and hospital can usually direct people to programs that provide nutrition and other health-related services.

In addition to programs designed specifically for older adults, the Supplemental Nutrition Assistance Program (SNAP) offers services to eligible people of all ages. As mentioned earlier, though, the participation rate for eligible seniors is only about 30 percent.

Meals for Singles Many older adults live alone, and singles of all ages face challenges in purchasing, storing, and preparing food. Large packages of meat and vegetables are often intended for families of four or more, and even a head of lettuce can spoil before one person can use it all. Many singles live in small dwellings and have little storage space for foods. A limited income presents additional obstacles. This section offers suggestions that can help to solve some of the problems singles face, beginning with a special note about the dangers of foodborne illnesses.

Foodborne Illnesses The risk of older adults getting a foodborne illness is greater than for other adults. The consequences of an upset stomach, diarrhea, fever, vomiting, abdominal cramps, and dehydration are oftentimes more severe, sometimes leading to paralysis, meningitis, or even death. For these reasons, older adults need to carefully follow the food safety suggestions presented in Chapter 19.

> DIETARY GUIDELINES FOR AMERICANS 2010

- Older adults should not eat or drink unpasteurized milk, milk products, or juices; raw or undercooked eggs, meat, poultry, fish, or shellfish; or raw sprouts.
- Older adults should eat deli meats and frankfurters only if they have been reheated to steaming hot.

congregate meals: nutrition programs that provide food for the elderly in conveniently located settings such as community centers.

Meals on Wheels: a nutrition program that delivers food for the elderly to their homes.

*Call Eldercare Locator at (800) 677-1116 or search www.eldercare.gov.

Spend Wisely People who have the means to shop and cook for themselves can cut their food bills simply by being wise shoppers. Large supermarkets are usually less expensive than convenience stores. A grocery list helps reduce impulse buying, and specials and coupons can save money when the items featured are those that the shopper needs and uses.

Buying the right amount so as not to waste any food is a challenge for people eating alone. They can buy fresh milk in the size best suited for personal needs. Boxes of milk that have been exposed to temperatures above those of pasteurization just long enough to sterilize the milk—a process called *ultrahigh temperature (UHT)*—are available and can be stored unopened on a shelf for as long as 3 months without refrigeration.

Foods in bulk are usually less expensive than packaged items. Staples such as rice, pastas, oatmeal, dry powdered milk, and dried legumes can be purchased in bulk and stored for months at room temperature.

A person who has ample freezer space can buy large packages of meat or whole chickens when they are on sale. Then the meat or chicken can be portioned and immediately wrapped into individual servings for the freezer. All the individual servings can be put in a bag marked appropriately with the contents and the date.

Frozen vegetables are more economical in large bags than in small boxes. After the amount needed is taken out, the bag can be closed tightly with a twist tie or rubber band. If the package is returned quickly to the freezer each time, the vegetables will stay fresh for a long time.

Finally, breads and cereals usually must be purchased in larger quantities. Again the amount needed for a few days can be taken out and the rest stored in the freezer. Consider buying day-old bread and baked goods for added savings.

Grocers will break open a package of wrapped meat and rewrap the portion needed. Similarly, eggs can be purchased by the half-dozen. Eggs do keep for long periods, though, if stored properly in the refrigerator.

Fresh fruits and vegetables generally cost less when they are in season. A person can buy individual pieces of fresh fruit at various stages of ripeness: a ripe one to eat right away, a semiripe one to eat soon after, and a green one to ripen on the windowsill. If vegetables are packaged in large quantities, the grocer can break open the package so that a smaller amount can be purchased. Small cans of fruits and vegetables, even though they are more expensive per unit, are a reasonable alternative, considering that it is expensive to buy a regular-size can and let the unused portion spoil.

Buy only what you will use.

Be Creative Creative chefs think of various ways to use foods when only large amounts are available. For example, a head of cauliflower can be divided into thirds. Then one-third is cooked and eaten hot. Another third is put into a vinegar and oil marinade for use in a salad. And the last third can be used in a casserole or stew.

A variety of vegetables and meats can be enjoyed stir-fried; inexpensive vegetables such as cabbage, celery, and onion are delicious when sautéed in a little oil with herbs or lemon added. Interesting frozen vegetable mixtures are available in larger grocery stores. Cooked, leftover vegetables can be dropped in at the last minute. A bonus of a stir-fried meal is that there is only one pan to wash. Similarly, a microwave oven allows a chef to use fewer pots and pans. Meals and leftovers can also be frozen or refrigerated in microwavable containers to reheat as needed.

Invite guests to share a meal.

Many frozen dinners offer nutritious options. Adding a fresh salad, a whole-wheat roll, and a glass of milk can make a nutritionally balanced meal.

Finally, single people might want to invite someone to share meals with them whenever there is enough food. It's likely that the person will return the invitation, and both parties will get to enjoy companionship and a meal prepared by others.

REVIEW IT Instruct an adult on how to shop for groceries and prepare healthy meals for one person on a tight budget.

Older people can benefit from both the nutrients provided and the social interaction available at congregate meals. Other government programs deliver meals to those who are home-bound. With creativity and careful shopping, those living alone can prepare nutritious, inexpensive meals.

Healthy meal patterns throughout adulthood support good health and long life. Physical activity, mental challenges, stress management, and social activities can also help people grow old comfortably. The next chapter describes how similar lifestyle choices help prevent chronic diseases as well.

Nutrition Portfolio

By eating a balanced diet, maintaining a healthy body weight, and engaging in a variety of physical, social, and mental activities, you can enjoy good health in later life. Visit older adults in your community and do the following:

- Consider whether they have the financial means, physical ability, and social support they need to eat adequately.
- Note whether they have experienced an unintentional loss of weight recently.
- Discuss how they occupy their time physically, socially, and mentally.
- Offer to analyze the diet of an older adult who you care about using Diet Analysis Plus. Have the person write down 1, 2, or even 3 days of their food and beverage intake, and then enter it into the program and print out a 3-day average report of the results. It will be fun and educational to go over it together. Remind the person that you are not a doctor and that this is a learning tool for an introductory nutrition course.

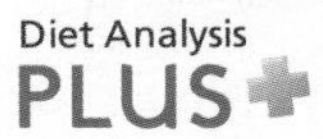

To complete this exercise, go to your Diet Analysis Plus at **www.cengagebrain.com.**

STUDY IT To review the key points of this chapter and take a practice quiz, go to the study cards at the end of the book.

REFERENCES

1. K. D. Kochanek and coauthors, Deaths: Preliminary data for 2009, *National Vital Statistics Reports* 59 (2011): 1–8.
2. H. Kung and coauthors, Deaths: Final data for 2005, *National Vital Statistics Reports* 56 (2008): 1–120.
3. E. S. Ford and coauthors, Low-risk lifestyle behaviors and all-cause mortality: Findings from the National Health and Nutrition Examination Survey III Mortality Study, *American Journal of Public Health* 101 (2011): 1922–1929.
4. A. L. Anderson and coauthors, Dietary patterns and survival of older adults, *Journal of the American Dietetic Association* 111 (2011): 84–91.
5. K. Khaw and coauthors, Combined impact of health behaviours and mortality in men and women: The EPIC-Norfolk Prospective Population Study, *PLoS Medicine* 5 (2008): e12.
6. L. B. Yates and coauthors, Exceptional longevity in men: Modifiable factors associated with survival and function to age 90 years, *Archives of Internal Medicine* 168 (2008): 284–290.
7. W. J. Chodzko-Zajko and coauthors, American College of Sports Medicine Position Stand: Exercise and physical activity for older adults, *Medicine and Science in Sports and Exercise* 41 (2009): 1510–1530; S. Mandic and coauthors, Characterizing differences in mortality at the low end of the fitness spectrum, *Medicine and Science in Sports and Exercise* 41 (2009): 1573–1579; P. Kokkinos and coauthors, Exercise capacity and mortality in black and white men, *Circulation* 117 (2008): 614–622.
8. S. Studenski and coauthors, Gait speed and survival in older adults, *Journal of the American Medical Association* 305 (2011): 50–58.
9. L. D. Baker and coauthors, Effects of aerobic exercise on mild cognitive impairment, *Archives of Neurology* 67 (2010): 71–79; T. Liu-Ambrose and M. G. Donaldson, Exercise and cognition in older adults: Is there a role for resistance training programmes? *British Journal of Sports Medicine* 43 (2009): 25–27.
10. Chodzko-Zajko and coauthors, 2009.
11. H. W. Park, Longevity, aging, and caloric restriction: Clive Maine McCay and the construction of a multidisciplinary research program, *Historical Studies in the Natural Sciences* 40 (2010): 79–124.
12. R. S. Sohal and coauthors, Life span extension in mice by food restriction depends on an energy imbalance, *Journal of Nutrition* 139 (2009): 533–539.
13. D. M. Huffman, Exercise as a calorie restriction mimetic: Implications for improving healthy aging and longevity, *Interdisciplinary Topics in Gerontology* 37 (2010): 157–174.
14. R. J. Colman and R. M. Anderson, Nonhuman primate calorie restriction, *Antioxidants and Redox Signaling* 14 (2011): 229–239.
15. R. K. Minor and coauthors, Dietary interventions to extend life span and health span based on calorie restriction, *Journal of Gerontology: Biological Sciences* 65 (2010): 695–703.
16. L. M. Redman and E. Ravussin, Caloric restriction in humans: Impact on physiological, psychological, and behavioral outcomes, *Antioxidants and Redox Signaling* 14 (2011): 275–287; L. Fontana, The scientific basis of caloric restriction leading to longer life, *Current Opinion in Gastroenterology* 25 (2009): 144–150.
17. O. Froy and R. Miskin, Effect of feeding regimens on circadian rhythms: Implications for aging and longevity, *Aging* 11 (2010): 7–27; Minor and coauthors, 2010; K. A. Varaday and M. K. Hellerstein, Do calorie restriction or alternate-day fasting regimens modulate adipose tissue physiology in a way that reduces chronic disease risk? *Nutrition Reviews* 66 (2008): 333–342.
18. F. Sofi and coauthors, Accruing evidence on benefits of adherence to the Mediterranean diet on health: An updated systematic review and meta-analysis, *American Journal of Clinical Nutrition* 92 (2010): 1189–1196.
19. M. E. Obrenovich and coauthors, Antioxidants in health, disease and aging, *CNS and Neurological Disorders Drug Targets* 10 (2011): 192–207; K. Chong-Han, Dietary lipophilic antioxidants: Implications and significance in the aging process, *Critical Reviews in Food Science and Nutrition* 50 (2010): 931–937; L. M. Willis, B. Shukitt-Hale, and J. A. Joseph, Recent advances in berry supplementation and age-related cognitive decline, *Current Opinion in Clinical Nutrition and Metabolic Care* 12 (2009): 91–94.
20. D. K. Childers and D. B. Allison, The "obesity paradox": A parsimonious explanation for relations among obesity, mortality rate and aging? *International Journal of Obesity* 34 (2010): 1231–1238.
21. O. Bouillanne and coauthors, Fat mass protects hospitalized elderly persons against morbidity and mortality, *American Journal of Clinical Nutrition* 90 (2009): 505–510.
22. P. Szulc and coauthors, Rapid loss of appendicular skeletal muscle mass is associated with higher all-cause mortality in older men: The prospective MINOS study, *American Journal of Clinical Nutrition* 91 (2010): 1227–1236.
23. G. A. Power and coauthors, Motor unit number estimates in masters runners: Use it or lose it? *Medicine and Science in Sports and Exercise* 42 (2010): 1644–1650; R. Koopman and L. J. van Loon, Aging, exercise, and muscle protein metabolism, *Journal of Applied Physiology* 106 (2009): 2040–2048; T. M. Manini and coauthors, Activity energy expenditure and change in body composition in late life, *American Journal of Clinical Nutrition* 90 (2009): 1336–1342; D. Paddon-Jones and B. B. Rasmussen, Dietary protein recommendations and the prevention of sarcopenia, *Current Opinion in Clinical Nutrition and Metabolic Care* 12 (2009): 86–90; D. K. Houston and coauthors, Dietary protein intake is associated with lean mass change in older, community-dwelling adults: The Health, Aging, and Body Composition (Health ABC) Study, *American Journal of Clinical Nutrition* 87 (2008): 150–155; D. Paddon-Jones and coauthors, Role of dietary protein in the sarcopenia of aging, *American Journal of Clinical Nutrition* 87 (2008): 1562S–1566S.
24. R. B. Ervin and B. A. Dye, The effect of functional dentition on Healthy Eating Index scores and nutrient intakes in a nationally representative sample of older adults, *Journal of Public Health Dentistry* 69 (2009): 207–216.
25. Position paper of the American Dietetic Association: Nutrition across the spectrum of aging, *Journal of the American Dietetic Association* 105 (2005): 616–633.
26. M. Hamer, C. J. Bates, and G. D. Mishra, Depression, physical function, and risk of mortality: National Diet and Nutrition Survey in adults older than 65 years, *American Journal of Geriatric Psychiatry* 19 (2011): 72–78.
27. H. Chang-Quan and coauthors, Health status and risk for depression among the elderly: A meta-analysis of published literature, *Age and Ageing* 39 (2010): 23–30; V. Colasanti and coauthors, Tests for the evaluation of depression in the elderly: A systematic review, *Archives of Gerontology and Geriatrics* 50 (2010): 227–230; D. R. Hoover and coauthors, Depression in the first year of stay for elderly long-term nursing home residents in the USA, *International Psychogeriatric Association* 22 (2010): 1161–1167.
28. R. S. Shim and coauthors, Prevalence, treatment, and control of depressive symptoms in the United States: Results from the National Health and Nutrition Examination Survey (NHANES), 2005-2008, *Journal of the American Board of Family Medicine* 24 (2011): 33–38.
29. J. Sarris, N. Schoendorfer, and D. J. Kavanagh, Major depressive disorder and nutritional medicine: A review of monotherapies and adjuvant treatments, *Nutrition Reviews* 67 (2009): 125–131.
30. M. F. Kuczmarski and coauthors, Higher Health Eating Index-2005 scores associated with reduced symptoms of depression in an urban population: Findings from the Healthy Aging in Neighborhoods of Diversity Across the Life Span (HANDLS) study, *Journal of the American Dietetic Association* 110 (2010): 383–389.
31. C. Denavas-Walt, B. D. Proctor, and J. C. Smith, US Census Bureau, *Income, Poverty, and Health Insurance Coverage in the United States: 2010* (Washington, D.C.: US Government Printing Office, 2011).
32. Paddon-Jones and Rasmussen, 2009; R. P. Heaney and D. K. Layman, Amount and type of protein influences bone health, *American Journal of Clinical Nutrition* 87 (2008): 1567S–1570S.
33. Position of the American Dietetic Association: Individualized nutrition approaches for older adults in health care communities, *Journal of the American Dietetic Association* 110 (2010): 1549–1553.
34. A. D. Smith and H. Refsum, Vitamin B-12 and cognition in the elderly, *American Journal of Clinical Nutrition* 89 (2009): 707S–711S.
35. Committee on Dietary Reference Intakes, *Dietary Reference Intakes for Thiamin, Riboflavin, Niacin, Vitamin B_6, Folate, Vitamin B_{12}, Pantothenic*

Acid, Biotin, and Choline, (Washington, D.C.: National Academies Press, 2000), p. 338.

36. Committee on Dietary Reference Intakes, *Dietary Reference Intakes for Calcium, Phosphorus, Magnesium, Vitamin D, and Fluoride* (Washington, D.C.: National Academies Press, 1997).
37. R. L. Bailey and coauthors, Dietary supplement use in the United States, 2003–2006, *Journal of Nutrition* 141 (2011): 261–266.
38. A. Weeden and coauthors, Vitamin and mineral supplements have a nutritionally significant impact on micronutrient intakes of older adults attending senior centers, *Journal of Nutrition for the Elderly* 29 (2010): 241–254.
39. M. Rhone and A. Basu, Phytochemicals and age-related eye diseases, *Nutrition Reviews* 66 (2008): 465–472.
40. W. G. Christen and coauthors, Dietary carotenoids, vitamins C and E, and risk of cataract in women. A Prospective Study, *Archives of Ophthalmology* 126 (2008): 102–109; A. G. Tan and coauthors, Antioxidant nutrient intake and the long-term incidence of age-related cataract: The Blue Mountains Eye Study, *American Journal of Clinical Nutrition* 87 (2008): 1899–1905.
41. S. Rautiainen and coauthors, Vitamin C supplements and the risk of age-related cataract: A population-based prospective cohort study in women, *American Journal of Clinical Nutrition* 91 (2010): 487–493.
42. L. S. Lim and coauthors, Relation of age-related cataract with obesity and obesity genes in an Asian population, *American Journal of Epidemiology* 169 (2009): 1267–1274.
43. R. D. Jager, W. F. Mieler, and J. W. Miller, Age-related macular degeneration, *New England Journal of Medicine* 358 (2008): 2606–2617.
44. J. K. Shen and coauthors, Oxidative damage in age-related macular degeneration, *Histology and Histopathology* 22 (2007): 1301–1308.
45. E. J. Johnson, Age-related macular degeneration and antioxidant vitamins: Recent findings, *Current Opinion in Clinical Nutrition and Metabolic Care* 13 (2010): 28–33; W. G. Christen and coauthors, Folic acid, pyridoxine, and cyanocobalamin combination treatment and age-related macular degeneration in women, *Archives of Internal Medicine* 169 (2009): 335–341; J. P. SanGiovanni and coauthors, Ω-3 long-chain polyunsaturated fatty acid intake and 12-y incidence of neovascular age-related macular degeneration and central geographic atrophy: AREDS report 30, a prospective cohort study from the Age-Related Eye Disease Study, *American Journal of Clinical Nutrition* 90 (2009): 1601–1607; C. Augood and coauthors, Oily fish consumption, dietary docosahexaenoic acid and eicosapentaenoic acid intakes, and associations with neovascular age-related macular degeneration, *American Journal of Clinical Nutrition* 88 (2008): 398–406; E. J. Johnson and coauthors, The influence of supplemental lutein and docosahexaenoic acid on serum, lipoproteins, and macular pigmentation, *American Journal of Clinical Nutrition* 87 (2008): 1521–1529; E. D. O'Connell and coauthors, Diet and risk factors for age-related maculopathy, *American Journal of Clinical Nutrition* 87 (2008): 712–722.
46. Y. J. Cheng and coauthors, Prevalence of doctor-diagnosed arthritis and arthritis-attributable activity limitation: United States, 2007–2009, *Morbidity and Mortality Weekly Report* 59 (2010): 1261–1265.
47. J. M. Hootman and C. G. Helmick, Projections of US prevalence of arthritis and associated activity limitations, *Arthritis and Rheumatism* 54 (2006): 226–229.
48. J. Bolen and coauthors, Arthritis as a potential barrier to physical activity among adults with heart disease: United States, 2005 and 2007, *Morbidity and Mortality Weekly Report* 58 (2009): 165–169.
49. T. Neogi, Gout, *New England Journal of Medicine* 364 (2011): 443–452.
50. N. G. Bazan, M. F. Molina, and W. C. Gordon, Docosahexaenoic acid signalolipidomics in nutrition: Significance in aging, neuroinflammation, macular degeneration, Alzheimer's, and other neurodegenerative diseases, *Annual Review of Nutrition* 31 (2011): 321–351.
51. Alzheimer's Association, 2011 Alzheimer's Disease Facts and Figures, *Alzheimer's and Dementia*, 2011.
52. R. Mayeux, Early Alzheimer's disease, *New England Journal of Medicine* 362 (2010): 2194–2201.
53. A. Burns and S. Iliffe, Clinical review: Alzheimer's disease, *British Medical Journal* 338 (2009): b158; H. W. Querfurth and F. M. LaFerla, Mechanisms of disease: Alzheimer's disease, *New England Journal of Medicine* 362 (2010): 329–344.
54. J. C. Lambert, Genome-wide association study identifies variants at CLU and CR1 associated with Alzheimer's disease, *Nature Genetics* 41 (2009): 1094–1099; National Institute on Aging, Alzheimer's Disease Genetics Fact Sheet, available at www.nia.nih.gov/alzheimers/publications/geneticsfs.htm, updated September 9, 2009.
55. P. I. Moreira and coauthors, Alzheimer disease and the role of free radicals in the pathogenesis of the disease, *CNS and Neurological Disorders Drug Targets* 7 (2008): 3–10.
56. J. A. Luchsinger and D. R. Gustafon, Adiposity and Alzheimer's disease, *Current Opinion in Clinical Nutrition and Metabolic Care* 12 (2009): 15–21; D. B. Miller and J. P. O'Callaghan, Do early-life insults contribute to the late-life development of Parkinson and Alzheimer disease? *Metabolism* 57 (2008): S44–S49.
57. A. Gella and N. Durany, Oxidative stress in Alzheimer disease, *Cell Adhesion and Migration* 13 (2009): 88–93.
58. R. J. Castellani and coauthors, Reexamining Alzheimer's disease: Evidence for a protective role for amyloid-beta protein precursor and amyloid-beta, *Journal of Alzheimer's Disease* 18 (2009): 447–452.
59. G. M. Cole and S. A. Frautschy, DHA may prevent age-related dementia, *Journal of Nutrition* 140 (2010): 869–874; Y. Gu and coauthors, Food combination and Alzheimer disease risk: A protective diet, *Archives of Neurology* 67 (2010): 699–706; E. Albanese and coauthors, Dietary fish and meat intake and dementia in Latin America, China, and India: A 10/66 Dementia Research Group population-based study, *American Journal of Clinical Nutrition* 90 (2009): 392–400; C. M. Milte, N. Sinn, and P. R. C. Howe, Polyunsaturated fatty acid status in attention deficit hyperactivity disorder, depression, and Alzheimer's disease: Towards an omega-3 index for mental health? *Nutrition Reviews* 67 (2009): 573–590; N. Scarmeas and coauthors, Physical activity, diet, and risk of Alzheimer disease, *Journal of the American Medical Association* 302 (2009): 627–637; L. J. Whalley and coauthors, n-3 Fatty acid erythrocyte membrane content, APOE ε4, and cognitive variation: An observational follow-up study in late adulthood, *American Journal of Clinical Nutrition* 87 (2008): 449–454.
60. Scarmeas and coauthors, 2009; E. B. Larson, Physical activity for older adults at risk for Alzheimer disease, *Journal of the American Medical Association* 300 (2008): 1077–1079; N. T. Lautenschlager and coauthors, Effect of physical activity on cognitive function in older adults at risk for Alzheimer disease, *Journal of the American Medical Association* 300 (2008): 1027–1037.
61. E. Sinforiani and coauthors, The effects of alcohol on cognition in the elderly: From protection to neurodegeneration, *Functional Neurology* 26 (2011): 103–106.
62. D. A. Dawson, A. J. Pulay, and B. F. Grant, A comparison of two single-item screeners for hazardous drinking and alcohol use disorder, *Alcoholism: Clinical and Experimental Research* 34 (2010): 364–374.
63. D. G. Blazer and L. T. Wu, The epidemiology of at-risk and binge drinking among middle-aged and elderly community adults: National Survey on Drug Use and Health, *American Journal of Psychiatry* 166 (2009): 1162–1169.
64. S. Matthews and D. W. Oslin, Alcohol misuse among the elderly: An opportunity for prevention, *American Journal of Psychiatry* 166 (2009): 1093–1095.
65. A. K. Hintzen and coauthors, Does alcohol craving decrease with increasing age? Results from a cross-sectional study, *Journal of Studies on Alcohol and Drugs* 72 (2011): 158–162; A. J. Barnes and coauthors, Prevalence and correlates of at-risk drinking among older adults: The Project SHARE Study, *Journal of General Internal Medicine* 25 (2010): 840–846.
66. T. Ahmed and N. Haboubi, Assessment and management of nutrition in older people and its importance to health, *Clinical Interventions in Aging* 5 (2010): 207–216.
67. X. Gao and coauthors, Food insecurity and cognitive function in Puerto Rican adults, *American Journal of Clinical Nutrition.*
68. C. A. Zizza, D. D. Arsiwalla, and K. J. Ellison, contribution of snacking to older adults' vitamin, carotenoid, and mineral intakes, *Journal of the American Dietetic Association* 110 (2010): 768–772; H. J. Silver, Oral strategies to supplement older adults' dietary intakes: Comparing the evidence, Nutrition Reviews 67 (2009): 21–31.
69. Position of the American Dietetic Association, American Society for Nutrition, and Society for Nutrition Education: Food and nutrition programs for community-residing older adults, *Journal of the American Dietetic Association* 110 (2010): 463–472.

LEARN IT Explain why certain nutrients and medications interact.

Nutrient-Drug Interactions

People older than the age of 65 take about one-third of all the over-the-counter and prescription drugs sold in the United States. They receive an average of 13 prescriptions a year and may take as many as 6 drugs at a time. They take a variety of non-vitamin-mineral supplements, such as glucosamine, as well. Most often, they take these drugs and supplements for heart disease, but also to treat arthritis, respiratory problems, and gastrointestinal disorders. They often go to different doctors for each condition and receive different prescriptions from each. Furthermore, physiologic changes associated with aging may alter drug metabolism and excretion, which may in turn, diminish drug effectiveness or create potential toxicities.[1] For all these reasons, physicians need to "start low and go slow" when prescribing for older adults.

An estimated 1 in 25 older adults takes prescription and nonprescription (over-the-counter) medications that pose a risk for a major drug–drug interaction.[2] To avoid harmful drug interactions, consumers need to inform all of their physicians and pharmacists of all the medicines being taken. These medicines enable people of all ages to enjoy better health, but they also bring side effects and risks.

This highlight focuses on some of the nutrition-related consequences of medical drugs, both prescription drugs and nonprescription drugs. Highlight 7 describes the relationships between nutrition and the drug alcohol, and Highlight 18 presents information on herbal supplements and other alternative therapies.

The Actions of Drugs

Most people think of drugs either as medicines that help them recover from illnesses or as illegal substances that lead to bodily harm and addiction. Actually, both uses of the term *drug* are correct because any substance that modifies one or more of the body's functions is, technically, a drug. Even medical drugs have both desirable and undesirable consequences within the body.

Consider aspirin. One action of aspirin is to limit the production of certain prostaglandins. Some prostaglandins help to produce fevers, some sensitize pain receptors, some cause contractions of the uterus, some stimulate digestive tract motility, some control nerve impulses, some regulate blood pressure, some promote blood clotting, and some cause inflammation. By interfering with prostaglandin actions, aspirin reduces fever and inflammation, relieves pain, and slows blood clotting, among other actions.

A person cannot use aspirin to produce one of its effects without producing all of its other effects. Someone who is prone to strokes and heart attacks might take aspirin to prevent blood clotting, but it will also dull that person's sense of pain. Another person who takes aspirin only for pain will also experience slow blood clotting. The anticlotting effect might be dangerous if it causes abnormal bleeding. A single two-tablet dose of aspirin doubles the bleeding time of wounds, an effect that lasts from 4 to 7 hours. For this reason, physicians instruct clients to refrain from taking aspirin before surgery.

The Interactions between Drugs and Nutrients

Hundreds of drugs and nutrients interact, and these interactions can lead to nutrient imbalances or interfere with drug effectiveness. Adverse nutrient–drug interactions are most likely if drugs are taken over long periods, if several drugs are taken, or if nutrition status is poor or deteriorating. Understandably, then, elderly people with chronic diseases are most vulnerable.

Nutrients and medications may interact in many ways:

- Drugs can alter food intake and the absorption, metabolism, and excretion of nutrients.
- Foods and nutrients can alter the absorption, metabolism, and excretion of drugs.
- Combinations can be toxic.

The following paragraphs describe these interactions, and Table H17-1 (p. 576) summarizes this information and provides specific examples.

Drugs Alter Food Intake

Some medications can make eating difficult or unpleasant. They may suppress appetite, alter taste sensations, induce nausea or vomiting, cause mouth dryness, or create inflammation or lesions in the mouth, stomach, or intestinal lining. Side effects, such as abdominal discomfort, constipation, and diarrhea may worsen when food is eaten. Medications that cause drowsiness may make a person too tired to eat. All of these complications limit food intake and can lead to weight loss and malnutrition if not resolved.

Some medications stimulate appetite and cause weight gain. Unintentional weight gain may result from the use of some antipsychotics, antidepressants, and corticosteroids (for example, prednisone). People using these drugs do not feel satiated and sometimes gain 40 to 60 pounds in just a few months. For many adults, fear of weight gain is a common reason for avoiding treatment.[3] For patients with diseases that

TABLE H17-1 **Examples of Diet-Drug interactions**

Drugs May Alter Food Intake by
• Altering the appetite (Amphetamines suppress appetite; corticosteroids increase appetite.)
• Interfering with taste or smell (Amphetamines change taste perceptions.)
• Inducing nausea or vomiting (Digitalis may do both.)
• Interfering with oral function (Some antidepressants may cause dry mouth.)
• Causing sores or inflammation in the mouth (Methotrexate may cause painful mouth ulcers.)
Drugs May Alter Nutrient Absorption by
• Changing the acidity of the digestive tract (Antacids may interfere with iron and folate absorption.)
• Damaging mucosal cells (Cancer chemotherapy may damage mucosal cells.)
• Binding to nutrients (Bile acid binders bind to fat-soluble vitamins.)
Foods and Nutrients May Alter Drug Absorption by
• Stimulating secretion of gastric acid (The antifungal agent ketoconazole is absorbed better with meals due to increased acid secretion.)
• Altering rate of gastric emptying (Intestinal absorption of drugs may be delayed when they are taken with food.)
• Binding to drugs (Calcium binds to tetracycline, reducing drug and calcium absorption.)
• Competing for absorption sites in the intestines (Dietary amino acids interfere with levodopa absorption.)
Drugs and Nutrients May Interact and Alter Metabolism by
• Acting as structural analogs (Warfarin and vitamin K are structural analogs.)
• Using similar enzyme systems (Phenobarbital induces liver enzymes that increase metabolism of folate, vitamin D, and vitamin K.)
• Competing for transport on plasma proteins (Fatty acids and drugs may compete for the same sites on the plasma protein albumin.)
Drugs May Alter Nutrient Excretion by
• Altering reabsorption in the kidneys (Some diuretics increase the excretion of sodium and potassium.)
• Causing diarrhea or vomiting (Diarrhea and vomiting may cause electrolyte losses.)
Foods May Alter Medication Excretion by
• Inducing activities of liver enzymes that metabolize drugs to allow their excretion (Components of charcoal-broiled meats increase metabolism of warfarin, theophylline, and acetominophen.)
Diet and Drug Interactions May Cause Toxicity
• Increasing side effects of the drug (Caffeine in beverages can increase adverse effects of stimulants.)
• Increasing drug action to excessive levels (Grapefruit components may block metabolism of drugs and enhance drugs' actions and side effects.)

cause wasting, such as cancer or AIDS, weight gain may be desirable. They may be prescribed appetite enhancers, such as megestrol acetate.

Drugs Alter Nutrient Absorption

Nutrient malabsorption is most likely to occur with medications that upset GI function or damage the intestinal mucosa. Antineoplastic and antiretroviral drugs are especially detrimental; nonsteroidal anti-inflammatory drugs (NSAIDS) and some antibiotics can have similar, though milder, effects.

Some medications bind nutrients in the GI tract, preventing their absorption. For example, bile acid binders, used to reduce cholesterol levels, also bind to the fat-soluble vitamins A, D, E, and K. Some antibiotics, notably tetracycline and ciprofloxacin, bind to the calcium in foods and supplements, which reduces the absorption of both the drug and the calcium. Other minerals, such as iron, magnesium, and zinc, may also bind to antibiotics. For this reason, pharmacists advise consumers to use dairy products and all mineral supplements at least 2 hours apart from these medications.

Medications that reduce stomach acidity may interfere with the absorption of vitamin B_{12}, folate, and iron. Examples include antacids, which neutralize stomach acid by acting as weak bases, and antiulcer drugs (such as proton pump inhibitors and H_2 blockers), which interfere with acid secretion.

Several drugs impede absorption by interfering with the intestinal metabolism or transport of nutrients into mucosal cells. For example, the antibiotics trimethoprim and pyrimethamine compete with folate for absorption into intestinal cells.

Diets Alter Drug Absorption

Most drugs are absorbed in the upper small intestine. Major influences on drug absorption include the stomach emptying rate, level of acidity, and direct interactions with dietary components. The drug's formulation also influences its absorption, and pharmacists often provide instructions advising whether food should be eaten or avoided when using a medication.

Drugs reach the small intestine more quickly when the stomach is empty. Therefore, taking a medication with meals may delay its absorption, even though the total amount absorbed may not be lower. As an example, aspirin works faster when taken on an empty stomach, but taking it with food is often encouraged to reduce stomach irritation.

Some drugs are better absorbed in an acid environment, whereas others are better absorbed in alkaline conditions. Drugs that are damaged by acid are available in coated forms that resist the action of the stomach's acidity.

Both nutrients and nonnutrients may bind to drugs and inhibit their absorption. For example, high-fiber diets may decrease the absorption of some tricyclic antidepressants. Phytates in foods can bind to digoxin, a drug prescribed for heart disease. As mentioned earlier, calcium and other minerals may bind to some antibiotics, reducing absorption of both the minerals and the drug.

Drugs Alter Nutrient Metabolism

Drugs and nutrients share many of the same enzyme systems in the small intestine and the liver. Consequently, some drugs may enhance or inhibit the activities of enzymes that are needed for nutrient metabolism. For example, the anticonvulsants phenobarbital and phenytoin increase levels of the liver enzymes that metabolize folate, vitamin D, and vitamin K; therefore, persons using these drugs may require supplements of these vitamins.

The drug methotrexate, used to treat cancer and inflammatory conditions, acts by interfering with folate metabolism and thus depriving rapidly dividing cancer cells of the folate they need to multiply. Methotrexate resembles folate in structure (see Figure H17-1) and competes with folate for the enzyme that converts folate to its active form.* The adverse effects of using methotrexate therefore include symptoms of folate deficiency. These adverse effects can be reduced by using a pre-activated form of folate (called leucovorin), which is often prescribed along with methotrexate to ensure that the body's rapidly dividing cells (such as cells of the digestive tract, skin cells, and red blood cells) receive adequate folate.

> **FIGURE H17-1 Folate and Methotrexate**

By competing for the enzyme that activates folate, methotrexate prevents cancer cells from obtaining the folate they need to multiply. This interference with folate metabolism creates a secondary deficiency of folate that deprives normal cells of the folate they need as well. Notice the similarities in their chemical structures.

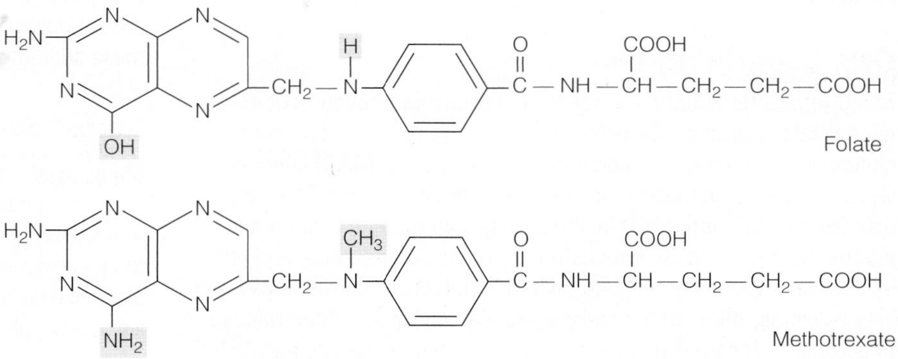

Folate

Methotrexate

Diet Alters Drug Metabolism

Some foods affect the activities of enzymes that metabolize drugs or may counteract the drugs' effects in other ways. For example, compounds in grapefruit and grapefruit juice interfere with enzymes that metabolize a number of drugs.[4] As a result of reduced enzyme action, blood concentrations of the drugs increase, leading to stronger physiological effects. The effect of the grapefruit juice lasts for a substantial period after the juice is consumed. Table H17-2 provides examples of drugs that interact with grapefruit juice, as well as some that are not affected.

A number of dietary factors affect the activity of the anticoagulant drug warfarin. The most important interaction is with vitamin K, which is structurally similar to warfarin. Warfarin acts by blocking the enzyme that activates vitamin K, thereby preventing the synthesis of blood-clotting factors. The amount of warfarin prescribed is dependent, in part, on how much vitamin K is in the diet. If vitamin K consumption from foods or supplements increases substantially, it can weaken the effect of the drug. Individuals using warfarin are advised to consume similar amounts of vitamin K daily to keep warfarin activity stable. The dietary sources highest in vitamin K are dark-green, leafy vegetables.

TABLE H17-2 Grapefruit Juice–Drug Interactions—Selected Examples

Drug Category	Drugs Affected by Grapefruit Juice	Drugs Unaffected by Grapefruit Juice
Cardiovascular drugs	Amiodarone Felodipine Nicardipine	Amlodipine Digoxin Diltiazem
Cholesterol-lowering drugs	Atorvastatin Lovastatin Simvastatin	Pravastatin Fluvastatin Rosuvastatin
Central nervous system drugs	Buspirone Carbamazepine Diazepam	Alprazolam Haloperidol Lorazapam
Anti-infective drugs	Saquinavir Erythromycin	Clarithromycin Quinine
Estrogens	Ethinylestradiol	17-β-estradiol
Anticoagulants	—	Acenocoumarol Warfarin
Immunosuppressants	Cyclosporine Tacrolimus	Prednisone

Drugs Alter Nutrient Excretion

Drugs that enhance urinary excretion may interfere with nutrient reabsorption in the kidneys, resulting in greater urinary losses. For example, some diuretics accelerate the excretion of calcium, potassium, and magnesium. Risk of mineral depletion is highest if multiple drugs with the same effect are used, if kidney function is impaired, or if medications are used for a long time. Note that some diuretics may cause mineral retention instead.

Some drugs increase the excretion of vitamin B_6. One example is isoniazid (INH), an antituberculosis drug similar in structure to vitamin B_6. By interfering with the vitamin's conversion to its active form, the drug accelerates vitamin B_6 excretion, thus inducing a deficiency. Because the drug must be taken for at least 6 months to treat infection, vitamin B_6 supplements are routinely given simultaneously to prevent deficiency.

Diets Alter Drug Excretion

Inadequate excretion of medications can cause toxicity, whereas excessive losses may reduce the amount available for therapeutic effect. Some food components can alter drug reabsorption by the kidneys. For example, the amount of the medication lithium that is reabsorbed by the kidneys correlates with the amount of sodium reabsorbed. Consequently, both dehydration and sodium depletion, which increase sodium reabsorption, may result in lithium retention. Similarly, a person with a high sodium intake will excrete more sodium in the urine, and therefore more lithium. Individuals using lithium are advised to maintain a consistent sodium intake from day to day in order to maintain a stable blood level of lithium.

*Other folate antagonists include aminopterin, sulfasalazine, pyrimethamine, trimethoprim, triamterene, carbamazepine, phenytoin, phenobarbital, and primidone.

Urine acidity can also affect drug excretion due to the effects of pH on a compound's chemical structure. The medication quinidine, used to treat arrhythmias, is excreted more readily in acidic urine. Foods or drugs that cause urine to become more alkaline (for example, sodium bicarbonate) may reduce quinidine excretion and raise its blood levels.

Diet-Drug Toxicities

Some interactions between foods and drugs can cause toxicity or exacerbate a drug's side effects. The combination of tyramine, a compound in some foods, and monoamine oxidase (MAO) inhibitors, which include some medications that treat depression and Parkinson's disease, can be fatal. MAO inhibitors block an enzyme that normally inactivates tyramine, as well as the hormones epinephrine and norepinephrine. When people who take MAO inhibitors consume excessive tyramine, the increase in tyramine can cause a sudden release of accumulated norepinephrine. This surge in norepinephrine results in severe headaches, rapid heartbeat, and a dangerous increase in blood pressure. For this reason, people taking MAO inhibitors are advised to restrict their intakes of foods with substantial amounts of tyramine (see Table H17-3).

TABLE H17-3 Examples of Foods with a High Tyramine Content

• Aged cheeses	• Mushrooms
• Aged or cured meats (sausage, salami)	• Prepared soy foods (miso, tempeh, tofu)
• Beer (draft or unpasteurized)	• Smoked or pickled fish (anchovies, caviar)
• Fava beans	• Soy sauce
• Fermented vegetables (sauerkraut, kimchi)	• Wine
	• Yeast extract (Marmite, Vegemite)

NOTE: The tyramine content of foods depends on storage conditions and processing; thus the amounts in similar products can vary substantially.

The Inactive Ingredients in Drugs

Besides the active ingredients, medicines may contain other substances such as sugar, sorbitol, lactose, and sodium. For most people who use medicines on occasion and in small amounts, such ingredients pose no problem. When medicines are taken regularly or in large doses, however, people on special diets may need to be aware of these additional ingredients and their effects.

Sugar, Sorbitol, and Lactose

Many liquid preparations contain sugar or sorbitol to make them taste better. For people who must regulate their intakes of carbohydrates, such as people with diabetes, the amount of sugar in these medicines may need to be considered. Large doses of liquids containing sorbitol may cause diarrhea. The lactose added as filler to some medications may cause problems for people who are lactose intolerant.

Sodium

Antibiotics and antacids often contain sodium. People who take Alka Seltzer, for example, may not realize that a single two-tablet dose may exceed their recommended sodium intake for a whole day. In addition, antacids neutralize stomach acid, and many nutrients depend on acid for their digestion. Taking any antacid regularly will reduce the absorption of many nutrients.

Nutrient interactions and risks are not unique to prescription drugs. People who buy over-the-counter drugs also need to protect themselves. The increasing availability of over-the-counter drugs allows people to treat themselves for many ailments from arthritis to yeast infections. Consumers need to ask their physicians about potential interactions and check with their pharmacists for instructions on taking drugs with foods. If problems arise, they should seek professional care without delay.

REFERENCES

1. A. Bhutto and J. E. Morley, The clinical significance of gastrointenstinal changes with aging, *Current Opinion in Clinical and Nutritional Metabolism* 11 (2008): 651–660; J. Tam-McDevitt, Polypharmacy, aging, and cancer, *Oncology* 22 (2008): 1052–1055.
2. D. M. Qato and coauthors, Use of prescription and over-the-counter medications and dietary supplements among older adults in the United States, *Journal of the American Medical Association* 300 (2008): 2867–2878.
3. K. Stein, When essential medications provoke new health problems: The metabolic effects of second-generation antipsychotics, *Journal of the American Medical Association* 110 (2009): 992–1001.
4. W. W. McCloskey, K. Zaiken, and R. R. Couris, Clinically significant grapefruit juice–drug interactions, *Nutrition Today* 43 (2008): 19–26; M. F. Paine and coauthors, Further characterization of a furanocoumarin-free grapefruit juice on drug disposition: Studies with cyclosporine, *American Journal of Clinical Nutrition* 87 (2008): 863–871.

CENGAGE LEARNING'S GLOBAL NUTRITION WATCH — GLOBAL NUTRITION RESOURCE CENTER

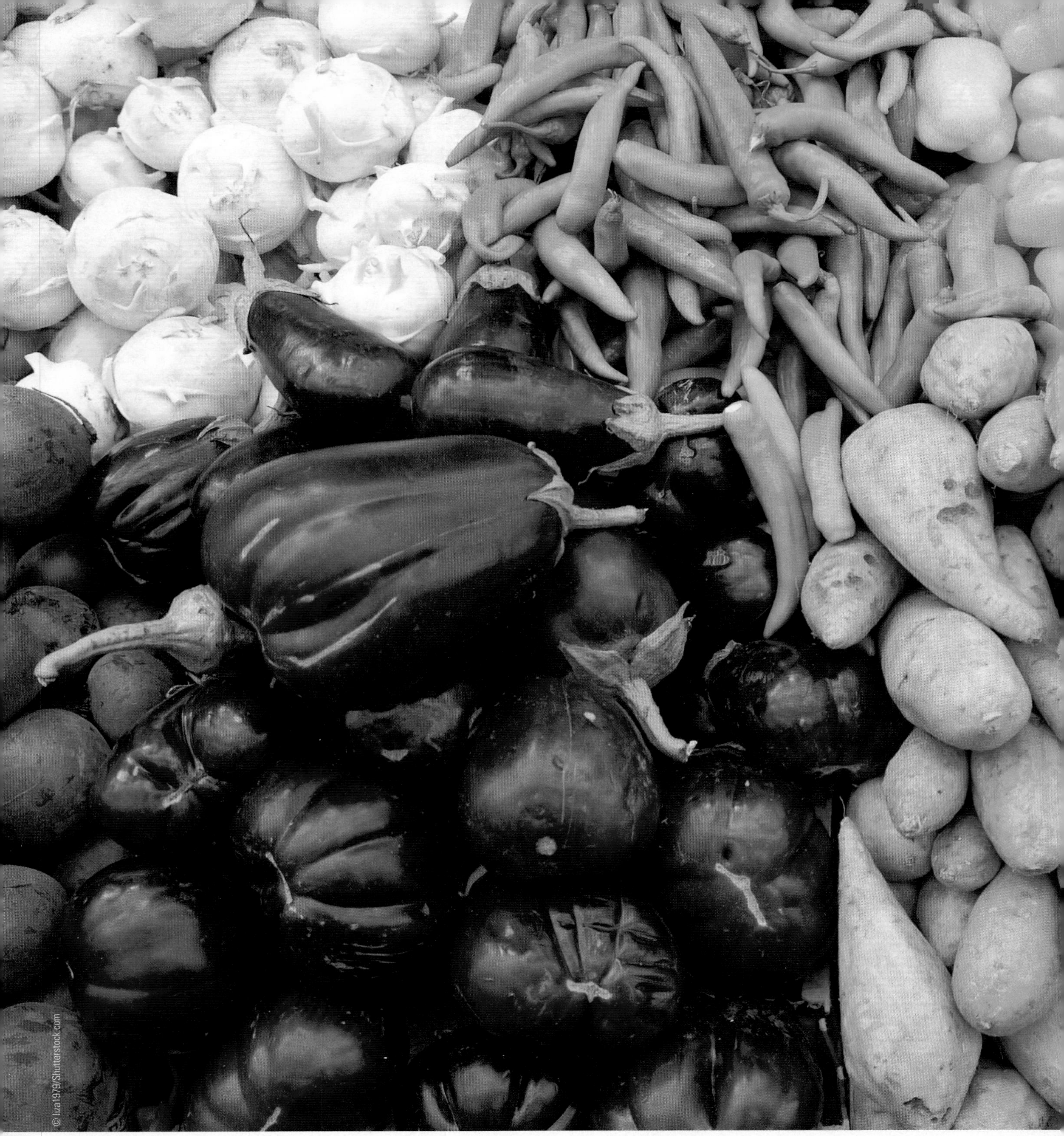

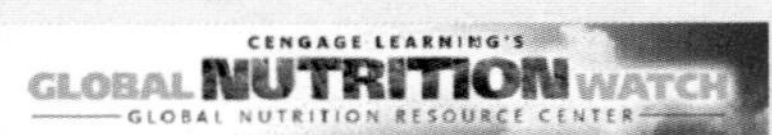

Dietary choices promote health, but they also have preventive and therapeutic measures. Go to Nutrition and Health/Disease → Nutrition and Cancer to learn how planning a proper diet can effectively reduce the risk of cancer and assist in the fight against the disease.

18

Diet and Health

LEARNING GPS

Nutrition in Your Life

No doubt, you're familiar with the recommendations. Eat more veggies. Eat more fiber. Eat more fish. Put down the saltshaker. Limit the fat. Be active. Don't smoke. And don't drink too much alcohol. What's the deal? If you follow this advice, will it really make a difference in how well or how long you live? In a word, yes. You can bet your life on it. If you could grow old in good health without having a heart attack or stroke, or getting diabetes, hypertension, or cancer, wouldn't you be willing to do just about anything—including improving your diet and activity habits? Of course, you would. And you can start today. In the Nutrition Portfolio at the end of this chapter, you can determine your risk factors for chronic diseases and learn how to minimize those risks.

Much of this text has described how good nutrition supports good health. This chapter examines some of the relationships between nutrition and disease—exploring how poor nutrition may promote the progression of diseases and how good nutrition may guard against the development of diseases. The bulk of this chapter focuses on the chronic diseases that pose the greatest threat to the lives of most people in developed countries, but it begins with a description of the immune system and its inflammatory response. As you will see, inflammation underlies the development of many chronic diseases. Chronic diseases develop over a lifetime as a result of metabolic abnormalities induced by such factors as genetics, age, gender, and lifestyle. As you have learned, diet is among the many lifestyle factors that influence the development of chronic diseases.[1]

18.1 Nutrition and Infectious Diseases

LEARN IT Identify factors that protect people from the spread of infectious diseases and describe the role of nutrition in immunity.

Infectious diseases such as smallpox once claimed the lives of many children and limited the average life expectancy of adults. Thanks to medical science's ability to identify disease-causing microorganisms and develop preventive strategies, most people now live well into their later years, and the average life expectancy far exceeds that of our ancestors. In developed nations, purification of water and safe handling of foods help prevent the spread of infection. Antibiotics and immunizations provide additional protection for individuals.[2]

Despite these advances, some infectious diseases still endanger many lives today. Disease strains such as tuberculosis and some foodborne infections, for example, have become resistant to antibiotics.[3] Nutrition cannot directly prevent or cure infectious diseases, but good nutrition can strengthen, and malnutrition can weaken, the body's defenses against them.

It is difficult to know exactly where infectious diseases fall among the leading causes of death. Compared with chronic diseases, infectious diseases pose a much greater challenge for public health officials who track disease prevalence. One physician might classify an ear infection as an infectious disease, whereas another calls it a disease of the ear. Trends change quickly as well. A disease, such as AIDS, that did not even exist until the early 1980s may suddenly appear and become one of the leading causes of death. A preventive strategy, such as food irradiation, may just as quickly eliminate hundreds of thousands of cases of foodborne infections each year. Public health strategies help the entire country defend against the spread of infection, and each individual's immune system provides a personal line of defense. A strong immune system depends on adequate nutrition. Poor nutrition weakens the immune system, which increases susceptibility to infections.

The Immune System The **immune system** defends the body so diligently and silently that people do not even notice the thousands of enemy attacks mounted against them every day (the accompanying glossary defines immune system terms). If the immune system fails, though, the body suddenly becomes vulnerable to every wayward disease-causing agent that comes its way. Infectious disease invariably follows.

The body's first lines of defense—the skin, mucous membranes, and GI tract—normally deter foreign substances. If these barriers fail, then the organs ♦ and cells of the immune system race into action. Foreign substances that gain entry into the body and elicit such a response—the **immune response**—are called **antigens.** Examples include bacteria, viruses, toxins, and food proteins that cause allergies.

♦ Organs of the immune system:
- Spleen
- Lymph nodes
- Thymus

infectious diseases: diseases caused by bacteria, viruses, parasites, or other microorganisms that can be transmitted from one person to another through air, water, or food; by contact; or through vector organisms such as mosquitoes.

GLOSSARY OF IMMUNE SYSTEM TERMS

antibodies: large proteins of the blood and body fluids, produced by the immune system in response to the invasion of the body by foreign molecules (usually proteins called *antigens*). Antibodies combine with and inactivate the foreign invaders, thus protecting the body.

antigens: substances that elicit the formation of antibodies or an inflammation reaction from the immune system. A bacterium, a virus, a toxin, and a protein in food that causes allergy are all examples of antigens.

B-cells: lymphocytes that produce antibodies. *B* stands for *bone marrow,* where the B-cells develop and mature.

cytokines (SIGH-toe-kines): special proteins that direct immune and inflammatory responses.

immune response: the body's reaction to foreign antigens, which neutralizes or eliminates them, thus preventing damage.

immune system: the body's natural defense against foreign materials that have penetrated the skin or mucous membranes.

immunoglobulins (IM-you-noh-GLOB-you-linz): proteins capable of acting as antibodies.

lymphocytes (LIM-foh-sites): white blood cells that participate in acquired immunity; B-cells and T-cells.

macrophages (MAK-roe-fay-jez): large phagocytic cells that serve as scavengers of the blood, clearing it of old or abnormal cells, cellular debris, and antigens.

neutrophils (NEW-tro-fills): the most common type of white blood cell. Neutrophils destroy antigens by phagocytosis.

phagocytes (FAG-oh-sites): white blood cells (neutrophils and macrophages) that have the ability to ingest and destroy foreign substances.

- **phagein** = to eat

phagocytosis (FAG-oh-sigh-TOH-sis): the process by which phagocytes engulf and destroy foreign materials.

T-cells: lymphocytes that attack antigens. *T* stands for the *thymus gland,* where the T-cells mature.

Of the 100 trillion cells that make up the human body, one in every hundred is a white blood cell. Two types of white blood cells, the phagocytes and lymphocytes, defend the body against infectious diseases.

Phagocytes: Neutrophils and Macrophages **Phagocytes,** the scavengers of the immune system, are the first to arrive at the scene if an invader, such as a microorganism, gains entry. Upon recognizing the foreign invader, the phagocyte engulfs and digests it, if possible, in a process called **phagocytosis.** Two types of immune system cells ingest and destroy foreign antigens by phagocytosis: **neutrophils** and **macrophages.** Neutrophils are the most common type of white blood cells and are responsible for much of the body's protection against infection. Macrophages move and kill bacteria more slowly than neutrophils, but they are larger and can engulf larger targets, including the body's dead and damaged cells. Phagocytes also secrete special proteins called **cytokines** that activate the metabolic and immune responses to infection.

Lymphocytes: B-cells There are two distinct types of **lymphocytes:** B-cells and T-cells. **B-cells** respond to infection by rapidly dividing and producing large proteins known as **antibodies.** Antibodies travel in the bloodstream to the site of infection. There they stick to the surfaces of antigens and kill or otherwise inactivate them, making them easy for phagocytes to ingest.

The antibodies are members of a class of proteins known as **immunoglobulins**—literally, large globular proteins that produce immunity. Antibodies react selectively to a specific foreign organism, and the B-cells retain a memory of how to make them. Consequently, the immune system can respond with greater speed the next time it encounters the same foreign organism. By doing so, B-cells play a major role in resistance to infection.

Lymphocytes: T-cells The **T-cells** travel directly to the invasion site to battle the invaders. T-cells recognize the antigens displayed on the surfaces of phagocyte cells and multiply in response. Then they release powerful chemicals to destroy all the foreign particles that have this antigen on their surfaces. As the T-cells begin to win the battle against infection, they release signals to slow down the immune response.

Unlike the phagocytes, which are capable of inactivating many different types of invaders, T-cells are highly specific. Each T-cell can attack only one type of antigen. This specificity is remarkable, for nature creates millions of antigens. After destroying a particular antigen, some T-cells retain the necessary information to serve as memory cells so that the immune system can rapidly produce the same type of T-cells again if the identical infection recurs.

T-cells actively defend the body against fungi, viruses, parasites, and a few types of bacteria; they can also destroy cancer cells. In organ transplant patients, T-cells participate in the rejection of newly transplanted tissues, which is why physicians prescribe immunosuppressive drugs following such surgery.

Each of the immune system cells ♦ plays a key role in fighting infectious disease. The differentiation and multiplication of these cells depends on a full array of nutrients.

♦ Cells of the immune system:
- Phagocytes
  - Neutrophils
  - Macrophages
- Lymphocytes
  - B-cells
  - T-cells

Nutrition and Immunity

Of all the body's systems, the immune system responds most sensitively to subtle changes in nutrition status. Malnutrition compromises both immune system tissues and immune responses.[4] From day to day, the immune system requires adequate amounts of protein, fat, carbohydrate, vitamins, and minerals to maintain its tissues and mount a response. Immune cells in tissues, lymph, and blood constantly survey their surroundings to detect foreign invaders and destroy those invaders before any symptoms of illness set in. If some of these foreign invaders escape destruction, then the immune system goes into action, initiating the immune response. The immune response demands greater amounts of nutrients for the synthesis of antibodies and signaling molecules such as cytokines, for cell multiplication, for free radical generation, and for the active process of ending the response.[5] Exactly which nutrients are needed in greater quantities, and how much, is not yet fully known, but the margin lists several nutrients ♦ known to play key roles in immunity.

♦ Nutrients known to affect immunity:
- Fatty acids
- Folate
- Iron
- Protein
- Selenium
- Vitamin A
- Vitamin B_6
- Vitamin C
- Vitamin D
- Vitamin E
- Zinc

> **FIGURE 18-1 Nutrition and Immunity**

Regardless of where a person enters the spiral, malnutrition, illness, and weakened immunity interact to compromise recovery and worsen malnutrition.

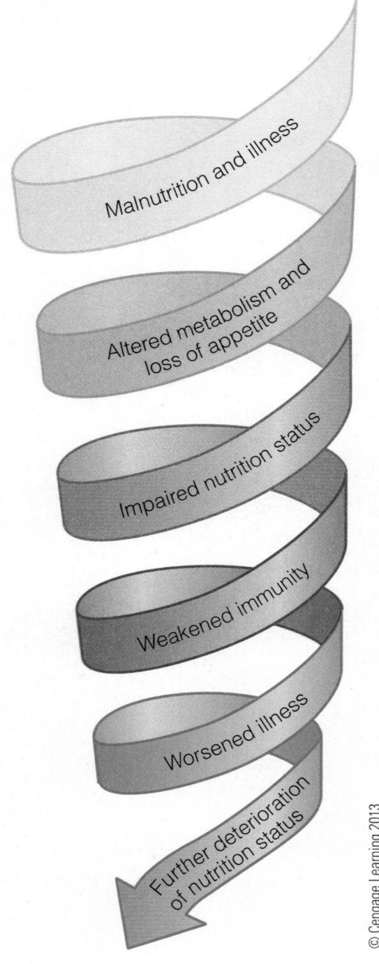

Impaired immunity opens the way for infectious diseases, which typically raise nutrient needs and reduce food intake. Consequently, nutrition status suffers further. Thus disease and malnutrition create a **synergistic** downward spiral that must be broken for recovery to occur (see Figure 18-1).

Quite simply, optimal immunity depends on optimal nutrition—enough, but not too much, of each of the nutrients. People with weakened immune systems, such as the elderly, benefit from a nutritious diet and supplements of selected nutrients.

HIV and AIDS Perhaps the most infamous infectious disease today is **AIDS (acquired immune deficiency syndrome).** AIDS develops from infection by **HIV (human immunodeficiency virus),** which is transmitted by direct contact with contaminated body fluids, including semen, vaginal secretions, and blood (but not saliva), or by passage of the infection from a mother to her infant during pregnancy, birth, or breastfeeding. HIV attacks the immune system and disables the body's defenses against other diseases. Then these diseases, which would produce only mild, if any, illness in people with healthy immune systems, destroy health and life.

Table 18-1 shows the impact of AIDS worldwide and in North America. For many years, the devastating effects of HIV infection seemed unstoppable. However, in the mid-to-late 1990s, the death rate in the United States from AIDS began to decline, and the progression from HIV to AIDS slowed dramatically. Even though remarkable progress has been made in understanding and treating HIV infection, the disease still has no cure. Without a cure, the best course is prevention. Unlike the chronic diseases featured in the remainder of this chapter, AIDS prevention does not in any way depend on good nutrition. Although good nutrition cannot prevent or cure AIDS, an adequate diet may improve responses to drugs, shorten hospital stays, promote independence, and improve the quality of life.[6] In addition, because common food bacteria can easily overwhelm a compromised immune system, attention to food safety is critical. (Chapter 19 provides food safety strategies.)

TABLE 18-1 The HIV and AIDS Epidemic at a Glance, 2009

Stage of Epidemic	World	North America
Individuals living with HIV infection or AIDS	33,300,000	1,500,000
Individuals newly infected with HIV	2,600,000	70,000
AIDS deaths	1,800,000	26,000

© Cengage Learning 2013

SOURCE: Joint United Nations Programme on HIV/AIDS, The 2010 UNA IDS *Report on the Global AIDS Epidemic,* www.unaids.org.

Inflammation and Chronic Diseases The immune system's response to infection or injury results in inflammation. The blood supply to the infected area increases and the blood vessels become permeable, which allows the white blood cells to rush to the site. As phagocytes engulf the offending microbes, they release oxidative products, such as hydrogen peroxide, that kill the microbes. In this acute phase, inflammation fights off the infection or injury, removes damaged tissue, heals wounds, and promotes recovery from external stressors.[7] In this way, acute inflammation and oxidative stress are beneficial.

When the inflammatory process persists, however, chronic inflammation is harmful. Cells of chronically inflamed tissues produce cytokines, oxidative products, blood clotting factors, and other bioactive chemicals that sustain the inflammatory response.[8] Such sustained inflammation threatens health and worsens the development of each of the chronic diseases discussed in the remainder of this chapter.

REVIEW IT Identify factors that protect people from the spread of infectious diseases and describe the role of nutrition in immunity.

Public health measures such as purification of water and safe handling of food help prevent the spread of infection in developed nations, and immunizations and antibiotics protect individuals. Nevertheless, some infectious diseases still endanger people today. Nutrition cannot prevent or cure infectious diseases, but adequate intakes of all the nutrients can help support the immune system as the body defends against disease-causing agents. If the immune system is impaired because of malnutrition or diseases such as AIDS, a person becomes vulnerable to infectious disease. Inflammation underlies many chronic diseases.

synergistic (SIN-er-JIS-tick): multiple factors operating together in such a way that their combined effects are greater than the sum of their individual effects.

AIDS (acquired immune deficiency syndrome): the late stage of HIV infection, in which severe complications of opportunistic infections and cancers develop.

HIV (human immunodeficiency virus): the virus that destroys lymphocytes and impairs immunity, eventually causing AIDS.

18.2 Nutrition and Chronic Diseases

LEARN IT List the leading nutrition-related causes of death in the United States.

Figure 18-2 shows the 10 leading causes of death in the United States.[9] Many of these deaths reflect chronic diseases that developed in response to lifestyle factors. ♦ Note that four of these causes of death, including the top two, have some relationship with diet. Taken together, these four conditions account for 60 percent of the nation's more than 2 million deaths each year. Worldwide, statistics are similar, with developing nations sharing many of the same chronic diseases as developed nations.[10]

♦ Lifestyle factors that contribute to the development of chronic diseases:
- Diet
- Physical inactivity
- Overweight
- Tobacco use
- Alcohol and drug abuse

This chapter explains how the major chronic diseases develop and summarizes their major links with nutrition. Earlier chapters that describe the connections between individual nutrients and diseases may have left the mistaken impression of "one disease–one nutrient" relationships. Indeed, valid links do exist between saturated fat and heart disease, calcium and osteoporosis, and antioxidant nutrients and cancer, but focusing only on these links oversimplifies the story. In reality, each nutrient may have connections with several diseases because its role in the body is not specific to a disease but to a body function. Furthermore, each of the chronic diseases develops in response to multiple risk factors, including many nondietary factors such as genetics, physical inactivity, and smoking. This chapter presents an integrated and balanced approach to disease prevention, paying careful attention to all of the factors involved. Table 18-2 (p. 586) presents some of the relationships between risk factors and chronic diseases. Figure 18-3 (p. 586) shows how many of the diseases themselves are risk factors for other chronic diseases. For example, a person with diabetes is likely to develop atherosclerosis and hypertension.

Notice that all of the diseases listed in Table 18-2 have a genetic component. A family history of a certain disease is a powerful indicator of a person's tendency to contract that disease. Still, lifestyle factors are often pivotal in determining whether that tendency will be expressed.[11] Genetics and lifestyle often work synergistically; for instance, cigarette smoking is especially likely to bring on heart disease in people who

Vegetables rich in fiber, phytochemicals, and the antioxidant nutrients (beta-carotene, vitamin C, and vitamin E) help to protect against chronic diseases.

> **FIGURE 18-2 The 10 Leading Causes of Death in the United States**

Many deaths have multiple causes, but diet influences the development of several chronic diseases—notably, heart disease, some types of cancer, stroke, and diabetes.

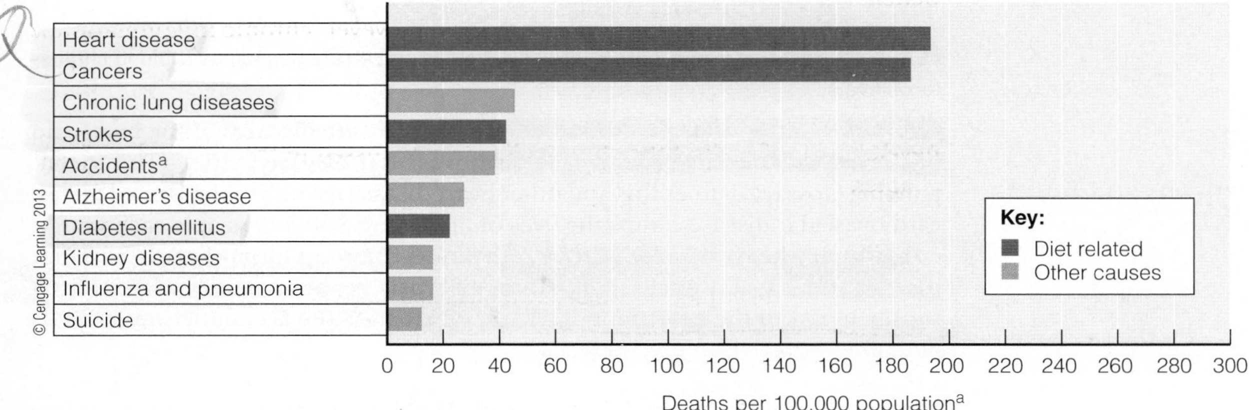

NOTE: Rates are age adjusted to allow relative comparisons of mortality among groups and over time.
[a]Motor vehicle and other accidents are the leading cause of death among people aged 15–24, followed by homicide, suicide, cancer, and heart disease. Alcohol contributes to about half of all accident fatalities.
SOURCE: Data from S. L. Murphy, J. Xu, and K. D. Kochanek and the Division of Vital Statistics, Deaths: Preliminary data for 2010, *National Vital Statistics Reports*, January 11, 2012.

TABLE 18-2 Risk Factors and Chronic Diseases

	Cancers	Hypertension	Diabetes (type 2)	Atherosclerosis	Obesity	Stroke
Dietary Risk Factors						
Diets high in added sugars (beverages)					✓	
Diets high in salty or pickled foods	✓	✓				
Diets high in saturated and/or *trans* fat	✓	✓	✓	✓	✓	✓
Diets low in fruits, vegetables, and other foods rich in fiber and phytochemicals	✓		✓	✓	✓	✓
Diets low in vitamins and/or minerals	✓	✓		✓		
Excessive alcohol intake	✓	✓		✓	✓	✓
Other Risk Factors						
Age	✓	✓	✓	✓		✓
Environmental contaminants	✓					
Genetics	✓	✓	✓	✓	✓	✓
Sedentary lifestyle	✓	✓	✓	✓	✓	✓
Smoking and tobacco use	✓	✓		✓		✓
Stress		✓		✓		✓

> FIGURE 18-3 Interrelationships among Chronic Diseases

Notice that many chronic diseases are themselves risk factors for other chronic diseases and that all of them are linked to obesity. The risk factors highlighted in blue define the metabolic syndrome.

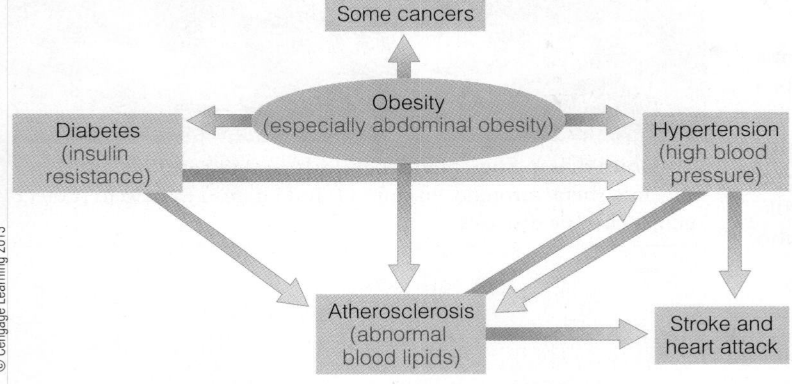

are genetically predisposed to develop it. Not smoking would benefit everyone's health, of course, regardless of genetic predisposition, but some recommendations to prevent chronic diseases best meet an individual's needs when family history is considered. For example, women with a family history of breast cancer might reduce their risks if they abstain from alcohol, whereas those with a family history of heart disease might benefit from one or two glasses of wine a week.

REVIEW IT List the leading nutrition-related causes of death in the United States.

Heart disease and cancers are the 2 leading causes of death in the United States, and strokes and diabetes also rank among the top 10. All four of these chronic diseases have significant links with nutrition. Other lifestyle risk factors and genetics are also important.

18.3 Cardiovascular Disease

LEARN IT Describe how atherosclerosis develops and strategies to lower blood cholesterol levels.

The major causes of death around the world today are diseases of the heart and blood vessels, collectively known as **cardiovascular disease (CVD)**. (The accompanying glossary defines this and other heart disease terms.) In the United States, cardiovascular disease claims the lives of more than 800,000 people each year.[12]

Coronary heart disease (CHD) is the most common form of cardiovascular disease and is usually caused by **atherosclerosis** in the **coronary arteries** that supply blood to the heart muscle. Atherosclerosis is the accumulation of lipids and other materials in the arteries.

How Atherosclerosis Develops As Highlight 16 points out, no one is free of the fatty streaks that may one day become the **plaques** of atherosclerosis.* For

*Plaque associated with atherosclerosis is known as *atheromatous* (ATH-er-OH-ma-tus) *plaque*.

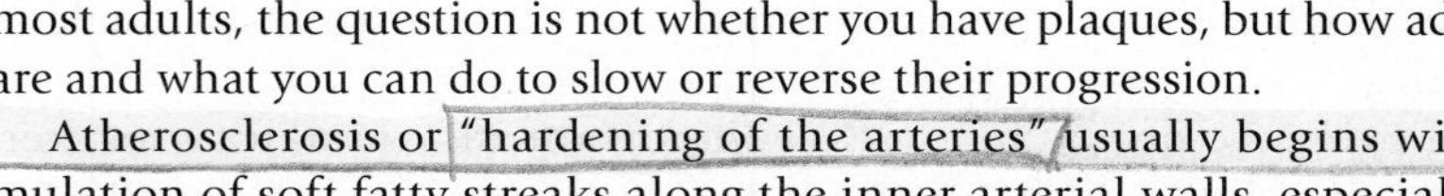

most adults, the question is not whether you have plaques, but how advanced they are and what you can do to slow or reverse their progression.

Atherosclerosis or "hardening of the arteries" usually begins with the accumulation of soft fatty streaks along the inner arterial walls, especially at branch points (review Figure H16-1 on p. 546). These fatty streaks gradually enlarge and harden as they fill with cholesterol, other lipids, and calcium, and they become encased in fibrous connective tissue, forming plaques. Plaques stiffen the arteries and narrow the passages through them. Most people have well-developed plaques by the age of 30. As Chapter 5 points out, a diet high in saturated fat is a major contributor to the development of plaques and the progression of atherosclerosis.[13] But atherosclerosis is much more than the simple accumulation of lipids within the artery wall—it is a complex inflammatory response to tissue damage. Indeed, extensive evidence confirms that inflammation is centrally involved in all stages of atherosclerosis.[14]

Inflammation The cells lining the blood vessels may incur damage from high LDL cholesterol, hypertension, toxins from cigarette smoking, obesity, elevated homocysteine, or some viral and bacterial infections.[15] Such damage increases the permeability of the blood vessel walls and elicits an inflammatory response. The immune system sends in macrophages (the large, phagocytic cells of the immune system mentioned earlier), and the smooth muscle cells of the artery wall try to repair the damage. Particles of LDL cholesterol become trapped in the blood vessel walls. Free radicals produced during inflammatory responses oxidize the LDL cholesterol, and the macrophages engulf it. The macrophages swell with large quantities of oxidized LDL cholesterol and eventually become the cells of plaque. Arterial damage and the inflammatory response also favor the formation of blood clots and allow minerals to harden plaque and form the fibrous connective tissue that encapsulates it.

The inflammatory response of atherosclerosis weakens the walls of the arteries and may cause an **aneurysm**—the abnormal bulging of a blood vessel wall. Aneurysms can rupture and lead to massive bleeding and death, particularly when a large blood vessel such as the aorta is affected. The central role of the inflammatory response in atherosclerosis has led researchers to look for signs or markers of inflammation in the blood vessel walls. One of the most promising of these markers is a protein known as **C-reactive protein (CRP).** High levels of CRP have proved to more accurately predict future heart attack than high

C-reactive protein (CRP): a protein released during the acute phase of infection or inflammation that enhances immunity by promoting phagocytosis and activating platelets. Its presence may be used to assess a person's risk of an impending heart attack or stroke.

GLOSSARY OF HEART DISEASE TERMS

aneurysm (AN-you-rizm): an abnormal enlargement or bulging of a blood vessel (usually an artery) caused by damage to or weakness in the blood vessel wall.

angina (an-JYE-nah or AN-ji-nah): a painful feeling of tightness or pressure in and around the heart, often radiating to the back, neck, and arms; caused by a lack of oxygen to an area of heart muscle.

atherosclerosis: a type of artery disease characterized by plaques along the inner walls of the arteries.

cardiovascular disease (CVD): a general term for all diseases of the heart and blood vessels.

CHD risk equivalents: disorders that raise the risk of heart attacks, strokes, and other complications associated with cardiovascular disease to the same degree as existing CHD. These disorders include symptomatic carotid artery disease, peripheral arterial disease, abdominal aortic aneurysm, and diabetes mellitus.

coronary arteries: blood vessels that supply blood to the heart.

coronary heart disease (CHD): the damage that occurs when the blood vessels carrying blood to the heart (the *coronary arteries*) become narrow and occluded.

embolism (EM-boh-lizm): the obstruction of a blood vessel by an *embolus* (EM-boh-luss), or traveling clot, causing sudden tissue death.

- **embol** = to insert, plug

heart attack: sudden tissue death caused by blockages of vessels that feed the heart muscle; also called *myocardial* (my-oh-KAR-dee-al) *infarction* (in-FARK-shun) or *cardiac arrest.*

- **myo** = muscle
- **cardial** = heart
- **infarct** = tissue death

hypertension: consistently higher-than-normal blood pressure. Hypertension that develops without an identifiable cause is known as *essential* or *primary hypertension*; hypertension that is caused by a specific disorder such as kidney disease is known as *secondary hypertension.*

plaques (PLACKS): the accumulation of fatty deposits, smooth muscle cells, and fibrous connective tissue that develops in the artery walls in atherosclerosis.

prehypertension: slightly higher than normal blood pressure, but not as high as hypertension (see Table 18-4, p. 589).

stroke: an event in which the blood flow to a part of the brain is cut off; also called *cerebrovascular accident (CVA).*

- **cerebro** = brain
- **vascular** = blood vessels

thrombosis (throm-BOH-sis): the formation of a *thrombus* (THROM-bus), or a blood clot, that may obstruct a blood vessel, causing gradual tissue death.

- **thrombo** = clot

transient ischemic (is-KEY-mik) **attack (TIA):** a temporary reduction in blood flow to the brain, which causes temporary symptoms that vary depending on the part of the brain affected. Common symptoms include light-headedness, visual disturbances, paralysis, staggering, numbness, and inability to swallow.

LDL cholesterol, which has a strong relationship with atherosclerosis, as a later section explains.[16]

Emerging evidence points to another important inflammatory marker, **lipoprotein-associated phospholipase A(2)** or **Lp-PLA(2)**. Lp-PLA(2) appears to be a highly specific marker of plaque inflammation and the formation of plaques that are most susceptible to rupture.[17] To better assess vascular inflammation and identify people at high risk of cardiovascular disease, Lp-LPA(2) is recommended as a diagnostic test in addition to traditional risk assessments.[18]

Plaques Once a plaque has formed, a sudden spasm or surge in blood pressure in an artery can tear away part of its fibrous coat, causing it to rupture. Some types of plaque are more unstable than others and are therefore more vulnerable to rupture.[19]* Such plaques have a thin fibrous cap, a large lipid core, and an abundance of macrophages—characteristics that undermine plaque stability. Researchers now know that the *composition* of plaques—rather than the *size* of plaques—is a key predictor of plaque rupture and subsequent clot formation. When plaques rupture, the immune system responds to the damage as it would to other tissue injuries.

Blood Clots **Platelets** are tiny disc-shaped bodies that cover an injured or damaged area. Together with other factors, platelets form blood clots. Abnormal blood clotting can trigger life-threatening events. For example, a blood clot may gradually grow large enough to restrict or close off a blood vessel **(thrombosis).** A *coronary thrombosis* blocks blood flow through an artery that feeds the heart muscle. A *cerebral thrombosis* blocks blood flow through an artery that feeds the brain. A clot may also break free from an artery wall and travel through the circulatory system until it lodges in a small artery and suddenly shuts off blood flow to the tissues **(embolism).**

The action of platelets is under the control of certain eicosanoids, known as prostaglandins and thromboxanes, which are made from the 20-carbon omega-6 and omega-3 fatty acids (introduced in Chapter 5). Each eicosanoid plays a specific role in helping to regulate ♦ many of the body's activities. Sometimes their actions oppose each other.[20] For example, one eicosanoid prevents clot formation, and another promotes it. Similarly, one dilates the blood vessels, and another constricts them. When omega-3 fatty acids are abundant in the diet, ♦ they make more of the kinds of eicosanoids that favor heart health.[21]

♦ Eicosanoids help to regulate:
- Blood pressure
- Blood clot formation
- Blood vessel contractions
- Immune response
- Nerve impulse transmissions

♦ Major sources of omega-3 fatty acids:
- Vegetable oils (canola, soybean, flaxseed)
- Walnuts, flaxseeds
- Fatty fish (mackerel, salmon, sardines)

Blood Pressure The stress of blood flow along artery walls can cause physical damage to arteries.[22] High blood pressure intensifies the stress of blood flow on arterial tissue, provoking a low-grade inflammatory state that may stimulate plaque formation and progression.[23]

The Result: Heart Attacks and Strokes When atherosclerosis in the coronary arteries becomes severe enough to restrict blood flow and deprive the heart muscle of oxygen, CHD develops. The person with CHD often experiences pain and pressure in the area around the heart **(angina).** A **heart attack** occurs when blood flow to the heart is cut off and that area of the heart muscle dies. Restricted blood flow to the brain causes a **transient ischemic attack (TIA)** or **stroke.** Coronary heart disease and strokes are the first and fourth leading causes of death, respectively, for adults in the United States.

Risk Factors for Coronary Heart Disease

Although atherosclerosis can develop in any blood vessel, the coronary arteries are most often affected, leading to CHD. Table 18-3 lists the major risk factors for CHD. Notice that some risk factors, such as diet and physical activity, are *modifiable,* meaning that they can be changed; others, such as age, gender, and family history are not modifiable. The

TABLE 18-3 Risk Factors for CHD

Major Risk Factors for CHD (not modifiable)
- Increasing age
- Male gender
- Family history of premature heart disease

Major Risk Factors for CHD (modifiable)
- High blood LDL cholesterol
- Low blood HDL cholesterol
- High blood pressure (hypertension)
- Diabetes
- Obesity (especially abdominal obesity)
- Physical inactivity
- Cigarette smoking
- An "atherogenic" diet (high in saturated fats and low in vegetables, fruits, and whole grains)

NOTE: Risk factors highlighted in yellow have relationships with diet.
SOURCE: Expert Panel on Detection, Evaluation, and Treatment of High Blood Cholesterol in Adults (Adult Treatment Panel III), *Third Report of the National Cholesterol Education Program (NCEP),* NIH publication no. 02-5215 (Bethesda, Md.: National Heart, Lung, and Blood Institute, 2002), pp. II-15–II-20.

lipoprotein-associated phospholipase A(2) or **(Lp-PLA(2)**: a lipoprotein-bound enzyme that generates potent proinflammatory and proatherogenic products such as oxidized free fatty acids and lysophosphatidylcholine. LpPLA(2) is a specific marker of plaque inflammation.

platelets: tiny, disc-shaped bodies in the blood, important in blood clot formation.

*Plaque that is susceptible to rupture because it has only a thin fibrous barrier between its lipid-rich core and the artery lining is called *vulnerable plaque.*

TABLE 18-4 Standards for CHD Risk Factors

Risk Factors	Desirable	Borderline	High Risk
Total blood cholesterol (mg/dL)	<200	200–239	≥240
LDL cholesterol (mg/dL)	<100[a]	130–159	160–189[b]
HDL cholesterol (mg/dL)	≥60	59–40	<40
Triglycerides, fasting (mg/dL)	<150	150–199	200–499[c]
Body mass index (BMI)[d]	18.5–24.9	25–29.9	≥30
Blood pressure (systolic and/or diastolic pressure)	<120/<80	120–139/80–89[e]	≥140/≥90[f]

[a]100–129 mg/dL LDL indicates a near or above optimal level.
[b]≥190 mg/dL LDL indicates a very high risk.
[c]≥500 md/dL triglycerides indicates a very high risk.
[d]Body mass index (BMI) is defined in Chapter 8; BMI standards are found on the inside back cover.
[e]These values indicate prehypertension.
[f]These values indicate stage one hypertension; ≥160/≥100 indicates stage two hypertension. Physicians use these classifications to determine medical treatment.

criteria for defining blood lipids, blood pressure, and obesity in relation to CHD risk are shown in Table 18-4; Table H16-1 on p. 547 presents cholesterol standards for children and adolescents.

By age 20, half of the adults in the United States have at least one major risk factor for CHD, and many have more than one.[24] Public health officials in both the United States and Canada recommend screening to identify risk factors in individuals and offer preventive advice for the population.[25] Regular screening and early detection have proved successful: since 1960, both blood cholesterol levels and deaths from cardiovascular disease among US adults have shown a continuous and substantial downward trend.[26] These trends also reflect behavior changes in individuals. As adults grow older, many of them stop smoking, limit alcohol consumption, and become mindful that their food choices can improve their cardiovascular health.

Age, Gender, and Family History A review of Table 18-3 shows that three of the major risk factors for CHD cannot be modified by diet or otherwise: age, gender, and family history. As men and women grow older, the risk of CHD rises. The increasing risk of CHD with advancing age reflects the steady progression of atherosclerosis.[27] On average, older people have more atherosclerosis than younger people do.

In men, aging becomes a significant risk factor at age 45 and older. CHD occurs about 10 to 15 years later in women than in men. Women younger than 45 tend to have lower LDL cholesterol than men of the same age, but women's blood cholesterol typically begins to rise between ages 45 and 55. Thus aging becomes a significant risk factor for women who are 55 and older. The gender difference has been attributed to a protective effect of estrogen in women, but CHD rates do not suddenly accelerate at menopause as naturally occurring estrogen levels taper off.[28] Rather, as in men, heart disease rates increase linearly with age. And, as in men, all of the major risk factors raise the risk of CHD in women. Ultimately, CHD kills as many women as men—and kills more women in the United States than any other disease.

Nonetheless, at every age, men have a greater risk of CHD than women do. The reasons for this gender difference are not completely understood, but they can be partly explained by the earlier onset of risk factors such as elevated LDL cholesterol and blood pressure in men. Levels of the amino acid homocysteine, which may damage artery walls and increase oxidative stress, rise with age and are generally higher in men. Researchers have not determined whether the damage is caused by homocysteine itself or by a factor associated with it.[29]

A history of early CHD in immediate family members is an independent risk factor even when other risk factors are considered. The more family members affected and the earlier the age of onset, the greater the risk.[30]

High LDL and Low HDL Cholesterol In population studies, the relationship between total blood cholesterol and atherosclerosis is strong—and most of the total cholesterol is made up of LDL cholesterol. The higher the LDL cholesterol, the greater the risk of CHD. In contrast, the lower the LDL cholesterol and blood pressure, the slower the progression of atherosclerosis.[31]

The LDL are clearly the most atherogenic lipoproteins. As Chapter 5 explains, HDL also carry cholesterol, but raised HDL represents cholesterol returning from the cells to the liver where it will be used to make bile and excreted in the GI tract. Thus high HDL indicates a *reduced* risk of atherosclerosis and heart attack. High LDL and low HDL correlate *directly* with heart disease, ♦ whereas low LDL and high HDL correlate *inversely* with risk.

♦ Cholesterol is carried in several lipoproteins, chief among them LDL and HDL (see Chapter 5 for details). Remember them this way:
- **L**DL = **L**ow-density lipoproteins = **L**ess healthy
- **H**DL = **H**igh-density lipoproteins = **H**ealthy

Any LDL cholesterol that remains in the blood after the body's cells take up the amount they need becomes vulnerable to oxidation. High blood levels of LDL cholesterol, especially oxidized LDL, trigger a series of events that promote plaque formation and contribute to plaque instability.[32] Oxidized LDL cholesterol stimulates production of atherogenic signaling molecules. These signaling molecules attract immune cells and allow them to adhere to, and penetrate, the innermost layer of the arterial wall. Within the arterial wall, these immune cells undergo transformation and replication into macrophages that form the lipid-rich foam cells characteristic of fatty streaks. Macrophages within the arterial wall perpetuate chronic inflammation by releasing inflammatory cytokines that contribute to plaque rupture.[33] When plaques rupture, a heart attack or stroke may occur. In the early stages of atherosclerosis, the goal of treatment is to slow plaque development. In the later stages, the goal of treatment is to stabilize plaques.

Research shows that atherosclerosis is reversible by removing the trapped macrophages from the arterial wall.[34] Key antioxidants such as resveratrol break the adhesion, releasing the macrophages from the arterial wall and promoting healing. (Highlight 11 discusses antioxidants and Highlight 13 describes the healthful actions of resveratrol and other phytochemicals.)

High Blood Pressure (Hypertension) Atherosclerosis is frequently accompanied by chronic high blood pressure (**hypertension**). The higher blood pressure is above normal, the greater the risk of heart disease. However, even values only slightly higher than desirable—classified as **prehypertension** in Table 18-4—increase the risk of heart attack and stroke.[35] This relationship between hypertension and heart disease risk holds true for men and women, young and old. High blood pressure injures the artery walls and accelerates plaque formation, thus initiating or worsening the progression of atherosclerosis. Then the plaques and reduced blood flow raise blood pressure further, and hypertension and atherosclerosis become mutually aggravating conditions.

Diabetes Diabetes—a major independent risk factor for all forms of cardiovascular disease—substantially increases the risk of death from CHD.[36] In diabetes, blood vessels often become blocked and circulation diminishes. Atherosclerosis progresses rapidly. For many people with diabetes, the risk of heart attack is similar to that of people with established CHD.[37] In fact, physicians describe diabetes and other disorders that have risks similar to CHD as **CHD risk equivalents.** Treatment to lower LDL cholesterol in diabetes follows the same recommendations as in CHD.

Obesity and Physical Inactivity Obesity, especially abdominal obesity, and physical inactivity significantly increase risk factors for CHD, contributing to high LDL cholesterol, low HDL cholesterol, hypertension, and diabetes.[38] Conversely, weight loss and physical activity protect against CHD by lowering LDL, raising HDL, improving insulin sensitivity, and lowering blood pressure.[39]

Cigarette Smoking Cigarette smoking is a powerful risk factor for CHD and other forms of cardiovascular disease. The risk increases the more a person smokes and is the same for men and women. Smoking damages the heart directly by increasing blood pressure and the heart's workload. It deprives the heart of oxygen and damages platelets, making blood clot formation likely. Toxins in cigarette smoke damage blood vessels, setting the stage for atherosclerosis. When people quit smoking, their risk of CHD begins to decline within a few months.[40]

Atherogenic Diet Diet also influences the risk of CHD. An "atherogenic" diet—high in saturated fats, *trans* fats, and cholesterol and low in vegetables, fruits, and whole grains—elevates LDL cholesterol. Conversely, diets rich in fruits, vegetables, and whole grains seem to lower the risk of CHD even more than might be expected based on risk factors such as LDL cholesterol alone.[41] Research shows that a diet rich in raw vegetables and fruits may protect against CHD even in people who are genetically susceptible to the disease.[42] Dietary strategies to reduce the risk of CHD are discussed fully in a later section.

Other Risk Factors The major risk factors for CHD listed in Table 18-3 (p. 588) and discussed in the previous sections have solid associations with the development of CHD. Nevertheless, other factors also seem to influence a person's risk of CHD. These factors, known as **emerging risk factors,** may be helpful in assessing an individual's risk of CHD. For example, some people with CHD, especially those with diabetes and those who are overweight, have elevated triglycerides. Elevated triglycerides are not directly atherogenic, and therefore do not represent an independent risk factor for CHD.[43] Rather, elevated triglycerides are considered an important marker of CHD risk because of the role they play in lipoprotein metabolism. When triglycerides are moderately high, remnants of very-low-density lipoproteins (VLDL) increase in the blood. These small, triglyceride-rich lipoproteins are highly atherogenic.

Metabolic Syndrome As Table 18-3 (p. 588) shows, most of the modifiable risk factors for CHD are directly related to diet. Several of these diet-related risk factors—low HDL, high blood pressure, elevated fasting blood glucose, and abdominal obesity—along with high blood triglycerides comprise a cluster of health risks known as **metabolic syndrome.** As Figure 18-3 (p. 586) shows, the risks that define metabolic syndrome ♦ underlie several chronic diseases and increase the risks of CHD and type 2 diabetes.[44] Abdominal obesity and **insulin resistance** are both hypothesized to be primary factors underlying the metabolic syndrome.[45] Metabolic syndrome, like the chronic diseases associated with it, also includes markers of inflammation and thrombosis.[46] Overeating and physical inactivity play a major role in the development of metabolic syndrome. The prevalence of metabolic syndrome among US adults is high, and treatment to reduce these risk factors for heart disease and diabetes should begin early and focus on changes in lifestyle.

♦ Metabolic syndrome includes any three of the following:
- Abdominal obesity: waist circumference >40 in (for men) or >35 in (for women)
- Triglycerides: ≥150 mg/dL
- HDL: <40 mg/dL (in men) or <50 mg/dL (in women)
- Blood pressure: ≥130/85 mm Hg
- Fasting glucose: ≥100 mg/dL

Recommendations for Reducing Coronary Heart Disease Risk

Recommendations to reduce CHD risk include both screening and intervention. The "How To" feature on p. 592 provides a tool to assess a person's 10-year heart disease risk. Notice that total cholesterol and HDL cholesterol are included in the assessment, but LDL cholesterol is not. LDL cholesterol is routinely estimated from measures of total cholesterol and HDL cholesterol and thus would not add information to this assessment.[47] Once a person's risks have been identified, treatment focuses on lowering LDL cholesterol. Lowering LDL cholesterol significantly reduces the incidence of CHD.[48] Treatment plans may include major lifestyle changes in diet, physical activity, and smoking cessation; medications; or combinations. The LDL cholesterol goals and treatment plans are specific to individuals, so they are best prescribed by a qualified health-care provider.

emerging risk factors: recently identified factors that enhance the ability to predict disease risk in an individual.

metabolic syndrome: a combination of risk factors—elevated fasting blood glucose, hypertension, abnormal blood lipids, and abdominal obesity—that greatly increase a person's risk of developing coronary heart disease; also called *Syndrome X, insulin resistance syndrome,* or *dysmetabolic syndrome.*

insulin resistance: the condition in which a normal amount of insulin produces a subnormal effect in muscle, adipose, and liver cells, resulting in an elevated fasting glucose; a metabolic consequence of obesity that precedes type 2 diabetes.

>How To

Assess Your Risk of Heart Disease

Do you know your heart disease risk score? Be aware that a high score does not mean that you *will* develop heart disease, but it should warn you of the possibility and prompt you to consult a physician about your health. You will need to know your blood cholesterol (ideally, the average of at least two recent measurements) and blood pressure (ideally, the average of several recent measurements). With this information in hand, find yourself in the five tables below and add the points for each risk factor.

Age (years)

	Men	Women
20–34	−9	−7
35–39	−4	−3
40–44	0	0
45–49	3	3
50–54	6	6
55–59	8	8
60–64	10	10
65–69	11	12
70–74	12	14
75–79	13	16

HDL (mg/dL)

	Men	Women
≥60	−1	−1
50–59	0	0
40–49	1	1
<40	2	2

Systolic Blood Pressure (mm Hg)

	Untreated		Treated	
	Men	Women	Men	Women
<120	0	0	0	0
120–129	0	1	1	3
130–139	1	2	2	4
140–159	1	3	2	5
≥160	2	4	3	6

Total Cholesterol (mg/dL)

	Age 20–39		Age 40–49		Age 50–59		Age 60–69		Age 70–79	
	Men	Women	Men	Women	Men	Women	Men	Women	Men	Women
<160	0	0	0	0	0	0	0	0	0	0
160–199	4	4	3	3	2	2	1	1	0	1
200–239	7	8	5	6	3	4	1	2	0	1
240–279	9	11	6	8	4	5	2	3	1	2
≥280	11	13	8	10	5	7	3	4	1	2

Smoking (any cigarette smoking in the past month)

	Age 20–39 Men	Age 20–39 Women	Age 40–49 Men	Age 40–49 Women	Age 50–59 Men	Age 50–59 Women	Age 60–69 Men	Age 60–69 Women	Age 70–79 Men	Age 70–79 Women
Smoker	8	9	5	7	3	4	1	2	1	1
Nonsmoker	0	0	0	0	0	0	0	0	0	0

Scoring Your Heart Disease Risk

Add up your total points: ________. Now find your total in the first column for your gender in the table at the right and then look to the next column for your approximate risk of developing heart disease within the next 10 years. Depending on your risk category, the following strategies can help reduce your risk:

- *>20% = High risk (CHD risk equivalent).*

  Try to lower LDL using all lifestyle changes and, most likely, lipid-lowering medications as well.

- *10–20% = Moderate risk*

  Try to lower LDL using all lifestyle changes and, possibly, lipid-lowering medications.

- *<10% = Low risk*

  Maintain or initiate lifestyle choices that help prevent elevation of LDL to prevent future heart disease.

Men		Women	
Total	Risk	Total	Risk
<0	<1%	<9	<1%
0–4	1%	9–12	1%
5–6	2%	13–14	2%
7	3%	15	3%
8	4%	16	4%
9	5%	17	5%
10	6%	18	6%
11	8%	19	8%
12	10%	20	11%
13	12%	21	14%
14	16%	22	17%
15	20%	23	22%
16	25%	24	27%
≥17	≥30%	≥25	≥30%

SOURCE: Adapted from Expert Panel on Detection, Evaluation, and Treatment of High Blood Cholesterol in Adults (Adult Treatment Panel III), *Third Report of the National Cholesterol Education Program (NCEP),* NIH publication no. 02-5216 (Bethesda, Md.: National Heart, Lung, and Blood Institute, 2002), section III.

TRY IT Estimate your 10-year risk for CHD using these charts from the Framingham Heart Study (or the American Heart Association website www.americanheart.org).

Cholesterol Screening To determine an individual's risk of CHD, health-care professionals review the person's health history and measure several blood lipids including total cholesterol, LDL cholesterol, HDL cholesterol, and triglycerides. Ideally, at least two measurements are taken at least 1 week apart and then compared to standards (shown earlier in Table 18-4 on p. 589). Single measurements may fail to identify those at risk or may misclassify them because blood cholesterol and other lipid concentrations vary significantly from day to day.

Lifestyle Changes Recommendations to reduce the risk of CHD focus first on lifestyle changes. To that end, people are encouraged to increase physical activity, lose weight (if necessary), implement dietary changes, and reduce exposure to tobacco smoke either by quitting smoking or by avoiding secondhand smoke.[49] Altering one's lifestyle is challenging, and instruction and counseling are critical for success. Health professionals can explain the reasons for change, set obtainable goals, and offer practical suggestions. If lifestyle changes fail to lower LDL or blood pressure to acceptable levels, then medications are prescribed. Table 18-5 summarizes strategies to reduce the risk of heart disease.[50] The "How To" feature on p. 594 offers suggestions for implementing a heart-healthy diet.

> DIETARY GUIDELINES FOR AMERICANS 2010

Consume less than 300 milligrams per day of dietary cholesterol. Individuals at high risk of cardiovascular disease should consume less than 200 milligrams of cholesterol daily.

REVIEW IT Describe how atherosclerosis develops and strategies to lower blood cholesterol levels.

Atherosclerosis is characterized by plaque build-up in artery walls. Plaques rupturing or blood clotting can cause heart attacks and strokes. Dietary recommendations to lower the risks of cardiovascular disease are summarized in Table 18-5. Quitting smoking and engaging in regular physical activity also improve heart health.

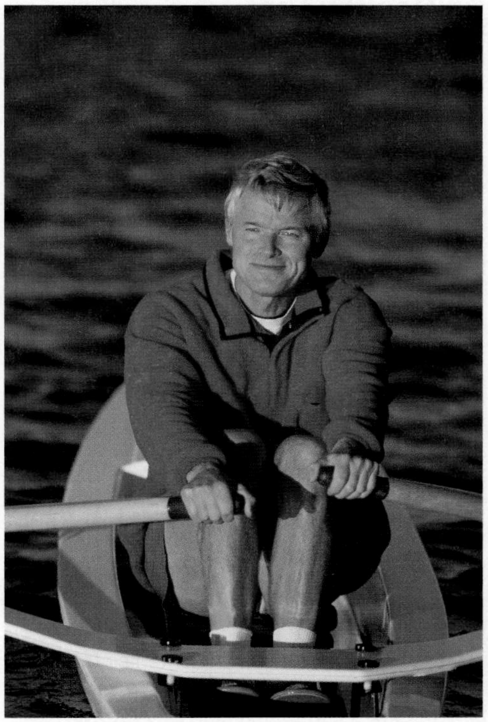

Regular aerobic exercise can help to defend against heart disease by strengthening the heart muscle, promoting weight loss, and improving blood lipids and blood pressure.

TABLE 18-5 Strategies to Reduce Risk of CHD

Dietary Strategies

- **Energy:** Balance energy intake and physical activity to prevent weight gain and to achieve or maintain a healthy body weight.
- **Saturated fat, *trans* fat, and cholesterol:** Choose lean meats, vegetables, and low-fat milk products; minimize intake of hydrogenated fats. Limit saturated fats to less than 7 percent of total kcalories, *trans* fat to less than 1 percent of total kcalories, and cholesterol to less than 300 milligrams a day.
- **Soluble fibers:** Choose a diet rich in vegetables, fruits, whole grains, and other foods high in soluble fibers.
- **Potassium and sodium:** Choose a diet high in potassium-rich fruits and vegetables, low-fat milk products, nuts, and whole grains. Choose and prepare foods with little or no salt (limit sodium intake to 1500 milligrams per day).
- **Added sugars:** Minimize intake of beverages and foods with added sugars.
- **Fish and omega-3 fatty acids:** Consume fatty fish rich in omega-3 fatty acids (salmon, tuna, sardines) at least twice a week.
- **Plant sterols and stanols:** Consume food products that contain added plant sterols or stanols.
- **Soy:** Consume soy foods to replace animal and dairy products that contain saturated fat and cholesterol.
- **Alcohol:** If alcohol is consumed, limit it to one drink daily for women and two drinks daily for men.

Lifestyle Choices

- **Physical activity:** Participate in at least 30 minutes of moderate-intensity endurance activity on most days of the week. The eventual goal should be an expenditure of at least 2000 kcalories weekly.
- **Smoking cessation:** Minimize exposure to any form of tobacco or tobacco smoke.

SOURCES: American Heart Association, www.heart.org, accessed April 2012; M. R. Flock and P. M. Kris-Etherton, Dietary Guidelines for Americans 2010: Implications for cardiovascular disease, *Current Atherosclerosis Reports* 13 (2011): 499–507.

>How To

Implement a Heart-Healthy Diet

Following a heart-healthy diet can require major changes in dietary choices. It helps to make a few changes at a time and to focus on positive choices (what to eat) first, rather than negative ones (what not to eat).

Grains

- Choose whole-grain breads and cereals that list "whole wheat" as the first ingredient on labels.
- Limit foods that list any *trans* fat in the Nutrition Facts panel or "hydrogenated oil" in the ingredients list.
- Limit products that contain tropical oils (coconut, palm, and palm kernel oil), which are high in saturated fat.

Fruits and Vegetables

- Consume fruits and vegetables frequently. Keep the refrigerator stocked with a variety of colorful fruits and vegetables (baby carrots, grapes, blueberries, melon).
- Incorporate at least one or two servings of fruits and vegetables into each meal.
- Choose canned products carefully. Canned vegetables (especially tomato-based products) may be high in sodium. Canned fruits may be high in added sugars.
- Limit high-sodium foods such as pickles, olives, sauerkraut, and kimchee.
- Limit french fries from fast-food restaurants, which are often loaded with *trans* fats.

Lunch and Dinner Entrées

- Limit meat, fish, and poultry servings to 5 ounces per day.
- Select lean cuts of beef, such as sirloin tip, round steak, and arm roast, and lean cuts of pork, such as center-cut ham, loin chops, and tenderloin. Trim visible fat before cooking.
- Select extra-lean ground meat and drain well after cooking. Use lean ground turkey, without skin added, in place of ground beef.
- Limit cholesterol-rich organ meats (liver, brain, sweetbreads).
- Limit egg yolks to no more than two per week; replace whole eggs in recipes with egg whites or commercial egg substitutes.
- Include more vegetarian entrées or legume dishes to reduce meat intake and boost vegetable, soluble fiber, and soy protein intakes.
- Restrict these high-sodium foods
  - Cured or smoked meats such as beef jerky, bologna, corned or chipped beef, frankfurters, ham, luncheon meats, salt pork, and sausage
  - Salty or smoked fish, such as anchovies, caviar, salted or dried cod, herring, sardines, and smoked salmon
  - Packaged, canned, or frozen soups, sauces, and entrées

Milk Products

- To obtain two to three servings of milk daily, include a portion of fat-free or low-fat milk, yogurt, or cottage cheese in each meal.
- Use yogurt or fat-free sour cream to make dips or salad dressings. Substitute evaporated fat-free milk for heavy cream.
- Limit foods high in saturated fat or sodium, such as cheese, processed cheeses, ice cream, and other milk-based desserts.

Fats and Oils

- Add nuts (not salted) and avocados to meals to increase monounsaturated fat intakes and make meals more appetizing.
- Include unsaturated vegetable oils in salad dressings and recipes, such as canola, corn, olive, peanut, saflower, sesame, soybean, and sunflower oils.
- Use margarines with added plant sterols or stanols regularly.
- Select soft margarines in tubs or liquid form; limit stick margarines and solid vegetable shortenings.

Spices and Seasonings

- Use salt only at the end of cooking, and you will need to add much less. Use salt substitutes at the table.
- Spices and herbs improve the flavor of foods without adding sodium. Try using more garlic, ginger, basil, curry or chili powder, cumin, pepper, lemon, mint, oregano, rosemary, and thyme.
- Check the sodium content on labels. Flavorings and sauces that are usually high in sodium include bouillon cubes, soy sauce, steak and barbecue sauces, relishes, mustard, and catsup.

Snacks and Desserts

- Select low-sodium and low-saturated fat choices such as unsalted pretzels, nuts, popcorn, chips, and crackers.
- Choose canned or dried fruits and some raw vegetables to boost fruit and vegetable intake.
- Enjoy angel food cake, which is made without egg yolks and added fat.
- Select low-fat frozen desserts such as sherbet, sorbet, fruit bars, and some low-fat ice creams.

TRY IT Plan heart-healthy meals for a day and analyze them using your personal profile. Discuss whether these meals meet the strategies listed in Table 18-5 (p. 593) and how to improve any shortcomings.

18.4 Hypertension

LEARN IT Present strategies to lower blood pressure.

Anyone concerned about atherosclerosis and the risk it presents must also be concerned about hypertension. Together, the two are a life-threatening combination. The higher the blood pressure is above normal, ♦ the greater the risk. For each 20 point increase in systolic blood pressure and 10 point increase in diastolic blood pressure, the risk of death from CVD doubles.[51] Low blood pressure, on the other hand, is generally a sign of long life expectancy and low heart disease risk.[52]

Hypertension affects about one out of three adults in the United States.[53] It contributes to an estimated 1 million heart attacks and more than 795,000 strokes each year. People usually do not feel the physical effects of high blood pressure, but it can impair life's quality and end life prematurely.

♦ Optimal resting blood pressure for adults is <120 over <80 mm Hg (120/80). Blood pressure is measured in millimeters of mercury (mm Hg). Blood pressure is measured both when the heart muscle contracts (*systolic* blood pressure) and when it relaxes (*diastolic* blood pressure).

How Hypertension Develops The underlying causes of most cases of hypertension are not fully understood, but much is known about the physiological factors that affect blood pressure. Blood pressure arises from contractions in the heart muscle that pump blood away from the heart (**cardiac output**) and the resistance blood encounters in the arterioles (**peripheral resistance**). When either cardiac output or peripheral resistance increases, blood pressure rises. ♦ Cardiac output is raised when heart rate or blood volume increases; peripheral resistance is affected mostly by the diameters of the arterioles. Blood pressure is therefore influenced by the nervous system, which regulates heart muscle contractions and the arteriole's diameters, and hormonal signals, which may cause fluid retention or blood vessel constriction. The kidneys also play a role in regulating blood pressure by controlling the secretion of the hormones involved in vasoconstriction and retention of sodium and water (review Figure 12-3 on p. 372).

♦ The equation describing this relationship is blood pressure = cardiac output × peripheral resistance.

Risk Factors for Hypertension Several major risk factors predicting the development of hypertension have been identified, including:

- *Aging.* Hypertension risk increases with age. An estimated two-thirds of persons 65 and older have hypertension.[54] Individuals who have normal blood pressure at age 55 still have a 90 percent risk of developing high blood pressure during their lifetimes.
- *Genetics.* Hypertension risk is similar among family members. It is also more prevalent and severe in certain ethnic groups: for African Americans in the United States, the prevalence of high blood pressure is among the highest in the world.[55] Compared with others, African Americans typically develop high blood pressure earlier in life, and their average blood pressure is much higher.[56]
- *Obesity.* Most people with hypertension—an estimated 60 percent—are obese. Obesity raises blood pressure in part by altering kidney function and promoting fluid retention.[57]
- *Salt sensitivity.* Approximately 30 to 50 percent of those with hypertension have blood pressure that is sensitive to salt.[58] These people can improve their blood pressure by reducing salt in their diets.
- *Alcohol.* Alcohol consumption, especially if consumed regularly in amounts greater than two drinks per day, ♦ is strongly associated with hypertension.[59] Alcohol is also associated with strokes independently of hypertension and its use may interfere with medications.

♦ One drink delivers ½ oz of pure ethanol:
- 5 oz wine
- 10 oz wine cooler
- 12 oz beer
- 1½ oz liquor (80 proof whiskey, scotch, rum, or vodka)

cardiac output: the volume of blood discharged by the heart each minute; determined by multiplying the stroke volume by the heart rate. The stroke volume is the amount of oxygenated blood the heart ejects toward the tissues at each beat. Cardiac output (volume/minute) = stroke volume (volume/beat) × heart rate (beats/minute).

peripheral resistance: the resistance to pumped blood in the small arterial branches (arterioles) that carry blood to the tissues.

Treatment of Hypertension The single most effective step people can take against hypertension is to find out whether they have it. At checkup time, a health-care professional can provide an accurate resting blood pressure reading.

TABLE 18-6 Lifestyle Modifications to Reduce Blood Pressure

Modification	Recommendation	Expected Reduction in Systolic Blood Pressure
Weight reduction	Maintain healthy body weight (BMI 18.5–24.9).	5–20 mm Hg/10 kg lost
DASH eating plan	Adopt a diet rich in fruits, vegetables, and low-fat milk products to reduce saturated fat intake.	8–14 mm Hg
Sodium restriction	Reduce dietary sodium intake to less than 2300 mg sodium (less than 6 g salt) per day, and further reduce intake to 1500 mg among persons who are 51 and older and those of any age who are African American or have hypertension, diabetes, or chronic kidney disease.	2–8 mm Hg
Physical activity	Perform aerobic physical activity for at least 30 minutes per day, most days of the week.	4–9 mm Hg
Moderate alcohol consumption	Men: Limit to two drinks per day. Women and lighter-weight men: Limit to one drink per day.	2–4 mm Hg

SOURCE: Adapted from US Department of Agriculture, US Department of Health and Human Services, *Dietary Guidelines for Americans*, 2010, www.dietaryguidelines.gov, and *Reference Card from the Seventh Report of the Joint National Committee on Prevention, Detection, Evaluation, and Treatment of High Blood Pressure (JNC 7)*, NIH publication no. 03-5231 (Bethesda, Md.: National Institutes of Health, National Heart, Lung, and Blood Institute, and National High Blood Pressure Education Program, May 2003).

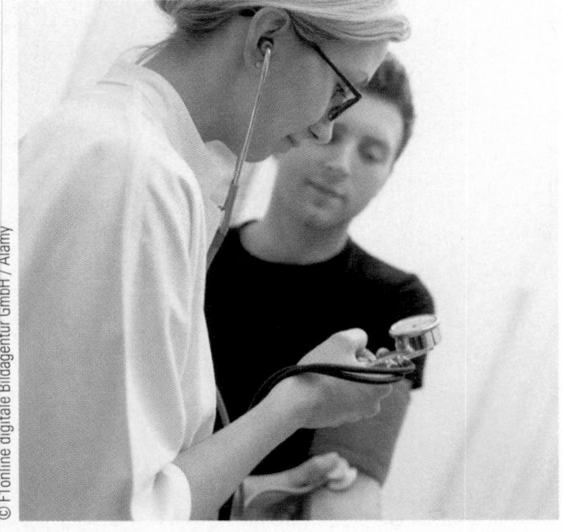

To guard against hypertension, have your blood pressure checked regularly.

Under normal conditions, blood pressure fluctuates continuously in response to a variety of factors including stress and such actions as talking or shifting position. Some people react emotionally to the procedure, which raises the blood pressure reading. For these reasons, if the resting blood pressure is above normal, the reading should be repeated before confirming the diagnosis of hypertension (see Table 18-4 on p. 589 for blood pressure standards). Thereafter, the blood pressure should be checked regularly. Both lifestyle modifications and medications are used to treat hypertension. Table 18-6 describes the lifestyle changes that reduce blood pressure and the expected reduction in systolic blood pressure for each change.

Weight Control Efforts to lower blood pressure focus on weight control. Weight loss alone is one of the most effective nondrug treatments for hypertension.[60] Those who are using drugs to control their blood pressure can often reduce or discontinue the drugs when they lose weight. Even a modest weight loss of 5 to 10 percent of body weight can lower blood pressure significantly.

Physical Activity The higher the blood pressure and the less active a person is to begin with, the greater the effect physical activity has in reducing blood pressure. Physical activity helps with weight control, of course, but moderate aerobic activity, such as 30 to 60 minutes of brisk walking most days, also helps to lower blood pressure directly.[61] Those who engage in regular aerobic activity may be able to control mild hypertension without medication.[62]

The DASH Diet Results of the Dietary Approaches to Stop Hypertension (DASH) trials show that an eating pattern rich in fruits, vegetables, low-fat milk products, whole grains, and nuts, and low in total fat and saturated fat can significantly lower blood pressure. In addition to lowering blood pressure, the DASH diet lowers total cholesterol and LDL cholesterol, and reduces inflammation.[63] Compared to the typical American diet, the DASH eating plan provides more fiber, potassium, magnesium, and calcium and less red meat, sweets, and sugar-containing beverages. Table 18-7 shows how the DASH eating plan compares to the USDA Food Pattern (introduced in Chapter 2). Both eating plans meet the goals specified in the *Dietary Guidelines for Americans*.[64]

> **DIETARY GUIDELINES FOR AMERICANS 2010**
African Americans and individuals with hypertension should increase their intakes of potassium.

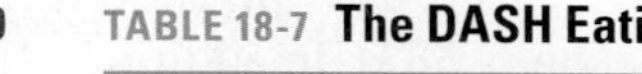

TABLE 18-7 **The DASH Eating Plan and the USDA Food Pattern Compared**

Food Group	DASH	USDA Food Pattern
Grains	6–8 oz	6 oz
Vegetables	2–2½ c	2½ c
Fruits	2–2½ c	2 c
Milk (fat-free/low-fat)	2–3 c	3 c
Lean meats, poultry, fish	6 oz or less	5½ oz
Nuts, seeds, legumes	4–5 oz per week	—[a]

NOTE: These diet plans are based on 2000 kcalories per day. Both DASH and the USDA Food Patterns recommend that fats and sugars be used sparingly and with discretion.
[a]The USDA Food Patterns combine nuts, seeds, and legumes with meat, poultry, and fish.

Salt/Sodium Intake Strong evidence supports the important role restricting sodium and/or salt plays in preventing and reducing hypertension. Lowering sodium intake reduces blood pressure regardless of gender or race, presence or absence of hypertension, or whether people follow the DASH diet or a typical American diet. The combination of the DASH diet with a limited intake of sodium, however, improves blood pressure better than either strategy alone.[65] Furthermore, the lower the sodium intake, the greater the drop in blood pressure. (See the "How To" feature in Chapter 12 on p. 380 for suggestions to limit sodium intake.) Research shows salt restriction provides additional protection against heart disease, beyond lowering blood pressure.[66]

> **DIETARY GUIDELINES FOR AMERICANS 2010**
Reduce daily sodium intake to less than 2300 milligrams and further reduce intake to 1500 milligrams among persons who are 51 and older and those of any age who are African American or have hypertension, diabetes, or chronic kidney disease.

Medications When diet and physical activity fail to reduce blood pressure, diuretics and antihypertensive drugs may be prescribed. Diuretics lower blood pressure by increasing fluid loss and lowering blood volume.

Some diuretics can lead to a potassium deficiency. People taking these diuretics need to include rich food sources of potassium or supplements daily and watch for signs of potassium imbalances such as weakness (particularly of the legs), unexplained numbness or tingling sensation, cramps, irregular heartbeats, and excessive thirst and urination. Blood potassium should be monitored regularly. Although some diuretics can lead to a potassium deficiency, others spare potassium. A combination of these two types of diuretics may be prescribed to prevent potassium deficiency.

The richest sources of potassium are fresh foods of all kinds.

Most people with hypertension use two or more medications to meet their blood pressure goals. Using a combination of drugs with different modes of action can reduce the doses needed and minimize side effects. In addition to diuretics, other medications ♦ commonly prescribed include ACE inhibitors, beta-blockers, and calcium channel blockers; these drugs are also used to treat various heart conditions.

♦ Hypertension medications:
- ACE inhibitors interfere with the conversion of angiotensin I to angiotensin II, a peptide that constricts blood vessels and releases hormones that increase blood pressure (review Figure 12-3 on p. 372). (*ACE* stands for *angiotensin-converting enzyme.*)
- Beta-blockers cause the heart to beat more slowly and with less force and the blood vessels to open up, which improves blood flow.
- Calcium channel blockers relax blood vessels and decrease the heart's pumping strength.

REVIEW IT Present strategies to lower blood pressure.
The most effective dietary strategy for preventing hypertension is weight control. Also beneficial are diets rich in fruits, vegetables, nuts, and low-fat milk products and low in fat, saturated fat, and sodium.

18.5 Diabetes Mellitus

LEARN IT Compare the dietary strategies to manage type 1 diabetes with those to prevent and treat type 2 diabetes.

The incidence of diabetes among children and adults has risen dramatically in the last decade (see Figure 18-4, p. 598). More than 25 million people in the United States have been diagnosed with diabetes.[67] As many as 79 million US adults 20 years of age and older have **prediabetes**—their blood glucose is elevated but not to such an extent as to be classified as diabetes.[68] ♦ People with prediabetes have a high risk of developing diabetes. The glossary on p. 599 defines diabetes terms.

♦ Fasting plasma glucose
- Normal: < 100 mg/dL
- Prediabetes: 100–125 mg/dL
- Diabetes: ≥125 mg/dL

> **FIGURE 18-4 Prevalence of Diabetes among Adults in the United States**

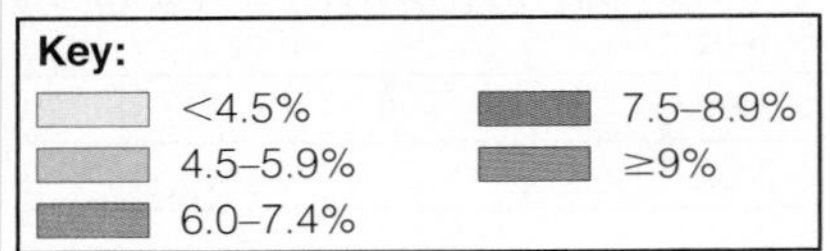

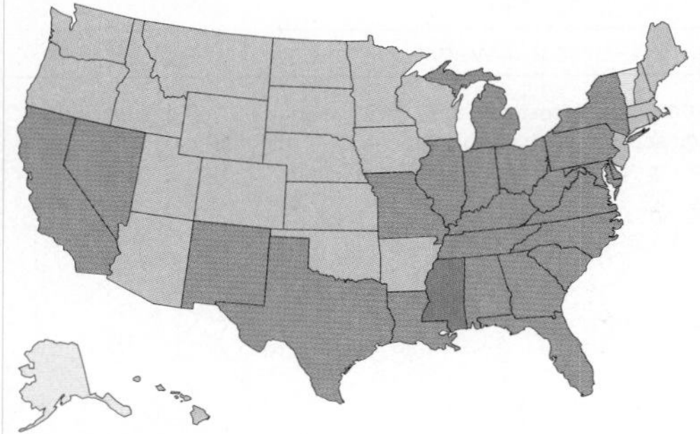

2000: Only two states had a prevalence of diabetes less than 4.5%, most had a prevalence between 4.5% and 7.4%, and only one state had a prevalence of 7.5% or greater.

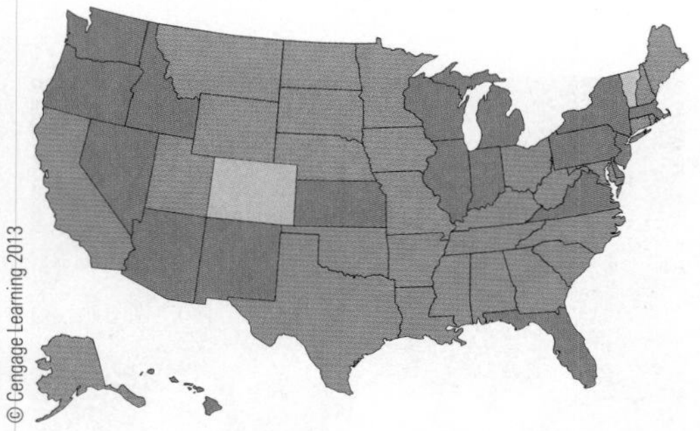

2009: No state had a prevalence of diabetes less than 4.5%, only two states had a prevalence rate less than 6%, and fifteen states had a prevalence of 9% or greater.

SOURCE: Centers for Disease Control and Prevention, www.cdc.gov/diabetes/statistics.

Diabetes ranks seventh among the leading causes of death (review Figure 18-2 on p. 585). In addition, diabetes underlies, or contributes to, several other major diseases, including heart disease, stroke, hypertension, blindness, and kidney failure. Heart disease is the leading cause of diabetes-related deaths. In fact, people with diabetes are twice as likely to develop cardiovascular problems as those without diabetes.

How Diabetes Develops **Diabetes mellitus** includes several metabolic disorders characterized by high blood glucose concentrations and disturbed insulin metabolism. People with diabetes may have insufficient insulin, ineffective insulin, or a combination of the two. The result is **hyperglycemia,** a marked elevation in blood glucose that can ultimately damage blood vessels, nerves, and tissues.

Table 18-8 shows the distinguishing features of its two main forms, type 1 diabetes and type 2 diabetes. As the following pages explain, the development of type 1 and type 2 diabetes differs, but some of their complications are similar.

To appreciate the problems presented by an absolute or relative lack of insulin, consider insulin's normal action. After a meal, insulin signals the body's cells to receive the energy nutrients from the blood—amino acids, glucose, and fatty acids. Insulin helps to maintain blood glucose within normal limits and stimulates protein synthesis, glycogen synthesis in liver and muscle, and fat synthesis. Without insulin, glucose regulation falters, and metabolism of the energy-yielding nutrients changes.

Type 1 Diabetes In **type 1 diabetes,** the less common type of diabetes (about 5 to 10 percent of all diagnosed cases), the pancreas loses its ability to synthesize the hormone insulin. Type 1 diabetes is an **autoimmune disorder.**[69] In most cases, the individual inherits a defect in which immune cells mistakenly attack and destroy the insulin-producing beta cells of the pancreas. The rate of beta cell destruction in type 1 diabetes varies. In some people (mainly infants and children), destruction is rapid; in others (mainly adults), it is slow. Type 1 diabetes commonly occurs in childhood and adolescence, but it can occur at any age, even late in life.

Without insulin, the body's energy metabolism changes, with such severe consequences as to threaten survival. The cells must have insulin to take up the needed fuels from the blood. People with type 1 diabetes must receive insulin either by injection or external pumps; insulin cannot be taken orally because it is a protein, and the enzymes of the GI tract would digest it.

Type 2 Diabetes **Type 2 diabetes** is the most prevalent form of diabetes, accounting for 90 to 95 percent of cases.[70] The primary defect in type 2 diabetes is insulin resistance, a reduced sensitivity to insulin. Consequently, muscle and adipose

TABLE 18-8 Features of Type 1 and Type 2 Diabetes

	Type 1	Type 2
Prevalence in diabetic population	5–10% of cases	90–95% of cases
Age of onset	<30 years	>40 years[a]
Associated conditions	Autoimmune diseases, viral infections, inherited factors	Obesity, aging, inherited factors
Major defect	Destruction of pancreatic beta cells; insulin deficiency	Insulin resistance; insulin deficiency (relative to needs)
Insulin secretion	Little or none	Varies; may be normal, increased, or decreased
Requirement for insulin therapy	Always	Sometimes
Older names	Juvenile-onset diabetes Insulin-dependent diabetes mellitus (IDDM)	Adult-onset diabetes Noninsulin-dependent diabetes mellitus (NIDDM)

[a]Incidence of type 2 diabetes is increasing in children and adolescence; in more than 90 percent of these cases, it is associated with overweight or obesity and a family history of type 2 diabetes.

hyperglycemia: elevated blood glucose concentrations.

autoimmune disorder: a condition in which the body develops antibodies to its own proteins and then proceeds to destroy cells containing these proteins. In type 1 diabetes, the body develops antibodies to its insulin and destroys the pancreatic cells that produce the insulin, creating an insulin deficiency.

GLOSSARY OF DIABETES TERMS

diabetes (DYE-uh-BEET-eez) **mellitus** (MELL-ih-tus or mell-EYE-tus): a group of metabolic diseases characterized by hyperglycemia resulting from defects in insulin secretion, insulin action, or both.

- **mellitus** = honey-sweet (sugar in urine)

prediabetes: condition in which blood glucose levels are higher than normal but not high enough to be diagnosed as diabetes; considered a major risk factor for future diabetes and cardiovascular diseases; formerly called *impaired glucose tolerance.*

type 1 diabetes: the less common type of diabetes in which the pancreas produces little or no insulin. Type 1 diabetes usually results from autoimmune destruction of pancreatic beta cells.

type 2 diabetes: the more common type of diabetes in which the cells fail to respond to insulin. Type 2 diabetes usually accompanies obesity and results from insulin resistance coupled with insufficient insulin secretion.

cells cannot remove glucose from the blood, and liver cells continue to make glucose. To compensate, the pancreas secretes larger amounts of insulin, and plasma insulin concentrations can rise to abnormally high levels (hyperinsulinemia). Over time, the pancreas becomes less able to compensate for the cells' reduced sensitivity to insulin, and hyperglycemia worsens. The high demand for insulin can eventually exhaust the beta cells of the pancreas and lead to impaired insulin secretion and reduced plasma insulin concentrations. Type 2 diabetes is therefore associated both with insulin resistance and with relative insulin deficiency; that is, the amount of insulin is insufficient to compensate for its diminished effect in cells.

Although the actual causes of type 2 diabetes are unknown, the risk is substantially increased by obesity (especially abdominal obesity), poor dietary habits, smoking, excessive alcohol consumption, aging, and physical inactivity. When people take action to control their lifestyle choices, prevention of type 2 diabetes is not only possible, but likely.[71] ♦ Even older adults can lower their diabetes risk by changing their lifestyles.[72]

♦ Lifestyle factors that lower diabetes risk:
- Healthy body weight
- Diet that follows the *Dietary Guidelines for Americans*
- Never smoking
- Limited alcohol intake
- Physical activity

Most people with type 2 diabetes are obese, and obesity itself can directly cause some degree of insulin resistance.[73] As discussed in Highlight 16, obesity has led to a dramatic rise in the incidence of type 2 diabetes among children and adolescents during the past two decades.[74] Inherited factors also strongly influence risk, and type 2 diabetes is more common in certain ethnic populations, including Native Americans, Hispanic Americans, Mexican Americans, African Americans, Asian Americans, and Pacific Islanders.

Inflammation contributes to insulin resistance, and its many links with obesity, metabolic syndrome, and CVD make inflammation central to the development of diabetes as well.[75] Chronic inflammation correlates with increases in blood glucose and decreases in insulin effectiveness.

Complications of Diabetes In both types of diabetes, glucose fails to gain entry into the cells and consequently accumulates in the blood. These two problems lead to both acute and chronic complications. Figure 18-5 (p. 600) summarizes the metabolic changes and acute complications that can arise in uncontrolled diabetes. Notice that when some glucose enters the cells, as in type 2 diabetes, many of the symptoms of type 1 diabetes do not occur.

Over the long term, the person with diabetes suffers not only from the acute complications shown in Figure 18-5, but also from its chronic effects. Chronically elevated blood glucose alters glucose metabolism in virtually every cell of the body. Some cells begin to convert excess glucose to sugar alcohols, for example, causing toxicity and cell distention—distended cells in the lenses of the eyes, for example, cause blurry vision. Some cells produce glycoproteins by attaching excess glucose to an amino acid in a protein; the altered proteins cannot function normally, which leads to a host of other problems. The structures of the blood vessels and nerves become damaged, leading to loss of circulation and nerve function. Infections occur due to poor circulation coupled with glucose-rich blood and urine. People with diabetes must pay special attention to hygiene and keep alert for early signs of infection. Early, aggressive treatment to control blood glucose significantly reduces the risk of long-term diabetes-related complications.

> **FIGURE 18-5 Metabolic Consequences of Untreated Diabetes**

The metabolic consequences of type 1 diabetes differ from those of type 2. In type 1, no insulin is available to allow any glucose to enter the cells. When glucose cannot enter the cells, a cascade of metabolic changes quickly follows. In type 2 diabetes, some glucose enters the cells. Because the cells are not "starved" for glucose, the body does not shift into the metabolism of fasting (losing weight and producing ketones).

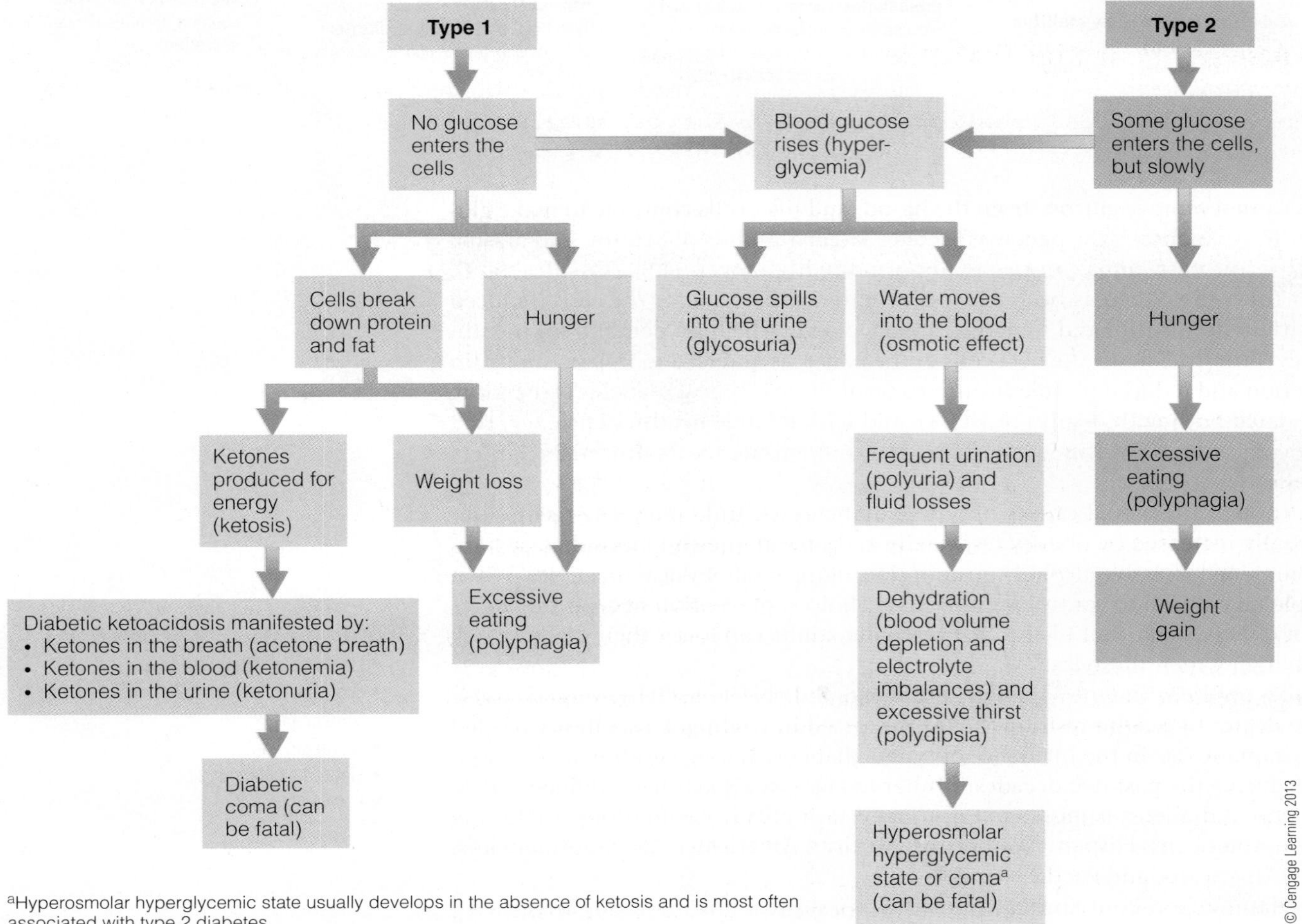

[a]Hyperosmolar hyperglycemic state usually develops in the absence of ketosis and is most often associated with type 2 diabetes.

Diseases of the Large Blood Vessels As mentioned, atherosclerosis tends to develop early, progress rapidly, and be more severe in people with diabetes. The interrelationships among insulin resistance, obesity, hypertension, and atherosclerosis help explain why about 75 percent of people with diabetes die as a consequence of cardiovascular diseases, especially heart attacks. Intensive diabetes treatment that keeps blood glucose levels tightly controlled can reduce the risk of cardiovascular disease among those with diabetes.*[76] Although intensive therapy is associated with some risks, including an increased risk of abnormally low blood sugar (hypoglycemia), benefits usually outweigh these disadvantages.

Diseases of the Small Blood Vessels For people with diabetes, disorders of the small blood vessels (capillaries)—called **microangiopathies**—may also develop and lead to loss of kidney function and retinal degeneration with accompanying loss of vision. About 85 percent of people with diabetes have impaired kidney function, loss of vision, or both. Consequently, diabetes is a leading cause of both kidney failure and blindness.[77]

Diseases of the Nerves Nerve tissues may also deteriorate with diabetes, expressed at first as a painful prickling sensation, often in the arms and legs. Later, the person loses sensation in the hands and feet. Injuries to these areas may go

*Intensive treatment may be inappropriate for some individuals with diabetes; examples include individuals with limited life expectancy or a history of hypoglycemia and middle-aged and older adults with previous heart disease or multiple heart disease risk factors.

microangiopathies: disorders of the small blood vessels.
- **micro** = small
- **angeion** = vessel
- **pathos** = disease

unnoticed, and infections can progress rapidly. With loss of both circulation and nerve function, undetected injury and infection may lead to death of tissue (gangrene), necessitating amputation of the limbs (most often the legs or feet). People with diabetes are advised to take conscientious care of their feet and visit a podiatrist regularly.

Recommendations for Diabetes

Diet is an important component of diabetes treatment. To maintain near-normal blood glucose levels, the diet is designed to deliver the same amount of carbohydrate each day, spaced evenly throughout the day. Several approaches can be used to plan such diets, but many people with diabetes learn to count carbohydrates using the exchange system that is presented in Appendix G.

Total Carbohydrate Intake Providing a consistent carbohydrate intake spaced throughout the day helps to maintain appropriate blood glucose levels and maximize the effectiveness of drug therapy. Eating too much carbohydrate at one time can raise blood glucose too high, stressing the already-compromised insulin-producing cells. Eating too little carbohydrate can lead to hypoglycemia. Low carbohydrate diets (less than 130 grams of carbohydrate per day) are not recommended.[78]

Carbohydrate Sources Different carbohydrate-containing foods have varying effects on blood glucose levels; for example, consuming a portion of white rice may cause blood glucose to rise higher and quicker than would a similar portion of barley. As Chapter 4 describes, this *glycemic effect* of foods is influenced by a food's fiber content, the preparation method, the other foods included in a meal, and individual tolerances. For individuals with diabetes, using the glycemic index may provide some additional benefit for achieving glucose control compared with that obtained by considering only the amount of carbohydrate consumed.[79] In addition, high-fiber, minimally processed foods—which typically have more moderate effects on blood glucose than do highly processed, starchy foods—are among foods frequently recommended for people with diabetes.

A common misconception is that people with diabetes need to avoid all sugar and sugar-containing foods. Because moderate consumption of sugar has not been shown to adversely affect glucose control, sugar recommendations for people with diabetes are similar to those for the general population, which suggests limiting foods and beverages with added sugars. Of course, sugars and sugary foods must be counted as part of the daily carbohydrate allowance.

Dietary Fat As mentioned earlier, people with diabetes have a high risk of developing cardiovascular diseases, and their guidelines for dietary fat are similar to those for others with high risks. Saturated fat intake should be limited to less than 7 percent of kcalories, *trans* fat intake should be minimized, and cholesterol intake should be limited to less than 200 milligrams daily.[80] Dietary strategies for preventing cardiovascular disease were presented earlier in Table 18-5 (p. 593).

Protein Protein intakes in the United States generally range from 15 to 20 percent of total kcalories. Protein intakes in this range need not be modified for individuals with diabetes and normal kidney function.[81] Higher protein intakes are discouraged because they may be detrimental to kidney function.

Alcohol Adults with diabetes can drink alcohol in moderation. Guidelines are similar to those for the general population, which advise a daily limit of one drink for women and two drinks for men.[82]

Recommendations for Type 1 Diabetes Normally, the body secretes a constant baseline amount of insulin at all times and secretes more as blood glucose rises following meals. People with type 1 diabetes, however, produce little or no insulin. They must learn to adjust the amount and schedule of their insulin doses to accommodate meals,

© Gregg Matthews/The New York Times/Redux

Phil Southerland has raced his bicycle across America and checks his glucose meter regularly. With appropriate diet, regular physical activity, frequent glucose checks, and treatment as needed, people with diabetes can achieve their dreams (www.teamtype1.org).

physical activity, and health status. To maintain blood glucose within a fairly normal range requires a lifelong commitment to a carefully coordinated program of diet, physical activity, and insulin.

Nutrition therapy for type 1 diabetes focuses on maintaining optimal nutrition status, controlling blood glucose, achieving a desirable blood lipid profile, controlling blood pressure, and preventing and treating the complications of diabetes. In addition to meeting basic nutrient requirements, the diet must provide a fairly consistent carbohydrate intake from day to day and at each meal and snack to help minimize fluctuations in blood glucose. Further alterations in diet may be necessary for the person with chronic complications such as cardiovascular or kidney disease.

Participation in all levels of physical activity is possible for people with type 1 diabetes who have good blood glucose control and no complications, but they should check with their physician first. One potential problem is hypoglycemia, which can occur during, immediately after, or many hours after physical activity.[83] To avoid hypoglycemia, the person must monitor blood glucose before and after activity to identify when adjustments in insulin or food intake are needed. Carbohydrate-rich foods should be readily available during and after activity.

Recommendations for Type 2 Diabetes In overweight people with type 2 diabetes, even moderate weight loss (5 to 10 percent of body weight) can help improve insulin resistance, blood lipids, and blood pressure. Together with diet, a regular routine of moderate physical activity not only supports weight loss, but also improves blood glucose control, blood lipid profiles, and blood pressure. Thus the benefits of regular, long-term physical activity for the treatment and prevention of type 2 diabetes are substantial.[84]

REVIEW IT Compare the dietary strategies to manage type 1 diabetes with those to prevent and treat type 2 diabetes.

Diabetes is characterized by high blood glucose and either insufficient insulin, ineffective insulin, or a combination of the two. People with type 1 diabetes coordinate diet, insulin, and physical activity to help control their blood glucose. Those with type 2 diabetes benefit most from a diet and physical activity program that controls glucose fluctuations and promotes weight loss.

18.6 Cancer

LEARN IT Differentiate among cancer initiators, promoters, and antipromoters and describe how nutrients or foods might play a role in each category.

Cancer, the growth of **malignant** tissue, ranks just below cardiovascular disease as a cause of death in the United States. (See the accompanying glossary of cancer terms.) As with cardiovascular disease, the prognosis for cancer today is far brighter than in the past. Identification of risk factors, new detection techniques, and innovative therapies offer hope and encouragement.

Cancer is not a single disorder. There are many **cancers,** that is, many different kinds of malignant growths. They have different characteristics, occur in different locations in the body, take different courses, and require different treatments.

People with cancer take comfort from the support of others and from the knowledge that medical science is waging an unrelenting battle in their defense.

How Cancer Develops The development of cancer, called **carcinogenesis,** often proceeds slowly and continues for several decades. A cancer arises from mutations in the genes that control cell division in a single cell. These mutations may promote cellular growth, interfere with growth restraint, or prevent cellular death.[85] The affected cell thereby loses its built-in capacity for halting cell division, and it produces daughter cells with the same genetic defects. As the abnormal mass of cells, called a **tumor,** grows, a network of blood vessels develops to supply the tumor with the nutrients it needs to support its growth.* The tumor can disrupt the functioning of the normal tissue around it, and some tumor cells may **metastasize** to other regions of the body.[86] Figure 18-6 illustrates cancer

*An abnormal mass of cells that is noncancerous is called a *benign* tumor.

> FIGURE 18-6 **Cancer Development**

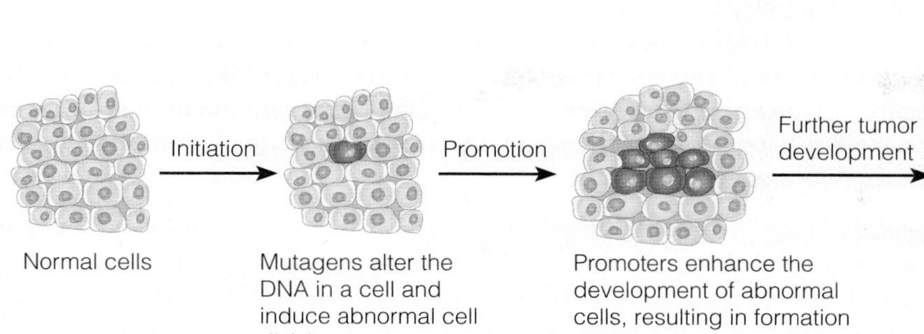
Initiation
Promotion
Further tumor development
Normal cells
Mutagens alter the DNA in a cell and induce abnormal cell division.
Promoters enhance the development of abnormal cells, resulting in formation of a tumor.

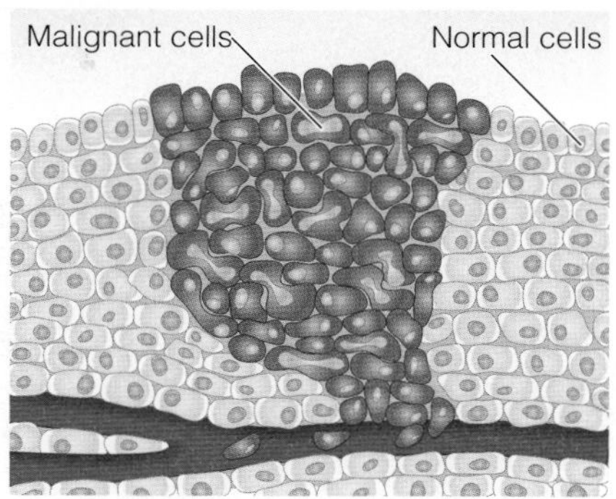
Malignant cells
Normal cells

The cancerous tumor releases cells into the bloodstream or lymphatic system (metastasis).

development. In leukemia (cancer affecting the white blood cells) the cells do not form a tumor, but rather accumulate in blood and other tissues.

The reasons cancers develop are numerous and varied. Vulnerability to cancer is sometimes inherited, such as when a person is born with a genetic defect that alters DNA structure, function, or repair. Certain metabolic processes may initiate carcinogenesis, as when phagocytes of the immune system produce oxidants that cause DNA damage or when chronic inflammation enhances the rate of cell division and the risk of a damaging mutation. More often, cancers are caused by interactions between a person's genes and the environment. Exposure to cancer-causing substances, or **carcinogens,** may either induce genetic mutations that lead to cancer or promote proliferation of cancerous cells.

Environmental Factors Among environmental factors, exposure to radiation and sun, water and air pollution, and smoking are known to cause cancer. Lack of physical activity may also play a role in the development of some cancers.[87] Men and women whose lifestyles include regular, vigorous physical activity have the lowest risk of colon cancer.[88] Physical activity may also protect against breast cancer by reducing body weight and by other mechanisms not related to body weight.[89]

Obesity itself is clearly a risk factor for certain cancers (such as colon, breast in postmenopausal women, endometrial, pancreas, kidney, and esophageal) and possibly for other types (such as gallbladder) as well.[90] Because different cancers have various causes, obesity's influence on cancer development depends on the site as well as other factors such as hormonal interactions. In the case of breast cancer in postmenopausal women, for example, the hormone estrogen is implicated. Obese postmenopausal women have much higher levels of estrogen than

GLOSSARY OF CANCER TERMS

antipromoters: factors that oppose the development of cancer.

cancers: malignant growths or tumors that result from abnormal and uncontrolled cell division.

carcinogenesis (CAR-sin-oh-JEN-eh-sis): the process of cancer development.

carcinogens (CAR-sin-oh-jenz or car-SIN-oh-jenz): substances that can cause cancer (the adjective is *carcinogenic*).

- **carcin** = cancer
- **gen** = gives rise to

initiators: factors that cause mutations that give rise to cancer, such as radiation and carcinogens.

malignant (ma-LIG-nant): describes a cancerous cell or tumor, which can injure healthy tissue and spread cancer to other regions of the body.

metastasize (me-TAS-tah-size): the spread of cancer from one part of the body to another.

promoters: factors that favor the development of cancers once they have begun.

tumor: an abnormal tissue mass with no physiological function; also called a *neoplasm* (NEE-oh-plazm).

Cancers are classified by the tissues or cells from which they develop:

Adenomas (ADD-eh-NOH-mahz) arise from glandular tissues.

Carcinomas (KAR-see-NOH-mahz) arise from epithelial tissues.

Gliomas (gly-OH-mahz) arise from glial cells of the central nervous system.

Leukemias (loo-KEE-mee-ahz) arise from white blood cell precursors.

Lymphomas (lim-FOH-mahz) arise from lymph tissue.

Melanomas (MEL-ah-NOH-mahz) arise from pigmented skin cells.

Sarcomas (sar-KOH-mahz) arise from connective tissues, such as muscle or bone.

lean women do because fat tissue produces estrogen. Researchers believe that the extended exposure to estrogen in obese women is linked to the increased risk of breast cancer after menopause.[91] The relationships between excessive body weight and certain cancers provide yet another reason to adopt a lifestyle that embraces physical activity and good nutrition.

As Table 18-9 shows, specific dietary constituents are associated with an increased risk of certain cancers. Some dietary factors may initiate cancer development (**initiators**), others may promote cancer development once it has started (**promoters**), and still others may protect against the development of cancer (**antipromoters**).

Dietary Factors—Cancer Initiators We do not know to what extent diet contributes to cancer development, although some experts estimate that diet may be linked to as many as one-third of all cases.[92] Consequently, many people think that certain foods are carcinogenic, especially those that contain additives or pesticides. As Chapter 19 explains, our food supply is one of the safest in the world. Additives that have been approved for use in foods are not carcinogens. Some pesticides are carcinogenic at high doses, but not at the concentrations commonly found on fruits and vegetables. The benefits of eating fruits and vegetables are far greater than the current risks.

Cancers of the head and neck correlate strongly with the combination of alcohol and tobacco use and with low intakes of fruits and vegetables. Alcohol intake alone is associated with cancers of the mouth, throat, and breast, and alcoholism often damages the liver and precedes the development of liver cancer.[93] These findings illustrate clearly why any potential benefit of moderate alcohol consumption on cardiovascular disease must be weighed against the potential dangers.

TABLE 18-9 Factors Associated with Cancer at Specific Sites

Cancer Sites	Risk Factors	Protective Factors
Breast (postmenopause)	Alcoholic drinks, body fatness, adult attained height,[a] abdominal fatness, adult weight gain	Lactation, physical activity
Breast (premenopause)	Alcoholic drinks, adult attained height,[a] greater birth weight	Lactation, body fatness
Colon and rectum	Red meat, processed meat, alcoholic drinks, body fatness, abdominal fatness, adult attained height[a]	Physical activity, foods containing dietary fiber, garlic, milk, calcium
Endometrium	Body fatness, abdominal fatness	Physical activity
Esophagus	Alcoholic drinks, body fatness, maté[b]	Nonstarchy vegetables, fruits, foods containing beta-carotene, foods containing vitamin C
Gallbladder	Body fatness	
Kidney	Body fatness	
Liver	Aflatoxins,[c] alcoholic drinks	
Lung	Arsenic in drinking water, beta-carotene supplements[d]	Fruits, foods containing carotenoids
Mouth, pharynx, and larynx	Alcoholic drinks	Nonstarchy vegetables, fruits, foods containing carotenoids
Nasopharynx	Cantonese-style salted fish	
Ovary	Adult attained height[a]	
Pancreas	Body fatness, abdominal fatness, adult attained height[a]	Foods containing folate
Prostate	Diets high in calcium	Foods containing lycopene, foods containing selenium, selenium[e]
Skin	Arsenic in drinking water	
Stomach	Salt, salty and salted foods	Nonstarchy vegetables, allium vegetables,[f] fruits

[a]Adult attained height is unlikely to directly modify the risk of cancer. It is a marker for genetic, environmental, hormonal, and also nutritional factors affecting growth during the period from preconception to completion of linear growth.
[b]As drunk traditionally in parts of South America, scalding hot through a metal straw. Any increased risk of cancer is judged to be caused by epithelial damage resulting from the heat, and not by the herb itself.
[c]Aflatoxins are toxins produced by molds or fungi. The main foods that may be contaminated are all types of grains (wheat, rye, rice, corn, barley, oats) and legumes, notably peanuts.
[d]The evidence is derived from studies using high-dose supplements (20 mg/day for beta-carotene; 25,000 international units/day for retinol) in smokers.
[e]The evidence is derived from studies using supplements at a dose of 200 μg/day. Selenium is toxic at higher doses.
[f]This includes vegetables such as garlic, onions, leeks, and shallots.
NOTE: Strength of evidence for all these factors is either "convincing" or "probable."
SOURCE: World Cancer Research Fund/American Institute for Cancer Research, *Food, Nutrition, Physical Activity and the Prevention of Cancer: A Global Perspective* (Washington, DC: AICR, 2007).

Cooking meats at high temperatures (frying, broiling) causes amino acids and creatine in the meats to react together and form carcinogens.* Grilling meat, ♦ fish, or other foods over a direct flame causes fat and added oils to splash on the fire and then vaporize, creating other carcinogens that rise and stick to the food.** Eating grilled food introduces these carcinogens to the digestive system, where they may damage the stomach and intestinal lining. Once these compounds are absorbed into the blood, however, they are detoxified by the liver.

♦ To minimize carcinogen formation during cooking:
- Roast or bake meats in the oven.
- When grilling, line the grill with foil or wrap the food in foil.
- Take care not to burn foods.
- Marinate meats beforehand.

Evidence from population studies spanning the globe for more than 30 years strongly suggests that diets high in red meat and processed meat (meat preserved by smoking, curing, or salting, or by the addition of preservatives) are a cause of colon cancer.[94] Based on such evidence, replacing most servings of red meat with poultry, fish, or legumes and choosing only occasional servings of grilled, fried, blackened, or smoked foods is in the best interest of health.

Another reason to moderate consumption of fried foods such as french fries and potato chips is the presence of acrylamide, a potential carcinogen. Acrylamide is produced when certain starches such as potatoes are fried or baked at high temperatures. Chapter 19 discusses acrylamide in foods.

Dietary Factors—Cancer Promoters Unlike carcinogens, which initiate cancers, some dietary components promote cancers. That is, once the initiating step has taken place, these components may accelerate tumor development.

Although studies of animals suggest that high-fat diets may promote cancer, studies of human beings have not proved that the effects of fat are independent of the effects of energy intake and physical activity. Overall, the evidence associating fats and oils with cancer risk is limited.[95]

The type of fat in the diet, however, may influence cancer promotion or prevention.[96] Studies of colon cancer implicate animal fats but not vegetable fat, and a number of studies suggest that omega-3 fatty acids from fish may protect against some cancers.[97] Thus the same dietary fat advice applies to cancer protection as to heart health: reduce saturated fat intake and increase omega-3 fatty acids.

Dietary Factors—Antipromoters Some foods may contain antipromoters—dietary compounds that defend against cancer. Table 18-9 (p. 604) includes these protective dietary factors. Research on dietary patterns of populations has led to recommendations aimed at reducing cancer risks.

Recommendations for Reducing Cancer Risks

A diet rich in fruits and vegetables may provide protection against the development of some cancers.[98] Fruits and vegetables contain both nutrients and phytochemicals with antioxidant activity, and these substances may prevent or reduce the oxidative reactions in cells that damage DNA. Phytochemicals may also help to enhance immune functions that protect against cancer development and promote enzyme reactions that inactivate carcinogens.[99] For example, the **cruciferous vegetables**—cabbage, cauliflower, broccoli, and brussels sprouts—contain a variety of phytochemicals that defend against cancers of the esophagus and endometrium.[100]

Cruciferous vegetables, such as cauliflower, broccoli, and brussels sprouts, contain nutrients and phytochemicals that may inhibit cancer development.

In addition, fruits and vegetables, as well as legumes and whole grains, are rich in fiber. As Chapter 4 explains, fiber may protect against cancer by binding, diluting, and rapidly removing potential carcinogens from the GI tract. High-fiber and whole-grain foods also help a person to maintain a healthy body weight—another preventive measure against cancer. Physical activity also helps maintain a healthy body weight and reduce the risk of some cancers. Table 18-10 (p. 606) summarizes dietary and lifestyle recommendations for reducing cancer risk.

*These carcinogens are *heterocyclic amines.*

**These carcinogens are *polycyclic aromatic hydrocarbons.*

cruciferous vegetables: vegetables of the cabbage family, including cauliflower, broccoli, and brussels sprouts.

TABLE 18-10 Recommendations for Reducing Cancer Risk

Body fatness: Be as lean as possible within the normal range of body weight.

- Ensure that body weight throughout childhood and adolescence projects toward the lower end of the normal adult BMI range by age 21.
- Maintain body weight within the normal range from age 21.
- Avoid weight gains and increases in waist circumference throughout adulthood.

Physical activity: Be physically active as part of everyday life.

- Be moderately physically active, equivalent to brisk walking, for at least 30 minutes every day.
- As fitness improves, aim for at least 60 minutes of moderate, or at least 30 minutes of vigorous, physical activity every day.
- Limit sedentary habits such as watching television.

Foods and drinks that promote weight gain: Limit consumption of energy-dense foods and avoid sugary drinks.

- Consume energy-dense foods (>225 kcalories/100 grams food), sparingly.
- Avoid drinks with added sugar and limit fruit juices.
- Consume "fast foods" sparingly, if at all.

Plant foods: Eat mostly foods of plant origin.

- Eat at least five servings of a variety of nonstarchy vegetables and fruits every day.
- Eat relatively unprocessed grains and/or legumes with every meal.
- Limit refined starchy foods.

Animal foods: Limit intake of red meat and avoid processed meat.

- Eat no more than 18 ounces of red meat a week, very little if any of which is processed.

Alcoholic drinks: Limit alcoholic drinks.

- If alcoholic drinks are consumed, limit consumption to no more than two drinks a day for men and one drink a day for women.

Preservation, processing, preparation: Limit consumption of salt and avoid moldy grains or legumes.

- Avoid salt-preserved, salted, or salty foods.
- Limit consumption of processed foods with added salt to ensure an intake of less than 6 grams of salt (2.4 grams of sodium) a day.
- Do not eat moldy grains or legumes.

Dietary supplements: Aim to meet nutritional needs through diet.

- Dietary supplements are not recommended for cancer prevention.

SOURCE: World Cancer Research Fund/American Institute for Cancer Research, *Food, Nutrition, Physical Activity and the Prevention of Cancer: A Global Perspective* (Washington, DC: AICR, 2007), pp. 373–390.

REVIEW IT Differentiate among cancer initiators, promoters, and antipromoters and describe how nutrients or foods might play a role in each category.

Some dietary factors, such as alcohol and heavily smoked foods, may initiate cancer development; others, such as animal fats, may promote cancer once it has gotten started; and still others, such as fiber, antioxidant nutrients, and phytochemicals, may act as antipromoters that protect against the development of cancer. By eating many fruits, vegetables, legumes, and whole grains and reducing saturated fat intake, people obtain the best possible nutrition at the lowest possible risk. Minimizing weight gain through regular physical activity and a healthy diet is also beneficial.

18.7 Recommendations for Chronic Diseases

LEARN IT Summarize dietary recommendations to prevent chronic diseases.

This chapter's discussion of chronic diseases began with the major cardiovascular diseases, described diabetes, and then went on to cancer—three different conditions with distinct sets of causes. Yet dietary excesses, particularly excess food energy and saturated fat intakes, increase the likelihood of all three diseases.[101] Similarly, all are responsive to diet, and in most cases, the beneficial foods are similar.

Not all diet recommendations apply equally to all of the diseases or to all people with a particular disease, but fortunately for the consumer, dietary recommendations do not contradict one another. In fact, they support one another. Most people can gain some disease-prevention benefits by making dietary changes. To that

TABLE 18-11 Dietary Guidelines and Recommendations for Chronic Diseases Compared

Dietary Guidelines 2010	Heart Disease	Hypertension	Diabetes	Cancer
Prevent and/or reduce overweight and obesity.	✓	✓	✓	✓
Increase physical activity and reduce time spent in sedentary behaviors.	✓	✓	✓	✓
Keep total fat 20 to 35 percent of kcalories.	✓			
Consume less than 10 percent of kcalories from saturated fatty acids and keep *trans*-fatty acid consumption as low as possible.	✓		✓	✓
Increase vegetable and fruit intake. Consume at least half of all grains as whole grains.	✓	✓	✓	✓
Reduce the intake of kcalories from added sugars.	✓		✓	
Reduce daily sodium intake to less than 2300 milligrams.[a]	✓	✓	✓	✓
If alcohol is consumed, it should be consumed in moderation.	✓	✓	✓	✓

[a]Further reduce daily sodium intake to 1500 milligrams among persons who are 51 and older and those of any age who are African American or have hypertension, diabetes, or chronic kidney disease.

end, the recommendations presented earlier for reducing the risks of heart disease (Table 18-5, p. 593), hypertension (Table 18-6, p. 596), and cancer (Table 18-10, p. 606) describe the kinds of foods people should include or limit. Table 18-11 compares the *Dietary Guidelines for Americans* with all these recommendations for chronic diseases. (A summary of the *Diet, Nutrition, and Prevention of Chronic Diseases* report from the World Health Organization (WHO) is presented in Appendix I.)

Several recommendations are aimed at weight control. Obesity is common in the United States, and it is linked with most of the chronic diseases that threaten life (review Figure 18-3 on p. 586). The problems of overweight multiply when medical conditions develop. For example, overweight people readily develop diabetes, which is often accompanied by high blood pressure and high blood cholesterol. Such a combination of problems may require only one treatment: adopting a healthful diet and regular exercise program.

Physical activity and a moderate weight loss of even 10 to 20 pounds can help improve blood glucose, blood lipids, and blood pressure.

> FIGURE 18-7 **The Healthy Eating Plate—An Alternative to USDA MyPlate**

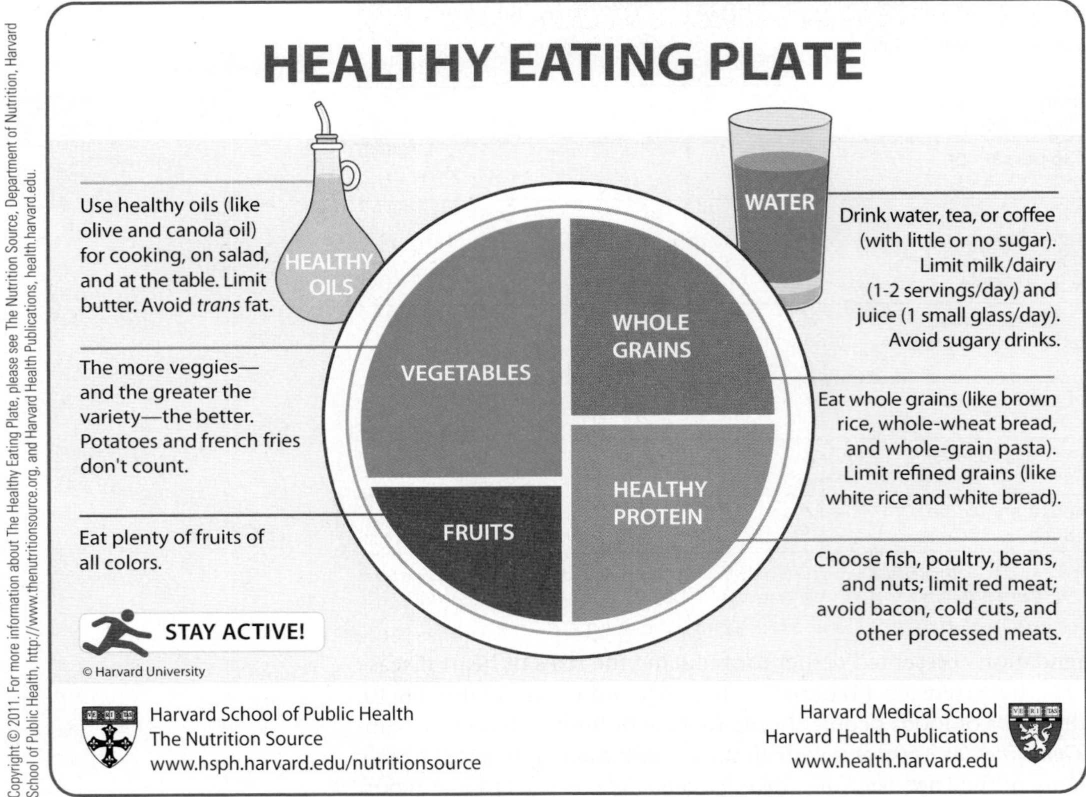

HEALTHY EATING PLATE
Use healthy oils (like olive and canola oil) for cooking, on salad, and at the table. Limit butter. Avoid *trans* fat.
HEALTHY OILS
WATER
Drink water, tea, or coffee (with little or no sugar). Limit milk/dairy (1-2 servings/day) and juice (1 small glass/day). Avoid sugary drinks.
The more veggies— and the greater the variety—the better. Potatoes and french fries don't count.
VEGETABLES
WHOLE GRAINS
Eat whole grains (like brown rice, whole-wheat bread, and whole-grain pasta). Limit refined grains (like white rice and white bread).
HEALTHY PROTEIN
Eat plenty of fruits of all colors.
FRUITS
Choose fish, poultry, beans, and nuts; limit red meat; avoid bacon, cold cuts, and other processed meats.
STAY ACTIVE!
© Harvard University
Harvard School of Public Health
The Nutrition Source
www.hsph.harvard.edu/nutritionsource
Harvard Medical School
Harvard Health Publications
www.health.harvard.edu

> **DIETARY GUIDELINES FOR AMERICANS 2010**
Increase physical activity and reduce time spent in sedentary behaviors.

The USDA Food Patterns and MyPlate offer guidance on selecting a healthy diet, but as Chapter 2 discusses, critics have noted several shortcomings. An alternative plate was created by the nutrition faculty from the Harvard School of Public Health.[102] The *Healthy Eating Plate* (Figure 18-7) is based on their critical review of links between diet and health.[103]

Recommendations for the Population The recommendations to prevent chronic diseases address the general population in the hope that all people at all levels of risk may benefit. Such a strategy is similar to national efforts to vaccinate to prevent measles, fluoridate water to prevent dental caries, and fortify grains with folate to prevent neural tube defects. These recommendations take a *preventive* or *population approach* and urge all people to make dietary changes believed to forestall or prevent diseases. Alternatively, recommendations that take a *medical* or *individual approach* urge dietary changes only for people who are known to need them.

Recommendations for Individuals People's hereditary susceptibility to diseases and their responsiveness to dietary measures vary. Unlike nutrient-deficiency diseases, which develop when nutrients are lacking and disappear when the nutrients are provided, chronic diseases are neither caused nor prevented by diet alone. Many people have followed dietary advice and developed heart disease or cancer anyway; others have ignored all advice and lived long and healthy lives. For many people, though, diet does influence the time of onset and course of some chronic diseases, and many health-care professionals urge dietary measures as part of a disease-prevention strategy.

To determine whether dietary recommendations are important to you personally, look at your family history to see which diseases are common to your relatives. In addition, examine your personal history, taking note of your body weight, blood pressure, blood lipid profile, and lifestyle habits such as smoking and physical activity.

Recommendations for Each Individual Even when recommendations are made "for individuals," they apply to large groups of people—those with hypertension or those with diabetes, for example. But that's expected to change in the next decade or so as research on the human genome provides the knowledge needed to create *specific* recommendations for *each* individual (as Highlight 6 explains).[104]

REVIEW IT Summarize dietary recommendations to prevent chronic diseases.
Clearly, optimal nutrition plays a key role in keeping people healthy and reducing the risk of chronic diseases. To have the greatest impact possible, dietary recommendations are aimed at the entire population, not just at the individuals who might benefit most. Recommendations focus on weight control and urge people to limit saturated and *trans* fat; increase fiber-rich fruits, vegetables, and whole grains; and balance food intake with physical activity. A person can do no better than to incorporate those suggestions into his or her daily life.

Not all diseases can be prevented through good nutrition, but many can. This chapter has presented the eating and activity patterns that can help people reduce the risk of chronic diseases and promote overall health. In addition to promoting health and preventing chronic diseases, a healthy eating pattern should also prevent foodborne illnesses, as described in the next chapter.

Nutrition Portfolio

Identifying your risk factors is the first step in taking action to defend yourself against heart attack, stroke, hypertension, diabetes, and cancer.

- Review your personal and family history of heart disease, hypertension, diabetes, and cancer.

Go to Diet Analysis Plus and choose one of the days on which you tracked your diet and activity for an entire day. Select the Energy Balance report to help you answer the following questions:

- In this report, did you expend more energy (kcalories) than you consumed or vice versa? Consider whether you are sedentary or overweight and how you might become more physically active and achieve a healthy body weight. If you smoke cigarettes, develop a reasonable plan for quitting.
- Learn whether you have high blood cholesterol or high blood pressure.

Go to Diet Analysis Plus and choose the Intake vs. Goals report to help you answer the following questions:

- In this report many dietary factors related to heart disease are assessed: sodium intake, fat intake, cholesterol intake, vitamin and mineral intake, and more. Please comment on how your Intake vs. Goals report relates to your risk factors. Is your diet helping to protect you or could it be putting you at higher risk?

Finally, go to Diet Analysis Plus and choose the Fat Breakdown report.

- We know from Chapter 5 that some types of fats appear to be protective against heart disease (monounsaturated and polyunsaturated) and some appear to put us more at risk (saturated fats and cholesterol). How was your intake for each type of fat?

Diet Analysis PLUS

To complete this exercise, go to your Diet Analysis Plus at **www.cengagebrain.com**.

STUDY IT To review the key points of this chapter and take a practice quiz, go to the study cards at the end of the book.

REFERENCES

1. A. Oliveira and coauthors, Major habitual dietary patterns are associated with acute myocardial infarction and cardiovascular risk markers in a Southern European population, *Journal of the American Dietetic Association* 111 (2011): 241–250; S. C. Larsson, J. Virtamo, and A. Wolk, Red meat consumption and risk of stroke in Swedish men, *American Journal of Clinical Nutrition* 94 (2011): 417–421; D. J. A. Jenkins and coauthors, Effect of a dietary portfolio of cholesterol-lowering foods given at 2 levels of intensity of dietary advice on serum lipids in hyperlipidemia, *Journal of the American Medical Association* 306 (2011): 831–839; E. S. Ford and coauthors, Healthy living is the best revenge: Findings from the European Prospective Investigation into Cancer and Nutrition-Potsdam Study, *Archives of Internal Medicine* 169 (2009): 1355–1362; A. Flood and coauthors, Dietary patterns as identified by factor analysis and colorectal cancer among middle-aged Americans, *American Journal of Clinical Nutrition* 88 (2008): 176–184; P. L. Lutsey, L. M. Steffen, and J. Stevens, Dietary intake and the development of the metabolic syndrome: The Atherosclerosis Risk in Communities Study, *Circulation* 117 (2008): 754–761; World Cancer Research Fund/American Institute for Cancer Research, *Summary: Food, Nutrition, Physical Activity, and the Prevention of Cancer: A Global Perspective* (Washington, DC: AICR, 2007).
2. Centers for Disease Control and Prevention, Ten great public health achievements: United States, 2001–2010, *Morbidity and Mortality Weekly Report*, www.cdc.gov/preview/mmwrhtm/mm6019a5.htm?s_cid=mm6019a5_w?s_, accessed November 17, 2011.
3. Centers for Disease Control and Prevention, Plan to combat extensively drug-resistant tuberculosis: Recommendations of the Federal Tuberculosis Task Force, *Morbidity and Mortality Weekly Report* 58 (2009): 1–43.
4. S. S. Percival, Nutrition and immunity: Balancing diet and immune function, *Nutrition Today* 46 (2011): 12-17; C. E. Taylor and C. A. Camargo Jr., Impact of micronutrients on respiratory infections, *Nutrition Reviews* 69 (2011): 259–269; R. Raqib and A. Carvioto, Nutrition, immunology, and genetics: Future perspectives, *Nutrition Reviews* 67 (2009): S227–S236.
5. Percival, 2011.
6. Position of the American Dietetic Association: Nutrition intervention and human immunodeficiency virus infection, *Journal of the American Dietetic Association* 110 (2010): 1105–1119.
7. B. Bistrian, Systemic response to inflammation, *Nutrition Reviews* 65 (2007): S170–S172.
8. B. K. Surmi and A. H. Hasty, The role of chemokines in recruitment of immune cells to the artery wall and adipose tissue, *Vascular Pharmacology* 52 (2010): 27–36; P. Libby, Inflammatory mechanisms: The molecular basis of inflammation and disease, *Nutrition Reviews* 65 (2007): S140–S146.
9. S. L. Murphy, J. Xu, and K. D. Kochanek and the Division of Vital Statistics, Deaths: Preliminary data for 2010, *National Vital Statistics Reports*, January 11, 2012.
10. World Health Organization, Chronic diseases and health promotion, www.who.int/chp/en, accessed October 27, 2011.
11. P. Perez-Martinez and coauthors, Calpain-10 interacts with plasma saturated fatty acid concentrations to influence insulin resistance in individuals with the metabolic syndrome, *American Journal of Clinical Nutrition* 93 (2011): 1136–1141; L. Qi and coauthors, Genetic predisposition, Western dietary pattern, and the risk of type 2 diabetes in men, *American Journal of Clinical Nutrition* 89 (2009): 1453–1458; A. Trichopoulou and coauthors, Genetic predisposition, nongenetic risk factors, and coronary infarct, *Archives of Internal Medicine* 168 (2008): 891–896.
12. V. L. Roger and coauthors, on behalf of the American Heart Association Statistics Committee and Stroke Statistics Subcommittee, Heart disease and stroke statistics—2012 update: A report from the American Heart Association, *Circulation* 125 (2012): 188–197.
13. A. Astrup and coauthors, The role of reducing intakes of saturated fat in the prevention of cardiovascular disease: Where does the evidence stand in 2010? *American Journal of Clinical Nutrition* 93 (2011): 684–688; R. J. Berlin and coauthors, Diet quality and the risk for cardiovascular disease: The Women's Health Initiative (WHI), *American Journal of Clinical Nutrition* 94 (2011): 49–57; M. U. Jakobsen and coauthors, Major types of dietary fat and risk of coronary heart disease: A pooled analysis of 11 cohort studies, *American Journal of Clinical Nutrition* 89 (2009): 1425–1432; E. Warensjo and coauthors, Markers of dietary fat quality and fatty acid desaturation as predictors of total and cardiovascular mortality: A population-based prospective study, *American Journal of Clinical Nutrition* 88 (2008): 203–209.
14. M. Drechsler and coauthors, Neutrophilic granulocytes—promiscuous accelerators of atherosclerosis, *Thrombosis and Haemostasis* 106 (2011): 839–848; G. K. Hansson and A. Hermansson, The immune system in atherosclerosis, *Nature Immunology* 12 (2011): 204–212; P. J. Murray and T. A. Wynn, Protective and pathogenic functions of macrophage subsets, *Nature Reviews. Immunology* 11 (2011): 723–737.
15. W. Insull, The pathology of atherosclerosis: Plaque development and plaque responses to medical treatment, *The American Journal of Medicine* 122 (2009): S3–S14.
16. W. Bekwelem and coauthors, White blood cell count, C-reactive protein, and incident heart failure in the Atherosclerosis Risk in Communities (ARIC) Study, *Annals of Epidemiology* 21 (2011): 739–748; S. Devaraj, U. Singh, and I. Jialal, The evolving role of C-reactive protein in atherothrombosis, *Clinical Chemistry* 55 (2009): 229–238.
17. I. Ikonomidis and coauthors, the role of lipoprotein-associated phospholipase A2 (Lp-PLA$_2$) in cardiovascular disease, *Reviews on Recent Clinical Trials* 6 (2011): 108–113; J. Y. Kim and coauthors, Lipoprotein-associated phospholipase A$_2$ activity is associated with coronary artery disease and markers of oxidative stress: A case-control study, *American Journal of Clinical Nutrition* 88 (2008): 630–637; H. S. Weintraub, Identifying the vulnerable patient with rupture-prone plaque, *American Journal of Cardiology* (12A) 101 (2008): 3F–10F.
18. M. H. Davidson and coauthors, Consensus panel recommendation for incorporating lipoprotein-associated phospholipase A$_2$ testing into cardiovascular disease risk assessment guidelines, *American Journal of Cardiology* (12A) 101 (2008): 51F–57F.
19. Insull, 2009; V. E. Friedewald and coauthors, The Editor's Roundtable: The vulnerable plaque, *American Journal of Cardiology* 102 (2008): 1644–1653.
20. W. S. Harris and coauthors, Omega-6 fatty acids and risk for cardiovascular disease, A Science Advisory from the American Heart Association Nutrition Subcommittee of the Council on Nutrition, Physical Activity, and Metabolism; Council on Cardiovascular Nursing; and Council on Epidemiology and Prevention, *Circulation* 119 (2009): 902–907.
21. A. C. Skulas-Ray and coauthors, Dose-response effects of omega-e fatty acids on triglycerides, inflammation, and endothelial function in healthy persons with moderate hypertriglyceridemia, *American Journal of Clinical Nutrition* 93 (2011): 243–252; F. Dangardt and coauthors, Omega-3 fatty acid supplementation improves vascular function and reduces inflammation in obese adolescents, *Atherosclerosis* 12 (2010): 580–585; M. N. Di Minno and coauthors, Exploring newer cardioprotective strategies: ω-3 fatty acids in perspective, *Thrombosis and Haemostasis* 104 (2010): 664–680.
22. J. Ando and K. Yamamoto, Effects of shear stress and stretch on endothelial function, *Antioxidants and Redox Signaling* 15 (2011): 1389–1403; J. J. Chiu, S. Usami, and S. Chien, Vascular endothelial responses to altered shear stress: Pathologic implications for atherosclerosis, *Annals of Medicine* 41 (2009): 19–28.
23. V. Cachofeiro and coauthors, Inflammation: A link between hypertension and atherosclerosis, *Current Hypertension Reviews* 5 (2009): 40–48.
24. A. L. Valderrama and coauthors, Million Hearts: Strategies to reduce the prevalence of leading cardiovascular disease risk factors—United States, 2011, *Morbidity and Mortality Weekly Report* 60 (2011): 1248–1251; N. T. Artinian and coauthors, Interventions to promote physical activity and dietary lifestyle changes for cardiovascular risk factor reduction in adults: A Scientific Statement from the American Heart Association, *Circulation* 122 (2010): 406–441; N. T. Nguyen and coauthors, Association of hypertension, diabetes, dyslipidemia, and metabolic syndrome with obesity: Findings from the National Health and Nutrition Examination Survey, 1999 to 2004, *Journal of the American College of Surgeons* 207 (2008): 928–934.
25. S. H. Woolf, A closer look at the economic argument for disease prevention, *Journal of the American Medical Association* 301 (2009): 536–538.
26. J. Fang and coauthors, Prevalence of Coronary Heart Disease: United States, 2006–2010, *Morbidity and Mortality Weekly Report* 60 (2011): 1377–1381.

27. Expert Panel on Detection, Evaluation, and Treatment of High Blood Cholesterol in Adults (Adult Treatment Panel III), *Third Report of the National Cholesterol Education Program (NCEP)*, NIH publication no. 02-5215 (Bethesda, Md.: National Heart, Lung, and Blood Institute, 2002), p. II-18.
28. Expert Panel on Detection, Evaluation, and Treatment of High Blood Cholesterol in Adults (Adult Treatment Panel III), 2002, p. VIII-2.
29. J. Durga and coauthors, Effect of 3 y of folic acid supplementation on the progression of carotid intima-media thickness and carotid arterial stiffness in older adults, *American Journal of Clinical Nutrition* 93 (2011): 941–949; M. Hoffman, Hypothesis: Hyperhomocysteinemia is an indicator of oxidant stress, *Medical Hypotheses* 77 (2011): 1088–1093; L. L. Humphrey and coauthors, Homocysteine level and coronary heart disease incidence: A systematic review and meta-analysis, *Mayo Clinic Proceedings* 83 (2008): 3–16.
30. Expert Panel on Detection, Evaluation, and Treatment of High Blood Cholesterol in Adults (Adult Treatment Panel III), 2002, p. II-19.
31. A. K. Chhatriwalla and coauthors, Low levels of low-density lipoprotein cholesterol and blood pressure and progression of coronary atherosclerosis, *Journal of the American College of Cardiology* 53 (2009): 1110–1115.
32. Insull, 2009; Expert Panel on Detection, Evaluation, and Treatment of High Blood Cholesterol in Adults (Adult Treatment Panel III), 2003, pp. II-2–II-3.
33. Murray and Wynn, 2011; Surmi and Hasty, 2010.
34. L. K. Curtiss, Reversing atherosclerosis? *New England Journal of Medicine* 360 (2009): 1144–1146; Y. M. Prak, M. Febbraio, and R. L. Silverstein, CD36 modulates migration of mouse and human macrophages in response to oxidized LDL and may contribute to macrophage trapping in the arterial intima, *Journal of Clinical Investigation* 119 (2009): 136–145.
35. Joint National Committee, *Prevention, Detection, Evaluation, and Treatment of High Blood Pressure, Seventh Report*, NIH publication no. 03-5233 (Bethesda, Md.: National Heart, Lung, and Blood Institute, 2003), pp. 1–3.
36. E. S. Ford, Trends in the risk for coronary heart disease among adults with diagnosed diabetes in the US: Findings from the National Health and Nutrition Examination Survey, 1999–2008, *Diabetes Care* 34 (2011): 1337–1343; R. A. DeFronzo and M. Abdul-Ghani, Assessment and treatment of cardiovascular risk in prediabetes: Impaired glucose tolerance and impaired fasting glucose, *American Journal of Cardiology* 108 (2011): 3B–24B.
37. Expert Panel on Detection, Evaluation, and Treatment of High Blood Cholesterol in Adults (Adult Treatment Panel III), 2002, pp. II-16, 11-50–11-53.
38. T. Y. Warren and coauthors, Sedentary behaviors increase risk of cardiovascular disease mortality in men, *Medicine and Science in Sports and Exercise* 42 (2010): 879–885; J. M. Sacheck, J. F. Kuder, and C. D. Economos, Physical fitness, adiposity, and metabolic risk factors in young college students, *Medicine and Science in Sports and Exercise* 42 (2010): 1039–1044; M. Hamer and E. Stamatakis, Physical activity and risk of cardiovascular disease events: Inflammatory and metabolic mechanisms, *Medicine and Science in Sports and Exercise* 41 (2009): 1206–1211; N. Orsini and coauthors, Combined effects of obesity and physical activity in predicting mortality among men, *Journal of Internal Medicine* 264 (2008): 442–451.
39. H. M. Ahmed and coauthors, Effects of physical activity on cardiovascular disease, *American Journal of Cardiology* 109 (2012): 288–295; R. Silvestre and coauthors, Effects of exercise at different times on postprandial lipemia and endothelial function, *Medicine and Science in Sports and Exercise* 40 (2008): 264–274; P. T. Williams, Reduced diabetic, hypertensive, and cholesterol medication use with walking, *Medicine & Science in Sports & Exercise* 40 (2008): 433–443.
40. S. A. Kenfield, Smoking and smoking cessation in relation to mortality in women, *Journal of the American Medical Association* 299 (2008): 2037–2047; Expert Panel on Detection, Evaluation, and Treatment of High Blood Cholesterol in Adults (Adult Treatment Panel III), 2002, p. II-16.
41. A. Mente and coauthors, A systematic review of the evidence supporting a causal link between dietary factors and coronary heart disease, *Archives of Internal Medicine* 169 (2009): 659–669.
42. R. Do and coauthors, The effect of chromosome 9p21 variants on cardiovascular disease may be modified by dietary intake: Evidence from a case/control and a prospective study, *PLoS Medicine* 8 (2011): e1001106.
43. M. Miller and coauthors, Triglycerides and cardiovascular disease: A Scientific Statement from the American Heart Association, *Circulation* 123 (2011): 2292–2333.
44. E. J. Gallagher, D. Leroith, and E. Karnieli, Insulin resistance in obesity as the underlying cause for the metabolic syndrome, *Mount Sinai Journal of Medicine* 77 (2010): 511–523; D. M. Minich and J. S. Bland, Dietary management of the metabolic syndrome beyond macronutrients, *Nutrition Reviews* 66 (2008): 429–444.
45. F. R. Jornayvaz, V. T. Samuel, and G. I. Shulman, The role of muscle insulin resistance in the pathogenesis of atherogenic dyslipidemia and nonalcoholic fatty liver disease associated with the metabolic syndrome, *Annual Review of Nutrition* 30 (2010): 273–290.
46. J. S. Chae and coauthors, Association of Lp-PLA(2) activity and LDL size with interleukin-6, an inflammatory cytokine and oxidized LDL, a marker of oxidative stress, in women with metabolic syndrome, *Atherosclerosis* 218 (2011): 499–506; Minich and Bland, 2008.
47. Expert Panel on Detection, Evaluation, and Treatment of High Blood Cholesterol in Adults (Adult Treatment Panel III), 2002, p. III-6.
48. Centers for Disease Control and Prevention, Vital signs: Prevalence, treatment, and control of high levels of low-density lipoprotein cholesterol—United States, 1999–2002 and 2005–2008, *Morbidity and Mortality Weekly Report* 60 (2011): 109–114; Expert Panel on Detection, Evaluation, and Treatment of High Blood Cholesterol in Adults (Adult Treatment Panel III), 2002, pp. II-1–II-4.
49. S. S. Gidding and coauthors, Implementing American Heart Association pediatric and adult nutrition guidelines: A Scientific Statement from the American Heart Association Nutrition Committee of the Council on Nutrition, Physical Activity and Metabolism, Council on Cardiovascular Disease in the Young, Council on Arteriosclerosis, Thrombosis and Vascular Biology, Council on Cardiovascular Nursing, Council on Epidemiology and Prevention, and Council for High Blood Pressure Research, *Circulation* 119 (2009): 1161–1175.
50. AHA Scientific Statement: Diet and lifestyle recommendations revision 2006, *Circulation* 114 (2006): 82–96; Expert Panel on Detection, Evaluation, and Treatment of High Blood Cholesterol in Adults (Adult Treatment Panel III), 2002, pp. V-1–V-28.
51. C. Rosendorff and coauthors, Treatment of hypertension in the prevention and management of ischemic heart disease: A Scientific Statement from the American Heart Association Council for High Blood Pressure Research and the Councils on Clinical Cardiology and Epidemiology and Prevention, *Circulation* 115 (2007): 2761–2788.
52. Chhatriwalla and coauthors, 2009.
53. Centers for Disease Control and Prevention, High blood pressure facts, www.cdc.gov/bloodpressure/facts.htm, accessed October 24, 2011.
54. Roger and coauthors, 2012.
55. Roger and coauthors, 2012.
56. K. C. Ferdinand and A. M. Armani, The management of hypertension in African Americans, *Critical Pathways in Cardiology* 6 (2007): 67–71.
57. J. Redon and coauthors, Mechanisms of hypertension in the cardiometabolic syndrome, *Journal of Hypertension* 27 (2009): 441–451; F. W. Visser and coauthors, Rise in extracellular fluid volume during high sodium depends on BMI in healthy men, *Obesity (Silver Spring)* 17 (2009): 1684–1688.
58. V. Savica, G. Bellinghieri, and J. D. Kopple, The effect of nutrition on blood pressure, *Annual Review of Nutrition* 30 (2010): 365–401.
59. Savica, Bellinghieri, and Kopple, 2010.
60. L. J. Appel on behalf of the American Society of Hypertension Writing Group, ASH Position paper: dietary approaches to lower blood pressure, *Journal of Clinical Hypertension* 11 (2009): 358–368.
61. E. G. Ciolac and coauthors, Effects of high-intensity aerobic interval training vs. moderate exercise on hemodynamic, metabolic and neuro-humoral abnormalities of young normotensive women at high familial risk for hypertension, *Hypertension Research* 33 (2010): 836–843; N. L. Chase and coauthors, The association of cardiorespiratory fitness and physical activity with incidence of hypertension in men, *American Journal of Hypertension* 22 (2009): 417–424.

62. Williams, 2008.
63. L. Azadbakht and coauthors, The Dietary Approaches to Stop Hypertension Eating Plan affects C-reactive protein, coagulation abnormalities, and hepatic function tests among type 2 diabetic patients, *Journal of Nutrition* 141 (2011): 1083–1088; J. F. Swain and coauthors, Characteristics of the diet patterns tested in the optimal macronutrient intake trial to prevent heart disease (OmniHeart): Options for a heart-healthy diet, *Journal of the American Dietetic Association* 108 (2008): 257–265; National Institutes of Health, National Heart, Lung, and Blood Institute, *Your Guide to Lowering Your Blood Pressure with DASH* (NIH Publication No. 06-4082, 2006).
64. US Department of Agriculture and US Department of Health and Human Services, *Dietary Guidelines for Americans 2005* (Washington, DC: Government Printing Office, January 2005).
65. F. Dumier, Dietary sodium intake and arterial blood pressure, *Journal of Renal Nutrition* 19 (2009): 57–60; National Institutes of Health, National Heart, Lung, and Blood Institute, 2006.
66. K. M. Dickinson, J. B. Keogh, and P. M. Clifton, Effects of a low-salt diet on flow-mediated dilation in humans, *American Journal of Clinical Nutrition* 89 (2009): 485–490.
67. Centers for Disease Control and Prevention, *Diabetes: Successes and opportunities for population-based prevention and control, At a glance, 2011*, www.cdc.gov/chronicdisease/resources/publications/AAG/ddt.htm, accessed November 9, 2011.
68. Centers for Disease Control, *National Diabetes Fact Sheet, 2011*, www.cdc.gov/diabetes/pubs/factsheets.htm, accessed October 24, 2011.
69. American Diabetes Association, Position Statement: Diagnosis and classification of diabetes mellitus, *Diabetes Care* 34 (2011): S62–S69.
70. American Diabetes Association, 2011.
71. J. P. Reis and coauthors, Lifestyle factors and risk for new-onset diabetes, *Annals of Internal Medicine* 155 (2011): 292–299.
72. D. Mozaffarian and coauthors, Lifestyle risk factors and new-onset diabetes mellitus in older adults: the Cardiovascular Health Study, *Archives of Internal Medicine* 169 (2009): 798–807.
73. American Diabetes Association, 2011.
74. B. Bennett and coauthors, Impaired insulin sensitivity and elevated ectopic fat in healthy obese vs. nonobese prepubertal children, *Obesity (Silver Spring)* 20 (2012): 371–375; K. L. Jones, Role of obesity in complicating and confusing the diagnosis and treatment of diabetes in children, *Pediatrics* 121 (2008): 361–368; J. A. Morrison and coauthors, Pre-teen insulin resistance predicts weight gain, impaired fasting glucose, and type 2 diabetes at age 18–19 y: A 10-y prospective study of black and white girls, *American Journal of Clinical Nutrition* 88 (2008): 778–788.
75. A. Das and S. Mukhopadhyay, The evil axis of obesity, inflammation and type-2 diabetes, *Endocrine Metabolic and Immune Disorders Drug Targets* 11 (2011): 23–31; J. M. Olefsky and C. K. Glass, Macrophages, inflammation, and insulin resistance, *Annual Review of Physiology* 17 (2010): 219–246.
76. A. Brown, L. R. Reynolds, and D. Bruemmer, Intensive glycemic control and cardiovascular disease: An update, *Nature Reviews. Cardiology* 7 (2010): 369–375; D. Hill and M. Fisher, The effect of intensive glycaemic control on cardiovascular outcomes, *Diabetes, Obesity and Metabolism* 12 (2010): 641–647; D. M. Nathan and coauthors, Modern-day clinical course of type 1 diabetes mellitus after 30 years' duration, *Archives of Internal Medicine* 169 (2009): 1307–1316; The Juvenile Diabetes Research Foundation Continuous Glucose Monitoring Study Group, Continuous glucose monitoring and intensive treatment of type 1 diabetes, *New England Journal of Medicine* 359 (2008): 1464–1476.
77. Department of Health and Human Services, Centers for Disease Control and Prevention, National Diabetes Fact Sheet, 2011, www.cdc.gov/diabetes/pubs/pdf/ndfs_2011.pdf, accessed November 9, 2011; N. R. Burrows and coauthors, Incidence of end-stage renal disease attributed to diabetes among persons with diagnosed diabetes: United States and Puerto Rico, 1996–2007, *Morbidity and Mortality Weekly Report* 59 (2010): 1361–1366; A. Whaley-Connell and coauthors, Diabetes mellitus and CKD awareness: The Kidney Early Evaluation Program (KEEP) and National Health and Nutrition Examination Survey (NHANES), *American Journal of Kidney Diseases* 53 (2009): S11–S21.
78. American Diabetes Association, Standards of medical care in diabetes: 2011, *Diabetes Care* 34 (2011): S11–S61.
79. American Diabetes Association, Standards of medical care in diabetes: 2011, 2011.
80. American Diabetes Association, Nutrition recommendations and interventions for diabetes, *Diabetes Care* 31 (2008): S61–S78.
81. American Diabetes Association, 2008.
82. American Diabetes Association, 2008.
83. American Diabetes Association, Standards of medical care in diabetes: 2011, 2011.
84. American College of Sports Medicine, American Diabetes Association, Joint position statement: Exercise and type 2 diabetes, *Medicine and Science in Sports and Exercise* 42 (2010): 2282–2303.
85. J. H. J. Hoeijmakers, DNA damage, aging, and cancer, *New England Journal of Medicine* 361 (2009): 1475–1485; C. M. Croce, Oncogenes and cancer, *New England Journal of Medicine* 358 (2008): 502–511.
86. A. C. Chiang and J. Massague, Molecular basis of metastasis, *New England Journal of Medicine* 359 (2008): 2814–2823.
87. M. M. Heinen and coauthors, Physical activity, energy restriction, and the risk of pancreatic cancer: A prospective study in the Netherlands, *American Journal of Clinical Nutrition* 94 (2011): 1314–1323; X. Sui and coauthors, Influence of cardiorespiratory fitness on lung cancer mortality, *Medicine and Science in Sports and Exercise* 42 (2010): 872–878; J. B. Peel and coauthors, A prospective study of cardiorespiratory fitness and breast cancer mortality, *Medicine and Science in Sports and Exercise* 41 (2009): 742–748; World Cancer Research Fund/American Institute for Cancer Research, *Policy and Action for Cancer Prevention. Food, Nutrition, and Physical Activity: A Global Perspective* (Washington, DC: AICR, 2009), pp. 12–28.
88. S. Y. Pan and M. DesMeules, Energy intake, physical activity, energy balance, and cancer: Epidemiologic evidence, *Methods in Molecular Biology* 472 (2009): 191–215; J. B. Peel and coauthors, Cardiorespiratory fitness and digestive cancer mortality: Findings from the Aerobics Center Longitudinal Study, *Cancer Epidemiology Biomarkers Prevention* 18 (2009): 1111–1117; K. Y. Wolin and coauthors, Physical activity and colon cancer prevention: A meta-analysis, *British Journal of Cancer* 100 (2009): 611–616.
89. H.-K. Na, and S. Oliynyk, Effects of physical activity on cancer prevention, *Annals of the New York Academy of Sciences* 1229 (2011): 176–183; Peel and coauthors, A prospective study of cardiorespiratory fitness and breast cancer mortality, 2009.
90. C. Eheman and coauthors, Annual Report to the Nation on the Status of Cancer, 1975–2008, featuring cancers associated with excess weight and lack of sufficient physical activity, *Cancer* (2012): doi10.1002/cncr.27514; S. Pendyala and coauthors, Diet-induced weight loss reduces colorectal inflammation: Implications for colorectal carcinogenesis, *American Journal of Clinical Nutrition* 93 (2011): 234–242; A. Vrieling and E. Kampman, The role of body mass index, physical activity, and diet in colorectal cancer recurrence and survival: A review of the literature, *American Journal of Clinical Nutrition* 92 (2010): 471–490; D. Li and coauthors, Body mass index and risk, age of onset, and survival in patients with pancreatic cancer, *Journal of the American Medical Association* 301 (2009): 2553–2562; Pan and DesMeules, 2009; World Cancer Research Fund/American Institute for Cancer Research, 2007, pp. 211–242.
91. N. H. Rod and coauthors, Low-risk factor profile, estrogen levels, and breast cancer risk among postmenopausal women, *International Journal of Cancer* 124 (2009): 1935–1940; World Cancer Research Fund/American Institute for Cancer Research, 2007, pp. 30–46.
92. L. H. Kushi and coauthors, American Cancer Society guidelines on nutrition and physical activity for cancer prevention, *CA: Cancer Journal for Clinicians* 62 (2012): 30–67; World Cancer Research Fund/American Institute for Cancer Research, 2007, pp. xiv–xxiv.
93. W. Y. Chen and coauthors, Moderate alcohol consumption during adult life, drinking patterns, and breast cancer risk, *Journal of the American Medical Association* 306 (2011): 1884–1890; World Cancer Research Fund/American Institute for Cancer Research, 2007, pp. 277–280.
94. A. Brevik and coauthors, Polymorphisms in base excision repair genes as colorectal cancer risk factors and modifiers of the effect of diets high in red meat, *Cancer Epidemiology, Biomarkers, and Prevention* 19 (2010): 3167–3173; S. Rohrmann, S. Hermann, and J. Linseisen, Heterocyuclic aromatic amine intake increases colorectal adenoma risk: Findings from a prospective European cohort study, *American Journal of Clinical Nutrition* 89 (2009): 1418–1424; R. Sinha and coauthors, Meat intake and mortality: A prospective study of over half a million people, *Archives of Internal Medicine* 169 (2009): 562–571; R. L. Santarelli, F. Pierre, and D. E. Corpet, Processed meat and colorectal cancer: A review of

epidemiologic and experimental evidence, *Nutrition and Cancer* 60 (2008): 131–144; World Cancer Research Fund/American Institute for Cancer Research, 2007, pp. 280–288.

95. World Cancer Research Fund/American Institute for Cancer Research, 2007, pp. 135–140.
96. M. Solanas and coauthors, Dietary olive oil and corn oil differentially affect experimental breast cancer through distinct modulation of the p21Ras signaling and the proliferation-apoptosis balance, *Carcinogenesis* 31 (2010): 871–879; J. Y. Lee, L. Zhao, and D. H. Hwang, Modulation of pattern recognition receptor-meditated inflammation and risk of chronic diseases by dietary fatty acids, *Nutrition Reviews* 68 (2009): 38–61.
97. T. M. Brasky and coauthors, Specialty supplements and breast cancer risk in the VITamins And Lifestyle (VITAL) Cohort, *Cancer Epidemiology, Biomarkers, and Prevention* 19 (2010): 1696–1708; World Cancer Research Fund/American Institute for Cancer Research, 2007, pp. 280–288; R. S. Chapkin, D. N. McMurray, and J. R. Lupton, Colon cancer, fatty acids and anti-inflammatory compounds, *Current Opinion in Gastroenterology* 23 (2007): 48–54; J. Shannon and coauthors, Erythrocyte fatty acids and breast cancer risk: A case-control study in Shanghai, China, *American Journal of Clinical Nutrition* 85 (2007): 1090–1097; E. Theodoratou and coauthors, Dietary fatty acids and colorectal cancer: A case-control study, *American Journal of Epidemiology* 166 (2007): 181–195.
98. T. J. Key, Fruit and vegetables and cancer risk, *British Journal of Cancer* 104 (2010): 6–11; L. B. Sansbury and coauthors, The effect of strict adherence to a high-fiber, high-fruit and -vegetable, and low-fat eating pattern on adenoma recurrence, *American Journal of Epidemiology* 170 (2009): 576–584.
99. World Cancer Research Fund/American Institute for Cancer Research, 2007, pp. 75–115, 182.
100. World Cancer Research Fund/American Institute for Cancer Research, 2007, pp. 75–115.
101. A. Galimanis and coauthors, Lifestyle and stroke risk: A review, *Current Opinion in Neurology* 22 (2009): 60–68; Gidding and coauthors, 2009; American Diabetes Association, 2008; Lutsey, Steffen, and Stevens, 2008; World Cancer Research Fund/American Institute for Cancer Research, 2007, pp. 373–390.
102. W. C. Willett and D. S. Ludwig, The 2010 Dietary Guidelines: The best recipe for health? *New England Journal of Medicine* 365 (2011): 1563–1565.
103. Harvard School of Public Health, The Nutrition Source, Healthy Eating Plate, www.hsph.harvard.edu/nutritionsource, accessed September 14, 2011.
104. G. C. Burdge and K. A. Lillycrop, Nutrition, epigenetics, and developmental plasticity: Implications for understanding human disease, *Annual Review of Nutrition* 30 (2010): 315–339; J. M. Ordovas, Genetic influences on blood lipids and cardiovascular disease risk: Tools for primary prevention, *American Journal of Clinical Nutrition* 89 (2009): 1509S–1517S; L. R. Ferguson and M. Philpott, Nutrition and mutagenesis, *Annual Review of Nutrition* 28 (2008): 313–329.

 LEARN IT Present arguments for and against the use of complementary and alternative medicine.

Complementary and Alternative Medicine

If you suffered from migraine headaches or severe joint pain, where would you turn for relief? Would you visit a physician? Or are you more likely to go to an herbalist or an acupuncturist? Most physicians diagnose and treat medical conditions in ways that are accepted by the established medical community; herbalists and acupuncturists, among others, offer alternatives to standard medical practice. Instead of taking two aspirin, for example, you might be advised to chew two fresh leaves of the herb feverfew or to swallow a tincture of white willow bark. Or you might receive a massage and several acupuncture needles.

Complementary and alternative medicine (CAM) has become increasingly popular in recent decades (see the accompanying glossary for this and related terms).[1] People use these therapies for a variety of reasons. Some want to take more responsibility for both maintaining their own health and finding cures for their own diseases, especially when traditional medical therapies prove ineffective. Others have become distrustful of, and feel overwhelmed by, the high-tech diagnostic tests and costly treatments that **conventional medicine** offers. This highlight explores alternative therapies in search of their possible benefits and with an awareness of their potential harms.

Defining Complementary and Alternative Medicine

By definition, complementary and alternative medicine is not conventional medicine because there is insufficient evidence that it is safe and effective. It includes a variety of approaches, philosophies, and treatments, some of which are defined in the accompanying glossary of alternative therapies. When these therapies are used instead of conventional medicine, they are called *alternative;* when used together with conventional medicine, they are called *complementary.* **Integrative medicine** combines conventional medicine and CAM treatments for which there is some high-quality scientific evidence of safety and effectiveness.[2]

A growing number of health-care professionals are learning about alternative therapies; more than half of US medical schools now offer elective courses in alternative medicine, and even more include discussions of these therapies in their required courses. More than ever before, health-care professionals are incorporating some of the beneficial alternative therapies into their practices, thus the gap between conventional medicine and CAM is narrowing.[3]

For some alternative therapies, preliminary and limited scientific evidence suggests some effectiveness; but for most, well-designed scientific studies have yet to determine safety and effectiveness. If proved safe and effective, an alternative therapy may be adopted by conventional medicine. Cancer radiation therapy, for example, was once considered an unconventional therapy, but it proved its clinical value and became part of accepted medical practice. In some cases, a therapy that is accepted by conventional medicine for a specific ailment is used for a different purpose in an alternative therapy. For example, chelation therapy, the preferred medical treatment for lead poisoning, is a common alternative therapy for cardiovascular disease.

Sound Research, Loud Controversy

Much information on alternative therapies comes from folklore, tradition, and testimonial accounts. Relatively few clinical trials have been conducted. Consequently, scientific evidence proving the safety and effectiveness of many alternative therapies is lacking. Some say that alternative therapies simply do not work; others suggest that these therapies have not been given a fair trial. In an effort to "explore complementary and alternative healing practices through

GLOSSARY

complementary and alternative medicine (CAM): diverse medical and health-care systems, practices, and products that are not currently considered part of conventional medicine; also called *adjunctive, unconventional,* or *unorthodox therapies.*

conventional medicine: diagnosis and treatment of diseases as practiced by medical doctors (M.D.), doctors of osteopathy (D.O.), and allied health professionals such as physical therapists and registered nurses; also called *allopathy; Western, mainstream, orthodox,* or *regular medicine;* and *biomedicine.*

integrative medicine: care that combines conventional and complementary therapies for which there is some high-quality scientific evidence of safety and effectiveness. Integrative medicine emphasizes the importance of the relationship between the practitioner and the patient and focuses on wellness, healing, and the whole person.

vigorous science," the National Center for Complementary and Alternative Medicine supports clinical trials of these therapies. Articles reporting the results of these clinical trials are available at their web site (**nccam.nih.gov/research**).

Sound research would answer two important questions. First, does the treatment offer better results than either doing nothing or giving a placebo? Second, do the benefits clearly outweigh the risks? Each of these points is worthy of elaboration.

Placebo Effect

Stories abound that credit alternative therapies with miraculous cures. Without scientific research to determine effectiveness, however, one is left to wonder whether it is the therapies or the placebo effect that produces the cure. Recall from Chapter 1 that giving a placebo often brings about a healing effect in people who believe they are receiving the treatment.[4] Conventional medicine tends to neglect this powerful remedy, whereas many alternative therapies embrace it.

Risks versus Benefits

Ideally, a therapy provides benefits with little or no risk. Figure H18-1 (p. 616) presents several examples of herbal remedies that appear to be generally safe and possibly effective in treating various conditions.[5] Such findings, if replicated, hold promise that these alternative therapies may one day be integrated into conventional medicine.

Some alternative therapies are innocuous, providing little or no benefit for little or no risk. Sipping a cup of warm tea with a pleasant aroma, for example, won't cure heart disease, but it may improve one's mood and help relieve tension. Given no physical hazard and little financial risk, such therapies are acceptable.

In contrast, other products and procedures are downright dangerous, posing great risks while providing no benefits. One example is the folk practice of geophagia (eating earth or clay), which can cause GI impaction and impair iron absorption. Another is the taking of laetrile to treat cancer, which can cause cyanide poisoning. Clearly, such therapies are too harmful to be used.

Perhaps most controversial are alternative therapies that may provide benefits, but also carry significant, unknown, or debatable risks. Smoking marijuana is an example of such an alternative therapy.[6] The compounds in marijuana seem to provide relief from symptoms such as nausea, vomiting, and pain that commonly accompany cancer, AIDS, and other diseases, but marijuana use also poses risks that some people consider acceptable whereas others deem intolerable. Figure H18-2 (p. 616) summarizes the relationships between risks and benefits.

GLOSSARY OF ALTERNATIVE THERAPIES

acupuncture (AK-you-PUNK-cher): a technique that involves piercing the skin with long thin needles at specific anatomical points to relieve pain or illness. Acupuncture sometimes uses heat, pressure, friction, suction, or electromagnetic energy to stimulate the points.

aroma therapy: a technique that uses oil extracts from plants and flowers (usually applied by massage or baths) to enhance physical, psychological, and spiritual health.

ayurveda (AH-your-VAY-dah): a traditional Hindu system of improving health by using herbs, diet, meditation, massage, and yoga to stimulate the body, mind, and spirit to prevent and treat disease.

bioelectromagnetic medical applications: the use of electrical energy, magnetic energy, or both to stimulate bone repair, wound healing, and tissue regeneration.

biofeedback: the use of special devices to convey information about heart rate, blood pressure, skin temperature, muscle relaxation, and the like to enable a person to learn how to consciously control these medically important functions.

biofield therapeutics: a manual healing method that directs a healing force from an outside source (commonly God or another supernatural being) through the practitioner and into the client's body; commonly known as "laying on of hands."

cartilage therapy: the use of cleaned and powdered connective tissue, such as collagen, to improve health.

chelation (kee-LAY-shun) **therapy:** the use of ethylene diamine tetraacetic acid (EDTA) to bind with metallic ions, thus healing the body by removing toxic metals.

chiropractic (KYE-roh-PRAK-tik): a manual healing method of manipulating the spine to restore health.

faith healing: healing by invoking divine intervention without the use of medical, surgical, or other traditional therapy.

herbal (ERB-al) **medicine:** the use of plants to treat disease or improve health; also known as *botanical medicine* or *phytotherapy*.

homeopathy (hoh-me-OP-ah-thee): a practice based on the theory that "like cures like," that is, that substances that cause symptoms in healthy people can cure those symptoms when given in very dilute amounts.

- **homeo** = like
- **pathos** = suffering

hydrotherapy: the use of water (in whirlpools, as douches, or packed as ice, for example) to promote relaxation and healing.

hypnotherapy: a technique that uses hypnosis and the power of suggestion to improve health behaviors, relieve pain, and heal.

imagery: a technique that guides clients to achieve a desired physical, emotional, or spiritual state by visualizing themselves in that state.

iridology: the study of changes in the iris of the eye and their relationships to disease.

macrobiotic diet: a philosophical approach of eating mostly plant-based foods such as whole grains, legumes, and vegetables, with small amounts of fish, fruits, nuts, and seeds.

- **macro** = large, great
- **biotic** = life

massage therapy: a healing method in which the therapist manually kneads muscles to reduce tension, increase blood circulation, improve joint mobility, and promote healing of injuries.

meditation: a self-directed technique of relaxing the body and calming the mind.

naturopathic (nay-chur-oh-PATH-ick) **medicine:** a system that taps the natural healing forces within the body by integrating several practices, including traditional medicine, herbal medicine, clinical nutrition, homeopathy, acupuncture, East Asian medicine, hydrotherapy, and manipulative therapy.

orthomolecular medicine: the use of large doses of vitamins to treat chronic disease.

ozone therapy: the use of ozone gas to enhance the body's immune system.

qi gong (chée GUNG): a Chinese system that combines movement, meditation, and breathing techniques to enhance the flow of qi (vital energy) in the body.

> FIGURE H18-1 **Examples of Herbal Remedies**

Ginger may relieve nausea and vomiting due to motion sickness or pregnancy.

Ginkgo may slow the loss of cognitive function associated with aging.

St. John's wort may be effective in treating mild depression.

American ginseng may improve glucose control in people with type 2 diabetes.

Saw palmetto may improve the symptoms associated with an enlarged prostate.

The gel of an aloe vera plant soothes a minor burn.

> FIGURE H18-2 **Risk-Benefit Relationships**

RISK: No (or little) → Much
BENEFIT: No (or little) → Much

BENEFIT \ RISK	No (or little) risk	Much risk
Much benefit	**Ideal situation** Benefits with little or no risk (Accept)	**Cautionary situation** Possible benefits with great or unknown risks (Consider carefully)
No (or little) benefit	**Neutral situation** Little or no benefit with little or no risk (Accept or reject as preferred)	**Dangerous situation** No benefits with great risks (Reject)

Nutrition-Related Alternative Therapies

Most alternative therapies fall outside the field of nutrition, but nutrition itself can be an alternative therapy. Furthermore, many alternative therapies prescribe specific dietary regimens even though most practitioners are not registered dietitians (see Highlight 1). Nutrition-related alternative therapies include the use of foods, vitamin and mineral supplements, and herbs to prevent and treat illnesses.

Foods

The many dietary recommendations presented throughout this text are based on scientific evidence and do *not* fall into the alternative therapies category; strategies that are still experimental, however, do. For example, alternative therapists may recommend macrobiotic diets to help prevent chronic diseases, whereas most registered dietitians would advise people to eat a balanced diet that includes servings from each of the five food groups. Similarly, enough scientific evidence is available to recommend including soy foods in the diet to protect against heart disease—but not to determine whether the phytoestrogens of soy supplements are safe or beneficial in managing the symptoms of menopause.

Highlight 13 explores the potential health benefits of soy and many other functional foods and concludes that no one food is magical. As part of a balanced diet, these foods can support good health and protect against disease. Importantly, the benefits derive from a variety of *foods.* More research is needed to determine the safety and effectiveness of taking supplements of the phytochemicals found in these foods.

Vitamin and Mineral Supplements

Like foods, vitamin and mineral supplements may fall into either the conventional or the alternative realm of medicine. For example, conventional advice recommends consuming 400 micrograms of folate to prevent neural tube defects, but not the taking of 1000 milligrams of vitamin C to prevent the common cold. Highlight 10 examines the appropriate use of supplements and potential dangers of excessive intakes.

As research on nutrition and chronic diseases has revealed many of the roles played by the vitamins and minerals in supporting health, conventional medicine has warmed up to the possibility that vitamin and mineral supplements might be an appropriate preventive therapy.[7] Some vitamin and mineral supplements appear to be in transition from alternative medicine to conventional medicine; that is, they have begun to prove their safety and effectiveness. Herbal remedies, however, still remain clearly in the realm of complementary and alternative medicine.

Herbal Remedies

From earliest times, people have used myriad herbs and other plants to cure aches and ills with varying degrees of success (review Figure H18-1). In fact, today's pharmaceutical industry originated from the use of plant-derived products for human health.[8] Upon scientific study, dozens of these folk remedies reveal their secrets. For example, myrrh, a plant resin used as a painkiller in ancient times, does indeed have an analgesic effect. The herb valerian, which has long been used as a tranquilizer, contains oils that have a sedative effect. Senna leaves, brewed as a laxative tea, produce compounds that act as a potent cathartic drug. Green tea, brewed from the dried leaves of *Camellia sinensis,* contains phytochemicals that induce cancer cells to self-destruct. Naturally occurring salicylates provide the same protective effects as low doses of aspirin. Salicylates are found in spices such as curry, paprika, and thyme; fruits; vegetables; teas; and candies flavored with wintergreen (methylsalicylate).

Beneficial compounds from wild species contribute to about half of our modern medicines. By analyzing these compounds, pharmaceutical labs can synthesize pure forms of the drugs. Unlike herbs and wild species, which vary from batch to batch, synthesized medicines deliver exact dosages. By synthesizing drugs, we are also able to conserve endangered species. Consider that it took all of the bark from one 40-foot-tall, 100-year-old Pacific yew tree to produce one 300 milligram dose of the anticancer drug paclitaxel (Taxol), until scientists learned how to synthesize it. Many yet undiscovered cures may be forever lost as wild species are destroyed, long before their secrets are revealed to medicine.

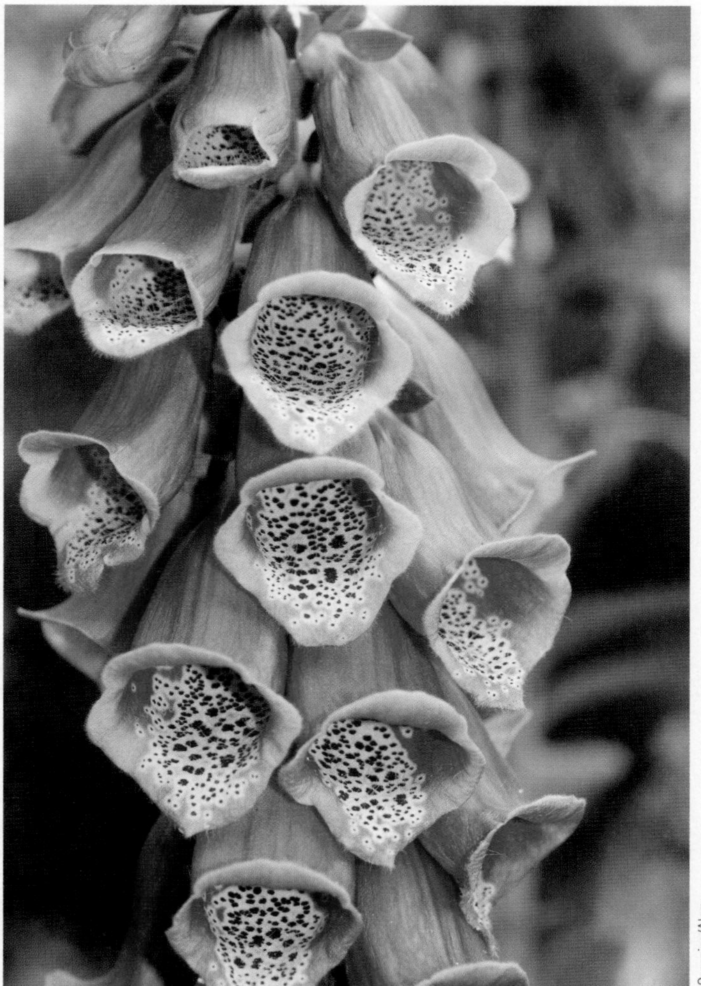

Digoxin, the most commonly prescribed heart medication, derives from the leaves of the foxglove plant *(Digitalis purpurea).*

Herbal Precautions

Plants are "natural," but that does not mean all plants are beneficial or even safe. Nothing could be more natural—and deadly—than the poisonous herb hemlock. Several herbal remedies have toxic effects. The popular Chinese herbal potion jin bu huan, which is used as a pain and insomnia remedy, has been linked with several cases of acute hepatitis. Germanium, a nonessential mineral commonly found in many herbal products, has been associated with chronic kidney failure. Paraguay tea produces symptoms of agitation, confusion, flushed skin, and fever. Kombucha tea, commonly used in the hopes of preventing cancer, relieving arthritis, curing insomnia, and stimulating hair regrowth, can cause severe metabolic acidosis. Table H18-1 (p. 618) provides an overview of selected herbs.[9]

TABLE H18-1 Overview of Selected Herbs

Common and Scientific Names	Claims	Research Findings	Risks[a]
Aloe (gel) *Aloe vera*	Promote wound healing	May help heal minor burns and abrasions; may cause infections in severe wounds	Generally considered safe
Black cohosh (stems and roots) *Actaea racemosa, Cimifuga racemosa*	Ease menopause symptoms	Conflicting evidence	May cause headaches, stomach discomfort, liver damage
Chamomile (flowers) *Matricaria recutita, Chamomilla recutita*	Relieve indigestion	Little evidence available	Generally considered safe
Chaparral (leaves and twigs) *Larrea tridentata*	Slow aging, "cleanse" blood, heal wounds, cure cancer, treat acne	No evidence available	Acute, toxic hepatitis; liver damage
Cinnamon (bark) *Cinnamomum zeylanicum, C. cassia*	Relieve indigestion, lower blood glucose and blood lipids	May lower blood glucose in type 2 diabetes	May have a blood-thinning effect; not safe for pregnant women or those taking diabetes medication
Comfrey (leafy plant) *Symphytum officinale, S. asperum, S. x uplandicum*	Soothe nerves	No evidence available	Liver damage
Echinacea (roots) *Enchinacea angustifolia, E. pallida, E. purpurea*	Alleviate symptoms of colds, flus, and infections; promote wound healing; boost immunity	Ineffective in preventing colds or other infections	Generally considered safe; may cause headache, dizziness, nausea
Ephedra (stems) *Ephedra sinica*	Promote weight loss	Little evidence available; FDA has banned the sale of ephedra-containing products	Rapid heart rate, tremors, seizures, insomnia, headaches, hypertension
Feverfew (leaves) *Tanacetum parthenium*	Prevent migraine headaches	May prevent migraine headaches	Generally considered safe; may cause mouth irritation, swelling, ulcers, and GI distress
Garlic (bulbs) *Allium sativum*	Lower blood lipids and blood pressure	May lower blood cholesterol slightly; conflicting evidence on blood pressure	Generally considered safe; may cause garlic breath, body odor, gas, and GI distress; inhibits blood clotting
Ginger (roots) *Zingiber officinale*	Prevent motion sickness, nausea	May relieve pregnancy-induced nausea; conflicting evidence on nausea caused by motion, chemotherapy, or surgery	Generally considered safe
Ginkgo (tree leaves) *Ginkgo biloba*	Improve memory, relieve vertigo	Little evidence available	Generally considered safe; may cause headache, GI distress, dizziness; may inhibit blood clotting
Ginseng (roots) *Panax ginseng* (Asian), *P. quinquefolius* (American)	Boost immunity, increase endurance	Little evidence available	Generally considered safe; may cause insomnia, headaches, and high blood pressure
Goldenseal (roots) *Hydrastis canadensis*	Relieve indigestion, treat urinary infections	Little evidence available	Generally considered safe; not safe for people with hypertension or heart disease
Kava (roots) *Piper methysticum*	Relieves anxiety, promotes relaxation	Little evidence available	Liver failure
Saw palmetto (ripe fruits) *Serenoa repens*	Relieve symptoms of enlarged prostate; diuretic; enhance sexual vigor	Little evidence available	Generally considered safe; may cause nausea, vomiting, diarrhea
St. John's wort (leaves and tops) *Hypericum perforatum*	Relieve depression and anxiety	May relieve mild depression	Generally considered safe; may cause fatigue, increased sensitivity to sunlight, and GI distress
Turmeric (roots) *Curcuma longa*	Reduces inflammation; relieves heartburn; prevents or treats cancer	No evidence available	Generally considered safe; may cause indigestion; not safe for people with gallbladder disease
Valerian (roots) *Valeriana officinalis*	Calm nerves, improve sleep	Little evidence available	Long-term use associated with liver damage
Yohimbe (tree bark) *Pausinystalia yohimbe*	Enhance "male performance"	No evidence available	Kidney failure, seizures

[a]Allergies are always a possible risk; see Table H18-2 (p. 620) for drug interactions. Pregnant women should not use herbal supplements.

Although some people use herbs to treat or prevent disease, herbs are not regulated as drugs; they are considered dietary supplements. The Food and Drug Administration (FDA) does not evaluate dietary supplements for safety or effectiveness, nor does it monitor their contents. Under the Dietary Supplement Health and Education Act, rather than the herb manufacturers having to prove the safety of their products, the FDA has the burden of proving that a product is not safe. Consequently, consumers may lack information about or find discrepancies regarding:

- *True identification of herbs.* Most mint teas are safe, for instance, but some varieties contain the highly toxic pennyroyal oil. Mistakenly used to soothe a colicky baby, mint tea laden with pennyroyal has been blamed for the liver and neurological injuries of at least two infants, one of whom died.
- *Purity of herbal preparations.* A young child diagnosed with lead poisoning had been given an herbal vitamin that contained large quantities of lead and mercury for 4 years. Twelve cases of lead poisoning among adults using Ayurvedic remedies were reported to the Centers for Disease Control and Prevention in recent years. One-fifth of Ayurvedic medicines purchased via the Internet for analyses contained detectable levels of lead, mercury, or arsenic.[10] Another herbal supplement, sold worldwide, is implicated in 10 cases of severe liver toxicity.[11] The suspected cause of the toxicity is unintended contamination or overdose of an ingredient that can lead to liver toxicity.
- *Appropriate uses and contraindications of herbs.* Herbal remedies alone may be appropriate for minor ailments—a cup of chamomile tea to ease gastric discomfort or the gel of an aloe vera plant to soothe a sunburn, for example—but not for major health problems such as cancer or AIDS.
- *Effectiveness of herbs.* Herbal remedies may claim to work wonders without having to prove effectiveness. Research studies often report conflicting findings, with some suggesting a benefit and others indicating no effectiveness.[12]
- *Variability of herbs.* Not all species are created equal. The various species of coneflower provide an example. *Echinacea purpurea,* for example, may help in the early treatment of colds, but *Echinacea augustifolia* may not.[13] Similarly, not all parts of a plant provide the same compounds. Leaves, roots, and oils contain different compounds and extracts, and the temperatures used during manufacturing may affect their potency. Consumers are not always aware of such differences, and manufacturers do not always make such distinctions when preparing and labeling supplements.
- *Accuracy of labels.* Supplements may contain none of an herb or mixed species, and labels are often inaccurate. More often than not, supplements do not contain the species or the quantities of active ingredients stated on their labels.[14] In several cases, supplements did not even contain herbs, but drugs that are known to interact with prescription medicines and lower blood pressure to dangerous levels. Such discrepancies in the contents of supplements can be dangerous to the consumer, interfere with scientific research, and make it difficult to interpret the findings. Consumers may want to shop for supplements bearing a logo from either US Pharmacopeia or Consumer Lab indicating that the contents have been analyzed and found to contain the ingredients and quantities listed on the label.
- *Safe dosages of herbs.* Herbs may contain active ingredients—compounds that affect the body. Each of these active ingredients has a different potency, time of onset, duration of activity, and consequent effects, making the plant itself too unpredictable to be useful. Foxglove leaves, for example, contain dozens of compounds that have an effect on the heart; digoxin, a drug derived from foxglove, offers a standard dosage that allows for a more predictable cardiac response. Even when herbs are manufactured into capsules or liquids, their concentrations of active ingredients differ dramatically from batch to batch and from the quantities stated on the labels.[15]
- *Interactions of herbs with medicines and other herbs.* Like drugs, herbs may interfere with, or potentiate, the effects of other herbs and drugs (see Table H18-2, p. 620).[16] A person taking both cardiac medication and the herb foxglove may be headed for disaster from the combined effect on the heart. Similarly, taking St. John's wort with medicines used to treat heart disease, depression, seizures, and certain cancers might diminish or exaggerate the intended effects.[17] Because *Ginkgo biloba* impairs blood clotting, it can cause bleeding problems for people taking aspirin or other blood-thinning medicines regularly.[18]
- *Adverse reactions and toxicity levels of herbs.* Herbs may produce undesirable reactions. The herbal root kava, commonly used to treat anxiety and insomnia, can cause liver abnormalities and may have such a sedating effect as to impair driving. Chinese herbal treatments containing *Aristolochia fangchi* are known to cause kidney damage and cancers.[19] Table H18-1 (p. 618) includes risks associated with commonly used herbs. To ensure the safety and standardization of herbal remedies, Congress may need to establish new regulations.[20]

Because herbal medicines are sold as dietary supplements, their labels cannot claim to cure a disease, but they can make various other claims. Not surprisingly, when a label claims that an herbal product may strengthen immunity, improve memory, support eyesight, or maintain heart health, consumers believe that taking the product will provide those benefits. Beware. Manufacturers need not prove effectiveness; they need only state on the product label that this claim "Has not been evaluated by the FDA." Consumers who decide to use herbs need to become informed of the possible risks.

Internet Precautions

As Highlight 1 points out, just because something appears on the Internet, "it ain't necessarily so." Keep in mind that the thousands of websites touting the benefits of herbal medicines and other dietary supplements are marketing their products. Most product advertisements

TABLE H18-2 Herb and Drug Interactions

Herb	Drug	Interaction
American ginseng	Estrogens, corticosteroids	Enhances hormonal response
American ginseng	Breast cancer therapeutic agent	Synergistically inhibits cancer cell growth
American ginseng, karela	Blood glucose regulators	Affect blood glucose levels
Echinacea (possible immunostimulant)	Cyclosporine and corticosteroids (immunosuppressants)	May reduce drug effectiveness
Evening primrose oil, borage	Anticonvulsants	Lower seizure threshold
Feverfew	Aspirin, ibuprofen, and other nonsteroidal anti-inflammatory drugs	Negates the effect of the herb in treating migraine headaches
Feverfew, garlic, ginkgo, ginger, and Asian ginseng	Warfarin, coumarin (anticlotting drugs, "blood thinners")	Prolong bleeding time; increase likelihood of hemorrhage
Garlic	Protease inhibitor (HIV drug)	May reduce drug effectiveness
Kava, valerian	Anesthetics	May enhance drug action
Kelp (iodine source)	Synthroid or other thyroid hormone replacers	Interferes with drug action
Kyushin, licorice, plantain, uzara root, hawthorn, Asian ginseng	Digoxin (cardiac antiarrhythmic drug derived from the herb foxglove)	Interfere with drug action and monitoring
St. John's wort, saw palmetto, black tea	Iron supplement	Tannins in herbs inhibit iron absorption
St. John's wort	Protease inhibitors (HIV drugs), warfarin (anticlotting drug), digoxin (cardiac antiarrhythmic drug), oral contraceptives, tamoxifen (breast cancer drug)	May enhance or reduce drug effectiveness
Valerian	Barbiturates	Causes excessive sedation

claim to prevent or treat specific diseases, but few include the FDA disclaimer statement. Many of the websites promote products by quoting researchers or physicians. Such quotations lend an air of authority to advertisements, but be aware that these sources may not even exist—and if they do, their comments may have been taken out of context. When asked, they may not agree at all with the claims attributed to them by the manufacturer.

Other deceits and dangers lurk in cyberspace as well. Potentially toxic substances, illegal and unavailable in many countries, are now easy to obtain via the Internet. Electronic access to products such as absinthe and oil of wormwood could be deadly. When the FDA discovers websites selling unapproved drugs, such as laetrile, it can order the business to shut down. But consumers need to remain vigilant because other similar businesses pop up quickly.

The Consumer's Perspective

Some health-care professionals may dismiss alternative therapies as ineffective and perhaps even dangerous, but many consumers think otherwise. In a survey of more than 20,000 people, almost 4 out of 10 adults had used at least one alternative therapy for a variety of medical complaints ranging from back pain and anxiety to heart disease and cancer.[21] Interestingly, those who seek alternative therapies seem to do so not so much because they are dissatisfied with conventional medicine as because they find these alternatives more in line with their beliefs about health and life.

Most often, people use alternative therapies in addition to, rather than in place of, conventional therapies. Few consult an alternative therapist without also seeing a physician. In fact, most people seek alternative therapies for nonserious medical conditions or for health promotion. They simply want to feel better and access is easy. Sometimes their symptoms are chronic and subjective, such as pain and fatigue, and difficult to treat. In these cases, the chances of finding relief are often as good with a placebo, standard medical intervention, or even nonintervention.

Consumers spend billions of dollars on alternative health services and related products such as herbs, crystals, and aromas.[22] As Highlight 1 points out, selecting a reliable practitioner depends on finding out about training, qualifications, and licenses. (To review how a person can identify health fraud and quackery, turn to p. 31. For a list of credible sources of nutrition information, see p. 32.)

In addition, consumers should inform their physicians about the use of any alternative therapies so that a comprehensive treatment plan can be developed and potential problems can be averted. When considering herbal products, remember to include supplements, teas, and garden plants. Sometimes herbal products may need to be discontinued, especially before surgery when interactions with anesthesia or normal blood clotting can be life-threatening.

Alternative therapies come in a variety of shapes and sizes. Both their benefits and their risks may be small, none, or great. Wise consumers and health-care professionals accept the beneficial, or even neutral, practices with an open mind and reject those practices known to cause harm. Making healthful choices requires understanding all the choices.

REFERENCES

1. R. E. Wells and coauthors, Complementary and alternative medicine use among adults with migraines/severe headaches, *Headache* 51 (2011): 1087–1097; P. M. Barnes, B. Bloom, and R. L. Nahin, Complementary and alternative medicine use among adults and children: United States, 2007, *National Health Statistics Reports*, December 10, 2008; K. J. Kemper, S. Vohra, R. Walls, and the Task Force on Complementary and Alternative Medicine, the Provisional Section on Complementary, Holistic, and Integrative Medicine, *Pediatrics* 122 (2008): 1374–1386.
2. R. Y. Teets, S. Dahmer, and E. Scott, Integrative medicine approach to chronic pain, *Primary Care* 37 (2010): 407–421; National Institutes of Health, National Center for Complementary and Alternative Medicine, Acupuncture for pain, http://nccam.nih.gov/health/acupuncture/acupuncture-for-pain.htm, created May 2009; National Institutes of Health, National Center for Complementary and Alternative Medicine, The use of complementary and alternative medicine in the United States, http://nccam.nih.gov/news/camstats/2007/camsurvey_fs1.htm, updated December 2008.
3. A. Nerurkar and coauthors, When conventional medical providers recommend unconventional medicine: Results of a national study, *Archives of Internal Medicine* 171 (2011): 862–864.
4. M. E. Wechsler and coauthors, Active albuterol or placebo, sham acupuncture, or no intervention in asthma, *New England Journal of Medicine* 365 (2011): 119–126.
5. M. S. Baliga and coauthors, Update on the chemopreventive effects of ginger and its phytochemicals, *Critical Reviews in Food Science and Nutrition* 51 (2011): 499–523; R. Nahas, and O. Sheikh, Complementary and alternative medicine for the treatment of major depressive disorder, *Canadian Family Physician* 57 (2011): 659–663; M. P. Freeman and coauthors, Complementary and alternative medicine in major depressive disorder: The American Psychiatric Association Task Force report, *Journal of Clinical Psychiatry* 71 (2010): 669–681; J. Sarris and D. J. Kavanagh, Kava and St. John's Wort: Current evidence for use in mood and anxiety disorders, *Journal of Alternative and Complementary Medicine* 15 (2009): 827–836.
6. L. Leung, Cannabis and its derivatives: Review of medical use, *Journal of the American Board of Family Medicine* 24 (2011): 452–462.
7. A. M. Hill, J. A. Fleming and P. M. Kris-Etherton, The role of diet and nutritional supplements in preventing and treating cardiovascular disease, *Current Opinion in Cardiology* 24 (2009): 433–441; W. G. Christen and coauthors, Folic acid, pyridoxine, and cyanocobalamin combination treatment and age-related macular degeneration in women, *Archives of Internal Medicine* 169 (2009): 335–341; National Institutes of Health State-of-the-Science Panel, National Institutes of Health State-of-the-Science Conference Statement: Multivitamin/mineral supplements and chronic disease prevention, *Annals of Internal Medicine* 145 (2006): 364–371.
8. D. M. Ribnicky and coauthors, Evaluation of botanicals for human health, *American Journal of Clinical Nutrition* 87 (2008): 472S–475S.
9. National Institutes of Health, National Center for Complementary and Alternative Medicine, Herbs at a glance, http://nccam.nih.gov/health/herbsataglance.htm, accessed October 11, 2011.
10. R. B. Saper and coauthors, Lead, mercury, and arsenic in US-and Indian-manufactured Ayurvedic medicines sold via the Internet, *Journal of the American Medical Association* 300 (2008): 915–923.
11. A. M. Schoepfer and coauthors, Herbal does not mean innocuous: Ten cases of severe hepatotoxicity associated with dietary supplements from Herbalife® products, *Journal of Hepatology* 47 (2007): 521–526.
12. B. E. Snitz and coauthors, *Ginkgo biloba* for preventing cognitive decline in older adults, *Journal of the American Medical Association* 302 (2009): 2663–2670; S. T. Dekosky and coauthors, *Ginkgo biloba* for prevention of dementia: A randomized controlled trial, *Journal of the American Medical Association* 300 (2008): 2253–2262; S. T. Dekosky and C. D. Furberg, Turning a new leaf: Ginkgo biloba in prevention of dementia? *Neurology* 70 (2008): 1730–1731; W. Weber and coauthors, *Hypericum perforatum* (St John's Wort) for attention-deficit/hyperactivity disorder in children and adolescents: A randomized controlled trial, *Journal of the American Medical Association* 299 (2008): 2633–2641.
13. M. Sharma and coauthors, Induction of multiple pro-inflammatory cytokines by respiratory viruses and reversal by standardized Echinacea, a potent antiviral herbal extract, *Antiviral Research* 83 (2009): 165–170; D. F. Birt and coauthors, *Echinacea* in infection, *American Journal of Clinical Nutrition* 87 (2008): 488S–492S.
14. Ribnicky and coauthors, 2008.
15. Ribnicky and coauthors, 2008.
16. National Institutes of Health, National Center for Complementary and Alternative Medicine, *Herbs at a Glance: A Quick Guide to Herbal Supplements* (2009): NIH Publication No. 09-6248.
17. National Institutes of Health, National Center for Complementary and Alternative Medicine, Herbs at a glance, St. John's wort, http://nccam.nih.gov/health/stjohnswort/ataglance.htm, updated July 2010; S. F. Zhou and X. Lai, An update on clinical drug interactions with the herbal antidepressant St. John's wort, *Current Drug Metabolism* 9 (2008): 394–409.
18. National Institutes of Health, National Center for Complementary and Alternative Medicine, Herbs at a glance: Ginkgo, http://nccam.nih.gov/health/ginkgo/ataglance.htm, updated July 2010.
19. F. D. Debelle, J. L. Vanherweghem, and J. L. Nortier, Aristolochic acid nephropathy: A worldwide problem, *Kidney International* 74 (2008): 158–169.
20. K. A. Clauson, M. L. Santamarina, and J. C. Rutledge, Clinically relevant safety issues associated with St. John's wort product labels, *BMC Complementary and Alternative Medicine* 8 (2008): 42.
21. Barnes, Bloom, and Nahin, 2008.
22. M. A. Alsawaf and A. Jatoi, Shopping for nutrition-based complementary and alternative medicine on the Internet: How much money might cancer patients be spending online? *Journal of Cancer Education* 22 (2007): 174–176.

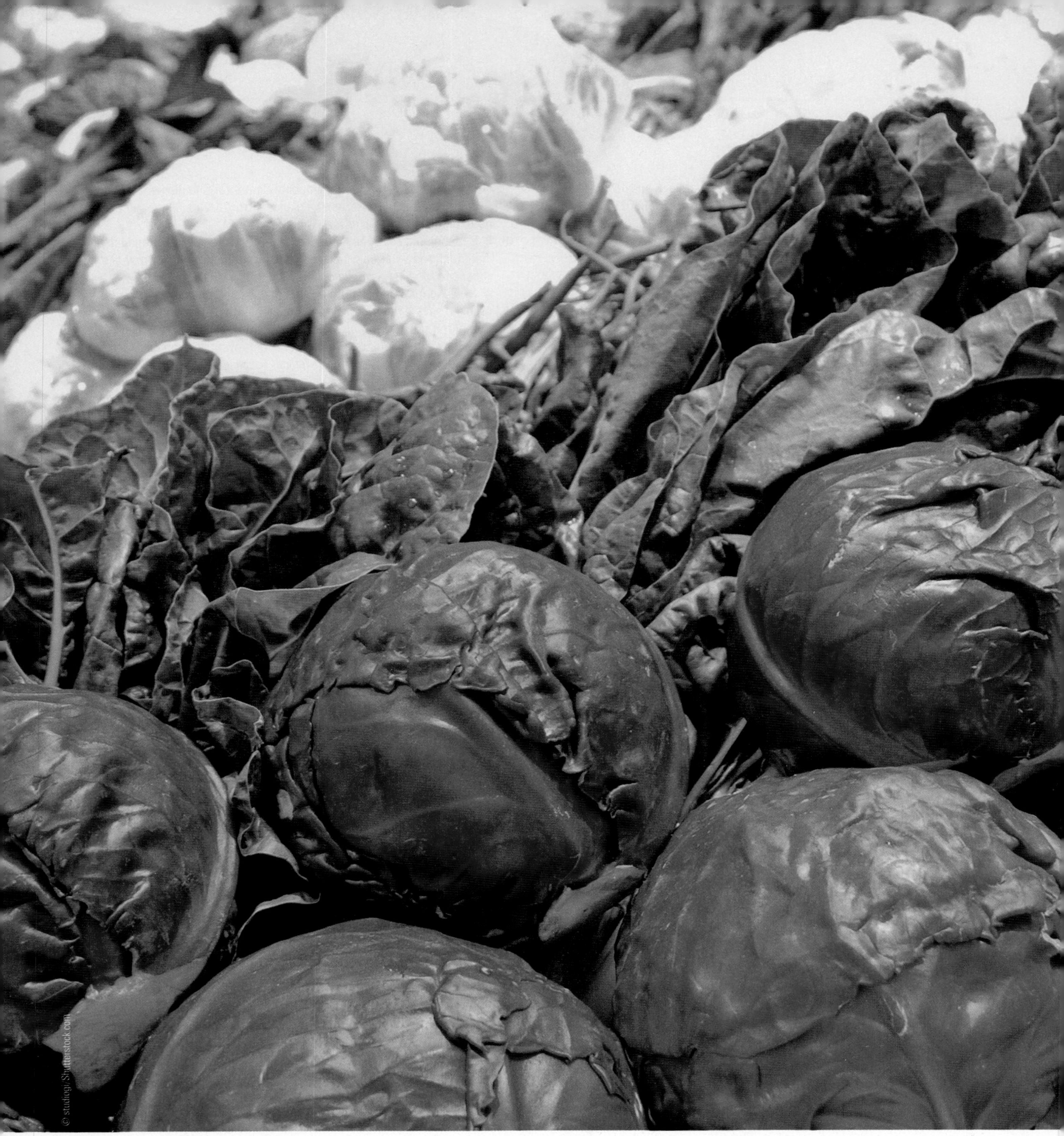

The use of pesticides helps ensure the survival of crops, but at what expense? Go to Food Safety/Food Sustainability → Harmful Residues & Toxins to review some of the effects pesticides have on consumers and the environment. After reading through some of the articles, describe a few low-impact alternatives.

19 Consumer Concerns about Foods and Water

LEARNING GPS

Nutrition in Your Life

Do you know what causes food poisoning and how to protect yourself against it? Were you alarmed to learn that french fries contain acrylamide or that fish contain mercury? Are you concerned about the pesticides that might linger on fruits and vegetables—or the hormones and antibiotics that remain in beef and chicken? Do you wonder whether foods contain enough nutrients—or too many additives? Making informed choices and practicing a few food safety tips will allow you to enjoy a variety of foods while limiting your risks of experiencing food-related illnesses. In the Nutrition Portfolio at the end of this chapter, you can review your food-handling practices.

Take a moment to consider the task of supplying food to more than 300 million people in the United States (and millions more in all corners of the world). To feed this nation, farmers grow and harvest crops; dairy producers supply milk products; ranchers raise livestock; shippers deliver foods to manufacturers by land, sea, and air; manufacturers prepare, process, preserve, and package products for refrigerated food cases and grocery-store shelves; and grocers store the food and supply it to consumers. After much time, much labor, and extensive transport, an abundant supply of a large variety of safe foods finally reaches consumers at reasonable market prices.

The **FDA** and other government and international agencies monitor this huge system using a network of people and sophisticated equipment. More than 1300 FDA inspectors process more than 24 million shipments from more than 300,000 manufacturers representing more than 150 countries. (The glossary on p. 624 identifies the various food regulatory agencies by their abbreviations.)

Government agencies focus on the potential **hazard** of foods, which differs from the **toxicity** of a substance—a distinction worth understanding. Anything can be toxic. Toxicity simply means that a substance *can* cause harm *if* enough is consumed. We consume many substances that are toxic, without **risk**,

hazard: a source of danger; used to refer to circumstances in which harm is possible under normal conditions of use.

toxicity: the ability of a substance to harm living organisms. All substances are toxic if high enough concentrations are used.

risk: a measure of the probability and severity of harm.

© Markus Moellenberg/Corbis

With the benefits of a safe and abundant food supply comes the responsibility to select, prepare, and store foods safely.

because the amounts are so small. The term *hazard,* on the other hand, is more relevant to our daily lives because it refers to the harm that is *likely* under real-life conditions. Consumers rely on these monitoring agencies to set **safety** standards and can learn to protect themselves from food hazards by taking a few preventive measures.

After the events of September 11, 2001, the threat of deliberate contamination of the US food supply became a pressing issue. The FDA works with the **USDA** and other government agencies to protect agriculture and other aspects of the food supply, but details of the war against domestic bioterrorism are beyond the scope of this discussion.

This chapter focuses on actions of individuals to promote food safety. It addresses the following food safety concerns:

- Foodborne illnesses
- Nutritional adequacy of foods
- Environmental contaminants
- Naturally occurring toxicants
- Pesticides
- Food additives
- Water safety

The chapter begins with the FDA's highest priority—the serious and prevalent threat of foodborne illnesses. The highlight that follows looks at genetically engineered foods.

safety: the condition of being free from harm or danger.

GLOSSARY OF FOOD REGULATORY AGENCIES

CDC (Centers for Disease Control): a branch of the Department of Health and Human Services that is responsible for, among other things, monitoring foodborne diseases.
www.cdc.gov

EPA (Environmental Protection Agency): a federal agency that is responsible for, among other things, regulating pesticides and establishing water quality standards.
www.epa.gov

FAO (Food and Agriculture Organization): an international agency (part of the United Nations) that has adopted standards to regulate pesticide use among other responsibilities.
www.fao.org

FDA (Food and Drug Administration): the federal agency responsible for ensuring the safety and wholesomeness of all dietary supplements and foods processed and sold in interstate commerce except meat, poultry, and eggs (which are under the jurisdiction of the USDA); inspecting food plants and imported foods; and setting standards for food composition and product labeling.
www.fda.gov

USDA (US Department of Agriculture): the federal agency responsible for enforcing standards for the wholesomeness and quality of meat, poultry, and eggs produced in the United States; conducting nutrition research; and educating the public about nutrition.
www.usda.gov

WHO (World Health Organization): an international agency concerned with promoting health and eradicating disease.
www.who.int

19.1 Foodborne Illnesses

LEARN IT Describe how foodborne illnesses can be prevented.

The FDA lists **foodborne illnesses** as the leading food safety concern because **outbreaks** of food poisoning far outnumber episodes of any other kind of food contamination. The CDC estimates 48 million cases of foodborne illnesses occur each year in the United States.[1] More than 100,000 people become so sick as to need hospitalization. For some 3000 people each year, the symptoms ♦ are so severe as to cause death. Most vulnerable are pregnant women; very young, very old, sick, or malnourished people; and those with a weakened immune system (as in AIDS). By taking the proper precautions, people can minimize their chances of contracting foodborne illnesses.

♦ Get medical help for these symptoms:

- Bloody diarrhea
- Diarrhea lasting more than 3 days
- Difficulty breathing
- Difficulty swallowing
- Double vision
- Fever lasting more than 24 hours
- Headache, muscle stiffness, and fever
- Numbness, muscle weakness, and tingling sensations in the skin
- Rapid heart rate, fainting, and dizziness

foodborne illnesses: illnesses transmitted to human beings through food and water, caused by either an infectious agent (foodborne infection) or a poisonous substance (food intoxication); commonly known as *food poisoning.*

outbreaks: two or more cases of a similar illness resulting from the ingestion of a common food.

Foodborne Infections and Food Intoxications Foodborne illness can be caused by either an infection or an intoxication. Table 19-1 summarizes the foodborne illnesses responsible for 90 percent of illnesses, hospitalizations, and deaths, along with their food sources, general symptoms, and prevention methods.

TABLE 19-1 Foodborne Illnesses

Common Organism Name	Most Frequent Food Sources	Onset and General Symptoms	Prevention Methods[a]
Foodborne Infections			
Campylobacter (KAM-pee-loh-BAK-ter) bacterium	Raw and undercooked poultry, unpasturized milk, contaminated water	Onset: 2 to 5 days. Diarrhea, vomiting, abdominal cramps, fever; sometimes bloody stools; lasts 2 to 10 days.	Cook foods thoroughly; use pasteurized milk; use sanitary food-handling methods.
E.coli: 0157[b] bacterium	Undercooked ground beef, unpasteurized milk and juices, raw fruits and vegetables, contaminated water, and person-to-person contact	Onset: 1 to 8 days. Severe bloody diarrhea, abdominal cramps, vomiting; lasts 5 to 10 days.	Cook ground beef thoroughly; use pasteurized milk; use sanitary food-handling methods; use treated, boiled, or bottled water.
Norovirus	Person-to-person contact; raw foods, salads, sandwiches	Onset: 1 to 2 days. Vomiting; lasts 1 to 2 days.	Use sanitary food-handling methods.
Listeria *(lis-TER-ee-AH)* bacterium	Unpasteurized milk; fresh soft cheeses; luncheon meats, hot dogs	Onset: 1 to 21 days. Fever, muscle aches; nausea, vomiting, blood poisoning, complications in pregnancy, and meningitis (stiff neck, severe headache, and fever).	Use sanitary food-handling methods; cook foods thoroughly; use pasteurized milk.
Clostridium (klo-STRID-ee-um) **perfringens** (per-FRINGE-enz) bacterium	Meats and meat products stored at between 120°F and 130°F	Onset: 8 to 16 hours. Abdominal pain, diarrhea, nausea; lasts 1 to 2 days.	Use sanitary food-handling methods; use pasteurized milk; cook foods thoroughly; refrigerate foods promptly and properly.
Salmonella (sal-moh-NEL-ah) bacteria (>2300 types)	Raw or undercooked eggs, meats, poultry, raw milk and other dairy products, shrimp, frog legs, yeast, coconut, pasta, and chocolate	Onset: 1 to 3 days. Fever, vomiting, abdominal cramps, diarrhea; lasts 4 to 7 days; can be fatal.	Use sanitary food-handling methods; use pasteurized milk; cook foods thoroughly; refrigerate foods promptly and properly.
Food Intoxications			
Botulism (BOT-chew-lizm) Botulinum toxin produced by *Clostridium botulinum* bacterium, which grows without oxygen, in low-acid foods, and at temperatures between 40°F and 120°F; the ***botulinum*** (BOT-chew-line-um) ***toxin*** responsible for botulism is called ***botulin*** (BOT-chew-lin).	Anaerobic environment of low acidity (canned corn, peppers, green beans, soups, beets, asparagus, mushrooms, ripe olives, spinach, tuna, chicken, chicken liver, liver pâté, luncheon meats, ham, sausage, stuffed eggplant, lobster, and smoked and salted fish)	Onset: 4 to 36 hours. Nervous system symptoms, including double vision, inability to swallow, speech difficulty, and progressive paralysis of the respiratory system; often fatal; leaves prolonged symptoms in survivors.	Use proper canning methods for low-acid foods; refrigerate homemade garlic and herb oils; avoid commercially prepared foods with leaky seals or with bent, bulging, or broken cans. Do not give infants honey because it may contain spores of *Clostridium botulinum,* which is a common source of infection for infants.
Staphylococcal (STAF-il-oh-KOK-al) **food poisoning** Staphylococcal toxin (produced by *Staphylococcus aureus* bacterium)	Toxin produced in improperly refrigerated meats; egg, tuna, potato, and macaroni salads; cream-filled pastries	Onset: 1 to 6 hours. Diarrhea, nausea, vomiting, abdominal cramps, fever; lasts 1 to 2 days.	Use sanitary food-handling methods; cook food thoroughly; refrigerate foods promptly and properly; use proper home-canning methods.
Toxoplasma (TOK-so-PLAZ-ma) parasite	Raw or undercooked meat; unwashed fruits and vegetables; contaminated water	Onset: 7 to 21 days. Swollen glands, fever, headache, muscle pain, stiff neck.	Use sanitary food-handling methods; cook foods thoroughly.

NOTE: Travelers' diarrhea is most commonly caused by *E. coli, Campylobacter jejuni, Shigella,* and *Salmonella.*

[a]The "How To" on pp. 628–629 provides more details on the proper handling, cooking, and refrigeration of foods.

[b]The most serious strain is *E. coli* STEC 0157.

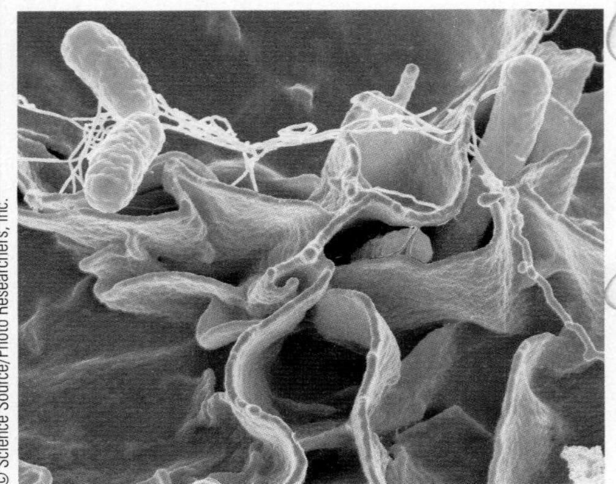

© Science Source/Photo Researchers, Inc.

An infection with *Salmonella* bacteria typically causes diarrhea, fever, and abdominal cramps for 12 to 72 hours.

© Polara Studios, Inc.

To prevent food intoxication from homemade flavored oils, wash and dry the herbs before adding them to the oil and keep the oil refrigerated.

Foodborne Infections Foodborne infections are caused by eating foods contaminated by infectious microbes. Among foodborne infections, norovirus and *Salmonella* are the leading causes of hospitalizations and deaths.[2] **Pathogens** commonly enter the GI tract in contaminated foods such as undercooked poultry and unpasteurized milk. Symptoms generally include abdominal cramps, fever, vomiting, and diarrhea.

Food Intoxications Food intoxications are caused by eating foods containing natural toxins or, more likely, microbes that produce toxins. The most common food toxin is produced by *Staphylococcus aureus*; it affects more than 1 million people each year. Less common, but more infamous, is *Clostridium botulinum*, an organism that produces a deadly toxin in anaerobic conditions such as improperly canned (especially home-canned) foods and homemade garlic or herb-flavored oils stored at room temperature. The botulinum toxin paralyzes muscles, making it difficult to see, speak, swallow, and breathe. Because death can occur within 24 hours of onset, botulism demands immediate medical attention. Even then, survivors may suffer the effects for months or years.

Other microbial toxins—called aflatoxins—are not common in the United States, but threaten the health of more than half the world's population.[3] Aflatoxins contaminate corn, grains, and nuts in tropical countries where foods are stored in warm, humid conditions that promote fungal growth. In humans, aflatoxins cause cancer. Strategies to reduce exposure in vulnerable populations need to become a worldwide priority.

Food Safety in the Marketplace

Transmission of foodborne illness has changed as our food supply and lifestyles have changed.[4] In the past, foodborne illness was caused by one person's error in a small setting, such as improperly refrigerated egg salad at a family picnic, and affected only a few victims. Today, we eat more foods that have been prepared and packaged by others. Consequently, when a food manufacturer or cruise ship chef makes an error, foodborne illness can quickly affect many people. An estimated 80 percent of reported foodborne illnesses are caused by errors in a commercial setting, such as the improper **pasteurization** of milk at a large dairy.

In the 2006 *E. coli* outbreak caused by contaminated fresh spinach, nearly 200 people became sick, and 2 elderly women and a 2-year-old boy died before consumers got the FDA message to not eat fresh spinach. In 2009, *Salmonella* was found in peanut butter that had been used in more than 2100 products made by more than 200 companies; another *Salmonella* outbreak in 2010 led to the recall of 500 million eggs from two farms. In 2011, a cantaloupe farm had to recall more than 300,000 cases of fruit when *Listeria* poisoning killed 29 people and made 139 others sick. These incidents and others focus the national spotlight on two important safety issues: disease-causing organisms are commonly found in foods, and safe food-handling practices can minimize harm from most of these foodborne pathogens.

Industry Controls The USDA, the FDA, and food-processing industries have developed and implemented programs to control foodborne illnesses. For example, USDA inspectors examine meat-processing plants every day to ensure that these facilities meet government standards. Facilities handling seafood, eggs, produce, and processed foods are inspected less often, but all food producers use a **Hazard Analysis Critical Control Point (HACCP)** plan to help prevent foodborne illnesses at their source. Each slaughterhouse, packer, distributor, and transporter of susceptible foods must identify "critical control points" that pose a risk of contamination and then devise and implement verifiable ways to eliminate or minimize the risk. The HACCP system has proved a remarkable success for domestic products, but such programs do not apply to imported foods.

pathogens (PATH-oh-jenz): microorganisms capable of producing disease.

pasteurization: heat processing of food that inactivates some, but not all, microorganisms in the food; not a sterilization process. Bacteria that cause spoilage are still present.

Hazard Analysis Critical Control Points (HACCP): a systematic plan to identify and correct potential microbial hazards in the manufacturing, distribution, and commercial use of food products; commonly referred to as "HASS-ip."

An estimated $2 trillion worth of products are imported into the United States from more than 150 countries each year. Many countries cooperate with the FDA and have adopted many of the safe food-handling practices used in the United States, but some imported foods come from countries with little or no regulatory oversight. To help consumers distinguish between imported and domestic foods, certain foods—including fish, shellfish, meats, fruits, vegetables, and some nuts—must display a Country of Origin Label specifying where they were produced.[5]

Consumer Awareness Canned and packaged foods sold in grocery stores are easily controlled, but rare accidents do happen. Batch numbering makes it possible to recall contaminated foods through public announcements via Internet, newspapers, television, and radio. In the grocery store, consumers can buy items before the "sell by" date and inspect the safety seals and wrappers of packages. A broken seal, bulging can lid, or mangled package fails to protect the consumer against microbes, insects, spoilage, or even vandalism.

State and local health regulations provide guidelines on the cleanliness of facilities and the safe preparation of foods for restaurants, cafeterias, and fast-food establishments. Even so, consumers can also take these actions to help prevent foodborne illnesses when dining out:

- Wash hands with hot, soapy water before meals.
- Expect clean tabletops, dinnerware, utensils, and food preparation areas.
- Expect cooked foods to be served piping hot and salads to be fresh and cold.
- Refrigerate take-home items within 2 hours and use leftovers within 3 to 4 days.

Improper handling of foods can occur anywhere along the line from commercial farms and manufacturers to supermarkets and restaurants to private homes. Maintaining a safe food supply requires everyone's efforts (see Figure 19-1).

Food Safety in the Kitchen

Whether microbes multiply and cause illness depends, in part, on a few key food-handling behaviors in the kitchen—whether the kitchen is in your home, a school cafeteria, a gourmet restaurant, or a canning manufacturer. Figure 19-2 (p. 628) summarizes the four simple things that can help most to prevent foodborne illness:

- *Clean.* Keep a clean, safe kitchen by washing hands and surfaces often. Wash countertops, cutting boards, sponges, and utensils in hot, soapy

> FIGURE 19-1 Food Safety from Farm to Table

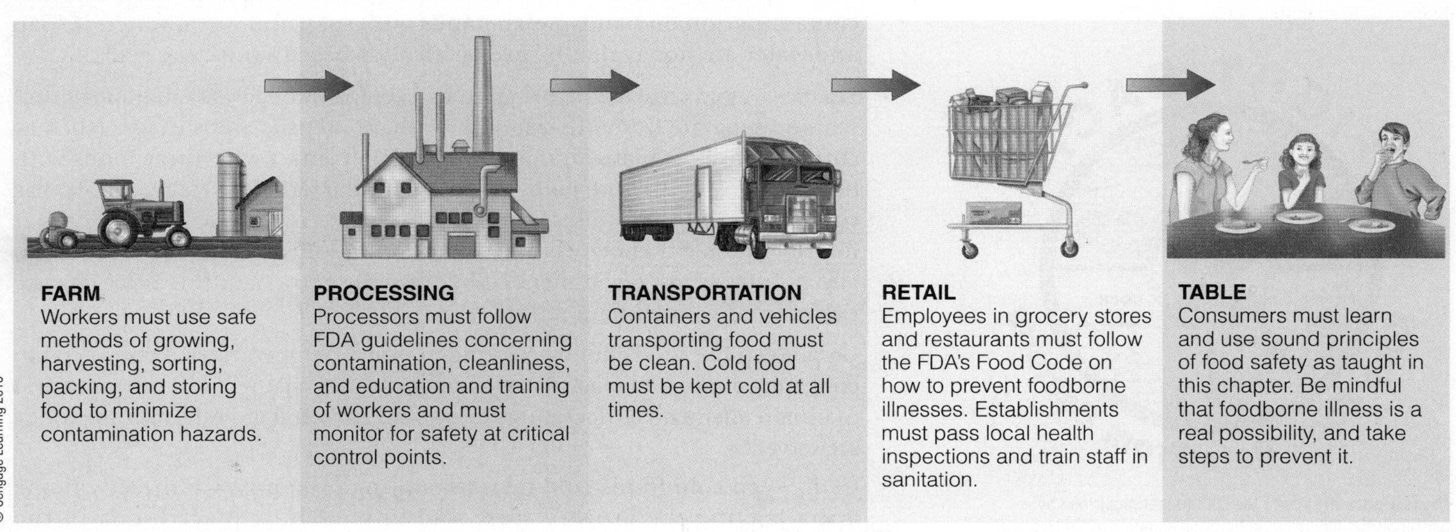

FARM
Workers must use safe methods of growing, harvesting, sorting, packing, and storing food to minimize contamination hazards.
PROCESSING
Processors must follow FDA guidelines concerning contamination, cleanliness, and education and training of workers and must monitor for safety at critical control points.
TRANSPORTATION
Containers and vehicles transporting food must be clean. Cold food must be kept cold at all times.
RETAIL
Employees in grocery stores and restaurants must follow the FDA's Food Code on how to prevent foodborne illnesses. Establishments must pass local health inspections and train staff in sanitation.
TABLE
Consumers must learn and use sound principles of food safety as taught in this chapter. Be mindful that foodborne illness is a real possibility, and take steps to prevent it.

>How To

Prevent Foodborne Illnesses

Most foodborne illnesses can be prevented by following four simple rules: clean, separate, cook, and chill.

Clean

- Wash fruits and vegetables in a clean sink with a scrub brush and warm water; store washed and unwashed produce separately.
- Use hot, soapy water to wash hands, utensils, dishes, nonporous cutting boards, and countertops before handling food and between tasks when working with different foods. Use a bleach solution on cutting boards (one capful per gallon of water).
- Cover cuts with clean bandages before food preparation; dirty bandages carry harmful microorganisms.
- Mix foods with utensils, not hands; keep hands and utensils away from mouth, nose, and hair.
- Anyone may be a carrier of bacteria and should avoid coughing or sneezing over food. A person with a skin infection or infectious disease should not prepare food.
- Wash or replace sponges and towels regularly.
- Clean up food spills and crumb-filled crevices.

Separate

- Wash all surfaces that have been in contact with raw meats, poultry, eggs, fish, and shellfish before reusing.
- Serve cooked foods on a clean plate with a clean utensil. Separate raw foods from those that have been cooked.
- Don't use marinade that was in contact with raw meat for basting or sauces.

Cook

- When cooking meats or poultry, use a thermometer to test the internal temperature. Insert the thermometer between the thigh and the body of a turkey or into the thickest part of other meats, making sure the tip of the thermometer is not in contact with bone or the pan. Cook to the temperature indicated for that particular meat (see Figure 19-5 on p. 630); cook hamburgers to at least medium well done. If you have safety questions, call the USDA Meat and Poultry Hotline: (800) 535-4555.
- Cook stuffing separately, or stuff poultry just prior to cooking.
- Do not cook large cuts of meat or turkey in a microwave oven; it leaves some parts undercooked while overcooking others.
- Cook eggs before eating them (soft-boiled for at least 3½ minutes; scrambled until set, not runny; fried for at least 3 minutes on one side and 1 minute on the other).
- Cook seafood thoroughly. If you have safety questions about seafood, call the FDA hotline: (800) FDA-4010.
- When serving foods, maintain temperatures at 140°F or higher.
- Heat leftovers thoroughly to at least 165°F.

Chill

- When running errands, stop at the grocery store last. When you get home, refrigerate the perishable groceries (such as meats and dairy products) immediately. Do not leave perishables in the car any longer than it takes for ice cream to melt.
- Put packages of raw meat, fish, or poultry on a plate before refrigerating to prevent juices from dripping on food stored below.
- Buy only foods that are solidly frozen in store freezers.
- Keep cold foods at 40° F or less; keep frozen foods at 0°F or less (keep a thermometer in the refrigerator).
- Marinate meats in the refrigerator, not on the counter.

(*Continued*)

> FIGURE 19-2 Fight Bac!
The FightBac! website (www.fightbac.org) describes four ways to keep food safe.

cross-contamination: the contamination of food by bacteria that occurs when the food comes into contact with surfaces previously touched by raw meat, poultry, or seafood.

water before and after each step of food preparation. To reduce bacterial contamination on hands, wash hands with soap and warm water; if soap and water are not available, use an alcohol-based sanitizing gel.[6]

- *Separate.* Avoid cross-contamination by keeping raw eggs, meat, poultry, and seafood separate from other foods. Wash all utensils and surfaces (such as cutting boards or platters) that have been in contact with these foods with hot, soapy water before using them again. Bacteria inevitably left on the surfaces from the raw meat can recontaminate the cooked meat or other foods—a problem known as **cross-contamination**. Washing raw eggs, meat, and poultry is not recommended because the extra handling increases the risk of cross-contamination.
- *Cook.* Keep hot foods hot by cooking to proper temperatures. Foods need to cook long enough to reach internal temperatures that will kill microbes and maintain adequate temperatures to prevent bacterial growth until the foods are served.
- *Chill.* Keep cold foods cold by refrigerating promptly. Go directly home upon leaving the grocery store and immediately place foods in the refrigerator or freezer. After a meal, refrigerate any leftovers immediately.

- Look for "Keep Refrigerated" or "Refrigerate After Opening" on food labels.
- Refrigerate leftovers promptly; use shallow containers to cool foods faster; use leftovers within 3 to 4 days.
- Thaw meats or poultry in the refrigerator, not at room temperature. If you must hasten thawing, use cool water (changed every 30 minutes) or a microwave oven.
- Freeze meat, fish, or poultry immediately if not planning to use within a few days.

In General

- Do not reuse disposable containers; use nondisposable containers or recycle instead.
- Do not taste food that is suspect. "If in doubt, throw it out."
- Throw out foods with danger-signaling odors. Be aware, though, that most food-poisoning bacteria are odorless, colorless, and tasteless.
- Do not buy or use items that have broken seals or mangled packaging; such containers cannot protect against microbes, insects, spoilage, or even vandalism. Check safety seals, buttons, and expiration dates.
- Follow label instructions for storing and preparing packaged and frozen foods; throw out foods that have been thawed or refrozen.
- Discard foods that are discolored, moldy, or decayed or that have been contaminated by insects or rodents.

For Specific Food Items

- *Canned goods.* Carefully discard food from cans that leak or bulge so that other people and animals will not accidentally ingest it; before canning, seek professional advice from the USDA Extension Service (check your phone book under US government listings, or ask directory assistance).
- *Milk and cheeses.* Use only pasteurized milk and milk products. Aged cheeses, such as cheddar and Swiss, do well for an hour or two without refrigeration, but they should be refrigerated or stored in an ice chest for longer periods.
- *Eggs.* Use clean eggs with intact shells. Do not eat eggs, even pasteurized eggs, raw; raw eggs are commonly found in Caesar salad dressing, eggnog, cookie dough, hollandaise sauce, and key lime pie. Cook eggs until whites are firmly set and yolks begin to thicken.
- *Honey.* Honey may contain dormant bacterial spores, which can awaken in the human body to produce botulism. In adults, this poses little hazard, but infants younger than 1 year of age should never be fed honey. Honey can accumulate enough toxin to kill an infant; it has been implicated in several cases of sudden infant death. (Honey can also be contaminated with environmental pollutants picked up by the bees.)
- *Mayonnaise.* Commercial mayonnaise may actually help a food to resist spoilage because of the acid content. Still, keep it refrigerated after opening.
- *Mixed salads.* Mixed salads of chopped ingredients spoil easily because they have extensive surface area for bacteria to invade, and they have been in contact with cutting boards, hands, and kitchen utensils that easily transmit bacteria to food (regardless of their mayonnaise content). Chill them well before, during, and after serving.
- *Picnic foods.* Choose foods that last without refrigeration, such as fresh fruits and vegetables, breads and crackers, and canned spreads and cheeses that can be opened and used immediately. Pack foods cold, layer ice between foods, and keep foods out of water.
- *Seafood.* Buy only fresh seafood that has been properly refrigerated or iced. Cooked seafood should be stored separately from raw seafood to avoid cross-contamination.

TRY IT After cutting the fat from a pork loin, you rinse the wooden cutting board under warm water before using it to chop vegetables. Discuss whether this precaution is adequate to protect against cross-contamination.

Unfortunately, consumers commonly fail to follow these simple food-handling recommendations. See the accompanying "How To" feature for additional food safety tips.

Safe Handling of Meats and Poultry Figure 19-3 (p. 630) presents label instructions for the safe handling of meat and poultry and two types of USDA seals. Meats and poultry contain bacteria and provide a moist, nutrient-rich environment that favors microbial growth. Ground meat is especially susceptible because it receives more handling than other kinds of meat and has more surface area exposed to bacterial contamination. Consumers cannot detect the harmful bacteria in or on meat. For safety's sake, cook meat thoroughly, using a thermometer to test the internal temperature (see Figure 19-4 on p. 630).

Unrelated to safe handling practices, **bovine spongiform encephalopathy (BSE)** is a slowly progressive, fatal disease that affects the central nervous system of cattle and wild game such as deer and elk.[7] A similar disease develops in people who have eaten contaminated beef from infected cows (milk products appear to be safe).* The USDA has taken numerous steps to prevent the transmission of BSE in cattle, and consequently, the risks from US cattle are extremely low.

*The human form of BSE is called *variant Creutzfeldt-Jakob Disease (vCJD)*.

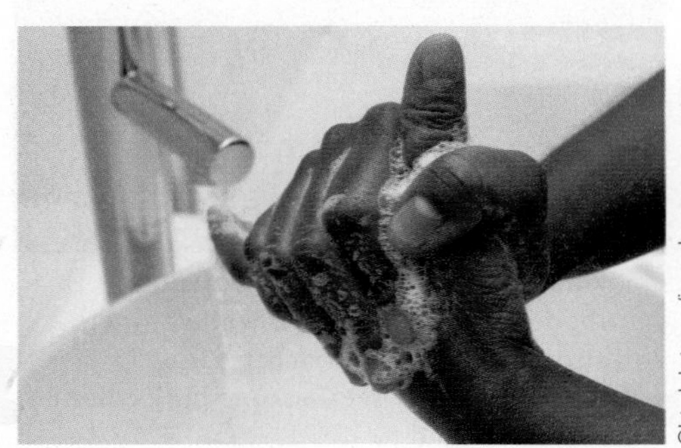

Wash your hands with warm water and soap for at least 20 seconds before preparing or eating food to reduce the chance of microbial contamination.

bovine spongiform encephalopathy (BOH-vine SPON-jih-form in-SEF-eh-LOP-eh-thee) or **BSE:** an often fatal illness of cattle and wild game that affects the nervous system and is transmitted to people by eating infected meats; commonly called *mad cow disease*.

> FIGURE 19-3 Meat and Poultry Safety, Grading, and Inspection Seals

Neither inspection nor grading guarantees that the product will not cause foodborne illnesses, but consumers can help to prevent foodborne illnesses by following the safe handling instructions.

The mandatory "Inspected and Passed by the USDA" seal ensures that meat and poultry products are safe, wholesome, and correctly labeled. Inspection does not guarantee that the meat is free of potentially harmful bacteria.

The voluntary "Graded by USDA" seal indicates that the product has been graded for tenderness, juiciness, and flavor. Beef is graded Prime (abundant marbling of the meat muscle), Choice (less marbling), or Select (lean). Similarly, poultry is graded A, B, or C.

Safe Handling Instructions

This product was prepared from inspected and passed meat and/or poultry. Some food products may contain bacteria that could cause illness if the product is mishandled or cooked improperly. For your protection, follow these safe handling instructions.

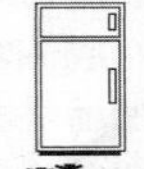

Keep refrigerated or frozen.
Thaw in refrigerator or microwave.

Keep raw meat and poultry separate from other foods. Wash working surfaces (including cutting boards), utensils, and hands after touching raw meat or poultry.

Cook thoroughly.

Keep hot foods hot. Refrigerate leftovers immediately or discard.

The USDA requires that safe handling instructions appear on all packages of meat and poultry.

Cook hamburgers to 160°F; color alone cannot determine doneness. Some burgers will turn brown before reaching 160°F, whereas others may retain some pink color, even when cooked to 175°F.

> FIGURE 19-4 Recommended Safe Temperatures (Fahrenheit)

Bacteria multiply rapidly at temperatures between 40°F and 140°F. Cook foods to the minimum internal temperatures shown on this thermometer and hold them at 140°F or higher.

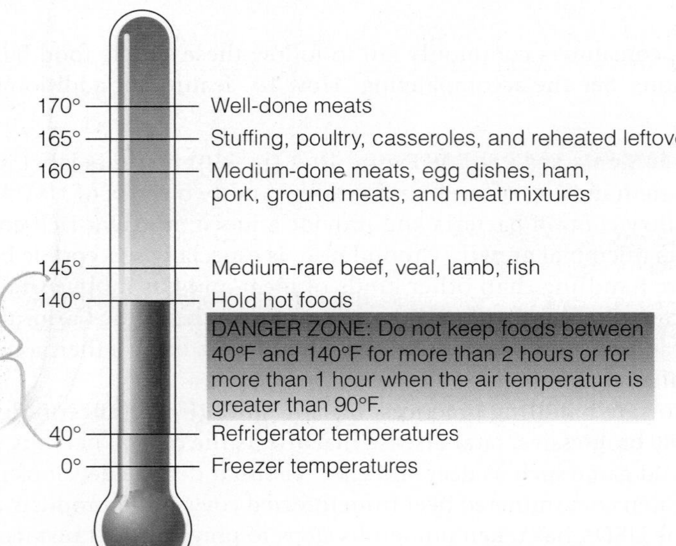

Safe Handling of Seafood Most seafood available in the United States and Canada is safe, but eating it undercooked or raw can cause severe illnesses—hepatitis, worms, parasites, viral intestinal disorders, and other diseases.* Rumor has it that freezing fish will make it safe to eat raw, but this is only partly true. Commercial freezing kills mature parasitic worms, but only cooking can kill all worm eggs and other microorganisms that can cause illness. For safety's sake, all seafood should be cooked until it is opaque.

As for **sushi,** even a master chef cannot detect harmful microbes that may occur in even the best-quality, freshest fish. The marketing term *sushi grade* implies wholesomeness, but is not legally defined and does not guarantee quality, purity, or freshness. Sushi can be safe to eat when chefs combine cooked seafood and other ingredients into these delicacies.

Eating raw oysters can be dangerous for anyone, but people with liver disease and weakened immune systems are most vulnerable. At least 10 species of bacteria found in raw oysters can cause serious illness and even death. Raw oysters may also carry the hepatitis A virus, which can cause liver disease. Some hot sauces can kill many of these bacteria, but not the virus; alcohol inactivates some bacteria, but not enough to guarantee protection (or to recommend drinking alcohol). Pasteurization of raw oysters—holding them at a specified temperature for a specified time—holds promise for killing bacteria without cooking the oyster or altering its texture or flavor.

As population density increases along the shores of seafood-harvesting waters, pollution inevitably invades the sea life there. Preventing seafood-borne illness is in large part a task of controlling water pollution. To help ensure a safe seafood market, the FDA requires processors to adopt food safety practices based on the HACCP system mentioned earlier.

Chemical pollution and microbial contamination lurk not only in the water, but also in the boats and warehouses where seafood is cleaned, prepared, and refrigerated. Because seafood is one of the most perishable foods, time and temperature are critical to its freshness, flavor, and safety. To keep seafood as fresh as possible, people in the industry must "keep it cold, keep it clean, and keep it moving." Wise consumers eat it cooked.

Other Precautions and Procedures Fresh food generally smells fresh. Not all types of food poisoning are detectable by odor, but some bacterial wastes produce "off" odors. If an abnormal odor exists, the food is spoiled. Throw it out or, if it was recently purchased, return it to the grocery store. Do not taste it. Table 19-2 lists safe refrigerator storage times for selected foods.

Local health departments and the USDA Extension Service can provide additional information about food safety. If precautions fail and a mild foodborne illness develops, drink clear liquids to replace fluids lost through vomiting and diarrhea. If serious foodborne illness is suspected, first call a physician. Then wrap the remainder of the suspected food and label the container so that the food cannot be mistakenly eaten, place it in the refrigerator, and hold it for possible inspection by health authorities.

> DIETARY GUIDELINES FOR AMERICANS 2010

Follow food safety recommendations when preparing and eating foods to reduce the risk of foodborne illnesses. To avoid microbial foodborne illness:

- Clean hands, food contact surfaces, and fruits and vegetables.
- Separate raw, cooked, and ready-to-eat foods while shopping, preparing, or storing foods.

*Diseases caused by toxins from the sea include ciguatera poisoning, scombroid poisoning, and paralytic and neurotoxic shellfish poisoning.

Eating raw seafood is a risky proposition.

TABLE 19-2 Safe Refrigerator Storage Times (≤40°F)

1 to 2 Days	Raw ground meats, breakfast or other raw sausages, raw fish or poultry; gravies
3 to 5 Days	Raw steaks, roasts, or chops; cooked meats, poultry, vegetables, and mixed dishes; lunchmeats (packages opened); mayonnaise salads (chicken, egg, pasta, tuna); fresh vegetables (spinach, green beans, tomatoes)
1 Week	Hard-cooked eggs, bacon or hot dogs (opened packages); smoked sausages or seafood; milk, cottage cheese
1 to 2 Weeks	Yogurt; carrots, celery, lettuce
2 to 4 Weeks	Fresh eggs (in shells); lunchmeats, bacon, or hot dogs (packages unopened); dry sausages (pepperoni, hard salami); most aged and processed cheeses (Swiss, brick)
2 Months	Mayonnaise (opened jar); most dry cheeses (Parmesan, Romano)

sushi: vinegar-flavored rice and seafood, typically wrapped in seaweed and stuffed with colorful vegetables. Some sushi is stuffed with raw fish; other varieties contain cooked seafood.

- Cook foods to a safe temperature to kill microorganisms.
- Chill (refrigerate) perishable foods promptly and defrost foods properly.
- Do *not* wash or rinse meat or poultry.
- Avoid raw (unpasteurized) milk or any products made from unpasteurized milk, raw or partially cooked eggs or foods containing raw eggs, raw or undercooked meat and poultry, unpasteurized juices, and raw sprouts.

Food Safety while Traveling People who travel to other countries have a 50–50 chance of contracting a foodborne illness, commonly described as **travelers' diarrhea.** Like many other foodborne illnesses, travelers' diarrhea is a sometimes serious, always annoying bacterial infection of the digestive tract. The risk is high because, for one thing, some countries' cleanliness standards for food and water are lower than those in the United States and Canada. For another, every region's microbes are different, and although people are immune to the microbes in their own neighborhoods, they have had no chance to develop immunity to the pathogens in places they are visiting for the first time. In addition to the food safety tips outlined on pp. 628–629, precautions while traveling include:

- Wash hands frequently with soap and hot water, especially before handling food or eating. Use sanitizing gel or hand wipes regularly.
- Eat only well-cooked and hot or canned foods. Eat raw fruits or vegetables only if washed in purified water and peeled with clean hands.
- Use purified, bottled water for drinking, making ice cubes, and brushing teeth. Alternatively, use disinfecting tablets or boil water.
- Refuse dairy products that have not been pasteurized and refrigerated properly.
- Travel with antidiarrheal medication in case efforts to avoid illness fail.

To sum up these recommendations, "Boil it, cook it, peel it, or forget it."

Advances in Food Safety Advances in technology have dramatically improved the quality and safety of foods available on the market. From pasteurization in the early 1900s to irradiation in the early 2000s, these advances offer numerous benefits, but they also raise consumer concerns.*

Irradiation The use of low-dose **irradiation** protects consumers from foodborne illnesses by:[8]

- Controlling mold in grains
- Sterilizing spices and teas for storage at room temperature
- Controlling insects and extending shelf life in fresh fruits and vegetables (inhibits the growth of sprouts on potatoes and onions and delays ripening in some fruits such as strawberries and mangoes)
- Destroying harmful bacteria in fresh and frozen beef, poultry, lamb, and pork

Some foods, however, are not candidates for irradiation. For example, when irradiated, high-fat meats develop off-odors, egg whites turn milky, grapefruits become mushy, and milk products change flavor. Incidentally, the milk in those boxes kept at room temperature on grocery-store shelves is *not* irradiated; it is sterilized with an **ultrahigh temperature (UHT) treatment.**

The use of food irradiation has been extensively evaluated over the past 50 years; approved for use in more than 40 countries; and supported by numerous health agencies, including the **FAO, WHO,** and the American Medical Association. Irradiation does not make foods radioactive, nor does it noticeably change

travelers' diarrhea: nausea, vomiting, and diarrhea caused by consuming food or water contaminated by any of several organisms, most commonly, *E. coli, Shigella, Campylobacter jejuni,* and *Salmonella.*

irradiation: sterilizing a food by exposure to energy waves, similar to ultraviolet light and microwaves; sometimes called *ionizing radiation.*

ultrahigh temperature (UHT) treatment: sterilizing a food by brief exposure to temperatures above those normally used.

*During the last century, pasteurization of milk helped to control typhoid fever, tuberculosis, scarlet fever, diphtheria, and other infectious diseases.

the taste, texture, or appearance of approved foods. ♦ Vitamin loss is minimal and comparable to amounts lost in other food-processing methods such as canning. Because irradiation kills bacteria without the use of heat, it is sometimes called "cold pasteurization."

Consumer Concerns about Irradiation Many consumers associate the term *radiation* with cancer, birth defects, and mutations, and consequently have strong negative emotions about using irradiation on foods. Some may mistakenly fear that irradiated food has been contaminated by radioactive particles, such as occurs in the aftermath of a nuclear accident. Some balk at the idea of irradiating, and thus sterilizing, contaminated foods and prefer instead the elimination of unsanitary slaughtering and food preparation conditions. Food producers, on the other hand, are eager to use irradiation, but they hesitate to do so until consumers are ready to accept it and willing to pay for it. Once consumers understand the benefits of irradiation, about half are willing to use irradiated foods, but most are not willing to pay more.

Regulation of Irradiation The FDA has established regulations governing the specific uses of irradiation and allowed doses. Each food that has been treated with irradiation must say so on its label. ♦ Labels can be misleading, however. Products that use irradiated foods as ingredients are not required to say so on the label. Furthermore, consumers may interpret the *absence* of the irradiation symbol to mean that the food was produced without any kind of treatment. This is not true; it is just that the FDA does not require label statements for other treatments used for the same purpose, such as postharvest fumigation with pesticides.

Other Pasteurizing Systems Other technologies using high-intensity pulsed light or electron beams have also been approved by the FDA. Like irradiation, these technologies kill microorganisms and extend the shelf life of foods without diminishing their nutrient content.

♦ Foods approved for irradiation:

- Eggs
- Fresh fruit (strawberries, citrus, papaya)
- Oysters, clams, mussels, scallops
- Raw beef, lamb, poultry, pork
- Spices, tea
- Vegetables (iceberg lettuce, fresh spinach, potatoes, tomatoes, onions)
- Wheat

♦ This international symbol, called the *radura*, identifies retail foods that have been irradiated. The phrases "Treated by irradiation" or "Treated with radiation" must accompany the symbol. The irradiation label is not required on commercially prepared foods that contain irradiated ingredients, such as spices.

REVIEW IT Describe how foodborne illnesses can be prevented.
Millions of people suffer mild to life-threatening symptoms caused by foodborne illnesses (review Table 19-1, p. 625). As the "How To" feature on pp. 628–629 describes, most of these illnesses can be prevented by storing and cooking foods at their proper temperatures and by preparing them in sanitary conditions. Irradiation of certain foods protects consumers from foodborne illnesses, but it also raises some concerns.

19.2 Nutritional Adequacy of Foods and Diets

LEARN IT Explain how to minimize nutrient losses in the kitchen.

In years past, when most foods were whole and farm fresh, the task of meeting nutrient needs primarily involved balancing servings from the various food groups. Today, however, foods have changed. Many "new" foods are available to appeal to consumers' demands for convenience and flavor, but not necessarily to deliver a balanced assortment of needed nutrients. Advertisers spend much effort and money encouraging consumers to buy their products quickly, frequently, and abundantly, not on promoting healthy eating habits—unless that would increase sales too.[9]

Obtaining Nutrient Information

To help consumers find their way among the abundance of available foods, the FDA has developed extensive nutrition labeling regulations, as Chapter 2 describes. In addition, the USDA's *Dietary Guidelines for Americans* help consumers combine foods into healthful eating patterns, and MyPlate helps them to put those guidelines into practice (see Chapter 2).

Minimizing Nutrient Losses

In addition to selecting nutritious foods and preparing them safely, consumers can improve their nutrition health by learning to store and cook foods in ways that minimize nutrient losses. Water-soluble vitamins are the most vulnerable of the nutrients, but both vitamins and minerals can be lost when they dissolve in water that is then discarded.

Fruits and vegetables contain enzymes that both synthesize and degrade vitamins. After a fruit or vegetable has been picked, vitamin synthesis stops, but degradation continues. To slow the degradation of vitamins, most fruits and vegetables should be kept refrigerated until used.* (Degradative enzymes are most active at warmer temperatures.)

Water-soluble vitamins readily dissolve in water. To prevent losses during washing, rinse fruits and vegetables before cutting. To minimize losses during cooking, steam or microwave vegetables. Alternatively, use the cooking water when preparing meals such as casseroles and soups.

Finally, keep in mind that most vitamin losses are not catastrophic and that a law of diminishing returns operates. Do not fret over small losses or waste time that may be valuable in improving your health in other ways. Be assured that if you start with plenty of fruits and vegetables and are reasonably careful in their storage and preparation, you will receive a sufficient supply of all the nutrients they provide.

REVIEW IT Explain how to minimize nutrient losses in the kitchen.

In the marketplace, food labels, the *Dietary Guidelines for Americans,* and MyPlate all help consumers learn about nutrition and how to plan healthy diets. At home, consumers can minimize nutrient losses from fruits and vegetables by refrigerating them, washing them before cutting them, storing them in airtight containers, and cooking them for short times in minimal water.

19.3 Environmental Contaminants

LEARN IT Explain how environmental contaminants get into foods and how people can protect themselves against contamination.

Concern about environmental contamination of foods is growing as the world becomes more populated and more industrialized. Industrial processes pollute the air, water, and soil. Plants absorb the **contaminants,** and people consume the plants (grains, vegetables, legumes, and fruits) or the meat and milk products from livestock that have eaten the plants. Similarly, polluted water contaminates the fish and other seafood that people eat. Environmental contaminants in air, water, and foods find their way into our bodies and have the potential to cause numerous health problems.

Harmfulness of Environmental Contaminants

The potential harmfulness of a contaminant depends in part on its **persistence**—the extent to which it lingers in the environment or in the body. Some contaminants in the environment are short-lived because microorganisms or agents such as sunlight or oxygen can break them down. Some contaminants in the body may linger for only a short time because the body rapidly excretes them or metabolizes them to harmless compounds. These contaminants present little cause for concern. Some contaminants, however, resist breakdown and can accumulate. Each level of the **food chain,** then, has a greater concentration than the one below **(bioaccumulation).** Figure 19-5 shows how bioaccumulation leads to high concentrations of toxins in animals and in people at the top of the food chain.

Contaminants enter the environment in various ways. Accidental spills are rare but can have devastating effects. More commonly, small amounts are released over long periods. The following paragraphs describe how three contaminants found their way into the food supply in the past. The first example involves a **heavy metal;** the others involve **organic halogens.**

Methylmercury

A classic example of acute contamination occurred in 1953 when a number of people in Minamata, Japan, became ill with a disease no one had seen before. By 1960, 121 cases had been reported, including 23 in infants. Mortality was high; 46 died, and the survivors suffered blindness, deafness,

contaminants: substances that make a food impure and unsuitable for ingestion.

persistence: stubborn or enduring continuance; with respect to food contaminants, the quality of persisting, rather than breaking down, in the bodies of animals and human beings.

food chain: the sequence in which living things depend on other living things for food.

bioaccumulation: the accumulation of contaminants in the flesh of animals high on the food chain.

heavy metal: a mineral ion such as mercury or lead, so called because of its relatively high atomic weight; many heavy metals are poisonous.

organic halogens: an organic compound containing one or more atoms of a halogen—fluorine, chlorine, iodine, or bromine.

*Some vitamins are easily destroyed by oxygen. To minimize the destruction of vitamins, store fruits and vegetables that have been cut and juice that has been opened in airtight containers and refrigerate them.

> FIGURE 19-5 **Bioaccumulation of Toxins in the Food Chain**

This example features fish as the food for human consumption, but bioaccumulation of toxins occurs on land as well when cows, pigs, and chickens eat or drink contaminated foods or water.

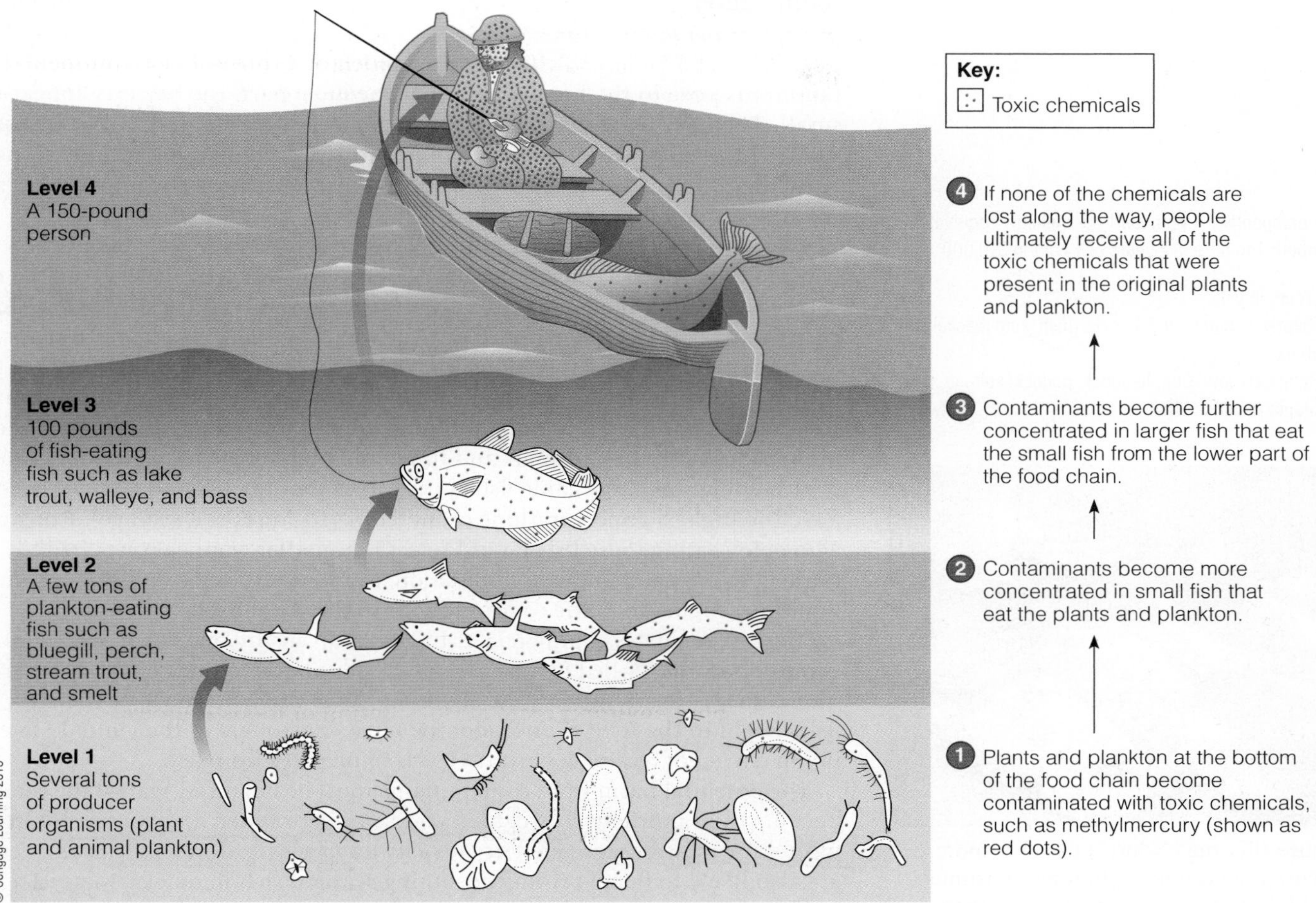
Key:
Toxic chemicals
Level 4
A 150-pound person
Level 3
100 pounds of fish-eating fish such as lake trout, walleye, and bass
Level 2
A few tons of plankton-eating fish such as bluegill, perch, stream trout, and smelt
Level 1
Several tons of producer organisms (plant and animal plankton)
4 If none of the chemicals are lost along the way, people ultimately receive all of the toxic chemicals that were present in the original plants and plankton.
3 Contaminants become further concentrated in larger fish that eat the small fish from the lower part of the food chain.
2 Contaminants become more concentrated in small fish that eat the plants and plankton.
1 Plants and plankton at the bottom of the food chain become contaminated with toxic chemicals, such as methylmercury (shown as red dots).
© Cengage Learning 2013

lack of coordination, and intellectual deterioration. The cause was ultimately revealed to be methylmercury contamination of fish from the bay where these people lived. The infants who contracted the disease had not eaten any fish, but their mothers had, and even though the mothers exhibited no symptoms during their pregnancies, the poison affected their unborn babies. Manufacturing plants in the region were discharging mercury-containing waste into the waters of the bay, the mercury was turning into methylmercury, and the fish in the bay were accumulating this poison in their bodies. Some of the affected families had been eating fish from the bay every day.

PBB and PCB In 1973, half a ton of **PBB (polybrominated biphenyls)**, toxic organic halogens, were accidentally mixed into some livestock feed that was distributed throughout the state of Michigan. The PBB found its way into millions of animals and then into the people who ate the meat. The seriousness of the accident came to light when dairy farmers reported that their cows were going dry, aborting their calves, and developing abnormal growths on their hooves. Although more than 30,000 cattle, sheep, and swine and more than a million chickens were destroyed, an estimated 97 percent of Michigan's residents had been exposed to PBB. Some of the exposed farm residents suffered nervous system aberrations and liver disorders.

A similar accident occurred in 1979 when **PCB (polychlorinated biphenyls)** contaminated rice oil in Taiwan. Women who had eaten the tainted rice oil gave birth to children with developmental problems. Decades later,

PBB (polybrominated biphenyl) and **PCB (polychlorinated biphenyl)**: toxic organic halogens used in pesticides, paints, and flame retardants.

young men who were exposed to PCB during gestation had reduced fertility. The interactive effects of PCB and mercury—two environmental contaminants found in fish—are especially damaging to brain functions such as balance and coordination.

Guidelines for Consumers How much of a threat do environmental contaminants pose to the food supply? For the most part, the hazards appear to be small. The FDA regulates the presence of contaminants in foods and requires foods with unsafe amounts to be removed from the market. Similarly, health agencies may issue advisories informing consumers about the potential dangers of eating contaminated foods.

♦ For perspective, 1 ppm (part per million) is equivalent to about 1 minute in 2 years or 1 cent in $10,000.

♦ Mercury in fish
- Relatively high: Tilefish, swordfish, king mackerel, shark
- Relatively low: Cod, haddock, pollock, salmon, sole, tilapia; most shellfish

Most recently, mercury poisoning has aroused concerns—even at levels one-tenth of those in the Minamata catastrophe. Fish and other seafood are the main sources of dietary mercury. Virtually all fish have at least trace amounts of mercury (median, 0.17 parts per million).[10] ♦ Mercury, PCB, chlordane, dioxins, and DDT are the toxins most responsible for fish contamination, but mercury leads the list by threefold. Chronic mercury exposure increases blood levels over time.[11]

Review Figure 19-5 (p. 635) and notice how toxins such as mercury become more concentrated in animals and in people high in the food chain. Because of bioaccumulation, large game fish at the top of the aquatic food chain ♦ generally have the highest concentrations of mercury (10 times the average). Consumers who enjoy eating these fish should select the smaller, younger ones (within legal limits). Also because of bioaccumulation, the concentrations in fish may be a million times higher than the concentrations in the water itself.

© Matt Farruggio

Because albacore ("white") tuna has more mercury than canned light tuna, consumers should limit their intake to no more than 6 ounces of albacore tuna per week.

The EPA regulates commercial fishing to help ensure that fish destined for consumption in the United States meet safety standards for mercury and other contaminants. Farm-raised fish usually have lower concentrations of mercury than fish caught in the wild. Consequently, most consumers in the United States are not in danger of receiving harmful levels of mercury from fish.

The potential harm from contaminants must be balanced against the potential benefits from nutrients.[12] Pregnant and lactating women and young children are most vulnerable because mercury toxicity damages the developing brain. Yet they are also likely to benefit from consuming seafood rich in omega-3 fatty acids. To receive the benefits and minimize the risks, pregnant and lactating women and young children can safely consume up to 12 ounces of seafood per week. In addition, they should limit their intake of albacore tuna and avoid eating large predatory fish altogether. ♦

♦ Pregnant and lactating women and young children should
- Avoid: tilefish (also called golden snapper or golden bass), swordfish, king mackerel, shark
- Limit weekly consumption to: 12 oz (cooked or canned) commercial fish and shellfish (such as shrimp, canned light tuna, salmon, pollock, and catfish) and 6 oz (cooked or canned) "white" albacore tuna

What about the noncommercial fish a person catches from a local lake, river, or ocean? After all, it's almost impossible to tell whether water is contaminated without sophisticated equipment. Each state monitors its waters and issues advisories to inform the public if chemical contaminants have been found in the local fish. To find out whether a fish advisory has been posted in your region, call the local or state environmental health department.

All things considered, fish continue to support a healthy diet, providing valuable protein, omega-3 fatty acids, and minerals. For most adults, the benefits of protecting against heart disease outweigh the risks of consuming seafood regularly. Ideally, consumers would select fish ♦ with high levels of omega-3 fatty acids and low levels of mercury.[13] In addition, they should select a variety of seafood to reduce the risk of exposure to contaminants from a single source.

♦ Fish relatively high in omega-3 fatty acids and low in mercury
- Anchovies, herring, lake trout, mackerel, pollock, salmon, sardines, smelt, tilapia

REVIEW IT Explain how environmental contaminants get into foods and how people can protect themselves against contamination.

Foods may become contaminated as pollutants enter the air, land, and sea. So far, the hazards appear relatively small. In all cases, two principles apply. First, remain alert to the possibility of contamination of foods and keep an ear open for public health announcements and advice. Second, eat a variety of foods. Varying food choices is an effective defensive strategy against the accumulation of toxins in the body. Each food eaten dilutes contaminants that may be present in other components of the diet.

19.4 Natural Toxicants in Foods

LEARN IT Identify natural toxicants and determine whether they are hazardous.

Consumers concerned about food contamination may think that they can eliminate all poisons from their diets by eating only "natural" foods. On the contrary, nature has provided plants with an abundant array of toxicants. A few examples will show how even "natural" foods may contain potentially harmful substances. They also show that although the *potential* for harm exists, *actual* harm rarely occurs.

Poisonous mushrooms are a familiar example of plants that can be harmful when eaten. Few people know, though, that other commonly eaten foods contain substances that can cause illnesses. Cabbage, bok choy, turnips, mustard greens, kale, brussels sprouts, cauliflower, broccoli, kohlrabi, and radishes contain small quantities of goitrogens—compounds that can enlarge the thyroid gland. Eating exceptionally large amounts of goitrogen-containing vegetables can aggravate a preexisting thyroid problem, but it usually does not initiate one. Problems may develop when exceptionally large amounts (2 to 3 pounds a day) of these vegetables are eaten raw; cooking deactivates the enzyme that normally inhibits the uptake of iodine in the thyroid.[14]

Lima beans and fruit seeds such as apricot pits contain cyanogens—inactive compounds that produce the deadly poison cyanide upon activation by a specific plant enzyme. For this reason, many countries restrict commercially grown lima beans to those varieties with the lowest cyanogen contents. As for fruit seeds, they are seldom deliberately eaten. An occasional swallowed seed or two presents no danger, but a couple of dozen seeds can be fatal to a small child. Perhaps the most infamous cyanogen in seeds is laetrile—a compound erroneously represented as a cancer cure. True, laetrile kills cancer, but only at doses that kill the person too. The combination of cyanide poisoning and lack of medical attention is life-threatening.

The humble potato contains many natural poisons including **solanine,** a powerful narcotic-like substance. ♦ Most of a potato's solanine is found in the sprouts and in the green layer that develops just beneath the skin. Solanine poisoning is extremely rare, however, because the small amounts of solanine normally found in potatoes are harmless—even when the potato skin is eaten. Solanine can be toxic, however, and presents a hazard when consumed in large quantities. Cooking does not destroy solanine, but it can be removed by peeling the potato. Symptoms of solanine poisoning include gastrointestinal disturbances and neurological disorders.

♦ Average solanine content: 8 mg/100 g potato
Toxic solanine dose: 20–25 mg/100 g potato

REVIEW IT Identify natural toxicants and determine whether they are hazardous.
Natural toxicants include the goitrogens in cabbage, cyanogens in lima beans, and solanine in potatoes. These examples of naturally occurring toxicants illustrate two familiar principles. First, any substance can be toxic when consumed in excess. Second, poisons are poisons, whether made by people or by nature. Remember, it is not the source of a chemical that makes it hazardous, but its chemical structure and the quantity consumed.

19.5 Pesticides

LEARN IT Debate the risks and benefits of using pesticides.

The use of **pesticides** in agriculture is controversial. They help to ensure the survival of crops, but they leave **residues** in the environment and on some of the foods we eat.

Hazards and Regulation of Pesticides Ideally, a pesticide destroys the pest and quickly degenerates to nontoxic products without accumulating in the food chain. Then, by the time consumers eat the food, no harmful residues remain. Unfortunately, no such perfect pesticide exists. As new pesticides are developed, government agencies assess their risks and benefits and vigilantly monitor their use.

solanine (SOH-lah-neen): a poisonous narcotic-like substance present in potato peels and sprouts.

pesticides: chemicals used to control insects, weeds, fungi, and other pests on plants, vegetables, fruits, and animals. Used broadly, the term includes herbicides (to kill weeds), insecticides (to kill insects), and fungicides (to kill fungi).

residues: whatever remains. In the case of pesticides, those amounts that remain on or in foods when people buy and use them.

Hazards of Pesticides Pesticides applied in the field may linger on foods. Health risks from pesticide exposure are probably small for healthy adults, but children, the elderly, and people with weakened immune systems may be vulnerable to some types of pesticide poisoning. To protect infants and children, government agencies set a **tolerance level** for each pesticide by first identifying foods that children commonly eat in large amounts and then considering the effects of pesticide exposure during each developmental stage.

Regulation of Pesticides Consumers depend on the EPA and the FDA to keep pesticide use within safe limits. These agencies evaluate the risks and benefits of a pesticide's use by asking such questions as: How dangerous is it? How much residue is left on the crop? How much harm does the pesticide do to the environment? How necessary is it? What are the alternatives to its use?

If the pesticide is approved, the EPA establishes a tolerance level for its presence in foods, well below the level at which it could cause any conceivable harm. Tolerance regulations also state the specific crops to which each pesticide can be applied. If a pesticide is misused, growers risk fines, lawsuits, and destruction of their crops.

Once tolerances are set, the FDA enforces them by monitoring foods and livestock feeds for the presence of pesticides. Over the past several decades of testing, the FDA has seldom found residues above tolerance levels, so it appears that pesticides are generally used according to regulations. Minimal pesticide use means lower costs for growers. In addition to costs, many farmers are also concerned about the environment, the quality of their farmland, and a safe food supply. Where violations are found, they are usually due to unusual weather conditions, use of unapproved pesticides, or misuse—for example, application of a particular pesticide to a crop for which it has not been approved.

As many as 400 varieties of fruits and vegetables are imported from other countries.

Pesticides from Other Countries An estimated two-thirds of the fruits and vegetables consumed in the United States are imported from other countries. Because other countries may not have the same pesticide regulations as the United States and Canada, imported foods may contain both pesticides that have been banned and permitted pesticides at concentrations higher than are allowed in domestic foods. A loophole in federal regulations allows US companies to manufacture and sell, to other countries, pesticides that are banned in this country. The banned pesticides then return to the United States on imported foods—a circuitous route that concerned consumers have called the "circle of poison." Federal inspectors sample imported foods and refuse entry if they are found to contain illegal pesticide residues. The United States, Mexico, and Canada work together, using a pesticide policy for all of North America.

Monitoring Pesticides

The FDA collects and analyzes samples of both domestic and imported foods. If the agency finds samples in violation of regulations, it can seize the products or order them destroyed. The FDA may also invoke a **certification** requirement that forces manufacturers, at their own expense, to have their foods periodically inspected and certified safe by an independent testing agency. Individual states also scan for pesticides (as well as for industrial chemicals) and provide information to the FDA.

Food in the Fields In addition to its ongoing surveillance, the FDA conducts selective surveys to determine the presence of particular pesticides in specific crops.

tolerance level: the maximum amount of a residue permitted in a food when a pesticide is used according to label directions.

certification: the process in which a private laboratory inspects shipments of a product for selected chemicals and then, if the product is found to be within acceptable levels of those chemicals, issues a guarantee to that effect.

For example, one year the agency searched for aldicarb in potatoes, captan in cherries, and diaminozide (the chemical name for Alar) in apples, among others. The actions taken that year required several certifications. Thus one shipper in Australia had to certify apples; one in Canada, peppers; one in Costa Rica, chayotes. All grapes from Mexico had to be certified and so did all mangoes from anywhere. This shows, incidentally, how many foods come from abroad—not only those already named, but hundreds more.

Food on the Plate In addition to monitoring foods in the field for pesticides, the FDA also monitors people's actual intakes. The agency conducts the Total Diet Study (sometimes called the "Market Basket Survey") to estimate the dietary intakes of pesticide residues by eight age and gender groups from infants to senior citizens. Four times a year, FDA surveyors buy almost 300 foods from US grocery stores, each time in several cities. They prepare the foods table ready and then analyze them not only for pesticides, but also for essential minerals, industrial chemicals, heavy metals, and radioactive materials. In all, the survey reports on more than 6000 samples a year, including samples imported from 100 countries. Most heavily sampled are fresh vegetables, fruits, and dairy products.

The Total Diet Study provides a direct estimate of the amounts of pesticide residues that remain in foods as they are usually eaten—after they have been washed, peeled, and cooked. Analyses reveal that 2 percent of domestic and 4 percent of imported samples exceed the amounts considered acceptable. The amount considered acceptable is "the daily intake of a chemical, which, if ingested over a lifetime, appears to be without appreciable risk." *Without appreciable risk* means "practical certainty that injury will not result even after a lifetime of exposure." All in all, these findings confirm the safety of the US food supply.

Consumer Concerns Despite these reassuring reports, consumers still worry that food monitoring may not be adequate. For one thing, manufacturers develop new pesticides all the time. For another, as described earlier, other countries use pesticides that are illegal for use here. For still another, although the regulations may protect US foods adequately, they may not necessarily protect the environment or the people who work in the fields. Concerns over poisoning of soil, waterways, wildlife, and workers may well be valid.

The FDA does not sample *all* food shipments or test for *all* pesticides in each sample. The FDA is a *monitoring* agency, and as such, it cannot, nor can it be expected to, guarantee 100 percent safety in the food supply. Instead, it sets standards so that substances do not become a hazard, checks enough samples to adequately assess average food safety, and acts promptly when problems or suspicions arise.

Minimizing Risks Whether consumers ingest pesticide residues depends on a number of factors. How much of a given food does the consumer eat? What pesticide was used on it? How much was used? How long ago was the food last sprayed? Did environmental conditions promote pest growth or pesticide breakdown? How well was the produce washed? Was it peeled or cooked? With so many factors, consumers cannot know for sure whether pesticide residues remain on foods, but they can minimize their risks by following the guidelines offered in Table 19-3 (p. 640); washing, peeling, and cooking fruits and vegetables reduces pesticide residue levels.[15] The food supply is protected well enough that consumers who take these precautions can feel secure that the foods they eat are safe.

Alternatives to Pesticides The use of pesticides has helped to generate higher crop yields that feed the world and protect against diseases transmitted by insects. Still, many consumers are leery. To feed a nation while using fewer pesticides requires creative farming methods. Highlight 19 describes how scientists can genetically alter plants to enhance their production of natural pesticides, and Chapter 20 presents alternative, or sustainable, agriculture methods. These methods include such practices as rotating crops, releasing organisms into fields to destroy pests, and planting nonfood crops nearby to kill pests or attract them

Washing fresh fruits and vegetables removes most, if not all, of the pesticide residues that might have been present.

TABLE 19-3 Minimize Pesticide Residues

When Shopping for Foods
• Select fruits and vegetables that do not have holes. • Select a variety of foods to minimize exposure to any one pesticide. • Consider buying certified organic foods when shopping for produce most likely to be contaminated (see Table 19-4, p. 641).
When Preparing Foods
• Trim the fat from meat, and remove the skin from poultry and fish; discard fats and oils in broths and pan drippings (pesticide residues concentrate in the animal's fat). • Wash fresh produce in warm running water, use a scrub brush, and rinse thoroughly. • Use a knife to peel an orange and grapefruit; do not bite into the peel. • Discard the outer leaves of leafy vegetables such as cabbage and lettuce. • Peel waxed fruits and vegetables; waxes don't wash off and can seal in pesticide residues. • Peel vegetables such as carrots and fruits such as apples when possible (peeling removes dirt, bacteria, and pesticides that remain in or on the peel, but also removes fibers, vitamins, and minerals).

♦ Organic foods that have met USDA standards may use this seal on their labels.

away from the food crops. For example, releasing sterile male fruit flies into orchards helps to curb the population growth of these pests; some flowers, such as marigolds, release natural insecticides and are often planted near crops such as tomatoes. Such alternative farming methods are more labor-intensive and may produce smaller yields than conventional methods, at least initially. Over time, though, by eliminating expensive pesticides, fertilizers, and fuels, these alternatives may actually cut costs more than they cut yields.

Organically Grown Crops Alternative methods are especially useful for farmers who want to produce and market **organic** crops that are grown and processed according to USDA regulations defining the use of synthetic fertilizers, herbicides, insecticides, fungicides, preservatives, and other chemical ingredients. ♦ Similarly, meat, poultry, eggs, and dairy products may be called organic if the livestock has been raised according to USDA regulations defining the grazing conditions and the use of organic feed, hormones, and antibiotics. In addition, producers may *not* claim products are organic if they have been irradiated, genetically engineered, or grown with fertilizer made from sewer sludge. Figure 19-6 shows examples of food labels for products using organic ingredients.

People can grow organic crops when their gardens or farms are relatively small.

organic: in agriculture, crops grown and processed according to USDA regulations defining the use of fertilizers, herbicides, insecticides, fungicides, preservatives, and other chemical ingredients.

> FIGURE 19-6 **Food Labels for Organic Products**

Foods made with 100 percent organic ingredients may claim "100% organic" and use the seal.	Foods made with at least 95 percent organic ingredients may claim "organic" and use the seal.	Foods made with at least 70 percent organic ingredients may list up to three of those ingredients on the front panel.	Foods made with less than 70 percent organic ingredients may list them on the side panel, but cannot make any claims on the front.

Most organic foods are marked as such, but consumers can also determine whether fruits and vegetables are organic by reading the product code on produce stickers. Codes for conventionally grown produce are four digits. Regular bananas, for example, have the code 4011. Codes for organic produce are five digits and begin with 9. (Thus the product code for organic bananas is 94011.) Codes for genetically modified produce are also five digits and begin with 8. (Genetically modified bananas are given the product code 84011.)

Consumers spend more than $20 billion a year on organic foods.[16] Reasons for buying organic include avoiding pesticides, benefiting the environment, protecting animals, improving worker safety, and obtaining safer and more nutritious foods. That organic products are safer or healthier for consumers than those grown using other methods, however, may not be the case. Using unprocessed animal manure as an organic fertilizer, for example, may transmit bacteria, such as *E. coli*, to human beings. For this reason, animal manure must be aged or composted before being used as fertilizer. Both organic and conventional methods may have advantages and disadvantages, and consumers must remain informed.

Each year, an environmental advocacy group publishes a list of the most popular fruits and vegetables that are most and least likely to have pesticide residues (see Table 19-4).[17] The suggestion is that because pesticide residues in conventionally grown foods are higher than in organic foods, consumers may want to pay attention to these lists when considering whether to make organic purchases. Further research contradicts such advice and indicates that the pesticide residues on even the most contaminated fruits and vegetables pose negligible risks and that using organic products does not reduce risks.[18] Whether buying conventionally grown or organically grown produce, consumers benefit most from eating at least five servings of fruits and vegetables daily.

Are organic foods nutritionally superior to conventional foods? For the most part, any nutrient differences reported have been small and within the range that normally occurs in crops.[19] Some research suggests that organic crops may have longer shelf life and better flavor, perhaps due to differences in soil type, soil nutrients, or environmental conditions.[20] Limited research suggests foods produced organically have increased amounts of some phytochemicals. Even so, there does not appear to be any real health benefits from eating organic foods.[21]

Many consumers are willing to pay a little more for organic produce.

TABLE 19-4 **Most and Least Pesticide-Contaminated Fruits and Vegetables**

Most Contaminated	Least Contaminated
Apples	Onions
Celery	Corn
Strawberries	Pineapples
Peaches	Avocados
Spinach	Asparagus
Nectarines (imported)	Peas
Grapes (imported)	Mangoes
Bell peppers	Eggplant
Potatoes	Cantaloupe
Blueberries	Kiwi
Lettuce	Cabbage
Kale	Watermelon

NOTE: These fruits and vegetables are ranked in order of their pesticide load.

REVIEW IT Debate the risks and benefits of using pesticides.
Pesticides can safely improve crop yields when used according to regulations, but they can also be hazardous when used inappropriately. The FDA tests both domestic and imported foods for pesticide residues in the fields and in market basket surveys of foods prepared table ready. Consumers can minimize their ingestion of pesticide residues on foods by following the suggestions in Table 19-3 on p. 640. Alternative farming methods may allow farmers to grow crops with few or no pesticides.

19.6 Food Additives

LEARN IT List common food additives, their purposes, and examples.

Additives confer many benefits on foods. Some reduce the risk of foodborne illness (for example, nitrites used in curing meat prevent poisoning from the botulinum toxin). Others enhance nutrient quality (as in vitamin D–fortified milk). Most additives are **preservatives** that help prevent spoilage during the time it takes to deliver foods long distances to grocery stores and then to kitchens. Some additives simply make foods look and taste good.

Without additives, bread would quickly get moldy, and salad dressing would go rancid.

Intentional additives are put into foods on purpose, whereas indirect additives may get in unintentionally before or during processing. This discussion begins with the regulations that govern additives, then presents intentional additives class by class, and finally says a word about indirect additives.

Regulations Governing Additives The FDA's regulation of additives focuses primarily on safety.[22] To receive permission to use a new additive in food products, a manufacturer must satisfy the FDA that the additive is:

- Effective (it does what it is supposed to do)
- Detectable and measurable in the final food product
- Safe (when fed in large doses to animals under strictly controlled conditions, it causes no cancer, birth defects, or other injury)

On approving an additive's use, the FDA writes a regulation stating in what amounts and in what foods the additive may be used. No additive receives permanent approval, and all must undergo periodic review.

The GRAS List Many familiar substances are exempted from complying with the FDA's approval process because they are **generally recognized as safe (GRAS),** based either on their extensive, long-term use in foods or on current scientific evidence. Several hundred substances are on the GRAS list, including such items as salt, sugar, caffeine, and many spices. Whenever substantial scientific evidence or public outcry has questioned the safety of any substance on the GRAS list, it has been reevaluated. If a legitimate question has been raised about a substance, it has been removed or reclassified. Meanwhile, the entire GRAS list is subjected to ongoing review.

The Delaney Clause One risk that the US law on additives refuses to tolerate at any level is the risk of cancer. To remain on the GRAS list, an additive must not have been found to be a **carcinogen** in any test on animals or human beings. The **Delaney Clause** (the part of the law that states this criterion) is uncompromising in addressing carcinogens in foods and drugs; in fact, it has been under fire for many years for being too strict and inflexible.

The Delaney Clause is best understood as a product of a different historical era. It was adopted decades ago at a time when scientists knew less about the relationships between carcinogens and cancer development. At that time, most substances were detectable in foods only in relatively large amounts, such as parts per thousand. Today, scientific understanding of cancer has progressed, and technology has advanced so that carcinogens in foods can be detected even when they are present only in parts per billion or even per trillion. ♦ Earlier, "zero risk" may

♦ For perspective, one part per trillion is equivalent to about 1 inch in 16 million miles; or 1 second in 32,000 years.

additives: substances not normally consumed as foods but added to food either intentionally or by accident.

preservatives: antimicrobial agents, antioxidants, and other additives that retard spoilage or maintain desired qualities, such as softness in baked goods.

generally recognized as safe (GRAS): food additives that have long been in use and are believed to be safe. First established by the FDA in 1958, the GRAS list is subject to revision as new facts become known.

carcinogen: a substance that can cause cancer.

Delaney Clause: a clause in the Food Additive Amendment to the Food, Drug, and Cosmetic Act that states that no substance that is known to cause cancer in animals or human beings at any dose level shall be added to foods.

have seemed attainable, but today we know it is not: all substances, no matter how pure, can be shown to be contaminated at some level with one carcinogen or another. For these reasons, the FDA prefers to deem additives (and pesticides and other contaminants) safe if lifetime use presents no more than a one-in-a-million risk of cancer to human beings. Thus, instead of the "zero-risk" policy of the Delaney Clause, the FDA uses a "negligible-risk" standard, sometimes referred to as the ***de minimis* rule.**

Margin of Safety Whatever risk level is permitted, actual risks must be determined by research. To determine risks posed by an additive, researchers feed test animals the additive at several concentrations throughout their lives. The additive is then permitted in foods in amounts 100 times *below* the lowest level that is found to cause any harmful effect—that is, at a 1/100 **margin of safety.** In many foods, *naturally* occurring substances occur with narrower margins of safety. Even nutrients pose risks at dose levels above those recommended and normally consumed: for older adults, the RDA for vitamin D is only 1/5 of the Upper Level.

Risks versus Benefits Of course, additives would not be added to foods if they only presented risks. In general, additives are used in foods when they offer benefits that outweigh the risks or make the risks worth taking. No amount of risk may be worth taking in the case of color additives that only enhance appearance but do not improve health or safety. In contrast, the FDA finds it worth taking the small risks associated with the use of nitrites on meat products, for example, because nitrites inhibit the formation of the deadly botulinum toxin. The choice involves a compromise between the risks of using additives and the risks of doing without them.

It is the manufacturers' responsibility to use only the amounts of additives that are necessary to achieve the needed effect, and no more. The FDA also requires that additives *not* be used:

- To disguise faulty or inferior products
- To deceive the consumer
- If use would significantly destroy nutrients
- If effects can be achieved by economical, sound manufacturing processes instead

© Polara Studios, Inc.

Both salt and sugar act as preservatives by withdrawing water from food; microbes cannot grow without water.

Intentional Food Additives

Intentional food additives are added to foods to give them some desirable characteristic: resistance to spoilage, color, flavor, texture, stability, or nutritional value. This section describes additives people most often ask about.

Foods can go bad in two ways. One way is relatively harmless: by losing their flavor and attractiveness. The other way is by becoming contaminated with microbes that cause foodborne illnesses, a hazard that justifies the use of antimicrobial agents.

Antimicrobials The most widely used antimicrobial agents are ordinary salt and sugar. Salt has been used throughout history to preserve meat and fish; sugar serves the same purpose in canned and frozen fruits and in jams and jellies. Both exert their protective effect primarily by capturing water and making it unavailable to microbes.

Other antimicrobial agents, the **nitrites,** are added to foods for three main purposes: to preserve color, especially the pink color of hot dogs and other cured meats; to enhance flavor by inhibiting rancidity, especially in cured meats and poultry; and to protect against bacterial growth. In amounts smaller than those needed to confer color, nitrites prevent the growth of the bacteria that produce the deadly botulinum toxin.

Nitrites clearly serve a useful purpose, but their use has been controversial. During the curing process and in the human body, nitrites can be converted to **nitrosamines.** Because some nitrosamines are known to cause cancer, the USDA and FDA regulate and monitor the use of nitrites in foods and beverages. Limited

***de minimis* rule:** a guideline that defines risk as a cancer rate of less than one cancer per million people exposed to a contaminant over a 70-year lifetime.

margin of safety: when speaking of food additives, a zone between the concentration normally used and that at which a hazard exists. For common table salt, for example, the margin of safety is 1/5 (five times the amount normally used would be hazardous).

intentional food additives: additives intentionally added to foods, such as nutrients, colors, and preservatives.

nitrites (NYE-trites): salts added to food to prevent botulism. One example is sodium nitrite, which is used to preserve meats.

nitrosamines (nye-TROHS-uh-meens): derivatives of nitrites that may be formed in the stomach when nitrites combine with amines. Nitrosamines are carcinogenic in animals.

evidence suggests that naturally occurring nitrites in fruits and vegetables may actually be beneficial to human health.[23]

Another food additive used in ready-to-eat meat and poultry products—such as sausages, hot dogs, and bologna—is a mixture of viruses known as **bacteriophages.** Bacteriophages destroy the bacterium *Listeria monocytogenes,* thus protecting consumers from the potentially life-threatening foodborne illness listeriosis. These additives are included in the ingredients list on food labels as a "bacteriophage preparation."

◆ Sulfites appear on food labels as:
- Sulfur dioxide
- Sodium sulfite
- Sodium bisulfite
- Potassium bisulfite
- Sodium metabisulfite
- Potassium metabisulfite

Antioxidants Another way food can go bad is by exposure to oxygen (oxidation). Often, these changes involve no hazard to health, but they damage the food's appearance, flavor, and nutritional quality. Oxidation is easy to detect when sliced apples or potatoes turn brown or when oil goes rancid. Antioxidants prevent these reactions. Among the antioxidants approved for use in foods are vitamin C (ascorbate) and vitamin E (tocopherol).

Another group of antioxidants, the **sulfites,** ◆ cost less than the vitamins. Sulfites prevent oxidation in many processed foods and alcoholic beverages (especially wine). Because some people experience adverse reactions, the FDA prohibits sulfite use on foods intended to be consumed raw, with the exception of grapes, and requires foods and drugs that contain sulfite additives to declare it on their labels. For most people, sulfites pose no hazard in the amounts used in products, but there is one more consideration—sulfites destroy the B vitamin thiamin. For this reason, the FDA prohibits their use in foods that are important sources of the vitamin, such as enriched grain products.

Two other antioxidants in wide use are **BHA** and **BHT,** which prevent rancidity in baked goods and snack foods.* Several tests have shown that animals fed large amounts of BHT develop *less* cancer when exposed to carcinogens and live *longer* than controls. Apparently, BHT protects against cancer through its antioxidant effect, which is similar to that of the antioxidant nutrients. The amount of BHT ingested daily from the US diet, however, contributes little to the body's antioxidant defense system. A caution: at intakes higher than those that protect against cancer, BHT *causes* cancer. Vitamins E and C remain the most important dietary antioxidants to strengthen defenses against cancer. (See Highlight 11 for a full discussion.)

Color additives not only make foods attractive, but they identify flavors as well. Everyone agrees that yellow jellybeans should taste lemony and black ones should taste like licorice.

Colors Only a few artificial colors remain on the FDA's list of additives approved for use in foods—a highly select group that has survived considerable testing. Colors derived from the natural pigments of plants must also meet standards of purity and safety. Examples of natural pigments commonly used by the food industry are the caramel that tints cola beverages and baked goods and the carotenoids that color margarine, cheeses, and pastas. Carotenoids are also added to the feed for farm-raised salmon, which deepens the pink flesh color.

Flavors Myriad natural flavors, artificial flavors, and flavor enhancers are among the most often used food additives. Many foods taste delicious because manufacturers have added the natural flavors of spices, herbs, essential oils, fruits, and fruit juices. Some spices, notably those used in Mediterranean cooking, provide antioxidant protection as well as flavors. Often, natural flavors are used in combination with artificial flavors.

One of the best-known flavor enhancers is **monosodium glutamate,** or MSG—a sodium salt of the amino acid glutamic acid. MSG is used widely in a number of foods, especially Asian foods, canned vegetables, soups, and processed meats. Besides enhancing the well-known sweet, salty, bitter, and sour tastes, MSG itself may possess a unique flavor. Adverse reactions to MSG—known as

bacteriophages (bak-TIR-ee-oh-fayjz): viruses that infect bacteria.
- **bacterio** = bacteria
- **phage** = eat

sulfites: salts containing sulfur that are added to foods to prevent spoilage.

BHA and **BHT:** preservatives commonly used to slow the development of off-flavors, odors, and color changes caused by oxidation.

monosodium glutamate (MSG): a sodium salt of the amino acid glutamic acid commonly used as a flavor enhancer. The FDA classifies MSG as a "generally recognized as safe" ingredient.

*BHA is butylated hydroxyanisole; BHT is butylated hydroxytoluene.

the **MSG symptom complex**—may occur in people with asthma and in sensitive individuals who consume large amounts of MSG, especially on an empty stomach. Otherwise, MSG is considered safe for adults. It is not allowed in foods designed for infants, however. Food labels require ingredient lists to itemize all additives, including MSG.

Sugar Alternatives The sugar alternatives, introduced in Chapter 4, are among the most widely used artificial flavor additives. Table 4-4 (p. 114) provides a summary of alternative sweeteners. This section presents safety issues surrounding a few of the most controversial ones.

Questions about saccharin's safety surfaced in 1977, when experiments suggested that large doses of saccharin (equivalent to hundreds of cans of diet soda daily for a lifetime) increased the risk of bladder cancer in rats. As a result, the FDA proposed banning saccharin. Public outcry in favor of saccharin was so loud, however, that Congress imposed a moratorium on the ban while additional safety studies were conducted. Products containing saccharin were required to carry a warning label until 2001, when studies concluded that saccharin did not cause cancer in humans. Common sense dictates that consuming large amounts of any substance is probably not wise, but at current, moderate intake levels, saccharin appears to be safe for most people.

Aspartame—a simple chemical compound made of two amino acids (phenylalanine and aspartic acid) and a methyl group (CH_3)—must bear a warning label for people with the inherited disease phenylketonuria (PKU). People with PKU are unable to dispose of any excess phenylalanine. The accumulation of phenylalanine and its by-products is toxic to the developing nervous system, causing irreversible brain damage. The little extra phenylalanine from aspartame poses only a small risk, even in heavy aspartame users, but people with PKU need to get all their required phenylalanine from protein- and nutrient-rich foods instead of from an artificial sweetener.

During metabolism in the body, the methyl group of aspartame momentarily becomes methyl alcohol (methanol)—a potentially toxic compound. This breakdown also occurs in aspartame-sweetened beverages when they are stored at warm temperatures over time. The amount of methanol produced may be safe to consume, but a person may not want to, considering that the beverage has lost its sweetness. In the body, enzymes convert methanol to formaldehyde, another toxic compound. Finally, formaldehyde is broken down to carbon dioxide. Before aspartame could be approved, the quantities of these products generated during metabolism had to be determined, and they were found to fall below the threshold at which they would cause harm. In fact, ounce for ounce, tomato juice yields six times as much methanol as a diet soda.

The amount of artificial sweetener considered safe is called the **Acceptable Daily Intake (ADI)** and represents the amount of consumption that, if maintained every day throughout a person's life, would still be considered safe by a wide margin. It usually reflects an amount 100 times less than the level at which no observed effects occur in animal research studies. The ADI for aspartame, for example, is 50 milligrams per kilogram of body weight. For a 150-pound adult, the ADI is equivalent to 97 packets of Equal or 20 cans of soft drinks sweetened only with aspartame every day for a lifetime. Most people who use aspartame consume less than 5 milligrams per kilogram of body weight per day. Table 4-4 (p. 114) includes the ADI for approved sweeteners.

Texture and Stability Some additives help to maintain a desirable consistency in foods. Emulsifiers keep mayonnaise stable, control crystallization in syrups, disperse spices in salad dressings, and allow powdered coffee creamer to dissolve easily. Gums are added to thicken foods and help form gels. Yeast may be added to provide leavening, and bicarbonates and acids may be used to control acidity.

Nutrients As mentioned earlier, nutrients are sometimes added as antioxidants (vitamins C and E) or for color (beta-carotene and other carotenoids). In addition, manufacturers sometimes add nutrients to fortify or maintain the nutritional

MSG symptom complex: an acute, temporary intolerance reaction that may occur after the ingestion of the additive MSG (monosodium glutamate). Symptoms include burning sensations, chest and facial flushing and pain, and throbbing headaches.

Acceptable Daily Intake (ADI): the estimated amount of a sweetener that individuals can safely consume each day over the course of a lifetime without adverse effect.

TABLE 19-5 Intentional Food Additives

Food Additive	Purpose	Common Examples
Antimicrobials	Prevent food spoilage from microorganisms	Salt, sugar, nitrites and nitrates (such as sodium nitrate), bacteriophages
Antioxidants	Prevent oxidative changes in color, flavor, or texture and delay rancidity and other damage to foods caused by oxygen	Vitamin C (erythorbic acid, sodium ascorbate), vitamin E (tocopherol), sulfites, BHA and BHT
Colors	Enhance appearance	Artificial: indigotine, erythrosine, tartrazine Natural: annatto (yellow), caramel (yellowish brown), carotenoids (yellowish orange), dehydrated beets (reddish brown), grape skins (red, green)
Flavors	Enhance taste	Salt, sugar, spices, artificial sweeteners, MSG
Emulsifiers and gums	Thicken, stabilize, or otherwise improve consistency and texture	Emulsifiers: lecithin, alginates, mono- and diglycerides Gums: agar, alginates, carrageenan, guar, locust bean, psyllium, pectin, xanthan gum, gum arabic, cellulose derivatives
Nutrients (vitamins and minerals)	Improve the nutritive value by replacing vitamins and minerals lost in processing (enrichment) or adding vitamins or minerals that may be lacking in the diet (fortification)	Thiamin, niacin, riboflavin, folate, iron (in grain products); iodine (in salt); vitamins A and D (in milk); vitamin C and calcium (in fruit drinks); vitamin B_{12} (in vegetarian foods)

quality of foods. Included among nutrient additives are the five nutrients added to grains (thiamin, riboflavin, niacin, folate, and iron), the iodine added to salt, the vitamins A and D added to milk, and the nutrients added to fortified breakfast cereals. Appropriate uses of nutrient additives are to:

- Correct dietary deficiencies known to result in diseases
- Restore nutrients to levels found in the food before storage, handling, and processing
- Balance the vitamin, mineral, and protein contents of a food in proportion to the energy content
- Correct nutritional inferiority in a food that replaces a more nutritious traditional food

A nutrient-poor food with nutrients added may appear to be nutrient-rich, but it is rich only in those nutrients chosen for addition. Table 19-5 summarizes intentional food additives.

Indirect Food Additives **Indirect** or **incidental additives** find their way into foods during harvesting, production, processing, storage, or packaging. Incidental additives may include tiny bits of plastic, glass, paper, tin, and other substances from packages as well as chemicals from processing, such as the solvent used to decaffeinate coffee. The following paragraphs discuss six different types of indirect additives that sometimes make headline news.

◆ Common foods containing acrylamide:
- French fries
- Potato chips
- Breakfast cereals
- Cookies

Acrylamide Raw potatoes don't have it, but french fries do—acrylamide, a compound that forms when carbohydrate-rich foods ◆ containing sugars and the amino acid asparagine are cooked at high temperatures. Apparently, acrylamide has been in foods ever since we started baking, frying, and roasting, but only recently has its presence been analyzed. At high doses, acrylamide causes cancer in animals and nerve damage in people. As such, scientists classify it as both a carcinogen and a **genotoxicant.** Quantities commonly found in foods, however, appear to be well below the amounts that cause such damage. The FDA is investigating how acrylamide is formed in foods, how its formation can be limited, and whether its presence is harmful.

Food Packaging The FDA ensures the safety of food packaging and assesses whether packaging materials might migrate into foods. These materials include coatings on can interiors, plastics, papers, and sealants.

Some microwave products are sold in "active packaging" that helps to cook the food; for example, pizzas are often heated on a metalized film laminated to paperboard. This film absorbs the microwave energy in the oven and reaches

indirect or **incidental additives:** substances that can get into food as a result of contact during growing, processing, packaging, storing, cooking, or some other stage before the foods are consumed; sometimes called *accidental additives.*

genotoxicant: a substance that mutates or damages genetic material.

temperatures as high as 500°F. At such temperatures, packaging components migrate into the food. For this reason, manufacturers must perform specific tests to determine whether materials are migrating into foods. If they are, their safety must be confirmed by strict procedures similar to those governing intentional additives.

Most microwave products are sold in "passive packaging" that is transparent to microwaves and simply holds the food as it cooks. These containers don't get much hotter than the foods, but materials still migrate at high temperatures. Consumers should not reuse these containers in the microwave oven. Instead they should use only glass or ceramic containers labeled as microwave safe; ♦ tiny air bubbles in some glass may expand when microwaved, causing the glass to break and glazes on some ceramics to leach, contaminating the food. In the United States, these ceramic containers cannot be sold without a permanent marking stating "Not for food use." Similarly, use only plastic wraps labeled as microwave-safe. Avoid using disposable styrofoam or plastic containers such as those used for carryout or margarine.

Similarly, a chemical known as bisphenol A can leach from hard-plastic bottles into water and other beverages. Drinking cold beverages from these bottles increases concentrations of bisphenol A in the body.[24] The FDA has some concern about the potential health effects of bisphenol A and is taking steps to reduce exposure in the food supply.[25]

♦ Quick test for using glass or ceramic containers in a microwave oven: Microwave the empty container for 1 minute.

- If it's hot, it's unsafe for the microwave.
- If it's warm, it's safe for short-term heating in the microwave.
- If it's cool, it's safe for long-term cooking in the microwave.

Dioxins Coffee filters, milk cartons, paper plates, and frozen food packages, if made from bleached paper, can contaminate foods with small quantities of **dioxins**—compounds formed during chlorine treatment of wood pulp during paper manufacture. Dioxin contamination of foods from such products appears only in trace quantities—in the parts-per-trillion range (recall, for perspective, that one part per trillion is equal to 1 second in 32,000 years). Such levels appear to present no health risks to people, but scientists recognize that dioxins are extremely toxic and are likely to cause cancer in humans. Accordingly, the paper industry has reduced its use of chlorine to cut dioxin exposure. In the meantime, the FDA has concluded that drinking milk from bleached-paper cartons presents no health hazard. Contrary to e-mail warnings, plastics do not yield dioxins when broken down, and dioxins are not released from plastic wrap when microwaved. Human exposure to dioxins comes primarily from foods such as beef, milk products, pork, fish, and shellfish.

Decaffeinated Coffee Many consumers have tried to eliminate caffeine from their diets by selecting decaffeinated coffee. To remove caffeine from coffee beans, manufacturers often use methylene chloride in a process that leaves traces of the chemical in the final product. The FDA estimates that the average cup of coffee decaffeinated this way contains about 0.1 part per million of methylene chloride, which seems to pose no significant threat. A person drinking decaffeinated coffee containing 100 times as much methylene chloride every day for a lifetime has a one-in-a-million chance of developing cancer from it. People are exposed to much more methylene chloride from other sources such as hair sprays and paint-stripping solutions. Still, some consumers prefer either to return to caffeine or to select coffee decaffeinated in another way, perhaps by steam. Unfortunately, manufacturers are not required to state on their labels the type of decaffeination process used in their products. Many labels provide consumer-information telephone numbers for those who have such questions.

Hormones Hormones are a unique type of incidental additive in that their use is intentional, but their presence in the final food product is not. The FDA has approved about a dozen hormones for use in food-producing animals, and the USDA has established limits for residues allowed in meat products.

Some ranchers in the United States treat cattle with **bovine growth hormone (BGH).** All cows make BGH naturally, but when given higher doses, hormone-treated animals produce leaner meats, and dairy cows produce more milk. Scientists can genetically alter bacteria to produce BGH, which allows laboratories to harvest huge quantities of the hormone and sell it to farmers as a drug.

dioxins (dye-OCK-sins): a class of chemical pollutants created as by-products of chemical manufacturing, incineration, chlorine bleaching of paper pulp, and other industrial processes. Dioxins persist in the environment and accumulate in the food chain.

bovine growth hormone (BGH): a hormone produced naturally in the pituitary gland of a cow that promotes growth and milk production; now produced for agricultural use by bacteria.

- **bovine** = of cattle

Indeed, traces of BGH do remain in the meat and milk of both hormone-treated and untreated cows. BGH residues have not been tested for safety in human beings because residues of the natural hormone have always been present in milk and meat, and the amount found in treated cows is within the range that can occur naturally. Furthermore, BGH, being a peptide hormone, is denatured by the heat used in processing milk and cooking meat, and it is also digested by enzymes in the GI tract. The FDA has determined that BGH absorption does not occur in humans and that BGH is biologically inactive in humans even if injected. According to the National Institutes of Health, "As currently used in the United States, meat and milk from [hormone] treated cows are as safe as those from untreated cows." Whether hormones that have passed through the animals into feces and then contaminated the soil and water interfere with plants or animals in the environment remains controversial.

When shopping for milk, consumers find options for conventional, hormone-free, and organic milk. They may expect key differences in the antibiotic and bacterial counts, nutritional value, and hormone composition—but that is not the case. A survey of more than 300 samples found the composition of all these milks to be similar.[26]

Antibiotics Like hormones, antibiotics are also intentionally given to livestock, and residues may remain in the meats and milks. Consequently, people consuming these foods receive tiny doses of antibiotics regularly, and those with sensitivity to antibiotics may suffer allergic reactions. To minimize drug residues in foods, the FDA requires a specified time between the time of medication and the time of slaughter to allow for drug metabolism and excretion.

Of greater concern to the public's health is the development of antibiotic resistance, which occurs when antibiotics are overused. Physicians and veterinarians use an estimated 5 million pounds of antibiotics to treat infections in people and animals, but farmers add five times as much to livestock feed to enhance growth. Not surprisingly, meat from these animals contains resistant bacteria. Such indiscriminate use of antibiotics can be catastrophic to the treatment of disease in human beings. Antibiotics are less effective in treating people who are infected with resistant bacteria. The FDA continues to monitor the use of antibiotics in the food industry with the goal of ensuring that antibiotics remain effective in treating human disease.

REVIEW IT List common food additives, their purposes, and examples.
On the whole, the benefits of food additives seem to justify the risks associated with their use. The FDA regulates the use of the intentional additives (summarized in Table 19-5, p. 646). Incidental additives sometimes get into foods during processing, but rarely present a hazard, although some processes such as treating livestock with hormones and antibiotics raise consumer concerns.

19.7 Consumer Concerns about Water

LEARN IT Discuss consumer concerns about water.

Foods are not alone in transmitting diseases; water is guilty too.[27] In fact, *Cryptosporidium* and *Cyclospora,* commonly found in fresh fruits and vegetables, and *Vibrio vulnificus,* found in raw oysters, are commonly transmitted through contaminated water. In addition to microorganisms, water may contain many of the same impurities that foods do: environmental contaminants, pesticides, and additives such as chlorine used to kill pathogenic microorganisms and fluoride used to protect against dental caries. A glass of "water" is more than just H_2O. This discussion examines the sources of drinking water, harmful contaminants, and ways to ensure water safety.

Sources of Drinking Water Water that is suitable for drinking is called **potable.** Only 1 percent of all the earth's water is potable. Drinking water comes from two sources—surface water and groundwater. In the United States, each source supplies water for about half of the population.

potable (POH-tah-bul): suitable for drinking.

Most major cities obtain their drinking water from surface water—the water in lakes, rivers, and reservoirs. Surface water is readily contaminated because it is directly exposed to acid rain, runoff from highways and urban areas, pesticide runoff from agricultural areas, and industrial wastes that are dumped directly into it. Surface water contamination is reversible, however, because fresh rain constantly replaces the water. It is also cleansed to some degree by aeration, sunlight, and plants and microorganisms that live in it.

Groundwater is the water in underground aquifers—rock formations that are saturated with and yield usable water. People who live in rural areas rely mostly on groundwater pumped up from private wells. Groundwater is contaminated more slowly than surface water, but also more permanently. Contaminants deposited on the ground migrate slowly through the soil before reaching groundwater. Once there, the contaminants break down less rapidly than in surface water due to the lack of aeration, sunlight, and aerobic microorganisms. The slow replacement of groundwater also helps contaminants remain for a long time. Groundwater is especially susceptible to contamination from hazardous waste sites, dumps and landfills, underground tanks storing gasoline and other chemicals, and improperly discarded household chemicals and solvents.

© Joe Cornish/Getty Images

Clean rivers represent irreplaceable water resources.

Water Systems and Regulations Public water systems treat water to remove contaminants that have been detected above acceptable levels. During treatment, a disinfectant (usually, chlorine) is added to kill bacteria. The addition of chlorine to public water is an important public health measure that appears to offer great benefits and small risks. On the one hand, chlorinated water has eliminated such waterborne diseases as typhoid fever, which once ravaged communities, killing thousands of people. On the other hand, it has been associated with an increase in bladder cancer and dioxin contamination of the environment. The EPA is responsible for ensuring that public water systems meet minimum standards for protecting public health.*

Even safe water may have characteristics that some consumers find unpleasant. Most of these problems reflect the mineral content of the water. For example, manganese and copper give water a metallic taste, and sulfur produces a "rotten egg" odor. Iron leaves a rusty brown stain on plumbing fixtures and laundry. Calcium and magnesium (commonly found in "hard water") build up in coffeemakers and hot water heaters. Similarly, soap is not easily rinsed away in hard water, leaving bathtubs and laundry looking dingy. For these and other reasons, some consumers have adopted alternatives to the public water system.

Home Water Treatments To ease concerns about the quality of drinking water, some people purchase home water-treatment systems. Because the EPA does not certify or endorse these water-treatment systems, consumers must shop carefully. Manufacturers offer a variety of units for removing contaminants from drinking water. None of them removes all contaminants, and each has its own advantages and disadvantages. Choosing the right treatment unit depends on the kinds of contaminants in the water. For example, activated carbon filters are particularly effective in removing chlorine, heavy metals such as mercury, and organic contaminants from sediment. Reverse osmosis forces pressurized water through a membrane, flushing out minerals such as sodium and some microorganisms such

*The EPA's safe drinking water hotline: (800) 426–4791.

as *Giardia.* Ozonation uses ozone gas to disinfect water. And distillation systems, which boil water and condense the steam to water, kill microorganisms but leave behind minerals such as lead. Therefore, before purchasing a home water-treatment unit, a consumer must first determine the quality of the water. In some cases, a state or county health department will test water samples or can refer the consumer to a certified laboratory. Consumers need to be aware that unscrupulous vendors may use scare tactics during home inspections to prompt sales.

Bottled Water Despite the higher cost, many people turn to bottled water as an alternative to tap water. More than 8 billion gallons of bottled water are sold in the United States annually—an average of more than 25 gallons per person.[28] The FDA regulates bottled drinking water and has established quality and safety standards ♦ compatible with those set by the EPA for public water systems. In addition, all bottled waters must be processed, packaged, and labeled in accordance with FDA regulations. Water quality may vary among brands because of variations in the source water used and company practices, but bottled water is neither safer nor healthier than tap water.

♦ The FDA:

- Approves and verifies the water source
- Inspects sanitizing procedures and bottling operations
- Examines the results of companies' analyses of source water and final product for bacteria, chemicals, and other contaminants

As Chapter 14 discusses, some bottled waters are marketed as "enhanced water"—water that has been enhanced with sweeteners, juices, coloring, flavors, vitamins, minerals, protein, or extra oxygen. Consumers perceive these bottled waters as healthful and sales have skyrocketed.

Labels on bottled water must identify the water's source. Approximately 75 percent of bottled waters derive from protected groundwater (from springs or wells)—the same as tap water. This water is usually treated before being bottled. For example, it may be disinfected with ozone gas rather than chlorine. Ozone kills microorganisms, then disintegrates spontaneously into water and oxygen, leaving behind no toxic by-products. Other bottled waters may be treated by filtration to remove pathogens and other particles. Bottled waters may also be treated by reverse osmosis or ion exchange to remove minerals. Alternatively, the water may be distilled into a vapor and then condensed again into water, thus removing any dissolved solids. These processes allow the bottle to be labeled "purified water." Most bottled waters do not contain fluoride; consequently, they do not provide the tooth protection of fluoridated water from community public water systems.

Realizing that fossil fuels and many gallons of water are used to create and transport plastic water bottles, concerned consumers use refillable bottles to save money and the environment.

Despite government regulations, some contamination has been detected in some bottled waters. Although the amounts of most contaminants found in bottled waters are probably insignificant, consumers should be aware that bottled water is not necessarily purer than the water from their taps. As a safeguard, the FDA recommends that bottled water be handled like other foods and be refrigerated after opening.

Protection of drinking water is a subject of ongoing concern and controversy. It may soon become a source of conflict between the world's nations as the population continues to grow and the renewable water supply remains constant. Estimates are that within the next 50 years, half of the world's people will not have enough clean water to meet their needs. To avert this potential calamity, we must take active steps to conserve water, clean polluted water, desalinate seawater, and curb population growth.

REVIEW IT Discuss consumer concerns about water.
Like foods, water may contain infectious microorganisms, environmental contaminants, pesticide residues, and additives. The EPA monitors the safety of the public water system, but many consumers choose home water-treatment systems or bottled water instead of tap water.

As this chapter said at the start, supplying food safely to hundreds of millions of people is an incredible challenge—one that is met, for the most part, with incredible efficiency. The following chapter describes a contrasting situation—that of the food supply not reaching the people.

Nutrition Portfolio

Practicing food safety allows you to eat a variety of foods, with little risk of foodborne illnesses.

- Review your food-handling practices and describe how effectively you wash your hands, utensils, and kitchen surfaces when preparing foods.
- Describe the steps you take to separate raw and cooked foods while storing and preparing them.
- Describe how you can ensure that you cook foods to a safe temperature and refrigerate perishable foods promptly.

Go to Diet Analysis Plus and choose one of the days on which you tracked your diet and activity for an entire day. Select the Intake Spreadsheet report to help you answer the following questions:

- Imagine for a moment that you got a foodborne illness on this particular day. Which food would you most suspect to have contained the illness-causing organism or toxin? Why would you suspect that food more than the other foods you ate that day?

To complete this exercise, go to your Diet Analysis Plus at **www.cengagebrain.com**.

STUDY IT To review the key points of this chapter and take a practice quiz, go to the study cards at the end of the book.

REFERENCES

1. M. T. Osterholm, Foodborne disease in 2011: The rest of the story, *New England Journal of Medicine* 364 (2011): 889–891; Centers for Disease Control and Prevention, Press release: New estimates more precise, December 15, 2010.
2. L. H. Gould and coauthors, Surveillance for foodborne disease outbreaks: United States, 2008, *Morbidity and Mortality Weekly Report* 60 (2011): 1197–1202.
3. C. P. Wild and Y. Y. Gong, Mycotoxins and human disease: A largely ignored global health issue, *Carcinogenesis* 31 (2010): 71–82.
4. Position of the American Dietetic Association: Food and water safety, *Journal of the American Dietetic Association* 109 (2009): 1449–1460.
5. Department of Agriculture, Mandatory Country of Origin Labeling of beef, pork, lamb, chicken, goat meat, wild and farm-raised fish and shellfish, perishable agricultural commodities, peanuts, pecans, ginseng, and macadamia nuts, *Federal Register* 74 (2009): 2658–2707.
6. Centers for Disease Control and Prevention, Handwashing: Clean hands save lives, www.cdc.gov/handwashing.
7. US Food and Drug Administration, All about BSE, www.fda.gov/animalveterinary, April 20, 2010.
8. Position of the American Dietetic Association: Food and water safety, *Journal of the American Dietetic Association* 109 (2009): 1449–1460.
9. C. Hawkes, Sales promotions and food consumption, *Nutrition Reviews* 67 (2009): 333–342.
10. B. C. Scudder and coauthors, Mercury in fish, bed sediment, and water from streams across the United States, 1998–2005: US Geological Survey Scientific Investigations Report 2009-5109, 2009.
11. D. R. Laks, Assessment of chronic mercury exposure within the US population, National Health and Nutrition Examination Survey, 1999–2006, *BioMetals* 22 (2009): 1103–1114.
12. A. Tsuchiya and coauthors, Fish intake guidelines: Incorporating n-3 fatty acid intake and contaminant exposure in the Korean and Japanese communities, *American Journal of Clinical Nutrition* 87 (2008): 1867–1875; A. L. Yaktine, M. C. Nesheim, and C. A. James, Nutrient and contaminant tradeoffs: Exchanging meat, poultry, or seafood for dietary protein, *Nutrition Reviews* 66 (2008): 113–122.
13. K. R. Mahaffey and coauthors, Balancing the benefits of n-3 polyunsaturated fatty acids and the risks of methylmercury exposure from fish consumption, *Nutrition Reviews* 69 (2011): 493–508.
14. M. Chu and T. F. Seltzer, Myxedema coma induced by ingestion of raw bok choy, *New England Journal of Medicine* 362 (2010): 1945–1946.
15. B. M. Keikotlhaile, P. Spanoghe, and W. Steurbaut, Effects of food processing on pesticide residues in fruits and vegetables: A meta-analysis approach, *Food and Chemical Toxicology* 48 (2010): 1–6.
16. C. Dimitri and L. Oberholtzer, *Marketing US organic foods: Recent trends from farms to consumers*. Economic Information Bulletin No. 58, US Department of Agriculture, Economic Research Service, September 2009.
17. Environmental Working Group, "EWG's Shoppers Guides to Pesticides," Environmental Working Group, Washington, DC, 2010.
18. C. K. Winter and J. M. Katz, Dietary exposure to pesticide residues from commodities alleged to contain the highest contamination levels, *Journal of Toxicology* 2011 (2011): 589–674.
19. A. D. Dangour and coauthors, Nutritional quality of organic foods: A systematic review, *American Journal of Clinical Nutrition* 90 (2009): 680–685.
20. J. P. Reganold and coauthors, Fruit and soil quality of organic and conventional strawberry agroecosystems, *PLoS ONE* 5 (2010): e12346.
21. A. D. Dangour and coauthors, Nutrition-related health effects of organic foods: A systematic review, *American Journal of Clinical Nutrition* 92 (2010): 203–210.
22. T. G. Neltner and coauthors, Navigating the US food additive regulatory program, *Comprehensive Reviews in Food Science and Food Safety* 10 (2011): 342–368.
23. N. G. Hord, Y. Tang, and N. S. Bryan, Food sources of nitrates and nitrites: The physiologic context for potential health benefits, *American Journal of Clinical Nutrition* 90 (2009): 1–10.
24. J. L. Carwile and coauthors, Use of polycarbonate bottles and urinary bisphenol A concentrations, *Environmental Health Perspectives* 117 (2009): 1368–1372.
25. Food and Drug Administration, Update on bisphenol A for use in food contact applications, www.fda.gov, revised March 2010.
26. J. Vicini and coauthors, Survey of retail milk composition as affected by label claims regarding farm-management practices, *Journal of the American Dietetic Association* 108 (2008): 1198–1203.
27. J. Yoder and coauthors, Surveillance for waterborne disease and outbreaks associated with drinking water and water not intended for drinking: United States, 2005–2006, *Morbidity and Mortality Weekly Report* 57 (2008): 39–68.
28. US Food and Drug Administration, Bottled water everywhere: Keeping it safe, August 25, 2008, www.fda.gov/consumer/updates/bottledwater082508.html.

LEARN IT Debate the pros and cons surrounding genetically engineered foods.

Food Biotechnology

© Tony Freeman/Photo Edit

Advances in food **biotechnology** promise just about everything from the frivolous (a tear-free onion) to the profound (a hunger-free world). Already biotechnology has produced leaner meats, longer shelf lives, better nutrient composition, and greater crop yields grown with fewer pesticides. Overall, biotechnology offers numerous opportunities to overcome food shortages, improve the environment, and eliminate disease. But it also raises concerns about possible risks to the environment and human health. Critics assert that biotechnology will exacerbate world hunger, destroy the environment, and endanger health. This highlight presents some of the many issues surrounding genetically engineered foods, and the accompanying glossary defines key terms.

The Promises of Genetic Engineering

For centuries, farmers have been selectively breeding plants and animals to shape the characteristics of their crops and livestock. They have created prettier flowers, hardier vegetables, and leaner animals. Consider the success of selectively breeding corn. Early farmers in Mexico began with a wild, native plant called teosinte (tay-oh-SEEN-tay) that bears only five or six kernels on each small spike. Many years of patient selective breeding have produced large ears filled with hundreds of plump kernels aligned in perfect formation, row after row.

Such genetic improvements, together with the use of irrigation, fertilizers, and pesticides, were responsible for more than half of the increases in US crop yields in the 20th century. Farmers still use selective breeding, but now, in the 21st century, advances in **genetic engineering** have brought rapid and dramatic changes to agriculture and food production.

Although selective breeding works, it is slow and imprecise because it involves mixing thousands of genes from two plants and hoping for the best. With genetic engineering, scientists can improve crops (or livestock) by introducing a copy of the specific gene needed to produce the desired trait. Figure H19-1 illustrates the difference. Once introduced, the selected gene acts like any other gene—it provides instructions for making a protein. The protein then determines a characteristic in the genetically modified plant or animal. In short, the process is now faster and more refined. Farmers no longer need to wait patiently for breeding to yield improved crops and animals, nor must they even respect natural lines of reproduction among species. Laboratory scientists can copy genes from one organism and insert them into almost any other organism—plant, animal, or microbe. Their work is changing not only the way farmers plant, fertilize, and harvest their crops, but also the ways the food industry processes food and consumers receive nutrients, phytochemicals, and drugs.

© STRI Antonio Mortaner

This wild predecessor of corn, with its sparse five or six kernels, bears little resemblance to today's large, full, sweet ears.

GLOSSARY

biotechnology: the use of biological systems or organisms to create or modify products. Examples include the use of bacteria to make yogurt, yeast to make beer, and cross-breeding to enhance crop production.

clone: a genetic copy of an animal, similar to identical twins but born at different times.

genetic engineering: the use of biotechnology to modify the genetic material of living cells so that they will produce new substances or perform new functions. Foods produced via this technology are called *genetically modified (GM)* or *genetically engineered (GE) foods.*

plant-pesticides: pesticides made by the plants themselves.

rennin: an enzyme that coagulates milk; found in the gastric juice of cows, but not human beings.

> FIGURE H19-1 **Selective Breeding and Genetic Engineering Compared**

Traditional Selective Breeding

Traditional selective breeding combines many genes from two varieties of the same species to produce one with the desired characteristics.

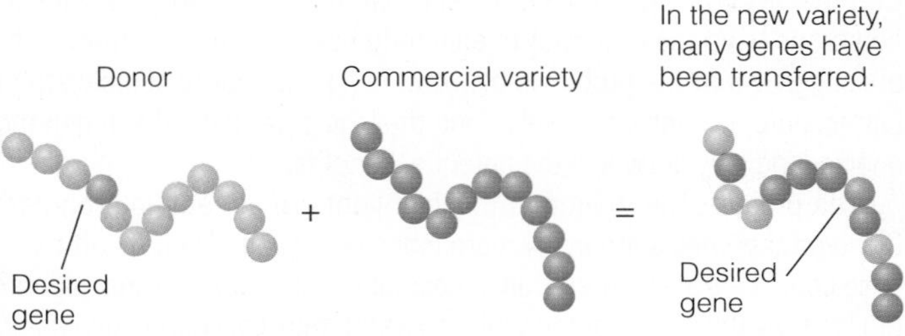

Genetic Engineering

Through genetic engineering, a single gene is (or several are) transferred from the same or different species to produce one with the desired characteristics.

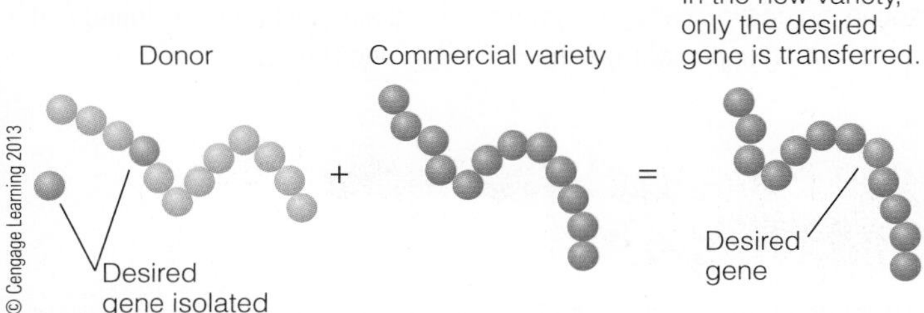

Extended Shelf Life

Among the first products of genetic engineering to hit the market were tomatoes that stay firm and ripe longer than regular tomatoes that are typically harvested green and ripened in the stores. These genetically modified tomatoes promise less waste and higher profits. Normally, tomatoes produce a protein that softens them after they have been picked. Scientists can now introduce into a tomato plant a gene that is a mirror image of the one that codes for the "softening" enzyme. This gene fastens itself to the RNA of the native gene and blocks synthesis of the softening protein. Without this protein, the genetically altered tomato softens more slowly than a regular tomato, allowing growers to harvest it at its most flavorful and nutritious vine-ripe stage.

Improved Nutrient Composition

Genetic engineering can also improve the nutrient composition of foods.[1] Instead of manufacturers adding nutrients to foods during processing, plants can be genetically altered to do their own fortification work—a strategy called *biofortification*.[2] Biofortification of staple crops with key vitamins and minerals can effectively combat the nutrient deficiency diseases that claim so many lives worldwide.[3] Genetically modifying wheat can improve its protein, zinc, and iron content. Soybeans may be implanted with a gene that upgrades soy

Genetically modified cauliflower is orange, reflecting a change in a single gene that increases its production of beta-carotene 100-fold.

protein to a quality approaching that of milk. Corn has been modified to contain twice the amount of lysine and tryptophan, its two limiting amino acids.[4] Soybean and canola plants can be genetically modified to alter the composition of their oils, making them richer in the heart-healthy monounsaturated fatty acids. "Golden rice," which has received genes from a daffodil and a bacterium that enable it to make beta-carotene, offers promise in treating vitamin A deficiency worldwide.[5] (Chapter 11 describes how vitamin A deficiency contributes to the deaths of 2 million children and the blindness of a half million each year.) Of course, increasing nutrients in crops may have unintended consequences as well. For example, when broccoli is manipulated to increase its selenium content, production of the cancer-fighting phytochemical sulforaphane declines.

In addition to enhancing the nutrient composition, genetically modified crops can also produce more of the phytochemicals that help maintain health and reduce the risks of chronic diseases (see Highlight 13). They can also be coaxed to produce less phytate, which allows more zinc to be absorbed. The possibilities seem endless.

Efficient Food Processing

Genetic engineering also helps to process foods more efficiently, which saves money. For example, the protein **rennin**, which is used to coagulate milk in the production of cheese, has traditionally been harvested from the stomachs of calves, a costly process. Now scientists can insert a copy of the rennin gene into bacteria and then use bacterial cultures to mass-produce rennin—saving time, money, space, and animals.

Genetic engineering can also help to bypass costly food-processing steps. At present, people who are lactose intolerant can buy milk that has been treated with the lactase enzyme. Wouldn't it be more convenient, and less expensive, if scientists could induce cows to make lactose-free milk directly? They're working on it. They have already successfully inserted into mice the genetic material needed to make lactase in their mammary glands, thereby producing low-lactose milk. Decaffeinated coffee beans are another real possibility.

Genetic research today has progressed well beyond tweaking a gene here and there to produce a desired trait. Scientists can now **clone** animals. By cloning animals, scientists have the ability to produce both needed food and pharmaceutical products. The FDA has declared that food from cloned livestock is safe to eat, but the USDA has asked farmers to keep cloned animals off the market. Cloned animals are used primarily for breeding; their offspring are used primarily for food. The percentage of consumers who have a favorable impression of eating food from cloned animals or their offspring is small, but increasing. The industry does not track cloned offspring entering the market. Because FDA does not distinguish between foods from cloned animals or their offspring and foods from conventional animals, food labels are not required to provide this information.[6]

Efficient Drug Delivery

Using cloned animals and other organisms in the development of pharmaceuticals is whimsically called "biopharming." For example, a cow cloned with the genetic equipment to make a vaccine in its milk could provide both nourishment and immunization to a whole village of people now left unprotected because they lack food and medical help. Similarly, researchers have figured out how to induce hens to produce eggs with a drug to treat multiple sclerosis. Bananas and potatoes have been designed to make hepatitis vaccines, and tobacco leaves to make AIDS drugs. Researchers can also harvest vaccines by genetically altering hydroponically grown tomato plants to secrete a protein through their root systems into the water. Using foods to deliver drugs is only a small part of the promise and potential that biotechnology offers the field of medicine.

Genetically Assisted Agriculture

Genetic engineering has helped farmers to increase yields, extend growing seasons, and grow crops that resist herbicides. About half of the soybean crops in the United States have been genetically engineered to withstand a potent herbicide. As a result, farmers can spray whole fields with this herbicide and kill the weeds without harming the soybeans.

Similarly, farmers can grow crops that produce their own pesticides—substances known as **plant-pesticides.** Corn, broccoli, and potatoes have received a gene from a bacterium that produces a protein that is toxic to leaf-chewing caterpillars (but not to humans). Yellow squash has been given two genes that confer resistance to the most common viral diseases. Potatoes can now produce a beetle-killing toxin in their leaves. These crops and many others like them are currently being grown or tested in fields around the United States. Growing crops that make their own pesticides allows farmers to save time, increase yields, and use fewer, or less harmful, pesticides. Genetically modified crops have decreased the environmental impact associated with pesticide use.

Other Possibilities

Many other biotechnology possibilities are envisioned for the near future. Shrimp may be empowered to fight diseases with genetic ammunition borrowed from sea urchins. Peanuts may have their allergens removed.[7] Plants may be given special molecules to help them grow in polluted soil. With these and other advances, farmers may reliably produce bumper crops of food every year on far fewer acres of land, with less loss of water and topsoil, and far less use of toxic pesticides and herbicides. Supporters of biotechnology predict that these efforts will enhance food production and help meet the challenge of feeding an ever-increasing world population. They contend that genetically modified crops have the potential to eliminate hunger and starvation. Others suggest that the problems of world hunger are more complex than biotechnology alone can resolve and that the potential risks of genetic engineering may outweigh the potential benefits.

The projects mentioned in this highlight are already in progress. Close on their heels are many more ingenious ideas. What if salt tolerance could be transplanted from a coastal marsh plant into crop plants? Could crops then be irrigated with seawater, thus conserving dwindling freshwater supplies? Or could crops be genetically designed to use less water? Would the world food supply increase if rice farmers could grow plants that were immune to disease? What if consumers could dictate which traits scientists insert into food plants? Would they choose to add phytochemicals to fight cancer or reduce the risk of heart disease? These and other possibilities seem unlimited, and though they may sound incredible, many such products have already been developed and are awaiting approval from the FDA, EPA, and USDA.

The Potential Problems and Concerns

Although many scientists hail biotechnology with confidence, others have reservations.[8] Most consumers know little about biotechnology or the extent to which their foods contain genetically modified foods. Some consumers have concerns about what they call "Frankenfoods." Those who oppose biotechnology fear for the safety of a world where genetic tampering produces effects that are not yet fully understood. They suspect that the food industry may be driven by potential profits, without ethical considerations or laws to harness the effects. They point out that even the scientists who developed the techniques cannot predict the ultimate outcomes of their discoveries. These consumers don't want to eat a scientific experiment or interfere with natural systems. Genetic decisions, they say, are best left to the powers of nature.

If science and the marketplace are allowed to drive biotechnology without restraint, critics fear that these problems may result:

- *Disruption of natural ecosystems.* New, genetically unique organisms that have no natural place in the food chain or evolutionary biological systems could escape into the environment and reproduce.
- *Introduction of diseases.* Newly created viruses may mutate to cause deadly diseases that may attack plants, animals, or human beings. Genetically modified bacteria may develop resistance to antibiotics, making the drugs useless in fighting infections.
- *Introduction of allergens and toxins.* Genetically modified crops may contain new substances that have consequences, such as causing allergies.[9]

Many consumers believe that genetically modified foods should be labeled as such.

- *Creation of biological weapons.* Fatal bacterial and viral diseases may be developed for use as weapons.
- *Ethical dilemmas.* Critics pose the question "How many human genes does an organism have to contain before it is considered human? For instance, how many human genes would a green pepper have to contain before one would have qualms about eating it?"

Proponents of biotechnology respond that evidence to date does not justify these concerns. Opponents counter that the lack of evidence showing harm does not provide evidence showing safety. These opposing views illustrate the tension between the forward thrust of science and the hesitation of consumers. Some would argue for more research on the safety of genetically modified food, while others assert that more research is a waste of resources and that it is time to embrace biotechnology with enthusiasm.[10] In addition to evaluating the potential risks and benefits, genetically modified crops need international oversight. Table H19-1, p. 656 summarizes the issues.

From another perspective, some argue that the concerns expressed by those protesting genetically engineered foods reflect prejudices acquired in an elitist world of fertile land and abundant food. Those living in poverty-stricken areas of the world do not have the luxury of determining how to grow crops and process foods. They cannot afford the delays created when protesters destroy test crops and disrupt scientific meetings. They need solutions now. People are starving, and genetic engineering holds great promise for increasing crop yields and providing those people with food.[11]

The work of Nobel Peace Prize Laureate Dr. Norman Borlaug and his team over the past four decades attests to the benefits of using technology to defend against hunger. By developing grains that resist pests and diseases, they have been able to increase yields and provide real solutions to global hunger problems. When Mexico used Borlaug's special breed of dwarf wheat, yields increased threefold compared with traditional varieties. India increased wheat production tenfold and became self-sufficient in its grain production. Pakistan increased its wheat production fivefold and sub-Saharan Africa more than tripled crop yields by changing farming practices. The combination of improved conventional systems and biotechnology produces more food to meet the nutritional needs of more people.[12]

At a minimum, critics of biotechnology have made a strong case for rigorous safety testing and labeling of new products. They contend, for example, that when a new gene has been introduced into a food, tests should ensure that other, unwanted genes have not accompanied it. If a disease-producing microorganism has donated genetic material, scientists must prove that no dangerous characteristic from the microorganism has also entered the food. If the inserted genetic material comes from a source to which some people develop allergies, such as nuts, then the new product should be labeled to alert them. Furthermore, if the newly altered genetic material creates proteins that have never before been encountered by the human body, their effects should be studied to ensure that people can eat them safely.

FDA Regulations

The FDA has taken the position that foods produced through biotechnology and cloning are not substantially different from others and require no special testing, regulations, or labeling. After all, most foods available today have already been genetically altered by years of selective breeding. The new vegetable broccoflower, a product of sophisticated cross-breeding of broccoli with cauliflower, met no testing or approval barriers on its way to the dinner plate. When the vegetable became available on the market, scientists studied its nutrient contents, but they did not question its safety.

In most cases, the new genetically modified food differs from the old conventional one only by a gene or two. The rennin produced by bacteria is structurally and functionally the same as the rennin produced by calves, for example. For that reason, the FDA considers it and other genetically engineered foods "generally recognized as safe (GRAS)."

A product such as the tomato described earlier need not be tested because its new gene *prevents* synthesis of a protein and adds nothing but a tiny fragment of genetic material. Nor does this tomato require special labeling because it is not significantly different from the many other varieties of tomatoes on the market. On the other hand, any substances introduced into a food (such as a hormone or protein) by way of bioengineering must meet the same safety standards applied to all additives. A tomato plant with a gene that, for example, produces a pesticide cannot be marketed until tests prove it safe for consumption. The FDA assures consumers that all bioengineered foods on the market today are as safe as their traditional counterparts.

Foods produced through biotechnology that are substantially different from others must be labeled to identify that difference. For example, if the nutrient composition of the new product differs from its traditional counterpart, as in the soybean and canola oils mentioned earlier, then labeling is required. Similarly, if an allergy-causing protein has been introduced to a nonallergenic food, then labeling must warn consumers.

TABLE H19-1 Food Biotechnology: Point, Counterpoint

Arguments in Opposition to Genetic Engineering	Arguments in Support of Genetic Engineering
1. **Ethical and moral issues.** It's immoral to "play God" by mixing genes from organisms unable to do so naturally. Religious and vegetarian groups object to genes from prohibited species occurring in their allowable foods.	1. **Ethical and moral issues.** Scientists throughout history have been persecuted and even put to death by fearful people who accuse them of playing God. Yet, today many of the world's citizens enjoy a long and healthy life of comfort and convenience due to once-feared scientific advances put to practical use.
2. **Imperfect technology.** The technology is young and imperfect—genes rarely function in just one way, their placement is imprecise ("shotgun"), and all of their potential effects are impossible to predict. Toxins are as likely to be produced as the desired trait. More than 95 percent of DNA is called "junk" because scientists have not yet determined its function.	2. **Advanced technology.** Recombinant DNA technology is precise and reliable. Many of the most exciting recent advances in medicine, agriculture, and technology were made possible by the application of this technology.
3. **Environmental concerns.** Environmental side effects are unknown. The power of a genetically modified organism to change the world's environments is unknown until such changes actually occur—then the "genie is out of the bottle." Once out, insects, birds, and the wind distribute genetically altered seed and pollen to points unknown.	3. **Environmental protection.** Genetic engineering may be the only hope of saving rain forest and other habitats from destruction. Through genetic engineering, farmers can make use of previously unproductive lands such as salt-rich soils and arid areas.
4. **"Genetic pollution."** Other kinds of pollution can often be cleaned up with money, time, and effort. Once genes are spliced into living things, those genes forever bear the imprint of human tampering.	4. **Genetic improvements.** Genetic side effects are more likely to benefit the environment than to harm it.
5. **Crop vulnerability.** Pests and diseases can quickly adapt to overtake genetically identical plants or animals around the world. Diversity is key to defense.	5. **Improved crop resistance.** Pests and diseases can be specifically fought on a case-by-case basis. Biotechnology is the key to defense.
6. **Loss of gene pool.** Loss of genetic diversity threatens to deplete valuable gene banks from which scientists can develop new agricultural crops.	6. **Gene pool preserved.** Thanks to advances in genetics, laboratories around the world are able to stockpile the genetic material of millions of species that, without such advances, would have been lost forever.
7. **Profit motive.** Genetic engineering will profit industry more than the world's poor and hungry.	7. **Everyone profits.** Industries benefit from genetic engineering, and a thriving food industry benefits the nation and its people, as witnessed by countries lacking such industries. Genetic engineering promises to provide adequate nutritious food for millions who lack such food today. Developed nations gain cheaper, more attractive, more delicious foods with greater variety and availability year round.
8. **Unproven safety for people.** Human safety testing of genetically altered products is generally lacking. The population is an unwitting experimental group in a nationwide laboratory study for the benefit of industry.	8. **Safe for people.** Human safety testing of genetically altered products is unneeded because the products are essentially the same as the original foodstuffs.
9. **Increased allergens.** Allergens can unwittingly be transferred into foods.	9. **Control of allergens.** A few allergens can be transferred into foods, but these are known. Also, foods likely to contain them are clearly labeled to warn consumers.
10. **Decreased nutrients.** A fresh-looking tomato or other produce held for several weeks may have lost substantial nutrients.	10. **Increased nutrients.** Genetic modifications can easily enhance the nutrients in foods.
11. **No product tracking.** Without labeling, the food industry cannot track problems to the source.	11. **Excellent product tracking.** The identity and location of genetically altered foodstuffs are known, and they can be tracked should problems arise.
12. **Overuse of herbicides.** Farmers, knowing that their crops resist herbicide effects, will use them liberally.	12. **Conservative use of herbicides.** Farmers will not waste expensive herbicides in second or third applications when the prescribed amount gets the job done the first time.
13. **Increased consumption of pesticides.** When a pesticide is produced by the flesh of produce, consumers cannot wash it off the skin of the produce with running water as they can with ordinary sprays.	13. **Reduced pesticides on foods.** Pesticides produced by plants in tiny amounts known to be safe for consumption are more predictable than applications by agricultural workers who make mistakes. Because other genetic manipulations will eliminate the need for postharvest spraying, fewer pesticides will reach the dinner table.
14. **Lack of oversight.** Government oversight is run by industry people for the benefit of industry—no one is watching out for the consumer.	14. **Sufficient regulation and rapid response.** Government agencies are efficient in identifying and correcting problems as they occur in the industry.

Most consumers want all genetically altered products clearly labeled. Consumer advocacy groups claim that by not requiring such labeling, the FDA forces millions of consumers to be guinea pigs, unwittingly testing genetically engineered foods. Additionally, they say, people who have religious objections to consuming foods to which genes of prohibited organisms have been added have no way of identifying those foods. For example, someone keeping a kosher kitchen may unknowingly use a food containing genes from a pig. Currently, labeling is voluntary. Manufacturers may state that a product has been "genetically engineered." Those who do would be wise to explain its purpose and benefit. When consumers recognize a personal health benefit, most tend to accept genetically engineered foods.

Speaking in defense of the FDA's position are the FDA itself, recognized as the nation's leading expert and advocate for food safety, and the Academy of Nutrition and Dietetics, which represents current scientific thinking in nutrition.[13] Many other scientific organizations agree, contending that biotechnology can deliver an improved food supply if we give it a fair chance to do so.

Will these new technologies provide foods to meet the needs of the future? Some would say yes. Biotechnology holds a world of promise, and with proper safeguards and controls, it may yield products that meet the needs of consumers almost perfectly.

REFERENCES

1. M. N. McGloughlin, Modifying agricultural crops for improved nutrition, *Nature Biotechnology* 30 (2010): 494–504.
2. P. J. White and M. R. Broadley, Biofortification of crops with seven mineral elements often lacking in human diets: Iron, zinc, copper, calcium, magnesium, selenium and iodine, *New Phytologist* 182 (2009): 49–84.
3. K. D. Hirschi, Nutrient biofortification of food crops, *Annual Review of Nutrition* 29 (2009): 401–421.
4. E. T. Nuss and S. A. Tanumihardjo, Quality protein maize for Africa: Closing the protein inadequacy gap in vulnerable populations, *Advanced Nutrition* 2 (2011): 217–224.
5. G. Tang and coauthors, Golden rice is an effective source of vitamin A, *American Journal of Clinical Nutrition* 89 (2009): 1776–1783.
6. US Food and Drug Administration, Animal cloning and food safety, January 15, 2008, www.fda.gov/consumer/updates/cloning011508.html.
7. M. Gallo and R. Sayre, Removing allergens and reducing toxins from food crops, *Current Opinion in Biotechnology* 20 (2009): 191–196.
8. P. G. Lemaux, Genetically engineered plants and foods: A scientist's analysis of the issues (part I), *Annual Review of Plant Biology* 59 (2008): 771–812.
9. A. Catani, Benefits and concerns associated with biotechnology-derived foods: Can additional research reduce children health risks? *European Review for Medical and Pharmacological Science* 13 (2009): 41–50; M. K. Selgrade and coauthors, Safety assessment of biotechnology products for potential risk of food allergy: Implications of new research, *Toxicological Sciences* 110 (2009): 31–39.
10. H. I. Miller, The regulation of agricultural biotechnology: Science shows a better way, *Nature Biotechnology* 27 (2010): 628–634; A. C. Dubock, Crop conundrum, *Nutrition Reviews* 67 (2009): 17–20; J. A. Magaña-Gómez and A. M. Calderón de la Barca, Risk assessment of genetically modified crops for nutrition and health, *Nutrition Reviews* 67 (2009): 1–16.
11. M. A. Parry and M. J. Hawkesford, Food security: Increasing yield and improving resource use efficiency, *Proceedings of the Nutrition Society* 69 (2010): 592–600.
12. M. S. Swaminathan, Achieving food security in times of crisis, *Nature Biotechnology* 27 (2010): 453–460.
13. Position of the American Dietetic Association: Agricultural and food biotechnology, *Journal of the American Dietetic Association* 106 (2006): 285–293.

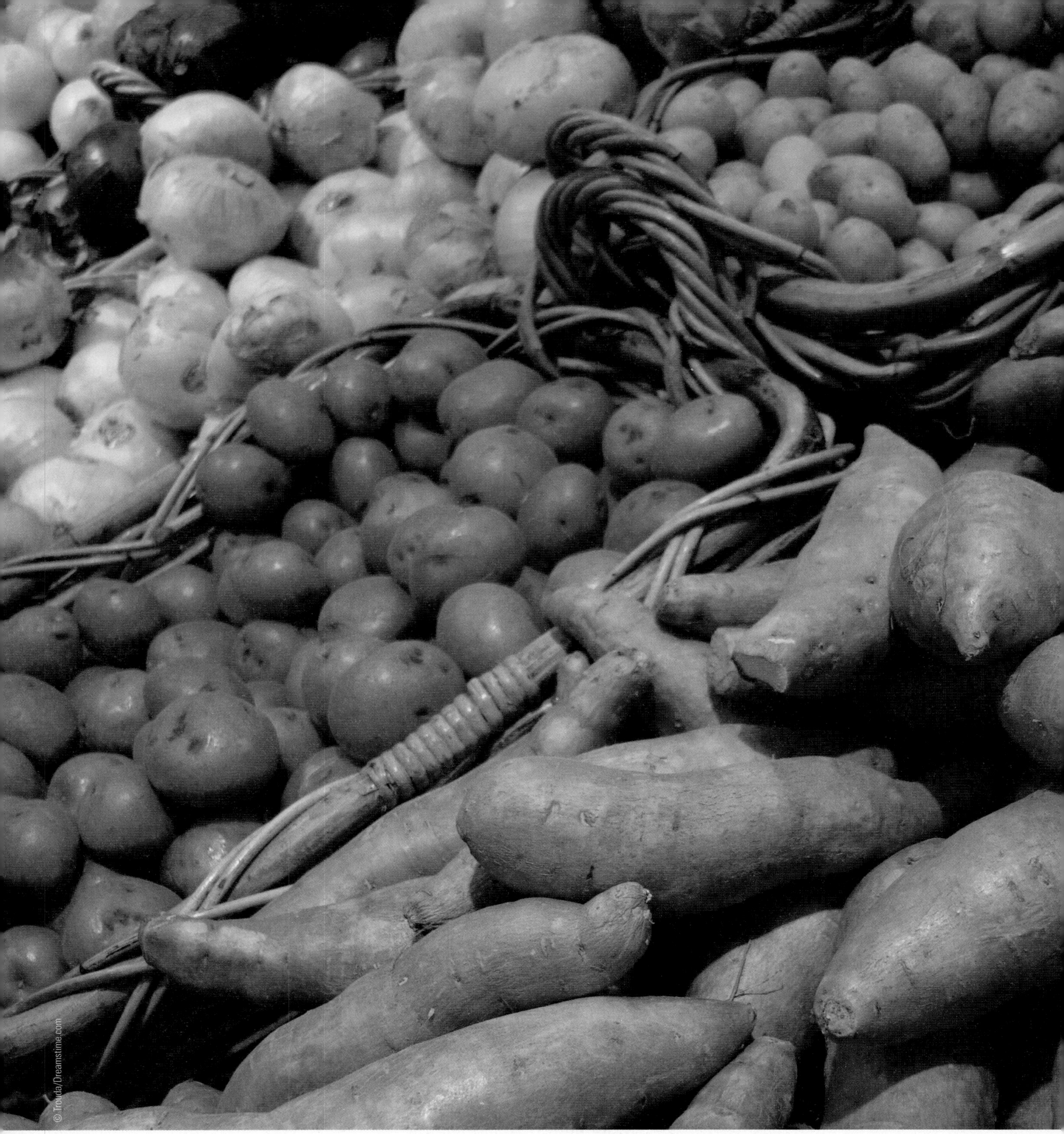

© Trouda/Dreamstime.com

CENGAGE LEARNING'S GLOBAL NUTRITION WATCH — GLOBAL NUTRITION RESOURCE CENTER

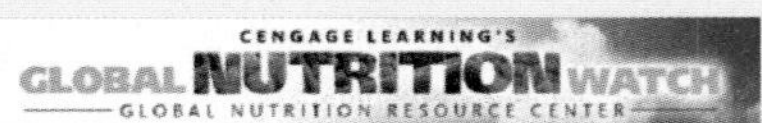

Environmental factors create food shortages around the globe. For the past decade, however, war and unrest have been the dominant causes of famine worldwide. Go to Food Safety/Food Sustainability → World Hunger/Sustainable Agriculture to review how politics have an increasing impact on global nutrition and access to resources. What other sociological elements can influence a person's access to food worldwide?

20 Hunger and the Global Environment

LEARNING GPS

Nutrition in Your Life

Imagine living with hunger from the moment you wake up until the time you thankfully fall asleep—and all through your dreams as well. Meal after meal, day after day, you have little or no food to eat. You know you need food, but you have no money. Would you beg on the street corner or go "dumpster diving" at the nearest fast-food restaurant? And then where would you find your next meal? How will you ever get enough to eat as long as you live in poverty? Resolving the hunger problem—whether in your community or on the other side of the world—depends on alleviating poverty and using resources wisely. In the Nutrition Portfolio at the end of this chapter, you can consider ways to get involved in hunger relief.

Worldwide, one person in every eight ♦ experiences persistent hunger—not the healthy appetite triggered by anticipation of a hearty meal, but the painful sensation caused by a lack of food.[1] In this chapter, **hunger** takes on the greater meaning—hunger that develops from prolonged, recurrent, and involuntary lack of food and results in discomfort, illness, weakness, or pain that exceeds the usual uneasy sensation. Such hunger deprives a person of the physical and mental energy needed to enjoy a full life and often leads to severe malnutrition and death. Tens of thousands of people die of hunger-related causes each day—one child every 5 seconds.

♦ 925 million of the world's 7 billion people (13%)

The enormity of the world hunger problem is reflected not only by huge numbers, but also by major challenges. As people populate and pollute the earth, resources become depleted, making food less available. Hunger and poverty, population growth, and environmental degradation are linked together; thus they tend to worsen one another. Because their causes overlap, so do their solutions; any initiative a person takes to help solve one problem will help solve others. Eliminating hunger requires a balance

hunger: consequence of food insecurity that, because of prolonged, involuntary lack of food, results in discomfort, illness, weakness, or pain that goes beyond the usual uneasy sensation.

among the distribution of food, the numbers of people, and the care of the environment.

Resolving the hunger problem may seem at first beyond the influence of one person. Can one person's choice to limit family size or to recycle a bottle or to volunteer at a food recovery program make a difference? ♦ In truth, such choices produce several benefits. For one, a person's action may influence many other people over time. For another, a repeated action becomes a habit, with compounded benefits. For still another, making choices with an awareness of the consequences gives a person a sense of personal control, hope, and effectiveness. The daily actions of many concerned people can help solve the problems of hunger in their own communities or on the other side of the world.

♦ "Never doubt that a small group of thoughtful, committed people can change the world. Indeed, it is the only thing that ever has."
—Margaret Mead

20.1 Hunger in the United States

LEARN IT Identify some reasons why hunger is present in a country as wealthy as the United States.

Ideally, all people at all times would have access to enough food to support an active, healthy life. In other words, they would experience **food security.** ♦ Unfortunately, more than 48 million people in the United States, including more than 16 million children, live in poverty and cannot afford to buy enough food to maintain good health.[2] Said another way, one out of seven households experiences hunger or the threat of hunger. Given the agricultural bounty and enormous wealth in this country, do these numbers surprise you? The limited or uncertain availability of nutritionally adequate and safe foods is known as **food insecurity** ♦ and is a major social problem in our nation today. Inadequate diets lead to poor health in adults and impaired physical, psychological, and cognitive development in children.

Table 20-1 presents the questions used in national surveys to identify food insecurity in the United States, and Figure 20-1 shows the most recent findings.[3]

♦ Food security categories:
- *High food security:* no indications of food-access problems or limitations
- *Marginal food security:* one or two indications of food-access problems but with little or no change in food intake

♦ Food insecurity categories:
- *Low food security:* reduced quality of life with little or no indication of reduced food intake; formerly known as *food insecurity without hunger*
- *Very low food security:* multiple indications of disrupted eating patterns and reduced food intake; formerly known as *food insecurity with hunger*

food security: access to enough food to sustain a healthy and active life.

food insecurity: limited or uncertain access to foods of sufficient quality or quantity to sustain a healthy and active life.

TABLE 20-1 Questions to Identify Food Insecurity in a US Household

To determine the extent of food insecurity in a household, surveys ask questions about behaviors and conditions known to characterize households having difficulty meeting basic food needs during the past 12 months. Most often, adults tend to protect their children from hunger. In the most severe cases, children also suffer from hunger and eat less.

1. Did you worry whether food would run out before you got money to buy more?
2. Did you find that the food you bought just didn't last and you didn't have money to buy more?
3. Were you unable to afford to eat balanced meals?
4. Did you or other adults in your household ever cut the size of your meals or skip meals because there wasn't enough food?
5. Did this happen in 3 or more months during the previous year?
6. Did you ever eat less than you felt you should because there wasn't enough money for food?
7. Were you ever hungry but didn't eat because you couldn't afford enough food?
8. Did you ever lose weight because you didn't have enough money to buy food?
9. Did you or other adults in your household ever not eat for a whole day because you were running out of money to buy food?
10. Did this happen in 3 or more months during the previous year?
11. Did you rely on only a few kinds of low-cost food to feed your children because you were running out of money to buy food?
12. Were you unable to feed your children a balanced meal because you couldn't afford it?
13. Were your children not eating enough because you just couldn't afford enough food?
14. Did you ever cut the size of your children's meals because there wasn't enough money for food?
15. Were your children ever hungry but you just couldn't afford enough food?
16. Did your children ever skip a meal because there wasn't enough money for food?
17. Did this happen in 3 or more months during the previous year?
18. Did your children ever not eat for a whole day because there wasn't enough money for food?

The more positive responses, the greater the food insecurity. Households with children answer all of the questions and are categorized as follows:

≤2 positive responses	=	food secure
3–7 positive responses	=	low food security
≥8 positive responses	=	very low food security

Households without children answer the first 10 questions and are categorized as follows:

≤2 positive responses	=	food secure
3–5 positive responses	=	low food security
≥6 positive responses	=	very low food security

Figure 20-1 shows the results of the 2010 surveys.

SOURCE: US Department of Agriculture, *Household Food Security in the United States, 2008,* www.ers.usda.gov.

> FIGURE 20-1 **Prevalence of Food Security in US Households, 2010**

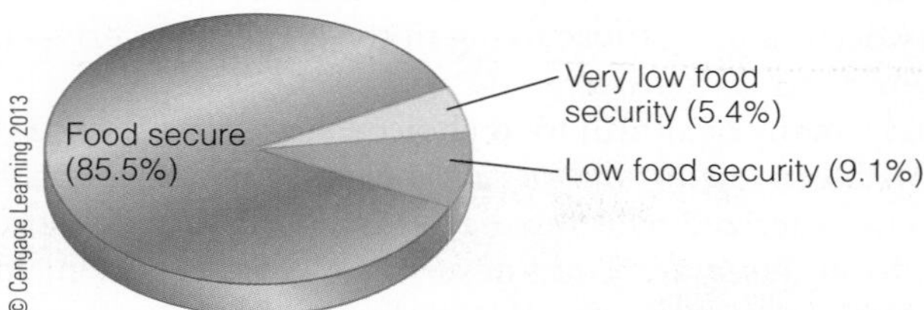

SOURCE: Economic Research Service, U.S. Department of Agriculture, www.ers.usda.gov, updated September 2011.

Responses to these questions provide crude, but necessary, data to estimate the degree of hunger in this country. Specific questions and measures focus on food insecurity in children. Even short-term adult food insecurity can interfere with a young child's development.[4]

Defining Hunger in the United States At its most extreme, people experience hunger because they have absolutely no food. More often, they have too little food (**food insufficiency**) and try to stretch their limited resources by eating small meals or skipping meals—often for days at a time. Sometimes hungry people obtain enough food to satisfy their hunger, perhaps by seeking food assistance or finding food through socially unacceptable ways—begging from strangers, stealing from markets, or scavenging through garbage cans, for example. Sometimes obtaining food raises concerns for food safety—for example, when rot, slime, mold, or insects have damaged foods or when people eat others' leftovers or meat from roadkill.

Food Poverty Hunger has many causes, but in developed countries, the primary cause is **food poverty.** People are hungry not because there is no food nearby to purchase, but because they lack money. The rate and severity of US poverty has increased over the past decade. An estimated 15 percent of the people in the United States lives in poverty.[5] Even those above the poverty line ♦ may not have food security. Physical and mental illnesses and disabilities, unemployment, low-paying jobs, unexpected or ongoing medical expenses, and high living expenses threaten financial stability. When money is tight, people are forced to choose between food and life's other necessities—utilities, housing, and medical care. Food costs are more variable and flexible; people can purchase fewer groceries to lower the monthly food bill, but they usually can't pay only a portion of the bills for electricity, rent, or medication. Other problems further contribute to food poverty, such as abuse of alcohol and other drugs; lack of awareness of available food assistance programs; and the reluctance of people, particularly the elderly, to accept what they perceive as "welfare" or "charity." Lack of resources remains the major cause of food poverty in developed countries, and solving this problem would do a lot to relieve hunger.

♦ The US poverty line for an individual is about $11,000 a year.

In the United States, poverty and hunger reach across various segments of society, touching some more than others—notably, single parents living in households with their children, Hispanics and African Americans, and those living in the inner cities. People living in poverty are simply unable to buy sufficient amounts of nourishing foods, even if they are wise shoppers. Consequently, their diets tend to be inadequate. For many of the children in these families, school lunch (and breakfast, where available) may be the only nourishment for the day. Otherwise they go hungry, waiting for an adult to find money for food. Not surprisingly, these children are more likely to have health problems and iron-deficiency anemia than those who eat regularly.[6] They also tend to perform poorly in school and in social situations. For adults, the risk of developing chronic diseases increases.[7] For pregnant women, the risk of developing gestational diabetes more than doubles.[8]

Feeding the hungry—in the United States.

food insufficiency: an inadequate amount of food due to a lack of resources.

food poverty: hunger resulting from inadequate access to available food for various reasons, including inadequate resources, political obstacles, social disruptions, poor weather conditions, and lack of transportation.

Obesity Paradox Ironically, hunger and obesity often exist side by side—sometimes within the same household or even the same person.[9] That hunger

> **FIGURE 20-2 The Poverty-Obesity Paradox**

Poverty → Hunger; Food insecurity

Hunger → Inadequate intake of energy, protein, vitamins, and minerals → Malnutrition

Food insecurity → Inadequate intake of energy, protein, vitamins, and minerals; Excessive intake of energy, fat, and sugar → Obesity

reflects an inadequate food intake and obesity implies an excessive intake seems paradoxical, but research studies have confirmed the association.[10] The highest rates of obesity occur among those living in the greatest poverty—the same people who live with food insecurity.[11]

Unfortunately, many healthful food choices, such as fruits and vegetables, are not readily available in low-income or rural neighborhoods.[12] Neighborhoods and communities characterized by limited access to nutritious and affordable foods are known as **food deserts.**[13] Food deserts are more prevalent in low-income and African American communities.[14] With limited access to grocery stores carrying varieties of fruits and vegetables, residents in these neighborhoods fall short of meeting dietary guidelines.[15] Furthermore, fruits and vegetables tend to cost more than the energy-dense foods that foster weight gain. Foods such as doughnuts, pizzas, and hamburgers provide the most energy and satiety for the least cost. Quite simply, poor-quality diets deliver more kcalories, but fewer nutrients, for less money; high-quality diets deliver fewer kcalories, but more nutrients, for more money.[16]

Economic uncertainty and stress greatly influence the prevalence of obesity.[17] People who are unsure about their next meal may overeat when food or money are available. Interestingly, food insecure people who do *not* participate in food assistance programs have a *greater* risk of obesity than those who do participate—illustrating that providing food actually helps to prevent obesity.[18] Figure 20-2 shows how poverty and food insecurity can lead to both malnutrition and obesity.

Relieving Hunger in the United States

The Academy of Nutrition and Dietetics calls for aggressive action to bring an end to domestic food insecurity and hunger and to achieve food and nutrition security for everybody living in the United States.[19] Many federal and local programs aim to prevent or relieve malnutrition and hunger in the United States.

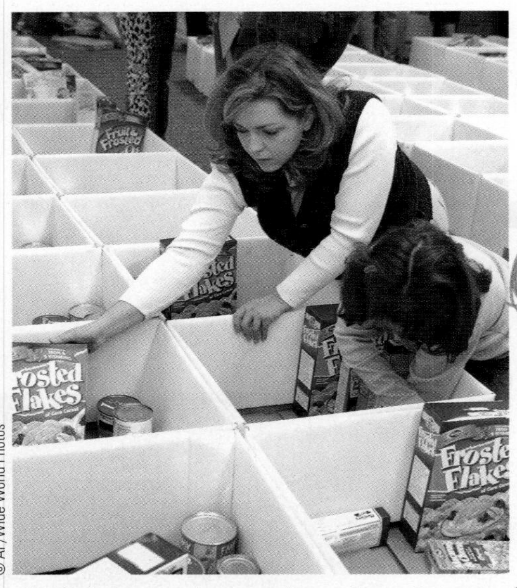

The fight against hunger depends on the helping hands of caring volunteers.

Federal Food Assistance Programs

Adequate nutrition and food security are essential to supporting good health and achieving the public health goals of the United States. To that end, an extensive network of federal assistance programs provides life-giving food to millions of US citizens daily. An estimated one out of every five Americans receives some kind of food assistance at a total cost of more than $90 billion per year. Even so, the programs are not fully successful in preventing hunger, but they do seem to improve the nutrient intakes of those who participate. Programs described in earlier chapters include the WIC program for low-income pregnant women, breastfeeding mothers, and their young children (Chapter 15); the school lunch, breakfast, and child-care food programs for children (Chapter 16); and the food assistance programs for older adults such as congregate meals and Meals on Wheels (Chapter 17).

The Supplemental Nutrition Assistance Program (SNAP), ♦ administered by the US Department of Agriculture (USDA), is the largest of the federal food assistance programs, both in amount of money spent and in number of people served. It provides assistance to more than 40 million people at a cost of more than $68 billion per year; about half of the recipients are children. The USDA issues debit cards through state agencies to households—people who buy and prepare food together. The amount a household receives depends on its size, resources, and income. The average monthly benefit is about $134 per person. Recipients may use the cards to purchase food and food-bearing plants and seeds, but not to buy tobacco, cleaning items, alcohol, or other nonfood items. The accompanying "How To" feature offers shopping tips for those on a limited budget.

♦ SNAP is the new name for the Food Stamp Program.

Food assistance programs improve nutrient intakes significantly, but hunger continues to plague the United States. Of the estimated 3.6 million homeless people in the United States who are eligible, only 15 percent of single adults and 50 percent of families receive food assistance. For some, reading, understanding, and completing the application can be difficult. For others, having to show identification and proof of homelessness can be frustrating. For many, accepting hunger is simply easier than meeting these challenges.

food deserts: neighborhoods and communities characterized by limited access to nutritious and affordable foods.

>How To

Plan Healthy, Thrifty Meals

Chapter 2 introduces the principles for planning a healthy diet. Meeting that goal on a limited budget adds to the challenge. To save money and spend wisely, plan and shop for healthy meals with the following tips in mind:

Planning

- Make a grocery list before going to the store to avoid expensive "impulse" items.
- Do not shop when hungry.
- Use leftovers.
- Center meals on rice, noodles, and other grains.
- Use small quantities of meat, poultry, fish, or eggs.
- Use legumes instead of meat, poultry, fish, or eggs several times a week.
- Use cooked cereals such as oatmeal instead of ready-to-eat breakfast cereals.
- Cook large quantities when time and money allow.
- Check for sales and use coupons for products you need; plan meals to take advantage of sale items.

Shopping

- Look for bargains on day-old bread and other bakery products.
- Select whole foods instead of convenience foods (potatoes instead of instant mashed potatoes, for example).
- Try store brands.
- Buy fresh produce that is in season; buy canned or frozen items at other times.
- Buy only the amount of fresh foods that you will eat before it spoils.
- Buy large bags of frozen items or dry goods; when cooking, take out the amount needed and store the remainder.
- Buy fat-free dry milk; mix and refrigerate quantities needed for a day or two. Buy fresh milk by the gallon or half-gallon.
- Buy less expensive cuts of meat. Chuck and bottom round roast are usually inexpensive; cover during cooking and cook long enough to make meat tender. Buy whole chickens instead of pieces.
- Compare the unit price (cost per ounce, for example) of similar foods so that you can select the least expensive brand or size.
- Buy nonfood items such as toilet paper and laundry detergent at discount stores instead of grocery stores.

For daily menus and recipes for healthy, thrifty meals, visit the USDA Center for Nutrition Policy and Promotion: **www.cnpp.usda.gov.**

TRY IT Search for "thrifty meals" at the USDA website (**www.cnpp.usda.gov**) and select a day's meals to analyze using your personal profile in a diet analysis program.

National Food Recovery Programs Efforts to resolve the problem of hunger in the United States do not depend solely on federal assistance programs. National **food recovery** ♦ programs have made a dramatic difference. The largest program, Feeding America, coordinates the efforts of more than 61,000 **food pantries, emergency shelters,** and **soup kitchens** that feed an estimated 37 million people a year.

♦ Four common methods of food recovery:
- *Field gleaning:* collecting crops from fields that either have already been harvested or are not profitable to harvest
- *Perishable food rescue or salvage:* collecting perishable produce from wholesalers and markets
- *Prepared food rescue:* collecting prepared foods from commercial kitchens
- *Nonperishable food collection:* collecting processed foods from wholesalers and markets

Each year, an estimated one-third of the world's food supply is wasted along the way from farm to final consumption. Consumers in wealthy nations such as the United States and Canada waste an estimated 200 to 250 pounds of food per person per year.[20] Food recovery programs collect and distribute good food that would otherwise go to waste. Volunteers might pick corn left in an already harvested field, a grocer might deliver ripe bananas to a local **food bank,** and a caterer might take leftover chicken salad to a community shelter, for example. All of these efforts help to feed the hungry in the United States.

Community Efforts Food recovery programs depend on volunteers. Concerned citizens work through local agencies and churches to feed the hungry. Community-based food pantries provide groceries, and soup kitchens serve prepared meals. Meals often deliver adequate nourishment, but most homeless people receive fewer than one and a half meals a day, so many are still inadequately nourished. A combination of various strategies helps to build food security in a community.

REVIEW IT Identify some reasons why hunger is present in a country as wealthy as the United States.

Food insecurity and hunger are widespread in the United States among those living in poverty. Ironically, hunger and poverty coexist with obesity. Government assistance programs help to relieve poverty and hunger. Food recovery programs and other community efforts also provide some relief.

food recovery: collecting wholesome food for distribution to low-income people who are hungry.

food pantries: programs that provide groceries to be prepared and eaten at home.

emergency shelters: facilities that are used to provide temporary housing.

soup kitchens: programs that provide prepared meals to be eaten on site.

food bank: a facility that collects and distributes food donations to authorized organizations feeding the hungry.

20.2 World Hunger

LEARN IT Identify some reasons why hunger is present in the developing countries of the world.

As distressing as hunger is in the United States, the prevalence is far greater and the consequences more severe in developing countries. Although hunger in developing countries has diverse causes, once again, the primary cause is poverty, and the poverty is far more extreme than in the United States. ♦ Most people cannot grasp the severity of poverty in the developing world. Of the world's 7 billion people, 25 percent have no land and no possessions *at all.* They are the "poorest poor." They survive on little more than $1 a day each, and they lack safe housing, clean water, and health care. They cannot read or write. The average US housecat receives twice as much protein every day as one of these people, and the cost of keeping that cat is greater than such a person's annual income.

♦ The international poverty line for an individual is $1.25 a day or about $456 a year.

The "poorest poor" are usually female. Many societies around the world undervalue females, providing girls with poorer diets and fewer opportunities than boys. Malnourished girls become malnourished mothers who give birth to low-birthweight infants—and the cycle of hunger, malnutrition, and poverty continues.

Not only does poverty cause hunger, but tragically, hunger worsens poverty by robbing a person of the good health and the physical and mental energy needed to be active and productive. Hungry people simply cannot work hard enough to get themselves out of poverty. Providing nourishment is a necessary investment in the well-being of both individuals and nations.[21] Economists calculate that cutting world hunger and malnutrition in half by 2015 would generate a value of more than $120 billion in longer, healthier, and more productive lives.

Food Shortages World hunger brings to mind victims of **famine,** a severe food shortage in an area that causes widespread starvation and death. In recent years, the natural causes of famine—drought, flood, and pests—have become less important than the political and economic crises created by people.[22] Figure 20-3 shows the hunger hotspots in the world.

Political Turbulence A sudden increase in food prices, a drop in workers' incomes, or a change in government policy can quickly leave millions hungry. An estimated 30 million people died during the Chinese famine of 1959 through 1961, the worst famine of the 20th century. The main cause was government policies associated with the Great Leap Forward, a government initiative that was intended to transform China's economy. However, the poorly planned communal farm system and the widespread waste of resources devastated the Chinese agricultural system.

The famine Somalia is currently experiencing has left an estimated 12 million people starving and more than 30,000 children younger than the age of 5 dead. Drought has clearly contributed to the crisis, but political factors are the primary cause of this famine. Without an effective government for the past decade, Somalia is plagued by political unrest and terrorist activities.

International efforts cannot fully relieve the hunger and poverty in Somalia and other parts of the world.

Armed Conflicts Armed conflict is another cause of famine worldwide. In times of war, farmers become warriors, their agricultural fields become battlegrounds, the citizens go hungry, and the warring factions often block famine relief. The world continues to struggle to find a middle ground between respecting the sovereignty of nations and insisting that all nations allow humanitarian assistance to reach the people. When supplementary food programs reach the people in war-torn countries, the children benefit.

Natural Disasters Natural disasters and other poor weather conditions create food shortages. In 2011, an earthquake and flooding in Japan dramatically reduced food supplies. During such natural disasters, emergency food relief

famine: widespread and extreme scarcity of food in an area that causes starvation and death in a large portion of the population.

> FIGURE 20-3 Hunger Hot Spots

Hunger is prevalent in the developing world, with some countries reporting hunger and malnutrition in more than half of their population.

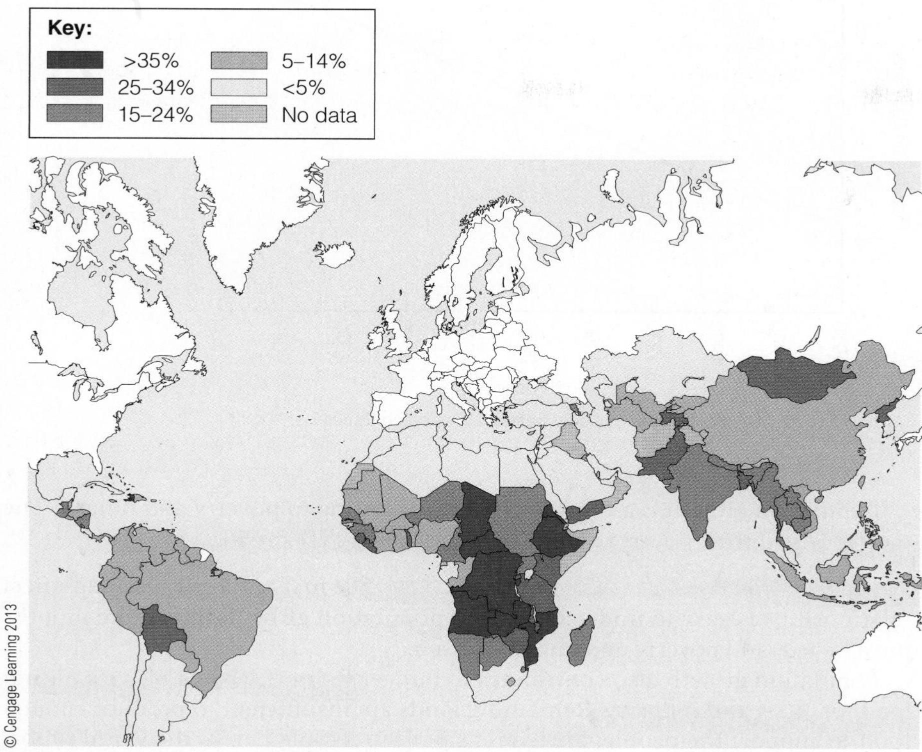

SOURCE: Food and Agriculture Organization of the United Nations, Food Security Statistics, 2010, www.fao.org/economic/ess/ess-fs/en.

from countries around the world provides a safety net for countries in need. International food relief programs also provide ongoing food assistance to countries such as South Sudan that are chronically short of food because of ongoing drought and poverty.

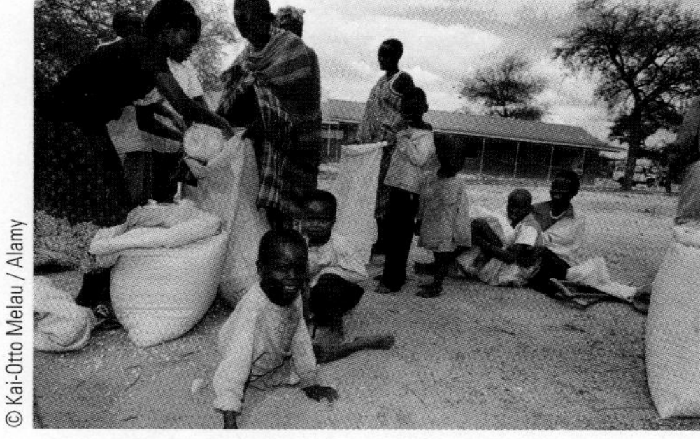

Feeding the hungry—in Kenya.

Poverty and Overpopulation Future demands for food require multiple solutions, including controlling overpopulation. The world's population is rising at an alarming rate, as Figure 20-4 (p. 666) shows. Skyrocketing numbers threaten earth's capacity to provide safe water and adequate food for its inhabitants.* Contaminated water and food shortages are responsible for much of the world's disease and death.

The sheer magnitude of the world's annual population increase of almost 83 million people is difficult to comprehend. Every minute, the world's population increases by another 158 people. In less than 6 months, the world adds the equivalent of another California.

As the world's population continues to grow, much of the increase is occurring in developing countries where hunger and malnutrition are already widespread. More people sharing the little food available can only worsen the problem. Stabilizing the population is essential if food production is ever going to be able to keep up with demands. Without population stabilization, the world can neither support the lives of people nor halt environmental degradation. Before the population problem can be resolved, it may be necessary to remedy the poverty problem. In countries around the world, economic growth has been accompanied by slowed population growth.

---

*The maximum number of people earth can support over time is its *human carrying capacity.*

> FIGURE 20-4 World Population

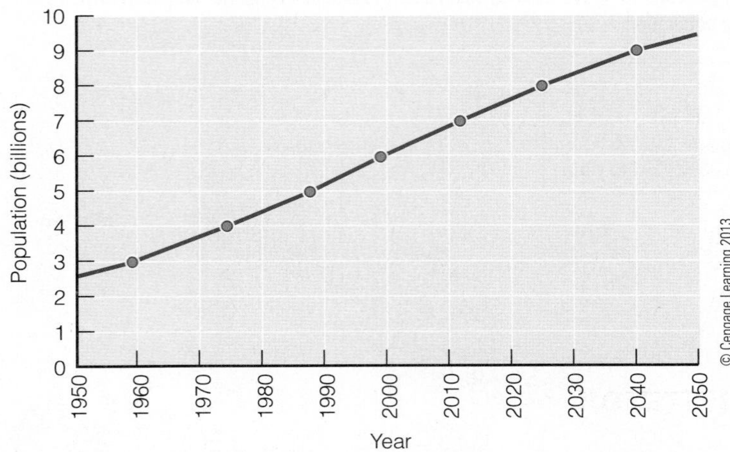

SOURCE: US Census Bureau, International Data Base, updated December 2008.

Population growth is a central factor contributing to poverty and hunger. The reverse is also true: poverty and hunger contribute to population growth.

Population Growth Leads to Hunger and Poverty The first of these cause-and-effect relationships is easy to understand. As a population grows larger, more mouths must be fed, and poverty and hunger worsen.

Population growth also contributes to hunger indirectly by using agricultural land for cities and industry. Remaining lands are insufficient to produce enough food to support the people. The world's poorest people live in the world's most damaged and inhospitable environments.

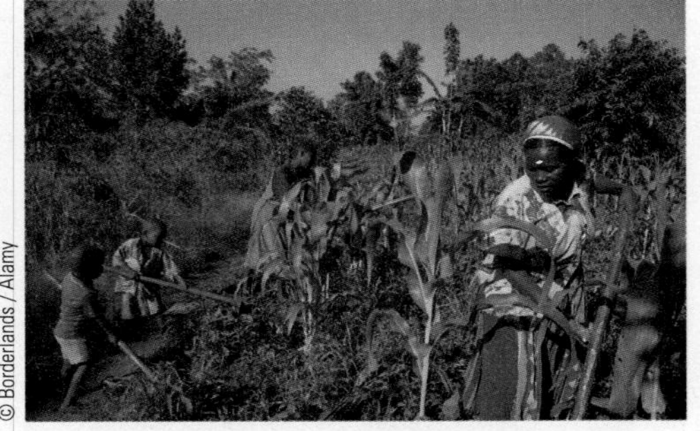

Families in developing countries depend on their children to help provide for daily needs.

Hunger and Poverty Lead to Population Growth How does poverty lead to overpopulation? Poverty and its consequences—inadequate food and shelter—leave women vulnerable to physical abuse, forced marriages, and prostitution. Furthermore, they lack access to reproductive health care and family planning counseling. Also, in some regions of the world, families depend on children to farm the land, haul water, and care for adults in their old age. For these families, children are an economic asset. Poverty claims many young children, who are among the most likely to die from malnutrition and disease. If a family faces ongoing poverty, the parents often choose to have many children to ensure that some will survive to adulthood. People are willing to risk having fewer children only if they are sure that their children will survive to adulthood and that the family can develop other economic assets (such as skills, businesses, and land).

Breaking the Cycle Relieving poverty and hunger may be a necessary first step in curbing population growth. When people attain better access to health care, education, and family planning, the death rate falls. At first, births outnumber deaths, but as the standard of living continues to improve, families become willing to risk having fewer children. Then the birth rate falls. Thus improvements in living standards help stabilize the population.

The link between improved economic status and slowed population growth has been demonstrated in several countries. Central to achieving this success is sustainable development that includes not only economic growth, but a sharing of resources among all groups. Where this has happened, population growth has slowed the most: in parts of Sri Lanka, Taiwan, Malaysia, and Costa Rica, for example. Where economic growth has occurred but only the rich have grown richer, population growth has remained high. Examples include Brazil, the Philippines, and Thailand, where large families continue to be a major economic asset for the poor.

As a society gains economic footing, education also becomes a higher priority. A society that educates its children, both males and females, experiences a drop in birth rates. Education, particularly for girls and women, brings improvements in family life, including improved nutrition, better sanitation, effective birth control, and elevated social and economic status. With improved conditions, more infants live to adulthood, making smaller families feasible.

REVIEW IT Identify some reasons why hunger is present in the developing countries of the world.

Natural causes such as drought, flood, and pests and political causes such as armed conflicts and government policies all contribute to the extreme hunger and poverty seen in the developing countries. In addition, overpopulation means more mouths to feed, which worsens the problems of poverty and hunger. Poverty and hunger, in turn, encourage parents to have more children to help support the family. Breaking this cycle requires improving the economic status of the people and providing them with health care, education, and family planning.

20.3 Malnutrition

LEARN IT Describe the consequences of nutrient and energy inadequacies.

Persistent hunger inevitably leads to malnutrition. Although malnutrition touches many adult lives, it most often strikes early in childhood and results in specific nutrient deficiencies or overall growth failure.

Nutrient Deficiencies A child suffering from a nutrient deficiency may continue to grow, but without adequate nourishment, body functions begin to fail and signs of deficiency diseases become apparent. Problems resulting from nutrient deficiencies include birth defects, learning disabilities, mental retardation, impaired immunity, blindness, incapacity to work, and premature deaths. Nutrients most likely to be deficient are iron, iodine, vitamin A, and zinc.[23]* The prevalence and consequences of these nutrient deficiencies stagger the mind. More than 30 percent of the world's population have iron-deficiency anemia, a leading cause of maternal deaths, preterm births, low birthweights, infections, and premature deaths. Iodine deficiency affects one out of seven, resulting in stillbirths and irreversible mental retardation (cretinism) in 40 million newborns every year. More than 100 million children (younger than age 5) suffer from symptoms of vitamin A deficiency—blindness, growth retardation, and poor resistance to common childhood infections such as measles. An estimated 20 percent of the world's population risk zinc deficiency, which contributes to growth failure, diarrhea, and pneumonia. (The deficiency symptoms of these nutrients and those of the other vitamins and minerals are presented in Chapters 10 through 13.)

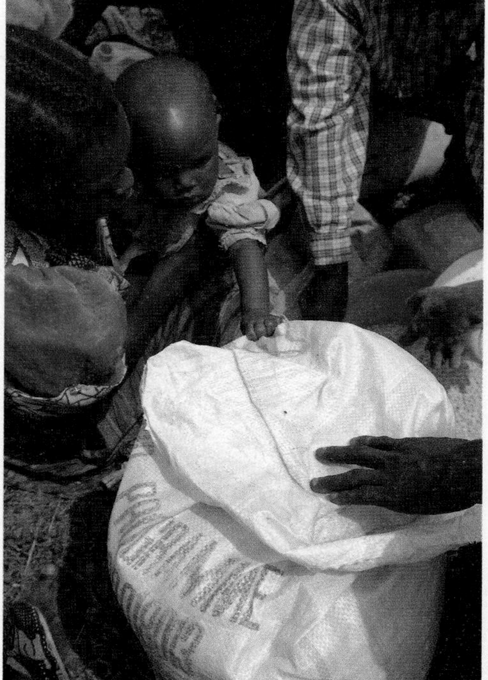

Donated food may temporarily ease hunger for some, but it is usually insufficient to prevent nutrient deficiencies or support growth.

The consequences of nutrient deficiencies are felt not only by individuals, but by entire nations. When people suffer from mental retardation, blindness, infections, and other consequences of malnutrition, the economy of their country declines as productivity decreases and health-care costs increase. The dramatic signs of malnutrition are most evident at each end of the life span in a nation's high infant mortality rate and short life expectancy.

Growth Failure In addition to specific nutrient deficiencies, inadequate food intake leads to poor growth in children. One child in six worldwide is born underweight, and one in four is underweight by the age of 5. These underweight children are malnourished and readily develop the diseases of poverty: parasitic and infectious diseases that cause diarrhea (dysentery and cholera), acute respiratory illnesses (pneumonia and whooping cough), measles, and malaria. The synergistic combination of infectious disease and malnutrition dramatically increases the likelihood of early death.[24] Compared with adequately nourished children, the risk of death is 2.5 times greater for children with mild malnutrition, 4.6 times greater for children with moderate malnutrition, and 8.4 times greater for children with severe malnutrition. Each year, 7.6 million children

*Nutrients associated with deficiencies that have characteristic clinical symptoms are sometimes referred to as Type I nutrients. Examples include iron (anemia), vitamin C (scurvy), and iodine (cretinism).

TABLE 20-2 Acute and Chronic Malnutrition Compared

	Acute Malnutrition	Chronic Malnutrition
Food deprivation	Current or recent	Long-term
Physical features	Rapid weight loss Wasting (underweight for height) Edema	Minimal height gains Stunting (short for age)
World prevalence of children younger than age 5	5 to 15%	20 to 50%
Clinical types	Kwashiorkor	Marasmus

younger than the age of 5 die; as many as 5 children *every minute* die as a result of hunger and malnutrition.[25] Most of them do not starve to death—they die from the diarrhea and dehydration that accompany infections.

Poor growth due to malnutrition is easy to overlook because a small child may look quite normal, but it is the most common sign of malnutrition. Growth may be impaired in two ways. Children who are suffering from **acute malnutrition** (recent severe food deprivation) may be underweight for their height (described as *wasting*). By comparison, children who have experienced **chronic malnutrition** (long-term food deprivation) are short for their age (described as *stunting*). An estimated 25 percent of the world's children younger than the age of 5 are stunted and about 10 percent are wasted.[26] Table 20-2 compares key features of acute malnutrition with those of chronic malnutrition.

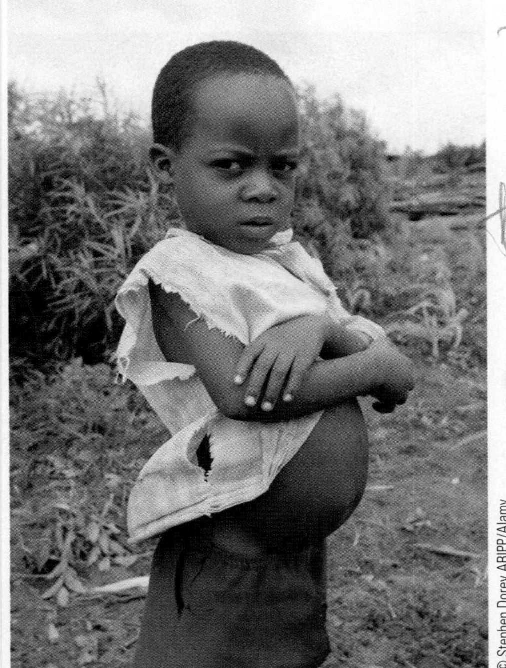

The edema characteristic of kwashiorkor is apparent in this child's swollen belly. Malnourished children commonly have an enlarged abdomen from parasites as well.

Historically, the wasting form of malnutrition, known as **kwashiorkor,** was attributed to protein deficiency. The stunting form of malnutrition, known as **marasmus,** was attributed to energy deficiency. Together, they were named *protein-energy malnutrition*. Traditional thinking about protein-energy malnutrition was challenged by findings that the diets of children with marasmus or kwashiorkor do not differ. Furthermore, treatment based on the concepts of protein-energy malnutrition failed to restore physiological homeostasis or healthy body composition.[27]

Current understanding of these two types of malnutrition—kwashiorkor and marasmus—recognizes that the primary cause of wasting and stunting is a deficiency of *many* nutrients, including amino acids, potassium, magnesium, zinc, and phosphorus.* Such poor-quality diets result in loss of appetite, diminished growth, and inability to resist infection or respond to environmental stresses. The following paragraphs present descriptions of the wasting associated with kwashiorkor and acute malnutrition and the stunting associated with marasmus and chronic malnutrition.

Wasting of Kwashiorkor Kwashiorkor typically reflects a sudden and recent deprivation of food (acute malnutrition). *Kwashiorkor* is a Ghanaian word that refers to the birth position of a child and is used to describe the illness a child develops when the next child is born. When a mother who has been nursing her first child bears a second child, she weans the first child and puts the second one on the breast. The first child, suddenly switched from nutrient-dense, protein-rich breast milk to a starchy, protein-poor cereal, soon begins to sicken and die. The child appears withdrawn or irritable and obviously ill. Loss of appetite interferes with any attempts to provide nourishment.

In kwashiorkor, some muscle wasting may occur, but it may not be apparent because the child's face, limbs, and abdomen become swollen with edema—a distinguishing feature of kwashiorkor. Fluid balance shifts in response to decreased concentrations of the blood protein albumin. A fatty liver develops due to a lack of the protein carriers to transport lipids out of the liver. The fatty liver lacks enzymes to clear metabolic toxins from the body, so their harmful effects

acute malnutrition: malnutrition caused by recent severe food restriction; characterized in children by underweight for height (*wasting*).

chronic malnutrition: malnutrition caused by long-term food deprivation; characterized in children by short height for age (*stunting*).

kwashiorkor (kwash-ee-OR-core or kwash-ee-or-CORE): severe malnutrition characterized by failure to grow and develop, edema, changes in the pigmentation of hair and skin, fatty liver, anemia, and apathy.

marasmus (ma-RAZ-mus): severe malnutrition characterized by poor growth, dramatic weight loss, loss of body fat and muscle, and apathy.

*Nutrients associated with deficiencies that impair growth, but have no characteristic clinical symptoms, are sometimes referred to as Type II nutrients. Examples include essential amino acids, potassium, sodium, magnesium, zinc, and phosphorus.

are prolonged. Inflammation in response to these toxins and to infections further contributes to the edema that accompanies kwashiorkor. Without sufficient tyrosine to make melanin, hair loses its color, and inadequate protein synthesis leaves the skin patchy and scaly, often with sores that fail to heal. The lack of proteins to carry or store iron leaves iron free. Free iron is common in kwashiorkor and may contribute to edema by increasing secretion of ADH—the antidiuretic hormone responsible for water retention.[28] In addition, iron may contribute to illness and death by promoting bacterial growth and free-radical damage. (Free-radical damage is discussed fully in Highlight 11.)

Stunting of Marasmus Appropriately named from the Greek word meaning "dying away," marasmus reflects a severe deprivation of food over a long time (chronic malnutrition). Marasmus occurs most commonly in children in all the overpopulated and impoverished areas of the world. Children living in poverty simply do not have enough to eat. They subsist on diluted cereal drinks that supply scant energy and protein of low quality; such food can barely sustain life, much less support growth. The loose skin on the buttocks and thighs sags down and looks as if the child is wearing baggy pants. Sadly, children with marasmus are often described as just "skin and bones."

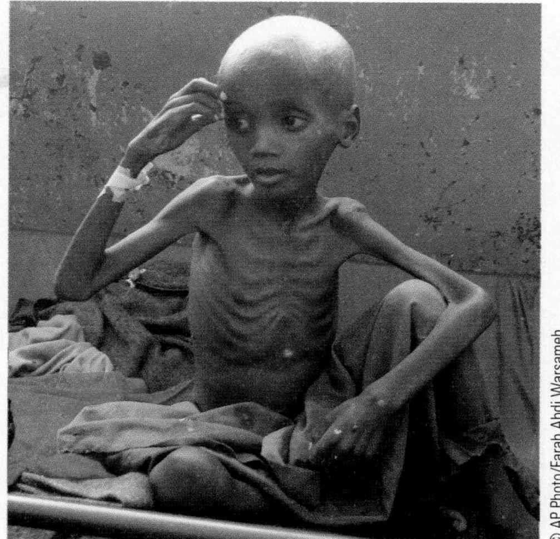
© AP Photo/Farah Abdi Warsameh

The severe wasting characteristic of marasmus is apparent in this child's "matchstick" arms.

Because the brain normally grows to almost its full adult size within the first 2 years of life, marasmus impairs brain development and learning ability. Reduced synthesis of key hormones slows metabolism and lowers body temperature. There is little or no fat under the skin to insulate against cold. Hospital workers find that children with marasmus need to be clothed, covered, and kept warm. Because these children often suffer delays in their mental and behavioral development, they also need loving care, a stimulating environment, and parental attention.

The starving child faces this threat to life by engaging in as little activity as possible—not even crying for food. The body musters all its forces to meet the crisis, so it cuts down on any expenditure of energy not needed for the functioning of the heart, lungs, and brain. Growth ceases; the child is no larger at age 4 than at age 2. Enzymes are in short supply, and the GI tract lining deteriorates. Consequently, what little food is eaten can't be digested and absorbed.

Rehabilitation

Ideally, optimal breastfeeding and improved complementary feedings would prevent malnutrition and save the lives of children. When mild to moderate malnutrition did occur, it might be quickly remedied with supplemental foods in the community. Severe acute malnutrition, on the other hand, requires hospitalization, which demands intensive nursing care, diet, and medication.[29]

Because the causes of malnutrition extend beyond insufficient food to include an individual's genetics, environmental stresses, and gastrointestinal bacteria, treatment must be individualized.[30] Providing foods fortified with multiple vitamins and minerals protects against some deficiency diseases; whether it improves illnesses, growth, or cognition is less clear.[31] To support physical growth, mental development, metabolic balance, and recovery from illnesses, malnourished children need specially formulated diets.[32]

To ensure rapid weight gain and correct nutrient deficiencies, children suffering from uncomplicated, but severe acute malnutrition may be given ready-to-use therapeutic food (RUTF)—a paste made of peanut butter and powdered milk and fortified with vitamins and minerals. Because it does not need to be mixed with water, the risk of bacterial contamination is greatly minimized. Another benefit is that it can be stored for 3 to 4 months without refrigeration. Whether these therapeutic foods should be used for the prevention of childhood malnutrition is currently being considered.[33] Limited research suggests that providing RUTF to young children threatened by malnutrition reduces the incidence of wasting.[34]

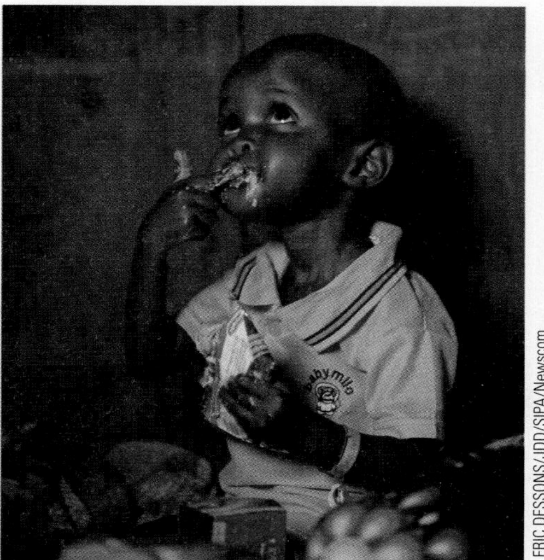
© ERIC DESSONS/JDD/SIPA/Newscom

Ready-to-use therapeutic food (RUTF)—made of peanut butter and powdered milk fortified with vitamins and minerals—nourishes hundreds of thousands of malnourished children each year.

In addition to nutrition intervention, children suffering from diarrhea commonly need rehydration. In severe cases, diarrhea will have incurred dramatic fluid and mineral losses that need to be replaced immediately to help raise the blood pressure and strengthen the heartbeat. Health-care workers around the world save millions of lives each year by effectively reversing dehydration and

◆ To prevent death from diarrheal disease, provide:

- Adequate sanitation
- Safe water
- Oral rehydration therapy

correcting the diarrhea ◆ with **oral rehydration therapy (ORT).** ORT is a simple, inexpensive, and effective treatment that consists of giving a sugar and salt solution orally.

REVIEW IT Describe the consequences of nutrient and energy inadequacies.
Hunger leads to malnutrition, which appears most evident in nutrient deficiencies and growth failure. Children suffering from acute malnutrition (recent severe food deprivation) may be underweight for their height, while those experiencing chronic malnutrition (long-term food deprivation) are short for their age. Problems resulting from nutrient deficiencies include preterm births and low birth weights (iron), stillbirths and cretinism (iodine), blindness (vitamin A), and growth failure (zinc). Treatment should be individualized to ensure rapid weight gain and correct nutrient deficiencies.

20.4 The Global Environment

LEARN IT Explain why relieving environmental problems will also help to alleviate hunger and poverty.

◆ World agriculture produces enough food to provide each person with 2720 kcal/day. Food consumption and waste in the United States averages 3747 kcal/day/person.

The world currently produces enough food to feed all of its people, ◆ and the problem of hunger today remains a problem of unequal distribution of land to grow crops or income to purchase foods. If present trends continue, however, the time is fast approaching when there will be an absolute deficit of food. This conclusion seems inescapable. The world's increasing population threatens the world's capacity to produce adequate food. Until the nations of the world resolve the population problem, they can neither support the lives of people already born nor remedy global trends toward environmental deterioration. And to resolve the population problem, a necessary first step is to remedy the poverty problem, for reasons already discussed. Of the almost 83 million people being added to the population each year, more than 95 percent are born in the most poverty-stricken areas of the world.

In recent years, even as crop yields set new record highs, more people needed more food and reserves dwindled. Food prices skyrocketed as the costs of energy and fertilizer increased dramatically. Hunger worsened as the world experienced its worst **food crisis** in a generation.[35] Efforts to identify some of the causes of, and possible solutions for, hunger have focused on the environment.

Hunger and Environment Connections Hunger interacts with the environment in two major ways:

- Food production to feed billions of people around the world damages the environment.
- A damaged environment cannot support food production to feed billions of people around the world.

Without concerted efforts to improve food production in ways that will protect the environment, the vicious cycle of hunger and environmental degradation will continue, and poverty and population growth will escalate.

Planting Crops Producing food costs the earth dearly. To grow food, we clear land—prairie, wetland, and forest—losing native ecosystems and wildlife. Then we plough the fields and plant crops. The soil loses nutrients as each crop is harvested, so fertilizer is applied. Some fertilizer runs off, polluting the waterways and stimulating algae growth. Some soil erodes into the waterways and interferes with the growth of aquatic plants and animals. By the time the water reaches the seas, it is unsuitable for most marine life. Agriculture is the largest single source of **nonpoint water pollution.** Pollution from "point sources," such as sewage plants or factories, is relatively easy to control, but runoff from fields and pastures enters waterways from so many broad regions that it is nearly impossible to control.

To protect crops against weeds and pests, farmers apply herbicides and pesticides. These chemicals also pollute the water and, wherever the wind carries them, the air. Most herbicides and pesticides injure more than weeds and pests; they also injure native plants, native insects, and animals that eat those plants and insects. Ironically,

oral rehydration therapy (ORT): the administration of a simple solution of sugar, salt, and water, taken by mouth, to treat dehydration caused by diarrhea. A simple ORT recipe:

- ½ L boiling water
- 4 tsp sugar
- ½ tsp salt

food crisis: a sharp rise in the rates of hunger and malnutrition, usually set off by a shock to either the supply of, or demand for, food and a sudden spike in food prices.

nonpoint water pollution: water pollution caused by runoff from all over an area rather than from discrete "point" sources. An example is the pollution caused by runoff from agricultural fields.

widespread use of pesticides and herbicides causes pests and weeds to evolve, becoming more resistant. Consequently, farmers must use more pesticides and herbicides. These chemicals pose hazards for farm workers who handle them, and the residues can create health problems for consumers as well (as Chapter 19 discusses).

Finally, when fields are irrigated, the water evaporates, but the salts do not. Consequently, salts accumulate on the soil surface. As the surface soil becomes increasingly salty, plant growth suffers. Irrigation can also deplete the water supply over time as it drains water from surface waters or from underground; then the water evaporates or runs off. Excessive irrigation can dry up rivers and lakes and lower the water table of a whole region. A vicious cycle develops. The drier the region becomes, the more farmers irrigate, and the more they irrigate, the drier the region becomes.

Without water, croplands become deserts.

Raising Livestock Raising livestock is also a major stressor on the environment.[36] Like plant crops, herds of livestock occupy land that once maintained itself in a natural state. The land suffers the losses of native plants and animals, soil erosion, water depletion, and desert formation. Alternatively, animals in large concentrated areas such as cattle feedlots create environmental problems when huge masses of animal wastes are produced. To prevent contamination of local soils and water supplies, the Environmental Protection Agency suggests several strategies for managing livestock, poultry, and horse waste. In addition to manure, cows produce large quantities of methane—a greenhouse gas that may contribute to climate change.[37]

In addition to the waste problems, animals must be fed; grain is grown for them on other land. That land may require fertilizers, herbicides, pesticides, and irrigation too. In the United States, more cropland is used to produce grains for livestock than to produce grains for people. Figure 20-5 compares the grain required to produce various foods.

> FIGURE 20-5 Pounds of Grain Needed to Produce 1 Pound of Bread and 1 Pound of Animal Weight Gain

To gain 1 pound, animals raised for food have to eat many more pounds of grain than it takes to make a pound of bread.

SOURCE: Idea and data from T. R. Reid, Feeding the planet, *National Geographic*, October 1998, pp. 58–74.

Fishing Fishing also incurs environmental costs. On the sea, we harvest fish with little thought of the dwindling supplies or the environmental damage incurred. Excessive fishing—catching fish at a faster rate than they can reproduce—diminishes the availability of seafood for people to eat, upsets the balance of marine life, and reduces water quality. Some fishing methods, such as nets and filament line, kill nonfood species and deplete large populations of aquatic animals such as dolphins. Some fishing methods damage the ocean floor. Table H20-1 (p. 681) in the highlight that follows this chapter presents lists of seafood sorted from an environmental perspective.

Catching wild fish cannot keep pace with increases in consumer consumption. The shortfall has spurred the rapid growth of aquaculture, which now provides almost half of the world's food fish and shellfish.[38] It continues to grow more rapidly than any other kind of food animal production.

Some aquaculture "farms" consist of vast net cages that enclose fish in ocean water or freshwater lakes, where natural water flow refreshes the cages. Other types of farms house fish in artificial ponds of various shapes positioned inland close to natural water or farther inland. On the coast, natural water is diverted through the ponds, bringing fresh water in and carrying out wastes. Farther inland, pond water is continuously filtered and cleansed, recycling through the ponds. All farmed fish must be fed, and fish chow consists of grains and fish harvested from wild species, diminishing their stocks for human consumption. Adequate environmental safeguards are a must to prevent environmental degradation from aquaculture.[39] Aquaculture can be sustainable, however, with the appropriate technologies and practices.

Fishing is also energy-intensive, requiring fuel for boats, refrigeration, processing, packing, and transport. Water pollution incurs health risks when people eat contaminated fish. Bioaccumulation of toxins in fish may rule out fish consumption altogether in some areas.

Energy Overuse The entire food industry, whether based on growing crops, raising livestock, or fishing, requires energy, which primarily entails burning **fossil fuels.** Massive fossil fuel use threatens the environment by causing air and water pollution, changing climate patterns, depleting the ozone layer, and more. In the United States, the food industry consumes about 20 percent of all the energy the

fossil fuels: coal, oil, and natural gas.

About half of the seafood consumed in the United States comes from aquaculture.

♦ Gallons of water to produce 1 serving of:
- Lettuce: 6
- Milk: 49
- Steak: 2600

water stress: intense demands on water resources by human activities such as municipal water supplies, industries, power plants, and agriculture.

nation uses. Most of this energy is used to run farm machinery and to produce fertilizers and pesticides. Energy is also used to process, package, transport, refrigerate, store, and prepare foods. Suggested changes that might reduce fossil fuel energy use by about 50 percent include:[40]

- Using small machinery and less fuel
- Replacing fertilizer with cover crops and manure
- Tilling to reduce soil erosion
- Reducing consumption of meat and dairy products (to levels of nutrient adequacy without excess)
- Limiting transportation distances
- Combining efficient practices with renewable systems (such as biomass and photovoltaic cells)

Water Misuse Food production also uses an enormous amount of water. It takes an estimated 1.5 million gallons ♦ of water to produce the food for one US consumer for 1 year.[41] According to the Environmental Protection Agency, current farming practices are responsible for an estimated 70 percent of the pollution in US rivers and streams. Growing crops adds sediment, nutrients, and pesticides to the water. Irrigating crops depletes groundwater supplies, causing the land to become desert, which ironically can lead to flooding.

As human populations grow, so does the demand for fresh water. In areas of high **water stress,** natural and manmade influences converge to limit access to safe drinking water (see Figure 20-6).[42] Poor water management causes many of the world's water problems.[43] Each day, people dump 2 million tons of waste into the world's rivers, lakes, and streams. By 2025, if present patterns continue, two of every three persons on earth will live in water-stressed conditions.

Biodiversity Biodiversity is declining.[44] By the year 2050, some 40,000 plant species may become extinct. Traditional agricultural practices and the increasing uniformity of global food habits have failed to conserve species diversity. Wheat, rice, soybeans, and maize provide 75 percent of the food energy around the world.[45] Only two dozen other crops provide the remainder. As people everywhere eat the same

> FIGURE 20-6 Water-Stress Hot Spots

"Water stress" reflects the amount of pressure placed on water resources by users such as municipal water systems, industries, power plants, and agricultural users. As human population increases in an area, water stress worsens.

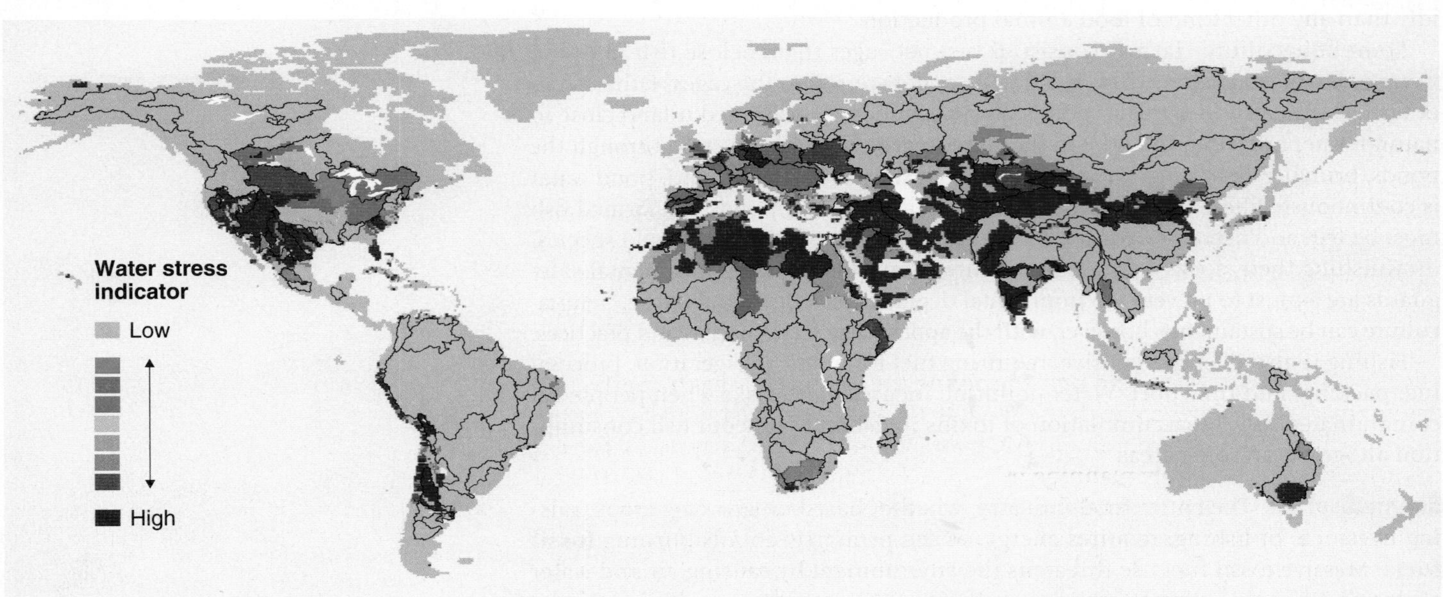

SOURCE: *World Water Assessment Programme*, The United Nations World Water Development Report 3: Water in a Changing World (*Paris: UNESCO, and London: Earthscan, 2009*). *p. 92.*

limited array of foods, local regions' native, genetically diverse plants no longer seem worth preserving. Yet, in the future, as the climate and environment change, those may be the very plants that people will need for food sources. A wild species of corn that grows in a dry climate, for example, might contain the genetic information necessary to help make domestic corn resistant to drought. (Highlight 19 offers several examples of how biotechnology is being used to improve food crops.)

Food Waste Food waste has a major environmental impact—both by squandering resources and by contributing to greenhouse gases.[46] More than 1 billion tons of food is wasted each year.[47] As a consequence, the many resources used in the production of these foods are also wasted. Each year, food waste accounts for more than 25 percent of the water used and about 300 million barrels of oil. In addition, the methane and carbon dioxide from decomposing food contributes to greenhouse gases and climate change.

In short, food production takes a tremendous toll on the environment. And environmental problems are reducing the world's ability to feed its people and keep them healthy. For the most part, our food production systems are not **sustainable.**

Sustainable Solutions

Can advances in agriculture compensate for the increases in population growth and the losses caused by environmental degradation? Historically, agricultural yields improved with advances in irrigation systems, fertilizers, and genetic strains. Today, however, the contributions these measures can make are reaching their limits, in part because they have also created environmental problems.

Adding more irrigation would not increase crop yields because almost all the land that can benefit from irrigation is already receiving it. In fact, rising concentrations of salt in the soil—a by-product of irrigation—are *lowering* yields on many of the world's irrigated croplands.

Nor can fertilizer use significantly enhance agricultural production. Much of the fertilizing that can be done is being done—and with great effect; fertilizers support some 40 percent of the world's total crop yields. Using more fertilizer, however, would not increase yields further and would add to the pollution of nearby waterways.

As for the development of high-yielding strains of crops, recent advances have been dramatic, but even they may be inadequate to change the overall trends. Furthermore, the raw materials necessary for developing new crops have become less available as genetic variation for many plant species is lost. Of the 5000 food plants grown throughout the world a few centuries ago, only 150 are cultivated in commercial agriculture today. Most of the world's population relies on only five cereals, three legumes, and three root crops to meet their energy needs. Even among these, valuable strains are vanishing.

Sustainable Agriculture For each environmental problem, agricultural solutions are being considered. Many farmers are implementing **sustainable agriculture** practices that can be adapted to meet the particular needs of a local area. The crop yields from farms that employ these practices often compare favorably with those from farms using less sustainable methods. Table 20-3 (p. 674) contrasts low-input, sustainable agriculture methods with high-input, unsustainable methods. Many sustainable practices are not really new, incidentally; they would be familiar to our great-grandparents. Farmers today are rediscovering the benefits of traditional techniques as they adapt and experiment with them in search of sustainable methods.

Sustainable Development The keys to solving the world's hunger, poverty, and environmental problems are in the hands of both the poor and the rich nations but require different efforts from them. The poor nations need to provide contraceptive technology and family planning information to their citizens, develop better programs to assist the poor, and slow and reverse the destruction of environmental resources. The rich nations need to stem their wasteful and polluting uses of resources and energy, which are contributing to global environmental degradation. They also must become willing to ease the debt burden that many poor nations face. Relieving poverty will help relieve environmental degradation and hunger.

sustainable: able to continue indefinitely; using resources at such a rate that the earth can keep on replacing them and producing pollutants at a rate with which the environment and human cleanup efforts can keep pace, so that no net accumulation of pollution occurs.

sustainable agriculture: ability to produce food indefinitely, with little or no harm to the environment.

TABLE 20-3 Agricultural Methods Compared

Environmental Issues	Unsustainable Methods	Sustainable Methods
Soil	Growing the same crop repeatedly on the same land takes nutrients out of the soil, making fertilizer use necessary; favors soil erosion; and invites weeds and pests to become established, making pesticide use necessary. Plowing the same way everywhere, allows water runoff and erosion.	Rotating crops increases nitrogen in the soil so there is less need to use fertilizers. Using appropriate plowing methods reduces soil erosion problems caused by weeds and pests. Using cover crops, crop rotation, no-till planting, contour planting, ridge till, mulch, terraces, and grass strips conserves both soil and water.
Fertilizer	Using fertilizers pollutes ground and surface water and increases costs as fossil fuels become scarce.	Reducing the use of fertilizers and using livestock manure more effectively lowers costs. Planting cover crops, such as legumes, after harvest restores nutrients and reduces erosion. Composting all plant residues not harvested into the soil improves its nutrient content and water-holding capacity.
Livestock	Feeding livestock in feedlots concentrates manure that pollutes water and releases methane, a global warming gas. Injecting animals with antibiotics prevents diseases.	Feeding livestock or buffalo on the open range allows their manure to fertilize the ground and releases no methane. Collecting feedlot animals' manure enables it to be used as fertilizer or treated before it is released. Maintaining animals' health can prevent disease.
Herbicide/ pesticide	Spraying herbicides and pesticides over large areas wipes out weeds, pests, and other plants and insects.	Using crop rotations, cover crops, and mechanical cultivation can control weeds. Using resistant crops and rotating crops foils pests that lay their eggs in the soil where last year's crop was grown. Using biological controls such as predators can destroy pests.
Water	Irrigating on a large scale depletes water supplies and concentrates salts in the soil.	Irrigating only during dry spells and applying only spot irrigation conserves water.
Energy	Using only fossil fuels depletes resources.	Using renewable energy technologies such as hydroelectric, biomass, photovoltaics, wind power, solar thermal, geothermal, biogas, and methanol conserves resources. Using machinery scaled to the job at hand, and operating it at efficient speeds conserves energy. Combining operations such as harrowing, planting, and fertilizing in the same operation conserves energy.

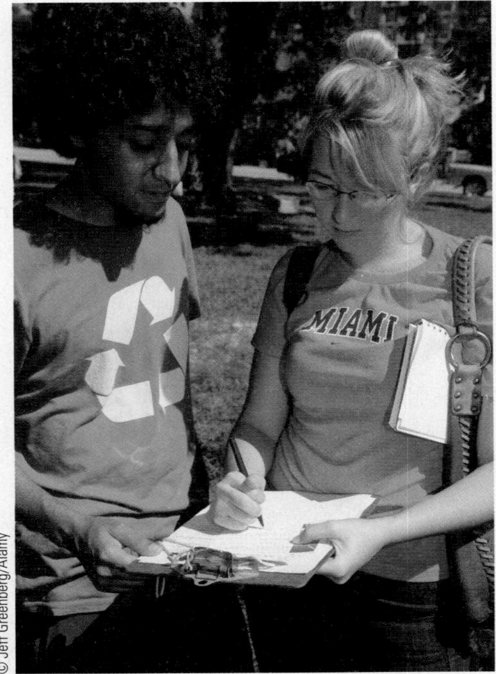

Each person's choice to get involved and be heard can help lead to needed change.

To rephrase a well-known adage: If you give a man a fish, he will eat for a day. If you teach him to fish and enable him to buy and maintain his own gear and bait, he will eat for a lifetime and help to feed others. Unlike food giveaways and money doles, which are only stop-gap measures, social programs that permanently improve the lives of the poor can permanently solve the hunger problem.

Sustainable Actions Every segment of our society can join in the fight against hunger, poverty, and environmental degradation. The federal government, the states, local communities, big business and small companies, educators, and all individuals have many opportunities to resolve these problems.

Dietitians and foodservice managers have a special role to play, and their efforts can make an impressive difference. Their professional organization, the Academy of Nutrition and Dietetics, urges members to conserve resources and minimize waste in both their professional and their personal lives.[48] In addition, members can educate themselves and others on hunger, its consequences, and programs to fight it; conduct research on the effectiveness and benefits of programs; and serve as advocates on the local, state, and national levels to help end hunger in the United States.[49] Globally, these professionals support programs that combat malnutrition, provide food security, promote self-sufficiency, respect local cultures, protect the environment, and sustain the economy.

Individuals can assist the global community in solving its poverty and hunger problems by joining and working for hunger-relief organizations (see Table 20-4). They can also support the needed changes in economic policies that influence food availability and price volatility both at home and in developing countries.[50]

TABLE 20-4 Hunger-Relief Organizations

Organization	Mission Statement
Action without Borders **www.idealist.org**	International organization seeking to connect people, organizations, and resources to help build a world where all people can live free and dignified lives.
Bread for the World **www.bread.org**	Non-partisan, Christian citizens' movement seeking to influence reform in policies, programs, and conditions that allow hunger and poverty to persist globally.
Catholic Relief Services **www.crs.org**	Humanitarian service agency assisting the impoverished and disadvantaged through community-based, sustainable development initiatives.
Community Food Security Coalition **www.foodsecurity.org**	North American coalition of diverse people and organizations working from the local to the international levels to build community food security.
Congressional Hunger Center **www.hungercenter.org**	Bipartisan organization training and inspiring leaders with the intent to end hunger, and advocating public policies to create a food-secure world.
Feeding America **www.feedingamerica.org**	Domestic charity organization providing food assistance through a nationwide network of member food banks and facilitating education to end hunger nationally.
Food and Agriculture Organization (FAO) of the United Nations **www.fao.org**	International organization leading efforts to defeat hunger by helping to develop and modernize countries' agriculture, forestry, and fishery practices.
Oxfam America **www.oxfamamerica.org**	International relief and development organization aiming to create lasting solutions to poverty, hunger, and injustice.
Pan American Health Organization **www.paho.org**	International public health agency aiming to strengthen national and local health systems with the purpose of improving the quality of, and lengthening, the lives of peoples in the Americas.
Society of St. Andrew **www.endhunger.org**	Ecumenical Christian ministry salvaging and redirecting large amounts of fresh produce to hunger agencies for distribution to the poor.
The Hunger Project **www.thp.org**	International relief organization emphasizing sustainable solutions such as rural development and self-reliance to facilitate food security.
United Nations Children's Fund (UNICEF) **www.unicef.org**	International organization advocating for the protection of children's rights, to help meet their basic needs and to expand their opportunities to reach their full potentials.
World Food Program **www.wfp.org**	Food aid branch of the United Nations aiming to prepare for, protect during, and provide assistance after, emergencies, as well as reducing hunger and undernutrition.
World Health Organization (WHO) **www.who.int**	United Nations agency acting as the authority on international public health by influencing policy, setting research agendas, establishing standards, and providing technical support to monitor and assess health trends.
World Hunger Year (WHY) **www.whyhunger.org**	Domestic organization supporting and funding community-based organizations intent on empowering individuals and building self-reliance to provide long-term solutions to hunger and poverty.

Most importantly, all individuals can try to make lifestyle choices ♦ that consider the environmental consequences. Many small decisions each day have major consequences for the environment. Highlight 20 describes how consumers can conserve resources and minimize waste when making food-related choices.

♦ A popular adage urges us to "Think globally, act locally."

The personal rewards of making environmentally friendly food choices are many, from saving money to the satisfaction of knowing that you are treading lightly on the earth. But do they really help? They do, if enough people join in. Because we number more than 7 billion, individual actions can add up to exert an immense impact.

"Be part of the solution, not part of the problem" another adage says. In other words, don't waste time or energy moaning and groaning about how bad things are: do something to improve them. They are our problems: human beings created them, and human beings must solve them.

REVIEW IT Explain why relieving environmental problems will also help to alleviate hunger and poverty.

Environmental degradation reduces our ability to produce enough food to feed the world's people. The rapid increase in the world's population exacerbates the situation. The global environment, which supports all life, is deteriorating, largely because of our irresponsible use of resources and energy. Governments, businesses, and all individuals have many opportunities to make environmentally conscious choices, which may help solve the hunger problem, improve quality of life, and generate jobs. Personal choices, made by many people, can have a great impact.

"We do not inherit the earth from our ancestors, we borrow it from our children." —ascribed to Chief Seattle, a 19th-century Native American leader

Nutrition Portfolio

Your choice to get involved in the fight against hunger—whether in your community or across the globe—can make a big difference in the health and survival of others.

- Find out about the hunger-relief programs in your area.
- Write to your legislators and voice your opinions on issues such as food assistance programs, environmental degradation, and international debt relief.
- Consider which environmentally friendly behaviors your are willing to adopt when making food-related choices.

Diet Analysis PLUS To complete this exercise, go to your Diet Analysis Plus at **www.cengagebrain.com.**

STUDY IT To review the key points of this chapter and take a practice quiz, go to the study cards at the end of the book.

REFERENCES

1. Food and Agriculture Organization of the United Nations, *The State of Food Insecurity in the World, 2010*, www.fao.org.
2. US Department of Agriculture, *Food Security in the United States: Key Statistics and Graphics*, ERS Research Briefs, www.ers.usda.gov/briefing/foodsecurity/stats_graphs.htm.
3. US Department of Agriculture, *Food Security in the United States: Key Statistics and Graphics*, ERS Research Briefs, www.ers.usda.gov/briefing/foodsecurity/stats_graphs.htm.
4. D. C. Hernandez and A. Jacknowitz, Transient, but not persistent, adult food insecurity influences toddler development, *Journal of Nutrition* 139 (2009): 1517–1524.
5. US Census Bureau, *Social, Economic, and Housing Statistics Division: Poverty* (Washington, D.C.: US Government Printing Office, 2011).
6. H. A. Eicher-Miller and coauthors, Food insecurity is associated with iron deficiency anemia in US adolescents, *American Journal of Clinical Nutrition* 90 (2009): 1358–1371; R. Rose-Jacobs and coauthors, Household food insecurity: Associations with at-risk infant and toddler development, *Pediatrics* 121 (2008): 65–72.
7. H. K. Seligman and D. Schillinger, Hunger and socioeconomic disparities in chronic disease, *New England Journal of Medicine* 363 (2010): 6–9.
8. C. M. Olson, Food insecurity and maternal health during pregnancy, *Journal of the American Dietetic Association* 110 (2010): 690–691.
9. J. C. Eisenmann and coauthors, Is food insecurity related to overweight and obesity in children and adolescents? A summary of studies, 1995–2009, *Obesity Reviews* 12 (2011): e73–e83.
10. B. J. Lohman and coauthors, Adolescent overweight and obesity: Links to food insecurity and individual, maternal, and family stressors, *Journal of Adolescent Health* 45 (2009): 230–237; E. Metallinos-Katsaras, B. Sherry, and J. Kallio, Food insecurity is associated with overweight in children younger than 5 years of age, *Journal of the American Dietetic Association* 109 (2009): 1790–1794.
11. A. Drewnowski, Obesity, diets, and social inequalities, *Nutrition Reviews* 67 (2009): S36–S39.
12. M. Franco and coauthors, Neighborhood characteristics and availability of healthy foods in Baltimore, *American Journal of Preventive Medicine* 35 (2008): 561–567.
13. Institute of Medicine and National Research Council, *The Public Health Effects of Food Deserts: Workshop summary* (Washington, DC: The National Academies Press, 2009).
14. J. Beaulac, E. Kristjansson, and S. Cummins, A systematic review of food deserts, 1966–2007, *Preventing Chronic Disease* 6 (2009): A105.
15. B. T. Izumi and coauthors, Associations between neighborhood availability and individual consumption of dark-green and orange vegetables among ethnically diverse adults in Detroit, *Journal of the American Dietetic Association* 111 (2011): 274–279.
16. A. Drewnowski, The cost of US foods as related to their nutritive value, *American Journal of Clinical Nutrition* 92 (2010):1181–1188; P. Monsivais and A. Drewnowski, Lower-energy-density diets are associated with higher monetary costs per kilocalorie and are consumed by women of higher socioeconomic status, *Journal of the American Dietetic Association* 109 (2009): 814–822; M. S. Townsend and coauthors, Less-energy-dense diets of low-income women in California are associated with higher energy-adjusted diet costs, *American Journal of Clinical Nutrition* 89 (2009): 1220–1226.
17. A. Offer, R. Pechey, and S. Ulijaszek, Obesity under affluence varies by welfare regimes: The effect of fast food, insecurity, and inequality, *Economics and Human Biology* 8 (2010): 297–308.
18. A. Karnik and coauthors, Food insecurity and obesity in New York City primary care clinics, *Medical Care* 49 (2011): 658–661; N. I. Larson and M. T. Story, Food insecurity and weight status among US children and families: A review of the literature, *American Journal of Preventative Medicine* 40 (2011): 166–173.
19. Position of the American Dietetic Association: Food insecurity in the United States, *Journal of the American Dietetic Association* 110 (2010): 1368–1377.
20. Food and Agriculture Organization of the United Nations, *Global Food Losses and Food Waste: Extent, Causes and* Prevention, 2011, www.fao.org.
21. S. de Pee and coauthors, How to ensure nutrition security in the global economic crisis to protect and enhance development of young children and our common future, *Journal of Nutrition* 140 (2010): 138S–142S; T. Atinmo and coauthors, Breaking the poverty/malnutrition cycle in Africa and the Middle East, *Nutrition Reviews* 67 (2009): 540–546.
22. M. W. Bloem, R. D. Semba, and K. Kraemer, Castel Gandolfo workshop: An introduction to the impact of climate change, the economic crisis, and the increase in the food prices on malnutrition, *Journal of Nutrition* 140 (2010): 132S–135S.
23. E. Boy and coauthors, Achievements, challenges, and promising new approaches in vitamin and mineral deficiency control, *Nutrition Reviews* 67 (2009): S24–S30.
24. R. L. Guerrant and coauthors, Malnutrition as an enteric infectious disease with long-term effects on child development, *Nutrition Reviews* 66 (2008): 487–505.
25. World Health Organization, 10 Facts on Child Health, October 2011, www.who.int/features/factfiles/child_health2/en/index.html.
26. P. Svedberg, How many people are malnourished, *Annual Review of Nutrition* 31 (2011): 263–283.

27. M. H. Golden, Evolution of nutritional management of acute malnutrition, *Indian Pediatrics* 47 (2010): 667–678.
28. T. Ahmed, S. Rahman, and A. Cravioto, Oedematous malnutrition, *Indian Journal of Medical Research* 130 (2009): 651–654.
29. D. R. Brewster, Inpatient management of severe malnutrition: Time for a change in protocol and practice, *Annals of Tropical Paediatrics* 31 (2011): 97–107.
30. W. A. Petri Jr. and coauthors, Association of malnutrition with amebiasis, *Nutrition Reviews* 67 (2009): S207–S215; T. Ahmed and coauthors, Use of metagenomics to understand the genetic basis of malnutrition *Nutrition Reviews* 67 (2009): S201–S206.
31. C. Best and coauthors, Can multi-micronutrient food fortification improve the micronutrient status, growth, health, and cognition of schoolchildren? A systematic review, *Nutrition Reviews* 69 (2011): 186–204.
32. M. H. Golden, Proposed recommended nutrient densitites for moderately malnourished children, *Food and Nutrition Bulletin* 30 (2009): S267–S342.
33. K. M. Hendricks, Ready-to-use therapeutic food for prevention of childhood undernutrition, *Nutrition Reviews* 68 (2010): 429–435.
34. S. Isanaka and coauthors, Effect of preventive supplementation with ready-to-use therapeutic food on the nutritional status, mortality, and morbidity of children aged 6 to 60 months in Niger: A cluster randomized trial, *Journal of the American Medical Association* 301 (2009): 277–285.
35. C. P. Timmer, Preventing food crises using a food policy approach, *Journal of Nutrition* 140 (2010): 224S–228S.
36. Food and Agriculture Organization of the United Nations, *Global Food Losses and Food Waste: Extent, Causes and Prevention,* 2011, www.fao.org.
37. A. Carlsson-Kanyama and A. D. González, Potential contributions of food consumption patterns to climate change, *American Journal of Clinical Nutrition* 89 (2009): 1704S–1709S.
38. Food and Agricultural Organization of the United Nations, *State of the World Fisheries and Aquaculture, 2008,* www.fao.org.
39. R. L. Naylor, Environmental safeguards for open-ocean aquaculture, *Issues in Science and Technology,* Spring 2006, pp. 53–58.
40. D. Pimentel and coauthors, Reducing energy inputs in the US food system, *Human Ecology* 36 (2008): 459–471.
41. D. Hinrichsen, Water pressure, www.nwf.org/nationalwildlife/article.cfm, accessed March 2009.
42. World Water Assessment Programme, *The United Nations World Water Development Report 3: Water in a Changing World* (Paris: UNESCO, and London: Earthscan, 2009).
43. The Role of Science in Solving the World's Emerging Water Problems, Arthur M. Sackler Colloquia of the National Academy of Sciences held in Irvine, California, October 8–10, 2004.
44. S. Milius, Losing life's variety, *Science News* 177 (2010): 20–25.
45. D. B. Lobell, W. Schlenker, and J. Costa-Roberts, Climate trends and global crop production since 1980, *Science* 333 (2011): 616–620.
46. K. D. Hall and coauthors, The progressive increase of food waste in America and its environmental impact, *PloSOne* 4 (2009): e7940.
47. Food and Agriculture Organization of the United States, *Global Food Losses and Food Waste: Extent, Causes and Prevention, 2011,* www.fao.org.
48. Position of the American Dietetic Association: Food and nutrition professionals can implement practices to conserve natural resources and support ecological sustainability, *Journal of the American Dietetic Association* 107 (2007): 1033–1043.
49. Position of the American Dietetic Association, 2010.
50. Food and Agriculture Organization of the United Nations, *The State of Food Insecurity in the World, 2011,* www.fao.org.

HIGHLIGHT > 20

LEARN IT Discuss ways consumers can conserve resources when making food-related choices.

Environmentally Friendly Food Choices

Chapter 20 concludes its examination of US and world hunger by focusing on the environment. It explains how producing enough food to feed billions of people around the world damages the environment—and how a damaged environment cannot adequately feed billions of people around the world. Efforts to resolve hunger and protect the environment demand improvements in our ways of producing, processing, packaging, transporting, storing, and preparing foods.[1] Because sustainable food strategies have social, economic, and environmental impacts, their success depends on pleasing the people, providing reasonably priced products, and protecting the planet.

The United Nations describes a nation's impact on the environment as its "ecological footprint"—a measure of the resources used to support a nation's consumption of food, materials, and energy. This measure takes into account the two most challenging aspects of sustainability—per capita resource consumption and population growth. As Figure H20-1 shows, the people of North America are the world's greatest consumers on a per capita basis. Some have estimated that it would take four more planet earths to accommodate every person in the world using resources at the level currently used in the United States. This highlight explores ways that consumers can conserve resources when making food-related choices. Such conscientious food choices can reduce pollution production and resource use.[2] The accompanying "How To" feature asks questions to determine your "food footprint"—a measure of the environmental impacts of food production, stated in terms of land usage.

Choice: Animal or Vegetable?

Some foods require more water, more fertilizer, more pesticides, and more energy for their production than others. One way to reduce these costs of food production is for consumers to eat low on the food chain. For the most part, that means eating more foods derived from plants and fewer foods derived from animals.[3] One study reports that compared with vegetarian diets, meat-based diets use 2.9 times more water, 2.5 times more energy, 13 times more fertilizer, and 1.4 times more pesticides.[4]

Consider that relatively little energy is needed to produce grains: it takes less than 1 kcalorie of fuel to produce each kcalorie of grain, whereas most animal-derived foods require from 10 to 90 kcalories of fuel per kcalorie of edible food; kcalories to produce fruits and vegetables are generally greater than grains, but less than animal-derived foods. In general, meat-based diets require much more energy, as well as more land and water, than do plant-based diets. An exception is livestock raised on the open range; these animals require about as much energy as most plant foods. Because we raise so much more grain-fed, than range-fed, livestock, however, the average energy requirement for meat production is high. Figure H20-2 (p. 680) shows how much less fuel vegetarian diets require than meat-based diets and shows that vegan diets require the least fuel of all.

To support our meat intake, we maintain several billion livestock, about four times our own weight in animals. Livestock consume 10 times as much grain each day as we do. We could use much of that grain to make grain products for ourselves and for others around the world. Making this shift could free up enough grain to feed 400 million people while using less fuel, water, and land.

> FIGURE H20-1 **Ecological Footprints**

The width of a bar represents the region's population, and the height represents per capita consumption (in terms of area of productive land or sea required to produce the natural resources consumed). Thus the footprint of the bar represents the region's total consumption. For example, Asia's population is more than 10 times greater than North America's, but because its consumption is only one-sixth as large, their footprints are similar in size.

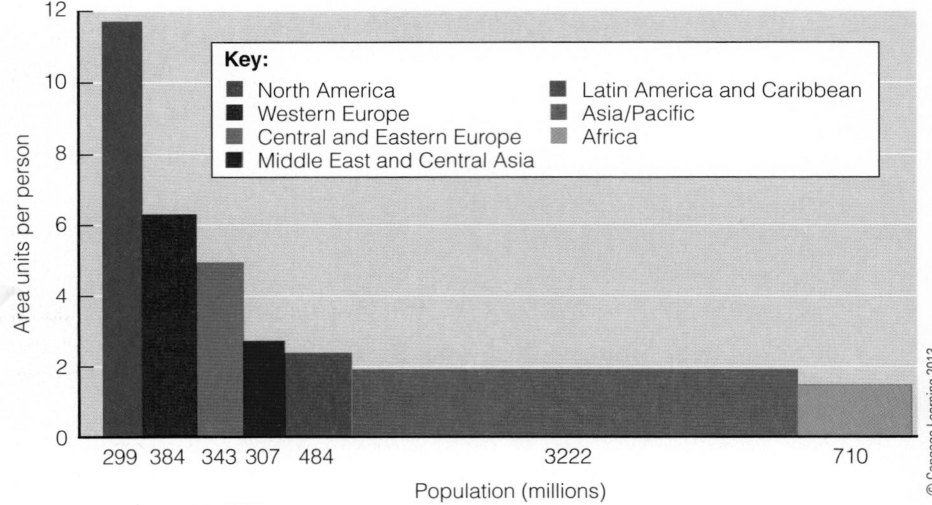

>How To

Determine Your Food Footprint

What's your food footprint? Answer the following questions, add up your score, and use the key below to determine how many acres are used to support your food habits.

Directions: Answer the questions below. See how you did by using the key below.

1. How often do you eat meat?
   a. I'm vegan (I eat no animal products.)
   b. I'm a vegetarian (I don't eat any meat.)
   c. I eat meat 1 to 4 days a week.
   d. I eat meat every day.

2. If you eat meat, which type of meat do you eat most often?
   a. I told you already I don't eat meat!
   b. Turkey or chicken
   c. Lamb or pork
   d. Beef

3. How often do you eat fast food?
   a. I never eat fast food.
   b. Sometimes; 2–3 times a month.
   c. I eat fast food a lot. But I would rather eat healthy!
   d. I love fast food. I would eat it everyday if I could (and sometimes I do).

4. How often do you buy food from local farmers' markets?
   a. I go to the farmers' market every week.
   b. I go to the farmers' market sometimes; maybe once a month.
   c. I don't know of any farmers' markets but I am interested in going!
   d. Never; I am not really interested in starting.

5. How much of your own food do you grow?
   a. I grow lots of my own food! I have a garden or a farm.
   b. I grow some food. I have a few plants in my yard or window garden.
   c. I don't have a garden but I would like to grow some food.
   d. I have never grown any food; I'm not that interested.

6. Do you try to eat food that is in season?
   a. Yes, I only eat food that is in season.
   b. Sometimes if I remember.
   c. I don't know what food is in season when, but I would like to learn.
   d. No, I eat what I want when I want.

7. How often do you eat home-cooked food?
   a. I eat home-cooked food almost every night.
   b. I eat home-cooked food 3-5 times a week.
   c. I eat out a lot, but I would eat more home-cooked food if I knew how.
   d. I never eat home-cooked food.

8. Do you try to buy fruit and vegetables that were grown locally or in your state?
   a. I always check to see where my food is grown. If it isn't grown locally, I don't buy it.
   b. I try to buy locally grown fruits and vegetables usually but not always.
   c. I never thought about where my food was grown. I will try to eat more local food.
   d. I don't care where my food comes from.

1 point for every **a** answer
2 points for every **b** answer
3 points for every **c** answer
5 points for every **d** answer

Your Score

8–14 Congratulations, you have awesome eating habits! Your next challenge: work to make sure ALL people have access to healthy, affordable food.

15–24 Not too bad. Seems like you have some good eating habits and aspire to have even better ones. 10–24 acres are used to support your food habits. Unfortunately we would still need at least one more planet to support your consumption.

25–33 Yikes, big foot! It seems some of your habits are damaging to the environment. 24–40 acres are used to support your food habits. We would need 3–4 more planets to sustain your lifestyle into the future.

Try It Review the questions with high scores and list steps you can take to reduce your food footprint.

SOURCE: *What's On Your Plate*? Curriculum Second Edition 2010 ed. Catherine Gund, Aubin Pictures.

> FIGURE H20-2 Amounts of Fuel Required to Feed People Eating at Different Points on the Food Chain

Three people who eat differently are compared here. Each has the same energy intake: 3300 kcalories a day. The fossil fuel amounts necessary to produce these different diets are calculated based on US conditions.

The meat eater consumes a typical US diet of meat, other animal products, and plant foods:

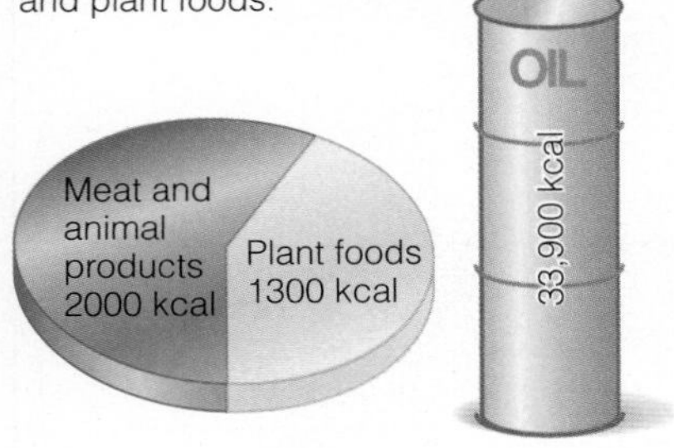

Fuel required to produce this food

The lacto-ovo-vegetarian eats a diet that excludes meats, but includes milk products and eggs:

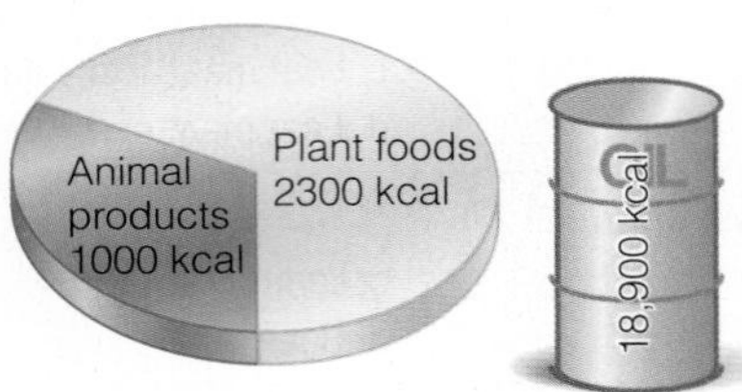

Fuel required to produce this food

The vegan eats a diet of plant foods only:

Fuel required to produce this food

Part of the solution to the livestock problem may be to cease feeding grain to animals and return to grazing them on the open range, which can be a sustainable practice. Ranchers have to manage the grazing carefully to hold the cattle's numbers to what the land can support without environmental degradation. To accomplish this, the economic benefits of traditional livestock and feed-growing operations would have to end. If producers were to pay the true costs of the environmental damage incurred by irrigation water, fertilizers, pesticides, and fuels, the price of meats might double or triple. According to classic economic theory, people would then buy less meat (reducing demand), and producers would respond by producing less meat (reducing supply). Meat production would then fall to a sustainable level.

Some consumers are trying to help solve some of these problems by choosing smaller portions of meat or selecting range-fed beef or buffalo only. Livestock on the range eat grass, which people cannot eat. "Rangeburger" buffalo also offers nutrient advantages over grain-fed beef because it is lower in fat and because the fat has more polyunsaturated fatty acids, including the omega-3 type.

Some consumers are opting for vegetarian, and even vegan, diets—at least occasionally. Choosing fish instead of meat may be a practical alternative if seafood is selected with an awareness of the environmental consequences. Table H20-1 presents seafood options grouped according to their impact on the environment.

Choice: Global or Local?

Plant-centered diets have an environmental advantage over meat-based diets, but some would argue that they don't go far enough. A more ecologically responsible diet is also based on locally grown products. On average, an item of food is transported 1500 miles before it is eaten—when the crops are grown and the final destination are both in the United States. That "our foods now travel more than we do" has several costly ramifications:

- *Energetically costly.* Foods must be refrigerated and transported thousands of miles to provide a full array of all produce all year round.
- *Socially unjust.* Farmers in impoverished countries, where the people are malnourished, are paid meager wages to grow food for wealthy nations.
- *Economically unwise.* It supports agribusinesses that buy land and labor cheaply in foreign countries instead of supporting local farmers raising crops in our communities.
- *Biologically risky.* Highly perishable foods are shipped from countries with unsafe drinking water and sanitation practices.

For all these reasons, consumers may want to improve the global environment by buying locally.

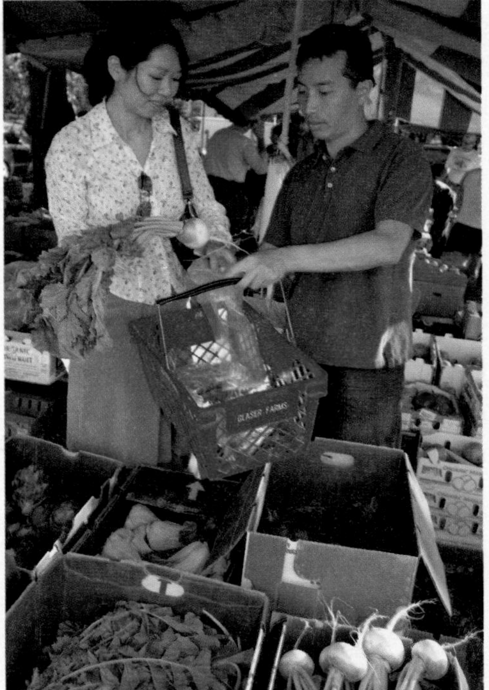

Locally grown foods offer benefits to both the local economy and the global environment.

TABLE H20-1 Environmental Impact of Commonly Eaten Seafood Choices

Seafood	Abundant, Well-Managed, and Fished or Farmed Using Methods that Are Environmentally Friendly	Fished or Farmed Using Methods that Raise Some Environmental Concerns	Overfished and/or Fished or Farmed Using Methods that Harm Other Marine Life or the Environment
Catfish	US farmed	—	—
Clams	Worldwide farmed, US Pacific and Atlantic wild-caught	US Atlantic wild-caught	—
Cod	US Pacific longline, jig, and trap-caught	US Pacific trawl-caught	US Atlantic wild-caught
Crab, blue	—	US trap-caught	—
Crab, Dungeness	US and Canada trap-caught	—	—
Crab, Jonah	—	US Atlantic wild-caught	—
Crab, king	—	US trap-caught	Imported trap-caught
Crab, kona	Australia wild-caught	Hawaii wild-caught	—
Crab, snow	—	Alaska and Canada wild-caught	—
Crab, stone	US Atlantic and Gulf of Mexico trap-caught	—	—
Flatfish, flounder	—	—	Atlantic wild-caught
Flatfish, halibut	US and Canadian Pacific wild-caught	US Pacific hook-and-line or bottom trawl, wild-caught	US Atlantic wild-caught, US Pacific set gillnet
Flatfish, plaice	—	US Pacific wild-caught	US Atlantic wild-caught
Pollock	Alaska wild-caught	—	—
Salmon	Alaska wild-caught	Washington wild-caught	Worldwide farmed
Scallops, bay	Worldwide farmed	—	—
Scallops, sea	—	Atlantic US and Canada wild-caught	—
Shrimp	Oregon wild-caught	US and Canada wild-caught, US Gulf of Mexico and South Atlantic wild-caught, US farmed	Imported farmed and wild-caught
Tilapia	US farmed	Central and South America farmed	China and Taiwan farmed
Tuna, albacore	British Columbia US troll/pole, Hawaii troll/pole or handline	Hawaii longline	Worldwide except Hawaii longline
Tuna, bigeye[a]	—	Hawaii pole/troll or handline, worldwide troll/poll, US Atlantic longline	Worldwide except US Atlantic longline
Tuna, bluefin	—	—	Worldwide wild-caught
Tuna, canned[a]	—	Worldwide wild-caught	—
Tuna, skipjack	Worldwide troll/pole, Hawaii troll/pole or handline	Hawaii longline	Imported longline
Tuna, yellowfin[a]	US Atlantic troll/pole	Hawaii troll/pole or handline, worldwide troll/pole, US Atlantic longline	Worldwide except US Atlantic longline

[a]Canned tuna is predominantly sold as either "chunk white" (albacore) or "chunk light" (yellowfin and skipjack tuna).
SOURCE: Adapted from the Seafood Guide of the Monterey Bay Aquarium Foundation, available at www.montereybayaquarium.org, accessed March 2009.

Buying locally offers other benefits as well. Families buying home-grown produce tend to eat a greater quantity and selection of fruits and vegetables.[5]

Adopting a local diet presents a bit of a challenge at first, especially when local fruits and vegetables are "out of season." But a nutritionally balanced diet of delicious foods is quite possible with a little creative planning.

Defining *Local*

Consumers shopping for "locally grown" produce are often trying to find fresh foods that will both help support small farms and protect the environment. Grocers tapping into this "local" marketing concept have seen significant increases in sales. But what exactly does *local* mean? How far is "locally grown" from the farm to the market? Given that *local* is not defined by the FDA, consumers and grocers do not always share the same understanding of the term. Some would say *local* means reasonably nearby—say, within 10 to 50 miles, or up to 200 miles. Others say it means within a day's drive, or within the state or a geographic region. Without agreement, consumers are left to identify which goods meet their limited definition, and grocers are free to market their goods with a broad definition.

Consumers who think of *local* as "fresh from a small farm" may be surprised to learn that is not always the case. "Local" may be a corporate-managed 1000-acre tomato farm or a family-run 3-acre blueberry farm. Because large retail stores need large volumes, they tend to deal with a few large producers. Shipments typically go through a distribution center, which can delay deliveries for several hours or even days. In contrast, small producers tend to sell to farmers' markets and small stores. Deliveries often arrive within hours of harvest.

Eco-Friendly Miles

Consumers who support the concept of purchasing local foods may believe that the fewer miles a food travels, the better for the environment—but this is not always the case. The type of transportation also contributes to environmental costs. In general, transporting by ship is cleaner than by airplane; by train is cleaner than by truck. It may be obvious that the eco-friendly choice for a person living in San Francisco

>How To

Make Environmentally Friendly Food-Related Choices

Food production taxes environmental resources and causes pollution. Consumers can make environmentally friendly choices at every step from food shopping to cooking and use of kitchen appliances to serving, cleanup, and waste disposal.

Food Shopping

Transportation

- Whenever possible, walk or ride a bicycle; use carpools and mass transit.
- Shop only once a week, share trips, or take turns shopping with others.
- When buying a car, choose an energy-efficient one.

Food Choices

- Choose foods low on the food chain; that is, eat more plants and fewer animals that eat plants (this suggestion also complements the *Dietary Guidelines for Americans* for eating for good health).
- Eat small portions of meat; select range-fed beef, buffalo, and poultry.
- Shop at farmers' markets and roadside stands for local foods; they require less transportation, packaging, and refrigeration.
- Limit use of imported canned beef products such as stews, chili, and corned beef that frequently come from cleared rain forest land.
- Choose seafood that has been farmed or fished in environmentally responsible ways (see Table H20-1, p. 681).
- Choose chickens from local farms.
- Plan wisely and buy smart to reduce the amount of food wasted.
- Donate excess food to food banks or shelters.
- Give food scraps to farmers for animal feed.

Food Packages

- Whenever possible, select foods with no packages; next best are minimal, reusable, or recyclable ones.
- Buy juices and sodas in large glass or recyclable plastic bottles (not small individual cans or cartons); grains, legumes, and nuts in bulk (not separate little packages); and eggs in pressed fiber cartons (not foam, unless it is recycled locally).
- Carry reusable string or cloth shopping bags; alternatively, reuse plastic bags.

Gardening

- Grow some of your own food, even if it is only herbs planted in pots on your kitchen windowsill.
- Compost all vegetable scraps, fruit peelings, and leftover plant foods.

Cooking Food

- Cook foods quickly in a stir-fry, pressure cooker, or microwave oven.
- When using the oven, bake a lot of food at one time and keep the door closed tightly.
- Use nondisposable utensils, dishes, and pans.
- Use pumps instead of spray products.
- Prepare smaller recipes to minimize waste.
- Eat leftovers.

Kitchen Appliances

- Use fewer small electrical appliances; open cans, mix batters, sharpen knives, and chop vegetables by hand.
- When buying a large appliance, choose an energy-efficient one.
- Consider solar power to meet home electrical needs.
- Set the water heater at $<130°F$ (54°C), no hotter; put it on a timer; wrap it and the hot-water pipes in insulation; install water-saving faucets.
- Set the refrigerator at 37°F to 40°F and the freezer to 0°F.
- Keep all appliances clean and in good repair.

Food Serving, Dish Washing, and Waste Disposal

- Use "real" plates, cups, and glasses instead of disposable ones.
- Use cloth towels and napkins, reusable storage containers with lids, reusable pans, and dishcloths instead of paper towels, plastic wrap, plastic storage bags, aluminum foil, and sponges.
- Run the dishwasher only when it is full.
- Recycle all paper, glass, plastic, and aluminum.

These suggested lifestyle changes can easily be extended from food to other areas.

Try It Find out if you are living a sustainable life by taking the food and health quizzes at **www.planetgreen.discovery.com/games-quizzes.**

is a California wine, but what about a person living in Atlanta? Is it "greener" to buy wine trucked across the country from California or shipped across the ocean from France?

Other Food-Related Choices

Eating more foods derived from plants and produced locally are two major trends that can influence the food industry. Consumers can also make dozens of other smaller decisions every day to help conserve resources and protect the environment. The accompanying "How To" feature lists some of the ways consumers can "tread lightly on the earth" through their daily food choices.

Chapter 20 and this highlight have presented many problems and have suggested that, although many of the problems are global in scope, the solutions depend on the actions of individual people at the local level. On learning of this, concerned people may take a perfectionist attitude, believing that they should be doing more than they realistically can, and so feel defeated. Keep in mind that striving for perfection even while falling short is progress. A positive attitude can bring about improvement, and sometimes improvement is enough. Celebrate the changes that are possible today by making them a permanent part of your life; do the same with changes that become possible tomorrow and every day thereafter. The results may surprise you.

REFERENCES

1. D. Pimentel and coauthors, Reducing energy inputs in the US food system, *Human Ecology* 36 (2008): 459–461.
2. A. Carlsson-Kanyama and A. D. González, Potential contributions of food consumption patterns to climate change, *American Journal of Clinical Nutrition* 89 (2009): 1704S–1709S.
3. H. J. Marlow and coauthors, Diet and the environment: Does what you eat matter? *American Journal of Clinical Nutrition* 89 (2009): 1699S–1703S.
4. Marlow and coauthors, 2009.
5. M. S. Nanney and coauthors, Frequency of eating homegrown produce is associated with higher intake among parents and their preschool-aged children in rural Missouri, *Journal of the American Dietetic Association* 107 (2007): 577–584.

Appendixes

© Kim D. French/Shutterstock.com

Appendix A Cells, Hormones, and Nerves

CONTENTS

This appendix is offered as an optional chapter for readers who want to enhance their understanding of how the body coordinates its activities. It presents a brief summary of the structure and function of the body's basic working unit (the cell) and of the body's two major regulatory systems (the hormonal system and the nervous system).

Cells

The body's organs are made up of millions of cells and of materials produced by them. Each **cell** is specialized to perform its organ's functions, but all cells have common structures (see the accompanying glossary and Figure A-l). Every cell is contained within a **cell membrane.** The cell membrane assists in moving materials into and out of the cell, and some of its special proteins act as "pumps" (described in Chapter 6). Some features of cell membranes, such as microvilli (Chapter 3), permit cells to interact with other cells and with their environments in highly specific ways.

Inside the membrane lies the **cytoplasm,** which is filled with **cytosol,** a jelly-like fluid. The cytoplasm contains much more than just cytosol, though. It is a highly organized system of fibers, tubes, membranes, particles, and subcellular **organelles** as complex as a city. These parts intercommunicate, manufacture and exchange materials, package and prepare materials for export, and maintain and repair themselves.

Within each cell is another membrane-enclosed body, the **nucleus.** Inside the nucleus are the **chromosomes,** which contain the genetic material, DNA. The DNA encodes all the instructions for carrying out the cell's activities. The role of DNA in coding for cell proteins is summarized in Figure 6-7 on p. 174. Chapter 6 also describes the variety of proteins produced by cells and some of the ways they perform the body's work.

Among the organelles within a cell are ribosomes, mitochondria, and lysosomes. Figure 6-7 briefly refers to the **ribosomes;** they assemble amino acids into proteins, following directions conveyed to them by RNA.

GLOSSARY OF CELL STRUCTURES

cell: the basic structural unit of all living things.

cell membrane: the thin layer of tissue that surrounds the cell and encloses its contents, made primarily of lipid and protein.

chromosomes: structures within the nucleus of a cell made of DNA and associated proteins. Human beings have 46 chromosomes in 23 pairs. Each chromosome has many genes.

cytoplasm (SIGH-toh-plazm): the cell contents, except for the nucleus.

cytosol: the fluid of cytoplasm that contains water, ions, nutrients, and enzymes.

endoplasmic reticulum (en-doh-PLAZ-mic reh-TIC-you-lum): a complex network of intracellular membranes. The *rough endoplasmic reticulum* is dotted with ribosomes, where protein synthesis takes place. The *smooth endoplasmic reticulum* bears no ribosomes.

Golgi (GOAL-gee) **apparatus:** a set of membranes within the cell where secretory materials are packaged for export.

lysosomes (LYE-so-zomes): cellular organelles; membrane-enclosed sacs of degradative enzymes.

mitochondria (my-toh-KON-dree-uh); singular *mitochondrion:* the cellular organelles that are made of membranes (lipid and protein) with enzymes mounted on them responsible for producing ATP aerobically.

nucleus: a major membrane-enclosed body within cells, which contains the cell's genetic material (DNA) embedded in chromosomes.

organelles: subcellular structures such as ribosomes, mitochondria, and lysosomes.

ribosomes (RYE-boh-zomes): protein-making organelles in cells that are composed of RNA and protein.

> FIGURE A-1 **The Structure of a Typical Cell**

The cell shown might be one in a gland (such as the pancreas) that produces secretory products (enzymes) for export (to the intestine). The rough endoplasmic reticulum with its ribosomes produces the enzymes; the smooth reticulum conducts them to the Golgi region; the Golgi membranes merge with the cell membrane, where the enzymes can be released into the extracellular fluid.

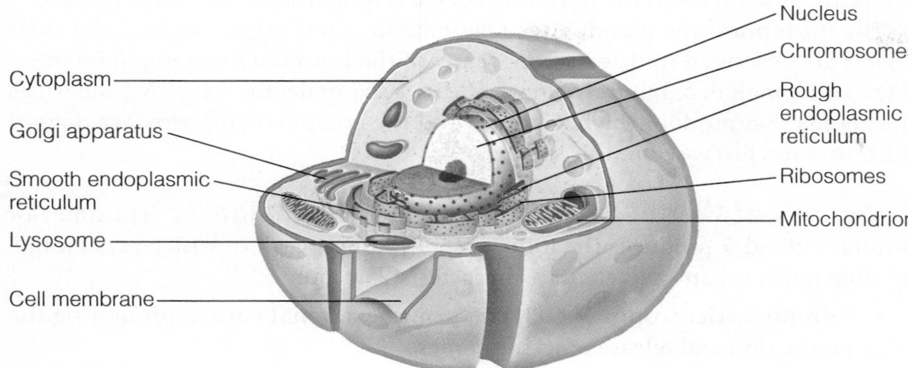

The **mitochondria** are made of intricately folded membranes that bear thousands of highly organized sets of enzymes on their inner and outer surfaces. Mitochondria are crucial to energy metabolism (described in Chapter 7) and muscles conditioned to work aerobically are packed with them. Their presence is implied whenever the TCA cycle and electron transport chain are mentioned because the mitochondria house the needed enzymes.*

The **lysosomes** are membranes that enclose degradative enzymes. When a cell needs to self-destruct or to digest materials in its surroundings, its lysosomes free their enzymes. Lysosomes are active when tissue repair or remodeling is taking place—for example, in cleaning up infections, healing wounds, shaping embryonic organs, and remodeling bones.

Besides these and other cellular organelles, the cell's cytoplasm contains a highly organized system of membranes, the **endoplasmic reticulum.** The ribosomes may either float free in the cytoplasm or be mounted on these membranes. A membranous surface dotted with ribosomes looks speckled under the microscope and is called "rough" endoplasmic reticulum; such a surface without ribosomes is called "smooth." Some intracellular membranes are organized into tubules that collect cellular materials, merge with the cell membrane, and discharge their contents to the outside of the cell; these membrane systems are named the **Golgi apparatus,** after the scientist who first described them. The rough and smooth endoplasmic reticula and the Golgi apparatus are continuous with one another, so secretions produced deep in the interior of the cell can be efficiently transported to the outside and released. These and other cell structures enable cells to perform the multitudes of functions for which they are specialized.

The actions of cells are coordinated by both hormones and nerves, as the next sections show. Among the types of cellular organelles are receptors for the hormones delivering instructions that originate elsewhere in the body. Some hormones penetrate the cell and its nucleus and attach to receptors on chromosomes, where they activate certain genes to initiate, stop, speed up, or slow down synthesis of certain proteins as needed. Other hormones attach to receptors on the cell surface and transmit their messages from there. The hormones ♦ are described in the next section; the nerves, in the one following.

♦ The study of hormones and their effects is *endocrinology*.

*For the reactions of glycolysis, the TCA cycle, and the electron transport chain, see Chapter 7 and Appendix C. The reactions of glycolysis take place in the cytoplasm; the conversion of pyruvate to acetyl CoA takes place in the mitochondria, as do the TCA cycle and electron transport chain reactions. The mitochondria then release carbon dioxide, water, and ATP as their end products.

Hormones

Hormones are chemical messengers secreted by a variety of glands in response to altered conditions in the body. Each hormone travels in the blood to all parts of the body, but only its specific target tissues or organs possess receptors to accept it. Only then can the hormone elicit a response to restore homeostasis.

The hormones, the glands they originate in, their target organs, and their effects are described in this section. Many of the hormones you might be interested in are included, but only a few are discussed in detail. Figure A-2 identifies the glands that produce the hormones, and the accompanying glossary defines the hormones discussed in this section.

APPENDIX A

♦ The pituitary gland in the brain has two parts—the *anterior* (front) and the *posterior* (hind).

Hormones of the Pituitary Gland and Hypothalamus The anterior pituitary gland ♦ produces the following hormones, each of which acts on one or more target organs and elicits a characteristic response:

- **Adrenocorticotropin (ACTH)** acts on the adrenal cortex, promoting the production and release of its hormones.
- **Thyroid-stimulating hormone** (TSH) acts on the thyroid gland, promoting the production and release of thyroid hormones.
- **Growth hormone (GH)** or **somatotropin** acts on all tissues, promoting growth, fat breakdown, and the formation of antibodies.
- **Follicle-stimulating hormone (FSH)** acts on the ovaries in the female, promoting their maturation, and on the testicles in the male, promoting sperm formation.
- **Luteinizing hormone (LH)** also acts on the ovaries, stimulating their maturation, the production and release of progesterone and estrogens, and ovulation; and on the testicles, promoting the production and release of testosterone.
- **Prolactin,** secreted in the female during pregnancy and lactation, acts on the mammary glands to stimulate their growth and the production of milk.

hormones: chemical messengers. Hormones are secreted by a variety of endocrine glands in response to altered conditions in the body. Each hormone travels to one or more specific target tissues or organs, where it elicits a specific response to maintain homeostasis.

GLOSSARY OF HORMONES

adrenocorticotropin (ad-REE-noh-KORE-tee-koh-TROP-in) or **ACTH:** a hormone, so named because it stimulates *(trope)* the adrenal cortex. The adrenal gland, like the pituitary, has two parts, in this case an outer portion *(cortex)* and an inner core *(medulla)*. The release of ACTH is mediated by *corticotropin-releasing hormone (CRH)*.

aldosterone: a hormone secreted by the adrenal glands that regulates blood pressure by increasing the reabsorption of sodium by the kidneys.

angiotensin: a hormone involved in blood pressure regulation.

antidiuretic hormone (ADH): a hormone produced by the pituitary gland in response to dehydration (or a high sodium concentration in the blood) that stimulates the kidneys to reabsorb more water and therefore to excrete less. In addition to its antidiuretic effect, ADH elevates blood pressure and so is also called *vasopressin* (VAS-oh-PRES-in).

calcitonin (KAL-see-TOH-nin): a hormone secreted by the thyroid gland that regulates calcium by lowering it when levels rise too high.

erythropoietin (eh-RITH-ro-POY-eh-tin): a hormone that stimulates red blood cell production.

estrogens: hormones responsible for the menstrual cycle and other female characteristics.

follicle-stimulating hormone (FSH): a hormone that stimulates maturation of the ovarian follicles in females and the production of sperm in males. (The ovarian follicles are part of the female reproductive system where the eggs are produced.) The release of FSH is mediated by *follicle-stimulating hormone releasing hormone (FSH–RH)*.

glucocorticoids: hormones from the adrenal cortex that affect the body's management of glucose.

growth hormone (GH): a hormone secreted by the pituitary that regulates the cell division and protein synthesis needed for normal growth (also called *somatotropin)*. The release of GH is mediated by *GH-releasing hormone (GHRH)* and *GH-inhibiting hormone (GHIH)*.

luteinizing (LOO-tee-in-EYE-zing) **hormone (LH):** a hormone that stimulates ovulation and the development of the corpus luteum (the small tissue that develops from a ruptured ovarian follicle and secretes hormones). In men, LH stimulates testosterone secretion. The release of LH is mediated by *luteinizing hormone-releasing hormone (LH–RH)*.

oxytocin (OCK-see-TOH-sin): a hormone that stimulates the mammary glands to eject milk during lactation and the uterus to contract during childbirth.

progesterone: the hormone of gestation (pregnancy).

prolactin (proh-LAK-tin): a hormone secreted from the anterior pituitary gland that acts on the mammary glands to promote the production of milk. The release of prolactin is mediated by *prolactin-inhibiting hormone (PIH)*.

relaxin: the hormone of late pregnancy.

renin (REN-in): an enzyme from the kidneys that hydrolyzes the protein angiotensinogen to angiotensin, which results in the kidneys reabsorbing sodium.

somatostatin (GHIH): a hormone that inhibits the release of growth hormone; the opposite of *somatotropin (GH)*.

testosterone: a steroid hormone from the testicles, or testes. The steroids, as explained in Chapter 5, are chemically related to, and some are derived from, the lipid cholesterol.

thyroid-stimulating hormone (TSH): a hormone secreted by the pituitary that stimulates the thyroid gland to secrete its hormones— thyroxin and triiodothyronine. The release of TSH is mediated by *TSH-releasing hormone (TRH)*.

> FIGURE A-2 The Endocrine System

These organs and glands release hormones that regulate body processes. An *endocrine gland* secretes its product directly into *(endo)* the blood; for example, the pancreas cells that produce insulin. An *exocrine gland* secretes its product(s) out *(exo)* to an epithelial surface either directly or through a duct; the sweat glands of the skin and the enzyme-producing glands of the pancreas are both examples. The pancreas is therefore both an endocrine and an exocrine gland.

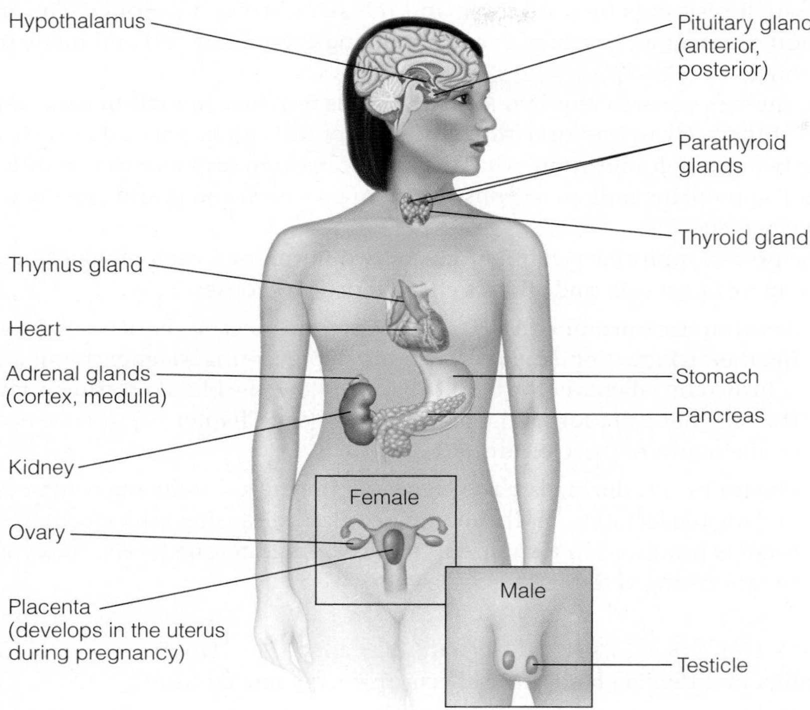

Each of these hormones has one or more signals that turn it on and another (or others) that turns it off. ♦ Among the controlling signals are several hormones from the hypothalamus:

♦ Hormones that are turned off by their own effects are said to be regulated by *negative feedback* (see Figure 3-12 on p. 84).

- **Corticotropin-releasing hormone (CRH),** which promotes release of ACTH, is turned on by stress and turned off by ACTH when enough has been released.
- **TSH-releasing hormone (TRH),** which promotes release of TSH, is turned on by large meals or low body temperature.
- **GH-releasing hormone (GHRH),** which stimulates the release of growth hormone, is turned on by insulin.
- **GH-inhibiting hormone (GHIH** or **somatostatin),** which inhibits the release of GH and interferes with the release of TSH, is turned on by hypoglycemia and/or physical activity and is rapidly destroyed by body tissues so that it does not accumulate.
- **FSH/LH-releasing hormone (FSH/LH-RH)** is turned on in the female by nerve messages or low estrogen and in the male by low testosterone.
- **Prolactin-inhibiting hormone (PIH)** is turned on by high prolactin levels and off by estrogen, testosterone, and suckling (by way of nerve messages).

Let's examine some of these controls. PIH, for example, responds to high prolactin levels (remember, prolactin promotes milk production). High prolactin levels ensure that milk is made and—by calling forth PIH—ensure that prolactin levels don't get too high. But when the infant is suckling—and creating a demand for milk—PIH is not allowed to work (suckling turns off PIH). The consequence: prolactin remains high, and milk production continues. Demand from the infant thus directly adjusts the supply of milk. The need is met through the interaction of the nerves and hormones.

As another example, consider CRH. Stress, perceived in the brain and relayed to the hypothalamus, switches on CRH. On arriving at the pituitary, CRH switches on ACTH. Then ACTH acts on its target organ, the adrenal cortex, which responds by producing and releasing stress hormones. The stress hormones trigger a cascade of events involving every body cell and many other hormones.

The numerous steps required to set the stress response in motion make it possible for the body to fine-tune the response; control can be exerted at each step. These two examples illustrate what the body can do in response to two different stimuli—producing milk in response to an infant's need and gearing up for action in an emergency.

The posterior pituitary gland produces two hormones, each of which acts on one or more target cells and elicits a characteristic response:

- **Antidiuretic hormone (ADH),** or **vasopressin,** acts on the arteries, promoting their contraction, and on the kidneys, preventing water excretion. ADH is turned on whenever the blood volume is low, the blood pressure is low, or the salt concentration of the blood is high (see Chapter 12). It is turned off by the return of these conditions to normal.
- **Oxytocin** acts during late pregnancy on the uterus, inducing contractions, and during lactation on the mammary glands, causing milk ejection. Oxytocin is produced in response to reduced progesterone levels, suckling, or the stretching of the cervix.

Hormones that Regulate Energy Metabolism Hormones produced by a number of different glands have effects on energy metabolism:

- Insulin from the pancreas beta cells is turned on by many stimuli, including high blood glucose. It acts on cells to increase glucose and amino acid uptake into them and to promote the secretion of GHRH.
- Glucagon from the pancreas alpha cells responds to low blood glucose and acts on the liver to promote the breakdown of glycogen to glucose, the conversion of amino acids to glucose, and the release of glucose into the blood.
- Thyroxin from the thyroid gland responds to TSH and acts on many cells to increase their metabolic rate, growth, and heat production.
- Norepinephrine and epinephrine ♦ from the adrenal medulla respond to stimulation by sympathetic nerves and produce reactions in many cells that facilitate the body's readiness for fight or flight: increased heart activity, blood vessel constriction, breakdown of glycogen and glucose, raised blood glucose levels, and fat breakdown. Norepinephrine and epinephrine also influence the secretion of the many hormones from the hypothalamus that exert control on the body's other systems.
- Growth hormone (GH) from the anterior pituitary (already mentioned).
- **Glucocorticoids** from the adrenal cortex become active during times of stress and carbohydrate metabolism.

♦ Norepinephrine and epinephrine were formerly called *noradrenalin* and *adrenalin*, respectively.

Every body part is affected by these hormones. Each different hormone has unique effects; and hormones that oppose each other are produced in carefully regulated amounts, so each can respond to the exact degree that is appropriate to the condition.

Hormones that Adjust Other Body Balances Hormones are involved in moving calcium into and out of the body's storage deposits in the bones:

- **Calcitonin** from the thyroid gland acts on the bones, which respond by storing calcium from the bloodstream whenever blood calcium rises above the normal range. It also acts on the kidneys to increase excretion of both calcium and phosphorus in the urine. Calcitonin plays a major role in infants and young children, but is less active in adults.

- Parathyroid hormone (parathormone or PTH) from the parathyroid gland responds to the opposite condition—lowered blood calcium—and acts on three targets: the bones, which release stored calcium into the blood; the kidneys, which slow the excretion of calcium; and the intestine, which increases calcium absorption.
- Vitamin D from the skin and activated in the kidneys acts with parathyroid hormone and is essential for the absorption of calcium in the intestine.

Figure 12-12 on p. 386 diagrams the ways vitamin D and the hormones calcitonin and parathyroid hormone regulate calcium homeostasis.

Another hormone has effects on blood-making activity:

- **Erythropoietin** from the kidneys is responsive to oxygen depletion of the blood and to anemia. It acts on the bone marrow to stimulate the making of red blood cells.

Another hormone is special for pregnancy:

- **Relaxin** from the ovaries is secreted in response to the raised progesterone and estrogen levels of late pregnancy. This hormone acts on the cervix and pelvic ligaments to allow them to stretch so that they can accommodate the birth process without strain.

Other agents help regulate blood pressure:

- **Renin** (an enzyme), from the kidneys, in cooperation with **angiotensin** in the blood responds to a reduced blood supply experienced by the kidneys and acts in several ways to increase blood pressure. Renin and angiotensin also stimulate the adrenal cortex to secrete the hormone aldosterone.
- **Aldosterone,** a hormone from the adrenal cortex, targets the kidneys, which respond by reabsorbing sodium. The effect is to retain more water in the bloodstream—thus, again, raising the blood pressure. Figure 12-3 (on p. 372) in Chapter 12 provides more details.

The Gastrointestinal Hormones Several hormones are produced in the stomach and intestines in response to the presence of food or the components of food:

- Gastrin from the stomach and duodenum stimulates the production and release of gastric acid and other digestive juices and the movement of the GI contents through the system.
- Cholecystokinin from the duodenum signals the gallbladder and pancreas to release their contents into the intestine to aid in digestion.
- Secretin from the duodenum calls forth acid-neutralizing bicarbonate from the pancreas into the intestine and slows the action of the stomach and its secretion of acid and digestive juices.
- Gastric-inhibitory peptide from the duodenum and jejunum inhibits the secretion of gastric acid and slows the process of digestion.

These hormones are defined and discussed in Chapter 3.

The Sex Hormones There are three major sex hormones:

- **Testosterone** from the testicles is released in response to LH (described earlier) and acts on all the tissues that are involved in male sexuality, promoting their development and maintenance.
- **Estrogens** from the ovaries are released in response to both FSH and LH and act similarly in females.
- **Progesterone** from the ovaries' corpus luteum and from the placenta acts on the uterus and mammary glands, preparing them for pregnancy and lactation.

This brief description of the hormones and their functions should suffice to provide an awareness of the enormous impact these compounds have on body processes. The body's other overall regulating agency is the nervous system.

Nerves

The nervous system has a central control system that can evaluate information about conditions within and outside the body, and a vast system of wiring that receives information and sends instructions. The control unit is the brain and spinal cord, called the **central nervous system;** and the vast complex of wiring between the center and the parts is the **peripheral nervous system.** The smooth functioning that results from the systems' adjustments to changing conditions is homeostasis.

The nervous system has two general functions: it controls voluntary muscles in response to sensory stimuli from them, and it controls involuntary, internal muscles and glands in response to nerve-borne and chemical signals about their status. In fact, the nervous system is best understood as two systems that use the same or similar pathways to receive and transmit their messages. The **somatic nervous system** controls the voluntary muscles; the **autonomic nervous system** controls the internal organs.

When scientists were first studying the autonomic nervous system, they noticed that when something hurt one organ of the body, some of the other organs reacted as if in sympathy for the afflicted one. They therefore named the nerve network they were studying the sympathetic nervous system. The term is still used today to refer to that branch of the autonomic nervous system that responds to pain and stress. The other branch is called the parasympathetic nervous system. (Think of the sympathetic branch as the responder when homeostasis needs restoring and the parasympathetic branch as the commander of function during normal times.) Both systems transmit their messages through the brain and spinal cord. Nerves of the two branches travel side by side along the same pathways to transmit their messages, but they oppose each other's actions (see Figure A-3).

An example will show how the sympathetic and parasympathetic nervous systems work to maintain homeostasis. When you go outside in cold weather, your skin's temperature receptors send "cold" messages to the spinal cord and brain.

Your conscious mind may intervene at this point to tell you to zip your jacket, but let's say you have no jacket. Your sympathetic nervous system reacts to the external stressor, the cold. It signals your skin-surface capillaries to shut down so that your blood will circulate deeper in your tissues, where it will conserve heat. Your sympathetic nervous system also signals involuntary contractions of the small muscles just under the skin surface. The product of these muscle contractions is heat, and the visible result is goose bumps. If these measures do not raise your body temperature enough, then the sympathetic nerves signal your large muscle groups to shiver; the contractions of these large muscles produce still more heat. All of this activity helps to maintain your homeostasis (with respect to temperature) under conditions of external extremes (cold) that would throw it off balance. The cold was a stressor; the body's response was resistance.

GLOSSARY OF NERVOUS SYSTEM

autonomic nervous system: the division of the nervous system that controls the body's automatic responses. Its two branches are the *sympathetic* branch, which helps the body respond to stressors from the outside environment, and the *parasympathetic* branch, which regulates normal body activities between stressful times.

central nervous system: the central part of the nervous system; the brain and spinal cord.

peripheral (puh-RIFF-er-ul) **nervous system:** the peripheral (outermost) part of the nervous system; the vast complex of wiring that extends from the central nervous system to the body's outermost areas. It contains both *somatic* and *autonomic* components.

somatic (so-MAT-ick) **nervous system:** the division of the nervous system that controls the voluntary muscles, as distinguished from the autonomic nervous system, which controls involuntary functions.

> FIGURE A-3 The Organization of the Nervous System

The brain and spinal cord evaluate information about conditions within and outside the body, and the peripheral nerves receive information and send instructions.

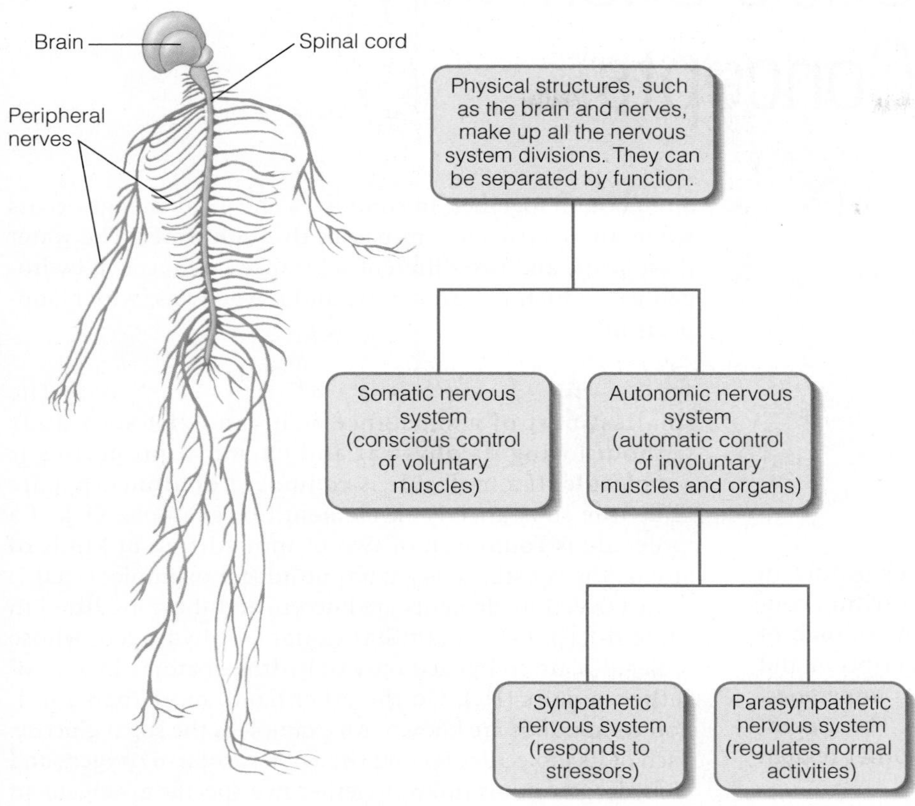

Now let's say you come in and sit by a fire and drink hot cocoa. You are warm and no longer need all that sympathetic activity. At this point, your parasympathetic nerves take over; they signal your skin-surface capillaries to dilate again, your goose bumps to subside, and your muscles to relax. Your body is back to normal. This is recovery.

Putting It Together

The hormonal and nervous systems coordinate body functions by transmitting and receiving messages. The point-to-point messages of the nervous system travel through a central switchboard (the spinal cord and brain), whereas the messages of the hormonal system are broadcast over the airways (the bloodstream), and any organ with the appropriate receptors can pick them up. Nerve impulses travel faster than hormonal messages do—although both are remarkably swift. Whereas your brain's command to wiggle your toes reaches the toes within a fraction of a second and stops as quickly, a gland's message to alter a body condition may take several seconds or minutes to get started and may fade away equally slowly.

Together, the two systems possess every characteristic a superb communication network needs: varied speeds of transmission, along with private communication lines or public broadcasting systems, depending on the needs of the moment. The hormonal system, together with the nervous system, integrates the whole body's functioning so that all parts act smoothly together.

Appendix B Basic Chemistry Concepts

CONTENTS

This appendix is intended to provide the background in basic chemistry you need to understand the nutrition concepts presented in this book. Chemistry is the branch of natural science that is concerned with the description and classification of **matter**, the changes that matter undergoes, and the **energy** associated with these changes. The accompanying glossary defines matter, energy, and other related terms.

Matter: The Properties of Atoms

Every substance has physical and chemical properties that distinguish it from all other substances and thus give it a unique identity. The physical properties include such characteristics as color, taste, texture, and odor, as well as the temperatures at which a substance changes its state (from a solid to a liquid or from a liquid to a gas) and the weight of a unit volume (its density). The chemical properties of a substance have to do with how it reacts with other substances or responds to a change in its environment so that new substances with different sets of properties are produced.

A physical change does not change a substance's chemical composition. The three physical states—ice, water, and steam—all consist of two hydrogen atoms and one oxygen atom bound together. In contrast, a chemical change occurs when an electric current passes through water. The water disappears, and two different substances are formed: hydrogen gas, which is flammable, and oxygen gas, which supports life.

Substances: Elements and Compounds The smallest part of a substance that can exist separately without losing its physical and chemical properties is a **molecule**. If a molecule is composed of **atoms** that are alike, the substance is an **element** (for example, O_2). If a molecule is **composed** of two or more different kinds of atoms, the substance is a **compound** (for example, H_2O).

Just over 100 elements are known, and these are listed in Table B-1 (p. B-1). A familiar example is hydrogen, whose molecules are composed only of hydrogen atoms linked together in pairs (H_2). On the other hand, more than a million compounds are known. An example is the sugar glucose. Each of its molecules is composed of 6 carbon, 6 oxygen, and 12 hydrogen atoms linked together in a specific arrangement (as described in Chapter 4).

The Nature of Atoms Atoms themselves are made of smaller particles. Within an atom's nucleus are protons (positively charged particles), and surrounding the nucleus are an equal number of electrons (negatively charged particles). The number of protons in the nucleus of an atom determines the atomic number. The positive charge on a proton is equal to the negative charge on an electron, so the charges cancel each other out and leave the atom neutral to its surroundings.

The nucleus may also include neutrons, subatomic particles that have no charge. Protons and neutrons are of equal mass, and together they give an atom its atomic mass. Electrons bond atoms together to make molecules, and they are involved in chemical reactions.

GLOSSARY

anion: a negatively changed ion.

atoms: the smallest components of an element that have all of the properties of the element.

cation: a positively changed ion.

compound: a substance composed of two or more different atoms—for example, water (H_2O).

covalent bonds: strong chemical bonds formed between atoms by sharing electrons.

element: a substance composed of atoms that are alike—for example, iron (Fe).

energy: the capacity to do work.

ion: an atom or group of atoms that have gained or lost one or more electrons and therefore have a negative or positive electrical charge.

matter: anything that takes up space and has mass.

molecule: two or more atoms of the same or different elements joined by chemical bonds. Examples are molecules of the element oxygen, composed of two oxygen atoms (O_2), and molecules of the compound water, composed of two hydrogen atoms and one oxygen atom (H_2O).

TABLE B-1 The Elements

Number of Protons (Atomic Number)	Element	Number of Electrons in Outer Shell
1	Hydrogen (H)	1
2	Helium (He)	2
3	Lithium (Li)	1
4	Beryllium (Be)	2
5	Boron (B)	3
6	Carbon (C)	4
7	Nitrogen (N)	5
8	Oxygen (O)	6
9	Fluorine (F)	7
10	Neon (Ne)	8
11	Sodium (Na)	1
12	Magnesium (Mg)	2
13	Aluminum (Al)	3
14	Silicon (Si)	4
15	Phosphorus (P)	5
16	Sulfur (S)	6
17	Chlorine (Cl)	7
18	Argon (Ar)	8
19	Potassium (K)	1
20	Calcium (Ca)	2
21	Scandium (Sc)	2
22	Titanium (Ti)	2
23	Vanadium (V)	2
24	Chromium (Cr)	1
25	Manganese (Mn)	2
26	Iron (Fe)	2
27	Cobalt (Co)	2
28	Nickel (Ni)	2
29	Copper (Cu)	1
30	Zinc (Zn)	2
31	Gallium (Ga)	3
32	Germanium (Ge)	4
33	Arsenic (As)	5
34	Selenium (Se)	6
35	Bromine (Br)	7
36	Krypton (Kr)	8
37	Rubidium (Rb)	1
38	Strontium (Sr)	2
39	Yttrium (Y)	2
40	Zirconium (Zr)	2
41	Niobium (Nb)	1
42	Molybdenum (Mo)	1
43	Technetium (Tc)	1
44	Ruthenium (Ru)	1
45	Rhodium (Rh)	1
46	Palladium (Pd)	—
47	Silver (Ag)	1
48	Cadmium (Cd)	2
49	Indium (In)	3
50	Tin (Sn)	4
51	Antimony (Sb)	5
52	Tellurium (Te)	6
53	Iodine (I)	7
54	Xenon (Xe)	8
55	Cesium (Cs)	1
56	Barium (Ba)	2
57	Lanthanum (La)	2
58	Cerium (Ce)	2
59	Praseodymium (Pr)	2
60	Neodymium (Nd)	2
61	Promethium (Pm)	2
62	Samarium (Sm)	2
63	Europium (Eu)	2
64	Gadolinium (Gd)	2
65	Terbium (Tb)	2
66	Dysprosium (Dy)	2
67	Holmium (Ho)	2
68	Erbium (Er)	2
69	Thulium (Tm)	2
70	Ytterbium (Yb)	2
71	Lutetium (Lu)	2
72	Hafnium (Hf)	2
73	Tantalum (Ta)	2
74	Tungsten (W)	2
75	Rhenium (Re)	2
76	Osmium (Os)	2
77	Iridium (Ir)	2
78	Platinum (Pt)	1
79	Gold (Au)	1
80	Mercury (Hg)	2
81	Thallium (Tl)	3
82	Lead (Pb)	4
83	Bismuth (Bi)	5
84	Polonium (Po)	6
85	Astatine (At)	7
86	Radon (Rn)	8
87	Francium (Fr)	1
88	Radium (Ra)	2
89	Actinium (Ac)	2
90	Thorium (Th)	2
91	Protactinium (Pa)	2
92	Uranium (U)	2
93	Neptunium (Np)	2
94	Plutonium (Pu)	2
95	Americium (Am)	2
96	Curium (Cm)	2
97	Berkelium (Bk)	2
98	Californium (Cf)	2
99	Einsteinium (Es)	2
100	Fermium (Fm)	2
101	Mendelevium (Md)	2
102	Nobelium (No)	2
103	Lawrencium (Lr)	2
104	Rutherfordium (Rf)	2
105	Dubnium (Db)	2
106	Seaborgium (Sg)	2
107	Bohrium (Bh)	2
108	Hassium (Hs)	2
109	Meitnerium (Mt)	2
110	Darmstadtium (Ds)	2

Key
- Elements found in energy-yielding nutrients, vitamins, and water
- Major minerals
- Trace minerals

Each element has a characteristic number of protons in its atom's nucleus. For example, the hydrogen atom (the simplest of all) possesses a single proton, with a single electron associated with it:

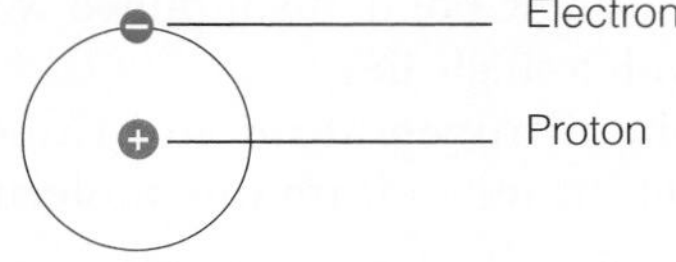

Hydrogen atom (H), atomic number 1

Just as hydrogen always has one proton, helium always has two, lithium three, and so on. The atomic number of each element is the number of protons in the nucleus of that atom, and this never changes in a chemical reaction; it gives the atom its identity. The atomic numbers for the known elements are listed in Table B-1.

In addition to hydrogen, the atoms most common in living things are carbon (C), nitrogen (N), and oxygen (O), whose atomic numbers are 6, 7, and 8, respectively. Their structures are more complicated than that of hydrogen, but each of them possesses the same number of electrons as there are protons in the nucleus. These electrons are found in orbits, or shells (shown below).

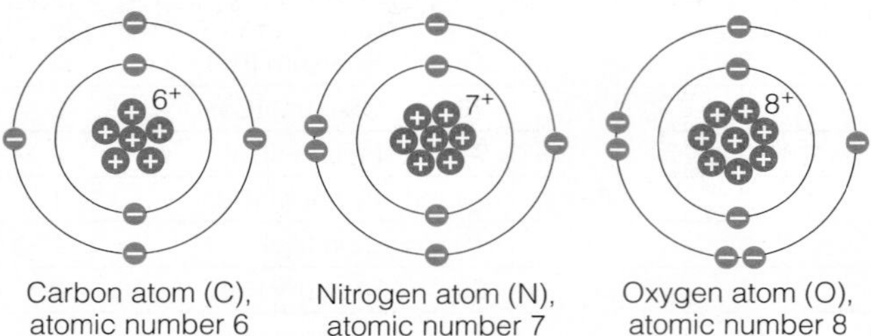

In these and all diagrams of atoms that follow, only the protons and electrons are shown. The neutrons, which contribute only to atomic weight, not to charge, are omitted.

The most important structural feature of an atom for determining its chemical behavior is the number of electrons in its outermost shell. The first, or innermost, shell is full when it is occupied by two electrons; so an atom with two or more electrons has a filled first shell. When the first shell is full, electrons begin to fill the second shell.

The second shell is completely full when it has eight electrons. A substance that has a full outer shell tends not to enter into chemical reactions. Atomic number 10, neon, is a chemically inert substance because its outer shell is complete. Fluorine, atomic number 9, has a great tendency to draw an electron from other substances to complete its outer shell, and thus it is highly reactive. Carbon has a half-full outer shell, which helps explain its great versatility; it can combine with other elements in a variety of ways to form a large number of compounds.

Atoms seek to reach a state of maximum stability or of lowest energy in the same way that a ball will roll down a hill until it reaches the lowest place. An atom achieves a state of maximum stability:

- By gaining or losing electrons to either fill or empty its outer shell.
- By sharing its electrons with other atoms and thereby completing its outer shell.

The number of electrons determines how the atom will chemically react with other atoms.

Chemical Bonding

Atoms often complete their outer shells by sharing electrons with other atoms. In order to complete its outer shell, a carbon atom requires four electrons. A hydrogen atom requires one. Thus, when a carbon atom shares electrons with four hydrogen atoms, each completes its outer shell (as shown below). Electron sharing binds the atoms together and satisfies the conditions of maximum stability for the molecule. The outer shell of each atom is complete because hydrogen effectively has the required 2 electrons in its first (outer) shell, and carbon has 8 electrons in its second (outer) shell; and the molecule is electrically neutral, with a total of 10 protons and 10 electrons.

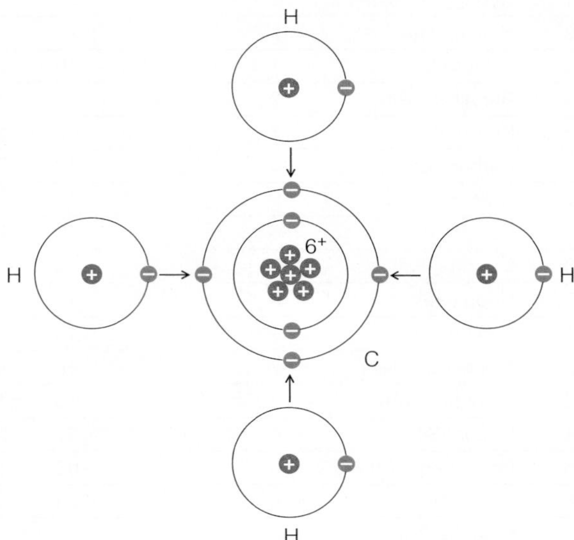

When a carbon atom shares electrons with four hydrogen atoms, a methane molecule is made.

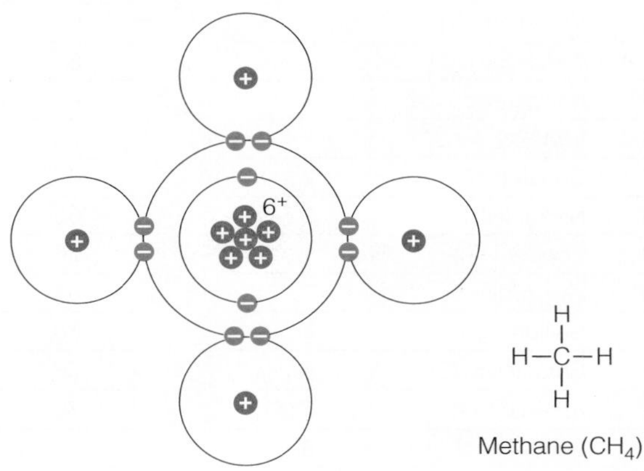

The chemical formula for methane is CH_4. Note that by sharing electrons, every atom achieves a filled outer shell.

Bonds that involve the sharing of electrons, like the bonds between carbon and the four hydrogens, are the most stable kind of association that atoms can form with one another. These bonds are called **covalent bonds,** and the resulting combination of atoms is called a molecule. A single pair of shared electrons forms a single bond. A simplified way to represent a single bond is with a single line.

Similarly, one nitrogen atom and three hydrogen atoms can share electrons to form one molecule of ammonia (NH_3):

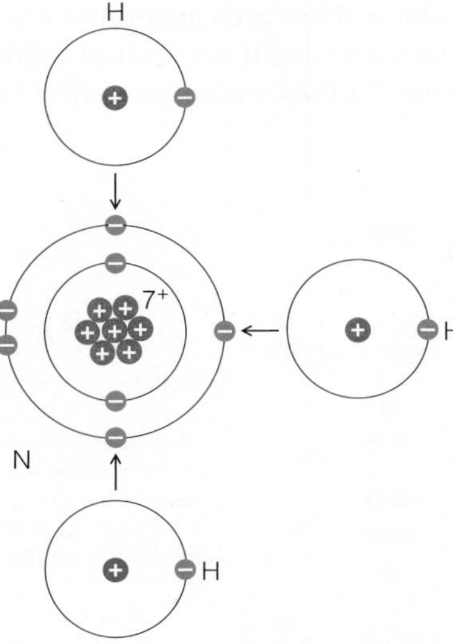

When a nitrogen atom shares electrons with three hydrogen atoms, an ammonia molecule is made.

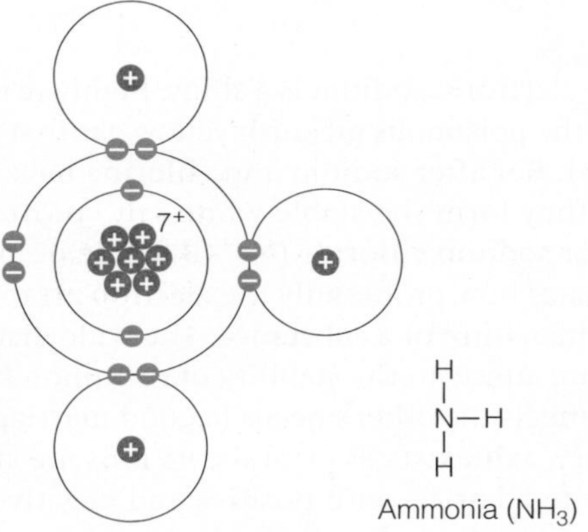

Ammonia (NH_3)

The chemical formula for ammonia is NH_3. Count the electrons in each atom's outer shell to confirm that it is filled.

One oxygen atom may be bonded to two hydrogen atoms to form one molecule of water (H_2O):

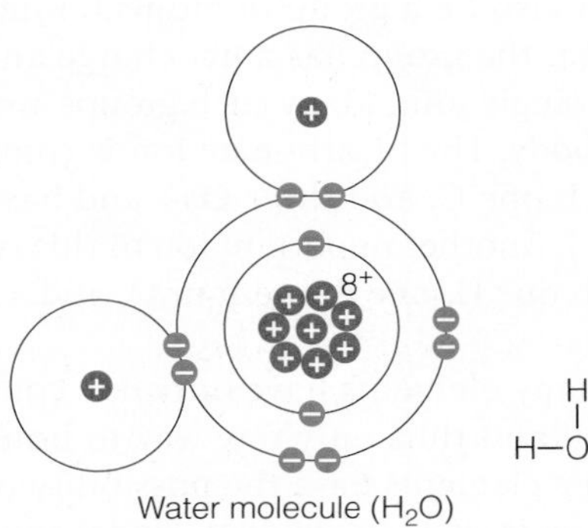

Water molecule (H_2O)

When two oxygen atoms form a molecule of oxygen, they must share two pairs of electrons. This double bond may be represented as two single lines:

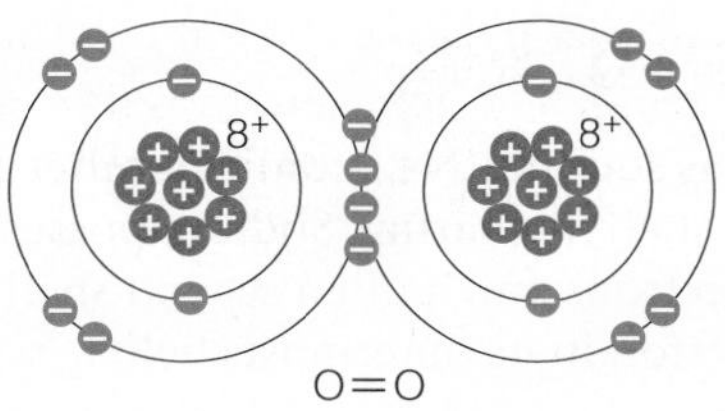

Oxygen molecule (O_2)

Small atoms form the tightest, most stable bonds. H, O, N, and C are the smallest atoms capable of forming one, two, three, and four electron-pair bonds, respectively. This is the basis for the statement in Chapter 4 that in drawings of compounds containing these atoms, hydrogen must always have one, oxygen two, nitrogen three, and carbon four bonds radiating to other atoms:

H—   —O—   —N— (with one bond up)   —C— (with one bond up and one bond down)

The stability of the associations between these small atoms and the versatility with which they can combine make them very common in living things. Interestingly all cells—whether they come from animals, plants, or bacteria—contain the same elements in very nearly the same proportions. The elements commonly found in living things are shown in Table B-2.

TABLE B-2 Elemental Composition of the Human Body

Element	Chemical Symbol	By Weight (%)
Oxygen	O	65.0
Carbon	C	18.0
Hydrogen	H	10.0
Nitrogen	N	3.0
Calcium	Ca	1.5
Phosphorus	P	1.0
Potassium	K	0.4
Sulfur	S	0.3
Sodium	Na	0.2
Chloride	Cl	0.1
Magnesium	Mg	0.1
Total		99.6[a]

[a]The remaining 0.4 percent by weight is contributed by the trace elements: chromium (Cr), copper (Cu), zinc (Zn), selenium (Se), molybdenum (Mo), fluorine (F), iodine (I), manganese (Mn), and iron (Fe). Cells may also contain variable traces of some of the following: boron (B), cobalt (Co), Lithium (Li), strontium (Sr), aluminum (AL), silicon (Si), Lead (Pb), vanadium (V), arsenic (As), bromine (Br), and others.

Formation of Ions

An atom such as sodium (Na, atomic number 11) cannot easily fill its outer shell by sharing. Sodium possesses a filled first shell of two electrons and a filled second shell of eight; there is only one electron in its outermost shell:

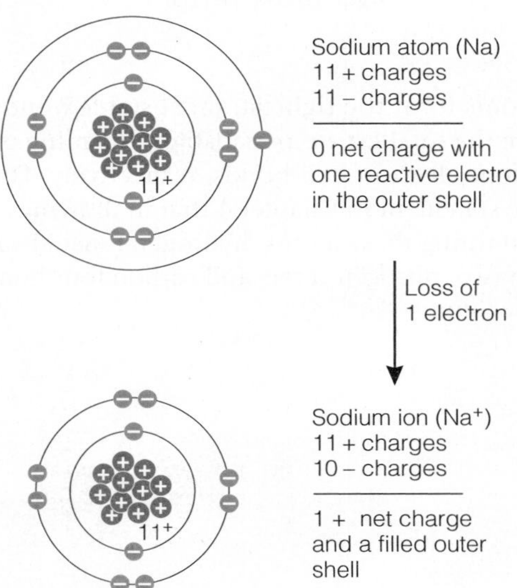

If sodium loses this electron, it satisfies one condition for stability: a filled outer shell (now its second shell counts as the outer shell). However, it is not electrically neutral. It has 11 protons (positive) and only 10 electrons (negative). It therefore has a net positive charge. An atom or molecule that has lost or gained one or more electrons and so is electrically charged is called an **ion**.

An atom such as chlorine (Cl, atomic number 17), with seven electrons in its outermost shell, can share electrons to fill its outer shell, or it can gain one electron to complete its outer shell and thus give it a negative charge:

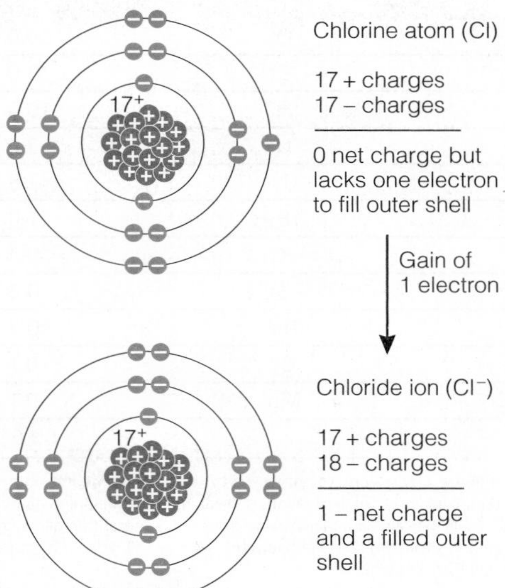

A positively charged ion such as sodium ion (Na^+) is called a **cation**; a negatively charged ion such as a chloride ion (Cl^-) is called an **anion**. Cations and anions attract one another to form salts:

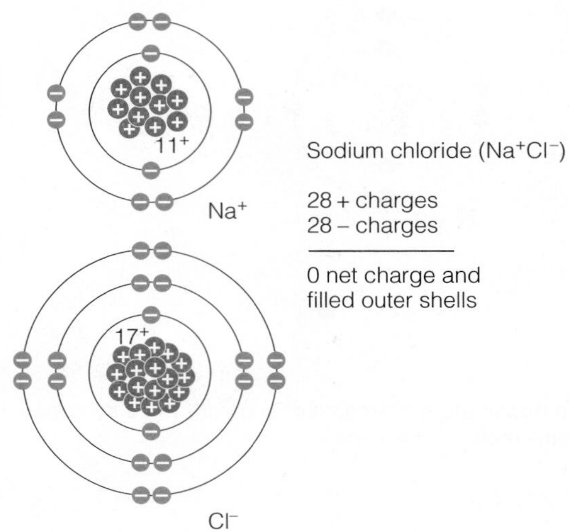

With all its electrons, sodium is a shiny, highly reactive metal; chlorine is the poisonous greenish yellow gas that was used in World War I. But after sodium and chlorine have transferred electrons, they form the stable white salt familiar to you as table salt, or sodium chloride (Na^+Cl^-). The dramatic difference illustrates how profoundly the electron arrangement can influence the nature of a substance. The wide distribution of salt in nature attests to the stability of the union between the ions. Each meets the other's needs (a good marriage).

When dry, salt exists as crystals; its ions are stacked very regularly into a lattice, with positive and negative ions alternating in a three-dimensional checkerboard structure. In water, however, the salt quickly dissolves, and its ions separate from one another, forming an electrolyte solution in which they move about freely. Covalently bonded molecules rarely dissociate like this in a water solution. The most common exception is when they behave like acids and release H^+ ions, as discussed in the next section.

An ion can also be a group of atoms bound together in such a way that the group has a net charge and enters into reactions as a single unit. Many such groups are active in the fluids of the body. The bicarbonate ion is composed of five atoms—one H, one C, and three Os—and has a net charge of −1 (HCO_3^-). Another important ion of this type is a phosphate ion with one H, one P, and four O, and a net charge of −2 (HPO_4^{-2}).

Whereas many elements have only one configuration in the outer shell and thus only one way to bond with other elements, some elements have the possibility of varied configurations. Iron is such an element. Under some conditions iron loses two electrons, and under other circumstances it loses three. If iron loses two electrons, it then has a net charge of +2, and we call it ferrous iron (Fe^{++}). If it donates three electrons to another atom, it becomes the +3 ion, or ferric iron (Fe^{+++}).

Ferrous iron (Fe^{++})	Ferric iron (Fe^{+++})
(had 2 outer-shell electrons but has lost them)	(had 3 outer-shell electrons but has lost them)
26 + charges	26 + charges
24 − charges	23 − charges
2 + net charge	3 + net charge

Remember that a positive charge on an ion means that negative charges—electrons—have been lost and not that positive charges have been added to the nucleus.

Water, Acids, and Bases

Water The water molecule is electrically neutral, having equal numbers of protons and electrons. When a hydrogen atom shares its electron with oxygen, however, that electron will spend most of its time closer to the positively charged oxygen nucleus. This leaves the positive proton (nucleus of the hydrogen atom) exposed on the outer part of the water molecule. We know, too, that the two hydrogens both bond toward the same side of the oxygen. These two facts explain why water molecules are polar: they have regions of more positive and more negative charge.

Polar molecules like water are drawn to one another by the attractive forces between the positive polar areas of one and the negative poles of another. These attractive forces, sometimes known as polar bonds or hydrogen bonds, occur among many molecules and also within the different parts of single large molecules. Although very weak in comparison with covalent bonds, polar bonds may occur in such abundance that they become exceedingly important in determining the structure of such large molecules as proteins and DNA.

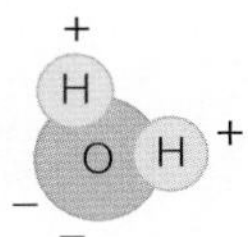

This diagram of the polar water molecule shows displacement of electrons toward the O nucleus; thus the negative region is near the O and the positive regions are near the H atoms.

Water molecules have a slight tendency to ionize, separating into positive (H^+) and negative (OH^-) ions. In pure water, a small but constant number of these ions is present, and the number of positive ions exactly equals the number of negative ions.

Acids An acid is a substance that releases H^+ ions (protons) in a water solution. Hydrochloric acid (HCl^-) is such a substance because it dissociates in a water solution into H^+ and Cl^- ions. Acetic acid is also an acid because it dissociates in water to acetate ions and free H^+:

```
   H  O             H  O
   |  ||            |  ||
H—C—C—O—H  ⟶  H—C—C—O⁻ + H⁺
   |                |
   H                H
```

Acetic acid dissociates into an acetate ion and a hydrogen ion.

The more H^+ ions released, the stronger the acid.

Chemists define degrees of acidity by means of the pH scale, which runs from 0 to 14. The pH expresses the concentration of H^+ ions: a pH of 1 is extremely acidic, 7 is neutral, and 13 is very basic. There is a tenfold difference in the concentration of H^+ ions between points on this scale. A solution with pH 3, for example, has 10 times as many H^+ ions as a solution with pH 4. At pH 7, the concentrations of free H^+ and OH^- are exactly the same—1/10,000,000 moles per liter.* At pH 4, the concentration of free H^+ ions is 1/10,000 moles per liter. This is a higher concentration of H^+ ions, and the solution is therefore acidic. Figure 3-6 on p. 75 presents the pH scale.

Bases A base is a substance that can combine with H^+ ions, thus reducing the acidity of a solution. The compound ammonia is such a substance. The ammonia molecule has two electrons that are not shared with any other atom; a hydrogen ion (H^+) is just a proton with no shell of electrons at all. The proton readily combines with the ammonia molecule to form an ammonium ion; thus a free proton is withdrawn from the solution and no longer contributes to its acidity. Many compounds containing nitrogen are important bases in living systems. Acids and bases neutralize each other to produce substances that are neither acid nor base.

```
  H                H
  |                |
:N—H + H⁺ ⟶  H—N⁺—H
  |                |
  H                H
```

Ammonia captures a hydrogen ion from water. The two dots here represent the two electrons not shared with another atom. These dots are ordinarily not shown in chemical structure drawings. Compare this drawing with the earlier diagram of an ammonia molecule (p. B-3).

Chemical Reactions

A chemical reaction results in the breakdown or formation of substances. Almost all such reactions involve a change in the bonding of atoms. Old bonds are broken, and new ones are formed. The nuclei of atoms are never involved in chemical reactions—only their outer-shell electrons participate. At the end of a chemical reaction, the number of atoms of each type is always the same as at the beginning. For example, two hydrogen molecules ($2H_2$) can react with one oxygen molecule (O_2) to form two water molecules ($2H_2O$). In this reaction two substances (hydrogen and oxygen) disappear, and a new one (water) is formed, but at the end of the reaction there are still four H atoms and two O atoms, just as there were at the beginning. Because the atoms are now linked in a different way, their characteristics or properties have changed.

In many instances chemical reactions involve not the relinking of molecules but the exchanging of electrons or protons among them. In such reactions the molecule that gains one or more electrons (or loses one or more hydrogen ions) is said to be reduced; the molecule that loses electrons (or gains protons) is oxidized. A hydrogen ion is equivalent to a proton.

*A mole is a certain number (about 6×10^{23}) of molecules. The pH of a solution is defined as the negative logarithm of the hydrogen ion concentration of the solution. Thus, if the concentration is 10^{-2} (moles per liter), the pH is 2; if 10^{-8}, the pH is 8; and so on.

Oxidation and reduction reactions take place simultaneously because an electron or proton that is lost by one molecule is accepted by another. The addition of an atom of oxygen is also oxidation because oxygen (with six electrons in the outer shell) accepts two electrons in becoming bonded. Oxidation, then, is loss of electrons, gain of protons, or addition of oxygen (with six electrons); reduction is the opposite—gain of electrons, loss of protons, or loss of oxygen. The addition of hydrogen atoms to oxygen to form water can thus be described as the reduction of oxygen *or* the oxidation of hydrogen.

If a reaction results in a net increase in the energy of a compound, it is called an endergonic, or "uphill," reaction (energy *erg*, is added into, *endo*, the compound). An example is the chief result of photosynthesis, the making of sugar in a plant from carbon dioxide and water using the energy of sunlight. Conversely, the oxidation of sugar to carbon dioxide and water is an exergonic, or "downhill," reaction because the end products have less energy than the starting products. Oftentimes, but not always, reduction reactions are endergonic, resulting in an increase in the energy of the products. Oxidation reactions often, but not always, are exergonic.

Chemical reactions tend to occur spontaneously if the end products are in a lower energy state and therefore are more stable than the reacting compounds. These reactions often give off energy in the form of heat as they occur. The generation of heat by wood burning in a fireplace and the maintenance of human body warmth both depend on energy-yielding chemical reactions. These downhill reactions occur easily although they may require some activation energy to get them started, just as a ball requires a push to start rolling.

Uphill reactions, in which the products contain more energy than the reacting compounds started with, do not occur until an energy source is provided. An example of such an energy source is the sunlight used in photosynthesis, where carbon dioxide and water (low-energy compounds) are combined to form the sugar glucose (a higher-energy compound). Another example is the use of the energy in glucose to combine two low-energy compounds in the body into the high-energy compound ATP (see Chapter 7). The energy in ATP may be used to power many other energy-requiring, uphill reactions. Clearly, any of many different molecules can be used as a temporary storage place for energy.

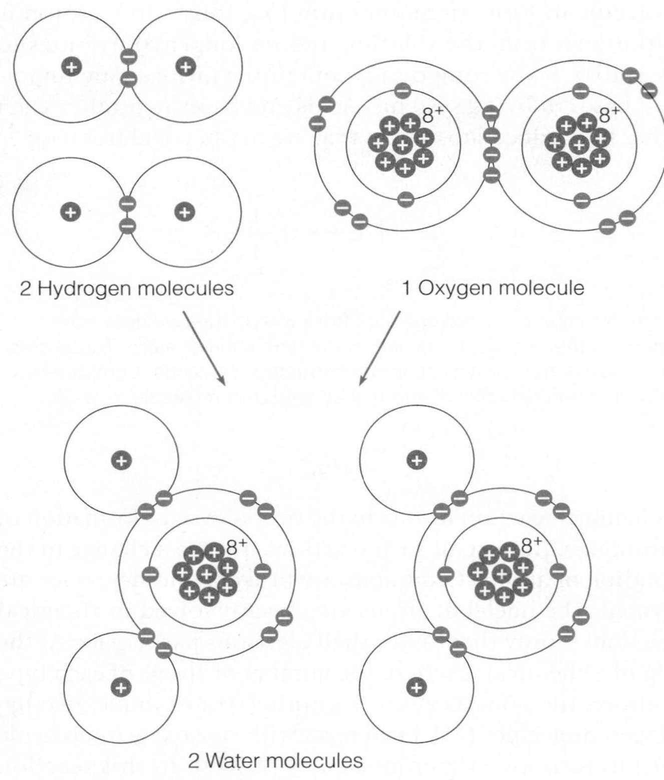

Hydrogen and oxygen react to form water.

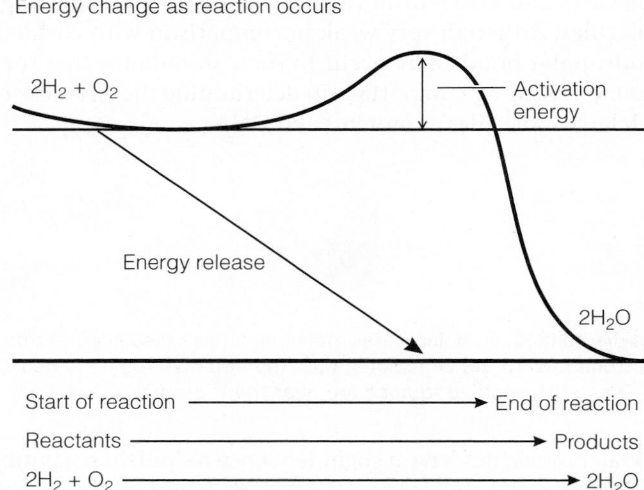

Formation of Free Radicals

Normally, when a chemical reaction takes place, bonds break and re-form with some redistribution of atoms and rearrangement of bonds to form new, stable compounds. Normally, bonds don't split in such a way as to leave a molecule with an odd, unpaired electron. When they do, free radicals are formed. Free radicals are highly unstable and quickly react with other compounds, forming more free radicals in a chain reaction. A cascade may ensue in which many highly reactive radicals are generated, resulting finally in the disruption of a living structure such as a cell membrane.

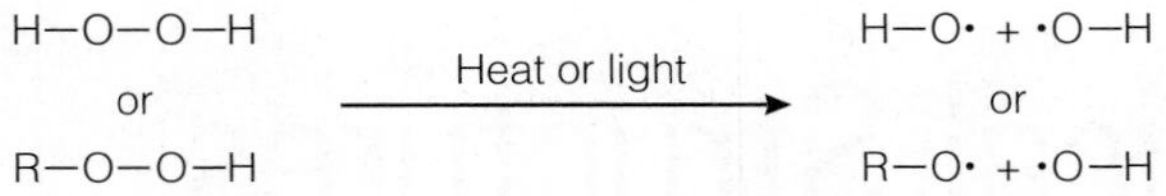

$$\text{H–O–O–H or R–O–O–H} \xrightarrow{\text{Heat or light}} \text{H–O}\cdot + \cdot\text{O–H or R–O}\cdot + \cdot\text{O–H}$$

Hydrogen peroxide or any hydroperoxide (R is any carbon chain with appropriate numbers of H) → Free radical

Free radicals are formed. The dots represent single electrons that are available for sharing (the atom needs another electron to fill its outer shell).

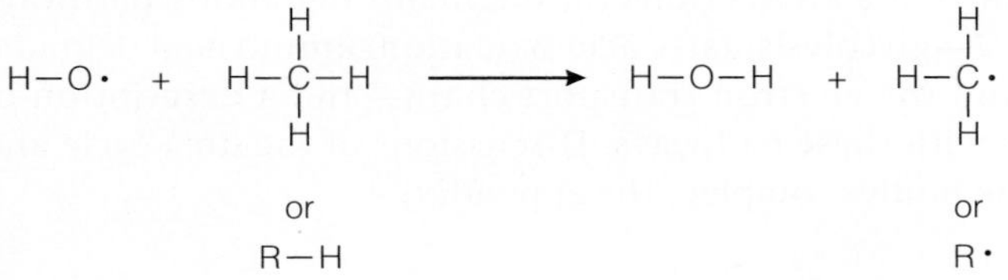

$$\text{H–O}\cdot + \text{H–CH}_2\text{–H (or R–H)} \longrightarrow \text{H–O–H} + \text{H–}\dot{\text{C}}\text{H}_2 \text{ (or R}\cdot\text{)}$$

Free radical + Compound with weak bond (perhaps an unsaturated fatty acid) → New stable compound (water or an alcohol) + Free radical

Free radicals destroy biological compounds. The free radical attacks a weak bond in a biological compound, disrupting it and forming a new stable molecule and another free radical. This free radical can attack another biological compound, and so on.

Oxidation of some compounds can be induced by air at room temperature in the presence of light. Such reactions are thought to take place through the formation of compounds called peroxides:

Peroxides:

H–O–O–H	Hydrogen peroxide
R–O–O–H	Hydroperoxides (R is any carbon chain with appropriate numbers of H)
R–O–O–R	Peroxide

Some peroxides readily disintegrate into free radicals, initiating chain reactions like those just described.

Free radicals are of special interest in nutrition because the antioxidant properties of vitamins C and E as well as beta-carotene and the mineral selenium are thought to protect against the destructive effects of these free radicals (see Highlight 11). For example, vitamin E on the surface of the lungs reacts with, and is destroyed by, free radicals, thus preventing the radicals from reaching underlying cells and oxidizing the lipids in their membranes.

APPENDIX B

Appendix C Biochemical Structures and Pathways

CONTENTS

The diagrams of nutrients presented here are meant to enhance your understanding of the most important organic molecules in the human diet. Following the diagrams of nutrients are sections on the major metabolic pathways mentioned in Chapter 7—glycolysis, fatty acid oxidation, amino acid degradation, the TCA cycle, and the electron transport chain—and a description of how alcohol interferes with these pathways. Discussions of the urea cycle and the formation of ketone bodies complete the appendix.

Carbohydrates

Monosaccharides

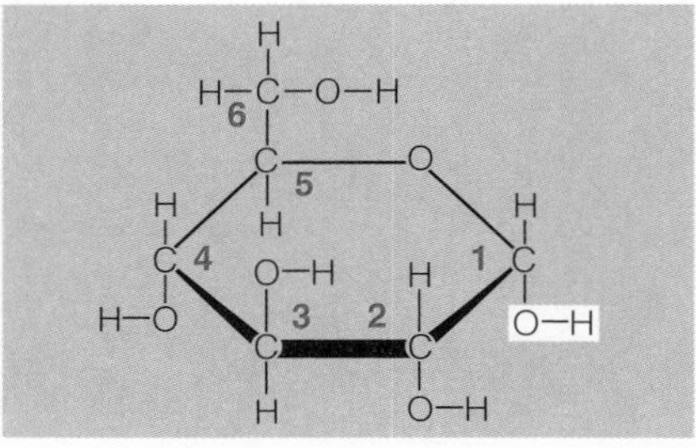

Glucose (alpha form). The ring would be at right angles to the plane of the paper. The bonds directed upward are above the plane; those directed downward are below the plane. This molecule is considered an alpha form because the OH on carbon 1 points downward.

Glucose (beta form). The OH on carbon 1 points upward.
Fructose, galactose: see Chapter 4.

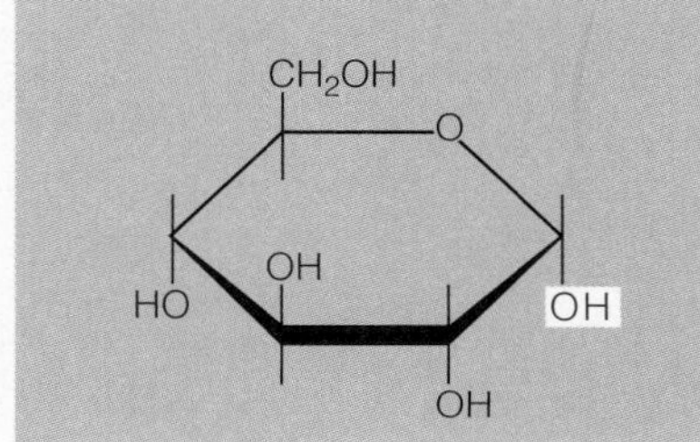

Glucose (alpha form) shorthand notation. This notation, in which the carbons in the ring and single hydrogens have been eliminated, will be used throughout this appendix.

Disaccharides

Maltose.

Lactose (alpha form).

Sucrose.

Polysaccharides As described in Chapter 4, starch, glycogen, and cellulose are all long chains of glucose molecules covalently linked together.

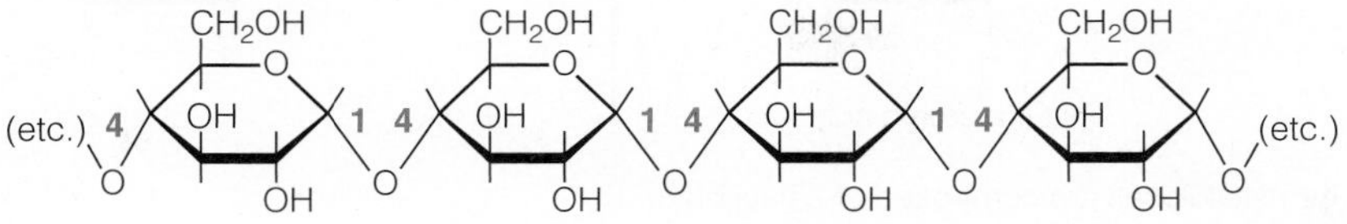

Starch. Two kinds of covalent linkages occur between glucose molecules in starch, giving rise to two kinds of chains. Amylose is composed of straight chains, with carbon 1 of one glucose linked to carbon 4 of the next (α-1,4 linkage). Amylopectin is made up of straight chains like amylose but has occasional branches arising where the carbon 6 of a glucose is also linked to the carbon 1 of another glucose (α-1,6 linkage).

Glycogen. The structure of glycogen is like amylopectin but with many more branches.

Cellulose. Like starch and glycogen, cellulose is also made of chains of glucose units, but there is an important difference: in cellulose, the OH on carbon 1 is in the beta position (see p. C-1). When carbon 1 of one glucose is linked to carbon 4 of the next, it forms a β-1,4 linkage, which cannot be broken by digestive enzymes in the human GI tract.

Amylose (unbranched starch)

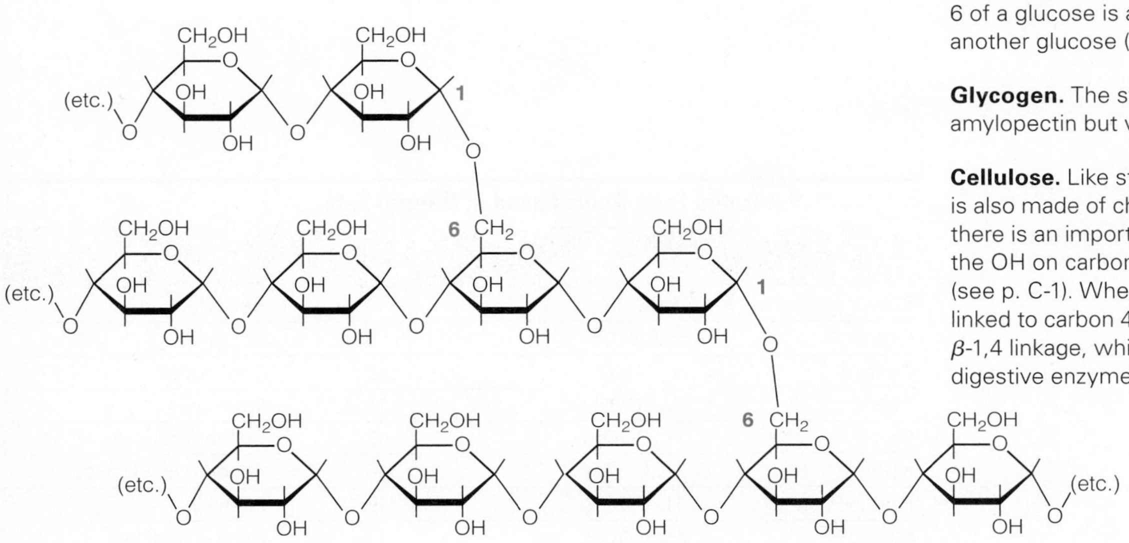

Amylopectin (branched starch)

Fibers, such as hemicelluloses, consist of long chains of various monosaccharides.

Monosaccharides common in the backbone chain of hemicelluloses:*

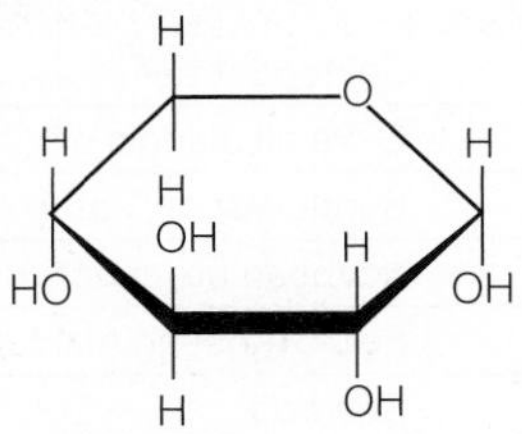

Xylose

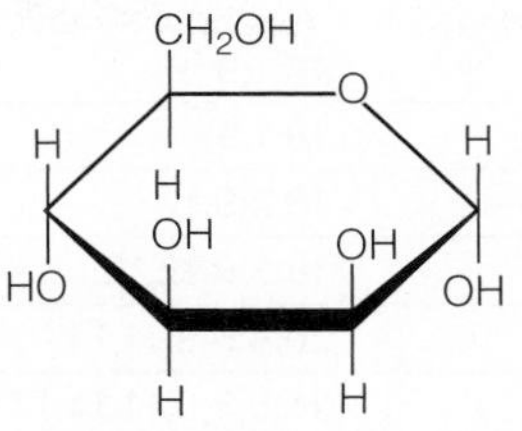

Mannose

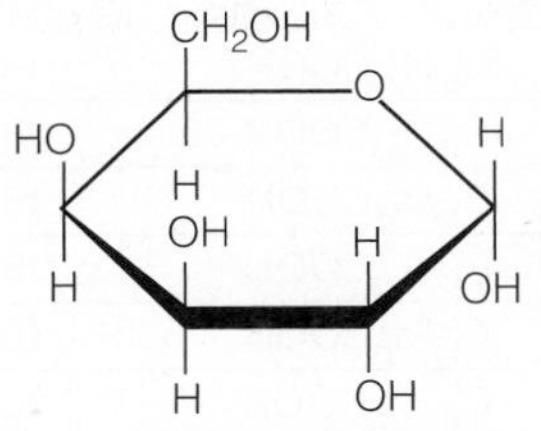

Galactose

*These structures are shown in the alpha form with the H on the carbon pointing upward and the OH pointing downward, but they may also appear in the beta form with the H pointing downward and the OH upward.

Monosaccharides common in the side chains of hemicelluloses:

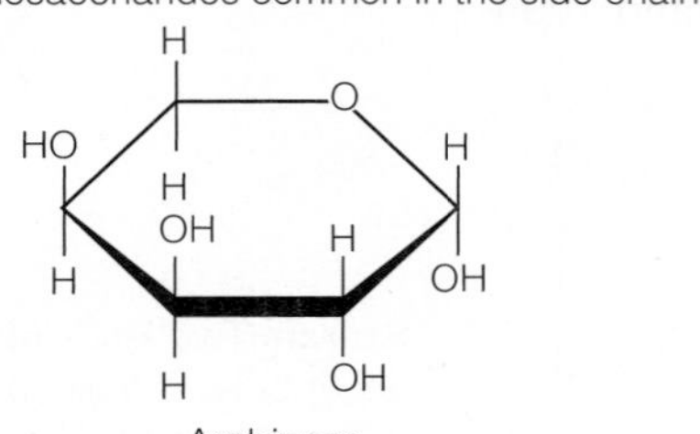

Arabinose

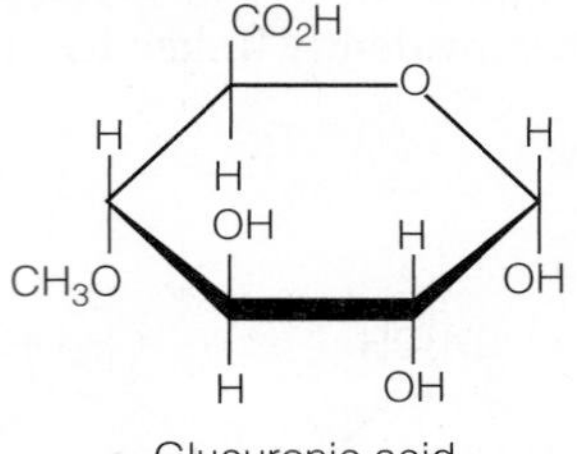

Glucuronic acid

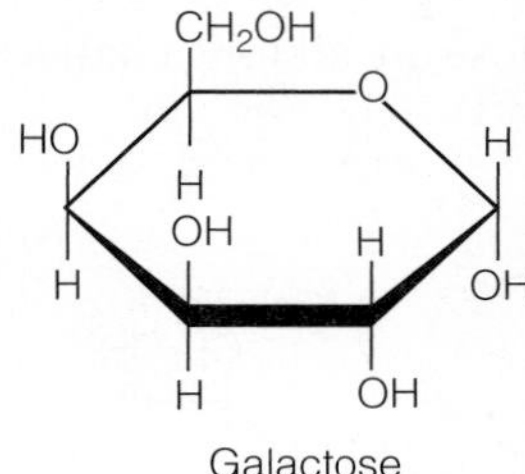

Galactose

Hemicelluloses. The most common hemicelluloses are composed of a backbone chain of xylose, mannose, and galactose, with branching side chains of arabinose, glucuronic acid, and galactose.

Lipids

TABLE C-1 Saturated Fatty Acids Found in Natural Fats

Saturated Fatty Acids	Chemical Formulas	Number of Carbons	Major Food Sources
Butyric	C_3H_7COOH	4	Butterfat
Caproic	$C_5H_{11}COOH$	6	Butterfat
Caprylic	$C_7H_{15}COOH$	8	Coconut oil
Capric	$C_9H_{19}COOH$	10	Palm oil
Laurie	$C_{11}H_{23}COOH$	12	Coconut oil, palm oil
Myristic[a]	$C_{13}H_{27}COOH$	14	Coconut oil, palm oil
Palmitic[a]	$C_{15}H_{31}COOH$	16	Palm oil
Stearic[c]	$C_{17}H_{35}COOH$	18	Most animal fats
Arachidic	$C_{19}H_{39}COOH$	20	Peanut oil
Behenic	$C_{21}H_{43}COOH$	22	Seeds
Lignoceric	$C_{23}H_{47}COOH$	24	Peanut oil

[a]Most common saturated fatty acids.

TABLE C-2 Unsaturated Fatty Acids Found in Natural Fats

Unsaturated Fatty Acids	Chemical Formulas	Number of Carbons	Number of Double Bonds	Standard Notation[a]	Omega Notation[b]	Major Food Sources
Palmitoleic	$C_{15}H_{29}COOH$	16	1	16:1;9	16:1ω7	Seafood, beef
Oleic	$C_{17}H_{33}COOH$	18	1	18:1;9	18:1ω9	Olive oil, canola oil
Linoleic	$C_{17}H_{31}COOH$	18	2	18:2;9,12	18:2ω6	Sunflower oil, safflower oil
Linolenic	$C_{17}H_{29}COOH$	18	3	18:3;9,12,15	18:3ω3	Soybean oil, canola oil
Arachidonic	$C_{19}H_{31}COOH$	20	4	20:4;5,8,11,14	20:4ω6	Eggs, most animal fats
Eicosapentaenoic	$C_{19}H_{29}COOH$	20	5	20:5;5,8,11,14,17	20:5ω3	Seafood
Docosahexaenoic	$C_{21}H_{31}COOH$	20	6	22:6;4,7,10,13,16,19	22:6ω3	Seafood

NOTE: A fatty acid has two ends; designated the methyl (CH_3) end and the carboxyl, or acid (COOH), end.

[a]Standard chemistry notation begins counting carbons at the acid end. The number of carbons the fatty acid contains comes first, followed by a colon and another number that indicates the number of double bonds; next comes a semicolon followed by a number or numbers indicating the positions of the double bonds. Thus the notation for linoleic acid, an 18-carbon fatty acid with two double bonds between carbons 9 and 10 and between carbons 12 and 13, is 18:2;9,12.

[b]Because fatty acid chains are lengthened by adding carbons at the acid end of the chain, chemists use the omega system of notation to ease the task of identifying them. The omega system begins counting carbons at the methyl end. The number of carbons the fatty acid contains comes first, followed by a colon and the number of double bonds; next come the omega symbol (ω) and a number indicating the position of the double bond nearest the methyl end. Thus linoleic acid with its first double bond at the sixth carbon from the methyl end would be noted 18:2ω6 in the omega system.

Protein: Amino Acids

The common amino acids may be classified into the seven groups listed below. Amino acids marked with an asterisk (*) are essential.

1. Amino acids with aliphatic side chains, which consist of hydrogen and carbon atoms (hydrocarbons):

$H-C(H)(NH_2)-C(=O)-OH$ **Glycine (Gly)**

$H_3C-C(H)(NH_2)-C(=O)-OH$ **Alanine (Ala)**

$(H_3C)_2CH-C(H)(NH_2)-C(=O)-OH$ **Valine* (Val)**

$(H_3C)_2CH-CH_2-C(H)(NH_2)-C(=O)-OH$ **Leucine* (Leu)**

$H_3C-CH_2-CH(CH_3)-C(H)(NH_2)-C(=O)-OH$ **Isoleucine* (Ile)**

2. Amino acids with hydroxyl (OH) side chains:

$HO-CH_2-C(H)(NH_2)-C(=O)-OH$ **Serine (Ser)**

$H_3C-CH(OH)-C(H)(NH_2)-C(=O)-OH$ **Threonine* (Thr)**

3. Amino acids with side chains containing acidic groups or their amides, which contain the group NH_2:

$HO-C(=O)-CH_2-C(H)(NH_2)-C(=O)-OH$ **Aspartic acid (Asp)**

$HO-C(=O)-CH_2-CH_2-C(H)(NH_2)-C(=O)-OH$ **Glutamic acid (Glu)**

$NH_2-C(=O)-CH_2-C(H)(NH_2)-C(=O)-OH$ **Asparagine (Asn)**

$NH_2-C(=O)-CH_2-CH_2-C(H)(NH_2)-C(=O)-OH$ **Glutamine (Gln)**

4. Amino acids with basic side chains:

$NH_2-CH_2-CH_2-CH_2-CH_2-C(H)(NH_2)-C(=O)-OH$ **Lysine* (Lys)**

$NH_2-C(=NH)-NH-CH_2-CH_2-CH_2-C(H)(NH_2)-C(=O)-OH$ **Arginine (Arg)**

$H-C=C-CH_2-C(H)(NH_2)-C(=O)-OH$ (imidazole ring: N, $N-H$, $C-H$) **Histidine* (His)**

5. Amino acids with aromatic side chains, which are characterized by the presence of at least one ring structure:

$C_6H_5-CH_2-C(H)(NH_2)-C(=O)-OH$ **Phenylalanine* (Phe)**

$HO-C_6H_4-CH_2-C(H)(NH_2)-C(=O)-OH$ **Tyrosine (Tyr)**

(indole ring)$-CH_2-C(H)(NH_2)-C(=O)-OH$ **Tryptophan* (Trp)**

6. Amino acids with side chains containing sulfur atoms:

$HS-CH_2-C(H)(NH_2)-C(=O)-OH$ **Cysteine (Cys)**

$CH_3-S-CH_2-CH_2-C(H)(NH_2)-C(=O)-OH$ **Methionine* (Met)**

7. Imino acid:

(pyrrolidine ring: $H-C-C$, $H-C-N-H$)$-C(H)-C(=O)-OH$ **Proline (Pro)**

Proline has the same chemical structure as the other amino acids, but its amino group has given up a hydrogen to form a ring.

Vitamins and Coenzymes

Vitamin A: retinol. This molecule is the alcohol form of vitamin A.

Vitamin A: retinal. This molecule is the aldehyde form of vitamin A.

Vitamin A: retinoic acid. This molecule is the acid form of vitamin A.

Vitamin A precursor: beta-carotene. This molecule is the carotenoid with the most vitamin A activity.

Thiamin. This molecule is part of the coenzyme thiamin pyrophosphate (TPP).

Thiamin pyrophosphate (TPP). TPP is a coenzyme that includes the thiamin molecule as part of its structure.

Riboflavin. This molecule is a part of two coenzymes—flavin mononucleotide (FMN) and flavin adenine dinucleotide (FAD).

Flavin mononucleotide (FMN). FMN is a coenzyme that includes the riboflavin molecule as part of its structure.

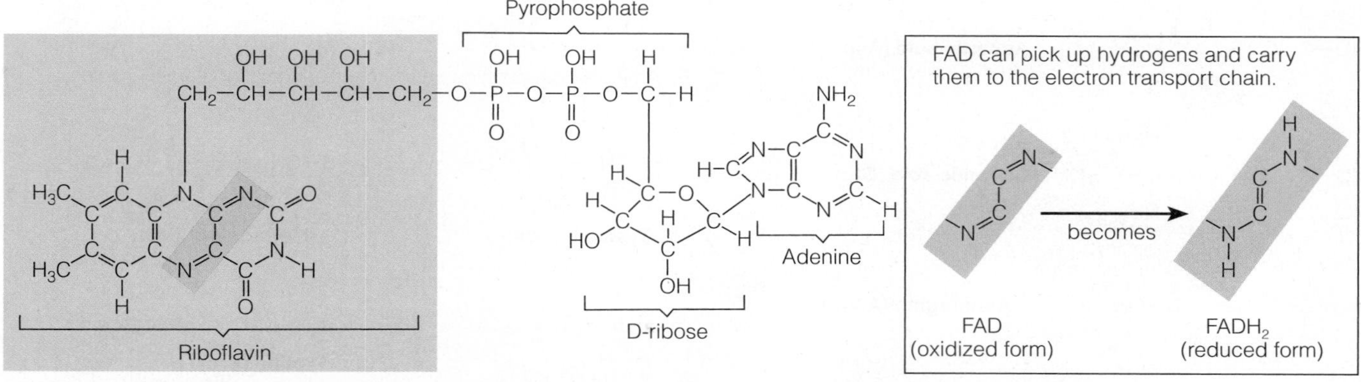

Flavin adenine dinucleotide (FAD). FAD is a coenzyme that includes the riboflavin molecule as part of its structure.

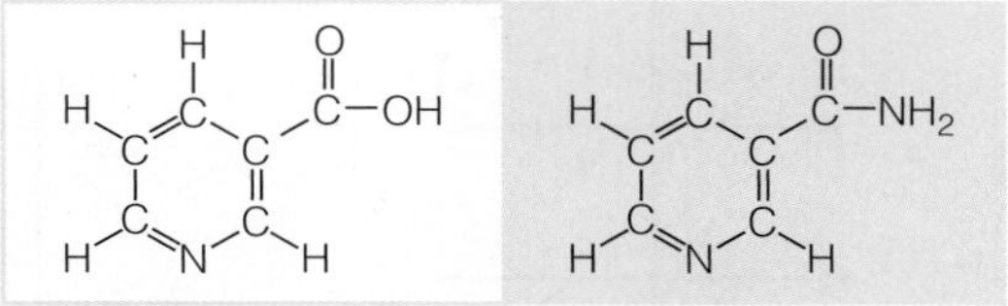

Nicotinic acid

Nicotinamide

Niacin (nicotinic acid and nicotinamide). These molecules are a part of two coenzymes—nicotinamide adenine dinucleotide (NAD^+) and nicotinamide adenine dinucleotide phosphate ($NADP^+$).

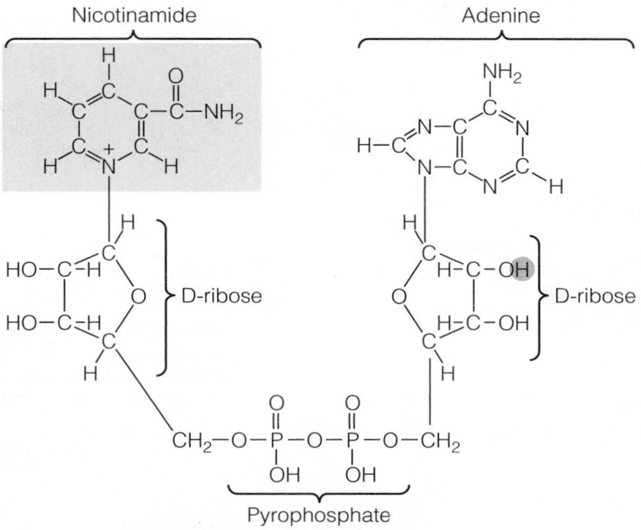

Nicotinamide

Adenine

D-ribose

D-ribose

Pyrophosphate

Nicotinamide adenine dinucleotide (NAD^+) and nicotinamide adenine dinucleotide phosphate ($NADP^+$). NADP has the same structure as NAD but with a phosphate group attached to the O instead of the H.

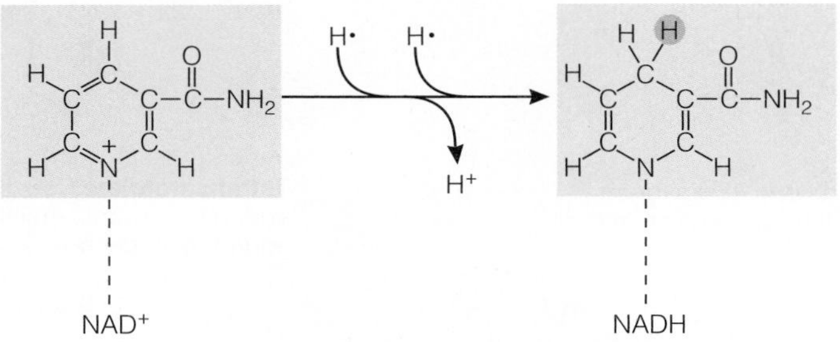

NAD^+

NADH

Reduced NAD^+ (NADH). When NAD^+ is reduced by the addition of H^+ and two electrons, it becomes the coenzyme NADH. (The dots on the H entering this reaction represent electrons—see Appendix B.)

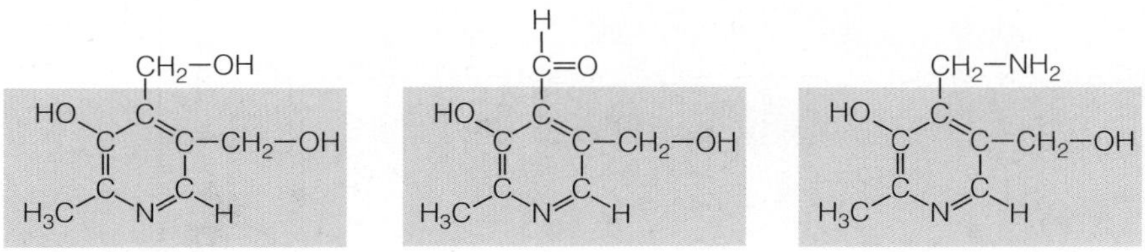

Pyridoxine

Pyridoxal

Pyridoxamine

Vitamin B_6 (a general name for three compounds—pyridoxine, pyridoxal, and pyridoxamine). These molecules are a part of two coenzymes—pyridoxal phosphate and pyridoxamine phosphate.

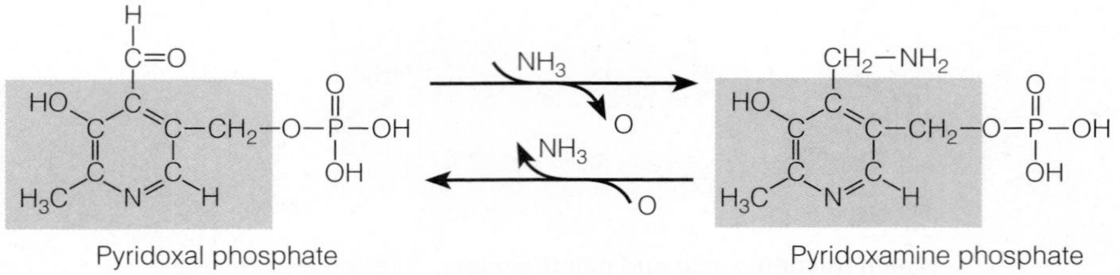

Pyridoxal phosphate

Pyridoxamine phosphate

Pyridoxal phosphate (PLP) and pyridoxamine phosphate. These coenzymes include vitamin B_6 as part of their structures.

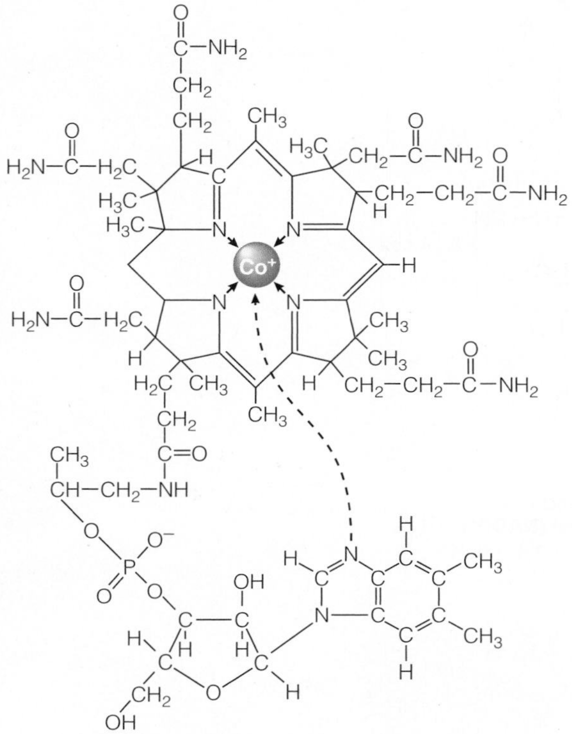

Vitamin B_{12} (cyanocobalamin). The arrows in this diagram indicate that the spare electron pairs on the nitrogens attract them to the cobalt.

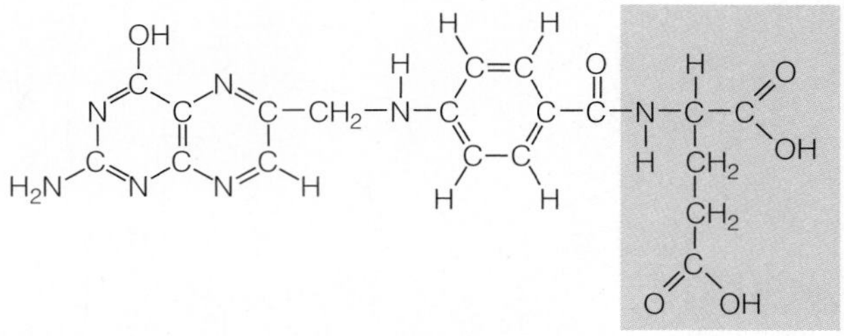

Folate (folacin or folic acid). This molecule consists of a double ring combined with a single ring and at least one glutamate (a nonessential amino acid marked in the box). Folate's biologically active form is tetrahydrofolate.

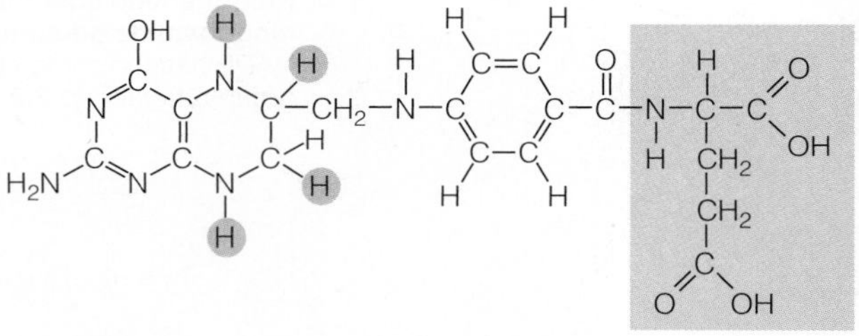

Tetrahydrofolate. This active coenzyme form of folate has four added hydrogens. An intermediate form, dihydrofolate, has two added hydrogens.

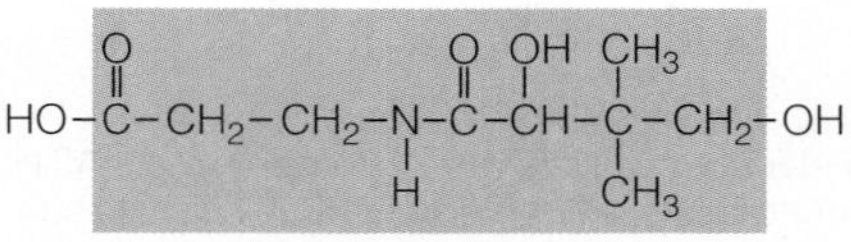

Pantothenic acid. This molecule is part of coenzyme A (CoA).

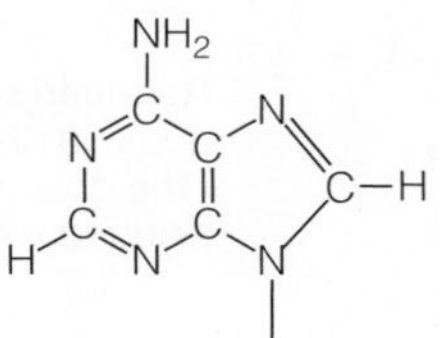

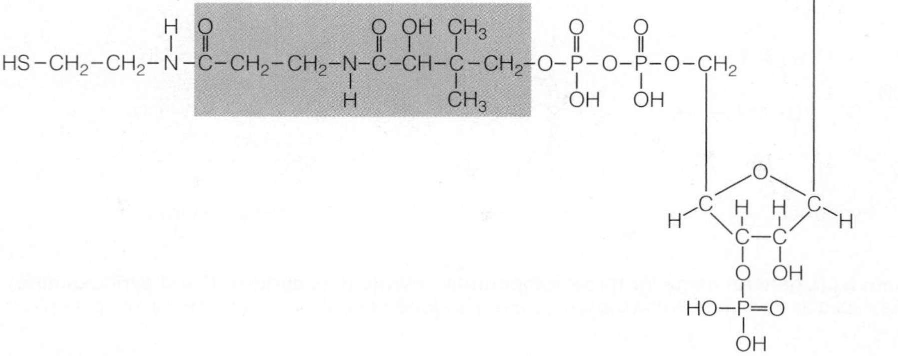

Coenzyme A (CoA). Coenzyme A is a coenzyme that includes pantothenic acid as part of its structure.

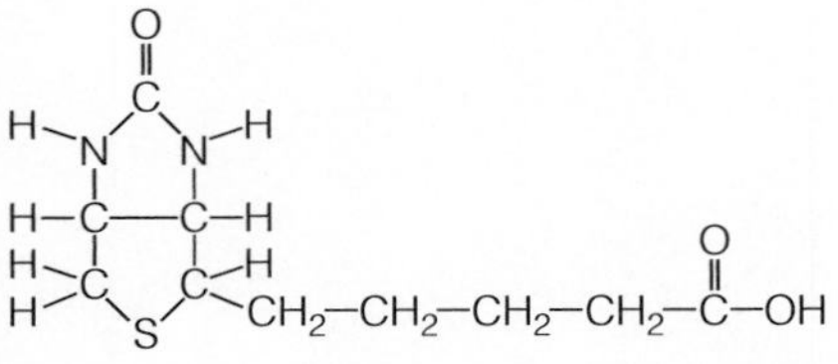

Biotin.

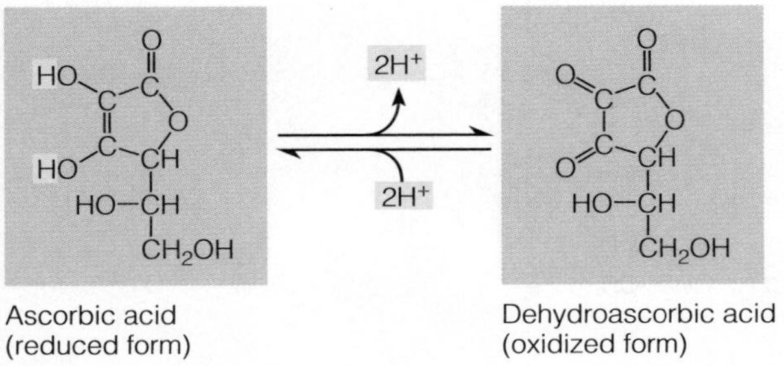

Ascorbic acid
(reduced form)

Dehydroascorbic acid
(oxidized form)

Vitamin C. Two hydrogen atoms with their electrons are lost when ascorbic acid is oxidized and gained when it is reduced again.

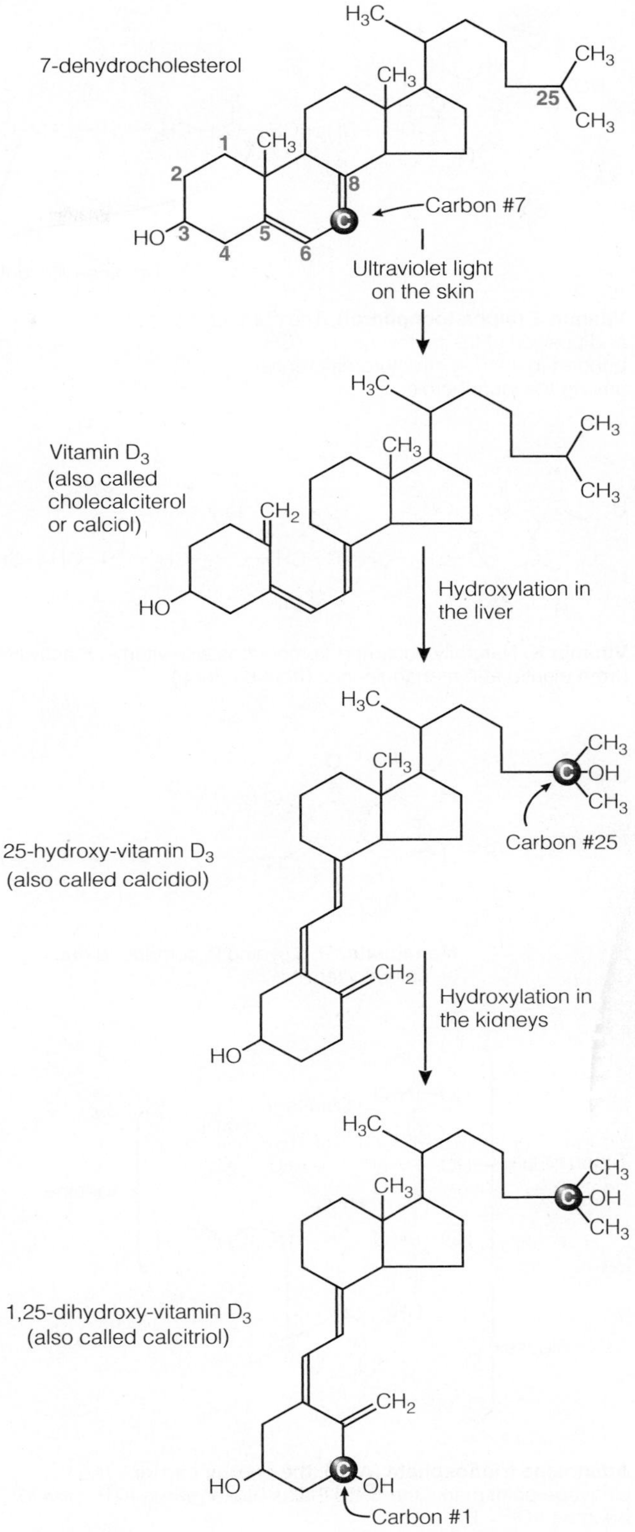

7-dehydrocholesterol

Carbon #7

Ultraviolet light on the skin

Vitamin D_3 (also called cholecalciterol or calciol)

Hydroxylation in the liver

25-hydroxy-vitamin D_3 (also called calcidiol)

Carbon #25

Hydroxylation in the kidneys

1,25-dihydroxy-vitamin D_3 (also called calcitriol)

Carbon #1

Vitamin D. The synthesis of active vitamin D begins with 7-dehydrocholesterol. (The carbon atoms at which changes occur are numbered.)

Tocotrienols contain double bonds here.

Vitamin E (alpha-tocopherol). The number and position of the methyl groups (CH_3) bonded to the ring structure differentiate among the tocopherols.

Vitamin K. Naturally occurring compounds with vitamin K activity include phylloquinones (from plants) and menaquinones (from bacteria).

Menadione. This synthetic compound has the same activity as natural vitamin K.

Triphosphate

Cleavage

Adenine

Ribose

Adenosine triphosphate (ATP), the energy carrier. The cleavage point marks the bond that is broken when ATP splits to become ADP + P.

Cleavage

+ H—O—H (Water)

Phosphate + ADP

H^+

Adenosine diphosphate (ADP).

Glycolysis

Figure C-1 depicts glycolysis. The following text describes key steps as numbered on the figure.

> **FIGURE C-1 Glycolysis**

Notice that galactose and fructose enter at different places but continue on the same pathway.

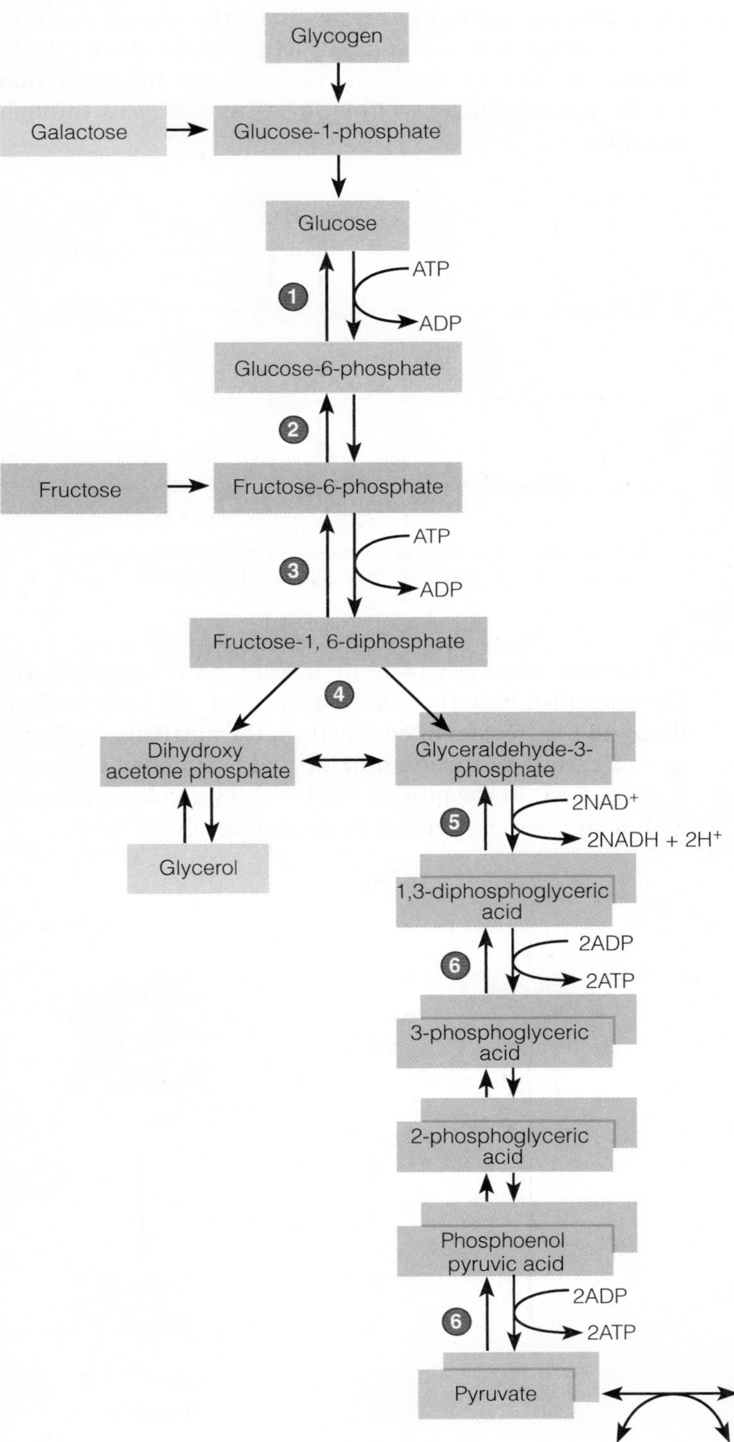

1. A phosphate is attached to glucose at the carbon that chemists call number 6 (review the first diagram of glucose on p. C-1 to see how chemists number the carbons in a glucose molecule). The product is called, logically enough, glucose-6-phosphate. One ATP molecule is used to accomplish this.

2. Glucose-6-phosphate is rearranged by an enzyme.

3. A phosphate is added in another reaction that uses another molecule of ATP. The product this time is fructose-1,6-diphosphate. At this point the 6-carbon sugar has a phosphate group on its first and sixth carbons and is ready to break apart.

4. When fructose-1,6-diphosphate breaks in half, the two 3-carbon compounds are not identical. Each has a phosphate group attached, but only glyceraldehyde-3-phosphate converts directly to pyruvate. The other compound, however, converts easily to glyceraldehyde-3-phosphate.

5. In the next step, enough energy is released to convert NAD^+ to NADH + H^+.

6. In two of the following steps ATP is regenerated.

Remember that in effect two molecules of glyceraldehyde-3-phosphate are produced from glucose; therefore, four ATP molecules are generated from each glucose molecule. Two ATP were needed to get the sequence started, so the net gain at this point is two ATP and two molecules of NADH + H^+. As you will see later, each NADH + H^+ moves to the electron transport chain to unload its hydrogens, producing more ATP.

Fatty Acid Oxidation

Figure C-2 presents fatty acid oxidation. The sequence is as follows.

1. The fatty acid is activated by combining with coenzyme A (CoA). In this reaction, ATP loses two phosphorus atoms (PP, or pyrophosphate) and becomes AMP (adenosine monophosphate)—the equivalent of a loss of two ATP.

2. In the next reaction, two H with their electrons are removed and transferred to FAD, forming $FADH_2$.

3. In a later reaction, two H are removed and go to NAD^+ (forming NADH + H^+).

4. The fatty acid is cleaved at the "beta" carbon, the second carbon from the carboxyl (COOH) end. This break results in a fatty acid that is two carbons shorter than the previous one and a 2-carbon molecule of acetyl CoA.

> FIGURE C-2 **Fatty Acid Oxidation**

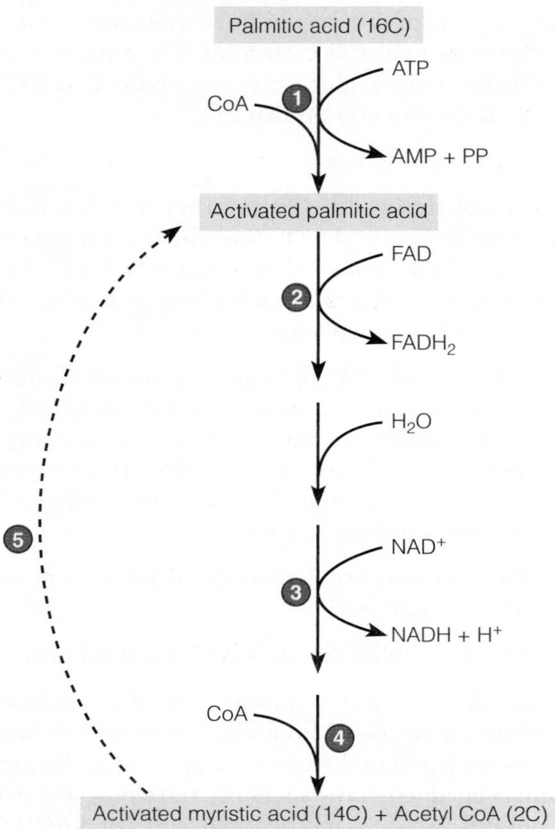

At the same time, another CoA is attached to the fatty acid, thus activating it for its turn through the series of reactions.

5. The sequence is repeated with each cycle producing an acetyl CoA and a shorter fatty acid until only a 2-carbon fatty acid remains—acetyl CoA.

In the example shown in Figure C-2, palmitic acid (a 16-carbon fatty acid) will go through this series of reactions seven times, using the equivalent of two ATP for the initial activation and generating seven $FADH_2$, seven $NADH + H^+$, and eight acetyl CoA. As you will see later, each of the seven $FADH_2$ will enter the electron transport chain, yielding two ATP (for a total of 14). Similarly, each $NADH + H^+$ will enter the electron transport chain, yielding three ATP (for a total of 21). Thus the oxidation of a 16-carbon fatty acid uses 2 ATP and generates 35 ATP. When the eight acetyl CoA enter the TCA cycle, even more ATP will be generated, as a later section describes.

Amino Acid Degradation

The first step in amino acid degradation is the removal of the nitrogen-containing amino group through either deamination (Figure 6-11 on p. 179) or transamination (Figure 6-12 on p. 179) reactions. Then the remaining carbon skeletons may enter the metabolic pathways at different places, as shown in Figure C-3 (p. C-12).

The TCA Cycle

The tricarboxylic acid, or TCA, cycle is the set of reactions that break down acetyl CoA to carbon dioxide and hydrogens. Pyruvate derived from glycolysis does not enter the TCA cycle directly; instead pyruvate enters the mitochondrion, loses a carbon group, and bonds with a molecule of CoA to become acetyl CoA. The TCA cycle uses any substance that can be converted to acetyl CoA directly or indirectly through pyruvate.

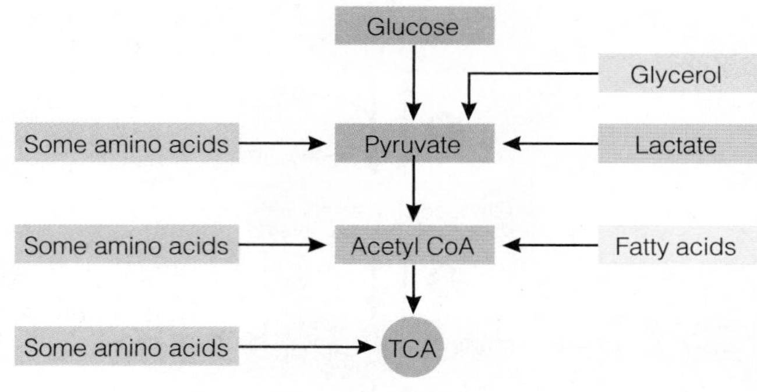

Any substance that can be converted to acetyl CoA directly, or indirectly through pyruvate, may enter the TCA cycle.

The step from pyruvate to acetyl CoA is complex. We have included only those substances that will help you understand the transfer of energy from the nutrients. Pyruvate loses a carbon to carbon dioxide and is attached to a molecule of CoA. In the process, NAD^+ picks up two hydrogens with their associated electrons, becoming $NADH + H^+$.

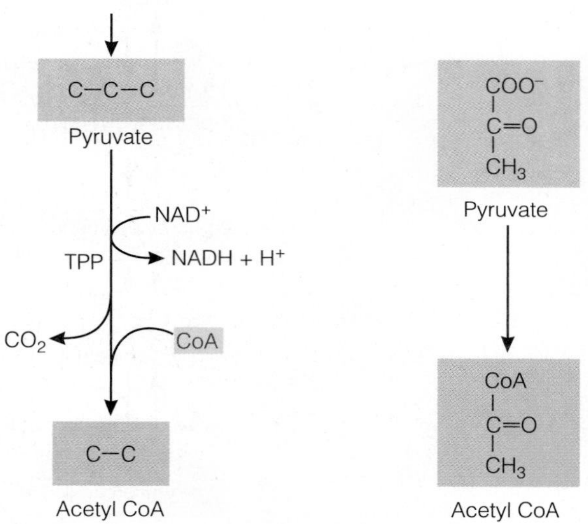

The step from pyruvate to acetyl CoA. (TPP and NAD are coenzymes containing the B vitamins thiamin and niacin, respectively.)

> FIGURE C-3 **Amino Acids Enter the Metabolic Pathways**

After losing their amino groups, carbon skeletons can be converted to one of seven molecules that can enter the TCA cycle (presented in Figure C-4).

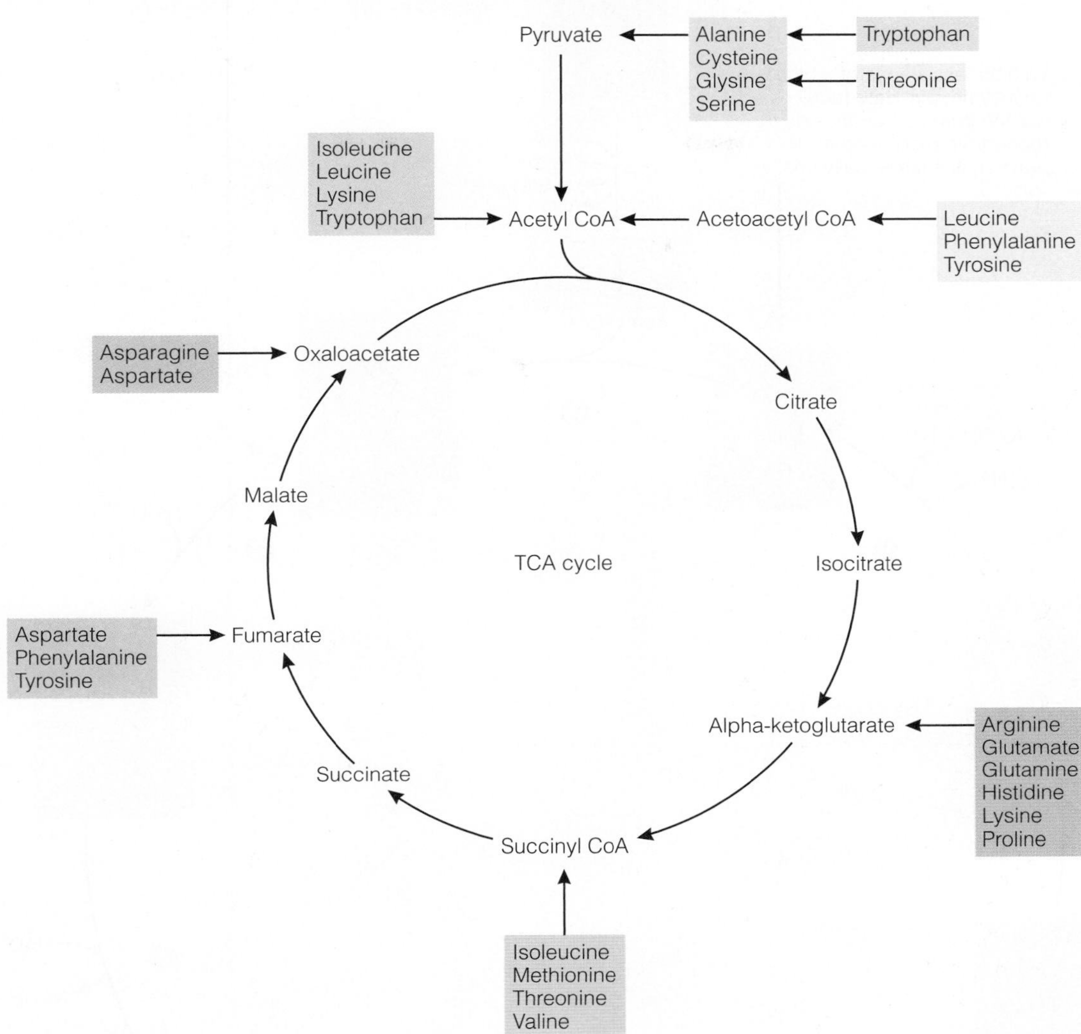

Let's follow the steps of the TCA cycle (see the corresponding numbers in Figure C-4 on the next page).

1. The 2-carbon acetyl CoA combines with a 4-carbon compound, oxaloacetate. The CoA comes off, and the product is a 6-carbon compound, citrate.
2. The atoms of citrate are rearranged to form isocitrate.
3. Now two H (with their two electrons) are removed from the isocitrate. One H becomes attached to the NAD^+ with the two electrons; the other H is released as H^+. Thus NAD^+ becomes NADH + H^+. (Remember this NADH + H^+, but let's follow the carbons first.) A carbon is combined with two oxygens, forming carbon dioxide (which diffuses away into the blood and is exhaled). What is left is the 5-carbon compound alpha-ketoglutarate.
4. Now two compounds interact with alpha-ketoglutarate—a molecule of CoA and a molecule of NAD^+. In this complex reaction, a carbon and two oxygens are removed (forming carbon dioxide); two hydrogens are removed and go to NAD^+ (forming NADH + H^+); and the remaining 4-carbon compound is attached to the CoA, forming succinyl CoA. (Remember this NADH + H^+ also. You will see later what happens to it.)
5. Now two molecules react with succinyl CoA—a molecule called GDP and one of phosphate (P). The CoA comes off, the GDP and P combine to form the high-energy compound GTP (similar to ATP), and succinate remains. (Remember this GTP.)
6. In the next reaction, two H with their electrons are removed from succinate and are transferred to a molecule of FAD (a coenzyme like NAD^+) to form $FADH_2$. The product that remains is fumarate. (Remember this $FADH_2$.)
7. Next a molecule of water is added to fumarate, forming malate.

> FIGURE C-4 The TCA Cycle

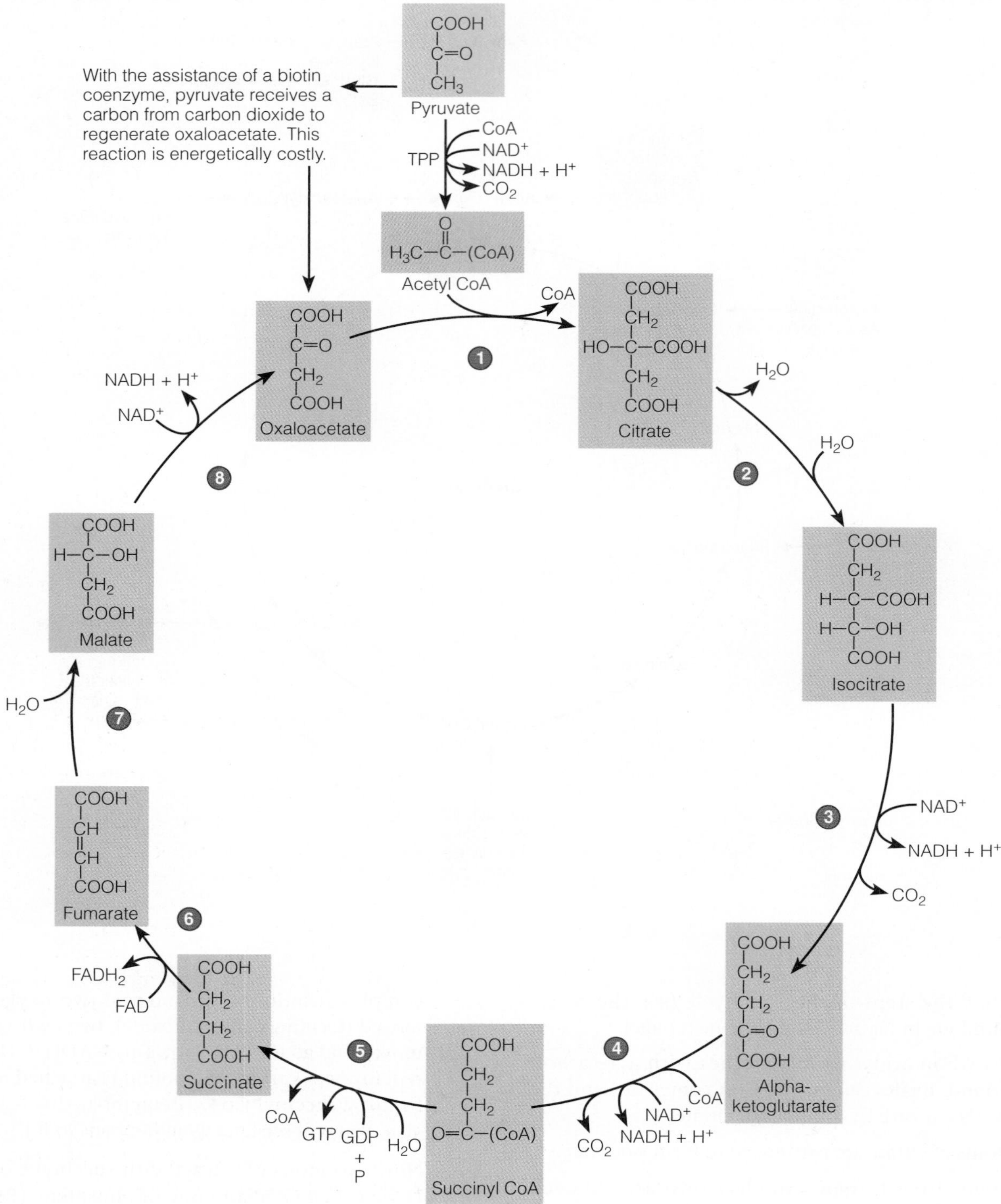

8. A molecule of NAD^+ reacts with the malate; two H with their associated electrons are removed from the malate and form $NADH + H^+$. The product that remains is the 4-carbon compound oxaloacetate. (Remember this $NADH + H^+$.)

We are back where we started. The oxaloacetate formed in this process can combine with another molecule of acetyl CoA (step 1), and the cycle can begin again, as shown in Figure C-4.

So far, we have seen two carbons brought in with acetyl CoA and two carbons ending up in carbon dioxide. But where are the energy and the ATP that were promised?

A review of the eight steps of the TCA cycle shows that the compounds $NADH + H^+$ (three molecules), $FADH_2$, and GTP capture energy originally found in acetyl CoA. To see how this energy ends up in ATP, we must follow the electrons further—into the electron transport chain.

The Electron Transport Chain

The six reactions described here are those of the electron transport chain, which is shown below the TCA cycle in Figure C-5. Since oxygen is required for these reactions, and ADP and P are combined to form ATP in several of them (ADP is phosphorylated), the reactions of the electron transport chain are also called *oxidative phosphorylation*.

An important concept to remember at this point is that an electron is not a fixed amount of energy. The electrons that bond the H to NAD^+ in NADH have a relatively large amount of energy. In the series of reactions that follow, they release this energy in small amounts, until at the end they are attached (with H) to oxygen (O) to make water (H_2O). In some of the steps, the energy they release is captured into ATP in coupled reactions.

1. In the first step of the electron transport chain, NADH reacts with a molecule called a flavoprotein, losing its electrons (and their H). The products are NAD^+ and reduced flavoprotein. A little energy is released as heat in this reaction.
2. The flavoprotein passes on the electrons to a molecule called coenzyme Q. Again they release some energy as heat, but ADP and P bond together and form ATP, storing much of the energy. This is a coupled reaction: ADP + P → ATP.
3. Coenzyme Q passes the electrons to cytochrome *b*. Again the electrons release energy.
4. Cytochrome *b* passes the electrons to cytochrome *c* in a coupled reaction in which ATP is formed: ADP + P → ATP.
5. Cytochrome *c* passes the electrons to cytochrome *a*.
6. Cytochrome *a* passes them (with their H) to an atom of oxygen (O), forming water (H_2O). This is a coupled reaction in which ATP is formed: ADP + P → ATP.

> FIGURE C-5 **The Electron Transport Chain**

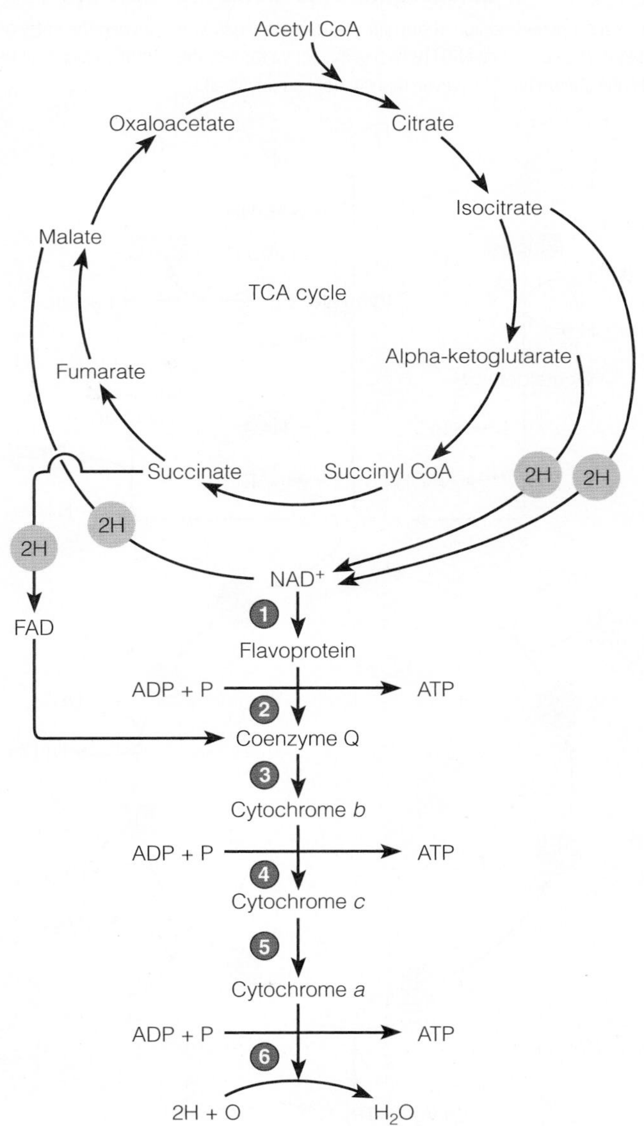

As Figure C-5 shows, each time NADH is oxidized (loses its electrons) by this means, the energy it releases is captured into three ATP molecules. When the electrons are passed on to water at the end, they are much lower in energy than they were originally. This completes the story of the electrons from NADH.

As for $FADH_2$, its electrons enter the electron transport chain at coenzyme Q. From coenzyme Q to water, ATP is generated in only two steps. Therefore, $FADH_2$ coming out of the TCA cycle yields just two ATP molecules.

One energy-receiving compound of the TCA cycle (GTP) does not enter the electron transport chain but gives its energy directly to ADP in a simple phosphorylation reaction. This reaction yields one ATP.

It is now possible to draw up a balance sheet of glucose metabolism (see Table C-3). Glycolysis has yielded 4 NADH + H^+ and 4 ATP molecules and has spent 2 ATP. The 2 acetyl CoA going through the TCA cycle have yielded 6 NADH + H^+, 2 $FADH_2$, and 2 GTP molecules. After the NADH + H^+ and $FADH_2$ have gone through the electron transport chain, there are 28 ATP. Added to these are the 4 ATP from glycolysis and the 2 ATP from GTP, making the total 34 ATP generated from one molecule of glucose. After the expense of 2 ATP is subtracted, there is a net gain of 32 ATP.*

*The total may sometimes be 30 ATP. The NADH + H^+ generated in the cytoplasm during glycolysis pass their electrons on to shuttle molecules, which move them into the mitochondria. One shuttle, malate, contributes its electrons to the electron transport chain before the first site of ATP synthesis, yielding 5 ATP. Another, glycerol phosphate, adds its electrons into the chain beyond that first site, yielding 3 ATP. Thus sometimes 5, and sometimes 3, ATP result from the NADH + H^+ that arise from glycolysis. The amount depends on the cell.

TABLE C-3 Balance Sheet for Glucose Metabolism

		ATP
Glycolysis:	4 ATP − 2 ATP	2
1 glucose to 2 pyruvate	2 NADH + H^+	3–5[a]
2 pyruvate to 2 acetyl CoA	2 NADH + H^+	5
TCA cycle and electron transport chain:		
2 isocitrate	2 NADH + H^+	5
2 alpha-ketoglutarate	2 NADH + H^+	5
2 succinyl CoA	2 GTP	2
2 succinate	2 $FADH_2$	3
2 malate	2 NADH + H^+	5
Total ATP collected from one molecule glucose:		30–32

[a]Each NADH + H^+ from glycolysis can yield 1.5 or 2.5 ATP. See the accompanying text.

A similar balance sheet from the complete breakdown of one 16-carbon fatty acid would show a net gain of 129 ATP. As mentioned earlier, 35 ATP were generated from the 7 $FADH_2$ and 7 NADH + H^+ produced during fatty acid oxidation. The 8 acetyl CoA produced will each generate 12 ATP as they go through the TCA cycle and the electron transport chain, for a total of 96 more ATP. After subtracting the 2 ATP needed to activate the fatty acid initially, the net yield from one 16-carbon fatty acid: 35 + 96 − 2 = 129 ATP.

These calculations help explain why fat yields more energy (measured as kcalories) per gram than carbohydrate or protein. The more hydrogen atoms a fuel contains, the more ATP will be generated during oxidation. The 16-carbon fatty acid molecule, with its 32 hydrogen atoms, generates 129 ATP, whereas glucose, with its 12 hydrogen atoms, yields only 32 ATP.

The TCA cycle and the electron transport chain are the body's major means of capturing the energy from nutrients in ATP molecules. Other means, such as anaerobic glycolysis, contribute energy quickly, but the aerobic processes are the most efficient.

Alcohol's Interference with Energy Metabolism

Highlight 7 provides an overview of how alcohol interferes with energy metabolism. With an understanding of the TCA cycle, a few more details may be appreciated. During alcohol metabolism, the enzyme alcohol dehydrogenase oxidizes alcohol to acetaldehyde while it simultaneously reduces a molecule of NAD^+ to NADH + H^+. The related enzyme acetaldehyde dehydrogenase reduces another NAD^+ to NADH + H^+ while it oxidizes acetaldehyde to acetyl CoA, the compound that enters the TCA cycle to generate energy. Thus, whenever alcohol is being metabolized in the body, NAD^+ diminishes, and NADH + H^+ accumulates, thus altering the body's "redox state." NAD^+ can oxidize, and NADH + H^+ can reduce, many other compounds as well. During alcohol metabolism, however, NAD^+ becomes unavailable for the multitude of reactions for which it is required.

As the previous sections just explained, for glucose to be completely metabolized, the TCA cycle must be operating, and NAD^+ must be present. If these conditions are not met (and when alcohol is present, they may not be), the pathway will be blocked, and traffic will back up—or an alternate route will be taken. Think about this as you follow the pathway shown in Figure C-6.

In each step of alcohol metabolism in which NAD^+ is converted to NADH + H^+, hydrogen ions accumulate, resulting in a dangerous shift of the acid-base balance toward acid (Chapter 12 explains acid-base balance). The accumulation of NADH + H^+ slows TCA cycle activity, so pyruvate and acetyl CoA build up. This condition favors the conversion of

> FIGURE C-6 Ethanol Enters the Metabolic Pathways

This is a simplified version of the glucose-to-energy pathway showing the entry of ethanol. The coenzyme NAD (which is the active form of the B vitamin niacin) is the only one shown here; however, many others are involved.

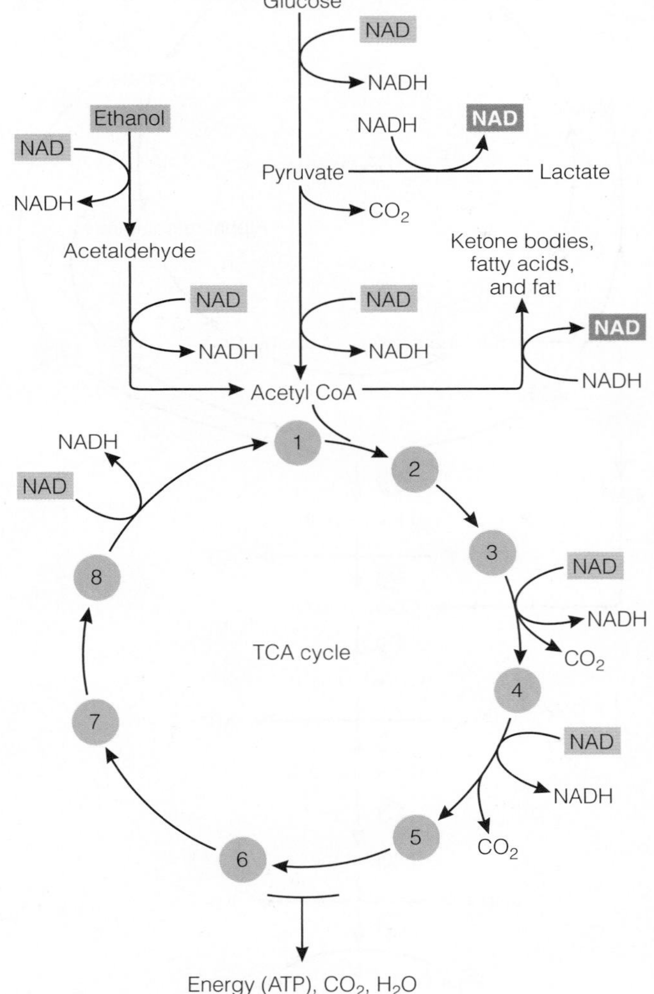

pyruvate to lactate, which serves as a temporary storage place for hydrogens from NADH + H^+. The conversion of pyruvate to lactate restores some NAD^+, but a lactate buildup has serious consequences of its own. It adds to the body's acid burden and interferes with the excretion of uric acid, causing goutlike symptoms. Molecules of acetyl CoA become building blocks for fatty acids or ketone bodies. The making of ketone bodies consumes acetyl CoA and generates NAD^+; but some ketone bodies are acids, so they push the acid-base balance further toward acid.

Thus alcohol cascades through the metabolic pathways, wreaking havoc along the way. These consequences have physical effects, as Highlight 7 describes.

The Urea Cycle

Chapter 6 sums up the process by which waste nitrogen is eliminated from the body by stating that ammonia molecules combine with carbon dioxide to produce urea. This is true, but it is not the whole story. Urea is produced in a multistep process within the cells of the liver.

Ammonia, freed from an amino acid or other compound during metabolism anywhere in the body, arrives at the liver by way of the bloodstream and is taken into a liver cell. There, it is first combined with carbon dioxide and a phosphate group from ATP to form carbamyl phosphate:

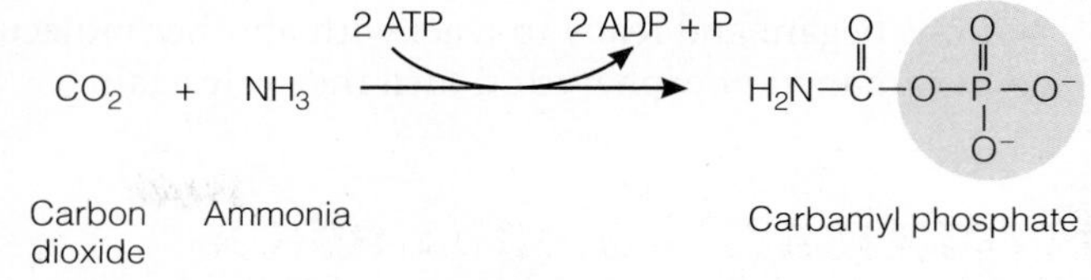

$$\underset{\text{Carbon dioxide}}{CO_2} + \underset{\text{Ammonia}}{NH_3} \xrightarrow[\text{2 ADP + P}]{\text{2 ATP}} \underset{\text{Carbamyl phosphate}}{H_2N{-}C({=}O){-}O{-}P({=}O)(O^-){-}O^-}$$

Figure C-7 shows the cycle of four reactions that follow.

1. Carbamyl phosphate combines with the amino acid ornithine, losing its phosphate group. The compound formed is citrulline.
2. Citrulline combines with the amino acid aspartic acid, to form argininosuccinate. The reaction requires energy from ATP. (ATP was shown earlier losing one phosphorus atom in a phosphate group, P, to become ADP. In this reaction, it loses two phosphorus atoms joined together, PP, and becomes adenosine monophosphate, AMP.)
3. Argininosuccinate is split, forming another acid, fumarate, and the amino acid arginine.
4. Arginine loses its terminal carbon with two attached amino groups and picks up an oxygen from water. The

> **FIGURE C-7 The Urea Cycle**

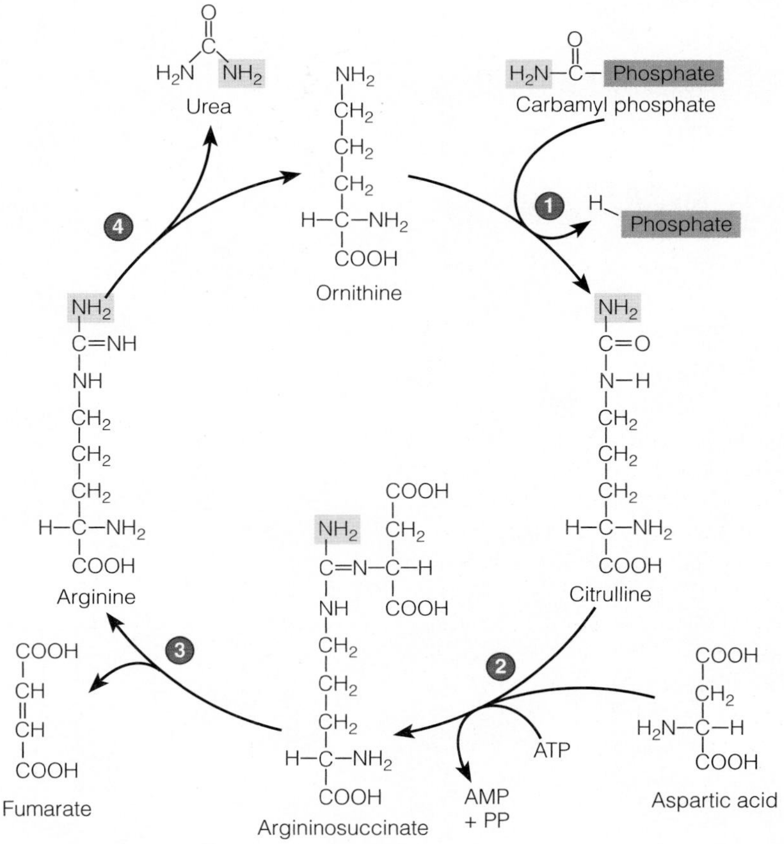

end product is urea, which the kidneys excrete in the urine. The compound that remains is ornithine, identical to the ornithine with which this series of reactions began, and ready to react with another molecule of carbamyl phosphate and turn the cycle again.

Formation of Ketone Bodies

Normally, fatty acid oxidation proceeds all the way to carbon dioxide and water. However, in ketosis (discussed in Chapter 7), an intermediate is formed from the condensation of two molecules of acetyl CoA: acetoacetyl CoA. Figure C-8 shows the formation of ketone bodies from that intermediate.

1. Acetoacetyl CoA condenses with acetyl CoA to form a 6-carbon intermediate, beta-hydroxy-beta-methylglutaryl CoA.
2. This intermediate is cleaved to acetyl CoA and acetoacetate.
3. Acetoactate can be metabolized either to beta-hydroxybutyrate acid (step 3a) or to acetone (3b).

Acetoacetate, beta-hydroxybutyrate, and acetone are the ketone bodies of ketosis. Two are real ketones (they have a C=O group between two carbons); the other is an alcohol that has been produced during ketone formation—hence the term *ketone bodies,* rather than ketones, to describe the three of them. There are many other ketones in nature; these three are characteristic of ketosis in the body.

> FIGURE C-8 The Formation of Ketone Bodies

$H_3C{-}C(=O){-}CH_2{-}C(=O){-}CoA$ + $H_3C{-}C(=O){-}CoA$ + H_2O

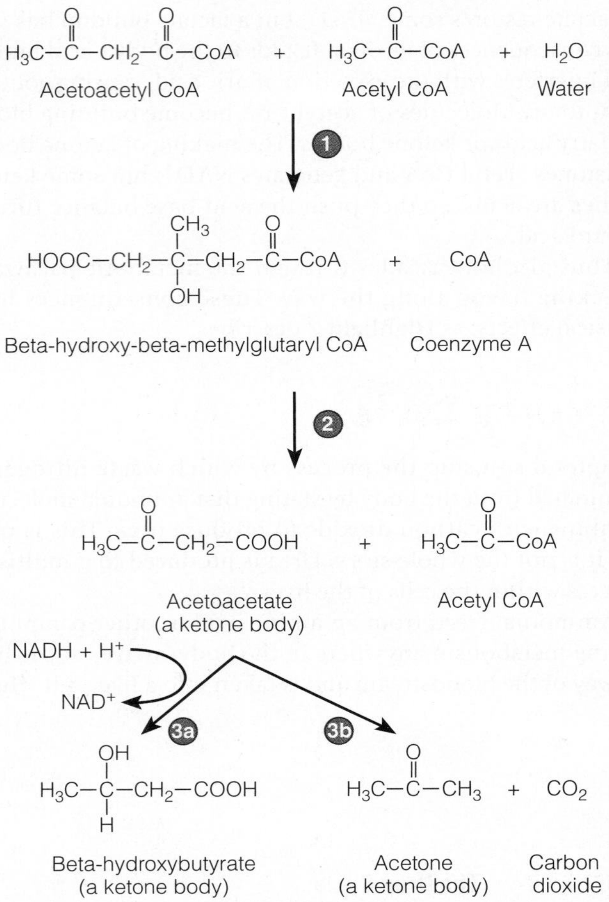

Acetoacetyl CoA + Acetyl CoA + Water

↓ 1

$HOOC{-}CH_2{-}C(CH_3)(OH){-}CH_2{-}C(=O){-}CoA$ + CoA

Beta-hydroxy-beta-methylglutaryl CoA + Coenzyme A

↓ 2

$H_3C{-}C(=O){-}CH_2{-}COOH$ + $H_3C{-}C(=O){-}CoA$

Acetoacetate (a ketone body) + Acetyl CoA

3a: NADH + H^+ → NAD^+

$H_3C{-}C(OH)(H){-}CH_2{-}COOH$

Beta-hydroxybutyrate (a ketone body)

3b:

$H_3C{-}C(=O){-}CH_3$ + CO_2

Acetone (a ketone body) + Carbon dioxide

Appendix D Measures of Protein Quality

In a world where food is scarce and many people's diets contain marginal or inadequate amounts of protein, it is important to know which foods contain the highest-quality protein. Chapter 6 describes protein quality, and this appendix presents different measures researchers use to assess the quality of a food protein. The accompanying glossary defines related terms.

CONTENTS

Amino Acid Scoring

Amino acid scoring evaluates a protein's quality by determining its amino acid composition and comparing it with that of a reference protein. The advantages of amino acid scoring are that it is simple and inexpensive, it easily identifies the limiting amino acid, and it can be used to score mixtures of different proportions of two or more proteins mathematically without having to make up a mixture and test it. Its chief weaknesses are that it fails to estimate the digestibility of a protein, which may strongly affect the protein's quality; it relies on a chemical procedure in which certain amino acids may be destroyed, making the pattern that is analyzed inaccurate; and it is blind to other features of the protein (such as the presence of substances that may inhibit the digestion or utilization of the protein) that would only be revealed by a test in living animals.

Table D-1 shows the reference pattern for the nine essential amino acids. To interpret the table, read, "For every 3210 units of essential amino acids, 145 must be histidine, 340 must be isoleucine, 540 must be leucine," and so on. To compare a test protein with the reference protein, the experimenter first obtains a chemical analysis of the test protein's amino acids. Then, taking 3210 units of the amino acids, the experimenter compares the amount of each amino acid to the amount found in 3210 units of essential amino acids in egg protein. For example, suppose the test protein contained (per 3210 units) 360 units of isoleucine; 500 units of leucine; 350 of lysine; and for each of the other amino acids, more units than egg protein contains. The two amino acids that are low are leucine (500 as compared with 540 in egg) and lysine (350 versus 440 in egg). The ratio—amino acid in the test protein divided by amino acid in egg—is 500/540 (or about 0.93) for leucine and 350/440 (or about 0.80) for lysine. Lysine is the

TABLE D-1 Reference Pattern for Amino Acid Scoring of Proteins

Essential Amino Acids	Reference Protein—Whole Egg (mg amino acid/g nitrogen)
Histidine	145
Isoleucine	340
Leucine	540
Lysine	440
Methionine + cystine[a]	355
Phenylalanine + tyrosine[b]	580
Threonine	294
Tryptophan	106
Valine	410
Total	3210

[a]Methionine is essential and is also used to make cystine. Thus the methionine requirement is lower if cystine is supplied.

[b]Phenylalanine is essential and is also used to make tyrosine if not enough of the latter is available. Thus the phenylalanine requirement is lower if tyrosine is also supplied.

APPENDIX D

GLOSSARY

amino acid scoring: a measure of protein quality assessed by comparing a protein's amino acid pattern with that of a reference protein; sometimes called *chemical scoring*.

biological value (BV): a measure of protein quality assessed by measuring the amount of protein nitrogen that is retained from a given amount of protein nitrogen absorbed.

net protein utilization (NPU): a measure of protein quality assessed by measuring the amount of protein nitrogen that is retained from a given amount of protein nitrogen eaten.

PDCAAS (protein digestibility-corrected amino acid score): a measure of protein quality assessed by comparing the amino acid score of a food protein with the amino acid requirements of preschool-age children and then correcting for the true digestibility of the protein; recommended by the FAO/WHO and used to establish protein quality of foods for Daily Value percentages on food labels.

protein efficiency ratio (PER): a measure of protein quality assessed by determining how well a given protein supports weight gain in growing rats; used to establish the protein quality for infant formulas and baby foods.

limiting amino acid (the one that falls shortest compared with egg). If the protein's limiting amino acid is 80 percent of the amount found in the reference protein, it receives a score of 80.

TABLE D-2 PDCAAS Values of Selected Foods

Food	PDCAAS
Casein (milk protein)	1.00
Egg white	1.00
Soybean (isolate)	.99
Beef	.92
Pea flour	.69
Kidney beans (canned)	.68
Chickpeas (canned)	.66
Pinto beans (canned)	.66
Rolled oats	.57
Lentils (canned)	.52
Peanut meal	.52
Whole wheat	.40

NOTE: 1.0 is the maximum PDCAAS a food protein can receive.

PDCAAS

The **protein digestibility-corrected amino acid score,** or **PDCAAS,** compares the amino acid composition of a protein with human amino acid requirements and corrects for digestibility. First the protein's amino acid composition is determined, and then it is compared against the amino acid requirements of preschool-aged children. This comparison reveals the most limiting amino acid—the one that falls shortest compared with the reference. If a food protein's limiting amino acid is 70 percent of the amount found in the reference protein, it receives a score of 70. The amino acid score is multiplied by the food's protein digestibility percentage to determine the PDCAAS. The accompanying "How To" provides an example of how to calculate the PDCAAS, and Table D-2 lists the PDCAAS values of selected foods.

>How To

Measure Protein Quality Using PDCAAS

To calculate the PDCAAS (protein digestibility-corrected amino acid score), researchers first determine the amino acid profile of the test protein (in this example, pinto beans). The second column of the table below presents the essential amino acid profile for pinto beans. The third column presents the amino acid reference pattern.

To determine how well the food protein meets human needs, researchers calculate the ratio by dividing the second column by the third column (for example, $30 \div 18 = 1.67$). The amino acid with the lowest ratio is the most limiting amino acid—in this case, methionine. Its ratio is the amino acid score for the protein—in this case, 0.84.

The amino acid score alone, however, does not account for digestibility. Protein digestibility, as determined by rat studies, yields a value of 79 percent for pinto beans. Together, the amino acid score and the digestibility value determine the PDCAAS:

$$\text{PDCAAS} = \text{protein digestibility} \times \text{amino acid score}$$

$$\text{PDCAAS for pinto beans} = 0.79 \times 0.84 = 0.66$$

Thus the PDCAAS for pinto beans is 0.66 as Table D-2 confirms.

The PDCAAS is used to determine the % Daily Value on food labels. To calculate the % Daily Value for protein for canned pinto beans, multiply the number of grams of protein in a standard serving (in the case of pinto beans, 7 grams per ½ cup) by the PDCAAS:

$$7 \text{ g} \times 0.66 = 4.62$$

This value is then divided by the recommended standard for protein (for children over age four and adults, 50 grams):

$$4.62 \div 50 = 0.09 \text{ (or 9\%)}$$

The food label for this can of pinto beans would declare that one serving provides 7 grams protein, and if the label included a % Daily Value for protein (which is optional), the value would be 9 percent.

Essential Amino Acids	Amino Acid Profile of Pinto Beans (mg/g protein)	Amino Acid Reference Pattern (mg/g protein)	Amino Acid Score
Histidine	30.0	18	1.67
Isoleucine	42.5	25	1.70
Leucine	80.4	55	1.46
Lysine	69.0	51	1.35
Methionine (+ cystine)	21.1	25	0.84
Phenylalanine (+ tyrosine)	90.5	47	1.93
Threonine	43.7	27	1.62
Tryptophan	8.8	7	1.26
Valine	50.1	32	1.57

Biological Value

The **biological value (BV)** of a protein measures its efficiency in supporting the body's needs. In a test of biological value, two nitrogen balance studies are done. In the first, no protein is fed, and nitrogen (N) excretions in the urine and feces are measured. It is assumed that under these conditions, N lost in the urine is the amount the body always necessarily loses by filtration into the urine each day, regardless of what protein is fed (endogenous N). The N lost in the feces (called metabolic N) is the amount the body invariably loses into the intestine each day, whether or not food protein is fed. (To help you remember the terms: endogenous N is "urinary N on a zero-protein diet"; metabolic N is "fecal N on a zero-protein diet.")

In the second study, an amount of protein slightly below the requirement is fed. Intake and losses are measured; then the BV is derived using this formula:

$$\text{BV} = \frac{\text{N retained}}{\text{N absorbed}} \times 100$$

The denominator of this equation expresses the amount of nitrogen *absorbed:* food N minus fecal N (excluding the metabolic N the body would lose in the feces anyway, even without food). The numerator expresses the amount of N *retained* from the N absorbed: absorbed N (as in the denominator) minus the N excreted in the urine (excluding the endogenous N the body would lose in the urine anyway even without food). The more nitrogen retained, the higher the protein quality. (Recall that when an essential amino acid is missing, protein synthesis stops, and the remaining amino acids are deaminated and the nitrogen excreted.)

Egg protein has a BV of 100, indicating that 100 percent of the nitrogen absorbed is retained. Supplied in adequate quantity, a protein with a BV of 70 or greater can support human growth as long as energy intake is adequate. Table D-3 presents the BV for selected foods.

This method has the advantages of being based on experiments with human beings (it can be done with animals, too, of course) and of measuring actual nitrogen retention. But it is also cumbersome, expensive, and often impractical, and it is based on several assumptions that may not be valid. For example, the physiology normal environment, or typical food intake of the subjects used for testing may not be similar to those for whom the test protein may ultimately be used. For another example, the retention of protein in the body does not necessarily mean that it is being well utilized. Considerable exchange of protein among tissues (protein turnover) occurs, but is hidden from view when only N intake and output are measured. The test of biological value wouldn't detect if one tissue were shorted.

TABLE D-3 Biological Values (BV) of Selected Foods

Egg	100
Milk	93
Beef	75
Fish	75
Corn	72

NOTE: 100 is the maximum BV a food protein can receive.

Net Protein Utilization

Like BV, **net protein utilization (NPU)** measures how efficiently a protein is used by the body and involves two balance studies. The difference is that NPU measures retention of food nitrogen rather than food nitrogen absorbed (as in BV). The formula for NPU is:

$$\text{NPU} = \frac{\text{N retained}}{\text{N intake}} \times 100$$

The numerator is the same as for BV, but the denominator represents food N intake only—not N absorbed.

This method offers advantages similar to those of BV determinations and is used more frequently, with animals as the test subjects. A drawback is that if a low NPU is obtained, the test results offer no help in distinguishing between two possible causes: a poor amino acid composition of the test protein or poor digestibility. There is also a limit to the extent to which animal test results can be assumed to be applicable to human beings.

TABLE D-4 Protein Efficiency Ratio (PER) Values of Selected Proteins

Casein (milk)	2.8
Soy	2.4
Glutein (wheat)	0.4

Protein Efficiency Ratio

The **protein efficiency ratio (PER)** measures the weight gain of a growing animal and compares it to the animal's protein intake. Until recently, the PER was generally accepted in the United States and Canada as the official method for assessing protein quality and it is still used to evaluate proteins for infants.

Young rats are fed a measured amount of protein and weighed periodically as they grow. The PER is expressed as:

$$\text{PER} = \frac{\text{weight gain (g)}}{\text{protein intake (g)}}$$

This method has the virtues of economy and simplicity but it also has many drawbacks. The experiments are time-consuming; the amino acid needs of rats are not the same as those of human beings; and the amino acid needs for growth are not the same as for the maintenance of adult animals (growing animals need more lysine, for example). Table D-4 presents PER values for selected foods.

APPENDIX D

Appendix E Nutrition Assessment

Nutrition assessment evaluates a person's health from a nutrition perspective. Many factors influence or reflect nutrition status. Consequently, the assessor, usually a registered dietitian assisted by other qualified health-care professionals, gathers information from many sources, including:

- Historical information.
- Anthropometric measurements.
- Physical examinations.
- Biochemical analyses (laboratory tests).

Each of these methods involves collecting data in a variety of ways and interpreting each finding in relation to the others to create a total picture.

The accurate gathering of this information and its careful interpretation are the basis for a meaningful evaluation. The more information gathered about a person, the more accurate the assessment will be. Gathering information is a time-consuming process, however, and time is often a rare commodity in the health care setting. Nutrition care is only one part of total care. It may not be practical or essential to collect detailed information on each person.

A strategic compromise is to screen clients by collecting preliminary data. Data such as height-weight and hematocrit are easy to obtain and can alert health-care workers to potential problems. **Nutrition screening** identifies clients who will require additional nutrition assessment. This appendix provides a sample of the procedures, standards, and charts commonly used in nutrition assessment.

CONTENTS

Historical Information

Clues about present nutrition status become evident with a careful review of a person's historical data (see Table E-1). Even when the data are subjective, they reveal important facts about a person. A thorough history identifies risk factors associated with poor nutrition status (see Table E-2) and provides a sense of the whole person. As you can see, many aspects of a person's life influence nutrition status and provide clues to possible problems.

TABLE E-1 Historical Data Used in Nutrition Assessments

Type of History	What It Identifies
Health history	Current and previous health problems and family health history that affect nutrient needs, nutrition status, or the need for intervention to prevent or alleviate health problems
Socioeconomic history	Personal, cultural, financial, and environmental influences on food intake, nutrient needs, and diet therapy options
Drug history	Medications (prescription and over-the-counter), illicit drugs, dietary supplements, and alternative therapies that affect nutrition status
Diet history	Nutrient intake excesses or deficiencies and reasons for imbalances

nutrition screening: the use of preliminary nutrition assessment techniques to identify people who are malnourished or are at risk for malnutrition.

TABLE E-2 Risk Factors for Poor Nutrition Status

Health History		
• Acquired immune deficiency syndrome (AIDS) • Alcoholism • Anorexia (lack of appetite) • Anorexia nervosa • Bulimia nervosa • Burns • Cancer • Chewing or swallowing difficulties (including poorly fitted dentures, dental caries, missing teeth, and mouth ulcers) • Chronic obstructive pulmonary disease • Circulatory problems • Constipation • Crohn's disease • Cystic fibrosis • Decubitus ulcers (pressure sores) • Dementia • Depleted blood proteins • Depression • Diabetes mellitus	• Diarrhea, prolonged or severe • Drug addiction • Dysphagia • Failure to thrive • Feeding disabilities • Fever • GI tract disorders or surgery • Heart disease • HIV infection • Hormonal imbalance • Hyperlipidemia • Hypertension • Infections • Kidney disease • Liver disease • Lung disease • Malabsorption • Mental illness • Mental retardation • Multiple pregnancies	• Nausea • Neurologic disorders • Organ failure • Overweight • Pancreatic insufficiency • Paralysis • Physical disability • Pneumonia • Pregnancy • Radiation therapy • Recent major illness • Recent major surgery • Recent weight loss or gain • Tobacco use • Trauma • Ulcerative colitis • Ulcers • Underweight • Vomiting, prolonged or severe
Socioeconomic History		
• Access to groceries • Activities • Age • Education	• Ethnic identity • Income • Kitchen facilities • Number of people in household	• Occupation • Religious affiliation
Drug History		
• Amphetamines • Analgesics • Antacids • Antibiotics • Anticonvulsant agents • Antidepressant agents • Antidiabetic agents	• Antidiarrheals • Antifungal agents • Antihyperlipemics • Antihypertensives • Antineoplastics • Antiulcer agents • Antiviral agents	• Catabolic steroids • Diuretics • Hormonal agents • Immunosuppressive agents • Laxatives • Oral contraceptives • Vitamin and other dietary supplements
Diet History		
• Deficient or excessive food intakes • Frequently eating out • Intravenous fluids (other than total parenteral nutrition) for 7 or more days	• Monotonous diet (lacking variety) • No intake for 7 or more days • Poor appetite	• Restricted or fad diets • Unbalanced diet (omitting any food group) • Recent weight gains or losses

An adept history taker uses the interview both to gather facts and to establish a rapport with the client. This section briefly reviews the major areas of nutrition concern in a person's history: health, socioeconomic factors, drugs, and diet.

Health History The assessor can obtain a **health history** from records completed by the attending physician, nurse, or other health-care professional. In addition, conversations with the client can uncover valuable information previously overlooked because no one thought to ask or because the client was not thinking clearly when asked.

An accurate, complete health history can reveal conditions that increase a client's risk for malnutrition (review Table E-2). Diseases and their therapies can have either immediate or long-term effects on nutrition status by interfering with ingestion, digestion, absorption, metabolism, or excretion of nutrients.

Socioeconomic History A **socioeconomic history** reveals factors that profoundly affect nutrition status. The ethnic background and educational level of

health history: an account of a client's current and past health status and disease risks.

socioeconomic history: a record of a person's social and economic background, including such factors as education, income, and ethnic identity.

both the client and the other members of the household influence food availability and food choices. An understanding of the community environment is also important in assessing nutrition status. For example, the interviewer should be familiar with the food habits of the major ethnic groups within the locale, regional food preferences, and nutrition resources and programs available in the community. Local health departments and social agencies often can provide such information.

Level of income also influences the diet. In general, the quality of the diet declines as income falls. At some point, the ability to purchase the foods required to meet nutrient needs is lost; an inadequate income puts an adequate diet out of reach. Agencies use poverty indexes to identify people at risk for poor nutrition and to qualify people for government food assistance programs.

Low income affects not only the power to purchase foods but also the ability to shop for, store, and cook them. A skilled assessor will note whether a person has transportation to a grocery store that sells a sufficient variety of low-cost foods, and whether the person has access to a refrigerator and stove.

Drug History The many interactions of foods and drugs require that health-care professionals take a **drug history** and pay special attention to any client who takes drugs routinely. If a person is taking any drug, the assessor records the name of the drug; the dose, frequency, and duration of intake; the reason for taking the drug; and signs of any adverse effects.

The interactions of drugs and nutrients may take many forms:

- Drugs can alter food intake and the absorption, metabolism, and excretion of nutrients.
- Foods and nutrients can alter the absorption, metabolism, and excretion of drugs.

Highlight 17 discusses nutrient-drug interactions in more detail, and Table H17-1 (p. 576) summarizes the mechanisms by which these interactions occur and provides specific examples.

Diet History A **diet history** provides a record of a person's eating habits and food intake and can help identify possible nutrient imbalances. Food choices are an important part of lifestyle and often reflect a person's philosophy. The assessor who asks nonjudgmental questions about eating habits and food intake encourages trust and enhances the likelihood of obtaining accurate information.

Assessors evaluate food intake using various tools such as the 24-hour recall, the usual intake record, the food record, and the food frequency questionnaire. Food models or photos and measuring devices can help clients identify the types of foods and quantities consumed. The assessor also needs to know how the foods are prepared and when they are eaten. In addition to asking about foods, assessors will ask about beverage consumption, including beverages containing alcohol or caffeine.

Besides identifying possible nutrient imbalances, diet histories provide valuable clues about how a person will accept diet changes should they be necessary. Information about what and how a person eats provides the background for realistic and attainable nutrition goals.

24-Hour Recall The **24-hour recall** provides data for one day only and is commonly used in nutrition surveys to obtain estimates of the typical food intakes for a population. The assessor asks the client to recount everything eaten or drunk in the past 24 hours or for the previous day.

An advantage of the 24-hour recall is that it is easy to obtain. It is also more likely to provide accurate data, at least about the past 24 hours, than estimates of average intakes over long periods. It does not, however, provide enough information to allow accurate generalizations about an individual's usual food intake. The previous day's intake may not be typical, for example, or the person may be unable to report portion sizes accurately or may conceal or forget information

drug history: a record of all the drugs, over-the-counter and prescribed, that a person takes routinely.

diet history: a record of eating behaviors and the foods a person eats.

24-hour recall: a record of foods eaten by a person for one 24-hour period.

APPENDIX E

about foods eaten. This limitation is partially overcome when 24-hour recalls are collected on several nonconsecutive days.

Usual Intake To obtain data about a person's usual intake, an inquiry might begin with "What is the first thing you usually eat or drink during the day?" Similar questions follow until a typical daily intake pattern emerges. This method can be useful, especially in verifying food intake when the past 24 hours have been atypical. It also helps the assessor verify food habits. For example, one person may always eat an afternoon snack; another may never eat breakfast. A person whose intake varies widely from day to day, however, may find it difficult to answer such general questions, and in that case, another food intake tool should be used to estimate nutrient intake.

Food Record Another tool for history taking is the **food record,** in which the person records food eaten, including the quantity and method of preparation. Chapter 9 (p. 284) provides an example. A food record can help both the assessor and the client to determine factors associated with eating that may affect dietary balance and adequacy.

Food records work especially well with cooperative people but require considerable time and effort on their part. A prime advantage is that the record keeper assumes an active role and may for the first time become aware of personal food habits and assume responsibility for them. It also provides the assessor with an accurate picture of the person's lifestyle and factors that affect food intake. For these reasons, a food record can be particularly useful in outpatient counseling for such nutrition problems as overweight, underweight, or food allergy. The major disadvantages stem from poor compliance in recording the data and conscious or unconscious changes in eating habits that may occur while the person is keeping the record.

Food Frequency Questionnaire An assessor uses a **food frequency questionnaire** to compare a client's food intake with the Daily Food Patterns. Clients may be asked how many servings of each of the following they eat in a typical day: grains; vegetables; fruits; protein foods; and solid fats and added sugars. This information helps pinpoint food groups, and therefore nutrients, that may be excessive or deficient in the diet. That a person ate no vegetables yesterday may not seem particularly significant, but never eating vegetables is a warning of possible nutrient deficiencies. When used with the usual intake or 24-hour recall approach, the food frequency questionnaire enables the assessor to double-check the accuracy of the information obtained.

Analysis of Food Intake Data After collecting food intake data, the assessor estimates nutrient intakes, either informally by using food guides or formally by using diet analysis programs. The assessor compares intakes with standards, usually nutrient recommendations or dietary guidelines, to determine how closely the person's diet meets the standards. Are the types and amounts of proteins, carbohydrates (including fiber), and fats (including cholesterol) appropriate? Are all food groups included in appropriate amounts? Is caffeine or alcohol consumption excessive? Are intakes of any vitamins or minerals (including sodium and iron) excessive or deficient? An informal evaluation is possible only if the assessor has enough prior experience with formal calculations to "see" nutrient amounts in reported food intakes without calculations. Even then, such an informal analysis is best followed by a spot check for key nutrients by actual calculation.

Formal calculations can be performed either manually (by looking up each food in a table of food composition, recording its nutrients, and adding them up) or by using a diet analysis program. The assessor then compares the intakes with standards such as the RDA.

Limitations of Food Intake Analysis Diet histories can be superbly informative, but the skillful assessor also keeps their limitations in mind. For example, a diet analysis program tends to imply greater accuracy than is possible to obtain from

food record: an extensive, accurate log of all foods eaten over a period of several days or weeks. A food record that includes associated information such as when, where, and with whom each food is eaten is sometimes called a *food diary*.

food frequency questionnaire: a checklist of foods on which a person can record the frequency with which he or she eats each food.

data as uncertain as the starting information. Nutrient contents of foods listed in tables of food composition or stored in computer databases are averages and, for some nutrients, incomplete. In addition, the available data on nutrient contents of foods do not reflect the amounts of nutrients a person actually absorbs. Iron is a case in point: its availability from a given meal may vary depending on the person's iron status; the relative amounts of heme iron, nonheme iron, vitamin C, meat, fish, and poultry eaten at the meal; and the presence of inhibitors of iron absorption such as tea, coffee, and nuts. (Chapter 13 describes the many factors that influence iron absorption from a meal.)

Furthermore, reported portion sizes may not be correct. The person who reports eating "a serving" of greens may not distinguish between ¼ cup and 2 whole cups; only individuals who have practice measuring food quantities can accurately report serving sizes. Children tend to remember the serving sizes of foods they like as being larger than serving sizes of foods they dislike.

An estimate of nutrient intakes from a diet history, combined with other sources of information, allows the assessor to confirm or eliminate the possibility of suspected nutrition problems. The assessor must constantly remember that nutrient intakes in adequate amounts do not guarantee adequate nutrient status for an individual. Likewise, insufficient intakes do not always indicate deficiencies, but instead alert the assessor to possible problems. Each person digests, absorbs, metabolizes, and excretes nutrients in a unique way; individual needs vary. Intakes of nutrients identified by diet histories are only pieces of a puzzle that must be put together with other indicators of nutrition status in order to extract meaning.

Anthropometric Measurements

Anthropometrics are physical measurements that reflect body composition and development (see Table E-3). They serve three main purposes: first, to evaluate the progress of growth in pregnant women, infants, children, and adolescents; second, to detect undernutrition and overnutrition in all age groups; and third, to measure changes in body composition over time.

Health-care professionals compare anthropometric measurements taken on an individual with population standards specific for gender and age or with previous measures of the individual. Measurements taken periodically and compared with previous measurements reveal changes in an individual's status.

Mastering the techniques for taking anthropometric measurements requires proper instruction and practice to ensure reliability. Once the correct techniques are learned, taking measurements is easy and requires minimal equipment.

Height and weight are well-recognized anthropometrics; other anthropometrics include skinfold measurements and various measures of lean tissue. Other measures are useful in specific situations. For example, a head circumference measurement may help to assess brain development in an infant, and an abdominal girth measurement supplies information about abdominal fluid retention in individuals with liver disease.

Measures of Growth and Development

Height and weight are among the most common and useful anthropometric measurements. Length measurements for infants and children up to age 3 and height measurements for children

TABLE E-3 Anthropometric Measurements Used in Nutrition Assessments

Type of Measurement	What It Reflects
Abdominal girth measurement	Abdominal fluid retention and abdominal organ size
Height-weight	Overnutrition and undernutrition; growth in children
Head circumference	Brain growth and development in infants and children under age 2
Skinfold	Subcutaneous and total body fat
Waist circumference	Body fat distribution

anthropometrics: measurements of the physical characteristics of the body, such as height and weight.
- **anthropos** = human
- **metric** = measuring

APPENDIX E

> FIGURE E-1 **Length Measurement of an Infant**

An infant is measured lying down on a measuring board with a fixed headboard and a movable footboard. Note that two people are needed to measure the infant's length.

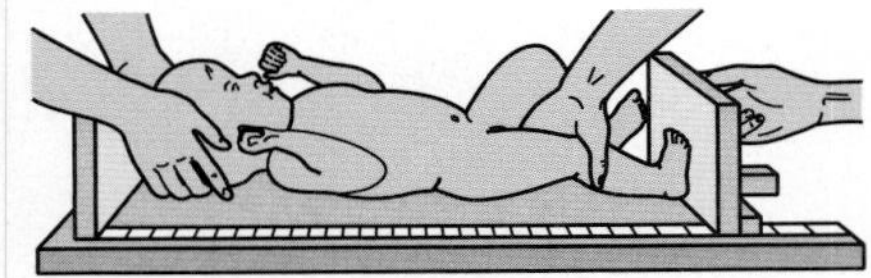

> FIGURE E-2 **Height Measurement of an Older Child or Adult**

Height is measured most accurately when the person stands against a flat wall to which a measuring tape has been affixed.

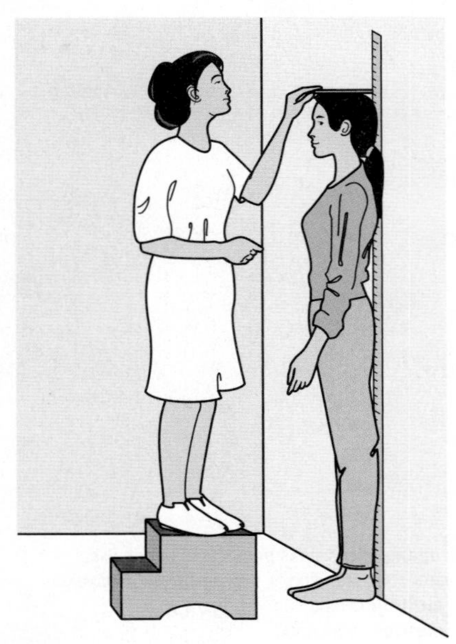

over 3 are particularly valuable in assessing growth and therefore nutrition status. For adults, height measurements alone are not critical, but help to estimate healthy weight and to interpret other assessment data. Once adult height has been reached, changes in body weight provide useful information in assessing overnutrition and undernutrition.

Height For infants and children younger than 3, health-care professionals may use special equipment to measure length. The assessor lays the barefoot infant on a measuring board that has a fixed headboard and movable footboard attached at right angles to the surface (see Figure E-1). Often two people are needed to obtain an accurate measurement: one to hold the infant's head against the headboard, and the other to keep the legs straight and do the measuring. This method provides the most accurate measure possible, but many health-care professionals use a less exacting method. They may simply hold the infant straight with its head against the headboard or other vertical support, mark the blanket with a chalk or pen at the infant's heel, and then measure the distance from the headboard to the mark. Even more informally and less accurately, they may lay the infant on a flat surface and extend a nonstretchable measuring tape along the side of the infant from the top of the head to the heel of the foot.

The procedure for measuring a child who can stand erect and cooperate is the same as for an adult. The best way to measure standing height is with the person's back against a flat wall to which a nonstretchable measuring tape or stick has been fixed (see Figure E-2). The person stands erect, without shoes, with heels together. The person's line of sight should be horizontal, with the heels, buttocks, shoulders, and head touching the wall. The assessor places a ruler, book, or other inflexible object on top of the head at a right angle to the wall; carefully checks the height measurement; and records it immediately in either inches or centimeters so that the correct measurement will not be forgotten.

The measuring rod of a scale is commonly used, but is less accurate because it bends easily. The assessor follows the same general procedure, asking the person to face away from the scale and to take extra care to stand erect.

Unfortunately, many health-care professionals merely ask clients how tall they are rather than measuring their height. Self-reported height is often inaccurate and should be used only as a last resort when measurement is impractical (in the case of an uncooperative client, an emergency admission, or the like).

Weight Valid weight measurements require scales that have been carefully maintained, calibrated, and checked for accuracy at regular intervals. Beam balance and electronic scales are the most accurate types of scales. To measure infants' weight, assessors use special scales that allow infants to lie or sit (see Figure E-3). Weighing infants naked, without diapers, is standard procedure. Children who can stand are weighed in the same way as adults (see Figure E-4). To make repeated measures useful, standardized conditions are necessary. Each weighing should take place at the same time of day (preferably before breakfast), in the same amount of clothing (without shoes), after the person has voided, and on the same scale. Special scales and hospital beds with built-in scales are available for weighing people who are bedridden. Bathroom scales are inaccurate and inappropriate in a professional setting. As with all measurements, the assessor records the observed weight immediately in either pounds or kilograms.

Head Circumference Assessors may also measure head circumference to confirm that infant growth is proceeding normally or to help detect protein malnutrition and evaluate the extent of its impact on brain size. To measure head circumference, the assessor places a nonstretchable tape so that it encircles the largest part of the infant's or child's head: just above the eyebrow ridges, just above the point where the ears attach, and around the occipital prominence at the back of the head. To ensure accurate recording, the assessor immediately notes the measure in either inches or centimeters.

Analysis of Measures in Infants and Children Growth retardation is a sign of poor nutrition status. Obesity is also a sign that dietary intervention may be needed.

Health professionals generally evaluate physical development by monitoring the growth rate of a child and comparing this rate with standard charts. Standard charts compare weight to age, height to age, and weight to height; ideally, height and weight are in roughly the same percentile. Although individual growth patterns may vary, a child's growth curve will generally stay at about the same percentile throughout childhood. In children whose growth has been retarded, nutrition rehabilitation will ideally induce height and weight to increase to higher percentiles. In overweight children, the goal is for weight to remain stable as height increases, until weight becomes appropriate for height.

To evaluate growth in infants, an assessor uses charts such as those in Figures E-5 (A and B) through E-10 (A and B). The assessor follows these steps to plot a weight measurement on a percentile graph:

- Select the appropriate chart based on age and gender.
- Locate the child's age along the horizontal axis on the bottom of the chart.
- Locate the child's weight in pounds or kilograms along the vertical axis.
- Mark the chart where the age and weight lines intersect, and read off the percentile.

To assess length, height, or head circumference, the assessor follows the same procedure, using the appropriate chart. (When length is measured, use the chart for birth to 36 months; when height is measured, use the chart for 2 to 20 years.) Head circumference percentile should be similar to the child's height and weight percentiles. With height, weight, and head circumference measures plotted on growth percentile charts, a skilled clinician can begin to interpret the data.

Percentile charts divide the measures of a population into 100 equal divisions. Thus half of the population falls above the 50th percentile, and half falls below. The use of percentile measures allows for comparisons among people of the same age and gender. For example, a 6-month-old female infant whose weight is at the 75 percentile weighs more than 75 percent of the female infants her age.

Head circumference is generally measured in children under 2 years of age. Because the brain grows rapidly before birth and during early infancy, extreme and chronic malnutrition during these times can impair brain development, curtailing the number of brain cells and the size of head circumference. Nonnutritional factors, such as certain disorders and genetic variation, can also influence head circumference.

Analysis of Measures in Adults For adults, health-care professionals typically compare weights with weight-for-height standards. One such standard is the body mass index (BMI), ♦ described in Chapter 8, which is useful for estimating the risk to health associated with overnutrition. The inside back cover shows BMI for various heights and weights.

Measures of Body Fat and Lean Tissue

Significant weight changes in both children and adults can reflect overnutrition and undernutrition with respect to energy and protein. To estimate the degree to which fat stores or lean tissues are affected by overnutrition or malnutrition, several anthropometric measurements are useful (review Table E-3 on p. E-5).

Skinfold Measures Skinfold measures provide a good estimate of total body fat and a fair assessment of the fat's location. Approximately half the fat in the body lies directly beneath the skin, and the thickness of this subcutaneous fat reflects total body fat. In some parts of the body, such as the back and the back of the arm over the triceps muscle, this fat is loosely attached; ♦ a person can pull it up between the thumb and forefinger to obtain a measure of skinfold thickness. To measure skinfold, a skilled assessor follows a standard procedure using reliable calipers (illustrated in Figure E-11, p. E-14) and then compares the measurement with standards. Triceps skinfold measures greater than 15 millimeters in men or 25 millimeters in women suggest excessive body fat.

Skinfold measurements correlate directly with the risk of heart disease. They assess central obesity and its associated risks better than do weight measures

> FIGURE E-3 Weight Measurement of an Infant

Infants sit or lie down on scales that are designed to hold them while they are being weighed.

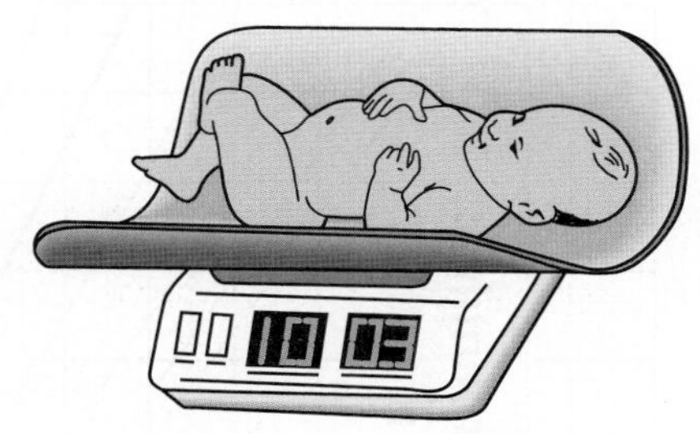

♦ The *body mass index (BMI)* is an index of a person's weight in relation to height, determined by dividing the weight in kilograms by the square of the height in meters:

$$\text{BMI} = \frac{\text{Weight (kg)}}{\text{Height (m)}^2}$$

♦ Common sites for skinfold measures:
- Triceps
- Biceps
- Subscapular (below shoulder blade)
- Suprailiac (above hip bone)
- Abdomen
- Upper thigh

> FIGURE E-4 Weight Measurement of an Older Child or Adult

Whenever possible, children and adults are measured on beam balance or electronic scales to ensure accuracy.

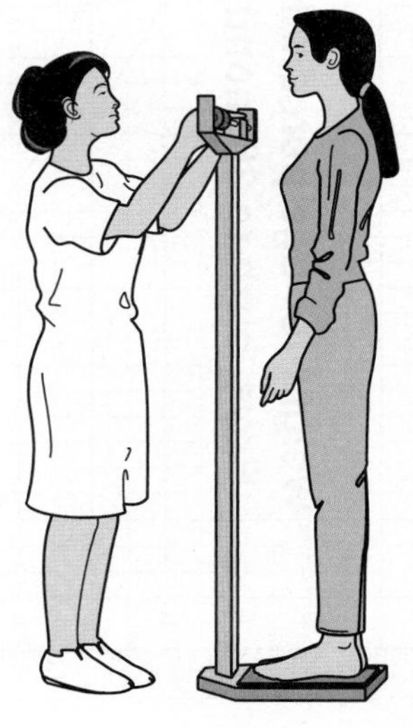

APPENDIX E

> FIGURE E-5A Weight-for-Age Percentiles: Boys, Birth to 36 Months

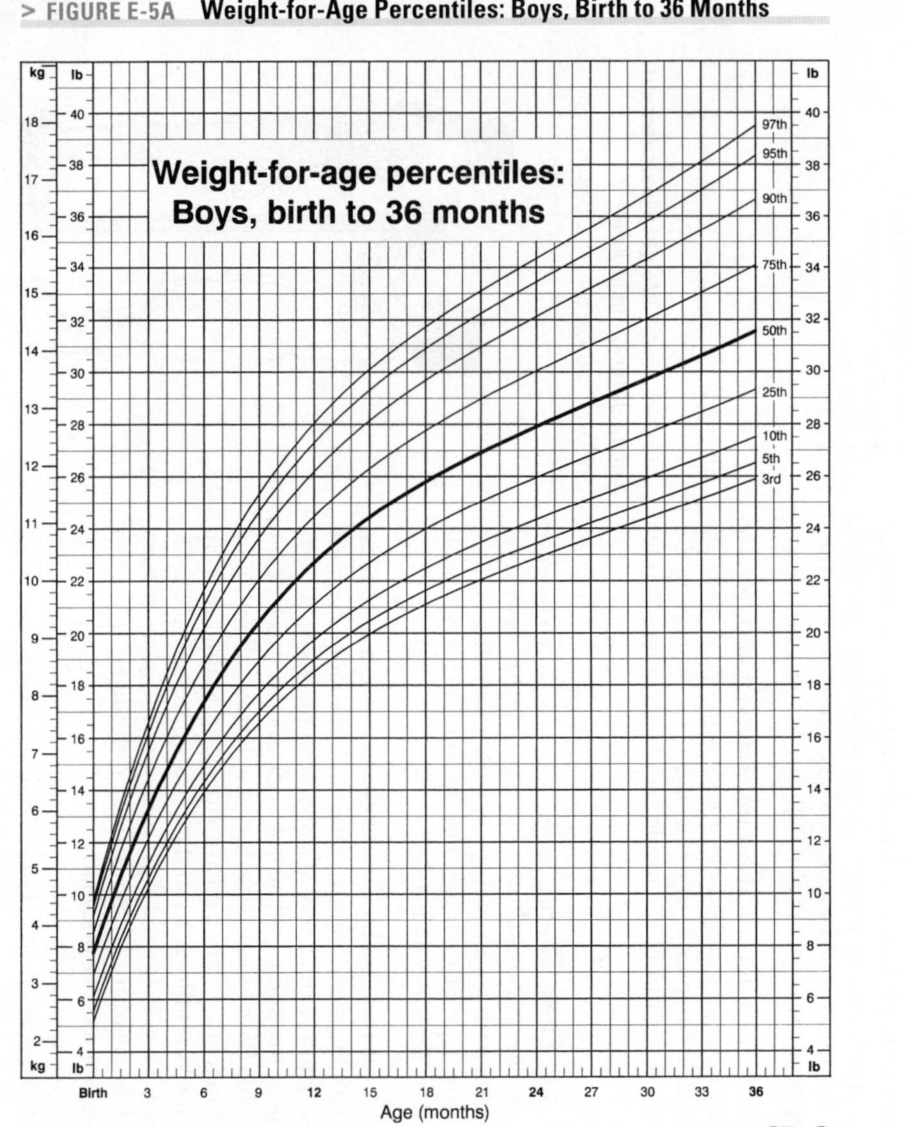

Figure 1. Weight-for-age percentiles, boys, birth to 36 months, CDC growth charts: United States

> FIGURE E-5B Weight-for-Age Percentiles: Girls, Birth to 36 Months

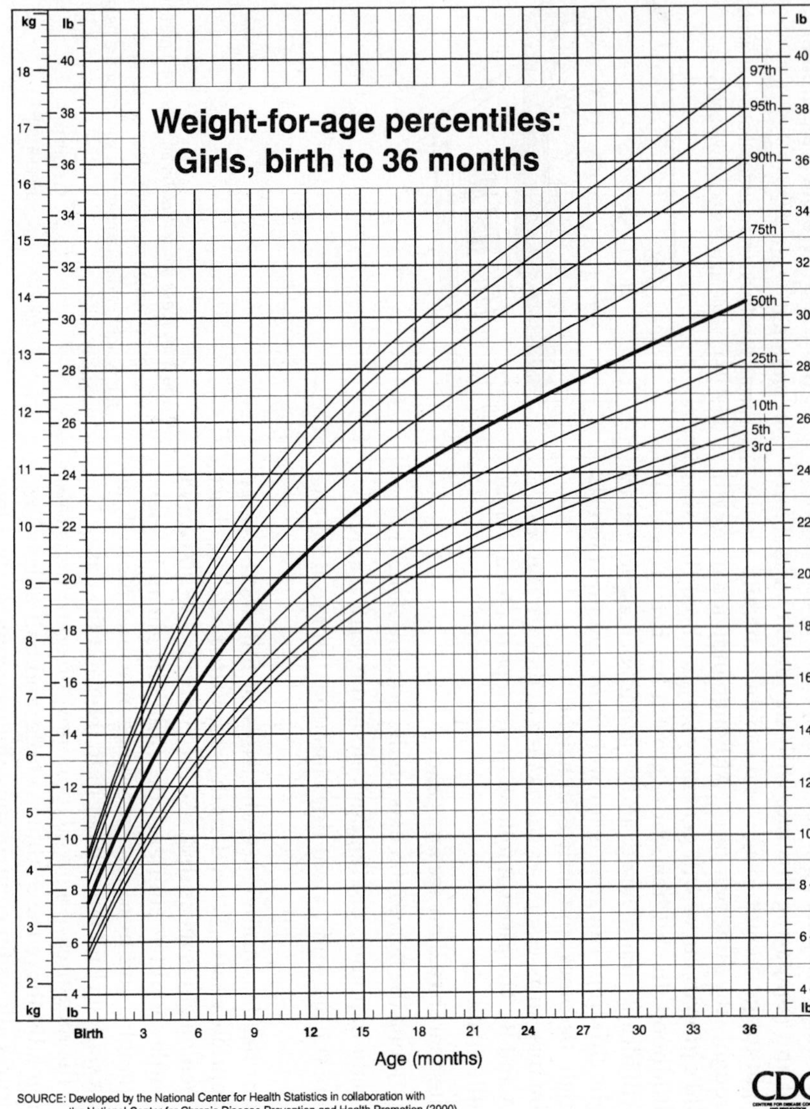

Figure 2. Weight-for-age percentiles, girls, birth to 36 months, CDC growth charts: United States

> FIGURE E-6A Length-for-Age Percentiles: Boys, Birth to 36 Months

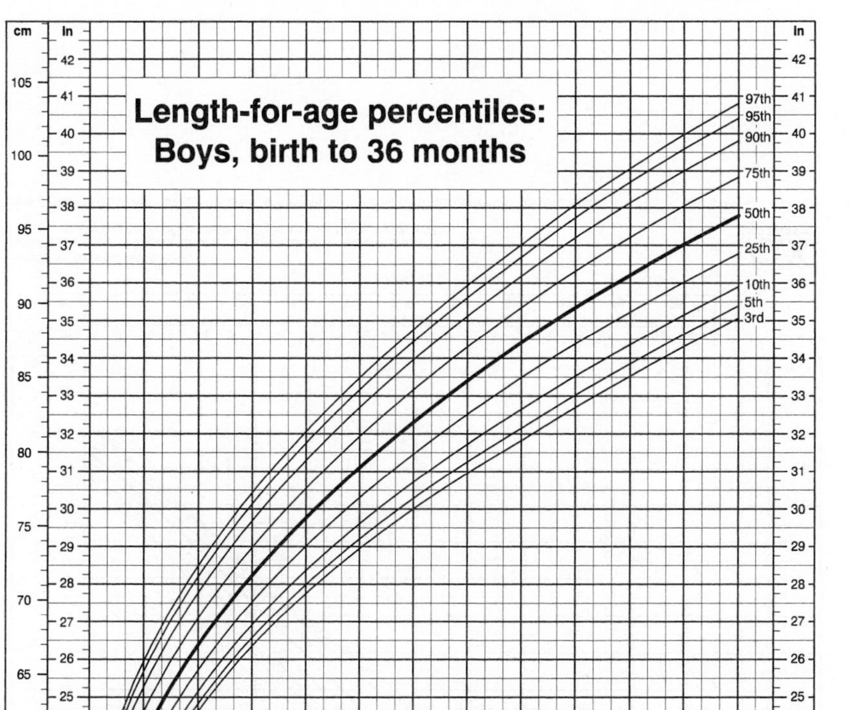

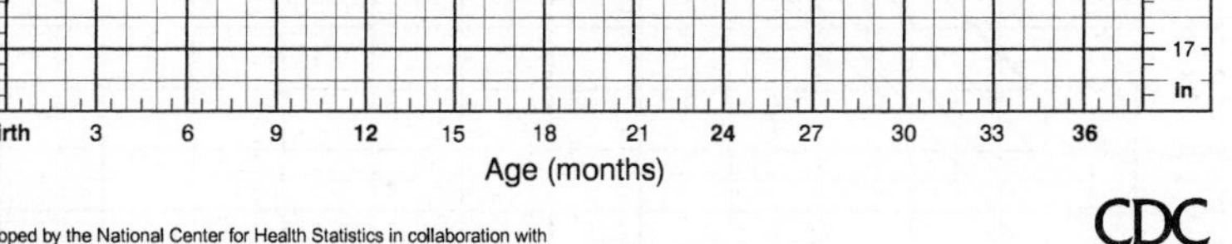

SOURCE: Developed by the National Center for Health Statistics in collaboration with the National Center for Chronic Disease Prevention and Health Promotion (2000).

Figure 3. Length-for-age percentiles, boys, birth to 36 months, CDC growth charts: United States

> FIGURE E-6B Length-for-Age Percentiles: Girls, Birth to 36 Months

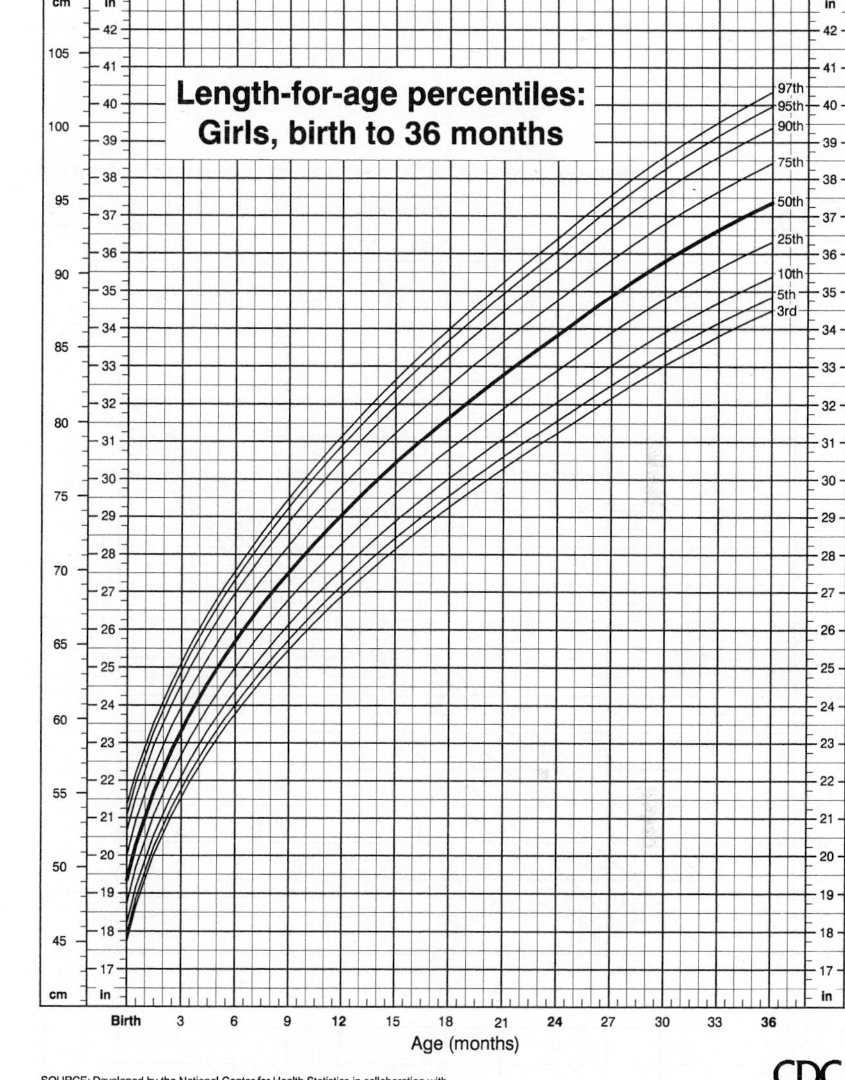

SOURCE: Developed by the National Center for Health Statistics in collaboration with the National Center for Chronic Disease Prevention and Health Promotion (2000).

Figure 4. Length-for-age percentiles, girls, birth to 36 months, CDC growth charts: United States

> FIGURE E-7A Weight-for-Length Percentiles: Boys, Birth to 36 Months

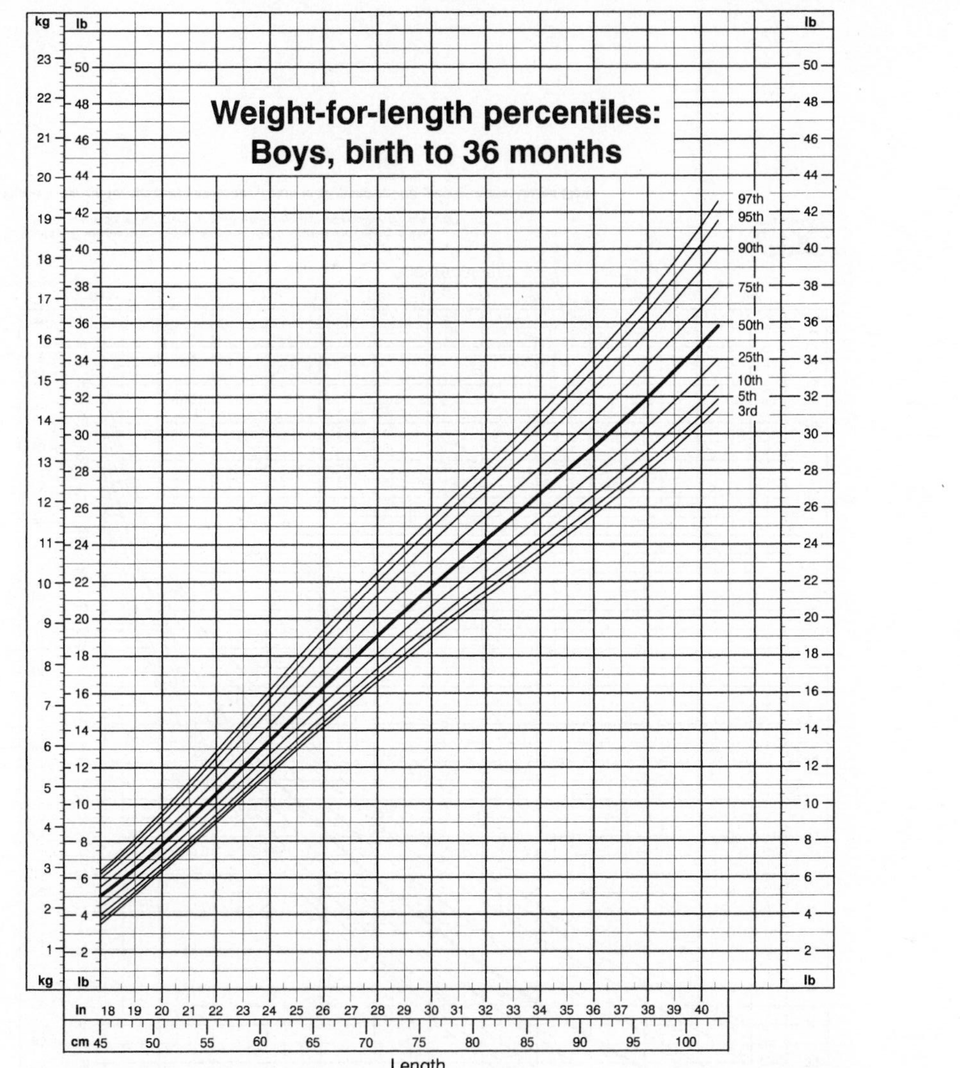

Figure 5. Weight-for-length percentiles, boys, birth to 36 months, CDC growth charts: United States

> FIGURE E-7B Weight-for-Length Percentiles: Girls, Birth to 36 Months

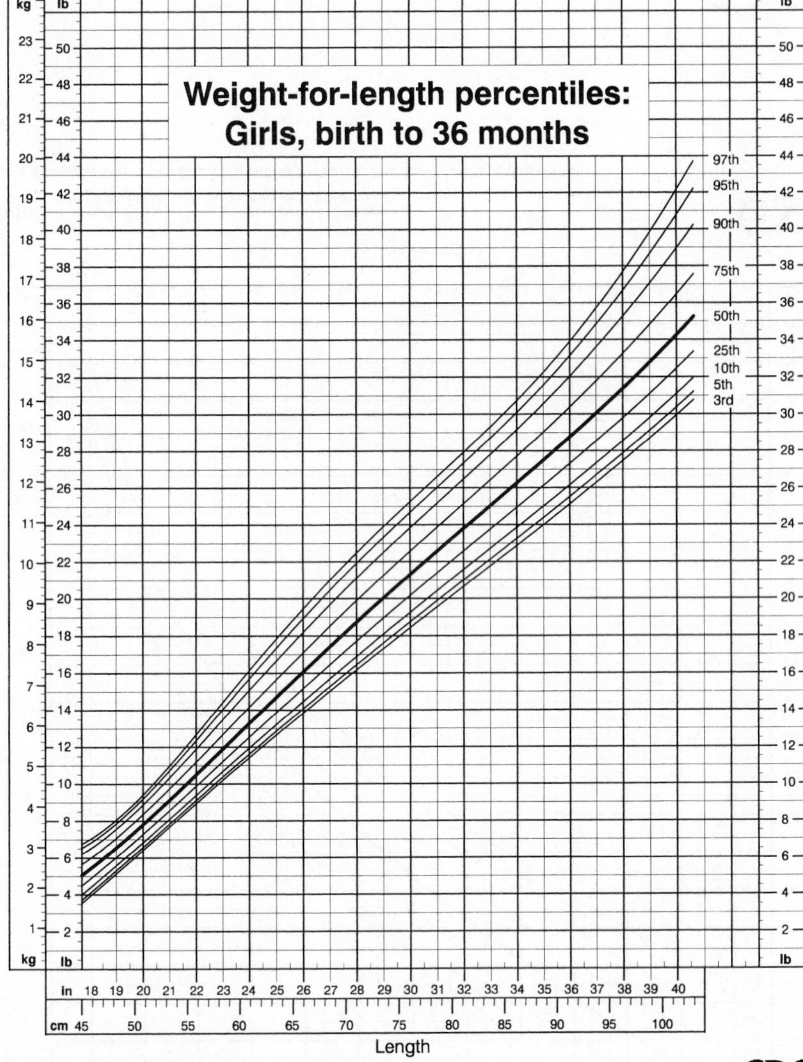

Figure 6. Weight-for-length percentiles, girls, birth to 36 months, CDC growth charts: United States

> FIGURE E-8A Weight-for-Age Percentiles: Boys, 2 to 20 Years

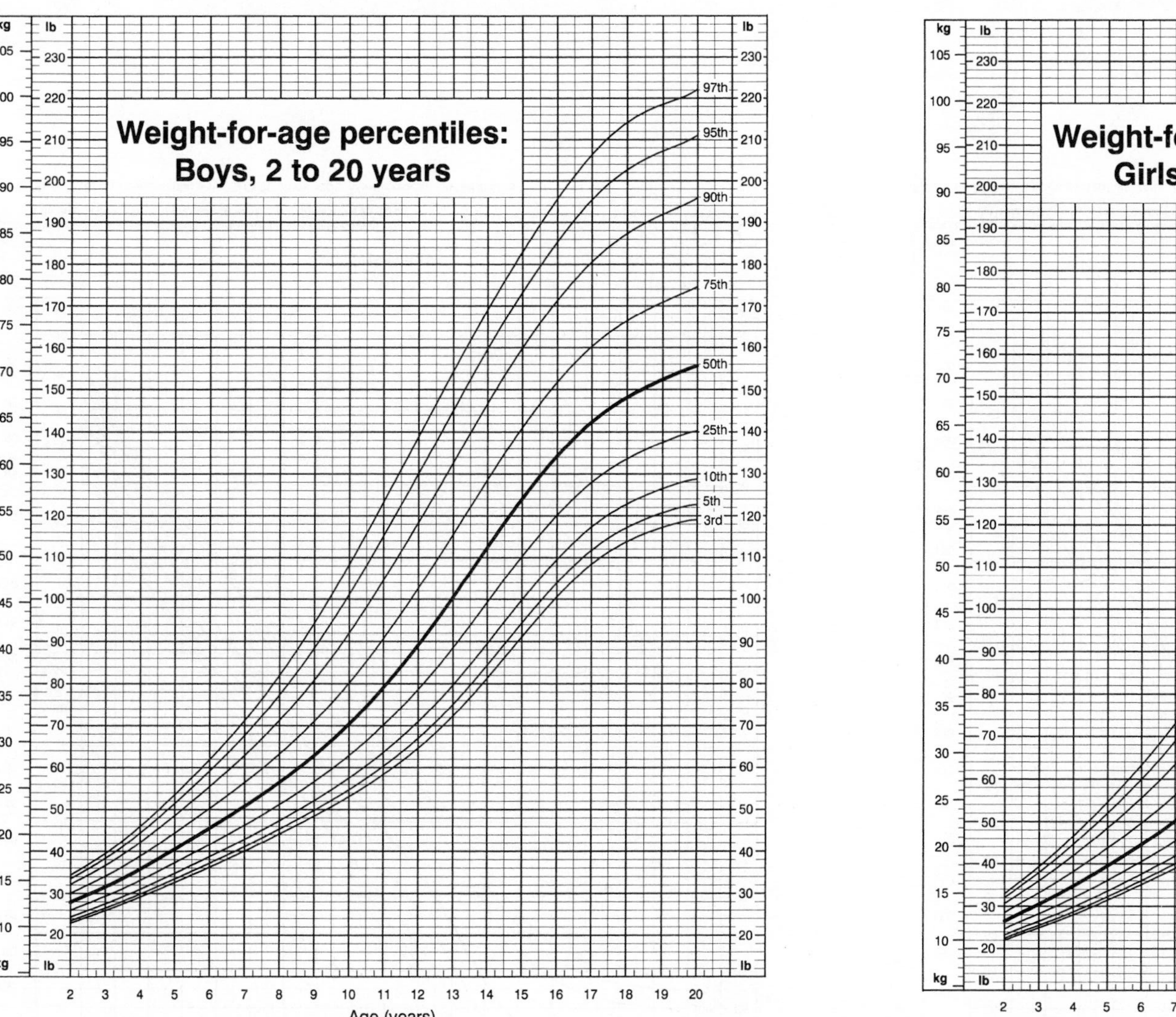

SOURCE: Developed by the National Center for Health Statistics in collaboration with the National Center for Chronic Disease Prevention and Health Promotion (2000).

Figure 9. Weight-for-age percentiles, boys, 2 to 20 years, CDC growth charts: United States

> FIGURE E-8B Weight-for-Age Percentiles: Girls, 2 to 20 Years

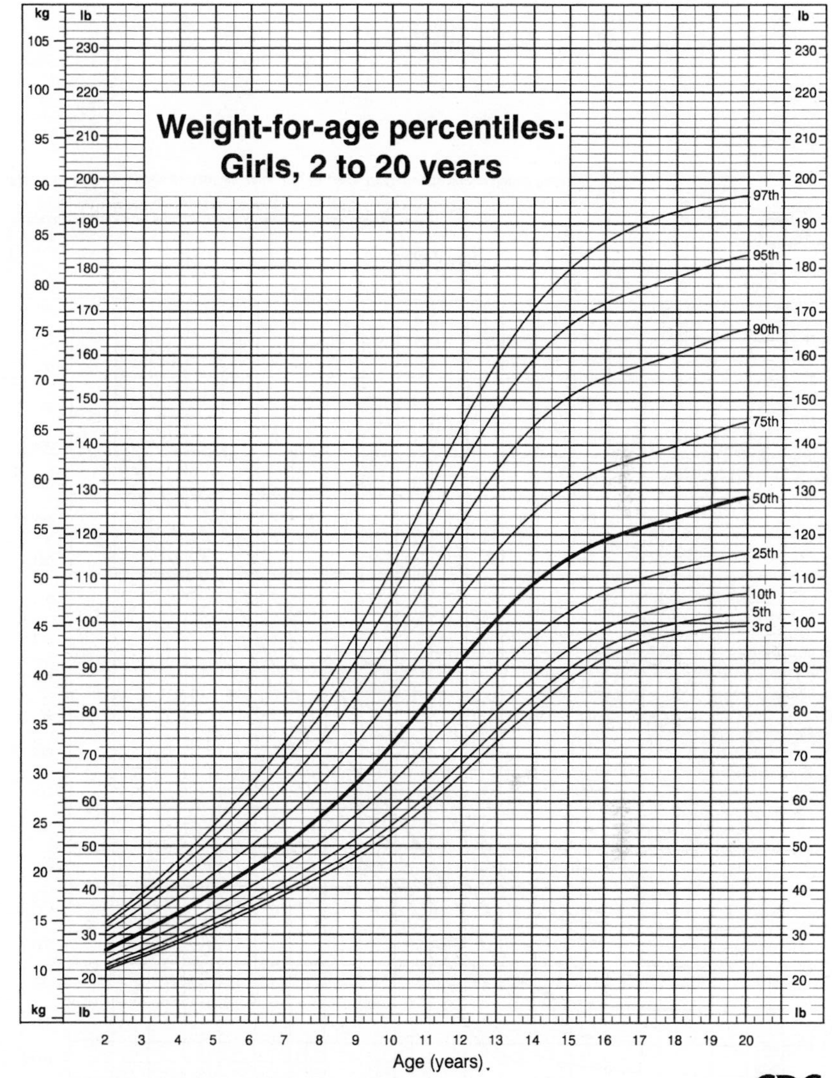

Weight-for-age percentiles:
Girls, 2 to 20 years

kg
lb
97th
95th
90th
75th
50th
25th
10th
5th
3rd
Age (years).
CDC

SOURCE: Developed by the National Center for Health Statistics in collaboration with the National Center for Chronic Disease Prevention and Health Promotion (2000).

Figure 10. Weight-for-age percentiles, girls, 2 to 20 years, CDC growth charts: United States

> FIGURE E-9A Stature-for-Age Percentiles: Boys, 2 to 20 Years

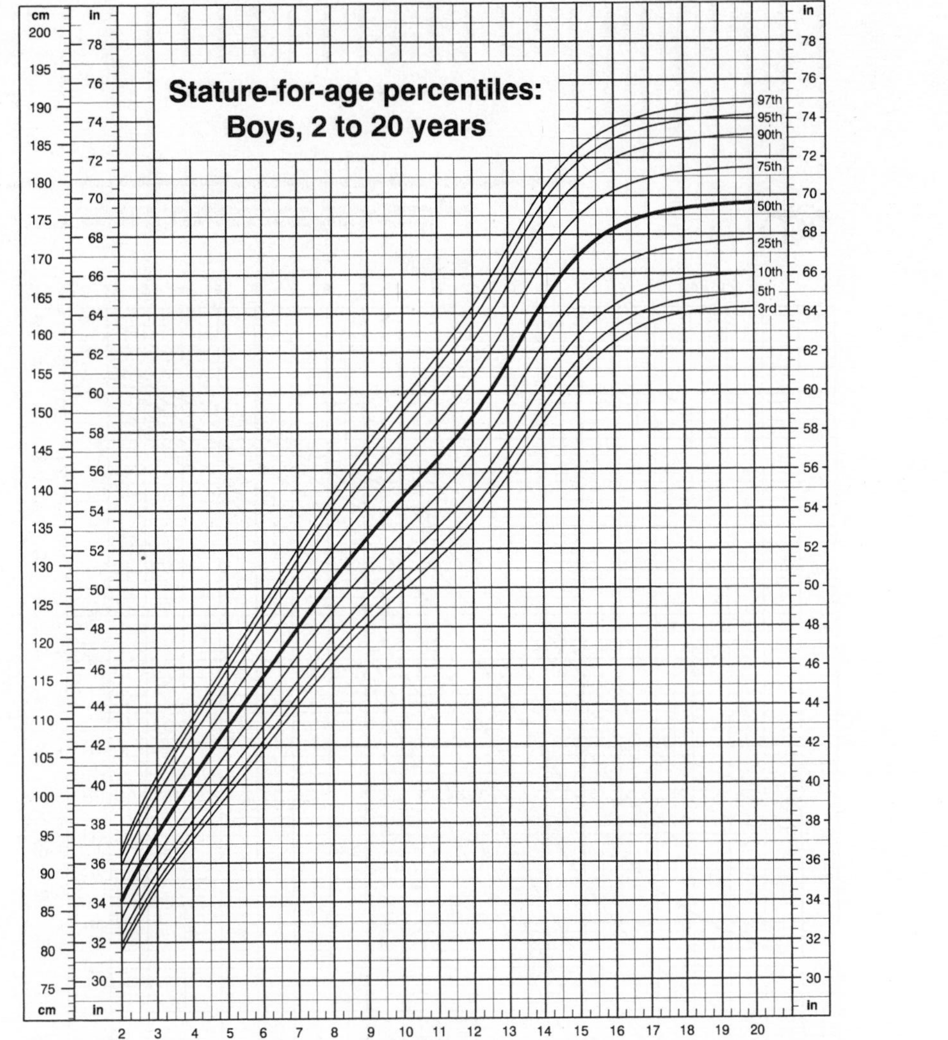

SOURCE: Developed by the National Center for Health Statistics in collaboration with the National Center for Chronic Disease Prevention and Health Promotion (2000).

Figure 11. Stature-for-age percentiles, boys, 2 to 20 years, CDC growth charts: United States

> FIGURE E-9B Stature-for-Age Percentiles: Girls, 2 to 20 Years

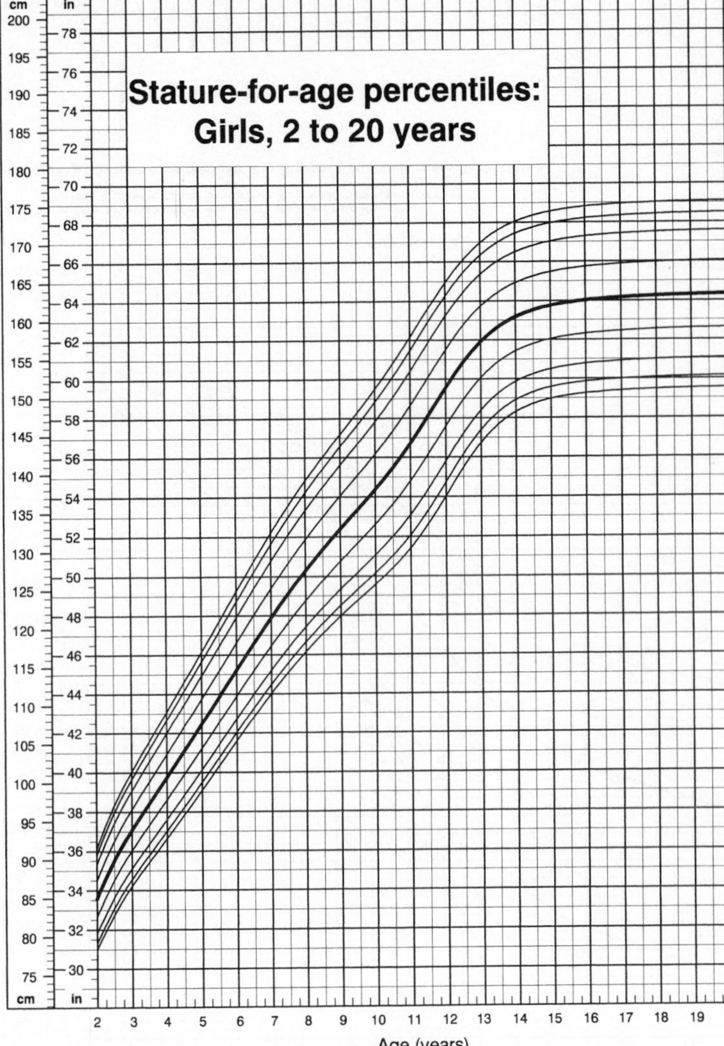

SOURCE: Developed by the National Center for Health Statistics in collaboration with the National Center for Chronic Disease Prevention and Health Promotion (2000).

Figure 12. Stature-for-age percentiles, girls, 2 to 20 years, CDC growth charts: United States

> FIGURE E-10A Weight-for-Stature Percentiles: Boys, 2 to 20 Years

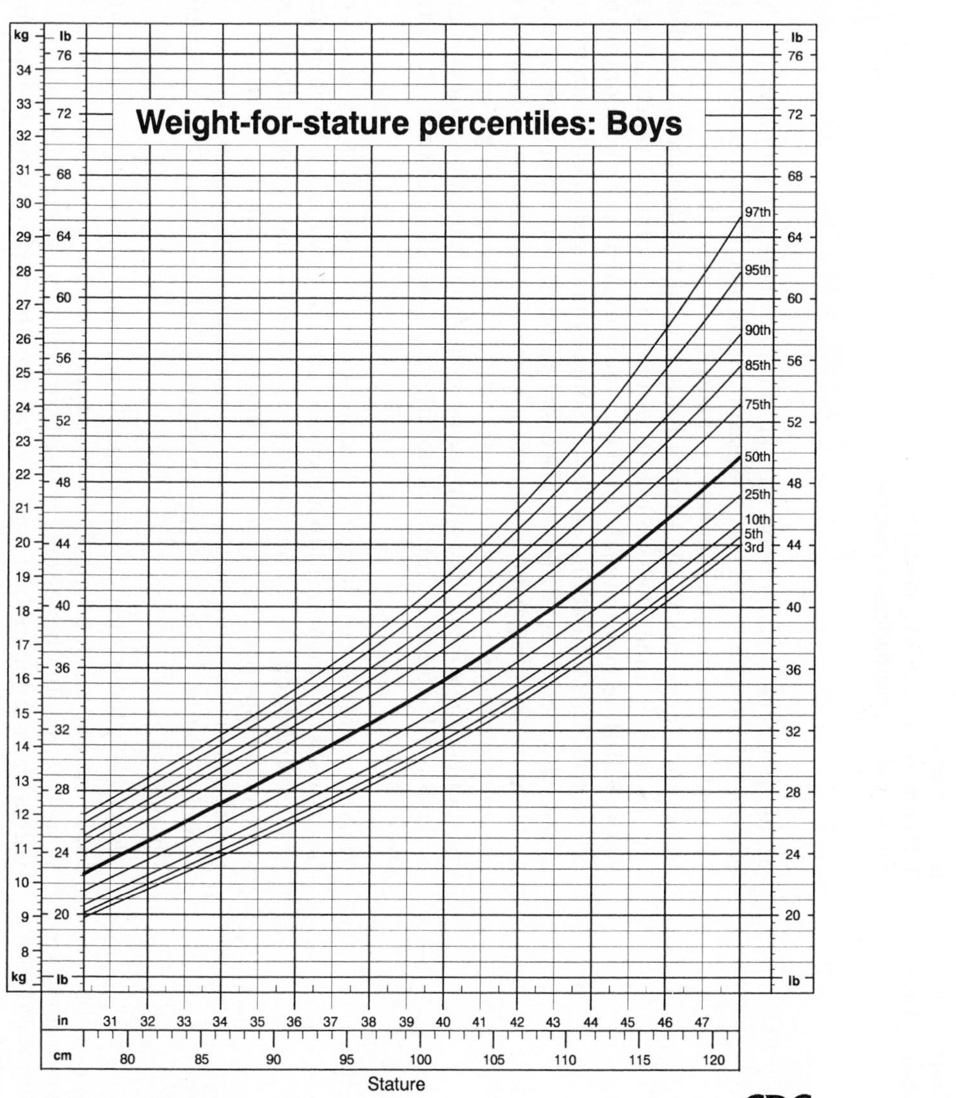

> FIGURE E-10B Weight-for-Stature Percentiles: Girls, 2 to 20 Years

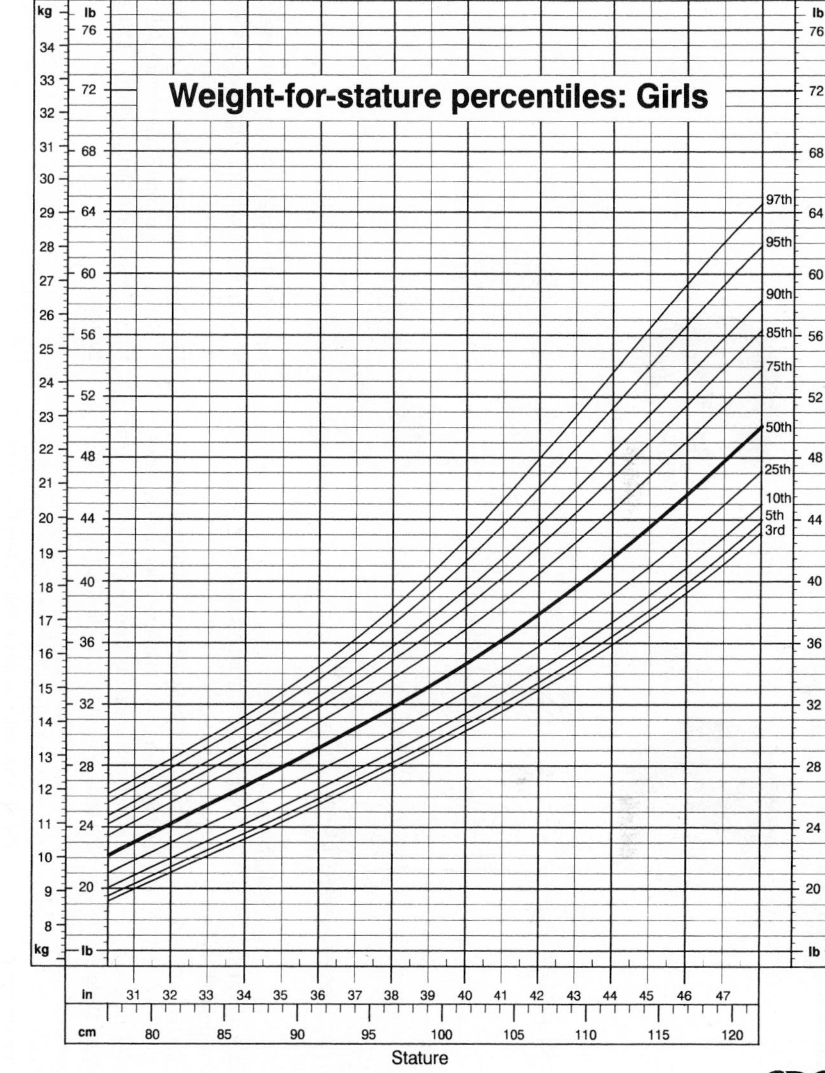

> FIGURE E-11 **How to Measure the Triceps Skinfold**

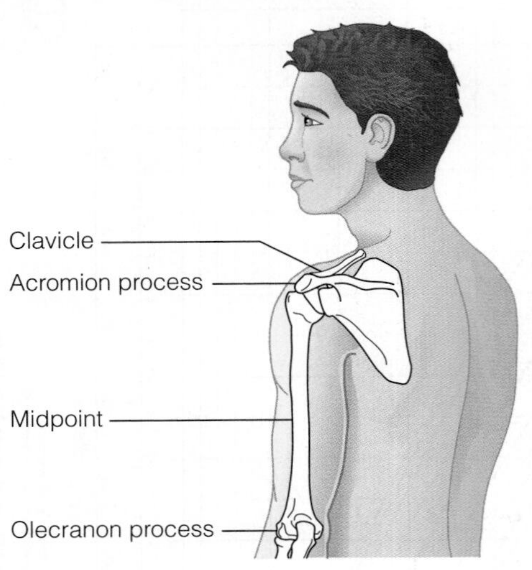

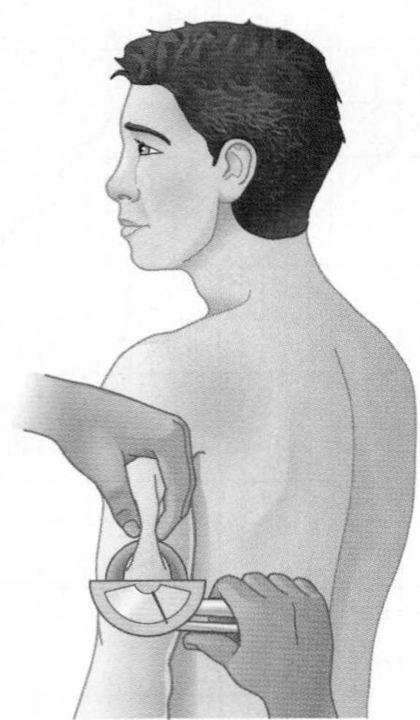

A. Find the midpoint of the arm:
1. Ask the subject to bend his or her arm at the elbow and lay the hand across the stomach. (If he or she is right-handed, measure the left arm, and vice versa.)
2. Feel the shoulder to locate the acromion process. It helps to slide your fingers along the clavicle to find the acromion process. The olecranon process is the tip of the elbow.
3. Place a measuring tape from the acromion process to the tip of the elbow. Divide this measurement by 2, and mark the midpoint of the arm with a pen.

B. Measure the skinfold:
1. Ask the subject to let his or her arm hang loosely to the side.
2. Grasp a fold of skin and subcutaneous fat between the thumb and forefinger slightly above the midpoint mark. Gently pull the skin away from the underlying muscle. (This step takes a lot of practice. To be sure you don't have muscle as well as fat, ask the subject to contract and relax the muscle. You should be able to feel if you are pinching muscle.)
3. Place the calipers over the skinfold at the midpoint mark, and read the measurement to the nearest 1.0 millimeter in two to three seconds. (If using plastic calipers, align pressure lines, and read the measurement to the nearest 1.0 millimeter in two to three seconds.)
4. Repeat steps 2 and 3 twice more. Add the three readings, and then divide by 3 to find the average.

alone. If a person gains body fat, the skinfold increases proportionately; if the person loses fat, it decreases. Measurements taken from central-body sites (around the abdomen) better reflect changes in fatness than those taken from upper sites (arm and back). A major limitation of the skinfold test is that fat may be thicker under the skin in one area than in another. A pinch at the side of the waistline may not yield the same measurement as a pinch on the back of the arm. This limitation can be overcome by taking skinfold measurements at several (often three) different places on the body (including upper-, central-, and lower-body sites) and comparing each measurement with standards for that site. Multiple measures are not always practical in clinical settings, however, and most often, the triceps skinfold measurement alone is used because it is easily accessible. Skinfold measures are not useful in assessing changes in body fat over time.

Waist Circumference Chapter 8 describes how fat distribution correlates with health risks and mentioned that the waist circumference is a valuable indicator of fat distribution. To measure waist circumference, the assessor places a nonstretchable tape around the person's body, crossing just above the upper hip bones and making sure that the tape remains on a level horizontal plane on all sides (see Figure E-12). The tape is tightened slightly, but without compressing the skin.

Waist-to-Hip Ratio Alternatively, some clinicians measure both the waist and the hips. The waist-to-hip ratio ♦ also assesses abdominal obesity, but provides no more information than using the waist circumference alone. In general, women with a waist-to-hip ratio of 0.80 or greater and men with a waist-to-hip ratio of 0.90 or greater have a high risk of health problems.

♦ To calculate the waist-to-hip ratio, divide the waistline measurement by the hip measurement. For example, a woman with a 28-inch waist and 38-inch hips would have a ratio of 28 ÷ 38 = 0.74.

> FIGURE E-12 **How to Measure Waist Circumference**

Place the measuring tape around the waist just above the bony crest of the hip. The tape runs parallel to the floor and is snug (but does not compress the skin). The measurement is taken at the end of normal expiration.

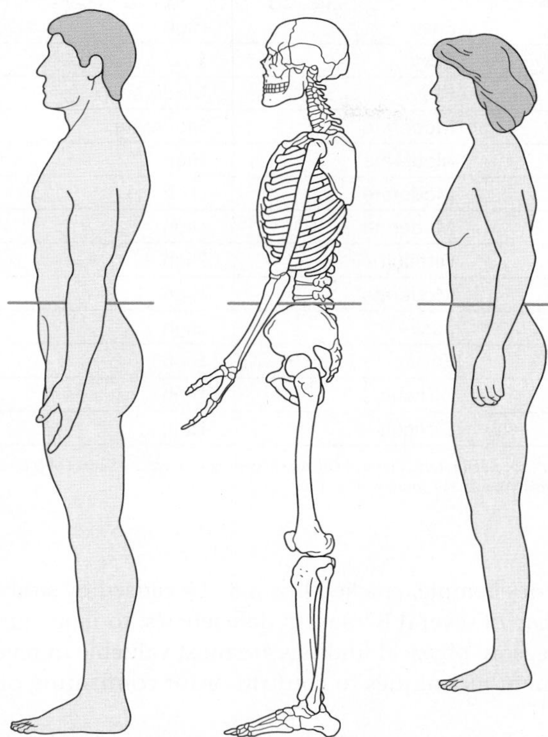

SOURCE: National Institutes of Health Obesity Education Initiative, *Clinical Guidelines on the Identification, Evaluation, and Treatment of Overweight and Obesity in Adults* (Washington, D.C.: U.S. Department of Health and Human Services, 1998), p. 59.

Hydrodensitometry To estimate body density using hydrodensitometry, the person is weighed twice—first on land and then again when submerged under water. Underwater weighing usually generates a good estimate of body fat and is useful in research, although the technique has drawbacks: it requires bulky, expensive, and nonportable equipment. Furthermore, submerging some people (especially those who are very young, very old, ill, or fearful) under water is not always practical.

Bioelectric Impedance To measure body fat using the bioelectric impedance technique, a very-low-intensity electrical current is briefly sent through the body by way of electrodes placed on the wrist and ankle. As is true of other anthropometric techniques, bioelectrical impedance requires standardized procedures and calibrated instruments to provide reliable results. Recent food intake and hydration status, for example, influence results. Bioelectrical impedance is most accurate for people within a normal fat range; it tends to overestimate fat in lean people and underestimate fat in obese people.

Chapter 8 (p. 246) illustrates several anthropometric measures of body fat and lean tissue. Clinicians use other methods to estimate body fat and its distribution as well. Each has its advantages and disadvantages as Table E-4 (p. E-16) summarizes.

Physical Examinations

An assessor can use a physical examination to search for signs of nutrient deficiency or toxicity. Like the other assessment methods, such an examination requires knowledge and skill. Many physical signs are nonspecific; they can reflect any of several nutrient deficiencies as well as conditions not related to nutrition

TABLE E-4 Methods of Estimating Body Fat and Its Distribution

Method	Cost	Ease of Use	Accuracy	Measures Fat Distribution
Height and weight	Low	Easy	High	No
Skinfolds	Low	Easy	Low	Yes
Circumferences	Low	Easy	Moderate	Yes
Ultrasound	Moderate	Moderate	Moderate	Yes
Hydrodensitometry	Low	Moderate	High	No
Heavy water tritiated	Moderate	Moderate	High	No
Deuterium oxide, or heavy oxygen	High	Moderate	High	No
Potassium isotope (^{40}K)	Very high	Difficult	High	No
Total body electrical conductivity (TOBEC)	High	Moderate	High	No
Bioelectric impedance (BIA)	Moderate	Easy	High	No
Dual-energy X-ray absorptiometry (DEXA)	High	Easy	High	No
Computed tomography (CT)	Very high	Difficult	High	Yes
Magnetic resonance imaging (MRI)	Very high	Difficult	High	Yes

SOURCE: Adapted with permisssion from G. A. Bray, a handout presented at the North American Association for the Study of Obesity and Emory University School of Medicine Conference on Obesity. Update: Pathophysiology, Clinical Consequences, and Therapeutic Options, Atlanta, Georgia, August 31–September 2, 1992.

(see Table E-5). For example, cracked lips may be caused by sunburn, windburn, dehydration, or any of several B vitamin deficiencies, to name just a few possible causes. For this reason, physical findings are most valuable in revealing problems for other assessment techniques to confirm or for confirming other assessment measures.

With this limitation understood, physical symptoms can be most informative and communicate much information about nutrition health. Many tissues and organs can reflect signs of malnutrition. The signs appear most rapidly in parts of the body where cell replacement occurs at a high rate, such as in the hair, skin, and digestive tract (including the mouth and tongue). The summary tables in Chapters 10, 11, 12, and 13 list additional physical signs of vitamin and mineral malnutrition.

TABLE E-5 Physical Findings Used in Nutrition Assessments

Body System	Healthy Findings	Malnutrition Findings	What the Findings Reflect
Hair	Shiny, firm in the scalp	Dull, brittle, dry, loose; falls out	Protein malnutrition
Eyes	Bright, clear pink membranes; adjust easily to light	Pale membranes; spots; redness; adjust slowly to darkness	Vitamin A, B vitamin, zinc, and iron status
Teeth and gums	No pain or caries, gums firm, teeth bright	Missing, discolored, decayed teeth; gums bleed easily and are swollen and spongy	Mineral and vitamin C status
Glands	No lumps	Swollen at front of neck	Protein malnutrition and iodine status
Tongue	Red, bumpy, rough	Sore, smooth, purplish, swollen	B vitamin status
Skin	Smooth, firm, good color	Off-color, scaly, flaky, cracked, dry, rough, spotty; "sandpaper" feel or sores; lack of fat under skin	Protein malnutrition, essential fatty acid, vitamin A, B vitamin, and vitamin C status
Nails	Firm, pink	Spoon-shaped, brittle, ridged, pale	Iron status
Internal systems	Regular heart rhythm, heart rate, and blood pressure; no impairment of digestive function, reflexes, or mental status	Abnormal heart rate, heart rhythm, or blood pressure; enlarged liver, spleen; abnormal digestion; burning, tingling of hands, feet; loss of balance, coordination; mental confusion, irritability, fatigue	Protein malnutrition and mineral status
Muscles and bones	Muscle tone; posture, long bone development appropriate for age	"Wasted" appearance of muscles; swollen bumps on skull or ends of bones; small bumps on ribs; bowed legs or knock-knees	Protein malnutrition, mineral, and vitamin D status

Biochemical Analyses

All of the approaches to nutrition assessment discussed so far are external approaches. Biochemical analyses or laboratory tests help to determine what is happening to the body internally. Common tests are based on analysis of blood and urine samples, which contain nutrients, enzymes, and metabolites that reflect nutrition status. Other tests, such as blood glucose, help pinpoint disease-related problems with nutrition implications. Tests that define fluid and electrolyte balance, acid-base balance, and organ function also have nutrition implications. Table E-6 lists biochemical tests most useful for assessing vitamin and mineral status.

The interpretation of biochemical data requires skill. Long metabolic sequences lead to the production of the end-products and metabolites seen in blood and urine. No single test can reveal nutrition status because many factors influence test results. The low blood concentration of a nutrient may reflect a primary deficiency of that nutrient, but it may also be secondary to the deficiency of one or several other nutrients or to a disease. Taken together with other assessment data,

TABLE E-6 Biochemical Tests Useful for Assessing Vitamin and Mineral Status

Nutrient	Assessment Tests
Vitamins	
Vitamin A	Serum retinol, retinol-binding protein
Thiamin[a]	Erythrocyte (red blood cell) transketolase activity, erythrocyte thiamin pyrophosphate
Riboflavin[a]	Erythrocyte glutathione reductase activity
Vitamin B_6[a]	Urinary xanthurenic acid excretion after tryptophan load test, erythrocyte transaminase activity, plasma pyridoxal 5′-phosphate (PLP)
Niacin	Plasma or urinary metabolites NMN (N-methyl nicotinamide) or 2-pyridone, or preferably both expressed as a ratio
Folate[b]	Serum folate, erythrocyte folate (reflects liver stores)
Vitamin B_{12}[b]	Serum vitamin B_{12}, serum and urinary methylmalonic acid, Schilling test
Biotin	Urinary biotin, urinary 3-hydroxyisovaleric acid
Vitamin C	Plasma vitamin C[c], leukocyte vitamin C
Vitamin D	Serum vitamin D
Vitamin E	Serum α-tocopherol, erythrocyte hemolysis
Vitamin K	Serum vitamin K, plasma prothrombin; blood-clotting time (prothrombin time) is not an adequate indicator
Minerals	
Phosphorus	Serum phosphate
Sodium	Serum sodium
Chloride	Serum chloride
Potassium	Serum potassium
Magnesium	Serum magnesium, urinary magnesium
Iron	Hemoglobin, hematocrit, serum ferritin, total iron-binding capacity (TIBC), erythrocyte protoporphyrin, serum iron, transferrin saturation
Iodine	Serum thyroxine or thyroid-stimulating hormone (TSH), urinary iodine
Zinc	Plasma zinc, hair zinc
Copper	Erythrocyte superoxide dismutase, serum copper, serum ceruloplasmin
Selenium	Erythrocyte selenium, glutathione peroxidase activity

[a]Urinary measurements for these vitamins are common, but may be of limited use. Urinary measurements reflect recent dietary intakes and may not provide reliable information concerning the severity of a deficiency.

[b]Folate assessments should always be conducted in conjunction with vitamin B_{12} assessments (and vice versa) to help distinguish the cause of common deficiency symptoms.

[c]Vitamin C shifts between the plasma and the white blood cells known as leukocytes; thus a plasma determination may not accurately reflect the body's pool. A measurement of leukocyte vitamin C can provide information about the body's stores of vitamin C. A combination of both tests may be more reliable than either one alone.

SOURCE: Adapted from H. E. Sauberlich, *Laboratory Tests for the Assessment of Nutritional Status* (Boca Raton, FIA.: CRC Press, 1999).

TABLE E-7 Normal Values for Serum Proteins

Indicator	Normal
Albumin (g/dL)	3.5–5.0
Transferrin (mg/dL)	200–400
Transthyretin (mg/dL)	16–40
Retinol-binding protein (mg/dL)	3–7
IGF-1 (μg/L)	300

NOTE: Levels less than normal suggest compromised protein status.

♦ A *subclinical deficiency* is a nutrient deficiency in the early stages before the outward signs have appeared.

♦ Transthyretin is also known as *prealbumin* or *thyroxine-binding prealbumin.*

however, laboratory test results help to create a picture that becomes clear with careful interpretation. They are especially useful in helping to detect subclinical malnutrition by uncovering early signs of malnutrition before the clinical signs of a classic deficiency disease appear.

Laboratory tests used to assess vitamin and mineral status (review Table E-6) are particularly useful when combined with diet histories and physical findings. Vitamin and mineral levels present in the blood and urine sometimes reflect recent rather than long-term intakes. This makes detecting subclinical deficiencies ♦ difficult. Furthermore, many nutrients interact; therefore, the amounts of other nutrients in the body can affect a lab value for a particular nutrient. It is also important to remember that nonnutrient conditions such as diseases influence biochemical measures.

It is beyond the scope of this text to describe all lab tests and their relations to nutrition status. Instead, the emphasis is on lab tests used to detect protein malnutrition and nutritional anemias.

Protein Malnutrition

No single biochemical analysis can adequately evaluate protein malnutrition. This discussion focuses on the measures commonly used today—transthyretin, retinol-binding protein, serum transferrin, and IGF-1 (insulin-like growth factor 1). Table E-7 provides standards for these indicators. Although serum albumin is easily and routinely measured, it lacks the sensitivity to assess protein malnutrition because of its long turnover rate.*

Transthyretin and Retinol-Binding Protein Transthyretin ♦ and retinol-binding protein occur as a complex in the plasma. They have a rapid turnover and thus respond quickly to dietary protein inadequacy and therapy.** Conditions other than malnutrition that lower transthyretin include metabolic stress, hemodialysis, and hypothyroidism; those that raise transthyretin include kidney disease and corticosteroid use. Conditions other than protein malnutrition that lower retinol-binding protein include vitamin A deficiency, metabolic stress, hyperthyroidism, liver disease, and cystic fibrosis; kidney disease raises retinol-binding protein levels.

Serum Transferrin Serum transferrin transports iron; consequently, its concentrations reflect both protein and iron status. Using transferrin as an indicator of protein status is complicated when an iron deficiency is present. Transferrin rises as iron deficiency grows worse and falls as iron status improves. Markedly reduced transferrin levels indicate severe protein malnutrition; in mild-to-moderate protein malnutrition, transferrin levels may vary, limiting their usefulness. Conditions other than protein malnutrition that lower transferrin include liver disease, kidney disease, and metabolic stress; those that raise transferrin include pregnancy, iron deficiency, hepatitis, blood loss, and oral contraceptive use. Although transferrin breaks down in the body more quickly than albumin, it is still relatively slow to respond to changes in protein intake and is not a sensitive indicator of the response to therapy.***

IGF-1 (Insulin-like Growth Factor 1) IGF-1 (insulin-like growth factor 1) declines in protein malnutrition. IGF-1 has a relatively short half-life and responds specifically to dietary protein rather than energy.**** For these reasons, it is a sensitive indicator of protein status and response to therapy. Conditions that decrease IGF-1 include anorexia nervosa, inflammatory bowel disease, celiac disease, HIV infection, and fasting.

Nutritional Anemias

Anemia, a symptom of a wide variety of nutrition- and nonnutrition-related disorders, is characterized by a reduced number of red blood cells. Iron, folate, and vitamin B_{12} deficiencies caused by inadequate intake, poor absorption, or abnormal metabolism of these nutrients are the most

*The half-life of albumin is 14 to 20 days, an indication of a slow degradation rate.

**The half-lives of transthyretin and retinol-binding protein are 2 days and 12 hours, respectively.

***The half-life of transferrin is 8 to 10 days.

****The half-life of IGF-1 is 12 to 15 hours.

common nutritional anemias. Some nonnutrition-related causes of anemia include massive blood loss, infections, hereditary blood disorders such as sickle-cell anemia, and chronic liver or kidney disease.

Assessment of Iron-Deficiency Anemia Iron deficiency, a common mineral deficiency, develops in stages. ♦ Chapter 13 describes iron deficiency in detail. This section describes tests used to uncover iron deficiency as it progresses. Table E-8 shows which laboratory tests detect various nutrition-related anemias, and Table E-9 (p. E-20) provides values used for assessing iron status. Although other tests are more specific in detecting early deficiencies, hemoglobin and hematocrit are the most commonly available tests.

♦ Stages of iron deficiency:
1. Iron stores diminish.
2. Transport iron decreases.
3. Hemoglobin production falls.

Hemoglobin Iron forms an integral part of the hemoglobin molecule that transports oxygen to the cells. In iron deficiency, the body cannot synthesize hemoglobin. Low hemoglobin values signal depleted iron stores. Table E-9 provides hemoglobin values used in nutrition assessment. Hemoglobin's usefulness in evaluating iron status is limited, however, because hemoglobin concentrations drop fairly late in the development of iron deficiency, and other nutrient deficiencies and medical conditions can also alter hemoglobin concentrations.

Hematocrit Hematocrit is commonly used to diagnose iron deficiency, even though it is an inconclusive measure of iron status. To measure the hematocrit, a clinician spins a volume of blood in a centrifuge to separate the red blood cells from the plasma. The hematocrit is the percentage of red blood cells in the total blood volume. Table E-9 includes values used to assess hematocrit status. Low values indicate incomplete hemoglobin formation, which is manifested by microcytic (abnormally small-celled), hypochromic (abnormally lacking in color) red blood cells.

TABLE E-8 Laboratory Tests Useful in Evaluating Nutrition-Related Anemias

Test or Test Result	What It Reflects
For Anemia (general)	
Hemoglobin (Hg)	Total amount of hemoglobin in the red blood cells (RBC)
Hematocrit (Hct)	Percentage of RBC in the total blood volume
Red blood cell (RBC) count	Number of RBC
Mean corpuscular volume (MCV)	RBC size; helps to determine if anemia is microcytic (iron deficiency) or macrocytic (folate or vitamin B_{12} deficiency)
Mean corpuscular hemoglobin concentration (MCHC)	Hemoglobin concentration within the average RBC; helps to determine if anemia is hypochromic (iron deficiency) or normochromic (folate or vitamin B_{12} deficiency)
Bone marrow aspiration	The manufacture of blood cells in different developmental states
For Iron-Deficiency Anemia	
↓ Serum ferritin	Early deficiency state with depleted iron stores
↓ Transferrin saturation	Progressing deficiency state with diminished transport iron
↑ Erythrocyte protoporphyrin	Later deficiency state with limited hemoglobin production
For Folate-Deficiency Anemia	
↓ Serum folate	Progressing deficiency state
↓ RBC folate	Later deficiency state
For Vitamin B_{12}–Deficiency Anemia	
↓ Serum vitamin B_{12}	Progressing deficiency state
Schilling test	Absorption of vitamin B_{12}

APPENDIX E

TABLE E-9 Criteria for Assessing Iron Status

Test	Age (yr)	Gender	Deficiency Value
Hemoglobin (g/dL)	0.5–10	M–F	<11
	11–15	M	<12
		F	<11.5
	>15	M	<13
		F	<12
	Pregnancy		<11
Hematocrit (%)	0.5–4	M–F	<32
	5–10	M–F	<33
	11–15	M	<35
		F	<34
	>15	M	<40
		F	<36
Serum ferritin (μg/L)	0.5–15	M–F	<10
	>15	M–F	<12
Total iron-binding capacity (μg/dL)	>15	M–F	>400
Serum iron (μg/dL)	>15	M–F	<60
Transferrin saturation (%)	0.5–4	M–F	<12
	5–10	M–F	<14
	>10	M–F	<16
Erythrocyte protoporphyrin (μg/dL RBC)	0.5–4	M–F	>80
	>4	M–F	>70

Low hemoglobin and hematocrit values alert the assessor to the possibility of iron deficiency. However, many nutrients and other conditions can affect hemoglobin and hematocrit. The other tests of iron status help pinpoint true iron deficiency.

Serum Ferritin In the first stage of iron deficiency, iron stores diminish. Measures of serum ferritin provide an estimate of iron stores. Such information is most valuable to iron assessment. Table E-9 shows serum ferritin cutoff values that indicate iron store depletion in children and adults. Serum ferritin is not reliable for diagnosing iron deficiency in infants, since normal serum ferritin values are often present in conjunction with iron-responsive anemia.

A decrease in transport iron characterizes the second stage of iron deficiency. This is revealed by an increase in the iron-binding capacity of the protein transferrin and a decrease in serum iron. These changes are reflected by the transferrin saturation, which is calculated from the ratio of the other two values as described in the following paragraphs.

Total Iron-Binding Capacity (TIBC) Iron travels through the blood bound to the protein transferrin. TIBC is a measure of the total amount of iron that transferrin can carry. Lab technicians measure iron-binding capacity directly. Table E-9 includes the cutoff for TIBC.

Serum Iron Lab technicians can also measure serum iron directly. Elevated values indicate iron overload; reduced values indicate iron deficiency. Table E-9 shows the deficient value for serum iron.

Transferrin Saturation The percentage of transferrin that is saturated with iron is an indirect measure that is derived from the serum iron and total iron-binding capacity measures as follows:

$$\% \text{ Transferrin} = \frac{\text{serum iron}}{\text{total iron-binding capacity}} \times 100$$

Table E-9 shows deficient transferrin saturation values for various age groups.

The third stage of iron deficiency occurs when the supply of transport iron diminishes to the point that it limits hemoglobin production. It is characterized by increases in erythrocyte protoporphyrin, a decrease in mean corpuscular volume, and decreased hemoglobin and hematocrit.

Erythrocyte Protoporphyrin The iron-containing portion of the hemoglobin molecule is heme. Heme is a combination of iron and protoporphyrin. Protoporphyrin accumulates in the blood when iron supplies are inadequate for the formation of heme. Lab technicians can measure erythrocyte protoporphyrin directly in a blood sample. The cutoffs for abnormal values of erythrocyte protoporphyrin are shown in Table E-9.

Mean Corpuscular Volume (MCV) A direct or calculated measure of the mean corpuscular volume (MCV) determines the average size of a red blood cell. Such a measure helps to classify the type of nutrient anemia. In iron deficiency, the red blood cells are smaller than average.

Assessment of Folate and Vitamin B_{12} Anemias

Folate deficiency and vitamin B_{12} deficiency present a similar clinical picture—an anemia characterized by abnormally large red blood cell precursors (megaloblasts) in the bone marrow and abnormally large, mature red blood cells (macrocytic cells) in the blood. Distinguishing between these two deficiencies is particularly important because their treatments differ. Giving folate to a person with vitamin B_{12} deficiency improves many of the lab test results indicative of vitamin B_{12} deficiency, but this is a dangerous error because vitamin B_{12} deficiency causes nerve damage that folate cannot correct. Thus inappropriate folate administration masks vitamin B_{12}–deficiency anemia, and nerve damage worsens. For this reason, it is critical to determine whether the anemia results from a folate deficiency or from a vitamin B_{12} deficiency. The following biochemical assessment techniques help to make this distinction.

Mean Corpuscular Volume (MCV) As previously mentioned, the MCV is a measure of red blood cell size. In folate and vitamin B_{12} deficiencies, the red blood cells are larger than average (macrocytic). Additional tests must be performed to differentiate folate deficiency from vitamin B_{12} deficiency.

Folate Levels Serum folate levels fluctuate with changes in folate intake and metabolism. Thus serum folate concentrations reflect current status, but provide little information about folate stores. As folate deficiency progresses and low serum levels persist, folate stores decline, resulting in folate depletion. Folate depletion is characterized by a fall in the folate concentrations of red blood cells (erythrocytes). As erythrocyte folate levels diminish, folate-deficiency anemia develops. Because low erythrocyte folate concentrations also occur with vitamin B_{12} deficiency, serum vitamin B_{12} concentrations must also be measured. Table E-10 shows standards for folate assessment.

Vitamin B_{12} Levels Serum and urinary methylmalonic acid are elevated in vitamin B_{12} deficiency, but not in folate deficiency. Thus this measure is useful in distinguishing between the two. Vitamin B_{12} deficiency usually arises from malabsorption. To determine whether malabsorption is the cause, a small oral dose of

TABLE E-10 Criteria for Assessing Folate and Vitamin B_{12}

	Deficient	Borderline	Acceptable
Serum folate (ng/mL)[a]	<3.0	3.0–5.9	>6.0
Erythrocyte folate (ng/mL)[a]	<140	140–159	>160
Serum vitamin B_{12} (pg/mL)	<150	150–200	≥201
Serum methylmalonic acid (nmol/L)	<376	—	—

NOTE: A nanogram (ng) is one-billionth of a gram; a picogram (pg) is one-trillionth of a gram.
[a]To convert folate values (ng/mL) to international standard units (nmol/L), multiply by 2.266.

vitamin B_{12} is given, and urinary excretion is measured. This procedure measures vitamin B_{12} absorption and is called a Schilling test.

Early stages of vitamin B_{12} deficiency can be detected by a low percentage saturation of its transport protein, a measure similar to iron's transferrin saturation. As the deficiency progresses, serum vitamin B_{12} concentrations fall. Table E-10 shows standards for vitamin B_{12} assessment.

Cautions about Nutrition Assessment

To give all the details of nutrition assessment procedures would entail writing another textbook. Nevertheless, any student of nutrition should know the basics of a proper nutrition assessment procedure for two reasons.

First, competent medical care includes attention to nutrition. Physicians should either employ a person skilled in nutrition assessment techniques or refer all clients to such a person to ensure the sound nutrition health of their clients. Health care facilities should make nutrition assessment a routine part of the initial workup on every client so that poor nutrition will not hinder the response to medical treatment and the recovery from illness.

Second, because nutrition is such a popular subject today, fraudulent practices are even more abundant than they have been in the past (and they have always been rampant). The knowledgeable consumer needs to know what procedures to expect in a nutrition assessment and what kinds of information they yield. This appendix has presented the basics of nutrition assessment for these reasons.

This caution is added: the tests outlined here yield information that becomes meaningful only when integrated into a whole picture by a skilled, experienced, and educated interpreter. Potential sources of error are many, from the taking of the initial data to their reporting and analysis. Each assessment method and measure is useful only as a part of the whole to confirm or eliminate the possibility of suspected nutrition problems. For example, the assessor must constantly remember that a sufficient intake of a nutrient does not guarantee adequate nutrient status for an individual. Conversely, the apparent inadequate intake of a nutrient does not, by itself, establish that a deficiency exists.

Similarly, many uncertainties, such as the calibration of the equipment, the skills of the measurer, and the perspective of the interpreter, limit the accuracy and value of anthropometric measures. This is also true of the results of a physical examination. Physical signs suggestive of malnutrition are nonspecific: they can reflect nutrient deficiencies or may be totally unrelated to nutrition. Assessors must interpret physical findings in light of other assessment findings. Finally, the usefulness of biochemical tests is also limited; the assessor must use caution in interpreting results. Vitamin and mineral blood concentrations may reflect disease processes, abnormal hormone levels, or aberrations other than dietary intake. Even if concentrations do reflect dietary intake, they may reflect what the person has been eating recently and not give a true picture of the person's nutrient status. Such complications sometimes make it difficult to detect a subclinical deficiency. Furthermore, many nutrients interact. The assessor has to keep in mind that an abnormal lab value for one nutrient may reflect abnormal status of other nutrients. The final diagnosis is therefore appropriately tentative, and its confirmation comes only after careful remedial steps successfully alleviate the observed problems.

Appendix F Physical Activity and Energy Requirements

Chapter 8 describes how to calculate estimated energy requirements (EER) for adults by using an equation that accounts for gender, age, weight, height, and physical activity level. Table F-1 presents additional equations to determine the EER for infants, children, adolescents, and pregnant and lactating women.

This appendix helps you determine the correct physical activity (PA) factor to use in the equations, either by calculating the physical activity level or by estimating it. For those who prefer to bypass these steps, the appendix presents tables that provide a shortcut to estimating total energy expenditure.*

CONTENTS

Calculating Physical Activity Level

To calculate your physical activity level, record all of your activities for a typical 24-hour day, noting the type of activity, the level of intensity, and the duration. Then, using a copy of Table F-2 (p. F-2), find your activity in the first column (or an activity that is reasonably similar) and multiply the number of minutes spent on that activity by the factor in the third column. Put your answer in the last column and total the accumulated values for the day. Now

TABLE F-1 Equations to Determine Estimated Energy Requirement (EER)

Infants	
0–3 months	EER = (89 × weight − 100) + 175
4–6 months	EER = (89 × weight − 100) + 56
7–12 months	EER = (89 × weight − 100) + 22
13–15 months	EER = (89 × weight − 100) + 20
Children and Adolescents	
Boys	
3–8 years	EER = 88.5 − (61.9 × age) + PA × [(26.7 × weight) + (903 × height)] + 20
9–18 years	EER = 88.5 − (61.9 × age) + PA × [(26.7 × weight) + (903 × height)] + 25
Girls	
3–8 years	EER = 135.3 − (30.8 × age) + PA × [(10.0 × weight) + (934 × height)] + 20
9–18 years	EER = 135.3 − (30.8 × age) + PA × [(10.0 × weight) + (934 × height)] + 25
Adults	
Men	EER = 662 − (9.53 × age) + PA × [(15.91 × weight) + (539.6 × height)]
Women	EER = 354 − (6.91 × age) + PA × [(9.36 × weight) + (726 × height)]
Pregnancy	
1st trimester	EER = nonpregnant EER + 0
2nd trimester	EER = nonpregnant EER + 340
3rd trimester	EER = nonpregnant EER + 452
Lactation	
0–6 months postpartum	EER = nonpregnant EER + 500 − 170
7–12 months postpartum	EER = nonpregnant EER + 400 − 0

NOTE: Select the appropriate equation for gender and age and insert weight in kilograms, height in meters, and age in years. See the text and Table F-3 to determine PA.

*This appendix, including the tables, is adapted from Committee on Dietary Reference Intakes, *Dietary Reference Intakes for Energy, Carbohydrate, Fiber, Fat, Fatty Acids, Cholesterol, Protein, and Amino Acids* (Washington, D.C.: National Academies Press, 2005).

add the subtotal of the last column to 1.1 (to account for basal energy and the thermic effect of food) as shown. This score indicates your physical activity level. Using Table F-3 (p. F-3), find the PA factor for your age and gender that correlates with your physical activity level and use it in the energy equations presented in Table F-1.

Estimating Physical Activity Level

As an alternative to recording your activities for a day, you can use the third column of Table F-3 to decide if your daily activity is sedentary, low active, active, or very active. Find the PA factor for your age and gender that correlates with your typical physical activity level and use it in the energy equations presented in Table F-1.

Using a Shortcut to Estimate Total Energy Expenditure

The DRI Committee has developed estimates of total energy expenditure based on the equations for adults presented in Table F-1. These estimates are presented in Table F-4 (p. F-4) for women and Table F-5 (p. F-5) for men. You can use these tables to estimate your energy requirement—that is, the number of kcalories needed to maintain your current body weight. On the table appropriate for your gender, find your height in meters (or inches) in the left-hand column. Then follow the row across to find your weight in kilograms (or pounds). (If you can't find your exact height and weight, choose a value between the two closest ones.) Look down the column to find the number of kcalories that corresponds to your activity level.

Importantly the values given in the tables are for 30-year-old people. Women 19 to 29 should add 7 kcalories per day for each year younger than age 30; older women should subtract 7 kcalories per day for each year older than age 30. Similarly men 19 to 29 should add 10 kcalories per day for each year younger than age 30; older men should subtract 10 kcalories per day for each year older than age 30.

TABLE F-2 Physical Activities and Their Scores

If your activity was equivalent to this . . .	Then list the number of minutes here and . . .	Multiply by this factor . . .	Add this column to get your physical activity level score:
Activities of Daily Living			
Gardening (no lifting)		0.0032	
Household tasks (moderate effort)		0.0024	
Lifting items continuously		0.0029	
Loading/unloading car		0.0019	
Lying quietly		0.0000	
Mopping		0.0024	
Mowing lawn (power mower)		0.0033	
Raking lawn		0.0029	
Riding in a vehicle		0.0000	
Sitting (idle)		0.0000	
Sitting (doing light activity)		0.0005	
Taking out trash		0.0019	
Vacuuming		0.0024	
Walking the dog		0.0019	
Walking from house to car or bus		0.0014	
Watering plants		0.0014	

TABLE F-2 Physical Activities and Their Scores (*continued*)

If your activity was equivalent to this . . .	Then list the number of minutes here and . . .	Multiply by this factor . . .	Add this column to get your physical activity level score:
Additional Activities			
Billiards		0.0013	
Calisthenics (no weight)		0.0029	
Canoeing (leisurely)		0.0014	
Chopping wood		0.0037	
Climbing hills (carrying 11 lb load)		0.0061	
Climbing hills (no load)		0.0056	
Cycling (leisurely)		0.0024	
Cycling (moderately)		0.0045	
Dancing (aerobic or ballet)		0.0048	
Dancing (ballroom, leisurely)		0.0018	
Dancing (fast ballroom or square)		0.0043	
Golf (with cart)		0.0014	
Golf (without cart)		0.0032	
Horseback riding (walking)		0.0012	
Horseback riding (trotting)		0.0053	
Jogging (6 mph)		0.0088	
Music (playing accordion)		0.0008	
Music (playing cello)		0.0012	
Music (playing flute)		0.0010	
Music (playing piano)		0.0012	
Music (playing violin)		0.0014	
Rope skipping		0.0105	
Skating (ice)		0.0043	
Skating (roller)		0.0052	
Skiing (water or downhill)		0.0055	
Squash		0.0106	
Surfing		0.0048	
Swimming (slow)		0.0033	
Swimming (fast)		0.0057	
Tennis (doubles)		0.0038	
Tennis (singles)		0.0057	
Volleyball (noncompetitive)		0.0018	
Walking (2 mph)		0.0014	
Walking (3 mph)		0.0022	
Walking (4 mph)		0.0033	
Walking (5 mph)		0.0067	
Subtotal			
Factor for basal energy and the thermic effect of food			**1.1**
Your physical activity level score			

TABLE F-3 Physical Activity Equivalents and Their PA Factors

Physical Activity Level	Description	Physical Activity Equivalents	Men, 19+ yr PA Factor	Women, 19+ yr PA Factor	Boys, 3–18 yr PA Factor	Girls, 3–18 yr PA Factor
1.0 to 1.39	Sedentary	Only those physical activities required for typical daily living	1.0	1.0	1.0	1.0
1.4 to 1.59	Low active	Daily living + 30–60 min moderate activity[a]	1.11	1.12	1.13	1.16
1.6 to 1.89	Active	Daily living + ⩾ 60 min moderate activity	1.25	1.27	1.26	1.31
1.9 and above	Very active	Daily living + ⩾ 60 min moderate activity *and* ≥ 60 min vigorous activity *or* ⩾ 120 min moderate activity	1.48	1.45	1.42	1.56

[a]Moderate activity is equivalent to walking at a pace of 3 to 4½ miles per hour.

TABLE F-4 Total Energy Expenditure (TEE in kCalories per Day) for Women 30 Years of Age[a] at Various Levels of Activity and Various Heights and Weights

Heights m (in)	Physical Activity Level	Weight[b] kg (lb)					
1.45 (57)		38.9 (86)	45.2 (100)	52.6 (116)	63.1 (139)	73.6 (162)	84.1 (185)
		kCalories					
	Sedentary	1564	1623	1698	1813	1927	2042
	Low active	1734	1800	1912	2043	2174	2304
	Active	1946	2021	2112	2257	2403	2548
	Very active	2201	2287	2387	2553	2719	2886
1.50 (59)		41.6 (92)	48.4 (107)	56.3 (124)	67.5 (149)	78.8 (174)	90.0 (198)
		kCalories					
	Sedentary	1625	1689	1771	1894	2017	2139
	Low active	1803	1874	1996	2136	2276	2415
	Active	2025	2105	2205	2360	2516	2672
	Very active	2291	2382	2493	2671	2849	3027
1.55 (61)		44.4 (98)	51.7 (114)	60.1 (132)	72.1 (159)	84.1 (185)	96.1 (212)
		kCalories					
	Sedentary	1688	1756	1846	1977	2108	2239
	Low active	1873	1949	2081	2230	2380	2529
	Active	2104	2190	2299	2466	2632	2798
	Very active	2382	2480	2601	2791	2981	3171
1.60 (63)		47.4 (104)	55.0 (121)	64.0 (141)	76.8 (169)	89.6 (197)	102.4 (226)
		kCalories					
	Sedentary	1752	1824	1922	2061	2201	2340
	Low active	1944	2025	2168	2327	2486	2645
	Active	2185	2276	2396	2573	2750	2927
	Very active	2474	2578	2712	2914	3116	3318
1.65 (65)		50.4 (111)	58.5 (129)	68.1 (150)	81.7 (180)	95.3 (210)	108.9 (240)
		kCalories					
	Sedentary	1816	1893	1999	2148	2296	2444
	Low active	2016	2102	2556	2425	2594	2763
	Active	2267	2364	2494	2682	2871	3059
	Very active	2567	2678	2824	3039	3254	3469
1.70 (67)		53.5 (118)	62.1 (137)	72.3 (159)	86.7 (191)	101.2 (223)	115.6 (255)
		kCalories					
	Sedentary	1881	1963	2078	2235	2393	2550
	Low active	2090	2180	2345	2525	2705	2884
	Active	2350	2453	2594	2794	2994	3194
	Very active	2662	2780	2938	3166	3395	3623
1.75 (69)		56.7 (125)	65.8 (145)	76.6 (169)	91.9 (202)	107.2 (236)	122.5 (270)
		kCalories					
	Sedentary	1948	2034	2158	2325	2492	2659
	Low active	2164	2260	2437	2627	2817	3007
	Active	2434	2543	2695	2907	3119	3331
	Very active	2758	2883	3054	3296	3538	3780
1.80 (71)		59.9 (132)	69.7 (154)	81.0 (178)	97.2 (214)	113.4 (250)	129.6 (285)
		kCalories					
	Sedentary	2015	2106	2239	2416	2593	2769
	Low active	2239	2341	2529	2731	2932	3133
	Active	2519	2634	2799	3023	3247	3472
	Very active	2855	2987	3172	3428	3684	3940

[a]For each year younger than 30, add 10 kcalories/day to TEE. For each year older than 30, subtract 10 kcalories/day from TEE.
[b]These columns represent a BMI of 18.5, 22.5, 25, 30, 35, and 40, respectively.

TABLE F-4 Total Energy Expenditure (TEE in kCalories per Day) for Women 30 Years of Age[a] at Various Levels of Activity and Various Heights and Weights (*continued*)

Heights m (in)	Physical Activity Level	Weight[b] kg (lb)					
1.85 (73)		63.3 (139)	73.6 (162)	85.6 (189)	102.7 (226)	119.8 (264)	136.9 (302)
				kCalories			
	Sedentary	2083	2179	2322	2509	2695	2882
	Low active	2315	2422	2624	2836	3049	3262
	Active	2605	2727	2904	3141	3378	3615
	Very active	2954	3093	3292	3562	3833	4103
1.90 (75)		66.8 (147)	77.6 (171)	90.3 (199)	108.3 (239)	126.4 (278)	144.4 (318)
				kCalories			
	Sedentary	2151	2253	2406	2603	2800	2996
	Low active	2392	2505	2720	2944	3168	3393
	Active	2693	2821	3011	3261	3511	3760
	Very active	3053	3200	3414	3699	3984	4270
1.95 (77)		70.3 (155)	81.8 (180)	95.1 (209)	114.1 (251)	133.1 (293)	152.1 (335)
				kCalories			
	Sedentary	2221	2328	2492	2699	2906	3113
	Low active	2470	2589	2817	3053	3290	3526
	Active	2781	2917	3119	3383	3646	3909
	Very active	3154	3309	3538	3838	4139	4439

[a]For each year younger than 30, add 7 kcalories/day to TEE. For each year older than 30, subtract 7 kcalories/day from TEE.
[b]These columns represent a BMI of 18.5, 22.5, 25, 30, 35, and 40, respectively.

TABLE F-5 Total Energy Expenditure (TEE in kCalories per Day) for Men 30 Years of Age[a] at Various Levels of Activity and Various Heights and Weights

Heights m (in)	Physical Activity Level	Weight[b] kg (lb)					
1.45 (57)		38.9 (86)	47.3 (100)	52.6 (116)	63.1 (139)	73.6 (163)	84.1 (185)
				kCalories			
	Sedentary	1777	1911	2048	2198	2347	2496
	Low active	1931	2080	2225	2393	2560	2727
	Active	2127	2295	2447	2636	2826	3015
	Very active	2450	2648	2845	3075	3305	3535
1.50 (59)		41.6 (92)	50.6 (107)	56.3 (124)	67.5 (149)	78.8 (174)	90.0 (198)
				kCalories			
	Sedentary	1848	1991	2126	2286	2445	2605
	Low active	2009	2168	2312	2491	2670	2849
	Active	2215	2394	2545	2748	2951	3154
	Very active	2554	2766	2965	3211	3457	3703
1.55 (61)		44.4 (98)	54.1 (114)	60.1 (132)	72.1 (159)	84.1 (185)	96.1 (212)
				kCalories			
	Sedentary	1919	2072	2205	2376	2546	2717
	Low active	2089	2259	2401	2592	2783	2974
	Active	2305	2496	2646	2862	3079	3296
	Very active	2660	2887	3087	3349	3612	3875

[a]For each year younger than 30, add 10 kcalories/day to TEE. For each year older than 30, subtract 10 kcalories/day from TEE.
[b]These columns represent a BMI of 18.5, 22.5, 25, 30, 35, and 40, respectively.

(continued)

TABLE F-5 Total Energy Expenditure (TEE in kCalories per Day) for Men 30 Years of Age[a] at Various Levels of Activity and Various Heights and Weights (*continued*)

Heights m (in)	Physical Activity Level	Weight[b] kg (lb)					
1.60 (63)		47.4 (104)	57.6 (121)	64.0 (141)	76.8 (169)	89.6 (197)	102.4 (226)
				kCalories			
	Sedentary	1993	2156	2286	2468	2650	2831
	Low active	2171	2351	2492	2695	2899	3102
	Active	2397	2601	2749	2980	3210	3441
	Very active	2769	3010	3211	3491	3771	4051
1.65 (65)		50.4 (111)	61.3 (129)	68.1 (150)	81.7 (180)	95.3 (210)	108.9 (240)
				kCalories			
	Sedentary	2068	2241	2369	2562	2756	2949
	Low active	2254	2446	2585	2801	3017	3234
	Active	2490	2707	2854	3099	3345	3590
	Very active	2880	3136	3339	3637	3934	4232
1.70 (67)		53.5 (118)	65.0 (137)	72.3 (159)	86.7 (191)	101.2 (223)	115.6 (255)
				kCalories			
	Sedentary	2144	2328	2454	2659	2864	3069
	Low active	2338	2542	2679	2909	3139	3369
	Active	2586	2816	2961	3222	3483	3743
	Very active	2992	3265	3469	3785	4101	4417
1.75 (69)		56.7 (125)	68.9 (145)	76.6 (169)	91.9 (202)	107.2 (236)	122.5 (270)
				kCalories			
	Sedentary	2222	2416	2540	2757	2975	3192
	Low active	2425	2641	2776	3020	3263	3507
	Active	2683	2927	3071	3347	3623	3900
	Very active	3108	3396	3602	3937	4272	4607
1.80 (71)		59.9 (132)	72.9 (154)	81.0 (178)	97.2 (214)	113.4 (250)	129.6 (285)
				kCalories			
	Sedentary	2301	2507	2628	2858	3088	3318
	Low active	2513	2741	2875	3132	3390	3648
	Active	2782	3040	3183	3475	3767	4060
	Very active	3225	3530	3738	4092	4447	4801
1.85 (73)		63.3 (139)	77.0 (162)	85.6 (189)	102.7 (226)	119.8 (264)	136.9 (302)
				kCalories			
	Sedentary	2382	2599	2718	2961	3204	3447
	Low active	2602	2844	2976	3248	3520	3792
	Active	2883	3155	3297	3606	3915	4223
	Very active	3344	3667	3877	4251	4625	4999
1.90 (75)		66.8 (147)	81.2 (171)	90.3 (199)	108.3 (239)	126.4 (278)	144.4 (318)
				kCalories			
	Sedentary	2464	2693	2810	3066	3322	3579
	Low active	2693	2948	3078	3365	3652	3939
	Active	2986	3273	3414	3739	4065	4390
	Very active	3466	3806	4018	4413	4807	5202
1.95 (77)		70.3 (155)	85.6 (180)	95.1 (209)	114.1 (251)	133.1 (293)	152.1 (335)
				kCalories			
	Sedentary	2547	2789	2903	3173	3443	3713
	Low active	2786	3055	3183	3485	3788	4090
	Active	3090	3393	3533	3875	4218	4561
	Very active	3590	3948	4162	4578	4993	5409

[a]For each year younger than 30, add 10 kcalories/day to TEE. For each year older than 30, subtract 10 kcalories/day from TEE.

[b]These columns represent a BMI of 18.5, 22.5, 25, 30, 35, and 40, respectively.

Appendix G Exchange Lists for Diabetes

Chapter 2 introduces the exchange system, and this appendix provides details from the *2008 Choose Your Foods: Exchange Lists for Diabetes.* Exchange lists can help people with diabetes to manage their blood glucose levels by controlling the amount and kinds of carbohydrates they consume. These lists can also help in planning diets for weight management by controlling kcalorie and fat intake.*

CONTENTS

The Exchange System

The exchange system sorts foods into groups by their proportions of carbohydrate, fat, and protein (Table G-1). These groups may be organized into several exchange lists of foods (Tables G-2 through G-12 on pp. G-3–G-16). For example, the carbohydrate group includes these exchange lists:

- Starch
- Fruits
- Milk (fat-free, reduced-fat, and whole)
- Sweets, Desserts, and Other Carbohydrates
- Nonstarchy Vegetables

TABLE G-1 The Food Lists

Lists	Typical Item/Portion Size	Carbohydrate (g)	Protein (g)	Fat (g)	Energy[a] (kcal)
Carbohydrates					
Starch[b]	1 slice bread	15	0–3	0–1	80
Fruits	1 small apple	15	—	—	60
Milk					
Fat-free, low-fat, 1%	1 c fat-free milk	12	8	0-3	100
Reduced-fat, 2%	1 c reduced-fat milk	12	8	5	120
Whole	1 c whole milk	12	8	8	160
Sweets, desserts, and other carbohydrates[c]	2 small cookies	15	varies	varies	varies
Nonstarchy vegetables	½ c cooked carrots	5	2	—	25
Meat and Meat Substitutes					
Lean	1 oz chicken (no skin)	—	7	0–3	45
Medium-fat	1 oz ground beef	—	7	4–7	75
High-fat	1 oz pork sausage	—	7	8+	100
Plant-based proteins	½ c tofu	varies	7	varies	varies
Fats	1 tsp butter	—	—	5	45
Alcohol	12 oz beer	varies	—	—	100

[a]The energy value for each exchange list represents an approximate average for the group and does not reflect the precise number of grams of carbohydrate, protein, and fat. For example, a slice of bread contains 15 grams of carbohydrate (60 kcalories), 3 grams protein (12 kcalories), and a little fat—rounded to 80 kcalories for ease in calculating. A ½ cup of vegetables (not including starchy vegetables) contains 5 grams carbohydrate (20 kcalories) and 2 grams protein (8 more), which has been rounded down to 25 kcalories.
[b]The Starch list includes cereals, grains, breads, crackers, snacks, starchy vegetables (such as corn, peas, and potatoes), and legumes (dried beans, peas, and lentils).
[c]The Sweets, Desserts, and Other Carbohydrates list includes foods that contain added sugars and fats such as sodas, candy, cakes, cookies, doughnuts, ice cream, pudding, syrup, and frozen yogurt.

*The Exchange Lists are the Basis of a meal planning system designed by a committee of the American Diabetes Association and The American Dietetic Association. While designed primary for people with diabetes and others who must follow special diets, the Exchange Lists are based on principles of good nutrition that apply to everyone.

Then any food on a list can be "exchanged" for any other on that same list. Another group for alcohol has been included as a reminder that these beverages often deliver substantial carbohydrate and kcalories, and therefore warrant their own list.

Serving Sizes

The serving sizes have been carefully adjusted and defined so that a serving of any food on a given list provides roughly the same amount of carbohydrate, fat, and protein, and, therefore, total energy. Any food on a list can thus be exchanged, or traded, for any other food on the same list without significantly affecting the diet's energy-nutrient balance or total kcalories. For example, a person may select 17 small grapes or ½ large grapefruit as one fruit exchange, and either choice would provide roughly 15 grams of carbohydrate and 60 kcalories. A whole grapefruit, however, would count as 2 fruit exchanges.

To apply the system successfully, users must become familiar with the specified serving sizes. A convenient way to remember the serving sizes and energy values is to keep in mind a typical item from each list (review Table G-1).

The Foods on the Lists

Foods do not always appear on the exchange list where you might first expect to find them. They are grouped according to their energy-nutrient contents rather than by their source (such as milks), their outward appearance, or their vitamin and mineral contents. For example, cheeses are grouped with meats (not milk) because, like meats, cheeses contribute energy from protein and fat but provide negligible carbohydrate.

For similar reasons, starchy vegetables such as corn, green peas, and potatoes are found on the Starch list with breads and cereals, not with the vegetables. Likewise, bacon is grouped with the fats and oils, not with the meats.

Diet planners learn to view mixtures of foods, such as casseroles and soups, as combinations of foods from different exchange lists. They also learn to interpret food labels with the exchange system in mind.

Controlling Energy, Fat, and Sodium

The exchange lists help people control their energy intakes by paying close attention to serving sizes. People wanting to lose weight can limit foods from the Sweets, Desserts, and Other Carbohydrates and Fats lists, and they might choose to avoid the Alcohol list altogether. The Free Foods list provide low-kcalorie choices.

By assigning items like bacon to the Fats list, the exchange lists alert consumers to foods that are unexpectedly high in fat. Even the Starch list specifies which grain products contain added fat (such as biscuits, cornbread, and waffles) by marking them with a symbol to indicate added fat (the symbols are explained in the table keys). In addition, the exchange lists encourage users to think of fat-free milk as milk and of whole milk as milk with added fat, and to think of lean meats as meats and of medium-fat and high-fat meats as meats with added fat. To that end, foods on the milk and meat lists are separated into categories based on their fat contents (review Table G-1). The Milk list is subdivided for fat-free, reduced fat, and whole; the meat list is subdivided for lean, medium fat, and high fat. The meat list also includes plant-based proteins, which tend to be rich in fiber. Notice that many of these foods (p. G-11) bear the symbol for "high fiber."

People wanting to control the sodium in their diets can begin by eliminating any foods bearing the "high sodium" symbol. In most cases, the symbol identifies foods that, in one serving, provide 480 milligrams or more of sodium. Foods on the Combination Foods or Fast Foods lists that bear the symbol provide more than 600 milligrams of sodium. Other foods may also contribute substantially to sodium (consult Chapter 12 for details).

Planning a Healthy Diet

To obtain a daily variety of foods that provide healthful amounts of carbohydrate, protein, and fat, as well as vitamins, minerals, and fiber, the meal plan for adults and teenagers should include at least:

- Two to three servings of nonstarchy vegetables
- Two servings of fruits
- Six servings of grains (at least three of whole grains), beans, and starchy vegetables
- Two servings of low-fat or fat-free milk
- About 6 ounces of meat or meat substitutes
- *Small* amounts of fat and sugar

The actual amounts are determined by age, gender, activity levels, and other factors that influence energy needs. Refer to Chapter 8 as you read through these sections to get an idea of how exchange lists can be useful in planning a diet.

TABLE G-2 Starch

The Starch list includes bread, cereals and grains, starchy vegetables, crackers and snacks, and legumes (dried beans, peas, and lentils).
1 starch choice = 15 grams carbohydrate, 0–3 grams protein, 0–1 grams fat, and 80 kcalories.

NOTE: In general, one starch exchange is ½ cup cooked cereal, grain, or starchy vegetable; ⅓ cup cooked rice or pasta; 1 ounce of bread product; ¾ ounce to 1 ounce of most snack foods.

Bread

Food	Serving Size
Bagel, large (about 4 oz)	¼ (1 oz)
⚠ Biscuit, 2½ inches across	1
Bread	
☺ reduced-kcalorie	2 slices (1½ oz)
white, whole-grain, pumpernickel, rye, unfrosted raisin	1 slice (1 oz)
Chapatti, small, 6 inches across	1
⚠ Cornbread, 1¾ inch cube	1 (1½ oz)
English muffin	½
Hot dog bun or hamburger bun	½ (1 oz)
Naan, 8 inches by 2 inches	¼
Pancake, 4 inches across, ¼ inch thick	1
Pita, 6 inches across	½
Roll, plain, small	1 (1 oz)
⚠ Stuffing, bread	⅓ cup
⚠ Taco shell, 5 inches across	2
Tortilla, corn, 6 inches across	1
Tortilla, flour, 6 inches across	1
Tortilla, flour, 10 inches across	⅓
⚠ Waffle, 4-inch square or 4 inches across	1

Cereals and Grains

Food	Serving Size
Barley, cooked	⅓ cup
Bran, dry	
☺ oat	¼ cup
☺ wheat	½ cup
☺ Bulgur (cooked)	½ cup
Cereals	
☺ bran	½ cup
cooked (oats, oatmeal)	½ cup
puffed	1½ cups
shredded wheat, plain	½ cup
sugar-coated	½ cup
unsweetened, ready-to-eat	¾ cup
Couscous	⅓ cup
Granola	
low-fat	¼ cup
⚠ regular	¼ cup
Grits, cooked	½ cup
Kasha	½ cup
Millet, cooked	⅓ cup

KEY

☺ = More than 3 grams of dietary fiber per serving.

⚠ = Extra fat, or prepared with added fat. (Count as 1 starch + 1 fat.)

🧂 = 480 milligrams or more of sodium per serving.

(*continued*)

TABLE G-2 Starch (*continued*)

Cereals and Grains—continued

Food	Serving Size
Muesli	¼ cup
Pasta, cooked	⅓ cup
Polenta, cooked	⅓ cup
Quinoa, cooked	⅓ cup
Rice, white or brown, cooked	⅓ cup
Tabbouleh (tabouli), prepared	½ cup
Wheat germ, dry	3 Tbsp
Wild rice, cooked	½ cup

Starchy Vegetables

Food	Serving Size
Cassava	⅓ cup
Corn	½ cup
on cob, large	½ cob (5 oz)
☺ Hominy, canned	¾ cup
☺ Mixed vegetables with corn, peas, or pasta	1 cup
☺ Parsnips	½ cup
☺ Peas, green	½ cup
Plantain, ripe	⅓ cup
Potato	
baked with skin	¼ large (3 oz)
boiled, all kinds	½ cup or ½ medium (3 oz)
⚠ mashed, with milk and fat	½ cup
french fried (oven-baked)[a]	1 cup (2 oz)
☺ Pumpkin, canned, no sugar added	1 cup
Spaghetti/pasta sauce	½ cup
☺ Squash, winter (acorn, butternut)	1 cup
☺ Succotash	½ cup
Yam, sweet potato, plain	½ cup

Crackers and Snacks[b]

Food	Serving Size
Animal crackers	8
Crackers	
⚠ round-butter type	6
saltine-type	6
⚠ sandwich-style, cheese or peanut butter filling	3
⚠ whole-wheat regular	2–5 (¾ oz)
☺ whole-wheat lower fat or crispbreads	2–5 (¾ oz)
Graham cracker, 2½-inch square	3
Matzoh	¾ oz
Melba toast, about 2-inch by 4-inch piece	4
Oyster crackers	20
Popcorn	3 cups
⚠ ☺ with butter	3 cups
☺ no fat added	3 cups
☺ lower fat	3 cups
Pretzels	¾ oz
Rice cakes, 4 inches across	2
Snack chips	
fat-free or baked (tortilla, potato), baked pita chips	15–20 (¾ oz)
⚠ regular (tortilla, potato)	9–13 (¾ oz)

Beans, Peas, and Lentils[c]

The choices on this list count as 1 starch + 1 lean meat.

Food	Serving Size
☺ Baked beans	⅓ cup
☺ Beans, cooked (black, garbanzo, kidney, lima, navy, pinto, white)	½ cup
☺ Lentils, cooked (brown, green, yellow)	½ cup
☺ Peas, cooked (black-eyed, split)	½ cup
🧂 ☺ Refried beans, canned	½ cup

KEY

☺ = More than 3 grams of dietary fiber per serving.

⚠ = Extra fat, or prepared with added fat. (Count as 1 starch + 1 fat.)

🧂 = 480 milligrams or more of sodium per serving.

[a]Restaurant-style french fries are on the Fast Foods list.

[b]For other snacks, see the Sweets, Desserts, and Other Carbohydrates list. For a quick estimate of serving size, an open handful is equal to about 1 cup or 1 to 2 ounces of snack food.

[c]Beans, peas, and lentils are also found on the Meat and Meat Substitutes list.

TABLE G-3 Fruits

Fruit[a]

The Fruits list includes fresh, frozen, canned, and dried fruits and fruit juices. 1 fruit choice = 15 grams carbohydrate, 0 grams protein, 0 grams fat, and 60 kcalories.

NOTE: In general, one fruit exchange is ½ cup canned or fresh fruit or unsweetened fruit juice; 1 small fresh fruit (4 ounces); 2 tablespoons dried fruit.

Food	Serving Size
Apple, unpeeled, small	1 (4 oz)
Apples, dried	4 rings
Applesauce, unsweetened	½ cup
Apricots	
canned	½ cup
dried	8 halves
☺ fresh	4 whole (5½ oz)
Banana, extra small	1 (4 oz)
☺ Blackberries	¾ cup
Blueberries	¾ cup
Cantaloupe, small	⅓ melon or 1 cup cubed (11 oz)
Cherries	
sweet, canned	½ cup
sweet fresh	12 (3 oz)
Dates	3
Dried fruits (blueberries, cherries, cranberries, mixed fruit, raisins)	2 Tbsp
Figs	
dried	1½
☺ fresh	1½ large or 2 medium (3½ oz)
Fruit cocktail	½ cup
Grapefruit	
large	½ (11 oz)
sections, canned	¾ cup
Grapes, small	17 (3 oz)
Honeydew melon	1 slice or 1 cup cubed (10 oz)
☺ Kiwi	1 (3½ oz)
Mandarin oranges, canned	¾ cup
Mango, small	½ (5½ oz) or ½ cup
Nectarine, small	1 (5 oz)
☺ Orange, small	1 (6½ oz)
Papaya	½ or 1 cup cubed (8 oz)
Peaches	
canned	½ cup
fresh, medium	1 (6 oz)
Pears	
canned	½ cup
fresh, large	½ (4 oz)
Pineapple	
canned	½ cup
fresh	¾ cup
Plums	
canned	½ cup
dried (prunes)	3
small	2 (5 oz)
☺ Raspberries	1 cup
☺ Strawberries	1¼ cup whole berries
☺ Tangerines, small	2 (8 oz)
Watermelon	1 slice or 1¼ cups cubes (13½ oz)

Fruit Juice

Food	Serving Size
Apple juice/cider	½ cup
Fruit juice blends, 100% juice	⅓ cup
Grape juice	⅓ cup
Grapefruit juice	½ cup
Orange juice	½ cup
Pineapple juice	½ cup
Prune juice	⅓ cup

KEY

☺ = More than 3 grams of dietary fiber per serving.

⚠ = Extra fat, or prepared with added fat. (Count as 1 starch + 1 fat.)

🧂 = 480 milligrams or more of sodium per serving.

[a]The weight listed includes skin, core, seeds, and rind.

TABLE G-4 Milk

The Milk list groups milks and yogurts based on the amount of fat they have (fat-free/low fat, reduced fat, and whole). Cheeses are found on the Meat and Meat Substitutes list and cream and other dairy fats are found on the Fats list.

NOTE: In general, one milk choice is 1 cup (8 fluid ounces or ½ pint) milk or yogurt.

Milk and Yogurts	
Food	**Serving Size**
Fat-free or low-fat (1%)	
1 fat-free/low-fat milk choice = 12 g carbohydrate, 8 g protein, 0–3 g fat, and 100 kcal.	
Milk, buttermilk, acidophilus milk, Lactaid	1 cup
Evaporated milk	½ cup
Yogurt, plain or flavored with an artificial sweetener	⅔ cup (6 oz)
Reduced-fat (2%)	
1 reduced-fat milk choice = 12 g carbohydrate, 8 g protein, 5 g fat, and 120 kcal.	
Milk, acidophilus milk, kefir, Lactaid	1 cup
Yogurt, plain	⅔ cup (6 oz)
Whole	
1 whole milk choice = 12 g carbohydrate, 8 g protein, 8 g fat, and 160 kcal.	
Milk, buttermilk, goat's milk	1 cup
Evaporated milk	½ cup
Yogurt, plain	8 oz

Dairy-Like Foods		
Food	**Serving Size**	**Count as**
Chocolate milk		
fat-free	1 cup	1 fat-free milk + 1 carbohydrate
whole	1 cup	1 whole milk + 1 carbohydrate
Eggnog, whole milk	½ cup	1 carbohydrate + 2 fats
Rice drink		
flavored, low fat	1 cup	2 carbohydrates
plain, fat-free	1 cup	1 carbohydrate
Smoothies, flavored, regular	10 oz	1 fat-free milk + 2½ carbohydrates
Soy milk		
light	1 cup	1 carbohydrate + ½ fat
regular, plain	1 cup	1 carbohydrate + 1 fat
Yogurt		
and juice blends	1 cup	1 fat-free milk + 1 carbohydrate
low carbohydrate (less than 6 grams carbohydrate per choice)	⅔ cup (6 oz)	½ fat-free milk
with fruit, low-fat	⅔ cup (6 oz)	1 fat-free milk + 1 carbohydrate

TABLE G-5 Sweets, Desserts, and Other Carbohydrates

1 other carbohydrate choice = 15 grams carbohydrate, variable grams protein, variable grams fat, and variable kcalories.

NOTE: In general, one choice from this list can substitute for foods on the Starch, Fruits, or Milk lists.

Beverages, Soda, and Energy/Sports Drinks		
Food	**Serving Size**	**Count as**
Cranberry juice cocktail	½ cup	1 carbohydrate
Energy drink	1 can (8.3 oz)	2 carbohydrates
Fruit drink or lemonade	1 cup (8 oz)	2 carbohydrates

TABLE G-5 Sweets, Desserts, and Other Carbohydrates (*continued*)

Beverages, Soda, and Energy/Sports Drinks—continued

Food	Serving Size	Count as
Hot chocolate		
regular	1 envelope added to 8 oz water	1 carbohydrate + 1 fat
sugar-free or light	1 envelope added to 8 oz water	1 carbohydrate
Soft drink (soda), regular	1 can (12 oz)	2½ carbohydrates
Sports drink	1 cup (8 oz)	1 carbohydrate

Brownies, Cake, Cookies, Gelatin, Pie, and Pudding

Food	Serving Size	Count as
Brownie, small, unfrosted	1¼-inch square, ⅞ inch high (about 1 oz)	1 carbohydrate + 1 fat
Cake		
angel food, unfrosted	1/12 of cake (about 2 oz)	2 carbohydrates
frosted	2-inch square (about 2 oz)	2 carbohydrates + 1 fat
unfrosted	2-inch square (about 2 oz)	1 carbohydrate + 1 fat
Cookies		
chocolate chip	2 cookies (2¼ inches across)	1 carbohydrate + 2 fats
gingersnap	3 cookies	1 carbohydrate
sandwich, with crème filling	2 small (about ⅔ oz)	1 carbohydrate + 1 fat
sugar-free	3 small or 1 large (¾-1 oz)	1 carbohydrate + 1–2 fats
vanilla wafer	5 cookies	1 carbohydrate + 1 fat
Cupcake, frosted	1 small (about 1¾ oz)	2 carbohydrates + 1–1½ fats
Fruit cobbler	½ cup (3½ oz)	3 carbohydrates + 1 fat
Gelatin, regular	½ cup	1 carbohydrate
Pie		
commercially prepared fruit, 2 crusts	⅙ of 8-inch pie	3 carbohydrates + 2 fats
pumpkin or custard	⅛ of 8-inch pie	1½ carbohydrates + 1½ fats
Pudding		
regular (made with reduced-fat milk)	½ cup	2 carbohydrates
sugar-free or sugar- and fat-free (made with fat-free milk)	½ cup	1 carbohydrate

Candy, Spreads, Sweets, Sweeteners, Syrups, and Toppings

Food	Serving Size	Count as
Candy bar, chocolate/peanut	2 "fun size" bars (1 oz)	1½ carbohydrates + 1½ fats
Candy, hard	3 pieces	1 carbohydrate
Chocolate "kisses"	5 pieces	1 carbohydrate + 1 fat
Coffee creamer		
dry, flavored	4 tsp	½ carbohydrate + ½ fat
liquid, flavored	2 Tbsp	1 carbohydrate
Fruit snacks, chewy (pureed fruit concentrate)	1 roll (¾ oz)	1 carbohydrate
Fruit spreads, 100% fruit	1½ Tbsp	1 carbohydrate
Honey	1 Tbsp	1 carbohydrate
Jam or jelly, regular	1 Tbsp	1 carbohydrate
Sugar	1 Tbsp	1 carbohydrate
Syrup		
chocolate	2 Tbsp	2 carbohydrates
light (pancake type)	2 Tbsp	1 carbohydrate
regular (pancake type)	1 Tbsp	1 carbohydrate

(*continued*)

APPENDIX G

TABLE G-5 Sweets, Desserts, and Other Carbohydrates (*continued*)

Condiments and Sauces[a]

Food	Serving Size	Count as
Barbeque sauce	3 Tbsp	1 carbohydrate
Cranberry sauce, jellied	¼ cup	1½ carbohydrates
🧂 Gravy, canned or bottled	½ cup	½ carbohydrate + ½ fat
Salad dressing, fat-free, low-fat, cream-based	3 Tbsp	1 carbohydrate
Sweet and sour sauce	3 Tbsp	1 carbohydrate

Doughnuts, Muffins, Pastries, and Sweet Breads

Food	Serving Size	Count as
Banana nut bread	1-inch slice (1 oz)	2 carbohydrates + 1 fat
Doughnut		
cake, plain	1 medium (1½ oz)	1½ carbohydrates + 2 fats
yeast type, glazed	3¾ inches across (2 oz)	2 carbohydrates + 2 fats
Muffin (4 oz)	¼ muffin (1 oz)	1 carbohydrate + ½ fat
Sweet roll or Danish	1 (2½ oz)	2½ carbohydrates + 2 fats

Frozen Bars, Frozen Desserts, Frozen Yogurt, and Ice Cream

Food	Serving Size	Count as
Frozen pops	1	½ carbohydrate
Fruit juice bars, frozen, 100% juice	1 bar (3 oz)	1 carbohydrate
Ice cream		
fat-free	½ cup	1½ carbohydrates
light	½ cup	1 carbohydrate + 1 fat
no sugar added	½ cup	1 carbohydrate + 1 fat
regular	½ cup	1 carbohydrate + 2 fats
Sherbet, sorbet	½ cup	2 carbohydrates
Yogurt, frozen		
fat-free	⅓ cup	1 carbohydrate
regular	½ cup	1 carbohydrate + 0–1 fat

Granola Bars, Meal Replacement Bars/Shakes, and Trail Mix

Food	Serving Size	Count as
Granola or snack bar, regular or low-fat	1 bar (1 oz)	1½ carbohydrates
Meal replacement bar	1 bar (1⅓ oz)	1½ carbohydrates + 0–1 fat
Meal replacement bar	1 bar (2 oz)	2 carbohydrates + 1 fat
Meal replacement shake, reduced kcalorie	1 can (10–11 oz)	1½ carbohydrates + 0–1 fat
Trail mix		
candy/nut-based	1 oz	1 carbohydrate + 2 fats
dried fruit-based	1 oz	1 carbohydrate + 1 fat

KEY

🧂 = 480 milligrams or more of sodium per serving.

[a]You can also check the Fats list and Free Foods list for other condiments.

TABLE G-6 Nonstarchy Vegetables

The Nonstarchy Vegetables list includes vegetables that have few grams of carbohydrates or kcalories; starchy vegetables are found on the Starch list. 1 nonstarchy vegetable choice = 5 grams carbohydrate, 2 grams protein, 0 grams fat, and 25 kcalories.

NOTE: In general, one nonstarchy vegetable choice is ½ cup cooked vegetables or vegetable juice or 1 cup raw vegetables. Count 3 cups of raw vegetables or 1½ cups of cooked vegetables as one carbohydrate choice.

Nonstarchy Vegetables[a]
Amaranth or Chinese spinach
Artichoke
Artichoke hearts
Asparagus
Baby corn
Bamboo shoots
Beans (green, wax, Italian)
Bean sprouts
Beets
🧂 Borscht
Broccoli
☺ Brussels sprouts
Cabbage (green, bok choy, Chinese)
☺ Carrots
Cauliflower
Celery
☺ Chayote
Coleslaw, packaged, no dressing
Cucumber
Eggplant
Gourds (bitter, bottle, luffa, bitter melon)
Green onions or scallions
Greens (collard, kale, mustard, turnip)
Hearts of palm
Jicama
Kohlrabi
Leeks
Mixed vegetables (without corn, peas, or pasta)
Mung bean sprouts
Mushrooms, all kinds, fresh
Okra
Onions
Oriental radish or daikon
Pea pods
☺ Peppers (all varieties)
Radishes
Rutabaga
🧂 Sauerkraut
Soybean sprouts
Spinach
Squash (summer, crookneck, zucchini)
Sugar pea snaps
☺ Swiss chard
Tomato
Tomatoes, canned
🧂 Tomato sauce
🧂 Tomato/vegetable juice
Turnips
Water chestnuts
Yard-long beans

KEY

☺ = More than 3 grams of dietary fiber per serving.

🧂 = 480 milligrams or more of sodium per serving.

[a]Salad greens (like chicory, endive, escarole, lettuce, romaine, spinach, arugula, radicchio, watercress) are on the Free Foods list.

TABLE G-7 Meat and Meat Substitutes

The Meat and Meat Substitutes list groups foods based on the amount of fat they have (lean meat, medium-fat meat, high-fat meat, and plant-based proteins).

Lean Meats and Meat Substitutes

1 lean meat choice = 0 grams carbohydrate, 7 grams protein, 0–3 grams fat, and 100 kcalories.

Food	Amount
Beef: Select or Choice grades trimmed of fat: ground round, roast (chuck, rib, rump), round, sirloin, steak (cubed, flank, porterhouse, T-bone), tenderloin	1 oz
Beef jerky	1 oz
Cheeses with 3 grams of fat or less per oz	1 oz
Cottage cheese	¼ cup
Egg substitutes, plain	¼ cup
Egg whites	2
Fish, fresh or frozen, plain: catfish, cod, flounder, haddock, halibut, orange roughy, salmon, tilapia, trout, tuna	1 oz
Fish, smoked: herring or salmon (lox)	1 oz
Game: buffalo, ostrich, rabbit, venison	1 oz
Hot dog with 3 grams of fat or less per oz (8 dogs per 14 oz package) *Note: May be high in carbohydrate.*	1
Lamb: chop, leg, or roast	1 oz
Organ meats: heart, kidney, liver *Note: May be high in cholesterol.*	1 oz
Oysters, fresh or frozen	6 medium
Pork, lean	
Canadian bacon	1 oz
rib or loin chop/roast, ham, tenderloin	1 oz
Poultry, without skin: Cornish hen, chicken, domestic duck or goose (well-drained of fat), turkey	1 oz
Processed sandwich meats with 3 grams of fat or less per oz: chipped beef, deli thin-sliced meats, turkey ham, turkey kielbasa, turkey pastrami	1 oz
Salmon, canned	1 oz
Sardines, canned	2 medium
Sausage with 3 grams of fat or less per oz	1 oz
Shellfish: clams, crab, imitation shellfish, lobster, scallops, shrimp	1 oz
Tuna, canned in water or oil, drained	1 oz
Veal, lean chop, roast	1 oz

Medium-Fat Meat and Meat Substitutes

1 medium-fat meat choice = 0 grams carbohydrate, 7 grams protein, 4–7 grams fat, and 130 kcalories.

Food	Amount
Beef: corned beef, ground beef, meatloaf, Prime grades trimmed of fat (prime rib), short ribs, tongue	1 oz
Cheeses with 4–7 grams of fat per oz: feta, mozzarella, pasteurized processed cheese spread, reduced-fat cheeses, string	1 oz
Egg *Note: High in cholesterol, so limit to 3 per week.*	1
Fish, any fried product	1 oz
Lamb: ground, rib roast	1 oz
Pork: cutlet, shoulder roast	1 oz
Poultry: chicken with skin; dove, pheasant, wild duck, or goose; fried chicken; ground turkey	1 oz
Ricotta cheese	2 oz or ¼ cup
Sausage with 4–7 grams of fat per oz	1 oz
Veal, cutlet (no breading)	1 oz

High-Fat Meat and Meat Substitutes

1 high-fat meat choice = 0 grams carbohydrate, 7 grams protein, 8+ grams fat, and 150 kcalories. These foods are high in saturated fat, cholesterol, and kcalories and may raise blood cholesterol levels if eaten on a regular basis. Try to eat 3 or fewer servings from this group per week.

Food	Amount
Bacon	
pork	2 slices (16 slices per lb or 1 oz each, before cooking)
turkey	3 slices (½ oz each before cooking)
Cheese, regular: American, bleu, brie, cheddar, hard goat, Monterey jack, queso, and Swiss	1 oz
Hot dog: beef, pork, or combination (10 per lb-sized package)	1
Hot dog: turkey or chicken (10 per lb-sized package)	1
Pork: ground, sausage, spareribs	1 oz
Processed sandwich meats with 8 grams of fat or more per oz: bologna, pastrami, hard salami	1 oz
Sausage with 8 grams fat or more per oz: bratwurst, chorizo, Italian, knockwurst, Polish, smoked, summer	1 oz

(continued)

KEY

= More than 3 grams of dietary fiber per serving.

= Extra fat, or prepared with added fat. (Count as 1 starch + 1 fat.)

= 480 milligrams or more of sodium per serving.

TABLE G-7 Meat and Meat Substitutes (*continued*)

Plant-Based Proteins[a]

1 plant-based protein choice = variable grams carbohydrate, 7 grams protein, variable grams fat, and variable kcalories.
Because carbohydrate content varies among plant-based proteins, you should read the food label.

Food	Serving Size	Count as
"Bacon" strips, soy-based	3 strips	1 medium-fat meat
Baked beans	⅓ cup	1 starch + 1 lean meat
Beans, cooked: black, garbanzo, kidney, lima, navy, pinto, white[a]	½ cup	1 starch + 1 lean meat
"Beef" or "sausage" crumbles, soy-based	2 oz	½ carbohydrate + 1 lean meat
"Chicken" nuggets, soy-based	2 nuggets (1½ oz)	½ carbohydrate + 1 medium-fat meat
Edamame	½ cup	½ carbohydrate + 1 lean meat
Falafel (spiced chickpea and wheat patties)	3 patties (about 2 inches across)	1 carbohydrate + 1 high-fat meat
Hot dog, soy-based	1 (1½ oz)	½ carbohydrate + 1 lean meat
Hummus	⅓ cup	1 carbohydrate + 1 high-fat meat
Lentils, brown, green, or yellow	½ cup	1 carbohydrate + 1 lean meat
Meatless burger, soy-based	3 oz	½ carbohydrate + 2 lean meats
Meatless burger, vegetable- and starch-based	1 patty (about 2½ oz)	1 carbohydrate + 2 lean meats
Nut spreads: almond butter, cashew butter, peanut butter, soy nut butter	1 Tbsp	1 high-fat meat
Peas, cooked: black-eyed and split peas	½ cup	1 starch + 1 lean meat
Refried beans, canned	½ cup	1 starch + 1 lean meat
"Sausage" patties, soy-based	1 (1½ oz)	1 medium-fat meat
Soy nuts, unsalted	¾ oz	½ carbohydrate + 1 medium-fat meat
Tempeh	¼ cup	1 medium-fat meat
Tofu	4 oz (½ cup)	1 medium-fat meat
Tofu, light	4 oz (½ cup)	1 lean meat

KEY

= More than 3 grams of dietary fiber per serving.

= Extra fat, or prepared with added fat. (Add an additional fat choice to this food.)

= 480 milligrams or more of sodium per serving (based on the sodium content of a typical 3-oz serving of meat, unless 1 or 2 oz is the normal serving size).

[a]Beans, peas, and lentils are also found on the Starch list; nut butters in smaller amounts are found in the Fats list.

TABLE G-8 **Fats**

Fats and oils have mixtures of unsaturated (polyunsaturated and monounsaturated) and saturated fats. Foods on the Fats list are grouped together based on the major type of fat they contain. 1 fat choice = 0 grams carbohydrate, 0 grams protein, 5 grams fat, and 45 kcalories.

NOTE: In general, one fat exchange is 1 teaspoon of regular margarine, vegetable oil, or butter; 1 tablespoon of regular salad dressing.

When used in large amounts, bacon and peanut butter are counted as high-fat meat choices (see Meat and Meat Substitutes list). Fat-free salad dressings are found on the Sweets, Desserts, and Other Carbohydrates list. Fat-free products such as margarines, salad dressings, mayonnaise, sour cream, and cream cheese are found on the Free Foods list.

Monounsaturated Fats

Food	Serving Size
Avocado, medium	2 Tbsp (1 oz)
Nut butters (trans fat-free): almond butter, cashew butter, peanut butter (smooth or crunchy)	1½ tsp
Nuts	
almonds	6 nuts
Brazil	2 nuts
cashews	6 nuts
filberts (hazelnuts)	5 nuts
macadamia	3 nuts
mixed (50% peanuts)	6 nuts
peanuts	10 nuts
pecans	4 halves
pistachios	16 nuts
Oil: canola, olive, peanut	1 tsp
Olives	
black (ripe)	8 large
green, stuffed	10 large

Polyunsaturated Fats

Food	Serving Size
Margarine: lower-fat spread (30%–50% vegetable oil, *trans* fat-free)	1 Tbsp
Margarine: stick, tub (*trans* fat-free) or squeeze (*trans* fat-free)	1 tsp
Mayonnaise	
reduced-fat	1 Tbsp
regular	1 tsp
Mayonnaise-style salad dressing	
reduced-fat	1 Tbsp
regular	2 tsp
Nuts	
Pignolia (pine nuts)	1 Tbsp
walnuts, English	4 halves
Oil: corn, cottonseed, flaxseed, grape seed, safflower, soybean, sunflower	1 tsp
Oil: made from soybean and canola oil—Enova	1 tsp
Plant stanol esters	
light	1 Tbsp
regular	2 tsp

Polyunsaturated Fats—continued

Food	Serving Size
Salad dressing	
🧂 reduced-fat *Note: May be high in carbohydrate.*	2 Tbsp
🧂 regular	1 Tbsp
Seeds	
flaxseed, whole	1 Tbsp
pumpkin, sunflower	1 Tbsp
sesame seeds	1 Tbsp
Tahini or sesame paste	2 tsp

Saturated Fats

Food	Serving Size
Bacon, cooked, regular or turkey	1 slice
Butter	
reduced-fat	1 Tbsp
stick	1 tsp
whipped	2 tsp
Butter blends made with oil	
reduced-fat or light	1 Tbsp
regular	1½ tsp
Chitterlings, boiled	2 Tbsp (½ oz)
Coconut, sweetened, shredded	2 Tbsp
Coconut milk	
light	⅓ cup
regular	1½ Tbsp
Cream	
half and half	2 Tbsp
heavy	1 Tbsp
light	1½ Tbsp
whipped	2 Tbsp
whipped, pressurized	¼ cup
Cream cheese	
reduced-fat	1½ Tbsp (¾ oz)
regular	1 Tbsp (½ oz)
Lard	1 tsp
Oil: coconut, palm, palm kernel	1 tsp
Salt pork	¼ oz
Shortening, solid	1 tsp
Sour cream	
reduced-fat or light	3 Tbsp
regular	2 Tbsp

KEY

🧂 = 480 milligrams or more of sodium per serving.

TABLE G-9 Free Foods

A "free" food is any food or drink choice that has less than 20 kcalories and 5 grams or less of carbohydrate per serving.

- Most foods on this list should be limited to 3 servings (as listed here) per day. Spread out the servings throughout the day. If you eat all 3 servings at once, it could raise your blood glucose level.
- Food and drink choices listed here without a serving size can be eaten whenever you like.

Low Carbohydrate Foods

Food	Serving Size
Cabbage, raw	½ cup
Candy, hard (regular or sugar-free)	1 piece
Carrots, cauliflower, or green beans, cooked	¼ cup
Cranberries, sweetened with sugar substitute	½ cup
Cucumber, sliced	½ cup
Gelatin	
dessert, sugar-free	
unflavored	
Gum	
Jam or jelly, light or no sugar added	2 tsp
Rhubarb, sweetened with sugar substitute	½ cup
Salad greens	
Sugar substitutes (artificial sweeteners)	
Syrup, sugar-free	2 Tbsp

Modified Fat Foods with Carbohydrate

Food	Serving Size
Cream cheese, fat-free	1 Tbsp (½ oz)
Creamers	
nondairy, liquid	1 Tbsp
nondairy, powdered	2 tsp
Margarine spread	
fat-free	1 Tbsp
reduced-fat	1 tsp
Mayonnaise	
fat-free	1 Tbsp
reduced-fat	1 tsp
Mayonnaise-style salad dressing	
fat-free	1 Tbsp
reduced-fat	1 tsp
Salad dressing	
fat-free or low-fat	1 Tbsp
fat-free, Italian	2 Tbsp
Sour cream, fat-free or reduced-fat	1 Tbsp
Whipped topping	
light or fat-free	2 Tbsp
regular	1 Tbsp

Condiments

Food	Serving Size
Barbecue sauce	2 tsp
Catsup (ketchup)	1 Tbsp
Honey mustard	1 Tbsp
Horseradish	
Lemon juice	
Miso	1½ tsp
Mustard	
Parmesan cheese, freshly grated	1 Tbsp
Pickle relish	1 Tbsp
Pickles	
🧂 dill	1½ medium
sweet, bread and butter	2 slices
sweet, gherkin	¾ oz
Salsa	¼ cup
🧂 Soy sauce, light or regular	1 Tbsp
Sweet and sour sauce	2 tsp
Sweet chili sauce	2 tsp
Taco sauce	1 Tbsp
Vinegar	
Yogurt, any type	2 Tbsp

Drinks/Mixes

Any food on the list—without a serving size listed—can be consumed in any moderate amount.

- 🧂 Bouillon, broth, consommé
- Bouillon or broth, low-sodium
- Carbonated or mineral water
- Club soda
- Cocoa powder, unsweetened (1 Tbsp)
- Coffee, unsweetened or with sugar substitute
- Diet soft drinks, sugar-free
- Drink mixes, sugar-free
- Tea, unsweetened or with sugar substitute
- Tonic water, diet
- Water
- Water, flavored, carbohydrate free

Seasonings

Any food on this list can be consumed in any moderate amount.

- Flavoring extracts (for example, vanilla, almond, peppermint)
- Garlic
- Herbs, fresh or dried
- Nonstick cooking spray
- Pimento
- Spices
- Hot pepper sauce
- Wine, used in cooking
- Worcestershire sauce

KEY

🧂 = 480 milligrams or more of sodium per serving.

TABLE G-10 Combination Foods

Many foods are eaten in various combinations, such as casseroles. Because "combination" foods do not fit into any one choice list, this list of choices provides some typical combination foods.

Entrees

Food	Serving Size	Count as
Casserole type (tuna noodle, lasagna, spaghetti with meatballs, chili with beans, macaroni and cheese)	1 cup (8 oz)	2 carbohydrates + 2 medium-fat meats
Stews (beef/other meats and vegetables)	1 cup (8 oz)	1 carbohydrate + 1 medium-fat meat + 0–3 fats
Tuna salad or chicken salad	½ cup (3½ oz)	½ carbohydrate + 2 lean meats + 1 fat

Frozen Meals/Entrees

Food	Serving Size	Count as
Burrito (beef and bean)	1 (5 oz)	3 carbohydrates + 1 lean meat + 2 fats
Dinner-type meal	generally 14–17 oz	3 carbohydrates + 3 medium-fat meats + 3 fats
Entrée or meal with less than 340 kcalories	about 8–11 oz	2–3 carbohydrates + 1–2 lean meats
Pizza		
cheese/vegetarian, thin crust	¼ of a 12 inch (4½–5 oz)	2 carbohydrates + 2 medium-fat meats
meat topping, thin crust	¼ of a 12 inch (5 oz)	2 carbohydrates + 2 medium-fat meats + 1½ fats
Pocket sandwich	1 (4½ oz)	3 carbohydrates + 1 lean meat + 1–2 fats
Pot pie	1 (7 oz)	2½ carbohydrates + 1 medium-fat meat + 3 fats

Salads (Deli-Style)

Food	Serving Size	Count as
Coleslaw	½ cup	1 carbohydrate + 1½ fats
Macaroni/pasta salad	½ cup	2 carbohydrates + 3 fats
Potato salad	½ cup	1½–2 carbohydrates + 1–2 fats

Soups

Food	Serving Size	Count as
Bean, lentil, or split pea	1 cup	1 carbohydrate + 1 lean meat
Chowder (made with milk)	1 cup (8 oz)	1 carbohydrate + 1 lean meat + 1½ fats
Cream (made with water)	1 cup (8 oz)	1 carbohydrate + 1 fat
Instant	6 oz prepared	1 carbohydrate
with beans or lentils	8 oz prepared	2½ carbohydrates + 1 lean meat
Miso soup	1 cup	½ carbohydrate + 1 fat
Oriental noodle	1 cup	2 carbohydrates + 2 fats
Rice (congee)	1 cup	1 carbohydrate
Tomato (made with water)	1 cup (8 oz)	1 carbohydrate
Vegetable beef, chicken noodle, or other broth-type	1 cup (8 oz)	1 carbohydrate

KEY

= More than 3 grams of dietary fiber per serving.

= Extra fat, or prepared with added fat.

= 600 milligrams or more of sodium per serving (for combination food main dishes/meals).

TABLE G-11 Fast Foods

The choices on the Fast Foods list are not specific fast-food meals or items, but are estimates based on popular foods. Ask the restaurant or check its website for nutrition information about your favorite fast foods.

Food	Serving Size	Count as
Breakfast Sandwiches		
🧂 Egg, cheese, meat, English muffin	1 sandwich	2 carbohydrates + 2 medium-fat meats
🧂 Sausage biscuit sandwich	1 sandwich	2 carbohydrates + 2 high-fat meats + 3½ fats
Main Dishes/Entrees		
🧂 ☺ Burrito (beef and beans)	1 (about 8 oz)	3 carbohydrates + 3 medium-fat meats + 3 fats
🧂 Chicken breast, breaded and fried	1 (about 5 oz)	1 carbohydrate + 4 medium-fat meats
Chicken drumstick, breaded and fried	1 (about 2 oz)	2 medium-fat meats
🧂 Chicken nuggets	6 (about 3½ oz)	1 carbohydrate + 2 medium-fat meats + 1 fat
🧂 Chicken thigh, breaded and fried	1 (about 4 oz)	½ carbohydrate + 3 medium-fat meats + 1½ fats
🧂 Chicken wings, hot	6 (5 oz)	5 medium-fat meats + 1½ fats
Oriental		
🧂 Beef/chicken/shrimp with vegetables in sauce	1 cup (about 5 oz)	1 carbohydrate + 1 lean meat + 1 fat
🧂 Egg roll, meat	1 (about 3 oz)	1 carbohydrate + 1 lean meat + 1 fat
Fried rice, meatless	½ cup	1½ carbohydrates + 1½ fats
🧂 Meat and sweet sauce (orange chicken)	1 cup	3 carbohydrates + 3 medium-fat meats + 2 fats
🧂 ☺ Noodles and vegetables in sauce (chow mein, lo mein)	1 cup	2 carbohydrates + 1 fat
Pizza		
Pizza		
🧂 cheese, pepperoni, regular crust	⅛ of a 14 inch (about 4 oz)	2½ carbohydrates + 1 medium-fat meat + 1½ fats
🧂 cheese/vegetarian, thin crust	¼ of a 12 inch (about 6 oz)	2½ carbohydrates + 2 medium-fat meats + 1½ fats
Sandwiches		
🧂 Chicken sandwich, grilled	1	3 carbohydrates + 4 lean meats
🧂 Chicken sandwich, crispy	1	3½ carbohydrates + 3 medium-fat meats + 1 fat
Fish sandwich with tartar sauce	1	2½ carbohydrates + 2 medium-fat meats + 2 fats
Hamburger		
🧂 large with cheese	1	2½ carbohydrates + 4 medium-fat meats + 1 fat
regular	1	2 carbohydrates + 1 medium-fat meat + 1 fat
🧂 Hot dog with bun	1	1 carbohydrate + 1 high-fat meat + 1 fat
Submarine sandwich		
🧂 less than 6 grams fat	6-inch sub	3 carbohydrates + 2 lean meats
🧂 regular	6-inch sub	3½ carbohydrates + 2 medium-fat meats + 1 fat
Taco, hard or soft shell (meat and cheese)	1 small	1 carbohydrate + 1 medium-fat meat + 1½ fats

(continued)

KEY

☺ = More than 3 grams of dietary fiber per serving.

⚠ = Extra fat, or prepared with added fat.

🧂 = 600 milligrams or more of sodium per serving (for fast-food main dishes/meals).

TABLE G-11 Fast Foods (*continued*)

Salads		
Food	**Serving Size**	**Count as**
Salad, main dish (grilled chicken type, no dressing or croutons)		1 carbohydrate + 4 lean meats
Salad, side, no dressing or cheese	Small (about 5 oz)	1 vegetable

Sides/Appetizers		
Food	**Serving Size**	**Count as**
French fries, restaurant style	small	3 carbohydrates + 3 fats
	medium	4 carbohydrates + 4 fats
	large	5 carbohydrates + 6 fats
Nachos with cheese	small (about 4½ oz)	2½ carbohydrates + 4 fats
Onion rings	1 serving (about 3 oz)	2½ carbohydrates + 3 fats

Desserts		
Food	**Serving Size**	**Count as**
Milkshake, any flavor	12 oz	6 carbohydrates + 2 fats
Soft-serve ice cream cone	1 small	2½ carbohydrates + 1 fat

KEY

= More than 3 grams of dietary fi ber per serving.

= Extra fat, or prepared with added fat.

= 600 milligrams or more of sodium per serving (for fast-food main dishes/meals).

TABLE G-12 Alcohol

1 alcohol equivalent = variable grams carbohydrate, 0 grams protein, 0 grams fat, and 100 kcalories.

NOTE: In general, one alcohol choice (½ ounce absolute alcohol) has about 100 kcalories. For those who choose to drink alcohol, guidelines suggest limiting alcohol intake to 1 drink or less per day for women, and 2 drinks or less per day for men. To reduce your risk of low blood glucose (hypoglycemia), especially if you take insulin or a diabetes pill that increases insulin, always drink alcohol with food. While alcohol, by itself, does not directly affect blood glucose, be aware of the carbohydrate (for example, in mixed drinks, beer, and wine) that may raise your blood glucose.

Alcoholic Beverage	Serving Size	Count as
Beer		
light (4.2%)	12 fl oz	1 alcohol equivalent + ½ carbohydrate
regular (4.9%)	12 fl oz	1 alcohol equivalent + 1 carbohydrate
Distilled spirits: vodka, rum, gin, whiskey, 80 or 86 proof	1½ fl oz	1 alcohol equivalent
Liqueur, coffee (53 proof)	1 fl oz	1 alcohol equivalent + 1 carbohydrate
Sake	1 fl oz	½ alcohol equivalent
Wine		
dessert (sherry)	3½ fl oz	1 alcohol equivalent + 1 carbohydrate
dry, red or white (10%)	5 fl oz	1 alcohol equivalent

Appendix H Table of Food Composition

This edition of the table of food composition includes a wide variety of foods. It is updated with each edition to reflect current nutrient data for foods, to remove outdated foods, and to add foods that are new to the marketplace.* The nutrient database for this appendix is compiled from a variety of sources, including the USDA Nutrient Database and manufacturers' data. The USDA database provides data for a wider variety of foods and nutrients than other sources. Because laboratory analysis for each nutrient can be quite costly, manufacturers tend to provide data only for those nutrients mandated on food labels. Consequently, data for their foods are often incomplete; any missing information on this table is designated as a dash. Keep in mind that a dash means only that the information is unknown and should not be interpreted as a zero. A zero means that the nutrient is not present in the food.

Whenever using nutrient data, remember that many factors influence the nutrient contents of foods. These factors include the mineral content of the soil, the diet fed to the animal or the fertilizer used on the plant, the season of harvest, the method of processing, the length and method of storage, the method of cooking, the method of analysis, and the moisture content of the sample analyzed. With so many influencing factors, users should view nutrient data as a close approximation of the actual amount.

For updates, corrections, and a list of more than 22,000 foods and codes found in the diet analysis software that accompanies this text, visit **www.cengagebrain.com/shop/ISBN0538495081**.

- ***Fats*** Total fats, as well as the breakdown of total fats to saturated, monounsaturated, polyunsaturated, and *trans* fats, are listed in the table. The fatty acids seldom add up to the total in part due to rounding but also because values may include some non-fatty acids, such as glycerol, phosphate, or sterols.
- ***Trans Fats*** *Trans* fat data have been listed in the table. Because food manufacturers have been required to report only *trans* fats on food labels since January 2006, much of the data is incomplete. Missing *trans* fat data are designated with a dash. As additional *trans* fat data become available, the table will be updated.
- ***Vitamin A and Vitamin E*** In keeping with the 2001 RDA for vitamin A, this appendix presents data for vitamin A in micrograms (µg) RAE. Similarly because the 2000 RDA for vitamin E is based only on the alpha-tocopherol form of vitamin E, this appendix reports vitamin E data in milligrams (mg) alpha-tocopherol, listed on the table as Vit E (mg α).
- ***Bioavailability*** Keep in mind that the availability of nutrients from foods depends not only on the quantity provided by a food, but also on the amount absorbed and used by the body—the bioavailability. The bioavailability of folate from fortified foods, for example, is greater than from naturally occurring sources. Similarly, the body can make niacin from the amino acid tryptophan, but niacin values in this table (and most databases) report preformed niacin only. Chapter 10 provides conversion factors and additional details.

*This food composition table has been prepared by Cengage Learning. The nutritional data are supplied by Axxya Systems.

- *Using the Table* The foods and beverages in this table are organized into several categories, which are listed at the head of each right-hand page. Page numbers are provided, and each group is color-coded to make it easier to find individual foods.
- *Caffeine Sources* Caffeine occurs in several plants, including the familiar coffee bean, the tea leaf, and the cocoa bean from which chocolate is made. Most human societies use caffeine regularly, most often in beverages, for its stimulant effect and flavor. Caffeine contents of beverages vary depending on the plants they are made from, the climates and soils where the plants are grown, the grind or cut size, the method and duration of brewing, and the amounts served. The accompanying table shows that, in general, a cup of coffee contains the most caffeine; a cup of tea, less than half as much; and cocoa or chocolate, less still. As for cola beverages, they are made from kola nuts, which contain caffeine, but most of their caffeine is added, using the purified compound obtained from decaffeinated coffee beans. The FDA lists caffeine as a multipurpose GRAS substance ♦ that may be added to foods and beverages. Drug manufacturers also use caffeine in many products.

♦ A GRAS substance is one that is "generally recognized as safe."

TABLE Caffeine Content of Selected Beverages, Foods, and Medications

Beverages and Foods	Serving Size	Average (mg)
Coffee		
Brewed	8 oz	95
Decaffeinated	8 oz	2
Instant	8 oz	64
Tea		
Brewed, green	8 oz	30
Brewed, herbal	8 oz	0
Brewed, leaf or bag	8 oz	47
Instant	8 oz	26
Lipton Brisk iced tea	12 oz	7
Nestea Cool iced tea	12 oz	12
Snapple iced tea (all flavors)	16 oz	42
Soft Drinks		
A&W Creme Soda	12 oz	29
Barq's Root Beer	12 oz	18
Coca-Cola	12 oz	30
Dr. Pepper, Mr. Pibb, Sunkist Orange	12 oz	36
A&W Root Beer, club soda, Fresca, ginger ale, 7-Up, Sierra Mist, Sprite, Squirt, tonic water, caffeine-free soft drinks	12 oz	0
Mello Yello	12 oz	51
Mountain Dew	12 oz	45
Pepsi	12 oz	32

TABLE Caffeine Content of Selected Beverages, Foods, and Medications (*continued*)

Beverages and Foods	Serving Size	Average (mg)
Energy Drinks		
Amp	8.4 oz	70
Aqua Blast	0.5 L	90
Aqua Java	0.5 L	55
E Maxx	8.4 oz	74
Java Water	0.5 L	125
KMX	8.4 oz	33
Krank	0.5 L	100
Red Bull	8.3 oz	67
Red Devil	8.4 oz	42
Sobe Adrenaline Rush	8.3 oz	77
Sobe No Fear	16 oz	141
Water Joe	0.5 L	65
Other Beverages		
Chocolate milk or hot cocoa	8 oz	5
Starbucks Frappuccino Mocha	9.5 oz	72
Starbucks Frappuccino Vanilla	9.5 oz	64
Yoohoo chocolate drink	9 oz	3
Candies		
Baker's chocolate	1 oz	26
Dark chocolate covered coffee beans	1 oz	235
Dark chocolate, semisweet	1 oz	18
Milk chocolate	1 oz	6
Milk chocolate covered coffee beans	1 oz	224
White chocolate	1 oz	0

Beverages and Foods	Serving Size	Average (mg)
Foods		
Frozen yogurt, Ben & Jerry's coffee fudge	1 cup	85
Frozen yogurt, Häagen-Dazs coffee	1 cup	40
Ice cream, Starbucks coffee	1 cup	50
Ice cream, Starbucks Frappuccino bar	1 bar	15
Yogurt, Dannon coffee flavored	1 cup	45

Drugs[a]	Serving Size	Average (mg)
Cold Remedies		
Coryban-D, Dristan	1 tablet	30
Diuretics		
Aqua-Ban	1 tablet	100
Pre-Mens Forte	1 tablet	100
Pain Relievers		
Anacin, BC Fast Pain Reliever	1 tablet	32
Excedrin, Midol, Midol Max Strength	1 tablet	65
Stimulants		
Awake, NoDoz	1 tablet	100
Awake Maximum Strength, Caffedrine, NoDoz Maximum Strength, Stay Awake, Vivarin	1 tablet	200
Weight-Control Aids		
Dexatrim	1 tablet	200

[a]A pharmacologically active dose of caffiene is defined as 200 milligrams.

NOTE: The FDA suggests a maximum of 65 milligrams per 12-ounce cola beverage but does not regulate the caffeine contents of other beverages. Because products change, contact the manufacturer for an update on products you use regularly.

SOURCE: Adapted from USDA database Release 18 (www.nal.usda.gov/fnic/foodcomp/Data/), Caffeine content of foods and drugs, Center for Science and the Public Interest (www.cspinet.org/new/cafchart.htm), and R. R. McCusker, B. A. Goldberger, and E. J. Cone, Caffeine content of energy drinks, carbonated sodas, and other beverages, *Journal of Analytical Toxicology* 30 (2006): 112–114.

TABLE H-1 Table of Food Composition (Computer code is for Cengage Diet Analysis program) (For purposes of calculations, use "0" for t, <1, <0.1, 0.01)

DA+ Code	Food Description	Quantity	Measure	Wt (g)	H_2O (g)	Ener (kcal)	Prot (g)	Carb (g)	Fiber (g)	Fat (g)	Fat Breakdown (g) Sat	Mono	Poly	*Trans*
Breads, Baked Goods, Cakes, Cookies, Crackers, Chips, Pies														
	Bagels													
8534	Cinnamon and raisin	1	item(s)	71	22.7	194	7.0	39.2	1.6	1.2	0.2	0.1	0.5	—
14395	Multi-grain	1	item(s)	61	—	170	6.0	35.0	1.0	1.5	0.5	0.1	0.4	—
8538	Oat bran	1	item(s)	71	23.4	181	7.6	37.8	2.6	0.9	0.1	0.2	0.3	—
4910	Plain, enriched	1	item(s)	71	25.8	182	7.1	35.9	1.6	1.2	0.3	0.4	0.5	0
4911	Plain, enriched, toasted	1	item(s)	66	18.7	190	7.4	37.7	1.7	1.1	0.2	0.3	0.6	0
	Biscuits													
25008	Biscuits	1	item(s)	41	15.8	121	2.6	16.4	0.5	4.9	1.4	1.4	1.8	—
16729	Scone	1	item(s)	42	11.5	148	3.8	19.1	0.6	6.2	2.0	2.5	1.3	—
25166	Wheat biscuits	1	item(s)	55	21.0	162	3.6	21.9	1.4	6.7	1.9	1.9	2.5	—
	Bread													
325	Boston brown, canned	1	slice(s)	45	21.2	88	2.3	19.5	2.1	0.7	0.1	0.1	0.3	—
8716	Bread sticks, plain	4	item(s)	24	1.5	99	2.9	16.4	0.7	2.3	0.3	0.9	0.9	—
25176	Cornbread	1	piece(s)	55	25.9	141	4.7	18.3	0.9	5.4	2.1	1.4	1.5	0
327	Cracked wheat	1	slice(s)	25	9.0	65	2.2	12.4	1.4	1.0	0.2	0.5	0.2	—
9079	Croutons, plain	¼	cup(s)	8	0.4	31	0.9	5.5	0.4	0.5	0.1	0.2	0.1	—
8582	Egg	1	slice(s)	40	13.9	113	3.8	19.1	0.9	2.4	0.6	0.9	0.4	—
8585	Egg, toasted	1	slice(s)	37	10.5	117	3.9	19.5	0.9	2.4	0.6	1.1	0.4	—
329	French	1	slice(s)	32	8.9	92	3.8	18.1	0.8	0.6	0.2	0.1	0.3	—
8591	French, toasted	1	slice(s)	23	4.7	73	3.0	14.2	0.7	0.5	0.1	0.1	0.2	—
42096	Indian fry, made with lard (Navajo)	3	ounce(s)	85	26.9	281	5.7	41.0	—	10.4	3.9	3.8	0.9	—
332	Italian	1	slice(s)	30	10.7	81	2.6	15.0	0.8	1.1	0.3	0.2	0.4	—
1393	Mixed grain	1	slice(s)	26	9.6	69	3.5	11.3	1.9	1.1	0.2	0.2	0.5	0
8604	Mixed grain, toasted	1	slice(s)	24	7.6	69	3.5	11.3	1.9	1.1	0.2	0.2	0.5	0
8605	Oat bran	1	slice(s)	30	13.2	71	3.1	11.9	1.4	1.3	0.2	0.5	0.5	—
8608	Oat bran, toasted	1	slice(s)	27	10.4	70	3.1	11.8	1.3	1.3	0.2	0.5	0.5	—
8609	Oatmeal	1	slice(s)	27	9.9	73	2.3	13.1	1.1	1.2	0.2	0.4	0.5	—
8613	Oatmeal, toasted	1	slice(s)	25	7.8	73	2.3	13.2	1.1	1.2	0.2	0.4	0.5	—
1409	Pita	1	item(s)	60	19.3	165	5.5	33.4	1.3	0.7	0.1	0.1	0.3	—
7905	Pita, whole wheat	1	item(s)	64	19.6	170	6.3	35.2	4.7	1.7	0.3	0.2	0.7	—
338	Pumpernickel	1	slice(s)	32	12.1	80	2.8	15.2	2.1	1.0	0.1	0.3	0.4	—
334	Raisin, enriched	1	slice(s)	26	8.7	71	2.1	13.6	1.1	1.1	0.3	0.6	0.2	—
8625	Raisin, toasted	1	slice(s)	24	6.7	71	2.1	13.7	1.1	1.2	0.3	0.6	0.2	—
10168	Rice, white, gluten free, wheat free	1	slice(s)	38	—	130	1.0	18.0	0.5	6.0	0	—	—	0
8653	Rye	1	slice(s)	32	11.9	83	2.7	15.5	1.9	1.1	0.2	0.4	0.3	—
8654	Rye, toasted	1	slice(s)	29	9.0	82	2.7	15.4	1.9	1.0	0.2	0.4	0.3	—
336	Rye, light	1	slice(s)	25	9.3	65	2.0	12.0	1.6	1.0	0.2	0.3	0.3	—
8588	Sourdough	1	slice(s)	25	7.0	72	2.9	14.1	0.6	0.5	0.1	0.1	0.2	—
8592	Sourdough, toasted	1	slice(s)	23	4.7	73	3.0	14.2	0.7	0.5	0.1	0.1	0.2	—
491	Submarine or hoagie roll	1	item(s)	135	40.6	400	11.0	72.0	3.8	8.0	1.8	3.0	2.2	—
8596	Vienna, toasted	1	slice(s)	23	4.7	73	3.0	14.2	0.7	0.5	0.1	0.1	0.2	—
8670	Wheat	1	slice(s)	25	8.9	67	2.7	11.9	0.9	0.9	0.2	0.2	0.4	—
8671	Wheat, toasted	1	slice(s)	23	5.6	72	3.0	12.8	1.1	1.0	0.2	0.2	0.4	—
340	White	1	slice(s)	25	9.1	67	1.9	12.7	0.6	0.8	0.2	0.2	0.3	—
1395	Whole wheat	1	slice(s)	46	15.0	128	3.9	23.6	2.8	2.5	0.4	0.5	1.4	—
	Cakes													
386	Angel food, prepared from mix	1	piece(s)	50	16.5	129	3.1	29.4	0.1	0.2	0	0	0.1	—
8772	Butter pound, ready to eat, commercially prepared	1	slice(s)	75	18.5	291	4.1	36.6	0.4	14.9	8.7	4.4	0.8	—
28517	Carrot	1	slice(s)	131	56.6	339	4.8	56.5	1.9	11.1	1.0	5.7	3.8	—
4931	Chocolate with chocolate icing, commercially prepared	1	slice(s)	64	14.7	235	2.6	34.9	1.8	10.5	3.1	5.6	1.2	—
8756	Chocolate, prepared from mix	1	slice(s)	95	23.2	352	5.0	50.7	1.5	14.3	5.2	5.7	2.6	—
393	Devil's food cupcake with chocolate frosting	1	item(s)	35	8.4	120	2.0	20.0	0.7	4.0	1.8	1.6	0.6	—
8757	Fruitcake, ready to eat, commercially prepared	1	piece(s)	43	10.9	139	1.2	26.5	1.6	3.9	0.5	1.8	1.4	—
1397	Pineapple upside down, prepared from mix	1	slice(s)	115	37.1	367	4.0	58.1	0.9	13.9	3.4	6.0	3.8	—
411	Sponge, prepared from mix	1	slice(s)	63	18.5	187	4.6	36.4	0.3	2.7	0.8	1.0	0.4	—
8817	White with coconut frosting, prepared from mix	1	slice(s)	112	23.2	399	4.9	70.8	1.1	11.5	4.4	4.1	2.4	—
8819	Yellow with chocolate frosting, ready to eat, commercially prepared	1	slice(s)	64	14.0	243	2.4	35.5	1.2	11.1	3.0	6.1	1.4	—

GE KEY: H-4 = Breads/Baked Goods H-10 = Cereal/Rice/Pasta H-14 = Fruit H-18 = Vegetables/Legumes H-28 = Nuts/Seeds H-30 = Vegetarian H-32 = Dairy H-40 = Eggs H-40 = Seafood H-42 = Meats H-46 = Poultry
46 = Processed Meats H-48 = Beverages H-52 = Fats/Oils H-54 = Sweets H-56 = Spices/Condiments/Sauces H-60 = Mixed Foods/Soups/Sandwiches H-64 = Fast Food H-84 = Convenience H-86 = Baby Foods

Chol (mg)	Calc (mg)	Iron (mg)	Magn (mg)	Pota (mg)	Sodi (mg)	Zinc (mg)	Vit A (µg)	Thia (mg)	Vit E (mg α)	Ribo (mg)	Niac (mg)	Vit B_6 (mg)	Fola (µg)	Vit C (mg)	Vit B_{12} (µg)	Sele (µg)
0	13	2.69	19.9	105.1	228.6	0.80	14.9	0.27	0.22	0.19	2.18	0.04	78.8	0.5	0	22.0
0	60	1.08	—	—	310.0	—	0	—	—	—	—	—	—	0	—	—
0	9	2.18	22.0	81.7	360.0	0.63	0.7	0.23	0.23	0.24	2.10	0.03	69.6	0.1	0	24.3
0	63	4.29	15.6	53.3	318.1	1.34	0	0.42	0.07	0.18	2.82	0.04	103.0	0.7	0	16.2
0	65	2.97	15.8	56.1	316.8	0.85	0	0.39	0.07	0.17	2.88	0.04	86.5	0	0	16.6
0	38	0.94	6.0	47.4	206.0	0.20	—	0.16	0.01	0.12	1.20	0.01	31.7	0.1	0.1	7.1
49	79	1.35	7.1	48.7	277.2	0.29	64.7	0.14	0.42	0.15	1.19	0.02	32.3	0	0.1	10.9
0	57	1.21	16.1	81.0	321.1	0.42	—	0.19	0.01	0.14	1.65	0.03	35.3	0.1	0.1	0
0	32	0.94	28.4	143.1	284.0	0.22	11.3	0.01	0.14	0.05	0.50	0.03	5.0	0	0	9.9
0	5	1.02	7.7	29.8	157.7	0.21	0	0.14	0.24	0.13	1.26	0.01	38.9	0	0	9.0
21	94	0.91	10.5	71.5	209.8	0.48	—	0.14	0.32	0.15	1.03	0.04	34.6	1.7	0.2	6.2
0	11	0.70	13.0	44.3	134.5	0.31	0	0.09	—	0.06	0.92	0.08	15.3	0	0	6.3
0	6	0.30	2.3	9.3	52.4	0.06	0	0.04	—	0.02	0.40	0.00	9.9	0	0	2.8
20	37	1.21	7.6	46.0	196.8	0.31	25.2	0.17	0.10	0.17	1.93	0.02	42.0	0	0	12.0
21	38	1.23	7.8	46.6	199.8	0.31	25.5	0.14	0.10	0.16	1.77	0.02	36.3	0	0	12.2
0	14	1.16	9.0	41.0	208.0	0.29	0	0.13	0.05	0.09	1.52	0.03	47.4	0.1	0	8.7
0	11	0.89	7.1	32.2	165.6	0.24	0	0.10	0.04	0.09	1.24	0.02	32.2	0	0	6.8
6	48	3.43	15.3	65.5	279.8	0.29	0	0.36	0.00	0.18	3.91	0.03	103.8	—	0	15.8
0	23	0.88	8.1	33.0	175.2	0.25	0	0.14	0.08	0.08	1.31	0.01	57.3	0	0	8.2
0	27	0.65	20.3	59.8	109.2	0.44	0	0.07	0.09	0.03	1.05	0.06	19.5	0	0	8.6
0	27	0.65	20.4	60.0	109.7	0.44	0	0.06	0.10	0.03	1.05	0.07	16.8	0	0	8.6
0	20	0.93	10.5	44.1	122.1	0.26	0.6	0.15	0.13	0.10	1.44	0.02	24.3	0	0	9.0
0	19	0.92	9.2	33.2	121.0	0.28	0.5	0.12	0.13	0.09	1.29	0.01	18.6	0	0	8.9
0	18	0.72	10.0	38.3	161.7	0.27	1.4	0.10	0.13	0.06	0.84	0.01	16.7	0	0	6.6
0	18	0.74	10.3	38.5	162.8	0.28	1.3	0.09	0.13	0.06	0.77	0.02	13.3	0.1	0	6.7
0	52	1.57	15.6	72.0	321.6	0.50	0	0.35	0.18	0.19	2.77	0.02	64.2	0	0	16.3
0	10	1.95	44.2	108.8	340.5	0.97	0	0.21	0.39	0.05	1.81	0.17	22.4	0	0	28.2
0	22	0.91	17.3	66.6	214.7	0.47	0	0.10	0.13	0.09	0.98	0.04	29.8	0	0	7.8
0	17	0.75	6.8	59.0	101.4	0.18	0	0.08	0.07	0.10	0.90	0.01	27.6	0	0	5.2
0	17	0.76	6.7	59.0	101.8	0.19	0	0.07	0.07	0.09	0.81	0.02	23.5	0.1	0	5.2
0	100	1.08	—	—	140	—	—	0.15	—	0.10	1.20	—	32.0	0	—	—
0	23	0.90	12.8	53.1	211.2	0.36	0	0.13	0.10	0.10	1.21	0.02	35.2	0.1	0	9.9
0	23	0.89	12.5	53.1	210.3	0.36	0	0.11	0.10	0.09	1.09	0.02	29.9	0.1	0	9.9
0	20	0.70	3.9	51.0	175.0	0.18	0	0.10	—	0.08	0.80	0.01	5.3	0	0	8.0
0	11	0.91	7.0	32.0	162.5	0.23	0	0.11	0.05	0.07	1.19	0.03	37.0	0.1	0	6.8
0	11	0.89	7.1	32.2	165.6	0.24	0	0.10	0.04	0.09	1.24	0.02	32.2	0	0	6.8
0	100	3.80	—	128.0	683.0	—	0	0.54	—	0.33	4.50	0.04	—	0	—	42.0
0	11	0.89	7.1	32.2	165.6	0.24	0	0.10	0.04	0.09	1.24	0.02	32.2	0	0	6.8
0	36	0.87	12.0	46.0	130.3	0.30	0	0.09	0.05	0.08	1.30	0.03	21.3	0.1	0	7.2
0	38	0.94	13.6	51.3	140.5	0.34	0	0.10	0.06	0.09	1.44	0.04	19.8	0	0	7.7
0	38	0.94	5.8	25.0	170.3	0.19	0	0.11	0.06	0.08	1.10	0.02	27.8	0	0	4.3
0	15	1.42	37.3	144.4	159.2	0.69	0	0.13	0.35	0.10	1.83	0.09	29.9	0	0	17.8
0	42	0.11	4.0	67.5	254.5	0.06	0	0.04	0.01	0.10	0.08	0.00	9.5	0	0	7.7
166	26	1.03	8.3	89.3	298.5	0.34	111.8	0.10	—	0.17	0.98	0.03	30.8	0	0.2	6.6
0	65	2.18	23.0	279.6	367.7	0.44	—	0.25	0.01	0.19	1.73	0.10	43.4	4.6	0	14.7
27	28	1.40	21.8	128.0	213.8	0.44	16.6	0.01	0.62	0.08	0.36	0.02	10.9	0.1	0.1	2.1
55	57	1.53	30.4	133.0	299.3	0.65	38.0	0.13	—	0.20	1.08	0.03	25.7	0.2	0.2	11.3
19	21	0.70	—	46.0	92.0	—	—	0.04	—	0.05	0.30	—	2.1	0	—	2.0
2	14	0.89	6.9	65.8	116.1	0.11	3.0	0.02	0.38	0.04	0.34	0.02	8.6	0.2	0	0.9
25	138	1.70	15.0	128.8	366.9	0.35	71.3	0.17	—	0.17	1.36	0.03	29.9	1.4	0.1	10.8
107	26	0.99	5.7	88.8	143.6	0.37	48.5	0.10	—	0.19	0.75	0.03	24.6	0	0.2	11.7
1	101	1.29	13.4	110.9	318.1	0.37	13.4	0.14	0.13	0.21	1.19	0.03	34.7	0.1	0.1	12.0
35	24	1.33	19.2	113.9	215.7	0.39	21.1	0.07	—	0.10	0.79	0.02	14.1	0	0.1	2.2

TABLE H-1 **Table of Food Composition (*continued*)** (Computer code is for Cengage Diet Analysis program) (For purposes of calculations, use "0" for t, <1, <0.1, 0.01)

DA+ Code	Food Description	Quantity	Measure	Wt (g)	H_2O (g)	Ener (kcal)	Prot (g)	Carb (g)	Fiber (g)	Fat (g)	Fat Breakdown (g) Sat	Mono	Poly	*Trans*
	Breads, Baked Goods, Cakes, Cookies, Crackers, Chips, Pies—*continued*													
8822	Yellow with vanilla frosting, ready to eat, commercially prepared	1	slice(s)	64	14.1	239	2.2	37.6	0.2	9.3	1.5	3.9	3.3	—
	Snack cakes													
8791	Chocolate snack cake, creme filled, with frosting	1	item(s)	50	9.3	200	1.8	30.2	1.6	8.0	2.4	4.3	0.9	—
25010	Cinnamon coffee cake	1	piece(s)	72	22.6	231	3.6	35.8	0.7	8.3	2.2	2.6	3.0	—
16777	Funnel cake	1	item(s)	90	37.6	276	7.3	29.1	0.9	14.4	2.7	4.7	6.1	—
8794	Sponge snack cake, creme filled	1	item(s)	43	8.6	155	1.3	27.2	0.2	4.8	1.1	1.7	1.4	—
	Snacks, chips, pretzels													
29428	Bagel chips, plain	3	item(s)	29	—	130	3.0	19.0	1.0	4.5	0.5	—	—	—
29429	Bagel chips, toasted onion	3	item(s)	29	—	130	4.0	20.0	1.0	4.5	0.5	—	—	—
38192	Chex traditional snack mix	1	cup(s)	45	—	197	3.0	33.3	1.5	6.1	0.8	—	—	—
654	Potato chips, salted	1	ounce(s)	28	0.6	155	1.9	14.1	1.2	10.6	3.1	2.8	3.5	—
8816	Potato chips, unsalted	1	ounce(s)	28	0.5	152	2.0	15.0	1.4	9.8	3.1	2.8	3.5	—
5096	Pretzels, plain, hard, twists	5	item(s)	30	1.0	114	2.7	23.8	1.0	1.1	0.2	0.4	0.4	—
4632	Pretzels, whole wheat	1	ounce(s)	28	1.1	103	3.1	23.0	2.2	0.7	0.2	0.3	0.2	—
4641	Tortilla chips, plain	6	item(s)	11	0.2	53	0.8	7.1	0.6	2.5	0.3	0.8	0.5	0.3
	Cookies													
8859	Animal crackers	12	item(s)	30	1.2	134	2.1	22.2	0.3	4.1	1.0	2.3	0.6	—
8876	Brownie, prepared from mix	1	item(s)	24	3.0	112	1.5	12.0	0.5	7.0	1.8	2.6	2.3	—
25207	Chocolate chip cookies	1	item(s)	30	3.7	140	2.0	16.2	0.6	7.9	2.1	3.3	2.1	—
8915	Chocolate sandwich cookie with extra creme filling	1	item(s)	13	0.2	65	0.6	8.9	0.4	3.2	0.7	2.1	0.3	1.1
14145	Fig Newtons cookies	1	item(s)	16	—	55	0.5	11.0	0.5	1.3	0	—	—	0
8920	Fortune cookie	1	item(s)	8	0.6	30	0.3	6.7	0.1	0.2	0.1	0.1	0	—
25208	Oatmeal cookies	1	item(s)	69	12.3	234	5.7	45.1	3.1	4.2	0.7	1.3	1.8	—
25213	Peanut butter cookies	1	item(s)	35	4.1	163	4.2	16.9	0.9	9.2	1.7	4.7	2.3	—
33095	Sugar cookies	1	item(s)	16	4.1	61	1.1	7.4	0.1	3.0	0.6	1.3	0.9	—
9002	Vanilla sandwich cookie with creme filling	1	item(s)	10	0.2	48	0.5	7.2	0.2	2.0	0.3	0.8	0.8	—
	Crackers													
9012	Cheese cracker sandwich with peanut butter	4	item(s)	28	0.9	139	3.5	15.9	1.0	7.0	1.2	3.6	1.4	—
9008	Cheese crackers (mini)	30	item(s)	30	0.9	151	3.0	17.5	0.7	7.6	2.8	3.6	0.7	—
33362	Cheese crackers, low sodium	1	serving(s)	30	0.9	151	3.0	17.5	0.7	7.6	2.9	3.6	0.7	—
8928	Honey graham crackers	4	item(s)	28	1.2	118	1.9	21.5	0.8	2.8	0.4	1.1	1.1	—
9016	Matzo crackers, plain	1	item(s)	28	1.2	112	2.8	23.8	0.9	0.4	0.1	0	0.2	—
9024	Melba toast	3	item(s)	15	0.8	59	1.8	11.5	0.9	0.5	0.1	0.1	0.2	—
9028	Melba toast, rye	3	item(s)	15	0.7	58	1.7	11.6	1.2	0.5	0.1	0.1	0.2	—
14189	Ritz crackers	5	item(s)	16	0.5	80	1.0	10.0	0	4.0	1.0	—	—	0
9014	Rye crispbread crackers	1	item(s)	10	0.6	37	0.8	8.2	1.7	0.1	0	0	0.1	—
9040	Rye wafer	1	item(s)	11	0.6	37	1.1	8.8	2.5	0.1	0	0	0	—
432	Saltine crackers	5	item(s)	15	0.8	64	1.4	10.6	0.5	1.7	0.2	1.1	0.2	0.5
9046	Saltine crackers, low salt	5	item(s)	15	0.6	65	1.4	10.7	0.5	1.8	0.4	1.0	0.3	—
9052	Snack cracker sandwich with cheese filling	4	item(s)	28	1.1	134	2.6	17.3	0.5	5.9	1.7	3.2	0.7	—
9054	Snack cracker sandwich with peanut butter filling	4	item(s)	28	0.8	138	3.2	16.3	0.6	6.9	1.4	3.9	1.3	—
9048	Snack crackers, round	10	item(s)	30	1.1	151	2.2	18.3	0.5	7.6	1.1	3.2	2.9	—
9050	Snack crackers, round, low salt	10	item(s)	30	1.1	151	2.2	18.3	0.5	7.6	1.1	3.2	2.9	—
9044	Soda crackers	5	item(s)	15	0.8	64	1.4	10.6	0.5	1.7	0.2	1.1	0.2	0.5
9059	Wheat cracker sandwich with cheese filling	4	item(s)	28	0.9	139	2.7	16.3	0.9	7.0	1.2	2.9	2.6	—
9061	Wheat cracker sandwich with peanut butter filling	4	item(s)	28	1.0	139	3.8	15.1	1.2	7.5	1.3	3.3	2.5	—
9055	Wheat crackers	10	item(s)	30	0.9	142	2.6	19.5	1.4	6.2	1.6	3.4	0.8	—
9057	Wheat crackers, low salt	10	item(s)	30	0.9	142	2.6	19.5	1.4	6.2	1.6	3.4	0.8	—
9022	Whole wheat crackers	7	item(s)	28	0.8	124	2.5	19.2	2.9	4.8	1.0	1.6	1.8	—
	Pastry													
16754	Apple fritter	1	item(s)	17	6.4	61	1.0	5.5	0.2	3.9	0.9	1.7	1.1	—
41565	Cinnamon rolls with icing, refrigerated dough	1	serving(s)	44	12.3	145	2.0	23.0	0.5	5.0	1.5	—	—	2.0
4945	Croissant, butter	1	item(s)	57	13.2	231	4.7	26.1	1.5	12.0	6.6	3.1	0.6	—
9096	Danish, nut	1	item(s)	65	13.3	280	4.6	29.7	1.3	16.4	3.8	8.9	2.8	—
9115	Doughnut with creme filling	1	item(s)	85	32.5	307	5.4	25.5	0.7	20.8	4.6	10.3	2.6	—

E KEY: H-4 = Breads/Baked Goods H-10 = Cereal/Rice/Pasta H-14 = Fruit H-18 = Vegetables/Legumes H-28 = Nuts/Seeds H-30 = Vegetarian H-32 = Dairy H-40 = Eggs H-40 = Seafood H-42 = Meats H-46 = Poultry
6 = Processed Meats H-48 = Beverages H-52 = Fats/Oils H-54 = Sweets H-56 = Spices/Condiments/Sauces H-60 = Mixed Foods/Soups/Sandwiches H-64 = Fast Food H-84 = Convenience H-86 = Baby Foods

Chol (mg)	Calc (mg)	Iron (mg)	Magn (mg)	Pota (mg)	Sodi (mg)	Zinc (mg)	Vit A (μg)	Thia (mg)	Vit E (mg α)	Ribo (mg)	Niac (mg)	Vit B_6 (mg)	Fola (μg)	Vit C (mg)	Vit B_{12} (μg)	Sele (μg)
35	40	0.68	3.8	33.9	220.2	0.16	12.2	0.06	—	0.04	0.32	0.01	17.3	0	0.1	3.5
0	58	1.80	18.0	88.0	194.5	0.52	0.5	0.01	0.54	0.03	0.46	0.07	13.0	1.0	0	1.7
26	55	1.36	9.9	91.9	277.6	0.30	—	0.17	0.23	0.16	1.29	0.02	36.1	0.3	0.1	9.6
62	126	1.90	16.2	152.1	269.1	0.65	49.5	0.23	1.54	0.32	1.86	0.04	50.4	0	0.3	17.7
7	19	0.54	3.4	37.0	155.1	0.12	2.1	0.06	0.50	0.05	0.52	0.01	17.0	0	0	1.3
0	0	0.72	—	45.0	70.0	—	0	—	—	—	—	—	—	0	0	—
0	0	0.72	—	50.0	300.0	—	0	—	—	—	—	—	—	0	0	—
0	0	0.55	—	75.8	621.2	—	0	0.09	—	0.05	1.21	—	12.1	0	—	—
0	7	0.45	19.8	465.5	148.8	0.67	0	0.01	1.91	0.06	1.18	0.20	21.3	5.3	0	2.3
0	7	0.46	19.0	361.5	2.3	0.30	0	0.04	2.58	0.05	1.08	0.18	12.8	8.8	0	2.3
0	11	1.29	10.5	43.8	514.5	0.25	0	0.13	0.10	0.18	1.57	0.03	51.3	0	0	1.7
0	8	0.76	8.5	121.9	57.6	0.17	0	0.12	—	0.08	1.85	0.07	15.3	0.3	0	—
0	19	0.25	15.8	23.2	45.5	0.26	0	0.00	0.46	0.01	0.13	0.02	2.2	0	0	0.7
0	13	0.82	5.4	30.0	117.9	0.19	0	0.10	0.03	0.09	1.04	0.01	30.9	0	0	2.1
18	14	0.44	12.7	42.2	82.3	0.23	42.2	0.03	—	0.05	0.24	0.02	7.0	0.1	0	2.8
13	11	0.69	12.4	62.1	108.8	0.24	—	0.08	0.54	0.06	0.87	0.01	17.8	0	0	4.1
0	2	1.01	4.7	17.8	45.6	0.10	0	0.02	0.25	0.02	0.25	0.00	6.0	0	0	1.1
0	10	0.36	—	—	57.5	—	0	—	—	—	—	—	—	0	—	—
0	1	0.12	0.6	3.3	21.9	0.01	0.1	0.01	0.00	0.01	0.15	0.00	5.3	0	0	0.2
0	26	1.93	48.8	176.7	311.1	1.42	—	0.26	0.23	0.13	1.35	0.09	34.9	0.3	0	17.4
13	27	0.65	21.1	112.8	154.1	0.46	—	0.08	0.73	0.09	1.85	0.05	23.5	0.1	0.1	4.8
18	5	0.30	1.7	12.2	49.4	0.08	—	0.04	0.28	0.05	0.31	0.01	9.5	0	0	3.1
0	3	0.22	1.4	9.1	34.9	0.04	0	0.02	0.16	0.02	0.27	0.00	5.0	0	0	0.3
0	14	0.76	15.7	61.0	198.8	0.29	0.3	0.15	0.66	0.08	1.63	0.04	26.3	0	0.1	2.3
4	45	1.43	10.8	43.5	298.5	0.33	8.7	0.17	0.01	0.12	1.40	0.16	45.6	0	0.1	2.6
4	45	1.43	10.8	31.8	137.4	0.33	5.1	0.17	0.09	0.12	1.40	0.16	26.7	0	0.1	2.6
0	7	1.04	8.4	37.8	169.4	0.22	0	0.06	0.09	0.08	1.15	0.01	12.9	0	0	2.9
0	4	0.89	7.1	31.8	0.6	0.19	0	0.11	0.01	0.08	1.10	0.03	4.8	0	0	10.5
0	14	0.55	8.9	30.3	124.4	0.30	0	0.06	0.06	0.04	0.61	0.01	18.6	0	0	5.2
0	12	0.55	5.9	29.0	134.9	0.20	0	0.07	—	0.04	0.70	0.01	12.8	0	0	5.8
0	20	0.72	—	10.0	135.0	—	—	—	—	—	—	—	—	0	—	—
0	3	0.24	7.8	31.9	26.4	0.23	0	0.02	0.08	0.01	0.10	0.02	4.7	0	0	3.7
0	4	0.65	13.3	54.5	87.3	0.30	0	0.04	0.08	0.03	0.17	0.03	5.0	0	0	2.6
0	10	0.84	3.3	23.1	160.8	0.12	0	0.01	0.14	0.06	0.78	0.01	20.9	0	0	1.5
0	18	0.81	4.1	108.6	95.4	0.11	0	0.08	0.01	0.06	0.78	0.01	18.6	0	0	2.9
1	72	0.66	10.1	120.1	392.3	0.17	4.8	0.12	0.06	0.19	1.05	0.01	28.0	0	0	6.0
0	23	0.77	15.4	60.2	201.0	0.31	0.3	0.13	0.57	0.07	1.71	0.04	24.1	0	0	3.0
0	36	1.08	8.1	39.9	254.1	0.20	0	0.12	0.60	0.10	1.21	0.01	27.0	0	0	2.0
0	36	1.08	8.1	106.5	111.9	0.20	0	0.12	0.60	0.10	1.21	0.01	27.0	0	0	2.0
0	10	0.84	3.3	23.1	160.8	0.12	0	0.01	0.14	0.06	0.78	0.01	20.9	0	0	1.5
2	57	0.73	15.1	85.7	255.6	0.24	4.8	0.10	—	0.12	0.89	0.07	17.9	0.4	0	6.8
0	48	0.74	10.6	83.2	226.0	0.23	0	0.10	—	0.08	1.64	0.03	19.6	0	0	6.1
0	15	1.32	18.6	54.9	238.5	0.48	0	0.15	0.15	0.09	1.48	0.04	35.1	0	0	1.9
0	15	1.32	18.6	60.9	84.9	0.48	0	0.15	0.15	0.09	1.48	0.04	15.0	0	0	10.1
0	14	0.86	27.7	83.2	184.5	0.60	0	0.05	0.24	0.02	1.26	0.05	7.8	0	0	4.1
14	9	0.26	2.2	22.4	6.8	0.09	7.1	0.03	0.07	0.04	0.23	0.01	6.3	0.2	0.1	2.6
0	—	0.72	—	—	340.1	—	0	—	—	—	—	—	—	—	—	—
38	21	1.15	9.1	67.3	424.1	0.42	117.4	0.22	0.47	0.13	1.24	0.03	50.2	0.1	0.1	12.9
30	61	1.17	20.8	61.8	236.0	0.56	5.9	0.14	0.53	0.15	1.49	0.06	54.0	1.1	0.1	9.2
20	21	1.55	17.0	68.0	262.7	0.68	9.4	0.28	0.24	0.12	1.90	0.05	59.5	0	0.1	9.2

TABLE H-1 Table of Food Composition (*continued*) (Computer code is for Cengage Diet Analysis program) (For purposes of calculations, use "0" for t, <1, <0.1, 0.01)

DA+ Code	Food Description	Quantity	Measure	Wt (g)	H_2O (g)	Ener (kcal)	Prot (g)	Carb (g)	Fiber (g)	Fat (g)	Fat Breakdown (g) Sat	Mono	Poly	*Trans*
	Breads, Baked Goods, Cakes, Cookies, Crackers, Chips, Pies—*continued*													
9117	Doughnut with jelly filling	1	item(s)	85	30.3	289	5.0	33.2	0.8	15.9	4.1	8.7	2.0	—
4947	Doughnut, cake	1	item(s)	47	9.8	198	2.4	23.4	0.7	10.8	1.7	4.4	3.7	—
9105	Doughnut, cake, chocolate glazed	1	item(s)	42	6.8	175	1.9	24.1	0.9	8.4	2.2	4.7	1.0	—
437	Doughnut, glazed	1	item(s)	60	15.2	242	3.8	26.6	0.7	13.7	3.5	7.7	1.7	—
10617	Toaster pastry, brown sugar cinnamon	1	item(s)	50	5.3	210	3.0	35.0	1.0	6.0	1.0	4.0	1.0	—
30928	Toaster pastry, cream cheese	1	item(s)	54	—	200	3.0	23.0	0	11.0	4.5	—	—	1.5
	Muffins													
25015	Blueberry	1	item(s)	63	29.7	160	3.4	23.0	0.8	6.0	0.9	1.5	3.3	—
9189	Corn, ready to eat	1	item(s)	57	18.6	174	3.4	29.0	1.9	4.8	0.8	1.2	1.8	—
9121	English muffin, plain, enriched	1	item(s)	57	24.0	134	4.4	26.2	1.5	1.0	0.1	0.2	0.5	—
29582	English muffin, toasted	1	item(s)	50	18.6	128	4.2	25.0	1.5	1.0	0.1	0.2	0.5	—
9145	English muffin, wheat	1	item(s)	57	24.1	127	5.0	25.5	2.6	1.1	0.2	0.2	0.5	—
8894	Oat bran	1	item(s)	57	20.0	154	4.0	27.5	2.6	4.2	0.6	1.0	2.4	—
	Granola bars													
38161	Kudos milk chocolate granola bars w/fruit and nuts	1	item(s)	28	—	90	2.0	15.0	1.0	3.0	1.0	—	—	—
38196	Nature Valley banana nut crunchy granola bars	2	item(s)	42	—	190	4.0	28.0	2.0	7.0	1.0	—	—	—
38187	Nature Valley fruit 'n' nut trail mix bar	1	item(s)	35	—	140	3.0	25.0	2.0	4.0	0.5	—	—	—
1383	Plain, hard	1	item(s)	25	1.0	115	2.5	15.8	1.3	4.9	0.6	1.1	3.0	—
4606	Plain, soft	1	item(s)	28	1.8	126	2.1	19.1	1.3	4.9	2.1	1.1	1.5	—
	Pies													
454	Apple pie, prepared from home recipe	1	slice(s)	155	73.3	411	3.7	57.5	2.3	19.4	4.7	8.4	5.2	—
470	Pecan pie, prepared from home recipe	1	slice(s)	122	23.8	503	6.0	63.7	—	27.1	4.9	13.6	7.0	—
33356	Pie crust mix, prepared, baked	1	slice(s)	20	2.1	100	1.3	10.1	0.4	6.1	1.5	3.5	0.8	—
9007	Pie crust, ready to bake, frozen, enriched, baked	1	slice(s)	16	1.8	82	0.7	7.9	0.2	5.2	1.7	2.5	0.6	—
472	Pumpkin pie, prepared from home recipe	1	slice(s)	155	90.7	316	7.0	40.9	—	14.4	4.9	5.7	2.8	—
	Rolls													
8555	Crescent dinner roll	1	item(s)	28	9.7	78	2.7	13.8	0.6	1.2	0.3	0.3	0.6	—
489	Hamburger roll or bun, plain	1	item(s)	43	14.9	120	4.1	21.3	0.9	1.9	0.5	0.5	0.8	—
490	Hard roll	1	item(s)	57	17.7	167	5.6	30.0	1.3	2.5	0.3	0.6	1.0	—
5127	Kaiser roll	1	item(s)	57	17.7	167	5.6	30.0	1.3	2.5	0.3	0.6	1.0	—
5130	Whole wheat roll or bun	1	item(s)	28	9.4	75	2.5	14.5	2.1	1.3	0.2	0.3	0.6	—
	Sport bars													
37026	Balance original chocolate bar	1	item(s)	50	—	200	14.0	22.0	0.5	6.0	3.5	—	—	—
37024	Balance original peanut butter bar	1	item(s)	50	—	200	14.0	22.0	1.0	6.0	2.5	—	—	—
36580	Clif Bar chocolate brownie energy bar	1	item(s)	68	—	240	10.0	45.0	5.0	4.5	1.5	—	—	0
36583	Clif Bar crunchy peanut butter energy bar	1	item(s)	68	—	250	12.0	40.0	5.0	6.0	1.5	—	—	0
36589	Clif Luna Nutz over Chocolate energy bar	1	item(s)	48	—	180	10.0	25.0	3.0	4.5	2.5	—	—	0
12005	PowerBar apple cinnamon	1	item(s)	65	—	230	9.0	45.0	3.0	2.5	0.5	1.5	0.5	0
16078	PowerBar banana	1	item(s)	65	—	230	9.0	45.0	3.0	2.5	0.5	1.0	0.5	0
16080	PowerBar chocolate	1	item(s)	65	6.4	230	10.0	45.0	3.0	2.0	0.5	0.5	1.0	0
29092	PowerBar peanut butter	1	item(s)	65	—	240	10.0	45.0	3.0	3.5	0.5	—	—	0
	Tortillas													
1391	Corn tortillas, soft	1	item(s)	26	11.9	57	1.5	11.6	1.6	0.7	0.1	0.2	0.4	—
1669	Flour tortilla	1	item(s)	32	9.7	100	2.7	16.4	1.0	2.5	0.6	1.2	0.5	—
	Pancakes, waffles													
8926	Pancakes, blueberry, prepared from recipe	3	item(s)	114	60.6	253	7.0	33.1	0.8	10.5	2.3	2.6	4.7	—
5037	Pancakes, prepared from mix with egg and milk	3	item(s)	114	60.3	249	8.9	32.9	2.1	8.8	2.3	2.4	3.3	—
1390	Taco shells, hard	1	item(s)	13	1.0	62	0.9	8.3	0.6	2.8	0.6	1.6	0.5	0.6
30311	Waffle, 100% whole grain	1	item(s)	75	32.3	200	6.9	25.0	1.9	8.4	2.3	3.3	2.1	—
9219	Waffle, plain, frozen, toasted	2	item(s)	66	20.2	206	4.7	32.5	1.6	6.3	1.1	3.2	1.5	—
500	Waffle, plain, prepared from recipe	1	item(s)	75	31.5	218	5.9	24.7	1.7	10.6	2.1	2.6	5.1	—

E KEY: H-4 = Breads/Baked Goods H-10 = Cereal/Rice/Pasta H-14 = Fruit H-18 = Vegetables/Legumes H-28 = Nuts/Seeds H-30 = Vegetarian H-32 = Dairy H-40 = Eggs H-40 = Seafood H-42 = Meats H-46 = Poultry -6 = Processed Meats H-48 = Beverages H-52 = Fats/Oils H-54 = Sweets H-56 = Spices/Condiments/Sauces H-60 = Mixed Foods/Soups/Sandwiches H-64 = Fast Food H-84 = Convenience H-86 = Baby Foods

Chol (mg)	Calc (mg)	Iron (mg)	Magn (mg)	Pota (mg)	Sodi (mg)	Zinc (mg)	Vit A (µg)	Thia (mg)	Vit E (mg α)	Ribo (mg)	Niac (mg)	Vit B_6 (mg)	Fola (µg)	Vit C (mg)	Vit B_{12} (µg)	Sele (µg)
22	21	1.49	17.0	67.2	249.1	0.63	14.5	0.26	0.36	0.12	1.81	0.08	57.8	0	0.2	10.6
17	21	0.91	9.4	59.7	256.6	0.25	17.9	0.10	0.90	0.11	0.87	0.02	24.4	0.1	0.1	4.4
24	89	0.95	14.3	44.5	142.8	0.23	5.0	0.01	0.08	0.02	0.19	0.01	18.9	0	0	1.7
4	26	0.36	13.2	64.8	205.2	0.46	2.4	0.53	—	0.04	0.39	0.03	13.2	0.1	0.1	5.0
0	0	1.80	—	70.0	190.0	—	—	0.15	—	0.17	2.00	0.20	40.0	0	0	—
10	100	1.80	—	—	220.0	—	—	0.15	—	0.17	2.00	—	40.0	0	0.6	—
20	56	1.02	7.8	70.2	289.4	0.28	—	0.17	0.75	0.15	1.25	0.02	34.0	0.4	0.1	8.8
15	42	1.60	18.2	39.3	297.0	0.30	29.6	0.15	0.45	0.18	1.16	0.04	45.6	0	0.1	8.7
0	30	1.42	12.0	74.7	264.5	0.39	0	0.25	—	0.16	2.21	0.02	42.2	0	0	—
0	95	1.36	11.0	71.5	252.0	0.38	0	0.19	0.16	0.14	1.90	0.02	43.5	0.1	0	13.5
0	101	1.63	21.1	106.0	217.7	0.61	0	0.24	0.25	0.16	1.91	0.05	36.5	0	0	16.6
0	36	2.39	89.5	289.0	224.0	1.04	0	0.14	0.37	0.05	0.23	0.09	50.7	0	0	6.3
0	200	0.36	—	—	60.0	—	0	—	—	—	—	—	—	0	0	—
0	20	1.08	—	120.0	160.0	—	0	—	—	—	—	—	—	0	—	—
0	0	0.00	—	—	95.0	—	0	—	—	—	—	—	—	0	—	—
0	15	0.72	23.8	82.3	72.0	0.50	0	0.06	—	0.03	0.39	0.02	5.6	0.2	0	4.0
0	30	0.72	21.0	92.3	79.0	0.42	0	0.08	—	0.04	0.14	0.02	6.8	0	0.1	4.6
0	11	1.73	10.9	122.5	327.1	0.29	17.1	0.22	—	0.16	1.90	0.05	37.2	2.6	0	12.1
106	39	1.80	31.7	162.3	319.6	1.24	100.0	0.22	—	0.22	1.03	0.07	31.7	0.2	0.2	14.6
0	12	0.43	3.0	12.4	145.8	0.07	0	0.06	—	0.03	0.47	0.01	14.0	0	0	4.4
0	3	0.36	2.9	17.6	103.5	0.05	0	0.04	0.42	0.06	0.39	0.01	8.8	0	0	0.5
65	146	1.96	29.5	288.3	348.8	0.71	660.3	0.14	—	0.31	1.21	0.07	32.6	2.6	0.1	11.0
0	39	0.93	5.9	26.3	134.1	0.18	0	0.11	0.02	0.09	1.16	0.02	31.1	0	0.1	5.5
0	59	1.42	9.0	40.4	206.0	0.28	0	0.17	0.03	0.13	1.78	0.03	47.7	0	0.1	8.4
0	54	1.87	15.4	61.6	310.1	0.53	0	0.27	0.23	0.19	2.41	0.02	54.2	0	0	22.3
0	54	1.86	15.4	61.6	310.1	0.53	0	0.27	0.23	0.19	2.41	0.01	54.2	0	0	22.3
0	30	0.69	24.1	77.1	135.5	0.57	0	0.07	0.26	0.04	1.04	0.06	8.5	0	0	14.0
3	100	4.50	40.0	160.0	180.0	3.75	—	0.37	—	0.42	5.00	0.50	100.0	60.0	1.5	17.5
3	100	4.50	40.0	130.0	230.0	3.75	—	0.37	—	0.42	5.00	0.50	100.0	60.0	1.5	17.5
0	250	4.50	100.0	370.0	150.0	3.00	—	0.37	—	0.25	3.00	0.40	80.0	60.0	0.9	14.0
0	250	4.50	100.0	230.0	250.0	3.00	—	0.37	—	0.25	3.00	0.40	80.0	60.0	0.9	14.0
0	350	5.40	80.0	190.0	190.0	5.25	—	1.20	—	1.36	16.00	2.00	400.0	60.0	6.0	24.5
0	300	6.30	140.0	125.0	100.0	5.25	—	1.50	—	1.70	20.00	2.00	400.0	60.0	6.0	—
0	300	6.30	140.0	190.0	100.0	5.25	0	1.50	—	1.70	20.00	2.00	400.0	60.0	6.0	—
0	300	6.30	140.0	200.0	95.0	5.25	0	1.50	—	1.70	20.00	2.00	400.0	60.0	6.0	5.1
0	300	6.30	140.0	130.0	120.0	5.25	0	1.50	—	1.70	20.00	2.00	400.0	60.0	6.0	—
0	21	0.32	18.7	48.4	11.7	0.34	0	0.02	0.07	0.02	0.39	0.06	1.3	0	0	1.6
0	41	1.06	7.0	49.6	203.5	0.17	0	0.17	0.06	0.08	1.14	0.01	33.3	0	0	7.1
64	235	1.96	18.2	157.3	469.7	0.61	57.0	0.22	—	0.31	1.73	0.05	41.0	2.5	0.2	16.0
81	245	1.48	25.1	226.9	575.7	0.85	82.1	0.22	—	0.35	1.40	0.12	104.6	0.7	0.4	—
0	13	0.25	11.3	29.7	51.7	0.21	0.1	0.03	0.09	0.01	0.25	0.03	9.2	0	0	0.6
71	194	1.60	28.5	171.0	371.3	0.87	48.8	0.15	0.32	0.25	1.47	0.08	28.5	0	0.4	20.0
10	203	4.56	15.8	95.0	481.8	0.35	262.7	0.34	0.64	0.46	5.86	0.68	49.5	0	1.9	8.3
52	191	1.73	14.3	119.3	383.3	0.51	48.8	0.19	—	0.26	1.55	0.04	34.5	0.3	0.2	34.7

DA+ Code	Food Description	Quantity	Measure	Wt (g)	H_2O (g)	Ener (kcal)	Prot (g)	Carb (g)	Fiber (g)	Fat (g)	Fat Breakdown (g)			
											Sat	Mono	Poly	*Trans*
Cereal, Flour, Grain, Pasta, Noodles, Popcorn														
	Grain													
2861	Amaranth, dry	½	cup(s)	98	9.6	365	14.1	64.5	9.1	6.3	1.6	1.4	2.8	—
1953	Barley, pearled, cooked	½	cup(s)	79	54.0	97	1.8	22.2	3.0	0.3	0.1	0	0.2	—
1956	Buckwheat groats, cooked, roasted	½	cup(s)	84	63.5	77	2.8	16.8	2.3	0.5	0.1	0.2	0.2	—
1957	Bulgur, cooked	½	cup(s)	91	70.8	76	2.8	16.9	4.1	0.2	0	0	0.1	—
1963	Couscous, cooked	½	cup(s)	79	57.0	88	3.0	18.2	1.1	0.1	0	0	0.1	—
1967	Millet, cooked	½	cup(s)	120	85.7	143	4.2	28.4	1.6	1.2	0.2	0.2	0.6	—
1969	Oat bran, dry	½	cup(s)	47	3.1	116	8.1	31.1	7.2	3.3	0.6	1.1	1.3	—
1972	Quinoa, dry	½	cup(s)	85	11.3	313	12.0	54.5	5.9	5.2	0.6	1.4	2.8	—
	Rice													
129	Brown, long grain, cooked	½	cup(s)	98	71.3	108	2.5	22.4	1.8	0.9	0.2	0.3	0.3	—
2863	Brown, medium grain, cooked	½	cup(s)	98	71.1	109	2.3	22.9	1.8	0.8	0.2	0.3	0.3	—
37488	Jasmine, saffroned, cooked	½	cup(s)	280	—	340	8.0	78.0	0	0	0	0	0	0
30280	Pilaf, cooked	½	cup(s)	103	74.0	129	2.1	22.2	0.6	3.3	0.6	1.5	1.0	—
28066	Spanish, cooked	½	cup(s)	244	184.2	241	5.7	50.2	3.3	1.9	0.4	0.6	0.7	0
2867	White glutinous, cooked	½	cup(s)	87	66.7	84	1.8	18.3	0.9	0.2	0	0.1	0.1	—
484	White, long grain, boiled	½	cup(s)	79	54.1	103	2.1	22.3	0.3	0.2	0.1	0.1	0.1	—
482	White, long grain, enriched, instant, boiled	½	cup(s)	83	59.4	97	1.8	20.7	0.5	0.4	0	0.1	0	—
486	White, long grain, enriched, parboiled, cooked	½	cup(s)	79	55.6	97	2.3	20.6	0.7	0.3	0.1	0.1	0.1	—
1194	Wild brown, cooked	½	cup(s)	82	60.6	83	3.3	17.5	1.5	0.3	0	0	0.2	—
	Flour and grain fractions													
505	All purpose flour, self-rising, enriched	½	cup(s)	63	6.6	221	6.2	46.4	1.7	0.6	0.1	0	0.2	—
503	All purpose flour, white, bleached, enriched	½	cup(s)	63	7.4	228	6.4	47.7	1.7	0.6	0.1	0	0.2	—
1643	Barley flour	½	cup(s)	56	5.5	198	4.2	44.7	2.1	0.8	0.2	0.1	0.4	—
383	Buckwheat flour, whole groat	½	cup(s)	60	6.7	201	7.6	42.3	6.0	1.9	0.4	0.6	0.6	—
504	Cake wheat flour, enriched	½	cup(s)	69	8.6	248	5.6	53.5	1.2	0.6	0.1	0.1	0.3	—
426	Cornmeal, degermed, enriched	½	cup(s)	69	7.8	255	5.0	54.6	2.8	1.2	0.1	0.2	0.5	0
424	Cornmeal, yellow whole grain	½	cup(s)	61	6.2	221	4.9	46.9	4.4	2.2	0.3	0.6	1.0	—
1978	Dark rye flour	½	cup(s)	64	7.1	207	9.0	44.0	14.5	1.7	0.2	0.2	0.8	—
1644	Masa corn flour, enriched	½	cup(s)	57	5.1	208	5.3	43.5	5.5	2.1	0.3	0.6	1.0	—
1976	Rice flour, brown	½	cup(s)	79	9.4	287	5.7	60.4	3.6	2.2	0.4	0.8	0.8	—
1645	Rice flour, white	½	cup(s)	79	9.4	289	4.7	63.3	1.9	1.1	0.3	0.3	0.3	—
1980	Semolina, enriched	½	cup(s)	84	10.6	301	10.6	60.8	3.2	0.9	0.1	0.1	0.4	—
2827	Soy flour, raw	½	cup(s)	42	2.2	185	14.7	14.9	4.1	8.8	1.3	1.9	4.9	—
1990	Wheat germ, crude	2	tablespoon(s)	14	1.6	52	3.3	7.4	1.9	1.4	0.2	0.2	0.9	—
506	Whole wheat flour	½	cup(s)	60	6.2	203	8.2	43.5	7.3	1.1	0.2	0.1	0.5	—
	Breakfast bars													
39230	Atkins Morning Start apple crisp breakfast bar	1	item(s)	37	—	170	11.0	12.0	6.0	9.0	4.0	—	—	—
10571	Nutri-Grain apple cinnamon cereal bar	1	item(s)	37	—	140	2.0	27.0	1.0	3.0	0.5	2.0	0.5	—
10647	Nutri-Grain blueberry cereal bar	1	item(s)	37	5.4	140	2.0	27.0	1.0	3.0	0.5	2.0	0.5	—
10648	Nutri-Grain raspberry cereal bar	1	item(s)	37	5.4	140	2.0	27.0	1.0	3.0	0.5	2.0	0.5	—
10649	Nutri-Grain strawberry cereal bar	1	item(s)	37	5.4	140	2.0	27.0	1.0	3.0	0.5	2.0	0.5	—
	Breakfast cereals, hot													
1260	Cream of Wheat, instant, prepared	½	cup(s)	121	—	388	12.9	73.3	4.3	0	0	0	0	0
365	Farina, enriched, cooked w/water and salt	½	cup(s)	117	102.4	56	1.7	12.2	0.3	0.1	0	0	0	—
363	Grits, white corn, regular and quick, enriched, cooked w/water and salt	½	cup(s)	121	103.3	71	1.7	15.6	0.4	0.2	0	0.1	0.1	—
8636	Grits, yellow corn, regular and quick, enriched, cooked w/salt	½	cup(s)	121	103.3	71	1.7	15.6	0.4	0.2	0	0.1	0.1	—
8657	Oatmeal, cooked w/water	½	cup(s)	117	97.8	83	3.0	14.0	2.0	1.8	0.4	0.5	0.7	0
5500	Oatmeal, maple and brown sugar, instant, prepared	1	item(s)	198	150.2	200	4.8	40.4	2.4	2.2	0.4	0.7	0.8	—
5510	Oatmeal, ready to serve, packet, prepared	1	item(s)	186	158.7	112	4.1	19.8	2.7	2.0	0.4	0.7	0.8	—
	Breakfast cereals, ready to eat													
1197	All-Bran	1	cup(s)	62	1.3	160	8.1	46.0	18.2	2.0	0.4	0.4	1.3	0
1200	All-Bran Buds	1	cup(s)	91	2.7	212	6.4	72.7	39.1	1.9	0.4	0.5	1.2	0

GE KEY: H-4 = Breads/Baked Goods H-10 = Cereal/Rice/Pasta H-14 = Fruit H-18 = Vegetables/Legumes H-28 = Nuts/Seeds H-30 = Vegetarian H-32 = Dairy H-40 = Eggs H-40 = Seafood H-42 = Meats H-46 = Poultry
46 = Processed Meats H-48 = Beverages H-52 = Fats/Oils H-54 = Sweets H-56 = Spices/Condiments/Sauces H-60 = Mixed Foods/Soups/Sandwiches H-64 = Fast Food H-84 = Convenience H-86 = Baby Foods

Chol (mg)	Calc (mg)	Iron (mg)	Magn (mg)	Pota (mg)	Sodi (mg)	Zinc (mg)	Vit A (µg)	Thia (mg)	Vit E (mg α)	Ribo (mg)	Niac (mg)	Vit B_6 (mg)	Fola (µg)	Vit C (mg)	Vit B_{12} (µg)	Sele (µg)
0	149	7.40	259.3	356.8	20.5	3.10	0	0.06	—	0.20	1.24	0.20	47.8	4.1	0	—
0	9	1.04	17.3	73.0	2.4	0.64	0	0.06	0.01	0.04	1.61	0.09	12.6	0	0	6.8
0	6	0.67	42.8	73.9	3.4	0.51	0	0.03	0.07	0.03	0.79	0.06	11.8	0	0	1.8
0	9	0.87	29.1	61.9	4.6	0.51	0	0.05	0.01	0.02	0.91	0.07	16.4	0	0	0.5
0	6	0.30	6.3	45.5	3.9	0.20	0	0.05	0.10	0.02	0.77	0.04	11.8	0	0	21.6
0	4	0.75	52.8	74.4	2.4	1.09	0	0.12	0.02	0.09	1.59	0.13	22.8	0	0	1.1
0	27	2.54	110.5	266.0	1.9	1.46	0	0.55	0.47	0.10	0.43	0.07	24.4	0	0	21.2
0	40	3.88	167.4	478.5	4.2	2.62	0.8	0.30	2.06	0.26	1.28	0.40	156.4	0	0	7.2
0	10	0.41	41.9	41.9	4.9	0.61	0	0.09	0.02	0.02	1.49	0.14	3.9	0	0	9.6
0	10	0.51	42.9	77.0	1.0	0.60	0	0.09	—	0.01	1.29	0.14	3.9	0	0	38.0
0	—	2.16	—	—	780.0	—	—	—	—	—	—	—	—	—	—	—
0	11	1.16	9.3	54.6	390.4	0.37	33.0	0.13	0.28	0.02	1.23	0.06	44.3	0.4	0	4.3
0	37	1.52	95.4	330.5	97.1	1.40	—	0.27	0.12	0.05	3.24	0.38	19.0	22.6	0	14.3
0	2	0.12	4.4	8.7	4.4	0.35	0	0.01	0.03	0.01	0.25	0.02	0.9	0	0	4.9
0	8	0.94	9.5	27.7	0.8	0.38	0	0.12	0.03	0.01	1.16	0.07	45.8	0	0	5.9
0	7	1.46	4.1	7.4	3.3	0.40	0	0.06	0.01	0.01	1.43	0.04	57.8	0	0	4.0
0	15	1.43	7.1	44.2	1.6	0.29	0	0.16	0.01	0.01	1.82	0.12	64.0	0	0	7.3
0	2	0.49	26.2	82.8	2.5	1.09	0	0.04	0.19	0.07	1.05	0.11	21.3	0	0	0.7
0	211	2.90	11.9	77.5	793.7	0.38	0	0.42	0.02	0.24	3.64	0.02	122.5	0	0	21.5
0	9	2.90	13.7	66.9	1.2	0.42	0	0.48	0.02	0.30	3.68	0.02	114.4	0	0	21.2
0	16	0.70	45.4	185.9	4.5	1.04	0	0.06	—	0.02	2.56	0.14	12.9	0	0	2.0
0	25	2.42	150.6	346.2	6.6	1.86	0	0.24	0.18	0.10	3.68	0.34	32.4	0	0	3.4
0	10	5.01	11.0	71.9	1.4	0.42	0	0.61	0.01	0.29	4.65	0.02	127.4	0	0	3.4
0	2	2.98	24.1	104.9	4.8	0.48	7.6	0.42	0.10	0.28	3.66	0.12	148.3	0	0	8.0
0	4	2.10	77.5	175.1	21.3	1.10	6.7	0.22	0.24	0.12	2.20	0.18	15.2	0	0	9.4
0	36	4.12	158.7	467.2	0.6	3.58	0.6	0.20	0.90	0.16	2.72	0.28	38.4	0	0	22.8
0	80	4.10	62.7	169.9	2.8	1.00	0	0.80	0.08	0.42	5.60	0.20	132.8	0	0	8.5
0	9	1.56	88.5	228.3	6.3	1.92	0	0.34	0.94	0.06	5.00	0.58	12.6	0	0	—
0	8	0.26	27.6	60.0	0	0.62	0	0.10	0.08	0.02	2.04	0.34	3.2	0	0	11.9
0	14	3.64	39.2	155.3	0.8	0.86	0	0.66	0.20	0.46	5.00	0.08	152.8	0	0	74.6
0	87	2.70	182.0	1067.0	5.5	1.65	2.5	0.24	0.82	0.48	1.83	0.18	146.4	0	0	3.2
0	6	0.90	34.4	128.2	1.7	1.76	0	0.27	—	0.07	0.97	0.18	40.4	0	0	11.4
0	20	2.32	82.8	243.0	3.0	1.74	0	0.26	0.48	0.12	3.82	0.20	26.4	0	0	42.4
0	200	—	—	90.0	70.0	—	—	0.22	—	0.25	3.00	—	—	9.0	—	—
0	200	1.80	8.0	75.0	110.0	1.50	—	0.37	—	0.42	5.00	0.50	40.0	0	—	—
0	200	1.80	8.0	75.0	110.0	1.50	—	0.37	—	0.42	5.00	0.50	40.0	0	0	—
0	200	1.80	8.0	70.0	110.0	1.50	—	0.37	—	0.42	5.00	0.50	40.0	0	0	—
0	200	1.80	8.0	55.0	110.0	1.50	—	0.37	—	0.42	5.00	0.50	40.0	0	0	—
0	862	34.91	21.4	150.8	732.6	0.86	—	1.59	—	1.47	21.55	2.15	431.0	0	0	—
0	5	0.58	2.3	15.1	383.3	0.09	0	0.07	0.01	0.05	0.57	0.01	39.6	0	0	10.6
0	4	0.73	6.1	25.4	269.8	0.08	0	0.10	0.02	0.07	0.87	0.03	39.9	0	0	3.8
0	4	0.73	6.1	25.4	269.8	0.08	2.4	0.10	0.02	0.07	0.87	0.03	39.9	0	0	3.3
0	11	1.05	31.6	81.9	4.7	1.17	0	0.09	0.09	0.02	0.26	0.01	7.0	0	0	6.3
0	26	6.83	49.9	126.4	403.5	1.03	0	1.02	—	0.05	1.56	0.30	42.2	0	0	11.1
0	21	3.96	44.7	112.4	240.9	0.92	0	0.60	—	0.04	0.77	0.18	18.7	0	0	3.8
0	241	10.90	224.4	632.4	150.0	3.00	300.1	1.40	—	1.68	9.16	7.44	800.0	12.4	12.0	5.8
0	57	13.64	186.4	909.1	614.5	4.55	464.5	1.09	1.42	1.27	15.45	6.09	1222.7	18.2	18.2	26.3

DA+ Code	Food Description	Quantity	Measure	Wt (g)	H_2O (g)	Ener (kcal)	Prot (g)	Carb (g)	Fiber (g)	Fat (g)	Fat Breakdown (g) Sat	Mono	Poly	*Trans*
	Cereal, Flour, Grain, Pasta, Noodles, Popcorn—*continued*													
1199	Apple Jacks	1	cup(s)	33	0.9	130	1.0	30.0	0.5	0.5	0	—	—	0
1204	Cap'n Crunch	1	cup(s)	36	0.9	147	1.3	30.7	1.3	2.0	0.5	0.4	0.3	—
1205	Cap'n Crunch Crunchberries	1	cup(s)	35	0.9	133	1.3	29.3	1.3	2.0	0.5	0.4	0.3	—
1206	Cheerios	1	cup(s)	30	1.0	110	3.0	22.0	3.0	2.0	0	0.5	0.5	—
3415	Cocoa Puffs	1	cup(s)	30	0.6	120	1.0	26.0	0.2	1.0	—	—	—	—
1207	Cocoa Rice Krispies	1	cup(s)	41	1.0	160	1.3	36.0	1.3	1.3	0.7	0	0	—
5522	Complete wheat bran flakes	1	cup(s)	39	1.4	120	4.0	30.7	6.7	0.7	—	—	—	0
1211	Corn Flakes	1	cup(s)	28	0.9	100	2.0	24.0	1.0	0	0	0	0	0
1247	Corn Pops	1	cup(s)	31	0.9	120	1.0	28.0	0.3	0	0	0	0	0
1937	Cracklin' Oat Bran	1	cup(s)	65	2.3	267	5.3	46.7	8.0	9.3	4.0	4.7	1.3	0
1220	Froot Loops	1	cup(s)	32	0.8	120	1.0	28.0	1.0	1.0	0.5	0	0	—
38214	Frosted Cheerios	1	cup(s)	37	—	149	2.5	31.1	1.2	1.2	—	—	—	—
372	Frosted Flakes	1	cup(s)	41	1.1	160	1.3	37.3	1.3	0	0	0	0	0
38215	Frosted Mini Chex	1	cup(s)	40	—	147	1.3	36.0	0	0	0	0	0	0
10268	Frosted Mini-Wheats	1	cup(s)	59	3.1	208	5.8	47.4	5.8	1.2	0	0	0.6	0
38216	Frosted Wheaties	1	cup(s)	40	—	147	1.3	36.0	0.3	0	0	0	0	0
1223	Granola, prepared	½	cup(s)	61	3.3	298	9.1	32.5	5.5	14.7	2.5	5.8	5.6	0
2415	Honey Bunches of Oats honey roasted	1	cup(s)	40	0.9	160	2.7	33.3	1.3	2.0	0.7	1.2	0.1	—
1227	Honey Nut Cheerios	1	cup(s)	37	0.9	149	3.7	29.9	2.5	1.9	0	0.6	0.6	—
2424	Honeycomb	1	cup(s)	22	0.3	83	1.5	19.5	0.8	0.4	0	—	—	—
10286	Kashi whole grain puffs	1	cup(s)	19	—	70	2.0	15.0	1.0	0.5	0	—	—	0
41142	Kellogg's Mueslix	1	cup(s)	83	7.2	298	7.6	60.8	6.1	4.6	0.7	2.4	1.5	0
1231	Kix	1	cup(s)	24	0.5	96	1.6	20.8	0.8	0.4	—	—	—	—
30569	Life	1	cup(s)	43	1.7	160	4.0	33.3	2.7	2.0	0.3	0.6	0.6	—
1233	Lucky Charms	1	cup(s)	24	0.6	96	1.6	20.0	0.8	0.8	—	—	—	—
38220	Multi Grain Cheerios	1	cup(s)	30	—	110	3.0	24.0	3.0	1.0	—	—	—	—
1201	Multi-Bran Chex	1	cup(s)	63	1.3	216	4.3	52.9	8.6	1.6	0	0	0.5	0
13633	Post Bran Flakes	1	cup(s)	40	1.5	133	4.0	32.0	6.7	0.7	0	—	—	—
1241	Product 19	1	cup(s)	30	1.0	100	2.0	25.0	1.0	0	0	0	0	0
32432	Puffed rice, fortified	1	cup(s)	14	0.4	56	0.9	12.6	0.2	0.1	0	—	—	—
32433	Puffed wheat, fortified	1	cup(s)	12	0.4	44	1.8	9.6	0.5	0.1	0	—	—	—
13334	Quaker 100% natural granola oats and honey	½	cup(s)	48	—	220	5.0	31.0	3.0	9.0	3.8	4.1	1.2	—
13335	Quaker 100% natural granola oats, honey, and raisins	½	cup(s)	51	—	230	5.0	34.0	3.0	9.0	3.6	3.8	1.1	—
2420	Raisin Bran	1	cup(s)	59	5.0	190	4.0	46.0	8.0	1.0	0	0.1	0.4	—
1244	Rice Chex	1	cup(s)	31	0.8	120	2.0	27.0	0.3	0	0	0	0	0
1245	Rice Krispies	1	cup(s)	26	0.8	96	1.6	23.2	0	0	0	0	0	0
5593	Shredded Wheat	1	cup(s)	49	0.4	177	5.8	40.9	6.9	1.1	0.1	0	0.2	0
1248	Smacks	1	cup(s)	36	1.1	133	2.7	32.0	1.3	0.7	—	—	—	—
1246	Special K	1	cup(s)	31	0.9	110	7.0	22.0	0.5	0	0	0	0	0
3428	Total corn flakes	1	cup(s)	23	0.6	83	1.5	18.0	0.6	0	0	0	0	0
1253	Total whole grain	1	cup(s)	40	1.1	147	2.7	30.7	4.0	1.3	—	—	—	—
1254	Trix	1	cup(s)	30	0.6	120	1.0	27.0	1.0	1.0	—	—	—	—
382	Wheat germ, toasted	2	tablespoon(s)	14	0.8	54	4.1	7.0	2.1	1.5	0.3	0.2	0.9	—
1257	Wheaties	1	cup(s)	36	1.2	132	3.6	28.8	3.6	1.2	—	—	—	—
	Pasta, noodles													
449	Chinese chow mein noodles, cooked	½	cup(s)	23	0.2	119	1.9	12.9	0.9	6.9	1.0	1.7	3.9	—
1995	Corn pasta, cooked	½	cup(s)	70	47.8	88	1.8	19.5	3.4	0.5	0.1	0.1	0.2	—
448	Egg noodles, enriched, cooked	½	cup(s)	80	54.2	110	3.6	20.1	1.0	1.7	0.3	0.5	0.4	0
1563	Egg noodles, spinach, enriched, cooked	½	cup(s)	80	54.8	106	4.0	19.4	1.8	1.3	0.3	0.4	0.3	—
440	Macaroni, enriched, cooked	½	cup(s)	70	43.5	111	4.1	21.6	1.3	0.7	0.1	0.1	0.2	0
2000	Macaroni, tricolor vegetable, enriched, cooked	½	cup(s)	67	45.8	86	3.0	17.8	2.9	0.1	0	0	0	—
1996	Plain pasta, fresh-refrigerated, cooked	½	cup(s)	64	43.9	84	3.3	16.0	—	0.7	0.1	0.1	0.3	—
1725	Ramen noodles, cooked	½	cup(s)	114	94.5	104	3.0	15.4	1.0	4.3	0.2	0.2	0.2	—
2878	Soba noodles, cooked	½	cup(s)	95	69.4	94	4.8	20.4	—	0.1	0	0	0	—
2879	Somen noodles, cooked	½	cup(s)	88	59.8	115	3.5	24.2	—	0.2	0	0	0.1	—
493	Spaghetti, al dente, cooked	½	cup(s)	65	41.6	95	3.5	19.5	1.0	0.5	0.1	0.1	0.2	—
2884	Spaghetti, whole wheat, cooked	½	cup(s)	70	47.0	87	3.7	18.6	3.2	0.4	0.1	0.1	0.1	—
	Popcorn													
476	Air popped	1	cup(s)	8	0.3	31	1.0	6.2	1.2	0.4	0	0.1	0.2	—
4619	Caramel	1	cup(s)	35	1.0	152	1.3	27.8	1.8	4.5	1.3	1.0	1.6	—

E KEY: H-4 = Breads/Baked Goods H-10 = Cereal/Rice/Pasta H-14 = Fruit H-18 = Vegetables/Legumes H-28 = Nuts/Seeds H-30 = Vegetarian H-32 = Dairy H-40 = Eggs H-40 = Seafood H-42 = Meats H-46 = Poultry
6 = Processed Meats H-48 = Beverages H-52 = Fats/Oils H-54 = Sweets H-56 = Spices/Condiments/Sauces H-60 = Mixed Foods/Soups/Sandwiches H-64 = Fast Food H-84 = Convenience H-86 = Baby Foods

Chol (mg)	Calc (mg)	Iron (mg)	Magn (mg)	Pota (mg)	Sodi (mg)	Zinc (mg)	Vit A (µg)	Thia (mg)	Vit E (mg α)	Ribo (mg)	Niac (mg)	Vit B_6 (mg)	Fola (µg)	Vit C (mg)	Vit B_{12} (µg)	Sele (µg)
0	0	4.50	8.0	30.0	130.0	1.50	150.2	0.37	—	0.42	5.00	0.50	100.0	15.0	1.5	2.4
0	5	6.80	20.0	73.3	266.7	5.00	2.5	0.51	—	0.57	6.68	0.67	133.5	0	0	6.7
0	7	6.53	18.7	73.3	240.0	5.13	2.4	0.51	—	0.57	6.68	0.67	133.7	0	0	6.7
0	100	8.10	40.0	95.0	280.0	3.75	150.3	0.37	—	0.42	5.00	0.50	200.0	6.0	1.5	11.3
0	100	4.50	8.0	50.0	170.0	3.75	0	0.37	—	0.42	5.00	0.50	100.0	6.0	1.5	2.0
0	53	6.00	10.7	66.7	253.3	2.00	200.1	0.49	—	0.56	6.67	0.67	133.3	20.0	2.0	5.8
0	0	24.00	53.3	226.7	280.0	20.00	300.1	2.00	—	2.27	26.67	2.67	533.3	80.0	8.0	4.1
0	0	8.10	3.4	25.0	200.0	0.16	149.8	0.37	—	0.42	5.00	0.50	100.0	6.0	1.5	1.4
0	0	1.80	2.5	25.0	120.0	1.50	150.0	0.37	—	0.42	5.00	0.50	100.0	6.0	1.5	2.0
0	27	2.40	80.0	293.3	200.0	2.00	299.9	0.49	—	0.56	6.67	0.67	133.3	20.0	2.0	14.4
0	0	4.50	8.0	35.0	150.0	1.50	150.1	0.37	—	0.42	5.00	0.50	100.0	15.0	1.5	2.3
0	124	5.60	19.9	68.4	261.3	4.67	—	0.46	—	0.52	6.22	0.62	124.4	7.5	1.9	—
0	0	6.00	3.7	26.7	200.0	0.20	200.1	0.49	—	0.56	6.67	0.67	133.3	8.0	2.0	1.8
0	133	12.00	—	33.3	266.7	4.00	—	0.49	—	0.56	6.67	0.67	266.7	8.0	2.0	—
0	0	16.66	69.4	196.7	5.8	1.74	0	0.43	—	0.49	5.78	0.58	115.7	0	1.7	2.4
0	133	10.80	0	46.7	266.7	10.00	—	1.00	—	1.13	13.33	1.33	533.3	8.0	4.0	—
0	48	2.58	106.8	329.4	15.3	2.45	0.6	0.44	6.77	0.17	1.30	0.17	50.0	0.7	0	17.0
0	0	10.80	21.3	0	253.3	0.40	—	0.49	—	0.56	6.67	0.67	133.0	0	2.0	—
0	124	5.60	39.8	112.0	336.0	4.67	—	0.46	—	0.52	6.22	0.62	248.9	7.5	1.9	8.8
0	0	2.03	6.0	26.3	165.4	1.13	—	0.28	—	0.32	3.74	0.37	75.0	0	1.1	—
0	0	0.36	—	60.0	0	—	0	0.03	—	0.03	0.80	0.00	—	0	—	—
0	48	6.83	74.2	363.3	257.5	5.67	136.7	0.67	6.00	0.67	8.33	3.08	615.0	0.3	9.2	14.4
0	120	6.48	6.4	28.0	216.0	3.00	120.2	0.30	—	0.34	4.00	0.40	160.0	4.8	1.2	4.8
0	149	11.87	41.3	120.0	213.3	5.33	0.9	0.53	—	0.60	7.12	0.71	142.4	0	0	10.7
0	80	3.60	12.8	48.0	168.0	3.00	—	0.30	—	0.34	4.00	0.40	160.0	4.8	1.2	4.8
0	100	18.00	24.0	85.0	200.0	15.00	—	1.50	—	1.70	20.00	2.00	400.0	15.0	6.0	—
0	108	17.50	64.8	237.7	410.6	4.05	171.1	0.40	—	0.45	5.40	0.54	432.2	6.5	1.6	4.9
0	0	10.80	80.0	266.7	280.0	2.00	—	0.49	—	0.56	6.67	0.67	133.3	0	2.0	—
0	0	18.00	16.0	50.0	210.0	15.00	225.3	1.50	—	1.70	20.00	2.00	400.0	60.0	6.0	3.6
0	1	4.43	3.5	15.8	0.4	0.14	0	0.36	—	0.25	4.94	0.01	2.7	0	0	1.5
0	3	3.80	17.4	41.8	0.5	0.28	0	0.31	—	0.21	4.23	0.02	3.8	0	0	14.8
0	61	1.20	51.0	220.0	20.0	1.05	0.5	0.13	—	0.12	0.82	0.07	15.0	0.2	0.1	8.3
0	59	1.20	49.0	250.0	20.0	0.99	0.5	0.13	—	0.12	0.80	0.08	14.1	0.4	0.1	8.8
0	20	10.80	80.0	360.0	360.0	2.25	—	0.37	—	0.42	5.00	0.50	100.0	0	2.1	—
0	100	9.00	9.3	35.0	290.0	3.75	—	0.37	—	0.42	5.00	0.50	200.0	6.0	1.5	1.2
0	0	1.44	12.8	32.0	256.0	0.48	120.1	0.30	—	0.34	4.80	0.40	80.0	4.8	1.2	4.1
0	18	2.90	60.3	179.3	1.1	1.37	0	0.14	—	0.12	3.47	0.18	21.1	0	0	2.0
0	0	0.48	10.7	53.3	66.7	0.40	200.2	0.49	—	0.56	6.67	0.67	133.3	8.0	2.0	17.5
0	0	8.10	16.0	60.0	220.0	0.90	225.1	0.52	—	0.59	7.00	2.00	400.0	21.0	6.0	7.0
0	752	13.53	0	22.6	157.9	11.28	112.8	1.13	22.56	1.28	15.04	1.50	300.8	45.1	4.5	1.2
0	1333	24.00	32.0	120.0	253.3	20.00	200.4	2.00	31.32	2.27	26.67	2.67	533.3	80.0	8.0	1.9
0	100	4.50	0	15.0	190.0	3.75	150.3	0.37	—	0.42	5.00	0.50	100.0	6.0	1.5	6.0
0	6	1.28	45.2	133.8	0.6	2.35	0.7	0.23	2.25	0.11	0.79	0.13	49.7	0.8	0	9.2
0	24	9.72	38.4	126.0	264.0	9.00	180.4	0.90	—	1.02	12.00	1.20	240.0	7.2	3.6	1.7
0	5	1.06	11.7	27.0	98.8	0.31	0	0.13	0.78	0.09	1.33	0.02	20.3	0	0	9.7
0	1	0.18	25.2	21.7	0	0.44	2.1	0.04	—	0.02	0.39	0.04	4.2	0	0	2.0
23	10	1.17	16.8	30.4	4.0	0.52	4.8	0.23	0.13	0.11	1.66	0.03	67.2	0	0.1	19.1
26	15	0.87	19.2	29.6	9.6	0.50	8.0	0.19	0.46	0.09	1.17	0.09	51.2	0	0.1	17.4
0	5	0.90	12.6	30.8	0.7	0.36	0	0.19	0.04	0.10	1.18	0.03	51.1	0	0	18.5
0	7	0.33	12.7	20.8	4.0	0.30	3.4	0.08	0.14	0.04	0.72	0.02	43.6	0	0	13.3
21	4	0.73	11.5	15.4	3.8	0.36	3.8	0.13	—	0.10	0.64	0.02	41.0	0	0.1	—
18	9	0.89	8.5	34.5	414.5	0.30	—	0.08	—	0.04	0.71	0.03	4.0	0.1	0	—
0	4	0.45	8.5	33.2	57.0	0.11	0	0.09	—	0.02	0.48	0.03	6.6	0	0	—
0	7	0.45	1.8	25.5	141.7	0.19	0	0.01	—	0.03	0.08	0.01	1.8	0	0	—
0	7	1.00	12.4	51.5	0.5	0.35	0	0.12	0.04	0.07	0.90	0.04	7.8	0	0	40.0
0	11	0.74	21.0	30.8	2.1	0.57	0	0.08	0.21	0.03	0.50	0.06	3.5	0	0	18.1
0	1	0.25	11.5	26.3	0.6	0.25	0.8	0.01	0.02	0.01	0.18	0.01	2.5	0	0	0
2	15	0.61	12.3	38.4	72.5	0.20	0.7	0.02	0.42	0.02	0.77	0.01	1.8	0	0	1.3

DA+ Code	Food Description	Quantity	Measure	Wt (g)	H_2O (g)	Ener (kcal)	Prot (g)	Carb (g)	Fiber (g)	Fat (g)	Fat Breakdown (g)			
											Sat	Mono	Poly	*Trans*
Cereal, Flour, Grain, Pasta, Noodles, Popcorn—*continued*														
4620	Cheese flavored	1	cup(s)	36	0.9	188	3.3	18.4	3.5	11.8	2.3	3.5	5.5	—
477	Popped in oil	1	cup(s)	11	0.1	64	0.8	5.0	0.9	4.8	0.8	1.1	2.6	—
Fruit and Fruit Juices														
	Apples													
952	Juice, prepared from frozen concentrate	½	cup(s)	120	105.0	56	0.2	13.8	0.1	0.1	0	0	0	—
225	Juice, unsweetened, canned	½	cup(s)	124	109.0	58	0.1	14.5	0.1	0.1	0	0	0	—
224	Slices	½	cup(s)	55	47.1	29	0.1	7.6	1.3	0.1	0	0	0	—
946	Slices without skin, boiled	½	cup(s)	86	73.1	45	0.2	11.7	2.1	0.3	0	0	0.1	—
223	Raw medium, with peel	1	item(s)	138	118.1	72	0.4	19.1	3.3	0.2	0	0	0.1	—
948	Dried, sulfured	¼	cup(s)	22	6.8	52	0.2	14.2	1.9	0.1	0	0	0	—
226	Applesauce, sweetened, canned	½	cup(s)	128	101.5	97	0.2	25.4	1.5	0.2	0	0	0.1	—
227	Applesauce, unsweetened, canned	½	cup(s)	122	107.8	52	0.2	13.8	1.5	0.1	0	0	0	—
38492	Crabapples	1	item(s)	35	27.6	27	0.1	7.0	0.9	0.1	0	0	0	—
	Apricot													
228	Fresh without pits	4	item(s)	140	120.9	67	2.0	15.6	2.8	0.5	0	0.2	0.1	—
229	Halves with skin, canned in heavy syrup	½	cup(s)	129	100.1	107	0.7	27.7	2.1	0.1	0	0	0	—
230	Halves, dried, sulfured	¼	cup(s)	33	10.1	79	1.1	20.6	2.4	0.2	0	0	0	—
	Avocado													
233	California, whole, without skin or pit	½	cup(s)	115	83.2	192	2.2	9.9	7.8	17.7	2.4	11.3	2.1	—
234	Florida, whole, without skin or pit	½	cup(s)	115	90.6	138	2.5	9.0	6.4	11.5	2.2	6.3	1.9	—
2998	Pureed	⅛	cup(s)	28	20.2	44	0.5	2.4	1.8	4.0	0.6	2.7	0.5	—
	Banana													
4580	Dried chips	¼	cup(s)	55	2.4	285	1.3	32.1	4.2	18.5	15.9	1.1	0.3	—
235	Fresh whole, without peel	1	item(s)	118	88.4	105	1.3	27.0	3.1	0.4	0.1	0	0.1	—
	Blackberries													
237	Raw	½	cup(s)	72	63.5	31	1.0	6.9	3.8	0.4	0	0	0.2	—
958	Unsweetened, frozen	½	cup(s)	76	62.1	48	0.9	11.8	3.8	0.3	0	0	0.2	—
	Blueberries													
959	Canned in heavy syrup	½	cup(s)	128	98.3	113	0.8	28.2	2.0	0.4	0	0.1	0.2	—
238	Raw	½	cup(s)	73	61.1	41	0.5	10.5	1.7	0.2	0	0	0.1	—
960	Unsweetened, frozen	½	cup(s)	78	67.1	40	0.3	9.4	2.1	0.5	0	0.1	0.2	—
	Boysenberries													
961	Canned in heavy syrup	½	cup(s)	128	97.6	113	1.3	28.6	3.3	0.2	0	0	0.1	—
962	Unsweetened, frozen	½	cup(s)	66	56.7	33	0.7	8.0	3.5	0.2	0	0	0.1	—
35576	**Breadfruit**	1	item(s)	384	271.3	396	4.1	104.1	18.8	0.9	0.2	0.1	0.3	—
	Cherries													
967	Sour red, canned in water	½	cup(s)	122	109.7	44	0.9	10.9	1.3	0.1	0	0	0	—
3000	Sour red, raw	½	cup(s)	78	66.8	39	0.8	9.4	1.2	0.2	0.1	0.1	0.1	—
3004	Sweet, canned in heavy syrup	½	cup(s)	127	98.2	105	0.8	26.9	1.9	0.2	0	0.1	0.1	—
969	Sweet, canned in water	½	cup(s)	124	107.9	57	1.0	14.6	1.9	0.2	0	0	0	—
240	Sweet, raw	½	cup(s)	73	59.6	46	0.8	11.6	1.5	0.1	0	0	0	—
	Cranberries													
3007	Chopped, raw	½	cup(s)	55	47.9	25	0.2	6.7	2.5	0.1	0	0	0	—
1717	Cranberry apple juice drink	½	cup(s)	123	102.6	77	0	19.4	0	0.1	0	0	0.1	—
1638	Cranberry juice cocktail	½	cup(s)	127	109.0	68	0	17.1	0	0.1	0	0	0.1	—
241	Cranberry juice cocktail, low calorie, with saccharin	½	cup(s)	119	112.8	23	0	5.5	0	0	0	0	0	—
242	Cranberry sauce, sweetened, canned	¼	cup(s)	69	42.0	105	0.1	26.9	0.7	0.1	0	0	0	—
	Dates													
244	Domestic, chopped	¼	cup(s)	45	9.1	125	1.1	33.4	3.6	0.2	0	0	0	—
243	Domestic, whole	¼	cup(s)	45	9.1	125	1.1	33.4	3.6	0.2	0	0	0	—
	Figs													
975	Canned in heavy syrup	½	cup(s)	130	98.8	114	0.5	29.7	2.8	0.1	0	0	0.1	—
974	Canned in water	½	cup(s)	124	105.7	66	0.5	17.3	2.7	0.1	0	0	0.1	—
973	Raw, medium	2	item(s)	100	79.1	74	0.7	19.2	2.9	0.3	0.1	0.1	0.1	—
	Fruit cocktail and salad													
245	Fruit cocktail, canned in heavy syrup	½	cup(s)	124	99.7	91	0.5	23.4	1.2	0.1	0	0	0	—
978	Fruit cocktail, canned in juice	½	cup(s)	119	103.6	55	0.5	14.1	1.2	0	0	0	0	—
977	Fruit cocktail, canned in water	½	cup(s)	119	107.6	38	0.5	10.1	1.2	0.1	0	0	0	—
979	Fruit salad, canned in water	½	cup(s)	123	112.1	37	0.4	9.6	1.2	0.1	0	0	0	—

E KEY: H-4 = Breads/Baked Goods H-10 = Cereal/Rice/Pasta H-14 = Fruit H-18 = Vegetables/Legumes H-28 = Nuts/Seeds H-30 = Vegetarian H-32 = Dairy H-40 = Eggs H-40 = Seafood H-42 = Meats H-46 = Poultry
6 = Processed Meats H-48 = Beverages H-52 = Fats/Oils H-54 = Sweets H-56 = Spices/Condiments/Sauces H-60 = Mixed Foods/Soups/Sandwiches H-64 = Fast Food H-84 = Convenience H-86 = Baby Foods

Chol (mg)	Calc (mg)	Iron (mg)	Magn (mg)	Pota (mg)	Sodi (mg)	Zinc (mg)	Vit A (µg)	Thia (mg)	Vit E (mg α)	Ribo (mg)	Niac (mg)	Vit B_6 (mg)	Fola (µg)	Vit C (mg)	Vit B_{12} (µg)	Sele (µg)
4	40	0.79	32.5	93.2	317.4	0.71	13.6	0.04	—	0.08	0.52	0.08	3.9	0.2	0.2	4.3
0	0	0.22	8.7	20.0	116.4	0.34	0.9	0.01	0.27	0.00	0.13	0.01	2.8	0	0	0.2
0	7	0.31	6.0	150.6	8.4	0.05	0	0.00	0.01	0.02	0.05	0.04	0	0.7	0	0.1
0	9	0.46	3.7	147.6	3.7	0.04	0	0.03	0.01	0.02	0.12	0.04	0	1.1	0	0.1
0	3	0.06	2.7	58.8	0.5	0.02	1.6	0.01	0.10	0.01	0.05	0.02	1.6	2.5	0	0
0	4	0.16	2.6	75.2	0.9	0.03	1.7	0.01	0.04	0.01	0.08	0.04	0.9	0.2	0	0.3
0	8	0.16	6.9	147.7	1.4	0.05	4.1	0.02	0.24	0.03	0.12	0.05	4.1	6.3	0	0
0	3	0.30	3.4	96.8	18.7	0.04	0	0.00	0.11	0.03	0.20	0.03	0	0.8	0	0.3
0	5	0.44	3.8	77.8	3.8	0.05	1.3	0.01	0.26	0.03	0.24	0.03	1.3	2.2	0	0.4
0	4	0.14	3.7	91.5	2.4	0.03	1.2	0.01	0.25	0.03	0.22	0.03	1.2	1.5	0	0.4
0	6	0.12	2.5	67.9	0.4	—	0.7	0.01	0.20	0.01	0.03	—	2.0	2.8	0	—
0	18	0.54	14.0	362.6	1.4	0.28	134.4	0.04	1.24	0.05	0.84	0.07	12.6	14.0	0	0.1
0	12	0.38	9.0	180.6	5.2	0.14	80.0	0.02	0.77	0.02	0.48	0.07	2.6	4.0	0	0.1
0	18	0.87	10.5	381.5	3.3	0.12	59.1	0.00	1.42	0.02	0.85	0.05	3.3	0.3	0	0.7
0	15	0.66	33.3	583.0	9.2	0.78	8.0	0.08	2.23	0.16	2.19	0.31	102.3	10.1	0	0.4
0	12	0.19	27.6	403.6	2.3	0.45	8.0	0.02	3.03	0.04	0.76	0.08	40.3	20.0	0	—
0	3	0.14	8.0	134.1	1.9	0.17	1.9	0.01	0.57	0.03	0.48	0.07	22.4	2.8	0	0.1
0	10	0.69	41.8	294.8	3.3	0.40	2.2	0.04	0.13	0.01	0.39	0.14	7.7	3.5	0	0.8
0	6	0.30	31.9	422.4	1.2	0.17	3.5	0.03	0.11	0.08	0.78	0.43	23.6	10.3	0	1.2
0	21	0.45	14.4	116.6	0.7	0.38	7.9	0.01	0.84	0.02	0.47	0.02	18.0	15.1	0	0.3
0	22	0.60	16.6	105.7	0.8	0.19	4.5	0.02	0.88	0.03	0.91	0.05	25.7	2.3	0	0.3
0	6	0.42	5.1	51.2	3.8	0.09	2.6	0.04	0.49	0.07	0.14	0.05	2.6	1.4	0	0.1
0	4	0.20	4.4	55.8	0.7	0.12	2.2	0.03	0.41	0.03	0.30	0.04	4.4	7.0	0	0.1
0	6	0.14	3.9	41.9	0.8	0.05	1.6	0.03	0.37	0.03	0.40	0.05	5.4	1.9	0	0.1
0	23	0.55	14.1	115.2	3.8	0.24	2.6	0.03	—	0.04	0.29	0.05	43.5	7.9	0	0.5
0	18	0.56	10.6	91.7	0.7	0.15	2.0	0.04	0.57	0.02	0.51	0.04	41.6	2.0	0	0.1
0	65	2.07	96.0	1881.6	7.7	0.46	0	0.42	0.38	0.11	3.45	0.38	53.8	111.4	0	2.3
0	13	1.67	7.3	119.6	8.5	0.09	46.4	0.02	0.28	0.05	0.22	0.05	9.8	2.6	0	0
0	12	0.25	7.0	134.1	2.3	0.08	49.6	0.02	0.05	0.03	0.31	0.03	6.2	7.8	0	0
0	11	0.44	11.4	183.4	3.8	0.12	10.1	0.02	0.29	0.05	0.50	0.03	5.1	4.6	0	0
0	14	0.45	11.2	162.4	1.2	0.10	9.9	0.03	0.29	0.05	0.51	0.04	5.0	2.7	0	0
0	9	0.26	8.0	161.0	0	0.05	2.2	0.02	0.05	0.02	0.11	0.04	2.9	5.1	0	0
0	4	0.13	3.3	46.8	1.1	0.05	1.7	0.01	0.66	0.01	0.05	0.03	0.6	7.3	0	0.1
0	4	0.09	1.2	20.8	2.5	0.02	0	0.00	0.15	0.00	0.00	0.00	0	48.4	0	0
0	4	0.13	1.3	17.7	2.5	0.04	0	0.00	0.28	0.00	0.05	0.00	0	53.5	0	0.3
0	11	0.05	2.4	29.6	3.6	0.02	0	0.00	0.06	0.00	0.00	0.00	0	38.2	0	0
0	3	0.15	2.1	18.0	20.1	0.03	1.4	0.01	0.57	0.01	0.06	0.01	0.7	1.4	0	0.2
0	17	0.45	19.1	291.9	0.9	0.12	0	0.02	0.02	0.02	0.56	0.07	8.5	0.2	0	1.3
0	17	0.45	19.1	291.9	0.9	0.12	0	0.02	0.02	0.02	0.56	0.07	8.5	0.2	0	1.3
0	35	0.36	13.0	128.2	1.3	0.14	2.6	0.03	0.16	0.05	0.55	0.09	2.6	1.3	0	0.3
0	35	0.36	12.4	127.7	1.2	0.15	2.5	0.03	0.10	0.05	0.55	0.09	2.5	1.2	0	0.1
0	35	0.36	17.0	232.0	1.0	0.14	7.0	0.06	0.10	0.04	0.40	0.10	6.0	2.0	0	0.2
0	7	0.36	6.2	109.1	7.4	0.09	12.4	0.02	0.49	0.02	0.46	0.06	3.7	2.4	0	0.6
0	9	0.25	8.3	112.6	4.7	0.11	17.8	0.01	0.47	0.02	0.48	0.06	3.6	3.2	0	0.6
0	6	0.30	8.3	111.4	4.7	0.11	15.4	0.02	0.47	0.01	0.43	0.06	3.6	2.5	0	0.6
0	9	0.37	6.1	95.6	3.7	0.10	27.0	0.02	—	0.03	0.46	0.04	3.7	2.3	0	1.0

TABLE H-1 Table of Food Composition (*continued*) (Computer code is for Cengage Diet Analysis program) (For purposes of calculations, use "0" for t, <1, <0.1, 0.01)

DA+ Code	Food Description	Quantity	Measure	Wt (g)	H_2O (g)	Ener (kcal)	Prot (g)	Carb (g)	Fiber (g)	Fat (g)	Fat Breakdown (g) Sat	Mono	Poly	*Trans*
Fruit and Fruit Juices—*continued*														
	Gooseberries													
982	Canned in light syrup	½	cup(s)	126	100.9	92	0.8	23.6	3.0	0.3	0	0	0.1	—
981	Raw	½	cup(s)	75	65.9	33	0.7	7.6	3.2	0.4	0	0	0.2	—
	Grapefruit													
251	Juice, pink, sweetened, canned	½	cup(s)	125	109.1	57	0.7	13.9	0.1	0.1	0	0	0	—
249	Juice, white	½	cup(s)	124	111.2	48	0.6	11.4	0.1	0.1	0	0	0	—
3022	Pink or red, raw	½	cup(s)	114	100.8	48	0.9	12.2	1.8	0.2	0	0	0	—
248	Sections, canned in light syrup	½	cup(s)	127	106.2	76	0.7	19.6	0.5	0.1	0	0	0	—
983	Sections, canned in water	½	cup(s)	122	109.6	44	0.7	11.2	0.5	0.1	0	0	0	—
247	White, raw	½	cup(s)	115	104.0	38	0.8	9.7	1.3	0.1	0	0	0	—
	Grapes													
255	American, slip skin	½	cup(s)	46	37.4	31	0.3	7.9	0.4	0.2	0.1	0	0	—
256	European, red or green, adherent skin	½	cup(s)	76	60.8	52	0.5	13.7	0.7	0.1	0	0	0	—
3159	Grape juice drink, canned	½	cup(s)	125	106.6	71	0	18.2	0.1	0	0	0	0	0
259	Grape juice, sweetened, with added vitamin C, prepared from frozen concentrate	½	cup(s)	125	108.6	64	0.2	15.9	0.1	0.1	0	0	0	—
3060	Raisins, seeded, packed	¼	cup(s)	41	6.8	122	1.0	32.4	2.8	0.2	0.1	0	0.1	—
987	**Guava, raw**	1	item(s)	55	44.4	37	1.4	7.9	3.0	0.5	0.2	0	0.2	—
35593	**Guavas, strawberry**	1	item(s)	6	4.8	4	0	1.0	0.3	0	0	0	0	—
3027	**Jackfruit**	½	cup(s)	83	60.4	78	1.2	19.8	1.3	0.2	0	0	0.1	—
990	**Kiwi fruit or Chinese gooseberries**	1	item(s)	76	63.1	46	0.9	11.1	2.3	0.4	0	0	0.2	—
	Lemon													
262	Juice	1	tablespoon(s)	15	13.8	4	0.1	1.3	0.1	0	0	0	0	0
993	Peel	1	teaspoon(s)	2	1.6	1	0	0.3	0.2	0	0	0	0	—
992	Raw	1	item(s)	108	94.4	22	1.3	11.6	5.1	0.3	0	0	0.1	—
	Lime													
269	Juice	1	tablespoon(s)	15	14.0	4	0.1	1.3	0.1	0	0	0	0	—
994	Raw	1	item(s)	67	59.1	20	0.5	7.1	1.9	0.1	0	0	0	—
995	**Loganberries, frozen**	½	cup(s)	74	62.2	40	1.1	9.6	3.9	0.2	0	0	0.1	—
	Mandarin orange													
1038	Canned in juice	½	cup(s)	125	111.4	46	0.8	11.9	0.9	0	0	0	0	—
1039	Canned in light syrup	½	cup(s)	126	104.7	77	0.6	20.4	0.9	0.1	0	0	0	—
999	**Mango**	½	cup(s)	83	67.4	54	0.4	14.0	1.5	0.2	0.1	0.1	0	—
1005	**Nectarine, raw, sliced**	½	cup(s)	69	60.4	30	0.7	7.3	1.2	0.2	0	0.1	0.1	—
	Melons													
271	Cantaloupe	½	cup(s)	80	72.1	27	0.7	6.5	0.7	0.1	0	0	0.1	—
1000	Casaba melon	½	cup(s)	85	78.1	24	0.9	5.6	0.8	0.1	0	0	0	—
272	Honeydew	½	cup(s)	89	79.5	32	0.5	8.0	0.7	0.1	0	0	0	—
318	Watermelon	½	cup(s)	76	69.5	23	0.5	5.7	0.3	0.1	0	0	0	—
	Orange													
14412	Juice with calcium and vitamin D	½	cup(s)	120	—	55	1.0	13.0	0	0	0	0	0	0
29630	Juice, fresh squeezed	½	cup(s)	124	109.5	56	0.9	12.9	0.2	0.2	0	0	0	—
14411	Juice, not from concentrate	½	cup(s)	120	—	55	1.0	13.0	0	0	0	0	0	0
278	Juice, unsweetened, prepared from frozen concentrate	½	cup(s)	125	109.7	56	0.8	13.4	0.2	0.1	0	0	0	—
3040	Peel	1	teaspoon(s)	2	1.5	2	0	0.5	0.2	0	0	0	0	—
273	Raw	1	item(s)	131	113.6	62	1.2	15.4	3.1	0.2	0	0	0	—
274	Sections	½	cup(s)	90	78.1	42	0.8	10.6	2.2	0.1	0	0	0	—
	Papaya, raw													
16830	Dried, strips	2	item(s)	46	12.0	119	1.9	29.9	5.5	0.4	0.1	0.1	0.1	—
282	Papaya	½	cup(s)	70	62.2	27	0.4	6.9	1.3	0.1	0	0	0	—
35640	**Passion fruit, purple**	1	item(s)	18	13.1	17	0.4	4.2	1.9	0.1	0	0	0.1	—
	Peach													
285	Halves, canned in heavy syrup	½	cup(s)	131	103.9	97	0.6	26.1	1.7	0.1	0	0	0.1	—
286	Halves, canned in water	½	cup(s)	122	113.6	29	0.5	7.5	1.6	0.1	0	0	0	—
290	Slices, sweetened, frozen	½	cup(s)	125	93.4	118	0.8	30.0	2.3	0.2	0	0.1	0.1	—
283	Raw, medium	1	item(s)	150	133.3	59	1.4	14.3	2.3	0.4	0	0.1	0.1	—
	Pear													
8672	Asian	1	item(s)	122	107.7	51	0.6	13.0	4.4	0.3	0	0.1	0.1	—
293	D'Anjou	1	item(s)	200	168.0	120	1.0	30.0	5.2	1.0	0	0.2	0.2	—
294	Halves, canned in heavy syrup	½	cup(s)	133	106.9	98	0.3	25.5	2.1	0.2	0	0	0	—
1012	Halves, canned in juice	½	cup(s)	124	107.2	62	0.4	16.0	2.0	0.1	0	0	0	—
291	Raw	1	item(s)	166	139.0	96	0.6	25.7	5.1	0.2	0	0	0	—

E KEY: H-4 = Breads/Baked Goods H-10 = Cereal/Rice/Pasta H-14 = Fruit H-18 = Vegetables/Legumes H-28 = Nuts/Seeds H-30 = Vegetarian H-32 = Dairy H-40 = Eggs H-40 = Seafood H-42 = Meats H-46 = Poultry
6 = Processed Meats H-48 = Beverages H-52 = Fats/Oils H-54 = Sweets H-56 = Spices/Condiments/Sauces H-60 = Mixed Foods/Soups/Sandwiches H-64 = Fast Food H-84 = Convenience H-86 = Baby Foods

Chol (mg)	Calc (mg)	Iron (mg)	Magn (mg)	Pota (mg)	Sodi (mg)	Zinc (mg)	Vit A (µg)	Thia (mg)	Vit E (mg α)	Ribo (mg)	Niac (mg)	Vit B_6 (mg)	Fola (µg)	Vit C (mg)	Vit B_{12} (µg)	Sele (µg)
0	20	0.42	7.6	97.0	2.5	0.14	8.8	0.03	—	0.07	0.19	0.02	3.8	12.6	0	0.5
0	19	0.23	7.5	148.5	0.8	0.09	11.3	0.03	0.28	0.02	0.23	0.06	4.5	20.8	0	0.5
0	10	0.45	12.5	202.2	2.5	0.08	0	0.05	0.05	0.03	0.40	0.03	12.5	33.6	0	0.1
0	11	0.25	14.8	200.1	1.2	0.06	1.2	0.05	0.27	0.02	0.25	0.05	12.4	46.9	0	0.1
0	25	0.09	10.3	154.5	0	0.07	66.4	0.04	0.14	0.03	0.23	0.06	14.9	35.7	0	0.1
0	18	0.50	12.7	163.8	2.5	0.10	0	0.04	0.11	0.02	0.30	0.02	11.4	27.1	0	1.1
0	18	0.50	12.2	161.0	2.4	0.11	0	0.05	0.11	0.03	0.30	0.02	11.0	26.6	0	1.1
0	14	0.07	10.4	170.2	0	0.08	2.3	0.04	0.15	0.02	0.30	0.05	11.5	38.3	0	1.6
0	6	0.13	2.3	87.9	0.9	0.02	2.3	0.04	0.09	0.02	0.14	0.05	1.8	1.8	0	0
0	8	0.27	5.3	144.2	1.5	0.05	2.3	0.05	0.14	0.05	0.14	0.07	1.5	8.2	0	0.1
0	9	0.16	7.5	41.3	11.3	0.04	0	0.28	0.00	0.44	0.18	0.04	1.3	33.1	0	0.1
0	5	0.13	5.0	26.3	2.5	0.05	0	0.02	0.00	0.03	0.16	0.05	1.3	29.9	0	0.1
0	12	1.06	12.4	340.3	11.6	0.07	0	0.04	—	0.07	0.46	0.07	1.2	2.2	0	0.2
0	10	0.14	12.1	229.4	1.1	0.12	17.1	0.03	0.40	0.02	0.59	0.06	27.0	125.6	0	0.3
0	1	0.01	1.0	17.5	2.2	—	0.3	0.00	—	0.00	0.03	0.00	—	2.2	0	
0	28	0.49	30.5	250.0	2.5	0.35	12.4	0.02	—	0.09	0.33	0.09	11.5	5.5	0	0.5
0	26	0.23	12.9	237.1	2.3	0.10	3.0	0.02	1.11	0.01	0.25	0.04	19.0	70.5	0	0.2
0	1	0.00	0.9	18.9	0.2	0.01	0.2	0.00	0.02	0.00	0.02	0.01	2.0	7.0	0	0
0	3	0.01	0.3	3.2	0.1	0.01	0.1	0.00	0.01	0.00	0.01	0.00	0.3	2.6	0	0
0	66	0.75	13.0	156.6	3.2	0.10	2.2	0.05	—	0.04	0.21	0.11	—	83.2	0	1.0
0	2	0.02	1.2	18.0	0.3	0.01	0.3	0.00	0.03	0.00	0.02	0.01	1.5	4.6	0	0
0	22	0.40	4.0	68.3	1.3	0.07	1.3	0.02	0.14	0.01	0.13	0.02	5.4	19.5	0	0.3
0	19	0.47	15.4	106.6	0.7	0.25	1.5	0.04	0.64	0.03	0.62	0.05	19.1	11.2	0	0.1
0	14	0.34	13.7	165.6	6.2	0.64	53.5	0.10	0.12	0.04	0.55	0.05	6.2	42.6	0	0.5
0	9	0.47	10.1	98.3	7.6	0.30	52.9	0.07	0.13	0.06	0.56	0.05	6.3	24.9	0	0.5
0	8	0.10	7.4	128.7	1.7	0.03	31.4	0.05	0.92	0.04	0.48	0.11	11.6	22.8	0	0.5
0	4	0.19	6.2	138.7	0	0.12	11.7	0.02	0.53	0.02	0.78	0.02	3.5	3.7	0	0
0	7	0.17	9.6	213.6	12.8	0.14	135.2	0.03	0.04	0.01	0.59	0.05	16.8	29.4	0	0.3
0	9	0.29	9.4	154.7	7.7	0.06	0	0.01	0.04	0.03	0.20	0.14	6.8	18.5	0	0.3
0	5	0.15	8.8	201.8	15.9	0.07	2.7	0.03	0.01	0.01	0.37	0.07	16.8	15.9	0	0.6
0	5	0.18	7.6	85.1	0.8	0.07	21.3	0.02	0.04	0.01	0.13	0.03	2.3	6.2	0	0.3
0	175	0.00	12.0	225.0	0	—	0	0.08	—	0.03	0.40	0.06	30.0	36.0	0	—
0	14	0.25	13.6	248.0	1.2	0.06	12.4	0.11	0.05	0.04	0.50	0.05	37.2	62.0	0	0.1
0	10	0.00	12.5	225.0	0	0.06	0	0.08	—	0.03	0.40	0.06	30.0	36.0	0	0.1
0	11	0.12	12.5	236.6	1.2	0.06	6.2	0.10	0.25	0.02	0.25	0.06	54.8	48.4	0	0.1
0	3	0.01	0.4	4.2	0.1	0.01	0.4	0.00	0.01	0.00	0.01	0.00	0.6	2.7	0	0
0	52	0.13	13.1	237.1	0	0.09	14.4	0.11	0.23	0.05	0.36	0.07	39.3	69.7	0	0.7
0	36	0.09	9.0	162.9	0	0.06	9.9	0.07	0.16	0.03	0.25	0.05	27.0	47.9	0	0.4
0	73	0.30	30.4	782.9	9.2	0.21	83.7	0.06	2.22	0.08	0.93	0.05	58.0	37.7	0	1.8
0	17	0.07	7.0	179.9	2.1	0.05	38.5	0.02	0.51	0.02	0.24	0.01	26.6	43.3	0	0.4
0	2	0.28	5.2	62.6	5.0	0.01	11.5	0.00	0.00	0.02	0.27	0.01	2.5	5.4	0	0.1
0	4	0.35	6.6	120.5	7.9	0.11	22.3	0.01	0.64	0.03	0.80	0.02	3.9	3.7	0	0.4
0	2	0.39	6.1	120.8	3.7	0.11	32.9	0.01	0.59	0.02	0.63	0.02	3.7	3.5	0	0.4
0	4	0.46	6.3	162.5	7.5	0.06	17.5	0.01	0.77	0.04	0.81	0.02	3.8	117.8	0	0.5
0	9	0.37	13.5	285.0	0	0.25	24.0	0.03	1.09	0.04	1.20	0.03	6.0	9.9	0	0.2
0	5	0.00	9.8	147.6	0	0.02	0	0.01	0.14	0.01	0.26	0.02	9.8	4.6	0	0.1
0	22	0.50	12.0	250.0	0	0.24	—	0.04	1.00	0.08	0.20	0.03	14.6	8.0	0	1.0
0	7	0.29	5.3	86.5	6.7	0.10	0	0.01	0.10	0.02	0.32	0.01	1.3	1.5	0	0
0	11	0.36	8.7	119.0	5.0	0.11	0	0.01	0.10	0.01	0.25	0.02	1.2	2.0	0	0
0	15	0.28	11.6	197.5	1.7	0.16	1.7	0.02	0.19	0.04	0.26	0.04	11.6	7.0	0	0.2

TABLE H-1 Table of Food Composition (*continued*) (Computer code is for Cengage Diet Analysis program) (For purposes of calculations, use "0" for t, <1, <0.1, 0.01)

DA+ Code	Food Description	Quantity	Measure	Wt (g)	H_2O (g)	Ener (kcal)	Prot (g)	Carb (g)	Fiber (g)	Fat (g)	Fat Breakdown (g) Sat	Mono	Poly	*Trans*
	Fruit and Fruit Juices—*continued*													
1017	**Persimmon**	1	item(s)	25	16.1	32	0.2	8.4	—	0.1	0	0	0	—
	Pineapple													
3053	Canned in extra heavy syrup	½	cup(s)	130	101.0	108	0.4	28.0	1.0	0.1	0	0	0	—
1019	Canned in juice	½	cup(s)	125	104.0	75	0.5	19.5	1.0	0.1	0	0	0	—
296	Canned in light syrup	½	cup(s)	126	108.0	66	0.5	16.9	1.0	0.2	0	0	0.1	—
1018	Canned in water	½	cup(s)	123	111.7	39	0.5	10.2	1.0	0.1	0	0	0	—
299	Juice, unsweetened, canned	½	cup(s)	125	108.0	66	0.5	16.1	0.3	0.2	0	0	0.1	—
295	Raw, diced	½	cup(s)	78	66.7	39	0.4	10.2	1.1	0.1	0	0	0	—
1024	**Plantain, cooked**	½	cup(s)	77	51.8	89	0.6	24.0	1.8	0.1	0.1	0	0	—
300	**Plum, raw, large**	1	item(s)	66	57.6	30	0.5	7.5	0.9	0.2	0	0.1	0	—
1027	**Pomegranate**	1	item(s)	154	124.7	105	1.5	26.4	0.9	0.5	0.1	0.1	0.1	—
	Prunes													
5644	Dried	2	item(s)	17	5.2	40	0.4	10.7	1.2	0.1	0	0	0	—
305	Dried, stewed	½	cup(s)	124	86.5	133	1.2	34.8	3.8	0.2	0	0.1	0	—
306	Juice, canned	1	cup(s)	256	208.0	182	1.6	44.7	2.6	0.1	0	0.1	0	—
	Raspberries													
309	Raw	½	cup(s)	62	52.7	32	0.7	7.3	4.0	0.4	0	0	0.2	—
310	Red, sweetened, frozen	½	cup(s)	125	90.9	129	0.9	32.7	5.5	0.2	0	0	0.1	—
311	**Rhubarb, cooked with sugar**	½	cup(s)	120	81.5	140	0.5	37.5	2.7	0.1	0	0	0.1	—
	Strawberries													
313	Raw	½	cup(s)	72	65.5	23	0.5	5.5	1.4	0.2	0	0	0.1	—
315	Sweetened, frozen, thawed	½	cup(s)	128	99.5	99	0.7	26.8	2.4	0.2	0	0	0.1	—
16828	**Tangelo**	1	item(s)	95	82.4	45	0.9	11.2	2.3	0.1	0	0	0	—
	Tangerine													
1040	Juice	½	cup(s)	124	109.8	53	0.6	12.5	0.2	0.2	0	0	0	—
316	Raw	1	item(s)	88	74.9	47	0.7	11.7	1.6	0.3	0	0.1	0.1	—
	Vegetables, Legumes													
	Amaranth													
1043	Leaves, boiled, drained	½	cup(s)	66	60.4	14	1.4	2.7	—	0.1	0	0	0.1	—
1042	Leaves, raw	1	cup(s)	28	25.7	6	0.7	1.1	—	0.1	0	0	0	—
8683	**Arugula leaves, raw**	1	cup(s)	20	18.3	5	0.5	0.7	0.3	0.1	0	0	0.1	—
	Artichoke													
1044	Boiled, drained	1	item(s)	120	100.9	64	3.5	14.3	10.3	0.4	0.1	0	0.2	—
2885	Hearts, boiled, drained	½	cup(s)	84	70.6	45	2.4	10.0	7.2	0.3	0.1	0	0.1	—
	Asparagus													
566	Boiled, drained	½	cup(s)	90	83.4	20	2.2	3.7	1.8	0.2	0	0	0.1	—
568	Canned, drained	½	cup(s)	121	113.7	23	2.6	3.0	1.9	0.8	0.2	0	0.3	—
565	Tips, frozen, boiled, drained	½	cup(s)	90	84.7	16	2.7	1.7	1.4	0.4	0.1	0	0.2	—
	Bamboo shoots													
1048	Boiled, drained	½	cup(s)	60	57.6	7	0.9	1.2	0.6	0.1	0	0	0.1	—
1049	Canned, drained	½	cup(s)	66	61.8	12	1.1	2.1	0.9	0.3	0.1	0	0.1	—
	Beans													
1801	Adzuki beans, boiled	½	cup(s)	115	76.2	147	8.6	28.5	8.4	0.1	0	—	—	—
511	Baked beans with franks, canned	½	cup(s)	130	89.8	184	8.7	19.9	8.9	8.5	3.0	3.7	1.1	—
513	Baked beans with pork in sweet sauce, canned	½	cup(s)	127	89.3	142	6.7	26.7	5.3	1.8	0.6	0.6	0.5	0
512	Baked beans with pork in tomato sauce, canned	½	cup(s)	127	93.0	119	6.5	23.6	5.1	1.2	0.5	0.7	0.3	—
1805	Black beans, boiled	½	cup(s)	86	56.5	114	7.6	20.4	7.5	0.5	0.1	0	0.2	—
14597	Chickpeas, garbanzo beans or bengal gram, boiled	½	cup(s)	82	49.4	134	7.3	22.5	6.2	2.1	0.2	0.5	0.9	—
569	Fordhook lima beans, frozen, boiled, drained	½	cup(s)	85	62.0	88	5.2	16.4	4.9	0.3	0.1	0	0.1	—
1806	French beans, boiled	½	cup(s)	89	58.9	114	6.2	21.3	8.3	0.7	0.1	0	0.4	—
2773	Great northern beans, boiled	½	cup(s)	89	61.1	104	7.4	18.7	6.2	0.4	0.1	0	0.2	—
2736	Hyacinth beans, boiled, drained	½	cup(s)	44	37.8	22	1.3	4.0	—	0.1	0.1	0.1	0	—
570	Lima beans, baby, frozen, boiled, drained	½	cup(s)	90	65.1	95	6.0	17.5	5.4	0.3	0.1	0	0.1	—
515	Lima beans, boiled, drained	½	cup(s)	85	57.1	105	5.8	20.1	4.5	0.3	0.1	0	0.1	—
579	Mung beans, sprouted, boiled, drained	½	cup(s)	62	57.9	13	1.3	2.6	0.5	0.1	0	0	0	—
510	Navy beans, boiled	½	cup(s)	91	58.1	127	7.5	23.7	9.6	0.6	0.1	0.1	0.4	0
32816	Pinto beans, boiled, drained, no salt added	½	cup(s)	63	58.8	14	1.2	2.6	—	0.2	0	0	0.1	—

E KEY: H-4 = Breads/Baked Goods H-10 = Cereal/Rice/Pasta H-14 = Fruit H-18 = Vegetables/Legumes H-28 = Nuts/Seeds H-30 = Vegetarian H-32 = Dairy H-40 = Eggs H-40 = Seafood H-42 = Meats H-46 = Poultry
6 = Processed Meats H-48 = Beverages H-52 = Fats/Oils H-54 = Sweets H-56 = Spices/Condiments/Sauces H-60 = Mixed Foods/Soups/Sandwiches H-64 = Fast Food H-84 = Convenience H-86 = Baby Foods

Chol (mg)	Calc (mg)	Iron (mg)	Magn (mg)	Pota (mg)	Sodi (mg)	Zinc (mg)	Vit A (µg)	Thia (mg)	Vit E (mg α)	Ribo (mg)	Niac (mg)	Vit B_6 (mg)	Fola (µg)	Vit C (mg)	Vit B_{12} (µg)	Sele (µg)
0	7	0.62	—	77.5	0.3	—	0	—	—	—	—	—	—	16.5	0	—
0	18	0.49	19.5	132.6	1.3	0.14	1.3	0.11	—	0.03	0.36	0.09	6.5	9.5	0	—
0	17	0.35	17.4	151.9	1.2	0.12	2.5	0.12	0.01	0.02	0.35	0.09	6.2	11.8	0	0.5
0	18	0.49	20.2	132.3	1.3	0.15	2.5	0.11	0.01	0.03	0.36	0.09	6.3	9.5	0	0.5
0	18	0.49	22.1	156.2	1.2	0.15	2.5	0.11	0.01	0.03	0.37	0.09	6.2	9.5	0	0.5
0	16	0.39	15.0	162.5	2.5	0.14	0	0.07	0.03	0.03	0.25	0.13	22.5	12.5	0	0.1
0	10	0.22	9.3	84.5	0.8	0.09	2.3	0.06	0.02	0.03	0.39	0.09	14.0	37.0	0	0.1
0	2	0.45	24.6	358.1	3.9	0.10	34.7	0.04	0.10	0.04	0.58	0.19	20.0	8.4	0	1.1
0	4	0.11	4.6	103.6	0	0.06	11.2	0.02	0.17	0.02	0.27	0.02	3.3	6.3	0	0
0	5	0.46	4.6	398.9	4.6	0.18	7.7	0.04	0.92	0.04	0.46	0.16	9.2	9.4	0	0.9
0	7	0.16	6.9	123.0	0.3	0.07	6.6	0.01	0.07	0.03	0.32	0.03	0.7	0.1	0	0
0	24	0.51	22.3	398.0	1.2	0.24	21.1	0.03	0.24	0.12	0.90	0.27	0	3.6	0	0.1
0	31	3.02	35.8	706.6	10.2	0.53	0	0.04	0.30	0.17	2.01	0.55	0	10.5	0	1.5
0	15	0.42	13.5	92.9	0.6	0.26	1.2	0.02	0.54	0.02	0.37	0.03	12.9	16.1	0	0.1
0	19	0.81	16.3	142.5	1.3	0.22	3.8	0.02	0.90	0.05	0.28	0.04	32.5	20.6	0	0.4
0	174	0.25	16.2	115.0	1.0	—	—	0.02	—	0.03	0.25	—	—	4.0	0	—
0	12	0.30	9.4	110.2	0.7	0.10	0.7	0.02	0.21	0.02	0.28	0.03	17.3	42.3	0	0.3
0	14	0.59	7.7	125.0	1.3	0.06	1.3	0.01	0.30	0.09	0.37	0.03	5.1	50.4	0	0.9
0	38	0.09	9.5	172.0	0	0.06	10.5	0.08	0.17	0.03	0.26	0.05	28.5	50.5	0	0.5
0	22	0.25	9.9	219.8	1.2	0.04	16.1	0.07	0.16	0.02	0.12	0.05	6.2	38.3	0	0.1
0	33	0.13	10.6	146.1	1.8	0.06	29.9	0.05	0.18	0.03	0.33	0.07	14.1	23.5	0	0.1
0	138	1.49	36.3	423.1	13.9	0.58	91.7	0.01	—	0.09	0.37	0.12	37.6	27.1	0	0.6
0	60	0.65	15.4	171.1	5.6	0.25	40.9	0.01	—	0.04	0.18	0.05	23.8	12.1	0	0.3
0	32	0.29	9.4	73.8	5.4	0.09	23.8	0.01	0.09	0.02	0.06	0.01	19.4	3.0	0	0.1
0	25	0.73	50.4	343.2	72.0	0.48	1.2	0.06	0.22	0.10	1.33	0.09	106.8	8.9	0	0.2
0	18	0.51	35.3	240.2	50.4	0.33	0.8	0.04	0.16	0.07	0.93	0.06	74.8	6.2	0	0.2
0	21	0.81	12.6	201.6	12.6	0.54	45.0	0.14	1.35	0.12	0.97	0.07	134.1	6.9	0	5.5
0	19	2.21	12.1	208.1	347.3	0.48	49.6	0.07	1.47	0.12	1.15	0.13	116.2	22.3	0	2.1
0	16	0.50	9.0	154.8	2.7	0.36	36.0	0.05	1.08	0.09	0.93	0.01	121.5	22.0	0	3.5
0	7	0.14	1.8	319.8	2.4	0.28	0	0.01	—	0.03	0.18	0.06	1.2	0	0	0.2
0	5	0.21	2.6	52.4	4.6	0.43	0.7	0.02	0.41	0.02	0.09	0.09	2.0	0.7	0	0.3
0	32	2.30	59.8	611.8	9.2	2.03	0	0.13	—	0.07	0.82	0.11	139.2	0	0	1.4
8	62	2.24	36.3	304.3	556.9	2.42	5.2	0.08	0.21	0.07	1.17	0.06	38.9	3.0	0.4	8.4
9	75	2.08	41.7	326.4	422.5	1.73	0	0.05	0.03	0.07	0.44	0.07	10.1	3.5	0	6.3
9	71	4.09	43.0	373.2	552.8	6.93	5.1	0.06	0.12	0.05	0.62	0.08	19.0	3.8	0	5.9
0	23	1.80	60.2	305.3	0.9	0.96	0	0.21	—	0.05	0.43	0.05	128.1	0	0	1.0
0	40	2.36	39.4	238.6	5.7	1.25	0.8	0.09	0.28	0.05	0.43	0.11	141.0	1.1	0	3.0
0	26	1.54	35.7	258.4	58.7	0.62	8.5	0.06	0.24	0.05	0.90	0.10	17.9	10.9	0	0.5
0	56	0.95	49.6	327.5	5.3	0.56	0	0.11	—	0.05	0.48	0.09	66.4	1.1	0	1.1
0	60	1.88	44.3	346.0	1.8	0.77	0	0.14	—	0.05	0.60	0.10	90.3	1.2	0	3.6
0	18	0.33	18.3	114.0	0.9	0.16	3.0	0.02	—	0.03	0.20	0.01	20.4	2.2	0	0.7
0	25	1.76	50.4	369.9	26.1	0.49	7.2	0.06	0.57	0.04	0.69	0.10	14.4	5.2	0	1.5
0	27	2.08	62.9	484.5	14.5	0.67	12.8	0.11	0.11	0.08	0.88	0.16	22.1	8.6	0	1.7
0	7	0.40	8.7	62.6	6.2	0.29	0.6	0.03	0.04	0.06	0.51	0.03	18.0	7.1	0	0.4
—	63	2.14	48.2	354.0	0	0.93	0	0.21	0.01	0.06	0.59	0.12	127.4	0.8	0	2.6
0	9	0.41	11.3	61.7	32.1	0.10	0	0.04	—	0.03	0.45	0.03	18.3	3.8	0	0.4

TABLE H-1 Table of Food Composition (*continued*) (Computer code is for Cengage Diet Analysis program) (For purposes of calculations, use "0" for t, <1, <0.1, 0.01)

DA+ Code	Food Description	Quantity	Measure	Wt (g)	H_2O (g)	Ener (kcal)	Prot (g)	Carb (g)	Fiber (g)	Fat (g)	Fat Breakdown (g)			
											Sat	Mono	Poly	*Trans*
	Vegetables, Legumes—*continued*													
1052	Pinto beans, frozen, boiled, drained	½	cup(s)	47	27.3	76	4.4	14.5	4.0	0.2	0	0	0.1	—
514	Red kidney beans, canned	½	cup(s)	128	99.0	108	6.7	19.9	6.9	0.5	0.1	0.2	0.2	—
1810	Refried beans, canned	½	cup(s)	127	96.1	119	6.9	19.6	6.7	1.6	0.6	0.7	0.2	—
1053	Shell beans, canned	½	cup(s)	123	111.1	37	2.2	7.6	4.2	0.2	0	0	0.1	—
1670	Soybeans, boiled	½	cup(s)	86	53.8	149	14.3	8.5	5.2	7.7	1.1	1.7	4.4	—
1108	Soybeans, green, boiled, drained	½	cup(s)	90	61.7	127	11.1	9.9	3.8	5.8	0.7	1.1	2.7	—
1807	White beans, small, boiled	½	cup(s)	90	56.6	127	8.0	23.1	9.3	0.6	0.1	0.1	0.2	—
575	Yellow snap, string or wax beans, boiled, drained	½	cup(s)	63	55.8	22	1.2	4.9	2.1	0.2	0	0	0.1	—
576	Yellow snap, string or wax beans, frozen, boiled, drained	½	cup(s)	68	61.7	19	1.0	4.4	2.0	0.1	0	0	0.1	—
	Beets													
584	Beet greens, boiled, drained	½	cup(s)	72	64.2	19	1.9	3.9	2.1	0.1	0	0	0.1	—
2730	Pickled, canned with liquid	½	cup(s)	114	92.9	74	0.9	18.5	3.0	0.1	0	0	0	—
581	Sliced, boiled, drained	½	cup(s)	85	74.0	37	1.4	8.5	1.7	0.2	0	0	0.1	—
583	Sliced, canned, drained	½	cup(s)	85	77.3	26	0.8	6.1	1.5	0.1	0	0	0	—
580	Whole, boiled, drained	2	item(s)	100	87.1	44	1.7	10.0	2.0	0.2	0	0	0.1	—
585	**Cowpeas or black-eyed peas, boiled, drained**	½	cup(s)	83	62.3	80	2.6	16.8	4.1	0.3	0.1	0	0.1	—
	Broccoli													
588	Chopped, boiled, drained	½	cup(s)	78	69.6	27	1.9	5.6	2.6	0.3	0.1	0	0.1	—
590	Frozen, chopped, boiled, drained	½	cup(s)	92	83.5	26	2.9	4.9	2.8	0.1	0	0	0.1	—
587	Raw, chopped	½	cup(s)	46	40.6	15	1.3	3.0	1.2	0.2	0	0	0	—
16848	**Broccoflower, raw, chopped**	½	cup(s)	32	28.7	10	0.9	1.9	1.0	0.1	0	0	0	—
	Brussels sprouts													
591	Boiled, drained	½	cup(s)	78	69.3	28	2.0	5.5	2.0	0.4	0.1	0	0.2	—
592	Frozen, boiled, drained	½	cup(s)	78	67.2	33	2.8	6.4	3.2	0.3	0.1	0	0.2	—
	Cabbage													
595	Boiled, drained, no salt added	1	cup(s)	150	138.8	35	1.9	8.3	2.8	0.1	0	0	0	—
35611	Chinese (pak choi or bok choy), boiled with salt, drained	1	cup(s)	170	162.4	20	2.6	3.0	1.7	0.3	0	0	0.1	—
16869	Kim chee	1	cup(s)	150	137.5	32	2.5	6.1	1.8	0.3	0	0	0.2	—
594	Raw, shredded	1	cup(s)	70	64.5	17	0.9	4.1	1.7	0.1	0	0	0	—
596	Red, shredded, raw	1	cup(s)	70	63.3	22	1.0	5.2	1.5	0.1	0	0	0.1	—
597	Savoy, shredded, raw	1	cup(s)	70	63.7	19	1.4	4.3	2.2	0.1	0	0	0	—
35417	**Capers**	1	teaspoon(s)	4	—	2	0	0	0	0	0	0	0	0
	Carrots													
8691	Baby, raw	8	item(s)	80	72.3	28	0.5	6.6	2.3	0.1	0	0	0.1	—
601	Grated	½	cup(s)	55	48.6	23	0.5	5.3	1.5	0.1	0	0	0.1	0
1055	Juice, canned	½	cup(s)	118	104.9	47	1.1	11.0	0.9	0.2	0	0	0.1	—
600	Raw	½	cup(s)	61	53.9	25	0.6	5.8	1.7	0.1	0	0	0.1	0
602	Sliced, boiled, drained	½	cup(s)	78	70.3	27	0.6	6.4	2.3	0.1	0	0	0.1	—
32725	**Cassava or manioc**	½	cup(s)	103	61.5	165	1.4	39.2	1.9	0.3	0.1	0.1	0	—
	Cauliflower													
606	Boiled, drained	½	cup(s)	62	57.7	14	1.1	2.5	1.4	0.3	0	0	0.1	—
607	Frozen, boiled, drained	½	cup(s)	90	84.6	17	1.4	3.4	2.4	0.2	0	0	0.1	—
605	Raw, chopped	½	cup(s)	50	46.0	13	1.0	2.6	1.2	0	0	0	0	—
	Celery													
609	Diced	½	cup(s)	51	48.2	8	0.3	1.5	0.8	0.1	0	0	0	—
608	Stalk	2	item(s)	80	76.3	13	0.6	2.4	1.3	0.1	0	0	0.1	—
	Chard													
1057	Swiss chard, boiled, drained	½	cup(s)	88	81.1	18	1.6	3.6	1.8	0.1	0	0	0	—
1056	Swiss chard, raw	1	cup(s)	36	33.4	7	0.6	1.3	0.6	0.1	0	0	0	—
	Collard greens													
610	Boiled, drained	½	cup(s)	95	87.3	25	2.0	4.7	2.7	0.3	0	0	0.2	—
611	Frozen, chopped, boiled, drained	½	cup(s)	85	75.2	31	2.5	6.0	2.4	0.3	0.1	0	0.2	—
	Corn													
29614	Yellow corn, fresh, cooked	1	item(s)	100	69.2	107	3.3	25.0	2.8	1.3	0.2	0.4	0.6	—
615	Yellow creamed sweet corn, canned	½	cup(s)	128	100.8	92	2.2	23.2	1.5	0.5	0.1	0.2	0.3	—
612	Yellow sweet corn, boiled, drained	½	cup(s)	82	57.0	89	2.7	20.6	2.3	1.1	0.2	0.3	0.5	—

E KEY: H-4 = Breads/Baked Goods H-10 = Cereal/Rice/Pasta H-14 = Fruit H-18 = Vegetables/Legumes H-28 = Nuts/Seeds H-30 = Vegetarian H-32 = Dairy H-40 = Eggs H-40 = Seafood H-42 = Meats H-46 = Poultry
6 = Processed Meats H-48 = Beverages H-52 = Fats/Oils H-54 = Sweets H-56 = Spices/Condiments/Sauces H-60 = Mixed Foods/Soups/Sandwiches H-64 = Fast Food H-84 = Convenience H-86 = Baby Foods

Chol (mg)	Calc (mg)	Iron (mg)	Magn (mg)	Pota (mg)	Sodi (mg)	Zinc (mg)	Vit A (µg)	Thia (mg)	Vit E (mg α)	Ribo (mg)	Niac (mg)	Vit B_6 (mg)	Fola (µg)	Vit C (mg)	Vit B_{12} (µg)	Sele (µg)
0	24	1.27	25.4	303.6	39.0	0.32	0	0.12	—	0.05	0.29	0.09	16.0	0.3	0	0.7
0	32	1.62	35.8	327.7	330.2	2.09	0	0.13	0.02	0.11	0.57	0.10	25.6	1.4	0	0.6
10	44	2.10	41.7	337.8	378.2	1.48	0	0.03	0.00	0.02	0.39	0.18	13.9	7.6	0	1.6
0	36	1.21	18.4	133.5	409.2	0.33	13.5	0.04	0.04	0.07	0.25	0.06	22.1	3.8	0	2.6
0	88	4.42	74.0	442.9	0.9	0.98	0	0.13	0.30	0.24	0.34	0.20	46.4	1.5	0	6.3
0	131	2.25	54.0	485.1	12.6	0.82	7.2	0.23	—	0.14	1.13	0.05	99.9	15.3	0	1.3
0	65	2.54	60.9	414.4	1.8	0.97	0	0.21	—	0.05	0.24	0.11	122.6	0	0	1.2
0	29	0.80	15.6	186.9	1.9	0.23	2.5	0.05	0.28	0.06	0.38	0.04	20.6	6.1	0	0.3
0	33	0.59	16.2	85.1	6.1	0.32	4.1	0.02	0.03	0.06	0.26	0.04	15.5	2.8	0	0.3
0	82	1.36	49.0	654.5	173.5	0.36	275.8	0.08	1.30	0.20	0.35	0.09	10.1	17.9	0	0.6
0	12	0.46	17.0	168.0	299.6	0.29	1.1	0.01	—	0.05	0.28	0.05	30.6	2.6	0	1.1
0	14	0.67	19.6	259.3	65.5	0.30	1.7	0.02	0.03	0.03	0.28	0.05	68.0	3.1	0	0.6
0	13	1.54	14.5	125.8	164.9	0.17	0.9	0.01	0.02	0.03	0.13	0.04	25.5	3.5	0	0.4
0	16	0.79	23.0	305.0	77.0	0.35	2.0	0.02	0.04	0.04	0.33	0.06	80.0	3.6	0	0.7
0	106	0.92	42.9	344.9	3.3	0.85	33.0	0.08	0.18	0.12	1.15	0.05	104.8	1.8	0	2.1
0	31	0.52	16.4	228.5	32.0	0.35	60.1	0.04	1.13	0.09	0.43	0.15	84.2	50.6	0	1.2
0	30	0.56	12.0	130.6	10.1	0.25	46.9	0.05	1.21	0.07	0.42	0.12	51.5	36.9	0	0.6
0	21	0.33	9.6	143.8	15.0	0.19	14.1	0.03	0.36	0.05	0.29	0.08	28.7	40.6	0	1.1
0	11	0.23	6.4	96.0	7.4	0.20	2.6	0.02	0.01	0.03	0.23	0.07	18.2	28.2	0	0.2
0	28	0.93	15.6	247.3	16.4	0.25	30.4	0.08	0.33	0.06	0.47	0.13	46.8	48.4	0	1.2
0	20	0.37	14.0	224.8	11.6	0.18	35.7	0.08	0.39	0.08	0.41	0.22	78.3	35.4	0	0.5
0	72	0.24	22.5	294.0	12.0	0.30	6.0	0.08	0.20	0.04	0.36	0.16	45.0	56.2	0	0.9
0	158	1.76	18.7	630.7	459.0	0.28	360.4	0.04	0.14	0.10	0.72	0.28	69.7	44.2	0	0.7
0	144	1.26	27.0	379.5	996.0	0.36	288.0	0.06	0.36	0.10	0.80	0.32	88.5	79.6	0	1.5
0	28	0.33	8.4	119.0	12.6	0.12	3.5	0.04	0.10	0.02	0.16	0.08	30.1	25.6	0	0.2
0	31	0.56	11.2	170.1	18.9	0.15	39.2	0.04	0.07	0.05	0.29	0.14	12.6	39.9	0	0.4
0	24	0.28	19.6	161.0	19.6	0.18	35.0	0.05	0.11	0.02	0.21	0.13	56.0	21.7	0	0.6
0	0	0.00	—	—	140	—	0	—	—	—	—	—	—	0	—	—
0	26	0.71	8.0	189.6	62.4	0.13	552.0	0.02	—	0.02	0.44	0.08	21.6	2.1	0	0.7
0	18	0.16	6.6	176.0	37.9	0.13	459.2	0.03	0.36	0.03	0.54	0.07	10.4	3.2	0	0.1
0	28	0.54	16.5	344.6	34.2	0.21	1128.1	0.11	1.37	0.07	0.46	0.26	4.7	10.0	0	0.7
0	20	0.18	7.3	195.2	42.1	0.15	509.4	0.04	0.40	0.04	0.60	0.08	11.6	3.6	0	0.1
0	23	0.26	7.8	183.3	45.2	0.15	664.6	0.05	0.80	0.03	0.50	0.11	10.9	2.8	0	0.5
0	16	0.27	21.6	279.1	14.4	0.35	1.0	0.08	0.19	0.04	0.87	0.09	27.8	21.2	0	0.7
0	10	0.19	5.6	88.0	9.3	0.10	0.6	0.02	0.04	0.03	0.25	0.10	27.3	27.5	0	0.4
0	15	0.36	8.1	125.1	16.2	0.11	0	0.03	0.05	0.04	0.27	0.07	36.9	28.2	0	0.5
0	11	0.22	7.5	151.5	15.0	0.14	0.5	0.03	0.04	0.03	0.26	0.11	28.5	23.2	0	0.3
0	20	0.10	5.6	131.3	40.4	0.07	11.1	0.01	0.14	0.03	0.16	0.04	18.2	1.6	0	0.2
0	32	0.16	8.8	208.0	64.0	0.10	17.6	0.01	0.21	0.04	0.25	0.05	28.8	2.5	0	0.3
0	51	1.98	75.3	480.4	156.6	0.29	267.8	0.03	1.65	0.08	0.32	0.07	7.9	15.8	0	0.8
0	18	0.64	29.2	136.4	76.7	0.13	110.2	0.01	0.68	0.03	0.14	0.03	5.0	10.8	0	0.3
0	133	1.10	19.0	110.2	15.2	0.21	385.7	0.03	0.83	0.10	0.54	0.12	88.4	17.3	0	0.5
0	179	0.95	25.5	213.4	42.5	0.22	488.8	0.04	1.06	0.09	0.54	0.09	64.6	22.4	0	1.3
0	2	0.61	32.0	248.0	242.0	0.48	13.0	0.20	0.09	0.07	1.60	0.06	46.0	6.2	0	0.2
0	4	0.48	21.8	171.5	364.8	0.67	5.1	0.03	0.09	0.06	1.22	0.08	57.6	5.9	0	0.5
0	2	0.36	21.3	173.8	0	0.50	10.7	0.17	0.07	0.05	1.32	0.04	37.7	5.1	0	0.2

TABLE H-1 Table of Food Composition (*continued*) (Computer code is for Cengage Diet Analysis program) (For purposes of calculations, use "0" for t, <1, <0.1, 0.01)

DA+ Code	Food Description	Quantity	Measure	Wt (g)	H_2O (g)	Ener (kcal)	Prot (g)	Carb (g)	Fiber (g)	Fat (g)	Fat Breakdown (g)			
											Sat	Mono	Poly	*Trans*
Vegetables, Legumes—*continued*														
614	Yellow sweet corn, frozen, boiled, drained	½	cup(s)	82	63.2	66	2.1	15.8	2.0	0.5	0.1	0.2	0.3	—
618	**Cucumber**	¼	item(s)	75	71.7	11	0.5	2.7	0.4	0.1	0	0	0	—
16870	**Cucumber, kim chee**	½	cup(s)	75	68.1	16	0.8	3.6	1.1	0.1	0	0	0	—
	Dandelion greens													
620	Chopped, boiled, drained	½	cup(s)	53	47.1	17	1.1	3.4	1.5	0.3	0.1	0	0.1	—
2734	Raw	1	cup(s)	55	47.1	25	1.5	5.1	1.9	0.4	0.1	0	0.2	—
1066	**Eggplant, boiled, drained**	½	cup(s)	50	44.4	17	0.4	4.3	1.2	0.1	0	0	0	—
621	**Endive or escarole, chopped, raw**	1	cup(s)	50	46.9	8	0.6	1.7	1.5	0.1	0	0	0	—
8784	**Jicama or yambean**	½	cup(s)	65	116.5	49	0.9	11.4	6.3	0.1	0	0	0.1	—
	Kale													
623	Frozen, chopped, boiled, drained	½	cup(s)	65	58.8	20	1.8	3.4	1.3	0.3	0	0	0.2	—
29313	Raw	1	cup(s)	67	56.6	33	2.2	6.7	1.3	0.5	0.1	0	0.2	—
	Kohlrabi													
1072	Boiled, drained	½	cup(s)	83	74.5	24	1.5	5.5	0.9	0.1	0	0	0	—
1071	Raw	1	cup(s)	135	122.9	36	2.3	8.4	4.9	0.1	0	0	0.1	—
	Leeks													
1074	Boiled, drained	½	cup(s)	52	47.2	16	0.4	4.0	0.5	0.1	0	0	0	—
1073	Raw	1	cup(s)	89	73.9	54	1.3	12.6	1.6	0.3	0	0	0.1	—
	Lentils													
522	Boiled	¼	cup(s)	50	34.5	57	4.5	10.0	3.9	0.2	0	0	0.1	—
1075	Sprouted	1	cup(s)	77	51.9	82	6.9	17.0	—	0.4	0	0.1	0.2	—
	Lettuce													
625	Butterhead leaves	11	piece(s)	83	78.9	11	1.1	1.8	0.9	0.2	0	0	0.1	—
624	Butterhead, Boston or Bibb	1	cup(s)	55	52.6	7	0.7	1.2	0.6	0.1	0	0	0.1	—
626	Iceberg	1	cup(s)	55	52.6	8	0.5	1.6	0.7	0.1	0	0	0	—
628	Iceberg, chopped	1	cup(s)	55	52.6	8	0.5	1.6	0.7	0.1	0	0	0	—
629	Looseleaf	1	cup(s)	36	34.2	5	0.5	1.0	0.5	0.1	0	0	0	—
1665	Romaine, shredded	1	cup(s)	56	53.0	10	0.7	1.8	1.2	0.2	0	0	0.1	—
	Mushrooms													
15585	Crimini (about 6)	3	ounce(s)	85	—	28	3.7	2.8	1.9	0	0	0	0	0
8700	Enoki	30	item(s)	90	79.7	40	2.3	6.9	2.4	0.3	0	0	0.1	—
1079	Mushrooms, boiled, drained	½	cup(s)	78	71.0	22	1.7	4.1	1.7	0.4	0	0	0.1	—
1080	Mushrooms, canned, drained	½	cup(s)	78	71.0	20	1.5	4.0	1.9	0.2	0	0	0.1	—
630	Mushrooms, raw	½	cup(s)	48	44.4	11	1.5	1.6	0.5	0.2	0	0	0.1	—
15587	Portabella, raw	1	item(s)	84	—	30	3.0	3.9	3.0	0	0	0	0	0
2743	Shiitake, cooked	½	cup(s)	73	60.5	41	1.1	10.4	1.5	0.2	0	0.1	0	—
	Mustard greens													
2744	Frozen, boiled, drained	½	cup(s)	75	70.4	14	1.7	2.3	2.1	0.2	0	0.1	0	—
29319	Raw	1	cup(s)	56	50.8	15	1.5	2.7	1.8	0.1	0	0	0	—
	Okra													
16866	Batter coated, fried	11	piece(s)	83	55.6	156	2.1	12.7	2.0	11.2	1.5	3.7	5.5	—
32742	Frozen, boiled, drained, no salt added	½	cup(s)	92	83.8	26	1.9	5.3	2.6	0.3	0.1	0	0.1	—
632	Sliced, boiled, drained	½	cup(s)	80	74.1	18	1.5	3.6	2.0	0.2	0	0	0	—
	Onions													
635	Chopped, boiled, drained	½	cup(s)	105	92.2	46	1.4	10.7	1.5	0.2	0	0	0.1	—
2748	Frozen, boiled, drained	½	cup(s)	106	97.8	30	0.8	7.0	1.9	0.1	0	0	0	—
1081	Onion rings, breaded and pan fried, frozen, heated	10	piece(s)	71	20.2	289	3.8	27.1	0.9	19.0	6.1	7.7	3.6	—
633	Raw, chopped	½	cup(s)	80	71.3	32	0.9	7.5	1.4	0.1	0	0	0	—
16850	Red onions, sliced, raw	½	cup(s)	57	50.7	24	0.5	5.8	0.8	0	0	0	0	—
636	Scallions, green or spring onions	2	item(s)	30	26.9	10	0.5	2.2	0.8	0.1	0	0	0	—
16860	**Palm hearts, cooked**	½	cup(s)	73	50.7	84	2.0	18.7	1.1	0.1	0	0	0.1	—
637	**Parsley, chopped**	1	tablespoon(s)	4	3.3	1	0.1	0.2	0.1	0	0	0	0	—
638	**Parsnips, sliced, boiled, drained**	½	cup(s)	78	62.6	55	1.0	13.3	2.8	0.2	0	0.1	0	—
	Peas													
639	Green peas, canned, drained	½	cup(s)	85	69.4	59	3.8	10.7	3.5	0.3	0.1	0	0.1	—
641	Green peas, frozen, boiled, drained	½	cup(s)	80	63.6	62	4.1	11.4	4.4	0.2	0	0	0.1	—
35694	Pea pods, boiled with salt, drained	½	cup(s)	80	71.1	32	2.6	5.2	2.2	0.2	0	0	0.1	—

E KEY: H-4 = Breads/Baked Goods H-10 = Cereal/Rice/Pasta H-14 = Fruit H-18 = Vegetables/Legumes H-28 = Nuts/Seeds H-30 = Vegetarian H-32 = Dairy H-40 = Eggs H-40 = Seafood H-42 = Meats H-46 = Poultry
6 = Processed Meats H-48 = Beverages H-52 = Fats/Oils H-54 = Sweets H-56 = Spices/Condiments/Sauces H-60 = Mixed Foods/Soups/Sandwiches H-64 = Fast Food H-84 = Convenience H-86 = Baby Foods

Chol (mg)	Calc (mg)	Iron (mg)	Magn (mg)	Pota (mg)	Sodi (mg)	Zinc (mg)	Vit A (μg)	Thia (mg)	Vit E (mg α)	Ribo (mg)	Niac (mg)	Vit B_6 (mg)	Fola (μg)	Vit C (mg)	Vit B_{12} (μg)	Sele (μg)
0	2	0.38	23.0	191.1	0.8	0.51	8.2	0.02	0.05	0.05	1.07	0.08	28.7	2.9	0	0.6
0	12	0.20	9.8	110.6	1.5	0.14	3.8	0.01	0.01	0.01	0.07	0.03	5.3	2.1	0	0.2
0	7	3.61	6.0	87.8	765.8	0.38	—	0.02	—	0.02	0.34	0.08	17.3	2.6	0	—
0	74	0.95	12.6	121.8	23.1	0.15	179.6	0.07	1.28	0.09	0.27	0.08	6.8	9.5	0	0.2
0	103	1.70	19.8	218.3	41.8	0.22	279.4	0.10	1.89	0.14	0.44	0.13	14.8	19.2	0	0.3
0	3	0.12	5.4	60.9	0.5	0.06	1.0	0.04	0.20	0.01	0.30	0.04	6.9	0.6	0	0
0	26	0.41	7.5	157.0	11.0	0.39	54.0	0.04	0.22	0.03	0.20	0.01	71.0	3.2	0	0.1
0	16	0.78	15.5	194.0	5.2	0.20	1.3	0.02	0.59	0.04	0.25	0.05	15.5	26.1	0	0.9
0	90	0.61	11.7	208.7	9.8	0.11	477.8	0.02	0.59	0.07	0.43	0.05	9.1	16.4	0	0.6
0	90	1.14	22.8	299.5	28.8	0.29	515.2	0.07	—	0.08	0.66	0.18	19.4	80.4	0	0.6
0	21	0.33	15.7	280.5	17.3	0.26	1.7	0.03	0.43	0.02	0.32	0.13	9.9	44.6	0	0.7
0	32	0.54	25.7	472.5	27.0	0.04	2.7	0.06	0.64	0.02	0.54	0.20	21.6	83.7	0	0.9
0	16	0.56	7.3	45.2	5.2	0.02	1.0	0.01	—	0.01	0.10	0.04	12.5	2.2	0	0.3
0	53	1.86	24.9	160.2	17.8	0.10	73.9	0.05	0.81	0.02	0.35	0.20	57.0	10.7	0	0.9
0	9	1.65	17.8	182.7	1.0	0.63	0	0.08	0.05	0.04	0.52	0.09	89.6	0.7	0	1.4
0	19	2.47	28.5	247.9	8.5	1.16	1.5	0.17	—	0.09	0.86	0.14	77.0	12.7	0	0.5
0	29	1.02	10.7	196.4	4.1	0.16	137.0	0.04	0.14	0.05	0.29	0.06	60.2	3.1	0	0.5
0	19	0.68	7.1	130.9	2.7	0.11	91.3	0.03	0.09	0.03	0.19	0.04	40.1	2.0	0	0.3
0	10	0.22	3.8	77.5	5.5	0.08	13.7	0.02	0.09	0.01	0.07	0.02	15.9	1.5	0	0.1
0	10	0.22	3.8	77.5	5.5	0.08	13.7	0.02	0.09	0.01	0.07	0.02	15.9	1.5	0	0.1
0	13	0.31	4.7	69.8	10.1	0.06	133.2	0.02	0.10	0.02	0.13	0.03	13.7	6.5	0	0.2
0	18	0.54	7.8	138.3	4.5	0.13	162.4	0.04	0.07	0.03	0.17	0.04	76.2	13.4	0	0.2
0	0	0.67	—	—	32.6	—	0	—	—	—	—	—	—	0	0	—
0	1	0.98	14.4	331.2	2.7	0.54	0	0.16	0.01	0.14	5.31	0.07	46.8	0	0	2.0
0	5	1.35	9.4	277.7	1.6	0.67	0	0.05	0.01	0.23	3.47	0.07	14.0	3.1	0	9.3
0	9	0.61	11.7	100.6	331.5	0.56	0	0.06	0.01	0.01	1.24	0.04	9.4	0	0	3.2
0	1	0.24	4.3	152.6	2.4	0.25	0	0.04	0.01	0.19	1.73	0.05	7.7	1.0	0	4.5
0	39	0.35	—	—	9.9	—	0	—	—	—	—	—	—	0	0	—
0	2	0.31	10.2	84.8	2.9	0.96	0	0.02	0.02	0.12	1.08	0.11	15.2	0.2	0	18.0
0	76	0.84	9.8	104.3	18.8	0.15	265.5	0.03	1.01	0.04	0.19	0.08	52.5	10.4	0	0.5
0	58	0.81	17.9	198.2	14.0	0.11	294.0	0.04	1.12	0.06	0.45	0.10	104.7	39.2	0	0.5
2	54	1.13	32.2	170.8	109.7	0.44	14.0	0.16	1.50	0.12	1.29	0.11	39.6	9.2	0	3.6
0	88	0.61	46.9	215.3	2.8	0.57	15.6	0.09	0.29	0.11	0.72	0.04	134.3	11.2	0	0.6
0	62	0.22	28.8	108.0	4.8	0.34	11.2	0.10	0.21	0.04	0.69	0.15	36.8	13.0	0	0.3
0	23	0.24	11.5	174.3	3.1	0.21	0	0.03	0.02	0.02	0.17	0.12	15.7	5.5	0	0.6
0	17	0.32	6.4	114.5	12.7	0.06	0	0.02	0.01	0.02	0.14	0.06	13.8	2.8	0	0.4
0	22	1.20	13.5	91.6	266.3	0.29	7.8	0.19	—	0.09	2.56	0.05	46.9	1.0	0	2.5
0	18	0.16	8.0	116.8	3.2	0.13	0	0.03	0.01	0.02	0.09	0.09	15.2	5.9	0	0.4
0	13	0.10	5.7	82.4	1.7	0.09	0	0.02	0.01	0.01	0.04	0.08	10.9	3.7	0	0.3
0	22	0.44	6.0	82.8	4.8	0.11	15.0	0.01	0.16	0.02	0.15	0.01	19.2	5.6	0	0.2
0	13	1.23	7.3	1318.4	10.2	2.72	2.2	0.03	0.36	0.12	0.62	0.53	14.6	5.0	0	0.5
0	5	0.23	1.9	21.1	2.1	0.04	16.0	0.00	0.02	0.00	0.05	0.00	5.8	5.1	0	0
0	29	0.45	22.6	286.3	7.8	0.20	0	0.06	0.78	0.04	0.56	0.07	45.2	10.1	0	1.3
0	17	0.80	14.5	147.1	214.2	0.60	23.0	0.10	0.02	0.06	0.62	0.05	37.4	8.2	0	1.4
0	19	1.21	17.6	88.0	57.6	0.53	84.0	0.22	0.02	0.08	1.18	0.09	47.2	7.9	0	0.8
0	34	1.57	20.8	192.0	192.0	0.29	41.6	0.10	0.31	0.06	0.43	0.11	23.2	38.3	0	0.6

DA+ Code	Food Description	Quantity	Measure	Wt (g)	H_2O (g)	Ener (kcal)	Prot (g)	Carb (g)	Fiber (g)	Fat (g)	Fat Breakdown (g)			
											Sat	Mono	Poly	*Trans*
Vegetables, Legumes—*continued*														
1082	Peas and carrots, canned with liquid	½	cup(s)	128	112.4	48	2.8	10.8	2.6	0.3	0.1	0	0.2	—
1083	Peas and carrots, frozen, boiled, drained	½	cup(s)	80	68.6	38	2.5	8.1	2.5	0.3	0.1	0	0.2	—
2750	Snow or sugar peas, frozen, boiled, drained	½	cup(s)	80	69.3	42	2.8	7.2	2.5	0.3	0.1	0	0.1	—
640	Snow or sugar peas, raw	½	cup(s)	32	28.0	13	0.9	2.4	0.8	0.1	0	0	0	—
29324	Split peas, sprouted	½	cup(s)	60	37.4	77	5.3	16.9	—	0.4	0.1	0	0.2	—
	Peppers													
644	Green bell or sweet, boiled, drained	½	cup(s)	68	62.5	19	0.6	4.6	0.8	0.1	0	0	0.1	—
643	Green bell or sweet, raw	½	cup(s)	75	69.9	15	0.6	3.5	1.3	0.1	0	0	0	—
1664	Green hot chili	1	item(s)	45	39.5	18	0.9	4.3	0.7	0.1	0	0	0	—
1663	Green hot chili, canned with liquid	½	cup(s)	68	62.9	14	0.6	3.5	0.9	0.1	0	0	0	—
1086	Jalapeno, canned with liquid	½	cup(s)	68	60.4	18	0.6	3.2	1.8	0.6	0.1	0	0.3	—
8703	Yellow bell or sweet	1	item(s)	186	171.2	50	1.9	11.8	1.7	0.4	0.1	0	0.2	—
1087	**Poi**	½	cup(s)	120	86.0	134	0.5	32.7	0.5	0.2	0	0	0.1	—
	Potatoes													
1090	Au gratin mix, prepared with water, whole milk and butter	½	cup(s)	124	97.7	115	2.8	15.9	1.1	5.1	3.2	1.5	0.2	—
1089	Au gratin, prepared with butter	½	cup(s)	123	90.7	162	6.2	13.8	2.2	9.3	5.8	2.6	0.3	—
5791	Baked, flesh and skin	1	item(s)	202	151.3	188	5.1	42.7	4.4	0.3	0.1	0	0.1	—
645	Baked, flesh only	½	cup(s)	61	46.0	57	1.2	13.1	0.9	0.1	0	0	0	—
1088	Baked, skin only	1	item(s)	58	27.4	115	2.5	26.7	4.6	0.1	0	0	0	—
5795	Boiled in skin, flesh only, drained	1	item(s)	136	104.7	118	2.5	27.4	2.1	0.1	0	0	0.1	—
5794	Boiled, drained, skin and flesh	1	item(s)	150	115.9	129	2.9	29.8	2.5	0.2	0	0	0.1	—
647	Boiled, flesh only	½	cup(s)	78	60.4	67	1.3	15.6	1.4	0.1	0	0	0	—
648	French fried, deep fried, prepared from raw	14	item(s)	70	32.8	187	2.7	23.5	2.9	9.5	1.9	4.2	3.0	—
649	French fried, frozen, heated	14	item(s)	70	43.7	94	1.9	19.4	2.0	3.7	0.7	2.3	0.2	—
1091	Hashed brown	½	cup(s)	78	36.9	207	2.3	27.4	2.5	9.8	1.5	4.1	3.7	—
652	Mashed with margarine and whole milk	½	cup(s)	105	79.0	119	2.1	17.7	1.6	4.4	1.0	2.0	1.2	0.7
653	Mashed, prepared from dehydrated granules with milk, water, and margarine	½	cup(s)	105	79.8	122	2.3	16.9	1.4	5.0	1.3	2.1	1.4	—
2759	Microwaved	1	item(s)	202	145.5	212	4.9	49.0	4.6	0.2	0.1	0	0.1	—
2760	Microwaved in skin, flesh only	½	cup(s)	78	57.1	78	1.6	18.1	1.2	0.1	0	0	0	—
5804	Microwaved, skin only	1	item(s)	58	36.8	77	2.5	17.2	4.2	0.1	0	0	0	—
1097	Potato puffs, frozen, heated	½	cup(s)	64	38.2	122	1.3	17.8	1.6	5.5	1.2	3.9	0.3	—
1094	Scalloped mix, prepared with water, whole milk and butter	½	cup(s)	124	98.4	116	2.6	15.9	1.4	5.3	3.3	1.5	0.2	—
1093	Scalloped, prepared with butter	½	cup(s)	123	99.2	108	3.5	13.2	2.3	4.5	2.8	1.3	0.2	—
	Pumpkin													
1773	Boiled, drained	½	cup(s)	123	114.8	25	0.9	6.0	1.3	0.1	0	0	0	—
656	Canned	½	cup(s)	123	110.2	42	1.3	9.9	3.6	0.3	0.2	0	0	—
	Radicchio													
8731	Leaves, raw	1	cup(s)	40	37.3	9	0.6	1.8	0.4	0.1	0	0	0	—
2498	Raw	1	cup(s)	40	37.3	9	0.6	1.8	0.4	0.1	0	0	0	—
657	**Radishes**	6	item(s)	27	25.7	4	0.2	0.9	0.4	0	0	0	0	—
1099	**Rutabaga, boiled, drained**	½	cup(s)	85	75.5	33	1.1	7.4	1.5	0.2	0	0	0.1	—
658	**Sauerkraut, canned**	½	cup(s)	118	109.2	22	1.1	5.1	3.4	0.2	0	0	0.1	—
	Seaweed													
1102	Kelp	½	cup(s)	40	32.6	17	0.6	3.8	0.5	0.2	0.1	0	0	—
1104	Spirulina, dried	½	cup(s)	8	0.4	22	4.3	1.8	0.3	0.6	0.2	0.1	0.2	—
1106	**Shallots**	3	tablespoon(s)	30	23.9	22	0.8	5.0	—	0	0	0	0	—
	Soybeans													
1670	Boiled	½	cup(s)	86	53.8	149	14.3	8.5	5.2	7.7	1.1	1.7	4.4	—
2825	Dry roasted	½	cup(s)	86	0.7	388	34.0	28.1	7.0	18.6	2.7	4.1	10.5	—
2824	Roasted, salted	½	cup(s)	86	1.7	405	30.3	28.9	15.2	21.8	3.2	4.8	12.3	—
8739	Sprouted, stir fried	½	cup(s)	63	42.3	79	8.2	5.9	0.5	4.5	0.6	1.0	2.5	0
	Soy products													
1813	Soy milk	1	cup(s)	240	211.3	130	7.8	15.1	1.4	4.2	0.5	1.0	2.3	0
2838	Tofu, dried, frozen (koyadofu)	3	ounce(s)	85	4.9	408	40.8	12.4	6.1	25.8	3.7	5.7	14.6	—

E KEY: H-4 = Breads/Baked Goods H-10 = Cereal/Rice/Pasta H-14 = Fruit H-18 = Vegetables/Legumes H-28 = Nuts/Seeds H-30 = Vegetarian H-32 = Dairy H-40 = Eggs H-40 = Seafood H-42 = Meats H-46 = Poultry
6 = Processed Meats H-48 = Beverages H-52 = Fats/Oils H-54 = Sweets H-56 = Spices/Condiments/Sauces H-60 = Mixed Foods/Soups/Sandwiches H-64 = Fast Food H-84 = Convenience H-86 = Baby Foods

Chol (mg)	Calc (mg)	Iron (mg)	Magn (mg)	Pota (mg)	Sodi (mg)	Zinc (mg)	Vit A (µg)	Thia (mg)	Vit E (mg α)	Ribo (mg)	Niac (mg)	Vit B_6 (mg)	Fola (µg)	Vit C (mg)	Vit B_{12} (µg)	Sele (µg)
0	29	0.96	17.9	127.5	331.5	0.74	368.5	0.09	—	0.07	0.74	0.11	23.0	8.4	0	1.1
0	18	0.75	12.8	126.4	54.4	0.36	380.8	0.18	0.41	0.05	0.92	0.07	20.8	6.5	0	0.9
0	47	1.92	22.4	173.6	4.0	0.39	52.8	0.05	0.37	0.09	0.45	0.13	28.0	17.6	0	0.6
0	14	0.65	7.6	63.0	1.3	0.08	17.0	0.04	0.12	0.02	0.19	0.05	13.2	18.9	0	0.2
0	22	1.34	33.6	228.6	12.0	0.62	4.8	0.12	—	0.08	1.84	0.14	86.4	6.2	0	0.4
0	6	0.31	6.8	112.9	1.4	0.08	15.6	0.04	0.34	0.02	0.32	0.15	10.9	50.6	0	0.2
0	7	0.25	7.5	130.4	2.2	0.09	13.4	0.04	0.27	0.02	0.35	0.16	7.5	59.9	0	0
0	8	0.54	11.3	153.0	3.2	0.13	26.6	0.04	0.31	0.04	0.42	0.12	10.4	109.1	0	0.2
0	5	0.34	9.5	127.2	797.6	0.10	24.5	0.01	0.46	0.02	0.54	0.10	6.8	46.2	0	0.2
0	16	1.28	10.2	131.2	1136.3	0.23	57.8	0.03	0.47	0.03	0.27	0.13	9.5	6.8	0	0.3
0	20	0.85	22.3	394.3	3.7	0.31	18.6	0.05	—	0.04	1.65	0.31	48.4	341.3	0	0.6
0	19	1.06	28.8	219.6	14.4	0.26	3.6	0.16	2.76	0.05	1.32	0.33	25.2	4.8	0	0.8
19	103	0.39	18.6	271.0	543.3	0.29	64.4	0.02	—	0.10	1.16	0.05	8.7	3.8	0	3.3
28	146	0.78	24.5	485.1	530.4	0.85	78.4	0.08	—	0.14	1.22	0.21	13.5	12.1	0	3.3
0	30	2.18	56.6	1080.7	20.2	0.72	2.0	0.12	0.08	0.09	2.84	0.62	56.6	19.4	0	0.8
0	3	0.21	15.3	238.5	3.1	0.18	0	0.06	0.02	0.01	0.85	0.18	5.5	7.8	0	0.2
0	20	4.08	24.9	332.3	12.2	0.28	0.6	0.07	0.02	0.06	1.77	0.35	12.8	7.8	0	0.4
0	7	0.42	29.9	515.4	5.4	0.40	0	0.14	0.01	0.02	1.95	0.40	13.6	17.7	0	0.4
0	13	1.27	34.1	572.0	7.4	0.46	0	0.14	0.01	0.03	2.13	0.44	15.0	18.4	0	—
0	6	0.24	15.6	255.8	3.9	0.21	0	0.07	0.01	0.01	1.02	0.21	7.0	5.8	0	0.2
0	16	1.05	30.8	567.0	8.4	0.39	0	0.08	0.09	0.03	1.34	0.37	16.1	21.2	0	0.4
0	8	0.51	18.2	315.7	271.6	0.26	0	0.09	0.07	0.02	1.55	0.12	19.6	9.3	0	0.1
0	11	0.43	27.3	449.3	266.8	0.37	0	0.13	0.01	0.03	1.80	0.37	12.5	10.1	0	0.4
1	23	0.27	19.9	344.4	349.6	0.31	43.0	0.09	0.44	0.04	1.23	0.25	9.4	11.0	0.1	0.8
2	36	0.21	21.0	164.8	179.5	0.26	49.3	0.09	0.53	0.09	0.90	0.16	8.4	6.8	0.1	5.9
0	22	2.50	54.5	902.9	16.2	0.72	0	0.24	—	0.06	3.46	0.69	24.2	30.5	0	0.8
0	4	0.31	19.4	319.0	5.4	0.25	0	0.10	—	0.01	1.26	0.25	9.3	11.7	0	0.3
0	27	3.44	21.5	377.0	9.3	0.29	0	0.04	0.01	0.04	1.28	0.28	9.9	8.9	0	0.3
0	9	0.41	10.9	199.7	307.2	0.21	0	0.08	0.15	0.02	0.97	0.08	9.0	4.0	0	0.4
14	45	0.47	17.4	252.2	423.7	0.31	43.5	0.02	—	0.06	1.28	0.05	12.4	4.1	0	2.0
15	70	0.70	23.3	463.1	410.4	0.49	0	0.08	—	0.11	1.29	0.22	13.5	13.0	0	2.0
0	18	0.69	11.0	281.8	1.2	0.28	306.3	0.03	0.98	0.09	0.50	0.05	11.0	5.8	0	0.2
0	32	1.70	28.2	252.4	6.1	0.20	953.1	0.02	1.29	0.06	0.45	0.06	14.7	5.1	0	0.5
0	8	0.23	5.2	120.8	8.8	0.25	0.4	0.01	0.90	0.01	0.10	0.02	24.0	3.2	0	0.4
0	8	0.23	5.2	120.8	8.8	0.25	0.4	0.01	0.90	0.01	0.10	0.02	24.0	3.2	0	0.4
0	7	0.09	2.7	62.9	10.5	0.07	0	0.00	0.00	0.01	0.06	0.01	6.8	4.0	0	0.2
0	41	0.45	19.6	277.1	17.0	0.30	0	0.07	0.27	0.04	0.61	0.09	12.8	16.0	0	0.6
0	35	1.73	15.3	200.6	780.0	0.22	1.2	0.03	0.17	0.03	0.17	0.15	28.3	17.3	0	0.7
0	67	1.12	48.4	35.6	93.2	0.48	2.4	0.02	0.32	0.04	0.16	0.00	72.0	1.2	0	0.3
0	9	2.14	14.6	102.2	78.6	0.15	2.2	0.18	0.38	0.28	0.96	0.03	7.1	0.8	0	0.5
0	11	0.36	6.3	100.2	3.6	0.12	18.0	0.02	—	0.01	0.06	0.09	10.2	2.4	—	0.4
0	88	4.42	74.0	442.9	0.9	0.98	0	0.13	0.30	0.24	0.34	0.20	46.4	1.5	0	6.3
0	120	3.39	196.1	1173.0	1.7	4.10	0	0.36	—	0.64	0.90	0.19	176.3	4.0	0	16.6
0	119	3.35	124.7	1264.2	140.2	2.70	8.6	0.08	0.78	0.12	1.21	0.17	181.5	1.9	0	16.4
0	52	0.25	60.4	356.6	8.8	1.32	0.6	0.26	—	0.12	0.69	0.10	79.9	7.5	0	0.4
0	60	1.53	60.0	283.2	122.4	0.28	0	0.14	0.26	0.16	1.23	0.18	43.2	0	0	11.5
0	310	8.27	50.2	17.0	5.1	4.16	22.1	0.42	—	0.27	1.01	0.24	78.2	0.6	0	46.2

DA+ Code	Food Description	Quantity	Measure	Wt (g)	H_2O (g)	Ener (kcal)	Prot (g)	Carb (g)	Fiber (g)	Fat (g)	Fat Breakdown (g)			
											Sat	Mono	Poly	*Trans*
Vegetables, Legumes—*continued*														
13844	Tofu, extra firm	3	ounce(s)	85	—	86	8.6	2.2	1.1	4.3	0.5	0.9	2.8	—
13843	Tofu, firm	3	ounce(s)	85	—	75	7.5	2.2	0.5	3.2	0	0.9	2.3	—
1816	Tofu, firm, with calcium sulfate and magnesium chloride (nigari)	3	ounce(s)	85	72.2	60	7.0	1.4	0.8	3.5	0.7	1.0	1.5	—
1817	Tofu, fried	3	ounce(s)	85	43.0	230	14.6	8.9	3.3	17.2	2.5	3.8	9.7	—
13841	Tofu, silken	3	ounce(s)	85	—	42	3.7	1.9	0	2.3	0.5	—	—	—
13842	Tofu, soft	3	ounce(s)	85	—	65	6.5	1.1	0.5	3.2	0.5	1.1	2.2	—
1671	Tofu, soft, with calcium sulfate and magnesium chloride (nigari)	3	ounce(s)	85	74.2	52	5.6	1.5	0.2	3.1	0.5	0.7	1.8	—
	Spinach													
663	Canned, drained	½	cup(s)	107	98.2	25	3.0	3.6	2.6	0.5	0.1	0	0.2	—
660	Chopped, boiled, drained	½	cup(s)	90	82.1	21	2.7	3.4	2.2	0.2	0	0	0.1	—
661	Chopped, frozen, boiled, drained	½	cup(s)	95	84.5	32	3.8	4.6	3.5	0.8	0.1	0	0.4	—
662	Leaf, frozen, boiled, drained	½	cup(s)	95	84.5	32	3.8	4.6	3.5	0.8	0.1	0	0.4	—
659	Raw, chopped	1	cup(s)	30	27.4	7	0.9	1.1	0.7	0.1	0	0	0	—
8470	Trimmed leaves	1	cup(s)	32	27.5	3	0.9	0	2.8	0.1	—	—	—	—
	Squash													
1662	Acorn winter, baked	½	cup(s)	103	85.0	57	1.1	14.9	4.5	0.1	0	0	0.1	—
29702	Acorn winter, boiled, mashed	½	cup(s)	123	109.9	42	0.8	10.8	3.2	0.1	0	0	0	—
29451	Butternut, frozen, boiled	½	cup(s)	122	106.9	47	1.5	12.2	1.8	0.1	0	0	0	—
1661	Butternut winter, baked	½	cup(s)	102	89.5	41	0.9	10.7	3.4	0.1	0	0	0	—
32773	Butternut winter, frozen, boiled, mashed, no salt added	½	cup(s)	121	106.4	47	1.5	12.2	—	0.1	0	0	0	—
29700	Crookneck and straightneck summer, boiled, drained	½	cup(s)	65	60.9	12	0.6	2.6	1.2	0.1	0	0	0.1	—
29703	Hubbard winter, baked	½	cup(s)	102	86.8	51	2.5	11.0	—	0.6	0.1	0	0.3	—
1660	Hubbard winter, boiled, mashed	½	cup(s)	118	107.5	35	1.7	7.6	3.4	0.4	0.1	0	0.2	—
29704	Spaghetti winter, boiled, drained, or baked	½	cup(s)	78	71.5	21	0.5	5.0	1.1	0.2	0	0	0.1	—
664	Summer, all varieties, sliced, boiled, drained	½	cup(s)	90	84.3	18	0.8	3.9	1.3	0.3	0.1	0	0.1	—
665	Winter, all varieties, baked, mashed	½	cup(s)	103	91.4	38	0.9	9.1	2.9	0.4	0.1	0	0.2	—
1112	Zucchini summer, boiled, drained	½	cup(s)	90	85.3	14	0.6	3.5	1.3	0	0	0	0	—
1113	Zucchini summer, frozen, boiled, drained	½	cup(s)	112	105.6	19	1.3	4.0	1.4	0.1	0	0	0.1	—
	Sweet potatoes													
666	Baked, peeled	½	cup(s)	100	75.8	90	2.0	20.7	3.3	0.2	0	0	0.1	—
667	Boiled, mashed	½	cup(s)	164	131.4	125	2.2	29.1	4.1	0.2	0.1	0	0.1	—
668	Candied, home recipe	½	cup(s)	91	61.1	132	0.8	25.4	2.2	3.0	1.2	0.6	0.1	—
670	Canned, vacuum pack	½	cup(s)	100	76.0	91	1.7	21.1	1.8	0.2	0	0	0.1	—
2765	Frozen, baked	½	cup(s)	88	64.5	88	1.5	20.5	1.6	0.1	0	0	0	—
1136	Yams, baked or boiled, drained	½	cup(s)	68	47.7	79	1.0	18.7	2.7	0.1	0	0	0	—
32785	**Taro shoots, cooked, no salt added**	½	cup(s)	70	66.7	10	0.5	2.2	—	0.1	0	0	0	—
	Tomatillo													
8774	Raw	2	item(s)	68	62.3	22	0.7	4.0	1.3	0.7	0.1	0.1	0.3	—
8777	Raw, chopped	½	cup(s)	66	60.5	21	0.6	3.9	1.3	0.7	0.1	0.1	0.3	—
	Tomato													
16846	Cherry, fresh	5	item(s)	85	80.3	15	0.7	3.3	1.0	0.2	0	0	0.1	—
671	Fresh, ripe, red	1	item(s)	123	116.2	22	1.1	4.8	1.5	0.2	0	0	0.1	—
675	Juice, canned	½	cup(s)	122	114.1	21	0.9	5.2	0.5	0.1	0	0	0	—
75	Juice, no salt added	½	cup(s)	122	114.1	21	0.9	5.2	0.5	0.1	0	0	0	—
1699	Paste, canned	2	tablespoon(s)	33	24.1	27	1.4	6.2	1.3	0.2	0	0	0.1	—
1700	Puree, canned	¼	cup(s)	63	54.9	24	1.0	5.6	1.2	0.1	0	0	0.1	—
1118	Red, boiled	½	cup(s)	120	113.2	22	1.1	4.8	0.8	0.1	0	0	0.1	—
3952	Red, diced	½	cup(s)	90	85.1	16	0.8	3.5	1.1	0.2	0	0	0.1	—
1120	Red, stewed, canned	½	cup(s)	128	116.7	33	1.2	7.9	1.3	0.2	0	0	0.1	—
1125	Sauce, canned	¼	cup(s)	61	55.6	15	0.8	3.3	0.9	0.1	0	0	0	—
8778	Sun dried	½	cup(s)	27	3.9	70	3.8	15.1	3.3	0.8	0.1	0.1	0.3	—
8783	Sun dried in oil, drained	¼	cup(s)	28	14.8	59	1.4	6.4	1.6	3.9	0.5	2.4	0.6	—
	Turnips													
678	Turnip greens, chopped, boiled, drained	½	cup(s)	72	67.1	14	0.8	3.1	2.5	0.2	0	0	0.1	—

E KEY: H-4 = Breads/Baked Goods H-10 = Cereal/Rice/Pasta H-14 = Fruit H-18 = Vegetables/Legumes H-28 = Nuts/Seeds H-30 = Vegetarian H-32 = Dairy H-40 = Eggs H-40 = Seafood H-42 = Meats H-46 = Poultry
6 = Processed Meats H-48 = Beverages H-52 = Fats/Oils H-54 = Sweets H-56 = Spices/Condiments/Sauces H-60 = Mixed Foods/Soups/Sandwiches H-64 = Fast Food H-84 = Convenience H-86 = Baby Foods

Chol (mg)	Calc (mg)	Iron (mg)	Magn (mg)	Pota (mg)	Sodi (mg)	Zinc (mg)	Vit A (µg)	Thia (mg)	Vit E (mg α)	Ribo (mg)	Niac (mg)	Vit B_6 (mg)	Fola (µg)	Vit C (mg)	Vit B_{12} (µg)	Sele (µg)
0	65	1.16	84.1	—	0	—	0	—	—	—	—	—	—	0	0	—
0	108	1.16	56.1	—	0	—	0	—	—	—	—	—	—	0	0	—
0	171	1.36	31.5	125.9	10.2	0.70	0	0.05	0.01	0.05	0.08	0.06	16.2	0.2	0	8.4
0	316	4.14	51.0	124.2	13.6	1.69	0.9	0.14	0.03	0.04	0.08	0.08	23.0	0	0	24.2
0	56	0.34	33.1	—	4.7	—	0	—	—	—	—	—	—	0	1.7	—
0	108	1.16	35.5	—	0	—	0	—	—	—	—	—	—	0	1.9	—
0	94	0.94	23.0	102.1	6.8	0.54	0	0.04	0.01	0.03	0.45	0.04	37.4	0.2	0	7.6
0	136	2.45	81.3	370.2	28.9	0.48	524.3	0.02	2.08	0.14	0.41	0.11	104.8	15.3	0	1.5
0	122	3.21	78.3	419.4	63.0	0.68	471.6	0.08	1.87	0.21	0.44	0.21	131.4	8.8	0	1.4
0	145	1.86	77.9	286.9	92.2	0.46	572.9	0.07	3.36	0.16	0.41	0.12	115.0	2.1	0	5.2
0	145	1.86	77.9	286.9	92.2	0.46	572.9	0.07	3.36	0.16	0.41	0.12	115.0	2.1	0	5.2
0	30	0.81	23.7	167.4	23.7	0.16	140.7	0.02	0.61	0.06	0.22	0.06	58.2	8.4	0	0.3
0	25	2.13	25.5	134.1	38.0	0.18	—	0.03	—	0.05	0.18	0.07	0	7.5	0	—
0	45	0.95	44.1	447.9	4.1	0.17	21.5	0.17	—	0.01	0.90	0.19	19.5	11.1	0	0.7
0	32	0.68	31.9	322.2	3.7	0.13	50.2	0.12	—	0.01	0.65	0.14	13.5	8.0	0	0.5
0	23	0.70	10.9	161.9	2.4	0.14	203.3	0.06	0.14	0.05	0.56	0.08	19.5	4.3	0	0.6
0	42	0.61	29.6	289.6	4.1	0.13	569.1	0.07	1.31	0.01	0.99	0.12	19.4	15.4	0	0.5
0	23	0.70	10.9	161.2	2.4	0.14	202.4	0.06	—	0.05	0.56	0.08	19.4	4.2	0	0.6
0	14	0.31	13.6	137.1	1.3	0.19	5.2	0.03	—	0.02	0.29	0.07	14.9	5.4	0	0.1
0	17	0.48	22.4	365.1	8.2	0.15	308.0	0.07	—	0.04	0.57	0.17	16.3	9.7	0	0.6
0	12	0.33	15.3	252.5	5.9	0.11	236.0	0.05	0.14	0.03	0.39	0.12	11.8	7.7	0	0.4
0	16	0.26	8.5	90.7	14.0	0.15	4.7	0.02	0.09	0.01	0.62	0.07	6.2	2.7	0	0.2
0	24	0.32	21.6	172.8	0.9	0.35	9.9	0.04	0.12	0.03	0.46	0.05	18.0	5.0	0	0.2
0	23	0.45	13.3	247.0	1.0	0.23	267.5	0.02	0.12	0.07	0.51	0.17	20.5	9.8	0	0.4
0	12	0.32	19.8	227.7	2.7	0.16	50.4	0.04	0.11	0.04	0.39	0.07	15.3	4.1	0	0.2
0	19	0.54	14.5	216.3	2.2	0.22	10	0.05	0.13	0.04	0.43	0.05	8.9	4.1	0	0.2
0	38	0.69	27.0	475.0	36.0	0.32	961.0	0.10	0.71	0.10	1.48	0.28	6.0	19.6	0	0.2
0	44	1.18	29.5	377.2	44.3	0.33	1290.7	0.09	1.54	0.08	0.88	0.27	9.8	21.0	0	0.3
7	24	1.03	10.0	172.6	63.9	0.13	0	0.01	—	0.03	0.36	0.03	10.0	6.1	0	0.7
0	22	0.89	22.0	312.0	53.0	0.18	399.0	0.04	1.00	0.06	0.74	0.19	17.0	26.4	0	0.7
0	31	0.47	18.4	330.1	7.0	0.26	913.3	0.05	0.67	0.04	0.49	0.16	19.3	8.0	0	0.5
0	10	0.35	12.2	455.6	5.4	0.13	4.1	0.06	0.23	0.01	0.37	0.15	10.9	8.2	0	0.5
0	10	0.28	5.6	240.8	1.4	0.37	2.1	0.02	—	0.03	0.56	0.07	2.1	13.2	0	0.7
0	5	0.42	13.6	182.2	0.7	0.15	4.1	0.03	0.25	0.02	1.25	0.03	4.8	8.0	0	0.3
0	5	0.41	13.2	176.9	0.7	0.15	4.0	0.03	0.25	0.02	1.22	0.04	4.6	7.7	0	0.3
0	9	0.22	9.4	201.5	4.3	0.14	35.7	0.03	0.45	0.01	0.50	0.06	12.8	10.8	0	0
0	12	0.33	13.5	291.5	6.2	0.20	51.7	0.04	0.66	0.02	0.73	0.09	18.5	15.6	0	0
0	12	0.52	13.4	278.2	326.8	0.18	27.9	0.06	0.39	0.04	0.82	0.14	24.3	22.2	0	0.4
0	12	0.52	13.4	278.2	12.2	0.18	27.9	0.06	0.39	0.04	0.82	0.14	24.3	22.2	0	0.4
0	12	0.97	13.8	332.6	259.1	0.20	24.9	0.02	1.41	0.05	1.00	0.07	3.9	7.2	0	1.7
0	11	1.11	14.4	274.4	249.4	0.22	16.3	0.01	1.23	0.05	0.91	0.07	6.9	6.6	0	0.4
0	13	0.82	10.8	261.6	13.2	0.17	28.8	0.04	0.67	0.03	0.64	0.10	15.6	27.4	0	0.6
0	9	0.24	9.9	213.3	4.5	0.15	37.8	0.03	0.48	0.01	0.53	0.07	13.5	11.4	0	0
0	43	1.70	15.3	263.9	281.8	0.22	11.5	0.06	1.06	0.04	0.91	0.02	6.4	10.1	0	0.8
0	8	0.62	9.8	201.9	319.6	0.12	10.4	0.01	0.87	0.04	0.59	0.06	6.7	4.3	0	0.1
0	30	2.45	52.4	925.3	565.7	0.53	11.9	0.14	0.00	0.13	2.44	0.09	18.4	10.6	0	1.5
0	13	0.73	22.3	430.4	73.2	0.21	17.6	0.05	—	0.10	0.99	0.08	6.3	28.0	0	0.8
0	99	0.58	15.8	146.2	20.9	0.10	274.3	0.03	1.35	0.05	0.30	0.13	85.0	19.7	0	0.6

TABLE H-1 Table of Food Composition (*continued*)

(Computer code is for Cengage Diet Analysis program) (For purposes of calculations, use "0" for t, <1, <0.1, 0.01)

DA+ Code	Food Description	Quantity	Measure	Wt (g)	H_2O (g)	Ener (kcal)	Prot (g)	Carb (g)	Fiber (g)	Fat (g)	Fat Breakdown (g) Sat	Mono	Poly	*Trans*
	Vegetables, Legumes—*continued*													
679	Turnip greens, frozen, chopped, boiled, drained	½	cup(s)	82	74.1	24	2.7	4.1	2.8	0.3	0.1	0	0.1	—
677	Turnips, cubed, boiled, drained	½	cup(s)	78	73.0	17	0.6	3.9	1.6	0.1	0	0	0	—
	Vegetables, mixed													
1132	Canned, drained	½	cup(s)	82	70.9	40	2.1	7.5	2.4	0.2	0	0	0.1	—
680	Frozen, boiled, drained	½	cup(s)	91	75.7	59	2.6	11.9	4.0	0.1	0	0	0.1	—
7489	V8 100% vegetable juice	½	cup(s)	120	—	25	1.0	5.0	1.0	0	0	0	0	0
7490	V8 low sodium vegetable juice	½	cup(s)	120	—	25	0	6.5	1.0	0	0	0	0	0
7491	V8 spicy hot vegetable juice	½	cup(s)	120	—	25	1.0	5.0	0.5	0	0	0	0	0
	Water chestnuts													
31073	Sliced, drained	½	cup(s)	75	70.0	20	0	5.0	1.0	0	0	0	0	0
31087	Whole	½	cup(s)	75	70.0	20	0	5.0	1.0	0	0	0	0	0
1135	**Watercress**	1	cup(s)	34	32.3	4	0.8	0.4	0.2	0	0	0	0	—
	Nuts, Seeds, and Products													
	Almonds													
32940	Almond butter with salt added	1	tablespoon(s)	16	0.2	101	2.4	3.4	0.6	9.5	0.9	6.1	2.0	—
1137	Almond butter, no salt added	1	tablespoon(s)	16	0.2	101	2.4	3.4	0.6	9.5	0.9	6.1	2.0	—
32886	Blanched	¼	cup(s)	36	1.6	211	8.0	7.2	3.8	18.3	1.4	11.7	4.4	—
32887	Dry roasted, no salt added	¼	cup(s)	35	0.9	206	7.6	6.7	4.1	18.2	1.4	11.6	4.4	—
29724	Dry roasted, salted	¼	cup(s)	35	0.9	206	7.6	6.7	4.1	18.2	1.4	11.6	4.4	—
29725	Oil roasted, salted	¼	cup(s)	39	1.1	238	8.3	6.9	4.1	21.7	1.7	13.7	5.3	—
508	Slivered	¼	cup(s)	27	1.3	155	5.7	5.9	3.3	13.3	1.0	8.3	3.3	0
1138	**Beechnuts, dried**	¼	cup(s)	57	3.8	328	3.5	19.1	5.3	28.5	3.3	12.5	11.4	—
517	**Brazil nuts, dried, unblanched**	¼	cup(s)	35	1.2	230	5.0	4.3	2.6	23.3	5.3	8.6	7.2	—
1166	**Breadfruit seeds, roasted**	¼	cup(s)	57	28.3	118	3.5	22.8	3.4	1.5	0.4	0.2	0.8	—
1139	**Butternuts, dried**	¼	cup(s)	30	1.0	184	7.5	3.6	1.4	17.1	0.4	3.1	12.8	—
	Cashews													
32931	Cashew butter with salt added	1	tablespoon(s)	16	0.5	94	2.8	4.4	0.3	7.9	1.6	4.7	1.3	—
32889	Cashew butter, no salt added	1	tablespoon(s)	16	0.5	94	2.8	4.4	0.3	7.9	1.6	4.7	1.3	—
1140	Dry roasted	¼	cup(s)	34	0.6	197	5.2	11.2	1.0	15.9	3.1	9.4	2.7	—
518	Oil roasted	¼	cup(s)	32	1.1	187	5.4	9.6	1.1	15.4	2.7	8.4	2.8	—
	Coconut, shredded													
32896	Dried, not sweetened	¼	cup(s)	23	0.7	152	1.6	5.4	3.8	14.9	13.2	0.6	0.2	—
1153	Dried, shredded, sweetened	¼	cup(s)	23	2.9	116	0.7	11.1	1.0	8.3	7.3	0.4	0.1	—
520	Shredded	¼	cup(s)	20	9.4	71	0.7	3.0	1.8	6.7	5.9	0.3	0.1	—
	Chestnuts													
1152	Chinese, roasted	¼	cup(s)	36	14.6	87	1.6	19.0	—	0.4	0.1	0.2	0.1	—
32895	European, boiled and steamed	¼	cup(s)	46	31.3	60	0.9	12.8	—	0.6	0.1	0.2	0.2	—
32911	European, roasted	¼	cup(s)	36	14.5	88	1.1	18.9	1.8	0.8	0.1	0.3	0.3	—
32922	Japanese, boiled and steamed	¼	cup(s)	36	31.0	20	0.3	4.5	—	0.1	0	0	0	—
32923	Japanese, roasted	¼	cup(s)	36	18.1	73	1.1	16.4	—	0.3	0	0.1	0.1	—
4958	**Flax seeds or linseeds**	¼	cup(s)	43	3.3	225	8.4	12.3	11.9	17.7	1.7	3.2	12.6	0
32904	**Ginkgo nuts, dried**	¼	cup(s)	39	4.8	136	4.0	28.3	—	0.8	0.1	0.3	0.3	—
	Hazelnuts or filberts													
32901	Blanched	¼	cup(s)	30	1.7	189	4.1	5.1	3.3	18.3	1.4	14.5	1.7	—
32902	Dry roasted, no salt added	¼	cup(s)	30	0.8	194	4.5	5.3	2.8	18.7	1.3	14.0	2.5	—
1156	**Hickory nuts, dried**	¼	cup(s)	30	0.8	197	3.8	5.5	1.9	19.3	2.1	9.8	6.6	—
	Macadamias													
32905	Dry roasted, no salt added	¼	cup(s)	34	0.5	241	2.6	4.5	2.7	25.5	4.0	19.9	0.5	—
32932	Dry roasted, with salt added	¼	cup(s)	34	0.5	240	2.6	4.3	2.7	25.5	4.0	19.9	0.5	—
1157	Raw	¼	cup(s)	34	0.5	241	2.6	4.6	2.9	25.4	4.0	19.7	0.5	—
	Mixed nuts													
1159	With peanuts, dry roasted	¼	cup(s)	34	0.6	203	5.9	8.7	3.1	17.6	2.4	10.8	3.7	—
32933	With peanuts, dry roasted, with salt added	¼	cup(s)	34	0.6	203	5.9	8.7	3.1	17.6	2.4	10.8	3.7	—
32906	Without peanuts, oil roasted, no salt added	¼	cup(s)	36	1.1	221	5.6	8.0	2.0	20.2	3.3	11.9	4.1	—
	Peanuts													
2807	Dry roasted	¼	cup(s)	37	0.6	214	8.6	7.9	2.9	18.1	2.5	9.0	5.7	—
2806	Dry roasted, salted	¼	cup(s)	37	0.6	214	8.6	7.9	2.9	18.1	2.5	9.0	5.7	—
1763	Oil roasted, salted	¼	cup(s)	36	0.5	216	10.1	5.5	3.4	18.9	3.1	9.4	5.5	—
1884	Peanut butter, chunky	1	tablespoon(s)	16	0.2	94	3.8	3.5	1.3	8.0	1.3	3.9	2.4	—
30303	Peanut butter, low sodium	1	tablespoon(s)	16	0.2	95	4.0	3.1	0.9	8.2	1.8	3.9	2.2	—
30305	Peanut butter, reduced fat	1	tablespoon(s)	18	0.2	94	4.7	6.4	0.9	6.1	1.3	2.9	1.8	—

KEY: H-4 = Breads/Baked Goods H-10 = Cereal/Rice/Pasta H-14 = Fruit H-18 = Vegetables/Legumes H-28 = Nuts/Seeds H-30 = Vegetarian H-32 = Dairy H-40 = Eggs H-40 = Seafood H-42 = Meats H-46 = Poultry
= Processed Meats H-48 = Beverages H-52 = Fats/Oils H-54 = Sweets H-56 = Spices/Condiments/Sauces H-60 = Mixed Foods/Soups/Sandwiches H-64 = Fast Food H-84 = Convenience H-86 = Baby Foods

Chol (mg)	Calc (mg)	Iron (mg)	Magn (mg)	Pota (mg)	Sodi (mg)	Zinc (mg)	Vit A (μg)	Thia (mg)	Vit E (mg α)	Ribo (mg)	Niac (mg)	Vit B_6 (mg)	Fola (μg)	Vit C (mg)	Vit B_{12} (μg)	Sele (μg)
0	125	1.59	21.3	183.7	12.3	0.34	441.2	0.04	2.18	0.06	0.38	0.06	32.0	17.9	0	1.0
0	26	0.14	7.0	138.1	12.5	0.09	0	0.02	0.02	0.02	0.23	0.05	7.0	9.0	0	0.2
0	22	0.86	13.0	237.2	121.4	0.33	475.1	0.04	0.24	0.04	0.47	0.06	19.6	4.1	0	0.2
0	23	0.74	20.0	153.8	31.9	0.44	194.7	0.06	0.34	0.10	0.77	0.06	17.3	2.9	0	0.3
0	20	0.36	12.9	260.0	310.0	0.24	100.0	0.05	—	0.03	0.87	0.17	—	30.0	0	—
0	20	0.36	—	450.0	70.0	—	100.0	0.02	—	0.02	0.75	—	—	30.0	0	—
0	20	0.36	12.9	240.0	360.0	0.24	50.0	0.05	—	0.03	0.88	0.17	—	15.0	0	—
0	7	0.00	—	—	5.0	—	0	—	—	—	—	—	—	2.0	—	—
0	7	0.00	—	—	5.0	—	0	—	—	—	—	—	—	2.0	—	—
0	41	0.06	7.1	112.2	13.9	0.03	54.4	0.03	0.34	0.04	0.06	0.04	3.1	14.6	0	0.3
0	43	0.59	48.5	121.3	72.0	0.49	0	0.02	4.16	0.10	0.46	0.01	10.4	0.1	0	0.8
0	43	0.59	48.5	121.3	1.8	0.48	0	0.02	—	0.09	0.46	0.01	10.4	0.1	0	—
0	78	1.34	99.7	249.0	10.2	1.13	0	0.07	8.95	0.20	1.32	0.04	10.9	0	0	1.0
0	92	1.55	98.7	257.4	0.3	1.22	0	0.02	8.97	0.29	1.32	0.04	11.4	0	0	1.0
0	92	1.55	98.7	257.4	117.0	1.22	0	0.02	8.97	0.29	1.32	0.04	11.4	0	0	1.0
0	114	1.44	107.5	274.4	133.1	1.20	0	0.03	10.19	0.30	1.43	0.04	10.6	0	0	1.1
0	71	1.00	72.4	190.4	0.3	0.83	0	0.05	7.07	0.27	0.91	0.03	13.5	0	0	0.7
0	1	1.39	0	579.7	21.7	0.20	0	0.16	—	0.20	0.48	0.38	64.4	8.8	0	4.0
0	56	0.85	131.6	230.7	1.1	1.42	0	0.21	2.00	0.01	0.10	0.03	7.7	0.2	0	671.0
0	49	0.50	35.3	616.7	15.9	0.58	8.5	0.22	—	0.12	4.20	0.22	33.6	4.3	0	8.0
0	16	1.21	71.1	126.3	0.3	0.94	1.8	0.12	—	0.04	0.31	0.17	19.8	1.0	0	5.2
0	7	0.81	41.3	87.4	98.2	0.83	0	0.05	0.15	0.03	0.26	0.04	10.9	0	0	1.8
0	7	0.81	41.3	87.4	2.4	0.83	0	0.05	—	0.03	0.26	0.04	10.9	0	0	1.8
0	15	2.06	89.1	193.5	5.5	1.92	0	0.07	0.32	0.07	0.48	0.09	23.6	0	0	4.0
0	14	1.95	88.0	203.8	4.2	1.73	0	0.12	0.30	0.07	0.56	0.10	8.1	0.1	0	6.5
0	6	0.76	20.7	125.2	8.5	0.46	0	0.01	0.10	0.02	0.13	0.07	2.1	0.3	0	4.3
0	3	0.45	11.6	78.4	60.9	0.42	0	0.01	0.09	0.00	0.11	0.06	1.9	0.2	0	3.9
0	3	0.48	6.4	71.2	4.0	0.21	0	0.01	0.04	0.00	0.11	0.01	5.2	0.7	0	2.0
0	7	0.54	32.6	173.0	1.4	0.33	0	0.05	—	0.03	0.54	0.15	26.1	13.9	0	2.6
0	21	0.80	24.8	328.9	12.4	0.11	0.5	0.06	—	0.03	0.32	0.10	17.5	12.3	0	—
0	10	0.32	11.8	211.6	0.7	0.20	0.4	0.08	0.18	0.05	0.48	0.18	25.0	9.3	0	0.4
0	4	0.19	6.5	42.8	1.8	0.14	0.4	0.04	—	0.01	0.19	0.03	6.1	3.4	0	—
0	13	0.75	23.2	154.8	6.9	0.51	1.4	0.15	—	—	0.24	0.14	21.4	10.1	0	—
0	142	2.13	156.1	354.0	11.9	1.83	0	0.06	0.14	0.06	0.59	0.39	118.4	0.5	0	2.3
0	8	0.62	20.7	390.2	5.1	0.26	21.5	0.17	—	0.07	4.58	0.25	41.4	11.4	0	—
0	45	0.98	48.0	197.4	0	0.66	0.6	0.14	5.25	0.03	0.46	0.17	23.4	0.6	0	1.2
0	37	1.31	51.9	226.5	0	0.74	0.9	0.10	4.58	0.03	0.61	0.18	26.4	1.1	0	1.2
0	18	0.64	51.9	130.8	0.3	1.29	2.1	0.26	—	0.04	0.27	0.06	12.0	0.6	0	2.4
0	23	0.88	39.5	121.6	1.3	0.43	0	0.23	0.19	0.02	0.76	0.12	3.4	0.2	0	3.9
0	23	0.88	39.5	121.6	88.8	0.43	0	0.23	0.19	0.02	0.76	0.12	3.4	0.2	0	3.9
0	28	1.24	43.6	123.3	1.7	0.44	0	0.40	0.18	0.05	0.83	0.09	3.7	0.4	0	1.2
0	24	1.27	77.1	204.5	4.1	1.30	0.3	0.07	—	0.07	1.61	0.10	17.1	0.1	0	1.0
0	24	1.26	77.1	204.5	229.1	1.30	0	0.06	3.74	0.06	1.61	0.10	17.1	0.1	0	2.6
0	38	0.92	90.4	195.8	4.0	1.67	0.4	0.18	—	0.17	0.70	0.06	20.2	0.2	0	—
0	20	0.82	64.2	240.2	2.2	1.20	0	0.16	2.52	0.03	4.93	0.09	52.9	0	0	2.7
0	20	0.82	64.2	240.2	296.7	1.20	0	0.16	2.84	0.03	4.93	0.09	52.9	0	0	2.7
0	22	0.54	63.4	261.4	115.2	1.18	0	0.03	2.49	0.03	4.97	0.16	43.2	0.3	0	1.2
0	7	0.30	25.6	119.2	77.8	0.45	0	0.02	1.01	0.02	2.19	0.07	14.7	0	0	1.3
0	6	0.29	25.4	107.0	2.7	0.47	0	0.01	1.23	0.02	2.14	0.07	11.8	0	0	1.2
0	6	0.34	30.6	120.4	97.2	0.50	0	0.05	1.20	0.01	2.63	0.06	10.8	0	0	1.4

DA+ Code	Food Description	Quantity	Measure	Wt (g)	H_2O (g)	Ener (kcal)	Prot (g)	Carb (g)	Fiber (g)	Fat (g)	Fat Breakdown (g)			
											Sat	Mono	Poly	*Trans*
	Nuts, Seeds, and Products—*continued*													
524	Peanut butter, smooth	1	tablespoon(s)	16	0.3	94	4.0	3.1	1.0	8.1	1.7	3.9	2.3	—
2804	Raw	¼	cup(s)	37	2.4	207	9.4	5.9	3.1	18.0	2.5	8.9	5.7	—
	Pecans													
32907	Dry roasted, no salt added	¼	cup(s)	28	0.3	198	2.6	3.8	2.6	20.7	1.8	12.3	5.7	—
32936	Dry roasted, with salt added	¼	cup(s)	27	0.3	192	2.6	3.7	2.5	20.0	1.7	11.9	5.6	—
1162	Oil roasted	¼	cup(s)	28	0.3	197	2.5	3.6	2.6	20.7	2.0	11.3	6.5	—
526	Raw	¼	cup(s)	27	1.0	188	2.5	3.8	2.6	19.6	1.7	11.1	5.9	—
12973	**Pine nuts or pignolia, dried**	1	tablespoon(s)	9	0.2	58	1.2	1.1	0.3	5.9	0.4	1.6	2.9	—
	Pistachios													
1164	Dry roasted	¼	cup(s)	31	0.6	176	6.6	8.5	3.2	14.1	1.7	7.4	4.3	—
32938	Dry roasted, with salt added	¼	cup(s)	32	0.6	182	6.8	8.6	3.3	14.7	1.8	7.7	4.4	—
1167	**Pumpkin or squash seeds, roasted**	¼	cup(s)	57	4.0	296	18.7	7.6	2.2	23.9	4.5	7.4	10.9	—
	Sesame													
32912	Sesame butter paste	1	tablespoon(s)	16	0.3	94	2.9	3.8	0.9	8.1	1.1	3.1	3.6	—
32941	Tahini or sesame butter	1	tablespoon(s)	15	0.5	89	2.6	3.2	0.7	8.0	1.1	3.0	3.5	—
1169	Whole, roasted, toasted	3	tablespoon(s)	10	0.3	54	1.6	2.4	1.3	4.6	0.6	1.7	2.0	—
	Soy nuts													
34173	Deep sea salted	¼	cup(s)	28	—	119	11.9	8.9	4.9	4.0	1.0	—	—	—
34174	Unsalted	¼	cup(s)	28	—	119	11.9	8.9	4.9	4.0	0	—	—	—
	Sunflower seeds													
528	Kernels, dried	1	tablespoon(s)	9	0.4	53	1.9	1.8	0.8	4.6	0.4	1.7	2.1	—
29721	Kernels, dry roasted, salted	1	tablespoon(s)	8	0.1	47	1.5	1.9	0.7	4.0	0.4	0.8	2.6	—
29723	Kernels, toasted, salted	1	tablespoon(s)	8	0.1	52	1.4	1.7	1.0	4.8	0.5	0.9	3.1	—
32928	Sunflower seed butter with salt added	1	tablespoon(s)	16	0.2	93	3.1	4.4	—	7.6	0.8	1.5	5.0	—
	Trail mix													
4646	Trail mix	¼	cup(s)	38	3.5	173	5.2	16.8	2.0	11.0	2.1	4.7	3.6	—
4647	Trail mix with chocolate chips	¼	cup(s)	38	2.5	182	5.3	16.8	—	12.0	2.3	5.1	4.2	—
4648	Tropical trail mix	¼	cup(s)	35	3.2	142	2.2	23.0	—	6.0	3.0	0.9	1.8	—
	Walnuts													
529	Dried black, chopped	¼	cup(s)	31	1.4	193	7.5	3.1	2.1	18.4	1.1	4.7	11.0	—
531	English or Persian	¼	cup(s)	29	1.2	191	4.5	4.0	2.0	19.1	1.8	2.6	13.8	—
	Vegetarian Foods													
	Prepared													
34222	Brown rice and tofu stir-fry (vegan)	8	ounce(s)	227	244.4	302	16.5	18.0	3.2	21.0	1.7	4.7	13.4	0
34368	Cheese enchilada casserole (lacto)	8	ounce(s)	227	80.3	385	16.6	38.4	4.1	17.8	9.5	6.1	1.1	—
34247	Five bean casserole (vegan)	8	ounce(s)	227	175.8	178	5.9	26.6	6.0	5.8	1.1	2.5	1.9	0
34261	Lentil stew (vegan)	8	ounce(s)	227	227.9	188	11.5	35.9	11.0	0.7	0.1	0.1	0.3	0
34397	Macaroni and cheese (lacto)	8	ounce(s)	227	352.1	391	18.1	37.1	1.0	18.7	9.8	6.0	1.8	0
34238	Steamed rice and vegetables (vegan)	8	ounce(s)	227	222.9	587	11.2	87.9	5.8	23.1	4.1	8.7	9.1	0
34308	Tofu rice burgers (ovo-lacto)	1	piece(s)	218	77.6	435	22.4	68.6	5.6	8.4	1.7	2.4	3.5	—
34276	Vegan spinach enchiladas (vegan)	1	piece(s)	82	59.2	93	4.9	14.5	1.8	2.4	0.3	0.6	1.3	—
34243	Vegetable chow mein (vegan)	8	ounce(s)	227	163.3	166	6.5	22.1	2.0	6.4	0.7	2.7	2.5	0
34454	Vegetable lasagna (lacto)	8	ounce(s)	227	178.9	208	13.7	29.9	2.6	4.1	2.3	1.1	0.3	—
34339	Vegetable marinara (vegan)	8	ounce(s)	252	200.7	104	3.0	16.7	1.4	3.1	0.4	1.4	1.0	0
34356	Vegetable rice casserole (lacto)	8	ounce(s)	227	178.9	238	9.7	24.4	4.0	12.5	4.9	3.5	3.1	—
34311	Vegetable strudel (ovo-lacto)	8	ounce(s)	227	63.1	478	12.0	32.4	2.5	33.8	11.5	16.7	3.9	0
34371	Vegetable taco (lacto)	1	item(s)	85	46.5	117	4.2	13.6	2.9	5.6	2.1	1.9	1.3	—
34282	Vegetarian chili (vegan)	8	ounce(s)	227	191.4	115	5.6	21.4	7.1	1.5	0.2	0.3	0.7	0
34367	Vegetarian vegetable soup (vegan)	8	ounce(s)	227	257.9	111	3.2	16.0	3.2	5.0	1.0	2.1	1.6	0
	Boca burger													
32067	All American flamed grilled patty	1	item(s)	71	—	90	14.0	4.0	3.0	3.0	1.0	—	—	0
32074	Boca chik'n nuggets	4	item(s)	87	—	180	14.0	17.0	3.0	7.0	1.0	—	—	0
32075	Boca meatless ground burger	½	cup(s)	57	—	60	13.0	6.0	3.0	0.5	0	—	—	0
32072	Breakfast links	2	item(s)	45	—	70	8.0	5.0	2.0	3.0	0.5	—	—	0
32071	Breakfast patties	1	item(s)	38	—	60	7.0	5.0	2.0	2.5	0	—	—	0
35780	Cheeseburger meatless burger patty	1	item(s)	71	—	100	12.0	5.0	3.0	5.0	1.5	—	—	0
33958	Original meatless chik'n patties	1	item(s)	71	—	160	11.0	15.0	2.0	6.0	1.0	—	—	0

Chol (mg)	Calc (mg)	Iron (mg)	Magn (mg)	Pota (mg)	Sodi (mg)	Zinc (mg)	Vit A (μg)	Thia (mg)	Vit E (mg α)	Ribo (mg)	Niac (mg)	Vit B_6 (mg)	Fola (μg)	Vit C (mg)	Vit B_{12} (μg)	Sele (μg)
0	7	0.30	24.6	103.8	73.4	0.47	0	0.01	1.44	0.02	2.14	0.09	11.8	0	0	0.9
0	34	1.67	61.3	257.3	6.6	1.19	0	0.23	3.04	0.04	4.40	0.12	87.6	0	0	2.6
0	20	0.78	36.8	118.3	0.3	1.41	1.9	0.12	0.35	0.03	0.32	0.05	4.5	0.2	0	1.1
0	19	0.75	35.6	114.5	103.4	1.36	1.9	0.11	0.34	0.03	0.31	0.05	4.3	0.2	0	1.1
0	18	0.68	33.3	107.8	0.3	1.23	1.4	0.13	0.70	0.03	0.33	0.05	4.1	0.2	0	1.7
0	19	0.69	33.0	111.7	0	1.23	0.8	0.18	0.38	0.04	0.32	0.06	6.0	0.3	0	1.0
0	1	0.47	21.6	51.3	0.2	0.55	0.1	0.03	0.80	0.02	0.37	0.01	2.9	0.1	0	0.1
0	34	1.29	36.9	320.4	3.1	0.71	4.0	0.26	0.59	0.05	0.44	0.39	15.4	0.7	0	2.9
0	35	1.34	38.4	333.4	129.6	0.73	4.2	0.26	0.61	0.05	0.45	0.40	16.0	0.7	0	3.0
0	24	8.48	303.0	457.4	10.2	4.22	10.8	0.12	0.00	0.18	0.99	0.05	32.3	1.0	0	3.2
0	154	3.07	57.9	93.1	1.9	1.17	0.5	0.04	—	0.03	1.07	0.13	16.0	0	0	0.9
0	21	0.66	14.3	68.9	5.3	0.69	0.5	0.24	—	0.02	0.85	0.02	14.7	0.6	0	0.3
0	94	1.40	33.8	45.1	1.0	0.68	0	0.07	—	0.02	0.43	0.07	9.3	0	0	0.5
0	59	1.07	—	—	148.1	—	0	—	—	—	—	—	—	0	—	—
0	59	1.07	—	—	9.9	—	0	—	—	—	—	—	—	0	—	—
0	7	0.47	29.3	58.1	0.8	0.45	0.3	0.13	2.99	0.03	0.75	0.12	20.4	0.1	0	4.8
0	6	0.30	10.3	68.0	32.8	0.42	0	0.01	2.09	0.02	0.56	0.06	19.0	0.1	0	6.3
0	5	0.57	10.8	41.1	51.3	0.44	0	0.03	—	0.02	0.35	0.07	19.9	0.1	0	5.2
0	20	0.76	59.0	11.5	83.2	0.85	0.5	0.05	—	0.05	0.85	0.13	37.9	0.4	0	—
0	29	1.14	59.3	256.9	85.9	1.20	0.4	0.17	—	0.07	1.76	0.11	26.6	0.5	0	—
2	41	1.27	60.4	243.0	45.4	1.17	0.8	0.15	—	0.08	1.65	0.09	24.4	0.5	0	—
0	20	0.92	33.6	248.2	3.5	0.41	0.7	0.15	—	0.04	0.51	0.11	14.7	2.7	0	—
0	19	0.97	62.8	163.4	0.6	1.05	0.6	0.01	0.56	0.04	0.14	0.18	9.7	0.5	0	5.3
0	29	0.85	46.2	129.0	0.6	0.90	0.3	0.10	0.20	0.04	0.32	0.15	28.7	0.4	0	1.4
0	353	6.34	118.3	501.4	142.2	2.03	—	0.23	0.07	0.14	1.49	0.36	51.8	24.8	0	14.8
39	441	2.44	34.6	191.2	1139.7	1.84	—	0.31	0.05	0.35	2.23	0.11	87.1	20.4	0.4	20.0
0	48	1.78	40.8	364.1	613.6	0.61	—	0.10	0.52	0.07	0.93	0.11	39.4	8.3	0	3.3
0	34	3.23	50.0	548.8	436.5	1.42	—	0.24	0.14	0.16	2.31	0.29	91.4	26.4	0	12.1
43	415	1.71	45.4	267.8	1641.0	2.32	—	0.32	0.27	0.48	2.18	0.13	74.2	0.9	0.8	33.3
0	91	3.31	153.1	810.1	3117.8	2.04	—	0.37	3.03	0.21	6.16	0.64	61.7	35.2	0	18.8
52	467	9.01	89.7	455.6	2449.5	2.06	—	0.27	0.12	0.26	3.43	0.29	106.6	2.0	0.1	43.0
0	117	1.13	40.4	170.5	134.2	0.68	—	0.07	—	0.07	0.53	0.10	52.1	1.8	0	5.1
0	189	3.70	28.0	310.3	372.7	0.76	—	0.13	0.05	0.11	1.43	0.14	45.6	8.0	0	6.5
10	176	1.86	41.9	470.0	759.4	1.14	—	0.26	0.05	0.25	2.49	0.22	64.7	19.0	0.4	21.8
0	17	0.94	19.1	189.9	439.6	0.42	—	0.15	0.55	0.08	1.36	0.12	40.6	23.5	0	10.8
17	190	1.28	29.3	414.2	626.0	1.24	—	0.16	0.35	0.29	2.00	0.19	72.6	56.0	0.2	5.8
29	200	2.15	24.5	181.0	512.1	1.24	—	0.28	0.20	0.31	2.88	0.11	88.3	17.4	0.2	19.7
7	77	0.88	26.3	174.1	280.7	0.59	—	0.08	0.04	0.06	0.49	0.08	38.7	4.6	0	3.0
0	65	1.98	41.0	543.1	390.7	0.74	—	0.14	0.15	0.10	1.31	0.18	59.3	20.3	0	4.4
0	46	1.87	34.9	550.3	729.5	0.56	—	0.13	0.55	0.09	1.99	0.27	47.8	29.9	0	1.4
5	150	1.80	—	—	280.0	—	0	—	—	—	—	—	—	0	—	—
0	40	1.44	—	—	500.0	—	—	—	—	—	—	—	—	0	—	—
0	60	1.80	—	—	270.0	—	0	—	—	—	—	—	—	0	—	—
0	20	1.44	—	—	330.0	—	0	—	—	—	—	—	—	0	—	—
0	20	1.08	—	—	280.0	—	0	—	—	—	—	—	—	0	—	—
5	80	1.80	—	—	360.0	—	—	—	—	—	—	—	—	0	—	—
0	40	1.80	—	—	430.0	—	—	—	—	—	—	—	—	0	—	—

APPENDIX H

DA+ Code	Food Description	Quantity	Measure	Wt (g)	H_2O (g)	Ener (kcal)	Prot (g)	Carb (g)	Fiber (g)	Fat (g)	Fat Breakdown (g)			
											Sat	Mono	Poly	*Trans*
	Vegetarian Foods—*continued*													
32066	Original patty	1	item(s)	71	—	70	13.0	6.0	4.0	0.5	0	—	—	0
32068	Roasted garlic patty	1	item(s)	71	—	70	12.0	6.0	4.0	1.5	0	—	—	0
37814	Roasted onion meatless burger patty	1	item(s)	71	—	70	11.0	7.0	4.0	1.0	0	—	—	0
	Gardenburger													
37810	BBQ chik'n with sauce	1	item(s)	142	—	250	14.0	30.0	5.0	8.0	1.0	—	—	0
39661	Black bean burger	1	item(s)	71	—	80	8.0	11.0	4.0	2.0	0	—	—	0
39666	Buffalo chik'n wing	3	item(s)	95	—	180	9.0	8.0	5.0	12.0	1.5	—	—	0
39665	Country fried chicken with creamy pepper gravy	1	item(s)	142	—	190	9.0	16.0	2.0	9.0	1.0	—	—	0
37808	Flamed grilled chik'n	1	item(s)	71	—	100	13.0	5.0	3.0	2.5	0	—	—	0
37803	Garden vegan	1	item(s)	71	—	100	10.0	12.0	2.0	1.0	—	—	—	0
39663	Homestyle classic burger	1	item(s)	71	—	110	12.0	6.0	4.0	5.0	0.5	—	—	0
37807	Meatless breakfast sausage	1	item(s)	43	—	50	5.0	2.0	2.0	3.5	0	—	—	0
37809	Meatless meatballs	6	item(s)	85	—	110	12.0	8.0	4.0	4.5	1.0	—	—	0
37806	Meatless riblets with sauce	1	item(s)	142	—	160	17.0	11.0	4.0	5.0	0	—	—	0
29913	Original	1	item(s)	71	—	90	10.0	8.0	3.0	2.0	0.5	—	—	0
39662	Sun-dried tomato basil burger	1	item(s)	71	—	80	10.0	11.0	3.0	1.5	0.5	—	—	0
29915	Veggie medley	1	item(s)	71	—	90	9.0	11.0	4.0	2.0	0	—	—	0
	Loma Linda													
9311	Big franks, canned	1	item(s)	51	—	110	11.0	3.0	2.0	6.0	1.0	1.5	3.5	0
9323	Fried chik'n with gravy	2	piece(s)	80	45.9	150	12.0	5.0	2.0	10	1.5	2.5	5.0	0
9326	Linketts, canned	1	item(s)	35	21.0	70	7.0	1.0	1.0	4.0	0.5	1.0	2.5	0
9336	Redi-Burger patties, canned	1	slice(s)	85	50.5	120	18.0	7.0	4.0	2.5	0.5	0.5	1.5	0
9350	Swiss Stake pattie with gravy, frozen	1	piece(s)	92	65.7	130	9.0	9.0	3.0	6.0	1.0	1.5	3.5	0
9354	Tender Rounds meatball substitute, canned in gravy	6	piece(s)	80	53.9	120	13.0	6.0	1.0	4.5	0.5	1.5	2.5	0
	Morningstar Farms													
33707	America's Original Veggie Dog links	1	item(s)	57	—	80	11.0	6.0	1.0	0.5	0	—	—	0
9362	Better'n Eggs egg substitute	¼	cup(s)	57	50.3	20	5.0	0	0	0	0	0	0	0
9371	Breakfast bacon strips	2	item(s)	16	6.8	60	2.0	2.0	0.5	4.5	0.5	1.0	3.0	0
9368	Breakfast sausage links	2	item(s)	45	26.8	80	9.0	3.0	2.0	3.0	0.5	1.5	1.0	0
33705	Chik'n nuggets	4	piece(s)	86	—	190	12.0	18.0	2.0	7.0	1.0	2.0	4.0	0
11587	Chik patties	1	item(s)	71	36.3	150	9.0	16.0	2.0	6.0	1.0	1.5	2.5	0
2531	Garden veggie patties	1	item(s)	67	40.1	100	10.0	9.0	4.0	2.5	0.5	0.5	1.5	0
33702	Spicy black bean veggie burger	1	item(s)	78	—	140	12.0	15.0	3.0	4.0	0.5	1.0	2.5	0
9412	Vegetarian chili, canned	1	cup(s)	230	172.6	180	16.0	25.0	10.0	1.5	0.5	0.5	0.5	0
	Worthington													
9424	Chili, canned	1	cup(s)	230	167.0	280	24.0	25.0	8.0	10.0	1.5	1.5	7.0	0
9436	Diced chik, canned	¼	cup(s)	55	42.7	50	9.0	2.0	1.0	0	0	0	0	0
9440	Dinner roast, frozen	1	slice(s)	85	53.2	180	14.0	6.0	3.0	11.0	1.5	4.5	5.0	0
9420	Meatless chicken slices, frozen	3	slice(s)	57	38.9	90	9.0	2.0	0.5	4.5	1.0	1.0	2.5	0
36702	Meatless chicken style roll, frozen	1	slice(s)	55	—	90	9.0	2.0	1.0	4.5	1.0	1.0	2.5	0
9428	Meatless corned beef, sliced, frozen	3	slice(s)	57	31.2	140	10.0	5.0	0	9.0	1.0	2.0	5.0	0
9470	Meatless salami, sliced, frozen	3	slice(s)	57	32.4	120	12.0	3.0	2.0	7.0	1.0	1.0	5.0	0
9480	Meatless smoked turkey, sliced	3	slice(s)	57	—	140	10.0	4.0	0	9.0	1.5	2.0	5.0	0
9462	Prosage links	2	item(s)	45	26.8	80	9.0	3.0	2.0	3.0	0.5	0.5	2.0	0
9484	Stakelets patty beef steak substitute, frozen	1	piece(s)	71	41.5	150	14.0	7.0	2.0	7.0	1.0	2.5	3.5	0
9486	Stripples bacon substitute	2	item(s)	16	6.8	60	2.0	2.0	0.5	4.5	0.5	1.0	3.0	0
9496	Vegetable Skallops meat substitute, canned	½	cup(s)	85	—	90	17.0	4.0	3.0	1.0	0	0	0.5	0
	Dairy													
	Cheese													
1433	Blue, crumbled	1	ounce(s)	28	12.0	100	6.1	0.7	0	8.1	5.3	2.2	0.2	—
884	Brick	1	ounce(s)	28	11.7	105	6.6	0.8	0	8.4	5.3	2.4	0.2	—
885	Brie	1	ounce(s)	28	13.7	95	5.9	0.1	0	7.8	4.9	2.3	0.2	—
34821	Camembert	1	ounce(s)	28	14.7	85	5.6	0.1	0	6.9	4.3	2.0	0.2	—
5	Cheddar, shredded	¼	cup(s)	28	10.4	114	7.0	0.4	0	9.4	6.0	2.7	0.3	—
888	Cheddar or colby	1	ounce(s)	28	10.8	112	6.7	0.7	0	9.1	5.7	2.6	0.3	—
32096	Cheddar or colby, low fat	1	ounce(s)	28	17.9	49	6.9	0.5	0	2.0	1.2	0.6	0.1	—
889	Edam	1	ounce(s)	28	11.8	101	7.1	0.4	0	7.9	5.0	2.3	0.2	—

KEY: H-4 = Breads/Baked Goods H-10 = Cereal/Rice/Pasta H-14 = Fruit H-18 = Vegetables/Legumes H-28 = Nuts/Seeds H-30 = Vegetarian H-32 = Dairy H-40 = Eggs H-40 = Seafood H-42 = Meats H-46 = Poultry
= Processed Meats H-48 = Beverages H-52 = Fats/Oils H-54 = Sweets H-56 = Spices/Condiments/Sauces H-60 = Mixed Foods/Soups/Sandwiches H-64 = Fast Food H-84 = Convenience H-86 = Baby Foods

Chol (mg)	Calc (mg)	Iron (mg)	Magn (mg)	Pota (mg)	Sodi (mg)	Zinc (mg)	Vit A (µg)	Thia (mg)	Vit E (mg α)	Ribo (mg)	Niac (mg)	Vit B_6 (mg)	Fola (µg)	Vit C (mg)	Vit B_{12} (µg)	Sele (µg)
0	60	1.80	—	—	280.0	—	0	—	—	—	—	—	—	0	—	—
0	60	1.80	—	—	370.0	—	0	—	—	—	—	—	—	0	—	—
0	100	2.70	—	—	300.0	—	—	—	—	—	—	—	—	0	—	—
0	150	1.08	—	—	890.0	—	—	—	—	—	—	—	—	0	—	—
0	40	1.44	—	—	330.0	—	—	—	—	—	—	—	—	0	—	—
0	40	0.72	—	—	1000.0	—	—	—	—	—	—	—	—	0	—	—
5	40	1.44	—	—	550.0	—	—	—	—	—	—	—	—	0	—	—
0	60	3.60	—	—	360.0	—	—	—	—	—	—	—	—	0	—	—
0	40	4.50	—	—	230.0	—	—	—	—	—	—	—	—	0	—	—
0	80	1.44	—	—	380.0	—	—	—	—	—	—	—	—	0	—	—
0	20	0.72	—	—	120.0	—	—	—	—	—	—	—	—	0	—	—
0	60	1.80	—	—	400.0	—	—	—	—	—	—	—	—	0	—	—
0	60	1.80	—	—	720.0	—	—	—	—	—	—	—	—	3.6	—	—
0	80	1.08	30.4	193.4	490.0	0.89	—	0.10	—	0.15	1.08	0.08	10.1	1.2	0.1	7.0
5	60	1.44	—	—	260.0	—	—	—	—	—	—	—	—	3.6	—	—
0	40	1.44	27.0	182.0	290.0	0.46	—	0.07	—	0.08	0.90	0.09	10.6	9.0	0	4.0
0	0	0.77	—	50.0	220.0	—	0	0.22	—	0.10	2.00	0.70	—	0	2.4	—
0	20	1.80	—	70.0	430.0	0.33	0	1.05	—	0.34	4.00	0.30	—	0	2.4	—
0	0	0.36	—	20.0	160.0	0.46	0	0.12	—	0.20	0.80	0.16	—	0	0.9	—
0	0	1.06	—	140.0	450.0	—	0	0.15	—	0.25	4.00	0.40	—	0	1.2	—
0	0	0.72	—	200.0	430.0	—	0	0.45	—	0.25	10.00	1.00	—	0	5.4	—
0	20	1.08	—	80.0	340.0	0.66	0	0.75	—	0.17	2.00	0.16	—	0	1.2	—
0	0	0.72	—	60.0	580.0	—	0	—	—	—	—	—	—	0	—	—
0	20	0.72	—	75.0	90.0	0.60	37.5	0.03	—	0.34	0.00	0.08	24.0	—	0.6	—
0	0	0.36	—	15.0	220.0	0.05	0	0.75	—	0.04	0.40	0.07	—	0	0.2	—
0	0	1.80	—	50.0	300.0	—	0	0.37	—	0.17	7.00	0.50	—	0	3.0	—
0	20	2.70	—	320.0	490.0	—	0	0.52	—	0.25	5.00	0.30	—	0	1.5	—
0	0	1.80	—	210.0	540.0	—	0	1.80	—	0.17	2.00	0.20	—	0	1.2	—
0	40	0.72	—	180.0	350.0	—	—	—	—	—	—	—	—	0	—	—
0	40	1.80	—	320.0	470.0	—	0	—	—	—	0.00	—	—	0	—	—
0	40	3.60	—	660.0	900.0	—	—	—	—	—	—	—	—	0	—	—
0	40	3.60	—	330.0	1130.0	—	0	0.30	—	0.13	2.00	0.70	—	0	1.5	—
0	0	1.08	—	100.0	220.0	0.24	0	0.06	—	0.10	4.00	0.08	—	0	0.2	—
0	20	1.80	—	120.0	580.0	0.64	0	1.80	—	0.25	6.00	0.60	—	0	1.5	—
0	250	1.80	—	250.0	250.0	0.26	0	0.37	—	0.13	4.00	0.30	—	0	1.8	—
0	100	1.08	—	240.0	240.0	—	0	0.37	—	0.13	4.00	0.30	—	0	1.8	—
0	0	1.80	—	130.0	460.0	0.26	0	0.45	—	0.17	5.00	0.30	—	0	1.8	—
0	0	1.08	—	95.0	800.0	0.30	0	0.75	—	0.17	4.00	0.20	—	0	0.6	—
0	60	2.70	—	60.0	450.0	0.23	0	1.80	—	0.17	6.00	0.40	—	0	3.0	—
0	0	1.44	—	50.0	320.0	0.36	0	1.80	—	0.17	2.00	0.30	—	0	3.0	—
0	40	1.08	—	130.0	480.0	0.50	0	1.20	—	0.13	3.00	0.30	—	0	1.5	—
0	0	0.36	—	15.0	220.0	0.05	0	0.75	—	0.03	0.40	0.08	—	0	0.2	—
0	0	0.36	—	10.0	390.0	0.67	0	0.03	—	0.03	0.00	0.01	—	0	0	—
21	150	0.08	6.5	72.6	395.5	0.75	56.1	0.01	0.07	0.10	0.28	0.04	10.2	0	0.3	4.1
27	191	0.12	6.8	38.6	158.8	0.73	82.8	0.00	0.07	0.10	0.03	0.01	5.7	0	0.4	4.1
28	52	0.14	5.7	43.1	178.3	0.67	49.3	0.02	0.06	0.14	0.10	0.06	18.4	0	0.5	4.1
20	110	0.09	5.7	53.0	238.7	0.67	68.3	0.01	0.06	0.14	0.18	0.06	17.6	0	0.4	4.1
30	204	0.19	7.9	27.7	175.4	0.87	74.9	0.01	0.08	0.10	0.02	0.02	5.1	0	0.2	3.9
27	194	0.21	7.4	36.0	171.2	0.87	74.8	0.00	0.07	0.10	0.02	0.02	5.1	0	0.2	4.1
6	118	0.11	4.5	18.7	173.5	0.51	17.0	0.00	0.01	0.06	0.01	0.01	3.1	0	0.1	4.1
25	207	0.12	8.5	53.3	273.6	1.06	68.9	0.01	0.06	0.11	0.02	0.02	4.5	0	0.4	4.1

TABLE H-1 Table of Food Composition (*continued*) (Computer code is for Cengage Diet Analysis program) (For purposes of calculations, use "0" for t, <1, <0.1, 0.01)

DA+ Code	Food Description	Quantity	Measure	Wt (g)	H_2O (g)	Ener (kcal)	Prot (g)	Carb (g)	Fiber (g)	Fat (g)	Fat Breakdown (g) Sat	Mono	Poly	*Trans*
Dairy—*continued*														
890	Feta	1	ounce(s)	28	15.7	75	4.0	1.2	0	6.0	4.2	1.3	0.2	—
891	Fontina	1	ounce(s)	28	10.8	110	7.3	0.4	0	8.8	5.4	2.5	0.5	—
8527	Goat cheese, soft	1	ounce(s)	28	17.2	76	5.3	0.3	0	6.0	4.1	1.4	0.1	—
893	Gouda	1	ounce(s)	28	11.8	101	7.1	0.6	0	7.8	5.0	2.2	0.2	—
894	Gruyere	1	ounce(s)	28	9.4	117	8.5	0.1	0	9.2	5.4	2.8	0.5	—
895	Limburger	1	ounce(s)	28	13.7	93	5.7	0.1	0	7.7	4.7	2.4	0.1	—
896	Monterey jack	1	ounce(s)	28	11.6	106	6.9	0.2	0	8.6	5.4	2.5	0.3	—
13	Mozzarella, part skim milk	1	ounce(s)	28	15.2	72	6.9	0.8	0	4.5	2.9	1.3	0.1	—
12	Mozzarella, whole milk	1	ounce(s)	28	14.2	85	6.3	0.6	0	6.3	3.7	1.9	0.2	—
897	Muenster	1	ounce(s)	28	11.8	104	6.6	0.3	0	8.5	5.4	2.5	0.2	—
898	Neufchatel	1	ounce(s)	28	17.6	74	2.8	0.8	0	6.6	4.2	1.9	0.2	—
14	Parmesan, grated	1	tablespoon(s)	5	1.0	22	1.9	0.2	0	1.4	0.9	0.4	0.1	—
17	Provolone	1	ounce(s)	28	11.6	100	7.3	0.6	0	7.5	4.8	2.1	0.2	—
19	Ricotta, part skim milk	¼	cup(s)	62	45.8	85	7.0	3.2	0	4.9	3.0	1.4	0.2	—
18	Ricotta, whole milk	¼	cup(s)	62	44.1	107	6.9	1.9	0	8.0	5.1	2.2	0.2	—
20	Romano	1	tablespoon(s)	5	1.5	19	1.6	0.2	0	1.3	0.9	0.4	0	—
900	Roquefort	1	ounce(s)	28	11.2	105	6.1	0.6	0	8.7	5.5	2.4	0.4	—
21	Swiss	1	ounce(s)	28	10.5	108	7.6	1.5	0	7.9	5.0	2.1	0.3	—
	Imitation cheese													
42245	Imitation American cheddar cheese	1	ounce(s)	28	15.1	68	4.7	3.3	0	4.0	2.5	1.2	0.1	—
53914	Imitation cheddar	1	ounce(s)	28	15.1	68	4.7	3.3	0	4.0	2.5	1.2	0.1	—
	Cottage cheese													
9	Low fat, 1% fat	½	cup(s)	113	93.2	81	14.0	3.1	0	1.2	0.7	0.3	0	—
8	Low fat, 2% fat	½	cup(s)	113	89.6	102	15.5	4.1	0	2.2	1.4	0.6	0.1	—
	Cream cheese													
11	Cream cheese	2	tablespoon(s)	29	15.6	101	2.2	0.8	0	10.1	6.4	2.9	0.4	—
17366	Fat-free cream cheese	2	tablespoon(s)	30	22.7	29	4.3	1.7	0	0.4	0.3	0.1	0	—
10438	Tofutti Better than Cream Cheese	2	tablespoon(s)	30	—	80	1.0	1.0	0	8.0	2.0	—	6.0	—
	Processed cheese													
24	American cheese food, processed	1	ounce(s)	28	12.3	94	5.2	2.2	0	7.1	4.2	2.0	0.3	—
25	American cheese spread, processed	1	ounce(s)	28	13.5	82	4.7	2.5	0	6.0	3.8	1.8	0.2	—
22	American cheese, processed	1	ounce(s)	28	11.1	106	6.3	0.5	0	8.9	5.6	2.5	0.3	—
9110	Kraft deluxe singles pasteurized process American cheese	1	ounce(s)	28	—	108	5.4	0	0	9.5	5.4	—	—	—
23	Swiss cheese, processed	1	ounce(s)	28	12.0	95	7.0	0.6	0	7.1	4.5	2.0	0.2	—
	Soy cheese													
10437	Galaxy Foods vegan grated parmesan cheese alternative	1	tablespoon(s)	8	—	23	3.0	1.5	0	0	0	0	0	0
10430	Nu Tofu cheddar flavored cheese alternative	1	ounce(s)	28	—	70	6.0	1.0	0	4.0	0.5	2.5	1.0	—
	Cream													
26	Half and half cream	1	tablespoon(s)	15	12.1	20	0.4	0.6	0	1.7	1.1	0.5	0.1	—
32	Heavy whipping cream, liquid	1	tablespoon(s)	15	8.7	52	0.3	0.4	0	5.6	3.5	1.6	0.2	—
28	Light coffee or table cream, liquid	1	tablespoon(s)	15	11.1	29	0.4	0.5	0	2.9	1.8	0.8	0.1	—
30	Light whipping cream, liquid	1	tablespoon(s)	15	9.5	44	0.3	0.4	0	4.6	2.9	1.4	0.1	—
34	Whipped cream topping, pressurized	1	tablespoon(s)	3	1.8	8	0.1	0.4	0	0.7	0.4	0.2	0	—
	Sour cream													
30556	Fat-free sour cream	2	tablespoon(s)	32	25.8	24	1.0	5.0	0	0	0	0	0	0
36	Sour cream	2	tablespoon(s)	24	17.0	51	0.8	1.0	0	5.0	3.1	1.5	0.2	—
	Imitation cream													
3659	Coffeemate nondairy creamer, liquid	1	tablespoon(s)	15	—	20	0	2.0	0	1.0	0	0.5	0	—
40	Cream substitute, powder	1	teaspoon(s)	2	0	11	0.1	1.1	0	0.7	0.7	0	0	—
904	Imitation sour cream	2	tablespoon(s)	29	20.5	60	0.7	1.9	0	5.6	5.1	0.2	0	—
35972	Nondairy coffee whitener, liquid, frozen	1	tablespoon(s)	15	11.7	21	0.2	1.7	0	1.5	0.3	1.1	0	—
35976	Nondairy dessert topping, frozen	1	tablespoon(s)	5	2.4	15	0.1	1.1	0	1.2	1.0	0.1	0	—
35975	Nondairy dessert topping, pressurized	1	tablespoon(s)	4	2.7	12	0	0.7	0	1.0	0.8	0.1	0	—

KEY: H-4 = Breads/Baked Goods H-10 = Cereal/Rice/Pasta H-14 = Fruit H-18 = Vegetables/Legumes H-28 = Nuts/Seeds H-30 = Vegetarian H-32 = Dairy H-40 = Eggs H-40 = Seafood H-42 = Meats H-46 = Poultry = Processed Meats H-48 = Beverages H-52 = Fats/Oils H-54 = Sweets H-56 = Spices/Condiments/Sauces H-60 = Mixed Foods/Soups/Sandwiches H-64 = Fast Food H-84 = Convenience H-86 = Baby Foods

Chol (mg)	Calc (mg)	Iron (mg)	Magn (mg)	Pota (mg)	Sodi (mg)	Zinc (mg)	Vit A (μg)	Thia (mg)	Vit E (mg α)	Ribo (mg)	Niac (mg)	Vit B_6 (mg)	Fola (μg)	Vit C (mg)	Vit B_{12} (μg)	Sele (μg)
25	140	0.18	5.4	17.6	316.4	0.81	35.4	0.04	0.05	0.23	0.28	0.12	9.1	0	0.5	4.3
33	156	0.06	4.0	18.1	226.8	0.99	74.0	0.01	0.07	0.05	0.04	0.02	1.7	0	0.5	4.1
13	40	0.53	4.5	7.4	104.3	0.26	81.6	0.02	0.05	0.10	0.12	0.07	3.4	0	0.1	0.8
32	198	0.06	8.2	34.3	232.2	1.10	46.8	0.01	0.06	0.09	0.01	0.02	6.0	0	0.4	4.1
31	287	0.04	10.2	23.0	95.3	1.10	76.8	0.01	0.07	0.07	0.03	0.02	2.8	0	0.5	4.1
26	141	0.03	6.0	36.3	226.8	0.59	96.4	0.02	0.06	0.14	0.04	0.02	16.4	0	0.3	4.1
25	211	0.20	7.7	23.0	152.0	0.85	56.1	0.00	0.07	0.11	0.02	0.02	5.1	0	0.2	4.1
18	222	0.06	6.5	23.8	175.5	0.78	36.0	0.01	0.04	0.08	0.03	0.02	2.6	0	0.2	4.1
22	143	0.12	5.7	21.5	177.8	0.82	50.7	0.01	0.05	0.08	0.02	0.01	2.0	0	0.6	4.8
27	203	0.11	7.7	38.0	178.0	0.79	84.5	0.00	0.07	0.09	0.02	0.01	3.4	0	0.4	4.1
22	21	0.07	2.3	32.3	113.1	0.14	84.5	0.00	—	0.05	0.03	0.01	3.1	0	0.1	0.9
4	55	0.04	1.9	6.3	76.5	0.19	6.0	0.00	0.01	0.02	0.01	0.00	0.5	0	0.1	0.9
20	214	0.14	7.9	39.1	248.3	0.91	66.9	0.01	0.06	0.09	0.04	0.02	2.8	0	0.4	4.1
19	167	0.27	9.2	76.9	76.9	0.82	65.8	0.01	0.04	0.11	0.04	0.01	8.0	0	0.2	10.3
31	127	0.23	6.8	64.6	51.7	0.71	73.8	0.01	0.06	0.12	0.06	0.02	7.4	0	0.2	8.9
5	53	0.03	2.1	4.3	60	0.12	4.8	0.00	0.01	0.01	0.00	0.00	0.4	0	0.1	0.7
26	188	0.15	8.5	25.8	512.9	0.59	83.3	0.01	—	0.16	0.20	0.03	13.9	0	0.2	4.1
26	224	0.05	10.8	21.8	54.4	1.23	62.4	0.01	0.10	0.08	0.02	0.02	1.7	0	0.9	5.2
10	159	0.08	8.2	68.6	381.3	0.73	32.3	0.01	0.07	0.12	0.03	0.03	2.0	0	0.1	4.3
10	159	0.09	8.2	68.6	381.3	0.73	32.3	0.01	0.07	0.12	0.04	0.03	2.0	0	0.1	4.3
5	69	0.15	5.7	97.2	458.8	0.42	12.4	0.02	0.01	0.18	0.14	0.07	13.6	0	0.7	10.2
9	78	0.18	6.8	108.5	458.8	0.47	23.7	0.02	0.02	0.20	0.16	0.08	14.7	0	0.8	11.5
32	23	0.34	1.7	34.5	85.8	0.15	106.1	0.01	0.08	0.05	0.02	0.01	3.8	0	0.1	0.7
2	56	0.05	4.2	48.9	163.5	0.26	83.7	0.01	0.00	0.05	0.04	0.01	11.1	0	0.2	1.5
0	0	0.00	—	—	135.0	—	0	—	—	—	—	—	—	0	—	—
23	162	0.16	8.8	82.5	358.6	0.90	57.0	0.01	0.06	0.14	0.04	0.02	2.0	0	0.4	4.6
16	159	0.09	8.2	68.6	381.3	0.73	49.0	0.01	0.05	0.12	0.03	0.03	2.0	0	0.1	3.2
27	156	0.05	7.7	47.9	422.1	0.80	72.0	0.01	0.07	0.10	0.02	0.02	2.3	0	0.2	4.1
27	338	0.00	0	33.8	459.0	1.22	114.0	—	—	0.14	—	—	—	0	0.2	—
24	219	0.17	8.2	61.2	388.4	1.02	56.1	0.00	0.09	0.07	0.01	0.01	1.7	0	0.3	4.5
0	60	0.00	—	75.0	97.5	—	—	—	—	—	—	—	—	—	—	—
0	200	0.36	—	—	190.0	—	—	—	—	—	—	—	—	0	—	—
6	16	0.01	1.5	19.5	6.2	0.08	14.6	0.01	0.05	0.02	0.01	0.01	0.5	0.1	0	0.3
21	10	0.00	1.1	11.3	5.7	0.03	61.7	0.00	0.15	0.01	0.01	0.00	0.6	0.1	0	0.1
10	14	0.01	1.4	18.3	6.0	0.04	27.2	0.01	0.08	0.02	0.01	0.01	0.3	0.1	0	0.1
17	10	0.00	1.1	14.6	5.1	0.03	41.9	0.00	0.13	0.01	0.01	0.00	0.6	0.1	0	0.1
2	3	0.00	0.3	4.4	3.9	0.01	5.6	0.00	0.01	0.00	0.00	0.00	0.1	0	0	0
3	40	0.00	3.2	41.3	45.1	0.16	23.4	0.01	0.00	0.04	0.02	0.01	3.5	0	0.1	1.7
11	28	0.01	2.6	34.6	12.7	0.06	42.5	0.01	0.14	0.03	0.01	0.00	2.6	0.2	0.1	0.5
0	0	0.00	—	30.0	0	—	0	0.01	—	0.01	0.20	—	—	0	—	—
0	0	0.02	0.1	16.2	3.6	0.01	0	0.00	0.01	0.00	0.00	0.00	0	0	0	0
0	1	0.11	1.7	46.3	29.3	0.34	0	0.00	0.21	0.00	0.00	0.00	0	0	0	0.7
0	1	0.00	0	28.9	12.0	0.00	0.2	0.00	0.12	0.00	0.00	0.00	0	0	0	0.2
0	0	0.00	0.1	0.9	1.2	0.00	0.3	0.00	0.05	0.00	0.00	0.00	0	0	0	0.1
0	0	0.00	0	0.8	2.8	0.00	0.2	0.00	0.04	0.00	0.00	0.00	0	0	0	0.1

DA+ Code	Food Description	Quantity	Measure	Wt (g)	H_2O (g)	Ener (kcal)	Prot (g)	Carb (g)	Fiber (g)	Fat (g)	Fat Breakdown (g)			
											Sat	Mono	Poly	*Trans*
	Dairy—*continued*													
	Fluid milk													
60	Buttermilk, low fat	1	cup(s)	245	220.8	98	8.1	11.7	0	2.2	1.3	0.6	0.1	—
54	Low fat, 1%	1	cup(s)	244	219.4	102	8.2	12.2	0	2.4	1.5	0.7	0.1	—
55	Low fat, 1%, with nonfat milk solids	1	cup(s)	245	220.0	105	8.5	12.2	0	2.4	1.5	0.7	0.1	—
57	Nonfat, skim or fat free	1	cup(s)	245	222.6	83	8.3	12.2	0	0.2	0.1	0.1	0	—
58	Nonfat, skim or fat free with nonfat milk solids	1	cup(s)	245	221.4	91	8.7	12.3	0	0.6	0.4	0.2	0	—
51	Reduced fat, 2%	1	cup(s)	244	218.0	122	8.1	11.4	0	4.8	3.1	1.4	0.2	—
52	Reduced fat, 2%, with nonfat milk solids	1	cup(s)	245	217.7	125	8.5	12.2	0	4.7	2.9	1.4	0.2	—
50	Whole, 3.3%	1	cup(s)	244	215.5	146	7.9	11.0	0	7.9	4.6	2.0	0.5	—
	Canned milk													
62	Nonfat or skim evaporated	2	tablespoon(s)	32	25.3	25	2.4	3.6	0	0.1	0	0	0	—
63	Sweetened condensed	2	tablespoon(s)	38	10.4	123	3.0	20.8	0	3.3	2.1	0.9	0.1	—
61	Whole evaporated	2	tablespoon(s)	32	23.3	42	2.1	3.2	0	2.4	1.4	0.7	0.1	—
	Dried milk													
64	Buttermilk	¼	cup(s)	30	0.9	117	10.4	14.9	0	1.8	1.1	0.5	0.1	—
65	Instant nonfat with added vitamin A	¼	cup(s)	17	0.7	61	6.0	8.9	0	0.1	0.1	0	0	—
5234	Skim milk powder	¼	cup(s)	17	0.7	62	6.1	9.1	0	0.1	0.1	0	0	—
907	Whole dry milk	¼	cup(s)	32	0.8	159	8.4	12.3	0	8.5	5.4	2.5	0.2	—
909	**Goat milk**	1	cup(s)	244	212.4	168	8.7	10.9	0	10.1	6.5	2.7	0.4	—
	Chocolate milk													
33155	Chocolate syrup, prepared with milk	1	cup(s)	282	227.0	254	8.7	36.0	0.8	8.3	4.7	2.1	0.5	—
33184	Cocoa mix with aspartame, added sodium and vitamin A, no added calcium or phosphorus, prepared with water	1	cup(s)	192	177.4	56	2.3	10.8	1.2	0.4	0.3	0.1	0	—
908	Hot cocoa, prepared with milk	1	cup(s)	250	206.4	193	8.8	26.6	2.5	5.8	3.6	1.7	0.1	0.2
69	Low fat	1	cup(s)	250	211.3	158	8.1	26.1	1.3	2.5	1.5	0.8	0.1	—
68	Reduced fat	1	cup(s)	250	205.4	190	7.5	30.3	1.8	4.8	2.9	1.1	0.2	—
67	Whole	1	cup(s)	250	205.8	208	7.9	25.9	2.0	8.5	5.3	2.5	0.3	—
70	**Eggnog**	1	cup(s)	254	188.9	343	9.7	34.4	0	19.0	11.3	5.7	0.9	—
	Breakfast drinks													
10093	Carnation Instant Breakfast classic chocolate malt, prepared with skim milk, no sugar added	1	cup(s)	243	—	142	11.1	21.3	0.7	1.3	0.7	—	—	—
10092	Carnation Instant Breakfast classic French vanilla, prepared with skim milk, no sugar added	1	cup(s)	273	—	150	12.9	24.0	0	0.4	0.4	—	—	—
10094	Carnation Instant Breakfast stawberry sensation, prepared with skim milk, no sugar added	1	cup(s)	243	—	142	11.1	21.3	0	0.4	0.4	—	—	—
10091	Carnation Instant Breakfast strawberry sensation, prepared with skim milk	1	cup(s)	273	—	220	12.5	38.8	0	0.4	0.4	—	—	—
1417	Ovaltine rich chocolate flavor, prepared with skim milk	1	cup(s)	258	—	170	8.5	31.0	0	0	0	0	0	0
8539	**Malted milk, chocolate mix, fortified, prepared with milk**	1	cup(s)	265	215.8	223	8.9	28.9	1.1	8.6	5.0	2.2	0.5	—
	Milkshakes													
73	Chocolate	1	cup(s)	227	164.0	270	6.9	48.1	0.7	6.1	3.8	1.8	0.2	—
3163	Strawberry	1	cup(s)	226	167.8	256	7.7	42.8	0.9	6.3	3.9	—	—	—
74	Vanilla	1	cup(s)	227	169.2	254	8.8	40.3	0	6.9	4.3	2.0	0.3	—
	Ice cream													
4776	Chocolate	½	cup(s)	66	36.8	143	2.5	18.6	0.8	7.3	4.5	2.1	0.3	—
12137	Chocolate fudge, no sugar added	½	cup(s)	71	—	100	3.0	16.0	2.0	3.0	1.5	—	—	0
16514	Chocolate, soft serve	½	cup(s)	87	49.9	177	3.2	24.1	0.7	8.4	5.2	2.4	0.3	—
16523	Sherbet, all flavors	½	cup(s)	97	63.8	139	1.1	29.3	3.2	1.9	1.1	0.5	0.1	—
4778	Strawberry	½	cup(s)	66	39.6	127	2.1	18.2	0.6	5.5	3.4	—	—	—
76	Vanilla	½	cup(s)	72	43.9	145	2.5	17.0	0.5	7.9	4.9	2.1	0.3	—
12146	Vanilla chocolate swirl, fat-free, no sugar added	½	cup(s)	71	—	100	3.0	14.0	2.0	3.0	2.0	—	—	0
82	Vanilla, light	½	cup(s)	76	48.3	125	3.6	19.6	0.2	3.7	2.2	1.0	0.2	—
78	Vanilla, light, soft serve	½	cup(s)	88	61.2	111	4.3	19.2	0	2.3	1.4	0.7	0.1	—

KEY: H-4 = Breads/Baked Goods H-10 = Cereal/Rice/Pasta H-14 = Fruit H-18 = Vegetables/Legumes H-28 = Nuts/Seeds H-30 = Vegetarian H-32 = Dairy H-40 = Eggs H-40 = Seafood H-42 = Meats H-46 = Poultry = Processed Meats H-48 = Beverages H-52 = Fats/Oils H-54 = Sweets H-56 = Spices/Condiments/Sauces H-60 = Mixed Foods/Soups/Sandwiches H-64 = Fast Food H-84 = Convenience H-86 = Baby Foods

Chol (mg)	Calc (mg)	Iron (mg)	Magn (mg)	Pota (mg)	Sodi (mg)	Zinc (mg)	Vit A (µg)	Thia (mg)	Vit E (mg α)	Ribo (mg)	Niac (mg)	Vit B_6 (mg)	Fola (µg)	Vit C (mg)	Vit B_{12} (µg)	Sele (µg)
10	284	0.12	27.0	370.0	257.3	1.02	17.2	0.08	0.12	0.37	0.14	0.08	12.3	2.5	0.5	4.9
12	290	0.07	26.8	366.0	107.4	1.02	141.5	0.04	0.02	0.45	0.22	0.09	12.2	0	1.1	8.1
10	314	0.12	34.3	396.9	127.4	0.98	144.6	0.09	—	0.42	0.22	0.11	12.3	2.5	0.9	5.6
5	306	0.07	27.0	382.2	102.9	1.02	149.5	0.11	0.02	0.44	0.23	0.09	12.3	0	1.3	7.6
5	316	0.12	36.8	419.0	129.9	1.00	149.5	0.10	0.00	0.42	0.22	0.11	12.3	2.5	1.0	5.4
20	285	0.07	26.8	366.0	100.0	1.04	134.2	0.09	0.07	0.45	0.22	0.09	12.2	0.5	1.1	6.1
20	314	0.12	34.3	396.9	127.4	0.98	137.2	0.09	—	0.42	0.22	0.11	12.3	2.5	0.9	5.6
24	276	0.07	24.4	348.9	97.6	0.97	68.3	0.10	0.14	0.44	0.26	0.08	12.2	0	1.1	9.0
1	93	0.09	8.6	105.9	36.7	0.28	37.6	0.01	0.00	0.09	0.05	0.01	2.9	0.4	0.1	0.8
13	109	0.07	9.9	141.9	48.6	0.36	28.3	0.03	0.06	0.16	0.08	0.02	4.2	1.0	0.2	5.7
9	82	0.06	7.6	95.4	33.4	0.24	20.5	0.01	0.04	0.10	0.06	0.01	2.5	0.6	0.1	0.7
21	359	0.09	33.3	482.5	156.7	1.21	14.9	0.11	0.03	0.48	0.27	0.10	14.2	1.7	1.2	6.2
3	209	0.05	19.9	289.9	93.3	0.75	120.5	0.07	0.00	0.30	0.15	0.06	8.5	1.0	0.7	4.6
3	214	0.05	20.3	296.0	95.3	0.76	123.1	0.07	0.00	0.30	0.15	0.06	8.7	1.0	0.7	4.7
31	292	0.15	27.2	425.6	118.7	1.06	82.2	0.09	0.15	0.38	0.20	0.09	11.8	2.8	1.0	5.2
27	327	0.12	34.2	497.8	122.0	0.73	139.1	0.11	0.17	0.33	0.67	0.11	2.4	3.2	0.2	3.4
25	251	0.90	50.8	408.9	132.5	1.21	70.5	0.11	0.14	0.46	0.38	0.09	14.1	0	1.1	9.6
0	92	0.74	32.6	405.1	138.2	0.51	0	0.04	0.00	0.20	0.16	0.04	1.9	0	0.2	2.5
20	263	1.20	57.5	492.5	110.0	1.57	127.5	0.09	0.07	0.45	0.33	0.10	12.5	0.5	1.1	6.8
8	288	0.60	32.5	425.0	152.5	1.02	145.0	0.09	0.05	0.41	0.31	0.10	12.5	2.3	0.9	4.8
20	273	0.60	35.0	422.5	165.0	0.97	160.0	0.11	0.10	0.45	0.41	0.06	5.0	0	0.8	8.5
30	280	0.60	32.5	417.5	150.0	1.02	65.0	0.09	0.15	0.40	0.31	0.10	12.5	2.3	0.8	4.8
150	330	0.50	48.3	419.1	137.2	1.16	116.8	0.08	0.50	0.48	0.26	0.12	2.5	3.8	1.1	10.7
9	444	4.00	88.9	631.1	195.6	3.38	—	0.33	—	0.45	4.44	0.44	4.0	26.7	1.3	8.0
9	500	4.50	100.0	665.0	192.0	3.75	—	0.37	—	0.51	5.00	0.49	100.0	30.0	1.5	9.0
9	444	4.00	88.9	568.9	186.7	3.38	—	0.33	—	0.45	4.44	0.44	88.9	26.7	1.3	8.0
9	500	4.47	100.0	665.0	288.0	3.75	—	0.37	—	0.51	5.07	0.50	100.0	30.0	1.5	8.8
5	350	3.60	100.0	—	270.0	3.75	—	0.37	—	—	4.00	0.40	—	12.0	1.2	—
27	339	3.76	45.1	577.7	230.6	1.16	903.7	0.75	0.15	1.31	11.08	1.01	18.6	31.8	1.1	12.5
25	300	0.70	36.4	508.9	252.2	1.09	40.9	0.10	0.11	0.50	0.28	0.05	11.4	0	0.7	4.3
25	256	0.24	29.4	412.0	187.9	0.81	58.9	0.10	—	0.44	0.39	0.10	6.8	1.8	0.7	4.8
27	332	0.22	27.3	415.8	215.8	0.88	56.8	0.06	0.11	0.44	0.33	0.09	15.9	0	1.2	5.2
22	72	0.61	19.1	164.3	50.2	0.38	77.9	0.02	0.19	0.12	0.14	0.03	10.6	0.5	0.2	1.7
10	100	0.36	—	—	65.0	—	—	—	—	—	—	—	—	0	—	—
22	103	0.32	19.0	192.0	43.3	0.45	66.6	0.03	0.22	0.13	0.11	0.03	4.3	0.5	0.3	2.5
0	52	0.13	7.7	92.6	44.4	0.46	9.7	0.02	0.02	0.08	0.07	0.02	6.8	5.6	0.1	1.3
19	79	0.13	9.2	124.1	39.6	0.22	63.4	0.03	—	0.16	0.11	0.03	7.9	5.1	0.2	1.3
32	92	0.06	10.1	143.3	57.6	0.49	85.0	0.03	0.21	0.17	0.08	0.03	3.6	0.4	0.3	1.3
10	100	0.00	—	—	65.0	—	—	—	—	—	—	—	—	0	—	—
21	122	0.14	10.6	158.1	56.2	0.55	97.3	0.04	0.09	0.19	0.10	0.03	4.6	0.9	0.4	1.5
11	138	0.05	12.3	194.5	61.6	0.46	25.5	0.04	0.05	0.17	0.10	0.04	4.4	0.8	0.4	3.2

DA+ Code	Food Description	Quantity	Measure	Wt (g)	H_2O (g)	Ener (kcal)	Prot (g)	Carb (g)	Fiber (g)	Fat (g)	Fat Breakdown (g) Sat	Mono	Poly	*Trans*
Dairy—*continued*														
	Soy desserts													
10694	Tofutti low fat vanilla fudge nondairy frozen dessert	½	cup(s)	70	—	140	2.0	24.0	0	4.0	1.0	—	—	—
15721	Tofutti premium chocolate supreme nondairy frozen dessert	½	cup(s)	70	—	180	3.0	18.0	0	11.0	2.0	—	—	—
15720	Tofutti premium vanilla nondairy frozen dessert	½	cup(s)	70	—	190	2.0	20.0	0	11.0	2.0	—	—	—
	Ice milk													
16517	Chocolate	½	cup(s)	66	42.9	94	2.8	16.9	0.3	2.1	1.3	0.6	0.1	—
16516	Flavored, not chocolate	½	cup(s)	66	41.4	108	3.5	17.5	0.2	2.6	1.7	0.6	0.1	—
	Pudding													
25032	Chocolate	½	cup(s)	144	109.7	155	5.1	22.7	0.7	5.4	3.1	1.7	0.2	0
1923	Chocolate, sugar free, prepared with 2% milk	½	cup(s)	133	—	100	5.0	14.0	0.3	3.0	1.5	—	—	—
1722	Rice	½	cup(s)	113	75.6	151	4.1	29.9	0.5	1.9	1.1	0.5	0.1	—
4747	Tapioca, ready to eat	1	item(s)	142	102.0	185	2.8	30.8	0	5.5	1.4	3.6	0.1	—
25031	Vanilla	½	cup(s)	136	109.7	116	4.7	17.6	0	2.8	1.6	0.9	0.2	0
1924	Vanilla, sugar free, prepared with 2% milk	½	cup(s)	133	—	90	4.0	12.0	0.2	2.0	1.5	—	—	—
	Frozen yogurt													
4785	Chocolate, soft serve	½	cup(s)	72	45.9	115	2.9	17.9	1.6	4.3	2.6	1.3	0.2	—
1747	Fruit varieties	½	cup(s)	113	80.5	144	3.4	24.4	0	4.1	2.6	1.1	0.1	—
4786	Vanilla, soft serve	½	cup(s)	72	47.0	117	2.9	17.4	0	4.0	2.5	1.1	0.2	—
	Milk substitutes													
	Lactose free													
16081	Fat-free, calcium fortified [milk]	1	cup(s)	240	—	80	8.0	13.0	0	0	0	0	0	0
36486	Low fat milk	1	cup(s)	240	—	110	8.0	13.0	0	2.5	1.5	—	—	—
36487	Reduced fat milk	1	cup(s)	240	—	130	8.0	12.0	0	5.0	3.0	—	—	—
36488	Whole milk	1	cup(s)	240	—	150	8.0	12.0	0	8.0	5.0	—	—	—
	Rice													
10083	Rice Dream carob rice beverage	1	cup(s)	240	—	150	1.0	32.0	0	2.5	0	—	—	—
17089	Rice Dream original rice beverage, enriched	1	cup(s)	240	—	120	1.0	25.0	0	2.0	0	—	—	—
10087	Rice Dream vanilla enriched rice beverage	1	cup(s)	240	—	130	1.0	28.0	0	2.0	0	—	—	—
	Soy													
34750	Soy Dream chocolate enriched soy beverage	1	cup(s)	240	—	210	7.0	37.0	1.0	3.5	0.5	—	—	—
34749	Soy Dream vanilla enriched soy beverage	1	cup(s)	240	—	150	7.0	22.0	0	4.0	0.5	—	—	—
13840	Vitasoy light chocolate soymilk	1	cup(s)	240	—	100	4.0	17.0	0	2.0	0.5	0.5	1.0	—
13839	Vitasoy light vanilla soymilk	1	cup(s)	240	—	70	4.0	10.0	0	2.0	0.5	0.5	1.0	—
13836	Vitasoy rich chocolate soymilk	1	cup(s)	240	—	160	7.0	24.0	1.0	4.0	0.5	1.0	2.5	—
13835	Vitasoy vanilla delite soymilk	1	cup(s)	240	—	120	7.0	13.0	1.0	4.0	0.5	1.0	2.5	—
	Yogurt													
3615	Custard style, fruit flavors	6	ounce(s)	170	127.1	190	7.0	32.0	0	3.5	2.0	—	—	—
3617	Custard style, vanilla	6	ounce(s)	170	134.1	190	7.0	32.0	0	3.5	2.0	0.9	0.1	—
32101	Fruit, low fat	1	cup(s)	245	184.5	243	9.8	45.7	0	2.8	1.8	0.8	0.1	—
29638	Fruit, nonfat, sweetened with low-calorie sweetener	1	cup(s)	241	208.3	123	10.6	19.4	1.2	0.4	0.2	0.1	0	—
93	Plain, low fat	1	cup(s)	245	208.4	154	12.9	17.2	0	3.8	2.5	1.0	0.1	—
94	Plain, nonfat	1	cup(s)	245	208.8	137	14.0	18.8	0	0.4	0.3	0.1	0	—
32100	Vanilla, low fat	1	cup(s)	245	193.6	208	12.1	33.8	0	3.1	2.0	0.8	0.1	—
5242	Yogurt beverage	1	cup(s)	245	199.8	172	6.2	32.8	0	2.2	1.4	0.6	0.1	—
38202	Yogurt smoothie, nonfat, all flavors	1	item(s)	325	—	290	10.0	60.0	6.0	0	0	0	0	0
	Soy yogurt													
34617	Stonyfield Farm O'Soy strawberry-peach pack organic cultured soy yogurt	1	item(s)	113	—	100	5.0	16.0	3.0	2.0	0	—	—	0
34616	Stonyfield Farm O'Soy vanilla organic cultured soy yogurt	1	item(s)	170	—	150	7.0	26.0	4.0	2.0	0	—	—	0
10453	White Wave plain silk cultured soy yogurt	8	ounce(s)	227	—	140	5.0	22.0	1.0	3.0	0.5	—	—	0

KEY: H-4 = Breads/Baked Goods H-10 = Cereal/Rice/Pasta H-14 = Fruit H-18 = Vegetables/Legumes H-28 = Nuts/Seeds H-30 = Vegetarian H-32 = Dairy H-40 = Eggs H-40 = Seafood H-42 = Meats H-46 = Poultry = Processed Meats H-48 = Beverages H-52 = Fats/Oils H-54 = Sweets H-56 = Spices/Condiments/Sauces H-60 = Mixed Foods/Soups/Sandwiches H-64 = Fast Food H-84 = Convenience H-86 = Baby Foods

Chol (mg)	Calc (mg)	Iron (mg)	Magn (mg)	Pota (mg)	Sodi (mg)	Zinc (mg)	Vit A (µg)	Thia (mg)	Vit E (mg α)	Ribo (mg)	Niac (mg)	Vit B_6 (mg)	Fola (µg)	Vit C (mg)	Vit B_{12} (µg)	Sele (µg)
0	0	0.00	—	8.0	90.0	—	0	—	—	—	—	—	—	0	—	—
0	0	0.00	—	7.0	180.0	—	0	—	—	—	—	—	—	0	—	—
0	0	0.00	—	2.0	210.0	—	0	—	—	—	—	—	—	0	—	—
6	94	0.15	13.1	155.2	40.6	0.36	15.7	0.03	0.05	0.11	0.08	0.02	3.9	0.5	0.3	2.2
16	76	0.05	9.2	136.2	48.5	0.47	90.4	0.02	0.05	0.11	0.06	0.01	3.3	0.1	0.2	1.3
35	149	0.46	31.3	226.7	137.0	0.71	—	0.05	0.00	0.22	0.15	0.06	8.3	1.2	0.5	4.9
10	150	0.72	—	330.0	310.0	—	—	0.06	—	0.26	—	—	—	0	—	—
7	113	0.28	15.8	201.4	66.4	0.52	41.6	0.03	0.05	0.17	0.34	0.06	4.5	0.2	0.2	4.8
1	101	0.15	8.5	130.6	205.9	0.31	0	0.03	0.21	0.13	0.09	0.03	4.3	0.4	0.3	0
35	146	0.17	17.2	188.9	136.4	0.52	—	0.04	0.00	0.22	0.10	0.05	8.0	1.2	0.5	4.6
10	150	0.00	—	190.0	380.0	—	—	0.03	—	0.17	—	—	—	0	—	—
4	106	0.90	19.4	187.9	70.6	0.35	31.7	0.02	—	0.15	0.22	0.05	7.9	0.2	0.2	1.7
15	113	0.52	11.3	176.3	71.2	0.31	55.4	0.04	0.10	0.20	0.07	0.04	4.5	0.8	0.1	2.1
1	103	0.21	10.1	151.9	62.6	0.30	42.5	0.02	0.07	0.16	0.20	0.05	4.3	0.6	0.2	2.4
3	500	0.00	—	—	125.0	—	100.0	—	—	—	—	—	—	0	0	—
10	300	0.00	—	—	125.0	—	100.0	—	—	—	—	—	—	0	—	—
20	300	0.00	—	—	125.0	—	98.2	—	—	—	—	—	—	0	—	—
35	300	0.00	—	—	125.0	—	58.1	—	—	—	—	—	—	0	—	—
0	20	0.72	—	82.5	100.0	—	—	—	—	—	—	—	—	1.2	—	—
0	300	0.00	13.3	60.0	90.0	0.24	—	0.06	—	0.00	0.84	0.07	—	0	1.5	—
0	300	0.00	—	53.0	90.0	—	—	—	—	—	—	—	—	0	1.5	—
0	300	1.80	60.0	350.0	160.0	0.60	33.3	0.15	—	0.06	0.80	0.12	60.0	0	3.0	—
0	300	1.80	40.0	260.0	140.0	0.60	33.3	0.15	—	0.06	0.80	0.12	60.0	0	3.0	—
0	300	0.72	24.0	200.0	140.0	0.90	—	0.09	—	0.34	—	—	24.0	0	0.9	—
0	300	0.72	24.0	200.0	120.0	0.90	—	0.09	—	0.34	—	—	24.0	0	0.9	—
0	300	1.08	40.0	320.0	150.0	0.90	—	0.15	—	0.34	—	—	60.0	0	0.9	—
0	40	0.72	—	320.0	115.0	—	0	—	—	—	—	—	—	0	—	—
15	300	0.00	16.0	310.0	100.0	—	—	—	—	0.25	—	—	—	0	—	—
15	300	0.00	16.0	310.0	100.0	—	—	—	—	0.25	—	—	—	0	—	—
12	338	0.14	31.9	433.7	129.9	1.64	27.0	0.08	0.04	0.39	0.21	0.09	22.1	1.5	1.1	6.9
5	369	0.62	41.0	549.5	139.8	1.83	4.8	0.10	0.16	0.44	0.49	0.10	31.3	26.5	1.1	7.0
15	448	0.19	41.7	573.3	171.5	2.18	34.3	0.10	0.07	0.52	0.27	0.12	27.0	2.0	1.4	8.1
5	488	0.22	46.6	624.8	188.7	2.37	4.9	0.11	0.00	0.57	0.30	0.13	29.4	2.2	1.5	8.8
12	419	0.17	39.2	536.6	161.7	2.03	29.4	0.10	0.04	0.49	0.26	0.11	27.0	2.0	1.3	12.0
13	260	0.22	39.2	399.4	98.0	1.10	14.7	0.11	0.00	0.51	0.30	0.14	29.4	2.1	1.5	—
5	300	2.70	100.0	580.0	290.0	2.25	—	0.37	—	0.42	5.00	0.50	100.0	15.0	1.5	—
0	100	1.08	24.0	5.0	20.0	—	0	0.22	—	0.10	—	0.04	—	0	0	—
0	150	1.44	40.0	15.0	40.0	—	—	0.30	—	0.13	—	0.08	—	0	0	—
0	400	1.44	—	0	30.0	—	0	—	—	—	—	—	—	0	—	—

APPENDIX H

TABLE H-1 Table of Food Composition (*continued*)

(Computer code is for Cengage Diet Analysis program) (For purposes of calculations, use "0" for t, <1, <0.1, 0.01)

DA+ Code	Food Description	Quantity	Measure	Wt (g)	H_2O (g)	Ener (kcal)	Prot (g)	Carb (g)	Fiber (g)	Fat (g)	Fat Breakdown (g)			
											Sat	Mono	Poly	*Trans*
Eggs														
	Eggs													
99	Fried	1	item(s)	46	31.8	90	6.3	0.4	0	7.0	2.0	2.9	1.2	—
100	Hard boiled	1	item(s)	50	37.3	78	6.3	0.6	0	5.3	1.6	2.0	0.7	—
101	Poached	1	item(s)	50	37.8	71	6.3	0.4	0	5.0	1.5	1.9	0.7	—
97	Raw, white	1	item(s)	33	28.9	16	3.6	0.2	0	0.1	0	0	0	—
96	Raw, whole	1	item(s)	50	37.9	72	6.3	0.4	0	5.0	1.5	1.9	0.7	—
98	Raw, yolk	1	item(s)	17	8.9	54	2.7	0.6	0	4.5	1.6	2.0	0.7	—
102	Scrambled, prepared with milk and butter	2	item(s)	122	89.2	204	13.5	2.7	0	14.9	4.5	5.8	2.6	0.7
	Egg substitute													
4028	Egg Beaters	¼	cup(s)	61	—	30	6.0	1.0	0	0	0	0	0	0
920	Frozen	¼	cup(s)	60	43.9	96	6.8	1.9	0	6.7	1.2	1.5	3.7	—
918	Liquid	¼	cup(s)	63	51.9	53	7.5	0.4	0	2.1	0.4	0.6	1.0	—
Seafood														
	Cod													
6040	Atlantic cod or scrod, baked or broiled	3	ounce(s)	85	64.6	89	19.4	0	0	0.7	0.1	0.1	0.2	—
1573	Atlantic cod, cooked, dry heat	3	ounce(s)	85	64.6	89	19.4	0	0	0.7	0.1	0.1	0.2	—
2905	**Eel, raw**	3	ounce(s)	85	58.0	156	15.7	0	0	9.9	2.0	6.1	0.8	—
	Fish fillets													
25079	Baked	3	ounce(s)	84	79.9	99	21.7	0	0	0.7	0.1	0.1	0.3	—
8615	Batter coated or breaded, fried	3	ounce(s)	85	45.6	197	12.5	14.4	0.4	10.5	2.4	2.2	5.3	—
25082	Broiled fish steaks	3	ounce(s)	85	68.1	128	24.2	0	0	2.6	0.4	0.9	0.8	—
25083	Poached fish steaks	3	ounce(s)	85	67.1	111	21.1	0	0	2.3	0.3	0.8	0.7	—
25084	Steamed	3	ounce(s)	85	72.2	79	17.2	0	0	0.6	0.1	0.1	0.2	—
25089	**Flounder, baked**	3	ounce(s)	85	64.4	113	14.8	0.4	0.1	5.5	1.1	2.2	1.4	—
1825	**Grouper, cooked, dry heat**	3	ounce(s)	85	62.4	100	21.1	0	0	1.1	0.3	0.2	0.3	—
	Haddock													
6049	Baked or broiled	3	ounce(s)	85	63.2	95	20.6	0	0	0.8	0.1	0.1	0.3	—
1578	Cooked, dry heat	3	ounce(s)	85	63.1	95	20.6	0	0	0.8	0.1	0.1	0.3	—
1886	**Halibut, Atlantic and Pacific, cooked, dry heat**	3	ounce(s)	85	61.0	119	22.7	0	0	2.5	0.4	0.8	0.8	—
1582	**Herring, Atlantic, pickled**	4	piece(s)	60	33.1	157	8.5	5.8	0	10.8	1.4	7.2	1.0	—
1587	**Jack mackerel, solids, canned, drained**	2	ounce(s)	57	39.2	88	13.1	0	0	3.6	1.1	1.3	0.9	—
8580	**Octopus, common, cooked, moist heat**	3	ounce(s)	85	51.5	139	25.4	3.7	0	1.8	0.4	0.3	0.4	—
1831	**Perch, mixed species, cooked, dry heat**	3	ounce(s)	85	62.3	100	21.1	0	0	1.0	0.2	0.2	0.4	—
1592	**Pacific rockfish, cooked, dry heat**	3	ounce(s)	85	62.4	103	20.4	0	0	1.7	0.4	0.4	0.5	—
	Salmon													
2938	Coho, farmed, raw	3	ounce(s)	85	59.9	136	18.1	0	0	6.5	1.5	2.8	1.6	—
1594	Broiled or baked with butter	3	ounce(s)	85	53.9	155	23.0	0	0	6.3	1.2	2.3	2.3	—
29727	Smoked chinook (lox)	2	ounce(s)	57	40.8	66	10.4	0	0	2.4	0.5	1.1	0.6	—
154	**Sardine, Atlantic with bones, canned in oil**	3	ounce(s)	85	50.7	177	20.9	0	0	9.7	1.3	3.3	4.4	—
	Scallops													
155	Mixed species, breaded, fried	3	item(s)	47	27.2	100	8.4	4.7	—	5.1	1.2	2.1	1.3	—
1599	Steamed	3	ounce(s)	85	64.8	90	13.8	2.0	0	2.6	0.4	1.0	0.8	—
1839	**Snapper, mixed species, cooked, dry heat**	3	ounce(s)	85	59.8	109	22.4	0	0	1.5	0.3	0.3	0.5	—
	Squid													
1868	Mixed species, fried	3	ounce(s)	85	54.9	149	15.3	6.6	0	6.4	1.6	2.3	1.8	—
16617	Steamed or boiled	3	ounce(s)	85	63.3	89	15.2	3.0	0	1.3	0.4	0.1	0.5	—
1570	**Striped bass, cooked, dry heat**	3	ounce(s)	85	62.4	105	19.3	0	0	2.5	0.6	0.7	0.9	—
1601	**Sturgeon, steamed**	3	ounce(s)	85	59.4	111	17.0	0	0	4.3	1.0	2.0	0.7	—
1840	**Surimi, formed**	3	ounce(s)	85	64.9	84	12.9	5.8	0	0.8	0.2	0.1	0.4	—
1842	**Swordfish, cooked, dry heat**	3	ounce(s)	85	58.5	132	21.6	0	0	4.4	1.2	1.7	1.0	—
1846	**Tuna, yellowfin or ahi, raw**	3	ounce(s)	85	60.4	92	19.9	0	0	0.8	0.2	0.1	0.2	—
	Tuna, canned													
159	Light, canned in oil, drained	2	ounce(s)	57	33.9	112	16.5	0	0	4.6	0.9	1.7	1.6	—
355	Light, canned in water, drained	2	ounce(s)	57	42.2	66	14.5	0	0	0.5	0.1	0.1	0.2	—
33211	Light, no salt, canned in oil, drained	2	ounce(s)	57	33.9	112	16.5	0	0	4.7	0.9	1.7	1.6	—
33212	Light, no salt, canned in water, drained	2	ounce(s)	57	42.6	66	14.5	0	0	0.5	0.1	0.1	0.2	—

KEY: H-4 = Breads/Baked Goods H-10 = Cereal/Rice/Pasta H-14 = Fruit H-18 = Vegetables/Legumes H-28 = Nuts/Seeds H-30 = Vegetarian H-32 = Dairy H-40 = Eggs H-40 = Seafood H-42 = Meats H-46 = Poultry = Processed Meats H-48 = Beverages H-52 = Fats/Oils H-54 = Sweets H-56 = Spices/Condiments/Sauces H-60 = Mixed Foods/Soups/Sandwiches H-64 = Fast Food H-84 = Convenience H-86 = Baby Foods

Chol (mg)	Calc (mg)	Iron (mg)	Magn (mg)	Pota (mg)	Sodi (mg)	Zinc (mg)	Vit A (µg)	Thia (mg)	Vit E (mg α)	Ribo (mg)	Niac (mg)	Vit B_6 (mg)	Fola (µg)	Vit C (mg)	Vit B_{12} (µg)	Sele (µg)
210	27	0.91	6.0	67.6	93.8	0.55	91.1	0.03	0.56	0.23	0.03	0.07	23.5	0	0.6	15.7
212	25	0.59	5.0	63.0	62.0	0.52	84.5	0.03	0.51	0.25	0.03	0.06	22.0	0	0.6	15.4
211	27	0.91	6.0	66.5	147.0	0.55	69.5	0.02	0.48	0.20	0.03	0.06	17.5	0	0.6	15.8
0	2	0.02	3.6	53.8	54.8	0.01	0	0.00	0.00	0.14	0.03	0.00	1.3	0	0	6.6
212	27	0.91	6.0	67.0	70.0	0.55	70.0	0.03	0.48	0.23	0.03	0.07	23.5	0	0.6	15.9
210	22	0.46	0.9	18.5	8.2	0.39	64.8	0.03	0.43	0.09	0.00	0.06	24.8	0	0.3	9.5
429	87	1.46	14.6	168.4	341.6	1.22	174.5	0.06	1.33	0.53	0.09	0.14	36.6	0.2	0.9	27.5
0	20	1.08	4.0	85.0	115.0	0.60	112.5	0.15	—	0.85	0.20	0.08	60.0	0	1.2	—
1	44	1.18	9.0	127.8	119.4	0.58	6.6	0.07	0.95	0.23	0.08	0.08	9.6	0.3	0.2	24.8
1	33	1.32	5.6	207.1	111.1	0.82	11.3	0.07	0.17	0.19	0.07	0.00	9.4	0	0.2	15.6
47	12	0.41	35.7	207.5	66.3	0.49	11.9	0.07	0.68	0.06	2.13	0.24	6.8	0.8	0.9	32.0
47	12	0.41	35.7	207.5	66.3	0.49	11.9	0.07	0.68	0.06	2.13	0.24	6.8	0.9	0.9	32.0
107	17	0.42	17.0	231.3	43.4	1.37	887.0	0.13	3.40	0.03	2.97	0.05	12.8	1.5	2.6	5.5
44	8	0.31	29.1	489.0	86.1	0.48	—	0.02	—	0.05	2.47	0.46	8.1	3.0	1.0	44.3
29	15	1.79	20.4	272.2	452.5	0.37	9.4	0.09	—	0.09	1.78	0.08	14.5	0	0.9	7.7
37	55	0.97	96.7	524.3	62.9	0.49	—	0.05	—	0.08	6.47	0.36	12.6	0	1.2	42.5
32	48	0.85	84.0	455.6	54.7	0.42	—	0.05	—	0.07	5.92	0.33	11.5	0	1.1	37.0
41	12	0.29	24.7	319.3	41.7	0.34	—	0.06	—	0.06	1.89	0.21	6.1	0.8	0.8	32.0
44	19	0.34	47.3	224.7	280.2	0.20	—	0.06	0.40	0.07	2.02	0.18	7.4	2.8	1.6	33.5
40	18	0.96	31.5	404.0	45.1	0.43	42.5	0.06	—	0.01	0.32	0.29	8.5	0	0.6	39.8
63	36	1.15	42.5	339.4	74.0	0.40	16.2	0.03	0.42	0.03	3.94	0.29	6.8	0	1.2	34.4
63	36	1.14	42.5	339.3	74.0	0.40	16.2	0.03	—	0.03	3.93	0.29	11.1	0	1.2	34.4
35	51	0.91	91.0	489.9	58.7	0.45	45.9	0.05	—	0.07	6.05	0.33	11.9	0	1.2	39.8
8	46	0.73	4.8	41.4	522.0	0.31	154.8	0.02	1.02	0.08	1.98	0.10	1.2	0	2.6	35.1
45	137	1.15	21.0	110.0	214.9	0.57	73.7	0.02	0.58	0.12	3.50	0.11	2.8	0.5	3.9	21.4
82	90	8.11	51.0	535.8	391.2	2.85	76.5	0.04	1.02	0.06	3.21	0.55	20.4	6.8	30.6	76.2
98	87	0.98	32.3	292.6	67.2	1.21	8.5	0.06	—	0.10	1.61	0.11	5.1	1.4	1.9	13.7
37	10	0.45	28.9	442.3	65.5	0.45	60.4	0.03	1.32	0.07	3.33	0.22	8.5	0	1.0	39.8
43	10	0.29	26.4	382.7	40.0	0.36	47.6	0.08	—	0.09	5.79	0.56	11.1	0.9	2.3	10.7
40	15	1.02	26.9	376.6	98.6	0.56	—	0.13	1.14	0.05	8.33	0.18	4.2	1.8	2.3	41.0
13	6	0.48	10.2	99.2	1134.0	0.17	14.7	0.01	—	0.05	2.67	0.15	1.1	0	1.8	21.6
121	325	2.48	33.2	337.6	429.5	1.10	27.2	0.04	1.70	0.18	4.43	0.14	10.2	0	7.6	44.8
28	20	0.38	27.4	154.8	215.8	0.49	10.7	0.02	—	0.05	0.70	0.06	17.2	1.1	0.6	12.5
27	20	0.22	45.9	238.0	358.7	0.78	32.3	0.01	0.16	0.05	0.84	0.11	10.2	2.0	1.1	18.2
40	34	0.20	31.5	444.0	48.5	0.37	29.8	0.04	—	0.00	0.29	0.39	5.1	1.4	3.0	41.7
221	33	0.85	32.3	237.3	260.3	1.48	9.4	0.04	—	0.39	2.21	0.04	11.9	3.6	1.0	44.1
227	31	0.62	28.9	192.1	356.2	1.49	8.5	0.01	1.17	0.32	1.69	0.04	3.4	3.2	1.0	43.7
88	16	0.91	43.4	279.0	74.8	0.43	26.4	0.09	—	0.03	2.17	0.29	8.5	0	3.8	39.8
63	11	0.59	29.8	239.7	388.5	0.35	198.9	0.06	0.52	0.07	8.30	0.19	14.5	0	2.2	13.3
26	8	0.22	36.6	95.3	121.6	0.28	17.0	0.01	0.53	0.01	0.18	0.02	1.7	0	1.4	23.9
43	5	0.88	28.9	313.8	97.8	1.25	34.9	0.03	—	0.09	10.02	0.32	1.7	0.9	1.7	52.5
38	14	0.62	42.5	377.6	31.5	0.44	15.3	0.37	0.42	0.04	8.33	0.77	1.7	0.8	0.4	31.0
10	7	0.79	17.6	117.3	200.6	0.51	13.0	0.02	0.49	0.07	7.03	0.06	2.8	0	1.2	43.1
17	6	0.87	15.3	134.3	191.5	0.43	9.6	0.01	0.19	0.04	7.52	0.19	2.3	0	1.7	45.6
10	7	0.78	17.6	117.4	28.3	0.51	0	0.02	—	0.06	7.03	0.06	2.8	0	1.2	43.1
17	6	0.86	15.3	134.4	28.3	0.43	0	0.01	—	0.04	7.52	0.19	2.3	0	1.7	45.6

TABLE H-1 Table of Food Composition (*continued*) (Computer code is for Cengage Diet Analysis program) (For purposes of calculations, use "0" for t, <1, <0.1, 0.01)

DA+ Code	Food Description	Quantity	Measure	Wt (g)	H_2O (g)	Ener (kcal)	Prot (g)	Carb (g)	Fiber (g)	Fat (g)	Fat Breakdown (g)			
											Sat	Mono	Poly	*Trans*
	Seafood—*continued*													
2961	White, canned in oil, drained	2	ounce(s)	57	36.3	105	15.0	0	0	4.6	0.7	1.8	1.7	—
351	White, canned in water, drained	2	ounce(s)	57	41.5	73	13.4	0	0	1.7	0.4	0.4	0.6	—
33213	White, no salt, canned in oil, drained	2	ounce(s)	57	36.3	105	15.0	0	0	4.6	0.9	1.4	1.9	—
33214	White, no salt, canned in water, drained	2	ounce(s)	57	42.0	73	13.4	0	0	1.7	0.4	0.4	0.6	—
	Yellowtail													
8548	Mixed species, cooked, dry heat	3	ounce(s)	85	57.3	159	25.2	0	0	5.7	1.4	2.2	1.5	—
2970	Mixed species, raw	2	ounce(s)	57	42.2	83	13.1	0	0	3.0	0.7	1.1	0.8	—
	Shellfish, meat only													
1857	Abalone, mixed species, fried	3	ounce(s)	85	51.1	161	16.7	9.4	0	5.8	1.4	2.3	1.4	—
16618	Abalone, steamed or poached	3	ounce(s)	85	40.7	177	28.8	10.1	0	1.3	0.3	0.2	0.2	—
	Crab													
1851	Blue crab, canned	2	ounce(s)	57	43.2	56	11.6	0	0	0.7	0.1	0.1	0.2	—
1852	Blue crab, cooked, moist heat	3	ounce(s)	85	65.9	87	17.2	0	0	1.5	0.2	0.2	0.6	—
8562	Dungeness crab, cooked, moist heat	3	ounce(s)	85	62.3	94	19.0	0.8	0	1.1	0.1	0.2	0.3	—
1860	**Clams, cooked, moist heat**	3	ounce(s)	85	54.1	126	21.7	4.4	0	1.7	0.2	0.1	0.5	—
1853	**Crayfish, farmed, cooked, moist heat**	3	ounce(s)	85	68.7	74	14.9	0	0	1.1	0.2	0.2	0.4	—
	Oysters													
8720	Baked or broiled	3	ounce(s)	85	68.6	89	5.6	3.2	0	5.8	1.3	2.1	1.9	—
152	Eastern, farmed, raw	3	ounce(s)	85	73.3	50	4.4	4.7	0	1.3	0.4	0.1	0.5	—
8715	Eastern, wild, cooked, moist heat	3	ounce(s)	85	59.8	117	12.0	6.7	0	4.2	1.3	0.5	1.6	—
8584	Pacific, cooked, moist heat	3	ounce(s)	85	54.5	139	16.1	8.4	0	3.9	0.9	0.7	1.5	—
1865	Pacific, raw	3	ounce(s)	85	69.8	69	8.0	4.2	0	2.0	0.4	0.3	0.8	—
1854	**Lobster, northern, cooked, moist heat**	3	ounce(s)	85	64.7	83	17.4	1.1	0	0.5	0.1	0.1	0.1	—
1862	**Mussel, blue, cooked, moist heat**	3	ounce(s)	85	52.0	146	20.2	6.3	0	3.8	0.7	0.9	1.0	—
	Shrimp													
158	Mixed species, breaded, fried	3	ounce(s)	85	44.9	206	18.2	9.8	0.3	10.4	1.8	3.2	4.3	—
1855	Mixed species, cooked, moist heat	3	ounce(s)	85	65.7	84	17.8	0	0	0.9	0.2	0.2	0.4	—
	Beef, Lamb, Pork													
	Beef													
4450	Breakfast strips, cooked	2	slice(s)	23	5.9	101	7.1	0.3	0	7.8	3.2	3.8	0.4	—
174	Corned beef, canned	3	ounce(s)	85	49.1	213	23.0	0	0	12.7	5.3	5.1	0.5	—
33147	Cured, thin siced	2	ounce(s)	57	32.9	100	15.9	3.2	0	2.2	0.9	1.0	0.1	—
4581	Jerky	1	ounce(s)	28	6.6	116	9.4	3.1	0.5	7.3	3.1	3.2	0.3	—
	Ground beef													
5898	Lean, broiled, medium	3	ounce(s)	85	50.4	202	21.6	0	0	12.2	4.8	5.3	0.4	—
5899	Lean, broiled, well done	3	ounce(s)	85	48.4	214	23.8	0	0	12.5	5.0	5.7	0.3	0.4
5914	Regular, broiled, medium	3	ounce(s)	85	46.1	246	20.5	0	0	17.6	6.9	7.7	0.6	—
5915	Regular, broiled, well done	3	ounce(s)	85	43.8	259	21.6	0	0	18.4	7.5	8.5	0.5	0.6
	Beef rib													
4241	Rib, small end, separable lean, 0″ fat, broiled	3	ounce(s)	85	53.2	164	25.0	0	0	6.4	2.4	2.6	0.2	—
4183	Rib, whole, lean and fat, ¼″ fat, roasted	3	ounce(s)	85	39.0	320	18.9	0	0	26.6	10.7	11.4	0.9	—
	Beef roast													
16981	Bottom round, choice, separable lean and fat, ⅛″ fat, braised	3	ounce(s)	85	46.2	216	27.9	0	0	10.7	4.1	4.6	0.4	—
16979	Bottom round, separable lean and fat, ⅛″ fat, roasted	3	ounce(s)	85	52.4	185	22.5	0	0	9.9	3.8	4.2	0.4	—
16924	Chuck, arm pot roast, separable lean and fat, ⅛″ fat, braised	3	ounce(s)	85	42.9	257	25.6	0	0	16.3	6.5	7.0	0.6	—
16930	Chuck, blade roast, separable lean and fat, ⅛″ fat, braised	3	ounce(s)	85	40.5	290	22.8	0	0	21.4	8.5	9.2	0.8	—
5853	Chuck, blade roast, separable lean, 0″ trim, pot roasted	3	ounce(s)	85	47.4	202	26.4	0	0	9.9	3.9	4.3	0.3	—
4296	Eye of round, choice, separable lean, 0″ fat, roasted	3	ounce(s)	85	56.5	138	24.4	0	0	3.7	1.3	1.5	0.1	—
16989	Eye of round, separable lean and fat, ⅛″ fat, roasted	3	ounce(s)	85	52.2	180	24.2	0	0	8.5	3.2	3.6	0.3	—

APPENDIX H

KEY: H-4 = Breads/Baked Goods H-10 = Cereal/Rice/Pasta H-14 = Fruit H-18 = Vegetables/Legumes H-28 = Nuts/Seeds H-30 = Vegetarian H-32 = Dairy H-40 = Eggs H-40 = Seafood H-42 = Meats H-46 = Poultry = Processed Meats H-48 = Beverages H-52 = Fats/Oils H-54 = Sweets H-56 = Spices/Condiments/Sauces H-60 = Mixed Foods/Soups/Sandwiches H-64 = Fast Food H-84 = Convenience H-86 = Baby Foods

Chol (mg)	Calc (mg)	Iron (mg)	Magn (mg)	Pota (mg)	Sodi (mg)	Zinc (mg)	Vit A (µg)	Thia (mg)	Vit E (mg α)	Ribo (mg)	Niac (mg)	Vit B_6 (mg)	Fola (µg)	Vit C (mg)	Vit B_{12} (µg)	Sele (µg)
18	2	0.36	19.3	188.8	224.5	0.26	2.8	0.01	1.30	0.04	6.63	0.24	2.8	0	1.2	34.1
24	8	0.55	18.7	134.3	213.6	0.27	3.4	0.00	0.48	0.02	3.28	0.12	1.1	0	0.7	37.2
18	2	0.36	19.3	188.8	28.3	0.26	0	0.01	—	0.04	6.63	0.24	2.8	0	1.2	34.1
24	8	0.54	18.7	134.4	28.3	0.27	3.4	0.00	—	0.02	3.28	0.12	1.1	0	0.7	37.3
60	25	0.53	32.3	457.6	42.5	0.56	26.4	0.14	—	0.04	7.41	0.15	3.4	2.5	1.1	39.8
31	13	0.28	17.0	238.1	22.1	0.29	16.4	0.08	—	0.02	3.86	0.09	2.3	1.6	0.7	20.7
80	31	3.23	47.6	241.5	502.6	0.80	1.7	0.18	—	0.11	1.61	0.12	11.9	1.5	0.6	44.1
144	50	4.84	68.9	295.0	980.1	1.38	3.4	0.28	6.74	0.12	1.89	0.21	6.0	2.6	0.7	75.6
50	57	0.47	22.1	212.1	188.8	2.27	1.1	0.04	1.04	0.04	0.77	0.08	24.4	1.5	0.3	18.0
85	88	0.77	28.1	275.6	237.3	3.58	1.7	0.08	1.56	0.04	2.80	0.15	43.4	2.8	6.2	34.2
65	50	0.36	49.3	347.0	321.5	4.65	26.4	0.04	—	0.17	3.08	0.14	35.7	3.1	8.8	40.5
57	78	23.78	15.3	534.1	95.3	2.32	145.4	0.12	—	0.36	2.85	0.09	24.7	18.8	84.1	54.4
117	43	0.94	28.1	202.4	82.5	1.25	12.8	0.03	—	0.06	1.41	0.11	9.4	0.4	2.6	29.1
43	36	5.30	37.4	125.0	403.8	72.22	60.4	0.07	0.98	0.06	1.04	0.04	7.7	2.8	14.7	50.7
21	37	4.91	28.1	105.4	151.3	32.23	6.8	0.08	—	0.05	1.07	0.05	15.3	4.0	13.8	54.1
89	77	10.19	80.8	239.0	358.9	154.45	45.9	0.16	—	0.15	2.11	0.10	11.9	5.1	29.8	60.9
85	14	7.82	37.4	256.8	180.3	28.27	124.2	0.10	0.72	0.37	3.07	0.07	12.8	10.9	24.5	131.0
43	7	4.34	18.7	142.9	90.1	14.13	68.9	0.05	—	0.20	1.70	0.04	8.5	6.8	13.6	65.5
61	52	0.33	29.8	299.4	323.2	2.48	22.1	0.01	0.85	0.05	0.91	0.06	9.4	0	2.6	36.3
48	28	5.71	31.5	227.9	313.8	2.27	77.4	0.25	—	0.35	2.55	0.08	64.6	11.6	20.4	76.2
150	57	1.07	34.0	191.3	292.4	1.17	0	0.11	—	0.11	2.60	0.08	15.3	1.3	1.6	35.4
166	33	2.62	28.9	154.8	190.5	1.32	57.8	0.02	1.17	0.02	2.20	0.10	3.4	1.9	1.3	33.7
27	2	0.71	6.1	93.1	509.2	1.44	0	0.02	0.06	0.05	1.46	0.07	1.8	0	0.8	6.1
73	10	1.76	11.9	115.7	855.6	3.03	0	0.01	0.12	0.12	2.06	0.11	7.7	0	1.4	36.5
23	6	1.53	10.8	243.2	815.9	2.25	0	0.04	0.00	0.10	2.98	0.19	6.2	0	1.5	16.0
14	6	1.53	14.5	169.2	627.4	2.29	0	0.04	0.13	0.04	0.49	0.05	38.0	0	0.3	3.0
58	6	2.00	17.9	266.2	59.5	4.63	0	0.05	—	0.23	4.21	0.23	7.6	0	1.8	16.0
69	12	2.21	18.4	250.0	62.4	5.86	0	0.08	—	0.23	5.10	0.16	9.4	0	1.7	19.0
62	9	2.07	17.0	248.3	70.6	4.40	0	0.02	—	0.16	4.90	0.23	7.6	0	2.5	16.2
71	12	2.30	18.5	242.4	72.4	5.18	0	0.08	—	0.23	4.93	0.17	8.5	0	1.6	18.0
65	16	1.59	21.3	319.8	51.9	4.64	0	0.06	0.34	0.12	7.15	0.53	8.5	0	1.4	29.2
72	9	1.96	16.2	251.7	53.6	4.45	0	0.06	—	0.14	2.85	0.19	6.0	0	2.1	18.7
68	6	2.29	17.9	223.7	35.7	4.59	0	0.05	0.41	0.15	5.05	0.36	8.5	0	1.7	29.3
64	5	1.83	14.5	182.0	29.8	3.76	0	0.05	0.34	0.12	3.92	0.29	6.8	0	1.3	23.0
67	14	2.15	17.0	205.8	42.5	5.93	0	0.05	0.45	0.15	3.63	0.25	7.7	0	1.9	24.1
88	11	2.66	16.2	198.2	55.3	7.15	0	0.06	0.17	0.20	2.06	0.22	4.3	0	1.9	20.9
73	11	3.12	19.6	223.7	60.4	8.73	0	0.06	—	0.23	2.27	0.24	5.1	0	2.1	22.7
49	5	2.16	16.2	200.7	32.3	4.28	0	0.05	0.30	0.15	4.69	0.34	8.5	0	1.4	28.0
54	5	1.98	15.3	193.1	31.5	3.95	0	0.05	0.34	0.13	4.37	0.31	7.7	0	1.5	25.2

TABLE H-1 Table of Food Composition (*continued*) (Computer code is for Cengage Diet Analysis program) (For purposes of calculations, use "0" for t, <1, <0.1, 0.01)

DA+ Code	Food Description	Quantity	Measure	Wt (g)	H_2O (g)	Ener (kcal)	Prot (g)	Carb (g)	Fiber (g)	Fat (g)	Fat Breakdown (g)			
											Sat	Mono	Poly	*Trans*
Beef, Lamb, Pork—*continued*														
	Beef steak													
4348	Short loin, t-bone steak, lean and fat, ¼″ fat, broiled	3	ounce(s)	85	43.2	274	19.4	0	0	21.2	8.3	9.6	0.8	—
4349	Short loin, t-bone steak, lean, ¼″ fat, broiled	3	ounce(s)	85	52.3	174	22.8	0	0	8.5	3.1	4.2	0.3	—
4360	Top loin, prime, lean and fat, ¼″ fat, broiled	3	ounce(s)	85	42.7	275	21.6	0	0	20.3	8.2	8.6	0.7	—
	Beef variety													
188	Liver, pan fried	3	ounce(s)	85	52.7	149	22.6	4.4	0	4.0	1.3	0.5	0.5	0.2
4447	Tongue, simmered	3	ounce(s)	85	49.2	242	16.4	0	0	19.0	6.9	8.6	0.6	0.7
	Lamb chop													
3275	Loin, domestic, lean and fat, ¼″ fat, broiled	3	ounce(s)	85	43.9	269	21.4	0	0	19.6	8.4	8.3	1.4	—
	Lamb leg													
3264	Domestic, lean and fat, ¼″ fat, cooked	3	ounce(s)	85	45.7	250	20.9	0	0	17.8	7.5	7.5	1.3	—
	Lamb rib													
182	Domestic, lean and fat, ¼″ fat, broiled	3	ounce(s)	85	40.0	307	18.8	0	0	25.2	10.8	10.3	2.0	—
183	Domestic, lean, ¼″ fat, broiled	3	ounce(s)	85	50.0	200	23.6	0	0	11.0	4.0	4.4	1.0	—
	Lamb shoulder													
186	Shoulder, arm and blade, domestic, choice, lean and fat, ¼″ fat, roasted	3	ounce(s)	85	47.8	235	19.1	0	0	17.0	7.2	6.9	1.4	—
187	Shoulder, arm and blade, domestic, choice, lean, ¼″ fat, roasted	3	ounce(s)	85	53.8	173	21.2	0	0	9.2	3.5	3.7	0.8	—
3287	Shoulder, arm, domestic, lean and fat, ¼″ fat, braised	3	ounce(s)	85	37.6	294	25.8	0	0	20.4	8.4	8.7	1.5	—
3290	Shoulder, arm, domestic, lean, ¼″ fat, braised	3	ounce(s)	85	41.9	237	30.2	0	0	12.0	4.3	5.2	0.8	—
	Lamb variety													
3375	Brain, pan fried	3	ounce(s)	85	51.6	232	14.4	0	0	18.9	4.8	3.4	1.9	—
3406	Tongue, braised	3	ounce(s)	85	49.2	234	18.3	0	0	17.2	6.7	8.5	1.1	—
	Pork, cured													
29229	Bacon, Canadian style, cured	2	ounce(s)	57	37.9	89	11.7	1.0	0	4.0	1.3	1.8	0.4	—
161	Bacon, cured, broiled, pan fried or roasted	2	slice(s)	16	2.0	87	5.9	0.2	0	6.7	2.2	3.0	0.7	0
35422	Breakfast strips, cured, cooked	3	slice(s)	34	9.2	156	9.8	0.4	0	12.5	4.3	5.6	1.9	—
189	Ham, cured, boneless, 11% fat, roasted	3	ounce(s)	85	54.9	151	19.2	0	0	7.7	2.7	3.8	1.2	—
29215	Ham, cured, extra lean, 4% fat, canned	2	2 ounce(s)	57	41.7	68	10.5	0	0	2.6	0.9	1.3	0.2	—
1316	Ham, cured, extra lean, 5% fat, roasted	3	ounce(s)	85	57.6	123	17.8	1.3	0	4.7	1.5	2.2	0.5	—
16561	Ham, smoked or cured, lean, cooked	1	slice(s)	42	27.6	66	10.5	0	0	2.3	0.8	1.1	0.3	—
	Pork chop													
32671	Loin, blade, chops, lean and fat, pan fried	3	ounce(s)	85	42.5	291	18.3	0	0	23.6	8.6	10	2.6	—
32672	Loin, center cut, chops, lean and fat, pan fried	3	ounce(s)	85	45.1	236	25.4	0	0	14.1	5.1	6.0	1.6	—
32682	Loin, center rib, chops, boneless, lean and fat, braised	3	ounce(s)	85	49.5	217	22.4	0	0	13.4	5.2	6.1	1.1	—
32603	Loin, center rib, chops, lean, broiled	3	ounce(s)	85	55.4	158	21.9	0	0	7.1	2.4	3.0	0.8	0.1
32478	Loin, whole, lean and fat, braised	3	ounce(s)	85	49.6	203	23.2	0	0	11.6	4.3	5.2	1.0	—
32481	Loin, whole, lean, braised	3	ounce(s)	85	52.2	174	24.3	0	0	7.8	2.9	3.5	0.6	—
	Pork leg or ham													
32471	Pork leg or ham, rump portion, lean and fat, roasted	3	ounce(s)	85	48.3	214	24.6	0	0	12.1	4.5	5.4	1.2	—
32468	Pork leg or ham, whole, lean and fat, roasted	3	ounce(s)	85	46.8	232	22.8	0	0	15.0	5.5	6.7	1.4	—
	Pork ribs													
32693	Loin, country style, lean and fat, roasted	3	ounce(s)	85	43.3	279	19.9	0	0	21.6	7.8	9.4	1.7	—
32696	Loin, country style, lean, roasted	3	ounce(s)	85	49.5	210	22.6	0	0	12.6	4.5	5.5	0.9	—

KEY: H-4 = Breads/Baked Goods H-10 = Cereal/Rice/Pasta H-14 = Fruit H-18 = Vegetables/Legumes H-28 = Nuts/Seeds H-30 = Vegetarian H-32 = Dairy H-40 = Eggs H-40 = Seafood H-42 = Meats H-46 = Poultry = Processed Meats H-48 = Beverages H-52 = Fats/Oils H-54 = Sweets H-56 = Spices/Condiments/Sauces H-60 = Mixed Foods/Soups/Sandwiches H-64 = Fast Food H-84 = Convenience H-86 = Baby Foods

Chol (mg)	Calc (mg)	Iron (mg)	Magn (mg)	Pota (mg)	Sodi (mg)	Zinc (mg)	Vit A (µg)	Thia (mg)	Vit E (mg α)	Ribo (mg)	Niac (mg)	Vit B_6 (mg)	Fola (µg)	Vit C (mg)	Vit B_{12} (µg)	Sele (µg)
58	7	2.56	17.9	233.9	57.8	3.56	0	0.07	0.18	0.17	3.29	0.27	6.0	0	1.8	10.0
50	5	3.11	22.1	278.1	65.5	4.34	0	0.09	0.11	0.21	3.93	0.33	6.8	0	1.9	8.5
67	8	1.88	19.6	294.3	53.6	3.85	0	0.06	—	0.15	3.96	0.31	6.0	0	1.6	19.5
324	5	5.24	18.7	298.5	65.5	4.44	6586.3	0.15	0.39	2.91	14.86	0.87	221.1	0.6	70.7	27.9
112	4	2.22	12.8	156.5	55.3	3.47	0	0.01	0.25	0.25	2.96	0.13	6.0	1.1	2.7	11.2
85	17	1.53	20.4	278.1	65.5	2.96	0	0.08	0.11	0.21	6.03	0.11	15.3	0	2.1	23.3
82	14	1.59	19.6	263.7	61.2	3.79	0	0.08	0.11	0.21	5.66	0.11	15.3	0	2.2	22.5
84	16	1.59	19.6	229.5	64.6	3.40	0	0.07	0.10	0.18	5.95	0.09	11.9	0	2.2	20.3
77	14	1.87	24.7	266.1	72.3	4.47	0	0.08	0.15	0.21	5.56	0.12	17.9	0	2.2	26.4
78	17	1.67	19.6	213.4	56.1	4.44	0	0.07	0.11	0.20	5.22	0.11	17.9	0	2.2	22.3
74	16	1.81	21.3	225.3	57.8	5.13	0	0.07	0.15	0.22	4.89	0.12	21.3	0	2.3	24.2
102	21	2.03	22.1	260.3	61.2	5.17	0	0.06	0.12	0.21	5.66	0.09	15.3	0	2.2	31.6
103	22	2.29	24.7	287.5	64.6	6.20	0	0.06	0.15	0.23	5.38	0.11	18.7	0	2.3	32.1
2130	18	1.73	18.7	304.5	133.5	1.70	0	0.14	—	0.31	3.87	0.19	6.0	19.6	20.5	10.2
161	9	2.23	13.6	134.4	57.0	2.54	0	0.06	—	0.35	3.13	0.14	2.6	6.0	5.4	23.8
28	5	0.38	9.6	195.0	798.9	0.78	0	0.42	0.11	0.09	3.53	0.22	2.3	0	0.4	14.2
18	2	0.22	5.3	90.4	369.6	0.56	1.8	0.06	0.04	0.04	1.76	0.04	0.3	0	0.2	9.9
36	5	0.67	8.8	158.4	713.7	1.25	0	0.25	0.08	0.12	2.58	0.11	1.4	0	0.6	8.4
50	7	1.13	18.7	347.7	1275.0	2.09	0	0.62	0.26	0.28	5.22	0.26	2.6	0	0.6	16.8
22	3	0.53	9.6	206.4	711.6	1.09	0	0.47	0.09	0.13	3.00	0.25	3.4	0	0.5	8.2
45	7	1.25	11.9	244.1	1023.1	2.44	0	0.64	0.21	0.17	3.42	0.34	2.6	0	0.6	16.6
23	3	0.39	9.2	132.7	557.3	1.07	0	0.28	0.10	0.10	2.10	0.19	1.7	0	0.3	10.7
72	26	0.74	17.9	282.4	57.0	2.71	1.7	0.52	0.17	0.25	3.35	0.28	3.4	0.5	0.7	29.7
78	23	0.77	24.7	361.5	68.0	1.96	1.7	0.96	0.21	0.25	4.76	0.39	5.1	0.9	0.6	33.2
62	4	0.78	14.5	329.1	34.0	1.76	1.7	0.44	—	0.20	3.66	0.26	3.4	0.3	0.4	28.4
56	22	0.57	21.3	291.7	48.5	1.91	0	0.48	0.08	0.18	6.68	0.57	0	0	0.4	38.6
68	18	0.91	16.2	318.1	40.8	2.02	1.7	0.53	0.20	0.21	3.75	0.31	2.6	0.5	0.5	38.5
67	15	0.96	17.0	329.1	42.5	2.10	1.7	0.56	0.17	0.22	3.90	0.32	3.4	0.5	0.5	41.0
82	10	0.89	23.0	318.1	52.7	2.39	2.6	0.63	0.18	0.28	3.95	0.26	2.6	0.2	0.6	39.8
80	12	0.85	18.7	299.4	51.0	2.51	2.6	0.54	0.18	0.26	3.89	0.34	8.5	0.3	0.6	38.5
78	21	0.90	19.6	292.6	44.2	2.00	2.6	0.75	—	0.29	3.67	0.37	4.3	0.3	0.7	31.6
79	25	1.09	20.4	296.8	24.7	3.24	1.7	0.48	—	0.29	3.96	0.37	4.3	0.3	0.7	36.0

DA+ Code	Food Description	Quantity	Measure	Wt (g)	H_2O (g)	Ener (kcal)	Prot (g)	Carb (g)	Fiber (g)	Fat (g)	Fat Breakdown (g) Sat	Mono	Poly	*Trans*
	Beef, Lamb, Pork—*continued*													
	Pork shoulder													
32626	Shoulder, arm picnic, lean and fat, roasted	3	ounce(s)	85	44.3	270	20.0	0	0	20.4	7.5	9.1	2.0	—
32629	Shoulder, arm picnic, lean, roasted	3	ounce(s)	85	51.3	194	22.7	0	0	10.7	3.7	5.1	1.0	—
	Rabbit													
3366	Domesticated, roasted	3	ounce(s)	85	51.5	168	24.7	0	0	6.8	2.0	1.8	1.3	—
3367	Domesticated, stewed	3	ounce(s)	85	50.0	175	25.8	0	0	7.2	2.1	1.9	1.4	—
	Veal													
3391	Liver, braised	3	ounce(s)	85	50.9	163	24.2	3.2	0	5.3	1.7	1.0	0.9	0.3
3319	Rib, lean only, roasted	3	ounce(s)	85	55.0	151	21.9	0	0	6.3	1.8	2.3	0.6	—
1732	Deer or venison, roasted	3	ounce(s)	85	55.5	134	25.7	0	0	2.7	1.1	0.7	0.5	—
	Poultry													
	Chicken													
29562	Flaked, canned	2	ounce(s)	57	39.3	97	10.3	0.1	0	5.8	1.6	2.3	1.3	—
	Chicken, fried													
29632	Breast, meat only, breaded, baked or fried	3	ounce(s)	85	44.3	193	25.3	6.9	0.2	6.6	1.6	2.7	1.7	—
35327	Broiler breast, meat only, fried	3	ounce(s)	85	51.2	159	28.4	0.4	0	4.0	1.1	1.5	0.9	—
36413	Broiler breast, meat and skin, flour coated, fried	3	ounce(s)	85	48.1	189	27.1	1.4	0.1	7.5	2.1	3.0	1.7	—
36414	Broiler drumstick, meat and skin, flour coated, fried	3	ounce(s)	85	48.2	208	22.9	1.4	0.1	11.7	3.1	4.6	2.7	—
35389	Broiler drumstick, meat only, fried	3	ounce(s)	85	52.9	166	24.3	0	0	6.9	1.8	2.5	1.7	—
35406	Broiler leg, meat only, fried	3	ounce(s)	85	51.5	177	24.1	0.6	0	7.9	2.1	2.9	1.9	—
35484	Broiler wing, meat only, fried	3	ounce(s)	85	50.9	179	25.6	0	0	7.8	2.1	2.6	1.8	—
29580	Patty, fillet or tenders, breaded, cooked	3	ounce(s)	85	40.2	256	14.5	12.2	0	16.5	3.7	8.4	3.7	—
	Chicken, roasted, meat only													
35409	Broiler leg, meat only, roasted	3	ounce(s)	85	55.0	162	23.0	0	0	7.2	1.9	2.6	1.7	—
35486	Broiler wing, meat only, roasted	3	ounce(s)	85	53.4	173	25.9	0	0	6.9	1.9	2.2	1.5	—
35138	Roasting chicken, dark meat, meat only, roasted	3	ounce(s)	85	57.0	151	19.8	0	0	7.4	2.1	2.8	1.7	—
35136	Roasting chicken, light meat, meat only, roasted	3	ounce(s)	85	57.7	130	23.1	0	0	3.5	0.9	1.3	0.8	—
35132	Roasting chicken, meat only, roasted	3	ounce(s)	85	57.3	142	21.3	0	0	5.6	1.5	2.1	1.3	—
	Chicken, stewed													
1268	Gizzard, simmered	3	ounce(s)	85	57.8	124	25.8	0	0	2.3	0.6	0.4	0.3	0.1
1270	Liver, simmered	3	ounce(s)	85	56.8	142	20.8	0.7	0	5.5	1.8	1.2	1.7	0.1
3174	Meat only, stewed	3	ounce(s)	85	56.8	151	23.2	0	0	5.7	1.6	2.0	1.3	—
	Duck													
1286	Domesticated, meat and skin, roasted	3	ounce(s)	85	44.1	287	16.2	0	0	24.1	8.2	11.0	3.1	—
1287	Domesticated, meat only, roasted	3	ounce(s)	85	54.6	171	20.0	0	0	9.5	3.5	3.1	1.2	—
	Goose													
35507	Domesticated, meat and skin, roasted	3	ounce(s)	85	44.2	259	21.4	0	0	18.6	5.8	8.7	2.1	—
35524	Domesticated, meat only, roasted	3	ounce(s)	85	48.7	202	24.6	0	0	10.8	3.9	3.7	1.3	—
1297	Liver pate, smoked, canned	4	tablespoon(s)	52	19.3	240	5.9	2.4	0	22.8	7.5	13.3	0.4	—
	Turkey													
3256	Ground turkey, cooked	3	ounce(s)	85	50.5	200	23.3	0	0	11.2	2.9	4.2	2.7	—
3263	Patty, batter coated, breaded, fried	1	item(s)	94	46.7	266	13.2	14.8	0.5	16.9	4.4	7.0	4.4	—
219	Roasted, dark meat, meat only	3	ounce(s)	85	53.7	159	24.3	0	0	6.1	2.1	1.4	1.8	—
222	Roasted, fryer roaster breast, meat only	3	ounce(s)	85	58.2	115	25.6	0	0	0.6	0.2	0.1	0.2	—
220	Roasted, light meat, meat only	3	ounce(s)	85	56.4	134	25.4	0	0	2.7	0.9	0.5	0.7	—
1303	Turkey roll, light and dark meat	2	slice(s)	57	39.8	84	10.3	1.2	0	4.0	1.2	1.3	1.0	—
1302	Turkey roll, light meat	2	slice(s)	57	42.5	56	8.4	2.9	0	0.9	0.2	0.2	0.1	0
	Processed Meats													
	Beef													
1331	Corned beef loaf, jellied, sliced	2	slice(s)	57	39.2	87	13.0	0	0	3.5	1.5	1.5	0.2	—

KEY: H-4 = Breads/Baked Goods H-10 = Cereal/Rice/Pasta H-14 = Fruit H-18 = Vegetables/Legumes H-28 = Nuts/Seeds H-30 = Vegetarian H-32 = Dairy H-40 = Eggs H-40 = Seafood H-42 = Meats H-46 = Poultry = Processed Meats H-48 = Beverages H-52 = Fats/Oils H-54 = Sweets H-56 = Spices/Condiments/Sauces H-60 = Mixed Foods/Soups/Sandwiches H-64 = Fast Food H-84 = Convenience H-86 = Baby Foods

Chol (mg)	Calc (mg)	Iron (mg)	Magn (mg)	Pota (mg)	Sodi (mg)	Zinc (mg)	Vit A (µg)	Thia (mg)	Vit E (mg α)	Ribo (mg)	Niac (mg)	Vit B_6 (mg)	Fola (µg)	Vit C (mg)	Vit B_{12} (µg)	Sele (µg)
80	16	1.00	14.5	276.4	59.5	2.93	1.7	0.44	—	0.25	3.33	0.29	3.4	0.2	0.6	28.6
81	8	1.20	17.0	298.5	68.0	3.46	1.7	0.49	—	0.30	3.66	0.34	4.3	0.3	0.7	32.7
70	16	1.93	17.9	325.7	40.0	1.93	0	0.07	—	0.17	7.17	0.40	9.4	0	7.1	32.7
73	17	2.01	17.0	255.1	31.5	2.01	0	0.05	0.37	0.14	6.09	0.28	7.7	0	5.5	32.7
435	5	4.34	17.0	279.8	66.3	9.55	8026	0.15	0.57	2.43	11.18	0.78	281.5	0.9	72.0	16.4
98	10	0.81	20.4	264.5	82.5	3.81	0	0.05	0.30	0.24	6.37	0.23	11.9	0	1.3	9.4
95	6	3.80	20.4	284.9	45.9	2.33	0	0.15	—	0.51	5.70	—	—	0	—	11.0
35	8	0.89	6.8	147.4	408.2	0.79	19.3	0.01	—	0.07	3.58	0.19	2.3	0	0.2	—
67	19	1.05	24.7	222.6	450.2	0.84	—	0.08	—	0.09	10.97	0.46	4.3	0	0.3	—
77	14	0.96	26.4	234.7	67.2	0.91	6.0	0.06	0.35	0.10	12.57	0.54	3.4	0	0.3	22.3
76	14	1.01	25.5	220.3	64.6	0.93	12.8	0.06	0.39	0.11	11.68	0.49	5.1	0	0.3	20.3
77	10	1.13	19.6	194.8	75.7	2.45	21.3	0.06	0.65	0.19	5.13	0.29	8.5	0	0.3	15.6
80	10	1.12	20.4	211.8	81.6	2.73	15.3	0.06	—	0.20	5.22	0.33	7.7	0	0.3	16.7
84	11	1.19	21.3	216.0	81.6	2.53	17.0	0.07	0.38	0.21	5.68	0.33	7.7	0	0.3	16.0
71	13	0.96	17.9	176.9	77.4	1.80	15.3	0.03	0.40	0.10	6.15	0.50	3.4	0	0.3	21.6
49	11	0.75	19.6	244.8	411.4	0.79	4.3	0.09	1.04	0.12	5.99	0.24	24.7	0	0.2	13.9
80	10	1.11	20.4	205.8	77.4	2.43	16.2	0.06	0.22	0.19	5.37	0.31	6.8	0	0.3	18.8
72	14	0.98	17.9	178.6	78.2	1.82	15.3	0.03	0.22	0.10	6.21	0.50	3.4	0	0.3	21.0
64	9	1.13	17.0	190.5	80.8	1.81	13.6	0.05	—	0.16	4.87	0.26	6.0	0	0.2	16.7
64	11	0.91	19.6	200.7	43.4	0.66	6.8	0.05	0.22	0.07	8.90	0.45	2.6	0	0.3	21.9
64	10	1.02	17.9	194.8	63.8	1.29	10.2	0.05	—	0.12	6.70	0.34	4.3	0	0.2	20.9
315	14	2.71	2.6	152.2	47.6	3.75	0	0.02	0.17	0.17	2.65	0.06	4.3	0	0.9	35.0
479	9	9.89	21.3	223.7	64.6	3.38	3385.8	0.24	0.69	1.69	9.39	0.64	491.6	23.7	14.3	70.1
71	12	0.99	17.9	153.1	59.5	1.69	12.8	0.04	0.22	0.13	5.20	0.22	5.1	0	0.2	17.8
71	9	2.29	13.6	173.5	50.2	1.58	53.6	0.14	0.59	0.22	4.10	0.15	5.1	0	0.3	17.0
76	10	2.29	17.0	214.3	55.3	2.21	19.6	0.22	0.59	0.39	4.33	0.21	8.5	0	0.3	19.1
77	11	2.40	18.7	279.8	59.5	2.22	17.9	0.06	1.47	0.27	3.54	0.31	1.7	0	0.3	18.5
82	12	2.44	21.3	330.0	64.6	2.69	10.2	0.07	—	0.33	3.47	0.39	10.2	0	0.4	21.7
78	36	2.86	6.8	71.8	362.4	0.47	520.5	0.04	—	0.15	1.30	0.03	31.2	0	4.9	22.9
87	21	1.64	20.4	229.6	91.0	2.43	0	0.04	0.28	0.14	4.09	0.33	6.0	0	0.3	31.6
71	13	2.06	14.1	258.5	752.0	1.35	9.4	0.09	0.87	0.17	2.16	0.18	38.5	0	0.2	20.8
72	27	1.98	20.4	246.6	67.2	3.79	0	0.05	0.54	0.21	3.10	0.30	7.7	0	0.3	34.8
71	10	1.30	24.7	248.3	44.2	1.48	0	0.03	0.07	0.11	6.37	0.47	5.1	0	0.3	27.3
59	16	1.14	23.8	259.4	54.4	1.73	0	0.05	0.07	0.11	5.81	0.45	5.1	0	0.3	27.3
31	18	0.76	10.2	153.1	332.3	1.13	0	0.05	0.19	0.16	2.72	0.15	2.8	0	0.1	16.6
19	4	0.21	10.8	242.1	590.8	0.50	0	0.01	0.07	0.08	4.05	0.23	2.3	0	0.2	7.4
27	6	1.15	6.2	57.3	540.4	2.31	0	0.00	—	0.06	0.99	0.06	4.5	0	0.7	9.8

TABLE H-1 Table of Food Composition (*continued*) (Computer code is for Cengage Diet Analysis program) (For purposes of calculations, use "0" for t, <1, <0.1, 0.01)

DA+ Code	Food Description	Quantity	Measure	Wt (g)	H_2O (g)	Ener (kcal)	Prot (g)	Carb (g)	Fiber (g)	Fat (g)	Fat Breakdown (g)			
											Sat	Mono	Poly	*Trans*
Processed Meats—*continued*														
	Bologna													
13459	Beef	1	slice(s)	28	15.1	90	3.0	1.0	0	8.0	3.5	4.3	0.3	—
13461	Light, made with pork and chicken	1	slice(s)	28	18.2	60	3.0	2.0	0	4.0	1.0	2.0	0.4	—
13458	Made with chicken and pork	1	slice(s)	28	15.0	90	3.0	1.0	0	8.0	3.0	4.1	1.1	—
13565	Turkey bologna	1	slice(s)	28	19.0	50	3.0	1.0	0	4.0	1.0	1.1	1.0	0
	Chicken													
7125	Breast, smoked	1	slice(s)	10	—	10	1.8	0.3	0	0.2	0	—	—	—
	Ham													
7127	Deli-sliced, honey	1	slice(s)	10	—	10	1.7	0.3	0	0.3	0.1	—	—	—
7126	Deli-sliced, smoked	1	slice(s)	10	—	10	1.7	0.2	0	0.3	0.1	—	—	—
8614	**Beef and pork mortadella, sliced**	2	slice(s)	46	24.1	143	7.5	1.4	0	11.7	4.4	5.2	1.4	—
1323	**Pork olive loaf**	2	slice(s)	57	33.1	133	6.7	5.2	0	9.4	3.3	4.5	1.1	—
1324	**Pork pickle and pimento loaf**	2	slice(s)	57	34.2	128	6.4	4.8	0.9	9.1	3.0	4.0	1.6	—
	Sausages and frankfurters													
37296	Beerwurst beef, beer salami (bierwurst)	1	slice(s)	29	16.6	74	4.1	1.2	0	5.7	2.5	2.7	0.2	—
37257	Beerwurst pork, beer salami	1	slice(s)	21	12.9	50	3.0	0.4	0	4.0	1.3	1.9	0.5	—
35338	Berliner, pork and beef	1	ounce(s)	28	17.3	65	4.3	0.7	0	4.9	1.7	2.3	0.4	—
37298	Bratwurst pork, cooked	1	piece(s)	74	42.3	181	10.4	1.9	0	14.3	5.1	6.7	1.5	—
37299	Braunschweiger pork liver sausage	1	slice(s)	15	8.2	51	2.0	0.3	0	4.5	1.5	2.1	0.5	—
1329	Cheesefurter or cheese smokie, beef and pork	1	item(s)	43	22.6	141	6.1	0.6	0	12.5	4.5	5.9	1.3	—
1330	Chorizo, beef and pork	2	ounce(s)	57	18.1	258	13.7	1.1	0	21.7	8.2	10.4	2.0	—
8600	Frankfurter, beef	1	item(s)	45	23.4	149	5.1	1.8	0	13.3	5.3	6.4	0.5	—
202	Frankfurter, beef and pork	1	item(s)	45	25.2	137	5.2	0.8	0	12.4	4.8	6.2	1.2	—
1293	Frankfurter, chicken	1	item(s)	45	28.1	100	7.0	1.2	0.2	7.3	1.7	2.7	1.7	0.1
3261	Frankfurter, turkey	1	item(s)	45	28.3	100	5.5	1.7	0	7.8	1.8	2.6	1.8	0.4
37275	Italian sausage, pork, cooked	1	item(s)	68	32.0	234	13.0	2.9	0.1	18.6	6.5	8.1	2.2	—
37307	Kielbasa or kolbassa, pork and beef	1	slice(s)	30	18.5	67	5.0	1.0	0	4.7	1.7	2.2	0.5	—
1333	Knockwurst or knackwurst, beef and pork	2	ounce(s)	57	31.4	174	6.3	1.8	0	15.7	5.8	7.3	1.7	—
37285	Pepperoni, beef and pork	1	slice(s)	11	3.4	51	2.2	0.4	0.2	4.4	1.8	2.1	0.3	—
37313	Polish sausage, pork	1	slice(s)	21	11.4	60	2.8	0.7	0	5.0	1.8	2.3	0.5	—
206	Salami, beef, cooked, sliced	2	slice(s)	52	31.2	136	6.5	1.0	0	11.5	5.1	5.5	0.5	—
37272	Salami, pork, dry or hard	1	slice(s)	13	4.6	52	2.9	0.2	0	4.3	1.5	2.0	0.5	—
40987	Sausage, turkey, cooked	2	ounce(s)	57	36.9	111	13.5	0	0	5.9	1.3	1.7	1.5	0.2
8620	Smoked sausage, beef and pork	2	ounce(s)	57	30.6	181	6.8	1.4	0	16.3	5.5	6.9	2.2	0
8619	Smoked sausage, pork	2	ounce(s)	57	32.0	178	6.8	1.2	0	16.0	5.3	6.4	2.1	0.1
37273	Smoked sausage, pork link	1	piece(s)	76	29.8	295	16.8	1.6	0	24.0	8.6	11.1	2.8	—
1336	Summer sausage, thuringer, or cervelat, beef and pork	2	ounce(s)	57	25.6	205	9.9	1.9	0	17.3	6.5	7.4	0.7	—
37294	Vienna sausage, cocktail, beef and pork, canned	1	piece(s)	16	10.4	37	1.7	0.4	0	3.1	1.1	1.5	0.2	—
	Spreads													
1318	Ham salad spread	¼	cup(s)	60	37.6	130	5.2	6.4	0	9.3	3.0	4.3	1.6	—
32419	Pork and beef sandwich spread	4	tablespoon(s)	60	36.2	141	4.6	7.2	0.1	10.4	3.6	4.6	1.5	—
	Turkey													
13604	Breast, fat free, oven roasted	1	slice(s)	28	—	25	4.0	1.0	0	0	0	0	0	0
13606	Breast, hickory smoked fat free	1	slice(s)	28	—	25	4.0	1.0	0	0	0	0	0	0
16049	Breast, hickory smoked slices	1	slice(s)	56	—	50	11.0	1.0	0	0	0	0	0	0
16047	Breast, honey roasted slices	1	slice(s)	56	—	60	11.0	3.0	0	0	0	0	0	0
16048	Breast, oven roasted slices	1	slice(s)	56	—	50	11.0	1.0	0	0	0	0	0	0
7124	Breast, oven roasted	1	slice(s)	10	—	10	1.8	0.3	0	0.1	0	0	0	—
13567	Turkey ham, 10% water added	2	slice(s)	56	40.9	70	10.0	2.0	0	3.0	0	0.4	0.6	0
37270	Turkey pastrami	1	slice(s)	28	20.3	35	4.6	1.0	0	1.2	0.3	0.4	0.3	—
3262	Turkey salami	2	slice(s)	57	39.1	98	10.9	0.9	0.1	5.2	1.6	1.8	1.4	0
37318	Turkey salami, cooked	1	slice(s)	28	20.4	43	4.3	0.1	0	2.7	0.8	0.9	0.7	—
Beverages														
	Beer													
866	Ale, mild	12	fluid ounce(s)	360	332.3	148	1.1	13.3	0.4	0	0	0	0	0
686	Beer	12	fluid ounce(s)	356	327.7	153	1.6	12.7	0	0	0	0	0	0
16886	Beer, non alcoholic	12	fluid ounce(s)	360	328.1	133	0.8	29.0	0	0.4	0.1	0	0.2	0

KEY: H-4 = Breads/Baked Goods H-10 = Cereal/Rice/Pasta H-14 = Fruit H-18 = Vegetables/Legumes H-28 = Nuts/Seeds H-30 = Vegetarian H-32 = Dairy H-40 = Eggs H-40 = Seafood H-42 = Meats H-46 = Poultry = Processed Meats H-48 = Beverages H-52 = Fats/Oils H-54 = Sweets H-56 = Spices/Condiments/Sauces H-60 = Mixed Foods/Soups/Sandwiches H-64 = Fast Food H-84 = Convenience H-86 = Baby Foods

Chol (mg)	Calc (mg)	Iron (mg)	Magn (mg)	Pota (mg)	Sodi (mg)	Zinc (mg)	Vit A (µg)	Thia (mg)	Vit E (mg α)	Ribo (mg)	Niac (mg)	Vit B_6 (mg)	Fola (µg)	Vit C (mg)	Vit B_{12} (µg)	Sele (µg)
20	0	0.36	3.9	47.0	310.0	0.56	0	0.01	—	0.03	0.67	0.04	3.6	0	0.4	—
20	40	0.36	5.6	45.6	300.0	0.45	0	—	—	—	—	—	—	0	—	—
30	20	0.36	5.9	43.1	300.0	0.39	0	—	—	—	—	—	—	0	—	—
20	40	0.36	6.2	42.6	270.0	0.51	0	—	—	—	—	—	—	0	—	—
4	0	0.00	—	—	100.0	—	0	—	—	—	—	—	—	0	—	—
4	0	0.12	—	—	100.0	—	0	—	—	—	—	—	—	0.6	—	—
4	0	0.12	—	—	103.3	—	0	—	—	—	—	—	—	0.6	—	—
26	8	0.64	5.1	75.0	573.2	0.96	0	0.05	0.10	0.07	1.23	0.06	1.4	0	0.7	10.4
22	62	0.30	10.8	168.7	842.9	0.78	34.1	0.16	0.14	0.14	1.04	0.13	1.1	0	0.7	9.3
33	62	0.75	19.3	210.7	740.7	0.95	44.3	0.22	0.22	0.06	1.41	0.23	21.0	4.4	0.3	4.5
18	3	0.44	3.5	66.5	264.9	0.71	0	0.02	0.05	0.03	0.98	0.04	0.9	0	0.6	4.7
12	2	0.15	2.7	53.3	261.0	0.36	0	0.11	0.03	0.04	0.68	0.07	0.6	0	0.2	4.4
13	3	0.32	4.3	80.2	367.7	0.70	0	0.10	—	0.06	0.88	0.05	1.4	0	0.8	4.0
44	33	0.95	11.1	156.9	412.2	1.70	0	0.37	0.01	0.13	2.36	0.15	1.5	0.7	0.7	15.7
24	1	1.42	1.7	27.5	131.5	0.42	641.0	0.03	0.05	0.23	1.27	0.05	6.7	0	3.1	8.8
29	25	0.46	5.6	88.6	465.3	0.96	20.2	0.10	0.10	0.06	1.24	0.05	1.3	0	0.7	6.8
50	5	0.90	10.2	225.7	700.2	1.93	0	0.35	0.12	0.17	2.90	0.30	1.1	0	1.1	12.0
24	6	0.67	6.3	70.2	513.0	1.10	0	0.01	0.09	0.06	1.06	0.04	2.3	0	0.8	3.7
23	5	0.51	4.5	75.2	504.0	0.82	8.1	0.09	0.11	0.05	1.18	0.05	1.8	0	0.6	6.2
43	33	0.52	9.0	90.9	379.8	0.50	0	0.02	0.09	0.11	2.10	0.14	3.2	0	0.2	10.4
35	67	0.66	6.3	176.4	485.1	0.82	0	0.01	0.27	0.08	1.65	0.06	4.1	0	0.4	6.8
39	14	0.97	12.2	206.7	820.8	1.62	6.8	0.42	0.17	0.15	2.83	0.22	3.4	0.1	0.9	15.0
20	13	0.44	4.9	84.4	283.0	0.61	0	0.06	0.06	0.06	0.87	0.05	1.5	0	0.5	5.4
34	6	0.37	6.2	112.8	527.3	0.94	0	0.19	0.32	0.07	1.55	0.09	1.1	0	0.7	7.7
13	2	0.15	2.0	34.7	196.7	0.30	0	0.05	0.00	0.02	0.59	0.04	0.7	0.1	0.2	2.4
15	2	0.29	2.9	37.3	199.3	0.40	0	0.10	0.04	0.03	0.71	0.03	0.4	0.2	0.2	3.7
37	3	1.14	6.8	97.8	592.8	0.92	0	0.04	0.08	0.08	1.68	0.08	1.0	0	1.6	7.6
10	2	0.16	2.8	48.4	289.3	0.53	0	0.11	0.02	0.04	0.71	0.07	0.3	0	0.4	3.3
52	12	0.84	11.9	169.0	377.1	2.19	7.4	0.04	0.10	0.14	3.24	0.18	3.4	0.4	0.7	0
33	7	0.42	7.4	101.5	516.5	0.71	7.4	0.10	0.07	0.06	1.66	0.09	1.1	0	0.3	0
35	6	0.33	6.2	273.9	468.9	0.74	0	0.12	0.14	0.10	1.59	0.10	0.6	0	0.4	10.4
52	23	0.87	14.4	254.6	1136.6	2.13	0	0.53	0.18	0.19	3.43	0.26	3.8	1.5	1.2	16.4
42	5	1.15	7.9	147.4	737.1	1.45	0	0.08	0.12	0.18	2.44	0.14	1.1	9.4	3.1	11.5
14	2	0.14	1.1	16.2	155.0	0.25	0	0.01	0.03	0.01	0.25	0.01	0.6	0	0.2	2.7
22	5	0.35	6.0	90.0	547.2	0.66	0	0.26	1.04	0.07	1.25	0.09	0.6	0	0.5	10.7
23	7	0.47	4.8	66.0	607.8	0.61	15.6	0.10	1.04	0.08	1.03	0.07	1.2	0	0.7	5.8
10	0	0.00	—	—	340.0	—	0	—	—	—	—	—	—	0	—	—
10	0	0.00	—	—	300.0	—	0	—	—	—	—	—	—	0	—	—
25	0	0.72	—	—	720.0	—	0	—	—	—	—	—	—	0	—	—
20	0	0.72	—	—	660.0	—	0	—	—	—	—	—	—	0	—	—
20	0	0.72	—	—	660.0	—	0	—	—	—	—	—	—	0	—	—
4	0	0.06	—	—	103.3	—	0	—	—	—	—	—	—	0	—	—
40	0	0.72	12.3	162.4	700.0	1.44	0	—	—	—	—	—	—	0	—	—
19	3	1.19	4.0	97.8	278.1	0.61	1.1	0.01	0.06	0.07	1.00	0.07	1.4	4.6	0.1	4.6
43	23	0.70	12.5	122.5	569.3	1.31	1.1	0.24	0.13	0.17	2.25	0.24	5.7	0	0.6	15.0
22	11	0.35	6.2	61.2	284.6	0.65	0.6	0.12	0.06	0.08	1.12	0.12	2.8	0	0.3	7.5
0	18	0.07	21.6	90.0	14.4	0.03	0	0.03	0.00	0.10	1.62	0.18	21.6	0	0.1	2.5
0	14	0.07	21.4	96.2	14.3	0.03	0	0.01	0.00	0.08	1.82	0.16	21.4	0	0.1	2.1
0	25	0.21	25.2	28.8	46.8	0.07	—	0.07	0.00	0.18	3.99	0.10	50.4	1.8	0.1	4.3

APPENDIX H

DA+ Code	Food Description	Quantity	Measure	Wt (g)	H_2O (g)	Ener (kcal)	Prot (g)	Carb (g)	Fiber (g)	Fat (g)	Fat Breakdown (g) Sat	Mono	Poly	*Trans*
Beverages—*continued*														
31609	Bud Light beer	12	fluid ounce(s)	355	335.5	110	0.9	6.6	0	0	0	0	0	0
31608	Budweiser beer	12	fluid ounce(s)	355	327.7	145	1.3	10.6	0	0	0	0	0	0
869	Light beer	12	fluid ounce(s)	354	335.9	103	0.9	5.8	0	0	0	0	0	0
31613	Michelob beer	12	fluid ounce(s)	355	323.4	155	1.3	13.3	0	0	0	0	0	0
31614	Michelob Light beer	12	fluid ounce(s)	355	329.8	134	1.1	11.7	0	0	0	0	0	0
	Gin, rum, vodka, whiskey													
857	Distilled alcohol, 100 proof	1	fluid ounce(s)	28	16.0	82	0	0	0	0	0	0	0	0
687	Distilled alcohol, 80 proof	1	fluid ounce(s)	28	18.5	64	0	0	0	0	0	0	0	0
688	Distilled alcohol, 86 proof	1	fluid ounce(s)	28	17.8	70	0	0	0	0	0	0	0	0
689	Distilled alcohol, 90 proof	1	fluid ounce(s)	28	17.3	73	0	0	0	0	0	0	0	0
856	Distilled alcohol, 94 proof	1	fluid ounce(s)	28	16.8	76	0	0	0	0	0	0	0	0
	Liqueurs													
33187	Coffee liqueur, 53 proof	1	fluid ounce(s)	35	10.8	113	0	16.3	0	0.1	0	0	0	—
3142	Coffee liqueur, 63 proof	1	fluid ounce(s)	35	14.4	107	0	11.2	0	0.1	0	0	0	—
736	Cordials, 54 proof	1	fluid ounce(s)	30	8.9	106	0	13.3	0	0.1	0	0	0	—
	Wine													
861	California red wine	5	fluid ounce(s)	150	133.4	125	0.3	3.7	0	0	0	0	0	0
858	Domestic champagne	5	fluid ounce(s)	150	—	105	0.3	3.8	0	0	0	0	0	0
690	Sweet dessert wine	5	fluid ounce(s)	147	103.7	235	0.3	20.1	0	0	0	0	0	0
1481	White wine	5	fluid ounce(s)	148	128.1	121	0.1	3.8	0	0	0	0	0	0
1811	Wine cooler	10	fluid ounce(s)	300	267.4	159	0.3	20.2	0	0.1	0	0	0	—
	Carbonated													
31898	7 Up	12	fluid ounce(s)	360	321.0	140	0	39.0	0	0	0	0	0	0
692	Club soda	12	fluid ounce(s)	355	354.8	0	0	0	0	0	0	0	0	0
12010	Coca-Cola Classic cola soda	12	fluid ounce(s)	360	319.4	146	0	40.5	0	0	0	0	0	0
693	Cola	12	fluid ounce(s)	368	332.7	136	0.3	35.2	0	0.1	0	0	0	—
2391	Cola or pepper-type soda, low calorie with saccharin	12	fluid ounce(s)	355	354.5	0	0	0.3	0	0	0	0	0	0
9522	Cola soda, decaffeinated	12	fluid ounce(s)	372	333.4	153	0	39.3	0	0	0	0	0	0
9524	Cola, decaffeinated, low calorie with aspartame	12	fluid ounce(s)	355	354.3	4	0.4	0.5	0	0	0	0	0	0
1415	Cola, low calorie with aspartame	12	fluid ounce(s)	355	353.6	7	0.4	1.0	0	0.1	0	0	0	—
1412	Cream soda	12	fluid ounce(s)	371	321.5	189	0	49.3	0	0	0	0	0	0
31899	Diet 7 Up	12	fluid ounce(s)	360	—	0	0	0	0	0	0	0	0	0
12031	Diet Coke cola soda	12	fluid ounce(s)	360	—	2	0	0.2	0	0	0	0	0	0
29392	Diet Mountain Dew soda	12	fluid ounce(s)	360	—	0	0	0	0	0	0	0	0	0
29389	Diet Pepsi cola soda	12	fluid ounce(s)	360	—	0	0	0	0	0	0	0	0	0
12034	Diet Sprite soda	12	fluid ounce(s)	360	—	4	0	0	0	0	0	0	0	0
695	Ginger ale	12	fluid ounce(s)	366	333.9	124	0	32.1	0	0	0	0	0	0
694	Grape soda	12	fluid ounce(s)	372	330.3	160	0	41.7	0	0	0	0	0	0
1876	Lemon lime soda	12	fluid ounce(s)	368	330.8	147	0.2	37.4	0	0.1	0	0	0	—
29391	Mountain Dew soda	12	fluid ounce(s)	360	314.0	170	0	46.0	0	0	0	0	0	0
3145	Orange soda	12	fluid ounce(s)	372	325.9	179	0	45.8	0	0	0	0	0	0
1414	Pepper-type soda	12	fluid ounce(s)	368	329.3	151	0	38.3	0	0.4	0.3	0	0	—
29388	Pepsi regular cola soda	12	fluid ounce(s)	360	318.9	150	0	41.0	0	0	0	0	0	0
696	Root beer	12	fluid ounce(s)	370	330.0	152	0	39.2	0	0	0	0	0	0
12044	Sprite soda	12	fluid ounce(s)	360	321.0	144	0	39.0	0	0	0	0	0	0
	Coffee													
731	Brewed	8	fluid ounce(s)	237	235.6	2	0.3	0	0	0	0	0	0	0
9520	Brewed, decaffeinated	8	fluid ounce(s)	237	234.3	5	0.3	1.0	0	0	0	0	0	0
16882	Cappuccino	8	fluid ounce(s)	240	224.8	79	4.1	5.8	0.2	4.9	2.3	1.0	0.2	—
16883	Cappuccino, decaffeinated	8	fluid ounce(s)	240	224.8	79	4.1	5.8	0.2	4.9	2.3	1.0	0.2	—
16880	Espresso	8	fluid ounce(s)	237	231.8	21	0	3.6	0	0.4	0.2	0	0.2	0
16881	Espresso, decaffeinated	8	fluid ounce(s)	237	231.8	21	0	3.6	0	0.4	0.2	0	0.2	0
732	Instant, prepared	8	fluid ounce(s)	239	236.5	5	0.2	0.8	0	0	0	0	0	0
	Fruit drinks													
29357	Crystal Light sugar-free lemonade drink	8	fluid ounce(s)	240	—	5	0	0	0	0	0	0	0	0
6012	Fruit punch drink with added vitamin C, canned	8	fluid ounce(s)	248	218.2	117	0	29.7	0.5	0	0	0	0	0
31143	Gatorade Thirst Quencher, all flavors	8	fluid ounce(s)	240	—	50	0	14.0	0	0	0	0	0	0
260	Grape drink, canned	8	fluid ounce(s)	250	210.5	153	0	39.4	0	0	0	0	0	0
17372	Kool-Aid (lemonade/punch/fruit drink)	8	fluid ounce(s)	248	220.0	108	0.1	27.8	0.2	0	0	0	0	—

KEY: H-4 = Breads/Baked Goods H-10 = Cereal/Rice/Pasta H-14 = Fruit H-18 = Vegetables/Legumes H-28 = Nuts/Seeds H-30 = Vegetarian H-32 = Dairy H-40 = Eggs H-40 = Seafood H-42 = Meats H-46 = Poultry
= Processed Meats H-48 = Beverages H-52 = Fats/Oils H-54 = Sweets H-56 = Spices/Condiments/Sauces H-60 = Mixed Foods/Soups/Sandwiches H-64 = Fast Food H-84 = Convenience H-86 = Baby Foods

Chol (mg)	Calc (mg)	Iron (mg)	Magn (mg)	Pota (mg)	Sodi (mg)	Zinc (mg)	Vit A (μg)	Thia (mg)	Vit E (mg α)	Ribo (mg)	Niac (mg)	Vit B_6 (mg)	Fola (μg)	Vit C (mg)	Vit B_{12} (μg)	Sele (μg)
0	18	0.14	17.8	63.9	9.0	0.10	0	0.03	—	0.10	1.39	0.12	14.6	0	0	4.0
0	18	0.10	21.3	88.8	9.0	0.07	0	0.02	—	0.09	1.60	0.17	21.3	0	0.1	4.0
0	14	0.10	17.7	74.3	14.2	0.03	0	0.01	0.00	0.05	1.38	0.12	21.2	0	0.1	1.4
0	18	0.10	21.3	88.8	9.0	0.07	0	0.02	—	0.09	1.60	0.17	21.3	0	0.1	4.0
0	18	0.14	17.8	63.9	9.0	0.10	0	0.03	—	0.10	1.39	0.12	14.6	0	0	4.0
0	0	0.01	0	0.6	0.3	0.01	0	0.00	—	0.00	0.00	0.00	0	0	0	0
0	0	0.01	0	0.6	0.3	0.01	0	0.00	0.00	0.00	0.00	0.00	0	0	0	0
0	0	0.01	0	0.6	0.3	0.01	0	0.00	0.00	0.00	0.00	0.00	0	0	0	0
0	0	0.01	0	0.6	0.3	0.01	0	0.00	0.00	0.00	0.00	0.00	0	0	0	0
0	0	0.01	0	0.6	0.3	0.01	0	0.00	—	0.00	0.00	0.00	0	0	0	0
0	0	0.02	1.0	10.4	2.8	0.01	0	0.00	0.00	0.00	0.05	0.00	0	0	0	0.1
0	0	0.02	1.0	10.4	2.8	0.01	0	0.00	—	0.00	0.05	0.00	0	0	0	0.1
0	0	0.02	0.6	4.5	2.1	0.01	0	0.00	0.00	0.00	0.02	0.00	0	0	0	0.1
0	12	1.43	16.2	170.6	15.0	0.14	0	0.01	0.00	0.04	0.11	0.05	1.5	0	0	—
0	—	—	—	—	—	—	—	—	—	—	—	—	—	—	0	—
0	12	0.34	13.2	135.4	13.2	0.10	0	0.01	0.00	0.01	0.30	0.00	0	0	0	0.7
0	13	0.39	14.8	104.7	7.4	0.18	0	0.01	0.00	0.01	0.15	0.06	1.5	0	0	0.1
0	18	0.75	15.0	129.0	24.0	0.18	—	0.01	0.03	0.03	0.13	0.03	3.0	5.4	0	0.6
0	—	—	—	0.6	75.0	—	—	—	—	—	—	—	—	—	—	—
0	18	0.03	3.5	7.1	74.6	0.35	0	0.00	0.00	0.00	0.00	0.00	0	0	0	0
0	—	—	—	0	49.5	—	0	—	—	—	—	—	—	0	—	—
0	7	0.41	0	7.4	14.7	0.06	0	0.00	0.00	0.00	0.00	0.00	0	0	0	0.4
0	14	0.06	3.5	14.2	56.8	0.11	0	0.00	0.00	0.00	0.00	0.00	0	0	0	0.3
0	7	0.08	0	11.2	14.9	0.03	0	0.00	0.00	0.00	0.00	0.00	0	0	0	0.4
0	11	0.06	0	24.9	14.2	0.03	0	0.02	0.00	0.08	0.00	0.00	0	0	0	0.3
0	11	0.39	3.5	28.4	28.4	0.03	0	0.02	0.00	0.08	0.00	0.00	0	0	0	0
0	19	0.18	3.7	3.7	44.5	0.26	0	0.00	0.00	0.00	0.00	0.00	0	0	0	0
0	—	—	—	77.0	45.0	—	—	—	—	—	—	—	—	—	—	—
0	—	—	—	18.0	42.0	—	0	—	—	—	—	—	—	0	—	—
0	—	—	—	70.0	35.0	—	—	—	—	—	—	—	—	—	—	—
0	—	—	—	30.0	35.0	—	—	—	—	—	—	—	—	—	—	—
0	—	—	—	109.5	36.0	—	0	—	—	—	—	—	—	0	—	—
0	11	0.65	3.7	3.7	25.6	0.18	0	0.00	0.00	0.00	0.00	0.00	0	0	0	0.4
0	11	0.29	3.7	3.7	55.8	0.26	0	0.00	—	0.00	0.00	0.00	0	0	0	0
0	7	0.41	3.7	3.7	33.2	0.14	0	0.00	0.00	0.00	0.05	0.00	0	0	0	0
0	—	—	—	0	70.0	—	—	—	—	—	—	—	—	—	—	—
0	19	0.21	3.7	7.4	44.6	0.36	0	0.00	—	0.00	0.00	0.00	0	0	0	0
0	11	0.14	0	3.7	36.8	0.14	0	0.00	—	0.00	0.00	0.00	0	0	0	0.4
0	—	—	—	0	35.0	—	—	—	—	—	—	—	—	—	—	—
0	18	0.18	3.7	3.7	48.0	0.26	0	0.00	0.00	0.00	0.00	0.00	0	0	0	0.4
0	—	—	—	0	70.5	—	0	—	—	—	—	—	—	0	—	—
0	5	0.02	7.1	116.1	4.7	0.04	0	0.03	0.02	0.18	0.45	0.00	4.7	0	0	0
0	7	0.14	11.8	108.9	4.7	0.00	0	0.00	0.00	0.03	0.66	0.00	0	0	0	0.5
12	144	0.19	14.4	232.8	50.4	0.50	33.6	0.04	0.09	0.27	0.13	0.04	7.2	0	0.4	4.6
12	144	0.19	14.4	232.8	50.4	0.50	33.6	0.04	0.09	0.27	0.13	0.04	7.2	0	0.4	4.6
0	5	0.30	189.6	272.6	33.2	0.11	0	0.00	0.04	0.42	12.34	0.00	2.4	0.5	0	0
0	5	0.30	189.6	272.6	33.2	0.11	0	0.00	0.04	0.42	12.34	0.00	2.4	0.5	0	0
0	10	0.09	9.5	71.6	9.5	0.01	0	0.00	0.00	0.00	0.56	0.00	0	0	0	0.2
0	0	0.00	—	160.0	40.0	—	0	—	—	—	—	—	—	0	—	—
0	20	0.22	7.4	62.0	94.2	0.02	5.0	0.05	0.04	0.05	0.05	0.02	9.9	89.3	0	0.5
0	0	0.00	—	30.0	110.0	—	0	—	—	—	—	—	—	0	—	—
0	130	0.17	2.5	30.0	40.0	0.30	0	0.00	0.00	0.01	0.02	0.01	0	78.5	0	0.3
0	14	0.45	5.0	49.6	31.0	0.19	—	0.03	—	0.05	0.04	0.01	4.3	41.6	0	1.0

DA+ Code	Food Description	Quantity	Measure	Wt (g)	H_2O (g)	Ener (kcal)	Prot (g)	Carb (g)	Fiber (g)	Fat (g)	Fat Breakdown (g) Sat	Mono	Poly	*Trans*
Beverages—*continued*														
17225	Kool-Aid sugar free, low calorie tropical punch drink mix, prepared	8	fluid ounce(s)	240	—	5	0	0	0	0	0	0	0	0
266	Lemonade, prepared from frozen concentrate	8	fluid ounce(s)	248	221.6	99	0.2	25.8	0	0.1	0	0	0	—
268	Limeade, prepared from frozen concentrate	8	fluid ounce(s)	247	212.6	128	0	34.1	0	0	0	0	0	—
14266	Odwalla strawberry C monster smoothie blend	8	fluid ounce(s)	240	—	160	2.0	38.0	0	0	0	0	0	0
10080	Odwalla strawberry lemonade quencher	8	fluid ounce(s)	240	—	110	0	28.0	0	0	0	0	0	0
10099	Snapple fruit punch fruit drink	8	fluid ounce(s)	240	—	110	0	29.0	0	0	0	0	0	0
10096	Snapple kiwi strawberry fruit drink	8	fluid ounce(s)	240	211.2	110	0	28.0	0	0	0	0	0	0
	Slim Fast ready-to-drink shake													
16054	French vanilla ready to drink shake	11	fluid ounce(s)	325	—	220	10.0	40.0	5.0	2.5	0.5	1.5	0.5	—
40447	Optima rich chocolate royal ready-to-drink shake	11	fluid ounce(s)	330	—	180	10.0	24.0	5.0	5.0	1.0	3.5	0.5	0
16055	Strawberries n cream ready to drink shake	11	fluid ounce(s)	325	—	220	10.0	40.0	5.0	2.5	0.5	1.5	0.5	—
	Tea													
33179	Decaffeinated, prepared	8	fluid ounce(s)	237	236.3	2	0	0.7	0	0	0	0	0	0
1877	Herbal, prepared	8	fluid ounce(s)	237	236.1	2	0	0.5	0	0	0	0	0	0
735	Instant tea mix, lemon flavored with sugar, prepared	8	fluid ounce(s)	259	236.2	91	0	22.3	0.3	0.2	0	0	0	—
734	Instant tea mix, unsweetened, prepared	8	fluid ounce(s)	237	236.1	2	0.1	0.4	0	0	0	0	0	0
733	Tea, prepared	8	fluid ounce(s)	237	236.3	2	0	0.7	0	0	0	0	0	0
	Water													
1413	Mineral water, carbonated	8	fluid ounce(s)	237	236.8	0	0	0	0	0	0	0	0	0
33183	Poland spring water, bottled	8	fluid ounce(s)	237	237.0	0	0	0	0	0	0	0	0	0
1821	Tap water	8	fluid ounce(s)	237	236.8	0	0	0	0	0	0	0	0	0
1879	Tonic water	8	fluid ounce(s)	244	222.3	83	0	21.5	0	0	0	0	0	0
Fats and Oils														
	Butter													
104	Butter	1	tablespoon(s)	14	2.3	102	0.1	0	0	11.5	7.3	3.0	0.4	—
2522	Butter Buds, dry butter substitute	1	teaspoon(s)	2	—	5	0	2.0	0	0	0	0	0	0
921	Unsalted	1	tablespoon(s)	14	2.5	102	0.1	0	0	11.5	7.3	3.0	0.4	—
107	Whipped	1	tablespoon(s)	9	1.5	67	0.1	0	0	7.6	4.7	2.2	0.3	—
944	Whipped, unsalted	1	tablespoon(s)	11	2.0	82	0.1	0	0	9.2	5.9	2.4	0.3	—
	Fats, cooking													
2671	Beef tallow, semisolid	1	tablespoon(s)	13	0	115	0	0	0	12.8	6.4	5.4	0.5	—
922	Chicken fat	1	tablespoon(s)	13	0	115	0	0	0	12.8	3.8	5.7	2.7	—
5454	Household shortening with vegetable oil	1	tablespoon(s)	13	0	115	0	0	0	13.0	3.4	5.5	2.7	2.2
111	Lard	1	tablespoon(s)	13	0	115	0	0	0	12.8	5.0	5.8	1.4	—
	Margarine													
114	Margarine	1	tablespoon(s)	14	2.3	101	0	0.1	0	11.4	2.1	5.5	3.4	2.1
5439	Soft	1	tablespoon(s)	14	2.3	103	0.1	0.1	0	11.6	1.7	4.4	2.1	3.0
32329	Soft, unsalted, with hydrogenated soybean and cottonseed oils	1	tablespoon(s)	14	2.5	101	0.1	0.1	0	11.3	2.0	5.4	3.5	—
928	Unsalted	1	tablespoon(s)	14	2.6	101	0.1	0.1	0	11.3	2.1	5.2	3.5	—
119	Whipped	1	tablespoon(s)	9	1.5	64	0.1	0.1	0	7.2	1.2	3.2	2.5	—
	Spreads													
54657	I Can't Believe It's Not Butter!, tub, soya oil (non-hydrogenated)	1	tablespoon(s)	14	2.3	103	0.1	0.1	0	11.6	2.8	2.0	5.1	0.1
2708	Mayonnaise with soybean and safflower oils	1	tablespoon(s)	14	2.1	99	0.2	0.4	0	11.0	1.2	1.8	7.6	—
16157	Promise vegetable oil spread, stick	1	tablespoon(s)	14	4.2	90	0	0	0	10.0	2.5	2.0	4.0	—
	Oils													
2681	Canola	1	tablespoon(s)	14	0	120	0	0	0	13.6	1.0	8.6	3.8	0.1
120	Corn	1	tablespoon(s)	14	0	120	0	0	0	13.6	1.8	3.8	7.4	0
122	Olive	1	tablespoon(s)	14	0	119	0	0	0	13.5	1.9	9.9	1.4	—

KEY: H-4 = Breads/Baked Goods H-10 = Cereal/Rice/Pasta H-14 = Fruit H-18 = Vegetables/Legumes H-28 = Nuts/Seeds H-30 = Vegetarian H-32 = Dairy H-40 = Eggs H-40 = Seafood H-42 = Meats H-46 = Poultry
= Processed Meats H-48 = Beverages H-52 = Fats/Oils H-54 = Sweets H-56 = Spices/Condiments/Sauces H-60 = Mixed Foods/Soups/Sandwiches H-64 = Fast Food H-84 = Convenience H-86 = Baby Foods

Chol (mg)	Calc (mg)	Iron (mg)	Magn (mg)	Pota (mg)	Sodi (mg)	Zinc (mg)	Vit A (µg)	Thia (mg)	Vit E (mg α)	Ribo (mg)	Niac (mg)	Vit B_6 (mg)	Fola (µg)	Vit C (mg)	Vit B_{12} (µg)	Sele (µg)
0	0	0.00	—	10.1	10.1	—	0	—	—	—	—	—	—	6.0	—	—
0	10	0.39	5.0	37.2	9.9	0.05	0	0.01	0.02	0.05	0.04	0.01	2.5	9.7	0	0.2
0	5	0.00	4.9	24.7	7.4	0.02	0	0.01	0.00	0.01	0.02	0.01	2.5	7.7	0	0.2
0	20	0.72	—	0	20.0	—	0	—	—	—	—	—	—	600.0	0	—
0	0	0.00	—	70.0	10.0	—	0	—	—	—	—	—	—	54.0	0	—
0	0	0.00	—	20.0	10.0	—	0	—	—	—	—	—	—	0	0	—
0	0	0.00	—	40.0	10.0	—	0	—	—	—	—	—	—	0	0	—
5	400	2.70	140.0	600.0	220.0	2.25	—	0.52	—	0.59	7.00	0.70	120.0	60.0	2.1	17.5
5	1000	2.70	140.0	600.0	220.0	2.25	—	0.52	—	0.59	7.00	0.70	120.0	30.0	2.1	17.5
5	400	2.70	140.0	600.0	220.0	2.25	—	0.52	—	0.59	7.00	0.70	120.0	60.0	2.1	17.5
0	0	0.04	7.1	87.7	7.1	0.04	0	0.00	0.00	0.03	0.00	0.00	11.9	0	0	0
0	5	0.18	2.4	21.3	2.4	0.09	0	0.02	0.00	0.01	0.00	0.00	2.4	0	0	0
0	5	0.05	2.6	38.9	5.2	0.02	0	0.00	0.00	0.00	0.02	0.00	0	0	0	0.3
0	7	0.02	4.7	42.7	9.5	0.02	0	0.00	0.00	0.01	0.07	0.00	0	0	0	0
0	0	0.04	7.1	87.7	7.1	0.04	0	0.00	0.00	0.03	0.00	0.00	11.9	0	0	0
0	33	0.00	0	0	2.4	0.00	0	0.00	—	0.00	0.00	0.00	0	0	0	0
0	2	0.02	2.4	0	2.4	0.00	0	0.00	—	0.00	0.00	0.00	0	0	0	0
0	7	0.00	2.4	2.4	7.1	0.00	0	0.00	0.00	0.00	0.00	0.00	0	0	0	0
0	2	0.02	0	0	29.3	0.24	0	0.00	0.00	0.00	0.00	0.00	0	0	0	0
31	3	0.00	0.3	3.4	81.8	0.01	97.1	0.00	0.32	0.01	0.01	0.00	0.4	0	0	0.1
0	0	0.00	0	1.6	120.0	0.00	0	0.00	0.00	0.00	0.00	0.00	0	0	0	—
31	3	0.00	0.3	3.4	1.6	0.01	97.1	0.00	0.32	0.01	0.01	0.00	0.4	0	0	0.1
21	2	0.01	0.2	2.4	77.7	0.01	64.3	0.00	0.21	0.00	0.00	0.00	0.3	0	0	0.1
25	3	0.00	0.2	2.7	1.3	0.01	78.0	0.00	0.26	0.00	0.00	0.00	0.3	0	0	0.1
14	0	0.00	0	0	0	0.00	0	0.00	0.34	0.00	0.00	0.00	0	0	0	0
11	0	0.00	0	0	0	0.00	0	0.00	0.34	0.00	0.00	0.00	0	0	0	0
0	0	0.00	0	0	0	0.00	0	0.00	—	0.00	0.00	0.00	0	0	0	—
12	0	0.00	0	0	0	0.01	0	0.00	0.07	0.00	0.00	0.00	0	0	0	0
0	4	0.01	0.4	5.9	133.0	0.00	115.5	0.00	1.26	0.01	0.00	0.00	0.1	0	0	0
0	4	0.00	0.3	5.5	155.4	0.00	142.7	0.00	1.00	0.00	0.00	0.00	0.1	0	0	0
0	4	0.00	0.3	5.4	3.9	0.00	103.1	0.00	0.98	0.00	0.00	0.00	0.1	0	0	0
0	2	0.00	0.3	3.5	0.3	0.00	115.5	0.00	1.80	0.00	0.00	0.00	0.1	0	0	0
0	2	0.00	0.2	3.4	97.1	0.00	73.7	0.00	0.45	0.00	0.00	0.00	0.1	0	0	0
0	4	0.00	0.3	5.5	155.3	0.00	142.6	0.00	0.72	0.00	0.00	0.00	0.1	0	0	0
8	2	0.06	0.1	4.7	78.4	0.01	11.6	0.00	3.03	0.00	0.00	0.08	1.1	0	0	0.2
0	10	0.18	—	8.7	90.0	—	—	0.00	—	0.00	0.00	—	—	0.6	—	
0	0	0.00	0	0	0	0.00	0	0.00	2.37	0.00	0.00	0.00	0	0	0	0
0	0	0.00	0	0	0	0.00	0	0.00	1.94	0.00	0.00	0.00	0	0	0	0
0	0	0.07	0	0.1	0.3	0.00	0	0.00	1.93	0.00	0.00	0.00	0	0	0	0

DA+ Code	Food Description	Quantity	Measure	Wt (g)	H_2O (g)	Ener (kcal)	Prot (g)	Carb (g)	Fiber (g)	Fat (g)	Fat Breakdown (g)			
											Sat	Mono	Poly	*Trans*
	Fats and Oils—*continued*													
124	Peanut	1	tablespoon(s)	14	0	119	0	0	0	13.5	2.3	6.2	4.3	—
2693	Safflower	1	tablespoon(s)	14	0	120	0	0	0	13.6	0.8	10.2	2.0	—
923	Sesame	1	tablespoon(s)	14	0	120	0	0	0	13.6	1.9	5.4	5.7	—
128	Soybean, hydrogenated	1	tablespoon(s)	14	0	120	0	0	0	13.6	2.0	5.8	5.1	—
130	Soybean, with soybean and cottonseed oil	1	tablespoon(s)	14	0	120	0	0	0	13.6	2.4	4.0	6.5	—
2700	Sunflower	1	tablespoon(s)	14	0	120	0	0	0	13.6	1.8	6.3	5.0	—
357	**Pam original no stick cooking spray**	1	serving(s)	0	0.2	0	0	0	0	0	0	0	0	—
	Salad dressing													
132	Blue cheese	2	tablespoon(s)	30	9.7	151	1.4	2.2	0	15.7	3.0	3.7	8.3	—
133	Blue cheese, low calorie	2	tablespoon(s)	32	25.4	32	1.6	0.9	0	2.3	0.8	0.6	0.8	—
1764	Caesar	2	tablespoon(s)	30	10.3	158	0.4	0.9	0	17.3	2.6	4.1	9.9	—
29654	Creamy, reduced calorie, fat-free, cholesterol-free, sour cream and/or buttermilk and oil	2	tablespoon(s)	32	23.9	34	0.4	6.4	0	0.9	0.2	0.2	0.5	—
29617	Creamy, reduced calorie, sour cream and/or buttermilk and oil	2	tablespoon(s)	30	22.2	48	0.5	2.1	0	4.2	0.6	1.0	2.4	—
134	French	2	tablespoon(s)	32	11.7	146	0.2	5.0	0	14.3	1.8	2.7	6.7	—
135	French, low fat	2	tablespoon(s)	32	17.4	74	0.2	9.4	0.4	4.3	0.4	1.9	1.6	—
136	Italian	2	tablespoon(s)	29	16.6	86	0.1	3.1	0	8.3	1.3	1.9	3.8	—
137	Italian, diet	2	tablespoon(s)	30	25.4	23	0.1	1.4	0	1.9	0.1	0.7	0.5	—
139	Mayonnaise-type	2	tablespoon(s)	29	11.7	115	0.3	7.0	0	9.8	1.4	2.6	5.3	—
942	Oil and vinegar	2	tablespoon(s)	32	15.2	144	0	0.8	0	16.0	2.9	4.7	7.7	—
1765	Ranch	2	tablespoon(s)	30	11.6	146	0.1	1.6	0	15.8	2.3	5.2	7.6	—
3666	Ranch, reduced calorie	2	tablespoon(s)	30	20.5	62	0.1	2.2	0	6.1	1.1	1.8	2.9	—
940	Russian	2	tablespoon(s)	30	11.6	107	0.5	9.3	0.7	7.8	1.2	1.8	4.4	—
939	Russian, low calorie	2	tablespoon(s)	32	20.8	45	0.2	8.8	0.1	1.3	0.2	0.3	0.7	—
941	Sesame seed	2	tablespoon(s)	30	11.8	133	0.9	2.6	0.3	13.6	1.9	3.6	7.5	—
142	Thousand Island	2	tablespoon(s)	32	14.9	118	0.3	4.7	0.3	11.2	1.6	2.5	5.8	—
143	Thousand Island, low calorie	2	tablespoon(s)	30	18.2	61	0.3	6.7	0.4	3.9	0.2	1.9	0.8	—
	Sandwich spreads													
138	Mayonnaise with soybean oil	1	tablespoon(s)	14	2.1	99	0.1	0.4	0	11.0	1.6	2.7	5.8	0
140	Mayonnaise, low calorie	1	tablespoon(s)	16	10.0	37	0	2.6	0	3.1	0.5	0.7	1.7	—
141	Tartar sauce	2	tablespoon(s)	28	8.7	144	0.3	4.1	0.1	14.4	2.2	3.8	7.7	—
	Sweets													
4799	**Butterscotch or caramel topping**	2	tablespoon(s)	41	13.1	103	0.6	27.0	0.4	0	0	0	0	—
	Candy													
1786	Almond Joy candy bar	1	item(s)	45	4.3	220	2.0	27.0	2.0	12.0	8.0	3.3	0.7	0
1785	Bit-O-Honey candy	6	item(s)	40	—	190	1.0	39.0	0	3.5	2.5	—	—	—
33375	Butterscotch candy	2	piece(s)	12	0.6	47	0	10.8	0	0.4	0.2	0.1	0	—
1701	Chewing gum, stick	1	item(s)	3	0.1	7	0	2.0	0.1	0	0	0	0	—
33378	Chocolate fudge with nuts, prepared	2	piece(s)	38	2.9	175	1.7	25.8	1.0	7.2	2.5	1.5	2.9	0.1
1787	Jelly beans	15	item(s)	43	2.7	159	0	39.8	0.1	0	0	0	0	—
1784	Kit Kat wafer bar	1	item(s)	42	0.8	210	3.0	27.0	0.5	11.0	7.0	3.5	0.3	0
4674	Krackel candy bar	1	item(s)	41	0.6	210	2.0	28.0	0.5	10.0	6.0	3.9	0.4	0
4934	Licorice	4	piece(s)	44	7.3	154	1.1	35.1	0	1.0	0	0.1	0	—
1780	Life Savers candy	1	item(s)	2	—	8	0	2.0	0	0	0	0	0	0
1790	Lollipop	1	item(s)	28	—	108	0	28.0	0	0	0	0	0	0
4679	M & Ms peanut chocolate candy, small bag	1	item(s)	49	0.9	250	5.0	30.0	2.0	13.0	5.0	5.4	2.1	—
1781	M & Ms plain chocolate candy, small bag	1	item(s)	48	0.8	240	2.0	34.0	1.0	10.0	6.0	3.3	0.3	—
4673	Milk chocolate bar, Symphony	1	item(s)	91	0.9	483	7.7	52.8	1.5	27.8	16.7	7.2	0.6	—
1783	Milky Way bar	1	item(s)	58	3.7	270	2.0	41.0	1.0	10.0	5.0	3.5	0.3	—
1788	Peanut brittle	1½	ounce(s)	43	0.3	207	3.2	30.3	1.1	8.1	1.8	3.4	1.9	—
1789	Reese's peanut butter cups	2	piece(s)	51	0.8	280	6.0	19.0	2.0	15.5	6.0	7.2	2.7	0
4689	Reese's pieces candy, small bag	1	item(s)	43	1.1	220	5.0	26.0	1.0	11.0	7.0	0.9	0.4	0
33399	Semisweet chocolate candy, made with butter	½	ounce(s)	14	0.1	68	0.6	9.0	0.8	4.2	2.5	1.4	0.1	—
1782	Snickers bar	1	item(s)	59	3.2	280	4.0	35.0	1.0	14.0	5.0	6.1	2.9	—
4694	Special Dark chocolate bar	1	item(s)	41	0.4	220	2.0	25.0	3.0	12.0	8.0	4.6	0.4	0
4695	Starburst fruit chews, original fruits	1	package(s)	59	3.9	240	0	48.0	0	5.0	1.0	2.1	1.8	—

KEY: H-4 = Breads/Baked Goods H-10 = Cereal/Rice/Pasta H-14 = Fruit H-18 = Vegetables/Legumes H-28 = Nuts/Seeds H-30 = Vegetarian H-32 = Dairy H-40 = Eggs H-40 = Seafood H-42 = Meats H-46 = Poultry
= Processed Meats H-48 = Beverages H-52 = Fats/Oils H-54 = Sweets H-56 = Spices/Condiments/Sauces H-60 = Mixed Foods/Soups/Sandwiches H-64 = Fast Food H-84 = Convenience H-86 = Baby Foods

Chol (mg)	Calc (mg)	Iron (mg)	Magn (mg)	Pota (mg)	Sodi (mg)	Zinc (mg)	Vit A (µg)	Thia (mg)	Vit E (mg α)	Ribo (mg)	Niac (mg)	Vit B_6 (mg)	Fola (µg)	Vit C (mg)	Vit B_{12} (µg)	Sele (µg)
0	0	0.00	0	0	0	0.00	0	0.00	2.11	0.00	0.00	0.00	0	0	0	0
0	0	0.00	0	0	0	0.00	0	0.00	4.63	0.00	0.00	0.00	0	0	0	0
0	0	0.00	0	0	0	0.00	0	0.00	0.19	0.00	0.00	0.00	0	0	0	0
0	0	0.00	0	0	0	0.00	0	0.00	1.10	0.00	0.00	0.00	0	0	0	0
0	0	0.00	0	0	0	0.00	0	0.00	1.64	0.00	0.00	0.00	0	0	0	0
0	0	0.00	0	0	0	0.00	0	0.00	5.58	0.00	0.00	0.00	0	0	0	0
0	0	0.00	0	0.3	1.5	0.01	0.1	0.00	0.00	0.00	0.00	0.00	0	0	0	0
5	24	0.06	0	11.1	328.2	0.08	20.1	0.00	1.80	0.03	0.03	0.01	7.8	0.6	0.1	0.3
0	28	0.16	2.2	1.6	384.0	0.08	—	0.01	0.08	0.03	0.01	0.01	1.0	0.1	0.1	0.5
1	7	0.05	0.6	8.7	323.4	0.03	0.6	0.00	1.56	0.00	0.01	0.00	0.9	0	0	0.5
0	12	0.08	1.6	42.6	320.0	0.05	0.3	0.00	0.21	0.01	0.01	0.01	1.9	0	0	0.5
0	2	0.03	0.6	10.8	306.9	0.01	—	0.00	0.71	0.00	0.01	0.01	0	0.1	0	0.5
0	8	0.25	1.6	21.4	267.5	0.09	7.4	0.01	1.60	0.01	0.06	0.00	0	0	0	0
0	4	0.27	2.6	34.2	257.3	0.06	8.6	0.01	0.09	0.01	0.14	0.01	0.6	0	0	0.5
0	2	0.18	0.9	14.1	486.3	0.03	0.6	0.00	1.47	0.01	0.00	0.01	0	0	0	0.6
2	3	0.19	1.2	25.5	409.8	0.05	0.3	0.00	0.06	0.00	0.00	0.02	0	0	0	2.4
8	4	0.05	0.6	2.6	209.0	0.05	6.2	0.00	0.60	0.01	0.00	0.01	1.8	0	0.1	0.5
0	0	0.00	0	2.6	0.3	0.00	0	0.00	1.46	0.00	0.00	0.00	0	0	0	0.5
1	4	0.03	1.2	8.4	354.0	0.01	5.4	0.00	1.84	0.01	0.00	0.00	0.3	0.1	0	0.1
0	5	0.01	1.5	8.4	413.7	0.01	0.9	0.00	0.72	0.01	0.00	0.00	0.3	0.1	0	0.1
0	6	0.20	3.0	51.9	282.3	0.06	13.2	0.01	0.98	0.01	0.16	0.02	1.5	1.4	0	0.5
2	6	0.18	0	50.2	277.8	0.02	0.6	0.00	0.12	0.00	0.00	0.00	1.0	1.9	0	0.5
0	6	0.18	0	47.1	300.0	0.02	0.6	0.00	1.50	0.00	0.00	0.00	0	0	0	0.5
8	5	0.37	2.6	34.2	276.2	0.08	4.5	0.46	1.28	0.01	0.13	0.00	0	0	0	0.5
0	5	0.27	2.1	60.6	249.3	0.05	4.8	0.01	0.30	0.01	0.13	0.00	0	0	0	0
5	1	0.03	0.1	1.7	78.4	0.02	11.2	0.01	0.72	0.01	0.00	0.08	0.7	0	0	0.2
4	0	0.00	0	1.6	79.5	0.01	0	0.00	0.32	0.00	0.00	0.00	0	0	0	0.3
8	6	0.20	0.8	10.1	191.5	0.05	20.2	0.00	0.97	0.00	0.01	0.07	2.0	0.1	0.1	0.5
0	22	0.08	2.9	34.4	143.1	0.07	11.1	0.01	—	0.03	0.01	0.01	0.8	0.1	0	0
0	18	0.33	30.3	126.5	65.0	0.36	0	0.01	—	0.06	0.21	—	—	0	—	—
0	20	0.00	—	—	150.0	—	0	—	—	—	—	—	—	0	—	—
1	0	0.00	0	0.4	46.9	0.01	3.4	0.00	0.01	0.00	0.00	0.00	0	0	0	0.1
0	0	0.00	0	0.1	0	0.00	0	0.00	0.00	0.00	0.00	0.00	0	0	0	0
5	22	0.74	20.9	69.5	14.8	0.54	14.4	0.02	0.09	0.03	0.12	0.03	6.1	0.1	0	1.1
0	1	0.05	0.9	15.7	21.3	0.02	0	0.00	0.00	0.01	0.00	0.00	0	0	0	0.5
3	60	0.36	16.4	126.0	30.0	0.51	0	0.07	—	0.22	1.07	0.05	59.6	0	0.1	2.0
3	40	0.36	—	168.8	50.0	—	0	—	—	—	—	—	—	0	—	—
0	0	0.22	2.6	28.2	126.3	0.07	0	0.01	0.07	0.01	0.04	0.00	0	0	0	—
0	0	0.00	—	0	0	—	0	0.00	—	0.00	0.00	—	—	0	—	0
0	0	0.00	—	—	10.8	—	0	0.00	—	0.00	0.00	—	—	0	—	1.0
5	40	0.36	36.5	170.6	25.0	1.13	14.8	0.03	—	0.06	1.60	0.04	17.3	0.6	0.1	1.9
5	40	0.36	19.6	127.4	30.0	0.46	14.8	0.02	—	0.06	0.10	0.01	2.9	0.6	0.1	1.4
22	228	0.82	61.0	398.6	91.9	1.00	0	0.06	—	0.25	0.14	0.10	10.9	2.0	0.4	—
5	60	0.18	19.8	140.1	95.0	0.41	15.1	0.02	—	0.06	0.20	0.02	5.8	0.6	0.2	3.3
5	11	0.51	17.9	71.4	189.2	0.37	16.6	0.05	1.08	0.01	1.12	0.03	19.6	0	0	1.1
3	40	0.72	45.4	217.4	180.0	0.93	0	0.12	—	0.08	2.35	0.07	28.1	0	0.1	2.3
0	20	0.00	18.9	169.9	80.0	0.32	0	0.04	—	0.06	1.22	0.03	12.0	0	0.1	0.8
3	5	0.44	16.3	51.7	1.6	0.23	0.4	0.01	—	0.01	0.06	0.01	0.4	0	0	0.5
5	40	0.36	42.3	—	140.0	1.37	15.3	0.03	—	0.06	1.60	0.05	23.5	0.6	0.1	2.7
0	0	1.80	45.5	136.0	50.0	0.59	0	0.01	—	0.02	0.16	0.01	0.8	0	0	1.2
0	10	0.18	0.6	1.2	0	0.00	—	0.00	—	0.00	0.00	0.00	0	30	0	0.5

TABLE H-1 Table of Food Composition (*continued*) (Computer code is for Cengage Diet Analysis program) (For purposes of calculations, use "0" for t, <1, <0.1, 0.01)

DA+ Code	Food Description	Quantity	Measure	Wt (g)	H_2O (g)	Ener (kcal)	Prot (g)	Carb (g)	Fiber (g)	Fat (g)	Fat Breakdown (g)			
											Sat	Mono	Poly	*Trans*
	Sweets—*continued*													
4698	Taffy	3	piece(s)	45	2.2	179	0	41.2	0	1.5	0.9	0.4	0.1	0.1
4699	Three Musketeers bar	1	item(s)	60	3.5	260	2.0	46.0	1.0	8.0	4.5	2.6	0.3	—
4702	Twix caramel cookie bars	2	item(s)	58	2.4	280	3.0	37.0	1.0	14.0	5.0	7.7	0.5	—
4705	York peppermint pattie	1	item(s)	39	3.9	160	0.5	32.0	0.5	3.0	1.5	1.2	0.1	0
	Frosting, icing													
4760	Chocolate frosting, ready to eat	2	tablespoon(s)	31	5.2	122	0.3	19.4	0.3	5.4	1.7	2.8	0.6	—
4771	Creamy vanilla frosting, ready to eat	2	tablespoon(s)	28	4.2	117	0	19.0	0	4.5	0.8	1.4	2.2	0
17291	Dec-A-Cake variety pack candy decoration	1	teaspoon(s)	4	—	15	0	3.0	0	0.5	0	—	—	—
536	White icing	2	tablespoon(s)	40	3.6	162	0.1	31.8	0	4.2	0.8	2.0	1.2	—
	Gelatin													
13697	Gelatin snack, all flavors	1	item(s)	99	96.8	70	1.0	17.0	0	0	0	0	0	0
2616	Sugar free, low calorie mixed fruit gelatin mix, prepared	½	cup(s)	121	—	10	1.0	0	0	0	0	0	0	0
548	**Honey**	1	tablespoon(s)	21	3.6	64	0.1	17.3	0	0	0	0	0	0
	Jams, jellies													
550	Jam or preserves	1	tablespoon(s)	20	6.1	56	0.1	13.8	0.2	0	0	0	0	—
42199	Jams, preserves, dietetic, all flavors, w/sodium saccharin	1	tablespoon(s)	14	6.4	18	0	7.5	0.4	0	0	0	0	—
552	Jelly	1	tablespoon(s)	21	6.3	56	0	14.7	0.2	0	0	0	0	—
545	**Marshmallows**	4	item(s)	29	4.7	92	0.5	23.4	0	0.1	0	0	0	—
4800	**Marshmallow cream topping**	2	tablespoon(s)	40	7.9	129	0.3	31.6	0	0.1	0	0	0	—
555	**Molasses**	1	tablespoon(s)	20	4.4	58	0	14.9	0	0	0	0	0	—
4780	**Popsicle or ice pop**	1	item(s)	59	47.5	47	0	11.3	0	0.1	0	0	0	—
	Sugar													
559	Brown sugar, packed	1	teaspoon(s)	5	0.1	17	0	4.5	0	0	0	0	0	0
563	Powdered sugar, sifted	⅓	cup(s)	33	0.1	130	0	33.2	0	0	0	0	0	—
561	White granulated sugar	1	teaspoon(s)	4	0	16	0	4.2	0	0	0	0	0	0
	Sugar substitute													
1760	Equal sweetener, packet size	1	item(s)	1	—	0	0	0.9	0	0	0	0	0	0
13029	Splenda granular no calorie sweetener	1	teaspoon(s)	1	—	0	0	0.5	0	0	0	0	0	0
1759	Sweet N Low sugar substitute, packet	1	item(s)	1	0.1	4	0	0.5	0	0	0	0	0	0
	Syrup													
3148	Chocolate syrup	2	tablespoon(s)	38	11.6	105	0.8	24.4	1.0	0.4	0.2	0.1	0	—
29676	Maple syrup	¼	cup(s)	80	25.7	209	0	53.7	0	0.2	0	0.1	0.1	—
4795	Pancake syrup	¼	cup(s)	80	30.4	187	0	49.2	0	0	0	0	0	0
	Spices, Condiments, Sauces													
	Spices													
807	Allspice, ground	1	teaspoon(s)	2	0.2	5	0.1	1.4	0.4	0.2	0	0	0	—
1171	Anise seeds	1	teaspoon(s)	2	0.2	7	0.4	1.1	0.3	0.3	0	0.2	0.1	—
729	Bakers' yeast, active	1	teaspoon(s)	4	0.3	12	1.5	1.5	0.8	0.2	0	0.1	0	—
683	Baking powder, double acting with phosphate	1	teaspoon(s)	5	0.2	2	0	1.1	0	0	0	0	0	0
1611	Baking soda	1	teaspoon(s)	5	0	0	0	0	0	0	0	0	0	0
8552	Basil	1	teaspoon(s)	1	0.8	0	0	0	0	0	0	0	0	—
34959	Basil, fresh	1	piece(s)	1	0.5	0	0	0	0	0	0	0	0	—
808	Basil, ground	1	teaspoon(s)	1	0.1	4	0.2	0.9	0.6	0.1	0	0	0	—
809	Bay leaf	1	teaspoon(s)	1	0	2	0	0.5	0.2	0.1	0	0	0	—
11720	Betel leaves	1	ounce(s)	28	—	17	1.8	2.4	0	0	—	—	—	—
730	Brewers' yeast	1	teaspoon(s)	3	0.1	8	1.0	1.0	0.8	0	0	0	0	0
11710	Capers	1	teaspoon(s)	5	—	0	0	0	0	0	0	0	0	0
1172	Caraway seeds	1	teaspoon(s)	2	0.2	7	0.4	1.0	0.8	0.3	0	0.2	0.1	—
1173	Celery seeds	1	teaspoon(s)	2	0.1	8	0.4	0.8	0.2	0.5	0	0.3	0.1	—
1174	Chervil, dried	1	teaspoon(s)	1	0	1	0.1	0.3	0.1	0	0	0	0	—
810	Chili powder	1	teaspoon(s)	3	0.2	8	0.3	1.4	0.9	0.4	0.1	0.1	0.2	—
8553	Chives, chopped	1	teaspoon(s)	1	0.9	0	0	0	0	0	0	0	0	—
51420	Cilantro (coriander)	1	teaspoon(s)	0	0.3	0	0	0	0	0	0	0	0	—
811	Cinnamon, ground	1	teaspoon(s)	2	0.2	6	0.1	1.9	1.2	0	0	0	0	0
812	Cloves, ground	1	teaspoon(s)	2	0.1	7	0.1	1.3	0.7	0.4	0.1	0	0.1	—
1175	Coriander leaf, dried	1	teaspoon(s)	1	0	2	0.1	0.3	0.1	0	0	0	0	—
1176	Coriander seeds	1	teaspoon(s)	2	0.2	5	0.2	1.0	0.8	0.3	0	0.2	0	—
1706	Cornstarch	1	tablespoon(s)	8	0.7	30	0	7.3	0.1	0	0	0	0	—

KEY: H-4 = Breads/Baked Goods H-10 = Cereal/Rice/Pasta H-14 = Fruit H-18 = Vegetables/Legumes H-28 = Nuts/Seeds H-30 = Vegetarian H-32 = Dairy H-40 = Eggs H-40 = Seafood H-42 = Meats H-46 = Poultry
= Processed Meats H-48 = Beverages H-52 = Fats/Oils H-54 = Sweets H-56 = Spices/Condiments/Sauces H-60 = Mixed Foods/Soups/Sandwiches H-64 = Fast Food H-84 = Convenience H-86 = Baby Foods

Chol (mg)	Calc (mg)	Iron (mg)	Magn (mg)	Pota (mg)	Sodi (mg)	Zinc (mg)	Vit A (µg)	Thia (mg)	Vit E (mg α)	Ribo (mg)	Niac (mg)	Vit B_6 (mg)	Fola (µg)	Vit C (mg)	Vit B_{12} (µg)	Sele (µg)
4	4	0.00	0	1.4	23.4	0.09	12.2	0.01	0.04	0.01	0.00	0.00	0	0	0	0.3
5	20	0.36	17.5	80.3	110.0	0.33	14.5	0.01	—	0.03	0.20	0.01	0	0.6	0.1	1.5
5	40	0.36	18.5	116.8	115.0	0.45	15.0	0.09	—	0.13	0.69	0.01	13.9	0.6	0.1	1.2
0	0	0.33	23.4	66.1	10.0	0.28	0	0.01	—	0.03	0.31	0.01	1.5	0	0	—
0	2	0.44	6.4	60.0	56.1	0.09	0	0.00	0.48	0.00	0.03	0.00	0.3	0	0	0.2
0	1	0.04	0.3	9.5	51.5	0.01	0	0.00	0.43	0.08	0.06	0.00	2.2	0	0	0
0	0	0.00	—	—	15.0	—	0	—	—	—	—	—	—	0	—	—
0	4	0.01	0.4	5.6	76.4	0.01	44.4	0.00	0.32	0.01	0.00	0.00	0	0	0	0.3
0	0	0.00	—	0	40.0	—	0	—	—	—	—	—	—	0	—	—
0	0	0.00	0	0	50.0	0.00	0	0.00	0.00	0.00	0.00	0.00	0	0	0	—
0	1	0.08	0.4	10.9	0.8	0.04	0	0.00	0.00	0.01	0.02	0.01	0.4	0.1	0	0.2
0	4	0.10	0.8	15.4	6.4	0.01	0	0.00	0.02	0.02	0.01	0.00	2.2	1.8	0	0.4
0	1	0.56	0.7	9.7	0	0.01	0	0.00	0.01	0.00	0.00	0.00	1.3	0	0	0.2
0	1	0.04	1.3	11.3	6.3	0.01	0	0.00	0.00	0.01	0.01	0.00	0.4	0.2	0	0.1
0	1	0.06	0.6	1.4	23.0	0.01	0	0.00	0.00	0.00	0.02	0.00	0.3	0	0	0.5
0	1	0.08	0.8	2.0	32.0	0.01	0	0.00	0.00	0.00	0.03	0.00	0.4	0	0	0.7
0	41	0.94	48.4	292.8	7.4	0.05	0	0.01	0.00	0.00	0.18	0.13	0	0	0	3.6
0	0	0.31	0.6	8.9	4.1	0.08	0	0.00	0.00	0.00	0.00	0.00	0	0.4	0	0.1
0	4	0.03	0.4	6.1	1.3	0.00	0	0.00	0.00	0.00	0.01	0.00	0	0	0	0.1
0	0	0.01	0	0.7	0.3	0.00	0	0.00	0.00	0.00	0.00	0.00	0	0	0	0.2
0	0	0.00	0	0.1	0	0.00	0	0.00	0.00	0.00	0.00	0.00	0	0	0	0
0	0	0.00	0	0	0	0.00	0	0.00	0.00	0.00	0.00	0.00	0	0	0	0
0	0	0.00	—	—	0	—	—	0.00	—	0.00	0.00	—	—	0	0	—
0	0	0.00	—	—	0	—	0	—	0.00	—	—	—	—	0	—	—
0	5	0.79	24.4	84.0	27.0	0.27	0	0.00	0.01	0.01	0.12	0.00	0.8	0.1	0	0.5
0	54	0.96	11.2	163.2	7.2	3.32	0	0.01	0.00	0.01	0.02	0.00	0	0	0	0.5
0	2	0.02	1.6	12.0	65.6	0.06	0	0.01	0.00	0.01	0.00	0.00	0	0	0	0
0	13	0.13	2.6	19.8	1.5	0.01	0.5	0.00	—	0.00	0.05	0.00	0.7	0.7	0	0.1
0	14	0.77	3.6	30.3	0.3	0.11	0.3	0.01	—	0.01	0.06	0.01	0.2	0.4	0	0.1
0	3	0.66	3.9	80.0	2.0	0.25	0	0.09	0.00	0.21	1.59	0.06	93.6	0	0	1.0
0	339	0.51	1.8	0.2	363.1	0.00	0	0.00	0.00	0.00	0.00	0.00	0	0	0	0
0	0	0.00	0	0	1258.6	0.00	0	0.00	0.00	0.00	0.00	0.00	0	0	0	0
0	2	0.02	0.6	2.6	0	0.01	2.3	0.00	0.01	0.00	0.01	0.00	0.6	0.2	0	0
0	1	0.01	0.4	2.3	0	0.00	1.3	0.00	—	0.00	0.00	0.00	0.3	0.1	0	0
0	30	0.58	5.9	48.1	0.5	0.08	6.6	0.00	0.10	0.00	0.09	0.03	3.8	0.9	0	0
0	5	0.25	0.7	3.2	0.1	0.02	1.9	0.00	—	0.00	0.01	0.01	1.1	0.3	0	0
0	110	2.29	—	155.9	2.0	—	—	0.04	—	0.07	0.19	—	—	0.9	0	—
0	6	0.46	6.1	50.7	3.3	0.21	0	0.41	—	0.11	1.00	0.06	104.3	0	0	0
0	—	—	—	—	105.0	—	—	—	—	—	—	—	—	—	0	—
0	14	0.34	5.4	28.4	0.4	0.11	0.4	0.01	0.05	0.01	0.07	0.01	0.2	0.4	0	0.3
0	35	0.89	8.8	28.0	3.2	0.13	0.1	0.01	0.02	0.01	0.06	0.01	0.2	0.3	0	0.2
0	8	0.19	0.8	28.4	0.5	0.05	1.8	0.00	—	0.00	0.03	0.01	1.6	0.3	0	0.2
0	7	0.37	4.4	49.8	26.3	0.07	38.6	0.01	0.75	0.02	0.20	0.09	2.6	1.7	0	0.2
0	1	0.01	0.4	3.0	0	0.01	2.2	0.00	0.00	0.00	0.01	0.00	1.1	0.6	0	0
0	0	0.01	0.1	1.7	0.2	0.00	1.1	0.00	0.01	0.00	0.00	0.00	0.2	0.1	0	0
0	23	0.19	1.4	9.9	0.2	0.04	0.3	0.00	0.05	0.00	0.03	0.00	0.1	0.1	0	0.1
0	14	0.18	5.5	23.1	5.1	0.02	0.6	0.00	0.17	0.01	0.03	0.01	2.0	1.7	0	0.1
0	7	0.25	4.2	26.8	1.3	0.02	1.8	0.01	0.01	0.01	0.06	0.00	1.6	3.4	0	0.2
0	13	0.29	5.9	22.8	0.6	0.08	0	0.00	—	0.01	0.03	—	0	0.4	0	0.5
0	0	0.03	0.2	0.2	0.7	0.01	0	0.00	0.00	0.00	0.00	0.00	0	0	0	0.2

DA+ Code	Food Description	Quantity	Measure	Wt (g)	H_2O (g)	Ener (kcal)	Prot (g)	Carb (g)	Fiber (g)	Fat (g)	Fat Breakdown (g)			
											Sat	Mono	Poly	*Trans*
Spices, Condiments, Sauces—*continued*														
1177	Cumin seeds	1	teaspoon(s)	2	0.2	8	0.4	0.9	0.2	0.5	0	0.3	0.1	—
11729	Cumin, ground	1	teaspoon(s)	5	—	11	0.4	0.8	0.8	0.4	—	—	—	—
1178	Curry powder	1	teaspoon(s)	2	0.2	7	0.3	1.2	0.7	0.3	0	0.1	0.1	—
1179	Dill seeds	1	teaspoon(s)	2	0.2	6	0.3	1.2	0.4	0.3	0	0.2	0	—
1180	Dill weed, dried	1	teaspoon(s)	1	0.1	3	0.2	0.6	0.1	0	0	0	0	—
34949	Dill weed, fresh	5	piece(s)	1	0.9	0	0	0.1	0	0	0	0	0	—
4949	Fennel leaves, fresh	1	teaspoon(s)	1	0.9	0	0	0.1	0	0	—	—	—	—
1181	Fennel seeds	1	teaspoon(s)	2	0.2	7	0.3	1.0	0.8	0.3	0	0.2	0	—
1182	Fenugreek seeds	1	teaspoon(s)	4	0.3	12	0.9	2.2	0.9	0.2	0.1	—	—	—
11733	Garam masala, powder	1	ounce(s)	28	—	107	4.4	12.8	0	4.3	—	—	—	—
1067	Garlic clove	1	item(s)	3	1.8	4	0.2	1.0	0.1	0	0	0	0	—
813	Garlic powder	1	teaspoon(s)	3	0.2	9	0.5	2.0	0.3	0	0	0	0	—
1068	Ginger root	2	teaspoon(s)	4	3.1	3	0.1	0.7	0.1	0	0	0	0	—
1183	Ginger, ground	1	teaspoon(s)	2	0.2	6	0.2	1.3	0.2	0.1	0	0	0	—
35497	Leeks, bulb and lower-leaf, freeze-dried	¼	cup(s)	1	0	3	0.1	0.6	0.1	0	0	0	0	—
1184	Mace, ground	1	teaspoon(s)	2	0.1	8	0.1	0.9	0.3	0.6	0.2	0.2	0.1	—
1185	Marjoram, dried	1	teaspoon(s)	1	0	2	0.1	0.4	0.2	0	0	0	0	—
1186	Mustard seeds, yellow	1	teaspoon(s)	3	0.2	15	0.8	1.2	0.5	0.9	0	0.7	0.2	—
814	Nutmeg, ground	1	teaspoon(s)	2	0.1	12	0.1	1.1	0.5	0.8	0.6	0.1	0	—
2747	Onion flakes, dehydrated	1	teaspoon(s)	2	0.1	6	0.1	1.4	0.2	0	0	0	0	—
1187	Onion powder	1	teaspoon(s)	2	0.1	7	0.2	1.7	0.1	0	0	0	0	—
815	Oregano, ground	1	teaspoon(s)	2	0.1	5	0.2	1.0	0.6	0.2	0	0	0.1	—
816	Paprika	1	teaspoon(s)	2	0.2	6	0.3	1.2	0.8	0.3	0	0	0.2	—
817	Parsley, dried	1	teaspoon(s)	0	0	1	0.1	0.2	0.1	0	0	0	0	—
818	Pepper, black	1	teaspoon(s)	2	0.2	5	0.2	1.4	0.6	0.1	0	0	0	—
819	Pepper, cayenne	1	teaspoon(s)	2	0.1	6	0.2	1.0	0.5	0.3	0.1	0	0.2	—
1188	Pepper, white	1	teaspoon(s)	2	0.3	7	0.3	1.6	0.6	0.1	0	0	0	—
1189	Poppy seeds	1	teaspoon(s)	3	0.2	15	0.5	0.7	0.3	1.3	0.1	0.2	0.9	—
1190	Poultry seasoning	1	teaspoon(s)	2	0.1	5	0.1	1.0	0.2	0.1	0	0	0	—
1191	Pumpkin pie spice, powder	1	teaspoon(s)	2	0.1	6	0.1	1.2	0.3	0.2	0.1	0	0	—
1192	Rosemary, dried	1	teaspoon(s)	1	0.1	4	0.1	0.8	0.5	0.2	0.1	0	0	—
11723	Rosemary, fresh	1	teaspoon(s)	1	0.5	1	0	0.1	0.1	0	0	0	0	—
2722	Saffron powder	1	teaspoon(s)	1	0.1	2	0.1	0.5	0	0	0	0	0	—
11724	Sage	1	teaspoon(s)	1	—	1	0	0.1	0	0	—	—	—	—
1193	Sage, ground	1	teaspoon(s)	1	0.1	2	0.1	0.4	0.3	0.1	0	0	0	—
30189	Salt substitute	¼	teaspoon(s)	1	—	0	0	0	0	0	0	0	0	0
30190	Salt substitute, seasoned	¼	teaspoon(s)	1	—	1	0	0.1	0	0	0	—	—	—
822	Salt, table	¼	teaspoon(s)	2	0	0	0	0	0	0	0	0	0	0
1194	Savory, ground	1	teaspoon(s)	1	0.1	4	0.1	1.0	0.6	0.1	0	—	—	—
820	Sesame seed kernels, toasted	1	teaspoon(s)	3	0.1	15	0.5	0.7	0.5	1.3	0.2	0.5	0.6	—
11725	Sorrel	1	teaspoon(s)	3	—	1	0.1	0.1	0	0	0	—	—	—
11721	Spearmint	1	teaspoon(s)	2	1.6	1	0.1	0.2	0.1	0	0	0	0	—
35498	Sweet green peppers, freeze-dried	¼	cup(s)	2	0	5	0.3	1.1	0.3	0	0	0	0	—
11726	Tamarind leaves	1	ounce(s)	28	—	33	1.6	5.2	0	0.6	—	—	—	—
11727	Tarragon	1	ounce(s)	28	—	14	1.0	1.8	0	0.3	—	—	—	—
1195	Tarragon, ground	1	teaspoon(s)	2	0.1	5	0.4	0.8	0.1	0.1	0	0	0.1	—
11728	Thyme, fresh	1	teaspoon(s)	1	0.5	1	0	0.2	0.1	0	0	0	0	—
821	Thyme, ground	1	teaspoon(s)	1	0.1	4	0.1	0.9	0.5	0.1	0	0	0	—
1196	Turmeric, ground	1	teaspoon(s)	2	0.3	8	0.2	1.4	0.5	0.2	0.1	0	0	—
11995	Wasabi	1	tablespoon(s)	14	10.7	10	0.7	2.3	0.2	0	—	—	—	—
	Condiments													
674	Catsup or ketchup	1	tablespoon(s)	15	10.4	15	0.3	3.8	0	0	0	0	0	—
703	Dill pickle	1	ounce(s)	28	26.7	3	0.2	0.7	0.3	0	0	0	0	—
138	Mayonnaise with soybean oil	1	tablespoon(s)	14	2.1	99	0.1	0.4	0	11.0	1.6	2.7	5.8	0
140	Mayonnaise, low calorie	1	tablespoon(s)	16	10.0	37	0	2.6	0	3.1	0.5	0.7	1.7	—
1682	Mustard, brown	1	teaspoon(s)	5	4.1	5	0.3	0.3	0	0.3	—	—	—	—
700	Mustard, yellow	1	teaspoon(s)	5	4.1	3	0.2	0.3	0.2	0.2	0	0.1	0	0
706	Sweet pickle relish	1	tablespoon(s)	15	9.3	20	0.1	5.3	0.2	0.1	0	0	0	—
141	Tartar sauce	2	tablespoon(s)	28	8.7	144	0.3	4.1	0.1	14.4	2.2	3.8	7.7	—
	Sauces													
685	Barbecue sauce	2	tablespoon(s)	31	18.9	47	0	11.3	0.2	0.1	0	0	0.1	0
834	Cheese sauce	¼	cup(s)	63	44.4	110	4.2	4.3	0.3	8.4	3.8	2.4	1.6	—
32123	Chili enchilada sauce, green	2	tablespoon(s)	57	53.0	15	0.6	3.1	0.7	0.3	0	0	0.1	0
32122	Chili enchilada sauce, red	2	tablespoon(s)	32	24.5	27	1.1	5.0	2.1	0.8	0.1	0	0.4	0

E KEY: H-4 = Breads/Baked Goods H-10 = Cereal/Rice/Pasta H-14 = Fruit H-18 = Vegetables/Legumes H-28 = Nuts/Seeds H-30 = Vegetarian H-32 = Dairy H-40 = Eggs H-40 = Seafood H-42 = Meats H-46 = Poultry
5 = Processed Meats H-48 = Beverages H-52 = Fats/Oils H-54 = Sweets H-56 = Spices/Condiments/Sauces H-60 = Mixed Foods/Soups/Sandwiches H-64 = Fast Food H-84 = Convenience H-86 = Baby Foods

Chol (mg)	Calc (mg)	Iron (mg)	Magn (mg)	Pota (mg)	Sodi (mg)	Zinc (mg)	Vit A (μg)	Thia (mg)	Vit E (mg α)	Ribo (mg)	Niac (mg)	Vit B_6 (mg)	Fola (μg)	Vit C (mg)	Vit B_{12} (μg)	Sele (μg)
0	20	1.39	7.7	37.5	3.5	0.10	1.3	0.01	0.07	0.01	0.09	0.01	0.2	0.2	0	0.1
0	20	—	—	43.6	4.8	—	—	—	—	—	—	—	—	—	—	—
0	10	0.59	5.1	30.9	1.0	0.08	1.0	0.01	0.44	0.01	0.06	0.02	3.1	0.2	0	0.3
0	32	0.34	5.4	24.9	0.4	0.10	0.1	0.01	—	0.01	0.05	0.01	0.2	0.4	0	0.3
0	18	0.48	4.5	33.1	2.1	0.03	2.9	0.00	—	0.00	0.02	0.01	1.5	0.5	0	—
0	2	0.06	0.6	7.4	0.6	0.01	3.9	0.00	0.01	0.00	0.01	0.00	1.5	0.9	0	—
0	1	0.02	—	4.0	0.1	—	—	0.00	—	0.00	0.01	0.00	—	0.3	0	—
0	24	0.37	7.7	33.9	1.8	0.07	0.1	0.01	—	0.01	0.12	0.01	—	0.4	0	—
0	7	1.24	7.1	28.5	2.5	0.09	0.1	0.01	—	0.01	0.06	0.02	2.1	0.1	0	0.2
0	215	9.24	93.6	411.1	27.5	1.07	—	0.09	—	0.09	0.70	—	0	0	0	—
0	5	0.05	0.8	12.0	0.5	0.03	0	0.01	0.00	0.00	0.02	0.03	0.1	0.9	0	0.4
0	2	0.07	1.6	30.8	0.7	0.07	0	0.01	0.01	0.00	0.01	0.08	0.1	0.5	0	1.1
0	1	0.02	1.7	16.6	0.5	0.01	0	0.00	0.01	0.00	0.02	0.01	0.4	0.2	0	0
0	2	0.20	3.3	24.2	0.6	0.08	0.1	0.00	0.32	0.00	0.09	0.01	0.7	0.1	0	0.7
0	3	0.06	1.3	19.2	0.3	0.01	0.1	0.01	—	0.00	0.02	0.01	2.9	0.9	0	0
0	4	0.23	2.8	7.9	1.4	0.03	0.7	0.01	—	0.01	0.02	0.00	1.3	0.4	0	0
0	12	0.49	2.1	9.1	0.5	0.02	2.4	0.00	0.01	0.00	0.02	0.01	1.6	0.3	0	0
0	17	0.32	9.8	22.5	0.2	0.18	0.1	0.01	0.09	0.01	0.26	0.01	2.5	0.1	0	4.4
0	4	0.06	4.0	7.7	0.4	0.04	0.1	0.01	0.00	0.00	0.02	0.00	1.7	0.1	0	0
0	4	0.02	1.5	27.1	0.4	0.03	0	0.01	0.00	0.00	0.01	0.02	2.8	1.3	0	0.1
0	8	0.05	2.6	19.8	1.1	0.04	0	0.01	0.01	0.00	0.01	0.02	3.5	0.3	0	0
0	24	0.66	4.1	25.0	0.2	0.06	5.2	0.01	0.28	0.01	0.09	0.01	4.1	0.8	0	0.1
0	4	0.49	3.9	49.2	0.7	0.08	55.4	0.01	0.62	0.03	0.32	0.08	2.2	1.5	0	0.1
0	4	0.29	0.7	11.4	1.4	0.01	1.5	0.00	0.02	0.00	0.02	0.00	0.5	0.4	0	0.1
0	9	0.60	4.1	26.4	0.9	0.03	0.3	0.00	0.01	0.01	0.02	0.01	0.2	0.4	0	0.1
0	3	0.14	2.7	36.3	0.5	0.04	37.5	0.01	0.53	0.01	0.15	0.04	1.9	1.4	0	0.2
0	6	0.34	2.2	1.8	0.1	0.02	0	0.00	—	0.00	0.01	0.00	0.2	0.5	0	0.1
0	41	0.26	9.3	19.6	0.6	0.28	0	0.02	0.03	0.01	0.02	0.01	1.6	0.1	0	0
0	15	0.53	3.4	10.3	0.4	0.04	2.0	0.00	0.02	0.00	0.04	0.02	2.1	0.2	0	0.1
0	12	0.33	2.3	11.3	0.9	0.04	0.2	0.00	0.01	0.00	0.03	0.01	0.9	0.4	0	0.2
0	15	0.35	2.6	11.5	0.6	0.03	1.9	0.01	—	0.01	0.01	0.02	3.7	0.7	0	0.1
0	2	0.04	0.6	4.7	0.2	0.01	1.0	0.00	—	0.00	0.01	0.00	0.8	0.2	0	—
0	1	0.07	1.8	12.1	1.0	0.01	0.2	0.00	—	0.00	0.01	0.01	0.7	0.6	0	0
0	4	—	1.1	2.7	0	0.01	—	0.00	—	—	—	—	—	—	0	—
0	12	0.19	3.0	7.5	0.1	0.03	2.1	0.01	0.05	0.00	0.04	0.01	1.9	0.2	0	0
0	7	0.00	0	603.6	0.1	—	0	—	—	—	—	—	—	0	—	—
0	0	0.00	—	476.3	0.1	—	0	—	—	—	—	—	—	0	—	—
0	0	0.01	0	0.1	581.4	0.00	0	0.00	0.00	0.00	0.00	0.00	0	0	0	0
0	30	0.53	5.3	14.7	0.3	0.06	3.6	0.01	—	—	0.05	0.02	—	0.7	0	0.1
0	3	0.21	9.2	10.8	1.0	0.27	0.1	0.03	0.01	0.01	0.15	0.00	2.6	0	0	0
0	—	—	—	—	0.1	—	—	—	—	—	—	—	—	—	—	—
0	4	0.22	1.2	8.7	0.6	0.02	3.9	0.00	—	0.00	0.01	0.00	2.0	0.3	0	—
0	2	0.16	3.0	50.7	3.1	0.03	4.5	0.01	0.06	0.01	0.11	0.03	3.7	30.4	0	0.1
0	85	1.48	20.2	—	—	—	—	0.06	—	0.02	1.16	—	—	0.9	0	—
0	48	—	14.5	128.1	2.6	0.17	—	0.04	—	—	—	—	—	0.6	0	—
0	18	0.51	5.6	48.3	1.0	0.06	3.4	0.00	—	0.02	0.14	0.03	4.4	0.8	0	0.1
0	3	0.14	1.3	4.9	0.1	0.01	1.9	0.00	—	0.00	0.01	0.00	0.4	1.3	0	—
0	26	1.73	3.1	11.4	0.8	0.08	2.7	0.01	0.10	0.01	0.06	0.01	3.8	0.7	0	0.1
0	4	0.91	4.2	55.6	0.8	0.09	0	0.00	0.06	0.01	0.11	0.04	0.9	0.6	0	0.1
0	13	0.11	—	—	—	—	—	0.02	—	0.01	0.07	—	—	11.2	0	—
0	3	0.07	2.9	57.3	167.1	0.03	7.1	0.00	0.21	0.02	0.21	0.02	1.5	2.3	0	0
0	12	0.10	2.0	26.1	248.1	0.03	2.6	0.01	0.02	0.01	0.03	0.01	0.3	0.2	0	0
5	1	0.03	0.1	1.7	78.4	0.02	11.2	0.01	0.72	0.01	0.00	0.08	0.7	0	0	0.2
4	0	0.00	0	1.6	79.5	0.01	0	0.00	0.32	0.00	0.00	0.00	0	0	0	0.3
0	6	0.09	1.0	6.8	68.1	0.01	0	0.00	0.09	0.00	0.01	0.00	0.2	0.1	0	—
0	3	0.07	2.5	6.9	56.8	0.03	0.2	0.01	0.01	0.00	0.02	0.00	0.4	0.1	0	1.6
0	0	0.13	0.8	3.8	121.7	0.02	9.2	0.00	0.08	0.01	0.03	0.00	0.2	0.2	0	0
8	6	0.20	0.8	10.1	191.5	0.05	20.2	0.00	0.97	0.00	0.01	0.07	2.0	0.1	0.1	0.5
0	4	0.06	3.8	65.0	349.7	0.04	3.8	0.00	0.20	0.01	0.15	0.01	0.6	0.2	0	0.4
18	116	0.13	5.7	18.9	521.6	0.61	50.4	0.00	—	0.07	0.01	0.01	2.5	0.3	0.1	2.0
0	5	0.36	9.5	125.7	61.9	0.11	—	0.02	0.00	0.02	0.63	0.06	5.7	43.9	0	0
0	7	1.05	11.1	231.3	113.8	0.14	—	0.01	0.00	0.21	0.61	0.34	6.6	0.3	0	0.3

TABLE H-1 Table of Food Composition (*continued*) (Computer code is for Cengage Diet Analysis program) (For purposes of calculations, use "0" for t, <1, <0.1, 0.01)

DA+ Code	Food Description	Quantity	Measure	Wt (g)	H_2O (g)	Ener (kcal)	Prot (g)	Carb (g)	Fiber (g)	Fat (g)	Fat Breakdown (g)			
											Sat	Mono	Poly	*Trans*
Spices, Condiments, Sauces—*continued*														
29688	Hoisin sauce	1	tablespoon(s)	16	7.1	35	0.5	7.1	0.4	0.5	0.1	0.2	0.3	—
1641	Horseradish sauce, prepared	1	teaspoon(s)	5	3.3	10	0.1	0.2	0	1.0	0.6	0.3	0	—
16670	Mole poblano sauce	½	cup(s)	133	102.7	156	5.3	11.4	2.7	11.3	2.6	5.1	3.0	—
29689	Oyster sauce	1	tablespoon(s)	16	12.8	8	0.2	1.7	0	0	0	0	0	—
1655	Pepper sauce or Tabasco	1	teaspoon(s)	5	4.8	1	0.1	0	0	0	0	0	0	—
347	Salsa	2	tablespoon(s)	32	28.8	9	0.5	2.0	0.5	0.1	0	0	0	—
52206	Soy sauce, tamari	1	tablespoon(s)	18	12.0	11	1.9	1.0	0.1	0	0	0	0	—
839	Sweet and sour sauce	2	tablespoon(s)	39	29.8	37	0.1	9.1	0.1	0	0	0	0	—
1613	Teriyaki sauce	1	tablespoon(s)	18	12.2	16	1.1	2.8	0	0	0	0	0	0
25294	Tomato sauce	½	cup(s)	150	132.8	63	2.2	11.9	2.6	1.8	0.2	0.4	0.9	0
728	White sauce, medium	¼	cup(s)	63	46.8	92	2.4	5.7	0.1	6.7	1.8	2.8	1.8	—
1654	Worcestershire sauce	1	teaspoon(s)	6	4.5	4	0	1.1	0	0	0	0	0	0
	Vinegar													
30853	Balsamic	1	tablespoon(s)	15	—	10	0	2.0	0	0	0	0	0	0
727	Cider	1	tablespoon(s)	15	14.0	3	0	0.1	0	0	0	0	0	0
1673	Distilled	1	tablespoon(s)	15	14.3	2	0	0.8	0	0	0	0	0	0
12948	Tarragon	1	tablespoon(s)	15	13.8	2	0	0.1	0	0	0	0	0	0
Mixed Foods, Soups, Sandwiches														
	Mixed dishes													
16652	Almond chicken	1	cup(s)	242	186.8	281	21.8	15.8	3.4	14.7	1.8	6.3	5.6	—
25224	Barbecued chicken	1	serving(s)	177	99.3	327	27.1	15.7	0.5	17.1	4.8	6.8	3.8	0
25227	Bean burrito	1	item(s)	149	81.8	326	16.1	33.0	5.6	14.8	8.3	4.7	0.9	—
9516	Beef and vegetable fajita	1	item(s)	223	143.9	397	22.4	35.3	3.1	18.0	5.9	8.0	2.5	—
16796	Beef or pork egg roll	2	item(s)	128	85.2	225	9.9	18.4	1.4	12.4	2.9	6.0	2.6	—
177	Beef stew with vegetables, prepared	1	cup(s)	245	201.0	220	16.0	15.0	3.2	11.0	4.4	4.5	0.5	—
30233	Beef stroganoff with noodles	1	cup(s)	256	190.1	343	19.7	22.8	1.5	19.1	7.4	5.7	4.4	—
16651	Cashew chicken	1	cup(s)	242	186.8	281	21.8	15.8	3.4	14.7	1.8	6.3	5.6	—
30274	Cheese pizza with vegetables, thin crust	2	slice(s)	140	76.6	298	12.7	35.4	2.5	12.0	4.9	4.7	1.6	—
30330	Cheese quesadilla	1	item(s)	54	18.3	190	7.7	15.3	1.0	10.8	5.2	3.6	1.3	—
215	Chicken and noodles, prepared	1	cup(s)	240	170.0	365	22.0	26.0	1.3	18.0	5.1	7.1	3.9	—
30239	Chicken and vegetables with broccoli, onion, bamboo shoots in soy based sauce	1	cup(s)	162	125.5	180	15.8	9.3	1.8	8.6	1.7	3.0	3.1	—
25093	Chicken cacciatore	1	cup(s)	244	175.7	284	29.9	5.7	1.3	15.3	4.3	6.2	3.3	0
28020	Chicken fried turkey steak	3	ounce(s)	492	276.2	706	77.1	68.7	3.6	12.0	3.4	2.9	3.9	—
218	Chicken pot pie	1	cup(s)	252	154.6	542	22.6	41.4	3.5	31.3	9.8	12.5	7.1	—
30240	Chicken teriyaki	1	cup(s)	244	158.3	364	51.0	15.2	0.7	7.0	1.8	2.0	1.7	—
25119	Chicken waldorf salad	½	cup(s)	100	67.2	179	14.0	6.8	1.0	10.8	1.8	3.1	5.2	—
25099	Chili con carne	¾	cup(s)	215	174.4	198	13.7	21.4	7.5	6.9	2.5	2.8	0.5	0
1062	Coleslaw	¾	cup(s)	90	73.4	70	1.2	11.2	1.4	2.3	0.3	0.6	1.2	—
1574	Crab cakes, from blue crab	1	item(s)	60	42.6	93	12.1	0.3	0	4.5	0.9	1.7	1.4	—
32144	Enchiladas with green chili sauce (enchiladas verdes)	1	item(s)	144	103.8	207	9.3	17.6	2.6	11.7	6.4	3.6	1.0	0
2793	Falafel patty	3	item(s)	51	17.7	170	6.8	16.2	—	9.1	1.2	5.2	2.1	—
28546	Fettuccine alfredo	1	cup(s)	244	88.7	279	13.1	46.1	1.4	4.2	2.2	1.0	0.4	0
32146	Flautas	3	item(s)	162	78.0	438	24.9	36.3	4.1	21.6	8.2	8.8	2.3	—
29629	Fried rice with meat or poultry	1	cup(s)	198	128.5	333	12.3	41.8	1.4	12.3	2.2	3.5	5.7	—
16649	General Tso chicken	1	cup(s)	146	91.0	296	18.7	16.4	0.9	17.0	4.0	6.3	5.3	—
1826	Green salad	¾	cup(s)	104	98.9	17	1.3	3.3	2.2	0.1	0	0	0	—
1814	Hummus	½	cup(s)	123	79.8	218	6.0	24.7	4.9	10.6	1.4	6.0	2.6	—
16650	Kung pao chicken	1	cup(s)	162	87.2	434	28.8	11.7	2.3	30.6	5.2	13.9	9.7	—
16622	Lamb curry	1	cup(s)	236	187.9	257	28.2	3.7	0.9	13.8	3.9	4.9	3.3	—
25253	Lasagna with ground beef	1	cup(s)	237	158.4	284	16.9	22.3	2.4	14.5	7.5	4.9	0.8	—
442	Macaroni and cheese, prepared	1	cup(s)	200	122.3	390	14.9	40.6	1.6	18.6	7.9	6.4	2.9	—
29637	Meat filled ravioli with tomato or meat sauce, canned	1	cup(s)	251	198.7	208	7.8	36.5	1.3	3.7	1.5	1.4	0.3	—
25105	Meat loaf	1	slice(s)	115	84.5	245	17.0	6.6	0.4	16.0	6.1	6.9	0.9	0
16646	Moo shi pork	1	cup(s)	151	76.8	512	18.9	5.3	0.6	46.4	6.9	15.8	21.2	—
16788	Nachos with beef, beans, cheese, tomatoes and onions	1	serving(s)	551	253.5	1576	59.1	137.5	20.4	90.8	32.6	41.9	9.4	—
6116	Pepperoni pizza	2	slice(s)	142	66.1	362	20.2	39.7	2.9	13.9	4.5	6.3	2.3	—
29601	Pizza with meat and vegetables, thin crust	2	slice(s)	158	81.4	386	16.5	36.8	2.7	19.1	7.7	8.1	2.2	—
655	Potato salad	½	cup(s)	125	95.0	179	3.4	14.0	1.6	10.3	1.8	3.1	4.7	—

E KEY: H-4 = Breads/Baked Goods H-10 = Cereal/Rice/Pasta H-14 = Fruit H-18 = Vegetables/Legumes H-28 = Nuts/Seeds H-30 = Vegetarian H-32 = Dairy H-40 = Eggs H-40 = Seafood H-42 = Meats H-46 = Poultry
5 = Processed Meats H-48 = Beverages H-52 = Fats/Oils H-54 = Sweets H-56 = Spices/Condiments/Sauces H-60 = Mixed Foods/Soups/Sandwiches H-64 = Fast Food H-84 = Convenience H-86 = Baby Foods

Chol (mg)	Calc (mg)	Iron (mg)	Magn (mg)	Pota (mg)	Sodi (mg)	Zinc (mg)	Vit A (µg)	Thia (mg)	Vit E (mg α)	Ribo (mg)	Niac (mg)	Vit B_6 (mg)	Fola (µg)	Vit C (mg)	Vit B_{12} (µg)	Sele (µg)
0	5	0.16	3.8	19.0	258.4	0.05	0	0.00	0.04	0.03	0.18	0.01	3.7	0.1	0	0.3
2	5	0.00	0.5	6.7	14.6	0.01	8.0	0.00	0.02	0.01	0.00	0.00	0.5	0.1	0	0.1
1	38	1.81	58.3	280.9	304.8	1.15	13.3	0.06	1.72	0.08	1.84	0.09	15.9	3.4	0.1	1.1
0	5	0.02	0.6	8.6	437.3	0.01	0	0.00	0.00	0.02	0.23	0.00	2.4	0	0.1	0.7
0	1	0.05	0.6	6.4	31.7	0.01	4.1	0.00	0.00	0.00	0.01	0.01	0.1	0.2	0	0
0	9	0.14	4.8	95.0	192.0	0.11	4.8	0.01	0.37	0.01	0.02	0.05	1.3	0.6	0	0.3
0	4	0.43	7.3	38.7	1018.9	0.08	0	0.01	0.00	0.03	0.72	0.04	3.3	0	0	0.1
0	5	0.20	1.2	8.2	97.5	0.01	0	0.00	—	0.01	0.11	0.03	0.2	0	0	—
0	5	0.30	11.0	40.5	689.9	0.01	0	0.01	0.00	0.01	0.22	0.01	1.4	0	0	0.2
0	23	1.24	28.9	536.8	268.6	0.36	—	0.08	0.52	0.08	1.64	0.20	23.2	32.0	0	1.0
4	74	0.20	8.8	97.5	221.3	0.25	—	0.04	—	0.11	0.25	0.02	3.1	0.5	0.2	—
0	6	0.30	0.7	45.4	55.6	0.01	0.3	0.00	0.00	0.01	0.03	0.00	0.5	0.7	0	0
0	0	0.00	—	—	0	—	0	—	—	—	—	—	—	0	—	—
0	1	0.03	0.7	10.9	0.7	0.01	0	0.00	0.00	0.00	0.00	0.00	0	0	0	0
0	1	0.09	0	2.3	0.1	0.00	0	0.00	0.00	0.00	0.00	0.00	0	0	0	5.0
0	0	0.07	—	2.3	0.7	—	—	0.07	—	0.07	0.07	—	—	0.3	0	—
41	68	1.86	58.1	539.7	510.6	1.50	31.5	0.07	4.11	0.22	9.57	0.43	26.6	5.1	0.3	13.6
120	26	1.70	32.4	419.7	500.9	2.67	—	0.09	0.01	0.24	6.87	0.40	15.0	7.9	0.3	19.5
38	333	3.01	52.5	447.6	510.6	1.98	—	0.28	0.01	0.30	1.92	0.19	122.9	8.2	0.3	15.9
45	85	3.65	37.9	475.0	756.0	3.52	17.8	0.38	0.80	0.29	5.33	0.39	69.1	23.4	2.1	28.3
74	31	1.68	20.5	248.3	547.8	0.89	25.6	0.32	1.28	0.24	2.55	0.18	38.4	4.0	0.3	17.5
71	29	2.90	—	613.0	292.0	—	—	0.15	0.51	0.17	4.70	—	—	17.0	0	15.0
74	69	3.25	35.8	391.7	816.6	3.63	69.1	0.21	1.25	0.30	3.80	0.21	48.6	1.3	1.8	27.9
41	68	1.86	58.1	539.7	510.6	1.50	31.5	0.07	4.11	0.22	9.57	0.43	26.6	5.1	0.3	13.6
17	249	2.78	28.0	294.0	739.2	1.42	47.6	0.29	1.05	0.33	2.84	0.14	61.6	15.3	0.4	18.6
23	190	1.04	13.5	75.6	469.3	0.86	58.3	0.11	0.43	0.15	0.89	0.02	21.6	2.4	0.1	9.2
103	26	2.20	—	149.0	600.0	—	—	0.05	—	0.17	4.30	—	—	0	—	29.0
42	28	1.19	22.7	299.7	620.5	1.32	81.0	0.07	1.11	0.14	5.28	0.36	16.2	22.5	0.2	12.0
109	47	1.97	40.0	489.3	492.1	2.13	—	0.11	0.00	0.20	9.81	0.57	16.1	14.0	0.3	22.6
156	423	8.79	110.3	1182.9	880.4	6.18	—	0.72	0.00	1.05	20.16	1.20	113.9	2.7	1.3	97.6
68	66	3.32	37.8	390.6	652.7	1.94	259.6	0.39	1.05	0.39	7.25	0.23	80.6	10.3	0.2	27.0
156	51	3.26	68.3	588.0	3208.6	3.75	31.7	0.15	0.58	0.36	16.68	0.88	24.4	2.0	0.5	36.1
42	20	0.82	23.9	202.5	246.5	1.13	—	0.05	0.62	0.09	4.06	0.25	15.8	2.5	0.2	10.7
27	42	2.83	50.6	636.8	864.8	2.36	—	0.15	0.01	0.22	3.18	0.19	58.1	10.3	0.6	7.3
7	41	0.53	9.0	162.9	20.7	0.18	47.7	0.06	—	0.05	0.24	0.11	24.3	29.4	0	0.6
90	63	0.64	19.8	194.4	198.0	2.45	34.2	0.05	—	0.04	1.74	0.10	31.8	1.7	3.6	24.4
27	266	1.07	38.5	251.4	276.3	1.26	—	0.07	0.02	0.16	1.27	0.17	44.6	59.3	0.2	6.0
0	28	1.74	41.8	298.4	149.9	0.76	0.5	0.07	—	0.08	0.53	0.06	47.4	0.8	0	0.5
9	218	1.83	38.4	163.9	472.9	1.24	—	0.41	0.00	0.35	2.85	0.09	96.2	1.6	0.4	38.4
73	146	2.66	61.3	222.9	885.7	3.43	0	0.10	0.10	0.16	3.00	0.26	95.7	0	1.2	36.7
103	38	2.77	33.7	196.0	833.6	1.34	41.6	0.33	1.60	0.18	4.17	0.27	97.0	3.4	0.3	22.0
66	26	1.46	23.4	248.2	849.7	1.40	29.2	0.10	1.62	0.18	6.28	0.28	23.4	12.0	0.2	19.9
0	13	0.65	11.4	178.0	26.9	0.21	59.0	0.03	—	0.05	0.56	0.08	38.3	24.0	0	0.4
0	60	1.91	35.7	212.8	297.7	1.34	0	0.10	0.92	0.06	0.49	0.49	72.6	9.7	0	3.0
65	50	1.96	63.2	427.7	907.2	1.50	38.9	0.15	4.32	0.14	13.22	0.58	42.1	7.5	0.3	23.0
90	38	2.95	40.1	493.2	495.6	6.60	—	0.08	1.29	0.28	8.03	0.21	28.3	1.4	2.9	30.4
68	233	2.22	40.1	420.1	433.6	2.70	—	0.21	0.21	0.29	3.06	0.22	50.4	15.0	0.8	21.4
34	310	2.06	40.0	258.0	784.0	2.06	180.0	0.27	0.72	0.43	2.18	0.08	64.0	0	0.5	30.6
15	35	2.10	20.1	283.6	1352.9	1.28	27.6	0.19	0.70	0.16	2.77	0.14	42.7	21.6	0.4	13.3
85	59	1.87	21.8	300.8	411.7	3.40	—	0.08	0.00	0.27	3.72	0.13	18.7	0.9	1.6	17.9
172	32	1.57	25.7	333.7	1052.5	1.82	49.8	0.49	5.39	0.36	2.88	0.31	21.1	8.0	0.8	30.0
154	948	7.32	242.4	1201.2	1862.4	10.68	259.0	0.29	7.71	0.81	6.39	1.09	148.8	16.0	2.6	44.1
28	129	1.87	17.0	305.3	533.9	1.03	105.1	0.26	—	0.46	6.09	0.11	73.8	3.3	0.4	26.1
36	258	3.14	31.6	352.3	971.7	2.02	49.0	0.37	1.13	0.37	3.77	0.19	64.8	15.6	0.6	22.8
85	24	0.81	18.8	317.5	661.3	0.38	40.0	0.09	—	0.07	1.11	0.17	8.8	12.5	0	5.1

TABLE H-1 Table of Food Composition (*continued*) (Computer code is for Cengage Diet Analysis program) (For purposes of calculations, use "0" for t, <1, <0.1, 0.01)

DA+ Code	Food Description	Quantity	Measure	Wt (g)	H_2O (g)	Ener (kcal)	Prot (g)	Carb (g)	Fiber (g)	Fat (g)	Fat Breakdown (g) Sat	Mono	Poly	*Trans*
Mixed Foods, Soups, Sandwiches—*continued*														
25109	Salisbury steaks with mushroom sauce	1	serving(s)	135	101.8	251	17.1	9.3	0.5	15.5	6.0	6.7	0.8	0
16637	Shrimp creole with rice	1	cup(s)	243	176.6	309	27.0	27.7	1.2	9.2	1.7	3.6	2.9	—
497	Spaghetti and meatballs with tomato sauce, prepared	1	cup(s)	248	174.0	330	19.0	39.0	2.7	12.0	3.9	4.4	2.2	—
28585	Spicy thai noodles (pad thai)	8	ounce(s)	227	73.3	221	8.9	35.7	3.0	6.4	0.8	3.3	1.8	—
33073	Stir fried pork and vegetables with rice	1	cup(s)	235	173.6	348	15.4	33.5	1.9	16.3	5.6	6.9	2.6	0
28588	Stuffed shells	2½	item(s)	249	157.5	243	15.0	28.0	2.5	8.1	3.1	3.0	1.3	—
16821	Sushi with egg in seaweed	6	piece(s)	156	116.5	190	8.9	20.5	0.3	7.9	2.2	3.2	1.5	—
16819	Sushi with vegetables and fish	6	piece(s)	156	101.6	218	8.4	43.7	1.7	0.6	0.2	0.1	0.2	—
16820	Sushi with vegetables in seaweed	6	piece(s)	156	110.3	183	3.4	40.6	0.8	0.4	0.1	0.1	0.1	—
25266	Sweet and sour pork	¾	cup(s)	249	205.9	265	29.2	17.1	1.0	8.1	2.6	3.5	1.5	0
16824	Tabouli, tabbouleh or tabuli	1	cup(s)	160	123.7	198	2.6	15.9	3.7	14.9	2.0	10.9	1.6	—
25276	Three bean salad	½	cup(s)	99	82.2	95	1.9	9.7	2.6	5.9	0.8	1.4	3.5	0
160	Tuna salad	½	cup(s)	103	64.7	192	16.4	9.6	0	9.5	1.6	3.0	4.2	—
25241	Turkey and noodles	1	cup(s)	319	228.5	270	24.0	21.2	1.0	9.2	2.4	3.5	2.3	—
16794	Vegetable egg roll	2	item(s)	128	89.8	201	5.1	19.5	1.7	11.6	2.5	5.7	2.6	—
16818	Vegetable sushi, no fish	6	piece(s)	156	99.0	226	4.8	49.9	2.0	0.4	0.1	0.1	0.1	—
	Sandwiches													
1744	Bacon, lettuce and tomato with mayonnaise	1	item(s)	164	97.2	341	11.6	34.2	2.3	17.6	3.8	5.5	6.7	—
30287	Bologna and cheese with margarine	1	item(s)	111	45.6	345	13.4	29.3	1.2	19.3	8.1	7.0	2.4	—
30286	Bologna with margarine	1	item(s)	83	33.6	251	8.1	27.3	1.2	12.1	3.7	5.0	2.1	—
16546	Cheese	1	item(s)	83	31.0	261	9.1	27.6	1.2	12.7	5.4	4.2	2.1	—
8789	Cheeseburger, large, plain	1	item(s)	185	78.9	564	32.0	38.5	2.6	31.5	12.5	10.2	1.0	1.8
8624	Cheeseburger, large, with bacon, vegetables, and condiments	1	item(s)	195	91.4	550	30.8	36.8	2.5	30.9	11.9	10.6	1.3	1.5
1745	Club with bacon, chicken, tomato, lettuce, and mayonnaise	1	item(s)	246	137.5	546	31.0	48.9	3.0	24.5	5.3	7.5	9.4	—
1908	Cold cut submarine with cheese and vegetables	1	item(s)	228	131.8	456	21.8	51.0	2.0	18.6	6.8	8.2	2.3	—
30247	Corned beef	1	item(s)	130	74.9	265	18.2	25.3	1.6	9.6	3.6	3.5	1.0	—
25283	Egg salad	1	item(s)	126	72.1	278	10.7	28.0	1.4	13.5	2.9	4.2	5.0	—
16686	Fried egg	1	item(s)	96	49.7	226	10.0	26.2	1.2	8.6	2.3	3.2	1.9	—
16547	Grilled cheese	1	item(s)	83	27.5	291	9.2	27.9	1.2	15.8	6.0	5.7	3.0	—
16659	Gyro with onion and tomato	1	item(s)	105	68.3	163	12.0	20.0	1.1	3.5	1.3	1.3	0.5	—
1906	Ham and cheese	1	item(s)	146	74.2	352	20.7	33.3	2.0	15.5	6.4	6.7	1.4	—
31890	Ham with mayonnaise	1	item(s)	112	56.3	271	13.0	27.9	1.9	11.6	2.8	4.0	4.0	—
756	Hamburger, double patty, large, with condiments and vegetables	1	item(s)	226	121.5	540	34.3	40.3	—	26.6	10.5	10.3	2.8	—
8793	Hamburger, large, plain	1	item(s)	137	57.7	426	22.6	31.7	1.5	22.9	8.4	9.9	2.1	—
8795	Hamburger, large, with vegetables and condiments	1	item(s)	218	121.4	512	25.8	40.0	3.1	27.4	10.4	11.4	2.2	—
25134	Hot chicken salad	1	item(s)	98	48.4	242	15.2	23.8	1.3	9.2	2.9	2.5	3.0	—
25133	Hot turkey salad	1	item(s)	98	50.1	224	15.6	23.8	1.3	6.9	2.3	1.6	2.5	—
1411	Hotdog with bun, plain	1	item(s)	98	52.9	242	10.4	18.0	1.6	14.5	5.1	6.9	1.7	—
30249	Pastrami	1	item(s)	134	71.2	328	13.4	27.8	1.6	17.7	6.1	8.3	1.2	—
16701	Peanut butter	1	item(s)	93	23.6	345	12.2	37.6	3.3	17.4	3.4	7.7	5.2	—
30306	Peanut butter and jelly	1	item(s)	93	24.2	330	10.3	41.9	2.9	14.7	2.9	6.5	4.4	—
1909	Roast beef submarine with mayonnaise and vegetables	1	item(s)	216	127.4	410	28.6	44.3	—	13.0	7.1	1.8	2.6	—
1910	Roast beef, plain	1	item(s)	139	67.6	346	21.5	33.4	1.2	13.8	3.6	6.8	1.7	—
1907	Steak with mayonnaise and vegetables	1	item(s)	204	104.2	459	30.3	52.0	2.3	14.1	3.8	5.3	3.3	—
25288	Tuna salad	1	item(s)	179	102.2	415	24.5	28.4	1.6	22.4	3.5	6.2	11.4	—
30283	Turkey submarine with cheese, lettuce, tomato, and mayonnaise	1	item(s)	277	168.0	529	30.4	49.4	3.0	22.8	6.8	6.0	8.6	—
31891	Turkey with mayonnaise	1	item(s)	143	74.5	329	28.7	26.4	1.3	11.2	2.6	2.6	4.8	—
	Soups													
25296	Bean	1	cup(s)	301	253.1	191	13.8	29.0	6.5	2.3	0.7	0.8	0.5	0
711	Bean with pork, condensed, prepared with water	1	cup(s)	253	215.9	159	7.3	21.0	7.3	5.5	1.4	2.0	1.7	—

…E KEY: H-4 = Breads/Baked Goods H-10 = Cereal/Rice/Pasta H-14 = Fruit H-18 = Vegetables/Legumes H-28 = Nuts/Seeds H-30 = Vegetarian H-32 = Dairy H-40 = Eggs H-40 = Seafood H-42 = Meats H-46 = Poultry …6 = Processed Meats H-48 = Beverages H-52 = Fats/Oils H-54 = Sweets H-56 = Spices/Condiments/Sauces H-60 = Mixed Foods/Soups/Sandwiches H-64 = Fast Food H-84 = Convenience H-86 = Baby Foods

Chol (mg)	Calc (mg)	Iron (mg)	Magn (mg)	Pota (mg)	Sodi (mg)	Zinc (mg)	Vit A (µg)	Thia (mg)	Vit E (mg α)	Ribo (mg)	Niac (mg)	Vit B_6 (mg)	Fola (µg)	Vit C (mg)	Vit B_{12} (µg)	Sele (µg)
60	74	1.94	23.8	314.5	360.5	3.45	—	0.10	0.00	0.27	3.95	0.13	20.8	0.7	1.6	17.4
180	102	4.68	63.2	413.1	330.5	1.72	94.8	0.29	2.06	0.11	4.75	0.21	75.3	12.9	1.2	49.3
89	124	3.70	—	665.0	1009.0	—	81.5	0.25	—	0.30	4.00	—	—	22.0	—	22.0
37	31	1.56	49.4	181.3	591.6	1.05	—	0.18	0.35	0.13	1.82	0.17	44.6	22.6	0.1	3.2
46	38	2.71	33.0	396.9	569.5	2.08	—	0.51	0.38	0.20	5.07	0.30	103.0	18.8	0.4	22.8
30	188	2.26	49.4	403.0	471.5	1.41	—	0.26	0.00	0.26	3.83	0.24	80.0	17.9	0.2	28.9
214	45	1.84	18.7	135.7	463.3	0.98	106.1	0.13	0.67	0.28	1.35	0.13	62.4	1.9	0.7	20.3
11	23	2.15	25.0	202.8	340.1	0.78	45.2	0.26	0.24	0.07	2.76	0.14	76.4	3.6	0.3	13.9
0	20	1.54	18.7	96.7	152.9	0.68	25.0	0.19	0.12	0.03	1.86	0.13	73.3	2.3	0	9.8
74	40	1.76	35.6	619.9	621.8	2.53	—	0.81	0.20	0.37	6.69	0.66	14.7	11.9	0.7	49.6
0	30	1.21	35.2	249.6	796.8	0.48	54.4	0.07	2.43	0.04	1.11	0.11	30.4	26.1	0	0.5
0	26	0.96	15.5	144.8	224.2	0.30	—	0.02	0.88	0.04	0.26	0.04	32.2	10.0	0	2.7
13	17	1.02	19.5	182.5	412.1	0.57	24.6	0.03	—	0.07	6.86	0.08	8.2	2.3	1.2	42.2
77	69	2.56	33.2	400.8	577.1	2.51	—	0.23	0.28	0.30	6.41	0.29	61.4	1.4	1.1	33.4
60	29	1.65	17.9	193.3	549.1	0.48	25.6	0.15	1.28	0.20	1.59	0.09	46.1	5.5	0.2	11.3
0	23	2.38	21.8	157.6	369.7	0.82	48.4	0.28	0.15	0.05	2.44	0.12	85.8	3.7	0	8.1
21	79	2.36	27.9	351.0	944.6	1.08	44.3	0.32	1.16	0.24	4.36	0.21	73.8	9.7	0.2	27.1
40	258	2.38	24.4	215.3	941.3	1.88	102.1	0.31	0.55	0.35	2.97	0.14	59.9	0.2	0.8	20.4
17	100	2.24	16.6	138.6	579.3	1.02	44.8	0.29	0.49	0.21	2.92	0.12	58.1	0.2	0.5	16.0
22	233	2.04	19.9	127.0	733.7	1.24	97.1	0.25	0.47	0.30	2.25	0.05	58.1	0	0.3	13.1
104	309	4.47	44.4	401.5	986.1	5.75	0	0.35	—	0.77	8.26	0.49	109.2	0	2.8	38.9
98	267	4.03	44.9	464.1	1314.3	5.20	0	0.33	—	0.67	8.25	0.47	95.6	1.4	2.4	6.6
71	157	4.57	46.7	464.9	1087.3	1.82	41.8	0.54	1.52	0.40	12.82	0.61	118.1	6.4	0.4	42.3
36	189	2.50	68.4	394.4	1650.7	2.57	70.7	1.00	—	0.79	5.49	0.13	86.6	12.3	1.1	30.8
46	81	3.04	19.5	127.4	1206.4	2.26	2.6	0.23	0.20	0.24	3.42	0.11	59.8	0.3	0.9	31.2
219	85	2.25	18.8	159.0	423.1	0.87	—	0.27	0.12	0.43	2.06	0.15	73.6	0.9	0.6	29.9
206	104	2.79	17.3	117.1	438.7	0.92	89.3	0.26	0.66	0.40	2.26	0.10	79.7	0	0.6	24.2
22	235	2.05	19.9	128.7	763.6	1.26	129.5	0.19	0.72	0.28	2.05	0.05	38.2	0	0.2	13.2
28	47	1.77	22.1	218.4	235.2	2.33	9.5	0.23	0.26	0.20	3.12	0.13	44.1	3.2	0.9	18.3
58	130	3.24	16.1	290.5	770.9	1.37	96.4	0.30	0.29	0.48	2.68	0.20	75.9	2.8	0.5	23.1
34	91	2.47	23.5	210.6	1097.6	1.13	5.6	0.57	0.50	0.26	3.80	0.25	60.5	2.2	0.2	20.2
122	102	5.85	49.7	569.5	791.0	5.67	0	0.36	—	0.38	7.57	0.54	76.8	1.1	4.1	25.5
71	74	3.57	27.4	267.2	474.0	4.11	0	0.28	—	0.28	6.24	0.23	60.3	0	2.1	27.1
87	96	4.92	43.6	479.6	824.0	4.88	0	0.41	—	0.37	7.28	0.32	82.8	2.6	2.4	33.6
39	115	1.88	20.0	172.4	505.1	1.19	—	0.23	0.28	0.22	4.84	0.19	46.4	0.5	0.2	19.9
37	114	1.99	21.3	189.0	494.5	1.07	—	0.22	0.28	0.20	4.26	0.22	46.4	0.5	0.2	23.4
44	24	2.31	12.7	143.1	670.3	1.98	0	0.23	—	0.27	3.64	0.04	48.0	0.1	0.5	26.0
51	80	3.02	22.8	182.2	1364.1	2.70	2.7	0.28	0.26	0.26	4.97	0.14	60.3	0.3	1.0	14.3
0	110	2.92	66.0	226.0	580.3	1.32	0	0.31	2.39	0.23	6.72	0.18	92.1	0	0	13.2
0	94	2.50	56.7	198.1	492.9	1.12	—	0.26	2.01	0.20	5.66	0.15	78.1	0.1	0	11.2
73	41	2.80	67.0	330.5	844.6	4.38	30.2	0.41	—	0.41	5.96	0.32	71.3	5.6	1.8	25.7
51	54	4.22	30.6	315.5	792.3	3.39	11.1	0.37	—	0.30	5.86	0.26	57.0	2.1	1.2	29.2
73	92	5.16	49.0	524.3	797.6	4.52	0	0.40	—	0.36	7.30	0.36	89.8	5.5	1.6	42.0
59	78	2.97	35.9	316.3	724.7	1.02	—	0.27	0.34	0.26	12.07	0.46	60.8	1.8	2.4	76.9
64	307	4.59	49.9	534.6	1759.0	2.74	74.8	0.49	1.19	0.65	3.91	0.31	121.9	10.5	0.6	42.1
67	100	3.46	34.3	304.6	564.9	3.00	5.7	0.28	0.74	0.32	6.84	0.46	64.4	0	0.3	40
5	79	3.05	61.8	588.8	689.0	1.41	—	0.27	0.02	0.15	3.63	0.23	140.1	3.6	0.2	7.9
3	78	1.89	43.0	371.9	883.0	0.96	43.0	0.08	1.08	0.03	0.52	0.03	30.4	1.5	0	7.8

TABLE H-1 Table of Food Composition (*continued*) (Computer code is for Cengage Diet Analysis program) (For purposes of calculations, use "0" for t, <1, <0.1, 0.01)

DA+ Code	Food Description	Quantity	Measure	Wt (g)	H_2O (g)	Ener (kcal)	Prot (g)	Carb (g)	Fiber (g)	Fat (g)	Fat Breakdown (g)			
											Sat	Mono	Poly	*Trans*
Mixed Foods, Soups, Sandwiches—*continued*														
713	Beef noodle, condensed, prepared with water	1	cup(s)	244	224.9	83	4.7	8.7	0.7	3.0	1.1	1.2	0.5	—
825	Cheese, condensed, prepared with milk	1	cup(s)	251	206.9	231	9.5	16.2	1.0	14.6	9.1	4.1	0.5	—
826	Chicken broth, condensed, prepared with water	1	cup(s)	244	234.1	39	4.9	0.9	0	1.4	0.4	0.6	0.3	—
25297	Chicken noodle soup	1	cup(s)	286	258.4	117	10.8	10.9	0.9	2.9	0.8	1.1	0.7	—
827	Chicken noodle, condensed, prepared with water	1	cup(s)	241	226.1	60	3.1	7.1	0.5	2.3	0.6	1.0	0.6	0
724	Chicken noodle, dehydrated, prepared with water	1	cup(s)	252	237.3	58	2.1	9.2	0.3	1.4	0.3	0.5	0.4	—
823	Cream of asparagus, condensed, prepared with milk	1	cup(s)	248	213.3	161	6.3	16.4	0.7	8.2	3.3	2.1	2.2	—
824	Cream of celery, condensed, prepared with milk	1	cup(s)	248	214.4	164	5.7	14.5	0.7	9.7	3.9	2.5	2.7	—
708	Cream of chicken, condensed, prepared with milk	1	cup(s)	248	210.4	191	7.5	15.0	0.2	11.5	4.6	4.5	1.6	—
715	Cream of chicken, condensed, prepared with water	1	cup(s)	244	221.1	117	3.4	9.3	0.2	7.4	2.1	3.3	1.5	—
709	Cream of mushroom, condensed, prepared with milk	1	cup(s)	248	215.0	166	6.2	14.0	0	9.6	3.3	2.0	1.8	0.1
716	Cream of mushroom, condensed, prepared with water	1	cup(s)	244	224.6	102	1.9	8.0	0	7.0	1.6	1.3	1.7	—
25298	Cream of vegetable	1	cup(s)	285	250.7	165	7.2	15.2	1.9	8.6	1.6	4.6	1.9	—
16689	Egg drop	1	cup(s)	244	228.9	73	7.5	1.1	0	3.8	1.1	1.5	0.6	—
25138	Golden squash	1	cup(s)	258	223.9	145	7.6	20.4	0.4	4.1	0.8	2.2	0.9	—
16663	Hot and sour	1	cup(s)	244	209.7	161	15.0	5.4	0.5	7.9	2.7	3.4	1.1	—
28054	Lentil chowder	1	cup(s)	244	202.8	153	11.4	27.7	12.6	0.5	0.1	0.1	0.2	0
28560	Macaroni and bean	1	cup(s)	246	138.8	146	5.8	22.9	5.1	3.7	0.5	2.2	0.6	0
714	Manhattan clam chowder, condensed, prepared with water	1	cup(s)	244	225.1	73	2.1	11.6	1.5	2.1	0.4	0.4	1.2	—
28561	Minestrone	1	cup(s)	241	185.4	103	4.5	16.8	4.8	2.3	0.3	1.4	0.4	0
717	Minestrone, condensed, prepared with water	1	cup(s)	241	220.1	82	4.3	11.2	1.0	2.5	0.6	0.7	1.1	—
28038	Mushroom and wild rice	1	cup(s)	244	199.7	86	4.7	13.2	1.7	0.3	0	0	0.2	0
828	New England clam chowder, condensed, prepared with milk	1	cup(s)	248	212.2	151	8.0	18.4	0.7	5.0	2.1	0.7	0.6	0
28036	New England style clam chowder	1	cup(s)	244	227.5	61	3.8	8.8	1.8	0.2	0.1	0	0	0
28566	Old country pasta	1	cup(s)	252	183.3	146	6.5	18.3	3.6	4.5	2.0	2.4	0.9	0
725	Onion, dehydrated, prepared with water	1	cup(s)	246	235.7	30	0.8	6.8	0.7	0	0	0	0	—
16667	Shrimp gumbo	1	cup(s)	244	207.2	166	9.5	18.2	2.4	6.7	1.3	2.9	2.0	—
28037	Southwestern corn chowder	1	cup(s)	244	217.8	98	4.9	17.0	2.4	0.5	0.1	0.1	0.2	0
30282	Soybean (miso)	1	cup(s)	240	218.6	84	6.0	8.0	1.9	3.4	0.6	1.1	1.4	—
25140	Split pea	1	cup(s)	165	119.7	72	4.5	16.0	1.6	0.3	0.1	0	0.2	0
718	Split pea with ham, condensed, prepared with water	1	cup(s)	253	206.9	190	10.3	28.0	2.3	4.4	1.8	1.8	0.6	—
726	Tomato vegetable, dehydrated, prepared with water	1	cup(s)	253	238.4	56	2.0	10.2	0.8	0.9	0.4	0.3	0.1	0
710	Tomato, condensed, prepared with milk	1	cup(s)	248	213.4	136	6.2	22.0	1.5	3.2	1.8	0.9	0.3	—
719	Tomato, condensed, prepared with water	1	cup(s)	244	223.0	73	1.9	16.0	1.5	0.7	0.2	0.2	0.2	—
28595	Turkey noodle	1	cup(s)	244	216.9	114	8.1	15.1	1.9	2.4	0.3	1.1	0.7	—
28051	Turkey vegetable	1	cup(s)	244	220.8	96	12.2	6.6	2.0	1.1	0.3	0.2	0.3	0
25141	Vegetable	1	cup(s)	252	228.1	82	5.2	16.5	4.5	0.3	0	0	0.1	0
720	Vegetable beef, condensed, prepared with water	1	cup(s)	244	224.0	76	5.4	9.9	2.0	1.9	0.8	0.8	0.1	—
28598	Vegetable gumbo	1	cup(s)	252	184.5	170	4.4	28.9	3.6	4.7	0.7	3.2	0.5	0
721	Vegetarian vegetable, condensed, prepared with water	1	cup(s)	241	222.7	67	2.1	11.8	0.7	1.9	0.3	0.8	0.7	—
Fast Food														
	Arby's													
36094	Au jus sauce	1	serving(s)	85	—	43	1.0	7.0	0	1.3	0.4	—	—	0.4
751	Beef 'n cheddar sandwich	1	item(s)	195	—	445	22.0	44.0	2.0	21.0	6.0	—	—	1.0
9279	Cheddar curly fries	1	serving(s)	198	—	631	8.0	73.0	7.0	37.4	6.8	—	—	5.7

Chol (mg)	Calc (mg)	Iron (mg)	Magn (mg)	Pota (mg)	Sodi (mg)	Zinc (mg)	Vit A (µg)	Thia (mg)	Vit E (mg α)	Ribo (mg)	Niac (mg)	Vit B_6 (mg)	Fola (µg)	Vit C (mg)	Vit B_{12} (µg)	Sele (µg)
5	20	1.07	7.3	97.6	929.6	1.51	12.2	0.06	1.22	0.05	1.03	0.03	19.5	0.5	0.2	7.3
48	289	0.80	20.1	341.4	1019.1	0.67	358.9	0.06	—	0.33	0.50	0.07	10.0	1.3	0.4	7.0
0	10	0.51	2.4	209.8	775.9	0.24	0	0.01	0.04	0.07	3.34	0.02	4.9	0	0.2	0
24	25	1.38	16.4	340.1	774.5	0.77	—	0.15	0.02	0.16	5.57	0.13	37.2	1.8	0.3	10.2
12	14	1.59	9.6	53.0	638.7	0.38	26.5	0.13	0.07	0.10	1.30	0.04	19.3	0	0	11.6
10	5	0.50	7.6	32.8	577.1	0.20	2.5	0.20	0.12	0.07	1.08	0.02	17.6	0	0.1	9.6
22	174	0.86	19.8	359.6	1041.6	0.91	62.0	0.10	—	0.27	0.88	0.06	29.8	4.0	0.5	8.0
32	186	0.69	22.3	310.0	1009.4	0.19	114.1	0.07	—	0.24	0.43	0.06	7.4	1.5	0.5	4.7
27	181	0.67	17.4	272.8	1046.6	0.67	178.6	0.07	—	0.25	0.92	0.06	7.4	1.2	0.5	8.0
10	34	0.61	2.4	87.8	985.8	0.63	163.5	0.02	—	0.06	0.82	0.01	2.4	0.2	0.1	7.0
10	164	1.36	19.8	267.8	823.4	0.79	81.8	0.10	1.01	0.29	0.62	0.05	7.4	0.2	0.6	6.0
0	17	1.31	4.9	73.2	775.9	0.24	9.8	0.05	0.97	0.05	0.50	0.00	2.4	0	0	2.9
1	80	1.20	17.5	340.7	787.9	0.56	—	0.12	1.05	0.18	3.32	0.12	39.5	10.7	0.3	4.3
102	22	0.75	4.9	219.6	729.6	0.48	41.5	0.02	0.29	0.19	3.02	0.05	14.6	0	0.5	7.6
4	262	0.78	42.4	542.2	515.6	0.88	—	0.16	0.52	0.30	1.14	0.16	32.7	12.5	0.7	6.0
34	29	1.24	19.5	373.3	1561.6	1.43	—	0.26	0.12	0.24	4.96	0.20	14.6	0.5	0.4	19.3
0	48	4.38	59.3	626.2	26.7	1.57	—	0.24	0.06	0.12	1.87	0.32	176.3	16.1	0	3.5
0	59	1.90	32.4	275.9	531.0	0.51	—	0.16	0.37	0.12	1.44	0.10	58.2	9.1	0	8.8
2	27	1.56	9.8	180.6	551.4	0.87	48.8	0.02	1.22	0.03	0.77	0.09	9.8	3.9	3.9	9.0
0	62	1.70	29.9	287.5	442.7	0.42	—	0.09	0.23	0.09	0.70	0.07	47.3	13.3	0	3.5
2	34	0.91	7.2	313.3	911.0	0.74	118.1	0.05	—	0.04	0.94	0.09	36.2	1.2	0	8.0
0	30	1.38	27.3	376.9	283.8	1.00	—	0.06	0.07	0.23	3.21	0.14	18.1	3.7	0.1	4.7
17	169	3.00	29.8	456.3	887.8	0.99	91.8	0.20	0.54	0.43	1.96	0.17	22.3	5.2	11.9	10.9
3	89	1.30	29.2	503.7	256.8	0.52	—	0.06	0.02	0.10	1.30	0.15	24.2	11.8	3.0	3.8
5	57	2.43	51.9	500.4	355.7	0.78	—	0.21	0.01	0.14	2.64	0.20	80.0	22.0	0.1	10.3
0	22	0.12	9.8	76.3	851.2	0.12	0	0.03	0.02	0.03	0.15	0.06	0	0.2	0	0.5
51	105	2.85	48.8	461.2	441.6	0.90	80.5	0.18	1.90	0.12	2.52	0.19	85.4	17.6	0.3	14.9
1	83	1.03	26.3	434.4	217.4	0.57	—	0.08	0.09	0.13	1.81	0.21	32.1	39.6	0.2	1.7
0	65	1.87	36.0	362.4	988.8	0.86	232.8	0.06	0.96	0.16	2.61	0.15	57.6	4.6	0.2	1.0
0	28	1.26	29.8	328.1	602.4	0.52	—	0.10	0.00	0.07	1.50	0.16	49.5	8.1	0	0.7
8	23	2.27	48.1	399.7	1006.9	1.31	22.8	0.14	—	0.07	1.47	0.06	2.5	1.5	0.3	8.0
0	20	0.60	10.1	169.5	334.0	0.20	10.1	0.06	0.43	0.09	1.26	0.06	12.7	3.0	0.1	2.0
10	166	1.36	29.8	466.2	711.8	0.84	94.2	0.09	0.44	0.31	1.35	0.15	5.0	15.6	0.6	9.2
0	20	1.31	17.1	273.3	663.7	0.29	24.4	0.04	0.41	0.07	1.23	0.10	0	15.4	0	6.1
26	28	1.40	24.1	223.8	395.9	0.75	—	0.21	0.01	0.12	2.85	0.15	45.0	7.1	0.1	13.7
21	38	1.48	25.0	423.4	348.7	0.99	—	0.09	0.01	0.09	3.64	0.27	23.9	10.6	0.2	10.3
0	40	2.08	39.1	681.5	670.3	0.67	—	0.16	0.00	0.09	2.70	0.26	36.3	22.4	0	2.2
5	20	1.09	7.3	168.4	773.5	1.51	190.3	0.03	0.58	0.04	1.00	0.07	9.8	2.4	0.3	2.7
0	56	1.85	38.9	360.2	518.4	0.64	—	0.18	0.64	0.08	1.77	0.17	57.6	21.9	0	4.1
0	24	1.06	7.2	207.3	814.6	0.45	171.1	0.05	1.39	0.04	0.90	0.05	9.6	1.4	0	4.3
0	0	—	—	—	1510.0	—	—	—	—	—	—	—	—	—	—	—
51	80	3.96	—	—	1274.0	—	—	—	—	—	—	—	—	1.8	—	—
0	80	3.24	—	—	1476.0	—	—	—	—	—	—	—	—	9.6	—	—

TABLE H-1 Table of Food Composition (*continued*) (Computer code is for Cengage Diet Analysis program) (For purposes of calculations, use "0" for t, <1, <0.1, 0.01)

DA+ Code	Food Description	Quantity	Measure	Wt (g)	H_2O (g)	Ener (kcal)	Prot (g)	Carb (g)	Fiber (g)	Fat (g)	Fat Breakdown (g) Sat	Mono	Poly	*Trans*
Fast Food—*continued*														
34770	Chicken breast fillet sandwich, grilled	1	item(s)	233	—	414	32.0	36.0	3.0	17.0	3.0	—	—	0
36131	Chocolate shake, regular	1	serving(s)	397	—	507	13.0	83.0	0	13.0	8.0	—	—	0
36045	Curly fries, large size	1	serving(s)	198	—	631	8.0	73.0	7.0	37.0	7.0	—	—	6.0
36044	Curly fries, medium size	1	serving(s)	128	—	406	5.0	47.0	5.0	24.0	4.0	—	—	4.0
752	Ham 'n cheese sandwich	1	item(s)	167	—	304	23.0	35.0	1.0	7.0	2.0	—	—	0
36048	Homestyle fries, large size	1	serving(s)	213	—	566	6.0	82.0	6.0	37.0	7.0	—	—	5.0
36047	Homestyle fries, medium size	1	serving(s)	142	—	377	4.0	55.0	4.0	25.0	4.0	—	—	4.0
33465	Homestyle fries, small size	1	serving(s)	113	—	302	3.0	44.0	3.0	20.0	4.0	—	—	3.0
9249	Junior roast beef sandwich	1	item(s)	125	—	272	16.0	34.0	2.0	10.0	4.0	—	—	0
9251	Large roast beef sandwich	1	item(s)	281	—	547	42.0	41.0	3.0	28.0	12.0	—	—	2.0
39640	Market Fresh chicken salad with pecans sandwich	1	item(s)	322	—	769	30.0	79.0	9.0	39.0	10.0	—	—	0
39641	Market Fresh Martha's Vineyard salad, without dressing	1	serving(s)	330	—	277	26.0	24.0	5.0	8.0	4.0	—	—	0
34769	Market Fresh roast turkey and Swiss sandwich	1	serving(s)	359	—	725	45.0	75.0	5.0	30.0	8.0	—	—	1.0
9267	Market Fresh roast turkey ranch and bacon sandwich	1	serving(s)	382	—	834	49.0	75.0	5.0	38.0	11.0	—	—	1.0
39642	Market Fresh Santa Fe salad, without dressing	1	serving(s)	372	—	499	30.0	42.0	7.0	23.0	8.0	—	—	2.0
39650	Market Fresh Southwest chicken wrap	1	serving(s)	251	—	567	36.0	42.0	4.0	29.0	9.0	—	—	1.0
37021	Market Fresh Ultimate BLT sandwich	1	item(s)	294	—	779	23.0	75.0	6.0	45.0	11.0	—	—	1.0
750	Roast beef sandwich, regular	1	item(s)	154	—	320	21.0	34.0	2.0	14.0	5.0	—	—	1.0
36132	Strawberry shake, regular	1	serving(s)	397	—	498	13.0	81.0	0	13.0	8.0	—	—	0
2009	Super roast beef sandwich	1	item(s)	198	—	398	21.0	40.0	2.0	19.0	6.0	—	—	1.0
36130	Vanilla shake, regular	1	serving(s)	369	—	437	13.0	66.0	0	13.0	8.0	—	—	0
	Auntie Anne's													
35371	Cheese dipping sauce	1	serving(s)	35	—	100	3.0	4.0	0	8.0	4.0	—	—	0
35353	Cinnamon sugar soft pretzel	1	item(s)	120	—	350	9.0	74.0	2.0	2.0	0	—	—	0
35354	Cinnamon sugar soft pretzel with butter	1	item(s)	120	—	450	8.0	83.0	3.0	9.0	5.0	—	—	0
35372	Marinara dipping sauce	1	serving(s)	35	—	10	0	4.0	0	0	0	0	0	0
35357	Original soft pretzel	1	serving(s)	120	—	340	10.0	72.0	3.0	1.0	0	—	—	0
35358	Original soft pretzel with butter	1	item(s)	120	—	370	10.0	72.0	3.0	4.0	2.0	—	—	0
35359	Parmesan herb soft pretzel	1	item(s)	120	—	390	11.0	74.0	4.0	5.0	2.5	—	—	—
35360	Parmesan herb soft pretzel with butter	1	item(s)	120	—	440	10.0	72.0	9.0	13.0	7.0	—	—	—
35361	Sesame soft pretzel	1	item(s)	120	—	350	11.0	63.0	3.0	6.0	1.0	—	—	0
35362	Sesame soft pretzel with butter	1	item(s)	120	—	410	12.0	64.0	7.0	12.0	4.0	—	—	0
35364	Sour cream and onion soft pretzel	1	item(s)	120	—	310	9.0	66.0	2.0	1.0	0	—	—	0
35366	Sour cream and onion soft pretzel with butter	1	item(s)	120	—	340	9.0	66.0	2.0	5.0	3.0	—	—	0
35373	Sweet mustard dipping sauce	1	serving(s)	35	—	60	0.5	8.0	0	1.5	1.0	—	—	0
35367	Whole wheat soft pretzel	1	item(s)	120	—	350	11.0	72.0	7.0	1.5	0	—	—	0
35368	Whole wheat soft pretzel with butter	1	item(s)	120	—	370	11.0	72.0	7.0	4.5	1.5	—	—	0
	Boston Market													
34978	Butternut squash	¾	cup(s)	143	—	140	2.0	25.0	2.0	4.5	3.0	—	—	0
35006	Caesar side salad	1	serving(s)	71	—	40	3.0	3.0	1.0	20.0	2.0	—	—	1.5
35013	Chicken Carver sandwich with cheese and sauce	1	item(s)	321	—	700	44.0	68.0	3.0	29.0	7.0	—	—	0
34979	Chicken gravy	4	ounce(s)	113	—	15	1.0	4.0	0	0.5	0	—	—	0
35053	Chicken noodle soup	¾	cup(s)	283	—	180	13.0	16.0	1.0	7.0	2.0	—	—	0
34973	Chicken pot pie	1	item(s)	425	—	800	29.0	59.0	4.0	49.0	18.0	—	—	7.0
35054	Chicken tortilla soup with toppings	¾	cup(s)	227	—	340	12.0	24.0	1.0	22.0	7.0	—	—	0
35007	Cole slaw	¾	cup(s)	125	—	170	2.0	21.0	2.0	9.0	2.0	—	—	0
35057	Cornbread	1	item(s)	45	—	130	1.0	21.0	0	3.5	1.0	—	—	1.0
34980	Creamed spinach	¾	cup(s)	191	—	280	9.0	12.0	4.0	23.0	15.0	—	—	0
34998	Fresh vegetable stuffing	1	cup(s)	136	—	190	3.0	25.0	2.0	8.0	1.0	—	—	0
34991	Garlic dill new potatoes	¾	cup(s)	156	—	140	3.0	24.0	3.0	3.0	1.0	—	—	0
34983	Green bean casserole	¾	cup(s)	170	—	60	2.0	9.0	2.0	2.0	1.0	—	—	0
34982	Green beans	¾	cup(s)	91	—	60	2.0	7.0	3.0	3.5	1.5	—	—	0

KEY: H-4 = Breads/Baked Goods H-10 = Cereal/Rice/Pasta H-14 = Fruit H-18 = Vegetables/Legumes H-28 = Nuts/Seeds H-30 = Vegetarian H-32 = Dairy H-40 = Eggs H-40 = Seafood H-42 = Meats H-46 = Poultry
6 = Processed Meats H-48 = Beverages H-52 = Fats/Oils H-54 = Sweets H-56 = Spices/Condiments/Sauces H-60 = Mixed Foods/Soups/Sandwiches H-64 = Fast Food H-84 = Convenience H-86 = Baby Foods

Chol (mg)	Calc (mg)	Iron (mg)	Magn (mg)	Pota (mg)	Sodi (mg)	Zinc (mg)	Vit A (µg)	Thia (mg)	Vit E (mg α)	Ribo (mg)	Niac (mg)	Vit B_6 (mg)	Fola (µg)	Vit C (mg)	Vit B_{12} (µg)	Sele (µg)
9	90	3.06	—	—	913.0	—	—	—	—	—	—	—	—	10.8	—	—
34	510	0.54	—	—	357.0	—	—	—	—	—	—	—	—	5.4	—	—
0	80	3.24	—	—	1476.0	—	—	—	—	—	—	—	—	9.6	—	—
0	50	1.98	—	—	949.0	—	—	—	—	—	—	—	—	6.0	—	—
35	160	2.70	—	—	1420.0	—	—	—	—	—	—	—	—	1.2	—	—
0	50	1.62	—	—	1029.0	—	—	—	—	—	—	—	—	12.6	—	—
0	30	1.08	—	—	686.0	—	—	—	—	—	—	—	—	8.4	—	—
0	30	0.90	—	—	549.0	—	—	—	—	—	—	—	—	6.6	—	—
29	60	3.06	—	—	740.0	—	0	—	—	—	—	—	—	0	—	—
102	70	6.30	—	—	1869.0	—	0	—	—	—	—	—	—	0.6	—	—
74	180	4.32	—	—	1240.0	—	—	—	—	—	—	—	—	30.0	—	—
72	200	1.62	—	—	454.0	—	—	—	—	—	—	—	—	33.6	—	—
91	360	5.22	—	—	1788.0	—	—	—	—	—	—	—	—	10.2	—	—
109	330	5.40	—	—	2258.0	—	—	—	—	—	—	—	—	11.4	—	—
59	420	3.60	—	—	1231.0	—	—	—	—	—	—	—	—	36.6	—	—
88	240	4.50	—	—	1451.0	—	—	—	—	—	—	—	—	7.8	—	—
51	170	4.68	—	—	1571.0	—	—	—	—	—	—	—	—	16.8	—	—
44	60	3.60	—	—	953.0	—	0	—	—	—	—	—	—	0	—	—
34	510	0.72	—	—	363.0	—	—	—	—	—	—	—	—	6.6	—	—
44	70	3.78	—	—	1060.0	—	—	—	—	—	—	—	—	6.0	—	—
34	510	0.36	—	—	350.0	—	—	—	—	—	—	—	—	5.4	—	—
10	100	0.00	—	—	510.0	—	—	—	—	—	—	—	—	0	—	—
0	20	1.98	—	—	410.0	—	0	—	—	—	—	—	—	0	—	—
25	30	2.34	—	—	430.0	—	—	—	—	—	—	—	—	0	—	—
0	0	0.00	—	—	180.0	—	0	—	—	—	—	—	—	0	—	—
0	30	2.34	—	—	900.0	—	0	—	—	—	—	—	—	0	—	—
10	30	2.16	—	—	930.0	—	—	—	—	—	—	—	—	0	—	—
10	80	1.80	—	—	780.0	—	—	—	—	—	—	—	—	1.2	—	—
30	60	1.80	—	—	660.0	—	—	—	—	—	—	—	—	1.2	—	—
0	20	2.88	—	—	840.0	—	0	—	—	—	—	—	—	0	—	—
15	20	2.70	—	—	860.0	—	—	—	—	—	—	—	—	0	—	—
0	30	1.98	—	—	920.0	—	—	—	—	—	—	—	—	0	—	—
10	40	2.16	—	—	930.0	—	—	—	—	—	—	—	—	0	—	—
40	0	0.00	—	—	120.0	—	0	—	—	—	—	—	—	0	—	—
0	30	1.98	—	—	1100.0	—	0	—	—	—	—	—	—	0	—	—
10	30	2.34	—	—	1120.0	—	—	—	—	—	—	—	—	0	—	—
10	59	0.80	—	—	35.0	—	—	—	—	—	—	—	—	22.2	—	—
0	60	0.43	—	—	75.0	—	—	—	—	—	—	—	—	5.4	—	—
90	211	2.85	—	—	1560.0	—	—	—	—	—	—	—	—	15.8	—	—
0	0	0.00	—	—	570.0	—	0	—	—	—	—	—	—	0	—	—
55	0	1.07	—	—	220.0	—	—	—	—	—	—	—	—	1.8	—	—
115	40	4.50	—	—	800.0	—	—	—	—	—	—	—	—	1.2	—	—
45	123	1.32	—	—	1310.0	—	—	—	—	—	—	—	—	18.4	—	—
10	41	0.48	—	—	270.0	—	—	—	—	—	—	—	—	24.5	—	—
5	0	0.71	—	—	220.0	—	0	—	—	—	—	—	—	0	—	—
70	264	2.84	—	—	580.0	—	—	—	—	—	—	—	—	9.5	—	—
0	41	1.48	—	—	580.0	—	—	—	—	—	—	—	—	2.5	—	—
0	0	0.85	—	—	120.0	—	0	—	—	—	—	—	—	14.3	—	—
5	20	0.72	—	—	620.0	—	—	—	—	—	—	—	—	2.4	—	—
0	43	0.38	—	—	180.0	—	—	—	—	—	—	—	—	5.1	—	—

DA+ Code	Food Description	Quantity	Measure	Wt (g)	H_2O (g)	Ener (kcal)	Prot (g)	Carb (g)	Fiber (g)	Fat (g)	Fat Breakdown (g)			
											Sat	Mono	Poly	*Trans*
Fast Food—*continued*														
34984	Homestyle mashed potatoes	¾	cup(s)	221	—	210	4.0	29.0	3.0	9.0	6.0	—	—	0
34985	Homestyle mashed potatoes and gravy	1	cup(s)	334	—	225	5.0	33.0	3.0	9.5	6.0	—	—	0
34988	Hot cinnamon apples	¾	cup(s)	145	—	210	0	47.0	3.0	3.0	0	—	—	0
34989	Macaroni and cheese	¾	cup(s)	221	—	330	14.0	39.0	1.0	12.0	7.0	—	—	0.5
51193	Market chopped salad with dressing	1	item(s)	563	—	580	11.0	31.0	9.0	48.0	9.0	—	—	1.0
34970	Meatloaf	1	serving(s)	218	—	480	29.0	23.0	2.0	33.0	13.0	—	—	0
39383	Nestle Toll House chocolate chip cookie	1	item(s)	78	—	370	4.0	49.0	2.0	19.0	9.0	—	—	0
34965	Quarter chicken, dark meat, no skin	1	item(s)	134	—	260	30.0	2.0	0	13.0	4.0	—	—	0
34966	Quarter chicken, dark meat, with skin	1	item(s)	149	—	280	31.0	3.0	0	15.0	4.5	—	—	0
34963	Quarter chicken, white meat, no skin or wing	1	item(s)	173	—	250	41.0	4.0	0	8.0	2.5	—	—	0
34964	Quarter chicken, white meat, with skin and wing	1	item(s)	110	—	330	50.0	3.0	0	12.0	4.0	—	—	0
34968	Roasted turkey breast	5	ounce(s)	142	—	180	38.0	0	0	3.0	1.0	—	—	0
35011	Seasonal fresh fruit salad	1	serving(s)	142	—	60	1.0	15.0	1.0	0	0	0	0	0
51192	Spinach with garlic butter sauce	1	serving(s)	170	—	130	5.0	9.0	5.0	9.0	6.0	—	—	0
34969	Spiral sliced holiday ham	8	ounce(s)	227	—	450	40.0	13.0	0	26.0	10.0	—	—	0
35003	Steamed vegetables	1	cup(s)	136	—	50	2.0	8.0	3.0	2.0	0	—	—	0
35005	Sweet corn	¾	cup(s)	176	—	170	6.0	37.0	2.0	4.0	1.0	—	—	0
35004	Sweet potato casserole	¾	cup(s)	198	—	460	4.0	77.0	3.0	17.0	6.0	—	—	0
	Burger King													
29731	Biscuit with sausage, egg, and cheese	1	item(s)	191	—	610	20.0	33.0	1.0	45.0	15.0	—	—	1.0
14249	Cheeseburger	1	item(s)	133	—	330	17.0	31.0	1.0	16.0	7.0	—	—	0.5
14251	Chicken sandwich	1	item(s)	219	—	660	24.0	52.0	4.0	40.0	8.0	—	—	2.5
3808	Chicken Tenders, 8 pieces	1	serving(s)	123	—	340	19.0	21.0	0.5	20.0	5.0	—	—	3.0
14259	Chocolate shake, small	1	item(s)	315	—	470	8.0	75.0	1.0	14.0	9.0	—	—	0
29732	Croissanwich with sausage and cheese	1	item(s)	106	37.2	370	14.0	23.0	0.5	25.0	9.0	12.7	3.3	2.0
14261	Croissanwich with sausage, egg, and cheese	1	item(s)	159	71.4	470	19.0	26.0	0.5	32.0	11.0	15.8	6.1	2.5
3809	Double cheeseburger	1	item(s)	189	—	500	30.0	31.0	1.0	29.0	14.0	—	—	1.5
14244	Double Whopper sandwich	1	item(s)	373	—	900	47.0	51.0	3.0	57.0	19.0	—	—	2.0
14245	Double Whopper with cheese sandwich	1	item(s)	398	—	990	52.0	52.0	3.0	64.0	24.0	—	—	2.5
14250	Fish Filet sandwich	1	item(s)	250	—	630	24.0	67.0	4.0	30.0	6.0	—	—	2.5
14255	French fries, medium, salted	1	serving(s)	116	—	360	4.0	41.0	4.0	20.0	4.5	—	—	4.5
14262	French toast sticks, 5 pieces	1	serving(s)	112	37.6	390	6.0	46.0	2.0	20.0	4.5	10.6	2.9	4.5
14248	Hamburger	1	item(s)	121	—	290	15.0	30.0	1.0	12.0	4.5	—	—	0
14263	Hash brown rounds, small	1	serving(s)	75	27.1	230	2.0	23.0	2.0	15.0	4.0	—	—	5.0
14256	Onion rings, medium	1	serving(s)	91	—	320	4.0	40.0	3.0	16.0	4.0	—	—	3.5
39000	Tendercrisp chicken sandwich	1	item(s)	286	—	780	25.0	73.0	4.0	43.0	8.0	—	—	4.0
37514	TenderGrill chicken sandwich	1	item(s)	258	—	450	37.0	53.0	4.0	10.0	2.0	—	—	0
14258	Vanilla shake, small	1	item(s)	296	—	400	8.0	57.0	0	15.0	9.0	—	—	0
1736	Whopper sandwich	1	item(s)	290	—	670	28.0	51.0	3.0	39.0	11.0	—	—	1.5
14243	Whopper with cheese sandwich	1	item(s)	315	—	760	33.0	52.0	3.0	47.0	16.0	—	—	1.5
	Carl's Jr													
33962	Carl's bacon Swiss crispy chicken sandwich	1	item(s)	268	—	750	31.0	91.0	—	28.0	28.0	—	—	—
10801	Carl's Catch fish sandwich	1	item(s)	215	—	560	19.0	58.0	2.0	27.0	7.0	—	1.9	—
10862	Carl's Famous Star hamburger	1	item(s)	254	—	590	24.0	50.0	3.0	32.0	9.0	—	—	—
10785	Charbroiled chicken club sandwich	1	item(s)	270	—	550	42.0	43.0	4.0	23.0	7.0	—	2.9	—
10866	Charbroiled chicken salad	1	item(s)	437	—	330	34.0	17.0	5.0	7.0	4.0	—	1.0	—
10855	Charbroiled Santa Fe chicken sandwich	1	item(s)	266	—	610	38.0	43.0	4.0	32.0	8.0	—	—	—
10790	Chicken stars, 6 pieces	1	serving(s)	85	—	260	13.0	14.0	1.0	16.0	4.0	—	1.6	—
34864	Chocolate shake, small	1	serving(s)	595	—	540	15.0	98.0	0	11.0	7.0	—	—	—
10797	Crisscut fries	1	serving(s)	139	—	410	5.0	43.0	4.0	24.0	5.0	—	—	—
10799	Double Western Bacon cheeseburger	1	item(s)	308	—	920	51.0	65.0	2.0	50	21.0	—	6.6	—

Chol (mg)	Calc (mg)	Iron (mg)	Magn (mg)	Pota (mg)	Sodi (mg)	Zinc (mg)	Vit A (µg)	Thia (mg)	Vit E (mg α)	Ribo (mg)	Niac (mg)	Vit B_6 (mg)	Fola (µg)	Vit C (mg)	Vit B_{12} (µg)	Sele (µg)
25	51	0.46	—	—	660.0	—	—	—	—	—	—	—	—	19.2	—	—
25	100	0.59	—	—	1230.0	—	—	—	—	—	—	—	—	24.9	—	—
0	16	0.28	—	—	15.0	—	—	—	—	—	—	—	—	0	—	—
30	345	1.65	—	—	1290.0	—	—	—	—	—	—	—	—	0	—	—
10	—	—	—	—	2010.0	—	—	—	—	—	—	—	—	—	—	—
125	140	3.77	—	—	970.0	—	—	—	—	—	—	—	—	1.8	—	—
20	0	1.32	—	—	340.0	—	—	—	—	—	—	—	—	0	—	—
155	0	1.52	—	—	260.0	—	0	—	—	—	—	—	—	0	—	—
155	0	2.14	—	—	660.0	—	0	—	—	—	—	—	—	0	—	—
125	0	0.89	—	—	480.0	—	0	—	—	—	—	—	—	0	—	—
165	0	0.78	—	—	960.0	—	0	—	—	—	—	—	—	0	—	—
70	20	1.80	—	—	620.0	—	0	—	—	—	—	—	—	0	—	—
0	16	0.29	—	—	20.0	—	—	—	—	—	—	—	—	29.5	—	—
20	—	—	—	—	200.0	—	—	—	—	—	—	—	—	—	—	—
140	0	1.73	—	—	2230.0	—	0	—	—	—	—	—	—	0	—	—
0	53	0.46	—	—	45.0	—	—	—	—	—	—	—	—	24.0	—	—
0	0	0.43	—	—	95.0	—	—	—	—	—	—	—	—	5.8	—	—
20	44	1.18	—	—	210.0	—	—	—	—	—	—	—	—	9.8	—	—
210	250	2.70	—	—	1620.0	—	89.9	—	—	—	—	—	—	0	—	—
55	150	2.70	—	—	780.0	—	—	0.24	—	0.31	4.17	—	—	1.2	—	—
70	64	2.89	—	—	1440.0	—	—	0.50	—	0.32	10.29	—	—	0	—	—
55	20	0.72	—	—	960.0	—	—	0.14	—	0.11	10.93	—	—	0	—	—
55	333	0.79	—	—	350.0	—	—	0.11	—	0.61	0.26	—	—	2.7	0	—
50	99	1.78	20.1	217.3	810.0	1.51	—	0.34	1.03	0.33	4.33	—	—	0	0.6	22.2
180	146	2.63	28.6	313.2	1060.0	2.08	—	0.38	1.66	0.51	4.72	0.28	—	0	1.1	38.0
105	250	4.50	—	—	1030.0	—	—	0.26	—	0.44	6.37	—	—	1.2	—	—
175	150	8.07	—	—	1090.0	—	—	0.39	—	0.59	11.05	—	—	9.0	—	—
195	299	8.08	—	—	1520.0	—	—	0.39	—	0.66	11.03	—	—	9.0	—	—
60	101	3.62	—	—	1380.0	—	—	—	—	—	—	—	—	3.6	—	—
0	20	0.71	—	—	590.0	—	0	0.15	—	0.48	2.30	—	—	8.9	—	—
0	60	1.80	21.3	124.3	440.0	0.57	—	0.31	0.98	0.19	2.88	0.05	—	0	0	13.7
40	80	2.70	—	—	560.0	—	—	0.25	—	0.28	4.25	—	—	1.2	—	—
0	0	0.36	—	—	450.0	—	0	0.11	0.83	0.06	1.35	0.17	—	1.2	—	—
0	100	0.00	—	—	460.0	—	0	0.14	—	0.09	2.32	—	—	0	—	—
75	79	4.43	—	—	1730.0	—	—	—	—	—	—	—	—	8.9	—	—
75	57	6.82	—	—	1210.0	—	—	—	—	—	—	—	—	5.7	—	—
60	348	0.00	—	—	240.0	—	—	0.11	—	0.63	0.21	—	—	2.4	0	—
51	100	5.38	—	—	1020.0	—	—	0.38	—	0.43	7.30	—	—	9.0	—	—
115	249	5.38	—	—	1450.0	—	—	0.38	—	0.51	7.28	—	—	9.0	—	—
80	200	5.40	—	—	1900.0	—	—	—	—	—	—	—	—	2.4	—	—
80	150	2.70	—	—	990.0	—	60.0	—	—	—	—	—	—	2.4	—	—
70	100	4.50	—	—	910.0	—	—	—	—	—	—	—	—	6.0	—	—
95	200	3.60	—	—	1330.0	—	—	—	—	—	—	—	—	9.0	—	—
75	200	1.80	—	—	880.0	—	—	—	—	—	—	—	—	30.0	—	—
100	200	3.60	—	—	1440.0	—	—	—	—	—	—	—	—	9.0	—	—
35	19	1.02	—	—	470.0	—	0	—	—	—	—	—	—	0	—	—
45	600	1.08	—	—	360.0	—	0	—	—	—	—	—	—	0	—	—
0	20	1.80	—	—	950.0	—	0	—	—	—	—	—	—	12.0	—	—
155	300	7.20	—	—	1730.0	—	—	—	—	—	—	—	—	1.2	—	—

DA+ Code	Food Description	Quantity	Measure	Wt (g)	H_2O (g)	Ener (kcal)	Prot (g)	Carb (g)	Fiber (g)	Fat (g)	Fat Breakdown (g) Sat	Mono	Poly	*Trans*
	Fast Food—*continued*													
14238	French fries, small	1	serving(s)	92	—	290	5.0	37.0	3.0	14.0	3.0	—	—	—
10798	French toast dips without syrup, 5 pieces	1	serving(s)	155	—	370	8.0	49.0	0	17.0	5.0	—	1.4	—
10802	Onion rings	1	serving(s)	128	—	440	7.0	53.0	3.0	22.0	5.0	—	0.8	—
34858	Spicy chicken sandwich	1	item(s)	198	—	480	14.0	48.0	2.0	26.0	5.0	—	—	—
34867	Strawberry shake, small	1	serving(s)	595	—	520	14.0	93.0	0	11.0	7.0	—	—	—
10865	Super Star hamburger	1	item(s)	348	—	790	41.0	52.0	3.0	47.0	14.0	—	—	—
38925	The Six Dollar burger	1	item(s)	429	—	1010	40.0	60.0	3.0	66.0	26.0	—	—	—
10818	Vanilla shake, small	1	item(s)	398	—	314	10.0	51.5	0	7.4	4.7	—	—	—
10770	Western Bacon cheeseburger	1	item(s)	225	—	660	32.0	64.0	2.0	30.0	12.0	—	4.8	—
	Chick Fil-A													
38746	Biscuit with bacon, egg, and cheese	1	item(s)	163	—	470	18.0	39.0	1.0	26.0	9.0	—	—	3.0
38747	Biscuit with egg	1	item(s)	135	—	350	11.0	38.0	1.0	16.0	4.5	—	—	3.0
38748	Biscuit with egg and cheese	1	item(s)	149	—	400	14.0	38.0	1.0	21.0	7.0	—	—	3.0
38753	Biscuit with gravy	1	item(s)	192	—	330	5.0	43.0	1.0	15.0	4.0	—	—	4.0
38752	Biscuit with sausage, egg, and cheese	1	item(s)	212	—	620	22.0	39.0	2.0	42.0	14.0	—	—	3.0
38771	Carrot and raisin salad	1	item(s)	113	—	170	1.0	28.0	2.0	6.0	1.0	—	—	0
38761	Chargrilled chicken Cool Wrap	1	item(s)	245	—	390	29.0	54.0	3.0	7.0	3.0	—	—	0
38766	Chargrilled chicken garden salad	1	item(s)	275	—	180	22.0	9.0	3.0	6.0	3.0	—	—	0
38758	Chargrilled chicken sandwich	1	item(s)	193	—	270	28.0	33.0	3.0	3.5	1.0	—	—	0
38742	Chicken biscuit	1	item(s)	145	—	420	18.0	44.0	2.0	19.0	4.5	—	—	3.0
38743	Chicken biscuit with cheese	1	item(s)	159	—	470	21.0	45.0	2.0	23.0	8.0	—	—	3.0
38762	Chicken Caesar Cool Wrap	1	item(s)	227	—	460	36.0	52.0	3.0	10.0	6.0	—	—	0
38757	Chicken deluxe sandwich	1	item(s)	208	—	420	28.0	39.0	2.0	16.0	3.5	—	—	0
38764	Chicken salad sandwich on wheat bun	1	item(s)	153	—	350	20.0	32.0	5.0	15.0	3.0	—	—	0
38756	Chicken sandwich	1	item(s)	170	—	410	28.0	38.0	1.0	16.0	3.5	—	—	0
38768	Chick-n-Strip salad	1	item(s)	327	—	400	34.0	21.0	4.0	20.0	6.0	—	—	0
38763	Chick-n-Strips	4	item(s)	127	—	300	28.0	14.0	1.0	15.0	2.5	—	—	0
38770	Cole slaw	1	item(s)	128	—	260	2.0	17.0	2.0	21.0	3.5	—	—	0
38776	Diet lemonade, small	1	cup(s)	255	—	25	0	5.0	0	0	0	0	0	0
38755	Hashbrowns	1	serving(s)	84	—	260	2.0	25.0	3.0	17.0	3.5	—	—	1.0
38765	Hearty breast of chicken soup	1	cup(s)	241	—	140	8.0	18.0	1.0	3.5	1.0	—	—	0
38741	Hot buttered biscuit	1	item(s)	79	—	270	4.0	38.0	1.0	12.0	3.0	—	—	3.0
38778	IceDream, small cone	1	item(s)	135	—	160	4.0	28.0	0	4.0	2.0	—	—	0
38774	IceDream, small cup	1	serving(s)	227	—	240	6.0	41.0	0	6.0	3.5	—	—	0
38775	Lemonade, small	1	cup(s)	255	—	170	0	41.0	0	0.5	0	—	—	0
38777	Nuggets	8	item(s)	113	—	260	26.0	12.0	0.5	12.0	2.5	—	—	0
38769	Side salad	1	item(s)	108	—	60	3.0	4.0	2.0	3.0	1.5	—	—	0
38767	Southwest chargrilled salad	1	item(s)	303	—	240	25.0	17.0	5.0	8.0	3.5	—	—	0
40481	Spicy chicken cool wrap	1	serving(s)	230	—	380	30.0	52.0	3.0	6.0	3.0	—	—	0
38772	Waffle potato fries, small, salted	1	serving(s)	85	—	270	3.0	34.0	4.0	13.0	3.0	—	—	1.5
	Cinnabon													
39572	Caramellata Chill w/whipped cream	16	fluid ounce(s)	480	—	406	10.0	61.0	0	14.0	8.0	—	—	0
39571	Cinnabon Bites	1	serving(s)	149	—	510	8.0	77.0	2.0	19.0	5.0	—	—	5.0
39570	Cinnabon Stix	5	item(s)	85	—	379	6.0	41.0	1.0	21.0	6.0	—	—	4.0
39567	Classic roll	1	item(s)	221	—	813	15.0	117.0	4.0	32.0	8.0	—	—	5.0
39568	Minibon	1	item(s)	92	—	339	6.0	49.0	2.0	13.0	3.0	—	—	2.0
39573	Mochalatta Chill w/whipped cream	16	fluid ounce(s)	480	—	362	9.0	55.0	0	13.0	8.0	—	—	0
39569	Pecanbon	1	item(s)	272	—	1100	16.0	141.0	8.0	56.0	10.0	—	—	5.0
	Dairy Queen													
1466	Banana split	1	item(s)	369	—	510	8.0	96.0	3.0	12.0	8.0	—	—	0
38552	Brownie Earthquake®	1	serving(s)	304	—	740	10.0	112.0	0	27.0	16.0	—	—	0.5
38561	Chocolate chip cookie dough blizzard,® small	1	item(s)	319	—	720	12.0	105.0	0	28.0	14.0	—	—	2.5
1464	Chocolate malt, small	1	item(s)	418	—	640	15.0	111.0	1.0	16.0	11.0	—	—	0.5
38541	Chocolate shake, small	1	item(s)	397	—	560	13.0	93.0	1.0	15.0	10.0	—	—	0.5
17257	Chocolate soft serve	½	cup(s)	94	—	150	4.0	22.0	0	5.0	3.5	—	—	0
1463	Chocolate sundae, small	1	item(s)	163	—	280	5.0	49.0	0	7.0	4.5	—	—	0

KEY: H-4 = Breads/Baked Goods H-10 = Cereal/Rice/Pasta H-14 = Fruit H-18 = Vegetables/Legumes H-28 = Nuts/Seeds H-30 = Vegetarian H-32 = Dairy H-40 = Eggs H-40 = Seafood H-42 = Meats H-46 = Poultry
= Processed Meats H-48 = Beverages H-52 = Fats/Oils H-54 = Sweets H-56 = Spices/Condiments/Sauces H-60 = Mixed Foods/Soups/Sandwiches H-64 = Fast Food H-84 = Convenience H-86 = Baby Foods

Chol (mg)	Calc (mg)	Iron (mg)	Magn (mg)	Pota (mg)	Sodi (mg)	Zinc (mg)	Vit A (µg)	Thia (mg)	Vit E (mg α)	Ribo (mg)	Niac (mg)	Vit B_6 (mg)	Fola (µg)	Vit C (mg)	Vit B_{12} (µg)	Sele (µg)
0	0	1.08	—	—	170.0	—	0	—	—	—	—	—	—	21.0	—	—
3	0	0.00	—	—	470.0	—	0	0.25	—	0.23	2.00	—	—	0	—	—
0	20	0.72	—	—	700.0	—	0	—	—	—	—	—	—	3.6	—	—
40	100	3.60	—	—	1220.0	—	—	—	—	—	—	—	—	6.0	—	—
45	600	0.00	—	—	340.0	—	0	—	—	—	—	—	—	0	—	—
130	100	7.20	—	—	980.0	—	—	—	—	—	—	—	—	9.0	—	—
145	279	4.29	—	—	1960.0	—	—	—	—	—	—	—	—	16.7	—	—
30	401	0.00	—	—	234.0	—	0	—	—	—	—	—	—	0	—	—
85	200	5.40	—	—	1410.0	—	60.0	—	—	—	—	—	—	1.2	—	—
270	150	2.70	—	—	1190.0	—	—	—	—	—	—	—	—	0	—	—
240	80	2.70	—	—	740.0	—	—	—	—	—	—	—	—	0	—	—
255	150	2.70	—	—	970.0	—	—	—	—	—	—	—	—	0	—	—
5	60	1.80	—	—	930.0	—	0	—	—	—	—	—	—	0	—	—
300	200	3.60	—	—	1360.0	—	—	—	—	—	—	—	—	0	—	—
10	40	0.36	—	—	110.0	—	—	—	—	—	—	—	—	4.8	—	—
65	200	3.60	—	—	1020.0	—	—	—	—	—	—	—	—	6.0	—	—
65	150	0.72	—	—	620.0	—	—	—	—	—	—	—	—	30	—	—
65	80	2.70	—	—	940.0	—	—	—	—	—	—	—	—	6.0	—	—
35	60	2.70	—	—	1270.0	—	0	—	—	—	—	—	—	0	—	—
50	150	2.70	—	—	1500.0	—	—	—	—	—	—	—	—	0	—	—
80	500	3.60	—	—	1350.0	—	—	—	—	—	—	—	—	1.2	—	—
60	100	2.70	—	—	1300.0	—	—	—	—	—	—	—	—	2.4	—	—
65	150	1.80	—	—	880.0	—	—	—	—	—	—	—	—	0	—	—
60	100	2.70	—	—	1300.0	—	—	—	—	—	—	—	—	0	—	—
80	150	1.44	—	—	1070.0	—	—	—	—	—	—	—	—	6.0	—	—
65	40	1.44	—	—	940.0	—	—	—	—	—	—	—	—	0	—	—
25	60	0.36	—	—	220.0	—	—	—	—	—	—	—	—	36.0	—	—
0	0	0.36	—	—	5.0	—	0	—	—	—	—	—	—	15.0	—	—
5	20	0.72	—	—	380.0	—	—	—	—	—	—	—	—	0	—	—
25	40	1.08	—	—	900.0	—	—	—	—	—	—	—	—	0	—	—
0	60	1.80	—	—	660.0	—	0	—	—	—	—	—	—	0	—	—
15	100	0.36	—	—	80.0	—	—	—	—	—	—	—	—	0	—	—
25	200	0.36	—	—	105.0	—	—	—	—	—	—	—	—	0	—	—
0	0	0.36	—	—	10.0	—	0	—	—	—	—	—	—	15.0	—	—
70	40	1.08	—	—	1090.0	—	0	—	—	—	—	—	—	0	—	—
10	100	0.00	—	—	75.0	—	—	—	—	—	—	—	—	15.0	—	—
60	200	1.08	—	—	770.0	—	—	—	—	—	—	—	—	24.0	—	—
60	200	3.60	—	—	1090.0	—	—	—	—	—	—	—	—	3.6	—	—
0	20	1.08	—	—	115.0	—	0	—	—	—	—	—	—	1.2	—	—
46	—	—	—	—	187.0	—	—	—	—	—	—	—	—	—	—	—
35	—	—	—	—	530.0	—	—	—	—	—	—	—	—	—	—	—
16	—	—	—	—	413.0	—	—	—	—	—	—	—	—	—	—	—
67	—	—	—	—	801.0	—	—	—	—	—	—	—	—	—	—	—
27	—	—	—	—	337.0	—	—	—	—	—	—	—	—	—	—	—
46	—	—	—	—	252.0	—	—	—	—	—	—	—	—	—	—	—
63	—	—	—	—	600.0	—	—	—	—	—	—	—	—	—	—	—
30	250	1.80	—	—	180.0	—	—	—	—	—	—	—	—	15.0	—	—
50	250	1.80	—	—	350.0	—	—	—	—	—	—	—	—	0	—	—
50	350	2.70	—	—	370.0	—	—	—	—	—	—	—	—	1.2	—	—
55	450	1.80	—	—	340.0	—	—	—	—	—	—	—	—	2.4	—	—
50	450	1.44	—	—	280.0	—	—	—	—	—	—	—	—	2.4	—	—
15	100	0.72	—	—	75.0	—	—	—	—	—	—	—	—	0	—	—
20	200	1.08	—	—	140.0	—	—	—	—	—	—	—	—	0	—	—

DA+ Code	Food Description	Quantity	Measure	Wt (g)	H_2O (g)	Ener (kcal)	Prot (g)	Carb (g)	Fiber (g)	Fat (g)	Fat Breakdown (g) Sat	Mono	Poly	*Trans*
	Fast Food—*continued*													
1462	Dipped cone, small	1	item(s)	156	—	340	6.0	42.0	1.0	17.0	9.0	4.0	3.0	1.0
38555	Oreo cookies blizzard, small	1	item(s)	283	—	570	11.0	83.0	0.5	21.0	10.0	—	—	2.5
38547	Royal Treats Peanut Buster® Parfait	1	item(s)	305	—	730	16.0	99.0	2.0	31.0	17.0	—	—	0
17256	Vanilla soft serve	½	cup(s)	94	—	140	3.0	22.0	0	4.5	3.0	—	—	0
	Domino's													
31606	Barbeque buffalo wings	1	item(s)	25	—	50	6.0	2.0	0	2.5	0.5	—	—	—
31604	Breadsticks	1	item(s)	30	—	115	2.0	12.0	0	6.3	1.1	—	—	—
37551	Buffalo Chicken Kickers	1	item(s)	24	—	47	4.0	3.0	0	2.0	0.5	—	—	—
37548	CinnaStix	1	item(s)	30	—	123	2.0	15.0	1.0	6.1	1.1	—	—	—
37549	Dot, cinnamon	1	item(s)	28	7.6	99	1.9	14.9	0.7	3.7	0.7	—	—	—
31605	Double cheesy bread	1	item(s)	35	—	123	4.0	13.0	0	6.5	1.9	—	—	—
31607	Hot buffalo wings	1	item(s)	25	—	45	5.0	1.0	0	2.5	0.5	—	—	—
	Domino's Classic hand tossed pizza													
31573	America's favorite feast, 12″	1	slice(s)	102	—	257	10.0	29.0	2.0	11.5	4.5	—	—	—
31574	America's favorite feast, 14″	1	slice(s)	141	—	353	14.0	39.0	2.0	16.0	6.0	—	—	—
37543	Bacon cheeseburger feast, 12″	1	slice(s)	99	—	273	12.0	28.0	2.0	13.0	5.5	—	—	—
37545	Bacon cheeseburger feast, 14″	1	slice(s)	137	—	379	17.0	38.0	2.0	18.0	8.0	—	—	—
37546	Barbeque feast, 12″	1	slice(s)	96	—	252	11.0	31.0	1.0	10.0	4.5	—	—	—
37547	Barbeque feast, 14″	1	slice(s)	131	—	344	14.0	43.0	2.0	13.5	6.0	—	—	—
31569	Cheese, 12″	1	slice(s)	55	—	160	6.0	28.0	1.0	3.0	1.0	—	—	0
31570	Cheese, 14″	1	slice(s)	75	—	220	8.0	38.0	2.0	4.0	1.0	—	—	0
37538	Deluxe feast, 12″	1	slice(s)	201	101.8	465	19.5	57.4	3.5	18.2	7.7	—	—	—
37540	Deluxe feast, 14″	1	slice(s)	273	138.4	627	26.4	78.3	4.7	24.1	10.2	—	—	—
31685	Deluxe, 12″	1	slice(s)	100	—	234	9.0	29.0	2.0	9.5	3.5	—	—	—
31694	Deluxe, 14″	1	slice(s)	136	—	316	13.0	39.0	2.0	12.5	5.0	—	—	—
31686	Extravaganzza, 12″	1	slice(s)	122	—	289	13.0	30.0	2.0	14.0	5.5	—	—	—
31695	Extravaganzza, 14″	1	slice(s)	165	—	388	17.0	40.0	3.0	18.5	7.5	—	—	—
31575	Hawaiian feast, 12″	1	slice(s)	102	—	223	10.0	30.0	2.0	8.0	3.5	—	—	—
31576	Hawaiian feast, 14″	1	slice(s)	141	—	309	14.0	41.0	2.0	11.0	4.5	—	—	—
31687	Meatzza, 12″	1	slice(s)	108	—	281	13.0	29.0	2.0	13.5	5.5	—	—	—
31696	Meatzza, 14″	1	slice(s)	146	—	378	17.0	39.0	2.0	18.0	7.5	—	—	—
31571	Pepperoni feast, extra pepperoni and cheese, 12″	1	slice(s)	98	—	265	11.0	28.0	2.0	12.5	5.0	—	—	—
31572	Pepperoni feast, extra pepperoni and cheese, 14″	1	slice(s)	135	—	363	16.0	39.0	2.0	17.0	7.0	—	—	—
31577	Vegi feast, 12″	1	slice(s)	102	—	218	9.0	29.0	2.0	8.0	3.5	—	—	—
31578	Vegi feast, 14″	1	slice(s)	139	—	300	13.0	40.0	3.0	11.0	4.5	—	—	—
	Domino's thin crust pizza													
31583	America's favorite, 12″	1	slice(s)	72	—	208	8.0	15.0	1.0	13.5	5.0	—	—	—
31584	America's favorite, 14″	1	slice(s)	100	—	285	11.0	20.0	2.0	18.5	7.0	—	—	—
31579	Cheese, 12″	1	slice(s)	49	—	137	5.0	14.0	1.0	7.0	2.5	—	—	—
31580	Cheese, 14″	1	slice(s)	68	27.0	214	8.8	19.0	1.4	11.4	4.6	2.9	2.5	—
31688	Deluxe, 12″	1	slice(s)	70	—	185	7.0	15.0	1.0	11.5	4.0	—	—	—
31697	Deluxe, 14″	1	slice(s)	94	—	248	10.0	20.0	2.0	15.0	5.5	—	—	—
31689	Extravaganzza, 12″	1	slice(s)	92	—	240	11.0	16.0	1.0	15.5	6.0	—	—	—
31698	Extravaganzza, 14″	1	slice(s)	123	—	320	14.0	21.0	2.0	20.5	8.0	—	—	—
31585	Hawaiian, 12″	1	slice(s)	71	—	174	8.0	16.0	1.0	9.5	3.5	—	—	—
31586	Hawaiian, 14″	1	slice(s)	100	—	240	11.0	21.0	2.0	13.0	5.0	—	—	—
31690	Meatzza, 12″	1	slice(s)	78	—	232	11.0	15.0	1.0	15.0	6.0	—	—	—
31699	Meatzza, 14″	1	slice(s)	104	—	310	14.0	20.0	2.0	20	8.0	—	—	—
31581	Pepperoni, extra pepperoni and cheese, 12″	1	slice(s)	68	—	216	9.0	14.0	1.0	14.0	5.5	—	—	—
31582	Pepperoni, extra pepperoni and cheese, 14″	1	slice(s)	93	—	295	13.0	20.0	1.0	19.0	7.5	—	—	—
31587	Vegi, 12″	1	slice(s)	71	—	168	7.0	15.0	1.0	9.5	3.5	—	—	—
31588	Vegi, 14″	1	slice(s)	97	—	231	10.0	21.0	2.0	13.5	5.0	—	—	—
	Domino's Ultimate deep dish pizza													
31596	America's favorite, 12″	1	slice(s)	115	—	309	12.0	29.0	2.0	17.0	6.0	—	—	—
31702	America's favorite, 14″	1	slice(s)	162	—	433	17.0	42.0	3.0	23.5	8.0	—	—	—
31590	Cheese, 12″	1	slice(s)	90	—	238	9.0	28.0	2.0	11.0	3.5	—	—	—
31591	Cheese, 14″	1	slice(s)	128	53.9	351	14.5	41.0	2.9	13.2	5.2	3.8	2.5	—
31589	Cheese, 6″	1	item(s)	215	—	598	22.9	68.4	3.9	27.6	9.9	—	—	—
31691	Deluxe, 12″	1	slice(s)	122	—	287	11.0	29.0	2.0	15.0	5.0	—	—	—

Chol (mg)	Calc (mg)	Iron (mg)	Magn (mg)	Pota (mg)	Sodi (mg)	Zinc (mg)	Vit A (µg)	Thia (mg)	Vit E (mg α)	Ribo (mg)	Niac (mg)	Vit B_6 (mg)	Fola (µg)	Vit C (mg)	Vit B_{12} (µg)	Sele (µg)
20	200	1.08	—	—	130.0	—	—	—	—	—	—	—	—	1.2	—	—
40	350	2.70	—	—	430.0	—	—	—	—	—	—	—	—	1.2	—	—
35	300	1.80	—	—	400.0	—	—	—	—	—	—	—	—	1.2	—	—
15	150	0.72	—	—	70.0	—	—	—	—	—	—	—	—	0	—	—
26	10	0.36	—	—	175.5	—	—	—	—	—	—	—	—	0	—	—
0	0	0.72	—	—	122.1	—	—	—	—	—	—	—	—	0	—	—
9	0	0.00	—	—	162.5	—	—	—	—	—	—	—	—	0	—	—
0	0	0.72	—	—	111.4	—	—	—	—	—	—	—	—	0	—	—
0	6	0.59	—	—	85.7	—	—	—	—	—	—	—	—	0	—	—
6	40	0.72	—	—	162.3	—	—	—	—	—	—	—	—	0	—	—
26	10	0.36	—	—	254.5	—	—	—	—	—	—	—	—	1.2	—	—
																—
22	100	1.80	—	—	625.5	—	—	—	—	—	—	—	—	0.6	—	—
31	140	2.52	—	—	865.5	—	—	—	—	—	—	—	—	0.6	—	—
27	140	1.80	—	—	634.0	—	—	—	—	—	—	—	—	0	—	—
38	190	2.52	—	—	900.0	—	—	—	—	—	—	—	—	0	—	—
20	140	1.62	—	—	600.0	—	—	—	—	—	—	—	—	0.6	—	—
27	190	2.16	—	—	831.5	—	—	—	—	—	—	—	—	0.6	—	—
0	0	1.80	—	—	110.0	—	0	—	—	—	—	—	—	0	—	—
0	0	2.70	—	—	150.0	—	0	—	—	—	—	—	—	0	—	—
40	199	3.56	—	—	1063.1	—	—	—	—	—	—	—	—	1.4	—	—
53	276	4.84	—	—	1432.2	—	—	—	—	—	—	—	—	1.8	—	—
17	100	1.80	—	—	541.5	—	—	—	—	—	—	—	—	0.6	—	—
23	130	2.34	—	—	728.5	—	—	—	—	—	—	—	—	1.2	—	—
28	140	1.98	—	—	764.0	—	—	—	—	—	—	—	—	0.6	—	—
37	190	2.70	—	—	1014.0	—	—	—	—	—	—	—	—	1.2	—	—
16	130	1.62	—	—	546.5	—	—	—	—	—	—	—	—	1.2	—	—
23	180	2.34	—	—	765.0	—	—	—	—	—	—	—	—	1.2	—	—
28	130	1.80	—	—	739.5	—	—	—	—	—	—	—	—	0	—	—
37	190	2.52	—	—	983.5	—	—	—	—	—	—	—	—	0	—	—
24	130	1.62	—	—	670.0	—	70.9	—	—	—	—	—	—	0	—	—
33	180	2.34	—	—	920.0	—	104.7	—	—	—	—	—	—	0	—	—
13	130	1.62	—	—	489.0	—	—	—	—	—	—	—	—	0.6	—	—
18	180	2.34	—	—	678.0	—	—	—	—	—	—	—	—	0.6	—	—
23	100	0.90	—	—	533.0	—	—	—	—	—	—	—	—	2.4	—	—
32	140	1.26	—	—	736.5	—	—	—	—	—	—	—	—	3.0	—	—
10	90	0.54	—	—	292.5	—	60.0	—	—	—	—	—	—	1.8	—	—
14	151	0.48	17.7	125.1	338.0	0.02	64.6	0.05	1.01	0.07	0.69	—	—	2.4	0.5	24.1
19	100	0.90	—	—	449.0	—	—	—	—	—	—	—	—	2.4	—	—
24	130	1.08	—	—	601.0	—	—	—	—	—	—	—	—	3.6	—	—
29	140	1.08	—	—	671.5	—	—	—	—	—	—	—	—	2.4	—	—
38	190	1.44	—	—	886.5	—	—	—	—	—	—	—	—	3.6	—	—
17	130	0.72	—	—	454.0	—	—	—	—	—	—	—	—	3.0	—	—
24	180	0.90	—	—	637.5	—	—	—	—	—	—	—	—	3.6	—	—
29	140	0.90	—	—	647.0	—	—	—	—	—	—	—	—	1.8	—	—
38	190	1.26	—	—	865.5	—	—	—	—	—	—	—	—	2.4	—	—
26	130	0.72	—	—	577.0	—	80.0	—	—	—	—	—	—	1.8	—	—
35	80	1.08	—	—	792.5	—	105.8	—	—	—	—	—	—	2.4	—	—
14	130	0.72	—	—	396.5	—	—	—	—	—	—	—	—	2.4	—	—
19	180	1.08	—	—	550.5	—	—	—	—	—	—	—	—	3.0	—	—
25	120	2.34	—	—	796.5	—	—	—	—	—	—	—	—	0.6	—	—
34	170	3.24	—	—	1110.0	—	—	—	—	—	—	—	—	0.6	—	—
11	110	1.98	—	—	555.5	—	70.0	—	—	—	—	—	—	0	—	—
18	189	3.78	32.0	209.9	718.1	1.75	99.8	0.29	1.13	0.31	5.44	—	—	0	0.6	45.6
36	295	4.67	—	—	1341.4	—	174.0	—	—	—	—	—	—	0.5	—	—
20	120	2.16	—	—	712.0	—	—	—	—	—	—	—	—	1.2	—	—

APPENDIX H

DA+ Code	Food Description	Quantity	Measure	Wt (g)	H_2O (g)	Ener (kcal)	Prot (g)	Carb (g)	Fiber (g)	Fat (g)	Fat Breakdown (g)			
											Sat	Mono	Poly	*Trans*
Fast Food—*continued*														
31700	Deluxe, 14″	1	slice(s)	156	—	396	15.0	42.0	3.0	20.0	7.0	—	—	—
31692	Extravaganzza, 12″	1	slice(s)	136	—	341	14.0	30.0	2.0	19.0	7.0	—	—	—
31701	Extravaganzza, 14″	1	slice(s)	186	—	468	20.0	43.0	3.0	25.5	9.5	—	—	—
31599	Hawaiian, 12″	1	slice(s)	114	—	275	12.0	30.0	2.0	13.0	5.0	—	—	—
31600	Hawaiian, 14″	1	slice(s)	162	—	389	17.0	43.0	3.0	18.0	6.5	—	—	—
31693	Meatzza, 12″	1	slice(s)	121	—	333	14.0	29.0	2.0	19.0	7.0	—	—	—
31703	Meatzza, 14″	1	slice(s)	167	—	458	19.0	42.0	3.0	25.0	9.5	—	—	—
31593	Pepperoni, extra pepperoni and cheese, 12″	1	slice(s)	110	—	317	13.0	29.0	2.0	17.5	6.5	—	—	—
31594	Pepperoni, extra pepperoni and cheese, 14″	1	slice(s)	155	—	443	18.0	42.0	3.0	24.0	9.0	—	—	—
31602	Vegi, 12″	1	slice(s)	114	—	270	11.0	30.0	2.0	13.5	5.0	—	—	—
31603	Vegi, 14″	1	slice(s)	159	—	380	15.0	43.0	3.0	18.0	6.5	—	—	—
31598	With ham and pineapple tidbits, 6″	1	item(s)	430	—	619	25.2	69.9	4.0	28.3	10.2	—	—	—
31595	With Italian sausage, 6″	1	item(s)	430	—	642	24.8	69.6	4.2	31.1	11.3	—	—	—
31592	With pepperoni, 6″	1	item(s)	430	—	647	25.1	68.5	3.9	32.0	11.7	—	—	—
31601	With vegetables, 6″	1	item(s)	430	—	619	23.4	70.8	4.6	28.7	10.1	—	—	—
	In-n-Out Burger													
34391	Cheeseburger with mustard and ketchup	1	serving(s)	268	—	400	22.0	41.0	3.0	18.0	9.0	—	—	0.5
34374	Cheeseburger	1	serving(s)	268	—	480	22.0	39.0	3.0	27.0	10.0	—	—	0.5
34390	Cheeseburger, lettuce leaves instead of buns	1	serving(s)	300	—	330	18.0	11.0	3.0	25.0	9.0	—	—	0
34377	Chocolate shake	1	serving(s)	425	—	690	9.0	83.0	0	36.0	24.0	—	—	1.0
34375	Double-Double cheeseburger	1	serving(s)	330	—	670	37.0	39.0	3.0	41.0	18.0	—	—	1.0
34393	Double-Double cheeseburger with mustard and ketchup	1	serving(s)	330	—	590	37.0	41.0	3.0	32.0	17.0	—	—	1.0
34392	Double-Double cheese-burger, lettuce leaves instead of buns	1	serving(s)	362	—	520	33.0	11.0	3.0	39.0	17.0	—	—	1.0
34376	French fries	1	serving(s)	125	—	400	7.0	54.0	2.0	18.0	5.0	—	—	0
34373	Hamburger	1	item(s)	243	—	390	16.0	39.0	3.0	19.0	5.0	—	—	0
34389	Hamburger with mustard and ketchup	1	serving(s)	243	—	310	16.0	41.0	3.0	10.0	4.0	—	—	0
34388	Hamburger, lettuce leaves instead of buns	1	serving(s)	275	—	240	13.0	11.0	3.0	17.0	4.0	—	—	0
34379	Strawberry shake	1	serving(s)	425	—	690	9.0	91.0	0	33.0	22.0	—	—	0.5
34378	Vanilla shake	1	serving(s)	425	—	680	9.0	78.0	0	37.0	25.0	—	—	1.0
	Jack in the Box													
30392	Bacon ultimate cheeseburger	1	item(s)	338	—	1090	46.0	53.0	2.0	77.0	30.0	—	—	3.0
1740	Breakfast Jack	1	item(s)	125	—	290	17.0	29.0	1.0	12.0	4.5	—	—	0
14074	Cheeseburger	1	item(s)	131	—	350	18.0	31.0	1.0	17.0	8.0	—	—	1.0
14106	Chicken breast strips, 4 pieces	1	serving(s)	201	—	500	35.0	36.0	3.0	25.0	6.0	—	—	6.0
37241	Chicken club salad, plain, without salad dressing	1	serving(s)	431	—	300	27.0	13.0	4.0	15.0	6.0	—	—	0
14064	Chicken sandwich	1	item(s)	145	—	400	15.0	38.0	2.0	21.0	4.5	—	—	2.5
14111	Chocolate ice cream shake, small	1	serving(s)	414	—	880	14.0	107.0	1.0	45.0	31.0	—	—	2.0
14073	Hamburger	1	item(s)	118	—	310	16.0	30.0	1.0	14.0	6.0	—	—	1.0
14090	Hash browns	1	serving(s)	57	—	150	1.0	13.0	2.0	10.0	2.5	—	—	3.0
14072	Jack's Spicy Chicken sandwich	1	item(s)	270	—	620	25.0	61.0	4.0	31.0	6.0	—	—	3.0
1468	Jumbo Jack hamburger	1	item(s)	261	—	600	21.0	51.0	3.0	35.0	12.0	—	—	1.5
1469	Jumbo Jack hamburger with cheese	1	item(s)	286	—	690	25.0	54.0	3.0	42.0	16.0	—	—	1.5
14099	Natural cut french fries, large	1	serving(s)	196	—	530	8.0	69.0	5.0	25.0	6.0	—	—	7.0
14098	Natural cut french fries, medium	1	serving(s)	133	—	360	5.0	47.0	4.0	17.0	4.0	—	—	5.0
1470	Onion rings	1	serving(s)	119	—	500	6.0	51.0	3.0	30.0	6.0	—	—	10
33141	Sausage, egg, and cheese biscuit	1	item(s)	234	—	740	27.0	35.0	2.0	55.0	17.0	—	—	6.0
14095	Seasoned curly fries, medium	1	serving(s)	125	—	400	6.0	45.0	5.0	23.0	5.0	—	—	7.0
14077	Sourdough Jack	1	item(s)	245	—	710	27.0	36.0	3.0	51.0	18.0	—	—	3.0
37249	Southwest chicken salad, plain, without salad dressing	1	serving(s)	488	—	300	24.0	29.0	7.0	11.0	5.0	—	—	0
14112	Strawberry ice cream shake, small	1	serving(s)	417	—	880	13.0	105.0	0	44.0	31.0	—	—	2.0

KEY: H-4 = Breads/Baked Goods H-10 = Cereal/Rice/Pasta H-14 = Fruit H-18 = Vegetables/Legumes H-28 = Nuts/Seeds H-30 = Vegetarian H-32 = Dairy H-40 = Eggs H-40 = Seafood H-42 = Meats H-46 = Poultry
6 = Processed Meats H-48 = Beverages H-52 = Fats/Oils H-54 = Sweets H-56 = Spices/Condiments/Sauces H-60 = Mixed Foods/Soups/Sandwiches H-64 = Fast Food H-84 = Convenience H-86 = Baby Foods

Chol (mg)	Calc (mg)	Iron (mg)	Magn (mg)	Pota (mg)	Sodi (mg)	Zinc (mg)	Vit A (µg)	Thia (mg)	Vit E (mg α)	Ribo (mg)	Niac (mg)	Vit B_6 (mg)	Fola (µg)	Vit C (mg)	Vit B_{12} (µg)	Sele (µg)
26	170	3.06	—	—	974.5	—	—	—	—	—	—	—	—	1.2	—	—
31	160	2.52	—	—	934.5	—	—	—	—	—	—	—	—	1.2	—	—
40	220	3.42	—	—	1260.0	—	—	—	—	—	—	—	—	1.2	—	—
19	150	1.98	—	—	717.0	—	—	—	—	—	—	—	—	1.2	—	—
26	210	2.88	—	—	1011.0	—	—	—	—	—	—	—	—	1.8	—	—
31	160	2.34	—	—	910.5	—	—	—	—	—	—	—	—	0	—	—
40	220	3.24	—	—	1230.0	—	—	—	—	—	—	—	—	0.6	—	—
27	150	2.16	—	—	840.5	—	86.5	—	—	—	—	—	—	0	—	—
37	220	3.06	—	—	1166.0	—	115.4	—	—	—	—	—	—	0.6	—	—
15	150	2.16	—	—	659.5	—	—	—	—	—	—	—	—	0.6	—	—
21	220	3.06	—	—	924.0	—	—	—	—	—	—	—	—	1.2	—	—
43	298	4.84	—	—	1497.8	—	—	—	—	—	—	—	—	1.5	—	—
45	302	4.89	—	—	1478.1	—	—	—	—	—	—	—	—	0.6	—	—
47	299	4.81	—	—	1523.7	—	167.9	—	—	—	—	—	—	0.6	—	—
36	307	5.10	—	—	1472.5	—	—	—	—	—	—	—	—	4.7	—	—
60	200	3.60	—	—	1080.0	—	—	—	—	—	—	—	—	12.0	—	—
60	200	3.60	—	—	1000.0	—	—	—	—	—	—	—	—	9.0	—	—
60	200	2.70	—	—	720.0	—	—	—	—	—	—	—	—	12.0	—	—
95	300	0.72	—	—	350.0	—	—	—	—	—	—	—	—	0	—	—
120	350	5.40	—	—	1440.0	—	—	—	—	—	—	—	—	9.0	—	—
115	350	5.40	—	—	1520.0	—	—	—	—	—	—	—	—	12.0	—	—
120	350	4.50	—	—	1160.0	—	—	—	—	—	—	—	—	12.0	—	—
0	20	1.80	—	—	245.0	—	0	—	—	—	—	—	—	0	—	—
40	40	3.60	—	—	650.0	—	—	—	—	—	—	—	—	9.0	—	—
35	40	3.60	—	—	730.0	—	—	—	—	—	—	—	—	12.0	—	—
40	40	2.70	—	—	370.0	—	—	—	—	—	—	—	—	12.0	—	—
85	300	0.00	—	—	280.0	—	—	—	—	—	—	—	—	0	—	—
90	300	0.00	—	—	390.0	—	—	—	—	—	—	—	—	0	—	—
140	308	7.38	—	540.0	2040.0	—	—	—	—	—	—	—	—	0.6	—	—
220	145	3.48	—	210.0	760.0	—	—	—	—	—	—	—	—	3.5	—	—
50	151	3.61	—	270.0	790.0	—	40.2	—	—	—	—	—	—	0	—	—
80	18	1.60	—	530.0	1260.0	—	—	—	—	—	—	—	—	1.1	—	—
65	280	3.35	—	560.0	880.0	—	—	—	—	—	—	—	—	50.4	—	—
35	100	2.70	—	240.0	730.0	—	—	—	—	—	—	—	—	4.8	—	—
135	460	0.47	—	840.0	330.0	—	—	—	—	—	—	—	—	0	—	—
40	100	3.60	—	250.0	600.0	—	0	—	—	—	—	—	—	0	—	—
0	10	0.18	—	190.0	230.0	—	0	—	—	—	—	—	—	0	—	—
50	150	1.80	—	450.0	1100.0	—	—	—	—	—	—	—	—	9.0	—	—
45	164	4.92	—	380.0	940.0	—	—	—	—	—	—	—	—	9.8	—	—
70	234	4.20	—	410.0	1310.0	—	—	—	—	—	—	—	—	8.4	—	—
0	20	1.42	—	1240.0	870.0	—	0	—	—	—	—	—	—	8.9	—	—
0	19	1.01	—	840.0	590.0	—	0	—	—	—	—	—	—	5.6	—	—
0	40	2.70	—	140.0	420.0	—	40.0	—	—	—	—	—	—	18.0	—	—
280	88	2.36	—	310.0	1430.0	—	—	—	—	—	—	—	—	0	—	—
0	40	1.80	—	580.0	890.0	—	—	—	—	—	—	—	—	0	—	—
75	200	4.50	—	430.0	1230.0	—	—	—	—	—	—	—	—	9.0	—	—
55	274	4.10	—	670.0	860.0	—	—	—	—	—	—	—	—	43.8	—	—
135	466	0.00	—	750.0	290.0	—	—	—	—	—	—	—	—	0	—	—

DA+ Code	Food Description	Quantity	Measure	Wt (g)	H_2O (g)	Ener (kcal)	Prot (g)	Carb (g)	Fiber (g)	Fat (g)	Fat Breakdown (g) Sat	Mono	Poly	*Trans*
	Fast Food—*continued*													
14078	Ultimate cheeseburger	1	item(s)	323	—	1010	40.0	53.0	2.0	71.0	28.0	—	—	3.0
14110	Vanilla ice cream shake, small	1	serving(s)	379	—	790	13.0	83.0	0	44.0	31.0	—	—	2.0
	Jamba Juice													
31645	Aloha Pineapple smoothie	24	fluid ounce(s)	730	—	500	8.0	117.0	4.0	1.5	1.0	—	—	—
31646	Banana Berry smoothie	24	fluid ounce(s)	719	—	480	5.0	112.0	4.0	1.0	0	—	—	—
31656	Berry Lime Sublime smoothie	24	fluid ounce(s)	728	—	460	3.0	106.0	5.0	2.0	1.0	—	—	—
31647	Carribean Passion smoothie	24	fluid ounce(s)	730	—	440	4.0	102.0	4.0	2.0	1.0	—	—	—
38422	Carrot juice	16	fluid ounce(s)	472	—	100	3.0	23.0	0	0.5	0	—	—	—
31648	Chocolate Moo'd smoothie	24	fluid ounce(s)	634	—	720	17.0	148.0	3.0	8.0	5.0	—	—	—
31649	Citrus Squeeze smoothie	24	fluid ounce(s)	727	—	470	5.0	110.0	4.0	2.0	1.0	—	—	—
31651	Coldbuster smoothie	24	fluid ounce(s)	724	—	430	5.0	100.0	5.0	2.5	1.0	—	—	—
31652	Cranberry Craze smoothie	24	fluid ounce(s)	793	—	460	6.0	104.0	4.0	0.5	0	—	—	—
31654	Jamba Powerboost smoothie	24	fluid ounce(s)	738	—	440	6.0	105.0	6.0	1.0	0	—	—	—
38423	Lemonade	16	fluid ounce(s)	483	—	300	1.0	75.0	0	0	0	0	0	0
31657	Mango-a-go-go smoothie	24	fluid ounce(s)	690	—	440	3.0	104.0	4.0	1.5	0.5	—	—	—
38424	Orange juice, freshly squeezed	16	fluid ounce(s)	496	—	220	3.0	52.0	0.5	1.0	0	—	—	—
38426	Orange/carrot juice	16	fluid ounce(s)	484	—	160	3.0	37.0	0	1.0	0	—	—	—
31660	Orange-a-peel smoothie	24	fluid ounce(s)	726	—	440	8.0	102.0	5.0	1.5	0	—	—	—
31662	Peach Pleasure smoothie	24	fluid ounce(s)	720	—	460	4.0	108.0	4.0	2.0	1.0	—	—	—
31665	Protein Berry Pizzaz smoothie	24	fluid ounce(s)	710	—	440	20.0	92.0	5.0	1.5	0	—	—	—
31668	Razzmatazz smoothie	24	fluid ounce(s)	730	—	480	3.0	112.0	4.0	2.0	1.0	—	—	—
31669	Strawberries Wild smoothie	24	fluid ounce(s)	725	—	450	6.0	105.0	4.0	0.5	0	—	—	—
38421	Strawberry Tsunami smoothie	24	fluid ounce(s)	740	—	530	4.0	128.0	4.0	2.0	1.0	—	—	—
38427	Vibrant C juice	16	fluid ounce(s)	448	—	210	2.0	50.0	1.0	0	0	0	0	0
38428	Wheatgrass juice, freshly squeezed	1	ounce(s)	28	—	5	0.5	1.0	0	0	0	0	0	0
	Kentucky Fried Chicken (KFC)													
31850	BBQ baked beans	1	serving(s)	136	—	220	8.0	45.0	7.0	1.0	0	—	—	0
31853	Biscuit	1	item(s)	57	—	220	4.0	24.0	1.0	11.0	2.5	—	—	3.5
51223	Boneless Fiery Buffalo Wings	6	item(s)	211	—	530	30.0	44.0	3.0	26.0	5.0	—	—	2.5
39386	Boneless Honey BBQ Wings	6	item(s)	213	—	570	30.0	54.0	5.0	26.0	5.0	—	—	2.5
51224	Boneless Sweet & Spicy Wings	6	item(s)	203	—	550	30.0	50.0	3.0	26.0	5.0	—	—	2.5
31851	Cole slaw	1	serving(s)	130	—	180	1.0	22.0	3.0	10.0	1.5	—	—	0
31842	Colonel's Crispy Strips	3	item(s)	151	—	370	28.0	17.0	1.0	20.0	4.0	—	—	2.5
31849	Corn on the cob	1	item(s)	162	—	150	5.0	26.0	7.0	3.0	1.0	—	—	0
51221	Double Crunch sandwich	1	item(s)	213	—	520	27.0	39.0	3.0	29.0	5.0	—	—	1.5
3761	Extra Crispy chicken, breast	1	item(s)	162	—	370	33.0	10.0	2.0	22.0	5.0	—	—	1.5
3762	Extra Crispy chicken, drumstick	1	item(s)	60	—	150	12.0	4.0	0	10.0	2.5	—	—	1.0
3763	Extra Crispy chicken, thigh	1	item(s)	114	—	290	17.0	16.0	1.0	18.0	4.0	—	—	1.5
3764	Extra Crispy chicken, whole wing	1	item(s)	52	—	150	11.0	11.0	1.0	7.0	1.5	—	—	0
51218	Famous Bowls mashed potatoes with gravy	1	serving(s)	531	—	720	26.0	79.0	6.0	34.0	9.0	—	—	3.5
51219	Famous Bowls rice with gravy	1	serving(s)	384	—	610	25.0	67.0	5.0	27.0	8.0	—	—	2.5
31841	Honey BBQ chicken sandwich	1	item(s)	147	—	290	23.0	40.0	2.0	4.0	1.0	—	—	0
31833	Honey BBQ wing pieces	6	item(s)	157	—	460	27.0	26.0	3.0	27.0	6.0	—	—	2.0
10859	Hot wings pieces	6	piece(s)	134	—	450	26.0	19.0	2.0	30.0	7.0	—	—	2.0
42382	KFC Snacker sandwich	1	serving(s)	119	—	320	14.0	29.0	2.0	17.0	3.0	—	—	1.0
31848	Macaroni and cheese	1	serving(s)	136	—	180	8.0	18.0	0	8.0	3.5	—	—	1.0
31847	Mashed potatoes with gravy	1	serving(s)	151	—	140	2.0	20.0	1.0	5.0	1.0	—	—	0.5
10825	Original Recipe chicken, breast	1	item(s)	161	—	340	38.0	9.0	2.0	17.0	4.0	—	—	1.0
10826	Original Recipe chicken, drumstick	1	item(s)	59	—	140	13.0	3.0	0	8.0	2.0	—	—	0.5
10827	Original Recipe chicken, thigh	1	item(s)	126	—	350	19.0	7.0	1.0	27.0	7.0	—	—	1.0
10828	Original Recipe chicken, whole wing	1	item(s)	47	—	140	10.0	4.0	0	9.0	2.0	—	—	0.5
51222	Oven roasted Twister chicken wrap	1	item(s)	269	—	520	30.0	46.0	4.0	23.0	3.5	—	—	0
31844	Popcorn chicken, small or individual	1	item(s)	114	—	370	19.0	21.0	2.0	24.0	4.5	—	—	2.5
31852	Potato salad	1	serving(s)	128	—	180	2.0	22.0	2.0	9.0	1.5	—	—	0
10845	Potato wedges, small	1	serving(s)	102	—	250	4.0	32.0	3.0	12.0	2.0	—	—	1.5
31839	Tender Roast chicken sandwich with sauce	1	item(s)	236	—	430	37.0	29.0	2.0	18.0	3.5	—	—	0

EKEY: H-4 = Breads/Baked Goods H-10 = Cereal/Rice/Pasta H-14 = Fruit H-18 = Vegetables/Legumes H-28 = Nuts/Seeds H-30 = Vegetarian H-32 = Dairy H-40 = Eggs H-40 = Seafood H-42 = Meats H-46 = Poultry
6 = Processed Meats H-48 = Beverages H-52 = Fats/Oils H-54 = Sweets H-56 = Spices/Condiments/Sauces H-60 = Mixed Foods/Soups/Sandwiches H-64 = Fast Food H-84 = Convenience H-86 = Baby Foods

Chol (mg)	Calc (mg)	Iron (mg)	Magn (mg)	Pota (mg)	Sodi (mg)	Zinc (mg)	Vit A (µg)	Thia (mg)	Vit E (mg α)	Ribo (mg)	Niac (mg)	Vit B_6 (mg)	Fola (µg)	Vit C (mg)	Vit B_{12} (µg)	Sele (µg)
125	308	7.39	—	480.0	1580.0	—	—	—	—	—	—	—	—	0.6	—	—
135	532	0.00	—	750.0	280.0	—	—	—	—	—	—	—	—	0	—	—
5	200	1.80	60.0	1000.0	30.0	0.30	—	0.37	—	0.34	2.00	0.60	60.0	102.0	0	1.4
0	200	1.44	40.0	1010.0	115.0	0.60	—	0.09	—	0.25	0.80	0.70	24.0	15.0	0.2	1.4
5	200	1.80	16.0	510.0	35.0	0.30	—	0.06	—	0.25	6.00	0.70	140.0	54.0	0	1.4
5	100	1.80	24.0	810.0	60.0	0.30	—	0.09	—	0.25	5.00	0.50	100.0	78.0	0	1.4
0	150	2.70	80.0	1030.0	250.0	0.90	—	0.52	—	0.25	5.00	0.70	80.0	18.0	0	5.6
30	500	1.08	60.0	810.0	380.0	1.50	—	0.22	—	0.76	0.40	0.16	16.0	6.0	1.5	4.2
5	100	1.80	80.0	1170.0	35.0	0.30	—	0.37	—	0.34	1.90	0.60	100.0	180.0	0	1.4
5	100	1.08	60.0	1260.0	35.0	16.50	—	0.37	—	0.34	3.00	0.40	121.5	1302.0	0	1.4
0	250	1.44	16.0	500.0	50.0	0.30	—	0.03	—	0.25	5.00	0.60	120.0	54.0	0	1.4
0	1200	1.80	480.0	1070.0	45.0	16.50	—	5.55	—	6.12	68.00	7.40	640.0	288.0	10.8	77.0
0	20	0.00	8.0	200.0	10.0	0.00	—	0.03	—	0.17	14.00	1.80	320.0	36.0	0	0
5	100	1.08	24.0	780.0	50.0	0.30	—	0.15	—	0.25	5.00	0.70	120.0	72.0	0	1.4
0	60	1.08	60.0	990.0	0	0.30	—	0.45	—	0.13	2.00	0.20	160.0	246.0	0	0
0	100	1.80	60.0	1010.0	125.0	0.60	—	0.45	—	0.25	3.00	0.50	120.0	132.0	0	2.8
0	250	1.80	80.0	1380.0	160.0	0.90	—	0.45	—	0.42	2.00	0.50	140.0	240.0	0.6	1.4
5	100	0.72	32.0	740.0	60.0	0.30	—	0.06	—	0.25	4.00	0.60	80.0	18.0	0	1.4
0	1100	2.62	60.0	650.0	240.0	0.58	—	0.08	—	0.17	1.20	0.70	58.3	60.0	0	5.6
5	150	1.80	32.0	810.0	70.0	0.30	—	0.09	—	0.34	6.00	1.00	160.0	60.0	0	1.4
5	250	1.80	40.0	1050.0	180.0	0.90	—	0.12	—	0.34	0.80	0.40	40.0	60.0	0.6	1.4
5	100	1.08	24.0	480.0	10.0	0.30	—	0.06	—	0.34	14.00	1.80	320.0	90.0	0	1.4
0	20	1.08	40.0	720.0	0	0.30	—	0.30	—	0.10	1.60	0.40	80.0	678.0	0	0
0	0	1.80	8.0	80.0	0	0.00	0	0.03	—	0.03	0.40	0.04	16.0	3.6	0	2.8
0	100	2.70	—	—	730.0	—	—	—	—	—	—	—	—	1.2	—	—
0	40	1.80	—	—	640.0	—	—	—	—	—	—	—	—	0	—	—
65	40	1.80	—	—	2670.0	—	—	—	—	—	—	—	—	1.2	—	—
65	40	1.80	—	—	2210.0	—	—	—	—	—	—	—	—	1.2	—	—
65	60	1.80	—	—	2000.0	—	—	—	—	—	—	—	—	1.2	—	—
5	40	0.72	—	—	270.0	—	—	—	—	—	—	—	—	12.0	—	—
65	40	1.44	—	—	1220.0	—	0	—	—	—	—	—	—	1.2	—	—
0	60	1.08	—	—	10.0	—	—	—	—	—	—	—	—	6.0	—	—
55	100	2.70	—	—	1220.0	—	—	—	—	—	—	—	—	6.0	—	—
85	20	2.70	—	—	1020.0	—	—	—	—	—	—	—	—	1.2	—	—
55	0	1.44	—	—	300.0	—	0	—	—	—	—	—	—	0	—	—
95	20	2.70	—	—	700.0	—	—	—	—	—	—	—	—	—	—	—
45	20	1.08	—	—	340.0	—	—	—	—	—	—	—	—	0	—	—
35	200	5.40	—	—	2330.0	—	—	—	—	—	—	—	—	6.0	—	—
35	200	4.50	—	—	2130.0	—	—	—	—	—	—	—	—	6.0	—	—
60	80	2.70	—	—	710.0	—	—	—	—	—	—	—	—	2.4	—	—
140	40	1.80	—	—	970.0	—	—	—	—	—	—	—	—	21.0	—	—
115	40	1.44	—	—	990.0	—	—	—	—	—	—	—	—	1.2	—	—
25	60	2.70	—	—	690.0	—	—	—	—	—	—	—	—	2.4	—	—
15	150	0.72	—	—	800.0	—	—	—	—	—	—	—	—	1.2	—	—
0	40	1.44	—	—	560.0	—	—	—	—	—	—	—	—	1.2	—	—
135	20	2.70	—	—	960.0	—	—	—	—	—	—	—	—	6.0	—	—
70	20	1.08	—	—	340.0	—	—	—	—	—	—	—	—	0	—	—
110	20	2.70	—	—	870.0	—	—	—	—	—	—	—	—	1.2	—	—
50	20	1.44	—	—	350.0	—	0	—	—	—	—	—	—	1.2	—	—
60	40	6.30	—	—	1380.0	—	—	—	—	—	—	—	—	15.0	—	—
25	40	1.80	—	—	1110.0	—	0	—	—	—	—	—	—	0	—	—
5	0	0.36	—	—	470.0	—	—	—	—	—	—	—	—	6.0	—	—
0	20	1.08	—	—	700.0	—	0	—	—	—	—	—	—	0	—	—
80	80	2.70	—	—	1180.0	—	—	—	—	—	—	—	—	9.0	—	—

TABLE H-1 Table of Food Composition (*continued*) (Computer code is for Cengage Diet Analysis program) (For purposes of calculations, use "0" for t, <1, <0.1, 0.01)

DA+ Code	Food Description	Quantity	Measure	Wt (g)	H_2O (g)	Ener (kcal)	Prot (g)	Carb (g)	Fiber (g)	Fat (g)	Fat Breakdown (g)			
											Sat	Mono	Poly	*Trans*
Fast Food—*continued*														
	Long John Silver													
39392	Baked cod	1	serving(s)	101	—	120	22.0	1.0	0	4.5	1.0	—	—	0
3777	Batter dipped fish sandwich	1	item(s)	177	—	470	18.0	48.0	3.0	23.0	5.0	—	—	4.5
37568	Battered fish	1	item(s)	92	—	260	12.0	17.0	0.5	16.0	4.0	—	—	4.5
37569	Breaded clams	1	serving(s)	85	—	240	8.0	22.0	1.0	13.0	2.0	—	—	2.5
37566	Chicken plank	1	item(s)	52	—	140	8.0	9.0	0.5	8.0	2.0	—	—	2.5
39404	Clam chowder	1	item(s)	227	—	220	9.0	23.0	0	10.0	4.0	—	—	1.0
39398	Cocktail sauce	1	ounce(s)	28	—	25	0	6.0	0	0	0	0	0	0
3770	Coleslaw	1	serving(s)	113	—	200	1.0	15.0	3.0	15.0	2.5	1.8	4.1	0
39400	French fries, large	1	item(s)	142	—	390	4.0	56.0	5.0	17.0	4.0	—	—	5.0
3774	Fries, regular	1	serving(s)	85	—	230	3.0	34.0	3.0	10.0	2.5	—	—	3.0
3779	Hushpuppy	1	piece(s)	23	—	60	1.0	9.0	1.0	2.5	0.5	—	—	1.0
3781	Shrimp, batter-dipped, 1 piece	1	piece(s)	14	—	45	2.0	3.0	0	3.0	1.0	—	—	1.0
39399	Tartar sauce	1	ounce(s)	28	—	100	0	4.0	0	9.0	1.5	—	—	—
39395	Ultimate Fish sandwich	1	item(s)	199	—	530	21.0	49.0	3.0	28.0	8.0	—	—	5.0
	McDonald's													
50828	Asian salad with grilled chicken	1	item(s)	362	—	290	31.0	23.0	6.0	10.0	1.0	—	—	0
2247	Barbecue sauce	1	item(s)	28	—	45	0	11.0	0	0	0	0	0	0
737	Big Mac hamburger	1	item(s)	219	—	560	25.0	47.0	3.0	30.0	10.0	—	—	1.5
29777	Caesar salad dressing	1	package(s)	44	—	150	1.0	5.0	0	13.0	2.5	—	—	—
38391	Caesar salad with grilled chicken, no dressing	1	serving(s)	278	230.6	181	26.4	10.5	3.1	6.0	2.9	1.7	0.8	0.2
38393	Caesar salad without chicken, no dressing	1	serving(s)	190	170.4	84	6.0	8.1	3.0	3.9	2.2	0.9	0.3	0.1
738	Cheeseburger	1	item(s)	119	—	310	15.0	35.0	1.0	12.0	6.0	—	—	1.0
29775	Chicken McGrill sandwich	1	item(s)	213	—	400	27.0	38.0	3.0	16.0	3.0	—	—	0
1873	Chicken McNuggets, 6 piece	1	serving(s)	96	—	250	15.0	15.0	0	15.0	3.0	—	—	1.5
3792	Chicken McNuggets, 4 piece	1	serving(s)	64	—	170	10.0	10.0	0	10.0	2.0	—	—	1.0
29774	Crispy chicken sandwich	1	item(s)	232	121.8	500	27.0	63.0	3.0	16.0	3.0	5.7	7.4	1.5
743	Egg McMuffin	1	item(s)	139	76.8	300	17.0	30.0	2.0	12.0	4.5	3.8	2.5	0
742	Filet-O-Fish sandwich	1	item(s)	141	—	400	14.0	42.0	1.0	18.0	4.0	—	—	1.0
2257	French fries, large	1	serving(s)	170	—	570	6.0	70.0	7.0	30.0	6.0	—	—	8.0
1872	French fries, small	1	serving(s)	74	—	250	2.0	30.0	3.0	13.0	2.5	—	—	3.5
33822	Fruit 'n Yogurt Parfait	1	item(s)	149	111.2	160	4.0	31.0	1.0	2.0	1.0	0.2	0.1	0
739	Hamburger	1	item(s)	105	—	260	13.0	33.0	1.0	9.0	3.5	—	—	0.5
2003	Hash browns	1	item(s)	53	—	140	1.0	15.0	2.0	8.0	1.5	—	—	2.0
2249	Honey sauce	1	item(s)	14	—	50	0	12.0	0	0	0	0	0	0
38397	Newman's Own creamy caesar salad dressing	1	item(s)	59	32.5	190	2.0	4.0	0	18.0	3.5	4.6	9.6	0
38398	Newman's Own low fat balsamic vinaigrette salad dressing	1	item(s)	44	29.1	40	0	4.0	0	3.0	0	1.0	1.2	0
38399	Newman's Own ranch salad dressing	1	item(s)	59	30.1	170	1.0	9.0	0	15.0	2.5	9.0	3.7	0
1874	Plain Hotcakes with syrup and margarine	3	item(s)	221	—	600	9.0	102.0	2.0	17.0	4.0	—	—	4.0
740	Quarter Pounder hamburger	1	item(s)	171	—	420	24.0	40.0	3.0	18.0	7.0	—	—	1.0
741	Quarter Pounder hamburger with cheese	1	item(s)	199	—	510	29.0	43.0	3.0	25.0	12.0	—	—	1.5
2005	Sausage McMuffin with egg	1	item(s)	165	82.4	450	20.0	31.0	2.0	27.0	10.0	10.9	4.6	0.5
50831	Side salad	1	item(s)	87	—	20	1.0	4.0	1.0	0	0	0	0	0
	Pizza Hut													
39009	Hot chicken wings	2	item(s)	57	—	110	11.0	1.0	0	6.0	2.0	—	—	0.3
14025	Meat Lovers hand tossed pizza	1	slice(s)	118	—	300	15.0	29.0	2.0	13.0	6.0	—	—	0.5
14026	Meat Lovers pan pizza	1	slice(s)	123	—	340	15.0	29.0	2.0	19.0	7.0	—	—	0.5
31009	Meat Lovers stuffed crust pizza	1	slice(s)	169	—	450	21.0	43.0	3.0	21.0	10.0	—	—	1.0
14024	Meat Lovers thin 'n crispy pizza	1	slice(s)	98	—	270	13.0	21.0	2.0	14.0	6.0	—	—	0.5
14031	Pepperoni Lovers hand tossed pizza	1	slice(s)	113	—	300	15.0	30.0	2.0	13.0	7.0	—	—	0.5
14032	Pepperoni Lovers pan pizza	1	slice(s)	118	—	340	15.0	29.0	2.0	19.0	7.0	—	—	0.5
31011	Pepperoni Lovers stuffed crust pizza	1	slice(s)	163	—	420	21.0	43.0	3.0	19.0	10.0	—	—	1.0
14030	Pepperoni Lovers thin 'n crispy pizza	1	slice(s)	92	—	260	13.0	21.0	2.0	14.0	7.0	—	—	0.5
10834	Personal Pan pepperoni pizza	1	slice(s)	61	—	170	7.0	18.0	0.5	8.0	3.0	—	—	1.0
10842	Personal Pan supreme pizza	1	slice(s)	77	—	190	8.0	19.0	1.0	9.0	3.5	—	—	1.0

Chol (mg)	Calc (mg)	Iron (mg)	Magn (mg)	Pota (mg)	Sodi (mg)	Zinc (mg)	Vit A (µg)	Thia (mg)	Vit E (mg α)	Ribo (mg)	Niac (mg)	Vit B_6 (mg)	Fola (µg)	Vit C (mg)	Vit B_{12} (µg)	Sele (µg)
90	20	0.72	—	—	240.0	—	—	—	—	—	—	—	—	0	—	—
45	60	2.70	—	—	1210.0	—	—	—	—	—	—	—	—	2.4	—	—
35	20	0.72	—	—	790.0	—	—	—	—	—	—	—	—	4.8	—	—
10	20	1.08	—	—	1110.0	—	0	—	—	—	—	—	—	0	—	—
20	0	0.72	—	—	480.0	—	0	—	—	—	—	—	—	2.4	—	—
25	150	0.72	—	—	810.0	—	—	—	—	—	—	—	—	0	—	—
0	0	0.00	—	—	250.0	—	—	—	—	—	—	—	—	0	—	—
20	40	0.36	—	222.7	340.0	0.70	—	0.07	—	0.08	2.34	—	—	18.0	—	—
0	0	0.00	—	—	580.0	—	0	—	—	—	—	—	—	24.0	—	—
0	0	0.00	—	370.0	350.0	0.30	0	0.09	—	0.01	1.60	—	—	15.0	—	—
0	20	0.36	—	—	200.0	—	0	—	—	—	—	—	—	0	—	—
15	0	0.00	—	—	160.0	—	0	—	—	—	—	—	—	1.2	—	—
15	0	0.00	—	—	250.0	—	0	—	—	—	—	—	—	0	—	—
60	150	2.70	—	—	1400.0	—	—	—	—	—	—	—	—	4.8	—	—
65	150	3.60	—	—	890.0	—	—	—	—	—	—	—	—	54.0	—	—
0	0	0.00	—	55.0	260.0	—	—	—	—	—	—	—	—	0	—	—
80	250	4.50	—	400.0	1010.0	—	—	—	—	—	—	—	—	1.2	—	—
10	40	0.18	—	30.0	400.0	—	—	—	—	—	—	—	—	0.6	—	—
67	178	1.77	—	708.9	767.3	—	—	0.15	—	0.19	10.62	—	127.9	29.2	0.2	—
10	163	1.15	17.1	410.4	157.7	—	—	0.08	—	0.07	0.40	—	102.6	26.8	0	0.4
40	200	2.70	—	240.0	740.0	—	60.0	—	—	—	—	—	—	1.2	—	—
70	150	2.70	—	510.0	1010.0	—	—	—	—	—	—	—	—	6.0	—	—
35	20	0.72	—	240.0	670.0	—	—	—	—	—	—	—	—	1.2	—	—
25	0	0.36	—	160.0	450.0	—	—	—	—	—	—	—	—	1.2	—	—
60	80	3.60	62.6	526.6	1380.0	1.53	41.8	0.46	2.27	0.39	12.85	—	104.4	6.0	0.4	—
230	300	2.70	26.4	218.2	860.0	1.59	—	0.36	0.82	0.51	4.31	0.20	109.8	1.2	0.9	—
40	150	1.80	—	250.0	640.0	—	36.2	—	—	—	—	—	—	0	—	—
0	20	1.80	—	—	330.0	—	0	—	—	—	—	—	—	9.0	—	—
0	20	0.72	—	—	140.0	—	0	—	—	—	—	—	—	3.6	—	—
5	150	0.67	20.9	248.8	85.0	0.53	0	0.06	—	0.17	0.35	—	19.4	9.0	0.3	—
30	150	2.70	—	210.0	530.0	—	5.0	—	—	—	—	—	—	1.2	—	—
0	0	0.36	—	210.0	290.0	—	0	—	—	—	—	—	—	1.2	—	—
0	0	0.00	—	0	0	—	0	—	—	—	—	—	—	0	—	—
20	61	0.00	3.0	16.0	500.0	0.20	—	0.01	15.43	0.02	0.01	0.64	2.4	0	0.1	0.1
0	4	0.00	1.3	8.8	730.0	0.01	—	0.00	0.00	0.00	0.00	0.00	0	2.4	0	0
0	40	0.00	1.8	70.4	530.0	0.03	0	0.01	—	0.08	0.01	0.02	0.6	0	0	0.2
20	150	2.70	—	280.0	620.0	—	—	—	—	—	—	—	—	0	—	—
70	150	4.50	—	390.0	730.0	—	10.0	—	—	—	—	—	—	1.2	—	—
95	300	4.50	—	440.0	1150.0	—	100.0	—	—	—	—	—	—	1.2	—	—
255	300	3.60	29.7	282.2	950.0	2.01	—	0.43	0.82	0.56	4.83	0.24	—	0	1.2	—
0	20	0.72	—	—	10.0	—	—	—	—	—	—	—	—	15.0	—	—
70	0	0.36	—	—	450.0	—	—	—	—	—	—	—	—	0	—	—
35	150	1.80	—	—	760.0	—	—	—	—	—	—	—	—	6.0	—	—
35	150	2.70	—	—	750.0	—	—	—	—	—	—	—	—	6.0	—	—
55	250	2.70	—	—	1250.0	—	—	—	—	—	—	—	—	9.0	—	—
35	150	1.44	—	—	740.0	—	—	—	—	—	—	—	—	6.0	—	—
40	200	1.80	—	—	710.0	—	57.7	—	—	—	—	—	—	2.4	—	—
40	200	2.70	—	—	700.0	—	57.7	—	—	—	—	—	—	2.4	—	—
55	300	2.70	—	—	1120.0	—	—	—	—	—	—	—	—	3.6	—	—
40	200	1.44	—	—	690.0	—	58.0	—	—	—	—	—	—	2.4	—	—
15	80	1.44	—	—	340.0	—	38.5	—	—	—	—	—	—	1.4	—	—
20	80	1.86	—	—	420.0	—	—	—	—	—	—	—	—	3.6	—	—

 (Computer code is for Cengage Diet Analysis program) (For purposes of calculations, use "0" for t, <1, <0.1, 0.01)

DA+ Code	Food Description	Quantity	Measure	Wt (g)	H_2O (g)	Ener (kcal)	Prot (g)	Carb (g)	Fiber (g)	Fat (g)	Fat Breakdown (g) Sat	Mono	Poly	*Trans*
Fast Food—*continued*														
39013	Personal Pan Veggie Lovers pizza	1	slice(s)	69	—	150	6.0	19.0	1.0	6.0	2.0	—	—	0.5
14028	Veggie Lovers hand tossed pizza	1	slice(s)	118	—	220	10.0	31.0	2.0	6.0	3.0	—	—	0.3
14029	Veggie Lovers pan pizza	1	slice(s)	119	—	260	10.0	30.0	2.0	12.0	4.0	—	—	0.3
31010	Veggie Lovers stuffed crust pizza	1	slice(s)	172	—	360	16.0	45.0	3.0	14.0	7.0	—	—	0.5
14027	Veggie Lovers thin 'n crispy pizza	1	slice(s)	101	—	180	8.0	23.0	2.0	7.0	3.0	—	—	0.5
39012	Wing blue cheese dipping sauce	1	item(s)	43	—	230	2.0	2.0	0	24.0	5.0	—	—	1.0
39011	Wing ranch dipping sauce	1	item(s)	43	—	210	0.5	4.0	0	22.0	3.5	—	—	0.5
	Starbucks													
38052	Cappuccino, tall	12	fluid ounce(s)	360	—	120	7.0	10.0	0	6.0	4.0	—	—	—
38053	Cappuccino, tall nonfat	12	fluid ounce(s)	360	—	80	7.0	11.0	0	0	0	0	0	0
38054	Cappuccino, tall soymilk	12	fluid ounce(s)	360	—	100	5.0	13.0	0.5	2.5	0	—	—	—
38059	Cinnamon spice mocha, tall nonfat w/o whipped cream	12	fluid ounce(s)	360	—	170	11.0	32.0	0	0.5	—	—	—	—
38057	Cinnamon spice mocha, tall w/whipped cream	12	fluid ounce(s)	360	—	320	10.0	31.0	0	17.0	11.0	—	—	—
38051	Espresso, single shot	1	fluid ounce(s)	30	—	5	0	1.0	0	0	0	0	0	0
38088	Flavored syrup, 1 pump	1	serving(s)	10	—	20	0	5.0	0	0	0	0	0	0
32562	Frappuccino bottled coffee drink, mocha	9½	fluid ounce(s)	298	—	190	6.0	39.0	3.0	3.0	2.0	—	—	—
32561	Frappuccino coffee drink, all bottled flavors	9½	fluid ounce(s)	281	—	190	7.0	35.0	0	3.5	2.5	—	—	—
38073	Frappuccino, mocha	12	fluid ounce(s)	360	—	220	5.0	44.0	0	3.0	1.5	—	—	—
38067	Frappuccino, tall caramel w/o whipped cream	12	fluid ounce(s)	360	—	210	4.0	43.0	0	2.5	1.5	—	—	—
38070	Frappuccino, tall coffee	12	fluid ounce(s)	360	—	190	4.0	38.0	0	2.5	1.5	—	—	—
39894	Frappuccino, tall coffee, light blend	12	fluid ounce(s)	360	—	110	5.0	22.0	2.0	1.0	0	—	—	—
38071	Frappuccino, tall espresso	12	fluid ounce(s)	360	—	160	4.0	33.0	0	2.0	1.5	—	—	—
39897	Frappuccino, tall mocha, light blend	12	fluid ounce(s)	360	—	140	5.0	28.0	3.0	1.5	0	—	—	—
39887	Frappuccino, tall Strawberries and Creme, w/o whipped cream	12	fluid ounce(s)	360	—	330	10.0	65.0	0	3.5	1.0	—	—	—
38063	Frappuccino, tall Tazo chai creme w/o whipped cream	12	fluid ounce(s)	360	—	280	10.0	52.0	0	3.5	1.0	—	—	—
38066	Frappuccino, tall Tazoberry	12	fluid ounce(s)	360	—	140	0.5	36.0	0.5	0	0	0	0	0
38065	Frappuccino, tall Tazoberry Crème	12	fluid ounce(s)	360	—	240	4.0	54.0	0.5	1.0	0	—	—	—
38080	Frappuccino, tall vanilla w/o whipped cream	12	fluid ounce(s)	360	—	270	10.0	51.0	0	3.5	1.0	—	—	—
39898	Frappuccino, tall white chocolate mocha, light blend	12	fluid ounce(s)	360	—	160	6.0	32.0	2.0	2.0	1.0	—	—	—
38074	Frappuccino, tall white chocolate w/o whipped cream	12	fluid ounce(s)	360	—	240	5.0	48.0	0	3.5	2.5	—	—	—
39883	Java Chip Frappuccino, tall w/o whipped cream	12	fluid ounce(s)	360	—	270	5.0	51.0	1.0	7.0	4.5	—	—	—
33111	Latte, tall w/nonfat milk	12	fluid ounce(s)	360	335.3	120	12.0	18.0	0	0	0	0	0	0
33112	Latte, tall w/whole milk	12	fluid ounce(s)	360	—	200	11.0	16.0	0	11.0	7.0	—	—	—
33109	Macchiato, tall caramel w/nonfat milk	12	fluid ounce(s)	360	—	170	11.0	30.0	0	1.0	0	—	—	—
33110	Macchiato, tall caramel w/whole milk	12	fluid ounce(s)	360	—	240	10.0	28.0	0	10.0	6.0	—	—	—
33107	Mocha coffee drink, tall nonfat, w/o whipped cream	12	fluid ounce(s)	360	—	170	11.0	33.0	1.0	1.5	0	—	—	—
38089	Mocha syrup	1	serving(s)	17	—	25	1.0	6.0	0	0.5	0	—	—	—
33108	Mocha, tall mocha w/whole milk	12	fluid ounce(s)	360	—	310	10.0	32.0	1.0	17.0	10.0	—	—	—
38042	Steamed apple cider, tall	12	fluid ounce(s)	360	—	180	0	45.0	0	0	0	0	0	0
38087	Tazo chai black tea, soymilk, tall	12	fluid ounce(s)	360	—	190	4.0	39.0	0.5	2.0	0	—	—	—
38084	Tazo chai black tea, tall	12	fluid ounce(s)	360	—	210	6.0	36.0	0	5.0	3.5	—	—	—
38083	Tazo chai black tea, tall nonfat	12	fluid ounce(s)	360	—	170	6.0	37.0	0	0	0	0	0	0
38076	Tazo iced tea, tall	12	fluid ounce(s)	360	—	60	0	16.0	0	0	0	0	0	0
38077	Tazo tea, grande lemonade	16	fluid ounce(s)	480	—	120	0	31.0	0	0	0	0	0	0

E KEY: H-4 = Breads/Baked Goods H-10 = Cereal/Rice/Pasta H-14 = Fruit H-18 = Vegetables/Legumes H-28 = Nuts/Seeds H-30 = Vegetarian H-32 = Dairy H-40 = Eggs H-40 = Seafood H-42 = Meats H-46 = Poultry
6 = Processed Meats H-48 = Beverages H-52 = Fats/Oils H-54 = Sweets H-56 = Spices/Condiments/Sauces H-60 = Mixed Foods/Soups/Sandwiches H-64 = Fast Food H-84 = Convenience H-86 = Baby Foods

Chol (mg)	Calc (mg)	Iron (mg)	Magn (mg)	Pota (mg)	Sodi (mg)	Zinc (mg)	Vit A (µg)	Thia (mg)	Vit E (mg α)	Ribo (mg)	Niac (mg)	Vit B_6 (mg)	Fola (µg)	Vit C (mg)	Vit B_{12} (µg)	Sele (µg)
10	80	1.80	—	—	280.0	—	—	—	—	—	—	—	—	3.6	—	—
15	150	1.80	—	—	490.0	—	—	—	—	—	—	—	—	9.0	—	—
15	150	2.70	—	—	470.0	—	—	—	—	—	—	—	—	9.0	—	—
35	250	2.70	—	—	980.0	—	—	—	—	—	—	—	—	9.0	—	—
15	150	1.44	—	—	480.0	—	—	—	—	—	—	—	—	9.0	—	—
25	20	0.00	—	—	550.0	—	0	—	—	—	—	—	—	0	—	—
10	0	0.00	—	—	340.0	—	0	—	—	—	—	—	—	0	—	—
25	250	0.00	—	—	95.0	—	—	—	—	—	—	—	—	1.2	0	—
3	200	0.00	—	—	100.0	—	—	—	—	—	—	—	—	0	0	—
0	250	0.72	—	—	75.0	—	—	—	—	—	—	—	—	0	0	—
5	300	0.72	—	—	150.0	—	—	—	—	—	—	—	—	0	0	—
70	350	1.08	—	—	140.0	—	—	—	—	—	—	—	—	2.4	0	—
0	0	0.00	—	—	0	—	0	—	—	—	—	—	—	0	0	—
0	0	0.00	—	—	0	—	0	—	—	—	—	—	—	0	0	—
12	219	1.08	—	530.0	110.0	—	—	—	—	—	—	—	—	0	—	—
15	250	0.36	—	510.0	105.0	—	—	—	—	—	—	—	—	0	—	—
10	150	0.72	—	—	180.0	—	—	—	—	—	—	—	—	0	0	—
10	150	0.00	—	—	180.0	—	—	—	—	—	—	—	—	0	0	—
10	150	0.00	—	—	180.0	—	—	—	—	—	—	—	—	0	0	—
0	150	0.00	—	—	220.0	—	—	—	—	—	—	—	—	0	—	—
10	100	0.00	—	—	160.0	—	—	—	—	—	—	—	—	0	0	—
0	150	0.72	—	—	220.0	—	—	—	—	—	—	—	—	0	—	—
3	350	0.00	—	—	270.0	—	—	—	—	—	—	—	—	21.0	—	—
3	350	0.00	—	—	270.0	—	—	—	—	—	—	—	—	3.6	0	—
0	0	0.00	—	—	30.0	—	0	—	—	—	—	—	—	0	0	—
0	150	0.00	—	—	125.0	—	0	—	—	—	—	—	—	1.2	0	—
3	350	0.00	—	—	370.0	—	—	—	—	—	—	—	—	3.6	0	—
3	150	0.00	—	—	250.0	—	—	—	—	—	—	—	—	0	—	—
10	150	0.00	—	—	210.0	—	—	—	—	—	—	—	—	0	0	—
10	150	1.44	—	—	220.0	—	—	—	—	—	—	—	—	0	—	—
5	350	0.00	39.8	—	170.0	1.35	—	0.12	—	0.47	0.36	0.13	17.5	0	1.3	—
45	400	0.00	46.6	—	160.0	1.28	—	0.12	—	0.54	0.34	0.14	16.8	2.4	1.2	—
5	300	0.00	—	—	160.0	—	—	—	—	—	—	—	—	1.2	—	—
30	300	0.00	—	—	135.0	—	—	—	—	—	—	—	—	2.4	—	—
5	300	2.70	—	—	135.0	—	—	—	—	—	—	—	—	0	—	—
0	0	0.72	—	—	0	—	0	—	—	—	—	—	—	0	0	—
55	300	2.70	—	—	115.0	—	—	—	—	—	—	—	—	0	—	—
0	0	1.08	—	—	15.0	—	0	—	—	—	—	—	—	0	0	—
0	200	0.72	—	—	70.0	—	—	—	—	—	—	—	—	0	0	—
20	200	0.36	—	—	85.0	—	—	—	—	—	—	—	—	1.2	0	—
5	200	0.36	—	—	95.0	—	—	—	—	—	—	—	—	0	0	—
0	0	0.00	—	—	0	—	0	—	—	—	—	—	—	0	0	—
0	0	0.00	—	—	15.0	—	0	—	—	—	—	—	—	4.8	0	—

DA+ Code	Food Description	Quantity	Measure	Wt (g)	H_2O (g)	Ener (kcal)	Prot (g)	Carb (g)	Fiber (g)	Fat (g)	Fat Breakdown (g) Sat	Mono	Poly	*Trans*
Fast Food—*continued*														
38045	Vanilla crème steamed nonfat milk, tall w/whipped cream	12	fluid ounce(s)	360	—	260	11.0	33.0	0	8.0	5.0	—	—	—
38046	Vanilla crème steamed soymilk, tall w/whipped cream	12	fluid ounce(s)	360	—	300	8.0	37.0	1.0	12.0	6.0	—	—	—
38044	Vanilla crème steamed whole milk, tall w/whipped cream	12	fluid ounce(s)	360	—	330	10.0	31.0	0	18.0	11.0	—	—	—
38090	Whipped cream	1	serving(s)	27	—	100	0	2.0	0	9.0	6.0	—	—	—
38062	White chocolate mocha, tall nonfat w/o whipped cream	12	fluid ounce(s)	360	—	260	12.0	45.0	0	4.0	3.0	—	—	—
38061	White chocolate mocha, tall w/whipped cream	12	fluid ounce(s)	360	—	410	11.0	44.0	0	20.0	13.0	—	—	—
38048	White hot chocolate, tall nonfat w/o whipped cream	12	fluid ounce(s)	360	—	300	15.0	51.0	0	4.5	3.5	—	—	—
38050	White hot chocolate, tall soymilk w/whipped cream	12	fluid ounce(s)	360	—	420	11.0	56.0	1.0	16.0	9.0	—	—	—
38047	White hot chocolate, tall w/whipped cream	12	fluid ounce(s)	360	—	460	13.0	50.0	0	22.0	15.0	—	—	—
	Subway													
15842	Cheese steak sandwich, 6″, wheat bread	1	item(s)	250	—	360	24.0	47.0	5.0	10.0	4.5	—	—	0
40478	Chicken and bacon ranch sandwich, 6″, white or wheat bread	1	serving(s)	297	—	540	36.0	47.0	5.0	25.0	10.0	—	—	0.5
38622	Chicken and bacon ranch wrap with cheese	1	item(s)	257	—	440	41.0	18.0	9.0	27.0	10.0	—	—	0.5
32045	Chocolate chip cookie	1	item(s)	45	—	210	2.0	30.0	1.0	10.0	6.0	—	—	0
32048	Chocolate chip M&M cookie	1	item(s)	45	—	210	2.0	32.0	0.5	10.0	5.0	—	—	0
32049	Chocolate chunk cookie	1	item(s)	45	—	220	2.0	30.0	0.5	10.0	5.0	—	—	0
4024	Classic Italian B.M.T. sandwich, 6″, white bread	1	item(s)	236	—	440	22.0	45.0	2.0	21.0	8.5	—	—	0
15838	Classic tuna sandwich, 6″, wheat bread	1	item(s)	250	—	530	22.0	45.0	4.0	31.0	7.0	—	—	0.5
15837	Classic tuna sandwich, 6″, white bread	1	item(s)	243	—	520	21.0	43.0	2.0	31.0	7.5	—	—	0.5
16397	Club salad, no dressing and croutons	1	item(s)	412	—	160	18.0	15.0	4.0	4.0	1.5	—	—	0
3422	Club sandwich, 6″, white bread	1	item(s)	250	—	310	23.0	45.0	2.0	6.0	2.5	—	—	0
4030	Cold cut combo sandwich, 6″, white bread	1	item(s)	242	—	400	20.0	45.0	2.0	17.0	7.5	—	—	0.5
34030	Ham and egg breakfast sandwich	1	item(s)	142	—	310	16.0	35.0	3.0	13.0	3.5	—	—	0
3885	Ham sandwich, 6″, white bread	1	item(s)	238	—	310	17.0	52.0	2.0	5.0	2.0	—	—	0
3888	Meatball marinara sandwich, 6″, wheat bread	1	item(s)	377	—	560	24.0	63.0	7.0	24.0	11.0	—	—	1.0
4651	Meatball sandwich, 6″, white bread	1	item(s)	370	—	550	23.0	61.0	5.0	24.0	11.5	—	—	1.0
15839	Melt sandwich, 6″, white bread	1	item(s)	260	—	410	25.0	47.0	4.0	15.0	5.0	—	—	—
32046	Oatmeal raisin cookie	1	item(s)	45	—	200	3.0	30.0	1.0	8.0	4.0	—	—	0
16379	Oven-roasted chicken breast sandwich, 6″, wheat bread	1	item(s)	238	—	330	24.0	48.0	5.0	5.0	1.5	—	—	0
32047	Peanut butter cookie	1	item(s)	45	—	220	4.0	26.0	1.0	12.0	5.0	—	—	0
4655	Roast beef sandwich, 6″, wheat bread	1	item(s)	224	—	290	19.0	45.0	4.0	5.0	2.0	—	—	0
3957	Roast beef sandwich, 6″, white bread	1	item(s)	217	—	280	18.0	43.0	2.0	5.0	2.5	—	—	0
16378	Roasted chicken breast, 6″, white bread	1	item(s)	231	—	320	23.0	46.0	3.0	5.0	2.0	—	—	0
34028	Southwest steak and cheese sandwich, 6″, Italian bread	1	item(s)	271	—	450	24.0	48.0	6.0	20.0	6.0	—	—	0
4032	Spicy Italian sandwich, 6″, white bread	1	item(s)	220	—	470	20.0	43.0	2.0	25.0	9.5	—	—	0
4031	Steak and cheese sandwich, 6″, white bread	1	item(s)	243	—	350	23.0	45.0	3.0	10.0	5.0	—	—	0
32050	Sugar cookie	1	item(s)	45	—	220	2.0	28.0	0.5	12.0	6.0	—	—	0
40477	Sweet onion chicken teriyaki sandwich, 6″, white or wheat bread	1	serving(s)	281	—	370	26.0	59.0	4.0	5.0	1.5	—	—	0
38623	Turkey breast and bacon melt wrap with chipotle sauce	1	item(s)	228	—	380	31.0	20.0	9.0	24.0	7.0	—	—	0

KEY: H-4 = Breads/Baked Goods H-10 = Cereal/Rice/Pasta H-14 = Fruit H-18 = Vegetables/Legumes H-28 = Nuts/Seeds H-30 = Vegetarian H-32 = Dairy H-40 = Eggs H-40 = Seafood H-42 = Meats H-46 = Poultry
6 = Processed Meats H-48 = Beverages H-52 = Fats/Oils H-54 = Sweets H-56 = Spices/Condiments/Sauces H-60 = Mixed Foods/Soups/Sandwiches H-64 = Fast Food H-84 = Convenience H-86 = Baby Foods

Chol (mg)	Calc (mg)	Iron (mg)	Magn (mg)	Pota (mg)	Sodi (mg)	Zinc (mg)	Vit A (µg)	Thia (mg)	Vit E (mg α)	Ribo (mg)	Niac (mg)	Vit B_6 (mg)	Fola (µg)	Vit C (mg)	Vit B_{12} (µg)	Sele (µg)
35	350	0.00	—	—	170.0	—	—	—	—	—	—	—	—	0	0	—
30	400	1.44	—	—	130.0	—	—	—	—	—	—	—	—	0	0	—
65	350	0.00	—	—	140.0	—	—	—	—	—	—	—	—	0	0	—
40	0	0.00	—	—	10.0	—	—	—	—	—	—	—	—	0	0	—
5	400	0.00	—	—	210.0	—	—	—	—	—	—	—	—	0	0	—
70	400	0.00	—	—	210.0	—	—	—	—	—	—	—	—	2.4	0	—
10	450	0.00	—	—	250.0	—	—	—	—	—	—	—	—	0	0	—
35	500	1.44	—	—	210.0	—	—	—	—	—	—	—	—	0	0	—
75	500	0.00	—	—	250.0	—	—	—	—	—	—	—	—	3.6	0	—
35	150	8.10	—	—	1090.0	—	—	—	—	—	—	—	—	18.0	—	—
90	250	4.50	—	—	1400.0	—	—	—	—	—	—	—	—	21.0	—	—
90	300	2.70	—	—	1680.0	—	—	—	—	—	—	—	—	9.0	—	—
15	0	1.08	—	—	150.0	—	—	—	—	—	—	—	—	0	—	—
10	20	1.00	—	—	100.0	—	—	—	—	—	—	—	—	0	—	—
10	0	1.00	—	—	100.0	—	—	—	—	—	—	—	—	0	—	—
55	150	2.70	—	—	1770.0	—	—	—	—	—	—	—	—	16.8	—	—
45	100	5.40	—	—	1030.0	—	—	—	—	—	—	—	—	21.0	—	—
45	100	3.60	—	—	1010.0	—	—	—	—	—	—	—	—	16.8	—	—
35	60	3.60	—	—	880.0	—	—	—	—	—	—	—	—	30.0	—	—
35	60	3.60	—	—	1290.0	—	—	—	—	—	—	—	—	13.8	—	—
60	150	3.60	—	—	1530.0	—	—	—	—	—	—	—	—	16.8	—	—
190	80	4.50	—	—	720.0	—	66.7	—	—	—	—	—	—	3.6	—	—
25	60	2.70	—	—	1375.0	—	—	—	—	—	—	—	—	13.8	—	—
45	200	7.20	—	—	1610.0	—	—	—	—	—	—	—	—	36.0	—	—
45	200	5.40	—	—	1590.0	—	—	—	—	—	—	—	—	31.8	—	—
45	150	5.40	—	—	1720.0	—	—	—	—	—	—	—	—	24.0	—	—
15	20	1.08	—	—	170.0	—	—	—	—	—	—	—	—	0	—	—
45	60	4.50	—	—	1020.0	—	—	—	—	—	—	—	—	18.0	—	—
15	20	0.72	—	—	200.0	—	—	—	—	—	—	—	—	0	—	—
20	60	6.30	—	—	920.0	—	—	—	—	—	—	—	—	18.0	—	—
20	60	4.50	—	—	900.0	—	—	—	—	—	—	—	—	13.8	—	—
45	60	2.70	—	—	1000.0	—	—	—	—	—	—	—	—	13.8	—	—
45	150	8.10	—	—	1310.0	—	—	—	—	—	—	—	—	21.0	—	—
55	60	2.70	—	—	1650.0	—	—	—	—	—	—	—	—	16.8	—	—
35	150	6.30	—	—	1070.0	—	—	—	—	—	—	—	—	13.8	—	—
15	0	0.72	—	—	140.0	—	—	—	—	—	—	—	—	0	—	—
50	80	4.50	—	—	1220.0	—	—	—	—	—	—	—	—	24.0	—	—
50	200	2.70	—	—	1780.0	—	—	—	—	—	—	—	—	6.0	—	—

APPENDIX H

TABLE H-1 Table of Food Composition (*continued*) (Computer code is for Cengage Diet Analysis program) (For purposes of calculations, use "0" for t, <1, <0.1, 0.01)

DA+ Code	Food Description	Quantity	Measure	Wt (g)	H_2O (g)	Ener (kcal)	Prot (g)	Carb (g)	Fiber (g)	Fat (g)	Fat Breakdown (g) Sat	Mono	Poly	*Trans*
Fast Food—*continued*														
15834	Turkey breast and ham sandwich, 6″, white bread	1	item(s)	227	—	280	19.0	45.0	2.0	5.0	2.0	—	—	0
16376	Turkey breast sandwich, 6″, white bread	1	item(s)	217	—	270	17.0	44.0	2.0	4.5	2.0	—	—	0
15841	Veggie Delite sandwich, 6″, wheat bread	1	item(s)	167	—	230	9.0	44.0	4.0	3.0	1.0	—	—	0
16375	Veggie Delite, 6″, white bread	1	item(s)	160	—	220	8.0	42.0	2.0	3.0	1.5	—	—	0
32051	White chip macadamia nut cookie	1	item(s)	45	—	220	2.0	29.0	0.5	11.0	5.0	—	—	0
	Taco Bell													
29906	7-Layer burrito	1	item(s)	283	—	490	17.0	65.0	9.0	18.0	7.0	—	—	1.0
744	Bean burrito	1	item(s)	198	—	340	13.0	54.0	8.0	9.0	3.5	—	—	0.5
749	Beef burrito supreme	1	item(s)	248	—	410	17.0	51.0	7.0	17.0	8.0	—	—	1.0
33417	Beef Chalupa Supreme	1	item(s)	153	—	380	14.0	30.0	3.0	23.0	7.0	—	—	0.5
34474	Beef Gordita Baja	1	item(s)	153	—	340	13.0	29.0	4.0	19.0	5.0	—	—	0
29910	Beef Gordita Supreme	1	item(s)	153	—	310	14.0	29.0	3.0	16.0	6.0	—	—	0.5
2014	Beef soft taco	1	item(s)	99	—	200	10.0	21.0	3.0	9.0	4.0	—	—	0
10860	Beef soft taco supreme	1	item(s)	135	—	250	11.0	23.0	3.0	13.0	6.0	—	—	0.5
34472	Chicken burrito supreme	1	item(s)	248	—	390	20.0	49.0	6.0	13.0	6.0	—	—	0.5
33418	Chicken Chalupa Supreme	1	item(s)	153	—	360	17.0	29.0	2.0	20.0	5.0	—	—	0
34475	Chicken Gordita Baja	1	item(s)	153	—	320	17.0	28.0	3.0	16.0	3.5	—	—	0
29909	Chicken quesadilla	1	item(s)	184	—	520	28.0	40.0	3.0	28.0	12.0	—	—	0.5
29907	Chili cheese burrito	1	item(s)	156	—	390	16.0	40.0	3.0	18.0	9.0	—	—	1.5
10794	Cinnamon twists	1	serving(s)	35	—	170	1.0	26.0	1.0	7.0	0	—	—	0
29911	Grilled chicken Gordita Supreme	1	item(s)	153	—	290	17.0	28.0	2.0	12.0	5.0	—	—	0
14463	Grilled chicken soft taco	1	item(s)	99	—	190	14.0	19.0	1.0	6.0	2.5	—	—	—
29912	Grilled Steak Gordita Supreme	1	item(s)	153	—	290	15.0	28.0	2.0	13.0	5.0	—	—	0
29904	Grilled steak soft taco	1	item(s)	128	—	270	12.0	20.0	2.0	16.0	4.5	—	—	0
29905	Grilled steak soft taco supreme	1	item(s)	135	—	235	13.0	21.0	1.0	11.0	6.0	—	—	—
2021	Mexican pizza	1	serving(s)	216	—	530	20.0	42.0	7.0	30.0	8.0	—	—	1.0
29894	Mexican rice	1	serving(s)	131	—	170	6.0	23.0	1.0	11.0	3.0	—	—	0
10772	Meximelt	1	serving(s)	128	—	280	15.0	22.0	3.0	14.0	7.0	—	—	0.5
2011	Nachos	1	serving(s)	99	—	330	4.0	32.0	2.0	21.0	3.5	—	—	2.0
2012	Nachos Bellgrande	1	serving(s)	308	—	770	19.0	77.0	12.0	44.0	9.0	—	—	3.0
2023	Pintos 'n cheese	1	serving(s)	128	—	150	9.0	19.0	7.0	6.0	3.0	—	—	0.5
34473	Steak burrito supreme	1	item(s)	248	—	380	18.0	49.0	6.0	14.0	7.0	—	—	0.5
33419	Steak Chalupa Supreme	1	item(s)	153	—	360	15.0	28.0	2.0	21.0	6.0	—	—	0
747	Taco	1	item(s)	78	—	170	8.0	13.0	3.0	10.0	3.5	—	—	0
2015	Taco salad with salsa, with shell	1	serving(s)	548	—	840	30.0	80.0	15.0	45.0	11.0	—	—	1.5
14459	Taco supreme	1	item(s)	113	—	210	9.0	15.0	3.0	13.0	6.0	—	—	0
748	Tostada	1	item(s)	170	—	230	11.0	27.0	7.0	10.0	3.5	—	—	0.5
Convenience Meals														
	Banquet													
29961	Barbeque chicken meal	1	item(s)	281	—	330	16.0	37.0	2.0	13.0	3.0	—	—	—
14788	Boneless white fried chicken meal	1	item(s)	286	—	310	10.0	21.0	4.0	20.0	5.0	—	—	—
29960	Fish sticks meal	1	item(s)	207	—	470	13.0	58.0	1.0	20.0	3.5	—	—	—
29957	Lasagna with meat sauce meal	1	item(s)	312	—	320	15.0	46.0	7.0	9.0	4.0	—	—	—
14777	Macaroni and cheese meal	1	item(s)	340	—	420	15.0	57.0	5.0	14.0	8.0	—	—	—
1741	Meatloaf meal	1	item(s)	269	—	240	14.0	20.0	4.0	11.0	4.0	—	—	—
39418	Pepperoni pizza meal	1	item(s)	191	—	480	11.0	56.0	5.0	23.0	8.0	—	—	—
33759	Roasted white turkey meal	1	item(s)	255	—	230	14.0	30.0	5.0	6.0	2.0	—	—	—
1743	Salisbury steak meal	1	item(s)	269	196.9	380	12.0	28.0	3.0	24.0	12.0	—	—	—
	Budget Gourmet													
1914	Cheese manicotti with meat sauce entrée	1	item(s)	284	194.0	420	18.0	38.0	4.0	22.0	11.0	6.0	1.3	—
1915	Chicken with fettucini entrée	1	item(s)	284	—	380	20.0	33.0	3.0	19.0	10.0	—	—	—
3986	Light beef stroganoff entrée	1	item(s)	248	177.0	290	20.0	32.0	3.0	7.0	4.0	—	—	—
3996	Light sirloin of beef in herb sauce entrée	1	item(s)	269	214.0	260	19.0	30.0	5.0	7.0	4.0	2.3	0.3	—
3987	Light vegetable lasagna entrée	1	item(s)	298	227.0	290	15.0	36.0	4.8	9.0	1.8	0.9	0.6	—

E KEY: H-4 = Breads/Baked Goods H-10 = Cereal/Rice/Pasta H-14 = Fruit H-18 = Vegetables/Legumes H-28 = Nuts/Seeds H-30 = Vegetarian H-32 = Dairy H-40 = Eggs H-40 = Seafood H-42 = Meats H-46 = Poultry
6 = Processed Meats H-48 = Beverages H-52 = Fats/Oils H-54 = Sweets H-56 = Spices/Condiments/Sauces H-60 = Mixed Foods/Soups/Sandwiches H-64 = Fast Food H-84 = Convenience H-86 = Baby Foods

Chol (mg)	Calc (mg)	Iron (mg)	Magn (mg)	Pota (mg)	Sodi (mg)	Zinc (mg)	Vit A (µg)	Thia (mg)	Vit E (mg α)	Ribo (mg)	Niac (mg)	Vit B_6 (mg)	Fola (µg)	Vit C (mg)	Vit B_{12} (µg)	Sele (µg)
25	60	2.70	—	—	1210.0	—	—	—	—	—	—	—	—	13.8	—	—
20	60	2.70	—	—	1000.0	—	—	—	—	—	—	—	—	13.8	—	—
0	60	4.50	—	—	520.0	—	—	—	—	—	—	—	—	18.0	—	—
0	60	2.70	—	—	500.0	—	—	—	—	—	—	—	—	13.8	—	—
15	20	0.72	—	—	160.0	—	—	—	—	—	—	—	—	0	—	—
25	250	5.40	—	—	1350.0	—	—	—	—	—	—	—	—	15.0	—	—
5	200	4.50	—	—	1190.0	—	5.9	—	—	—	—	—	—	4.8	—	—
40	200	4.50	—	—	1340.0	—	9.9	—	—	—	—	—	—	6.0	—	—
40	150	2.70	—	—	620.0	—	—	—	—	—	—	—	—	3.6	—	—
35	100	2.70	—	—	780.0	—	—	—	—	—	—	—	—	2.4	—	—
40	150	2.70	—	—	620.0	—	—	—	—	—	—	—	—	3.6	—	—
25	100	1.80	—	—	630.0	—	—	—	—	—	—	—	—	1.2	—	—
40	150	2.70	—	—	650.0	—	—	—	—	—	—	—	—	3.6	—	—
45	200	4.50	—	—	1360.0	—	—	—	—	—	—	—	—	9.0	—	—
45	100	2.70	—	—	650.0	—	—	—	—	—	—	—	—	4.8	—	—
40	100	1.80	—	—	800.0	—	—	—	—	—	—	—	—	3.6	—	—
75	450	3.60	—	—	1420.0	—	—	—	—	—	—	—	—	1.2	—	—
40	300	1.80	—	—	1080.0	—	—	—	—	—	—	—	—	0	—	—
0	0	0.37	—	—	200.0	—	0	—	—	—	—	—	—	0	—	—
45	150	1.80	—	—	650.0	—	—	—	—	—	—	—	—	4.8	—	—
30	100	1.08	—	—	550.0	—	14.6	—	—	—	—	—	—	1.2	—	—
40	100	2.70	—	—	530.0	—	—	—	—	—	—	—	—	3.6	—	—
35	100	2.70	—	—	660.0	—	—	—	—	—	—	—	—	3.6	—	—
35	120	1.44	—	—	565.0	—	29.2	—	—	—	—	—	—	3.6	—	—
40	350	3.60	—	—	1000.0	—	—	—	—	—	—	—	—	4.8	—	—
15	100	1.44	—	—	790.0	—	—	—	—	—	—	—	—	3.6	—	—
40	250	2.70	—	—	880.0	—	—	—	—	—	—	—	—	2.4	—	—
3	80	0.71	—	—	530.0	—	0	—	—	—	—	—	—	0	—	—
35	200	3.60	—	—	1280.0	—	—	—	—	—	—	—	—	4.8	—	—
15	150	1.44	—	—	670.0	—	—	—	—	—	—	—	—	3.6	—	—
35	200	4.50	—	—	1250.0	—	9.9	—	—	—	—	—	—	9.0	—	—
40	100	2.70	—	—	530.0	—	—	—	—	—	—	—	—	3.6	—	—
25	80	1.08	—	—	350.0	—	—	—	—	—	—	—	—	1.2	—	—
65	450	7.20	—	—	1780.0	—	—	—	—	—	—	—	—	12.0	—	—
40	100	1.08	—	—	370.0	—	—	—	—	—	—	—	—	3.6	—	—
15	200	1.80	—	—	730.0	—	—	—	—	—	—	—	—	4.8	—	—
																—
50	40	1.08	—	—	1210.0	—	0	—	—	—	—	—	—	4.8	—	—
45	80	1.44	—	—	1200.0	—	—	—	—	—	—	—	—	18.0	—	—
55	20	1.44	—	—	710.0	—	—	—	—	—	—	—	—	0	—	—
20	100	2.70	—	—	1170.0	—	—	—	—	—	—	—	—	0	—	—
20	150	1.44	—	—	1330.0	—	0	—	—	—	—	—	—	0	—	—
30	0	1.80	—	—	1040.0	—	0	—	—	—	—	—	—	0	—	—
35	150	1.80	—	—	870.0	—	0	—	—	—	—	—	—	0	—	—
25	60	1.80	—	—	1070.0	—	—	—	—	—	—	—	—	3.6	—	—
60	40	1.44	—	—	1140.0	—	0	—	—	—	—	—	—	0	—	—
85	300	2.70	45.4	484.0	810.0	2.29	—	0.45	—	0.51	4.00	0.22	30.7	0	0.7	—
85	100	2.70	—	—	810.0	—	—	0.15	—	0.42	6.00	—	—	0	—	—
35	40	1.80	38.9	280.0	580.0	4.71	—	0.17	—	0.36	4.28	0.27	18.9	2.4	2.5	—
30	40	1.80	57.7	540.0	850.0	4.81	—	0.15	—	0.29	5.53	0.37	38.4	6.0	1.6	—
15	283	3.03	78.5	420.0	780.0	1.39	—	0.22	—	0.45	3.13	0.32	74.8	59.1	0.2	—

DA+ Code	Food Description	Quantity	Measure	Wt (g)	H_2O (g)	Ener (kcal)	Prot (g)	Carb (g)	Fiber (g)	Fat (g)	Fat Breakdown (g) Sat	Mono	Poly	*Trans*
	Convenience Meals—*continued*													
	Healthy Choice													
9425	Cheese French bread pizza	1	item(s)	170	—	340	22.0	51.0	5.0	5.0	1.5	—	—	—
9306	Chicken enchilada suprema meal	1	item(s)	320	251.5	360	13.0	59.0	8.0	7.0	3.0	2.0	2.0	—
3821	Familiar Favorites lasagna bake with meat sauce entrée	1	item(s)	255	—	270	13.0	38.0	4.0	7.0	2.5	—	—	—
13744	Familiar Favorites sesame chicken with vegetables and rice entrée	1	item(s)	255	—	260	17.0	34.0	4.0	6.0	2.0	2.0	2.0	—
9316	Lemon pepper fish meal	1	item(s)	303	—	280	11.0	49.0	5.0	5.0	2.0	1.0	2.0	—
9322	Traditional salisbury steak meal	1	item(s)	354	250.3	360	23.0	45.0	5.0	9.0	3.5	4.0	1.0	—
9359	Traditional turkey breasts meal	1	item(s)	298	—	330	21.0	50.0	4.0	5.0	2.0	1.5	1.5	—
	Stouffers													
2313	Cheese French bread pizza	1	serving(s)	294	—	380	15.0	43.0	3.0	16.0	6.0	—	—	—
11138	Cheese manicotti with tomato sauce entrée	1	item(s)	255	—	360	18.0	41.0	2.0	14.0	6.0	—	—	—
2366	Chicken pot pie entrée	1	item(s)	284	—	740	23.0	56.0	4.0	47.0	18.0	12.4	10.5	—
11116	Homestyle baked chicken breast with mashed potatoes and gravy entrée	1	item(s)	252	—	270	21.0	21.0	2.0	11.0	3.5	—	—	—
11146	Homestyle beef pot roast and potatoes entrée	1	item(s)	252	—	260	16.0	24.0	3.0	11.0	4.0	—	—	—
11152	Homestyle roast turkey breast with stuffing and mashed potatoes entrée	1	item(s)	273	—	290	16.0	30.0	2.0	12.0	3.5	—	—	—
11043	Lean Cuisine Comfort Classics baked chicken and whipped potatoes and stuffing entrée	1	item(s)	245	—	240	15.0	34.0	3.0	4.5	1.0	2.0	1.0	0
11046	Lean Cuisine Comfort Classics honey mustard chicken with rice pilaf entrée	1	item(s)	227	—	250	17.0	37.0	1.0	4.0	1.0	1.0	1.0	0
9479	Lean Cuisine Deluxe French bread pizza	1	item(s)	174	—	310	16.0	44.0	3.0	9.0	3.5	0.5	0.5	0
360	Lean Cuisine One Dish Favorites chicken chow mein with rice	1	item(s)	255	—	190	13.0	29.0	2.0	2.5	0.5	1.0	0.5	0
11054	Lean Cuisine One Dish Favorites chicken enchilada Suiza with Mexican-style rice	1	serving(s)	255	—	270	10.0	47.0	3.0	4.5	2.0	1.5	1.0	0
9467	Lean Cuisine One Dish Favorites fettucini alfredo entrée	1	item(s)	262	—	270	13.0	39.0	2.0	7.0	3.5	2.0	1.0	0
11055	Lean Cuisine One Dish Favorites lasagna with meat sauce entrée	1	item(s)	298	—	320	19.0	44.0	4.0	7.0	3.0	2.0	0.5	0
	Weight Watchers													
11164	Smart Ones chicken enchiladas suiza entrée	1	item(s)	255	—	340	12.0	38.0	3.0	10.0	4.5	—	—	—
39763	Smart Ones chicken oriental entrée	1	item(s)	255	—	230	15.0	34.0	3.0	4.5	1.0	—	—	—
11187	Smart Ones pepperoni pizza	1	item(s)	198	—	400	22.0	58.0	4.0	9.0	3.0	—	—	—
39765	Smart Ones spaghetti bolognese entrée	1	item(s)	326	—	280	17.0	43.0	5.0	5.0	2.0	—	—	—
31512	Smart Ones spicy szechuan style vegetables and chicken	1	item(s)	255	—	220	11.0	34.0	4.0	5.0	1.0	—	—	—
	Baby Foods													
787	Apple juice	4	fluid ounce(s)	127	111.6	60	0	14.8	0.1	0.1	0	0	0	—
778	Applesauce, strained	4	tablespoon(s)	64	56.7	26	0.1	6.9	1.1	0.1	0	0	0	—
779	Bananas with tapioca, strained	4	tablespoon(s)	60	50.4	34	0.2	9.2	1.0	0	0	0	0	—
604	Carrots, strained	4	tablespoon(s)	56	51.7	15	0.4	3.4	1.0	0.1	0	0	0	—
770	Chicken noodle dinner, strained	4	tablespoon(s)	64	54.8	42	1.7	5.8	1.3	1.3	0.4	0.5	0.3	—
801	Green beans, strained	4	tablespoon(s)	60	55.1	16	0.7	3.8	1.3	0.1	0	0	0	—
910	Human milk, mature	2	fluid ounce(s)	62	53.9	43	0.6	4.2	0	2.7	1.2	1.0	0.3	—
760	Mixed cereal, prepared with whole milk	4	ounce(s)	113	84.6	128	5.4	18.0	1.5	4.0	2.2	1.2	0.4	—
772	Mixed vegetable dinner, strained	2	ounce(s)	57	50.3	23	0.7	5.4	0.8	0	—	—	0	—
762	Rice cereal, prepared with whole milk	4	ounce(s)	113	84.6	130	4.4	18.9	0.1	4.1	2.6	1.0	0.2	—
758	Teething biscuits	1	item(s)	11	0.7	44	1.0	8.6	0.2	0.6	0.2	0.2	0.1	—

Chol (mg)	Calc (mg)	Iron (mg)	Magn (mg)	Pota (mg)	Sodi (mg)	Zinc (mg)	Vit A (µg)	Thia (mg)	Vit E (mg α)	Ribo (mg)	Niac (mg)	Vit B_6 (mg)	Fola (µg)	Vit C (mg)	Vit B_{12} (µg)	Sele (µg)
10	350	3.60	—	—	600.0	—	—	—	—	—	—	—	—	0	—	—
30	40	1.44	—	—	580.0	—	—	—	—	—	—	—	—	3.6	—	—
20	100	1.80	—	—	600.0	—	—	—	—	—	—	—	—	0	—	—
35	18	0.72	—	—	580.0	—	—	—	—	—	—	—	—	12.0	—	—
35	20	0.36	—	—	580.0	—	—	—	—	—	—	—	—	30.0	—	—
45	80	2.70	—	—	580.0	—	—	—	—	—	—	—	—	21.0	—	—
35	40	1.80	—	—	600.0	—	—	—	—	—	—	—	—	0	—	—
30	200	1.80	—	230.0	660.0	—	—	—	—	—	—	—	—	2.4	—	—
70	250	1.44	—	550.0	920.0	—	—	—	—	—	—	—	—	6.0	—	—
65	150	2.70	—	—	1170.0	—	—	—	—	—	—	—	—	2.4	—	—
55	20	0.72	—	490.0	770.0	—	0	—	—	—	—	—	—	0	—	—
35	20	1.80	—	800.0	960.0	—	—	—	—	—	—	—	—	6.0	—	—
45	40	1.08	—	490.0	970.0	—	—	—	—	—	—	—	—	3.6	—	—
25	40	1.16	—	500.0	650.0	—	—	—	—	—	—	—	—	3.6	—	—
30	64	0.38	—	370.0	650.0	—	—	—	—	—	—	—	—	0	—	—
20	150	2.70	—	300.0	700.0	—	—	—	—	—	—	—	—	15.0	—	—
25	40	0.72	—	380.0	650.0	—	—	—	—	—	—	—	—	2.4	—	—
20	150	0.72	—	350.0	510.0	—	—	—	—	—	—	—	—	2.4	—	—
15	200	0.72	—	290.0	690.0	—	0	—	—	—	—	—	—	0	—	—
30	250	1.47	—	610.0	690.0	—	—	—	—	—	—	—	—	2.4	—	—
40	200	0.72	—	—	800.0	—	—	—	—	—	—	—	—	2.4	—	—
35	40	0.72	—	—	790.0	—	—	—	—	—	—	—	—	6.0	—	—
15	200	1.08	—	401.0	700.0	—	69.1	—	—	—	—	—	—	4.8	—	—
15	150	3.60	—	—	670.0	—	—	—	—	—	—	—	—	9.0	—	—
10	40	1.44	—	—	890.0	—	—	—	—	—	—	—	—	0	—	—
0	5	0.72	3.8	115.4	3.8	0.03	1.3	0.01	0.76	0.02	0.10	0.03	0	73.4	0	0.1
0	3	0.12	1.9	45.4	1.3	0.01	0.6	0.01	0.36	0.02	0.04	0.02	1.3	24.5	0	0.2
0	3	0.12	6.0	52.8	5.4	0.04	1.2	0.01	0.36	0.02	0.08	0.04	3.6	10.0	0	0.4
0	12	0.20	5.0	109.8	20.7	0.08	320.9	0.01	0.29	0.02	0.25	0.04	8.4	3.2	0	0.1
10	17	0.40	9.0	89.0	14.7	0.32	70.4	0.03	0.12	0.04	0.44	0.04	7.0	0	0	2.4
0	23	0.40	12.0	87.6	3.0	0.12	10.8	0.02	0.04	0.04	0.20	0.02	14.4	0.2	0	0
9	20	0.02	1.8	31.4	10.5	0.10	37.6	0.01	0.04	0.02	0.10	0.01	3.1	3.1	0	1.1
12	249	11.82	30.6	225.7	53.3	0.80	28.4	0.49	—	0.65	6.54	0.07	12.5	1.4	0.3	
—	12	0.18	6.2	68.6	4.5	0.08	77.1	0.01	—	0.02	0.28	0.04	4.5	1.6	0	0.4
12	271	13.82	51.0	215.5	52.2	0.72	24.9	0.52	—	0.56	5.90	0.12	9.1	1.4	0.3	4.0
0	11	0.39	3.9	35.5	28.4	0.10	3.1	0.02	0.02	0.05	0.47	0.01	5.4	1.0	0	2.6

Appendix I WHO Nutrition Recommendations

The World Health Organization (WHO) has assessed the relationships between diet and the development of chronic diseases. This appendix presents its nutrition recommendations:

- Energy: sufficient to support growth, physical activity, and a healthy body weight (BMI between 18.5 and 24.9) and to avoid weight gain greater than 11 pounds (5 kilograms) during adult life
- Total fat: 15 to 30 percent of total energy
- Saturated fatty acids: <10 percent of total energy
- Polyunsaturated fatty acids: 6 to 10 percent of total energy
- Omega-6 polyunsaturated fatty acids: 5 to 8 percent of total energy
- Omega-3 polyunsaturated fatty acids: 1 to 2 percent of total energy
- *Trans*-fatty acids: <1 percent of total energy
- Total carbohydrate: 55 to 75 percent of total energy
- Sugars: <10 percent of total energy
- Protein: 10 to 15 percent of total energy
- Cholesterol: <300 mg per day
- Salt (sodium): <5 g salt per day (<2 g sodium per day), appropriately iodized
- Fruits and vegetables: ≥400 g per day (about 1 pound)
- Total dietary fiber: >25 g per day from foods
- Physical activity: one hour of moderate-intensity activity, such as walking, on most days of the week

Appendix J Healthy People 2020

Table 1-4 (p. 25) lists the objectives from the Nutrition and Weight Status section of the Healthy People 2020 initiative. Table J-1 presents additional nutrition-related objectives from other topic areas.

TABLE J-1 Nutrition-Related Objectives from Other Topic Areas

Access to Health Services

- Increase the proportion of persons who receive appropriate evidence-based clinical preventive services.

Adolescent Health

- Increase the proportion of schools with a school breakfast program.

Arthritis, Osteoporosis, and Chronic Back Conditions

- Reduce hip fractures among older adults.
- Reduce the proportion of adults with osteoporosis.

Cancer

- Reduce the cancer death rate.
- Increase the mental and physical health-related quality of life of cancer survivors.

Diabetes

- Reduce the annual number of new cases of diagnosed diabetes in the population.
- Reduce the death rate among the population with diabetes.
- Reduce the diabetes death rate.
- Improve glycemic control among the population with diagnosed diabetes.
- Improve lipid control among persons with diagnosed diabetes.
- Increase the proportion of the population with diagnosed diabetes whose blood pressure is under control.
- Increase the proportion of persons with diagnosed diabetes who receive formal diabetes education.
- Increase prevention behaviors in persons at high risk for diabetes with prediabetes.

Early and Middle Childhood

- Increase the proportion of elementary, middle, and senior high schools that require school health education.

Educational and Community-Based Programs

- Increase the proportion of preschool Early Head Start and Head Start programs that provide health education to prevent health problems in the following areas: unintentional injury; violence; tobacco use and addiction; alcohol and drug use, unhealthy dietary patterns; and inadequate physical activity, dental health, and safety.
- Increase the proportion of elementary, middle, and senior high schools that provide comprehensive school health education to prevent health problems in the following areas: unintentional injury; violence; suicide; tobacco use and addiction; alcohol or other drug use; unintended pregnancy, HIV/AIDS, and STD infection; unhealthy dietary patterns; and inadequate physical activity.
- Increase the proportion of college and university students who receive information from their institution on each of the priority health risk behavior areas (all priority areas; unintentional injury; violence; suicide; tobacco use and addiction; alcohol and other drug use; unintended pregnancy, HIV/AIDS, and STD infection; unhealthy dietary patterns; and inadequate physical activity).
- Increase the proportion of worksites that offer an employee health promotion program to their employees.
- Increase the number of community-based organizations (including local health departments, tribal health services, nongovernmental organizations, and state agencies) providing population-based primary prevention services.

Environmental Health

- Reduce blood lead levels in children.
- Reduce the number of US homes that are found to have lead-based paint or related hazards.

(continued)

TABLE J-1 Nutrition-Related Objectives from Other Topic Areas (*continued*)

Food Safety
• Reduce infections caused by key pathogens transmitted commonly through food. • Reduce the number of outbreak-associated infections due to Shiga toxin-producing *E. coli* 0157, or *Campylobacter, Listeria*, or *Salmonella* species associated with food commodity groups. • Reduce severe allergic reactions to food among adults with a food allergy diagnosis. • Increase the proportion of consumers who follow key food safety practices. • Improve food safety practices associated with foodborne illness in foodservice and retail establishments.
Heart Disease and Stroke
• Increase overall cardiovascular health in the US population. • Reduce coronary heart disease deaths. • Reduce stroke deaths. • Reduce the proportion of persons in the population with hypertension. • Reduce the proportion of adults with high total blood cholesterol levels. • Reduce the mean total blood cholesterol levels among adults. • Increase the proportion of adults with prehypertension who meet the recommended guidelines. • Increase the proportion of adults with hypertension who meet the recommended guidelines. • Increase the proportion of adults with elevated LDL cholesterol who have been advised by a health-care provider regarding cholesterol-lowering management including lifestyle changes and, if indicated, medication. • Increase the proportion of adults with elevated LDL-cholesterol who adhere to the prescribed LDL cholesterol–lowering management lifestyle changes and, if indicated, medication.
Maternal, Infant, and Child Health
• Reduce low birth weight (LBW) and very low birth weight (VLBW). • Reduce preterm births. • Increase the proportion of pregnant women who receive early and adequate prenatal care. • Increase the proportion of mothers who achieve a recommended weight gain during their pregnancies. • Increase the proportion of women of childbearing potential with intake of at least 400 micrograms of folic acid from fortified foods or dietary supplements. • Reduce the proportion of women of childbearing potential who have low red blood cell folate concentrations. • Increase the proportion of women delivering a live birth who received preconception care services and practiced key recommended preconception health behaviors. • Increase the proportion of infants who are breastfed. • Increase the proportion of employers that have worksite lactation support programs. • Reduce the proportion of breastfed newborns who receive formula supplementation within the first 2 days of life. • Reduce the occurrence of fetal alcohol syndrome (FAS). • Reduce occurrence of neural tube defects.
Mental Health and Mental Disorders
• Reduce the proportion of adolescents who engage in disordered eating behaviors in an attempt to control their weight.
Older Adults
• Increase the proportion of the health-care workforce (including dietitians) with geriatric certification.
Oral Health
• Increase the proportion of the US population served by community water systems with optimally fluoridated water.
Physical Activity
• Reduce the proportion of adults who engage in no leisure-time physical activity. • Increase the proportion of adolescents and adults who meet current federal physical activity guidelines for aerobic physical activity and for muscle-strengthening activity. • Increase the proportion of the nation's public and private schools that require daily physical education for all students. • Increase the proportion of adolescents who participate in daily school physical education. • Increase regularly scheduled elementary school recess in the United States. • Increase the proportion of children and adolescents who do not exceed recommended limits for screen time.

SOURCE: Adapted from Healthy people 2020: www.healthypeople.gov.

Appendix K Aids to Calculation

Many mathematical problems have been worked out in the "How To" features of the text. This appendix offers additional help and examples.

Conversions

A conversion factor is a fraction that converts a measurement expressed in one unit to another unit—for example, from pounds to kilograms or from feet to meters. To create a conversion factor, an equality (such as 1 kilogram = 2.2 pounds) is expressed as a fraction:

$$\frac{1\text{ kg}}{2.2\text{ lb}} \text{ and } \frac{2.2\text{ lb}}{1\text{ kg}}$$

To convert the units of a measurement, use the fraction with the desired unit in the numerator.

Example 1 Convert a weight of 130 pounds to kilograms.
Multiply 130 pounds by the conversion factor that includes both pounds and kilograms, with the desired unit (kilograms) in the numerator:

$$130\text{ lb} \times \frac{1\text{ kg}}{2.2\text{ lb}} = \frac{130\text{ kg}}{2.2} = 59\text{ kg}$$

Alternatively, to convert a measurement from one unit of measure to another, multiply the given measurement by the appropriate equivalent found on the next page of weights and measures.

Example 2 Convert 64 fluid ounces to liters.
Locate the equivalent measure from the volume section on the next page (1 ounce = 0.03 liter) and multiply the number of ounces by 0.03:

$$64\text{ oz} \times 0.03\text{ oz/L} = 1.9\text{ L}$$

Percentages

A percentage is a fraction whose denominator is 100. For example:

$$50\% = \frac{50}{100}$$

Like other fractions, percentages are used to express a portion of a quantity. Fractions whose denominators are numbers other than 100 can be converted to percentages by first dividing the numerator by the denominator and then multiplying the result by 100.

Example 3 Express 5/8 as a percent.

$$\frac{5}{8} = 5 \div 8 = 0.625$$

$$0.625 \times 100 = 62.5\%$$

The following examples show how to calculate specific percentages.

Example 4 Suppose your energy intake for the day is 2000 kcalories (kcal) and your recommended energy intake is 2400 kcalories. What percent of the recommended energy intake did you consume?

Divide your intake by the recommended intake.
2000 kcal (intake) ÷ 2400 kcal (recommended) = 0.83
Multiply by 100 to express the decimal as a percent.
0.83 × 100 = 83%

Example 5 Suppose a man's intake of vitamin C is 120 milligrams and his RDA is 90 milligrams. What percent of the RDA for vitamin C did he consume?

Divide the intake by the recommended intake.
120 mg (intake) ÷ 90 mg (RDA) = 1.33
Multiply by 100 to express the decimal as a percent.
1.33 × 100 = 133%

Example 6 Dietary recommendations suggest that carbohydrates provide 45 to 65 percent of the day's energy intake. If your energy intake is 2000 kcalories, how much carbohydrate should you eat?

Because this question has a range of acceptable answers, work the problem twice. First, use 45% to find the least amount you should eat.

Divide 45 by 100 to convert to a decimal.
45 ÷ 100 = 0.45
Multiply kcalories by 0.45.
2000 kcal × 0.45 = 900 kcal
Divide kcalories by 4 to convert carbohydrate kcal to grams.
900 kcal ÷ 4 kcal/g = 225 g

Now repeat the process using 65% to find the maximum number of grams of carbohydrates you should eat.

Divide 65 by 100 to convert it to a decimal.
65 ÷ 100 = 0.65
Multiply kcalories by 0.65.
2000 kcal × 0.65 = 1300 kcal
Divide kcalories by 4 to convert carbohydrate kcal to grams.
1300 kcal ÷ 4 kcal/g = 325 g

If you plan for between 45% and 65% of your 2000-kcalorie intake to be from carbohydrates, you should eat between 225 grams and 325 grams of carbohydrates.

Weights and Measures

Length

1 centimeter (cm) = 0.39 inches (in)
1 foot (ft) = 30 centimeters (cm)
1 inch (in) = 2.54 centimeters (cm)
1 meter (m) = 39.37 inches (in)

Weight

1 gram (g) = 0.001 kilograms (kg)
= 1000 milligrams (mg)
= 0.035 ounces (oz)
1 kilogram (kg) = 1000 grams (g)
= 2.2 pounds (lb)
1 microgram (μg) = 0.001 milligrams (mg)
1 milligram (mg) = 0.001 grams (g)
= 1000 micrograms (μg)
1 ounce (oz) = 28 grams (g)
= 0.03 kilograms (kg)
= 1/16 or 0.0625 pound (lb)
1 pound (lb) = 454 grams (g)
= 0.45 kilograms (kg)
= 16 ounces (oz)

Volume

1 cup = 16 tablespoons (tbs or T)
= 0.25 liters (L)
= 236 milliliters (mL, commonly rounded to 250 mL)
= 8 ounces (oz)
1 liter (L) = 33.8 fluid ounces (fl oz)
= 0.26 gallons (gal)
= 2.1 pints (pt)
= 1.06 quarts (qt)
= 1000 milliliters (mL)
1 milliliter (mL) = 0.001 liters (L)
= 0.03 fluid ounces (fl oz)
= 1/5 teaspoon (tsp)
1 ounce (oz) = 0.03 liters (L)
= 30 milliliters (mL)
= 2 tablespoons (tbs)
1 pint (pt) = 2 cups (c)
= 0.47 liters (L)
= 16 ounces (oz)
= 0.5 quarts (qt)
1 quart (qt) = 4 cups (c)
= 0.95 liters (L)
= 32 ounces (oz)
= 1/4 or 0.25 gallon (gal)
= 2 pints (pt)
1 tablespoon (tbs or T) = 3 teaspoons (tsp)
= 15 milliliters (mL)
1 teaspoon (tsp) = 5 milliliters (mL)
1 gallon (gal) = 16 cups (c)
= 3.8 liters (L)
= 128 ounces (oz)
= 8 pints (pt)
= 4 quarts (qt)
1 cup (c) = 8 ounces (oz)
= 16 tablespoons (tbs)
= 250 milliliters (mL)

Energy

1 megajoule (MJ) = 240 kcalories (kcal)
1 kilojoule (kJ) = 0.24 kcalories (kcal)
1 kcalorie (kcal) = 4.2 kilojoule (kJ)
1 g alcohol = 7 kcal = 29 kJ
1 g carbohydrate = 4 kcal = 17 kJ
1 g fat = 9 kcal = 37 kJ
1 g protein = 4 kcal = 17 kJ

Temperature

To change from Fahrenheit (°F) to Celsius (°C), subtract 32 from the Fahrenheit measure and then multiply that result by 0.56.

To change from Celsius (°C) to Fahrenheit (°F), multiply the Celsius measure by 1.8 and add 32 to that result.

A comparison of some useful temperatures is given below.

	Celsius	Fahrenheit
Boiling point	100°C	212°F
Body temperature	37°C	98.6°F
Freezing point	0°C	32°F

Glossary

Many medical terms have their origins in Latin or Greek. By learning a few common derivations, you can glean the meaning of words you have never heard of before. For example, once you know that "hyper" means above normal, "glyc" means glucose, and "emia" means blood, you can easily determine that "hyperglycemia" means high blood glucose. The derivations **below** will help you to learn many terms presented in this glossary.

General

a- or **an-** = not or without
ana- = up
ant- or **anti-** = against
ante- or **pre-** or **pro-** = before
bi- or **di-** = two, twice
cata- = down
co- = with or together
dys- or **mal-** = bad, difficult, painful
endo- = inner or within
epi- = upon
exo- = outside of or without
extra- = outside of, beyond, or in addition
gen- or **-gen** = gives rise to, producing
homeo- = like, similar, constant unchanging state
hyper- = over, above, excessive
hypo- = below, under, beneath
in- = not
inter- = between, in the midst
intra- = within
-itis = infection or inflammation
-lysis = break
macro- = large or long
micro- = small
mono- = one, single
neo- = new, recent
oligo- = few or small
-osis or **-asis** = condition
para- = near
peri- = around, about
poly- = many or much
semi- = half
-stat or **-stasis-** = stationary
tri- = three

Body

angi- or **vaso-** = vessel
arterio- = artery
cardiac or **cardio-** = heart
-cyte = cell
enteron = intestine
gastro- = stomach
hema- or **-emia** = blood
hepatic = liver
myo- or **sarco-** = muscle
nephr- or **renal** = kidney
neuro- = nerve
osteo- = bone
pulmo- = lung
ure- or **-uria** = urine
vena = vein

Chemistry

-al = aldehyde
-ase = enzyme
-ate = salt
glyc- or **gluc-** = sweet (glucose)
hydro- or **hydrate** = water
lipo- = lipid
-ol = alcohol
-ose = carbohydrate
saccha- = sugar

24-hour recall: a record of foods eaten by a person for one 24-hour period.

A

absorption: the uptake of nutrients by the cells of the small intestine for transport into either the blood or the lymph.

Academy of Nutrition and Dietetics: the professional organization of dietitians in the United States; formerly the American Dietetic Association. The Canadian equivalent is Dietitians of Canada, which operates similarly.

Acceptable Daily Intake (ADI): the estimated amount of a sweetener that individuals can safely consume each day over the course of a lifetime without adverse effect.

Acceptable Macronutrient Distribution Ranges (AMDR): ranges of intakes for the energy nutrients that provide adequate energy and nutrients and reduce the risk of chronic diseases.

accredited: approved; in the case of medical centers or universities, certified by an agency recognized by the US Department of Education.

acesulfame (AY-sul-fame) **potassium:** an artificial sweetener composed of an organic potassium salt that has been approved for use in both the United States and Canada; also known as *acesulfame-K* because K is the chemical symbol for potassium.

acetaldehyde (ass-et-AL-duh-hide): an intermediate in alcohol metabolism.

acetyl CoA (ASS-eh-teel, or ah-SEET-il, coh-AY): a 2-carbon compound (*acetate,* or *acetic acid*) to which a molecule of CoA is attached.

acid controllers: medications used to prevent or relieve indigestion by suppressing production of acid in the stomach; also called *H2 blockers.*

acid-base balance: the equilibrium in the body between acid and base concentrations.

acidosis (assi-DOE-sis): above-normal acidity in the blood and body fluids.

acids: compounds that release hydrogen ions in a solution.

acne: a chronic inflammation of the skin's follicles and oil-producing glands, which leads to an accumulation of oils inside the ducts that surround hairs; usually associated with the maturation of young adults.

acupuncture (AK-you-PUNK-cher): a technique that involves piercing the skin with long thin needles at specific anatomical points to relieve pain or illness. Acupuncture sometimes uses heat, pressure, friction, suction, or electromagnetic energy to stimulate the points.

acute malnutrition: malnutrition caused by recent severe food restriction; characterized in children by underweight for height (*wasting*).

adaptive thermogenesis: adjustments in energy expenditure related to changes in environment such as extreme cold and to physiological events such as overfeeding, trauma, and changes in hormone status.

added sugars: sugars and other kcaloric sweeteners that are added to foods during processing, preparation, or at the table. Added sugars do not include the naturally occurring sugars found in fruits and milk products.

additives: substances not normally consumed as foods but added to food either intentionally or by accident.

adequacy (dietary): providing all the essential nutrients, fiber, and energy in amounts sufficient to maintain health.

Adequate Intake (AI): the average daily amount of a nutrient that appears sufficient to maintain a specified criterion; a value used as a guide for nutrient intake when an RDA cannot be determined.

adipokines (ADD-ih-poe-kines): proteins synthesized and secreted by adipose cells.

adiponectin: a protein produced by the fat cells that inhibits inflammation and protects against insulin resistance, type 2 diabetes, and cardiovascular disease.

adipose (ADD-ih-poce) **tissue:** the body's fat tissue; consists of masses of triglyceride-storing cells.

adolescence: the period from the beginning of puberty until maturity.

adrenal glands: glands adjacent to, and just above, each kidney.

adrenocorticotropin (ad-REE-noh-KORE-tee-koh-TROP-in) or **ATCH:** a hormone, so named because it stimulates *(trope)* the adrenal cortex. The adrenal gland, like the pituitary, has two parts, in this case the outer portion *(cortex)* and an inner core *(medulla)*. The release of ACTH is mediated by *corticotropin-releasing hormone (CRH).*

adverse reactions: unusual responses to food (including intolerances and allergies).

aerobic (air-ROE-bic): requiring oxygen.

aerobic physical activity: activity in which the body's large muscles move in a rhythmic manner for a sustained period of time. Aerobic activity, also called *endurance activity,* improves cardiorespiratory fitness. Brisk walking, running, swimming, and bicycling are examples.

AIDS (acquired immune deficiency syndrome): the late stage of HIV infection, in which severe complications of opportunistic infections and cancers develop.

alcohol: a class of organic compounds containing hydroxyl (OH) groups.

alcohol abuse: a pattern of drinking that includes failure to fulfill work, school, or home responsibilities; drinking in situations that are physically dangerous (as in driving while intoxicated); recurring alcohol-related legal problems (as in aggravated assault charges); or continued drinking despite ongoing social problems that are caused by or worsened by alcohol.

alcohol dehydrogenase (dee-high-DROJ-eh-nayz): an enzyme active in the stomach and the liver that converts ethanol to acetaldehyde.

alcoholism: a pattern of drinking that includes a strong craving for alcohol, a loss of control and an inability to stop drinking once begun, withdrawal symptoms (nausea, sweating, shakiness, and anxiety) after heavy drinking, and the need for increasing amounts of alcohol to feel "high."

alcohol-related birth defects (ARBD): malformations in the skeletal and organ systems (heart, kidneys, eyes, ears) associated with prenatal alcohol exposure.

alcohol-related neurodevelopmental disorder (ARND): abnormalities in the central nervous system and cognitive development associated with prenatal alcohol exposure.

aldosterone (al-DOS-ter-own): a hormone secreted by the adrenal glands that regulates blood pressure by increasing the reabsorption of sodium by the kidneys. Aldosterone also regulates chloride and potassium concentrations.

alkalosis (alka-LOE-sis): above-normal alkalinity (base) in the blood and body fluids.

alpha-lactalbumin (lact-AL-byoo-min): a major protein in human breast milk, as opposed to *casein* (CAY-seen), a major protein in cow's milk.

alpha-tocopherol: the active vitamin E compound.

Alzheimer's disease: a degenerative disease of the brain involving memory loss and major structural changes in neuron networks; also known as *senile dementia of the Alzheimer's type (SDAT), primary degenerative dementia of senile onset,* or *chronic brain syndrome.*

amenorrhea (ay-MEN-oh-REE-ah): the absence of or cessation of menstruation. *Primary amenorrhea* is menarche delayed beyond 16 years of age. *Secondary amenorrhea* is the absence of three to six consecutive menstrual cycles.

amino (a-MEEN-oh) **acids:** building blocks of proteins. Each contains an amino group, an acid group, a hydrogen atom, and a distinctive side group, all attached to a central carbon atom.

amino acid pool: the supply of amino acids derived from either food proteins or body proteins that collect in the cells and circulating blood and stand ready to be incorporated in proteins and other compounds or used for energy.

amino acid scoring: a measure of protein quality assessed by comparing a protein's amino acid pattern with that of a reference protein; sometimes called *chemical scoring.*

ammonia: a compound with the chemical formula NH_3, produced during the deamination of amino acids.

amniotic (am-nee-OTT-ic) **sac:** the "bag of waters" in the uterus, in which the fetus floats.

amylase (AM-ih-lace): an enzyme that hydrolyzes amylose (a form of starch). Amylase is a *carbohydrase,* an enzyme that breaks down carbohydrates.

anabolic steroids: drugs related to the male sex hormone, testosterone, that stimulate the development of lean body mass.

anabolism (an-AB-o-lism): reactions in which small molecules are put together to build larger ones. Anabolic reactions require energy.

anaerobic (AN-air-ROE-bic): not requiring oxygen.

anaphylactic (ana- fill-LAC-tic) **shock:** a life-threatening, whole-body allergic reaction to an offending substance.

androstenedione: hormones made in the adrenal glands that serve as precursors to the male hormone testosterone; falsely promoted as burning fat, building muscle, and slowing aging.

anecdote: a personal account of an experience or event; not reliable scientific information.

anemia (ah-NEE-me-ah): literally, "too little blood." Anemia is any condition in which too few red blood cells are present, or the red blood cells are immature (and therefore large) or too small or contain too little hemoglobin to carry the normal amount of oxygen to the tissues. Anemia is not a disease itself but can be a symptom of many different disease conditions, including many nutrient deficiencies, bleeding, excessive red blood cell destruction, and defective red blood cell formation.

anencephaly (AN-en-SEF-a-lee): an uncommon and always fatal type of neural tube defect characterized by the absence of a brain.

aneurysm (AN-you-rizm): an abnormal enlargement or bulging of a blood vessel (usually an artery) caused by damage to or weakness in the blood vessel wall.

angina (an-JYE-nah or AN-ji-nah): a painful feeling of tightness or pressure in and around the heart, often radiating to the back, neck, and arms; caused by a lack of oxygen to an area of heart muscle.

angiotensin I (AN-gee-oh-TEN-sin): an inactive precursor that is converted by an enzyme to yield active angiotensin II.

angiotensin II: a hormone involved in blood pressure regulation.

angiotensinogen: a precursor protein that is hydrolyzed to angiotensin I by renin.

anions (AN-eye-uns): negatively charged ions.

anorexia (an-oh-RECK-see-ah) **nervosa:** an eating disorder characterized by a refusal to maintain a minimally normal body weight and a distortion in perception of body shape and weight.

antacids: medications used to relieve indigestion by neutralizing acid in the stomach.

antagonist: a competing factor that counteracts the action of another factor. When a drug displaces a vitamin from its site of action, the drug renders the vitamin ineffective and thus acts as a vitamin antagonist.

anthropometric (AN-throw-poe-MET-rick): relating to measurement of the physical characteristics of the body, such as height and weight.

antibodies: large proteins of the blood and body fluids, produced by the immune system in response to the invasion of the body by foreign molecules (usually proteins called *antigens*). Antibodies combine with and inactivate the foreign invaders, thus protecting the body.

antidiuretic hormone (ADH): a hormone produced by the pituitary gland in response to dehydration (or a high sodium concentration in the blood) that stimulates the kidneys to reabsorb more water and therefore to excrete less. In addition to its antidiuretic effect, ADH elevates blood pressure and so is also called *vasopressin* (VAS-oh-PRES-in).

antigens: substances that elicit the formation of antibodies or an inflammation reaction from the immune system. A bacterium, a virus, a toxin, and a protein in food that causes allergy are all examples of antigens.

antioxidants: in the body, substances that significantly decrease the adverse effects of free radicals on normal physiological functions.

antioxidants: as a food additive, preservatives that delay or prevent rancidity of fats in foods and other damage to food caused by oxygen.

antipromoters: factors that oppose the development of cancer.

antiscorbutic (AN-tee-skor-BUE-tik) **factor:** the original name for vitamin C.

anus (AY-nus): the terminal outlet of the GI tract.

aorta (ay-OR-tuh): the large, primary artery that conducts blood from the heart to the body's smaller arteries.

apoptosis: cell death.

appendix: a narrow blind sac extending from the beginning of the colon that contains bacteria and lymph cells.

appetite: the integrated response to the sight, smell, thought, or taste of food that initiates or delays eating.

aquaculture: the practice of fish farming.

arachidonic (a-RACK-ih-DON-ic) **acid:** an omega-6 polyunsaturated fatty acid with 20 carbons and four double bonds; present in small amounts in meat and other animal products and synthesized in the body from linoleic acid.

ariboflavinosis (ay-RYE-boh-FLAY-vin-oh-sis): riboflavin deficiency.

aroma therapy: a technique that uses oil extracts from plants and flowers (usually applied by massage or baths) to enhance physical, psychological, and spiritual health.

arteries: vessels that carry blood from the heart to the tissues.

artesian water: water drawn from a well that taps a confined aquifer in which the water is under pressure.

arthritis: inflammation of a joint, usually accompanied by pain, swelling, and structural changes.

artificial fats: zero-energy fat replacers that are chemically synthesized to mimic the sensory and cooking qualities of naturally occurring fats but are totally or partially resistant to digestion.

artificial sweeteners: sugar substitutes that provide negligible, if any, energy; sometimes called *nonnutritive sweeteners.*

ascorbic acid: one of the two active forms of vitamin C. Many people refer to vitamin C by this name.

-ase (ACE): a word ending denoting an enzyme. The word beginning often identifies the compounds the enzyme works on.

aspartame (ah-SPAR-tame or ASS-par-tame): an artificial sweetener composed of two amino acids (phenylalanine and aspartic acid); approved for use in both the United States and Canada.

atherosclerosis (ATH-er-oh-scler-OH-sis): a type of artery disease characterized by plaques (accumulations of lipid-containing material) on the inner walls of the arteries.

atoms: the smallest components of an element that have all of the properties of the element.

ATP or **adenosine** (ah-DEN-oh-seen) **triphosphate** (try-FOS-fate): a common high-energy compound composed of a purine (adenine), a sugar (ribose), and three phosphate groups.

atrophic (a-TRO-fik) **gastritis** (gas-TRY-tis): chronic inflammation of the stomach accompanied by a diminished size and functioning of the mucous membrane and glands. This condition is also characterized by inadequate hydrochloric acid and intrinsic factor—two substances needed for vitamin B_{12} absorption.

atrophy (AT-ro-fee): becoming smaller; with regard to muscles, a decrease in size (and strength) because of disuse, undernutrition, or wasting diseases.

autoimmune disorder: a condition in which the body develops antibodies to its own proteins and then proceeds to destroy cells containing these proteins. In type 1 diabetes, the body develops antibodies to its insulin and destroys the pancreatic cells that produce the insulin, creating an insulin deficiency.

autonomic nervous system: the division of the nervous system that controls the body's automatic responses. Its two branches are the *sympathetic* branch, which helps the body respond to stressors from the outside environment, and the *parasympathetic* branch, which regulates normal body activities between stressful times.

avidin (AV-eh-din): the protein in egg whites that binds biotin.

ayurveda (AH-your-VAY-dah): a traditional Hindu system of improving health by using herbs, diet, meditation, massage, and yoga to stimulate the body, mind, and spirit to prevent and treat disease.

B

bacteriophages (bak-TIR-ee-oh-fayjz): viruses that infect bacteria.

balance (dietary): providing foods in proportion to one another and in proportion to the body's needs.

bariatrics: the field of medicine that specializes in treating obesity.

basal metabolic rate (BMR): the rate of energy use for metabolism under specified conditions: after a 12-hour fast and restful sleep, without any physical activity or emotional excitement, and in a comfortable setting. It is usually expressed as kcalories per kilogram body weight per hour.

basal metabolism: the energy needed to maintain life when a body is at complete digestive, physical, and emotional rest.

bases: compounds that accept hydrogen ions in a solution.

B-cells: lymphocytes that produce antibodies. *B* stands for *bone marrow,* where the B-cells develop and mature.

beer: an alcoholic beverage traditionally brewed by fermenting malted barley and adding hops for flavor.

behavior modification: the changing of behavior by the manipulation of antecedents (cues or environmental factors that trigger behavior), the behavior itself, and consequences (the penalties or rewards attached to the behavior).

belching: the release of air or gas from the stomach through the mouth.

benign: an abnormal mass of cells that is noncancerous.

beriberi: the thiamin-deficiency disease.

beta-carotene (BAY-tah KARE-oh-teen): one of the carotenoids; an orange pigment and vitamin A precursor found in plants.

beta-hydroxymethylbutryate (HMB): a metabolite of the amino acid leucine promoted to increase muscle mass and strength.

BHA and **BHT:** preservatives commonly used to slow the development of off-flavors, odors, and color changes caused by oxidation.

bicarbonate: an alkaline compound with the formula HCO_3 that is secreted from the pancreas as part of the pancreatic juice. Bicarbonate is also produced in all cell fluids from the dissociation of carbonic acid to help maintain the body's acid-base balance.

bifidus (BIFF-id-us, by-FEED-us) **factors:** factors in colostrum and breast milk that favor the growth of the "friendly" bacterium *Lactobacillus* (lack-toh-ba-SILL-us) *bifidus* in the infant's intestinal tract, so that other, less desirable intestinal bacteria will not flourish.

bile: an emulsifier that prepares fats and oils for digestion; an exocrine secretion made by the liver, stored in the gallbladder, and released into the small intestine when needed.

binders: chemical compounds in foods that combine with nutrients (especially minerals) to form complexes the body cannot absorb. Examples include *phytates* (FYE-tates) and *oxalates* (OCK-sa-lates).

binge drinking: pattern of drinking that raises blood alcohol concentration to 0.08 percent or higher; usually corresponds to four or more drinks for women and five or more drinks for men on a single occasion, generally within hours.

binge-eating disorder: an eating disorder with criteria similar to those of bulimia nervosa, excluding purging or other compensatory behaviors.

bioaccumulation: the accumulation of contaminants in the flesh of animals high on the food chain.

bioavailability: the rate at and the extent to which a nutrient is absorbed and used.

bioelectromagnetic medical applications: the use of electrical energy, magnetic energy, or both to stimulate bone repair, wound healing, and tissue regeneration.

biofeedback: the use of special devices to convey information about heart rate, blood pressure, skin temperature, muscle relaxation, and the like to enable a person to learn how to consciously control these medically important functions.

biofield therapeutics: a manual healing method that directs a healing force from an outside source (commonly God or another supernatural being) through the practitioner and into the client's body; commonly known as "laying on of hands."

biological value (BV): a measure of protein quality assessed by measuring the amount of protein nitrogen that is retained from a given amount of protein nitrogen absorbed.

biotechnology: the use of biological systems or organisms to create or modify products. Examples include the use of bacteria to make yogurt, yeast to make beer, and cross-breeding to enhance crop production.

biotin (BY-oh-tin): a B vitamin that functions as a coenzyme in metabolism.

blastocyst (BLASS-toe-sist): the developmental stage of the zygote when it is about 5 days old and ready for implantation.

blind experiment: an experiment in which the subjects do not know whether they are members of the experimental group or the control group.

bloating: uncomfortable abdominal fullness or distention.

blood lipid profile: results of blood tests that reveal a person's total cholesterol, triglycerides, and various lipoproteins.

body composition: the proportions of muscle, bone, fat, and other tissue that make up a person's total body weight.

body mass index (BMI): an index of a person's weight in relation to height; determined by dividing the weight (in kilograms) by the square of the height (in meters).

bolus (BOH-lus): a portion; with respect to food, the amount swallowed at one time.

bomb calorimeter (KAL-oh-RIM-eh-ter): an instrument that measures the heat energy released when foods are burned, thus providing an estimate of the potential energy (kcalories) of the foods.

bone density: a measure of bone strength. When minerals fill the bone matrix (making it dense), they give it strength.

bone meal: crushed or ground bone preparations intended to supply calcium to the diet. Calcium from bone is not well absorbed and is often contaminated with toxic minerals such as arsenic, mercury, lead, and cadmium.

bottled water: drinking water sold in bottles.

botulism (BOT-chew-lism): an often fatal foodborne illness caused by the ingestion of foods containing a toxin produced by bacteria that grow without oxygen.

bovine growth hormone (BGH): a hormone produced naturally in the pituitary gland of a cow that promotes growth and milk production; now produced for agricultural use by bacteria.

bran: the protective coating around the kernel of grain, rich in nutrients and fiber.

branched-chain amino acids: the essential amino acids leucine, isoleucine, and valine, which are present in large amounts in skeletal muscle tissue; falsely promoted as fuel for exercising muscles.

breast milk bank: a service that collects, screens, processes, and distributes donated human milk.

brown adipose tissue: masses of specialized fat cells packed with pigmented mitochondria that produce heat instead of ATP.

brown sugar: refined white sugar crystals to which manufacturers have added molasses syrup with natural flavor and color; 91 to 96 percent pure sucrose.

buffers: compounds that keep a solution's pH constant when acids or bases are added.

bulimia (byoo-LEEM-ee-ah) **nervosa:** an eating disorder characterized by repeated episodes of binge eating usually followed by self-induced vomiting, misuse of laxatives or diuretics, fasting, or excessive exercise.

C

caffeine: a natural stimulant found in many common foods and beverages, including coffee, tea, and chocolate; may enhance endurance by stimulating fatty acid release. High doses cause headaches, trembling, rapid heart rate, and other undesirable side effects.

calbindin: a calcium-binding transport protein that requires vitamin D for its synthesis.

calcidiol: vitamin D found in the blood that is made from the hydroxylation of calciol in the liver; also called *25-hydroxyvitamin D.*

calciferol (kal-SIF-er-ol): vitamin D.

calciol: vitamin D derived from animals in the diet or made in the skin from 7-dehydrocholesterol, a precursor of cholesterol, with the help of sunlight; also called *cholecalciferol* or *vitamin* D_3.

calcitonin (KAL-seh-TOE-nin): a hormone secreted by the thyroid gland that regulates blood calcium by lowering it when levels rise too high.

calcitriol: vitamin D that is made from the hydroxylation of calcidiol in the kidneys; the biologically active hormone; also called *1,25-dihydroxyvitamin D* or *active vitamin D.*

calcium: the most abundant mineral in the body; found primarily in the body's bones and teeth.

calcium rigor: hardness or stiffness of the muscles caused by high blood calcium concentrations.

calcium tetany (TET-ah-nee): intermittent spasm of the extremities due to nervous and muscular excitability caused by low blood calcium concentrations.

calcium-binding protein: a protein in the intestinal cells, made with the help of vitamin D, that facilitates calcium absorption.

calmodulin (cal-MOD-you-lin): a calcium-binding protein that regulates such cell activities as muscle contractions.

calories: units by which energy is measured. Food energy is measured in kilocalories (1000 calories equal 1 kilocalorie), abbreviated **kcalories** or **kcal**. One kcalorie is the amount of heat necessary to raise the temperature of 1 kilogram (kg) of water 1°C. The scientific use of the term *kcalorie* is the same as the popular use of the term *calorie.*

cancers: malignant growths or tumors that result from abnormal and uncontrolled cell division.

capillaries (CAP-ill-aries): small vessels that branch from an artery. Capillaries connect arteries to veins. Exchange of oxygen, nutrients, and waste materials takes place across capillary walls.

carbohydrase (KAR-boe-HIGH-drase): an enzyme that hydrolyzes carbohydrates.

carbohydrate loading: a regimen of moderate exercise followed by the consumption of a high-carbohydrate diet that enables muscles to store glycogen beyond their normal capacities; also called *glycogen loading* or *glycogen super compensation.*

carbohydrates: compounds composed of carbon, oxygen, and hydrogen arranged as monosaccharides or multiples of monosaccharides. Most, but not all, carbohydrates have a ratio of one carbon molecule to one water molecule: $(CH_2O)_n$.

carbonated water: water that contains carbon dioxide gas, either naturally occurring or added, that causes bubbles to form in it; also called *bubbling* or *sparkling water.* The FDA defines seltzer, soda, and tonic waters as soft drinks; they are not regulated as water.

carbonic acid: a compound with the formula H_2CO_3 that results from the combination of carbon dioxide (CO_2) and water (H_2O); of particular importance in maintaining the body's acid-base balance.

carcinogenesis (CAR-sin-oh-JEN-eh-sis): the process of cancer development.

carcinogens (CAR-sin-oh-jenz or car-SIN-oh-jenz): substances that can cause cancer; the adjective is *carcinogenic.*

cardiac output: the volume of blood discharged by the heart each minute; determined by multiplying the stroke volume by the heart rate. The stroke volume is the amount of oxygenated blood the heart ejects toward the tissues at each beat. Cardiac output (volume/minute) = stroke volume (volume/beat) × heart rate (beats/minute).

cardiorespiratory conditioning: improvements in heart and lung function and increased blood volume, brought about by aerobic training.

cardiorespiratory endurance: the ability to perform large-muscle, dynamic exercise of moderate to high intensity for prolonged periods.

cardiovascular disease (CVD): a general term for all diseases of the heart and blood vessels. Atherosclerosis is the main cause of CVD. When the arteries that carry blood to the heart muscle become blocked, the heart suffers damage known as *coronary heart disease (CHD).*

carnitine (CAR-neh-teen): a nonessential, nonprotein amino acid made in the body from lysine that helps transport fatty acids across the mitochondrial membrane. As a supplement, carnitine supposedly "burns" fat and spares glycogen during endurance events, but in reality it does neither.

carotenoids (kah-ROT-eh-noyds): pigments commonly found in plants and animals, some of which have vitamin A activity. The carotenoid with the greatest vitamin A activity is beta-carotene.

cartilage therapy: the use of cleaned and powdered connective tissue, such as collagen, to improve health.

catabolism (ca-TAB-o-lism): reactions in which large molecules are broken down to smaller ones. Catabolic reactions release energy.

catalyst (CAT-uh-list): a compound that facilitates chemical reactions without itself being changed in the process.

cataracts (KAT-ah-rakts): clouding of the eye lenses that impairs vision and can lead to blindness.

cathartic (ka-THAR-tik): a strong laxative.

cations (CAT-eye-uns): positively charged ions.

CDC (Centers for Disease Control): a branch of the Department of Health and Human Services that is responsible for, among other things, monitoring foodborne diseases.

celiac disease: an intestinal disorder in which the inability to absorb the protein portion of gluten results in an immune response that damages intestinal cells; also called *celiac sprue* or *gluten-sensitive enteropathy.*

cell: the basic structural unit of all living things.

cell differentiation (DIF-er-EN-she-AY-shun): the process by which immature cells develop specific functions different from those of the original that are characteristic of their mature cell type.

cell membrane: the thin layer of tissue that surrounds the cell and encloses its contents; made primarily of lipid and protein.

central nervous system: the central part of the nervous system; the brain and spinal cord.

central obesity: excess fat around the trunk of the body; also called *abdominal fat* or *upper-body fat.*

cerebral thrombosis: a clot that blocks blood flow through an artery that feeds the brain.

certification: the process in which a private laboratory inspects shipments of a product for selected chemicals and then, if the product is found to be within acceptable levels of those chemicals, issues a guarantee to that effect.

certified lactation consultants: health-care providers who specialize in helping new mothers establish a healthy breastfeeding relationship with their newborn. These consultants are often registered nurses with specialized training in breast and infant anatomy and physiology.

certified nutritionist or **certified nutritional consultant** or **certified nutrition therapist:** a person who has been granted a document declaring his or her authority as a nutrition professional.

cesarean section: a surgically assisted birth involving removal of the fetus by an incision into the uterus, usually by way of the abdominal wall.

chaff: the outer inedible part of a grain; also called the *husk.*

CHD risk equivalents: disorders that raise the risk of heart attacks, strokes, and other complications associated with cardiovascular disease to the same degree as existing CHD. These disorders include symptomatic carotid artery disease, peripheral arterial disease, abdominal aortic aneurysm, and diabetes mellitus.

cheilosis (kye-LOH-sis or kee-LOH-sis): a condition of reddened lips with cracks at the corners of the mouth.

chelate (KEY-late): a substance that can grasp the positive ions of a mineral.

chelation (kee-LAY-shun) **therapy:** the use of ethylene diamine tetraacetic acid (EDTA) to bind with metallic ions, thus healing the body by removing toxic metals.

chiropractic (KYE-roh-PRAK-tik): a manual healing method of manipulating the spine to restore health.

chloride (KLO-ride): the major anion in the extracellular fluids of the body. Chloride is the ionic form of chlorine, Cl^-.

chlorophyll (KLO-row-fil): the green pigment of plants, which absorbs light and transfers the energy to other molecules, thereby initiating photosynthesis.

cholecalciferol (KO-lee-kal-SIF-er-ol): vitamin D derived from animals in the diet and made in the skin from 7-dehydrocholesterol, a precursor of cholesterol, with the help of sunlight; also called *vitamin D_3.*

cholecystokinin (COAL-ee-SIS-toe-KINE-in), or **CCK:** a hormone produced by cells of the intestinal wall. Target organ: the gallbladder. Response: release of bile and slowing of GI motility.

cholesterol (koh-LESS-ter-ol): one of the sterols containing a four-ring carbon structure with a carbon side chain.

cholesterol-free: less than 2 milligrams cholesterol per serving and 2 grams or less saturated fat and *trans* fat combined per serving.

choline (KOH-leen): a nitrogen-containing compound found in foods and made in the body from the amino acid methionine. Choline is part of the phospholipid lecithin and the neurotransmitter acetylcholine.

chromium: an essential trace mineral that enhances the activity of insulin.

chromium picolinate (CROW-mee-um pick-oh-LYNate): a trace mineral supplement; falsely promoted as building muscle, enhancing energy, and burning fat. *Picolinate* is a derivative of the amino acid tryptophan, which seems to enhance chromium absorption.

chromosomes: structures within the nucleus of a cell made of DNA and associated proteins. Human beings have 46 chromosomes in 23 pairs. Each chromosome has many genes.

chronic diseases: diseases characterized by a slow progression and long duration. Examples include heart disease, cancer, and diabetes.

chronic malnutrition: malnutrition caused by long-term food deprivation; characterized in children by short height for age (*stunting*).

chronological age: a person's age in years from his or her date of birth.

chylomicrons (kye-lo-MY-cronz): the class of lipoproteins that transport lipids from the intestinal cells to the rest of the body.

chyme (KIME): the semiliquid mass of partly digested food expelled by the stomach into the duodenum.

cirrhosis (seer-OH-sis): advanced liver disease in which liver cells turn orange, die, and harden, permanently losing their function; often associated with alcoholism.

cis: on the near side of; refers to a chemical configuration in which the hydrogen atoms are located on the same side of a double bond.

citric acid cycle: a series of metabolic reactions that break down molecules of acetyl CoA to carbon dioxide and hydrogen atoms; also called the *TCA cycle, tricarboxylic acid cycle,* or the *Kreb's cycle.*

clinically severe obesity: a BMI of 40 or greater or a BMI of 35 or greater with additional medical problems. A less preferred term used to describe the same condition is *morbid obesity.*

clone: a genetic copy of an animal, similar to identical twins but born at different times.

CoA (coh-AY): coenzyme A; the coenzyme derived from the B vitamin pantothenic acid and central to energy metabolism.

coenzymes: complex organic molecules that work with enzymes to facilitate the enzymes' activity. Many coenzymes have B vitamins as part of their structures.

cofactor: a small, inorganic or organic substance that facilitates the action of an enzyme.

colitis (ko-LYE-tis): inflammation of the colon.

collagen (KOL-ah-jen): the structural protein from which connective tissues such as scars, tendons, ligaments, and the foundations of bones and teeth are made.

colonic irrigation: the popular, but potentially harmful practice of "washing" the large intestine with a powerful enema machine.

colostrum (ko-LAHS-trum): a milklike secretion from the breast, present during the first few days after delivery before milk appears; rich in protective factors.

complementary and alternative medicine (CAM): diverse medical and health-care systems, practices, and products that are not currently considered part of conventional medicine; also called *adjunctive, unconventional,* or *unorthodox therapies.*

complementary proteins: two or more dietary proteins whose amino acid assortments complement each other in such a way that the essential amino acids missing from one are supplied by the other.

complex carbohydrates: polysaccharides (starches and fibers).

compound: a substance composed of two or more different atoms—for example, water (H_2O).

conception: the union of the male sperm and the female ovum; fertilization.

condensation: a chemical reaction in which water is released as two molecules combine to form one larger product.

conditionally essential amino acid: an amino acid that is normally nonessential, but must be supplied by the diet in special circumstances when the need for it exceeds the body's ability to produce it.

conditionally essential nutrient: a nutrient that is normally nonessential, but must be supplied by the diet in special circumstances when the need for it exceeds the body's ability to produce it.

conditioning: the physical effect of training; improved flexibility, strength, and endurance.

confectioners' sugar: finely powdered sucrose, 99.9 percent pure.

congregate meals: nutrition programs that provide food for the elderly in conveniently located settings such as community centers.

conjugated linoleic acid: several fatty acids that have the same chemical formula as linoleic acid (18 carbons, two double bonds) but with different configurations (the double bonds occur on adjacent carbons).

constipation: the condition of having infrequent or difficult bowel movements.

contaminants: substances that make a food impure and unsuitable for ingestion.

contamination iron: iron found in foods as the result of contamination by inorganic iron salts from iron cookware, iron-containing soils, and the like.

control group: a group of individuals similar in all possible respects to the experimental group except for the treatment. Ideally, the control group receives a placebo while the experimental group receives a real treatment.

convenient dietary supplements: liquid mean replacers, energy drinks, energy bars, and energy gels that athletes and active people use to replenish energy and nutrients when time is limited.

conventional medicine: diagnosis and treatment of diseases as practiced by medical doctors (M.D.), doctors of osteopathy (D.O.), and allied health professionals such as physical therapists and registered nurses; also called *allopathy; Western, mainstream, orthodox,* or *regular medicine;* and *biomedicine.*

cool-down: 5 to 10 minutes of light activity, such as walking or stretching, following a vigorous workout to gradually return the body's core to near-normal temperature.

copper: an essential trace mineral that is part of many enzymes.

Cori cycle: the pathway in which glucose is metabolized to lactate (by anaerobic glycolysis) in the muscle, lactate is converted back to glucose in the liver, then glucose is returned to the muscle; named after the scientist who elucidated this pathway.

corn sweeteners: corn syrup and sugars derived from corn.

corn syrup: a syrup made from cornstarch that has been treated with acid, high temperatures, and enzymes to produce glucose, maltose, and dextrins. It may be dried and used as corn syrup solids. See also *high-fructose corn syrup (HFCS).*

cornea (KOR-nee-uh): the transparent membrane covering the outside of the eye.

coronary arteries: blood vessels that supply blood to the heart.

coronary heart disease (CHD): the damage that occurs when the blood vessels carrying blood to the heart (the *coronary arteries*) become narrow and occluded.

coronary thrombosis: a clot that blocks blood flow through an artery that feeds the heart muscle.

correlation (CORE-ee-LAY-shun): the simultaneous increase, decrease, or change in two variables. If A increases as B increases, or if A decreases as B decreases, the correlation is positive. (This does not mean that A causes B or vice versa.) If A increases as B decreases, or if A decreases as B increases, the correlation is negative. (This does not mean that A prevents B or vice versa.) Some third factor may account for both A and B.

cortical bone: the very dense bone tissue that forms the outer shell surrounding trabecular bone and comprises the shaft of a long bone.

coupled reactions: pairs of chemical reactions in which some of the energy released from the breakdown of one compound is used to create a bond in the formation of another compound.

covert (KOH-vert) hidden, as if under covers.

CP, or **creatine phosphate:** a high-energy compound in muscle cells that acts as a reservoir of energy that can maintain a steady supply of ATP. CP provides the energy for short bursts of activity; also called *phosphocreatine*.

C-reactive protein (CRP): a protein released during the acute phase of infection or inflammation that enhances immunity by promoting phagocytosis and activating platelets. Its presence may be used to assess a person's risk of an impending heart attack or stroke.

creatine (KREE-ah-tin): a nitrogen-containing compound that combines with phosphate to form the high-energy compound creatine phosphate (or phosphocreatine) in muscles.

cretinism (CREE-tin-ism): a congenital disease characterized by mental and physical retardation and commonly caused by maternal iodine deficiency during pregnancy.

critical periods: finite periods during development in which certain events occur that will have irreversible effects on later developmental stages; usually a period of rapid cell division.

cross-contamination: the contamination of food by bacteria that occurs when the food comes into contact with surfaces previously touched by raw meat, poultry, or seafood.

cruciferous vegetables: vegetables of the cabbage family, including cauliflower, broccoli, and brussels sprouts.

crypts (KRIPTS): tubular glands that lie between the intestinal villi and secrete intestinal juices into the small intestine.

cultural competence: having an awareness and acceptance of cultures and the ability to interact effectively with people of diverse cultures.

cyclamate (SIGH-kla-mate): an artificial sweetener composed of a sodium or calcium salt of cyclamic acid that is being considered for approval in the United States and is available in Canada as a tabletop sweetener, but not as an additive.

cytokines (SIGH-toe-kines): special proteins that direct immune and inflammatory responses.

cytoplasm (SIGH-toh-plazm): the cell contents, except for the nucleus.

cytosol: the fluid of cytoplasm that contains water, ions, nutrients, and enzymes.

D

Daily Values (DV): reference values developed by the FDA specifically for use on food labels.

deamination (dee-AM-ih-NAY-shun): removal of the amino (NH_2) group from a compound such as an amino acid.

defecate (DEF-uh-cate): to move the bowels and eliminate waste.

deficient: inadequate; a nutrient amount that fails to meet the body's needs and eventually results in deficiency symptoms.

dehydration: the condition in which body water output exceeds water input. Symptoms include thirst, dry skin and mucous membranes, rapid heartbeat, low blood pressure, and weakness.

Delaney Clause: a clause in the Food Additive Amendment to the Food, Drug, and Cosmetic Act that states that no substance that is known to cause cancer in animals or human beings at any dose level shall be added to foods.

***de minimus* rule:** a guideline that defines risk as a cancer rate of less than one cancer per million people exposed to a contaminant over a 70-year lifetime.

denaturation (dee-NAY-chur-AY-shun): the change in a protein's shape and consequent loss of its function brought about by heat, agitation, acid, base, alcohol, heavy metals, or other agents.

dental caries: decay of teeth.

dental plaque: a gummy mass of bacteria that grows on teeth and can lead to dental caries and gum disease.

dextrose: the name food manufacturers use for the sugar that is chemically the same as glucose.

DHEA (dehydroepiandrosterone) and **androstenedione:** hormones made in the adrenal glands that serve as precursors to the male hormone testosterone; falsely promoted as burning fat, building muscle, and slowing aging.

DHF (dihydrofolate): a coenzyme form of folate.

diabetes (DYE-uh-BEET-eez): chronic disorders of carbohydrate metabolism, usually characterized by hyperglycemia resulting from insufficient or ineffective insulin; clinically called *diabetes mellitus* (MELL-ih-tus or MELL-eye-tus).

diarrhea: the frequent passage of watery bowel movements.

diet: the foods and beverages a person eats and drinks.

diet history: a record of eating behaviors and the foods a person eats.

dietary fibers: in plant foods, the *nonstarch polysaccharides* that are not digested by human digestive enzymes, although some are digested by GI tract bacteria.

dietary folate equivalents (DFE): the amount of folate available to the body from naturally occurring sources, fortified foods, and supplements, accounting for differences in the bioavailability from each source.

Dietary Reference Intakes (DRI): a set of nutrient intake values for healthy people in the United States and Canada. These values are used for planning and assessing diets and include: Estimated Average Requirements (EAR), Recommended Dietary Allowances (RDA), Adequate Intakes (AI), and Tolerable Upper Intake Levels (UL).

dietary supplement: any pill, capsule, tablet, liquid, or powder that contains vitamins, minerals, herbs, or amino acids; intended to increase dietary intake of these substances.

dietetic technician: a person who has completed a minimum of an associate's degree from an accredited university or college and an approved dietetic technician program that includes a supervised practice experience. See also *dietetic technician, registered (DTR)*.

dietetic technician, registered (DTR): a dietetic technician who has passed a national examination and maintains registration through continuing professional education.

dietitian: a person trained in nutrition, food science, and diet planning. See also *registered dietitian*.

digestion: the process by which food is broken down into absorbable units.

digestive enzymes: proteins found in digestive juices that act on food substances, causing them to break down into simpler compounds.

digestive system: all the organs and glands associated with the ingestion and digestion of food.

dioxins (dye-OCK-sins): a class of chemical pollutants created as by-products of chemical manufacturing, incineration, chlorine bleaching of paper pulp, and other industrial processes. Dioxins persist in the environment and accumulate in the food chain.

dipeptide (dye-PEP-tide): two amino acids bonded together.

diploma mills: entities without valid accreditation that provide worthless degrees.

direct calorimetry: a means of estimating energy expenditure by measuring the amount of heat released.

disaccharides (dye-SACK-uh-rides): pairs of monosaccharides linked together. See Appendix C for the chemical structures of the disaccharides.

discretionary kcalories: the kcalories remaining in a person's energy allowance after consuming enough nutrient-dense foods to meet all nutrient needs for a day.

disordered eating: eating behaviors that are neither normal nor healthy, including restrained eating, fasting, binge eating, and purging.

dispensable amino acids: nonessential amino acids.

dissociates (dis-SO-see-aites): physically separates.

distilled water: water that has been vaporized and recondensed, leaving it free of dissolved minerals.

diverticula (dye-ver-TIC-you-la): sacs or pouches that develop in the weakened areas of the intestinal wall (like bulges in an inner tube where the tire wall is weak).

diverticulitis (DYE-ver-tic-you-LYE-tis): infected or inflamed diverticula.

diverticulosis (DYE-ver-tic-you-LOH-sis): the condition of having diverticula.

DNA (deoxyribonucleic acid): the double helix molecules of which genes are made.

docosahexaenoic (DOE-cossa-HEXA-ee-NO-ick) **acid (DHA):** an omega-3 polyunsaturated fatty acid with 22 carbons and six double bonds; present in fatty fish and synthesized in limited amounts in the body from linolenic acid.

dolomite: a compound of minerals (calcium magnesium carbonate) found in limestone and marble. Dolomite is powdered and is sold as a calcium-magnesium supplement. However, it may be contaminated with toxic minerals, is not well absorbed, and interferes with absorption of other essential minerals.

double-blind experiment: an experiment in which neither the subjects nor the researchers know which subjects are members of the experimental group and which are serving as control subjects, until after the experiment is over.

Down syndrome: a genetic abnormality that causes mental retardation, short stature, and flattened facial features.

drink: a dose of any alcoholic beverage that delivers ½ ounce of pure ethanol: 5 ounces of wine, 10 ounces of wine cooler, 12 ounces of beer, or 1½ ounces of liquor.

drug: a substance that can modify one or more of the body's functions.

drug history: a record of all the drugs, over-the-counter and prescribed, that a person takes routinely.

DTR: see *dietetic technician, registered.*

duodenum (doo-oh-DEEN-um, doo-ODD-num): the top portion of the small intestine (about "12 fingers' breadth" long in ancient terminology).

duration: length of time (for example, the time spent in each activity session).

dysentery (DISS-en-terry): an infection of the digestive tract that causes diarrhea.

dysphagia (dis-FAY-jah): difficulty swallowing.

E

eating disorders: disturbances in eating behavior that jeopardize a person's physical or psychological health.

eating pattern: customary intake of foods and beverages over time.

eclampsia (eh-KLAMP-see-ah): a condition characterized by extremely high blood pressure, elevated protein in the urine, seizures, and possibly coma.

edema (eh-DEEM-uh): the swelling of body tissue caused by excessive amounts of fluid in the interstitial spaces; seen in protein deficiency (among other conditions).

edentulous (ee-DENT-you-lus): lack of teeth.

eicosanoids (eye-COSS-uh-noyds): derivatives of 20-carbon fatty acids; biologically active compounds that help to regulate blood pressure, blood clotting, and other body functions. They include *prostaglandins* (PROS-tah-GLAND-ins), *thromboxanes* (throm-BOX-ains), and *leukotrienes* (LOO-ko-TRY-eens).

eicosapentaenoic (EYE-cossa-PENTA-ee-NO-ick) **acid (EPA):** an omega-3 polyunsaturated fatty acid with 20 carbons and five double bonds; present in fatty fish and synthesized in limited amounts in the body from linolenic acid.

electrolyte solutions: solutions that can conduct electricity.

electrolytes: salts that dissolve in water and dissociate into charged particles called ions.

electron transport chain: the final pathway in energy metabolism that transports electrons from hydrogen to oxygen and captures the energy released in the bonds of ATP; also called the *respiratory chain.*

element: a substance composed of atoms that are alike—for example, iron (Fe).

embolism (EM-boh-lizm): the obstruction of a blood vessel by an *embolus* (EM-boh-luss), or traveling clot, causing sudden tissue death.

embryo (EM-bree-oh): the developing infant from 2 to 8 weeks after conception.

emergency shelters: facilities that are used to provide temporary housing.

emerging risk factors: recently identified factors that enhance the ability to predict disease risk in an individual.

emetic (em-ETT-ic): an agent that causes vomiting.

empty-kcalorie foods: a popular term used to denote foods that contribute energy but lack protein, vitamins, and minerals.

emulsifier (ee-MUL-sih-fire): substances with both water-soluble and fat-soluble portions that promotes the mixing of oils and fats in a watery solution.

endogenous (en-DODGE-eh-nus): from within the body.

endoplasmic reticulum (en-doh-PLAZ-mic reh-TIC-you-lum): a complex network of intracellular membranes. The *rough endoplasmic reticulum* is dotted with ribosomes, where protein synthesis takes place. The *smooth endoplasmic reticulum* bears no ribosomes.

endosperm: the inner edible part of a kernel of grain, rich in starch and proteins.

enema: solution inserted into the rectum and colon to stimulate a bowel movement and empty the lower large intestine.

energy: the capacity to do work. The energy in food is chemical energy. The body can convert this chemical energy to mechanical, electrical, or heat energy.

energy balance: the energy (kcalories) consumed from foods and beverages compared with the energy expended through metabolic processes and physical activities.

energy density: a measure of the energy a food provides relative to the amount of food (kcalories per gram).

energy-yielding nutrients: the nutrients that break down to yield energy the body can use (carbohydrate, fat, and protein).

enhanced water: water that is fortified with ingredients such as vitamins, minerals, protein, oxygen, or herbs. Enhanced water is marketed as *vitamin water, sports water, oxygenated water,* and *protein water.*

enriched: the addition to a food of nutrients that were lost during processing so that the food will meet a specified standard.

enteropancreatic (EN-ter-oh-PAN-kree-AT-ik) **circulation:** the circulatory route from the pancreas to the intestine and back to the pancreas.

enzymes: proteins that facilitate chemical reactions without being changed in the process; protein catalysts.

EPA (Environmental Protection Agency): a federal agency that is responsible for, among other things, regulating pesticides and establishing water quality standards.

epidemic (ep-ih-DEM-ick): the appearance of a disease (usually infectious) or condition that attacks many people at the same time in the same region.

epigenetics: the study of heritable changes in gene function that occur without a change in the DNA sequence.

epiglottis (epp-ih-GLOTT-iss): cartilage in the throat that guards the entrance to the trachea and prevents fluid or food from entering it when a person swallows.

epinephrine (EP-ih-NEFF-rin): a hormone of the adrenal gland that modulates the stress response; formerly called *adrenaline*. When administered by injection, epinephrine counteracts anaphylactic shock by opening the airways and maintaining heartbeat and blood pressure.

epithelial (ep-i-THEE-lee-ul) **cells:** cells on the surface of the skin and mucous membranes.

epithelial tissue: the layer of the body that serves as a selective barrier between the body's interior and the environment. Examples are the cornea of the eyes, the skin, the respiratory lining of the lungs, and the lining of the digestive tract.

ergocalciferol (ER-go-kal-SIF-er-ol): vitamin D derived from plants in the diet; also called *vitamin* D_2.

ergogenic (ER-go-JEN-ick) **aids:** substances or techniques used in an attempt to enhance physical performance.

erythrocyte (eh-RITH-ro-cite) **hemolysis** (he-MOLL-uh-sis): the breaking open of red blood cells (erythrocytes); a symptom of vitamin E–deficiency disease in human beings.

erythrocyte protoporphyrin (PRO-toe-PORE-fe-rin): a precursor to hemoglobin.

erythropoietin (eh-RITH-ro-POY-eh-tin): a hormone that stimulates red blood cell production.

esophageal (ee-SOF-ah-GEE-al) **sphincter:** a sphincter muscle at the upper or lower end of the esophagus. The lower esophageal sphincter is also called the *cardiac sphincter.*

esophagus (ee-SOFF-ah-gus): the food pipe; the conduit from the mouth to the stomach.

essential amino acids: amino acids that the body cannot synthesize in amounts sufficient to meet physiological needs.

essential fatty acids: fatty acids that the body cannot synthesize in amounts sufficient to meet physiological needs.

essential nutrients: nutrients a person must obtain from food because the body cannot synthesize them in amounts sufficient to meet physiological needs; also called *indispensable nutrients.* About 40 nutrients are currently known to be essential for human beings.

Estimated Average Requirement (EAR): the average daily amount of a nutrient that will maintain a specific biochemical or physiological function in half the healthy people of a given age and gender group.

Estimated Energy Requirement (EER): the average dietary energy intake that maintains energy balance and good health in a person of a given age, gender, weight, height, and level of physical activity.

estrogens: hormones responsible for the menstrual cycle and other female characteristics.

ethanol: a particular type of alcohol found in beer, wine, and liquor; also called *ethyl alcohol.*

ethnic foods: foods associated with particular cultural groups.

excessive drinking: heavy drinking, binge drinking, or both.

exchange lists: diet-planning tools that organize foods by their proportions of carbohydrate, fat, and protein. Foods on any single list can be used interchangeably.

exercise: planned, structured, and repetitive body movements that promote or maintain physical fitness.

exogenous (eks-ODGE-eh-nus): from outside the body, usually from foods.

experimental group: a group of individuals similar in all possible respects to the control group except for the treatment. The experimental group receives the real treatment.

extra lean: less than 5 grams of fat, 2 grams of saturated fat and *trans* fat combined, and 95 milligrams cholesterol per serving and per 100 grams of meat, poultry, and seafood.

extracellular fluid: fluid outside the cells. Extracellular fluid includes two main components—the interstitial fluid between cells and the intravascular fluid of plasma. Extracellular fluid accounts for approximately one-third of the body's water.

F

FAD (flavin adenine dinucleotide): a coenzyme form of riboflavin.

fad diets: popular eating plans that promise quick weight loss. Most fad diets severely limit certain foods or overemphasize others (for example, never eat potatoes or pasta or eat cabbage soup daily).

faith healing: healing by invoking divine intervention without the use of medical, surgical, or other traditional therapy.

false negative: a test result indicating that a condition is not present (negative) when in fact it is present (therefore false).

false positive: a test result indicating that a condition is present (positive) when in fact it is not (therefore false).

famine: widespread and extreme scarcity of food in an area that causes starvation and death in a large portion of the population.

FAO (Food and Agriculture Organization): an international agency (part of the United Nations) that has adopted standards to regulate pesticide use among other responsibilities.

fat replacers: ingredients that replace some or all of the functions of fat and may or may not provide energy.

fat-free: less than 0.5 gram of fat per serving (and no added fat or oil); synonyms include *zero-fat, no fat,* and *nonfat.*

fats: lipids that are solid at room temperature (77°F or 25°C).

fatty acid: an organic compound composed of a carbon chain with hydrogens attached and an acid group (COOH) at one end and a methyl group (CH_3) at the other end.

fatty acid oxidation: the metabolic breakdown of fatty acids to acetyl CoA; also called *beta oxidation.*

fatty liver: an early stage of liver deterioration seen in several diseases, including kwashiorkor and alcoholic liver disease. Fatty liver is characterized by an accumulation of fat in the liver cells.

fatty streaks: accumulations of cholesterol and other lipids along the walls of the arteries.

FDA (Food and Drug Administration): a part of the Department of Health and Human Services' Public Health Service that is responsible for ensuring the safety and wholesomeness of all dietary supplements and food processed and sold in interstate commerce except meat, poultry, and eggs (which are under the jurisdiction of the USDA); inspecting food plants and imported foods; and setting standards for food composition and product labeling.

female athlete triad: a potentially fatal combination of three medical problems—disordered eating, amenorrhea, and osteoporosis.

fermentable: the extent to which bacteria in the GI tract can break down fibers to fragments that the body can use.

ferritin (FAIR-ih-tin): the iron storage protein.

fertility: the capacity of a woman to produce a normal ovum periodically and of a man to produce normal sperm; the ability to reproduce.

fetal alcohol spectrum disorder: a range of physical, behavioral, and cognitive abnormalities caused by prenatal alcohol exposure.

fetal alcohol syndrome (FAS): a cluster of physical, behavioral, and cognitive abnormalities associated with prenatal alcohol exposure, including facial malformations, growth retardation, and central nervous disorders.

fetal programming: the influence of substances during fetal growth on the development of diseases in later life.

fetus (FEET-us): the developing infant from 8 weeks after conception until term.

fibrocystic (FYE-bro-SIS-tik) **breast disease:** a harmless condition in which the breasts develop lumps, sometimes associated with caffeine consumption. In some, it responds to abstinence from caffeine; in others, it can be treated with vitamin E.

fibrosis (fye-BROH-sis): an intermediate stage of liver deterioration seen in several diseases, including viral hepatitis and alcoholic liver disease. In fibrosis, the liver cells lose their function and assume the characteristics of connective tissue cells (fibers).

field gleaning: collecting crops from fields that either have already been harvested or are not profitable to harvest.

filtered water: water treated by filtration, usually through *activated carbon filters* that reduce the lead in tap water, or by *reverse osmosis* units that force pressurized water across a membrane removing lead, arsenic, and some microorganisms from tap water.

fitness: the characteristics that enable the body to perform physical activity; more broadly, the ability to meet routine physical demands with enough reserve energy to rise to a physical challenge; or the body's ability to withstand stress of all kinds.

flatulence: passage of excessive amounts of intestinal gas.

flavonoids (FLAY-von-oyds): yellow pigments in foods; phytochemicals that may exert physiological effects on the body.

flaxseeds: the small brown seeds of the flax plant; valued in nutrition as a source of fiber, lignans, and omega-3 fatty acids.

flexibility: the capacity of the joints to move through a full range of motion; the ability to bend and recover without injury.

flora: bacteria in the intestines.

fluid balance: maintenance of the proper types and amounts of fluid in each compartment of the body fluids.

fluorapatite (floor-APP-uh-tite): the stabilized form of bone and tooth crystal, in which fluoride has replaced the hydroxyl groups of hydroxyapatite.

fluoride: an essential trace mineral that makes teeth stronger and more resistant to decay.

fluorosis (floor-OH-sis): discoloration and pitting of tooth enamel caused by excess fluoride during tooth development.

FMN (flavin mononucleotide): a coenzyme form of riboflavin.

folate (FOLE-ate): a B vitamin; also known as folic acid, folacin, or pteroylglutamic (tare-o-EEL-glue-TAM-ick) acid (PGA). The coenzyme forms are *DHF (dihydrofolate)* and *THF (tetrahydrofolate).*

follicle-stimulating hormone (FSH): a hormone that stimulates maturation of the ovarian follicles in females and the production of sperm in males. (The ovarian follicles are part of the female reproductive system where the eggs are produced.) The release of FSH is mediated by *follicle-stimulating hormone releasing hormone (FSH–RH).*

food allergy: an adverse reaction to food that involves an immune response; also called *food-hypersensitivity reaction.*

food aversions: strong desires to avoid particular foods.

food bank: a facility that collects and distributes food donations to authorized organizations feeding the hungry.

food chain: the sequence in which living things depend on other living things for food.

food cravings: strong desires to eat particular foods.

food crisis: a sharp rise in the rates of hunger and malnutrition, usually set off by a shock to either the supply of, or demand for, food and a sudden spike in food prices.

food deserts: neighborhoods and communities characterized by limited access to nutritious and affordable foods.

food frequency questionnaire: a checklist of foods on which a person can record the frequency with which he or she eats each food.

food group plans: diet-planning tools that sort foods into groups based on nutrient content and then specify that people should eat certain amounts of foods from each group.

food insecurity: limited or uncertain access to foods of sufficient quality or quantity to sustain a healthy and active life.

food insufficiency: an inadequate amount of food due to a lack of resources.

food intolerances: adverse reactions to foods that do not involve the immune system.

food pantries: programs that provide groceries to be prepared and eaten at home.

food poverty: hunger resulting from inadequate access to available food for various reasons, including inadequate resources, political obstacles, social disruptions, poor weather conditions, and lack of transportation.

food record: an extensive, accurate log of all foods eaten over a period of several days or weeks. A food record that includes associated information such as when, where, and with whom each food is eaten is sometimes called a *food diary.*

food recovery: collecting wholesome food for distribution to low-income people who are hungry.

food security: access to enough food to sustain a healthy and active life.

food substitutes: foods that are designed to replace other foods.

foodborne illness: illness transmitted to human beings through food and water, caused by either an infectious agent (foodborne infection) or a poisonous substance (food intoxication); commonly known as *food poisoning.*

foods: products derived from plants or animals that can be taken into the body to yield energy and nutrients for the maintenance of life and the growth and repair of tissues.

fortified: the addition to a food of nutrients that were either not originally present or present in insignificant amounts. Fortification can be used to correct or prevent a widespread nutrient deficiency or to balance the total nutrient profile of a food.

fossil fuels: coal, oil, and natural gas.

fraudulent: the promotion, for financial gain, of devices, treatments, services, plans, or products (including diets and supplements) that alter or claim to alter a human condition without proof of safety or effectiveness.

free: "nutritionally trivial" and unlikely to have a physiological consequence; synonyms include *without, no,* and *zero.* A food that does not contain a nutrient naturally may make such a claim, but only as it applies to all similar foods (for example, "applesauce, a fat-free food").

free radical: an unstable molecule with one or more unpaired electrons.

frequency: the number of occurrences per unit of time (for example, the number of activity sessions per week).

fructose (FRUK-tose or FROOK-tose): a monosaccharide; sometimes known as *fruit sugar* or *levulose.* Fructose is found abundantly in fruits, honey, and saps.

fuel: compounds that cells can use for energy. The major fuels include glucose, fatty acids, and amino acids; other fuels include ketone bodies, lactate, glycerol, and alcohol.

full term: between the thirty-eighth and forty-second week of pregnancy.

functional foods: whole, fortified, or modified foods that contain physiologically active compounds that provide health benefits beyond their nutrient contributions; sometimes called *designer foods* or *nutraceuticals.*

G

g: grams; a unit of weight equivalent to about 0.03 ounces.

galactose (ga-LAK-tose): a monosaccharide; part of the disaccharide lactose.

gallbladder: the organ that stores and concentrates bile. When it receives the signal that fat is present in the duodenum, the gallbladder contracts and squirts bile through the bile duct into the duodenum.

gangrene (GANG-green): death of tissue, usually due to insufficient blood supply.

gastric glands: exocrine glands in the stomach wall that secrete gastric juice into the stomach.

gastric juice: the digestive secretion of the gastric glands of the stomach.

gastrin: a hormone secreted by cells in the stomach wall. Target organ: the glands of the stomach. Response: secretion of gastric acid.

gastroesophageal reflux: the backflow of stomach acid into the esophagus, causing damage to the cells of the esophagus and the sensation of heartburn; commonly known as *heartburn* or *acid indigestion.*

gastroesophageal reflux disease (GERD): a condition characterized by the backflow of stomach acid into the esophagus two or more times a week.

gastrointestinal (GI) tract: the digestive tract. The principal organs are the stomach and intestines.

gatekeepers: with respect to nutrition, key people who control other people's access to foods and thereby exert profound impacts on their nutrition. Examples are the spouse who buys and cooks the food, the parent who feeds the children, and the caregiver in a day-care center.

gene expression: the process by which a cell converts the genetic code into RNA and protein.

gene pool: all the genetic information of a population at a given time.

generally recognized as safe (GRAS): food additives that have long been in use and are believed to be safe. First established by the FDA in 1958, the GRAS list is subject to revision as new facts become known.

genes: sections of chromosomes that contain the instructions needed to make one or more proteins.

genetic engineering: the use of biotechnology to modify the genetic material of living cells so that they will produce new substances or perform new functions. Foods produced via this technology are called *genetically modified (GM)* or *genetically engineered (GE) foods.*

genetics: the study of genes and inheritance.

genome (GEE-nome): the complete set of genetic material (DNA) in an organism or a cell. The study of genomes is called *genomics.*

genomics: the study of all the genes in an organism and their interactions with environmental factors.

genotoxicant: a substance that mutates or damages genetic material.

geophagia: the specific craving for nonfood items such as clay, baby powder, chalk, ash, ceramics, paper, paint chips, charcoal, or dirt.

germ: the seed that grows into a mature plant, especially rich in vitamins and minerals.

gestation (jes-TAY-shun): the period from conception to birth. For human beings, the average length of a healthy gestation is 40 weeks. Pregnancy is often divided into 3-month periods, called *trimesters.*

gestational diabetes: glucose intolerance with onset or first recognition during pregnancy.

gestational hypertension: high blood pressure that develops in the second half of pregnancy and resolves after childbirth, usually without affecting the outcome of the pregnancy.

ghrelin (GRELL-in): a protein produced by the stomach cells that enhances appetite and decreases energy expenditure.

glands: cells or groups of cells that secrete materials for special uses in the body. Glands may be *exocrine* (EKS-oh-crin) *glands,* secreting their materials "out" (into the digestive tract or onto the surface of the skin), or *endocrine* (EN-doe-crin) *glands,* secreting their materials "in" (into the blood).

glossitis (gloss-EYE-tis): an inflammation of the tongue.

glucagon (GLOO-ka-gon): a hormone secreted by special cells in the pancreas in response to low blood glucose concentration. Glucagon elicits release of glucose from liver glycogen stores.

glucocorticoids: hormones from the adrenal cortex that affect the body's management of glucose.

glucogenic amino acids: amino acids that can make glucose via either pyruvate or TCA cycle intermediates.

gluconeogenesis (gloo-ko-nee-oh-JEN-ih-sis): the making of glucose from a noncarbohydrate source such as amino acids or glycerol.

glucose (GLOO-kose): a monosaccharide; sometimes known as *blood sugar* in the body or *dextrose* in foods.

glucose polymers: compounds that supply glucose, not as single molecules, but linked in chains somewhat like starch. The objective is to attract less water from the body into the digestive tract (osmotic attraction depends on the number, not the size, of particles).

glucose tolerance factors (GTF): small organic compounds that enhance insulin's actions.

glycemic (gly-SEEM-ic) **index:** a method of classifying foods according to their potential for raising blood glucose.

glycemic response: the extent to which a food raises the blood glucose concentration and elicits an insulin response.

glycerol (GLISS-er-ol): an alcohol composed of a three-carbon chain, which can serve as the backbone for a triglyceride.

glycogen (GLY-ko-jen): an animal polysaccharide composed of glucose; a storage form of glucose manufactured and stored in the liver and muscles. Glycogen is not a significant food source of carbohydrate and is not counted as a dietary carbohydrate in foods.

glycolysis (gly-COLL-ih-sis): the metabolic breakdown of glucose to pyruvate. Glycolysis does not require oxygen (anaerobic).

goblet cells: cells of the GI tract (and lungs) that secrete mucus.

goiter (GOY-ter): an enlargement of the thyroid gland due to an iodine deficiency, malfunction of the gland, or overconsumption of a goitrogen. Goiter caused by iodine deficiency is *simple goiter.*

goitrogen (GOY-troh-jen): a substance that enlarges the thyroid gland and causes *toxic goiter.* Goitrogens occur naturally in such foods as cabbage, kale, brussels sprouts, cauliflower, broccoli, and kohlrabi.

Golgi (GOAL-gee) **apparatus:** a set of membranes within the cell where secretory materials are packaged for export.

good source of: the product provides between 10 and 19 percent of the Daily Value for a given nutrient per serving.

gout (GOWT): a common form of arthritis characterized by deposits of uric acid crystals in the joints.

growth hormone (GH): a hormone secreted by the pituitary that regulates the cell division and protein synthesis needed for normal growth; also called *somatotropin.* The release of GH is mediated by *GH-releasing hormone (GHRH)* and *GH-inhibiting hormone.*

H

hard water: water with a high calcium and magnesium content.

Hazard Analysis Critical Control Points (HACCP): a systematic plan to identify and correct potential microbial hazards in the manufacturing, distribution, and commercial use of food products; commonly referred to as "HASS-ip."

hazard: a source of danger; used to refer to circumstances in which harm is possible under normal conditions of use.

HDL (high-density lipoprotein): the type of lipoprotein that transports cholesterol back to the liver from the cells; composed primarily of protein.

health claims: statements that characterize the relationship between a nutrient or other substance in a food and a disease or health-related condition.

health history: an account of a client's current and past health status and disease risks.

healthy: on food labels, a food that is low in fat, saturated fat, cholesterol, and sodium and that contains at least 10 percent of the Daily Values for vitamin A, vitamin C, iron, calcium, protein, or fiber.

Healthy Eating Index: a measure that assesses how well a diet meets the recommendations of the *Dietary Guidelines for Americans* and MyPlate.

Healthy People: a national public health initiative under the jurisdiction of the US Department of Health and Human Services (DHHS) that identifies the most significant preventable threats to health and focuses efforts toward eliminating them.

heart attack: sudden tissue death caused by blockages of vessels that feed the heart muscle; also called *myocardial* (my-oh-KAR-dee-al) *infarction* (in-FARK-shun) or *cardiac arrest.*

heartburn: a burning sensation in the chest area caused by backflow of stomach acid into the esophagus; medically known as *gastroesophageal reflux.*

heat stroke: a dangerous accumulation of body heat with accompanying loss of body fluid.

heavy drinking: consuming an average of more than one drink per day for women and more than two drinks per day for men.

heavy metals: mineral ions such as mercury and lead, so called because they are of relatively high atomic weight. Many heavy metals are poisonous.

Heimlich (HIME-lick) **maneuver (abdominal thrusts):** a technique for dislodging an object from the trachea of a choking person; named for the physician who developed it.

hematocrit (hee-MAT-oh-krit): the percentage of total blood volume that consists of red blood cells.

heme (HEEM) **iron:** the iron in foods that is bound to the hemoglobin and myoglobin proteins; found only in meat, fish, and poultry.

hemochromatosis (HE-moh-KRO-ma-toe-sis): a genetically determined failure to prevent absorption of unneeded dietary iron that is characterized by iron overload and tissue damage.

hemoglobin (HE-moh-GLO-bin): the oxygen-carrying protein of the red blood cells that transports oxygen from the lungs to tissues throughout the body; hemoglobin accounts for 80 percent of the body's iron.

hemolytic (HE-moh-LIT-ick) **anemia:** the condition of having too few red blood cells as a result of erythrocyte hemolysis.

hemophilia (HE-moh-FEEL-ee-ah): a hereditary disease in which the blood is unable to clot because it lacks the ability to synthesize certain clotting factors.

hemorrhagic (hem-oh-RAJ-ik) **disease:** a disease characterized by excessive bleeding.

hemorrhoids (HEM-oh-royds): painful swelling of the veins surrounding the rectum.

hemosiderin (heem-oh-SID-er-in): an iron-storage protein primarily made in times of iron overload.

hepatic portal vein: the vein that collects blood from the GI tract and conducts it to the liver.

hepatic vein: the vein that collects blood from the liver and returns it to the heart.

hepcidin: a hormone produced by the liver that regulates iron balance.

herbal (ERB-al) **medicine:** the use of plants to treat disease or improve health; also known as *botanical medicine* or *phytotherapy.*

hGH (human growth hormone): a hormone produced by the brain's pituitary gland that regulates normal growth and development; also called *somatotropin.*

high: on food labels, 20 percent or more of the Daily Value for a given nutrient per serving; synonyms include *rich in* or *excellent source.*

high fiber: 5 grams or more of fiber per serving. A high-fiber claim made on a food that contains more than 3 grams of fat per serving and per 100 grams of food must also declare total fat.

high food security: no indications of food-access problems or limitations.

high potency: One hundred percent or more of the Daily Value for the nutrient in a single supplement and for at least two-thirds of the nutrients in a multinutrient supplement.

high-fructose corn syrup (HFCS): a syrup made from cornstarch that has been treated with an enzyme that converts some of the glucose to the sweeter fructose; made especially for use in processed foods and beverages, where it is the predominant sweetener.

high-quality proteins: dietary proteins containing all the essential amino acids in relatively the same amounts that human beings require. They may also contain nonessential amino acids.

high-risk pregnancy: a pregnancy characterized by indicators that make it likely the birth will be surrounded by problems such as premature delivery, difficult birth, restricted growth, birth defects, and early infant death.

histamine (HISS-tah-mean or HISS-tah-men): a substance produced by cells of the immune system as part of a local immune reaction to an antigen.

HIV (human immunodeficiency virus): the virus that destroys lymphocytes and impairs immunity, eventually causing AIDS.

HMB (beta-hydroxybetamethylbutyrate): a metabolite of the branched-chain amino acid leucine promoted to increase muscle mass and strength.

homeopathy (hoh-me-OP-ah-thee): a practice based on the theory that "like cures like," that is, that substances that cause symptoms in healthy people can cure those symptoms when given in very dilute amounts.

homeostasis (HOME-ee-oh-STAY-sis): the maintenance of constant internal conditions (such as blood chemistry, temperature, and blood pressure) by the body's control systems. A homeostatic system is constantly reacting to external forces to maintain limits set by the body's needs.

honey: sugar (mostly sucrose) formed from nectar gathered by bees. Composition and flavor vary, but honey always contains a mixture of sucrose, fructose, and glucose.

hormones: chemical messengers. Hormones are secreted by a variety of glands in response to altered conditions in the body. Each hormone travels to one or more specific target tissues or organs, where it elicits a specific response to maintain homeostasis.

hormone-sensitive lipase: an enzyme inside adipose cells that responds to the body's need for fuel by hydrolyzing triglycerides so that their parts (glycerol and fatty acids) enter the general circulation and thus become available to other cells for fuel. The signals to which this enzyme responds include epinephrine and glucagon, which oppose insulin.

hourly sweat rate: the amount of weight lost plus fluid consumed during exercise per hour. One pound equals roughly 2 cups (500 milliliters) of fluid.

human carrying capacity: the maximum number of people the earth can support over time.

human genome (GEE-nome): the complete set of genetic material (DNA) in a human being.

hunger: the painful sensation caused by a lack of food that initiates food-seeking behavior; a consequence of food insecurity that, because of prolonged, involuntary lack of food, results in discomfort, illness, weakness, or pain that goes beyond the usual uneasy sensation.

husk: the outer inedible part of a grain; also called the *chaff.*

hydrochloric acid: an acid composed of hydrogen and chloride atoms (HCl) that is normally produced by the gastric glands.

hydrogenation (HIGH-dro-jen-AY-shun or high-DROJ-eh-NAY-shun): a chemical process by which hydrogens are added to monounsaturated or polyunsaturated fatty acids to reduce the number of double bonds, making the fats more saturated (solid) and more resistant to oxidation (protecting against rancidity). Hydrogenation produces *trans*-fatty acids.

hydrolysis (high-DROL-ih-sis): a chemical reaction in which one molecule is split into two molecules, with hydrogen (H) added to one and a hydroxyl group (OH) added to the other (from water, H_2O).

hydrophilic (high-dro-FIL-ick): a term referring to water-loving, or water-soluble, substances.

hydrophobic (high-dro-FOE-bick): a term referring to water-fearing, or non-water-soluble, substances; also known as *lipophilic* (fat loving).

hydrotherapy: the use of water (in whirlpools, as douches, or packed as ice, for example) to promote relaxation and healing.

hydroxyapatite (high-drox-ee-APP-ah-tite): crystals made of calcium and phosphorus.

hyperactivity: inattentive and impulsive behavior that is more frequent and severe than is typical of others a similar age; professionally called *attention-deficit/hyperactivity disorder (ADHD).*

hypercalcemia: high blood calcium that may develop from a variety of disorders, including vitamin D toxicity. It does *not* develop from a high calcium intake.

hyperglycemia: elevated blood glucose concentrations.

hyperplastic obesity: obesity due to an increase in the *number* of fat cells.

hypertension: consistently higher-than-normal blood pressure. Hypertension that develops without an identifiable cause is known as *essential* or *primary hypertension;* hypertension that is caused by a specific disorder such as kidney disease is known as *secondary hypertension.*

hyperthermia: an above-normal body temperature.

hypertrophic obesity: obesity due to an increase in the *size* of fat cells.

hypertrophy (high-PER-tro-fee): growing larger; with regard to muscles, an increase in size (and strength) in response to use.

hypnotherapy: a technique that uses hypnosis and the power of suggestion to improve health behaviors, relieve pain, and heal.

hypoglycemia (HIGH-po-gly-SEE-me-ah): an abnormally low blood glucose concentration.

hyponatremia (HIGH-poe-na-TREE-mee-ah): a decreased concentration of sodium in the blood.

hypothalamus (high-po-THAL-ah-mus): a brain center that controls activities such as maintenance of water balance, regulation of body temperature, and control of appetite.

hypothermia: a below-normal body temperature.

hypothesis (hi-POTH-eh-sis): an unproven statement that tentatively explains the relationships between two or more variables.

hypothyroidism: underactivity of the thyroid gland that may be caused by iodine deficiency or any number of other causes.

I

ileocecal (ill-ee-oh-SEEK-ul) **valve:** the sphincter separating the small and large intestines.

ileum (ILL-ee-um): the last segment of the small intestine.

imagery: a technique that guides clients to achieve a desired physical, emotional, or spiritual state by visualizing themselves in that state.

imitation foods: foods that substitute for and resemble another food, but are nutritionally inferior to it with respect to vitamin, mineral, or protein content. If the substitute is not inferior to the food it resembles and if its name provides an accurate description of the product, it need not be labeled "imitation."

immune response: the body's reaction to foreign antigens, which neutralizes or eliminates them, thus preventing damage.

immune system: the body's natural defense against foreign materials that have penetrated the skin or mucous membranes.

immunity: the body's ability to defend itself against diseases.

immunoglobulins (IM-you-noh-GLOB-you-linz): proteins capable of acting as antibodies.

implantation (IM-plan-TAY-shun): the embedding of the blastocyst in the inner lining of the uterus.

indigestion: incomplete or uncomfortable digestion, usually accompanied by pain, nausea, vomiting, heartburn, intestinal gas, or belching.

indirect calorimetry: a means of estimating energy expenditure by measuring the amount of oxygen consumed.

indirect or **incidental additives:** substances that can get into food as a result of contact during growing, processing, packaging, storing, cooking, or some other stage before the foods are consumed; sometimes called *accidental additives.*

indispensable amino acids: essential amino acids.

indispensable nutrients: nutrients a person must obtain from food because the body cannot synthesize them in amounts sufficient to meet physiological needs; also called *essential nutrients.*

infectious diseases: diseases caused by bacteria, viruses, parasites, or other microorganisms that can be transmitted from one person to another through air, water, or food; by contact; or through vector organisms such as mosquitoes.

inflammation: an immunological response to cellular injury characterized by an increase in white blood cells.

initiators: factors that cause mutations that give rise to cancer, such as radiation and carcinogens.

inorganic: not containing carbon or pertaining to living things.

inositol (in-OSS-ih-tall): a nonessential nutrient that can be made in the body from glucose. Inositol is a part of cell membrane structures.

insoluble fibers: nonstarch polysaccharides that do not dissolve in water. Examples include the tough, fibrous structures found in the strings of celery and the skins of corn kernels.

insulin (IN-suh-lin): a hormone secreted by special cells in the pancreas in response to (among other things) elevated blood glucose concentration. Insulin controls the transport of glucose from the bloodstream into the muscle and fat cells.

insulin resistance: the condition in which a normal amount of insulin produces a subnormal effect in muscle, adipose, and liver cells, resulting in an elevated fasting glucose; a metabolic consequence of obesity that precedes type 2 diabetes.

integrative medicine: care that combines conventional and complementary therapies for which there is some high-quality scientific evidence of safety and effectiveness. Integrative medicine emphasizes the importance of the relationship between the practitioner and the patient and focuses on wellness, healing, and the whole person.

intensity: the degree of exertion while exercising (for example, the amount of weight lifted or the speed of running).

intentional food additives: additives intentionally added to foods, such as nutrients, colors, and preservatives.

intermittent claudication (klaw-dih-KAY-shun): severe calf pain caused by inadequate blood supply. It occurs when walking and subsides during rest.

Internet (the Net): a worldwide network of millions of computers linked together to share information.

interstitial (IN-ter-STISH-al) **fluid:** fluid between the cells (intercellular), usually high in sodium and chloride. Interstitial fluid is a large component of extracellular fluid.

intestinal ischemia (is-KEY-me-ah): a diminished blood flow to the intestines that is characterized by abdominal pain, forceful bowel movements, and blood in the stool.

intra-abdominal fat: fat stored within the abdominal cavity in association with the internal abdominal organs, as opposed to the fat stored directly under the skin (subcutaneous fat); also called *visceral fat.*

intracellular fluid: fluid within the cells, usually high in potassium and phosphate. Intracellular fluid accounts for approximately two-thirds of the body's water.

intrinsic factor: a glycoprotein (a protein with short polysaccharide chains attached) secreted by the stomach cells that binds with vitamin B_{12} in the small intestine to aid in the absorption of vitamin B_{12}.

invert sugar: a mixture of glucose and fructose formed by the hydrolysis of sucrose in a chemical process; sold only in liquid form and sweeter than sucrose. Invert sugar is used as a food additive to help preserve freshness and prevent shrinkage.

iodide: the ion form of iodine.

iodine: an essential trace mineral that is needed for the synthesis of thyroid hormones.

ions (EYE-uns): atoms or molecules that have gained or lost electrons and therefore have electrical charges. Examples include the positively charged sodium ion (Na^+) and the negatively charged chloride ion (Cl^-).

iridology: the study of changes in the iris of the eye and their relationships to disease.

iron: an essential trace mineral that is needed for the transport of oxygen and the metabolism of energy nutrients.

iron deficiency: the state of having depleted iron stores.

iron overload: toxicity from excess iron.

iron-deficiency anemia: severe depletion of iron stores that results in low hemoglobin and small, pale red blood cells. Anemias that impair hemoglobin synthesis are *microcytic* (small cell).

irradiation: sterilizing a food by exposure to energy waves, similar to ultraviolet light and microwaves; sometimes called *ionizing radiation.*

irritable bowel syndrome: an intestinal disorder of unknown cause. Symptoms include abdominal discomfort and cramping, diarrhea, constipation, or alternating diarrhea and constipation.

IU: international units; a measure of vitamin activity used before direct chemical analysis was available.Many fortified foods and supplements use IU on their labels.

J

jejunum (je-JOON-um): the first two-fifths of the small intestine beyond the duodenum.

joule: a measure of *work* energy; the amount of energy expended when 1 kilogram is moved 1 meter by a force of 1 newton.

K

kcal: abbreviation of kcalories; a unit by which energy is measured.

kcalorie: a unit by which energy is measured. One kcalorie is the amount of heat necessary to raise the temperature of 1 kilogram (kg) of water 1°C. The scientific use of the term *kcalorie* is the same as the popular use of the term *calorie.*

kcalorie (energy) control: management of food energy intake.

kcalorie-free: fewer than 5 kcalories per serving.

kefir (keh-FUR): a fermented milk created by adding *Lactobacillus acidophilus* and other bacteria that break down lactose to glucose and galactose, producing a sweet, lactose-free product.

keratin (KARE-uh-tin): a water-insoluble protein; the normal protein of hair and nails.

keratinization: accumulation of keratin in a tissue; a sign of vitamin A deficiency.

keratomalacia (KARE-ah-toe-ma-LAY-shuh): softening of the cornea that leads to irreversible blindness; a sign of severe vitamin A deficiency.

Keshan (KESH-an or ka-SHAWN) **disease:** the heart disease associated with selenium deficiency; named for one of the provinces of China where it was first studied. Keshan disease is characterized by heart enlargement and insufficiency; fibrous tissue replaces the muscle tissue that normally composes the middle layer of the walls of the heart.

keto (KEY-toe) **acid:** an organic acid that contains a carbonyl group (C=O).

ketogenic amino acids: amino acids that are degraded to acetyl CoA.

ketone (KEE-tone) **bodies:** compounds produced during the incomplete breakdown of fat when glucose is not available in the cells.

ketosis (kee-TOE-sis): an undesirably high concentration of ketone bodies in the blood and urine.

Kreb's cycle: named after the scientist who elucidated this biochemistry, a series of metabolic reactions that break down molecules of acetyl CoA to carbon dioxide and hydrogen atoms; also called the *citric acid cycle* or the *TCA cycle.*

kwashiorkor (kwash-ee-OR-core or kwash-ee-or-CORE): severe malnutrition characterized by failure to grow and develop, edema, changes in the pigmentation of hair and skin, fatty liver, anemia, and apathy.

L

lactadherin (lack-tad-HAIR-in): a protein in breast milk that attacks diarrhea-causing viruses.

lactase deficiency: a lack of the enzyme required to digest the disaccharide lactose into its component monosaccharides (glucose and galactose).

lactase: an enzyme that hydrolyzes lactose.

lactate: a 3-carbon compound produced from pyruvate during anaerobic metabolism.

lactation: production and secretion of breast milk for the purpose of nourishing an infant.

lacteals (LACK-tee-als): the lymphatic vessels of the intestine that take up nutrients and pass them to the lymph circulation.

lactoferrin (lack-toh-FERR-in): a protein in breast milk that binds iron and keeps it from supporting the growth of the infant's intestinal bacteria.

lacto-ovo-vegetarians: people who include milk, milk products, and eggs, but exclude meat, poultry, and seafood from their diets.

lactose (LAK-tose): a disaccharide composed of glucose and galactose; commonly known as *milk sugar.*

lactose intolerance: a condition that results from the inability to digest the milk sugar lactose; characterized by bloating, gas, abdominal discomfort, and diarrhea. Lactose intolerance differs from milk allergy, which is caused by an immune reaction to the protein in milk.

lactovegetarians: people who include milk and milk products, but exclude meat, poultry, seafood, and eggs from their diets.

large intestine or **colon** (COAL-un): the lower portion of intestine that completes the digestive process. Its segments are the *ascending colon*, the *transverse colon*, the *descending colon*, and the *sigmoid colon*.

larynx: the upper part of the air passageway that contains the vocal cords; also called the *voice box*.

laxatives: substances that loosen the bowels and thereby prevent or treat constipation.

LDL (low-density lipoprotein): the type of lipoprotein derived from very-low-density lipoproteins (VLDL) as triglycerides are removed and broken down; composed primarily of cholesterol.

lean: less than 10 grams of fat, 4.5 grams of saturated fat and *trans* fat combined, and 95 milligrams of cholesterol per serving and per 100 grams of meat, poultry, and seafood.

lean body mass: the body minus its fat.

lecithin (LESS-uh-thin): one of the phospholipids. Both nature and the food industry use lecithin as an emulsifier to combine water-soluble and fat-soluble ingredients that do not ordinarily mix, such as water and oil.

legumes (lay-GYOOMS or LEG-yooms): plants of the bean and pea family, with seeds that are rich in protein compared with other plant-derived foods.

leptin: a protein produced by fat cells under direction of the *ob* gene that decreases appetite and increases energy expenditure.

less: on food labels, at least 25 percent less of a given nutrient or kcalories than the comparison food; synonyms include *fewer* and *reduced*.

less cholesterol: 25 percent or less cholesterol than the comparison food (reflecting a reduction of at least 20 milligrams per serving), and 2 grams or less saturated fat and *trans* fat combined per serving.

less fat: 25 percent or less fat than the comparison food.

less saturated fat: 25 percent or less saturated fat and *trans* fat combined than the comparison food.

let-down reflex: the reflex that forces milk to the front of the breast when the infant begins to nurse.

levulose: an older name for fructose.

license to practice: permission under state or federal law, granted on meeting specified criteria, to use a certain title (such as dietitian) and offer certain services. Licensed dietitians may use the initials LD after their names.

life expectancy: the average number of years lived by people in a given society.

life span: the maximum number of years of life attainable by a member of a species.

light or **lite:** one-third fewer kcalories than the comparison food; 50 percent or less of the fat or sodium than the comparison food; any use of the term other than as defined must specify what it is referring to (for example, "light in color" or "light in texture").

lignans: phytochemicals present in flaxseed that are converted to phytosterols by intestinal bacteria and are under study as possible anticancer agents.

limiting amino acid: the essential amino acid found in the shortest supply relative to the amounts needed for protein synthesis in the body. Four amino acids are most likely to be limiting: lysine, methionine, threonine, and tryptophan.

lingual: pertaining to the tongue.

linoleic (lin-oh-LAY-ick) **acid:** an essential fatty acid with 18 carbons and two double bonds.

linolenic (lin-oh-LEN-ick) **acid:** an essential fatty acid with 18 carbons and three double bonds.

lipase (LYE-pase): an enzyme that hydrolyzes lipids (fats). *Lingual lipase* refers to the fat-digesting enzyme secreted from the salivary gland at the base of the tongue.

lipids: a family of compounds that includes triglycerides, phospholipids, and sterols. Lipids are characterized by their insolubility in water. (Lipids also include the fat-soluble vitamins, described in Chapter 11.)

lipoprotein lipase (LPL): an enzyme that hydrolyzes triglycerides passing by in the bloodstream and directs their parts into the cells, where they can be metabolized for energy or reassembled for storage.

lipoprotein-associated phospholipase A(2) or **Lp-PLA(2):** a lipoprotein-bound enzyme that generates potent proinflammatory and proatherogenic products such as oxidized free fatty acids and lysophosphatidylcholine. Lp-PLA(2) is a specific marker of plaque inflammation.

lipoproteins (LIP-oh-PRO-teenz): clusters of lipids associated with proteins that serve as transport vehicles for lipids in the lymph and blood.

lipotoxicity: the adverse effects of fat in nonadipose tissues.

liquor or **distilled spirits:** an alcoholic beverage traditionally made by fermenting and distilling a carbohydrate source such as molasses, potatoes, rye, beets, barley, or corn.

listeriosis: an infection caused by eating food contaminated with the bacterium *Listeria monocytogenes*, which can be killed by pasteurization and cooking but can survive at refrigerated temperatures; certain ready-to-eat foods, such as hot dogs and deli meats, may become contaminated after cooking or processing, but before packaging.

liver: the organ that manufactures bile. (The liver's many other functions are described in Chapter 7.)

longevity: long duration of life.

low: on food labels, an amount that would allow frequent consumption of a food without exceeding the Daily Value for the nutrient. A food that is naturally low in a nutrient may make such a claim, but only as it applies to all similar foods (for example, "fresh cauliflower, a low-sodium food"); synonyms include *little*, *few*, and *low source of*.

low birthweight (LBW): a birthweight of 5½ pounds (2500 grams) or less; indicates probable poor health in the newborn and poor nutrition status in the mother during pregnancy, before pregnancy, or both. Optimal birthweight for a full-term baby is 6.8 to 7.9 pounds (about 3100 to 3600 grams).

low cholesterol: 20 milligrams or less cholesterol per serving and 2 grams or less saturated fat and *trans* fat combined per serving.

low fat: 3 grams or less fat per serving.

low food security: reduced quality of life with little or no indication of reduced food intake; formerly known as *food insecurity without hunger*.

low kcalorie: 40 kcalories or less per serving.

low saturated fat: 1 gram or less saturated fat and less than 0.5 gram of *trans* fat per serving.

low sodium: 140 milligrams or less per serving.

low-risk pregnancy: a pregnancy characterized by factors that make it likely the birth will be normal and the infant healthy.

lumen (LOO-men): the space within a vessel, such as the intestine.

lutein (LOO-teen): a plant pigment of yellow hue; a phytochemical believed to play roles in eye functioning and health.

luteinizing (LOO-tee-in-EYE-zing) **hormone (LH):** a hormone that stimulates ovulation and the development of the corpus luteum (the small tissue that develops from a ruptured ovarian follicle and secretes hormones); so called because the follicle turns yellow as it matures. In men, LH stimulates testosterone secretion. The release of LH is mediated by *luteinizing hormone–releasing hormone (LH–RH)*.

lycopene (LYE-koh-peen): a pigment responsible for the red color of tomatoes and other red-hued vegetables; a phytochemical that may act as an antioxidant in the body.

lymph (LIMF): a clear yellowish fluid that is similar to blood except that it contains no red blood cells or platelets. Lymph from the GI tract transports fat and fat-soluble vitamins to the bloodstream via lymphatic vessels.

lymphatic (lim-FAT-ic) **system:** a loosely organized system of vessels and ducts that convey fluids toward the heart. The GI part of the lymphatic system carries the products of fat digestion into the bloodstream.

lymphocytes (LIM-foh-sites): white blood cells that participate in acquired immunity; B-cells and T-cells.

lysosomes (LYE-so-zomes): cellular organelles; membrane-enclosed sacs of degradative enzymes.

M

macrobiotic diet: a philosophical approach of eating mostly plant-based foods such as whole grains, legumes, and vegetables, with small amounts of fish, fruits, nuts, and seeds.

macrocytic: abnormally large blood cells.

macronutrients: carbohydrate, fat, and protein; the nutrients the body requires in relatively large amounts (many grams daily).

macrophages (MAK-roe-fay-jez): large phagocytic cells that serve as scavengers of the blood, cleaning it of old or abnormal cells, cellular debris, and antigens.

macrosomia (mak-roh-SO-me-ah): abnormally large body size. In the case of infants, a birthweight at the 90th percentile or higher for gestational age (roughly 9 lb—or 4000 g—or more); macrosomia results from prepregnancy obesity, excessive weight gain during pregnancy, or uncontrolled gestational diabetes.

macular (MACK-you-lar) **degeneration:** deterioration of the macular area of the eye that can lead to loss of central vision and eventual blindness. The *macula* is a small, oval, yellowish region in the center of the retina that provides the sharp, straight-ahead vision so critical to reading and driving.

magnesium: a cation within the body's cells, active in many enzyme systems.

major minerals: essential mineral nutrients the human body requires in relatively large amounts (greater than 100 milligrams per day); sometimes called *macrominerals.*

malignant (ma-LIG-nant): describes a cancerous cell or tumor, which can injure healthy tissue and spread cancer to other regions of the body.

malnutrition: any condition caused by excess or deficient food energy or nutrient intake or by an imbalance of nutrients.

maltase: an enzyme that hydrolyzes maltose.

maltose (MAWL-tose): a disaccharide composed of two glucose units; sometimes known as *malt sugar.*

mammary glands: glands of the female breast that secrete milk.

manganese: an essential trace mineral that acts as a cofactor for many enzymes.

maple sugar: a sugar (mostly sucrose) purified from the concentrated sap of the sugar maple tree.

marasmus (ma-RAZ-mus): severe malnutrition characterized by poor growth, dramatic weight loss, loss of body fat and muscle, and apathy.

margin of safety: when speaking of food additives, a zone between the concentration normally used and that at which a hazard exists. For common table salt, for example, the margin of safety is 1/5 (five times the amount normally used would be hazardous).

marginal food security: one or two indications of food-access problems but with little or no change in food intake.

massage therapy: a healing method in which the therapist manually kneads muscles to reduce tension, increase blood circulation, improve joint mobility, and promote healing of injuries.

mastication: the process of chewing.

matrix (MAY-tricks): the basic substance that gives form to a developing structure; in the body, the formative cells from which teeth and bones grow.

matter: anything that takes up space and has mass.

Meals on Wheels: a nutrition program that delivers food for the elderly to their homes.

meat replacements: products formulated to look and taste like meat, fish, or poultry; usually made of textured vegetable protein.

meditation: a self-directed technique of relaxing the body and calming the mind.

megaloblastic: abnormally large blood cells.

menadione (men-uh-DYE-own): the synthetic form of vitamin K.

Menkes disease: a genetic disorder of copper transport that creates a copper deficiency and results in mental retardation, poor muscle tone, seizures, brittle kinky hair, and failure to thrive.

MEOS or **microsomal** (my-krow-SO-mal) **ethanol oxidizing system:** a system of enzymes in the liver that oxidize not only alcohol but also several classes of drugs.

metabolic syndrome: a combination of risk factors—insulin resistance, hypertension, abnormal blood lipids, and abdominal obesity—that greatly increase a person's risk of developing coronary heart disease; also called *Syndrome X, insulin resistance syndrome,* or *dysmetabolic syndrome.*

metabolic water: water generated during metabolism.

metabolism: the sum total of all the chemical reactions that go on in living cells. *Energy metabolism* includes all the reactions by which the body obtains and expends the energy from food.

metalloenzymes (meh-TAL-oh-EN-zimes): enzymes that contain one or more minerals as part of their structures.

metallothionein (meh-TAL-oh-THIGH-oh-neen): a sulfur-rich protein that avidly binds with and transports metals such as zinc.

metastasize (me-TAS-tah-size): the spread of cancer from one part of the body to another.

methylation: the addition of a methyl group (CH_3).

MFP factor: a peptide released during the digestion of meat, fish, and poultry that enhances nonheme iron absorption.

mg NE: milligrams niacin equivalents; a measure of niacin activity.

mg: milligrams; one-thousandth of a gram.

micelles (MY-cells): tiny spherical complexes of emulsified fat that arise during digestion; most contain bile salts and the products of lipid digestion, including fatty acids, monoglycerides, and cholesterol.

microangiopathies: disorders of the small blood vessels.

microarray technology: research tools that analyze the expression of thousands of genes simultaneously and search for particular gene changes associated with a disease. DNA microarrays are also called *DNA chips.*

microcytic (my-cro-SIT-ic) **hypochromic** (high-po-KROME-ic) **anemia:** small, pale red blood cells that develop in iron-deficiency anemia.

microgram (μg): one-millionth of a gram.

microgram DFE (μg DFE): micrograms dietary folate equivalents; a measure of folate activity.

microgram (μg RAE): micrograms retinol activity equivalents; a measure of vitamin A activity.

micronutrients: vitamins and minerals; the nutrients the body requires in relatively small amounts (milligrams or micrograms daily).

microvilli (MY-cro-VILL-ee or MY-cro-VILL-eye): tiny, hairlike projections on each cell of every villus that can trap nutrient particles and transport them into the cells; singular *microvillus.*

milk anemia: iron-deficiency anemia that develops when an excessive milk intake displaces iron-rich foods from the diet.

milliequivalents per liter (mEq/L): the concentration of electrolytes in a volume of solution. Milliequivalents are a useful measure when considering ions because the number of charges reveals characteristics about the solution that are not evident when the concentration is expressed in terms of weight.

mineral oil: a purified liquid derived from petroleum and used to treat constipation.

mineral water: water from a spring or well that naturally contains at least 250 to 500 parts per million (ppm) of minerals. Minerals give water a distinctive flavor. Many mineral waters are high in sodium.

mineralization: the process in which calcium, phosphorus, and other minerals crystallize on the collagen matrix of a growing bone, hardening the bone.

minerals: inorganic elements. Some minerals are essential nutrients required in small amounts by the body for health.

misinformation: false or misleading information.

mitochondria (my-toh-KON-dree-uh): the cellular organelles made of membranes (lipid and protein) with enzymes mounted on them that are responsible for producing ATP aerobically; the singular is *mitochondrion*.

mmol: millimoles; one-thousandth of a mole, the molecular weight of a substance. To convert mmol to mg, multiply by the atomic weight of the substance.

moderate-intensity physical activity: physical activity that requires some increase in breathing and/or heart rate and expends 3.5 to 7 kcalories per minute. Walking at a speed of 3 to 4.5 miles per hour (about 15 to 20 minutes to walk 1 mile) is an example.

moderation (alcohol): up to one drink per day for women with no more than three drinks on any single day and up to two drinks per day for men with no more than four drinks on any single day.

moderation (dietary): providing enough but not too much of a substance.

molasses: the thick brown syrup produced during sugar refining. Molasses retains residual sugar and other by-products and a few minerals; blackstrap molasses contains significant amounts of calcium and iron.

molecule: two or more atoms of the same or different elements joined by chemical bonds. Examples are molecules of the element oxygen, composed of two oxygen atoms (O_2), and molecules of the compound water, composed of two hydrogen atoms and one oxygen atom (H_2O).

molybdenum (mo-LIB-duh-num): an essential trace mineral that acts as a cofactor for many enzymes.

monoglycerides: molecules of glycerol with one fatty acid attached. A molecule of glycerol with two fatty acids attached is a *diglyceride*.

monosaccharides (mon-oh-SACK-uh-rides): carbohydrates of the general formula $C_nH_{2n}O_n$ that typically form a single ring. The monosaccharides important in nutrition are *hexoses*, sugars with six atoms of carbon and the formula $C_6H_{12}O_6$.

monosodium glutamate (MSG): a sodium salt of the amino acid glutamic acid commonly used as a flavor enhancer. The FDA classifies MSG as a "generally recognized as safe" ingredient.

monounsaturated fatty acid (MUFA): a fatty acid that lacks two hydrogen atoms and has one double bond between carbons—for example, oleic acid. A *monounsaturated fat* is composed of triglycerides in which most of the fatty acids are monounsaturated.

more: on food labels, at least 10 percent more of the Daily Value for a given nutrient than the comparison food; synonyms include *added* and *extra*.

motility: the ability of the GI tract muscles to move.

mouth: the oral cavity containing the tongue and teeth.

MSG symptom complex: an acute, temporary intolerance reaction that may occur after the ingestion of the additive MSG (monosodium glutamate). Symptoms include burning sensations, chest and facial flushing and pain, and throbbing headaches.

mucous (MYOO-kus) **membranes:** the membranes, composed of mucus-secreting cells, that line the surfaces of body tissues.

mucus (MYOO-kus): a slippery substance secreted by cells of the GI lining (and other body linings) that protects the cells from exposure to digestive juices (and other destructive agents). The lining of the GI tract with its coat of mucus is a *mucous membrane*. (The noun is *mucus*; the adjective is *mucous*.)

muscle dysmorphia (dis-MORE-fee-ah): a psychiatric disorder characterized by a preoccupation with building body mass.

muscle endurance: the ability of a muscle to contract repeatedly without becoming exhausted.

muscle power: the product of force generation (strength) and movement velocity (speed); the speed at which a given amount of exertion is completed.

muscle strength: the ability of muscles to work against resistance.

mutations: permanent changes in the DNA that can be inherited.

myoglobin: the oxygen-holding protein of the muscle cells.

N

NAD (nicotinamide adenine dinucleotide): the main coenzyme form of the vitamin niacin. Its reduced form is NADH.

NADP (the phosphate form of NAD): a coenzyme form of niacin.

nanoceuticals: substances with extremely small particles that have been manufactured by nanotechnology.

nanotechnology: a manufacturing technology that manipulates atoms to change the structure of matter.

narcotic (nar-KOT-ic): a drug that dulls the senses, induces sleep, and becomes addictive with prolonged use.

natural water: water obtained from a spring or well that is certified to be safe and sanitary. The mineral content may not be changed, but the water may be treated in other ways such as with ozone or by filtration.

naturopathic (nay-chur-oh-PATH-ick) **medicine:** a system that taps the natural healing forces within the body by integrating several practices, including traditional medicine, herbal medicine, clinical nutrition, homeopathy, acupuncture, East Asian medicine, hydrotherapy, and manipulative therapy.

neotame (NEE-oh-tame): an artificial sweetener composed of two amino acids (phenylalanine and aspartic acid); approved for use in the United States.

net protein utilization (NPU): a measure of protein quality assessed by measuring the amount of protein nitrogen that is retained from a given amount of protein nitrogen eaten.

neural tube: the embryonic tissue that forms the brain and spinal cord. The two main types of neural tube defects are *spina bifida* (literally "split spine") and *anencephaly* ("no brain").

neural tube defects: malformations of the brain, spinal cord, or both during embryonic development that often result in lifelong disability or death.

neurofibrillary tangles: snarls of the threadlike strands that extend from the nerve cells, commonly found in the brains of people with Alzheimer's dementia.

neurons: nerve cells; the structural and functional units of the nervous system. Neurons initiate and conduct nerve impulse transmissions.

neuropeptide Y: a chemical produced in the brain that stimulates appetite, diminishes energy expenditure, and increases fat storage.

neurotransmitters: chemicals that are released at the end of a nerve cell when a nerve impulse arrives there. They diffuse across the gap to the next cell and alter the membrane of that second cell to either inhibit or excite it.

neutrophils (NEW-tro-fills): the most common of white blood cell. Neutrophils destroy antigens by phagocytosis.

niacin (NIGH-a-sin): a B vitamin. The coenzyme forms are *NAD (nicotinamide adenine dinucleotide)* and *NADP (the phosphate form of NAD)*. Niacin can be eaten preformed or made in the body from its precursor, tryptophan, an essential amino acid.

niacin equivalents (NE): the amount of niacin present in food, including the niacin that can theoretically be made from its precursor, tryptophan.

niacin flush: a temporary burning, tingling, and itching sensation that occurs when a person takes a large dose of nicotinic acid; often accompanied by a headache and reddened face, arms, and chest.

night blindness: slow recovery of vision after flashes of bright light at night or an inability to see in dim light; an early symptom of vitamin A deficiency.

nitrites (NYE-trites): salts added to food to prevent botulism. One example is sodium nitrite, which is used to preserve meats.

nitrogen balance: the amount of nitrogen consumed (N in) as compared with the amount of nitrogen excreted (N out) in a given period of time.

nitrosamines (nye-TROHS-uh-meens): derivatives of nitrites that may be formed in the stomach when nitrites combine with amines. Nitrosamines are carcinogenic in animals.

nonessential amino acids: amino acids that the body can synthesize.

nonexercise activity thermogenesis (NEAT): energy expended in everyday spontaneous activities.

nonheme iron: the iron in foods that is not bound to proteins; found in both plant-derived and animal-derived foods.

nonnutritive sweeteners: sweeteners that yield no energy (or insignificant energy in the case of aspartame).

nonperishable food collection: collecting processed foods from wholesalers and markets.

nonpoint water pollution: water pollution caused by runoff from all over an area rather than from discrete "point" sources. An example is the pollution caused by runoff from agricultural fields.

nucleotide bases: the nitrogen-containing building blocks of DNA and RNA—cytosine (C), thymine (T), uracil (U), guanine (G), and adenine (A). In DNA, the base pairs are A–T and C–G and in RNA, the base pairs are A–U and C–G.

nucleotides: the subunits of DNA and RNA molecules, composed of a phosphate group, a 5-carbon sugar (deoxyribose for DNA and ribose for RNA), and a nitrogen-containing base.

nucleus: a major membrane-enclosed body within cells, which contains the cell's genetic material (DNA), embedded in chromosomes.

nursing bottle tooth decay: extensive tooth decay due to prolonged tooth contact with formula, milk, fruit juice, or other carbohydrate-rich liquid offered to an infant in a bottle.

nutrient claims: statements that characterize the quantity of a nutrient in a food.

nutrient density: a measure of the nutrients a food provides relative to the energy it provides. The more nutrients and the fewer kcalories, the higher the nutrient density.

nutrient profiling: ranking foods based on their nutrient composition.

nutrients: chemical substances obtained from food and used in the body to provide energy, structural materials, and regulating agents to support growth, maintenance, and repair of the body's tissues. Nutrients may also reduce the risks of some diseases.

nutrigenetics: the science of how genes affect the activities of nutrients.

nutrigenomics: the science of how nutrients affect the activities of genes.

nutrition: the science of foods and the nutrients and other substances they contain, and of their actions within the body (including ingestion, digestion, absorption, transport, metabolism, and excretion). A broader definition includes the social, economic, cultural, and psychological implications of food and eating.

nutrition assessment: a comprehensive analysis of a person's nutrition status that uses health, socioeconomic, drug, and diet histories; anthropometric measurements; physical examinations; and laboratory tests.

nutrition screening: the use of preliminary nutrition assessment techniques to identify people who are malnourished or are at risk for malnutrition.

nutritional genomics: the science of how nutrients affect the activities of genes *(nutrigenomics)* and how genes affect the activities of nutrients *(nutrigenetics)*.

nutritionist: a person who specializes in the study of nutrition. Note that this definition does not specify qualifications and may apply not only to registered dietitians but also to self-described experts whose training is questionable. Most states have licensing laws that define the scope of practice for those calling themselves nutritionists.

nutritive sweeteners: sweeteners that yield energy, including both sugars and sugar alcohols.

O

obese: overweight with adverse health effects; BMI 30 or more.

obesogenic (oh-BES-oh-JEN-ick) **environment:** all the factors surrounding a person that promote weight gain, such as increased food intake, especially of unhealthy choices, and decreased physical activity.

obligatory (ah-BLIG-ah-TORE-ee) **water excretion:** the amount of water the body has to excrete each day to dispose of its wastes—about 500 milliliters (about 2 cups, or 1 pint).

oils: lipids that are liquid at room temperature (77°F or 25°C).

olestra: a synthetic fat made from sucrose and fatty acids that provides 0 kcalories per gram; also known as *sucrose polyester.*

oligopeptide (OL-ee-go-PEP-tide): string of four to nine amino acids.

omega: the last letter of the Greek alphabet (ω), used by chemists to refer to the position of the closest double bond to the methyl (CH_3) end of a fatty acid.

omega-3 fatty acid: a polyunsaturated fatty acid in which the closest double bond to the methyl (CH_3) end of the carbon chain is three carbons away.

omega-6 fatty acid: a polyunsaturated fatty acid in which the closest double bond to the methyl (CH_3) end of the carbon chain is six carbons away.

omnivorous: an eating pattern that includes foods derived from both animals and plants.

opsin (OP-sin): the protein portion of visual pigment molecules.

oral rehydration therapy (ORT): the administration of a simple solution of sugar, salt, and water, taken by mouth, to treat dehydration caused by diarrhea.

organelles: subcellular structures such as ribosomes, mitochondria, and lysosomes.

organic: in agriculture, crops grown and processed according to USDA regulations defining the use of fertilizers, herbicides, insecticides, fungicides, preservatives, and other chemical ingredients.

organic: in chemistry, a substance or molecule containing carbon-carbon bonds or carbon-hydrogen bonds. This definition excludes coal, diamonds, and a few carbon-containing compounds that contain only a single carbon and no hydrogen, such as carbon dioxide (CO_2), calcium carbonate ($CaCO_3$), magnesium carbonate ($MgCO_3$), and sodium cyanide (NaCN).

organic: on food labels, that at least 95 percent of the product's ingredients have been grown and processed according to USDA regulations defining the use of fertilizers, herbicides, insecticides, fungicides, preservatives, and other chemical ingredients.

organic halogens: an organic compound containing one or more atoms of a halogen—fluorine, chlorine, iodine, or bromine.

orlistat (OR-leh-stat): a drug used in the treatment of obesity that inhibits pancreatic lipase activity in the GI tract, thus blocking digestion and absorption of dietary fat and limiting energy intake.

orthomolecular medicine: the use of large doses of vitamins to treat chronic disease.

osmosis: the movement of water across a membrane *toward* the side where the solutes are more concentrated.

osmotic pressure: the amount of pressure needed to prevent the movement of water across a membrane.

osteoarthritis: a painful, degenerative disease of the joints that occurs when the cartilage in a joint deteriorates; joint structure is damaged, with loss of function; also called *degenerative arthritis*.

osteoblasts: cells that build bone during growth.

osteocalcin (os-teo-KAL-sen): a calcium-binding protein in bones, essential for normal mineralization.

osteoclasts: cells that destroy bone during growth.

osteomalacia (OS-tee-oh-ma-LAY-shuh): a bone disease characterized by softening of the bones. Symptoms include bending of the spine and bowing of the legs. The disease occurs most often in adult women.

osteoporosis (OS-tee-oh-pore-OH-sis): a disease in which the bones become porous and fragile due to a loss of minerals; also called *adult bone loss*.

outbreaks: two or more cases of a similar illness resulting from the ingestion of a common food.

overnutrition: excess energy or nutrients.

overt (oh-VERT): out in the open and easy to observe.

overweight: body weight greater than the weight range that is considered healthy; BMI 25 to 29.9.

ovum (OH-vum): the female reproductive cell, capable of developing into a new organism upon fertilization; commonly referred to as an egg.

oxaloacetate (OKS-ah-low-AS-eh-tate): a carbohydrate intermediate of the TCA cycle.

oxidants (OKS-ih-dants): compounds (such as oxygen itself) that oxidize other compounds. Compounds that prevent oxidation are called *antioxidants*, whereas those that promote it are called *prooxidants*.

oxidation (OKS-ee-day-shun): the process of a substance combining with oxygen; oxidation reactions involve the loss of electrons.

oxidative stress: a condition in which the production of oxidants and free radicals exceeds the body's ability to handle them and prevent damage.

oxytocin (OCK-see-TOH-sin): a hormone that stimulates the mammary glands to eject milk during lactation and the uterus to contract during childbirth.

oyster shell: a product made from the powdered shells of oysters that is sold as a calcium supplement, but it is not well absorbed by the digestive system.

ozone therapy: the use of ozone gas to enhance the body's immune system.

P

pancreas: a gland that secretes digestive enzymes and juices into the duodenum. (The pancreas also secretes hormones into the blood that help to maintain glucose homeostasis.)

pancreatic (pank-ree-AT-ic) **juice:** the exocrine secretion of the pancreas that contains enzymes for the digestion of carbohydrate, fat, and protein as well as bicarbonate, a neutralizing agent. The juice flows from the pancreas into the small intestine through the pancreatic duct. (The pancreas also has an endocrine function, the secretion of insulin and other hormones.)

pantothenic (PAN-toe-THEN-ick) **acid:** a B vitamin. The principal active form is part of coenzyme A, called "CoA" throughout Chapter 7.

parathyroid hormone: a hormone from the parathyroid glands that regulates blood calcium by raising it when levels fall too low; also known as *parathormone* (PAIR-ah-THOR-moan).

pasteurization: heat processing of food that inactivates some, but not all, microorganisms in the food; not a sterilization process. Bacteria that cause spoilage are still present.

pathogens (PATH-oh-jenz): microorganisms capable of producing disease.

PBB (polybrominated biphenyl) and **PCB (polychlorinated biphenyl):** toxic organic halogens used in pesticides, paints, and flame retardants.

PDCAAS (protein digestibility–corrected amino acid score): a measure of protein quality assessed by comparing the amino acid score of a food protein with the amino acid requirements of preschool-age children and then correcting for the true digestibility of the protein; recommended by the FAO/WHO and used to establish protein quality of foods for Daily Value percentages on food labels.

peak bone mass: the highest attainable bone density for an individual, developed during the first three decades of life.

peer review: a process in which a panel of scientists rigorously evaluates a research study to assure that the scientific method was followed.

pellagra (pell-AY-gra): the niacin-deficiency disease.

pepsin: a gastric enzyme that hydrolyzes protein. Pepsin is secreted in an inactive form, *pepsinogen*, which is activated by hydrochloric acid in the stomach.

pepsinogen: an inactive compound that is activated by hydrochloric acid in the stomach to form pepsin.

peptic ulcer: a lesion in the mucous membrane of either the stomach (a *gastric ulcer*) or the duodenum (a *duodenal ulcer*).

peptidase: a digestive enzyme that hydrolyzes peptide bonds. *Tripeptidases* cleave tripeptides; *dipeptidases* cleave dipeptides. *Endopeptidases* cleave peptide bonds within the chain to create smaller fragments, whereas *exopeptidases* cleave bonds at the ends to release free amino acids.

peptide bond: a bond that connects the acid end of one amino acid with the amino end of another, forming a link in a protein chain.

percent Daily Value (%DV): the percentage of a Daily Value recommendation found in a specified serving of food for key nutrients based on a 2000-kcalorie diet.

percent fat-free: may be used only if the product meets the definition of *low fat* or *fat-free* and must reflect the amount of fat in 100 grams (for example, a food that contains 2.5 grams of fat per 50 grams can claim to be "95 percent fat-free").

perinatal: referring to the time between the twenty-eighth week of gestation and 1 month after birth.

peripheral (puh-RIFF-er-ul) **nervous system:** the peripheral (outermost) part of the nervous system; the vast complex of wiring that extends from the central nervous system to the body's outermost areas. It contains both *somatic* and *autonomic* components.

peripheral resistance: the resistance to pumped blood in the small arterial branches (arterioles) that carry blood to the tissues.

perishable food rescue or **salvage:** collecting perishable produce from wholesalers and markets.

peristalsis (per-ih-STALL-sis): wavelike muscular contractions of the GI tract that push its contents along.

pernicious (per-NISH-us) **anemia:** a blood disorder that reflects a vitamin B_{12} deficiency caused by lack of intrinsic factor and characterized by abnormally large and immature red blood cells. Other symptoms include muscle weakness and irreversible neurological damage.

persistence: stubborn or enduring continuance; with respect to food contaminants, the quality of persisting, rather than breaking down, in the bodies of animals and human beings.

pesticides: chemicals used to control insects, weeds, fungi, and other pests on plants, vegetables, fruits, and animals. Used broadly, the term includes herbicides (to kill weeds), insecticides (to kill insects), and fungicides (to kill fungi).

pH: the unit of measure expressing a substance's acidity or alkalinity. The lower the pH, the higher the H^+ ion concentration and the stronger the acid. A pH above 7 is alkaline, or base (a solution in which OH^- ions predominate).

phagocytes (FAG-oh-sites): white blood cells (neutrophils and macrophages) that have the ability to ingest and destroy foreign substances.

phagocytosis (FAG-oh-sigh-TOH-sis): the process by which phagocytes engulf and destroy foreign materials.

pharmacological effect: the body's response to a large dose of a nutrient (levels commonly available only from supplements) that overwhelms some body system and acts like a drug.

pharynx (FAIR-inks): the passageway leading from the nose and mouth to the larynx and esophagus, respectively.

phenylketonuria (FEN-il-KEY-toe-NEW-ree-ah) or **PKU:** an inherited disorder characterized by failure to metabolize the amino acid phenylalanine to tyrosine.

phosphocreatine (PC): a high-energy compound in muscle cells that acts as a reservoir of energy that can maintain a steady supply of ATP and provides the energy for short bursts of activity; also called *creatine phosphate (CP).*

phospholipid (FOS-foe-LIP-id): a compound similar to a triglyceride but having a phosphate group (a phosphorus-containing salt) and choline (or another nitrogen-containing compound) in place of one of the fatty acids.

phosphorus: a major mineral found mostly in the body's bones and teeth.

photosynthesis: the process by which green plants use the sun's energy to make carbohydrates from carbon dioxide and water.

phylloquinone (FILL-oh-KWIN-own): the naturally occurring form of vitamin K.

physical activity: bodily movement produced by muscle contractions that substantially increase energy expenditure.

physiological age: a person's age as estimated from her or his body's health and probable life expectancy.

physiological effect: the body's response to a normal dose of a nutrient (levels commonly found in foods) that provides a normal blood concentration.

physiological fuel value: the number of kcalories that the body derives from a food, in contrast to the number of kcalories determined by calorimetry.

phytic (FYE-tick) **acid:** a nonnutrient component of plant seeds; also called *phytate* (FYE-tate). Phytic acid occurs in the husks of grains, legumes, and seeds and is capable of binding minerals such as zinc, iron, calcium, magnesium, and copper in insoluble complexes in the intestine, which the body excretes unused.

phytochemicals (FIE-toe-KEM-ih-cals): nonnutrient compounds found in plants that confer taste, color, and other characteristics. Some phytochemicals have biological activity in the body.

phytoestrogens: phytochemicals structurally similar to human estrogen that weakly mimic or modulate estrogen's action in the body. Phytoestrogens include the isoflavones *genistein, daidzein,* and *glycitein.*

pica (PIE-ka): the general term for eating nonfood items. The specific craving for nonfood items that come from the earth, such as clay or dirt, is known as *geophagia.*

pigment: a molecule capable of absorbing certain wavelengths of light so that it reflects only those that we perceive as a certain color.

placebo (pla-see-bo): an inert, harmless medication given to provide comfort and hope; a sham treatment used in controlled research studies.

placebo effect: a change that occurs in response to expectations in the effectiveness of a treatment that actually has no pharmaceutical effects.

placenta (plah-SEN-tuh): the organ that develops inside the uterus early in pregnancy, through which the fetus receives nutrients and oxygen and returns carbon dioxide and other waste products to be excreted.

plant-based diets: an eating pattern that derives most of its protein from plant products (although some animal products may be included).

plant-pesticides: pesticides made by the plants themselves.

plant sterols: phytochemicals that have structural similarities to cholesterol and lower blood cholesterol by interfering with cholesterol absorption. Plant sterols include *sterol esters* and *stanol esters.*

plaques (PLACKS): the accumulation of fatty deposits, smooth muscle cells, and fibrous connective tissue that develops in the artery walls in atherosclerosis.

platelets: tiny, disc-shaped bodies in the blood, important in blood clot formation.

PLP (pyridoxal phosphate): the primary active coenzyme form of vitamin B_6.

point of unsaturation: the double bond of a fatty acid, where hydrogen atoms can easily be added to the structure.

polar: characteristic of a neutral molecule, such as water, that has opposite charges spatially separated within the molecule.

polypeptide: many (10 or more) amino acids bonded together.

polysaccharides: compounds composed of many monosaccharides linked together. An intermediate string of 3 to 10 monosaccharides is an *oligosaccharide.*

polyunsaturated fatty acid (PUFA): a fatty acid that lacks four or more hydrogen atoms and has two or more double bonds between carbons—for example, linoleic acid (two double bonds) and linolenic acid (three double bonds). A *polyunsaturated fat* is composed of triglycerides in which most of the fatty acids are polyunsaturated.

portion sizes: the quantity of a food served or eaten at one meal or snack; *not* a standard amount.

post term: after the forty-second week of pregnancy.

postpartum amenorrhea: the normal temporary absence of menstrual periods immediately following childbirth.

potable (POT-ah-bul): water that is suitable for drinking.

potassium: the principal cation within the body's cells; critical to the maintenance of fluid balance, nerve impulse transmissions, and muscle contractions.

powdered bone: crushed or ground bone preparations intended to supply calcium to the diet. Calcium from bone is not well absorbed and is often contaminated with toxic minerals such as arsenic, mercury, lead, and cadmium.

prebiotics: food components (such as fibers) that are not digested by the human body but are used as food by the GI bacteria to promote their growth and activity.

precursors: substances that precede others; with regard to vitamins, compounds that can be converted into active vitamins; also known as *provitamins.*

prediabetes: condition in which blood glucose levels are higher than normal but not high enough to be diagnosed as diabetes; considered a major risk factor for future diabetes and cardiovascular diseases; formerly called *impaired glucose tolerance.*

preeclampsia (PRE-ee-KLAMP-see-ah): a condition characterized by high blood pressure and some protein in the urine.

preformed vitamin A: dietary vitamin A in its active form.

prehypertension: slightly higher-than-normal blood pressure, but not as high as hypertension.

prenatal alcohol exposure: subjecting a fetus to a pattern of excessive alcohol intake characterized by substantial regular use or heavy episodic drinking.

prepared food rescue: collecting prepared foods from commercial kitchens.

preservatives: antimicrobial agents, antioxidants, and other additives that retard spoilage or maintain desired qualities, such as softness in baked goods.

pressure ulcers: damage to the skin and underlying tissues as a result of compression and poor circulation; commonly seen in people who are bedridden or chair-bound.

preterm (premature): prior to the thirty-eighth week of pregnancy.

primary deficiency: a nutrient deficiency caused by inadequate dietary intake of a nutrient.

probiotics: living microorganisms found in foods that, when consumed in sufficient quantities, are beneficial to health.

processed foods: foods that have been treated to change their physical, chemical, microbiological, or sensory properties.

proenzyme: the inactive form of an enzyme; also called a *zymogen*.

progesterone: the hormone of gestation (pregnancy).

progressive overload principle: the training principle that a body system, in order to improve, must be worked at frequencies, durations, or intensities that gradually increase physical demands.

prolactin (pro-LAK-tin): a hormone secreted from the anterior pituitary gland that acts on the mammary glands to promote the production of milk. The release of prolactin is mediated by *prolactin-inhibiting hormone (PIH)*.

promoters: factors that favor the development of cancers once they have begun.

proof: a way of stating the percentage of alcohol in distilled liquor. Liquor that is 100 proof is 50 percent alcohol; 90 proof is 45 percent, and so forth.

prooxidants: substances that significantly induce oxidative stress.

protease (PRO-tee-ace): an enzyme that hydrolyzes proteins.

protein digestibility: a measure of the amount of amino acids absorbed from a given protein intake.

protein efficiency ratio (PER): a measure of protein quality assessed by determining how well a given protein supports weight gain in growing rats; used to establish the protein quality for infant formulas and baby foods.

protein turnover: the degradation and synthesis of protein.

proteins: compounds composed of carbon, hydrogen, oxygen, and nitrogen atoms, arranged into amino acids linked in a chain. Some amino acids also contain sulfur atoms.

protein-sparing action: the action of carbohydrate (and fat) in providing energy that allows protein to be used for other purposes.

proteome: all proteins in a cell. The study of all proteins produced by a species is called *proteomics*.

proteomics: the study of the all proteins produced by a species.

puberty: the period in life in which a person becomes physically capable of reproduction.

public health dietitians: dietitians who specialize in providing nutrition services through organized community efforts.

public water: water from a municipal or county water system that has been treated and disinfected.

purified water: water that has been treated by distillation or other physical or chemical processes that remove dissolved solids. Because purified water contains no minerals or contaminants, it is useful for medical and research purposes.

purines: compounds of nitrogen-containing bases such as adenine, guanine, and caffeine. Purines that originate from the body are *endogenous* and those that derive from foods are *exogenous*.

pyloric (pie-LORE-ic) **sphincter:** the circular muscle that separates the stomach from the small intestine and regulates the flow of partially digested food into the small intestine; also called *pylorus* or *pyloric valve*.

pyruvate (PIE-roo-vate): a 3-carbon compound that plays a key role in energy metabolism.

Q

qi gong (chée GUNG): a Chinese system that combines movement, meditation, and breathing techniques to enhance the flow of qi (vital energy) in the body.

quality of life: a person's perceived physical and mental well-being.

R

rachitic (ra-KIT-ik) **rosary:** the poorly formed rib attachments that may develop in a vitamin D deficiency; literally, "the rosary of rickets."

radura: an international symbol used to identify retail foods that have been irradiated.

randomization (ran-dom-ih-zay-shun): a process of choosing the members of the experimental and control groups without bias.

raw sugar: the first crop of crystals harvested during sugar processing. Raw sugar cannot be sold in the United States because it contains too much filth (dirt, insect fragments, and the like). Sugar sold as "raw sugar" domestically has actually gone through more than half of the refining steps.

RD: see *registered dietitian*.

Recommended Dietary Allowance (RDA): the average daily amount of a nutrient considered adequate to meet the known nutrient needs of practically all healthy people; a goal for dietary intake by individuals.

rectum: the muscular terminal part of the intestine, extending from the sigmoid colon to the anus.

reduced kcalorie: at least 25 percent fewer kcalories per serving than the comparison food.

reference protein: a standard against which to measure the quality of other proteins.

refined: the process by which the coarse parts of a food are removed. When wheat is refined into flour, the bran, germ, and husk are removed, leaving only the endosperm.

refined flour: finely ground endosperm that is usually enriched with nutrients and bleached for whiteness; sometimes called *white flour*.

reflux: a backward flow.

registered dietitian (RD): a person who has completed a minimum of a bachelor's degree from an accredited university or college, has completed approved course work and a supervised practice program, has passed a national examination, and maintains registration through continuing professional education.

registration: listing; with respect to health professionals, listing with a professional organization that requires specific course work, experience, and passing of an examination.

relaxin: the hormone of late pregnancy.

remodeling: the dismantling and re-formation of a structure.

renin (REN-in): an enzyme from the kidneys that hydrolyzes the protein angiotensinogen to angiotensin I, which results in the kidneys reabsorbing sodium.

rennin: an enzyme that coagulates milk; found in the gastric juice of cows, but not human beings.

replication (REP-lih-KAY-shun): repeating an experiment and getting the same results.

requirement: the lowest continuing intake of a nutrient that will maintain a specified criterion of adequacy.

residues: whatever remains. In the case of pesticides, those amounts that remain on or in foods when people buy and use them.

resistance training: the use of free weights or weight machines to provide resistance for developing muscle strength, power, and endurance; also called *weight training*. A person's own body weight may also be used to provide resistance such as when a person does push-ups, pull-ups, or abdominal crunches.

resistant starches: starches that escape digestion and absorption in the small intestine of healthy people.

respiratory chain: the final pathway in energy metabolism that transports electrons from hydrogen to oxygen and captures the energy released in the bonds of ATP; also called the *electron transport chain.*

resting metabolic rate (RMR): similar to the basal metabolic rate (BMR), a measure of the energy use of a person at rest in a comfortable setting, but with less stringent criteria for recent food intake and physical activity. Consequently, the RMR is slightly higher than the BMR.

retina (RET-in-uh): the innermost membrane of the eye, composed of several layers including one that contains the rods and cones.

retinal (RET-ih-nal): the aldehyde form of vitamin A.

retinoic (RET-ih-NO-ick) **acid:** the acid form of vitamin A.

retinoids (RET-ih-noyds): chemically related compounds with biological activity similar to that of retinol; metabolites of retinol.

retinol (RET-ih-nol): the alcohol form of vitamin A.

retinol activity equivalents (RAE): a measure of vitamin A activity; the amount of retinol that the body will derive from a food containing preformed retinol or its precursor beta-carotene.

retinol-binding protein (RBP): the specific protein responsible for transporting retinol.

rheumatoid (ROO-ma-toyd) **arthritis:** a disease of the immune system involving painful inflammation of the joints and related structures.

rhodopsin (ro-DOP-sin): a light-sensitive pigment of the retina that contains the retinal form of vitamin A and the protein opsin.

riboflavin (RYE-boh-flay-vin): a B vitamin. The coenzyme forms are *FMN (flavin mononucleotide)* and *FAD (flavin adenine dinucleotide).*

ribose: a naturally occurring 5-carbon sugar needed for the synthesis of ATP.

ribosomes (RYE-boh-zomes): protein-making organelles in cells that are composed of RNA and protein.

rickets: the vitamin D–deficiency disease in children characterized by inadequate mineralization of bone (manifested in bowed legs or knock-knees, outward-bowed chest, and "beads" on ribs). A rare type of rickets, not caused by vitamin D deficiency, is known as *vitamin D–refractory rickets.*

risk: a measure of the probability and severity of harm.

risk factor: a condition or behavior associated with an elevated frequency of a disease but not proved to be causal. Leading risk factors for chronic diseases include obesity, cigarette smoking, high blood pressure, high blood cholesterol, physical inactivity, and a diet high in saturated fats and low in vegetables, fruits, and whole grains.

RNA (ribonucleic acid): a compound similar to DNA, but RNA is a single strand with a ribose sugar instead of a deoxyribose sugar and uracil instead of thymine as one of its bases.

S

saccharin (SAK-ah-ren): an artificial sweetener that has been approved for use in the United States. In Canada, saccharin can only be used as a tabletop sweetener, not as an additive.

safety: the condition of being free from harm or danger.

saliva: the secretion of the salivary glands. Its principal enzyme begins carbohydrate digestion.

salivary glands: exocrine glands that secrete saliva into the mouth.

salt: a compound composed of a positive ion other than H^+ and a negative ion other than OH^-. An example is sodium chloride (Na^+ Cl^-).

salt sensitivity: a characteristic of individuals who respond to a high salt intake with an increase in blood pressure or to a low salt intake with a decrease in blood pressure.

sarcopenia (SAR-koh-PEE-nee-ah): loss of skeletal muscle mass, strength, and quality.

satiating: having the power to suppress hunger and inhibit eating.

satiation (say-she-AY-shun): the feeling of satisfaction and fullness that occurs during a meal and halts eating. Satiation determines how much food is consumed during a meal.

satiety (sah-TIE-eh-tee): the feeling of fullness and satisfaction that occurs after a meal and inhibits eating until the next meal. Satiety determines how much time passes between meals.

saturated fat-free: less than 0.5 gram of saturated fat and 0.5 gram of *trans* fat per serving.

saturated fatty acid: a fatty acid carrying the maximum possible number of hydrogen atoms—for example, stearic acid. A *saturated fat* is composed of triglycerides in which most of the fatty acids are saturated.

scurvy: the vitamin C–deficiency disease.

secondary deficiency: a nutrient deficiency caused by something other than an inadequate intake such as a disease condition or drug interaction that reduces absorption, accelerates use, hastens excretion, or destroys the nutrient.

secretin (see-CREET-in): a hormone produced by cells in the duodenum wall. Target organ: the pancreas. Response: secretion of bicarbonate-rich pancreatic juice.

sedentary: physically inactive (literally, "sitting down a lot").

segmentation (SEG-men-TAY-shun): a periodic squeezing or partitioning of the intestine at intervals along its length by its circular muscles.

selenium (se-LEEN-ee-um): an essential trace mineral that is part of an antioxidant enzyme.

senile dementia: the loss of brain function beyond the normal loss of physical adeptness and memory that occurs with aging.

senile plaques: clumps of the protein fragment beta-amyloid on the nerve cells, commonly found in the brains of people with Alzheimer's dementia.

serotonin (SER-oh-TONE-in): a neurotransmitter important in sleep regulation, appetite control, and sensory perception, among other roles. Serotonin is synthesized in the body from the amino acid tryptophan with the help of vitamin B_6.

set point: the point at which controls are set (for example, on a thermostat). The set-point theory that relates to body weight proposes that the body tends to maintain a certain weight by means of its own internal controls.

serving sizes: the standardized quantity of a food; such information allows comparisons when reading food labels and consistency when following the *Dietary Guidelines.*

sickle-cell anemia: a hereditary form of anemia characterized by abnormal sickle- or crescent-shaped red blood cells. Sickled cells interfere with oxygen transport and blood flow. Symptoms are precipitated by dehydration and insufficient oxygen (as may occur at high altitudes) and include hemolytic anemia (red blood cells burst), fever, and severe pain in the joints and abdomen.

simple carbohydrates: monosaccharides and disaccharides (the sugars).

small for gestational age (SGA): term describing an infant whose birth weight is low compared with the number of weeks in utero, often reflecting growth failure.

small intestine: a 10-foot length of small-diameter intestine that is the major site of digestion of food and absorption of nutrients. Its segments are the *duodenum, jejunum,* and *ileum.*

socioeconomic history: a record of a person's social and economic background, including such factors as education, income, and ethnic identity.

sodium: the principal cation in the extracellular fluids of the body; critical to the maintenance of fluid balance, nerve impulse transmissions, and muscle contractions.

sodium bicarbonate (baking soda): a white crystalline powder that is used to buffer acid that accumulates in the muscles and blood during high-intensity exercise.

sodium-free and **salt-free:** less than 5 milligrams of sodium per serving.

soft water: water with a high sodium or potassium content.

solanine (SOH-lah-neen): a poisonous narcotic-like substance present in potato peels and sprouts.

solid fats: fats that are not usually liquid at room temperature; commonly found in most foods derived from animals and vegetable oils that have been hydrogenated. Solid fats typically contain more saturated and *trans* fats than most oils (Chapter 5 provides more details).

soluble fibers: nonstarch polysaccharides that dissolve in water to form a gel. An example is pectin from fruit, which is used to thicken jellies.

solutes (SOLL-yutes): the substances that are dissolved in a solution. The number of molecules in a given volume of fluid is the *solute concentration*.

somatic (so-MAT-ick) **nervous system:** the division of the nervous system that controls the voluntary muscles, as distinguished from the autonomic nervous system, which controls involuntary functions.

somatostatin (GHIH): a hormone that inhibits the release of growth hormone; the opposite of *somatotropin (GH)*.

soup kitchens: programs that provide prepared meals to be eaten on site.

sperm: the male reproductive cell, capable of fertilizing an ovum.

sphincter (SFINK-ter): a circular muscle surrounding, and able to close, a body opening. Sphincters are found at specific points along the GI tract and regulate the flow of food particles.

spina (SPY-nah) **bifida** (BIFF-ih-dah): one of the most common types of neural tube defects; characterized by the incomplete closure of the spinal cord and its bony encasement.

sports anemia: a transient condition of low hemoglobin in the blood, associated with the early stages of sports training or other strenuous activity.

spring water: water originating from an underground spring or well. It may be bubbly (carbonated), or "flat" or "still," meaning not carbonated. Brand names such as "Spring Pure" do not necessarily mean that the water comes from a spring.

starches: plant polysaccharides composed of many glucose molecules.

sterile: free of microorganisms, such as bacteria.

sterols (STARE-ols or STEER-ols): compounds containing a four-ring carbon structure with any of a variety of side chains attached.

stevia (STEE-vee-ah): glycosides found in the leaves of a South American shrub that are used as an herbal sweetener; sold in the United States as a dietary supplement that provides sweetness without kcalories.

stomach: a muscular, elastic, saclike portion of the digestive tract that grinds and churns swallowed food, mixing it with acid and enzymes to form chyme.

stools: waste matter discharged from the colon; also called *feces* (FEE-seez).

stress: any threat to a person's well-being; a demand placed on the body to adapt.

stress eating: eating in response to arousal.

stress fractures: bone damage or breaks caused by stress on bone surfaces during exercise.

stress response: the body's response to stress, mediated by both nerves and hormones.

stressors: environmental elements, physical or psychological, that cause stress.

stroke: an event in which the blood flow to a part of the brain is cut off; also called *cerebrovascular accident (CVA)*.

structure-function claims: statements that characterize the relationship between a nutrient or other substance in a food and its role in the body.

subclavian (sub-KLAY-vee-an) **vein:** the vein that provides passageway from the lymphatic system to the vascular system.

subclinical deficiency: a deficiency in the early stages, before the outward signs have appeared.

subcutaneous fat: fat stored directly under the skin.

subjects: the people or animals participating in a research project.

successful weight-loss maintenance: achieving a weight loss of at least 10 percent of initial body weight and maintaining the loss for at least one year.

sucralose (SUE-kra-lose): an artificial sweetener composed of sucrose with Cl atoms instead of OH groups; approved for use in the United States and Canada.

sucrase: an enzyme that hydrolyzes sucrose.

sucrose (SUE-krose): a disaccharide composed of glucose and fructose; commonly known as *table sugar, beet sugar,* or *cane sugar.* Sucrose also occurs in many fruits and some vegetables and grains.

sudden infant death syndrome (SIDS): the unexpected and unexplained death of an apparently well infant; the most common cause of death of infants between the second week and the end of the first year of life; also called *crib death.*

sugar alcohols: sugarlike compounds that can be derived from fruits or commercially produced from dextrose; also called *polyols.* Sugar alcohols are absorbed more slowly than other sugars and metabolized differently in the human body; they are not readily utilized by ordinary mouth bacteria. Examples are maltitol, mannitol, sorbitol, xylitol, isomalt, and lactitol.

sugar-free: less than 0.5 gram of sugar per serving.

sugars: simple carbohydrates composed of monosaccharides and disaccharides.

sulfate: a salt produced from the oxidation of sulfur.

sulfites: salts containing sulfur that are added to foods to prevent spoilage.

sulfur: a mineral present in the body as part of some proteins.

sushi: vinegar-flavored rice and seafood, typically wrapped in seaweed and stuffed with colorful vegetables. Some sushi is stuffed with raw fish; other varieties contain cooked seafood.

sustainable: able to continue indefinitely; using resources at such a rate that the earth can keep on replacing them and producing pollutants at a rate with which the environment and human cleanup efforts can keep pace, so that no net accumulation of pollution occurs.

sustainable agriculture: ability to produce food indefinitely, with little or no harm to the environment.

symptomatic allergy: an immune response that produces antibodies and symptoms.

synergistic (SIN-er-JIS-tick): multiple factors operating together in such a way that their combined effects are greater than the sum of their individual effects.

T

tagatose (TAG-ah-tose): a monosaccharide structurally similar to fructose that is incompletely absorbed and thus provides only 1.5 kcalories per gram; approved for use as a "generally recognized as safe" ingredient.

TCA cycle or **tricarboxylic** (try-car-box-ILL-ick) **acid cycle:** a series of metabolic reactions that break down molecules of acetyl CoA to carbon dioxide and hydrogen atoms; also called the *citric acid cycle* or the *Kreb's cycle* after the biochemist who elucidated its reactions.

T-cells: lymphocytes that attack antigens. *T* stands for the *thymus gland,* where the T-cells mature.

tempeh (TEM-pay): a fermented soybean food, rich in protein and fiber.

teratogen (ter-AT-oh-jen): a substance that causes abnormal fetal development and birth defects.

teratogenic (ter-AT-oh-jen-ik): causing abnormal fetal development and birth defects.

testosterone: a steroid hormone from the testicles, or testes. The steroids are chemically related to, and some are derived from, the lipid cholesterol.

textured vegetable protein: processed soybean protein used in vegetarian products such as soy burgers.

theory: a tentative explanation that integrates many and diverse findings to further the understanding of a defined topic.

thermic effect of food (TEF): an estimation of the energy required to process food (digest, absorb, transport, metabolize, and store ingested nutrients); also called the *specific dynamic effect (SDE) of food* or the *specific dynamic activity (SDA) of food.* The sum of the TEF and any increase in the metabolic rate due to overeating is known as *diet-induced thermogenesis (DIT).*

thermogenesis: the generation of heat; used in physiology and nutrition studies as an index of how much energy the body is expending.

THF (tetrahydrofolate): a coenzyme form of folate.

thiamin (THIGH-ah-min): a B vitamin. The coenzyme form is *TPP (thiamin pyrophosphate).*

thirst: a conscious desire to drink.

thoracic (thor-ASS-ic) **duct:** the main lymphatic vessel that collects lymph and drains into the left subclavian vein.

thrombosis (throm-BOH-sis): the formation of a *thrombus* (THROM-bus), or a blood clot, that may obstruct a blood vessel, causing gradual tissue death.

thyroid-stimulating hormone (TSH): a hormone secreted by the pituitary that stimulates the thyroid gland to secrete its hormones—thyroxin and triiodothyronine. The release of TSH is mediated by *TSH-releasing hormone (TRH).*

thyrotropin: another name for thyroid-stimulating hormone (TSH).

tocopherol (tuh-KOFF-er-ol): a general term for several chemically related compounds, one of which has vitamin E activity.

tofu (TOE-foo): a curd made from soybeans, rich in protein and often fortified with calcium; used in many Asian and vegetarian dishes in place of meat.

Tolerable Upper Intake Level (UL): the maximum daily amount of a nutrient that appears safe for most healthy people and beyond which there is an increased risk of adverse health effects.

tolerance level: the maximum amount of residue permitted in a food when a pesticide is used according to the label directions.

toxicity: the ability of a substance to harm living organisms. All substances are toxic if high enough concentrations are used.

TPP (thiamin pyrophosphate): the coenzyme form of thiamin.

trabecular (tra-BECK-you-lar) **bone:** the lacy inner structure of calcium crystals that supports the bone's structure and provides a calcium storage bank.

trace minerals: essential mineral nutrients the human body requires in relatively small amounts (less than 100 milligrams per day); sometimes called *microminerals.*

trachea (TRAKE-ee-uh): the air passageway from the larynx to the lungs; also called the *windpipe.*

training: practicing an activity regularly, which leads to conditioning. (Training is what you do; conditioning is what you get.)

trans: on the other side of; refers to a chemical configuration in which the hydrogen atoms are located on opposite sides of a double bond.

***trans* fat-free:** less than 0.5 gram of *trans* fat and less than 0.5 gram of saturated fat per serving.

transamination (TRANS-am-ih-NAY-shun): the transfer of an amino group from one amino acid to a keto acid, producing a new nonessential amino acid and a new keto acid.

transcription: the process of messenger RNA being made from a template of DNA.

***trans*-fatty acids:** fatty acids with hydrogens on opposite sides of the double bond.

transferrin (trans-FAIR-in): the iron transport protein.

transient ischemic (is-KEY-mik) **attack (TIA):** a temporary reduction in blood flow to the brain, which causes temporary symptoms that vary depending on the part of the brain affected. Common symptoms include light-headedness, visual disturbances, paralysis, staggering, numbness, and inability to swallow.

translation: the process of messenger RNA directing the sequence of amino acids and synthesis of proteins.

travelers' diarrhea: nausea, vomiting, and diarrhea caused by consuming food or water contaminated by any of several organisms, most commonly, *E. coli, Shigella, Campylobacter jejuni,* and *Salmonella.*

triglycerides (try-GLISS-er-rides): the chief form of fat in the diet and the major storage form of fat in the body; composed of a molecule of glycerol with three fatty acids attached; also called *triacylglycerols* (try-ay-seel-GLISS-er-ols).

tripeptide: three amino acids bonded together.

tumor: an abnormal tissue mass with no physiological function; also called a *neoplasm* (NEE-oh-plazm).

turbinado (ter-bih-NOD-oh) **sugar:** sugar produced using the same refining process as white sugar, but without the bleaching and anti-caking treatment. Traces of molasses give turbinado its sandy color.

type 1 diabetes: the less common type of diabetes in which the pancreas produces little to no insulin. Type 1 diabetes usually results from autoimmune destruction of pancreatic beta cells.

type 2 diabetes: the more common type of diabetes in which the cells fail to respond to insulin. Type 2 diabetes usually accompanies obesity and results from insulin resistance coupled with insufficient insulin secretion.

U

ulcer: a lesion of the skin or mucous membranes characterized by inflammation and damaged tissues.

ultrahigh temperature (UHT) treatment: sterilizing a food by brief exposure to temperatures above those normally used.

umbilical (um-BILL-ih-cul) **cord:** the ropelike structure through which the fetus's veins and arteries reach the placenta; the route of nourishment and oxygen to the fetus and the route of waste disposal from the fetus. The scar in the middle of the abdomen that marks the former attachment of the umbilical cord is the *umbilicus* (um-BILL-ih-cus), commonly known as the "belly button."

uncoupled reactions: chemical reactions in which energy is released as heat.

undernutrition: deficient energy or nutrients.

underweight: body weight lower than the weight range that is considered healthy; generally defined as BMI <18.5.

unsaturated fatty acid: a fatty acid that lacks hydrogen atoms and has at least one double bond between carbons (includes monounsaturated and polyunsaturated fatty acids). An *unsaturated fat* is composed of triglycerides in which most of the fatty acids are unsaturated.

unspecified eating disorders: eating disorders that do not meet the defined criteria for specific eating disorders.

urea (you-REE-uh): the principal nitrogen-excretion product of protein metabolism. Two ammonia fragments are combined with carbon dioxide to form urea.

USDA (US Department of Agriculture): the federal agency responsible for enforcing standards for the wholesomeness and quality of meat, poultry, and eggs produced in the United States; conducting nutrition research; and educating the public about nutrition.

uterus (YOU-ter-us): the muscular organ within which the infant develops before birth.

V

validity (va-lid-ih-tee): having the quality of being founded on fact or evidence.

variables: factors that change. A variable may depend on another variable (for example, a child's height depends on his age), or it may be independent (for example, a child's height does not depend on the color of her eyes). Sometimes both variables correlate with a third variable (a child's height and eye color both depend on genetics).

variety (dietary): eating a wide selection of foods within and among the major food groups.

vasoconstrictor (VAS-oh-kon-STRIK-tor): a substance that constricts or narrows the blood vessels.

vasopressin (VAS-oh-PRES-in): another name for antidiuretic hormone, so called because it elevates blood pressure.

vegans (VEE-gans): people who exclude all animal-derived foods (including meat, poultry, fish, eggs, and milk and milk products) from their diets; also called *pure vegetarians, strict vegetarians,* or *total vegetarians.*

vegetarians: a general term used to describe people who exclude meat, poultry, fish, or other animal-derived foods from their diets.

veins (VANES): vessels that carry blood to the heart.

very low food security: multiple indications of disrupted eating patterns and reduced food intake; formerly known as *food insecurity with hunger.*

very low sodium: 35 milligrams or less per serving.

vigorous-intensity physical activity: physical activity that requires a large increase in breathing and/or heart rate and expends more than 7 kcalories per minute. Walking at a very brisk pace (>4.5 miles per hour) or running at a pace of at least 5 miles per hour are examples.

villi (VILL-ee, VILL-eye): fingerlike projections from the folds of the small intestine; singular *villus.*

visceral fat: fat stored within the abdominal cavity in association with the internal abdominal organs; also called *intra-abdominal fat.*

viscous: a gel-like consistency.

vitamin A: all naturally occurring compounds with the biological activity of *retinol* (RET-ih-nol), the alcohol form of vitamin A.

vitamin A activity: a term referring to both the active forms of vitamin A and the precursor forms in foods without distinguishing between them.

vitamin B_6: a family of compounds—pyridoxal, pyridoxine, and pyridoxamine. The primary active coenzyme form is *PLP (pyridoxal phosphate).*

vitamin B_{12}: a B vitamin characterized by the presence of cobalt. The active forms of coenzyme B_{12} are methylcobalamin and deoxyadenosylcobalamin.

vitamin D_2: vitamin D derived from plants in the diet; also called *ergocalciferol.*

vitamin D_3: vitamin D derived from animals in the diet and made in the skin from 7-dehydrocholesterol, a precursor of cholesterol, with the help of sunlight; also called *cholecalciferol.*

vitamins: organic, essential nutrients required in small amounts by the body for health. Vitamins regulate body processes that support growth and maintain life.

VLDL (very-low-density lipoprotein): the type of lipoprotein made primarily by liver cells to transport lipids to various tissues in the body; composed primarily of triglycerides.

VO_2max: the maximum rate of oxygen consumption by an individual at sea level.

vomiting: expulsion of the contents of the stomach up through the esophagus to the mouth.

vulnerable plaque: plaque that is susceptible to rupture because it has only a thin fibrous barrier between its lipid-rich core and the artery lining.

W

waist circumference: an anthropometric measurement used to assess a person's abdominal fat.

warm-up: 5 to 10 minutes of light activity, such as easy jogging or cycling, prior to a workout to prepare the body for more vigorous activity.

water balance: the balance between water intake and output (losses).

water intoxication: the rare condition in which body water contents are too high in all body fluid compartments.

wean: to gradually replace breast milk with infant formula or other foods appropriate to an infant's diet.

websites: Internet resources composed of text and graphic files, each with a unique URL (Uniform Resource Locator) that names the site (for example, www.usda.gov).

weight management: maintaining body weight in a healthy range by preventing gradual weight gain over time and losing weight if overweight, and by preventing weight losses and gaining weight if underweight.

well water: water drawn from ground water by tapping into an aquifer.

Wernicke-Korsakoff (VER-nee-key KORE-sah-kof) **syndrome:** a neurological disorder typically associated with chronic alcoholism and caused by a deficiency of the B vitamin thiamin; also called *alcohol-related dementia.*

wheat flour: any flour made from the endosperm of the wheat kernel.

whey protein: a by-product of cheese production; falsely promoted as increasing muscle mass. Whey is the watery part of milk that separates from the curds.

white sugar: granulated sucrose or "table sugar," produced by dissolving, concentrating, and recrystallizing raw sugar.

WHO (World Health Organization): an international agency concerned with promoting health and eradicating disease.

whole grain: a grain that maintains the same relative proportions of starchy endosperm, germ, and bran as the original (all but the husk); not refined.

whole-wheat flour: any flour made from the entire wheat kernel.

Wilson's disease: a genetic disorder of copper metabolism that creates a copper toxicity and results in neurologic symptoms such as tremors, impaired speech, inappropriate behaviors, and personality changes.

wine: an alcoholic beverage traditionally made by fermenting a sugar source such as grape juice.

without appreciable risk: practical certainty that injury will not result even after a lifetime of exposure.

X

xanthophylls (ZAN-tho-fills): pigments found in plants responsible for the color changes seen in autumn leaves.

xerophthalmia (zer-off-THAL-mee-uh): progressive blindness caused by inadequate mucus production due to severe vitamin A deficiency.

xerosis (zee-ROW-sis): abnormal drying of the skin and mucous membranes; a sign of vitamin A deficiency.

Y

yields: production per acre.

yogurt: milk product that results from the fermentation of lactic acid in milk by *Lactobacillus bulgaricus* and *Streptococcus thermophilus.*

Z

zinc: an essential trace mineral that is part of many enzymes and a constituent of insulin.

zygote (ZY-goat): the initial product of the union of ovum and sperm; a fertilized ovum.

zymogen (ZYE-mo-jen): the inactive precursor of an enzyme; sometimes called a *proenzyme.*

Index

- Page numbers in bold indicate definition of term.
- Page numbers followed by the letter "f" indicate figures or photographs.
- Page numbers followed by the letter "n" indicate footnotes.
- Page numbers followed by the letter "t" indicate tables.

A

E

F

O

P

W

X

Y

Z

eview the key points of this chapter below, then take the practice quiz on the back of this card.

.1 Food Choices

- People select foods based on such factors as taste and convenience, but selections based on nutrition knowledge may better support good health.
- Individual foods are neither "good" nor "bad"; daily food choices made over a lifetime may improve or impair a person's health.

.2 The Nutrients

- Foods provide nutrients—substances that provide energy, structural materials, and regulating agents to support the growth, maintenance, and repair of the body's tissues. Essential nutrients *must* be obtained from foods.
- The six classes of nutrients include carbohydrates, lipids (fats), proteins, vitamins, minerals, and water. Carbohydrates, lipids, proteins, and vitamins are organic, meaning they contain carbon; minerals and water are inorganic, meaning they do not contain carbon.
- Energy is measured in kcalories—a measure of heat energy. One kcalorie is the amount of heat necessary to raise the temperature of 1 kg water 1°C.
- The energy-yielding nutrients are carbohydrate (4 kcal/g), fat (9 kcal/g), and protein (4 kcal/g). Alcohol provides 7 kcal/g, but it is not considered a nutrient.
- Vitamins, minerals, and water do not provide energy; instead, they facilitate a variety of activities in the body.

.3 The Science of Nutrition

- The science of nutrition is the study of nutrients and other substances in foods and the body's handling of them.
- Researchers follow the scientific method (review Figure 1-3, p. 12). They randomly assign control and experimental groups, use large sample sizes, provide placebos, and are blind to treatments. Their findings are reviewed and replicated by other scientists before being accepted as valid.
- Correlations indicate an association between variables, not a cause.

.4 Dietary Reference ntakes

- Dietary Reference Intakes (DRI) are a set of nutrient intake values used to plan and evaluate diets for healthy people.
- Estimated Average Requirement (EAR) defines the amount of a nutrient that supports a specific function in the body for half of the population. Recommended Dietary Allowance (RDA) is based on the EAR and establishes a goal for dietary intake that will meet the needs of almost all healthy people. Adequate Intake (AI) serves a similar purpose when an RDA cannot be determined.
- Estimated Energy Requirement (EER) defines the average amount of energy intake needed to maintain energy balance, and Acceptable Macronutrient Distribution Ranges (AMDR) define the proportions contributed by carbohydrate, fat, and protein to a healthy diet. Carbohydrates should consist of 45 to 65 percent of kcalories, fat 20 to 35 percent, and protein 10 to 35 percent.
- Tolerable Upper Intake Level (UL) establishes the highest amount that appears safe for regular consumption.

1.5 Nutrition Assessment

- Malnutrition develops when people get too little, too much, or an imbalance of energy or nutrients.
- Four nutrition assessment methods include historical information on diet and health, anthropometric measurements, physical examinations, and laboratory tests. Together, these methods reveal the stages of a nutrient deficiency (review Figure 1-7, p. 23).
- A primary deficiency is caused by an inadequate intake of a nutrient; a secondary deficiency is caused by a condition that reduces absorption, accelerates use, increases excretion, or destroys the nutrient.

1.6 Diet and Health

- Risk factors such as obesity and cigarette smoking increase the likelihood of disease development (review Table 1-6).
- Some risk factors, such as genetics, are important but cannot be changed. Recommendations focus on changeable, personal life choices such as diet and activity habits.
- Diet has no influence on some diseases but is linked closely to others (review Table 1-5).

TABLE 1-5 Leading Causes of Death in the United States

	Percentage of Total Deaths
1. **Heart disease**	24.1
2. **Cancers**	23.3
3. Chronic lung diseases	5.6
4. **Strokes**	5.2
5. Accidents	4.8
6. Alzheimer's disease	3.4
7. **Diabetes mellitus**	2.8
8. Pneumonia and influenza	2.0
9. Kidney disease	2.0
10. Suicide	1.5

NOTE: The diseases highlighted in bold have relationships with diet.

SOURCE: Deaths: Preliminary data for 2010, *National Vital Statistics Reports*, January 11, 2012, Centers for Disease Control and Prevention, www.cdc.gov/nchs.

TABLE 1-6 Factors Contributing to Deaths in the United States

Factors	Percentage of Deaths
Tobacco	18
Poor diet/inactivity	15
Alcohol	4
Microbial agents	3
Toxic agents	2
Motor vehicles	2
Firearms	1
Sexual behavior	1
Illicit drugs	1

SOURCE: A. H. Mokdad and coauthors, Actual causes of death in the United States, 2000, *Journal of the American Medical Association* 291 (2004): 1238–1245, with corrections from *Journal of the American Medical Association* 293 (2005): 298.

Take the quiz below to test your mastery of the key chapter concepts.

1.1 Food Choices (pp. 4–6)

LEARN IT Describe how various factors influence personal food choices.

1. When people eat the foods typical of their families or geographic region, their choices are influenced by:
   a. habit. c. personal preference.
   b. nutrition. d. heritage or tradition.

1.2 The Nutrients (pp. 6–12)

LEARN IT Name the six major classes of nutrients and identify which are organic and which yield energy.

2. What is the difference between organic and inorganic?
3. How much energy do carbohydrates, fats, and proteins yield per gram? How is energy measured?
4. Describe how alcohol resembles nutrients. Why is alcohol not considered a nutrient?
5. The nutrient found most abundantly in both the human body and most foods is:
   a. fat. b. water. c. minerals. d. proteins.
6. The inorganic nutrients are:
   a. proteins and fats. c. minerals and water.
   b. vitamins and minerals. d. vitamins and proteins.
7. The energy-yielding nutrients are:
   a. fats, minerals, and water.
   b. minerals, proteins, and vitamins.
   c. carbohydrates, fats, and vitamins.
   d. carbohydrates, fats, and proteins.

1.3 The Science of Nutrition (pp. 12–17)

LEARN IT Explain the scientific method and how scientists use various types of research studies and methods to acquire nutrition information.

8. What is the science of nutrition?
9. Explain how variables might be correlational but not causal.
10. Studies of populations that reveal correlations between dietary habits and disease incidence are:
    a. clinical trials. c. case-control studies.
    b. laboratory studies. d. epidemiological studies.
11. An experiment in which neither the researchers nor the subjects know who is receiving the treatment is known as:
    a. double blind. c. blind variable.
    b. double control. d. placebo control.

1.4 Dietary Reference Intakes (pp. 17–21)

LEARN IT Define the four categories of the DRI and explain the purposes.

12. What judgment factors are involved in setting the energy ar nutrient recommendations?
13. An RDA represents the:
    a. highest amount of a nutrient that appears safe for mo healthy people.
    b. lowest amount of a nutrient that will maintain a specifie criterion of adequacy.
    c. average amount of a nutrient considered adequate to me the known nutrient needs of practically all healthy people.
    d. average amount of a nutrient that will maintain a speci biochemical or physiological function in half the people.

1.5 Nutrition Assessment (pp. 21–25)

LEARN IT Explain how the four assessment methods are used detect energy and nutrient deficiencies and excesses.

14. What methods are used in nutrition surveys? What kinds information can these surveys provide?
15. Historical information, physical examinations, laboratory test and anthropometric measurements are:
    a. techniques used in diet planning.
    b. steps used in the scientific method.
    c. approaches used in disease prevention.
    d. methods used in a nutrition assessment.
16. A deficiency caused by an inadequate dietary intak is called a(n):
    a. overt deficiency. c. primary deficiency.
    b. covert deficiency. d. secondary deficiency.

1.6 Diet and Health (pp. 25–27)

LEARN IT Identify several risk factors and explain their relationshi to chronic diseases.

17. Behaviors such as smoking, dietary habits, physical activit and alcohol consumption that influence the development disease are known as:
    a. risk factors. c. preventive agents.
    b. chronic causes. d. disease descriptors.

ultiple Choice Answers

d 5.b 6.c 7.d 10.d 11.a 13.c 15.d 16.c 17.a

2 Planning a Healthy Diet

eview the key points of this chapter below, then take the practice quiz on the back of this card.

.1 Principles and Guidelines

- A well-planned diet delivers *adequate* nutrients, a *balanced* array of nutrients, and an appropriate amount of *energy* (*kcalories*). It is based on *nutrient-dense* foods, *moderate* in substances that can be detrimental to health, and *varied* in its selections.
- The *Dietary Guidelines for Americans* offer practical advice on how to eat for good health (review Table 2-1, p. 39).

.2 Diet-Planning Guides

- The USDA Food Patterns help consumers select the types and amounts of foods to provide adequacy, balance, and variety in the diet. It makes it easier to plan a diet that includes a balance of grains, vegetables, fruits, protein foods, and milk products. In making any food choice, remember to view the food in the context of the total diet.
- The combination of many different foods provides the array of nutrients that are essential to a healthy diet (review Figure 2-2, pp. 42–43).
- MyPlate reminds consumers to make healthy choices from the five food groups (review Figure 2-4).

> FIGURE 2-4 **MyPlate**

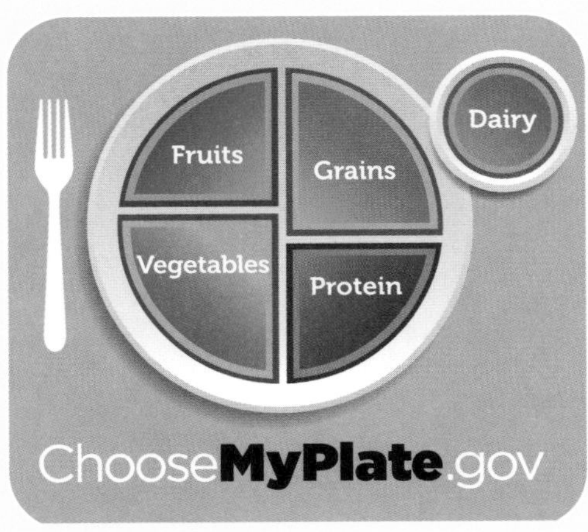

SOURCE: USDA, www.choosemyplate.gov.

2.3 Food Labels

- Food labels list ingredients in descending order of predominance by weight, nutrition facts based on standard serving sizes, and Daily Values based on a 2000-kcalorie diet (review Figure 2-8).
- Nutrient claims reflect the quantity of a nutrient (high or low), health claims reflect relationships between a nutrient and a disease (potassium reduces risk of hypertension), and structure-function claims reflect relationships between a nutrient and its function in the body (calcium builds bones).

> FIGURE 2-8 **Example of a Food Label**

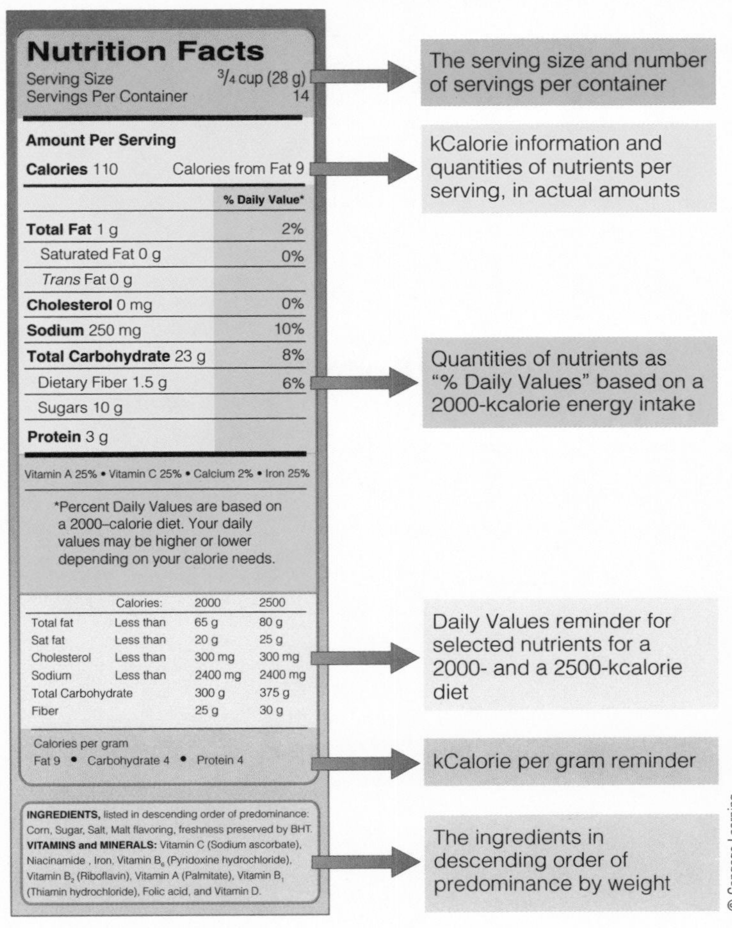

Take the quiz below to test your mastery of the key chapter concepts.

2.1 Principles and Guidelines (pp. 35–40)

LEARN IT Explain how each of the diet-planning principles can be used to plan a healthy diet.

1. What recommendations appear in the *Dietary Guidelines for Americans*?

2. The diet-planning principle that provides all the essential nutrients in sufficient amounts to support health is:
   a. balance.
   b. variety.
   c. adequacy.
   d. moderation.

3. A person who chooses a chicken leg that provides 0.5 milligram of iron and 95 kcalories instead of 2 tablespoons of peanut butter that also provides 0.5 milligram of iron but 188 kcalories is using the principle of nutrient:
   a. control.
   b. density.
   c. adequacy.
   d. moderation.

4. Which of the following is consistent with the *Dietary Guidelines for Americans*?
   a. Choose a diet restricted in fat and cholesterol.
   b. Balance the food you eat with physical activity.
   c. Choose a diet with plenty of milk products and meats.
   d. Eat an abundance of foods to ensure nutrient adequacy.

2.2 Diet-Planning Guides (pp. 40–53)

LEARN IT Use the USDA Food Patterns to develop a meal plan within a specified energy allowance.

5. Review the *Dietary Guidelines*. What types of grocery selections would you make to achieve those recommendations?

6. According to the USDA Food Patterns, cheese is grouped as a:
   a. meat.
   b. protein food.
   c. milk product.
   d. miscellaneous fat.

7. Foods within a given food group of the USDA Food Patterns are similar in their contents of:
   a. energy.
   b. proteins and fibers.
   c. vitamins and minerals.
   d. carbohydrates and fats.

8. In the exchange system, each portion of food on any given li... provides about the same amount of:
   a. energy.
   b. satiety.
   c. vitamins.
   d. minerals.

9. Enriched grain products are fortified with:
   a. fiber, folate, iron, niacin, and zinc.
   b. thiamin, iron, calcium, zinc, and sodium.
   c. iron, thiamin, riboflavin, niacin, and folate.
   d. folate, magnesium, vitamin B_6, zinc, and fiber.

2.3 Food Labels (pp. 53–60)

LEARN IT Compare and contrast the information on food labels t... make selections that meet specific dietary and health goals.

10. What information can you expect to find on a food label? Ho... can this information help you choose between two simil... products?

11. What are the Daily Values? How can they help you meet healt... recommendations?

12. Describe the differences between nutrient claims, healt... claims, and structure-function claims.

13. Food labels list ingredients in:
    a. alphabetical order.
    b. ascending order of predominance by weight.
    c. descending order of predominance by weight.
    d. manufacturer's order of preference.

14. "Milk builds strong bones" is an example of a:
    a. health claim.
    b. nutrition fact.
    c. nutrient content claim.
    d. structure-function claim.

15. Daily Values on food labels are based on a:
    a. 1500-kcalorie diet.
    b. 2000-kcalorie diet.
    c. 2500-kcalorie diet.
    d. 3000-kcalorie diet.

ultiple Choice Answers
c 3. b 4. b 6. c 7. c 8. a 9. c 13. c 14. d 15. b

STUDY IT

3 Digestion, Absorption, and Transport

Review the key points of this chapter below, then take the practice quiz on the back of this card.

3.1 Digestion

- Digestion breaks down foods into nutrients. Absorption brings the nutrients into the cells of the small intestine for transport to the body's cells.
- Food enters the mouth and travels down the esophagus and through the upper and lower esophageal sphincters to the stomach, then through the pyloric sphincter to the small intestine, on through the ileocecal valve to the large intestine, past the appendix to the rectum, ending at the anus (review Figure 3-1).
- The wavelike contractions of peristalsis and the periodic squeezing of segmentation keep things moving at a reasonable pace. Along the way, secretions from the salivary glands, stomach, pancreas, liver (via the gallbladder), and small intestine deliver fluids and digestive enzymes (review Table 3-1).

3.2 Absorption

- The many folds and villi of the small intestine increase its surface area, making nutrient absorption efficient.
- Nutrients pass through the cells of the intestinal villi and enter either the blood (if they are water soluble or small fat fragments) or the lymph (if they are fat soluble).

3.3 The Circulatory Systems

- Nutrients leaving the digestive system via the blood are routed directly to the liver before being transported to the body's cells.
- Nutrients leaving via the lymphatic system bypass the liver at first, but eventually enter the vascular system via the thoracic duct, which opens into the subclavian vein.

3.4 The Health and Regulation of the GI Tract

- A diverse and abundant bacteria population supports GI health.
- The regulation of GI processes depends on the coordinated efforts of the hormonal system and the nervous system.
- Together, digestion and absorption break down foods into nutrients for the body's use.
- To function optimally, a healthy GI tract needs a balanced diet, adequate rest, and regular physical activity.

FIGURE 3-1 The Gastrointestinal Tract

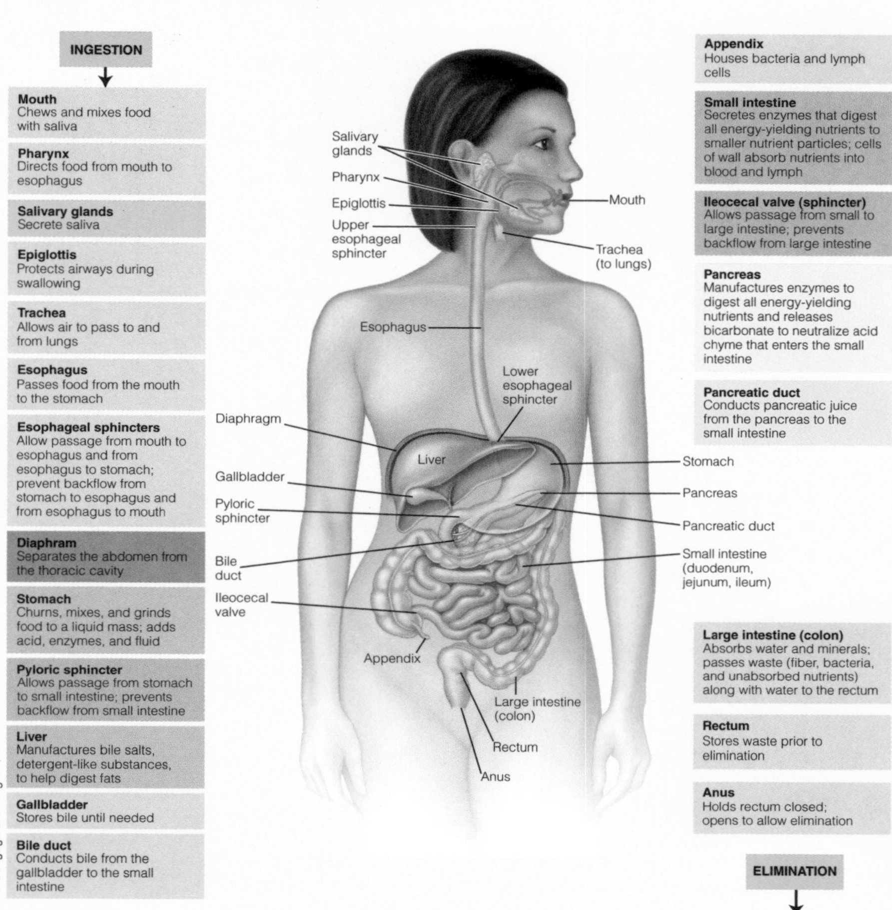

INGESTION
Mouth
Chews and mixes food with saliva
Pharynx
Directs food from mouth to esophagus
Salivary glands
Secrete saliva
Epiglottis
Protects airways during swallowing
Trachea
Allows air to pass to and from lungs
Esophagus
Passes food from the mouth to the stomach
Esophageal sphincters
Allow passage from mouth to esophagus and from esophagus to stomach; prevent backflow from stomach to esophagus and from esophagus to mouth
Diaphram
Separates the abdomen from the thoracic cavity
Stomach
Churns, mixes, and grinds food to a liquid mass; adds acid, enzymes, and fluid
Pyloric sphincter
Allows passage from stomach to small intestine; prevents backflow from small intestine
Liver
Manufactures bile salts, detergent-like substances, to help digest fats
Gallbladder
Stores bile until needed
Bile duct
Conducts bile from the gallbladder to the small intestine
© Cengage Learning
Salivary glands
Pharynx
Epiglottis
Upper esophageal sphincter
Mouth
Trachea (to lungs)
Esophagus
Lower esophageal sphincter
Diaphragm
Liver
Gallbladder
Pyloric sphincter
Bile duct
Ileocecal valve
Appendix
Stomach
Pancreas
Pancreatic duct
Small intestine (duodenum, jejunum, ileum)
Large intestine (colon)
Rectum
Anus
Appendix
Houses bacteria and lymph cells
Small intestine
Secretes enzymes that digest all energy-yielding nutrients to smaller nutrient particles; cells of wall absorb nutrients into blood and lymph
Ileocecal valve (sphincter)
Allows passage from small to large intestine; prevents backflow from large intestine
Pancreas
Manufactures enzymes to digest all energy-yielding nutrients and releases bicarbonate to neutralize acid chyme that enters the small intestine
Pancreatic duct
Conducts pancreatic juice from the pancreas to the small intestine
Large intestine (colon)
Absorbs water and minerals; passes waste (fiber, bacteria, and unabsorbed nutrients) along with water to the rectum
Rectum
Stores waste prior to elimination
Anus
Holds rectum closed; opens to allow elimination
ELIMINATION

TABLE 3-1 Summary of Digestive Secretions and Their Major Actions

Organ or Gland	Target Organ	Secretion	Action
Salivary glands	Mouth	Saliva	Fluid eases swallowing; salivary enzyme breaks down some **carbohydrate.***
Gastric glands	Stomach	Gastric juice	Fluid mixes with bolus; hydrochloric acid uncoils **proteins;** enzymes break down proteins; mucus protects stomach cells.*
Pancreas	Small intestine	Pancreatic juice	Bicarbonate neutralizes acidic gastric juices; pancreatic enzymes break down **carbohydrates, fats,** and **proteins.**
Liver	Gallbladder	Bile	Bile is stored until needed.
Gallbladder	Small intestine	Bile	Bile emulsifies **fat** so that enzymes can have access to break it down.
Intestinal glands	Small intestine	Intestinal juice	Intestinal enzymes break down **carbohydrate, fat,** and **protein** fragments; mucus protects the intestinal wall.

*Saliva and gastric juice also contain lipases, but most fat breakdown occurs in the small intestine.

TEST IT

Take the quiz below to test your mastery of the key chapter concepts.

3.1 Digestion (pp. 69–77)

LEARN IT Explain how foods move through the digestive system, describing the actions of the organs, muscles, and digestive secretions along the way.

1. Describe the challenges associated with digesting food and the solutions offered by the human body.
2. Name five organs that secrete digestive juices. How do the juices and enzymes facilitate digestion?
3. The semiliquid, partially digested food that travels through the intestinal tract is called:
   a. bile.
   b. lymph.
   c. chyme.
   d. secretin.
4. The muscular contractions that move food through the GI tract are called:
   a. hydrolysis.
   b. sphincters.
   c. peristalsis.
   d. bowel movements.
5. The main function of bile is to:
   a. emulsify fats.
   b. catalyze hydrolysis.
   c. slow protein digestion.
   d. neutralize stomach acidity.
6. The pancreas neutralizes stomach acid in the small intestine by secreting:
   a. bile.
   b. mucus.
   c. enzymes.
   d. bicarbonate.
7. Which nutrient passes through the GI tract mostly undigested and unabsorbed?
   a. fat
   b. fiber
   c. protein
   d. carbohydrate

3.2 Absorption (pp. 77–79)

LEARN IT Describe the anatomical details of the intestinal cells that facilitate nutrient absorption.

8. The fingerlike projections on the small intestine that dramatically increase its surface area are called:
   a. villi.
   b. crypts.
   c. goblet cells.
   d. chylomicrons.
9. Absorption occurs primarily in the:
   a. mouth.
   b. stomach.
   c. small intestine.
   d. large intestine.

3.3 The Circulatory Systems (pp. 80–82)

LEARN IT Explain how nutrients are routed in the circulatory system from the GI tract into the body and identify which nutrients enter th blood directly and which must first enter the lymph.

10. All blood leaving the GI tract travels first to the:
    a. heart.
    b. liver.
    c. kidneys.
    d. pancreas.
11. Which nutrients leave the GI tract by way of the lymphat system?
    a. water and minerals
    b. proteins and minerals
    c. all vitamins and minerals
    d. fats and fat-soluble vitamins

3.4 The Health and Regulation of the GI Tract (pp. 82–86)

LEARN IT Describe how bacteria, hormones, and nerves influenc the health and activities of the GI tract.

12. How does the composition of the diet influence the functionin of the GI tract?
13. What steps can you take to help your GI tract function a its best?
14. Digestion and absorption are coordinated by the:
    a. pancreas and kidneys.
    b. liver and gallbladder.
    c. hormonal system and the nervous system.
    d. vascular system and the lymphatic system.
15. Gastrin, secretin, and cholecystokinin are examples of:
    a. crypts.
    b. enzymes.
    c. hormones.
    d. goblet cells.

ultiple Choice Answers

c 4.c 5.a 6.d 7.b 8.a 9.c 10.b 11.d 14.c 15.c

4 The Carbohydrates: Sugars, Starches, and Fibers

Review the key points of this chapter below, then take the practice quiz on the back of this card.

4.1 The Chemist's View of Carbohydrates

- Carbohydrates include monosaccharides, disaccharides, and polysaccharides (review Table 4-1).
- Carbohydrates are made of carbon (C), oxygen (O), and hydrogen (H); each atom forms a specified number of chemical bonds: carbon forms four, oxygen forms two, and hydrogen forms one (review Figure 4-1, p. 96).
- Monosaccharides (glucose, fructose, and galactose) all have the same chemical formula ($C_6H_{12}O_6$), but their structures differ. Disaccharides (maltose, sucrose, and lactose) each contain a glucose paired with one of the three monosaccharides.
- A condensation reaction can bond two monosaccharides together to form a disaccharide and water (review Figure 4-4, p. 98). A hydrolysis reaction can use water to split a disaccharide into its two monosaccharides (review Figure 4-5, p. 98).
- Chains of monosaccharides are called polysaccharides and include glycogen, starches, and dietary fibers. Both glycogen and starch are storage forms of glucose—glycogen in the body, and starch in plants—and both yield energy.
- Dietary fibers contain glucose (and other monosaccharides), but their bonds cannot be broken by human digestive enzymes; they yield little, if any, energy.
- Soluble fibers dissolve in water to form gels and are easily digested by bacteria in the colon. Insoluble fibers do not dissolve in water or form gels and are less readily fermented.

TABLE 4-1 **The Carbohydrate Family**

Monosaccharides
Glucose
Fructose
Galactose
Disaccharides
Maltose (glucose + glucose)
Sucrose (glucose + fructose)
Lactose (glucose + galactose)
Polysaccharides
Glycogen[a]
Starches (amylose and amylopectin)
Fibers (soluble and insoluble)

[a]Glycogen is a polysaccharide, but not a common dietary source of carbohydrate.

4.2 Digestion and Absorption of Carbohydrates

- The body digests starches into the disaccharide maltose. Maltose and the other disaccharides (lactose and sucrose) from foods are broken down into monosaccharides, which are absorbed (review Figure 4-9, p. 103)
- Fibers help to regulate the passage of food through the GI tract and slow the absorption of glucose.
- Lactose intolerance occurs when there is insufficient lactase to digest the disaccharide lactose found in milk and milk products. Symptoms include GI distress.

4.3 Glucose in the Body

- Dietary carbohydrates provide glucose that can be used by the cells for energy, stored by the liver and muscles as glycogen, or converted into fat if intakes exceed needs.
- All of the body's cells depend on glucose; those of the central nervous system are especially dependent on it.
- Without glucose, the body is forced to break down its protein tissues to make glucose and to alter energy metabolism to make ketone bodies from fats.
- Blood glucose regulation depends on two pancreatic hormones: insulin to move glucose from the blood into the cells when levels are high and glucagon to free glucose from glycogen stores and release it into the blood when levels are low (review Figure 4-10, p. 107).

4.4 Health Effects and Recommended Intakes of Sugars

- Excessive intakes of sugars may increase the risk of dental caries, displace needed nutrients and fiber, and contribute to obesity when energy intake exceeds needs.
- Concentrated sweets are relatively low in nutrients, high in kcalories, and may need to be limited; sugars that occur naturally in fruits, vegetables, and milk are acceptable.
- To control weight gain, blood glucose, and dental caries, consumers may use alternative sweeteners (artificial sweeteners, herbal products, and sugar alcohols) to limit kcalories and minimize sugar intake (review Table 4-4, p. 114).

4.5 Health Effects and Recommended Intakes of Starch and Fibers

- Adequate intake of fiber fosters weight management, lowers blood cholesterol, and may help prevent colon cancer, diabetes, hemorrhoids, appendicitis, and diverticulosis.
- Excessive intake of fiber displaces energy- and nutrient-dense foods, causes intestinal discomfort and distention, and may interfere with mineral absorption.
- Because starches and fibers help control body weight and prevent heart disease, cancer, diabetes, and GI disorders, the *Dietary Guidelines* suggest plenty of whole grains, vegetables, legumes, and fruits—enough to provide 45 to 65 percent of the daily energy intake from carbohydrate.

Take the quiz below to test your mastery of the key chapter concepts.

4.1 The Chemist's View of Carbohydrates (pp. 96–101)

LEARN IT Identify the monosaccharides, disaccharides, and polysaccharides common in nutrition by their chemical structures and major food sources.

1. What happens in a condensation reaction? In a hydrolysis reaction?
2. How are starch and glycogen similar, and how do they differ? How do the fibers differ from the other polysaccharides?
3. Disaccharides include:
   a. starch, glycogen, and fiber.
   b. amylose, pectin, and dextrose.
   c. sucrose, maltose, and lactose.
   d. glucose, galactose, and fructose.
4. The making of a disaccharide from two monosaccharides is an example of:
   a. digestion.
   b. hydrolysis.
   c. condensation.
   d. gluconeogenesis.
5. The significant difference between starch and cellulose is that:
   a. starch is a polysaccharide, but cellulose is not.
   b. animals can store glucose as starch, but not as cellulose.
   c. hormones can make glucose from cellulose, but not from starch.
   d. digestive enzymes can break the bonds in starch, but not in cellulose.

4.2 Digestion and Absorption of Carbohydrates (pp. 101–104)

LEARN IT Summarize carbohydrate digestion and absorption.

6. What role does fiber play in digestion and absorption?
7. Describe lactose intolerance and its symptoms.
8. The ultimate goal of carbohydrate digestion and absorption is to yield:
   a. fibers.
   b. glucose.
   c. enzymes.
   d. amylase.
9. The enzyme that breaks a disaccharide into glucose and galactose is:
   a. amylase.
   b. maltase.
   c. sucrase.
   d. lactase.

4.3 Glucose in the Body (pp. 104–109)

LEARN IT Explain how the body maintains its blood glucose concen tration and what happens when blood glucose rises too high or fall too low.

10. What are the possible fates of glucose in the body? What i the protein-sparing action of carbohydrate?
11. The storage form of glucose in the body is:
    a. insulin.
    b. maltose.
    c. glucagon.
    d. glycogen.
12. With insufficient glucose in metabolism, fat fragment combine to form:
    a. dextrins.
    b. mucilages.
    c. phytic acids.
    d. ketone bodies.
13. What does the pancreas secrete when blood glucose rises When blood glucose falls?
    a. insulin; glucagon
    b. glucagon; insulin
    c. insulin; glycogen
    d. glycogen; epinephrine

4.4 Health Effects and Recommended Intakes of Sugars (pp. 109–115)

LEARN IT Describe how added sugars can contribute to healt problems.

14. What are the dietary recommendations regarding concentrate sugar intakes?
15. Describe the risks and benefits of using alternative sweetener

4.5 Health Effects and Recommended Intakes of Starch and Fibers (pp. 115–121)

LEARN IT Identify the health benefits of, and recommendations fo starches and fibers.

16. What foods provide starches and fibers?
17. Carbohydrates are found in virtually all foods except:
    a. milks.
    b. meats.
    c. breads.
    d. fruits.
18. What percentage of the daily energy intake should come fro carbohydrates?
    a. 15 to 20
    b. 25 to 30
    c. 45 to 50
    d. 45 to 65

ultiple Choice Answers

c 4.c 5.d 8.b 9.d 11.d 12.d 13.a 17.b 18.d

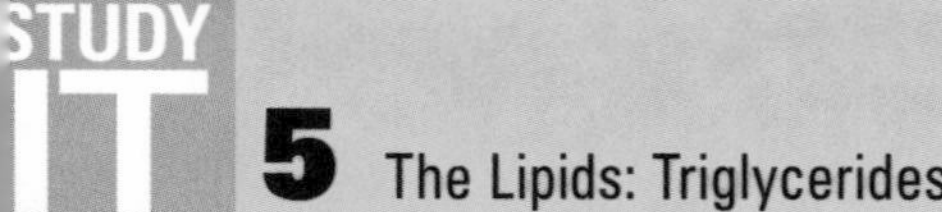

5 The Lipids: Triglycerides, Phospholipids, and Sterols

Review the key points of this chapter below, then take the practice quiz on the back of this card.

5.1 The Chemist's View of Fatty Acids and Triglycerides

- The predominant lipids both in foods and in the body are triglycerides: glycerol with three fatty acids attached by way of condensation reactions (review Figure 5-3, p. 133). Other lipids include phospholipids and sterols (review Table 5-2).
- Fatty acids vary in the length of their carbon chains, their degrees of unsaturation, and the location of their double bond(s). Saturated fatty acids are fully loaded with hydrogens; unsaturated (monounsaturated or polyunsaturated) fatty acids are missing hydrogens and have double bonds.
- Hydrogenation protects against oxidation (thereby promoting shelf-life) and alters the texture of foods by making liquid vegetable oils more solid. This process makes polyunsaturated fats more saturated and creates *trans*-fatty acids.

5.2 The Chemist's View of Phospholipids and Sterols

- The chemical structure of phospholipids, including lecithin, allows them to be soluble in both water and fat. In the body, phospholipids are part of cell membranes; in foods, phospholipids act as emulsifiers to mix fats with water.
- Sterols have a multiple-ring structure that differs from the structure of other lipids. In the body, sterols include cholesterol, bile, vitamin D, and some hormones. Animal-derived foods contain cholesterol.

5.3 Digestion, Absorption, and Transport of Lipids

- Bile emulsifies fats, making them accessible to the lipases that dismantle triglycerides to monoglycerides and fatty acids for absorption (review Figure 5-12, p. 140).
- Four types of lipoproteins carry triglycerides, phospholipids, and cholesterol throughout the body: *chylomicrons* are the largest and contain mostly dietary triglycerides, *VLDL* are smaller and are about half triglycerides, *LDL* are smaller still and contain mostly cholesterol, and *HDL* are the densest and are rich in protein (review Figure 5-16, p. 142).

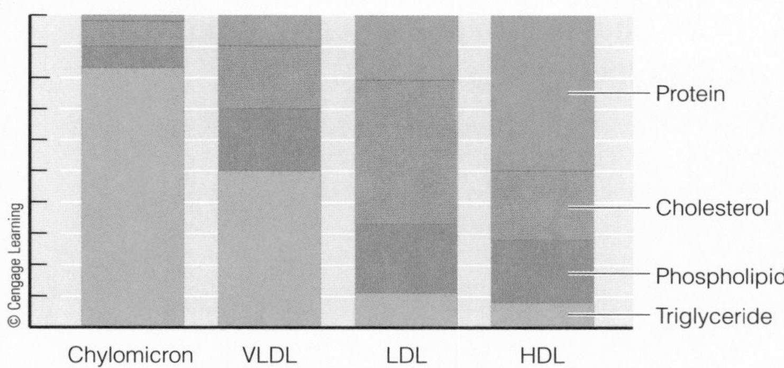

TABLE 5-2 The Lipid Family

Triglycerides

- 1 Glycerol (per triglyceride) and
- 3 Fatty acids (per triglyceride); depending on the number of double bonds, fatty acids may be:
  - *Saturated* (no double bonds)
  - *Monounsaturated* (one double bond)
  - *Polyunsaturated* (more than one double bond); depending on the location of the double bonds, polyunsaturated fatty acids may be:
    - *Omega-3* (double bond closest to methyl end is 3 carbons away)
    - *Omega-6* (double bond closest to methyl end is 6 carbons away)

Phospholipids (such as lecithin)

Sterols (such as cholesterol)

5.4 Lipids in the Body

- In the body, triglycerides provide energy, insulate against temperature extremes, protect against shock, and help the body use carbohydrate and protein efficiently.
- Linoleic acid (18 carbons, omega-6) and linolenic acid (18 carbons, omega-3) are essential fatty acids, serving as structural parts of cell membranes and as precursors to the longer fatty acids that can make eicosanoids.
- The body stores fat if given excesses, and uses body fat for energy when needed. (The liver can also convert excess carbohydrate and protein into fat.) Fat breakdown requires carbohydrate for maximum efficiency; without carbohydrate, fatty acids break down to ketone bodies.

5.5 Health Effects and Recommended Intakes of Saturated Fats, *Trans* Fats, and Cholesterol

- Some fat in the diet is necessary, but too much fat provides energy (kcalories) without nutrients, which leads to obesity and nutrient inadequacies.
- Too much saturated fat, *trans* fat, and cholesterol increases the risk of heart disease and possibly cancer.
- Recommendations advise that a diet be moderate in total fat and low in saturated fat, *trans* fat, and cholesterol.

5.6 Health Effects and Recommended Intakes of Monounsaturated and Polyunsaturated Fats

- Some fat in the diet has health benefits, especially the monounsaturated and polyunsaturated fats that protect against heart disease and possibly cancer.
- The *Dietary Guidelines* recommend replacing saturated fats with monounsaturated and polyunsaturated fats, particularly omega-3 fatty acids from foods such as fatty fish, not from supplements.
- Many selection and preparation strategies can help bring these goals within reach, and food labels help to identify foods consistent with these guidelines.

Take the quiz below to test your mastery of the key chapter concepts.

5.1 The Chemist's View of Fatty Acids and Triglycerides (pp. 129–136)

LEARN IT Recognize the chemistry of fatty acids and triglycerides and differences between saturated and unsaturated fats.

1. Name three classes of lipids found in the body and in foods. What are some of their functions in the body? What features do fats bring to foods?
2. What features distinguish fatty acids from each other?
3. What does the term *omega* mean with respect to fatty acids? Describe the roles of the omega fatty acids in disease prevention.
4. Describe the structure of a triglyceride.
5. What does hydrogenation do to fats? What are *trans*-fatty acids, and how do they influence heart disease?
6. Saturated fatty acids:
   a. are always 18 carbons long.
   b. have at least one double bond.
   c. are fully loaded with hydrogens.
   d. are always liquid at room temperature.
7. A triglyceride consists of:
   a. three glycerols attached to a lipid.
   b. three fatty acids attached to a glucose.
   c. three fatty acids attached to a glycerol.
   d. three phospholipids attached to a cholesterol.
8. The difference between *cis*- and *trans*-fatty acids is:
   a. the number of double bonds.
   b. the length of their carbon chains.
   c. the location of the first double bond.
   d. the configuration around the double bond.

5.2 The Chemist's View of Phospholipids and Sterols (pp. 136–137)

LEARN IT Describe the chemistry, major food sources, and roles of phospholipids and sterols.

9. Which of the following is *not* true? Lecithin is:
   a. an emulsifier.
   b. a phospholipid.
   c. an essential nutrient.
   d. a constituent of cell membranes.

5.3 Digestion, Absorption, and Transport of Lipids (pp. 137–144)

LEARN IT Summarize fat digestion, absorption, and transport.

10. What do lipoproteins do? What are the differences among the chylomicrons, VLDL, LDL, and HDL?
11. Chylomicrons are produced in the:
    a. liver. c. gallbladder.
    b. pancreas. d. small intestine.
12. Transport vehicles for lipids are called:
    a. micelles. c. blood vessels.
    b. lipoproteins. d. monoglycerides.

5.4 Lipids in the Body (pp. 144–147)

LEARN IT Outline the major roles of fats in the body, including a discussion of essential fatty acids and the omega fatty acids.

13. Which of the following is *not* true? Fats:
    a. contain glucose.
    b. provide energy.
    c. protect against organ shock.
    d. carry vitamins A, D, E, and K.
14. The essential fatty acids include:
    a. stearic acid and oleic acid.
    b. oleic acid and linoleic acid.
    c. palmitic acid and linolenic acid.
    d. linoleic acid and linolenic acid.

5.5 Health Effects and Recommended Intakes of Saturated Fats, *Trans* Fats, and Cholesterol (pp. 147–149)

LEARN IT Explain the relationships among saturated fats, *trans* fat and cholesterol and chronic diseases, noting recommendations.

15. How does excessive fat intake influence health? What factor influence LDL, HDL, and total blood cholesterol?
16. What are the dietary recommendations regarding fat and cho lesterol intake? List ways to reduce intake.
17. The lipoprotein most associated with a high risk of heart dis ease is:
    a. CHD. b. HDL. c. LDL. d. LPL

5.6 Health Effects and Recommended Intakes of Monounsaturated and Polyunsaturated Fats (pp. 149–157)

LEARN IT Explain the relationships between monounsaturated an polyunsaturated fats and health, noting recommendations.

18. List foods that are high in saturated fats and those that ar high in unsaturated fats.
19. What is the Daily Value for fat (for a 2000-kcalorie diet)?
20. A person consuming 2200 kcalories a day who wants to mee health recommendations should limit daily fat intake to:
    a. 20 to 35 grams. c. 75 to 100 grams.
    b. 50 to 85 grams. d. 90 to 130 grams.

ultiple Choice Answers
c 7.c 8.d 9.c 11.d 12.b 13.a 14.d 17.c 20.b

STUDY IT

6 Protein: Amino Acids

Review the key points of this chapter below, then take the practice quiz on the back of this card.

6.1 The Chemist's View of Proteins

- Proteins are more chemically complex than carbohydrates or lipids; they are made of 20 different amino acids, 9 of which the body cannot make. These 9 are the essential amino acids—histidine, isoleucine, leucine, lysine, methionine, phenylalanine, threonine, tryptophan, and valine.
- Each amino acid contains an amino group, an acid group, a hydrogen atom, and a distinctive side group, all attached to a central carbon atom.
- Cells link amino acids together in a series of condensation reactions to create proteins (review Figure 6-3, p. 169). The distinctive sequence of amino acids in each protein determines its unique shape and function.

6.2 Digestion and Absorption of Proteins

- The stomach's hydrochloric acid first denatures dietary proteins, then enzymes cleave them into smaller polypeptides and some amino acids.
- Pancreatic and intestinal enzymes split polypeptides further, to oligo-, tri-, and dipeptides, and then split most of these to single amino acids that can be absorbed into the intestinal cells (review Figure 6-6, p. 172).

6.3 Proteins in the Body

- Cells synthesize proteins according to the genetic information provided by the DNA in the nucleus of each cell (review Figure 6-7, p. 174). This information dictates the sequence in which amino acids are linked together to form a given protein. Sequencing errors occasionally occur, sometimes with significant consequences.
- A sampling of protein functions are summarized in Table 6-3.
- Proteins are constantly being synthesized and broken down as needed.
- The body's assimilation of amino acids into proteins and its release of amino acids via protein degradation and excretion can be tracked by measuring nitrogen balance, which should be positive during growth and steady in adulthood. An energy deficit or an inadequate protein intake may force the body to break down lean body tissue and use amino acids as fuel, creating a negative nitrogen balance.
- Protein eaten in excess of need is degraded and stored as body fat.

6.4 Protein in Foods

- High-quality proteins deliver all of the essential amino acids in adequate amounts, which ensures protein synthesis. Mixtures of foods containing complementary proteins can each supply the amino acids missing in the other.
- In addition to its amino acid content, the quality of a protein is measured by its digestibility and its ability to support growth.

6.5 Health Effects and Recommended Intakes of Protein

- Protein deficiency impairs the body's ability to grow and function optimally.
- Excess protein offers no advantage and may incur health problems as well.
- The optimal diet is adequate in energy from carbohydrate and fat and delivers 0.8 grams of protein per kilogram of healthy body weight each day.
- Healthy people do not need protein or amino acid supplements.

TABLE 6-3 Protein Functions in the Body

Structural materials	Proteins form integral parts of most body tissues and provide strength and shape to skin, tendons, membranes, muscles, organs, and bones.
Enzymes	Proteins facilitate chemical reactions.
Hormones	Proteins regulate body processes. (Some, but not all, hormones are proteins.)
Fluid balance	Proteins help to maintain the volume and composition of body fluids.
Acid-base balance	Proteins help to maintain the acid-base balance of body fluids by acting as buffers.
Transportation	Proteins transport substances, such as lipids, vitamins, minerals, and oxygen, around the body.
Antibodies	Proteins inactivate foreign invaders, thus protecting the body against diseases.
Energy and glucose	Proteins provide some fuel, and glucose if needed, for the body's energy needs.
Other	The protein fibrin creates blood clots; the protein collagen forms scars; the protein opsin participates in vision.

Take the quiz below to test your mastery of the key chapter concepts.

6.1 The Chemist's View of Proteins

(pp. 168–171)

LEARN IT Recognize the chemical structures of amino acids and proteins.

1. Explain how the sequence of amino acids affects protein shape.
2. What are essential amino acids?
3. Which part of its chemical structure differentiates one amino acid from another?
   a. its side group
   b. its acid group
   c. its amino group
   d. its double bonds
4. Isoleucine, leucine, and lysine are:
   a. proteases.
   b. polypeptides.
   c. essential amino acids.
   d. complementary proteins.

6.2 Digestion and Absorption of Proteins (pp. 171–172)

LEARN IT Summarize protein digestion and absorption.

5. In the stomach, hydrochloric acid:
   a. denatures proteins and activates pepsin.
   b. hydrolyzes proteins and denatures pepsin.
   c. emulsifies proteins and releases peptidase.
   d. condenses proteins and facilitates digestion.

6.3 Proteins in the Body (pp. 173–180)

LEARN IT Describe how the body makes proteins and uses them to perform various roles.

6. What are enzymes? What roles do they play in chemical reactions? Describe the differences between enzymes and hormones.
7. How does the body use amino acids? What is deamination? Define *nitrogen balance*. What conditions are associated with zero, positive, and negative balance?
8. Proteins that maintain the acid-base balance of the blood and body fluids by accepting and releasing hydrogen ions are:
   a. buffers.
   b. enzymes.
   c. hormones.
   d. antigens.
9. If an essential amino acid that is needed to make a protein i unavailable, the cells must:
   a. deaminate another amino acid.
   b. substitute a similar amino acid.
   c. break down proteins to obtain it.
   d. synthesize the amino acid from glucose and nitrogen.
10. Protein turnover describes the amount of protein:
   a. found in foods and the body.
   b. absorbed from the diet.
   c. synthesized and degraded.
   d. used to make glucose.

6.4 Protein in Foods (pp. 181–182)

LEARN IT Explain the differences between high-quality and low quality proteins, including notable food sources of each.

11. How can vegetarians meet their protein needs without eatin meat?
12. Which of the following foods provides the highest qualit protein?
   a. egg
   b. corn
   c. gelatin
   d. whole grains

6.5 Health Effects and Recommended Intakes of Protein (pp. 182–188)

LEARN IT Identify the health benefits of, and recommendations fo protein.

13. How might protein excess, or the type of protein eaten, infl ence health?
14. What factors are considered in establishing recommende protein intakes?
15. What are the benefits and risks of taking protein and amir acid supplements?
16. The protein RDA for a healthy adult who weighs 180 pounds i
   a. 50 milligrams/day.
   b. 65 grams/day.
   c. 180 grams/day.
   d. 2000 milligrams/day.
17. Which of these foods has the least protein per ½ cup?
   a. rice
   b. broccoli
   c. pinto beans
   d. orange juice

ultiple Choice Answers

a 4.c 5.a 8.a 9.c 10.c 12.a 16.b 17.d

Review the key points of this chapter below, then take the practice quiz on the back of this card.

7.1 Chemical Reactions in the Body

- During digestion, the energy-yielding nutrients—carbohydrates, lipids, and proteins—are broken down to glucose (and other monosaccharides), glycerol, fatty acids, and amino acids. These compounds may enter metabolic pathways to yield energy (review Figure 7-5).
- Enzymes with their coenzymes help cells use nutrients to build compounds (anabolism) or break them down to release energy (catabolism)—review Figure 7-2, p. 200.
- ATP—a high-energy compound—captures the energy released during catabolism (review Figure 7-4, p. 201).

7.2 Breaking Down Nutrients for Energy

- Glucose breakdown begins with glycolysis, a pathway that produces pyruvate (review Figure 7-6, p. 204).
- Pyruvate may be converted to lactate anaerobically (without oxygen) or to acetyl CoA aerobically (with oxygen).
- Pyruvate can make glucose; acetyl CoA cannot make glucose (review Figure 7-9, p. 206).
- The glycerol part of a triglyceride can make either pyruvate (and then glucose) or acetyl CoA. The fatty acids of a triglyceride *cannot* make glucose; they can provide abundant acetyl CoA (review Figure 7-11, p. 208).
- Some amino acids can be used to make glucose; others can be used either to provide energy or to make fat. Before an amino acid enters these metabolic pathways, its nitrogen-containing amino group must be removed through deamination.
- The digestion of carbohydrate yields glucose (and other monosaccharides); some glucose is stored as glycogen, and some is broken down to pyruvate and acetyl CoA.
- The digestion of fat yields glycerol and fatty acids; some are reassembled and stored as body fat, and others are broken down to acetyl CoA.
- The digestion of protein yields amino acids; most amino acids are used to build body protein or other nitrogen-containing compounds, some are broken down to acetyl CoA, and others enter the TCA cycle directly.
- Acetyl CoA may enter the TCA cycle to release energy (review Figure 7-16, p. 212) or combine with other molecules of acetyl CoA to make body fat.

> FIGURE 7-5 Simplified Overview of Energy-Yielding Pathways

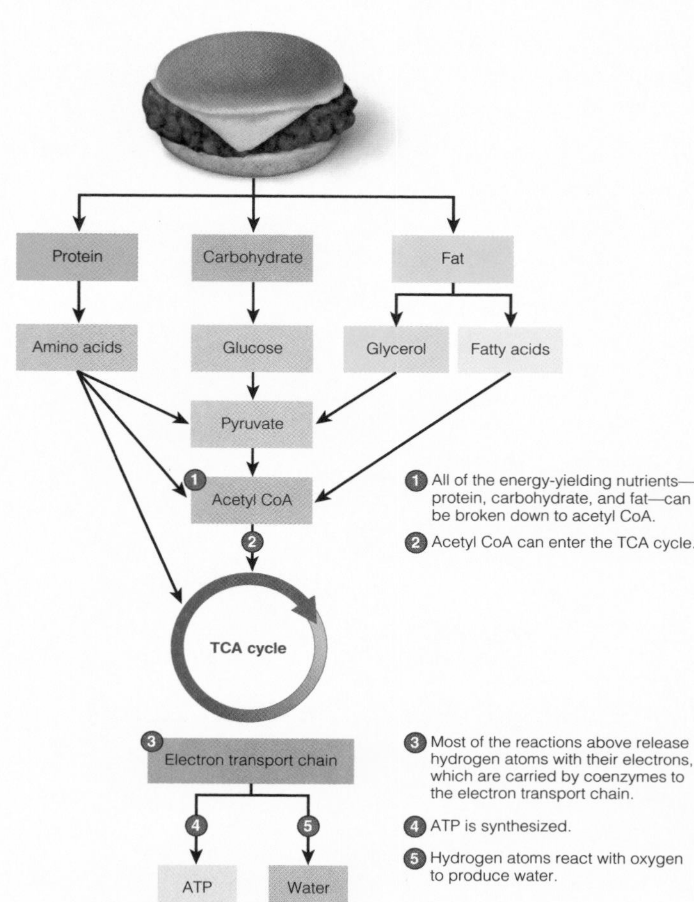
Protein
Carbohydrate
Fat
Amino acids
Glucose
Glycerol
Fatty acids
Pyruvate
Acetyl CoA
TCA cycle
Electron transport chain
ATP
Water
1 All of the energy-yielding nutrients—protein, carbohydrate, and fat—can be broken down to acetyl CoA.
2 Acetyl CoA can enter the TCA cycle.
3 Most of the reactions above release hydrogen atoms with their electrons, which are carried by coenzymes to the electron transport chain.
4 ATP is synthesized.
5 Hydrogen atoms react with oxygen to produce water.
© Cengage Learning

7.3 Feasting and Fasting

- If energy intake exceeds energy needs, the result will be weight gain—regardless of whether the excess is from protein, carbohydrate, or fat (review Table 7-2). The body is most efficient at storing excess energy from dietary fat.
- When fasting, the body adapts to conserve energy and minimize losses by increasing fat breakdown to fuel most cells, using glycerol and amino acids to make glucose for the brain and red blood cells, producing ketones for the brain, suppressing appetite, and slowing metabolism.

TABLE 7-2 Review of Energy-Yielding Nutrient Endpoints

Nutrient	Yields energy?	Yields glucose?	Yields amino acids and body proteins?	Yields fat stores?
Carbohydrates (glucose)	Yes	Yes	Yes—when nitrogen is available, can yield *nonessential* amino acids	Yes
Lipids (fatty acids)	Yes	No	No	Yes
Lipids (glycerol)	Yes	Yes—when carbohydrate is unavailable	Yes—when nitrogen is available, can yield *nonessential* amino acids	Yes
Proteins (amino acids)	Yes	Yes—when carbohydrate is unavailable	Yes	Yes

Take the quiz below to test your mastery of the key chapter concepts.

7.1 Chemical Reactions in the Body

(pp. 198–201)

LEARN IT Identify the nutrients involved in energy metabolism and the high-energy compound that captures the energy released during their breakdown.

1. Define *metabolism*, *anabolism*, and *catabolism*; give an example of each.
2. What are coenzymes, and what service do they provide in metabolism?
3. Hydrolysis is an example of a(n):
   a. coupled reaction.
   b. anabolic reaction.
   c. catabolic reaction.
   d. synthesis reaction.
4. During metabolism, released energy is captured and transferred by:
   a. enzymes.
   b. pyruvate.
   c. acetyl CoA.
   d. adenosine triphosphate.

7.2 Breaking Down Nutrients for Energy

(pp. 201–213)

LEARN IT Summarize the main steps in the energy metabolism of glucose, glycerol, fatty acids, and amino acids.

5. Name the four basic units, derived from foods, that are used by the body in energy metabolism. How many carbons are in the "backbones" of each?
6. Define *aerobic* and *anaerobic*. How does insufficient oxygen influence metabolism?
7. How does the body dispose of excess nitrogen?
8. The body derives most of its energy from:
   a. proteins and fats.
   b. vitamins and minerals.
   c. glucose and fatty acids.
   d. glycerol and amino acids.
9. Glycolysis:
   a. requires oxygen.
   b. generates abundant energy.
   c. converts glucose to pyruvate.
   d. produces ammonia as a by-product.
10. The pathway from pyruvate to acetyl CoA:
    a. produces lactate.
    b. is known as gluconeogenesis.
    c. is metabolically irreversible.
    d. requires more energy than it produces.
11. For complete oxidation, acetyl CoA enters:
    a. glycolysis.
    b. the TCA cycle.
    c. the Cori cycle.
    d. the electron transport chain.
12. Deamination of an amino acid produces:
    a. vitamin B_6 and energy.
    b. pyruvate and acetyl CoA.
    c. ammonia and a keto acid.
    d. carbon dioxide and water.
13. Before entering the TCA cycle, each of the energy-yieldin nutrients is broken down to:
    a. ammonia.
    b. pyruvate.
    c. electrons.
    d. acetyl CoA.

7.3 Feasting and Fasting (pp. 213–219)

LEARN IT Explain how an excess of any of the three energy-yieldin nutrients contributes to body fat and how an inadequate intake of ar of them shifts metabolism.

14. What adaptations does the body make during a fast? What ar ketone bodies? Define *ketosis*.
15. Distinguish between a loss of *fat* and a loss of *weight*, an describe how each might happen.
16. The body stores energy for future use in:
    a. proteins.
    b. acetyl CoA.
    c. triglycerides.
    d. ketone bodies.
17. During a fast, when glycogen stores have been depleted, th body begins to synthesize glucose from:
    a. acetyl CoA.
    b. amino acids.
    c. fatty acids.
    d. ketone bodies.
18. During a fast, the body produces ketone bodies by:
    a. hydrolyzing glycogen.
    b. condensing acetyl CoA.
    c. transaminating keto acids.
    d. converting ammonia to urea.

ultiple Choice Answers

c 4. d 8. c 9. c 10. c 11. b 12. c 13. d 16. c 17. b 18. b

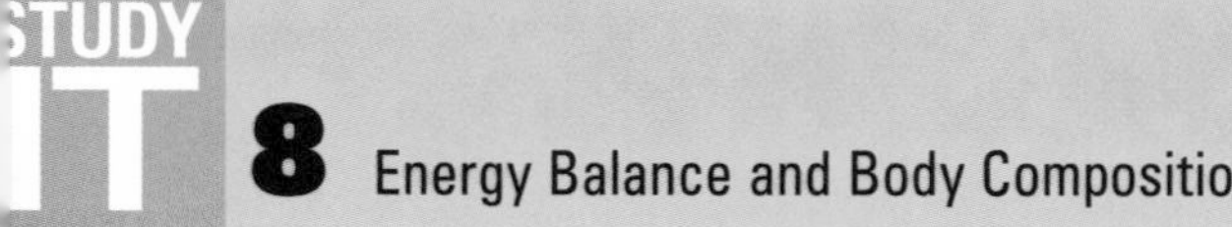

Energy Balance and Body Composition

Review the key points of this chapter below, then take the practice quiz on the back of this card.

8.1 Energy Balance

- When energy consumed equals energy expended, a person is in energy balance and body weight is stable.
- If more energy is taken in than is expended, a person gains weight. If more energy is expended than is taken in, a person loses weight.

8.2 Energy In: The kCalories Foods Provide

- Scientists use a bomb calorimeter to estimate the potential energy of foods by measuring the heat energy released when foods are burned.
- Hunger and appetite initiate eating, whereas satiation and satiety stop and delay eating, respectively (review Figure 8-2, p. 234). Each responds to messages from the nervous and hormonal systems. Superimposed on these signals are complex factors involving emotions, habits, and other aspects of human behavior.

8.3 Energy Out: The kCalories the Body Expends

- A person in energy balance takes in energy from food and expends much of it on basal metabolic activities, some of it on physical activities, and a little on the thermic effect of food (review Figure 8-4).
- Energy requirements vary from person to person based on such factors as gender, age, weight, and height as well as the intensity and duration of physical activity.

8.4 Body Weight and Body Composition

- Body weight standards are based on a person's weight in relation to height, called the body mass index (BMI), and reflect disease risks. BMI does not identify body fat or its distribution, and it may misclassify muscular people as overweight.

> FIGURE 8-4 Components of Energy Expenditure

The amount of energy expended in voluntary physical activities has the greatest variability, depending on a person's activity patterns. For a sedentary person, physical activities may account for less than half as much energy as basal metabolism, whereas an extremely active person may expend as much on activity as for basal metabolism.

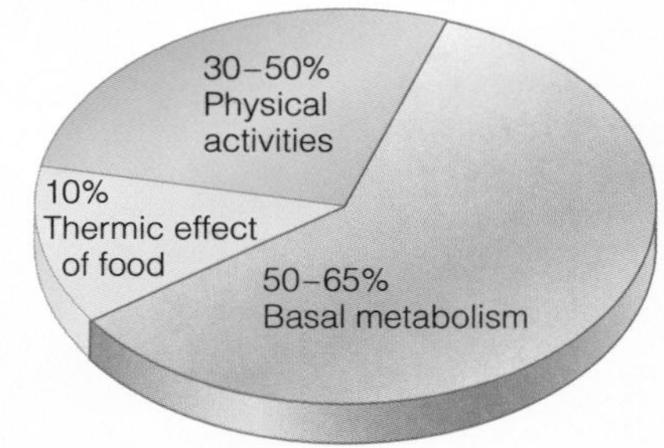

The amount of energy expended in a day differs for each individual, but in general, basal metabolism is the largest component of energy expenditure and thermic effect of food is the smallest.

> FIGURE 8-7 Distribution of Body Weights in US Adults

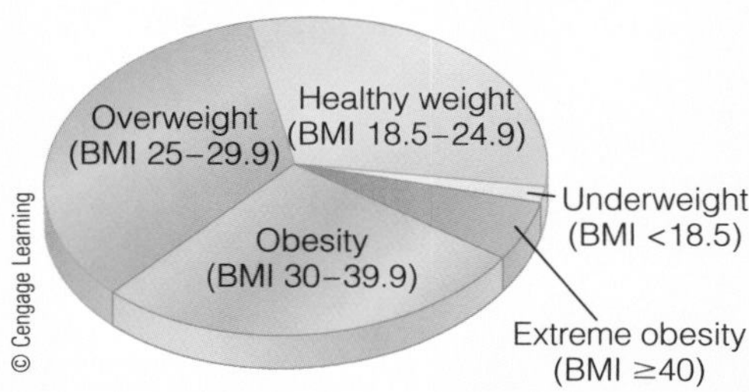

- $\text{BMI} = \dfrac{\text{weight (kg)}}{\text{height (m)}^2}$ or $\dfrac{\text{weight (lb)}}{\text{height (in)}^2} \times 703.$

  Underweight: BMI <18.5
  Healthy: BMI 18.5 to 24.9
  Overweight: BMI 25 to 29.9
  Obese: BMI ≥30
- Two-thirds of US adults have a BMI greater than 25 (review Figure 8-7).
- The ideal amount of body fat varies from person to person, but excess body fat poses health risks.
- Central obesity—excess abdominal fat distributed around the trunk of the body—presents greater health risks than excess fat distributed on the lower body (review Figure 8-8 and Figure 8-9, p. 245).
- Women with a waist circumference greater than 35 inches and men with a waist circumference greater than 40 inches have a high risk of central obesity–related health problems.

8.5 Health Risks Associated with Body Weight and Body Fat

- The healthiest weight for an individual depends on personal factors such as body fat distribution, family health history, and current health status. At the extremes, both overweight and underweight impose health risks (review Figure 8-11).

> FIGURE 8-11 BMI and Mortality

This J-shaped curve describes the relationship between body mass index (BMI) and mortality and shows that both underweight and overweight present risks of a premature death.

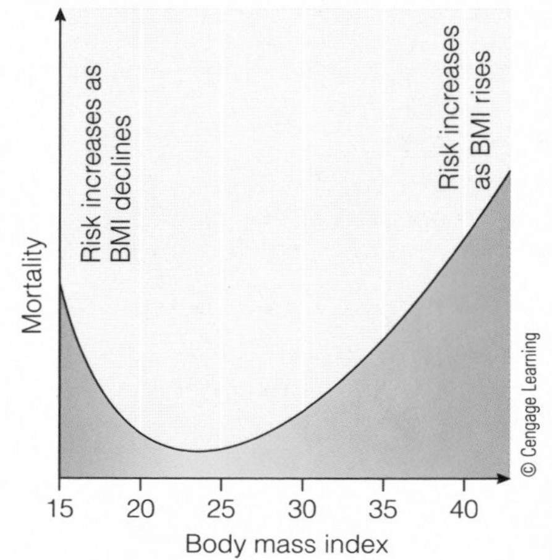

Take the quiz below to test your mastery of the key chapter concepts.

8.1 Energy Balance (p. 232)

LEARN IT Describe energy balance and the consequences of not being in balance.

1. A person who consistently consumes 1700 kcalories a day and expends 2200 kcalories a day for a month would be expected to:
   a. lose ½ to 1 pound.
   b. gain ½ to 1 pound.
   c. lose 4 to 5 pounds.
   d. gain 4 to 5 pounds.

8.2 Energy In: The kCalories Foods Provide (pp. 232–236)

LEARN IT Discuss some of the physical, emotional, and environmental influences on food intake.

2. A bomb calorimeter measures:
   a. physiological fuel.
   b. energy available from foods.
   c. kcalories a person derives from foods.
   d. heat a person releases in basal metabolism.
3. The psychological desire to eat that accompanies the sight, smell, or thought of food is known as:
   a. hunger.
   b. satiety.
   c. appetite.
   d. palatability.
4. A person watching television after dinner reaches for a snack during a commercial in response to:
   a. external cues.
   b. hunger signals.
   c. stress arousal.
   d. satiety factors.

8.3 Energy Out: The kCalories the Body Expends (pp. 236–240)

LEARN IT List the components of energy expenditure and factors that might influence each.

5. The largest component of energy expenditure is:
   a. basal metabolism.
   b. physical activity.
   c. indirect calorimetry.
   d. thermic effect of food.
6. A major factor influencing BMR is:
   a. hunger.
   b. food intake.
   c. body composition.
   d. physical activity.
7. The thermic effect of an 800-kcalorie meal is about:
   a. 8 kcalories.
   b. 80 kcalories.
   c. 160 kcalories.
   d. 200 kcalories.

8.4 Body Weight and Body Composition (pp. 241–246)

LEARN IT Distinguish between body weight and body compositio including methods to assess each.

8. What problems are involved in defining "ideal" body weight?
9. What is central obesity, and what is its relationship to disease
10. For health's sake, a person with a BMI of 21 might want to:
    a. lose weight.
    b. maintain weight.
    c. gain weight.
11. Which of the following reflects height and weight?
    a. body mass index
    b. central obesity
    c. waist circumference
    d. body composition

8.5 Health Risks Associated with Body Weight and Body Fat (pp. 246–249)

LEARN IT Identify relationships between body weight and chroni diseases.

12. Which of the following increases disease risks?
    a. BMI 19–21
    b. BMI 22–25
    c. lower-body fat
    d. central obesity

ultiple Choice Answers

c 2.b 3.c 4.a 5.a 6.c 7.b 10.b 11.a 12.d

9 Weight Management: Overweight, Obesity, and Underweight

Review the key points of this chapter below, then take the practice quiz on the back of this card.

9.1 Overweight and Obesity

- Fat cells develop by increasing in number and size (review Figure 9-2, p. 263).
- Preventing weight gain depends on limiting the number of fat cells; weight loss depends on decreasing the size of fat cells.
- Lipoprotein lipase (LPL) removes triglycerides from the blood for storage in both adipose tissue and muscle cells. LPL activity is influenced by weight and gender.
- With weight gains or losses, the body adjusts in an attempt to return to its previous weight (set point theory).

9.2 Causes of Overweight and Obesity

- Obesity has multiple causes and different combinations of causes in different people.
- Some environmental causes, such as overeating and physical inactivity, may be within a person's control, and some, such as genetics, may be beyond it.
- Proteins such as ghrelin and leptin regulate food intake and energy homeostasis.

9.3 Problems of Overweight and Obesity

- Whether a person should lose weight depends on factors such as the extent of overweight, age, health risks, and genetics.
- Not all obesity will cause disease or shorten life expectancy. Just as there are unhealthy, normal-weight people, there are healthy, overweight people.
- Some people risk more in the process of losing weight than in remaining overweight.
- Fad diets and weight-loss supplements can be as physically and psychologically damaging as excess body weight.

9.4 Aggressive Treatments for Obesity

- Obese people with high risks of medical problems may need aggressive treatment, including drugs or surgery.
- Others may benefit most from improving eating and physical activity habits.

9.5 Weight-Loss Strategies

- A person who adopts a lifelong "eating plan for good health" rather than a "diet for weight loss" will be more likely to keep the lost weight off.
- Table 9-3 (p. 277) provides several tips for successful weight management.

> FIGURE 9-8 *Animated* **Influence of Physical Activity on Discretionary kCalories**

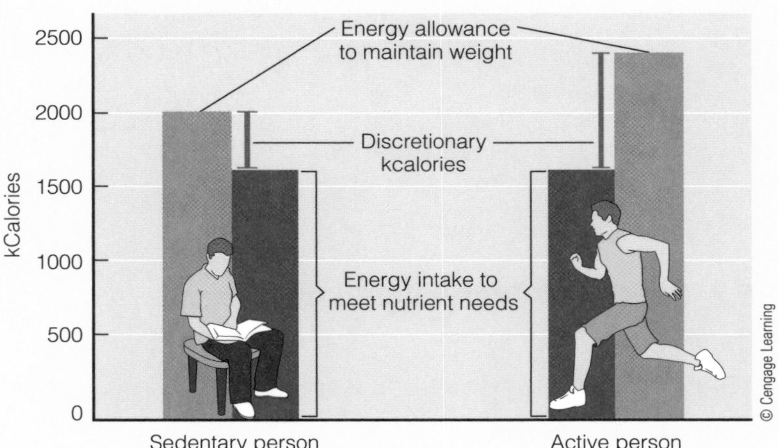

- Physical activity can increase energy expenditure (review Figure 9-8), improve body composition, help control appetite, reduce stress and stress eating, and enhance physical and psychological well-being.
- A surefire remedy for obesity has yet to be found; a combination of approaches is most effective.
- Weight loss depends on adjusting diet and physical activity so that more energy is expended than is taken in.
- For weight loss, energy intake should be reduced by 500 to 1000 kcalories per day, depending on starting body weight and usual food intake.
- Safe rate for weight loss is ½ to 2 pounds per week or 10 percent body weight per 6 months.
- Behavior modification and cognitive restructuring retrain habits to support a healthy eating and activity plan.
- Treatment requires time, individualization, and sometimes the assistance of a registered dietitian or support group.
- Preventing weight gains and maintaining weight losses require vigilant attention to diet and physical activity; taking care of oneself is a lifelong responsibility.

9.6 Underweight

- Both the incidence of underweight and the health problems associated with it are less prevalent than overweight and its associated problems.
- To gain weight, a person must train physically and increase energy intake by selecting energy-dense foods, eating regular meals, taking larger portions, and consuming extra snacks and beverages.
- Table 9-5 (p. 287) includes a summary of weight-gain strategies.

Take the quiz below to test your mastery of the key chapter concepts.

9.1 Overweight and Obesity (pp. 261–263)

LEARN IT Describe how body fat develops and why it can be difficult to maintain weight gains and losses.

1. With weight loss, fat cells:
   a. decrease in size only.
   b. decrease in number only.
   c. decrease in both number and size.
   d. decrease in number, but increase in size.
2. Describe the role of lipoprotein lipase (LPL).

9.2 Causes of Overweight and Obesity (pp. 264–267)

LEARN IT Review some of the causes of obesity.

3. Obesity is caused by:
   a. overeating.
   b. inactivity.
   c. defective genes.
   d. multiple factors.
4. The protein produced by the fat cells under the direction of the *ob* gene is called:
   a. leptin.
   b. serotonin.
   c. sibutramine.
   d. phentermine.

9.3 Problems of Overweight and Obesity (pp. 268–270)

LEARN IT Discuss the physical, social, and psychological consequences of overweight and obesity.

5. Which of the following is *not* used to evaluate the risks to health from obesity?
   a. body mass index
   b. blood leptin levels
   c. waist circumference
   d. disease risk profiles

9.4 Aggressive Treatments for Obesity (pp. 270–272)

LEARN IT Explain the risks and benefits, if any, of several aggressive ways to treat obesity.

6. Gastric bypass surgery:
   a. is the best noninvasive treatment for obesity.
   b. limits food intake by reducing the capacity of the stomach.
   c. allows a person to eat unlimited amounts of food without weight gain.
   d. suppresses hunger by increasing production of gastrointestinal hormones.

9.5 Weight-Loss Strategies (pp. 272–285)

LEARN IT Outline reasonable strategies for achieving and maintaining a healthy body weight.

7. What are the benefits of increased physical activity in a weight-loss program?
8. Describe the behavioral strategies for changing an individual's dietary habits. What role does personal attitude play?
9. A realistic goal for weight loss is to reduce body weight:
   a. down to the weight a person was at age 25.
   b. down to the ideal weight in the weight-for-height tables.
   c. by 10 percent over 6 months.
   d. by 15 percent over 3 months.
10. A nutritionally sound weight-loss diet might restrict daily energy intake to create a:
    a. 1000-kcalorie-per-month deficit.
    b. 500-kcalorie-per-month deficit.
    c. 500-kcalorie-per-day deficit.
    d. 3500-kcalorie-per-day deficit.
11. Successful weight loss depends on:
    a. avoiding fats and limiting water.
    b. taking supplements and drinking water.
    c. increasing proteins and restricting carbohydrates.
    d. reducing energy intake and increasing physical activity.
12. Physical activity does *not* help a person to:
    a. lose weight.
    b. retain muscle.
    c. maintain weight loss.
    d. lose fat in trouble spots.
13. Which strategy would *not* help an overweight person to lose weight?
    a. Exercise.
    b. Eat slowly.
    c. Limit high-fat foods.
    d. Eat energy-dense foods regularly.

9.6 Underweight (pp. 285–287)

LEARN IT Summarize strategies for gaining weight.

14. Which strategy would *not* help an underweight person to gain weight?
    a. Exercise.
    b. Drink plenty of water.
    c. Eat snacks between meals.
    d. Eat large portions of foods.

Multiple Choice Answers
a 3. d 4. a 5. b 6. b 9. c 10. c 11. d 12. d 13. d 14. b

TUDY IT 10 The Water-Soluble Vitamins: B Vitamins and Vitamin C

eview the key points of this chapter below, then take the practice quiz on the back of this card.

0.1 The Vitamins—An Overview

- The vitamins are organic, essential nutrients needed in tiny amounts in the diet both to prevent deficiency diseases and to support optimal health. The body handles the vitamins differently depending on whether they are water- or fat-soluble (review Table 10-2).
- The water-soluble vitamins are the B vitamins and vitamin C; the fat-soluble vitamins are vitamins A, D, E, and K. The B vitamins include thiamin, niacin, riboflavin, vitamin B_6, folate, vitamin B_{12}, pantothenic acid, and biotin.

0.2 The B Vitamins

- The B vitamins serve as coenzymes—small organic molecules closely associated with enzymes that facilitate the work of cells (review Figure 10-2).
- Thiamin is part of the coenzyme TPP, which assists in energy metabolism. Deficiency can result in beriberi. Thiamin occurs in small quantities in many nutritious foods; pork is an exceptionally good source.
- Riboflavin is part of the coenzymes FMN and FAD that accept and donate hydrogens during energy metabolism. Milk and milk products are good sources.
- Niacin is part of the coenzymes NAD and NADP that participate in many metabolic reactions. The deficiency disease, pellagra, causes diarrhea, dermatitis, dementia, and eventually death ("the 4 Ds"). Toxicity produces "niacin flush"—a tingling, painful sensation. The amino acid tryptophan can be converted to niacin in the body: 60 mg tryptophan = 1 NE (niacin equivalent). Good sources of niacin are protein-rich foods.
- Biotin plays a critical role in energy metabolism, replenishing oxaloacetate in the TCA cycle. Biotin is widespread in foods; deficiencies and toxicities are rare.
- Pantothenic acid is part of coenzyme A that forms acetyl CoA in many metabolic pathways. Pantothenic acid is widespread in foods; deficiencies and toxicities are rare.
- Vitamin B_6 occurs as pyridoxal, pyridoxine, and pyridoxamine; all can become part of the coenzyme PLP, which is active in amino acid metabolism. Deficiency causes convulsions; toxicity causes nerve damage.
- Folate is part of the coenzyme THF that activates vitamin B_{12}, synthesizes DNA, and regenerates the amino acid methionine from homocysteine. Folate helps prevent neural tube defects. Excessive folate can mask the anemia of a vitamin B_{12} deficiency, but it will not prevent the associated nerve damage. Folate is abundant in legumes, fruits, and vegetables.

TABLE 10-2 Water-Soluble and Fat-Soluble Vitamins Compared

	Water-Soluble Vitamins: B Vitamins and Vitamin C	Fat-Soluble Vitamins: Vitamins A, D, E, and K
Absorption	Directly into the blood	First into the lymph, then the blood
Transport	Travel freely	Many require transport proteins
Storage	Circulate freely in water-filled parts of the body	Stored in the cells associated with fat
Excretion	Kidneys detect and remove excess in urine	Less readily excreted; tend to remain in fat-storage sites
Toxicity	Possible to reach toxic levels when consumed from supplements	Likely to reach toxic levels when consumed from supplements
Requirements	Needed in frequent doses (perhaps 1 to 3 days)	Needed in periodic doses (perhaps weeks or even months)

NOTE: Exceptions occur, but these differences between the water-soluble and fat-solule vitamins are valid generalizations.

- Vitamin B_{12} activates folate, synthesizes DNA, regenerates methionine from homocysteine, and maintains the sheath that protects nerve fibers. Deficiencies typically occur when either hydrochloric acid or intrinsic factor is lacking. Vitamin B_{12} is found primarily in foods derived from animals.
- Many substances that people claim as B vitamins are not. Fortunately, a variety of foods from each food group provides an adequate supply of all B vitamins.

10.3 Vitamin C

- Vitamin C acts as an antioxidant—a substance that decreases the adverse effects of free radicals in the body.
- Vitamin C works as a cofactor in the synthesis of collagen, neurotransmitters (serotonin and norepinephrine), hormones (thyroxin), and other compounds.
- Vitamin C deficiency causes scurvy.

> FIGURE 10-2 Coenzyme Action

Some vitamins form part of the coenzymes that enable enzymes either to synthesize compounds (as illustrated by the lower enzymes in this figure) or to dismantle compounds (as illustrated by the upper enzymes).

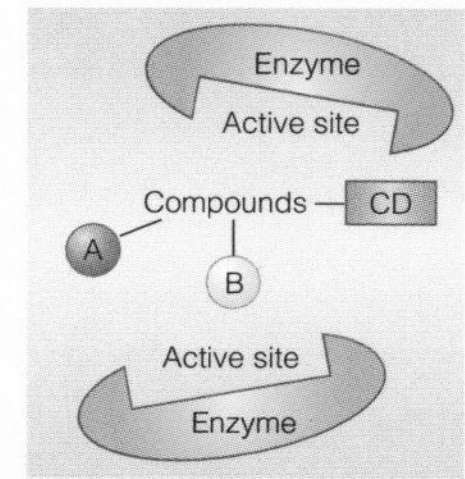

Without coenzymes, compounds A, B, and CD don't respond to their enzymes.

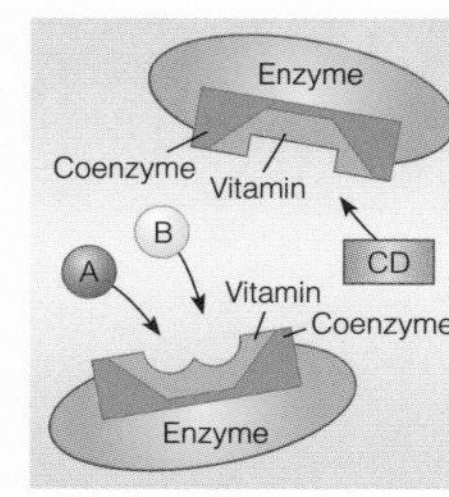

With the coenzymes in place, compounds are attracted to their sites on the enzymes . . .

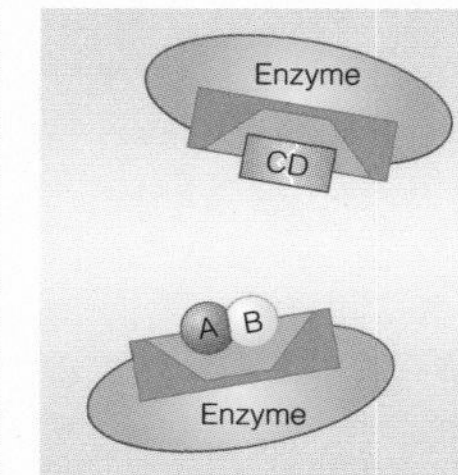

. . . and the reactions proceed instantaneously. The coenzymes often donate or accept electrons, atoms, or groups of atoms.

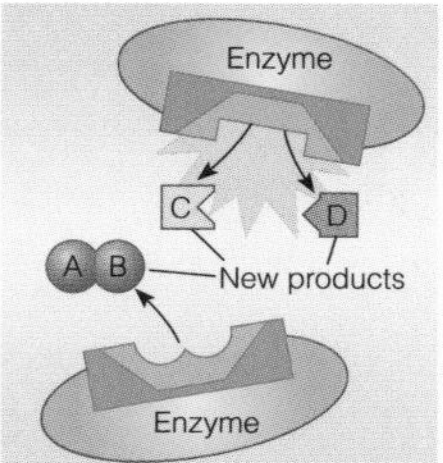

The reactions are completed with either the formation of a new product, AB, or the breaking apart of a compound into two new products, C and D, and the release of energy.

Take the quiz below to test your mastery of the key chapter concepts.

10.1 The Vitamins—An Overview (pp. 297–300)

LEARN IT Describe how vitamins differ from the energy nutrients and how fat-soluble vitamins differ from water-soluble vitamins.

1. Vitamins:
   a. are inorganic compounds.
   b. yield energy when broken down.
   c. are soluble in either water or fat.
   d. perform best when linked in long chains.

2. The rate at and the extent to which a vitamin is absorbed and used in the body is known as its:
   a. bioavailability.
   b. intrinsic factor.
   c. physiological effect.
   d. pharmacological effect.

10.2 The B Vitamins (pp. 300–322)

LEARN IT Identify the main roles, deficiency symptoms, and food sources for each of the B vitamins (thiamin, riboflavin, niacin, biotin, pantothenic acid, vitamin B_6, folate, and vitamin B_{12}).

3. Which B vitamins are involved in energy metabolism? Protein metabolism? Cell division?

4. What is the relationship of tryptophan to niacin?

5. Describe the relationship between folate and vitamin B_{12}.

6. What risks are associated with high doses of niacin? Vitamin B_6? Vitamin C?

7. Many of the B vitamins serve as:
   a. coenzymes.
   b. antagonists.
   c. antioxidants.
   d. serotonin precursors.

8. With respect to thiamin, which of the following is the most nutrient dense?
   a. 1 slice whole-wheat bread (69 kcalories and 0.1 milligram thiamin)
   b. 1 cup yogurt (144 kcalories and 0.1 milligram thiamin)
   c. 1 cup snow peas (69 kcalories and 0.22 milligram thiamin)
   d. 1 chicken breast (141 kcalories and 0.06 milligram thiamin)

9. The body can make niacin from:
   a. tyrosine.
   b. serotonin.
   c. carnitine.
   d. tryptophan.

10. The vitamin that protects against neural tube defects is:
    a. niacin.
    b. folate.
    c. riboflavin.
    d. vitamin B_{12}.

11. A lack of intrinsic factor may lead to:
    a. beriberi.
    b. pellagra.
    c. pernicious anemia.
    d. atrophic gastritis.

12. Which of the following is a B vitamin?
    a. inositol
    b. carnitine
    c. vitamin B_{15}
    d. pantothenic acid

10.3 Vitamin C (pp. 322–327)

LEARN IT Identify the main roles, deficiency symptoms, and foo sources for vitamin C.

13. Vitamin C serves as a(n):
    a. coenzyme.
    b. antagonist.
    c. antioxidant.
    d. intrinsic factor.

14. The requirement for vitamin C is highest for:
    a. smokers.
    b. athletes.
    c. alcoholics.
    d. the elderly.

See p. 328 for a summary of the water-soluble vitamins.

ultiple Choice Answers

c 2.a 7.a 8.c 9.d 10.b 11.c 12.d 13.c 14.a

11 The Fat-Soluble Vitamins: A, D, E, and K

eview the key points of this chapter below, then take the practice quiz on the back of this card.

1.1 Vitamin A and Beta-Carotene

- Vitamin A is found in the body in three forms: retinol, retinal, and retinoic acid. Together, they are essential to vision, healthy epithelial tissues, and growth.
- Vitamin A deficiency is a major health problem worldwide, leading to infections, blindness, and keratinization.
- Toxicity can also cause problems and is most often associated with supplement abuse.
- Animal-derived foods such as liver and whole or fortified milk provide retinoids, whereas brightly colored plant-derived foods such as spinach, carrots, and pumpkins provide beta-carotene and other carotenoids.
- In addition to serving as a precursor for vitamin A, beta-carotene may act as an antioxidant in the body.

1.2 Vitamin D

- Vitamin D can be synthesized in the body with the help of sunlight (see Figure 11-9) or obtained from fortified milk.
- Vitamin D sends signals to three primary target sites: the GI tract to absorb more calcium and phosphorus, the bones to release more, and the kidneys to retain more. These actions maintain blood calcium concentrations and support bone formation.
- A vitamin D deficiency causes rickets in childhood and osteomalacia in later life.

11.3 Vitamin E

- Vitamin E (alpha-tocopherol) acts as an antioxidant, defending lipids and other components of the cells against oxidative damage.
- Deficiencies are rare, but they do occur in premature infants, the primary symptom being erythrocyte hemolysis (red blood cell breakage).
- Vitamin E is found predominantly in vegetable oils and appears to be one of the least toxic of the fat-soluble vitamins.

11.4 Vitamin K

- Vitamin K helps with blood clotting (review Figure 11-12, p. 355), and its deficiency causes hemorrhagic disease (uncontrolled bleeding).
- Bacteria in the GI tract can make vitamin K; people typically receive about half of their requirements from bacterial synthesis and half from foods such as green vegetables and vegetable oils.
- Because people depend on bacterial synthesis for vitamin K, deficiency is most likely in newborn infants and in people taking antibiotics.

> FIGURE 11-9 **Vitamin D Synthesis and Activation**

The final activation step in the kidneys is tightly regulated by hormones.

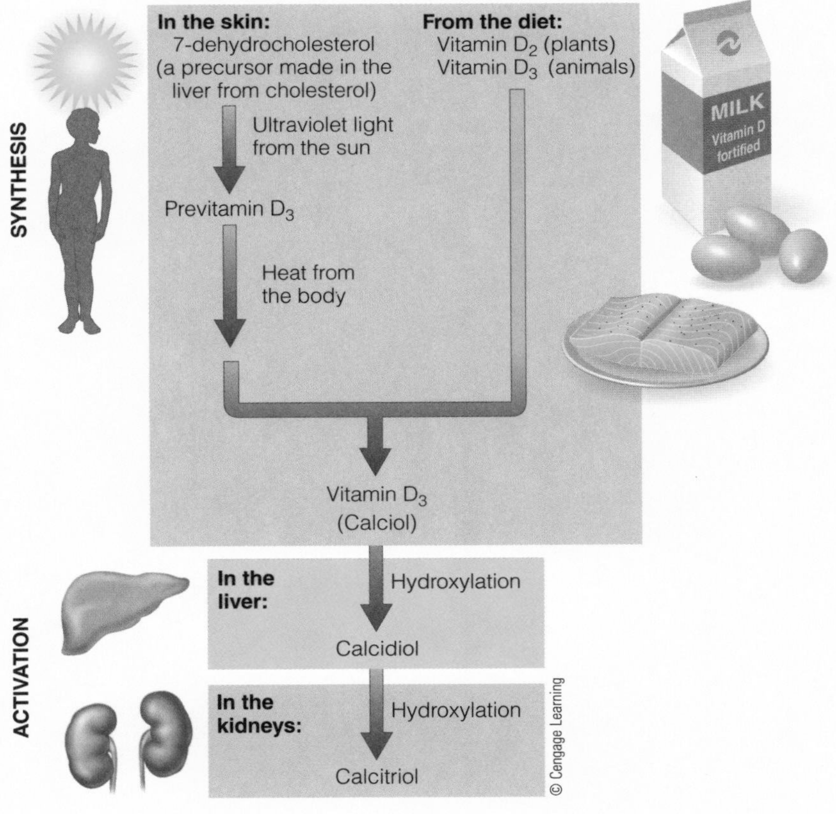

Take the quiz below to test your mastery of the key chapter concepts.

11.1 Vitamin A and Beta-Carotene

(pp. 340–347)

LEARN IT Identify the main roles, deficiency symptoms, and food sources for vitamin A.

1. What are vitamin precursors? Name the precursors of vitamin A, and tell in what classes of foods they are located. Give examples of foods with high vitamin A activity.
2. The form of vitamin A active in vision is:
   a. retinal.
   b. retinol.
   c. rhodopsin.
   d. retinoic acid.
3. Vitamin A–deficiency symptoms include:
   a. rickets and osteomalacia.
   b. hemorrhaging and jaundice.
   c. night blindness and keratomalacia.
   d. fibrocystic breast disease and erythrocyte hemolysis.
4. Good sources of vitamin A include:
   a. oatmeal, pinto beans, and ham.
   b. apricots, turnip greens, and liver.
   c. whole-wheat bread, green peas, and tuna.
   d. corn, grapefruit juice, and sunflower seeds.

11.2 Vitamin D (pp. 347–352)

LEARN IT Identify the main roles, deficiency symptoms, and sources for vitamin D.

5. How is vitamin D unique among the vitamins?
6. To keep minerals available in the blood, vitamin D targets:
   a. the skin, the muscles, and the bones.
   b. the kidneys, the liver, and the bones.
   c. the intestines, the kidneys, and the bones.
   d. the intestines, the pancreas, and the liver.
7. Vitamin D can be synthesized from a precursor that the body makes from:
   a. bilirubin.
   b. tocopherol.
   c. cholesterol.
   d. beta-carotene.

11.3 Vitamin E (pp. 353–354)

LEARN IT Identify the main roles, deficiency symptoms, and foc sources for vitamin E.

8. Vitamin E's most notable role is to:
   a. protect lipids against oxidation.
   b. activate blood-clotting proteins.
   c. support protein and DNA synthesis.
   d. enhance calcium deposits in the bones.
9. The classic sign of vitamin E deficiency is:
   a. rickets.
   b. xeropthalmia.
   c. muscular dystrophy.
   d. erythrocyte hemolysis.

11.4 Vitamin K (pp. 354–356)

LEARN IT Identify the main roles, deficiency symptoms, and sourc for vitamin K.

10. What conditions may lead to vitamin K deficiency?
11. Without vitamin K:
    a. muscles atrophy.
    b. bones become soft.
    c. skin rashes develop.
    d. blood fails to clot.
12. A significant amount of vitamin K comes from:
    a. vegetable oils.
    b. sunlight exposure.
    c. bacterial synthesis.
    d. fortified grain products.

See p. 357 for a summary of the fat-soluble vitamins.

ultiple Choice Answers

a 3.c 4.b 6.c 7.c 8.a 9.d 11.d 12.c

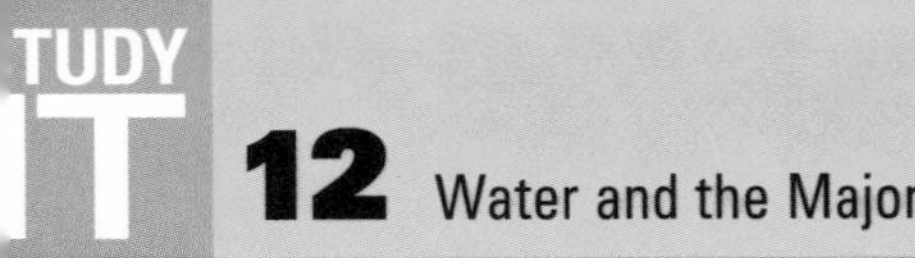

12 Water and the Major Minerals

eview the key points of this chapter below, then take the practice quiz on the back of this card.

2.1 Water and the Body Fluids

- Water makes up about 60 percent of an adult's body weight.
- Water assists with the transport of nutrients and waste products throughout the body, participates in chemical reactions, acts as a solvent, serves as a shock absorber, and regulates body temperature.
- To maintain water balance, intake from liquids, foods, and metabolism must equal losses from the kidneys, skin, lungs, and GI tract (review Table 12-3, p. 370).
- Whenever the body experiences low blood volume, low blood pressure, or highly concentrated body fluids, the actions of ADH, renin, angiotensin, and aldosterone restore homeostasis (review Figure 12-3).
- Electrolytes (charged minerals) in the fluids help distribute the fluids inside and outside the cells, thus ensuring the appropriate water balance and acid-base balance to support all life processes.
- Excessive losses of fluids and electrolytes upset these balances, and the kidneys play a key role in restoring homeostasis.

FIGURE 12-3 ***Animated*** **How the Body Regulates Blood 'olume and Blood Pressure**

ne renin-angiotensin-aldosterone system helps regulate blood volume and ierefore blood pressure.

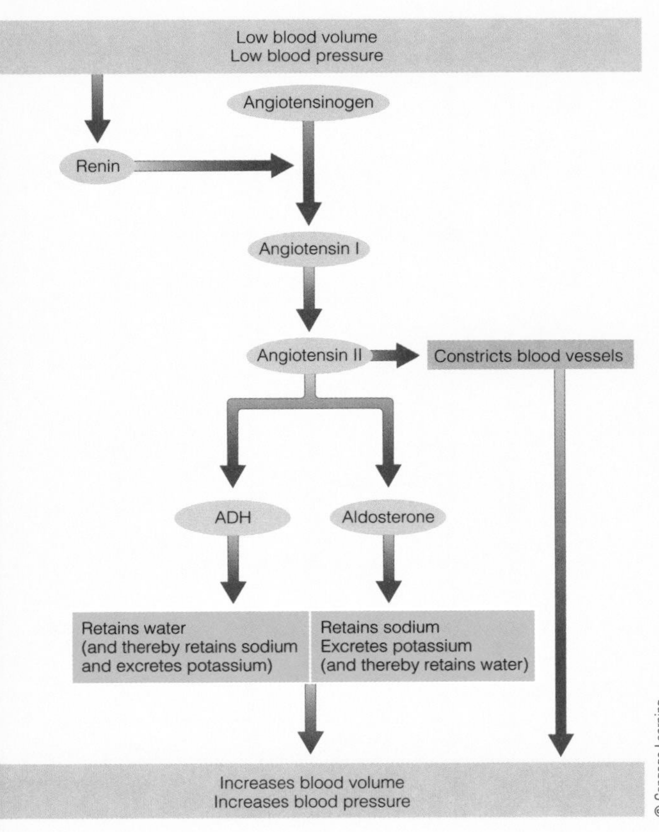

12.2 The Minerals—An Overview

- The major minerals are needed in the diet and found in the body in larger quantities than the trace minerals (review Figure 12-9, p. 378).
- Minerals are inorganic elements that retain their chemical identities; receive special handling and regulation in the body; and may bind with other substances or interact with other minerals, thus limiting their absorption.
- The major minerals, especially sodium, chloride, and potassium, influence the body's fluid balance; whenever an anion moves, a cation moves—always maintaining homeostasis.
- Sodium, chloride, potassium, calcium, and magnesium are key members of the team of nutrients that direct nerve impulse transmission and muscle contraction. They are also the primary nutrients involved in regulating blood pressure.
- Phosphorus and magnesium participate in many reactions involving glucose, fatty acids, amino acids, and the vitamins. Calcium, phosphorus, and magnesium combine to form the structure of the bones and teeth. Each major mineral also plays other specific roles in the body.

12.3 The Major Minerals

- Sodium is the main cation outside cells and one of the primary electrolytes responsible for maintaining fluid balance. Dietary deficiency is rare; excesses seem to aggravate hypertension, and so health professionals advise a diet moderate in salt and sodium.
- Chloride is the major anion outside cells, and it associates closely with sodium. In addition to its role in fluid balance, chloride is part of the stomach's hydrochloric acid.
- Potassium is the primary cation inside cells and plays an important role in maintaining fluid balance. Fresh foods, notably fruits and vegetables, are its best sources.
- Calcium is found primarily in the bones where it provides a rigid structure and a reservoir of calcium for the blood. Blood calcium participates in muscle contraction, blood clotting, and nerve impulses, and it is closely regulated by a system of hormones and vitamin D (review Figure 12-12, p. 386). Milk and milk products are good sources of calcium, but certain vegetables and tofu also provide calcium. Even when calcium intake is inadequate, blood calcium remains normal, but at the expense of bone loss, which can lead to osteoporosis.
- Phosphorus accompanies calcium both in the crystals of bone and in many foods such as milk. Phosphorus is also important in energy metabolism, as part of phospholipids, and as part of the genetic materials DNA and RNA.
- Magnesium supports bone mineralization and participates in numerous enzyme systems and in heart function. It is found abundantly in legumes and dark green, leafy vegetables and, in some areas, in water.
- Sulfate is found in all protein-containing foods. Its primary role in amino acids is to stabilize proteins by forming disulfide bridges.

Take the quiz below to test your mastery of the key chapter concepts.

12.1 Water and the Body Fluids (pp. 368–377)

LEARN IT Explain how the body regulates fluid balance.

1. List the roles of water in the body.
2. List the sources of water intake and routes of water excretion.
3. What is ADH? Where does it exert its action? What is aldosterone? How does it work?
4. How does the body use electrolytes to regulate fluid balance?
5. The body generates water during the:
   a. buffering of acids.
   b. dismantling of bone.
   c. metabolism of minerals.
   d. oxidation of energy nutrients through the electron transport chain.
6. Regulation of fluid and electrolyte balance and acid-base balance depends primarily on the:
   a. kidneys.
   b. intestines.
   c. sweat glands.
   d. specialized tear ducts.

12.2 The Minerals—An Overview (pp. 377–378)

LEARN IT List some of the ways minerals differ from vitamins and other nutrients.

7. What do the terms *major* and *trace* mean when describing the minerals in the body?
8. Describe some characteristics of minerals that distinguish them from vitamins.
9. The distinction between the major and trace minerals reflects the:
   a. ability of their ions to form salts.
   b. amounts of their contents in the body.
   c. importance of their functions in the body.
   d. capacity to retain their identity after absorption.

12.3 The Major Minerals (pp. 379–394)

LEARN IT Identify the main roles, deficiency symptoms, and food sources for each of the major minerals (sodium, chloride, potassium, calcium, phosphorus, magnesium, and sulfate).

10. What is the major function of sodium in the body? Describe how the kidneys regulate blood sodium. Is a dietary deficiency of sodium likely? Why or why not?
11. List calcium's roles in the body. How does the body keep blood calcium constant regardless of intake?
12. Name significant food sources of calcium. What are the co sequences of inadequate intakes?
13. List the roles of phosphorus in the body. Discuss the relationshi between calcium and phosphorus. Is a dietary deficiency phosphorus likely? Why or why not?
14. State the major functions of chloride, potassium, magnesiu and sulfur in the body. Are deficiencies of these nutrients like to occur in your own diet? Why or why not?
15. The principal cation in extracellular fluids is:
    a. sodium.
    b. chloride.
    c. potassium.
    d. phosphorus.
16. The role of chloride in the stomach is to help:
    a. support nerve impulses.
    b. convey hormonal messages.
    c. maintain a strong acidity.
    d. assist in muscular contractions.
17. Which would provide the most potassium?
    a. bologna
    b. potatoes
    c. pickles
    d. whole-wheat bread
18. Calcium homeostasis depends on:
    a. vitamin K, aldosterone, and renin.
    b. vitamin K, parathyroid hormone, and renin.
    c. vitamin D, aldosterone, and calcitonin.
    d. vitamin D, calcitonin, and parathyroid hormone.
19. Calcium absorption is hindered by:
    a. lactose.
    b. oxalates.
    c. vitamin D.
    d. stomach acid.
20. Phosphorus assists in many activities in the body, but *not*:
    a. energy metabolism.
    b. the clotting of blood.
    c. the transport of lipids.
    d. bone and teeth formation.
21. Most of the body's magnesium can be found in the:
    a. bones.
    b. nerves.
    c. muscles.
    d. extracellular fluids.

See p. 394 for a summary of the major minerals.

ultiple Choice Answers

d 6. a 9. b 15. a 16. c 17. b 18. d 19. b 20. b 21. a

eview the key points of this chapter below, then take the practice quiz on the back of this card.

3.1 The Trace Minerals—An Overview

- The body needs tiny amounts of the trace minerals.
- The trace minerals can be toxic at levels not far above estimated requirements—a consideration for supplement users (review Figure 13-1, p. 404).
- Like the other nutrients, the trace minerals are best obtained by eating a variety of foods.

3.2 The Trace Minerals—As Individuals

- Iron is found in hemoglobin and myoglobin where it carries oxygen for energy metabolism; iron also acts as a cofactor for some enzymes. Iron deficiency is most common among infants and young children, teenagers, women of childbearing age, and pregnant women. Symptoms include fatigue and anemia. Iron overload is most common in men. Heme iron, which is found only in meat, fish, and poultry, is better absorbed than nonheme iron, which occurs in most foods (review Figure 13-4). Nonheme iron absorption is improved by eating iron-containing foods with foods containing the MFP factor and vitamin C; absorption is limited by phytates and oxalates.
- Zinc-requiring enzymes participate in reactions affecting growth, vitamin A activity, and pancreatic digestive enzyme synthesis. Both dietary zinc and zinc-rich pancreatic secretions (via enteropancreatic circulation) are available for absorption. Absorption is monitored by a special binding protein (metallothionein) in the small intestine. Protein-rich foods derived from animals are the best sources of bioavailable zinc. Fiber and phytates in cereals bind zinc, limiting absorption. Symptoms of deficiency include growth retardation and sexual immaturity.
- Iodide, the iodine ion, is an essential component of the thyroid hormone. A deficiency can lead to goiter (enlargement of the thyroid gland) and can impair fetal development, causing cretinism. Iodization of salt has largely eliminated iodine deficiency in the United States and Canada.
- Selenium is an antioxidant nutrient that works closely with the glutathione peroxidase enzyme and vitamin E. Selenium is found with protein in foods. Deficiencies are associated with a predisposition to a type of heart abnormality known as Keshan disease.
- Copper is a component of several enzymes, all of which are involved with oxygen or oxidation. Some act as antioxidants; others are essential to iron metabolism. Legumes, whole grains, and shellfish are good sources of copper.
- Manganese-dependent enzymes are involved in bone formation and various metabolic processes. Because manganese is widespread in plant foods, deficiencies are rare, although regular use of calcium and iron supplements may limit manganese absorption.
- Fluoride makes teeth more resistant to decay. Fluoridation of public water reduces the incidence of dental caries; excess fluoride during tooth development can discolor and pit tooth enamel (fluorosis).
- Chromium enhances insulin's action. Deficiency can result in a diabetes-like condition. Chromium is widely available in unrefined foods including brewer's yeast, whole grains, and liver.
- Molybdenum is a part of many metalloenzymes. Deficiencies are unknown and toxicity is rare. It is found in a variety of foods.

13.3 Contaminant Minerals

- Contaminant minerals include the heavy metals lead, mercury, and cadmium that enter the food supply by way of soil, water, and air pollution.
- Lead typifies the ways all heavy metals behave in the body: they interfere with nutrients that are trying to do their jobs. The "good guy" nutrients are shoved aside by the "bad guy" contaminants. Then, when the contaminants cannot perform the role of the nutrients, health diminishes.
- To safeguard our health, we must defend ourselves against contamination by eating nutrient-rich foods and preserving a clean environment.

> **FIGURE 13-4 Heme and Nonheme Iron in Foods**

About 40% of the iron in meat, fish, and poultry is bound into heme; the other 60% is nonheme iron.

Key:
Heme
Nonheme

All of the iron in foods derived from plants is nonheme iron.

Heme accounts for about 10% of the average daily iron intake, but it is well absorbed (about 25%).

Nonheme iron accounts for the remaining 90%, but it is less well absorbed (about 17%).

Take the quiz below to test your mastery of the key chapter concepts.

13.1 The Trace Minerals—An Overview

(pp. 403–405)

LEARN IT Summarize key factors unique to the trace minerals.

1. Discuss the importance of balanced and varied diets in obtaining the essential minerals and avoiding toxicities.
2. Describe some of the ways trace minerals interact with one another and with other nutrients.

13.2 The Trace Minerals (pp. 405–424)

LEARN IT Identify the main roles, deficiency symptoms, and food sources for each of the trace minerals (iron, zinc, iodine, selenium, copper, manganese, fluoride, chromium, and molybdenum).

3. Distinguish between heme and nonheme iron. Discuss the factors that enhance iron absorption.
4. Iron absorption is impaired by:
   a. heme.
   b. phytates.
   c. vitamin C.
   d. MFP factor.
5. Which of these people is *least* likely to develop an iron deficiency?
   a. 3-year-old boy
   b. 52-year-old man
   c. 17-year-old girl
   d. 24-year-old woman
6. Which of the following would *not* describe the blood cells of a severe iron deficiency?
   a. anemic
   b. microcytic
   c. pernicious
   d. hypochromic
7. Which provides the most absorbable iron?
   a. 1 apple
   b. 1 cup milk
   c. 3 ounces steak
   d. ½ cup spinach
8. What causes iron overload? What are its symptoms?
9. Describe the similarities and differences in the absorption and regulation of iron and zinc.
10. Discuss possible reasons for a low intake of zinc. What factors affect the bioavailability of zinc?
11. The intestinal protein that helps to regulate zinc absorption is:
    a. albumin.
    b. ferritin.
    c. hemosiderin.
    d. metallothionein.
12. A classic sign of zinc deficiency is:
    a. anemia.
    b. goiter.
    c. mottled teeth.
    d. growth retardation.
13. What public health measure has been used in preventir simple goiter?
14. Cretinism is caused by a deficiency of:
    a. iron.
    b. zinc.
    c. iodine.
    d. selenium.
15. The mineral best known for its role as an antioxidant is:
    a. copper.
    b. selenium.
    c. manganese.
    d. molybdenum.
16. What measure has been recommended for protection again tooth decay?
17. Fluorosis occurs when fluoride:
    a. is excessive.
    b. is inadequate.
    c. binds with phosphorus.
    d. interacts with calcium.
18. Which mineral enhances insulin activity?
    a. zinc
    b. iodine
    c. chromium
    d. manganese

See p. 425 for a summary of the trace minerals.

13.3 Contaminant Minerals (p. 424)

LEARN IT Describe how contaminant minerals disrupt body process and impair nutrition status.

19. Which of the following does lead *not* compete with?
    a. iron
    b. zinc
    c. fluoride
    d. calcium

ultiple Choice Answers

b 5.b 6.c 7.c 11.d 12.d 14.c 15.b 17.a 18.c 19.c

14 Fitness: Physical Activity, Nutrients, and Body Adaptations

eview the key points of this chapter below, then take the practice quiz on the back of this card.

4.1 Fitness

- Physical activity promotes good health and reduces the risk of developing a number of diseases.
- The components of fitness are flexibility, muscle strength and endurance, and cardiorespiratory endurance; these are obtained by conditioning the body, through training, to adapt to the activity performed.
- Participating in cardiorespiratory, strength, and flexibility activities improves fitness and benefits health (review Table 14-2, p. 440).

4.2 Energy Systems and Fuels o Support Activity

- ATP and CP are high-energy compounds used by the body to provide energy to the muscle cells.

  During rest: ATP + creatine → CP

  During activity: CP → ATP + creatine
- Fuel mixture during physical activity depends on diet, the intensity and duration of activity, and training (review Table 14-4).
- During intense activity, the fuel mix is mostly glucose; during less intense activity, fat makes a greater contribution.
- With endurance training, muscle cells adapt to store more glycogen and to rely less on glucose and more on fat for energy.
- Active athletes may need more protein than sedentary people do; they typically eat more food and therefore obtain enough protein without supplements (review Table 14-5, p. 451).

4.3 Vitamins and Minerals o Support Activity

- With the possible exception of iron for women, well-nourished active people and athletes do not need dietary supplements.

14.4 Fluids and Electrolytes to Support Activity

- Active people need to drink plenty of water (review Table 14-6, p. 454).
- Symptoms of dehydration and heat stroke:

  Headache

  Nausea

  Dizziness

  Clumsiness

  Stumbling

  Sudden cessation of sweating (hot, dry skin)

  Confusion or other mental changes
- Endurance athletes need to drink both water and carbohydrate-containing beverages, especially during training and competition.
- During events lasting longer than 3 hours, athletes may need to replace sodium losses to prevent hyponatremia.

14.5 Diets for Physically Active People

- To enhance athletic performance and overall health, a diet should provide ample fluid and a variety of nutrient-dense foods in quantities to meet energy needs.
- Carbohydrate-rich foods that are light and easy to digest are best for pregame and postgame meals.

'ABLE 14-4 **Primary Fuels Used for Activities of Different Intensities and Durations**

Activity Intensity	Activity Duration	Preferred Fuel Source	Oxygen Needed?	Activity Example
:xtreme[a]	8 to 10 sec	ATP-CP (immediate availability)	No (anaerobic)	100-yard dash, shot put
'ery high	20 sec to 3 min	ATP from carbohydrate (lactate)	No (anaerobic)	¼-mile run at maximal speed
ligh	3 min to 20 min	ATP from carbohydrate	Yes (aerobic)	Cycling, swimming, or running
Moderate	More than 20 min	ATP from fat	Yes (aerobic)	Hiking

All levels of activity intensity use the ATP-CP system initially; extremely intense short-term activities rely solely on the ATP-CP system.

Take the quiz below to test your mastery of the key chapter concepts.

14.1 Fitness (pp. 438–444)

LEARN IT Describe the health benefits of being physically fit and explain how to develop the components of fitness.

1. Explain the overload principle.
2. Define *cardiorespiratory conditioning* and list some of its benefits.
3. Physical inactivity is linked to all of the following diseases except:
   a. cancer.
   b. diabetes.
   c. emphysema.
   d. hypertension.
4. The progressive overload principle can be applied by performing:
   a. an activity less often.
   b. an activity with more intensity.
   c. an activity in a different setting.
   d. a different activity each day of the week.

14.2 Energy Systems and Fuels to Support Activity (pp. 444–451)

LEARN IT Identify the factors that influence fuel use during physical activity and the types of activities that depend more on glucose or fat, respectively.

5. What types of activity are anaerobic? Which are aerobic? Describe the relationships among energy expenditure, type of activity, and oxygen use.
6. The process that regenerates glucose from lactate is known as the:
   a. Cori cycle.
   b. ATP-CP cycle.
   c. adaptation cycle.
   d. cardiac output cycle.
7. "Hitting the wall" is a term runners sometimes use to describe:
   a. dehydration.
   b. competition.
   c. indigestion.
   d. glucose depletion.
8. The technique endurance athletes use to maximize glycogen stores is called:
   a. aerobic training.
   b. muscle conditioning.
   c. carbohydrate loading.
   d. progressive overloading.
9. Conditioned muscles rely less on ________ and more on ________ for energy.
   a. protein; fat
   b. fat; protein
   c. glycogen; fat
   d. fat; glycogen

14.3 Vitamins and Minerals to Support Activity (pp. 451–453)

LEARN IT List which vitamins and mineral supplements, if an athletes may need and why.

10. Why are some athletes likely to develop iron-deficienc anemia? Compare iron-deficiency anemia and sports anemi explaining the differences.
11. Vitamin or mineral supplements taken just before an event a useless for improving performance because the:
    a. athlete sweats the nutrients out during the event.
    b. stomach can't digest supplements during physical activity.
    c. nutrients are diluted by all the fluids the athlete drinks.
    d. body needs hours or days for the nutrients to do their work
12. Physically active young women, especially those who a endurance athletes, are prone to:
    a. energy excess.
    b. iron deficiency.
    c. protein overload.
    d. vitamin A toxicity.

14.4 Fluids and Electrolytes to Support Activity (pp. 453–457)

LEARN IT Identify the factors that influence an athlete's fluid nee and describe the differences between water and sports drinks.

13. Discuss the importance of hydration during training, and li recommendations to maintain fluid balance.
14. The body's need for ________ far surpasses its need for a other nutrient.
    a. water
    b. protein
    c. vitamins
    d. carbohydrate

14.5 Diets for Physically Active People (pp. 457–460)

LEARN IT Discuss an appropriate daily eating pattern for athlete and list one example of a recommended pre- or post-game meal.

15. A recommended pregame meal includes plenty of fluids an provides between:
    a. 300 and 800 kcalories, mostly from fat-rich foods.
    b. 50 and 100 kcalories, mostly from fiber-rich foods.
    c. 1000 and 2000 kcalories, mostly from protein-rich foods.
    d. 300 and 800 kcalories, mostly from carbohydrate-rich food

ultiple Choice Answers

c 4. b 6. a 7. d 8. c 9. c 11. d 12. b 14. a 15. d

Review the key points of this chapter below, then take the practice quiz on the back of this card.

15.1 Nutrition prior to Pregnancy

- Prior to pregnancy, the health and behaviors of both men and women can influence fertility and fetal development. In preparation, they can achieve and maintain a healthy body weight, choose an adequate and balanced diet, be physically active, receive regular medical care, manage chronic conditions, and avoid harmful influences.

15.2 Growth and Development during Pregnancy

- The infant develops through three stages—the zygote, embryo, and fetus (review Figure 15-2, p. 472). Each organ and tissue grows on its own schedule (review Figure 15-4, p. 473).
- Times of intense development are critical periods that depend on nutrients (review Figure 15-3, p. 473).
- Without folate, the neural tube fails to develop completely during the first month of pregnancy; all women of childbearing age should take folate daily.

15.3 Maternal Weight

- A healthy pregnancy depends on a sufficient weight gain.
- Women who begin their pregnancies at a healthy weight need to gain about 30 pounds to cover the growth and development of the placenta, uterus, blood, breasts, and infant (review Figure 15-8, p. 479).
- Physical activity throughout pregnancy can help a woman develop the strength she needs to carry the extra weight and maintain habits that will help her lose it after the birth.

15.4 Nutrition during Pregnancy

- Energy and nutrient needs are high during pregnancy; the diet should include an extra serving from each of the five food groups.
- Supplements of iron and folate are recommended.
- Nausea, constipation, and heartburn commonly accompany pregnancy and can usually be alleviated with a few simple strategies (review Table 15-2, p. 483). Food cravings do not typically reflect physiological needs.

15.5 High-Risk Pregnancies

- High-risk pregnancies (review Table 15-3) threaten the life and health of both mother and infant and are likely to produce an infant with a low birth weight.
- Proper nutrition and abstinence from smoking, alcohol, and other drugs improve the outcome as can prenatal care to monitor for gestational diabetes and preeclampsia.

15.6 Nutrition during Lactation

- Lactating women need extra fluid and enough energy and nutrients to produce about 25 ounces of milk a day.
- Breastfeeding is contraindicated for those with HIV/AIDS.
- Alcohol, other drugs, smoking, and contaminants may reduce milk production or enter breast milk and impair infant development.

TABLE 15-3 High-Risk Pregnancy Factors

Factor	Condition that Raises Risk
Maternal weight	
• Prior to pregnancy	Prepregnancy BMI either <18.5 or ≥25
• During pregnancy	Insufficient or excessive pregnancy weight gain (see Table 15-1, p. 477)
Maternal nutrition	Nutrient deficiencies or toxicities; eating disorders
Socioeconomic status	Poverty, lack of family support, low level of education, limited food available
Lifestyle habits	Smoking, alcohol or other drug use
Age	Teens, especially 15 years or younger; women 35 years or older
Previous pregnancies	
• Number	Many previous pregnancies (3 or more to mothers younger than age 20; 4 or more to mothers age 20 or older)
• Interval	Short or long intervals between pregnancies (<18 months or >59 months)
• Outcomes	Previous history of problems
• Multiple births	Twins or triplets
• Birthweight	Low- or high-birthweight infants
Maternal health	
• High blood pressure	Development of gestational hypertension
• Diabetes	Development of gestational diabetes
• Chronic diseases	Diabetes; heart, respiratory, and kidney disease; certain genetic disorders; special diets and medications

Take the quiz below to test your mastery of the key chapter concepts.

15.1 Nutrition prior to Pregnancy (pp. 469–470)

LEARN IT List the ways men and women can prepare for a healthy pregnancy.

1. The health and behaviors of men and women prior to pregnancy can influence all of the following except:
   a. fertility. c. placenta development.
   b. infant's gender. d. infant's mental development.

15.2 Growth and Development during Pregnancy (pp. 470–475)

LEARN IT Describe fetal development from conception to birth and explain how maternal malnutrition can affect critical periods.

2. Describe the placenta and its function.
3. Explain why women of childbearing age need folate in their diets. How much is recommended, and how can women ensure that these needs are met?
4. The spongy structure that delivers nutrients to the fetus and returns waste products to the mother is called the:
   a. embryo. c. placenta.
   b. uterus. d. amniotic sac.

15.3 Maternal Weight (pp. 475–479)

LEARN IT Explain how both underweight and overweight can interfere with a healthy pregnancy and how weight gain and physical activity can support maternal health and infant growth.

5. Which of these strategies is *not* a healthy option for an overweight woman?
   a. Limit weight gain during pregnancy.
   b. Postpone weight loss until after pregnancy.
   c. Follow a weight-loss diet during pregnancy.
   d. Try to achieve a healthy weight before becoming pregnant.
6. A reasonable weight gain during pregnancy for a normal-weight woman is about:
   a. 10 pounds. c. 30 pounds.
   b. 20 pounds. d. 40 pounds.

15.4 Nutrition during Pregnancy (pp. 479–484)

LEARN IT Summarize the nutrient needs of women during pregnancy.

7. Which nutrients are needed in the greatest amounts during pregnancy? Why are they so important? Describe wise food choices for the pregnant woman.

15.5 High-Risk Pregnancies (pp. 484–491)

LEARN IT Identify factors predicting low-risk and high-risk pregnancies and describe ways to manage them.

8. What is the significance of infant birthweight in terms of the child's future health?
9. Describe some of the special problems of the pregnant adolescent. Which nutrients are needed in increased amounts?
10. What practices should be avoided during pregnancy? Why?
11. To help prevent neural tube defects, grain products are now fortified with:
   a. iron. c. protein.
   b. folate. d. vitamin C.
12. Pregnant women should *not* take supplements of:
   a. iron. c. vitamin A.
   b. folate. d. vitamin C.
13. The combination of high blood pressure, protein in the urine and edema signals:
   a. jaundice.
   b. preeclampsia.
   c. gestational diabetes.
   d. gestational hypertension.

15.6 Nutrition during Lactation (pp. 492–497)

LEARN IT Summarize the nutrient needs of women during lactation

14. To facilitate lactation, a mother needs:
   a. about 5000 kcalories a day.
   b. adequate nutrition and rest.
   c. vitamin and mineral supplements.
   d. a glass of wine or beer before each feeding.
15. A breastfeeding woman should drink plenty of water to:
   a. produce more milk.
   b. suppress lactation.
   c. prevent dehydration.
   d. dilute nutrient concentration.
16. A woman may need iron supplements during lactation:
   a. to enhance the iron in her breast milk.
   b. to provide iron for the infant's growth.
   c. to replace the iron in her body's stores.
   d. to support the increase in her blood volume.

ultiple Choice Answers

b 4. c 5. c 6. c 11. b 12. c 13. b 14. b 15. c 16. c

eview the key points of this chapter below, then take the practice quiz on the back of this card.

6.1 Nutrition during Infancy

- The primary food for infants during the first 12 months is either breast milk or iron-fortified formula.
- Breast milk offers both nutrients and immunological protection.
- Protective factors in breast milk:
  - Antibodies
  - Oligosaccharides
  - Bifidus factors
  - Lactoferrin
  - Lactadherin
  - Growth factor
  - Lipase enzyme
- At 4 to 6 months of age, infants should gradually begin eating solid foods, including iron-fortified cereals and vitamin-C rich fruits and vegetables.
- By 1 year, infants are drinking from a cup and eating a variety of foods.
- Infants may benefit from supplements containing vitamin D, iron, and fluoride (review Table 16-2, p. 510).

6.2 Nutrition during Childhood

- Children's appetites and nutrient needs reflect their stage of growth.
- Children who are chronically hungry and malnourished suffer growth retardation; temporary hunger and mild nutrient deficiencies produce more subtle problems—such as poor academic performance.
- Iron deficiency is widespread and has many physical and behavioral consequences.
- "Hyper" behavior is not caused by poor nutrition; misbehavior may be due to lack of sleep, too little physical activity, or too much television, among other factors.
- Some children have food allergies—adverse reactions that involve an immune response—most often caused by peanuts, tree nuts, milk, eggs, wheat, soybeans, fish, or shellfish.
- Childhood obesity is a major health problem (review Figure 16-9, p. 526).
- Children need to eat nutrient-dense foods and learn how to make healthful diet and activity choices.

16.3 Nutrition during Adolescence

- The need for iron increases during adolescence for both males and females; blood losses incurred through menstruation increase iron needs for females further.
- Sufficient calcium intake during adolescence supports optimal bone growth and density.
- The adolescent growth spurt increases the need for energy and nutrients.
- Adolescents who drink soft drinks regularly have a higher energy intake and a lower calcium intake; they are also more likely to be overweight.

TABLE 16-7 Recommended Eating and Physical Activity Behaviors to Prevent Obesity

The Expert Committee of the American Medical Association recommends the following healthy habits for children 2 to 18 years of age to help prevent childhood obesity:

- Limit consumption of sugar-sweetened beverages, such as soft drinks and fruit-flavored punches.
- Eat the recommended amounts of fruits and vegetables every day (2 to 4.5 cups per day based on age).
- Learn to eat age-appropriate portions of foods.
- Eat foods low in energy density such as those high in fiber and/or water and moderate in fat.
- Eat a nutritious breakfast every day.
- Eat a diet rich in calcium.
- Eat a diet balanced in recommended proportions for carbohydrate, fat, and protein.
- Eat a diet high in fiber.
- Eat together as a family as often as possible.
- Limit the frequency of restaurant meals.
- Limit television watching or other screen time to no more than 2 hours per day and do not have televisions or computers in bedrooms.
- Engage in at least 60 minutes of moderate to vigorous physical activity every day.

SOURCE: S. E. Barlow, Expert Committee recommendations regarding the prevention, assessment, and treatment of child and adolescent overweight and obesity: Summary report, *Pediatrics* 120 (2007): S164–S192.

Take the quiz below to test your mastery of the key chapter concepts.

16.1 Nutrition during Infancy (pp. 505–516)

LEARN IT List some of the immune factors in breast milk and describe the appropriate foods for infants during the first year of life.

1. What are the appropriate uses of formula feeding? What criteria would you use in selecting an infant formula?
2. Why are solid foods not recommended for an infant during the first few months of life? When is an infant ready to start eating solid food?
3. Identify foods that are inappropriate for infants and explain why they are inappropriate.
4. A reasonable weight for a healthy 5-month-old infant who weighed 8 pounds at birth might be:
   a. 12 pounds.
   b. 16 pounds.
   c. 20 pounds.
   d. 24 pounds.
5. Dehydration can develop quickly in infants because:
   a. much of their body water is extracellular.
   b. they lose a lot of water through urination and tears.
   c. only a small percentage of their body weight is water.
   d. they drink lots of breast milk or formula, but little water.
6. An infant should begin eating solid foods between:
   a. 2 and 4 weeks.
   b. 1 and 3 months.
   c. 4 and 6 months.
   d. 8 and 10 months.

16.2 Nutrition during Childhood (pp. 517–536)

LEARN IT Explain how children's appetites and nutrient needs reflect their stage of growth and why iron deficiency and obesity are often concerns during childhood.

7. What nutrition problems are most common in children? What strategies can help prevent these problems?
8. Describe the relationships between nutrition and behavior. How does television influence nutrition?
9. Describe a true food allergy. Which foods most often cause allergic reactions? How do food allergies influence nutrition status?
10. Describe the problems associated with childhood obesity and the strategies for prevention and treatment.
11. List strategies for introducing nutritious foods to children.
12. What impact do school meal programs have on the nutrition status of children?
13. Among US and Canadian children, the most prevalent nutrient deficiency is of:
    a. iron.
    b. folate.
    c. protein.
    d. vitamin D.
14. A true food allergy always:
    a. elicits an immune response.
    b. causes an immediate reaction.
    c. creates an aversion to the offending food.
    d. involves symptoms such as headaches or hives.
15. Which of the following strategies is *not* effective?
    a. Play first, eat later.
    b. Provide small portions.
    c. Encourage children to help prepare meals.
    d. Use dessert as a reward for eating vegetables.

16.3 Nutrition during Adolescence (pp. 536–540)

LEARN IT Discuss some of the challenges in meeting the nutrie needs of adolescents.

16. Describe the changes in nutrient needs from childhoo to adolescence. Why is an adolescent girl more likely t develop an iron deficiency than an adolescent boy?
17. How do adolescents' eating habits influence their nutrie intakes?
18. To help teenagers consume a balanced diet, parents can:
    a. monitor the teens' food intake.
    b. give up—parents can't influence teenagers.
    c. keep the pantry and refrigerator well stocked.
    d. forbid snacking and insist on regular, well-balanced meals.
19. During adolescence, energy and nutrient needs:
    a. reach a peak.
    b. fall dramatically.
    c. rise, but do not peak until adulthood.
    d. fluctuate so much that generalizations can't be made.
20. The nutrients most likely to fall short in the adolescer diet are:
    a. sodium and fat.
    b. folate and zinc.
    c. iron and calcium.
    d. protein and vitamin A.
21. To balance the day's intake, an adolescent who ea a hamburger, fries, and cola at lunch might benefit most from dinner of:
    a. fried chicken, rice, and banana.
    b. ribeye steak, baked potato, and salad.
    c. pork chop, mashed potatoes, and apple juice.
    d. spaghetti with meat sauce, broccoli, and milk.

ultiple Choice Answers

b 5. a 6. c 13. a 14. a 15. d 18. c 19. a 20. c 21. d

17 Life Cycle Nutrition: Adulthood and the Later Years

eview the key points of this chapter below, then take the practice quiz on the back of this card.

7.1 Nutrition and Longevity

- Life expectancy in the United States increased dramatically in the 20th century.
- Factors that enhance longevity include limited or no alcohol use, regular balanced meals, weight control, abstinence from smoking, regular physical activity, and adequate sleep.
- Energy restriction in animals seems to lengthen their lives; whether such dietary intervention in human beings is beneficial remains unknown.
- Nutrition—especially when combined with regular physical activity—can influence aging and longevity by supporting good health and preventing disease.

7.2 The Aging Process

- Changes that accompany aging can impair nutrition status. Among physiological changes, hormone activity alters body composition, immune system changes raise the risk of infections, atrophic gastritis interferes with digestion and absorption, and tooth loss limits food choices.
- Psychological changes such as depression, economic changes such as loss of income, and social changes such as loneliness contribute to poor food intake.

7.3 Energy and Nutrient Needs of Older Adults

- Some nutrients need special attention in the diet, but supplements are not routinely recommended.
- Table 17-5 helps determine the risk of malnutrition in older adults.

7.4 Nutrition-Related Concerns of Older Adults

- Senile dementia and cognitive losses afflict many older adults; some face loss of vision due to cataracts or macular degeneration or cope with the pain of arthritis.

TABLE 17-5 Risk Factors for Malnutrition in Older Adults

	These questions help *determine* the risk of malnutrition in older adults:
Disease	• Do you have an illness or condition that changes the types or amounts of foods you eat?
Eating poorly	• Do you eat fewer than two meals a day? Do you eat fruits, vegetables, and milk products daily?
Tooth loss or mouth pain	• Is it difficult or painful to eat?
Economic hardship	• Do you have enough money to buy the food you need?
Reduced social contact	• Do you eat alone most of the time?
Multiple medications	• Do you take three or more different prescribed or over-the-counter medications daily?
Involuntary weight loss or gain	• Have you lost or gained 10 pounds or more in the last 6 months?
Needs assistance	• Are you physically able to shop, cook, and feed yourself?
Elderly person	• Are you older than 80?

NOTE: A complete description of DETERMINE and its scoring system are available online from the American Academy of Family Physicians: www.aafp.org/afp/980301ap/edits.html.

- Table 17-2 provides a summary of the nutrient concerns of aging.
- Some problems may be inevitable, but others are preventable and good nutrition may play a key role.

17.5 Food Choices and Eating Habits of Older Adults

- Congregate meals provide older adults with both nutrients and social interactions; those who are homebound receive delivered meals.
- Adults living alone can prepare nutritious, inexpensive meals with a little creativity and careful shopping.

ABLE 17-2 Nutrient Concerns of Aging

Nutrient	Effect of Aging	Comments
/ater	Lack of thirst and decreased total body water make dehydration likely.	Mild dehydration is a common cause of confusion. Difficulty obtaining water or getting to the bathroom may compound the problem.
nergy	Need decreases as muscle mass decreases (sarcopenia).	Physical activity moderates the decline.
ber	Likelihood of constipation increases with low intakes and changes in the GI tract.	Inadequate water intakes and lack of physical activity, along with some medications, compound the problem.
otein	Needs may stay the same or increase slightly.	Low-fat, high-fiber legumes and grains meet both protein and other nutrient needs.
tamin B_{12}	Atrophic gastritis is common.	Deficiency causes neurological damages; supplements may be needed.
tamin D	Increased likelihood of inadequate intake; skin synthesis declines.	Daily sunlight exposure in moderation or supplements may be beneficial.
alcium	Intakes may be low; osteoporosis is common.	Stomach discomfort commonly limits milk intake; calcium substitutes or supplements may be needed.
on	In women, status improves after menopause; deficiencies are linked to chronic blood losses and low stomach acid output.	Adequate stomach acid is required for absorption; antacid or other medicine use may aggravate iron deficiency; vitamin C and meat increase absorption.
nc	Intakes are often inadequate and absorption may be poor, but needs may also increase.	Medications interfere with absorption; deficiency may depress appetite and sense of taste.

Take the quiz below to test your mastery of the key chapter concepts.

17.1 Nutrition and Longevity (pp. 553–556)

LEARN IT Describe the role nutrition plays in longevity.

1. Life expectancy in the United States is about:
   a. 48 to 60 years.
   b. 58 to 70 years.
   c. 68 to 80 years.
   d. 78 to 90 years.
2. The human life span is about:
   a. 85 years.
   b. 100 years.
   c. 115 years.
   d. 130 years.
3. A 72-year-old person whose physical health is similar to that of people 10 years younger has a(n):
   a. chronological age of 62.
   b. physiological age of 72.
   c. physiological age of 62.
   d. absolute age of minus 10.
4. Rats live longest when given diets that:
   a. eliminate all fat.
   b. provide lots of protein.
   c. allow them to eat freely.
   d. restrict their energy intakes.

17.2 The Aging Process (pp. 556–560)

LEARN IT Summarize how nutrition interacts with the physical, psychological, economic, and social changes involved in aging.

5. Which characteristic is *not* commonly associated with atrophic gastritis?
   a. inflamed stomach
   b. vitamin B_{12} toxicity
   c. bacterial overgrowth
   d. lack of intrinsic factor

17.3 Energy and Nutrient Needs of Older Adults (pp. 560–563)

LEARN IT Explain why the needs for some nutrients increase or decrease during aging.

6. Why does the risk of dehydration increase as people age?
7. Why do energy needs usually decline with advancing age?
8. Identify some factors that complicate the task of setting nutrient standards for older adults.
9. On average, adult energy needs:
   a. decline 5 percent per year.
   b. decline 5 percent per decade.
   c. remain stable throughout life.
   d. rise gradually throughout life.

17.4 Nutrition-Related Concerns of Older Adults (pp. 564–568)

LEARN IT Identify how nutrition might contribute to, or prevent, th development of age-related problems associated with vision, arthrit the brain, and alcohol use.

10. Which nutrients seem to protect against cataract developmen
    a. minerals
    b. lecithins
    c. antioxidants
    d. amino acids
11. The best dietary advice for a person with osteoarthritis mig be to:
    a. avoid milk products.
    b. take fish oil supplements.
    c. take vitamin E supplements.
    d. lose weight, if overweight.

17.5 Food Choices and Eating Habits of Older Adults (pp. 568–572)

LEARN IT Instruct an adult on how to shop for groceries and prepa healthy meals for one person on a tight budget.

12. What characteristics contribute to malnutrition in old people?
13. Congregate meal programs are preferable to Meals on Whee because they provide:
    a. nutritious meals.
    b. referral services.
    c. social interactions.
    d. financial assistance.
14. The OAA Nutrition Program is available to:
    a. all people 65 years and older.
    b. all people 60 years and older.
    c. homebound people only, 60 years and older.
    d. low-income people only, 60 years and older.

ultiple Choice Answers

c 2. d 3. c 4. d 5. b 9. b 10. c 11. d 13. c 14. b

18 Diet and Health

eview the key points of this chapter below, then take the practice quiz on the back of this card.

8.1 Nutrition and Infectious Diseases

- Public health measures such as purification of water and safe handling of food help prevent the spread of infection in populations; immunizations and antibiotics protect individuals.
- Nutrition cannot prevent or cure infectious diseases, but adequate intakes of all the nutrients can help support the immune system as the body defends against disease-causing agents.
- If the immune system is impaired because of malnutrition or diseases such as AIDS, a person becomes vulnerable to infectious disease.
- Inflammation underlies many chronic diseases.

8.2 Nutrition and Chronic Diseases

- Heart disease, cancers, and strokes are the three leading causes of death in the United States, and diabetes also ranks among the top 10.
- These four chronic diseases have significant links with nutrition and with one another (review Figure 18-3); other lifestyle risk factors and genetics also play a role.

8.3 Cardiovascular Disease

- Atherosclerosis is characterized by a build-up of plaque in an artery wall.
- Rupture of plaque or abnormal blood clotting can cause heart attacks and strokes.
- Dietary recommendations to lower the risks of cardiovascular disease are summarized in Table 18-5 (p. 593); quitting smoking and engaging in regular physical activity also improve heart health.

8.4 Hypertension

- The most effective dietary strategy for preventing hypertension is weight control.
- The DASH eating plan is rich in fruits, vegetables, nuts, and low-fat milk products and low in fat, saturated fat, and sodium (review Table 18-7, p. 597).

FIGURE 18-3 Interrelationships among Chronic Diseases

otice that many chronic diseases are themselves risk factors for other chronic seases and that all of them are linked to obesity. The risk factors highhted in blue define the metabolic syndrome.

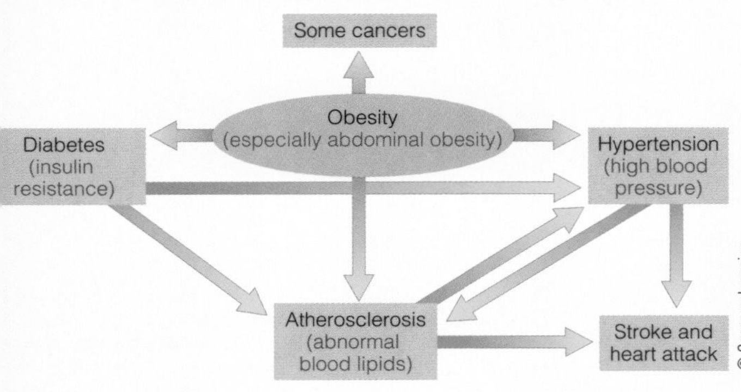

TABLE 18-11 Dietary Guidelines and Recommendations for Chronic Diseases Compared

Dietary Guidelines 2010	Heart Disease	Hypertension	Diabetes	Cancer
Prevent and/or reduce overweight and obesity.	✓	✓	✓	✓
Increase physical activity and reduce time spent in sedentary behaviors.	✓	✓	✓	✓
Keep total fat 20 to 35 percent of kcalories.	✓			
Consume less than 10 percent of kcalories from saturated fatty acids and keep *trans*-fatty acid consumption as low as possible.	✓		✓	✓
Increase vegetable and fruit intake. Consume at least half of all grains as whole grains.	✓	✓	✓	✓
Reduce the intake of kcalories from added sugars.	✓		✓	
Reduce daily sodium intake to less than 2300 milligrams.[a]	✓	✓	✓	✓
If alcohol is consumed, it should be consumed in moderation.	✓	✓	✓	✓

[a]Further reduce daily sodium intake to 1500 milligrams among persons who are 51 and older and those of any age who are African American or have hypertension, diabetes, or chronic kidney disease.

18.5 Diabetes Mellitus

- Diabetes is characterized by high blood glucose and either insufficient insulin, ineffective insulin, or a combination of the two.
- People with type 1 diabetes coordinate diet, insulin, and physical activity to help control their blood glucose.
- People with type 2 diabetes benefit most from a diet and physical activity program that controls glucose fluctuations and promotes weight loss.

18.6 Cancer

- Some dietary factors, such as alcohol and heavily smoked foods, may initiate cancer development; others, such as animal fats, may promote cancer once it develops; and still others, such as fiber, antioxidant nutrients, and phytochemicals, may serve as antipromoters that protect against cancer development.
- To obtain the best possible nutrition at the lowest possible cancer risk, eat many fruits, vegetables, legumes, and whole grains and little saturated fat; minimize weight gain through regular physical activity and a healthy diet (review Table 18-10, p. 606).

18.7 Recommendations for Chronic Diseases

- Optimal nutrition keeps people healthy and reduces the risk of chronic diseases.
- Dietary recommendations are aimed at populations and focus on controlling weight; limiting saturated and *trans* fat; increasing fiber-rich fruits, vegetables, and whole grains; and balancing food intake with physical activity (review Table 18-11).

TEST IT

Take the quiz below to test your mastery of the key chapter concepts.

18.1 Nutrition and Infectious Diseases (pp. 582–584)

LEARN IT Identify factors that protect people from the spread of infectious diseases and describe the role of nutrition in immunity.

1. The immune cells of the body do *not* include:
   a. B-cells.
   b. T-cells.
   c. antigens.
   d. phagocytes.
2. Which of the following produce antibodies?
   a. phagocytes
   b. T-cells
   c. antigens
   d. B-cells

18.2 Nutrition and Chronic Diseases (pp. 585–586)

LEARN IT List the leading nutrition-related causes of death in the United States.

3. The leading cause of death in the United States is:
   a. AIDS.
   b. cancer.
   c. diabetes.
   d. heart disease.

18.3 Cardiovascular Disease (pp. 586–594)

LEARN IT Describe how atherosclerosis develops and strategies to lower blood cholesterol levels.

4. Plaques in the arteries contribute to the development of:
   a. cancer.
   b. diabetes.
   c. atherosclerosis.
   d. infectious diseases.
5. Which blood lipid correlates directly with heart disease?
   a. HDL
   b. LDL
   c. VLDL
   d. triglycerides
6. Weight loss and physical activity may protect against heart disease by:
   a. raising LDL and insulin levels.
   b. raising HDL and lowering blood pressure.
   c. lowering LDL and increasing clot formation.
   d. improving insulin sensitivity and raising blood pressure.

18.4 Hypertension (pp. 595–597)

LEARN IT Present strategies to lower blood pressure.

7. What is the most effective strategy for most people to lower their blood pressure?
   a. lose weight
   b. restrict salt
   c. monitor glucose
   d. supplement protein

18.5 Diabetes Mellitus (pp. 597–602)

LEARN IT Compare the dietary strategies to manage type 1 diabet with those to prevent and treat type 2 diabetes.

8. Describe the differences between type 1 diabetes and type diabetes.
9. All of the following factors increase the risk of type 2 diabet *except*:
   a. aging.
   b. inactivity.
   c. obesity.
   d. smoking.
10. The most important dietary strategy in diabetes is to:
    a. provide for a consistent carbohydrate intake.
    b. restrict fat to 30 percent of daily kcalories.
    c. limit carbohydrate intake to 300 milligrams a day.
    d. take multiple vitamin and mineral supplements daily.

18.6 Cancer (pp. 602–606)

LEARN IT Differentiate among cancer initiators, promoters, and a tipromoters and describe how nutrients or foods might play a role each category.

11. Describe the characteristics of a diet that might offer the be protection against the onset of cancer.
12. Which of the following help(s) to protect against cancer?
    a. alcohol
    b. pickled foods
    c. phytochemicals
    d. omega-6 fatty acids

18.7 Recommendations for Chronic Disease (pp. 606–609)

LEARN IT Summarize dietary recommendations to prevent chron diseases.

13. Describe the dietary choices that best protect against t development of most chronic diseases.

ultiple Choice Answers

c 2. d 3. d 4. c 5. b 6. b 7. a 9. d 10. a 12. c

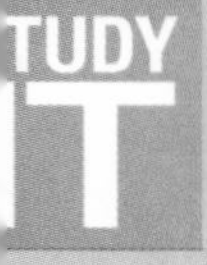

19 Consumer Concerns about Foods and Water

eview the key points of this chapter below, then take the practice quiz on the back of this card.

9.1 Foodborne Illnesses

- Maintaining a safe food supply requires everyone's efforts (review Figure 19-1).
- Millions of people suffer mild to life-threatening symptoms caused by foodborne illnesses (review Table 19-1, p. 625).
- The "How To" (pp. 628–629) describes how these illnesses can be prevented by storing and cooking foods at their proper temperatures and by preparing them in sanitary conditions.
- Irradiation of certain foods protects consumers from foodborne illness, but also raises some concerns.

9.2 Nutritional Adequacy of Foods and Diets

- In the marketplace, food labels, the *Dietary Guidelines for Americans,* and MyPlate all help consumers learn about nutrition and how to plan healthy diets.
- At home, consumers can minimize nutrient losses from fruits and vegetables by refrigerating them, washing them before cutting them, storing them in airtight containers, and cooking them for short times in minimal water.

9.3 Environmental Contaminants

- Environmental contamination of foods is a concern, but so far, the hazards appear relatively small.
- Remain alert to the possibility of contamination, and keep an ear open for public health announcements and advice.
- Eat a variety of foods to protect against the accumulation of toxins in the body. Each food eaten dilutes contaminants that may be present in other components of the diet.

9.4 Natural Toxicants in Foods

- Natural toxicants include the goitrogens in cabbage, cyanogens in lima beans, and solanine in potatoes.
- Any substance can be toxic when consumed in excess.
- Poisons are poisons, whether made by people or by nature. The source of a chemical does not make it hazardous; its chemical structure and the quantity consumed do.

19.5 Pesticides

- Pesticides can safely improve crop yields when used according to regulations, but they can also be hazardous when used inappropriately.
- The FDA tests both domestic and imported foods for pesticide residues in the fields and in market basket surveys of foods prepared table ready.
- Consumers can minimize their ingestion of pesticide residues on foods by following the suggestions in Table 19-4 (p. 641).
- Alternative farming methods may allow farmers to grow crops with few or no pesticides.

19.6 Food Additives

- The benefits of food additives seem to justify the risks associated with their use.
- The FDA regulates the use of these intentional additives: antimicrobial agents (such as nitrites) to prevent microbial spoilage; antioxidants (such as vitamins C and E, sulfites, and BHA and BHT) to prevent oxidative changes; colors (such as tartrazine) and flavor enhancers (such as MSG) to appeal to senses; and nutrients (such as iodine in salt) to enrich or fortify foods (see Table 19-5, p. 646).
- Incidental additives sometimes get into foods during processing, but rarely present a hazard; some processes such as treating livestock with hormones and antibiotics raise consumer concerns.

19.7 Consumer Concerns about Water

- Like foods, water may contain infectious microorganisms, environmental contaminants, pesticide residues, and additives.
- The EPA monitors the safety of the public water system, but many consumers choose home water treatment systems or bottled water instead of tap.

FIGURE 19-1 Food Safety from Farm to Table

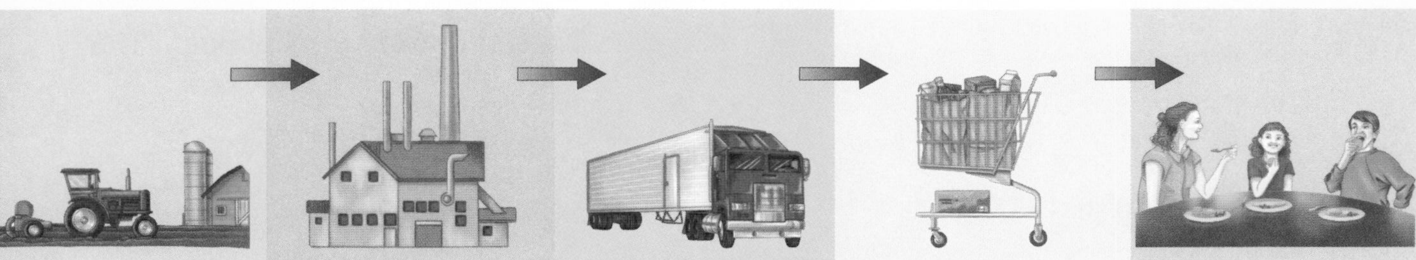

FARM
Workers must use safe methods of growing, harvesting, sorting, packing, and storing food to minimize contamination hazards.

PROCESSING
Processors must follow FDA guidelines concerning contamination, cleanliness, and education and training of workers and must monitor for safety at critical control points.

TRANSPORTATION
Containers and vehicles transporting food must be clean. Cold food must be kept cold at all times.

RETAIL
Employees in grocery stores and restaurants must follow the FDA's Food Code on how to prevent foodborne illnesses. Establishments must pass local health inspections and train staff in sanitation.

TABLE
Consumers must learn and use sound principles of food safety as taught in this chapter. Be mindful that foodborne illness is a real possibility, and take steps to prevent it.

Take the quiz below to test your mastery of the key chapter concepts.

19.1 Foodborne Illnesses (pp. 625–633)

LEARN IT Describe how foodborne illnesses can be prevented.

1. Distinguish between the two types of foodborne illnesses and provide an example of each.
2. What special precautions apply to meats? To seafood?
3. Eating a contaminated food such as undercooked poultry or unpasteurized milk might cause a:
   a. food allergy.
   b. food infection.
   c. food intoxication.
   d. botulinum reaction.
4. The temperature danger zone for foods ranges from:
   a. −20°F to 120°F.
   b. 0°F to 100°F.
   c. 20°F to 120°F.
   d. 40°F to 140°F.
5. Examples of foods that frequently cause foodborne illness are:
   a. canned foods.
   b. steaming-hot foods.
   c. fresh fruits and vegetables.
   d. raw milk, seafood, meat, and eggs.
6. Irradiation can help improve our food supply by:
   a. cooking foods quickly.
   b. killing microorganisms.
   c. minimizing the use of preservatives.
   d. improving the nutrient content of foods.

19.2 Nutritional Adequacy of Foods and Diets (pp. 633–634)

LEARN IT Explain how to minimize nutrient losses in the kitchen.

7. Describe which nutrients are most vulnerable to destruction.

19.3 Environmental Contaminants (pp. 634–636)

LEARN IT Explain how environmental contaminants get into foods and how people can protect themselves against contamination.

8. How does bioaccumulation of toxins occur and how does that influence the safety of foods?

19.4 Natural Toxicants in Foods (p. 637)

LEARN IT Identify natural toxicants and determine whether they are hazardous.

9. Solanine is an example of a(n):
   a. heavy metal.
   b. artificial color.
   c. natural toxicant.
   d. animal hormone.

19.5 Pesticides (pp. 637–642)

LEARN IT Debate the risks and benefits of using pesticides.

10. How do pesticides become a hazard to the food supply, a[nd] how are they monitored? In what ways can people reduce t[he] concentrations of pesticides in and on foods that they prepar[e]

19.6 Food Additives (pp. 642–648)

LEARN IT List common food additives, their purposes, and example[s]

11. What is the difference between a GRAS substance and regulated food additive? Give examples of each.
12. The standard that deems additives safe if lifetime u[se] presents no more than a one-in-a-million risk of cancer known as the:
    a. Delaney Clause.
    b. zero-risk policy.
    c. GRAS list of standards.
    d. negligible-risk policy.
13. Common antimicrobial additives include:
    a. salt and nitrites.
    b. carrageenan and MSG.
    c. dioxins and sulfites.
    d. vitamin C and vitamin E.
14. Common antioxidants include:
    a. BHA and BHT.
    b. tartrazine and MSG.
    c. sugar and vitamin E.
    d. nitrosamines and salt.
15. Incidental additives that may enter foods during processi[ng] include:
    a. dioxins and BGH.
    b. dioxins and folate.
    c. beta-carotene and agar.
    d. nitrites and irradiation.

19.7 Consumer Concerns about Water (pp. 648–651)

LEARN IT Discuss consumer concerns about water.

16. Chlorine is added to water to:
    a. protect against dental caries.
    b. destroy harmful minerals such as lead and mercury.
    c. kill pathogenic microorganisms.
    d. remove the sulfur that produces a "rotten egg" odor.

ultiple Choice Answers

b 4.d 5.d 6.b 9.c 12.d 13.a 14.a 15.a 16.c

eview the key points of this chapter below, then take the practice quiz on the back of this card.

0.1 Hunger in the United States

- Food insecurity and hunger are widespread in the United States among those living in poverty (review Figure 20-1).
- Ironically, hunger and malnutrition can occur among obese people (review Figure 20-2).
- Government assistance programs help to relieve poverty and hunger; food recovery programs and other community efforts are equally important.

0.2 World Hunger

- Natural causes such as drought, flood, and pests and political causes such as armed conflicts and government policies all contribute to the extreme hunger and poverty seen in the developing countries.
- More people means more mouths to feed, which worsens the problems of poverty and hunger.
- Poverty and hunger encourage parents to have more children; breaking this cycle requires improving the economic status of the people and providing them with health care, education, and family planning.
- Hunger leads to malnutrition, which appears most evident in nutrient deficiencies and growth failure.

0.3 Malnutrition

- Inadequate food intake in children leads to poor growth and nutrient deficiencies.
- Children suffering from acute malnutrition (recent severe food deprivation) may be underweight for their height, whereas those experiencing chronic malnutrition (long-term food deprivation) are short for their age.
- Problems resulting from nutrient deficiencies include preterm births and low birth weights (iron), stillbirths and cretinism (iodine), blindness (vitamin A), and growth failure (zinc).
- Treatment for malnutrition should be individualized to ensure rapid weight gain and to correct nutrient deficiencies.

20.4 The Global Environment

- The global environment is deteriorating, largely because of our irresponsible use of resources and energy.
- Environmental degradation reduces our ability to produce enough food to feed the world's people.
- The rapid increase in the world's population exacerbates the situation.
- Governments, businesses, and all individuals can make environmentally conscious choices to help solve the hunger problem, improve the quality of life, and generate jobs.
- Personal choices, made by many people, can have a great impact.

FIGURE 20-1 **Prevalence of Food Security in US Households, 2010**

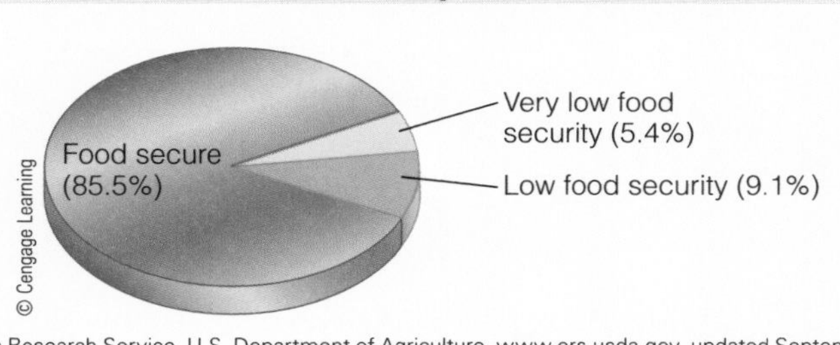

)URCE: Economic Research Service, U.S. Department of Agriculture, www.ers.usda.gov, updated September 2011.

> FIGURE 20-2 **The Poverty-Obesity Paradox**

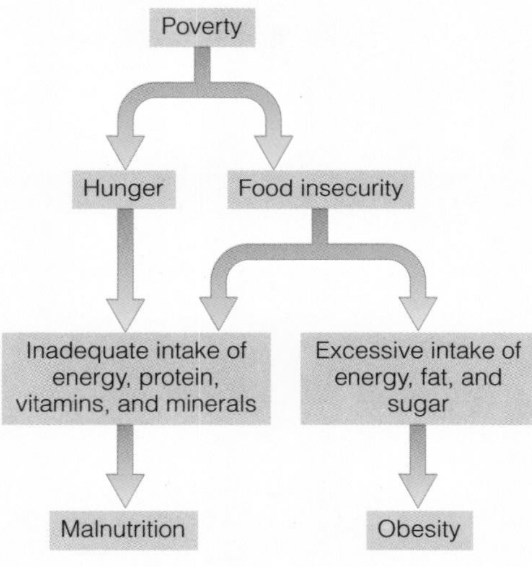

Take the quiz below to test your mastery of the key chapter concepts.

20.1 Hunger in the United States

(pp. 660–663)

LEARN IT Identify some reasons why hunger is present in a country as wealthy as the United States.

1. Food insecurity refers to the:
   a. uncertainty of foods' safety.
   b. fear of eating too much food.
   c. limited availability of foods.
   d. reliability of food production.

2. The most common cause of hunger in the United States is:
   a. poverty.
   b. alcohol abuse.
   c. mental illness.
   d. lack of education.

3. SNAP debit cards cannot be used to purchase:
   a. tomato plants.
   b. birthday cakes.
   c. cola beverages.
   d. laundry detergent.

4. Which action is not typical of a food recovery program?
   a. gathering potatoes from a harvested field
   b. collecting overripe tomatoes from a wholesaler
   c. offering SNAP debit cards to low-income people
   d. delivering restaurant leftovers to a community shelter

20.2 World Hunger (pp. 664–667)

LEARN IT Identify some reasons why hunger is present in the developing countries of the world.

5. The primary cause of the worst famine in the 20th century was:
   a. armed conflicts.
   b. natural disasters.
   c. food contaminations.
   d. government policies.

6. Which of the following is most critical in providing food to a the world's people?
   a. decreasing air pollution
   b. increasing water supplies
   c. decreasing population growth
   d. increasing agricultural land

20.3 Malnutrition (pp. 667–670)

LEARN IT Describe the consequences of nutrient and energ inadequacies.

7. The most likely cause of death in malnourished children is:
   a. growth failure.
   b. diarrheal disease.
   c. simple starvation.
   d. vitamin A deficiency.

20.4 The Global Environment (pp. 670–675)

LEARN IT Explain why relieving environmental problems will als help to alleviate hunger and poverty.

8. Discuss the different paths by which rich and poor cou tries can attack the problems of world hunger and th environment.

ultiple Choice Answers
c 2.a 3.d 4.c 5.d 6.c 7.b

Daily Values for Food Labels

The Daily Values are standard values developed by the Food and Drug Administration (FDA) for use on food labels. The values are based on 2000 kcalories a day for adults and children over 4 years old. Chapter 2 provides more details.

Nutrient	Amount
Protein[a]	50 g
Thiamin	1.5 mg
Riboflavin	1.7 mg
Niacin	20 mg NE
Biotin	300 μg
Pantothenic acid	10 mg
Vitamin B_6	2 mg
Folate	400 μg
Vitamin B_{12}	6 μg
Vitamin C	60 mg
Vitamin A	5000 IU[b]
Vitamin D	400 IU[b]
Vitamin E	30 IU[b]
Vitamin K	80 μg
Calcium	1000 mg
Iron	18 mg
Zinc	15 mg
Iodine	150 μg
Copper	2 mg
Chromium	120 μg
Selenium	70 μg
Molybdenum	75 μg
Manganese	2 mg
Chloride	3400 mg
Magnesium	400 mg
Phosphorus	1000 mg

[a]The Daily Values for protein vary for different groups of people: pregnant women, 60 g; nursing mothers, 65 g; infants under 1 year, 14 g; children 1 to 4 years, 16 g.
[b]Equivalent values for nutrients expressed as IU are: vitamin A, 1500 RAE (assumes a mixture of 40% retinol and 60% beta-carotene); vitamin D, 10 μg; vitamin E, 20 mg.

Food Component	Amount	Calculation Factors
Fat	65 g	30% of kcalories
Saturated fat	20 g	10% of kcalories
Cholesterol	300 mg	Same regardless of kcalories
Carbohydrate (total)	300 g	60% of kcalories
Fiber	25 g	11.5 g per 1000 kcalories
Protein	50 g	10% of kcalories
Sodium	2400 mg	Same regardless of kcalories
Potassium	3500 mg	Same regardless of kcalories

GLOSSARY OF NUTRIENT MEASURES

kcal: kcalories; a unit by which energy is measured (Chapter 1 provides more details).

g: grams; a unit of weight equivalent to about 0.03 ounces.

mg: milligrams; one-thousandth of a gram.

μg: micrograms; one-millionth of a gram.

IU: international units; an old measure of vitamin activity determined by biological methods (as opposed to new measures that are determined by direct chemical analyses). Many fortified foods and supplements use IU on their labels.

- For vitamin A, 1 IU = 0.3 μg retinol, 3.6 μg β-carotene, or 7.2 μg other vitamin A carotenoids
- For vitamin D, 1 IU = 0.02 μg cholecalciferol
- For vitamin E, 1 IU = 0.67 natural α-tocopherol (other conversion factors are used for different forms of vitamin E)

mg NE: milligrams niacin equivalents; a measure of niacin activity (Chapter 10 provides more details).

- 1 NE = 1 mg niacin
  = 60 mg tryptophan (an amino acid)

μg DFE: micrograms dietary folate equivalents; a measure of folate activity (Chapter 10 provides more details).

- 1 μg DFE = 1 μg food folate
  = 0.6 μg fortified food or supplement folate taken with food
  = 0.5 μg supplement folate taken on an empty stomach

μg RAE: micrograms retinol activity equivalents; a measure of vitamin A activity (Chapter 11 provides more details).

- 1 μg RAE = 1 μg retinol
  = 12 μg β-carotene
  = 24 μg other vitamin A carotenoids

mmol: millimoles; one-thousanth of a mole, the molecular weight of a substance. To convert mmol to mg, multiply by the atomic weight of the substance.

- For sodium, mmol × 23 = mg Na
- For chloride, mmol × 35.5 = mg Cl
- For sodium chloride, mmol × 58.5 = mg NaCl

Diet Analysis PLUS+ Student Guide

Table of Contents

Getting Started

Diet Analysis Plus is a health and nutrition management program that allows you to create personal profiles, estimate activity level, track diet and exercise, and create detailed reports. The Diet Analysis Plus online system (also referred to as DA+) makes it easy for you to evaluate the types and serving sizes of the foods you consume.

*Tip:*Diet Analysis Plus includes a**Tutorial** which includes useful step-by-step information, such as how to create your primary profile and other basic features. This tutorial is always available from the **Tutorial** link at the top of the page.

Accessing Helpful Information

When you're working in a specific report from the **Reports** page, you can click the Report Title to bring up an **Info box** which explains the functions on the screen.

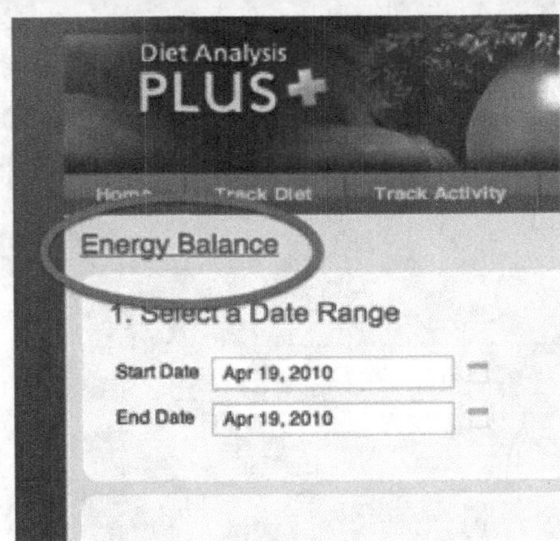

When you're working from the **Track Diet** page, you can click the Search Tips link to bring up a short **Help note** which contains an explanation of the report.

Entering a New Course Access Code

Your instructor can provide you with a **Course Access Code** that allows you to register for new or additional courses.

To register with a **Course Access Code**:.

Step 1. After you have received the **Course Access Code** from your instructor, you can log into DA+.

Step 2. Enter your Course Access Code in the **Course Access Code** text box.

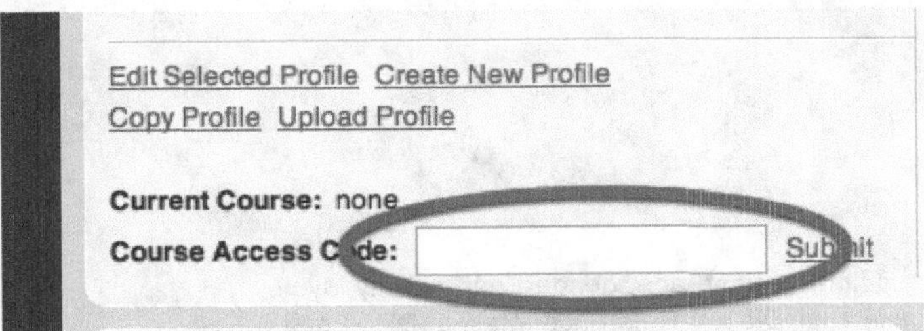

Step 3. Enter your new code exactly as it appears and click the Submit link. A confirmation text box will appear at the top of the page.

System Setup

To ensure the best experience with DA Plus and enjoy all of its features, please use the information in this section to optimize your computer system and browser settings. System requirements, browser configuration, and software conflicts are important issues when troubleshooting. Due to the constantly shifting nature of Web browsers, it is important to follow these recommendations, to take advantage of the most compatible configuration.

This section lists the basic hardware, software, and system settings you need to run Diet Analysis Plus. If your system meets the basic hardware needs, you can download any of the free software you need from the links in the following sections.

Windows System Requirements

- **Microsoft Windows** XP, Vista, or 7

- **Web browsers**: Firefox 3+, Chrome, and Internet Explorer 6, 7, or 8
- **Internet Connection**: Use of the Diet Analysis Plus web site requires an Internet connection speed of **56k** or higher
- To print tests, **Adobe Acrobat Reader** 4.05b or higher.
(Get Reader from get.adobe.com/reader/.)
- **Adobe Flash Player** 9 or higher is required.
(Get Player from http://get.adobe.com/flashplayer/.)

Macintosh System Requirements

- **Mac OS X** 10.4 or higher
- **Web browsers:** Firefox 3+ and Safari 3+
- **Adobe Flash Player** 9 or higher is required.
(Get Player from http://get.adobe.com/flashplayer/.)
- **Internet Connection**: Use of the Diet Analysis Plus web site requires an Internet connection speed of **56k** or higher

---

Linux System Requirements

- **Red Hat Linux 9.0** (or similar), X Windows System.
- **Web browser**: Firefox 3+
- **Adobe Acrobat Reader** 4.05b or higher is required to print tests.
(Get Reader from get.adobe.com/reader/.)
- **Adobe Flash Player** 9 or higher is required.
(Get Player from http://get.adobe.com/flashplayer/.)
- **Internet Connection**: Use of the Diet Analysis Plus web site requires an Internet connection speed of **56k** or higher

Setting Up Profiles

What is a Profile?

Nutritional requirements vary depending on an individual's height, weight, age, gender, and activity level. Your profile records this information and uses it to determine your Dietary Reference Intakes (DRIs) and to create custom reports. Before using Diet Analysis Plus, you must first create a profile.

Tip: Also see Using Multiple Profiles for more details on profile management.

Creating a Profile

Step 1. When you log into DA Plus for the first time, you will be asked to crate a primary profile. You can create multiple profiles for different analyses.

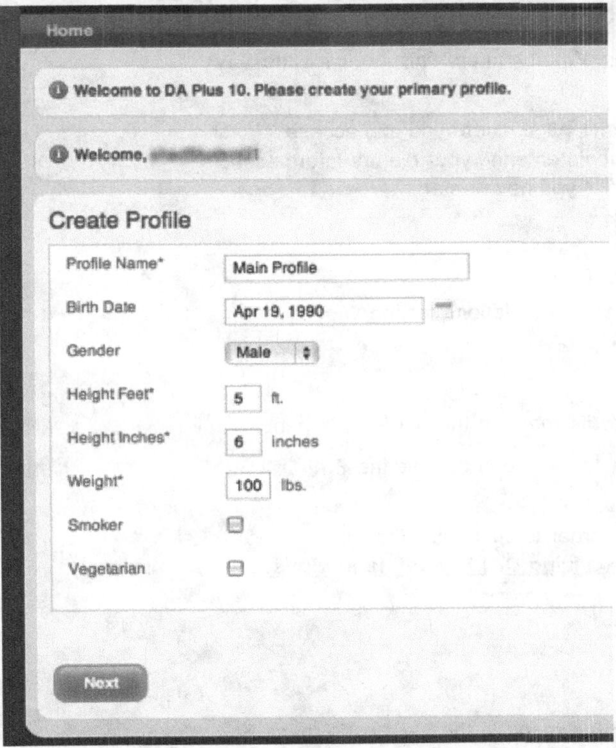

Step 2. Enter the information for your primary profile.

Step 3. When you are finished, click the **Next** button to see the **Activity Questionnaire** page.

Note: The **Long Activity Questionnaire** is the default when setting up a primary profile; you can use a shorter version when setting up additional profiles.

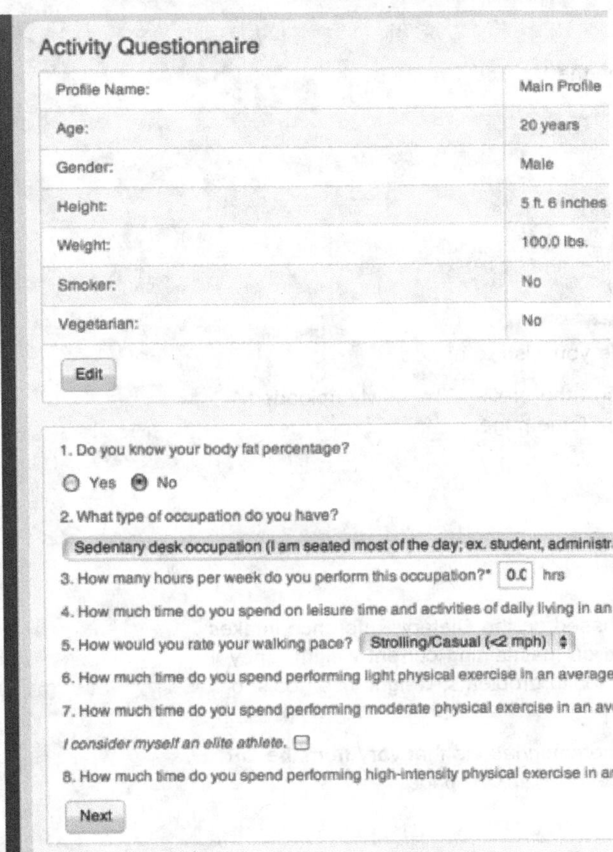

Step 4. Answer the questions on the questionnaire. The lifestyle related questions on this page will help the DA Plus system determine your activity level.

Step 5. When you are finished, click the **Next** button to see the **Confirm Profile** page.

Step 6. Click the **Edit** button(s) to change any of your information. When you are finished, click the **Save** button to save your profile.

After your primary profile is complete, you are directed to the **Home** page where you can access information on your profiles, food intakes, reports, and labs. To begin entering your dietary information, click the **Track Diet** tab on the main menu to find and list the foods you have eaten.

Viewing DRI Recommendations

After you have created a profile, Diet Analysis Plus determines recommendations for the **Dietary Reference Intakes** (DRI).

To see a profile's DRI recommendations:

Step 1. From the Home page, select the profile from the **Main Profile** menu in the **My Profile** pane.

Step 2. Click the **View Full DRI** button in the **Dietary Reference Intakes** pane to see the **Full DRI Page**.

Tip: You can edit the values in many of the fields on this page in order to customize the recommendations to your needs. For more information, see **Customizing DRI Recommendations**.

Managing an Existing Profile

Use the steps outlined below to view or edit an existing profile:

Step 1. From the Home page, select the profile from the **Main Profile** menu in the **My Profile** pane.

Step 2. Click the **Edit Selected Profile** link to makes changes to the profile and/or questionnaire.

Step 3. When you are finished, click **Save** to be returned to the **Home** page.

Importing a DA+ 9.0 Profile

To import a DA+ 9.0 profile:

Step 1. From the Home page, click the **Upload Profile** link to see the **Upload Profile** page.

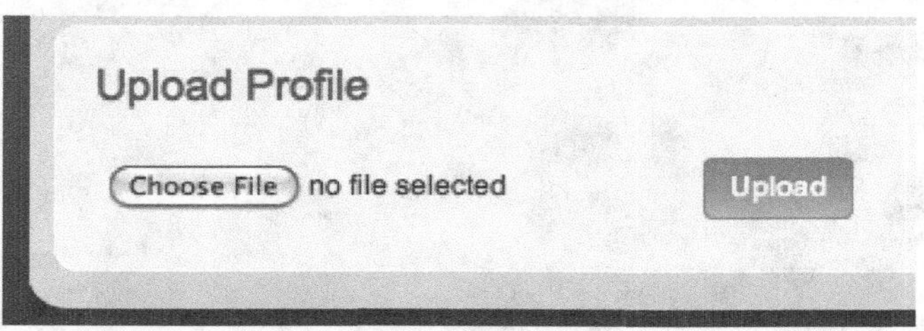

Step 4. Click the **Choose File** button and navigate to the **.dap** file you wish to upload.

Step 5. Once you selected the file, click the **Upload** button to import the profile. The newly uploaded profile will now appear in the **Active Profile** menu on the Home page.

Advanced Tips

+ Customizing DRI Recommendations

Diet Analysis Plus determines recommendations for each profile based on the Dietary Reference Intakes (DRI) published by the USDA. These recommendations are based on maintaining current weight. They do not take into account special needs based on athletic activity, health problems, weight loss goals, or other factors that may require variations for individuals.

Diet Analysis Plus is not a clinical tool, and will not create any recommendations that vary from the DRI. However, you may change many of the recommendations on the **View Full DRI** page.

To change the DRIs for a Profile:

Step 1. From the Home page, select the profile from the **Activity Profile** menu in the My Profile pang.

Step 2. Dietary Reference Intakes pane, click the **View Full DRI** button.

Step 3. Click the **Edit** button.

Step 4. Change the quantities in any editable field, then click **Save**. Your modified DRIs will now be used for the Recommendations reflected in all reports.

+ Using Multiple Profiles

A profile is two important things: a description of an individual, and the record of foods and activities recorded for that individual. You may need to use more than one more profile for yourself for several different reasons:

- Profile information changes (weight loss, pregnancy, age, etc.)
- Multiple class assignments with differing profile information for either yourself or a hypothetical profile
- Various profile settings and reports to show the effects of changes in diet and activity

---

+ Deleting Profiles: CAUTION!

A profile is not just the name and information in the **Profile** menu. A profile is also the record of all the foods recorded for that profile name in the **Track Diet** page, and all the activities recorded in the **Track Activities** page.

If you delete a profile, you will be deleting all the foods and activities as well as the name. Deleted profiles cannot be restored.

Tracking Your Diet

You can track your diet by recording everything you eat and drink during the day. Make sure you record everything you consume, including water, drinks, condiments, cooking oils, and alcoholic beverages.

Getting Started Tracking Your Diet

Step 1. On the main menu, click the **Track Diet** tab. The **Track Diet** page will open.

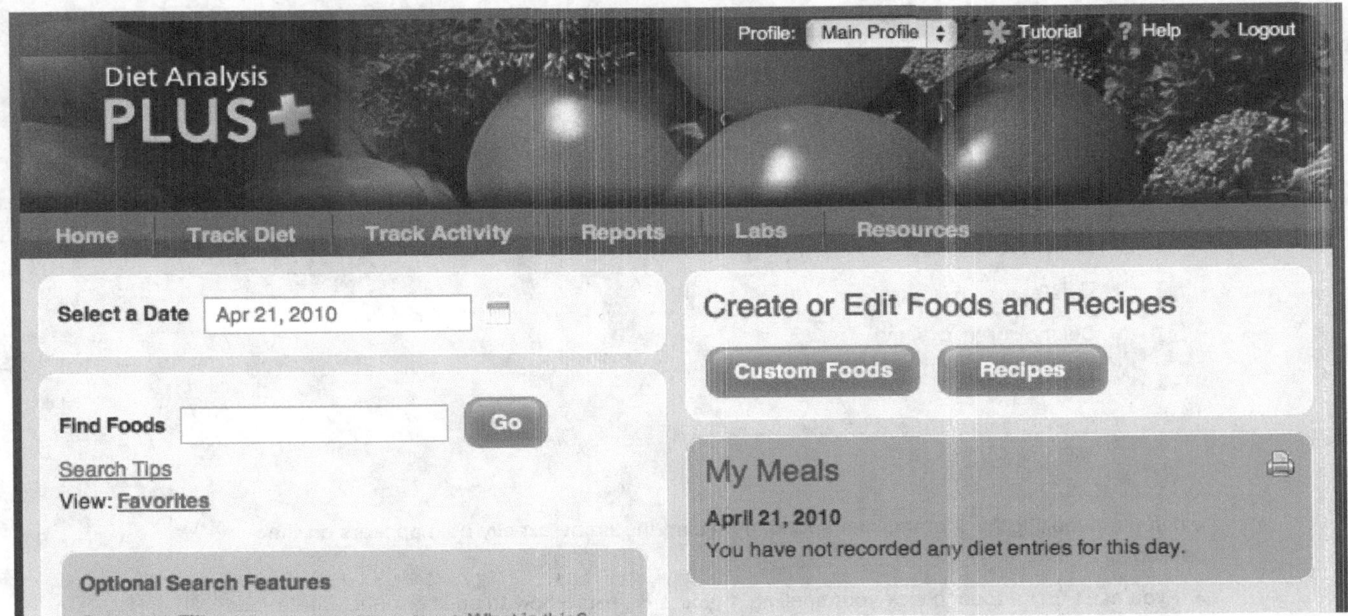

Step 2. Click the calendar icon next to the **Select a Date** field to open a drop-down calendar.

<< < April, 2010 > >> x

Sun	Mon	Tue	Wed	Thu	Fri	Sat
28	29	30	31	1	2	3
4	5	6	7	8	9	10
11	12	13	14	15	16	17
18	19	20	21	22	23	24
25	26	27	28	29	30	1
2	3	4	5	6	7	8

Apr 21, 2010 Clean Today

Step 3. Click a date to select it. The date you select will be highlighted.

Step 4. Use the arrows at the top to change the month or year.

Finding a Food

Step 1. In the **Find Foods** field, type the name of a food.

Step 2. Click the **Go** button.

A **Food List** will appear which provides a list of foods that contain the word or words you typed.

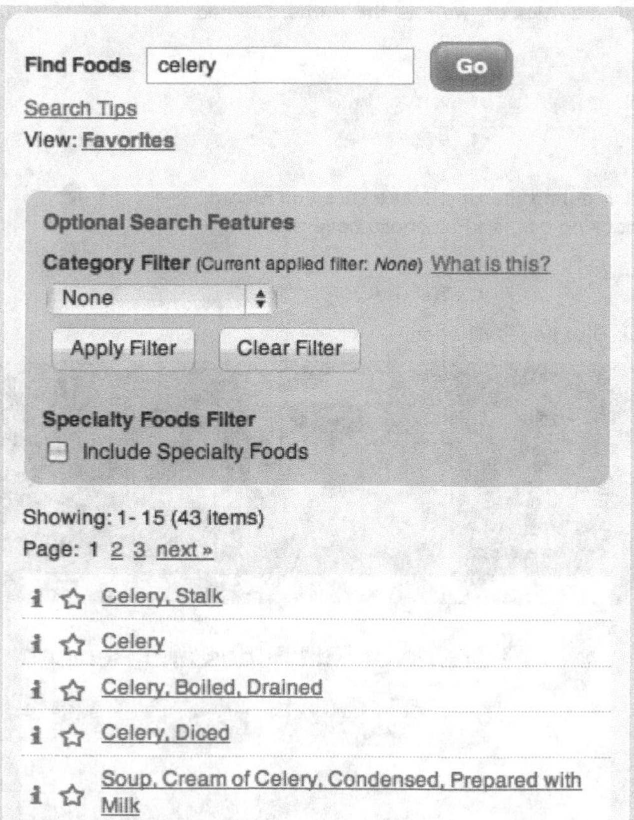

Search Tips:

• If you are looking for a brand-name food, try typing the name exactly as it appears on the packaging.

• If you can't find a food, check your spelling. If you are unsure how to spell or punctuate a food name, use just the first few letters of each word in the name. For example, "kell min whe" will find **"Kellogg's Mini-Wheats."**

• Ue the Category Filter and/or Specialty Foods Filter to fine-tune your search.

• If the list of foods is very long, try adding specifics. For example, "fried chicken" finds a shorter, more specific list than "chicken."

• Common foods come up first in a search, then generic foods, then brand names. To find your food, you may have to scroll through the list, using the page numbers and arrows that appear inside the **Food List** field.

• To scroll through a long list, use the page numbers and arrows that appear inside the Food List box.

• The **i** button to the left of each item opens a pop-up showing the nutrient information about the food. Sometimes this information can help you choose the right food.

Selecting Portions

Step 1. In the **Food List**, select the food name to open the **Serving Size** window for adding the food to a meal.

Step 2. Check the serving size of the food, the units of measurement, and the meal it was consumed.

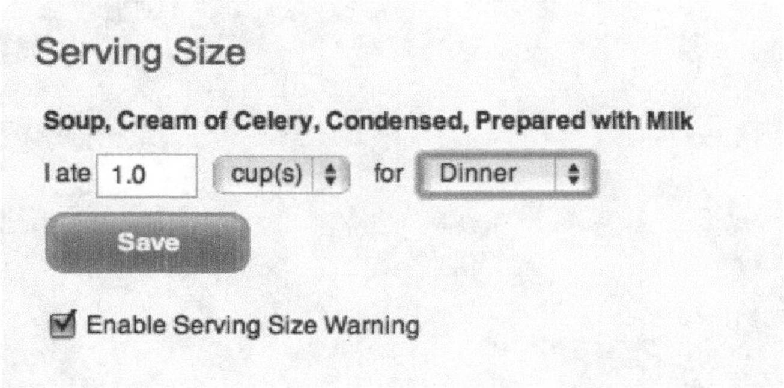

Serving Size

Soup, Cream of Celery, Condensed, Prepared with Milk

I ate 1.0 cup(s) for Dinner

Save

☑ Enable Serving Size Warning

Careful! By default, all foods open showing the amount for one "serving," but servings are expressed in different units. Some food servings are "1 item." Others are "8 ounces" or "1 pint," or even "500 grams." If you are unsure what the serving size unit means, use the drop-down to change the unit of measurement. The quantity will automatically recalculate.

Step 3. Carefully estimate the amount of food you ate.

• **If you measured all the food in your meals** before you ate them and kept a record, this is an excellent practice. This is the most accurate way to track your diet.

• **If you did not measure all the food you ate**, estimate the amounts you ate. All the reports used in your assignments will be based on the information you enter about each food. Do your best to be realistic about your portion sizes. If you underestimate or overestimate, the reports you use for your assignments will be less reliable.

Step 4. Click the meal to which you want to add the food: **Breakfast**, **Lunch**, **Dinner**, or **Snack**.

Step 5. Click the **Save** button and you will see the food added to your **Food List**.

Step 6. Go to **Step 1**, and **Search** for your next food. Continue with these steps until you have listed everything you ate each day. Be as thorough as possible for best results.

Editing the My Meals Food List

Step 1. To **change** the amount, serving size, or meal of a food item, click the **Edit** button in the **My Meals** window.

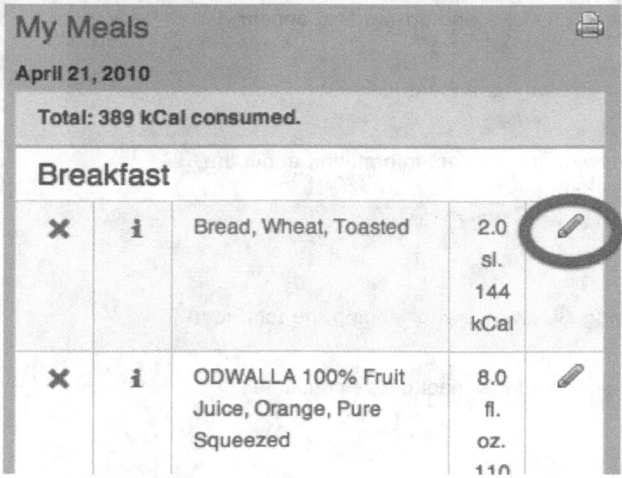

The **Serving Size** window will open.

Step 2. Make your changes as needed and click **Save**.

Step 3. To **delete** an item from the food list, click the ✖ **button** in the **My Meals** window.

Advanced Tips

Customize your list of foods by making a list of favorites, adding a food, or adding a recipe.

+ Making a Favorites List

You can create a "short list" of foods you eat frequently by adding them to your **Favorites** list.

Step 1. To **add** a food to your **Favorites** list, click the ☆ **button**.

Step 2. To **see** your **Favorites** list, click the **View: Favorites** link.

Tip: When you create a new food or recipe, it is automatically added to your **Favorites** list.

+ Creating Custom Foods

A **"Food"** is any food item for which nutritional information is available. To add a customized food to your account, you need to at least know the food's calories per serving -- and it is helpful to have nutrient information as well.

When adding a new food, it is helpful to have a Nutrition Facts label from a processed food. If you need to add an unprocessed food to your list of **Favorites**, you can research its nutritional information yourself.

To **add** a new food:

Step 1. On the **Track Diet** menu, click the **Custom Foods** button. The **Custom Food List** page will open.

Step 2. Click the **Create New Food** button. The **Nutrition Facts** page will open.

Step 3. Enter the information available for the food.

- For **packaged foods**, you can use the Daily Value for some nutrients on the **Nutrition Facts label**. You can also use grams or micrograms if the label lists nutrients in those units.
- For **supplements**, you can use grams, micrograms, or the Daily Value, if the label lists

nutrients in those units.

Step 4. When you are done, click the **Save** button at the bottom of the window. The **Custom Food List** will open with the new food listed.

Step 5. Click on any food name to do further edits or click **Close** to return to the **Track Diet** page.

Tip: The new Food is added to your **Favorites** list automatically.

+ *Creating Recipes*

A **"Recipe"** is a group of Foods. To create a Recipe, you don't need to know any nutrient information. The nutrient information comes from the Foods that are part of the Recipe.

Important Recipes that you create do not have My Pyramid or Exchange values. Those values cannot be calculated from the foods in the recipe.

To **add** a new recipe:

Step 1. On the Track Diet menu, click the **Recipes** button. The **Recipe List** page will open.

Step 2. Click the **Create New Recipe** button. Type the recipe name, and specify a number of portions. Then click the **Save** button to be returned to the Recipe List page.

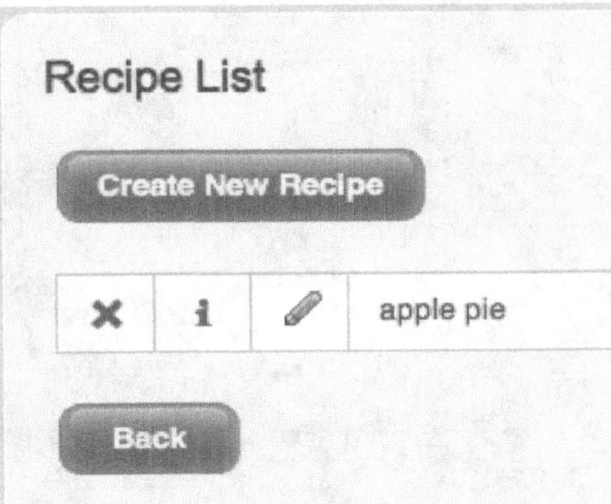

Tip: The **Number of Servings** is important. Diet Analysis Plus will divide the Recipe by this number, in order to calculate a serving size and nutrient content.

Step 3. In the **Recipe** box, click the **edit** button and then the **Add Ingredient** button to search for and choose a food to add to the recipe.

Step 4. Type the name of an ingredient and click the **Go** button to see a list of foods from which you can select.

Step 5. Click the **name** of the food and the **Add/Edit Ingredient** dialog will open.

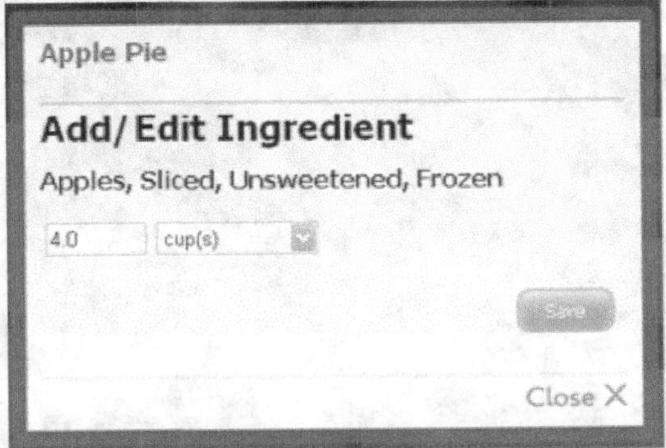

Step 6. Type in the amount to add to the recipe, and check the units of measurement in the drop-down menu. When you are finished, click the **Save** button to be returned to the **Recipe** page.

Step 7. Click the **Add Ingredient** button to add additional food items until the ingredient list for that recipe is complete.

Step 8. When you are finished adding food items to the recipe, click the **Back** button to be returned to the **Recipe List** page.

Step 8. Click the **Back** button to be returned to the **Track Diet** page.

Tip: The new Recipe is automatically added to your **Favorites** list.

Tracking Your Activities

You can track your activities by recording everything you do for 24 hours each day. Record everything, including activities that don't seem like "exercise," such as working on a computer or cooking. The calculation of calories burned will be more accurate if the activity list is as complete as possible.

Any time that is not accounted for in a 24 hour period will be calculated as "resting."

Getting Started Tracking Your Activities

Step 1. Click the **Track Activity** tab on the main menu.

Step 2. Click the calendar icon next to the **Select a Date** field to open a drop-down **calendar**.

<< < April, 2010 > >> x

Sun	Mon	Tue	Wed	Thu	Fri	Sat
28	29	30	31	1	2	3
4	5	6	7	8	9	10
11	12	13	14	15	16	17
18	19	20	21	22	23	24
25	26	27	28	29	30	1
2	3	4	5	6	7	8

Apr 21, 2010 Clean Today

Step 3. Click a date to select it. The date you select will be highlighted.

Step 4. Use the arrows at the top to change the month or year.

Adding an Activity

Step 1. In the **Find Activities** box, type the name of an activity and click the **Go** button.

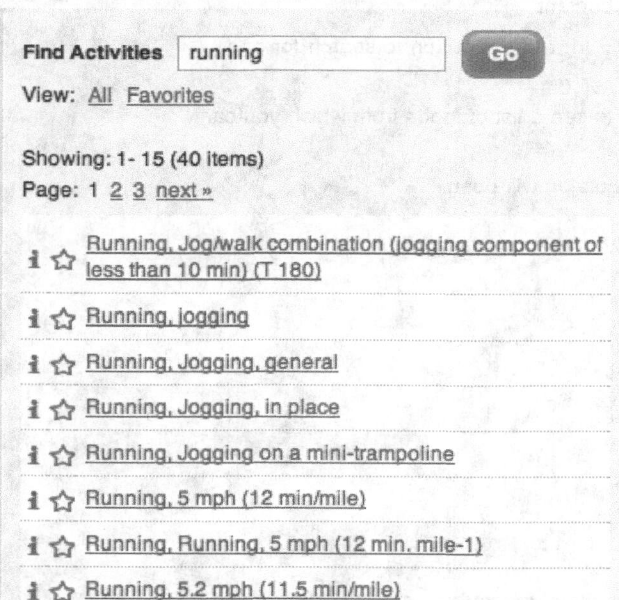

In the activity list that appears you will see a list of activities that contain the word or words you typed.

Search Tips:

- If you can't find an activity, try different or partial words. For example, instead of "bicycling," try "cycl."
- To browse the entire list of activities, click the **All** link.
- To scroll through a long list, use the page numbers.

Step 2. Select the activity that is the best match.

Tip: To the left of each item there is an **information** button. Click the **i** to see details about the energy burned by an activity, or MET value.

Step 3. Click a **name** to choose that activity. The **Activity Duration** window will open.

Activity Duration

Running, 5.2 mph (11.5 min/mile)

0 hour(s) 0 minute(s)

Save

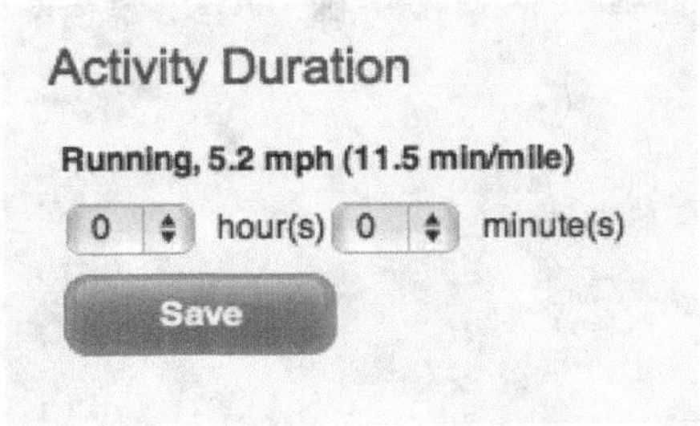

Step 4. Type in the amount of time you spent on the activity, then click **Save.**

Step 5. Go back to Step 1, and Search for the next activity. Continue with these steps until you have listed everything you did for 24 hours each day. Be as thorough as possible for best results.

Editing Activities

To edit the **Activity Duration**:

Step 1. Click the **Edit** icon to choose an activity in the My Activities window. The **Activity Duration** window will open.

Step 2. Make the necessary changes and click **Save**.

Step 3. Click the ✖ **button** to the left of the activity name to delete it.

Viewing Reports

Diet Analysis Plus allows you to generate a variety of reports showing analyses of the diet and activity information you have recorded for a profile. These reports can be part of an assignment and they are necessary when you are working with some of the Labs. You can easily print any of the reports for a profile in either PDF or RTF formats. These can then be printed, saved to disk, or emailed.

Use the **Reports** page to examine report information and to observe how reports are affected by changing variables such as dates viewed and meals included.

Getting Started Viewing Reports

Click the **Reports** tab on the main menu to open the **Reports** page. All available reports are shown under **Nutrients**, **Spreadsheets**, and **Advanced**.

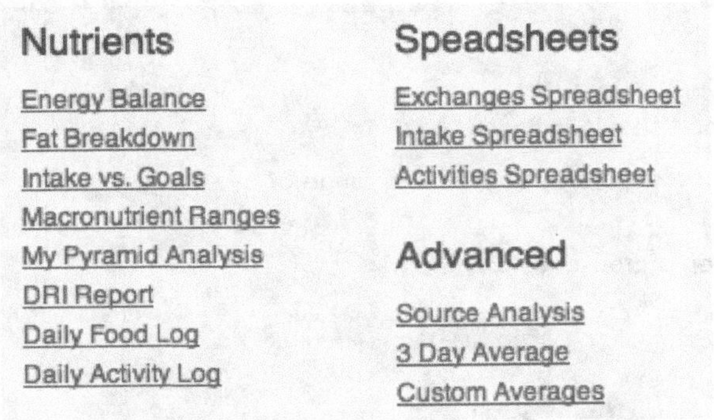

Nutrients

Energy Balance
Fat Breakdown
Intake vs. Goals
Macronutrient Ranges
My Pyramid Analysis
DRI Report
Daily Food Log
Daily Activity Log

Speadsheets

Exchanges Spreadsheet
Intake Spreadsheet
Activities Spreadsheet

Advanced

Source Analysis
3 Day Average
Custom Averages

Each report includes options for setting parameters that define the data to include. Below is an example of the **Macronutrient Ranges** report showing some of the report options.

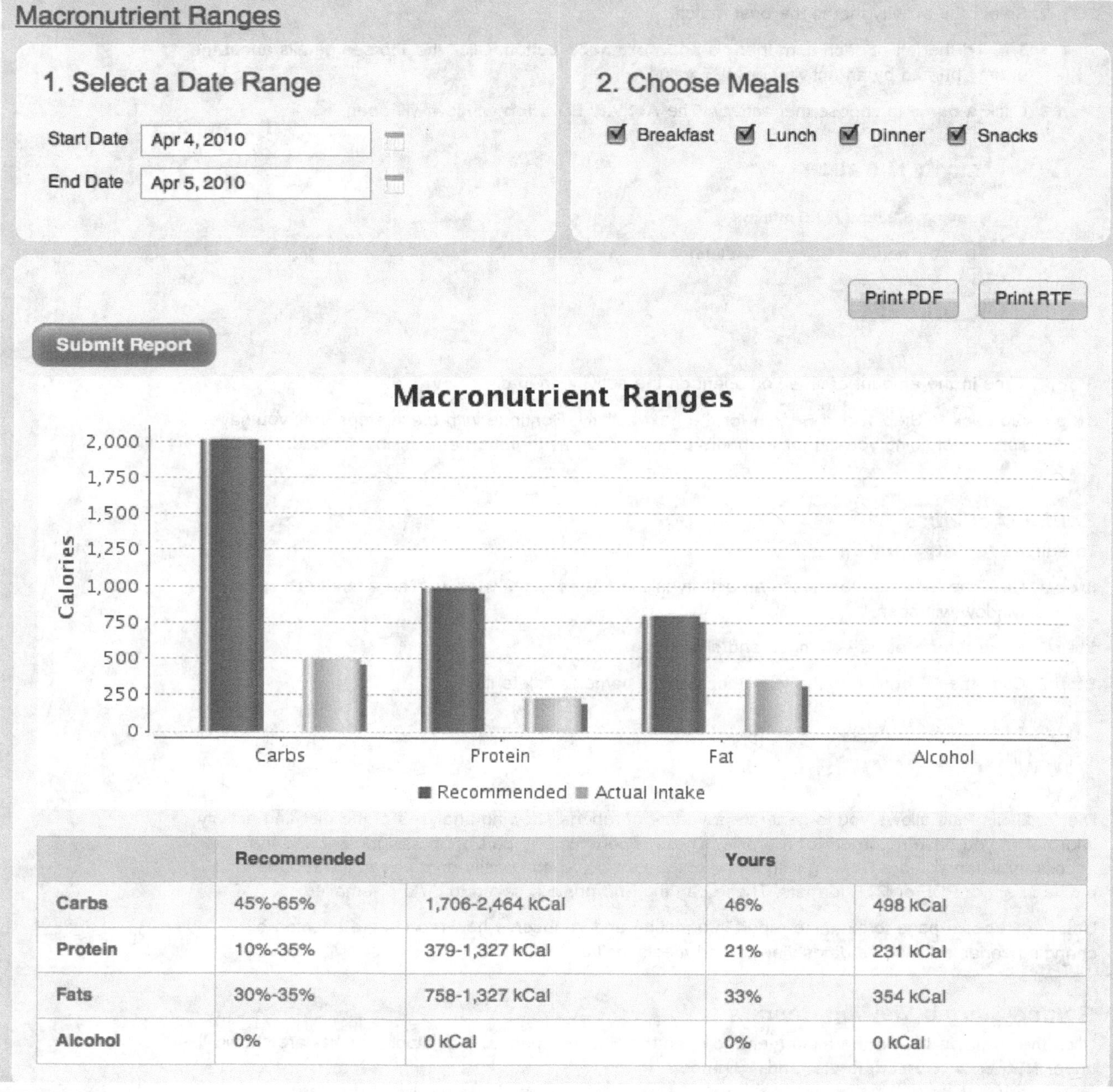

	Recommended		Yours	
Carbs	45%-65%	1,706-2,464 kCal	46%	498 kCal
Protein	10%-35%	379-1,327 kCal	21%	231 kCal
Fats	30%-35%	758-1,327 kCal	33%	354 kCal
Alcohol	0%	0 kCal	0%	0 kCal

Choosing Dates

To create a report, you must first choose the date range that the report should include.

Tip: Some reports use information from only one day at a time. Other reports let you choose a range of dates to include in a report.

Step 1. Click the calendar icon next to the **Select a Date** field to open a drop-down **calendar**.

If you have listed foods or activities for a day, it is highlighted.

<< < April, 2010 > >>						x
Sun	Mon	Tue	Wed	Thu	Fri	Sat
28	29	30	31	1	2	3
4	5	6	7	8	9	10
11	12	13	14	15	16	17
18	19	20	21	22	23	24
25	26	27	28	29	30	1
2	3	4	5	6	7	8
Apr 21, 2010	Clean				Today	

Step 2. Click a date to select it.

Step 3. Use the arrows at the top to change the month or year.

Step 4. For many reports you will need to repeat this process for both the **Start Date** and **End Date**.

Tip: Generating reports may be slow if you select a range of dates that includes many days. If it is taking a very long time, try specifying fewer days.

Choosing Content

The **Choose Meals** pane allows you to choose the meals to include in a report.

Step 1. Click the check box to change the selections.

2. Choose Meals

☑ Breakfast ☑ Lunch ☑ Dinner ☑ Snacks

Step 2. In some reports, you may also be asked to choose a nutrient from a menu.

Using Reports

Diet Analysis Plus generates a variety of reports showing analyses of the diet and activity information recorded for a profile.

It is important to understand the data in the reports — what the scales measure, what conditions are being compared, and what the limitations are of the information provided.

Note: The **Custom Average** and **3-Day Average** reports are not generated on-screen and are only available as PDF's. See the section on Printing Reports for more details

---

IMPORTANT!: A Note about Nutrient Data

On the **Track Diet** page, you can search for foods that you ate. This search uses a database of thousands of foods, which includes information about the nutrients in the foods.

The Diet Analysis Plus database derives information from many sources. The data for brand name foods often comes directly from the companies that make them. Some of those foods have data for many nutrients. Others have data for only the nutrients the companies are required by regulation to report.

Sometimes you may see an asterisk (*) indicating that the nutrient content shown on the report may be lower than the actual nutrient content. This means that you may be getting more of a nutrient than the report shows.

Keep this in mind when reading reports, and consider the implications when drawing conclusions about actual dietary intake. Your instructor can provide more in-depth guidance on interpreting report data.

Profile DRI Goals Report

The **Profile DRI Goals** report shows the Dietary Reference Intakes (DRI) recommended for the profile. All the other reports use these DRI Goals as the baseline for comparisons with actual dietary intake.

The report shows the basic information from the profile that affects the recommendations, and then lists all the nutrients. Some nutrients have a dash next to them instead of a number. These nutrients don't have a recommendation.

You don't need to choose a date for this report, because it is based on the static personal information in the profile.

Intake vs. Goals Report

The **Intake vs. Goals** report compares actual dietary intake with the DRI goals for your profile.

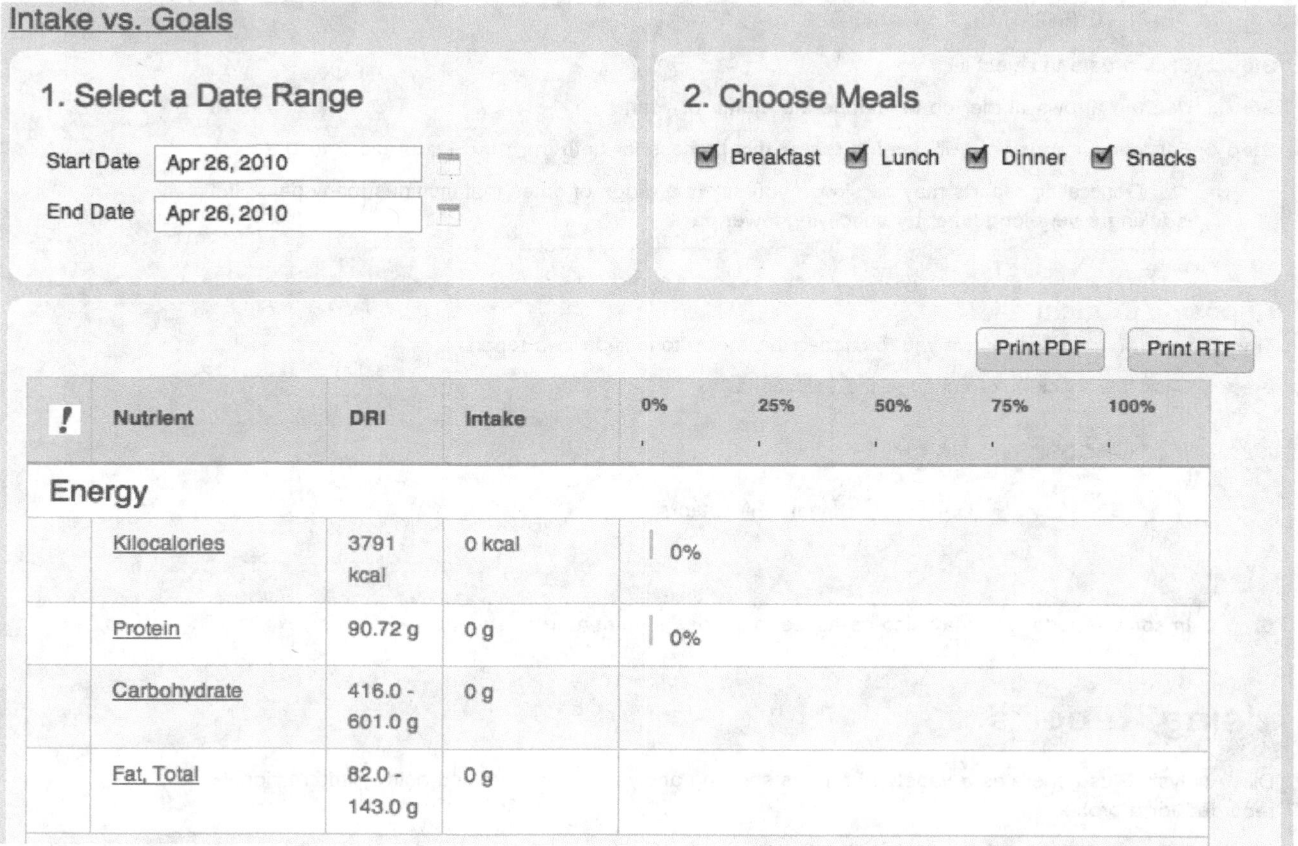

Intake vs. Goals

1. Select a Date Range

Start Date Apr 26, 2010

End Date Apr 26, 2010

2. Choose Meals

☑ Breakfast ☑ Lunch ☑ Dinner ☑ Snacks

Print PDF Print RTF

!	Nutrient	DRI	Intake	0% 25% 50% 75% 100%
Energy				
	Kilocalories	3791 kcal	0 kcal	0%
	Protein	90.72 g	0 g	0%
	Carbohydrate	416.0 - 601.0 g	0 g	
	Fat, Total	82.0 - 143.0 g	0 g	

The **DRI** column shows the recommended amount of each nutrient you should have consumed based on the data in the profile. The **Intake** column shows the amount of each nutrient that you actually consumed. The graph shows the **percentage of the recommended DRI you consumed** of each nutrient.

To create an **Intake vs. Goals** report:

Step 1. Select a **Start Date** and **End Date** by using the **Calendar**.

Step 2. Use the check boxes under **Choose Meals** to select the meals to include. This limits the comparison by meal.

Step 3. When you are finished making selections, click **Print PDF** or **Print RTF**.

Macronutrient Ranges Report

The calories you eat come from the three **macronutrients: protein**, **carbohydrates**, and **fat**. (Calories can also come from alcohol, which is not a nutrient.)

The Dietary Reference Intakes (DRI) for these energy yielding nutrients are called the **Acceptable Macronutrient Distribution Ranges (AMDR)**. These ranges are the minimum and maximum percentages of each macronutrient that should comprise total caloric intake. Those percentages are:

- 45 to 65 percent from carbohydrate
- 10 to 35 percent from protein

- 20 to 35 percent from fat

The **Macronutrient Ranges** report compares the recommended percentage ranges with your actual dietary intake.

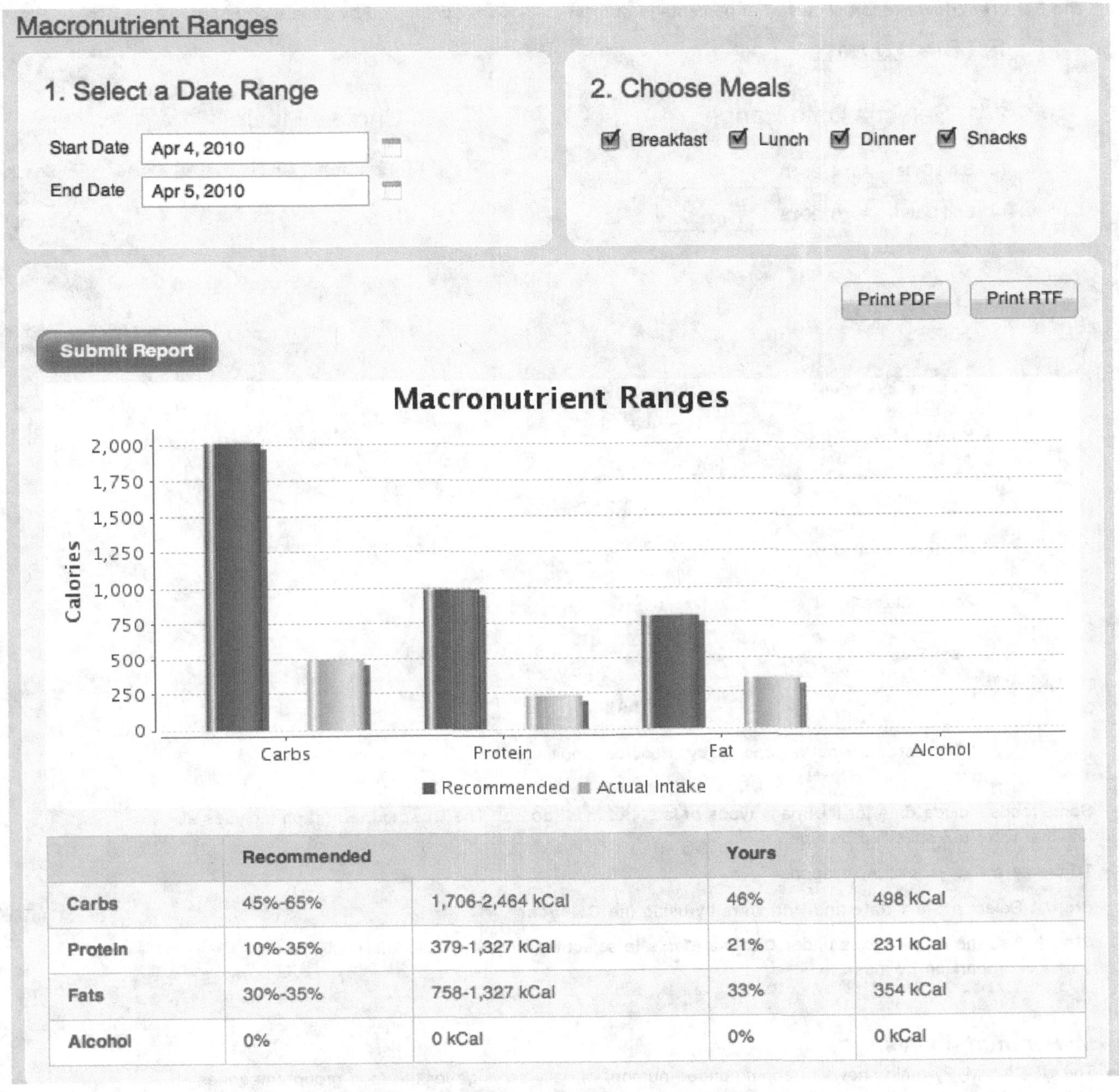

	Recommended		Yours	
Carbs	45%-65%	1,706-2,464 kCal	46%	498 kCal
Protein	10%-35%	379-1,327 kCal	21%	231 kCal
Fats	30%-35%	758-1,327 kCal	33%	354 kCal
Alcohol	0%	0 kCal	0%	0 kCal

The first graph shows the **Recommended** percentages for each macronutrient. The second graph shows **Actual** caloric (**kcal**) intake. It also shows calories from **alcohol**.

The table shows the recommended percentages and calories or carbohydrates, protein, and fat, and the percentages and calories that you ate.

Note that it is important to consider the total number of calories consumed as well as the percentage. If you eat twice as many calories as is recommended, with 35% of those calories coming from fat, you will have eaten twice as much fat as the recommended maximum.

To create a **Macronutrient Ranges** report:

Step 1. Select a **Start Date** and **End Date** by using the **Calendar**.

Step 2. Use the check boxes under **Choose Meals** to select the meals to include. This limits the comparison by meal.

Fat Breakdown Report

There is no DRI recommendation for many fats, but nutritionists and other standards bodies often describe recommendations for different fats as percentages of total caloric intake.

The **Fat Breakdown** report shows what percentage of **total calories** is contributed by each type of fat.

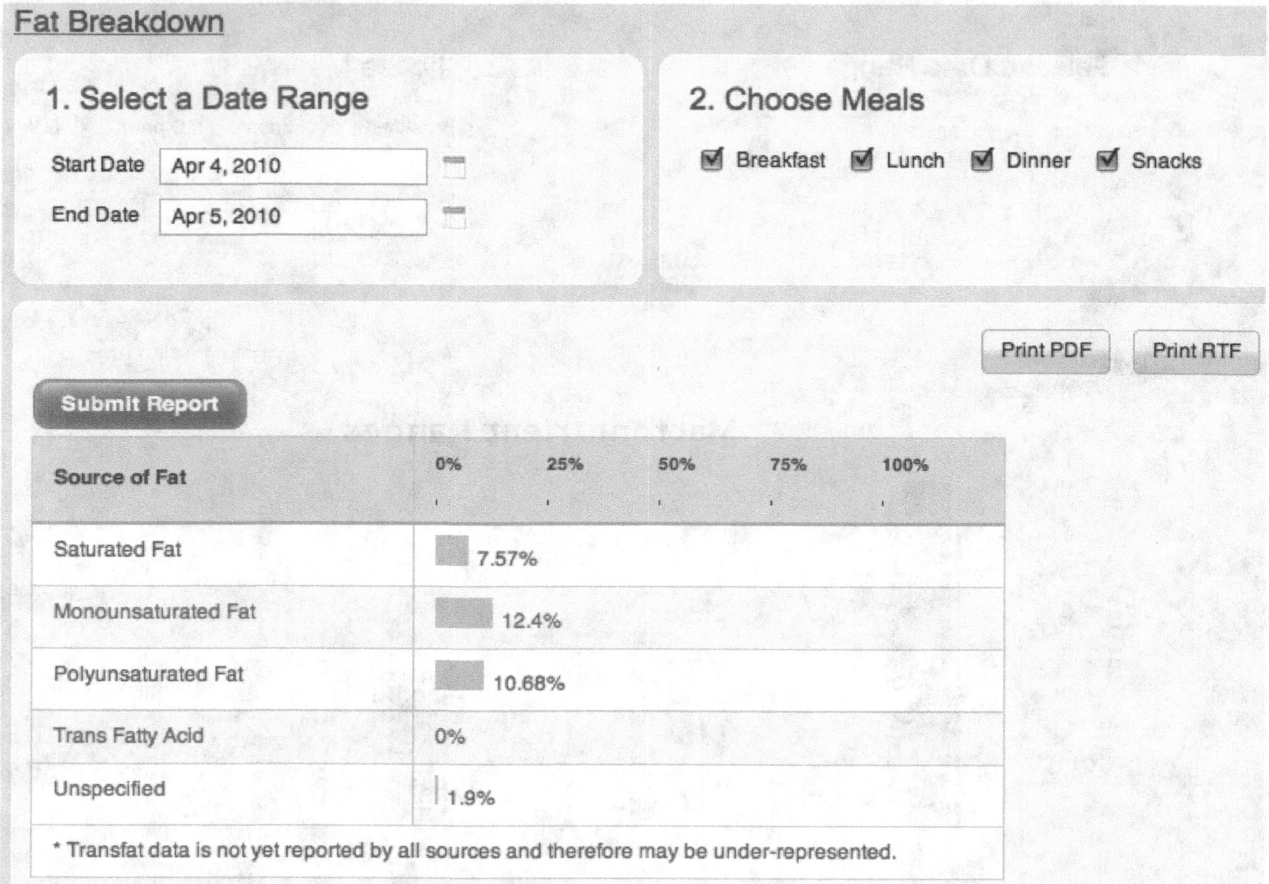

Fat Breakdown

1. Select a Date Range

Start Date Apr 4, 2010

End Date Apr 5, 2010

2. Choose Meals

☑ Breakfast ☑ Lunch ☑ Dinner ☑ Snacks

Print PDF Print RTF

Submit Report

Source of Fat	0% 25% 50% 75% 100%
Saturated Fat	7.57%
Monounsaturated Fat	12.4%
Polyunsaturated Fat	10.68%
Trans Fatty Acid	0%
Unspecified	1.9%

* Transfat data is not yet reported by all sources and therefore may be under-represented.

Some foods include data for the many types of fats, but most do not. The **Unspecified** graph includes all fats that were not specified by type.

To create a **Fat Breakdown** report:

Step 1. Select a **Start Date** and **End Date** by using the **Calendar**.

Step 2. Use the check boxes under **Choose Meals** to select the meals to include. This limits the comparison by meal.

MyPyramid Analysis

The USDA's MyPyramid shows a recommended number of daily servings for the food group categories of Grains, Vegetables, Fruits, Milk, Meats and Beans, and Discretionary Calories. It also includes recommendations for eating whole grains, various types of vegetables, and for oils.

For detailed explanations of the MyPyramid recommendations, click the links to specific items on the report screen, or click the MyPyramid logo to go to the www.mypyramid.gov home page.

The **MyPyramid Analysis** report graphically illustrates the comparison between the recommendations of MyPyramid and the number of servings from each food groups you ate.

My Pyramid Analysis

1. Select a Date Range

Start Date Apr 4, 2010

End Date Apr 5, 2010

2. Choose Meals

☑ Breakfast ☑ Lunch ☑ Dinner ☑ Snacks

Print PDF Print RTF

Submit Report

	Goal*		Actual	% Goal
Grains	10.0 oz. eq.	tips	5.9 oz. eq.	59.5 %
Vegetables	4.0 cup eq.	tips	0.2 cup eq.	6.2 %
Fruits	2.5 cup eq.	tips	0.4 cup eq.	15 %
Milk	3.0 cup eq.	tips	0.8 cup eq.	27.7 %
Meat & Beans	7.0 oz. eq.	tips	4.1 oz. eq.	58.9 %
Discretionary	648.0		252.5	39 %

MyPyramid.gov
STEPS TO A HEALTHIER YOU

Your results are based on a 3200 calorie pattern.

Make Half Your Grains Whole! Aim for at least 5.0 oz. eq. whole grains.

Vary Your Veggies! Aim for this much every week:

Dark Green Vegetables = 3.0 cups weekly

Important! Foods added using the **Create a Food** and **Create a Recipe** feature of Diet Analysis Plus do not have MyPyramid values, and are not included in the MyPyramid Analysis report.

To create a **Food Pyramid Analysis** report:

Step 1. Select a **Start Date** and **End Date** by using the **Calendar**.

Step 2. Use the check boxes under **Choose Meals** to select the meals to include. This limits the comparison by meal.

Intake Spreadsheet Report

The **Intake Spreadsheet** report shows all the foods eaten on the selected date, and the amount of each nutrient contained in each food.

This report can be saved to a file that can be imported into a spreadsheet program.

To create an **Intake Spreadsheet** report:

Step 1. Use the **Calendar** to select a date for the **Select a Date** field.

Step 2. Use the check boxes under **Choose Meals** to select the meals to include. This limits the comparison by meal.

(*Optional*) Click the **Export Excel File** button to export the report as a text file that can be opened in a spreadsheet program.

Source Analysis Report

You can influence your intake of various nutrients using the foods you choose to eat. You can raise your intake of a nutrient by choosing foods that are rich in that nutrient. Similarly, you can lower your intake of a nutrient by limiting foods that contain a lot of that nutrient.

The **Source Analysis** report allows you to choose one nutrient, then see all the foods you ate that contain that nutrient.

The **Amount** column shows how much you ate of each food.

To see the nutrient amounts in each food you ate, select the nutrient from the **Nutrient** drop-down list (in this example, "Vitamin C)."

The **graph** shows how each food contributes to the total intake of the nutrient selected. The graph does *not* show the percentage of the DRI, or the percentage of the Daily Value.

Note: "100%" on the graph is equal to "all that you actually ate." It does not express anything about how much you "should" have eaten.

You may see stars next to some of the foods on the list. A **orange star** marks foods containing more than 20% of the Daily Value for the nutrient. A **silver star** marks foods containing 10% to 20% of the Daily Value.

Tip: Daily Value is not the same as the DRI of a nutrient, which is based on a profile. Your instructor can provide guidance on the definition and uses of Daily Values.

To create a **Source Analysis** report:

Step 1. Select a day using the **Calendar**.

Step 2. Use the check boxes under **Choose Meals** to select the meals to include. This limits the comparison by meal.

Step 3. From the **Nutrient** drop-down list, choose the nutrient you want to analyse.

Energy Balance Report

Calories are different from nutrients in an important way. Their consumption is affected not just by food eaten, but also by energy expended in activity.

The **Energy Balance** report shows net caloric intake over time. It lets you see whether you retained more calories than you needed, fewer calories than you needed, or used about the same amount as you ate.

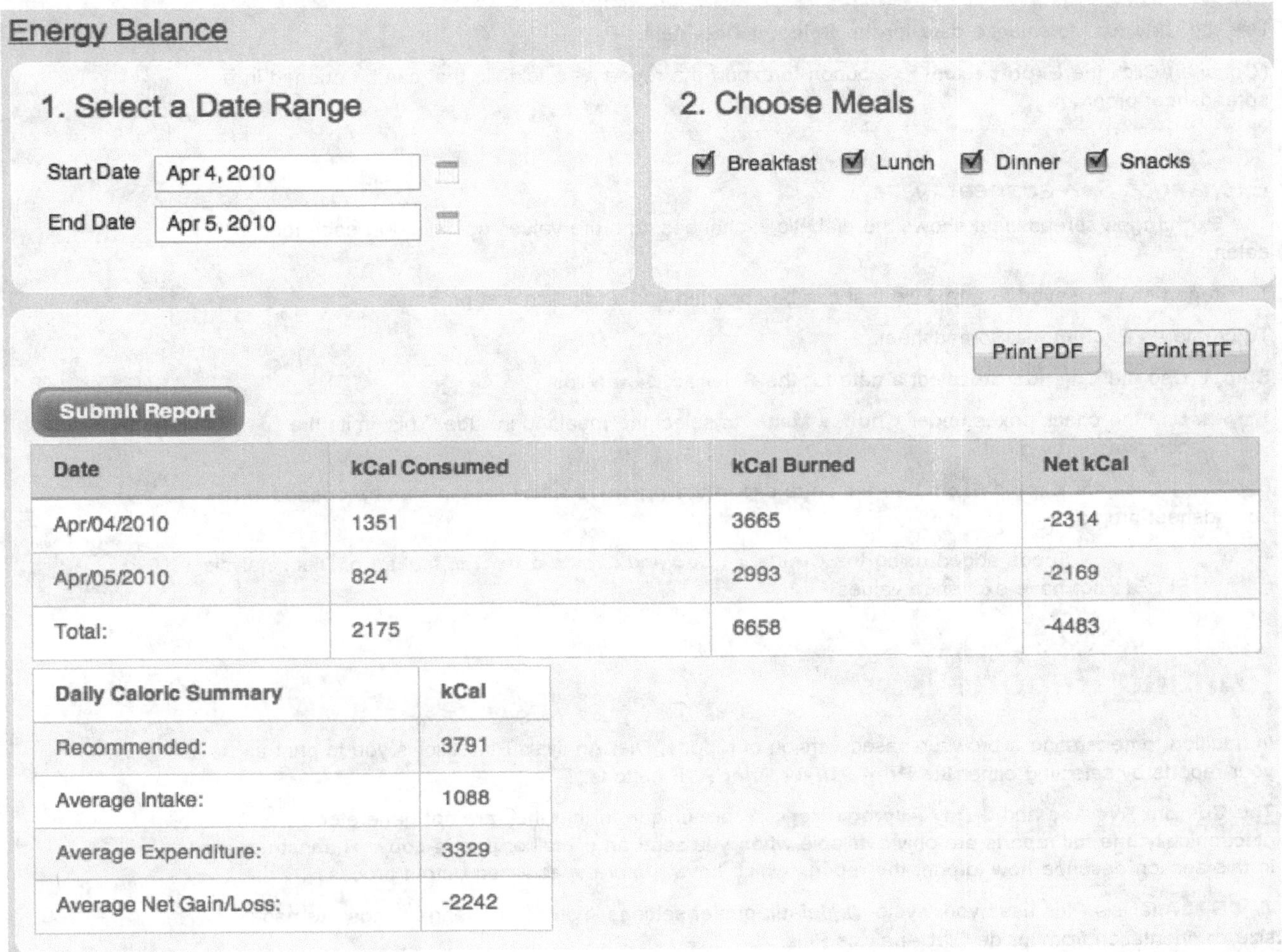

Energy Balance

1. Select a Date Range

Start Date Apr 4, 2010

End Date Apr 5, 2010

2. Choose Meals

☑ Breakfast ☑ Lunch ☑ Dinner ☑ Snacks

Print PDF Print RTF

Submit Report

Date	kCal Consumed	kCal Burned	Net kCal
Apr/04/2010	1351	3665	-2314
Apr/05/2010	824	2993	-2169
Total:	2175	6658	-4483

Daily Caloric Summary	kCal
Recommended:	3791
Average Intake:	1088
Average Expenditure:	3329
Average Net Gain/Loss:	-2242

The **kcal Consumed** column shows how many calories you ate. The **kcal Burned** column shows how many calories you used performing activities and calories burned at rest. The calories burned at rest are calculated by multiplying the amount of time (in hours) left in the day not performing other activities by your profile weight in kilograms and again by 1.1(an estimated MET for resting). The **Net Calories** column shows whether you ate more than you burned, or burned more than you ate. The **Net kcal** column shows how many calories you needed per day according to the profile.

If the **total Net kcal** is a positive number, you ate more calories than you used. This might result in weight gain.

If the **total Net kcal** is a negative number, then you used more than you ate. This might result in weight loss.

Careful! Energy metabolism is complex and varies for each person. Your instructor can provide more detailed information about caloric expenditure and weight change.

To create an **Energy Balance Analysis** report:

Step 1. Select a **Start Date** and **End Date** by using the **Calendar**.

Step 2. Use the check boxes under **Choose Meals** to select the meals to include. This limits the comparison by meal.

Activities Spreadsheet

The **Activities Spreadsheet** report shows all the activities recorded for a day. It lists the rate at which calories were burned by the activity (in calories per kilogram per hour), the duration of the activity, and the total number of calories burned.

The report will account for all 24 hours in the day. If you did not record 24 hours of activity, Diet Analysis Plus will calculate unaccounted for time as having been spent resting.

This report can be saved to a text file that can be imported into a spreadsheet program.

To create an **Activities Spreadsheet** report:

Use the **Calendar** to select a date for the **Select a Date** field.

(*Optional*) Click the **Export Excel File** button to export the report as a text file that can be opened in a spreadsheet program.

Exchanges Spreadsheet

The **Exchanges** spreadsheet shows the diabetic exchanges, and the values contained in each food eaten.

This report can be saved to a text file that can be imported into a spreadsheet program.

To create an **Exchanges** Spreadsheet:

Step 1. Use the **Calendar** to select a date for the **Select a Date** field.

Step 2. Use the check boxes under **Choose Meals** to select the meals to include. This limits the comparison by meal.

(*Optional*) Click the **Export Excel File** button to export the report as a text file that can be opened in a spreadsheet program.

> *Important!* Foods added using the **Create a Food** and **Create a Recipe** feature of Diet Analysis Plus do not have exchange values.

Printing Reports

In addition to generating a browser-based version of reports, Diet Analysis Plus allows you to print all your reports by selecting either the **Print PDF** or **Print RTF** buttons.

The **Custom Average** and **3-Day Average Reports** are unique in that they are not generated electronically; the full reports are only available when you send an e-mail or print a copy. The instructions in this section describe how to print the reports which have a more specialized output process.

Tip: Diet Analysis Plus uses your system's default printer settings. You can't change things like page size or orientation from inside Diet Analysis Plus.

Printing the Intake Spreadsheet

The **Intake Spreadsheet** lists all the foods you have eaten and shows all the nutrient details for each food. For this type of report Diet Analysis Plus prints a spreadsheet for each day individually.

Step 1. Before generating your report, verify you have the correct information selected in the **Select a Date** and **Choose Meals** options. Make any necessary changes.

Intake Spreadsheet

1. Select a Date

Select a Date Apr 4, 2010

2. Choose Meals

☑ Breakfast ☑ Lunch ☑ Dinner ☑ Snacks

Print PDF Print RTF Export Excel File

Submit Report

Item Name	Meal	Quantity		
Cereal, Granola, Prepared	Breakfast	0.5 cup(s)		
Milk, Non Fat Skim or Fat Free	Breakfast	8.0 fluid ounce(s)		
Orange	Breakfast	1.0 item(s) - 2 5/8 in. diameter, sphere		

Step 2. Click **Print PDF** to open a PDF preview in a separate browser window. Click **Print RTF** to download an .rtf file to disk. Click **Export Excel File** to download a .csv file to disk.

Step 3. Once downloaded, you can email the file(s).

Printing the 3-Day Average Report

Most professors assign a 3-day Average assignment.

Diet Analysis Plus selects the first day on which you entered any food, and will average that day followed by the next two days with data. You can change the any day of the average by clicking on the Calendar next to the day you would like to change.

To print the 3-Day Average Report:

Step 1. Choose three dates from the **Calendar** icons to the right of the three date text boxes.

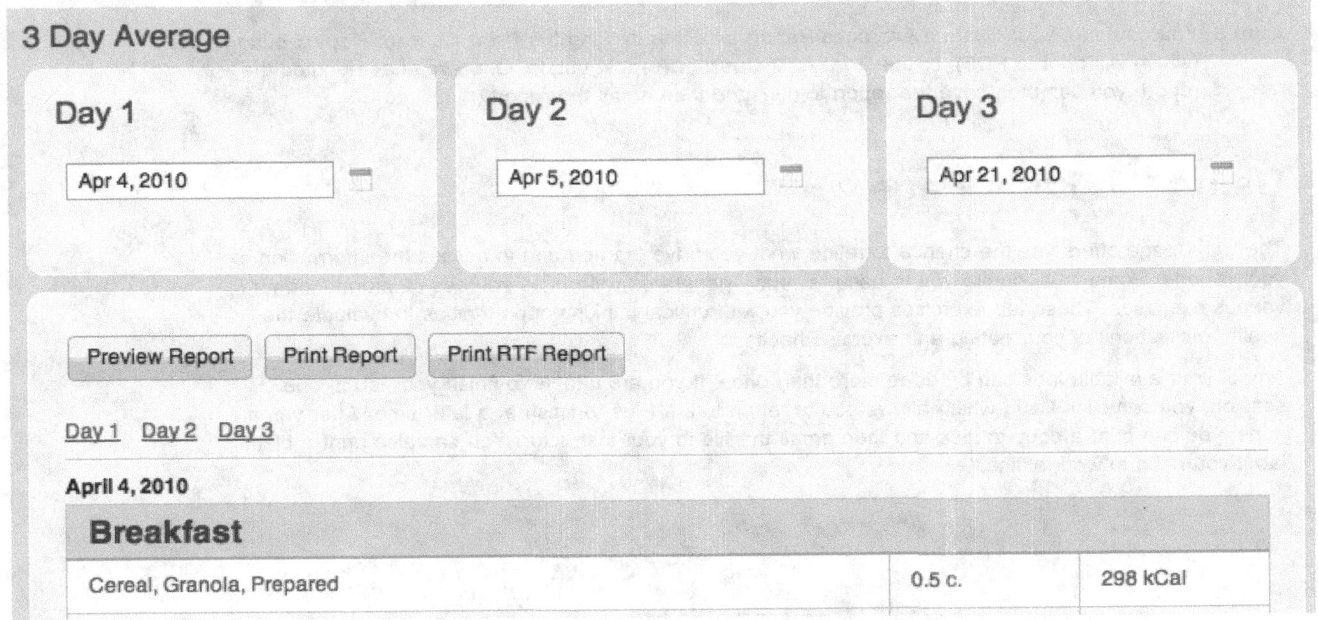

Step 2. Click the **Preview Report** button to preview your 3 Day Report. Press the **Back** button on your

browser to return to the main menu.

Step 3. You can print your 3 Day Report by either clicking the **Print Report** button (which will print to PDF), or the **Print RTF** button. Once you have printed the report, you can then save the report to disk and then email the report (if desired).

Printing a Custom Average Report

Use the **Custom Average** menu to choose your own date range and meals, and to choose which reports to generate.

To print a **Custom Average** report:

Step 1. Choose a **Start Date** and an **End Date**.

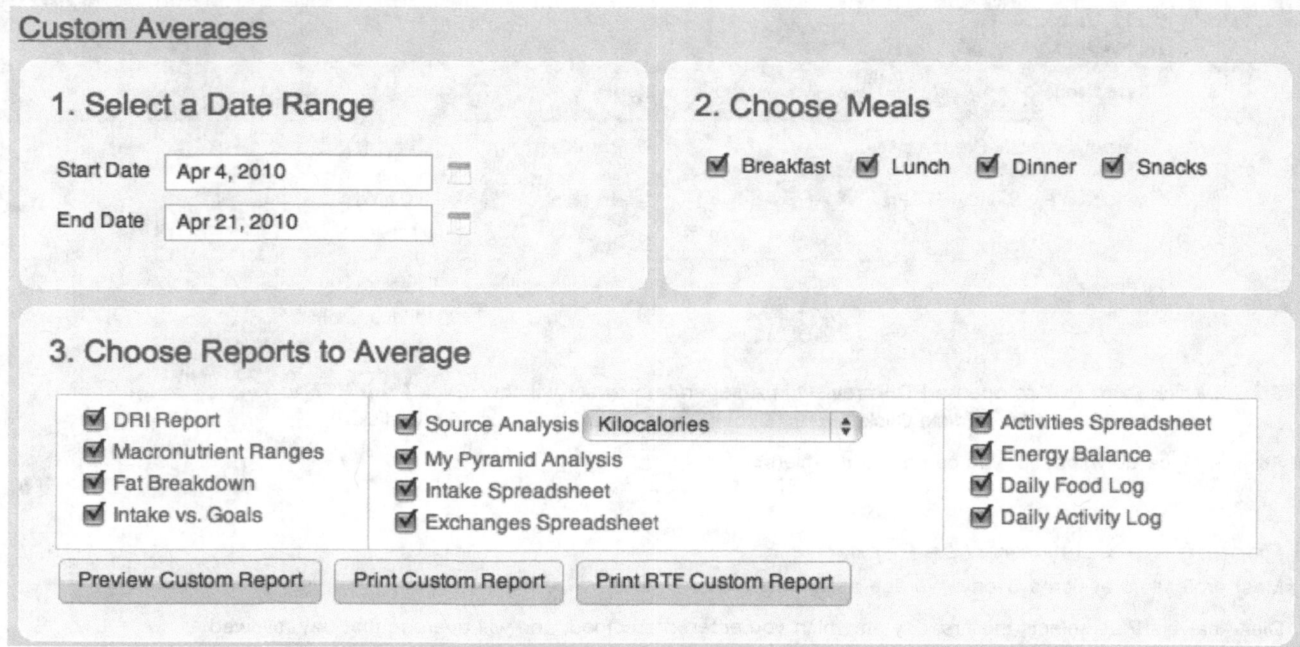

Step 2. Click the check boxes under **Choose Meals** to choose which meals to average.

Step 3. Click the check boxes under **Choose Reports to Average** to choose which reports to print.

Step 4. Click the **Preview Custom Report** button. A **Print Preview** window will open. Press the **Back** button on your browser to return to the main menu.

Step 5. You can print your Custom Averages Report by either clicking the **Print Custom Report** button (which will print to PDF), or the **Print RTF Custom Report** button. Once you have printed the report, you can then save the report to disk and then email the report (if desired).

Using the Labs Page

The **Labs** page offers you the chance to refine what you have learned and to assess the information in your reports. Some labs require you to use previously generated reports as sources of information for various exercises. These lab exercises provide you with critical thinking opportunities to evaluate the health implications of your eating and exercise habits.

Any of your available labs can be done more than once. If you are unable to finish your lab in one session, you can click **Save** which allows you to return to DA Plus to finish at a later time. After you are done, you can print a copy to disk and then email the file to your instructor. You can also print a blank lab if you need to work offline.

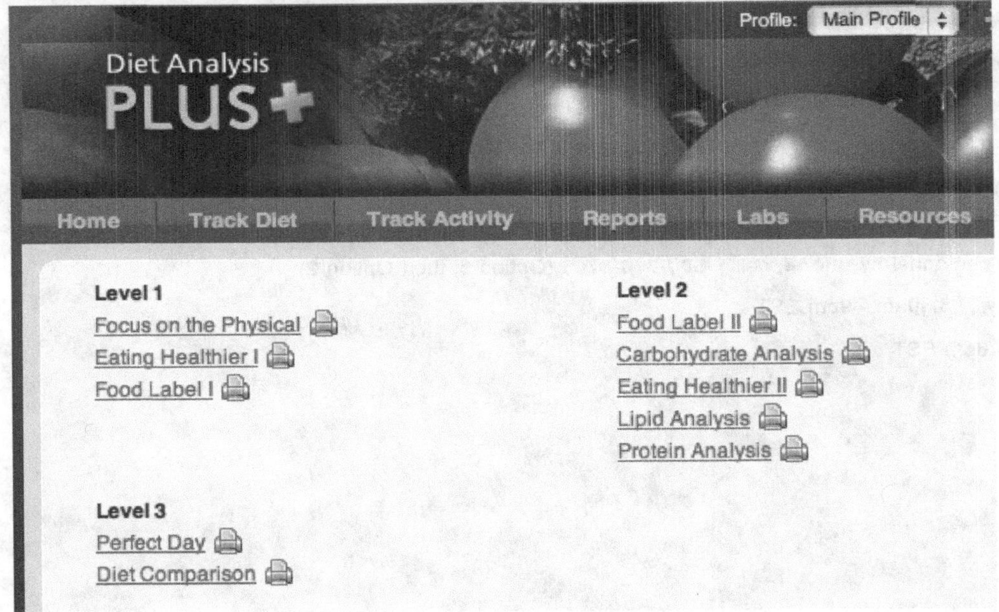

Lab Basics

Step 1. Click the **Labs** tab to see the labs that have been assigned to you.

Step 2. Click a lab **name** to open that activity.

Step 3. Review the **Objective** and **Lab Resources** before beginning.

Step 4. Enter your answers as required.

Step 5. Click **Next** to continue to the next page, or click **Save** to resume the lab at later time.

Step 6. When you have completed the lab, click the **Save and Submit** button to submit the lab results to your instructor.

Printing Labs

To print your lab:

Step 1. Click the **Labs** tab to see the labs that have been assigned to you.

Step 2. Click the print icon next to the lab you would like to print. A PDF version of your lab will open in a new window.

Step 3. Print or save the PDF from your browser toolbar.

E-mailing Labs

Step 1. Click the **Labs** tab to see the labs that have been assigned to you.

Step 2. Follow the steps in **Printing Labs** above to print a copy of your lab in PDF format. Use your favorite email program to attach this PDF file to your email.

Technical Support

If you have trouble using or accessing DA Plus, you can contact support to get further assistance. From the **Cengage Learning Technical Support** site you can get information on how to contact our experts by phone, e-mail form, or online chat.

To access **Online Chat/Customer Service Support** , you can either click the **Technical Support** link at the bottom of any page in DA Plus or direct your browser to www.cengage.com/support/.

Once you are at the Cengage Learning support site, you can follow these steps:

Step 1. Select **Diet Analysis Plus** from the **Student** drop-down menu.

Step 2. Click **Go** and the **Technical Support** page will open.

Step 3. Choose from any of the following technical support resources:

- Select **Chat Online** to chat with a technical support representative.
- Select **Submit your questions under Contact Us to access the online e-mail form.**

Note: Chat and e-mail support are available 24/7. When using an e-mail form, support requests are usually responded to within 48 hours.

To contact technical support personnel by **phone**, call **1-800-354-9706** (Option 5, then Option 2).

- Monday - Thursday: 8:30am - 9pm EST
- Friday: 8:30am - 6pm EST

Personal Dietary Assesment

A Project Guide for Diet Analysis

Maureen A. Reidenauer R.D. | Michele Fisher Ph.D., R.D.

Australia • Brazil • Japan • Korea • Mexico • Singapore • Spain • United Kingdom • United States

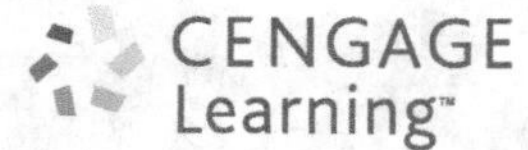

Personal Dietary Assesment: A Project Guide for Diet Analysis

Maureen A. Reidenauer R.D.
Michele Fisher Ph.D., R.D.

Executive Editors:
Michele Baird
Maureen Staudt
Michael Stranz

Project Development Manager:
Linda deStefano

Senior Marketing Coordinators:
Sara Mercurio
Lindsay Shapiro

Senior Production / Manufacturing Manager:
Donna M. Brown

PreMedia Services Supervisor:
Rebecca A. Walker

Rights & Permissions Specialist:
Kalina Hintz

Cover Image:
Getty Images*

ISBN-13: 978-0-759-35524-8

ISBN-10: 0-759-35524-X

Cengage Learning
5191 Natorp Boulevard
Mason, Ohio 45040
USA

Cengage Learning is a leading provider of customized learning solutions with office locations around the globe, including Singapore, the United Kingdom, Australia, Mexico, Brazil, and Japan. Locate your local office at: **international.cengage.com/region**

Cengage Learning products are represented in Canada by Nelson Education, Ltd.

For your lifelong learning solutions, visit **custom.cengage.com**

Visit our corporate website at **cengage.com**

Printed in the United States of America

Acknowledgements

We would like to take this opportunity to thank our husbands and families for their continued love and support throughout this project.

This guide is dedicated to Jordan Pollock a boy whose sense of awe and wonder with learning is an inspiration to us all.

PERSONAL DIETARY ASSESSMENT

Contents List

Contents continued.

PERSONAL DIETARY ASSESSMENT
Project Introduction

Personal Dietary Assessment [PDA] is designed for you to use with **Diet Analysis Plus** software application and can be used in any of your Nutrition or Health and Wellness courses. PDA has been used by students who agreed that this project was extremely appropriate and useful in their lives and studies.

Using the ten analytic reports generated by Diet Analysis Plus , you will be able to apply the concepts presented in the classroom to evaluate your current eating and lifestyle habits. The sequential flow of assignments encourages critical thinking and allows you to develop strong assessment skills. You will be able to identify areas of deficiency and excess of nutrients within current eating habits and you will be able to identify ways to adjust your diet to better meet your nutritional needs and attain your personal health goals.

Throughout this PDA assignment guide, nutrition and wellness principles will be reinforced to aid with your studies. PDA encourages you to apply newly learned guidelines and principles to your life. Learning objectives for each portion of PDA will highlight competencies that you should be comfortable with and be able to demonstrate upon successful completion.

PERSONAL DIETARY ASSESSMENT LEARNING OBJECTIVES

- Students will review their personalized nutrition profile, based on their age, weight, and gender and activity level.
- Students will gain an understanding of nutrient composition of their current food intake.
- Students will be able to compare their diets to their identified needs by analysis of graphic reports generated by **Diet Analysis Plus** software.
- Students will identify areas of deficiency and excess of nutrients with their current eating habits.
- Students will identify ways to adjust diet to better meet nutritional needs and attain personal health goals.

Getting Started

1. **REGISTER AND LOG IN:**
   * Go to the web address that is found on the pin code card for the software.
   * Type in the ACCESS CODE that is found on Pin Card you purchased.
   * Follow directions to register your e-mail and password.
2. **SET-UP PROFILE:**
   * Click "Start Here" to create your personal nutritional needs profile.
   * Name your profile and enter age, gender, height, weight and activity level.
   * Review definitions on activity levels prior to selection to ensure accurate calorie calculations.
   * Print out your personalized PROFILE.
3. **RECORD INTAKE FOR THREE DAYS:**
   * Using PDA-Form 1 record all food and beverage intake for each of the three days to be used for evauation. Use one form for each individual day you are recording intake. Ideally you should use two typical weekdays and one adjacent weekend.
   * Record as much information as possible regarding brand name of food or beverage, the closest approximation of portion ingested and the emotions that were being experienced around that intake.
   * Do not change your eating habits during this recording period.
4. **INPUT INTAKE:**
   * Go to TRACK DIET to input the foods that you recorded.
   * In SEARCH box, enter the food that you recorded.
   * Choose the closest correct food item from the list which results from search.
   * Fill in the amount of that food that you ate.
   * Indicate which meal the food was included.
   * Continue adding foods until all foods for Day 1 are represented.
   * Complete steps for Day 2 and Day 3.
5. **FAMILIARIZE YOURSELF WITH THE REPORTS:**
   * Click on CREATE Reports to obtain the analysis of your diet.
   * View all 10 Report Types at this time as you will use these reports throughout the course of this assessment effort.
6. **PRINT AVERAGE REPORT:**
   * Attempt to print the INTAKE vs. GOALS report and submit to instructor.
7. **COMPLETE ASSIGNMENTS AS DIRECTED:**
   * PDA requires you to find and record data found on various reports that the software generates.
   * You will need to refer to your course textbook to formulate responses that will completely meet requirements of assignment.

PERSONAL DIETARY ASSESSMENT

Daily Food Intake Record

PDA-FORM 1

Name: ______________________________ *Day/Date:* ____________________

MEAL	TIME	FOOD	AMOUNT	LOCATION	EMOTION	COMMENTS
BREAKFAST						
SNACK						
LUNCH						
SNACK						
DINNER						
SNACK						

PART I

Food Ways

PART I OVERVIEW:

Decisions about what to eat, when to eat and even where we will eat are the result of the many influences in our lives. Our culture, region, religion, foods available and finances influence food choices. How we make our food selections can greatly impact our overall health and sense of well-being. As consumers, we need to become more aware of how nutrition claims and advertising can influence our eating patterns.

LEARNING OBJECTIVES:

Upon completing assignment:

1. You will be able to identify factors that influence daily food choices.
2. You will be able to evaluate the nutritional claims made in advertising.

REPORTS NEEDED TO COMPLETE ASSIGNMENT:

* PDA-Form1
* Printout from Website

ASSIGNMENT DUE DATE: ______________________________

COMMENTS:

____

____

____

____

____

____

____

PERSONAL DIETARY ASSESSMENT

Daily Food Intake Record

PDA-FORM 1

Name: ____________________ *Day/Date:* ____________________

MEAL	TIME	FOOD	AMOUNT	LOCATION	EMOTION	COMMENTS
BREAKFAST						
SNACK						
LUNCH						
SNACK						
DINNER						
SNACK						

Name: ______________________ Course: ______________

PART I
Assignment

BRIEFLY ANSWER THE FOLLOWING QUESTIONS.

1. List your three favorite foods:
   *
   *
   *

   Why are these your favorite foods?

2. List your three least preferred foods:
   *
   *
   *

   Why are these your least preferred foods?

3. Using food records recorded for three days, detail how often you consumed your favorite foods—if you did not consume, *detail* why these foods were not included in the 72-hour period.

4. During the three day recording period, how many times did you eat solely because you were hungry?

5. What factors influenced your food choices during the three day recording period?

6. What claims, advertisements or food labels influenced your intake?

7. Which of the five nutrition principles/factors/building blocks appear to be missing in your current eating pattern as highlighted by three day recording? *Explain your answer.*
Missing:

Explain:

8. What two lifestyle factors could you change in order to improve eating habits? Explain how this will improve your eating habits.
*
*
Explain:

9. Go to the following website and complete the assignment as detailed. http://webquiz. iLrn.com/tutorials/B_holp07/P_hoeger1A/public. Detail your results:

10. What did you learn about your lifestyle that was most interesting? From the results of the quiz from #9, identify health behaviors you would like to change and explain why it is important for you to address these factors.

PART II

Nutrition Tools

PART II OVERVIEW:

When trying to meet nutritional needs there are many tools available to assist you. Nutrient recommendations detail a healthy person's need for energy and other important nutrients. Dietary Reference Intakes [DRI] and the Daily Values [DV] are the most common nutrition standards used in the U.S. The Food Guide Pyramid graphically demonstrates the nutritional concepts of adequacy, balance, moderation and variety, but lacks in attention to calorie control and water intake. The exchange system allows for careful attention to the macronutrient content of a person's diet and is often used for diabetics and those on weight programs.

LEARNING OBJECTIVES:

Upon completing assignment:

1. You will be able to identify your daily nutrition needs.
2. You will know how a food label provides nutrient composition.
3. You will be able to assess composition of daily intake using Food Guide Pyramid.
4. You will be able to assess diet in terms of the exchange system.

REPORTS NEEDED TO COMPLETE ASSIGNMENT:

- Profile
- Food Guide Analysis
- Exchange Spreadsheet

ASSIGNMENT DUE DATE: ______________________________

COMMENTS:

____

____

____

____

____

____

Name: ______________________________ Course: ______________

PART II
Assignment

1. Using the ***Profile*** report, detail your daily recommended intakes for the following nutrients:
   * Calories:
   * Protein:
   * Fat:
   * Fiber:
   * Water:
   * Sodium:
   * Vitamin C:

2. Examine your average Food Guide Pyramid servings for each of the food groups and respond to the following:
   * What food groups were in the preferred range? ***Detail.***
   * What food groups were in the less than preferred range? ***Detail.***
   * What food groups exceeded range? ***Detail.***

3. Detail why it is important to meet the correct number of servings from each food group represented on the Food Guide Pyramid.

4. Discuss ways that you can change current eating habits to better meet pyramid guidelines.

5. Using Exchanges Spreadsheet answer the following:
   * Total number of vegetable servings:
   * Total number of bread/starch servings:
   * Total number of fruit servings:
   * Total number of meat servings:
   * Total number of fat servings:
   * Total number of milk servings:

6. Explain why the spreadsheet highlights servings of meat from different categories—very lean, med fat, high fat.

- ***Very Lean:***

- ***Medium Fat:***

- ***High Fat:***

7. Which food groups were in the preferred range and which ones are in excess? ***Explain.***

- ***Preferred Range:***

- ***Excess Range:***

8. Which food group was your biggest concern? Why?

9. Were there any surprises in completing this assignment? ***Detail.***

10. Are there any changes that you want to make in your diet? Discuss how likely you are to make these changes.

PART III

Digestion

PART III OVERVIEW:

Digestion is the process that transforms the foods that you eat into the nutrients for your body. Proper digestion and absorption of nutrients from the foods you eat depends on optimal function of the gastrointestinal [GI] tract. Many lifestyle issues such as illness, physical activity, stress and environmental contaminants may impact the health of the GI tract.

LEARNING OBJECTIVES:

Upon completing assignment:

1. You will be able to identify foods high in fiber.
2. You will be able to suggest changes that can improve diet.
3. You will be able to evaluate lifestyle factors that affect health.

REPORTS NEEDED TO COMPLETE ASSIGNMENT:

- Macronutrient Ranges
- Source Analysis

ASSIGNMENT DUE DATE: ______________________________

COMMENTS:

____

____

____

____

____

____

____

Name: ______________________ Course: ______________

PART III
Assignment

1. List foods in your diet that required the most mastication and nutrient release upon ingestion.

2. Did you experience any episodes of indigestion during recording period?
* If 'yes' explain foods that caused symptoms.
* If 'no' detail foods that have caused symptoms in past.

3. What causes symptoms from foods identified in question #2?

4. How much alcohol did you consume during the three day recording period?
Types of Alcohol:
* Servings of alcohol
* Grams of alcohol

Based on this intake would you be identified as which of the following:
* social drinker
* moderate drinker
* problem drinker/alcohol abuser

Explain:

5. How did the foods in your diet three day recording period influence how your GI tract functioned? ***Explain.***

6. What does Diet Guidelines suggest as recommended intake of alcohol?

✷ How does your intake compare to recommendations? ***Explain.***

7. Alcohol molecules diffuse through stomach walls and reach the brain within a minute, in the presence of food this process is slowed.

✷ Based on this concept, how quickly did the alcohol in your diet reach your brain?

8. Alcohol generates 7 calories per gram – how many calories did alcohol contribute to your diet?

_____ gm × 7cal/gm = _______ calories ______ % of calories from alcohol

Comments:

9. Which of the building blocks, balance moderation variety and adequacy helps prevent deficiencies associated with digestion difficulties? ***Explain.***

10. Discuss how you can improve your diet and lifestyle habits that could help GI tract function more effectively.

PART IV
Carbohydrates

PART IV OVERVIEW:

Carbohydrates in foods are known as the sugars, starches, and fibers. Simple carbohydrates include the sugars such as glucose, fructose, sucrose and lactose. Complex carbohydrates are the starches and fibers. The main function of sugars and starches are to provide energy to the body. Fiber has many health benefits which include keeping your digestive tract healthy, helping reduce your risk of heart disease and diabetes and helping you maintain a healthy body weight. Diets that are high in simple sugars and low in fiber can lead to many health concerns.

LEARNING OBJECTIVES:

Upon completing assignment:

1. You will be able to identify foods that are high in simple carbohydrates.
2. You will be able to identify foods that are high in complex carbohydrates.
3. You will be able to use guidelines / records to evaluate your diet.
4. You will be able to suggest menu changes to improve diet.

REPORTS NEEDED TO COMPLETE ASSIGNMENT:

- Macronutrient Ranges
- Intake Vs Goals
- Source Analysis

ASSIGNMENT DUE DATE: ______________________________

COMMENTS:

____

____

____

____

____

____

Name: ______________________ Course: ______________

PART IV
Assignment

Briefly answer the following questions.

1. List the three foods that contributed the most carbohydrates to your daily intake.

*
*
*

Comments:

2. List the three foods that contributed the most fiber to your daily intake.

*
*
*

Comments:

3. List three foods that were low in fiber or did not provide any fiber to your daily intake.

*
*
*

Comments:

4. What % of your calories on average came from carbohydrates? The Acceptable Macronutrient Distribution Range (AMDR) states that 45 to 65% of your calories should come from carbohydrates. How did your intake compare to this?

Percentage:

Comparison:

5. How many grams of carbohydrate on average did you consume? The latest Dietary Reference Intakes (DRIs) suggest that you should have a minimum of 130 grams/day of carbohydrate. How did your intake compare to recommendation?

Grams:

Comparison:

6. What % of your Recommended Daily Nutrient (RDN) for carbohydrates did you consume? If you exceeded 100% of your RDN for carbohydrates, what 3 foods contributed to this?

Percentage:

✷

✷

✷

7. How many grams of fiber did you consume on average? How did this value compare to the fiber recommendations of 25 grams per day for women and 38 grams per day for males?

Grams:

Comparison:

8. Did your average daily intake provide the minimum number of servings of foods from each of the fiber-containing food group in the FOOD GUIDE PYRAMID? Which food group(s) fell short of the recommendations?

Food Guide Minimums:

Food Groups that fell short:

✷

✷

✷

Comments:

9. What changes might you make among your vegetable, fruit and grain choices to increase the fiber content of your diet?

10. Did you use any sugar substitutes in your diet? What are the benefits and concerns associated with use of sugar alternatives?

Sugar substitutes used:

Benefits of sugar substitutes:

Concerns with use of sugar substitutes:

PART V
Lipids

PART V OVERVIEW:

Most of the lipids in foods are fats and oils. Fats are a concentrated source of energy and provide essential fatty acids. In addition they contribute flavor, texture, and mouth feel to the foods you eat. In the body fats are important because they are the primary form of stored energy provide insulation against temperature changes and help protect vital organs. Diets high in fat and cholesterol can lead to numerous health issues such as heart disease, obesity, certain cancers and gallbladder disease.

LEARNING OBJECTIVES:

Upon completing assignment:

1. You will be able to identify foods high in fat and cholesterol.
2. You will be able to use guidelines and nutrient recommendations to evaluate diet for fat content.
3. You will be able to suggest dietary changes that could improve fat and cholesterol intake.

REPORTS NEEDED TO COMPLETE ASSIGNMENT:

- Macronutrient Ranges
- Intake Vs Goals
- Source Analysis
- Fat Breakdown

ASSIGNMENT DUE DATE: ______________________________

COMMENTS:

____

____

____

____

____

____

Name: ______________________ Course: ______________

PART V
Assignment

1. How many grams of total fat on average did you consume?
   What percent of your Recommended Daily Nutrient (RDN) did you consume?

   Explain.

2. List the top three foods that contributed the most total fat to your intake.
   *
   *
   *
   Detail:

3. What percentage of your total calories came from fat?
   The Acceptable Macronutrient Distribution Range states that 20 – 35% of your total daily calories should come from fat.
   How did your intake compare to this for total fat?

4. What percentage of calories came from saturated fat?
   Health recommendations suggest < 10% of your calories should come from saturated fats.
   How did your intake compare to this recommendation?

5. List the top three foods that contributed the most saturated fat to your daily intake.
   *
   *
   *
   Detail:

6. What percentage of your total calories came from monounsaturated fats?
Health recommendations suggest that > *10% of your calories should come from monounsaturated fats.*
How did your intake compare to this recommendation?

7. What percentage of your total calories came from polyunsaturated fats?
Health recommendations suggest that about 10% of your calories should come from polyunsaturated fats.
How did your intake compare to this recommendation?

8. How many milligrams of cholesterol did you consume on average?
The Dietary Guidelines recommend an intake of dietary cholesterol of less than 300 milligrams per day.
How did your intake compare to this recommendation?

9. List the top three foods that contributed the most cholesterol to your daily intake.

-
-
-

Detail:

10. List three health issues associated with high fat, high saturated fat and/or high cholesterol intakes.

Recommend ways that you can or will change your diet to improve your fat and cholesterol intakes.

PART VI

Proteins

PART VI OVERVIEW:

All cells need protein to function properly. Protein has many important roles within the body such as supporting growth and repair of the body's tissues, helping to create enzymes and antibodies and maintaining your body's fluid balance. Protein intake in excess or deficit can lead to serious health concerns. The typical diet provides two types of proteins, animal and vegetable. Vegetarian diets can meet nutritional needs as long as there is adequate intake and variety of protein rich foods.

LEARNING OBJECTIVES:

Upon completing assignment:

1. You will be able to identify quantity and quality of protein in diet.
2. You will be able to evaluate adequacy of protein in diet.
3. You will be able to suggest food choices that are high in protein and fiber and low in fat.

REPORTS NEEDED TO COMPLETE ASSIGNMENT:

- Macronutrient Ranges
- Intake Vs Goals
- Source Analysis

ASSIGNMENT DUE DATE: ______________________________

COMMENTS:

____

____

____

____

____

____

Name: ______________________ *Course:* ______________________

PART VI
Assignment

1. Compare your daily recommended intake of protein (in grams) to your actual intake (in grams).
Recommended:
Actual:

What percent of your Dietary Reference Intake [DRI] did you consume in protein?
Percentage:
Comments:

2. Select one day and identify two significant sources of animal protein (those foods that supplied more than 4 grams/ serving) and your significant sources of plant protein (those foods that supplied more than 2 grams/serving). Specify the serving size consumed and the grams of protein provided.
Animal Protein:
✱
✱
Vegetable Protein:
✱
✱
Comments:

3. How many servings from each of the protein food groups in the Food Guide Pyramid did you consume on average?

Were they within the recommended number of servings? ***Details.***

How can you improve your intake? ***Details.***

4. What percentage of your total daily calories came from protein?
The Acceptable Macronutrient Distribution Range (AMDR) states that 10–35% of your calories should come from protein.

How did your intake compare to this for protein?

5. Using the number of calories you consumed on average, calculate the number of grams of protein you should consume if 20% of your calories come from protein.
Show calculations:

6. Using the Recommended Dietary Allowance (RDA) of 0.8 gram protein per kilogram body weight per day, how much protein (grams) should you consume daily?
Show calculations:

7. Compare the values obtained in questions #5 and #6 and discuss why they are the same or different.

Which value is preferred? Why?

8. List and discuss three reasons that people are vegetarian.
*
*
*
Comments:

9. What are the benefits and concerns with vegetarian diets?
Benefits:

Concerns:

10. Discuss the two major forms of protein energy malnutrition that exist throughout the world.
*

*

PART VII

Vitamins

PART VII OVERVIEW:

Vitamins are compounds required by the body in small amounts to maintain health. Vitamins do not provide energy but assist in the release of energy from the foods you eat. Vitamin deficiencies and toxicities can cause health concerns. A well balanced diet following dietary guidelines can provide all necessary levels of vitamins. The use of dietary supplements needs to be carefully evaluated to prevent health issues.

LEARNING OBJECTIVES:

Upon completing assignment:

1. You will be able to identify food sources of vitamins.
2. You will be able to identify major functions of selected vitamins.
3. You will be able to suggest diet changes that could provide adequate intake of selected vitamins.

REPORTS NEEDED TO COMPLETE ASSIGNMENT:

* Intake Vs Goals
* Intake Spreadsheet

ASSIGNMENT DUE DATE: ______________________________

COMMENTS:

____

____

____

____

____

____

____

Name: ______________________ Course: ______________

PART VII
Assignment

1. List all of the fat soluble vitamins that were less than 75% of your % Recommended Daily Nutrient (RDN).
   ✱
   ✱
   ✱

2. What foods could you add to your diet to increase your intake of these vitamins (list each vitamin and 3 good food sources)?
   Vitamin ______:
   ✱
   ✱
   ✱
   Vitamin ______:
   ✱
   ✱
   ✱

3. List all fat soluble vitamins that exceeded 200% of your Recommended Daily Nutrient (RDN). What foods contributed most to your intake of these vitamins (list each vitamin and 3 foods from your diet)?
   Vitamin ______:
   ✱
   ✱
   Vitamin ______:
   ✱
   ✱

4. For those fat soluble vitamins that you consumed less than 75% of your RDN or exceeded 200% of your RDN, what effect could this have on your health? (Be specific—list each vitamin and a health issue associated with it).
   Vitamin ______:
   ✱
   ✱
   Vitamin ______:
   ✱
   ✱

5. List all of the water soluble vitamins that were less than 75% of your % Recommended Daily Nutrient (RDN).
   ✱
   ✱
   ✱
   ✱
   ✱

6. What foods could you add to your diet to increase your intake of these vitamins (list each vitamin and 3 good food sources)?

Vitamin ______:

*
*
*

Vitamin ______:

*
*
*

7. List all water soluble vitamins that exceeded 200% of your Recommended Daily Nutrient (RDN). What foods contributed most to your intake of these vitamins (list each vitamin and 3 foods from your diet)?

Vitamin ______:

*
*
*

Vitamin ______:

*
*
*

8. For those water soluble vitamins that you consumed less than 75% of your RDN or exceeded 200% of your RDN, what effect could this have on your health? (Be specific—list each vitamin and a health issue associated with it).

Vitamin ______:

*
*
*

Vitamin ______:

*
*
*

9. List the four antioxidant nutrients. Which foods are the best sources of these nutrients? What other compounds in foods have antioxidant activity?

* *
* *

10. Does your diet have a wide variety of antioxidant rich foods? List these foods. How can you add more antioxidant rich foods to your diet and what benefits may it have on your health?

PART VIII
Minerals

PART VIII OVERVIEW:

Minerals are inorganic compounds required by the body to maintain health. They can be classified as either major or trace depending on how much is present in your body. Like vitamins, mineral deficiencies and toxicities can cause health problems. The consumption of a variety of well balanced foods from all of the food groups will ensure that you receive adequate amounts of minerals from your diet.

LEARNING OBJECTIVES:

Upon completing assignment:

1. You will be able to identify food sources of selected minerals.
2. You will be able to use guidelines to evaluate intake of minerals.
3. You will be able to suggest diet change that could provide for adequate intake of identified minerals.
4. You will be able to evaluate use of and problems with dietary supplements.

REPORTS NEEDED TO COMPLETE ASSIGNMENT:

- Intake Vs Goals
- Intake Spreadsheet

ASSIGNMENT DUE DATE: ______________________

COMMENTS:

____

____

____

____

____

____

____

Name: ______________________________ *Course:* ____________

PART VIII
Assignment

1. Review your average intake of minerals. List the MAJOR minerals that were less than 75% of your % Dietary Reference Intake [DRI].

*
*
*
*

2. What foods could you add to your diet to increase your intake of these minerals (list each mineral and 3 good food sources)?

Mineral ____________:
*
*
*

Mineral ____________:
*
*
*

Comments:

3. List all major minerals that exceeded 200% of your Recommended Daily Nutrient (RDN). What foods contributed most to your intake of these minerals (list each mineral and 3 foods from your diet)?

Mineral ____________:
*
*
*

Mineral ____________:
*
*
*

Mineral ____________:
*
*
*

Comments:

4. For those major minerals that you consumed less than 75% of your Recommended Daily Nutrient (RDN) or exceeded 200% of your Recommended Daily Nutrient (RDN), what effect could this have on your health? (Be specific—list each mineral and a health issue associated with it)

Mineral _____________:

*

*

Mineral _____________:

*

*

Mineral _____________:

*

*

Comments:

5. List the TRACE minerals that were less than 75% of your % Recommended Daily Nutrient (RDN).

Mineral _____________:

*

*

Mineral _____________:

*

*

Mineral _____________:

*

*

6. What foods could you add to your diet to increase your intake of these minerals (list each mineral and 3 good food sources)?

Mineral _____________:

*

*

*

Mineral _____________:

*

*

*

Comments:

Name: ____ Course: ____________________

7. List any trace mineral that exceeded 200% of your Recommended Daily Nutrient (RDN). What foods contributed most to your intake of these minerals (list each mineral and 3 foods from your diet)?

Mineral ____________*:*

*

*

*

Mineral ____________*:*

*

*

*

Comments:

8. For those trace minerals that you consumed less than 75% of your Recommended Daily Nutrient (RDN) or exceeded 200% of your Recommended Daily Nutrient (RDN), what effect could this have on your health? (Be specific—list each mineral and a health issue associated with it).

Mineral ____________*:*

*

*

Mineral ____________*:*

*

*

Comments:

9. Do you take any dietary supplement? If yes, which one(s) and why?

After reviewing your intake of vitamins and minerals from the foods you consume do you feel you need a supplement? Explain.

10. Discuss the pros and cons of dietary supplements.

Pros:

Cons:

Comments:

PART IX

Physical Activity

PART IX OVERVIEW:

Physical activity can provide numerous health benefits such as: reducing the risk of chronic diseases including heart disease and diabetes; lowering blood pressure; improving one's emotional well being by relieving stress; and keeping bones, muscles and joints healthy. The type and amount of physical activity you engage in can improve cardio-respiratory endurance, strength and endurance, and the ratio of lean muscle to body fat. The physically fit athlete needs a diet that provides nutrient dense foods and provides for adequate hydration.

LEARNING OBJECTIVES:

Upon completing assignment:

1. You will be able to identify how exercise impacts daily energy equation.
2. You will be able to discuss how composition of diet can/does impact athletic performance.
3. You will be able to list benefits of water versus sports drinks in the diet.

REPORTS NEEDED TO COMPLETE ASSIGNMENT:

* Activities Spreadsheet
* Energy Balance
* Source Analysis

ASSIGNMENT DUE DATE: ______________________

COMMENTS:

____

____

____

____

____

____

____

Name: ______________________________ Course: ______________

PART IX
Assignment

1. During the recording period, on average, how long were you engaged in the following activities:
   * Stretching:
   * Warm up:
   * Aerobic activity:
   * Strength training:
   * Cool down:

2. What is your biggest obstacle to engaging in physical activity? Explain

3. What changes can you or did you make in regards to current physical activity level?

4. What health benefits do you expect as a result of these changes? Explain.

5. Using Source Analysis report, what was your average water intake?

   Was intake able to provide for needs? Explain.

6. How did your fluid intake change on days that you engaged in physical activity?

   What type of fluids did you select? Why these?

7. What was your average caffeine intake during the three day recording period?

Comments:

8. Most competitive sports organizations restrict caffeine intake to 5-6 cups intake two hours prior to participation in competition if you were a professional athlete would this rule affect you? Explain.

9. Professional athletes in training often use protein supplements. Explain why someone in training might think that they should be using these supplements.

Discuss potential risks associated with excessive protein intake.

10. Discuss health concerns associated with use of Erogenic agents

PART X

Energy Balance/Weight Management

PART X OVERVIEW:

The energy balance equation compares the calories from the foods you consume to the calories used to maintain your body's metabolism and physical activity to determine if you will maintain, lose or gain weight. Over consumption and under consumption of calories can cause health concerns. Obesity is directly linked to many chronic health diseases such as diabetes, heart disease and certain cancers. Disordered eating is caused by many emotional and psychosocial factors. Eating disorders have many short term and long term consequences on health.

LEARNING OBJECTIVES:

Upon completing assignment:

1. You will be able to assess if your BMI indicates health concerns.
2. You will be able to identify if you are at risk of health concerns associated with obesity.
3. You will be able to identify if you are at risk for disordered eating issues.

REPORTS NEEDED TO COMPLETE ASSIGNMENT:

* Personal Profile
* Energy Balance

ASSIGNMENT DUE DATE: ______________________________

COMMENTS:

____

____

____

____

____

____

____

Name: ______________________________ Course: ____________________

PART X
Assignment

1. What is your Body Mass Index [BMI] and how are you classified? Did this finding surprise you? Why?
BMI:
Classification:

Comments:

2. What was your average calorie intake? What percentage of your personal Recommended Daily Nutrient (RDN) for calories did you consume?
Average calorie intake:
Percentage:

Comments:

3. If you consumed the same amount of calories per day over a month (from question #2), would you lose, gain, or maintain your weight? Explain.

4. Discuss three health risks associated with elevated BMI/excessive body fat.
   *
   *
   *

Comments:

5. What are the limitations to using BMI? Discuss three methods used to measure body fat.
Limitations:
Other Methods:
*
*
*

6. Obesity is a complex issue with many causes. What are five factors that play a role in its development?
*
*
*
*
*

Are you at risk for developing obesity?

7. Discuss how you can achieve and maintain a healthy body weight. Be specific with regards to your current habits.

8. Give an example of a popular "fad" diet and briefly explain why it is considered a "fad" diet.

9. Discuss "CENTRAL OBESITY" and its impact on health.

Does this reflect your current weight status?

10. Discuss ways to combat, treat, and prevent eating disorders.